Introduction to Green Chemistry

Second Edition

Albert S. Matlack

CRC Press is an imprint of the
Taylor & Francis Group, an **informa** business

CRC Press
Taylor & Francis Group
6000 Broken Sound Parkway NW, Suite 300
Boca Raton, FL 33487-2742

Printed in the United States of America on acid-free paper
10 9 8 7 6 5 4 3 2 1

International Standard Book Number: 978-1-4200-7811-4 (Hardback)

Library of Congress Cataloging-in-Publication Data

Matlack, Albert S., 1923-
Introduction to green chemistry / Albert Matlack. -- 2nd ed.
p. cm.
Includes bibliographical references and index.
ISBN 978-1-4200-7811-4 (hard back : alk. paper)
1. Environmental chemistry--Industrial applications. 2. Environmental management. I. Title.

TP155.2.E58M38 2010
660.028'6--dc22
2009049175

Visit the Taylor & Francis Web site at
http://www.taylorandfrancis.com

and the CRC Press Web site at
http://www.crcpress.com

Contents

Preface to Second Edition

Three concepts of green chemistry have proven to be popular and are widely quoted. Anastas and Warner[1] devised 12 principles of green chemistry:

1. It is better to prevent waste formation than to treat it after it is formed.
2. Design synthetic methods to maximize incorporation of all material used in the process into the final product.
3. In synthetic methods, where practicable, use or generate materials of low human toxicity and environmental impact.
4. Chemical product design should aim to preserve efficacy while reducing toxicity.
5. Auxiliary materials (solvents, extractants, etc.) should be avoided if possible or otherwise made innocuous.
6. Energy requirements should be minimized: syntheses should be conducted at ambient temperature or pressure.
7. A raw material should, where practicable, be renewable.
8. Unnecessary derivatization (such as protection or deprotection) should be avoided, where possible.
9. Selectively catalyzed processes are superior to stoichiometric processes.
10. Chemical products should be designed to be degradable to innocuous products when disposed of and not be environmentally persistent.
11. Process monitoring should be used to avoid excursions leading to the formation of hazardous materials.
12. Materials used in a chemical process should be chosen to minimize hazard and risk.

Winterton[2] proposed 12 more green principles:

1. Identify and quantify by-products.
2. Report conversions, selectivities, and productivities.
3. Establish full mass-balance for the process.
4. Measure catalyst and solvent losses in air and aqueous effluent.
5. Investigate basic thermochemistry.
6. Anticipate heat and mass transfer limitations.
7. Consult a chemical or process engineer.
8. Consider the effect of overall process on choice of chemistry.
9. Help develop and apply sustainability measures.
10. Quantify and minimize the use of utilities.
11. Recognize where safety and waste minimization are incompatible.
12. Monitor, report, and minimize the laboratory waste emitted.

Anastas and Zimmerman[3] developed 12 principles of green engineering:

1. Designers need to strive to ensure that all material and energy inputs and outputs are as inherently nonhazardous as possible.
2. It is better to prevent waste than to treat or clean up waste after it is formed.
3. Separation and purification operations should be designed to minimize energy consumption and materials use.
4. Products, processes, and systems should be designed to maximize mass, energy, space, and time efficiency.
5. Products, processes, and systems should be "output pulled" rather than "input pushed" through the use of energy and materials.
6. Embedded entropy and complexity must be viewed as an investment when making design choices on recycle, reuse, and beneficial disposition.
7. Targeted durability, not immortality, should be a design goal.
8. Design for unnecessary or capability (e.g., "one size fits all") solutions should be considered a design flaw.
9. Material diversity in multicomponent products should be minimized to promote disassembly and value retention.
10. The design of products, processes, and systems must include integration and interconnectivity with available energy and materials flows.
11. Products, processes, and systems should be designed for performance in a commercial "afterlife."
12. Material and energy inputs should be renewable rather than depleting.

A group at the University of Nottingham has put these together into "IMPROVEMENTS PRODUCTIVELY."[4]

Barry Trost introduced the concept of atom economy, where all the reagents end up in the product of the reaction and no by-products form.[5]

Roger Sheldon introduced the concept of the E-factor, which is obtained by dividing the amount of waste by the amount of product.[6] A higher number means more waste and more environmental impact.

It has been about 10 years since the first edition was written. Comparison with current statistics will show whether any progress has been made toward cleaner processes. Many topics that were barely mentioned in the first edition have become major areas of research. These include fluorous biphasic catalysis, metal organic frameworks, process intensification, dynamic kinetic resolutions of optical isomers, and directed evolution of enzymes. Government funding has pushed the development of fuel cells and biofuels, perhaps at the expense of better technologies.

The second edition is still meant to be a combination text and reference that can lead the reader to the frontiers of green chemistry. The teacher will need to select what there is time for studying in the course. (For those who desire a shorter text, Mike Lancaster has written one.[7] There is also a text on green engineering.[8]) The author assigns about two papers for outside reading each week.

These are best if they cover both sides of an issue so that the student can make up his own mind. The recommended reading in the first edition is still good, but it has been supplemented by newer papers. The exercises at the ends of the chapters are meant to bring the general principles in the book to the specific locality where the student lives. A few students embrace them enthusiastically but many reports are superficial. The challenge is to devise a grading system that will improve this. A final point is that while the Internet can be extremely useful, it is not the best first place to go to for a general review of a class of compounds. The industrial chemical encyclopedias are better for this. It is fashionable for authors to speak of how green their process is. Readers should look at this critically since it may be true only in part or not at all.[9,10]

Albert S. Matlack

REFERENCES

1. P.T. Anastas and J.C. Warner, *Green Chemistry: Theory and Practice*, Oxford University Press, New York, 1998.
2. N. Winterton, Twelve more green chemistry principles, *Green Chem.*, 2001, *3*, G73.
3. P.T. Anastas and J.B. Zimmerman, Peer reviewed: Design through the 12 principles of green engineering, *Environ. Sci. Technol.*, 2003, *37*, 94A.
4. S.Y. Tang, R.A. Bourne, R.L. Smith, and M. Poliakoff, The 24 principles of green engineering and green chemistry: Improvements, productively, *Green Chem.*, 2008, *10*, 268.
5. B.M. Trost, The atom economy—A search for synthetic efficiency, *Science*, 1991, *254*, 1471; On inventing reactions for atom efficiency, *Acc. Chem. Res.*, 2002, *35*, 695.
6. R.A. Sheldon, The E factor, 15 years on, *Green Chem.*, 2007, *9*, 1273.
7. M. Lancaster, *Green Chemistry—An Introductory Text*, Royal Society of Chemistry, Cambridge, UK, 2002.
8. D.T. Allen and D.R. Shonnard, *Green Engineering—Environmentally Conscious Design of Chemical Processes*, Prentice Hall, Upper Saddle River, NJ, 2002.
9. G. Ondrey, Go green, but be realistic, *Chem. Eng.*, 2008, *115*(4), 5.
10. P. Nieuwenhuizen, T. Vanroelen, D. Lyon, H. Bos-Brouwers, and E. Croufer, Can it be easy to be "green"? *Chem. Ind.* (London), Apr. 7, 2008, 19.

Preface to First Edition

This book is intended for chemists, chemical engineers, and others who want to see a better world through chemistry and a transition from its present unsustainable course[1] to a sustainable future.[2] (A sustainable future is one that allows future generations as many options as we have today.) It is meant to serve as an introduction to the emerging field of green chemistry—of pollution prevention. It is based on a one-semester three-credit course[3] given at the senior—graduate level interface at the University of Delaware each year from 1995 to 1998.

Books and courses in environmental chemistry usually deal with contaminants that enter air, water, and soil as a result of human activities: how to analyze for them and what to add to the smokestack or tailpipe to eliminate them. They are also concerned with how to get the contaminants out of the soil once they are there. Because texts such as those by Andrews et al.,[4] Baird,[5] Crosby,[6] Gupta,[7] Macalady,[8] Manahan,[9] and Spiro and Stigliani[10] cover this area adequately, such material need not be repeated here.

Green chemistry[11] avoids pollution by utilizing processes that are "benign by design." (The industrial ecology[12] being studied by engineers and green chemistry are both parts of one approach to a sustainable future.) Ideally, these processes use nontoxic chemicals[13] and produce no waste, while saving energy and helping our society achieve a transition to a sustainable economy. It had its origins in programs such as 3M's "Pollution Prevention Pays." It was formalized in the United States by the Pollution Prevention Act of 1990. Since then, the U.S. Environmental Protection Agency and the National Science Foundation have been making small grants for research in the area. Some of the results have been summarized in symposia organized by these agencies.[14]

This book cuts across traditional disciplinary lines in an effort to achieve a holistic view. The material is drawn from inorganic chemistry, biochemistry, organic chemistry, chemical engineering, materials science, polymer chemistry, conservation, and so on. While the book is concerned primarily with chemistry, it is necessary to indicate how this fits into the larger societal problems. For example, in the discussion of the chemistry of low-emissivity windows and photovoltaic cells, it is also pointed out that enormous energy savings would result from incorporating passive solar heating and cooling in building design. Living close enough to walk or bike to work or to use public transportation instead of driving alone will save much more energy than would better windows. (Two common criticisms of scientists is that their training is too narrow and that they do not consider the social impacts of their work.)

This book begins with a chapter on the need for green chemistry, including the toxicity of chemicals and the need for minimization of waste. The next three chapters deal with the methods that are being studied to replace some especially noxious materials. Chapters 5–8 cover various ways to improve separations and to reduce waste. Chapter 9 continues this theme and switches to combinations of biology and chemistry. Chapter 10 discusses many optical resolutions done by enzymes or whole cells. Chapter 11, on agrochemicals, continues the biological theme. Chapters 12 through 15 cover various aspects of sustainability, such as where energy and materials will come from if not from petroleum, natural gas, or coal; how to promote sustainability by making things last longer; and the role of recycling in reducing demands on the natural resource base. The last three chapters try to answer the question that arises at nearly every meeting on green chemistry: Why is it taking so long for society to implement new knowledge of how to be green?

The topics within these chapters are not confined to specific areas. For example, cyclodextrins are discussed not only under supported reagents, but also under separations by inclusion compounds or under chemistry in water. Surfactants have been placed under doing chemistry in water. They are also mentioned as alternatives for cleaning with organic solvents and in the discussion on materials from renewable sources. An effort has been made to cross-reference such items. But for the reader who has any doubt about where to find an item, a comprehensive index is included.

The industrial chemistry on which much green chemistry is based may be foreign to many in academia, but many good references are available.[15] There is much more emphasis in industrial chemistry on catalysis and the organometallic mechanisms that often go with them. Improved catalysts are often the key to improved productivity, using less energy and generating less waste.[16] Again, many good sources of information are available.[17] In addition, there are two books on the chemistry and biology of water, air, and soil.[18]

Each chapter in *Introduction to Green Chemistry* lists recommended reading, consisting primarily of review papers and portions of books and encyclopedias. This allows detailed study of a subject. The introductions to current journal articles often contain valuable references on the status of a field and trends in research. However, the traditional grain of salt should be applied to some news items from trade journals, which may be little more than camouflaged advertisements.

The examples in the book are drawn from throughout the world. In the student exercises that accompany each chapter, readers are often asked to obtain data on their specific location. In the United States, one need look no further than the local newspaper for the results of the toxic substances release inventory. This data can also be found on Web sites of the U.S. Environmental Protection Agency (http://www.epa.gov/opptintr/tri) and state environmental agencies,[19] as well as on comparable Web sites of other countries (e.g., http://www.unweltbundesamt.de for the German environmental agency). Non-governmental organizations also post some of this data (e.g., the Environmental Defense "Chemical Scorecard," www.scorecard.org, and the Committee for the National Institute for the Environment, www.cnie.org). The data available on the Internet is growing rapidly.[20] The environmental compliance records of more than 600 U.S. companies can be found at http://es.epa.gov/oeca/sfi.index.html. A catalog of U.S. Environmental Protection Agency documents can be found at http://www.epa.gov/ncepihoni.catalog.html. Data on the toxic properties of chemicals can be found at toxnet.nlm.nih.gov and http://www.chemquik.com.

Some of the exercises sample student attitudes. Others call for student projects in the lab or in the community. Some of the questions are open-ended in the sense that society has yet to find a good answer for them, but they leave room for discussion.

Those using this volume as a textbook will find field trips helpful. These might include visits to a solar house, a farm using sustainable agriculture, a tannery, a plant manufacturing solar cells, and so on. Although the course at the University of Delaware had no laboratory, one would be useful to familiarize students with techniques of green chemistry that they might not encounter in the regular courses. These might include the synthesis, characterization, and evaluation of a zeolite, running a reaction in an extruder, using a catalytic membrane reactor, adding ultrasound or microwaves to a reaction, making a chemical by plant cell culture, doing biocatalysis, making a compound by organic electrosynthesis, running a reaction in super-critical carbon dioxide, and use of a heterogeneous catalyst in a hot tube. Ideally, students would run a known reaction first, then an unknown one of their own choice (with appropriate safety precautions). Such a lab would require the collaboration of several university departments.

There is a myth that green chemistry will cost more. This might be true if something was to be added at the smokestack or outlet pipe. However, if the whole process is examined and rethought, being green can save money. For example, if a process uses solvent that escapes into the air, there may be an air pollution problem. If the solvent is captured and recycled to the process, the savings from not having to buy fresh solvent may be greater than the cost of the equipment that recycles it. If the process is converted to a water-based one, there may be additional savings.

It is hoped that many schools will want to add green chemistry to their curricula. Sections of this book can be used in other courses or can be used by companies for in-house training. The large number of references makes the book a guide to the literature for anyone interested in a sustainable future.

Albert S. Matlack

REFERENCES

1. J. Lubchenko, *Science*, 1998, *279*: 491.
2. (a) R. Goodland and H. Daly. *Ecol. Appl.*, 1996, *6*: 1002. (b) A. Merkel, *Science*, 1998, *281*: 336.
3. A.S. Matlack, *Green Chem.*, 1999, *1*(1): Gl7.
4. J.E. Andrews, P. Brimblecombe, T.D. Jickells, P.S. Liss, eds. *Introduction to Environmental Chemistry*. Blackwen Science, Cambridge, MA 1995.
5. C.L. Baird, *Environmental Chemistry*, W.H. Freeman, New York, 1995.
6. D.G. Crosby, *Environmental Toxicology and Chemistry*, Oxford University Press, New York, 1998.
7. R.S. Gupta, *Environmental Engineering and Science: An Introduction*, Government Institutes, Rockville, MD, 1997.
8. D.L. Macalady, *Perspectives in Environmental Chemistry*, Oxford University Press, Oxford, England, 1997.
9. S.E. Manahan, *Environmental Chemistry*, 6th ed., Lewis Publishers, Boca Raton, FL, 1994; *Fundamentals of Environmental Chemistry*, Lewis Publishers, Boca Baton, FL, 1993.
10. T.C. Spiro, W.M. Stigliani, *Chemistry of the Environment*, Prentice Hall, Upper Saddle River, NJ, 1996.
11. J. Clark, *Chem. Br.*, 1998, *34*(10): 43.
12. (a) B. Hileman, *Chem. Eng. News*, luly 20, 1998, 41; (b) J. Darmstadter, *Chem. Eng. News*, August 10, 1998, 6; (c) N.E. Gallopoulos, *Chem. Eng. News*, August 10, 1988, 7; (d) T.E. Graedel, and B.R. Allenby, *Industrial Ecology*, 1995; *Design for Environment*, 1996; *Industrial Ecology and the Automobile*, 1996, all from Prentice Hall, Paramus, NJ. (e) B.R. Allenby, *Industrial Ecology: Policy Framework and Implementation*, Prentice Hall, Paramus, NJ, 1998.
13. A.W. Gessner, *Chem. Bag. Prog.*, 1998, *94*(12), 59.
14. (a) Anon., *Preprints ACS Div. Environ. Chem.*, 1994, *34*(2), pp. 175–431. (b) P.T. Anastas, C.A. Farris, eds. *Benign by Design: Alternative Synthetic Design for Pollution Prevention*, ACS Symp. 577, Washington, D.C., 1994; (c) P.T. Anastas, T.C. Williamson, eds. *Green Chemistry: Designing Chemistry for the Environment*, ACS Symp. 626, Washington, D.C., 1996; (d) S.C. DeVito,

R.L. Garrett, eds. *Designing Safer Chemicals: Green Chemistry for Pollution Prevention*, ACS Symp. 640, Washington, D.C., 1996; *Chemtech*, 1996, *26*(11): 34. (e) J.J. Breen, M.J. Dellarco, eds. *Pollution Prevention in Industrial Processes: The Role of Process Analytical Chemistry*, ACS Symp. 508, Washington, D.C., 1994. (f) P.T. Anastas, T.C. Williamson, eds., *Green Chemistry: Frontiers in Benign Chemical Syntheses and Processes*, Oxford University Press, Oxford, England, 1998.

15. (a) W. Buchner, R. Schliebs, C. Winter, K.H. Buchel, *Industrial inorganic Chemisty*, VCH, Weinheirn, 1989; (b) K. Weisserrnel, H.-J. Arpe, *Industrial Organic Chemistry*, 3rd ed., VCH, Weinheim, 1997. (c) P. J. Chenier, *Survey of Industrial Chemistry*, 2nd ed., VCH, Weinheim, 1992; (d) H. Wittcoff, B. Reuben, *Industrial Organic Chemicals*, 2nd ed., John Wiley, New York, 1996; (e) J.I. Kroschwitz, ed., *Kirk-Othmer Encyclo. Chem. Technol.* 4th ed., John Wiley, 1991; (f) J.I. Kroschwitz, ed., *Encyclo. Polymer Sci. Eng.*, 2nd ed., John Wiley, New York, 1985–1989; (g) W. Gerhartz, ed., *Ulimann's Encyclo. Ind. Chem.*, 5th ed., VCH, Weinheim, 1985.
16. J. Haber, *Pure. App. Chem.*, 1994, *66*: 1597
17. (a) J.N. Armor, *Environmental Catalysis*, A.C.S. Symp. 552, Washington, D.C., 1993; (b) O.W. Parshall, S.D. Ittel, *Homogeneous Catalysis*, 2nd ed., John Wiley, New York, 1992; (c) B.C. Gates, *Kirk-Othmer Encyclo. Chem. Technol.*, 4th ed., 1993, *5*, 320; (d) W.R. Moser, D.W. Slocum, eds., *Homogeneous Transition Metal-Catalyzed Reactions*, Adv Chem 230, American Chemical Society, Washington, D.C., 1992; (e) J. P. Colliman, L. S. Hegedus, *Principles and Applications of Organotransition Chemistry*, University Science Books, Mill Valley, CA, 1980; (f) C. Elschenbroich, A. Salzer, *Organometallics—A Concise Introduction*, VCH, Weinheim, 1992; (g) G. Braca, *Oxygenates by Homologation or CO Hydrogenation with Metal Complexes*, Kluwer Academic Publishers, Dordrect, the Netherlands, 1994; (h) F.P. Pruchnik, *Organometallic Chemistry of the Transition Elements*, Plenum, NY, 1993.
18. (a) J. Tolgyessy, ed., *Chemistry and Biology of Water, Air, Soil—Environmental Aspects*, Elsevier, Amsterdam, 1993; (b) B. Evangelou, *Environmental Soil and Water Chemistry*, Wiley, NY, 1998.
19. A. Kumar, K. Desai, K. Kumar, *Environ. Prog.*, 1998, *17*(2), S11. (gives a directory of World Wide Web sites)
20. (a) S.M. Bachrnch, ed., *The Internet: A Guide for Chemists*, American Chemical Society, Washington, D.C., 1996; (b) T. Murphy, C. Briggs-Erickson, *Environmental Guide to the Internet*, Government Institutes, Rockville, MD, 1998; (c) K. O'Donnell, L. Winger, *The Internet for Scientists*, Harwood, Amsterdam, 1997; (d) L.E.J. Lee, P. Chin, D.D. Mosser, *Biotechnol. Adv.*, 1998, *16*: 949; (e) B.J. Thomas, *The Wand Wide Web for Scientists and Engineers*, SAE Intemational, Warrendale, PA, 1998; (f) S. Lawrence, C.L. Giles, *Science*, 1998, *280*: 98; (g) R.E. Maizell, *How To Find Chemical Information: A Guide for Practicing Chemists, Educators and Students*, Wiley, NY, 1998.

CHAPTER 1

Introduction

This chapter will consider what is toxic, what is waste, why accidents occur, and how to reduce all of these.[1]

1.1 GENERAL BACKGROUND

In the glorious days of the 1950s and 1960s, chemists envisioned chemistry as the solution to a host of society's needs. Indeed, they created many of the things we use today and take for granted. The discovery of Ziegler–Natta catalysis of stereospecific polymerization alone resulted in major new polymers. The chemical industry grew by leaps and bounds until it employed about 1,027,000 workers in the United States in 1998.[2] By 2007, this number had dropped to 872,200. Some may remember the DuPont slogan, "Better things for better living through chemistry." In the Sputnik era, the scientist was a hero. At the same time, doctors aided by new chemistry and antibiotics felt that infectious diseases had been conquered.

Unfortunately, amid the numerous success stories were some adverse outcomes that chemists had not foreseen. It was not realized that highly chlorinated insecticides such as DDT [1,1-bis(4-chlorophenyl)-2,2,2-trichloroethane] (**1.1** Schematic), also known as dichlorodiphenyltrichloroethane (made by the reaction of chloral with chlorobenzene), would bioaccumulate in birds. This caused eggshell thinning and nesting failures, resulting in dramatic population declines in species such as peregrine falcon, bald eagle, osprey, and brown pelican. Rachel Carson[3] was one of the first to call attention to this problem. Now that these insecticides have been banned in the United States, the species are recovering. Some are still made and used in other countries, but may return to the United States by long-range aerial transport (e.g., from Mexico). (Compounds applied to plants, building surfaces, and such may evaporate and enter the atmosphere where they may remain until returned to the ground at distant points by rain or by cooling of the air.[4]) DDT is still made in Mexico, China, India, and Russia. A global treaty to ban these persistent pollutants is being sought.[5]

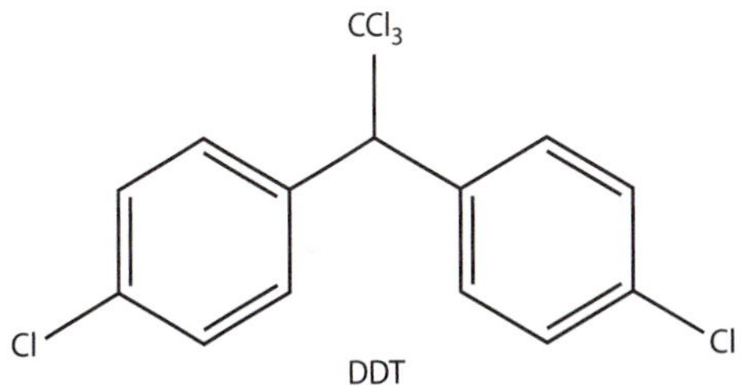

CCl_3

Cl Cl

DDT

1.1 Schematic

It was also not appreciated that these compounds and other persistent highly chlorinated compounds, such as polychlorinated biphenyls (PCBs), can act as estrogen mimics.[6,7] Surfactants such as those made from alkylphenols and ethylene oxide are also thought to do this, although perhaps to a lesser extent. The effects are now showing up in populations of native animals, raising questions about possible effects in humans. A program is being set up to screen 86,000 commercial pesticides and chemicals for this property.[8]

Thalidomide (**1.2** Schematic) was used to treat nausea in pregnant women from the late 1950s to 1962. It was withdrawn from the market after 8000 children in 46 countries were born with birth defects.[9]

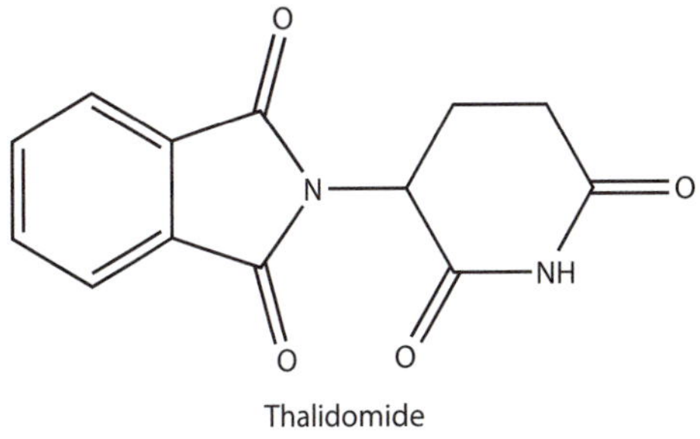

O N O NH O O

Thalidomide

1.2 Schematic

The compound has other uses as a drug as long as it is not given to pregnant women. In Brazil, it is used to treat leprosy. Unfortunately, some doctors there have not taken the warning seriously enough and several dozen deformed births have occurred.[10] The U.S. Food and Drug Administration (FDA) has approved its use for treating painful inflammation of leprosy.[11] It also inhibits human immunodeficiency virus (HIV) and can prevent the weight loss that often accompanies the acquired immunodeficiency syndrome (AIDS). Celgene is using it as a lead compound for an anti-inflammatory drug and is looking for analogues with reduced side effects.[12] The analogue below (**1.3** Schematic) is 400–500 times as active as thalidomide. Revlimid (**1.4** Schematic) is approved for treating multiple myeloma.[13]

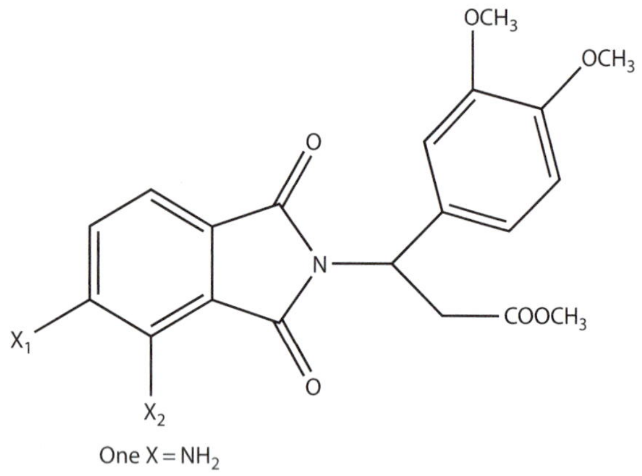

One X = NH_2

1.3 Schematic

Revlimid

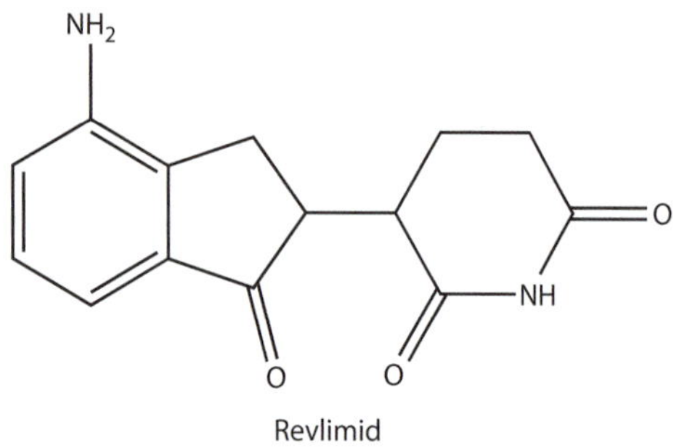

1.4 Schematic

Chlorofluorocarbons were developed as safer alternatives to sulfur dioxide and ammonia as refrigerants. Their role in the destruction of the ozone layer was not anticipated. Tetraethyl lead was used as an antiknock agent in gasoline until it was learned that it was causing lead poisoning and lowering the intelligence quotient (IQ) in children. We have still not decided what to do with the waste from nuclear power plants, which will remain radioactive for longer than the United States has been in existence. Critics still question the advisability of using the Yucca Mountain, Nevada, site.[14] They say that the finding of ^{36}Cl from atomic bomb tests in the 1940s at the depth of the repository indicates that surface water can get into this site.[15]

Doctors did not anticipate the development of drug-resistant malaria and tuberculosis. The emergence of Legionnaire's disease, Lyme disease, AIDS, Hantavirus, and Ebola virus was not anticipated. Most drug companies are still unwilling to tackle tropical diseases because they fear that poor people afflicted with diseases will not be able to pay for the drugs.[16]

Today, there is often public suspicion toward scientists.[17] Some picture a mad chemist with his stinks and smells. There is a notion among some people "that science is boring, conservative, close-minded, devoid of mystery, and a negative force in society."[18] Chemophobia has increased. Many people think that chemicals are bad and "all natural" is better, even though a number of them do not know what a chemical is. There is a feeling that scientists should be more responsible for the influence of their work on society. Liability suits have proliferated in the United States. This has caused at least three companies to declare bankruptcy: Johns Manville for asbestos in 1982, A. H. Robins for its "Dalkon Shield" contraceptive device in 1985, and Dow Corning for silicone breast implants in 1995.[19] Doctors used to be respected pillars of their communities. Today they are subjects of malpractice suits, some of which only serve to increase the cost of health care. Medical implant research is threatened by the unwillingness of companies such as DuPont and Dow Corning to sell plastics for the devices to implant companies.[20] The chemical companies fear liability suits. Not long ago drug companies became so concerned about lawsuits on childhood vaccines that many were no longer willing to make them. Now that the U.S. Congress has passed legislation limiting the liability, vaccine research is again moving forward. The lawsuits had not stimulated research into vaccines with fewer side effects, but instead had caused companies to leave the market.

1.2 TOXICITY OF CHEMICALS IN THE ENVIRONMENT

The public's perception of toxicity and risk often differs from that found by scientific testing.[21] The idea that "natural"[22] is better than "chemical" is overly simplistic. Many chemicals that are found in nature are extremely potent biologically. Mycotoxins are among these.[23] Aflatoxins (**1.5** Schematic) were discovered when turkeys fed moldy groundnut (peanut) meal became ill and died. They are among the most potent carcinogens known.

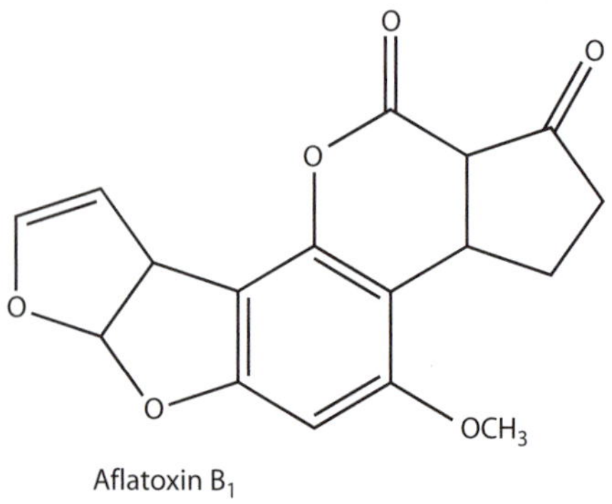

Aflatoxin B_1

1.5 Schematic

Vikings went berserk after eating derivatives of lysergic acid (**1.6** Schematic) made by the ergot fungus growing on rye.

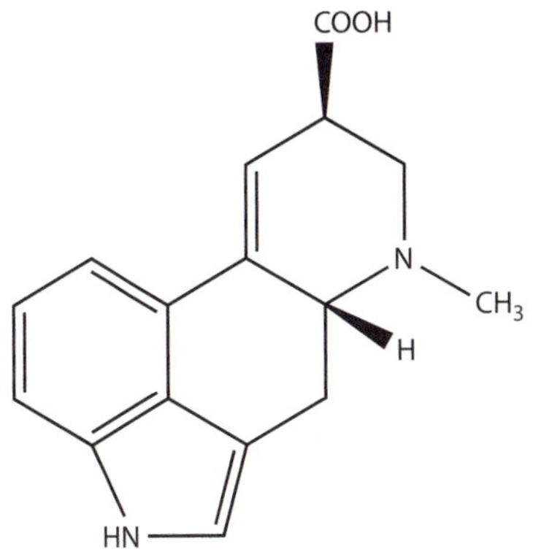

Lysergic acid

1.6 Schematic

Some *Amanita* mushrooms are notorious for the poisons that they contain. A Japanese fish delicacy of globefish or other fishes may contain potent poisons (such as tetrodotoxin; **1.7** Schematic)[24] if improperly prepared.[25] Tainted fish cause death in 6–24 h in 60% of those who consume it. They die from paralysis of the lungs. Oysters may also contain poisons acquired from their diet. The extract of the roots of the sassafras tree (*Sassafras albidum*) (**1.8** Schematic) is used to flavor the soft drink "root beer" but contains the carcinogen safrole, which must be removed before use.

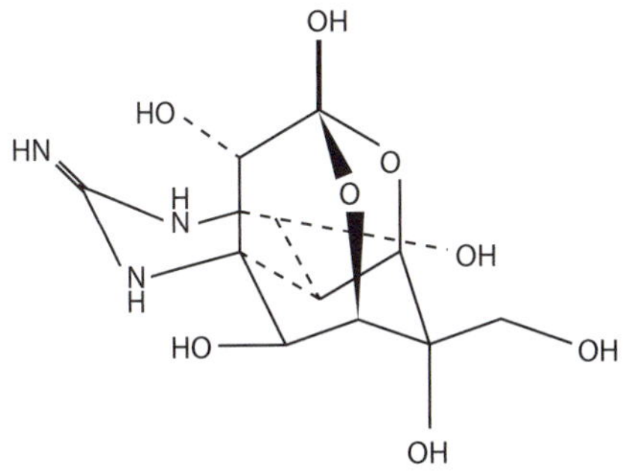

Tetrodotoxin

1.7 Schematic

Safrole

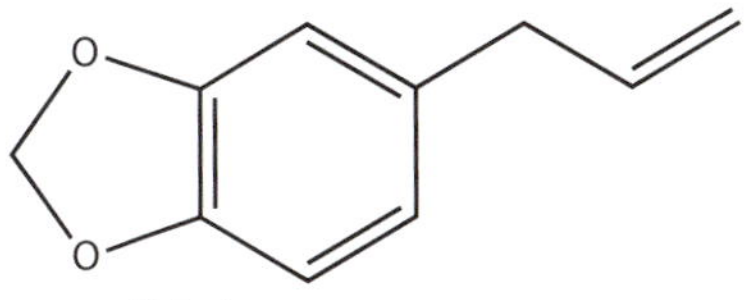

1.8 Schematic

The natural insecticide Sabadilla, which is popular with organic farmers, contains 30 alkaloids present at a level of 3–6% in the seeds of *Schoenocaulon officinale*.[26] It can affect the cardiovascular system, respiration, nerve fibers, and skeletal muscles of humans. Gastrointestinal symptoms and hypotension may also result from its ingestion.

The U.S. Congress added the Delaney clause to the Food, Drug, and Cosmetic Act in 1958.[27] The clause reads: "No additive shall be deemed to be safe if it is found to induce cancer when ingested by man or animal, or if it is found, after tests which are appropriate for the evaluation of the safety of food additives, to induce cancer in man or animal." It does not cover natural carcinogens in foods or environmental carcinogens, such as chlorinated dioxins and polychlorinated biphenyls.[3] Some Americans feel that food additives are an important source of cancer, and that the Delaney clause should have been retained. Bruce Ames, the father of the Ames test for mutagens, disagrees.[28] The Delaney clause was repealed on August 3, 1996.[29]

Ames feels that, instead of worrying so much about the last traces of contaminants in foods, we should focus our attention on the real killers.

Annual Preventable Deaths in the United States[30]

Active smoking	430,700
Overweight and sedentary	400,000
Alcoholic beverages	100,000
Passive smoking	53,000
Auto accidents	43,300
AIDS	37,500
Homicides	34,000
Suicides	30,575
Falls	14,900
Drowning	4400
Fires	3200
Cocaine	4202
Heroin and morphine	4175
Bee stings	3300
Radon, to nonsmokers	2500
Lightning	82
Recalcitrant farm animals	>20
Dog mauling	17
Snakebite	12

The World Health Organization (WHO) estimates that the number of smoking deaths worldwide is about 3 million/year.[31] Cigarette use is increasing among American college students, despite these statistics.[32] Indoor radon contributes to about 12% of the lung cancer deaths in the United States each year.[33] Infectious diseases cause 37% of all deaths worldwide.[34] Many of these could be prevented by improved sanitation. There are 3 million pesticide poisonings, including 220,000 fatalities and 750,000 chronic illnesses, in the world each year. Smoke from cooking with wood fires kills 4 million children in the world each year.

It is estimated that perhaps 80% of cancers are environmental in origin and related to lifestyle. There is "convincing" evidence of a connection between excess weight and cancers of the colon, rectum, esophagus, pancreas, kidney, and breast in postmenopausal women.[35] (There are genetic factors, now being studied by the techniques of molecular biology, that predispose some groups to heightened risk, e.g., breast cancer in women.[36]) In addition to cancers caused by tobacco and ethanol, there are those caused by being overweight,[37] too much sun, smoked foods, foods preserved with a lot of salt, and viruses (for cancers of the liver and cervix). Consumption of the blue-green alga, *Microcystis*, has increased liver cancer in China.[38] Many species of cyanobacteria produce the neurotoxin (**1.9** Schematic).[39]

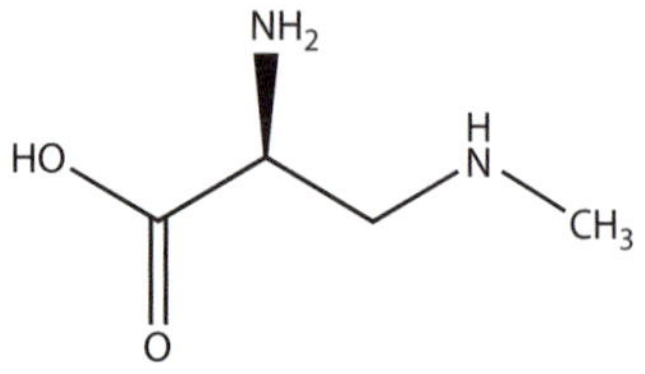

1.9 Schematic

Perhaps the biggest killer is sodium chloride, a compound necessary for life, that plays a role in the regulation of body fluids and blood pressure.[40] It raises the blood pressure of many of the 65 million Americans with hypertension, increasing the risk of osteoporosis, heart attack, and stroke.[41] The U.S. National Academy of Sciences suggests limiting the consumption of sodium chloride to 4 g/day (1.5 g of sodium).[42] This means cutting back on processed foods (source of 80% of the total), such as soups, frozen dinners, salted snacks, ham, soy sauce (18% salt), ready-to-eat breakfast cereals, and others. The Dietary Approaches to Stop Hypertension (DASH) diet, which reduces salt, has lots of fruits and vegetables, and minimizes saturated fat from meat and full-fat milk, is as effective as drugs in lowering blood pressure in many cases. However, this may not be the whole story. If there is an adequate intake of calcium, magnesium, and potassium, together with fruits and vegetables in a low-fat diet, the sodium may not need to be reduced, as shown in a 1997 study.[43] The DASH diet also reduces the risk of heart disease by 24% and of stroke by 18%.[44] People who do not smoke, are active physically, drink alcohol in moderation, and eat at least five servings of fruits and vegetables a day live 14 years longer on average than those who do none of these.[45] In the United States, snacks and soft drinks have tended to supplant nutrient-rich foods, such as fruits, vegetables, and milk. Not eating fruits and vegetables poses a greater cancer risk than traces of pesticides in foods.[46] Fruits and vegetables often contain natural antioxidants,[47] such as the resveratrol (**1.10** Schematic) found in grapes. Resveratrol inhibits tumor initiation, promotion, and progression.[48] Sirtus Pharmaceuticals was founded in 2004 to develop derivatives of resveratrol for age-related diseases.[49]

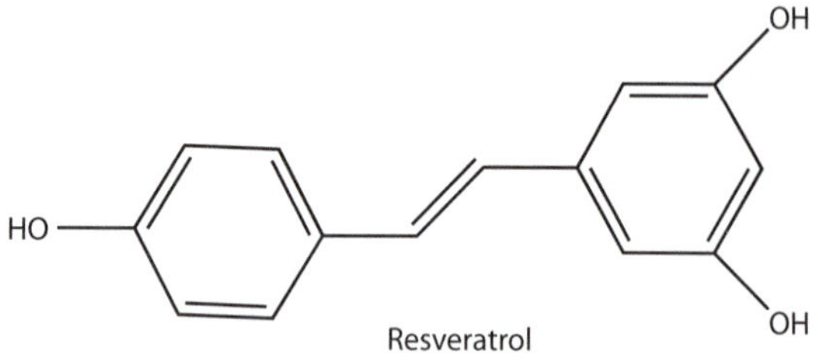

Resveratrol

1.10 Schematic

Antioxidants have also reduced atherosclerotic heart disease.[50] Thus, foods contain many protective substances, as well as some antinutrients, such as enzyme inhibitors and natural toxins.[51]

A U.S. National Research Council report concludes that natural and synthetic carcinogens are present in human foods at such low levels that they pose little threat.[52] It points out that consuming too many calories as fat, protein, carbohydrates, or ethanol is far more likely to cause cancer than consuming the synthetic or natural chemicals in the diet. However, it also mentions several natural substances linked to increased cancer risk: heterocyclic amines formed in the overcooking of meat; nitrosoamines, aflatoxins, and other mycotoxins.[53] Typical of the heterocyclic amine mutagens are compounds **1.11** Schematic and **1.12** Schematic, the first from fried beef and the second from broiled fish.[54]

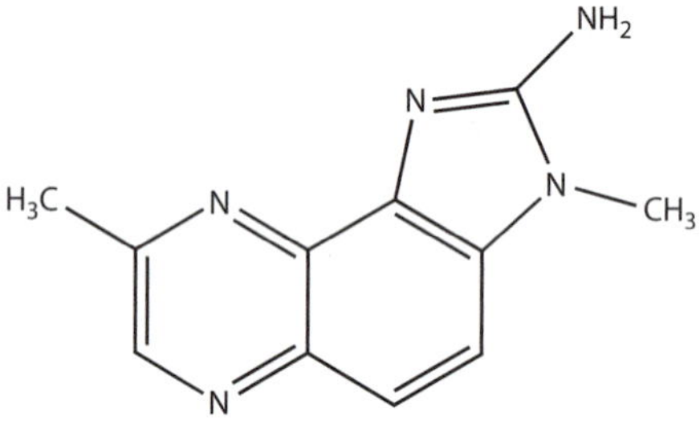

1.11 Schematic

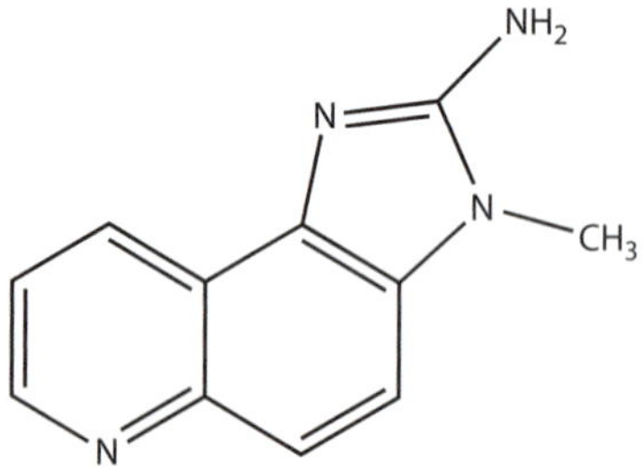

1.12 Schematic

Deep frying with soybean, sunflower, and corn oil allows air oxidation of linoleates to highly toxic (2E)-4-hydroxy-2-nonenal (**1.13** Schematic), which has been linked to Parkinson's and Alzheimer's diseases.[55]

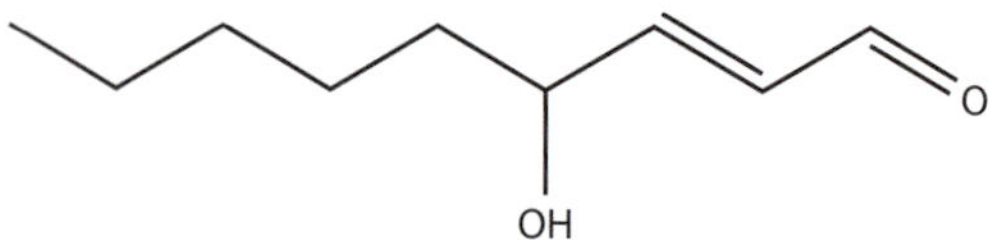

1.13 Schematic

Two reviews cover the incidence of cancer and its prevention by diet and other means.[56] Cancer treatments have had little effect on the death rates, so that prevention is the key.[57]

Being sedentary is a risk factor for diseases such as heart attack and late-onset diabetes. One-third of adult

Americans are obese, perhaps as much the result of cheap gasoline as the plentiful supply of food.[58] Obesity-related complications result in 400,000 premature deaths in the United States each year. This is less of a problem in most other countries. For example, the incidence of obesity in the United Kingdom is 20%.

Prevention of disease is under-used.[59] Needle exchange programs could prevent 17,000 AIDS infections in the United States each year. Vaccines are not used enough. For example, only 61% of the people in Massachusetts are fully vaccinated. Only 15–30% of elderly, immunocompromised persons, and those with pulmonary or cardiac conditions have been vaccinated against pneumonia. A 1998 study, in New Jersey and Quebec, of patients older than 65 who had been prescribed cholesterol-lowering drugs found that, on average, the prescription went unfilled 40% of the year. Good drug compliance lowered the cholesterol level by 39%, whereas poor compliance lowered it by only 11%.[60]

Many persons take unnecessary risks by using alternative, rather than conventional, medicine.[61] Although some of it works (e.g., a Chinese herbal medicine for irritable bowel syndrome), much of it is ineffective.[62] The use of herbal extracts and dietary supplements may help some people, but many such materials are ineffective and may be dangerous.[63] One lot of "Plantago" contained digitalis that sent two people to hospital emergency rooms with heart blockage. Cases of central nervous system depression and heavy metal poisoning have also been reported. One child died while being treated with herbal extracts in a case where conventional medicine might have saved her life. Of 260 Asian patent medicines bought in California stores, 83 contained undeclared pharmaceuticals or heavy metals, and 23 contained more than one adulterant. Tests of 10 brands of St. John's Wort, a popular herbal antidepressant, found that two had 20% of the potency listed on the label, six had 50–90%, and two had 30–40% more than that listed.[64] The problem is that the U.S. Dietary Supplement Health and Education Act of 1994, said to have been passed by strong industry lobbying, does not allow the FDA to regulate these materials. Herbal extracts are also used in many other countries.[65] The most research on them has been done in Germany, where the Bundesinstitut fur Arzneimittel and Medizinproduckte regulates them more effectively than is done in the United States. Many of the drugs used by conventional medicine originated in folk medicine. These were all tested carefully for safety and efficacy, and provided in consistent standardized amounts, before acceptance by conventional medicine. The National Institutes of Health Office of Dietary Supplements in the United States has an international database at http://dietary/supplements.info.nih.gov. See also www.amfoundation.org/herbmed.htm. Ephedra was banned in the United States in 2004 because of its dangers.[66] Asian governments are screening traditional medicines for safety and effectiveness in the search for new drugs.[67]

An additional note of caution must be added to this discussion. Tryptophan (**1.14** Schematic), an essential amino acid for humans, was sold in health food stores as a sleeping pill.[68] It was taken off the market in 1989 after a lot made by a new microorganism at Showa Denko caused an outbreak of a rare blood disease that killed 39 people.[69] Although the material was 99% pure, it contained over 60 trace contaminants.

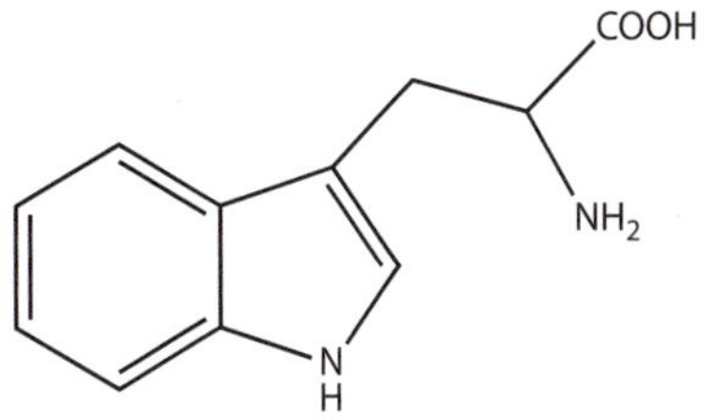

1.14 Schematic

The deadly contaminant compounds **1.15** Schematic and **1.16** Schematic were present at about 0.01%. The moral is to test products from new processes on animals first. Some samples of over-the-counter 5-hydroxy-L-tryptophan contained impurities that caused eosinophila-myalgia syndrome.[70]

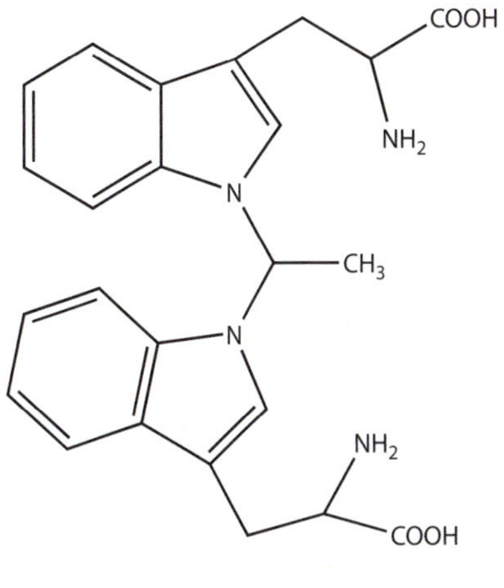

1.15 Schematic

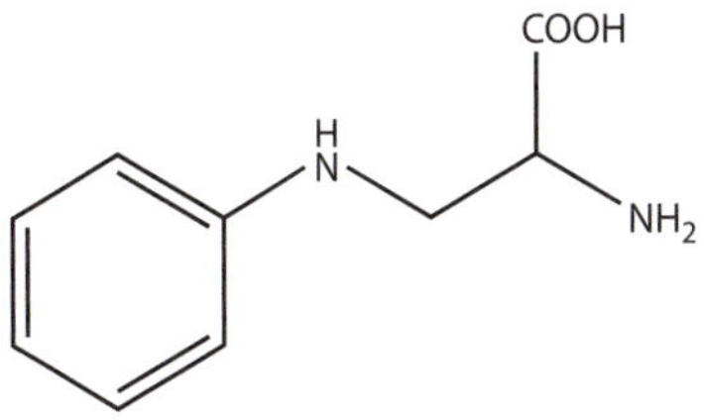

1.16 Schematic

Another sad saga resulted from the sale, by itinerant salesmen in Spain, of aniline-denatured rapeseed oil for cooking.[71] The toxic oil syndrome affected 20,000 people, killing 839. The toxic compound is thought to have been compound **1.17** Schematic.

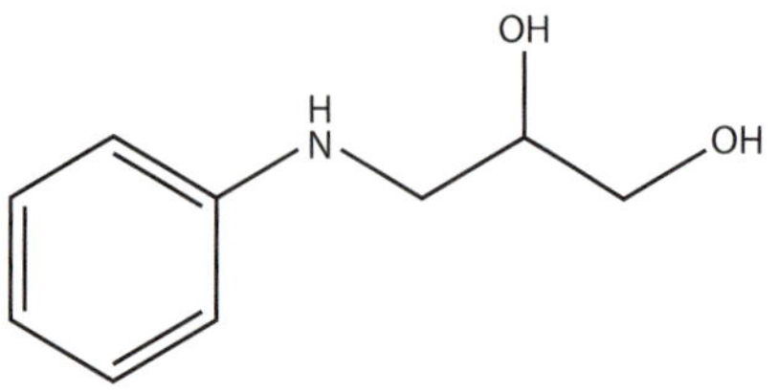

3-Phenylamino-1,2-propanediol

1.17 Schematic

Three cases of dangerous substitutions in China occurred in 2006–2008. The addition of melamine to pet food, probably to make it appear to have a higher protein content, killed dozens of dogs and cats.[72] A complex of melamine and cyanuric acid (**1.18** Schematic) caused the deaths.

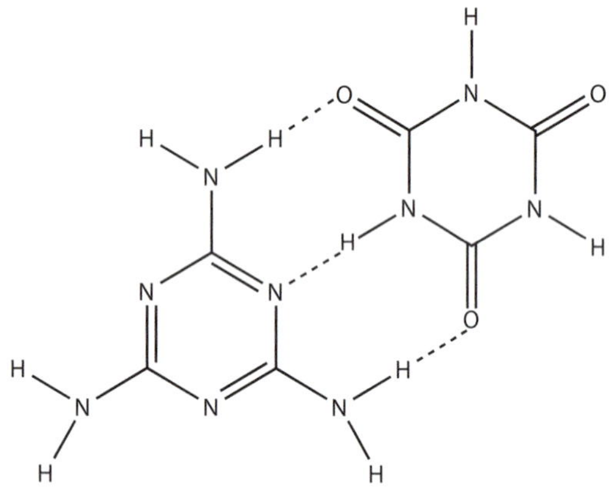

1.18 Schematic

The use of diethylene glycol in place of the more expensive 1,2-propanediol in cough syrup caused several deaths in China and Panama.[73] Heparin (**1.19** Schematic) contaminated with oversulfated heparin killed 66 patients.[74]

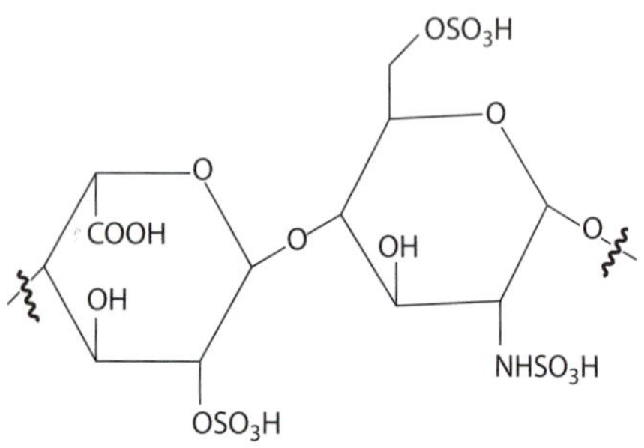

1.19 Schematic

Research on the long-term effects of low-level pollution continues. Air pollution by ozone, sulfur dioxide, and particulates in Britain kills 24,000 people annually.[75] It is estimated that exposure to diesel exhaust over a 70-year lifetime will cause 450 cases of cancer per million people in California.[76] Small particulates ($<2.5\ \mu m$) in air increase mortality, cardiovascular disease, pneumonia, and chronic obstructive pulmonary disease.[77] Long thin carbon nanoparticles exhibit asbestos-like effects in mice.[78] Epidemiologists continue to investigate clusters of diseases such as the two to four times higher incidence of neural tube and certain other defects in children born within ¼ mile of Superfund sites in California,[79] the lower birth weight and prematurity of infants born to women living next to the Lipari landfill in New Jersey,[80] the brain cancers in researchers at the Amoco laboratory in Napierville, Illinois and at the University of Maine,[81] and a cancer cluster near a Superfund site at Toms River, New Jersey. A U.S. General Accounting Office report suggests that the list of Superfund sites may have to be doubled because of ground and drinking water contamination.[82] Some persons may have multiple chemical sensitivities to low levels of pollutants.[83]

Children are more sensitive to chemicals than adults.[84] Childhood cancer, autism, attention deficit hyperactivity disorder, defects in reproductive tracts and asthma are increasing. The fetus may be the most sensitive.[85] A program to monitor a sample of children from conception to age 21 has been proposed, but had not started by 2008.[86] Pregnant women who smoke tend to have more low birth weight babies and those who drink alcoholic beverages risk fetal alcohol syndrome. Environmental standards are often set using a statistical approach.[87] The U.S. Environmental Protection Agency (EPA) sets pesticide limits at 1% of the level found to have no effects in animals. A 1996 law to protect children may lower this level another 10-fold. Some pesticide makers have gone to tests of pesticides on adult humans in the United Kingdom, presumably in an effort to obtain higher limits.[88] Some people consider this unethical. Chemical safety is an international challenge.[89] Complete health effects data are available for only about 7% of the chemicals produced in more than 1 million lb annually.[90] The U.S. EPA and the American Chemistry Council (formerly the Chemical Manufacturers Association) have reached an agreement to test 3000 chemicals in an effort to speed up the work being done by the Organization for Economic cooperation and Development in Europe. At first the Synthetic Organic Chemical Manufacturers Association (now Society of Chemical Manufactures and Affiliates) did not agree with the way in which this testing will be done,[91] but agreed to it later. Animal rights activists have objected to the tests because animals will be used in at least some of the testing.[92] Data on somewhat more than half of these had been received by 2008.[93] There was also the problem of about 270 orphan chemicals that no company was willing to sponsor. The EPA hopes to assess 9000 chemicals made in amounts greater than 25,000 lb per year by 2012.[94] Canada, Mexico, and the United States have agreed to cooperate in the assessment of chemicals.[95] The adoption of REACH (Registration, Evaluation and Authorization of Chemicals) by the European Union (EU) in December 2006 promises to help establish the toxicities of many chemicals to supplement what is being done in North America.[96]

The precautionary principle is applied much more in the EU than in the rest of the world. If a chemical substance appears to be causing harm, it may need to be regulated before the final proof of harm is obtained.[97] Environmentalists like this approach. Chemical companies dislike it. Green chemistry chooses less toxic materials over more toxic ones[98] and tries to minimize the use of flammable, explosive, or highly reactive materials.[99] It is not always easy to figure out which is least toxic. Toxicities can vary with the species as well as with the age and sex of the animal. The manner of application may also vary. This is illustrated in the following by some LD_{50} (50% of the animals die) data on chemicals that are not very toxic and some that are very toxic[100]:

Ethanol: 10.6 g/kg for young rats; 7.6 g/kg for old rats
Malathion: 1.0 g/kg for female rats; 1.375 g/kg for male rats
Glyphosate: 4.873 g/kg for rats; 1.568 g/kg for mice
HCN average fatal dose for a human: 50–60 mg
Aflatoxin: 18.2 mg/50 g body weight day-old ducking
Acetone: 10.7 mL/kg orally for rats

This test probably would not pick up long-term effects caused by bioaccumulation. Populations of animals can also be decimated by chemical effects that do not kill the animals (e.g., the eggshell thinning of birds at the top of the food chain, feminization of males, and behavioral changes, such as not feeding or protecting the young, or losing the ability to avoid predators).

Many chemists alive today have worked with compounds later found to be carcinogens, neurotoxins, and so on. All chemicals need to be treated with due respect.[101]

1.3 ACCIDENTS WITH CHEMICALS

Chemists take pride in their ability to tame dangerous chemicals in an effort to make the things that society needs. In fact, some companies seek business by advertising their ability to do custom syntheses with such chemicals. [102]Aerojet Fine Chemicals offers syntheses with azides and vigorous oxidations.[103] Carbolabs offers custom syntheses with phosgene, fluorinating agents, and nitration. Custom syntheses with phosgene are also offered by PPG Industries, Hatco, Rhone-Poulenc, and SPNE (see Chapter 2). Pressure Chemical, Dynamit Nobel, and Rutgers suggest that they can do the hazardous reactions for others.[104] A hazardous reagent may be attractive for fine chemical syntheses if it gives a cleaner product with less waste or saves two or three steps. It may also be used because it is the traditional way of doing the job. Methods for screening unknown reactions for hazards have been summarized.[105]

Chemistry is a relatively safe occupation. (Underground coal mining is one of the most dangerous in the United States. A total of 47 coal miners lost their lives in 1995 from surface and underground mining.[106]) In the United States in 1996, the nonfatal injury and illness rate for chemical manufacturing was 4.8:100 full-time workers, compared with 10.6 for all of manufacturing. There were 34 deaths in the chemical industry, about 5% of those for all manufacturing.[107] The injury rate of the chemical industry in the United Kingdom fell to an all time low in 1997, 0.37 accidents per 100,000 h. This was in the middle of those for manufacturing industries, worse than the textile industry, but better than the food, beverage, and tobacco industries.[108] However, despite countless safety meetings and inspections and safety prizes, accidents still happen.[109] In chemistry, just as in airline safety, some of the accidents can be quite dramatic. There were 23,000 accidents with toxic chemicals in the United States in 1993–1995 (i.e., 7670/year compared with 6900/year in 1988–1992). The accidents in 1993–1995 included 60 deaths and the evacuation of 41,000 people. Statistics for the United States compiled by the Chemical Safety and Hazard Investigation Board reveal an average of 1380 chemical accidents resulting in death, injury, or evacuation each year for the 10 years before 1999.[110] Each year, these accidents caused an average of 226 deaths and 2000 injuries. About 60,000 chemical accidents are reported annually in the United States. The American Chemistry Council in the United States reported 793 fires, explosions, and chemical releases in 1995.[111] American Chemistry Council members reported about 2000 accidents in 2003, about the same as a decade before.[112] About 50% of global chemical incidents are in the United States and 30% in the EU.[113] The United Kingdom is second only to the United States in recorded incidents.

There is usually an investigation to determine the cause of an accident. The Chemical Safety and Hazard Investigation Board looked at 900 chemical accidents in 2007 and chose a few to investigate. In its 10-year history, it has completed 43 investigations with nine more in the works.[114] It is independent, but cannot write regulations or levy fines. It has found that about half of the 167 major chemical accidents over a 20-year period were caused by runaway reactions, which caused 108 deaths. Neither the U.S. Occupational Safety and Health Administration nor the U.S. EPA was willing to write new regulations on reactive chemicals. There are 2000 batch reactors in the EU with 110 runaway reactions per year.[115] The Board found that 281 dust explosions killed 119 workers. The U.S. Occupational Safety and Health Administration (OSHA) declined making new regulations to cover this even though it has regulations on dust in grain elevators.[114] A dust explosion in a sugar refinery in Georgia killed eight workers and injured more.[116]

Knowing the cause should help eliminate similar accidents in the future. Then engineering steps may be taken to produce a fail-safe system. These may involve additional alarms, interlocks (such as turning off microwaves before the oven door can be opened), automatic shut-offs if any leaks occur, and secondary vessels that would contain a spill. A special sump contained a leak of nerve gas when an O-ring failed at an incinerator at Tooele, Utah.[117] Underground storage tanks (made of noncorroding materials) may be fitted with a catchment basin around the fill pipe, automatic shut-off devices to prevent overfilling, and a double wall complete with an interstitial monitor[118] Clearly, such methods can work, but they have not reduced the overall incidence of accidents, as shown in the foregoing data. The best solution will be to satisfy society's needs with a minimum of hazardous chemicals.[119] A few of the many accidents will be discussed in the following to show how and why they occurred, together with some green approaches that could eliminate them. (Most of these are the ones described in the first edition because the nature and causes of accidents since then are very similar.)

The chemical industry received a wake-up call for improved safety when 40 tons of methyl isocyanate escaped from a pesticide plant into a densely populated area of Bhopal, India on December 3, 1984. This resulted in 5000 deaths and 200,000 injuries.[120] According to one reviewer,[121] the accident was the ultimate outcome of faulty technological design, years of poor management of an unprofitable and highly dangerous facility, years of ignoring an outrageously bad safety record on the part of both the parent company and the Indian government, inadequate education and training of the workforce, uncontrolled growth of an industrial population center, a nonexistent emergency response system, and the community's ignorance about the dangers in its midst. At the time of the accident, a refrigeration system, a temperature indicator, and a flare tower were not functioning.[122] Curiously, the Maharashtra Development Council (in the state next to the one in which Bhopal is located) advertised in 1998. The strengthening of the green movement and the growing protests against environmental pollution in many Western countries provide an opportunity for India to emerge as a major player in the global dye market. Given the global tendency to take full advantage of lower labor costs and less stringent effluent legislation, India has a competitive advantage.[123]

The explosion of the nuclear reactor at Chernobyl (spelling changed recently to Chornobyl) in the Ukraine on April 26, 1986 sent radioactive material as far away as Sweden.[124] The current death toll is 65. There has been a huge increase in childhood thyroid cancer, with cases as far as 500 km away.[125] (U.S. bomb tests have also increased the incidence of thyroid cancers in the western United States.[126]) There is a 30-km exclusion zone around the plant where no one is allowed to live. This was created by the evacuation of 135,000 people.[127] The accident is said to have happened "because of combination of the physical characteristics of the reactor, the design of the control rods, human error and management shortcomings in the design, and implementation of the safety experiment."

The world's worst radioactive contamination (twice that of Chernobyl) was at Mayak in Russia. This resulted from the explosion of a radioactive waste storage unit on September 29, 1957 and deliberate dumping of liquid waste into the Techa River in 1949–1956.[128]

Although no nuclear plants in the United States have exploded, there have been some scary incidents. The near melt-down at Three Mile Island, near Harrisburg, Pennsylvania, involved a faulty valve.[129] At the Salem, New Jersey, nuclear plants, there have been "repetitive equipment problems and personnel errors," the latter including manually overriding safety alarms.[130] The Nuclear Regulatory Commission shut down all three units at the site in 1995 for "repeated failures in their preventive maintenance programs" according to an ex-employee.[131] Fines of more than 700,000 dollars have been imposed on the utility.[132] The corporate culture was said to be "production, production, production." A company official said, "We had a lackadaisical, casual approach."[133] They were restarted after 2 years, presumably after the problems had been corrected. Since then, there have been outages due to a leak when an operator exceeded the design pressure for a coolant system and problems with a water intake system.[134] In the last two decades, these power plants have operated only 52% of the time, putting them among the 10 worst of the 110 nuclear plants in the United States. The U.S. General Accountability Office has questioned the effectiveness of the Nuclear Regulatory Commission in regulating such plants.[135] This is the second largest commercial generating station in the United States.

At the Peach Bottom nuclear plant in southeastern Pennsylvania, an operator was found fast asleep. Years later, in 2008, Exelon discontinued security service with Wackenhut after a guard was found sleeping on duty.[136] Federal laboratories in the United States have also had problems: explosion of a solution of hydroxylamine nitrate in dilute nitric acid at the Hanford plant in Richland, Washington (nearly identical with earlier ones at other federal facilities);[137] a series of safety problems, including an explosion, at Los Alamos, New Mexico;[138] and a series of seven incidents, including equipment problems, leading to the closure of the High Flux Isotope Reactor Facility at Oak Ridge, Tennessee.[139] The Oak Ridge problems included inadequate and ineffective communication among all parties, inattention to detail, a significant lack of trust and respect, high levels of frustration, and too much paperwork. New nuclear plants[140] are no longer being built in the United States, but some have been proposed. Large numbers are being planned for East Asia.[141] France generates 75% of its electricity from nuclear energy. There has been an increase in the incidence of leukemia within a 35-km radius of a nuclear waste-reprocessing plant at La Hague on the Normandy coast.[142] First Energy Operating Co. was fined $5.45 million for knowing about but not repairing a corroded head on a nuclear reactor in Ohio.[143]

The realization of the seriousness of global warming has led to many proposals for more nuclear plants.[144] Electricity need not be generated by nuclear power. Generating it from fossil fuels contributes to global warming. Producing it from renewable sources, such as wind, wave power, hydropower, geothermal, and solar energy, does not (see Chapter 15). Sweden has voted to phase out nuclear energy. The German government has agreed to phase out the country's 19 nuclear reactors.[145] It has been estimated that offshore wind power sources could produce electricity 40% more cheaply than the nuclear power stations planned for Japan.[146] Energy conservation can help a great deal in reducing the amount of energy needed.

The flammable gases used by the petrochemical industry have been involved in many accidents.[147] A fire and explosion following a leak of ethylene and isobutane from a pipeline at a Phillips plant in Pasadena, Texas, in 1989, killed 23 people and injured 130.[148] OSHA fined the company 4 million dollars. Accidents of this type can happen anywhere in the world where petrochemical industries are located. There have been explosions at an ethylene plant in Beijing, China,[149] at a Shell air separations plant in Malaysia,[150] at a Shell ethylene and propylene plant in Deer Park, Texas,[151] at a Shell propylene plant in Norco, Louisiana,[152] at a BASF plant in Ludwigshafen, Germany, that used pentane to blow polystyrene,[153] and at a Texaco refinery at Milford Haven, England.[154] The problems at the BASF plant could have been avoided by blowing the polystyrene with nitrogen or carbon dioxide instead of pentane. Two weeks before this incident at the BASF plant, four people were injured by a fire from a benzene leak.[155] BASF had two other accidents in October 1995, a polypropylene fire in Wilton, England, and spraying of a heat transfer fluid over the plant and adjacent town in Ludwigshafen, Germany.[156] There was an explosion in the hydrogenation area of the company's 1,4-butanediol plant in Geismar, Louisiana, on April 15, 1997,[157] which resulted from internal corrosion of a hydrogen line. This corrosion might have been detected by periodic nondestructive testing with ultrasound. The explosion and fires at the Texaco refinery have been attributed to making modifications in the plant but not training people on how to use the modified equipment, having too many alarms (2040 in the plant), insufficient inspection of the corrosion of the equipment, not learning from past experience, and reduced operator staffing.[158] Investigation of an explosion and fire at Shell Chemical's Belpre, Ohio, thermoplastic elastomer plant revealed that at the time of the accident, roughly seven times the normal amount of butadiene had inadvertently been added to the reactor. Alarms indicated that the reactor had been overcharged, but interlocks were manually overridden to initiate the transfer of raw materials into the reactor vessel, contrary to established procedures.[159] The federal government fined the company 3 million dollars for the various citations in relation to the accident.[160]

Some companies have repeated problems. BP's Texas City refinery has had 43 deaths in 33 years.[161] The March 23, 2005 fire that killed 15 and injured 170 was said to be due to aging infrastructure, overzealous cost cutting, inadequate design, and risk blindness. BP has also been involved in major pipeline spills in Alaska.[162] There has been a failure to monitor the corrosion of pipelines.

The difference between a minor incident and a major accident with fatalities may depend on the amount of wind at the time and whether or not a spark or a welding torch, which can ignite the mixture, is nearby. A switch from a base of petroleum and natural gas to renewable resources would eliminate many of these types of accidents. The use of more paper and fewer plastics will help (e.g., paper bags at supermarkets). Plastics made by biocatalysis [e.g., poly(lactic acid) and poly(3-hydroxybutyrate)] can be used instead of polypropylene. Plastics made from proteins from corn, milk, or soybeans may be able to replace many of those derived from petrochemicals. Acetic acid can be produced by fermentation rather than by the reaction of methanol with carbon monoxide (a toxic gas).[163] It may be a nuisance to have to mop up a leak or spill in a fermentation plant, but it is unlikely that there will be any fire or explosion (see Chapters 9 and 12).

Investigation of the causes of accidents should help prevent recurrence of the same types, but this does not seem to work in some cases. An explosion killed four employees and injured 18 others at a Terra Industries, Port Neal, Iowa, fertilizer plant.[164] The Iowa fire marshall blamed an overheated pump, which recirculated ammonium nitrate solution that was left running during a shutdown of the unit. It caused water to evaporate, allowing the ammonium nitrate to crystallize.[165] The EPA investigation concluded that the explosion resulted from a lack of written, safe operation procedures. "In the days and weeks just prior to the explosion equipment failures and maintenance problems were chronic."[166] Ammonium nitrate has been involved in other major catastrophes.[167] An explosion at Oppau, Germany, in September 1921 killed 1000 people. The shock was felt 145 miles away. A "terrific explosion" of two freighters loading ammonium nitrate containing 1% mineral oil killed 512 people, including many in chemical plants adjacent to the dock.[168] An explosion of 300 tons of ammonium nitrate at an Atofina plant in Toulouse, France, killed 30 people in 2001.[169] Ammonium nitrate is used primarily as a fertilizer. It can be replaced by crop rotations involving nitrogen-fixing legumes, spreading animal manure on fields, and fewer lawns (see Chapter 11).

These examples show that even though companies have a great deal of experience in handling hazardous materials, accidents can still occur. This includes toxic gases as well. A leak in the hydrogen cyanide unit at a Rohm & Haas, Deer Park, Texas plant sent 32 workers to the hospital.[170] Presumably, the hydrogen cyanide was being used to react with acetone in the synthesis of methyl methacrylate. An alternative route that does not use hydrogen cyanide is available.[171] Isobutylene is oxidized catalytically to methacrolein and then to methacrylic acid, which is esterified with methanol to give methyl methacrylate. The methacrolein can also be made by the hydroformylation of propyne, although this does involve the use of toxic carbon monoxide and flammable hydrogen.[172] These processes also eliminate the ammonium bisulfate waste from the process using hydrogen cyanide. A leak of hydrogen fluoride at a Marathon Petroleum plant in Texas City, Texas, sent 140 people to the hospital for observation and treatment of inflamed eyes and lungs and caused the evacuation of 3000 more.[173] Replacement of the hydrogen fluoride with a nonvolatile solid acid would eliminate such a problem (see Chapter 6). Many companies in the United Kingdom do not store hazardous materials correctly, which has resulted in some fires and explosions.[174]

Stern government warnings are not enough to prevent such releases. ICI spilled 704 lb of ethylene dichloride, in July 1996, and 331,000 lb of chloroform, in April 1996, at its Runcorn plant near Liverpool, England. Other leaks at the plant involved mercury, trichloroethylene, and hexachlorobenzene.[175] The company promised the British Environment Agency that it would prevent spills in the future. Then, on June 4, 1997, a leak of titanium tetrachloride caused the closure of a nearby road for 2 h, and on June 5, 1997, an oil spill occurred at another site. The Environment Agency shut down the titanium dioxide plant saying, "It is outrageous that, within weeks of ICI being called to a meeting with the Agency, where it promised to clean up its act, its plants have been involved in two further leaks."[176]

Abnormal events (i.e., the unforseen) can lead to accidents. The Napp Industries Lodi, New Jersey plant was destroyed by an explosion and fire on April 21, 1995 while doing toll manufacturing. Five workers were killed, dozens injured, and 400 nearby residents evacuated.[177] A line feeding benzaldehyde into a mixture of aluminum powder and sodium hydrosulfite plugged. In trying to clear the blockage workers inadvertently introduced some water. They went home at 7:30 p.m. No more was done until 6:00 a.m. when the morning shift arrived. Although they blanketed the reaction with nitrogen, continuing build-up of heat led to the explosion at 8:00 a.m. The U.S. Occupational Safety and Health Administration fined the company 127,000 dollars for 18 alleged safety and health violations that involved "multiple mistakes and mismanagement."[178] An abnormal event in a biocatalysis plant would not cause such problems (see Chapter 9).

Human error is a factor in many accidents. At 5:00 a.m., fire at a poly(vinyl chloride) pipe plant in Samson, Alabama, resulted in evacuation of 2500 residents within a 2-mile radius.[179] It was caused by a 40-gal mixing vat overheating when the heater was left on overnight by mistake. This accident might have been avoided by putting the heater on a timer. A better solution would be to blow the polymer with an inert gas, such as nitrogen or carbon dioxide.

Add to these accidents the oil spills from ships, such as the Exxon Valdez in 1989, that lost 11 million gal of heavy crude oil into Prince William Sound in Alaska.[180] Major spills have also occurred off Ireland, France, Japan, Scotland, and Spain. In February 1996, a spill of 65,000 tons occurred off Wales.[181] This plus plant accidents at Hoechst in the same month and the problem of disposing of an obsolete oil platform have resulted in a proposed set of Integrated Pollution Prevention Control rules in Europe. Opinions differ on how long it takes these marine ecosystems to recover. Duck, murre, orcas (killer whales), and sea otter populations in Alaska may take years to recover.[182] Measures suggested by the U.S. Oil Pollution Act of 1990 to minimize future spills include double-hulled ships, tug escort zones for tight passages, and tanker-free zones for critical areas. To these might be added minimum training requirements for the crews. Exxon now employs the following preventive measures in Prince William Sound[183]:

Tugs escort tankers.
The U.S. Coast Guard follows traffic with radar.
Reduced speeds are used.
Traffic restrictions are tighter in bad weather.
Better equipment and training have made it safer to miss icebergs.
A harbor pilot assists navigation.
The use of alcohol and drugs is prohibited and random testing is done.

In 1991, 235 oil spills were reported in the Port of Philadelphia, Pennsylvania.[184] Canada has 12 spills a day, of which nine are due to oil.[185] A crewman's mistake on a tanker released 18,000 gal of tetrachloroethylene into the shrimp-fishing ground near Port O'Connor, Texas, in 1996.[186] There are 400 oil spills on the north slope of Alaska each year. Staff cuts in the 1990s are said to have undermined preventive maintenance.[187] The oil tanker "Prestige" sank off Vigo, Spain, in 2002. The ship knew it had a leak, but was refused permission to fix the leak for correct by France, Portugal, and Spain. This just made the spill worse.[188] Europe's largest peacetime fire at Buncefield, U.K., resulted from continuing to pump gasoline into a tank for >40 min after it was full.[189]

Accidents can be very expensive, not just in plant replacement costs, but also in compensation to victims (as well as lost profits from lost sales). When operators at General Chemical's Richmond, California, plant overheated a railroad tank car, a safety relief valve opened and sent fuming sulfuric acid over the area. More than 20,000 people sought medical treatment. Five freeways and several rapid-transit stations closed. A fund of 92.8 million dollars has been set up to compensate the victims.[190] A release of sulfur trioxide from General Chemical's Augusta, Georgia, plant on November 17, 1998, sent 90 people to the hospital with eye, nose, and lung irritation. This exceeded a worst-case scenario that had been prepared for the plant.[191] These reagents are often used to make detergents. The use of alkyl polyglycoside detergents, which are made from sugars and fats, would reduce or eliminate the need for these reagents.

An October 23, 1992, release of 2.5 tons of unburned hydrocarbons, mainly butadiene, from an extinguished flare at an Oxy Chem plant in Corpus Christi, Texas, resulted in an out-of-court settlement of a class action lawsuit for 65.7 million dollars by nearby residents with health complaints.[192] A jury in New Orleans, Louisiana, awarded damages of 3.4 billion dollars for a fire that resulted when a tank car carrying butadiene caught fire in September 1987. The fire caused the evacuation of more than 1000 nearby residents.[193] A fire occurred at the University of Texas on October 19, 1996, destroying a laboratory.[194] It was caused by pouring a solvent into which sodium had been cut, down the drain, the reaction of remaining traces of sodium with water igniting the hydrogen being evolved in place of igniting the mixture. The six-alarm fire was the 10th incident in the building that involved the Austin Fire Department since February 1992. Under pressure from the Fire Department, the university will spend 30.2 million dollars to bring the building up to current fire safety standards.

More than 44 million Americans live or work near places that pose risks from the storage or use of dangerous industrial chemicals.[195] The cost of accidents may be more than just a monetary one to the company. A fire and explosion occurred on July 4, 1993, in a Sumitomo Chemical plant in Niihama, Japan, that made over half of all the epoxy-encapsulation resin for semiconductor chips used in the entire world. Cutting off the supply would have been a serious inconvenience to the customers. The company took the unusual step of letting other companies use its technology until it could rebuild its own plant, so that a supply crisis never developed. The company still supplied 50% of the world's requirements for that resin in 1999.[196]

Why do these accidents continue to happen? One critic says, "Hourly workers struggle to maintain production in the face of disabled or ignored alarms, undocumented and often uncontrollable bypasses of established components, operating levels that exceed design limits, postponed and severely reduced turnaround maintenance and increasing maintenance on 'hot' units by untrained, temporary non-union contract workers."[197] Another source mentions "institutional realities that undercut corporate safety goals, such as incentives that promote safety violations in the interest of short-term profitability, shielding upper management from 'bad news' and turnover of management staff."[198] A third says that "Many chemical plant disasters have been precipitated by an unplanned change in process, a change in equipment or a change in personnel."[199]

It has been estimated that the U.S. petrochemical industry could save up to 10 billion dollars/year by avoiding abnormal situations or learning how to better deal with them.[200] A team studying abnormal situations management identified eight key issues:

- Lack of management leadership
- The significant role of human errors
- Inadequate design of the work environment
- Absence of procedures for dealing with abnormal operations (as opposed to emergencies)
- Loss of valuable information from earlier minor incidents
- The potential economic return
- Transferability of good abnormal situations performance to other plants
- The importance of teamwork and job design

The paper mentions 550 major accidents at U.S. petrochemical plants (each with more than 500,000 dollars lost) in the last 5 years, with 12.9 billion dollars total equipment damage. Learning from case histories is helpful, but is evidently not enough.[201]

It is easy to blame accidents on human error, but good design can often minimize this.[202] Avoid poor lighting or contrast. Provide a checklist so that the operator will find it easy to recall all of the necessary information. Most valves have right-handed threads. Do not mix in any that have left-handed threads. Mount them so that they are easy to access, and the labels are easy to see.

Experts in industrial safety recommend a shift to inherently safer chemistry.[203] This involves minimizing inventories of hazardous chemicals by generating on-site as needed.[204] Less toxic alternatives can be used, for example using dimethyl carbonate instead of dimethyl sulfate for methylations (see Chapter 2). Process intensification in continuous microreactors eliminates the possibility of runaway reactions and the need for heeding explosive limits while increasing yields and purity[205] (see Chapter 8). The Massachusetts 1989 Toxics Use Reduction Act requires companies to think through their complete processes with the idea of reducing the use of toxics, but does not require them to implement the plans.[206] However, many of them have done so with a total savings of $88 million.

New Jersey adopted a Toxic Catastrophe Prevention Act in 1986, which requires risk assessment by companies.[207] New Jersey enacted a pollution Prevention Act in 1991 and an inherently safer technology law to prevent terrorism at chemical plants in 2005.[208] The Kanawha Valley Hazardous Assessment Project in West Virginia developed worst-case scenarios for 12 chemical plants in the area.[209] The chemicals studied included acrylonitrile, vinylidene chloride, butyl isocyanate, methylene chloride, chlorine, phosphorous trichloride, hydrogen sulfide, methyl isocyanate, phosgene, ethylene oxide, sulfur trioxide, and others. Even though inherently safer chemistry will save lives and money and deter attacks by terrorists, the American Chemistry Council objects to it whenever it is proposed for new legislation.[210] The system of more fences, guards, and checkpoints ignores the fact that the terrorists who attacked the World Trade Center in New York and the Pentagon Building in Arlington, Virginia, came by air. The vulnerability of plants protected by fences was shown when two people cut through two fences and a locked gate at a liquid natural gas storage facility in Lynn, Massachusetts, in 2006. The incident went unnoticed for five days, despite routine perimeter inspections and a system to monitor intrusions.[211] Ethanol can be made by hydrating the flammable gas ethylene. It can also be made by fermentation of corn. Terrorists might consider the former as a target, but why would they attack the latter, which is the equivalent of a brewery?

In addition to the amounts of chemicals released to the environment through accidents, the U.S. Toxic Release Inventory shows the release of 2.43 billion lb into air, land, or water in 1996,[212] down from 3.21 billion lb in 1992.[213,214] (The data are available on a number of Internet sites, including www.epa.gov/opptintr/tri.) The total decline since the law became effective in 1987 has been 59%.[215] This right-to-know law was enacted in the aftermath of the Bhopal accident. It now covers nearly 650 chemicals out of about 72,000 in commerce. (It is credited with causing industries to reduce emissions more than the usual command-and-control regulations that can lead to cumbersome, adversarial legal proceedings.[216] Companies do not like being considered the worst polluter in their area. According to the Political Economy Research Institute, DuPont led the list of the top 100 air polluters in 2002.) Facilities are not required to report releases unless they manufacture or process more than 25,000 lb or handle more than 10,000 lb of the chemicals annually.[217] The U.S. EPA has proposed lower reporting amounts for some, especially toxic chemicals [e.g., 10 lb of chlordane (a persistent chlorinated insecticide), polychlorinated biphenyls, or mercury; and 0.1 g dioxins]. Some compounds have also been removed from the list. Acetone (129 million lb, released in 1993) and nonaerosol forms of sulfuric acid (106 million lb injected underground of 130 million lb, released in 1993) have been removed from the list recently because they are not likely to cause adverse effects to the environment or to human health under conditions of normal use.[218] However, the amount of production-related waste has remained relatively constant at about 37 billion lb since 1991, when data collection began.[219] About 250 million metric tons of hazardous waste are generated each year in the United States. This is about 1 ton/person.[220] The number of Superfund sites, where toxic waste was deposited in the past and that now need to be cleaned up, is expected to reach 2000, with an estimated cost of 26 million dollars to clean up each site.[221]

The numbers in the Toxic Release Inventory must not be taken out of context. Large reductions in the numbers for hydrogen chloride and sulfuric acid between 1987 and 1996 resulted from narrowing the reporting requirements to cover only airborne releases.[222] About half of the major reductions in waste generated by an 80-company sample resulted from redefining on-site recycling activities as in process recovery, which does not have to be reported.[223] A change in the production level of a plant can also change the numbers. Numbers can also change when a more accurate system of measurement is implemented. Coal-fired electric power plants were not required to report releases until the middle of 1999.

The Toxic Release Inventory for Delaware for 1993–2006 will illustrate how the system works (see below). For comparison, Kennecott's Utah copper mine released 96,991,889 lb in 2006.

A study of the 1993–1996 data illustrates some of the reasons for the emissions and their varying levels. The large releases from the first two companies were largely solvents used in painting cars. These might be reduced to near zero by the use of powder coating for finishing cars, and cleaning parts with aqueous detergents (see Chapter 8). Part of the reduction by Chrysler may have involved substitution of a nonreportable solvent for a reportable one (as suggested by a GM employee). The big reductions for both companies stem from major shutdowns to retool for new models. During this period, Chrysler reduced its use per unit of production by 6%, whereas GM increased its use per unit of production by 19%. Included in the Sun Oil releases was 150,000 lb of ethylene oxide. The nylon plant losses were mainly hydrochloric acid from traces of chloride in the coal burned to power the plant. The increased releases in 1995–1996 reflect a new requirement to report nitrate released by the wastewater treatment plant. A scrubber could be used to remove the hydrogen chlorine from stack gases. A metals company lost primarily trichloroethylene used in cleaning metal parts. Cleaning with aqueous detergent could reduce the amount lost to zero (see Chapter 8). (Elsewhere in Delaware, the town of Smyrna has to treat its water supply to remove trichloroethylene before distributing the water to its customers.) The releases from Formosa Plastics included 111,000 lb of the carcinogen, vinyl chloride. The substitution of polyolefins, including those made with new metallocene catalysts, for polyvinyl chloride might eliminate the need to make vinyl chloride. Other releases in the state included the carcinogens acrylonitrile, benzene, dichloromethane, and ethylene oxide. For comparison, the state of Delaware estimates that cars and trucks contribute 138,040 lb/day of volatile organic compounds into the air. This means that local traffic puts out in 2.5 days what the Valero refinery releases in a year. The poultry industry, in the same county as the nylon plant, probably releases more nitrates to streams than the nylon plant (by putting excess manure on fields). The total emissions in Delaware are less

Toxic Release Inventory for Delaware (Partial)

	Pounds Released per Year				
	1993	1996	1998	2000	2006
Chrysler	987,440	105,655	441,994	483,604	177,320
General Motors	826,311	581,039	421,327	278,115	133,218
Sun Oil refinery	463,010	92,500	—	34,270	85,737
Invista nylon	455,900	654,970	1,043,900	801,279	710,172
Valero	344,549	162,642	1,919,578	1,755,753	3,315,541
Formosa Plastics	146,621	132,862	131,977	134,526	102,478
Delmarva Power				1,868,578	1,591,913
NRG power plant				3,041,931	3,722,465
Total					11,171,515

than what some single companies release in other parts of the nation (e.g., Eastman Chemical, 29 million lb).[224]

One interesting approach to the loss of volatile liquids is the use of an evaporation suppressant for the styrene used to cure unsaturated polyester resisns.[225] The suppressant, made from a bisphenol. A epoxy resin, stearic acid, colophony, triethanolamine, glycerol monostearate, calcium stearate, and sorbitan monostearate, may coat the surface to retard evaporation.

Chemicals in lakes and streams caused 46 states to issue public health warnings to avoid or curtail the eating of fish. Mercury was the cause in 60% of the cases, polychlorinated biphenyls 22%, chlordane (a banned chlorinated insecticide) 7%, and DDT 2%, the remainder being spread over 25 chemicals.[226] Fish advisories went up 26% from 1995 to 1996 to the point where 15% of the lakes and 5% of the rivers in the United States were covered.[227] In fairness to industry, it should be pointed out that a study by the Lindsay Museum, in Walnut Creek, California, found that 70% of the chemicals in San Francisco Bay came from the daily activities of ordinary people (e.g., oil from leaky cars, copper dust from brake pads, garden fertilizers, and pesticides).[228] Nearly 50% of the oil in the world's waters comes from people carelessly discarding used oil on the ground or down drains.[229] Some reduction has been obtained by labeling storm drains with "Only rain down the drain" or "Drains to creek." Sixty-six percent of Delaware rivers and streams do not meet minimum standards for swimming, and 29% do not support normal aquatic life.[230]

Occupational illness and injury[231] cost 30–40 billion dollars/year in the United States.[232] In 1994, there were 6.8 million injuries and illness in private industry, amounting to 8.4 cases per 100 workers. Nearly two-thirds were disorders associated with repeated trauma, such as carpal tunnel syndrome.[233] The Occupational Safety and Health Act of 1970 set up the National Institute of Safety and Health (NIOSH) to study the problem and OSHA to deal with it through inspections and regulations. Both have received much criticism with respect to their effectiveness that they are still struggling to find more effective ways of dealing with the problem.[234] NIOSH is searching for practical ways of protecting workers, especially those in small businesses, from methylene chloride, tetrachloroethylene, diesel exhaust in coal mines, isocyanates, 2-methoxyethanol, and others. OSHA is expanding a plan that worked well in Maine, a state that used to have one of the worst accident and illness records in the United States.[235] The 200 firms with the worst records were asked to look for deficiencies and to correct them. They were also inspected. These measures cut injuries and illnesses over a two-year period.

1.4 WASTE AND ITS MINIMIZATION

Nearly everything made in the laboratory ends up as waste. After the materials are made, characterized, and tested, they may be stored for a while, but eventually they are discarded. In schools, the trend is to run experiments on a much smaller scale, which means that less material has to be purchased and less waste results.[236] (Even less material will be used if chemistry-on-a-chip becomes commonplace.[237]) Two things limit how small a scale industrial chemists can use. One is the relatively large amount of a polymer needed for the fabrication of molded pieces for testing physical properties. There is a need to develop smaller-scale tests that will give data that are just as good. The second factor is the tendency of salesmen to be generous in offering samples for testing by potential customers. A lot of the samples received by the potential customer may never be used.

Some industrial wastes result because it is cheaper to buy new material than to reclaim used material. Some catalysts fall into this category. The 1996 American Chemical Society National Chemical Technician Award went to a technician at Eastman Chemical who set up a program for recovering cobalt, copper, and nickel from spent catalysts for use by the steel industry.[238] This process for avoiding landfill disposal gave Eastman significant savings. Some waste metal salts can be put into fertilizer as trace elements that are essential for plant growth. However, this practice has been abused in some cases by putting in toxic waste (e.g., some that contain dioxins and heavy metals).[239] Even a waste as cheap as sodium chloride can be converted back to the sodium hydroxide and hydrochloric acid that it may have come from, by electrodialysis using bipolar membranes.[240] (Membrane separations are covered in Chapter 7.) Waste acid can be recovered by vacuum distillation in equipment made of fluoropolymers.[241]

Mixed solvents can be difficult to recover. If they are kept separate, they can be reclaimed by distillation on-site. The capital investment required is paid back by the reduced need to buy new solvent. In Germany "completely enclosed vapor cleaners" give 99% reduction in solvent emissions.[242] After cleaning and draining, the airtight chamber is evacuated and the solvent vapors are captured by chilling and adsorption on carbon. When a sensor shows that the solvent is down to 1 g/m^3, the vacuum is released so that the lid can be opened.

In some cases the amount of waste may be considered too small to justify the research needed to find a use for it or to improve the process to eliminate it. The staff may also be so busy with potentially profitable new ventures that it hesitates to take time to devise a new process for a mature product made in an established plant. This may be particularly true in the current period of downsizing and restructuring.

Improved process design can minimize waste. Deborah Luper suggests asking the following questions[243]:

Are you using raw materials that are the target of compound-specific regulations?
Can carriers and solvents be eliminated, reduced, or recycled?
Can you find riches in someone else's wastes?
Are there riches in your wastes?
Where does the product end up after use and what happens to its components?

What percentage of the incoming materials leaves the plant in the finished products?
What support equipment and processes generate waste?
How long is the waste treatment train?
Can process water be recycled?
What can be achieved with better process control?

Improved housekeeping can often lead to reduced emissions and waste.[244] The leaky valves and seals can be fixed or they can be replaced with new designs that minimize emissions.[245] These include diaphragm valves, double mechanical seals with interstitial* liquids, magnetic drives, better valve packings, filled fabric seals for floating roof tanks, and so on. Older plants may have some of the worst problems.[246] The U.S. EPA fined Eastman Kodak 8 million dollars for organic solvents leaking from the 31 miles of industrial sewers at its Rochester, New York, plant. This emphasizes the need for regular inspections and preventive maintenance, which in the long run is the cheapest method. Vessels with smooth interiors lined with non- or low-stick poly(tetrafluoroethylene) can be selected for batch tanks that require frequent cleaning. These might be cleaned with high-pressure water jets instead of solvents. Perhaps a vessel can be dedicated to a single product, instead of several, so that it does not have to be cleaned as often. If a product requires several rinses, the last one can be used as the first one for a new lot of product. Volatile organic compounds can be loaded with dip tubes instead of splash loading. Exxon (now Exxon-Mobil) has used such methods to cut emissions of volatile organic compounds by 50% since 1990.[247] Install automatic high-level shut-offs on tanks. Use wooden pallets over again, instead of considering them as throw-away items. The current ethic should be reduce, reuse, and recycle in that order.

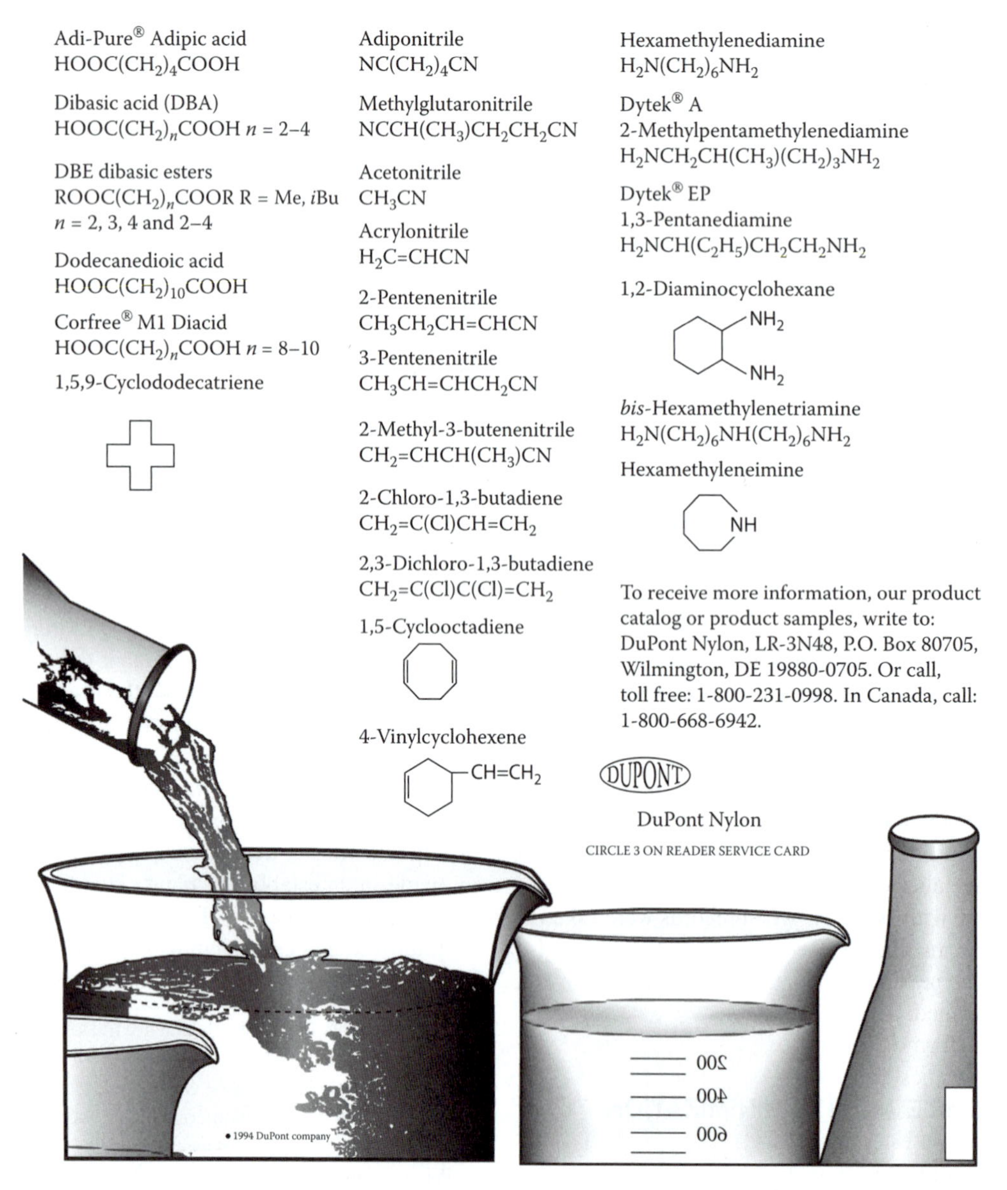

Adi-Pure® Adipic acid
HOOC(CH2)4COOH
Dibasic acid (DBA)
HOOC(CH2)nCOOH n = 2–4
DBE dibasic esters
ROOC(CH2)nCOOR R = Me, iBu
n = 2, 3, 4 and 2–4
Dodecanedioic acid
HOOC(CH2)10COOH
Corfree® M1 Diacid
HOOC(CH2)nCOOH n = 8–10
1,5,9-Cyclododecatriene
Adiponitrile
NC(CH2)4CN
Methylglutaronitrile
NCCH(CH3)CH2CH2CN
Acetonitrile
CH3CN
Acrylonitrile
H2C=CHCN
2-Pentenenitrile
CH3CH2CH=CHCN
3-Pentenenitrile
CH3CH=CHCH2CN
2-Methyl-3-butenenitrile
CH2=CHCH(CH3)CN
2-Chloro-1,3-butadiene
CH2=C(Cl)CH=CH2
2,3-Dichloro-1,3-butadiene
CH2=C(Cl)C(Cl)=CH2
1,5-Cyclooctadiene
4-Vinylcyclohexene
CH=CH2
Hexamethylenediamine
H2N(CH2)6NH2
Dytek® A
2-Methylpentamethylenediamine
H2NCH2CH(CH3)(CH2)3NH2
Dytek® EP
1,3-Pentanediamine
H2NCH(C2H5)CH2CH2NH2
1,2-Diaminocyclohexane
NH2
NH2
bis-Hexamethylenetriamine
H2N(CH2)6NH(CH2)6NH2
Hexamethyleneimine
NH
To receive more information, our product catalog or product samples, write to:
DuPont Nylon, LR-3N48, P.O. Box 80705, Wilmington, DE 19880-0705. Or call, toll free: 1-800-231-0998. In Canada, call: 1-800-668-6942.
DUPONT
DuPont Nylon
CIRCLE 3 ON READER SERVICE CARD
• 1994 DuPont company

1.20 Schematic (Reprinted from *Chem. Eng. News*, Oct. 24, 1994, with permission of E. I. DuPont de Nemours & Co., Inc.)

A waste is not a waste if it can be reused. For example, one steel manufacturer drops pickle liquor down a 100-ft-tall tower at 1200°F to recover iron oxide for magnetic oxide and hydrogen chloride for use again in pickle liquor.[248] The pomace left over from processing pears and kiwis can be dried and used to increase the dietary fiber in other foods.[249] Food-processing wastes and wastes from biocatalytic processes often become feed for animals. A refinery stream of ethane, methane, butane, and propane, which was formerly flared as waste, will be processed to recover propane for conversion to propylene and then to polypropylene.[250] Organic chemical wastes may end up as fuel for the site's power plant or for a cement kiln, but more valuable uses would be preferable. Waste exchanges are being set up. One company's waste may be another's raw material. For example, calcium sulfate from flue gas desulfurization in Denmark and Japan ends up in dry wall for houses. If the waste exchange merely pairs up an acid and a base so that the two can be neutralized, rather than reclaimed, the result is waste salts. Admittedly, these are probably not as toxic as the starting acid and base, but they still have to be disposed of somewhere. Several general references on waste minimization and pollution prevention are available.[251]

On-line continuous monitoring of reactions can reduce waste by better control of the reaction and elimination of many laboratory samples that are discarded later. This can be used in the laboratory to replace analyses by thin-layer or other chromatography. For example, an infrared probe was used to monitor a copolymerization of 2-ethylhexyl acrylate and styrene in an aqueous emulsion.[252] The yield in organic reactions is seldom 100%. If uses can be found for the by-products, then everything can be sold for more than the fuel value of the by-products. There may still be problems if the market sizes for the various products do not fit the volumes produced or that can be produced in a slightly altered process. A few examples of upgrading will be given.

Dimethyl terephthalate for the production of poly(ethylene terephthalate) is produced by the cobalt salt-catalyzed oxidation of *p*-xylene with oxygen (reaction **1.21**).[253] In this free radical process, some biphenyl derivatives are formed. In addition, triesters are formed from any trimethylbenzenes in the feed. Thus, the still bottoms contain several compounds, which are all methyl esters. Hercules (now Ashland) found that transesterification of this mixture with ethylene glycol led to a mixture of polyols that could be used with isocyanates to form rigid polyurethanes. For the price, the "Terate" product was hard to beat.

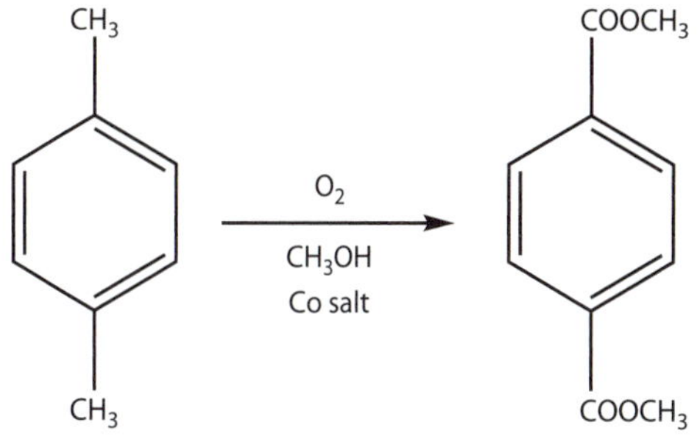

CH_3 — CH_3 →(O_2; CH_3OH, Co salt)→ $COOCH_3$ — $COOCH_3$

1.21 Schematic

DuPont markets a number of intermediates from its manufacture of fibers,[254] as shown by the accompanying advertisement often seen in *Chemical Engineering News* in 1994 (**1.20** Schematic). (Most of the DuPont fiber plants have been sold to Koch Industries.) Adipic acid is prepared by oxidation of a mixture of cyclohexanol and cyclohexanone, obtained, in turn, by the oxidation of cyclohexane (reaction **1.22**).[255]

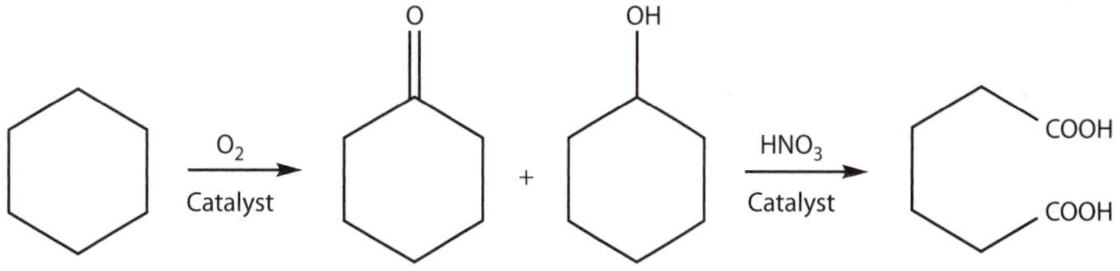

→(O_2, Catalyst)→ O + OH →(HNO_3, Catalyst)→ COOH, COOH

1.22 Schematic

Unfortunately, this process produces undesired nitrogen oxides at the same time. Some shorter dibasic acids are also formed. Trimerization of 1,3-butadiene is used to prepare 1,5,9-cyclododectriene, an intermediate for another nylon.[256] In the process, 1,5-cyclooctadiene and 4-vinylcyclohexene are formed as by-products (reaction **1.23** Schematic).

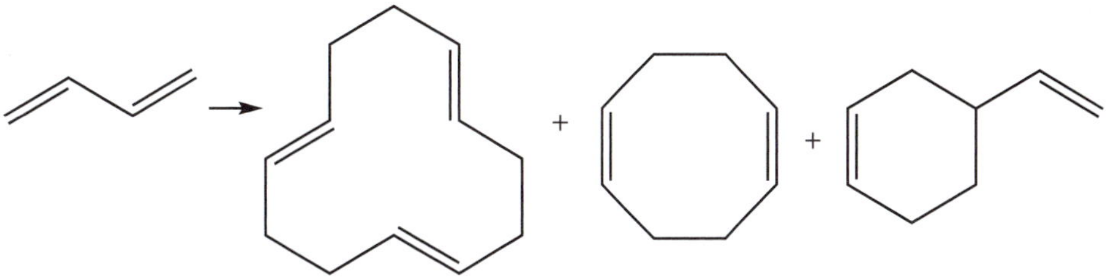

1.23 Schematic

The hexamethylenediamine used with adipic acid to make nylon-6,6 is made by reduction of adiponitrile, prepared, in turn, by the addition of hydrogen cyanide to 1,3-butadiene (reaction **1.24**).[257]

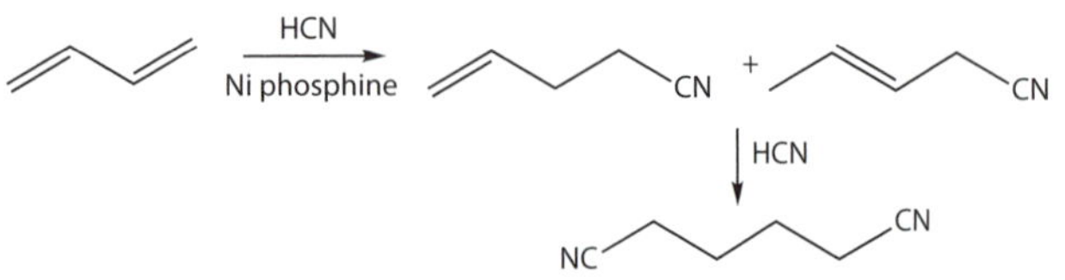

1.24 Schematic

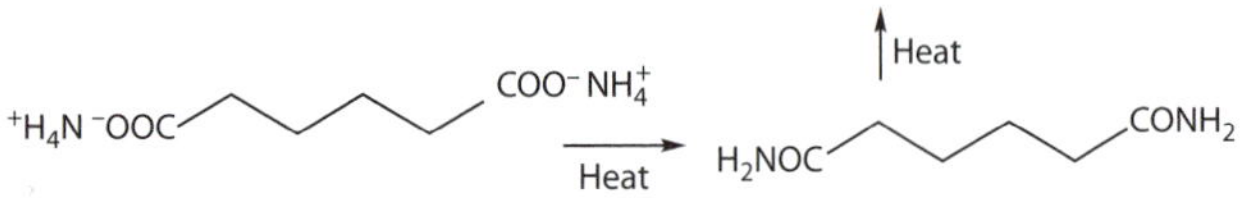

1.25 Schematic

An earlier process that did not use the toxic hydrogen cyanide was discarded because it produced more waste than the HCN route. In it, the adiponitrile was made by heating ammonium adipate (see reaction **1.25**).

The 2-pentenenitrile, 2-methyl-3-butenenitrile, and methylglutaronitrile in **1.20** Schematic are by-products of this reaction sequence. DuPont is still studying the phosphines used as ligands for the nickel in an effort to find one bulky enough to favor terminal addition only.[258] Reduction of the various nitriles leads to the amines in **1.20** Schematic, including the cyclic ones. The 2,3-dichloro-1,3-butadiene is probably a by-product in the synthesis of 2-chloro-1,3-butadiene used to make Neoprene rubber. DuPont also polymerized acrylonitrile to prepare poly(acrylonitrile) fiber (Orlon). Acetonitrile is obtained as a by-product of the ammoxidation of propylene to produce acrylonitrile (reaction **1.26**).

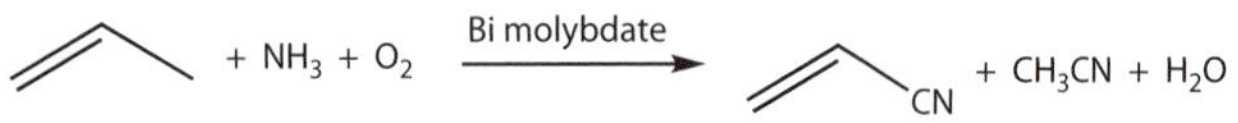

1.26 Schematic

Paul V. Tebo has described a DuPont goal of zero emissions and zero waste.[259]

The hydroformylation[260] of propene to form butyraldehyde invariably produces some isobutyraldehyde at the same time (reaction **1.27**).[261] One of the best processes uses a water-soluble rhodium phosphine complex to produce 94.5% of the former and 4.5% of the latter.[262] The products form a separate layer that is separated from the water. Rhodium is expensive so it is important to lose as little as possible. In 10 years of operation by Rhone-Poulenc-Ruhrchemie, 2 million metric tons of butyraldehyde have been made with the loss of only 2 kg of rhodium. The process is 10% cheaper than the usual one. Higher olefins are not soluble enough in water to work well in the process. The process does work for omega-alkenecarboxylic acids such as 10-undecenoic acid, where a 97:3 normal/iso compound is obtained in 99% conversion.[263] For higher alkenecarboxylic acids, a phase transfer catalyst, such as dodecyltrimethylammonium bromide, must be used. However, this lowers the normal/iso ratio. (Chapter 8 contains more on biphasic reactions.)

1.27 Schematic

Over the years a variety of uses have been found for isobutyraldehyde by Eastman Chemical and others.[264] It is converted to isobutyl alcohol, neopentyl glycol, isobutyl acetate, isobutyric acid, isobutylidene diurea, methylisoamyl ketone, and various hydrogenation and esterification products (**1.28** Schematic).

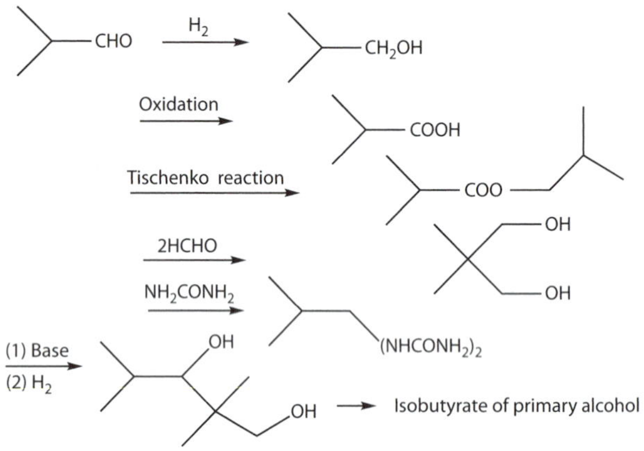

1.28 Schematic

Neopentyl glycol is used in the preparation of polyesters. Because there are no β-hydrogen atoms, the polymers are more stable. The self-condensation of isobutyraldehyde followed by reduction leads to 2,2,4-trimethyl-1,3-pentanediol, the monoisobutyrate of which is the most common coalescing agent (used at 0.5–2 vol.%) in latex paints. A few latex paints no longer require coalescing agents.[265] Isobutyl acetate is used as a solvent for nitrocellulose coatings. Isobutylidene diurea is used as a slow-release fertilizer. If there is a surplus of isobutyraldehyde, recent work has shown that it can be decarbonylated to propylene over a palladium on silica catalyst.[266] There is also the possibility of dehydrogenation to methacrolein for conversion to methacrylic acid and then to methyl methacrylate. Dehydration of isobutyl alcohol would produce isobutylene for conversion to methyl-*tert*-butyl ether, although this would probably be uneconomical.

Uses have been found for 2-methyl-1,3-propanediol (a by-product of the preparation of 1,4-butanediol) in the personal care industry.[267]

The preparation of rayon (regenerated cellulose) by the viscose process[268] is being phased out, owing to the environmental pollution of the process (**1.29** Schematic).

$$\text{CellOH} + CS_2 + \text{NaOH} \longrightarrow \text{CellOCSS}^- \text{Na}^+ \longrightarrow \text{CellOH} + \text{Na}^+$$

1.29 Schematic

In the new process (1.1), a solution of cellulose in *N*-methylmorpholine-*N*-oxide (**1.30** Schematic) is run into water to produce the fibers.[269] The solvent is recovered, so that there is no waste. The process not only eliminates the very flammable and odorous carbon disulfide, but also produces a much stronger fiber because the degradation of the molecular weight of the viscose process is avoided. A new process uses an ionic liquid as the solvent (see Chapter 8).

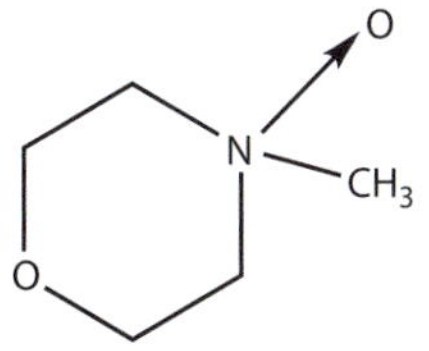

1.30 Schematic

1.5 CONCLUSIONS

The challenge is to reduce the incidence and severity of accidents, waste, the toxicity of chemicals, and the amount of energy used, while still providing the goods that society needs. Several provocative papers suggest some ways of doing this.[270] The key is in the preparation of more sophisticated catalysts. Thus, solid acids may be able to replace the risky hydrogen fluoride and sulfuric acid used in alkylation reactions in the refining of oil. Zeolites offer the promise of higher yields through size and shape selectivity (see Chapter 6). With the proper catalysts, oxidations with air and hydrogen peroxide may replace heavy metal-containing oxidants (see Chapter 4). Enantioselective catalysis may allow the preparation of the biologically active optical isomer without the unwanted one (see Chapter 10). It may be possible to run the reaction in water at or near room temperature using biocatalysis instead of in a solvent or at high temperature (see Chapter 9). Some processes yield more by-product salts than the desired product. Sheldon[271] recommends a salt-free diet by improved catalytic methods. These and other possibilities will be examined in the chapters that follow.

Bodor[272] has suggested the design of biologically safer chemicals through retrometabolic design. For example, the ethylene glycol used widely as an antifreeze in cars might be replaced with less hazardous propylene glycol. The former is converted by the body to glycolaldehyde, glyoxylic acid, and oxalic acid (**1.31** Schematic), whereas the latter gives the normal body metabolites lactic acid and pyruvic acid (**1.32** Schematic).[273]

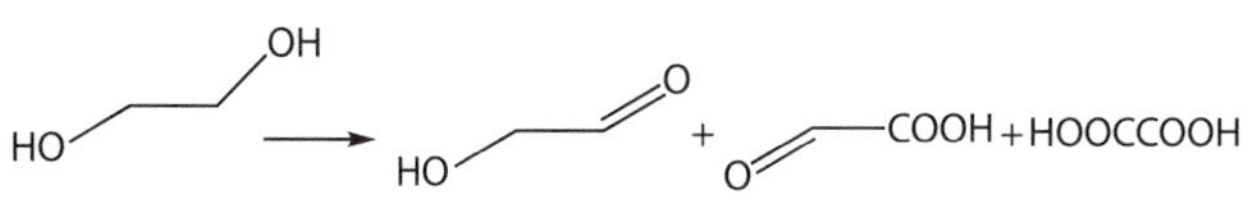

1.31 Schematic

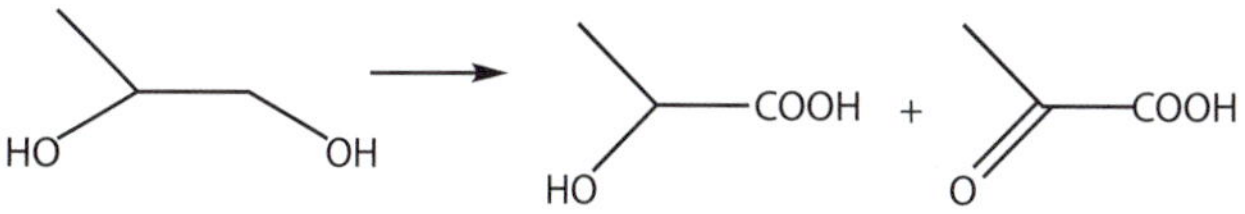

1.32 Schematic

A lethal dose of ethylene glycol for man is 1.4 mL/kg. The problem is that its sweet taste makes it attractive to children and pets. An alternative approach is to add a bittering agent to it. The estimated lethal dose of propylene glycol for man is 7 mL/kg.

The U.S. National Science and Technology Council has laid out a research and development strategy for toxic substances and waste.[274] Hirschhorn[275] has suggested ways of achieving prosperity without pollution.

Two examples will indicate improvements that have come about by simple economics of the marketplace. Olefins higher than ethylene are usually converted to the epoxides indirectly by adding hypochlorous acid, and then base to split out hydrogen chloride (**1.33** Schematic). However, an Arco process (**1.34** Schematic) does this without the formation of by-product salt.[276] Dow is switching to the new route.[277] (A newer route using titanium silicalite and hydrogen peroxide is covered in Chapter 4.)

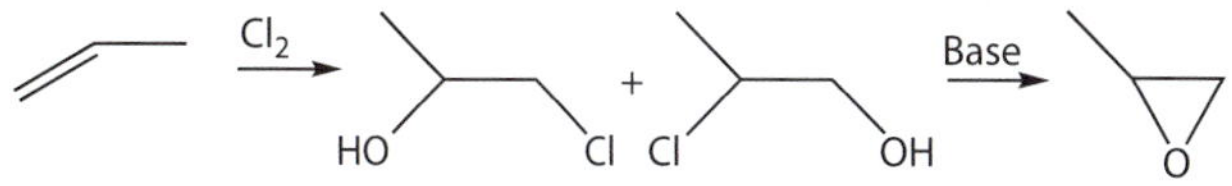

1.33 Schematic

The hydroperoxide formed by the air oxidation of ethylbenzene is used to convert propylene to its oxide. The by-product 1-phenylethanol is dehydrated to styrene, a second valuable product. The disadvantage of such a process is that the demand for such products has to be equal on a molar basis.

Propylene used to be converted to isotactic polypropylene with titanium tetrachloride and diethylaluminum chloride in a hydrocarbon solvent.[278] The atactic polypropylene obtained by evaporation of the solvent after filtration of the desired isotactic polymer was of little value, some going into adhesives. An acidic deashing step was necessary to remove residual titanium, aluminum, and chloride from the polymer, the metal-containing residues ending up in a landfill. This process was supplanted by high mileage catalysts during which the titanium chloride was supported on magnesium chloride.[279] These were activated by triethylaluminum in the presence of ligands that enhance the stereoselectivity of the catalysts. The result was a product that required no deashing and no removal of atactic polymer. In the next step in the evolution, the solvent was eliminated by polymerization in the gas phase or in liquid propylene. By the proper choice of catalyst, the polymer can be obtained in large enough granules so that the older practice of extruding molten polymer to form a strand that was chopped into "molding powder" is no longer necessary. The field is still

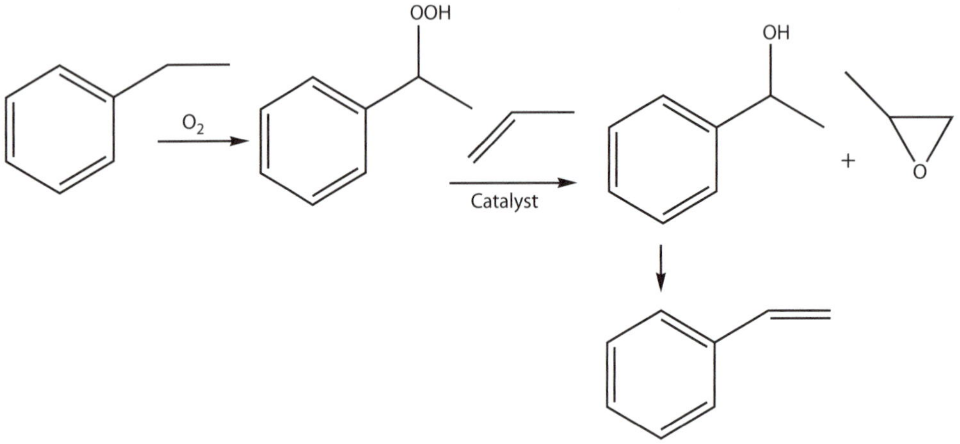

1.34 Schematic

evolving. Metallocene single-site catalysts[280] allow greater control of the product and have led to new products. Ethylene-α-olefin copolymers can be made from ethylene alone, the α-olefin being made *in situ*.[281] A typical catalyst is shown in **1.35** Schematic. It is activated by methyl alumoxane. Methods of preparing syndiotactic polypropylene in a practical way are now available. A new polypropylene made with such catalysts may be able to supply the properties now found only in plasticized polyvinyl chloride. All but the last type of catalyst show little tolerance to air, moisture, and polar groups. The catalyst in **1.36** Schematic is active in the presence of ethers, ketones, esters, and water.[282] Ethylene can even be polymerized in water by one of the palladium diimine catalysts.[283] The initial polymerization probably forms a shell of polyethylene that protects the catalyst from the water. To make the products even greener, consumer use must be reduced, the low level of reuse and recycle must be raised, and a renewable source of the propylene, rather than petroleum, must be used. Propylene could be made by reduction and then dehydration of acetone from fermentation (see Chapters 12 and 14).

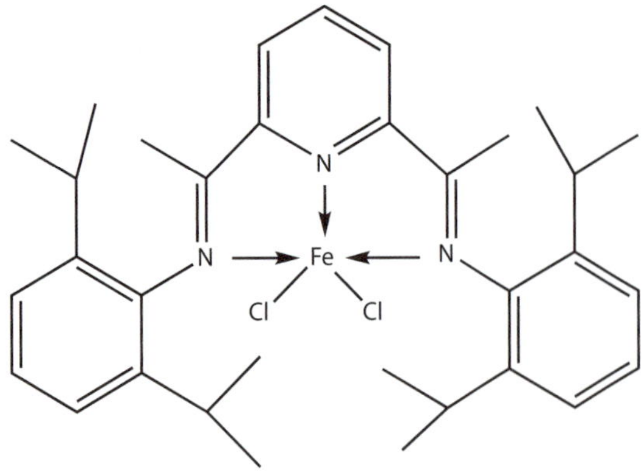

1.35 Schematic

Hirschhorn[284] feels that "An environmentally driven industrial revolution is beginning." Brain Rushton, President of the American Chemical Society in 1995, said that "We will gradually eliminate environmentally unsound processes and practices from our industry. We will build a better environment to work and live in. We will keep scientists and engineers both employed and at the cutting edge of technology as they serve the competitive needs of the nation. Last but by no means least, we will create a better image for chemistry and the profession."[285]

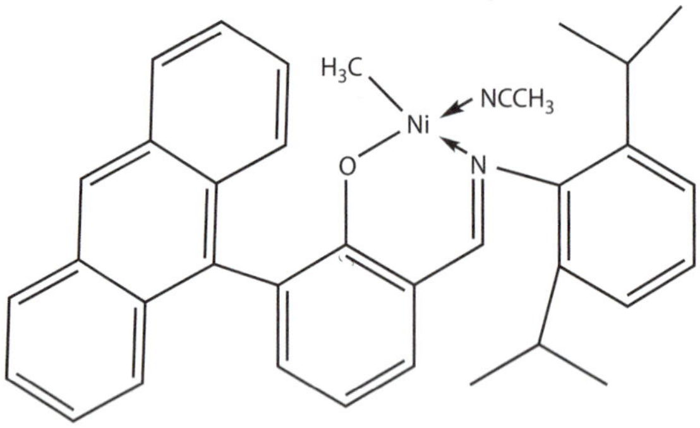

1.36 Schematic

Lastly, Gro Harlem Brundtland, formex head of the WHO and former prime minister of Norway and the secretary general of the World Commission on Environment and Development, has noted that "The obstacles to sustainability are not mainly technical. They are social, institutional and political."

1.6 SUMMARY OF SOME IMPORTANT POINTS

1. Scientists may not be able to anticipate the influence of their work on society.
2. Chemophobia is rampant at the same time that the real risks are often unappreciated.
3. "Natural" is often misinterpreted.
4. In the United States, oversight on herbal remedies and dietary supplements by the FDA would be desirable.
5. Chemical accidents will continue to happen. Human error will often be the cause. If the chemicals being used

are nontoxic, the severity of the accidents will be reduced greatly.

6. The Toxic Release Inventory is proving to be a good way of encouraging voluntary reductions of emissions.
7. A waste is not a waste if a valuable use can be found for it. However, it is better to find a better process that eliminates the waste.
8. While improved housekeeping can reduce releases, it is better to choose a new process with benign reagents.
9. If you do not use it, you cannot lose it, you do not have to buy it, and you do not have to treat it for disposal.
10. Green chemistry and sustainability are intertwined with the social and political fabric and the country and cannot be considered in isolation.
11. Green chemistry need not be more expensive. Pollution prevention is often cheaper than end-of-the-pipe treatments. Rethinking the entire process can often be helpful.
12. The problems are not all elsewhere.
13. There may be a hierarchy of approaches to the problem from the least change to the most change.
14. Be patient, since it may take a while to work the bugs out of the new process.
15. An expensive catalyst is not too expensive if none is lost. The lowest price per unit of performance is what counts. This must include all costs, even the diffuse costs of emissions.

REFERENCES

1. I.T. Horvath and P.T. Anastas, eds, *Chem. Rev.*, 2007, *107*, 2167 (issue on green chemistry).
2. Anon., *Chem. Eng. News*, May 18, 1998, 19; November 12, 2007.
3. R.L. Carson, *Silent Spring*, Houghton–Mifflin, Boston, 1962.
4. (a) E. Wilson, *Chem. Eng. News*, Oct. 12, 1998, 16; (b) D.A. Kurtz, *Long Range Transport of Pesticides*, Lewis Publishers, Boca Raton, FL, 1990.
5. (a) Anon., *Chem. Eng. News*, Nov. 11, 1995, 16; (b) B. Baker, *BioScience*, Mar. 1996, 183; (c) B. Hileman, *Chem. Eng. News*, July 6, 1998, 4.
6. www.endocrinedisruption.com.
7. (a) T. Colburn, J.P. Myers, and D. Dumanowski, *Our Stolen Future*, Dutton, New York, 1996; (b) B. Hileman, *Chem. Eng. News*, Oct. 9, 1995, 30.
8. Anon., *Chem. Eng. News*, Aug. 17, 1998, 33.
9. (a) T. Stephens and R. Brynner, *Dark Remedy: The Impact of Thalidomide and its Revival as a Vital Medicine*, Perseus, Cambridge, MA, 2001; (b) T. Stephens, *Chem. Br.*, 2001, *37*(11), 38.
10. Anon., *Harvard Health Lett.*, 1998, *23*(4), 4.
11. R. Hoffmann, *Chem. Eng. News*, Apr. 29, 1996.
12. Anon., *Chem. Ind.* (*London*), 1998, 591.
13. Anon., *Chem. Eng. News*, Dec. 18, 2006, 27.
14. I.J. Winograd and E.H. Roseboom, Jr., *Science*, 2008, *320*, 1428.
15. G.W. Muller, *Chemtech*, 1997, *27*(1), 2.
16. J. Johnson. *Chem. Eng. News*, Jan. 11, 1999, 28.
17. J. Schummer, B. Bensaude-Vincent, and B. van Tiggelen, eds, *The Public Image of Chemistry*, World Scientific Publishing, Singapore, 2007.
18. (a) Anon., *Chem. Ind.*, 1996, 396; (b) C. Djerassi, *Science*, 1996, *272*, 1858.
19. D.R. Hofstadter, *Science*, 1998, *281*, 512.
20. M. Reisch, *Chem. Eng. News*, May 22, 1995, 6.
21. R.F. Service, *Science*, 1994, *266*, 726.
22. (a) R.S. Stricoff, *Handbook of Laboratory Health and Safety*, 2nd ed., Wiley, New York, 1995; (b) National Research Council, *Prudent Practices in the Laboratory: Handing and Disposal of Chemicals*, National Academy of Sciences Press, Washington, DC, 1995; (c) R. Rawls, *Chem. Eng. News*, Aug. 14, 1995, 4; (d) R.E. Lenga, *The Sigma-Aldrich Library of Chemical Safety Data*, 2nd ed., Sigma Aldrich Corp, Milwaukee, WI, 1988; (e) N.I. Sax and R.A. Lewis, *Dangerous Properties of Industrial Materials*, 7th ed., van Nostrand–Reinhold, New York, 1989; (f) D.V. Sweet, R.L. Sweet, Sr., eds, *Registry of Toxic Effects of Chemicals Substances*, Diane Publishers, Upland, PA, 1994; (g) C. Maltoni and I.J. Selikoff, Living in a chemical world—occupational and environmental significance of industrial carcinogens, *Ann. NY Acad. Sci.*, 1988, *534*, whole issue; (h) J. Lynch, *Kirk–Othmer Encyclopedia of Chemical Technology*, 4th ed., Wiley, New York, 1995, *14*, 199; (i) B. Ballantyne, *Kirk–Othmer Encyclopedia of Polymer Science and Engineering*, 2nd ed., 1989, *16*, 878; (j) R.P. Pohanish and S.A. Greene, *Hazardous Materials Handbook*, Wiley, New York, 1996; (k) R.J. Lewis, Sr., *Hazardous Chemicals Desk Reference*, 4th ed., Wiley, New York, 1996; (l) G. Schuurmann, B. Market, *Ecotoxicology, Ecological Fundamentals, Chemical Exposure and Biological Effects*, Wiley, New York, 1997; (m) P. Calow, *Handbook of Ecotoxicology*, Blackwell Scientific, London, 1993; (n) M. Richards, *Environmental Xenobiotics*, Taylor & Francis, London, 1996.
23. B.M. Jacobson, *Chem. Eng. News*, Jan. 11, 1999, 2.
24. R.D. Coker, *Chem. Ind.* (*London*), 1995, 260.
25. (a) S. Budavaria, M.J. O'Neil, A. Smith, P.E. Heckelman, and J.F. Kinneary, eds, *Merck Index*, 12th ed., Merck & Co., Whitehouse Station, NJ, 1996, 1578; (b) Anon., *Tufts University Diet and Nutrition Letter*, 1996, *14*(6), 8.
26. T. Nishikawa, D. Urabe, and M. Isobe, *Angew. Chem. Int. Ed.*, 2004, *43*, 4782.
27. D. Hanson, *Chem. Eng. News*, Sept. 12, 1994, 16.
28. B.N. Ames and L.S. Gold, *Angew. Chem. Int. Ed.*, 1990, *29*, 1197; *258*, 261.
29. Anon., *Amicus J.*, 1996, *18*(3), 4.
30. (a) Anon., *ASH Smoking Health Rev.*, July–Aug. 1998, 7; (b) T. Levin, *On Earth*, 2008, *30*(2), 44; (c) Anon., *Harvard Health Lett.*, 2001, *26*(8), 6; (d) Anon., *World Watch*, 2002, *15*(6), 40; (e) Anon., *University of California Berkeley Wellness Lett.*, 2004, *20*(9), 1.
31. Anon., *New Sci.*, Apr. 25, 1998, 19.
32. H. Wechsler, N.A. Rigotti, J. Gledhill-Hoyt, and H. Lee, *JAMA*, 1998, *280*, 1673.
33. Anon., *Environ. Sci. Technol.*, 1998, *32*, 213A.
34. (a) D. Pimentel, M. Tort, L. D' Anna, A. Krawic, J. Berger, J. Rossman, F. Mugo, N. Doon, M. Shriberg, E. Howard, S. Lee, and J. Talbot, Ecology of increasing disease, population growth and environmental degradation, *BioScience*, 1998, *48*(10), 817; (b) Anon., *1998–1999 World Resources: A Guide to the Global Environmental*, World Resources Institute, Washington, DC; (c) Anon., *Chem. Eng. News*, May 11, 1998, 21.

35. www.dietandcancerreport.org.
36. F.P. Perera, *Science*, 1997, *278*, 1068.
37. (a) Anon., *Johns Hopkins Medical Lett., Health After 50*, 1996, *8*(6), 1; (b) M. Nestle, *Science*, 2003, *299*, 781, 846–859 (special section on obesity); (c) F as in Fat: How Obesity Policies Are Failing in America 2007, www.healthyamericans.org/reports/obesity2007; (d) C. Runyan, *World Watch*, 2001, *14*(2), 11.
38. R. Schuhmacher, *Chem. Ind (London)*, July 17, 2006, 24.
39. Anon., *Chem. Eng. News*, Apr. 11, 2005, 26.
40. Anon., *University California Berkeley Wellness Lett.*, Sept. 1998, *14*(12), 6.
41. (a) Anon., *Johns Hopkins Medical Lett., Health After 50*, 1996, *8*(2), 6; (b) Anon., *Tufts University Diet and Nutrition Lett.*, 1996, *14*(4), 1; (c) 1996, *14*(5), 4–6; (d) Anon., *Tufts Health and Nutrition Lett.*, Feb. 2006, special supplement on salt.
42. (a) Anon., *Duke Medicine Health News*, 2008, *14*(6), 12; (b) Anon., *University of California Berkeley Wellness Lett.*, 2007, *23*(6), 1.
43. (a) G. Taubes, *Science*, 1998, *281*, 898; (b) D.A. McCarron, *Science*, 1998, *281*, 933.
44. (a) Anon., *Tufts Health & Nutrition Lett.*, 2008, *26*, 1; (b) N.R. Cook, J. Cutler, E. Obarzanek, E. LBuring, K.M. Rexrode, S.K. Kumanyika, L.J. Appel, and P.K. Whelton, *Br. Med. J.*, 2007, *334*, 885.
45. Anon., *University of California Berkeley Wellness Lett.*, 2008, *24*(8), 1.
46. (a) Anon., *Environ. Sci. Technol.*, 1998, *32*, 81A; (b) Anon., *Chem. Eng. News*, Nov. 24, 1997, 60.
47. (a) S.J. Risch and C.-T. Ho, eds, *Spices: Flavor Chemistry and Antioxidant Properties*. A.C.S. Symp. 660, Washington, DC, 1997; (b) J.T. Kumpulainen and J.T. Salonen, *Natural Antioxidants and Anticarcinogens in Nutrition, Health and Disease*, Special Pub. 240, Royal Society of Chemistry, Cambridge, 1999.
48. M. Jang, L. Cai, G.O. Udeani, K.V. Slowing, C.F. Thomas, C.W.W. Beecher, H.H.S. Fong, N.R. Farnsworth, A.D. Kinghorn, R.G. Mehti, R.C. Moon, and J.M. Pezzuto, *Science*, 1997, *275*, 218.
49. Y. Bhattacharjee, *Science*, 2008, *320*, 593.
50. M.N. Diaz, B. Frei, J.A. Vita, and J.F. Keaney, Jr., *N. Engl. J. Med.*, 1997, *337*, 408.
51. F. Shahidi, Ed., *Antinutrients and Phytochemicals in Foods*, A.C.S. Symp. 662, Washington, DC, 1997.
52. National Research Council, *Carcinogens and Anticarcinogens in the Human Diet*, National Academy of Science Press, Washington, DC, 1996.
53. (a) Anon., *Chem. Ind. (London)*, 1996, 159; (b) J. Long, *Chem. Eng. News*, Feb. 26, 1996, 7; (c) Anon., *Environ. Sci. Technol.*, 1996, *30*(5), 199A.
54. H. Kasai, Z. Yamaizumi, T. Shiomi, S. Yokoyama, T. Miyazawa, K. Wakabayashi, M. Nagao, T. Sugimura, and S. Nishimura, *Chem. Lett.*, 1981, 485.
55. Anon., *Chem. Eng. News*, May 9, 2005, 35.
56. (a) C.M. Williams, *Chem. Ind. (London)*, 1993, 280; (b) W. Watson and B. Gloding, *Chem. Br.*, 1998, *34*(7), 45; (c) www.dietandcancerreport.org; (d) J. Ali, *Chem. Ind. (London)*, May 19, 2003, 15.
57. J.C. Bailar, III and H.L. Gornik, *N. Engl. J. Med.*, 1997, *336*, 1569.
58. (a) I. Wickelgren, *Science*, 1998, *280*, 1364; (b) G. Taubes, *Science*, 1998, *280*, 1367; (c) L.A. Campfield, F.J. Smith, and P. Burn, *Science*, 1998, *280*, 1383.
59. B.R. Jasny and F.E. Bloom, *Science*, 1998, *280*, 1507.
60. Anon., *University California Berkeley Wellness Lett.*, 1999, *15*(5), 8.
61. T. Murcott, *The Whole Story: Alternative Medicine on Trial*, Palgrave MacMillan, New York, 2005.
62. P.B. Fontanarosa, G.D. Lundberg, eds, *JAMA*, 1998; *280*, 1569–1618.
63. (a) R.S. diPaola, H. Zhang, G.H. Lambert, R. Meeker, E. Licitra, M.M. Rafe, B.T. Zhu, H. Spaulding, S. Goodin, M.B. Toledano, W.W. Hait, and M.A. Gallo, *N. Engl. J. Med.*, 1998, *339*, 785; (b) N.R. Slifman, W.R. Obermeyer, B.K. Aloi, S.M. Musser, W.A. Correll, Jr., S.M. Cichowicz, J.M. Betz, and L.A. Love, *N. Engl. J. Med.*, 1998, *339*, 806; (c) Y. Beigel, I. Ostfeld, and N. Schoenfeld, *N. Engl. J. Med.*, 1998, *339*, 827; (d) M. Angell and J.P. Kassirer, *N. Engl. J. Med.*, 1998, *339*, 839; (e) M.J. Coppes, R.A. Anderson, R.M. Egeler, and J.E.A. Wolf, *N. Engl. J. Med.*, 1998, *339*, 846; (f) R.J. Ko, F. LoVecchio, S.C. Curry, and T. Bagnasco, *N. Engl. J. Med.*, 1998, *339*, 847; (g) A. Bluming, *N. Engl. J. Med.*, 1998, *339*, 855; (h) P. Kurtzweil, *FDA Consumer*, 1998, *32*(5), 28.
64. Anon., *University of California at Berkeley Wellness Lett.*, 1998, *15*(3), 1.
65. L.D. Lawson and R. Bauer, *Phytomedicines of Europe: Chemical and Biological Activity*, A.C.S. Symp. 691, Washington, DC, 1998. (Regulatory efforts in European countries are covered in the first two chapters.)
66. (a) E. Stokstad, *Science*, 2004, *304*, 188; (b) D.M. Marcus and A.P. Grollman, *Science*, 2003, *301*, 1669.
67. (a) D. Normile, *Science*, 2003, *299*, 188; (b) R. Stone, *Science*, 2008, *319*, 709.
68. Anon., *University of California at Berkeley Wellness Lett.*, 1995, *11*(6), 5.
69. (a) A.N. Mayeno and G.J. Gleich, *Trends Biotechnol.*, 1994, *12*, 346; (b) L.R. Ember, *Chem. Eng. News*, June 17, 1996, 20.
70. Anon., *Chem. Eng. News*, Sept. 7, 1998, 30.
71. A.M.-C. Ibanez, *An. Quim. Ser. C.*, 1987, *83*(1), 107; *Chem. Abstr.*, 1987, *107*, 95547.
72. S.L. Rovner, *Chem. Eng. News*, May 12, 2008, 41.
73. (a) J.F. Tremblay, *Chem. Eng. News*, May 22, 2006, 11; (b) J. Kemsley, *Chem. Eng. News*, May 12, 2008, 37.
74. J. Kemsley, *Chem. Eng. News*, May 12, 2008, 38.
75. (a) M. Day, *New Sci.*, Jan. 24, 1998; 16; (b) G. Davison and C.N. Hewitt, eds, *Air Pollution in the United Kingdom*, Royal Society of Chemistry, Cambridge, 1997.
76. C.M. Cooney, *Environ. Sci. Technol.*, 1998, *32*, 250A.
77. K.A. Colburn and P.R.S. Johnson, *Science*, 2003, *299*, 665.
78. B. Halford, *Chem. Eng. News*, May 26, 2008, 9.
79. Anon., *Chem. Eng. News*, July 7, 1997, 30.
80. Anon., *Chem. Eng. News*, Sept. 22, 1997, 17.
81. P. Morse, *Chem. Eng. News*, Sept. 7, 1998, 12; Nov. 9, 1998, 42.
82. Anon., *Chem. Eng. News*, Jan. 4, 1999, 16.
83. (a) N.A. Ashford and C.S. Miller, *Chemical Exposures: Low Levels and High Stakes*, 2nd ed., Van Nostrand–Reinhold, New York, 1998; (b) J.D. Spengler and A.C. Rohr, *Chem. Eng. News*, Sept. 21, 1998, 105.

84. (a) B. Hileman, *Chem. Eng. News*, Mar. 10, 1997, 35; (b) P. Liese, *Chem. Ind. (London)*, 1998, 548; (c) B. Hileman, *Chem. Eng. News*, Apr. 7, 2003, 23; (d) P.J. Landrigan and L. Goldman, *Chem. Eng. News*, July 21, 2003, 3.
85. K.D. Sinclair, R.G. Lea, W.D. Rees, and L.E. Young, *Soc. Reprod. Fertil. Suppl.*, 2007, *64*, 425–443.
86. (a) B. Hileman, *Chem. Eng. News*, Aug. 15, 2005, 35; (b) Anon., *Chem. Eng. News*, Mar. 12, 2007, 28.
87. V. Barnett and A. O'Hagan, eds, *Setting Environmental Standards—The Statistical Approach to Handling Uncertainty and Variation*, Chapman & Hall/CRC Press, Boca Raton, FL, 1997.
88. (a) J. Kaiser, *Science*, 1999, *283*, 18; (b) B. Hileman, *Chem. Eng. News*, Aug. 3, 1998, 8.
89. B.-U. Hildebrandt and U. Schlottmann, *Angew. Chem. Int. Ed.*, 1998, *37*, 1317.
90. (a) J. Johnson, *Chem. Eng. News*, Apr. 27, 1998, 7; May 18, 1998, 7; (b) Anon., *Environ. Sci. Technol.*, 1999, *33*, 15A; (c) Anon., *Chem. Ind. (London)*, 1998, 827; (d) J.V. Rodricks, ed., *Calculated Risks—The Toxicity and Human Health Risk of Chemicals in Our Environment*, Cambridge University Press, UK, 2007.
91. J. Johnson, *Chem. Eng. News*, Jan. 18, 1999, 13; Mar. 8, 1999, 9.
92. J. Johnson, *Chem. Eng. News*, Jan. 25, 1999, 10.
93. C. Hogue, *Chem. Eng. News*, June 9, 2008, 33.
94. Anon., *Chem. Eng. News*, Aug. 27, 2007, 29.
95. C. Hogue, *Chem. Eng. News*, Oct. 1, 2007, 28.
96. (a) P.L. Short, *Chem. Eng. News*, Aug. 13, 2007, 33; (b) S. Boxerman, C. Bell, K.N. Nordlander, and N.K. Strickland, *Chem. Eng.*, 2008, *115*(3), 38, 42; (c) M. McMillan, *Chem. Ind. (London)*, March 10, 2008, 24; (d) M. Schapiro, *Exposed: The Toxic Chemistry of Everyday Products and What's at Stake for American Power*, Chelsea Green Publishing, White River Junction, VT, 2007.
97. K.H. Whiteside, *Precautionary Politics: Principle and Practice in Confronting Environmental Risk*, MIT Press, Cambridge, MA, 2006.
98. www.epa.gov/ecotox; http://toxtown.nim.nih.gov.
99. T. Kletz, *Chem. Br.*, 1999, *35*(1), 37.
100. S. Budavari, M.J. O'Neil, A. Smith, P.E. Heckelman, and J.F. Kinneary, eds, *The Merck Index*, 12th ed., Merck & Co., Whitehouse Station, NJ, 1996.
101. W.A. Burgess, *Recognition of Health Hazards in Industry*, 2nd ed., Wiley, New York, 1995.
102. K.J. Watkins, *Chem. Eng. News*, Aug. 13, 2001, 17.
103. (a) Advertisement, *Chem. Eng. News*, Jan. 11, 1999 (inside back cover); (b) S.C. Stinson, *Chem. Eng. News*, July 13, 1998, 57, 71; *Chem. Eng. News*, Jan. 27, 2003, 1.
104. Anon., *Chem. Eng. News*, July 10, 2000, 70; *Chem. Eng. News*, Nov. 22, 1999, 33; *Chem. Eng. News*, Apr. 21, 2003, 30; *Chem. Eng. News*, Feb. 14, 2000, 94.
105. G. Amery, *Aldrichchim. Acta*, 2001, *34*(2), 61.
106. K. Snyder, U.S. Mine Safety and Health Administration, 1997.
107. A.M. Thayer, *Chem. Eng. News*, Apr. 27, 1998, 15.
108. Anon., *Chem. Br.*, 1998, *34*(8), 13.
109. F.P. Lees, *Loss Prevention in the Process Industries*, Butterworth–Heinemann, Oxford, 1996.
110. (a) J. Johnson, *Chem. Eng. News*, Mar. 15, 1999, 12; (b) Anon., *Chem. Eng. News*, Nov. 2, 2009, 22.
111. G. Parkinson, *Chem. Eng.*, 1997, *104*(1), 21.
112. Anon., *Chem. Eng. News*, Apr. 19, 2004, 29.
113. J. Cheftel, *Chem. Ind. (London)*, Mar. 7, 2005, 12.
114. J. Johnson, *Chem. Eng. News*, Apr. 7, 2008, 44; Aug. 13, 2007, 38; June 16, 2003, 7.
115. K.R. Westerterp and E.J. Molga, *Ind. Eng. Chem. Res.*, 2004, *43*, 4585.
116. J. Johnson, *Chem. Eng. News*, Feb. 18, 2008, 5.
117. *Chem. Eng. News*, Dec. 1998, 23.
118. W. Stellmach, *Chem. Eng. Prog.*, 1998, *94*(8), 71.
119. (a) F.R. Spellman and N.E. Whiting, *Safety Engineering: Principles and Practices*, Government Institutes, Rockville, MD, 1999; (b) T.A. Kletz, *Process Safety: A Handbook of Inherently Safer Design*, 2nd ed., Taylor & Francis, Philadelphia, 1996; (c) D.A. Crowl, ed., *Inherently Safer Chemical Processes*, A.I.Ch.E., New York, 1996.
120. (a) M. Heylin, *Chem. Eng. News*, Dec. 19, 1994, 3; (b) S. Jasanoff, ed., *Learning from Disaster: Risk Management after Bhopal*, University of Pennsylvania Press, Philadelphia, PA, 1994; (c) W. Lepkowski, *Chem. Eng. News*, Dec. 19, 1994, 8; (d) D. Lapierre and J. Moro, *Five Past Midnight in Bhopal*, Warner Books, Clayton, Victoria, Australia, 2002; (e) C. Crabb, *Science*, 2004, *306*, 1670; (f) T. d'Silva, *The Black Box of Bhopal: A Closer Look at the World's Deadliest Industrial Disaster*, Trafford Publishing, Bloomington, Indiana, 2006; (g) J.-F. Tremblay, *Chem. Eng. News*, Jan. 24, 2005, 26, 28.
121. H.S. Brown, *Chem. Eng. News*, Oct. 10, 1994, 38.
122. P. Orum, *Chem. Ind. (London)*, 1998, 500.
123. J.-F. Tremblay, *Chem. Eng. News*, Aug. 17, 1998, 20.
124. (a) M. Freemantle, *Chem. Eng. News*, Apr. 29, 1996, 18; (b) C. Hohenemser, F. Warner, B. Segerstahl, and V.M. Novikov, *Environment*, 1996, *38*(3), 3; (c) D. Hanson, *Chem. Eng. News*, Sept. 12, 2005, 11; (d) K. Charman, *World Watch*, 2006, *19*(4), 12, 19; (e) J. Bohannon, *Science*, 2005, *309*, 1663; (f) R. Stone, *Science*, 2006, *312*, 180; (g) R.K. Chesser and R.J. Baker, *Am. Sci.*, 2006, *94*, 542.
125. (a) M. Balter. *Science*, 1996, *270*, 1758; (b) J. Webb. *New Sci.*, Apr. 1, 1995, 7.
126. J. Johnson, *Chem. Eng. News*, Aug. 17, 1997, 10.
127. R. Stone, *Science*, 1998, *281*, 623.
128. (a) B. Segerstahl, A. Akleyev, and V. Novikov, *Environment*, 1997, *39*(1), 12; (b) E. Marshall, *Science*, 1997, *275*, 1062; (c) Anon., *Environ. Sci. Technol.*, 1998, *32*, 80A; (d) R. Stone, *Science*, 1999, *283*, 158.
129. J.S. Walker, *Three Mile Island: A Nuclear Crisis in Historical Perspective*, University of California Press, Berkeley, CA, 2004.
130. P. Milford, *Wilmington Delaware News J.*, July 22, 29; Aug. 8, 1994; Jan. 5, 1995.
131. P. Silverberg, *Chem. Eng.*, 1996, *103*(5), 5.
132. P. Milford, *Wilmington Delaware News J.*, Dec. 14, 1995, B1.
133. P. Milford, *Wilmington Delaware News J.*, May 7, 1996, B1.
134. J. Montgomery, D. Thompson, Jr., and P. Milford, *Wilmington, Delaware News J.*, Apr. 2, 1998, B1; Apr. 11, 1998, B3; Aug. 11, 1998, B2; Sept. 22, 1998, B1; Dec. 4, 1998, B2; Dec. 10, 1998, B2; Dec. 12, 1998, B1.
135. Anon., *Chem. Eng. News*, July 21, 1997, 22.
136. (a) L.R. Ember, D.J. Hanson, G. Hess, B. Hileman, C. Hogue, J. Johnson, and S.R. Morrissey, *Chem. Eng. News*, Jan. 21, 2008, 35; (b) Anon., *Wall Street J.*, Jan. 11, 2008, B6.

137. J. Johnson, *Chem. Eng. News*, Aug. 4, 1997, 10.
138. Anon., *Chem. Eng. News*, Sept. 8, 1997, 25.
139. A Lawler, *Science*, 1998, *279*, 1444; 1998, *280*, 29.
140. C. Ramsey and M. Modarres, *Commercial Nuclear Power—Assuring Safety for the Future*, Wiley, New York, 1998.
141. P.H. Abelson, *Science*, 1996, *272*, 465.
142. M. Balter, *Science*, 1997, *275*, 610.
143. Anon., *Chem. Eng. News*, May 2, 2005, 25.
144. (a) D. Whitford, *Fortune*, Aug. 6, 207, 42; (b) M. Freemantle, *Chem. Eng. News*, Sept. 13, 2004, 31; (c) E. Marshall, D. Clery, G. Yidong, and D. Normile, *Science*, 2005, *309*, 1168, 1172, 1177; (d) M. Jacoby, *Chem. Eng. News*, Aug. 24, 2009, 14; (e) A. Kadak, *Technol. Rev.*, 2009, 112(2), 11.
145. Anon., *Amicus J.*, 1999, *21*(1), 16.
146. Anon., *Chem. Eng. News.*, Jan. 18, 33.
147. S.L. Wilkinson, *Chem. Eng. News*, Nov. 9, 1998, 71.
148. A.M. Thayer, *Chem. Eng. News*, Aug. 27, 1998, 15.
149. Anon., *Chem. Eng. News*, July 7, 1997, 22.
150. (a) Anon., *Chem. Eng. News*, June 29, 1998, 18; (b) P.M. Morse, *Chem. Eng. News*, Mar. 23, 1998, 17.
151. Anon., *Chem. Eng. News*, June 30, 1997, 19.
152. G. Peaff, *Chem. Eng. News*, Jan. 16, 1995, 16.
153. Anon., *Chem. Eng.*, 1998, *105*(10), 48.
154. Anon., *Chem. Eng. Prog.*, 1998, *94*(4), 86.
155. Anon., *Chem. Eng. News*, Aug. 17, 1998, 11.
156. Anon., *Chem. Eng. News*, Oct. 16, 1995, 9.
157. Anon., *Chem. Eng. News*, Apr. 28, 1997, 13; May 26, 1997, 12.
158. Anon., *The Explosion and Fires at the Texaco Refinery*, Milford Haven, 24 July, 1994, HSE Books, Sudbury, Suffolk, U.K.
159. Anon., *Chem. Eng. News*, Nov. 7, 1994, 9.
160. Anon., *Chem. Eng. News*, Dec. 5, 1994, 11.
161. J. Johnson, *Chem. Eng. News*, Nov. 13, 2006, 31; Jan. 22, 2007, 10; July 16, 2007, 7; (b) Chemical Safety and Hazard Investigation Board news release, Jan. 17, 2008; (c) G. Hess, *Chem. Eng. News*, Nov. 7, 2005, 10; (d) R. D'Aquino and S. Berger, *Chem. Eng. Prog.*, 2007, *103*(2), 14.
162. Natural Resources Defense Council, *Nature's Voice*, May/June 2006.
163. K. Weissermel and H.-J. Arpe, *Industrial Organic Chemistry*, 2nd ed., VCH, Weinheim, 1993, 17–176.
164. Anon., *Chem. Eng. News*, Jan. 30, 1995, 11.
165. Anon., *Chem. Eng. News*, July 31, 1995, 8.
166. Anon., *Chem. Eng. News*, Jan. 29, 1996, 11.
167. (a) T. Urbanski, *Chemistry and Technology of Explosives*, Pergamon Press, Oxford, 1965, *2*, 459–462; (b) B.S. Hopkins, *General Chemistry for Colleges*, DC Health, Boston, 1930, 560.
168. (a) W.H. Shearon, Jr., *Chem. Eng. News*, May 12, 1947, 1334; (b) *Chem. Eng. News*, Apr. 28, 1947, 1198; (c) M.S. Reisch, *Chem. Eng. News*, Jan. 12, 1998, 88; (d) A. Thayer, *Chem. Eng. News*, Apr. 21, 1997, 11.
169. (a) K. Watkins, *Chem. Eng. News*, Oct. 1, 2001, 14; (b) Anon., *Chem. Eng. News*, June 17, 2002, 9.
170. Anon., *Chem. Eng. News*, Oct. 24, 1994, 10.
171. D. Arntz, *Catal. Today*, 1993, *18*, 173.
172. R.A. Sheldon, *Chemtech*, 1994, *24*(3), 38.
173. W. Worthy, *Chem. Eng. News*, Nov. 9, 1997, 6.
174. G. Ondrey, *Chem. Eng.*, 1998, *105*(6), 29.
175. Anon., *Chem. Ind.* (*London*), 1998, 289.
176. (a) M. Reisch, *Chem. Eng. News*, June 23, 1997, 22; (b) Anon., *Chem. Br.*, 1997, *33*(7), 5.
177. (a) E. Kirschner, *Chem. Eng. News*, May 1, 1995, 6; (b) Anon., *Chem. Eng. News*, Oct. 27, 1997, 18; (c) C. Cooper, *Chem. Eng.*, 1997, *104*(11), 55.
178. Anon., *Chem. Eng. News*, Oct. 30, 1995, 10.
179. Anon., *Chem. Eng. News*, Dec. 12, 1994, 13.
180. Anon., *Chem. Eng. News*, Aug. 1, 1994, 17.
181. K. Fouhy, *Chem. Eng.*, 1996, *103*(4), 37.
182. (a) J.J. Adams, Natural Resources Defense Council, [letter], Oct. 1994; (b) D.A. Wolfe, M.J. Hamedi, J.A. Galt, G. Watabayashi, J. Short, C. O'Claire, S. Rice, J. Michel, J.R. Payne, J. Braddock, S. Hanna, and D. Sale, *Environ. Sci. Technol.*, 1994, *28*, 560A; (c) T.R. Loughlin, ed., *Marine Mammals and the Exxon Valdez*, Academic, San Diego, 1994; (d) D.L. Garshelis and C.B. Johnson, *Science*, 1999, *283*, 176; (e) J. Wheelwright, *Degrees of Disaster, Prince William Sound, How Nature Reels and Rebounds*, Yale University Press, New Haven, 1996; (f) F. Pearce, *New Sci.*, May 5, 2001, 4; (g) Anon., *Science*, 2008, *320*, 297.
183. Exxon Corp., Environment, Health and Safety Progress Report for 1995. 1996, p. 11.
184. K. Fouhy, *Chem. Eng.*, 1996, *103*(4), 37.
185. M. Fingas, *Chem. Ind.* (*London*), 1995, 1005.
186. Anon., *Chem. Ind.* (*London*), 1996, 275.
187. J. Pelley, *Environ. Sci. Technol.*, 2001, *35*, 276A.
188. (a) J. Bohannon, X. Bosch, and J. Withgott, *Science*, 2002, *298*, 1695; (b) J. Bohannon and X. Bosch, *Science*, 2003, *299*, 490; (c) S. Diez, E. Jover, J.M. Bayona, and J. Albaiges, *Environ. Sci. Technol.*, 2007, *41*, 3075.
189. (a) Anon., *Chem. Ind.* (*London*), Dec. 19, 2005, 4; (b) E. Dorey, *Chem. Ind.* (*London*), June 5, 2006, 12.
190. Anon., *Chem. Eng. News*, June 26, 1995, 13.
191. J. Johnson, *Chem. Eng. News*, Dec. 7, 1998, 36.
192. S. Ainsworth, *Chem. Eng. News*, July 3, 1995, 6.
193. P. Morse, *Chem. Eng. News*, Sept. 15, 1997, 11.
194. M.B. Brennan, *Chem. Eng. News*, June 23, 1997, 29.
195. (a) *Nowhere To Hide*, National Environmental Law Center, Boston, 1995; (b) *Chem. Ind.* (*London*), 1995, 677.
196. J.-F. Tremblay, *Chem. Eng. News*, Jan. 11, 1999, 17.
197. C. Bedford, *Chem. Ind.* (*London*), 1995, 36.
198. R. Baldini, *Chem. Eng. News*, May 8, 1995, 4.
199. E. Kirschner, *Chem. Eng. News*, Apr. 24, 1995, 23.
200. (a) I. Nimmo, *Chem. Eng. Prog.*, 1995, *91*(9), 36; (b) P. Bullemer and I. Nimmo, *Chem. Eng. Prog.*, 1998, *94*(1), 43.
201. (a) T. Kletz, *What Went Wrong: Case Histories of Process Plant Disasters*, Gulf Publishing, Houston, 1994; (b) *Loss Prevention*, vol. 18, American Institute Chemical Engineers Publ. T-93, New York, 1994; (c) C.R. Nelms, *What You Can Learn From Things That Go Wrong*, Failsafe Network, Richmond, VA, 1996.
202. (a) S.L. Wikinson, *Chem. Eng. News*, Nov. 9, 1998, 82; (b) H. Carmichael, *Chem. Br.*, 1998, *34*(4), 37; (c) H. Abbott and M. Tyler, *Safety by Design: A Guide to Management and Law of Designing for Product Safety*, 2nd ed., Glower Publishing Aldershot, UK, England, 1997.
203. (a) T. Kletz, *Learning from Accidents*, 3rd ed., Golf Professional Publishing, Oxford, 2001; (b) T. Kletz, *By Accident—A Life Preventing Them in Industry*, PFV

Publications, London, 2000; (c) D.C. Hendershot, *Process Safety Prog.*, 2006, *25*(2), 98. Additional references on inherently safer processes can be found by checking Dennis C. Hendershot in Google.
204. (a) P. Baybutt, *Chem. Eng. Prog.*, 2003, *99*(12), 35; (b) J. Johnson, *Chem. Eng. News*, Feb. 3, 2003, 23.
205. A.S. Matlack, *Green Chem.*, 2003, *5*, G3.
206. R.L. Reibstein and C.J. Cleveland, *Encyl. Earth*, 2006.
207. R. Baldini, *Chem. Eng. News*, May 8, 1995, 4.
208. L. Ember, *Chem. Eng. News*, Dec. 12, 2005, 24.
209. W. Lepkowski, *Chem. Eng. News*, June 20, 1994, 24.
210. (a) G. Hess, *Chem. Eng. News*, Mar. 17, 2008, 11; (b) D. Hanson, *Chem. Eng. News*, June 30, 2008, 27; (c) Anon., *Chem. Eng. News*, Oct. 26, 2009, 21.
211. Anon., *Chem. Eng. News*, Jan. 8, 2007, 39.
212. D.J. Hanson, *Chem. Eng. News*, July 6, 1998, 19.
213. Anon., *Chem. Ind.* (*London*), 1994, 320.
214. D. Hanson, *Chem. Eng. News*, Apr. 25, 1994, 8.
215. Anon., *Chem. Eng. Prog.*, 2008, *104*(3), 18.
216. C.Q. Jia, A. DiGuardo, and D. Mackay, *Environ. Sci. Technol.*, 1996, *30*(2), 86A.
217. B. Hileman, *Chem. Eng. News*, Jan. 1999, 4.
218. Anon., *Environ. Sci. Technol.*, 1995, *29*(9), 395A.
219. Anon., *Chem. Eng.*, 1995, *102*(4), 50.
220. B. Hileman, *Chem. Eng. News*, May 22, 1995, 43.
221. L. Ember, *Chem. Eng. News*, Sept. 19, 1994, 23.
222. G.W. Sage, *Chem. Eng. News*, Aug. 24, 1998, 8.
223. T.E. Natan, Jr. and C.G. Miller, *Environ. Sci. Technol.*, 1998, *32*, 368A.
224. Delaware Department of Natural Resources and Environmental Control [memo], Dover, DE, May 1996.
225. E. Kicko-Walczak and E. Grzywa, *Macromol. Symp.*, 1998, *127*, 265.
226. K. Miller, *Wilmington Delaware News J.*, Aug. 29, 1995.
227. Anon., *Environ. Sci. Technol.*, 1997, *31*, 451A.
228. Anon., *New Sci.*, June 3, 1995, 11.
229. M. Murray, *Wilmington, Delaware News J.*, May 1, 1996, A1.
230. C.A.G. Tulou, *Outdoor Delaware*, 1998, *7*(1), 2.
231. L. diBerardinis, ed., *Handbook of Occupational Safety and Health*, Wiley, New York, 1998.
232. (a) B. Hileman, *Chem. Eng. News*, Mar. 4, 1996, 16; (b) May 6, 1996, 9.
233. Anon., *Chem. Eng. News*, Jan. 22, 1996, 17.
234. D.J. Shanefield, *Chem. Eng. News*, Oct. 23, 1995, 4.
235. (a) R. Zanetti, *Chem. Eng.*, 1995, *102*(10), 5; (b) D. Hanson, *Chem. Eng. News*, May 22, 1995, 9.
236. (a) D. Mayo, R.M. Pike, and P.K. Trumper, *Microscale Organic Laboratory*, 3rd ed., Wiley, New York, 1994; (b) M.M. Singh, R.M. Pike, and Z. Szafran, *General Chemistry Micro- and Macroscale Laboratory*, Wiley, New York, 1994; (c) M.M. Singh, K.C. Swallow, R.M. Pike, and Z. Szafran, *J. Chem. Ed.*, 1993, *70*, A39; (d) M.M. Singh, R.M. Pike, and Z. Szafran, Preprints A.C.S. *Div. Environ. Chem.*, 1994, *34*(2), 194; (e) Z. Szafran, R.M. Pike, and J.C. Foster, *Microscale General Chemistry Laboratory with Selected Macroscale Experiments*, Wiley, New York, 1993; (f) Z. Szafran, R.M. Pike, and M.M. Singh, *Microscale Inorganic Chemistry: A Comprehensive Laboratory Experience*, Wiley, New York, 1991; (g) J.L. Skinner, *Microscale Chemistry—Experiments in Miniature*, Royal Society of Chemistry, Cambridge, 1998.
237. M. Freemantle, *Chem. Eng. News*, Feb. 22, 1999, 27.
238. Anon., *Chem. Eng. News*, June 3, 1996.
239. Anon., *Chem. Eng. News*, Mar. 30, 1998, 29.
240. S. Mazrou, H. Kerdjoudj, A.T. Cherif, A. Elmidaoui, and J. Molenat, *New J. Chem.*, 1998, *22*, 355.
241. Anon., *Environ. Sci. Technol.*, 1998, *32*, 119A.
242. Guide to Cleaner Technologies—Cleaning and Degreasing Process Changes, U.S. EPA/625/R-93/017 Washington, DC.
243. D. Luper, *Chem. Eng. Prog.*, 1996; *92*(6), 58.
244. (a) N. Chadha, *Chem. Eng. Prog.*, 1994, *90*(11), 32; (b) Office of Technology Assessment, *Industry, Technology and the Environment: Competitive Challenges and Business Opportunities*, Washington, DC, 1994, Chap. 8; (c) K. Martin and T.W. Bastock, eds, *Waste Minimisation: A Chemist's Approach*, Royal Society of Chemistry, Cambridge, 1994; (d) J.H. Siegell, *Chem. Eng.*, *103*(6), 92; (e) M. Venkatesh and C.W. Moores, *Chem. Eng. Prog.*, 1998, *94*(11), 26; (f) S.B. Billatos and N.A. Basaly, *Green Technology and Design for the Environment*, Taylor & Francis, Washington, DC, 1997; (g) M. Venkatesh, *Chem. Eng. Prog.*, 1997, *93*(5), 33. (h) P. Crumpler, *Chem. Eng.*, 1997, *104*(10), 102.
245. (a) K. Fouhy, *Chem. Eng.*, 1995, *102*(1), 41; (b) J. Jarosch, *Chem. Eng.*, 1996, *103*(5), 120; (c) *Chem. Eng.*, 1996, *103*(5), 127; (d) D.M. Carr, *Chem. Eng.*, 1995, *102*(8), 78; (e) B.J. Netzel, *Chem. Eng.*, 1995, *102*(8), 82B; (f) Anon., *Chem. Eng.*, 1995, *102*(11); (g) C. Brown and P. Dixon, *Chem. Eng. Prog.*, 1996, *92*, 42; (h) J.H. Siegell, *Chem. Eng. Prog.*, 1998, *94*(11), 33.
246. E.M. Kirschner, *Chem. Eng. News*, July 10, 1995, 14.
247. Environmental, Health and Safety Progress Report for 1995, Exxon Corp, New York, p. 11.
248. J. Szekely and G. Trapaga, *Technol. Rev.*, 1995, *98*(1), 30.
249. M.A. Martin-Cabrejas, R.M. Esteban, F.J. Lopez-Andreu, K. Waldron, and R.R. Selvendran, *J. Agric. Food Chem.*, 1995, *43*, 662.
250. M. Reisch, *Chem. Eng. News*, Feb. 10, 1997, 9.
251. (a) B. Crittenden and S.T. Kolaczkowski, *Waste Minimization Guide*, *Inst. Chem. Eng.*, Rugby, UK, 1994; (b) D.F. Ciambrone, *Waste Minimization as a Strategic Weapon*, Lewis Publishers, Boca Raton, FL, 1995; (c) T.E. Higgins, *Pollution Prevention Handbook*, Lewis Publishers, Boca Raton, FL, 1995; (d) D.T. Allen, *Adv. Chem. Eng.*, 1994, *19*, 251; (e) L. Theodore, *Pollution Prevention*, van Nostrand–Reinhold, New York, 1992; (f) L. Theodore, *Pollution Prevention—Problems and Solutions*, Gordon and Breach, Reading, 1994; (g) M.J. Healy, *Pollution Prevention Opportunity Assessments—A Practical Guide*, Wiley, New York, 1998; (h) G.F. Nalven, *Practical Engineering Perspectives—Environmental Management and Pollution Prevention*, American Institute of Chemical Engineers, New York, 1997; (i) D.T. Allen and K.S. Ross, *Pollution Prevention for Chemical Processes*, Wiley, New York, 1997; (j) J.H. Clark, ed., *Chemistry of Waste Minimisation*, Blackie, London, 1995; (k) EPA, *Pollution Prevention Guidance Manual for the Dye Manufacturing Industry*, U.S. Environmental Protection Agency, EPA/741/B-92-001.
252. (a) E.G. Chatzi, O. Kammona, and C. Kiparissides, *J. Appl. Polym. Chem.*, 1997, *63*, 799; (b) Such probes are available commercially. For example, from ASI Applied Systems, Annapolis, MD.

253. K. Weissermel and H.-J. Arpe, *Industrial Organic Chemistry*, 2nd ed., VCH, Weinheim, 1993, 388–390.
254. (a) L.E. Manzer in P.T. Anastas and C.A. Farris, *Benign By Design: Alternative Synthetic Design for Pollution Prevention*, A.C.S. Symp. 577, Washington, DC, 1994, 146–147; (b) M. Joucla, P. Marion, P. Grenouillet, and J. Jenck in J.R. Kosak, and T.A. Johnson, eds, *Catalysis of Organic Reactions*, Marcel Dekker, New York, 1994, p. 127.
255. K. Weissermel and H.-J. Arpe, *Industrial Organic Chemistry*, 2nd ed., VCH, Weinheim, 1993, 238–239.
256. K. Weissermel and H.-J. Arpe, *Industrial Organic Chemistry*, 2nd ed., VCH, Weinheim, 1993, 241–242.
257. K. Weissermel and H.-J. Arpe, *Industrial Organic Chemistry*, 2nd ed., VCH, Weinheim, 1993, 244–245.
258. K.A. Kreutzer and W. Tam, *Chem. Abstr.*, 1996, *124*, 118251, 147112.
259. (a) J.H. Krieger, *Chem. Eng. News*, July 8, 1996, 12; (b) P.V. Tebo, *Chemtech*, 1998, *28*(3), 8.
260. P.W.N.M. vanLeeuwen and C. Claver, eds, *Rhodium-Catalyzed Hydroformylation*, Kluwer Academic Publishers, Dordrecht, 2000.
261. K. Weissermel and H.-J. Arpe, *Industrial Organic Chemistry*, 2nd ed., VCH, Weinheim, 1993, 127–131.
262. B. Cornils and E. Wiebus, *Chemtech*, 1995, *25*(1), 33.
263. B. Fell, C. Schobben, and G. Papadogianakis, *J. Mol. Catal., A: Chem.*, 1995, *101*, 179.
264. (a) SRI International, *Chemical Economics Handbook*, Jan. 1991, under Oxo Chemicals, Menlo Park, CA, (b) H. Bach, R. Gartner, and B. Cornils, *Ullmann's Encyclopedia Industrial Chemistry*, 5th ed., VCH, Weinheim, 1985, *A4*, 450–452.
265. Anon., *Chem. Eng. News*, June 16, 2008, 62.
266. R. Song, D. Ostgard, and G.V. Smith. In: W.E. Pascoe, ed., *Catalysis of Organic Reactions*, Marcel Dekker, New York, 1992.
267. E.M. Kirschner, *Chem. Eng. News*, July 1, 1996, 18.
267. A. Turbak, *Encyclopedia of Polymer Science and Engineering*, 2nd ed., Wiley, New York, 1988, *14*, 45.
269. (a) M. Hirami, *J. Macromol. Sci. Pure Appl. Chem.*, 1996, *A33*, 1825; (b) C. O'Driscoll, *Chem, Br.*, 1996, *32*(12), 27; (c) G. Parkinson, *Chem. Eng.*, 1996, *103*(10), 19; (d) S. Dobson, *Chem. Ind.* (*London*), 1995, 870.
270. (a) R.A. Sheldon, *Chemtech*, 1994, *24*(3), 38; (b) C.B. Dartt and M.E. Davis, *Ind. Chem. Res.*, 1994, *33*, 2887; (c) J.A. Cusumano, *Chemtech*, 1992, *22*(8), 482; *Appl. Catal. A General*, 1994, *113*, 181; (d) J. Haber, *Pure Appl. Chem.*, 1994, *66*, 1597; (e) A. Mittelman and D. Lin, *Chem. Ind.* (*London*), 1995, 694.
271. R.A. Sheldon, *Chemtech*, 1994, *24*(3), 38.
272. (a) N. Bodor, *Chemtech*, 1995, *25*(10), 22; (b) N. Bodor. In: S.C. de Vito, and R.L. Garrett, eds, *Designing Safer Chemicals: Green Chemistry for Pollution Prevention.* A.C.S. Symp. 640, Washington, DC, 1996; (c) See also, S.C. de Vito, *Chemtech*, 1996, *26*(11), 34 for more ways to design safer chemicals.
273. L.J. Casarett and J. Doull, eds, *Toxicology—The Basic Science of Poisons*, Macmillan, New York, 1975, 195, 514, 516, 720.
274. National Science and Technology Council, A National R&D Strategy for Toxic Substances and Hazardous and Solid Waste, Sept. 1995, Available from U.S. EPA, Office of Research and Development, Washington, DC, 20460.
275. J.S. Hirschhorn, *Prosperity Without Pollution*, van Nostrand-Reinhold, New York, 1991.
276. K. Weissermel and H.-J. Arpe, *Industrial Organic Chemicals*, 2nd ed., VCH, Weinheim, 1993, 265–267.
277. M. McCoy, *Chem. Eng. News*, 1998, Aug. 3, 20.
278. J. Boor, Jr., *Ziegler-Natta Catalysts and Polymerization*, Academic Press, New York, 1979.
279. G. Fink, R. Mulhaupt, and H.H. Brintzinger, eds, *Ziegler Catalysis—Recent Scientific Innovations and Technological Improvements*, Springer, Heidelberg, 1995.
280. (a) J. Scheirs and W. Kaminsky, eds, *Metallocene-Based Polyolefins*, Wiley, New York, 2000; (b) T. Takahasha, ed., *Metallocenes in Regio- and Stereoselective Syntheses*, Springer, Heidelberg, 2005; (c) W. Kaminsky, ed., *Macromol. Symp.*, 2006, *236*, 1–258.
281. B. Rieger, L.S. Baugh, S. Kacker, and S. Striegler, eds, *Late Transition Metal Polymerization Catalysis*, Wiley, New York, 2003.
282. T.R. Younkin, E.F. Connor, J.I. Henderson, S.K. Friedrich, R.H. Grubbs, and D.A. Bansleben, *Science*, 2000, *287*, 460.
283. S.-M.Yu, A. Berkefeld, I. Gottker-Schnitmann, G. Muller, and S. Mecking, *Macromolecules*, 2007, *40*, 421.
284. J.S. Hirschhorn, *Chemtech*, 1995, *25*(4), 6.
285. B.M. Rushton, *Chem. News*, Jan. 2, 1995, 2.

It would also be good to read all or most of the following before finishing this book:

A. Johansson, *Clean Technology*, Lewis Publishers, Boca Raton, FL, 1992.

K. Martin and T.W. Bastock, *Waste Minimisation: A Chemist's Approach*, Royal Society of Chemistry, Cambridge, 1994.

For additional reading on a sustainable future, see

J.S. Hirschhorn, *Chemtech*, 1995; 25(4):6.

World Commission on Environment and Development, *Our Common Future*, Oxford University Press, Oxford, 1987.

G.G. Lebel and H. Kane, *Sustainable Development: A Guide to Our Common Future—The Report of the World Commission on Environment and Development*, Global Tomorrow Coalition, Washington, DC, 1989.

D.H. Meadows, D.L. Meadows, and J. Randers, *Beyond the Limits: Confronting Global Collapse, Envisioning a Sustainable Future*, Chelsea Green Publishers, Mills, VT, 1992.

RECOMMENDED READING

1. Waste Minimization

   D.T. Allen, Adv. Chem. Eng., 1994, 19, 251–289, 304–312, 318–323.

   N. Chadha, Chem. Eng. Prog., 1994, 90(11), 32.

   D. Luper, Chem. Eng. Prog., 1996, 92(6), 58.

2. Safer Chemicals and Pollution Prevention

   R.A. Sheldon, Chemtech, 1994, 24(3), 38.

   J.A. Cusumano, J. Chem. Ed., 1995, 72, 959; Chemtech, 1992, 22(8), 482.

   C.B. Dartt and M.E. Davis, Ind. Eng. Chem. Res., 1994, 33, 2887.

   J.F. Murphy, Chem. Eng. Prog., 2007, 103(8), 33.

http://home.att.net/~d.c.hendershot/papers/htm or put Dennis Hendershot into Google and look for the paper that he presented to the Green Chemistry Conference in Washington, DC in 2003.

3. Carcinogens in Foods
   B.N. Ames and L.S. Gold, Science, 1992, 258, 261; Angew, Chem. Int. Ed. Engl., 1990, 29, 1197.
4. Alternative Medicine
   M. Angell and J.P. Kassirer, N. Engl. J. Med., 1998, 339, 839.
5. Ecology of Increasing Disease
   D. Pimentel, M.L. Tort, L. D'Anna, A. Krawic, J. Berger, J. Rossman, F. Mugo, J. Doon, M. Shriberg, E. Howard, S. Lee, and J. Talbot, Bioscience, 1998, 48, 817.

EXERCISES

1. Check the companies in your area in the Toxic Release Inventory (or a comparable compilation; see Preface for websites) Compare their releases with comparable companies in other areas. Can you devise ways in which the releases might be reduced? Could an alternative process eliminate the release altogether? If possible, visit one of the plants and see what the employees tell you about possible reduction of the releases.
2. Do you look at nuclear energy as a way to avert global warming or as a potential problem?
3. Are the EPA and the FDA (or comparable agencies in your country) impediments to progress or our guardians? Are they antiquated and unrealistic?
4. Check a year of *Chem. Eng. News*, *Chem. Ind.* (*London*), or other chemical news magazine for fires, explosions, spills, and other accidents.
5. Were there any fires, explosions, spills, and accidents in your laboratory last year? If so, who goofed on what?
6. Do you contribute to air and water pollution?
7. Are there any Superfund or comparable sites near you? If so, what led to them?
8. How do you feel about the use of tobacco, alcohol, salt, chemical pesticides, and being overweight and sedentary?
9. Should herbal teas, alternative medicines, and health food supplements be regulated by the government?
10. Do oil spills heal quickly or is there a long-term effect on nature?
11. Should all rivers and lakes be fishable and swimmable?
12. Pick out some wastes of industry that have often been pollutants. How could they have been avoided or made into useful products? If you have trouble finding some, try lignin, fly ash, "red mud" (a waste from refining aluminum ore), spent sulfuric acid, chicken feathers, nitrogen oxides from making nylon, or calcium sulfite from scrubbing stack gases.
13. Too much salt can be deadly. How can the food industry be persuaded to put less in processed foods?
14. Inherently safer chemistry can save lives and money. Why is the chemical industry opposed to it?
15. How can chemical companies be persuaded to endorse the precautionary principle?

CHAPTER 2

Doing without Phosgene, Hydrogen Cyanide, and Formaldehyde

2.1 INTRODUCTION

The chemicals used as reagents are often highly reactive and are used as they are. As such, they may react readily with components of the human body, such as water, or the hydroxyl and amino groups of proteins and nucleic acids. The next chapters will deal with efforts to replace these toxic materials with inherently safer chemistry. It may not be possible to replace them completely. This leads to a hierarchy of approaches (a tiered approach follows) from least change to the most change:

- No change: Those using the process are confident that they can handle it. They may not have had any accidents recently.
- Go to a closed system and carefully avoid leaks.
- Do not ship the chemical or store it in large quantities. Preferably, generate it on site *in situ* as needed.
- Farm out the work to specialist companies that are experienced in handling it and are willing to do custom syntheses.
- Replace the reagent with a less but still toxic reagent that may be easier to handle.
- Replace the reagent with a nontoxic reagent.
- Make the product by a different route, using nontoxic reagents.
- Substitute other products for the ones made with the toxic reagent, preferably ones based on renewable natural resources.
- Ask whether the use is really necessary at all (e.g., drinking a 10% solution of sugar in water from a poly(ethylene terephthalate) soda bottle instead of just going to a water fountain).
- Ban or tax the reagent or the use. Tetraethyllead, formerly used as an antiknock agent in gasoline, was banned when the effect on children's IQ was found.
- Chlorofluorocarbons are taxed, as well as being phased out, because they destroy the ozone layer.

This scheme will be applied to phosgene (bp 7.56°C),[1] a reagent that is made by the reaction of carbon monoxide with chlorine over activated carbon at elevated temperatures (**2.1** Schematic).[2]

$$CO + Cl_2 \longrightarrow COCl_2$$

2.1 Schematic

Both the reagents and the product are quite toxic. Phosgene was used as a war gas in World War I. It causes irritation of the eyes and nose at 3 ppm or more. The effects are insidious because the major discomfort comes not at first, but later. Current practice is to use it in closed systems and not to store or ship it. Workers wear indicator film badges to monitor their exposure and to alert them to danger. Custom phosgenations are offered by PPG Industries,[3] Hatco,[4] Rhone-Phoulenc,[5] SNPE[6] (including its subsidiary Isochem[7]), and Sigma-Aldrich.[8] SNPE has a plant that makes the phosgene as needed, so that at any one time only a few kilograms are present in the reactor and piping, an amount that can be handled by a scrubber using 10% sodium hydroxide. This is fine unless something springs a leak. The company distributed a book on phosgene in 1998.[9] A handbook on phosgenations is available.[10] There has been some concern about the development of phosgene in chloroform that has been stored for a long time.[11] The best solution is to use solvents other than chloroform (a carcinogen). The largest industrial use of phosgene[12] is in the preparation of isocyanates, most of which are used for the preparation of polyurethanes. Some isocyanates are also used to make agricultural chemicals. The second major use is in the preparation of polycarbonates.

In the laboratory bis(trichloromethyl)carbonate (triphosgene; 2.2) (mp 79–83°C) can be used in place of phosgene for greater safety and convenience.[1,13] The Sigma-Aldrich Chemical Co. catalog describes it as a moisture-sensitive lachrymator that should be handled with gloves in a hood. The company offers a kit for its conversion to phosgene in the lab (**2.2** Schematic).

$$Cl_3COCOOCCl_3$$

2.2 Schematic

2.2 PREPARATION OF ISOCYANATES

The usual method of preparation can be illustrated by that of toluenediisocyanate, a reagent that is then reacted with diols to form polyurethanes[14] (see also reviews[15]) (**2.3** Schematic).

Most of the weight of the phosgene is lost as hydrogen chloride, which is probably neutralized to form salts that are discarded. Bayer is building a new plant in China that will carry out the phosgenation in the gas phase at over 300°C. It will use 80% less solvent, use 40–60% less energy and save 20% in investment costs.[16]

Other commonly used diisocyanates are also made by reacting the diamines with phosgene (**2.4** Schematic).

The diisocyanates are reacted with diols to produce polyurethanes that end up in consumer products, such as seat cushions, mattresses, insulation, car bumpers, swim suits, floor coatings, paints, and adhesives.[17] The aromatic diisocyanates are cheaper, but yield less light-stable polymers than the aliphatic ones.

2.2.1 Other Ways to Make Isocyanates from Amines

The tragedy at Bhopal[18] involved methyl isocyanate. It is reacted with 1-naphthol to make a carbamate insecticide (**2.5** Schematic).

The DuPont Co. has developed a method (**2.6** Schematic) to produce methyl isocyanate on demand so that it need not be stored. The company no longer needs to ship tank cars of it from West Virginia to the Gulf Coast, and the new process is cheaper.[19] (As mentioned in Chapter 1, integrated pest management involving *Bacillus thuringiensis* could replace this insecticide.)

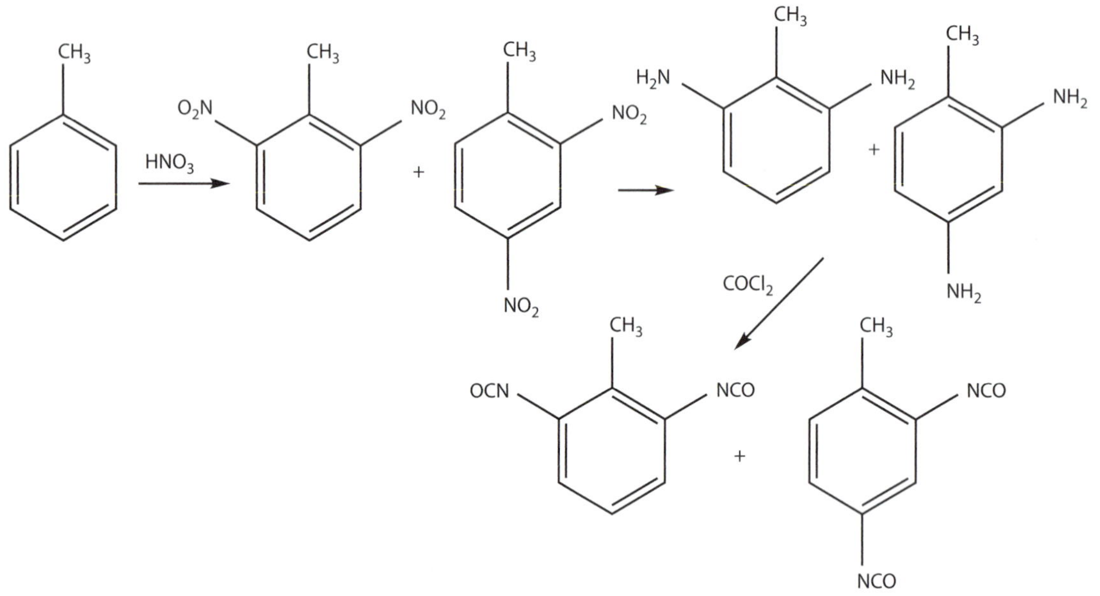

CH_3 HNO_3 O_2N CH_3 NO_2 + CH_3 NO_2 NO_2 H_2N CH_3 NH_2 + CH_3 NH_2 NH_2 $COCl_2$ OCN CH_3 NCO + CH_3 NCO NCO

2.3 Schematic

OCN NCO
4,4'-Methylenebis(phenyl isocyanate)

OCN NCO
4.4'-Methylenebis(cyclohexyl isocynate)

NCO NCO
Isophorone diisocyanate

OCN NCO
1,6-Diisocyanatohexane

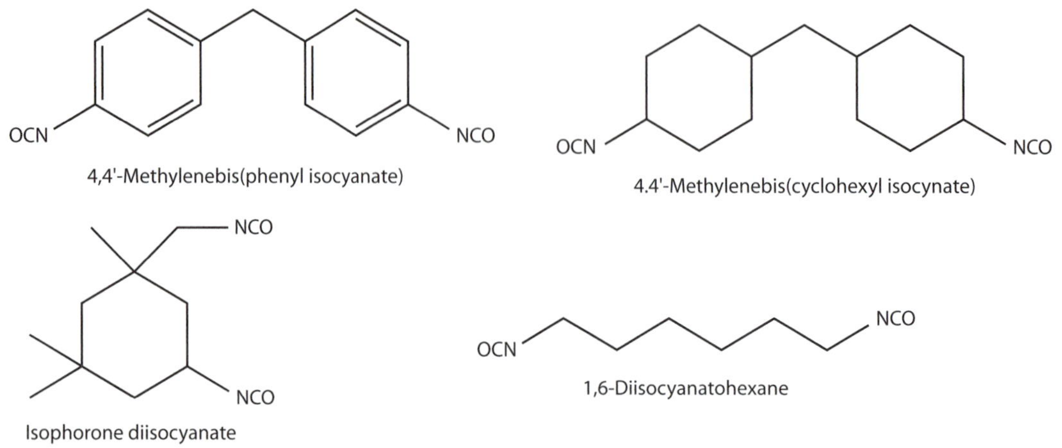

2.4 Schematic

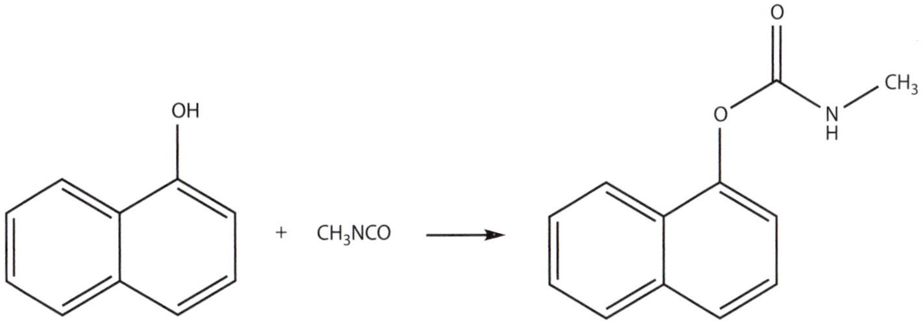

$$\text{1-naphthol} + CH_3NCO \longrightarrow \text{1-naphthyl } N\text{-methylcarbamate}$$

2.5 Schematic

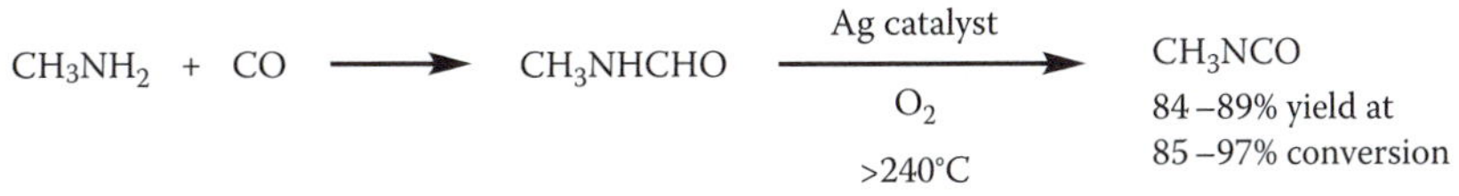

$$CH_3NH_2 + CO \longrightarrow CH_3NHCHO \xrightarrow[O_2,\ >240°C]{\text{Ag catalyst}} CH_3NCO$$

84–89% yield at
85–97% conversion

2.6 Schematic

The reaction can be run in a mini reactor with a silver catalyst in etched channels.[20] However, neighbors of the Rhone-Poulenc (now Bayer Crop Science) Institute Plant, West Virginia have been concerned about the 250,000 lb of methyl isocyanate stored there.[21]

A number of workers have treated both aliphatic and aromatic amines with dialkyl carbonates to form urethanes that can be pyrolyzed to isocyanates. Lead catalysts have given 95–98% selectivity (**2.7** Schematic).[22] (The percentage selectivity refers only to the amount of material reacted because the percentage conversion may not be 100%. In many cases, selectivity to the desired product drops as the conversion approaches 100%.)

The less toxic zinc acetate catalyst gave the product in 98% yield.[23] Angles et al.[24] obtained 70–95% yields using a sodium hydride catalyst. Aresta et al.[25] made urethanes with almost complete selectivity using phosphorus acid as a catalyst. This may involve an intermediate $(C_6H_5)_2P(O)OCOOR$. The phosphorus acid can be recovered and does not contaminate the product.

Tsujimoto et al.[26] obtained a 98% yield of bisurethane (**2.8** Schematic).

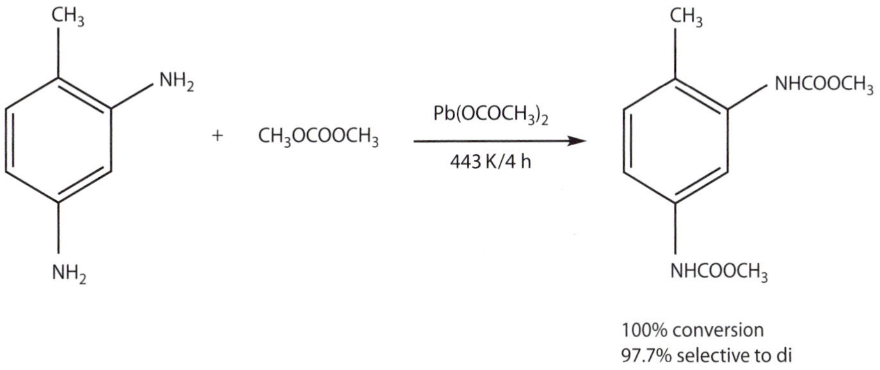

$$\text{2,4-}(NH_2)_2C_6H_3CH_3 + CH_3OCOOCH_3 \xrightarrow[443\ K/4\ h]{Pb(OCOCH_3)_2} \text{2,4-}(NHCOOCH_3)_2C_6H_3CH_3$$

100% conversion
97.7% selective to di

2.7 Schematic

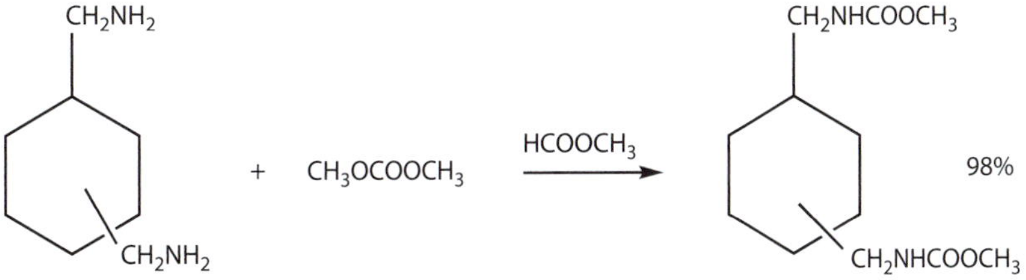

$$(CH_2NH_2)_2C_6H_{10} + CH_3OCOOCH_3 \xrightarrow{HCOOCH_3} (CH_2NHCOOCH_3)_2C_6H_{10}\quad 98\%$$

2.8 Schematic

The yields of isocyanates prepared by such routes can be good. Romano and co-workers made 1,6-diisocyanatohexane from hexamethylenediamine.[27] (Enichem built a commercial toluenediisocyanate plant in Qatar.[28] Dow has purchased Enichem's polyurethanes business.) Ookawa et al.[29] ran a similar conversion to produce the diisocyanate in 92.5% yield with 1.2% monoisocyanate, which could be recycled to the process (**2.9** Schematic).

These reactions require a carbonate that is usually made from phosgene (**2.10** Schematic).[30] Workers at Enichem have developed a commercial process for producing dimethyl carbonate by the oxidative carbonylation of methanol using cobalt or copper catalysts (**2.11** Schematic).[31] A selectivity of 96% was obtained when the reaction was carried out in the gas phase at 130°C with a palladium(II) chloride–copper(II) acetate–magnesium(II) chloride catalyst.[32] The reaction has also been run in a eutectic mixture of copper(I) chloride–potassium chloride at 150°C, with 94–96% selectivity to dimethyl carbonate and 2–5% selectivity to methyl ether.[33] After the product distils out, the molten salt can be used for the next run. Dimethyl carbonate has also been prepared from urea and methanol in triglyme in high yields using the technique of reactive distillation.[34] (Reactive distillation involves distilling out a product or by-product as it is formed. This can minimize exposure to the reaction conditions and shift equilibria in the desired direction.) The azeotrope of dimethyl carbonate and methanol that distils out is somewhat difficult to separate. A membrane separation might solve this problem. Note that this process is based on the less toxic carbon dioxide rather than carbon monoxide. The by-product can be reacted with carbon dioxide to make more urea.

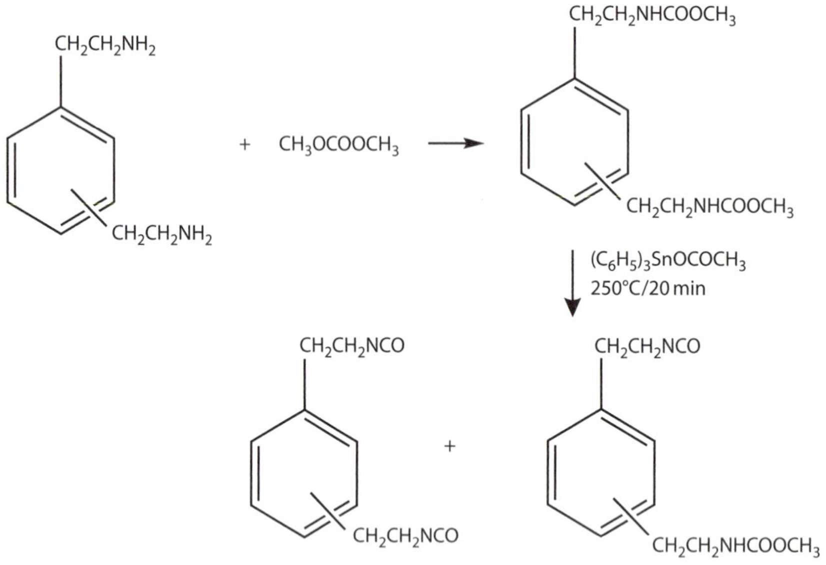

$C_6H_4(CH_2CH_2NH_2)_2 + CH_3OCOOCH_3 \longrightarrow C_6H_4(CH_2CH_2NHCOOCH_3)_2$

$\xrightarrow[250^\circ C/20\,min]{(C_6H_5)_3SnOCOCH_3} C_6H_4(CH_2CH_2NCO)_2 + C_6H_4(CH_2CH_2NCO)(CH_2CH_2NHCOOCH_3)$

2.9 Schematic

$$2CH_3OH + COCl_2 + 2NaOH \longrightarrow CH_3OCOOCH_3 + 2NaCl + 2H_2O$$

2.10 Schematic

$$2CH_3OH + CO + 0.5O_2 \xrightarrow{\text{Catalyst}} CH_3OCOOCH_3 + H_2O$$

$$CH_3OH + NH_2CONH_2 \xrightarrow[\text{Heat}]{\text{Triglyme}} (NH_2COOCH_3) + NH_3$$

$$(NH_2COOCH_3) \xrightarrow{CH_3OH} CH_3OCOOCH_3 + NH_3$$

98.2% selectivity at 98.3% conversion

2.11 Schematic

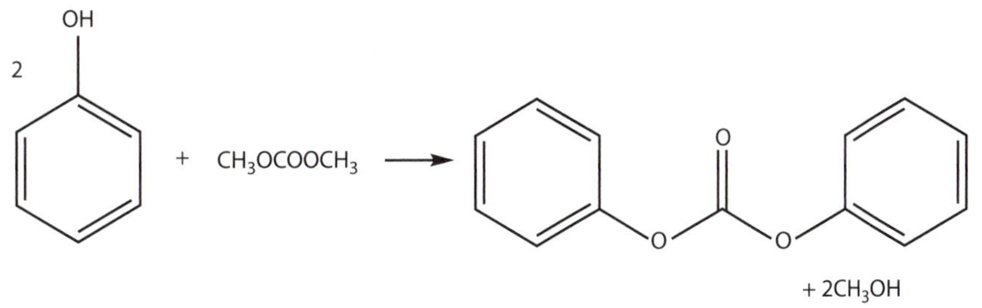

2 OH + $CH_3OCOOCH_3$ → (diphenyl carbonate) + $2CH_3OH$

2.12 Schematic

Another route based on carbon dioxide uses a nickel phenanthroline catalyst with 2,2-dimethoxypropane as a scavenger for the by-product water to give 81% selectivity at 83% conversion.[35] A similar reaction has been carried out in supercritical (sc) carbon dioxide with a tin catalyst.[36] The by-product acetone is reconverted to the acetal in a separate step.

Diphenyl carbonate can be made from dimethyl carbonate by exchange with phenol (**2.12** Schematic).[37] It can also be made directly from phenol, but the yields are not as high.[38] A combination of palladium and manganese catalysts is used. The water formed is sparged out with excess reaction gas to shift the reaction to the desired product. Diphenyl carbonate is preferred over dimethyl carbonate for the production of polycarbonates.

The dialkyl carbonates can also be used in benign alkylations (**2.13** Schematic), instead of toxic dialkyl sulfates and iodoalkanes, with improved selectivity.[39]

Excess dialkyl carbonate can be the solvent. No by-product salts are formed. The by-product alcohol and carbon dioxide could be recycled. Arylacetonitriles, phenols, aromatic amines, trialkylamines, acetylenes, silica, and titania can also be alkylated with dialkyl carbonates. Tetraalkoxysilanes for use in sol gel syntheses can be made without chlorine-containing intermediates.[40] The alkylation of titanium dioxide, followed by later hydrolysis, offers the possibility of a new manufacturing process that would eliminate the by-product chlorinated dioxins formed in the current titanium tetrachloride process. Methylation of ammonium chloride leads to tetramethylammonium chloride in 96% yield.[41] Alkylation of catechol with dimethyl carbonate in the presence of $LiOH/Al_2O_3$ at 553 K gave the monoether with 84% selectivity at 100% conversion. Some 3-methylcatechol also formed in the process.[42] When the reaction of aniline with dimethyl carbonate is carried out in the gas phase with zeolite KY, the reaction was 93.5% selective for *N*-methylaniline at 99.6% conversion. When zeolite NaX was used, the selectivity to *N,N*-dimethylaniline was 95.6% at 100% conversion. (For more on the size and shape selectivity of zeolite catalysts, see Chapter 6.) Benzene has been alkylated to a 34.7:1 toluene/xylene mixture over a zirconium–tungsten oxide catalyst at 100–200°C with 99.3% conversion of the dimethyl carbonate.[43] Methanol and methyl ether also formed. Zeolite catalysts have been used with dimethyl carbonate to prepare methyl esters from carboxylic acids in 93–99% yield in a process that could replace hazardous diazomethane.[44] Monoalkylation of arylacetonitriles can also be done with trimethyl*orthoformate* in 70% yield.[45]

Several variants of the dialkyl carbonate route to carbamates have been reported. Valli and Alper[46] may have made the carbonate *in situ* (**2.14** Schematic).

The process works for both aliphatic and aromatic amines. Harstock et al.[47] used anodic oxidation instead of oxygen (**2.15** Schematic).

n-Butylamine was converted with carbon monoxide, oxygen, and methanol in 100% yield using a copper

$$ArOCH_2X + CH_3OCOOCH_3 \xrightarrow[180-200^\circ C]{K_2CO_3} ArOCH(CH_3)X + CO_2 + CH_3OH \quad X = CN, COOCH_3$$

2.13 Schematic

$$RNH_2 + CO + 0.5O_2 + CH_3OH \xrightarrow[\text{Bipyridyl}]{\text{Pd on clay}} RNHCOOCH_3 \; (79\text{–}100\%) + H_2O$$

2.14 Schematic

$$ArNH_2 + CO + Pd(OCOCH_3)_2 + Cu(OCOCH_3)_2 + CH_3OH \rightarrow ArNHCOOCH_3$$

2.15 Schematic

chloride–palladium chloride catalyst in sc carbon dioxide.[48] A similar run with aniline using a palladium chloride–zirconyl sulfate catalyst gave the corresponding urethane with 99% selectivity at 100% conversion.[49]

A number of processes proceed through intermediate ureas (**2.16** Schematic).[50] With a palladium on charcoal catalyst, the selectivity to carbamate is 97% (**2.17** Schematic). This is done all in one pot at up to 96.5% conversion of aniline with 94% selectivity. The use of a $[Ru(CO)_3I_3]^-$ $(C_4H_9)_4N^+$ catalyst gave the urea with 99% selectivity at 59% conversion.[51] Reaction of diphenylurea with dimethyl carbonate using a sodium methoxide catalyst produced urethane with 99% selectivity at 75% conversion.[52] Ureas can also be made by reacting an amine with carbon dioxide with 99% selectivity.[53] Cyclohexylamine was converted to urea using a cesium hydroxide catalyst in an ionic liquid in 98% yield.[54]

Another method makes urea through the intermediate acetoacetanilide.[55] The catalyst was reused five times with no loss in activity. The by-product acetone might be pyrolyzed to ketene, which would dimerize to diketene that would react with more amine to form acetoacetanilide (**2.18** Schematic).

These processes that substitute carbon monoxide, oxygen, and an alcohol for phosgene appear to be on the verge of commercial viability. There may be questions of catalyst life and recyclability that are not mentioned in the papers and patents. Some reaction times need to be shortened through the use of improved catalysts or by microwave heating. Aresta et al.[56] do say that their palladium–copper catalyst system can be recycled. The use of supported catalysts that could be recovered by filtration could simplify workups and recycling. (Supported catalysts are described in Chapter 5.)

McGhee and co-workers at Monsanto have studied the preparation of isocyanates using nontoxic carbon dioxide (instead of the poisonous carbon monoxide) with the amine (**2.19** Schematic).[57]

The preferred bases include triethylamine, amidines, and guanidines. Dehydration of the carbamate has been done with phosphorus trichloride, phosphorus oxytrichloride, phosphorus pentoxide, acetic anhydride, benzenesulfonic acid anhydride, *o*-sulfobenzoic anhydride, or other similar compounds.

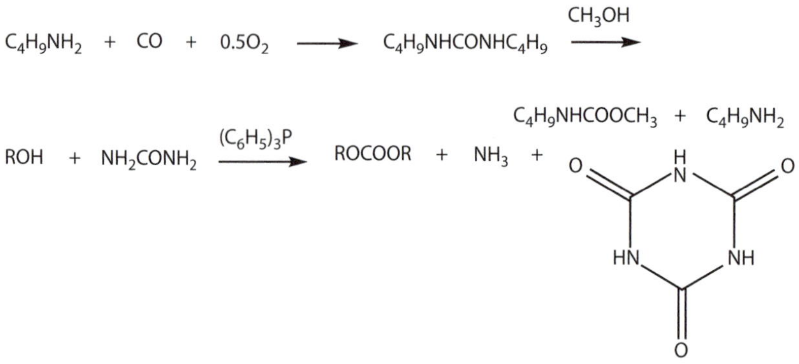

$$C_4H_9NH_2 + CO + 0.5O_2 \longrightarrow C_4H_9NHCONHC_4H_9 \xrightarrow{CH_3OH} C_4H_9NHCOOCH_3 + C_4H_9NH_2$$

$$ROH + NH_2CONH_2 \xrightarrow{(C_6H_5)_3P} ROCOOR + NH_3 +$$

2.16 Schematic

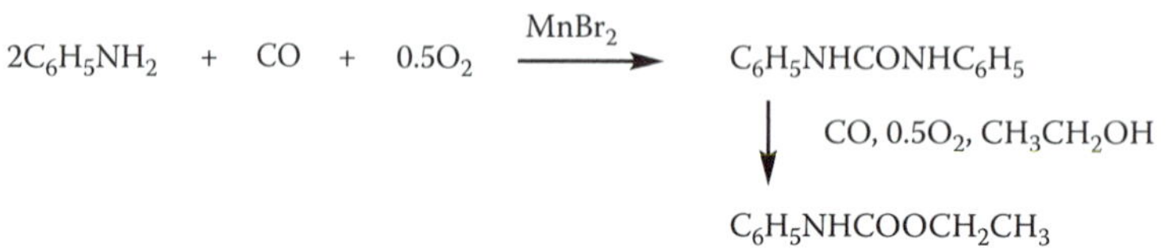

$$2C_6H_5NH_2 + CO + 0.5O_2 \xrightarrow{MnBr_2} C_6H_5NHCONHC_6H_5 \xrightarrow{CO,\ 0.5O_2,\ CH_3CH_2OH} C_6H_5NHCOOCH_2CH_3$$

2.17 Schematic

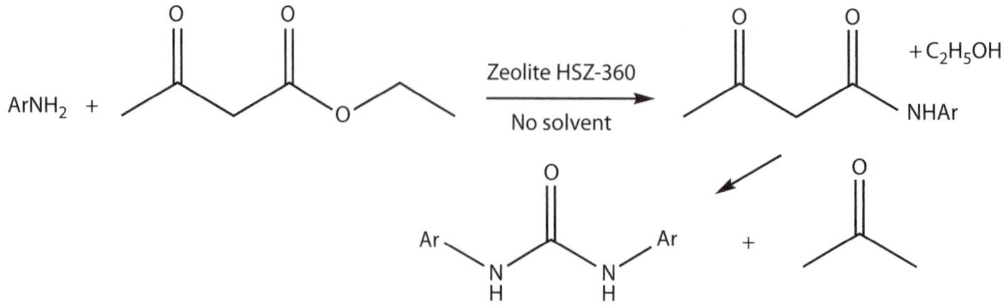

2.18 Schematic

$$RNH_2 + CO_2 + \text{Base} \longrightarrow RNHCOO^- \text{BaseH}^+ \xrightarrow{\text{Dehydrating agent}} \underset{94\text{–}98\%}{RNCO}$$

2.19 Schematic

2.2.2 Isocyanates from Nitro Compounds

Aromatic nitro compounds can be converted to carbamates by treatment with carbon monoxide and an alcohol, usually methanol, in the presence of palladium, rhodium, or ruthenium catalysts.[58] Wehman et al.[59] have reported the best yields, which can be close to quantitative (**2.20** Schematic).

$$C_6H_5NO_2 + 3CO \xrightarrow[CH_3OH]{\text{Pd catalyst}} C_6H_5NHCOOCH_3 + C_6H_5NHCONHC_6H_5 + 2CO_2$$

2.20 Schematic

For dinitro compounds, using Pd(*o*-phenanthroline)$_2$ $(OSO_2CF_3)_2$ plus 4-chlorobenzoic acid, both mono- and dicarbamates are formed. With the onset of time, all of the monocarbamate can be converted to the dicarbamate. It is possible that the by-product *N,N*-diphenylurea can be recycled to the next reaction and be converted *in situ* to more carbamate.

Such reactions may proceed by carbonylation of an intermediate palladium amide. Giannoccaro et al.[60] have carried out the stoichiometric reaction of amines with palladium and nickel compounds and carbon monoxide to prepare isocyanates (**2.21** Schematic).

$$((C_6H_5)_3P)_2PdClNHR + CO \longrightarrow ((C_6H_5)_3P)_2PdClCONHR \xrightarrow{CuCl_2} (RNHCOCl) \longrightarrow RNCO + HCl$$

2.21 Schematic

$$C_6H_5NO_2 + CO + CH_3OH \xrightarrow[\substack{CH_3OCH_2CH_2OCH_3 \\ 170°C/3\,h}]{\substack{PdCl_2 \\ H_5PV_2Mo_{10}O_{40}}} C_6H_5NHCOOCH_3 \quad \substack{98\% \text{ conversion} \\ 96\% \text{ selectivity}}$$

2.22 Schematic

The reaction of the nitro compound has also been carried out with a heteropolyacid as part of the catalyst (**2.22** Schematic).[61]

Under less favorable conditions with palladium catalysts, other workers have found aniline, azobenzene, and azoxybenzene as by-products.[62] The yields reported with ruthenium catalysts are lower than with palladium catalysts. Mukherjee et al.[63] reacted nitrobenzene with carbon monoxide and methanol with a sodium methoxide plus ruthenium catalyst to give 80% carbamate and 18% aniline. Based on the mechanism suggested by Giannoccaro et al., it should be possible to recycle the aniline to the next run. Gargulak et al.[64] used a $Ru((C_6H_5)_2PCH_2CH_2P(C_6H_5)_2)(CO)_3$ catalyst with nitrobenzene to prepare the carbamate. They felt that their catalyst is an improvement over the short lifetimes of earlier ones.

This approach, as exemplified by the work of Wehman et al., appears ripe for commercialization. The drawbacks of the reaction are the need for three equivalents of carbon monoxide and relatively large amounts of noble metal catalysts. The use of relatively high levels of catalyst is not a problem if the catalyst can be used over a long time period and has good activity, as in a column in a continuous process. A variant using a 1:1 molar ratio of nitrobenzene to aniline eliminates the need for the noble metal catalyst.[65] The catalyst is a mixture of sulfur, sodium methoxide, and ammonium metavanadate (**2.23** Schematic).

Further work on these various processes may be needed to optimize catalyst life, separation, and recycle, as well as reaction rates. The preferred solvent, if any is required, would be ethanol instead of methanol, for the former is less toxic and can be produced readily from renewable sources by fermentation (see Chapter 9).

$$C_6H_5NO_2 + 3CO + C_6H_5NH_2 \xrightarrow{NH_4VO_3} C_6H_5NHCONHC_6H_5 \xrightarrow{CH_3OH} C_6H_5NHCOOCH_3 \quad (96\% \text{ at } 100\% \text{ conversion})$$

2.23 Schematic

2.2.3 Isocyanates from Isocyanic Acid

Tertiary aliphatic isocyanates can be made by addition of isocyanic acid (a toxic material) to olefins. Cytec (formerly American Cyanamid) makes a diisocyanate in this way.[66] The reaction can also be controlled to produce the monoadduct, which can be used as a monomer. Tertiary isocyanates of this type react at lower rates than primary ones in addition reactions. The isocyanic acid can be produced by the pyrolysis of cyanuric acid, which, in turn, can be prepared from the urea (**2.24** Schematic and **2.25** Schematic).

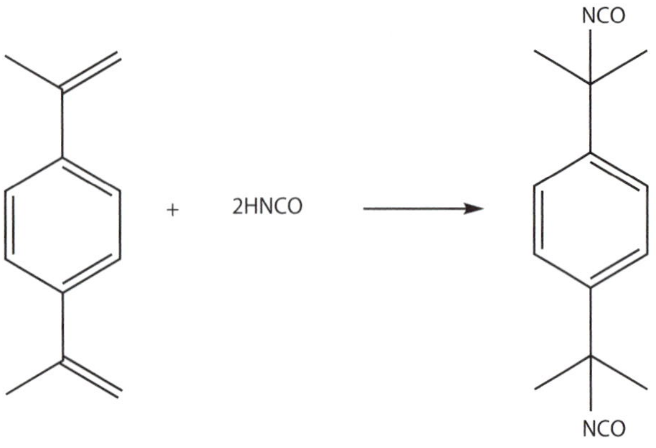

+ 2HNCO → NCO NCO

2.24 Schematic

H_2N O NH_2 → HNCO ↔ HO N OH N N OH

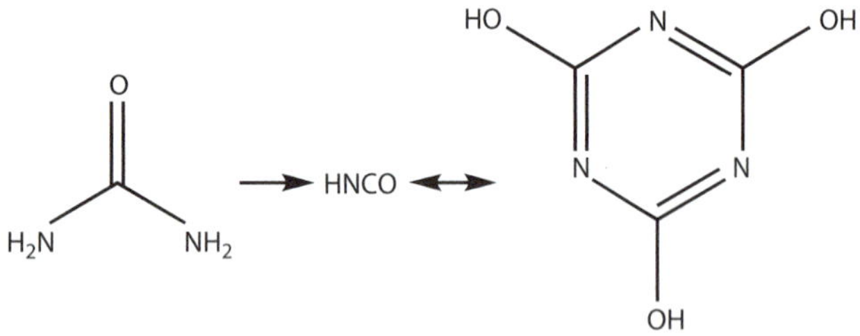

2.25 Schematic

The method is also applicable to α-pinene (from pine trees), which is cheaper than petroleum-based diisopropenylbenzenes.[67] The reaction (**2.26** Schematic) can be carried out to produce either the mono- or diadduct. Alternative routes to the diadduct might be based on 1,8-diamino-*p*-menthane (from Rohm & Haas, now Dow) using the methods in Section 2.2.1, as long as no high-temperature pyrolyses were used to recover the isocyanate from a carbamate.

Isocyanates can also be made by reaction of salts of isocyanic acid with alkyl halides.[68] A disadvantage is that this produces by-product salts (**2.27** Schematic).

This method has been applied to a monomer synthesis (**2.28** Schematic).[69] The quaternary ammonium salt is used as a phase transfer catalyst. The potassium iodide converts

$$C_2H_5Br + KNCO \longrightarrow C_2H_5NCO + KBr$$

2.27 Schematic

the starting chloride to a more reactive iodide *in situ*. The alcohol traps the intermediate isocyanate before it can react with the small amount of water that has to be present. If no methanol is present, the corresponding urea is formed in up to 87% yield. If only 0.3 equivalent of water is used, the product is isocyanurate, the cyclic trimer of isocyanate. Where $n = 1$, the isocyanatoethyl methacrylate is a useful monomer now made with phosgene (**2.29** Schematic).[70] Dow has patented a route from ethanolamine plus a dialkyl carbonate, followed by transesterification to a methacrylate carbamate, which is pyrolyzed to the isocyanate (**2.30** Schematic) in 50% yield.[71]

2.2.4 Isocyanates by Addition of Urethanes to Double Bonds

Ethyl urethane (a carcinogen) can be added to activated double bonds in the presence of palladium catalysts to give a mixture of *cis* and *trans* adducts (**2.31** Schematic).[72]

If such an addition reaction can be carried out cleanly in high yields, di- and polyisocyanates might be made from ethylenediacrylate, *N,N′*-methylenebis(acrylamide), or an unsaturated polyester made from maleic anhydride. However, it would be desirable to find a noncarcinogenic analogue of the ethyl urethane, preferably one for which the adduct would unblock at a lower temperature. It may also be necessary to hydrogenate the double bonds before unblocking.

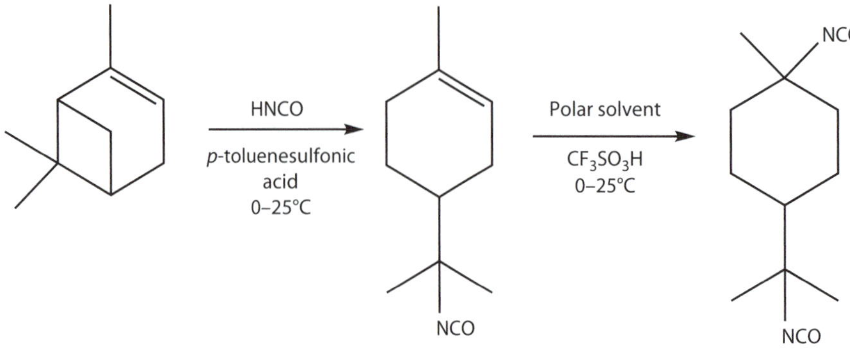

HNCO → *p*-toluenesulfonic acid 0–25°C NCO → Polar solvent CF_3SO_3H 0–25°C NCO NCO

2.26 Schematic

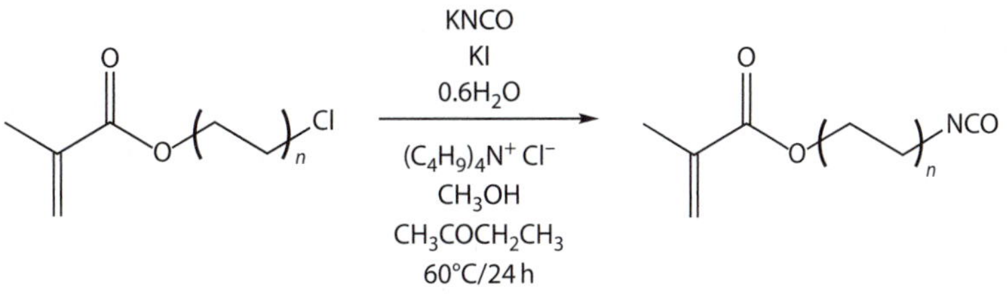

2.28 Schematic

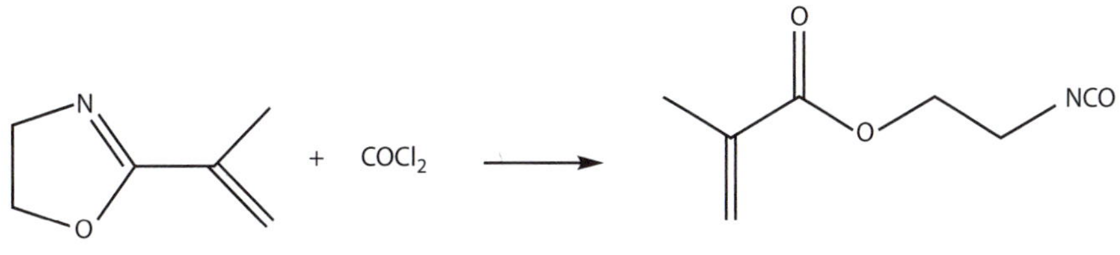

2.29 Schematic

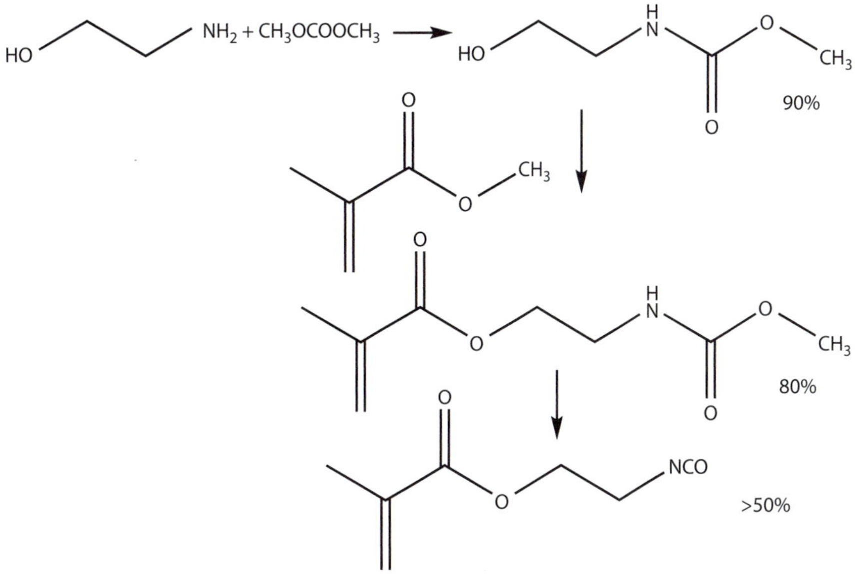

2.30 Schematic

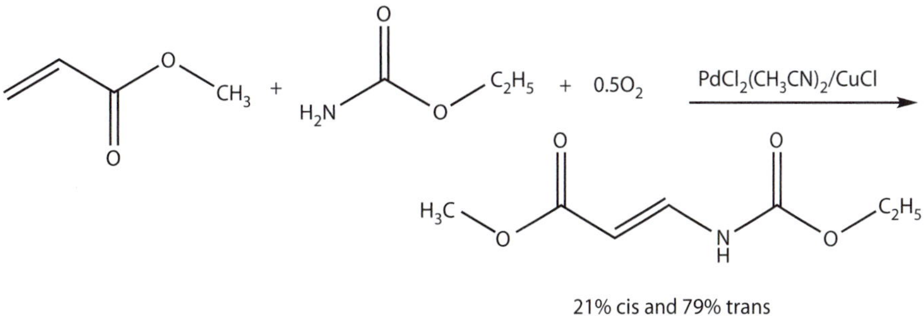

2.31 Schematic

2.2.5 Isocyanates by Rearrangements

Isocyanates can also be made by a number of rearrangements of carboxylic acid derivatives that proceed via nitrenes (**2.32** Schematic). Of these, the most practical commercially would be the one based on amides. Terephthalamide (easily made from inexpensive dimethyl terephthalate) and its cyclohexyl analogue have been converted to isocyanates via *N*-chloro compounds (**2.33** Schematic).[73]

A more direct route has been used to prepare carbamates from aliphatic carboxylic acid amides (**2.34** Schematic).[74] (A *hydrotrope* is a compound that helps other compounds normally not very soluble in water to dissolve in water.) The one that was used was compound **2.35** Schematic.

Pyrolysis of the carbamate (obtained in 90–96% yield) would lead to isocyanate. This type of reaction can also be carried out in 95% yield using *N*-bromosuccinimide and sodium methoxide in methanol or with bromine and sodium hydroxide.[75] This suggests the possibility of generating the bromine or iodine during the reaction by electrolysis of sodium bromide or sodium iodide so that no waste salts result. The starting amides can be made by heating the ammonium salts of the acids. This method does have the disadvantage of producing by-product salts (**2.36** Schematic).

It is conceivable that the Lossen rearrangement (Equation 2.25) might be practical commercially (**2.37** Schematic).

Hydroxylamine can be prepared from ammonia and hydrogen peroxide using a titanium silicalite catalyst in 83% yield.[76] The by-product acetic acid could be recycled to acetic anhydride by pyrolyzing part of it to ketene.

2.2.6 Reducing the Toxicity of Isocyanates

Isocyanates can be dangerous, as shown by the tragedy with methyl isocyanate at Bhopal. Several workers at W.L. Gore Associates (Newark, Delaware) suffered disabilities while using isocyanates, probably toluenediisocyanate, outside of a hood. Isocyanates are the most common class of chemicals causing occupational asthma.[77] The problems are greater in small businesses, such as decorating or small car spraying, for which safety precautions are less likely to be used. To enable the isocyanates to be used outside of a hood, it is necessary to reduce this toxicity. This can be done by increasing

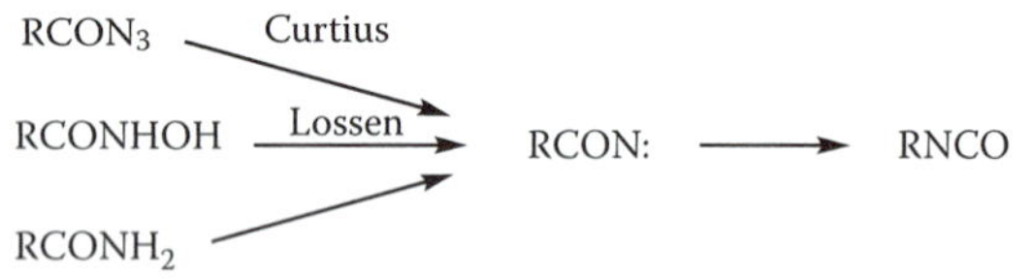

2.32 Schematic

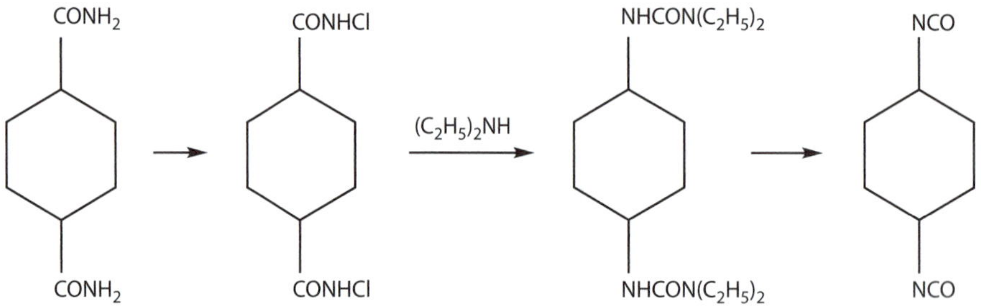

2.33 Schematic

$$RCONH_2 + NaOCl + NaOH + CH_3OH \xrightarrow[\text{Hydrotrope and surfactant}]{\text{With or without}} RNHCOOCH_3 \quad R = C_6\text{–}C_{11}$$

2.34 Schematic

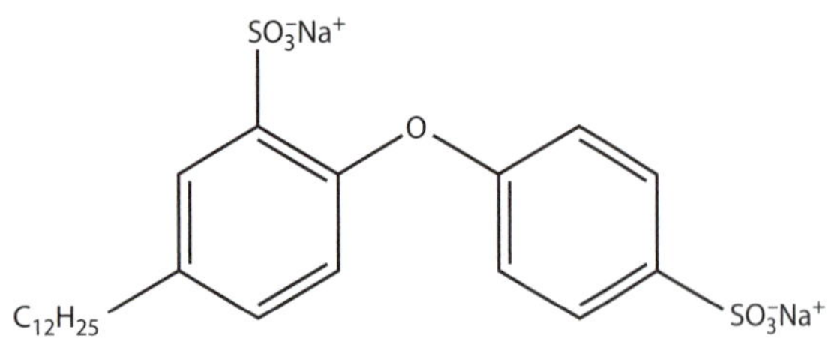

2.35 Schematic

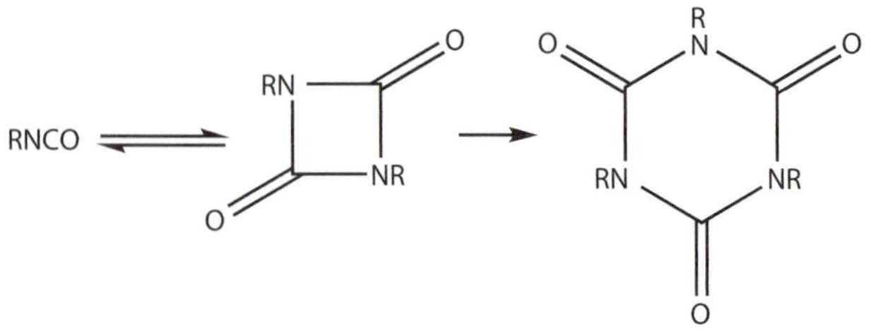

2.37 Schematic

$$RCOOCH_3 + NH_2OH \longrightarrow RCONHOH \xrightarrow{(CH_3CO)_2O} RNCO + CH_3COOH$$

2.36 Schematic

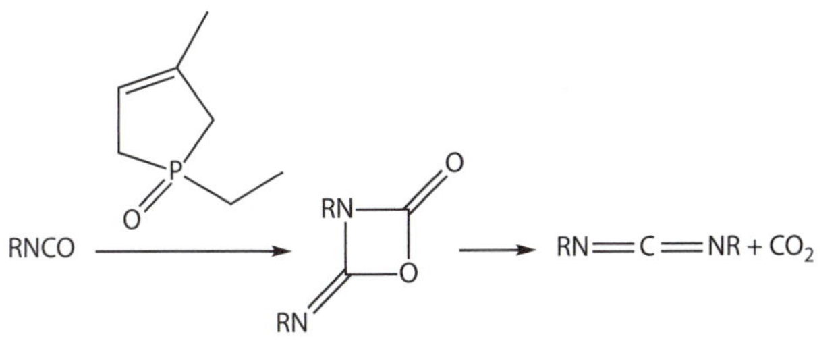

RNCO → RN=C=NR + CO_2

2.38 Schematic

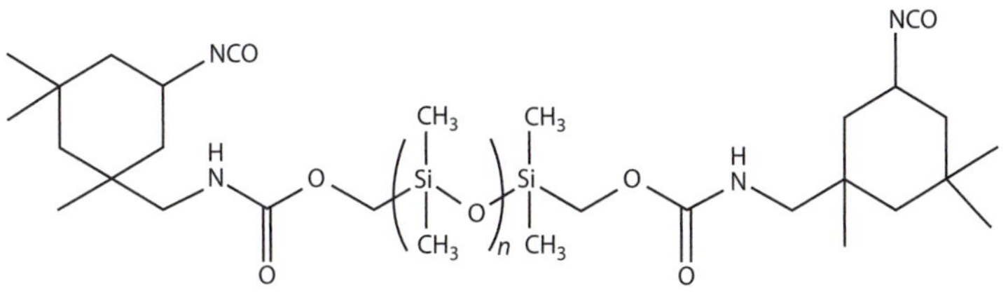

2.39 Schematic

their molecular weight, which decreases the vapor pressure, or by generating them *in situ*. A common way is to react one of two isocyanate groups selectively. An isocyanate may form a dimer or trimer (**2.38** Schematic).[78] The trimer is a useful stiffening structure in some polyurethanes. Phosphine oxides or phosphates can convert an isocyanate to a carbodiimide. This is used commercially to convert some 4,4′-diphenyldiisocyanate to the carbodiimide so that the resulting product is a liquid that may be easier to handle than the starting solid. If both isocyanate groups react, the result is a polycarbodiimide that is useful for cross-linking coatings.[79] Polycarbodiimides can be blocked with diethylamine so that they are inert until the amine released as a powder coating is cured by heating.[80] [Powder coatings (which contain no solvents) are discussed in Chapter 9.]

Another common method is to react an excess of diisocyanate with a diol (e.g., a polymeric diol), as in the following, which is used to repair deteriorating stone by curing with moisture (**2.39** Schematic).[81] In this compound, one isocyanate group is more reactive than the other. Another approach is to prepare a diisocyanate from an amine-terminated polymer, such as amine-terminated poly(ethylene oxide).

Various HA-type reagents can add to the isocyanate group (**2.40** Schematic).[82] (A way to remember the many reactions in this chapter is to use the rule that HA and AA reagents can add to multiple bonds, often reversibly. Among the multiple bonds are N=C and C=O.)

For polyurethanes HA is an alcohol; for polyureas it is an amine. This type of reaction can be used to block an isocyanate and make it unreactive to water until unblocking is done thermally with loss of a volatile blocking agent. For example, a polypropyleneglycol urethane prepolymer with isocyanate groups blocked with methylethylketoxime can

$$\text{RNCO} + \text{HA} \rightleftharpoons \text{RNHCOA}$$

2.40 Schematic

be used with 2,2-bis(hydroxylmethyl)propionic acid and tetraethylenepentamine in a one-package, cross-linkable, waterborne coating. Deblocking occurs at 110–140°C after the water has evaporated.[83] It is necessary to recover the methylethylketoxime from the exhaust air so that it can be recycled, instead of being let out to pollute the surrounding air.

Similarly, an aqueous treatment for wool uses an isocyanate blocked with a bisulfite. In this case, the by-product bisulfite salt can be removed, and possibly recycled, by washing the fabric with water at the end. The dissociation temperature decreases in the following order: alcohol > lactam > phenol > ketoxime > active methylene compound. ε-Caprolactam is a typical lactam and ethyl acetoacetate a typical active methylene compound. This thermal reversibility has been used to prepare a polyurethane from 1,4-butanediol, 4,4-diphenyldiisocyanate, and 4,4′-isopropylidenediphenol (bisphenol A) that can be recycled just by remolding (**2.41** Schematic).[84]

Another way to produce an isocyanate that is part of a polymer is to polymerize an isocyanate alone or with

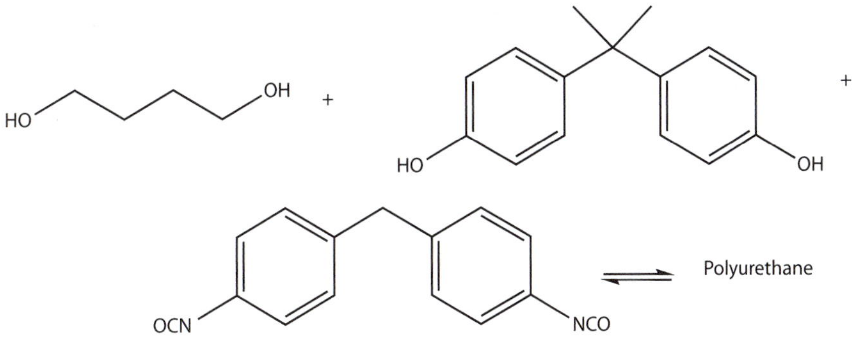

⇌ Polyurethane

2.41 Schematic

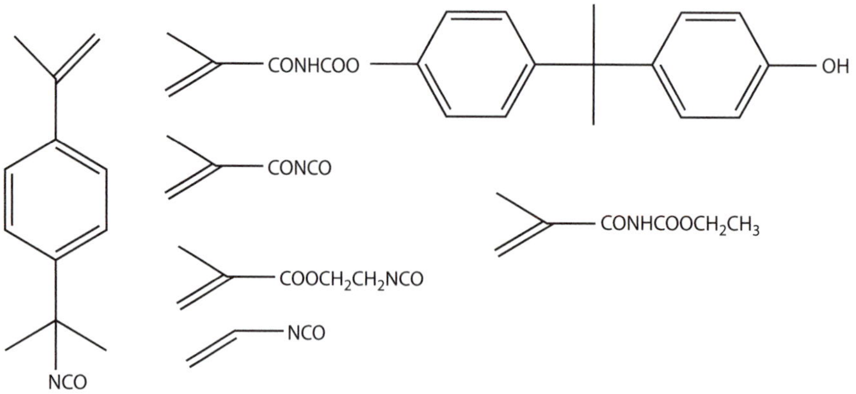

2.42 Schematic

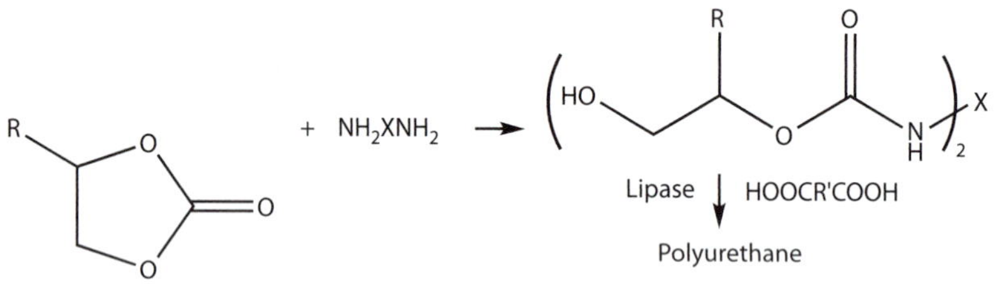

2.43 Schematic

comonomers. Typical isocyanate monomers are shown in the structure (**2.42** Schematic).[85]

Use of these comonomers in minor amounts can lead to cross-linkable polymers. For example, Yukawa et al. copolymerized isopropenylcumylisocyanate with butyl acrylate and then performed a series of blocking and further polymerization steps to prepare a waterborne coating for cars. Methacryloylisocyanate was reacted with bisphenol A to yield a new monomer.[86] It is possible that some of these isocyanate copolymers might be used to replace some resins now made with the carcinogen, formaldehyde (e.g., in crease-proofing cotton textiles). It may also be possible to use a blocked isocyanate monomer to make copolymers in which the isocyanate can be unblocked thermally for curing. A final way to put an isocyanate group into a polymer is to run a Hofmann reaction on a copolymer of acrylamide. The isocyanate groups formed could be trapped with a blocking agent or used directly.

2.2.7 Polyurethanes Made without Isocyanates

Cyclic carbonates, such as ethylene and propylene carbonates, have been reacted with diamines to produce diols, which have then been converted to polyurethanes by reaction with a dicarboxylic acid using a lipase.[87] Although the molecular weights have been low, for example, M_W 9350 and M_n 5345, it may be possible to increase these using chain extension agents such as bisanhydrides, bisepoxides, and bisoxazolines (**2.43** Schematic).

Bis(cyclic carbonates) reacted with diamines without solvent to give polyurethanes of low molecular weight, for example, M_n 3600–4200 (**2.44** Schematic).[88]

Copolymerization of a cyclic urea and a cyclic carbonate yielded a copolymer polyurethane (**2.45** Schematic).[89]

The cyclic carbonates needed for these reactions can be made by reaction of the corresponding epoxides[90] or by reaction of a diol with urea (**2.46** Schematic).[91]

There are many other uses for cyclic carbonates.[92] *N*-vinylcarbazole can be prepared using ethylene carbonate instead of the usual acetylene and potassium hydroxide (**2.47** Schematic).[93]

Huntsman sells glycerol carbonate.

Alpha, omega-aminoalcohols have been converted to polyurethanes with di-*tert*-butyltricarbonate.[94] Further study of the reactions and catalysts may make it possible to replace

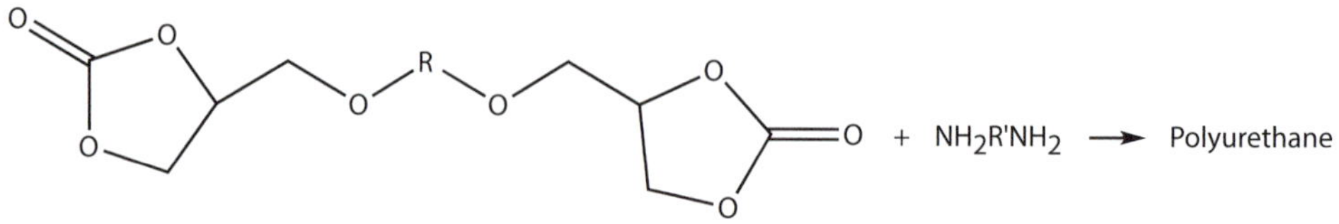

2.44 Schematic

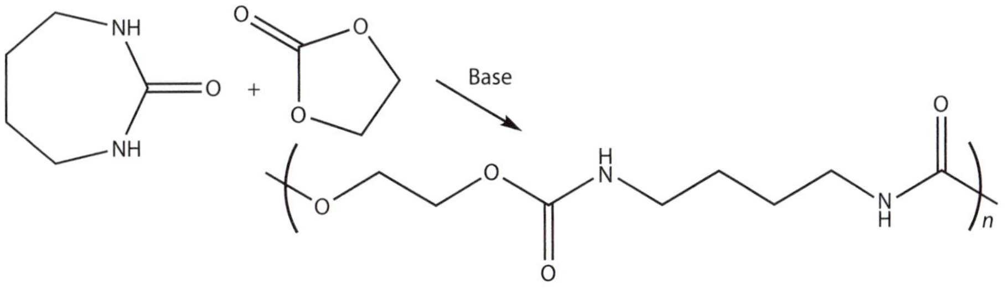

2.45 Schematic

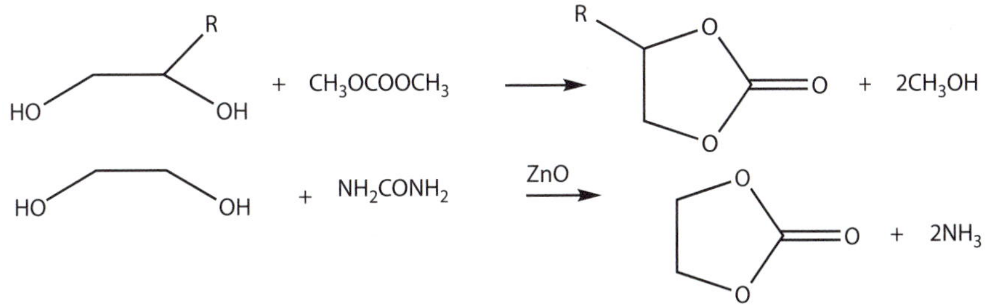

2.46 Schematic

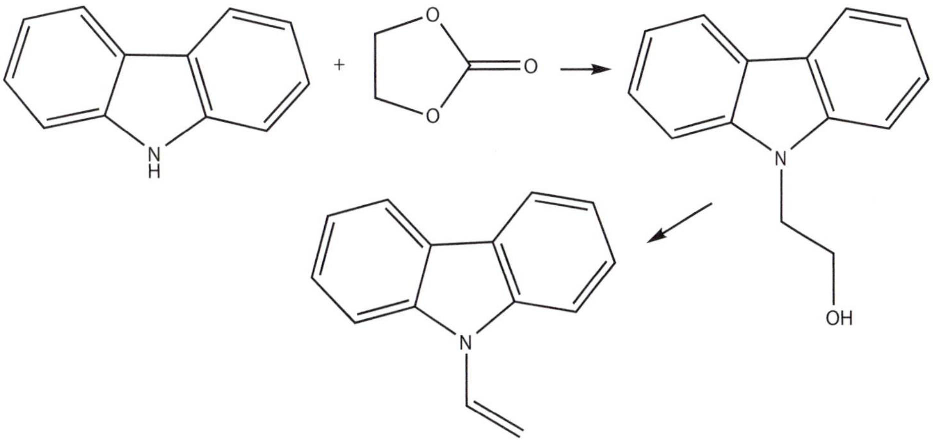

2.47 Schematic

the tricarbonate with a dialkyl or diphenyl carbonate. The intermediate hydroxyurethane could be heated to drive off the blocking alcohol or phenol, which could be trapped for recycle. M_W was 19,000–34,000. T_M was 155°C for $n = 6$ and 195°C for $n = 4$ (**2.48** Schematic).

2.3 POLYCARBONATES

The traditional preparation of polycarbonates uses phosgene (**2.49** Schematic).[95] This can be carried out in solution with pyridine as a base in an interfacial system with sodium hydroxide as a base and methylene chloride as a solvent. This method is being displaced by an ester exchange method (**2.50** Schematic) with diphenyl carbonate, which requires no solvent and produces no salts. It does require high temperatures and a good vacuum or nitrogen flow to extract out the last of the phenol (bp 182°C). Asahi Chemical uses a solid-state polymerization on a powdered prepolymer to extract the phenol more easily to give molecular weights of 60,000 or more.[96] Numerous patents and papers describe a variety of basic catalysts and conditions for this reaction with diphenyl carbonate.[97] Bis(2,2,2-trichloroethyl) carbonate has also been used to make this polymer at 200°C.[98]

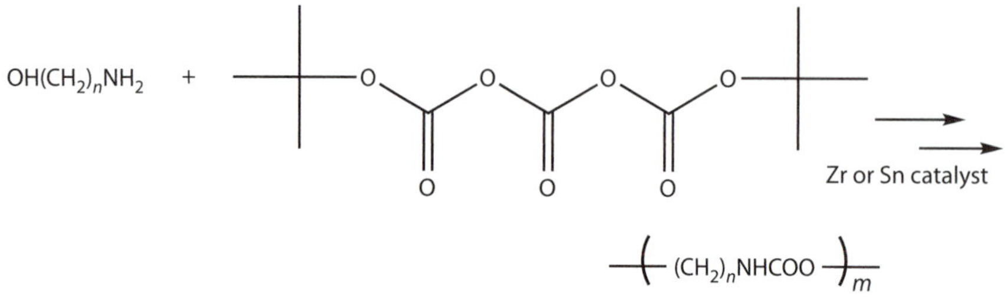

2.48 Schematic

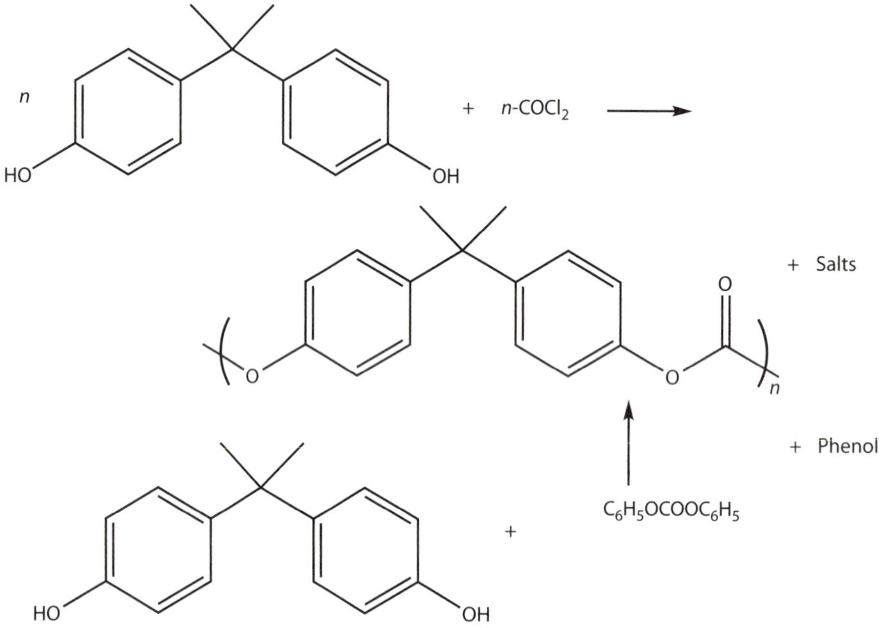

2.49 Schematic

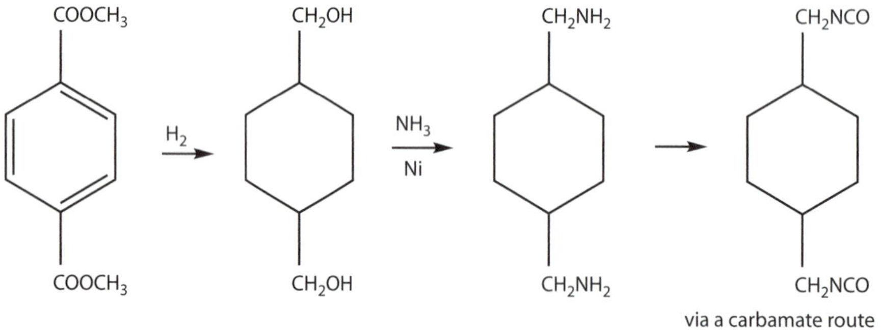

2.50 Schematic

Dimethyl carbonate has been used with 1,4-cyclohexanedimethanol to prepare a biodegradable, biocompatible polymer (**2.51** Schematic).[99]

Glucose can be reduced to sorbitol, which can then be heated to form isosorbide.[100] It could be converted to a polycarbonate with phosgene but not dimethyl carbonate. Further study may find conditions where a dialkyl carbonate or diphenyl carbonate would work. The bicyclic structure stiffens the polymer, M_n 50,000 and T_g 163.5°C.

Bisphenol polycarbonate has also been prepared by treating bisphenol with carbon monoxide and oxygen using a palladium catalyst to a polymer of M_W 12,900 and M_n 5600.[101] Chain extension with bisanhydride, bisepoxide, or bisoxazoline might raise the molecular weight to a useful range (**2.52** Schematic).

Epoxides can be copolymerized with carbon dioxide using a zinc diiminate catalyst (**2.53** Schematic).[102]

A typical copolymer has M_n 21,300 and a polydispersity of 1.07 with 95% carbonate linkages. Some catalysts produce the copolymers epoxides and carbonates. This may offer an interesting opportunity to eliminate the use of the toxic ethylene oxide in making polyethyleneglycol. International

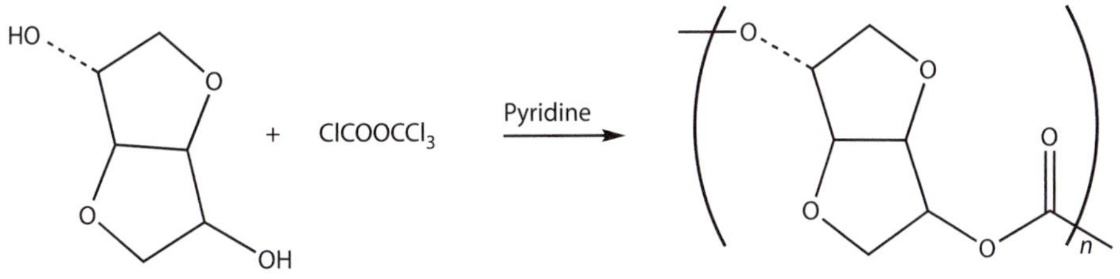

+ $ClCOOCCl_3$ Pyridine

2.51 Schematic

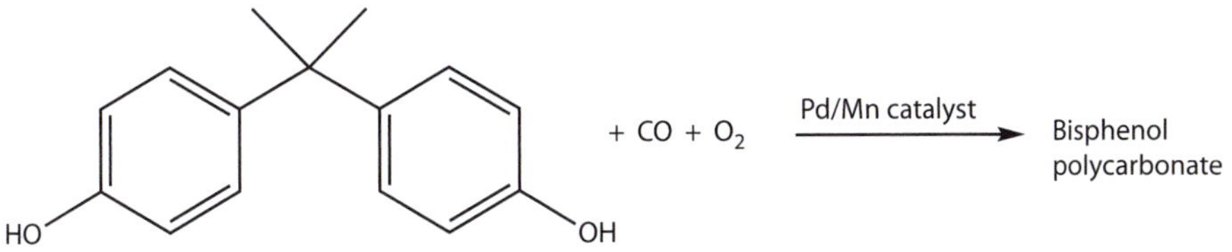

+ CO + O_2 Pd/Mn catalyst Bisphenol polycarbonate

2.52 Schematic

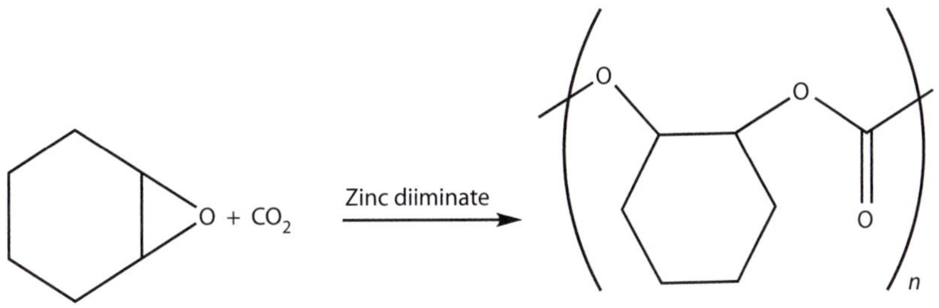

O + CO_2 Zinc diiminate

2.53 Schematic

Polyol Chemicals Inc. uses hydrogenolysis to convert sugars to mixed glycols.[103] Ethylene glycol can be converted to a carbonate by reaction with urea and then polymerized to polyethyleneglycol.

2.4 SUMMARY AND CONCLUSIONS FOR PHOSGENE

Polycarbonates and polyurethanes are so useful that they are here to stay. Elimination of phosgene in the preparation of the former is now commercial at SABIC (formerly General Electric Plastics), Asahi, and Mitsubishi. With recent improvements in the syntheses of isocyanates without phosgene, the industry is on the verge of a commercial process eliminating it here also. DuPont has developed a way to make methyl isocyanate on demand so that it does not need to be stored or shipped. Going one step further, a combination of integrated pest management, pyrethrin insecticides, and neem antifeedants might eliminate altogether the need to make carbamate insecticides. Thus, most of the steps in the hierarchy of approaches outlined in the Introduction to this chapter are being tried.

The challenge now is to make other isocyanates in closed systems on demand, so that they do not need to be stored or shipped before being converted to less volatile products or to blocked reagents. Even better, use systems that trap them in harmless forms *in situ*. Some of the reagents used to make the raw materials for isocyanates are dangerous in their own right (e.g., the reaction of two carcinogens to produce 4,4′-diaminodiphenylamine) (**2.54** Schematic).

Some rethinking may be needed to use less harmful materials to end up with similar properties in the final polyurethane. Polymers made from the foregoing reagent need stiffness. Because amines can be made from alcohols and ammonia with nickel or ruthenium catalysts, the following

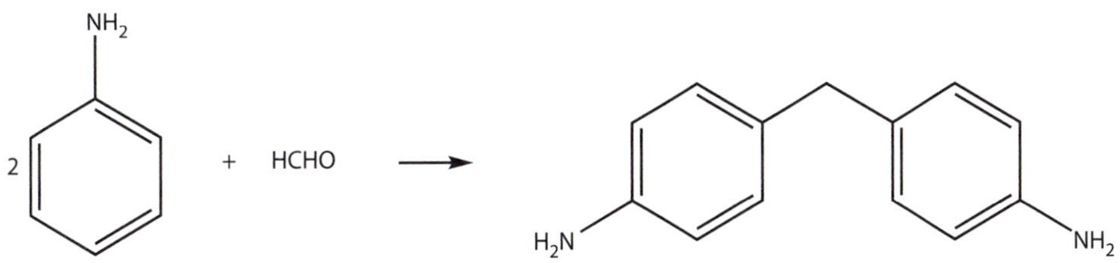

2 + HCHO →

2.54 Schematic

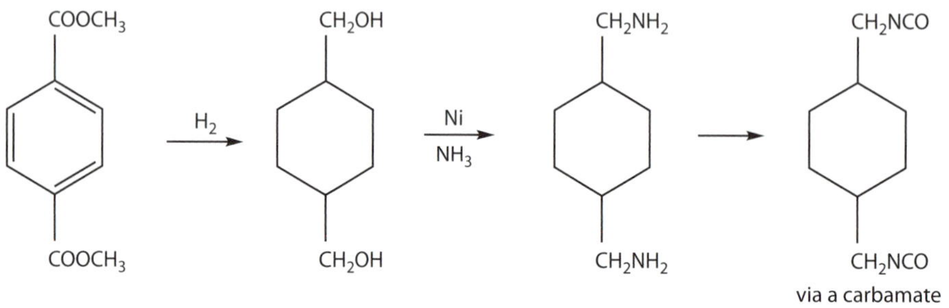

2.55 Schematic

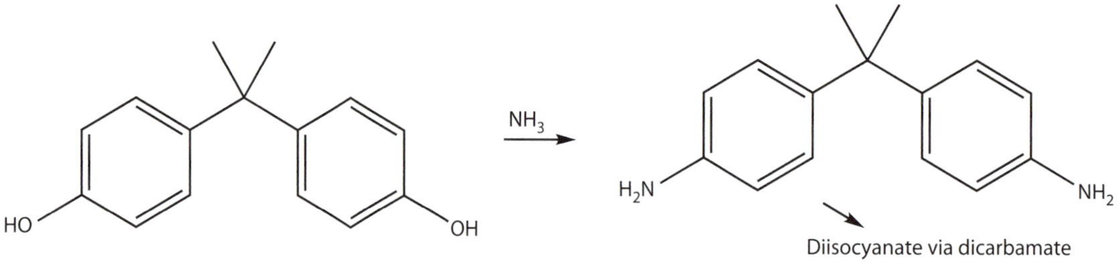

2.56 Schematic

stiffeners are possible. Eastman Chemical uses this *cis–trans* mixture of diols to make a polyester. Another stiffening possibility is made from bisphenol. The alternative is to use the foregoing cyclic diols as stiffeners with more flexible aliphatic diisocyanates. Such changes will depend on relative costs of extent of changes required in the plants, the perceived danger of the carcinogens, and the demand for light-stable polyurethanes (**2.55** Schematic and **2.56** Schematic).

There is the question of just how toxic one intermediate is relative to another. In the alternative methods that avoid phosgene, carbon dioxide is less toxic than carbon monoxide, and ethanol is less toxic than methanol, suggesting the use of diethyl carbonate instead of dimethyl carbonate; the aliphatic amine intermediates are less toxic than the aromatic amines and the nitro compounds from which the latter are made; routes to amines from diols are less dangerous than those going through nitriles; higher molecular weight analogues are less toxic (by inhalation) than those of lower molecular weight; routes via carboxylic acids may be less dangerous than those via nitro compounds and a solid might be less likely to give problems than a liquid that might penetrate the skin. If the material is not a problem in ordinary usage, or if the alternative costs a lot more to make, it probably will not be displaced. If it is a minor problem, then containment and good occupational hygiene may be the answer.

Some processes that claim replacements for phosgene still use dangerous material. Manada and Murakami[104] have made methyl chloroformate without phosgene as follows (**2.57** Schematic).

Items for further study include the following:

1. Use of an HA reagent that will make it easier to recover the isocyanate than is possible with $RNHCOOCH_3$.

   Could diphenyl carbonate replace dimethyl carbonate, at least for isocyanates that boil appreciably higher than phenol? Could methylethylketoxime or ethyl acetoacetate fit into some of the syntheses?
2. Will the DuPont method for methyl isocyanate work for higher aliphatic amines and aromatic amines? The patent claims amines up to 10 carbon atoms but gives no examples. Could this method be adapted for use with copolymers of *N*-vinylformamide?
3. Ureas can be prepared from amines and carbon dioxide.[105] They can also be prepared by heating amines with urea at 160°C.[106] Treatment with excess alcohol (or other hydroxyl compounds) would give a carbonate (or analogue) for pyrolysis to an isocyanate (**2.58** Schematic).

   The methanolysis of *N,N′*-dibutylurea to give the carbamate is known.[107] This would overcome some of the problems of the Monsanto synthesis.
4. Phenol can be converted to aniline by treatment with ammonia at high temperatures. Can a nickel, ruthenium, or other metal catalyst be found to do this for aromatic diols, such as 4,4′-sulfonyldiphenol?

   The product shown in structure **2.59** Schematic is a known compound that can be made from aniline hydrochloride and acetone in 74% yield.[108] It has also been

$$4NO + O_2 + 4CH_3OH \longrightarrow 4CH_3ONO + 2H_2O \xrightarrow[\text{CO, } 60^\circ C]{NOCl} ClCOOCH_3 + 2NO$$

2.57 Schematic

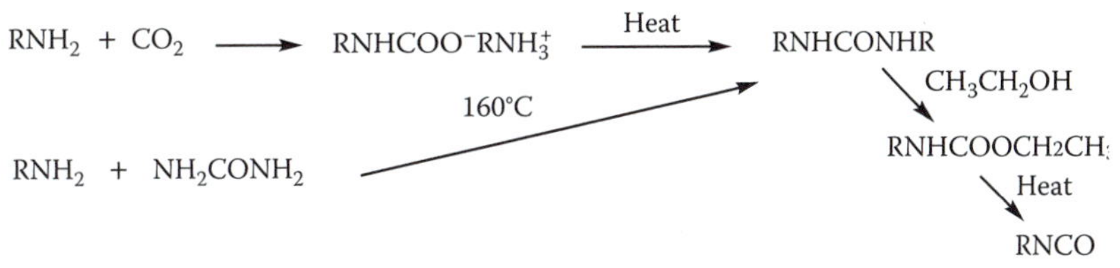

$RNH_2 + CO_2 \longrightarrow RNHCOO^-RNH_3^+ \xrightarrow{Heat} RNHCONHR$

$RNH_2 + NH_2CONH_2 \xrightarrow{160°C} RNHCONHR$

$RNHCONHR \xrightarrow{CH_3CH_2OH} RNHCOOCH_2CH_3 \xrightarrow{Heat} RNCO$

2.58 Schematic

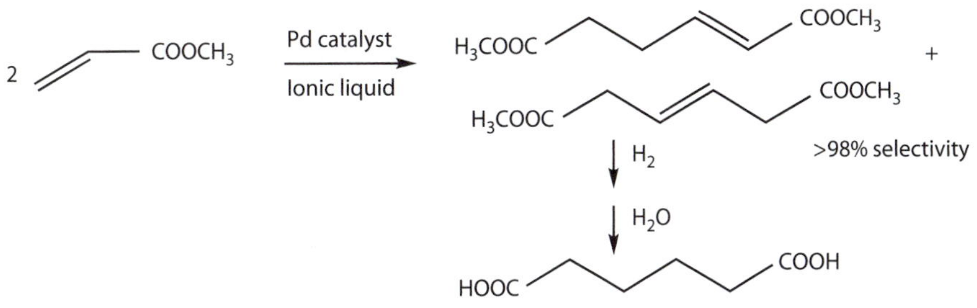

2 $CH_2=CHCOOCH_3$ —Pd catalyst, Ionic liquid→ H_3COOC…$COOCH_3$ + H_3COOC…$COOCH_3$ (>98% selectivity) —H_2→ —H_2O→ HOOC…COOH

2.59 Schematic

made by reduction of the corresponding dinitro compound.[109] Another possibility is to reduce the bisphenol to the corresponding dialchohol, which can then be treated with ammonia to produce a diamine.

5. Methyl *N*-phenylcarbamate has been reacted with formaldehyde to give a biscarbamate that was pyrolyzed to 4,4′-diisocyanatodiphenylmethane. Can a catalyst be found so that acetone can be substituted for formaldehyde?
6. Is there a way of oxidizing an amide to a hydroxamic acid that can then undergo a Lossen rearrangement *in situ*? This might be possible with hydrogen peroxide and titanium silicalite (or a related system) or with a dioxime.

2.5 REPLACEMENTS FOR HCN

Hydrogen cyanide and its salts are highly toxic. There are alternatives that can replace it in most of its uses. The heap leaching of gold ore has resulted in some disastrous spills, such as the one in Romania that contaminated a 50 km section of the river. Haber Inc. has a process that uses no cyanide, recovers more gold, and speeds up the extraction at no extra cost.[110] A Mexican process uses thiourea, which can be recovered and reused.[111] Australian scientists use chloride with a special grade of activated carbon.[112] Gold ore has also been extracted with thiourea plus ferric sulfate and an ionic liquid.[113]

The DuPont process for the preparation of adipic acid and hexamethylenediamine for the preparation of nylon 6,6 starting with the addition of HCN to 1,3-butadiene was described in Chapter 1. Both of these can be made from renewable raw materials. Petroselenic acid (*cis*-6-octadienoic acid) can be cleaved to adipic and lauric acids by ozonolysis (Chapter 9). John Frost uses genetically modified organisms to alter the aromatic amino acid synthesis to produce *cis*, *cis*-muconic acid, which is hydrogenated to adipic acid (Chapter 9). The ammonium salt of the adipic acid can be heated to produce adiponitrile, which is reduced to hexamethylenediamine. Reaction of 1,3-butadiene with carbon monoxide and methanol gave a mixture of unsaturated methyl esters, which were then treated with carbon monoxide and hydrogen to make the aldehyde esters in 99.2% conversion and 95.6% yield.[114] The aldehyde ester could be reacted with ammonia and hydrogen to produce caprolactam for nylon 6. It could also be oxidized to adipic acid for nylon 6,6. Methyl acrylate can be dimerized to unsaturated diesters with a palladium catalyst and an ionic liquid with or without sc carbon dioxide with >98% selectivity (**2.60** Schematic).[115]

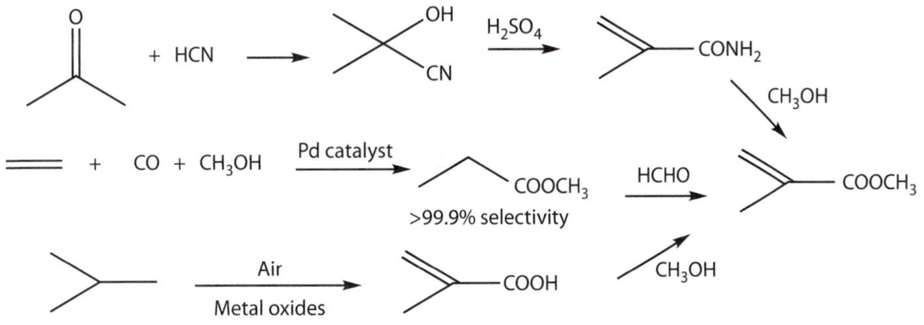

Acetone + HCN → acetone cyanohydrin (OH, CN) —H_2SO_4→ methacrylamide ($CONH_2$) —CH_3OH→ methyl methacrylate ($COOCH_3$)

Ethylene + CO + CH_3OH —Pd catalyst→ $CH_3CH_2COOCH_3$ (>99.9% selectivity) —HCHO→ methyl methacrylate ($COOCH_3$)

Isobutylene —Air, Metal oxides→ methacrylic acid (COOH) —CH_3OH→ methyl methacrylate

2.60 Schematic

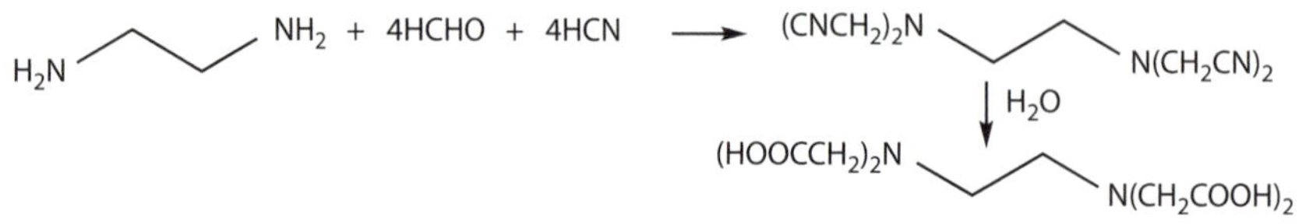

H_2N–CH_2CH_2–NH_2 + 4HCHO + 4HCN → $(CNCH_2)_2N$–CH_2CH_2–$N(CH_2CN)_2$ —(H_2O)→ $(HOOCCH_2)_2N$–CH_2CH_2–$N(CH_2COOH)_2$

2.61 Schematic

Methyl methacrylate has been manufactured from acetone and HCN via the cyanohydrin. It is also made by the hydroformylation of ethylene followed by reaction with formaldehyde.[116] The best way is the process of EverNu technology to make methacrylic acid by the oxidation of isobutane (**2.61** Schematic).[117]

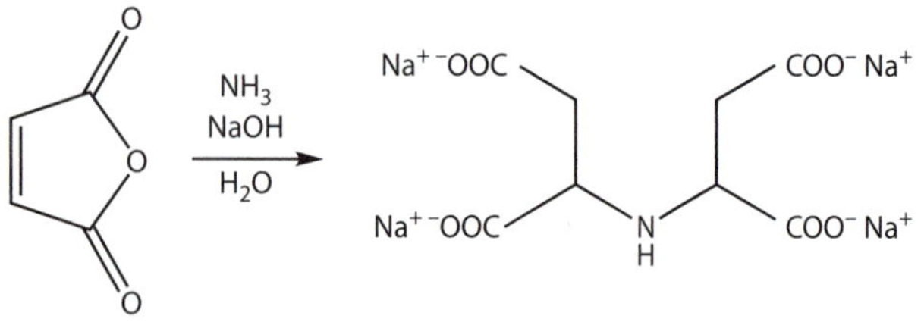

Maleic anhydride —(NH_3, NaOH, H_2O)→ Na^+ ^-OOC…N(H)…COO^- Na^+ (tetrasodium iminodisuccinate)

2.62 Schematic

The usual way to make an amino acetic acid is to treat the amine with formaldehyde and HCN, followed by hydrolysis of the nitrile (**2.62** Schematic). Bayer has used a different method to develop a chelating agent that can be substituted for ethylenediaminetetraacetic acid (EDTA) (**2.63** Schematic).[118]

Monsanto has eliminated its use of formaldehyde and HCN in the preparation of glyphosate by an oxidation of diethanolamine to iminodiacetic acid (Chapter 11).

The Strecker synthesis of alpha-amino acids proceeds through the cyanohydrin (**2.64** Schematic). An alternative is to replace the hydrogen cyanide with acetamide and carbon monoxide using a palladium or cobalt catalyst.[119] Alpha-ketoacids can be converted to alpha-amino acids with ammonium formate using an iridium catalyst in 91–97% yield.[120] A two-stage enzymatic process converted acetaldehyde to alanine (**2.65** Schematic).[121]

The food additive potassium ferrocyanide has been used to convert aryl bromides to aryl nitriles.[122] It should be tried in the Strecker reaction.

Tert-butylamine has been prepared by the Ritter reaction using HCN and sulfuric acid followed by hydrolysis of the formamide. BASF is now able to make it in 99% yield from isobutylene and ammonia (**2.66** Schematic).[123]

HCN is no longer needed to fumigate houses since the house can just be heated enough to kill the insects (Chapter 11).

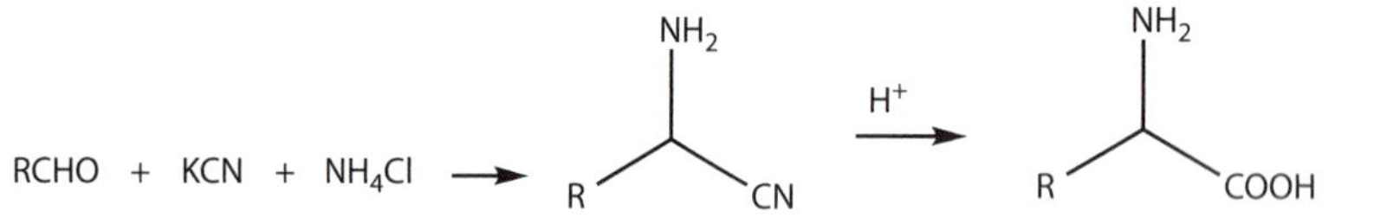

RCHO + KCN + NH_4Cl → R–CH(NH_2)–CN —(H^+)→ R–CH(NH_2)–COOH

2.63 Schematic

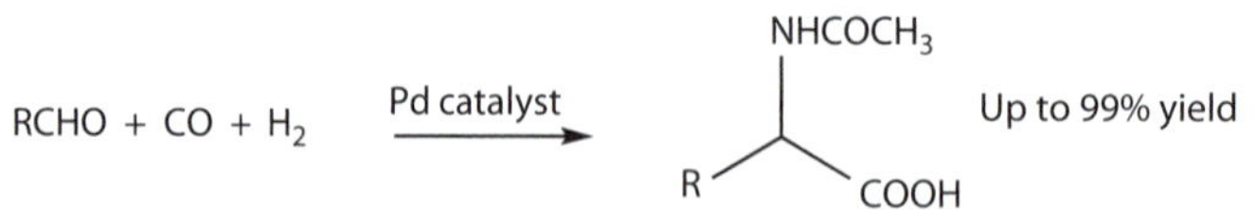

RCHO + CO + H_2 —(Pd catalyst)→ R–CH($NHCOCH_3$)–COOH   Up to 99% yield

2.64 Schematic

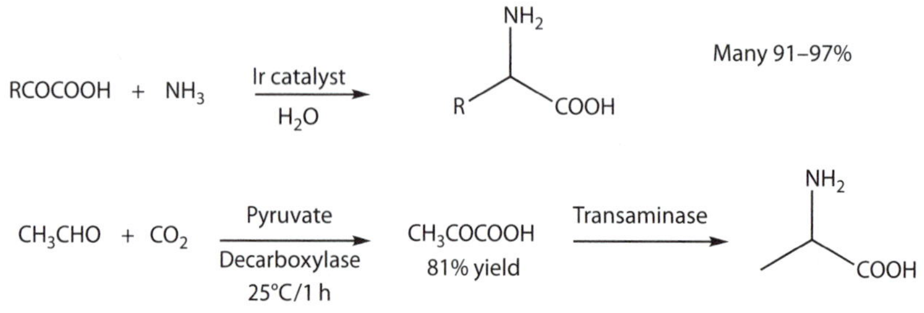

$RCOCOOH$ + NH_3 —(Ir catalyst, H_2O)→ R–CH(NH_2)–COOH   Many 91–97%

CH_3CHO + CO_2 —(Pyruvate Decarboxylase, 25°C/1 h)→ $CH_3COCOOH$ 81% yield —(Transaminase)→ CH_3–CH(NH_2)–COOH

2.65 Schematic

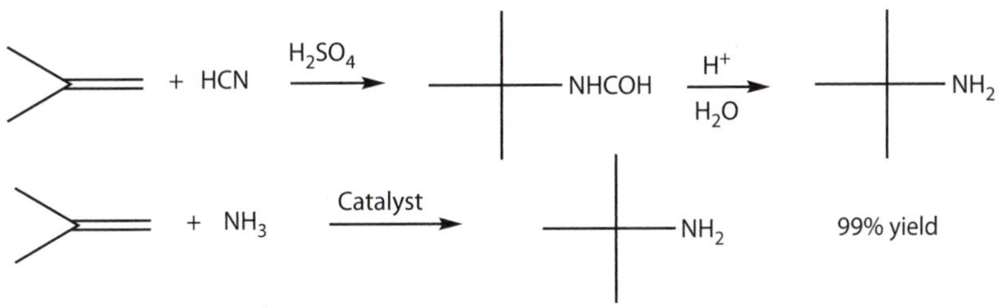

2.66 Schematic

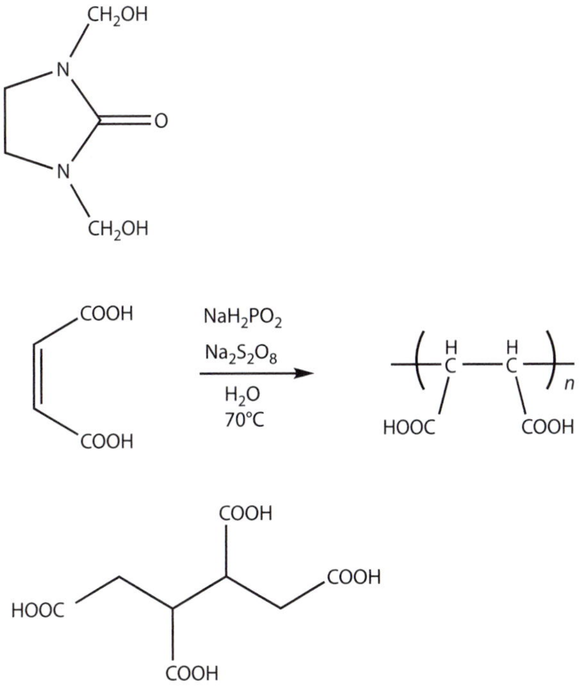

2.67 Schematic

2.6 ELIMINATING FORMALDEHYDE

Formaldehyde is a carcinogen that is a leading cause of the "sick building syndrome."[124] It is released by the phenol–formaldehyde and urea–formaldehyde adhesives used in the plywood and particle board. Columbia Forest Products has converted all its plywood and particleboard to an adhesive made from soy protein flour and an azetidinium salt (normally used as a wet-strength resin for paper).[125] The University of Southern Mississippi has developed a similar system.[126] Evertech LLC has developed a system of vacuum impregnation of softwoods with urea, glyoxal, and starch to form a dense wood to replace old growth hardwoods.[127] Rohm & Haas (now Dow) has worked with Johns Manville to develop a system using an aqueous acrylic thermosetting binder to replace the use of formaldehyde in fiberglass insulation.[128]

Another source of formaldehyde in the home is the use of dimethylolethyleneurea as a crease-proofing resin for cotton. This is being replaced by 1,2,3,4-butanetetracarboxylic acid, poly(maleic acid),[129] and many other reagents that will react with the hydroxyl groups of the cellulose to cross-link it (**2.67** Schematic).

REFERENCES

1. T.A. Ryan, C. Ryan, E.A. Seddon, and K. Seddon, *Phosgene and Related Carbonyl Halides*, Elsevier, Amsterdam, 1996.
2. W. Schneider and W. Diller, *Ullmann's Encyclopedia of Industrial Chemistry*, 5th ed., SNPE Paris, France, 1991, *A19*, 411.
3. Anon., *Chem. Eng. News*, Jan. 29, 1996, 17.
4. Anon., *Chem. Eng. News*, July 15, 1996, 13.
5. Anon., *Chem. Eng. News*, Dec. 11, 1995, 20.
6. (a) S.C. Stinson, *Chem. Eng. News*, Jan. 18, 1999, 69; (b) *Chem. Eng. News*, June 1, 1998, 26.
7. R. Mullin, *Chem. Eng. News*, March 1, 2004, 20.
8. M. McCoy, *Chem. Eng. News*, May 31, 2004, 17.
9. J.-P. Senet, *The Recent Advance in Phosgene Chemistry*, SNPE Paris, France, 1998.

10. L. Cotarca and H. Eckert, *Phosgenations—A Handbook*, Wiley-VCH, Weinheim, 2003.
11. E. Turk, *Chem. Eng. News*, Mar. 2, 1998, 6.
12. Anon., *Chem. Market Reporter*, February 16, 2004, 27.
13. (a) Anon., *Aldrichim. Acta*, 1988, *21*(2), 47; (b) L. Cotarca, P. Delogu, A. Nardelli, and V. Junjic, *Synthesis*, 1996, 553.
14. K. Weissermel and H.-J. Arpe, *Industrial Organic Chemistry*, 2nd ed., VCH, Weinheim, 1993, 373–379.
15. (a) R.H. Richter and R.D. Priester, Jr., *Kirk–Othmer Encyclopedia of Chemical Technology*, 4th ed., SNPE Paris, France, 1995, *14*, 902; (b) J.K. Backus, C.D. Blue, P.M. Boyd, F.J. Cama, J.H. Chapman, J.L. Eakin, S.J. Harasin, E.R. McAfee, C.G. McCarty, N.H. Nodelman, J.N. Rieck, H.-G.L. Schmelzer, and E.P. Squiller, *Encyclopedia Polymer Science Engineering*, 2nd ed., 1988, *13*, 243.
16. (a) D. Williams, *Chem. Ind.* (*London*), July 9, 2007, 7; (b) G. Ondrey, *Chem. Eng.*, 2007, *114*(5), 17; (c) Anon., *Chem. Eng. News*, Sept. 25, 2006, 47.
17. (a) K.J. Saunders, *Organic Polymer Chemistry*, 2nd ed., Chapman & Hall, London, 1988, 358; (b) M.P. Stevens. *Polymer Chemistry—An Introduction*, 3rd ed., Wiley, New York, 378–381.
18. M. Heylin and W. Lepkowski, *Chem. Eng. News*, Dec. 19, 1994, 3, 8.
19. (a) J.R. Thomen, *Chem. Eng. News*, Feb. 6, 1995, 2; (b) V.N.M. Rao and G.E. Heinsohn, U.S. patent 4,537, 726 (1985); (c) L.E. Manzer, *Catal. Today*, 1993, *18*(2), 199.
20. (a) J. Haggin, *Chem. Eng. News*, June 3, 1996, 38; (b) R.F. Service, *Science*, 1998, *282*, 400.
21. W. Lepkowski, *Chem. Eng. News*, Nov. 21, 1994, 11.
22. (a) Z.-H. Fu and Y. Ono, *J. Mol. Catal.*, 1994, *91*, 399; (b) S. Wang, G. Zhang, X. Ma, and J. Gong, *Ind. Eng. Chem. Res.*, 2007, *46*, 6858.
23. T. Baba, A. Kobayashi, Y. Kawanami, K. Inazu, A. Ishikawa, T. Echizenn, K. Murai, S. Aso, and M. Inomata, *Green Chem.*, 2005, *7*, 159.
24. E. Angles, A. Santillan, I. Martinez, A. Ramirez, E. Moreno, M. Salmon, and R. Martinez, *Synth. Commun.*, 1994, *24*, 2441.
25. M. Aresta, C. Barloco, and E. Quaranta, *Tetrahedron*, 1995, *51*, 8073.
26. (a) T. Tsujimoto and T. Okawa, U.S. patent 5,391,805 (1995); (b) Anon., *Chem. Abstr.*, 1995, *123*, 56828.
27. (a) F. Mizia, F. Rivetti, and U. Romano, *Chem. Abstr.*, 1994, *120*, 194501; (b) U.S. patent 5,315,034 (1994).
28. (a) F. Rivetti, personal communication, 1998; (b) Anon., *Chem. Eng. News*, Nov. 29, 1999, 13.
29. (a) T. Ookawa, H. Igarashi, and T. Tsujimoto, *Chem. Abstr.*, 1995, *123*, 199647; (b) T. Tsujimoto, T. Aoki, and H. Matsunaga, *Chem. Abstr.*, 1995, *123*, 341248.
30. A.G. Shaikh and S. Sivaram, *Chem. Rev.*, 1996, *96*, 951.
31. (a) F. Rivetti, U. Romano, and D. Delledonne. In: P.T. Anastas and T.C. Williamson, eds, *Green Chemistry—Designing Chemistry for the Environment*, A.C.S. Symp. 626, Washington, DC, 1996; Preprints A.C.S. *Div. Environ. Chem.*, 1994, *34*(2), 332; (b) F. Rivetti, D. Delledonne, and D. Dreoni, *Chem. Abstr.*, 1994; *120*, 76901; (c) D. Delledonne, F. Rivetti, and U. Romano, *J. Organomet. Chem.*, 1995, *488*, C15; (d) P. Tundo and M. Selva, *Chemtech*, 1995, *25*(5), 31; (e) P. Tundo, C.A. Marques, and M. Selva. In: P.T. Anastas and T.C. Williamson, eds, *Green Chemistry—Designing Chemistry for the Environment.* A.C.S. Symp. 626, Washington, DC, 1996; Preprints A.C.S. *Div. Environ. Chem.*, 1994, *34*(2), 313, 336; (f) N. diMuzio, C. Fusi, F. Rivetti, and G. Sasselli, U.S. patent 5,210,269 (1993).
32. W. Yanji, Z. Xinqiang, Y. Baoguo, Z. Bingchang, and C. Jinsheng, *Appl. Catal. A*, 1998, *171*, 255.
33. Z. Kricsfalussy, H. Waldman, and H.-J. Traenckner, *Ind. Eng. Chem. Res.*, 1998, *37*, 865.
34. J.Y. Ryer, U.S. patent 5,902,894 (1999); U.S. patent 6,010,976 (2000).
35. M. Alba, J.-C. Choi, and T. Sakakura, *Green Chem.*, 2004, *6*, 524.
36. G. Ondrey, *Chem. Eng.*, 1999, *106*(6), 23.
37. G.E. Harrison, A.J. Dennis, and M. Sharif, U.S. patent 5,426,207 (1995); H. Lee, S.-J. Kim, B.S. Ahn, W.K. Lee, and H.S. Kim, *Catal. Today*, 2003, *87*, 139.
38. (a) H.-J. Buysch, C. Hesse, J. Rechner, R. Schomaecker, P. Wagner, and D. Kaufman, U.S. patent 5,498,742 (1996); (b) M. Takagi, H. Miyagi, T. Yoneyama, and Y. Ohgomori, *J. Mol. Catal. A*, 1998, 129L1; (c) M. Goyal, R. Nagahata, J.-I. Sugiyama, M. Asai, M. Ueda, and K. Takeuchi, *J. Mol. Catal. A*, 1999, *137*, 147.
39. (a) A. Bomben, C.A. Marques, M. Selva, and P. Tundo, *Tetrahedron*, 1995, *51*, 11573; (b) M. Selva, C.A. Marques, and P. Tundo, *J. Chem. Soc. Perkin* 1995, *1*, 1889; (c) Y. Ono, *Pure Appl. Chem.*, 1996, *68*, 367; *Catal. Today*, 1997, *35*, 15; *Appl. Catal. A*, 1997, *155*, 133; (d) M. Selva and A. Perosa, *Green Chem.*, 2008, *10*, 457; (e) M. Selva, E. Militello, and M. Fabris, *Green Chem.*, 2008, *10*, 73; (f) P. Tundo and M. Selva, *Acc. Chem. Res.*, 2002, *35*, 706; (g) P. Tundo, *Pure Appl. Chem.*, 2001, *73*, 1117.
40. M. Okamoto, K. Miyazaki, A. Kado, and E. Suzuki, *Chem. Commun.*, 2001, 1839.
41. Z. Zheng, T. Wu, and X. Zhou, *Chem. Commun.*, 2006, 1864.
42. Y. Fu, T. Baba, and Y. Ono, *Appl. Catal. A*, 1998, *166*, 419, 425.
43. C.D. Chang and T.J. Huang, U.S. patent 5,804,690 (1998).
44. M. Selva, P. Tundo, D. Brunelli, and A. Perosa, *Green Chem.*, 2007, *9*, 463.
45. M. Selva and P. Tundo, *J. Org. Chem.*, 1998, *63*, 9540.
46. V.L.K. Valli and H. Alper, *Organometallics*, 1995, *14*, 80.
47. F.W. Harstock, D.G. Herrington, and L.B. McMahon, *Tetrahedron Lett.*, 1994, *35*, 8761.
48. J. Li, H. Jiang, and M. Chen, *Green Chem.*, 2001, *3*, 137.
49. F. Shi, Y. Deng, T.S. Ma, and H. Yang, *J. Catal.*, 2001, *203*, 525.
50. T. Jiang, X. Ma, Y. Zhou, S. Liang, J. Zhang, and B. Han, *Green Chem.*, 2008, *10*, 465.
51. (a) S. Kanagasabapathy, S.P. Gupta, and R.V. Chaudhari, *Ind. Eng. Chem. Res.*, 1994, *33*, 1; (b) K.-T. Li and Y.-J. Peng, *J. Catal.*, 1993, *143*, 631.
52. S.A.R. Mulla, C.V. Rode, A.A. Kelkar, and S.P. Gupte, *J. Mol. Catal. A*, 1997, *122*, 103.
53. J. Gao, H. Li, Y. Zhang, and Y. Zhang, *Green Chem.*, 2007, *9*, 572.
54. F. Shi, Y. Deng, T.S. Ma, J. Peng, Y. Gu, and B. Qiao, *Angew. Chem. Int. Ed.*, 2003, *42*, 3257.
55. F. Bigi, R. Maggi, G. Sartori, and E. Zambonin, *Chem. Commun.*, 1998, 513.
56. M. Aresta, C. Berloco, and E. Quaranta, *Tetrahedron*, 1995, *51*, 8073.

57. (a) W.D. McGhee, M. Paster, D. Riley, K. Ruettimann, J. Solodar, and T. Waldman. In: P.T. Anastas and T.C. Williamson, eds, *Green Chemistry—Designing Chemistry for the Environment*, A.C.S. Symp. 626, Washington, DC, 1996; Preprints A.C.S. *Div. Environ. Chem.*, 1994, *34*(2), 206; (b) T.E. Waldman and W.D. McGhee, *J. Chem. Soc., Chem. Commun.*, 1994, *699*, 957; (c) W.D. McGhee, Y. Pan, and J.J. Talley, *Tetrahedron Lett.*, 1994, *35*, 839; (d) W.D. McGhee and D. Riley, *J. Org. Chem.*, 1995, *60*, 6205; (e) W.D. McGhee, D. Riley, K. Christ, Y. Pan, and B. Parnas, *J. Org. Chem.*, 1995, *60*, 2820; (f) W.D. McGhee, M.D. Paster, D.P. Riley, K.W. Ruettimann, A.J. Solodar, and T.E. Waldman, *Chem. Abstr.*, 1995, *123*, 287146; (g) W.D. McGhee and D.R. Riley, U.S. patent 5,371,183 (1994); (h) D. Riley, W.D. McGhee, and T. Waldman. In: P.T. Anastas and C.A. Farris, eds, *Benign by Design,* A.C.S. Symp. 577, Washington, DC, 1994, 122.
58. (a) S. Cenini and F. Ragaini, *Catalytic Reductive Carbonylation of Organic Nitro Compounds,* Kluwer Academic, Dordrecht, 1997; (b) A.M. Tafesh and J. Weigung, *Chem. Rev.*, 1996, *96*, 2035; (c) F. Ragaini, M. Macchi, and S.S. Cenini, *J. Mol. Catal. A*, 1997, *127*, 33; (d) S.J. Skoog and W.L. Gladfelter, *J. Am. Chem. Soc.*, 1997, *119*, 11049; (e) E. Bolzacchini, R. Lucini, S. Meinardi, M. Orlandi, and B. Rindone, *J. Mol. Catal. A*, 1996, *110*, 227; (f) S.B. Halligudi, N.H. Khan, R.I. Kureshy, E. Suresh, and K. Venkatsubramanian, *J. Mol. Catal. A*, 1997, *124*, 147; (g) R. Santi, A.M. Romano, F. Panella, G. Mestroni, and A.S.O. Santii, *J. Mol. Catal.*, 1999, *144*, 41.
59. (a) P. Wehman, G.C. Dol, E.R. Moorman, P.C.J. Kamer, P.W.N.M. van Leeuwen, J. Fraanje, and K. Goubitz, *Organometallics*, 1994, *13*, 4856; (b) P. Wehman, V.E. Kaasjager, W.D.J. deLange, F. Hartl, P.C.J. Kamer, P.W.N.M. van Leeuwen, J. Fraanje, and K. Goubitz, *Organometallics*, 1995, *14*, 3751; (c) P. Wehman, P.C.J. Kamer, and P.W.N.M. van Leeuwen, *J. Chem. Soc., Chem. Commun.*, 1996, 217.
60. P. Giannoccaro, I. Tommasi, and M. Aresta, *J. Organomet. Chem.*, 1994, *476*, 13.
61. Y. Izumi, K. Urabe, and M. Onaka, *Zeolite, Clay and Heteropolyacid in Organic Reactions*, VCH, Weinheim, 1992, 157.
62. (a) N.P. Reddy, A.M. Masdeu, B. Elali, and H. Alper, *J. Chem. Soc., Chem. Commun*, 1994, 863; (b) R. Santi, A.M. Romano, F. Panella, and C. Santini, *J. Mol. Catal. A*, 1997, *127*, 95; (c) P. Wehman, L. Borst, P.C.J. Kamer, and P.W.N.M. van Leeuwen, *J. Mol. Catal. A*, 1996, *112*, 23.
63. D.K. Mukherjee, B.K. Palit and C.R. Saha, *J. Mol. Catal.*, 1994, *91*, 19.
64. J.D. Gargulak and W.L. Gladfelter, *J. Am. Chem. Soc.*, 1994, *116*, 3792; also in P.T. Anastas and C.A. Farris, *Benign by Design*, A.C.S. Symp. 577, Washington, DC, 1994, 46.
65. V. Macho, M.S. Kralik, and F. Halmo, *J. Mol. Catal. A*, 1996, *109*, 119.
66. R.H. Richter and R.D. Priester, Jr., *Kirk–Othmer Encyclopedia of Chemical Technology*, 4th ed., Wiley, New York, 1995, *14*, 902.
67. P.L. Brusky, J.H. Kyung, and R.A. Grimm. In: J.R. Kosak and T.A. Johnson, eds, *Catalysis of Organic Reactions,* Marcel Dekker, New York, 1994, 473.
68. (a) R.H. Richter and R.D. Priester, Jr., *Kirk–Othmer Encyclopedia of Chemical Technology*, 4th ed., Wiley, New York, 1995, *14*, 902; (b) J.K. Backus, C.D. Blue, P.M. Boyd, F.J. Cama, J.H. Chapman, J.L. Eakin, S.J. Harasin, E.R. McAfee, C.G. McCarty, N.H. Nodelman, J.N. Rieck, H.G. Schmelzer, and E.P. Squiller, *Encylopedia of Polymer Science and Engineering*, 2nd ed., Wiley, New York, 1988, *13*, 243.
69. C. Dubosclard–Gottardi, P. Caubere, and Y. Fort, *Tetrahedron*, 1995, *51*, 2561.
70. M. Wakasa, S. Watabe, and H. Yokoo, *Chem. Abstr.*, 1989, *110*, 231127.
71. Dow Chemical, *Chem. Abstr.*, 1988, *109*, 169865.
72. F. Ragaini, T. Longo, and S. Cenini, *J. Mol. Catal. A*, 1996, *110*, L171.
73. (a) H. Zengel and M. Bergfeld, U.S. patent 4,457,871 (1984); (b) D. Hentschel, H. Zengel, and M. Bergfeld, U.S. patent 4,223,145 (1979); (c) R.H. Richter and R.D. Priester, Jr., *Kirk–Othmer Encyclopedia of Chemical Technology*, 4th ed., Wiley, New York, 1995, *14*, 902.
74. D.S. Rane and M.M. Sharma, *J. Chem. Technol. Biotechnol.*, 1994, *59*, 271.
75. X. Huang and J.W. Keillor, *Tetrahedron Lett.*, 1997, *38*, 313.
76. M.A. Mantegozza, M. Padovan, G. Petrini, and P. Roffia, U.S. 5,320,819 patent (1994).
77. J. Jarvis, R. Aguis, and L. Sawyer, *Chem. Br.*, 1996, *32*(6), 51.
78. (a) J.W. Reisch, R.J. Blackwell, R.T. Wojcik, J.M. O'Connor, and K.B. Chandalia, *Surf. Coat. Int.*, 1995, *78*, 380; (b) R.H. Richter and R.D. Priester, Jr., *Kirk–Othmer Encyclopedia of Chemical Technology*, 4th ed., Wiley, New York, 1995, *14*, 902.
79. W. Brown, *Surf. Coat. Int.*, 1995, *78*, 238.
80. J.W. Taylor, M.J. Collins, and D.R. Bassett, *J. Coat. Technol.*, 1995, *67*, 43.
81. M. Puterman, B. Jansen, and H. Kober, *J. Appl. Polym. Sci.*, 1996, *59*, 1237.
82. R.H. Richter and R.D. Priester, Jr., *Kirk–Othmer Encyclopedia of Chemical Technology*, 4th ed., Wiley, New York, 1995, *14*, 902.
83. H. Xiao, H.X. Xiao, B. Suthar, and K.C. Frisch, *J. Coat. Technol.*, 1995, *67*, 19.
84. F.C. Onwumere and J.F. Pazos, U.S. patent 5,491,210 (1996).
85. (a) Y. Yukawa, M. Yabuta, and A. Tominaga, *Prog. Org. Coat.*, 1994, *24*, 359; (b) D.-J. Liaw and D.-L. Ou, *J. Appl. Polym. Sci.*, 1996, *59*, 1529.
86. (a) Y. Yukawa, M. Yabuta, and A. Tominaga, *Prog. Org. Coat.*, 1994, *24*, 359; (b) D.-J. Liaw and D.-L. Ou, *J. Appl. Polym. Sci.*, 1996, *59*, 1529.
87. R.W. Cabe and A. Taylor, *Chem. Commun.*, 2002, 934, *Green Chem.*, 2004, *6*, 151.
88. B. Ochiai, Y. Satoh, and T. Endo, *Green Chem.*, 2005, *7*, 765.
89. F. Schmitz, H. Keul, and H. Hocker, *Macromol. Rapid Commun.*, 1997, *18*, 699.
90. (a) R.L. Paddock and S.B.T. Nguyen, *Chem. Commun.*, 2004, 1622; (b) J.-S. Tian, C.-X. Miao, J.Q. Wang, C. Cai, Y. Zhao, and L.N. He, *Green Chem.*, 2007, *9*, 566; (c) C.-X. Miao, J.-Q. Wang, Y. Wu, Y. Du, and L.-N. He, *ChemSusChem*, 2008, *1*, 236; (d) J. Sun, S. Zhang, W. Cheng, and J. Ren,

Tetrahedron Lett., 2008, *49*, 3588; (e) T. Takahashi, T. Watahiki, S. Kitazume, H. Yasuda, and T. Sakakura, *Chem. Commun.*, 2006, 1664.
91. (a) X. Zhao, Y. Zhang, and Y. Wang, *Ind. Eng. Chem. Res.*, 2004, *43*, 4038; (b) B.M. Bhanage, S.-I. Fujita, Y. Ikushima, and M. Arai, *Green Chem.*, 2003, *5*, 429.
92. J.H. Clements, *Ind. Eng. Chem. Res.*, 2003, *42*, 663.
93. G. Ondrey, *Chem. Eng.*, 2005, *112*(12), 22.
94. R.M. Versteegen, R.P. Sijbesma, and E.W. Meijer, *Angew. Chem. Int. Ed.*, 1999, *38*, 2917.
95. (a) K.J. Saunders, *Organic Polymer Chemistry,* 2nd ed., Chapman & Hall, London, 1988, 269; (b) K. Komiya, M. Aminaka, K. Hasegawa, H. Hachiya, H. Okamoto, S. Fukuoka, H. Yoneda, I. Fukawa, and T. Dozono, *Green Chemistry—Designing Chemistry for the Environment*, A.C.S. Symp. 626, Washington, DC, 1996; Preprints A.C.S. *Div. Environ. Chem.*, 1994, *34*(2), 343.
96. K. Komya, K. Tomoyasu, and S. Fukuoka, *Chem. Abstr.*, 1993, *119*, 250, 783.
97. See for example (a) S. Kuze, U.S. patent 5,349,043 (1994); (b) M. Yokoyama, K. Takakura, and J. Takano, U.S. patent 5,391,691 (1995); (c) S. Kuehling, W. Alewelt, H. Kauth, and D. Freitag, U.S. patent 5,314,985 (1994); (d) S. Fukuoka, M. Kawamura, K. Komiya, M. Tojo, H. Hachiya, K. Hasegawa, M. Aminaka, H. Okamoto, I. Fukawa, and S. Konno, *Green Chem.*, 2003, *5*, 497; (e) W.B. Kim, U.A. Joshi, and J.S. Lee, *Ind. Eng. Chem. Res.*, 2004, *43*, 1897–1914; (f) G. Parkinson, *Chem. Eng.*, 1999, *106*(10), 21.
98. G. Parkinson, *Chem. Eng.*, 1996, *103*(8), 21.
99. V. Pokharkar and S. Sivaram, *Polymer*, 1995, *36*, 4851.
100. S. Chatti, G. Schwarz, and H.R. Kricheldorf, *Macromolecules*, 2006, *39*, 9064.
101. M. Ishu, I.M. Goyal, M. Ueda, K. Takeuchi, and M. Asai, *Macromol. Rapid Commun.*, 2001, *22,* 376.
102. M. Cheng, E.B. Lobkovsky, and G.W. Coates, *J. Am. Chem. Soc.*, 1998, *120*, 11018.
103. U.S. EPA, 2008. Green Chemistry Award Entries, 26.
104. (a) N. Manada and M. Murakami, *Bull. Chem. Soc. Japan*, 1994, *67*, 2856; (b) S.C. Stinson, *Chem. Eng. News*, Nov. 25, 1996, 41.
105. S. Coffey, ed., *Rodd's Chemistry of Carbon Compounds*, 2nd ed., Elsevier, Amsterdam, Washington, DC, 1964, *1C*, 312.
106. S.R. Sandler and W. Karo, *Organic Functional Group Preparations,* Academic, New York, 1971, *2*, 137.
107. (a) S. Kanagasabapathy, S.P. Gupta, and R.V. Chaudhari, *Ind. Eng. Chem. Res.*, 1994, *33*, 1; (b) K.-T. Li and Y.-J. Peng, *J. Catal.*, 1993, *143*, 631.
108. C.E. Hoyle, K.S. Ezzell, Y.G. No, K. Malone, and S.F. Thames, *Polym. Degrad. Stab.*, 1989, *25*, 325.
109. (a) A.B. Mossman and W.L. Chiang, U.S. patent 5,037,994 (1991); (b) Anon., *Chem. Abstr.*, 1991, *115*, 231840.
110. G. Ondrey, *Chem. Eng.*, 2004, *111*(2), 14.
111. www.ipsnews.net/news.asp?idnews=40932.
112. Anon., *New Sci.*, April 1, 2000, 11.
113. J.A. Whitehead, G.A. Lawrance, and A. McCluskey, *Green Chem.*, 2004, *6*, 313.
114. S. Lane, U.S. patent 5,840,959 (1998).
115. (a) M. Picquet, S. Stutzmann, I. Tkatchenko, I. Tommasi, J. Zimmermann, and P. Wasserscheid, *Green Chem.*, 2003, *5*, 153; (b) D. Ballivet, T. Tkatchenko, M. Picquet, M. Solins, G. Francio, P. Wasserscheid, and W. Leitner, *Green Chem.*, 2003, *5*, 232.
116. (a) K.J. Watkins, *Chem. Eng. News*, Mar. 19, 2001, 16; (b) P.L. Short, *Chem. Eng. News*, Aug. 27, 2001, 23; (c) G. Ondrey, *Chem. Eng.*, 2003, *110*(7), 19.
117. U.S. EPA, 2008 Green Chemistry Award Entries, Washington, DC, 23.
118. S.K. Ritter, *Chem. Eng. News*, July 2, 2001, 24.
119. (a) R.M. Gomez, P. Sharma, J.L. Arias, J. Perez-Flores, L. Velasco, and A. Cabrera, *J. Mol. Catal. A*, 2001, *170*, 271; (b) M. Beller and M. Eckert, *Angew. Chem. Int. Ed.*, 2000, *39*, 1010; (c) M. Beller, M. Eckert, W.A. Moradi, and H. Neumann, *Angew. Chem. Int. Ed.*, 1999, *38*, 1454.
120. S. Ogo, K. Uehara, T. Abura, and S. Fukuzumi, *J. Am. Chem. Soc.*, 2004, *126*, 3020.
121. M. Miyazaki, M. Shibue, K. Ogino, H. Nakamura, and H. Maeda, *Chem. Commun.*, 2001, 1800.
122. T. Schareina, A. Zapf, and M. Beller, *Chem. Commun.*, 2004, 1388.
123. Anon., *Green Chem.*, 2001, *3*, G22.
124. (a) Anon., *Chem. Eng. News*, April 21, 2008, 37; (b) Anon., *Chem. Eng. News*, July 14, 2008, 28.
125. U.S. EPA Presidential Green Chemistry Award 2007 to K. Li, Columbia Forest Products and Hercules (now Ashland), Washington, DC.
126. S.F. Thames and J.W. Rawlins, U.S. EPA Green Chemistry Award Entries, Washington, DC, 14.
127. EverTech LLC, U.S. EPA, 2008 Green Chemistry Award Entries, Washington, DC, 23.
128. L.R. Raber, *Chem. Eng. News*, October 16, 2006, 55.
129. D. Chen, C.Q. Wang, and X. Qiu, *Ind. Eng. Chem. Res.*, 2005, *44*, 7921.

RECOMMENDED READING

1. W. Schneider and W. Diller, *Ullmann's Encyclopedia of Industrial Chemistry*, 5th ed., VCH, Weinheim, 1991, A19, 411 (on phosgene).
2. K. Weissermel and H.-J. Arpe, *Industrial Organic Chemistry,* 4th ed., 1993, 379–385 (on isocyanates).
3. R.H. Richter and R.D. Priester, Jr., *Kirk–Othmer Encyclopedia of Chemical Technology*, 4th ed., Wiley, New York, 1995, *14*, 906–908, 918–924 (making isocyanates without phosgene).
4. K.J. Saunders, *Organic Polymer Chemistry,* 2nd ed., Chapman & Hall, London, 1988, 269–272 (preparation of polycarbonates).
5. J. Jarvis, R. Aguis, and L. Sawyer, *Chem. Br.*, 1996, *32*(6), 51 (chemicals and asthma).

EXERCISES

1. Methyl isocyanate was involved in the tragedy at Bhopal. What would our society have to do to get along without it altogether?

2. Devise a copolymer containing blocked isocyanate groups, such that on reaction with a difunctional reagent to cure it, no reblocking would occur, and no volatile compound would come off.
3. Dimethyl carbonate can be used to alkylate phenols, aromatic amines, acetylenes, and active methylene compounds. What other alkylations might be carried out with it that are now performed with iodomethane or dimethyl sulfate?
4. Go through the various syntheses for isocyanates and figure out how to recycle all the by-products to the processes, wherever possible.
5. Other dangerous, but widely used, chemicals that need to be replaced by less hazardous ones include formaldehyde, benzene, acrylonitrile, hydrogen cyanide, and hydrogen fluoride. Pick one and see what would be involved in making the same end products without it.

CHAPTER 3

The Chlorine Controversy

3.1 THE PROBLEM

Greenpeace and other environmental organizations have called for the phasing out of chlorine and the products made using it.[1] Other groups, including the Society of Toxicology, American Medical Association (AMA), the American Chemical Society (ACS),[2] Michigan Environmental Science Board,[3] American Chemical Council (formerly Chemical Manufacturers Association) Chemical Manufacturers Association Chlorine Chemistry Council,[4] American Industrial Health Council,[5] and United Kingdom (U.K.) Chemical Industries Association,[6] have said that this is unnecessary. The last calls the request the result of chlorophobia. It points out that there are about 1500 natural chlorine-containing compounds, including epibatidine (**3.1** Schematic), a painkiller 200 times more powerful than morphine, which is used by an Ecuadorian tree frog for defense. It is a highly poisonous nerve toxin used by natives in their blowpipes. Gribble[7] has also called attention to about 2000 natural chlorine compounds, including many from marine algae. (Marine algae contain vanadium haloperoxidases that catalyze halogenations.[8]) One group of fungi, basidomycetes, even produce chloromethane as well as chlorinated aromatic compounds.[9] Methyl bromide is emitted by *Brassica* plants (in the mustard family).[10] Bromophenols are part of the prized flavor of shrimp.[11] These natural halogen compounds include at least two that are essential for human life, hydrochloric acid in the stomach and thyroxine (**3.2** Schematic) in the thyroid gland.

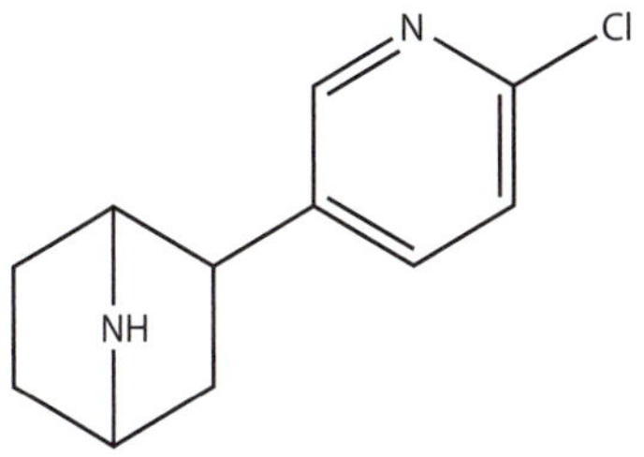

3.1 Schematic

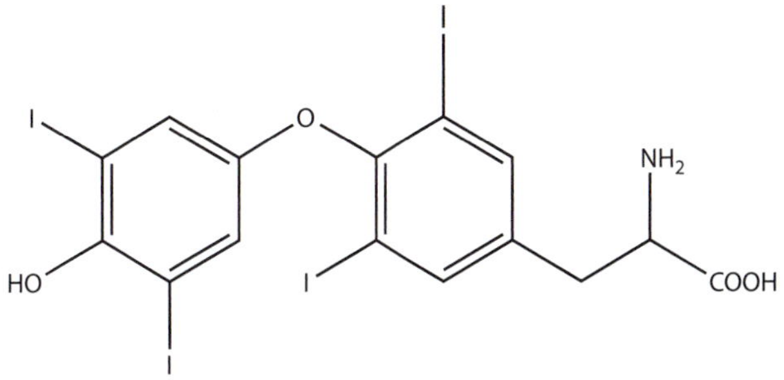

3.2 Schematic

(The status of the chlorine debate has been reviewed.[12])

Our society uses a great many products containing chlorine, as seen in Table 3.1, from *Chemical Engineering News*.[13] The chlorine industry in Europe used 10 million metric tons of chlorine in 1992. Of this, 20% went into polyvinyl chloride, 8% into intermediates, 6% into solvents, 5% into other organic compounds, and 2% into inorganic chemicals. About 36% ended up in hydrochloric acid. The remaining 23% was disposed of as salt. This amounted to 3.5 million tons of polyvinyl chloride; 761,000 tons of chlorinated paraffins and poly(chloroprene); 537,000 tons of solvents, such as perchloroethylene and methylene chloride; 561,000 tons of sodium hypochlorite; and other inorganic chemicals. Products made with chlorine, but not containing it, included 1.3 million tons of intermediates for polyurethanes, 517,000 tons of other resins and plastics, 447,000 tons of dyes and crop-protection chemicals, and 370,000 tons of inorganic compounds, such as titanium dioxide.[14] Chlorine is used in 98% of water disinfection, in the production of 96% of agricultural chemicals, and in making 85% of drugs.[15]

When phasing out the use of chlorine means major changes for a large industry, why would environmental groups ask for it? The answer is that halogen imparts significant biological activity to compounds. In the right place, this has been very useful. Unfortunately, they, and active metabolites from them, have collected in unintended places, such

Table 3.1 Chlorine Derivatives Are Used in a Huge Array of Products

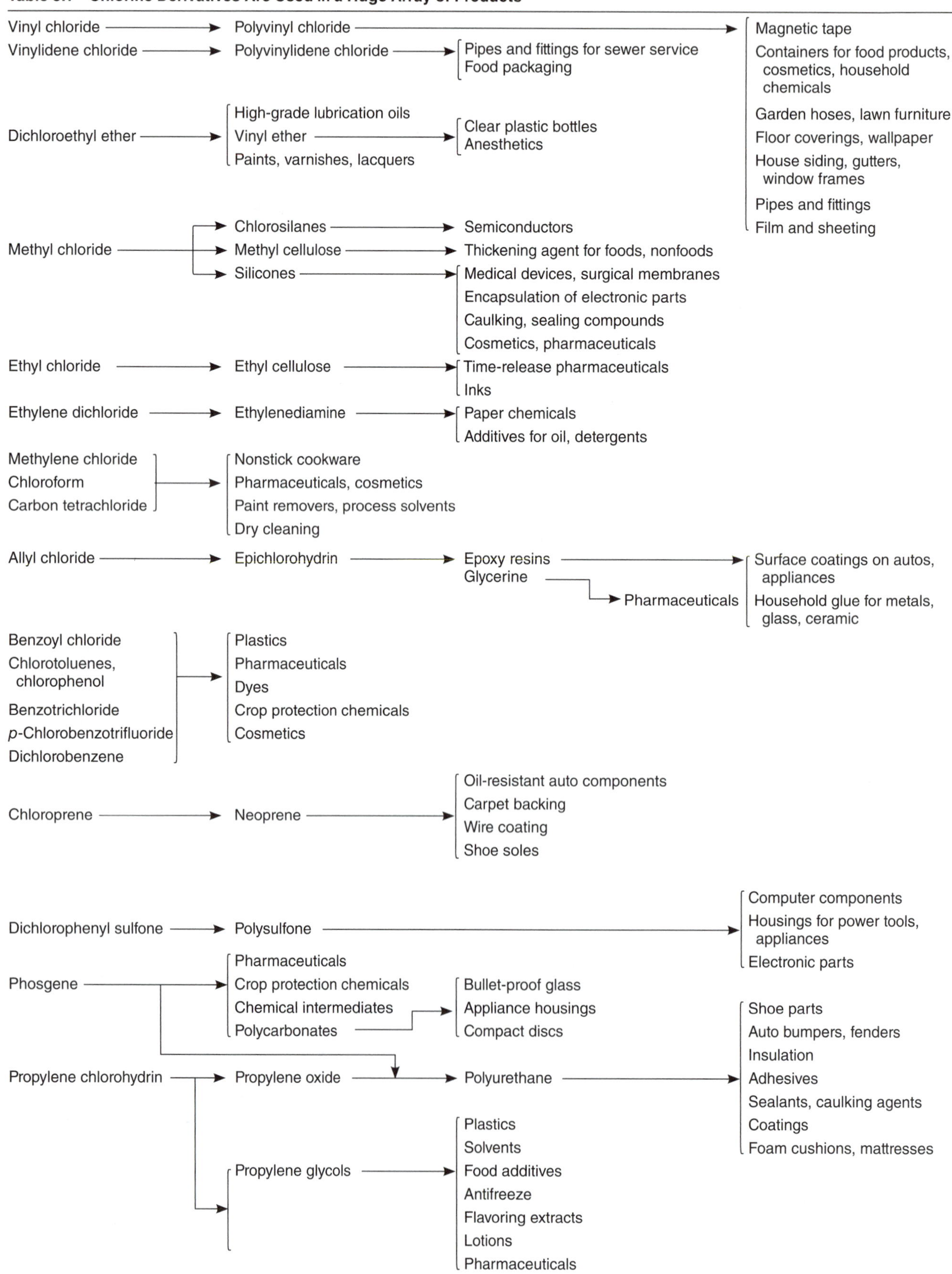

continued

Table 3.1 Chlorine Derivatives Are Used in a Huge Array of Products (Continued)

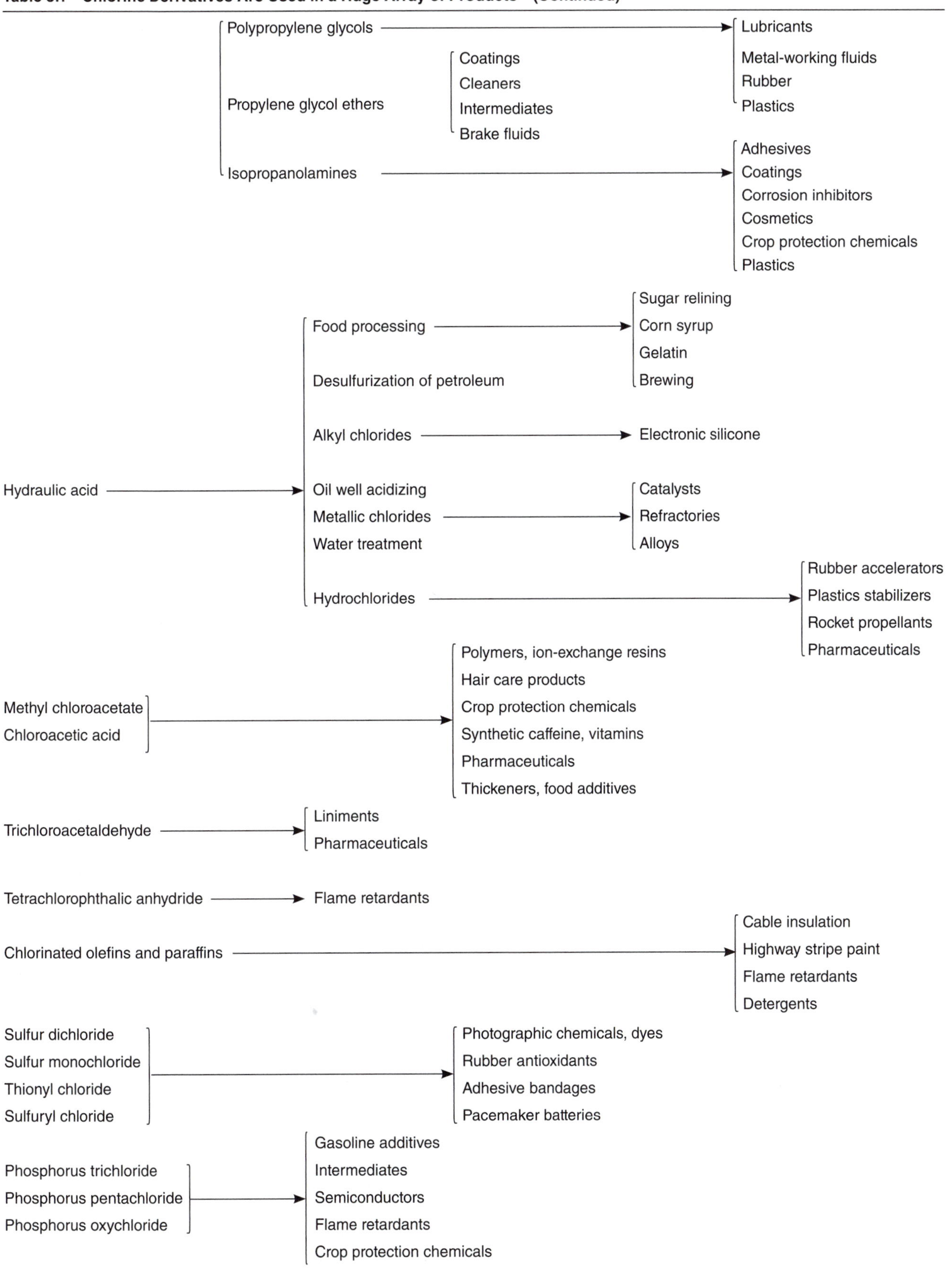

continued

Table 3.1 Chlorine Derivatives Are Used in a Huge Array of Products (Continued)

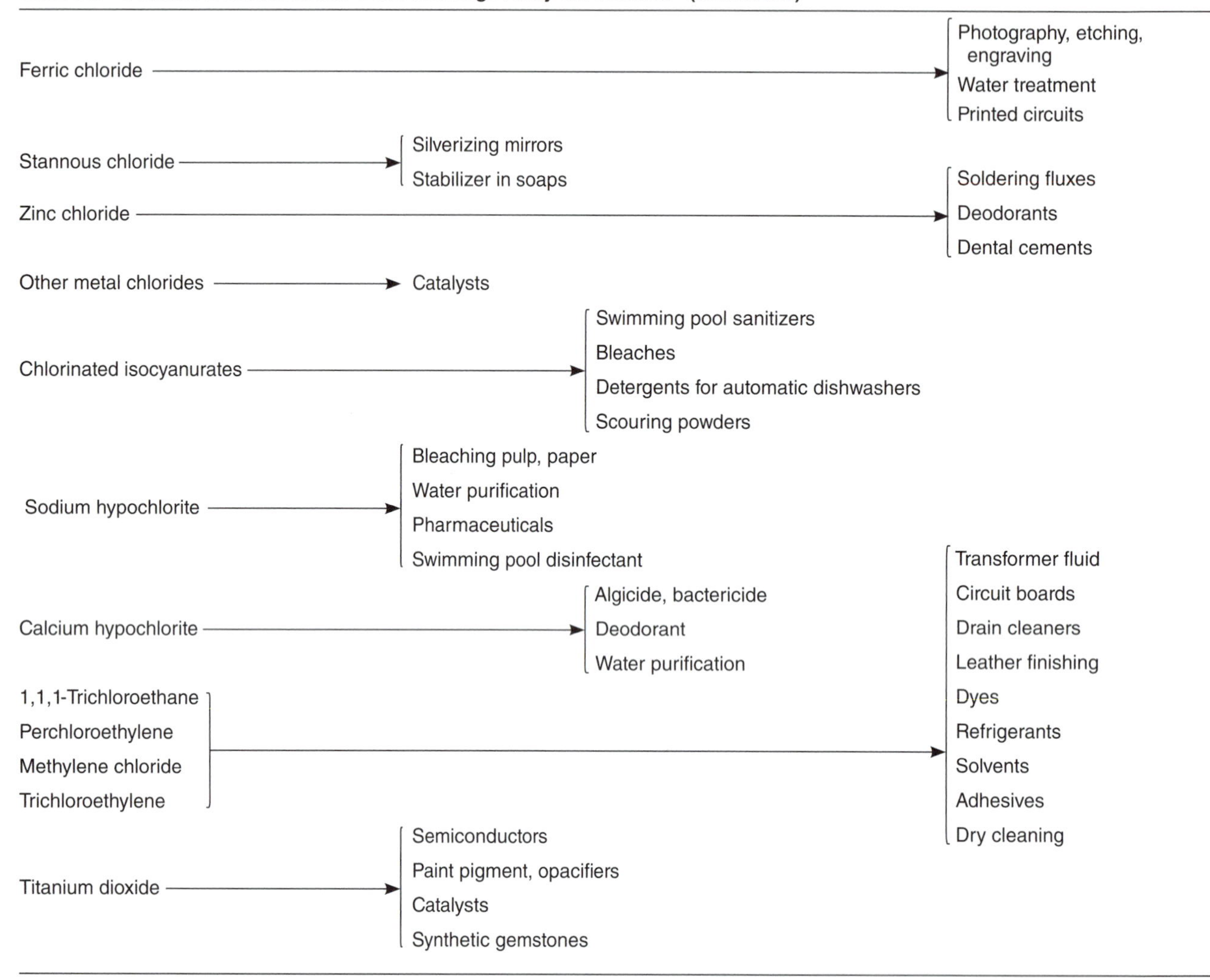

Source: Reprinted from B. Hileman, J.R. Long, and E.M. Kirschner, *Chem. Eng. News*, Nov. 21, 1994, 14–15. Copyright 1994. With permission.

as groundwater, rivers, and air.[16] Once in the air the compounds can travel long distances and come to earth again in rain, or just by cooling of the air.[17] There, some of them have created serious problems with nontarget organisms. Unfortunately, some environmental monitoring has focused only on the starting compound, excluding active metabolites derived from it, thus giving a distorted picture. Some of the metabolites are more toxic than the original compounds.

3.2 TOXICITY OF CHLORINE COMPOUNDS

Chlorine was used as a war gas in World War I and more recently in Iraq.[18] It continues to be a source of accidents. Nine people died when a tank car of chlorine ruptured in a derailment in South Carolina,[19] three in a train derailment in Texas,[20] and 29 in a truck collision in China.[21] There were serious leaks, but no fatalities, during transfers in Missouri,[22] Louisiana,[23] and Arizona.[24] The first was due to the use of an unsuitable hose that was mislabeled.

There have been proposals for stronger tank cars.[25] The Netherlands and Switzerland prohibit the shipment of chlorine by rail. Akzo Nobel sells a skid-mounted electrolysis unit that can be operated at a customer's plant to produce 15,000 metric tons of chlorine per year.[26] MIOX Corporation provides an electrolysis unit on-site, which means that the only compound shipped is sodium chloride.[27] Many water treatment facilities have switched to sodium hypochlorite, but this requires shipping a lot of water.

As mentioned in Chapter 1, persistent highly chlorinated compounds, such as DDT and PCBs, have caused eggshell thinning in birds, leading to reproductive failure. Running the standard toxicity tests on rodents did not pick this up. (The heterocyclic compound $C_{10}H_6N_2Br_4Cl_2$ has been found in the eggs of birds of the Pacific Ocean. It is probably of natural origin, because it is not found in the Great Lakes of the Untied States.[28]) Although useful in applications such as protection of houses from termites with chlordane, persistence in the environment has frequently turned out to be undesirable. Analysis of tree bark samples has shown that persistent organochlorine pollutants have global distribution.[29] The concentration in seabirds of Midway Island in the northern Pacific Ocean is almost as high as in birds of the Great Lakes between Canada and the United States.[30] Polybromobiphenyl flame retardants have been found in the blubber of sperm whales.[31] The concentration of the insecticide Toxaphene, a mixture made by chlorinating camphene, in fishes of Lake Laberge in the southern Yukon of Canada is high, presumably the result of atmospheric deposition and biomagnification.[32] This material and many other highly chlorinated compounds have been banned in the United States for many years. However, Toxaphene is still used in Central America and Eurasia. The Nordic countries and the Netherlands have called for a global ban on such persistent organic pollutants.[33] They point out that such insecticides appear to be cheap and easy to apply, but when their adverse effects on nontarget species are taken into account, they are as expensive, or more so, than the alternatives. (The treaty banning persistent organic pollutants went into effect in May 2004 and now has 134 parties. The United States has not ratified the treaty.[34]) DDT was banned in the United States in 1973, but its more toxic metabolite DDE, formed by dehydrohalogenation, is still being detected in well water in the Midwestern United States.[35] DDT may speed up the development of Parkinson's disease. Exposure of the fetus may slow down or stunt mental and physical growth.[36] The use of methoxychlor, which is 500 times less toxic and breaks down in the environment, has been proposed.[37] (For alternatives to pesticides see Chapter 11.)

Polychlorobiphenyls are responsible for many fish advisories (which advise fishermen not to eat too much fish caught in the body of water). Mohawk Indians near Massena, New York, who have eaten relatively large amounts of fish contaminated with polychlorinated biphenyls, have a "very high rate of hypothyroidism." Their children have a "striking" increase in diabetes and asthma.[26] Polychlorinated biphenyls were used as heat-transfer fluids in plants processing edible oils. Leakage of these into the oil caused more than 50 deaths of persons who ate the oil in Japan in 1968 and 1973.[27] Biphenyl–diphenyl ether fluids are much less toxic and their presence can be detected by odor at 1 ppm or more. Dredging of sediments has been used to remove polychlorinated biphenyls from a lake in Sweden,[38] the New Bedford, Massachusetts harbor,[39] and the Housatonic River in Massachusetts,[40] and has been suggested for the Hudson River in New York[41] and two rivers downstream from Dow's Midland, Michigan plant.[42] (The former regional administrator of the EPA, who left in May 2008, says that she was forced to resign because of her efforts to get Dow to remove the dioxins from the rivers.[43–45])

Henschler[46] has reviewed the toxicity of organochlorine compounds. Many are carcinogens (e.g., vinyl chloride, which is associated with liver and biliary tract cancers and angiosarcomas).[47] One of the most toxic is 2,3,7,8-tetrachlorodibenzo-*p*-dioxin (**3.3** Schematic).

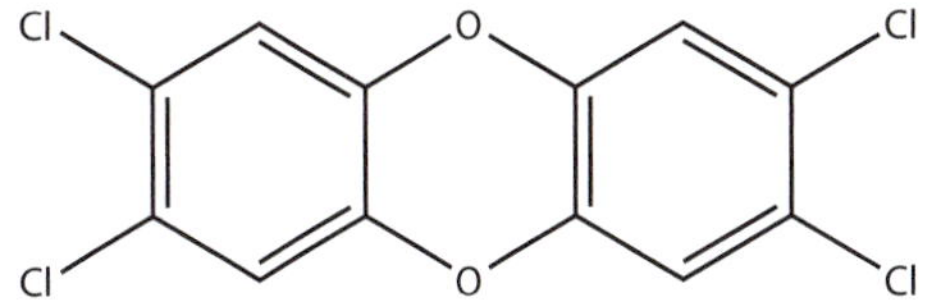

3.3 Schematic

Sixty-five horses died within a few weeks when their paddocks were sprayed with waste oil containing 33 ppm of this dioxin in 1971 in Missouri.[48] In 1953, a runaway reaction in a vat of trichlorophenol at BASF's Ludwigshafen, Germany, plant caused workers to be contaminated with this dioxin. In addition to the chloracne that this caused in some workers, there was an 18% increase in medical problems in the exposed group.[49] This included thyroid disease, intestinal and respiratory infections, and disorders of the peripheral nervous system, as well as appendicitis. A 1976 accident that released kilogram quantities of dioxin near Seveso, Italy, resulted in a shift in the sex ratio of babies born in the area until 1984. There were 26 males and 48 females. Normally 106 males are born for every 100 females.[50] According to the report of the U.S. EPA, dioxins, even in minute amounts, can cause disruption of regulatory hormones, reproductive and immune system disorders, and abnormal fetal development.[51] The dioxins present in "Agent Orange" used in the Vietnam War may have caused birth defects and deformities.[52] Physicians for Social Responsibility, Environmental Defense Fund, and others say that further regulation of dioxin sources is necessary, because effects on the fetus can occur at extremely low levels, levels that now occur in the tissues of some subpopulations.[53] Crummett[54] has written a history of the dioxin problem.

Waste combustion accounts for 95% of dioxin emissions in the United States, with 5100 g from hospital incinerators, 3000 g from municipal incinerators, and 35 g from hazardous waste incinerators.[55] (New regulations are designed to reduce the amount from medical incinerators.[56] Some dioxins originate from forest fires and other natural soruces.[57]) The annual global output of polychlorodibenzo-*p*-dioxins and related dibenzofurans is about 13,000 kg/year. Paper mills in the United States release 110 g of dioxins and related materials each year. When bleaching is done with chlorine dioxide instead of chlorine, dioxins drop below the detectable level. One of the problems in hospitals may be the use of gloves of polyvinyl chloride to avoid the dermatitis in some workers that is caused by the protein in natural rubber gloves. Guayule rubber is a possible alternative, as it is hypoallergenic.[58] BASF has offered a vanadium oxide-containing catalyst for the removal of dioxins from municipal waste incinerators, which is now being used in at least two commercial installations in Germany.[59] Other methods of dioxin control have been studied.[60] A study by Queens College of the State University of New York concluded that eliminating dioxin sources would not cause economic loss.[61] This would be done by a combination of replacing incineration by recycling and installing autoclaves to sterilize infectious waste at hospitals.

Combustion of poly(vinyl chloride), some 250,000 tons per year in the United States, is a source of dioxins.[62] The presence of even small amounts can ruin the recycling of poly(ethylene terephthalate). Removal of poly(vinyl chloride) from film, bottles, and other consumer items by regulation could help with the problem. Newer polyolefins[63] and thermoplastic polyurethanes[64] can be used instead. Other problems of poly(vinyl chloride) include carcinogenic monomer, plasticizers that are endocrine disruptors, and stabilizers that may contain cadmium and lead.

Some chloracnegens, such as **3.4** Schematic, are so potent that drug companies hesitate to work with them.[65] The widespread use of triclosan (**3.5** Schematic) as a bactericide in personal care products needs to be reduced now

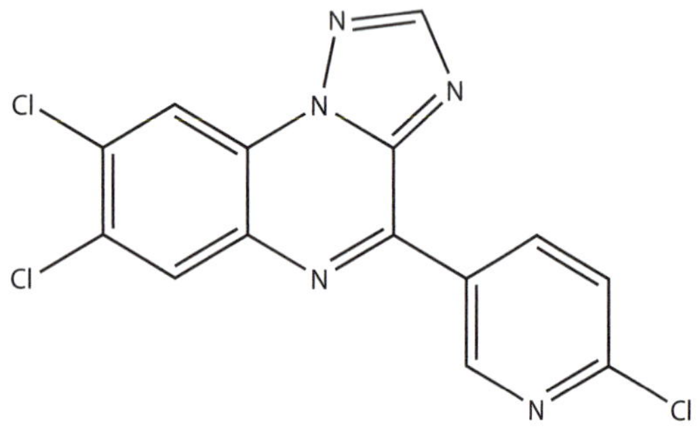

3.4 Schematic

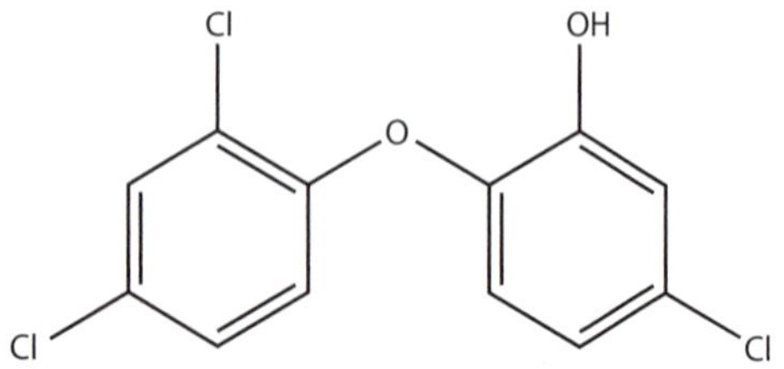

3.5 Schematic

that several strains of bacteria resistant to it have been isolated.[66] Soap and water alone will suffice in many of these applications. The use of dichlobenil (**3.6** Schematic) to kill roots penetrating underground pipes might be eliminated by designing tighter pipe joints.[67]

3.3 ESTROGEN MIMICS

Man-made chemicals are now ubiquitous in the environment.[68] Each of us carries large numbers of them in our

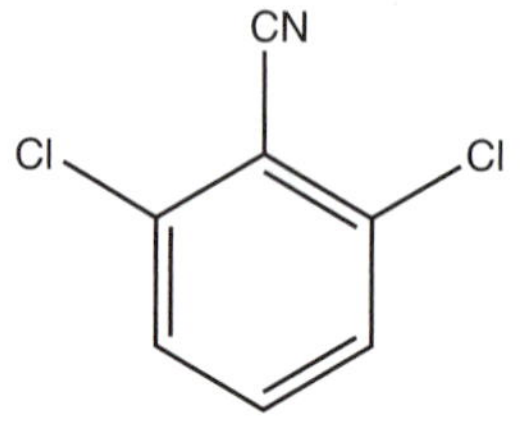

3.6 Schematic

bodies at very low levels. A reporter, Bill Moyers, found 84 synthetic chemicals in his blood.[69] Are they harmless or might they be implicated in some of the rising rates of attention deficit disorder, childhood asthma, juvenile cancer, autoimmune diseases, obesity, osteoporosis, early testicular cancer, breast cancer, prostate cancer, Parkinson's disease, and Alzheimer's disease? While some of these have genetic factors, these are no more than 10–20% of the total cases.

Colburn et al.[70] proposed that several synthetic chemicals, especially highly chlorinated ones such as polychlorinated biphenyls and insecticides, can act as hormone mimics in wild life. These include feminized young male alligators in a Florida lake, where a pesticide spill occurred in 1980, feminized male birds, and feminized fish.[71] On the other hand, feminization of fish in some rivers in the United Kingdom is due to human estrogens that are not removed by the wastewater treatment plants.[72] Aquatic hypoxia can also cause endocrine disruption in fish.[73] It has been suggested that estrogen mimics[74] may account for the declining ratio of male to female births in Canada, Denmark, the Netherlands, and the United States,[75] some of the increased levels of breast cancer[76] in developed countries, the two- to fourfold rise in testicular cancer in industrialized countries, and a drop in the sperm counts in some cities and countries.[77] Sperm counts in the United States and Europe have dropped over 50% since the late 1930s. They have dropped in Denmark but not in Finland.[78] An alternative explanation for the lower sperm counts is the wearing of tight underwear, which keeps the testes too warm to function well.[79]

Colburn points out that the first 38 weeks of life, that is, during gestation, are the most dangerous. Hair cells of the ear are formed at about two months after conception. Faulty development can lead to hearing impairment involved in attention deficit disorder. The male sex organs are formed at about the same time. Polychlorinated biphenyls and dioxins passed from the mother to the child during pregnancy and through breast milk weaken the immune system.[80] Studies on rodents find problems at levels not far above those in some people. At low levels of endocrine disruptors, the dose–response curve may not be linear, the response decreasing after reaching a certain level. This makes it impossible to extrapolate from high-dose experiments. The issue is highly contentious, with the manufacturers maintaining that their products are useful and harmless but with their critics saying that they used rodent strains that were insensitive and/or rat chow that would suppress the effect.[81]

A dose–response relation has been found for dieldrin (a highly chlorinated insecticide) and breast cancer.[82] On the other hand, the phytoestrogens of soy products may account for the low incidence of breast and prostate cancer in Japan.[83] Parkinson's disease is more likely with people who use pesticides.[84] At one time, diethylstilbestrol (DES; **3.7** Schematic)

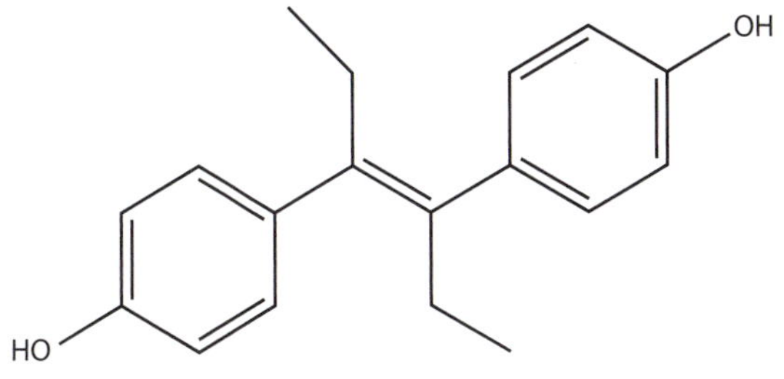

3.7 Schematic

was used to prevent miscarriages in pregnant women. It was banned in 1971 after it was linked to a rare vaginal cancer in the daughters of these women. Some of the grandsons of women who took the drug are being born with defects in their reproductive organs. The drug caused a mutation in the DNA or a change in methylation pattern ("epigenetics").

Other classes of commonly used compounds are also under suspicion as hormone mimics. These include alkylphenols and their reaction products with ethylene oxide, which are used as detergents, as well as phthalates, which are used as plasticizers.[85] These detergents might be replaced with alkyl polyglycosides that can be made from renewable sugars and fats (see Chapter 12). Di-*n*-butyl phthalate disrupts development in male rats, causing missing or displaced testes.[86] A Swedish study found a sevenfold greater incidence of testicular cancer among those who worked with polyvinyl chloride-containing phthalates.[87] There is concern about the use of dinonyl phthalate plasticizers in polyvinyl chloride toys used by children younger than the age of 3.[88] Bans have been enacted in the EU based on the precautionary principle, which states that if a risk is suspected, it is not necessary to wait until the final proof is shown.[89] The American Chemistry Council (a trade association of 134 U.S. chemical manufacturers) dislikes the use of this principle. There is also concern about the amount of di(2-ethylhexyl)phthalate that leaches from the polyvinyl chloride bags during intravenous administration of nutrients and drugs.[90] Adequate substitutes are already available. Vinyl chloride can be copolymerized with a plasticizing comonomer such as vinyl stearate or vinyl 2-ethylhexanoate. Grafting butyl acrylate to polyvinyl chloride also works.[91] The polyvinyl chloride can be replaced by ethylene–styrene copolymers[92] or polypropylene plastomer made with metallocene catalysts. Many other plasticizers are available, but many of them, including acetyltributyl citrate, will cost more.[93] BASF is marketing the diisononyl ester of cyclohexanedicarboxylic acid as safe, but it will not solve problems of leaching migration and evaporation.[94]

When pregnant mice are fed very small doses of bisphenol A, their male offspring have enlarged prostates as adults.[95] This has caused concern about the use of polycarbonates made from this material in reusable baby bottles, water carboys, and such.[96] This may be a problem only if the polymer is abused by heating it too long in water that is too hot. Styrene and its dimer and trimer may also be endocrine disruptors.[97] Most of the estrogenic compounds have potencies of 0.02–0.0001 that of estradiol or diethylstilbestrol. Many explanations, including estrogen mimics, have been advanced for the decline in amphibian populations and the increased incidence of deformities in frogs.[98] None has yet been proven.

Polybrominated diphenyl ether flame retardants, which are developmental toxins in animals, are now found in breast milk.[99] Decabromodiphenyl ether is converted to more toxic, less brominated analogs by sunlight and by soil microbes.[100]

Perfluorooctanesulfonate, used by 3M to make fabrics resistant to soiling, is a carcinogen that is now distributed globally.[101] The company has replaced it with a shorter perfluorosulfonamide (**3.8** Schematic).[102]

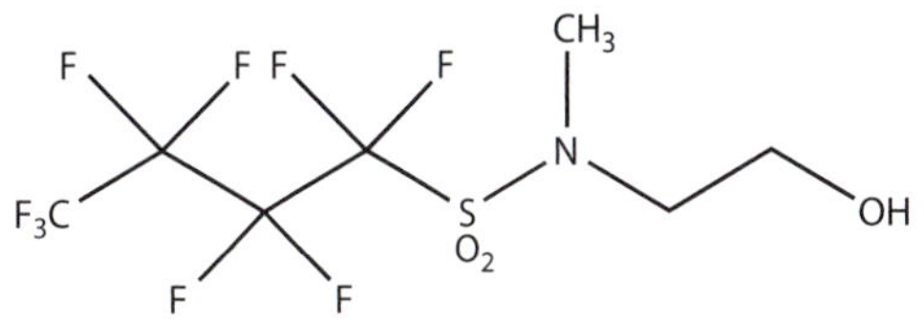

3.8 Schematic

The perfluorooctanoate used by DuPont in polymerizing tetrafluoroethylene has contaminated the drinking water near its plant in West Virginia.[103] DuPont is now doing the polymerization in sc carbon dioxide, which does not require perfluorooctanoate (see Chapter 8). Arkema is now polymerizing vinylidine fluoride with a non-fluorine-containing surfactant.[104] Most perfluorooctanoic acid is produced in the environment by oxidation of the corresponding alcohol, which is used on paper, carpet, and such.[105] Dow sells a polyolefin emulsion that can replace the fluorochemicals used on fast food packaging.[106] Despite many years of successful fluoridation of water to reduce the incidence of dental caries, there are still those who question the practice.[107]

The amount of perchlorate, which can interfere with iodide uptake, in water as the result of making rocket fuel is in a safe range in food.[108]

There is concern about pharmaceuticals in drinking water.[109] Many are not completely removed in water treatment plants. Their effects on the health of aquatic plants and animals as well as humans are uncertain. There are a number of ways to remove them: (1) subsurface flow constructed wetlands,[110] (2) estrogen-degrading bacteria,[111] (3) ozone,[112] (4) granular activated carbon,[113] and (5) ultrasound[114] and peroxides with an Fe-TAML catalyst (**3.9** Schematic).[115] Oxidation of acetaminophen, the most commonly used

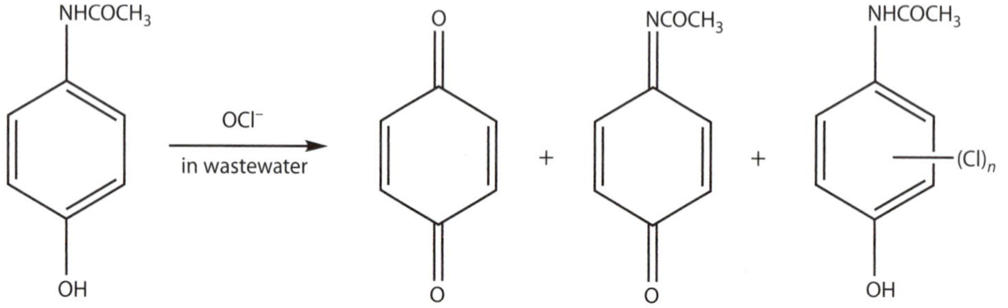

3.9 Schematic

painkiller, with hypochlorite led to two toxic compounds as well as chlorination products.[116]

The chemical industry is suspicious of the claims and is hesitant to make changes based on them.[117] However, increased testing[118] is being done. The U.S. EPA has finalized its testing procedure.[119] The American Chemistry Council has voted to spend 850,000 dollars to study the problem.[120] The European Modulators Steering Group of the European Chemical Industry Council has started a 7 million dollar testing program.[121] Other research is also being planned.[122]

3.4 BLEACHING PAPER

The traditional bleaching of paper with chlorine leads to an effluent containing 45–90 kg of organic waste per ton of pulp, of which 4–5 kg is organically bound chlorine.[123] These include various chlorinated furanones, cyclopentenediones, and other materials.[124] These are derived from lignin, humic acids, and such. Typical compounds are structures **3.10** Schematic and **3.11** Schematic. These are weakly

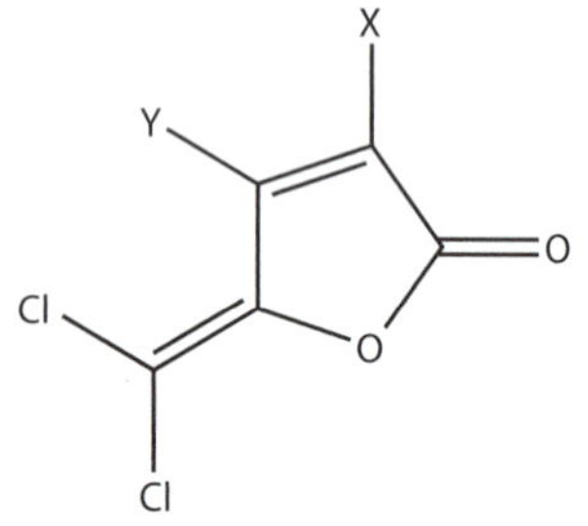

3.10 Schematic

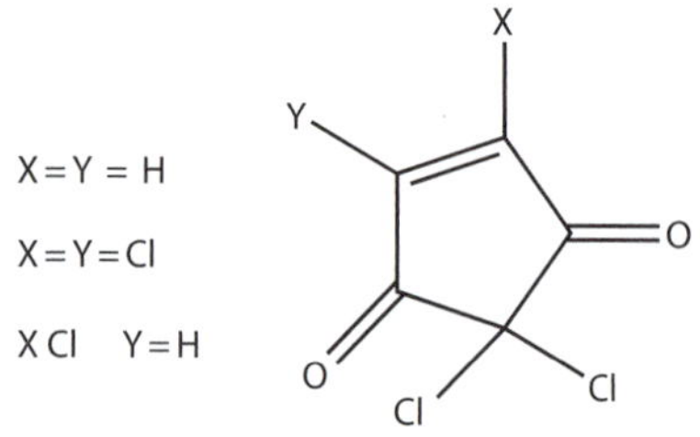

X=Y = H

X=Y=Cl

X Cl Y=H

3.11 Schematic

mutagenic in the Ames test. The amount of chlorine per ton drops to 46 g for recycled paper.[125]

Various other reagents are being tried in order to reduce these chlorinated materials as well as the chlorinated dioxins. These include chlorine dioxide, hydrogen peroxide, oxygen, peracids, ozone, dioxiranes, and enzymes.[126] Most mills switched to chlorine dioxide to reduce the levels of chlorinated dioxins.[127] Chloride dioxide must be used with care since it can explode when in the liquid state.[128] This reduced the other chlorinated compounds to 0.9–1.7 kg/ton.[129] When ozone is used in combination with chlorine dioxide, 1 kg of ozone can save up to 4 kg of chorine dioxide.[130] To eliminate all the chlorinated compounds in the mill effluent, it is necessary to go to a nonchlorine bleaching agent.[131] This is being done in some mills in Europe.[132] When this is done, the resin and fatty acids in the effluent still exhibit harmful sublethal effects on the fish in the stream. To avoid this, some mills have switched to closed systems with a total recycle of water. More will do so as they work out the technical problems involved (e.g., deposits that tend to build up).[133] Pitch deposits have been decreased up to 75% by the fungal pretreatment of eucalyptus wood.[134]

Pulp can be bleached with hydrogen peroxide.[135] Sodium perborate works just as well.[136] Polyoxometalates can be used with hydrogen peroxide or oxygen so that the capital cost is only 1% that of a comparable plant using ozone.[137] (For other uses of polyoxometalates, see Chapter 6.) Transition metal ions in the wrong form can decompose the hydrogen peroxide. Sodium silicate is used to prevent this, but scale deposits may build up. Dow Chemical is marketing a complexing agent that avoids this problem.[138] The bleaching by hydrogen peroxide works better if a prebleach with ozone or chlorine dioxide is used. It works even better if EDTA is added after the pretreatment, but before the hydrogen peroxide.[139] Collins and co-workers have studied the ways in which the organic ligand in a metal catalyst can be oxidized, and then they have altered the structure to avoid these. The result is a catalyst (**3.12** Schematic) that can be used with hydrogen peroxide to bleach paper pulp as well as dyes in wastewater.[140]

Some dyes in dye effluent can also be decolorized with hydrogen peroxide in the presence of ultraviolet (UV) light.[141] This facilitates reuse of the wastewater. Other

3.12 Schematic

transition catalysts for bleaching textiles and wood have been studied.[142] Cheaper hydrogen peroxide may become available if processes preparing it directly from hydrogen and oxygen with a palladium catalyst prove to be practical.[143] Headwaters & Evonik have such a system that produces hydrogen peroxide with up to 100% selectivity. A commercial plant is planned for 2009.[144] A system for generating hydrogen peroxide on-site electrochemically is also under development.[145] This would increase the incentive to replace both chlorine and chlorine dioxide. Peracids have also been used to bleach paper pulp[146] as well as jute and cotton.[147] Cotton can also be bleached with hydrogen peroxide, instead of the usual sodium hypochlorite,[148] light plus aqueous sodium borohydride,[149] and light plus sodium peroxocarbonate.[150]

One plant is Sweden produces 315,000 ton/year of pulp that is bleached with ozone. The pulp is just as strong as the standard pulp bleached with chlorine. Almost all (95%) of the effluent waste is recycled.[151] Ozone is also being used to bleach *Eucalyptus* in Brazil.[152] Dioxiranes work well in bleaching pulp. They can be generated *in situ* from a ketone and peroxymonosulfate ion (**3.13** Schematic).[153]

3.13 Schematic

Xylanases are being used in a prebleaching step to remove some of the hemicellulose present.[154] These reduce the amount of chemical bleaching agent needed. They are followed by chlorine dioxide, hydrogen peroxide, dimethyldioxirane, or some other bleaching agent. The enzymes must be free of cellulases. Most work at 37°C, but a few, derived from thermophilic organisms, are effective at 55°C and 65°C. One derived from *Streptomyces thermoviolaceus* is effective in a 3-h pretreatment at 65°C when followed by hydrogen peroxide. A combination of laccase and *N*-hydroxybenzotriazole has also been found to help the bleaching of wood pulp.[155] Hydrogen is abstracted first from the *N*-hydroxybenzotriazole oxidation mediator, which then abstracts hydrogen from the substrate lignin, which in turn reacts with oxygen.

Many items of paper that are used commonly in the bleached form do not really need to be bleached, for example towels, toilet paper, and facial tissues.

Organochlorine compounds are also found in the wastewater when wool is treated with chlorine in the "Hercosett" process for shrink-proofing wool. Several shrink-proofing processes that use no chlorine have been devised: (1) hydrogen peroxide pretreatment, followed by chitosan,[156] (2) cross-linking with glycerol triglycidyl ether,[157] (3) curing a mixture of a difunctional acrylate and an epoxysilicone with UV light,[158] and (4) 2-min pretreatment in a glow discharge followed by application of a cationic acrylic copolymer.[159]

3.5 DISINFECTING WATER

Many ways of disinfecting water are being studied.[160] Some do not kill all the pathogenic organisms. Some produce harmful by-products. The challenge is to have satisfactory disinfection at minimal cost and maximum rate.

Chlorine was first used to disinfect water[161] in Jersey City, New Jersey in 1908. It kills many microorganisms that are present in water, including those that cause cholera, typhoid, dysentery, and hepatitis.[162] It has saved a huge number of lives. Even so, there have been 116 disease outbreaks in the United States since 1986 owing to contaminated tap water.[163] In the process of chlorinating water, small amounts of chloroform,[164] a carcinogen, are formed by chlorination of natural organic matter in the water. Since 1979 the U.S. EPA has limited the level to 100 ppb. A new level of 0.3 mg/L has been proposed.[165] However, a study in California found an increase in the incidence of miscarriages when the level was higher than 75 μg/L.[166] Disinfection with chloroamines eliminates the chloroform, but sometimes forms some highly toxic cyanogen chloride as a by-product.[167] When used in Washington, DC, it caused leaching of lead from the pipes.[168] When chlorine was used, there was a protective coating of lead chloride in the pipes. Polymeric *N*-haloamines have been used to coat surfaces of fabrics, glass, and plastics.[169] In addition, many polymeric quaternary ammonium and phosphonium salts, often chemically attached to the substrate, have been used for this purpose.[170] Such materials may have use in hospitals and food processing plants. The use of sodium hypochlorite in food sterilization in Japan led to 250 ppb of chloroform in one product. When phosphoric acid was used instead, almost no chloroform was present.[171] Chlorination of the cold water around poultry being processed resulted in one of the most potent mutagens known (**3.14** Schematic).

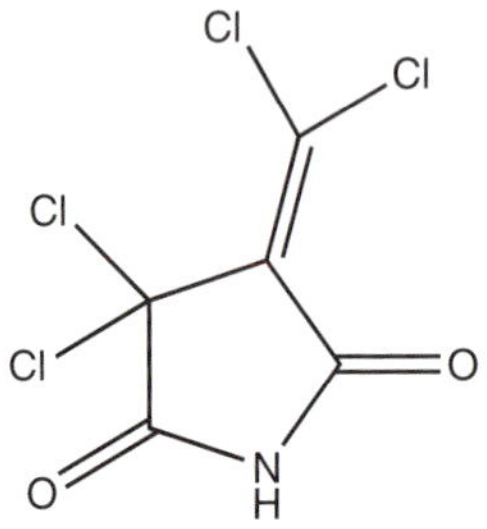

3.14 Schematic

It could be deactivated by sodium bisulfite or L-cysteine.[172] Substitution of chlorine dioxide for chlorine in the poultry chiller water gave no mutagenicity. No chloroform is formed when this reagent is used.[173] A mixture of 29 live, harmless bacteria can be put on chickens to prevent colonization by Salmonella.[174] Feeding chickens *Saccharomyces boulardii* a couple days before transport prevents an increase in *Campylobacter* and *Salmonella* during transport.[175]

Ozone can be used to sterilize water, but it has no residual action once the water gets into the distribution pipe.[176] It could reduce the amount of chlorine or sodium hypochlorite needed. It might also produce traces of formaldehyde, a carcinogen, in the water. Combinations of UV light and titanium dioxide are being studied as a means of disinfecting air and water.[177] The systems work, but the quantum yields may be somewhat low for an economical process.[178] It is said to destroy those cyanobacterial toxins that conventional treatment methods do not remove.[179] Pulsed UV light from a xenon flash bulb is said to destroy pathogens, including *Cryptosporidium*, which is not affected by chlorine, at a price comparable with chlorination.[180] (A 1993 outbreak of *Cryptosporidium* infection in Milwaukee, Wisconsin, sickened 400,000 persons and several died.) No chemicals are required and no chloroform or other by-products form. The system has also been supplemented with ultrasound and electromagnetic fields in Phoenix, Arizona.[181] A combination of UV light and hydrogen peroxide is also being used in plants today.[182] Campers use a battery-powered UV light to sterilize water.[183] Proctor and Gamble sells packets of calcium hypochlorite and iron sulfate to sterilize water in developing nations.[184]

Sterilization by purely physical means would eliminate questionable by-products. The classic method is to boil the water, but this is probably too energy-intensive for city water supplies. There have been proposals for dual water supplies in homes, with a small line for drinking water and a larger one for flushing the toilets and watering the lawn. While this would be very costly to retrofit communities, it would allow more options in how municipal water is treated. Campers use biological filters to clean up water for drinking. With proper prefiltration to remove larger particles, this is feasible on a commercial scale. The throughput has to be high and the membrane has to resist fouling or be easy to clean. Use of a membrane with built-in polymeric quaternary ammonium salts[185] would be even better. Food is sometimes sterilized by high pressure, with or without heat and ultrasound, a process that is in commercial use today.[186] Food has also been sterilized with pulsed electric fields.[187] High-intensity ultrasound is often used to disrupt bacterial cells. This should be feasible, although the energy costs would need to be studied carefully.[188] Low-cost disinfection of water by passing it through titanium alloy plates charged with direct current is being used in Brazil. It is said to cost one-fifth as much as traditional technologies.[189] It certainly should be studied for application elsewhere. High-energy electron beams are being used to disinfect water in a pilot facility in Miami, Florida.[190] Gamma rays from a ^{60}Co source offer another alternative. This could be a restricted option, because handling the radioactive source requires considerable care.

Treated wastewater from municipal sewage plants requires a treatment with chlorine before discharge into a river, at least in the warmer months of the year. Because residual chlorine can kill the flora and fauna of the river, treated wastewater is usually treated further with sulfur dioxide or sodium bisulfite to remove the chlorine.[191] This seems like an ideal place to use alternative methods of disinfection, because no residual activity is needed or desired. Disinfection with UV light is now a viable alternative.[192]

Scale in cooling towers and boilers can be prevented by ultrasound or electric fields, which alter the calcium carbonate crystals that form so that they can be removed. No chemicals are necessary.[193]

It is certainly possible to disinfect water without using chlorine or compounds containing it. Further study is needed on the need for residual activity in the distribution system of cities, the possible toxicity of the products of alternative reagents, and the relative costs of the various systems. These systems are also applicable to water reuse, which will increase in the future.[194] The purely physical systems have considerable merit, because no chemicals are required and no noxious by-products are formed. Invasive organisms can be removed from ballast water by such methods.[195] The use of silver on clothing to prevent bacterial contamination is a bad idea. The silver may wash off and end up in a wastewater treatment plant where it can inhibit microorganisms that purify the water. If it goes out with sludge that is spread on fields, it may interfere with beneficial bacteria such as Rhizobia.[196]

3.6 CHLOROFLUOROCARBONS AND OZONE DEPLETION

Chlorofluorocarbons were devised as nonflammable, nontoxic, noncorrosive alternatives to refrigerants such as sulfur dioxide, ammonia, and chloromethane.[197] Their uses grew to include blowing agents for foam insulation, aerosol propellants, and solvents for cleaning electronic parts.[198] Typical compounds are the following:

Compound	Designation[a]
$CFCl_3$	CFC 11
CF_2Cl_2	CFC 12
CF_2ClCCl_2F	CFC 113

[a] The first number for the chlorofluoromethanes is the number of hydrogen atoms plus one, the second is the number of fluorine atoms, and the remaining atoms are chlorine.[199]

Unfortunately, these compounds were found to destroy stratospheric ozone.[200] This is the layer that protects us from UV light. Without it, or with less of it, there will be more skin cancers and cataracts. The decline has been worldwide, but most significantly in the polar regions. The ozone hole in Antarctica was 25 million km^2 in September 1998 (i.e., 2.5 times the area of Europe).[201] Ozone levels at the North Pole have fallen by 40% since 1982.[202] A 15-year study shows an increase in the wavelengths that can damage DNA of 8% per decade at 40° north latitude, the latitude of Philadelphia, Pennsylvania, and Madrid, Spain.

An international agreement to phase out chlorofluorocarbons was reached in 1990 in Montreal.[203] It seems to be working, for tropospheric anthropogenic chloride has peaked and is now decreasing.[204]

The search for drop-in replacements for refrigerants has focused on hydrofluorocarbons (HFCs).[205] Typical ones are the following:

Compound	Designation
$CHClF_2$	HCFC 22
CH_2F_2	HFC 32
CH_2FCF_3	HFC 134a
CH_3CClF_2	HCFC 141b
CH_3CF_2H	HFC 152a

$$Cl_2C{=}CCl_2 + HF \longrightarrow Cl_2CHCCl_2F \xrightarrow{HF} Cl_2CHCClF_2 \xrightarrow{HF} Cl_2CHCF_3$$

3.16 Schematic

DuPont doubled its capacity for the last one.[206] ICl (now Akzo Nobel) has expanded its plant for HFC 134a.[207] Matsushita Electrical Industrial Co. opted for a 23:25:52 blend of difluoromethane–pentafluoroethane–1,1,1,2-tetrafluoroethane as a refrigerant.[208] If the ozone-depleting potential of CFC 11 is assigned a value of 1.0, then hydrochlorofluorocarbons (HCFCs) have values from 0.01 to 0.11, HFCs 0.0, methyl chloroform 0.1, and carbon tetrachloride 1.1.[203] HCFCs are now being phased out.[209] Although HFCs do not deplete the ozone layer, they are powerful greenhouse gases. (The most potent greenhouse gas known is sulfur hexafluoride, of which 85,7000 tons is now in the atmosphere.[210] Recommendations for decreasing its use include not using it to fill tires, sports shoes, and sound-insulating windows; to degas aluminum, to blanket magnesium; and to use thermal destruction for any that uses of sulfur hexafluoride is emitted by the electronics industry.) Since none of these uses of sulfur hexafluoride is completely innocent, they too will be phased out over a period of years.[211] HCFC-123 ($CHCl_2CF_3$) caused liver disease in nine workers when a leak developed at a smelting plant in Belgium.[212] Failure to remove a perfluorocarbon cleaning fluid from equipment caused 51 deaths from dialyzer units.[213] There is also concern about the trifluoroacetic acid that can be formed from their degradation in the atmosphere and its effect on wild life when it falls back to earth in rain.[214]

These compounds are usually prepared by the addition of hydrogen fluoride to a double bond or by the replacement of chlorine by fluorine in the presence of a metal fluoride catalyst.[215] The trick is finding the catalyst and conditions that will give the desired selectively in high yield. An outstanding example is **3.15** Schematic. Tri- and tetrachloroethylene can be used as starting materials.[216]

$$CH_2{=}CCl_2 + HF \xrightarrow{AlF_3} CH_3CFCl_2 \quad \text{99.5\% yield}$$

3.15 Schematic

The catalyst for the first set of reactions (**3.16** Schematic) was a mixture of antimony pentachloride and titanium tetrachloride. These sequences may involve addition–elimination reactions as well as nucleophilic substitution (**3.17** Schematic).

$$ClH{=}CCl_2 + HF \xrightarrow{\beta\text{-}CrF_3} CF_3CH_2F \quad \text{HFC 134a}$$

3.17 Schematic

The first use of chlorofluorocarbons to be banned was that of an aerosol propellant. These chlorofluorocarbons were used in cosmetics and paints. Other pressurized gases can be used, although propane is somewhat too flammable for this use. A fingertip pump or a squeeze bulb can also be used to provide the pressure to produce the aerosol spray. One can use a rub-on stick deodorant. A shaving brush can be used. Chlorofluorocarbons have also been used in preparing polymer foams, especially for insulation. Other gases, such as nitrogen, argon, carbon dioxide, pentane, or the like, can be used instead, but may not be as soluble in the polymers being foamed.[217] Dow Chemical replaced chlorofluorocarbons with carbon dioxide in the preparation of foamed sheets of polystyrene.[218] Another system prefroths a mixture of a di- or triacrylate and an acrylated urethane oligomer with air, and then cures it with an electron beam to form a cellular material.[219] Methyl formate is much cheaper and works just as well as fluorine compounds in the blowing of polyurethanes.[220] Arkema is selling 1,2-dichloroethylene for this purpose.[221] Chemical blowing agents, such as azodicarbonamide, can be used. In making polyurethanes, the presence of a little water will generate carbon dioxide from the isocyanate to foam the material. The insulating value of these gases may not be as high as that of the chlorofluorocarbons, so that thicker layers might have to be used. Eventually, the gas will diffuse out of the polymer and equilibrate with the surrounding air. If this happens in a short time period, then there is little advantage in using the chlorofluorocarbon. If this happens in 100 years, a system of recovering insulation from used appliances could be set up to prevent release of the gases to the atmosphere. The answer is probably somewhere in between. Another system to produce carbon dioxide for foaming polymers is to decarboxylate a maleic anhydride–styrene copolymer with a potassium hydroxide catalyst.[222]

Many options are being explored for cleaning electronic parts and such without using chlorofluorocarbons such as 1,1,2-trichloro-1,2,2-trifluoroethane (HCFC-113).[223] These include perfluorocarbons, aqueous detergents, terpenes, alcohols, ketones, plasma cleaning, laser cleaning, and chlorocarbons in zero-emission equipment. The alcohols and ketones would have to be used in explosion-proof equipment. The best approach is to use manufacturing methods that leave no residues in the first place (i.e., working under nitrogen or argon) and with no-residue soldering fluxes.[224] Propylene carbonate and γ-butyrolactone have been effective in removing cured photoresist in the printed circuit board industry, where chlorofluorocarbons were formerly used.[225]

The phasing out of chlorofluorocarbons means that there are stocks that need to be disposed of in a harmless way. Japan had 600,000 metric tons awaiting disposal in 1996.[226] Hitachi has a catalytic system for burning these and absorbing the acidic gases in a solution of lime. Burning in a lime kiln gives 99.9999% destruction.[227] Chlorocarbons can be burned completely using copper(I) chloride–potassium chloride on silica catalyst at 400°C.[228] The halogen atoms can also be removed by treatment with sodium in liquid ammonia at room temperature.[229] Sodium vapor in a flame at 1000°C also destroys chlorofluorocarbons.[230] Burdenic and Crabtree have devised a method that uses sodium oxalate at 270°C.[231]

$$CF_2Cl_2 + 2Na_2C_2O_4 \xrightarrow{270°C} 2NaF + 2NaCl + 4CO_2 + C$$

3.18 Schematic

The best system for dichlorodifluoromethane appears to be conversion to a useful HFC (**3.19** Schematic).[232] This

$$CF_2Cl_2 \text{ (CFC-12)} \xrightarrow[Pd/C]{H_2} CF_2H_2 \text{ (HFC-32)} \quad \text{70–90\% selectivity}$$

3.19 Schematic

can also be done by electrolysis using lead electrodes with 92.6% selectivity (**3.19** Schematic).[233]

Perfluorocarbons are 6000–25,000 times as potent as carbon dioxide in producing global warming.[234] The largest single source is the aluminum industry, which releases 28,0000 metric tons of perfluoromethane and 3200 metric tons of perfluoroethane into the world's air each year. Aluminum metal is produced by the electrolysis of trisodium hexafluoroaluminate.[235] The process uses 15 kg fluoride per metric ton of aluminum, of which 10–25% is lost. Research is being done on new types of electrodes in an effort to reduce this. A protective cathode coating made from titanium diboride and colloidal alumina reduces erosion of the cathode and creates higher current efficiency.[236] This saves energy and may reduce the production of perfluorocarbons. Electrolysis in a new electrolyte at 700°C uses aluminum–bronze electrodes and produces no perfluorocarbons.[237] Perfluoromethane can be converted to the useful monomer tetrafluoroethylene in a carbon arc, but the selectivity needs improvement.[238] Some perfluoroethane is used for etching semiconductor materials and in plasma cleaning of chemical vapor deposition chambers.[239] The material in the off gases can be destroyed (99.6%) in a microwave plasma with oxygen and natural gas in milliseconds.[240] This requires a scrubber because HF and COF_2 are among the compounds formed. An alternative method uses a membrane to remove nitrogen before burning the concentrated stream at 900°C.[241]

Bromine-containing compounds, called halons, were developed as nontoxic, noncorrosive fire extinguishers. The two most commonly used ones are bromochlorodifluoromethane (halon 1211) and bromotrifluoromethane (halon 1301), which have ozone-depleting potentials of 4 and 16, respectively.[242] They are used in airplanes, libraries, and computer installations, where the use of water could be damaging. Only about 5% of that released is actually used in fires. The rest is released in testing. Clearly, a better method of testing is needed. Atmospheric levels of halons were still rising in 1998.[243] Finding a drop-in replacement is proving to be difficult.[244] 3M sells $CF_3CF_2COCF(CF_3)_2$ as a replacement for halon 1301.[245] The best course of action is to blanket the fire with an inert gas, such as carbon dioxide or nitrogen, in cases where there is no danger of hurting people. One worker died and another 15 were injured when a fire suppression system that used carbon dioxide malfunctioned during maintenance at the Idaho National Engineering and Environmental Laboratory.[246] Another worker died of nitrogen asphyxiation at a Union Carbide (now Dow) plant in Louisiana.[247]

Methyl bromide is a major source of bromine in the atmosphere. It is about 50 times as efficient as chloride in depleting ozone.[248] It is used as a soil and crop fumigant.

It is degraded rapidly by soil bacteria, but much of it still escapes.[249] Adding organic wastes, such as composted manure, reduced losses by 12%.[250] To minimize volatilization, it should be injected at great depths in moist soil under a tarp with the soil surface packed before and after application.[251] One of the best barriers consists of two layers of polyethylene with a layer of an ethylene–vinyl alcohol copolymer in between. An ammonium thiosulfate fertilizer also reduces the emissions.[252] When used as a fumigant[253] for grain, it can be recovered on carbon for reuse, stripping being done by electrical heating.[254] Countries agreed to cut its usage by one-fourth by 2001 and by half by 2005.[255] Since then, the first date has been extended to 2005 and even past 2008 in the United States.[256] The timing has been the subject of considerable debate. Farmers say that there is no single economically viable alternative available. On the other hand, a United Nations study concluded that alternatives[257] are already available for nearly all of its uses.[258] Integrated pest management may reduce the need for it. For example, crop rotation can control some nematodes without the need for chemicals.[259] A new method kills insects in grain with microwaves as the grain flows from the elevator bucket into storage.[260] Lowering the temperature of stored grain to 65°F can control some pests and is less expensive than fumigation.[261] Raising the temperature to 122°F in combination with diatomaceous earth killed 98% of flour beetles.[262] Storage of grain under nitrogen or carbon dioxide also eliminates insects. Food irradiation is another alternative to the use of methyl bromide.[263] (For more on insect control, see Chapter 11.)

Most work on the replacement of chlorofluorocarbon refrigerants has focused on drop-in replacements for existing equipment.[264] With items for which the stock turns over every few years, it is reasonable to use nonfluorine compounds. Over 8 million refrigerators and freezers are discarded each year in the United States.[265] Ammonia is still used in large refrigeration units. It could be put into home refrigerators if a sensor to warn of leaks were included. Such sensors do exist. This would mean building a unit without copper tubing and with the use of seals resistant to ammonia. Small amounts of ammonia are not particularly harmful, as shown by the use of household ammonia to clean glass windows. Bosch-Siemens, the largest manufacturer of refrigerators in the world, is now using a propane–isobutane mixture as the refrigerant in Europe. Although the gas is flammable, it should pose no more of a problem than having a gas stove in a home. Five million such refrigerators were in use in 1996, with not a single fire having been reported. Presumably, a gas odorant is included or an instrumental sensor for the gas is built in.[266] It is illegal to use hydrocarbon refrigerants in home refrigerators in the United States.[267] Hydrocarbon refrigerants have been used in 200,000 cars in the United States without a single incident.[268]

Absorption refrigerating systems, such as gas refrigerators, use ammonia/water and lithium bromide/water as the refrigerants. These units are compact, noise- and vibration-free, and simple. They do require more energy. Perhaps this can be offset by more insulation.[269] Solar energy is one source of heat for them. Carbon dioxide has been suggested as a refrigerant but would require pressures three to five times higher than used now.[270] Cooling can be done by passing an electrical current through a junction of dissimilar metals, the electrocaloric or Peltier effect.[271] Refrigerators based on this principle are rugged, reliable, contain no moving parts, and are low cost, but they consume more energy. A figure of merit, ZT, of three or more is needed before they can be adopted widely. The highest ZT, 2.4, has been obtained with a complex mutilayered material.[272] Some vinylidene fluoride copolymers have electrocaloric effects.[273] Further research is needed to find simpler materials that work in the desired temperature range. Some metals become hot when magnetized and cool when demagnetized, the magnetocaloric effect. An alloy of aluminum, erbium, and dysprosium is an example. If cheaper materials can be found that do not require high magnetic fields, this could eliminate the need for refrigerants altogether.[274] A refrigerator using gadolinium, containing varying amounts of silicon and germanium, has been tested for a year.[275] These are also being tested in air conditioners for cars. Thermoacoustic refrigeration is possible using intense sound (200 dB) in a compressor that has no moving parts. The sound outside the device is not loud.[276] In countries in the temperate zone, it may be possible to cut the amount of energy needed for the household refrigerator by circulating outside air through its shell in the winter or even on cool summer evenings. Some hotels in Chicago used the air from the city's network of underground tunnels for cooling until visitors complained of musty odors. They should be able to go back to this by running the air over activated carbon to remove the odors. Passive solar heating and cooling could remove the need for a lot of air conditioning in homes. In cooler climates, such as in New England, it probably is not necessary to have an air conditioner in one's car. (Energy usage is discussed in Chapter 15.)

3.7 CHLORINATED SOLVENTS

Chlorinated solvents replaced more flammable hydrocarbon and other solvents. They have often been chosen for their stability, ease of drying, and ability to remove oils. Unfortunately, they can cause cirrhosis of the liver and in some cases cancer. Efforts are being made to find replacements for methylene chloride, chloroform, carbon tetrachloride, methyl chloroform, and others. Degreasing with trichloroethylene is still common and dry cleaning of clothing with perchloroethylene is still the standard. (The most common organic contaminants in ground water at hazardous waste sites are trichloroethylene and perchloroethylene.[277]) Better containment is one approach. As mentioned in Chapter 1, completely enclosed vapor cleaners are now available in Germany for use with trichloroethylene.[278] In Dover, Delaware, Capitol Cleaners reduced its use of perchloroethylene by 96% by putting in a new closed-loop system.[279]

A better approach is to find alternative methods of cleaning that do not use chlorinated solvents.[280] As mentioned in Section 3.6, under the use of chlorofluorocarbons for cleaning, these include aqueous detergent systems, sometimes augmented by high-pressure jets or ultrasound,[281] blasting with particles of ice or dry ice,[282] laser cleaning,[283] plasma cleaning, or other means.[284] Commercial units for cleaning clothes with supercritical carbon dioxide are now available.[285] (See Chapter 8 for more on the use of supercritical carbon dioxide.) In addition, reexamination of the process may lead to elimination of the contaminant oil in the first place. Aqueous-based cutting oils for machining would be one example.[286] A corrosion inhibitor may have to be present. Water remaining after rinsing can be removed with jets of air or by centrifugation. Coating a tungsten carbide cutting tool with a nanolayer of tungsten plus tungsten disulfide eliminates the need for any cutting oil.[287]

Ethyl lactate has been suggested as a relatively nontoxic biodegradable solvent that is inexpensive. It can be made from renewable materials by a new fermentation process.[288] Photoresists and organic contaminants can be removed with ozone in water in a process that uses no acids, hydrogen peroxide, or high temperatures while saving water.[289]

Dry cleaning with perchloroethylene is used with certain fabrics of silk, wool, or rayon that might shrink in the usual laundering in hot water. Recent studies have shown that it can be replaced with "multipurpose wet cleaning."[290] This uses soaps, gentle washing, steam treatment, and, sometimes, microwave drying. Bad stains are removed by a concentrated detergent pretreatment. Delicate fabrics may receive handwashing. The process is said to work on clothing of silk, wool, rayon, and cotton. Decamethylcyclopentasiloxane can clean more clothes per gallon than perchloroethylene.[291] Most fabrics of cotton or wool are treated with shrink-proofing reagents today anyway. Wash-and-wear garments that do not need dry cleaning are common today. They can also eliminate the need for energy- and time-intensive ironing of the garment.

Methylene chloride is a common ingredient of paint strippers. On a house, an alternative is to heat with a heat gun and then scrape off the paint with a putty knife. Other solvents are needed for stripping finishes from furniture. Epidemiological work suggests that methylene chloride can cause cancer of the liver, bile duct, and brain.[292] It is possible that a professional furniture refinisher could be hired to remove the finish by ultrasonic cleaning or bombardment with particles. Further work is needed.

It is now possible to polymerize isobutylene to butyl rubber in toluene or heptane instead of the usual chloromethane.[293] A new titanium sandwich catalyst does the trick. European manufacturers are phasing out some chloroparaffin lubricants.[294] In Japan, rubber is being chlorinated in water instead of carbon tetrachloride.[295] This still involves chlorine, but does get rid of the solvent. 1,2-Dimethoxyethane is being used as an alternative to chlorinated solvents for the Schmidt reaction for conversion of an ester to an amine through an acid azide and a carbamate.[296] This eliminated accidents from polyazides made *in situ* from methylene chloride or chloroform (**3.20** Schematic).

$$RCOOCH_3 + NaN_3 \xrightarrow{CH_3SO_3H} RCON_3 \xrightarrow{CH_3OH} RNHCOOCH_3 \longrightarrow RNH_2$$

3.20 Schematic

Photocatalytic titanium dioxide coatings can be self-cleaning, converting organic materials, such as bacteria and dirt, to carbon dioxide at practical rates at room temperature.[297] This offers the possibility of self-cleaning buildings and windows (made by PPG and Pilkington).

3.8 SYNTHESES WHERE CHLORINE IS NOT IN THE FINAL PRODUCT

Frequently, chlorine is not in the final product. Instead it is in by-product hydrogen chloride or salts. Workers at Flexsys (formerly Monsanto) have devised some syntheses that obviate the need for chlorine (**3.21** Schematic through **3.22** Schematic).[298] In each case, the traditional process is shown first.

The benzamide formed in the second example (see **3.22** Schematic) is recycled to the process. It is also possible to use aniline in place of the benzamide in a process that produces 74% less organic waste, 99% less inorganic waste, and 97% less wastewater.[299]

In one case, the halogen is generated *in situ* by electrolysis so that none appears in the final product and none is lost (**3.23** Schematic).[300] The yield with sodium bromide is 89%, with sodium chloride 6%. If this process is general for olefins and if the toxic acetonitrile can be eliminated, this could become an important way to eliminate by-product salts in the preparation of epoxides. As mentioned in Chapter 1, propylene oxide is often made by adding hypochlorous acid to propylene and then de-hydrohalogenating with base.[301] The alternative process that uses the oxidation of ethylbenzene with oxygen to form the hydroperoxide followed by transfer of an oxygen to propylene is gaining in popularity. The best method is the oxidation of propylene with hydrogen peroxide using a titanium silicalite catalyst.[302] Direct oxidation of

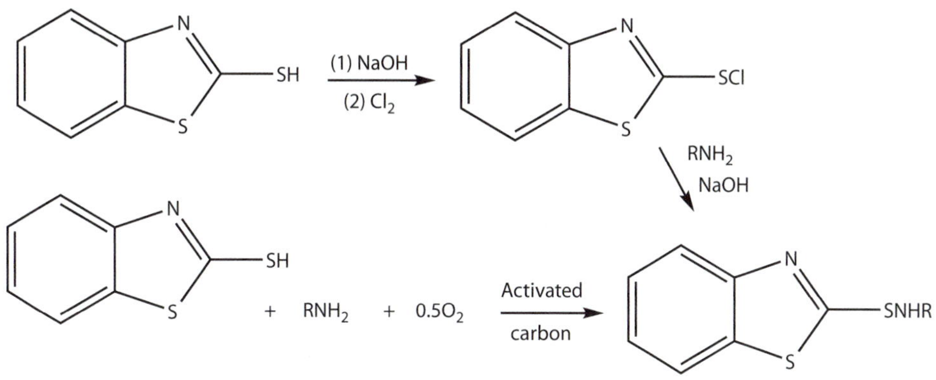

3.21 Schematic

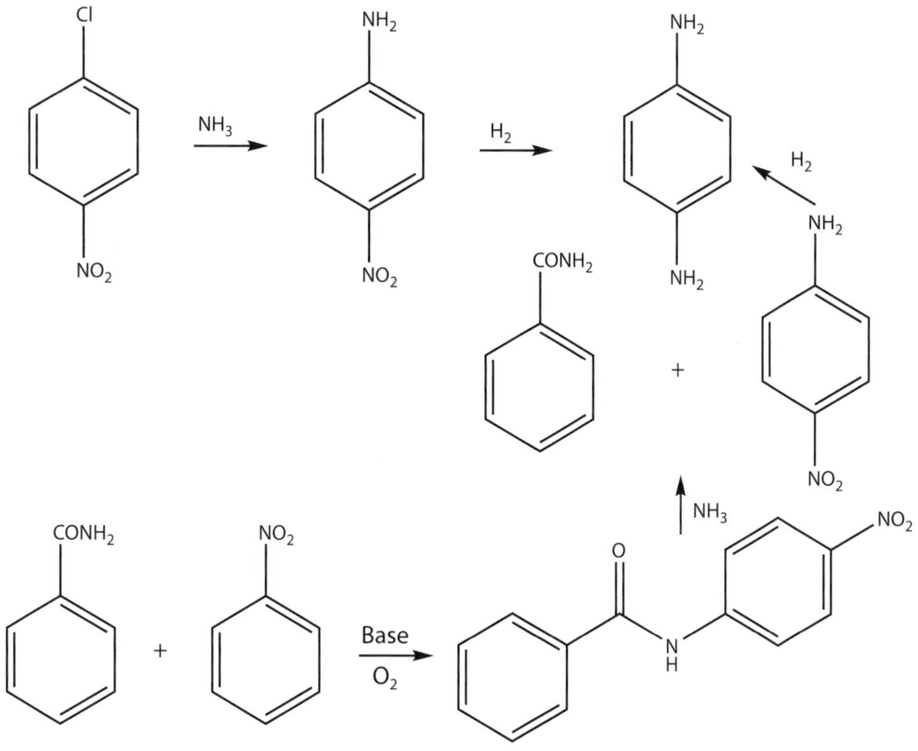

3.22 Schematic

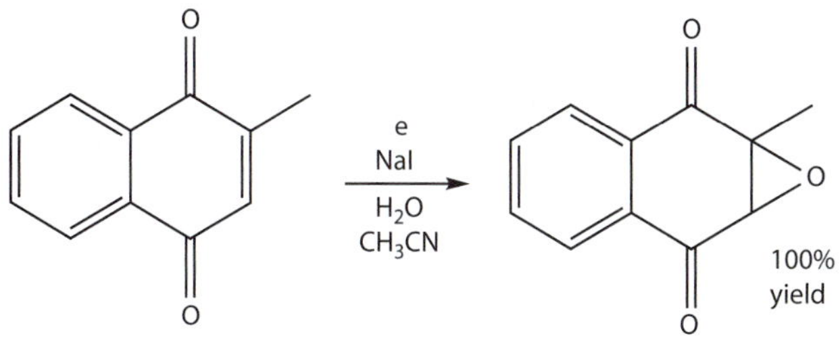

3.23 Schematic

propylene with oxygen to give propylene oxide can be highly selective at low conversions (e.g., >99% selective with a gold/titanium dioxide catalyst at 1–2% conversion).[303] A $EuCl_3$/Zn/acetic acid system was 94% selective at 3% conversion.[304] The challenge is to obtain high selectivity at nearly complete conversion. A membrane reactor that removed the propylene oxide continuously as formed might help.

Halogens can also be generated *in situ* by the action of an oxidizing agent on a sodium halide.[305] In this example (**3.24** Schematic), the oxidizing agent is potassium peroxymonosulfate (Oxone). The reactions were run in water diluted with acetone, ethyl acetate, or carbon tetrachloride. Bromine

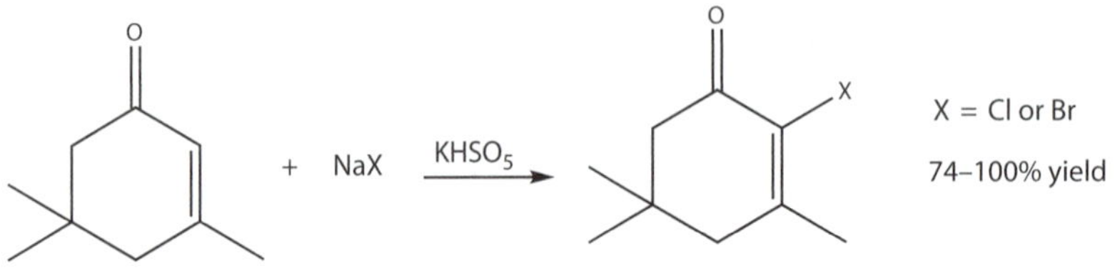

+ NaX $\xrightarrow{KHSO_5}$ X = Cl or Br; 74–100% yield

3.24 Schematic

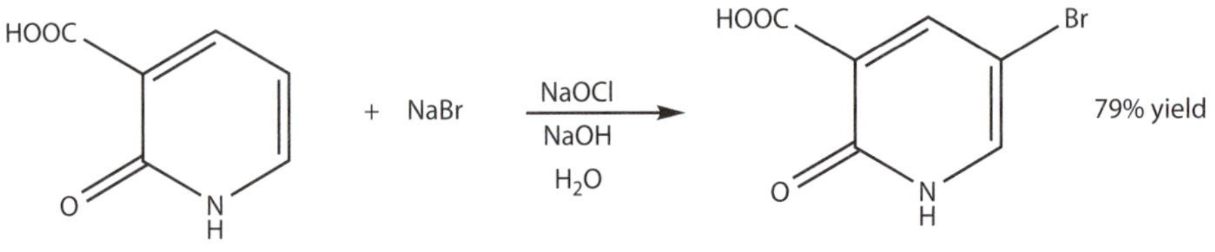

+ NaBr $\xrightarrow{NaOCl,\ NaOH,\ H_2O}$ 79% yield

3.25 Schematic

can also be generated *in situ* by the action of sodium perborate on sodium or potassium bromide.[306] It can also be made by the use of sodium bromide and sodium hypochlorite (**3.25** Schematic).[307] Such processes eliminate the dangers inherent in handling and storing the halogens, but not any toxicity associated with the halogen-containing products. It is also possible that the brominated compound will be used in a nucleophilic substitution that will produce sodium bromide.

The Heck reaction has been improved so that the usual halogen salts are not formed, and no phosphine ligands are required (**3.26** Schematic).[308] However, a toxic gas is formed.

Many chlorine compounds are biologically active. Some, such as the polychlorinated dibenzo-*p*-dioxins, are extremely potent. Many of the problems with some of these compounds are worldwide. Ozone depletion by chlorofluorocarbons is an example. Chlorine compounds do not necessarily stay where you put them. The widely used herbicide atrazine (**3.27** Schematic) is now widely distributed in ground water. Germany, Italy, and the Netherlands have limited its use.[1]

Long-range transport by winds is bringing chlorinated pesticides to places as remote as the Arctic island of Svalbard, Norway.[310] In a sample of 450 polar bears from this locality, seven had both male and female genitals.[311] It has been estimated that the use of chlorinated pesticides could be reduced by two-thirds with only a marginal influence on farmers, just by changing tilling and cultivating practices.[1] Although many highly chlorinated insecticides, such as DDT, have been outlawed in developed nations, they are still used in other parts of the world. Thus, there is a need for making the ban global. We have also come to realize that a chemical

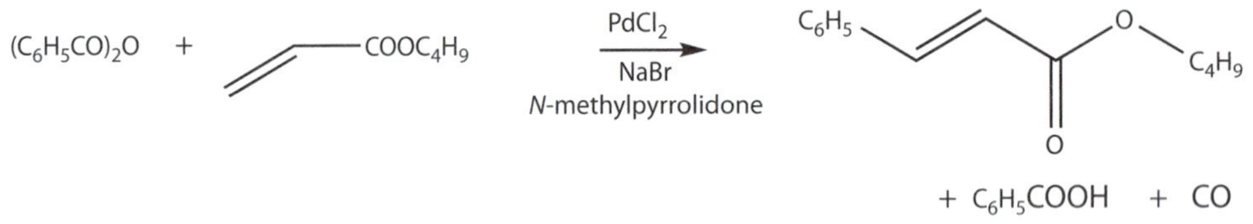

$(C_6H_5CO)_2O$ + $CH_2{=}CH{-}COOC_4H_9$ $\xrightarrow{PdCl_2,\ NaBr}$ (N-methylpyrrolidone) $C_6H_5CH{=}CHCOOC_4H_9$ + C_6H_5COOH + CO

3.26 Schematic

3.9 SUMMARY AND CONCLUSIONS

Greenpeace called for a ban on the use of chlorine and its compounds. The industry responded with "a combination of anger, indignation and denial."[309] Companies came together to mount a public information campaign. Why should the two sides take such extreme positions? Greenpeace has learned that a carefully thought-out analysis of a problem may attract little attention from the public or the media. However, taking an extreme position and carrying out highly visible stunts can. This group hung a banner high up on the multistory DuPont building in Wilmington, Delaware, at the time of the annual shareholders meeting. The banner protested DuPont's production of chlorofluorocarbons. The industry is large and profitable, with many paid-off plants. It does not want to see its business wiped out. It does not want to spend money to change unless there is compelling evidence forcing it to do so.

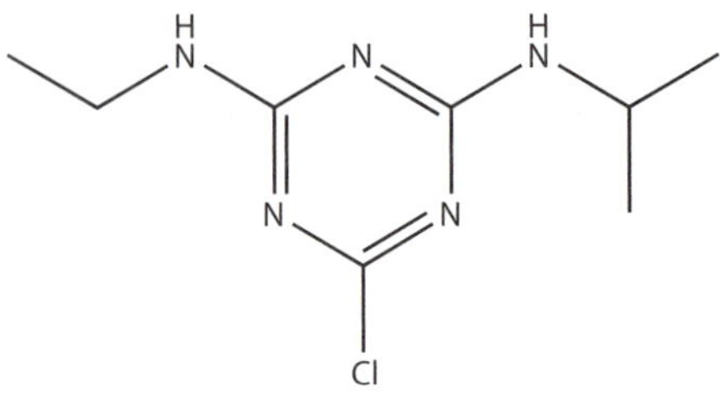

3.27 Schematic

need not kill an animal outright to wipe out a population. This can happen by altering behavior patterns or reproductive success. If male animals are feminized so that they do not breed successfully, or if females take no interest in their young and do not provide them with food, or if animals lose the ability to find and capture prey, the species will not survive. Nor will the species survive if it loses the behavior that helps it avoid predators. The thinning of eggshells threatened to wipe out species of several predatory birds, such as ospreys, eagles, peregrine falcons, and brown pelicans. All of this means that new chlorine compounds and others for release into the environment will have to be screened more thoroughly. There is also the nagging question of how much of what is seen in animals will be applicable to humans. This means finding ways to screen for cancers that may take 30 years to develop. Fortunately, the two sides in the controversy, plus government, are now in agreement that more research is needed and money is becoming available for it.

Chlorinated solvents might be handled by simple containment. The deadly dioxins might be removed from stack gases with the BASF device, or 2% nitrilotriacetic acid might be added to prevent their formation.[312] A membrane with a prefilter could be used to remove natural organic matter from drinking water before treatment with chlorine to reduce the formation of unwanted by-products, such as chloroform. Systems might be set up to segregate chlorine-containing plastics out of waste going into incinerators. This would be very difficult with the film used to package food. Chlorinated polymers could be banned from such applications. Their use might also be lessened by adding a tax. The possibility of loss of dangerous reagents, such as chlorine or phosgene during shipment or storage, could be reduced greatly by generation on-site as needed. About half the chlorine used ends up in by-product hydrogen chloride and salts that may present a disposal problem. Recent improvements to the Deacon process for converting hydrogen chloride back to chlorine with oxygen in the presence of copper chloride may help reduce this problem.[313] If every plant were independent, then, in theory at least, each could electrolyze aqueous solutions of its waste sodium chloride back to chlorine and sodium hydroxide for reuse. The real value of the current chlorine controversy may be in getting companies to rethink their entire process from start to finish.[314] As waste disposal costs rise, there should be greater incentive to devise chlorine-free processes.

A German study found that about half of the 2.9 million metric tons of chlorine used there annually could be eliminated at a net cost increase of only 1%.[315] Eighty percent of the chlorine used in Germany goes into propylene oxide, epoxy resins, polyvinyl chloride, and phosgene. Propylene oxide can be made by the oxidation of propylene with hydrogen peroxide or hydroperoxides at lower cost. Diphenyl carbonate can be used to replace phosgene in the preparation of polycarbonates (see Chapter 2). It may be possible to replace the use of phosgene in making isocyantes for polyurethane resins. Up to 50% of the polyvinyl chloride could be replaced, mainly with polyolefins. (Polypropylene with the properties of plasticized polyvinyl chloride is now available.[316]) Houses can be covered with wood, aluminum, brick, stucco, and so forth, as well as with polyvinyl chloride. Sewer pipes can be made of polyethylene as easily as from polyvinyl chloride. Some cities in Germany and Austria have banned the use of poly(vinyl chloride).[1] Epichlorohydrin is used to react with bisphenols in the preparation of epoxy resins. There are other ways of making bisepoxides (e.g., oxidation of olefins with hydroperoxides). Elf Atochem sells limonene dioxide, epoxidized poly(1,3-butadiene), and epoxidized vegetable oils that may be made by such a route and that might then be made into epoxy resins.[317] This would require some experimentation to develop epoxy resins with properties that matched or exceeded the properties of the current favorites. A titanium dioxide process using a dialkyl carbonate could eliminate chlorodioxins associated with the usual chloride process (Chapter 2). Alkyl ketenes for the treatment of cellulose are usually made by reaction of the carboxylic acid with phosphorus trichloride followed by treatment of the acid chloride with base. They can be made by dehydration of the carboxylic acid with silica at high temperature in 90% selectivity (**3.28** Schematic).[318]

$$C_7H_{15}COOH \xrightarrow[745\text{–}795K]{SiO_2} C_6H_{13}CH{=}C{=}O \quad \text{90\% selectivity}$$

3.28 Schematic

The controversy has not hurt business for the chlorine industry, which is booming.[319] There have been some shifts in the product line. Methylene chloride demand is dropping. More trichloroethylene is being produced as a feedstock for fluorocarbon production. More sodium chlorite is being produced as a precursor for chlorine dioxide to be used in bleaching paper. The trend should start downward in the future as chlorine-free alternatives are found and as ageing plants are retired.

Although many chlorinated compounds are destined to remain in the market for a long time, society does owe a debt of gratitude to environmental organizations such as Greenpeace and the World Wildlife Fund (with which Theo Colborn is associated) for calling attention to problems that may not have been evident to others. Their call has increased the funding related to these problems.

REFERENCES

1. (a) G. Graff, *Technol. Rev.*, 1995, *98*(1), 54; (b) E.J. Hoekstra and E.W.B. deLeer, *Chem. Br.*, 1995, 127.
2. B. Hileman, *Chem. Eng. News*, Oct. 17, 1994, 6.
3. E. Kirschner, *Chem. Eng. News*, July 11, 1994, 5.
4. B. Hileman, *Chem. Eng. News*, Aug. 29, 1994, 8.

5. Anon., *Chem. Eng. News*, Jan. 30, 1995, 18.
6. Anon., *Chem. Ind.* (London), 1995, 160.
7. (a) G.W. Gribble, *Environ. Sci. Technol.*, 1994, *28*, 310A; (b) *Acc. Chem. Res.*, 1998, *31*, 141; (c) *Am. Sci.*, 2004, *92*, 342.
8. A. Butler, *Science*, 1998, *281*, 207.
9. J.A. Field, F.J.M. Verhagen, and E. deJong, *Trends Biotechnol.*, 1995, *13*, 451.
10. Anon., *Chem. Eng. News*, Oct. 5, 1998, 41.
11. F.B. Whitfield, F. Helidoniotis, K.J. Shaw, and D. Svoronos, *J. Agric. Food Chem.*, 1997, *45*, 4398.
12. (a) G. Porter, T.E. Graedel, W.C. Keene, G.W. Gribble, J. Fauvarque, H. Galal-Gorchev, J. Miyamoto, M.J. Molina, H.W. Sidebottom, J.A. Franklin, K. Ballschmiter. C. Rappe, A. Hanberg, G. Menges, and A.E. Fischli, IUPAC white book on chlorine, *Pure Appl. Chem.*, 1996, *68*, 1683–1823; (b) R.C. Ahlert and F.C. Brown, *Environ. Prog.*, 1998, *17*(3), 161; (c) T.F. Yosie, *Environ. Sci. Technol.*, 1996, *30*, 499A; (d) C.T. Howlett, Jr. and T. Collins, *Chem. Eng. News*, Oct. 18, 2004, 40; (e) N. Winterton, *Green Chem.*, 2000, *2*, 173–225.
13. B. Hileman, J.R. Long, and E.M. Kirschner, *Chem. Eng. News*, Nov. 21, 1994, 12.
14. (a) Anon., *Chem. Ind. (London)*, 1995, 722; (b) P. Layman, *Chem. Eng. News*, Mar. 17, 1997, 17; (c) R.U. Ayres and L.W. Ayres, *J. Ind. Ecol.*, 1997, *1*(1), 81; 1997, *1*(2), 65.
15. D.R. Rea, private communication, Jan. 18, 1995.
16. R.E. Hester and R.M. Harrison, eds, *Chlorinated Organic Micropollutants*, Royal Society of Chemistry, Cambridge, UK, 1996.
17. J.M. Blais, D.W. Schindler, D.C.G. Muir, L.E. Kimpe, D.B. Donald, and B. Rosenberg, *Nature*, 1998, *395*, 585.
18. Anon., *Chem. Eng. News*, June 11, 2007, 24.
19. (a) G. Hess, *Chem. Eng. News*, Aug. 18, 2008, 28; (b) Anon., *Chem. Eng. News*, Mar. 9, 2009, 21.
20. M. Reisch, *Chem. Eng. News*, July 5, 2004, 5.
21. Anon., *Chem. Eng. News*, Apr. 11, 2005, 13.
22. Anon., *Chem. Eng. News*, Dec. 23, 2002.
23. Anon., *Chem. Eng. News*, July 28, 2003, 39.
24. Anon., *Chem. Eng. News*, Dec. 1, 2003, 27.
25. G. Hess, *Chem. Eng. News*, June 23, 2008, 23.
26. G. Ondrey, *Chem. Eng.*, 2008, *115*(5), 16.
27. U.S. EPA, *Presidential Green Chem. Award Entries*, 27 (to Miox Corp.), Washington, DC.
28. S.A. Tittlemier, M. Simon, W.M. Jarman, J.E. Elliott, and R.J. Norstrom, *Environ. Sci. Technol.*, 1999, *33*, 26.
29. S.L. Simonich and R.A. Hites, *Science*, 1995, *269*, 1851.
30. R. Renner, *Environ. Sci. Technol.*, 1996, *30*(1), 15A.
31. (a) D. MacKenzie, *New Sci.*, July 4, 1998, 6; (b) J. deBoer, P.G. Wester, H.J.C. Klamer, W.E. Lewis, and J.P. Boon, *Nature*, 1998, *394*, 28.
32. K.A. Kidd, D.W. Schindler, D.C.G. Muir, W.L. Lockhart, and R.H. Hesslein, *Science*, 1995, *269*, 240.
33. Anon., *Chem. Eng. News*, Nov. 6, 1995, 16.
34. Anon., *Chem. Eng. News*, Nov. 20, 2006, 58.
35. D.W. Kolpin, E.M. Thurman, and D.A. Goolsby, *Environ. Sci. Technol.*, 1996, *30*(1), 335.
36. B. Hileman, *Chem. Eng. News*, Sept. 25, 2006, 18.
37. J.L. Neumeyer, *Chem. Eng. News*, Sept. 25, 2006, 12.
38. B. Hileman, *Chem. Eng. News*, Feb. 1, 1999, 24.
39. C.J.M. Meershoek and R.M. Reeves, *Int. News Fats, Oils, Relat. Mater.*, 1998, *9*, 542, 543.
40. G. Bremle and P. Larsson, *Environ. Sci. Technol.*, 1998, *32*, 3491.
41. K.S. Betts, *Environ. Sci. Technol.*, 1998, *32*, 536A.
42. Anon., *Environ. Sci. Technol.*, 1998, *32*, 257A.
43. (a) R. Renner, *Environ. Sci. Technol.*, 1998, *32*, 360A; (b) M.A. Rivlin, *Amicus J.*, 1998, *19*(4), 30; (c) C.W. Schmidt, *Chem. Innovation*, 2001, *31*(12), 49.
44. C. Hogue, *Chem. Eng. News*, Aug. 11, 2008, 15.
45. Anon., *Chem. Eng. News*, July 28, 2008, 41.
46. D. Henschler, *Angew. Chem. Int. Ed.*, 1994, *33*, 1920.
47. Anon., *Chem. Eng. News*, Feb. 1, 1999, 27.
48. M. Heylin, *Chem. Eng. News*, Nov. 21, 1994, 5.
49. (a) D. MacKenzie, *New Sci.*, Sept. 10, 1994, 8; (b) *Chem. Ind. (London)*, 1994, 704.
50. (a) Anon., *Chem. Eng. News*, Aug. 19, 1996. (b) P. Mocarelli, P.L. Brambilla, P.M. Gerthoux, D.G. Patterson, and L.L. Needham, *Lancet*, 1996, *348*, 409.
51. (a) D. Hanson, *Chem. Eng. News*, May 30, 1994, 13; (b) B. Hileman, *Chem. Eng. News*, Sept. 19, 1994, 6; (c) *Environ. Sci. Technol*, 1994, *28*, 507A, 512A; (d) J. Johnson, R. Clapp, P. de Fur, E. Silbergeld, and P. Washburn, *Environ. Sci. Technol.*, 1995, *29*, 25A–33A.
52. C.M. Cooney, *Environ. Sci. Technol.*, 1999, *33*, 12A.
53. (a) B. Hileman, *Chem. Eng. News*, July 25, 1994, 21; (b) Anon., *Chem. Eng. News*, Nov. 24, 1997, 49.
54. W.R. Crummett, *Decades of Dioxin: Limelight on a Molecule*, Xlibris Bloomington, Indiana, 2002.
55. (a) R.A. Kerr, *Science*, 1994, *266*, 1162; (b) L.P. Brzuzy and R.A. Hites, *Environ. Sci. Technol.*, 1996, *30*, 1797; (c) B. Hileman, *Chem. Eng. News*, June 29, 1998, 25.
56. Anon., *Chem. Eng. News*, Aug. 25, 1997, 21.
57. G.W. Gribble, *Chem. Eng. News*, June 15, 1998, 4.
58. K. Cornish and D.J. Siler, *Chemtech*, 1996, *26*(8), 38.
59. (a) Anon., *Appl. Catal. B*, 1995, *6*, N17; (b) Anon., *Eur. Chem. News*, 1995, *63*(1650), 23.
60. J. Blanco, E. Alvarez, and C. Knapp, *Chem. Eng.*, 1999, *106*(11), 149.
61. Anon., *Chem. Eng. News*, July 18, 1996, 22.
62. P.W. McRandle, *World Watch*, 2006, *19*(2), 8.
63. A.H. Tullo, *Chem. Eng. News*, May 20, 2002, 13.
64. Anon., *Chem. Eng. News*, May 1, 2000, 18.
65. (a) A.R. MacKenzie and S. Brooks, *Chem. Eng. News*, Nov. 23, 1998, 8; (b) A.R. Mackenzie and S. Brooks, *Chem. Br.*, 1998, *34*(12), 18.
66. (a) L. McMurry, M. Oethinger, and S.B. Levy, *Nature*, 1998, *394*, 531; (b) P.M. Morse, *Chem. Eng. News*, Feb. 1, 1999, 40.
67. Sold by General Chemical Company as "RootX."
68. J. Thornton, *Pandora's Poison—Chlorine, Health and a New Environment*, MIT Press, Cambridge, 2000.
69. M. Reisch, *Chem. Eng. News*, April 2, 2001, 9.
70. (a) T. Colburn, J.P. Meyers, and D. Dumanowski, *Our Stolen Future*, Dutton, New York, 1996; (b) www.endrocrinedisruption.org.
71. (a) *Chem, Ind. (London)*, 1996, 364; (b) R. Stone, *Science*, 1994, *265*, 308; (c) M. Lee, *Chem. Br.*, 1996, *32*(6), 5; (d) E. Culotta, *Science*, 1995, *267*, 330; (e) *Chem. Ind. (London)*, 1994, 320.
72. (a) R. Renner, *Environ. Sci. Technol.*, 1997, *31*, 312A; 1998, *32*, 8A, 395A; (b) S. Jobling, M. Nolan, C.R. Tyler, G. Brighty, and J.P. Sumpter, *Environ. Sci. Technol.*, 1998, *32*, 2498; (c) *Chem. Ind. (London)*, 1998, 76.

73. R.S.S. Wu, B.S. Zhou, D.J. Randall, N.Y.S. Woo, and P.K.S. Lam, *Environ. Sci. Technol.*, 2003, *37*, 1137.
74. (a) H. Carmichael, *Chem. Br.*, 1998, *34*(10), 25; (b) D. Cadbury, *The Feminization of Nature*, Penguin, London, 1997; (c) L.H. Keith, *Environmental Endocrine Disruptors—A Handbook of Property Data.*, Wiley, New York, 1997; (d) T.E. Wiese and W.R. Kelce, *Chem. Ind.* (*London*), 1997, 648. (e) R.E. Hester and R.M. Harrison, eds, *Endocrine Disrupting Chemicals*, Royal Society Chemistry, Cambridge, 1999; (f) M. Burke, *Chem. Br.*, 2003, *39*(1), 30; (g) J. Miyamoto and J. Burger, eds, *Pure Appl. Chem.*, 2003, *75*(11–12), 1617–2615; (h) J. Lintelmann, A. Katayama, L. Shore, and A. Wenzel, *Pure Appl. Chem.*, 2003, *75*, 631–681; (i) S. Krimsky, *Hormonal Chaos: The Scientific and Social Origins of the Endocrine Hypothesis*, Johns Hopkins Press, Baltimore, MD, 2000; (j) T. Schettler, G.M. Solomon, M. Valente, and A. Huddle, *Generations at Risk: Reproductive Health and the Environment*, MIT Press, Cambridge, 1999; (k) D.O. Norris and J.A. Carr, *Endocrine Disruption: Biological Bases for Health Effects in Wildlife and Humans*, Oxford University Press, New York, 2005.
75. *Chem. Eng. News*, Apr. 6, 1998, 35.
76. M. Patlak, *Environ. Sci. Technol.*, 1996, *30*(5), 210A.
77. (a) S. Swan, *Environ. Health Perspect.*, 1997, *105*, 1228; (b) *Chem. Ind.* (*London*), 1997, 927.
78. B. Halwell, *Futurist*, 1999, *33*(9), 14.
79. L. Stoloff, *Chem. Eng. News*, Mar. 6, 1995, 5.
80. J. Kaiser, *Science*, 2000, *288*, 424.
81. (a) B. Hileman, *Chem. Eng. News*, May 5, 2003, 40; (b) *Chem. Eng. News*, Apr. 7, 2003, 7.
82. A.P. Hoyer, P. Granjean, T. Jorgensen, J.W. Brock, and H.B. Hartvig, *Lancet*, 1998, *352*, 1816.
83. R.L. Williams and T. Rutledge, *Chem. Ind.* (*London*), 1998, 14.
84. J.M. Hatcher, K.D. Pennell, and G.W. Miller, *Trends Pharm. Sci.*, 2008, *29*(6), 322.
85. (a) A.M. Warhurst, *Chem. Ind.* (*London*), 1995, 756; (b) R. Renner, *Environ. Sci. Technol.*, 1996, *30*, 19A; 1997, *31*, 316A; (c) G. Vines, *New Sci.*, Aug. 26, 1995, 22; (d) M. Bokern and H.H. Harms, *Environ. Sci. Technol.*, 1997, *31*, 1849; (e) B. Thiele, K. Gunther, and M.J. Schwuger, *Chem. Rev.*, 1997, *97*, 3247; (f) *Int. News Fats, Oils Relat. Mater.*, 1997, *8*, 1269.
86. *Chem. Eng. News*, July 13, 1998, 55.
87. (a) *Chem. Ind.* (*London*), 1998, 72; (b) W. MacDonald, *Chem. Eng.*, 1998, *105*(2), 52; (c) D. MacKenzie, *New Sci.*, Jan. 24, 1998, 13.
88. (a) B. Hileman, *Chem. Eng. News*, Nov. 30, 1998, 8; (b) *Chem. Eng. News*, Dec. 7, 1998, 35; (c) *Chem. Ind.* (*London*), 1998, 633; (d) *Chem. Br.*, 1998, *34*(6), 9; (e) *Chem. Eng. News*, June 1, 1998, 32; (f) C. Cray, *Chem. Eng. News*, Jan. 11, 1999, 2.
89. (a) B. Hileman, *Chem. Eng. News*, July 11, 2005, 11; (b) P. Mitchell, *Chem. Ind.* (*London*), Oct. 4, 2004, 7; (c) D. Hanson, *Chem. Eng. News*, Aug. 4, 2008, 8.
90. L. Ember, *Chem. Eng. News*, Mar. 15, 1999, 41.
91. S.C. Gaynor, *Chem. Eng. News*, Jan. 11, 1999, 2.
92. *Chem. Eng. News*, Dec. 21, 1998, 11.
93. A.H. Tullo, *Chem. Eng. News*, Nov. 14, 2005, 29.
94. M. Rahman and C.S. Brasel, *Prog. Polym. Sci.*, 2004, *29*, 1223.
95. B. Hileman, *Chem. Eng. News*, Mar. 24, 1997, 37.
96. (a) J.E. Biles, T.P. McNeal, T.H. Begley, and H.C. Hollifield, *J. Agric. Food Chem.*, 1997, *45*, 3541; (b) A.J. LaCovey, *Chem. Eng. News*, Sept. 28, 1998, 4; (c) *Chem. Ind.* (*London*), 1998, 872; (d) G. Hess, *Chem. Eng. News*, Sept. 22, 2008, 13; (e) B.E. Erickson, *Chem. Eng. News*, Nov. 17, 2008, 42; (f) M. Jacoby, *Chem. Eng. News*, Dec. 15, 2008, 31; (g) M. Voith, Chem. Eng. News, July 20, 2009, 28.
97. *Chem. Ind.* (*London*), 1998, 324.
98. (a) B. Hileman, *Chem. Eng. News*, May 25, 1998, 40; (b) C.W. Schmidt, *Environ. Sci. Technol.*, 1997, *31*, 324A.
99. (a) P. Webster, *Science*, 2004, *304*, 1730; (b) Anon., *World Watch*, 2004, *17*(1), 30.
100. (a) Anon., *Chem. Eng. News*, June 9, 2008, 39; (b) S. Everts, *Chem. Eng. News*, June 26, 2006, 8.
101. R. Renner, *Environ. Sci. Technol.*, 2001, *35*, 180A.
102. J.C. d'Eon, M.D. Hurley, T.J. Wallington, and S.A. Mabury, *Environ. Sci. Technol.*, 2006, *40*, 1862.
103. (a) Anon., *Wilmington, Delaware News J.*, June 10, 2008, A1; (b) Anon., *Chem. Eng. News*, Dec. 4, 2006, 44.
104. U.S. EPA, *Presidential Green Chem. Award Entries*, 2008, 33.
105. (a) C. Hogue, *Chem. Eng. News*, Apr. 21, 2003, 9; (b) D.A. Ellis, J.W. Martin, A.O. deSilva, S.A. Mabury, M.D. Hurley, M.P.S. Andersen, and T.J. Wallington, *Environ. Sci. Technol.*, 2004, *38*, 3316; (c) M.J.A. Dinglasan, Y. Ye, E.A. Edwards, and S.A. Mabury, *Environ. Sci. Technol.*, 2004, *38*, 2857; (d) S.K. Ritter, *Chem. Eng. News*, June 14, 2004, 44.
106. M. Reisch, *Chem. Eng. News*, Dec. 4, 2006, 40.
107. (a) M. McCoy, *Chem. Eng. News*, Apr. 16, 2001, 42; (b) B. Hileman, *Chem. Eng. News*, Mar. 27, 2006, 11; (c) C. Bryson, *The Fluoride Deception*, Seven Stories Press, New York, 2004; (d) B. Hileman, *Chem. Eng. News*, Sept. 4, 2006, 34.
108. (a) Anon., *Chem. Eng. News*, June 4, 2007, 27; (b) C. Hogue, *Chem. Eng. News*, May 14, 2007, 39.
109. (a) O.A. Jones, J. Lester, and N. Vouhvoulis, *Trends Biotechnol.*, 2005, *23*(4), 163; (b) B. Halford, *Chem. Eng. News*, Feb. 25, 2008, 13; (c) G.T. Ankley, B.W. Brooks, D.B. Huggett, and J.P. Sumpter, *Environ. Sci. Technol.*, 2007, *41*, 8211; (d) D.S. Aga, ed., *Fate of Pharmaceuticals in the Environment and in Water*, CRC Press, Boca Raton, FL; (e) P.K. Jjemba, *Pharma-Ecology: Occurrence and Fate of Pharmaceutical and Personal Care Products in the Environment*, Wiley, Hoboken, NJ, 2008.
110. V. Matamoros and J.M. Bayona, *Environ. Sci. Technol.*, 2006, *40*, 5811.
111. (a) Anon., *Chem. Eng. News*, Jan. 18, 2007, 44; (b) P. Walter, *Chem. Ind. (London)*, Sept. 8, 2008, 8.
112. M.M. Huber, A. Gobel, A. Joss, N. Herman, H. Siegrist, T.A. Ternes, and U. vonGunten, *Environ. Sci. Technol.*, 2005, *39*, 4290.
113. T.A. Ternes, M. Meisenheimer, D. McDowell, F. Sacher, H.-J. Brauch, B. Haist-Guldo, G. Preuss, U. Wilme, and N. Zulei-Seibert, *Environ. Sci. Technol.*, 2002, *36*, 3855.
114. H. Fu, R.P.S. Suri, R.F. Chimchirian, E. Helmig, and R. Constable, *Environ. Sci. Technol.*, 2007, *41*, 5869.
115. T.J. Colllins et al., *Environ. Sci. Technol.*, 2008, *42*, 1296.
116. A. Schaefer, *Environ. Sci. Technol.*, 2006, *40*, 412.
117. (a) *Chem. Ind.* (*London*), 1996, 316; (b) W.R. Moomaw, *Chem. Eng. News*, Apr. 1, 1996, 34; (c) B. Hileman, *Chem. Eng. News*, Mar. 18, 1996; (d) J. Johnson, *Environ. Sci. Technol.*, 1996, *30*, 168A.

118. (a) R. Kendall, R.L. Dickerson, G.P. Giesy, and W.P. Suk, eds, *Principles and Processes for Evaluating Endocrine Disruption*, SETAC Press, Pensacola, FL, 1998; (b) B. Hileman, *Chem. Eng. News*, Feb. 9, 1998, 7; (c) J. Johnson, *Chem. Eng. News*, Mar. 8, 1999, 9.
119. B. Hileman, *Chem. Eng. News*, Oct. 12, 1998, 18.
120. *Chem. Eng. News*, June 10, 1996.
121. *Chem. Ind.* (*London*), 1998, 415.
122. (a) B. Hileman, *Chem. Eng. News*, May 13, 1996, 28; (b) J. Johnson, *Environ. Sci. Technol.*, 1996, *30*, 242A; (c) *Chem. Eng. News*, Jan. 18, 1999, 33; (d) *Chem. Eng. News*, Oct. 26, 1998, 24.
123. E. Bolzacchini, A.M. Brambilla, M. Orlandi, and B. Rindone, Preprints A.C.S. *Div. Environ. Chem.*, 1994, *34*(2), 355.
124. (a) R. Franzen and L. Kornberg, *Tetrahedron Lett.*, 1995, *36*, 3905; (b) R. Franzen and L. Kornberg, *Environ. Sci. Technol.*, 1994, *28*, 2222; (c) A. Smeds, R. Franzen, and L. Kronberg, *Environ. Sci. Technol.*, 1995, *29*, 1839.
125. R. Mestel, *New Sci.*, Sept. 24, 1994, 10.
126. (a) T. McDonough, *Kirk–Othmer Encyclopedia of Chemical Technology*, 4th ed., Wiley, New York, 1992, *4*, 301; (b) V. Turoski, ed., Preprints A.C.S. *Div. Environ Chem.*, 1995, *35*(2), 249–339; (c) P.W. Hart, A.W. Rudie, and J.C. Joseph, eds, *Advances in Pulping and Papermaking*, AIChE Symp. 307 1995, *91*, 25, 38, 53, 92; (d) M.R. Servos, K.R. Munkittrick, J.H. Carey, and G.J. Van Der Kraak, eds, *Environmental Fate and Effects of Pulp and Paper Mill Effluents.*, St. Lucie Press, Delray Beach, FL, 1996; (e) E.M. Kirschner, *Chem. Eng. News*, Sept. 29, 1997, 15; (f) B.N. Brogdon, P.W. Hart, J.C. Ransdell, and B.L. Scheller, eds, *Innovative Advances in the Forest Products Industries*, AIChE. Symp., New York, 1998, *94*(319).
127. (a) E. Wilson, *Chem. Eng. News*, Sept. 18, 1995; (b) M. McCoy, *Chem. Eng. News*, Feb. 1, 1999, 18.
128. A.P. Sattelberger and R.P. Currier, *Chem. Eng. News*, Dec. 9, 2002, 4.
129. A.B. McKague and D.W. Reeve, Preprints A.C.S. *Div. Environ. Chem.*, 1995, *35*(2), 335.
130. C. Chirat and D. Lachenal, *TAPPI J.*, 1997, *80*(9), 209.
131. K.-E.L. Eriksson and R.B. Adolphson, *TAPPI J.*, 1997, *80*(6), 80.
132. (a) V. Peck and R. Daley, *Environ. Sci. Technol.*, 1994, *28*, 524A; (b) S. Blomback, J.-E. Eriksson, K. Idner, and B. Warnquist, *TAPPI J.*, 1999, *82*(10), 131.
133. C.-J. Alfthan, *Pap. Technol.* (*London*), 1995, *36*(6), 20.
134. A. Gutierrez, M.J. Martinez, J.C. Del Rio, J. Romero, J. Canaval, G. Lenon, and A.T. Martinez, *Environ. Sci. Technol.*, 2000, *34*, 3705.
135. P. Grundstrom and T. Granfeldt, *Paper Technol.* (*London*), 1997, *38*(9), 29.
136. S. Varennes, C. Daneault, and M. Parenteau, *TAPPI J.*, 1996, *79*(3), 245.
137. (a) G.S. Samdani, *Chem. Eng.*, 1994, *101*(8), 19; (b) I.A. Weinstock, R.H. Atalla, R.S. Reiner, M.A. Moen, K.E. Hammel, C.J. Houtman, C.L. Hill, and M.K. Harrup, *J. Mol. Catal. A*, 1997, *116*, 59; (c) I.A. Weinstock, C.L. Hill, R.H. Atalla, M.W. Wemple, R.S. Reiner, C.S. Sullivan, C.J. Houtman, S.E. Reichel, R. Heintz, E.L. Springer, J.S. Reddy, J.J. Cowan, and E.M. Barbuzzi, *Green Chemistry and Chemical Engineering Conference*, Washington, DC, June 30–July 2, 1998; (d) I.A. Weinstock, R.H. Atalla, R.S. Reiner, M.A. Moen, and K.E. Hammel, *New J. Chem.*, 1996, *20*, 269.
138. J. Haggin, *Chem. Eng. News*, Oct. 9, 1995, 7.
139. P. Axegard, E. Bergnor, M. Ek, and U.E. Ekholm, *TAPPI J.*, 1996, *79*(1), 113.
140. (a) T.J. Collins and C.P. Horwitz, *Chem. Abstr.*, 1998, *128*, 155, 724; (b) C.P. Horwitz, D.R. Fooksman, L.D. Vuocolo, S.W. Gordon-Wylie, N.J. Cox, and T.J. Collins, *J. Am. Chem. Soc.*, 1998, *120*, 4867; (c) R. Dagani, *Chem. Eng. News*, July 5, 1999, 30.
141. Y. Yang, D.T. Wyatt II, and M. Bahorsky, *Text Chem. Color* 1998, *30*(4), 27.
142. R. Hage, and A. Lienke, *Angew. Chem. Int.* Ed., 2006, *45*, 206.
143. (a) D. Hairston, *Chem. Eng.*, 1995, *102*(7), 67; (b) G. Parkinson, *Chem. Eng.*, 1997, *104*(8), 25; (c) A.P. Gelbein, *Chemtech*, 1998, *28*(12), 1.
144. U.S. EPA, Presidential Green Chemistry Challenge Award, Washington, DC, 2007.
145. (a) M.E. Fraser, A.S. Woodman, E.B. Anderson, and A.N. Pirri, Preprints A.C.S. *Div. Environ. Chem.*, 1996, *36*(1), 175; (b) G. Parkinson, *Chem. Eng.*, 1997, *104*(1), 17.
146. D.E. Fletcher, N.G. Johansson, J.J. Basta, A.-S. Holm, and E. Wackerberg, *TAPPI J.*, 1997, *80*(12), 143.
147. Y. Cai and S.K. David, *Text. Res. J.*, 1997, *67*, 459.
148. B.A. Evans, L. Boleslawski, and J.E. Boliek, *Text. Chem. Color.*, 1997, *29*(3), 28.
149. (a) A. Ouchi, T. Obata, H. Sakai, and M. Sakuragi, *Green Chem.*, 2002, *3*, 221; (b) A. Ouchi, T. Obata, T. Oishi, H. Isakai, T. Hayashi, W. Ando, and J. Ito, *Green Chem.*, 2004, *6*, 198.
150. (a) A. Ouchi, H. Sakai, T. Oishi T. Hayashi, W. Ando, and J. Ito, *Green Chem.*, 2003, *5*, 516; (b) A. Ouchi and H. Sakai, *Green Chem.*, 2003, *5*, 329.
151. Anon., *Appl. Catal. B*, 1994, *4*, N3.
152. K.L. Patrick, *Pulp Paper*, Oct. 1997, 61.
153. D. Santiago, A. Rodriguez, J. Swec, A.L. Baumstasrk, and A.J. Ragauskas, *Ind. Eng. Chem. Res.*, 1995, *34*, 400.
154. (a) G. Elegir, M. Sykes, and T.W. Jeffries, *Enzyme Microb. Technol.*, 1995, *17*, 954; (b) P. Bajpai and P.K. Bajpai, *TAPPI J.*, 1996, *79*(4), 225; (c) J. Hamilton, D.J. Senior, L.A. Rodriguez, L.D. Santiago, J. Swec, and A.J. Ragauskas, *TAPPI J.*, 1996, *79*(4), 231; (d) L.P. Christov and B.A. Prior, *Enzyme Microb. Technol.*, 1996, *18*, 244; (e) A.P. Garg, A.J. McCarthy, and J.C. Roberts, *Enzyme Microb. Technol.*, 1996, *18*, 260.
155. J. Sealy and A.J. Ragauskas, *Enzyme Microb. Technol.*, 1998, *23*, 422.
156. M.R. Julia, M. Cot, P. Erra, D. Jocic, and J.M. Canal, *Text. Chem. Color*, 1998, *30*(8), 78.
157. T.J. Kang and S.J. Moon, *Text. Res. J.*, 1998, *68*, 719.
158. K.J. Dodd, C.M. Carr, and K. Byrne, *Text. Res. J.*, 1998, *68*, 10.
159. I.M. Zuchairah, M.T. Pailthorpe, and S.K. David, *Text. Res. J.*, 1997, *67*, 69.
160. (a) P. Cartwright, *Chem. Eng.*, 2006, *113*(3), 50; (b) R. D'Aquino, *Chem. Eng. Prog.*, 2006, *102*(4), 8; (c) T. Oppenlander, *Photochemicall Purification of Water and Air—Advanced Oxidation Processes: Principles, Reaction*

Mechanisms and Reaction Concepts, Wiley-VCH, Weinheim, 2002; (d) D. Hairston, *Chem. Eng.*, 2003, *110*(7), 25; (e) M. Kimber, *Chem. Br.*, 2003, *39*(5), 26.
161. (a) G.C. White, *Handbook of Chlorination and Alternative Disinfectants*, 4th ed., Wiley, New York, 1998; (b) T. Studt, *R&D* (*Cahners*), Mar. 1998, 59.
162. *Univ. California Berkeley Wellness Lett.*, Sept. 1995, 5.
163. *Environ. Sci. Technol.*, 1995, *29*, 173A.
164. K.S. Betts, *Environ. Sci. Technol.*, 1998, *32*, 546A.
165. (a) L. Raber, *Chem. Eng. News*, June 28, 1998, 28; (b) *Chem. Eng. News*, Apr. 13, 1998, 36.
166. K.S. Betts, *Environ. Sci. Technol.*, 1998, *32*, 169A.
167. *Chem. Br.*, 1996, *32*(5), 14.
168. J.A. Switzer, V.V. Rajasekharan, S. Boonsalee, E.A. Kulp, and E.W. Bohannon, *Environ. Sci. Technol.*, 2006, *40*, 3384.
169. (a) M.W. Ekonian, J.H. Putnam, and S.D. Worley, *Ind. Eng. Chem. Res.*, 1998, *37*, 2873. (b) Y. Sun and G. Sun, *Ind. Eng. Chem. Res.*, 2004, *43*, 5015; (c) J. Liang, Y. Chen, X. Ren, R. Wu, K. Barnes, S.D. Worley, R.M. Broughton, U. Cho, H. Kocer, and T.S. Huang, *Ind. Eng. Chem. Res.*, 2007, *46*, 6425; (d) J. Liang, K. Barnes, A. Akdag, S.D. Worley, J. Lee, R.M. Broughton, and T.S. Huang, *Ind. Eng. Chem. Res.*, 2007, *46*, 1861.
170. (a) Y. Uemura, I. Moritake, S. Kurihara, and T. Nonaka, *J. Appl. Polym. Sci.*, 1999, *72*, 371; (b) S. Borman, *Chem. Eng. News*, May 28, 2001, 13; (c) Z. Shi, K.G. Neoh, and E.T. Kang, *Ind. Eng. Chem. Res.*, 2007, *46*, 439; (d) O. Bouloussa, F. Rondolez, and V. Semety, *Chem. Commun.*, 2008, 951; (e) V. Sambhy, B.R. Peterson, and A. Sen, *Angew. Chem. Int. Ed.*, 2008, *47*, 1250; (f) U.S. EPA, *Presidential Green Chem. Award Entries*, 2008, 22 to DuraBan International.
171. M. Miyahara, M. Toyoda, and Y. Saito, *J. Agric. Food Chem.*, 1995, *43*, 320.
172. (a) K.L. Stevens, R.E. Wilson, and M. Friedman, *J. Agric. Food Chem.*, 1995, *43*, 2424; (b) W.F. Haddon, R.G. Binder, R.Y. Wong, L.A. Harden, R.E. Wilson, M. Benson, and K.L. Stevens, *J. Agric. Food Chem.*, 1996, *44*, 256.
173. L.-S. Tsai, R. Wilson, and V. Randall, *J. Agric. Food Chem.*, 1997, *45*, 2267.
174. Anon., *Agric. Res.*, 1999, *47*(3), 27.
175. J. Lee, *Agric. Res.*, 1998, *46*(12), 18.
176. (a) *Univ. California Berkeley Wellness Lett.*, Sept. 1995, 5; (b) P.V. Shanbhag and K.K. Sirkar, *J. Appl. Polym. Sci.*, 1998, *69*, 1263; (c) M. Dore, *Chemistry of Oxidants and Treatment of Water*, Wiley, New York, 1997; (d) G. Parkinson, *Chem. Eng.*, 1998, *105*(2), 21.
177. (a) T. Matsunaga and M. Okochi, *Environ., Sci. Technol.*, 1995, *29*, 501; (b) K. Rajeshwar, *Chem. Ind.* (*London*), 1996, 454; (c) Y. Horie, D.A. David, M. Taya, and S. Tone, *Ind. Eng. Chem. Res.*, 1996, *35*, 3920; (d) J.K. Kim, H.S. Choi, E. Kim, Y.S. Lee, Y.H. Kim, and S.W. Kim, *Biotechnol. Lett.*, 2003, *24*, 1397.
178. E. Wilson, *Chem. Eng. News*, July 1, 1996, 29.
179. P.K.J. Robertson, L.A. Lawton, B. Munch, and J. Rouzade, *Chem. Commun*, 1997, 393.
180. (a) L.G. Parkinson, *Chem. Eng.*, 1998, *105*(2), 21; (b) Safe Water Solutions, Brown Deer, Wisconsin-trade literature.
181. *Environ. Sci. Technol.*, 1997, *31*, 120A.
182. R. Renner, *Environ. Sci. Technol.*, 1996, *30*, 284A.
183. SteriPEN, 2007 data sheet.
184. I. Amato, *Chem. Eng. News*, Apr. 17, 2006, 39.
185. S.D. Worley and G. Sun, *Trends Polym. Sci.*, 1996, *4*, 364; (b) G. Li, J. Shen, and Y. Zhu, *J. Appl. Polym. Sci.*, 1998, *67*, 1761; (c) E.V. Anufrieva, E.F. Panarin, V.D. Pautov, A.B. Kirpach, M.G. Krakovyak, V.B. Luschik, and M.V. Solovskij, *Macromol. Chem. Phys.*, 1997, *198*, 3871.
186. (a) P. Lopez and J. Burgos, *J. Agric. Food Chem.*, 1995, *43*, 620; (b) P. Butz, A. Fernandez, H. Fister, and B.L. Tauscher, *Agric. Food Chem.*, 1998, *45*, 302; (c) X. Felipe, M. Capellas, and A.J.R. Law, *J. Agric. Food Chem.*, 1998, *45*, 627. (d) S.L. Wilkinson, *Chem. Eng. News*, Nov. 10, 1997, 24. (e) S. Grant, M. Patterson, and D. Ledward, *Chem. Ind.* (*London*), 2000, 55; (f) K.B. Connolly, *Food Processing*, 2007, *68*(7), 49; (g) L. McGinnis, *Agric. Res.*, 2006, *54*(10), 4.
187. G. Ondrey, *Chem. Eng.*, 2006, *113*(13), 12.
188. U.S. EPA, *Presidential Green Chem. Award Entries*, 2008, 35 to Ashland.
189. U.S. Office of Science and Technology, *Technology for a Sustainable Future*, U.S. Government Printing Office, Washington, DC, 1994, 133.
190. (a) W.J. Cooper, M.J. Nickelsen, D.C. Kajdi, C.N. Kurucz, and T.D. Waite, Preprints A.C.S. *Div. Environ. Chem.*, 1995, *35*(2), 718–725; (b) *Chem. Br.*, *32*(5), 14.
191. (a) G.R. Helz and A.C. Nweke, *Environ. Sci. Technol.*, 1995, *29*, 1018; (b) W.A. Maccrechan, J.S. Jensen, and G.R. Helz, *Environ. Sci. Technol.*, 1998, *32*, 3640.
192. (a) *Environ. Sci. Technol.*, 1996, *30*, 277A; (b) *Environ. Sci. Technol.*, 1997, *31*, 213A.
193. (a) Anon., *Chem. Eng. Prog.*, 2002, *98*(9), 10; (b) Clearwater Enviro., *Pollution Equipment News*, 2008, *41*(3), 14.
194. (a) B. Durham and L. Patria, *Chem. Eng.*, 2006, *113*(10), 50; (b) T. Asano, F.L. Burton, H.L. Leverenz, R. Tsuchihashi, and G. Tchobanoglous, *Water Reuse—Issues, Technologies and Applications*, Metcalf and Eddy, McGraw-Hill, New York, NY, 2007; (c) J. Kemsley, *Chem. Eng. News*, Jan. 28, 2008, 71.
195. F.C. Dobbs and A. Rogerson, *Environ. Sci. Technol.*, 2005, *39*, 259A.
196. R.L. Rundle, *Wall Street J.*, June 6, 2006, D1.
197. B.A. Nagengast, *ASHRAE J.* 1995, *37*(3), 54.
198. E. Kirschner, *Chem. Eng. News*, July 4, 1994, 12.
199. K. Weissermel and H.-J. Arpe, *Industrial Organic Chemistry*, 2nd ed., VCH, Weinheim, 1993, 55.
200. (a) F.S. Rowland and M.J. Molina, *Chem. Eng. News*, Aug. 15, 1994, 8; (b) P. Zurer, *Chem. Eng. News*, Jan. 2, 1995, 9; (c) *Chem. Eng. News*, Feb. 12, 1996, 24; (d) D. Hinrichsen, *Amicus J.*, 1996, *18*(3), 35; (e) R. Luther and C. Wolz, *Environ. Sci. Technol.*, 1997, *31*, 142A; (e) E.A. Parson, *Protecting the Ozone Layer—Science and Strategy*, Oxford University Press, Oxford, 2003; (f) B. Hileman, *Chem. Eng. News*, May 2, 2005, 28.
201. L. Mastny, *World Watch*, 1999, *12*(1), 11.
202. (a) *Chem. Ind.* (*London*), 1997, 291; (b) For information on the polar stratospheric cloud particles involved in the destruction of ozone in the Arctic, see J. Schreiner, C. Voigt, A. Kohlmann, F. Arnold, K. Mauersberger, and N. Larsen, *Science*, 1999, *283*, 968.
203. (a) E.A. Parsons and O. Greene, *Environment*, 1995, *37*(2), 18; (b) P. Zurer, *Chem. Eng. News*, Mar. 13, 1995, 8.

204. (a) S.A. Montzka, J.H. Butler, R.C. Myers, T.M. Thompson, T.H. Swanson, A.D. Clarke, L.T. Lock, and J.W. Elkins, *Science*, 1996, *272*, 1318; (b) B. Hileman, *Chem. Eng. News*, Sept. 5, 2005, 13.
205. (a) D. Hairston, *Chem. Eng.*, 1995, *102*(1), 65; (b) *Environ. Sci. Technol.*, 1994, *28*, 509A; (c) Microinfo, Technological Development and Pollution Abatement: A Study of How Enterprises Are Finding Alternatives to Chlorofluorocarbons, Alton, Hampshire, UK, 1995; (d) K.B. Miller, C.W. Purcell, J.M. Matchett, and M.H. Turner, *Strategies for Managing Ozone-Depleting Refrigerants: Confronting the Future*, Battelle Press, Columbus, OH, 1995.
206. *Chem. Eng. News*, Oct. 2, 1995, 12.
207. *Chem. Eng. News*, Jan. 2, 1995, 11.
208. G.S. Samdani, *Chem. Eng.*, Feb., 1995, 15.
209. (a) A.H. Tullo *Chem. Eng. News*, Apr. 24, 2006, 24; (b) S. Rovner, *Chem. Eng. News*, Oct. 1, 2007, 11.
210. M. Maiss and C.A.M. Brenninkmeijer, *Environ. Sci. Technol.*, 1998, *32*, 3077.
211. (a) P.S. Zurer, *Chem. Eng. News*, Dec. 4, 1995, 26. (b) B. Hileman, *Chem. Eng. News*, July 20, 1998, 13; Dec. 21, 1998, 27.
212. P. Zurer, *Chem. Eng. News*, Aug. 25, 1997, 8.
213. P. Short, *Chem. Eng. News*, Oct. 12, 2001, 6.
214. (a) D. Zehavi, J.N. Seiber, and C. Wujcik, Preprints A.C.S. *Div. Environ. Chem.*, 1995, *35*(1), 190; (b) Anon., *Chem. Ind. (London)*, 1995, 581, (c) E. Wilson, *Chem. Eng. News*, July 31, 1995, 4; (d) C.E. Wujcik, D. Zehavi, and J.N. Seiber, Preprints A.C.S. *Div. Environ. Chem.*, 1997, *37*(1), 31.
215. (a) L.E. Manzer and V.N.M. Rao, *Adv. Catal.*, 1993, *39*, 329; (b) L.E. Manzer in, J.R. Kosak, and T.A. Johnson, eds, *Catalytic Organic Reactions*, Marcel Dekker, New York, 1994, 411; (c) B.E. Smart and R.E. Fernandez, *Kirk–Othmer Encyclopedia of Chemistry Technology*, 4th ed., Wiley, New York, 1994, *11*, 499; (d) L.E. Manzer. In: P.T. Anastas and C.A. Farris, eds, *Benign by Design*, A.C.S. Symp., 577, New York, Washington, DC, 1994, *144*, 152.
216. (a) S. Brunet, C. Batiot, and M. Calderon, *J. Mol. Catal. A*, 1996, *108*, 11; (b) C. Batiot, S. Brunet, J. Barrault, and M. Blanchard, *J. Chem. Soc. Chem. Commun.*, 1994, 867; (c) E. Kemnitz, A. Kohne, I. Grohmann, A. Lippitz, and WES Unger, *J. Catal.* 1996, *159*, 270; (d) E. Kemnitz, A. Hess, G. Rother, and S. Troyanov, *J. Catal.* 1996, *159*, 332.
217. (a) J.G. Lee and R.W. Flumerfelt, *J. Appl. Polym. Sci.*, 1995, *58*, 2213; (b) M. McCoy, *Chem. Eng. News*, Aug. 17, 1998, 18.
218. Presidential Green Chemistry Challenge Award Recepients, U.S. EPA 744-K-96-001, July 1996, 3 (see U.S. patents 5,250,577 and 5,266,605).
219. R.W. Greer and G.L. Wilkes, *J. Appl. Polym. Sci.*, 1996, *62*, 1115.
220. U.S.E.P.A., *Presidential Green Chem. Award Entries*, 2008, 43 Foam Supplies, Inc.
221. Anon., *Chem. Eng. News*, Nov. 1, 2004, 16.
222. A.A. Stroeks and L.W. Steelbakkers, *Chem. Abstr.*, 1995, *123*, 257946.
223. J.B. Durkee, *The Parts Cleaning Handbook Without CFC's—How to Manage The Change.* Hanser-Gardner, Cincinnati, OH, 1994.
224. S.S. Seelig, *Today's Chemist at Work*, 1996, *5*(6), 48.
225. K.I. Papathomas and A.C. Bhatt, *J. Appl. Polym. Sci.*, 1996, *59*, 2029.
226. G. Parkinson, *Chem. Eng.*, 1996, *103*(7), 17.
227. G. Ondrey, *Chem. Eng.*, 1997, *104*(7), 21.
228. M.L.H. Green, R.M. Lago, and S.C. Tsang, *J. Chem. Soc., Chem. Commun.*, 1995, 365.
229. J. de Angelis and A. Heyduk, *Chem. Eng. News*, Mar. 11, 1996, 4.
230. *Chem. Eng. News*, Aug. 11, 1997, 35.
231. J. Burdeniuc and R.H. Crabtree, *Science*, 1996, *271*, 340.
232. A. Wiersma, E.J.A.X. van de Sandt, M. Makkee, C.P. Luteijn, H. van Bekkum, and J.A. Moulijn, *Catal. Today*, 1996, *27*, 257.
233. N. Sonoyama and T. Sakata, *Environ. Sci. Technol.*, 1998, *32*, 375.
234. (a) G.S. Samdani, *Chem. Eng.*, 1995, *102*(3), 25; (b) *Chem. Eng. News*, Feb. 13, 1995, 25; (c) E. Cook, *Chem. Ind. (London)*, 1995, 204.
235. J.T. Staley and W. Haupin, *Kirk–Othmer Encyclopedia of Chemical Technology*, 4th ed., Wiley, New York, 1992, *2*, 191.
236. *R&D (Cahners)*, 1998, *40*(10), 159.
237. G. Ondrey, *Chem. Eng.*, 2005, *112*(7), 18.
238. D.T. Chen, M.M. David, G.V.D. Tiers, and J.N. Schroepfer, *Environ. Sci. Technol.*, 1998, *32*, 3237.
239. E. Kirschner, *Chem. Eng. News*, July 24, 1995, 28.
240. (a) C.L. Hartz, J.W. Bevan, M.W. Jackson, and B.A. Wofford, *Environ. Sci. Technol.*, 1998, *32*, 682; (b) G. Ondrey, *Chem. Eng.*, 2003, *110*(7), 22.
241. G. Ondrey, *Chem. Eng.*, 1999, *106*(1), 21.
242. (a) E.A. Parsons and O. Greene, *Environment*, 1995, *37*(2), 18; (b) P. Zurer, *Chem. Eng. News*, Mar. 13, 1995, 8.
243. *Chem. Eng. News*, Mar. 2, 1998, 27.
244. (a) M. Freemantle, *Chem. Eng. News*, Sept. 19, 1994, 29; Jan. 30, 1995; (b) A.W. Miziolek and W. Tsang, eds, *Halon Replacements: Technology and Science*, A.C.S. Symp. 611, Washington, DC, 1995; (c) A.E. Finnerty and following papers, Preprints A.C.S. *Div. Environ. Chem.*, 1994, *34*(2), 757–807.
245. R. Dagani, *Chem. Eng. News*, Mar. 3, 2003, 44.
246. *Chem. Eng. News*, Sept. 28, 1998, 19.
247. A.M. Thayer, *Chem. Eng. News*, Apr. 27, 1998, 15.
248. (a) P. Zurer, *Chem. Eng. News*, Nov. 14, 1994, 29; (b) C.H. Bell, N. Price, and B. Chakrabarti, eds, *The Methyl Bromide Issue*, Wiley, New York, 1996.
249. *Chem. Eng. News*, Oct. 30, 1995, 14.
250. J. Gan, S.R. Yates, S. Papiernik, and D. Crowley, *Environ. Sci. Technol.*, 1998, *32*, 3094.
251. (a) J. Gan, S.R. Yates D. Wang, and W.F. Spencer, *Environ. Sci. Technol.*, 1996, *30*, 1629; (b) D. Wang, S.R. Yates, F.F. Ernst, J. Gan, F. Gao, and J.O. Becker, *Environ. Sci. Technol.*, 1997, *31*, 3017; (c) D. Wang, S.R. Yates, F.F. Ernst, J. Gan, and W.A. Jury, *Environ. Sci. Technol.*, 1997, *31*, 3686; (d) S.R. Yates, D. Wang, F.F. Ernst, and J. Gan, *Environ. Sci. Technol.*, 1997, *31*, 1136.
252. J. Gan, S.R. Yates, J.O. Becker, and D. Wang, *Environ. Sci. Technol.*, 1998, *32*, 2438.
253. J.N. Seiber, J.A. Knuteson, J.E. Woodrow, N.L. Wolfe, M.V. Yates, and S.R. Yates, eds, *Fumigants: Environmental Behavior, Exposure and Analysis*, A.C.S. Symp. 652, Washington, DC, 1996.

254. G.S. Samdani, *Chem. Eng.*, 1994, *101*(10), 21.
255. *Chem. Ind. (London)*, 1995, 995.
256. (a) P. Morse, *Chem. Eng. News*, Oct. 26, 1998, 9; Nov. 9, 1998, 46; (b) B. Hileman, *Chem. Eng. News*, Jan. 17, 2005, 30.
257. *Chem. Ind. (London)*, 1997, 894.
258. (a) *Chem. Eng. News*, Aug. 7, 1995, 22; Jan. 29, 1996, 18; (b) P. Zurer, *Chem. Eng. News*, Dec. 18, 1995, 8.
259. G. Parkinson, *Chem. Eng.*, 1997, *104*(12), 46.
260. J. Beard, *New Sci.*, June 1, 1996, 24.
261. L. McGraw, *Agric. Res.*, 1998, *46*(7), 21.
262. A.K. Dowdy, *Agric. Res.*, 1998, *46*(7), 22.
263. P.J. Skerrett, *Technol. Rev.*, 1997, *100*(8), 28.
264. I.R. Shankland, Preprints A.C.S. *Div. Environ. Chem.* 1994, *34*(2), 743–755.
265. Anon., *Chem. Eng. News*, September 3, 2001, 68.
266. (a) *Environ. Sci. Technol.*, 1994, *28*, 171Aj; (b) C. Millais, *Chem. Ind. (London)*, 1994, 484; (c) E. Ayres and H. French, *World Watch*, 1996, *9*(5), 15.
267. B. Hileman, *Chem. Eng. News*, Aug. 3, 1998, 33.
268. (a) A. Shanley, *Chem. Eng.*, 1997, *104*(11), 63; (b) B. Hileman, *Chem. Eng. News*, Nov. 30, 1998, 21.
269. (a) S. Haaf and H. Henrici, *Ullmann's Encyclopedia of Industrial Chemistry*, 5th ed., VCH, Weinheim, 1988, *B3*, 19–21; (b) J. Beard, *New Sci.*, May 3, 1993, 17; (c) A. de Lucas, M. Donate, and J.F. Rodriquez, *Ind. Eng. Chem. Res.*, 2007, *46*, 345; (d) L.G. Gordeeva, A. Frent, G. Restuccia, and Y.I. Aristov, 2007, *46*, 2747.
270. (a) T. Kamiya, *Chem. Eng.*, 1999, *106*(11), 58; (b) M.S. Reisch, *Chem. Eng. News*, Nov. 17, 2008, 35.
271. (a) T.M. Tritt, *Science*, 1996, *272*, 1276; (b) S. Half and H. Henrici, *Ullmann's Encyclopedia of Industrial Chemistry*, 5th ed., VCH, Weinheim, 1988, *B3*, 19–25; (c) S.L. Wilkinson, *Chem. Eng. News*, Apr. 3, 2000, 31.
272. (a) R.F. Service, *Science*, 2007, *317*, 1318; (b) 2006, *311*, 1860; (b) L.E. Bell, *Science*, 2008, 321, 1457.
273. B. Neese, B. Chu, S.-G. Lu, Y. Wang, E. Furman, and Q.M. Zhang, *Science*, 2008, *321*, 821.
274. (a) D. Pendick, *New Sci.*, Sept. 10, 1994, 21. (b) S.L. Wilkinson, *Chem. Eng. News*, May 31, 1999, 5; (c) G.J. Miller, *Chem. Soc. Rev.*, 2006, *35*, 799; (d) V.K. Pecharsky and K.A. Gschneidner, Jr., *Mater. Matter. (Sigma-Aldrich)*, 2007, *2*(4), 4.
275. (a) J. Glanz, *Science*, 1998, *279*, 2045; (b) *R&D (Cahners)*, 1998, *40*(12), 7.
276. (a) D. Mackenzie, *Science*, 1997, *278*, 2060; (b) S.L. Wilkinson, *Chem. Eng. News*, May 31, 1999, 5; (c) S. Garrett and S. Backhaus, *Am. Sci.*, 2000, *88*, 516.
277. P.L. McCarthy, *Science*, 1997, *276*, 1521.
278. EPA, *Guide to Cleaner Technology—Cleaning and Degreasing Process Changes*, U.S. Environmental Protection Agency, Washington, DC, EPA/625/R-93/017, Feb. 1994.
279. A. Farrell, *Delaware Estuary News*, 1996, *6*(3), 6.
280. (a) EPA, *Guide to Cleaner Technologies—Alternatives to Chlorinated Solvents for Cleaning and Degreasing*, U.S. Environmental Protection Agency, Washington, DC, EPA/625/R-93/016, Feb. 1994; (b) D. Pendick, *New Sci.*, Sept. 10, 1994, 21; (c) M. McCoy, *Chem. Eng. News*, Nov. 14, 2005, 19.
281. (a) H. Stromberg, *Rubber World*, 1995, *212*(5), 14; (b) *Chem. Eng.*, 1997, *104*(11), 107.
282. (a) Anon., *Pollut. Equip. News*, Aug. 1998, 16; Equipment from Ice Cleaning Systems, Indianapolis, IN; (b) Anon., *Paper Technol.*, 2002, *43*(8), 116.
283. (a) V. Leclair, *Environ. Sci. Technol.*, 1997, *31*, 215A; (b) S.H. Lusi, *Chemtech*, 1998, *28*(6), 56.
284. P.M. Randall, *Engineers' Guide to Cleaner Production Technologies*, Technomic, Lancaster, PA, 1997.
285. (a) *R&D (Cahners)*, 1997, *39*(9), 37; (b) Global Technologies "Drywash", Bridgeport, CT.
286. D. Klamann, *Ullmann's Encyclopedia of Industrial Chemistry*, 5th ed., VCH, Weinheim, 1990, *A15*, 480.
287. R. Komanduri, *Green Chemistry and Chemical Engineering Conference*, Washington, DC, June 23, 1997.
288. J. Knight, *New Sci.*, Mar. 21, 1998, 7.
289. (a) Presidential Green Chemistry Challenge Award Recipients, U.S. EPA 744-K-97-003, Sept. 1997, 4; (b) R.R. Mathews, U.S. patent 5,464,480, 1995.
290. (a) EPA, *Summary of a Report on Multipurpose Wet Cleaning*, U.S. Environmental Protection Agency, Washington, DC, EPA744-S-94-001, June 1994; (b) H. Black, *Environ. Sci. Technol.*, 1996, *30*, 284A; (c) *Environment*, 1995, *37*(1), 22; (d) D. Sinsheimer, R. Gottlieb, and C. Farrar, *Environ. Sci. Technol.*, 2002, *36*, 1649.
291. F. Case, *Inform*, 2006, *17*(9), 559.
292. J. Huff, J. Bucher, and J.C. Barrett, *Science*, 1996, *272*, 1083.
293. (a) *Chem. Ind. (London)*, 1995, 362; (b) M. Baird and F. Barsan, *J. Chem. Soc., Chem. Commun.*, 1995, 1065; (c) V. Touchard, R. Spitz, C. Boissaon, and M.-F. Llauro, *Macromol. Rapid Commun.*, 2004, *25*, 1953; (d) S. Garratt, A. Guerrero, D.L. Hughes, and M. Bochmann, *Angew. Chem. Int., Ed.*, 2004, *43*, 2166.
294. *Chem. Eng. News*, June 26, 1995, 14.
295. G.S. Samdani, *Chem. Eng.*, 1995, *102*(5), 15.
296. N. Galvez, M. Moreno-Manas, R.M. Sebastian, and A. Vallribera, *Tetrahedron*, 1996, 52, 1609.
297. (a) R. Dagani, *Chem. Eng. News*, July 14, 1998, 14; (b) W.A. Jacoby, P.C. Maness, E.J. Wolfrum, D.M. Blake, and J.A. Fennell, *Environ. Sci. Technol.*, 1998, *32*, 2650.
298. (a) M.K. Stern, *J. Org. Chem.*, 1994, *59*, 5627; (b) R.A. Sheldon, *Chemtech*, 1994, *24*(3), 38.
299. (a) M.K. Stern, B.K. Cheng, and J. Clark, *New J. Chem.*, 1996, *20*, 259; (b) L.R. Raber, *Chem. Eng. News*, July 6, 1998, 25.
300. N. Takano, *Chem. Lett.*, 1996, 85.
301. *Chem. Eng. News*, June 22, 1998, 11.
302. A.H. Tullo and P.L. Short, *Chem. Eng. News*, Oct. 9, 2006, 22.
303. T. Hayashi, K. Tanaka, and M. Haruto, *J. Catal.*, 1998, *178*, 566.
304. I. Yamanaka, K. Nakagika, and K. Otsuka, *Appl. Catal. A*, 1998, *171*, 309.
305. R.K. Dieter, L.E. Nice, and S.E. Velu, *Tetrahedron Lett.*, 1996, *36*, 2377.
306. (a) G.W. Kalbalka and K. Yang, *Synth. Commun.*, 1998, *28*, 3807; (b) B.P. Bandgar and N.J. Nigal, *Synth. Commun.*, 1998, *28*, 3225.
307. B. Hache, J.-S. Duceppe, and P.L. Beaulieu, *Synthesis*, 2002, 528.
308. M.S. Stephan, A.J.J.M. Teunissen, G.K.M. Verzijl, and J.G. de Vries, *Angew. Chem. Int. Ed.*, 1998, *37*, 662.
309. M. Heylin, *Chem. Eng. News*, Nov. 21, 1994, 5.

310. M. Oehme, J.-E. Haugen, and M. Schlabach, *Environ. Sci. Technol.*, 1996, *30*, 2294.
311. C. Holden, *Science*, 1998, *280*, 2053.
312. R. Addink, R.H.W.L. Paulus, and K. Olie, *Environ. Sci. Technol.*, 1996, *30*, 2350.
313. (a) H.Y. Pan, R.G. Minet, S.W. Benson, and T.T. Tsotsis, *Ind. Eng. Chem. Res.*, 1994, *33*, 2996; (b) E. Wilson, *Chem. Eng. News*, Sept. 11, 1995, 9.
314. R. Lundquist, *Chem. Eng. News*, Nov. 28, 1994, 3.
315. G.S. Samdani, *Chem. Eng.*, 1995, *102*(2), 13.
316. (a) D.E. Mouzakis, M. Gahleitner, and J. Karger-Kocsis, *J. Appl. Polym. Sci.*, 1998, *70*, 873; (b) Y. Hu, M.T. Krejchi, C.D. Shah, C.L. Myers, and R.M. Waymouth, *Macromolecules*, 1998, *31*, 6908.
317. Elf Atochem (now Arkema), 1996 trade brochure, Bridgeport, CT.
318. A.J. Becerra, R. Martinez, M.C. Huff, and M.A. Barteau, *Catal. Today*, 2005, *107–108*, 244.
319. (a) E.M. Kirschner, *Chem. Eng. News*, Jan. 9, 1995, 11; Jan. 1, 1996, 12; (b) B. Hileman, J.R. Long, and E.M. Kirschner, *Chem. Eng. News*, Nov. 21, 1994, 12; (c) P.L. Layman, *Chem. Eng. News*, May 6, 1996, 22; (d) *Chem. Eng. News*, July 17, 1995, 30; (e) N. Botha, *Chem. Ind. (London)*, 1995, 832; (f) *Chem. Eng. News*, Oct. 16, 1995, 8; (g) *Chem. Eng. News*, Apr. 22, 1996, 19; (h) M.S. Reisch, *Chem. Eng. News*, Feb. 13, 1995, 15; (i) P.M. Morse, *Chem. Eng. News*, Oct. 20, 1997, 19; (j) M. McCoy, *Chem. Eng. News*, Sept. 7, 1998, 17; Nov. 23, 1998, 26; (k) C. Martin, *Chem. Br.*, 1997, *33*(5), 44; (1) M. Beal, *Chem. Ind.*, 1997, 434; (m) Anon., *Chem. Eng. News*, Sept. 8, 2008, 21.

RECOMMENDED READING

1. G. Graff, *Technol. Rev.*, 1995, *98*(1), 54; the chlorine controversy.
2. G.W. Gribble, *Environ. Sci. Technol.*, 1994, *28*, 310A; review of natural chlorine compounds.
3. D. Henschler, *Angew. Chem. Int. Ed.*, 1994, *33*, 1920; toxicity of chlorine compounds.
4. B. Hileman, J.R. Long, and E.M. Kirschner, *Chem. Eng. News*, Nov. 21, 1994, 12–26; uses of chlorine compounds and the controversy.
5. R. Stone, *Science*, 1994, *265*, 308; estrogen mimics and the chlorine controversy.
6. E. Ayres and H. French, *World Watch*, 1996, *9*(5), 15; propane/isobutane blends in European refrigerators.
7. H.Greim, *Angew. Chem. Int. Ed.*, 2005, *44*, 5568; endrocrine disruptors.

EXERCISES

1. Look up the fish advisories in your area to see how many are due to polychlorinated biphenyls and other chlorinated compounds.
2. In your laboratory work for the past month, how many times have you run substitution reactions on halides or produced waste halide salts? In looking back on the work, can you think of alternatives that would avoid the use of halogens?
3. How does your city disinfect its drinking water and its treated wastewater?
4. Are there any places in your areas where the groundwater is contaminated with trichloroethylene, perchloroethylene, chlorinated herbicides, or chlorinated insecticides? If so, do any of them require treatment?
5. Where is dioctyl phthalate used as a plasticizer? Devise a copolymer made with a plasticizing comonomer that would eliminate the need for the phthalate.
6. Compare the properties of the newer polyolefins made with metallocene (W. Kaminsky and M. Arndt, *Adv. Polym. Sci.*, 1997, *127*, 143) and Brookhart catalysts with those of plasticized polyvinyl chloride.
7. Compare the properties of epoxy resins based on bisphenol A–epichlorohydrin with those based on epoxides that can be made without the use of epichlorohydrin.
8. Devise a potentially commercial method of preparing aluminum by electrolysis that does not involve any fluoride.
9. Visit the local supermarket and hardware store. See how many items are marked with the polyvinyl chloride symbol.
10. How could polyvinylidene chloride barrier coatings in food packaging be replaced? If you have trouble coming up with an answer, see G. Ondrey. *Chem. Eng.*, 1998, *105*(1), 29; and H.C. Silvis. *Trends Polym. Sci.*, 1997, *5*(3), 75.

CHAPTER 4

Toxic Heavy-Metal Ions

4.1 THE PROBLEM

The toxicities of heavy metal ions are well known.[1] These include cadmium, chromium, cobalt, copper, lead, mercury, tin, and zinc. For convenience, nonmetals such as arsenic, asbestos, and selenium are also included here. Mercury-containing wastes dumped into Minamata Bay in Japan killed hundreds of people and sickened many more.[2] Hundreds of deaths also occurred in Iraq when people ate grain treated with an organomercury biocide intended to protect the young seedlings from fungi.[3] The expression "mad as a hatter" originated when hatters got poisoned with the mercury compounds used in the felting of hats. Some authors have postulated the poisoning of modern man from silver amalgam dental fillings, much as lead poisoning from the use of lead cooking and storage vessels may have led to the decline of the Roman empire.[4] Dentists are being encouraged to separate the mercury from waste amalgam before it goes down the drain.[5] In 1995, fishermen in the United States were advised not to eat the fish that they caught in 1306 different bodies of water because of the mercury in them.[6] This figure should be compared with 438 advisories for polychlorinated biphenyls, 122 for chlordane, 52 for dioxins, and 35 for DDT. Game fish may accumulate mercury to 225,000 times the concentration in the surrounding water.[7] There is enough mercury in the Florida Everglades to poison some wading birds that feed on fish.[8] The amount of mercury that people should be allowed to eat in fish is a matter of current debate.[9]

Hundreds of people in Japan suffered from a degenerative bone disease called itai-itai because they drank water containing cadmium.[10] They lived downstream from a mine and smelter that produced zinc and cadmium. In West Bengal, India, where 800,000 people drink well water containing over 50 μg/L arsenic, 200,000 people developed skin lesions from drinking the water.[11] In Bangladesh, 70 million people drink well water containing arsenic.[12] The problem arose when international agencies drilled deeper wells to avoid contaminated surface water, but did not test the water. Too much arsenic in drinking water is known to result in more deaths from cancer.[13] In a region in northern Chile, arsenic-containing water caused highly elevated rates of lung and bladder cancers.[14] Selenium can also be bioaccumulated. Waterfowl in refuges in the western United States have been poisoned by agricultural drainage from seleniferous soils.[15]

Contamination of the environment with toxic heavy metals began long ago. Evidence of the Roman use of copper and lead can be found in ice cores from Greenland and in cores of European bogs.[16] Mining of gold and silver has left a legacy of mercury at mining sites.[17] The Carson River Superfund site in Nevada is estimated to release 150–400 kg/year of mercury, which is comparable with the 300 kg/year released by a 1000-MW coal-fired power plant. Mercury vapor can persist in the atmosphere for as much as a year. Households can also be sources of metals. Tetraethyllead was used as an antiknock agent in gasoline for about 50 years, despite the fact that the dangers of lead were known at the time.[18] Both ethanol and tetraethyllead prevented engine knocking, but only the latter could be patented. It has now been banned in the European Union and in North America because it can lower the IQ of children.[19] Harmless titanium dioxide has replaced basic lead carbonate as a white pigment in paint. As a result of these two uses, unacceptable levels of lead persist in some soils and older houses. In England, roughly 60% of the copper in the sewage came from the plumbing in houses.[20] Galvanized tanks accounted for 50% of the zinc in the sewage system. Highway runoff in Ohio contains zinc and cadmium from tire crumbs; copper and lead from brake dust; zinc, chromium, copper, and magnesium from engines; as well as oil and grease.[21] Streets in Stockholm, Sweden, receive appreciable amounts of cadmium, copper, lead, antimony, and zinc from wear on brake linings and tire tread.[22]

The form or valence of the metal can make a big difference in its toxicity.[23] The tin plate on a food can is essentially harmless. Yet, tributyltin derivatives are used to kill barnacles on ships. Mercury amalgam fillings for teeth are relatively harmless, although there is some debate on their effects over long times (as mentioned earlier). On the other hand, a Dartmouth professor died after getting a couple of drops of dimethylmercury on gloves that allowed it to penetrate to her skin.[24] Chromium(VI) is carcinogenic, but chromium(III) is an essential element for human nutrition. Chromium, cobalt, copper, fluoride, manganese, and zinc are among the 27 elements essential for humans. The metal ions are often found in enzymes. The amounts required may be small. The average adult human contains about 2–3 g of zinc.[25] It is easy to be overdosed with materials such as zinc, fluoride, or even iron, if there is too much in the diet or in the drinking water. People who take large amounts of dietary supplements are at the most risk. If a metal compound has a very low solubility in water and in body fluids and is not volatile, then it may not be a problem at all. Much work has been done to determine the speciation of metals in soil, water, combustion gases, and others.[26] Nickel is also carcinogenic, apparently by catalyzing the oxidation of bisulfite ion to monoperoxysulfate ion, which is then converted to sulfate radical, which, in turn, attacks the DNA.[27] Nickel in jewelry is the most common cause of contact allergies in several European countries.[28] Arsenic, cadmium, lead, and mercury are more toxic than the other foregoing metals and are not essential for humans.

The reader would do well to check the extent of contamination by toxic heavy metal ions of the area where he or she lives. Delaware will be used as an example to illustrate common sources. The Delaware River Estuary is contaminated with arsenic, cadmium, chromium, copper, lead, mercury, and silver.[29] The arsenic, chromium, copper, and lead are largely from point sources. Dredge spoil disposal is a problem because it contains copper, mercury, and polycyclic aromatic hydrocarbons. The use of sewage sludge as a soil amendment is limited by the presence of toxic heavy metal ions. An example is Middletown, Delaware, where the lead in the sludge originates in a local battery plant. Zinc is a problem in Red Clay and White Clay Creeks below two former vulcanize fiber plants that used zinc chloride to soften cellulose. Although the releases have stopped, release of zinc from the sediments in Red Clay Creek keeps the stream contaminated. The east branch of the Maurice River above Millville, New Jersey, is contaminated with arsenic from an agricultural chemical plant upstream. Tetraethyllead was once made at the Chambers Works of the DuPont Co., across the river from Wilmington, Delaware. The company is studying phytoremediation by ragweed and other plants as a means of reducing the lead content of soils at the plant. Phytoremediation can work if the hyperaccumulating plant is collected and burned so that the reside can be sent to a metal refinery.[30] Chloramone Corp. is being taken to court for releases of mercury from its chlorine plant in Delaware City, Delaware.[31] Municipal solid waste from northern Delaware was burned one time at an incinerator in Chester, Pennsylvania. Incinerators operating at 1000–1100°C evaporate 98–100% of the cadmium, copper, lead, and zinc so that compounds of these elements end up in the fly ash, which must then be treated as hazardous waste.[32]

Materials containing these metals have been quite useful in many applications in the past. Before examining ways to prevent pollution from them, their uses must be considered. The following list has been put together from the relevant sections in the *Kirk-Othmer Encyclopedia of Chemical Technology*, *Ullmann's Encyclopedia of Industrial Chemistry*, and, for mercury, a paper by Randall:[33]

- Arsenic:
  - Chromium copper arsenate as a wood preservative
  - Calcium and sodium arsenate herbicides
  - Lead arsenate formerly used on fruit crops
  - Formerly in cotton defoliants
  - Sodium arsenite in cattle and sheep dips
  - Alloys with copper and lead for bearings
  - With aluminum, gallium, and indium in semiconductors
  - Gallium aluminum arsenide in solar cells
  - Gallium arsenide phosphide in light-emitting diodes
  - Gallium arsenide infrared detectors, lasers, and photocopiers
  - Arsanilic acid as a growth additive for poultry and swine

Arsenic compounds are also by-products of processing gold and lead ores and of coal combustion.

- Asbestos:
  - Matrix reinforcement in shingles for houses, vinyl floor tile, brake linings, and clutch facings
  - Insulation
  - Portland cement
  - Formerly in gloves and fire suits
- Cadmium:
  - Nickel–cadmium, silver–cadmium, and mercury–cadmium batteries
  - Plating
  - Pigments such as sulfide and selenide
  - Cadmium laurate with barium laurate as stabilizers for polyvinyl chloride
  - Semiconductors, such as sulfide and selenide, as in solar cells
  - Phosphor in television tubes
  - Brazing and low-melting alloys (e.g., in copper)

Some are also released from the burning of coal and the use of phosphate fertilizers.

- Cobalt:
  - With molybdenum in desulfurization catalysts
  - Catalyst for the oxo process, *p*-xylene oxidation, and others
  - Electroplating
  - Salts of fatty acids as driers for paints
  - Color for glass
  - Helps rubber adhere to a steel tire cord
  - Formerly a foam stabilizer in beer where it caused heart problems and some deaths

- Chromium:
  - Electroplating and chromizing steel.
  - Stainless steel.
  - Chromite grain in foundries.
  - Chromium dioxide in magnetic tape.
  - Pigments, such as chromium oxide and lead chromate.
  - Zinc chromate for corrosion protection of cooling towers, and others.
  - Tanning of leather.
  - Dyeing of textiles.
  - Chromium copper arsenate as a wood preservative (used 62% of the chromic acid in the United States).
  - Catalyst for polymerization of ethylene.
  - Copper chromite catalyst for hydrogenation.
  - Zinc chromate near the zinc anode gives batteries 50–80% more shelf life.
  - Coal combustion is the largest source of chromium in the environment.
- Copper:
  - Wiring, printed circuits
  - Plumbing
  - Roofing
  - Auto radiators
  - Coins
  - Algicide in swimming pools and reservoirs
  - Fungicides
  - Electroplating
  - Solar collectors
  - Chromium copper arsenate wood preservative
  - Cuprous oxide in antifouling coatings for ships
  - Pigments, such as copper phthalocyanine
- Lead:
  - Pipe and cable sheathing.
  - 60% of the total goes into lead–acid batteries.
  - 14% of the total goes into lead glazes for glass and china. (Lead can leach from pieces that are not fired properly.[34])
  - Stabilizers for polyvinyl chloride.
  - Pigments such as basic lead carbonate (in white house paint), red lead (in primer paint for steel), lead chromate.
  - Shielding against x-rays.
  - Cathode ray tubes.
  - Chemically resistant linings.
  - Solder.
  - Tetraethyllead antiknock agent for gasoline.
  - Lead shot, ammunition, yacht keels, wheel weights sinkers for fishing.
  - Lead azide is a detonator for explosives.
  - Lead shot and sinkers have poisoned waterfowl that have ingested them.
- Mercury:
  - Fungicide in seed treatment
  - Slimicide in paper mills
  - Biocide in latex paints
  - Dental amalgams
  - Batteries
  - Catalysts for transesterification of vinyl acetate as well as addition to triple bounds
  - Thermometers and leveling bulbs
  - Mercury vapor and fluorescent lamps
  - Electrical switches
  - Mercury cathode for electrolysis of sodium chloride in the manufacture of chorine and sodium hydroxide
  - Pigments, such cadmium mercury sulfide
  - Mercury fulminate detonator for explosives

Coal burning and incinerators are the largest sources.[35] Mercury metal boils at 357°C so that it escapes from the stacks. Even crematoria are sources, the mercury coming from the fillings in the teeth.[36]

- Nickel:
  - Stainless steel
  - Other alloys with steel and other metals
  - Electroplating
  - Coins and jewelry
  - Nickel–cadmium batteries
  - Catalysts for hydrogenation
  - Light stabilizer in polyolefins

Some also come from burning coal. The smelting of nickel and copper sulfide ores has resulted in tremendous air pollution from sulfur dioxide. At one time this could be detected on white pines (a sensitive species) at a distance of 10 miles from the smelter at Sudbury, Ontario. INCO now uses a flash smelter, which roasts the ore with oxygen in a closed vessel that gives a concentrated stream of sulfur dioxide and saves half the energy. Thus, 90% of the sulfur in the ore can be recovered after conversion to sulfuric acid at the rate of 2300 tons/day. Even at this rate of recovery the facility emitted 265,000 tons of sulfur dioxide in 1996 and 235,000 tons in 1998.[37]

- Zinc:
  - Zinc chromate and phosphate for corrosion protection
  - Galvanizing
  - Nickel oxide–zinc batteries
  - Flaked zinc pigment
  - Electroplating
  - Alloys
  - Zinc dithiocarbamate fungicides and stabilizers
  - Treating cellulose to make vulcanized fiber
  - Phosphors
  - Zinc dithiophosphates in lubricating oils
  - Zinc oxide in tires
  - Opaque agent in some sunscreens

Various approaches are being studied in an effort to eliminate or minimize the problems with the toxic heavy metals while not doing away with the end uses. These will be discussed in the following sections. They include (a) alternative nonaqueous methods for which loss in water has been a problem, (b) the use of other less harmful metal or organic coatings, (c) the use of catalytic, rather than stoichiometric, amounts of reagents in the chemical process, (d) the use of a reagent on a support or in a separate phase so that none is lost, (e) reagents that contain no metal, (f) development of sure-fire collection and recycling techniques that do not allow the item containing the metal to enter the general waste stream, (g) separation and stockpiling in the case where the material is an unwanted by-product, and (h) regulation, and even banning, if the harm exceeds the benefits. Cadmium is a by-product of zinc mining. Arsenic is a by-product of other mining.[38] Both may have to be stockpiled. Organomercury biocides have been banned in latex paints, seed treatments, and other uses.[39] A New Jersey law is trying to lower the mercury content of batteries to 1 ppm. It also sets up a collection program for spent batteries. California and Minnesota restrict the disposal of fluorescent lights (each of which can contain up to 25–50 mg of mercury). Technologies exist for the recovery of the mercury in lamps (such as heating broken lamps to a high temperature under a vacuum).[40] A federal law for the United States that would phase out the use of mercury in disposable batteries has been proposed.[41] The U.S. Environmental Protection Agency (EPA) has proposed new standards that will reduce the emission of mercury from incinerators by 80% and lead and cadmium emissions by 95%.[42] The EPA has limited the air discharge of chromium compounds from decorative electroplating ranks to 0.015 mg/m^3 of air.[43] The exposure limit for asbestos is 0.1 fiber per cubic centimeter of air.[44] The World Health Organization (WHO) has set the limits on drinking water contaminants at (μg/L) the following rates: lead 10, nickel 20, copper 2000, boron 300, antimony 5, arsenic 10, trihalomethanes 100, bromate 25, atrazine 2, lindane 2, and simazine 2.[45] Atrazine and simazine are commonly used herbicides for corn and other crops. Lindane is a highly chlorinated insecticide. (Agrochemicals will be discussed in Chapter 11.) Witco (now Chemtura) voluntarily stopped the sale of cadmium-containing heat stabilizers for polyvinyl chloride on June 30, 1994.[46]

4.2 END-OF-THE-PIPE TREATMENTS

It would be preferable to eliminate the sources of these toxic heavy metal ions. However, in some cases this will be difficult. The switch from burning coal to generate electrical power to renewable sources of energy to moderate global warming will take many years. Losses of metal ions from mining processes such as flotation and hydrometallurgy will continue although perhaps at a reduced rate.[47] Surplus cadmium from the mining of zinc and arsenic from other mining will have to be dealt with. The leaching of arsenic and selenium from soils will continue to be problems unless changes in land use and other practices can reduce them. It will be necessary to find ways to eliminate them from effluent air and water.

The gasification of oil in the Shell process leaves the vanadium and nickel from the oil in the soot ash. Lurgi has devised a process to recover the vanadium in a form that can be used by the metal alloys industry.[48] The International Metals Reclamation Company (a subsidiary of INCO) operates a facility for the recycling of nickel–cadmium batteries in Pennsylvania. "Since 1978, Inmetco has recycled nearly 1.5 billion lb of material that might otherwise have been disposed of in landfills." The recovered cadmium goes back into new batteries.[49] A process recycles batteries containing mercury by heating to 650°C. The mercury volatilizes and is condensed in a wet scrubber. The residue of zinc, iron, and others can be recycled to standard smelters.[50] Zinc can be removed from galvanized steel by vaporization at 397°C/10 Pa.[51] Efforts are being made to extend the technique to removal of tin from copper wire, copper from circuit boards, and metal from batteries.

Mercury can be removed from flue gases by adsorption on activated carbon, with or without iodine or sulfur.[52] This may require cooling the gases, which would be unnecessary otherwise. Domestic waste averages about 3–5 g of mercury per ton. Most of the mercury ends up in the ash fraction, which must be dealt with separately. Ten percent sulfur on sepiolite (a silicate) is said to compare well with sulfur on activated carbon for removal of mercury in gas streams at 320 K.[53] A 4-A molecular sieve containing 1% silver can be used to clean up natural gas containing up to 300 ppb of mercury.[54] Another method uses a hollow-fiber membrane containing an oxidizing liquid, the mercuric ion being precipitated later as mercury(II) sulfide.[55] An older method used sulfur on activated carbon.

Metal workers clean up their wastes by a variety of methods, which include chemical precipitation, electrowinning, and ion exchange.[56] Many electroplating firms are now reducing the amount of wastewater by using the last rinse of one batch as the first rinse of the next batch. Chemical precipitation has some problems. If the precipitate is sent to a landfill, the metals in it are thrown away and may possibly be leached from the landfill. The precipitate may also be sent off-site to a metals recovery firm. Although precipitation can lower the metals content to acceptable levels in favorable cases, some metal ions are still left in the wastewater. Precipitation of mercury as the sulfide leaves 11 ppb of mercury in the effluent.[57] This is the process used at the chlor-alkali plant in Delaware City, Delaware. Since there can be tremendous bioaccumulation of mercury, this may not be enough to allow the fish in the river to be eaten. It would be better for the plant to switch to the well-established "Nafion" (a perfluorinated polymeric sulfonic acid made by DuPont) membrane process that requires no mercury.[58] Most chlor-alkali plants had switched from the mercury cell process by 2008. Governments are considering stockpiling the mercury so that it does not fall into the hands of gold miners in developing nations and end up contaminating the land there.[59] Ion exchange does not remove all the metal ions either. It can be quite slow at low levels of metal ions. At least, the recovered metal ions can be recycled to the head of the plant. Electrowinning is hard at very low levels of metal ions. Use of a carbon fiber electrode can help some. Electrolytic fluidized beds are said to help and to be cheaper than ion exchange.[60] Concentration of the wastewater by reverse osmosis before electrowinning would reduce the amount lost. Combining this with total water recycle would help. If the rinses are concentrated by reverse osmosis, it may be possible to recycle them to the process without electrowinning or ion exchange. This might require analysis of each plating bath before use to assure the proper concentrations of ingredients.

Asarco has devised a method of treating the water from a lead–zinc mine with sulfate-reducing bacteria that generate hydrogen sulfide *in situ* to precipitate the metal ions.[61] Arsenic can be precipitated from the effluent of gold extraction as $FeAsO_4 \cdot 2H_2O$.[62] Aluminum-loaded Shirasu zeolites can also remove arsenic from water.[63] A system for poor villagers in Bangladesh uses a two-stage system with sand and processed cast iron followed by a second bucket containing sand, charcoal, and brick chips.[64] Carboxymethyl-β-cyclodextrin can complex cadmium simultaneously with organic compounds such as anthracene, trichlorobenzene, and DDT.[65] It is suggested for the cleanup of leachates from sites containing mixed hazardous waste. A zeolite containing a hexadecyltrimethylammonium salt has been used to pick chromate, molybdate, and selenate ions from water.[66] Chromate ions can be removed selectively in the presence of sulfate, chloride, bicarbonate, and nitrate ions, with a cross-linked resin containing dipicolylamine copper(II).[67] Nalco markets a polymeric chelating agent (NALMET) that removes copper, lead, nickel, and zinc ions from wastewater to lower levels than can be obtained by precipitation processes.[68] Presumably, the metal ions can be recovered for reuse. Lezzi et al. have devised some polymeric sulfur-containing reagents for nearly complete removal of lead and mercury ions from water (**4.1** Schematic and **4.2** Schematic).[69]

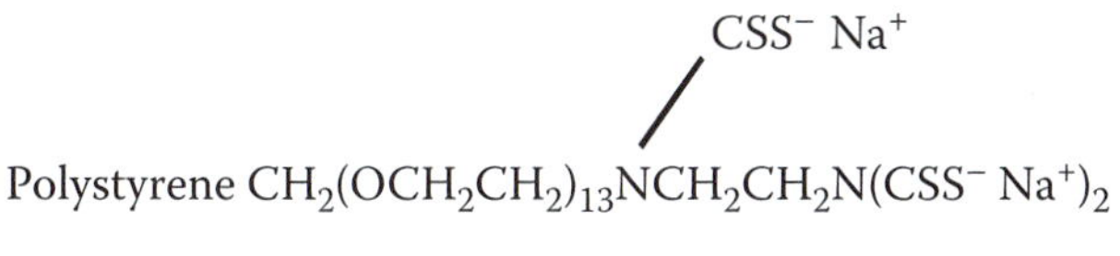

$$\text{Polystyrene } CH_2(OCH_2CH_2)_{13}N(CSS^- Na^+)CH_2CH_2N(CSS^- Na^+)_2$$

4.1 Schematic

$$\text{Polystyrene } CH_2(OCH_2CH_2)_{11}OCH_2CH_2SH$$

4.2 Schematic

The former can reduce lead and mercury ions from 18–19.5 ppm to 0.005 and 0.05 ppm, respectively, in 7 h. The latter can take mercury ion from 20 ppm to less than 10 ppb in 2 h. Unless much shorter times can be used, these may be impractical for commercial use. A dithiol can reduce mercury from 34.5 to 0.008 ppm in 15 min (**4.3** Schematic).[70]

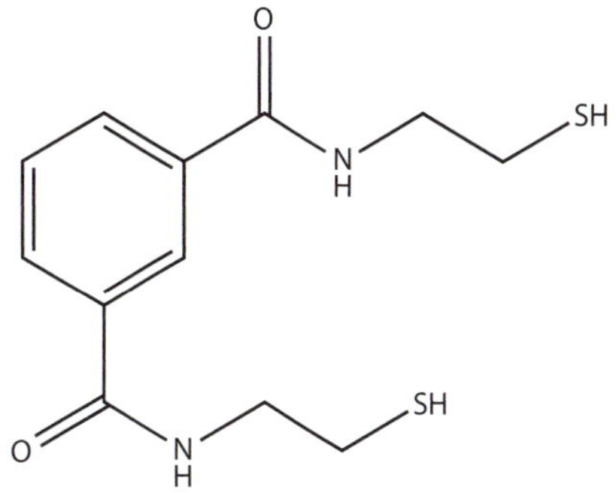

4.3 Schematic

More selective chelating agents will help improve separations of metal ions from each other and from waste streams. These could aid the recovery of metals from mining wastes and thereby reduce the presence of metal ions in the wrong places. For example, they may help separate the metal ions obtained by heap leaching of tailings, as done with sulfuric acid on copper ores. IBC Advanced Technologies, Provo, Utah, has attached macrocyclic ligands of the crown ether type to silica and titania with a spacer to produce "SuperLig."[71] These can separate lead from tin and zinc, mercury from sulfuric acid solutions, and lead from water. They are useful with trace amounts of metal ions and can be used over again. One such macrocyclic tetracarboxylic acid selects lead over zinc by a factor of 10^6 (**4.4** Schematic). (These metals are often found together in ores.[72])

Nickel and cobalt also occur together in ores. A new separation uses a mixture of trioctyl and tridecylamines in kerosene in the pores of microporous poly(tetrafluoroethylene).[73] Cobalt and copper ions pass through the membrane, but nickel ions do not. A long-sought goal of chemists is to find chelating agents that are specific for each metal ion. Progress is being made toward this goal. Tsukube et al. have a ligand that is "perfectly" selective for silver ion over lead, copper, cobalt, zinc, and nickel ions (**4.5** Schematic).[74]

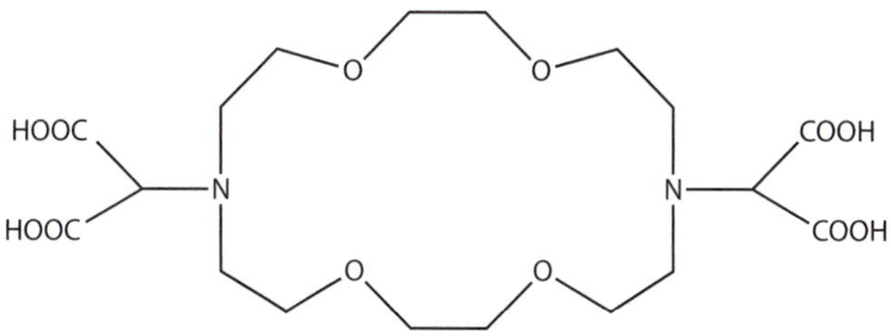

4.4 Schematic

A cross-linked polystyrene polymer containing thiacrown ether groups is able to remove over 99% of Hg(II) from dilute aqueous solutions in 30 min, selectively, in the presence of other metal ions (**4.6** Schematic).[75]

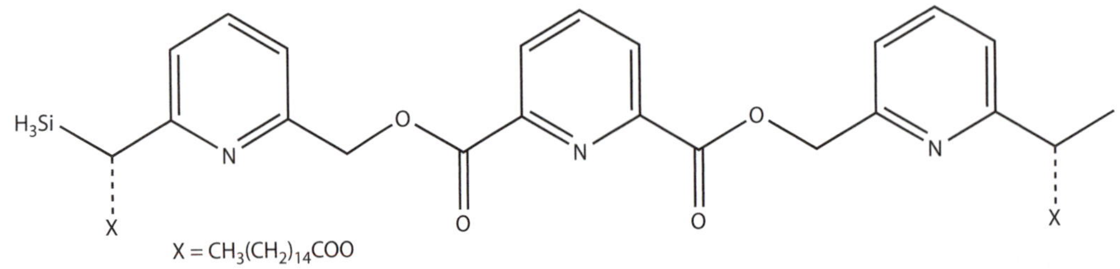

$X = CH_3(CH_2)_{14}COO$

4.5 Schematic

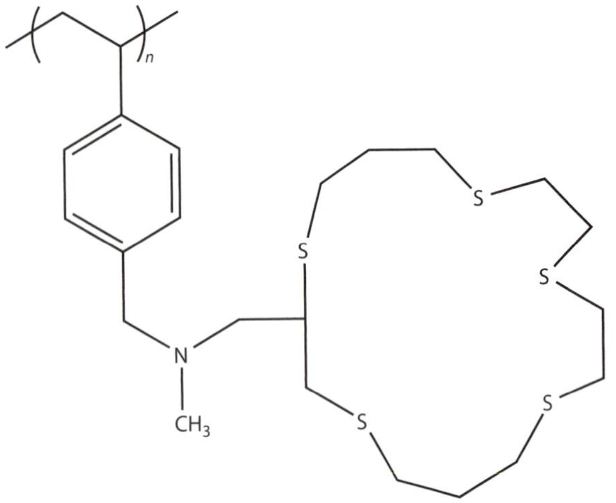

4.6 Schematic

A mercaptopropyl nanoporous silica extracts Hg(II) but not Cd(II), Pb(II), Zn(II), Co(II), Fe(III), Cu(II), or Ni(II).[76] The pore size may play a role in the selectivity.

A calixarene dithiocarbamate can separate mercury and gold in the presence of lead, cadmium, nickel, and platinium (**4.7** Schematic).[77]

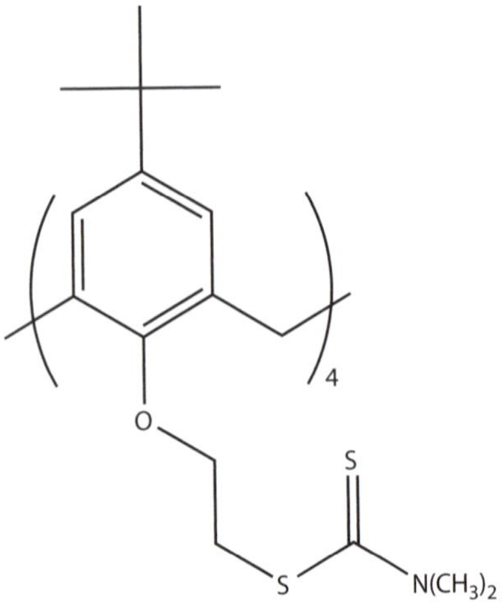

4.7 Schematic

Another calixarene, with long alkyl side chains and eight phosphate groups, has been used to recover uranium from seawater in 68–94% yields.[78] Seawater contains 3 ppb of uranium. The resin could be recycled several times. Reaxa sells a series of microporous and macroporous polymeric scavengers for traces of metal ions under the "QuadraPure" name. They contain thiourea, iminodiacetate, aminomethylphosphonic acid, *o*-aminobenzenethiol, ethylenediamine, and aminopropylimidazole groups.[79]

4.3 BIOCIDES

Biocides are used to prevent organisms from growing in unwanted places.[80] They help keep algae out of swimming pools and cooling water. They prevent the growth of bacteria and fungi in paints, paper mills, leather factories, and others. As more and more solvent-based systems are replaced by waterborne ones, their use will increase. They are also used to preserve wood[81] and to keep fouling organisms off ships. The use of organomercury compounds in seed treatments, latex paints, and paper mills has been outlawed. Pentachlorophenol is no longer used to preserve wood. The chromium copper arsenate that is used on wood is not entirely free of problems. A study of decks in Connecticut found that soil beneath the decks contained levels of arsenic that exceeded regulatory limits.[82] The amount of chromium approached levels of concern, but was within regulatory limits. It too has been phased out and replaced by an alkaline copper quaternary compound, copper boron azole, and copper azole. Decks made from wood flour and recycled plastics do not need these preservatives.[83] Wood has also been preserved by chemical reaction with carboxylic acid anhydrides, isocyanates, epoxides, formaldehyde, and acrylonitrile.[84] (The last two are carcinogens.) Supercritical fluids (described in Chapter 8) can be used to help insert these reagents into the wood.

Biocides that release formaldehyde, a known carcinogen, are also of some concern. These are often hydroxymethylamines or their ethers, such as **4.8** Schematic.

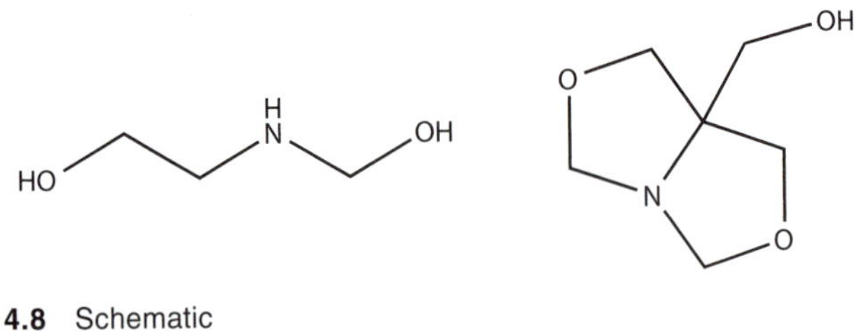

4.8 Schematic

N-Chloro compounds, such as **4.9** Schematic and **4.10** Schematic, are also being looked at as part of the reexamination of the general use of chlorine compounds. Isothiazolinones, such as **4.11** Schematic, are still acceptable. Quaternary ammonium salts, undecylenic acid, and sorbic acid are also acceptable. The last is used in the preservation of foods. Sodium and zinc salts of pyridine-2-thiol-*N*-oxide are also used in personal care products, architectural coatings and antifoulant paints.[80]

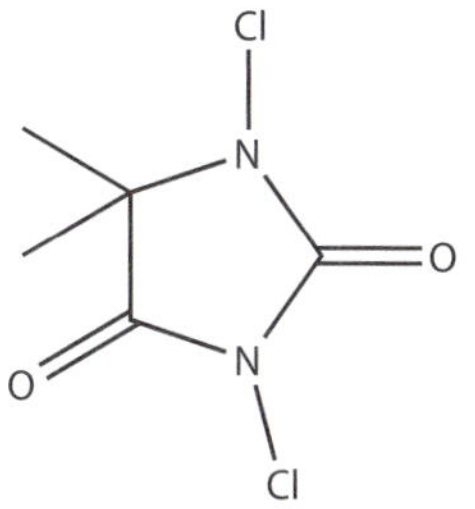

4.9 Schematic

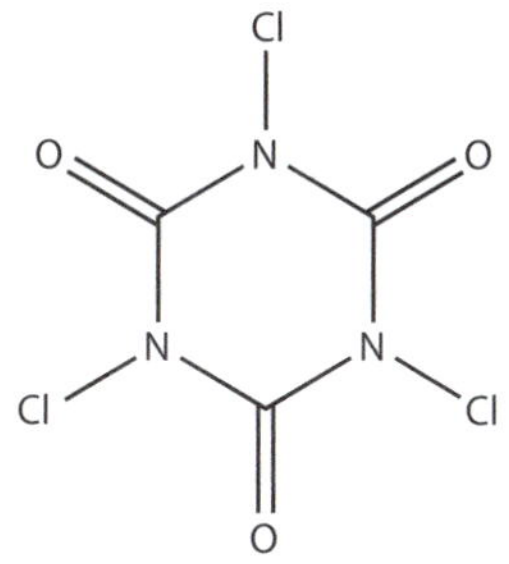

4.10 Schematic

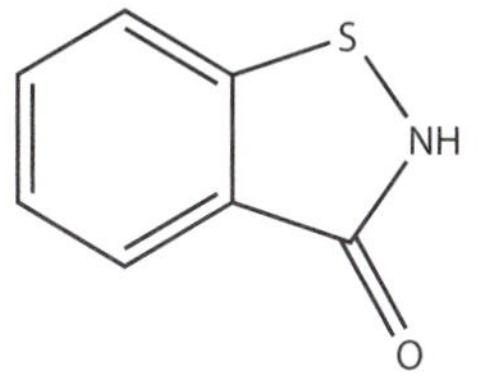

4.11 Schematic

Chlorine is often used to inhibit the growth of organisms in cooling water.[85] The problem is that it can still kill after it is put back into the river. (See Chapter 3 for ways of sterilizing water.) Chlorine dioxide can be used as a biocide for process water.[86] Ozone treatment is certified for use in cooling tower water in California.[87] A combination of low-toxicity hydrogen peroxide and peracetic acid controlled the microbial slime in a paper mill for a year at a reduced cost.[88] Tetrakis(hydroxymethyl) phosphonium sulfate ($[(HOCH_2)_4P^+]_2SO_4^{2-}$) is suitable for industrial water treatment.[89] It does not bioaccumulate and has nontoxic degradation products. The typical dose is lower than the lethal concentration for fish. When air-cooled heat exchangers are used, no chemicals are needed and no aquatic organisms are lost.[90]

Didecyldimethylammonium chloride is used to prevent stain and mildew on freshly cut lumber.[91] If a quaternary ammonium salt monomer could be impregnated into wood and then graft polymerized to it, the process might displace some of the chromium copper arsenate treatments. Some antibacterial polymers (e.g., **4.12** Schematic) work by gradual hydrolysis to release the active group.[92]

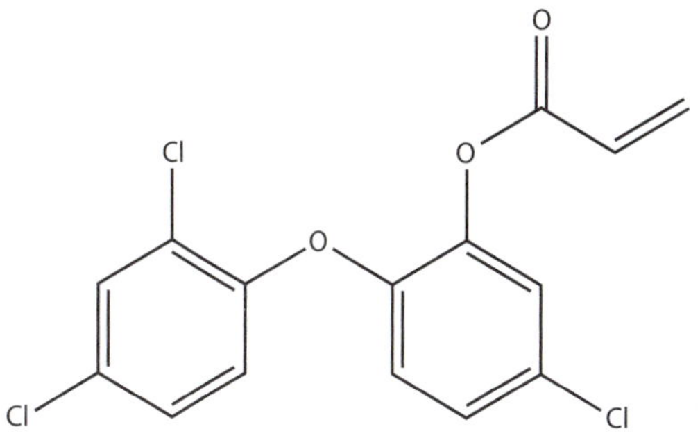

4.12 Schematic

Another polymeric biocide (**4.13** Schematic) has been found to adhere to cotton to give a "durable" antiodor finish to socks and towels.[93] Poly(hexamethylenebiguanide hydrochloride) is already used in swimming pools and in personal care products.

Antifoulant coatings for ships work by the slow leaching of the toxicant into the water. Copper oxide is used, as well as longer-lasting tributyltin compounds.[94] When the ships are in the deep ocean, the tin compound does not cause a problem. However, if they are in port most of the time, as recreational boats are, the metal compounds can kill desirable animals in the harbor. For example, elevated levels of tributyltin compounds have been found in stranded bottlenose dolphins.[95] The International Maritime Organization has agreed to ban the use of tributyltin compounds in antifouling paints on ships.[96] Thus, it is desirable to replace them with a system that is more environmentally friendly.[97]

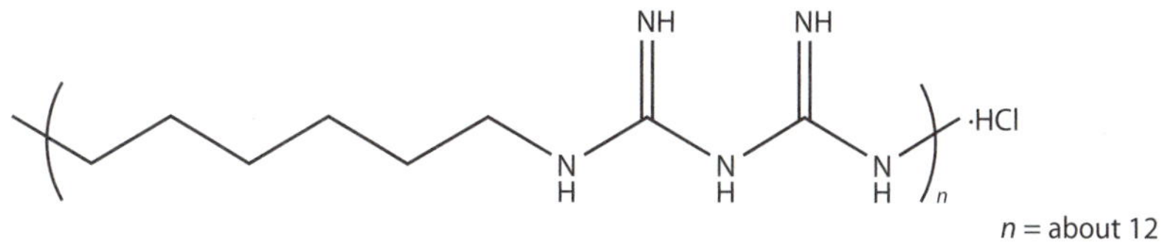

n = about 12

4.13 Schematic

Rohm & Haas (now Dow) has provided an isothiazolone (**4.14** Schematic) that can last for 3 years.[98] The second compound, the methacrylate monomer (**4.15** Schematic), is said to lead to polymers with antifoulant activity.[99]

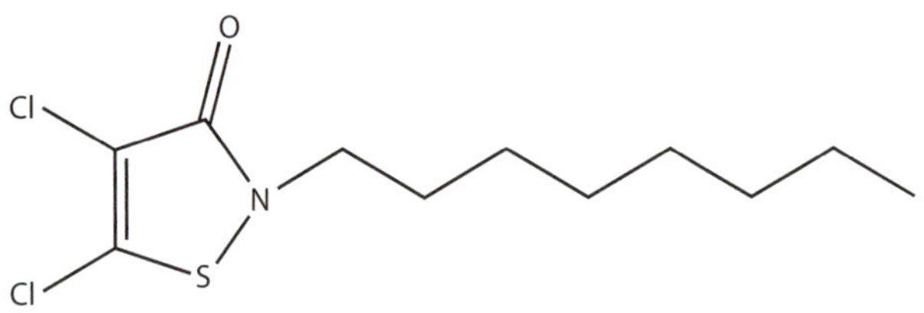

4.14 Schematic

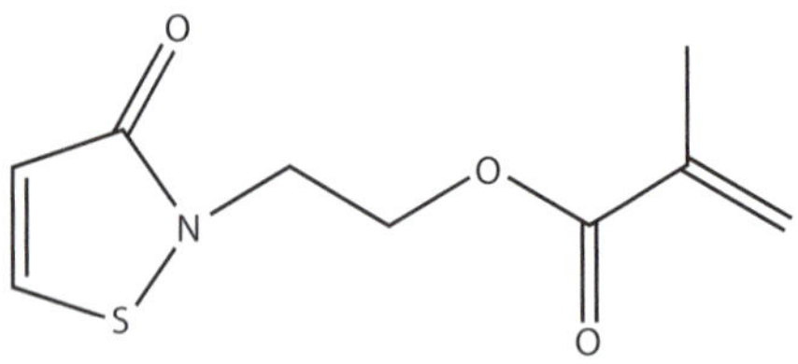

4.15 Schematic

cooling systems of power plants, and aquaculture cages. Numerous investigators are studying the chemical defenses of sponges, nudibranchs, gorgonians, and bryozoans that repel the usual fouling organisms. Many compounds with antifouling activity have been isolated, including isonitriles,[103] a pyridine nitro compound,[104] two of which are shown (**4.16** Schematic). It is hoped that simpler analogs that are more amenable to total synthesis will also prove to be effective. More needs to be known about the modes of action of these inhibitors. They may work by more than one mechanism. Bacteria (such as *Pseudomonas aeruginosa*[105]) settle first on a clean surface, form a biofilm, and release an attractant for the larger fouling organisms. (Individual bacteria respond to a quorum-sensing chemical that causes them to form a biofilm.[106]) Do the natural antifoulants inhibit bacterial growth or prevent the release of the quorum-sensing chemical or the attractant or repel the bacteria? Do they inhibit the curing of the adhesive secreted by the fouling organism? Do they act similar to insect antifeedants and repel the fouling organism? Would insect antifeedants have any effect on barnacle larvae? The ideal coating would be one that did not release anything, at least not until triggered by the bacteria or the fouling organism. It is possible that a polymeric coating containing quaternary ammonium salt groups (on spacers), in an amount such that the coating would swell somewhat, but not dissolve in seawater, would kill off bacteria that tried to grow on the ship.

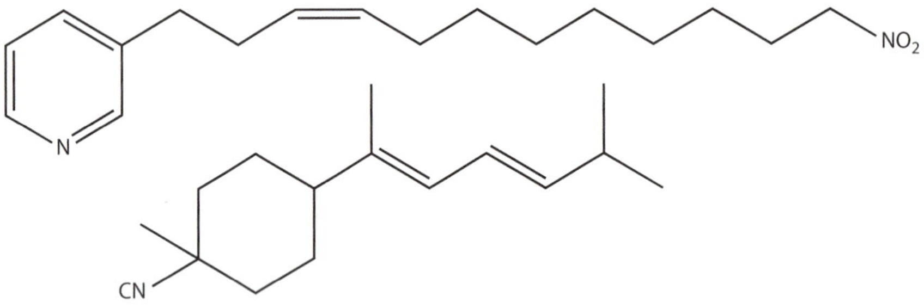

4.16 Schematic

An alternative to shedding antifoulant coatings would be a nontoxic surface to which marine organisms could not attach, or at least not attach tightly. Silicones and fluoropolymers have been suggested for this purpose.[100] At least one polydimethylsiloxane is sold for this purpose.[101] The fluoropolymers include polyurethanes based on fluoroether polyols and those made from fluorinated acrylates and methacrylates. Ideally, the motion of the ship though the water would dislodge the fouling organisms. If not, then gentle rubbing with brushes or scrubbing with jets of water could be used. The relatively high cost of the coating would be offset by the reduced need to repaint. Superhydrophobic coatings that mimic the microscopic roughness of the lotus leaf may also be suitable for the prevention of biofouling.[102]

Many sessile animals in the sea are not fouled by the algae, hydroids, mussels, and barnacles that foul ships,

Efforts are also under way to replace the tin compounds used as stabilizers in polyvinyl chloride[107] and in free radical reactions in organic chemistry.[108]

4.4 CATALYSTS FOR REACTIONS OTHER THAN OXIDATION

The search is on for catalysts to replace those containing toxic heavy metals. The addition of hydrogen chloride to acetylene to form vinyl chloride is catalyzed by mercuric chloride. Rhodium(III) chloride on activated carbon works just as well and is much less toxic.[109] It should be tried also in other addition reactions of acetylene as well as in *trans*-esterification reactions of vinyl acetate. The reduction of 2-ethyl-2-hexanal to 2-ethylhexanol can be catalyzed by a mixture of copper, zinc, manganese, and aluminum oxides in 100% yield.[110] This is said to be a replacement for carcinogenic copper chromite.

Spent catalysts to be reclaimed either on site or at a central processing facility, rather than being sent to a landfill. Johnson Matthey uses oxidation in sc water for the recovery of precious metals from spent catalysts.[111] However, with metals that are not too toxic, it is often cheaper to recycle them to other uses.[112] Spent nickel catalysts end up in stainless steel. Others, such as those containing copper, iron, magnesium, manganese, and zinc, can be treated with sulfuric acid and then used as micronutrients in fertilizer. Several thousand tons of spent catalysts are recycled in this way each year. There have been some abuses with arsenic getting into the fertilizer and leading to new regulations.[113]

4.5 DYES AND PIGMENTS

Chromium is important in mordant dyeing. Cadmium, chromium, copper, lead, and mercury have been important in inorganic pigments. These pigments provide good heat and light stability at relatively low cost. However, their manufacture can lead to release of toxic heavy metal ions into the environment. Other releases can occur when products made from them end up for disposal in an incinerator or landfill. Many efforts are under way to replace them with less harmful materials.

Red cerium sulfide is sold by Rhone-Poulenc (now Rhodia) as a replacement for pigments based on cadmium and lead.[114] Lead chromate is being replaced by nontoxic yellow bismuth vanadate, which has good hiding power and lightfastness.[115] The light stability and heat stability of red $CaTaO_2N$ and yellow $LaTaON_2$ are better than those of cadmium pigments.[116] More expensive organic pigments[117] are being used to replace oranges and reds. Many cheaper azo pigments lack the light stability needed for outdoor applications. Two light-stable types made by Ciba Specialities (BASF Corp.), diketopyrrolopyrroles (**4.17** Schematic) and quinacridones (**4.18** Schematic), are shown.[118] White basic lead carbonate has been replaced by titanium dioxide in house paints. Red lead chromate and yellow lead chromate are being phased out of paints.

Bird feathers are often brightly colored, but most contain no pigments. The colors are due to interference patterns.[119] The same principle is now being used in luster pigments.[120] Chemical vapor deposition, sputtering, vacuum deposition, as well as aqueous means, are being used to put thin layers of iron oxide, titanium dioxide, tin oxide, zirconium oxide, and others on mica, aluminum flakes, and others, to produce these colors. Thin films of brilliant blue boron nitride have been deposited on silicon by the titanium(IV) chloride-catalyzed decomposition of borazine.[121] Investigation of an ancient Mayan blue paint, which is resistant to acid, alkali, solvents, oxidation and reducing agents, biocorrosion, and moderate heat, has shown that it consists of indigo in a clay lattice that contains metal and metal oxide nanoparticles.[122] This suggests the intercalation of other dyes into clay lattices to produce stable pigments. Core–shell quantum dots have been suggested as pigments with a range of colors.[123]

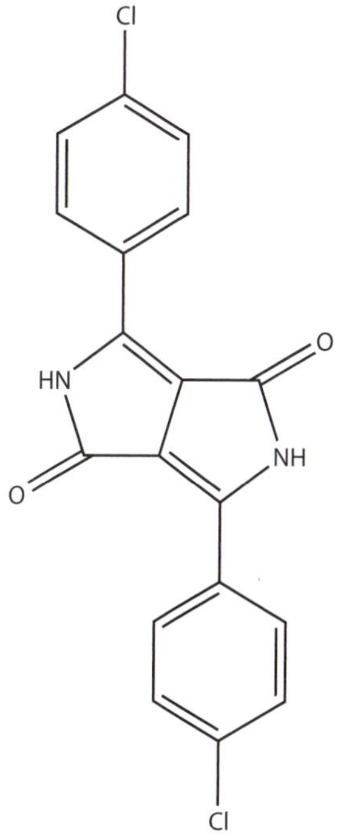

Cl
HN
O
O
NH
Cl

4.17 Schematic

H
N
O
O
N
H

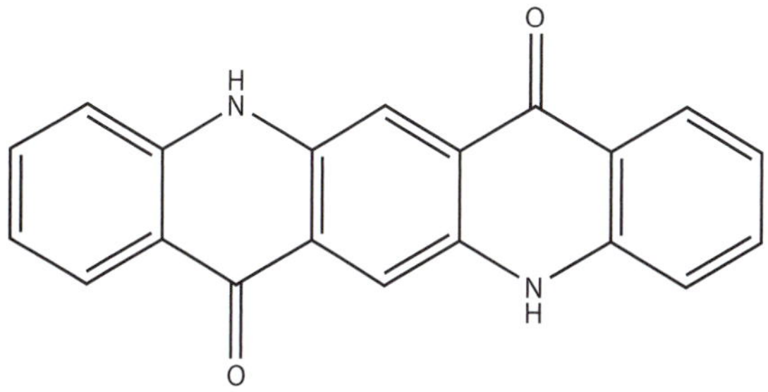

4.18 Schematic

Sewekow[124] has outlined the requirements for eco-textiles for "green garments." These include avoidance of harmful metals, formaldehyde-containing reagents, halogenated dye carriers (such as trichlorobenzene), and carcinogenic dyes (such as those based on benzidine and those azo dyes that could lead to carcinogenic aromatic amines on reduction). There is already a ban in Germany on azo dyes that could release *o*-toluidine, 2-naphthylamine, and *p*-chloroaniline on reduction.[125] Eco-friendly wet processing of cotton uses a pectinase, hydrogen peroxide bleaching, reactive dyeing, and enzymatic desizing.[126] Most (70%) wool is dyed with chromium mordant dyes. Iron has been used as a nontoxic replacement for chromium and cobalt to produce yellow, orange, red, blue, brown, and black dyes on protein and polyamide fibers.[127] Disperse dyes have also been applied to cotton and wool in supercritical carbon dioxide, after a pretreatment with a polyether, to give light shades.[128] This nontoxic, nonexplosive method reduces the amount of wastewater. Transfer printing done by the sublimation of dyes from paper to polyester fabrics also reduces the amount of wastewater. Ink jet printing of textiles is also used.[129] Exhausted dye baths can sometimes be used for the next batch by the addition of fresh reagents to reduce the volume of wastewater. The sodium dithionite and other reducing agents often used to solubilize dyes for application to fabrics, after which they are insolubilized by reoxidation, can be replaced by sugars or by electrolytic reduction.[130] This clean use of electricity eliminates waste salts.

Permanent-press fabrics are often cured with *N*-methylol compounds, such as **4.19** Schematic, and tend to release formaldehyde after the cure. [Formaldehyde (a carcinogen) from various formaldehyde resins is a problem in some homes.[131]] These can be replaced with polycarboxylic acids, such as the 1,2,3,4-butanetetracarboxylic acid (**4.20** Schematic), which cure via the cyclic anhydride.[132] Both dyeing and durable-press treatments can be combined in a single step, if desired.[133]

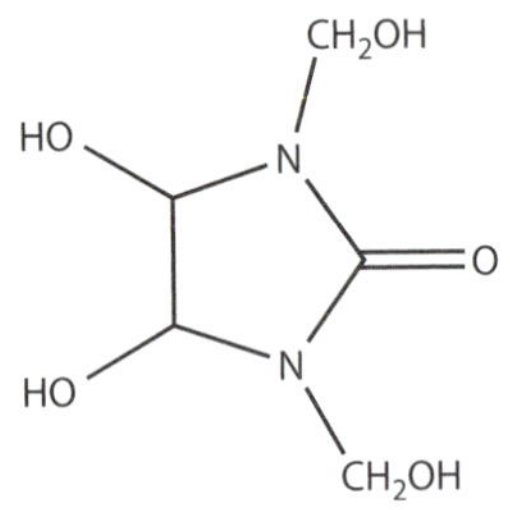

4.19 Schematic

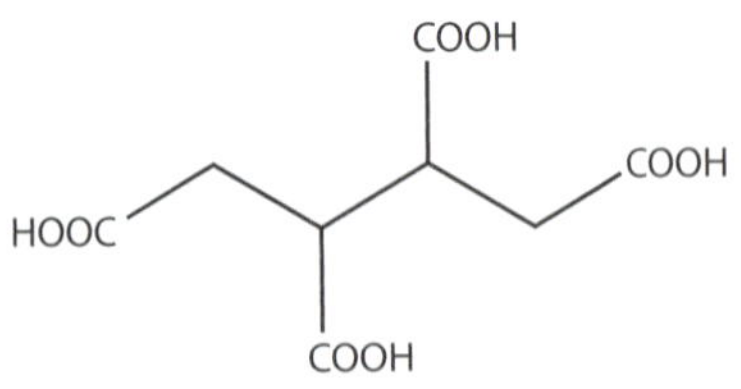

4.20 Schematic

No formaldehyde is used in the process, so that none is released. Copolymers of maleic anhydride should also be useful in this application and might be cheaper than the tetracarboxylic acid.[134] After exposure, a photographic film based on silver chloride must be processed by a wet process that uses hydroquinone. Digital cameras, which use no silver, are now widely available.[135]

4.6 ELECTRICAL USES

Mercury is often used in electrical switches. A nontoxic, cost-effective drop-in replacement is a gallium alloy containing indium, zinc, and copper.[136] The use of mercury switches is decreasing owing to a shift to solid-state devices.[137] Mercury switch thermostats are being replaced by fully electronic devices. There is still some mercury in every fluorescent lamp. There are other ways of generating light that approximates sunlight. Xenon gas discharge lamps[138] are used in accelerated testing of plastics because their output is similar to sunlight. They contain no mercury or other metals. If they can be fabricated to the size needed in domestic lighting and can operate as efficiently as a fluorescent tube, they could provide an alternative to the use of mercury. If only concentrated light can be obtained from current xenon lamps, then this might be distributed throughout the building through optical fibers or pipes. One fluorescent lamp uses xenon and a Eu(III)-doped $LiGdF_4$ phosphor to produce orange or red light without the need for mercury.[139] Additional phosphors are being studied in an effort to obtain white light. White light has been produced by the use of carbon-doped silica as a phosphor.[140] This avoids the need to use metals, such as silver, cadmium, geranium, and rare earths, which are present in the usual phosphor in a fluorescent lamp. No mercury is used. A GaN light-emitting diode with a "white" phosphor has been suggested for lighting.[141] LEDs are now used in traffic lights, automobile tail lights, and in lamps that can last up to 50,000 h.[142] Replacing conventional lighting in rooms by LEDs may happen in the future.

There is an effort to use methylcyclopentadienylmanganese tricarbonyl as an antiknock agent in gasoline in the United States, which is controversial.[143] Ethyl Corp. feels that the compound is harmless, for it has been used in Canada for 19 years. Critics say no systematic study has been done to measure its effects in Canada. The manganese salts come out as 0.2–10 μm particles.[144] Particles under 2.5 μm can lodge deep in the lungs and cause respiratory ailments. Companies that make automobiles say that it interferes with catalytic converters and sensors in cars. Ethanol works and contains no metal.

There is a problem with the widespread use of arsenic, cadmium, and selenium in electronic and photovoltaic devices. Cadmium mercury telluride is used in infrared-sensing night goggles. Cadmium sulfide, cadmium selenide, gallium arsenide, and analogs are used in solar cells. If their use becomes widespread, then an efficient system of collecting used cells for reprocessing will be needed. Some workers feel that it will be better to use nontoxic silicon cells wherever possible. (Solar cells are discussed in Chapter 15.)

The preparation of gallium arsenide is a dangerous process employing two pyrophoric reagents (**4.21** Schematic).[145]

$$Ga(CH_3)_3 + AsH_3 \longrightarrow GaAs + 3CH_4$$

4.21 Schematic

Chemical vapor deposition[146] of a single source precursor offers a safer alternative (**4.22** Schematic).[147]

$$Ga(As(tert\text{-}butyl)_2)_3 \longrightarrow GaAs$$

4.22 Schematic

Gallium nitride can be prepared by the pyrolysis of gallium dialkylamides. Wherever possible, it would be safer to substitute gallium nitride for gallium arsenide. Cadmium selenide can be made by the pyrolysis of cadmium 2-pyridineselenolate.[148] Thin films of cadmium telluride have been made by spraying a mixture of cadmium and tellurium compounds dissolved in a trialkylphosphine on to a substrate at temperatures higher than 250°C.[149] Cadmium phosphide can be made by pyrolysis of a cadmium phosphine complex.[150] Such methods are described as less toxic and easier to control. They avoid the use of the toxic gases hydrogen selenide and telluride.

4.7 LEATHER

Most leather is tanned with chromium(III) salts, which react with carboxyl groups in the collagen, so that it is no longer biodegradable.[151] A ton of raw hide produces 200 kg of good leather.[152] Untanned rejects end up as gelatin and glue. In a typical case, only 60% of the chromium is taken up by the leather, and the rest are discharged in waste liquors.[153] If the hide is pretanned with glutaraldehyde, the chromium usage can be reduced by 50%.[154] Various schemes have been devised to recover the chromium from the wastewater for reuse. In one, more than 98% of the chromium is recovered with a carboxylic ion-exchange resin.[155]

Iron, titanium, and zirconium salts can be substituted for the chromium ones.[156] Vegetable tannins, such as **4.23** Schematic, can also be used, but they slow down the tanning process. Similar phenols can be found in the residue from tea leaves left after the manufacture of instant tea. Perhaps, they could be used in making leather to eliminate or reduce the amount of another waste product. Getting reagents to penetrate the hide is a problem. Newer methods, such as ultrasonication and supercritical fluid extraction, may help reduce the time required to make leather, so that these alternative tanning agents can be used instead of the chromium. Enzymatic dehairing has reduced waste and increased yields.[157] A sulfonated melamine–urea–formaldehyde resin plus a vegetable tannin produced leather equivalent to that made with chromium.[158]

OH
HO
O
OH
OH
OH

4.23 Schematic

4.8 METAL FINISHING

Electroplating wastes have often contaminated waterways. A number of nonaqueous methods are available to avoid this problem. These include dipping (as in galvanizing steel with zinc), cladding, laser-assisted coatings, thermal spray coatings, vacuum metallization, chemical vapor deposition, sputtering, and in-mold metallization of plastics.[159] Sol–gel ceramic coatings are also used. Work has also been done at Los Alamos on laser-driven vapor deposition to deposit uniform coatings.[160] Recent work has been summarized in a U.S. EPA guide to alternative metal finishes, from which much of the following discussion has been taken.[161] The EPA has also set up a National Metal Finishing Center at Ann Arbor, Michigan.[162] The coatings are designed to prevent corrosion, increase wear, and to improve the appearance of objects.

Cadmium coatings impart corrosion resistance and natural lubricity. They are usually applied from an electroplating bath of cadmium cyanide. Wastewater from this process must be treated to remove both toxic ions. Other anions such as fluoroborate, sulfate, chloride, and pyrophosphate can sometimes replace cyanide. Zinc–nickel, zinc–cobalt, and zinc–tin alloys can be plated as replacements for cadmium. They have the necessary corrosion and wear resistance, but not the lubricity of cadmium coatings. Aluminum deposited by ion vapor deposition can replace cadmium coatings in some applications. In this process, there are no toxic materials and no liquid waste. The process involves evaporating aluminum and partially charging it in a vacuum before it is deposited on a substrate of the opposite charge. Such coatings may require the use of a lubricant to replace the natural lubricity of cadmium. They can be porous, in which case they may have to be peened (rubbed) with glass beads to eliminate the pores.

The use of chromium plating for decorative effects is subject to the whims of fashion. Thirty years ago, each car had chromium-plated bumpers and strips. Today, very few do. Plastic wheel covers are still shiny from what are probably aluminum coatings. Where the wear resistance of chromium coatings is important, as in machine parts, a coating of hard titanium nitride can be applied by physical vapor deposition. The chromium on hydraulic cylinders can be replaced by nitrocarburizing in ammonia followed by oxidation.[163] (Coatings of silicon dioxide and aluminum oxide can also be applied by vapor deposition.) The electroplating of nickel–tungsten–silicon carbide and nickel–tungsten–boron alloys is also being studied as a substitute for chromium coatings. Chromate is also used for conversion coatings for aluminum and zinc. The coatings provide corrosion resistance and may help adhesion of organic coatings. Zirconium deposited on the surface as zirconium oxide can substitute for the chromate. Titanates and molybdates have also been studied for this application. The etching of polyvinyl chloride surfaces before plating can be done with ultrasound instead of the CrO_3/H_2SO_4 that is often used.[164]

Cyanide is no longer needed in the electroplating of copper. There are commercially available alternatives. The formaldehyde used in the electroless deposition of copper coatings is carcinogenic. A nonhazardous reductant is needed to replace the formaldehyde. These methods still require treatment of the wastewater to remove copper ion.

Electrodialysis is said to be able to recover all the copper in electroplating wastewater.[165]

A new method is available that replaces silver in the silvering of mirrors.[166] A reflective multilayer coating of silicon and silica deposited by chemical vapor deposition replaces the usual metal films.

4.9 OXIDATION

4.9.1 Introduction

Organic chemists often use stoichiometric amounts of heavy metal compounds as oxidants. These reagents include chromium oxides, potassium dichromate, pyridinium chlorochromate, potassium ferricyanide, lead dioxide, lead tetraacetate, manganese dioxide, manganese(III) acetate, potassium permanganate, mercuric oxide, osmium tetroxide, silver oxide, and others. These result in large amounts of waste that must be disposed of or recycled. Traces of them may remain in the wastewater or the product. Many of these contain metal compounds that are quite toxic. The activation of C–H bonds with reagents, such mercuric acetate and selenium dioxide, has these problems. Even reagents, such as sodium hypochlorite and *m*-chloroperoxybenzoic acid, result in much waste, although it is much less toxic.

Schemes have been devised to substitute less toxic metals for more toxic ones. Potassium ferrate on K10 montmorillontie clay has been used to replace potassium chromate and potassium permanganate in the oxidation of alcohols to aldehydes and ketones in 54–100% yields.[167] The potassium ferrate is made by the action of sodium hypochlorite on iron(III) nitrate or by treatment of iron(III) sulfate with potassium peroxymonosulfate.[168] After the oxidation, any excess oxidizing agent, and its reduced form, are easy to recover by filtration or centrifugation. In another case, manganese-containing reagents have been substituted for more toxic ones containing chromium and selenium (**4.24** Schematic).[169] Selenium dioxide was used formerly in the first step and pyridinium chlorochromate in the second. This process still generates large amounts of waste.

$Mn(OCOCH_3)_3$, KBr → 71% (OH); MnO_2 → 100% (O)

4.24 Schematic

The trend is toward oxidants that are relatively harmless and that generate no noxious by-products. Oxygen, and materials derived from it, are examples. Ozone is quite toxic by itself, but produces no harmful by-products. It is used commercially for the cleavage of oleic acid to azelaic acid and pelargonic acid.[170]

Hydrogen peroxide is made from oxygen (**4.25** Schematic).[171] The anthraquinone is reduced back to the anthrahydroquinone with hydrogen using a nickel or palladium catalyst to complete the cycle. The anthraquinone used may also contain a 2-ethyl group. Headwaters-Evonik will have a plant producing hydrogen peroxide from hydrogen and oxygen in almost quantitative yield in 2009.[172] This should lower its cost.

Several other oxidizing agents can be made from hydrogen peroxide and thus be derived indirectly from oxygen. These include sodium perborate, sodium percarbonate, urea peroxide, peracids, potassium peroxymonosulfate, amine oxides, dioxiranes, and iodosobenzene (**4.26** Schematic).

Oxygen and many of these derived oxidizing agents are used with transition metal catalysts, which must be recovered and recycled. A typical mechanism is shown in **4.27** Schematic.

The ligands on these metal ions often tend to be oxidized at the same time as the substrate, leading to low turnover numbers.[173] Efforts to stabilize porphyrin ligands have involved replacement of the most oxidizable hydrogen atoms on the pyrrole rings with halogen atoms and putting halogenated aryl groups in the *meso*-positions.[174]

Oxidation of ligands can be avoided by the use of purely inorganic catalysts. If the catalysts are insoluble in the

4.25 Schematic

$$H_2O_2 + H_2SO_4 \longrightarrow H_2SO_5 \longrightarrow KOSO_2SOOH$$

$$H_2O_2 + RCOOH \longrightarrow RCOOOH$$

4.26 Schematic

$$RH + Co(III) \longrightarrow R^{\cdot} + Co(II) + H^+$$

$$R^{\cdot} + O_2 \longrightarrow ROO^{\cdot} \xrightarrow{RH} ROOH + R^{\cdot}$$

$$ROOH + Co(II) \longrightarrow RO^{\cdot} + Co(III) + OH^-$$

$$RO^{\cdot} + RH \longrightarrow ROH + R^{\cdot}$$

4.27 Schematic

medium, as in zeolites or heteropolyacids, the workup is much simpler. If the oxidation can be run in the gas phase or in a melt, no solvent has to be separated. If the oxidant is oxygen, a membrane reactor can sometimes be used to keep the oxygen concentration low and allow the product to be separated continuously, thereby avoiding overoxidation.

There are many reviews that cover various aspects of oxidation. These include the ones on alkane activation,[175] catalytic selective oxidation,[176] metal complexes of dioxygen,[177] metal-catalyzed oxidation,[178] biomimetic oxidations,[179] oxidation with peroxides,[180] catalytic oxidations with peroxides,[181] catalytic oxidations with oxygen,[182] oxidations with dioxiranes,[183] and oxidation of pollutants.[184]

The goals of the current work include

1. 100% selectivity and yield at 100% conversion.
2. Eliminate waste, especially toxic heavy metals.
3. Replace stoichiometric reactions with catalytic ones.
4. Replace very toxic metals such as mercury, lead, selenium, and osmium.
5. High turnover number and turnover frequency. Possible use of microwaves to speed up reactions.
6. Easy separation and reycle for 100 or more times. Eliminate oxidizable ligands on metals.
7. Combine reaction and separation, for example, by decanting a product.
8. Use only one equivalent of the oxidant with 100% utilization of it.
9. Use no solvent, especially not chlorinated ones or ones that can oxidize. Water in a biphasic system is all right.
10. Continuous methods, possibly in a microreactor, are preferred over batch ones.
11. Replace noble metal catalysts with iron or titanium ones.
12. Use no transition metal.

The rest of this chapter will show some of the best examples of progress toward these goals.

4.9.2 Oxidation by Oxygen

Although oxidations by oxygen are not common in the usual laboratory and in the manufacture of fine chemicals, they are often the standard methods of making large-volume commodity chemicals. Some typical ones are given in reactions **4.28** Schematic through **4.35** Schematic.[185]

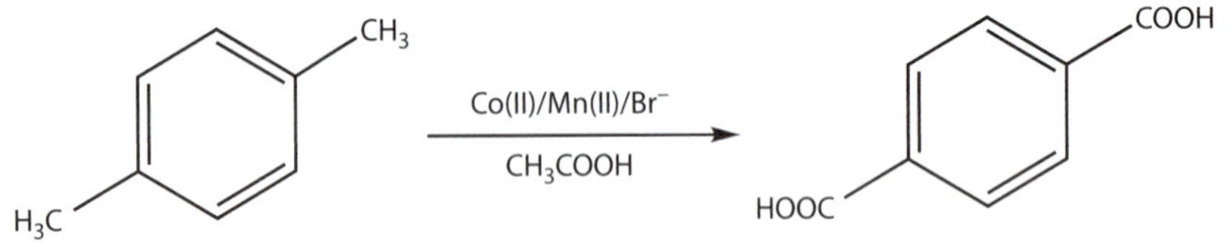

4.28 Schematic

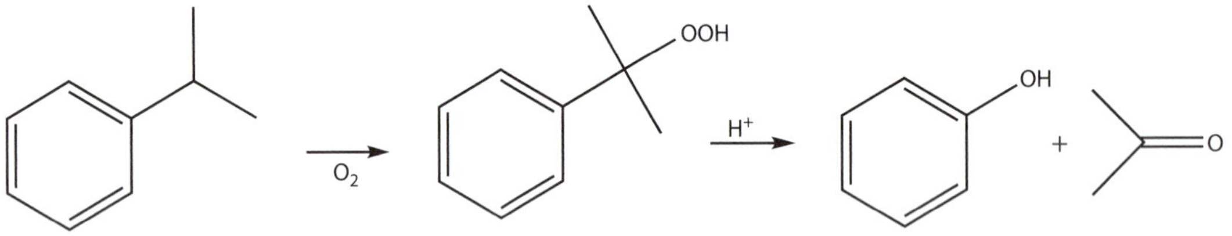

4.29 Schematic

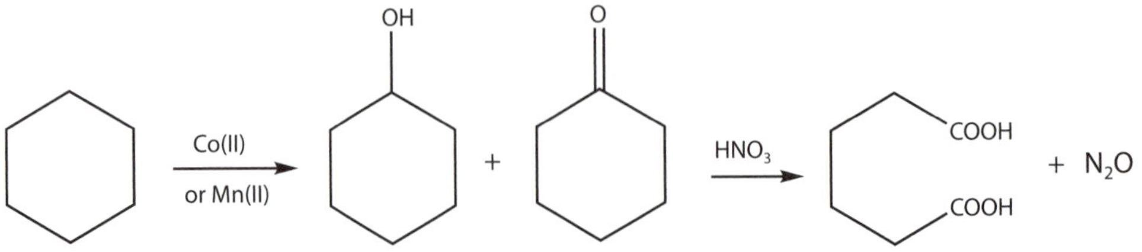

4.30 Schematic

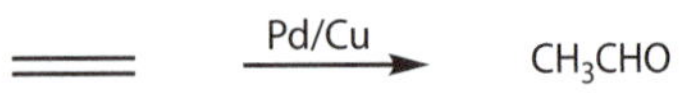

4.31 Schematic

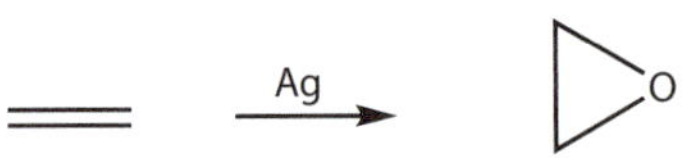

4.32 Schematic

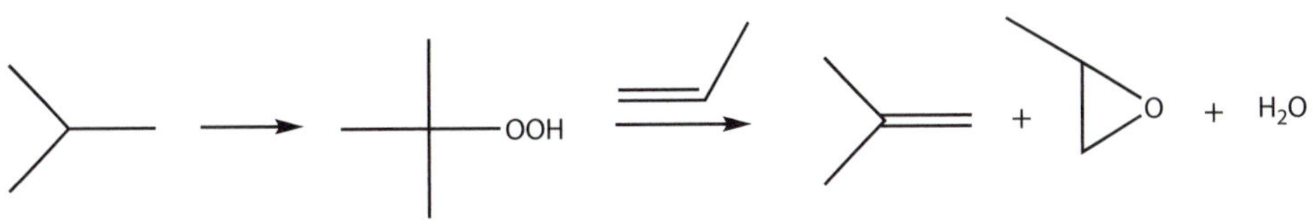

4.33 Schematic

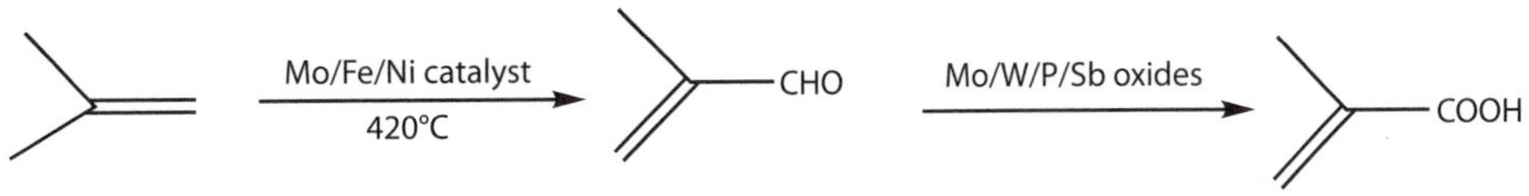

4.34 Schematic

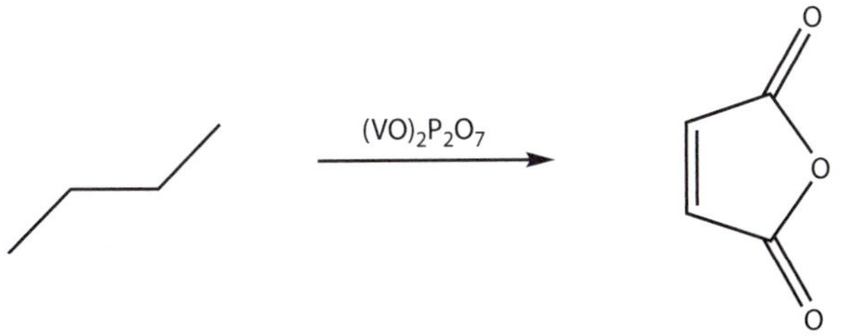

4.35 Schematic

m-Xylene can be oxidized to isophthalic acid by the method used to make terephthalic acid (**4.28** Schematic). Nylon plants that make adipic acid sometimes lose polluting nitrogen oxides (**4.30** Schematic). The synthesis of acetaldehyde from ethylene (**4.32** Schematic) employs a tandem oxidation in which the copper(II) chloride reoxidizes the palladium (O) and oxygen reoxidizes the copper(I) chloride formed in the process. A similar synthesis has been used to convert 1-decene to 2-decanone. Polyaniline was used with palladium(II), eliminating the need for copper(II) chloride and the corrosion caused by chloride ion.[186] The vanadium pyrophosphate (**4.35** Schematic) used in the oxidation of butane to maleic anhydride is encapsulated in silica to strengthen it for use in a recirculating fluidized solids reactor.[187] The use of *tert*-butylhydroperoxide to convert propylene to propylene oxide produces the by-product isobutylene, which can be converted to *tert*-butylmethyl ether, an antiknock agent for gasoline. If ethylbenzene is used instead of isobutene, the by-product is styrene, which can be polymerized to polystyrene. The only defect in such syntheses is that the market demand must be large enough to absorb all of the by-product.

The commercial oxidation of cyclohexane to cyclohexanol and cyclohexanone by oxygen for use in making nylon-6,6 is very inefficient. This oxidation can now be carried out in over 90% yield with a manganese porphyrin trapped in an indium imidazoledicarboxylate metal organic framework.[188]

An iron vanadate catalyst was used with oxygen to convert fluorene to fluorenone in 60% yield (**4.36** Schematic). However, if this catalyst was doped with cesium ion, the yield rose to over 95%.[189] The oxidation could be done with air using a protonated potassium manganese oxide molecular sieve with 100% selectivity at 97% conversion.[190]

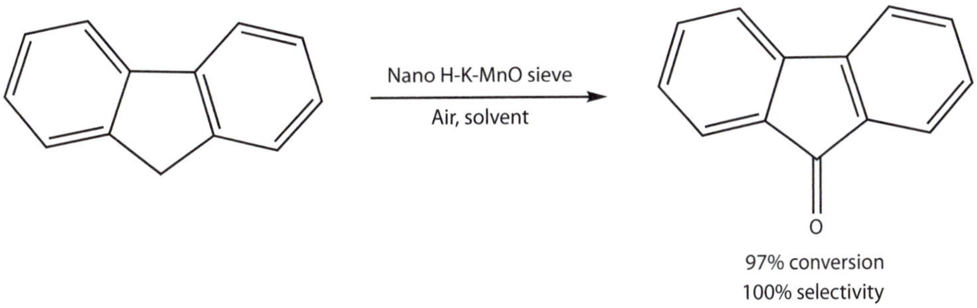

4.36 Schematic

Toluene has been oxidized to benzoic acid in 99% selectivity at 98% conversion in reversed micelles of aqueous sodium bromide in sc carbon dioxide with a cobalt perfluorocarboxylate catalyst (**4.37** Schematic).[191]

Xylenes have been oxidized with manganese bromide and oxygen in sc water in a continuous flow system using a

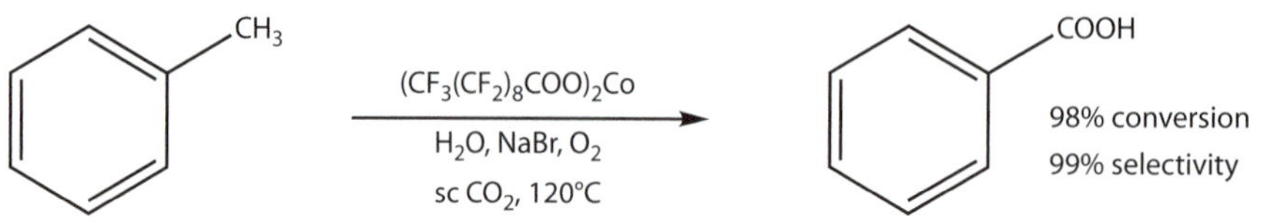

4.37 Schematic

contact time of <20 s to produce 94–97% of the dicarboxylic acid and 3–6% of benzoic acid.[192]

3-Picoline can be oxidized by oxygen in water at 250–400°C with a V_2O_5/TiO_2 catalyst to niacin in 96% yield (**4.38** Schematic).[193]

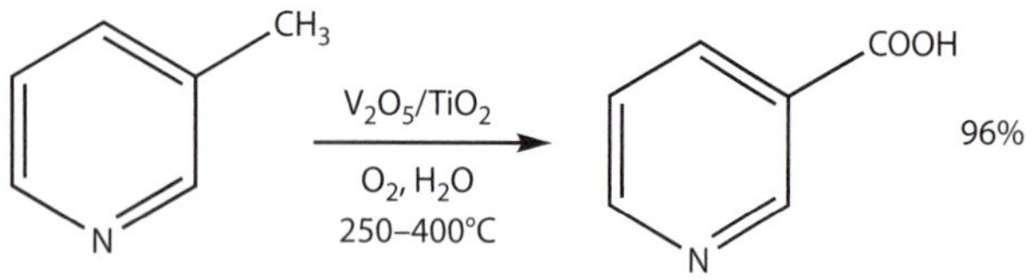

4.38 Schematic

Glucose can be oxidized to gluconic acid by oxygen (**4.39** Schematic) in the presence of palladium on alumina, a palladium–bismuth catalyst, or a platinum–bismuth catalyst in 99–100% yield (**4.39** Schematic).[194]

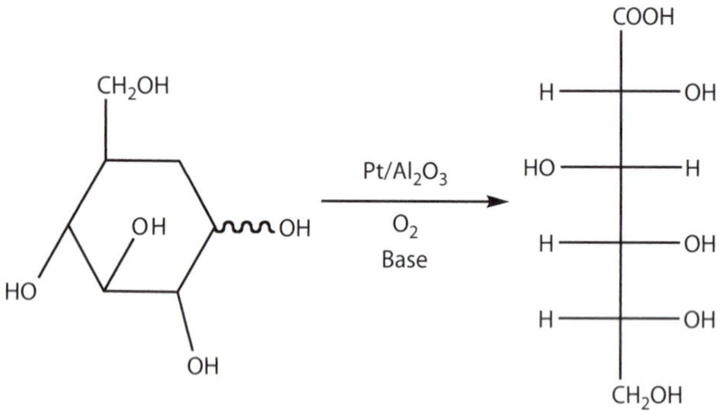

4.39 Schematic

Gluconic acid is a possible substitute for the sodium tripolyphosphate used in detergents to complex calcium and magnesium ions in the water. The phosphate can cause algal blooms in rivers and lakes receiving the wastewater from the washing process. This method can also be used for further conversion of the gluconic acid to 2-ketogluconic acid with 98% selectivity at 98% conversion.[195] The method allows the conversion of the hydroxyl group in 10-undecen-1-ol to a carboxylic acid without touching the double bond.[196] Gold on carbon catalysts have been used to oxidize ethylene glycol to glycolic acid with 98% selectivity[197] and 1,2-propyleneglycol to lactic acid with 100% selectivity at 78% conversion (**4.40** Schematic).[198]

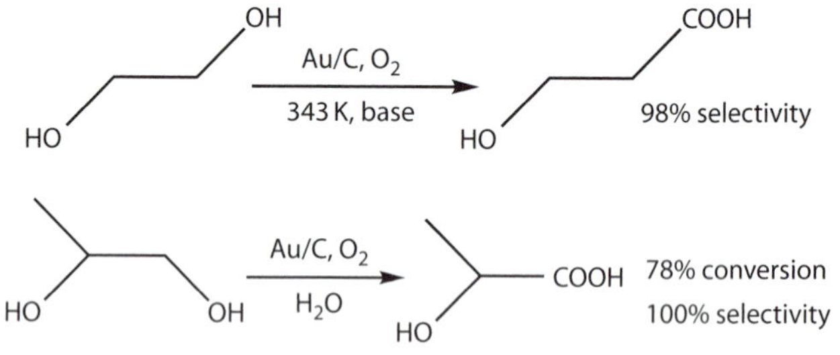

4.40 Schematic

Aliphatic aldehydes, such as butyraldehyde, have been converted to the corresponding carboxylic acid using air with a yttrium cluster catalyst in >99% yields.[199]

One molar aqueous ethylene glycol has been converted to glycolaldehyde in 90% yield using an alcohol oxidase, a method said to give better selectivity than chemical methods.[200] Laccase has been used with oxygen in the presence of an azine to convert benzyl alcohols to aldehydes in 87–100% yields.[201] (For more on biocatalysis, see Chapter 9.)

A few of the many methods for the oxidation of alcohols to aldehydes and ketones with oxygen will illustrate some of the techniques used:

a. In water with a dimeric ruthenium catalyst to convert benzyl alcohol to benzaldehyde with 99% conversion and 99% selectivity (**4.41** Schematic).[202]
b. In water with a palladium phenanthroline disulfonate catalyst plus sodium acetate to convert 2-pentanol to 2-pentanone in 90% yield with 100% selectivity.[203] The aqueous layer containing the catalyst could be used again in the next run.

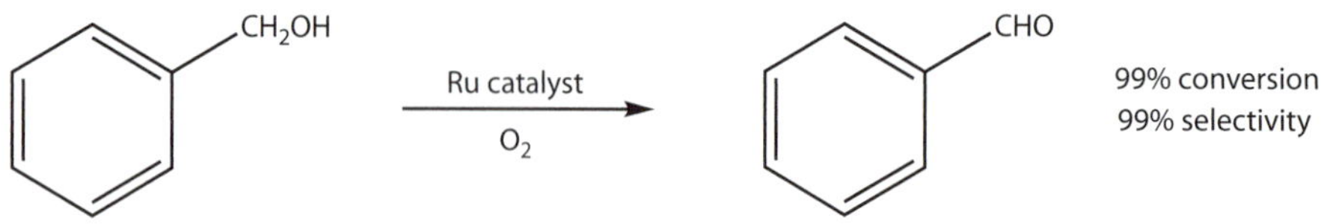

4.41 Schematic

c. With RuMn/hydrotalcite catalyst to convert benzyl alcohol to benzaldehyde in 1 h with 99% yield at 100% conversion.[204]

d. Use of a gold–palladium/titanium dioxide catalyst without solvent to convert benzyl alcohol to benzaldehyde with >96% selectivity and turnover frequencies of up to 270,000 h^{-1}.[205]

e. Conversion of benzyl alcohol to benzaldehyde in 100% yield using $H_5PV_2Mo_{10}O_{40}$ in sc carbon dioxide.[206] Note the absence of any organic ligand in the heteropolyacid catalyst, which can be filtered off and reused. No metal leaches. After removal of the carbon dioxide, only the product remains.

f. Use of fiber-supported 2,2,6,6-tetramethylpiperidine-1-oxyl (Fiber Cat) with manganese and cobalt nitrates to convert benzyl alcohol to benzaldehyde in >99% selectivity at >99% conversion.[207] The oxidation mediator can be filtered off and reused with no loss in activity. Oxidation mediators, such as TEMPO, abstract a hydrogen atom from the substrate, which then reacts with the oxygen. With *N*-hydroxyphthalimide, the first step is abstraction of a hydrogen atom to make the radical, which then abstracts the hydrogen atom from the substrate.

g. In refluxing water with a polymer-supported palladium catalyst for the conversion of benzyl alcohol to benzaldehyde in 97% yield.[208] After filtering off the catalyst, the benzaldehyde layer can be decanted off.

Two reactions illustrate the use of air oxidation where toxic selenium dioxide or bromination followed by hydrolysis and oxidation would normally be used. The first (**4.42** Schematic) oxidizes a methylphenol to a phenolaldehyde in 95% yield.[209] The second (**4.43** Schematic) uses air to oxidize α-pinene to verbenone in 77% yield.[210] This eliminates the need for the lead tetraacetate, sodium dichromate, sulfuric acid, and benzene in the two-step procedure in *Organic Syntheses*.[211]

Binaphthols are important as ligands for transition metal catalysts used in stereoselective syntheses. The oxidation of a suspension of powdered 2-naphthol in water with ferric chloride and air gives the corresponding binaphthol

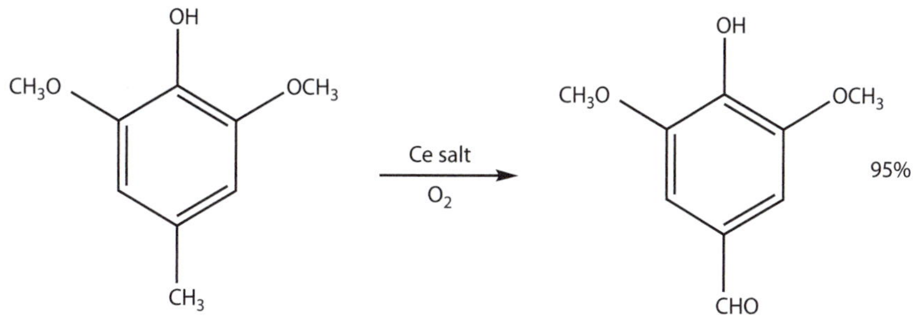

4.42 Schematic

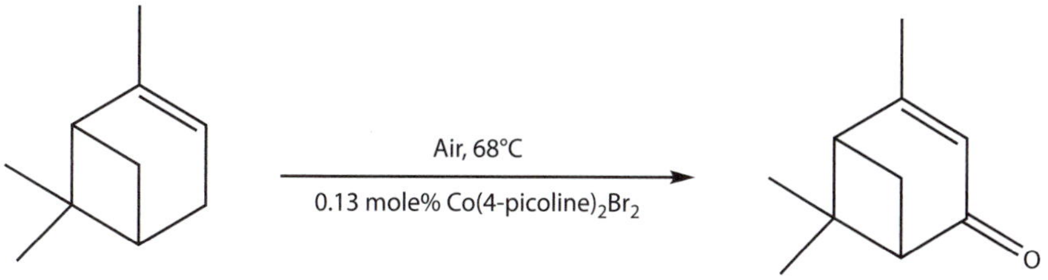

4.43 Schematic

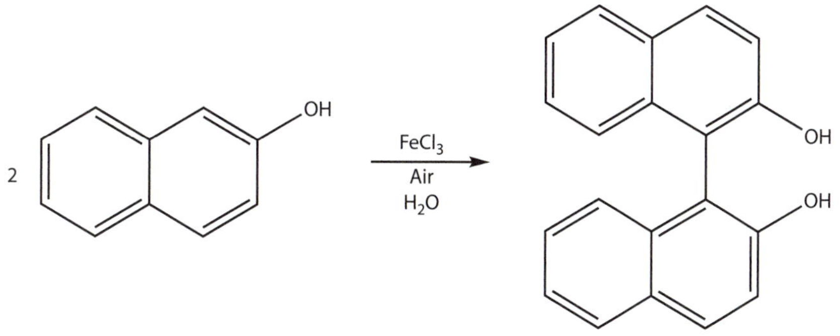

4.44 Schematic

(**4.44** Schematic) in 95% yield. This is an improvement over homogeneous syntheses, which are accompanied by quinone formation.[212] The workup consists of filtration, washing with water, drying, and recrystallizaiton from toluene. The reaction can also be run with a catalytic amount of inexpensive copper(II) sulfate on alumina to produce the binaphthol in 97% yield.[213]

2,3-Naphthalenediol can be oxidized with oxygen using a copper triflate catalyst to a polymer in >99% yield (**4.45** Schematic).[214]

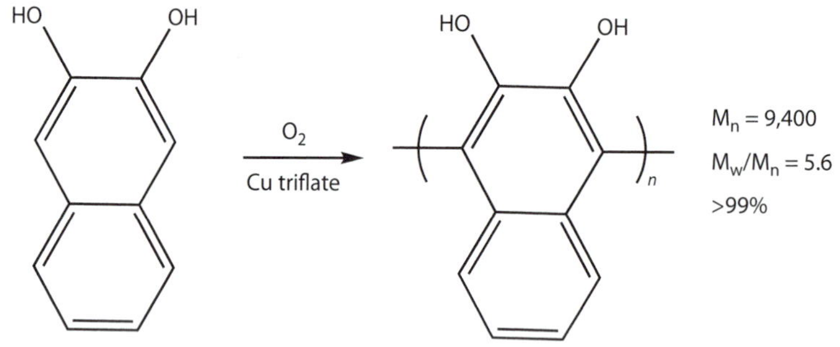

4.45 Schematic

High selectivity at high conversions has been observed in the photocatalyzed oxidation of hydrocarbons in zeolites using oxygen.[215] Even red light can be used. In some cases, dyes are exchanged for the alkali cations of the zeolites. The yield of one of several possible products from 2-methylpentene and 1-methylcyclopentane can be as high as 100%, compared with much lower selectivity when the reaction is run in solution. Scaling up these processes to a commercial scale poses significant challenges. Perhaps, a fluidized bed of powdered zeolite can be used in a tubular reactor. A solvent-free way to desorb the products from the zeolites is needed. Perhaps, simple heating will suffice. (For more on oxidations in zeolites, see Chapter 6.)

The oxidation of cyclooctene to its epoxide by oxygen using a potassium salt of an iron silicotungstate polyoxometalate was 98% selective with a turnover number of 10,000.[216] Nitrobenzene can be converted to 2-nitrophenol by oxygen in >99% selectivity using a heteropolyacid (**4.46** Schematic).[217]

4.46 Schematic

Triethylamine was oxidized to its oxide using a ruthenium catalyst in 98% yield (**4.47** Schematic).[218]

Care must be taken in all of these oxidations to avoid fires and explosions. This means operating outside the explosive limits of the system. This is not a factor if a continuous microreactor is used (see Chapter 8). It also means testing for peroxides before workup. A common method is to see if a sample liberates iodine from a solution of potassium iodide. Chemists at Praxair Inc. have devised a reactor that uses pure or nearly pure oxygen below a gas containment baffle with nitrogen above.[219] The oxygen efficiency is about 99%, which allows increased productivity and selectivity at lower operating pressures or temperatures. The rate of reaction is increased. The amounts of vent gas, lost solvent, and lost reactant are reduced. In the oxidation of *p*-xylene, 25–62% less loss to carbon monoxide and dioxide was found together with less color.

4.9.3 Oxidation with Hydrogen Peroxide without Titanium

The problem of nitrogen oxide emissions from nylon plants was mentioned previously. Mallinckrodt produces nitrogen oxides when it dissolves metals in nitric acid. By using hydrogen peroxide to oxidize the nitrogen oxides to nitric acid, it has eliminated 30 tons/year of the oxides and reduced its nitric acid use by 109 tons/year.[220]

The use of hydrogen peroxide to replace the chlorine and sodium hypochlorite used to bleach paper and textiles was discussed in Chapter 3.

Olefins can be converted to the corresponding epoxides with hydrogen peroxide under a variety of conditions.[221] A few of the many methods for the preparation of epoxides and for the oxidation of alcohols to aldehydes and ketones will be given to illustrate the variety of techniques employed. When a manganese ion was complexed with a tetraarylporphyrin–carboxylic acid, cyclooctene was oxidized to the corresponding epoxide in

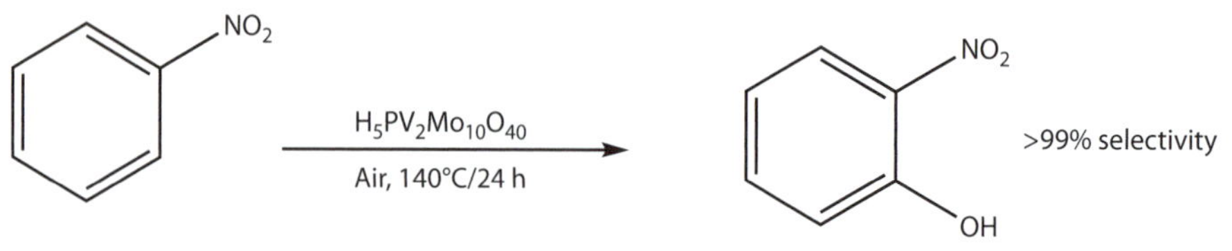

$$(CH_3CH_2)_3N \xrightarrow[\substack{20°C,\ 1\ atm\ O_2 \\ ClCH_2CH_2Cl}]{RuCl_3\ nH_2O} (CH_3CH_2)_3N \rightarrow O \quad 98\%$$

4.47 Schematic

100% yield in 3 min at 0°C.[222] Putting a manganese bipyridyl complex into a zeolite gave a catalyst good for 1000 cycles. In contrast to many catalysts, it showed no tendency to decompose hydrogen peroxide and underwent no self-oxidation.[223] By adjusting the acidity of the zeolite, the temperature, and time of the reaction, it was possible to favor epoxides, diols, or diacids. After 4 h (22% conversion) at 20°C, the product consisted of 81% epoxide and 14% diol. After 18 h at the same temperature, cyclohexene (62% conversion) gave 6% epoxide and 79% diol. At 100% conversion (40 h), the product was the diacid in 80% yield. Repeated regeneration of the catalyst was possible. This type of catalyst offers considerable promise, especially if the zeolite cavity can be chosen to favor a given product in higher selectivity. As a solid, it allows ease of separation from the mixture and the possibility of using it in a column if the rate can be increased by the use of higher temperatures or by addition of microwaves. It might be suitable as a replacement for the more toxic ozone in the cleavage of oleic acid to nonanoic acid and azelaic acid.

Solid catalysts can offer ease of separation as well as high selectivity. Cyclohexene has been converted to its epoxide with hydrogen peroxide and hydrotalcite [$Mg_{10}Al_2(OH)_{24}CO_3$] in methanol at 60°C in more than 99% yield at 100% conversion.[224] With 1-octene, the yield was 95% at 95% conversion. No transition metal is needed. The high yields may be at least partly due to the presence of the base, which may prevent the ring-opening side reactions often found in media lacking a base. The catalyst can be reused with no loss in activity. Solid catalysts with no organic ligands may last longer, because there is no organic ligand to be oxidized. Use of microwaves in the oxidation of cyclohexene gave a 100% yield in 1 min.[225] Oxidatively and solvolytically stable manganese-substituted heteropolyanions, such as the Keggin compound $[(Mn(II)(H_2O)_3)_2(SbW_9O_{33})_2]^{12}$, gave more than 99% regioselectivity in the oxidation of limonene if a biphasic system of ethylene dichloride–water was used (**4.48** Schematic).[226] However, there was some isomerization to the corresponding allylic alcohol during the reaction.

Cyclooctene was converted to its epoxide by hydrogen peroxide in water with an alkylated polyethyleneimine plus a polyoxometalate in 100% yield at 99% conversion.[227] Cyclohexene was converted to its epoxide in 98% yield in 45 min using a combination of hydrogen peroxide, sodium bicarbonate, and a molybdenum salicyloxime.[228] This may have involved a percarbonate anion. *Trans*-stilbene was converted to its epoxide with 97% selectivity at 100% conversion with an iron(III) chloride 2,6-pyridinedicarboxylic acid and base system.[229]

$2H_2O_2$

H_2O

$ClCH_2CH_2Cl$

O

4.48 Schematic

Cinnamyl alcohol was oxidized to cinnamaldehyde in 96–97% yield with 5% hydrogen peroxide using a platinum catalyst without solvent.[230] The yield with benzyl alcohol to benzaldehyde was >99%. This last reaction gave benzaldehyde in 99% yield when the catalyst was an iron–vanadium–molybdenum polyoxometalate.[231] Benzyl alcohol could be converted to benzoic acid in 100% yield at 100% conversion when more hydrogen peroxide was used with a zinc–tungsten polyoxometalate.[232]

A "green" route to adipic acid uses no solvent or halide ion and produces no nitrogen oxides.[233] Four equivalents of 30% aqueous hydrogen peroxide are used with air, sodium tungstate, and $CH_3(C_8H_7)_3N^+HSO_4^-$ (as a phase-transfer catalyst) to oxidize cyclohexene to adipic acid in 90–93% yield. After the adipic acid is filtered off, the aqueous phase can be reused by adding more phase-transfer catalyst and hydrogen peroxide. The reaction probably proceeds through the epoxide and glycol.

If the hydrogen peroxide is used with a catalytic amount of a ketone, such as hexafluoroacetone, olefins can be converted to epoxides, aldehydes to acids, sulfides to sulfoxides, tertiary amines to amine oxides, and so on.[234] The active reagent is the hydroperoxide resulting from the addition of hydrogen peroxide to the carbonyl group of the ketone or the dioxirane.

Methylrheniumtrioxide has been used as a catalyst for the oxidation of methylnaphthalenes, phenols, and phenol ethers to quinones with hydrogen peroxide (**4.49** Schematic).[235] The product ratio was 7:1 in favor of the desired vitamin. The current industrial preparation of the vitamin with chromium trioxide gives 38–60% yields. This results in 18 kg of chromium-containing waste per 1 kg of vitamin. Hydrogen peroxide of this concentration is hazardous to use. Water is an inhibitor for the reaction. However, it is possible to use 35% hydrogen peroxide with acetic anhydride to give a 10:1 selectivity for the desired vitamin. Rhenium oxides also work, but give lower yields. Further work is needed to find an active rhenium catalyst that is easy to separate and recycle to the next run. Rhenium compounds are not very toxic. Methylrheniumtrioxide can also be used with 30% hydrogen peroxide to convert olefins to epoxides with more than 98–99% selectivity.[236] A pyridine or bipyridine is used to accelerate the oxidation and prevent ring opening to by-products.

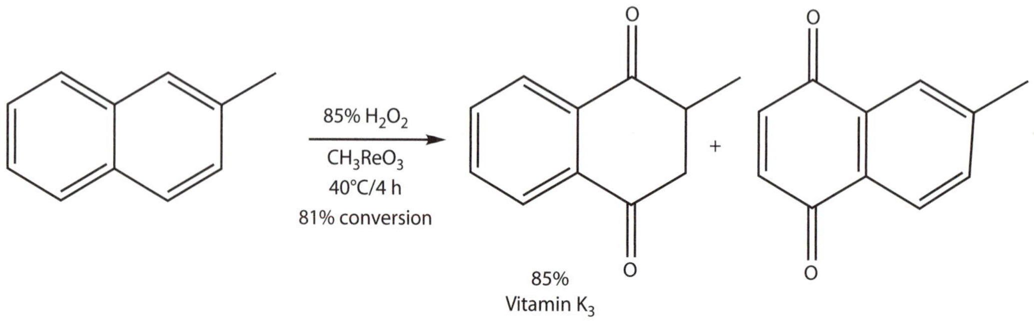

85% H_2O_2
CH_3ReO_3
40°C/4 h
81% conversion
+
85%
Vitamin K_3

4.49 Schematic

Soybean peroxidase can be used with hydrogen peroxide to oxidize alcohols to aldehydes and ketones.[237] The use of hydrogen peroxide with an immobilized lipase has allowed the oxidation of linoleic acid to a monoepoxide in 91% yield. The enzyme could be reused 15 times.[238] (Additional oxidations and other reactions catalyzed by enzymes are described in Chapter 9.)

Many efforts have been made to provide an osmium-free alternative to the Sharpless dihydroxylation of olefins. An iron catalyst gave a nearly quantitative yield of racemic dimethyl tartrate from tartaric acid (**4.50** Schematic).[239]

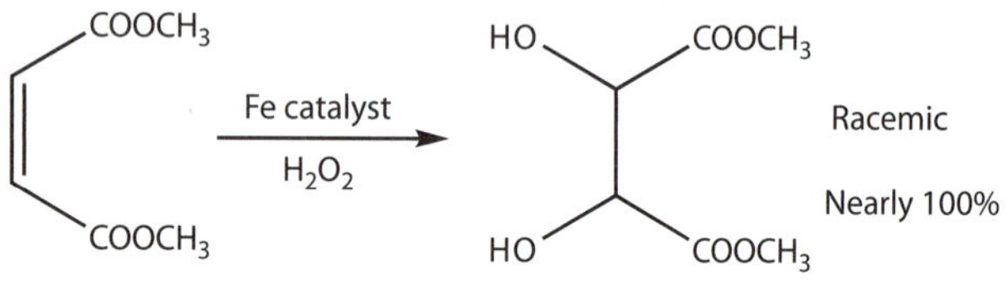

$COOCH_3$
$COOCH_3$
Fe catalyst
H_2O_2
HO $COOCH_3$
HO $COOCH_3$
Racemic
Nearly 100%

4.50 Schematic

Cyclooctene was converted to a mixture of diol and epoxide with a manganese oxide chelate plus 2,6-dichlorobenzoic acid (**4.51** Schematic).[240]

Cyclohexene was converted to the *trans* diol in 98% yield using a Nafion catalyst without solvent.[241] (Nafion is a polymeric perfluorosulfonic acid.)

Oxidation of phenol with hydrogen peroxide and titanium silicalite leads to a mixture of catechol and hydroquinone.

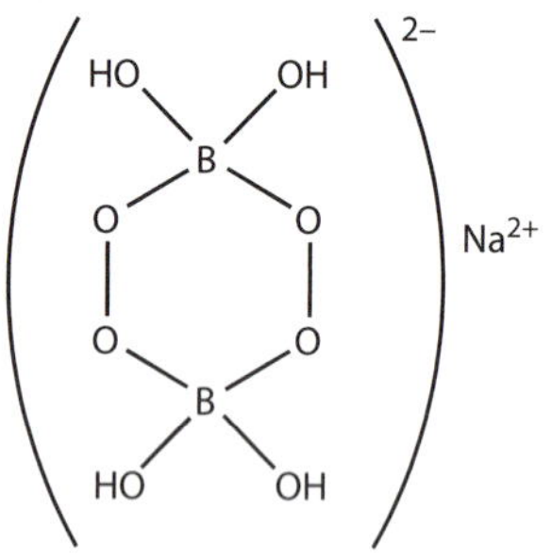

HO OH
B
O O
O O
B
HO OH
2−
Na^{2+}

4.52 Schematic

Various catalysts are being studied to be more selective for one or the other. A microporous niobium phosphate led to a 21:1 catechol–hydroquinone.[242] A potassium manganese–molybdenum polyoxometalate gave hydroquinone with no catechol with 86% selectivity at 45% conversion.[243]

A biphasic system of aqueous ammonia and hydrogen peroxide and benzaldehyde with a sodium zinc tungstate polyoxometalate produced benzaldehydeoxime in 99.9% selectivity at 95% conversion.[244] *N*-Methylmorpholine was converted quantitatively to its oxide with hydrogen peroxide using a layered double hydroxide containing tungstate ions.[245]

The hazards of concentrated hydrogen peroxide can be avoided by using sodium perborate (**4.52** Schematic), sodium percarbonate, and urea peroxide. The first two are inexpensive, stable, nontoxic, easily handled bleaching agents used in detergents. They are suitable for many types of oxidation, such as formation of epoxides from olefins, oxidation of

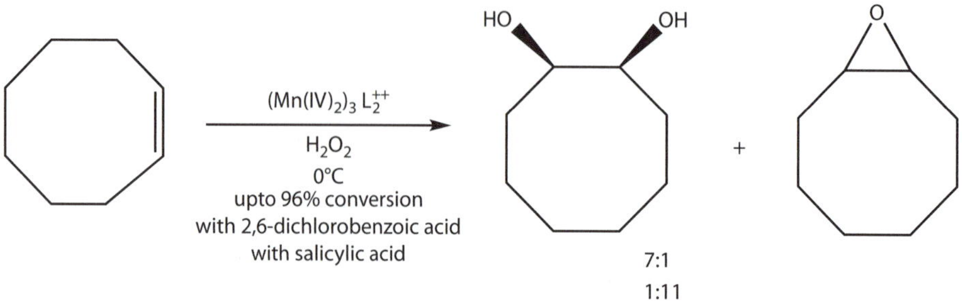

$(Mn(IV)_2)_3\ L_2^{++}$
H_2O_2
0°C
upto 96% conversion
with 2,6-dichlorobenzoic acid
with salicylic acid
HO OH
+
7:1
1:11

4.51 Schematic

$2\,Na_2CO_3 + H_2O_2$ $NH_2CONH_2 + H_2O_2$

4.53 Schematic

sulfides to sulfoxides, and sulfones, lactones from cyclic ketones, and so on.[246] Sodium perborate has a cyclic structure (**4.53** Schematic). The other two are just hydrogen-bonded complexes. The urea–hydrogen peroxide complex can be stored for a year at room temperature.[247]

Urea peroxide has been used with methylrheniumtrioxide to epoxidize allylic alcohols[248] and to oxidize 2,5-dimethylfuran to 3-hexene-2,5-dione (97% yield).[249] This avoids the need for the use of concentrated hydrogen peroxide with this catalyst. The urea also avoids acid-catalyzed secondary ring-opening reactions through its buffering action. The urea–hydrogen peroxide complex can also be used to oxidize secondary amines to nitroxides (**4.54** Schematic).[250]

Sodium percarbonate is used as an algaecide to avoid the use of copper salts.[251]

a molybdenum 2,4-pentanedionate[253] or a polyimidetriazole molybdenum complex.[254] The second catalyst could be recycled 10 times, although the yield fell to 92% in the sixth cycle and to 70% in the 10th cycle. Putting the molybdenum complex in a molecular sieve raised the conversion in the epoxidation of styrene from 25% to 96% (94% selectivity). Thus, both the rate and yield increased.[255] A molybdenum on silica catalyst converted cyclohexene to its epoxide with 98% selectively at 100% conversion.[256] A magnesium aluminum-layered double hydroxide pillared with a polyvanadate ion was used in the epoxidation of geraniol to its epoxide with 97% selectively at 95% conversion (**4.55** Schematic).[257]

Use of a silicotungstate polyoxometalate catalyst in the epoxidation of *cis*-2-octene and other olefins gave 99% conversion and selectivity with 99% utilization of the hydrogen peroxide.[258]

These methods provide ease of separation of the catalyst from the reaction mixture for reuse, as well as improved

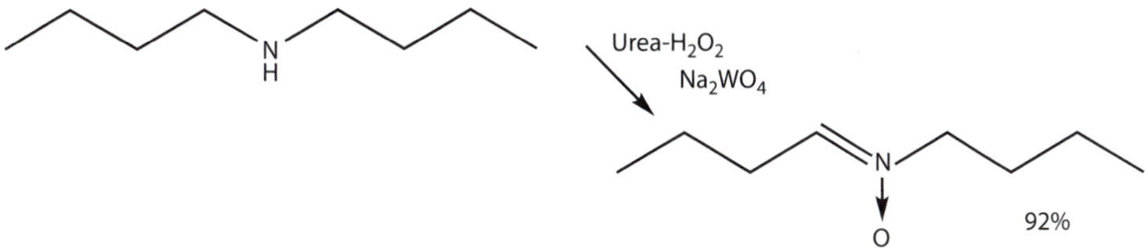

N H

Urea-H_2O_2

Na_2WO_4

N O

92%

4.54 Schematic

4.9.4 Oxidation by Hydroperoxides without Titanium

One of the commercial processes for the production of propylene oxide uses the transition metal-catalyzed epoxidation by *tert*-butylhydroperoxide. Various molybdenum complexes have been used to catalyze the epoxidation of other olefins by *tert*-butylhydroperoxide. Cyclooctene is converted to its epoxide quantitatively using a molybdenum complex of *N*-octadecylpyridylpyrazole.[252] Cyclohexene has been converted quantitatively to its epoxide using either

yields. Further work is needed to keep the yields up with the recycled catalysts. It may be necessary to use inorganic ligands or supports, rather than organic ones, to achieve this.

It may not be necessary to use a transition metal catalyst at all. A 96% yield has been obtained in the epoxidation of a cyclic ketone using *tert*-butylhydroperoxide with a potassium fluoride on an alumina catalyst (**4.56** Schematic).[259]

Another insoluble catalyst, made by exchanging the sodium ion in zeolite Y with a copper histidine complex, gave 100% selectivity in the oxidation of an alcohol to a carboxylic acid using *tert*-butylhydroperoxide.[260]

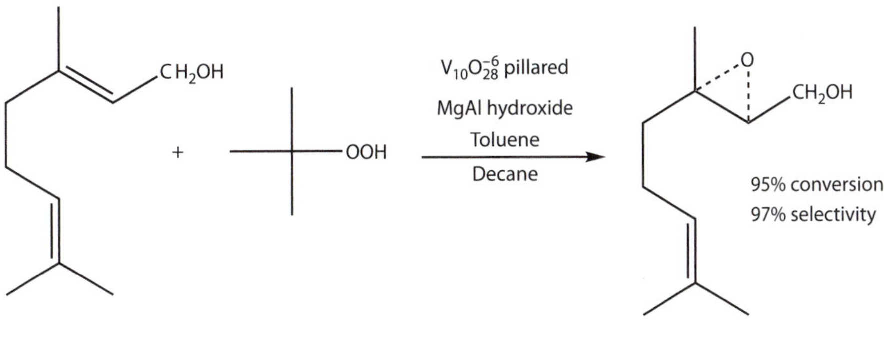

CH_2OH

+ OOH

$V_{10}O_{28}^{-6}$ pillared

MgAl hydroxide

Toluene

Decane

O

CH_2OH

95% conversion

97% selectivity

4.55 Schematic

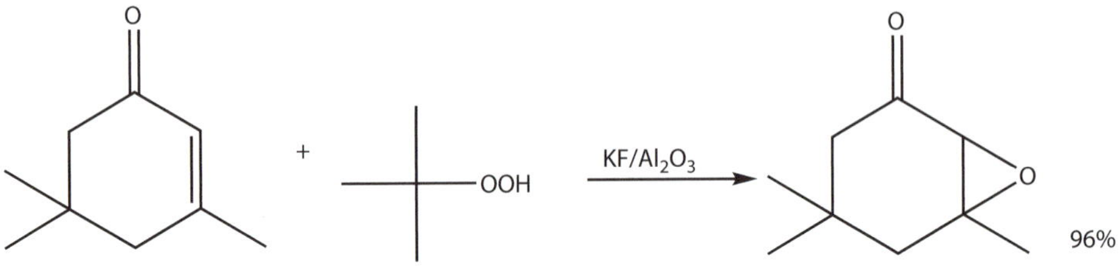

O + —OOH KF/Al_2O_3 → O 96%

4.56 Schematic

Secondary alcohols have been oxidized to ketones with excess *tert*-butylhydroperoxide in up to 93–99% yields using a zirconium catalyst.[261]

4.9.5 Oxidations by Hydrogen Peroxide and Hydroperoxides in the Presence of Titanium

A breakthrough came when Enichem in Italy discovered titanium silicalite (TS-I, a titanium-containing microporous silica molecular sieve) as a catalyst for oxidations with hydrogen peroxide.[262] The company uses this commercially for the preparation of cyclohexanone oxime (for nylon 6) (**4.57** Schematic) and the oxidation of phenol to a mixture of catechol and hydroquinone (**4.52** Schematic). It has built a 2000 metric tons/year plant to make propylene oxide by this process (**4.58** Schematic).[263] The selectivity to propylene oxide can be as high as 100% at 0°C, 98% at 25°C, and 95% at 40°C.[264] The reaction of cyclohexanone with ammonia and hydrogen peroxide can also be done with H-ZSM-5 zeolite coated with titanium silicalite giving 99% selectivity.[265]

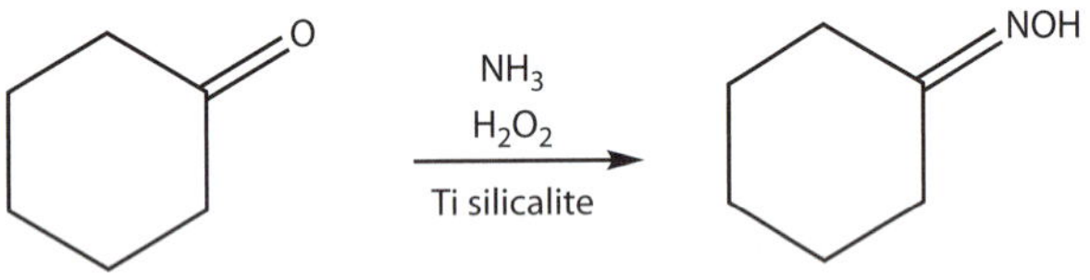

O NH_3 H_2O_2 → Ti silicalite NOH

4.57 Schematic

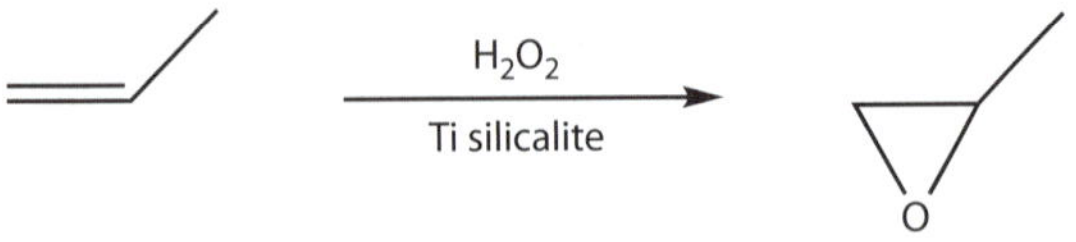

H_2O_2 → Ti silicalite O

4.58 Schematic

The substitution of other metals for titanium in the silicalite gives catalysts that usually do not work as well. In any event, titanium is less toxic than many others that could be substituted for it.

The major limitation of titanium silicalite is the small pore size, which restricts its use to small molecules. Its discovery has spawned a great deal of work in many laboratories to put the titanium into solids with larger pores. Corma and coworkers have prepared titanium-β and titanium MCM-41 as large-pore zeolites.[266] With these, the epoxidation of 1-hexene to its epoxide with hydrogen peroxide was 96–99% selective. With *tert*-butylhydroperoxide the epoxidation was nearly 100% selective. The highest activity and selectivity for epoxidation of olefins with organic peroxides was obtained when some of the silicon atoms were substituted with methyl groups. Titanium-β-fluoride was more selective to epoxide than titanium-β hydroxide in the epoxidation of methyl oleate with hydrogen peroxide (95% compared with 80%).

A catalyst made by treating silica with tetraisopropyltitanate was used with *tert*-butylhydroperoxide to produce a mixture of the two epoxides from limonene in 90% yield (**4.59** Schematic).[267]

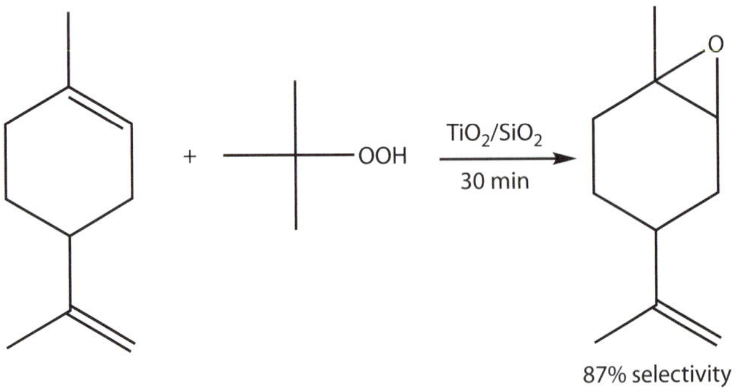

+ —OOH TiO_2/SiO_2 → 30 min O 87% selectivity

4.59 Schematic

It gave a 96% yield of the epoxide from 2-heptene. Thus, the limitation of TS-1 catalyst to small molecules is being overcome. Microporous amorphous solids appear to work as well as their crystalline molecular sieve counterparts.

4.9.6 Oxidation by Other Peroxy Compounds

Dioxiranes are good selective oxidants.[268] Some β-diketones have been oxidized to alcohols (**4.60** Schematic) in 95% or higher yield.[269] The dioxiranes are prepared and used in solution. The nickel catalyst speeds up reaction. Hydrocarbons can be functionalized (**4.61** Schematic) in up to 92% yield in this way.[270] A dioxirane phase-transfer catalyst, produced *in situ* with potassium peroxymonosulfate, has been used to epoxidize an olefin to an epoxide (**4.62** Schematic) in up to 92% yield.[271]

Dimethyldioxirane converts *N*,*N*-dimethylhydrazones to the corresponding nitriles in 94–98% yields in 2–3 min.[272] It has also been used to convert cyclooctatetraene to

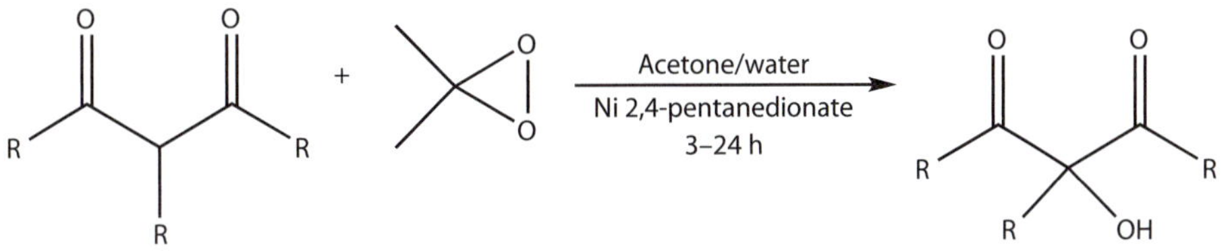

4.60 Schematic

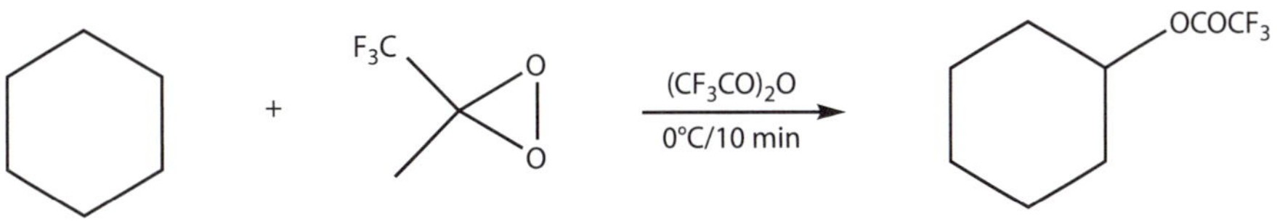

4.61 Schematic

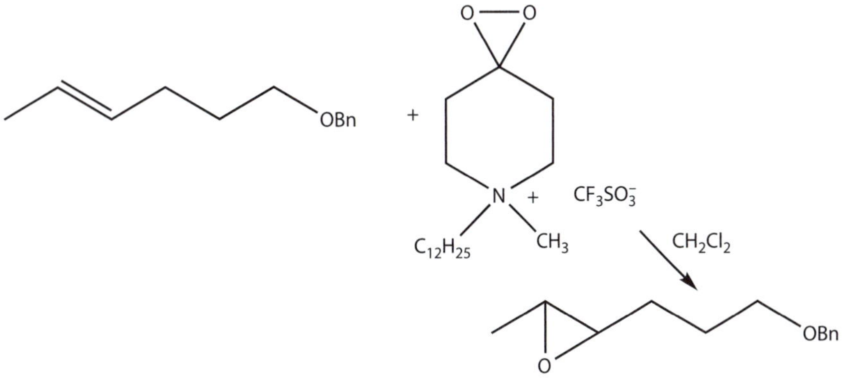

4.62 Schematic

tetraepoxides.[273] A dioxirane analog, a perfluorinated dialkyloxaziridine, has been used to oxidize alcohols to ketones.[274]

Potassium peroxymonosulfate has been used in the Nef reaction to convert 1-nitrohexane to caproic acid in 98% yield.[275] Replacement of the potassium ion by tetrabutylammonium ion allows the oxidant to be used in acetonitrile and methylene chloride. (Less toxic solvents would be desirable.) When this reagent was used with a manganese catalyst, 1-octene was epoxidized in more than 99% yield.[276] It also converted ethylbenzene to acetophenone with more than 99% selectivity.

Butylamine has been oxidized quantitatively to butyronitrile by 2-chloroperbenzoic acid using a ruthenium tetramesitylporphyrin catalyst.[277]

4.9.7 Oxidation by Amine Oxides and Iodoso Compounds

Methylcyclohexane has been oxidized to an alcohol using an amine oxide with a ruthenium porphyrin catalyst (**4.63** Schematic).[278] Cyclohexane gave a 95% yield of a 6.7:1.6 mixture of cyclohexanone and cyclohexanol.[279] (This suggests using hydrogen peroxide with a catalytic amount of 2,6-dichloropyridine and a ruthenium zeolite or heteropolymetallate.) Manganese[280] and iron[281] porphyrins gave lower yields. A similar system converted 1-phenyl-1,3-butadiene to an aldehyde (via epoxide isomerization) in 99% yield (**4.64** Schematic).[282] A PEGylated version of the catalyst with the same *N*-oxide converted cyclooctene to its epoxide in a 99% yield at 99% conversion.[283]

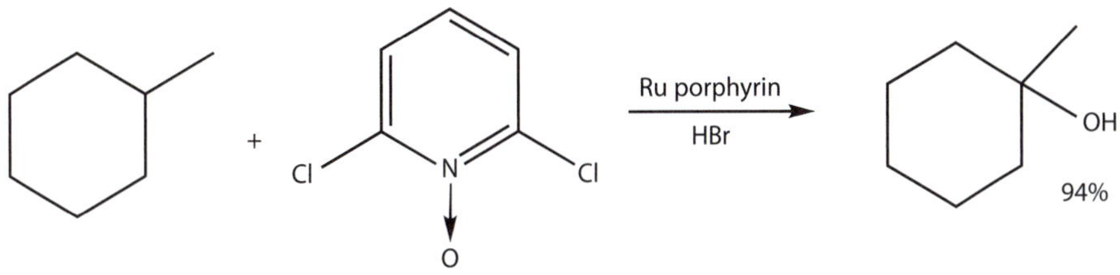

4.63 Schematic

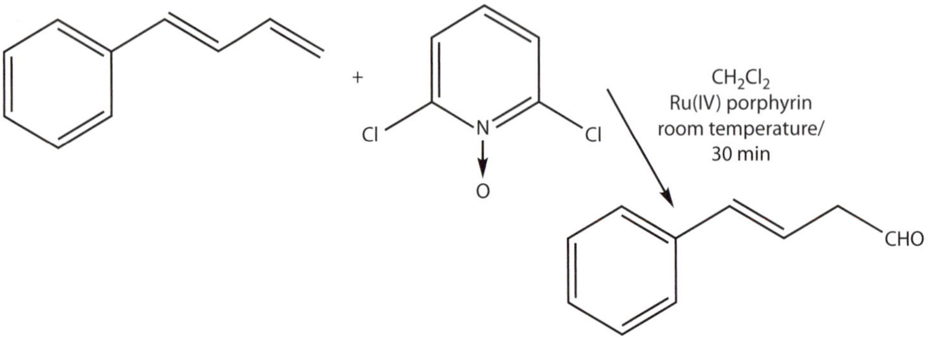

4.64 Schematic

Iodosobenzene and other hypervalent iodine compounds offer low toxicity and mild conditions for oxidation reactions.[284] Iodosobenzene can be made by oxidation of iodobenzene to iodobenzenediacetate, followed by treatment with the base.[285] It has been used with a manganese chelate and a trace of pyridine-*N*-oxide to convert olefins to their epoxides.[286] A manganese porphyrin on silica catalyst used with iodosobenzene converted cyclooctene to its epoxide in a 100% yield with no leaching of the metal.[287] 2-Iodoxybenzoic acid has been used in mild oxidations but has a tendency to explode. 2-Iodoxybenzoic acid esters are said to be safe to use.[288] The problem with both the amine oxides and the iodoso compounds is that they produce reduced by-products that will have to be isolated and recycled to avoid waste.

4.9.8 Electrically Assisted Oxidations

The use of electricity in reactions is clean and, at least in some cases, can produce no waste. Toxic heavy metal ions need not be involved in the reaction. Hazardous or expensive reagents, if needed, can be generated *in situ* where contact with them will not occur. The actual oxidant is used in catalytic amounts, with its reduced form being reoxidized continuously by the electricity. In this way, 1 mol% of ruthenium(III) chloride can be used in aqueous sodium chloride to oxidize benzyl alcohol to benzaldehyde at 25°C in 80% yield. The benzaldehyde can, in turn, be oxidized to benzoic acid by the same system in 90% yield.[289] The actual oxidant is ruthenium tetroxide. Naphthalene can be oxidized to naphthoquinone with 98% selectivity using a small amount of cerium salt in aqueous methanesulfonic acid when the cerium(III) that forms is reoxidized to cerium(IV) electrically.[290] Substituted aromatic compounds can be oxidized to the corresponding phenols electrically with a platinum electrode in trifluoroacetic acid, triethylamine, and methylene chloride.[291] With ethylbenzoate, the product is a mixture of 44:34:22 *o*/*m*/*p*-hydroxybenzoates in 89% yield. Water is used to hydrolyze the intermediate trifluoroacetate. In another case, electricity is used to generate sodium hypoiodite and sodium hydroxide continuously for the epoxidation of 2-methylanaphthoquinone (**4.65** Schematic).[292] The yield is 100%. No waste is produced and no hazardous halogen has to be handled or shipped. Unfortunately, the yield drops to 89% when sodium bromide is substituted for the iodide and to 6% when sodium chloride is used. One wonders whether such a process could be used to convert propylene to propylene oxide economically. It would also be desirable to use a solvent that is less toxic than acetonitrile. Propylene may be soluble enough in just water for the reaction to be used.

4.9.9 Oxidation by Halogens and Halogen Compounds

Iodine has been used to oxidize aldehydes to esters in alcohols in 91–98% yields (**4.66** Schematic).[293]

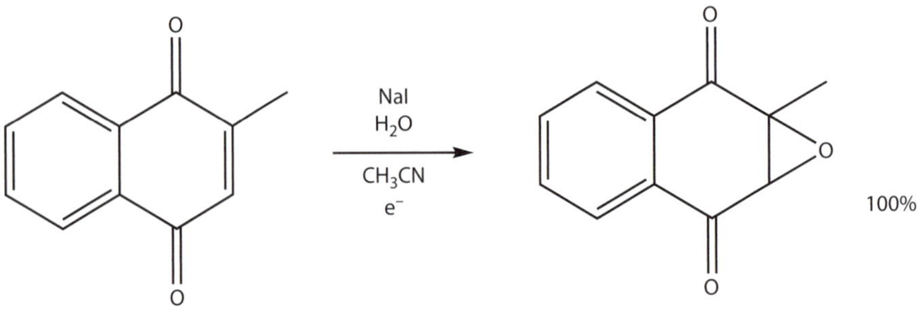

4.65 Schematic

$$\text{RCHO} \xrightarrow[\text{KOH in R'OH}]{I_2} \text{RCOOR'} \qquad 91\text{–}98\%$$

4.66 Schematic

It is possible that the iodine and base could be generated in the way cited earlier for the oxidation of 2-methylnaphthoquinone.

Sodium hypochlorite, as in household bleach, is an inexpensive oxidant. In use, it does produce the by-product sodium chloride, which, in theory at least, could be recycled by electrolysis. It has been used to oxidize a variety of olefins to their epoxides in the presence of manganese chelates as catalysts, with yields of 80–95%.[294] Quaternary ammonium salts can be used as phase-transfer catalysts. Amine oxides are also added in catalytic amounts in some cases. Manganese chelates of porphyrins and salicylidene amines are used as catalysts. A typical example is shown in reaction **4.67** Schematic. With optically active chelates, the oxidation can be highly stereospecific. Alcohols can be oxidized to aldehydes or ketones in 30 min at room temperature using aqueous sodium hypochlorite, ethyl acetate, and tetrabutylammonium bromide as a phase-transfer catalyst.[295] Benzaldehyde was obtained in 93% yield and octaldehyde in 86% yield. Alkyldimethylamines have also been oxidized to amine oxides using sodium hypochlorite with iron-halogenated porphyrin catalysts.[296] The presence of the 16 halogen atoms in the porphyrin stabilizes this ligand to oxidation. Ligand oxidation is a common problem limiting the lifetimes of oxidation catalysts.

An *ortho*-dimethylbenzene has been oxidized selectively to a toluic acid in a two-phase system in which the product is extracted into water as it is formed (**4.67** Schematic). Three equivalents of sodium hypochlorite are used with sodium hydroxide, 1% ruthenium(III) chloride, and 5% tetrabutylammonium bisulfate to produce the product in 98% yield.[297] The oxidation takes place in the organic phase. This example illustrates the principle of removing the product as it is formed, before further side reactions can take place, so that a high yield can be obtained.

Upjohn and Pharmacia (now Pfizer) have used the steroidal intermediate, bisnoraldehyde, to make progesterone and corticosteroids for many years. They developed an oxidation process for the conversion of the primary alcohol to bisnoraldehyde (**4.68** Schematic). This uses sodium hypochlorite bleach with 4-hydroxy-2,2,6,6-tetramethyl-1-piperidinyloxy radical in a two-phase system.[298]

The many advantages of the method over earlier ones include no heavy metals, no noxious emissions, no hazardous reagents or reaction mixtures, increased utilization of the soya sterol feedstock (from 15% to 100%), elimination of ethylene dichloride (a carcinogen) solvent, 89% less nonrecoverable solvent waste, 79% less aqueous waste, and the wastes being nontoxic. Presumably, a method for recycling the expensive catalyst has been developed.

Periodates have been used in a variety of oxidations. Tetrabutylammonium periodate can be used in nonaqueous media to oxidize alcohols to aldehydes (60–97% yields), mercaptans to disulfides (90–100% yields), and dialkylsulfides to sulfoxides (60–75% yields).[299] Sodium periodate has been used with ruthenium complexes as catalysts for the oxidation of olefins to epoxides (45–99% yields) and alcohols to ketones (75% yield). The epoxidation of the olefins is complicated by cleavage of the carbon–carbon bond.[300] The reduced iodine compounds can be converted back to periodate electrochemically. Sodium perbromate with a ruthenium(III) chloride catalyst has been used to oxidize primary alcohols to either the aldehyde or the carboxylic acid in 90% yield, depending on the reaction conditions.[301] Styrene has been converted to an aziridine with ammonia, ammonium iodide, sodium hypochlorite, and a surfactant in water with 99% selectivity and a 90% yield (**4.69** Schematic).[302]

4.9.10 Oxidation with Dinitrogenoxide

Dinitrogenoxide (nitrous oxide) may be a useful, relatively harmless oxidant if a good inexpensive source of it can be found. It can be obtained from the by-product gases from the oxidation of cyclohexanone by nitric acid in the manufacture of nylon. It can also be made from ammonium nitrate. It is said to offer a wider range of process safety than oxygen.[303] Benzene can be converted to phenol in 100% selectivity using dinitrogen oxide with FeZSM-5 zeolite at room temperature. Solutia considered using this process as

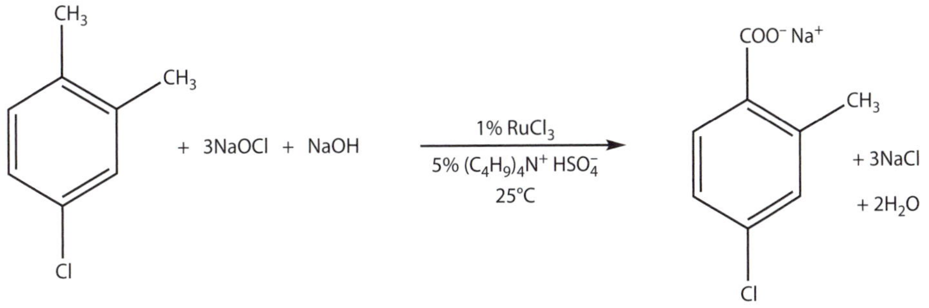

CH_3, CH_3, Cl (4-chloro-1,2-dimethylbenzene) + 3NaOCl + NaOH $\xrightarrow[5\%\ (C_4H_9)_4N^+\ HSO_4^-,\ 25°C]{1\%\ RuCl_3}$ $COO^-\ Na^+$, CH_3, Cl + 3NaCl + $2H_2O$

4.67 Schematic

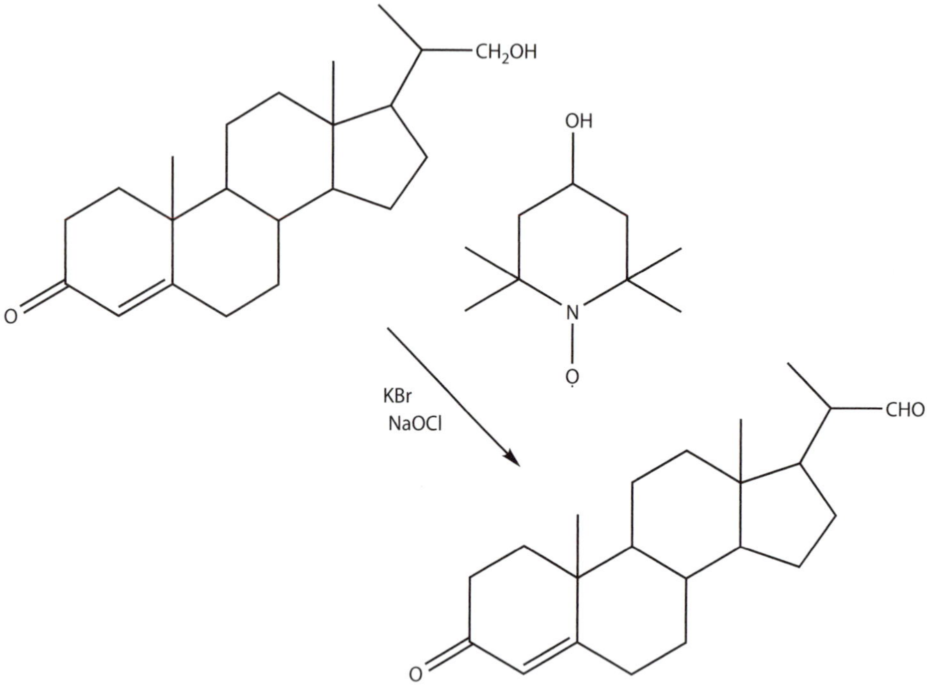

4.68 Schematic

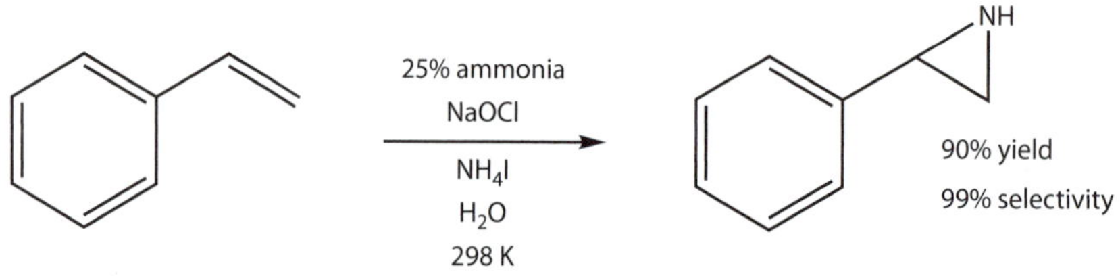

4.69 Schematic

part of a route to adipic acid for use in making nylon.[304] This transformation is also of considerable interest to makers of phenol, for it would eliminate the by-product acetone formed in the current process that is based on cumene hydroperoxide. (Adipic acid can also be made from glucose, without the need for cyclohexane, nitric acid, or benzene, as described in Chapter 9.)

4.10 MISCELLANEOUS: ASBESTOS AS A TOXIC MATERIAL

Asbestos fiber is a carcinogen that can cause lung cancer. In earlier years when this was not known, the material was used widely in insulation, fire suits, and so on, because it could not burn and was inexpensive. In friction materials, it shows a high affinity for resins used with it, good heat resistance, a high coefficient of friction, and low abrasion.[305] Many materials have been suggested as replacements. The cheaper ones are natural minerals, such as mica, talc, attapulgite, and vermiculite. Somewhat more expensive are glass and steel fibers. More expensive yet are fibers of alumina and silica. Fibers of organic polymers, such as polypropylene and polyacrylonitrile, are also suitable for some applications. Strong aramide, carbon, and polybenzimidazole fibers are too expensive for most applications. The replacement selected will depend on the application. Filled polymers should be relatively simple, although it may require a hydrophobically coated filler. Brake linings for cars and trucks pose more of a challenge. These may require glass or metal fibers or even aramide or graphite fibers. At one time, brake linings were the principal source of asbestos fibers on city streets. One method of disposal of asbestos is to dissolve it in phosphoric acid, neutralize it with lime and use the product as fertilizer.[306]

REFERENCES

1. (a) For example, see the Environmental Health Criteria published by the World Health Organization in Geneva, Switzerland: 1, mercury, 1976; 3, lead, 1977; 15, tin, 1980;

18, arsenic, 1981; 36, fluoride, 1984; 58, selenium, 1986; 61, chromium, 1988; 85, lead, 1989; 86, mercury, 1989; 101, methylmercury; 118, inorganic mercury, 1991; 106, beryllium, 1990; 108, nickel, 1991; 134 and 135, cadmium, 1992; (b) F.W. Ohme, ed., *Toxicity of Heavy Metals in the Environment*, Dekker, New York, 1978–1979; (c) S.A. Katz and H. Salem, *The Biological and Environmental Chemistry of Chromium*, VCH, Weinheim, 1994; (d) J.S. Thayer, *Environmental Chemistry of the Heavy Elements: Hydride and Organic Compounds*, VCH, Deerfield Beach, FL, 1995; (e) A. Sigel, H. Sigel, eds, *Metal Ions in Biological System, vol. 29: Biological Properties of Metal Alkyls*, Dekker, New York, 1993; (f) G.L. Fisher and M.A. Gallo, eds, *Asbestos Toxicity*, Dekker, New York, 1988; (g) C. Hogue, *Chem. Eng. News*, Aug. 9, 2004, 23.

2. (a) J.F. Tremblay, *Chem. Eng. News*, June 3, 1996, 8; Mar. 25, 1996, 21; (b) A. Sigel and H. Sigel, *Metal Ions in Biological Systems, vol. 34: Mercury and Its Effects on Environment and Biology*, Dekker, New York, 1997.
3. C. Baird, *Environmental Chemistry*, W.H. Freeman, New York, 1995, 361; pp. 347–394 cover As, Cd, Pb, and Hg.
4. (a) M.J. Vimy, *Chem. Ind. (Lond.)*, 1995, 14; (b) P. Belsky, *Chem. Eng. News*, Oct. 12, 1998, 11; (c) A. Cutler, *Chem. Eng. News*, July 13, 1998, 8; (d) T. Newton, *Chem. Br.*, 2002, *38*(10), 24.
5. B.E. Erickson, *Chem. Eng. News*, July 28, 2008, 47.
6. Anon., *Environ. Sci. Technol.*, 1996, *30*, 331A.
7. (a) B.S. Murdock, *Science*, 1996, *272*, 1247; (b) R. Macdonald, D. Mackay, and B. Hickie, *Environ. Sci. Technol.*, 2002, *36*, 457A.
8. P. Vaithiyanathan, C.J. Richardson, R.G. Kavanaugh, C.B. Craft, and T. Barkay, *Environ. Sci. Technol.*, 1996, *30*, 2591.
9. (a) J. Kaiser, *Science*, 1996, *271*, 1045; (b) J. Johnson, *Chem. Eng. News*, Jan. 5, 1998, 8; May 11, 1998, 22; (c) R. Renner, *Environ. Sci. Technol.*, 1998, *32*, 444A; (d) B. Baker, *Bioscience*, 1998, *48*, 900.
10. C. Baird, *Environ. Chem.*, W.H. Freeman, New York, 1995, 379.
11. (a) P. Bagla and J. Kaiser, *Science*, 1996, *274*, 174; (b) J.C. Hering, Preprints *A.C.S. Div. Environ. Chem.*, 1998, *38*(2), 289.
12. (a) Anon., *Science*, 1998, *281*, 1261; (b) W. Lepkowski, *Chem. Eng. News*, Nov. 9, 1998, 12; Nov. 16, 1998, 27; (c) B. Sarkar, *Chem. Eng. News*, Dec. 14, 1998, 8; (d) Y. Bhattacharjee, *Science*, 2007, *35*, 1659.
13. (a) Anon., *Environ. Sci. Technol.*, 1996, *30*, 65A; (b) P.L. Short, *Chem. Eng. News*, Apr. 23, 2007, 13; (c) S. Ahuja, *Arsenic Contamination of Groundwater*, Wiley, Hoboken, NJ, 2008; (d) A.A. Meharg, *Venomous Earth: How Arsenic Caused the World's Worst Mass Poisoning*, Macmillan, New York, 2005; (e) National Research Council, *Arsenic in Drinking Water—2001 Update*, National Academy Press, Washington, DC, 2001.
14. (a) B. Hileman, *Chem. Eng. News*, Feb. 1, 1999, 24; (b) M.L. Biggs, R. Haque, L. Moore, A. Smith, C. Ferreccio, and C. Hopenhayn-Rich, *Science*, 1998, *281*, 785; (c) H. Pringle, *Science*, 2009, *324*, 1130.
15. Y. Zhang and J.N. Moore, *Environ. Sci. Technol.*, 1996, *30*, 2613.
16. (a) S. Hong, J.-P. Candelone, C.C. Patterson, and C.F. Boutron, *Science*, 1996, *272*, 246; (b) J.O. Nriagu, *Science*, 1996, *272*, 223; 1998, *281*, 1622; (c) W. Shotyk, D. Weiss, P.G. Appleby, A.K. Chelburkin, R. Frei, M. Gloor, J.D. Kramers, S. Reese, and W.O. Van Der Knapp, *Science*, 1998, *281*, 1635; (d) K.J.R. Rosman, W. Chisholm, S. Hong, J.-P. Candelone, and C.F. Boutron, *Environ. Sci. Technol.*, 1997, *31*, 3413; (e) D. Weiss, W. Shotyk, A.K. Cheburkin, M. Gloor, and S. Reese, *Water Air Soil Pollut.*, 1997, *100*, 311–324; (f) R. Lutter and E. Irwin, *Environment*, 2002, *44*(9), 24.
17. (a) M.S. Gustin, G.E. Taylor, and T.L. Leonard, *Environ. Sci. Technol.*, 1996, *30*, 2572; (b) J.O. Nriagu, *Chem. Br.*, 1994, *30*, 650.
18. B. Hileman, *Chem. Eng. News*, Apr. 17, 2000, 29.
19. (a) H.W. Mielke, *Am. Sci.*, 1999, *87*(1), 62; (b) B.P. Lanphear, *Science*, 1998, *281*, 1617; (c) *Chem. Eng. News*, Aug. 3, 1998, 11.
20. Anon., *Environ. Sci. Technol.*, 1996, *30*, 375A.
21. Anon., *Environ. Sci. Technol.*, 1997, *31*, 261A; 1998, *32*, 175A.
22. D.S.T. Hjortenkrans, B.G. Bergback, and A.V. Haggerud, *Environ. Sci. Technol.*, 2007, *41*, 5224.
23. M. Saunders, *Chem. Eng. News*, Sep. 1, 2008, 6.
24. (a) J. Long, *Chem. Eng. News*, June 16, 1997, 11; (b) M.B. Blayney, J.S. Winn, and D.W. Nierenberg, *Chem. Eng. News*, May 12, 1997, 7; (c) T.Y. Toribara, T.W. Clarkson, and D.W. Nierenberg, *Chem. Eng. News*, June 16, 1997, 6; (d) K. Klug, *N. Engl. J. Med.*, 1998, *338*, 1692; (e) D.W. Nierenberg, R.E. Nordgren, M.B. Chang, R.W. Siegler, M.B. Blayney, F. Hochberg, T.Y. Toribara, E. Cernichiari, and T. Clarkson, *N. Engl. J. Med.*, 1998, *338*, 1672.
25. J.M. Berg and Y. Shi, *Science*, 1996, *271*, 1081.
26. K.C. Galbreath and C.J. Zygarlicke, *Environ. Sci. Technol.*, 1996, *30*, 2421 [mercury in coal combustion gases].
27. A.M. Rouhi, *Chem. Eng. News*, Oct. 21, 1996, 43.
28. A. Motluk, *New Sci.*, Jan. 30, 1999, 22.
29. D.E. Wilmington, *Delaware Estuary News*, 1995, *5*(3), 8.
30. (a) S. Cherian and M.M. Oliveira, *Environ. Sci. Technol.*, 2005, *39*, 9377; (b) D. van der Lelie, J.-P. Schwitzguebel, D.J. Glass, J. Vangroneveld, and A. Baker, *Environ. Sci. Technol.*, 2001, *35*, 446A; (c) T. Klaus-Joerger, R. Joerger, E. Olsson, and C.-G. Granqvist, *Trends Biotechnol.*, 2001, *19*, 15; (d) I. Raskin and B.E. Ensley, eds, *Phytoremediation of Toxic Metals: Using Plants to Clean Up the Environment*, Wiley, New York, 2000.
31. M. Murray, *Wilmington, Delaware News J.*, Mar. 9, 1995.
32. A. Jakob et al., *Environ. Sci. Technol.*, 1995, *29*, 242.
33. P.M. Randall, *Environ. Prog.*, 1995, *14*, 232.
34. M.J. Hynes and B. Jonson, *Chem. Soc. Rev.*, 1997, *26*, 133.
35. (a) I. Olmez and M.R. Ames, *Pure Appl. Chem.*, 1997, *69*(1), 35; (b) J. Johnson, *Environ. Sci. Technol.*, 1997, *31*, 218A; (c) C.M. Cooney, *Environ. Sci. Technol.*, 1998, *32*, 211A.
36. (a) C.A. Phillips, T. Gladding, and S. Maloney, *Chem. Br.*, 1994, *30*, 646; (b) K.M. Reese, *Chem. Eng. News*, Dec. 7, 1998, 80.
37. *INCO*, 1995 and 1998 Annual Reports, INCO Toronto, Canada.
38. V. Dutre and C. Vandecasteele, *Environ. Sci. Technol.*, 1998, *32*, 2782.
39. S.C. de Vito, *Kirk–Othmer Encyclopedia of Chemical Technology*, 4th ed., Wiley, New York, 1995, *16*, 212.

40. (a) K.A. O'Connell, *Waste Age's Recycling Times*, 1997, *9*(13), 6; *9*(25), 4; 1998, *10*(2), 5; (b) Bethlemen Lamp Recycling Co. literature Bethlehem Lamp Recycling, Bethlehem, PA.
41. Anon., *Chem. Eng. News*, June 6, 1994, 22.
42. (a) Anon., *Chem. Eng. News*, Mar. 25, 1996, 18; (b) J. Johnson, *Chem. Eng. News*, Nov. 23, 1998, 11.
43. D. Hanson, *Chem. Eng. News*, Feb. 27, 1995, 34.
44. Anon., *Chem. Eng. News*, Mar. 6, 1995, 20.
45. J. Rose, *Environ. Sci. Technol.*, 1995, *29*, 171A.
46. W.R. Toller and W.E. Mahoney, *Chem. Eng. News*, Apr. 25, 1994.
47. (a) R.E. Hester and R.M. Harrison, eds, *Mining and Its Environmental Impact*, Royal Society of Chemistry, Cambridge, UK, 1994; (b) C. Bodsworth, *The Extraction and Refining of Metals*, CRC Press, Boca Raton, FL, 1994; (c) A. Burkin, *Chemical Hydrometallurgy: Theory and Principles*, Imperial College Press, London, 2001.
48. J.H. Krieger and R.M. Baum, *Chem. Eng. News*, June 27, 1994, 14.
49. *INCO*, Annual Report, 1995, 26, INCO Toronto, Canada.
50. G. Parkinson, *Chem. Eng.*, 1996, *103*(4), 19.
51. G.S. Samdani, *Chem. Eng.*, 1995, *102*(6), 17.
52. (a) M. Simon, P. Jonk, G.L. Wuhl-Couturier, and M. Daunderer, *Ullmann's Encyclopedia Industrial Chemistry*, 5th ed., VCH, Weinheim, 1990, *A16*, 281; (b) G.E. Dunham, S.J. Miller, R. Chang, and P. Bergman, *Environ. Prog.*, 1998, *17*, 203; (c) S.V. Krishnan, B.K. Guellett, and W. Jozewicz, *Environ. Prog.*, 1997, *16*, 47; (d) J.A. Korpiel and R.D. Vidic, *Environ. Sci. Technol.*, 1997, *31*, 2319; (e) S.A. Benson, S.J. Miller, and E.S. Olson, Preprints *A.C.S. Div. Environ. Chem.*, 1998, *38*(2), 163; (f) R.V. Peltier, *Platts Power*, 2003, *147*(4), 40; (g) G. Ondrey, *Chem. Eng.*, 2008, *115*(4), 16.
53. M.I. Guijarro, S. Mendioroz, and V. Nunoz, *Ind. Eng. Chem. Res.*, 1998, *37*, 1088.
54. (a) G.S. Samdani, *Chem. Eng.*, 1995, *102*(4), 17; (b) P.G. Menon, *Appl. Catal. A*, 1995, *124*, N17.
55. G. Ondrey, *Chem. Eng.*, 1998, *105*(6), 27.
56. (a) U.S. Environmental Protection Agency, *Guides to Pollution Prevention: The Metal Finishing Industry*, Doc. EP 7.8, 1992, p. 76; (b) M. Valenti, *Mech. Eng.*, Oct. 1994, 68; (c) J.B. Kushner and A.S. Jushner, *Water and Waste Control for the Plating Shop*, 3rd ed., Hanser–Gardner, Cincinnati, OH, 1994; (d) L. Hartinger, *Handbook of Effluent Treatment and Recycling for the Metal Finishing Industry*, 2nd ed., ASM International & Finishing Publications, Novelty, OH, 1994.
57. M. Novak and W. Singer, *Kirk-Othmer Encyclopedia of Chemical Technology*, 4th ed., 1995, *16*, 228.
58. (a) W.G. Grot, *Macromol. Symp.*, 1994, *82*, 161; (b) C. Hogue, *Chem. Eng. News*, June 15, 2009, 24.
59. (a) C. Hogue, *Chem. Eng. News*, May 28, 2007, 26; July 2, 2007, 21; (b) Anon., *Chem. Eng. News*, Nov. 19, 2007, 36.
60. G. Parkinson, *Chem. Eng.*, 1998, *105*(8), 21.
61. Anon., *Presidential Green Chemistry Challenge*, EPA 744-K-96-001, July 1996, p. 20; (b) D.R. Baghurst, J. Barrett, and D.M.P. Mingos, *J. Chem. Soc. Chem. Commun.*, 1995, 323.
63. Y. Xu., A. Ohki, and S. Maeda, *Chem. Lett.*, 1998, 1015.
64. S. Ritter, *Chem. Eng. News*, Feb. 12, 2007, 19.
65. X. Wang and M.L. Brusseau, *Environ. Sci. Technol.*, 1995, *29*, 2632.
66. R.S. Bowman, G.M. Haggerty, R.G. Huddleston, D. Neel, and M.M. Flynn, Preprints A.C.S. *Div. Environ. Chem.*, 1994, *34*(1), 178; *Chem. Abstr.*, 1995, *123*, 92, 547.
67. D. Zhao, A.K. Sen Gupta, and L. Stewart, *Ind. Eng. Chem. Res.*, 1998, *37*, 4383.
68. Anon., *Presidential Green Chemistry Challenge*, EPA 744-K-96-001, July 1996, p. 33.
69. (a) A. Lezzi and S. Cobianco, *J. Appl. Polym. Chem.*, 1994, *54*, 889; (b) A. Lezzi, S. Cobianco, and A. Roggero, *J. Polym. Sci. A Polym. Chem.*, 1994, *32*, 1877.
70. M. Matlock and D.A. Atwood, ACS Orlando Meeting, Environ., Apr. 2002, 209.
71. (a) G.S. Samdani, *Chem. Eng.*, 1994, *101*(11), 19; (b) R.M. Izatt, K. Pawlak, and J.S. Bradshaw, *Chem. Rev.*, 1995, *95*, 2529; (c) R.M. Izatt, J.S. Bradshaw, R.L. Bruening, B.J. Tarbet, and M.L. Bruening, *Pure Appl. Chem.*, 1995, *67*, 1069; (d) R.M. Izatt, J.S. Bradshaw, and R.L. Bruening, *Pure Appl. Chem.*, 1996, *68*, 12373.
72. E. Brucher, B. Gyori, J. Emri, S. Jakab, Z Kovacs, P. Solymosi, and I. Toth, *J. Chem. Soc. Dalton Trans.*, 1995, 3353.
73. J. Masrchese, M. Campderros, and A Acosta, *J. Chem. Technol. Biotechnol.*, 1995, *64*, 293.
74. H. Tsukube, J.-I. Uenishi, N. Kojima, and O. Yonemitsu, *Tetrahedron Lett.*, 1995, *36*, 2257.
75. T.F. Baumann, J.G. Reynolds, and G.A. Fox, *Chem. Commun*, 1998, 1637.
76. J. Brown, L. Mercier, and T.J. Pinnavaia, *Chem. Commun*, 1999, 69.
77. A.T. Yordanov, J.T. Mague, and D.M. Roundhill, *Inorg. Chem.*, 1995, *34*, 5084.
78. Y. Koide, H. Terasaki, H. Sato. H. Shosenji, and K. Yamada, *Bull. Chem. Soc. Jpn.*, 1996, *69*, 785.
79. Reaxa and Aldrich literature, 2008, Reaxa, Manchester, UK and Aldrich, Milwaukee, WI.
80. (a) T. McEntee, *Kirk–Othmer Encyclopedia of Chemical Technology*, 4th ed., Wiley, New York, 1995, *14*, 174; (b) H.W. Rossmore, *Handbook of Biocide and Preservative Use*, Blackie Academic & Professional, Glasgow, Scotland, 1995; (c) W. Paulus, *Microbiocides for the Protection of Materials—A Handbook*, Chapman & Hall, London, 1994; (d) S.C. Stinson, *Chem. Eng. News*, Nov. 21, 1994, 46; (e) Anon., *Chem. Eng. News*, Aug. 5, 1996, 8; (f) M. McCoy, *Chem. Eng. News*, Nov. 9, 1998, 21; (g) B. Buecker and R. Post, *Chem. Eng. Prog.*, 1998, *94*(9), 45; (h) A. Shanley and T. Kamiya, *Chem. Eng.*, 1998, *105*(13), 69.
81. T.P. Schulz, H. Militz, M.H. Freeman, B. Goodell, and D.D. Nicholas, eds, *Development of Commercial Wood Preservatives*, ACS Symp. 982, Oxford University Press, New York, 2008.
82. D.E. Stilwell and K.D. Gorny–Piascik, Preprints A.C.S. *Div. Environ. Chem.*, 1996, *36*(2), 50.
83. (a) G. Monaghan, *Coat. Technol.*, 2007, *4*(7), 30; (b) M.S. Reisch, *Chem. Eng. News*, Aug. 9, 2004, 14.
84. (a) E.D. Suttie, *Chem. Ind.* (*Lond.*), 1997, 720; (b) Anon., *Chem. Eng. News*, Apr. 2, 2007, 26.
85. C. Frayner, *Modern Cooling Water Treatment Practice*, Chemical Publishing, New York, 1998.
86. D.W. Hairston, *Chem. Eng.*, 1995, *102*(10), 67.
87. J. Johnson, *Environ. Sci. Technol.*, 1995, *29*, 72A.
88. S. Bhattacharjee and R. Far, *TAPPI J.*, 1997, *80*(12), 43.

89. (a) Anon., *Presidential Green Chemistry Challenge*, EPA 744-K-96-001, July 1996, 39; (b) L. Raber, *Chem. Eng. News*, June 30, 1997, 7.
90. R. Mukherjee, *Chem. Eng. Prog.*, 1997, *93*(2), 26.
91. T. Chen, C. Dwyre-Gygax, R.S. Smith, and C. Brueil, *J. Agric. Food Chem.*, 1995, *43*, 1400.
92. S.T. Oh et al., *J. Appl. Polym. Sci.*, 1996, *59*, 1871.
93. J.D. Payne and D.W. Kudner, *Text Chem. Color*, 1996, *28*(5), 28.
94. (a) P.L. Layman, *Chem. Eng. News*, May 1, 1995, 23; (b) M.S. Reisch, *Chem. Eng. News*, Oct. 14, 1996; (c) S.J. deMora, *Tributyltin: Case Study of an Environmental Contaminant*, Cambridge University Press, Cambridge, UK 1996; (d) A.M. Rouhi, *Chem. Eng. News*, Apr. 27, 1998, 41; (e) M.R. Blumhorst, ed., *Tributyltin Compounds in the Aquatic Environment*; Preprints A.C.S. *Div. Environ. Chem.*, 1998, *38*(1), 91–138; (f) J.T. Walker and S. Surman, *Industrial Biofouling—Detection, Prevention and Control*, Wiley, Chichester, 2000.
95. K. Kannan, K. Senthilkumar, B.G. Loganathan, S. Takahashi, D.K. Odell, and S. Tanabe, *Environ. Sci. Technol.*, 1997, *31*, 296.
96. (a) K. Christen, *Environ. Sci. Technol.*, 1999, *33*, 11A; (b) Anon., *Chem. Ind.* (*Lond.*), 1999, 951.
97. I. Omae, *Chem. Rev.*, 2003, *103*, 3431.
98. *Presidential Green Chemistry Challenge*, EPA 744-K-96-001, July 1996, 4.
99. B.S. Kim, C.K. Seo, and C.J. You, U.S. patent 5,472,993 (1995).
100. (a) S. Stinson, *Chem. Eng. News*, Apr. 17, 1995, 23; (b) S. Borman, *Chem. Eng. News*, Mar. 7, 1994; (c) J.D. Adkins, A.E. Mera, M.A. Roe-Short, G.T. Pawlikowski, and R.F. Brady, Jr., *Prog. Org. Coat.*, 1996, *29*, 1; (d) J.R. Carroll, R.M. Bradley, and I. Kalmikoff, *Mod. Paint. Coat.*, 1993, *83*(10), 114.
101. (a) T. Provder, S. Malliprakash, S.H. Amin, A. Majid, and J. Texter, *Macromol. Symp.*, 2006, *242*, 279; (b) P.L. Short, *Chem. Eng. News*, Mar. 5, 2007, 37.
102. (a) X.-M. Li, D. Reinhoudt, and M. Crego-Calama, *Chem. Soc. Rev.*, 2007, *36*, 1350; (b) H.J. Lee and C. Wills, *Chem. Ind. (Lond.)*, Apr. 13, 2009, 21.
103. (a) T. Okino, E. Yoshimura, H. Hirota, and N. Fusetani, *Tetrahedron*, 1996, *52*, 9447; (b) N. Tomczak, D. Janczewski, M. Han, and G.J. Vansco, *Prog. Polym. Sci.*, 2009, 34, 393.
104. G.-Y.-S. Wang, M. Kuramoto, D. Uemura, A. Yamada, K. Yamaguchi, and K. Yazawa, *Tetrahedron Lett.*, 1996, *37*, 1813.
105. C. Potera, *Technol. Rev.*, 1998, *101*(4), 30.
106. L.M. Jarvis, *Chem. Eng. News*, June 9, 2008, 15.
107. N.A. Mohamed, *Polym. Degrad. Stabil.*, 1997, *56*, 317.
108. (a) P.A. Baguley and J.C. Walton, *Angew Chem. Int. Ed.*, 1998, *37*, 3073; (b) Anon., *Chem. Eng. News*, July 21, 2008, 47; (c) W.G. Schultz and L. Wang, *Chem. Eng. News*, Aug. 29, 2005; (d) K. Inoue, A. Sawada, I. Shibata, and A. Baba, *J. Am. Chem. Soc.*, 2002, *124*, 906; (e) A. Parsons, *Chem. Br.*, 2002, *38*(2), 42; (f) Anon., *Chem. Eng. News*, Nov. 13, 2006, 38.
109. S.A. Panova, G.K. Shestakov, and O.N. Temkin, *J. Chem. Soc. Chem. Commun.*, 1994, 977.
110. G. Horn and C.D. Frohning, U.S. patent 5,302,569 (1994).
111. J. Frankham, *Chem. Ind.* (*Lond.*), Mar. 1, 2004, 18.
112. (a) J.A. Lassner, L.B. Lasher, R.L. Koppel, and J.N. Hamilton, *Chem. Eng. Prog.*, 1994, *90*(8), 95; (b) *Chem. Eng. News*, June 3, 1996, 57.
113. (a) Anon., *Chem. Eng. News*, Sep. 13, 1999, 17; (b) B. Hileman, *Chem. Eng. News*, May 11, 1998, 24; (c) Anon., *Chem. Eng. News*, Dec. 4, 2000, 49.
114. (a) G.S. Samdani, *Chem. Eng.*, 1995, *102*(3), 19; (b) *Chem. Eng. News*, June 13, 1994, 27; (c) G. Parkinson, *Chem. Eng.*, 1996, *103*(10), 21; (d) A. Shanley, *Chem. Eng.*, 1999, *106*(2), 67.
115. M. Novotny, Z. Solc, and M. Trojan, *Kirk–Othmer Encyclopedia of Chemical Technology*, 4th ed., Wiley, New York, 1996, *19*, 29.
116. M. Jansen and H.P. Letschert, *Nature*, 2000, *404*, 980.
117. (a) W. Herbst and K. Hunger, *Industrial Organic Pigments*, 3rd ed., Wiley-VCH, New York, 2004; (b) H.M. Smith, *High Performance Pigments*, Wiley-VCH, Weinheim, 2002.
118. (a) D.W. Hairston, *Chem. Eng.*, 1996, *103*(6), 67; (b) D. Smock, *Plast World*, 1995, *53*(3), 24.
119. M. Srinivasarao, *Chem. Rev.*, 1999, *99*, 1935.
120. (a) W. Ostertag and N. Mronga, *Macromol. Symp.*, 1995, *100*, 163; (b) A. Stephens, U.S. patent 5,693,134 (1997); (c) R. Schmid, N. Mronga, C. Kaliba, W. Ostertag, and H. Schmidt, U.S. patent 5,693,135 (1997); (d) Anon., *Chem. Eng. News*, Sep. 22, 1997, 12; (e) G. Parkinson, *Chem. Eng.*, 1998, *105*(10), 21; (f) G. Pfaff and P. Reynders, *Chem. Rev.*, 1999, *99*, 1963; (g) S. Pellicori and M. Colton, *R&D* (*Cahners*), 1999, *41*(3), 39.
121. (a) M. Mokhtari, H.S. Park, H.W. Roesky, S.E. Johnson, W. Bolse, J. Conrad, and W. Plass, *Chem. Eur. J.*, 1996, *2*, 1269; (b) M. Mokhtari, H.S. Park, S.E. Johnson, W. Bolse, and H.W. Roesky, *Chem. Mater.*, 1997, *9*, 23.
122. (a) M. Jose-Yacaman, L. Rendon, J. Arenas, and M.C.S. Puche, *Science*, 1996, *273*, 223; (b) Anon., *Science*, 2008, *319*, 1315; (d) J. Kemsley, *Chem. Eng. News*, Mar. 3, 2008, 9.
123. N.L. Pickett, O. Masala, and J. Harris, *Mater. Matters* (*Aldrich*), 2008, *3*(1), 24; (b) M.A. Fierke, F. Li, and A. Stein, *Mater. Matters* (*Aldrich*), 2008, *3*(1), 10.
124. U. Sewekow, *Text Chem. Color*, 1996, *28*(1), 21.
125. Anon., *Text Chem. Color*, 1996, *28*(4), 11.
126. M.A. Ibrahim, M. El-Hossamy, M.S. Morsy, and B.M. Eid, *J. Appl. Polym. Sci.*, 2004, *93*, 1825.
127. (a) J. Sokolowska-Gajda, H.S. Freeman, and A. Reife, *Text Res. J.*, 1994, *64*, 388; (b) H.S. Freeman, J. Sokolowska-Gajda, A. Reife, L.D. Claxton, and V.S. Houk, *Text Chem. Color*, 1995, *27*(2), 13; (c) H.S. Freeman, A. Reife, and J. Sokolowska-Gajda, U.S. patent 5,376,151 (1994); (d) H.A. Bardole, H.S. Freeman, and A. Reife, *Text Res. J.*, 1998, *68*, 141.
128. B. Gebert, W. Saus, D. Knittel, H.-J. Buschmann, and E. Schollmeyer, *Text Res. J.*, 1994, *64*, 371.
129. Anon., *Chem. Eng. News*, Aug. 29, 2005, 14.
130. (a) R.S. Blackburn and A. Harvey, *Environ. Sci. Technol.*, 2004, *38*, 4034; (b) T. Bechtold, E. Burtscher, A. Turcanu, and O. Bobletter, *Text Res. J.*, 1997, *67*, 635; (c) T. Bechtold, E. Burtscher, and A. Turcanu, *Text Chem. Color*, 1998, *30*(8), 72.
131. T.J. Kelly, D.L. Smith, and J. Satola, *Environ. Sci. Technol.*, 1999, *33*, 81.
132. (a) C.Q. Yang, L. Xu, S. Li, and Y. Jiang, *Text Res. J.*, 1998, *68*, 457; (b) C.Q. Yang and X. Wang, *Text Res. J.*, 1996, *66*, 595; (c). C.M. Welch, *Text Chem. Color*, 1997, *29*(2), 21.

133. M. Raheel and C. Guo, *Text Res. J.*, 1998, *68*, 571.
134. G. Xu and C.Q.-X. Yang, *J. Appl. Polym. Chem.*, 1999, *74*, 907.
135. (a) R. Peters and R. Sikorski, *Science*, 1998, *280*, 457; (b) J. Dreyfuss, *Mod. Maturity*, 2000, *43R*(4), 68.
136. L.T. Taylor, *Presidential Green Chemistry Challenge*, EPA 744-K-96-001, July 1996, 9.
137. P.M. Randall, *Environ. Prog.*, 1995, *14*, 232.
138. S.-C. Hwang and W.R. Weltmer, Jr., *Kirk–Othmer Encyclopedia of Chemical Technology*, 4th ed., Wiley, New York, 1995, *13*, 1.
139. R.T. Wegh, H. Donker, K. Oskam, and A. Meijerink, *Science*, 1999, *283*, 663.
140. W.H. Green, K.P. Le, J. Grey, T.T. Au, and M.J. Sailor, *Science*, 1997, *276*, 1826.
141. (a) G. Fasol, *Science*, 1997, *278*, 1902; (b) M.J. Sailor, *Science*, 1997, *278*, 2036.
142. (a) E. Mills, *Science*, 2005, *308*, 1263; (b) E.E. Schubert and J.K. Kim, *Science*, 2005, *308*, 1276; R. Gaugan, *R&D (Cahners)*, 2005, *46*(11), 36.
143. (a) Anon., *Chem. Eng. News*, Apr. 10, 1995; Mar. 18, 1996; (b) H. Clewell, *Chem. Eng. News*, May 13, 1996; (c) B. Hileman, *Chem. Eng. News*, July 27, 1998, 13.
144. J. Zayed, B. Hong, and G. L'Esperance, *Environ. Sci. Technol.*, 1999, *33*, 3341.
145. S.M. Gates, *Chem. Rev.*, 1996, *96*, 1519, 1530.
146. T.T. Kodas and M. Hampden-Smith, *The Chemistry of Metal Chemical Vapor Deposition*, VCH, Weinheim, 1994.
147. D.A. Atwood, V.O. Atwood, A.H. Cowley, R.A. Jones, J.L. Atwood, and S.G. Bott, *Inorg. Chem.*, 1994, *33*, 3251.
148. Y. Cheng, T.J. Emge, and J.C. Brennan, *Inorg. Chem.*, 1994, *33*, 3711.
149. R.F. Service, *Science*, 1996, *271*, 922.
150. M.A. Beswick, P.R. Raithby, C.A. Russell, A. Steiner, K.L. Verhorevoort, G.N. Ward, and D.S. Wright, *Angew Chem. Int. Ed. Engl.*, 1995, *34*, 2662.
151. (a) D.G. Bailey, P.R. Buechler, A.L. Everett, and S.N. Feairheller, *Kirk–Othmer Encyclopedia of Chemical Technology*, 3rd ed., Wiley, New York, 1981, *14*, 208; (b) T.C. Thorstensen, *Kirk–Othmer Encyclopedia Chemical Technology*, 4th ed., Wiley, New York, 1995, *15*, 159; (c) E. Heidemann, *Ullmann's Encyclopedia of Industrial Chemistry*, 5th ed., VCH, Weinheim, 1990, *A15*, 259; (d) P.L. Kronick, *Chemtech*, 1995, *25*(7), 31; (e) A.D. Covington, *Chem. Soc. Rev.*, 1997, *26*, 111; (f) T. Covington, G. Lampard, and M. Pennington, *Chem. Br.*, 1998, *34*(4), 40.
152. N. Natchimuthu, G. Radhakishnan, K. Palanivel, K. Ramamurthy, and J.S. Anand, *Polym. Int.*, 1994, *33*, 329.
153. T.F. O'Dwyer and B.K. Hodnett, *J. Chem. Technol. Biotechnol*, 1995, *62*, 30.
154. R.C. deMesa, Preprints A.C.S. *Div. Environ. Chem.*, 1994, *34*(2), 309.
155. D. Petruzzelli, R Passino, and G. Tiravanti, *Ind. Eng. Chem. Res.*, 1995, *34*, 2612.
156. E. Heidemann, *Ullmann's Encyclopedia of Industrial Chemistry*, 5th ed., VCH, Weinheim, 1990, *A15*, 259.
157. (a) S. Saravanabhavan, P. Thanikaivelan, J. Rao, and B.U. Nair, *Environ. Sci. Technol.*, 2005, *39*, 3776; (b) S. Saravanabhavan, P. Thanikaivelan, J.R. Rao, B.U. Nair, and T. Ramasami, *Environ. Sci. Technol.*, 2008, *42*, 1731.
158. C. Simon and A. Pizzi, *J. Appl. Polym. Sci.*, 2003, *88*, 1889.
159. (a) L. Pawlowski, *The Science and Engineering of Thermal Spray Coatings*, John Wiley, New York, 1995; (b) K.H. Stern, ed., *Metallurgical and Ceramic Protective Coatings*, Chapman & Hall, London, 1996; (c) C.C. Berndt and S. Sampath, eds, *1995 Advances in Thermal Spray Science and Technology: Proceedings of the 8th National Thermal Spray Conference*, ASM International, Novelty, OH, 1995; (d) A. Ohmori, ed., *ITSC '95: Thermal Spraying—Current Status and Future Trends, Proceedings of the 1995 International Thermal Spray Conference*, ASM International, Novelty, OH, 1995; (e) W. Miller, *Encyclopedia of Polymer Science Engineering*, 2nd ed., 1987, *9*, 580; (f) T.T. Kodas and M. Hampden-Smith, *The Chemistry of Metal Chemical Vapor Deposition*, VCH, Weinheim, 1994; (g) W.S. Rees, Jr., *CVD of Nonmetals*, VCH, Weinheim, 1996; (h) V. Comello, *R&D (Cahners)*, 1998, *40*(6), 71; (i) S. Almeida, *Ind. Eng. Chem. Res.*, 2001, *40*, 15.
160. G.S. Samdani, *Chem. Eng.*, 1994, *101*(12), 23.
161. Guide to Cleaner Technologies—Alternative Metal Finishes. EPA/625/R-94/007, Cincinnati, OH, Sep. 1994.
162. (a) Anon., *Environ. Sci. Technol.*, 1995, *29*, 407A, 548A; (b) Anon., *Chem. Eng. News*, July 10, 1995, 19.
163. U.S.E.P.A. *Green Chemistry Award Entries*, 2008, *21*, Commercial Fluid Power LLC, Dover, OH.
164. Y. Zhao, C. Bao, R. Feng, and T.J. Mason, *J. Appl. Polym. Sci.*, 1998, *68*, 1411.
165. M.J. Semmens, D. Dillon, and B.C. Riley, *Environ. Prog.*, 2001, *20*, 251.
166. Anon., *Chem. Eng. News*, Jan. 22, 1996, 21.
167. (a) L. Delaude, P. Laszlo, and P. Lehance, *Tetrahedron Lett.*, 1995, *36*, 8505; (b) V.K. Sharma, Preprints A.C.S. *Div. Environ. Chem.*, 2000, *40*(1), 131.
168. M.D. Johnson, U.S. patent 5,746,944 (1998).
169. S. Takano, M. Moriya, K. Tanaka, and K. Ogasawara, *Synthesis*, 1994, 687.
170. K. Weissermel and H.-J. Arpe, *Industrial Organic Chemistry*, 2nd ed., VCH. Weinheim, 1993, 204.
171. K. Weissermel and H.-J. Arpe, *Industrial Organic Chemistry*, 2nd ed., VCH, Weinheim, 1993, 326.
172. S.K. Ritter, *Chem. Eng. News*, July 9, 2007, 35.
173. D. Ramprasad, A.G. Gilicinski, T.J. Markley, and G.P. Pez, *Inorg. Chem.*, 1994, *33*, 2841.
174. (a) J.E. Lyons, P.E. Ellis, Jr., and H.K. Myers, Jr., *J. Catal.*, 1995, *155*, 59; (b) J.G. Goll, K.T. Moore, A. Ghosh, and M.J. Therien, *J. Am. Chem. Soc.*, 1996, *118*, 8344; (c) D. Dolphin, T.G. Traylor, and L.Y. Xie, *Acc. Chem. Res.*, 1997, *30*, 251; (d) K.T. Moore, I.T. Horvath, and M.J. Therien, *J. Am. Chem. Soc.*, 1997, *119*, 1791; (e) A. Porhiel, A. Bondon, and J. Leroy, *Tetrahedron Lett.*, 1998, *39*, 4829.
175. (a) O. Reiser, *Angew Chem. Int. Ed. Engl.*, 1994, *33*, 69; (b) C.L. Hill, ed., *Activation and Functionalization of Alkanes*, Wiley-Interscience, New York, 1989; (c) M.T. Benson, T.R. Cundari, E.W. Moody, et al., *J. Organomet. Chem.*, 1995, *504*, 1–152 [whole issue]; (d) J.A. Davies, P.L. Watson, J.F. Leibman, and A. Greenberg, *Selective Hydrocarbon Activation*, VCH, Weinheim, 1990; (e) B.K. Warren and T. Oyama, eds, *Heterogeneous Hydrocarbon Oxidation*. ACS Symp. 638, Washington, DC, 1996; (f) A.E. Shilov and

G.B. Shulpin, *Chem. Rev.*, 1997, *97*, 2879; (g) S.S. Stahl, J.H. Labinger, and J.E. Bercaw, *Angew Chem. Int. Ed. Engl.*, 1998, *37*, 2181.

176. (a) S.T. Oyama and J.W. Hightower, *Catalytic Selective Oxidation*, ACS Symp. 523, Washington, DC, 1993; (b) V.C. Corberan and S.V. Bellon, *New Developments in Selective Oxidation II*, Elsevier, Amsterdam, 1994; (c) W.A. Herrmann, H. Heaney, R. Sheldon, W. Adam, L. Hadjiarapoglu, F. Hoft, S. Warwel, M. Sojka, M.R. Klass, J. Fossey, J. Sorba, D. Lefort, and K. Dear, *Top. Curr. Chem.*, 1993, *164*, 1–125; (d) G. Centi, M. Misono, and accompanying papers, *Catal. Today*, 1998, *41*, 287–457 [solid catalysts]; (e) W.W. Wckenfelder, A.R. Bowers, and J.A. Roth, *Chemical Oxidation: Technologies for the Nineties, vol. 6, Proceedings*, Technomic Publishing, Lancaster, PA, 1997; (f) Anon., *Adv. Synth. Catal.*, 2004, *346*, 107–375; (g) Anon., *Catal. Today*, 2004, 91–92 [whole issue]; (h) T. Katsuki, *Asymmetric Oxidation Reactions*, Oxford Universsity Press, Oxford, 2001; (i) G. Canti, F. Cavani, and F. Trifiro, *Selective Oxidation by Heterogeneous Catalysis*, Kluwer Academic/Plenum Publishers, New York, 2000; (j) B.K. Hodnet, *Heterogeneous Catalytic Oxidation—Fundamental and Technological Aspects of the Selective and Total Oxidation of Organic Compounds*, Wiley, 2000; (k) T. Mallat and A. Baiker, eds, *Catal. Today*, 2000, *57*, 1–166; (l) P. Arpentinier, F. Cavani, and F. Trifio, *The Technology of Catalytic Oxidations*, Editions Technip, Paris, 2001; (j) F. Cavani and J.H. Teles, *ChemSusChem*, 2009, 2, 508.
177. I.M. Klotz, D.M. Kurtz, Jr., M.H. Dickman, M.T. Pope, D.H. Busch, and N.W. Alcock, *Chem. Rev.*, 1994, *94*(3), 567–856 [whole issue].
178. (a) B. Meunier, *Chem. Rev.*, 1992, *92*, 1411; (b) R.A. Sheldon, ed., *Metalloporphyrins in Catalytic Oxidation Dekker*, New York, 1994; (c) G.A. Barf and R.A. Sheldon, *J. Mol. Catal.*, 1995, *102*, 23; (d) S.-I. Murahashi, *Angew Chem. Int. Ed. Engl.*, 1995, *34*, 2443; (e) T. Naota, H. Takaya, and S.-I. Murahashi, *Chem. Rev.*, 1998, *98*, 2607–2622 [using Ru]; (f) T. Hirao, *Chem. Rev.*, 1997, *97*, 2707 [using V]; (g) R.A. Sheldon, I.W.C.E. Arends, and H.E.B. Lempers, *Catal. Today*, 1998, *41*, 387 [catalyst on clays, zeolites, etc.]; (h) R.A. Sheldon, ed., *J. Mol. Catal. A: Chem.*, 1997, *117*, 1–471; (i) J.-E. Backvall, ed., *Modern Oxidation Methods*, Wiley-VCH, Weinheim, 2004; (j) C. Limberg, *Angew Chem. Int. Ed.*, 2003, *42*, 5932; (k) J.-M. Bregeault, *Dalton Trans.*, 2003, 3289; (l) J.M. Thomas, *J. Mol. Catal. A*, 1999, *146*, 77.
179. (a) B. Meunier, ed., *J. Mol. Catal. A: Chem.*, 1996, *113*, 1–422; (b) Y. Moro-oka and M. Akita, *Catal. Today*, 1998, *41*, 327; (c) W. Nam, *Acc. Chem. Res.*, 2007, *40*(7) [whole issue].
180. (a) A. Sobkowiak, H.-C. Tung, and D.T. Sawyer, *Prog. Inorg. Chem.*, 1992, *40*, 291; (b) K.M. Dear, *Top. Curr. Chem.*, 1993, *164*, 115; (c) S. Wilson, *Chem. Ind. (Lond.)*, 1994, 255; (d) C.W. Jones and J.H. Clark, eds, *Applications of Hydrogen Peroxide and Derivatives*, Royal Society of Chemistry, Cambridge, 1999; (e) W. Adam, ed., *Peroxide Chemistry Research Report—Mechanistic and Preparative Aspects of Oxygen Transfers*, Wiley-VCH, Weinheim, 2000.
181. (a) G. Strukul, ed., *Catalytic Oxidations with Hydrogen Peroxide*, Kluwer Academic, the Netherlands, 1993; (b) R.A. Sheldon, *Chemtech*, 1991, 566; *Top. Curr. Chem.*, 1993, *164*, 21.
182. (a) T. Mukaiyama and T. Yamada, *Bull. Chem. Soc. Jpn.*, 1995, *68*, 17; (b) W. Partenheimer, *Catal. Today*, 1995, *23*, 69.
183. (a) W. Adam and L. Hadjiarapoglu, *Top. Curr. Chem.*, 1993, *164*, 45; (b) R. Curci, A. Dinoi, and M.F. Rubino, *Pure Appl. Chem.*, 1995, *67*, 811.
184. B. Meunier and A. Sorokin, *Acc. Chem. Res.*, 1997, *30*, 470.
185. (a) K. Weissermel and H.-J. Arpe, *Industrial Organic Chemistry*, VCH, Weinheim, 1993, 88, 351, 238, 162, 142, 265, 281, 366, 302, 187; (b) F. Cavani and F. Trifiro, *Chemtech*, 1994, *24*(4), 18 [maleic anhydride].
186. T. Hirao, M. Higuchi, B. Hatano, and I. Ikeda, *Tetrahedron Lett.*, 1995, *36*, 5925.
187. (a) R.M. Contractor, H.S. Horowitz, G.M. Sisler, and E. Bordes, *Catal. Today*, 1997, *37*, 51; (b) B. Kubias, U. Rodemerck, H.-W. Zanthoff, and M. Meisel, *Catal. Today*, 1996, *32*, 243; (c) J. Haggin, *Chem. Eng. News*, Apr. 3, 1995, 20.
188. M. Jacoby, *Chem. Eng. News*, Sep. 15, 2008, 9.
189. A.P.E. York, A. Bruckner, G.-U. Wolf, P.-M. Wilde, and M. Meisel, *Chem. Commun.*, 1996, 239.
190. S.L. Suib, *Acc. Chem. Res.*, 2008, *41*, 479.
191. J. Zhu and S.C. Tsang, *Catal. Today*, 2003, *81*, 673.
192. E. Garcia-Verdugo, J. Fraga-Dubreuil, P.A. Hamley, W.B. Thomas, K. Whiston, and M. Poliakoff, *Green Chem.*, 2005, *7*, 294; M. Poliakoff, *Adv. Synth. Catal.*, 2004, *346*, 307.
193. G. Prinz, D. Heinz, and W.F. Holderich, Preprints A.C.S. *Div. Environ. Chem*, 2000, *40*(1), 179.
194. (a) I. Nikov and K. Paev, *Catal. Today*, 1995, *24*, 41; (b) M. Besson, F. Lahmer, P. Gallezot, P. Fuertes, and G. Fleche, *J. Catal.*, 1995, *152*, 116; (c) P. Gallezot, *Catal. Today*, 1997, *37*, 405.
195. A. Abbadi and H. van Bekkum, *Appl. Catal. A*, 1995, *124*, 409.
196. A.-B. Crozon, M. Besson, and P. Gallezot, *New J. Chem.*, 1998, *22*, 269.
197. S. Biella, G.L. Castiglioni, C. Fumigalli, L. Prati, and M. Rossi, *Catal. Today*, 2002, *72*, 143.
198. L. Prati and M. Rossi, *J. Catal.*, 2000, *176*, 552.
199. P.W. Roesky, G. Canseco-Melchor, and A. Zulys, *Chem. Commun.*, 2004, 738.
200. K. Isobe and H. Nishise, *J. Mol. Catal. B*, 1995, *1*, 37.
201. (a) A. Potthast, T. Rosenau, C.L. Chen, and J.S. Gratzl, *J. Mol. Catal. A*, 1996, *108*, 5; (b) T. Rosenau, A. Potthast, C.L. Chen, and J.S. Gratzl, *Synth. Commun.*, 1996, *26*, 315.
202. N. Komiya, T. Nakae, H. Sato, and T. Naota, *Chem. Commun.*, 2006, 4829.
203. (a) G.-J. ten Brink, I.W.C.E. Arends, and R.A. Sheldon, *Science*, 2000, *287*, 1636; (b) R.A. Sheldon, I.W.C.E. Arends, G.-J. Ten Brink, and A. Dijksman, *Acc. Chem. Res*, 2002, *35*, 774.
204. K. Ebitani, K. Motokura, T. Mizugaki, and K. Kaneda, *Angew Chem. Int. Ed.*, 2005, *44*, 3423.
205. M. Freemantle, *Chem. Eng. News*, Jan. 23, 2006, 8.
206. G. Maayan, B. Ganchegui, W. Leitner, and R. Neumann, *Chem. Commun.*, 2006, 2230.
207. M. Gilhespy, M. Lok, and X. Baucherel, *Chem. Commun.*, 2005, 1085.
208. Y. Uozumi and R. Nakao, *Angew Chem. Int. Ed.*, 2003, *43*, 194.
209. T. Yoshikuni, *J. Chem. Technol. Biotechnol.*, 1994, *59*, 353.

210. M. Lajunen and A.M.P. Koskinen, *Tetrahedron Lett.*, 1994, *35*, 4461.
211. M.R. Sivik, K.J. Stanton, and L.A. Paquette, *Org. Synth.*, 1993, *72*, 57.
212. K. Ding, Y. Wang, L. Zhang, Y. Wu, and T. Matsuura, *Tetrahedron*, 1996, *52*, 1005.
213. T. Sakamoto, H. Yonehara, and C. Pac, *J. Org. Chem.*, 1994, *59*, 6859.
214. S. Habaue, T. Sako, and Y. Okamoto, *Macromolecules*, 2003, *88*, 2604.
215. (a) H. Frei, F. Blatter, and H. Sun, *Chemtech*, 1996, *26*(6), 24; (b) X. Li and V. Ramamurthy, *J. Am. Chem. Soc.*, 1996, *118*, 10666.
216. Y. Nishiyama, Y. Nakagawa, and N. Mizuno, *Angew Chem. Int. Ed.*, 2001, *40*, 3639.
217. A. M. Khenkin, L. Weiner, and R. Neumann, *J. Am. Chem. Soc.*, 2005, *127*, 9988.
218. S.L. Jain and B. Sain, *Chem. Commun.*, 2002, 1040.
219. (a) A.K. Roby and J.P. Kingsley, *Chemtech*, 1996, *26*(2), 39; (b) Anon., *Presidential Green Chemistry Challenge*, EPA 744-K-96-001, July 1996, 30; (c) J. Kwamya and M. Greene, Preprints A.C.S. *Div. Environ. Chem.*, 1997, *37*(1), 356.
220. *Presidential Green Chemistry Challenge*, EPA 744-K-96-001, July 1996, 37.
221. (a) G. Grigoropoulou, J.H. Clark, and J.A. Elings, *Green Chem.*, 2003, *5*, 1; (b) C.W. Jones, *Applications of Hydrogen Peroxide and Derivatives*, Royal Society of Chemistry, Cambridge, 1999.
222. S. Banfi, F. Legramandi, F. Montanari, G. Pozzi, and S. Quici, *J. Chem. Soc. Chem. Commun.*, 1991, 1285.
223. P.-P. Knops-Gerrits, D. de Vos F. Thibault-Starzyk, and P.A. Jacobs, *Nature*, 1994, *369*, 543.
224. S. Ueno, K. Yamaguchi, K. Yoshida, K. Ebitani, and K. Koneda, *Chem. Commun.*, 1998, 295.
225. U.R. Pillai, E. Sahle-Demessie, and R.S. Varma, *Tetrahedron Lett.*, 2002, *43*, 2909.
226. M. Bosing, A Noh, I. Loose, and B. Krebs, *J. Am. Chem. Soc.*, 1998, *120*, 7552.
227. A. Haimov, H. Cohen, and R. Neumann, *J. Am. Chem. Soc.*, 2004, *126*, 11762.
228. N. Gharah, S. Chakraborty, A.K. Muklerjee, and R. Bhattacharyya, *Chem. Commun.*, 2004, 2630.
229. G. Anilkumar, B. Bitterlich, F.G. Gelalcha, M.K. Tse, and M. Beller, *Chem. Commun.*, 2007, 289.
230. Y. Kon, Y. Usui, and K. Sato, *Chem. Commun.*, 2007, 4399.
231. P. Nagaraju, N. Pasha, P.S.S. Prasad, and N. Lingaiah, *Green Chem.*, 2007, *9*, 1126.
232. D. Sloboda-Rozner, P.L. Alsters, and R. Neumann, *J. Am. Chem. Soc.*, 2003, *125*, 5280.
233. K. Sato, M. Aoki, and R. Noyori, *Science*, 1998, *281*, 1646.
234. P.A. Ganeshpure and W. Adam, *Synthesis*, 1996, 179.
235. (a) W. Adam, W.A. Herrmann, J. Lin, C.R. Saha-Moller, R.W. Fischer, D.G. Correia, and M. Shimizu, *Angew Chem. Int. Ed. Engl.*, 1994, *33*, 2475; *J. Org. Chem.*, 1994, *59*, 8281; *J. Mol. Catal. A*, 1995, *97*, 15; (b) W.A. Hermann, J.D. Correia, F.E. Kuhn, G.R.J. Artus, and C.C. Romao, *Chem. Eur. J.*, 1996, *2*(2), 168; (c) J.H. Espenson, *Chem. Commun.*, 1999, 479.
236. (a) J. Rudolph, K.L. Reddy, J.P. Chiang, and K.B. Sharpless, *J. Am. Chem. Soc.*, 1997, *119*, 6189; (b) C. Coperet, H. Adolfsson, and K.B. Sharpless, *Chem. Commun.*, 1997, 1565; (c) H. Rudler, J.R. Gregorio, B. Denise, J.-M.L. Bregeault, and A. Deloffre, *J. Mol. Catal. A: Chem.*, 1998, *133*, 255.
237. J.P. McEldoon, A.R. Pokora, and J.S. Dordick, *Enzyme Microb. Technol.*, 1995, *17*, 359.
238. S. Warwel and M.R. Klass, *J. Mol. Catal. B*, 1995, *1*, 29.
239. M. Fujita, M. Costas, and L. Que, Jr., *J. Am. Chem. Soc.*, 2003, *125*, 99012.
240. J.W. de Boer, J. Brinksma, W.R. Browne, A. Meetlllsma, P.L. Alsters, R. Hage, and B.L. Feringa, *J. Am. Chem. Soc.*, 2005, *127*, 7990.
241. Y. Usui, K. Sato, and M. Tanaka, *Angew Chem. Int. Ed.*, 2003, *42*, 5623.
242. N.K. Mal, A. Bhaumik, P. Kumar, and M. Fujiwara, *Chem. Commun.*, 2003, 872.
243. S. Lin, Y. Zhen, S.M. Wang, and Y.-M. Dei, *J. Mol. Catal. A*, 2000, *156*, 113.
244. D. Sloboda-Rozner and R. Neumann, *Green Chem.*, 2006, *8*, 679.
245. B.M. Choudary, B. Bharathi, C.V. Reddy, M.L. Kantam, and K.V. Raghavan, *Chem. Commun.*, 2001, 1736.
246. (a) J. Muzart, *Synthesis*, 1995, 1325; (b) A. McKillop and W.R. Sanderson, *Tetrahedron*, 1995, *51*, 6145; (c) D.T.C. Yang, Y.H. Cao, and G.W. Kabalka, *Synth. Commun.*, 1995, *25*, 3695.
247. H. Heaney, *Aldrichim. Acta*, 1993, *26*, 35.
248. (a) W. Adam, R. Kumar, T.I. Reddy, and M. Renz, *Angew Chem. Int. Ed. Engl.*, 1996, *35*, 880; (b) T.R. Boehlow and C.D. Spilling, *Tetrahedron Lett.*, 1996, *37*, 2717.
249. J. Finlay, M.A. McKervey, and H.Q.N. Gunaratne, *Tetrahedron Lett.*, 1998, *39*, 5651.
250. E. Marcantoni, M. Petrini, and O. Polimanti, *Tetrahedron Lett.*, 1995, *36*, 3561.
251. BioSafe Systems, Glastonbury, CT.
252. W.R. Thiel, M. Angstl, and N. Hansen, *J. Mol. Catal. A*, 1995, *103*, 5.
253. S.V. Kotov and E. Balbolov, *J. Mol. Catal. A*, 2001, *176*, 41.
254. J.-H. Ahn and D.C. Sherrington, *Chem. Commun.*, 1996, 643.
255. R. Clarke and D.J. Cole-Hamilton, *J. Chem. Soc. Dalton Trans.*, 1993, 1913.
256. U. Arnold, R.S. de Cruz, D. Mandelli, and U. Schuchardt, *J. Mol. Catal. A*, 2001, *165*, 149.
257. A.L. Villa, D.E. de Vos, F. Verpoort, B.F. Sels, and PA. Jacobs, *J. Catal.*, 2001, *198*, 223.
258. (a) K. Kamata, K. Yonehara, Y. Sumida, K. Yamaguchi, S. Hikicki, and N. Mizuno, *Science*, 2003, *300*, 964; (b) N. Mizuno, S. Hikichi, K. Yamaguchi, S. Uchida, Y. Nakagawa, K. Uehara, and K. Kamata, *Catal. Today*, 2006, *117*, 32.
259. J.M. Fraile, J.I. Garcia, J.A. Mayoral, and F. Figueras, *Tetrahedron Lett.*, 1996, *37*, 5995.
260. B.M. Weckjuysen, A.A. Verberckmoes, I.P. Vannijvel, J.A. Pelgrims, P.L. Buskins, P.A. Jacobs, and R.A. Schoonheydt, *Angew Chem. Int. Ed. Engl.*, 1995, *34*, 2652.
261. K. Krohn, *Synthesis*, 1997, 1115.
262. (a) M.G. Clerici and P. Ingallina, *J. Catal.*, 1993, *140*, 71; (b) U. Romano, M.G. Clerici, P. Ingallina, L. Rossi, G. Petrini, G. Leofanti, M.A. Mantegazza, and F. Pignataro, Preprints

A.C.S. *Div. Environ. Chem.*, 1994, *34*(2), 320, 325, 328; (c) C. Dartt and M.E. Davis, *Catal. Today*, 1994, *19*, 151; (d) A. Thangaraj, R. Kumar, and P. Ratnasamy, *J. Catal.*, 1991, *131*, 294; (e) J. Haggin, *Chem. Eng. News*, Aug. 5, 1996, 26; (f) M. Freemantle, *Chem. Eng. News*, July 29, 1996, 47; (g) I.W.C.E. Arends, R.A. Sheldon, M. Wallau, and U. Schuchardt, *Angew Chem. Int. Ed. Engl.*, 1997, *36*, 1144; (h) G.N. Vayssilov, *Catal. Rev. Sci. Eng.*, 1997, *39*, 209; (i) R. Murugavel and H.W. Roesky, *Angew Chem. Int. Ed. Engl.*, 1997, *36*, 477; (j) R.A. Sheldon, M. Wallau, I.W.C.E. Arends, and U. Schuchardt, *Acc. Chem. Res.*, 1998, *31*, 485.

263. F. Rivetti, personal communication, 1998.
264. L. Chen, G.K. Chuah, and S. Jaenicke, *J. Mol. Catal. A: Chem.*, 1998, *132*, 281.
265. G. Thiele and E. Roland, U.S. patent 5,525,563 (1996).
266. (a) A. Corma, M.T. Navarro, and J.P. Pariente, *J. Chem. Soc. Chem. Commun.*, 1994, 147; (b) T. Blasco, A. Corma, M.T. Navarro, and J.P. Pariente, *J. Catal.*, 1995, *156*, 65; (c) A. Corma, P. Esteve, A. Martinez, and S. Valencia, *J. Catal.*, 1995, *152*, 18; (d) A. Corma, P. Esteve, and A. Martinez, *J. Catal.*, 1996, *161*(11), 1339; (e) T. Blasco, M.A. Camblor, A. Corma, P. Esteve, A. Martinez, C. Prieto, and S. Valencia, *Chem. Commun.*, 1996, 2367; (f) M.A. Camblor, A. Corma, P. Esteve, A. Martinez, and S. Valencia, *Chem. Commun.*, 1997, 795; (g) A. Corma, J.L. Jorda, M.T. Navarro, and F. Rey, *Chem. Commun.*, 1998, 1899; (h) A. Corma, M. Domine, J.A. Gaona, J.L. Jorda, M.T. Navarro, F. Rey, J. Perez, Pariente, J. Tsuji, B. McCulloch, and L.T. Nemeth, *Chem Commun.*, 1998, 2211.
267. J.M. Frail, J.I. Garcia, J.A. Mayoral, S.C. de Menorval, and F. Rachdi, *J. Chem. Soc. Chem. Commun.*, 1995, 539.
268. (a) W. Adam, C.R. Saha-Moller, and C.-G. Zhao, *Org. Reactions*, 2002, *1*, 219–516; (b) W. Adam, C.R. Saha-Moller, and P.A. Ganeshpure, *Chem. Rev.*, 2001, *101*, 3499–3548; (c) R. Curci, L. D'Accolti, and C. Tusco, *Acc. Chem. Res.*, 2006, *39*, 1.
269. W. Adam and A.K. Smerz, *Tetrahedron*, 1996, *52*, 5799.
270. G. Asensio, R. Mello, M.E. Gonzalez-Nunez, G. Castellano, and J. Corral, *Angew Chem. Int. Ed. Engl.*, 1996, *35*, 217.
271. S.E. Denmark, D.C. Forbes, D.S. Hays, J.S. DePue, and R.G. Wilde, *J. Org. Chem.*, 1995, *60*, 1391.
272. A. Altamura, L. D'Accolti, A. Detomaso, A. Dinoi, M. Fiorentino, C. Fusco, and R. Curci, *Tetrahedron Lett.* 1998, *39*, 2009.
273. R.W. Murray, M. Singh, and N.P. Rath, *Tetrahedron Lett.*, 1998, *39*, 2899.
274. A. Arnone, R. Bernardi, M. Cavicchioli, and G. Resnati, *J. Org. Chem.*, 1995, *60*, 2314.
275. P. Ceccherelli, M. Curini, M.C. Marcotullio, F. Epifano, and O. Rosati, *Synth. Commun.*, 1998, *28*, 3057.
276. J. Wessel and R.H. Crabtree, *J. Mol. Catal. A: Chem.*, 1996, *113*, 13.
277. A.J. Bailey and B.R. James, *Chem. Commun.*, 1996, 2343.
278. H. Ohtake, T. Higuchi, and M. Hirobe, *J. Am. Chem. Soc.*, 1992, *114*, 10660.
279. J.T. Groves, M. Bonchio, T. Carofiglio, and K. Shalyaev, *J. Am. Chem. Soc.*, 1996, *118*, 8961.
280. Y. Iamamoto, M.D. Assis, K.J. Ciuffi, C.M.C. Prado, B.Z. Prellwitz, M. Moraes, O.R. Nascimento, and H.C. Sacco, *J. Mol. Catal. A: Chem.*, 1997, *116*, 365.
281. C.-C. Guo, *J. Catal.*, 1998, *178*, 182.
282. J. Chen and C.M. Che, *Angew Chem. Int. Ed.*, 2004, *43*, 4950.
283. J.-L. Zhang and C.-M. Che, *Org. Lett.*, 2002, *4*, 1911.
284. (a) T. Wirth, *Angew Chem. Int. Ed.*, 2005, *44*, 3656–3665; (b) R.M. Moriarty, *Hypervalent Iodine in Organic Chemistry*, Wiley-Interscience, Hoboken, NJ.
285. C.J. Mazac, *Kirk–Othmer Encyclopedia of Chemical Technology*, 3rd ed., Wiley, New York, 1981, *13*, 671.
286. T. Hamada, T. Fukuda, H. Imanishi, and T. Katsuki, *Tetrahedron*, 1996, *52*, 515.
287. F.S. Vinhado, C.M.C. Prado-Manso, H.C. Sacco, and Y. Iamamoto, *J. Mol. Catal A*, 2001, *174*, 279.
288. M. Rouhi, *Chem. Eng. News*, April 7, 2003, 28.
289. S. Rajendran and D.C. Trivedi, *Synthesis*, 1995, 153.
290. J. Utley, *Chem. Ind. (Lond.)*, 1994, 215.
291. K. Fujimoto, Y. Tokuda, H. Mackawa, Y. Matsubara, T. Mizuno, and I. Nishiguchi, *Tetrahedron*, 1996, *52*, 3889.
292. N. Takano, M. Ogata, and N. Takeno, *Chem. Lett.*, 1996, 85.
293. S. Yamada, D. Morizono, and K. Yamamoto, *Tetrahedron Lett.*, 1992, *33*, 4329.
294. (a) L.B. Chiavetto, G. Guglielmetti, C. Querci, and M. Ricci, *Tetrahedron Lett.*, 1996, *37*, 1091; (b) J. Skarzewski, A. Gupta, and A. Vogt, *J. Mol. Catal. A*, 1995, *103*, L63; (c) A.M. d'A.R. Gonsalves, M.M. Pereira, A.C. Serra, R.A.W. Johnstone, and M.L.P.G. Nunes, *J. Chem. Soc. Perkin Trans.* 1, 1994, 2053; (d) S.L. Vander Velde and E.N. Jacobsen, *J. Org. Chem.*, 1995, *60*, 5380.
295. G.A. Mirafzal and A.M. Lozeva, *Tetrahedron Lett.*, 1998, *39*, 7263.
296. D.R. Hill, J.E. Celebuski, R.J. Pariza, M.S. Chorghade, M. Levenberg, T. Pagano, G. Cleary, P. West, and D. Whittern, *Tetrahedron Lett.*, 1996, *37*, 787.
297. Y. Sasson, A.El A. Al Quntar, and A. Zoran, *Chem. Commun.*, 1998, 73.
298. (a) *Presidential Green Challenge US Environmental Protection Agency*, EPA 744-K-96-001, July 1996, 18; (b) B.D. Hewitt, *Chem. Abstr.*, 1995, *123*, 314, 263; (c) A.E.J. deNooy, A.C. Besemer, and H. van Berkkum, *Synthesis*, 1996, 1153. Review on oxidation with nitroxyl radicals.
299. H. Firouzabadi, A Sardarian, and H. Badparva, *Bull. Chem. Soc. Jpn.*, 1996, *69*, 685.
300. (a) A.J. Bailey, W.P. Griffith, A.J.P. White, and D.J. Williams, *J. Chem. Soc. Chem. Commun.*, 1994, 1833; (b) A.E.M. Boelrijk, M.M. van Velzen, T.X. Neenan, J. Reedijk, H. Kooijman, and A.L. Spek, *J. Chem. Soc. Chem. Commun.*, 1995, 2465; (c) G.A. Bard and R.A. Sheldon, *J. Mol. Catal.*, 1995, *98*, 143.
301. A. Behr and K. Eusterwiemann, *J. Organomet. Chem.*, 1991, *403*, 209.
302 C. Varszegi, M. Ernst, F. van Laar, B.F. Sels, E. Schwab, and D.E. de Vos, *Angew Chem. Int. Ed.*, 2008, *47*, 1477.
303. (a) J.T. Groves and J.S. Roman, *J. Am. Chem. Soc.*, 1995, *117*, 5594; (b) J. Ettedgui and R. Neumann, *J. Am. Chem. Soc.*, 2009, 131, 4.
304. (a) G.I. Panov, A.K. Uriarte, M.A. Rodkin, and V.I. Sobolev, *Catal. Today*, 1998, *41*, 365; (b) A.M. Thayer, *Chem. Eng. News*, Apr. 6, 1998, 21; (c) V.I. Sobolev, A.S. Kharitonov, Y.A. Paukshtis, and G.A. Panov, *J. Mol. Catal.*, 1993, *84*, 117; (d) S. Bordiga, R. Buzzoni, F. Geobaldo, C. Lamberti, E. Giamello, A. Zecchina, G. Leofanti, G. Petrini, G. Tozzola,

and G. Vlaic, *J. Catal.*, 1996, *158*, 486; (e) M.S. Reisch, *Chem. Eng. News*, Mar. 2, 1998, 19; (f) K. Fouhy, *Chem. Eng.*, 1997, *104*(2), 15.

305. C.R. Jolicoeur, *Kirk–Othmer Encyclopedia of Chemical Technology*, 4th ed., Wiley, New York, 1992, *3*, 683.
306. A. Pawelczyk and B. Szczygiel, *Pure Appl. Chem.*, 2009, 81, 113.

RECOMMENDED READING

1. K.M. Dear, *Top. Curr. Chem.*, 1993, *164*, 115.
2. R.A. Sheldon, *Top. Curr. Chem.*, 1993, *164*, 21.
3. S. Wilson, *Chem. Ind.* (*Lond.*), 1994, 255.
4. C.R. Jolicoeur, *Kirk–Othmer Encyclopedia of Chemical Technology*, 4th ed., Wiley, New York, 1992, *3*, 683–684.
5. A. McKillop, W.R. Sanderson, *Tetrahedron*, 1995, *51*, 6145.
6. J. Muzart, *Synthesis*, 1995, 132.
7. B. Hileman, *Chem. Eng. News*, Apr. 17, 2000, 29.
8. S. Almeida, *Ind. Eng. Chem. Res.*, 2001, *40*(3), 15.

EXERCISES

1. How is the disposal of used fluorescent lamps and batteries handled where you live? If they can release mercury and other heavy metal into the environment, devise a system that will not.
2. Check the *Toxic Release Inventory* (or a comparable compilation) for releases of heavy metal ions in your area. Suggest alternative chemistry that could eliminate the releases.
3. What are the fish advisories in your area due to?
4. Keep a list for a week of the toxic metal ions that can be found in your home, or car, or that you use in the laboratory. Could any of the metal ions end up contaminating the environment?
5. Check to see what biocides are in the products in and around your home.
6. Is the current removal of asbestos from buildings overkill?
7. Visit the physics or engineering departments on campus to see what nonaqueous methods they use to apply metal coatings.
8. What other reagents can you think of that might be regenerated electrochemically to reduce the amount of reagent needed?
9. Tabulate the oxidants used in an annual volume of *Organic Syntheses*. How might environmentally friendly oxidants be substituted for any that involves toxic metals or that produces large amounts of waste salts?
10. What might be substituted for Ni–Cd batteries?
11. Visit some local industries that use toxic heavy metal ions to see how they handle their wastes. Include a visit to the local boat showroom to find out what they recommend for antifouling paints.

CHAPTER 5

Solid Catalysts and Reagents for Ease of Workup

5.1 INTRODUCTION

The most obvious advantage of a solid catalyst or reagent is that it can be removed from the reaction mixture by simple filtration or centrifugation. This allows quick recovery for reuse in the next run. Alternatively, the solid can be put in a column with the reaction mixture flowing through it. There can be other advantages as well. If the catalyst is expensive or toxic, this provides a way to not lose it and to minimize exposure to it. If the bulk solid catalyst is expensive, less of it will be needed if it is spread over the surface of a solid support. The use of the solid material can minimize waste. Consider, for example, the use of a strong acid ion-exchange resin in place of *p*-toluenesulfonic acid to catalyze an esterification. The resin can be recovered for reuse by filtration. The *p*-toluenesulfonic acid has to be removed by washing with aqueous base, after which it is usually discarded as waste. The solid catalyst can be used at very high catalyst levels because it can be recovered and used again. If the reaction is being performed on a solid support, a large excess of another reagent can be used to drive the reaction to completion. The excess reagent and any by-products can easily be washed off the solid. The vigor of some reactions can be moderated by putting a reagent on a solid. However, this may involve the use of extra solvent to put it on to the solid and then to extract the product from the solid. Some solids, such as clays and zeolites, offer size and shape selectivity, so that higher yields and fewer by-products can be obtained. Some catalysts can catalyze not only the desired reaction, but also side reactions. Isolating catalytic sites on a support can sometimes eliminate these side reactions. The support may also limit the possible conformations of the catalytic site in a way that limits side reactions. As an example, the immobilization of some enzymes increases their thermal stability. Altus Biologics, Cambridge, Massachusetts, does this by cross-linking enzyme crystals.[1]

There are some disadvantages to the use of solid catalysts and regents. It may not be possible to obtain a high loading, so that reactions are slower and more catalyst may be needed. Particles with the high surface area needed for the support to obtain higher loadings may not be very mechanically stable, and the catalyst may suffer from attrition in use. DuPont has solved this problem for a vanadium pyrophosphate catalyst, used in the oxidation of *n*-butane to maleic anhydride, by putting a hard porous shell of silica around it.[2] The resulting particles are stable enough to be used in a circulating fluidized bed. This has increased the yield in the reaction. A polymeric support needs to swell in the medium to allow the reaction to proceed. It may not swell properly in some media. If the particles are too fine, flow rates through the catalyst bed may be too low. Groups on the surface may be active, whereas those inside the particle may not be accessible for reaction.

Table 5.1 compares the advantages and disadvantages of homogeneous and heterogeneous catalysts. Many books and reviews cover the nature, preparation, characterization,[4] and use of homogeneous[5] and heterogeneous[6] catalysts. Others describe supported metal catalysts,[7] other supported catalysts and reagents,[8] and their use in preparative chemistry,[9] the use of polymers as supports,[10] trends in industrial catalysis,[11] and the environmentally friendly nature of solid catalysts and reagents.[12]

The support may not be an innocent bystander. It may act as a ligand for any metal compound put on its surface and in doing so alter the steric and electronic properties of the metal complex. Groups on its surface, such as hydroxyl and carboxyl, may react with the metal complex added to produce a new compound. Binders to improve mechanical stability and promoters to improve activity are often added. An inert filler can also be used to dissipate heat in exothermic reactions, as in the vanadium pyrophosphate catalyst on a ceramic support used in the conversion of butane to maleic anhydride.[13] The promoter may be another metal ion, which can then react with the first metal ion to form a cluster.[14] It can also promote by preventing attrition of the solid. The surface of the solid will vary with the coordinately unsaturated active site being at edges, corners, and defects. Putting the active catalytic group right on the surface may not allow it to form the right conformation for the desired reaction. In this case, a spacer group of several atoms may have to be added. It is occasionally possible to boost activity, as with some polymers, by putting more than one functional group

Table 5.1 Characteristics of Homogeneous and Heterogeneous Catalysts

Characteristic	Homogeneous	Heterogeneous
Ease of separation	May be difficult	Just filter
Thermal stability	Often low	Usually high
Sensitivity to oxygen and moisture	May be high	Often low
Life	Low	Often high
Range of suitable solvents	Limited	Almost no limit
Corrosion and plating out	Sometimes	None
Selectivity	Often high and easy to modify	Less selective, several types of sites may be present
Reproducibility	High	Very dependent on mode of preparation
Efficiency	High (all sites are active)	May be low (only sites on the surface are active)

Source: Adapted from U. Schubert, *New J. Chem.*, 1994, 18, 1049. With permission of Gauthier-Villars/ESME-23 rue Linois-Paris cedex 15 and Elsevier.

on the surface to create a multidentate ligand. Enzymes are proteins that often have several groups in widely different parts of the molecule that curl around a metal ion to form an active site. A low loading can isolate active sites so that some deactivation reactions that involve two sites can be prevented. Considering the various reactions that can occur in putting the catalyst on the support, it is clear that more than one type of site can be present. Sometimes, selectivity to the desired product can be improved by poisoning one type of site, as in the use of ethyl *p*-anisate with magnesium chloride-supported catalysts for the polymerization of propylene to reduce the amount of atactic polymer formed (Chapter 1). Selectivity can be improved by using clays and zeolites that offer selectivity by size and shape between layers in the former and in porous channels in the latter.

Heterogeneous catalysts can be deactivated in several ways. A metal may be poisoned by a strong ligand (e.g., a sulfur compound on a hydrogenation catalyst). Metal particles may sinter and agglomerate so that surface area is reduced. This process may be aided by reaction with one of the reactants to form a different compound, as in palladium hydride formation in hydrogenations. Many metal-catalyzed reactions involve metal–carbon bonds that insert carbon monoxide, ethylene, and the like. Site isolation can reduce this aggregation. The active catalyst may undergo leaching of the active material, especially if strongly coordinating solvents or reactants are used. Coke often builds up when catalysts are used at elevated temperatures.[15] With inorganic supports, it is common to regenerate such catalysts by burning off the organic matter in oxygen at an elevated temperature. Some promoters help prevent this buildup of carbon. For example, the addition of some gold to a nickel catalyst used in the steam reforming of hydrocarbons eliminates the usual deactivation with time, apparently by preventing the formation of graphite.[16] The gold may catalyze the reaction of the coke with steam to form carbon monoxide and hydrogen. When this is a big problem, it is sometimes possible to cycle the catalyst through the reaction vessel, then to a regeneration vessel, and back to the reaction continuously. On the other hand, there are times when some coking can passivate a catalyst's surface, possibly by blocking undesirable sites, and lead to enhanced selectivity.[17] Attrition can be a problem, especially if fluidized beds are used to avoid channeling.

The catalysts can be placed on the support in a variety of ways. One common way is to coat the support with an aqueous solution of catalyst or its precursor and then dry and, if necessary, activate it by reduction, oxidation, or other such reactions. Vapor deposition can also be used. Such materials may not be held on strongly; thus, leaching can be a problem in strongly coordinating solvents. The ionic bonding that results when an ion-exchange resin picks up an anion or cation will be stronger as long as the pH is kept in an acceptable range. The bonding will be stronger yet if a stable chelate is formed. The most stable bonds for holding the catalyst are covalent. Of these, the best are those that are resistant to the reaction being run, as well as to hydrolysis and oxidation.

Green chemistry provides a strong motivation to heterogenize homogeneous catalysts. With the proper chemistry, it is possible to attach almost any catalyst to an inorganic support, a magnetic bead, or a polymer. The rest of this chapter will serve as a progress report on what has been done and the many challenges that remain. Numerous examples from the literature will be used to illustrate the range of possibilities. The use of inorganic and polymer supports, as well as ion-exchange resins, will be discussed in this chapter. Clays and zeolites will be covered in the following chapter. Immobilized, enzymatic catalysts will be discussed in Chapter 9.

5.2 THE USE OF INORGANIC SUPPORTS

When one places a reagent on a support, it spreads out into a thin accessible layer. If the reactant is now added, the local concentrations of reactants may be high, thus favoring a bimolecular reaction. This is also a way to bring together two reactants that may not be particularly soluble in the same solvent. For example, one compound can be placed on a support as a solution in water. Then, after drying, the solid can be treated with a reactant in a water-immiscible solvent. If the by-products of the reaction do not interfere with the reaction, the support may be reloaded with more of the reagent for use again. If they do, it may be necessary to purify the support before it can be used again. A few examples from the literature will serve to illustrate some of the possibilities.

Sodium has been dispersed on the surface of alumina, sodium chloride, or titanium dioxide by stirring with excess support at 180–190°C under argon. The first was the most effective for reducing titanium(III) chloride for the McMurray reaction **5.1**, being better than the commonly used reduction with zinc.[18]

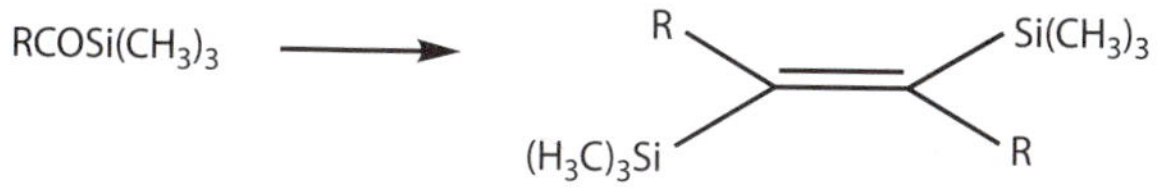

$RCOSi(CH_3)_3$ ⟶ $R(Si(CH_3)_3)C{=}C(R)Si(CH_3)_3$

5.1 Schematic

The reagent is described as inexpensive, readily prepared and nonpyrophoric, but air sensitive. Sodium dispersed on titanium dioxide in a similar way has been used to reduce zinc chloride to "active" zinc for reaction with RX to give RZnX. This method produces fewer by-products from Wurtz coupling.[19] SiGNa, Inc. encapsulates lithium and sodium in porous silica so that they can be used safely.[20]

Many organic reactions have been run on alumina.[21] The hydrolysis of nitriles can be performed quickly on excess alumina. Further hydrolysis to the carboxylic acid is slow (**5.2** Schematic).[22]

$$RCN \text{ on } Al_2O_3 \xrightarrow{60°C} RCONH_2 \quad 82\text{–}92\%$$

5.2 Schematic

Potassium permanganate on alumina has been used to oxidize benzylic hydrocarbons to ketones (**5.3** Schematic), often in high yields.[23]

Silica is another common support. Copper(II) nitrate on silica (**5.4** Schematic) has been used to regenerate carbonyl compounds from their oximes, tosylhydrazones, 1,3-dithiolanes, or 1,3-dithianes in 88–98% yields.[24]

Potassium fluoride on alumina is sometimes used as a base.[25] It may be that the fluoride ion displaces some surface hydroxyl from the alumina to produce some potassium hydroxide. Lithium chloride, bromide, and iodide on silica have been used to open epoxide rings (**5.5** Schematic) to give the corresponding halohydrins.[26]

Some Witting reactions (**5.6** Schematic) are much faster when the aldehyde is put on silica first.[27]

It is sometimes possible to eliminate some by-products and increase yields of the desired products by putting the dienophile in a Diels–Alder reaction on silica (**5.7** Schematic).[28] Under standard conditions, the reaction yields 51% of the desired product and 35% of the dimethylamine adduct of the starting quinone. The reaction of acetyl chloride with alkylene glycols on silica was 98–100% selective for the monoacetate at 52–66% conversion (**5.8** Schematic).[29]

There are also some cases where the support is coated with a homogeneous film that is insoluble in the medium. Dibutyl sulfide was oxidized in methylene chloride to the corresponding sulfoxide quantitatively with cerium ammonium nitrate in an aqueous film on silica.[30] There was no overoxidation to the sulfone. A polyethylene glycol film on silica has been used with a water-soluble rhodium catalyst in the hydroformylation of 1-hexene.[31] The activity was as good as in the comparable homogeneous system, and there was

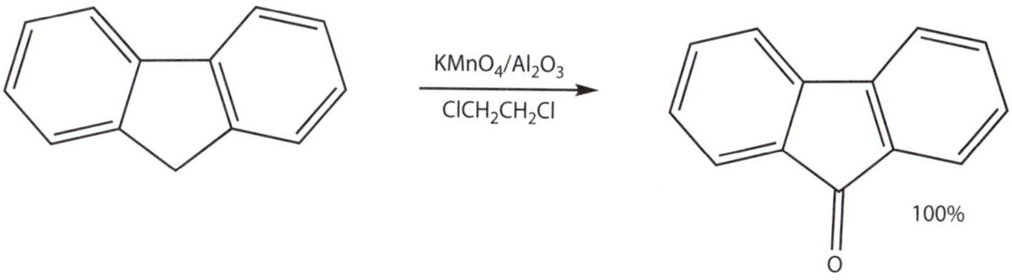

$KMnO_4/Al_2O_3$, $ClCH_2CH_2Cl$ ⟶ 100%

5.3 Schematic

$Cu(NO_3)_2/SiO_2$ ⟶ CHO 98%

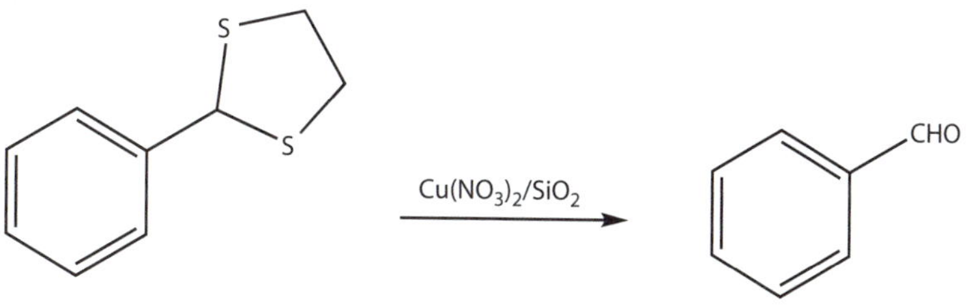

5.4 Schematic

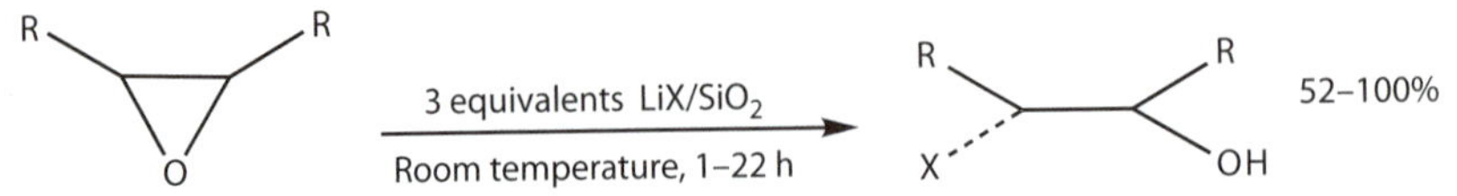

3 equivalents LiX/SiO_2, Room temperature, 1–22 h; 52–100%

5.5 Schematic

$$(C_6H_5)_3P^{+-}CHCOOCH_3 + RCHO \xrightarrow[\text{Hexane, 25°C}]{SiO_2} RCH{=}CHCOOCH_3 \quad (C_6H_5)_3PO$$

5.6 Schematic

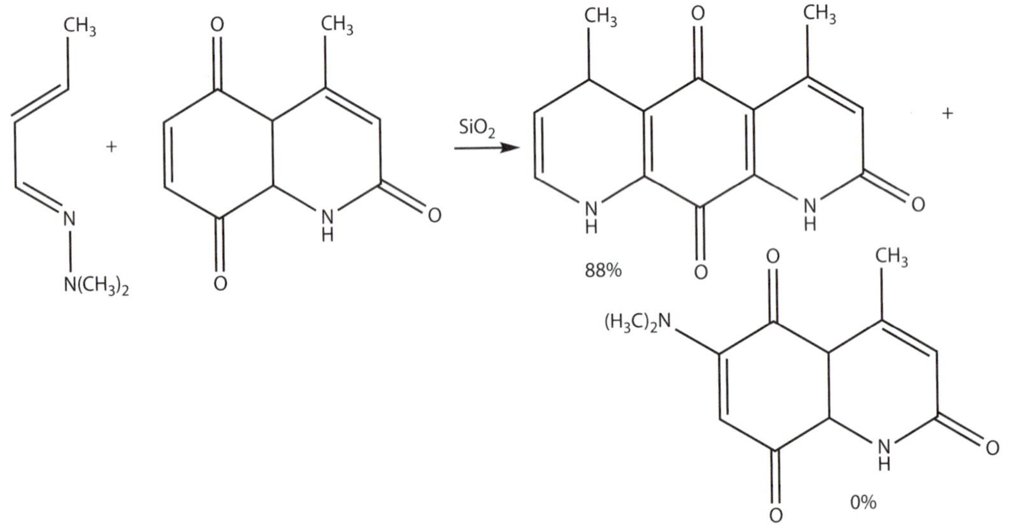

SiO_2; 88%; 0%

5.7 Schematic

$$HO(CH_2)_nOH \text{ on } SiO_2 + CH_3COCl \longrightarrow HO(CH_2)_nOCOCH_3$$

98–100% selectivity at 52–66% conversion

5.8 Schematic

little isomerization to 2-hexene. A 6:1 ratio of linear/branched isomers was obtained (**5.9** Schematic).

Palladium complexed with a water-soluble phosphine, the sodium salt of triphenylphosphine-*m*-sulfonic acids, in an aqueous film on silica catalyzed allylation in acetonitrile in 100% conversion (**5.10** Schematic).[32] No metal was lost. The catalyst could be recycled.

A latex-supported catalyst has been used to isolate sites. Styrene has been polymerized in the presence of an ionene diblock copolymer (a water-soluble cationic copolymer) to form a graft copolymer latex.[33] The cobalt phthalocyanine sulfonate catalyst $[CoPc(SO_3^- Na^+)_4]$ was added and became attached to the cationic polymer. When this catalyst was used for the oxidation of thiols to disulfides by oxygen, the activity was 15 times that in a polymer-free system. This site isolation suppresses the formation of inactive dioxygen-bridged μ-peroxo complexes.

Hydroxyl groups are present on the surfaces of oxides such as alumina, silica, titania, magnesium oxide, and others. These can be used to anchor catalysts to supports. In one case, the surface hydroxyl group was converted first to a potassium salt by treatment with potassium acetate at pH 8.4 and then to a cobalt salt with cobalt(II) chloride. This was used as catalyst for the oxidation of phenols to quinones by oxygen (**5.11** Schematic).[34]

An iridium complex was attached to a silica surface without the use of any base (**5.12** Schematic).[35] The silica behaves as a simple ligand would.

A rhodium phosphine was anchored to silica (**5.13** Schematic).[36]

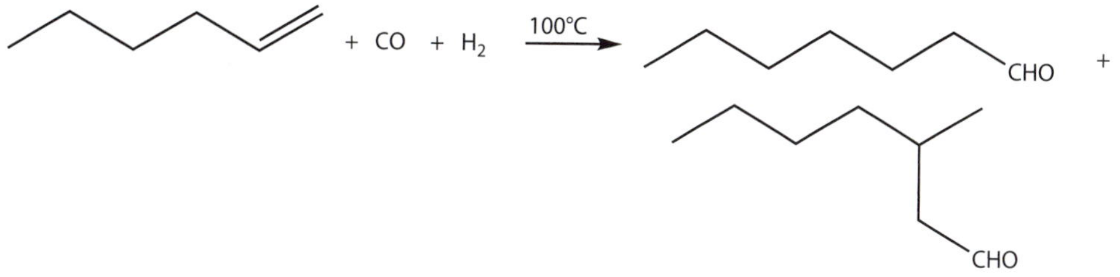

5.9 Schematic

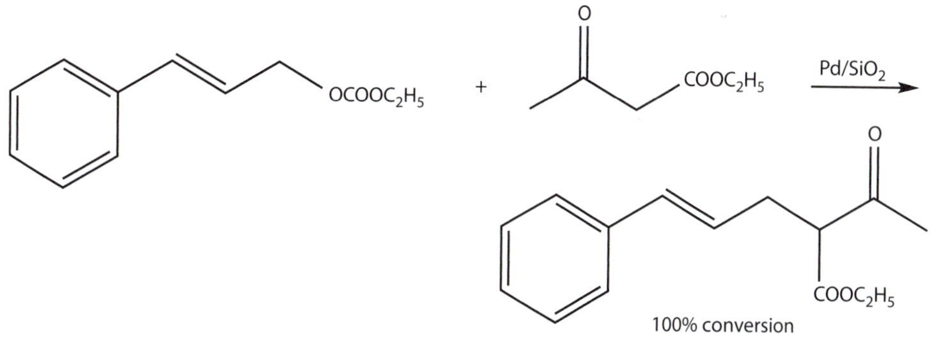

5.10 Schematic

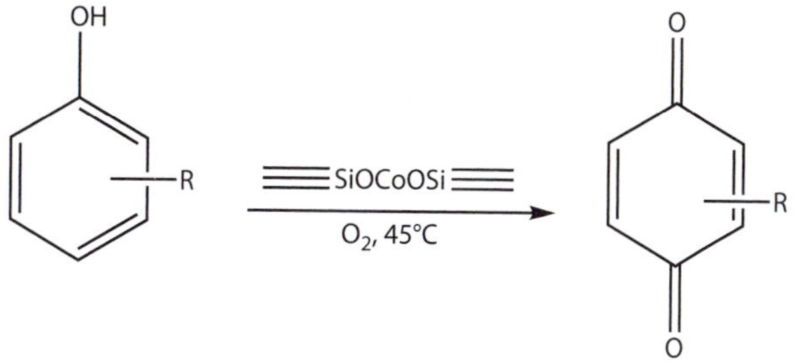

5.11 Schematic

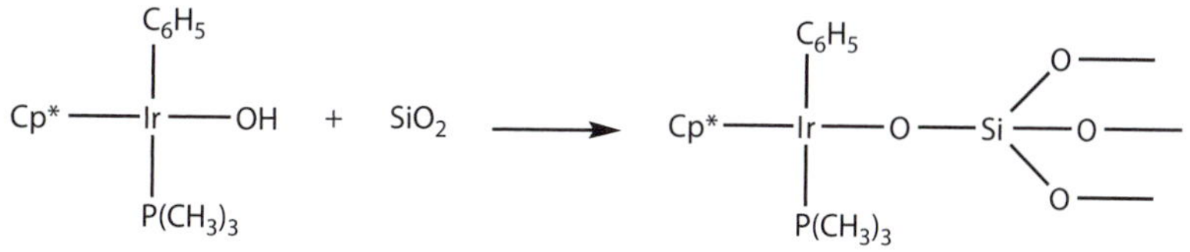

5.12 Schematic

$$\equiv Si{-}OH + CH_3Rh(P(CH_3)_3)_2(CO) \longrightarrow \equiv Si{-}ORh(P(CH_3)_3)_2(CO) + CH_4$$

5.13 Schematic

An allyltris (trimethylphosphine) rhodium complex with an RhOSi linkage has also been prepared.[37]

Metal chlorides can also be put on oxide supports. Zinc chloride was put on alumina in tetrahydrofuran, followed by dilution with methylene chloride, then boiling with a few drops of water and finally evaporation and drying. It was used as a catalyst to prepare a diketone (**5.14** Schematic).[38] This method prevented side reactions that occurred when zinc chloride was used in tetrahydrofuran. With no zinc chloride at all, the reaction was sluggish. Metal chlorides, as well as protonic acids, on such supports are sold by Contract Chemicals in England as "Envirocats" (short for environmentally friendly catalysts).[39] They are stable to air. One of them has been used to dehydrate an oxime to a nitrile (**5.15** Schematic).[40]

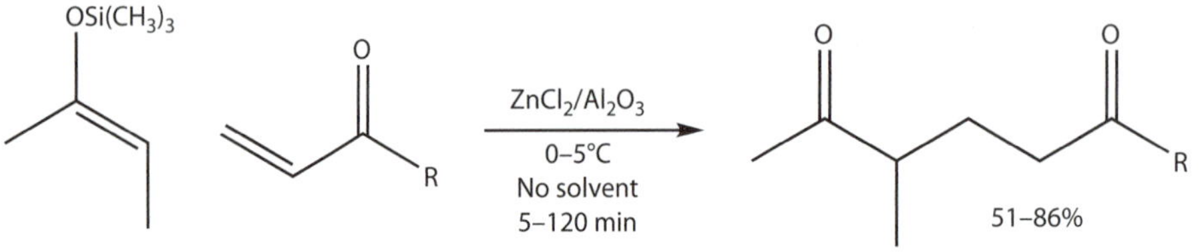

$OSi(CH_3)_3$ ... $ZnCl_2/Al_2O_3$, 0–5°C, No solvent, 5–120 min → 51–86%

5.14 Schematic

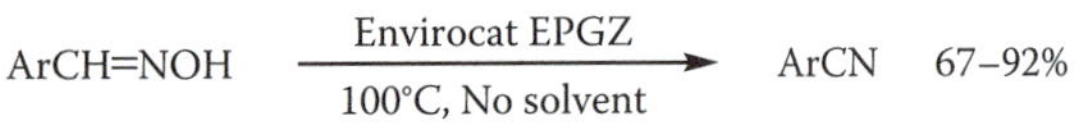

ArCH=NOH —(Envirocat EPGZ, 100°C, No solvent)→ ArCN 67–92%

5.15 Schematic

The catalysts are nontoxic powders that can be reused several times. (Solid acids and bases will be discussed in detail in Chapter 6.)

There is some question about the nature of the active sites in these supported metal chlorides. One can envision a number of structures that might result when zinc chloride on silica is treated with water. (Similar structures might be written for zinc chloride on alumina.) These might include the structures in **5.16** Schematic. Other hydroxyl and chloride-bridged structures might also be present.

A supported palladium catalyst for the Heck reaction (**5.17** Schematic) could be reused several times with little loss in activity.[41] Heterogeneous catalysts for the Heck reaction, such as palladium on silica and platinum on alumina, can perform just as well in the Heck reaction[42] as homogeneous catalysts.[43]

In some cases, the palladium dissolves and reprecipitates during the reaction with the support, such as NaY zeolite, acting as a reservoir for the palladium.[44] Use of a chelating group for the palladium on the support prevented any loss of the metal in a Suzuki reaction that gave 100% conversion.[45] The conversion was 97% after the seventh use (**5.18** Schematic).

DSM has reacted bromobenzene and butyl acrylate in a ligand-free Heck reaction on a kg scale with a 99% yield.[46] Chlorobenzene reacted with styrene to give *trans*-stilbene in >99% yield when cesium acetate was used with a palladium pincer ligand.[47] Aryl esters of phenols[48] and aryl phosphoric

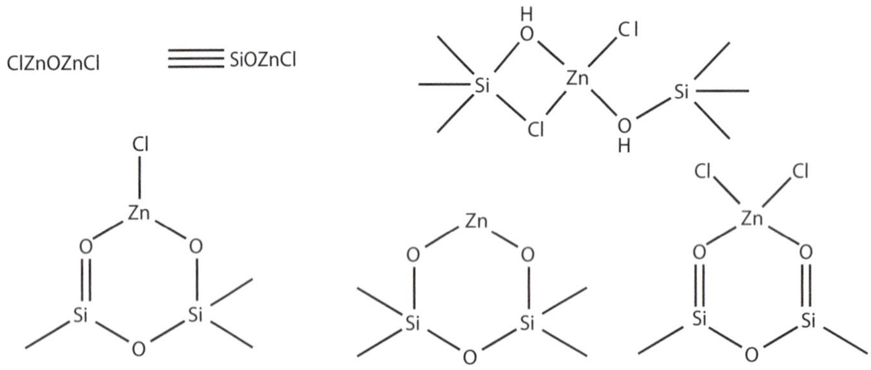

ClZnOZnCl ≡SiOZnCl

5.16 Schematic

+ $COOCH_3$ + $Pd(P(C_6H_5)_3)_2$ →

X = Sn, Al or Ti

$COOCH_3$ 100% selectivity

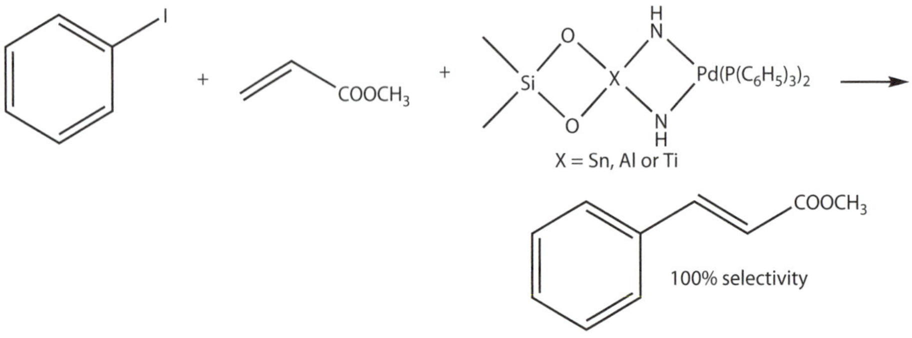

5.17 Schematic

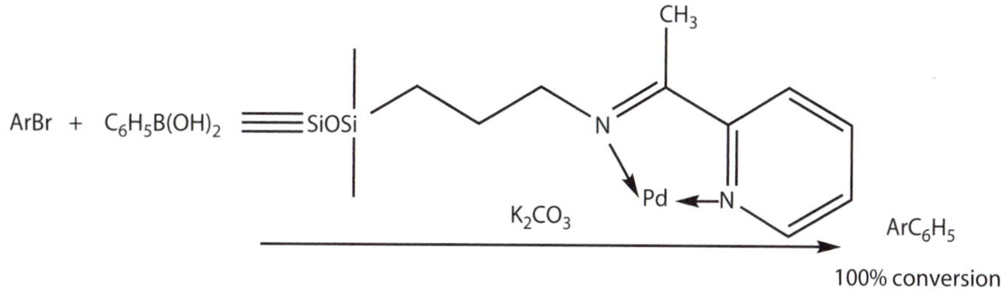

ArBr + $C_6H_5B(OH)_2$ ≡SiOSi–(CH₂)₃–N=C(CH_3)–(2-pyridyl), N→Pd←N; K_2CO_3 → ArC_6H_5 100% conversion

5.18 Schematic

acids[49] have been used instead of aryl halides in the Heck reaction.

Reduction of rhodium(III) chloride with zinc borohydride in the presence of silica produced a catalyst that gave nearly 100% chemoselectivity to 2-phenylpropionaldehyde (**5.19** Schematic).[50] No loss of the expensive rhodium occurred. (Rhodium is expensive enough so that industrial processes cannot afford to lose more than traces of it.)

The most common way to attach groups to metal oxide surfaces is to treat the surface with alkoxysilanes that contain other groups that can be reacted further to attach other things. Methoxides and ethoxides are used commonly (**5.20** Schematic).

The product of the first reaction was added to a double bond in a fullerene using a platinum catalyst.[51] The amino group in the second product has been used to support a peracid.[52] The products have possible use in the disinfection of gases and in the treatment of wastewater containing low levels of organic compounds. Another route to the peracid involves hydrolysis of the nitrile to the carboxylic acid, followed by conversion to the peracid with hydrogen peroxide. The peracid is then used to oxidize cyclic olefins to their epoxides, with 96–100% selectivity at 55–62% conversion.[53] The amino group in the aminopropylsilica has been used to attach a dye by first treating with thiophosgene to produce an isothiocyanate, which reacted with an amino group in the dye to form a thiourea.[54] It has also been used to attach a platinum catalyst that was used to catalyze the oxidation of cyclohexanol with oxygen to give cyclohexanone in 98% yield.[55] In this last example, magnesium oxide could be substituted for silica with equivalent results. A typical way to add a metal catalyst is to complex it with a chelating agent made from the aminopropylsilica and an aldehyde, such as salicylaldehyde or 2-pyridinealdehyde.[56] Rhodium and palladium were attached at the same time to silica treated with 3-isocyanopropyltriethoxysilane to produce a catalyst for the reduction of arenes at 40°C at 1 atm hydrogen.[57] The activity was higher than any yet reported for other homogeneous or heterogeneous catalysts. A rhodium catalyst attached to silica with a diphenylphosphinopropylsilyl group was also much more active than its homogeneous counterpart for the same reduction.[58]

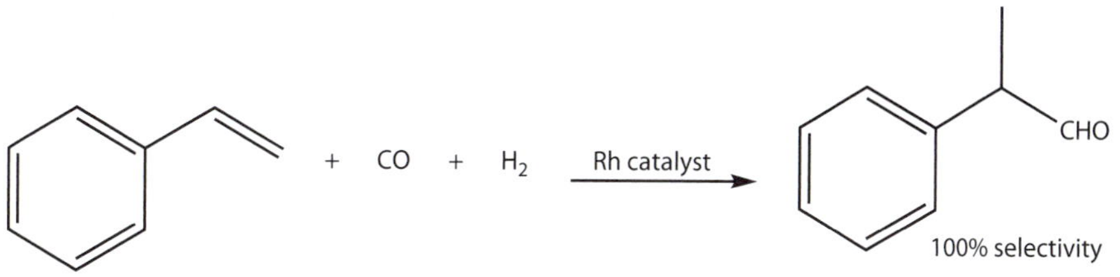

Styrene + CO + H_2 —Rh catalyst→ 2-phenylpropionaldehyde (CHO) 100% selectivity

5.19 Schematic

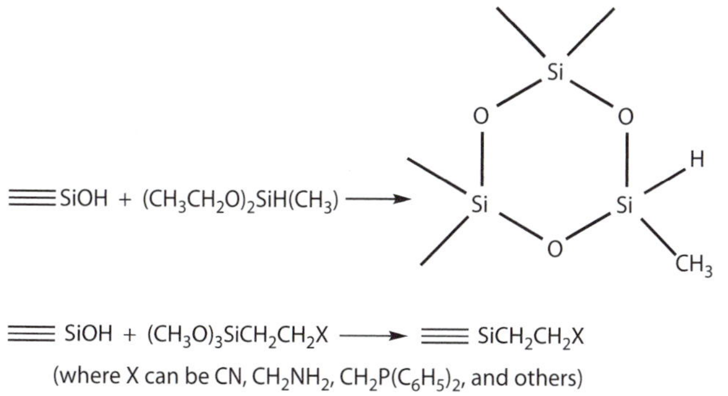

≡SiOH + $(CH_3CH_2O)_2SiH(CH_3)$ → cyclic Si–O–Si–O–Si–O (Si bearing H and CH_3)

≡ SiOH + $(CH_3O)_3SiCH_2CH_2X$ → ≡ $SiCH_2CH_2X$

(where X can be CN, CH_2NH_2, $CH_2P(C_6H_5)_2$, and others)

5.20 Schematic

The product where X contained phosphorus was used to complex nickel compounds.[59] In another instance, triethoxysilane was added to an allylcalixarene, and the product was then used to treat silica, thus anchoring the calixarene to the support.[60] Various chelating agents have been attached to silica by such routes.[61] When X was CH_2Cl, it was possible to attach a salicyclideneamino chelating agent, which chelated manganese(III).[62] The regular pore structure of the mesoporous silica used was not disrupted in the process. When X is CH_2Cl, a cyclopentadienyl group can be put on silica by reaction with cyclopentadienyllithium.[63] This has been used to pick up fullerenes, which could be recovered

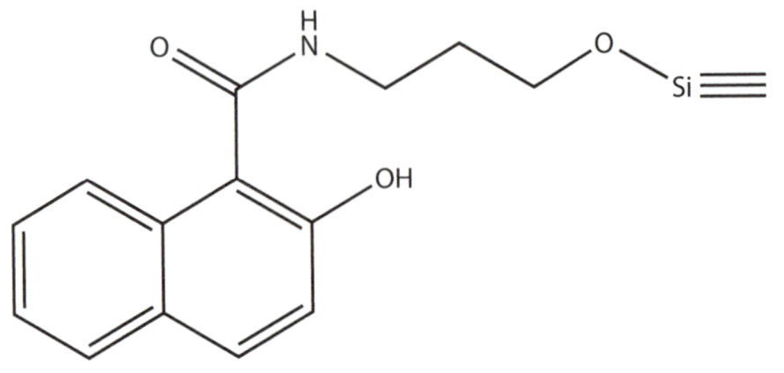

5.21 Schematic

later. It also provides a way to put in transition metal ions as metallocenes.

The process has also been used to attach photosensitizers to silica.[64] The product in structure **5.21** Schematic was made by treating aminopropylsilica with 2-hydroxy-1-naphthoic acid. It was more effective in the photochemical dechlorination of polychlorinated biphenyls than a homogeneous sensitizer.

Aminopropylsilica has also been used to attach enzymes (**5.22** Schematic). The aminopropylsilica is treated first with glutaraldehyde and then with the enzyme.[65] Galactosidase has been attached to glass in this way.[66] The immobilized enzyme is useful in removing lactose from milk, for the benefit of those persons who are intolerant of lactose. A similar technique was used to attach glucose oxidase to magnetite (Fe_3O_4) particles, starting with a trimethoxysilane.[67] This permits easy separation with a magnet. The immobilized enzyme was 50% less active than the native enzyme, but kept 95% of its activity after 9 months at 4°C. Another way to attach an enzyme to a support is to treat the metal oxide support first with 2,4,6-trichloro-1,3,5-triazine (cyanuric chloride) (**5.23** Schematic) and then with the enzyme.[68] This was done on both alumina and silica. A lipase immobilized in this way retained 80% of its activity after 336 h. The catalyst could be reused three times before its activity dropped to 50% of the original value and it was still yielding more than 90% enantiomeric excess of *S*(+)-2-arylpropionic acids. An enzyme from *Candida cylindracea* immobilized on silica in a similar way was 37 times more stable than the native enzyme. It retained 80% of its activity after 336 h. Loss of activity on aging or use is a problem with many enzymes. (Biocatalysis is discussed in more detail in Chapter 9.)

Another way to attach groups to silica involves sequential treatment of silica with tetrachlorosilane, diethylamine, and 3-hydroxypropyldiephenyphosphine (**5.24** Schematic).[69] This method was used to attach nickel and rhodium to glass, quartz, and silica surfaces.

Similar techniques have been used to incorporate silica into various polymers. (Mineral fillers are often given hydrophobic coatings to help in their dispersion in polymers such as polyolefins.[70]) Methacrylatopropyltrimethoxysilane has been used to add the methacrylate group to silica. The treated silica was used in the emulsion polymerization of ethyl acrylate.[71] Transparent films could be formed from the product. Living polystyrene was end-capped with a triethoxysilane (**5.25** Schematic), after which it was used to treat silica and alumina.[72]

The authors were interested in using this technique to disperse inorganic pigments. A living polymer has also been end-capped by treatment with chlorosilanes containing

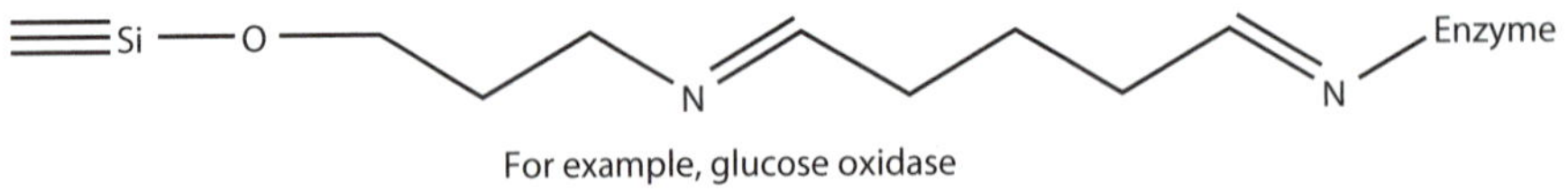

For example, glucose oxidase

5.22 Schematic

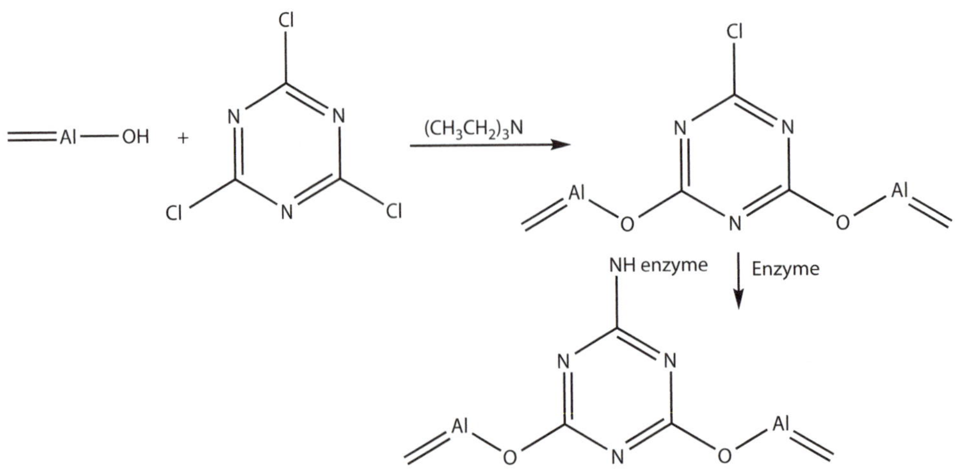

5.23 Schematic

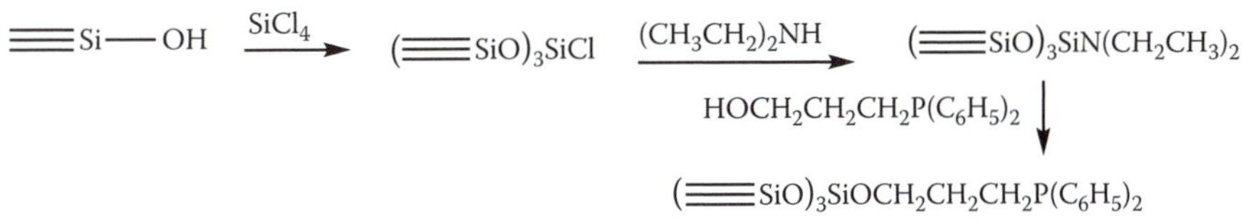

$$\equiv Si{-}OH \xrightarrow{SiCl_4} (\equiv SiO)_3SiCl \xrightarrow{(CH_3CH_2)_2NH} (\equiv SiO)_3SiN(CH_2CH_3)_2 \xrightarrow{HOCH_2CH_2CH_2P(C_6H_5)_2} (\equiv SiO)_3SiOCH_2CH_2CH_2P(C_6H_5)_2$$

5.24 Schematic

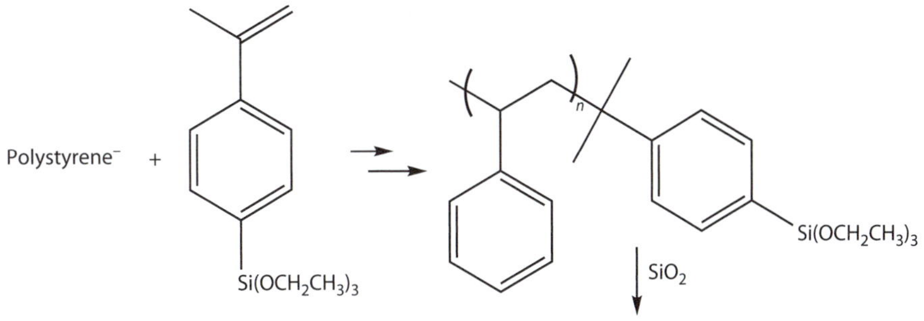

Polystyrene⁻ + Si(OCH2CH3)3 SiO2 Si(OCH2CH3)3 n

5.25 Schematic

protective groups that were later removed to give terminal hydroxyl and amino groups (**5.26** Schematic).[73] A ruthenium complex was made from the final amine.

A copolymerization of styrene, divinylbenzene, and maleic anhydride was carried out in the presence of silica. This was hydrolyzed in boiling water and then treated with chloroplatinic acid to add platinum on silica.[74] Presumably the anhydride reacted with the hydroxyl groups on the surface of the silica. Then the platinum(II) formed a salt with some of the carboxyl groups.

Another way to insert a noble metal catalyst on to silica is as follows (**5.27** Schematic). When used in the hydroformylation of propylene, this catalyst was as good as $Rh(H)(CO)(P(C_6H_5)_3)_3$. It gave a 1:1 mixture of *n*-butyraldehyde and isobutyraldehyde,[75] compared with 3:1 for the reaction in solution. The linear aldehyde is more valuable, but a number of uses have been developed for the isobutyraldehyde. If the market needed more of the latter, this might be an excellent way to make it.

The gradual evolution of catalysts for olefin polymerization was described in Chapter 1. The use of magnesium chloride as a support for titanium tetrachloride led to high mileages that allowed the catalyst residues to be left in the polyolefin.[76] The polymer particle is an enlarged replica of the catalyst particle. To make the desired compact granules of polymer with a high bulk density (0.50 g/cm^3 or higher), the catalyst must be compact at the same time that it has a high surface area. Newer metallocene catalysts are soluble in hydrocarbons. To obtain the desired polymer morphology, they must be put on supports, such as alumina, silica, or zeolites.[77]

The sol–gel route has been used to heterogenize homogeneous catalysts.[78] In the process, one or more metal alkoxides are gradually hydrolyzed to produce a metal oxide or a mixture of metal oxides. The advantage is that the second metal oxide is found throughout the first one, rather than just on the surface, as could result from treatment of a preformed metal oxide support.[79] In other cases, the transition metal catalyst is just entrapped physically by being present during the hydrolysis of the tetraalkoxysilane.[80] The method can be run to give high surface areas, which can increase catalytic activity. A recent example involves the cohydrolysis of two alkoxysilanes in the presence of alumina (**5.28** Schematic).[81] The resulting coating imparted "extremely

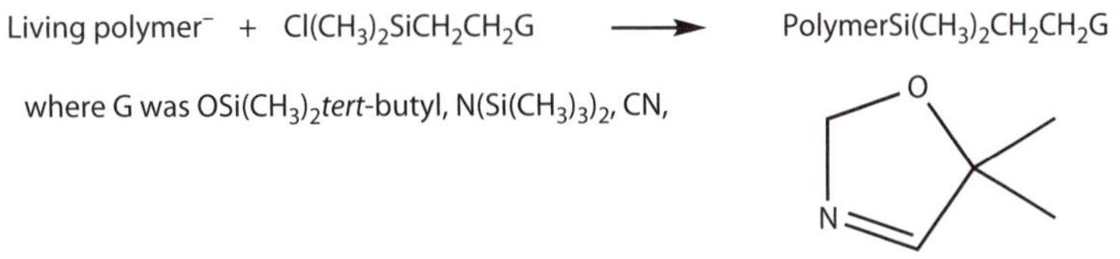

$$\text{Living polymer}^- + Cl(CH_3)_2SiCH_2CH_2G \longrightarrow PolymerSi(CH_3)_2CH_2CH_2G$$

where G was $OSi(CH_3)_2$*tert*-butyl, $N(Si(CH_3)_3)_2$, CN,

O N

5.26 Schematic

$$\equiv SiOH + P(CH_2OH)_3 \longrightarrow \equiv SiOCH_2P(CH_2OH)_2 \xrightarrow{Rh_4(CO)_{12}}$$

5.27 Schematic

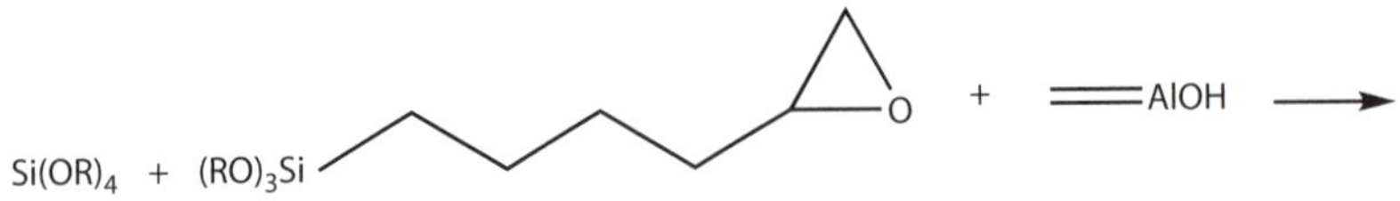

$Si(OR)_4$ + $(RO)_3Si$ + =AlOH →

5.28 Schematic

high" scratch resistance to plastics. This could result in longer-lasting plastic eye glass lenses and dinnerware. Current materials are limited by the tendency to scratch easily and, for dinnerware, to stain.

The sol–gel method has also been used to put an NHC(S)NHC(O)Ph on silica.[82] A rhodium catalyst made from this could be used at least five times in the hydroformylation of styrene. By including some methyltriethoxysilane in the cohydrolysis of tetraethoxysilane and titanium(IV) isopropoxide, it was possible to vary the surface polarity of the amorphous microporous mixed oxide catalysts used in oxidations with hydrogen peroxide.[83] The methyl groups slowed down the deactivation of the catalyst and made it possible to regenerate them thermally.

Colloidal metal clusters, which offer a high surface area for better activity, have been stabilized by polymers. Thus, a homogeneous dispersion of a cobalt-modified platinum cluster was stabilized by a coordinating polymer, poly(*N*-vinyl-2-pyrrolidone).[84] Addition of the cobalt(II) [or iron(III)] doubled the activity and increased the selectivity from 12% to 99% when the catalyst was used to reduce cinnamaldehyde (**5.29** Schematic).

The system was very sensitive to the transition metal ion added to the platinum. When it was nickel, hydrocinnamaldehyde was obtained in 97% selectivity.[85] Palladium supported on a carbon nanofiber gave hydrocinnamaldehyde as the only product.[86] The intermediates in these reductions are probably metal hydride clusters. When the colloidal platinum is supported on magnesium oxide, without another transition metal, the reduction produces the unsaturated

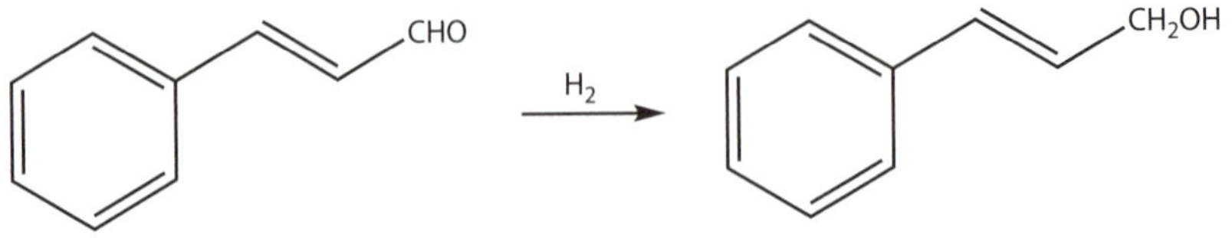

CHO $\xrightarrow{H_2}$ CH_2OH

5.29 Schematic

alcohol with 97% selectivity.[87] A rhodium colloid stabilized by the same polymer was used with a water-soluble phosphine in the hydroformylation of propylene to produce a 1:1 mixture of *n*-butyraldehyde and isobutyraldehyde in 99% yield.[88] It could be used at least seven times, as long as it was not exposed to air.

Colloidal palladium can also be stabilized with poly(*N*-vinyl-2-pyrrolidone).[89] Other polymeric stabilizers has been made as a model used with palladium, platinum, and other metals.[90] These include poly(2-ethyl-2-oxazoline), poly(methacrylic acid),[91] block copolymers containing COOH groups,[92] poly(*N*-isopropyl acrylamide),[93] and poly(styrene-*b*-4-vinylpyridine).[94] Colloidal iron has been stabilized not only by (*N*-2-pyrrolidone) but also by oleic acid.[95] Such ferrofluids are of interest for magnetofluid seals and bearings, as well as for other uses. Palladium and rhodium catalysts have also been supported on dendrimers containing amine and phosphine groups.[96] Dendrimers are spherical polymers made by starting with a multifunctional molecule that is reacted with a reagent that gives a product for which each new end becomes difunctional. This process is repeated several times to produce the spherical polymer.[97] A silicon-based dendrimer was made and fitted with dimethylamino groups that complexed nickel. The complex was used as a catalyst for the addition of carbon tetrachloride to methyl methacrylate.[98] The dendrimer offers a controlled dispersion of the catalytic sites on the surface of the nanoscopic particle with easy recovery for reuse. The isolation of the sites prevented deactivation. Copper can also be placed in the voids inside polyamidoamine dendrimers with 4–64 copper atoms per particle.[99] Soluble dendrimer catalysts can be recovered for reuse by precipitation with a nonsolvent or by ultrafiltration. Polymer-supported dendrimer catalysts can be recovered by filtration or centrifugation.[100]

Various workers have been making cluster compounds to bridge the gap between homogeneous catalysts and their heterogeneous counterparts. Platinum–rhenium clusters have been made as models for bimetallic catalysts.[101] It is hoped that this will shed light on some of the promoter effects found when a second metal is added to catalysts. The authors feel that the selective ligand displacement at rhenium is relevant to the mode of action of heterogeneous Pt/Re catalysts. The ligand in the clusters is $Ph_2PCH_2PPh_2$ (**5.30** Schematic).

Roesky and co-workers have used reactions of silanetriols to produce a variety of metallasiloxane cage structures.[102] Some of these are shown in **5.31** Schematic.

An aluminum silsesquioxane has been made a model for silica-supported catalysts. It catalyzed (1 mol%) a Diels–Alder

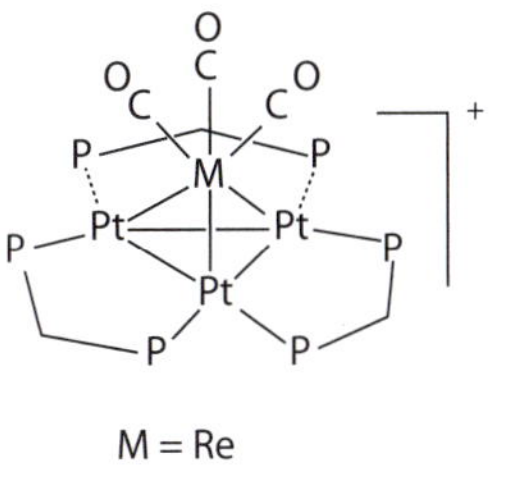

M = Re

5.30 Schematic

reaction between a diene and an unsaturated ketone (**5.32** Schematic).[103] Titanium analogues have also been made.[104]

Lithium–ytterbium and lithium–zirconium silsesquioxanes have also been prepared as models for catalysts.[105]

Silsesquisiloxanes are made by the hydrolysis of trialkoxysilanes or trichlorosilanes.[106] The octameric units are similar to structures in zeolites. They also model a silica particle. Perfluoroalkylsilsesquioxanes, prepared in nearly quantitative yields, are hydrophobic, thermally and hydrolytically stable.[107] Some have been inserted into polymers. In one case four methacrylate groups have been added to an octameric structure (**5.33** Schematic).[108] The product polymerized readily with heat or light to a polymer containing 65% silica. An octameric silsesquisiloxane has also been appended to polystyrene (**5.34** Schematic).[109]

5.3 ION-EXCHANGE RESINS

Ion-exchange resins[110] have been used for many years, mainly to remove unwanted ions from water. For example, calcium and magnesium ions in hard water can be removed in this way. A two-bed system of a strong acid cation exchanger followed by a strong base anion exchanger can be used to deionize water, as an alternative to distilled water. Electrodeionization of copper ion-containing wastewater using alternative cation- and anion-exchange membranes can be used to recover copper and very pure water for recycling to the electroplating plant so that there is no waste.[111] (Recovery of metal ions from wastewater from metal-finishing plants was discussed in Chapter 4.) A weak base ion exchanger has been used to recover acetic acid from a 1% aqueous solution.[112] Thus, the resins can be used to recover valuable chemicals from waste streams before the water is discharged from the plant. The resins can also be used as insoluble catalysts or reagents before or after further modification. Because of microenvironmental effects and site–site interactions, polymer-supported reactions may differ from their homogeneous counterparts in the course of the reaction, selectivity, rate, and stereochemical results.[113] This can sometimes be very advantageous. Before describing such uses, a brief summary of what the resins are and how

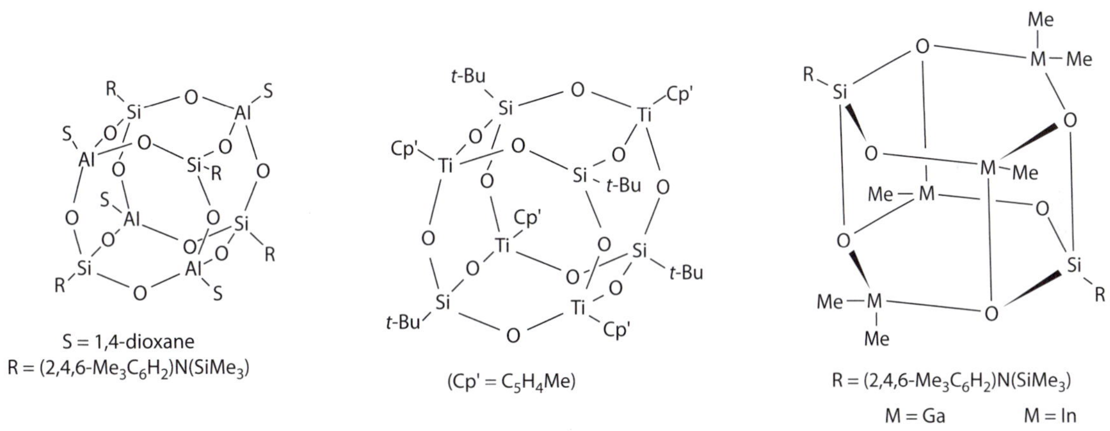

S = 1,4-dioxane
R = (2,4,6-$Me_3C_6H_2$)N($SiMe_3$)

(Cp' = C_5H_4Me)

R = (2,4,6-$Me_3C_6H_2$)N($SiMe_3$)
M = Ga M = In

5.31 Schematic

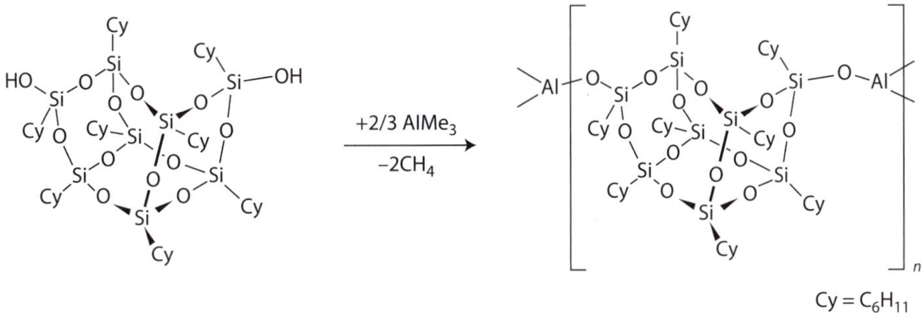

+2/3 $AlMe_3$
−2CH_4

Cy = C_6H_{11}

5.32 Schematic

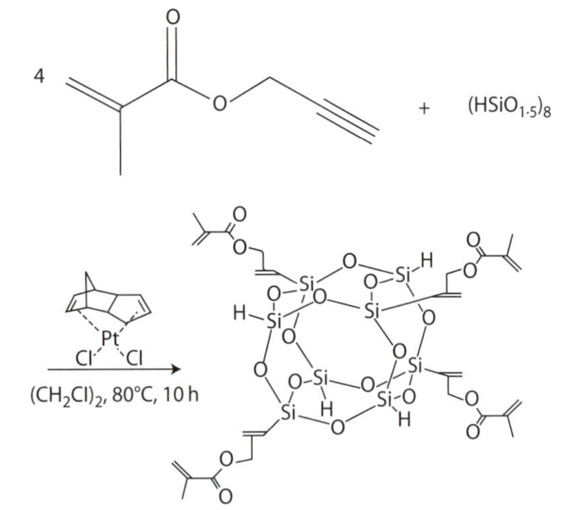

5.33 Schematic

they are made will be given. This will focus mainly on those derived from polystyrene.

Styrene containing some divinylbenzene and a water-immiscible solvent as a porogen are polymerized in aqueous suspension using a free radical initiator. When done properly, this results in a sturdy bead containing pores where the solvent used to be. This macroreticular resin has more surface area and, after derivatization, a much greater exchange capacity than a solid bead of the same size. Recent work has shown how to improve the beads by making them monodisperse.[114] The method uses monodisperse seed from an emulsion polymerization for the following suspension polymerization. The reactions (**5.35** Schematic) to form the actual ion-exchange resins from this support are shown.

The use of dimethylamine in place of trimethylamine would produce a tertiary amine. Pyridine has also been used in the quaternatization to produce anion-exchange membrances.[115] (Membranes are covered in Chapter 7.) The amount of divinylbenzene in the polymerization can be varied widely, but for use in water it is about 8%. Commercial beads are about 0.3–1.2 mm in diameter. Glass-fiber-supported resins provide higher rates of ion exchange and regeneration than beads.[116]

There are other ways to obtain some of these products. Efforts have been made to replace the formaldehyde–hydrogen chloride mixture, which is carcinogenic. One method uses

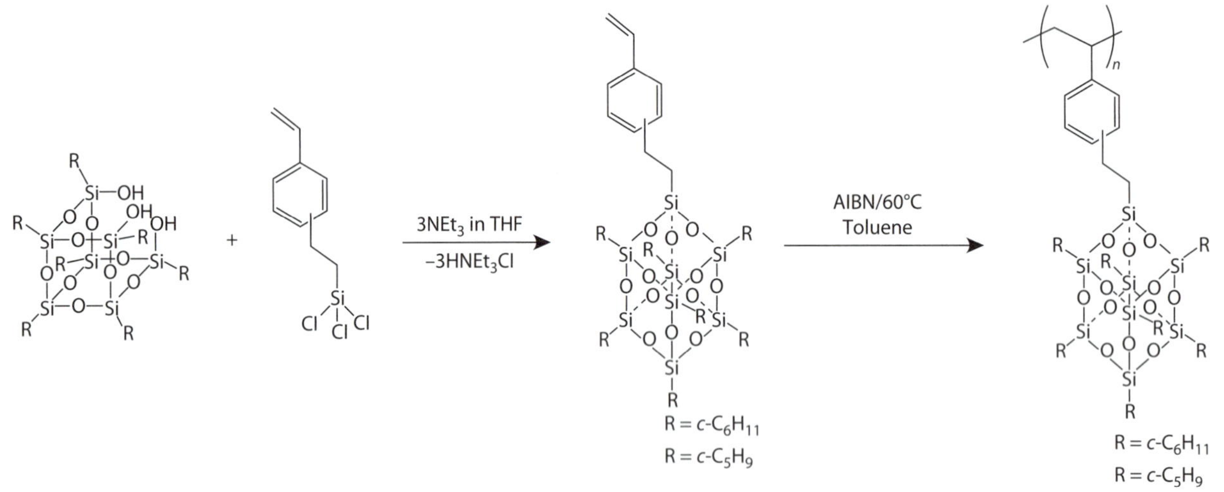

5.34 Schematic

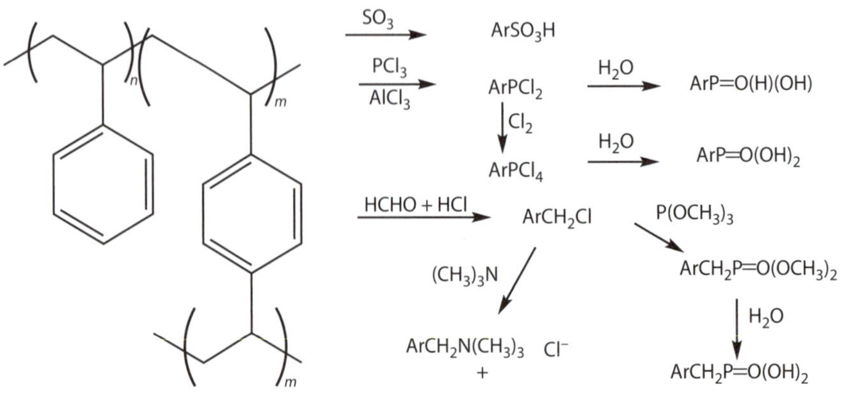

5.35 Schematic

trioxane with trimethylchlorosilane and tin(IV) chloride.[117] Another copolymerizes 4-vinylbenzyl chloride with divinylbenzene. Still another copolymerizes 4-vinyltoluene with divinylbenzene and then chlorinates the methyl group with sulfuryl chloride[118] or cobalt(III) acetate in the presence of lithium chloride.[119] Aminomethyl groups have been introduced into the styrene–divinylbenzene copolymer by treatment with *N*-chloromethylphthalimide and iron(III) chloride, followed by removal of the protecting phthaloyl group.[120] Acylation of cross-linked polystyrene with acetic anhydride, followed by reductive amination of the ketone, yielded a 1-aminoethyl-polystyrene.[121] Quaternization of this amine with iodomethane gave an anion-exchange resin. A greener method would be to quaternize with less toxic dimethyl carbonate.

The functional group can also be introduced in a monomer used in the original polymerization. Sulfonic acids can be introduced as their sodium salts with sodium 4-vinylbenzenesulfonate. In one case, this was used with *N,N*-dimethylacrylamide and methylenebis(acrylamide) as comonomers.[122] Sodium 2-sulfoethyl methacrylate was also used to add the sulfonic acid. Hollow polyethylene fibers grafted with sodium 4-vinylbenzenesulfonate were as effective (after conversion to the acid form) in acid-catalyzed reactions as the usual beads of ion-exchange resins.[123] Weak acid resins can be made by copolymerization of monomers, such as acrylic acid, methacrylic acid, and maleic anhydride, with a cross-linking monomer.

The chloromethyl group can be modified to introduce chelating groups for more specific ion exchange[124] (e.g., by reaction with ethylenediamine,[125] glycine,[126] or the sodium salt of iminodiacetic acid). The product from ethylenediamine could be used for complexation of copper(II) ion. It was also used to scavenge carbon dioxide, which could be recovered later by thermal desorption.[127] Some metal ion separations are helped by putting both sulfonic and phosphonic acid groups in the resin (e.g., europium(III) from nitric acid).

The acidities of the cation-exchange resins fall in the following order: $SO_3H > PO_3H_2 > PO_2H_2 > COOH$. The quaternary ammonium base is stronger than the tertiary amine base. The efficiency of columns of beads is increased by a small, uniform particle size, low degree of cross-linking, high temperature, low flow rate, low ionic concentration in the incoming liquid, and a high length/diameter ratio. In practice, these must be balanced against cost and time. It is desirable to have the resin denser than the liquid. The resin must swell to be active, so that the degree of cross-linking should not be too high. This swelling also means that some extra room has to be left in the column for this. There are several ways that the resins can wear out. These include the following: attrition in use, in part due to swelling and shrinking; oxidation by air, chlorine, hydrogen peroxide, and others; thermal degradation; radiation; desulfonation; and fouling. The use of quaternary ammonium hydroxide resins may be limited to 40–100°C. When radioactive solutions are involved, inorganic exchangers are used.[128]

Sulfonic acid resins can be used as solid catalysts for esterifications and other acid-catalyzed reactions. Amberlyst 15 was a more effective catalyst for the preparation of esters of phenethyl alcohol and cyclohexanol than sulfated zirconia, an acid clay, and dodecatungstophosphoric acid.[129] (Amberlyst and Amberlite are trademarks of Rohm & Haas.) (See Chapter 6 for more details on solid acids and bases.) The same catalyst gave 86–96% yields of hydroxyesters when a lactone was stored with a hydroxyacid.[130] Diols can be monoacylated in 58–91% yields by transesterification with ethyl propionate in the presence of Dowex 50W (a product of the Dow Chemical Co.).[131] Modification of the sulfonic acid resin with 2-mercaptoethylamine produced a catalyst for the reaction of phenol with acetone to produce bisphenol A (**5.36** Schematic) in 99.5% yield.[132] After 20 cycles, the yield was still 98.7%. When used as catalysts, ion-exchange resins can last for 6 months to 2 years.

Anion-exchange resins can be used as reagents.[133] They are commercially available with acetate, borohydride, bromide, chloride, dichromate, iodide, and periodate anions. Metal carbonyl-containing anions such as $Co(CO)_4^-$, $Rh(CO)I_2^-$, and $(Pt_{15}(CO)_{30})^{2-}$ have also been used.[134] A polystyrene-based anion-exchange resin in the chloride form reacted with rhodium(III) chloride to produce a catalyst for hydration of acetylenes and for double-bond migration in allyl compounds.[135] It was leach-proof and recyclable. It might replace the more dangerous mercury catalysts sometimes used in hydration of acetylenes. Copolymers of acrylamide and 3-(acryloylamino) propyltrimethylammonium chloride have been used to bind $(PtCl_6)^{2-}$ for later reduction to platinum metal to produce a hydrogenation catalyst.[136] Strong base-exchange resins have been used with sodium borohydride and nickel acetate to reduce aromatic aldehydes to arylmethyl compounds in 78–98% yields,[137] aryl oximes to benzyl

2 OH + O → HO OH

5.36 Schematic

amines,[138] and olefins to their saturated analogues.[139] Monosubstituted olefins were reduced quantitatively in 1 h at 0°C. Di- and trisubstituted olefins were not reduced under these conditions. When the reaction was carried out at 65°C for 1 h, the disubstituted olefins, but not the trisubstituted ones, were reduced. The resin was recycled by rinsing with dilute hydrochloric acid and then treating it with aqueous sodium borohydride. Thioacetates were reduced to thiols with a borohydride resin used with palladium acetate.[140] Amine oxides were reduced to the corresponding amines using a borohydride resin with a copper sulfate catalyst in 93–96% yields.[141] Polymeric ozonides have been reductively decomposed with a borohydride resin.[142] Zinc borohydride on a solid aluminum phosphate has been used to reduce epoxides to the corresponding alcohols in 97% yields.[143] Ziroconium borohydride on cross-linked poly(vinylpyridine) has been used to reduce carbonyl compounds without touching the olefinic double bond in yields of 80–96%.[144] The resin was regenerable by washing with acid and then with base, then treating again with zirconium borohydride. This illustrates the diversity of supports that can be used with borohydride ion. Each of these methods produces waste salts. It would be better to reduce catalytically with hydrogen using supported metal catalysts, such as palladium on charcoal, platinum on alumina, and others, if the desired selectivity can be obtained. In the reduction of thioacetates, it would be necessary to use a less active metal sulfide catalyst to avoid poisoning. Catalytic reduction would avoid not only the waste salts, but also the high cost of the reagents.

An anion-exchange resin in the fluoride form has been used to remove the *tert*-butyldiphenylsilyl protecting group from an alcohol.[145] A resin with bromate ion has been used as an oxidizing agent.[146] The resin was regenerated for reuse by washing with aqueous sodium bisulfite and sodium chloride and then exchange with sodium bromate. A resin with the sulfhydryl ion was used to produce mercaptans from alkyl halides.[147]

Nucleophilic substitution of the chloromethylated cross-linked polystyrene resin has been used to attach a wide variety of groups. Among these are those shown (**5.37** Schematic): Potassium diphenylphosphide was used to make the first phosphine.[148] The microenvironmental effects of this catalyst in the formation of benzyl benzoate by the Mitsunobu reaction **5.38** were studied at 18%, 40%, 67%, and 100% substitution. The yield and purity of the product were higher at the two lowest levels. The yield of 97% compared favorably with the 85% obtained with the homogeneous counterpart.

A palladium catalyst derived from this polymeric phosphine gave yields almost identical with its homogeneous counterpart in reaction **5.39**. There was no decrease in activity after five cycles of use.[149]

Alcohols were added to the dihydroprans.[150] The furan derivative was used in a Diels–Alder reaction with a fullerene.[151] Magnesium was added to the anthracene derivative, after which it was used to prepare benzyl and allyl magnesium chlorides. The ability to take up magnesium diminished over several cycles of reuse.[152] Polymeric nonlinear optical materials were made by reaction of chloromethylated polymer with the substituted pyridine to form quaternary ammonium salts, and also by reaction with variously substituted alkoxides.[153] The guanidine was used to catalyze the reaction of soybean oil with methanol to give the methyl ester.[154] The catalyst was almost as active as a homogeneous one. There was some leaching after nine catalytic cycles, perhaps owing to displacement of the guanidino group by methanol. An effort to avoid this by adding five more methylene groups between the aromatic ring and the nitrogen gave a less active catalyst. The cyclopentadiene derivative was used to prepare a rhodium metallocene for use as a catalyst in the carbonylation of a diene.[155] The diphosphine was complexed with palladium(II) chloride for use in the Heck reaction of iodobenzene with methyl cinnamate (as shown earlier in this chapter with a different supported catalyst).[156] The palladium had a much higher turnover number than that in a monomeric palladium counterpart, presumably owing to isolation of the active sites. The catalyst was easy to recycle without loss of activity. It could be handled in air and had high thermal stability. The supported titanium tartrate was used as a catalyst for the Diels–Alder reaction of cyclopentadiene with methacrolein.[157] It was far more active than a homogeneous catalyst owing to site isolation. It could be recovered and reused.

Polymeric counterparts of 4-dimethylaminopyridine have been prepared by several groups. Reilly Industries (now Vertellus) sells one as Reillex PolyDMAP (**5.40** Schematic).[158] When used in esterifications, the polymeric catalyst offers easy separation, reduced toxicity, and the ability to use it in excess. It can be reused. The monomeric reagent, which is often used in organic synthesis,[159] is highly toxic. The two other polymeric analogues of 4-dimethylaminopyridine (**5.41** Schematic) are based on other chemistries.[160] The first is a polyamide and the second is a polyurethane. In the second case, glycerol was also sometimes added to produce an insoluble product.

Another polymeric analog has been formed by the reaction of an alternating copolymer of ethylene and carbon monoxide and 4-aminopyridine to form a polymeric pyrrole.[161] Reaction of 4-aminopyridine with epichlorohydrin produced another analogue.[162] One has also been made by reaction of a polymeric acid chloride with a pyridine-substituted piperazine.[163] Reilly Industries also sells copolymers of 4-vinylpyridine as ion-exchange resins.[164]

A nonionic cross-linked polymethacrylate resin (Amberlite XAD7) was used in the reduction of the ketone (**5.42** Schematic) by an aqueous suspension of *Zygosaccharomyces rouxii* cells[165] Adsorption by the resin kept the concentrations of the substrate and product below levels that would be lethal to the cells, yet allowed enough to be in solution for the

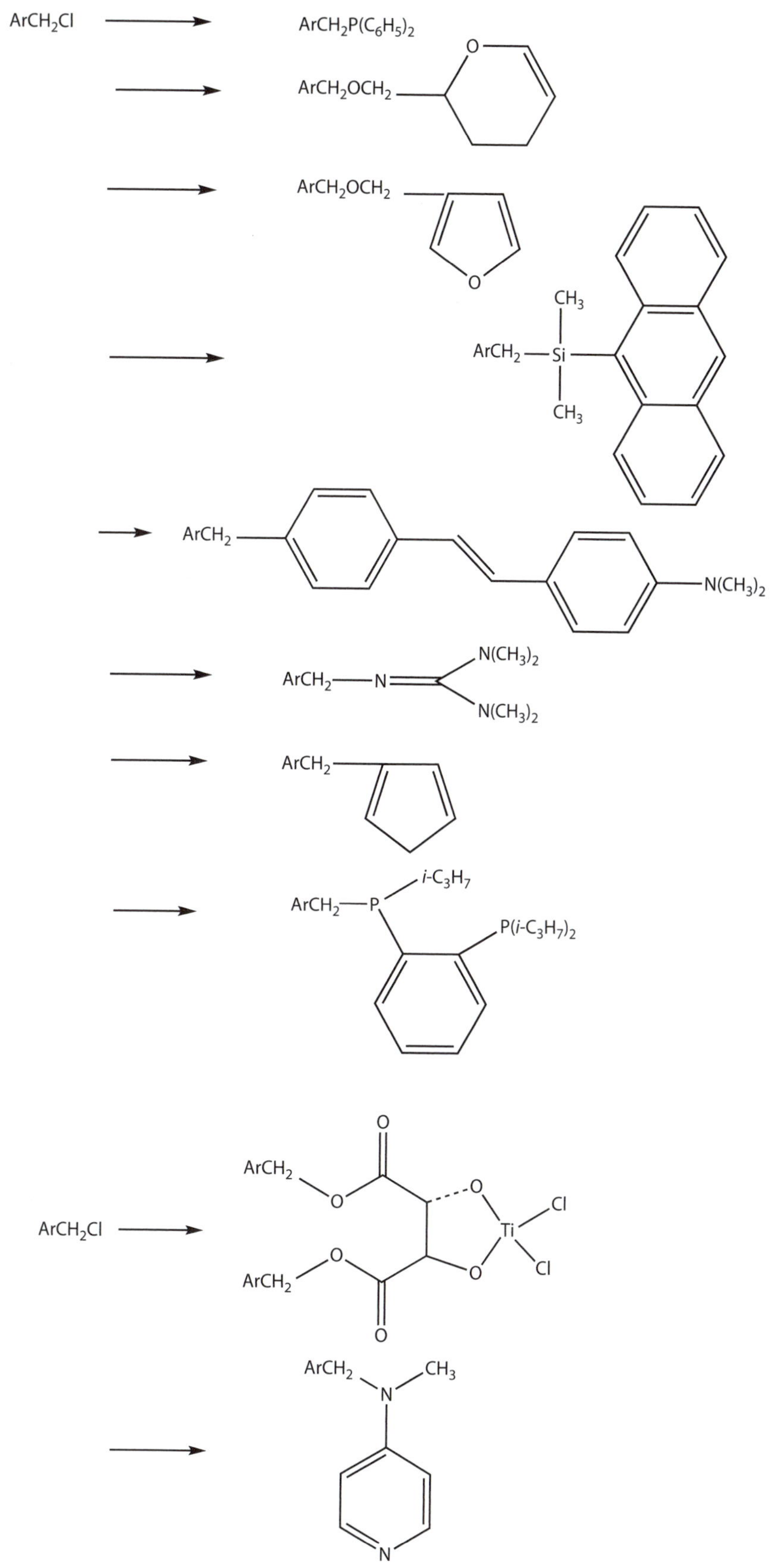

$ArCH_2Cl$ → $ArCH_2P(C_6H_5)_2$

→ $ArCH_2OCH_2$–

→ $ArCH_2OCH_2$–

→ $ArCH_2$–Si(CH_3)(CH_3)–

→ $ArCH_2$–

–$N(CH_3)_2$

→ $ArCH_2$–N=C($N(CH_3)_2$)($N(CH_3)_2$)

→ $ArCH_2$–

→ $ArCH_2$–P(i-C_3H_7)– $P(i\text{-}C_3H_7)_2$

$ArCH_2Cl$ → $ArCH_2$, O, O, Ti, Cl, Cl, $ArCH_2$, O, O

→ $ArCH_2$–N–CH_3, N

5.37 Schematic

$$C_6H_5COOH + C_6H_5CH_2OH + C_2H_5OOCN{=}NCOOC_2H_5 + \text{polymeric phosphine} \longrightarrow$$
$$C_6H_5COOCH_2C_6H_5 + C_2H_5OOCNHNHCOOC_2H_5 + \text{polymeric phosphine oxide}$$

5.38 Schematic

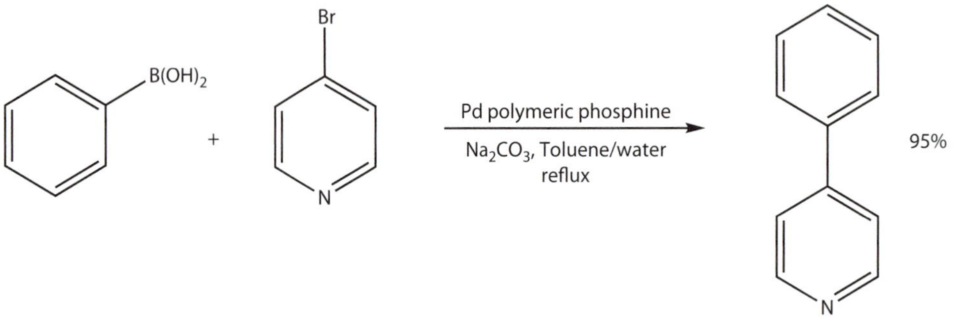

Br
B(OH)$_2$
+
Pd polymeric phosphine
Na_2CO_3, Toluene/water
reflux
N
N
95%

5.39 Schematic

n
CH_3
N
N
Reillex PolyDMAP

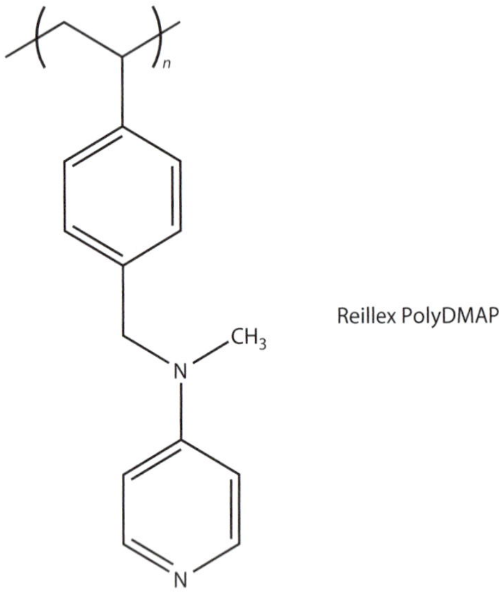

5.40 Schematic

reduction to take place in 96% yield with more than 99.9% enantiomeric excess (often abbreviated %ee).

A flow process using packed columns of immobilized reagents, catalysts, and scavengers has been used to prepare the alkaloid oxomaritidine in seven steps with an overall yield of 40% (**5.43** Schematic).[166] A similar system was used to prepare the alkaloid plicamine.[167]

5.4 COMBINATORIAL CHEMISTRY

Merrifield pioneered the solid-phase method for the preparation of polypeptides.[168] The typical support used is a chloromethylated polystyrene cross-linked with 1% divinylbenzene. A generalized scheme is shown in **5.44** Schematic.

R is typically 9-fluorenylmethyl or *tert*-butyl. In the former compound, the protecting group can be removed with piperidine and in the latter one with trifluoroacetic acid. The cycle of deprotection and coupling can be repeated many times to produce polypeptides. The general requirements are that the reagents used in the synthesis and deprotection steps should not cleave the tether or the polypeptide. The final cleavage removes both the support and the protecting group. Over the years, many variations have been devised to perform the cleavages under milder conditions. The big advantage of the method is the ease of removal of unreacted reagents and by-products by simple washing. Large excesses of reagents can be used to speed up the reactions and ensure completion of reactions. In a study of the preparation of peptide guanidines, the solid-phase method offered several advantages. The reaction was seven times as fast as in solution. The yields were improved greatly. Problems with the solubility of reagents were eliminated, because a wider choice of solvents for the guanylation reaction was now possible.[169] Another recent innovation is the use of ultrasound to assist the coupling of zinc carboxylates with the Merrifield resin.[170] Ultrasound also accelerated the cleavage of the final polypeptide from the resin by ethanolamines and aqueous sodium carbonate or sodium hydroxide. The process involves an initial transesterification, followed by hydrolysis of the new ester.

Combinatorial chemistry uses such methods to make mixtures of many possible compounds (called libraries) that can then be screened for the desired property to select those compounds that are active.[171] The methods are suitable not only for polypeptides, but also for oligonucelotides,[172] oligosaccharides,[173] and a variety of small molecules, including many heterocyclic compounds. The method's popularity is because our knowledge of the interactions of molecules with receptors is not sufficient to allow prediction of the optimum structure. The change in conformations of both the

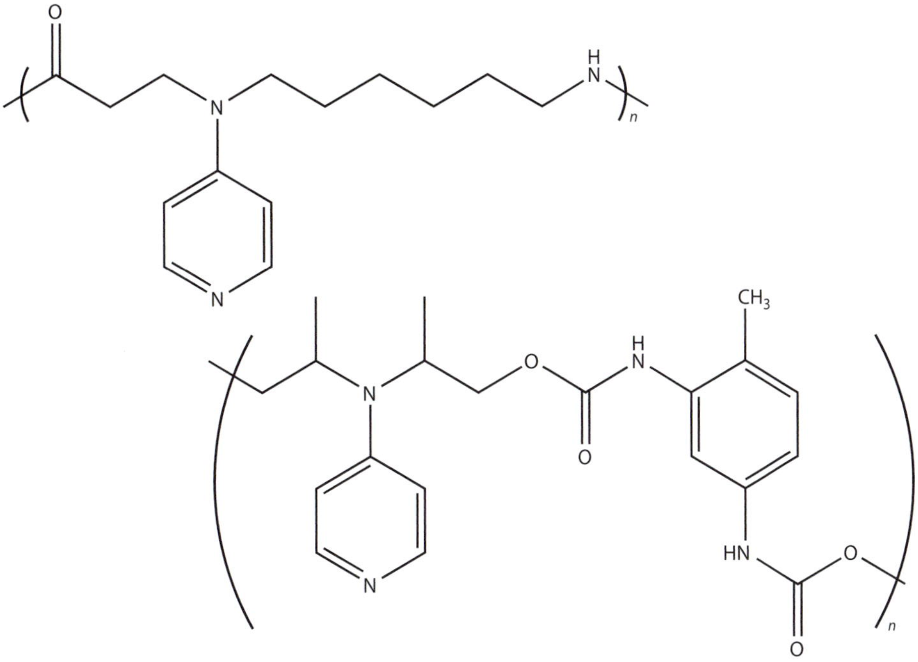

5.41 Schematic

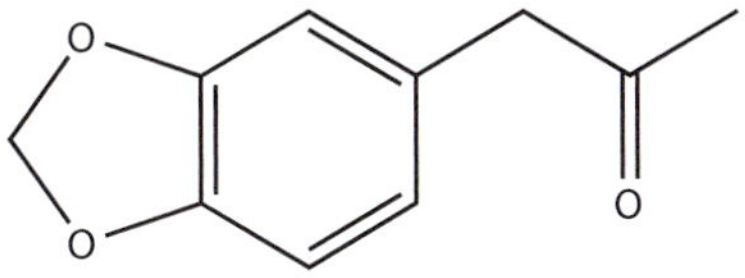

5.42 Schematic

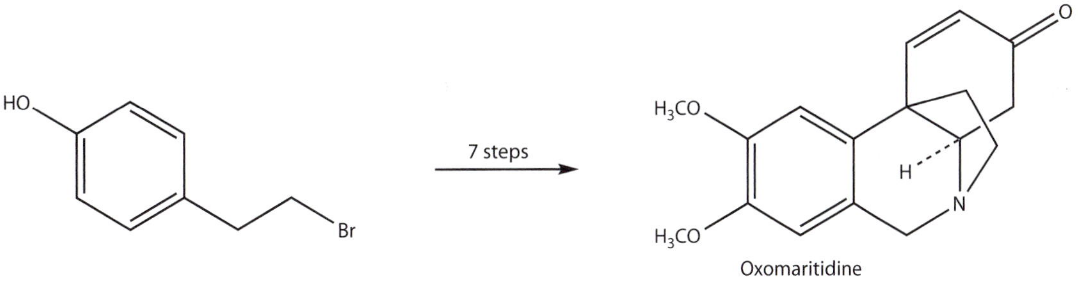

5.43 Schematic

ligand and the receptor during complexation can be hard to predict. In addition, several weak interactions can sometimes combine to provide strong binding. The method can also uncover new unnatural structures for biological targets. This is a shotgun method that works if there is a lead structure. Many leads are from natural products.[174] The drug industry has been very disappointed by libraries made without leads.

A popular method is the split-and-mix one. The first reagent is attached to the support. The beads are then split into several batches, each of which receives a different building block. All the batches are recombined and again split into several lots, each of which receives a different added building block. This process is continued until the desired library of hundreds or thousands of compounds is complete. This method is good for making small amounts of a large number of compounds. A second method, parallel synthesis, makes a smaller library, but with larger amounts of each compound. In this method, multiple reaction vessels (e.g., a multiwell plate) allow the preparation of a different compound in each well. Both methods have been automated.

The libraries are then screened against the desired target to pick out the active compounds. Sensitive methods are needed, for the amount of a compound is small. High-throughput robotic screening is common. For example, potential catalysts for exothermic reactions have been screened by infrared thermography.[175] In another example, catalysts for hydrosilylation were screened by the bleaching of unsaturated dyes.[176] Sometimes it is not necessary to remove the compound from the support before screening. When an active compound is found, it must be identified.[177]

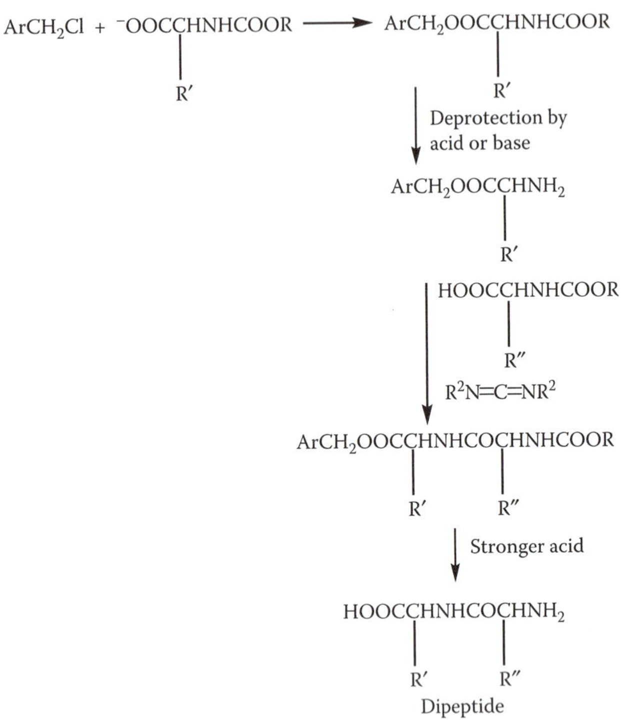

$ArCH_2Cl + {}^-OOCCH(R')NHCOOR \longrightarrow ArCH_2OOCCH(R')NHCOOR$

Deprotection by acid or base ↓

$ArCH_2OOCCH(R')NH_2$

$HOOCCH(R'')NHCOOR$

$R^2N{=}C{=}NR^2$ ↓

$ArCH_2OOCCH(R')NHCOCH(R'')NHCOOR$

Stronger acid ↓

$HOOCCH(R')NHCOCH(R'')NH_2$

Dipeptide

5.44 Schematic

In parallel synthesis, this is not a problem because the well number will have its history recorded. With split synthesis, it may be necessary to repeat the synthesis, checking smaller and smaller sublibraries until the active compound is found. Colored beads and vial caps have been used to keep track of the reactions.[178] An elegant way to identify the beads is to start with beads in a tiny porous polypropylene capsule that also contains a radiofrequency-encodable microchip that can record each step in the process.[179] Matrix-assisted laser desorption/ionization–time of flight (MALDI-TOF) mass spectroscopy has been used to identify compounds from single beads.[180] This method is especially useful because it can be used with very small amounts of compounds with molecular weights as high as several thousands.[181] If the compound is attached to the support by a photolabile link, the bead can be analyzed without prior cleavage of the compound from it. Solid-state nuclear magnetic resonance[182] and infrared microspectroscopy[183] have also been used for compounds on single beads.

Screening of the first library may not reveal a compound with strong activity. If one with weak activity is found, a new library can be made around it. This process may have to be repeated several times before a compound with strong activity is found.

The Merrifield resin does have some limitations. Because it does not swell in some solvents, reaction rates can be low. This has led to a search for other more polar resins. A variety of polyethylene glycol units[184] have been put in supports by different methods. One method is to attach a polyethylene glycol chain to the cross-linked polystyrene by graft polymerization or other means.[185] The polyethylene glycol chains are free to move about because all the cross-linking is in the polystyrene part. Such copolymers have also been used as solid-phase transfer catalysts, such that they can be recovered for reuse by simple filtration.[186] In another method, cross-linking of the polystyrene has been done with tetraethylene glycol diacrylate instead of divinylbenzene.[187] Another resin has been made by copolymerizing a polyethylene glycol end-capped with acrylamide groups with *N,N*-dimethylacrylamide and *N*-ethoxycarbonylethyl-*N*-methylacrylamide and then modifying the resin to incorporate a benzyl alcohol group to produce a highly porous, high-capacity resin (**5.45** Schematic).[188]

Cross-linked supports have also been made by copolymerization of polyethylene glycol terminated with methacrylate groups with cross-linking agents such as trimethylopropane trimethacrylate.[189] This type of resin has also been used as phase-transfer catalysts.[190]

Solid supports tend to have low loading. To avoid this problem, soluble polymeric supports have been used in solution.[191] The only requirement is that there be an easy way to remove excess reagents and by-products at each stage. Polyethylene glycol monomethyl ether has been used for this purpose.[192] Purification at each stage is achieved by ultrafiltration or by precipitation with a large amount of ethyl ether or ethanol.[193] Modified versions (**5.46** Schematic) have been used in the synthesis of oligosaccharides.[194]

Purification at each step was achieved by fast column chromatography in ethyl acetate. The polymer stays at the top of the column while the reagents and by-products pass through. The polymer is then eluted with 4:1 ethyl acetate/methanol.

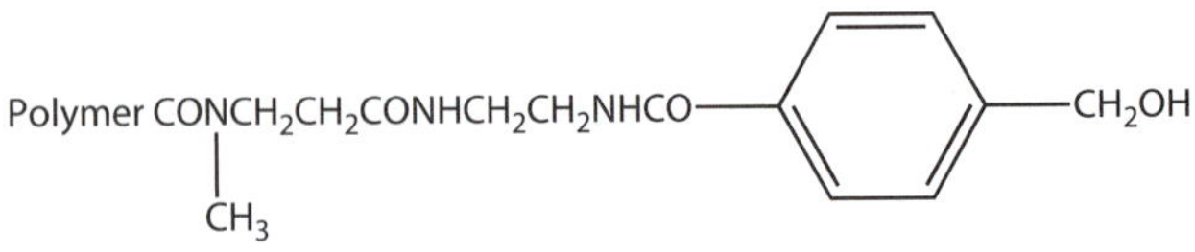

Polymer $CON(CH_3)CH_2CH_2CONHCH_2CH_2NHCO{-}C_6H_4{-}CH_2OH$

5.45 Schematic

$HOCH_2CH_2S(CH_2CH_2O)_nCH_3$ $\quad$ $HOCH_2{-}C_6H_4{-}CONH(CH_2CH_2O)_nH$

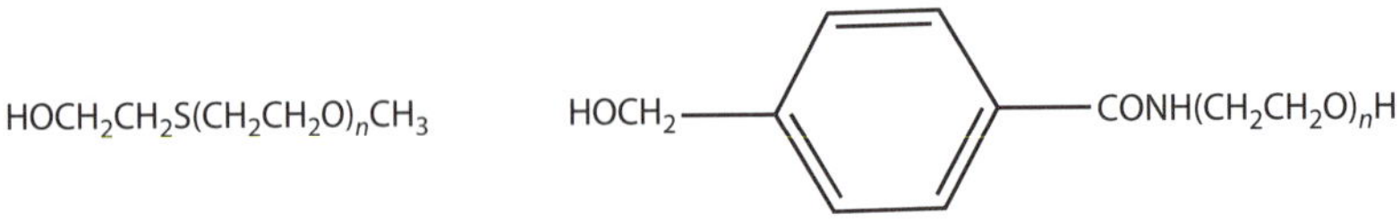

5.46 Schematic

Another system is to remove excess reagents and byproducts with solid polymeric reagents that react with them, which enables separation by filtration. For example, a polymeric amine can be used to remove isocyanates, acylchlorides, and sulfonyl chlorides. A polymeric base can be used to remove hydrochloric acid. A polymeric isocyanate can be used to remove amines.[195]

Ligands have also been attached to polyethylene glycol so that they can be recovered for recycle by the foregoing methods. A phosphine isocyanate has been reacted with polyethylene glycol for use in the Staudinger reaction with alkyl azides to form a phosphine imine.[196] An alkaloid has been attached to the monomethyl ether of polyethylene glycol for use in the Sharpless asymmetrical dihydroxylation of olefins. The reaction was complete in the same time period, with no decrease in yield or enantioselectivity, as occured when the alkaloid was used by itself.[197] (Asymmetrical reactions are covered in Chapter 10.)

Combinatorial chemistry has also been applied to the preparation of catalysts[198] and to inorganic compounds.[199] New catalysts for the oxidation of methanol in fuel cells,[200] the polymerization of olefins,[201] and asymmetrical organic reactions (or chiral recognition)[202] have been found by this method. A rate enhancement of 1 million for the ring-opening hydrolysis of an isoxazole (**5.47** Schematic) was found with a "synzyme" made by alkylating poly(ethylene imine) with three different groups.[203]

Among the inorganic materials under study by this method are phosphors,[204] superconductors, magnetic materials,[205] thin-film dielectric materials,[206] pigments,[207] and zeolites.[208] Many of these use magnetron sputtering of one or more elements at a time on a substrate through a series of masks. Combinatorial chemistry should also be applied to optimization of ligands (plain or polymeric) for the separation of metal ions. Better separations are needed by the mining, metal finishing, and other industries.

Combinatorial chemistry offers a good way to find a new material while operating on a small scale. Thus, it does not produce much waste. However, if the synthesis of the new compound is scaled up, there will be a problem with wastes. Ways will need to be found to recycle excess reagents, to minimize use of solvents, and, if possible, to develop ways to do without protecting groups.

Combinatorial methods are also being used in biology. Beetles have been making mixtures of large-ring polyamines as defensive secretions for a long time.[209] More recently, combinatorial antibody libraries have been made for primary amide hydrolysis.[210]

5.5 OTHER USES OF SUPPORTED REAGENTS

The problems in the use of chlorine and chlorine compounds for the disinfection of water were discussed in Chapter 3. One problem is the residual toxicity of chlorine to nontarget organisms, as in cooling tower effluents. Using ozone as an alternative involves a very toxic material, but does eliminate the residual toxicity problem. The use of antibacterial polymers is being investigated as an alternative way of disinfecting water and air. Several are shown in **5.48** Schematic through **5.51** Schematic.

A silane containing hydroxyl and quaternary ammonium salt groups was treated with a triisocyanate to prepare a biocidal film that was active against *Escherichia coli* for as long as 1 month.[211] Cellulose was treated with the appropriate trimethoxysilane to attach a phosphonium salt.[212] Polyphosphonium salts based on polystyrene have been prepared with spacers of varying lengths between the aromatic ring and the phosphorus atom.[213] Polymeric dichlorohydantoins were used to disinfect potable water and air. Less than 1 ppm of free chlorine was released into the water.[214] The bromine analogue had even higher activity. A macroreticular copolymer made from divinylbenzene and epithiopropyl methacrylate was treated first with triethylenetetramine and then complexed with silver ion to a polymer with high antibacterial activity. No silver ion dissolves in the water. There is no loss in activity after the polymer is reused several times. Such polymers may prove to be effective alternatives to the use of added chemicals to sterilize water and air. They could be used in films, columns, or other equipment.

Phase-transfer catalysts are used to facilitate reactions between reagents that are in two different phases (e.g., 1-bromooctane in toluene with aqueous potassium iodide to form 1-iodooctane). They are usually quaternary ammonium

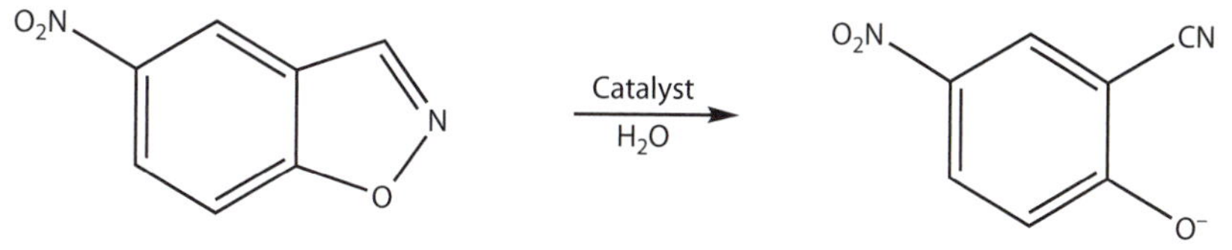

O_2N N O —Catalyst, H_2O→ O_2N CN O^-

5.47 Schematic

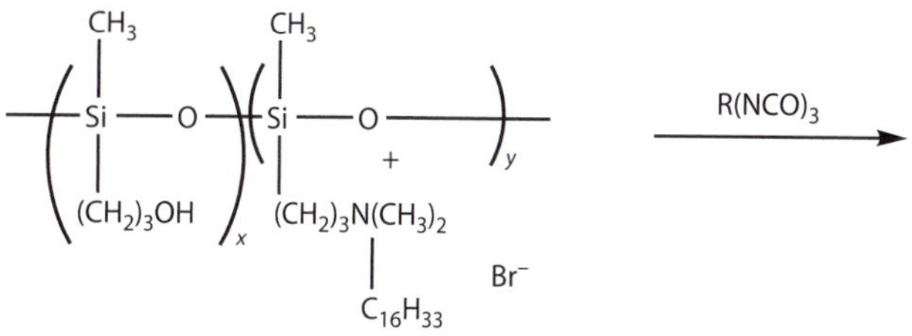

CH_3 Si O $(CH_2)_3OH$ x; CH_3 Si O $(CH_2)_3N(CH_3)_2$ + $C_{16}H_{33}$ Br^- y —$R(NCO)_3$→

5.48 Schematic

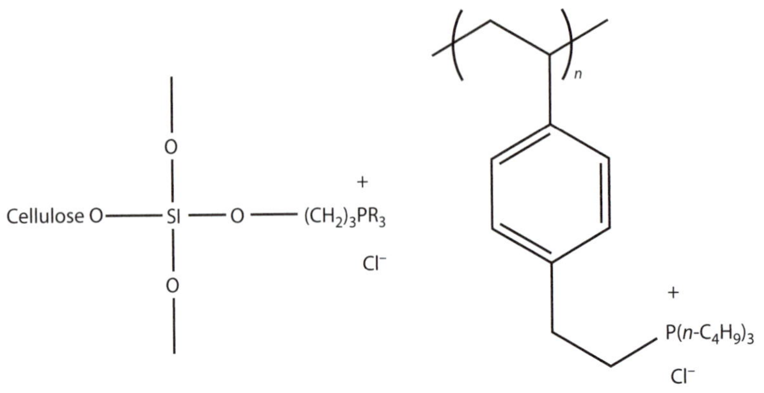

5.49 Schematic

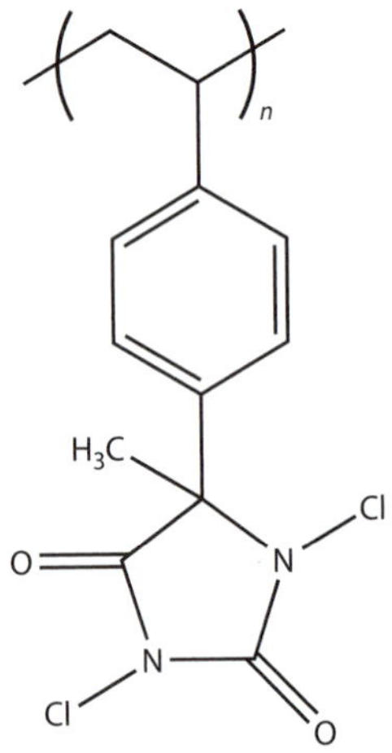

5.50 Schematic

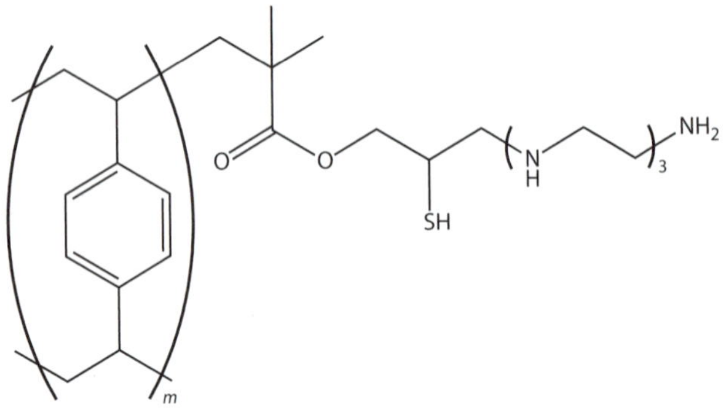

5.51 Schematic

or phosphonium salts or crown ethers. They can complicate the workup of the reaction and may be difficult to recover for reuse. When they are insoluble polymeric ones, workup and recycle can be done by simple filtration.[215] The process is called triphase catalysis. In favorable cases, their activity can be comparable with that of their lower molecular weight analogues. They are often based on cross-linked polystyrene, for which spacers between the aromatic ring and the quaternary onium salts can increase activity by two- to fourfold. Copolymerization of 4-vinylbenzyl chloride with styrene and *N,N*-dimethylacrylamide, followed by treatment with tri-*n*-butylphosphine, produced catalysts that were used in the reaction of benzyl chloride with solid potassium acetate (**5.52** Schematic).[216]

Increasing the density of the phosphonium salt increased the rate of reaction of potassium iodide with 1-bromooctane in a silica-supported catalyst (**5.53** Schematic).[217] The catalyst made from silica with 100-Å pores was 10 times as active as that from 60-Å pore silica. When used in toluene, the catalyst could be reused four times with no loss in activity.

A polymeric formamide was a catalyst for the reaction of alkyl bromides with potassium thiocyanate, even though the corresponding formamide was inactive. The catalyst was made by copolymerizing styrene and divinylbenzene with 4 mol% of the monomer shown in **5.54** Schematic.[218]

Presumably, the polymer works because of a multiplicity of functional groups that behave as weak ligands in the proper conformation. Further experiments of this type with other functional groups that might behave as ligands for inorganic salts should be tried. Perhaps, combinatorial chemistry can be applied here.

Supports can sometimes be used to prevent the loss of expensive or toxic reagents or catalysts. Cinchona alkaloids are used in the enantioselective dihydroxylation of olefins

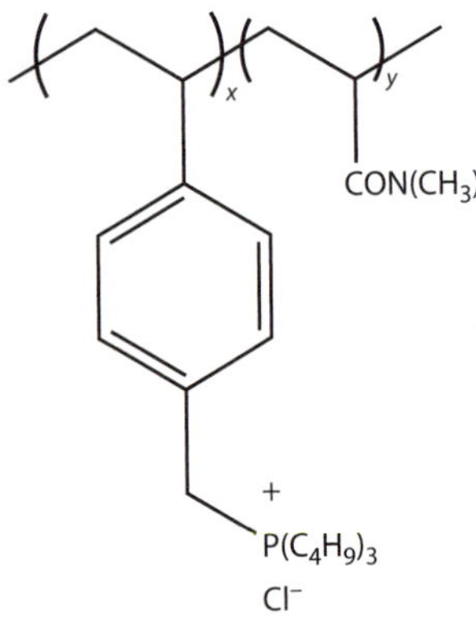

5.52 Schematic

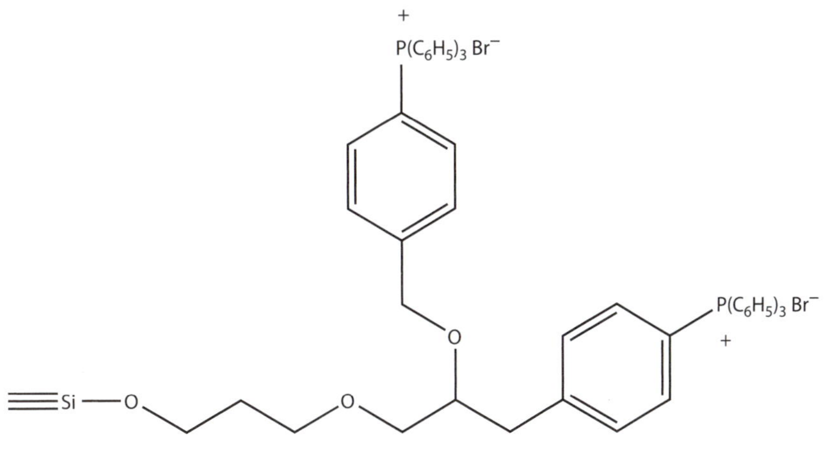

5.53 Schematic

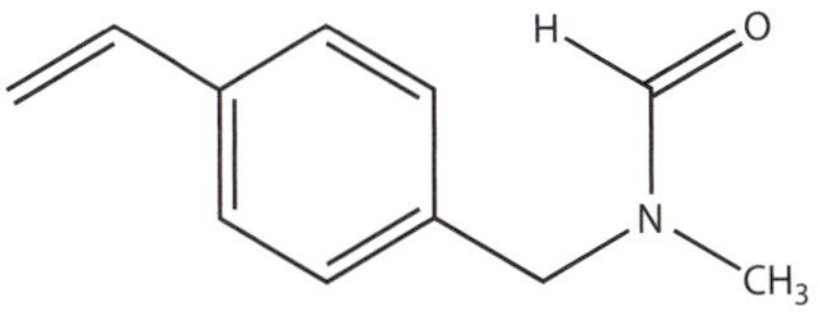

5.54 Schematic

with osmium tetroxide. Although the amount used may be low on a mole percent basis, it is an appreciable amount on a weight basis. One cannot afford to discard it. Terpolymers were prepared from an alkaloid derivative of 4-vinylbenzoic acid, 2-hydroxylethyl methacrylate, and ethylene glycol dimethacrylate. These were used with osmium tetroxide and *N*-methylmorpholine-*N*-oxide or potassium ferricyanide in the dihydroxylation. Stilbene gave the desired product with 95% enantiomeric excess and 70% yield, compared with 97%ee for a homogeneous reaction.[219] The polymer with the poisonous osmium tetroxide complexed to it was recovered by filtration. Further work is needed, because the enantioselectivity with stilbene and other olefins tends to be slightly lower than with the corresponding reactions in solution. Poly(vinylpyridine) has been used to support osmium tetroxide in the dihydroxylation of olefins with hydrogen peroxide.[220] The yields with cyclooctene and *trans*-4-octene were 99–100%. The polymeric complex could be recycled several times, but was slowly oxidized. Use of an inorganic support, such as a zeolite or heteropolymetalate, would avoid the oxidation. The use of rhodium supported on cross-linked poly(vinylpyridine) for the hydroformylation of propylene and 1-octene works well enough that it is said to "have commercial potential."[221]

A silver dichromate complex with poly(ethyleneimine) has been used to oxidize benzyl alcohol in toluene to benzaldehyde in 98% yield.[222] The polymeric reagent could be regenerated several times with no loss in activity. No chromium was lost in the reaction. Polypropylene fabric with vinylpyridine graft polymerized on to it has also been used to support dichromate in the oxidation of alcohols.[223] Less than 1 ppm of chromium was in solution after the reaction. This is one method of avoiding the loss of toxic heavy metal ions. There may be losses in mining the chromium ore and converting it to dichromate. A better method, as mentioned in Chapter 4, would be to select an insoluble catalyst that can be used in an oxidation with air. Even so, proof of the ability to recycle a catalyst repeatedly must show that no leaching occurs. A chromium aluminum phosphate molecular sieve catalyst used in the oxidation of α-pinene to verbenone by a hydroperoxide was active in reuse, but the filtrate was also able to catalyze the oxidation.[224] (This reaction was described in Chapter 4. For more on zeolite and molecular sieve catalysts, see Chapter 6.)

Polymeric ligands can offer advantages in the separation of metal ions. A copolymer of 4-vinylpyridine and divinylbenzene (**5.55** Schematic) separates iron(III) quantitatively.[225] A polymeric hydroxamic acid favors iron(III) over other ions.[226]

Sherrington and co-workers have used polybenzimidazoles and polyimides to complex metal ion catalysts and thus avoid or minimize losses of the metal ions, some of which are expensive. The first has been used to complex palladium(II) for the Wacker reaction [(**5.56** Schematic) of 1-octene and 1-decene to ketones].[227] (The other reagents for this reaction are copper(II) chloride and oxygen.)

In some cases with 1-decene, the activity was higher than the homogeneous counterpart. The catalyst was recycled seven times. There was some loss of palladium at first, but after about six cycles only 1 ppm was lost per cycle. No hydrogen chloride needed to be added. No copper(II) chloride needed to be added after the first cycle. The catalyst

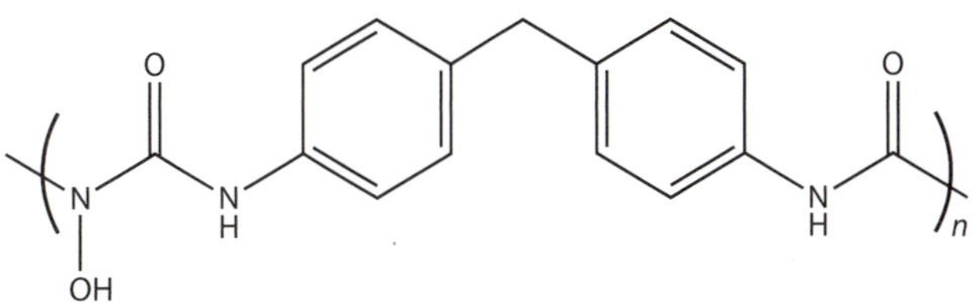

5.55 Schematic

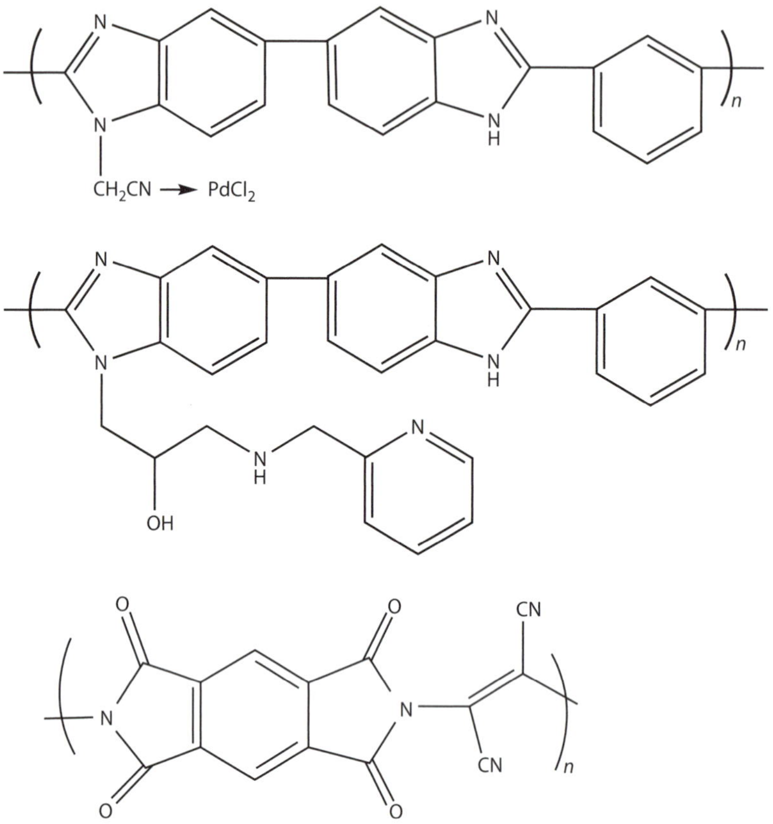

5.56 Schematic

was used at 120°C and was stable to about 400°C. The second polymer was used with molybdenum(VI) and *tert*-butylhydroperoxide to convert cyclohexene to its epoxide in quantitative yield.[228] The catalyst was recycled nine times with no detectable loss of molybdenum, but the activity did decline. The polymeric ligand may favor site isolation, whereas some other polymeric ligands allow formation of unwanted oxygen-bridged bimetallic species. The third polymer was used to complex palladium(II) for the Wacker reaction of 1-octene.[229] Both nitrile groups probably coordinate with the palladium. The material was used for six cycles. There was significant leaching in the first cycle, but none in later cycles. The challenge is to devise active supported catalysts that lose no metal and no activity on repeated use. This could allow their use in packed bed columns. The strategy of Sherrington and co-workers appears to be to use supporting polymers that contain bidentate or tridentate ligand groups to hold the metal ion more tightly. To be active, the metal ion must be coordinately unsaturated. This means that the multidentate ligand must leave an open site on the metal ion, or allow one or more bonding groups to be displaced by the incoming reactants. All excess metal ions must be washed out of the supported catalyst before being used, or it may appear to come out in the first use. Deactivation mechanisms deserve further study. Lower loading might favor site isolation and decrease any loss of activity owing to reactions that produce inactive binuclear species. If loss of activity is due to a buildup of tarry side products that tend to block pores, a new more selective catalyst may be needed or these by-products might be extracted in a regeneration step. Sometimes, ultrasonication can do this during the reaction and prolong catalyst life.

A palladium derivative or an oligo-*p*-phenyleneterephthamide performed better as a catalyst for the reduction of phenylacetylene to ethylbenzene than palladium on carbon or silica or alumina catalysts.[230] A rhodium(I) complex of the polyamide was used to catalyze the addition of silanes to 1,3-dienes with good regio- and stereoselectivity.[231] Rhodium is one of the metals that is so expensive that losses must be kept at an absolute minimum.

This polyamide was also used to support a palladium catalyst for the selective reduction of alkynes and dienes.[232] Phenylacetylene was reduced to styrene in 100% yield and 100% conversion. The reduction of 1-octyne was 80% selective to 1-octene at 100% conversion. There was no loss in activity or selectivity after 11 successive runs. A polymeric analogue of triphenylphosphine (**5.57** Schematic) has been used to immobilize rhodium dicarbonylacetylacetonate for the hydroformylation of 1-octene to the corresponding aldehyde (**5.58** Schematic).[233]

The linear/branched ratio was 7.5:1, which was greater than that found with the control catalyst in solution.

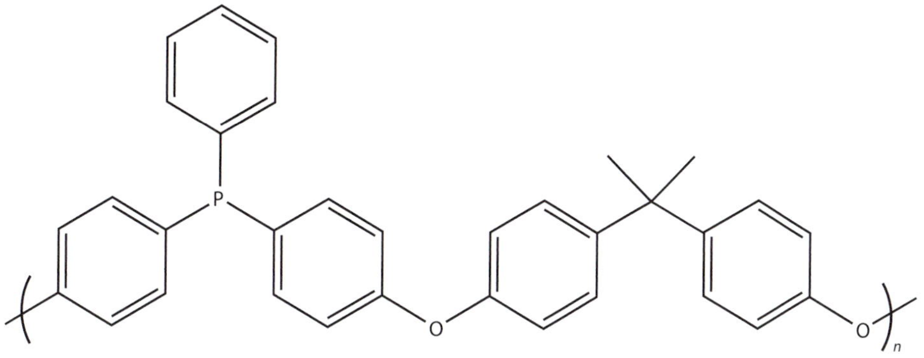

5.57 Schematic

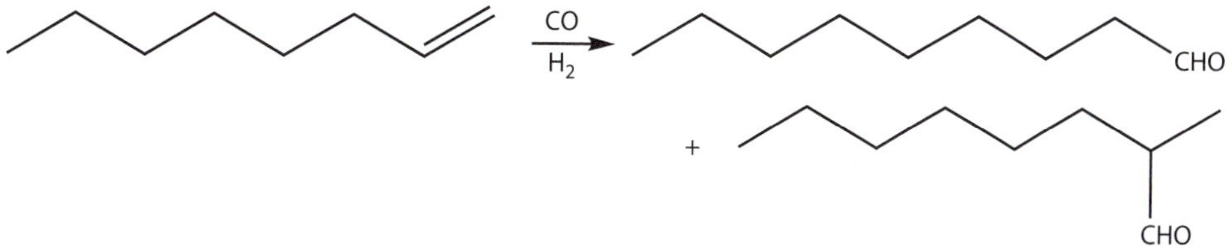

5.58 Schematic

Rhodium(III) chloride has been immobilized on a support made by polymerization of vinylpyridine and divinylbenzene in the presence of silica. The best activity for the conversion of methanol to acetic acid by carbon monoxide was obtained after 20% of the pyridine groups were quaternized with methyl iodide. This suggests ionic bonding of a tetrahalorhodate ion to the polymer.[234]

Frequently, supported catalysts show lower activity than homogeneous catalysts because of low loading. Leaching of the metal may be a problem if the complex is not strong enough. A polymer from the bipyridine monomer (**5.59** Schematic) has also been used to complex rhodium compounds.[235] These polymeric complexes should be more stable than those prepared from poly(vinylpyridine).

Bergbreiter et al.[236] have circumvented this problem of lower activity by complexing the rhodium salt on to poly(alkene oxide) oligomers containing diphenylphosphine groups (**5.60** Schematic). The compound was used as a catalyst for the hydrogenation of olefins. No rhodium was lost. This type of polymer shows inverse temperature solubility.

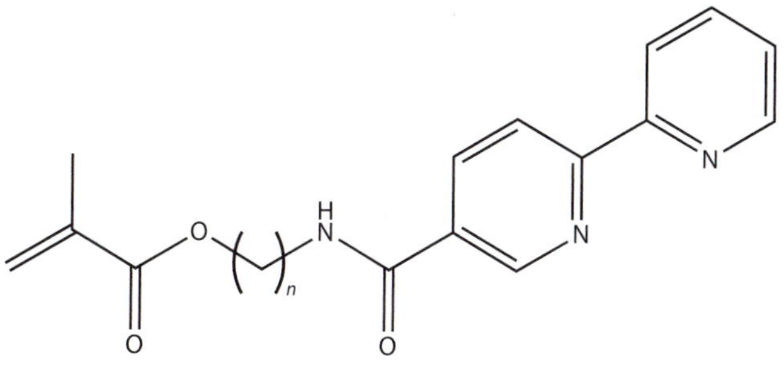

5.59 Schematic

$$((C_6H_5)_2PCH_2CH_2(OCH_2\underset{\displaystyle CH_3}{\underset{|}{CH}})_{59}(OCH_2CH_2)_{15}OCH_2CH_2P(C_6H_5)_2)_{1.5}\,RhCl$$

5.60 Schematic

When the temperature was raised, the polymeric catalyst separated from solution for easy recovery and reuse. This type of "smart" catalyst will separate from solution if the reaction is too exothermic. The catalytic activity ceases until the reaction cools down and the catalyst redissolves. Poly(*N*-isopropylacrylamide) also shows inverse temperature solubility in water. By varying the polymers and copolymers used, the temperature of phase separation could be varied (e.g., from 25 to 80°C).[237] A terpolymer of 2-isopropenylanthraquinone, *N*-isopropylacrylamide, and acrylamide has been used in the preparation of hydrogen peroxide instead of 2-ethylanthraquinone.[238] The polymer separates from solution when the temperature exceeds 33°C to allow recovery of the hydrogen peroxide. Some anthraquinone is always lost to oxidation in such processes. Such an inverse solubility is thought to be due to the release of water from the hydrated forms of the polymers to give the anhydrous forms. This method is used to isolate hydroxypropylcellulose in its commercial preparation.

Cheaper hydrogen peroxide can now be made directly from hydrogen and oxygen in almost 100% yield in a process being used commercially by Headwaters-Evonik.[239] The process uses 4 nm particles of a 50:1 Pd:Pt catalyst, where sodium polyacrylate has been added to prevent the growth of the unwanted crystal face. The arrangement of atoms and the catalytic activity can vary with the crystal face. Gold catalysts also show a size effect.[240] Although bulk gold is inert, 3–5 nm nanoparticles can catalyze a variety of reactions. One catalyst has active 0. 5nm sites that contain about 10 gold atoms.[241] It is now possible to monitor changes in catalysts during reactions using x-rays and transmission electron microscopy.[242] Rhodium on magnesium oxide particles flattened on exposure to oxygen and returned to their original shape when CO was used to reduce the oxide layer.

Efforts to heterogenize homogeneous enantioselective oxidation catalysts often give lower enantioselectivity. This may at least partly be due to the polymer not being able to achieve the proper conformation for the metal chelate catalyst with the substrate. Sometimes, insertion of a spacer group between the polymer and the chelate can help achieve the right conformation. By studying various supports, it has been possible to achieve 92%ee in the oxidation of 1-phenylcyclohexene to its epoxide in 72% yield, using an amine oxide–peracid combination. (This is the Jacobsen oxidation.) Thus, results comparable with those using monomolecular analogues can be obtained, combined with ease of separation for reuse.[243] (Other enantioselective reactions are described in Chapter 10.) One of the porous styrene-based resins used is shown in **5.61** Schematic.

The desire to convert benzene directly to phenol with 30% hydrogen peroxide was mentioned in Chapter 4. A polymer-supported salicylimine vanadyl complex (1 mol%) was used to catalyze this reaction. Phenol was obtained in 100% yield at 30% conversion.[244] There was no leaching of the metal. The catalyst was recycled 10 times after which it started to break up. Oxidation of ligands is often a problem with oxidation catalysts. Inorganic supports not subject to such oxidation need to be tried to extend the life of such catalytic agents.

The Swern oxidation of alcohols to aldehydes is carried out with dimethyl sulfoxide. A polymeric version (**5.62** Schematic) used poly(hexylene sulfoxide), obtained by the oxidation of the corresponding sulfide with hydrogen peroxide to facilitate recovery and reuse, and at the same time, the bad smell of dimethyl sulfide was avoided. Quantitative yields were obtained with 1-octanol and 6-undecanol.[245]

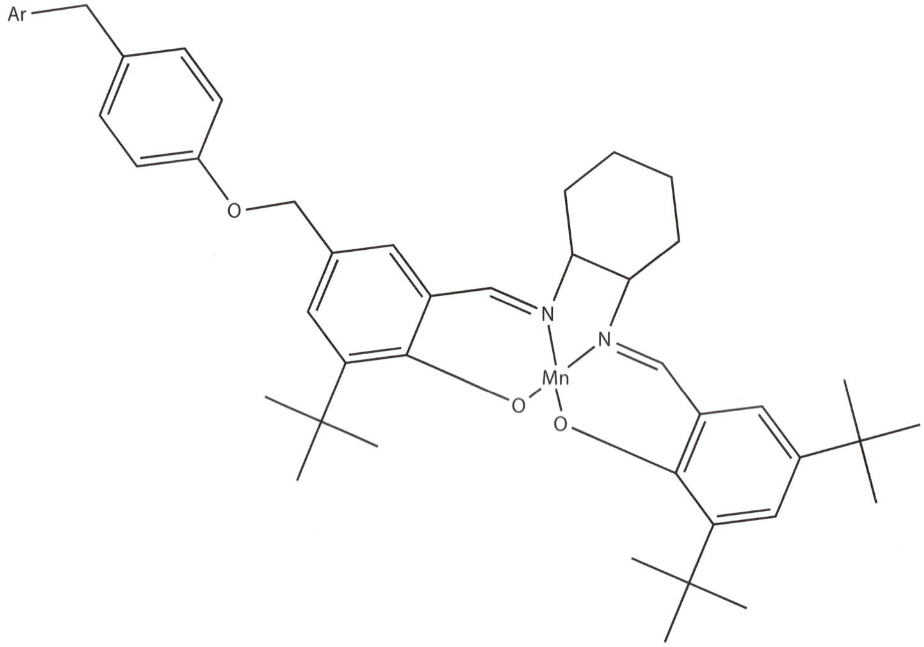

Ar
O
N
N
Mn
O
O

5.61 Schematic

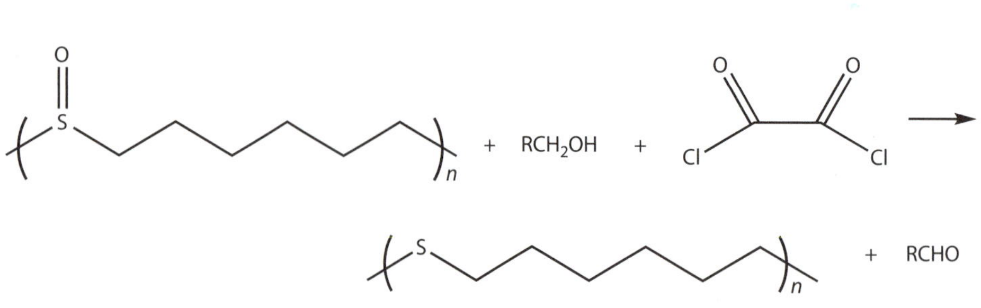

O
S
n
+ RCH_2OH +
O O
Cl Cl
S
n
+ RCHO

5.62 Schematic

However, it would be better to use a process that does not produce waste salts and that does not use methylene chloride in the workup.

A polymeric protecting group was used in a synthesis of pyridoxine by the Diels–Alder reaction (**5.63** Schematic).[246]

It is possible to put mutually incompatible reagents in the same flask if each is on a separate resin. The reactions in **5.64** Schematic could be run in a sequential fashion or simultaneously in cyclohexane. The reagent in the first step was poly(vinylpyridinium dichromate), in the second step, a perbromide on a strongly basic resin (Amberlyst A26), and in the third step, 4-chloro-3-hydroxy-1-methyl-5-(trifluoromethyl) pyrazole on another strongly basic resin (Amberlite IRA900).[247]

5.6 CYCLODEXTRINS

Cyclodextrins are included here as water-soluble supports. They will also be mentioned in Chapter 7 on separations and in Chapter 8 on running reactions in water instead of organic solvents. Cyclodextrins[248] are made by the enzymatic modification of starch.[249] They are made commercially by Cerestar and Wacker Chemie. They have conical structures

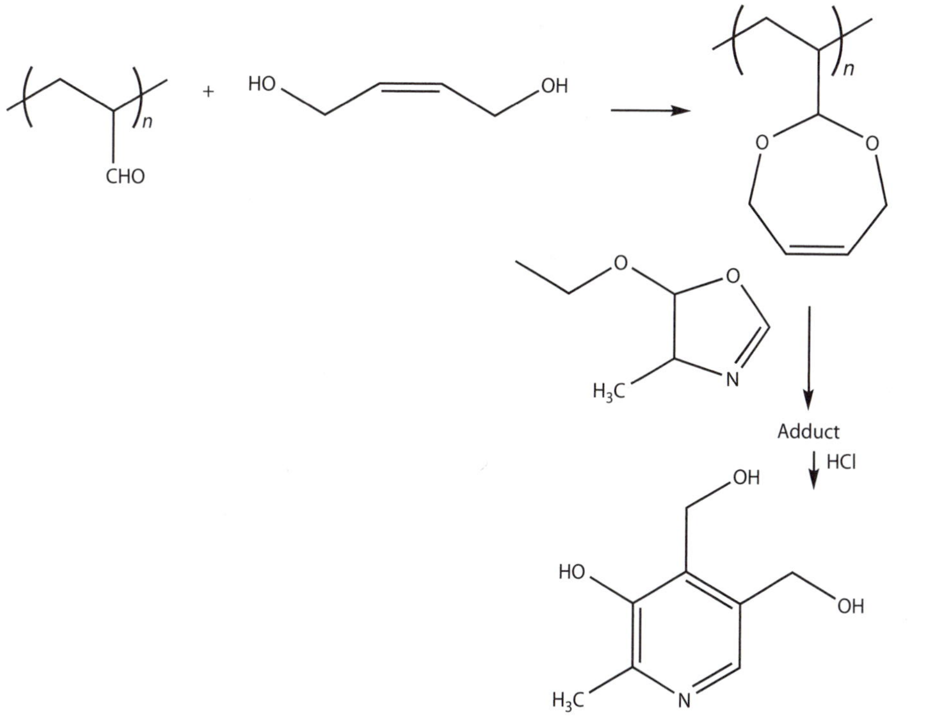

5.63 Schematic

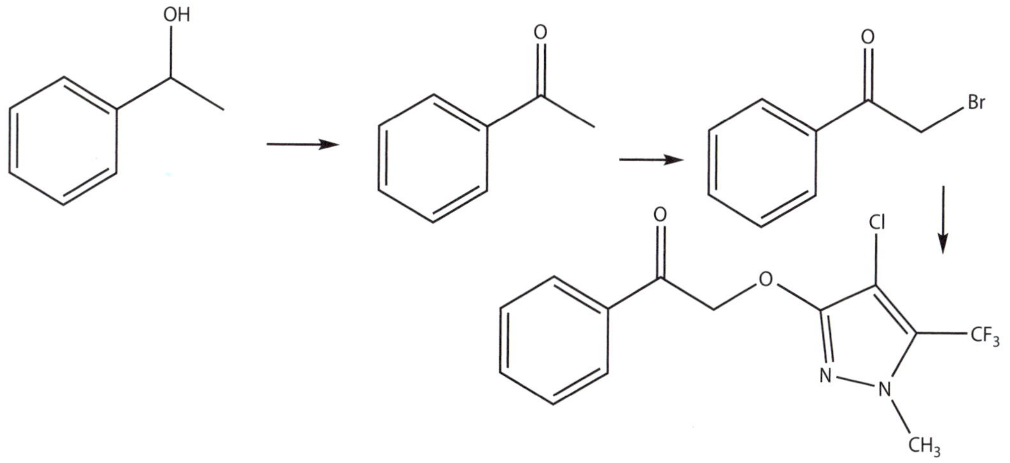

5.64 Schematic

Table 5.2 Properties of α-, β-, and γ-Cyclodextrins

Cyclodextrin Type	MW	Solubility in Water (g/100 mL)	Diameter (Å)	
			Cavity	Outside
α	972	14.5	4.7–5.2	14.6
β	1135	1.85	6.0–6.4	15.4
γ	1295	23.2	7.5–8.3	17.5

of six, seven, and eight glucose units in rings, denoted α-, β-, and γ-cyclodextrin, respectively. Mercian Corporation has a process for β-cyclodextrin, which is more selective than usual, that produces no α- and only a small amount of the γ-product.[250] Use of another cyclodextrin glycosyltransferase from a different organism gave 89% alpha, 9% beta, and 2% gamma.[251] A cyclic tetrasaccharide has also been made enzymatically from starch.[252] The insides are apolar and hydrophobic, whereas the outsides are hydrophilic (Table 5.2).

All are 7.9–8.0 Å in height. A wide variety of guests can fit into these hosts, both in solution and in the solid state. These include paraffins, alcohols, carboxylic acids, noble gases, and some polymers.[253] Sometimes, only part of the molecule needs to fit in. β- and γ-cyclodextrins form such inclusion compounds with polypropylene glycols with molecular weights up to 1000.[254] The guests have reduced volatility and enhanced stability to heat, light, and oxygen. Liquid drugs can be converted to solids that are not hygroscopic. Taste and odor problems can be overcome. Two substances that might react with one another can be packaged together if one of them is isolated in a cyclodextrin. These hosts can also be used to separate molecules based on their size and shape. The bitter flavors in citrus juices, naringen, and limonin can be separated in this way. Cyclodextrins form insoluble complexes with cholesterol in foods. β-Cyclodextrin prevents the enzymatic browning of apple juice by picking up the phenols present before enzymes can act on them.[255] At a ratio of two β-cyclodextrin to one linoleic acid (but not at a 1:1 ratio), oxidation by lipoxygenase is reduced.[256]

The ability to stabilize is also part of green chemistry, for, if things last longer, there will be less need to replace them, and overall consumption will go down. Perpropionic acid can be stabilized by α-cyclodextrin so that 97% remains after 50 days at room temperature.[257] Without the cyclodextrin, 5% decomposes in just 11 days. The material may be useful for bleaching and disinfection. Some artificial food colors are of questionable toxicity. Some have already been removed from the market. Natural colors offer alternatives, but often lack the necessary resistance to heat and low pH. If cyclodextrins can stabilize colors, such as the red in beet juice, there might be no further need for the artificial colors.

Cyclodextrins can be used as "reaction vessels." Chlorination, bromination,[258] carboxylation,[259] azo coupling, and others favor the *para*-isomers over the *meta*- and *ortho*-isomers when the compound being treated is inside a cyclodextrin. Pericyclic reactions, such as the Diels–Alder reaction, are facilitated when done in cyclodextrins. A typical carboxylation used to make a monomer for a high-melting polyester is shown in reaction **5.65**. The dicarboxylic acid was obtained in 65% yield with 79% selectivity. Fortunately, there are better ways (see Chapter 6) to make this compound without the use of toxic carbon tetrachloride. This method gives 100% selectivity at 71% conversion in the reaction of biphenyl-4-carboxylic acid to form the 4,4′-biphenyldicarboxylic acid.

β-Cyclodextrin accelerates the platinum-catalyzed addition of triethoxysilane to styrene.[260] The reaction is 100% complete in 30 min at 50°C with the cyclodextrin, but is only 45% complete without it. The Wacker reaction of 1-decene produces several ketones owing to double-bond isomerization. If the reaction is run in a dimethylcyclodextrin, isomerization is reduced by faster reoxidation of palladium (O).[261] The product 2-decanone is obtained with 98–99% selectivity (**5.66** Schematic). No organic solvent is needed; no chloride is needed. This is an important improvement because the chloride required in the usual Wacker reaction produces unwanted corrosion. A bis(imidazoylmethyl)-β-cyclodextrin (**5.67** Schematic) converts a messy aldehyde cyclization, which gives a mixture of products, to a clean reaction.[262]

Polymerization of phenyl or cyclohexyl methacrylate in 2,6-dimethyl-β-cyclodextrin in water using a potassium persulfate–potassium bisulfite initiator gave better yields and higher molecular weights than those obtained with an azo initiator in tetrahydrofuran.[263] Other polymerizations of

β-Cyclodextrin; Cu salt, Aqueous alkali, CCl_4; HOOC, COOH

5.65 Schematic

$PdSO_4$, H_2O; $CuSO_4$, O_2, $HPV_6Mo_6O_{40}$; O

5.66 Schematic

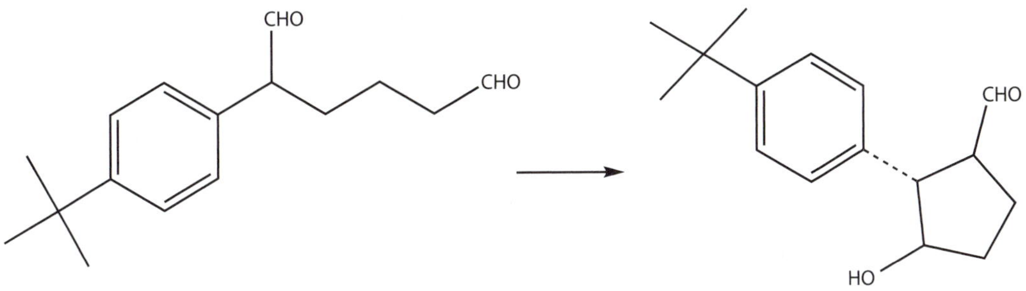

5.67 Schematic

hydrophobic monomers initiated by free radicals have been carried out with the aid of cyclodextrins as well.[264]

Many derivatives of cyclodextrins have been prepared. These include amines, hydroxylamines, hydrazines, ethers, and others.[265] A dimeric β-cyclodextrin sulfide (**5.68** Schematic) binds cholesterol 200–300 times as tightly as β-cyclodextrin itself.[266]

A polymer was made by treating first with methyl β-cyclodextrin and then with hexamethylenediisocyanate, then with 2-hydroxyethyl methacrylate, followed by polymerization with an azo initiator in the presence of cholesterol (i.e., molecularly imprinted with cholesterol). After extraction of the cholesterol, it was 30–45 times more effective in picking up cholesterol than β-cyclodextrin.[267] Many others have been prepared as "artificial enzymes" or enzyme mimics.[268] Some of these are chelating agents (**5.69** Schematic) that bind metal ions more strongly than the underivatized cyclodextrin.

Cyclodextrins have also been immobilized on silica. β-Cyclodextrin has also been incorporated into a water-soluble copolymer with a molecular weight of 373,000 (**5.70** Schematic).[269]

Presumably, other maleic anhydride copolymers could be modified in the same way. A β-cyclodextrin polymer cross-linked with epichlorohydrin scavenged di(2-ethylhexyl) phthalate from water with an equilibrium constant of 928. It could be used again after extraction of the phthalate with aqueous methanol.[270]

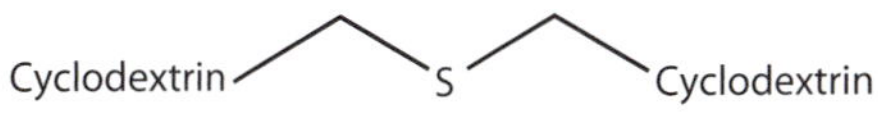

5.68 Schematic

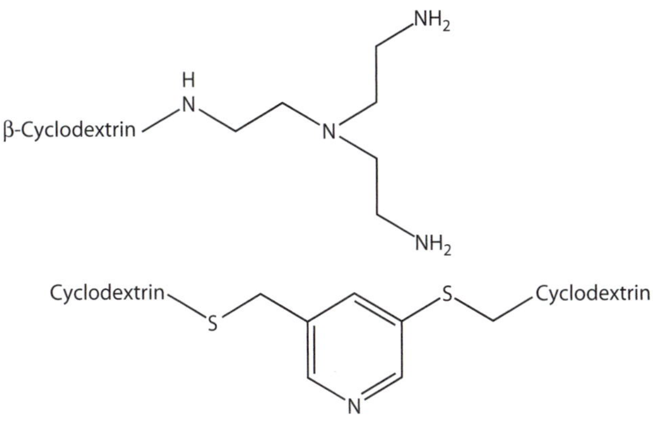

5.69 Schematic

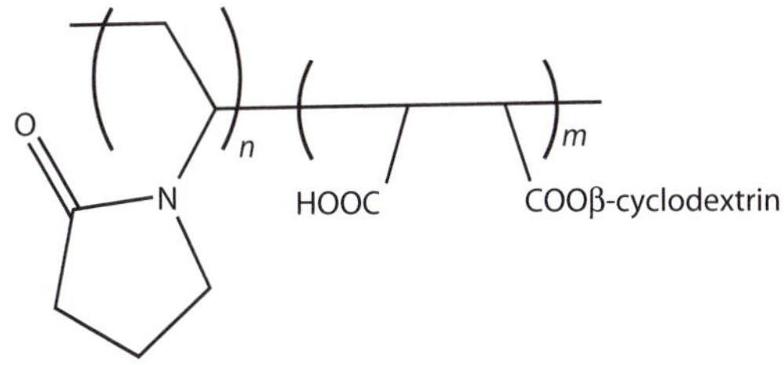

5.70 Schematic

Larger analogues of cyclodextrins are known. A cyclic decasaccharide of rhamnose with a 130-Å hole is known.[271]

REFERENCES

1. (a) A.L. Margolin, *Trends Biotechnol.*, 1996, 14; (b) N. Khalaf, C.P. Govardhan, J.J. Lalonde, R.A. Persichetti, Y.-F. Wang, and A.L. Margolin, *J. Am. Chem. Soc.* 1996, *118*, 5494; (c) J.J. Lalonde, *J. Am. Chem. Soc.* 1995, *117*, 6845; *Chemtech*, 1997, *27*(2), 38; (d) D. Hairston, *Chem. Eng.*, 1995 *102*(5), 56.
2. (a) J. Haggin, *Chem. Eng. News*, Apr. 3, 1995, 20; (b) R.M. Contractor, *Chem. Eng. Sci.*, 1999, *54*, 5627.
3. U. Schubert, *New J. Chem.*, 1994, *18*, 1049.
4. (a) J. Haber, J.H. Block, and B. Delmon, *Pure Appl. Chem.*, 1995, *67*, 1257–1306; (b) M. Baerns, *Basic Principles in Applied Catalysis*, Springer-Verlag, Heidelberg, 2003; (c) S.M. Roberts, G. Poignant, and I.V. Kozhevnikov, *Catalysts for Fine Chemical Synthesis*, 2 vols., Wiley, New York, 2001; (d) S.D. Jackson, J.S.J. Hargreaves, and D. Lennon, eds, *Catalysis in Application*, Royal Society of Chemistry, Cambridge, 2003; (e) M.E. Ford, *Catalysis of Organic Reactions*, Dekker, New York, 2000; (f) B. Cornils, W.A. Hermann, M. Muhler, and C.-H. Wong, eds, *Catalysis from A to Z*, Wiley-VCH, Weinheim, 2000; (g) P. Braunstein, L.A. Oro, and P.R. Raithby, eds, *Metal Clusters in Chemistry*, 3 vols., Wiley-VCH, Weinheim, 1999; (h) G. Ertl, H. Knozinger, and J. Weitkamp, eds, *Prepartion of Solid Catalysts*, Wiley-VCH, Weinheim, 1999; (i) K. Kirchner and W. Weissensteiner, *Organometallic Chemistry and Catalysis*, Springer-Verlag, New York, 2001; (j) B. Viswanathan, S. Sivasankar, and A.V. Ramaswamy, *Catalysis—Principles and Applications*, CRC Press, Boca Raton, FL, 2002.

5. (a) G.W. Parshall and S.D. Ittel, *Homogeneous Catalysis*, 2nd ed., Wiley, New York, 1992; (b) L. Brandsma, S.F. Vasilevsky, and H.D. Verkruijsse, *Applications of Transition Metal Catalysts in Organic Synthesis*, Springer-Verlag, Berlin, 1997; (c) B. Cornils and W.A. Herrmann, eds, *Applied Homogeneous Catalysis with Organometallic Compoounds*, 2nd ed., 3 vols., Wiley-VCH, Weinheim, 2002; (d) S.M. Roberts and G. Poignant, eds, *Catalysts for Fine Chemical Synthesis*, 2 vols., Wiley, New York, 2002; (e) B. Cornils, W.A. Herrmann, I.T. Horvath, W. Leitner, S. Mecking, H. Olivier-Bourbigou, and D. Vogt, *Multiphase Homogeneous Catalysis*, 2 vols., Wiley-VCH, Weinheim, 2005; (f) A. Kirschning, H. Monenschein, and R. Wittenberg, *Angew. Chem. Int. Ed.*, 2001, *40*, 651; (g) S. Bhaduri and D. Mukesh, *Homogeneous Catalysis*, Wiley, New York, 2000.
6. (a) B.C. Gates, *Encyclopedia of Chemical Technology*, 4th ed., Wiley, New York, 1993, *5*, 324 [homogeneous]; 340 [heterogeneous]; (b) J.A. Moulijin, P.W.N.M. van Leeuwen, and R.A. Van Santen, eds, *Catalysis—An Integrated Approach to Homogeneous, Heterogeneous and Industrial Catalysis*, Elsevier, Amsterdam, 1993; (c) J.M. Thomas and W.J. Thomas, *Principles and Practice of Heterogeneous Catalysis*, VCH, Weinheim, 1997; (d) G.V. Smith and F. Notheisz, *Heterogeneous Catalysis in Organic Chemistry*, Academic Press, San Diego, 1999; (e) G. Ertl and H. Knoezinger, eds, *Handbook of Heterogeneous Catalysis*, Wiley-VCH, Weinheim, 1997; (f) M. Bowker, *The Basis and Applications of Heterogeneous Catalysis*, Oxford University Press, Oxford, 1998; (g) A.F. Carley, P.R. Davies, G.J. Hutchings, and M.S. Spencer, *Surface Chemistry and Catalysis*, Kluwer Academic/Plenum, New York, 2002; (h) H.F. Rase, *Handbook of Commercial Catalysts—Heterogeneous Catalysts*, CRC Press, Boca Raton, FL, 200; (i) J.A. Gladysz, *Chem. Rev.*, 2002, *102*(10), 2002, "Recoverable Catalysts and Reagents"; (j) R.A. Sheldon and H. vanBekkum, eds, *Fine Chemicals Through Heterogeneous Catalysis*, Wiley-VCH, Weinheim, 2000.
7. (a) F.R. Hartley, ed., *Supported Metal Complexes—Catalysis by Metal Complexes*, D. Riedel, Dordrecht, the Netherlands, 1985; (b) J.-M. Basset, ed., *J. Mol. Catal.*, 1994, *86*, 51–343; (c) D.W. Robinson, *Encyclopedia of Chemical Technology*, 4th ed., 1993, *5*, 448; (d) W.A. Herrmann and B. Cornils, *Angew. Chem. Int. Ed.*, 1997, *36*, 1049; (e) E. Linder, T. Schneller, F. Auer, and H.A. Mayer, *Angew. Chem. Int. Ed.*, 1999, *38*, 2154.
8. (a) J.H. Clark, A.P. Kybett, and D.J. Macquarrie, *Supported Reagents: Preparation, Analysis, Applications*, VCH, New York, 1992; (b) R.P. Nielsen, P.H. Nielson, H. Malmos, T. Damhus, B. Diderichsen, H.K. Nielsen, M. Simonsen, H.E. Schiff, A. Oestergaard, H.S. Olsen, P. Eigtved, and T.K. Nielsen, *Kirk–Othmer Encyclopedia Chemical Technology*, 4th ed., Wiley, New York, 1994, *9*, 584; (c) R.P. Nielsen, *Kirk–Othmer Encyclopedia of Chemical Technology*, 4th ed., Wiley, New York, 1993, *5*, 383; (d) B.K. Hodnett, A.P. Kybett, J.H. Clark, and K. Smith, *Supported Reagents and Catalysts in Chemistry*, Royal Society Of Chemistry. Special Pub. 216, Cambridge, 1998; (e) M. Benaglia, A. Puglisi, and F. Cozzi, *Chem. Rev.*, 2003, *103*, 3401, "Polymer-supported organic catalysts".
9. (a) R.A. Sheldon and J. Dakka, *Catal. Today*, 1994, *19*, 215; (b) J.H. Clark, *Catalysis of Organic Reactions by Supported Inorganic reagents*, VCH, Weinheim, 1994; (c) R.L. Augustine, *Heterogeneous Catalysis for the Synthetic Chemist*, Dekker, New York, 1995; (d) P. Laszlo, ed., *Preparatory Chemistry Using Supported Reagents*, Academic Press, San Diego, 1987; (e) W.T. Ford, ed., *Polymeric Reagents and Catalysts*, American Chemical Society, Symp 308, Washington, DC, 1986; (f) J.S. Fruchtel and G. Jung, *Angew. Chem. Int. Ed.,* 1996, *35*, 17; (g) P. Hodge and D.C. Sherrington, eds, *Polymer-Supported Reactions in Organic Synthesis*, Wiley, New York, 1980; (h) K. Smith, ed., *Solid Supports and Catalysts in Organic Synthesis*, Ellis Horwood, Chichester, 1992; (i) D.C. Sherrington and P. Hodge, eds, *Syntheses and Separations Using Functionalized Polymers*, Wiley, New York, 1988; (j) E.A. Bekturov and S.E. Kudaibergenov, *Catalysis by Polymers*, Hutig and Wepf, Heidelberg, 1997; (k) F.Z. Dorwald, *Organic Synthesis on Solid Phase Supports, Linkers, Reactions*, 2nd ed., Wiley-VCH, Weinheim, 2002; (l) M.R. Buchmeister, ed., *Polymeric Materials in Organic Synthesis and Catalysis*, Wiley-VCH, Weinheim, 2003; (m) C. Song, J. Garces, and Y. Sugi, eds, *Shape-Selective Catalysis—Chemical Synthesis and Hydrocarbon Processing*, A.C.S. Symp. 738, Oxford University, New York, 1999; (n) J.H. Clark and C.N. Rhoads, *Clean Synthesis Using Porous Inorganic Solid Catalysts and Supported Reagents*, Royal Society of Chemistry, Cambridge, 2000; (o) F. Guillier, D. Orain, and M. Bradley, *Chem. Rev.*, 2000, *100*, 2091, "linkers and cleavage strategies in solid-phase organic syntheses and combinatorial chemistry"; (p) R.E. Sammelson and M.J. Kurth, *Chem. Rev.*, 2001, *101*, 137, "C–C bond forming solid phase reactions"; (q) P. Seneci, *Solid-Phase Synthesis and Combinatorial Technologies*, Wiley-VCH, Weinheim, 2000; (r) A.W. Czarnik, ed., *Solid-Phase Organic Synthesis*, Wiley, New York, 2001; (s) K. Burgess, ed., *Solid-Phase Organic Synthesis*, Wiley, New York, 1998; (t) D.C. Sherrington and A.P. Kybett, *Supported Catalysts and Their Applications*, Royal Society of Chemistry, Cambridge, 2001; (u) P. Blaney, R. Grigg, and Sridharan, *Chem. Rev.*, 2002, *102*, 2607, "traceless solid-phase organic synthesis"; (v) J. Tella-Puche and F. Albericio, eds, *The Power of Functional Group Resins*, Wiley-VCH, Weinheim, 2008.
10. (a) A. Akelah and A. Moet, *Functionalized Polymers and Their Applications*, Chapman & Hall, London, 1990; (b) D.E. Bergbreiter and C.R. Mason, eds, *Functional Polymers*, Plenum, New York, 1989; (c) C.U. Pitman, Jr., Polymer-supported catalysts. In: G. Wilkinson, ed., *Comprehensive Organometallic Chemistry*, Pergamon, Oxford, 1992, *8*, 553; (d) N.K. Mathur, C.K. Narang, and R.E. Williams, *Polymers as Aids in Organic Chemistry*, Academic Press, New York, 1980; (e) E.A. Bekturov and S.E. Kudaibergenov, *Polymeric Catalysis*, Huthig and Wepf, Heidelberg, 1994; (f) D. Bellus, *J. Macromol. Sci. Pure Appl. Chem.*, 1994, *A31*, 1355; (g) G. Odian, *Principles of Polymerization*, 4th ed., Wiley–Interscience, New York, 2004, 760–777; (h) P. Hodge, *Encyclopedia of Polymer Science Engineering*, 2nd ed., 1988, *12*, 618; (i) D.C. Sherrington, *Chem. Commun.*, 1998, 2275; (j) M.R. Buchmeiser, *Polymeric Materials in Organic Syntheses and Catalysis*, Wiley-VCH, Weinheim, 2003.

11. (a) D. Arntz, *Catal. Today*, 1993, *18*, 173; (b) J.M. Thomas, *Angew. Chem. Int. Engl.* 1994, *33*, 913; (c) M. Otake, *Chemtech*, 1995, *25*(9), 36; (d) J.F. Roth, *Chemtech*, 1991, *21*(6), 357; (e) M.E. Davis and S.L. Suib, eds, *Selectivity in Catalysis*, A.C.S. Symp. 517, Washington, DC, 1993; (f) J.R. Kosak and T.A. Johnson, *Catalysis of Organic Reactions*, Dekker, New York, 1993; (g) M.G. Scaros and M.L. Prunier, eds, *Catalysis of Organic Reactions*, Dekker, New York, 1995; (h) R.E. Malz, *Catalysis of Organic Reactions*, Dekker, New York, 1996; (i) R.J. Farrauto and C.H. Bartholomew, *Fundamentals of Industrial Catalytic Processes*, Chapman & Hall, London, 1997; (j) R.J. Farrauto and C.H. Bartholomew, *Fundamentals of Industrial Catalytic Processes*, Kluwer Academic, Norwell, MA, 1999; (k) R.J. Wijngearden, A.R. Kronberg, and K.R. Westerterp, *Industrial Catalysis, Optimizing Catalysts and Processes*, Wiley-VCH, Weinheim, 1998.
12. (a) J.A. Cusumano, *Appl. Catal. A*, 1994, *113*, 181; (b) J.F. Roth, *Appl. Catal. A*, 1994, *113*, 134; (c) D.C. Sherrington, *Chem. Ind. (London)*, 1991, *1*, 15; (d) J.H. Clark, S.R. Cullen, S.J. Barlow, and T.W. Bastock, *J. Chem. Soc. Perkin Trans. 2*, 1994, 1117; (e) J.H. Clark and D.J. Macquarrie, *Chem. Soc. Rev.*, 1996, *25*, 303, (f) R.A. Sheldon and I. Arends, *Green Chemistry and Catalysis*, Wiley-VCH, Weinheim, 2007; (g) R.H. Crabtree and P.A. Anastas, *The Handbook of Green Chemistry–Green Catalysis*, 12 vols., Wiley-VCH, Weinheim, 2009; (h) R.B. Subramanian, B.W.-L. Jang, and J.J. Spivey, eds, Environmental catalysis—green chemistry, *Catal. Today*, 2000, *55*, 1–204; (i) G. Rothenberg, *Catalysis—Concepts and Green Applications*, Wiley-VCH, Weinheim, 2008; (j) G. Rothenberg, *Catalysis: Concepts and Green Applications*, Wiley-VCH, Weinheim, 2008.
13. M.J. Ledoux, C. Crouzet, C. Pham-Hui, V. Turinew, K. Kourtakis, P.L. Mills, and J.J. Lerou, *J. Catal.*, 2001, *203*, 495.
14. (a) R.D. Adams and F.A. Cotton, eds, *Catalysis by Di- and Polynuclear Metal Cluster Compounds*, Wiley-VCH, New York, 1998; (b) E.R. Bernstein, ed., *Chemical Reactions in Clusters*, Oxford University Press, Oxford, England, 1996.
15. J.R. Rostrup-Nielsen, ed., *Catal. Today* 1997, *37*, 225–331.
16. F. Besenbacher, I. Chorkendorff, B.S. Clausen, B. Hammer, A.M. Molenbroek, J.K. Norskov, and I. Stensgaard, *Science*, 1998, *279*, 1913.
17. M. Jacoby, *Chem. Eng. News*, May 4, 1998, 41.
18. A. Furstner, G. Seidel, B. Gabor, C. Kopiske, C. Kruger, and R. Mynott, *Tetrahedron*, 1995, *51*, 8875.
19. H.S. Stadtmueller, B. Greve, K. Lennick, A. Chair, and P. Knochel, *Synthesis*, 1995, 69.
20. U.S. EPA, *Presidential Green Chemistry Challenge Awards Program*, 2008, 4.
21. (a) G.W. Kalbalka and R.M. Payne, *Tetrahedron*, 1997, *53*, 7999; (b) S.D. Jackson and J.S.J. Hargreaves, eds, *Metal Oxide Catalysis*, Wiley-VCH, Weinheim, 2008.
22. C.P. Wilgus, S. Downing, E. Molitor, S. Bains, R.M. Pagni, and G.W. Kabalka, *Tetrahedron Lett.*, 1995, *36*, 3469.
23. D. Zhao and D.G. Lee, *Synthesis*, 1994, 915.
24. J.G. Lee and J.P. Hwang, *Chem. Lett.*, 1995, 507.
25. V.K. Yadav and K.K. Kapoor, *Tetrahedron*, 1996, *52*, 3659; *Tetrahedron Lett.*, 1994, *35*, 9481.
26. H. Kotsuki and T. Shimanouchi, *Tetrahedron Lett.*, 1996, *37*, 1845.
27. V.J. Patil and U. Mavers, *Tetrahedron Lett.*, 1996, *37*, 1281.
28. J.M. Perez, *Tetrahedron Lett.*, 1996, *37*, 6955.
29. H. Ogawa, M. Amano, and T. Chihara, *Chem. Commun.*, 1998, 495.
30. M.H. Ali, D.R. Leach, and C.E. Schmitz, *Synth. Commun*, 1998, *28*, 2969.
31. M.J. Naughton and R.S. Drago, *J. Catal.*, 1995, *155*, 383. (For a similar reaction of methyl acrylate see G Fremy, E. Montflier, J.F. Carpenter, Y. Castanet, and A. Montreux, *J. Catal.*, 1996, *162*, 339.)
32. P. Schneider, F. Quignard, A. Choplin, and D. Sinou, *New J. Chem.*, 1996, *20*, 545.
33. E.T.W.M. Schipper, R.P.M. Pinckaers, P. Piet, and A.L. German, *Macromolecules*, 1995, *28*, 2194.
34. Z. Deng, G.R. Dieckmann, and S.H. Langer, *Chem. Commun*, 1996, 1789.
35. T.Y. Meyer, K.A. Woerpel, B.M. Novak, and R.G. Bergman, *J. Am. Chem. Soc.*, 1994, *116*, 10290.
36. S.L. Scott, M. Szpakowicz, A. Mills, and C.C. Santini, *J. Am. Chem. Soc.*, 1998, *120*, 1883.
37. S.L. Scott, P. Dufour, C.C. Santini, and J.-M. Basset, *Inorg. Chem.*, 1996, *35*, 869; *J. Chem. Soc. Chem. Commun.*, 1994, 2011.
38. B.C. Ranu, M. Saha, and S. Bhar, *J. Chem. Soc. Perkin Trans.*, 1994, *1*, 2197.
39. K. Tanabe, *Appl. Catal. A*, 1994, *113*, 147.
40. B.P. Bandgar, S.R. Jagtap, S.B. Ghodeshwar, and P.P. Wadgaonkar, *Synth Commun.*, 1995, *25*, 2993.
41. J. Kiviaho, T. Hanaoka, Y. Kubota, and Y. Sugi, *J. Mol. Catal. A Chem.*, 1995, *101*, 25.
42. L.F. Tietze, H. Ila, and H.P. Bell, *Chem. Rev.*, 2004, *104*, 3453.
43. (a) R.L. Augustine and S.T. O'Leary, *J. Mol. Catal.*, 1992, *72*, 229; (b) M.-Z. Cai, C.-S. Song, and X. Huang, *Synthesis*, 1997, 521; (c) J. Li, A.W.-H. Mau, and C.R. Strauss, *Chem. Commun*, 1997, 1275; (d) R. Dagani, *Chem. Eng. News*, Sept. 21, 1998, 70; (e) K. Kohler, M. Wagner, and L. Djakovitch, *Catal. Today*, 2001, *66*, 105.
44. S. SProckl, W. Kleist, M.A. Gruber, and K. Kohler, *Angew. Chem. Int. Ed.*, 2004, *43*, 1881.
45. S. Paul and J.H.Clark, *Green Chem.*, 2003, *5*, 635.
46. M.T. Reetz and J.G. deVries, *Chem. Commun.*, 2004, 1559.
47. D. Morales-Morales, R. Redon, C. Yung, and C.M. Jensen, *Chem. Commun.*, 2004, 1619.
48. L.J. Goossen and J. Paetzold, *Angew. Chem. Int. Ed.*, 2002, *41*, 1237.
49. A. Inoue, H. Shinokubo, and K. Oshima, *J. Am. Chem. Soc.*, 2003, *125*, 1484.
50. M. Lenarda, R. Ganzerla, S. Paganelli, L. Storaro, and R. Zanoni, *J. Mol. Catal. A Chem.*, 1996, *105*, 117.
51. J.-M. Planeix, B. Coq, L.-C. de Menorva, and P. Medina, *Chem. Commun.*, 1996, 2087.
52. (a) F. Cassidy, *Appl. Catal. B*, 1996, *8*, N23; (b) J.P. Sankey, S. Wilson, and J. McAdam, U.S. patent 5,698,326 (1997).
53. J.A. Elings, R. Ait–Meddour, J.H. Clark, and D.J. Macquarrie, *Chem. Commun.*, 1998, 2701.
54. D. Crowther and X. Liu, *J. Chem. Soc. Chem. Commun.*, 1995, 2445.
55. C.-G. Jia, F.-Y. Liang, L.-J. Tang, M.-Y. Huang, and Y.-Y. Jiang, *Macromol. Chem. Phys.*, 1994, *195*, 3225.

56. (a) D.M. Haddleton, D. Kukulj, and A.P. Radique, *Chem. Commun*, 1999, 99; (b) I.C. Chisem, J. Rafelt, M.T. Shieh, J. Chisem, J.H. Clark, R. Jachuck, D. Macquarrie, C. Ramshaw, and K. Scott, *Chem. Commun*, 1998, 1949.
57. H. Gao and R.J. Angelici, *J. Am. Chem. Soc.*, 1997, *119*, 6937.
58. H. Gao and R.J. Angelici, *J. Mol. Catal. A*, 1999, *149*, 63.
59. J. Blumel, *Inorg Chem.*, 1994, *33*, 5050.
60. G. Arena, A. Casnati, A. Contino, L. Mirone, D. Sciotto, and R. Ungaro, *Chem. Commun.*, 1996, 2277.
61. J.F. Biernat, P. Konieczka, B.J. Tarbet, J.S. Bradshaw, and R.M. Izatt, *Sep Purif Methods*, 1994, *23*(2), 77.
62. P. Sutra and D. Brunel, *Chem. Commun.*, 1996, 2485.
63. B. Nie and V.M. Rotello, *J. Org. Chem.*, 1996, *61*, 1870.
64. M. Ayadim and J.P. Soumillion, *Tetrahedron Lett.*, 1995, *36*, 4615; 1996, *37*, 381.
65. C. Bruning and J. Grobe, *J. Chem. Soc. Chem. Commun.*, 1995, 2323.
66. J. Rogalski, A. Dawidowicz, and A. Leonowicz, *J. Mol. Catal.*, 1994, *93*, 233.
67. M. Shimomura, H. Kikuchi, T. Yamauchi, and S. Miyauchi, *J. Macromol. Sci. Pure Appl. Chem.*, 1996, *A33*, 1687.
68. J.M. Moreno and J.V. Sinisterra, *J. Mol. Catal.*, 1994, *93*, 357; *J. Mol. Catal. A Chem.*, 1995, *98*, 171.
69. M.G.L. Petrucci and A.K. Kakkar, *J. Chem. Soc. Chem. Commun.*, 1995, 1577.
70. P. Dubois, M. Alexandre, F. Hindryckx, and R. Jerome, *J. Macromol. Sci., Rev. Macromol. Chem.*, 1998, *C38*, 511.
71. (a) E. Bourgeat–Lami, P. Espiard, and A. Guyot, *Polymer*, 1995, *36*, 4386; (b) P. Espiard and A. Guyot, *Polymer*, 1995, *36*, 4391; (c) P. Eslpiard, A. Guyot, J. Perez, G. Vigier, and L. David, *Polymer*, 1995, *36*, 4397.
72. M. Ohata, M. Yamamoto, A. Takano, and Y. Isono, *J. Appl. Polym. Sci.*, 1996, *59*, 399.
73. M.A. Peters, A.M. Belu, R.W. Linton, L. Dupray, T.J. Meyer, and J.M. de Simone, *J. Am. Chem. Soc.*, 1995, *117*, 3380.
74. C.-W. Chen, H. Yu, M.-Y. Huang, and Y.-Y. Jiang, *Macromol. Rep.*, 1995, *A32*, 1039.
75. T. Shido, T. Okazaki, and M. Ichikawa, *J. Catal.*, 1995, *157*, 436.
76. E. Albizzati and M. Galimberti, *Catal. Today*, 1998, *41*, 159.
77. F. Ciardelli, A. Altomare, and M. Michelotti, *Catal. Today*, 1998, *41*, 149.
78. U. Schubert, *New J. Chem.*, 1994, *18*, 1049.
79. P. Lengo, M. DiSerio, A. Sorrentino, and E. Santacesaria, *Appl. Catal. A*, 1998, *167*, 85.
80. (a) J. Blum, D. Avnir, and H. Schumann, *Chemtech*, 1999, *29*(2), 32; (b) B. Heinrichs, F. Noville, and J.-P. Pirard, *J. Catal.*, 1997, *170*, 366; (c) S. Bharathi and O. Lev, *Chem. Commun*, 1997, 2303.
81. H.K. Schmidt, *Macromol. Symp.*, 1996, *101*, 333, 340.
82. D. Cauzzi, M. Lanfranchi, G. Marzolini, G. Predieri, A. Tiripicchio, M. Costa, and R. Zanoni, *J Organomet Chem.*, 1995, *488*, 115.
83. S. Klein and W.F. Maier, *Angew. Chem. Int. Ed.*, 1996, *35*, 2230.
84. (a) W.-Y. Yu, H.-F. Liu, and Q. Tao, *Chem. Commun.*, 1996, 1773; (b) W. Yu, Y. Wang, H. Liu, and W. Zheng, *J. Mol. Catal. A*, 1996, *112*, 105.
85. H. Feng and H. Liu, *J. Mol. Catal. A Chem.*, 1997, *126*, L5.
86. C. Pham-Huu, N. Keller, G. Ehret, L.J. Charbonniere, R. Ziessel, and M.J. Ledoux, *J. Mol. Catal. A*, 2001, *170*, 155.
87. W. Yu, H. Liu, and X. An, *J. Mol. Catal. A Chem.*, 1998, *129*, L9.
88. M. Han and H. Liu, *Macromol. Symp.* 1996, *105*, 179.
89. (a) A.B.R. Mayer, J.E. Mark, and S.H. Hausner, *J. Appl. Polym. Sci.*, 1998, *70*, 1209; (b) N. Toshima and P. Lu, *Chem. Lett.*, 1996, 729.
90. (a) B.L. Cushing, V.L. Kolesnichenko, and C.J. O'Connor, *Chem. Rev.*, 2004, *104*, 3893, (b) S. Laurent, D. Forge, M. Port, A. Roch, C. Robic, L.V. Elat, and R.N. Muller, *Chem. Rev.*, 2008, *108*, 2064.
91. A.B.R. Mayer and J.E. Mark, *Polym. Bull.*, 1996, *37*, 683.
92. J.F. Cieben, R.T. Clay, B.H. Sohn, and R.E. Cohen, *New J. Chem.*, 1998, *22*, 685, 745.
93. C.-W. Chen and M. Akashi, *J. Polym Sci., Polym. Chem.*, 1997, *35*, 1329.
94. S. Klingelhofer, W. Heitz, A. Greiner, S. Oestreich, S. Forster, and M. Antonietta, *J. Am. Chem. Soc.*, 1997, *119*, 10116.
95. R. Dagani, *Chem. Eng. News*, Jan. 13, 1997, 26.
96. (a) R.T. Reetz, G. Lohmer, and R. Schwickardi, *Angew. Chem. Int. Ed.*, 1997, *36*, 1526; (b) M.A. Hearshaw and J.R. Moss, *Chem. Commun.*, 1999, 1.
97. (a) P.R. Dvornic and D.A. Tomalia, *Chem. Br.*, 1994, *30*, 641; (b) D.A. Tomalia and H.D. Durst, *Top. Curr. Chem.*, 1993, *165*, 193; (c) D.A. Tomalia and R. Esfand, *Chem. Ind. (London)*, 1997, 416; (d) J.M.J. Frechet, C.J. Hawker, I. Gitsov, and J.W. Leon, *J. Macromol. Sci. Pure Appl. Chem.*, 1996, *A33*, 1399; (e) G.R. Newkome, C.N. Moorefield, and F. Vogtle, *Dendritic Molecules*, VCH, Weinheim, 1996; (f) F. Zeng and S.C. Zimmerman, *Chem. Rev.*, 1997, *97*, 1681; (g) M. Fischer and F. Vogtle, *Angew. Chem. Int. Ed.*, 1999, *38*, 884; (h) L. Balogh, D.A. Tomalia, and G.L. Hagnauer, *Chem. Innovation*, 2000, *30*(3), 19.
98. (a) J.W.J. Knapen, A.W. van der Made, J.C. de Wilde, P.W.N.M. van Leeuwen, P. Wijkens, D.M. Grove, and G. van Koten, *Nature*, 1994, *372*, 659; (b) A.W. Kley, R.J. Gossage, J.T.B.H. Jastrebski, J. Boersma, and G. van Koten, *Angew. Chem. Int. Ed.*, 2000, *39*, 176.
99. (a) R.F. Service, *Science*, 1999, *283*, 165; (b) R. Dagani, *Chem. Eng. News*, Feb. 8, 1999, 33.
100. S. Lebreton, S. Monaghan, and M. Bradley, *Aldrichim. Acta*, 2001, *34*(3), 75.
101. J. Xiao, L. Hao, and R.J. Puddephatt, *Organometallics*, 1995, *14*, 4183.
102. (a) R. Murugavel, V. Chandrasekhar, and H.W. Roesky, *Acc. Chem. Res.*, 1996, *29*, 183; (b) A. Voigt, R. Murugavel, V. Chandrasekhar, N. Winkhofer, H.W. Roesky, H.-G. Schmidt, and I. Uson, *Organometallics*, 1996, *15*, 1610; (c) M. Montero, A. Voigt, M. Teichert, I. Uson, and H.W. Roesky, *Angew. Chem. Int. Ed.*, 1995, *34*, 2504; (d) R. Murugavel, A. Voigt, M.G. Walawalkar, and H.W. Roesky, *Chem. Rev.*, 1996, *96*, 2205; (e) C. Rennekamp, A. Gouzyr, A. Klemp, H.W. Roesky, C. Bronneke, J. Karcher, and R. Herbst–Irmer, *Angew. Chem. Int. Ed.*, 1997, *36*, 404; (f) A. Voigt, R. Murugavel, M.L. Montero, H. Wessel, F.-Q. Liu, H.W. Roesky, I. Uson, T. Albers, and El Parisini, *Angew. Chem. Int. Ed.*, 1997, *36*, 1001; (g) A. Voigt, M.G. Walawalkar, R. Murugavel, H.W. Roesky, E. Parisini, and P. Lubini, *Angew. Chem. Int. Ed.*, 1997, *36*, 2203.

103. H.C.L. Abbenhuis, H.W.G. van Herwijnen, and R.A. van Santen, *Chem. Commun.*, 1996, 1941.
104. K. Wada, M. Nakashita, M. Bundo, K. Ito, T. Kondo, and T.-A. Mitsudo, *Chem. Lett.*, 1998, 659.
105. V. Lorenz, S. Giessmann, Y.K. Gun'ko, K. Fischer, J.W. Gilje, and F.T. Edelmann, *Angew. Chem. Int. Ed.*, 2004, *43*, 4603.
106. A. Provatas and J.G. Matison, *Trends Polym. Sci.*, 1997, *5*, 327.
107. J.M. Mabry, A. Vij, S.T. Iacomo, and B.D. Viers, *Angew. Chem. Int. Ed.*, 2008, *47*, 4137.
108. A. Sellinger and R.M. Laine, *Macromolecules*, 1996, *29*, 2327.
109. T.S. Haddad and J.D. Lichtenhan, *Macromoleclues*, 1996, *29*, 7302.
110. (a) C.E. Harland, *Ion Exchange: Theory and Practice.*, 2nd ed., Royal Society of Chemistry, London, 1994; (b) M. Street, *Ind. Eng. Chem. Res.*, 1995, *34*, 2841; (c) R.L. Albright and P.A. Yarnell, *Encyclopedia of Polymer Science Engineering*, 2nd ed., Wiley, New York, 1987, *8*, 341; (d) F. de Darnell and T.V. Ardon, *Ullmann's Encyclopedia of Industrial Chemistry*, 5th ed., VCH, Weinheim, 1989, *A14*, 393; (e) C. Dickert, *Kirk–Othmer Encyclopedia of Chemical Technology*, 4th ed., Wiley, New York, 1995, *14*, 737–783; (f) R.M. Wheaton and L.J. Lefevre, *Kirk–Othmer Encyclopedia of Chemical Technology*, 3rd ed., Wiley, New York, 1981, *13*, 678; (g) F. de Silva, *Chem. Eng.*, 1994, *101*(7), 86; (h) J.T. McNulty, *Chem. Eng.*, 1997, *104*(6), 94; (i) R.A. Williams, A. Dyer, and M.J. Hudson, *Progress in Ion Exchange: Advances and Applications*, Royal Society Chemistry, Special Publ., 196, Cambridge, UK, 1997; (j) J.A. Grieg, ed., *Ion Exchange Developments and Applications*, Royal Society Chemistry/ Soc. Chem. Ind., Cambridge, UK, 1996; (k) A.K. Sengupta, *Ion Exchange Technology: Advances in Pollution Control.*, Technomic, Lancaster, PA, 1995; (l) P. Barbaro and F. Liguori, eds., *Chem. Rev., 2009, 109*, 515- 529; S.D. Alexandratos, *Ind. Eng. Chem. Res., 2009, 48*, 388 .
111. M.J. Semmens and C. Riley, *National Centre for Clean Industrial and Treatment Technologies—Activities Reports, Michigan Technological University*, Houghton, MI, 10/95–12/96, 51.
112. F.L.D. Cloete and A.P. Marais, *Ind. Eng. Chem. Res.*, 1995, *34*, 2464.
113. (a) P. Hodge, *Chem. Soc. Rev.*, 1997, *26*, 417; (b) S.J. Shuttleworth, A.M. Allin, and P.K. Sharma, *Synthesis*, 1997, 1217.
114. (a) F. Svec and J.M.J. Frechet, *Science*, 1996, *273*, 205; (b) Y.-C. Liang, F. Svec, and J.M.J. Frechet, *J. Polym. Sci. Polym. Chem.*, 1997, *35*, 2631; (c) K. Lewandowski, F. Svec, and J.M.J. Frechet, *J. Appl. Polym. Sci.*, 1998, *67*, 597; (d) Bayer Corp. Advertisement, *Chem. Eng. News*, Mar. 15, 1999, 17.
115. T. Sata, Y. Yamane, and K. Matsusaki, *J. Polym. Sci. Polym. Chem.*, 1998, *36*, 49.
116. J. Economy, L. Dominguez, and C.L. Mangun, *Ind. Eng. Chem. Res.*, 2002, *41*, 6436.
117. S. Itsuno, K. Uchikoshi, and K. Ito, *J. Am. Chem. Soc.*, 1990, *112*, 8187.
118. T. Balakrishman and V. Rajendran, *J. Macromol. Sci., Pure Appl. Chem.*, 1996, *A33*, 103.
119. Q. Sheng and H.D.H. Stover, *Macromolecules*, 1997, *30*, 6712.
120. C.C. Zikos and N.G. Ferderigos, *Tetrahedron Lett.*, 1995, *36*, 3741.
121. X. Hui and H. Xizhang, *Polym. Bull.*, 1998, *40*, 47.
122. M. Kralik, M. Hronec, S. Lora, G. Palma, M. Zecca. A. Biffs, and B. Corain, *J. Mol. Catal. A Chem.*, 1995, *97*, 145; 1995, *101*, 143.
123. T. Mizota, S. Tsuneda, K. Saito, and T. Sugo, *J. Catal.*, 1994, *149*, 243.
124. (a) J.H. Hodgkin, *Encyclopedia of Polymer Science Engineering*, 2nd ed., 1985, *3*, 363; (b) S.D. Alexandratos, *Sep. Purif. Methods*, 1992, *21*(1), 1; (c) S.D. Alexandratos and L.A. Hussain, *Ind. Eng. Chem. Res.*, 1995, *34*, 251; (d) S.D. Alexandratos, A.W. Trochimczuk, E.P. Horwitz, and R.C. Gatrone, *J. Appl. Polymer Sci.*, 1996, *61*, 273.
125. N. Ohtani, Y. Inoue, Y. Inagaki, K. Fukuda, and T. Nishiyama, *Bull. Chem. Soc. Japan*, 1995, *68*, 1669.
126. D.R. Patel, M.K. Dalal, and R.N. Ram, *J. Mol. Chem. A Chem.*, 1996, *109*, 141.
127. Y. Cohen and R.W. Peters, eds, *Novel Adsorbents and Their Environmental Applications*, AlChE Symp. 309, New York, 1995, *91*, 49.
128. (a) Z. Zheng, D. Gu, R.G. Anthony, and E. Klavetter, *Ind. Eng. Chem. Res.*, 1995, *34*, 2142; (b) *Chem. Eng. Progress*, 1995, *91*(2), 22.
129. G.D. Yadav and P.H. Mehta, *Int. Eng. Chem. Res.* 1994, *33*, 2198.
130. R.C. Anand and N. Selvapalam, *Synth. Commun.*, 1994, *24*, 2743.
131. T. Nishiguchi, S. Fujisaki, Y. Ishii, Y. Yano, and A. Nishida, *J. Org. Chem.*, 1994, *59*, 1191.
132. K. Berg, G. Fennhoff, R. Pakull, H.J. Buysch, B. Wehrle, A. Eitel, K. Wulff, and J. Kirsch, Eur. patent 567, *857*, 1993; *Chem. Abstr.*, 1994, *120*, 271, 422.
133. A. Chesney, *Green Chem.*, 1999, *1*, 209.
134. (a) L. Hong and E. Ruckenstein, *J. Mol. Catal. A Chem.*, 1995, *101*, 115; (b) S. Bhaduri and K. Sharma, *Chem. Commun.*, 1996, 207.
135. M. Setty–Fichman, Y. Sasson, and J. Blum, *J. Mol. Catal. A Chem.*, 1997, *126*, 27.
136. E. Baumgarten, A. Fiebes, and A. Stumpe, *J. Polym. Sci., Polym. Chem.*, 1996, *34*, 1889.
137. B.P. Bandgar, S.N. Kshirsagar, and P.P. Wadgaonkar, *Synth Commun.*, 1995, *25*, 941.
138. B.P. Bandgar, S.M. Nikat, and P.P. Wadgaonkar, *Synth Commun.*, 1995, *25*, 863.
139. J. Choi and N.M. Yoon, *Synthesis*, 1996, 597.
140. J. Choi and N.M. Yoon, *Synth. Commun.*, 1995, *25*, 2655.
141. T.B. Sim, J.H. Ahn, and N.M. Yoon, *Synthesis*, 1996, 324.
142. S. Rimmer and J.R. Ebdon, *J. Polym. Sci., Polym. Chem.*, 1996, *34*, 3573, 3591.
143. J.M. Campelo, R. Chakraborty, and J.M. Marinas, *Synth. Commun.*, 1996, *26*, 415.
144. B. Tamami and N. Goudarzian, *J. Chem. Soc. Chem. Commun.*, 1994, 1079.
145. E. Larsen, T. Kofoed, and E.B. Pedersen, *Synthesis*, 1995, 1121.
146. B. Tamami and M.A.K. Zarchi, *Eur. Polymer J* 1995, *31*, 715.
147. J. Choi and N.M. Yoon, *Synthesis*, 1995, 373.
148. (a) G. Braca, A.M.R. Galletti, M DiGirolamo, G. Sbrana, R. Silla, and P. Ferrarini, *J. Mol. Catal. A Chem.*, 1995, *96*, 203; (b) S.D. Alexandratos and D.H.J. Miller, *Macromolecules*, 1996, *29*, 8025.

149. I. Fenger and C. LeDrian, *Tetrahedron Lett.*, 1998, *39*, 4287.
150 (a) L.A. Thompson and J.A. Ellman, *Tetrahedron Lett.*, 1994, *35*, 9333; (b) G. Liu and J.A. Ellman, *J. Org. Chem.*, 1995, *60*, 7712.
151. B. Nie, *Tetrahedron Lett.*, 1995, *36*, 3617.
152. T.R. van den Ancker, S. Harvey, and C.L. Raston, *J. Organomet Chem.*, 1995, *502*, 35.
153. T.J. Marks and M.A. Ratner, *Angew. Chem. Int. Ed.*, 1995, *34*, 155.
154. U. Schuchardt, R.M. Vargas, and G. Gelbard, *J. Mol. Catal. A Chem.*, 1996, *109*, 37.
155. D.P. Dygutsch and P.E. Eilbracht, *Tetrahedron*, 1996, *52*, 5461.
156. P.-W. Wang and M.A. Fox, *J. Org. Chem.*, 1994, *59*, 5358.
157. J.M. Fraile, J.A. Mayoral, A.J. Royo, R.V. Salvador, B. Altava, S.V. Luis, and M.I. Burguete, *Tetrahedron*, 1996, *52*, 9853.
158. J.G. Keay and E.F.V. Scriven In: J.R. Kosak and T.A. Johnson, eds, *Catalysis of Organic Reactions*, Dekker, New York, 1994, 339.
159. (a) R. Murugan and E.F.V. Scriven, *Aldrichchimica Acta*, 2003, *36*(1), 21; (b) A.C. Spivey and S. Arseniyadis, *Angew. Chem. Int. Ed.*, 2004, *43*, 5436.
160. (a) S. Liu and W.K. Fife, *Macromolecules*, 1996, *29*, 3334; (b) S. Jingwu, H. Jitao, and S. Shixiang, *Macromol. Rep.*, 1995, *32*, 319.
161. Y. Feng, L. Zhang, J. Huang, and J. Sun, *J. Appl. Poly. Sci.*, 1998, *69*, 2303.
162. J. Huang, Q, Wang, J. Yao, L. Zhang, and J. Sun, *J. Appl. Poly. Sci.*, 1999, *71*, 1101.
163. J.-T. Huang, S.-H. Zheng, J.-A. Zhang and J.-W. Sun, *J. Appl. Polym. Sci.*, 2000, *75*, 593.
164. Anon., *Chem. Eng. News*, Apr. 17, 2000, 35.
165. B.A. Anderson, M.M. Hansen, A.R. Harkness. C.L. Henry, J.T. Vicenzi, and M.J. Zmijewski, *J. Am. Chem. Soc.*, 1995, *117*, 12358.
166. I.R. Baxendale, J. Deeley, C.M. Griffiths-Jones, S.V. Ley, S. Saaby, and G.K. Tranmer, *Chem. Commun.*, 2006, 2566.
167. Anon., *Chem. Eng. News*, December 16, 2002, 36.
168. (a) R.B. Merrifield, *Angew. Chem. Int. Ed.*, 1985, *24*, 799; (b) R.B. Merrifield, *Life During a Golden Age of Peptide Chemistry: The Concept and Development of Solid–Phase Peptide Synthesis*, American Chemical Society, Washington, DC, 1993; (c) G. Barany, N. Kneib-Cordonier, and D.G. Mullen, *Encyclopedia of Polymer Science and Engineering*, 2nd ed., Wiley, New York, 1988, *12*, 811–858; (d) R. Epton, ed., *Innovation and Perspectives in Solid-Phase Synthesis: Peptides, Proteins and Nucleic Acids, Biological and Biomedical Applications.*, Mayflower Worldwide, Birmingham, U.K., 1994.
169. J. Kowalski and M.A. Lipton, *Tetrahedron Lett.*, 1996, *37*, 5839.
170. M.V. Anuradha and B. Ravindranath, *Tetrahedron*, 1995, *51*, 5671, 5675.
171. (a) S. Borman, *Chem. Eng. News*, Apr. 6, 1998, 47; Mar. 8, 1999, 33; May 15, 2000, 53; Aug. 27, 2001, 49; Nov. 11, 2002, 43; Oct. 27, 2003, 45; Oct. 4, 2004, 32; (b) N.K. Terrett, *Combinatorial Chemistry*, Oxford University Press, New York, 1998; (c) D.J. Ecker, ed., *Combinatorial Chemistry—A Section of Biotechnology and Bioengineering*, Wiley, New York, 1998; (d) S.H. de Witt and A.W. Czarnik, *Acc. Chem. Res.*, 1996, *29*, 114–170; (e) I.M. Chaiken and K.D. Janda, eds, *Molecular Diversity and Combinatorial Chemistry: Libraries and Drug Discovery*, American Chemical Society, Washington, DC, 1996; (f) S. Wilson and A.W. Czarnik, *Combinatorial Chemistry—Synthesis and Application*, Wiley, New York, 1997; (g) E.M. Gordon and J.F. Kerwin, eds, *Combinatorial Chemistry and Molecular Diversity in Drug Discovery*, Wiley, New York, 1998; (h) A.W. Czarnik and S.H. Dewitt, *A Practical Guide to Combinatorial Chemistry*, American Chemical Society Professional Reference Book, Washington, DC, 1997; (i) L.A. Thompson and J.A. Ellman, *Chem. Rev.*, 1996, *96*, 555; (j) B.A. Bunin, *The Combinatorial Index*, Academic Press, San Diego, 1997; (k) G. Jung, *Combinational Peptide and Nonpeptide Libraries*, VCH, Weinheim, 1996; (l) G. Lowe, *Chem. Soc. Rev.*, 1995, *24*, 309; (m) S. Cabilly, ed., *Combinatorial Peptide Library Protocols*, Humana Press, Totowa, NJ, 1998; (n) J.A. Bristol, ed., *Tetrahedron*, 1997, *53*, 6573–6697; (o) J.W. Szostak, ed., *Chem. Rev.*, 1997, *97*, 347–509; (p) S. Kobayashi, *Chem. Soc. Rev.*, 1999, *28*, 1; (q) D.J. Tapolczay, R.J. Kobylecki, L.J. Payne, and B. Hall, *Chem. Ind. (London)*, 1998, 772; (r) K. Burgess, *Solid-Phase Organic Synthesis*, Wiley, New York, 1999; (s) B.A. Lorsbach and M.J. Kurth, *Chem. Rev.*, 1999, *99*, 1549; (t) A. Czarnik, *Solid Phase Organic Synthesis*, Wiley, New York, *1*, 2001; (u) H. Fenneri, ed., *Combinatorial Chemistry—A Practical Approach*, Oxford University Press, Oxford, 2000; (v) P. Senici, *Solid Phase and Combinatorial Technologies*, Wiley, New York, 2000; (w) S.A. Kates, *Solid-Phase Synthesis*, Dekker, New York, 2000.
172. (a) E. Leikauf, F. Barnekow, and H. Koster, *Tetrahedron*, 1996, *52*, 6913; (b) Y. Wang, H. Zhong, and W. Voelter, *Chem. Lett.*, 1995, 273; (c) N.N. Bhongle and J.Y. Tang, *Synth. Commun.*, 1995, *25*, 3671; (d) M.E. Schwartz, R.R. Breaker, G.T. Asteriadis, and G.R. Gough, *Tetrahedron Lett.*, 1995, *36*, 27.
173. (a) J.T. Randolph, K.F. McClure, and S.J. Danishefsky, *J. Am. Chem. Soc.*, 1995, *117*, 5712; (b) J.Y. Roberge, X. Beebe, and S.J. Danishefsky, *Science*, 1995, *269*, 202; (c) R. Liang, L. Yan, J. Loebach, M. Ge, Y. Uozumi, K. Sekanina, N. Horan, J. Gildersleeve, C. Thompson, A. Smith, K. Biswas, W.C. Still, and D. Kahne, *Science*, 1996, *274*, 1520; (d) A.M. Rouhi, *Chem. Eng. News*, Sept. 23, 1996, 62; Dec. 2, 1995, 4; (e) Y. Ito, O. Kanie, and T. Ogawa, *Angew. Chem. Int. Ed.*, 1996, *35*, 2510; (f) P. Arya and R.N. Ben, *Angew. Chem. Int. Ed.*, 1997, *36*, 1280; (g) S. Borman, *Chem. Eng. News*, July 20, 1998, 49; (h) O.J. Plante, E.R. Palmacci, and P.H. Seeberger, *Science*, 2001, *291*, 1523; (i) H.M.I. Osborn and T.H. Khan, *Tetrahedron*, 1999, *55*, 1807; (j) R.F. Service, *Science*, 2001, *291*, 805.
174. R. Breinbauer, I.R. Vetter, and H. Waldmann, *Angew. Chem. Int. Ed.*, 2002, *41*, 2879.
175. (a) S.J. Taylor and J.P. Morken, *Science*, 1998, *280*, 267; (b) A. Holzwarth, H.-W. Schmidt, and W.F. Maier, *Angew. Chem. Int. Ed.*, 1998, *37*, 2644.
176. R. Dagani, *Chem. Eng. News*, Sept. 21, 1998, 13.
177. A.W. Czarnik, *Anal. Chem.*, 1998, *70*, 378A.
178. (a) J.W. Guiles, C.L. Lanter, and R.A. Rivero, *Angew. Chem. Int. Ed.*, 1998, *37*, 926; (b) B.J. Egner, S. Rana, H. Smith, N.

Bouloc, J.G. Frey, W.S. Brocklesby, and M. Bradley, *Chem. Commun.*, 1997, 735.

179. (a) K.C. Nicolaou, X.-Y. Xiao, Z. Parankoosh, A. Senyei, and M.P. Nova, *Angew. Chem. Int. Ed.*, 1995, *34*, 2289; (b) E.J. Moran, S. Sarshar, J.F. Cargill, M.M. Shahbaz, A. Lio, A.M.M. Mjalli, and R.W. Armstrong, *J. Am. Chem. Soc.*, 1995, *117*, 10787; (c) R.F. Service, *Science*, 1995, *270*, 577; (d) T. Czarnik and M. Nova, *Chem. Br.*, 1997, *33*(10), 39.
180. (a) B.J. Egner, G.J. Langley, and M. Bradley, *J. Org. Chem.*, 1995, *60*, 2652; *J. Chem. Soc. Chem. Commun.*, 1995, 2163; (b) M. Cardno and M. Bradley, *Tetrahedron Lett.*, 1996, *37*, 135; (c) *Chem. Eng. News*, Apr. 29, 1996, 53; (d) J.N. Kyranos and J.C. Hogan, Jr., *Anal. Chem.*, 1998, *70*, 389A; (e) *Chem. Eng. News*, Sept. 22, 1997, 27; (f) D.C. Schriemer, D.R. Bundle, L. Li, and O. Hindgaul, *Angew. Chem. Int. Ed.*, 1998, *37*, 3383.
181. (a) M. Przybylski and M.O. Glocker, *Angew. Chem. Int. Ed.*, 1996, *35*, 806; (b) G. Montaudo, *Trends Polymer Sci.*, 1996, *4*(3), 81.
182. (a) R.C. Anderson, J.P. Stokes, and M.J. Shapiro, *Tetrahedron Lett.*, 1995, *36*, 5311; (b) S.K. Sarkar, R.S. Garigipati, J.L. Adams, and P.A. Keifer, *J. Am. Chem. Soc.*, 1996, *118*, 2305; (c) P.A. Keifer, *J. Org. Chem.*, 1996, *61*, 1558; (d) I.E. Pop, C.F. Dhalluin, B.P. Deprez, P.C. Melnyk, G.M. Lippens, and A.L. Tartar, *Tetrahedron*, 1996, *52*, 12209; (e) W.L. Fitch, G. Detre, C.P. Holmes, J.N. Shoolery, and P.A. Keifer, *J. Org. Chem.*, 1994, *59*, 7955.
183. (a) K. Russell, D.C. Cole, F.M. McLaren, and D.E. Pivonka, *J. Am. Chem. Soc.*, 1996, *118*, 7941; (b) B. Yan, G. Kumaravl, H. Anjaria, A. Wu, R.C. Petter, C.F. Jewell, Jr., and J.R. Wareing, *J. Org. Chem.*, 1995, *60*, 5736; (c) B. Yan and G. Kumaravel, *Tetrahedron*, 1996, *52*, 843; (d) B. Yan, *Acc. Chem. Res.*, 1998, *31*, 621; (e) W.J. Haap, T.B. Walk, and G. Jung, *Angew. Chem. Int. Ed.*, 1998, *37*, 3311.
184. J.M. Harris and S. Zalipsky, eds, *Poly(ethylene glycol): Chemistry and Biological Applications*, A.C.S. Symp. 680, Washington, DC, 1997.
185. (a) B.-D. Park, H.-I. Lee, S.-J. Ryoo, and Y.-S. Lee, *Tetrahedron Lett.*, 1997, *38*, 591; (b) P.H. Toy, T.S. Reger, and K.D. Janda, *Aldrichchim. Acta*, 2000, *33*(3), 87.
186. (a) M.L. Hallensleben and F. Lucarelli, *Polym. Bull.*, 1996, *37*, 759; (b) G.E. Totten, N.A. Clinton, and P.L. Matlock, *J. Macromol. Sci., Rev. Macromol. Chem.*, 1998, *C38*, 77, 126, 137.
187. M. Renil and V.N.R. Pillai, *J. Appl. Polym. Sci.*, 1996, *61*, 1585.
188. M. Renil and M. Meldal, *Tetrahedron Lett.*, 1995, *36*, 4647.
189. M. Kempe and G. Barany, *J. Am. Chem. Soc.*, 1996, *118*, 7083.
190. M. Teodorescu, M. Dimonie, and J. Languri, *Angew. Macromol. Chem.*, 1997, *251*, 81.
191. (a) D.L. Coffen, ed., *Tetrahedron*, 1998, *54*, 3955–4141; (b) P. Wentworth, Jr., *Chem. Commun.*, 1999, 1917.
192. K.E. Geckeler, *Adv. Polym. Sci.*, 1995, *121*, 31–62.
193. S. Borman, *Chem. Eng. News*, July 31, 1995, 25.
194. L. Jiang, R.C. Hartley, and T.-H. Chan, *Chem. Commun.*, 1996, 2193.
195. (a) A.G.M. Barrett, M.L. Smith, and F.J. Zecri, *Chem. Commun.*, 1998, 2317; (b) R.J. Booth and J.C. Hodges, *Acc. Chem. Res.*, 1999, *32*, 18; *J. Am. Chem. Soc.*, 1997, *119*, 4882; (c) D.L. Flynn, J.Z. Crich, R.V. Devraj, S.L. Hockermann, J.J. Parlow, M.S. South, and S. Woodard, *J. Am. Chem. Soc.*, 1997, *119*, 4874.
196. P. Wentworth, Jr., A.M. Vandersteen, and K.D. Janda, *Chem. Commun.*, 1997, 759.
197. H. Han and K.D. Janda, *J. Am. Chem. Soc.*, 1996, *118*, 7632.
198. (a) S. Borman, *Chem. Eng. News*, Nov. 4, 1996, 37; (b) *Chem. Eng. News*, Dec. 23, 1996, 17; (c) F.C. Moates, M. Somani, J. Annamalai, J.T. Richardson, D. Luss, and R.C. Willson, *Ind. Eng. Chem. Res.*, 1996, *35*, 4801; (d) B. Jandeleit and H. Weinberg, *Chem. Ind. (London)*, 1998, 795; (e) R. Dagani, *Chem. Eng. News*, June 15, 1998, 6; (f) R. Schlogl, *Angew. Chem. Int. Ed.*, 1998, *37*, 2333; (g) J.W. Saalfrank and W.F. Maier, *Angew. Chem. Int. Ed.*, 2004, *43*, 2028; (h) V.V. Guiliants, ed., *Catal. Today*, 2001, *67*, 307–409; (i) E.W. McFarland and W.H. Weinberg, *Trends Biotechnol.*, 1999, *17*, 107; (j) S. Dahmen and S. Brase, *Synthesis*, 2001, 1431; (k) S. Senkan, *Angew. Chem. Int. Ed.*, 2001, *40*, 312.
199. (a) X.-D. Xiang, X. San, G. Briceno, Y. Lou, K.-A. Wang, H. Chang, W.G. Wallace-Freedman, S.-W. Chen, and P.G. Schultz, *Science*, 1995, *268*, 1738; (b) G. Briceno, H. Chang, X. Sun, P.G. Schultz, and X.-D. Xiang, *Science*, 1995, *270*, 273; (c) X.-D. Xiang, *Chem. Ind. (London)*, 1998, 800; (d) R. Dagani, *Chem. Eng. News*, Mar. 8, 1999, 51; (e) L.L. Bellavance, *Chem. Eng.*, 1999, *106*(10), 76; (f) B. Jandeleit, D.J. Schaefer, T.S. Powers, H.W. Turner, and W.H. Weinberg, *Angew. Chem. Int. Ed.*, 1999, *38*, 2494.
200. E. Reddington, A. Sapienza, B. Gurau, R. Viswanathan, S. Sarangapani, E.S. Smotkin, and T.E. Mallouk, *Science*, 1998, *280*, 1735.
201. (a) T.R. Boussie, C. Coutard, H. Turner, V. Murphy, and T.S. Powers, *Angew. Chem. Int. Ed.*, 1998, *37*, 3272; (b) M. Stork, A. Herrmann, T. Nemnich, M. Klapper, and K. Mullen, *Angew. Chem. Int. Ed.*, 2000, *39*, 4367.
202. (a) M.T. Reetz, M.H. Becker, K.M. Kuhling, and A. Holzwarth, *Angew. Chem. Int. Ed.*, 1998, *37*, 2647; (b) A. Whiting, *Chem. Br.*, 1999, *35*(3), 31; (c) M.D. Weingarten, K. Sekanina, and W.C. Still, *J. Am. Chem. Soc.*, 1998, *120*, 9112; (d) P. Murer, K. Lewandowski, F. Svec, and J.M.J. Frechet, *Chem. Commun.*, 1998, 2559; (e) M.T. Reetz, *Angew. Chem. Int. Ed.*, 2001, *40*, 284.
203. F. Hollfelder, A.J. Kirby, and D.S. Tawfik, *J. Am. Chem. Soc.*, 1997, *119*, 9578.
204. (a) R.F. Service, *Science*, 1997, *227*, 474; *Technol. Rev.*, 1998, *101*(3), 34; (b) E. Danielson, M. Devenney, D.M. Giaquinta, J.H. Golden, R.C. Haushalter, E.W. McFarland, W.H. Weinberg, and X.D. Wu, *Science*, 1998, *279*, 837; (c) D.J. Tapolczay, R.J. Kobylecki, L.J. Payne, and B. Hall, *Chem. Ind. (London)*, 1998, 772.
205. P.G. Schultz, X. Xiang, and I. Goldwasser, U.S. patent 5,776,359 (1998).
206. E. Wilson, *Chem. Eng. News*, Mar. 16, 1998, 9.
207. Anon., *Chem. Eng. News*, Apr. 27, 1998, 12.
208. J. Klein, C.W. Lehmann, H.-W. Schmidt, and W.F. Maier, *Angew. Chem. Int. Ed.*, 1998, *37*, 3369.
209. (a) F.C. Schroeder, J.J. Farmer, A.B. Attygalle, S.R. Smedley, T. Eisner, and J. Meinwald, *Science*, 1998, *281*, 428; (b) S. Borman, *Chem. Eng. News*, July 20, 1998, 11.

210. C. Gao, B.J. Lavey, C.-HL. Lo, A. Datta. P. Wentworth, and K.D. Janda, *J. Am. Chem. Soc.*, 1998, *120*, 2211.
211. J. Hazziza-Laskar, G. Helary, and G. Sauvet, *J. Appl. Polym. Sci.*, 1995, *58*, 77.
212. A. Kanazawa, T. Ikeda, and T. Endo, *J. Appl. Polym. Sci.*, 1994, *54*, 1305.
213. A. Kanazawa, T. Ikeda, and T. Endo, *J. Polym. Sci. Polym. Chem.*, 1994, *32*, 1997.
214. (a) G. Sun, W.B. Wheatley, and S.D. Worley, *Ind. Eng. Chem. Res.*, 1994, *33*, 168; (b) G. Sun, L.C. Allen, E.P. Luckie, W.B. Wheatley, and S.D. Worley, *Ind. Eng. Chem. Res.*, 1995, *34*, 4106.
215. (a) C.M. Starks, C.L. Liotta, and M. Halpern, eds, *Phase Transfer Catalysis—Fundamentals, Applications and Industrial Perspectives*, Chapman & Hall, New York, 1994, 207–239; (b) M. Halpern, *Ullmann's Encyclopedia of Industrial Chemistry*, 5th ed., VCH, Weinheim, 1991, *A19*, 295; (c) E.V. Dehmlow and S. Dehmlow, *Phase Transfer Catalysis*, 3rd ed., VCH, Weinheim, 1993; (d) Y. Goldberg, *Phase Transfer Catalysis: Selected Problems and Applications*, Gordon and Breach, Newark, NJ, 1992, Chapter 5; (e) V. Percec and C. Pugh, *Encyclopedia of Polymer Science Engineering*, 2nd ed., 1987, *10*, 443; (f) B.C. Gates and J. Lieto, *Encyclopedia of Polymer Science Engineering*, 2nd ed., 1985, *2*, 708; (g) M.E. Halpern, *Phase-Transfer Catalysis: Mechanisms and Syntheses*, A.C.S. Symp. 659, Washington, DC, 1997; (h) S.J. Tavener and J.H. Clark, *Chem. Ind. (London)*, 1997, 22.
216. T. Iizawa, Y. Yamada, Y. Ogura, and Y. Sato, *J. Polym. Sci., A Polym. Chem.*, 1994, *32*, 2057.
217. J.H. Clark, S.J. Tavener, and S.J. Barlow, *Chem. Commun.*, 1996, 2429.
218. S. Kondo, M. Takesue, M. Suzuki, H. Kunisada, and Y. Yuki, *J. Macromol. Sci. Pure Appl. Chem.*, 1994, *A31*, 2033.
219. (a) D. Pini, A. Petri, and P. Salvadori, *Tetrahedron*, 1994, *50*, 11321; (b) A. Petri, D. Pini, and P. Salvadori, *Tetrahedron Lett.*, 1995, *36*, 1549.
220. W.A. Herrmann, R.M. Kratzer, J. Blumel, H.B. Friedrich, R.W. Fischer, D.C. Apperley, J. Mink, and O. Berkesi, *J. Mol. Catal. A Chem.*, 1997, *120*, 197.
221. N. Yoneda, Y. Nakagawa, and T. Mimami, *Catal. Today*, 1997, *36*, 357.
222. N. Goudarzian, P. Ghahramani, and S. Hossini, *Polym. Int.*, 1996, *39*, 61.
223. R.T. Peltonen, K.B. Ekman, and J.H. Nasman, *Ind. Eng. Chem. Res.*, 1994, *33*, 235.
224. R.A. Sheldon, M. Wallau, I.W.C.E. Arends, and U. Schuchardt, *Acc. Chem. Res.*, 1998, *31*, 485.
225. V. Gutanu, C. Luca, C. Turta, V. Neagu, V. Sofranschi, M. Cherdivarenco, and B.C. Simionescu, *J. Appl. Polym. Sci.*, 1996, *59*, 1371.
226. K. Kurita, S.-I. Watabe, S.-I. Nishimura, and S. Ishii, *J. Polym. Sci. Polym. Chem.*, 1996, *34*, 429.
227. (a) D.C. Sherrington and H.G. Tang, *Macromol. Symp.*, 1994, *80*, 193; (b) H.G. Tang and D.C. Sherrington, *J. Mol. Catal.*, 1994, *94*, 7; (c) G. Olason and D.C. Sherrington, *Macromol. Symp.*, 1998, *131*, 127.
228. M.M. Miller and D.C. Sherrington, *J. Catal.*, 1995, *152*, 368, 377.
229. J.-H. Ahn and D.C. Sherrington, *Macromolecules*, 1996, *29*, 4164.
230. F. Arena, G. Cum, R. Gallo, and A. Parmaliana, *J. Mol. Catal. A Chem.*, 1996, *110*, 235.
231. (a) Z.M. Michalska, B. Ostaszewski, and K. Strzelec, *J. Organomet. Chem.*, 1995, *496*, 19; (b) Z.M. Michalska and K. Strzlec, *J. Mol. Catal. A*, 2001, *177*, 71.
232. Z.M. Michalska, B. Ostaszewski, J. Zientarska, J. Zientarska, and J.W. Sobczak, *J. Mol. Chem. A Chem.*, 1998, *129*, 207.
233. E. Bonaplata, H. Ding, B.E. Hanson, and J.E. McGrath, *Polymer*, 1995, *36*, 3035.
234. D. Jiang X. Li, and E. Wang, *Macromol. Symp.*, 1996, *105*, 161.
235. J.H. van Esch, M.A.M. Hoffmann, and R.J.M. Nolte, *J. Org. Chem.*, 1995, *60*, 1599.
236. D.E. Bergbreiter, V.M. Mariagnaman, and L. Zhang, *Adv. Mater*, 1995, *7*(1), 69; *Macromol. Symp.*, 1996, *105*, 9.
237. (a) D.E. Bergbreiter, B.L. Case, Y.-S. Liu, and J.W. Caraway, *Macromolecules*, 1998, *31*, 6053; (b) D.J. Gravert, A. Datta, P. Wentworth, Jr., and K.D. Janda, *J. Am. Chem. Soc.*, 1998, *120*, 9481.
238. J.E. Guillet, K.C. Kohler, and G. Friedman, U.S. patent 5,624,543 (1997).
239. (a) G. Ondrey, *Chem. Eng.*, 2005, *112*(4), 15; (b) B. Zhou, *11th Annual Green Chemistry & Engineering Conference Abstracts*, Washington, DC, 2007, 30; (c) M. Rueter, B. Zhou, and S. Parasher, U.S. patent 7,144,565 (2006); (d) Z. Zhou, Z. Wu, C. Zhang, and B. Zhou, U.S. patent application 20080081017 (2008); (e) Z. Wu, Z. Zhou, M. Rueter, and B. Zhou, U.S. patent application 20080193368 (2008).
240. (a) J. Kemsley, *Chem. Eng. News*, Aug. 25, 2008, 9; (b) R. Burks, *Chem. Eng. News*, Sept. 24, 2007, 87.
241. A.A. Herzing, C.J. Kiely, A.F. Carley, P. Landon, and G.J. Hutchings, *Science*, 2008, *321*, 1331.
242. P. Nolte, A. Stierle, N.Y. Jin-Phillipp, N. Kasper, T.U. Schull, and H. Dosch, *Science*, 2008, *321*, 1654.
243. (a) L. Canali, E. Cowan, H. Deleuze, C.L. Gibson, and D.C. Sherrington, *Chem. Commun.*, 1998, 2561; (b) X.-D. Du and X.-D. Yu, *J. Polym. Sci., Polym. Chem.*, 1997, *35*, 3249; (c) M.D. Angelino and P.E. Laibinis, *Macromolecules*, 1998, *31*, 7581.
244. S.K. Das, A. Kumar, Jr., S. Nandrajog, and A. Kumar, *Tetrahedron Lett.*, 1995, *36*, 7909.
245. T. Oyama, T. Ozaki, and Y. Chujo, *Polym. Bull.*, 1997, *38*, 379.
246. H. Ritter and R. Sperber, *Macromolecules*, 1994, *27*, 5919.
247. J.J. Parlow, *Tetrahedron Lett.*, 1995, *36*, 1395.
248. (a) J. Szejtli, *Cyclodextrin Technology*, Kluwer Academic, Dordrecht, the Netherlands, 1988; (b) J.K. Poudrier, *Todays Chem. Work*, 1995, *4*(2), 25; (c) W.J. Shieh and A.R. Hedges, *J. Macromol. Sci. Pure Appl. Chem.*, 1996, *A33*, 673; (d) G. Wenz, *Angew. Chem. Int. Ed.*, 1994, *33*, 803; (e) E. Weber, *Kirk–Othmer Encyclopedia of Chemical Technology*, 4th ed., Wiley, New York, 1995, *14*, 129–131; (f) J.L. Atwood, *Ullmann's Encyclopedia of Industrial Chemistry*, VCH, Weinheim, 1989, *A14*, 119; (g) V.T. D'Souza and K.B. Lipowitz, eds, *Chem. Rev.*, 1998, *98*, 1741–2045; (h) M. McCoy, *Chem. Eng. News*, Mar. 1, 1999, 25; (i) J.-M. Lehn, ed., *Comprehensive Supramolecular Chemistry*, Pergamon, Oxford, *3*, 1996; (j) M.M. Conn and J. Rebek, Jr., *Chem. Rev.*, 1997, *97*, 1647; (k) K.A. Connors, *Chem. Rev.*, 1997, *97*, 1325.

249. T.-J. Kim, B.-C. Kim, and H.-S. Lee, *Enzyme Microb. Technol.*, 1995, *17*, 1057.
250. T. Kamiya, *Chem. Eng.*, 1998, *105*(8), 44.
251. B.N. Gewande and A.Y. Patkar, *Enzyme Microb. Technol.*, 2001, *28*, 735.
252. *Chem. Eng.*, 2002, *109*(4), 23.
253. (a) L. Huang and A.E. Tonelli, *J. Macromol. Sci., Rev. Macromol. Chem.*, 1998, *C38*, 781; (b) A. Harada, *Adv. Polym. Sci.*, 1997, *133*, 141; (c) G. Wenz, M.B. Steinbrunn, and R. Landfester, *Tetrahedron*, 1997, *53*, 15575.
254. A. Harada, *Macromolecules*, 1995, *28*, 8406.
255. K. Hicks, R.M. Haines, C.B.S. Tong, G.M. Sapers, Y. El-Atawy, P.L. Irwin, and P.A. Seib, *J. Agric. Food Chem.*, 1996, *44*, 2591.
256. (a) W.A. Reichenbach and D.B. Min, *J. Am. Oil Chem. Soc.*, 1997, *74*, 1329; (b) J.M. Lopez-Nicolas, R. Bru, and F. Garcia-Carmona, *J. Agric. Food Chem.*, 1997, *45*, 1144.
257. M. Granger, M. Dupont, and H. Ledon, U.S. patent 5,382,571 (1995).
258. P. Velusamy, K. Pitchumani, and C. Srinivasan, *Tetrahedron*, 1996, *52*, 3487.
259. (a) H. Hirai, Y. Shiraishi, and K. Saito, *Macromol. Rapid Commun.*, 1995, *16*, 31, 697; (b) H. Hirai, *J. Macromol. Sci. Pure Appl. Chem.*, 1994, *A31*, 1491.
260. L.N. Lewis and C.A. Sumpter, *J. Mol. Catal. A Chem.*, 1996, *104*, 293.
261. E. Monflier, E. Blouet, Y. Barbaux, and A. Mortreux, *Angew. Chem. Int. Ed.*, 1994, *33*, 2100.
262. R. Breslow, J. Desper, and Y. Huang, *Tetrahedron Lett.*, 1996, *37*, 2541.
263. J. Jeromin and H. Ritter, *Macromol. Rapid Commun.*, 1998, *19*, 377.
264. (a) W. Lau, *Macromol. Symp.*, 2002, *182*, 283; (b) M. Fischer and H. Ritter, *Macromol. Rapid Commun.*, 2000, *21*, 142; (c) S. Rimmer, *Macromol. Symp.*, 2000, *150*, 149; (d) H. Ritter and Tabatabai, *Prog. Polym. Sci.*, 2002, *27*, 1713.
265. (a) S. Hanessian, A. Benalil, and C. Laferriere, *J. Org. Chem.*, 1995, *60*, 4786; (b) K.A. Martin, M.A. Mortellaro, R.W. Sweger, L.E. Fikes, D.T. Winn, S. Clary, M.P. Johnson, and A.W. Czarnik, *J. Am. Chem. Soc.*, 1995, *117*, 10443; (c) P.R. Ashton, R. Koniger, J.F. Stoddart, D. Alker, and V.D. Harding, *J. Org. Chem.*, 1996, *61*, 903; (d) S.K. Young, P.L. Vadja, and E. Napadensky, *Polym. Preprints*, 2001, *42*(2), 162; (e) E. Engeldinger, D. Armsprach, and D. Matt, *Chem. Rev.*, 2003, *103*, 4147.
266. R. Breslow and B. Zhang, *J. Am. Chem. Soc.*, 1996, *118*, 8495.
267. (a) K. Sreenivasan, *J. Appl. Polym. Sci.*, 1998, *68*, 1857, 1863; (b) H. Asanuma, M. Kakazu, M. Shibata, T. Hishiya, and M. Komiyana, *Chem. Commun.*, 1997, 1971.
268. (a) R. Breslow, *Acc. Chem. Res.*, 1995, *28*, 146; (b) R. Breslow, S. Halfon, and B. Zhang, *Tetrahedron*, 1995, *51*, 377; (c) Y. Murakami, J.-I. Kikuchi. Y. Hisaeda, and O. Hayashida, *Chem. Rev.*, 1996, *96*, 721; (d) C.A. Haskard, C.J. Easton, B.L. May, and S.F. Lincoln, *Inorg. Chem.*, 1996, *35*, 1059; (e) Jiang and Lawrence, *J. Am. Chem. Soc.*, 1995, *117*, 1857.
269. M. Weickenmeier and G. Wenz, *Macromol. Rapid Commun.*, 1996, *17*, 731.
270. S. Murai, S. Imajo, Y. Takasu, K. Takahashi, and K. Hatori, *Environ. Sci. Technol.*, 1998, *32*, 782.
271. G. Gattuso, S. Menzer, S.A. Nepogodiev, J.F. Stoddart, and D.J. Williams, *Angew. Chem. Int. Ed.*, 1997, *36*, 1451.

RECOMMENDED READING

1. D. Arntz, *Catal. Today* 1993, *18*, 173 (trends in the chemical industry).
2. S. Borman, *Chem. Eng. News*, Feb. 12, 1996, 29–54.
3. D.C. Sherrington, *Chem. Ind.*, 1991, *1*, 15.
4. J.K. Poudrier, *Todays Chem. Work*, 1995, *4*(2), 25.
5. J.H. Clark and D.J. Macquarrie, *Chem. Soc. Rev.*, 1996, *25*, 303 (environmentally friendly catalytic methods using supported reagents and catalysts).
6. J.A. Gladysz, *Pure Appl. Chem.*, 2001, *73*, 1319 (ideal synthesis).
7. C.C. Tzschucke, *Angew. Chem. Int. Ed.*, 2002, *41*, 3965 (modern techniques for organic synthesis).
8. A.T. Bell, *Science*, 2003, *299*, 1688 (characterization of heterogeneous catalysts).
9. D.J. Cole-Hamilton, *Science*, 2003, *299*, 1702 (separation of homogeneous catalysts).
10. P. Hodge, *Ind. Eng. Chem. Res.*, 2005, *44*, 8542 (polymer-supported reagents catalysts and scavengers in flow systems).
11. P. Kundig, *Science*, 2006, *314*, 430 (trends in organic synthesis).

EXERCISES

1. Compare the sizes of the cavities of cyclodextrins with those of natural food colors, such as that of beets, to see if shelf-stable food colors could be produced using any of the cyclodextrins.
2. How could you immobilize a crown ether for recovery of a metal ion?
3. How might combinatorial chemistry be applied to optimizing ligands for catalysts?
4. How many times should a supported catalyst be reused to be sure that none is being lost and that the activity remains high?
5. Design an ion-exchange resin that can be used at 200°C.
6. Would it be better to put an oxidation catalyst on an organic or an inorganic support?
7. Compare the environmental effects of supported toxic metal oxidants with oxidations performed with hydrogen peroxide or oxygen.
8. Should each home be outfitted with a sterilizing filter for incoming water instead of chlorination at a central plant?
9. How would you put a shell of alumina or a transition metal oxide on a particle of magnetic iron oxide? (If you are puzzled, see P. Lengo, M. DiSerio, A. Sorrentino, V Solinas, and E. Santacesaria. *Appl. Catal. A*, 1998, *167*, 85.)
10. Why do so few students and faculty in university chemistry departments use supported reagents and catalysis?

CHAPTER 6

Solid Acids and Bases

6.1 INTRODUCTION

This chapter continues the discussion of the use of solid reagents to minimize exposure to hazardous reagents, to make workups easier, and to minimize waste. Liquid acids are also corrosive, may be difficult to recycle for repeated use, and may show low activity or selectivity in some reactions.[1] Large amounts of sulfuric acid and hydrogen fluoride are used in petroleum refining for the alkylation of isobutene with olefins to produce high-octane gasoline.[2] A typical reaction of this type (**6.1** Schematic) is shown.

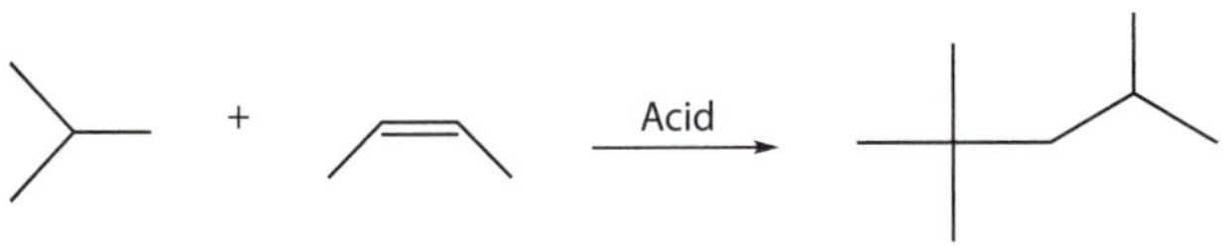

Acid

6.1 Schematic

Hydrogen fluoride (bp 19.5°C) is more potent biologically than other Bronsted acids[3] and could be a real problem if it were released into the environment.[4] Exposure to hydrogen fluoride over just 2% of the body or to 50 ppm in air can lead to death.[5] Olah has devised a way of lowering its volatility by the addition of a tertiary amine.[6] One refinery is using this method. Another company has patented a method for using far less hydrogen fluoride.[7] Mobil used 1300 ppm of a 2:1 hydrogen fluoride/boron trifluoride catalyst in the foregoing reaction. The 2-butene conversion was 100%. The ratio of desired trimethylpentane to dimethylhexane was 35:1. However, the 22:1 ratio of isobutane to 2-butene must be reduced to 12:1 or lower to be useful in a refinery.[8] A preliminary announcement by Hydrocarbon Technologies describes a benign solid superacid catalyst for the reaction that appears to be stable under normal-operating conditions in a continuous-flow laboratory system.[9] If this passes operation in the pilot plant, it could answer the problem. However, several other solid catalysts for the reaction have undergone rapid deactivation in use.

Solid catalysts for this reaction are now practical. ABB Lummus and Akzo Nobel have an "AlkyClean" process that uses a zeolite at 70–90°C to produce 96 octane gasoline.[10,11] It uses three reactors, of which two are in production cycling every 1–3 h between alkylation and mild catalyst regeneration by hydrogen. (Hydrogenation of the pore-plugging olefins releases them.) Every few weeks a stronger regeneration is done at 250°C. The process has been demonstrated in a refinery in Finland. The UOP "Alkylene" process uses a fluidized bed of Pt on $AlCl_3/Al_2O_3$ at 10–40°C together with regeneration by hydrogenation. It has been demonstrated in a refinery in Utah and is being used in a refinery in Azerbaijan.[11–13] Exelus uses a zeolite with two fixed bed reactors. After 10–12 h of alkylation at 60–90°C, the catalyst is regenerated with 2 h of hydrogenation.[14,15] It produces gasoline with >98 octane. It is being installed in a refinery in Europe.

An alternate process used in several refineries dimerizes isobutene with a solid acid catalyst, for example, with the ion-exchange resin Amberlyst 15, to dimers, which are then hydrogenated (**6.2** Schematic).[16,17]

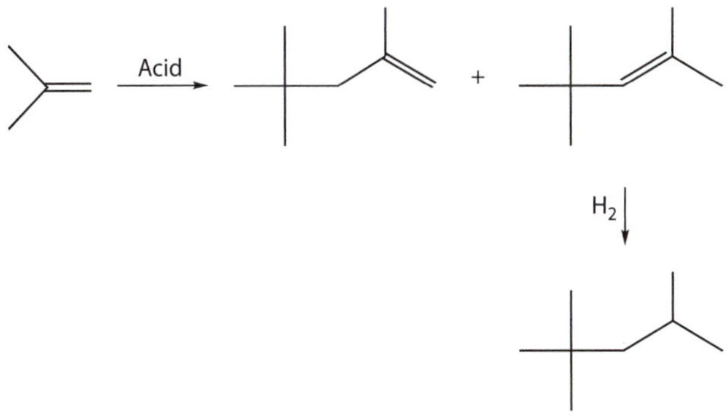

Acid

H_2

6.2 Schematic

Hydrogen fluoride is also used in the alkylation of benzene with linear olefins to produce detergent alkylate for sulfonation to produce detergents. This acid is now being replaced in some plants by fluorided silica–alumina catalysts in the UOP–Petresa process.[18]

The Hoechst–Celanese three-step process for the preparation of ibuprofen has replaced an older six-step process.[19] Although the new process produces much less waste, hydrogen fluoride is used as both solvent and catalyst in the acylation step (**6.3** Schematic). It should be possible to replace the dangerous hydrogen fluoride. Two possibilities are the UOP–Petresa catalyst and a high surface area polymeric persulfonic acid.[20] (These may require a film of liquid acid on the surface to prevent deactivation of the catalyst.)

Many types of solid acid and base catalysts are known.[21] Superacids are those that are at least as strong as 100% sulfuric acid.[22] The acid strengths are measured using basic indicators and are assigned a Hammett acidity function, H_{0-}. Table 6.1 lists some superacids, with the strongest at the top.

Only the first and last are liquids. The strongest Bronsted acid is $H(CHB_{11}Cl_{11})$.[23] A nonoxidizing Lewis superacid containing many fluorine atoms is strong enough to coordinate fluorobenzene (**6.4** Schematic).[24]

The stronger acids can catalyze some reactions that the weaker ones cannot (e.g., the alkylation of isobutene by 2-butene). As a rule of thumb, it is probably better to use the weakest acid that will do the job to avoid unwanted side reactions. Many solid acids, such as clays and molecular sieves, are shape- and size-selective, so that this also enters into the decision on which ones to use. The various types of

6.3 Schematic

Table 6.1 Comparison of Acid Strengths by $-H_0$

	$-H_0$
$HF–SbF_5$ (1:1)	20
$FSO_3H–SbF_5$ (1:0.2)	20
Sulfated ZrO_2	16.0
$SbF_5/SiO_2–Al_2O_3$	13.8
$AlCl_3–CuCl_2$	13.8
$H_3PW_{12}O_{40}$	13.3
$Cs_{2.5}H_{0.5}PW_{12}O_{40}$	13.3
Nafion	12.0
Sulfuric acid (100%)	12.0

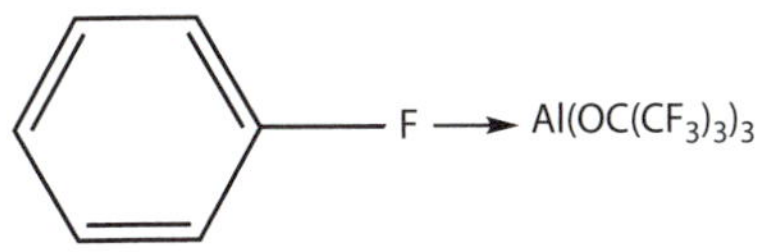

C_6H_5–F ⟶ $Al(OC(CF_3)_3)_3$

6.4 Schematic

solid acids and bases will be discussed in the following. Some soluble catalysts that have advantages over the current ones will also be included. Because many good reviews are available, only a few examples from the literature will be given to illustrate the diversity of possibilities, the trends in current research, and the needs.

6.2 POLYMERIC SULFONIC ACIDS

Sulfonic acid ion-exchange resins that can catalyze esterification, etherification, and addition of alcohols and water to olefins were covered in Chapter 5. Degussa (now Evonik) has made a polysiloxane analog with alkyl sulfonic acid groups on it.[25] It is stable up to 230°C, compared with 120°C for the polystyrene-based ones.

DuPont markets a polymeric perfluorinated sulfonic acid as Nafion H (**6.5** Schematic). Dow sells a similar resin.

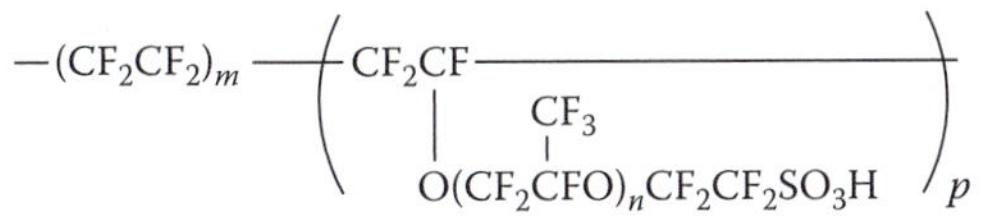

$$-(CF_2CF_2)_m-\left(CF_2CF(O(CF_2CF(CF_3)O)_nCF_2CF_2SO_3H)\right)_p$$

6.5 Schematic

The use of Nafion in electrolytic cells for the production of chlorine to eliminate the use of mercury was mentioned in Chapter 4. Nafion H has been used as an acid in many organic syntheses.[26] It is the polymeric equivalent of trifluoromethanesulfonic acid (triflic acid).[27] When used as a catalyst for the pinacol rearrangement of 1,2-diols, dehydration was avoided.[28] However, 1,4- and 1,5-diols did dehydrate with it.[29]

It was also useful in an alkylation of aniline (**6.6** Schematic).[30] The reaction could not be carried out with the usual sulfonic acid ion-exchange resin, because its maximum use temperature was 120°C. The product cannot be made directly from acetone and aniline owing to the formation of a dihydroquinoline by-product (from two molecules of acetone and one of aniline). A shape-selective zeolite might allow the reaction to take place without the formation of this by-product. An inexpensive way of making this diamine from acetone and aniline, similar to the preparation of bisphenol A from acetone and phenol, could lead to new families of polyamides, polyimides, polyureas, polyurethanes, and epoxy resins. A palladium catalyst supported on Nafion dimerized ethylene much faster in water than in organic solvents. The butane was easy to separate.[31]

The alkylation (**6.7** Schematic) was carried out by passing a solution of the epoxide through a 10-cm column of Nafion H powder.[32] Although the yield was comparable with those obtained with Sn(IV) chloride and with boron trifluoride ethereate, it would be desirable to replace the CH_2Cl_2/47 $CFCl_3$/3 CF_3CH_2OH solvent used to swell the polymer with one that is not toxic and not capable of destroying the ozone layer.

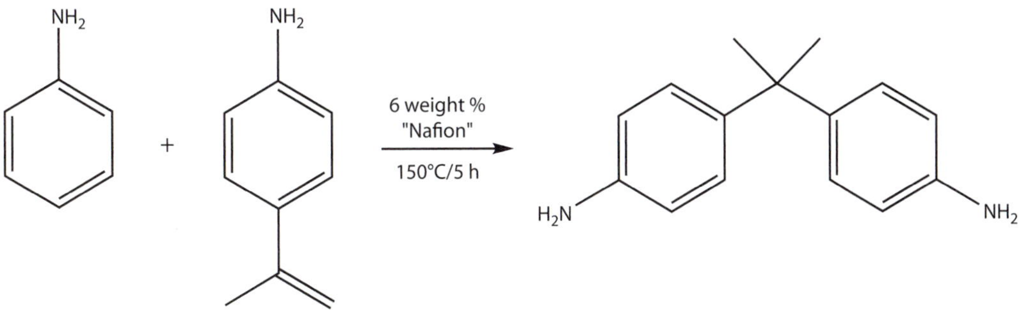

NH_2 NH_2 + 6 weight % "Nafion" 150°C/5 h H_2N NH_2

6.6 Schematic

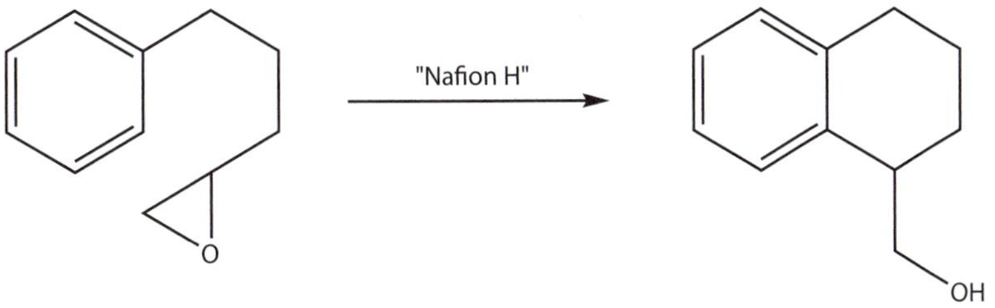

"Nafion H" O OH

6.7 Schematic

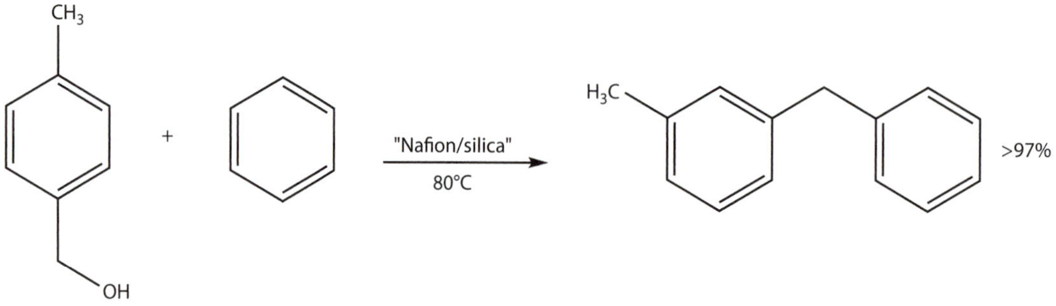

6.8 Schematic

The activity of Nafion as a catalyst has been increased by applying it on silica by the sol–gel method using tetramethoxysilane[33] or tetraethoxysilane and dimethyldiethoxysilane.[34] This increased its activity as a solid acid catalyst up to 100 times that of bulk Nafion. It performed better than when it was on carbon in the dimerization of α-methylstyrene. It was a much better catalyst (seven times on a weight basis) than Amberlyst 15, a typical sulfonic acid ion-exchange resin. It was more active in the benzylation of benzene (**6.8** Schematic) than trifluoromethylsulfonic acid, a reaction in which Amberlyst 15 and *p*-toluenesulfonic acid were inactive.[35]

Hybrid membranes of zirconia in Nafion have been made by the hydrolysis of zirconium tetrabutoxide by water in the pores of a film.[36]

A Nafion–silica analog, which was made by the sol–gel method, gave over 99% conversion in the alkylation of benzene with 1-dodecene to make detergent alkylate (**6.9** Schematic). It was also useful in acylations of aromatic compounds.[37]

It is also possible to put both sulfonic acid and amine groups in the same polymer without them neutralizing each other (**6.10** Schematic).[38]

6.3 POLYMER-SUPPORTED LEWIS ACIDS

The use of inorganic supports for Bronsted and Lewis acids[39] was described in Chapter 5. These included the metal chlorides, as well as protonic acids, on inorganic supports that are sold by Contract Chemicals as "Envirocats".[40] They are air-stable, nontoxic powders that can be reused several times.

Polypropylene containing hydroxyl groups was used with aluminum chloride and boron trifluoride to prepare

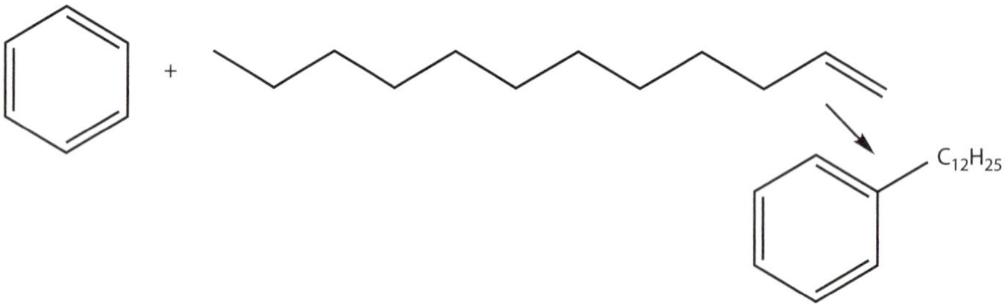

6.9 Schematic

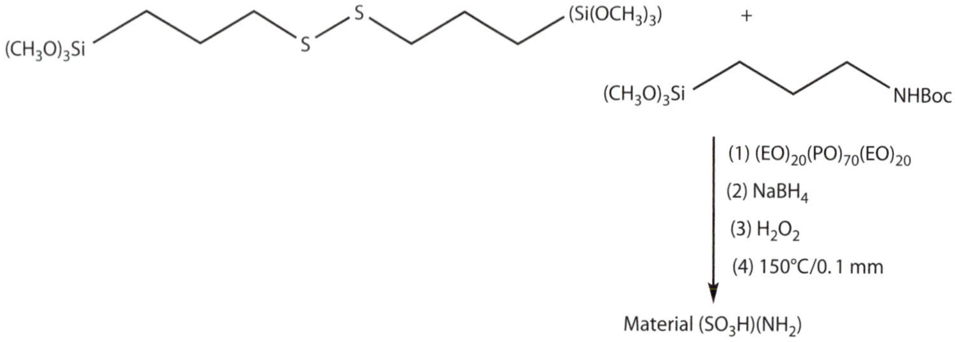

6.10 Schematic

PPOAlCl_2 and PPOBF_2.[41] These were used to polymerize isobutylene and vinyl isobutyl ether. They could be reused for this purpose many times. A styrene–divinylbenzene copolymer sulfonic acid was used with aluminum chloride as a catalyst for the preparation of esters, such as butyl lactate, in 99.5% yield.[42] This may have involved polymer SO_3AlCl_2. The protonic acid *p*-toluenesulfonic acid has been used as its salt with poly(4-vinylpyridine) in the hydrolysis of tetrahydropyranyl ethers (i.e., in the liberation of alcohols that have been protected by reaction with dihydropyran).[43]

6.4 SULFATED ZIRCONIA

Sulfated zirconia is the strongest solid acid known.[44] It can isomerize butanes and alkylate isobutene with *cis*-2-butene. Not everyone considers it to be a superacid.[45] It can be made by treatment of zirconium dioxide (zirconia) with sulfuric acid[46] or by sol–gel methods starting with zirconium alkoxides.[47] Sometimes, the gel is dried with supercritical carbon dioxide in an effort to maintain a large surface area. There is also a debate about what the material actually is. Infrared spectral studies of material prepared by pyrolysis of zirconium sulfate suggest disulfate and monosulfate groups on the surface (**6.11** Schematic).[48] Hydrogen sulfates and cyclic sulfates may also be possible. Some authors feel that sulfated tetragonal, but not monoclinic, zirconia is active.[49] When sulfur is lost on heating, the tetragonal to monoclinic transition can occur. Others believe that the monoclinic form is also active.[50]

The activity of sulfated zirconia can be enhanced by adding platinum, iridium, osmium, palladium, iron, or yttrium.[51] Manganese increased the activity 1000-fold,[52] but it decays rapidly.[53] With platinum as a promoter, the increase in activity is not as great, but it does not decay. Both catalysts have induction periods in the isomerization of butane. It is thought that the promoters work by producing intermediate butanes. Although the platinum one may act catalytically, the iron–manganese one may act irreversibly or may be deactivated by by-products. The platinum may be in Pt–O and Pt–Pt bonds.[54] Platinum-modified sulfated zirconia with hydrogen showed constant 90–100% selectivity in the isomerization of hexane to branched C_6 compounds.[55] (This isomerization can also be performed with 99% selectivity using Pt-H-β-zeolite.[56]) A new method for isomerizing petroleum naphtha, to increase the octane number, with platinum-modified sulfated zirconia is said to be cheaper than using aluminum chloride and to offer increased capacity over the conventional zeolite method.[57]

Because the sulfate in sulfated zirconia is subject to losses when the catalyst is calcined and under reducing conditions, other more stable oxyanions are being examined. Tungstated zirconia is being studied.[58] A platinum-promoted tungstated zirconia was less active than sulfated zirconia.[59] Molybdate has also been used (e.g., to produce an esterification catalyst that gave 95% yield).[60] Other oxides have been sulfated, including titanium dioxide and tin(IV) oxide.[61] The sulfated titanium dioxide was active in cracking cumene at 180°C, whereas a silica–alumina catalyst was inactive at this temperature.

Because sulfated zirconia contains both protonic and Lewis acid acidity,[62] it is possible that poisoning one type selectively would lead to improved lifetimes and yields. Another possibility would be to replace sulfuric acid with a sulfonic acid, such as triflic acid, to produce zirconium triflates on zirconia. This might require different promoter metal ions. The use of rare earth metal triflates as Lewis acid catalysts is described in Section 6.6.

6.5 SUPPORTED METAL OXIDES

Tantalum oxide on silica, made by treatment of silica with tantalum alkoxides, is a catalyst for the Beckman rearrangement of cyclohexanone to caprolactam (**6.12** Schematic).[63] The catalyst lasted at least 10 h in operation. This process eliminates the by-product ammonium sulfate from the present commercial rearrangement in sulfuric acid. Caprolactam is polymerized to produce nylon-6.

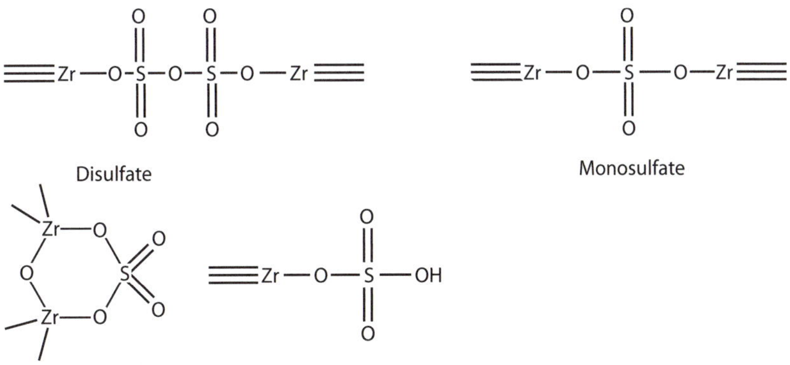

Disulfate

Monosulfate

6.11 Schematic

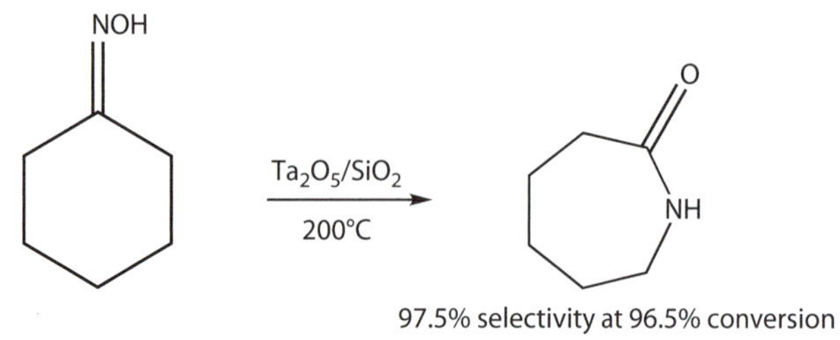

NOH

Ta_2O_5/SiO_2

200°C

O

NH

97.5% selectivity at 96.5% conversion

6.12 Schematic

Alkylations can also be performed with iron oxide catalysts (**6.13** Schematic).[64] Other alcohols, such as *n*-propyl and *n*-butyl alcohols, also worked. Highly acidic catalysts can be obtained by supporting tungsten oxide on iron, tin, and titanium oxides.[65]

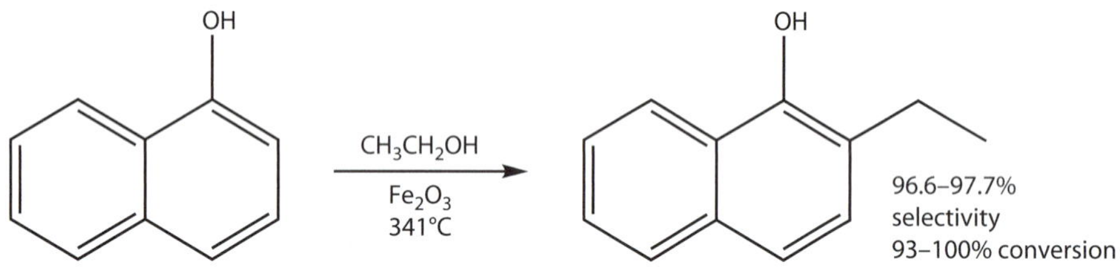

OH

CH_3CH_2OH

Fe_2O_3

341°C

OH

96.6–97.7% selectivity

93–100% conversion

6.13 Schematic

Benzylation of benzene with benzyl chloride using a $CuFeO_4$ catalyst gave diphenylmethane in 81.5% yield.[66] The catalyst was reused five times, with no loss in activity. $HNbMoO_6$ is a strong water-tolerant solid acid catalyst.[67] High surface area TiN has been used in alkylations with high yields (**6.14** Schematic).[68]

6.6 RARE EARTH TRIFLATES

Many Lewis acids, such as metal halides, must be used in the absence of water to avoid hydrolysis. Some reactions require them in stoichiometric amounts. Rare earth and a few other metal triflates are stable to water, alcohols, and carboxylic acids. They can be used as water-tolerant acids in catalytic amounts.[69] Although they usually are not solid catalysts, they are included here because they can be recycled for reuse with no loss in activity. This consists of adding water, extracting away the organic compound, concentrating the aqueous phase to allow the salt to recrystallize, and drying the salt. The aqueous phase may simply be evaporated and the residue dried to recover the catalyst quantitatively. For example, this method was used to recover the catalyst from esterification of alcohols with acetic acid to give 95–99% yields of acetates.[70] Although not strictly a rare earth, scandium triflate is included in this group.

However, a catalyst made by microencapsulating scandium triflate in polystyrene was easy to recover and reuse with no loss in activity by filtration.[71] (Wako sells these.) In the imino aldol reaction (**6.15** Schematic), its activity was greater than that of the unencapsulated counterpart, maintaining activity (90% yield) after seven cycles.[72] (Many Lewis acids do not work well in this reaction.)

Anisole can be acylated with acetic anhydride in 99% yield (**6.16** Schematic). Ytterbium triflate can also be used. The yields are low when there is no activating group in the ring. The rate is accelerated by the addition of lithium perchlorate.[73] Acylation of alcohols works well with 1 mol% of scandium triflate as a catalyst (**6.17** Schematic).[74] The less toxic toluene has also been used as the solvent in such acylations. The yields with 4-dimethylaminopyridine and tributylphosphine as catalysts were lower and the reaction times longer.

Scandium triflate also catalyzes carbonyl-ene reactions (**6.18** Schematic).[75]

Rare earth triflates can also be used in the reaction of aldehydes with aromatic rings (**6.19** Schematic).[76]

The catalysts are stable to amines (**6.20** Schematic).[77]

They can also act as catalysts for some reactions that are normally catalyzed by base (**6.21** Schematic).[78] They are suitable for aldol, Diels–Alder, pinacol, and ring-opening reactions. Yttrium triflate has been used as a catalyst for the polymerization of tetrahydrofuran[79] and dysprosium triflate for the polymerization of vinyl isobutyl ether.[80]

Hafnium triflate has been used for the acylation and alkylation of aromatic compounds.[81] It has also been used in aromatic nitration in a process (**6.22** Schematic) that eliminates the usual waste acid from such reactions.[82] The

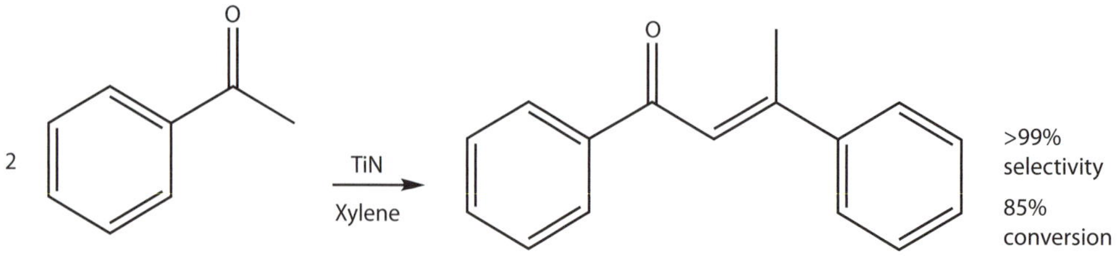

2

O

TiN

Xylene

O

>99% selectivity

85% conversion

6.14 Schematic

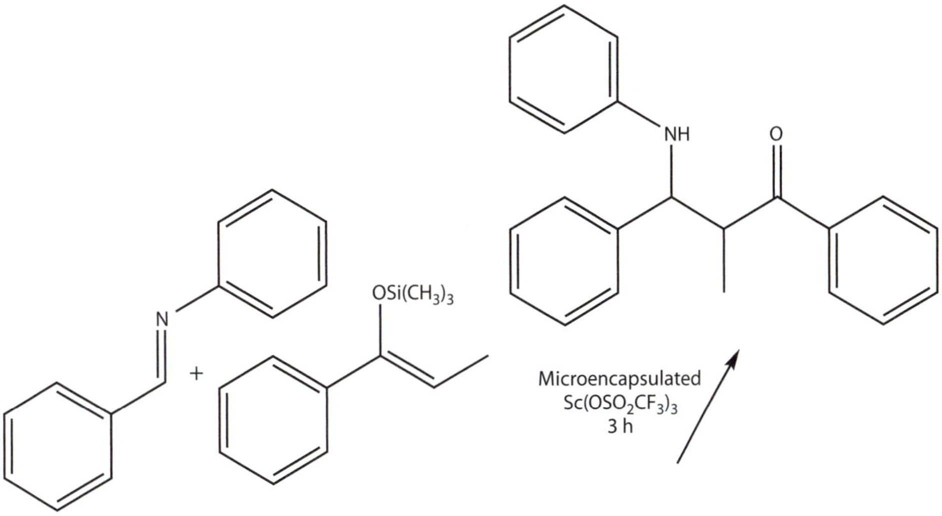

6.15 Schematic

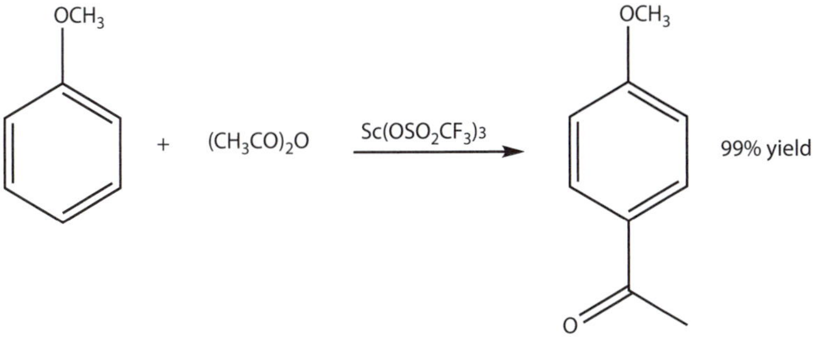

6.16 Schematic

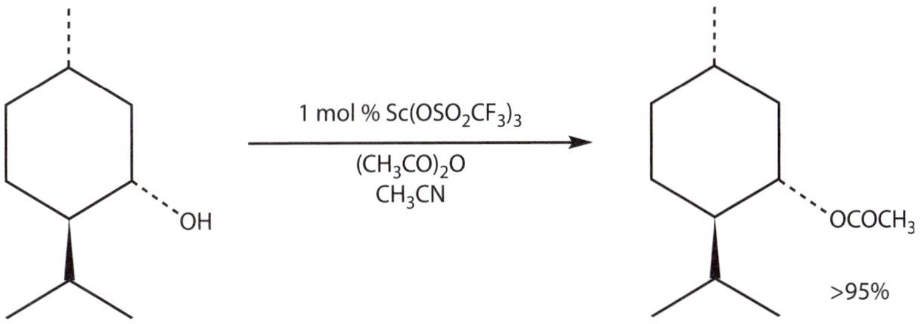

6.17 Schematic

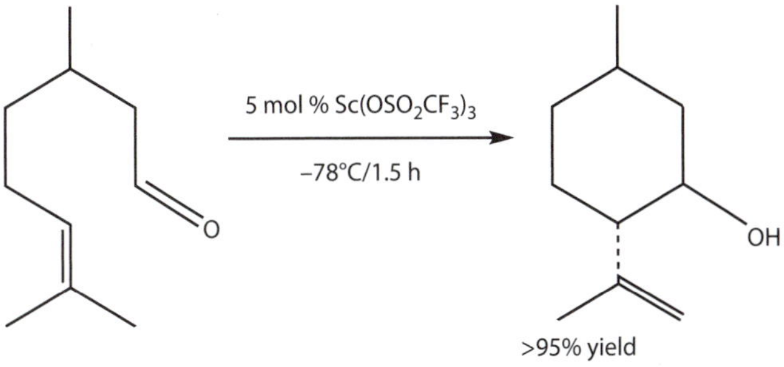

6.18 Schematic

products are intermediates in the synthesis of toluene diisocyanates used in making polyurethanes. The catalyst could be reused with little or no loss in activity or yield. (For more on isocyanates, see Chapter 2.)

Scandium dodecyl sulfate has been used as a catalyst in an aqueous dispersion (**6.23** Schematic).[83] The reaction was 5000 times faster in water than in methylene chloride. It was also slower in dimethyl sulfoxide, acetonitrile, ethyl ether, tetrahydrofuran, and methanol.

Scandium was used to make a polyester in which only the primary hydroxyl groups had reacted (**6.24** Schematic).[84]

Indium(III) chloride is another Lewis acid that is stable in water. (For the use of indium (III) chloride without organic solvents, see Chapter 8.) It has been used as a catalyst in Diels–Alder and other reactions, such as **6.25** Schematic.[85] After the product is extracted with an organic solvent, the aqueous layer can be used to catalyze another reaction. It should also be possible to just separate the organic layer without the use of a solvent.

Tris(pentafluorophenyl)boron is an air-stable, water-tolerant Lewis acid for aldol, Diels–Alder, and Michael reactions.[86] It is effective at 2 mol% in some reactions. Its recycle has not been demonstrated.

6.7 SOLID BASES

Potassium fluoride on alumina is a solid base.[87] (It forms potassium hydroxide by exchange of fluoride for hydroxyl groups on the surface of the alumina.) An example of its use

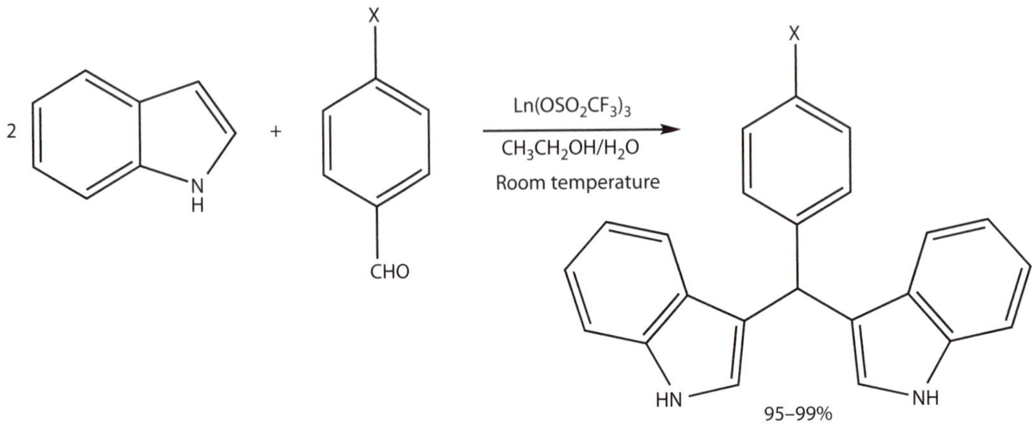

2 N H + X CHO $Ln(OSO_2CF_3)_3$ CH_3CH_2OH/H_2O Room temperature X HN NH 95–99%

6.19 Schematic

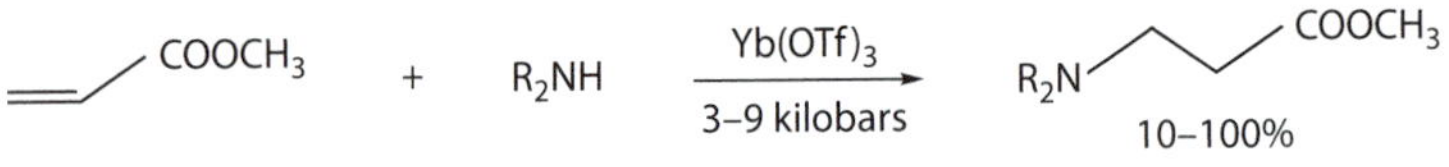

$COOCH_3$ + R_2NH $Yb(OTf)_3$ 3–9 kilobars R_2N $COOCH_3$ 10–100%

6.20 Schematic

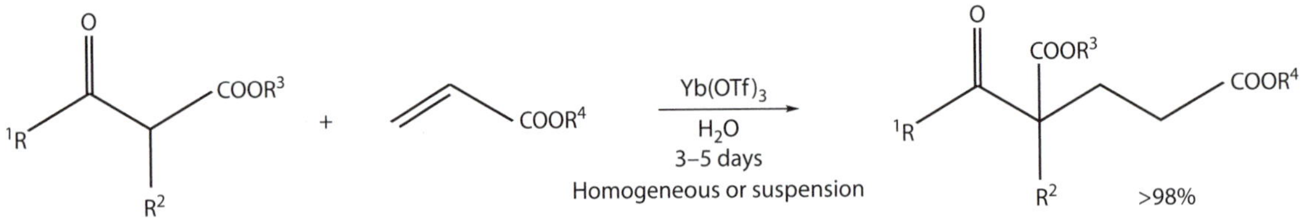

O ^1R $COOR^3$ R^2 + $COOR^4$ $Yb(OTf)_3$ H_2O 3–5 days Homogeneous or suspension O $COOR^3$ ^1R $COOR^4$ R^2 >98%

6.21 Schematic

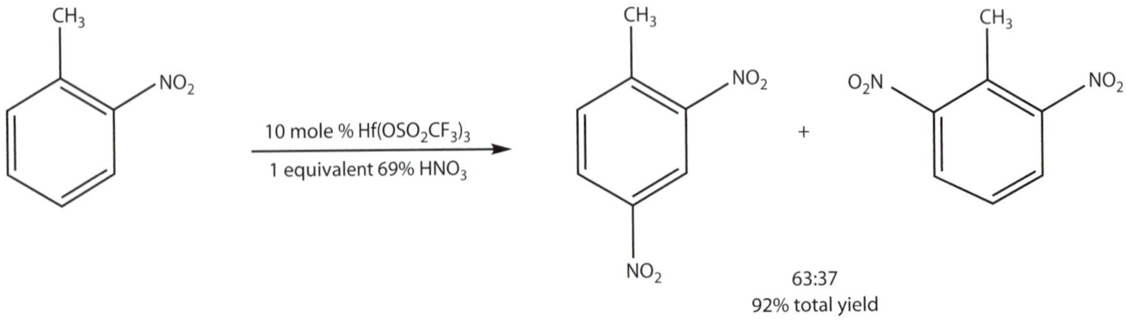

CH_3 NO_2 10 mole % $Hf(OSO_2CF_3)_3$ 1 equivalent 69% HNO_3 CH_3 NO_2 NO_2 + O_2N CH_3 NO_2

63:37
92% total yield

6.22 Schematic

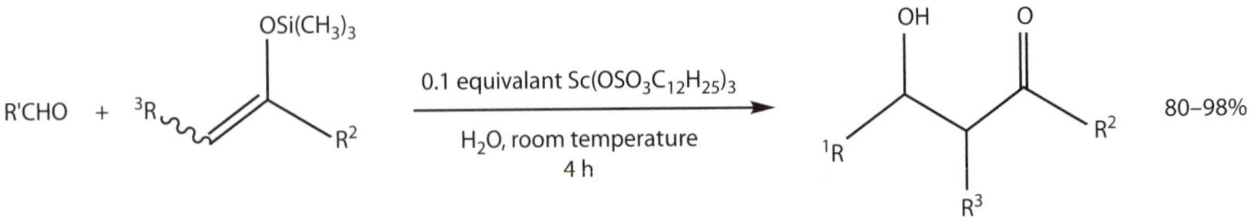

R'CHO + ^3R $OSi(CH_3)_3$ R^2 0.1 equivalant $Sc(OSO_3C_{12}H_{25})_3$ H_2O, room temperature 4 h OH O ^1R R^2 R^3 80–98%

6.23 Schematic

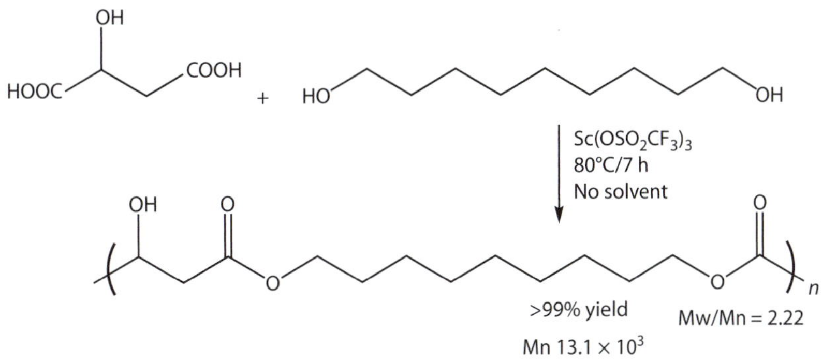

6.24 Schematic

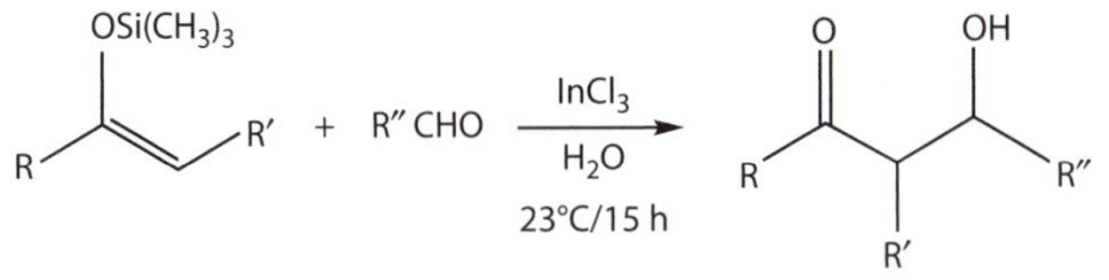

6.25 Schematic

is given by reaction **6.26** Schematic.[88] It has also been used without solvent in the Tishchenko reaction of benzaldehyde to give benzyl benzoate in 94% yield.[89] Potassium fluoride on an aluminum phosphate molecular sieve was a weaker base in the isomerization of 1-butene.[90]

Alkali and alkaline earth carbonates catalyzed the Knoevenagel reaction (**6.27** Schematic) of benzaldehyde with malononitrile.[91]

Layered double hydroxide carbonates of magnesium and aluminum are called hydrotalcites (e.g., $Mg_6Al_2(OH)_{16}CO_3$–H_2O). They can be used as solid bases before or after calcinations to produce mixed magnesium–aluminum oxides.[92] They have also been made inside mesoporous silica for additional size and shape selectivity in reactions.[93] They have been used to catalyze the addition of alcohols to acrylonitrile[94]; the reduction of ketones with isopropyl alcohol, with 92–95% selectivity (the Meerwein–Ponndorff–Verley reduction)[95]; the reaction of glycerol with glycerol trioleate to produce a monoglyceride[96]; and the aldol and Knoevenagel reactions, in 88–98% yield.[97] When the aldol reaction was run at 0°C, the hydroxyketone was obtained in 88–97% yield, with no dehydration to the unsaturated ketone. The Meerwein–Ponndorff–Verley reduction has also been run in the gas phase with aluminum-free titanium β-zeolite with 95% selectivity.[98] A calcined hydrotalcite has been used to alkylate aniline exclusively to the monoalkyl derivative (**6.28** Schematic).[99] Such selectivity is difficult to obtain by other methods.

They have also been used for the addition of trimethylsilyl cyanide to aldehydes and ketones in up to 99% yield[100] and for the condensation of aldehydes with active methyl groups.[101]

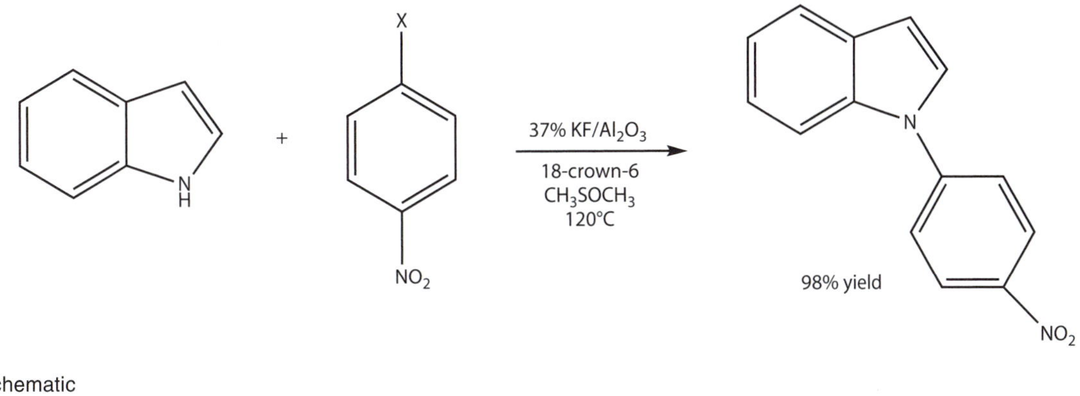

6.26 Schematic

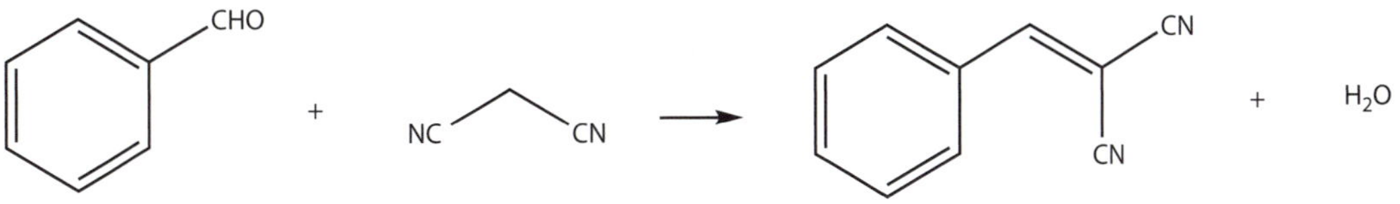

6.27 Schematic

NH_2 + CH_3OH —Calcined hydrotalcite, 425°C→ $NHCH_3$ (68% conversion)

6.28 Schematic

Hydrogenation of acetonitrile in a Co/Ni/Mg/Al layered hydroxide prevented the formation of di- and triamines.[102]

Magnesium oxide, calcium oxide, and hydroxyapatite [$Ca_{10}(PO_4)_6(OH)_2$] have been used to catalyze the opening of epoxide rings by trimethylsilyl cyanide, with 92–99% regioselectivity.[103] This is higher than that found in homogeneous systems. Trimethylsilyl cyanide is the synthetic equivalent of toxic hydrogen cyanide. Intercalation of cobalt phthalocyaninetetrasulfonate into a layered magnesium–aluminum double hydroxide increased its activity in the air oxidation of 2,6-dimethylphenol by a factor of 125.[104] Such layered double hydroxides can also serve as hosts for guests, such as carboxylates, sulfates, phosphates, porphyrins, and polyoxometalates that can be put in by anion exchange.[105] The heat and light stability of an anionic dye were improved by intercalation in a Zn/Al-layered double hydroxide.[106]

The *ortho*-alkylation of phenol (**6.29** Schematic) was carried out with a catalyst that was a solid solution of cerium(IV) and magnesium oxides.[107] At 32% conversion, the selectivity was 90% for *o*-cresol and 8.6% for 2,6-xylenol. [The latter is a monomer for a poly(phenylene oxide), an engineering plastic.] There was no decay of activity.

A silica-supported phenoxide gave 99% yield in the Michael reaction (**6.30** Schematic).[108]

Superbases for isomerization of alkene such as β-pinene were made by treating alumina or magnesium oxide successively with an alkali metal hydroxide and an alkali metal under nitrogen.[109] Potassium oxide in mesoporous alumina is also a superbase.[110] Large transparent sheets of Zn/Al-layered double hydroxide have been made and used for ion exchange.[111]

Quaternary ammonium hydroxides and amines made in zeolite MCM-41 have been used in Knoevenagel, Michael, and aldol reactions.[112]

6.8 ZEOLITES AND RELATED MATERIALS

Zeolites[113] are well known to chemical engineers for industrial separations and as catalysts; the average chemist in an organic chemical laboratory uses them only as molecular sieves for drying solvents. They offer the promise of more environmentally friendly catalysts that will improve selectivity and reduce waste.[114] Zeolites[115] are inorganic aluminosilicate polymers. They contain AlO_4 and SiO_4 tetrahedra, linked by shared oxygen atoms. No two aluminum units are adjacent. For each aluminum, there must be an additional cation, usually from an alkali or alkaline earth metal, for electrical neutrality. There are about 50 natural zeolites, with names such as chabazite, clinoptilolite, faujasite, and mordenite. (Mineral names usually end in *ite.*) In addition, many are synthetic.

The structures contain channels in one, two, or three directions. When one channel intersects another, there may be a larger cage. The channels contain the alkali or alkaline earth metal cations (i.e., the exchangeable cations and water). Diagram **6.31** Schematic[116] shows the channels of ZSM-5. Structure **6.32** Schematic[117] is in zeolites X and Y. Each corner represents a silicon or aluminum atom. There is an oxygen atom on each line between them. I and II indicate where cations might be.

The silicon-to-aluminum ratio also varies with typical values such as

- Zeolite A 1.0
- Zeolite X 1.1
- Zeolite Y 2.4
- Silicalite infinite (has no exchangeable cations)

Other zeolites may have ratios as high as 10–100.

After drying, some zeolites are about 50% voids. The pore diameters in angstroms (Å) and percentage pore volumes for several zeolites with sodium cations are shown in Table 6.2.

OH + CH_3OH —CeO_2/MgO, 450–550°C→ OH, CH_3 + H_3C, OH, CH_3 (98% total yield)

6.29 Schematic

CH_3CH_2OH

20°C/0.25 h

6.30 Schematic

The pore diameter varies with the cation in it. Data for zeolites A and X are shown in Table 6.3. Schematic 6.33 shows the dimensions of the pores with the sizes of some hydrocarbons.

The names of zeolites are confusing and relatively uninformative. Workers in the field keep an atlas of zeolites[118] nearby, but the ordinary chemical laboratory probably will not have one. These materials have been named in a variety of ways, but the most common one involves the place where the material was first made. ZSM-5 means zeolite SOCONY–Mobil No. 5; SOCONY reflects the origin of the company as Standard Oil of New York. Newer materials from the same place are designated as MCM, with a number following, which means Mobil Composition of Matter. VPI-5 was made at Virginia Polytechnic Institute and State University. The man who made it is now at the California Institute of

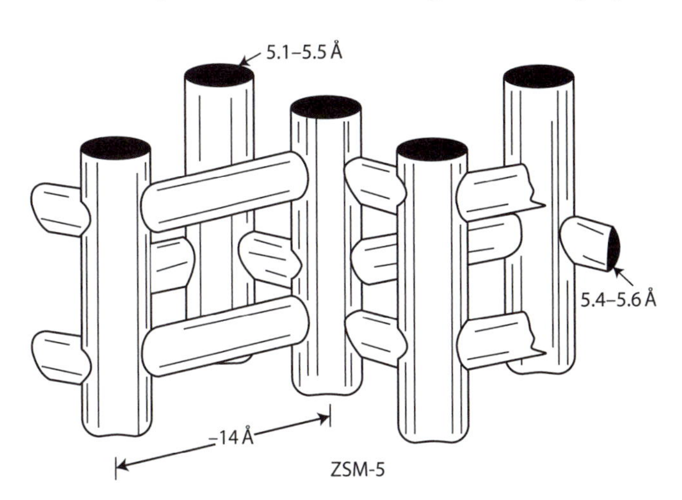

6.31 Schematic (Reprinted from V. Ramamurthy, D.F. Eaton, and J.V. Caspar, *Acc. Chem. Res.*, 1992, *25*, 299. Copyright 1992 American Chemical Society. With permission.)

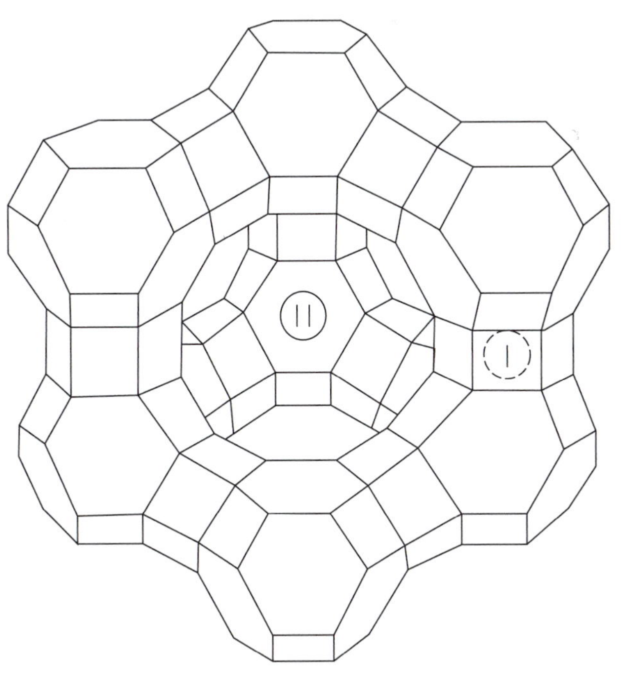

6.32 Schematic (Reprinted from V. Ramamurthy, D.F. Eaton, and J.V. Caspar, *Acc. Chem. Res.*, 1992, *25*, 299. Copyright 1992 American Chemical Society. With permission.)

Table 6.2 Comparison of Pore Diameters and Pore Volumes

Characteristic	4A	X	Y	Synthetic Mordenite	ZSM-5
Pore diameter	4.1	7.4	7.4	6.7–7	6
Pore volume (%)	47	50	48	28	—

Source: J. Huang, V. Subramanian, R. Agrawal, and R. Berry, *IEEE Journal on Selected Areas in Communications*, 2009, *27*(2), 226–234.

Table 6.3 Variation in Pore Diameters as a Function of Its Cation

Cation	Zeolite	Diameter (Å)
Zeolite A		
Calcium exchanged	5A	4.2–4.4
Sodium exchanged	4A	3.6–4.0
Potassium exchanged	3A	3.3
Zeolite X		
Sodium exchanged	13X	7.4
Calcium exchanged		7.8

Technology, where he makes CIT zeolites. TS-1 (titanium silicalite) and titanium-β are titanium-containing molecular sieves. USY is an ultrastable version of zeolite Y. For those who do not work in the field all of the time, it would be desirable to have a more informative system. It would be an improvement if each paper mentioning a zeolite would define it the first time the name is given. This could include the relative amounts of the elements present, whether it contains one-, two-, or three-dimensional channels, the pore size in angstroms, and the number of oxygen atoms in the major channel window. A simple one might be (10:1:0.25 Si/Al/Na, 3-D, 4.0 Å, 10 O atoms). If the information were available, the acidity or basicity could also be mentioned as well as whether the acid sites were Bronsted or Lewis acid.

Zeolites can be made by the formation of a gel from sodium aluminate, sodium silicate, and sodium hydroxide in the presence of a template, usually a quaternary ammonium hydroxide, followed by digestion in which the material crystallizes (**6.34** Schematic).[119] (This is sol–gel technology.[120]) Nonhydrolytic sol–gel technology uses a metal chloride plus a metal alkoxide, as in tetrachlorosilane plus tetraisopropylsilane plus tetraisopropyl titanate, to prepare a catalyst for oxidation.[121] Zeolites A and X have been made from fly ash found in coal-fired power plants.[122]

A wide variety of quaternary ammonium salts have been used in an effort to prepare new zeolitic structures.[123] Molecular modeling can sometimes help in the design of new structures.[124] Combinatorial synthesis was performed using a 100-well Teflon block.[125] Tetrapropylammonium hydroxide was used to make titanium silicalite (TS-1).[126] Tetrabutylammonium hydroxide was used to prepare ZSM-11.

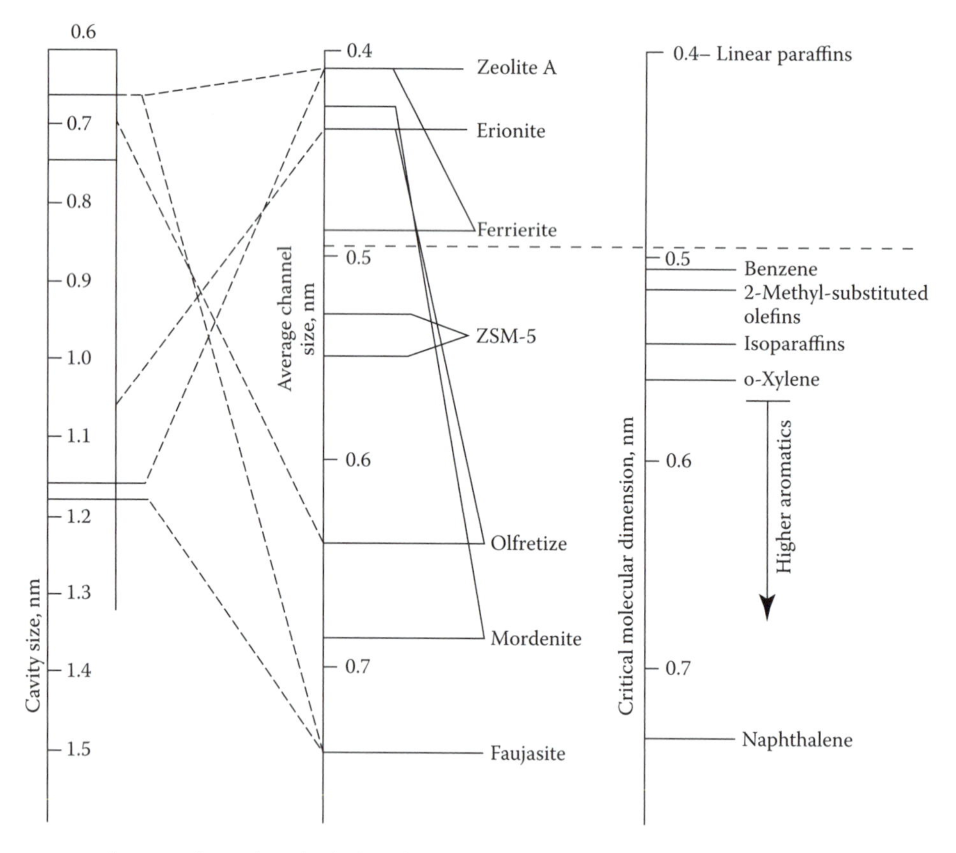

6.33 Schematic compares the pore dimensions (nm) of zeolites with the dimensions of hydrocarbons. (Reprinted from E.G. Derouane in M.S. Whittington and A.J. Jacobsen, eds, *Intercalation Chemistry*, Academic Press, 1982, 101. With permission.)

$$NaAlO_2 + Na_2SiO_3 + NaOH + R_4N^+OH^- \longrightarrow Gel \xrightarrow{100°C/6\,h} Zeolite$$

6.34 Schematic

Hexadecyltrimethylammonium chloride was used with tetramethylammonium silicate with sodium aluminate and sodium hydroxide to prepare mesoporous (pores larger than about 15 Å) zeolites.[127] Digestion was accomplished with microwave radiation in 1 min at 160°C to give MCM-41. (Mobil's MCM-41 series has uniform pores that can be varied from 16 to 100 Å.[128] One way to vary the pore size is to use mixtures of templates.[129]) The use of seed crystals can sometimes shorten times in such microwave digestions.[130] Application of a magnetic field during the preparation of MCM-41 produced aligned pores.[131] The syntheses are sometimes modified by inclusion of alkyltrialkoxysilanes, which may contain a variety of functional groups, such as methacryloyl, vinyl, mercaptan, amine, or others, to alter the hydrophobicity and reactivity of the final product.[132] Zeolites have also been made from alkylene- and phenylenebistrialkoxysilanes.[133] Although most templates are burned out, some can be recovered for reuse by extraction with supercritical carbon dioxide.[134] A tetraethylammonium fluoride template was removed with aqueous acetic acid.[135] Removal of the template with ozone at 250°C gave a more uniform pore-size distribution.[136] A tin silicate analog was made from tin(IV) chloride, hexadecyltrimethylammonium chloride, tetramethylammonium hydroxide, and tetramethylammonium silicate.[137]

Although the reactions are usually run in water, ethylene glycol, propyl alcohol, sulfolane, pyridine, and so on have also been used.[138] Occasionally when metal cations are provided for charge compensation, it is possible to prepare zeolites under nonbasic conditions. CoZSM-5 has been prepared from a 0.25 HF/0.75 NaF/0.25 NaCl/1.25 $(C_3H_7)_4N^+$ Br^- plus $[(C_2H_5)_4N^+]_2CoCl_2^{2-}$ mixture at 170°C.[139] CuMCM-41 can be made starting with a mixture of $C_{16}H_{33}N(CH_3)_3^+$ Br^-, ethylamine, and $AlCl_3 \cdot 6H_2O$ in water.[140]

Nonionic templates, such as amines, have also been used.[141] Mesoporous silica molecular sieves have been made from *n*-dodecylamine, tetraethoxysilane, ethanol, and water.[142] After the product dried in air, the template was recovered by extraction with hot ethanol. Usually, the template, which may occasionally be expensive, is burned out. Large-cage zeolites have been made with diamines as templates.[143] The use of $C_{12}H_{25}NHCH_2CH_2NH_2$ gave a mesoporous sieve stable up to 1000°C, and to boiling water for more than 150 h, which is much greater than that observed with most mesoporous sieves.[144] The mechanical and hydrothermal stability of MCM-41 has been improved by postsynthesis treatment with aluminum chloride hydrate.[145] There was no change in the x-ray pattern after 16 h in boiling water. Previous AlMCM-41 became amorphous after boiling in water for 6 h.

Nonionic surfactants such $C_{11}H_{23}$–$C_{15}H_{31}O(CH_2CH_2O)_nH$ can be used in place of the amine.[146] These can also be used with alumina, titania, and zirconia. These surfactants are less expensive and less toxic than the amines, and they are also biodegradable. Block copolymers of ethylene oxide have also been used.[147] A large-pore silica zeolite has been prepared using bis(pentamethylcyclopentadienyl) cobalt(III) hydroxide as a template.[148] An EMT zeolite has been made using 18-crown-6 as the template.[149] Much larger voids (320–360 nm) have been made in alumina, titanium dioxide, and zirconium oxide by use of the sol–gel method with polystyrene latex spheres (which are removed at the end of the synthesis).[150]

After the zeolite is made, it may be modified.[151] Burning out the quaternary ammonium salt template leaves hydroxyl groups that are acidic. These can be neutralized with a variety of ions. If the ion is cesium, the resulting zeolite is basic.[152] For zeolite X, the basicity is CsX > RbX > KX > NaX. Various ions can be exchanged for sodium and potassium and other ions in zeolites. Adding in some rare earth ions can make the zeolite more stable.[153] Aqueous exchange with copper(II) ions adds copper on two types of sites. However, the copper can be added on only one type of site selectively by using copper(II) hexafluoroacetylacetonate.[154] The hydroxyl group reacted with triisopropoxyaluminum to form a catalyst that was more active in the reduction of ketones with isopropyl alcohol than was triisopropoxyaluminum itself.[155] Boron trifluoride etherate was added to modify the acidity of the catalyst for etherification.[156]

Molybdenum sulfide clusters can be added with $Mo_3S_4^{4+}$ ion or by molybdenum carbonyl, followed by hydrogen sulfide.[157] Palladium ion can be added, and then reduced to palladium metal by isopropyl alcohol with ultrasound or by hydrogen.[158] Its activity in the Heck reaction was comparable with that of some of the best homogeneous catalysts. Ruthenium clusters were formed inside zeolite X by addition of the ruthenium ion and then reducing with hydrogen.[159] Sodium clusters were placed into zeolite Y by dissolving sodium in the zeolite.[160] These are inorganic electrides. Some organic cations formed inside zeolites can be stable for weeks.[161] A ruthenium dinitrogen complex made inside a zeolite was stable to 523 K.[162]

It is possible to form a complex in the supercage of a zeolite by introducing the metal ion first and then the ligand. If the complex is too large to enter the pore, it cannot escape, but can be used for catalysis in place.[163] This was done in zeolite Y with iron and manganese bipyridyl complexes.[164] Such encapsulation can offer steric protection against oxidation of the catalyst, giving longer life to oxidation catalysts.[165] It can also prevent the dimerization of ruthenium and rhodium ions that occurs when they are oxidized in solution.[166] Copper phthalocyanine has been made in zeolite Y by this method.[167]

Removal of some aluminum is often desirable to stabilize a zeolite. This can be done hydrothermally with steam, or with tetrachlorosilane or ammonium hexafluorosilicate.[168] Gaseous metal halides can be used to introduce aluminum, boron, gallium, and indium into the H or ammonium form of a zeolite. To improve selectivity, the pore openings can be made smaller by the chemical vapor deposition of silica from tetraethoxysilane.[169] Pore size can also be adjusted with tetrachlorosilane, titanium(IV) chloride, and antimony pentachloride.[170] Both pore size and activity in HY zeolites have been modified by sequential treatment with silane and NO.[171] To utilize the size- and shape-selectivity of zeolite catalysts in reactions, it may be necessary to deactivate exterior acid sites. This can be done with bulky amines, such as 4-methylquinoline, coking the catalyst with mesitylene, introducing silicon polymers, and so on. In one case, it was performed by grinding the zeolite with barium acetate supported on silica in an agate mortar.[172] Tetrabutyl and tetraphenylgermanium have also been used for this purpose.[173]

There are other microporous solids that are not aluminosilicates; hence, they are not zeolites.[174] The titanosilicates, such as titanium silicalite (TS-1), which are selective catalysts for oxidations, are discussed in Chapter 4. Much work is going into extending these findings to materials with larger pores to accommodate larger substrates for oxidation. Molecular sieves also include the aluminum phosphate (AlPO) family.[175] A wide variety of metal ions have been included in these. The MgAlPO-5 preparation involved a 20-min digestion with microwaves or a conventional digestion for 24 h.[176] Methylphosphonic acid can also be used in place of phosphoric acid.[177] Cage compounds that may serve as models for the aluminum phosphates have been made by treatment of trimethylaluminum with *tert*-butylphosphonic acid.[178] A molecular sieve containing zinc has been prepared from 2-aminoethylphosphonic acid.[179] Alkylenebisphosphonates have been used with copper, zinc, and zirconium ions to prepare layered structures.[180] Gallophosphate[181] and cobalt gallophosphate[182] have also been prepared. Mesoporous molecular sieves of alumina have been prepared starting with aluminum *sec*-butoxide, using nonionic surfactants as templates.[183]

Sacrificial template methods have been used to make a variety of porous materials.[184] Preparation of microporous silica in the presence of polystyrene spheres gave a hierarchical solid with both micropores and macropores after removal of the polystyrene by dissolving out or burning out.[185] Such materials have also been made with carbon black, carbon nanotubes, or nanofibers which are burned out.[186] Hierarchical materials are faster catalysts due to better transport and slower deactivation by pore blocking. Mesoporous carbon materials have been made by using MCM 48 as a template in a resorcinol/formaldehyde resin, the zeolite being removed by hydrogen fluoride after the product was carbonized.[187] Porous ceramics have also been made using organic templates which are removed by burning out.[188] Some organic polymers have intrinsic microporosity, for example, the polymer in **6.35** Schematic.[189]

Measuring the acidity of the Bronsted and Lewis acid sites is problematic. The adsorption and desorption of various amines have been used, but there is some disagreement about what it means.[190] Some workers prefer to use isopropylamine, which desorbs only from Bronsted sites. Solid-state nuclear magnetic resonance (NMR) has also been used in the study of reactions on zeolites.[191] Many zeolites crystallize into crystals that are too fine for conventional x-ray analysis. A method that uses synchrotron x-rays on microcrystalline powders promises to make it much easier to determine the structures of zeolites and related materials.[192] A solid-state NMR method using diphosphines has been used to locate acid sites.[193] Solid-state phosphorus NMR can distinguish Bronsted and Lewis acid sites in USY.[194] Various other methods have been used to study the structures and reactions of zeolites.[195]

There are many uses of zeolites and molecular sieves, based on their size and shape selectivity. As catalysts, they offer less or no corrosion, little waste, ready adaptation to continuous processes, high thermostability, and other attributes.[196] Their limitations include deactivation owing to occasional plugging of pores with secondary products, and the difficulty of using them with bulky molecules. The first limitation can sometimes be overcome with a continuous loop through the reactor and then through a regeneration cycle. The propylene oxide oligomers that can build up in the pores of titanium silicalite during the oxidation of propylene with hydrogen peroxide can be removed by heating with refluxing dilute hydrogen peroxide to reactivate the catalyst.[197] The limitation on the use of bulky molecules accounts for the tremendous amount of work on mesoporous solids being done today.[198] In cases where delaminated zeolites can be used, this limitation has been overcome.[199]

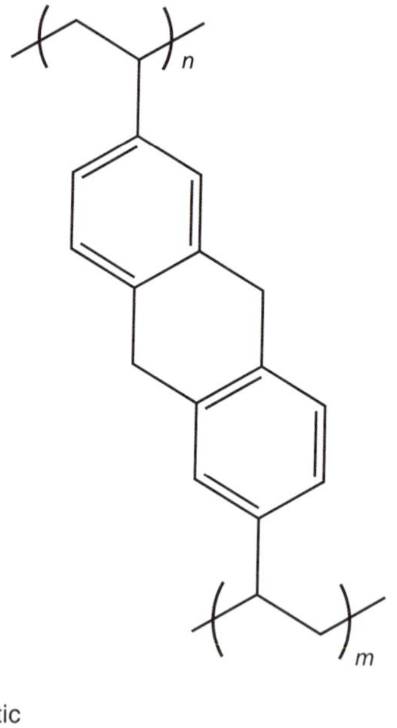

n

m

6.35 Schematic

The largest commercial use of zeolites is probably in detergents, where they pick up the calcium and magnesium ions in hard water. They replace sodium tripolyphosphate, which causes eutrophication of water bodies. (Phosphorus is the limiting nutrient in many lakes and rivers.) Zeolites are also used to separate linear from branched paraffins (e.g., by calcium zeolite A), *p*-xylene from mixed xylenes (e.g., by barium zeolite X), and 95% oxygen from air (for use in steel making and in wastewater treatment).[200] This method of enriching oxygen uses less energy than the conventional cryogenic distillation. Carbon dioxide, hydrogen sulfide, and mercaptans can be removed from gas streams with zeolites. The catalytic cracking of petroleum is done with zeolite X or Y, stabilized with rare earth cations. Natural gas has been converted first to methanol and then to gasoline with ZSM-5 in New Zealand. Clinoptilolite is used to remove ammonium ion (another nutrient) from wastewater. A few of the better examples of the use of zeolites as catalysts for organic reactions will be given in the following.[201]

Chiba and Arco (now BP) use a zeolite in their alkylation of benzene with ethylene to produce ethylbenzene for conversion to styrene.[202] The older process used aluminum chloride. The use of reactive distillation in this reaction can give 99.7% specificity at 100% conversion.[203] The ethylene is fed in at the bottom of the catalytic zone in the refluxing benzene. The ethylbenzene goes to the distillation pot. Any polyalkylated material is transalkylated later with benzene with the same molecular sieve catalyst. Cumene can be produced by the alkylation of benzene with propylene (**6.36** Schematic) in a higher than 99% yield with β-zeolite.[204] The catalyst gives a higher yield than the phosphoric acid on a support that it replaces. By using the catalytic distillation method mentioned for ethylbenzene, the yield of 99.95% pure cumene is 99.6%, which is 5–6% higher than in a conventional plant.[205]

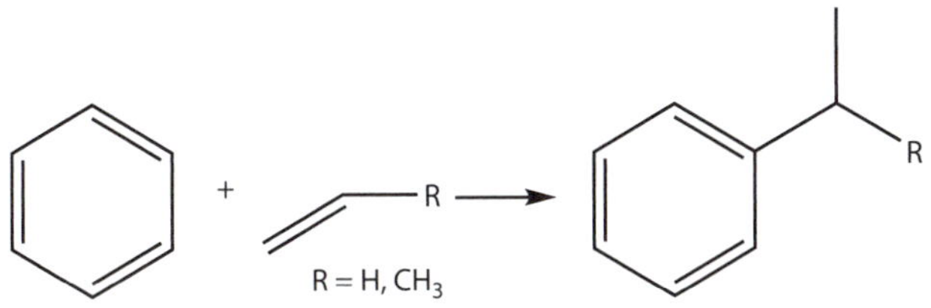

The principle of removing a product as soon as it is formed, and before any secondary reactions can occur, is one that should be applied more widely to increase yields. It may be possible to apply it to the reaction of isobutane with butenes.

Benzene is also alkylated with 1-olefins (e.g., where R is $C_{10}H_{21}$) to prepare "detergent alkylate" for sulfonation to make detergents. A UOP–Petresa process, using a fluorided silica–alumina catalyst, gave 92% linear alkylbenzene and 6% branched alkylbenzene at 97% conversion with a catalyst life of 182 h.[206] The catalyst could be regenerated by washing first with a paraffin, then with an alcohol, and then drying. The cost of the process is 30% lower than one using hydrogen fluoride. Liang et al. used zeolite HY, which tended to gum up and become deactivated.[207] They overcame this problem by using a circulating fluidized bed in which the catalyst passed through the reactor and then through the regenerator. The system worked, and it was easy to operate. Steam-treated mordenite gave 67% of the 2-phenyl product and 33% of the 3-phenyl product with 98% selectivity and 96% conversion.[208]

Zeolites are useful in favoring *p*-disubstituted benzene products, which have smaller diameters than the other isomers. The raw material for the terephthalic acid used in making poly(ethylene terephthalate) is *p*-xylene. It has been produced from *n*-pentane (**6.37** Schematic) using an MFI zeolite catalyst.[209] The significant finding was that the xylene fraction was 99% *para*. The other fractions are not lost. Toluene can be disproportionated to *p*-xylene and benzene with H-ZSM-5 treated with a little hexamethyldisiloxane to

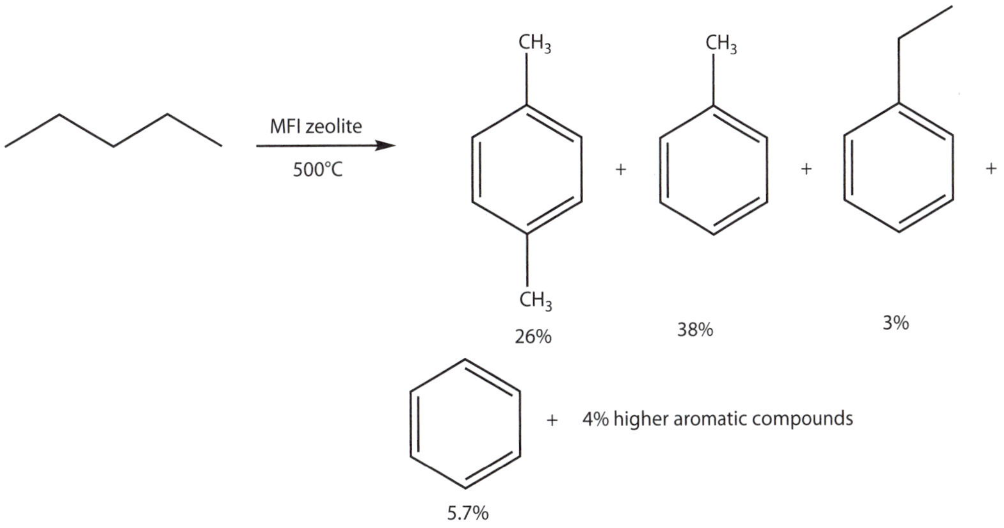

R = H, CH_3

6.36 Schematic

MFI zeolite, 500°C → 26% + 38% + 3% + 5.7% + 4% higher aromatic compounds

6.37 Schematic

give 99% *p*-xylene, so that the usual separation of the *ortho*- and *meta*-isomers with another zeolite would not be required.[210] Benzene can be transalkylated with the higher aromatics to give toluene. Ethylbenzene can be isomerized to *p*-xylene. Ethylbenzene can be alkylated with ethanol in the presence of a modified ZSM-5 catalyst to produce *p*-diethylbenzene with 97% selectivity.[211] A cobalt/manganese acetate catalyst was used with air to convert *p*-xylene to terephthalic acid in 99.4% yield at 100% conversion, but about 0.5% manganese leached out.[212]

Halogenation[213] and benzoylation of substituted benzenes also favor the *para*-products. Nitration to toluene with *n*-propyl nitrate in the presence of H-ZSM-5 gives 95% *para*-selectivity.[214] Benzene can be nitrated with 65% nitric acid in the vapor phase at 170°C over a mordenite catalyst.[215] (Most zeolites are not stable to protonic mineral acids. Mordenite is an exception.) The nitration of aromatic compounds can also be carried out with a sulfuric acid/silica catalyst that is slurried in methylene chloride to give 97–98% yield.[216] The catalyst could be recovered and reused with less than 3% loss in activity after three runs. These methods eliminate the problem of what to do with spent by-product sulfuric acid from conventional nitrations.

The acetylation of alcohols and phenols with acetic anhydride and HSZ-360 zeolite at 60°C used no solvent. The acetate of 1-dodecanol was obtained in 98% yield, and the acetate from 1-naphthol in 100% yield.[217] Acetamides were made by treatment of the amines with acetic acid in the presence of HY zeolite in 98–99% yield.[218] The catalyst could be reused three times with no loss in activity or yield. The acylation of anisole with acetic anhydride at 100°C without solvent using zeolite H-β gave 4-methoxyacetophenone (**6.38** Schematic) in 98% yield.[219] The catalyst could be recovered, regenerated, and reused with no decrease in yield. These reactions show that not all zeolites are used at high temperatures in the vapor phase.

Anisole has also been acylated with aliphatic or aromatic carboxylic acids, with 92–95.6% selectivity at 70–87% conversion, using zeolite Y9 at 155°C.[220] Acylation of benzene with acetic acid using H-ZSM-5 gave acetophenone with 91% selectivity at 43% conversion.[221] In one case, the *ortho*-isomer (**6.39** Schematic) is favored, presumably because the intramolecular cyclic transition state fits better in the ZSM-5 zeolite than whatever transition state leads to the *para*-isomer.[222] This is a Fries rearrangement. The *ortho*/*para* ratio was 98.7:1.3.

Naphthalene (**6.40** Schematic)[223] and biphenyl (**6.41** Schematic)[224] also alkylate selectively.

Industrially important olefins can be made with the help of molecular sieves. Methanol can be converted into 50% ethylene and 30% propylene with a SAPO catalyst at 350–500°C.[225] Isobutene can be made by isomerization of *n*-butenes over clinoptilolite at 450°C with 91.6% selectivity at 23.5% conversion,[226] or with H-ferrierite with 92%

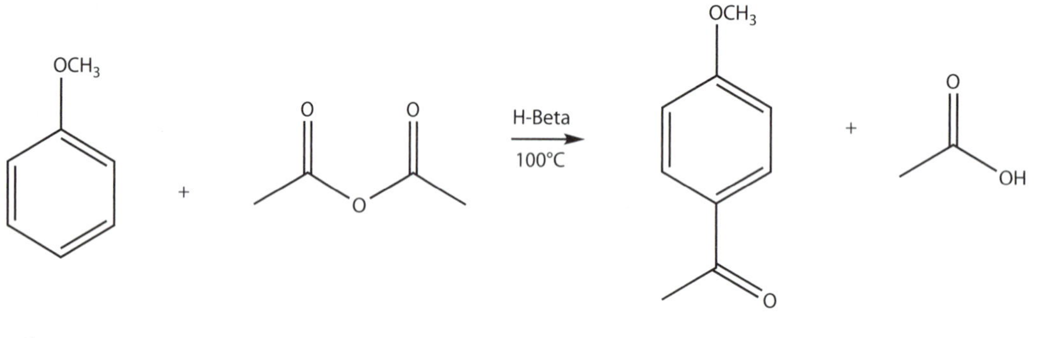

OCH_3 O O H-Beta 100°C OCH_3 O O O OH

6.38 Schematic

OH O O ZSM-5 250°C/30 min O O OH O 30% 70% + CH_3COOH

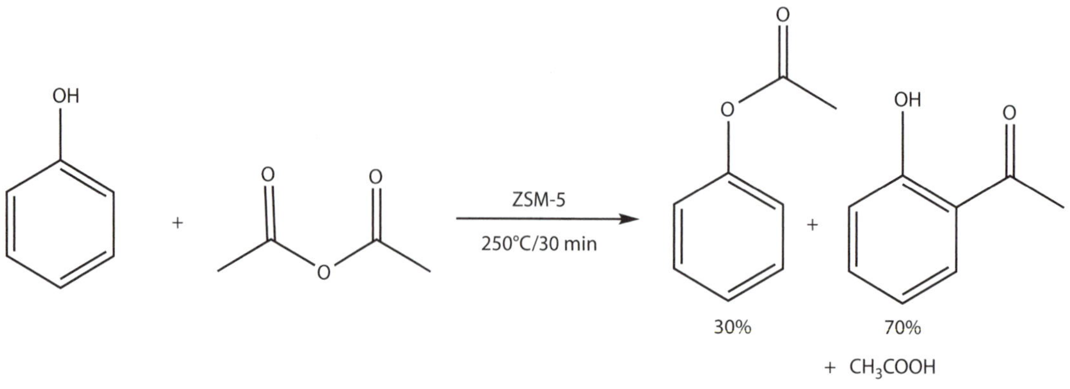

6.39 Schematic

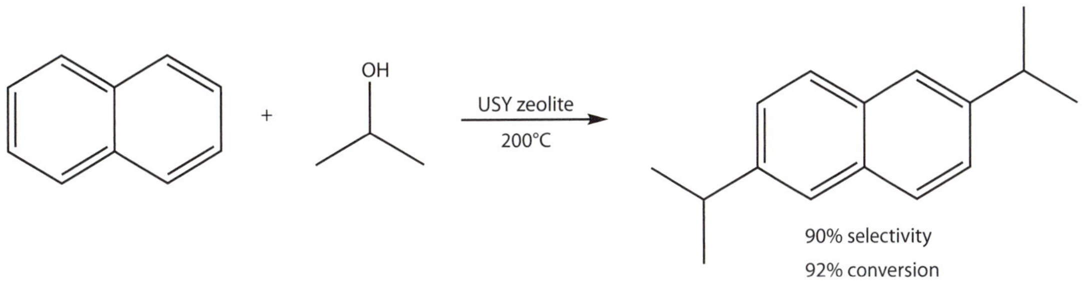

OH
+
USY zeolite
200°C
90% selectivity
92% conversion

6.40 Schematic

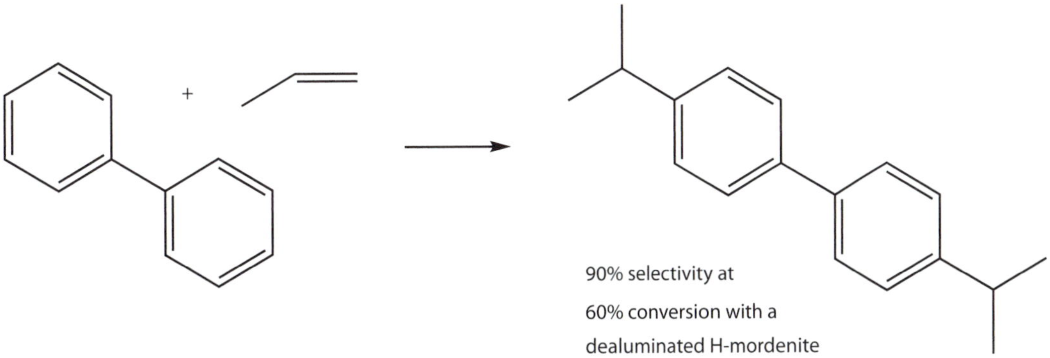

+
90% selectivity at
60% conversion with a
dealuminated H-mordenite

6.41 Schematic

selectivity.[227] Zeolites with 8-membered rings are too small, and 12-membered rings do not suppress char, but 10-membered ring zeolites let the isobutene out while suppressing dimer and oligomer formation.[228]

Acetonitrile can be produced from ethanol, ammonia, and oxygen in 99% yield using a SAPO catalyst at 350°C.[229] It can be reduced to ethylamine with 98% selectivity using a 1.1 Co/1.1 Ni/0.9 mg/1.0 Al-layered double hydroxide and hydrogen at 393 K.[230] Methylamine and dimethylamine are more valuable than trimethylamine. When methanol and ammonia are reacted in a zeolite, such as clinoptilolite, mordenite, or chabazite, the products are largely the desired monomethyl and dimethylamines, one of the best distributions being 73.1% mono-, 19.4% di-, and 1.4% tri-, with 97.7% conversion.[231] Alkylation of aniline with methanol over chromium phosphate produced the *N*-methylaniline with 94–95% selectivity.[232]

Sometimes the choice of zeolite can determine which product is obtained. This is true in the alkylation of 4-methylimidazole with methanol (**6.42** Schematic) in the vapor phase.[233] The product from a Meerwein–Ponndorff–Verley reduction (**6.43** Schematic) varied with the zeolite catalyst.[234] High selectivity for acrolein, propylene, and allyl ether from allyl alcohol (**6.44** Schematic) can be obtained by using different zeolites.[235] A combination of acid or base strength and cavity size may be responsible for these effects.

Hydrogenation is selective for size and shape.[236] Cyclohexene, but not cyclododecene, was reduced by hydrogen with a rhodium in NaY zeolite. The Beckmann rearrangement of cyclohexanone oxime to caprolactam (**6.45** Schematic) (for polymerization to nylon-6) is usually done with sulfuric acid, which ends up as low-value ammonium sulfate. No ammonium sulfate is formed when zeolites or molecular sieves are used. The rearrangement is 95% selective at 100% conversion with a ZSM-5 catalyst,[237] 93% selective at 99% conversion with a B-MFT zeolite,[238] and more than 98% selective at 68% conversion with H-β D.[239] Sumitomo has worked out a commercial process.[240]

The conversion of *tert*-butanol to pivalic acid with H-ZSM-5 (**6.46** Schematic) eliminates the usual sulfuric acid.[241]

MCM-41 has been used to replace sodium or potassium hydroxide in the synthesis of jasminaldehyde (**6.47** Schematic). Selectivity of 90% at more than 80% conversion was obtained using a 1.5:1 ratio of benzaldehyde/1-heptanal and 5% catalyst.[242] An AlPO catalyst gave 86% selectivity at 96% conversion.[243]

Photochemical reactions of materials enclosed in zeolites can lead to different proportions of products, or in some cases to different products than those run in solutions.[244] The distribution can vary with the zeolite. The enhanced selectivity in the oxidation of hydrocarbons with oxygen[245] was mentioned in Chapter 4. The oxidation of cyclohexane in NaY zeolite

H_3C N N H + CH_3OH → H_3C N N CH_3 + CH_3 H_3C N N

1,4-dimethyl
main product with zeolite beta

1,5-dimethyl
product with zeolite Y

6.42 Schematic

O

Zeolite CsX + isopropyl alcohol
92% selectivity

OH 77%

Zeolite NaX + isopropyl alcohol
86% selectivity

OH 87%

6.43 Schematic

OH H-CsY → O

LiNaY →

CsNaY → O

6.44 Schematic

NOH Zeolite → O NH

6.45 Schematic

with oxygen and visible light to yield cyclohexane hydroperoxide with complete selectivity at more than 40% conversion may have considerable industrial potential. Heating the hydroperoxide yields only cyclohexanone, which can be oxidized to adipic acid for use in making nylon-6,6.[246]

OH + CO → (10 min, 100°C) COOH 100%

6.46 Schematic

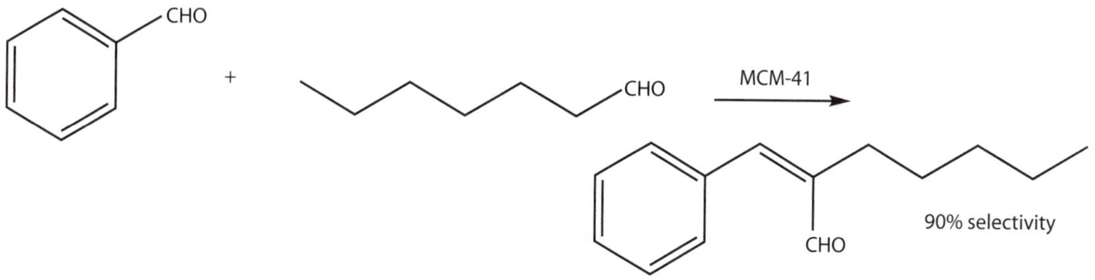

CHO + CHO MCM-41 → CHO 90% selectivity

6.47 Schematic

Photorearrangements of aryl esters in zeolites (**6.48** Schematic) give different product distributions.[247] Photosensitized oxidation of stilbene with oxygen (**6.49** Schematic) leads to different products in solution and in a zeolite.[248]

Selective photoreduction of the double bond in steroidal α,β-unsaturated ketones was carried out with MCM-41 and NaY in isopropyl alcohol/hexane (**6.50** Schematic).[249] No reduction occurs in the absence of zeolite.

Clearly, photochemical reactions in zeolites offer interesting possibilities in directing syntheses. Light is an environmentally clean reagent. The scale-up of such reactions for commercial use may require some innovative equipment. There is also the question of quantum yields, for not all the light may penetrate the zeolite. This will not be a problem if sunlight can be used. Any photosensitizers that are required will need to be recycled and reused repeatedly.

Zeolites can also be used in polymerizations. Tetrahydrofuran has been polymerized in high yield by acetic anhydride in combination with a dealuminated zeolite Y to a product with a narrow molecular weight distribution.[250] H-ZSM-5 also works in the presence of a trace of water.[251] Acrylonitrile has been polymerized in the mesoporous zeolite MCM-41,

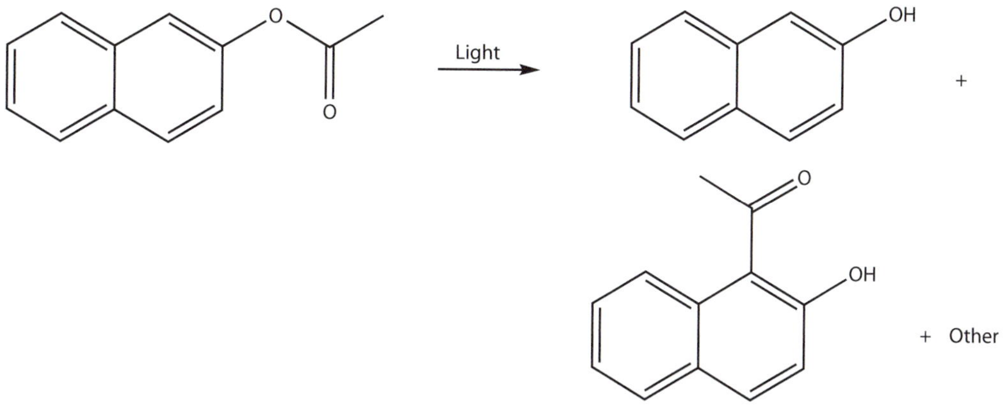

O O Light → OH + O OH + Other

6.48 Schematic

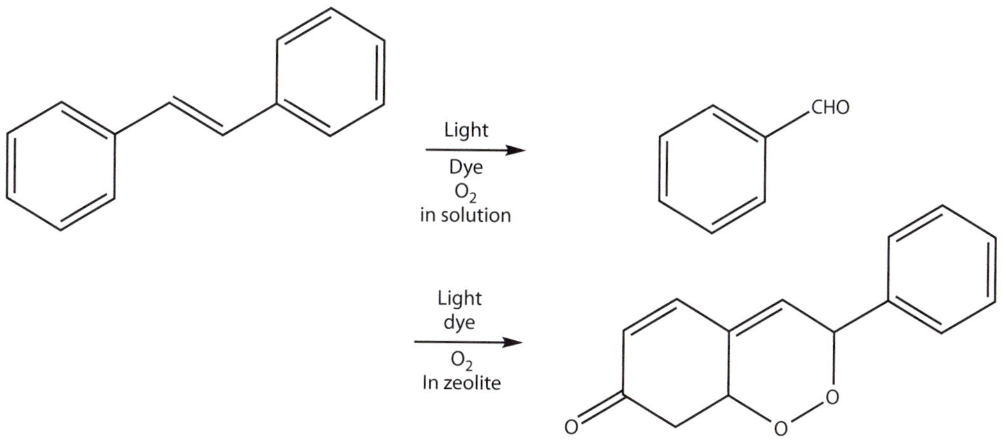

Light Dye O_2 in solution → CHO

Light dye O_2 In zeolite → O O O

6.49 Schematic

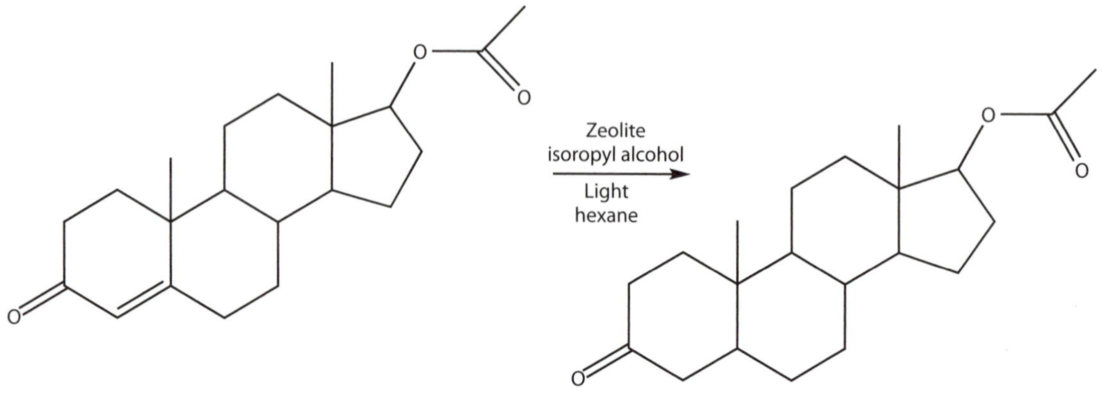

Zeolite
isoropyl alcohol
Light
hexane

6.50 Schematic

using potassium persulfate as a catalyst, and then converted to graphite.[252] Polymerization of methyl methacrylate in MCM-41 with an azo initiator gave a polymer with a number average molecular weight of 360,000 and a polydispersity of 1.7 compared with a control without the zeolite of 36,000 and 2.8.[253] The limited space in the zeolite hindered bimolecular termination.

Zeolites and molecular sieves may be useful in places where organic materials would decompose. Mordenite can be used to concentrate the sulfur dioxide in flue gas.[254] The higher concentration in the concentrate makes recovery by conversion to sulfuric acid easier. A silver–sodium zeolite A molecular sieve can be used to recover mercury from gas streams to prevent destruction of aluminum heat exchangers by mercury.[255] A synthetic dealuminized zeolite can pick up mercury and dioxins from incinerator flue gas without picking up sulfur dioxide.[256] It lasts for 3 years. Inorganic ion exchangers can be used with radioactive materials that cause the usual organic ion-exchange resins to deteriorate. Doped titanates can remove cesium and strontium preferentially over sodium, calcium, barium, and magnesium ions.[257]

It is now possible to prepare defect-free zeolite membranes for use in separations.[258] (More details on separations can be found in Chapter 7.) A mordenite membrane on a porous alumina support had a separation factor of more than 160 for benzene over *p*-xylene.[259] A ZSM-5 membrane on porous alumina had a separation factor of 31 for *n*-butane over isobutane at 185°C.[260] Similar results were obtained with silicalite on alumina tubes.[261] Free-standing membranes of titanium dioxide have been prepared.[262] Mesoporous silica films containing a variety of functional groups have been prepared.[263] The next step is to use the membranes in membrane reactors to separate products as they form.

In summary, zeolites and molecular sieves are versatile solid acids and bases that can be tailored to provide selectivity in reactions by size and shape. After estimating the sizes of the starting materials, products, and transition state, a series of sieves approximating the required sizes and acidity or basicity can be tested.[264] After finding one that works, it can be optimized by selective deactivation of unwanted sites, narrowing of the pore openings, if necessary, and stabilization by dealumination or addition of stabilizing ions. If deactivation in use is a problem, methods of regeneration will need to be studied.

6.9 METAL ORGANIC FRAMEWORKS

This is a very active area. Research is directed toward what kinds can be made and what applications they might have.[265] BASF offers some for sale.[266] A typical metal organic framework is made from a difunctional carboxylic acid,[267] sulfonic acid,[268] phosphonic acid,[269] hydroxamic acid (for Fe[III]),[270] boronic acid,[271] or diamine[272] with a divalent metal ion (**6.51** Schematic).

In addition to terephthalic acid, other multifunctional acids may be used (**6.52** Schematic).

Various diamines are used (**6.53** Schematic).

A variety of multifunctional phenols have been used with the boronic acids (**6.54** Schematic).

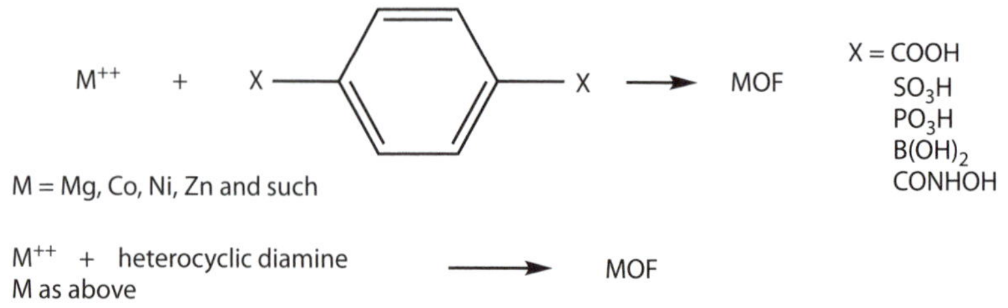

M^{++} + X–C_6H_4–X → MOF

X = COOH, SO_3H, PO_3H, $B(OH)_2$, CONHOH

M = Mg, Co, Ni, Zn and such

M^{++} + heterocyclic diamine → MOF

M as above

6.51 Schematic

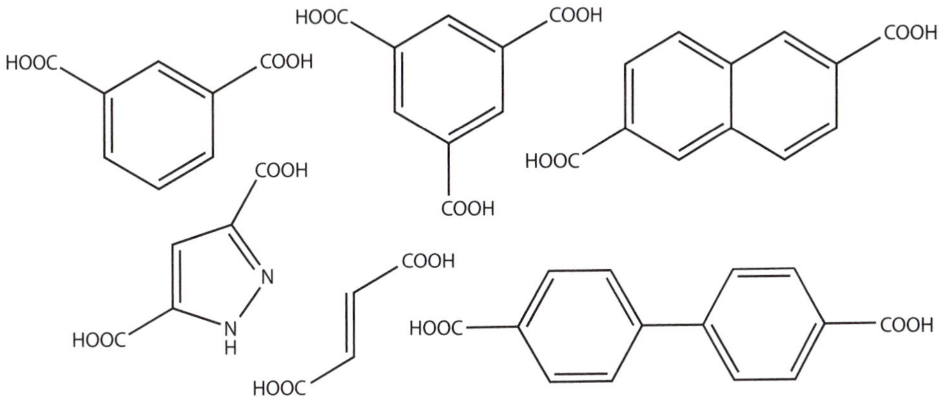

6.52 Schematic

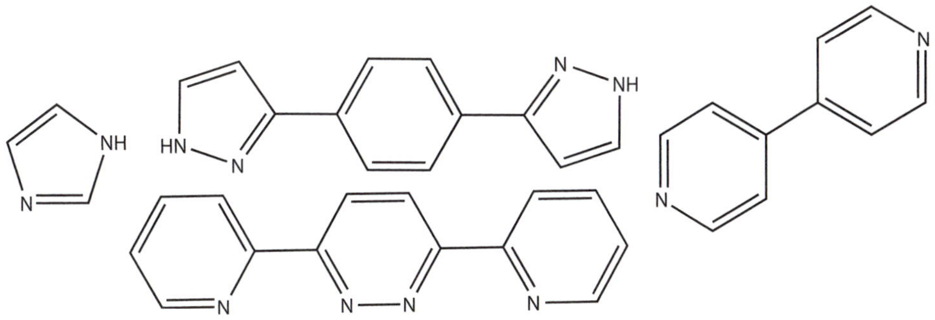

6.53 Schematic

Zn(II) is one of the most common ions used with the acids. Cu(II), Co(II), and Ni(II) are also used. Tri- and tetravalent ions such as Eu(III), La(III), Ce(III), Er(III), Lu(III), Yb(III), Cr(III), V(III), Fe(III), and Zr(IV) have also been used. Zn(II), Co(II), Pd(II), and Pt(II) have been used with the di- and triamines. The polyphenols have also been reacted with Sn, Mo, W, Cd, and Nb.

The covalent organic porous frameworks may have 60% open space, compared with up to 50% in zeolites. One containing boron has a surface area of 4210 m^2/g, a density of 0.17 g/cm^3, and is stable to 500°C. The open space can be used for the storage of gases. A framework made from magnesium ion and 2,5-dihydroxyterephthalic acid can hold 35% of its weight of carbon dioxide compared to 20.7% for zeolite 13X.[273] The storage of hydrogen in metal organic frameworks is being studied.[274] A zinc terephthalate can store 6 wt% at 77 K but only 2 wt% at room temperature.[275] Storage of methane, nitric oxide, carbon monoxide, oxygen,

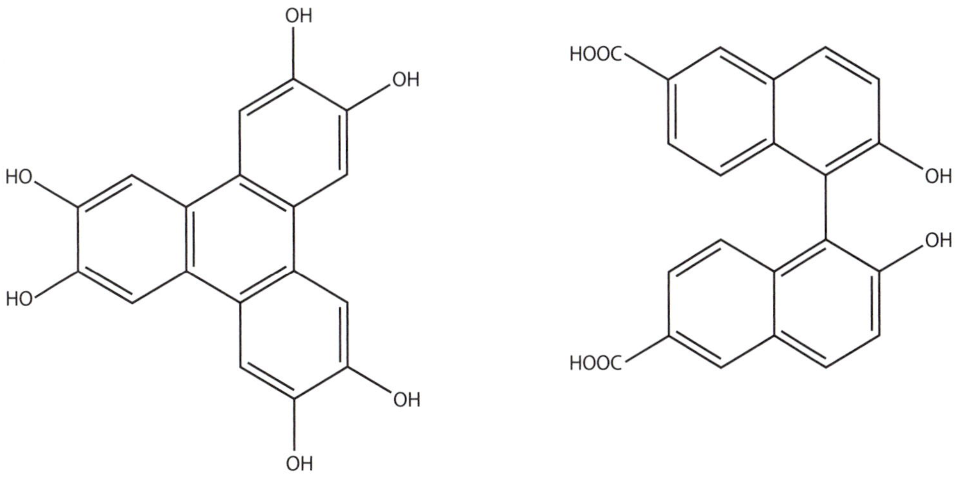

6.54 Schematic

sulfur dioxide, and ammonia in nanoporous materials has been investigated.[276] The use of metal organic framework materials has been suggested for the separation of mixtures of hydrocarbons.[277]

A variety of compounds have been inserted into the cavities.[278] Metal nanoparticles of Pd, Cu, and Au have been inserted into zinc terephthalate by chemical vapor deposition followed by reduction.[279] Heteropolymetalates, such as $K_2PW_{11}O_{40}$, have been inserted into chromium terephthalate.[280] A chromium terephthalate containing ethylenediamine or diethylenetriamine catalyzed the reaction of benzaldehyde with ethyl cyanoacetate to give ethyl 2-cyanocinnamate with 99.3% selectivity at 97.7% conversion.[281] Manganese benzenetristetrazolate, which has two kinds of Mn sites, catalyzed a similar reaction of benzaldehyde (**6.55** Schematic).[282]

Polymerization of styrene or methyl methacrylate in the pores of zinc or copper terephthalate led to higher molecular weights and increased isotacticity.[283] A cage from palladium and tris(4-pyridyl)-2,4,6-triazine altered the course of the light-catalyzed reaction of benzil.[284] Some 2 + 2 cycloadditions were also altered (**6.56** Schematic).

The cavities in a cage made from zinc iodide and tris(4-pyridyl)-2,4,6-triazine were large enough for reactions of bulky molecules with reagents such as phenyl isocyanate, octanoic anhydride, and maleic anhydride (**6.57** Schematic).[285]

The amino group in a framework made from gadolinium nitrate and 2-aminoterephthalic acid could be reacted with ethyl isocyanate to form a urea and with crotonic anhydride to produce a crotonamide in new frameworks.[286]

6.10 CLAYS

Clays are layered, hydrated silicates of aluminum, iron, and magnesium.[287] Most are crystalline inorganic polymers, but a few are amorphous. They consist of tetrahedral and octahedral layers, as shown in **6.58** Schematic. The tetrahedral layers contain silicon and oxygen, but sometimes with some substitution of silicon by aluminum or iron. The octahedral layers are mainly aluminum, iron(III), iron(II), or magnesium ions. The spaces between the layers contain water and any cations necessary to maintain electrical neutrality. The water can be taken out reversibly by heating. Montmorillonite (in the smectite family) has two tetrahedral layers for each octahedral one. Its overall formula is $[Al_{1.67}Mg_{0.33}(Na_{0.33})]Si_4O_{10}(OH)_2$. This is one of the more common clays used for the catalysis of organic reactions.

The layers in the smectites can be expanded by intercalation with water, alcohols, ethylene glycol, glycerol, and such. (For reviews on intercalation into layered lattices, see O'Hare[288] and Oriakhi.[289]) The initial spacing between layers is 12–14 Å when clay is intercalated with ethylene glycol. If the structure is heated to 550°C, the structure collapses to a 10 Å spacing. The swelling of sodium montmorillonite with water has been modeled.[290] The collapse of the structure on

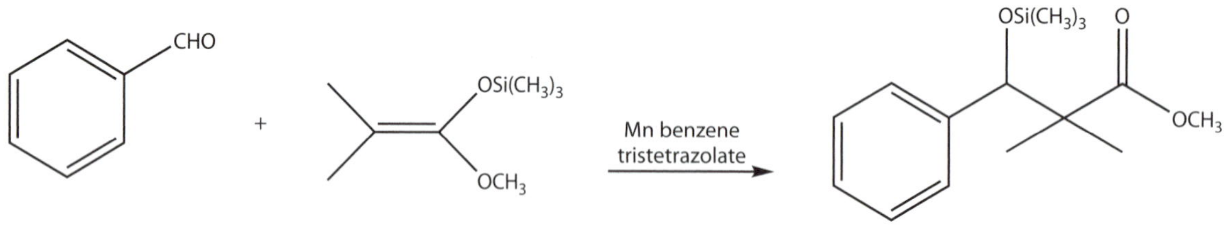

CHO + OSi(CH3)3 OCH3 — Mn benzene tristetrazolate → OSi(CH3)3 O OCH3

6.55 Schematic

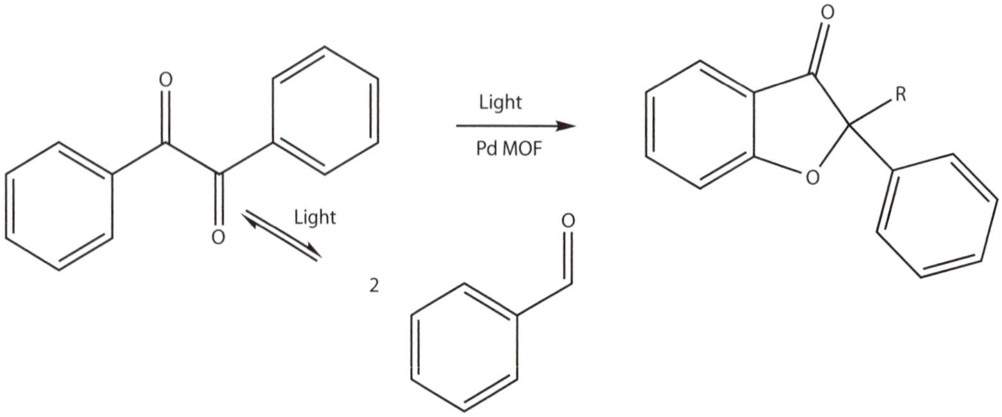

O O — Light / Pd MOF → O R O; Light ⇌ 2 O

6.56 Schematic

6.57 Schematic

heating to high temperatures has been a problem in some catalyses, because the surface area and, hence, the catalytic activity decrease greatly. To avoid this problem, the clay layers can be propped open.[291] Smectites can be pillared with oxides of aluminum, zirconium, cerium, iron, or other such elements to give 30 Å between layers. In the most common case, an aluminum hydroxy cluster cation $[(Al_{13}O_4(OH)_{24}(H_2O)_{12})]^{7+}$ is inserted by cation exchange and then heated to form the oxide pillars. The spaces between layers in montmorillonite pillared with iron are 76 Å.[292] Even with pillaring, many clays are not very stable to thermal and hydrothermal treatments, which detracts from their use as

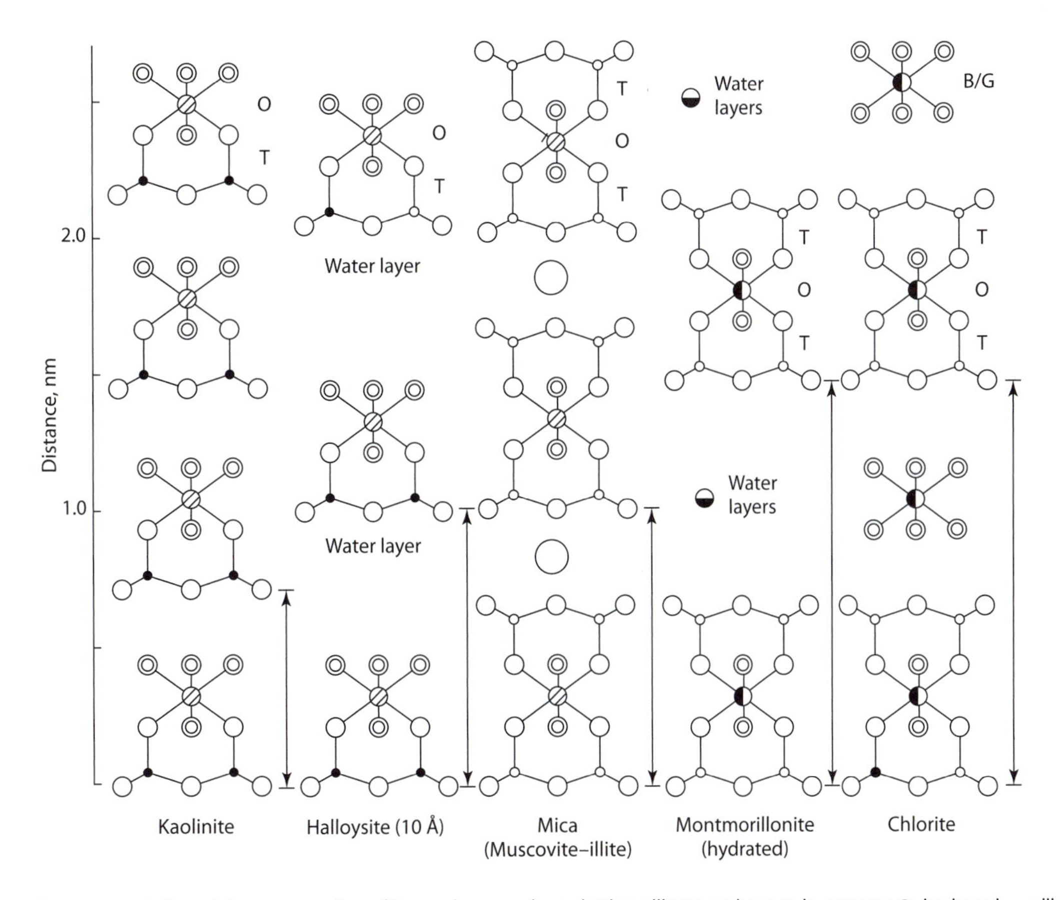

6.58 Schematic representation of the succession of layers in some layer lattice silicates where ○ is oxygen; ◎, hydroxyl; •, silicon; ∘ Si–Al; ⊘, aluminum; ◐, Al–Mg; ◯, Potassium; ◒, Na–Ca. Sample layers are designated as O, octahedral; T, tetrahedral; and B/G, brucite- or gibbsite-like. The distance (in nm) depicted by arrows between repeating layers are 0.72, kaolinite; 1.01, halloysite (10 Å); 1.00, mica; ca. 1.5, montmorillonite; and 1.41, chlorite. (Reprinted from T. Dombrowski, *Kirk–Othmer Encyclopedia of Chemical Technology*, 4th ed., Wiley, New York, 1993, *6*, 381. Copyright 1993 John Wiley & Sons. With permission.)

catalysts. A "highly stable" clay has been made by pillaring with $Al_2(OH)_3Cl$.[293] One pillared with Ce(III)/Al was stable to steam at 800°C.[294]

Clays have many uses.[295] These include making bricks, chinaware, filled paper, filled plastics, drilling muds, and so on. The use of clay in paint can make it thixotropic (i.e., the paint is thick until pushed with the brush, whereupon it becomes temporarily thinner). This is due to the plate-like structures of clays. Attapulgite can be used in brake linings for cars and trucks as a replacement of carcinogenic asbestos. Kaolin can be heated to 600°C to form molecular sieves. The cation-exchange capacity of clays is essential for the mineral nutrition of plants. Putting 4% exfoliated clay into nylon-6 raised the heat distortion temperature 50°C and increased the tensile strength with no loss in impact strength.[296] Putting 5% of exfoliated clay into poly(ethylene terephthalate) raised the heat distortion temperature 20–50°C and increased both modulus and crystallization rate three-fold.[297] (Slow crystallization of this polymer leads to uneconomically long cycle times in molding.) Similar techniques have been used to add clays into polyethylene,[298] polypropylene,[299] polyimides,[300] and epoxy resins.[301]

Cationic and anionic clays have been used as catalysts in many organic reactions, including electrophilic aromatic substitutions, addition and elimination reactions of olefins, cyclizations, isomerizations, rearrangements, and such.[302] As catalysts for the Friedel–Crafts reaction they offer several advantages over aluminum chloride and other typical Lewis acids. They are easy to handle, noncorrosive, and low in cost. Yields and conversions are high with reduced times and catalyst levels. Compared with some reactions that require stoichiometric amounts of the Lewis acid, zinc chloride on K10 montmorillonite can be used at a level 2000 times lower. Polyalkylation and rearrangements are often absent. They are regenerable. No toxic waste is produced. The acidity can be modified easily by varying the cation and any anions on it.

Alcohols, phenols, thiols, and amines can be acetylated with acetic anhydride in the presence of montmorillonite K10 under mild conditions.[303] Octyl acetate was obtained in 1 h from 1-octanol, in 96% yield, using K10 at room temperature. Peracetylation of sugars, such as glucose, has been done in similar fashion in 92–99% yield.[304]

Alcohols were used to alkylate aromatic compounds in 90–95% yield with K10 montmorillonite, but the cationic intermediate isomerized so that several isomers were obtained with different points of chain attachment.[305] Ultrasound accelerated the benzylation (**6.59** Schematic) of aromatic compounds (where X = Cl and R = H or Cl).[306]

Nonpoisonous and nonhydrolyzable neodymium chloride has been used on clay for the same reaction, giving 80% selectivity where X = Cl and R = H.[307] There was no loss in activity after 5000 turnovers. Zinc chloride on K10 montmorillonite can also be used with reagents in the vapor phase (**6.60** Schematic).[308] Experiments with the same catalyst in solution gave 3–50% yield. The use of boron trifluoride can be avoided by oligomerizing 1-decene to oils in 81% conversion using K10 montmorillonite.[309]

Reactions on clay catalysts show some selectivity by size and shape. Aluminum montmorillonite was used to rearrange phenyl ethers (**6.61** Schematic).[310] The bulkier *tert*-butyl group led to only the *para*-product. Allylation of aromatic compounds with octenols (**6.62** Schematic) gave only the unbranched isomer.[311] Aniline can be ethylated with ethanol

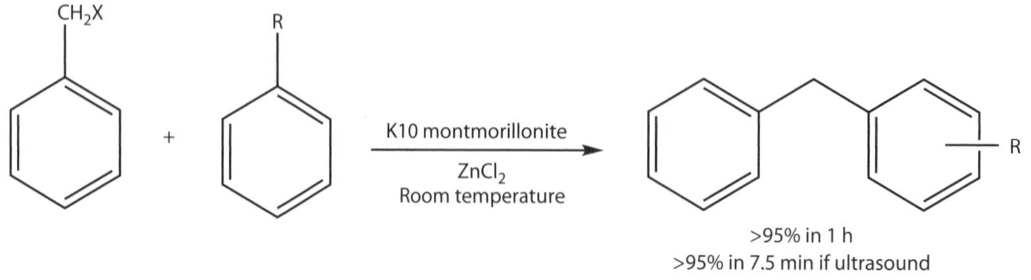

CH_2X + R → K10 montmorillonite, $ZnCl_2$, Room temperature → R

>95% in 1 h
>95% in 7.5 min if ultrasound

6.59 Schematic

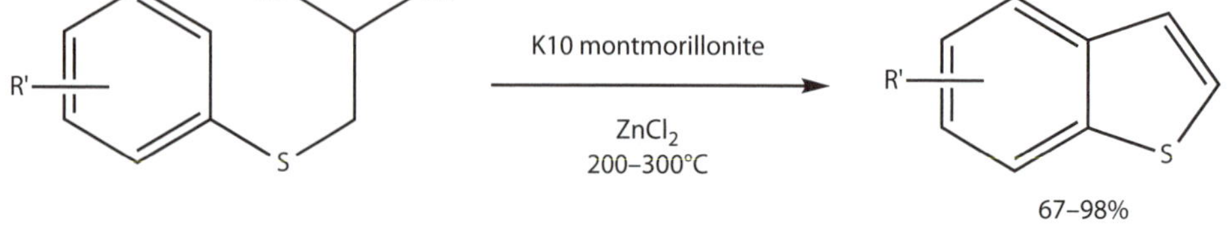

RO, OR, R', S → K10 montmorillonite, $ZnCl_2$, 200–300°C → R', S

67–98%

6.60 Schematic

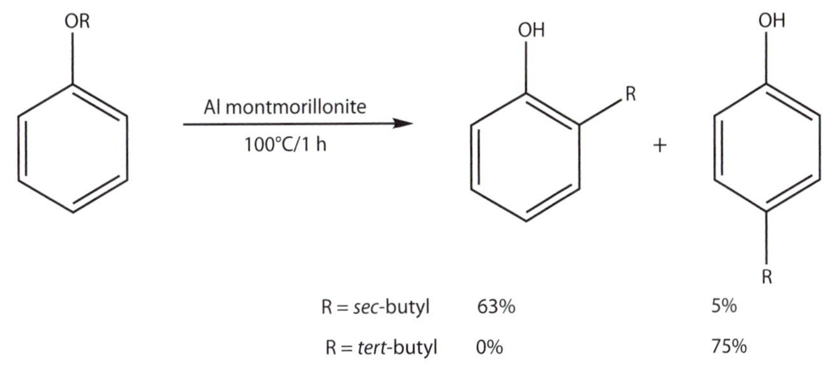

6.61 Schematic

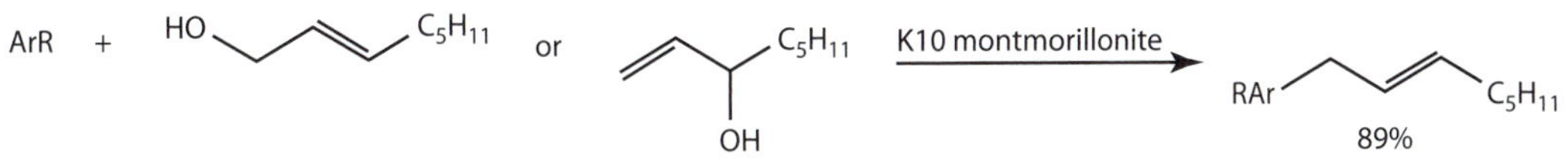

6.62 Schematic

to give 86% *N*-ethylaniline and 14% *N,N*-diethylaniline in 52% conversion with, or without, vanadium oxide at 400°C.[312]

Copper nitrate on montmorillonite (Claycop) was used with acetic anhydride to nitrate chlorobenzene in 100% yield, giving 13:85 *ortho*/*para*-isomers.[313] When an iron oxide pillared clay was used, 94% *para*-isomer was obtained.[314] Claycop and its iron analog were used to oxidize a pyrazoline to a pyrazole (**6.63** Schematic) in 64–97% yield.[315]

Kaolin was used as a catalyst for the preparation of thioketals from ketones in 90–95% yield.[316] Thioketals were hydrolyzed in 91–100% yield using a mixture of finely ground $Fe(NO_3)_3 \cdot 9H_2O$ and K10 montmorillonite in hexane.[317] Microwaves were used to accelerate the reaction of *ortho*-phenylenediamine with carboxylic acids (**6.64** Schematic) over K10 montmorillonite.[318] Microwaves gave better conversions in the reaction of imidazole with ethyl acrylate (**6.65** Schematic) than thermal activation.[319]

Selectivity in the dehydration of olefins is improved with pillared clays. Clays with aluminum oxide or mixed aluminum and iron pillars converted isopropyl alcohol to propylene with more than 90% selectivity.[320] A small amount of isopropyl ether was formed. When zeolite Y is used, the two products are formed in roughly equal amounts. A tantalum-pillared montmorillonite converted 1-butanol to butenes at 500°C with 100% selectivity at 41% conversion.[321] The product contained a 17:20:16 mixture of 1-butene/*cis*-2-butene/*trans*-2-butene. No butyraldehyde or butyl ether was formed. A pillared clay was used for the alkylation of benzene with 1-dodecene without the formation of dialkylated products.[322] The carbonylation of styrene proceeded in 100% yield (**6.66** Schematic).[323]

A pillared clay containing a quaternary ammonium bromide surfactant was used as an inexpensive phase transfer catalyst for the reaction of alkyl halides with sodium azide to form alkyl azides in 82–93% yield.[324] It could be reused at least twice, with little or no loss in yield.

Clays were also pillared with polyoxometalate ions, such as $[(PW_{11}VO_{40})^{4-}]$.[325] Other layered materials can also be pillared. Zirconium phosphate was pillared with chromium(III) oxide.[326] A layered titanate was pillared with silica.[327] The need

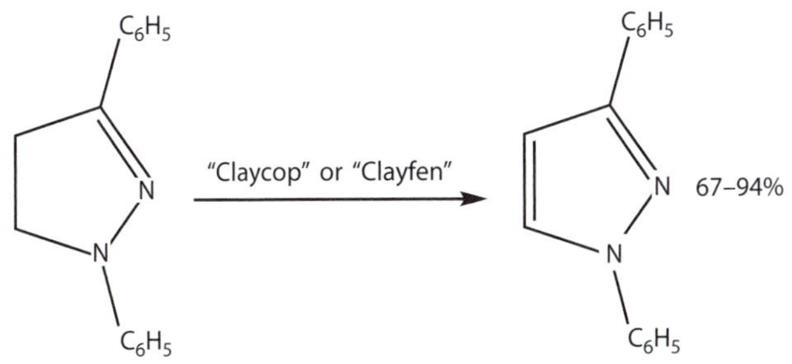

6.63 Schematic

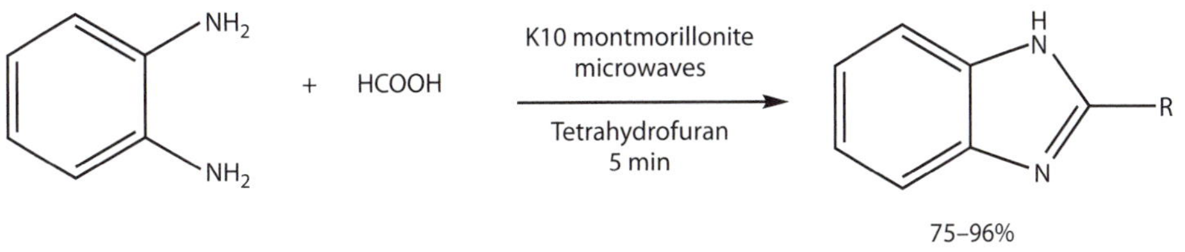

6.64 Schematic

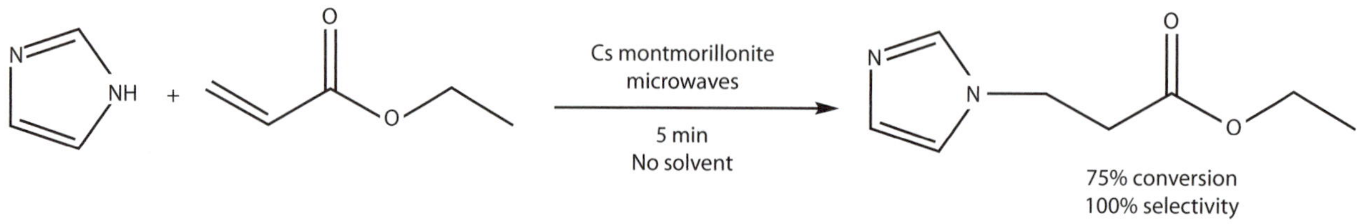

Cs montmorillonite microwaves

5 min No solvent

75% conversion 100% selectivity

6.65 Schematic

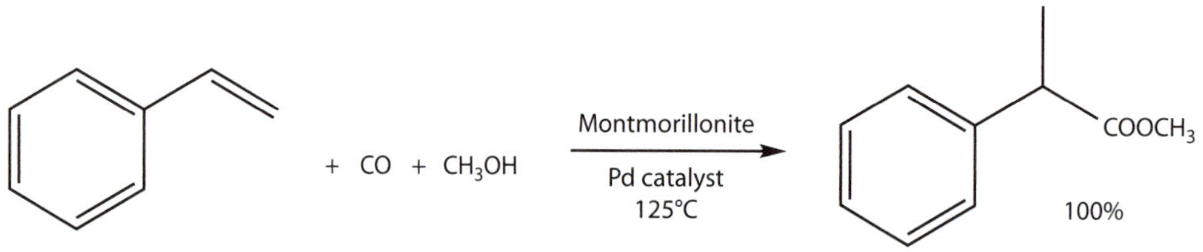

+ CO + CH_3OH Montmorillonite Pd catalyst 125°C

$COOCH_3$ 100%

6.66 Schematic

for a separate pillaring step has been avoided in the preparation of some porous lamellar silicas that are structurally similar to pillared clays.[328] Eight to twelve carbon diamines were used with tetraethoxysilane in ethanol with added water to make them in 18 h at room temperature. The template was recovered by solvent extraction before the silica was calcined.

A tandem reaction using a mixture of titanium montmorillonite and magnesium aluminum hydrotalcite in one pot gave 93% yield at >99% conversion (**6.67** Schematic).

A self-supporting film of montmorillonite was prepared in aqueous phosphoric acid, which immobilizes the clay by cross-linking.[329] The film remained intact in hot water and retained its ion-exchange capacity. Such films offer the potential of separating mixtures. Pillared, layered structures may be able to separate flat molecules from others. They may also be able to separate more highly branched structures from linear molecules. The first step is to estimate the sizes of the molecules in the mixture relative to the interlayer distances and the density of pillars. These properties of the membrane can be varied. Vegetable oils have to be winterized by crystallizing out the saturated triglycerides. Biodiesel fuel, which consists of methyl oleate from the transesterification of rapeseed oil with methanol, must be winterized by crystallizing out the methyl stearate so that the fuel can be used in cold weather. The proper pillared, layered solids might permit passage of the saturated esters but not the unsaturated ones. The process might also be used to remove the cholesterol-raising saturated fats from food oils. Separation of 2,6-diisopropylnaphthalene from its 2,7-isomer should also be tried. There should also be some reactions of flat molecules that could be more selective if run with a catalyst in a pillared clay. There is great interest in mesoporous analogs of titanium silicalite for the selective oxidation of larger molecules. A layered silicate pillared with titania should be tried in such oxidations. Clays pillared with polyoxometalates, containing vanadium, molybdenum, or tungsten, should also be tried in oxidations.

6.11 HETEROPOLYACIDS

Heteropolyacids[330] that may be useful catalysts are primarily of the Keggin ($[X^{n+}M_{12}O_{40}]^{(8-n)-}$) and Dawson ($[X_2M_{18}O_{62}]^{6-}$) type. In the Keggin ions, X can be P^{5+}, As^{5+}, B^{3+}, Si^{4+}, or Ge^{4+}, and M can be Mo, V, or W, where more than one valence state is possible. For the Dawson ions, X can be P^{5+}, with M as for

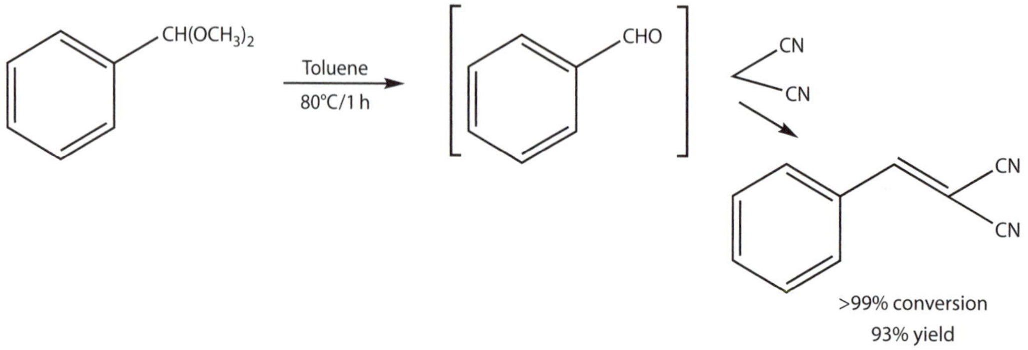

$CH(OCH_3)_2$ Toluene 80°C/1 h [CHO] CN CN

CN CN

>99% conversion 93% yield

6.67 Schematic

the Keggin ions. The structure of the $(PW_{12}O_{40})^{3-}$ ion is shown diagrammatically (**6.68** Schematic). The central heteroatom in the more or less spherical structure is surrounded by metal oxide tetrahedra. The materials are strong acids. Both Bronsted and Lewis acid acidities are present. Various ligands, such as methylpyridines, can coordinate with them.[331] Because the complex with ethyl ether is insoluble in ethyl ether, this can be used as a method of isolation. Transition metals, such as Mn(III), in the polyoxometalate ion can be oxidized or reduced without destroying the polyoxometalate.[332] Heteropolymetalates can serve as ligands for transition metal ions.[333] As an example, cyclopentadienyltitanium has been inserted in this way.[334]

The heteropolyacids are very soluble in water. They can form sparingly soluble or insoluble salts with ions such as ammonium, cerium, potassium, silver, and such. The acids are often soluble in organic solvents, such as alcohols, ketones, carboxylic acids, and carboxylic esters. Long-chain tetraalkylammonium salts can also be soluble in organic solvents. In a sense, heteropolyacids are soluble versions of insoluble metal oxide catalysts. They can be used as catalysts both in solution and as solids. A catalyst that is soluble in water would be a solid if used alone or in a hydrocarbon medium. They can also be placed on or in insoluble supports such as the zeolites MCM-41 and SBA-15.[335] In the insoluble forms at least, they offer the advantages of easy separation and recovery for reuse.

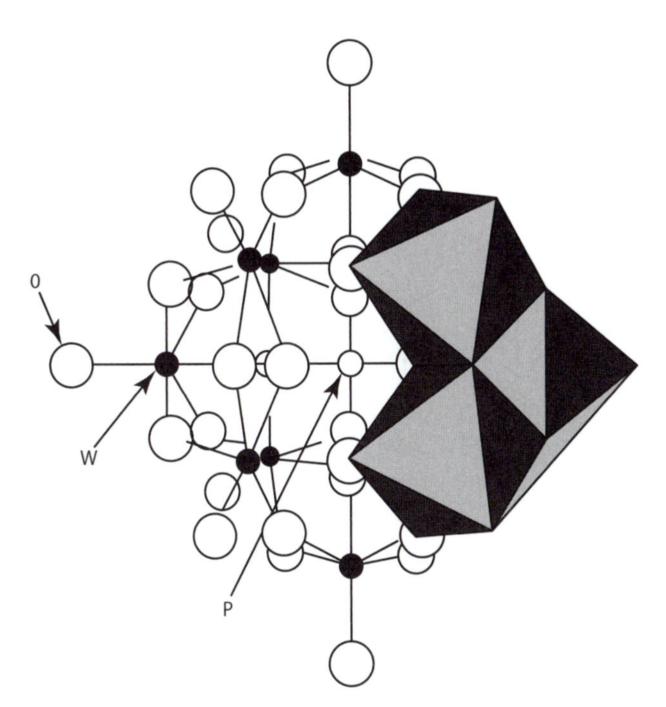

6.68 Schematic. (Reprinted from M. Misono, *Catal. Rev. Sci. Eng.*, 1987, *29*, 269. Copyright 1987 Marcel Dekker. With permission.)

Heteropolyacids can be prepared (**6.69** Schematic) in aqueous solutions at pH 1–2. Salts may be obtained by careful neutralization with an alkali carbonate. Although the heteropolyacids are stable in acidic aqueous solutions, they tend to decompose in alkaline aqueous solutions. The resistance to hydrolysis varies as $Si > Ge > P > As$ and $W > Mo > V$. The acids are relatively stable thermally, the decomposition temperatures following the order $H_3PW_{12}O_{40}$ (610°C) > $H_4SiW_{12}O_{40}$ (540°C) > $H_3PMo_{12}O_{40}$ (495°C) > $H_4SiMo_{12}O_{40}$ (375°C). Ammonium and cesium salts of heteropolyacids can sometimes have organized microporous structures.[336] The salts of Dawson acids and 1,6-diaminohexane are also microporous.[337] It is unclear whether these organized microporous structures depart from the generalized globular structure shown in **6.68** Schematic.

Polyoxometalates can be functionalized with triols, alkyltriethoxysilanes, and such.[338] Bridging O atoms can be replaced by aryl nitrido bridges.

Heteropolyacids have been used as catalysts in a variety of acid-catalyzed and oxidation reactions.[339] A few examples will be given to show the advantages of these catalysts. Showa Denko uses a heteropolyacid catalyst for the addition of acetic acid to ethylene to produce ethyl acetate.[340] Butyl acrylate can be made with 96% selectivity at 98% conversion in a flow system using $H_3PW_{12}O_{40}$ on carbon.[341] The activity of immobilized dodecatungstosilicic acid on carbon in vapor-phase esterification is higher than that of the zeolite and ion-exchange resins now used by industry.[342] In the esterification of ethyl alcohol with acetic acid with $Cs_{2.5}H_{0.5}PW_{12}O_{40}$ silica, the turnover frequency based on the acid sites present was higher than those with Amberlyst 15 (a sulfonic acid ion-exchange resin) and H-ZSM-5 (an acidic zeolite).[343] When the esterification of pyridine-2,6-dicarboxylic acid with 1-butanol was carried out homogeneously with $H_3PW_{12}O_{40}$, the diester was obtained in 100% yield.[344] When the heterogeneous catalyst $Ce_{0.87}H_{0.4}PW_{12}O_{40}$ was used, the activity was lower, but the catalyst could be recovered and reused. In view of the activity of scandium and rare earth triflates as acids, discussed earlier, further work should be done with scandium and rare earth salts of heteropolyacids, varying both the elements used and the stoichiometry.

tert-Butyl methyl ether was a common gasoline additive that raises the octane number and reduces the air pollution from cars. It can be made with nearly 100% selectivity at 71% conversion using $H_3PW_{12}O_{40}$ at 85°C.[345] $H_4GeW_{12}O_{40}$[346] and $H_3PW_{12}O_{40}$[347] are also effective catalysts. In current manufacturing, Amberlyst 15 (a sulfonic acid ion-exchange resin) has the disadvantages of thermal instability and loss

$$12Na_2MoO_4 + Na_2SiO_3 + 26HCl \longrightarrow H_4SiMo_{12}O_{40} + 26NaCl + 11H_2O$$

6.69 Schematic

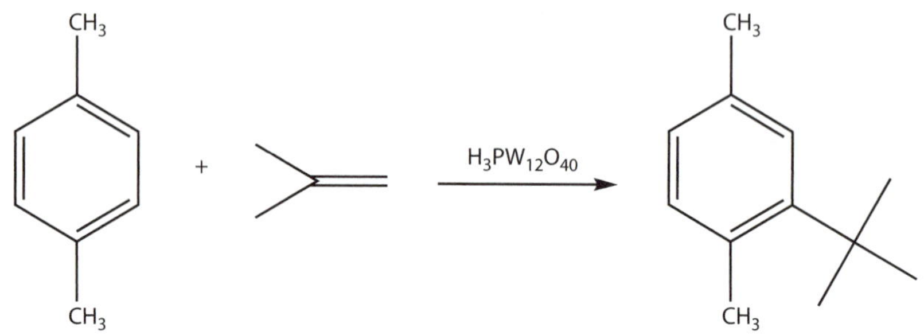

6.70 Schematic

of acid sites by leaching. Alkylation of *p*-xylene with isobutylene using $H_3PW_{12}O_{40}$ (**6.70** Schematic) yields the monoalkylated product with 75% selectivity (90% conversion).[348] With a sulfuric acid catalyst, the selectivity is only 7%.

Dihydromyrcene was hydrated in aqueous acetic acid with 90% selectivity (to dihydromyrcenol and its acetate) at 21% conversion with $H_3PW_{12}O_{40}$ (**6.71** Schematic) without the isomerization and cyclization that often accompany such reactions.[349] This catalyst was much more active than sulfuric acid and Amberlyst 15 (a sulfonic acid ion-exchange resin).

The amount of cesium can determine the pore size in a microporous heteropolyacid salt. The pore size is $Cs_{2.5} > Cs_{2.2} > Cs_{2.1}$ in $Cs_nH_{3-n}PW_{12}O_{40}$ with the best activity for larger substrates in the first.[350] Reaction **6.72** could be carried out with the first, but not with the other two. Typical microporous zeolites, zeolite Y and H-ZSM-5, were inactive.

Heteropolyacids can be very useful in oxidation reactions. In contrast with metal chelate catalysts that usually become oxidized and deactivated eventually, the heteropolyacids are extremely stable to oxidation. Acetaldehyde is produced commercially from ethylene by the Wacker reaction with a palladium(II) chloride catalyst, copper(II) chloride, oxygen, and water. The corrosive conditions are a disadvantage of the process. Catalytica Inc. has devised a process (**6.73** Schematic) that uses only 1% as much palladium and chloride as the usual process. It uses a small amount of palladium(II) chloride with a partial sodium salt of phosphomolybdovanadic acid.[351]

The same reaction (R = H) was run to 80% conversion in 50 h with palladium(II) chloride and a manganese phosphomolybdovanadate on silica.[352] Palladium(II) sulfate was used with a phosphomolybdovanadic acid to convert 1-butene ($R = C_2H_5$) to 2-butanone with more than 95% selectivity at 98% conversion.[353] Cyclohexene was converted to cyclohexanone with 97% selectivity using $Pd(NO_3)_2/CuSO_4/H_3PMo_{12}O_{40}/O_2$ in 1 h at 80°C.[354] Palladium(II) sulfate and copper(II) sulfate were used with a phosphomolybdovanadic

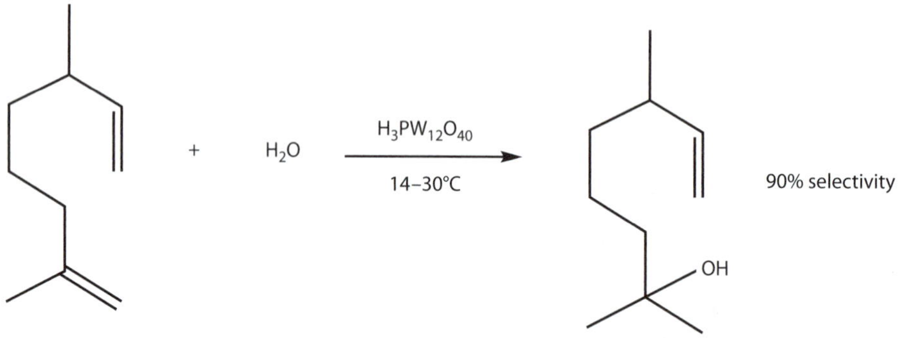

6.71 Schematic

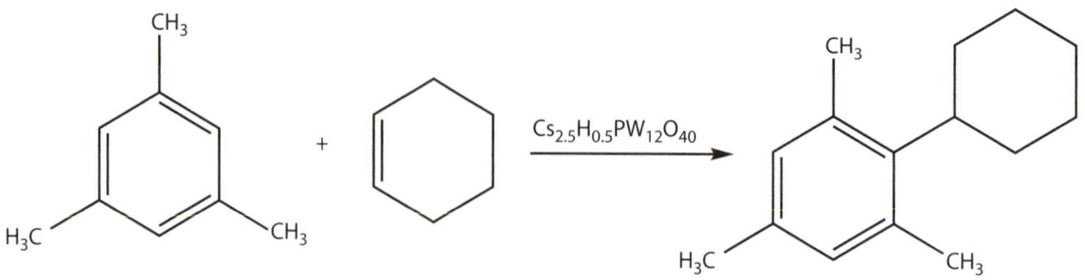

6.72 Schematic

$$\text{CH}_2{=}\text{CH-R} \xrightarrow[\text{Pd(II)}, O_2, H_2O]{\text{Heteropolyacid}} \text{CH}_3\text{C(=O)R} \quad R = H, n\text{-}C_2H_5, n\text{-}C_8H_{17}$$

6.73 Schematic

acid and a per (2,6-di-*O*-methyl) β-cyclodextrin in an aqueous system to convert 1-decene ($R = C_8H_{17}$) to 2-decanone with 98% selectivity at 100% conversion.[355] The cyclodextrin derivative was used as an inverse phase-transfer catalyst to take the olefin into the aqueous phase where it could react with the catalyst. The reaction has also been combined with a further oxidation to convert ethylene directly to acetic acid in 94% yield[356] using a combination of palladium and tellurium salts with magnesium silicotungstate on a silica support. A commercial plant is being built to run this reaction by this catalyst or a similar one.[357]

Butane has been isomerized to isobutane in 95% yield (at 24.5% conversion) with platinum/$Cs_{2.5}H_{0.5}PW_{12}O_{40}$ at 200–300°C under hydrogen at 0.05 atm.[358] The hydrogen reduced deactivation of the catalyst. The oxidation of isobutane to isobutylene was needed for the synthesis of *tert*-butyl methyl ether to put into gasoline. Isobutylene, in turn, can be oxidized to methacrylic acid for conversion to methyl methacrylate, an important monomer (**6.74** Schematic). Making methyl methacrylate in this way avoids the use of toxic hydrogen cyanide in the present commercial process.

The oxidative dehydrogenation of isobutane to isobutylene was carried out with a salt of a Dawson acid, $K_7P_2W_{17}MnO_{61}$, with 79% selectivity to isobutylene, 3% to propylene, and 18% to carbon monoxide and carbon dioxide.[359] Methacrolein was oxidized to methacrylic acid with 87% selectivity at 84% conversion using a heteropolyacid salt made with molybdenum, arsenic, copper, phosphorus, vanadium, and cesium compounds.[360] Isobutane can be oxidized directly to methacrylic acid using $Cs_{2.5}Ni_{0.08}H_{0.34}PVMo_{11}O_{40}$ and oxygen at 320°C, but with only 36% selectivity, much carbon monoxide and dioxide are formed at the same time.[361] Further work is needed to improve the yields in these reactions.

Heteropolyacids can be used as catalysts in oxidations with hydrogen peroxide, *tert*-butyl hydroperoxide, or others. Hydrogen peroxide was used with $[(C_6H_{13})_4N][(PO_4)\{WO(O_2)_2\}_4]$ to convert 1-nonene to its epoxide in 93–94% yield.[362] This system converted cyclooctene to its epoxide in 99% yield. The conversion of 1-octene to the corresponding epoxide took place with 95% selectivity at more than 85% conversion using hydrogen peroxide with $H_3PW_{12}O_{40}$ and a quaternary ammonium salt, such as hexadecylpyridinium chloride, as a phase-transfer agent.[363] Unfortunately, the chlorinated solvents that were necessary for the reaction gradually decomposed the catalyst. An insoluble catalyst for this reaction that required no organic solvent was made by the sol–gel method (**6.75** Schematic).[364] Cyclooctene was converted to its epoxide (80% conversion) with hydrogen peroxide using this catalyst. Oxidation of cyclohexane with hydrogen peroxide and a $SiW_{10}[Fe(H_2O)]_2O_{38}^{6-}$ catalyst at 83°C for 96 h gave a 55:45 cyclohexanol/cyclohexanone mixture in 66% conversion with 95% utilization of hydrogen peroxide.[365] If the time can be reduced and cheap hydrogen peroxide is available, this could replace the current relatively inefficient air oxidation of cyclohexane in the production of nylon.

Sulfides have been converted to the corresponding sulfoxides with *tert*-butyl hydroperoxide in 95–99% selectivity using $H_5PV_2Mo_{10}O_{40}$ on a carbon catalyst.[366] Combinatorial synthesis of polyoxometalates identified one with a PVW ratio of 1:2:10 that was used to oxidize tetrahydrothiophene to its sulfoxide at 95°C with 95% selectivity.[367] Oxidation of dibenzothiophene with hydrogen peroxide in the presence of phosphotungstic acid produced the corresponding sulfone in nearly 100% yield.[368] The sulfone can then be removed on silica. This process can be used to remove sulfur from petroleum, leaving only 0.005% in the oil. The sulfone could be burned off the silica and the sulfur dioxide converted to elemental sulfur in a Claus unit. Such a process is needed to make low-sulfur gasoline and diesel fuel in order to reduce air pollution.

Neumann and coworkers used sandwich-type polyoxometalates as catalysts for such reactions.[369] The sandwich consists of transition metal ions between two Keggin anions,

$$(CH_3)_2C{=}O + HCN \rightarrow (CH_3)_2C(OH)CN \xrightarrow{\text{Acid}} \xrightarrow{CH_3OH} CH_2{=}C(CH_3)COOCH_3$$

$$(CH_3)_3CH \xrightarrow{O_2} (CH_3)_2C{=}CH_2 \xrightarrow{O_2} CH_2{=}C(CH_3)CHO \xrightarrow{O_2} CH_2{=}C(CH_3)COOH \xrightarrow{CH_3OH} CH_2{=}C(CH_3)COOCH_3$$

6.74 Schematic

$$Si(OC_2H_5)_4 + C_6H_5Si(OC_2H_5)_3 + (RO)_3Si(CH_2)_nN^+Cl^- \xrightarrow{H_2O} \text{Support} \xrightarrow{\text{Polyoxometalate}} \text{Catalyst}$$

6.75 Schematic

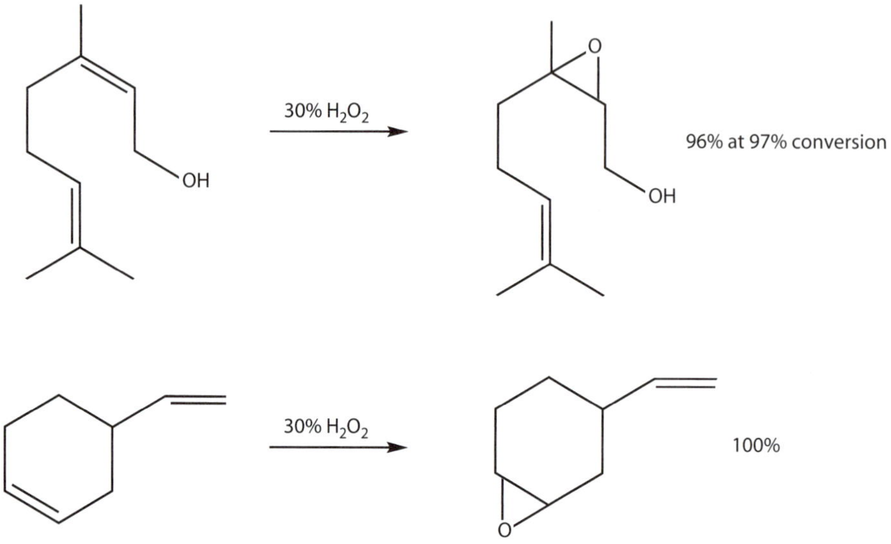

30% H_2O_2

OH

OH

96% at 97% conversion

30% H_2O_2

100%

6.76 Schematic

as in $K_{12}WZnMn(II)_2(ZnW_9O_{34})_2$. In some cases, quaternary ammonium salts were use as the cations to take the catalyst into organic solvents. Sulfides were oxidized to sulfoxides with 30% aqueous hydrogen peroxide in 85–90% yield, with some sulfone also being formed. The system was more than 99% selective in the conversion of cyclooctene to its epoxide. The system also shows good selectivity between double bonds (**6.76** Schematic), probably the result of the bulky anion and the increased electron density in alkyl-substituted double bonds.

This type of catalyst has also been used with ozone to oxidize cyclohexane to cyclohexanone with more than 98% selectivity at 41% conversion.[370] This appears to be an even better route to replace the current commercial oxidation of cyclohexane than its reaction above that used hydrogen peroxide. Because ozone is toxic, care will be needed to contain it. A Pd polyoxometalate has been used to hydrogenate an aromatic ring without touching a ketone (**6.77** Schematic).[371]

Polyoxometalates, such as $K_5SiVW_{11}O_{40}$, are used in an environmentally benign bleaching process for kraft wood pulp that is now said to be cost-competitive.[372] The process involves two steps: first, anaerobic delignification of the pulp; second, aerobic oxidation of the organic waste in the bleaching liquor to carbon dioxide and water. The recovery of the polyoxometalate for reuse is 99.99%. No chlorine or organic solvents are used. Oxygen is the terminal oxidant. This process can be used in an effluent-free mill.

Heteropolyacid catalysts are also sometimes used with light. Ethanol has been dehydrogenated by $H_4SiW_{12}O_{40}$ with light to acetaldehyde and hydrogen with 100% selectivity.[373] Photocatalysis is also useful in removing low concentrations of organic pollutants from wastewater.[374] Chlorophenols in water were photolyzed in water with oxygen and $H_3PW_{12}O_{40}$, with 97% removal in 120 min. The commonly used titanium dioxide gave 77% removal under these conditions. This could be a very practical process if it can be performed with a supported catalyst in sunlight.

Several ways have been used to place finely dispersed heteropolyacids on supports. Polyaniline has been used as a support for $H_4SiW_{12}O_{40}$.[375] Heteropolyacids have been placed on silica,[376] in mesoporous zeolites,[377] and even in a viral particle from which the RNA had been removed.[378] A polyoxometalate catalyst became more active and selective as an oxidation catalyst when it was supported on a silica that contained polyethyleneoxy chains [i.e., $SiO(CH_2CH_2O)_nCH_2CH_2OCH_3$].[379] Reuse was not mentioned. The carbon chain might oxidize gradually during use, in contrast with catalysts without carbon atoms. Heteropolyacids have also been embedded in polyethersulfone and poly(phenylene oxide) to increase their activity.[380] $H_3PMo_{12}O_{40}$ dispersed in a polyethersulfone had 10 times the

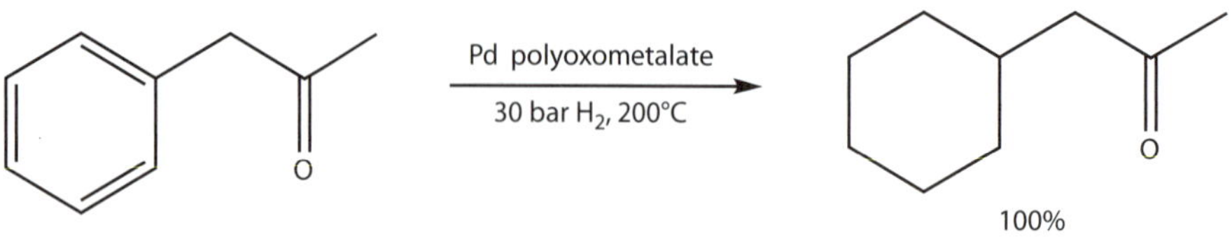

Pd polyoxometalate

30 bar H_2, 200°C

100%

6.77 Schematic

activity in the oxidation of ethanol to acetaldehyde by oxygen as the bulk heteropolyacid.[381] $K_8SiW_{11}O_{39}$ was reacted with a trichlorosilyl-terminated poly(dimethylsiloxane) to anchor the polyoxometalate to the polymer.[382] The use of polyoxometalates to pillar clays has been mentioned earlier.

Finally, antiviral activity has been found in some polyoxometalates, such as $K_7PTi_2W_{10}O_{40} \cdot 7H_2O$, against human immunodeficiency virus (HIV).[383]

REFERENCES

1. (a) K. Tanabe, *Appl. Catal. A*, 1994, *113*, 147; (b) H. Yamamoto and K. Ishihara, eds., *Acid Catalysis in Modern Organic Synthesis*, Wiley-VCH, Weinheim, 2008, 2 vol.
2. (a) J.F. Roth, *Chemtech*, 1991, *21*, 361; (b) L.F. Albright, *Chemtech*, 1998, *28*(6), 40.
3. R.A. Smith, *Kirk–Othmer Encyclopedia of Chemical Technology*, 4th ed., Wiley, New York, 1994, *11*, 374.
4. (a) E. Furimsky, *Catal. Today*, 1996, *30*, 244; (b) www.csb.gov, Aug. 26, 2009
5. Anon., *Chem. Eng. News*, Sep. 13, 1999, 40.
6. (a) G.A.S. Olah, *Chem. Abstr.*, 1992, *116*, 40893; (b) G. Parkinson, *Chem. Eng.*, 1995, *102*(12), 17; (c) *Chem. Eng. News*, Feb. 27, 1995.
7. C.D. Change and P.G. Rodewald, Jr, U.S. patent 5,457,257 (1995).
8. A. Gelbein and R. Piccolini, *Chemtech*, 1996, *26*(1), 29.
9. (a) Anon., *Hydrocarbon Processing*, 1996, *75*(2), 40; (b) G. Parkinson, *Chem. Eng.*, 1996, *103*(1), 23.
10. Anon., *Hydrocarbon Processing*, 2002, *81*(5), 35.
11. G. Parkinson, *Chem. Eng.*, 2005, *112*(4), 21.
12. G. Parkinson, *Chem. Eng.*, 2001, *108*(1), 27.
13. S.K. Ritter, *Chem. Eng. News*, Oct. 1, 2001, 63.
14. R. D'Aquino and L. Marvidis, *Chem. Eng. Prog.*, 2007, *103*(1), 8.
15. G. Ondrey, *Chem. Eng.*, 2007, *114*(1), 11.
16. A.K. Kolah, Q. Zhiven, and S.M. Mahajani, *Chem. Innov.*, 2001, *31*(3), 15.
17. M. Marchionna, M. diGirolamo, and R. Patrini, *Catal. Today*, 2001, *65*, 397.
18. J.A. Horsley, *Chemtech*, 1997, *27*(10), 45.
19. R.A. Sheldon, *Chemtech*, 1994, *24*(3), 38.
20. P. Botella, A. Corina, and J. Lopez-Nieto, 1999, *185*, 371.
21. (a) J.M. Thomas, *Angew. Chem. Int. Ed.*, 1994, *33*, 926, 931; (b) A. Corma, *Chem. Rev.*, 1995, *95*, 559; (c) Y. Izumi, *Catal. Today*, 1997, *33*, 371; (d) J.A. Horsley, *Chemtech*, 1997, *27*(10), 45; (e) I.E. Maxwell, *Cattech*, 1997, *1*(1), 5, 27; (f) E. Iglesia, D.G. Barton, J.A. Biscardi, M.J.L. Gines, and S.L. Soled, *Catal. Today*, 1997, *38*, 339.
22. (a) G.A. Olah, G.K.S. Prakash, and J. Sommer, *Superacids*, 2nd ed., Wiley-Interscience, New York, 2009; (b) T.A. O'Donnell, *Superacids and Acidic Melts as Inorganic Chemical Reaction Media*, VCH, Weinheim, 1993; (c) M. Misono and T. Okuhara, *Chemtech*, 1993, *23*(11), 23; (d) G.A. Olah, *Superacid Chemistry*, 2nd ed., Wiley, Hoboken, NJ, 2009.
23. (a) E.S. Stoyanov, S.P. Hoffmann, M. Juhasz, and C.A. Reed, *J. Am. Chem. Soc.*, 2006, *128*, 3160; (b) A. Avelar, F.S. Tham, and C.A. Reed, *Angew. Chem. Int. Ed.*, 2009, *48*, 3491.
24. L.O. Muller, D. Himmel, J. Stauffer, J. Slattery, G. Santiso-Quinones, V. Brecht, and I. Krossing, *Angew. Chem. Int. Ed.*, 2008, *47*, 7659.
25. G. Parkinson, *Chem. Eng.*, 1996, *103*(10), 25.
26. (a) G.A. Olah, P.S. Iyer, and G.K.S. Prakash, *Synthesis*, 1986, 513; (b) S.J. Sondheimer, N.J. Bunce, and C.A. Fyfe, *J. Macromol. Sci. Rev. Macromol. Chem.*, 1986, *C26*(3), 353.
27. J.B. Simonato, *Chem. Ind.* (*Lond.*), Dec. 19, 2005, 20.
28. I. Busci, A. Molnar, M. Bartok, and G.A. Olah, *Tetrahedron*, 1994, *50*, 8195.
29. I. Busci, A. Molnar, M. Bartok, and G.A. Olah, *Tetrahedron*, 1995, *51*, 3319.
30. J.R. Kosak and T.A. Johnson, eds, *Catalysis of Organic Reactions*, Dekker, New York, 1993, 487.
31. A.J. Seen, K.J. Cavell, A.M. Hodges, and A.W.-H. Mau, *J. Mol. Catal.*, 1994, *94*, 163.
32. S.K. Taylor, M.G. Dickinson, S.A. May, D.A. Pickering, and P.C. Sadek, *Synthesis*, 1998, 1133.
33. (a) Q. Sun, M.A. Harmer, and W.E. Farneth, *Chem. Commun.*, 1996, 1201; (b) Q. Sun, W.E. Farneth, and M.A. Harmer, *J. Catal.*, 1996, *164*, 62; (c) M.A. Harmer, W.E. Farneth, and Q. Sun, *J. Am. Chem. Soc.*, 1996, *118*, 7708; (d) B. Torok, I. Kiricsi, A. Molnar, and G.A. Olah, *J. Catal.*, 2000, *193*, 132.
34. Q. Deng, R.B. Moore, and K.A. Mauritz, *J. Appl. Polym. Sci.*, 1998, *68*, 747.
35. Q. Sun, M.A. Harmer, and W.E. Farneth, *Ind. Eng. Chem. Res.*, 1997, *36*, 5541.
36. W. Apichatachutapan, R.B. Moore, and K.A. Mauritz, *J. Appl. Polym. Chem.*, 1996, *62*, 417.
37. M.A. Harmer, Q. Sun, M.J. Michalczyk, and Z. Yang, *Chem. Commun.*, 1997, 1803.
38. J. Alauzun, I.A. Mehdi, C. Reye, and R.J.P. Corriu, *J. Am. Chem. Soc.*, 2006, *128*, 8718.
39. M. Santelli and J.-M. Pons, *Lewis Acids and Selectivity in Organic Synthesis*, CRC Press, Boca Raton, FL, 1996.
40. (a) K. Tanabe, *Appl. Catal. A*, 1994, *113*, 147; (b) S.J. Barlow, K. Martin, J.H. Clark, and A.J. Teasdale, *Chem. Abstr.*, 1996, *125*, 280, 521; (c) B. Trenbirth, T.W. Bastock, J.H. Clark, K. Martin, and P. Price, *Chem. Abstr.*, 1998, *129*, 262, 700; (d) A. Watson, J. Rafelt, D.J. Macquarrie, J.H. Clark, K. Wilson, J. Mdoe, and D.B. Jackson, Preprints A.C.S. *Div. Environ. Chem.*, 1999, *39*(1), 457; (e) D.J. Macquarrie, J.H. Clark, J.E.G. Mdoe, D.B. Jackson, and P.M. Price, Preprints A.C.S. *Div. Environ. Chem.*, 1999, *39*(1), 460.
41. (a) T.C. Chung and A. Kumar, *Polym. Bull.*, 1992, *28*, 123; (b) T.C. Chung, A. Kumar, and D. Rhubright, *Polym. Bull.*, 1993, *30*, 385; (c) F.J.-Y. Chen, T.-C. Chung, J.E. Stanat, and S.H. Lee, U.S. patent 5,770,539 (1998).
42. H. Yang, B. Li, and Y. Fang, *Synth. Commun.*, 1994, *24*, 3269.
43. Z. Li and A. Ganesan, *Synth. Commun.*, 1998, *28*, 3209.
44. (a) X. Song and A. Sayari, *Catal. Rev. Sci. Eng.*, 1996, *38*, 329; *Chemtech*, 1995, *25*(8), 27; (b) A. Corma, A. Martinez, and C. Martinez, *J. Catal.*, 1994, *149*, 52; (c) A. Corma, *Chem. Rev.*, 1995, *95*, 559; (d) J.M. Thomas, *Angew. Chem. Int. Ed.*, 1994, *33*, 926, 931; (e) A.S.C. Brown and J.S.J. Hargreaves, *Green Chem.*, 1999, *1*, 17; (f) B.M. Reddy and M.K. Patel, *Chem. Rev.*, 2009, *109*, 2185–2208.

45. (a) D. Tichit, D. el Alami, and F. Figueras, *J. Catal.*, 1996, *163*, 18; (b) K.T. Wan, C.B. Khouw, and M.E. Davis, *J. Catal.*, 1996, *158*, 311; (c) F. Babou, G. Coudurier, and J.C. Vedrina, *J. Catal.*, 1995, *152*, 341; (d) T.-K. Cheung, J.L. d'Itri, and B.C. Gates, *J. Catal.*, 1995, *153*, 344; (e) D. Farcasiu, A. Ghenciu, G. Marino, and K.D. Rose, *J. Am. Chem. Soc.*, 1997, *119*, 11826; (f) T.-K. Cheung and B.C. Gates, *Chemtech*, 1997, *27*(9), 28; (g) A.S.C. Brown and J.S.J. Hargreaves, *Green Chem.*, 1999, 17.
46. (a) D. Farcasiu and J.Q. Li, *Appl. Catal. A*, 1995, *125*, 97; (b) M.R. Gonzalez, J.M. Kobe, K.B. Fogash, and J.A. Dumesic, *J. Catal.*, 1996, *160*, 290; (c) D. Farcasiu, J.Q. Li, and S. Cameron, *Appl. Catal. A*, 1997, *154*, 173; (d) M.-T. Tran, N.S. Gnep, G. Szabo, and M. Guisnet, *Appl. Catal. A*, 1998, *171*, 207.
47. (a) T. Lopez, J. Navarrette, R. Gomez, O. Navaro, F. Figueras, and H. Armendariz, *Appl. Catal. A*, 1995, *125*, 217; (b) D.A. Ward and E.I. Ko, *J. Catal.*, 1994, *150*, 18; (c) H. Armendariz, B. Coq, D. Tichit, R. Dutartre, and F. Figueras, *J. Catal.*, 1998, *173*, 345.
48. E.E. Platero, M.P. Mentruit, C.O. Arean, and A. Zecchina, *J. Catal.*, 1996, *162*, 268.
49. (a) C. Morterra, G. Cerrato, F. Pima, and Signoretto, *J. Catal.*, 1995, *157*, 109; (b) R. Srinivasan, R.E. Keogh, D.R. Milburn, and B.H. Davis, *J. Catal.*, 1995, *153*, 123.
50. W. Stichert and F. Schuth, *J. Catal.*, 1998, *174*, 242.
51. (a) K. Arcata, *Appl. Catal. A*, 1996, *146*, 3; (b) J.C. Yori and J.M. Parera, *Appl. Catal. A*, 1996, *147*, 145; (c) S.N. Koyande, R.G. Jaiswal, and R.V. Jayaram, *Ind. Eng. Chem. Res.*, 1998, *37*, 908; (d) M.S. Wong and J.Y. Ying, Abstracts Boston A.C.S. Meeting, Aug. 1998, CATL 038, [for yttrium].
52. (a) R. Srinavasan, R.E. Keogh, and B.H. Davis, *Appl. Catal. A*, 1995, *130*, 135; (b) T.-K. Cheung, J.L. d'Itrl, and B.C. Gates, *J. Catal.*, 1995, *151*, 464.
53. J.E. Tabora and R.J. Davis, *J. Catal.*, 1996, *162*, 125.
54. T. Shishido, T. Tanaka, and H. Hattori, *J. Catal.*, 1997, *172*, 24.
55. (a) J.-M. Manoli, C. Potvin, M. Muhler, U. Wild, G. Resofski, T. Buchholz, and Z. Paal, *J. Catal.*, 1998, *178*, 338; (b) S.R. Vaudagna, R.A. Cornelli, S.A. Canavese, and N.S. Figoli, *J. Catal.*, 1997, *169*, 389.
56. H.Y. Chu, M.P. Rosynek, and J.H. Lunsford, *J. Catal.*, 1998, *178*, 352.
57. G. Parkinson, *Chem. Eng.*, 1999, *106*(3), 19.
58. (a) J.G. Santiesteban, J.C. Vartuli, S. Han, R.D. Bastian, and C.D.C. Hang, *J. Catal.*, 1997, *168*, 431; (b) R.A. Boyse and E.I. Ko, *J. Catal.*, 1997, *171*, 191; 1998, *179*, 100.
59. (a) G. Larsen and L.M. Petkovic, *J. Mol. Catal. A: Chem.*, 1996, *113*, 517; (b) G. Larsen, E. Lotero, S. Raghaven, R.D. Parra, and C.A. Querini, *Appl. Catal. A*, 1996, *139*, 201; (c) S.R. Vaudagna, R.A. Comelli, S.A. Canavese, and N.S. Figoli, *Appl. Catal. A*, 1998, *168*, 93; 1998, *164*, 265.
60. B. Manohar, V.R. Reddy, and B.M. Reddy, *Synth. Commun.*, 1998, *28*, 3183.
61. (a) A. Corma, A. Martinez, and C. Martinez, *Appl. Catal. A*, 1996, *144*, 249; (b) S.P. Chavan, P.K. Zubaidha, S.W. Dantale, A. Keshavaraja, A.V. Ramaswamy, and T. Ravindranathan, *Tetrahedron Lett.*, 1996, *37*, 233, 237; (c) Y.-Y. Huang, B.-Y. Zhao, and Y.-C. Xie, *Appl. Catal. A*, 1998, *171*, 65.
62. T. Reimer, D. Spielbauer, M. Hunger, G.A.H. Mekhemer, and H. Knozinger, *J. Chem. Soc., Chem. Commun.*, 1994, 1181.
63. T. Ushikubo and K. Wada, *J. Catal.*, 1994, *148*, 138.
64. (a) H. Grabowska, W. Mista, L. Syper, J. Wrzyszcz, and M. Zawadzki, *Angew. Chem. Int. Ed.*, 1996, *35*, 1512; *J. Catal.*, 1996, *160*, 134; (b) J. Wrzyszcz, H. Grabowska, W. Mista, L. Syper, and M. Zawadzki, *Appl. Catal. A*, 1998, *166*, L249.
65. M. Hino and K. Arata, *Bull. Chem. Soc. Jpn.*, 1994, *67*, 1472.
66. S.P. Ghorpade, V.S. Darshane, and S.G. Dixit, *Appl. Catal. A*, 1998, *166*, 135.
67. C. Tagusagawa, A. Takagaki, S. Hayashi, and K. Domen, *J. Am. Chem. Soc.*, 2008, *130*, 7230.
68. A. Fischer, P. Makowski, J.-O. Muller, M. Antonietti, A. Thomas, and F. Goettmann, *ChemSusChem*, 2008, *1*, 444.
69. (a) J. Inanaga, Y. Sugimoto, and T. Hanamoto, *New J. Chem.*, 1995, *19*, 707; (b) R.W. Marshman, *Aldrich Chim. Acta*, 1995, *28*(3), 77; (c) W. Xie, Y. Jin, and G. Wang, *Chemtech*, 1999, *29*(2), 23; (d) S. Kobayashi, M. Sugiura, H. Kitagawa, and W.W.-L. Lam, *Chem. Rev.*, 2002, *102*, 2227–2302; (e) S. Kobayashi and K. Manabe, *Acc. Chem. Res.*, 2002, *35*, 209.
70. A.G.M. Barett and D.C. Braddock, *Chem. Commun.*, 1997, 351.
71. (a) S. Kobayashi and R. Akiyama, *Chem. Commun.*, 2003, 449; (b) S. Kobayashi and R. Akiyama, *Pure Appl. Chem.*, 2001, *73*, 1103.
72. (a) S. Kobayashi and S. Nagasyama, *J. Am. Chem. Soc.*, 1998, *120*, 2985; 1997, *119*, 11049; (b) M. Rouhi, *Chem. Eng. News*, Mar. 30, 1998, 5.
73. A. Kawada, S. Mitamura, and S. Kobayashi, *Chem. Commun.*, 1996, 183.
74. K. Ishihara, M. Kubota, H. Kurihara, and H. Yamamoto, *J. Am. Chem. Soc.*, 1995, *117*, 4413.
75. V.K. Aggarwal, G.P. Vennall, P.N. Davey, and S.C. Newman, *Tetrahedron Lett.*, 1998, *39*, 1997.
76. (a) D. Chen, L. Yu, and P.G. Wang, *Tetrahedron Lett.*, 1996, *37*, 4467; (b) T. Tsuchimoto, T. Hiyama, and S.-I. Fukuzawa, *Chem. Commun.*, 1996, 2345.
77. G. Jenner, *Tetrahedron Lett.*, 1995, *36*, 233.
78. E. Keller and L. Feringa, *Tetrahedron Lett.*, 1996, *37*, 1879.
79. N.E. Drysdale and R.E. Bockrath, *Chem. Abstr.*, 1995, *123*, 33862.
80. N.E. Drysdale, U.S. patent 5,475,069 (1995).
81. (a) I. Hachiya, M. Moriwaki, and S. Kobayashi, *Bull. Chem. Soc. Jpn.*, 1995, *68*, 2053; (b) S. Kobayashi, M. Moriwaki, and I. Hachiya, *Tetrahedron Lett.*, 1996, *37*, 2053.
82. F.J. Waller, A.G.M. Barrett, D.C. Braddock, and D. Ramprasad, *Chem. Commun.*, 1997, 613; *Tetrahedron Lett.*, 1998, *39*, 1641.
83. S. Kobayashi and T. Wakabayashi, *Tetrahedron Lett.*, 1998, *39*, 5389.
84. A. Takasu, Y. Shibata, Y. Narukawa, and T. Hirabayashi, *Macromolecules*, 2007, *40*, 151.
85. (a) T.-P. Loh, J. Pei, and G.-Q. Cao, *Chem. Commun.*, 1996, 1819; (b) T.-P. Loh and X.-R. Li, *Chem. Commun.*, 1996, 1929; (c) T.-P. Loh, J. Pei, and M. Lin, *Chem. Commun.*, 1996, 2315.
86. K. Ishihara, N. Hanaki, M. Funahashi, M. Miyata, and H. Yamamoto, *Bull. Chem. Soc. Jpn.*, 1995, *68*, 1721.

87. Y. Ono and T. Baba, *Catal. Today*, 1997, *38*, 321.
88. W.J. Smith, III, and J.S. Sawyer, *Tetrahedron Lett.*, 1996, *37*, 299.
89. H. Handa, T. Baba, H. Sugisawa, and Y. Ono, *J. Mol. Chem. A: Chem.*, 1998, *134*, 171.
90. J.H. Zhu, Y. Wang, Q.H. Xu, and H. Hattori, *Chem. Commun.*, 1996, 1889.
91. M.A. Aramendia, V. Borau, C. Jimenez, J.M. Marinas, and F.J. Romero, *Chem. Lett.*, 1995, *24*, 279.
92. (a) A. Corma, V. Fomes, and F. Rey, *J. Catal.*, 1994, *148*, 205; (b) J.I. D Cosimo, V.K. Diez, M. Xu, E. Iglesia, and C.R. Apesteguia, *J. Catal.*, 1998, *178*, 499; (c) A.M. Auerbach, K.A. Carrado, and K.K. Dutta, *Handbook of Layered Materials: Layered Double Hydroxides*, Marcel Dekker, New York, 2004; (d) V.S. Shirure, B.P. Nikhade, and V.G. Pangarkar, *Ind. Eng. Chem. Res.*, 2007, *46*, 3086.
93. L. Li and J. Shi, *Chem. Commun.*, 2008, 996.
94. P.S. Kumbhar, J. Sanchez-Valente, and F. Figueras, *Chem. Commun.*, 1998, 1091.
95. P.S. Kumbhar, J. Sanchez-Valente, J. Lopez, and F. Figueras, *Chem. Commun.*, 1998, 535.
96. A. Corma, S. Iborra, S. Miquel, and J. Primo, *J. Catal.*, 1998, *173*, 315.
97. (a) M.L. Kantam, B.M. Choudary, C.V. Reddy, K.K. Rao, and F. Figueras, *Chem. Commun.*, 1998, 1033; (b) K.K. Rao, M. Gravelle, J.S. Valente, and F.L. Figueras, *J. Catal.*, 1998, *173*, 115; (c) B.M. Choudary, M.L. Kantam, B. Kavita, C.V. Reddy, K.K. Rao, and F. Figueras, *Tetrahedron Lett.*, 1998, *39*, 3555.
98. J.C. van der Waal, P.J. Kunkeler, K. Tan, and H. van Bekkum, *J. Catal.*, 1998, *173*, 74.
99. J. Santhanalakshmi and T. Raja, *Appl. Catal. A*, 1996, *147*, 69.
100. B.M. Choudary, N. Narender, and V. Bhuma, *Synth. Commun.*, 1995, *25*, 2829.
101. M.J. Climent, A. Corma, S. Iborra, and J. Primo, *J. Catal.*, 1995, *151*, 60.
102. B. Coq, D. Tichit, and S. Ribet, *J. Catal.*, 2000, *189*, 117.
103. M. Onaka, A. Ohta, K. Sugita, and Y. Izumi, *Appl. Catal. A*, 1995, *125*, 203.
104. M. Chibwe and T.J. Pinnavaia, *J. Chem. Soc., Chem. Commun.*, 1993, 278.
105. (a) S.P. Newman and W. Jones, *New J. Chem.*, 1998, *22*, 105; (b) S. Carlino, *Chem. Br.*, 1997, *33*(9), 59; (c) E.L. Crepaldi, P.C. Pavan, and J.J.B. Valim, *Chem. Commun.*, 1999, 155.
106. P. Tang, X. Xu, Y. Lin, and D. Li, *Ind. Eng. Chem. Res.*, 2008, *47*, 2478.
107. S. Sato, K. Koizumi, and F. Nozaki, *J. Catal.*, 1998, *178*, 264.
108. D.J. Macquarrie, *Tetrahedron Lett.*, 1998, *39*, 4125.
109. H. Gorzawski and W.F. Holderich, *J. Mol. Catal. A*, 1999, *144*, 181.
110. L.B. Sun, J. Yang, J.H. Kou, F.N. Gu, Y. Chun, Y. Wang, J.H. Zhu, and G. Zou, *Angew. Chem. Int. Ed.*, 2008, *47*, 3418.
111. L. Wang, C. Li, M. Liu, D.G. Evans, and X. Duan, *Chem. Commun.*, 2007, 123.
112. (a) I. Rodriguez, S. Iborra, A. Corma, F. Rey, and J.L. Gorda, *Chem. Commun.*, 1999, 593; (b) B.M. Chaudary, M.L. Kantam, P. Sreekanth, T. Bandopadhyay, F. Figueras, and A. Tuel, *J. Mol. Catal. A*, 1999, *142*, 361; (c) X. Lin, G.K. Chuah, and S. Jaenicke, *J. Mol. Catal. A*, 1999, *150*, 287.
113. (a) P.-S.E. Dai, *Catal. Today*, 1995, *26*, 3; (b) C.B. Dartt and M.E. Davis, *Ind. Eng. Chem. Res.*, 1994, *33*, 2887; (c) M.E. Davis and S.L. Suib, *Selectivity in Catalysis*, A.C.S. Symp. 517, Washington, DC, 1993; (d) R.A. Sheldon, *Chem. Ind. (Lond.)*, 1997, 12; (e) N.Y. Chen, W.E. Garwood, and F.G. Dwyer, *Shape Selective Catalysis in Industrial Applications*, 2nd ed., Dekker, Monticello, New York, 1996; (f) M.E. Davis, *Cattech*, 1997, *1*(1), 19; (g) A. Corma and H. Garcia, *Catal. Today*, 1997, *38*, 257; (h) R.S. Downing, H. van Bekkum and R.A. Sheldon, *Cattech*, 1997, *1*(2), 95; (i) H.K. Beyer, *Studies in Surface Science and Catalysis 94: Catalysis by Microporous Materials*, Elsevier, Amsterdam, 1995.
114. (a) W. Buchner, R. Schliebs, G. Winter, and K.H. Buchel, *Industrial Inorganic Chemistry*, VCH, Weinheim, 1989, 321; (b) B.C. Gates, *Kirk–Othmer Encyclopedia of Chemical Technology*, 4th ed., Wiley, New York, 1993, *5*, 358–363; (c) G.H. Kuhl and C.T. Kresge, *Kirk–Othmer Encyclopedia of Chemical Technology*, 4th ed., Wiley, New York, 1995, *16*, 888–925; (d) D.W. Breck and R.A. Anderson, *Kirk–Othmer Encyclopedia of Chemical Technology*, 3rd ed., Wiley, New York, 1981, *15*, 638–669; (e) M.E. Davis, *Ind. Eng. Chem. Res.*, 1991, *30*, 1675; (f) S.L. Suib, *Chem. Rev.*, 1993, *93*, 803; (g) J. Weitkamp, H.G. Karge, H. Pfeifer, and W. Holderich, eds, *Zeolites and Related Microporous Materials*, Elsevier, Amsterdam, 1994; (h) R. Szostak, *Molecular Sieves*, van Nostrand-Reinhold, New York, 1989; (i) T. Hattori and T. Yashima, *Zeolites and Microporous Crystals*, Elsevier, Amsterdam, 1994; (j) R.F. Lobo, J.S. Beck, S.L. Suib, D.R. Corbin, M.E. Davis, L.E. Ton, and S.I. Zones, eds, *Microporous and Macroporous Materials*, Materials Research Society, Pittsburgh, 1996, *431*; (k) A. Dyer, *An Introduction to Zeolite Molecular Sleves*, Wiley, New York, 1988; (l) H. van Bekkum, E.M. Flanigen, and J.C. Jansen, eds, *Introduction to Zeolite Science and Practice*, Elsevier, Amsterdam, 1991; (m) F.R. Ribeiro, A. Alvarez, C. Henriques, F. Lemos, J.M. Lopes, and M.F. Ribeiro, *J. Mol. Catal. A: Chem.*, 1995, *96*, 245; (n) C. Bremard and D. Bougeard, *Adv. Mater.*, 1995, *7*(1), 10; (o) J. Weitkamp, *Handbook of Heterogeneous Catalysis*, VCH, Weinheim, 1997; (p) J. Weitkamp, *Molecular Sieves: Science and Technology*, Springer, Berlin, 1998; (q) H. Chon, S.I. Woo, and S.E. Park, eds, *Recent Advances and New Horizons in Zeolite Science and Technology—Studies in Surface Science and Catalysis*, Elsevier, Amsterdam, 1996, *102*; (r) N. Herron and D.R. Corbin, eds, *Inclusion Chemistry with Zeolites: Nanoscale Materials by Design: Topics in Inclusion Science*, Kluwer Academic, Dordrecht, the Netherlands, 1995, *6*; (s) R.E. Morris and P.A. Wright, *Chem. Ind.*, 1998, 256; (t) M.E. Davis, *Chem. Eur. J.*, 1997, *3*, 1745–1750; (u) A. Corma, *Chem. Rev.*, 1997, *97*, 2373; (v) R. Richards, ed., *Surface and Nanomolecular Catalysis*, Taylor and Francis/CRC Press, Boca Raton, FL, 2006; (w) F. Schuth, K. Sing, and J. Weitkamp, eds, *Handbook of Porous Solids*, Wiley-VCH, Weinheim, 2002; (x) M. Guisnet and J.-P. Gilson, *Zeolites for Cleaner Technologies*, Imperial College Press, London, 2002; (y) J.H. Clark and C.N. Rhodes, *Clean Synthesis Using Porous Inorganic Solid Catalysts and Supported Reagents*, Royal Society of Chemistry, Cambridge, 2000; (z) R. Xu, W. Pang, J. Yu, Q. Huo, and J. Chen, *Chemistry of Zeolites and Related Porous Materials*, Synthesis and Structure, Wiley, Hoboken, NJ, 2007; (aa) J.H. Clark, *Acc.*

Chem. Res., 2002, *35*, 791; (bb) S.M. Auerbach, K.A. Carrado, and P.K. Dutta, *Handbook of Zeolite Science and Technology*, Marcel Dekker, New York, 2003.

115. V. Ramamurthy, D.F. Eaton, and J.V. Caspar, *Acc. Chem. Res.*, 1992, *25*, 299.
116. B.C. Gates, *Kirk–Othmer Encyclopedia of Chemical Technology*, Wiley, New York, 1993, *5*, 358–363.
117. B.C. Gates, *Kirk–Othmer Encyclopedia Chemical Technology*, Wiley, New York, 1993, *5*, 358–363.
118. C. Baerlocher, W.M. Meier, and D.H. Olsen, *Atlas of Zeolite Framework Types*, 5th ed., Elsevier, Amsterdam, 2001.
119. (a) M.L. Occelli and H.E. Robson, *Zeolite Synthesis*, A.C.S. Symp. 398, Washington, DC, 1989; (b) M.L. Occelli and H. Kessler, *Synthesis of Porous Materials: Zeolites, Clays and Nanostructures*, Dekker, New York, 1997; (c) S. Biz and M.L. Occelli, *Catal. Rev. Sci. Eng.*, 1998, *40*(3), 329; (d) S. Feng and R. Xu, *Acc. Chem. Res.*, 2001, *34*, 239; (e) C.S. Cundy and P.A. Cox, *Chem. Rev.*, 2003, *103*, 663–702; (f) F. Schuth, *Angew. Chem. Int. Ed.*, 2003, *42*, 3605; (g) M. Maple, *Chem. Ind. (Lond.)*, Apr. 17, 2006, 19.
120. (a) S.J. Monaco and E.I. Ko, *Chemtech*, 1998, *28*(6), 23; (b) A.C. Pierre, *Introduction to Sol–Gel Processing*, Kluwer Academic, Boston, 1998; (c) C.J.L. Brinker and G.W. Scherer, *The Physics and Chemistry of Sol–Gel Processes*, Academic, New York, 1990; (d) L.L. Hench and R. Orefice, *Krik–Othmer Encyclopedia of Chemical Technology*, 4th ed., 1997, *22*, 497; (e) G. Bellussi, ed., *Catal. Today*, 1998, *41*, 3–41; (f) E.I. Ko, ed., *Catal. Today*, 1997, *35*, 203–365; (g) L.C. Klein, E.J. Pope, S. Sakka, and J.L. Woolfrey, eds, *Sol–Gel Processing of Advanced Materials—Ceramic Transactions*, American Ceramic Society, Westerville, Ohio, 1998, *81*; (h) J.D. Wright and W.A.J.M. Sommerdijk, *Sol–Gel Materials, Chemistry and Applications*, Gordon and Breach, Amsterdam, 2001; (i) B. Dunn and J.I. Zink, *Acc. Chem. Res.*, 2007, *40*(9), 730–895 [issue on sol–gel technology].
121. V. Lafond, P.H. Mutin, and A. Vioux, *J. Mol. Catal. A*, 2002, *182/183*, 81.
122. (a) H.-L. Chang and W.-H. Shih, *Ind. Eng. Chem. Res.*, 2000, *39*, 4185; (b) C.L. Choi, M. Park, D.H. Keem, J.-E. Kim, B.-Y. Park, and J. Choi, *Environ. Sci. Technol.*, 2001, *35*, 2810; (c) N. Moreno, X. Querol, C. Ayora, C.F. Pereira, and M. Janssen-Jurkovicova, *Environ. Sci. Technol*, 2001, *35*, 3526.
123. (a) M.E. Davis, *Chemtech*, 1994, *24*(9), 22; (b) P. Wagner, M. Yoshikawa, M. Lovallo, K. Tusji, M. Taspatsis, and M.E. Davis, *Chem. Commun.*, 1997, 2179; (c) C.Y. Chen, L.W. Finger, R.C. Medrud, P.A. Crozier, I.Y. Chan, T.V. Harris, and I. Zones, *Chem. Commun.*, 1999, 1775; (d) M.A. Camblor, A. Corma, and L.A. Villaescusa, *Chem. Commun.*, 1997, 749.
124. (a) G.J. Hutchings, G.W. Watson, and D.J. Willock, *Chem. Ind. (Lond.)*, 1997, 603; (b) R. Millini, *Catal. Today*, 1998, *41*, 41.
125. D.E. Akporiaye, I.M. Dahl, A. Karisson, and R. Wendelbo, *Angew, Chem. Int. Ed. Engl.*, 1998, *37*, 609.
126. (a) M.A. Uguina, D.P. Serrano, G. Ovejero, R. van Grieken, and M. Camacho, *Appl. Catal. A*, 1995, 124, 391; (b) H. Gao, J. Suo, and S. Li, *J. Chem. Soc. Chem. Commun.*, 1995, 835; (c) C.B. Khouw and M.E. Davis, *J. Catal.*, 1995, *151*, 77.
127. (a) C.-G. Wu and T. Bein, *Chem. Commun.*, 1996, 925; (b) S.E. Park, D.S. Kim, J.-S. Chang, and W.Y. Kim, *Catal. Today*, 1998, *44*, 301; (c) S. van Donk, A.H. Jansson, J.H. Bitter, and K.P. deJong, *Catal. Rev. Sci. Eng.*, 2003, *45*, 297.
128. (a) N. Coustel, F.D. Renzo, and F. Fajula, *J. Chem. Soc. Chem. Commun.*, 1994, 967; (b) R. Schmidt, D. Akporiaye, M. Stoker, and O.H. Ellestad, *J. Chem. Soc. Chem. Commun.*, 1994, 1491; (c) J.Y. Ying, C.P. Mehnert, and M.S. Wong, *Angew. Chem. Int. Ed.*, 1999, *38*, 56–77; (d) P. Selvam, S.K. Bhatia, and C.G. Sonwane, *Ind. Eng. Chem. Res.*, 2001, *40*, 3237; (e) X. Tang, S. Liu, Y. Wang, W. Huang, E. Sominski, O. Palchik, Y. Koltypin, and A. Gedanken, *Chem. Commun.*, 2000, 2119.
129. S. Namba, A. Mochizuki, and M. Kito, *Chem. Lett.*, 1998, *27*, 569.
130. C.S. Cundy, R.J. Plaisted, and J.P. Zao, *Chem. Commun.*, 1998, 1465.
131. S.H. Tolbert, A. Firouzi, G.D. Stucky, and B.F. Chmelka, *Science,* 1997, *278*, 264.
132. (a) C.E. Fowler, S.L. Burkett, and S. Mann, *Chem. Commun.*, 1997, 1769; 1998, 1825; (b) C.E. Fowler, B. Lebeau, and S. Mann, *Chem. Commun.*, 1998, 1825; (c) D.L. Ou, A. Adamjee, and A.B. Seddon, *Surf. Coat. Int.*, 1996, *79*, 496; (d) M.H. Lim, C.F. Blanford, and A. Stein, *J. Am. Chem. Soc.*, 1997, *117*, 4090.
133. (a) K. Yamamoto, Y. Sakata, Y. Nohara, Y. Takahashi, and T. Tatsumi, *Science*, 2003, *300*, 470; (b) B. Boury, R.J.P. Corriu, V. Le Strat, P. Delord, and M. Nobili, *Angew. Chem. Int. Ed.*, 1999, *38*, 3172; (c) D.A. Loy, K. Rahimian, and M. Samara, *Angew. Chem. Int. Ed.*, 1999, *38*, 555; (d) G. Cerveau, R.J.P. Corriu, and E. Framery, *Chem. Commun.*, 1999, 2081; (e) C. Yoshima-Ishii, T. Assefa, N. Coombs, M.J. MacLachlan, and G.A. Ozin, *Chem. Commun.*, 1999, 2539; (f) M.P. Kapoor, Q. Yang, and S. Inagaki, *J. Am. Chem. Soc.*, 2002, *124*, 15176; (g) A. Sayari and Y. Yang, *Chem. Commun.*, 2002, 2582; (h) H. Muramatsu, R. Corriu, and B. Boury, *J. Am. Chem. Soc.*, 2003, *125*, 854.
134. (a) S. Kawi and M.W. Lai, *Chem. Commun.*, 1998, 1407; *Chemtech*, 1998, *28*(12), 26; (b) K.J.C. van Bommel, A. Figgeri, and S. Shinkai, *Angew. Chem. Int. Ed.*, 2003, *42*, 980.
135. C.W. Jones, K. Tsuji, and M.E. Davis, *Nature*, 1998, *393*, 52.
136. M.T.J. Keene, R. Denoyel, and P.L. Llewellyn, *Chem. Commun.*, 1998, 2203.
137. T.K. Das, K. Chaudhari, A.J. Chandwadkar, and S. Sivasanker, *J. Chem. Soc. Chem. Commun.*, 1995, 2495.
138. (a) R.E. Morris and S.J. Weigel, *Chem. Soc. Rev.*, 1997, *26*, 309; (b) R. Ryoo, S. Jun, J.M. Kim, and M.J. Kim, *Chem. Commun.*, 1997, 2225; (c) S.B. Hong, M.A. Camblor, and M.E. Davis, *J. Am. Chem. Soc.*, 1997, *119*, 761.
139. C.I. Round, C.D. Williams, and C.V.A. Duke, *Chem. Commun.*, 1997, 1849.
140. W. Lin, Q. Cai, W. Pang, and Y. Yue, *Chem. Commun.*, 1998, 2473.
141. (a) M.E. Davis, *Chemtech*, 1994, *24*(9), 22; (b) P.S. Singh and K. Kosuge, *Chem. Lett.*, 1998, 101; (c) R. Mokaya and W. Jones, *J. Catal.*, 1997, *172*, 211.
142. (a) P.T. Tanev and T.J. Pinnavaia, *Science*, 1995, *267*, 865; (b) N. Jlagappan, B.V.N. Raju, and C.N.R. Rao, *Chem. Commun.*, 1996, 2243.

143. (a) X. Bu, P. Feng, and G.D. Stucky, *Science*, 1997, *278*, 2080; (b) P.T. Tanev, Y. Liang, and T.J. Pinnavaia, *J. Am. Chem. Soc.*, 1997, *119*, 8616.
144. (a) S.S. Kim, W. Zhang, and T.J. Pinnavaia, *Science*, 1998, *282*, 1302; (b) Anon., *Chem. Eng. News*, Nov. 16, 1998, 30.
145. R. Mokaya and W. Jones, *Chem. Commun.*, 1998, 1839.
146. (a) S.A. Bagshaw, E. Prouzet, and T.J. Pinnavaia, *Science*, 1995, *269*, 1242; (b) R. Richer and L. Mercier, *Chem. Commun.*, 1998, 1775; (c) X. Zhang, Z. Zhang, J. Suo, and S. Li, *Chem. Lett.*, 1998, 755; (d) E. Prouzet and T.J. Pinnavaia, *Angew. Chem. Int. Ed.*, 1997, *36*, 516.
147. (a) D. Zhao, Q. Huo, J. Feng, B.F. Chmelka, and G.D. Stucky, *J. Am. Chem. Soc.*, 1998, *120*, 6024; (b) C.G. Goltner, *Chem. Commun.*, 1998, 2287; (c) C.G. Goltner, S. Henke, M.C. Weissenberger, and M. Antonietti, *Angew. Chem. Int. Ed.*, 1998, *37*, 613; (d) D. Zhao, J. Feng, Q. Huo, N. Melosh, G.H. Frederickson, B. Chmelka, and G.D. Stucky, *Science*, 1998, *279*, 548.
148. (a) C.C. Freyhardt, M. Tsapatsis, R.F. Lobo, K.J. Balkus, Jr, and M.E. Davis, *Nature*, 1996, *381*, 295; (b) J. Haggin, *Chem. Eng. News*, May 27, 1996, 5.
149. F. Dougnier and J.-L. Guth, *J. Chem. Soc. Chem. Commun.*, 1995, 1951.
150. (a) B.T. Holland, C.F. Blanchard, and A. Stein, *Science*, 1998, *281*, 538; (b) B.T. Holland, L. Abrams, and A. Stein, *J. Am. Chem Soc.*, 1999, *121*, 4308.
151. (a) E.F. Vansant, *J. Mol. Catal. A: Chem.*, 1997, *115*, 379; (b) J.H. Clark and D.J. Macquarrie, *Chem. Commun.*, 1998, 853; (c) A. Bleloch, B.F.G. Johnson, S.V. Ley, A.J. Price, D.S. Shephard, and A.V. Thomas, *Chem. Commun.*, 1999, 1907; (d) X.S. Zhao, C. Qs. Lu, and X. Hu, *Chem. Commun.*, 1999, 1391; (e) Y. Tao, H. Kanoh, L. Abrams, and K. Kaneko, *Chem. Rev.*, 2006, *106*, 896.
152. (a) D. Barthomeuf, *Catal. Rev. Sci. Eng.*, 1996, *38*, 521; (b) D.K. Murray, T. Howard, P.W. Gougen, T.R. Krawietz, and J.F. Haur, *J. Am. Chem. Soc.*, 1994, *116*, 6354; (c) K.R. Kloetstra and H. van Bekkum, *J. Chem. Soc. Chem. Commun.*, 1995, 1005; (d) M. Hunger, U. Schenk, B. Burger, and J. Weitkamp, *Angew. Chem. Int. Ed.*, 1997, *36*, 2504.
153. W. Zhang and T.J. Pinnavaia, *Chem. Commun.*, 1998, 1185.
154. Y. Kuroda, K. Yagi, Y. Yoshikawa, and M. Nagao, *Chem. Commun.*, 1997, 2241.
155. (a) R. Anwander, C. Palm, G. Gerstberger, O. Groeger, and G. Engelhardt, *Chem. Commun.*, 1998, 1811; (b) R. Mokaya and W. Jones, *Chem. Commun.*, 1997, 2185.
156. K. Wilson and J.H. Clark, *Chem. Commun.*, 1998, 2135.
157. (a) M. Taniguchi, Y. Ishii, T. Murata, T. Tatsumi, and M. Hidai, *J. Chem. Soc. Chem. Commun.*, 1995, 2533; (b) Y. Okamoto, *Catal. Today*, 1997, *39*, 45.
158. (a) S. Tanabe, H. Matsumoto, T. Mizushima, K. Okitsu, and Y. Maeda, *Chem. Lett.*, 1996, 327; (b) C.P. Mehnert and J.Y. Ying, *Chem. Commun.*, 1997, 2215.
159. (a) C.T. Fishel, R.J. Davis, and J.M. Garcia, *Chem. Commun.*, 1996, 649; (b) B. Moraweck, G. Bergeret, M. Cattenot, V. Kougionas, C. Geantet, J.-L. Portefaix, J.L. Zotin, and M. Breysse, *J. Catal.*, 1997, *165*, 45; (c) D.S. Shepard, T. Maschmeyer, G. Sankar, J.M. Thomas, D. Ozkaya, B.F.G. Johnson, R. Raja, R.D. Oldroyd, and R.G. Bell, *Chem. Eur. J.*, 1998, *4*, 1214.
160. P.P. Edwards, P.A. Anderson, and J.M. Thomas, *Acc. Chem. Res.*, 1996, *29*, 23.
161. (a) V.J. Rao, N. Prevost, V.L. Ramamurthy, M. Kojima, and L.J. Johnson, *Chem. Commun.*, 1997, 2209; (b) M.L. Cano, A. Corma, V. Fornes, H. Garcia, M.A. Miranda, C. Baerlocher, and C. Lengauer, *J. Am. Chem. Soc.*, 1996, *118*, 11006.
162. H. Miessner and K. Richter, *Angew. Chem. Int. Ed.*, 1998, *37*, 117.
163. (a) P.C.H. Mitchell, *Chem. Ind.*, 1991, 308; (b) M. Sykora, K. Maruszewski, S.M. Treffert-Ziemelis, and J.R. Kincaid, *J. Am. Chem. Soc.*, 1998, *120*, 3490.
164. (a) G. Vanko, Z. Homonnay, S. Nagy, A. Vertes, G. Pal-Borbely, and H.K. Beyer, *Chem. Commun.*, 1996, 785; (b) P.-P.H.J.M. Knops-Gerrits, F.C. De Schryver, M. van der Auweraer, H. Van Mingroot, X.-Y. Li, and P.A. Jacobs, *Chem. Eur. J.*, 1996, *2*, 592.
165. (a) D.E. de Vos, F. Thibault-Starzyk, P.P. Knops-Gerrits, R.F. Parton, and P.A. Jacobs, *Macromol. Symp.*, 1994, *80*, 175; (b) R.F. Parton, I.F.J. Vankelecom, M.J.A. Casselman, C.P. Bezoukhanova, J.B. Uytterhoeven, and P.A. Jacobs, *Nature*, 1994, *370*, 541; (c) S. Borman, *Chem. Eng. News*, Aug. 29, 1994, 34; (d) J. Ryczkowski, *Appl. Catal. A*, 1995, *125*, N19; (e) K.J. Balkus, Jr, M. Eissa, and R. Levado, *J. Am. Chem. Soc.*, 1995, *117*, 10753; (f) B.-Z. Zhan and X.-Y. Li, *Chem. Commun.*, 1998, 349.
166. (a) F. Bedioui, L. Roue, E. Briot, J. Devynck, K.J. Balkus, Jr, and J.F. Diaz, *New J. Chem.*, 1996, *20*, 1235; (b) P.K. Dutta and S.K. Das, *J. Am. Chem. Soc.*, 1997, *119*, 4311.
167. S. Seelan, A.K. Sinha, D. Srinvinas, and S. Sivasanker, *J. Mol. Catal. A*, 2000, *157*, 163.
168. (a) A. Corma, A. Martinez, and C. Martinez, *Appl. Catal. A*, 1996, *134*, 169; (b) P. Bartl, B. Zibrowius, and W. Holderich, *Chem. Commun.*, 1996, 1611.
169. (a) Y.S. Bhat, J. Das, and A.B. Halgeri, *J. Catal.*, 1995, *155*, 154; 1996, *159*, 368; *Bull. Chem. Soc. Jpn.*, 1996, *69*, 469; (b) J.-H. Kim, A. Ishida, M. Okajima, and M. Niwa, *J. Catal.*, 1996, *161*, 387.
170. Y.-H. Yue, Y. Tang, Y. Tang, Y. Liu, and Z. Gao, *Ind. Eng. Chem. Res.*, 1996, *35*, 430.
171. E. Klemm, M. Seitz, H. Scheldat, and G. Emig, *J. Catal.*, 1998, *173*, 177.
172. T. Inui, S.-B. Pu, and J.-I. Kugai, *Appl. Catal. A*, 1996, *146*, 285.
173. (a) C. Nedez, A. Choplin, J. Corker, J.-M. Basset, J.-F. Joly, and E. Benazzi, *J. Mol. Catal.*, 1994, *92*, L239; (b) E. Benazzi, S. de Tavernier, P. Beccat, J.F. Joly, C. Nedez, A. Choplin, and J.-M. Basset, *Chemtech*, 1994, *24*(10), 13.
174. (a) Y. Wan, H. Yang, and D. Zhao, *Acc. Chem. Res.*, 2006, *39*, 423; (b) Y. Lu, *Angew. Chem. Int. Ed.*, 2006, *45*, 7664; (c) X. He and D. Antonelli, *Angew. Chem. Int. Ed.*, 2002, *41*, 214.
175. (a) D.E. Akporiaye, H. Fjellvag, E.N. Halvorsen, T. Haug, A. Karlsson, and K.P. Lillerud, *Chem. Commun.*, 1996, 1553; (b) T. Kimura, Y. Sugahara, and K. Kuroda, *Chem. Commun.*, 1998, 559; (c) B.T. Holland, P.K. Isbester, C.F. Blanchard, E.J. Muson, and A. Stein, *J. Am. Chem. Soc.*, 1997, *119*, 6796; (d) P. Feng, Y. Xia, X. Bu, and G.D. Stucky, *Chem. Commun.*, 1997, 949; (e) J. Yu and R. Xu, *Chem. Soc. Rev.*, 2006, *35*, 593; (f) J.M. Thomas and J. Klinowski, *Angew. Chem. Int. Ed.*, 2007, *46*, 7160; (g) A.K. Cheetham, G. Ferey, and T. Loiseau, *Angew. Chem. Int. Ed.*, 1999, *38*, 3268.

176. S.L. Cresswell, J.H. Parsonage, P.G. Riby, and M.J.K. Thomas, *Chem. Commun.*, 1995, 2513.
177. (a) J. Rocha, *Chem. Commun.*, 1996, 2513; (b) K. Maeda, Y. Kiyozumi, and F. Mizukami, *Angew. Chem. Int. Ed.*, 1994, *33*, 2335.
178. Y. Yang, M.G. Walawalkar, J. Pinkas, H.W. Roesky, and H.-G. Schmidt, *Angew. Chem. Int. Ed.*, 1998, *37*, 96.
179. S. Drumel, P. Janvier, D. Deniaud, and B. Bujoli, *J. Chem. Soc. Chem. Commun.*, 1995, 1051.
180. (a) G. Alberti, S. Murcia-Mascaros, and R. Vivani, *J. Am. Chem. Soc.*, 1998, *120*, 9291; (b) M. Fang, D.M. Kaschak, A.C. Sutorik, and T.E. Mallouk, *J. Am. Chem. Soc.*, 1997, *119*, 12184; (c) D.M. Poojary, B. Zhang, and A. Clearfield, *J. Am. Chem. Soc.* 1997, *119*, 12550.
181. S. Feng, X. Xu, G. Yang, R. Xu, and F.P. Glasser, *J. Chem. Soc. Dalton Trans.*, 1995, 2147.
182. A.R. Cowley and A.M. Chippindale, *Chem. Commun.*, 1996, 673.
183. S.A. Bagshaw and T.J. Pinnavaia, *Angew. Chem. Int. Ed.*, 1996, *35*, 1102.
184. (a) K.J.C. van Bommel, A. Firiggeri, and S. Shinkai, *Angew. Chem. Int. Ed.*, 2003, *42*, 980; (b) Y. Wan, Y. Shi, and D. Zhao, *Chem. Commun.*, 2007, 897.
185. B.T. Holland, L. Abrams, and A. Stein, *J. Am. Chem. Soc.*, 1999, *121*, 4308.
186. M. Hartmann, *Angew. Chem. Int. Ed.*, 2004, *43*, 5880.
187. C. Liang, Z. Li, and S. Dai, *Angew. Chem. Int. Ed.*, 2008, *47*, 3696.
188. P. Colombo, *Science*, 2008, *322*, 381.
189. (a) C. Weder, *Angew. Chem. Int. Ed.*, 2008, *47*, 448; (b) N. Du, G.P. Robertson, I. Pinnau, S. Thomas, and M.D. Guiver, *Macromolecular Rapid Commun.*, 2009, *30*, 584; (c) K.J. Msayib et al., *Angew. Chem. Int. Ed.*, 2009, *48*, 3273; (d) M.G. Schwab, B. Fassbender, H.W. Spiess, A. Thomas, X. Feng, and K. Mullen, *J. Am. Chem. Soc.*, 2009, *131*, 7216.
190. (a) W.E. Farneth and R.J. Gorte, *Chem. Rev.*, 1995, *95*, 615; (b) C. Lee, D.J. Parrillo, R.J. Gorte, and W.E. Farneth, *J. Am. Chem. Soc.*, 1996, *118*, 3262; (c) J.A. Lercher, C. Grundling, and G. Eder-Mirth, *Catal. Today*, 1996, *27*, 353; (d) J.C. Lavalley, *Catal. Today*, 1996, *27*, 377; (e) A. Zecchina, C. Lamberti, and S. Bordiga, *Catal. Today*, 1998, *41*, 169; (f) V. Solinas and I. Ferino, *Catal. Today*, 1998, *41*, 179; (g) S. Savitz, A.L. Myers, R.J. Gorte, and D. White, *J. Am. Chem. Soc.*, 1998, *120*, 5701; (h) R.S. Drago, S.C. Dias, M. Torrealba, and L. De Lima, *J. Am. Chem. Soc.*, 1997, *119*, 4444; (i) R.S. Drago, J.A. Dias, and T.O. Maier, *J. Am. Chem. Soc.*, 1997, *119*, 7702.
191. (a) G. Engelhardt and D. Michel, *High-Resolution Solid-State NMR of Silicates and Zeolites*, Wiley, New York, 1987; (b) M.T. Janicke, C.C. Landry, S.C. Christiansen, D. Kumar, G.D. Stucky, and B.F. Chmelka, *J. Am. Chem. Soc.*, 1998, *120*, 6940; (c) K. Inumaru, N. Jin, S. Uchida, and M. Misono, *Chem. Commun.*, 1998, 1489; (d) A. Simon, L. Delmotte, J.-M. Chezeau, and L. Huve, *Chem. Commun.*, 1997, 263.
192. (a) T. Wessels, C. Baerlocher, and L.B. McCusker, *Science*, 1999, *284*, 477; (b) R. Dagani, *Chem. Eng. News*, Apr. 19, 1999, 11; (c) M. O'Keefe, *Angew. Chem. Int. Ed.*, 2009, *48*, 8182 .
193. C.P. Grey, L. Peng, and P.J. Chupas, *J. Am. Chem. Soc.*, 2004, *126*, 12254.
194. J. Sutovich, A.W. Peters, E.F. Rakiewicz, R.F. Wormsbecher, S.M. Mattingly, and K.T. Mueller, *J. Catal.*, 1999, *183*, 155.
195. (a) S.J.L. Billinge and I. Levin, *Science*, 2007, *316*, 561; (b) E. Stavitski, M.H.F. Kox, I. Swart, F.M.F. de Groot, and B.M. Weckhuysen, *Angew. Chem. Int. Ed.*, 2008, *47*, 3543.
196. (a) G. Perot and M. Guisnet, *J. Mol. Catal.*, 1990, *61*, 173. (b) M.E. Davis, *Acc. Chem. Res.*, 1993, *26*, 111; *Ind. Eng. Chem.*, 1991, *30*, 1675; (c) S. Bhatia, *Zeolite Catalysis: Principles and Applications*, CRC Press, Boca Raton, FL, 1990; (d) N.Y. Chen, W.E. Garwood, and F.G. Dwyer, *Shape-Selective Catalysis in Industrial Applications*, Dekker, New York, 1989; (e) Y. Izumi, K. Urabe, and M. Onaka, *Zeolite, Clay and Heteropoly Acid in Organic Reactions*, VCH, Weinheim, 1992; (f) G. Hutchings, *Chem. Br.*, 1992, *28*, 1006; (g) W. Holderich, M. Hesse, and F. Naumann, *Angew. Chem. Int. Ed.*, 1988, *27*, 226; (h) B.A. Lerner, *Chem. Ind. (Lond.)*, 1997, 16; (i) C. Crabb, *Chem. Eng.*, 2001, *108*(5), 59; (j) M. Pagliaro, R. Ciriminna, and G. Palmisano, *Chem. Soc. Rev.*, 2007, *36*, 932.
197. G.F. Thiele and E. Roland, *J. Mol. Catal. A: Chem.*, 1997, *117*, 351.
198. M.E. Davis, ed., *Catal. Today*, 1994, *19*, 1–209.
199. (a) A. Corma, V. Fornes, J. Martinez-Triguero, and S.B. Pergher, *J. Catal.*, 1999, *186*, 57; (b) A. Corma, U.K. Diaz, M.E. Domine, and V. Fornes, *Angew. Chem. Int. Ed.*, 2000, *39*, 1499; *Chem. Commun.*, 2000, 137; (c) I. Rodriguez, M.J. Climent, S. Ibrra, V. Fornes, and A.Corma, *J. Catal.*, 2000, *192*, 441.
200. K.S. Knaebel. *Chem. Eng.*, 1995, *102*(11), 92.
201. S.E. Sen, S.M. Smith, and K.A. Sullivan, *Tetrahedron*, 1999, *55*, 12657.
202. (a) G. Ondrey, *Chem. Eng.*, 1995, *102*(9), 21; (b) *Chem. Eng. News*, Apr. 22, 1996, 19.
203. L.A. Lawrence, U.S. patent 5,476,978 (1995).
204. (a) G. Parkinson, *Chem. Eng.*, 1996, *103*(1), 19; (b) *Chem. Eng. News*, Dec. 18, 1995, 12; (c) *Chem. Eng. News*, Oct. 2, 1995, 11; (d) M. Sasidharan, K.M. Reddy, and R. Kumar, *J. Catal.*, 1995, *154*, 216.
205. S. Samdani, *Chem. Eng.*, 1995, *102*(8), 23.
206. (a) J.L.G. de Almeida, M. Dufaux, Y.B. Taarit, and C. Naccache, *J. Am. Oil Chem. Soc.*, 1994, *71*, 675; (b) J.A. Kocal, U.S. patent 5,196,574 (1993); (c) M.F. Cox and D.L. Smith, *Int. News Fats Oils Relat. Mater.*, 1997, *8*, 19; (d) J.A. Horsley, *Chemtech*, 1997, *27*(10), 45.
207. W. Liang, Z. Yu, Y. Jin, Z. Wang, Y. Wang, Y. Wang, M. He, and E. Min, *J. Chem. Tech. Biotechnol.*, 1995, *62*, 98.
208. T.-C. Tsai, I. Wang, S.-J. Li, and J.-Y. Liu, *Green Chem.*, 2003, *5*, 404.
209. Y.S. Bhat, J. Das, and A.B. Halgeri, *Appl. Catal. A*, 1995, *130*, L1.
210. C.D. Chang and P.G. Rodewald, U.S. patent 5,498,814 (1996).
211. Y.S. Bhat, J. Das, and B. Halgeri, *J. Catal.*, 1995, *155*, 154; *Appl. Catal. A: Gen.*, 1994, *115*, 257.
212. S.A. Chavan, D. Srinivas, and P. Ratnasamy, *Chem. Commun.*, 2001, 1124.
213. (a) K. Smith and D. Bahzad, *Chem. Commun.*, 1996, 467; (b) A. Botta, H.-J. Buysch, and L. Puppe, *Angew. Chem. Int. Ed.*, 1991, *30*, 1689; (c) A.N. Singh, D. Bhattacharya, and S. Sharma, *J. Mol. Catal. A: Chem.*, 1995, *102*, 139.

214. T.J. Kwok, K. Jayasuriya, R. Dumavarapu, and B.W. Brodman, *J. Org. Chem.*, 1994, *59*, 4939.
215. L. Bertea, H.W. Kouwenhoven, and R. Prins, *Appl. Catal. A*, 1995, *129*, 229.
216. J.M. Riego, Z. Sedin, J.M. Zaldivar, N.C. Marziano, and C. Tortato, *Tetrahedron Lett.*, 1996, *37*, 513.
217. R. Ballini, G. Boscia, S. Carloni, L. Ciarelli, R. Maggi, and G. Sartori, *Tetrahedron Lett.*, 1998, *39*, 6049.
218. N. Narender, T. Srinvasu, S.J. Kulkarni, and K.V. Raghavan, *Green Chem.*, 2000, *2*, 104.
219. K. Smith, Z. Zhenhua, and P.K.G. Hodgson, *J. Mol. Catal. A: Chem.*, 1998, *134*, 121.
220. Y. Ma, Q.L. Wang, W. Jiang, and B. Zuo, *Appl. Catal. A*, 1997, *165*, 199.
221. A.P. Singh and A.K. Pandey, *J. Mol. Catal. A: Chem.*, 1997, *123*, 141.
222. Y.V. Subba Rao, S.J. Kulkarni, M. Subrahmanyam, and A.V.R. Rao, *Appl. Catal. A*, 1995, *133*, L1.
223. (a) S.-J. Chu and Y.-W. Chen, *Appl. Catal. A*, 1995, *123*, 51; (b) R. Brzozowski and W. Tecza, *Appl. Catal. A*, 1998, *166*, 21.
224. Y. Sugi, X.-L. Tu, T. Matsuzaki, T.-A. Hanaoka, Y. Kubota, J.-H. Kim, M. Matsumoto, K. Nakajima, and A. Igarashi, *Catal. Today*, 1996, *31*, 3.
225. G. Parkinson, *Chem. Eng.*, 1996, *103*(1), 17.
226. H.C. Woo, K.H. Lee, and J.S. Lee, *Appl. Catal. A*, 1996, *134*, 147.
227. P. Meriaudeau, V.A. Tuan, N.H. Lee, and G. Szabo, *J. Catal.*, 1997, *169*, 397.
228. J. Houzvicka, S. Hansildaar, and V. Ponec, *J. Catal.*, 1997, *167*, 273.
229. S.J. Kulkami, R.R. Rao, M. Subrahmanyam, and A.V.R. Rao, *J. Chem. Soc. Chem. Commun.*, 1994, 273.
230. B. Coq, D. Tichit, and S. Rilbit, *J. Catal.*, 2000, *189*, 117.
231. (a) G.J. Hutchings, T. Thermistocleous, and R.G. Copperthwaite, U.S. patent 5,488,165 (1996); (b) C. Grundling, G. Eder-Mirth, and J.A. Lercher, *J. Catal.*, 1996, *160*, 299; (c) F.C. Wilhelm, T.R. Gaffney, G.E. Parris, and B.A. Aufdembrink, U.S. patent 5,399,769 (1995); (d) D.R. Corbin, S. Schwarz, and G.C. Sonnichsen, *Catal. Today*, 1997, *37*, 71.
232. (a) F.M. Bautista, J.M. Campelo, A. Garcia, D. Luna, J.M. Marinas, A.A. Romero, and M.R. Urbano, *J. Catal.*, 1997, *172*, 103; *Appl. Catal. A*, 1998, *166*, 39; (b) L.F.M. Bautista, J.M. Campelo, A. Garcia, D. Luna, J.M. Marinas, and A.A. Romero, *Appl. Catal. A*, 1998, *166*, 39.
233. (a) Y. Ono, Y. Izawa, and Z.-H. Fu, *J. Chem. Soc. Chem. Commun.*, 1995; (b) Y. Ono, *Cattech*, 1997, *1*(1), 31.
234. C.F. de Graauw, J.A. Peters, H. van Bekkum, and J. Huskens, *Synthesis*, 1994, 1007.
235. G.J. Hutchings and D.F. Lee, *J. Chem. Soc. Chem. Commun.*, 1994, 2503.
236. E.J. Creyghton and R.S. Downing, *J. Mol. Catal. A: Chem.*, 1998, *134*, 47.
237. H. Sato, *Catal. Rev. Sci. Eng.*, 1997, *39*(4), 395.
238. W.F. Holderich, J. Roseler, G. Heitmann, and A.T. Liebens, *Catal. Today*, 1997, *37*, 353.
239. M.A. Camblor, A. Corma, H. Garcia, V. Semmer-Herledan, and S. Valencia, *J. Catal.*, 1998, *177*, 267.
240. J.-F. Tremblay, *Chem. Eng. News*, Jan. 11, 1999, 19.
241. A.G. Stepanov, M.V. Luzgin, V.N. Romannikov, V.N. Sidelnikov, and K.I. Zamaraev, *J. Catal.*, 1996, *164*, 411.
242. M.J. Climent, A. Corma, R. Guil-Lopez, S. Iborra, and J. Primo, *J. Catal.*, 1988, *175*, 70.
243. M.J. Climent, A. Corma, H. Garcia, R. Guil-Lopez, S. Iborra, and V. Fornes, *J. Catal.*, 2001, *197*, 385.
244. (a) V. Ramamurthy, D.F. Eaton, and J.V. Caspar, *Acc. Chem. Res.*, 1992, *25*, 299; (b) N.J. Turro, *Chem. Commun.*, 2002, 2279.
245. (a) H. Frei, F. Blatter, and H. Sun, *Chemtech*, 1996, *26*(6), 24; (b) H. Sun, F. Blatter, and H. Frei, *J. Am. Chem. Soc.*, 1996, *118*, 6873; (c) F. Blatter, H. Sun, S. Vasenkov, and H. Frei, *Catal. Today*, 1998, *41*, 297; (d) X. Li and V. Ramamurthy, *J. Am. Chem. Soc.*, 1996, *118*, 10666; (e) R.J. Robbins and V. Ramamurthy, *Chem. Commun.*, 1997, 1071; (f) V. Ramamurthy, P. Lakshminarasimhan, C.P. Grey, and L.J. Johnston, *Chem. Commun.*, 1998, 2411; (g) A. Pace and E.L. Clennan, *J. Am. Chem. Soc.*, 2002, *124*, 11236; (h) S. Uppili and V. Ramamurthy, *Org. Lett.*, 2002, *4*, 87; (i) V. Ramamurthy, J. Shailaja, L.S. Kaanumalle, R.B. Sunoj, and J. Chandrasekhar, *Chem. Commun.*, 2003, 1987; (j) J. Sivaguru, A. Natarajan, L.S. Kaanamalle, J. Shailaj, S. Uppili, A. Joy, and V. Ramamurthy, *Acc. Chem. Res.*, 2003, *36*, 509.
246. H. Sun, F. Blatter, and H. Frei, *J. Am. Chem. Soc.*, 1996, *118*, 6873.
247. (a) K. Pitchumani, M. Warrier, C. Cui, R.G. Weiss, and V. Ramamurthy, *Tetrahedron Lett.*, 1996, *37*, 6251; (b) V. Ramamurthy, K. Pitchumani, and M. Warrier, *J. Am. Chem. Soc.*, 1996, *118*, 9428; (c) M. Rouhi, *Chem. Eng. News*, Oct. 7, 1996, 7; (d) K. Pitchumani, M. Warrier, and V. Ramamurthy, *J. Am. Chem. Soc.*, 1996, *118*, 9428.
248. (a) X. Li and V. Ramamurthy, *Tetrahedron Lett.*, 1996, *37*, 5235; (b) H. Takeya, Y. Kuriyama, and M. Kojima, *Tetrahedron Lett.*, 1998, *39*, 5967; (c) C.H. Tung, H. Wang, and Y.-M. Ying, *J. Am. Chem. Soc.*, 1998, *120*, 5179.
249. V.J. Rao, S.R. Uppili, D.R. Corbin, S. Schwarz, S.R. Lustig, and V. Ramamurthy, *J. Am. Chem. Soc.*, 1998, *120*, 2480.
250. T.L. Lambert and J.F. Knifton, U.S. patent 5,466,778 (1995).
251. (a) R. Becker, C. Sigwart, C. Palm, and W. Franzischka, *Chem. Abstr.*, 1995, *122*, 266344; (b) R. Dostalek, R. Fischer, U. Mueller, and R. Becker, *Chem. Abstr.*, 1995, *122*, 266347.
252. C.-G. Wu and T. Bein, *Science*, 1994, *266*, 1013.
253. S.M. Ng, S.-I. Ogino, T. Aida, K.A. Koyano, and T. Tatsumi, *Macromol. Rapid Commun.*, 1997, *18*, 991.
254. J.P. Dunn, Y. Cai, L.S. Liebmann, H.G. Stenger, Jr, and D.R. Simpson, *Ind. Eng. Chem. Res.*, 1996, *35*, 1409.
255. T.Y. Yan, *Ind. Eng. Chem. Res.*, 1994, *33*, 3010.
256. G.S. Samdani, *Chem. Eng.*, 1994, *101*(10), 19.
257. (a) T.M. Nenoff, J.E. Miller, S.G. Thoma, and D.E. Trudell, *Environ. Sci. Technol.*, 1996, *30*, 3630; (b) R.G. Anthony, R.G. Dosch, D. Gu, and C.V. Philip, *Ind. Eng. Chem. Res.*, 1994, *33*, 2702; (c) G.S. Samdani, *Chem. Eng.*, 1994, *101*(11), 27.
258. (a) P.R.H.P. Rao and M. Matsukata, *Chem. Commun.*, 1996, 1441; (b) R. Lai and G.R. Gavalas, *Ind. Eng. Eng. Chem. Res.*, 1998, *37*, 4275; (c) J. Dong and Y.S. Lin, *Ind. Eng. Chem. Res.*, 1998, *37*, 2404; (d) S.J. Roser, H.M. Patel, M.R. Lovell, J.E. Mur, and S. Mann, *Chem. Commun.*, 1998, 829; (e) G. Cho, I.-S. Moon, Y.-G. Shul, K.-T. Jung, J.-S. Lee, and B.M. Fung,

Chem. Lett., 1998, 355; (f) M.E. Raimondi, T. Maschmeyer, R.H. Templer, and J.M. Seddon, *Chem. Commun.*, 1997, 1843; (g) D. Zhao, P. Yang, D.I. Margolese, B.F. Chmelka, and G.D. Stucky, *Chem. Commun.*, 1998, 2499; (h) R. Mokaya, W. Zhou, and W. Jones, *Chem. Commun.*, 1998, 51; (i) K.W. Terry, C.G. Lugmair, and T.D. Tilley, *J. Am. Chem. Soc.*, 1997, *119*, 9745; (j) A. Shimojima, Y. Sugahara, and K. Kuroda, *J. Am. Chem. Soc.*, 1998, *120*, 4528.

259. N. Nishiyama, K. Ueyama, and M. Matsukata, *J. Chem. Soc. Chem. Commun.*, 1995, 1967.
260. (a) Y. Yan, M.E. Davis, and G.R. Gavalas *Ind. Eng. Chem. Res.*, 1995, *34*, 1652; (b) Y. Yan, M. Tsapatisis, G.R. Gavalas, and M.E. Davis, *J. Chem. Soc. Chem. Commun.*, 1995, 227.
261. B.S. Kang and G.R. Gavalas, *Ind. Eng. Chem. Res.*, 2002, *41*, 3145.
262. (a) T.A. Ostomel and G.D. Stucky, *Chem. Commun.*, 2004, 1016; (b) C. Haseloh, S.Y. Choi, M. Mamak, N. Coombs, S. Petrov, N. Chopra, and G.A. Ozin, *Chem. Commun.*, 2004, 1460; (c) K.-S. Jang, M.-G. Song, S.-H. Cho, and J.-D. Kim, *Chem. Commun.*, 2004, 1514.
263. F. Cagnol, D. Grosso, and C. Sanchez, *Chem. Commun.*, 2004, 1742.
264. J. Haggin, *Chem. Eng. News*, Aug. 19, 1996, 8.
265. (a) M.J. Zaworotki, *Angew. Chem. Int. Ed.*, 2000, *39*, 3052; (b) C.N.R. Rao, S. Natarajan, and R. Vaidhyanathan, *Angew. Chem. Int. Ed.*, 2004, *43*, 1466; (c) F.A. Paz and J. Klinowski, *Chem. Ind. (Lond.)*, Mar. 20, 2006, 21; (d) F.A.A. Paz and J. Klinowski, *Pure Appl. Chem.*, 2007, *79*, 1097; (e) D. Maspoch, D. Ruiz-Molina, and J. Veciana, *Chem. Soc. Rev.*, 2007, *36*, 770; (f) P.A. Wright, *Microporous Framework Solids*, Royal Society of Chemistry, Cambridge, 2008; (g) S. Natarajan and S. Mandal, *Angew. Chem. Int. Ed.*, 2008, *47*, 4798; (h) G. Ferey, *Chem. Soc. Rev.*, 2008, *37*, 191; (i) D. Farruseng, S. Aguada, and C. Pinel, *Angew. Chem. Int. Ed.*, 2009, *48*, 7502; (j) J.R. Long and O.M. Yaghi, eds. (for issue), *Chem. Soc. Rev.*, 2009, *38*, 1213–1450.
266. M. Jacoby, *Chem. Eng. News*, Aug. 26, 2008, 13.
267. (a) P. Zurer, *Chem. Eng. News*, Nov. 22, 1999, 14; (b) S.Y. Yang, L.S. Long, R.B. Huang, and L.S. Zheng, *Chem. Commun.*, 2002, 472; (c) F. Millange, C. Serre, and G. Ferey, *Chem. Commun.*, 2002, 822; (d) Y.-C. Liao, F.-L. Liao, W.-K. Chang, and S.-L. Wang, *J. Am. Chem. Soc.*, 2004, *126*, 1320; (e) Y. Cui, H.L. Ngo, P.S. White, and W. Lin, *Chem. Commun.*, 2003, 994; (f) K. Barthelet, J. Marrot, G. Ferey, and D. Riou, *Chem. Commun.*, 2004, 520; (g) F.A.A. Paz and J. Klinowski, *Chem. Commun.*, 2003, 1484; (h) Z. Wang, V.C. Kravtsov, and M.J. Zaworotko, *Angew. Chem. Int. Ed.*, 2005, *44*, 2877; (i) D.T. de Lill and C.L. Cahill, *Chem. Commun.*, 2006, 4946; (j) C. Livage, P.M. Forster, N. Guillou, M.M. Tafoya, A.K. Cheetham, and G. Ferey, *Angew. Chem Int. Ed.*, 2007, *46*, 5877; (k) C. Serre, C. Mellot-Draznieks, S. Surble, N. Audebrand, Y. Filinchuk, and G. Ferey, *Science*, 2007, *315*, 1828; (l) K. Koh, A.G. Wong-Foy, and A.J. Matzger, *Angew. Chem. Int. Ed.*, 2008, *47*, 677; (m) L. Pan, X. Huang, J. Li, Y. Wu, and N. Zhang, *Angew. Chem. Int. Ed.*, 2000, *39*, 527.
268. (a) F. Gandara, J. Perles, N. Snejko, M. Iglesias, B. Gomez-Lor, E. Gutierrez-Puebla, and M.A. Monge, *Angew. Chem. Int. Ed.*, 2006, *45*, 7998; (b) P. Mal, D. Schultz, K. Beyeh, K. Rissalnen, and J.R. Nitschke, *Angew. Chem. Int. Ed.*, 2008, *47*, 8297.
269. A. Clearfield, Z. Wang, and J.M. Heising, Anaheim A.C.S. Meeting, Mar. 2004, INORG 1019.
270. Y. Bai, D. Guo, C.-Y. Duan, D.-B. Dang, K.-L. Pang, and Q.-J. Meng, *Chem. Commun.*, 2004, 186.
271. (a) R. Dagani, *Chem. Eng. News*, May 28, 2007, 32; (b) H.M. El-Kaderi, J.R. Hunt, J.L. Mendosa-Cortes, A.P. Cote, R.E. Taylor, M. O'Keefe, and O.M. Yaghi, *Science*, 2007, *316*, 268; (c) A.P. Cote, H.M. El-Kaderi, H. Furukawa, J.R. Hunt, and O.M. Yaghi, *J. Am. Chem. Soc.*, 2007, *129*, 12914; (d) M. Mastalerz, *Angew. Chem. Int. Ed.*, 2008, *47*, 445.
272. (a) P. Zurer, *Chem. Eng. News*, May 1, 2000, 14; (b) R. Baum, *Chem. Eng. News*, Feb. 13, 1995, 37; (c) Y.-Q. Tian, C.-X. Cai, Y. Ji, X.-Z. You, S.-M. Peng, and G.-H. Lee, *Angew. Chem. Int. Ed.*, 2002, *41*, 1384; (d) M. Ruben, J. Rojo, F.J. Romeroa-Salguero, L.H. Uppadine, and J.-M. Lehn, *Angew. Chem. Int. Ed.*, 2004, *43*, 3644; (e) K.S. Park, Z. Ni, A.P. Cote, J.Y. Choi, R. Huang, F.J. Uribe-Romo, H.K. Chae, M. O'Keefe, and O.M. Yaghi, *Proc. Nat. Acad. Science USA*, 2006, *103*, 10186.
273. (a) S.R. Caskey, A.G. Wong-Foy, and A.J. Matzger, *J. Am. Chem. Soc.*, 2008, *130*, 10870; (b) R. Banerjee, H. Furakawa, D. Britt, C. Knobler, M. O'Keefe, and O.M. Yaghi, *J. Am. Chem. Soc.*, 2009, *131*, 3875.
274. M. Dinca and J.R. Long, *Angew. Chem. Int. Ed.*, 2008, *47*, 6766.
275. (a) S.S. Kaye, A. Dailley, O.M. Yaghi, and J.R. Long, *J. Am. Chem. Soc.*, 2007, *129*, 14176; (b) S. Borman, *Chem. Eng. News*, Dec. 22, 2003, 39.
276. R.E. Morris and P.S. Wheatley, *Angew. Chem. Int. Ed.*, 2008, *47*, 4966.
277. (a) D. Dubbeldam, C.J. Galvin, K.S. Walton, D.E. Ellis, and R.Q. Snurr, *J. Am. Chem. Soc.*, 2008, *130*, 10884; (b) F. Stallmach, S. Groger, V. Kungel, J. Karger, O.M. Yaghi, M. Hesse, and U. Muller, *Angew. Chem. Int. Ed.*, 2006, *45*, 2123.
278. D.J. Tranchemontagne, Z. Ni, M. O'Keefe, and O.M. Yaghi, *Angew. Chem. Int. Ed.*, 2008, *47*, 5136.
279. S. Hermes, M.-K. Schhroter, R. Schmid, L. Khodeir, M. Muhler, A. Tissler, R.W. Fischer, and R.A. Fischer, *Angew. Chem. Int. Ed.*, 2005, *44*, 6237.
280. (a) G. Ferey, C. Mellot-Drazieks, C. Serre, F. Millange, J. Dutoir, S. Surble, and I. Margiolaki, *Science*, 2005, *309*, 2040; (b) R. Dagani, *Chem. Eng. News*, Sep. 26, 2005, 11.
281. Y.K. Hwang, D.-Y. Hong, J.-S. Chang, S.H. Jhung, Y.-K. Seo, J. Kim, A. Vimont, M. Daturi, C. Serre, and G. Ferey, *Angew. Chem. Int. Ed.*, 2008, *47*, 4144.
282. S. Ritter, *Chem. Eng. News*, Apr. 21, 2008, 8.
283. T. Uemura, Y. Ono, K. Kitagawa, and S. Kitagawa, *Macromolecules*, 2008, *41*, 87.
284. (a) T. Furusawa, M. Kawano, and M. Fujita, *Angew. Chem. Int. Ed.*, 2007, *46*, 5717; (b) M.Yoshizawa, Y. Takeyama, T. Kusukawa, and M. Fujita, *Angew. Chem. Int. Ed.*, 2002, *41*, 1347.
285. T. Kawamichi, T. Kodama, M. Kawano, and M. Fujita, *Angew. Chem. Int. Ed.*, 2008, *47*, 8030.
286. Y.-F. Song and L. Cronin, *Angew. Chem. Int. Ed.*, 2008, *47*, 4635.
287. (a) T. Dombrowski, *Kirk–Othmer Encyclopedia of Chemical Technology*, 4th ed., Wiley, New York, 1993, *6*, 381; (b) H.H. Murray, *Ullmann's Encyclopedia of Industrial Chemistry*,

5th ed., VCH, Weinheim, 1986, *A7*, 109; (c) Y. Izumi, K. Urabe, and M. Onaka, *Zeolite, Clay and Heteropoly Acid in Organic Reactions*, VCH, New York, 1992, 49.
288. D. O'Hare, *New J. Chem.*, 1994, *18*, 989.
289. C. Oriakhi, *Chem. Br.*, 1998, *34*(11), 59.
290. S. Karaborni, B. Smit, W. Heidug, J. Urai, and E. van Oort, *Science*, 1996, *271*, 1102.
291. (a) S.P. Kaldare, V. Ramaswamy, and A.V. Ramaswamy, *Catal. Today*, 1999, *49*, 313; (b) S. Cheng, *Catal. Today*, 1999, *49*, 303; (c) A. Gil, L.M. Gandia, and M.A. Vincente, *Catal. Rev. Sci. Eng.*, 2000, *42*, 145; (d) A. Baumgartner, K. Sattler, J. Thum, and J. Breu, *Angew. Chem. Int. Ed.*, 2008, *47*, 1640; (e) M. Stocker, W. Seidl, L. Seyfarth, J. Senker, and J. Breu, *Chem. Commun.*, 2008, 629; (f) A. Gil, S.A. Korilli, and M.A. Vincente, *Catal. Rev. Sci. Eng*., 2008, 50, 150.
292. T. Mandalia, M. Crespin, D. Messad, and F. Bergaya, *Chem. Commun.*, 1998, 2111.
293. T.J. Pinnavaia and J.G. Guan, U.S. patent 5,583,082 (1996).
294. J.M. Dominguez, J.C. Botello-Pozos, A. Lopez-Ortega, M.T. Ramirez, G. Sandoval-Flores, and A. Rojas-Hernandez, *Catal. Today*, 1998, *43*, 69.
295. P. Sennett, *Kirk–Othmer Encyclopedia of Chemical Technology*, 4th ed., Wiley, New York, 1993, *6*, 405.
296. L. Liu, Z. Qi, and X. Zhu, *J. Appl. Polym. Sci.*, 1999, *71*, 1133.
297. Y. Ke, C. Long, and Z. Qi, *J. Appl. Polym. Sci.*, 1999, *71*, 1139.
298. H.G. Jeon, H.-T. Jung, and S.D. Hudson, *Polym. Bull.*, 1998, *41*, 107.
299. (a) M. Kato, A. Usuki, and A. Okada, *J. Appl. Polym. Sci.*, 1997, *66*, 1781; (b) N. Hasegawa, M. Kawasumi, M. Kato, A. Usuki, and A. Okada, *J. Appl. Polym Sci.*, 1998, *67*, 87.
300. Z.-K. Zhu, Y. Yang, J. Yin, X.-Y. Wang, Y.-C. Ke, and Z.-N. Qi, *J. Appl. Polym. Sci.*, 1999, *73*, 2063.
301. D.C. Lee and L.W. Jang, *J. Appl. Polym. Sci.*, 1998, *68*, 1997.
302. (a) M. Balogh and P. Laszlo, *Organic Chemistry Using Clays*, Springer, Berlin, 1992; (b) P. Laszlo. In: J.R. Kosak and T.A. Johnson, eds, *Catalysis of Organic Reactions*, Dekker, New York, 1994, 429; (c) A. Vaccari, *Catal. Today*, 1998, *41*, 53; (d) R.S. Varma, *Tetrahedron*, 2002, *58*, 1235–1256.
303. A.-X. Li, T.-S. Li, and T.-H. Ding, *Chem. Commun.*, 1997, 1389.
304. P.M. Bhaskar and D. Loganathan, *Terahedron Lett.*, 1998, *39*, 2215.
305. O. Sieskind and P. Albrecht, *Terahedron Lett.*, 1993, *34*, 1171.
306. J.J. van der Eynde, A. Mayence, and Y. Van Haverbeke, *Tetrahedron Lett.*, 1995, 36, 3133.
307. D. Baudry, A. Dormond, and F. Montagne, *New J. Chem.*, 1994, *18*, 871.
308. P.D. Clark, A. Kirk, and J.G.K. Yee, *J. Org. Chem.*, 1995, *60*, 1936.
309. (a) S.M. Pillai and M. Ravindranathan, *J. Chem. Soc. Chem. Commun.*, 1994, 1813; (b) G.D. Yadav and N.S. Doshi, *Green Chem.*, 2002, *4*, 528.
310. J.-I. Tateiwa, T. Nishimura, H. Horiuchi, and S. Uemura, *J. Chem. Soc. Perkin Trans. I*, 1994, 3367.
311. K. Smith and G.M. Pollard, *J. Chem. Soc. Perkin Trans. I*, 1994, 3519.
312. S. Narayanan, K. Deshpande, and B.P. Prasad, *J. Mol. Catal.*, 1994, *88*, L271.
313. B. Gigante, A.O. Prazeres, M.J. Marcelo-Curto, A. Cornelis, and P. Laszlo, *J. Org. Chem.*, 1995, 60, 3445.
314. T. Mishra and K.M. Parida, *J. Mol. Catal. A: Chem.*, 1997, *121*, 91.
315. K. Bougrin, M. Soufiaoui, and M. El Yazidi, *Tetrahedron Lett.*, 1995, *36*, 4065.
316. D. Ponde, H.B. Borate, A. Sudalai, T. Ravindranathan, and V.H. Deshpande, *Tetrahedron Lett.*, 1996, *37*, 4605.
317. M. Hirano, K. Ukawa, S. Yakabe, J.H. Clark, and T. Morimoto, *Synthesis*, 1997, 858.
318. K. Bougrin and M. Soufiaoui, *Tetrahedron Lett.*, 1995, *36*, 3683.
319. R.M. Martin-Aranda, M.A. Vincente-Rodriguez, J.M. Lopez-Pestana, A.J. Lopez-Peinado, A. Jerez, J. de Lopez-Gonzalez, and M.A. Banares-Munoz, *J. Mol. Catal. A: Chem.*, 1997, *124*, 115.
320. A.K. Ladavos, P.N. Trikalitis, and R.J. Pomonis, *J. Mol. Catal. A: Chem.*, 1996, *106*, 241.
321. G. Guiu and P. Grange, *J. Chem. Soc. Chem. Commun.*, 1994, 1729.
323. R. Mokaya and W. Jones, *J. Chem. Soc. Chem. Commun.*, 1994, 929.
324. K. Nozaki, M.L. Kantam, T. Horiuchi, and H. Takaya, *J. Mol. Catal. A: Chem.*, 1997, *118*, 247.
325. R.S. Varma and K.P. Naicker, *Tetrahedron Lett.*, 1998, *39*, 2915.
326. (a) C.-W. Hu, Q.-I. He, Y.-H. Zhang, Y.-Y. Liu, Y.-F. Zhang, T.-D. Tang, J.-Y. Zhang, and E.-Y. Wang, *J. Chem. Soc. Chem. Commun.*, 1996, 121; (b) C. Hu, Q. He, Y. Zhang, E. Wang, T. Okuhara, and M. Misono, *Catal. Today*, 1996, *30*, 141.
327. D.J. Jones, J. Roziere, P. Maireles-Torres, A. Jimenez-Lopez, P. Olivera-Pastor, E. Rodriguez-Castellon, and A.A.G. Tomlinson, *Inorg. Chem.*, 1995, *34*, 4611.
328. S. Udomsak and R.G. Anthony, *Catal. Today*, 1994, *21*, 197.
329. P.T. Tanev and T.J. Pinnavaia, *Science*, 1996, *271*, 1267.
330. M. Isayama, K. Sakata, and T. Kunitake, *Chem. Lett.*, 1993, *22*, 1283.
331. (a) Y. Izumi, K. Urabe, and M. Onaka, *Zeolite, Clay and Heteropolyacid in Organic Reactions*, VCH, Weinheim, 1992, 99; (b) M.T. Pope and A. Muller, *Polyoxometalates: From Platonic Solids to Anti-Retroviral Activity*, Kluwer, Dordrecht, 1994; (c) M.T. Pope and A. Muller, *Angew. Chem. Int. Ed.*, 1991, *30*, 34; (d) A. Corma, *Chem. Rev.*, 1995, *95*, 588; (e) J.M. Thomas and K.I. Zamaraev, eds, *Perspectives in Catalysis*, Blackwell Scientific, Oxford, UK, 1992, 431; (f) M. Misono and T. Okuhara, *Chemtech*, 1993, *23*(11), 23; (g) C.L. Hill, ed., *Chem. Rev.*, 1998, *98*, 1–390.
332. H.Y. Woo, H. So, and M.T. Pope, *J. Am. Chem. Soc.*, 1996, *118*, 621.
333. (a) X.-Y. Zhang, C.J. O'Connor, G.B. Jameson, and M.T. Pope, *Inorg. Chem.*, 1996, *35*, 30; (b) R. Neier, C. Trojanowski, and R. Mates, *J. Chem. Soc. Dalton Trans.*, 1995, 2521.
334. C.J. Gomez-Garcia, J.J. Borras-Almenar, E. Coronado, and L. Ouahab, *Inorg. Chem.*, 1994, *33*, 4016.
335. B.M. Rapko, M. Pohl, and R.G. Finke, *Inorg. Chem.*, 1994, *33*, 3625.
335. (a) W. Kaleta and K. Nowinska, *Chem. Commun.*, 2001, 535; (b) K. Inumaru, T. Ishihara, Y. Kamiya, T. Okuhara, and S. Yamanaka, *Angew. Chem. Int. Ed.*, 2007, *46*, 7625.
336. J.L. Bonardet, J. Fraissard, G.B. McGarvey, and J.B. Moffat, *J. Catal.*, 1995, *151*, 147.

337. M. Holscher, U. Englert, B. Zibrowius, and W.F. Holderich, *Angew. Chem. Int. Ed.*, 1994, *33*, 2491.
338. (a) A. Proust, R. Thouvenot, and P. Gouzerh, *Chem. Commun.*, 2008, 1837; (b) J. Hsao, Y. Xia, L. Wang, L. Ruhlmann, Y. Zhu, Q. Li, P. Yin, Y. Wei, and H. Guo, *Angew. Chem. Int. Ed.*, 2008, *47*, 2626.
339. (a) C.L. Hill and C.M. Prosser-McCartha, *Coord. Chem. Rev.*, 1995, *143*, 407; (b) I.V. Kozhevnikov, *Catal. Rev. Sci. Eng.*, 1995, *37*, 311; *J. Mol. Catal. A: Chem.*, 1997, *117*, 151; (c) C.L. Hill, ed., *J. Mol. Catal. A: Chem.*, 1996, *114*, 1–359; (d) F. Cavani, *Catal. Today*, 1998, *41*, 73.
340. G. Parkinson, *Chem. Eng.*, 1996, *103*(8), 15.
341. P. du Pont, J.C. Vedrine, E. Paumard, G. Hecquet, and F. Lefebvre, *Appl. Catal. A*, 1995, *129*, 217.
342. W. Chu, X. Yang, X. Ye, and Y. Wu, *Appl. Catal. A*, 1996, *145*, 125.
343. Y. Izumi, M. Ono, M. Ogawa, and K. Urabe, *Chem. Lett.*, 1993, *22*, 825.
344. M.N. Timofeeva, R.I. Maksimovskaya, E.A. Paukshtis, and I.V. Kozhevnikov, *J. Mol. Catal. A: Chem.*, 1995, *102*, 73.
345. G.D. Yadov and N. Kirthivasan, *J. Chem. Soc. Chem. Commun.*, 1995, 203.
346. S. Shibata, S.-I. Nakata, T. Okuhara, and M. Misono, *J. Catal.*, 1997, *166*, 263.
347. S. Shibata and M. Misono, *Chem. Commun.*, 1998, 1293.
348. H. Soeda, T. Okuhara, and M. Misono, *Chem. Lett.*, 1994, *23*, 909.
349. I.V. Kozhevnikov, A. Sinnema, A.J.A. van der Weerdt, and H. van Bekkum, *J. Mol. Catal. A: Chem.*, 1997, *120*, 63.
350. T. Okuhara, T. Nishimura, and M. Misono, *Chem. Lett.*, 1995, *24*, 155.
351. J.H. Grate, D.R. Hamm, and S. Mahajan. In: J.R. Kosak and T.A. Johnson, eds, *Catalysis of Organic Reactions*, Dekker, New York, 1994, 213.
352. K. Nowinska, D. Dudko, and R. Golon, *Chem. Commun.*, 1996, 277.
353. A.W. Stobbe-Kreemers, G. van der Lans, M. Makkee, and J.J.F. Scholten, *J. Catal.*, 1995, *154*, 187.
354. Y. Kim, H. Kim, J. Lee, K. Sim, Y. Has, and H. Paik, *Appl. Catal. A*, 1997, *155*, 15.
355. E. Monflier, S. Tilloy, E. Blouet, Y. Barbaux, and A. Mortreux, *J. Mol. Catal. A: Chem.*, 1996, *109*, 27.
356. T. Suzuki, H. Yoshikawa, K. Abe, K. Sano, and Showa Denko, U.S. patent 5,405,996 (1995).
357. G. Parkinson, *Chem. Eng.*, 1996, *103*(8), 15.
358. K. Na, T. Okuhara, and M. Misono, *J. Catal.*, 1997, *170*, 96.
359. F. Cavani, C. Comuzzi, G. Dolcetti, E. Etienne, R.G. Finke, G. Selleri, F. Trifiro, and A. Trovarelli, *J. Catal.*, 1996, *160*, 317.
360. K. Nagai, S.H. Amano, Y. Nagaoka, T. Ui, and T. Yamamoto, *Chem. Abstr.*, 1994, *121*, 231601.
361. N. Mizuno, M. Tateishi, and M. Iwamoto, *Appl. Catal. A*, 1994, *118*, L1.
362. N.M. Gresley, W.P. Griffith, A.C. Laemmel, H.I.S. Nogueira, and B.C. Parkin, *J. Mol. Catal. A: Chem.*, 1997, *117*, 185.
363. D.C. Duncan, R.C. Chambers, E. Hecht, and C.L. Hill, *J. Am. Chem. Soc.*, 1995, *117*, 681.
364. R. Neumann and H. Miller, *J. Chem. Soc. Chem. Commun.*, 1995, 2277.
365. N. Mizuno, C. Nozaki, I. Kiyoto, and M. Misono, *J. Am. Chem. Soc.*, 1998, *120*, 9267.
366. R.D. Gall, C.L. Hills, and J.E. Walker, *J. Catal.*, 1996, *159*, 473.
367. C.L. Hill and R.D. Gall, *J. Mol. Catal A: Chem.*, 1996, *114*, 103.
368. F.M. Collins, A.R. Lucy, and C. Sharp, *J. Mol. Catal A: Chem.*, 1997, *117*, 397.
369. (a) R. Neumann and A.M. Khenkin, *Inorg. Chem.*, 1995, *34*, 5753; (b) R. Neumann and M. Gara, *J. Am. Chem. Soc.*, 1994, *116*, 5509; 1995, *117*, 5066; (c) R. Neumann and D. Juwiler, *Tetrahedron*, 1996, *52*, 8781; (d) R. Neumann, A.M. Khenkin, D. Juwiler, H. Miller, and M. Gara, *J. Mol. Catal. A: Chem.*, 1998, *117*, 169.
370. R. Neumann and A.M. Khenkin, *Chem. Commun.*, 1998, 1967.
371. V. Kogan, Z. Aizenshtat, and R. Neumann, *New J. Chem.*, 202, *26*, 272.
372. (a) I.A. Weinstock, R.H. Atalla, R.S. Reiner, M.A. Moen, K.E. Hammel, C.J. Hartman, and C.L. Hill, *New J. Chem.*, 1996, *20*, 269; (b) R.F. Service, *Science*, 1995, *268*, 500; (c) G.S. Samdani, *Chem. Eng.*, 1995, *102*(5), 19; (d) D.M. Sonnen, R.S. Reiner, R.H. Atalla, and I.A. Weinstock, *Ind. Eng. Chem. Res.*, 1997, *36*, 4134; (e) D.V. Evtuguin, C.P. Neto, J. Rocha, and J.D.P. de Jesus, *Appl. Catal. A*, 1998, *167*, 123; (f) K. Ruuttunan and T. Vuorinen, *Ind. Eng. Chem. Res.*, 2005, *44*, 4284; (g) A.R. Gaspar, J.A.F. Gamelas, D.V. Evtuguin, and C.P. Neto, *Green Chem.*, 2007, *9*, 717.
373. J. Haggin, *Chem. Eng. News*, Aug. 5, 1996, 26.
374. (a) A. Mylonas and E. Papaconstantinou, *J. Mol. Catal.*, 1994, *92*, 261; (b) E. Papaconstantinou, *Chem. Eng. News*, June 24, 1996; (c) J. Haggin, *Chem. Eng. News*, Jan. 8, 1996, 27.
375. (a) R. Dziembaj, et al., *J. Mol. Catal. A: Chem.*, 1996, *112*, 423; (b) M. Hasik, A. Pron, J.B. Raynor, and W. Luzny, *New J. Chem.*, 1995, *19*, 1155.
376. (a) Y. Izumi, K. Urabe, and M. Onaka, *Catal. Today*, 1997, *35*, 183; (b) S. Soled, S. Miseo, G. McVicker, W.E. Gates, A. Gutierrez, and J. Paes, *Catal. Today*, 1997, *36*, 441; (c) K. Nowinska, M. Sopa, D. Dudko, and M. Mocna, *Catal. Lett.*, 1997, *49*, 43; (d) T. Balsco, A. Corma, A. Martinez, and P. Martinez-Escolano, *J. Catal.*, 1998, *177*, 306.
377. (a) S.R. Mukai, T. Masuda, I. Ogino, and K. Hashimoto, *Appl. Catal. A*, 1997, *165*, 219; (b) I.V. Kozhevnikov, K.R. Kloetstra, A. Sinnema, H.W. Zandbergen, and H. van Bekkum, *J. Mol. Catal. A: Chem.*, 1996, *114*, 287.
378. T. Douglas and M. Young, *Nature*, 1998, *393*, 152.
379. R. Neumann and M. Cohen, *Angew. Chem. Int. Ed.*, 1997, *36*, 1669.
380. (a) G.I. Park, S.S. Lim, J.S. Choi, I.K. Song, and W.Y. Lee, *J. Catal.*, 1998, *178*, 378; (b) J.K. Lee, I.K. Song, and W.Y. Lee, *J. Mol. Catal. A: Chem.*, 1997, *120*, 207.
381. I.K. Song, J.K. Lee, and W.Y. Lee, *Appl. Catal. A*, 1994, *119*, 107.
382. D.E. Katsoulis, U.S. patent 5,391,638 (1995).
383. M. Inoue and T. Yamase, *Bull. Chem. Soc. Jpn.*, 1995, *68*, 3055.
384. E.G. Derouane in M.S. Whittington and A.J. Jacobsen, eds, *Intercalation Chemistry*, Academic Press, New York, 1982, 101.
385. M. Misono, "Heterogeneous catalysis by heteropoly compounds of molybdenum and tungsten." *Catal. Rev. Sci. Eng.*, 1987, *29*, 269.

RECOMMENDED READING

1. G. Perot and M. Guisnet, *J. Mol. Catal.*, 1990, *61*, 173.
2. K. Tanabe, *Appl. Catal. A*, 1994, *113*, 147.
3. M. Misono and T. Okuhara, *Chemtech*, 1993, *23*(11), 23.
4. J.M. Thomas, *Angew. Chem. Int. Ed.*, 1994, *33*, 931–933.
5. Y. Izumi, K. Urabe, and M. Onaka, *Zeolite Clay and Heteropoly Acid in Organic Reactions*, VCH, Weinheim, 1992, 1, 2, 21, 22, 99–103.
6. J.H. Clark and C.N. Rhodes, *Clean Synthesis Using Porous Inorganic Solid Catalysts and Supported Reagents*, Royal Society of Chemistry, Cambridge, 2000 [Compare to this text].
7. A. Proust, R. Thouvenot, and P. Gouzerh, *Chem. Commun.*, 2008, 1837.

EXERCISES

1. Compare clays, heteropolyacids, and zeolites as catalysts for organic reactions.
2. Pick out the acidic and basic catalysts in the reactions in a volume of *Organic Syntheses*. Which one might be replaced with solid acids and bases? What advantages might there be to doing this?
3. Devise a system of more meaningful names for zeolites.
4. Select a reaction that requires a stoichiometric amount of aluminum chloride as a catalyst. Try running it with a rare earth triflate.
5. Suggest some reactions that might be more selective if run in pillared clays. What separations might be run with such clays?
6. How would you characterize a new zeolite or pillared clay or heteropolyacid?
7. Having found a zeolite catalyst for a reaction, how would you optimize it?
8. Can you think of a reaction that might be improved by reaction distillation (i.e., one that has not already been run in this way)?
9. What is the potential of carbon molecular sieves?
10. If you have never run a catalytic vapor-phase reaction in a hot tube, pick one that could use a solid acid or base as a catalyst and try it.
11. Predict areas where high-surface-area Nafion might be able to replace the use of hydrogen fluoride and concentrated sulfuric acid as catalysis.

CHAPTER 7

Chemical Separations

7.1 THE GENERAL PICTURE

Separations[1] use 6% of the total energy consumed by industry in the United States.[2] They can account for up to 70% of the plant costs. The many types include (a) removal of a catalyst, (b) removal of unchanged starting material, (c) removal of by-products, and (d) recovery of solvent or others from a reaction mixture. Common separations involve (a) aromatic from aliphatic, (b) substituted from unsubstituted, (c) linear from branched, (d) positional isomers, (e) diastereoisomers and enantiomers, (f) small molecules from polymers, (g) one polymer from another, (h) charged from uncharged, (i) one metal ion from another, (j) one anion from another, and so forth. The traditional methods have not been entirely satisfactory. Distillation becomes more difficult with heat-sensitive materials, close-boiling mixtures, and azeotropes. Crystallization may not be as selective as desirable and may leave part of the product in the mother liquor as waste.[3] Extraction may not be completely selective and may involve the use of large volumes of solvents, frequently with pH changes that result in waste salts. Chromatography[4] is fine for small amounts and analytical purposes, but it becomes cumbersome on an industrial scale. However, simulated moving bed chromatography (a method in which much of the solvent is recirculated and the inlets and outlets are moved as the profile of the material moves through the column) is making the technique more practical and economical for fine chemicals production.[5] Supercritical fluid chromatography is also being scaled up for the separation of optical isomers.

The goals of improved separations[6] include (a) saving energy, (b) reducing waste, (c) avoiding the generation of waste salts, (d) minimizing recycle, (e) avoiding or minimizing the use of organic solvents, and (f) saving money. Current practice for preventing further reaction of the desired product may involve energy-intensive quenching and running to partial conversion with much recycle. An example is the oxidation of methanol to formaldehyde with oxygen over a silver catalyst at 700°C, followed by quick quenching to 150°C to prevent further oxidation to formic acid.[7] A common method of removing an acid catalyst from a reaction mixture is to wash the mixture (in a solvent) with a mild base in water, which produces waste salts. The use of supported reagents and solid catalysts, which can be recovered for reuse by centrifugation or filtration to avoid this, was covered in Chapters 5 and 6. To these must be added a third method, standing acoustical waves of ultrasound, which has been used in the separation of cells and blood samples.[8]

The use of catalysts based on polymers with inverse temperature solubility, often copolymers of *N*-isopropylacrylamide, to allow recovery by raising the temperature to precipitate the polymer for filtration,[9] was mentioned in Chapter 5. The opposite, if the catalyst is soluble hot, but not cold, has also been used in ruthenium-catalyzed additions to the triple bonds of acetylenes (**7.1** Schematic).[10] The long aliphatic tail of the phosphine ligand caused the catalyst to be insoluble at room temperature so that it could be recovered by filtration. There was no loss in yield or selectivity after seven cycles of use. A phosphine-modified poly(*N*-isopropylacrylamide) in 90% aqueous ethanol/heptane has been used in the hydrogenation of 1-olefins.[11] The mixture is biphasic at 22°C, but single phase at 70°C, the temperature at which the reaction takes place. This is still not ideal, because it takes energy to heat and cool, and it still uses flammable solvents.

where the catalyst is $[Ru(\mu\text{-}OCOH)(CO)_2((C_6H_5)_2P(CH_2CH_2)_{50}\,CH_2CH_3)]$

7.1 Schematic

These are two types of "smart" polymers that respond to temperature, pH, light, electric fields, chemicals, or ionic strength.[12]

Physical methods of separation are preferred to those using chemical reagents or organic solvents. These are the standard methods of separation of minerals. Iron oxides can be recovered magnetically.[13] The iron oxide in kaolin clay is removed by high-intensity magnetic separation. The treated clay is used as a filler for paper. Magnetite coated with poly(ethylene oxide) and poly(propylene oxide) have been used to remove organic compounds from hydrogen.[14] A cobalt catalyst in a carbon molecular sieve has been recovered magnetically.[15] Ores can be separated by a variety of methods involving radiation. If the pieces of ore are passed by a sensor one at a time, the desired type can be selected by infrared, ultraviolet, visible, or x-ray radiation. A jet of air is used to remove the selected pieces. This method is also applicable to the sorting of cans, bottles, and packages in mixed solid waste. It does have trouble when a container is made of more than one material, such as a steel can with an aluminum lid. Thus, there is an incentive to make containers, their caps, lids, and labels, all of the same material. This method is much easier than processes that dissolve the mixed plastics in a solvent and then try to separate them by differential precipitation. Machine vision that picks out a particular image pattern has been used to remove aflatoxin-contaminated nuts from a stream of pistachios passing at the rate of 163 kg per channel per hour with 97.8% selectivity.[16]

Pits in dried plums have been separated by pressing the "depitted" plum on a belt against a force transducer below the belt, allowing the plum with the pit to be removed.[17] The chemical industry is making increasing use of these and other sensing techniques for on-line monitoring of reaction mixtures in plants. They offer the advantage of instantaneous information, without having to wait for the laboratory to perform analyses. They also avoid the waste and disposal problems associated with the unused portions of materials taken for analysis in the laboratory. They should lead to higher yields. An electric field has been used to reduce energy consumption in the dewatering of sludge.[18] Sludge contained 85–88% water going in and 50–70% coming out.

Separation by density is often used in sink-float techniques. In minerals processing, the bath is often a slurry of magnetite (Fe_3O_4). The magnetite is easy to remove magnetically. Flotation is a common method of concentrating ores[19] and removing ink from paper during recycling. In the flotation of minerals, particles are made hydrophobic with a collector, and a frothing agent is added. Air is bubbled through the system and the ore-laden froth is collected. A depressant is used sometimes to make unwanted accompanying minerals hydrophilic so that they will sink. Hydrometallurgy, in which the minerals are dissolved, is used occasionally.[20] The metal ion is removed with a special hydrophobic chelating agent in a hydrocarbon such as kerosene. It is then taken into water again by treatment with acid for collection by electrolytic deposition. Physical separations on the usual chemical mixtures involve the use of micro- and mesoporous solids, such as the pillared clays, metal organic frameworks, and zeolites discussed in Chapter 6. Other such methods will be covered later in this chapter.

Magnetic beads are available for some bioseparations,[21] which are often more complicated than the usual chemical separations.[22] An antibody can be placed on them for an antigen that is to be separated. Compared with affinity chromatography in a column, the beards offer no plugging with cell debris or channeling. They have been used in immunoassays for pesticides and polychlorinated biphenyls.[23] Iron oxide-filled ion-exchange resins have been used to recover zinc ion from wastewater.[24] The waste stream flows through a slurry of the resin particles. After removal of the zinc by acid, the resin can be reused. Magnetite-impregnated styrene–divinylbenzene copolymer resins have also been used for combinatorial synthesis of polypeptides.[25] This simplified the splitting and mixing procedure. Resin beads containing a metal ion held by a tetradentate chelating agent are marketed for the separation of histidine-containing proteins.[26] The metal may not be iron. A process that uses magnetic particles (30 nm–25 μm) coated with selective extractants is said to be cheaper and faster than ion-exchange resins for the recovery of cadmium, zinc, transuranic elements, and such from dilute waste streams.[27]

The best approach for improving separations is to work toward reactions that achieve 100% yields at 100% conversions. Frequently, this will require more selective catalysts. The previous chapter contained an example moving in this direction. Toluene was disproportionated to benzene and xylenes using a silica-modified zeolite catalyst.[28] After removal of benzene and unchanged toluene by distillation, the xylene remaining was a 99% *para*-isomer. It was clean enough to put directly into the process of oxidation to terephthalic acid. This avoided the usual separation of xylenes by crystallization or by a molecular sieve. There are times when equilibrium can be shifted by removal of a product or by-product continuously to give 100% conversion. The familiar esterification with azeotropic removal of water or removal of water with a molecular sieve is an example.

Reactive distillation[29] is a method of removing the product from the reaction zone as it is formed, so that further reaction does not occur. An example cited in the preceding chapter involved running ethylene into a column of molecular sieve catalyst containing refluxing benzene. The product ethylbenzene (needed for the production of styrene) was collected in the distillation pot with minimal dialkylation.[30] Any diethylbenzene that did form was transalkylated with benzene to give more ethylbenzene, using the same catalyst. The result was that ethylbenzene was obtained with 99.7% specificity at 100% conversion. A second example is the preparation of *tert*-butyl methyl ether for use as an additive to gasoline, during which up to 99.99% conversion of the isobutylene is achieved.[31] The ether flows downward from the column of catalyst into the distillation pot, while the isobutylene–methanol azeotrope is removed continuously and recycled to the reaction section. Lactic acid has been recovered from an aqueous solution as its butyl ester by continuous reactive distillation using an ion-exchange resin catalyst, the ester accumulating in the pot.[32] Continuous extraction with methylisobutylketone has been used to achieve over 90% conversion of fructose to 5-hydroxymethylfurfural using an H-mordenite catalyst.[33]

It is also possible to separate some mixtures by reacting out a desired compound under conditions in which the others present will not react. The C_5 refinery stream can be heated gently until the cyclopentadiene present has dimerized to dicyclopentadiene. The other C_5 compounds can then be distilled off.[34] It is also possible to separate isobutene from mixed butenes by acid-catalyzed hydration to *tert*-butyl alcohol or addition of methanol to produce *tert*-butyl methyl ether.[35] The success of these methods hinges on the products being useful, as such, or being reconvertible to the original compounds. Dicyclopentadiene can be cracked to cyclopentadiene. The *tert*-butyl alcohol can be dehydrated to isobutylene. No matter what the separation technique is, chemical reaction or adsorption, it is important to be able to recover the material easily in an energy-efficient way. With adsorption, the usual methods are heating or evacuation or a combination of the two. For example, a zeolite that has been used for drying can be heated to remove the water so that the zeolite can be reused.

Another useful separation technique for gas–liquid–liquid systems[36] involves catalysts that are only soluble in one of two phases. An outstanding example was cited in Chapter 1. The hydroformylation of propylene to produce butyraldehydes (in the Ruhrchemie/Rhone–Poulenc process) is done by passing propylene gas into an aqueous solution of a ruthenium catalyst.[37] The product can be removed continuously as a separate layer. Union Carbide (now part of Dow Chemical) has devised a variant of this process with added *N*-methylpyrrolidone that extends it to higher olefins, such as octene, dodecene, styrene, and dienes.[38] Another group added ethanol to the water so that 1-octene could be hydroformylated.[39] Dodecene has been hydroformylated in a microemulsion using a rhodium catalyst and a polyglycol ether surfactant with 99% conversion.[40] This reaction has also been run in a water/heptane mixture with a water-soluble Rh catalyst giving a conversion of 96%.[41] Both the starting olefin and the product were in the heptane phase for easy recovery of the product. A further variation is seen in the carboxylation of styrene in 98% yield with carbon monoxide in toluene/water using a water-soluble palladium catalyst.[42] Another example is the Shell process for higher olefins in which ethylene is passed into a diol containing an organonickel catalyst.[43] The oligomeric α-olefins form a second phase that is easy to separate. The best examples are those that require no organic solvent.

Nonvolatile ionic liquids have considerable potential as media for reactions,[44] such as that shown (**7.2** Schematic).[45] The product was separated by decantation. The ionic liquid layer could be reused several times, with no loss in catalytic activity. The tetrafluoroborate and hexafluorophosphate salts are stable to air and to water. The tetrachloroaluminate,

BF_4^- H_3C C_4H_9 $PdCl_2$, $(C_6H_5)_3P$ 100% selectivity

7.2 Schematic

which can be used as both a catalyst and a medium in acylation and alkylation reactions, is sensitive to moisture. A wide variety of inorganic and organic materials, including some polymers, are soluble in ionic liquids. Ionic liquids can be used in liquid–liquid extractions instead of volatile organic solvents.[46] The product is recovered from the ionic liquid by evaporation. Examples of the use of such ionic liquids include the dimerization of propylene, using an ethylaluminum chloride salt and a nickel phosphine, during which the product separates as a separate layer,[47] the metathesis of 2-pentene to 2-butene and 3-hexene with a tungsten catalyst,[48] and catalytic hydrogenation of the double bonds, but not the nitrile groups, in poly(acrylonitrile-*co*-1,3-butadiene) using a ruthenium catalyst in toluene.[49] In the last example, the ionic liquid containing the catalyst could be separated by decantation and reused several times without any significant loss of activity or selectivity. If the organic product remains dissolved in the ionic liquid and is relatively nonvolatile, it can be recovered by extraction with supercritical carbon dioxide.[50] In some cases, just adding carbon dioxide will cause a second phase to form in the carbon dioxide-expanded ionic liquid. Three phase systems with heterogeneous catalysts have also been used.[51] For example, a system with the catalyst in the bottom ionic liquid layer, a middle layer of water and a top layer of cyclohexane has been used. (Other uses of supercritical fluids are given in Chapter 8.)

Phase-separable homogeneous catalysis[52] also includes fluorous biphasic systems.[53] The latter use a liquid fluorocarbon that is insoluble in organic solvents at 25°C, or more, the mixture becomes homogeneous on heating, and the reaction takes place. On cooling to 25°C, the product separates. The fluorocarbon phase, which contains the catalyst, can be reused. For the system to work, perfluoroalkyl groups have to be inserted into the ligands for the metal catalysts and, sometimes, into the reagents, A typical example is the hydroformylation of 1-decene with hydrogen and carbon monoxide in 50:50 toluene/perfluoro(methylcyclohexane) at 100°C using an HRh (CO)$\{P[CH_2CH_2(CF_2)_5CF_3]_3\}_3$ catalyst.[54] At 98% conversion, the selectivity to aldehydes was 90%, and the selectivity to 2-olefins, which did not react further, was 10%. In nine reaction-separation cycles, the total turnover was more than 35,000, with a loss of 1.18 ppm Rh per mole of undecanals. Such a system has been used in quantitative transesterification of a 1:1 mixture of alcohol and ester using a perfluoroalkyl tin catalyst (**7.3** Schematic).[55]

The perfluoroalkyl group can be put into reagents and, if desired, removed after the reaction to get the product. It can be used in scavengers, for example, a perfluoroalkyl

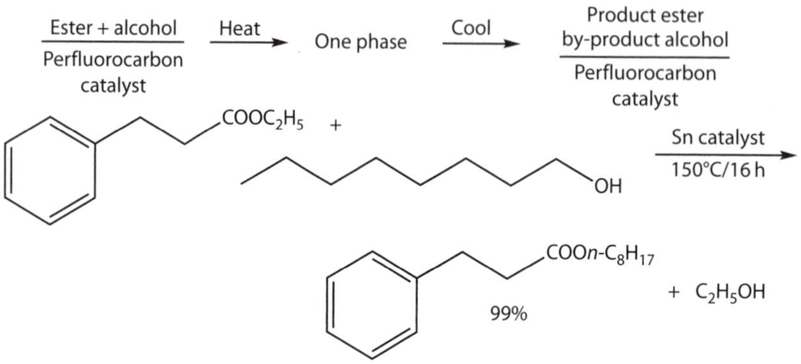

Ester + alcohol / Perfluorocarbon catalyst —Heat→ One phase —Cool→ Product ester by-product alcohol / Perfluorocarbon catalyst

$C_6H_5CH_2CH_2COOC_2H_5$ + n-octanol (OH) —Sn catalyst, 150°C/16 h→ $C_6H_5CH_2CH_2COO\text{n-}C_8H_{17}$ (99%) + C_2H_5OH

7.3 Schematic

isocyanate to remove an amine, with the resulting product being removed by passage of the reaction mixture through a column of fluorous silica. It can also be put on catalysts to give temperature-dependent phase miscibility in ordinary organic solvents, so that a perfluorocarbon solvent is not needed (**7.4** Schematic).[56]

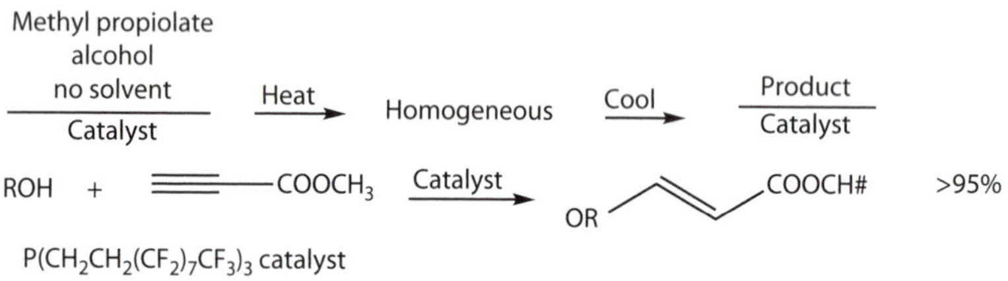

Methyl propiolate alcohol no solvent / Catalyst —Heat→ Homogeneous —Cool→ Product / Catalyst

ROH + ≡—$COOCH_3$ —Catalyst→ OR–CH=CH–COOCH# >95%

$P(CH_2CH_2(CF_2)_7CF_3)_3$ catalyst

7.4 Schematic

Fluorous Technologies Inc., Pittsburgh, PA can supply many of the compounds needed for such work.

Although such systems may be useful in specialized cases, they will probably remain too expensive for general industrial use. Putting on a fluorous tag and removing it later, plus the chromatography on fluorous silica, add to the cost. Perfluorocarbons are also greenhouse gases, so that precautions have to be taken not to lose any.

Systems with two aqueous phases have been used in the separation of proteins and metal ions.[57] These involve water-soluble polymers, such as poly(ethylene glycol), with or without salts, such as potassium phosphate.[58] Such systems eliminate the need for volatile organic solvents. A typical example is the separation of myoglobin with aqueous poly-(*N*-vinylacetamide)-dextran, the protein going to the bottom phase containing the dextran.[59]

Hydrotropes, such as sodium isopropylbenzenesulfonate, which increase the solubility of organic compounds in water, have been used to extract compounds from plants that might be sensitive to heat.[60] The products can be recovered by dilution of the extracts with water. After separation of the organic compound layer, the filtrate could be concentrated for reuse by reverse osmosis.

In theory, at least, there may be exceptions to the generalization that pH changes for causing separation produce waste salts. The sodium hydroxide used to extract a phenol into water could be recycled (**7.5** Schematic).

The reaction would have to be run such that a minimum of water would need to be removed by reverse osmosis or by evaporation. Carbon dioxide under pressure has been used to free 6-hydroxy-2-naphthoic acid from its potassium salt, to recover the acid and recycle potassium carbonate.[61] Ammonia or tertiary amines can be used to separate carboxylic acids. The ammonia or amine can be recovered by heating under a vacuum. This type of reaction is used in some floor polishes. The polymer containing the carboxylic acid groups is soluble in aqueous ammonia until applied. It loses its solubility in water when the ammonia evaporates. However, the polish can be stripped from the floor by the use of fresh aqueous ammonia. Trimethylamine, in ketones such as methyl ethyl or methyl isobutyl isobutyl ketone, has been used to recover lactic and succinic acids from their aqueous solutions. The free acids were obtained and the base recovered for recycle by heating the salts.[62] Trioctylamine plus Aliquat 336 (trioctylmethylammonium chloride) in decanol has also been used.[63] The effect of temperature on such extractions has been studied.[64] Amines have been separated

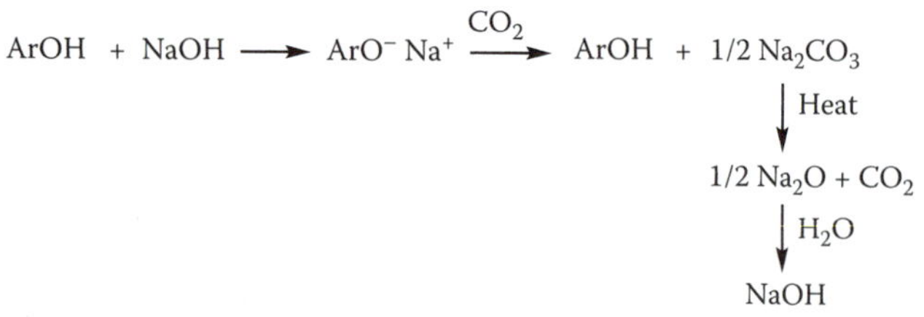

$$ArOH + NaOH \longrightarrow ArO^- Na^+ \xrightarrow{CO_2} ArOH + 1/2\,Na_2CO_3$$

$$1/2\,Na_2CO_3 \xrightarrow{Heat} 1/2\,Na_2O + CO_2$$

$$1/2\,Na_2O \xrightarrow{H_2O} NaOH$$

7.5 Schematic

by adding carbon dioxide to form an ammonium carbamate that separates from the organic solvent.[65]

There are also examples for which there is no need to separate a catalyst, because it can be left in the product without adverse effects. Magnesium chloride-supported catalysts for the polymerization of propylene attain such high mileage that they can be left in the polymer. Earlier less-efficient catalysts had to be removed by an acidic extraction process that produced titanium- and aluminum-containing wastes. The earlier processes also produced a heptane-soluble polymer that had to be removed and sometimes ended up as waste. The newer processes produce so little that it can be left in the product. Thus, improved catalysts have eliminated waste. (See Chapter 1 for more details.)

7.2 INCLUSION COMPOUNDS

The use of zeolites, molecular sieves, metal organic frameworks, and pillared clays to sort molecules by size and shape was described in Chapter 6. Some organic molecules also contain cavities or channels that guests can fit into and thus separate them from other compounds. The host–guest combinations are often referred to as inclusion compounds or clathrates.[66] The cyclodextrins—conical molecules containing six, seven, or eight glucose units in rings—are probably the most common. They were covered in Chapter 5. Others include crown ethers, cryptates, cyclophanes, and calixarenes, all of which are more expensive than cyclodextrins.[67] The price may not matter if they can be recovered and used repeatedly. The best systems might insert them into membranes for continuous use or in columns for cyclical use. The goal is to approach the exquisite selectivity of biological systems. This is part of current work in molecular recognition.

The guest must have a suitable size and shape to fit into the host. Cavities lined with nonpolar groups may favor nonpolar guests, whereas a polar lining may favor a polar guest. Host channels may favor linear molecules. Host cages may favor spherical molecules. Hydrogen bonding of amides with OH and NH may stabilize some structures.[68] π-Stacking may be important. The cavities are usually 4–8 Å in diameter. Inclusion compounds can be formed by putting the host into excess guest, or putting the two into a common solvent. Sometimes, the complex will form on exposure of the crystalline host to the vapor of a guest. Sometimes, the two can just be ground together. The stability of the inclusion compounds varies greatly. If the compound depends on the crystal structure of the host, the guest will be liberated if the inclusion compound is dissolved. This may be a simple way to recover the guest after a separation. Some inclusion compounds are stable only in the presence of excess guest. Others may be so stable that they must be heated strongly or placed in a substantial vacuum to recover the guest.

Urea is an inexpensive material that crystallizes with roughly 5-Å diameter channels in a structure held together by hydrogen bonding. Linear alkanes, usually with six or more carbon atoms, can fit into these channels.[69] As long as the linear alkyl group is present, other groups, such as aldehyde, ketone, ether, ester, carboxylic acid, hydroxyl, amine, nitrile, mercaptan, sulfide and aromatic ring, can also be present in the guest. If urea is crystallized from a methanolic solution containing fatty acids, the crystals are enriched in the saturated acids over the unsaturated ones.[70] Mono-unsaturated acids fit better than polyunsaturated ones. The crystals can be filtered off and dissolved in water to liberate the fatty acids. By using such techniques to eliminate saturated and polyunsaturated fatty acids, 97–99% oleic acid has been obtained from olive oil's fatty acids.[71] It is possible that this method could be applied to biodiesel fuel obtained by the methanolysis or ethanolysis of rapeseed oil, to eliminate the saturated esters that tend to crystallize out in cold winter weather.[72] An untried alternative might be to use a desaturase enzyme to convert the saturated ester in the original mixture into an unsaturated one.

Inserting the guest into the host can sometimes improve the selectivity of its reactions. The polymerization of 1,3-butadiene in urea crystals preferentially leads to the crystalline 1,4-*trans* polymer.[73] Such solid-state reactions can improve the selectivity in many reactions.[74] If performed with a chiral diol, the reaction can give products with high enantioselectivity (**7.6** Schematic).

Occasionally, a chiral diol can serve as host for one optical isomer (e.g., an epoxide), allowing the other one to be distilled out. The guest optical isomer can then be recovered by stronger

Light

Ph = phenyl

99% ee
48% yield

7.6 Schematic

7.7 Schematic

heating. A quaternary ammonium salt, derived from the amino acid leucine, has been used to resolve 1,1′-bi-2-naphthol (**7.7** Schematic) by formation of an inclusion compound.[75]

Hydrazine can be stabilized by formation of an inclusion compound with hydroquinone.[76] Uncomplexed anhydrous hydrazine can be explosive. Explosive acetylene can be stabilized by storage in strapped porphyrins[77] as well as in cucurbit[6]uril.[78] Highly toxic dimethyl sulfate can be handled more easily as an inclusion compound with toxic 18-crown-6. Both of these toxic compounds could be avoided through the use of dimethyl carbonate (as described in Chapter 2). Reactive intermediates, such as benzyne, have been stabilized by generating them inside hosts.[79] Even the noble gas xenon can be trapped reversibly by hosts such as a cryptophane, 4-*tert*-butyl-calix[4]arene, or α-cyclodextrin.[80]

A variety of substituted ureas have been used to form inclusion compounds.[81] Cucurbituril made from urea, glyoxal, and formaldehyde is a band-shaped molecule with a 4-Å hole at the top and the bottom connected to a 5.5-Å cavity (**7.8** Schematic). It forms strong complexes with some diamines (e.g., 1,6-diaminohexane) and with alkylammonium ions, such as protonated 1-aminopentane.[82]

$NH_2CONH_2 + OHCCHO + HCHO \rightarrow$

7.8 Schematic

Cucurbit[n]urils come in various sizes,[83] may contain functional groups,[84] and may be chiral.[85] Cucurbit[7]uril can distinguish between dipeptides based on the sequence of amino acids, such as tyrosinylglycine ($K = 3.6$ million) and glycyltyrosine ($K = 200$).[86] It has also been used as a host for drugs.[87]

A basket-shaped host that is a crown ether containing a bicyclic urea structure (**7.9** Schematic) (where **R** is phenyl) has been used to bind the herbicide Paraquat analog (**7.10** Schematic) (where the anion is hexafluorophosphate) and some polymeric derivatives of it with K up to 57,000.[88] Even stronger binding with a K of 7 million has been obtained with a cyclic urea porphyrin.[89]

7.9 Schematic

7.10 Schematic

Rebek and co-workers[90] have studied the self-assembly of molecules containing substituted ureas. Dimerization of two calixarenes (where Bn is benzyl and Ar is 4-fluorophenyl) (**7.11** Schematic) by hydrogen bonding produces a host that encapsulates benzene, fluorobenzene, *p*-difluorobenzene, and pyrazine, but not toluene. The strongest complex was that of difluorobenzene. Stability favors the best fit that leaves no empty space in the cavity. The self-assembled dimer of the urea in (**7.12** Schematic) (where R = ethoxycarbonyl) encapsulated methane but not ethane.

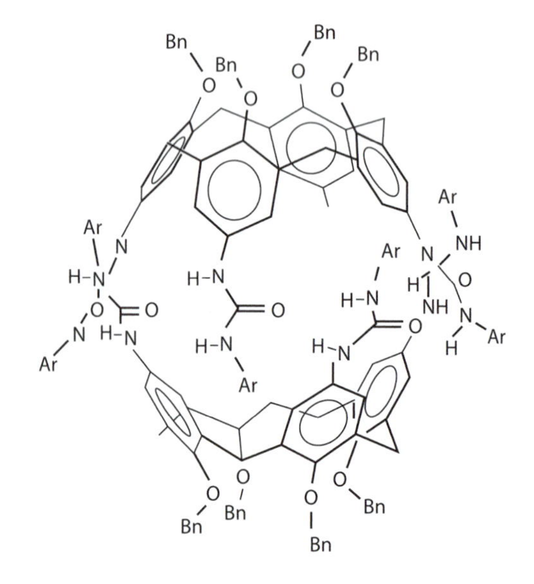

7.11 Schematic

Hydrogen bonding is also important in the sheet-like guanidinium sulfonate hosts, for which the gallery height can be varied from 3.0 to 11.5 Å by changing the sulfonate.[91] For example, the distance is 3 Å for dithionate, 5.5 Å for ethane-1,2-disulfonate, 9.5 Å for 2,6-naphthalene disulfonate, and

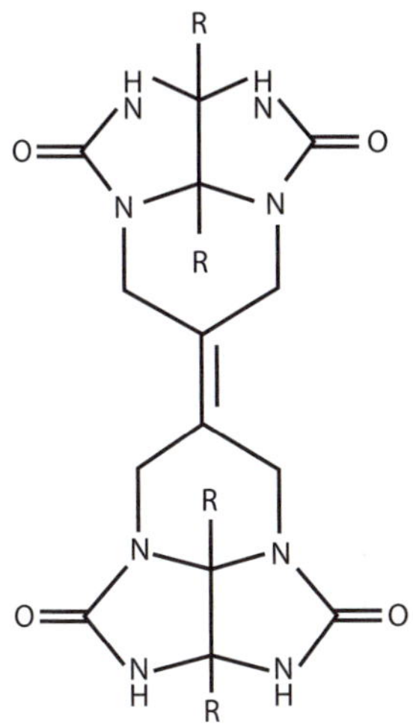

7.12 Schematic

11 Å for 4,4′-biphenyldisulfonate. A typical preparation involves crystallization from a 2:1 mixture of guanidinium hydrochloride and 4,4′-biphenyldisulfonic acid in methanol in the presence of a guest. Typical guests include acetonitrile, benzonitrile, *m*-xylene, styrene, and toluene.

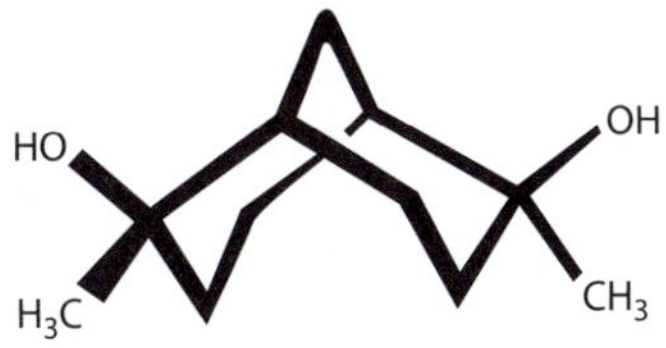

7.13 Schematic

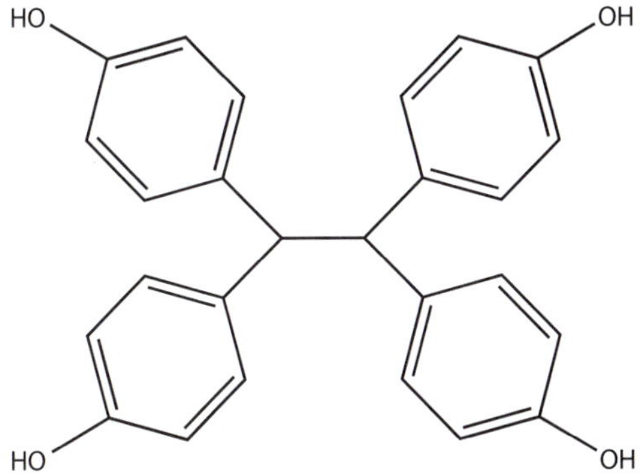

7.14 Schematic

Several diols form inclusion compounds.[92] The bicyclic one shown in **7.13** Schematic contains tubular canals that guests can fit into.[93] The crystals form inclusion compounds with the vapor of 1,2-dimethoxyethane, 1,2-dichloroethane, 1,4-dioxane, ethyl acetate, toluene, chlorobenzene, and others. The tetrols (**7.14** Schematic and **7.15** Schematic) also form inclusion compounds with various guests. Compound **7.14** favors acetone over isopropyl alcohol, acetonitrile over isopropyl alcohol, acetonitrile over methanol, and pyridine over methanol.[94] Compound **7.15** is described as an organic analog of zeolites with hydrogen-bonded molecular sheets.[95] Esters, such as ethyl benzoate, and ketones, such as 5-nonanone, can be guests.

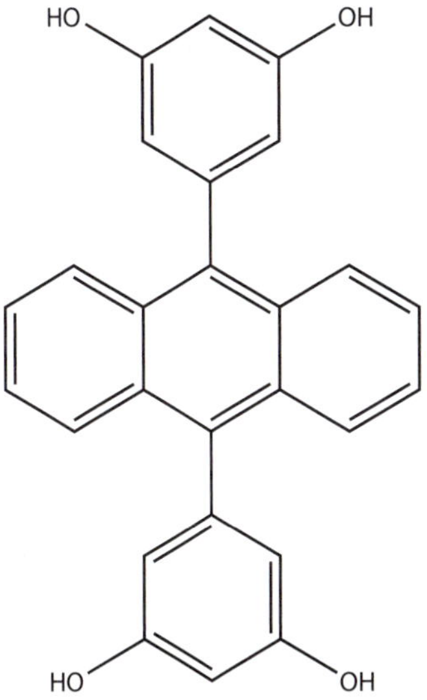

7.15 Schematic

Tris(*o*-phenylenedioxy)cyclotriphosphazene (**7.16** Schematic) can store iodine,[96] stilbene, and azobenzene.[97] It can store methane and carbon dioxide, but hydrogen, nitrogen, and oxygen are excluded.[98]

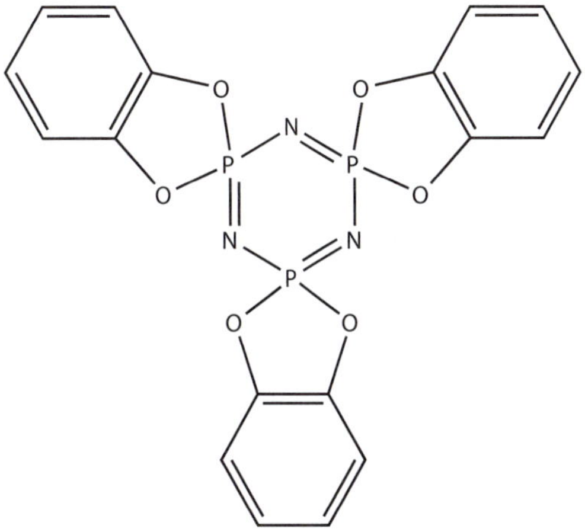

7.16 Schematic

The octasulfide (where Ar is 3,4-dimethylphenyl) (**7.17** Schematic) can act as a host for 1,4-dioxane, *N,N*-dimethylformamide, tetrahydrofuran, toluene, *tert*-butyl alcohol, ethyl acetate, limonene, and *o*-xylene.[99]

Some pyrazolones can also form inclusion compounds. Compound **7.18** (in which Ar is *meta*- or *para*-phenylene) forms inclusion compounds with methanol, ethanol, isopropyl alcohol, acetone, 2-butanone, tetrahydrofuran, and 1,4-dioxane, often with two guest molecules for one host molecule.[100]

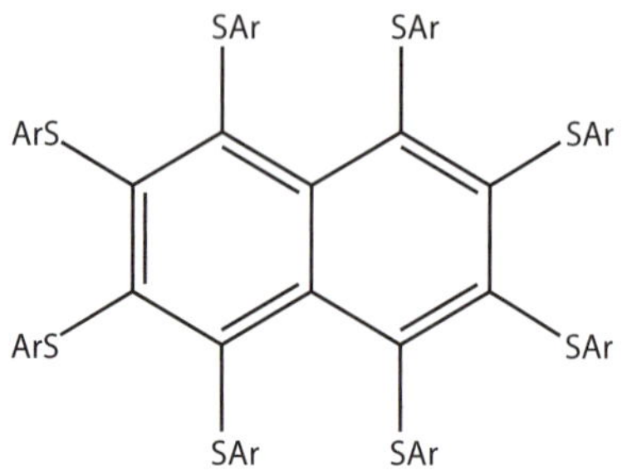

7.17 Schematic

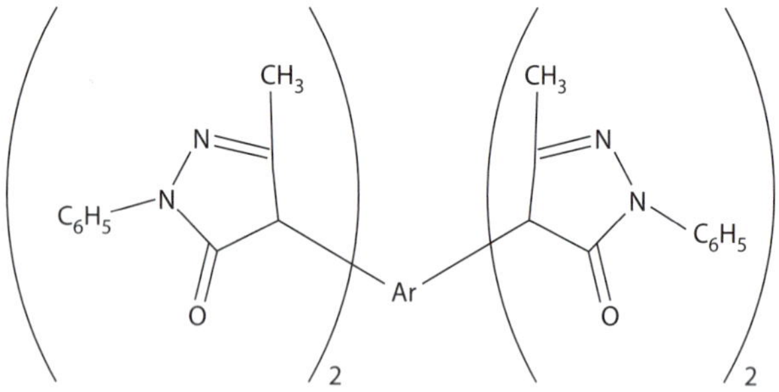

7.18 Schematic

Carboxylic acids can also form clathrates. Xylene isomers can be separated by the formation of clathrates with 1,1′-binaphthyl-2,2′-dicarboxylic acid.[101] *m*-Xylene does not form a complex. The *o*-xylene complex decomposes at 50–100°C, and the *p*-xylene complex at 100–120°C. Charge transfer is a factor in the host–guest complex of 4-methyl-3,5-dinitrobenzoic acid with 2,6-dimethylnaphthalene.[102] Charge transfer with tetranitrofluorenone has also been used to remove 60% of the dialkyldibenzothiophenes from petroleum that contains 1920 ppm sulfur, although no inclusion compound is involved.[103] As polynitro compounds are often explosive, a better charge acceptor is needed.

The host (**7.19** Schematic) was found by combinatorial synthesis (from a library of 100 salts). It forms host–guest complexes with several alcohols, ketones, xylenes, and others (e.g., 2:3 with methanol, 1:1 with ethanol, 2:2 with acetone, and 2:1 with acetonitrile).[104]

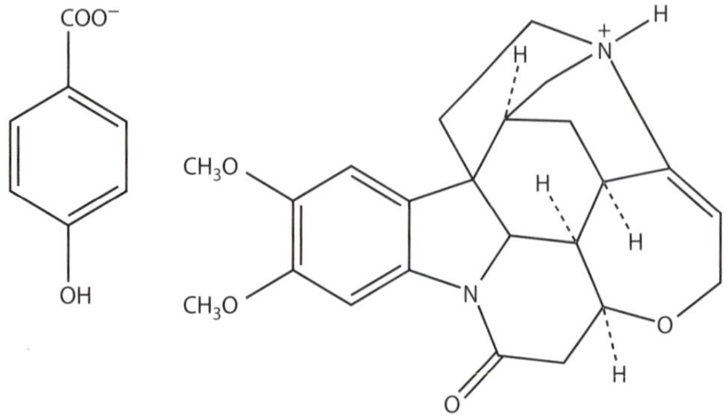

7.19 Schematic

Cobalt, nickel, and zinc salts of 1,3,5-benzenetricarboxylic acid are solids with 4–5-Å pores, which can often accommodate ammonia, ethanol, or water, but not larger molecules, as guests.[105] Cobalt and zinc terephthalates can also act as hosts for molecules such as pyrazine.[106] (For more on metal organic frameworks, see Chapter 6.) Crystals of $Co(en)_3Cl_3$ (where en = ethylenediamine) have holes that can pick up argon, oxygen, xenon, and methane.[107]

The purpose in showing these rather exotic structures is to indicate the many possibilities for selectivity for shapes,[108] some of which are not possible with zeolites, pillared clays, and others. Sometimes, the host structure adjusts to the guest to give an induced fit. By using this method, it is possible to select compounds based solely on size. For practical use, it may often be necessary to immobilize these structures on supports or place them in membranes.

High selectivity can be obtained by imprinting polymers with neutral molecules.[109] In this process, a cross-linked polymer is prepared in the presence of a template. Then the template is removed by solvent extraction. The extracted polymer is then used to pick up the template molecules from other sources. Among the examples in the literature are some that deal with atrazine (an herbicide), cholesterol, other sterols, dipeptides, *N*-acetyltryptophane resolution (L-isomer favored by a factor of 6), adenine, and barbiturates.[110] The polymerizations in the first two examples are shown in (**7.20** Schematic). (The cross-linking comonomer with the cholesterol-containing monomer was ethylenebismethacrylate. The cholesterol was cleaved from the polymer with sodium hydroxide in methanol.)

The technique has also been used with metal ions. One imprinted polymer separated Al(III) from Ni(II) by a factor of 1427.[111] The Swedish company MIP Technologies can supply imprinted polymers on a kilogram scale.[112]

Calixarenes[113] can also be used to separate gases and liquids by forming host–guest complexes with them. Calix[4]arene (**7.21** Schematic) can form complexes with acetone, carbon tetrachloride, and methane.[114] The methane complex is stable to 320°C. The analog with a *p-tert*-butyl group in each aromatic ring picks up carbon monoxide and carbon dioxide but not hydrogen.[115]

7.3 SEPARATION OF IONS

The desire to attain near-biological specificity in the separation of ions was mentioned in Chapter 4. Some progress is being made in this direction with ligands that form fairly specific complexes. These may not be inclusion compounds, but at least some involve encapsulation of the ion. These have been developed by "tuning" macrocycles by varying ring size, ring substitution, and the donor set of oxygen, nitrogen, and sulfur atoms.[116] For calix[*n*]arenes, it can involve varying the ring size, as well as the extent and type of substitution on both the aromatic rings and the hydroxyl groups.[117] Devising more

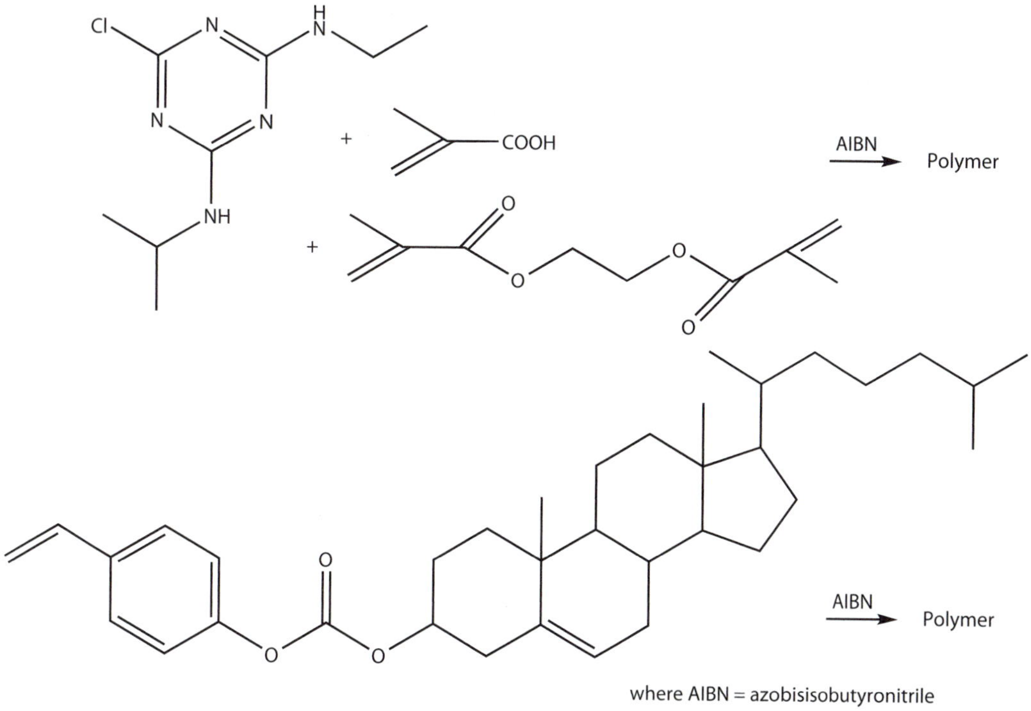

AIBN → Polymer

AIBN → Polymer

where AIBN = azobisisobutyronitrile

7.20 Schematic

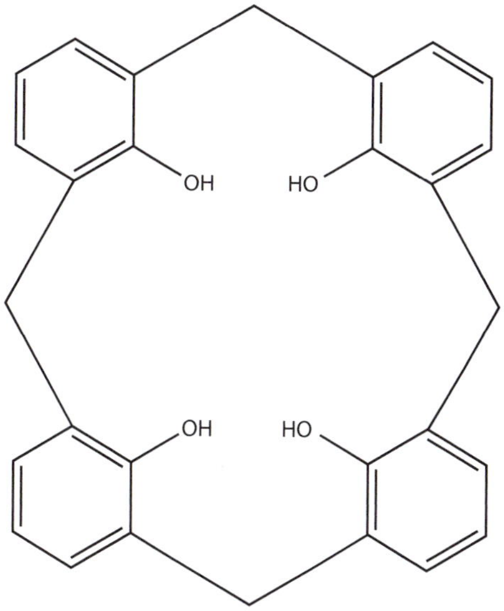

7.21 Schematic

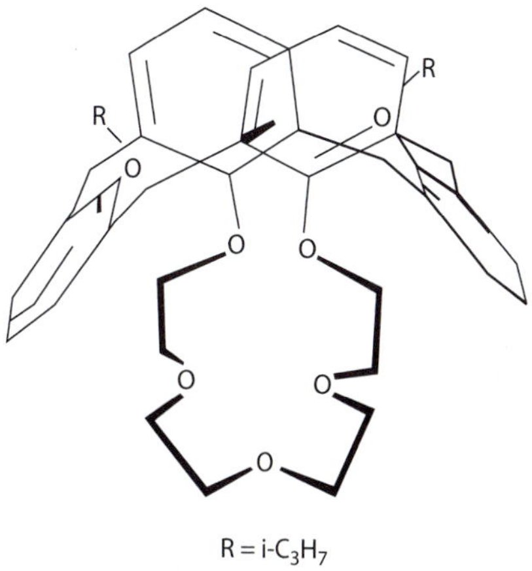

$R = i\text{-}C_3H_7$

7.22 Schematic

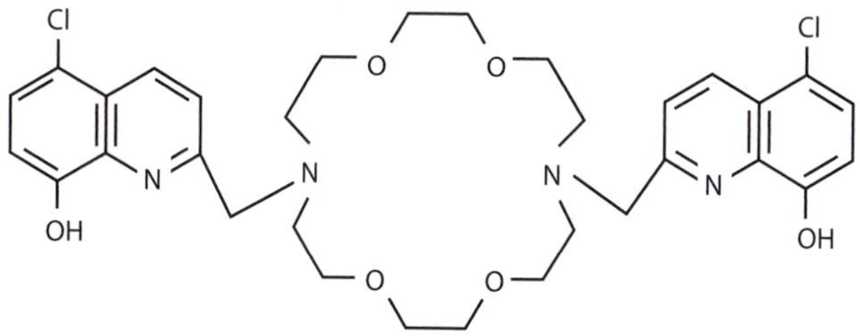

7.23 Schematic

specific receptors for anions may also involve guanidinium salts and other azonia compounds, as well as inclusion of a transition metal ion in the macrocycle.[118] Others may not have rings, but may just be di- or polyfunctional.[119]

Casnati and et al.[120] have devised a calixarene crown ether (**7.22** Schematic) that shows a preference for potassium over sodium of 22,000. A macrocyclic polyether with attached 8-hydroxyquinoline groups (**7.23** Schematic) favors barium over other alkaline earth cations by a factor of more than 10 million.[121] Compound **7.24** shows a preference of magnesium over calcium of 590.[122] Compound **7.25** has the highest known binding constant for Ag(I), log $K = 19.6$.[123]

Lithium chloride has been separated from seawater by using **7.26** Schematic despite the presence of sodium, potassium, magnesium, and calcium chlorides.[124]

Ammonium ion, often found in wastewater, can be collected in the presence of sodium and potassium ions using layered $HCa_2Nb_3O_{10}{\cdot}1.5H_2O$.[125]

There is also an active search for selective anion acceptors.[126] The cryptate (**7.27** Schematic) favors fluoride over chloride by a factor of 40 million.[127]

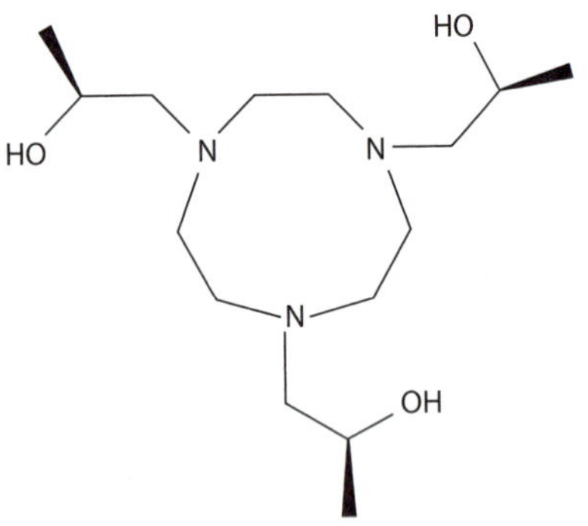

7.24 Schematic

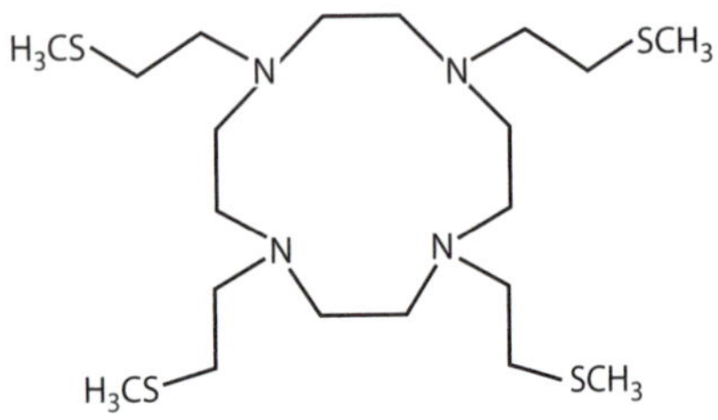

7.25 Schematic

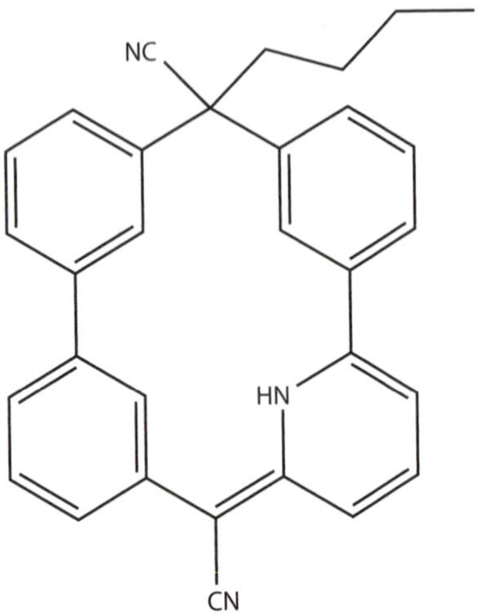

7.26 Schematic

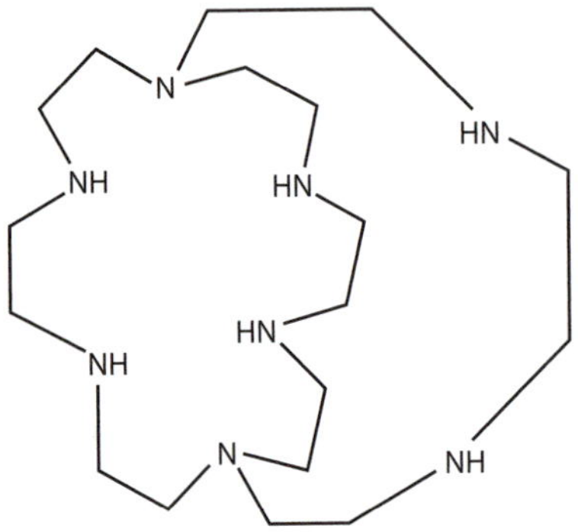

A boron compound (**7.28** Schematic) collects fluoride with $K = 5$ million.[128]

Calix[4]bipyrrole collects chloride selectively with $K = 2.9$ million.[129]

KCN complexes with the crown ether (**7.29** Schematic) with $K = 19$ million.[130]

7.27 Schematic

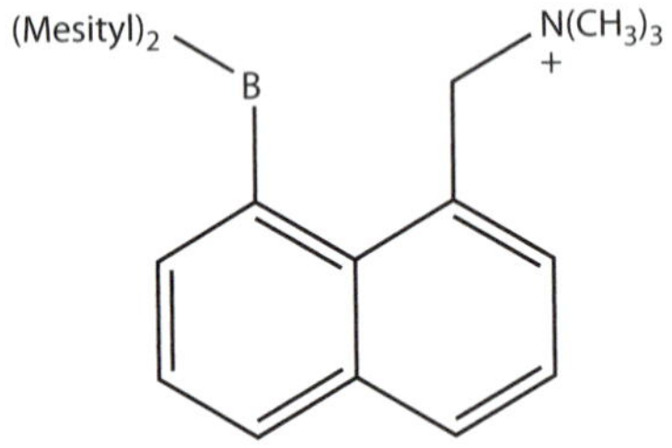

7.28 Schematic

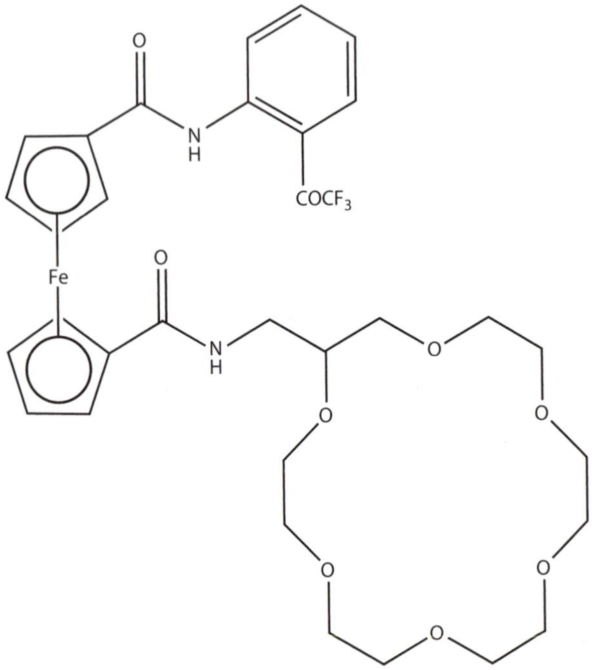

7.29 Schematic

The selective extraction and recovery of oxyanions is also being studied.[131] Many bodies of water become eutrophic when excess phosphate from detergents and fertilizer washes in. This overenrichment results in undesirable algal blooms. Agents that complex phosphate may allow it to be removed from treated wastewater, recovered and reused. The first complexing agent (**7.30** Schematic) (where X = S) complexes dihydrogen phosphate with a *K* of 820 and acetate with a *K* of 470; chloride, hydrogen sulfate, nitrate, and perchlorate are held much more weakly.[132] The *K* for the second one with phosphate (**7.31** Schematic) (where *R* is H) is 12,000.[133]

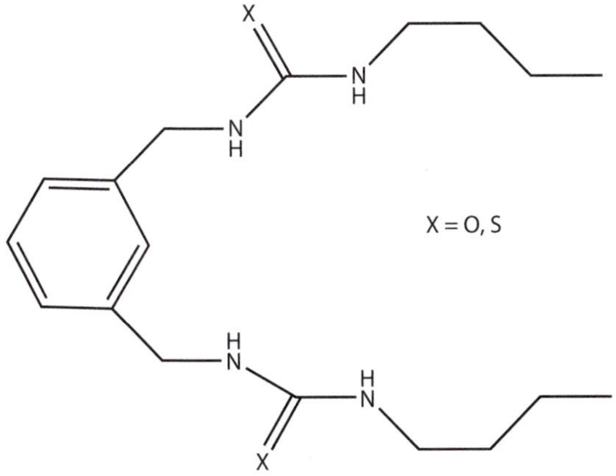

X = O, S

7.30 Schematic

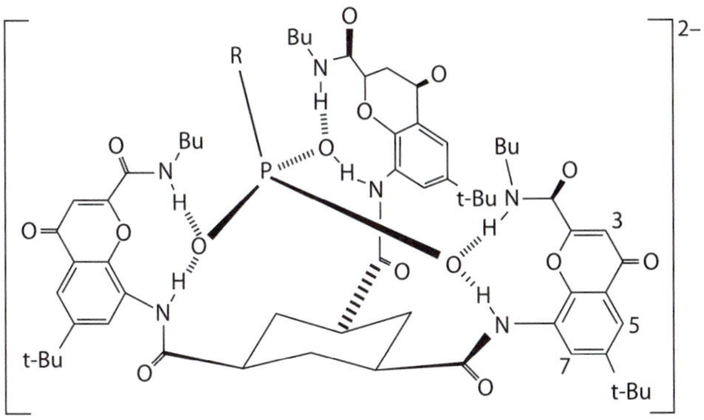

7.31 Schematic

Strong complexes of phosphates are known for **7.30** and compound **7.32** ($K = 190{,}000$ for $H_2PO_4^-$)[134] and the zinc(II) complex of compound **7.33** (log $K = 5.8$–7.9 for $ROPO_4^-$).[135]

The pyridinium salt (**7.34** Schematic) is a receptor for tricarboxylic acids with log $K = 4.5$–5.0.[136] The trisguanidinium salt (**7.35** Schematic) has a binding constant for citrate of 6.9×10^3.[137]

Sugars such as fructose and glucose can be complexed with the arylboronic acid (**7.36** Schematic), presumably through the formation of cyclic boronates.[138] (Although not ions, the sugars are included here because of the similarity of the complexing agent to those effective for ions.) A bisboronic acid of a μ-oxobis [porphinatoiron(III)] forms complexes with glucose and galactose, with association constants of 10^4–10^5.[139] Fructose can be separated from glucose by using a liquid membrane with a similar boronic acid in microporous polypropylene.[140] When glucose isomerase is included in the liquid, the output can be more than 80% fructose. High-fructose corn syrup is used widely as a sweetener.

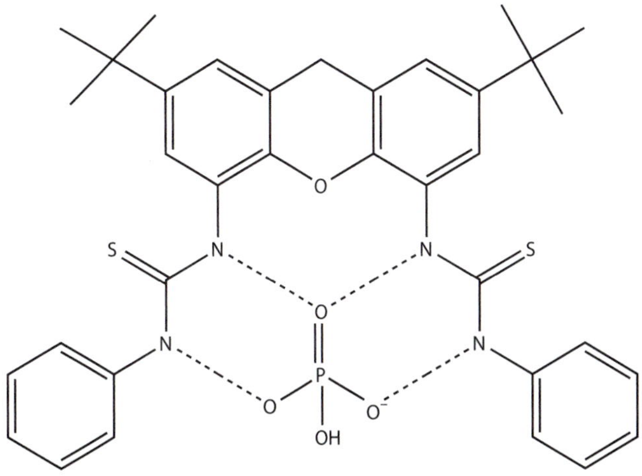

7.32 Schematic

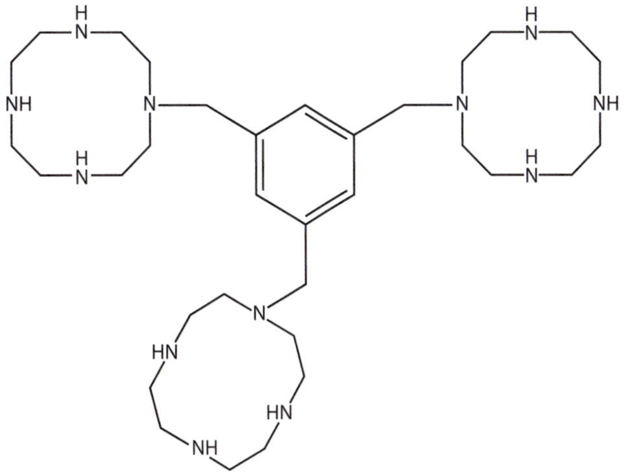

7.33 Schematic

7.4 MEMBRANE SEPARATIONS

Membranes can be used to separate molecules that differ in size, polarity, ionic character, hydrophilicity, and hydrophobicity.[141] Their use is less energy-intensive than distillation. They can often separate azeotropes and close-boiling mixtures. They can sometimes replace traditional methods (such as solvent extraction, precipitation and chromatography) that can be inefficient, expensive, or may result in the loss of substantial amounts of product. Thermally and chemically sensitive molecules can be handled. Membranes can be porous or nonporous, solid or liquid, organic, or inorganic.

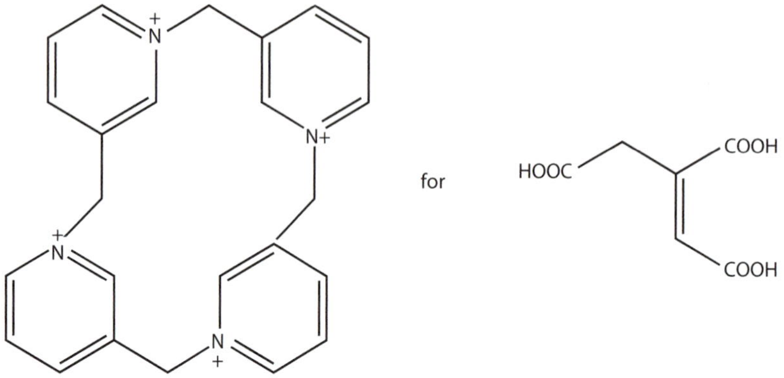

for

HOOC COOH COOH

7.34 Schematic

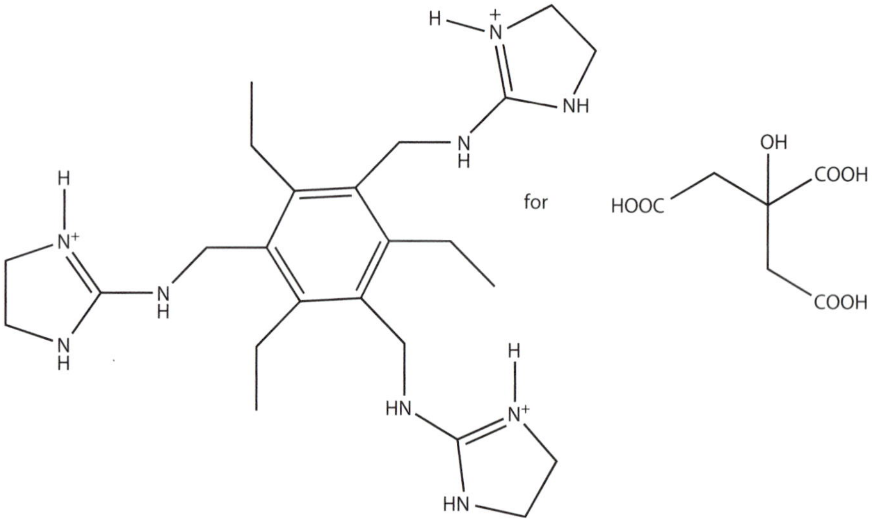

for

HOOC OH COOH COOH

7.35 Schematic

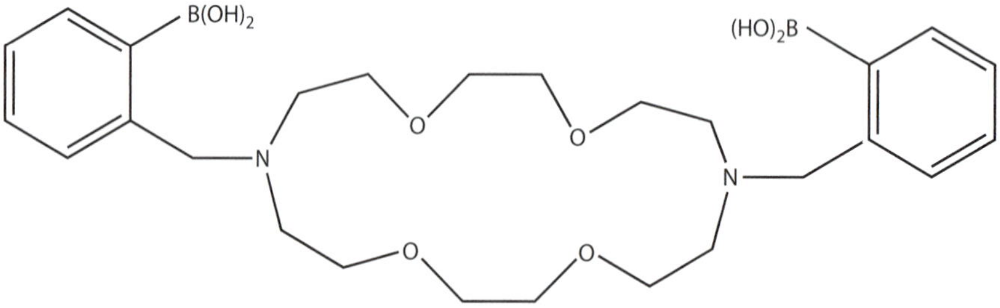

B(OH)$_2$ (HO)$_2$B

7.36 Schematic

Microfiltration[142] membranes have 0.1–10-μm pores. They can be used to remove bacteria and viruses from drinking water, as an alternative to chlorination.[143] The removal of viruses is most effective when the membrane is made of poly(*N*-benzyl-4-vinylpyridinium chloride).[144] Ultrafiltration members have 0.002–0.05-μm pores. They can remove macromolecules such as proteins. An example is the concentration of cheese whey, the liquid left when cheese is made. Ultrafiltration has also been used for the recovery of water-soluble paints in the rinse water from coating automobiles, oil from aqueous emulsions of metal-working fluids, polyvinyl alcohol from textile desizing,[145] and in the removal of polymeric materials from drugs, such as alkaloids. In the refining of vegetable oils, it removes 100% of the phospholipids, 80–85% of the free fatty acids, and most of the pigments.[146] This process eliminates water–acid treatment to remove the phospholipids, base to remove the free fatty acids, and deodorization by vacuum stripping. It can also be used for small organic molecules (e.g., 2-phenylethanol) held in micelles (in water) that are too large to pass through the pore.[147]

Reverse osmosis[148] uses membranes with 5–20-Å pores or no pores at all. Some authors add a category called nanofiltration[149] that is roughly between ultrafiltration and reverse osmosis. A membrane with 20-Å pores retained 90–98% of sugars and magnesium sulfate, but passed sodium chloride.[150] Acidic copper sulfate-containing wastewater was concentrated by reverse osmosis and then passed through a nanofiltration membrane to recover the copper sulfate. The recovered water was recycled to the process. This is an appropriate option for electroplating waste. At low pH, acetic acid and lactic acid are largely undissociated and pass through such a membrane. At higher pH, they do not go through it. These acids can also be separated by Nafion membranes,[151] weak base ion-exchange resins,[152] and by extraction with long-chain tertiary amines.[153] Cheese whey, which contains 4–6% sodium chloride, and soy sauce, which contains 18% salt, can be desalted in this way. Such a process can replace more expensive salting-out procedures, in which salts are added to cause a product to separate.[154] The fatty acids, oils, and fats left in the wastewater from oil-processing plants can be recovered in this way.[155] An aromatic polyamide membrane for desalination by reverse osmosis had improved flux, with no loss in ion rejection after treatment with hydrofluoric acid.[156]

Electrodialysis uses stacks of pairs of anion- and cation-exchange membranes in deionizing water and in recovery of formic, acetic, lactic, gluconic, citric, succinic, and glutamic acids from their sodium and potassium salts in fermentation broths.[157] This may have an advantage over processes that involve purification through calcium salts. Electrodialytic bipolar membranes have been used to recover concentrated mineral acids from dilute solution.[158] They can be used to convert sodium chloride into hydrogen chloride and sodium hydroxide in a process that avoids the use of chlorine.[159] Soy protein has been precipitated by electroacidification with bipolar membranes.[160] No chemical acids or bases were needed. The effluents could be reused. The product was just as acceptable as a conventional one. Membranes of perfluorinated sulfonic acid polymers, such as Nafion, pass water and cations freely, but are impermeable to anions.[161] They are used in electrochemical processes, such as the electrolysis of aqueous sodium chloride to produce chlorine and sodium hydroxide. They are replacing polluting mercury cells in this application. They are also used in fuel cells.

Membranes can be made in various shapes. The plate and frame types have less surface area per unit of space than the spirally wound ones; hollow-fiber ones[162] have the most. In commercial reverse osmosis, spirally wound ones are used in 75% of the cases and hollow fibers in 25%. Fouling can be a problem with membranes. A prefilter[163] is used to remove larger particles. The liquid can be flowed across a surface instead of directly at it to reduce fouling. High-frequency flow reversal and local shear enhancement (possibly by ultrasound) can reduce fouling.[164] Mechanical vibration at 60 Hz is said to increase the flow rate five-fold.[165] If the particles causing the fouling have a net surface charge, they can be repelled by a membrane with the same surface charge.[166] The use of an electromagnetic field with a conventional reverse osmosis membrane reduced fouling severalfold and doubled the flux.[167] Sulfonated polysulfone membranes are fouled less by proteins.[168] (Hydrophilic surfaces seem to be helpful in general.) The fouling of membranes by natural organic matter delayed the acceptance of the technology for purifying drinking water.[169] The problem appears to be solved. A system in San Marcos, Texas, removes 99.99% of Cryptosporidium, Giardia, and viruses. Once a month, the system is cleaned with sodium hypochlorite. Occasionally an aqueous citric acid treatment at pH 3 is used.[170] A system for wastewater uses a 1 min backwash after 10–20 min of operation. Once a quarter, the system is cleaned with 1% NaOH plus sodium hypochlorite followed by 0.5% citric acid.[171] An oxidizing biocide (0.2–0.5 ppm) is fed continuously. A clever system has been used to prepare a self-cleaning membrane for the concentration of a solution of casein. Trypsin is attached to the aminated polysulfone membrane with glutaraldehyde.[172] Any proteins that settle on the membrane are digested. When the fouling impurities in drinking water are better identified, such an enzymatic technique may be applicable. If an inorganic membrane becomes fouled with organic matter, the latter can be burned off. If the membrane fouled very often, this could be inconvenient for a water treatment plant.

Microporous membranes that are not wetted by the liquid on either side can be used to concentrate solutes to high levels at low temperature and pressure. The volatile component crosses the vapor gap to the receiving liquid. This technique of osmotic distillation has been used to concentrate fruit and vegetable juices, as well as pharmaceuticals, sugars, proteins and salts of carboxylic acids.[173]

The pores of microporous membranes can be filled with liquids to form liquid membranes that can be used in separations.[174] Transport through such membranes is often facilitated by the addition of a carrier to the liquid to improve the separation factor or the flux rate. A solution of a long-chain tertiary aliphatic amine in kerosene in the pores of a poly(tetrafluoroethylene) membrane was used to separate nickel(II) ion from cobalt(II) and copper(II) ions.[175] The cobalt and copper ions passed through, but the nickel ion did not. (The separation of nickel from cobalt in ores is a long-standing problem.) Visser et al.[176] have studied the carrier-mediated transport of ions through liquid membranes in polymer pores using salicylimine–uranyl complexes. Dihydrogenphosphate ion moved through 140 times as fast as chloride ion. Catecholamines have been extracted using a boronic acid calixarene in 2-nitrophenyl octyl ether (**7.37** Schematic).[177] (The analog with an octadecyloxy group *meta* to the boronic acid group was about nine times as effective as the one shown.) This is analogous to the extraction of sugars with boronic acids described earlier.

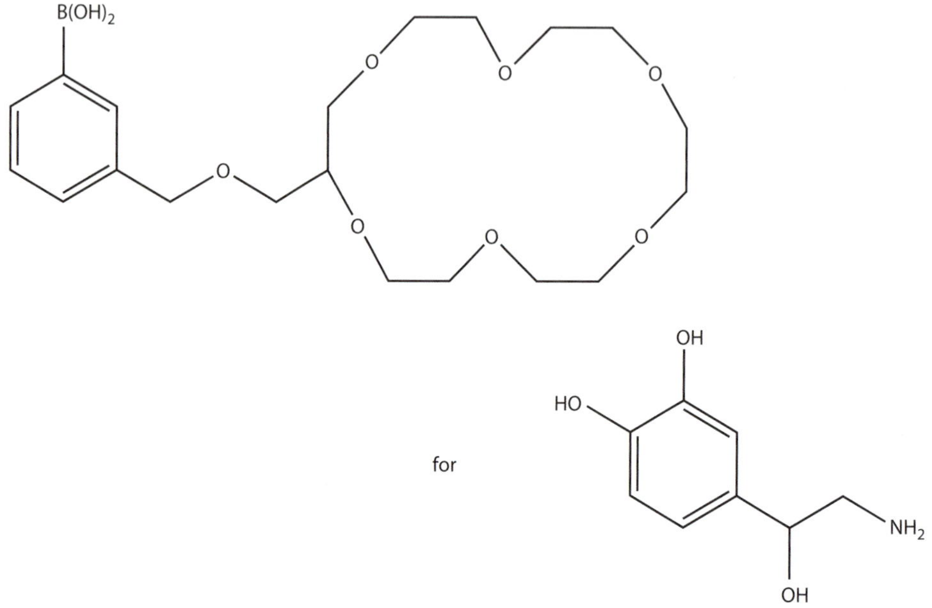

7.37 Schematic

The recovery of phenylalanine from a fermentation broth has been simplified by using a microporous poly(tetrafluoroethylene) membrane with tri-*n*-octylmethylammonium chloride in toluene in the pores.[178] Phenylalanine can also be separated using the quaternary ammonium salt with 2-nitrophenyl octyl ether in a cellulose triacetate membrane.[179] Kerosene flowing in hollow-fiber membranes can remove 99.9% of organic pollutants, such as benzene, *p*-dichlorobenzene, chloroform, and carbon tetrachloride, from wastewater outside the fibers.[180]

Anions can be separated by varying the applied potential across a porous polypropylene membrane with polypyrrole in the holes.[181] Chloride ion had high permeability, whereas sulfate and benzoate had low permeability.

Almost all gas separations are performed with nonporous membranes.[182] The rate of movement through the membrane depends on the diffusivity and solubility of the gas in the polymer.[183] Diffusion is higher for small molecules. It is faster in rubbery polymer than in a glassy polymer. The selection of nitrogen over pentane in natural rubber is 10, but in polyvinyl chloride, it is 100,000. The rate of movement of the gas through the polymer is inversely related to the thickness. For a rapid rate, the polymer film should be as thin as possible. For a mechanically strong defect-free film, the limit is about 20 μm. In practice, a very thin film is placed on a thicker porous structure of the same or a different material. To obtain a satisfactory flux rate, the surface area is made as large as possible, often by the use of hollow fibers. The flux rate can vary with the temperature and pressure. The driving force for movement through the membrane is a pressure, concentration, or pH gradient. The separation factor, α, usually becomes lower as the flux rate increases. If the separation factor is low, it may be necessary to use more than one stage. For practical use, the membrane must be mechanically, thermally, and chemically stable over a long period of time and have a high selectivity for the desired separation.

Common industrial applications include the following separations: (a) nitrogen and oxygen from air, (b) hydrogen from refinery gases, (c) hydrogen from ammonia plant off-gas, (d) carbon dioxide from methane (for natural gas and landfill gas), (e) carbon monoxide from hydrogen (for adjustment of the ratio for hydroformylation of olefins), (f) helium from methane, (g) hydrogen sulfide from methane (for natural gas), and (h) organic vapors from air.[184] Gas separation membranes offer low-energy consumption, reduced environmental impact, suitable cost-effectiveness at low gas volumes, low maintenance costs, and ease of operation, with space and weight efficiency.[185] Because the units are modular, it is easy to expand a plant. Some examples will be given to illustrate the diversity of approaches used. Nitrogen from commercial hollow-fiber membrane systems is available in 95–99.9% purity.[186] Membrane-based nitrogen costs one-third to one-half as much as liquid nitrogen. The best separation factors (23–32) are with carbon molecular sieves that contain silica.[187] Separation factors for oxygen over nitrogen (8–12, and in one case 26) are obtained with aromatic polyimides and polytriazoles.[188] The permeability was low where the α-value was 26 at 30°C and this dropped to 13 for the same membrane at 100°C. This membrane had a separation factor of 169 for hydrogen over nitrogen at the higher temperature. Polyaniline membranes separate oxygen over nitrogen with an α-value of 14.[189] The corresponding values of hydrogen over nitrogen, helium over nitrogen, and carbon dioxide over methane are 314, 366, and 78, respectively. Polyaniline can be prepared in the pores of microporous polypropylene by polymerization of aniline with ammonium persulfate.[190] This gives a strong flexible membrane. The inclusion of cobalt complexes in membranes for facilitated transport of oxygen often raises the α-value of oxygen over nitrogen to 7–15.[191] The flux rate can also go up. The problem with these systems is one of limited stability, partly owing to the oxidation of the cobalt complex. Zeolites and carbon molecular sieves are also used to separate oxygen from nitrogen.[192] The latter have an α-value of 9–16, but are vulnerable to moisture. To avoid this, hollow-fiber membranes of the molecular sieve have been coated with copolymers of tetrafluoroethylene.[193] The coated material gave an α-value of 200 for hydrogen over methane. Air products has used $Li_3Co(CN)_5 \cdot 2HCON(CH_3)_2$ to adsorb more oxygen reversibly than a carbon molecular sieve.[194] It lost 15% of its activity over 545 cycles in 17 days. It also degraded in moist air. For commercial use, the half-life must be greater than half a year.

There is also interest in separating carbon dioxide from nitrogen in flue gases. Both have important industrial uses. The separation factor for carbon dioxide over nitrogen with polycarbonate and polysulfone membranes[195] is 35–40; with polyethylene glycol in cellulose acetate[196] is 22; with polyether imides containing polyethylene oxide units[197] is 70; with an amine modified polyimide[198] is 814; with a copolymer of dimethylaminoethyl methacrylate and acrylonitrile[199] is 60–90; and with 40% 3-methylsulfolane in poly(trimethylsilylmethyl methacrylate)[200] is more than 40. This last combination also separated out the sulfur dioxide in stack gas with an α-value for sulfur dioxide over carbon dioxide of 29.[201] The 3-methylsulfolane is added as a plasticizer to increase the flux rate. For a commercial membrane, it would have to remain in the membrane and not be evaporated over a long time period. An alternative is to devise an equivalent comonomer for use with the trimethylsilylmethyl methacrylate. Poly(trimethylsilylmethyl methacrylate) (**7.38** Schematic) and poly(trimethylsilyl-1-propyne) (**7.39** Schematic) are used because they have higher gas permeabilities than many other polymers.[202] Work in this area is aimed at increasing the selectivity at high permeability and at improving the stability of the polymers under extended use.

The aromatic polyimides, cited earlier, separate carbon dioxide over methane with α-values of 150–160. Aromatic polyoxadiazoles have α-valves of 100–200.[203] A zeolite T membrane had a separation factor of 400.[204] The permeability

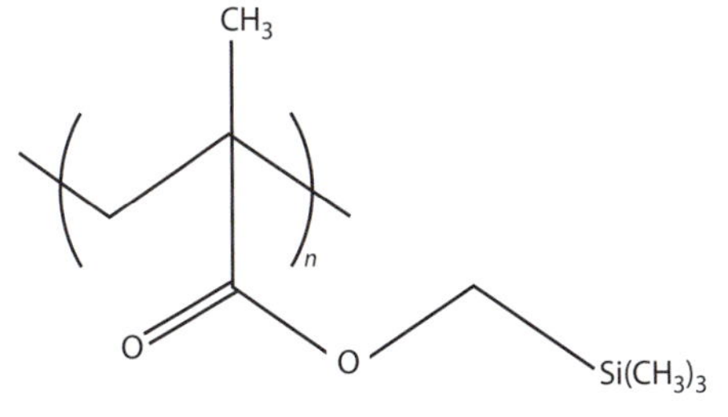

Poly(trimethylsilylmethyl methacrylate)

7.38 Schematic

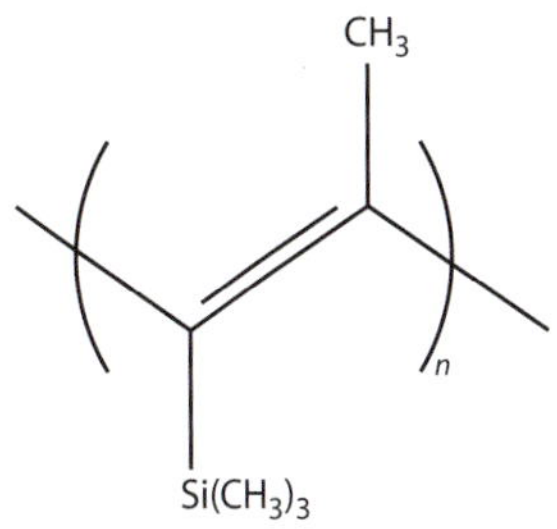

Poly(trimethylsilyl-1-propyne)

7.39 Schematic

of one aromatic polyimide was improved by two to four orders of magnitude by carbonizing it on porous alumina,[205] the final α-value being 100. In this case, the final actual membrane was probably a porous carbon molecular sieve. Facilitated transport has also been used to increase the separation factor. A porous poly(vinylidene difluoride) membrane with ethanolamine or diethanolamine in the pores gave a separation factor of carbon dioxide over methane of 2000.[206] Such a method is less energy intensive than when an amine is used in the usual solvent method.

Aromatic polyimides, containing trifluoromethyl groups, have been used to separate hydrogen over carbon monoxide, with α-values of 26–26 and hydrogen over methane with α-values of 73–380.[207] An α-value of 22 was obtained with a membrane of a copolymer of methyl methacrylate and tris(trimethylsiloxy) [γ-(methacryloxy)propyl] silane, in a process estimated to cost only 40% of the cost of the present cryogenic distillation.[208]

Air Products has devised a nanoporous carbon membrane on macroporous alumina that passes hydrocarbons, such as butane, 100 times faster than hydrogen.[209] The hydrocarbon molecules adsorb in the holes and diffuse through, while blocking the passage of the hydrogen. The process can recover hydrogen from refinery gases for use, rather than just sending it to the flare. Such a "reverse selective" membrane, made from poly(methyl-2-pentyne) containing fumed silica, removed hydrogen from benzene.[210] Olefins can be recovered from the nitrogen used in olefin polymerization plants with a membrane more permeable to the olefin.[211] Two stages are needed. Volatile organic compounds, at concentrations higher than 1% in air, can be recovered economically using rubbery membranes of silicone or polyurethane.[212] These include chlorofluorocarbons, hydrochlorofluorocarbons, tetrachloroethylene, carbon tetrachloride, benzene, toluene, *p*-xylene, and hexane, with separation factors of 30–210. This offers an alternative to carbon adsorption and cold traps for recovery of valuable compounds, at the same time avoiding air pollution.

Ethylene has been separated from ethane by a silver nitrate solution passing countercurrent in a hollow-fiber polysulfone.[213] This separation has also been performed with the silver nitrate solution between two sheets of a polysiloxane.[214] An α of 165 was obtained with a composite hollow-fiber membrane made from poly(ethylene oxide),

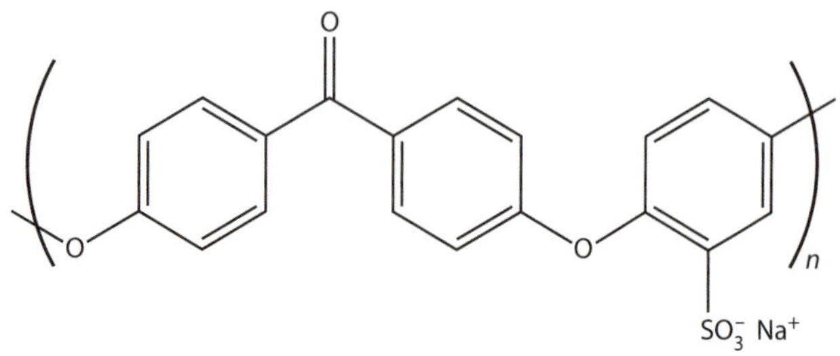

7.40 Schematic

poly(butylene terephthalate), and silver nitrate.[215] A hydrated silver ion-exchanged Nafion film separated 1,5-hexadiene from 1-hexene separation factors of 50–80.[216] Polyethylene, graft-polymerized with acrylic acid and then converted to its silver salt, favored isobutylene over isobutane by a factor of 10. A membrane of poly(vinylidene fluoride) with silver nitrate in glycerol in the pores separated 1-butene from butane with an α value of 850.[217] Olefins, such as ethylene, can be separated from paraffins by electroinduced facilitated transport using a Nafion membrane containing copper ions and platinum.[218] A carbon molecular sieve made by pyrolysis of a polyimide, followed by enlargement of the pores with water at 400°C, selected propylene over propane with an α-value greater than 100 at 35°C.[219] A membrane of poly(vinyl alcohol) cross-linked with glutaraldehyde containing silver hexafluoroantimonate separated propylene from propane with an α value of 125.[220] ITQ 32 zeolite can be used to separate linear olefins from paraffins without the need for silver.[221] The ratio of the rates of diffusion of propylene to propane was 1611 at 333 K.

Carbon molecular sieves can also be used to recover fluorine-containing gases for recycling at semiconductor plants.[222] (These are potent greenhouse gases.)

The polyetherketone (**7.40** Schematic) has a separation factor of 1.5×10^6 for water vapor over nitrogen and can be used to dehumidify gases.[223]

Chiral membrane separations are also possible. The modified poly(L-glutamate) (**7.41** Schematic) separated racemic tryptophane completely.[224]

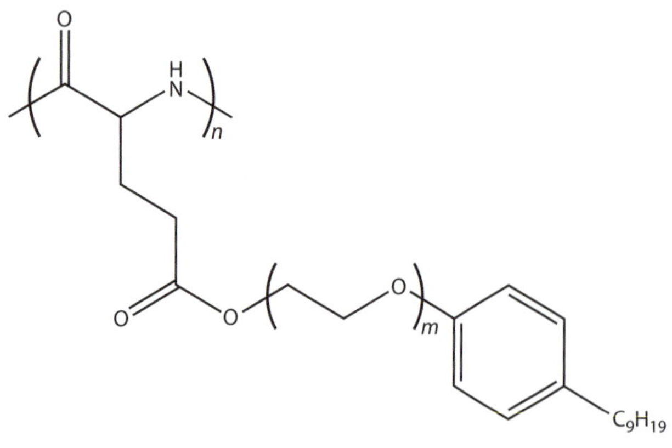

7.41 Schematic

Inorganic membranes[225] are much more expensive than organic ones, but they have certain advantages.[226] In processing foods, they offer ease of sterilization by heating and resistance to chlorine-containing sterilants. Organic fouling substances can be burned off. Stack gases do not have to be cooled before the membrane can be used. They are also resistant to organic solvents. Ceramic membranes for micro- and ultrafiltration can last 5–10 years.[227] Ceramic membranes can also be produced with pore sizes of 5–100 Å.[228] The sol–gel method can be used to place thin layers on a porous inorganic support.[229] Thin layers of silica can be placed on porous Vycor glass by alternating treatments with tetrachlorosilane and water.[230] The resulting membrane had a separation factor for hydrogen over nitrogen of 500–1000 at 600°C. It might be used on the tail gas of an ammonia plant. This same separation has been accomplished with an α-value of more than 1000 with a silica membrane made by chemical vapor deposition of tetraethoxysilane in the pores of an alumina tube at 400–600°C.[231] The membrane was stable to steam at 500°C. Similar ultrathin films have been made by plasma polymerization of hexamethyldisiloxane.[232] Volatile solvents can be removed from air using La_2O_3-modified γ-alumina membranes.[233] For removal of acetone from nitrogen, the separation factor was about 1000. Up to 99% of a red anionic dye was removed from wastewater with a titanium dioxide membrane with 0.2-µm pores.[234] This is a simpler process than trying to oxidize the dye in the wastewater and should allow reuse of the recovered dye. Membranes of carbon with 100-Å pores can be prepared by the pyrolysis of mixtures of polyacrylonitrile and other polymers.[235] This is similar to the preparation of graphite fibers.

Defect-free zeolite films are available for separations (also see Chapter 6). They are usually made on a porous inorganic support, often with seeding.[236] These have been used to separate benzene over *p*-xylene by an α-value of more than 160, to separate benzene over cyclohexane by $\alpha = 15–20(550)$, to separate *n*-butane over isobutene by $\alpha = 88$,[237] to separate water over propanol by $\alpha = 71$,[238] to separate hydrogen over isobutane by $\alpha = 151$, to separate hydrogen over methane by $\alpha = 500$,[239] and to separate hydrogen over argon by $\alpha = 5.3$. Zeolites have also been embedded in nickel foil[240] and in glassy polymers.[241] The latter gave a separation factor of carbon dioxide over methane of 34.[242] Silicalite molecular sieves have been applied on porous alumina and stainless steel.[243] Using these, the separation factor was >200 for hexane over benzene, 40 for *n*-octane over isooctane, and 200 for hexane over 2,2-dimethylhexane at 200°C. The silicalite on sintered stainless steel selectively permeated acetic acid over water from 5 to 40% solutions. Self-supporting films of montmorillonite clay have also been prepared.[244]

Pervaporation is a process in which compounds move selectively across a membrane in response to a vacuum.[245] This a less energetic process than the usual distillation. The most common use is in the dehydration of ethanol, which needs less than 20% of the energy of the azeotropic distillation process.[246]

(To remove the remaining water from the 95% ethanol/5% water azeotrope, a hydrocarbon such as benzene is added and distillation is continued until the water is removed as a new azeotrope. This is the process that is being displaced.) The separation factors for water over alcohol are high, such as 12,000 for a polymeric quaternary ammonium salt complex of κ-carrageenan (a polysaccharide potassium sulfonate from seaweed) with a flux rate of 600 $g/m^2/h$,[247] 6300 for a polycarbonate grafted with 4-vinylpyridine[248]; 3000–4500 for blends of polyvinyl alcohol with sodium alginate,[249] 3500 for a polyacrylic acid–polycationic polymer blend with a flux rate of 1.6 $kg/m^2/h$,[250] and 1000–2000 for a membrane of polyvinyl alcohol-poly(styrenesulfonic acid barium salt).[251]

The opposite process, the selective removal of ethanol, is not as easy. It requires a more hydrophobic membrane and the separation factors are much lower. It has the potential to be very useful. The yeasts that form ethanol during fermentation are inhibited by it. This limits the concentration of ethanol to 6–20% depending on the particular yeast used. If ethanol were removed continuously during fermentation, more alcohol could be produced from a given batch of yeast. Even if the ethanol permeate contained a significant amount of water, this water could be removed by one of the aforementioned membranes. This two-membrane process could reduce the price of fuel ethanol to make it more competitive with gasoline. This could be part of a switch to a sustainable future that could help moderate the rate of global warming. Separation factors for 2.5–10% ethanol in water include 60 for silicalite in stainless steel,[252] 89 at 1.81 $kg/m^2/h$ with a silicalite membrane on alumina,[253] about 30 for an aluminum-free analog of zeolite ZSM-5 in silicone rubber,[254] 17 for poly(trimethylsilyl-1-propyne),[255] and 12–15 (with a flux rate of 200 $g/m^2/h$ with no decline in flux over time) for 1% of poly(dimethylsiloxane) in poly(trimethylsilyl-1-propyne).[256] Poly(trimethylsilyl-1-propyne) oxidizes readily in air, so that membranes made of it tend to deteriorate with age.[257] The poly(dimethylsiloxane) may inhibit this reaction. It is possible that the addition of a polymeric antioxidant would delay aging of the membranes.

Water can be removed from methanol by a membrane of polyvinyl alcohol cross-linked with polyacrylic acid, with a separation factor of 465.[258] A polymeric hydrazone of 2,6-pyridinedialdehyde has been used to dehydrate azeotropes of water with *n*- and *i*-propyl alcohol, *s*- and *tert*-butyl alcohol, and tetrahydrofuran.[259] *Clostridium acetobutylicum*, which is used to produce 1-butanol, is inhibited by it. Pervaporation through a poly(dimethylsiloxane) membrane filled with cyclodextrins, zeolites or oleyl alcohol kept the concentration in the broth lower than 1% and removed the inhibition.[260] Acetic acid can be dehydrated with separation factors of 807 for poly(4-methyl-1-pentene) grafted with 4-vinylpyridine,[261] 150 for polyvinyl alcohol cross-linked with glutaraldehyde,[262] more than 1300 for a doped polyaniline film (4.1 $g/m^2/h$),[263] 125 for a nylon-polyacrylic acid membrane (5400 $g/m^2/h$), and 72 for a polysulfone.[264] Pyridine can be dehydrated with a membrane of a copolymer of acrylonitrile and 4-styrenesulfonic acid to give more than 99% pyridine.[265] A hydrophobic silicone rubber membrane removes acetone selectively from water. A hydrophilic cross-linked polyvinyl alcohol membrane removes water selectively from acetone. Both are more selective than distillation.[266] Glycerol can be dried by pervaporation through a poly(vinyl alcohol) cross-linked by maleic anhydride membrane with a separation factor of 5,598–26,000.[267]

Benzene can be separated over cyclohexane with an α-value of 26 with a polyvinyl alcohol–poly(allylamine) blend containing a cobalt(II) complex.[268] An α-value of 60 has been obtained by pervaporation with a poly(acrylonitrile-*co*-methyl methacrylate) membrane.[269] Membranes of porous polyethylene grafted with glycidyl methacrylate[270] and poly(*N,N*-dimethylacrylamide-*co*-methyl methacrylate)[271] have also been used in this separation with separation factors of 21–22. This is a separation that would be difficult to do by size and by distillation. The two boil only 2°C apart. The cyclohexane produced by the reduction of benzene is the starting material for nylon. The best solution to the problem is to run the reduction to 100% completion.

Pervaporation can also be used to remove organic compounds from wastewater. Elastomeric hydrophobic polymer membranes are used to remove compounds more hydrophobic than acetone present at levels of 200–50,000 ppm.[272] Compounds such as 2-butanone, benzene, and trichloroethylene can be recovered. A polyphosphazene–poly(dimethylsiloxane) graft copolymer membrane was able to concentrate 7% ethyl acetate in water up to 97%. The use of silicalite in poly(dimethylsiloxane) improves both selectivity and flux rate.[273]

Catalytic membrane reactors combine reaction and separation into a single step.[274] They can result in a cheaper and cleaner product that needs no further processing. The catalyst can be on one side of the membrane (fluidized[275] or not) or actually in the membrane. The reaction product can be removed from it as formed, as in the hydrolysis of a macromolecule, for which an ultrafiltration membrane keeps the polymer in the reaction vessel, but allows the hydrolysate to pass. Pervaporation of a liquid passing through can be used to remove the product from it. Continuous removal of a product can shift the equilibrium in the desired direction, as in dehydrogenations. It can eliminate inhibition of cells by the products of their reaction. By this process, unwanted secondary reactions of the primary products can be eliminated or reduced. Sometimes a membrane can prevent poisons from getting to a catalyst. Membranes can allow slow, steady addition of a reagent to avoid locally high concentration that could lead to hot spots. One reactor design uses mixing spheres in microporous tubes to eliminate hot spots and improve yields.[276] A few examples will be given to illustrate the diversity of reactions that can be run in membrane reactors. Many more can be found in the foregoing review references.

Removal of hydrogen through membranes has improved conversions and yields in some dehydrogenations.[277] The dehydrogenation of *n*-butane in a Pd/Ag membrane reactor increased conversions four- to six-fold over equilibrium values.[278] Conversions were up 1.5 times when a polyimide membrane was used.[279] One problem has been to obtain a high enough permeation rate of hydrogen through dense palladium membranes. To raise the flux rate, very thin layers of palladium have been placed on tantalum and niobium[280] or on porous ceramics.[281]

Ethylene and acetaldehyde are the main products when carbon monoxide is reduced by hydrogen permeating through a membrane of palladium.[282] The main products are methane and methanol when a supported palladium catalyst is used, instead of the membrane. Cyclododecatriene was reduced with hydrogen to cyclododecene with 88% selectivity at 96% conversion, and with 80% selectivity at 100% conversion, using a membrane with palladium in the pores of alumina.[283] A microporous ceramic membrane containing titanium dioxide and platinum was used in the reduction of 2-hexyne to *cis*-2-hexene with hydrogen, with 100% selectivity.[284] Vegetable oils reduced with hydrogen in a tubular ceramic membrane reactor with palladium in the membrane gave a product with 2–5% *trans*-content, about three to four times lower than in a conventional slurry reactor.[285] (*trans*-Fatty acids cause as much heart disease as saturated fats.)

Ceramic membranes of mixed oxides, formed by heating mixtures of lanthanum, strontium, cobalt, and iron carbonates or nitrates, allow oxygen, but not nitrogen, to pass.[286] This eliminates the need for an air separation plant. Air Products sells a unit for use in the laboratory so that an oxygen cylinder is not needed.[287] Such a membrane reactor was used to oxidize methane to carbon monoxide with 90% selectivity at 98% conversion.[288] This process is 25% cheaper than the usual method of making synthesis gas from natural gas.[289] Yttria-stabilized zirconia also permeates oxygen selectively over nitrogen, but the flux rate is lower.[290] The highest oxide ion mobility known is that of cubic Bi_2O_3 electrodeposited on gold.[291] The yield of C_2 compounds has been improved by feeding methane and oxygen on opposite sides of a mixed barium cerium and gadolinium oxide membrane.[292] The best selectivity was 70% at 8% conversion with the membrane, compared with 32% selectivity without it. The selectivity to C_2 products was 94% with a Y_2O_3/Bi_2O_3 membrane at higher than 800°C.[293] With a $Mn/Na_2WO_4/SiO_2$ membrane, the product contained 30% ethane, 40% ethylene, and 10% propylene at 20% conversion.[294] Another group obtained 85% selectivity to coupled products using a membrane reactor of Sr, La, and Bi oxides with silver on a yttria-stabilized zirconia tube.[295] A better solution to this problem of making C_2 compounds is to make the ethylene by dehydration of ethanol from fermentation. (For making monomers and other materials from renewable sources, see Chapter 12. This will help moderate global warming, as described in Chapter 15.)

The exotherm from the reaction of sulfur trioxide with toluene can be moderated by feeding the sulfur trioxide through a tube with 0.5-μm pores.[296] Iron phthalocyanine can be prepared in the cages in sodium Y zeolite. This is then protected by a cover of cross-linked poly(dimethylsiloxane). The cyclohexane on one side of this membrane has been oxidized by the *tert*-butylhydroperoxide on the other side at room temperature to a 95:5 mixture of cyclohexanone and cyclohexanol at 25% conversion.[297] The zeolite protected the catalyst from oxidation.[298] The membrane could be regenerated by washing with acetone and drying. This reaction is of interest because the oxidation of cyclohexane is used to make nylons. The challenge is to use a membrane reactor with oxygen to obtain high conversions to these products. Continuous separation of the products through a membrane might help.

Chiral catalytic membranes have been prepared by occluding homogeneous ruthenium, rhodium, and manganese catalysts in a polymerizing mixture that leads to a cross-linked poly(dimethylsiloxane) rubber.[299] The ruthenium catalyst was used to reduce methyl acetoacetate with up to 70% enantioselectivity and the rhodium one gave 90% ee. The manganese catalyst was used with styrene on one side and aqueous sodium hypochlorite on the other to produce the epoxide with 52% ee. The ruthenium and manganese catalysts were used three times, with no loss of activity or enantioselectivity. This appears to be a satisfactory way to heterogenize homogeneous catalysts.

Continuous membrane bioreactors can have advantages. One is the removal of inhibition of cells by the product. This was shown in the oxidation of naphthalene (**7.42** Schematic) by *Pseudomonas fluorescens*.[300] An ultrafiltration membrane kept the cells and insoluble naphthalene on one side, while the soluble product was removed continuously from the other side. The yield was three times that in a batch process. Membrane bioreactors are in common use for the treatment of wastewater.[301]

Inhibition of animal cell cultures by ammonia has been overcome either by passing oxygen through a porous poly(tetrafluoroethylene) membrane into the culture or by continuous removal through a membrane of the same

7.42 Schematic

material (0.2-μm pores) to aqueous phosphoric acid on the other side.[302] Urea was removed by an anion-exchange membrane clamped over an insolubilized polyvinyl alcohol membrane containing urease.[303] No ammonium ion returned back into the feed solution. A urease in poly(ethylene-*co*-vinyl acetate) converted urea on one side to ammonia for recovery on the other.[304]

A cheaper process for the environmentally friendly solvent ethyl lactate uses a membrane to separate ammonia and water by-products. The ammonium lactate made by fermentation of carbohydrates is cracked by heating to lactic acid, which is then esterified with ethanol. This process eliminates the large amounts of waste salts formed with other processes.[305] This solvent has been suggested as a replacement for 80% of the 3.8 million tons of solvent used each year in the United States for electronics, paints, coatings, textiles, cleaning adhesives, printing, and de-icing.

A membrane cell recycle reactor with continuous ethanol extraction by dibutyl phthalate increased the productivity four-fold with increased conversion of glucose from 45 to 91%.[306] The ethanol was then removed from the dibutyl phthalate with water. It would be better to do this second step with a membrane. In another process, microencapsulated yeast converted glucose to ethanol, which was removed by an oleic acid phase containing a lipase that formed ethyl oleate.[307] This could be used as a biodiesel fuel. Continuous ultrafiltration has been used to separate the propionic acid produced from glycerol by a *Propionibacterium*.[308] Whey proteins have been hydrolyzed enzymatically and continuously in an ultrafiltration reactor, with improved yields, productivity, and elimination of peptide coproducts.[309] Continuous hydrolysis of a starch slurry has been carried out with α-amylase immobilized in a hollow-fiber reactor.[310] Oils have been hydrolyzed by a lipase immobilized on an aromatic polyamide ultrafiltration membrane with continuous separation of one product through the membrane to shift the equilibrium toward the desired products.[311] Such a process could supplant the current energy-intensive industrial one that takes 3–24 h at 150–260°C. Lipases have also been used to prepare esters. A lipase–surfactant complex in hexane was used to prepare a wax ester found in whale oil, by the esterification of 1-hexadecanol with palmitic acid in a membrane reactor.[312] After 1 h, the yield was 96%. The current industrial process runs at 250°C for up to 20 h.

Some oxidoreductases require expensive cofactors, such as nicotine adenine dinucleotide (NADH). This requires continuous regeneration of the cofactor in an enzymatic reaction using a sacrificial substrate. Nanofiltration membranes have been used to separate the two enzyme systems.[313] A sulfonated polysulfone nanofiltration was used in the reduction of fructose to mannitol, using glucose to gluconolactone to regenerate the NADH.

The use of membrane reactors can be expected to expand dramatically in the future. The search for improved selectivity, permeability, stability, and resistance to fouling of membranes will continue.

REFERENCES

1. (a) N.N. Li, ed., *Separation and Purification Technology*, Dekker, New York, 1992; (b) S.D. Barnicki and J.J. Siirola, *Kirk–Othmer Encyclopedia of Chemical Technology*, 4th ed., Wiley, New York, 1997, *21*, 923–962; (c) J.D. Seader and E.J. Henley, *Separation Process Principles*, Wiley, New York, 1998; (d) J.L. Humphrey and G.E. Keller II, *Separation Process Technology*, McGraw-Hill, New York, Wiley, New York 1997; (e) C.E. Meloan, *Chemical Separations—Principles, Techniques and Experiments*, Wiley, New York, 1999.
2. H.S. Muralidhara, *Chemtech*, 1994, *24*(5), 36.
3. G.D Botsaris and K. Toyokura, eds, *Separation and Purification by Crystallization. Am. Chem. Soc. Symp., 667*, Washington, DC, 1997.
4. J. Gerstner, *Chem. Br.*, 1997, *33*(2), 40.
5. (a) D.E. Rearick, M Kearney, and D.D. Costesso, *Chemtech*, 1997, *27*(9), 36; (b) R. Wooley, Z. Ma, and N.-H.L. Wang, *Ind. Eng. Chem. Res.*, 1998, *37*, 3699; (c) A.M. Thayer, *Chem. Eng. News*, Sept. 5, 2005, 49; (d) T. Aida and P.L. Silveston, *Cyclic Separating Reactors*, Blackwell Publishing, Ames, IA, 2005.
6. (a) P. Radecki, J. Crittenden, D. Shonnard, and J. Bulloch, eds, *Emerging Separation and Separative Reaction Technologies for Process Waste Reduction: Adsorption and Membrane Systems*, AIChE, New York, 1999 (b) G. Karet, *R&D (Cahners)*, 1999, *41*(3), 35; (c) C.A.M. Afonso and J.P.G. Crespo, eds, *Green Separation Processes–Fundamentals and Applications*, Wiley-VCH, Weinheim, 2005; (d) C.C. Tzschucke, C. Markert, W. Bannwarth, S. Roller, A. Hebel, and R. Haag, *Angew. Chem Int. Ed.*, 2002, *41*, 3965; (e) D.J. Cole-Hamilton, *Science*, 2003, *299*, 1702; (f) K. Sundmacher, A. Kienle, and A. Seidel-Morgenstern, *Integrated Chemical Processes*, Wilely-VCH, Weinheim, 2005; (g) R.D. Noble and R. Agrawal, *Ind. Eng. Chem. Res.*, 2005, *44*, 2887; (h) J. Charpentier, *Ind. Eng. Chem. Res.*, 2007, *46*, 3465.
7. K. Weissermel and H.-J. Arpe., *Industrial Organic Chemistry*, 2nd ed., VCH, Weinheim, 1993, 36.
8. (a) C. Arthur. *New. Sci. Jan.* 28, 1995, 21; (b) P.W.S. Pui, F. Trampler, S.A. Sonderhoff, M. Groeschi, D.G. Kilburn, and J.M. Piret, *Biotechnol. Progr.*, 1995, *11*, 146; (c) J.J. Hawkes and W.T. Coakley, *Enzyme. Microb. Technol.*, 1996, *19*, 57; (d) W.T. Coakley, *Trends Biotechnol*, 1997, *15*, 506; (e) S.-H. Hwang and Y.-M. Koo, *Biotechnol. Lett.*, 2003, *25*, 345.
9. (a) H. Feil, Y.H. Bae, J. Feijen, and S.W. Kim, *Macromol. Rapid Commun.*, 1993, *14*, 465; (b) S. Kurihara, A Minagoshi, and T. Nonaka, *J. Appl. Polym. Sci.*, 1996, *62*, 153.
10. O. Lavastre P. Bebin, O. Marchaland, and P.H. Dixneuf, *J. Mol. Catal. A. Chem.*, 1996, *108*, 29.
11. D.E Bergbreiter, Y.-S. Liu, and P.L. Osborn, *J. Am. Chem. Soc.*, 1998, *120*, 4250.
12. B. Jeong and A. Gutowska, *Trends Biotechnol.*, 2002, *20*, 305.
13. H.E. Cohen, *Ullmann's Encyclopedia of Industrial Chemistry*, 5th ed., VCH, Weinheim, 1988, *B2*, 14–1.

14. G.D. Moeser, K.H. Roach, W.H. Green, P.E. Laibinis, and T.A. Hatton, *Ind. Eng. Chem. Res.*, 2002, *41*, 4739.
15. A.-H. Lu, W. Schmidt, N. Matoussevitch, H.L. Bonneman, B. Spliethoff, B. Tesche, E. Bill, W. Kiefer, and F. Schuth, *Angew. Chem. Int. Ed.*, 2004, *43*, 4303.
16. T.C. Pearson and T.F. Schatzki, *J. Agric. Food. Chem.*, 1998, *46*, 2248.
17. E.S. Jackson, *Agric. Res.*, 2007, *55*(8), 23.
18. G. Ondrey, *Chem. Eng.*, 2005, *112*(7), 14.
19. B. Yarar, *Ullmann's Encyclopedia of Industrial Chemistry*, 5th ed., VCH, Weinheim, 1988, *B2*, 23–1.
20. E. Muller, R. Berger, W.C.G. Kosters, and M. Cox, *Ullmann's Encylopedia of Industrial Chemistry*, 5th ed., 1988, *B3*, 6–43.
21. (a) Anon., *Science*, 1995, *267*, 1101; (b) Anon., *Science*, 1996, *274*, 1019; (c) P.D. Rye, *Biotechnol*, 1996, *14*, 155.
22. G. Street, ed., *Highly Selective Separations in Biotechnology*, Chapman & Hall, London, 1994.
23. (a) A. Dubina, A. Gascon, I. Ferrer, and D. Barcelo, *Environ. Sci. Technol.* 1996, *30*, 509; (b) A. Oubina, A. Gascon, and D. Barcelo, *Environ. Sci. Technol.*, 1996, *30*, 513; (c) T.S. Lawruk, C.E. Lachman, S.W. Jourdan, and J.R. Fleeker, *Environ. Sci. Technol.*, 1996, *30*, 695.
24. S. Samdani, *Chem. Eng.*, 1995, *102*(8), 25.
25. M.J. Szymonifka and K.T. Chapman, *Tetrahedron Lett.*, 1995, *36*, 1597.
26. Anon., *Science*, 1995, *270*, 385.
27. (a) L. Nunez and M.D. Kaminki, *Chemtech*, 1998, *28*(9), 41; (b) Anon., *Chem. Ind. (Lond.)*, 1998, 197.
28. C.D. Chang and P.G. Rodewald, U.S. patent 5,498,814 (1996).
29. (a) G.G. Podrebarac, F.T.T. Ng, and G.L. Rempel, *Chemtech*, 1997, 27(5), 37; (b) M.P. Dudkovic, *Catal. Today*, 1999, *48*, 5; (c) K. Sundmacher and A. Kienle, eds, *Reactive Distillation: Status and Future Directions*, Wiley-VCH, Weinheim, 2003; (d) W.L. Luyben and C-C. Yu, *Reactive Distillation Design and Control*, Wiley-AIChE, Hoboken, NJ, 2008.
30. L.A. Lawrence, U.S. patent 5,476,978 (1995).
31. (a) M.A. Isla and H.A. Irazoqui, *Ind. Eng. Chem. Res.*, 1996, *35*, 2696; (b) J.R. Fair, *Chem. Eng.*, 1998, *105*(11), 158.
32. R. Kumar and S.M. Mahajani, *Ind. Eng. Chem. Res.*, 2007, *46*, 6873.
33. (a) C. Moreau, R. Durand, S. Razigade, J. Duhamet, P. Faugeras, P. Rivalier, and P. Ros, *Appl. Catal. A*, 1996, *145*, 211; (b) P. Rivalier, M.C. Hayes, D.P. Herzog, and F.M. Rubio, *Catal Today*, 1995. *24*, 165.
34. K. Weissermel and H.-J. Arpe, *Industrial Organic Chemistry*, 2nd ed., VCH, Weinheim, 1993, 121.
35. K. Weissermel and H.-J. Arpe, *Industrial Organic Chemistry*, 2nd ed., VCH, Weinheim, 1993, 69.
36. R.V. Chaudhari, A. Bhattacharya, and B.M. Bhanage, *Catal. Today*, 1995, *24*, 123.
37. B. Cornils and E. Wiebus, *Chemtech*, 1995, *25*(1), 33.
38. J. Haggin, *Chem. Eng. News*, Apr. 17 1995, 25.
39. R.M. Deshpande, Purwanto, H. Dalmas, and R.V. Chaudhari, *J. Mol. Catal. A. Chem.*, 1997, *126*, 133.
40. M. Haumann, H. Koch, P. Hugo, and R. Schomacker, *Appl. Catal.*, 2002, *225*, 239.
41. Y. Wang, J. Jiang, Q. Miao, X. Wu, and Z. Jin, *Catal. Today*, 2002, *74*, 85.
42. S. Tilloy, E. Monflier, F. Bertoux, Y. Castanet, and A. Mortreux, *New J. Chem.*, 1997, *21*, 529.
43. J. Haggin, *Chem. Eng. News*, Oct. 2, 1995, 25.
44. M. Freemantle, *Chem. Eng. News*, Mar. 30, 1998, 32.
45. S.M. Silva, P.A.Z. Suarez, R.F. de Souza, and J. Dupont, *Polym. Bull.*, 1998, *40*, 401.
46. (a) J.G. Huddleston, H.D. Willauer, R.P. Swatloski, A.E. Visser, and R.D. Rogers, *Chem. Commun.*, 1998, 1765; (b) M. Freemantle, *Chem. Eng. News*, Aug. 24, 1998, 12.
47. Y. Chauvin, S. Einloft, and H. Olivier, *Ind. Eng. Chem. Res.*, 1995, *34*, 1149.
48. Y. Chauvin and F. diMarco-van Tiggelen, U.S. patent 5,675,051 (1997).
49. L.A. Muller, J. Dupont, and R.F. deSouza, *Macromol. Rapid Commun.*, 1998, *19*, 409.
50. (a) J.F. Brennecke, L.A. Blanchard, E.J. Beckman, and D. Hancu, *Nature*, 1999, *399*, 28; (b) M. Freemantle., *Chem. Eng. News*, May 10, 1999, 9.
51. P. Tundo and A. Perosa, *Chem. Soc. Rev.*, 2007, *36*, 532.
52. B.E. Hanson and J.R. Zoeller, eds, *Catal Today*, 1998, *42*(4), 371–471.
53. (a) I.T. Horvath., *Acc. Chem. Res.*, 1998, *31*, 641; (b) E. deWolf, G van Koten, and B.-J. Deelman, *Chem. Soc. Rev.*, 1999, *28*, 37; (c) D.P. Curran, *Angew. Chem. Int. Ed.*, 1998, *37*, 1175; (d) A. Studer, S. Hadida, R. Ferritto, S.-Y. Kim, P. Jeger, P. Wipf, and D.P. Curran, *Science*, 1997, *275*, 823, (e) B. Cornils, *Angew. Chem. Int. Ed.*, 1997, *36*, 2057; (f) D.P. Curran, S. Hanida, M. Hashino, A. Studer, P. Wipf, P. Jeger, S.-Y. Kim, and R. Ferritto, U.S. patent 5,777,121 (1998); (g) J.A. Gladysz, D.P. Curran, and I.T. Horvath, eds, *Handbook of Fluorous Chemistry*, Wiley-VCH, Weinheim, 2004; (h) W. Zhang, *Tetrahedron*, 2003, *59*, 4475; (i) J.A. Gladysz and D.P. Curran, eds, *Tetrahedron*, 2002, *58*, 3823–4131; (j) D.P. Curran, *Science*, 2008, *321*, 1645; (k) D.P. Curran, *Aldrichchim. Acta*, 2006, *39*(1), 3; W. Zhang, *Green Chem.*, 2009, *11*, 911; (m) S.K. Ritter, *Chem. Eng.* Oct. *12*, 2009, 45.
54. I.T. Rorvath, G. Kiss, R.A. Cook, J.E. Bond, P.A. Stevens, J. Rabai, and E.J. Mozeleski, *J. Am. Chem. Soc.*, 1998, *120*, 3133.
55. S. Ritter, *Chem. Eng. News*, Oct. 1, 2001, 15.
56. S. Ritter, *Chem. Eng. News*, Nov. 26, 2001, 9.
57. R.D. Rogers and M.A. Eiteman. eds, *Aqueous Biphasic Separations: Biomolecules to Metal Ions*. Plenum, New York, 1995.
58. (a) R.D. Rogers., *Green Chemistry and Chemical Engineering Conference Notes*, Washington, DC, June 1997, 17; (b) J.G. Huddleston, H.D. Willauer, S.T. Griffin, and R.D. Rogers, *Ind. Eng. Chem. Rev.*, 1999, *38*, 2523–2539.
59. A. Kishida, S. Nakano, Y. Kikunaga, and M. Akashi *J. Appl. Polym. Sci.*, 1998, *67*, 255.
60. S.P. Mishra and V.G. Gaikar, *Ind. Eng. Chem. Res.*, 2004, *43*, 5339.
61. J. Kulpe, H. Strutz, H-M Rueffer, and S. Rittner, U.S. patent 5,546,564 (1996).
62. S.M. Husson and C.J. King, *Ind. Eng. Chem. Res.*, 1998, *37*, 2996.
63. G. Kyuchoukov, M. Marinova, J. Molinier, J. Albet, and G. Malmary, *Ind. Eng. Chem. Res.*, 2001, *40*, 5635.
64. R. Canari and A.M. Eyal, *Ind. Eng. Chem. Res.*, 2004, *43*, 7608.
65. (a) X. Xie, C.L. Liotta and C.A. Eckert, *Ind. Eng. Chem. Res.*, 2004, *43*, 7907; (b) V. Stastny and D.M. Rudkevich, *J. Am. Chem. Soc.*, 2007, *129*, 1018.

66. (a) E. Weber, Kirk-Othmer, *Encyclopedia of Chemical Technology*, 4th ed., Wiley, New York, 1995, *14*, 122; (b) J.L. Atwood, J.E.D. Davies, and D.D. MacNicol, eds, *Inclusion Compounds*, Academic Press, London, 1984; (c) G. Venz, *Angew. Chem. Int. Ed.*, 1994, *33*, 803; (d) J.L. Atwood, *Ullmann's Encyclopedia Industrial Chemistry*, 5th ed., VCH, Weinheim, 1989, *A14*, 119; (e) D.J. Cram and J.M. Cram, *Container Molecules and Their Guests*, Royal Society Chemistry Cambridge, UK, 1998; (f) E.D. Sloan, *Clarthrate Hydrates of Natural Gases*, 2nd ed., Dekker, New York, 1998; (g) S.C. Zimmerman, *Science*, 1997, *276*, 543; (h) P.J. Langley and J. Hulliger, *Chem. Soc. Rev.*, 1999, 28, 279–291; (i) F. Laeri, F. Schuth, U. Simon, and M. Wark, eds, *Host–Guest-Systems Based on Nanoporous Crystals*, Wiley-VCH, Weinheim, 2003; (j) F. Hof, S.L. Craig, C. Nuckolls, and J. Rebek, *Angew. Chem. Int. Ed.*, 2003, *41*, 1488.
67. (a) J.-M. Lehn, ed., *Comprehensive Supramolecular Chemistry*, Pergamon, Oxford, 1996; (b) G.G. Chapman and J.C. Sherman, *Tetrahedron*, 1997, *53*, 15911; (c) A. Nangia and G.R. Desiraju, *Top. Curr. Chem.*, 1998, *198*, 58–95; (d) H-J. Schneider, *Angew. Chem. Int. Ed.*, 2009, *48*, 3924.
68. (a) J.L. Sessler and J. Jayawickramarajah, *Chem. Commun.*, 2005, 1939; (b) R.P. Sijbesma and E.W. Meijer, *Chem. Commun.*, 2003, 5; (c) G. Cooke and V.M. Rotello, *Chem. Soc. Rev.*, 2002, *31*, 275.
69. K.D.M. Harris, *Chem. Soc. Rev.*, 1997, *26*, 279.
70. D.G. Hayes, *Inform*, 2002, *13*, 832.
71. (a) R. Brockmann, G. Demmering, U. Kreutzer, M. Lindemann, J. Rlachenka, and U. Steinberner, *Ullmann's Encyclopedia of Industrial Chemistry*, 5th ed., VCH, Weinheim, 1987, *A10*, 267; (b) D.G. Hayes, Y.C. Bengtsson, J.M. Van Alstine, and F. Setterwall, *J. Am. Oil. Chem. Soc.*, 1998, *75*, 1403.
72. N.K. McGuire, *Today's Chemist at Work*, 2003, *12*(1), 10.
73. (a) H. Morawetz, *Encyclopedia of Polymer Science Engineering*, 2nd ed., Wiley, New York, 1987, *7*, 731; (b) M. Farina and G. DiSilvestro, *Encyclopedia of Polymer Science Engineering*, 2nd ed., Wiley, New York, 1988, *12*, 486; (c) I. Pasquon, L. Porri, and U. Giannini, *Encyclopedia of Polymer Science Engineering*, 2nd ed., Wiley, New York, 1989, *15*, 739.
74. F. Toda, *Acc. Chem. Res.*, 1995, *28*, 480.
75. F. Toda and K. Tanaka, *Chem. Commun.*, 1997, 1087.
76. F. Toda, S. Hyoda, K. Okada, and K. Hirotsu, *J. Chem. Soc. Chem. Commun.*, 1995, 1531.
77. J. Nakazawi, J. Hagawari, M. Mizuki, Y. Shimazaki, F. Tani, and Y. Naruta, *Angew. Chem. Int. Ed.*, 2005, *44*, 3744.
78. S. Lim, H. Kim, N. Selvapalam, K.-J. Kim, S.J. Cho, G. Seo, and K. Kim, *Angew. Chem. Int. Ed.*, 2008, *47*, 352.
79. A.M. Rouhi, *Chem. Eng. News*, May 12. 1997, 43.
80. (a) K. Bartik, M. Luhmer, J.-P. Dutasta, A. Collet, and J. Reisse, *J. Am. Chem. Soc.*, 1998, *120*, 784; (b) E.B. Breuwer, G.D. Enright, and J.A. Ripmeester, *Chem. Commun.*, 1997, 939.
81. (a) T. Martin, U. Obst, and J. Rebek Jr., *Science*, 1998, *281*, 1842; (b) R.J. Jansen, A.E. Rowan, R. de Gelder, H.W. Scheeren, and R.J.M. Nolte, *Chem. Commun.*, 1998, 121; (c) J.M. Rivera. T. Martin, and J. Rebek Jr., *Science*, 1998, *279*, 1021; (d) Y. Fokunaga and J. Rebek Jr., *J. Am. Chem. Soc.*, 1998, *120*, 66; (e) T. Szabo, B.M. O'Leary, and J. Rebek Jr., *Angew. Chem. Int. Ed.*, 1998, *37*, 3410; (f) D. Whang, J. Heo, J.H. Park, and K. Kim, *Angew. Chem. Int. Ed.*, 1998, *37*, 78; (g) T. Heinz, D.M. Rudkevich, and J. Rebek Jr., *Nature*, 1998, *394*, 764; (h) R. Dagani., *Chem. Eng. News*, Dec. 1, 1997, 4.
82. (a) E. Weber, ed., *Top. Curr. Chem.*, 1995, 1–156; (b) D. Whang, Y.-M. Jeon, J. Heo, and K. Kim, *J. Am. Chem. Soc.*, 1996, *118*, 11333; (c) Y.-M. Jeon, D. Whang, J. Kim, and K. Kim., *Chem. Lett.*, 1996, 503.
83. (a) J. Lagona, P. Mukhopadhyay, S. Chakrabarti, and L. Isaacs, *Angew. Chem. Int. Ed.*, 205, *44*, 4844; (b) J.W. Lee, S. Samal, N. Selvapalam, H.-J. Kim, and K. Kim, *Acc. Chem. Res.*, 2003, *36*, 621.
84. K.K. Kim, N. Selvapalam, Y.H. Ko, K.M. Park, D. Kim, and J. Kim, *Chem. Soc. Rev.*, 2007, *36*, 267.
85. W.-H. Huang, P.Y. Zavalij, and L. Isaacs, *Angew. Chem. Int. Ed.*, 2007, *46*, 7425.
86. M.V. Rekharsky, H. Yamamura, Y.K. Ko, N. Selvapalam, K. Kim, and Y. Inoue, *Chem. Commun.*, 2008, 2236.
87. N. Saleh, A.L. Koner, and W.M. Nau, *Angew. Chem. Int. Ed.*, 2008, *47*, 5398.
88. A.P.H.J. Schenning, B. deBruin, A.E. Rowan, H. Kooijman, A.L. Spek, and R.J.M. Nolte, *Angew. Chem. Int. Ed.*, 1995, *34*, 2132.
89. A.E. Rowan, P.P.M. Aarts, and K.W.M. Koutstaal, *Chem. Commun.*, 1998, 611.
90. (a) J. Rebek Jr., *Chem. Soc. Rev.*, 1996, *25*, 255; (b) B.C. Hamann, K.D. Shimizu, and J. Rebek Jr, *Angew. Chem. Int. Ed.*, 1996, *35*, 1326; (c) R.S. Meissner, J. Rebek Jr., and J. deMendoza, *Science*, 1995, *270*, 1485; (d) C. Valdes, U.P. Spitz, L.M. Toledo, S.W. Kubik, and J. Rebek Jr., *J. Am. Chem. Soc.*, 1995, *117*, 12733.
91. (a) M. Ward, *Chem. Br.*, 1998, *34*(9), 52; (b) V.A. Russell, C.C. Evans, W. Li, and M.D. Ward, *Science*, 1997; (c) J.A. Swift, A.M. Pivovar, A.M. Reynolds, and M.D. Ward, *J. Am. Chem. Soc.*, 1998, *120*, 5887; (d) R. Dagani, *Chem. Eng. News*, June 8, 1998, 38; (e) K.T. Holman and M.D. Ward, *Angew. Chem. Int. Ed.*, 2000, *39*, 1653; (f) A.T. Holman, A.M. Pinovar, and M.D. Ward, *Science*, 2001, *294*, 1907.
92. (a) F. Toda, K. Tanaka, H. Koshima, and S.I. Khan, *Chem. Commun.*, 1998, 2503; (b) Y. Miyake, J. Hirose, Y. Hasegawa, K. Sada, and M. Miyata, *Chem. Commun.*, 1998, 111; (c) F. Toda, *Pure. Appl. Chem.*, 2001, *73*, 1137.
93. (a) A.T. Ung, D. Gizachew, R. Bishop, M.L. Scudder, I.G. Dance, and D.C. Craig., *J. Am. Chem. Soc.*, 1995, *117*, 8745; (b) R. Bishop, D.O. Craig, A. Marougkas, and M.L. Scudder, *Tetrahedron*, 1994, *50*, 8749.
94. H. Suzuki, *Tetrahedron Lett.*, 1994, *35*, 5015.
95. K. Endo, T. Sawaki, M. Koyanagi, K. Kobayashi, H. Maruda, and Y. Aoyama, *J. Am. Chem. Soc.*, 1995, *117*, 8341.
96. T. Hertzsch, F. Budde, E. Weber, and J. Hulliger, *Angew. Chem. Int. Ed.*, 2002, *41*, 2282.
97. R. Sozzani, A. Commoti, S. Bracco, and R. Simonutti, *Angew. Chem. Int. Ed.*, 2004, *43*, 2792.
98. P. Sozzani, S. Bracco, A. Comotti, L. Ferretti, and R. Simonutti, *Angew. Chem. Int. Ed.*, 2005, *44*, 1816.
99. G.A. Downing, C.S. Frampton, J.H. Gall, and D.D. MacNicol, *Angew. Chem. Int. Ed.*, 1996, *35*, 1547.
100. K. Reiner, R. Richter, S. Hauptmann, J. Becher, and L. Hennig, *Tetrahedron*, 1995, *51*, 13291.

101. K. Beketov, E. Weber, J. Seidel, K. Kohnke, K. Makhkamov, and Bakhtiyar., *Chem. Commun.*, 1999, 91.
102. V.R. Pedireddi, W. Jones, A.P. Chorlton, and A. Docherty, *Tetrahedron Lett.*, 1998, *39*, 5409.
103. V. Meille, E. Schulz, M. Vrinat, and M. Lemaire, *Chem. Commun.*, 1998, 305.
104. K. Sada, K. Yoshikawa, and M. Miyata, *Chem. Commun.*, 1998, 1763.
105. (a) O.M. Yaghi, H. Li, and T.L. Groy, *J. Am. Chem. Soc.*, 1996, *118*, 9096; (b) O.M. Yaghi, H. Li, C. Davis, D. Richardson, and T.L. Groy, *Acc. Chem. Res.*, 1998, *31*, 474; (c) Y. Adyama, *Top. Curr. Chem.*, 1998, *198*, 131–162; (d) O.M. Yaghi, C.E. Davis, G. Li, and H. Li. *J. Am. Chem. Soc.*, 1997, *119*, 2861; (e) H.J. Choi and M.P. Suh, *J. Am. Chem. Soc.*, 1998, *120*, 10622.
106. (a) H. Li, M. Eddaoudi, T.L. Groy, and O.M. Yaghi, *J. Am. Chem. Soc.*, 1998, *120*, 8571; (b) R.H. Groeneman, L.R. MacGillivray, and J.L Atwood, *Chem. Commun.*, 1998, 2735.
107. S. Takamizawa, T. Akatsuka and T, Ueda, *Angew. Chem. Int. Ed.*, 2008, *47*, 1689.
108. R. Bishop, *Chem. Soc. Rev.*, 1996, *25*, 311.
109. (a) M.T. Muldoon and L.H. Stanker, *Chem. Ind.*, 1996, 204; (b) K. Mosbach and O. Ramstrom, *Biotechnology Chem. Ind. (Lond.),* 1996, *14*, 163; (c) R.A. Bartsch and M. Maeda, eds, *Molecular and Ionic Recognition with Imprinted Polymers. Am. Chem. Soc. Symp.* 703, Washington, DC, 1998; (d) G. Wulff, *Chemtech*, 1998, *28*(11), 19; (e) J. Steinke, D.C. Sherrington, and I. R. Dunkin, *Adv. Polymer. Sci.*, 1995, *123*, 81; (f) K. Haupt and K. Mosbach, *Trends Biotechnol.*, 1998, *16*, 468; (g) B. Sellergren, *Angew. Chem. Int. Ed.*, 2000, *39*, 1031; (h) R. A. Bartsch and M. Maeda, eds, *Molecular and Ionic Recognition with Imprinted Polymers, ACS Symp. 703*, Oxford University Press, New York, 1998; (i) S.A. Piletsky, S. Alcock, and A.P.F. Turner, *Trends. Biotechnol.*, 2001, *19*, 9; (j) G. Wulff, *Chem. Rev.*, 2002, *102*, 1; (k) B. Sellergren, *Molecularly Imprinted Polymers*, Elsevier, Amsterdam, 2000; (l) K. Haupt, *Chem. Commun.*, 2003, 171; (m) S.C. Zimmerman and N.G. Lemcoff, *Chem. Commun.*, 2004, 5; (n) M. Komiyama, T. Takeuchi, and H. Asanuma, *Molecular Imprinting*, Wiley-VCH, Weinheim, 2002; (o) J.D. Marty and M. Mauzac, *Adv. Polym. Sci.*, 2005, *172*, 1; (p) C. Alexander, L. Davidson, and W. Hayes, *Tetrahedron*, 2003, *59*, 2025; (q) M.Y. San and O. Ramstrom, eds, *Molecularly Imprinted Materials—Science and Technology*, CRC Press, Boca Raton, FL, 2004.
110. (a) J. Matsui, O. Doblhoff, and T. Takeuchi, *Chem. Lett.*, 1995, 489; (b) M.J. Whitcombe, M.E. Rodriguez, P. Villar, and E.N. Vulfson, *J. Am. Chem. Soc.*, 1995, *117*, 7105; (c) M. Kempe and K. Mosbach, *Tetrahedron. Lett.*, 1995, *36*, 3563; (d) M. Yoshikawa, J. Izumi, T. Kitao, *Chem. Lett.*, 1996, 611; (e) J. Mathew-Krotz and K.J. Shea, *J. Am. Chem. Soc.*, 1996, *118*, 8154; (f) K. Tanabe, T. Takeuchi, J. Matsui, K. Ikebukuro, K. Yano, and I. Karube, A.C.S. Symp., *757*, Washington, DC, 2000; (g) K. Sreenivasan, *J. Appl. Polym. Sci.*, 1998, *70*, 19; (h) S.-H. Cheong, A.E. Rachkov, J.-K. Park, K. Yano, and I. Karube, *J. Polym. Sci. A. Polym. Chem.*, 1998, *36*, 1725; (i) J. Haginaka, and H. Sanbe, *Chem. Lett.*, 1998, 1089.
111. M. Andac, E. Ozyapi, S. Senel, R. Say, and A. Denizli, *Ind. Eng. Chem. Res.*, 2006, *46*, 1780.
112. Anon., *Chem. Eng. News*, Nov. 13, 2006, 39.
113. L. Mandolini and R. Ungaro, *Calixarenes in Action*, Imperial College Press, London, 2000.
114. J.L. Atwood, L.J. Barbour, and A. Jerga, *Science*, 2002, *296*, 2367.
115. (a) S.Ritter, *Chem. Eng. News*, May 31, 2004, 7; (b) J.L. Atwood, L.J. Barbour, and A. Jerga, *Angew. Chem. Int. Ed.*, 2004, *43*, 2948.
116. (a) L.P. Lindoy, *Pure. Appl. Chem.*, 1997, *69*, 2179; (b) J.S. Bradshaw and R.M. Izatt, *Acc. Chem. Res.*, 1997, *30*, 338; (c) A.H. Bond, M.L. Dietz, and R.D. Rogers, eds, *Metal Ion Separation and Preconcentration: Progress and Opportunities, ACS Symp. 716*, Oxford University Press, Oxford, 1999; (d) G.W. Gokel, W.M. Leevy, and M.E. Weber, *Chem. Rev.*, 2004, *104*, 2723.
117. (a) A. Ikeda and S. Shinkai, *Chem. Rev.*, 1997, *97*, 1713; (b) C.D. Gutsche, *Aldrichim. Acta.* 1995, *28*(1), 3; (c) O.M. Falana, H.F. Koch, D.M. Roundhill, G.L. Lumetta, and B.P. Hay, *Chem Commun.*, 1998, 503; (d) F. Arnaud-Neu, S. Fuangswasdi, A. Notti, S. Pappalardo, and M.F. Parisi, *Angew. Chem. Int. Ed.*, 1998, 37, 112; (e) P. Schmitt, P.D. Beer, M.G.B. Drew, and P.D. Sheen, *Angew. Chem. Int. Ed.*, 1997, *36*, 1840; (f) G. Lumetta and R.D. Rogers, *Calixarene Molecules for Separations*, A.C.S. Symp., Washington, DC, 2000.
118. (a) F.P. Schmidtchen and M. Berger, *Chem. Rev.*, 1997, *97*, 1609; (b) P.D. Beer, *Acc. Chem. Res.*, 1998, *31*, 71; (c) M. Berger and F.P. Schmidtchen, *Angew. Chem. Int. Ed.*, 1998, *37*, 2694; (d) A. Bianchi, K. Bowman-James, and E. Garcia-Espana, *Supramolecular Chemistry of Anions*, Wiley-VCH, New York, 1997; (e) M.M.G. Antonisse and D.N. Reinhoudt, *Chem. Commun.*, 1998, 443; (f) P.D. Beer, M.G.B. Drew, D. Hesek, and K.C. Nam, *Chem. Commun.*, 1997, 107.
119. F.-G. Klarner and B. Kahlert, *Acc. Chem. Res.*, 2003, *36*, 919.
120. A. Casnati, A. Pochini, R. Ungaro, O. Bouchi, F. Ugozzoli, R.J.M. Egberink, H. Struijk, R. Lugtenberg, F. deJong, and D.N. Reinhoudt, *Chem. Eur. J.* 1996, *2*, 436.
121. X.X. Zhang, A.V. Bordunov, J.S. Bradshaw, N.K. Dalley, X. Kou, and R.M Izatt. *J. Am. Chem. Soc.*, 1995, *117*, 11507.
122. J. Huskens and A.D. Sherry, *Chem. Commun.*, 1997, 845.
123. T. Gyr, H-R. Macke, and M. Hening, *Angew. Chem. Int. Ed.*, 1997, 36, 2786.
124. S. Tsuchiya, Y. Nakatani, R. Ibrahim, and S. Ogawa, *J. Am. Chem. Soc.*, 2002, *124*, 4936.
125. R. Chitrakar, S. Tezuka, A. Sonoda, H. Kakita, K. Sakane, K. Ooi, and T. Hirotsu, *Ind. Eng. Chem. Res.*, 2008, *47*, 176.
126. (a) D.E. Kaufmann and A. Otten, *Angew. Chem. Int. Ed.*, 1994, *33*, 1832; (b) J.L. Atwood, K.T. Holman, and J.W. Steed, *Chem. Commun.*, 1996, 1401; (c) J.L. Sessler, P.A. Gale, and W.-S. Cho, eds, *Anion Receptor Chemistry*, Royal Society Chemistry, Cambridge, UK, 2006; (d) P.A. Gale, S.E. Garcia-Garrido, and *J. Garric, Chem. Soc. Rev.*, 2008, *37*, 151; (e) M.J. Chmielewski, T. Zielinski, and J. Jurczak, *Pure Appl. Chem.*, 2007, *79*, 1087; (f) S.O. Kang, R.A. Begum, and K. Bowman-Jones, *Angew. Chem. Int. Ed.*, 2006, *45*, 7894; (g) P.A. Gale, *Acc. Chem. Res.*, 2006, *39*, 465.
127. S.D. Reilly, G.R.K. Khalsa, D.K. Ford, J.R. Brainard, B.P. Hay, and P.H. Smith, *Inorg. Chem.*, 1995, *34*, 569.
128. C.-W. Chiu and F.P. Gabbai, *J. Am. Chem. Soc.*, 2006, *128*, 14248.

129. J. Sessler, D. An, W.-S. Cho, V. Lynch, and M. Marquez, *Chem. Commun.*, 2005, 540.
130. H. Miyaji, D.-S. Kim, B.Y. Chang, E. Park, S.-M. Park, and K.H. Ahn, *Chem. Commun.*, 2008, 753.
131. D.M. Roundhill and H.F. Koch, *Chem. Soc. Rev.*, 2002, *31*, 60.
132. S. Nishizawa, P. Buhlmann, M. Iwao, and Y. Umezawa, *Tetrahedron Lett.*, 1995, *36*, 6483.
133. (a) C. Raposo N. Perez, M. Almaraz, L. Mussons, C. Caballero, and J.R. Moran, *Tetrahedron Lett.*, 1995, *36*, 3255; (b) P. Buhlmann, S. Nishizawa, K. P. Xiao, and Y. Umezawa, *Tetrahedron*, 1997, *53*, 1647.
134. P. Buhlmann, S. Nishizawa, K.P. Xiao, and Y. Umezawa, *Tetrahedron* 1997, *53*, 1647.
135. E. Kimura, S. Aoki, T. Koike, and M. Shiro, *J. Am. Chem. Soc.*, 1997, *119*, 3068.
136. S. Shinoda, M. Tadokoro, H. Tsukube, and R. Arakawa, *Chem. Commun.*, 1998, 181.
137. (a) A. Metzger and E.V. Anslyn, *Angew. Chem. Int. Ed.*, 1998, *37*, 649; (b) A. Metzger, V.M. Lynch, and E.V. Anslyn, *Angew. Chem. Int. Ed.*, 1997, *36*, 862.
138. K. Nakaskima and S. Shinkai, *Chem. Lett.*, 1995, 443.
139. M. Takeuchi. T. Imada, and S. Shinkai, *J. Am. Chem. Soc.*, 1996, *118*, 10658.
140. M.-F. Paugam, J.A. Riggs, and B.D. Smith, *Chem. Commun.*, 1996, 2539.
141. (a) R. Singh, *Chemtech*, 1996, *26*(6), 46, 1998, *28*(4), 33; (b) R.W. Baker, *Kirk–Othmer. Encyclopedia Chemical Technology*, 4th ed., Wiley, New York, 1995, *16*, 135; (c) R.D. Noble and S.A. Stern, eds, *Membrane Separations Technology, Principles and Applications*, Elsevier, Amsterdam, 1995; (d) T. Matsuura, *Synthetic Membranes and Membrane Separation Processing*, CRC Press, Boca Raton, FL, 1994; (e) M. Kurihara, *J. Macromol. Sci. Pure. Appl. Chem.*, 1994, *A31*, 1791; (f) J. Howell, V. Sanchez, and R.W. Field, eds, *Membranes in Bioprocessing: Theory and Applications*, Chapman & Hall, London, 1994; (g) W.S.W. Ho and K.K. Sirkar, eds, *Membrane Handbook*, Chapman & Hall, London, 1992; (h) Y. Osada and T. Nakagawa, *Membrane Science and Technology*, Dekker, New York, 1992; (i) I. Cabasso, *Encyclopedia of Polymer Science in Engineering*, 2nd ed., Wiley, New York, 1987, 9, 509–579; (j) A. Johansson, *Clean Technology*, Lewis Publishers, Boca Raton, FL, 1992, 109; (k) H. Strathmann, *Ullmann's Encyclopedia of Industrial Chemistry*, 5th ed., VCH, Weinheim, 1990, *A16*, 187; (l) K. Scott and R. Hughes, eds, *Industrial Membrane Separation Technology*, Blackie Academic, London, 1996; (m) P. Cartwright, *Chem. Eng.*, 1994, *101*(9), 84; (n) M. Mulder, *Basic Principles of Membrane Technology*, 2nd ed., Kluwer Academic, Dordrecht, 1996; (o) R.A. Bartsch, and J.D. Way, *Chemical Separations with Liquid Membranes*, Am. Chem. Symp. 642, Washington, DC, 1996; (p) X. Wang and H.G. Spencer, *Trends Polym. Sci.*, 1997, *5*(2), 38. (polymer membranes for food processing); (q) R. Rogers, *Chem. Eng. News*, Aug. 24, 1998, 38; (r) D. Paul and K. Ohlrogge, *Environ. Prog.*, 1998, *17*(3), 137; (s) A.P. Nunes and K.-V. Peinemann, *Membrane Technology in the Chemical Industry*, 2nd ed., Wiley-VCH, Weinheim, 2006; (t) I. Pinnau and B.D. Freeman, eds, *Membrane Formation and Modification*, A.C.S. Symp. 744, Oxford University Press, Oxford, 2000; (u) R. Baker, *Membrane Technology and Applications*, McGraw Hill, New York, 2000; (v) E. Driolli and D. Paul, eds, *Ind. Eng. Chem. Res.*, 2005, *44*, 7609–7729; (w) N.N. Li, ed., *Advanced Membrane Technology and Applications*, Wiley-AlChE, Hoboken, NJ, 2008; (x) R. Baker, *Membrane Technology and Applications*, 2nd Ed., Wiley, Hoboken, NJ, 2004.
142. (a) W.J. Koros, Y.H. Ma, and T. Shimidzu, *Pure Appl. Chem.*, 1996, 68, 1479; (b) L.J. Zeman and A.L. Zydney, eds, *Microfiltration and Ultrafiltration: Principles and Applications*, Dekker, New York, 1996; (c) B. Elias and J. van Cleef, *Chem. Eng.*, 1998, *105*(9), 94; (d) K. Torzewski, *Chem. Eng.*, *115*(3), 33.
143. G. Pinholster, *Environ. Sci. Technol.*, 1995, *29*, 174A.
144. N. Kawabata, I. Fujita, and T. Inoue, *J. Appl. Polym. Sci.*, 1996, *60*, 911.
145. Anon., *Textile World*, March/April 2008, 38.
146. (a) R. Subramanian and M. Nakajima, *J. Am. Oil. Chem. Soc.* 1997, *74*, 971; (b) D. Hairston, *Chem. Eng.*, 1997, *104*(10), 34; (c) G. Parkinson, *Chem. Eng.*, 1997, *104*(5), 33; (d) R. Subramanian, M. Nakajima, K.S.M.S. Raghavarao, and T. Kimura, *J. Am. Oil Chem. Soc.*, 2004, *81*, 313.
147. (a) M. Abe and Y. Kondo, *Chemtech*, 1999, *93*(2), 54; (b) A. Lambert, P. Plucinski, and I. V. Kozhevnikov, *Chem. Commun.*, 2003, 714; (c) K. Materna, E. Goralska, A. Sobczynska, and J. Szymanowski, *Green Chem.*, 2004, *6*, 176.
148. (a) D. Bhattacharyya, W.C. Mangum, and M.E. Williams, *Kirk–Othmer Encyclopedia Chemical Technology*, 4th ed., Wiley, New York, 1997, *21*, 303; (b) J. Kucera, *Chem. Eng. Prog.*, 1997, *93*(2), 54; 2008, *104*(5), 30; (c) D.A. Musale and S.S. Kulkarni, *J. Macro. Sci. Rev. Macromol. Sci.*, 1998, *C38*, 615; (d) M. Cheryan, *Ultrafiltration and Microfiltration Handbook*, Technomic, Lancaster, PA, 1997; (e) K. Moftah, *Chem. Eng.*, 2003, *110*(9), 62; (f) M. Vioth, *Chem. Eng. News*, Oct. 19, 2009, 20; (g) Q. Yang, K.W. and T-S. Chung, *Environ. Sci. Technol.*, 2009, *43*, 2800.
149. P. Vandezande, L.E.M. Gevers, and I.F.J. Vankelecom, *Chem. Soc. Rev.*, 2008, *37*, 365.
150. L.P. Raman, M. Cheryan, and N. Rajagopalan., *Chem. Eng. Prog.*, 1994, *90*(3), 68.
151. R. Wodzki and J. Nowaczyk, *J. Appl. Polym. Sci.*, 1997, *63*, 355.
152. F.L.D. Cloete and A.P. Marais, *Ind. Eng. Chem. Res.*, 1995, *34*, 2464.
153. (a) M. San-Martin,C. Pazos, and J. Coca, *J. Chem. Technol. Biotechnol.* 1996, 65, 281; (b) Y.S. Mok, K.H. Lee, and W.K. Lee, *J. Chem. Technol. Biotechnol.*, 1996, 65, 309.
154. Anon., *Chem. Eng. News*, May 20, 1996, 34.
155. R.A. Dangel, D. Astraukis, and J. Palmateer, *Environ. Prog.*, 1995, *14*(1), 65.
156. A. Kulkami, D. Mukherjee, and W.N. Gill, *J. Appl. Polym. Sci.*, 199, *60*, 483.
157. (a) S.C. Stinson, *Chem. Eng. News*, Nov. 25, 1996, 47; (b) S. Huang, X Wu, C. Yuan, and T. Liu, *J. Chem. Technol. Biotechnol.*, 1995, *64*, 109; (c) C. Huang and T. Xu, *Environ. Sci. Technol.*, 2006, *40*, 5233; (d) M. Bailly and D.Bar, *Chem. Eng.*, 2002, *109*(7), 51; (e) A. Saxena, G.S. Gohil, and V.K. Shahi, *Ind. Eng. Chem. Res.*, 2007, *46*, 1270.
158. N.P. Chopey, *Chem. Eng. (Lond.)*,1989, *96*(12), 81.

159. S. Mazrou, H. Kerdjoudj, A.T. Cherif, A. Elmidaoui, and J. Molenat, *New J. Chem.*, 1998, *22*, 355.
160. L. Bazinet, F. Lamarche, R. Labresque, and D. Ippersiel, *J. Agric Food Chem.*, 1997, *45*, 3788.
161. C. Zaluski and G. Wu, *Macromolecules*, 1994, *2*, 6750.
162. (a) I. Moch Jr., *Kirk–Othmer Encyclopedia of Chemical Technology*, 4th ed., Wiley, New York, 1995, *13*, 312; (b) Anon., *Chem. Eng.*, 1997, *104*(2), 109.
163. Anon., *Chem. Eng.*, 1997, *104*(2), 109.
164. (a) M. Dekker and R. Boom, *Trends Biotechnol.*, 1995, *13*, 129; (b) B. Elias and J van Cleef, *Chem. Eng.*, 1998, *105*(10), 94; (c) B. Culkin, A. Plotkin, and M. Monroe, *Chem. Eng. Prog.*, 1998, *94*(1), 29; (d) S. Shelley, *Chem. Eng.*, 1997, *104*(2), 107.
165. (a) G. Parkinson, *Chem. Eng.* 1996, *103*(12), 21; (b) S. Shelly, *Chem. Eng.*, 1997, *104*(2), 107; (c) C. Crabb and C. Armesto, *Chem. Eng.*, 1999, *106*(11), 45; 166. M. Freemantle, *Chem. Eng. News*, Nov. 11, 1996, 33.
167. G. Parkinson, *Chem. Eng.*, 1999, *106*(3), 23.
168. (a) G. Pozniak, M. Bryjak, and W. Trochimczuk, *Angew. Makromol. Chem.*, 1995, *233*, 23; (b) M. Zhou, H. Liu, J.E. Kilduff, R.I. Langer, D.G. Anderson, and G. Balfort, *Environ. Sci. Technol.*, 2009, *43*, 3865.
169. M. Elimelech, G. Amy, and M. Clark, eds, *Preprints*, *ACS Div. Environ. Chem.* 1996, *36*(2), 87–135 (a symposium).
170. E. M. Scibelli, *Pollution Eng.*, Feb. 2007, 42.
171. B. Buecker, *Chem. Eng.*, 2007, *114*(5), 63.
172. M. Rucka, G. Pozniak, B. Turkiewicz, and W. Trochimczuk, *Enzyme Microb. Technol.*, 1996, *18*, 477.
173. P.A. Hogan, R.P. Canning, P.A Peterson, R.A. Johnson, and A.S. Michaels, *Chem. Eng. Prog.* 1998, *94*(7), 49.
174. R.A. Bartsch and J.D. Way, eds, *Chemical Separations with Liquid Membranes*, Am. Chem. Soc. Symp. 642, Washington, DC, 1996.
175. J. Marchese, M. Campderros, and A. Acosta, *J. Chem. Technol. Biotechnol.*, 1995, *64*, 293.
176. (a) H.C. Visser, D.N. Reinhoudt, and F. de Jong, *Chem. Soc. Rev.*, 1994, 23, 75; (b) H. Visser, D.M. Rudkevich, W. Verboom, F deJong, and D.N. Reinhoudt, *J. Am. Chem. Soc.*, 1994, *116*, 11554; (c) D.M. Rudkevich, J.D. Mercer-Chalmers, W. Verboom, R. Ungaro, F. de Jong, and D.N. Reinhoudt, 1995, *117*, 6124.
177. M.-F. Paugam, J.I. Bien, B.D. Smith, L.A.J. Chrisstoffels, F. de Jong, and D.N. Reinhoudt, *J. Am. Chem. Soc.*, 1996, *118*, 9820.
178. J.A. Adarkar, S.B. Sawant, J.B. Joshl, and V.G. Pangarkar, *Biotechnol. Prog.*, 1997, *13*, 493.
179. T.A. Munro and B.D. Smith, *Chem. Commun.*, 1997, 2167.
180. G. Samdani, *Chem. Eng.*, 1994, *101*(8), 21.
181. M. Morita, *J. Appl. Polym. Sci.*, 1998, *70*, 647.
182. (a) E.R. Hensema and J.P. Boom, *Macromol. Symp.*, 1996, *102*, 409; (b) D.R. Paul and Yampolskii, *Polymeric Gas Separation Membranes*, CRC Press, Boca Raton, FL, 1994; (c) R.L. Kesting and A.K.F. Fritzsche, *Polymeric Gas Separation Membrane*, Wiley, New York, 1996; (d) W.J. Koros, G.K. Fleming, S.M. Jordan, T.H. Kim, and H.H. Hoehn, *Prog. Polym. Sci.*, 1988, *13*, 339; (e) N. Toshima, *Polymers for Gas Separation*, VCH, Weinheim, 1992; (f) G. Maier, *Angew. Chem. Int. Ed.*, 1998, *37*, 2960–2974; (g) M. Freemantle, *Chem. Eng. News*, Oct. 3, 2005, 49.
183. (a) B. Freeman and I. Pinnau, *Trends Polym. Sci.*, 1997, *5*(5), 167; (b) B.D. Freeman and I. Pinnau, *Polymer Membranes for Gas and Vapor Separation*, A.C.S. Symp. 733, Washington, DC, 1999.
184. (a) J. Humphrey, *Chem. Eng. Prog.*, 1995, *91*(10), 31, 39; (b) W.J. Koros, *Chem. Eng. Prog.*, 1995, *91*(10), 68, 71, 76; (c) F. Bernardo, E. Driolli, and G. Golemme, *Ind. Eng. Chem. Res.*, 2009, *48*, 4638.
185. M.A. Henson and W.J. Koros, *Ind. Eng. Chem. Res.*, 1994, *33*, 1901.
186. (a) Anon., *Chen. Eng.*, 1994, *101*(9), 83; (b) K.P. Michael, *Chem. Eng.*, 1997, *104*(1), 72; (c) Anon., *Chem. Eng. News*, Aug. 26 1996, 16; (d) M.S. Reisch, *Chem. Eng. News*, Apr. 3 1995, 11; (e) A. Shanley, *Chem. Eng.*, 1998, *105*(2), 61.
187. (a) Y.M. Lee, H.B. Park, and I.Y. Suh, Chicago ACS Meeting, *Polym. Mater. Sci. Eng.*, 2001, *85*, 293; (b) Y.M. Lee and H.B. Park, Orlando ACS Meeting, April 2002, PMSE 90.
188. (a) Y. Li, X. Wang, M. Ding, and J. Xu, *J. Appl. Polym. Sci.*, 1996, *61*, 741; (b) L.M. Robeson, W.F. Burgoyne, M. Langsam, A.C. Savoca, and C.F. Tien, *Polymer*, 1994, *35*, 4970; (c) G. Xuesong, and L. Fengcai, *Polymer*, 1995, *36*, 1035; (d) H. Kawakami, M. Mikawa, and S. Nagaoka, *J. Appl. Polym. Sci.*, 1996, *62*, 965; (e) Y. Li, M. Ding, and J. Xu., *J. Macromol. Sci. Pure. Appl. Chem.*, 1997, *A34*, 461; (f) T-S. Chung and E.R Kafchinski, *J. Appl. Polym. Sci.*, 1997, 65, 1555; (g) Y. Li, M. Ding, and J. Xu, *Macromol. Chem. Phys.*, 1997, *198*, 2769.
189. (a) M.-J. Chang, Y.-H. Liao, A.S. Myerson, and T.K. Kwei, *J. Appl. Polym. Sci.*, 1996, *62*, 1427; (b) L. Rebattet, M. Escoubes, E. Genies, and M. Pimeri, *J. Appl. Polym. Sci.*, 1995, *57*, 1595; 1995, *58*, 923.
190. (a) J. Yang, C. Zhao, D. Cui, J. Hou, M. Wan, and M. Xu, *J. Appl. Polym. Sci.*, 1995, *56*, 831; (b) C.R. Martin, Preparation of conducting polymers in porous membranes. *Acc. Chem. Res.*, 1995, *28*, 61.
191. (a) Z. Zhang and S. Lin, *Macromol. Rapid Commun.*, 1995, *16*, 927; (b) M.J. Choi, C.N. Park, and Y.M. Lee, *J. Appl. Polym. Sci.*, 1995, *58*, 2373; (c) S.H. Chen and J. Y. La, *J. Appl. Polym. Sci.*, 1996, *59*, 1129; (d) H. Nishide, T. Suzuki, Y. Soejima, and E. Tsuchida, *Macromol. Symp.*, 1994, *80*, 145. (e) C.-K. Park, M.-J. Choi, and Y.M. Lee, *J. Appl. Polym. Sci.*, 1997, *66*, 483.
192. (a) E.M. Kirschner, *Chem. Eng. News*, Dec. 12, 1994, 16; (b) H. Kita, M. Yoshino, K. Tanaka, and K.-I. Okamoto, *Chem. Commun.*, 1997, 1051; (c) J.-I. Hayashi, M. Yamamoto, K. Kusakabe, and S. Morooka, *Ind. Eng. Chem. Res.*, 1997, *36*, 2134.
193. C.W. Jones and W.J. Koros, *Ind. Eng. Chem. Res.*, 1995, *34*, 158, 164.
194. (a) D. Ramprasad, G.P. Pez, B.H. Toby, T.J. Markley, and R.M. Pearlstein, *J. Am. Chem. Soc.*, 1995, *117*, 10694; (b) G. Parkinson, *Chem. Eng.*, 1996, *103*(1), 21; (c) J. Haggin, *Chem. Eng. News*, Feb. 3, 1996, 35; (d) D. Ramprasad, T.J. Markley, and G.P. Pez, *J. Mol. Catal. A. Chem.*, 1997, *117*, 273.
195. G. Parkinson, *Chem. Eng.*, 1996, *103*(8), 15.
196. J. Li, K. Nagai, T. Nakagawa, and S. Wang., *J. Appl. Polym. Sci.*, 1995, *58*, 1455.
197. K. Okamoto, M. Fujii, S. Okamyo, H. Suzuki, K. Tanaka, and H. Kita., *Macromolecules*, 1995, *28*, 6950.

198. K.-I. Okamoto, N. Yasugi, T. Kawabata, K. Tanaka, and H. Kita, *Chem. Lett.*, 1995, *24*, 613.
199. M. Yoshikawa, K. Fujimoto, H. Kinugawa, T. Kitao, Y. Kamiya, and N. Ogata, *J. Appl. Polym. Sci.*, 1995, *58*, 1771.
200. J. Li, K. Tachihara, K. Nagai, T. Nakagawa, and S. Wang, *J. Appl. Polym. Sci.*, 1996, *60*, 1645.
201. J. Li, K. Nagai, T. Nakagawa, and S. Wang, *J. Appl. Polym. Sci.*, 1996, *61*, 2467.
202. D.S. Pope, W.J. Koros, and H.B. Hopfenberg, *Macromolecules*, 1994, *27*, 5839.
203. M.E. Sena and C.T. Andrade, *Polym. Bull.*, 1995, *34*, 643.
204. Y. Cui, H. Kita, and K.-I. Okamoto, *Chem. Commun.*, 2003, 2154.
205. J.-I. Hayashi, M. Yamamoto, K. Kusakabe, and S. Morooka, *Ind. Eng. Chem. Res.*, 1995, *34*, 4364.
206. M. Teramoto, K. Nakai, N. Ohnishi, Q. Huang, T. Watari, and H. Matsuyama, *Ind. Eng. Chem. Res.*, 1996, *35*, 538.
207. K. Tanaka, M. Okano, H. Kita, K.-I. Okamoto, and S. Nishi, *Polym. J.*, 1994, *26*, 1186.
208. K.E. Porter, A.B. Hinchliffe, and B.J. Tighe, *Ind. Eng. Chem. Res.*, 1997, *36*, 830.
209. (a) G. Parkinson, *Chem. Eng.*, 1997, *104*(1), 15; (b) J. Haggin, *Chem. Eng. News*, Apr. 4 1994, 25.
210. T.C. Merkel, B.D. Freeman, R.J. Spontak, Z. He, I. Pinnau, P. Meakin, and A.J. Hill, *Science*, 2002, *296*, 519.
211. J.H. Krieger, *Chem. Eng. News*, Dec. 18, 1995, 25.
212. (a) R.W. Baker, J. Kaschemekat, and J.G. Wijmans, *Chemtech*, 1996, *26*(7), 37; (b) Anon., *Hydrocarbon Process*, 1996, *75*(8), 91–92; (c) R.P. Ponangi and P.N. Pintauro, *Ind. Eng. Chem. Res.*, 1996, *35*, 2756; (d) V. Simmons, *Chem. Eng.*, 1994, *101*(9), 92; (e) A. Alpers, B. Keil, O. Ludtke, and K. Ohbugge, *Ind. Eng. Chem. Res.*, 1999, 38, 3754; (f) B. Xia, S. Majumdar, and K.K. Sirkar, *Ind. Eng. Chem. Res.*,1999, *38*, 3462.
213. D.T. Tsou, M.W. Blachman, and J.C. Davis, *Ind. Eng. Chem. Res.*, 1994, 33, 3209.
214. D.G. Besserabov, R.D. Sanderson, E.P. Jacobs, and I.N. Beckman, *Ind. Eng. Chem. Res.*, 1995, *34*, 1769.
215. K. Nymeyer, T. Visser, R. Assen, and M. Wessling, *Ind. Eng. Chem. Res.*, 2004, *43*, 720.
216. R. Rabago, D.L. Bryant, O.A. Koval, and R.D. Noble, *Ind. Eng. Chem. Res.*, 1996, *35*, 1090.
217. A.S. Kovvali, H. Chen, and K.K. Sirkar, *Ind. Eng. Chem. Res.*, 2002, *41*, 347.
218. D.G. Bessarabov, R.D. Sanderson, V.V. Valuev, Y.M. Popkov, and S.F. Timashev, *Ind. Eng. Chem. Res.*, 1997, *36*, 2487.
219. H. Suda and K. Haraya, *Chem. Commun.*, 1997, 93.
220. J.H. Kim, B.R. Min, K.B. Lee, J.Won, and Y.S.Kang, *Chem. Commun.*, 2002, 2732.
221. M. Palomino, A. Cantin, A. Corma, S. Leiva, F. Rey, and S. Valencia, *Chem. Commun.*, 2007, 1233.
222. G. Parkinson, *Chem. Eng.*, 1998, *105*(8), 17.
223. F. Wang, T. Chen, and J. Xu, *Macromol. Rapid Commun.*, 1998, *19*, 135.
224. N. Ogata, *Macromol Symp.*, 1995, *98*, 543.
225. (a) P.R. Bhave, ed., *Inorganic Membranes: Synthesis, Characteristics and Applications*, van Nostrand Reinhold, New York, 1991; (b) A.J. Burggraaf, ed., *Fundamentals of Inorganic Membrane Science and Technology*, Ser 4, Elsevier, Amsterdam, 1996; (c) H.P. Tsieh, *Inorganic Membranes for Separation and Reaction*, *Membrane Science and Technology*, Ser. 3, Elsevier, Amsterdam, 1996; (d) A.F. Sammells and M.V. Mundschau, *Nonporous Inorganic Membranes for Chemical Processing*, Wiley-VCH, Weinheim, 2006.
226. K. Keizer and H. Verweij, *Chemtech*, 1996, *26*(1), 37.
227. (a) D. Hairston, *Chem. Eng.*, 1994, *101*(12), 63; (b) G. Parkinson, *Chem. Eng.*, 1995, *102*(12), 23.
228. M.A. Anderson, *Presidential Green Challenge*, EPA 744-K-96–001, July 1996, 9.
229. D.A. Ward and E.I. Ko, *Ind. Eng. Chem. Res.*, 1995, *34*, 421.
230. S. Kim and G. R. Gavalas, *Ind. Chem. Res.*, 1995. *34*, 168.
231. S. Yan, H. Maeda, K. Kusakabe, S. Morooka, *Ind. Eng. Chem. Res.*, 1994, *33*, 2096.
232. L. Zuri, M.S. Silverstein, M. Narkis, *J. Appl. Polym. Sci.*, 1996, *62*, 2147.
233. P. Huang, N. Xu, J. Shi, and Y.S. Lin, *Ind. Eng. Chem. Res.*, 1997, *36*, 3815.
234. J.J. Porter and S. Zhang, *Text Chem. Col.*, 1997, *29*(4), 29.
235. V.M. Linkov, R.D. Sanderson, and E.P. Jacobs, *Polym. Int.*, 1994, *35*, 239.
236. (a) N. Nishiyama, K. Ueyama, and M. Matsukata, *J. Chem. Soc. Chem. Commun.*, 1995, 1967. (b) P.R.H.P. Rao and M. Matsukata, *Chem. Commun.*, 1996, 1441; (c) Y. Yan, M. Tsapatsis, G.R. Gavalas, and M.E. Davis, *J. Chem. Soc. Chem. Commun.*, 1995, 227; (d) Yan, M.E. Davis and G.R. Gavalas, *Ind. Eng. Chem. Res.*, 1995, *34*, 1652; (e) Y. Yan and T. Bein, *J. Am. Chem. Soc.*, 1995, *117*, 9990; (f) J.E. Antia and R. Govind, *Appl. Catal. A.*, 1995, *131*, 107; (g) G.S. Samdani, *Chem. Eng.*, 1995, *102*(5), 15; *102*(7), 21; (h) S. Shimizu, Y. Kiyozumi, and P. Mizukami, *Chem. Lett.*, 1996, *25*, 403; (i) H. Kita, T. Inoue, H. Asamura, N. Tanaka, and K. Okamoto, *Chem. Commun.*, 1997, 45; (j) J. Hadlund, B. Schoeman, and J. Sterte, *Chem. Commun.*, 1997, 1193; (k) B.-K. Sea, E. Soewito, M. Watanabe, and K. Kusakabe, *Ind. Eng. Chem. Res.*, 1998, *37*, 2502; (l) N. Nishiyama, A. Koide, Y. Egashira, and K. Ueyama, *Chem. Commun.*, 1998, 2147; (m) K. Kusakabe, T. Kuroda, A. Murata, and S. Morooka, *Ind. Eng. Chem. Res.*, 1997, *36*, 649; (n) S. Mintova, J. Hedlund, B. Schoeman, Valtchev, and J. Sterte, *Chem. Commun.*, 1997, 15; (o) M.A. Snyder and M. Tsapatsis, *Angew. Chem. Int. Ed.*, 2007, *46*, 7560; (p) J. Choi, S. Ghosh, Z. Lai, and M. Tsapatsis, *Angew. Chem. Int. Ed.*, 2006, *45*, 1154; (q) S. Eslava, F. Iacopi, M.R. Baklanov, C.E.A. Kirschhock, K. Maex, and J.A. Martens, *J. Am. Chem. Soc.*, 2007, *129*, 9288.
237. Z. Lai and M. Tsapatsis, *Ind. Eng. Chem. Res.*, 2004, *43*, 3000.
238. M. Nomura. T. Yamaguchi, and S.-I. Nakao, *Ind. Eng. Chem. Res.*, 1997, *36*, 4217.
239. M.A. Salomon, J. Coronas, M. Menendez, and J. Santamaria, *Chem. Commun.*, 1998, 125.
240. R.M. de Vos and H. Verweij, *Science*, 1998, *279*, 1710.
241. P. Kolsch, D. Venzke, M. Noak, P. Toussaint, and J. Caro, *J. Chem. Soc. Chem. Commun.*, 1994, 2491.
242. J-M. Duval, A.J.B. Klempermann, B. Folkers. M.H. Mulder, G. Desgrandchamps, and C.A. Smolders, *J. Appl. Polym. Sci.*, 1994, *54*, 409.
243. (a) H.H. Funke, A.M. Argo, J. Falconer, and R.D. Noble, *Ind. Eng. Chem. Res.*, 1997, *36*, 137; (b) H.H. Funke, M.G. Kovalchick, J.L. Falconer, and R.D. Noble, *Ind. Eng. Chem. Res.*, 1996, *35*, 1575; (c) C. Zhang, J. Yin, S. Xiang, and

H. Li, *Chem. Commun.*, 1996, 1285; (d) K. Keizer, Z.A.L.P. Vroon, H. Verweij, and A.J. Burggraaf, *J. Chem. Technol. Biotechnol.*, 1996, *65*, 389; (e) T. Sano, S. Ejiri, M. Hasegawa, Y. Kawakami, N. Enomoto, Y. Tamai, and H. Yanagishita, *Chem. Lett.*, 1995, 153.
244. M. Isayama, K. Sakata, and T. Kunitake, *Chem. Lett.* 1993, 1283.
245. (a) T.M. Aminabhavi, R.S. Khinnavar, S.B. Harogoppad, U.S Althal, Q.T. Nguyen, and K.C. Hansen, *J. Macromol. Sci. Rev. Macromol. Chem.*, 1994, *C34*, 139; (b) J. Neel., *Makromol. Chem. Macromol. Symp.*, 1993, *70/71*, 327; (c) R.Y.M. Huang, ed., *Pervaporation Membrane Separation Processes*, Elsevier, Amsterdam, 1991; (d) A. Shanley, *Chem. Eng.*, 1994, *101*(9), 34; (e) X. Feng and R.Y.M. Huang, *Ind. Eng. Chem. Res.*, 1997, *36*, 1048; (f) N. Wynn, *Chem. Eng. Prog.*, 2001, *97*(10), 66.
246. R.W. Baker, *Kirk–Othmer Encylopedia of Chemical Technology*, 4th ed., Wiley, New York, 1995, *16*, 135.
247. J. Degal and K.-H. Lee, *J. Appl. Polym. Sci.*, 1996, *60*, 1177.
248. S-H. Chen and J-Y. Lai, *J. Appl. Polym. Sci.*, 1995, *56*, 1353.
249. J. Jegal and K.H. Lee, *J. Appl. Polym. Sci.*, 1996, *61*, 389.
250. H. Karakane, M. Tsuyumoto, Y. Maeda, and Z. Honda, *J. Appl. Polym. Sci.*, 1991, *42*, 3229.
251. S. Takegami, H. Yamada, and S. Tsuji, *Polym. J.*, 1992, *24*, 1239.
252. T. Sano, H. Yanagishita, Y. Kiyozumi, D. Kitamoto, and P. Mizukami, *Chem. Lett.*, 1992, *21*, 2413.
253. X. Lin, H. Kita, and K-I. Okamoto, Chicago A.C.S. Meeting, *Polym. Mater. Sci. Eng.*, 2001, *85*, 289.
254. X. Chen, Z. Ping and Y. Long, *J. Appl. Polym. Sci.*, 1998, *67*, 629.
255. T. Masuda, M. Takatsuka, B-Z. Tang, and T. Higashimura, *J. Membr. Sci.*, 1990, *49*, 69.
256. Y.S. Kang, E.M. Shin, B. Jung, and J.-J. Kim, *J. Appl. Polym. Sci.*, 1994, *53*, 317.
257. (a) K. Nagal, A. Higuchi, and T. Nakagawa, *J. Appl. Polym. Sci.*, 1994, *54*, 207, 1353. (b) K. Nagai and T. Nakagawa, *J. Appl. Polym. Sci.*, 1994, *54*, 1651.
258. J.-W. Rhim, H.-K. Kim, and K.-H Lee, *J. Appl. Polym. Sci.*, 1996, *61*, 1767.
259. E. Oikawa, S. Tamura, Y. Arai, and T. Aoki, *J. Appl. Polym. Sci.*, 1995, *58*, 1205.
260. E. Favre, Q.T. Nguyen, and S. Bruneau, *J. Chem. Technol. Biotechnol.*, 1996, *65*, 221.
261. J.-Y. Lai, Y.-L. Yin, and K.-R. Lee, *Polym. J.*, 1995, *27*, 813.
262. C.K. Yeom and K.-H. Lee, *J. Appl. Polym. Sci.*, 1996, *59*, 1271.
263. S-C. Huang, I.J. Ball, and R.B. Kaner, *Macromolecules*, 1998, *31*, 5456.
264. G.H. Koops, J.A.M. Nolten, M.H. Mulder, and C.A. Smolders, *J. Appl. Polym. Sci.*, 1994, *54*, 385.
265. (a) B.-K. Oh, W.J. Wang, and Y.M. Lee, *J. Appl. Polym. Sci.*, 1996, *59*, 227; (b) Y.M. Lee and B.-K. Oh, *Macromol. Symp.*, 1997, *118*, 425.
266. R.W. Baker, *Kirk–Othmer Encyclopedia of Chemical Technology*, 4th ed., Wiley, New York, 1995, *16*, 135.
267. D.B. Khairnar and V.G. Pangarkar, *J. Am. Oil Chem. Soc.*, 2004, *81*, 505.
268. C.K. Park, B.-K. Oh, M.J. Choi, and Y.M. Lee, *Polym. Bull.*, 1994, *33*, 591.
269. S.K. Ray, S.B. Sawant, J.B. Joshi, and V.G. Pangarkar, *Ind. Eng. Chem. Res.* 1997, *36*, 5265.
270. H. Wang, X. Lin, K. Tanaka, H. Kita, and K.-I. Okamoto, *J. Polym. Sci. A. Polym. Chem.*, 1998, *36*, 2247.
271. K. Inui, H. Okumura, T. Miyata, and T. Uragami, *Polym. Bull.*, 1997, *39*, 733.
272. (a) A. Bac, D. Roizard, P. Lochon, and J. Ghanbaja, *Macromol Symp.*, 1996, *102*, 225; (b) T.A. Barber and B.D. Miller, *Chem. Eng.*, 1994, *101*(9), 98; (c) A.L. Athayde, R. W. Baker, R. Daniels, M.H. Lee, and J.H. Ly, *Chemtech*, 1997, *27*(1), 34. (d) S. Mishima, H. Kaneoka, and T. Nakagawa, *J. Appl. Polym. Sci.*, 1999, 71, 273. (e) M. Hoshi, M. Kobayashi, T. Saitoh, A. Higuchi, and T. Nakagawa, *J. Appl. Polym. Sci.*, 1999, *71*, 273.
273. W. Ji and S.K. Sikdar, *Ind. Eng. Chem. Res.*, 1996, *35*, 1124.
274. (a) J.A. Dalmon, A. Giroir-Fendler, C. Mirodatos, and H.Mozzanega, eds, *Catal Today*, 1995, *25*(3–4), 199–129; (b) G. Saracco and V. Specchie, *Catal. Rev. Sci. Eng.*, 1994, *36*, 305; (c) D.M.F. Prazeres and J.M.S. Cabral, *Enzyme. Microbiol. Technol.*, 1994, *16*, 738; (d) E. Drioll and L. Giorno, *Chem. Ind.*, 1996, 19; (e) H.P. Hsieh, *Inorganic Membranes for Separation and Reaction, Membrane Science and Technology*, Ser 3, Elsevier, Amsterdam, 1996; (f) J. Coronas and J. Santamaria, *Catal. Today*, 1999, *51*, 377; (g) K.K. Sirkar, P.V. Shanbhag, and A.B. Kowali, *Ind. Eng. Chem.*, 1999, *38*, 3715–3737; (h) J.G.S. Masrcano and T.T. Tsotsis, *Catalytic Membranes and Membrane Reators*, Wiley-VCH, Weinheim, 2002; (i) J. Santamaria and M. Menendez, *Catal. Today*, 2001, *67*, 1–291; (j) I.F.J. Vankelecom, *Chem. Rev.*, 2002, *102*, 3779; (k) W. Yang and T.T. Tsotsis, *Catal. Today*, 2003, *82*, 1–281.
275. A-E.M. Adris, and J.R. Grace, *Ind. Eng. Chem. Res.*, 1997, *36*, 4549.
276. M. Jacoby, *Chem. Eng. News*, Apr. 7, 1997, 51.
277. (a) J.K. Ali and D.W.T. Rippin, *Ind. Eng. Chem. Res.*, 1995, *34*, 722; (b) F. Cavani and P. Trifiro, *Appl. Catal. A.*, 1995, *133*, 234; (c) J.P. Collins, R.W. Schwartz, R. Sehgal, T.L. Ward, O.J. Brinker, G.P. Hagen, and C.A. Udovich, *Ind. Eng. Chem. Res.*, 1996, *35*, 4398; (d) E. Kikuchi, *Cattech*, 1997, *1*(1), 67.
278. (a) E. Gobina and R. Hughes, *Appl. Catal. A*, 1996, *137*, 119; (b) J.P. Collins, R.W. Schwartz, R. Sehgal, T.L. Ward, O.J. Brinker, G.P. Hagen, and C.A. Udovich, *Ind. Eng. Chem. Res.*, 1996, *35*, 4398.
279. M.E. Rezac, W.J. Koros, and S.J. Miller, *Ind. Eng. Chem. Res.*, 1995, *34*, 862.
280. (a) R.E. Buxbaum and A.B. Kinney, *Ind. Eng. Chem. Res.*, 1996, *35*, 530; (b) G.S. Samdani, *Chem. Eng.*, 1995, *102*(8), 19.
281. D. Hairston, *Chem. Eng.*, 1996, *103*(2), 59.
282. S. Naito and N. Amemiya, *Chem. Lett.*, 1996, 893.
283. G. Weissmeier and D. Honicke, *Ind. Eng. Chem. Res.*, 1996, *35*, 4412.
284. C. Lange, S. Storck, B. Teschi, and W.H. Maier, *J. Catal.*, 1998, *175*, 280.
285. G. Parkinson, *Chem. Eng.*, 1997, *104*(5), 33.
286. S.J. Xu and W.J. Thompson, *Ind. Eng. Chem. Res.*, 1998, *37*, 1290; (b) T. Ishihara, J.A. Kilner, M. Honda, and Y. Takita, *J. Am. Chem. Soc.*, 1997, *119*, 2747; (c) J. Kilner, S. Benson, J. Lane, and D. Waller, *Chem. Ind. (Lond.)*, 1997, 907; (d) X. Qi, Y.S. Lin, and S.L. Swartz, *Ind. Eng. Chem. Res.*, 2000, *39*, 646; (e) A.C. vanVeen, M. Rebeilleau, D. Farrusseng, and

C. Mirodatos, *Chem. Commun.*, 2003, 32; (f) S. Guntuka, S. Banerjee, S. Farooq, and M.P. Srinivasan, *Ind. Eng. Chem. Res.*, 2008, *47*, 154; (g) G.Ondrey, *Chem. Eng.*, 2007, *114*(9), 20; (h) X.Tan, Y.Liu, and K.Li, *Ind. Eng. Chem. Res.*, 2005, *44*, 61.

287. G. Ondrey, *Chem. Eng.*, 2005, *112*(7), 17.
288. (a) G.J. Stiegel and R.D. Srivastava, *Chem. Ind. (Lond.)*, 1994, 854; (b) A.G. Dixon, W.R. Moser, and Y.H.M, *Ind. Eng. Chem. Res.*, 1994, *33*, 3015; (c) U. Alachandran, J.T. Dusek, P.S. Maiya, B.Ma. R.L. Mieville, M.S. Kleefisch and C.A. Udovich, *Catal. Today*, 1997, *36*, 265.
289. (a) G. Parkinson, *Chem. Eng.*, 1997, *104*(6), 19; (b) Anon., *Chem. Eng. News*, May 11, 1998, 10.
290. G. Xomeritakis and Y-S. Lin, *Eng. Chem. Res.*, 1994, *33*, 2607.
291. J.A. Switzer, M.G. Shumsky, and E.W. Bohannan, *Science*, 1999, *284*, 293.
292. T. Hibino, K-I. Ushiki, and Y. Kuwahara, *J. Chem. Soc. Chem. Commun.*, 1995, 1001.
293. Y. Zeng and Y.S. Lin, *Ind. Eng. Chem. Res.*, 1997, *36*, 277.
294. E.M. Cordi, S. Pak, M.P. Rosynek, and J.H. Lunsford, *Appl. Catal. A*, 1997, *155*, L1.
295. X-M. Guo, K. Hidajat, C.-B. Ching, and H.-F. Chen, *Ind. Eng. Chem. Res.*, 1997, *36*, 3576.
296. G.S. Samdani, *Chem. Eng*, 1995, *102*(8), 17.
297. (a) R.F. Parton, I.F.J. Vankelecom, M.J.A. Casselman, C.P. Bezoukhanova, J.B. Uytterhoeven, and P.A. Jacobs, *Nature*, 1994, *370*, 541; (b) R.F. Parton, I.F.J. Vankelecom, D. Fas, K.B.M. Janssen, P.-P. Knops-Gerrits, and P.A. Jacobs, *J. Mol. Catal. A. Chem.*, 1996, *113*, 283, 287.
298. D.E. de Vos, F. Thibault-Starzyk, P.-P. Knops-Gerrits, R.F. Parton, and P.A. Jacobs, *Macromol. Symp.*, 1994, *80*, 175, 185.
299. (a) I.F.J. Vankelecom, D. Tas, R.F. Parton, V. Van de Vyver, and P.A. Jacobs, *Angew. Chem. Int. Ed.*, 1996, *35*, 1346; (b) I. Vankelcom, A. Wolfson, S. Geresh, M. Landau, M. Gottlieb, and M. Hershkovitz, *Chem. Commun.*, 1999, 2407.
300. A. Bosetti, D. Bianchi, N. Andriollo, D. Cidaria, P. Cesti, G. Sello, and P. DiGennaro, *J. Chem. Technol. Biotechnol.*, 1996, *66*, 375.
301. (a) C. Butcher, *Chem. Eng.*, 2007, *114* (10), 33; (b) F. Fatone, P. Battistoni, P. Pavan, and F. Cecchi, *Ind. Eng. Chem. Res.*, 2007, *46*, 6688.
302. (a) M. Schneider, I.W. Marison, and U. von Stocker, *Enzyme Microb. Technol.*, 1994, *16*, 957; (b) M. Schneider, F. Reymond, I.W. Marison, and U. von Stocker, *Enzyme. Microb. Technol.*, 1995, *17*, 839.
303. D.-H. Chen, J.-C. Leu, and T.-C. Huang, *J. Chem. Technol. Biotechnol.*, 1994, *61*, 351.
304. T. Kamiya, *Chem. Eng.*, 1998, *105*(8), 44.
305. (a) G. Parkinson, *Chem. Eng.*, 1998, *105*(4), 27; (b) L.R. Raber, *Chem. Eng. News*, July 6, *1998*, 25.
306. H.N. Chang, J.W. Yang, Y.S. Park, D.J. Kim, and H.O. Han, *J. Biotechnol.*, 1992, *24*, 329.
307. A.C. Oliveira and J.M.S. Cabral, *J. Chem. Technol. Biotechnol.*, 1991, *52*, 219.
308. P. Boyaval, C. Corre, and M.-N. Madec, *Enzyme. Microb. Technol.*, 1994, *16*, 883.
309. A. Perea and U. Ugalde, *Enzyme. Microb. Technol.*, 1996, *18*, 29.
310. Y.-H. Ju, W.-J. Chen, and C.-K. Lee, *Enzyme. Microb. Technol.*, 1995, *17*, 685.
311. R. Molinari, M.E. Santoro, and E. Drioli, *Ind. Eng. Chem. Res.*, 1994, *33*, 2591.
312. Y. Isono, H. Nabetani, and M. Nakajima, *J. Am. Oil. Chem. Soc.*, 1995, *72*, 887.
313. (a) B. Nidetzky, D. Haltrich, and K.D. Kulbe, *Chemtech*, 1996, *26*(1), 31; (b) D. Gollhofer, B. Nidetzky, M. Fuerlinger, and K.D. Kulbe, *Enzyme. Microb. Technol.*, 1995, *17*, 235.

RECOMMENDED READING

1. A. Johansson, *Clean Technology*, Lewis Publishers, Boca Raton, FL, 1992, 109–125.
2. L.P. Raman, M. Cheryan, and N. Rajagopalan, *Chem. Eng. Prog.*, 1994, *90*(3), 68.
3. A. Shanley, *Chem. Eng.*, 1994, *101*(9), 34.
4. P. Cartwright, *Chem. Eng.* 1994, *101*(9), 84.
5. E. Driolo and L Giorno, *Chem. Ind. (Lond.)*, 1996, 19.
6. R. Singh, *Chemtech*, 1998, *28*(4), 33.
7. G. Karet, *R&D (Cahners)* 1999, *41*(3), 35.
8. For a more detailed review see, RW Baker, *Kirk–Othmer Encyclopedia of Chemical Technology*, 4th ed., Wiley, New York, 1995, *16*, 135.
9. J. Charpentier, *Ind. Eng. Chem. Res.*, 2007, *46*, 3465 (review of modern chemical engineering including separations).
10. D.J. Cole-Hamilton, *Science*, 2003, *299*, 1688 (separation of homogeneous catalysts).
11. R. Mullin, *Chem. Eng. News*, Apr. 5, 2004, 22 (nanofiltration).
12. M. Freemantle, *Chem. Eng. News*, June 16, 2003, 33; Oct. 3, 2005, 49 (membrane separations).
13. C.C.Tzschucke, *Angew. Chem. Int. Ed.*, 2002, *41*, 3965 (modern separation techniques).
14. R.D. Noble and R. Agrawal, *Ind. Eng. Chem Res.*, 2005, *44*, 2887 (needs of separation research).
15. K. Torzewski, *Chem. Eng.*, 2008, *115*(3), 33 (summary of membrane methods).

EXERCISES

1. If you have never used a membrane for a separation, have one of the biochemists near you show you how it is done.
2. Think of the most difficult separations you have done. Could any of them have been performed more easily by using the methods outlined in this chapter?
3. Design a ligand for gold that will allow recovery of the metal from ores without the use of cyanide ion.
4. If you have never made an inclusion compound, look up the separation of a long-chain aliphatic compound with urea and try it in the laboratory.
5. Reactive distillation and catalytic membrane reactors separate products of reactions before secondary reactions can occur. Think of some places where they can be applied to favor mono-, rather than di-additions or substitutions.
6. Look up the dimensions and coordinating ability of some common ions that are difficult to separate from others.

Compare these with the dimensions and conformations of typical crown ethers, cryptands, and calixarenes. Then extrapolate the data to the design of a new highly selective reagent.

7. Think of a difficult separation of small organic molecules. Match the sizes with those of the cavities of hosts that might allow a separation by the formation of an inclusion compound.
8. Compare the selectivity of the best synthetic membranes and ligands for separating ions with the specificities that are achieved by biological membranes. How might we mimic the biological membranes more successfully?

CHAPTER 8

Working without Organic Solvents

8.1 ADVANTAGES AND DISADVANTAGES OF SOLVENTS

Solvents are very useful. They can be used to[1]

1. Place reagents into a common phase where they can react
2. Dissolve solids so that they can be pumped from place to place
3. Lower viscosity and facilitate mixing
4. Regulate temperatures of reactions by heating at reflux
5. Moderate the vigor of exothermic reactions
6. Allow recovery of solids by filtration or centrifugation
7. Extract compounds from mixtures
8. Purify compounds by recrystallization
9. Convert pyrophoric materials to nonpyrophoric solutions (as for aluminum alkyls)
10. Remove azeotrope compounds from reactions
11. Clean equipment and clothing
12. Place a thin film of material on to a substrate

The United States uses 160 billion gallons of solvent each year.[2] The largest single use is for vapor degreasing, followed by dry cleaning and then by immersion cleaning of parts.

The disadvantages of solvents include the following:

1. Loss of 10–15 million tons of solvents (with a fuel value of 2 billion dollars) each year[3]
2. Reaction of lost solvents in air with nitrogen oxides in sunlight to produce ground level ozone
3. Destruction of upper atmosphere ozone by chlorofluorocarbons
4. Toxicity of chlorinated and other solvents to workers
5. Miscarriages caused by ethers of ethylene glycol
6. Birth defects from exposure to solvents[4]
7. Fires and explosions that may result from use
8. Monetary cost

Large point sources can account for much loss of solvents. For example, the largest sources in Delaware (see Chapter 1 for a table of sources) were until recently the painting of cars in two auto assembly plants. Nonpoint and smaller sources often produce more total emissions than the large sources. Delaware violates the Clean Air Act for ground level ozone several times each year, largely because of too many cars. There are many smaller commercial sources of solvent emissions, including dry-cleaning shops, printing establishments, metal-cutting fluids in machine shops, auto body repair shops, and many others. The consumers also emit solvents from their homes when they use paint remover, oil-based paints, adhesives, spot remover, charcoal lighter fluid, aerosol cans of personal care products, hair spray, nail polish, and gasoline-powered tools, among many. The use of chlorinated solvents minimizes the risk of fire, but causes toxicity problems, such as liver damage and cancer. If other volatile solvents are used, all equipment in the plant must be explosion-proof, which includes placing every electrical switch in a heavy metal container. The limits being set on these volatile organic compound (VOC) emissions from various products are the subject of some controversy in the United States.[5]

A tiered approach to this problem from the least change to the most change might be the following:

1. Place something on the plant outlet to destroy or recover the solvent.
2. Enclose the operation so that the solvent is not lost (see Chapter 3 for low-emission degreasing in a closed system).
3. Substitute a less harmful solvent.
4. Use supercritical carbon dioxide (also written as sc carbon dioxide) as a solvent.
5. Use water.
6. Use no solvent.
7. Switch to another process that eliminates the need for the solvent.
8. Can we do just as well without the product made with the solvent?

For laboratory use and, perhaps, for small-volume production, another first stage can be added. Microreactors, which require much less solvent, are being studied for fluorination,[6] DNA chips, and other uses.[7] (See a later section in this chapter for more details.) A variation on item 2 is the use of a polyhydroxyester base on aromatic epoxy resins and aliphatic carboxylic acids to reduce the loss of styrene from unsaturated polyester resins by 95–99%.[8] Lasco Bathware (eight plants) captures the emissions of styrene and burns them.[9]

Various methods are used for the recovery or destruction of solvent emissions.[10] Reichhold Chemical in Dover, Delaware, has added a gas-fired incineration unit to destroy emissions of 1,3-butadiene and acrylonitrile.[11] The energy use in such systems tends to be high. The energy use and the temperature can occasionally be reduced with catalysts.[12] In some cases the air containing the solvent can be run into the plant boiler. If the amount of the VOC is low, a biofilter may be suitable for removing it. Capitol Cleaners, a family-owned dry cleaning business in Dover, Delaware, has reduced its use of perchloroethylene by 96%, through the use of new tighter equipment and a system that filters and redistills used solvent. The pay-back period for the investment was 4 years.[13] Dow has designed closed system equipment for the use of chlorinated solvents.[14] Solvent emissions can also be recovered for reuse by adsorption and by refrigeration.[15] If a large volume of air is used to keep the levels of solvents in the air of the plant low, then cold traps are not particularly effective. If the concentration of the organic compound in air is higher than 1%, recovery by membranes (see Chapter 7) may be possible.[16]

Recovery of used solvents by distillation, on site or off site at a central facility, is easier if each solvent is stored separately. A single solvent for the whole plant would be desirable if a solvent must be used. Sandoz Pharmaceutical (now Novartis) has done this for a process, with the result that it now produces only 1.5 lb of waste per pound of product instead of the former 17.5 lb.[17] For an investment of 2.1 million dollars, they save 775,000 dollars/year. Processes that use solvent mixtures should be restudied to see if a single solvent can be used instead. Solvent recovery almost never reaches 100%. A plant extracting oil from soybeans recovered 99.92% by the use of desolventizer toasters with fully countercurrent stripping steam.[18] It still lost 0.3 gallons of solvent per ton of beans. It is hard to extract the last traces of solvent from a viscous polymer. If sc carbon dioxide extraction were used, the amount left in the beans would not matter. This is a big improvement over an incident that happened a few years ago in Louisville, Kentucky, in a similar plant. Hexane lost from the plant seeped into the sewer beneath a local street. A spark from a passing car that backfired ignited the hexane, causing the street to blow up.

Various substitutes have been suggested for objectionable solvents.[19] Alkanes and alcohols produce much less ground level ozone than equal amounts of alkenes and aldehydes.[20] Toluene and 2-butanone produce less than the xylenes. Ferro Corporation is selling Grignard reagents in the dibutyl ether of diethyleneglycol (butyl diglyme), which it says is a safer solvent owing to its higher boiling point and lower solubility in water than the usual ether solvents.[21] DuPont (now Koch Industries) is marketing a mixture of dimethyl succinate, dimethyl glutarate, and dimethyl adipate, a by-product of making nylon (see Chapter 1), as a less toxic solvent.[22] Monsanto (now Solutia) sells a similar mixture.[23] Eastman says that its diethoxymethane can be used as a replacement for tetrahydrofuran, methylene chloride, and the dimethyl ether of ethylene glycol.[24] *tert*-Butyl methyl ether can be used as a replacement for 1,2-dimethoxyethane, 1,4-dioxane, methylene chloride, and tetrahydrofuran.[25] Its solubility in water is small so that the workup of a reaction can involve the addition of water. *N*-Methylpyrrolidone is a low-toxicity solvent that can be used to clean silicon chips.[26] Toxic hexamethylphosphoramide can be replaced by 1,1,3,3-tetramethylguanidine in reactions with samarium (II) iodide.[27] "Methyl siloxane fluids" [probably poly(dimethylsiloxane) oils] have been suggested as alternatives to chlorofluorocarbons for precision cleaning.[28] Methyl soyate (the methyl ester of soybean fatty acids) is being sold as a replacement for trichloroethylene used in cleaning metal parts.[29] Ethyl lactate is a solvent based on renewable raw materials that may be suitable for many applications. (See Chapter 7 for its preparation.) Piperylene sulfone, which can be made by reacting piperylene (1,3-pentadiene) with sulfur dioxide, has been suggested as a substitute for dimethyl sulfoxide that is used as a solvent in reactions. After the reaction is complete, the mixture is heated to decompose the sulfone to piperylene and sulfur dioxide for recycling.[30]

Perfluorohexane has been used as a replacement for carbon tetrachloride in brominations and as a "universal" medium for suspension polymerization.[31] It can be used with monomers that contain water-sensitive groups, such as isocyanate, chlorocarbonyl, and trimethoxysilyl. Perfluorotriethylamine has been used as a solvent for Lewis acid-catalyzed reactions.[32] Oxychem sells 4-chlorobenzotrifluoride as a solvent. Benzotrifluoride itself has been used as a substitute for methylene chloride.[33] Before these materials are used on a large scale, their potential as greenhouse gases should be checked. As mentioned in Chapter 3, tetrafluoromethane is a very potent greenhouse gas. Ethyl lactate and propyleneglycol ethers have been suggested as replacements for ethyleneglycol ethers used by the semiconductor industry.[34] The use of 1-butyl-3-methylimidazolium hexafluorophosphate and other ionic liquids[35] to replace organic solvents is described later in this chapter.

Progress in using carbon dioxide, ionic liquids, or water as solvents, as well as reactions performed without solvent will now be examined. These approaches are better than simply substituting another organic solvent for the objectionable one. There are also some applications for which the use of solvent-containing materials can be avoided by a choice of other systems. If a building is made of wood, it has to be painted, sometimes with an oil-based paint. If it is covered with brick, stone, stucco, aluminum, or poly(vinyl chloride), no painting is needed. A room covered with wallpaper (put up with an aqueous adhesive) does not need to be painted. Wash-and-wear garments do not have to be dry-cleaned. If a nail polish free of solvents cannot be devised, it will not be a big problem for society.

8.2 WORKING WITHOUT SOLVENT

Many industrial processes for making commodity chemicals are carried out in the gas phase without solvent using heterogeneous catalysts. For example, methanol is made from synthesis gas (a mixture of carbon monoxide and hydrogen) using zinc oxide-containing catalysts.[36] The heat of the exothermic reaction is moderated by injecting cold synthesis gas at several places in the reactor. Oxidation of methanol to formaldehyde with oxygen is done in the gas phase, with an excess of methanol to keep outside the explosive limits. Water is used in the mixture to moderate the heat evolved. In some reactions, carbon dioxide is used as the inert heat-moderating gas. Fluorinations can be controlled by adding the fluorine in a dilute nitrogen stream. Ethylene and propylene can be polymerized in the gas phase using fluidized bed techniques.[37] Bayer uses a neodymium catalyst for the fluidized gas-phase polymerization of 1,3-butadiene in a new process that uses 50% less energy and costs 30% less.[38] Hydrocarbon solvents were formerly used for these processes. Propylene is also polymerized as a liquid without solvent. Showa Denko has a process for the oxidation of ethylene in the gas phase to acetic acid using a palladium catalyst.[39] It produces one-tenth as much effluent as the process for making acetaldehyde in solution. Arylacetonitriles can be methylated in the gas phase at 180°C, using gas–liquid phase-transfer catalysis with a film of potassium carbonate-containing polyethylene glycol on corundum (a form of aluminum oxide) spheres (**8.1** Schematic).[40] The nontoxic dimethyl carbonate replaces the highly toxic cancer-suspect dimethyl sulfate and iodomethane normally used in methylations. The by-product methanol can be converted to more dimethyl carbonate by oxidative carbonylation. The only waste is nontoxic carbon dioxide. (See Chapter 2 for more detail on reactions of dimethyl carbonate.)

General Electric Plastics (now SABIC) has replaced the methylene chloride used in recrystallizing bisphenol A with a fractional, falling-film, melt crystallization.[41] This simplified system has saved energy, at the same time eliminating the drying and distillation steps. The mechanical extraction of soybeans does not remove quite as much oil as extraction by hexane, but the press cake is a higher-energy animal feed.[42] This can also be done in an extruder with a residence time of 30 s at 135°C.[43] The pressed oil commands a higher price in niche markets.

There are times when excess starting material can serve as the solvent. Excess liquid chlorine has been used in the chlorination of natural rubber to eliminate the need for the usual carbon tetrachloride.[44] It is difficult to remove the last traces of carbon tetrachloride from the chlorinated rubber. The dimethyl terephthalate and terephthalic acid used in the preparation of poly(ethylene terephthalate) (**8.2** Schematic) are made by the oxidation of *p*-xylene.[45] The dimethyl terephthalate is recrystallized from the starting material, *p*-xylene. The terephthalic acid is recrystallized from water at 225–227°C.

Many reactions can be carried out without solvent[46] when the reagents are liquids or when the mixture can be melted to produce a liquid.[47] This includes cases in which a solid reagent gradually dissolves in the other as the reaction proceeds. Esters of rosin are made in melts, which solidify on cooling. Rosin is a mixture of resin acids obtained from pine trees. The one shown, abietic acid, can be esterified with pentaerythritol (**8.3** Schematic), as shown, or with other alcohols.

The current literature contains many reactions that are run without solvent. Frequently, this can offer improved yields and selectivity as well as simplicity. Almost all of them still use solvents, such as toxic methylene chloride and flammable ethyl ether, in the workup and sometimes involve chromatographic separations. These solvents might be avoided by filtering off a solid catalyst, and then using it in the next run without washing it with solvent to recover the last of the product. Other possibilities include distillation under vacuum, extraction with sc carbon dioxide, and extraction with water plus or minus a hydrotropic agent, possibly at high temperature. (Some of these are discussed later.) If the reaction has been run on a solid support, displacement of a less polar product by a more polar starting material (as in the oxidation of an alcohol to a ketone) may be possible. The goal is to find simple, convenient, and fast energy-efficient ways in which to isolate the product and recycle catalysts and supports. If the product is obtained in 100% yield at 100% conversion, no further treatment will be needed. In the much more likely case that unreacted starting materials and by-products are present, the product will have to be obtained by distillation, membrane

$$ArCH_2CN + CH_3OCOOCH_3 \xrightarrow{K_2CO_3} ArCH(CH_3)CN + CH_3OH$$

99% conversion
>99.5% selectivity

8.1 Schematic

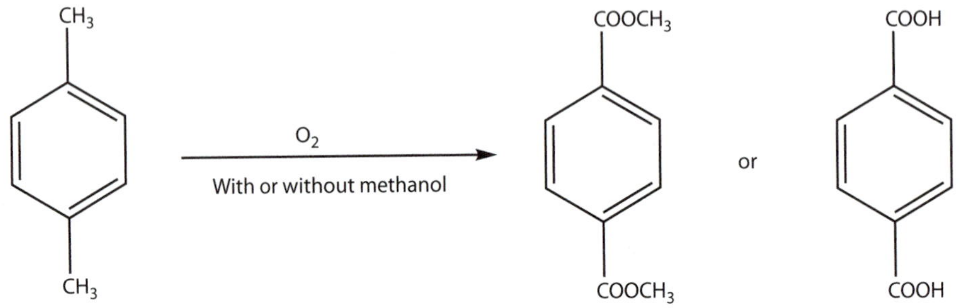

8.2 Schematic

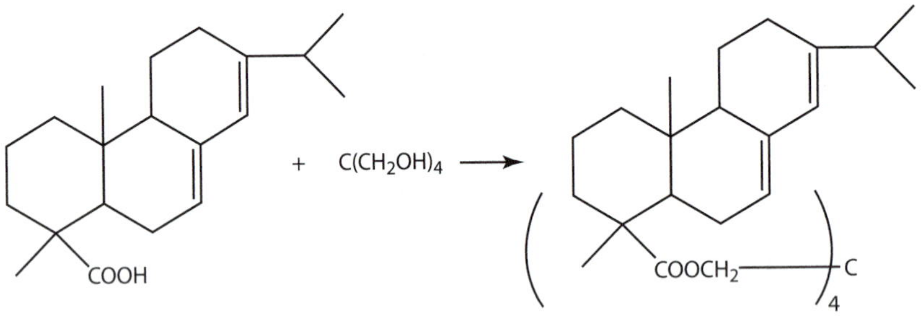

8.3 Schematic

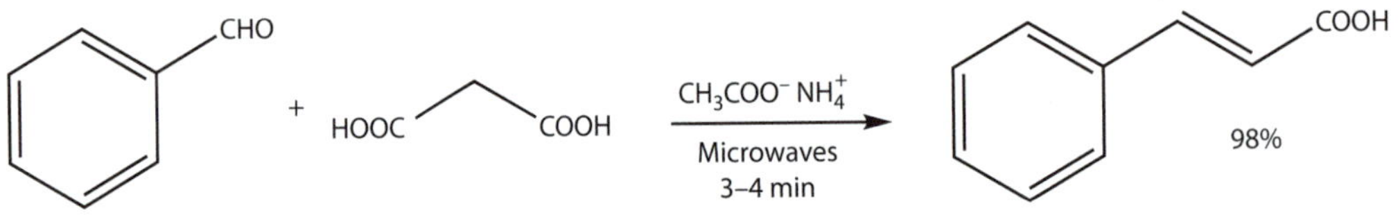

8.4 Schematic

separation, or other means. A few examples will illustrate the possibilities.

Varma has run many types of reactions on nontoxic supports, such as alumina, silica, and clay, using microwaves.[48] (For more on the use of microwaves, see Chapter 15.) The method offers selectivity under safe, fast, and mild reaction conditions in open vessels. The supports can be recycled. The reactions include oxidations, reductions, nitrations, aldol condensations, and others. Solvents were used in the workups. Other workers have devised a condensation (**8.4** Schematic) of aromatic aldehydes with malonic acid (without support) that uses no solvents at all.[49] The workup consisted of pouring the glassy reaction mixture from the Erlenmeyer flask into ice water, filtering off the solid, and drying it. The conventional procedure for this reaction requires heating at 120°C for 3 h to give the product in 96% yield.

Aldehydes can be condensed with aliphatic nitro compounds (**8.5** Schematic) using an ion-exchange resin[50] or a potassium-exchanged-layered zirconium phosphate as catalyst.[51]

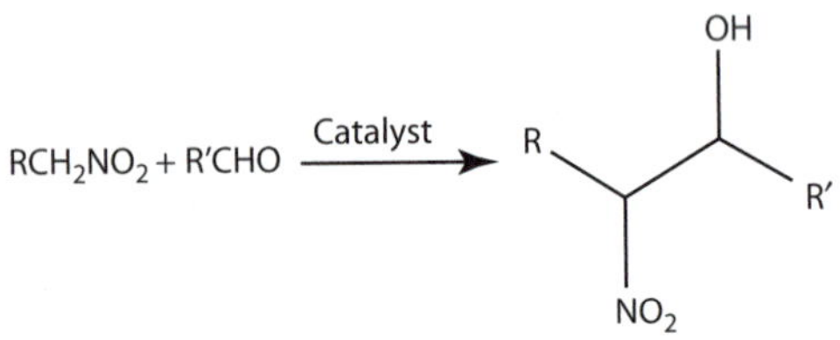

8.5 Schematic

The yields were 62–89%. The ion-exchange resin and the zirconium phosphate were washed with hazardous methylene chloride to recover the products. The catalyst could be reused with little loss in activity. Michael additions of β-diketones to α-,β-unsaturated ketones can be run without solvent with the help of microwave radiation (**8.6** Schematic) to give the adducts in 63–95% yield.[52] Cerium (III) chloride can also be used. No polymerization of the methyl vinyl ketone takes place.

Condensation and reduction have been combined in the preparation of methylisobutylketone from acetone (**8.7** Schematic).[53] For a reaction run at 160°C for 2 h, the selectivity to methylisobutylketone was 92.5% at 46% conversion. The by-products were 0.6% isopropyl alcohol and 3.6%

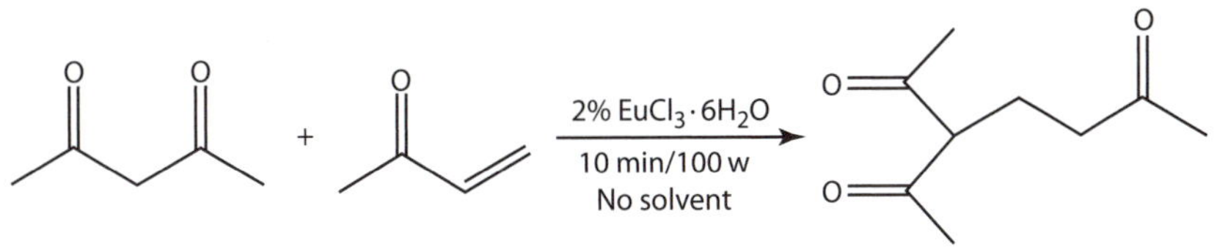

2% $EuCl_3 \cdot 6H_2O$
10 min/100 w
No solvent

8.6 Schematic

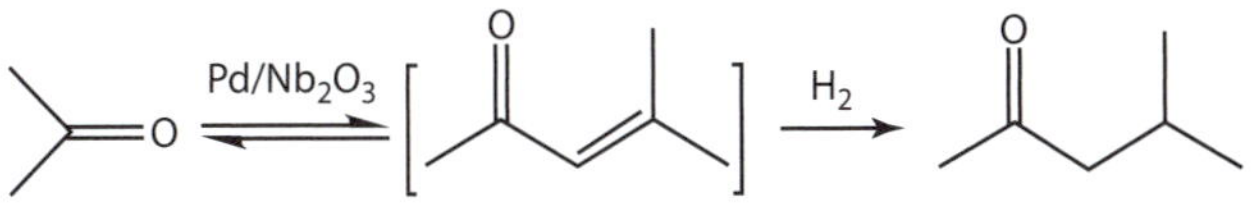

Pd/Nb_2O_3
H_2

8.7 Schematic

diisobutylketone. The activity and selectivity remained constant for 1500 h.

2-Hydroxypyridine has been *N*-alkylated (as the pyridone) with benzyl chloride (**8.8** Schematic) quantitatively without solvent.[54] However, methylene chloride was used in the workup.

Some reactions can be quite exothermic. Before running a reaction without solvent, it would be best to check the heat of reaction by thermal analytical methods. This will avoid runaway reactions. A thermoregulator should also be used to control the temperature.

Grinding a metal salt with a ligand can sometimes produce the metal complex faster and more conveniently than working in solvent.[59]

Thionocarbamates (**8.10** Schematic) were made by grinding the ingredients together with a mortar and pestle at room temperature in air.[60]

The Friedel–Crafts alkylation of benzene with isopropyl bromide (**8.11** Schematic) was run by grinding the ingredients in an agate mortar for 45 min, pouring onto crushed ice, and extracting with ether.[61]

The more environmentally benign solid acid $Cs_{2.5}H_{0.5}PW_{12}O_{40}$ was ground in a mortar with a diol for 5 min, and then left for 9 h at room temperature to give the pinacol rearrangement (**8.12** Schematic).[62]

Benzene can be nitrated with 70% nitric acid in the vapor phase over a molybdenum oxide/silica catalyst (**8.13** Schematic).[63]

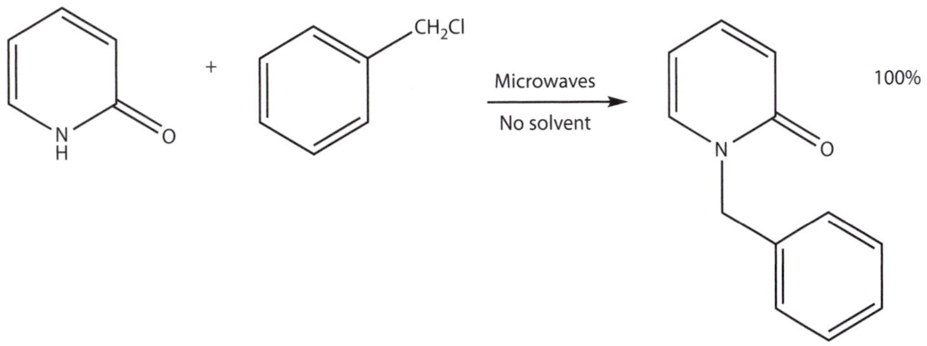

CH_2Cl
Microwaves
No solvent
100%

8.8 Schematic

Some reactions can be run in the solid state with the aid of ultrasound, grinding,[55] heat, host–guest complexes, and so forth.[56] These include Michael reactions, Baeyer–Villiger reactions, benzylic acid rearrangements, etherifications, and others. Calix[4]arene (MP 310°C) can be converted to its *p*-isopropyl derivative (**8.9** Schematic) by grinding it with a nickel sulfate on alumina catalyst, heating to 210°C, and then introducing propylene for 6 h.[57] Chloroform was used in the workup. This procedure resolved the problem of the low solubility of the starting material under the usual alkylation conditions.

Grinding pyrogallol and isovaleraldehyde with a mortar and pestle for 5 min gave the calix[4]pyrogallolarene, a faster process than methods using solvents.[58]

Organocatalysis has also been carried out in a ball mill (**8.14** Schematic).[64]

Solvents can be eliminated from some reactions by running them in ball mills.[65] Methyl methacrylate can be polymerized in a vibratory ball mill using a talc catalyst.[66] Ball mills are feasible for reactions on a commercial scale. Huge mills are used by the mining industry in grinding ores before flotation. If the mill is used repeatedly to make the same product, the problem of cleaning it out between runs is eliminated. It is also possible to use an inert liquid diluent in the milling, so that the product can be recovered as a slurry. The preferred diluent would be water if it is inert for the reaction in question.

Films of polyimides have been made (**8.15** Schematic) without solvent by vapor deposition of the diamine and

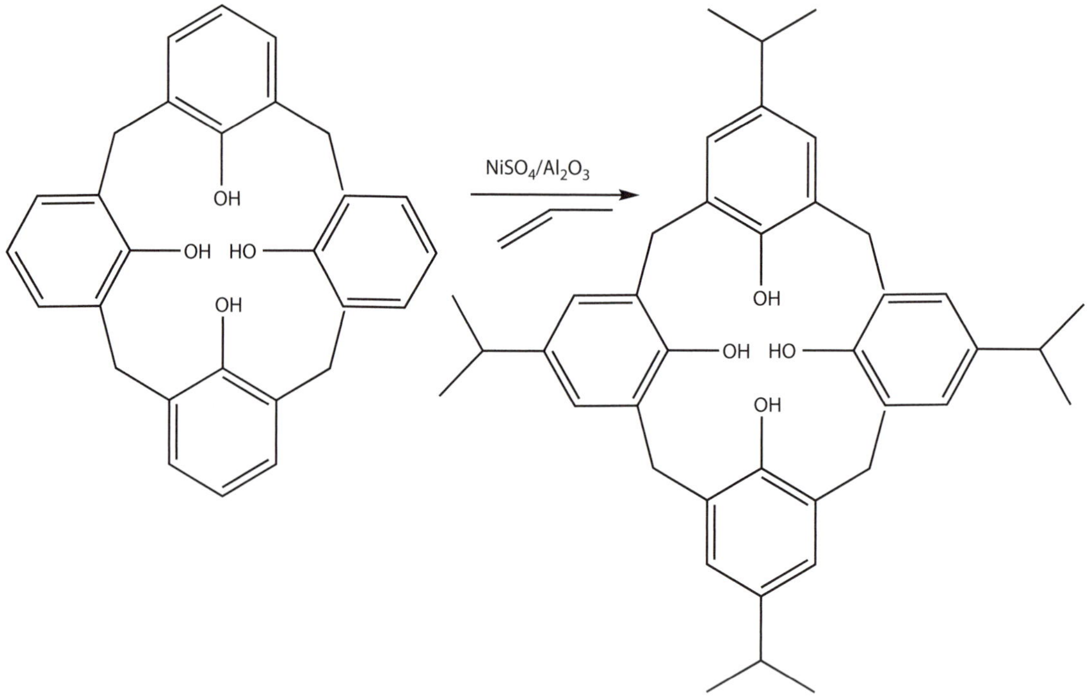

8.9 Schematic

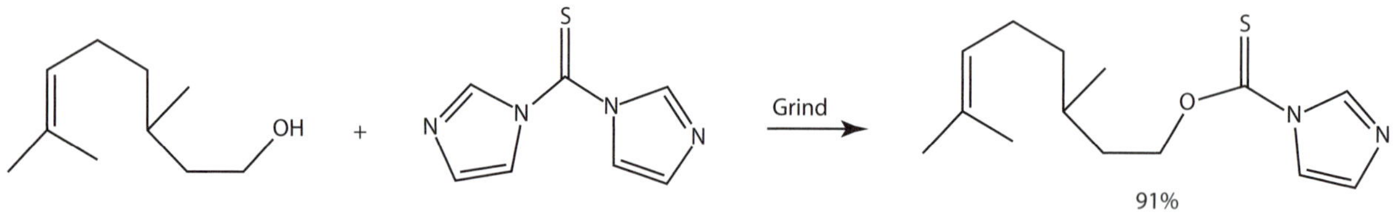

8.10 Schematic

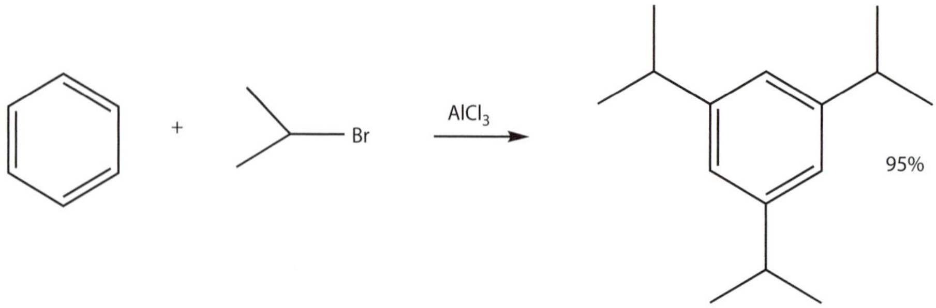

8.11 Schematic

dianhydride on to a cold surface, followed by heating.[67] The films were equivalent to those made from solution.

Inorganic compounds can also be made by solid-state synthesis.[68] The reactions are essentially controlled explosions. Two of these reactions are shown in Schematic **8.16**.

The analyses of traces of aromatic solvents and pesticides in water can now be made without the customary solvent extraction.[69] The materials are extracted on a fine silica fiber coated with poly(dimethylsiloxane), on a polyacrylate, or on Parafilm. The samples are then desorbed thermally in the inlet of the gas chromatograph, or with Parafilm, just examined by infrared light.

Flash vacuum pyrolysis is done without solvent.[70]

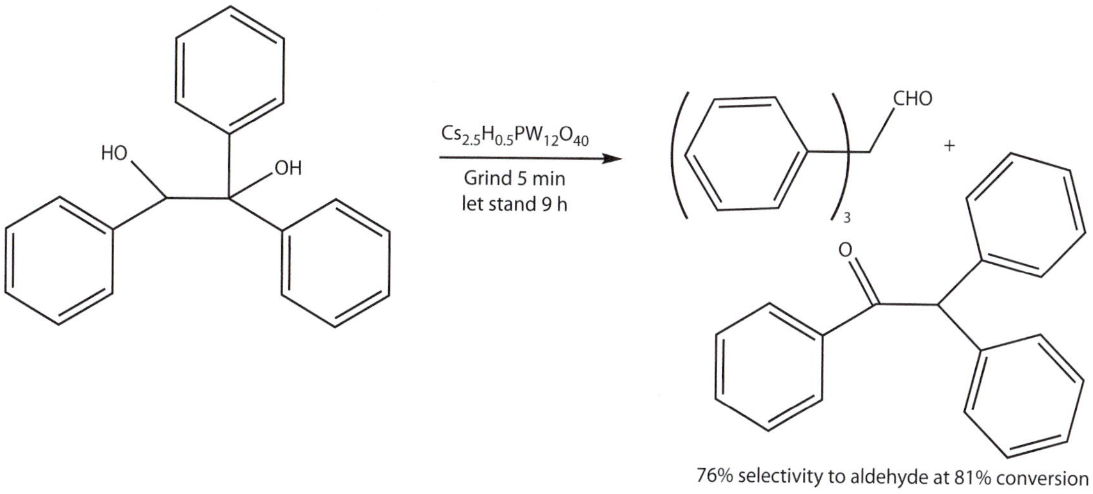

$Cs_{2.5}H_{0.5}PW_{12}O_{40}$
Grind 5 min
let stand 9 h

76% selectivity to aldehyde at 81% conversion

8.12 Schematic

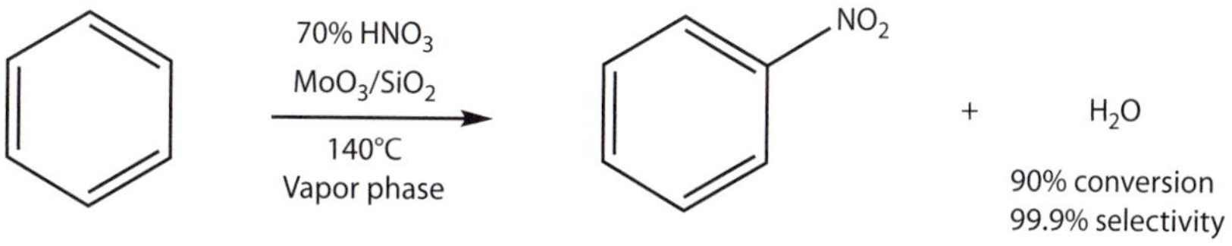

70% HNO_3
MoO_3/SiO_2
140°C
Vapor phase

90% conversion
99.9% selectivity

8.13 Schematic

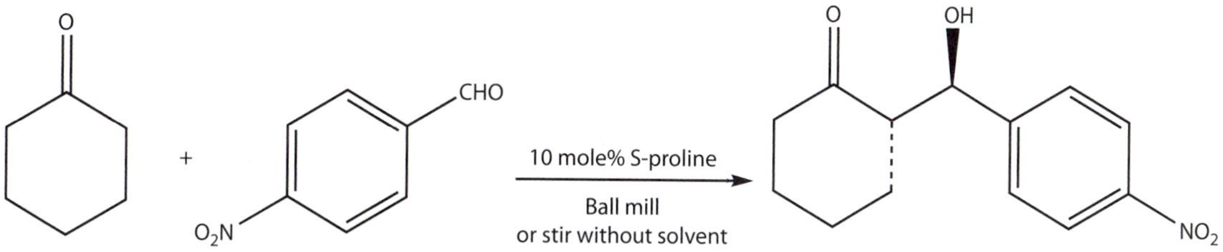

10 mole% S-proline
Ball mill
or stir without solvent

8.14 Schematic

→ Polyimide

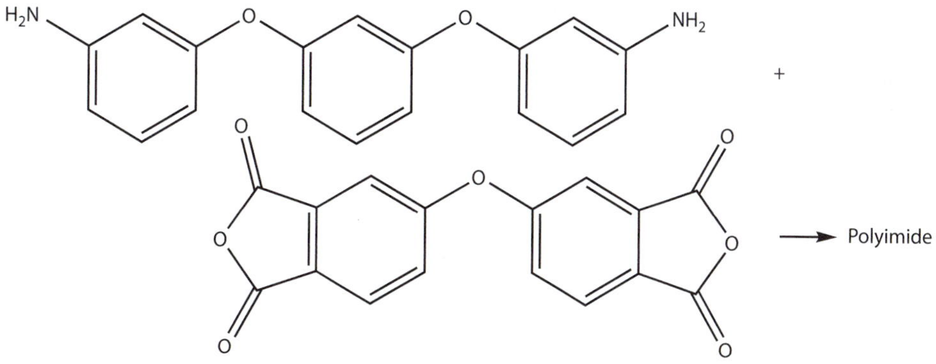

8.15 Schematic

8.3 PROCESS INTENSIFICATION

These methods[71] are continuous with short contact times, which minimize further reactions, for higher yields of purer products. Mixing and heat transfer are good. There is no need to worry about explosive limits of reactant mixtures. Reaction conditions can be optimized quickly. The need for a pilot plant is eliminated. To increase capacity, just add another unit, that is, "number up." Reactions can be run with or without solvent. Such units offer smaller, cheaper, and

$$TiCl_3 + Li_3N \longrightarrow TiN + 3LiCl$$

$$Fe_2O_3 + Al \longrightarrow Fe + Al_2O_3$$

8.16 Schematic

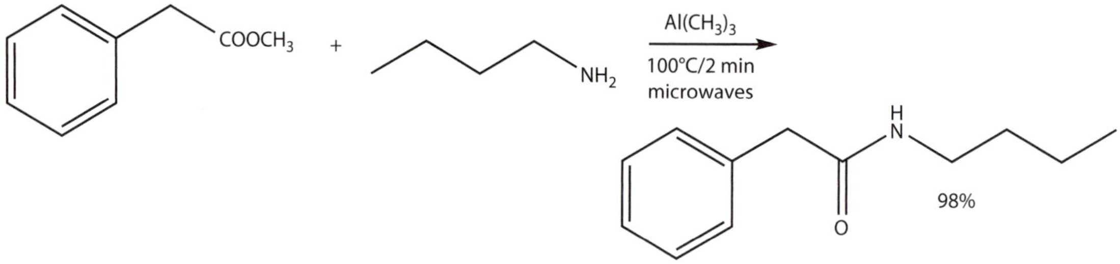

8.17 Schematic

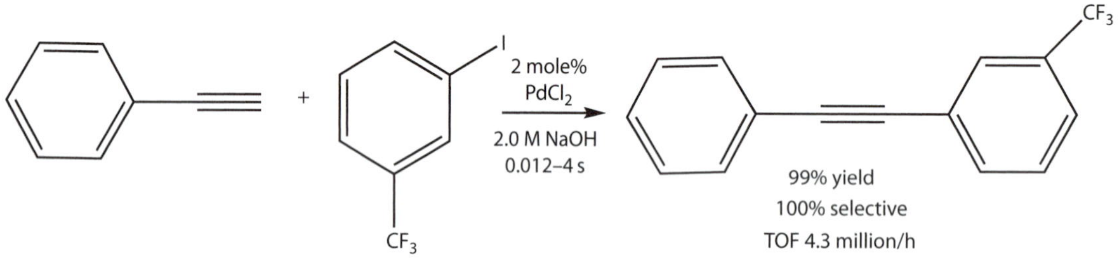

8.18 Schematic

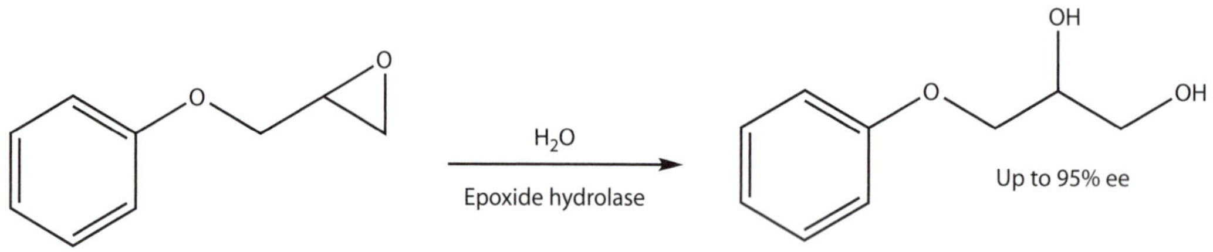

8.19 Schematic

safer chemical plants. Commercial units are available from Thales Nano, Corning Glass, Uniqsis, Protensive, Ehrfield Mikrotechnik, and others.

8.3.1 Microchannels and Tubes

The channels can be in plastic, metal, or glass. The tubes can be glass, Teflon, or poly(vinyl chloride). For longer lengths in small units, these can be in spiral or back and forth fashion. A Corning unit 30 cm × 30 cm × 60 cm can produce 40 metric tons of product a year.[72] Catalysts can be used in packed beds or attached to the surfaces. A few of the many reactions that have been run in such reactors will be given to illustrate the range of possibilities.

1-Methyl-3-butylimidazolium bromide has been prepared in 2–6 mm channels at a rate of 9.3 kg/day, a rate 20 times that of batch processes.[73] Ozone has been used to oxidize phosphites, amines, and 1-olefins with up to 100% selectivity in 1 s.[74] A three-stream system of ester, amine, and trimethylaluminum entering a mixing zone has been used to prepare amides in 98% yield in 2 min using microwaves at 100°C (Schematic **8.17**).[75]

Many reactions have been run using palladium catalysts. A Sonogashira reaction was run in 0.012–4 s to give nearly quantitative yields with a turnover frequency of 4.3 million/h (Schematic **8.18**).

Enzymatic reactions can also be carried out in microreactors, as illustrated by the hydrolysis of phenyl glycidyl ether (Schematic **8.19**).[76]

The short contact times allow some reactions to be run at higher temperatures than they would be in a batch reaction. The one below would be run at –78°C.[77] The mixture is in the first reactor for 3.4 s and in the second for 2.9 s (Schematic **8.20**).

The cationic polymerization of butyl vinyl ether gave just as high a molecular weight at 0°C as it did at –78°C.[78] The reaction was complete in <0.5 s.

Fine particles can also be made in microreactors. An ethanolic solution of danazole (a steroid drug) was put into water to form 364 nm particles.[79] A coaxial flow device with an inner diameter of 150 μm in the inner tube and of 1.7 mm in the outer tube was used to prepare <7 nm particles of Fe_3O_4 from Fe(II) and Fe(III) salts and tetramethylammonium hydroxide.[80] Such nanoparticles have also been

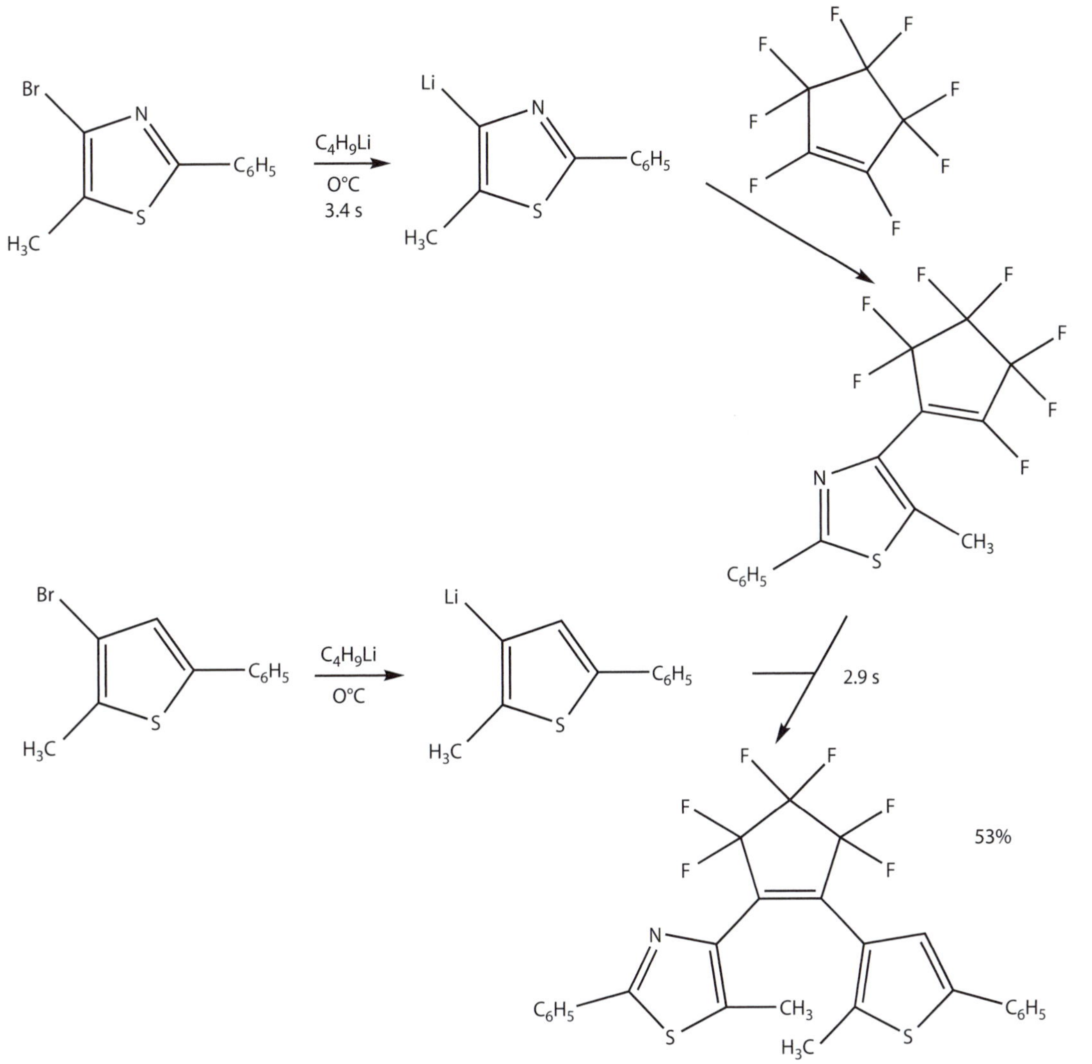

8.20 Schematic

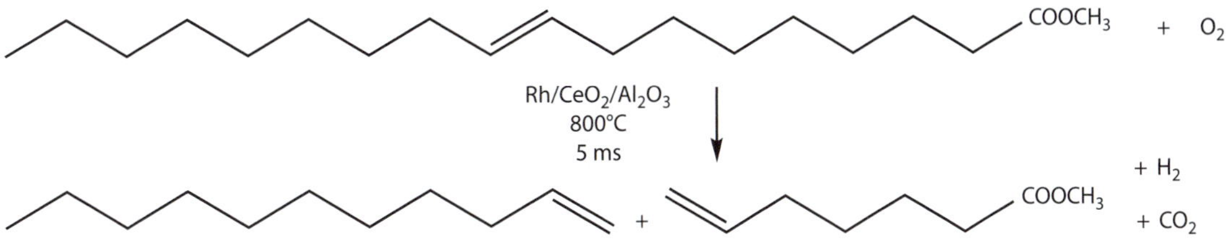

8.21 Schematic

prepared using an aqueous stream mixed with one of a perfluorocarbon for droplets in 25 μm channels.[81] Such systems are used to prevent clogging. Monodisperse quantum dots of cadmium selenide have been prepared by a similar method using gas–liquid segmented flow in a microreactor with multiple temperature zones.[82]

Segmented flow of aqueous droplets in mineral oil or hexane has been used to prepare polyamides and other insoluble products in poly(vinyl chloride) tubing with an inner diameter of 1.59 mm.[83] Polyethylene tubing has also been used.

Reactions in microreactors can be optimized in a short time, especially if one with multiple channels is used.[84]

Microreactors are being used commercially. Lonza is preparing kilogram quantities with them.[85] Nippoh Chemicals (Tokyo) is building a plant that will make 1000 metric tons of trimethyl orthoacetate per year using millimeter-sized tubes. It will occupy one-fifth the space at half the cost of a batch facility.[86]

Microreactors are not just for small-scale production. Velocys (now Oxford Catalysts) is designing a gas to liquid

$CH_2=C(CH_3)COOCH_3$ —Light→ $-[CH_2-C(CH_3)(COOCH_3)]_n-$ 18 cm disc for 40 kg/h

$RR'C=O$ + $ClCH_2CN$ —KOH, tetrahydrofuran, $C_6H_5CH_2N(C_2H_5)_3^+ Cl^-$, 20°C, <1 s→ epoxide (R, R', CN) 90% conversion 15 cm disc for 8 tons/year

8.22 Schematic

plant for stranded natural gas on offshore oil platforms. It will use steam reforming followed by a Fischer–Tropsch synthesis.[87] Dimethylbenzothiophenes have been removed from petroleum with 30% hydrogen peroxide, phosphotungstic acid, and octadecyltrimethylammonium bromide as a surfactant in a 1-mm Teflon capillary in 1.3 min at 70°C.[88] (This reaction is needed to lower the sulfur content of fuels.) Methyl oleate has been cleaved by oxygen at 800°C over a rhodium/cerium oxide/aluminum oxide catalyst in 5 ms (Schematic **8.21**).[89] A pound a day of product was made from a reactor 1 cm in length. The olefins may replace some of those made from petroleum.

8.3.2 Thin Film Reactors

Spinning disc reactors, in which the reagents drop on to a disc spinning at several hundred r.p.m., have been used for the polymerization of methyl methacrylate[90] and in the Darzens reaction.[91] In the Darzens reaction, the reaction time was reduced by 99.9%, the inventory by 99%, and the impurity level by 93%.

Magnesium hydroxide and magnesium oxide nanoparticles have been made in a reactor with a 12 cm disc.[92]

The spinning tube in a tube reactor of Holl Technologies consists of a rotor spinning at 3000 r.p.m. inside a stator with an annular gap of 0.25–1.5 mm, which ensures excellent mixing of the reagents. Sizes vary from desktop units that can process pounds per day to one about 15-ft-long that can handle a ton per hour. Biodiesel fuel is being made by the transesterification of vegetable oil in such an apparatus with a 4.5 in. stator tube with sodium hydroxide as catalyst in 1 s at 50°C to produce 1 million gallons per year at a plant in North Carolina.[93]

8.3.3 Extruders

Extruders are in common use for the processing of polymers, as in the spinning of synthetic fibers. They are also used for the extrusion cooking of foods, such as ready-to-eat breakfast cereals. They are not part of the standard equipment of the usual chemical laboratory, yet they offer considerable promise for running reactions without solvent on a continuous basis.

An extruder is essentially a pipe with a small hole at one end. Materials are put in at one end, as from a hopper, and pushed through by screws until they exit from the other end. Twin intermeshing screws can give good mixing.[94] Laboratory extruders can offer a variety of options. They may have zones heated to different temperatures, inlets at varying points along the tube where materials can be added, vents that allow volatile substances to exit, and a variety of hole sizes and shapes at the end. Thus, it is possible to melt one material, add a reagent later, and take off a volatile by-product farther down the barrel. The reaction time can be varied by the screw speed and the length of the barrel. The feed material can be blanketed with an inert gas, such as nitrogen, if needed.

Polymerization is often incomplete. Traces of residual monomer may adversely affect the properties of the polymer, such as lowering the glass transition temperature (the temperature at which a rubbery material becomes glassy). Residual monomer is very undesirable if it is a carcinogen such as acrylonitrile or vinyl chloride. Polymers for use in contact with food must not lose a toxic monomer into the food. Solvents and monomers are very difficult to remove completely from polymers, but can be removed in a vented extruder. Residual methyl methacrylate can be removed from poly(methyl methacrylate) in a vented extruder.[95] The molecular weights of polymers that are too large for the intended application can be reduced by passage through an extruder. The opposite can also be accomplished. The molecular weights of polymers with amino or hydroxyl groups can be increased by chain extension with blocked diisocyanates, the blocking agent coming out the vent. Poly(ethylene terephthalate) with carboxylic acid end groups can be reacted with bis(oxazolines) to increase the molecular weight and improve the physical properties.[96]

Poly(methyl methacrylate) can be converted to monomer for recycling by passage through a continuous twin-screw extruder at a little over 400°C.[97] The 95% yield of 95–99% purity compares favorably with the 60–80% yield of 60–80% purity of the batch process, which is 10 times slower.

Biodegradable plates, knives, forks, and leaf bags can be made by molding or extrusion of starch plasticized with 10% water.[98] After use, the materials could be composted. A mixture of benzylated wood and polystyrene can be extruded easily.[99] The properties of the product are comparable with those of polystyrene.

Most reactive extrusion has dealt with polymers that can be difficult to handle by other techniques.[100] Polymers can be cross-linked, grafted with other monomers, and so on. Extrusion cross-linking of polyethylene wire coating raises the maximum temperature of use. Polypropylene degrades in this process, but can be cross-linked successfully if multifunctional acrylate monomers are included with the peroxide used in the process.[101] Butyl rubber can be halogenated. Poly(vinyl chloride) can be reacted with sodium mercaptides. Polymers containing ester or carbonate groups can be transesterified.[102] Polycarbonate and poly(butylene terephthalate) can be melt-blended in the presence of a titanium alkoxide catalyst.[103] A small amount of transesterification can make dissimilar polymers compatible, even though without it they would be incompatible. Block copolymers have been prepared by melt-blending a mixture of poly(ethylene terephthalate) with nylon 6 or nylon 6,6.[104] Transamidation can also take place when a mixture of polyamides is run through an extruder.[105]

Coextrusion of L-lactide (a cyclic dimer lactone of lactic acid) with hydroxy-terminated poly(caprolactone) in the presence of stannous octoate gave a block copolymer.[106] Block copolymers can also be used as compatibilizers for the homopolymers, in addition to having interesting properties in their own right.[107] Copolymers of ethylene and vinyl acetate can be treated with methanol and a little sodium methoxide to remove some of the acetate groups from the copolymer.[108] The by-product methyl acetate comes out the vent of the extruder. *n*-Butyl alcohol can be used in the same way.[109] The product copolymer is useful as a gas barrier coating on films for packaging food.

Graft polymerization in an extruder is used to alter the properties of a polymer (e.g., make a polyolefin more polar so that it can stick to more polar surfaces). This is another way of making compatibilizers for dissimilar polymers. Maleic anhydride can be grafted to polyethylene in an extruder.[110] The anhydride groups in the graft copolymer can be reacted with a terminal group in a condensation polymer to produce a compatibilizer.[111] A polypropylene–maleic anhydride graft was reacted with a terminal amino group in a polyamide (polycaprolactam) (**8.23** Schematic) by reactive processing in an extruder.[112] Blends of polypropylene and nylons offer properties not available in either homopolymer.

The same technique can be used to dye a material that is otherwise difficult to dye. An ethylene–propylene copolymer rubber was reacted first with maleic anhydride and then with an aromatic amine dye in an extruder to produce a dyed rubber.[113] Dye sites can also be inserted into polyolefins by grafting them with dimethylaminoethyl methacrylate, using azo or peroxide catalysts in an extruder.[114] *N*-Vinylimidazole has been grafted to polyethylene in an extruder with the help of dicumylperoxide.[115] The product was mixed with an acrylic acid-modified polypropylene and used to compatibilize polyethylene and polypropylene. This could be helpful in the recycling of mixed polyolefins from municipal solid waste. Recycling of cross-linked (thermoset) polymers is more of a problem because they cannot be remelted in an extruder. However, they can be if thermally reversible cross-links have been used in their preparation.[116] These include polyurethanes, Diels–Alder reaction products, and so on.

Glycidyl methacrylate has also been grafted to polypropylene to insert an epoxy group for later reaction with something else.[117] The product was used for coupling with carboxy-terminated poly(butylene terephthalate) (**8.24** Schematic) in a twin-screw extruder.[118]

Glycidyl methacrylate has also been grafted to polyethylene and to poly(ethylene-*co*-propylene) using peroxides in extruders.[119] The reactive group need not be pendant. It can be part of a copolymer. Ethylene–methyl acrylate copolymers and styrene–maleic anhydride copolymers have been reacted with amines in extuders.[120]

The extruder can be used for a variety of polymerizations even if no preformed polymer is present.[121] These include the continuous anionic polymerization of caprolactam to produce nylon 6[122]; anionic polymerization of caprolactone[123]; anionic polymerization of styrene[124]; cationic copolymerization of 1,3-dioxolane and methylal[125]; free radical polymerization of methyl methacrylate[126]; addition of ammonia to maleic anhydride to form poly(succinimide)[127]; and preparation of an acrylated polyurethane from poly(caprolactone), 4,4′-methylenebis(phenyl isocyanate), and 2-hydroxyethyl acrylate.[128]

The technique of reaction injection molding to prepare molded parts is slightly different. Polyurethanes can be made in this way by injecting separate streams of diol and diisocyanate into the mold. In the metathesis polymerization of dicyclopentadiene,[129] the procatalyst is put in one stream and the catalyst activator in the other. The procatalyst is usually a tungsten or molybdenum complex and the catalyst activator is a metal alkyl. Polymerization is through opening the norbornene double bond with cross-linking by opening some of the cyclopentene double bonds (**8.25** Schematic).

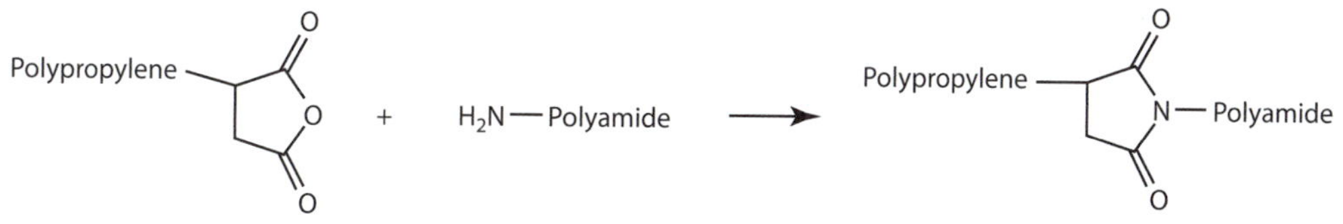

8.23 Schematic

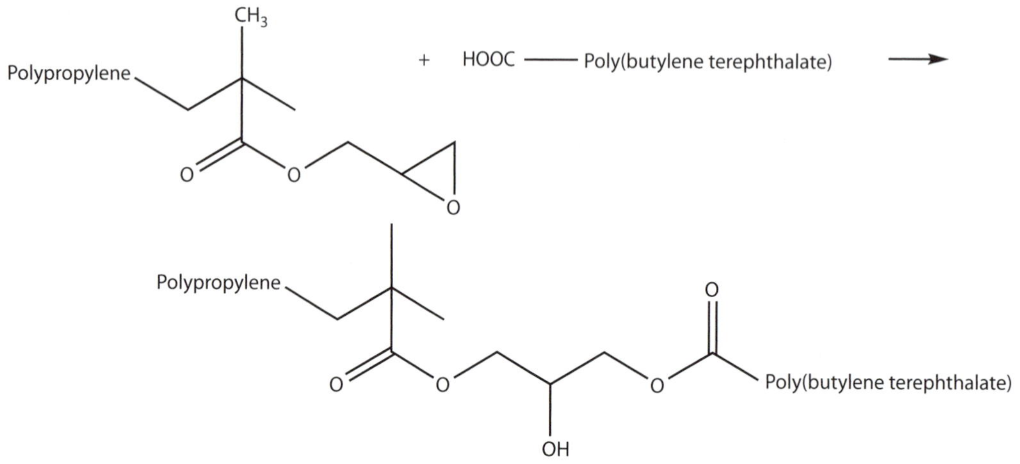

8.24 Schematic

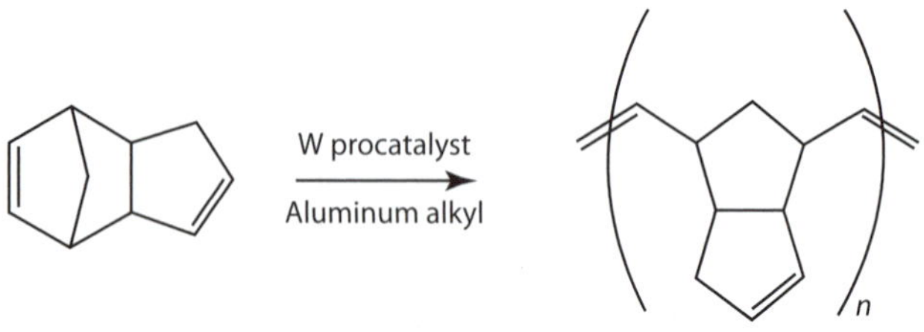

8.25 Schematic

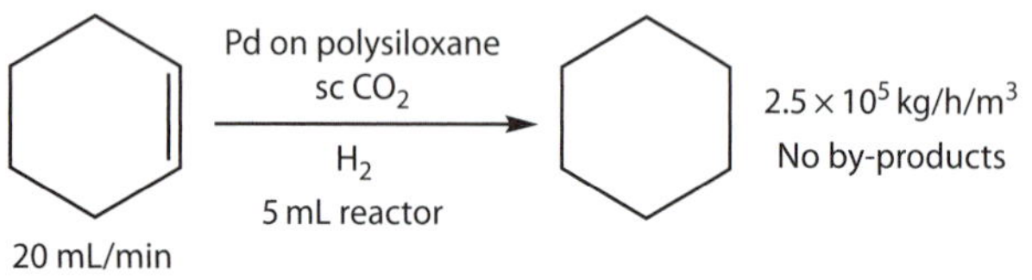

8.26 Schematic

The monomer usually contains a dissolved rubber to make the liquid more viscous and to improve the impact strength of the molding.

Hydrogen has been used to reduce vegetable oils at 150°C in a twin-screw extruder coated with a palladium catalyst.[130] This continuous process with 20% conversion per pass decreased catalyst consumption and eliminated a catalyst separation step. However, the hydrogenation rate was only one-tenth that when a palladium-on-carbon catalyst slurry was used. It might be increased by putting more catalyst on the extruder, by using a longer extruder barrel, or by using the catalyst slurry method in an extruder.

These examples show that it is possible to handle gases and liquids of low molecular weight in reactions in extruders. The presence of a polymer is optional. It is possible to remove volatile by-products through a vent. The catalyst, if any is needed, can be in a coating in the extruder. Viscous mixtures are not a problem. Adding ultrasound to the extruder can make the material seem less viscous.[131] Adding microwaves to speed up the extrusion will require more development. A jacketed extruder can be used to remove heat if necessary. Thus, it should be possible to run many of the usual reactions in extruders. The only deterrents for the chemist in the laboratory might be the need to work on a slightly larger scale and the increased time to clean the extruder between runs. It is possible that miniextruders exist or that the present ones can be scaled down. This underused apparatus offers great potential for the elimination of solvents.

8.3.4 Other Methods of Process Intensification

Reactions that take several hours under conventional conditions can be done in a few minutes with microwaves, as shown by many examples in earlier chapters. Examples that use ultrasound to speed up reactions involving solids can also be found in earlier chapters. (Chapter 15 contains more information on these techniques.[132])

The rates of hydrogenation are limited by the low solubility of hydrogen in the usual solvents. Much higher rates can be obtained in sc carbon dioxide with which hydrogen is miscible, as shown by the hydrogenation of cyclohexene (Schematic **8.26**).[133]

Gas phase biocatalysis in a tubular reactor offers possible reductions in plant size of up to 1000 times.[134] Enzymes are more stable thermally when restricted amounts of water are present (Schematic **8.27**).

8.4 CARBON DIOXIDE AS A SOLVENT

Carbon dioxide is nontoxic, nonflammable, relatively inert, abundant, and inexpensive. It does have some limitations. It has to be used under pressure and is not a very good solvent for many substances. As a solvent, it is used most commonly as a supercritical fluid (i.e., above its critical temperature of 31.1°C and critical pressure of 7.38 MPa). In this region the density of the material and its solvent power can be varied by changing the temperature and pressure. The system has properties between those of gases and liquids.

COOH + ROH —(Gas phase lipase, 60–100°C, N_2)→ COOR

97.8% conversion
2.2–2.5 kg/h with
2.0 kg enzyme in
a packed bed

8.27 Schematic

Diffusivity and mass transfer are better than in liquids, such as the usual organic solvents. The low critical temperature allows heat-sensitive materials to be processed without damage. Sc carbon dioxide is not the only supercritical fluid that can be used but it is among the few that offer the combination of properties given. Many have much higher critical temperatures or are flammable. Several reviews are available.[135]

That many chemical substances are not soluble in sc carbon dioxide permits selective extraction.[136] It is often used with foods, for which it eliminates the possibility of leaving toxic residues of solvents such as methylene chloride. It also avoids the hydrolysis that might occur when esters (for flavors or fragrances) are recovered by steam distillation. It has been used to extract the flavor from hops, the caffeine from coffee, fat and cholesterol from foods,[137] pecan oil,[138] lavender oil (for which hydrolysis of linalyl acetate could occur in steam distillation),[139] ginseng (from which it does not extract pesticide residues),[140] ginger,[141] microalgae,[142] cooked chicken,[143] ethanol from cider,[144] and many others. One method used with aromas and contaminants in water is to pass the liquid through polypropylene tubes immersed in liquid carbon dioxide.[145] The addition of a few percent of a solvent such as ethanol or hexane increases the number of things that can be extracted. The addition of ethanol (a relatively nontoxic solvent) to the sc carbon dioxide has been used in the extraction of oil from cottonseed,[146] oil from sunflower seed,[147] phospholipids from canola oil,[148] and antioxidants from the seed coat of *Tamarindus indica*.[149] By using sc carbon dioxide, the oil content of seeds can be extracted for measurement in minutes, compared with hours in a Soxhlet extractor.[150] (Oil seeds have also been extracted with aliphatic hydrocarbons, alkyl chlorides, alcohols, and water.[151]) Mono-, di-, and triglycerides can be fractionated to more than 90% purity with sc carbon dioxide.[152] Salicylic acid (*o*-hydroxybenzoic acid) of 99% purity can be obtained by extraction away from the much less soluble *p*-hydroxybenzoic acid.[153] The plastic contaminants in paper for recycling can be removed (55–79%) using sc carbon dioxide fortified with a little pentane or ethanol.[154] Motor oil can be recycled with a combination of sc carbon dioxide and a ceramic ultrafiltration membrane.[155] The oil passes through, leaving the metal-laden residue behind. Environmental samples for analysis (such as contaminated soils) can be extracted with sc carbon dioxide.[156] Supercritical fluids can be used to regenerate porous adsorbents at 50–60°C, which is an improvement over the 800°C used in a furnace or the 150°C used with steam.[157] Residual monomers in a copolyester of glycolic and lactic acids can be removed by extraction with sc carbon dioxide.[158] It can also be used with an anion-exchange resin to recover indole continuously.[159] This might be applied to coal tar. Sc carbon dioxide has also been used to extract organic compounds from ionic fluids, such as 1-butyl-3-methylimidazolium hexafluorophosphate, as a way of working up reactions in this medium.[160] In some cases, it is only necessary to add carbon dioxide to form a second phase that can be separated, that is, carbon dioxide expanded solvents.[161]

Heavy metal ions can be extracted with sc carbon dioxide containing chelating agents for them.[162] Uranium can be extracted from aqueous nitrates (such as might be obtained by dissolving spent fuel rods) by putting 1–3% tributyl phosphate in the carbon dioxide.[163] A hydrocarbon solvent, such as dodecane, is normally used in this process. Such extractions work better if the chelating agent contains many fluorine atoms to make it more soluble in the carbon dioxide.[164] A highly fluorinated dithiocarbamate allowed recovery of up to 98% of mercury, lead, chromium, and iron.[165] Copper (II) has been extracted from aqueous solutions with fluorinated β-diketones in carbon dioxide.[166]

The ease of penetration of materials by supercritical fluids offers a way of inserting something into them.[167] Sc carbon dioxide has been used to insert biocides into wood,[168] polyethylene glycol into leather,[169] and monomers and initiators, for subsequent graft polymerization, into polymers, such as polyethylene and bisphenol A polycarbonate.[170] Polymers swollen in this way are plasticized by the carbon dioxide. Polystyrene film has been made by drawing such a material.[171] This is also a way of foaming the polymers by releasing the pressure in the same way that some breakfast cereals are prepared.[172] Extrusion processing of foods from sc carbon dioxide has been considered for continuous making of bread, low-fat snacks, and similar food.[173] Highly oriented microfibrils of poly(acrylonitrile) have been prepared by spraying a solution of the polymer in dimethylformamide into sc carbon dioxide.[174] Drugs have been microencapsulated by the rapid expansion of sc carbon dioxide containing the drug and poly(lactic acid).[175] Microparticles of poly(methyl methacrylate) and polystyrene have been prepared by spraying solutions of the polymers in organic solvents into liquid carbon dioxide.[176] A coaxial nozzle can be used for the preparation of microparticles of polymers.[177] Microparticles of rare earth metal acetates have been prepared by passing a solution in dimethyl sulfoxide into sc carbon dioxide.[178] Nanoparticles of metal oxides and nitrides have been prepared in similar fashion.[179] Solutions of steroid drugs, such as progesterone in sc carbon dioxide, can be released through a micronozzle to produce 4.5-μm particles. These are finer than the 7.5-μm particles from jet milling. This may be cheaper, faster, and kinder to heat-sensitive compounds.[180] It is possible that the preparation of regenerated cellulose fibers from solutions in *N*-methylmorpholine-*N*-oxide might be improved by spinning into carbon dioxide, instead of into water, as is currently done. This would make it easier to recycle the solvent. If desired, foamed fibers for extra bulk could be prepared in this way.

Most polymers are insoluble in sc carbon dioxide, with the exception of highly fluorinated ones.[181] Even so, some perfluoropolymers, such as poly(tetrafluoroethylene), still require temperatures of 200–300°C to bring them into solution.[182] DuPont is now polymerizing tetrafluoroethylene in sc carbon dioxide.[183] This eliminated the toxic perfluorooctanesulfonate used in the earlier process. Silicone polymers are also somewhat soluble. Poly(dimethylsiloxane) has been fractionated into different molecular weights using sc carbon dioxide at 70°C.[184]

Many reactions in sc carbon dioxide require a perfluorosurfactant. Silicones are not as good. Poly(ether carbonates) from the polymerization of a cyclic carbonate, such as propylene carbonate, and acetate-functional polyethers are much cheaper and work well.[185] A variety of cationic and free radical-catalyzed polymerizations have been carried out in this solvent using special surfactants.[186] An example is a block copolymer of styrene and 1,1-dihydroperfluorooctyl acrylate, which can emulsify up to 20% of a hydrocarbon as core–shell-type micelles.[187] (The polymerizations have been done primarily by de Simone and coworkers.[188]) Some have involved the use of perfluoroazo- and perfluoroperoxide initiators. These include polymerization of methyl methacrylate,[189] acrylic acid,[190] butyl acrylate,[191] styrene,[192] divinylbenzene,[193] and vinyl acetate, with or without ethylene.[194] Methyl methacrylate has also been polymerized in a biphasic system of carbon dioxide and water to produce a stable latex without a surfactant.[195] Isobutylene has been polymerized by a 2-chloro-2,4,4-trimethylpentane–titanium tetrachloride initiator.[196] Vinyl ethers and oxetanes have been polymerized with initiators such as ethylaluminum dichloride and boron trifluoride etherate, without incorporation of carbon dioxide into the polymers.[197] The polymerization of propylene oxide with a zinc glutarate catalyst gave a copolymer of carbon dioxide.[198] Cyclohexene oxide behaves similarly.[199] Under other conditions propylene carbonate can be obtained in almost quantitative yield.[200] (It is a candidate for addition to diesel fuel to reduce particulate emissions.) The oxidative polymerization of 2,6-dimethylphenol to a polyether has also been performed in sc carbon dioxide.[201] The ring-opening metathesis polymerization of norbornene with ruthenium tosylate has been carried out in carbon dioxide.[202] The *cis*/*trans*-ratio could be controlled by traces of methanol. Bisphenol A polycarbonate has been made in sc carbon dioxide.[203] Many anionic polymerizations may be difficult to achieve because the anions react with carbon dioxide.

Sc carbon dioxide is suitable as a medium for reactions of small molecules as well, including hydrogenation, hydroformylation, Diels–Alder reactions, amination, alkylation, isomerization, oxidation, cyclizations, and others, as well as inorganic and organometallic reactions.[204] Free radical reactions, such as the oxidation of cumene with oxygen to give cumene hydroperoxide and the bromination of toluene to produce benzyl bromide, proceed normally.[205] This provides an alternative to the carbon tetrachloride often employed in halogenations. Tetraheptylammonium bromide was a satisfactory phase-transfer catalyst for the reaction of benzyl chloride with solid potassium bromide in sc carbon dioxide containing 5 mol% acetone.[206] The corresponding reaction with potassium cyanide has been carried without the acetone.[207] The reaction of soybean oil with glycerol to produce monoglycerides proceeded smoothly in carbon dioxide.[208] This eliminated the alkali catalyst and gave a better conversion to a lighter-colored product, with less odor. It might be possible to combine this reaction with the initial extraction of the oil from the beans.

Jessop and coworkers pointed out that homogeneous catalysis in supercritical fluids can offer high rates, improved selectivity, and elimination of mass-transfer problems.[209] They used a ruthenium phosphine catalyst to reduce sc carbon dioxide to formic acid using hydrogen.[210] The reaction might be used to recycle waste carbon dioxide from combustion. It also avoids the use of poisonous carbon monoxide to make formic acid and its derivatives. There is no need for the usual solvent for such a reaction, because the excess carbon dioxide is the solvent. If the reaction is run in the presence of dimethylamine, dimethylformamide is obtained with 100% selectivity at 92–94% conversion.[211] In this example, the ruthenium phosphine catalyst was supported on silica. Asymmetric catalytic hydrogenation of dehydroamino acid derivatives (**8.28** Schematic) can be performed in carbon dioxide using ruthenium or rhodium chiral phosphine catalysts.[212]

The solubility of the catalyst salt is improved by the use of 3,5-bis(trifluoromethyl)phenylborate or triflate anions. Unsaturated fatty acids can be reduced in the same way.[213] Hydrogenations in sc carbon dioxide can be more selective

R COOCH$_3$ NHCOCH$_3$ + H$_2$ —Ru catalyst→ R COOCH$_3$ NHCOCH$_3$ 91–99.7% ee

8.28 Schematic

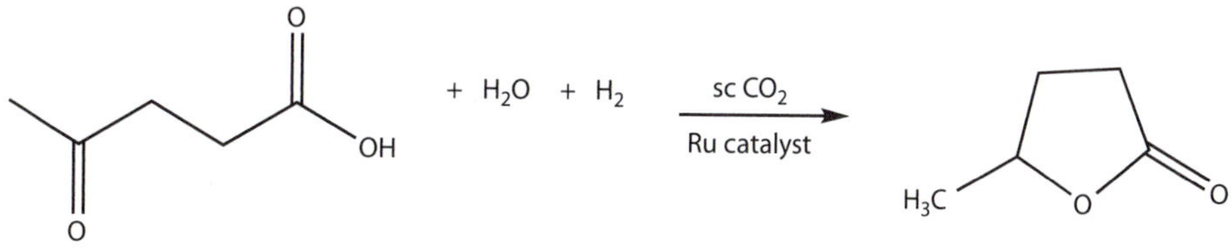

+ H_2O + H_2 $\xrightarrow[\text{Ru catalyst}]{\text{sc } CO_2}$

8.29 Schematic

than in the gas phase while using 35 times less catalyst.[214] Cyclohexene was reduced with hydrogen and a polysiloxane-supported palladium catalyst in a continuous-flow reactor in 95–98% yield. Various olefins and acetylenes were reduced in a microchannel reactor in >90% yield in <1 s.[215] A similar continuous process was used to prepare γ-valerolactone (Schematic **8.29**).[216]

Epoxides, oximes, nitriles, aldehydes, ketones, and nitro compounds can also be reduced.[217] By varying the temperature, the products from the reduction of nitrobenzene can be selected from aniline, cyclohexylamine, dicyclohexylamine, and cyclohexane. In the same way, acetophenone can be converted to 1-phenylethanol, 1-cyclohexylethanol, or ethylcyclohexane.

Hydroformylation of propylene with carbon monoxide and hydrogen can also be carried out in sc carbon dioxide using a dicobaltoctacarbonyl catalyst.[218] However, the method appears to offer no advantage over carrying out the reaction in water, as described in Chapter 1. It might offer advantages with higher olefins that are not soluble enough in water to use it as a solvent. Olefins have been converted to their epoxides with *tert*-butylhydroperoxide and vanadium, titanium, or molybdenum catalysts with more than 99% selectivity at more than 99% conversion.[219] Oxidation of a substituted cysteine with *tert*-butylhydroperoxide (**8.30** Schematic) in sc carbon dioxide showed a pronounced pressure-dependent diastereoselectivity, with up to 95% diastereoisomeric excess, whereas conventional solvents gave 0% excess.[220]

Mesitylene underwent monoalkylation with isopropyl alcohol (or prophylene) with 100% selectivity at 50% conversion using a polysiloxane-supported solid acid catalyst.[221]

A combination of sc carbon dioxide and light was used in flow reactors to make some metal complexes that could not be prepared by other methods, for example, cyclopentadienyl $Mn(CO)_2(H_2)$ and $Cr(CO)_5$(ethylene).[222] A manganese hydride was added to olefins in carbon dioxide.[223] A zeolite in sc carbon dioxide selectively adsorbs 2,7-dimethylnaphthalene so that it can be separated from the desired 2,6-dimethylnaphthalene.[224] The latter is oxidized to the dicarboxylic acid for conversion to a higher-melting analog of poly(ethylene terephthalate). A pillared clay made in sc carbon dioxide had a higher catalytic-cracking activity and selectivity than the usual pillared smectite.[225] Sc carbon dioxide is a better medium than organic solvents for inserting large amounts of guest into a pillared clay without contamination with organic solvent, as shown by intercalation of 4-aminoazobenzene into a montmorillonite pillared with tetramethylammonium ion.[226] Sc carbon dioxide is also used to dry aerogels, such as that from zirconium dioxide.[227] This prevents loss of surface area by partial collapse during drying.

Enzymes can also be used in sc carbon dioxide.[228] A *Pseudomonas* lipase immobilized on silica gel gave better conversions and enantioselectivity in the acetylation of racemic alcohols with acetic anhydride than when used in organic solvents.[229] The lipase-catalyzed esterification of glycidol gave 83% enantioselectivity, which is as favorable as when the reaction is run in organic solvents.[230] An immobilized lipase was used in the ethanolysis of cod-liver oil.[231] Another immobilized lipase was used to convert oleic acid to various esters.[232] The use of a lipase in sc carbon dioxide for analyses of fats in foods cuts solvent use by 98%.[233] Polyesters were made enzymatically in carbon dioxide.[234]

One of the limitations on the use of sc carbon dioxide as a reaction medium has been the difficulty of getting substances dissolved or finely dispersed in it. Johnston and coworkers overcame this difficulty for bovine serum albumin by making a microemulsion of water in carbon dioxide using the fluorinated surfactant (**8.31** Schematic).[235]

The biological activity of the protein is not lost in the process. The authors point out that "Carbon dioxide and

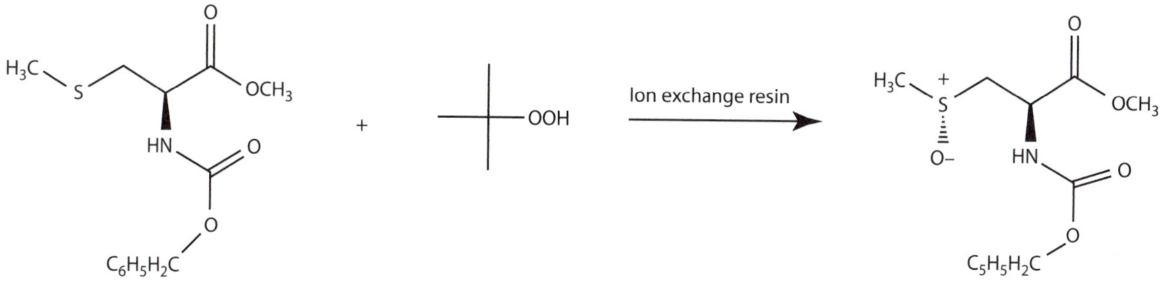

Ion exchange resin

8.30 Schematic

$$CF_3O[CF_2CF(CF_3)O]_3CF_2COO^-\ NH_4^+$$

8.31 Schematic

water are the least expensive and most abundant solvents on earth." This method has also been used to prepare CO_2/H_2O microemulsions of inorganic compounds, such as potassium permanganate and potassium dichromate.[236] Such micellar systems have been used in hydroformylations, hydrogenations, and carbon–carbon bond-forming reactions.[237] Similar CO_2/H_2O micelles have been formed using a hydrophilic dendrimer with a coating of perfluoroalkyl groups.[238] These were used to extract a dye from water into carbon dioxide. (Anionic dyes, i.e., xanthene carboxylic acids such as fluorescein, have also been extracted from water at pH 1–6 with an amine dendrimer with hexadecanoylamide groups on the surface [in a toluene solution] without the use of carbon dioxide.[239]) It may be possible to extend these approaches to other liquids that are immiscible with carbon dioxide.

Sc carbon dioxide is also useful in a variety of applications as a replacement for organic solvents. Union Carbide (now Dow Chemical) markets a process for spray-painting in which the level of emissions of VOCs is reduced by 80%.[240] Less paint has to be used because the microatomization caused by the evaporation of carbon dioxide eliminates pinholes so that thinner coatings can be used. It can be used as a replacement for tetrachloroethylene in the dry cleaning of clothes.[241] Micell Technologies (now Fred Butler, COOL Clean Technologies, Inc. Global Technologies, Wallingford, CT) sells and operates the equipment for the process. This process involves high pressures, which may be a disadvantage in neighborhood cleaning shops. If so, the local cleaner could send the clothes to a central cleaning facility. Water pollution can be avoided by dyeing cotton, wool, and poly(ethylene terephthalate) fabrics in sc carbon dioxide.[242] The cotton and wool require pretreatment with a polyether that dissolves the dye and swells the fibers. Energy that would be required to dry fabrics dyed in water is saved. Any dye that does not stick to the fabric can be reused in the next run. Textiles of poly(ethylene terephthalate)–cotton can be sized and desized in liquid carbon dioxide.[243] This is also a good way to put on fluoropolymers for abrasion resistance. Block copolymers of tetrahydropyranyl, heptafluorobutyl, and pentadecylfluorooctyl methacrylates have been used as photoresists in microlithography.[244] Photolysis in the presence of triarylsulfonium or diphenyliodinium salts produces acid that hydrolyzes the tetrapyranyl groups, making the image insoluble in sc carbon dioxide. Pulsed sc carbon dioxide containing less than 5% propylene carbonate has been used to remove conventional photoresists from silicon wafers.[245] This replaces the aqueous sulfuric acid or organic solvents used in conventional processes. (Ozone in cold water has also been used to do this without the carbon dioxide.[246] This eliminates the acid or organic solvents, but the ozone is much more toxic.) Sc carbon dioxide is being used to make improved pigmented power coatings and artificial bone, a combination of calcium hydroxyapatite and a biodegradable polymer.[247] Body tissues were sterilized with sc carbon dioxide.[248]

Hydroquinone can be monoalkylated cleanly with supercritical methanol at 350°C/120 bars for 1 h (**8.32** Schematic). Aniline can be monomethylated with 98% selectivity under the same conditions (**8.33** Schematic).[249]

This is an alternative for some of the methylations with dimethyl carbonate that were discussed in Chapter 2.

8.5 WATER AS A REACTION MEDIUM

Supercritical water, with or without oxygen or catalysts, can be used to decompose organic waste materials.[250] Nerve gases, such as VX, can be disposed of in this way.[251] Corrosion can be a problem, especially if a chlorine-containing compound is decomposed. This may require that the reactor be made of titanium or tantalum. The corrosion can be reduced by the use of sodium carbonate.[252] Use of this system at 380°C reduced that content of polychlorinated biphenyls in a sample from 20 mg/L to less than 0.5 µg/L. Trichloroethylene, 1,1,1-trichloroethane, and trichloroacetic

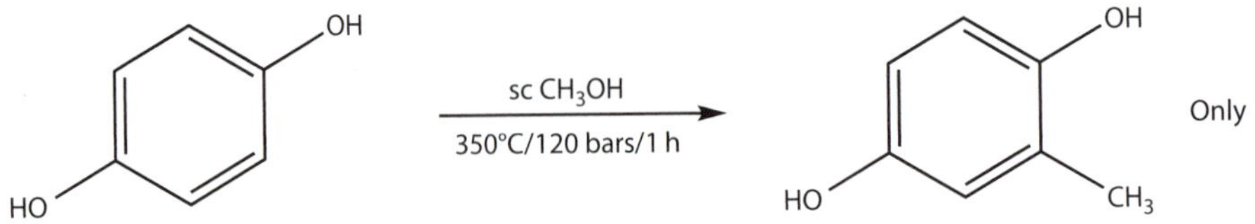

sc CH_3OH, 350°C/120 bars/1 h — Only

8.32 Schematic

sc CH_3OH, 350°C/120 bars/1 h — 98% selectivity

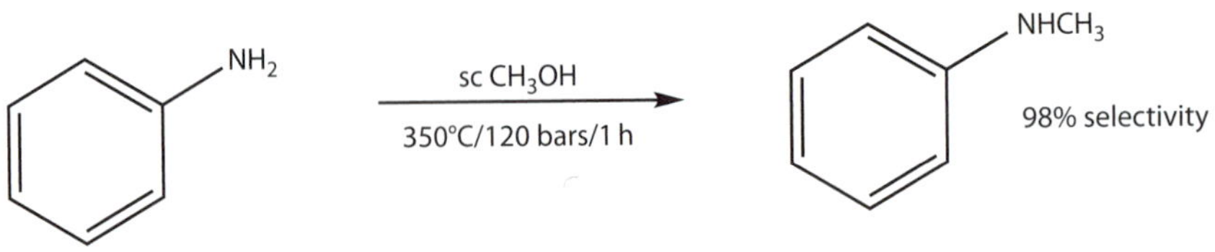

8.33 Schematic

acid were destroyed with 99.96% efficiency at 450°C for 60 s using 1.5% hydrogen peroxide plus sodium bicarbonate.[253] Sodium nitrate or sodium nitrite could be used in place of the hydrogen peroxide. Nitrates, ammonium hydroxide, and amines are all converted to nitrogen at 350–360°C. Emulsions of petroleum, water, and solids can be broken by heating to 350°C. Supercritical water has been used to recover 2,4-diaminotoluene from distillation residues from the manufacture of 2,4-toluenediisocyanate.[254]

The critical temperature of water is 374°C. This is too hot for most organic compounds. However, exploratory work has been done with supercritical and near-critical water.[255] At 300°C the polarity and density of water approach those of acetone at room temperature. Cyclohexanol (**8.34** Schematic) can be dehydrated to cyclohexene in 85% yield at 278°C in 18 h. Pinacol (**8.35** Schematic) can be rearranged quantitatively at 275°C in 60 min. Quantitative ring opening of 2,5-dimethylfuran (**8.36** Schematic) occurs at 250°C for 30 min. Acetals and esters (**8.37** Schematic) can be hydrolyzed under such conditions.

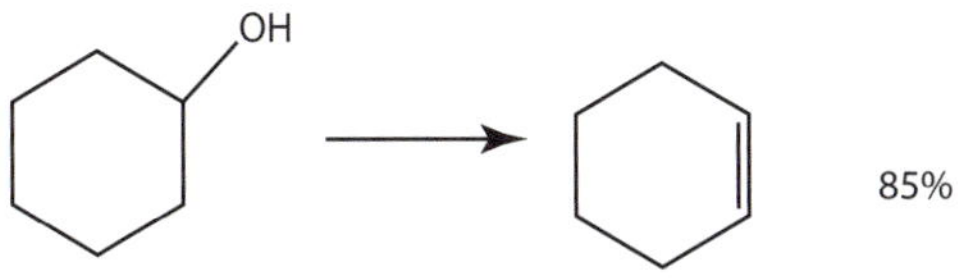

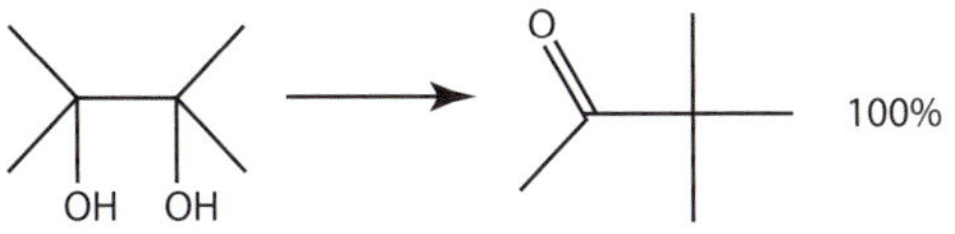

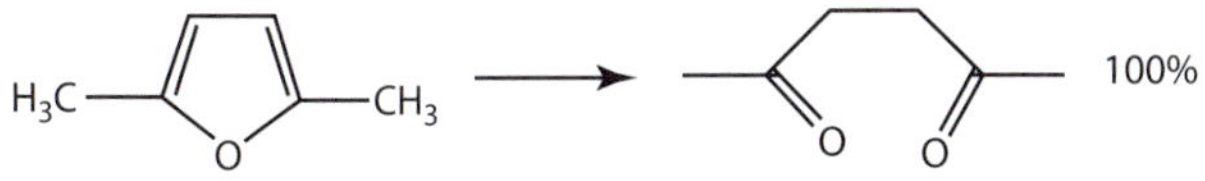

OH

85%

8.34 Schematic

O

OH OH

100%

8.35 Schematic

H_3C O CH_3

O O

100%

8.36 Schematic

Diaryl ethers are cleaved rapidly. Poly(ethylene terephthalate) is cleaved at 300°C. Vegetable oils can be hydrolyzed in more than 97% yield with water in 15–20 min at 260–280°C.[256] The nitrile, amide, and ester groups in various copolymers of butadiene and ethylene can be hydrolyzed to carboxylic acid groups with water at 250–300°C.[257]

p-Xylene can be oxidized to terephthalic acid in 90% yield plus 5–8% benzoic acid in less than 20 s using hydrogen peroxide with a manganese bromide catalyst at 400°C (Schematic **8.38**).[258]

A similar continuous-flow system was used to convert 5-cyanopentylamine to caprolactam at 400°C in 96 s in 90% yield at 94% conversion (Schematic **8.39**).[259]

1-Hexanol was converted to an amide using ammonium acetate at 400°C in 10 min (Schematic **8.40**).[260]

Acetylation of benzyl alcohol with an equivalent of acetic anhydride in a microreactor gave benzyl acetate in 99% yield and 99% conversion in 15 s at 200°C.[261]

Metal nitrates, such as those of aluminum, copper, iron, nickel, and zirconium, have been converted to metal oxide nanoparticles in a flow-through system in water at 673 K.[262] Nickel formate was converted to nickel particles in water at 673 K, the hydrogen for the reduction coming from the formate.[263]

Phenols can be alkylated with secondary and tertiary alcohols in water at 250–350°C, but the reaction is slow.[264] These reactions have no catalysts and no solvent, so that there is no need to recover either. Many products separate on cooling. Alkynes can be trimerized to the corresponding aromatic compounds with cyclopentadienylcobaltdicarbonyl in more than 95% yield.[265]

Not all yields are as good as those just quoted. The yields of the desired olefin in the Heck reaction (**8.41** Schematic) are only about 25%, several other products being formed at the same time.[266]

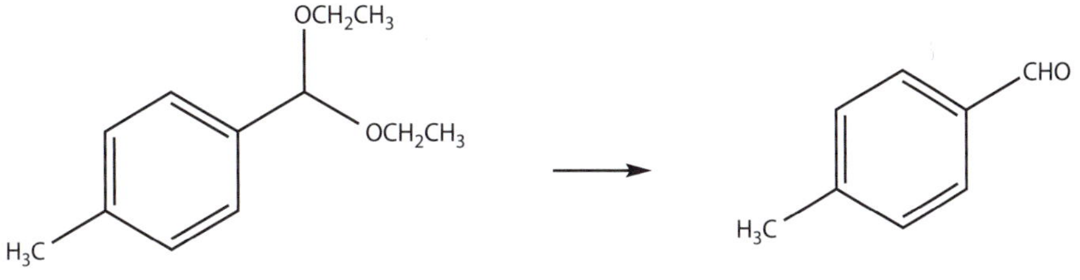

OCH_2CH_3

OCH_2CH_3

H_3C

CHO

H_3C

8.37 Schematic

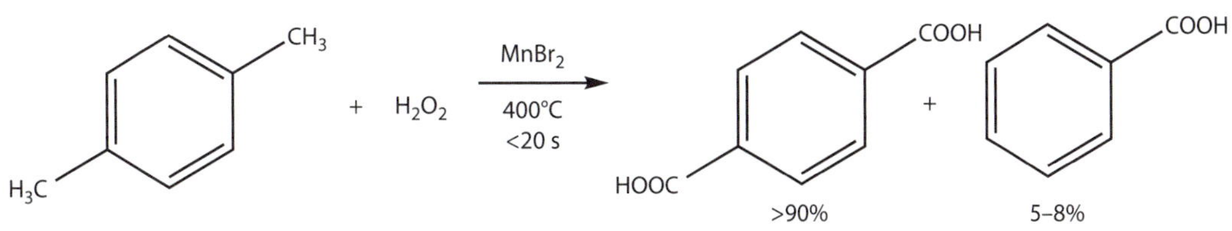

CH_3

H_3C

+ H_2O_2

$MnBr_2$

400°C

<20 s

COOH

HOOC

>90%

+

COOH

5–8%

8.38 Schematic

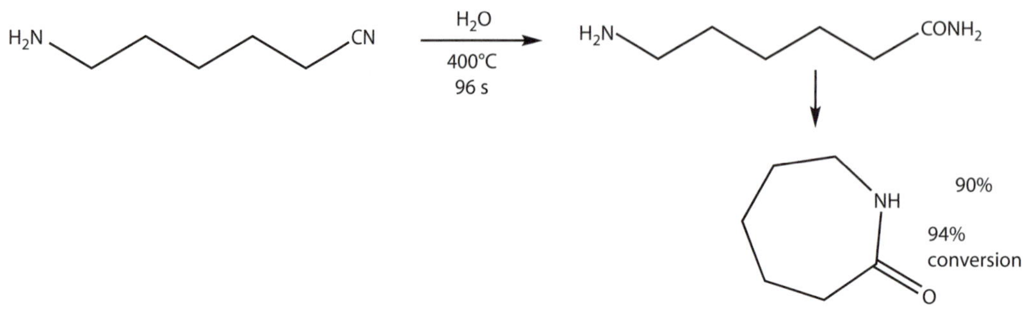

8.39 Schematic

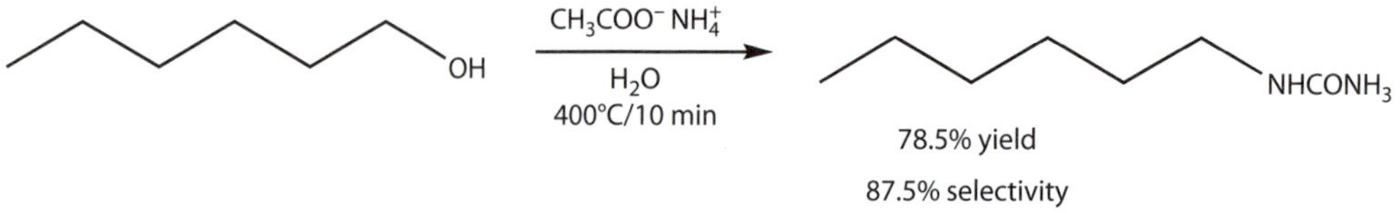

8.40 Schematic

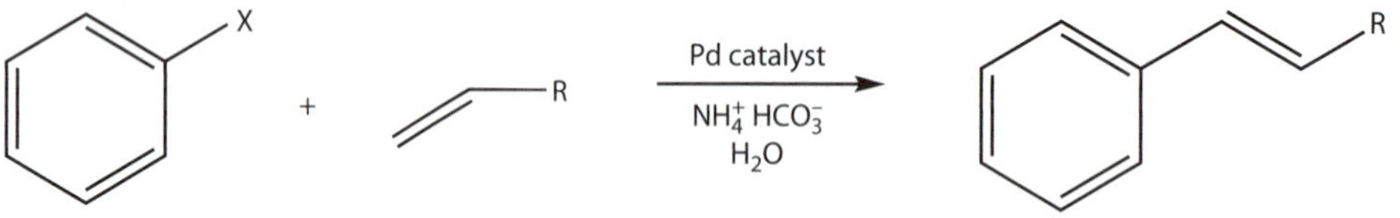

8.41 Schematic

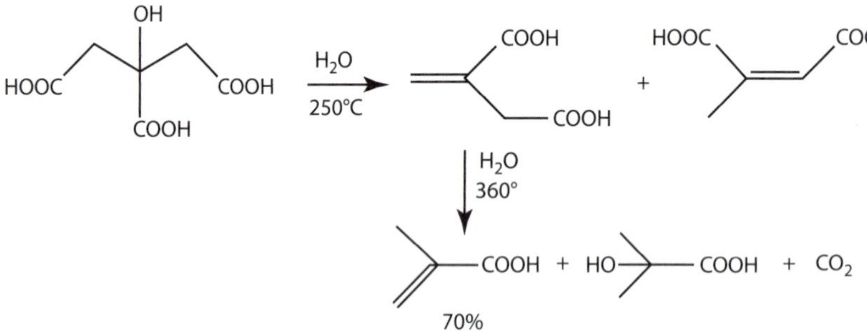

8.42 Schematic

Methacrylic acid, which is normally derived from petroleum, can be made from citric acid (**8.42** Schematic), which in turn is made by fermentation of glucose.[267] The combined yield of itaconic and citraconic acids (in about 3:1 ratio) in the first step was over 90%. These two interconvert rapidly. Decarboxylation of the itaconic acid gave methacrylic acid in 70% yield. Inclusion of 3-hydroxy-2-methylpropionic acid as product raised the yield to over 80%. D-Glucose can be converted to D-erythrose (**8.43** Schematic) in 50% yield using water at 400°C for 0.11 s.[268]

Water at 250–300°C can remove organic compounds from soil, catalysts, and some sludges just as effectively as organic solvents do.[269]

Water promotes many reactions at temperatures no higher than 100°C.[270] These include the Diels–Alder, Heck, Michael, and Mukaiyama reactions, dipolar cycloadditions, Claisen rearrangement, pinacol couplings, conjugated additions, hydroformylation, and many others. The rate enhancement may be due to a "hydrophobic effect" (i.e., the two organic reactants being closer together because they are not very soluble in water). As an example, the addition of phenylazide to norbornene (**8.44** Schematic) goes at the

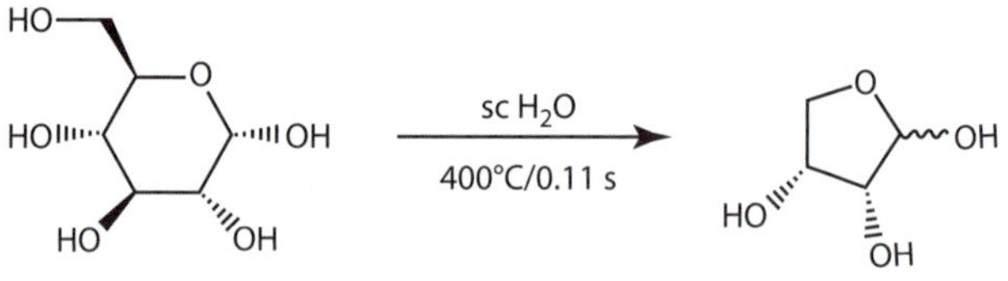

8.43 Schematic

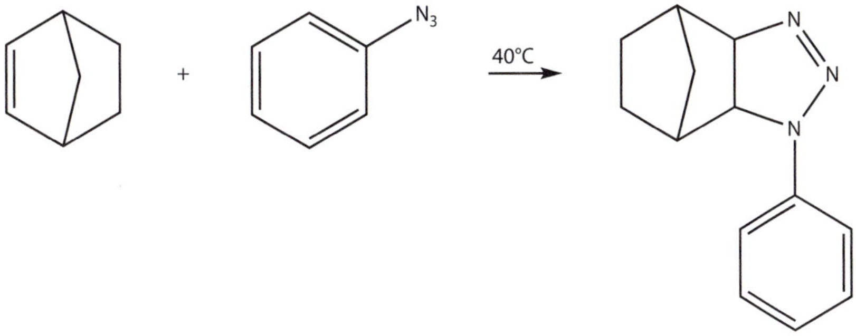

8.44 Schematic

relative rate of 5 in hexane, 7 in methanol, 72 in 94:6 water/*tert*-butyl alcohol, and 250 in 99:1 water/*N*-cyclohexylpyrrolidone.[271]

In one case, a different product is obtained in a condensation (**8.45** Schematic).[272] No dehydration of the product occurs.

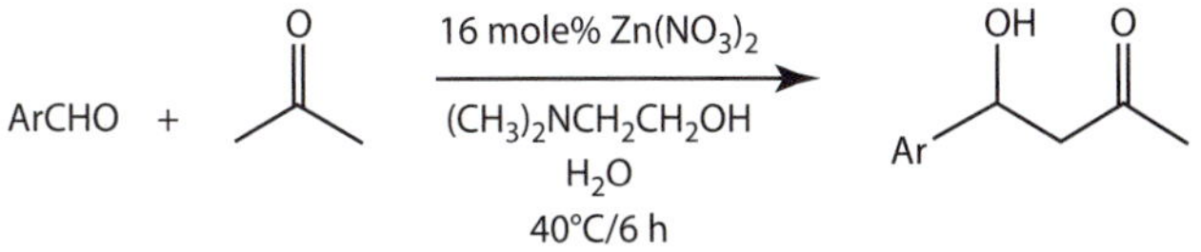

8.45 Schematic

Organic compounds that are insoluble in water can be handled sometimes in micelles (microscopically fine droplets) with or without the aid of ultrasound or surfactants.[273]

Some organometallic reactions can be carried out in water.[274] The Heck reaction, mentioned earlier, involves an organopalladium intermediate. The ring-opening metathesis polymerization of norbornenes containing polar groups, such as ethers, can be done in water using ruthenium salt catalysts. In water, the ruthenium must coordinate better with the olefin than the ether or ester. Aqueous Barbier–Grignard-type reactions can be carried out with indium, tin, and zinc metals in water.[275] Tin (II) chloride has also been used.[276] A typical reaction of this type is shown in **8.46** Schematic.[277]

When zinc is used in such reactions, ultrasound is often used to keep the surface clean and thus promote the reaction. Indium is nontoxic and resistant to oxidation by air. It can be recovered electrolytically and recycled. Such processes eliminate the need to protect and deprotect functional groups. A nitrogen atmosphere is unnecessary.[278] (The use of indium (III) chloride[279] and other Lewis acids as catalysts in water is discussed in Chapter 6.)

Many of the reactions are catalyzed by transition metal catalysts. A variety of water-soluble ligands have been devised to make the catalysts soluble in water. The most common is a sulfonated phosphine made by the sulfonation of triphenylphosphine (**8.47** Schematic) with fuming sulfuric acid.[280]

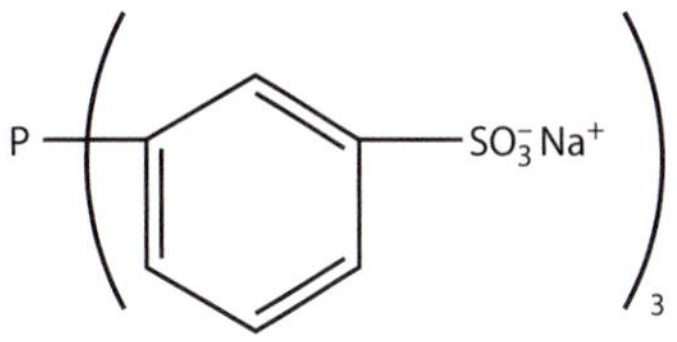

8.47 Schematic

A variety of other water-soluble ligands (**8.48** Schematic through **8.51** Schematic) have also been used.

The reaction of phosphine-containing alcohols with 2-sulfobenzoic acid cyclic anhydride (see **8.48** Schematic) has been used as a milder method of producing a water-soluble phosphine than the use of fuming sulfuric acid.[281] The second phosphine (see **8.49** Schematic) was made by the addition of lithium diphenylphosphide to the commercially available acrylamidesulfonate and then converted to the lithium salt.[282] (Many other phosphine sulfonates have also been made for this use.[283]) The guanidinium salts (**8.50** Schematic) used in aqueous Heck reactions are more stable to oxidation by air than the sulfonated phosphines.[284] Poly(acrylic acid) salts can be modified (see **8.51** Schematic) to incorporate phosphine ligands.[285] The one shown was used with rhodium in the hydrogenation of 2-acetamidocinnamic acid to produce the saturated acid in higher than 97% yield and 56% ee. (The literature reports up to 93% ee under other conditions; other phosphines containing –COONa groups have also been used.[286]) Derivatives from polyethylene

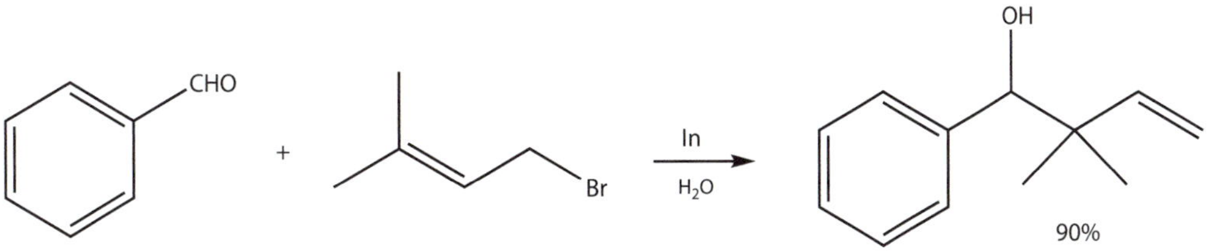

8.46 Schematic

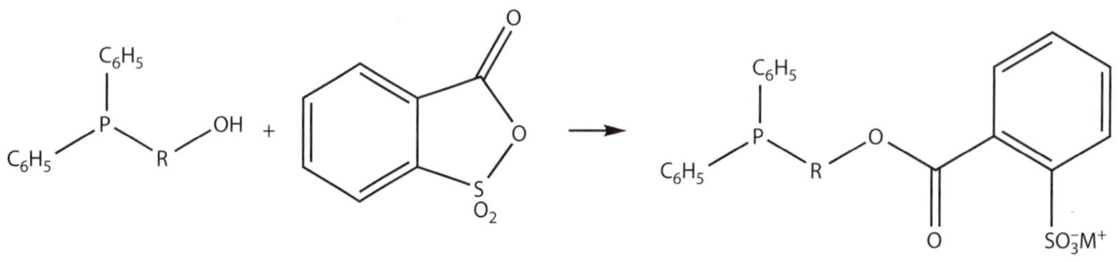

8.48 Schematic

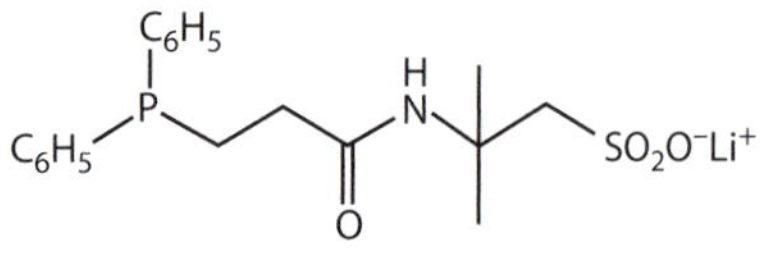

8.49 Schematic

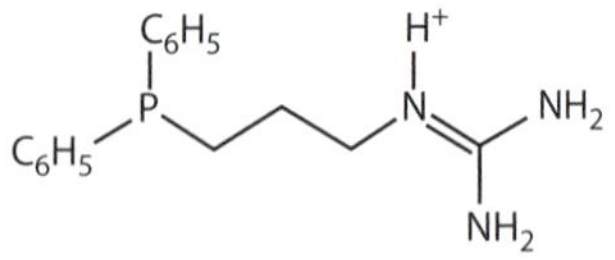

8.50 Schematic

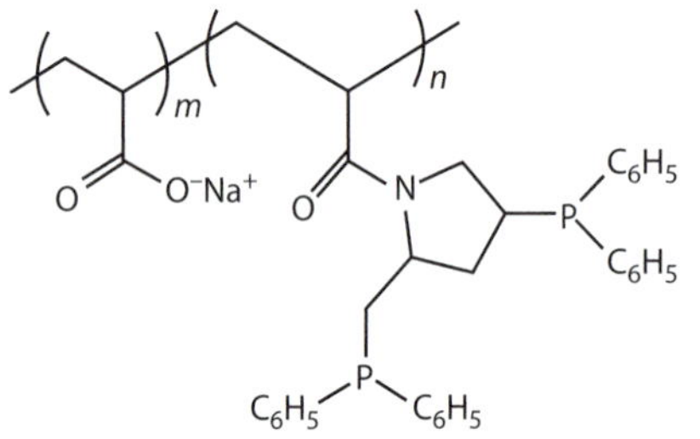

8.51 Schematic

oxide and polypropylene oxide have also been used as water-soluble ligands.[287] The ones shown in (**8.52** Schematic) were used with rhodium in the hydroformylation of 1-hexene, 1-octene, and 1-dodecene:

$$(C_6H_5)_{3-m}P[C_6H_4\text{-p-}(OCH_2CH_2)_nOH]_m \qquad m = 1, 2, \text{ or } 3$$

8.52 Schematic

The yield of aldehydes at 95–97% conversion was 90–91%. The catalyst goes into the organic phase at higher temperatures and returns to the aqueous phase on cooling. On the fifth cycle with 1-hexene, the conversion dropped to 92% and the yield of aldehyde to 87%. A phosphine containing a trimethylammonium group was used with ruthenium in a living ring-opening metathesis polymerization.[288]

These water-soluble catalysts are especially useful when the product is a separate organic phase that can be separated continuously.[289] Examples include the hydroformylation of propylene and the telomerization of dienes. The catalyst in the aqueous phase can be used over and over again without loss. This includes reactions during which the aqueous solution of the catalyst is supported on silica or other supports.[290] A rhodium catalyst in water on silica was 20 times more active in the hydroformylation of methyl acrylate than the homogeneous counterpart.[291] The hydroformylation of methyl 10-undecenoate gave a 94% yield of the corresponding 11-formyl compound.[292] For larger molecules, it was necessary to include a phase-transfer catalyst such as dodecyltrimethylammonium bromide. This lowered the normal/iso ratio below the 97:3 in the foregoing example. Similar quaternary ammonium phase-transfer catalysts[293] have been used in the condensation of aldehydes with ketones[294] and with aqueous Heck reactions.[295]

Reactions can also be carried out in microemulsions. (Microemulsions have finer droplets of the oil phase than the usual emulsions.) They have been used in the hydroxycarbonylation of iodobenzene (**8.53** Schematic).[296] When the surfactant was sodium dodecyl sulfate, some butanol had to be present to achieve the microemulsion. A typical run used 2 mmol aryl iodide, 0.5 g surfactant, 1 mL butanol, 10 mL water, and 2 mmol base. The butanol was unnecessary with nonionic surfactants, such as the one derived from 1-dodecanol with 23 eq of ethylene oxide. The advantages of such a system include the following: (a) No organic solvent is needed. The substrate is the oil phase. (b) The microemulsions form without the need for vigorous agitation. (c) No excess base is needed, in contrast with some reactions in which phase-transfer catalysis is used. (d) The surfactants

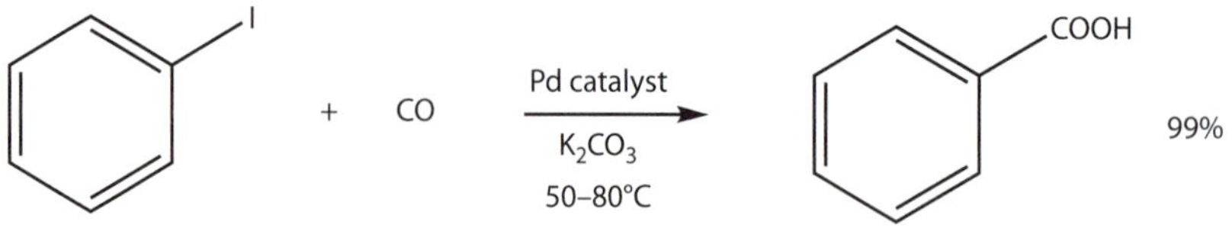

8.53 Schematic

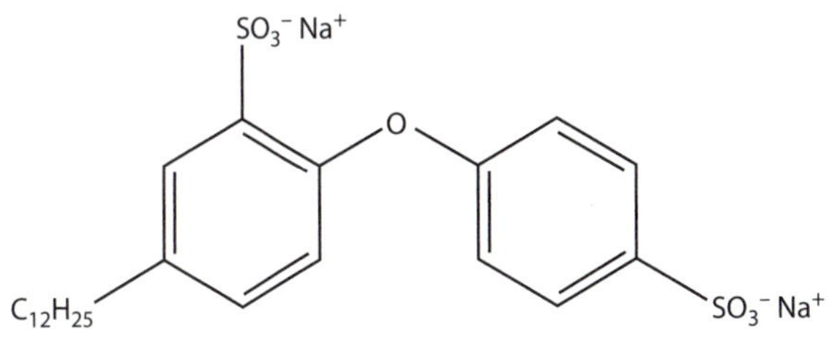

8.54 Schematic

can be recovered and recycled. They are inexpensive and biodegradable.

In one case, fine droplets were reacted in a column without the need for a surfactant. Cyclooctene was introduced through a sparger at the bottom of a column containing an aqueous solution of the oxidant, potassium peroxymonosulfate. The oxidation took place as the droplets rose to the surface. The reaction was nearly 99% selective.[297] The organic phase could be pumped around the loop repeatedly to increase conversion. The epoxide product was collected by filtration as the organic layer was pumped around the loop. Conversion was over 60% in 3 h.

Hydrotropic agents such as sodium xylenesulfonate increase the solubility of organic compounds in water. Dow sells a hydrotropic disulfonate diphenyl ether as Dowfax 2Al (**8.54** Schematic).

Aqueous sodium xylenesulfonate has been used to extract fragrances, such as 2-phenylethanol, as a mild replacement for the usual steam distillation.[298] The alcohol was recovered by dilution with additional water. The hydrotrope could be reused after concentration of the aqueous phase. A mixture of *ortho*- and *para*-chlorobenzoic acids was separated by dissolving the *ortho*-isomer in water containing sodium 2-butoxyethyl sulfate.[299] The *o*-chlorobenzoic acid was recovered by diluting the extract with more water to precipitate it. Base-catalyzed condensations (**8.55** Schematic) were carried out with the aid of sodium 2-butoxyethylsulfate or sodium *p*-toluenesulfonate as hydrotropes.[300]

The reagents stayed in the aqueous phase. The product precipitated as formed. The hydrotropic solution showed no problems after being recycled 4 times. This method has also been used in the Hantzsch condensation (**8.56** Schematic) where, again, the product was recovered by filtration.

Cyclodextrins can also be used to make some organic compounds that are insoluble in water go into aqueous solution. These were discussed in Chapter 5.

Water worked, whereas organic solvents did not, in the cyclization of polyepoxides to ladder polyether toxins of the type produced by "red tides."[301]

A relatively low solubility in water may be enough for the desired reaction to take place. Chapter 4 included an example where a suspension of powdered 2-naphthol in water was oxidized by air in the presence of an iron (III) chloride catalyst to produce the corresponding binaphthol in 95% yield.[302] The by-product quinones usually found when the reaction is carried out in solution with stoichiometric amounts of metal-containing oxidants were absent.

Polyethylene glycol has been used as an analog of water in some cases. Ethers have been made (**8.57** Schematic) in 68–91% yield using powdered potassium hydroxide (containing 12% water) in polyethylene glycol.[303]

Although organic solvents were used in the workup, it might be possible to use filtration plus ultrafiltration, followed by distillation instead. Hydroformylation of 1-hexene was achieved in polyethylene glycol by using a cobalt catalyst containing polyethyleneoxy substitutents.[304] The catalyst phase was separated for reuse.

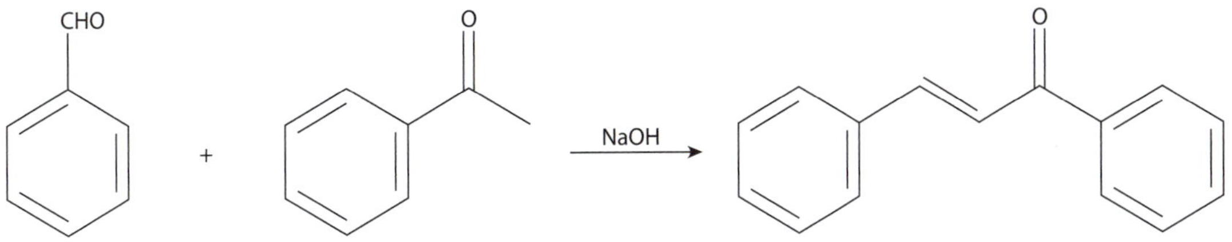

8.55 Schematic

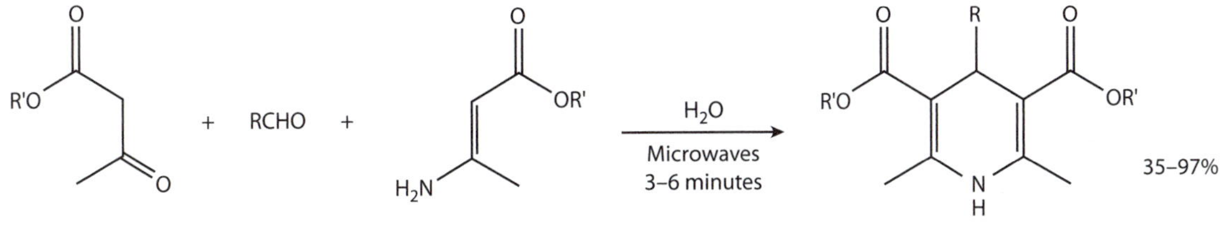

8.56 Schematic

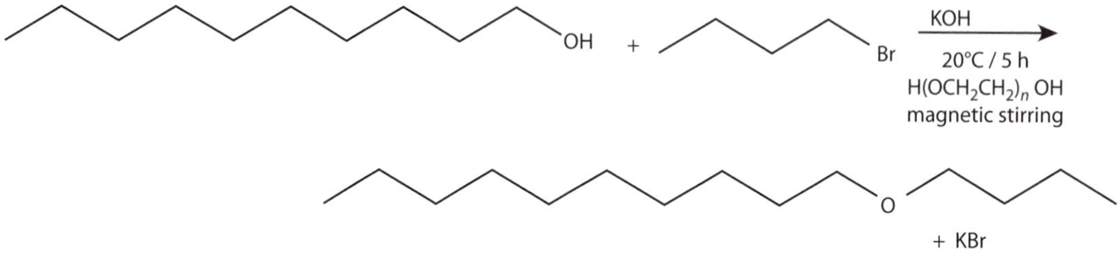

OH + Br KOH 20°C / 5 h $H(OCH_2CH_2)_nOH$ magnetic stirring

O + KBr

8.57 Schematic

Polymerization in aqueous emulsions is common industrial practice. The emulsion has a much lower viscosity than a comparable solution in a solvent. It is possible to obtain higher molecular weights in the emulsion because there is no solvent present to act as a chain-transfer agent. Frequently, the product emulsions can be used as such (e.g., as adhesives or coatings).

8.6 IONIC LIQUIDS

Ionic liquids are salts that melt below 100°C.[305] They have very low vapor pressures, can dissolve many organic and inorganic compounds, are stable up to 300°C, and can have low flammability. Their density, viscosity, solubility, miscibility, stability to water, thermal stability, and other properties can be varied by the choice of cation and anion. Some are hydrophobic and others are hydrophilic. Mixtures of these can lead to a biphasic system.[306] Such a system has been used to produce a tetraphasic system with a butylmethylimidazolium chloride below water below tetradecyltrihexylphosphonium chloride below pentane. Some typical structures are shown below, where *R* is alkyl and *X* can be halide, sulfonate, halometalate, carboxylate, bis(sulfonylimide), tetrafluoroborate, hexafluorophosphate, and the like (Schematic **8.58**). Some contain functional groups[307] and some are zwitterions.[308]

Research in the field is concerned with what can be made, their characterization, and possible applications.

Many syntheses have been carried out using ionic liquids.[309] These include aldol reactions,[310] cyclotrimerizations of alkynes,[311] conversion of epoxides to carbonates,[312] Grignard reactions,[313] esterification (Schematic **8.59**),[314] hydroformylation (Schematic **8.60**),[315] hydrogenation (Schematic **8.61**),[316] ketal formation,[317] oxidation,[318] and ring-closing metathesis (Schematic **8.62**).[319] Enzymatic reactions can also be carried out in ionic liquids (Schematic **8.63**).[320] Free radical, ionic, and metathesis polymerizations have been done in ionic liquids.[321] Workups vary. The product may separate as a separate phase. Addition of carbon dioxide may cause the separation of a separate phase. The product may be extracted by sc carbon dioxide[322] or another solvent for the product which does not dissolve in the ionic liquid. In some cases, the product can be distilled or sublimed out. These methods allow the ionic liquid to be reused in the next run.

Supported ionic liquids have been developed to minimize the amount needed and to aid separations.[323] Supports include silica,[324] carbon,[325] and chitosan.[326] Polymers containing imidazolium salts have also been used.[327]

A few examples will illustrate these points.

Industry is looking for applications for ionic liquids.[328] Relatively few have been found. BASF has found that 1-methylimidazole scavenges HCl much better than triethylamine (BASIL process) in the production of alkoxyphenylphosphines for use in making photoinitiators for coatings.[329] PetroChina has a 65,000 tons/year plant for the dimerization of isobutene to isooctene with an aluminum

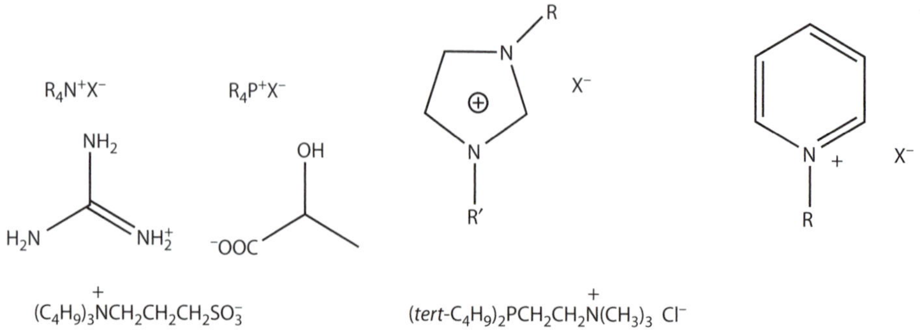

$R_4N^+X^-$ $R_4P^+X^-$ X^- X^-

NH_2 H_2N NH_2^+ OH ^-OOC

$(C_4H_9)_3\overset{+}{N}CH_2CH_2CH_2SO_3^-$ (*tert*-$C_4H_9)_2\overset{+}{P}CH_2CH_2N(CH_3)_3\ Cl^-$

8.58 Schematic

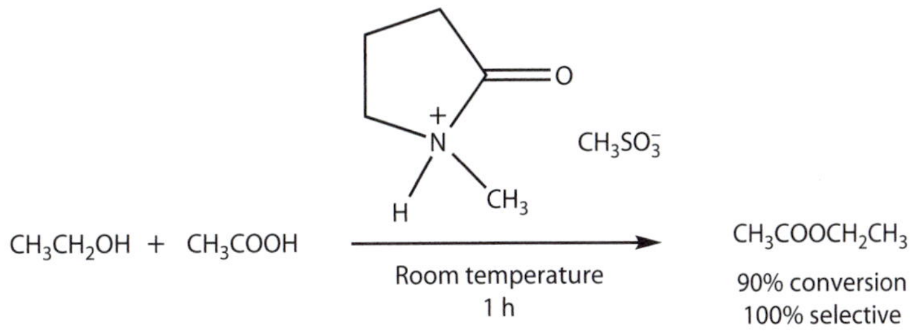

$CH_3SO_3^-$

$CH_3CH_2OH + CH_3COOH \rightarrow CH_3COOCH_2CH_3$

Room temperature
1 h

90% conversion
100% selective

8.59 Schematic

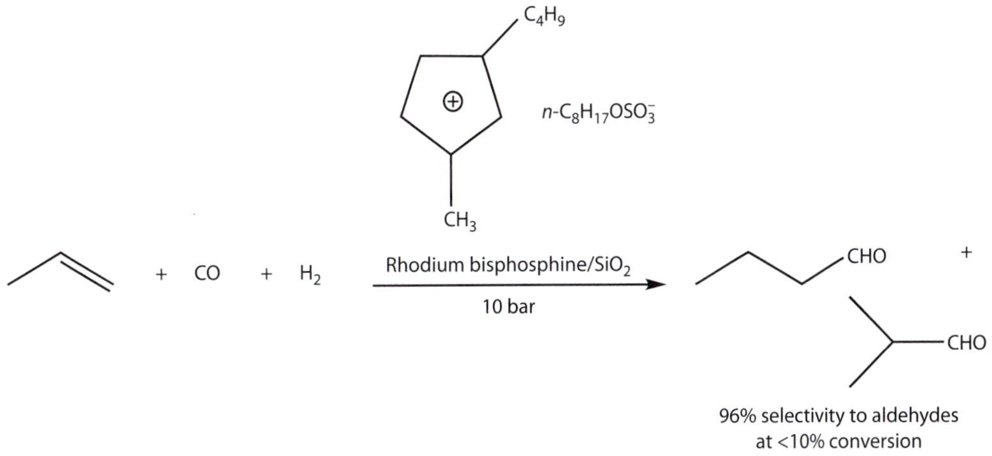

C_4H_9 n-$C_8H_{17}OSO_3^-$ CH_3

+ CO + H_2 → CHO +

Rhodium bisphosphine/SiO_2
10 bar

CHO

96% selectivity to aldehydes
at <10% conversion

8.60 Schematic

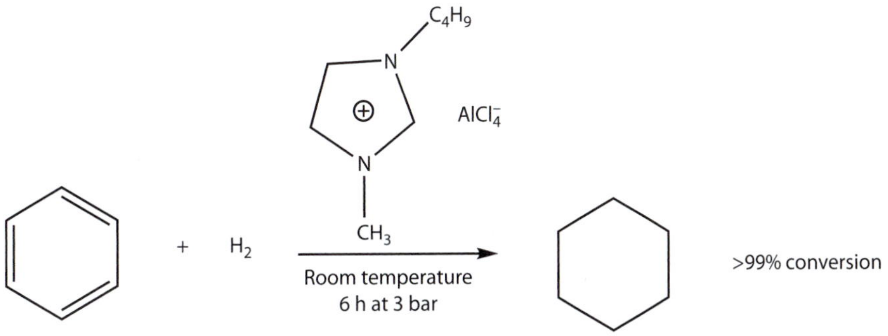

C_4H_9 N $AlCl_4^-$ N CH_3

+ H_2 →

Room temperature
6 h at 3 bar

>99% conversion

8.61 Schematic

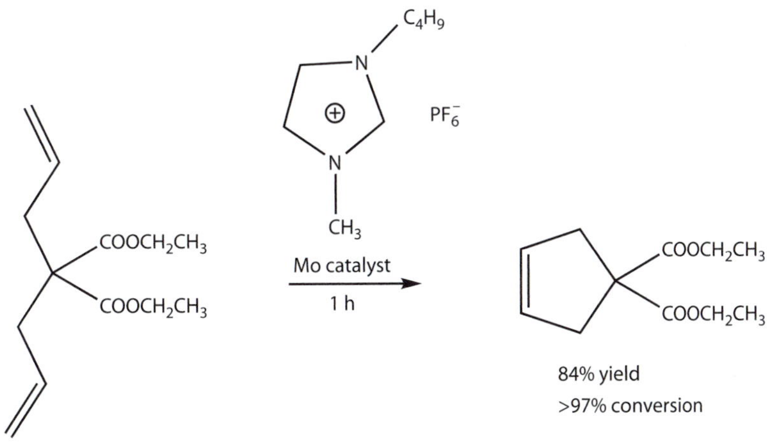

C_4H_9 N PF_6^- N CH_3

$COOCH_2CH_3$
$COOCH_2CH_3$

Mo catalyst
1 h

$COOCH_2CH_3$
$COOCH_2CH_3$

84% yield
>97% conversion

8.62 Schematic

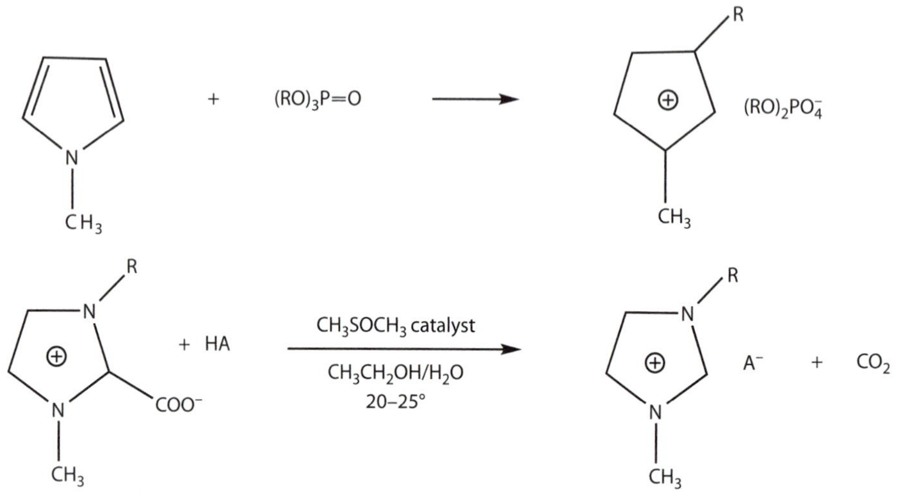

8.63 Schematic

chloride-based ionic liquid followed by hydrogenation to isooctane. The reaction is also run in France. BASF converts tetramethyleneglycol to 1,4-dichlorobutane in an ionic liquid Cellulose can be dissolved in 1-ethyl-3-methylimidazolium acetate and other ionic liquids and spun into fibers.[330] Cellulose-polymer blends can be made in this way by BASF. Cellulose can be reacted to give esters, ethers, and urethanes in ionic liquids.[331] Silk fibroin can also be spun from solutions in ionic liquids.[332] Ionic liquids are being investigated for lubricants, capacitors, electrolytes for batteries and dye-sensitized solar cells, and in the storage of toxic gases used in the semiconductor industry, such as boron trifluoride, phosphine, and arsine.[333] They can serve both as the template and as the medium for the synthesis of zeolites and metal organic frameworks.[334]

There are some problems with the use of ionic liquids. The presence of small amounts of chloride and water can alter the properties.[335] Two ways of avoiding chloride are shown below.

The high viscosity of ionic liquids interferes with some uses. Salts of amino acids have lower viscosities. Tetrabutylphosphonium prolinate[336] and tetraethylammonium L-alaninate[337] are examples. There have been questions about cost and toxicity. If none is lost and the liquid can be used over and over again, cost should not matter. The cost need not be high. Quaternary ammonium salts are in widespread use as cationic surfactants, fabric softeners, and bacteriocides in mouthwashes. These are not very toxic. Others can be quite toxic, depending on the length of side chains and the anions present.[338] Reactions in ionic liquids can be green, at least sometimes.[339]

A mixture of an alcohol and either an amidine or a guanidine can be switched to a high-polarity ionic liquid with carbon dioxide at atmospheric pressure (Schematic **8.64**).[340] The reaction can be reversed by passing nitrogen through the mixture or by heating to 50–60°C.

"Deep eutectic solvents" are related to ionic liquids. Mixing choline chloride and urea in a 1:2 molar ratio produces a liquid melting at 12°C (Schematic **8.65**).[341]

They can dissolve metal oxides and sugars. Enzymatic reactions can be run in them. Dibasic carboxylic acids, such as malonic acid, can be used instead of urea. They are easy to prepare, do not react with water, are biodegradable, and, presumably, have low toxicities, since choline and urea are both found in the human body.

8.7 SURFACTANTS AND CLEANING

Surfactants are often used when reactions are carried out in water. However, their presence in the final product may be undesirable. In polymerizations, it is possible to use a polymerizable surfactant that becomes part of the final copolymer.[342] A typical one (**8.66** Schematic) contains an acrylate group.

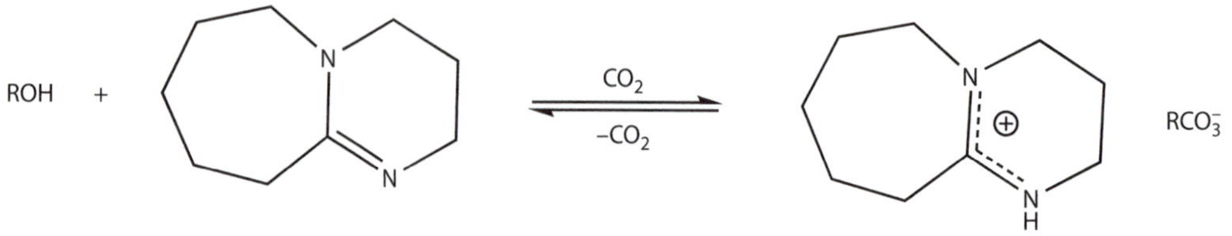

8.64 Schematic

$(CH_3)_3\overset{+}{N}CH_2CH_2OH\ Cl^- + NH_2CONH_2$

8.65 Schematic

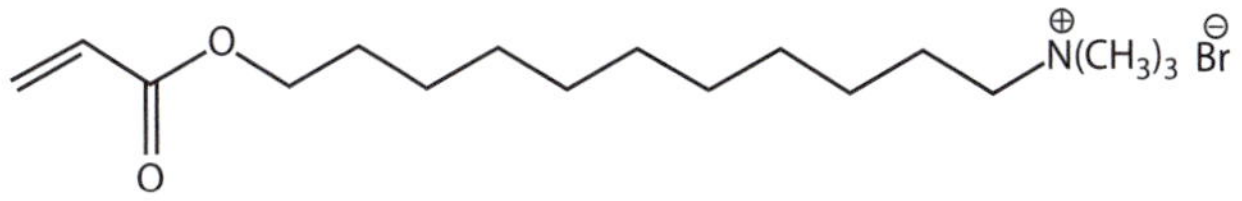

8.66 Schematic

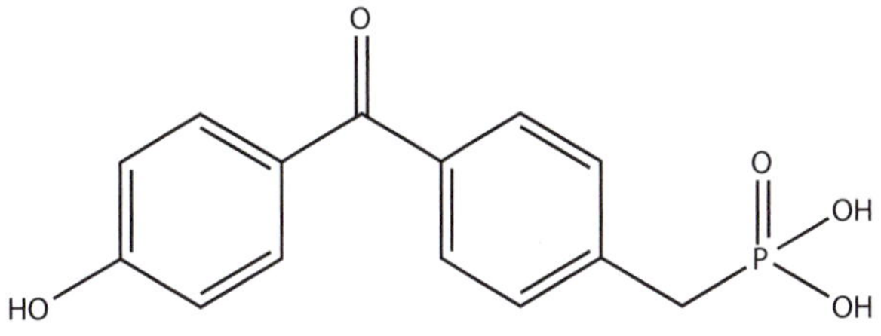

8.68 Schematic

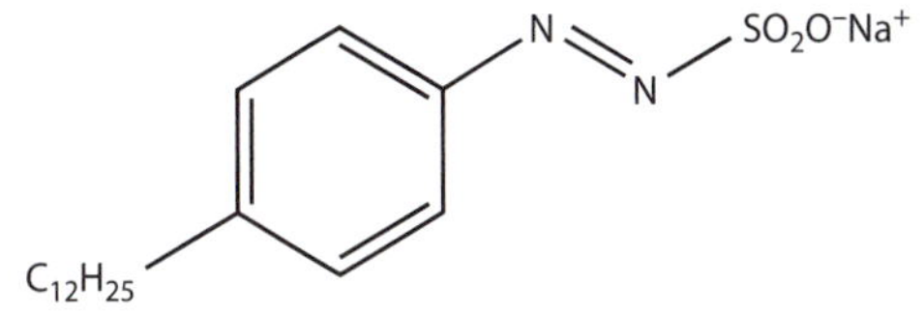

8.69 Schematic

Surfactants (i.e., surface-active agents) consist of molecules containing a hydrophobic unit and a hydrophilic unit. Soap, sodium stearate, is an example. The polar group can also be cationic, as in quaternary ammonium salts, or nonionic, as in polyethyleneoxy units. Destructible surfactants contain cleavable units so that after cleavage the parts no longer act as surface-active agents. The most common ones contain an acetal or ketal linkage that can be cleaved by acid.[343] Two typical ones are shown in **8.67** Schematic, where *R* is a long hydrocarbon chain.

The problem has been to obtain surfactants that cleave relatively quickly under mild conditions. Union Carbide (now Dow Chemical) is now marketing a proprietary one made from glycerol for use in treating wastewater. For example, the grease and oil left behind when industrial rags and clothing are cleaned can be separated by cleaving the surfactant with acid, which breaks the emulsion.[344] The two layers are then treated separately for recycle or disposal. The company suggests other uses in paper and textile processing, metal working fluids, and other commercial uses. This technique may also be of value in working up organic reactions run in water. Ideally, there should be some way of recovering the surfactant for recycling to the next run. Collecting the parts and making the acetal separately is a possibility.

Photodestructible surfactants have also been made.[345] When the first one (**8.68** Schematic) is irradiated with light in the presence of base, phosphoric acid is eliminated. In the second (**8.69** Schematic), nitrogen and sulfuric acid are lost.

A surfactant made from 2-dodecylfuran and *N*(4-hydroxyphenyl)maleimide decomposed by a reverse Diels–Alder reaction on heating at 95°C for 1 h.[346] A latex paint is an aqueous emulsion. After the paint dries, the surfactant is still present and may interfere with the resistance of the paint to washing to remove fingerprints or other marks. A destructible surfactant would be valuable in this application. It would have to destruct by the action of light or air. Latex paints are not entirely free of solvent. A few percent of a higher-boiling solvent, a coalescing agent, is usually, but not always, necessary to form a pinhole-free film from the particles of polymer after the water evaporates. A typical one (**8.70** Schematic) is derived from isobutyraldehyde.

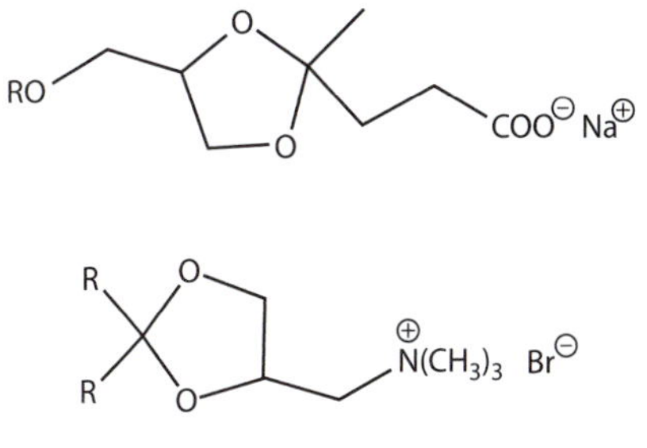

8.67 Schematic

Another approach is to try to develop more effective surfactants that can be used at lower levels. A combination of *N*-tetradecyldiethanolamine and copper (II) is effective at 16% of the usual amount of surfactant in forming a microemulsion.[347] The use of copper (II) is questionable for many applications after which the metal ion might end up in wastewater.

Soaps and detergents are surfactants that are used in cleaning.[348] Soaps are salts of fatty acids (e.g., sodium stearate). Detergents are salts of sulfonic acids, quaternary ammonium salts, tertiary amine oxides, ethylene oxide adducts of alcohols, and phenols.[349] Their use in cleaning to replace chlorinated solvents[350] was covered in Chapter 3, Section 3.7. (See also the use of hydrogen peroxide,[351] ethyl lactate,[352] and ultrasound[353] to clean process equipment.) Fluidized baths can be used to burn off organic residues from small machine parts.[354] Some other environmental aspects of their use will be covered here briefly.

Home laundry systems in Europe use 4 gallons of water per wash compared with 18–23 gallons in the United States.[355] The trend toward lower temperatures in washing has increased the use of enzymes in solid detergents.[356] Because much of the cost of doing the wash is in heating the water, the European system appears to be better.

The sodium alkylbenzenesulfonates are an important class of detergents. The original ones made by alkylating benzene with propylene tetramer did not biodegrade, resulting in foaming rivers and such.[357] The current

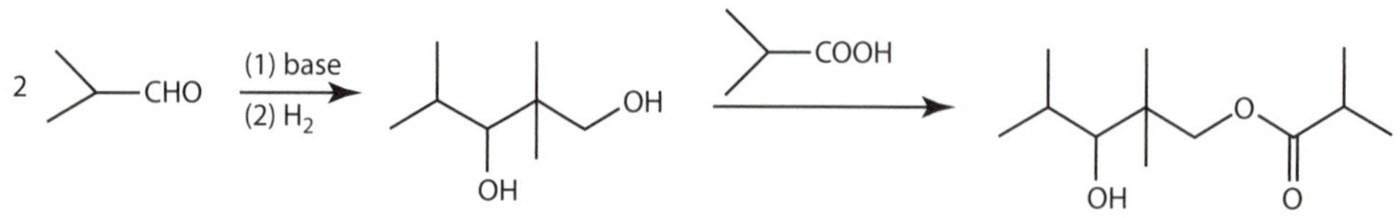

8.70 Schematic

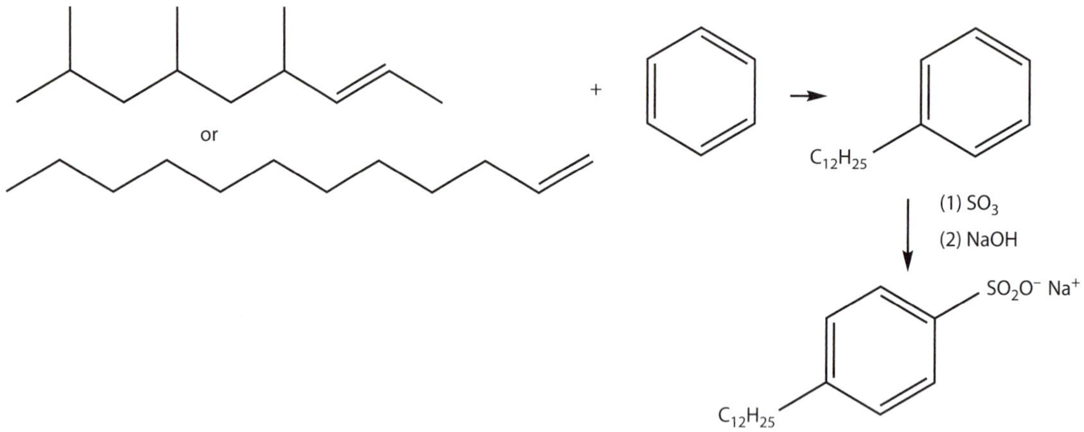

8.71 Schematic

linear alkylbenzenesulfonates (**8.71** Schematic) made from α-olefins are biodegradable.

The second problem with detergents has been the use of sodium tripolyphosphate (**8.72** Schematic) as a builder (i.e., an agent to sequester the calcium and magnesium ions in the water). This has caused the eutrophication (excessive growth of algae) of lakes and rivers receiving wastewater.[358]

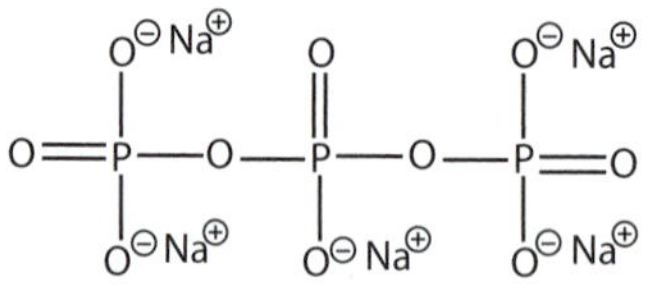

8.72 Schematic

Fish kills have also resulted from wastewater. Various replacements for phosphate are used in those places that have banned phosphate. The use of zeolites is common. Gluconic acid, prepared by the palladium-catalyzed air oxidation of glucose, can be used.[359] Poly(acrylic acid) can be used, but is not biodegradable.[360] Poly(aspartic acid) (**8.73** Schematic) made from maleic anhydride and ammonia solves this problem.[361]

A product made from renewable raw materials has also been active in this application. Oxidation of the amylopectin fraction of starch with periodic acid cleaves some vicinal diol structures to dialdehydes, which can be oxidized further to dicarboxylic acids.[362] For good activity, over 80% of the original glucopyranose units must remain. A 1:1 blend of a tertiary amine-*N*-oxide and an alcohol ethoxylate cleaned hard surfaces without the need for phosphate.[363] These detergents may not form complexes with calcium and magnesium ions. Thus, the use of nonionic detergents may obviate the need for a builder.

The alternative to not using a phosphate is to take it out of the wastewater at the treatment plant.[364] The recovered phosphate can be used as fertilizer. This method is used in Germany and Sweden.

Many surfactants are made from petroleum with hazardous reagents. Sulfur trioxide was used with the carcinogen benzene to make the alkybenzenesulfonates described earlier. The carcinogen ethylene oxide is used to make many nonionic surfactants from phenols and long-chain alcohols. A common surfactant, sodium dodecyl sulfate, is made from an alcohol derived from coconut oil by reduction followed by treatment with sulfur trioxide. Long-chain tertiary amines derived from natural fats and oils are quaternized with

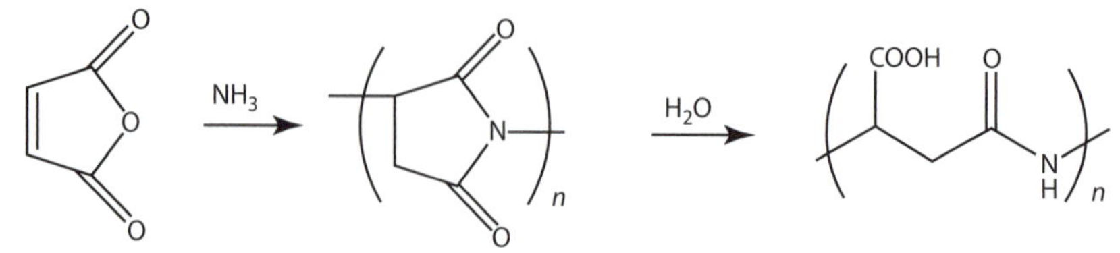

8.73 Schematic

$$(CH_3)_3\overset{+}{N}CH_2COO^- + ROH \xrightarrow[\text{No solvent}]{CH_3SO_3H} (CH_3)_3\overset{+}{N}CH_2COOR \quad CH_3SO_3^-$$

$$\downarrow C_{12}H_{25}NH_2$$

$$(CH_3)_3\overset{+}{N}CH_2CONHC_{12}H_{25} \quad CH_3SO_3^-$$

8.74 Schematic

methyl chloride. Perhaps this can be done with dimethyl carbonate. Soaps are made by hydrolysis of fats and oils with sodium or potassium hydroxide. They are biodegradable. The challenge is to obtain detergent activity from compounds made from renewable sources so that biodegradability will not be a problem.[365]

Henkel makes alkyl polyglycoside surfactants from alcohols derived from coconut and palm oils, and glucose derived from cornstarch.[366] The preparation can be done with β-D-glucosidase.[367] Sucrose monoesters of long-chain fatty acids can also be made enzymatically.[368] A variety of biosurfactants are produced by various organisms that can be grown in culture.[369] Dow Chemical sells "Ecosurf" detergents made by reaction of ethylene oxide and 1-hexadecanol from palm oil.[370] Kao Corporation sells alkyl glyceryl ethers made by hydrolysis of alkyl glycidyl ethers.[371] The hydrolysis is carried out in a tubular reactor in 3 min at 250–300°C to make 1000 metric tons/year. It would be better to make these directly from glycerol. Surfactants have also been made from glycine betaine (Schematic **8.79**).[372]

Environmentally friendly detergent ingredients are in demand to satisfy consumers.[373]

Thin films of titanium dioxide prepared by the sol–gel technique from titanium alkoxides catalyze the photooxidation of thin films of organic matter[374] (as mentioned in Chapter 3). The films are transparent and abrasion-resistant. They can photocatalyze the oxidation of about 1–5 μm/day of organic film. Consistent activity in the oxidation of stearic acid and benzene was observed over 300 h, as long as some water was present. This technique may be useful in self-cleaning windows, windshields, and architectural surfaces. The coating makes the windshield more hydrophilic so that it does not fog. Condensed water forms a film, rather than beads. Molds as well as dirt could be removed in this way. The photocatalytic activity is increased by inclusion of a little fluoride or platinum in the coating.[375] The coating of titanium dioxide is said to be a good adsorbent and photocatalyst for the destruction of formaldehyde in "sick" buildings.[376] Another technique for self-cleaning outside surfaces might be to apply a perfluorocarbon that dirt would not strongly adhere to, so that it would be washed off by the next rain. This is analogous to the treatment applied now to upholstery and carpets to repel dirt. Titanium dioxide has also been used with light to deactivate a pressure-sensitive adhesive so that materials held by it (e.g., a bandage) can be removed easily.[377]

8.8 COATINGS

Thin films of organic compounds are often applied from solvents. These films include adhesives, coatings for magnetic tapes, inks, paints, and varnishes. Over 11% of all products have some kind of coating. Worldwide sales of coating-based products are over 5 trillion dollars.[378] A number of reviews of surface coatings and paints are available.[379] Several other reviews cover ways of minimizing the use of solvents in these.[380] Much of the discussion that follows is taken from these latter reviews.

Waterborne coatings[381] can replace solvent-containing ones in many uses. Typical water-based paints are emulsions of polymers, such as poly(vinyl acetate-*co*-butyl acrylate), most of them containing a small amount of a higher-boiling solvent as a coalescing agent. Thus, they do emit some VOCs, but far less than if the usual solvent-based paint or varnish was used. Waterborne paints that are free of VOCs are being developed. They cause less sensitivity to asthmatics than regular waterborne paints.[382] Many water-based coatings contain polymers with carboxylic acid or secondary or tertiary amine groups that are solubilized with a volatile base, such as ammonia, or an acid, such as lactic acid, which are emitted and should be trapped as the film dries.[383] Monomers that can be used to make polymers of this type include acrylic acid, methacrylic acid, maleic anhydride, 2-dimethylaminoethyl methacrylate, and others. Condensation polymers, such as polyamides, based on diethylenetriamine have amine groups that enable the polymer to dissolve at low pH.

Many automobiles and appliances are coated with a waterborne primer applied by electrodeposition in a process called electrocoating.[384] The metal car or appliance is one electrode. Either anodic or cathodic electrodeposition can be used. This is better than a spray system with solvent in that small recesses and sharp edges can be coated evenly, for they, too, are part of the electrode. Paint utilization is over 95%, in contrast with a spray system for which much overspray would be lost or would have to be recovered. After the car is coated, it is dried and baked to remove the ammonia or small carboxylic acid to produce a water-insensitive coating. In addition, the system may contain reagents that undergo further reaction to cross-link the coating. If blocked isocyanates are included, the blocking agent that comes off with the baking will have to be trapped. If the isocyanate is present as a cyclic dimer, there will be no blocking agent to be trapped. (See Chapter 2 for this chemistry.) If a copolymer of

$$\text{polymer-COO}^{\ominus}\text{NH}_4^{\oplus} + \text{polymer-(oxazoline)} \xrightarrow{-\text{NH}_3} \text{polymer-COOCH}_2\text{CH}_2\text{NHCO-polymer}$$

8.75 Schematic

glycidyl methacrylate is used, it can react with amine groups in the formulation during the baking. There are some limitations to this method. The baked coating is usually an insulator, making it impossible to apply another coat by this method. However, there are a few systems that produce a conductive first coat so that a second coat can be applied by the same method. The vat that the car passes through is large. It would not be convenient to change colors very often if topcoats were applied in this way. The color of the primer coat does not matter. DuPont has a new system that eliminates the primer coat with a reduction of 45–50 kg of carbon dioxide emissions per car.[385]

Waterborne polyurethane coatings for wood and floors can be based on prepolymers made by reacting diisocyanate with 2,2-bis(hydroxymethyl)propionic acid.[386] These are usually two-component systems in which mixing occurs at the time of use (e.g., using a dual-spray nozzle). It is desirable to pick a polyisocyanate that will react faster with an alcohol group than with water. 2-Hydroxyethyl acrylate can be incorporated in the prepolymer to help cross-linking. Inclusion of isocyanate dimers (uretdiones) will allow later cross-linking. Water-based epoxy resins may use a dispersion of a liquid epoxide in water that is cured by a polyamine, or an acrylic polymer containing carboxylic acid groups.[387] Waterborne fluorochemical coatings are used for nonstick, dirt-resistant, and corrosion-resistant coatings.[388] One system (**8.75** Schematic) uses an ammonium carboxylate in one chain and an oxazoline in another. As the ammonia is lost from the film, the carboxylic acid formed reacts with the oxazoline to form an amide. The clear, colorless coating resists graffiti, marine-fouling organisms, solvents, acids, bases, and sodium chloride. It can also be used as a mold-release and ice-release coating.

Electropolymerization of monomers can produce pinhole-free coatings on steel at room temperature and atmospheric pressure in aqueous systems.[389] The coatings tend to be even because thin spots conduct electricity better and cause more polymerization. An alternating copolymer prepared in this way from styrene and the maleimide of 4-aminobenzoic acid had a weight average molecular weight of 146,000 and a glass transition temperature of 231°C. It was stable up to 350°C under oxygen and showed good adhesion to steel.

Powder coatings use neither solvent nor water.[390] They can give nearly zero emissions. Thick coats can be applied in contrast with the more than one thin coat needed with solvent-borne coatings. Overspray can be reused so that there is no waste, in contrast with the lost overspray with some sprayed coatings. A thermoplastic polymer can be used for the powder. Alternatively, cross-linking can be done after the powder is on the object being coated. One method is to dip the hot object in a fluidized bed of the powder. Another is to spray electrically charged powder on to the object, which is given the opposite charge. The object is then heated to melt the powder. The powder must not melt or sinter or react under normal shipping or storage conditions. Typically, the glass transition temperature of the powder is 50–55°C. Then, it is expected to cure to a film free of pinholes at the lowest possible temperature. It has been a challenge to lower the curing temperatures so that the technique can be used on plastics as well as with metals. A way around this is to melt the powder at a modest temperature and then cure it with light or an electron beam.[391] One method uses a methacrylate prepolymer containing bisphenol A dimethacrylate as a crystalline cross-linking agent, so that the viscosity of the initial melt is low.[392] Dow and Ferro both sell proprietary powder coatings for use on wood.[393] Another method is to apply the powder coating to the mold and then to mold the plastic into it. Flame spraying, in which the polymer is melted by brief contact with a flame and hits the object to be coated as molten droplets that form a pinhole-free film, is another possibility.[394] Traffic stripes on roads may be made of thermoplastic polymers applied in this way. Laser ablation of fluoropolymer films has been used to apply coatings of poly(tetrafluoroethylene).[395] The polymer depolymerizes and repolymerizes in the process. Powder coatings can be glossy, or flat (nonglossy) if a flatting agent has been added.

The process consumes less energy than a system using solvent spraying. A lot of air has to be passed through to prevent fires and explosions in the latter case, and all of it has to be heated, at least in winter. However, powder systems must be handled in a way that prevents dust explosions. The use of an inert gas, such as nitrogen, to prevent this is a possibility. Application of the powder as an aqueous dispersion is sometimes used to prevent it. This also allows polymers with lower glass transition temperatures and, possibly, lower baking temperatures to be used. The water has to be evaporated before the final film coalesces and cures. Another limitation is the work required to clean the equipment between color changes. If one apparatus can be dedicated to one color, then this is not a problem. It has also proved difficult to incorporate metal flake pigments into powder coatings.

A popular type of powder coating consists of a carboxy-terminated polyester mixed with a diepoxide.[396] The epoxides include bisphenol A diepoxides, triglycidylisocyanurate, and epoxidized vegetable oils. Blocked isocyanates are also used in powder coatings.[397] The use of oximes of hindered

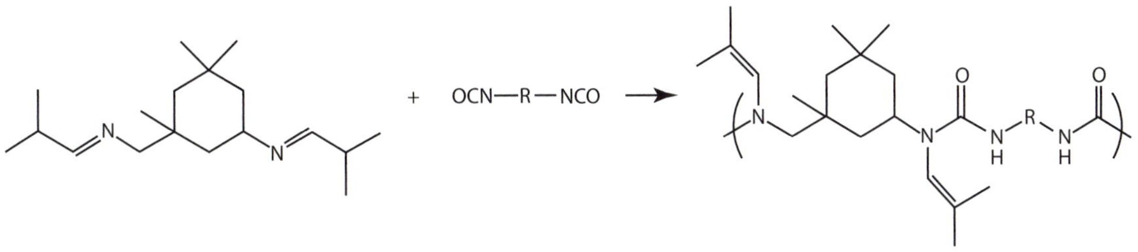

8.76 Schematic

ketones, such as diisopropyl ketone, allows deblocking at lower temperatures than with ordinary oximes, such as that from 2-butanone. This system cures at 129–140°C. Many current systems cure at 165–200°C. Removal of all of the diisopropylketone oxime out of the coating during curing could be a problem. The ketone oxime or other blocking agent would have to be trapped from the exhaust air for reuse. If the diisocyanate is present as a cyclic dimer (uretdione) that opens on heating, there will be no emissions to trap.[398] Another system that evolves nothing on curing uses ketimines, which react through their enamine isomers with isocyanates (**8.76** Schematic) to form polyureas.[399] It may be suitable for powder coatings, or for a 100% solids coating applied from a dual-spray nozzle.

Powder coatings are used on furniture, lawn mowers, appliances, auto fittings, trucks, store shelving, exercise equipment, aluminum extrusions, and so forth. They are useful on pipes when thick coatings are needed.[400] Pipeline joints can be coated in the field by this method. An alternative is to use a heat-shrinkable sleeve of polyethylene.

Coatings cured by light or by electron beam radiation[401] use no solvents and produce virtually no emissions. The cure can be fast (e.g., less than 1 s). Energy usage is less than that with solvent or waterborne coatings. They do have some limitations. Light does not go around corners. Pigments can interfere with the passage of light. Systems that cure by a free radical mechanism are inhibited by air. This means that an inert atmosphere must be used. An alternative is to use protective waxes that bloom to the surface to keep oxygen out. It is difficult to incorporate stabilizing antioxidants in systems that cure by free radical polymerization. The high viscosity of the prepolymer is lowered by dilution with a polymerizable monomer. If this monomer is an acrylate, skin irritation and sensitization are possible. Sometimes, methacrylates are used to avoid this. It is desirable to cure all of the monomer so that it will not leach out later. Shrinkage that interferes with adhesion is a problem with the polymerization of vinyl monomers. Thus, enough acrylate groups must be present for a fast cure, but not so many that shrinkage is excessive.

Polyester acrylates and urethane acrylates are in common use.[402] Cationic systems use vinyl ethers, such as triethyleneglycol divinyl ether, epoxides, and polyols. These avoid some of the problems of the acrylates. Skin irritation is much less of a problem. Shrinkages are less with ring-opening epoxides. Cures are not inhibited by oxygen, but they are slower and may require heat for completion. The presence of pigment is less of a problem. The photoinitiators[403] used, such as a triarylsulfonium tetrafluoroborate, leave traces of acid in the coating. The photoinitiators for curing the acrylates tend to be benzophenones and benzoin derivatives. Work is being done to find photoinitiators that work with visible light, which is absorbed less by pigments.[404] Two such initiators are shown in **8.77** Schematic.

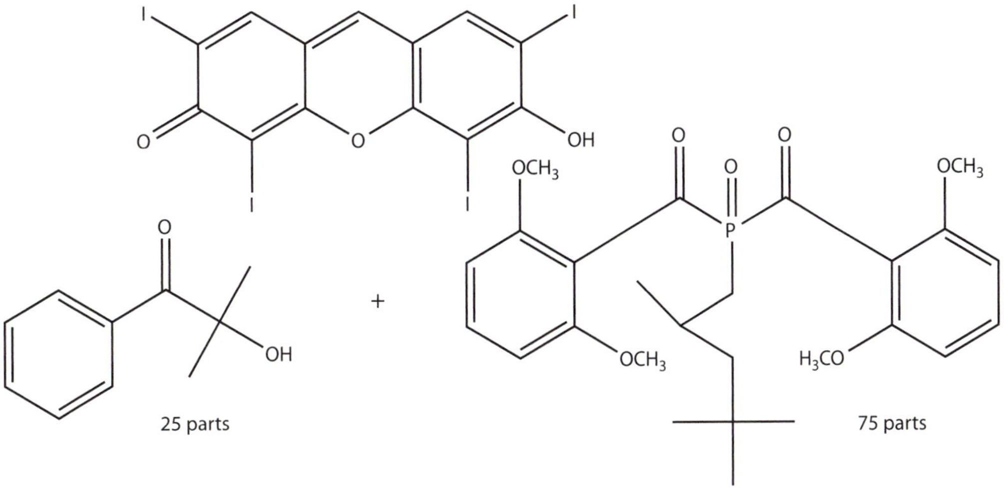

25 parts

75 parts

8.77 Schematic

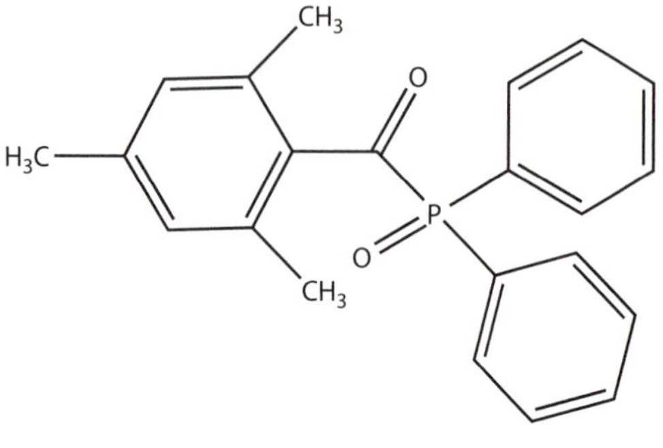

8.78 Schematic

Some cures can even be done in sunlight in 2 min using the acylphosphine initiator (**8.78** Schematic).[405]

Inhibition of the polymerization by air was controlled by using 1:1 diacrylate–poly(methyl methacrylate) mixtures or by using the acrylate in a laminate between two glass plates. These cost-effective cures use no external energy and no lamps. Composites of glass fiber and epoxides from vegetable oils have been cured in sunlight with diaryliodinium and triaryl sulfonium salts in 25 min.[406] Further work is needed to speed up this photocationic cure.

No photoinitiator is needed when the acrylates are cured by electron beams. The need to use a reactive monomer diluent to lower the viscosity is lessened by the use of hyperbranched or dentritic polyester methacrylates.[407] In one case, 15% of trimethylolpropane triacrylate was used as a cross-linking monomer. Compact molecules have lower viscosities than linear ones. Dendrimers are spherical molecules made by repeated branching from a multifunctional core.[408] In this case the hydroxyl groups on the surface have been converted to methacrylates. In addition to the low viscosity, these materials have higher curing rates. If water-based, radiation-curable latexes[409] are used, there is no need to use a reactive monomer to lower the viscosity. However, more energy and time are needed to remove the water before turning on the light.

Radiation-cured coatings are used mostly on flat surfaces. They work well on vinyl flooring and in coil coating. The latter involves coating sheet metal that is fabricated into a variety of shapes and structures after coating. Radiation curing is suitable for heat-sensitive materials such as paper, wood, and some plastics. It is common with printing inks, for which a fast cure is important. In Europe, furniture parts are coated before assembly of the furniture. In the United States, where the furniture is assembled first, solvent-based nitrocellulose lacquers are preferred as coatings. It should be possible to coat furniture after assembly with a system with enough lamps and mirrors so that all parts of the furniture receive adequate light. Photocationic initiation has the best chance of working. It should also be possible for the United States to switch to the European system. Another option would be to switch to more furniture molded from plastic. Cross-linked, scratch-resistant coatings are applied to eyeglasses, to plastic headlights for cars, and to windows by radiation curing. Optical fibers for communication are also coated by radiation curing.

There are some other ways of applying coatings without the use of solvents. Plastisols[410] involve the use of a paste of poly(vinyl chloride) particles in a plasticizer. Metal objects dipped in this can then be heated to harden the coating by dissolution of the plasticizer into the polymer. An example would be a wire rack for a dishwasher. Colored paint films can be laminated to plastics such as automobile parts and siding for houses by molding.[411] Ideally, the dry paint film would be prepared by extrusion or other means that avoided the use of solvent. It is also possible to produce a coating from two monomer ingredients that polymerize in flight. Polyurea coatings[412] can be made in this way using a dual-spray nozzle. Enough polymerization must occur in flight so that the coating will not sag, but not enough that leveling does not occur. It should be possible to extend this to single monomers or prepolymers that can be polymerized by cationic initiation by adding the catalyst at the spray nozzle. Ultrasonic spray nozzles might give the best results. The simplest systems to try would involve flat, horizontal surfaces, such as coil coating or flooring.

Silicon dioxide films can be applied to plastics by plasma or glow discharge polymerization of silanes, such as tetramethoxysilane and hexamethyldisiloxane.[413] These films can serve as oxygen barriers on polypropylene film to replace poly(vinylidene chloride) coatings. Aromatic ether amides have also been made that afford good oxygen barriers to replace poly(vinylidene chloride).[414] The sol–gel process has been used to apply hard scratch-resistant coatings to plastics.[415] Coatings of silicon dioxide are also oxygen barriers for films of polypropylene and poly(ethylene terephthalate). Application of such coatings to the insides of poly(ethylene terephthalate) soda bottles might allow them to be used as refillable containers, thus reducing plastics consumption and waste.

High scratch resistance on plastics has been obtained by reaction **8.79** Schematic. (This process does give off a lower alcohol, which will have to be recovered for recycle.)

A possible alternative is to graft polymerize a vinyltrialkoxysilane to the surface of the polymer for later cross-linking by moisture.[416] Mitsubishi produces a diamond-like coating on the inside of plastic beverage bottles using acetylene in a radio-frequency plasma.[417] This raises the barrier to oxygen and carbon dioxide 10-fold. Bottles can be

$Si(OR)_4$ + $(RO)_3Si$–(CH2)n–epoxide + AlOOH + HCl + H_2O ⟶ Coating

8.79 Schematic

coated at the rate of 12,000–18,000 per hour per machine. The U.S. Food and Drug Administration (FDA) has approved the process. It may be possible to prepare diamond coatings that are good scratch-resistant oxygen barriers using the three-laser technique.[418] In contrast with earlier methods, this technique uses carbon dioxide at room temperature without a vacuum. It has been used primarily on metals. This could also make plastic dinnerware more acceptable in the home. The dinnerware would be light in weight, unbreakable, and free of scratches and stains.

A pulsed plasma was used to prepare pinhole-free films from relatively nontoxic *N*-vinylpyrrolidone.[419] The pulsing reduced fragmentation of the monomer and cross-linking. This method should be tried with other monomers. Plasmas are often used for the modification of polymer surfaces.[420] These methods are relatively rapid and use no solvent. Decorative coatings of TiN and other inorganic compounds can be applied to metals and other inorganic substrates by sputtering, chemical vapor deposition, plasmas, and such, as described in Chapter 4.[421] Coatings can fail from poor surface preparation, pinholes, lack of resistance to light, or other causes.[422] Some coatings have been designed to assimilate oil and to displace water from surfaces so that surface preparation can be skipped or minimized.[423] One such polymer is **8.80** Schematic. These systems need further development before they can come into common use.

Car painting[424] used to be the largest source of solvent emissions in Delaware. It should be possible to finish a car using the methods outlined in the foregoing, especially as some of them are already used on trucks and lawn mowers. A typical finish for an automobile includes a zinc phosphate pretreatment for corrosion resistance, a primer put on by electrocoating, a second coating put on by powder coating, a waterborne colored basecoat, and finally a solvent-borne clear coat to protect the finish from acid rain, bird droppings, and such.[425] General Motors Buick uses a two-component polyurethane clear coat. Glamour finishes, such as those containing metal flakes, are popular. Luster pigments, based on very thin layers of metal oxides on mica (where the color comes from interference patterns), can be used in powder coatings to obtain the same effect.[426] These pigments are transparent to ultraviolet light so that radiation curing can be used. Occasionally, the clear topcoat has been waterborne.[427] A consortium of U.S. automakers opened a facility in Michigan to test powder coating clear coats in August 1996.[428] A clear coat from PPG (PPG Industries, Pittsburgh, PA) that contains a patented flow additive has been used on 50,000 BMW cars, saving up to 1.5 kg of solvent per vehicle.[429] Because trucks and lawn mowers can be finished by powder coating, part of the problem may be the public preference for glamour finishes. The chromium-plated bumper disappeared for economic reasons. It is possible that adding a 500 dollar surcharge for a car with a glamour finish might have reduced the solvent emissions in Delaware. A new system that uses a plasma to clean the steel and put on an alkyl silane finish can replace the zinc phosphate steps, providing superior corrosion performance, eliminating wastewater, and making it easier to recycle the car at the end of its useful life.[430]

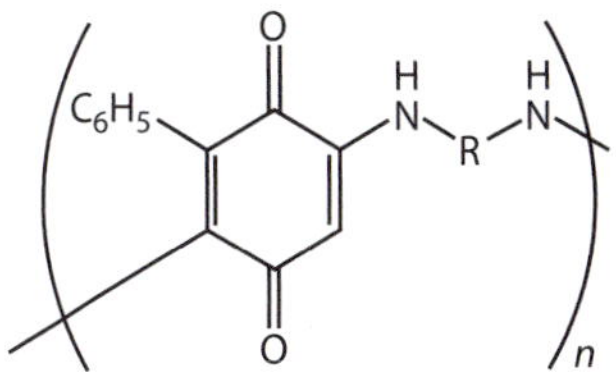

C_6H_5 O H N R H N O n

8.80 Schematic

In contrast to the localized emissions of auto assembly plants, printing[431] with solvents is done in many small, scattered plants. Eighty percent of printing plants have fewer than 20 employees. The U.S. Environmental Protection Agency (EPA) has set up a joint project to help them reduce emissions.[432] Even with solvent recovery and other emission control devices, publication gravure printing alone emits 26 million lb/year of toluene.[433] A waterborne ink for this use has been demonstrated in Europe. A good plant can capture up to 96% of the solvent from printing, but 80% to the low 90%s is more common.[434] Waterborne and radiation-cured inks are at the forefront of the effort to reduce these emissions.[435] The waterborne inks still contain small amounts of solvent. Overprint coatings contain 1–5% solvent, down from 10–12% 5 years earlier.[436] A typical water-based ink contains a polymeric carboxylic acid that is neutralized with ammonia or an amine.[437] These inks work well with porous materials, such as paper and cardboards, but are harder to use with plastics, which do not absorb moisture. An ink containing no solvent is used for the commercial printing of checks by Deluxe Corporation.[438] The plant has zero volatile emissions. The presence of carboxylic acid groups in the resin allows cleanup of equipment with aqueous base. Radiation-cured inks are finding increased use in spite of the fact that they cost 3 times as much as regular inks.

Although ballpoint pens are said to "write dry" on porous materials such as paper, the ink in them still contains some solvent.[439] A typical ink may contain 30% phenylglycol and 8% 1,2-propanediol.

There are about 40,000 screen printers in the United States, each with an average of 14 employees.[440] The screen is a stencil that must be cleaned with solvent between runs. Solvent recovery on-site by distillation is reducing the amount of new solvent needed. Less hazardous solvents are now being used (e.g., a mixture of propylene glycol monomethyl ether, propylene glycol monomethyl ether acetate, and cyclohexanone). High-pressure jets of aqueous emulsion cleaners are also being used. The next step in the evolution of the industry is to switch to nonimpact printing using ink jets, which will eliminate the screens.

Hot-melt[441] and waterborne adhesives and sealants are replacing those with solvent.[442] The most common hot-melt adhesives are based on copolymers of ethylene and

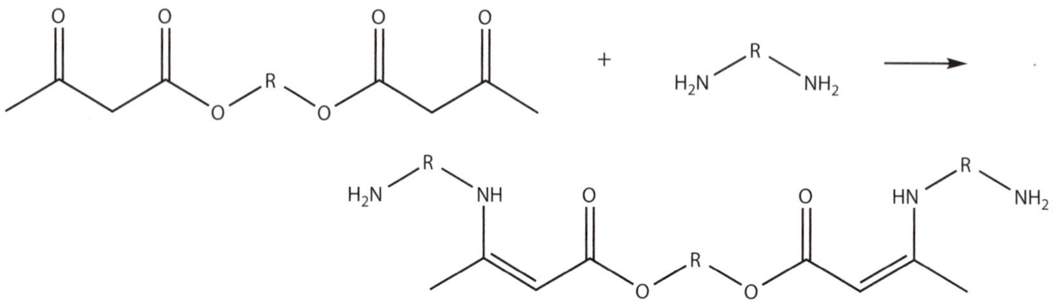

8.81 Schematic

vinyl acetate. They bond by solidifying on cooling, which can be fast if the substrates are at room temperature, as in book-binding or carton-sealing. Efforts are being made to find adhesives that will be easier to handle in the repulping of paper bonded with them. (Putting carboxylic acid groups into the hot-melt adhesive should help.) The iron-on patch for mending clothing is a hot-melt adhesive based on a polyamide. A hot-melt adhesive made from an isocyanate-terminated polyurethane is cured by moisture.[443] The most common aqueous emulsion adhesive is based on poly(vinyl acetate-*co*-vinyl alcohol). It is made by the emulsion polymerization of vinyl acetate followed by partial hydrolysis. Goodyear is using a water-based adhesive in all of its tire retreading centers.[444] Waterborne and hot-melt adhesives are preferred over solvent-based ones for cars, except where mechanical fasteners are needed for part removal.[445]

It is also possible to bond without an adhesive. Two pieces of plastic or two pieces of metal can be bonded by ultrasonic welding. Many plastic films are joined by heat sealing (i.e., heating with a hot metal bar until the films join together). Bags for food items are often made in this way. Mixtures of fibers, in which one is a lower-melting thermoplastic fiber, can be bonded by melting the lower-melting fiber while pressing against the other one. This is done sometimes with a pattern on a hot roll to make nonwoven fabrics. A variant is to spin a bicomponent fiber in which one component is lower melting. A high-powered diode laser has been used to bond polypropylene without any environmentally unfriendly pretreatments.[446]

Isocyanate-based sealants are in common use. The toxicity of the isocyanates has led to a search for less toxic alternatives. One is based on the reaction of bis(acetoacetates) with diamines (**8.81** Schematic).[447] The sealant cures when the bis(enamine) is mixed with more of the starting bis(acetoacetate) in the presence of a catalyst.

Magnetic tapes are usually coated with coatings from solvents, such as methyl ethyl ketone, methylisobutylketone, and toluene. Aqueous replacements can eliminate all or nearly all of the solvents.[448] One system used iron oxide in a blend of a polyurethane and ethylene–vinyl chloride copolymer emulsions thickened with hydroxyethyl cellulose, which was cross-linked with a melamine–formaldehyde resin. Coating was done at line speeds of 100 m/min. The whole system proved to be 15% cheaper than coating from solvent. In another system, traces of methanol are evolved on drying and would have to be captured. This replaces a line where 600 kg/h of solvent would have to be recovered and recycled. Most magnetic tapes have been replaced by CDs.

Hair sprays are used by many persons to hold their hair in place. A typical spray consists of a polymer in ethanol and a little water with dimethyl ether and *n*-butane as propellants.[449] These organic compounds contribute to ground level ozone. They are also a fire hazard. Efforts are underway to reduce these problems. Dilution with more water gives an unacceptable spray. The use of different polymers in high-solids formulations in ethanol containing more water is being explored. Eastman Chemical has prepared a water-dispersible sulfopolyester in a melt without solvent, with no toxic by-products.[450] This polymer can be used in 55% ethanol. The need for a propellant can be eliminated by using a finger or rubber bulb-actuated pump. If hair style changes so that ponytails and braids become popular again, the severity of the problem will decrease.

REFERENCES

1. (a) D. Mendenhall, *Chem. Eng. News*, Feb. 6, 1995, 3; (b) E.M. Kirschner, *Chem. Eng. News*, June 20, 1994, 13; (c) C. Reichardt, *Solvent and Solvent Effects in Organic Chemistry*, 3rd ed., Wiley-VCH, Weinheim, 2003; (d) G. Carrea and Sergio Riva, eds, *Organinc Synthesis in Non-Aqueous Media*, Wiley-VCH, Weinheim, 2008; (e) G. Wypych, *Chem. Eng.*, 2006, *113*(6), 54; (f) R. Gant, C. Jimenez-Gonzalez, A. Ten Kate, P.A. Crafts, M. Jones, L. Powell, J.H. Atherton, and J.L. Cordiner, *Chem. Eng.*, 2006, *113*(3), 30.
2. A.R. Gavaskar, R.F. Offenbuttel, L.A. Hernon–Kenny, J.A. Jones, M.A. Salem, J.R. Becker, and J.E. Tabor, *Onsite Solvent Recovery*, U.S. EPA/600/R-94/026, Cincinnati, Ohio, Sept. 1993.
3. S. Deng, A. Sourirajan, T. Matsuura, and B. Farnand, *Ind. Eng. Chem. Res.*, 1995, *34*, 4494.
4. (a) B. Hileman, *Chem. Eng. News*, Mar. 29, 1999, 6; (b) S. Khattak, G.K. Moghtaderk, K. McMartin, M. Barrera, D. Kennedy, and G. Koren, *J. Am. Med. Assoc.*, 1999, *281*, 1106.

5. (a) Anon., *Chem. Eng. News*, Aug. 24, 1998, 44; (b) D.J. Hanson, *Chem. Eng. News*, June 1, 1998, 33.
6. (a) Anon., *Chem. Eng. News*, May 31, 1999, 28; (b) R.D. Chambers and R.C.H. Spink, *Chem. Commun.*, 1999, 883.
7. (a) R.F. Service, *Science*, 1998, *282*, 396, 399; (b) A.I. Stankiewicz and J.A. Moulijn, *Chem. Eng. Prog.*, 2000, *96*(1), 22; (c) D.C. Hendershot, *Chem. Eng. Prog.*, 2000, *96*(1), 35.
8. E. Kicko–Walczak and E. Grzywa, *Macromol. Symp.*, 1998, *127*, 265.
9. (a) Anon., *Chem. Info.*, 2008, *46*(3), 22; (b) R. Levy, *Pollut. Equip. News*, Apr. 2008, 6.
10. B. Mills, *Surf. Coat. Int.*, 1998, *81*(5), 223.
11. M. Murray, *Wilmington, Delaware News J.*, May 1, 1996, Bl.
12. G. Makin and M. Langdon, *Surf. Coat. Int.*, 1995, *78*(6), 265.
13. A. Farrell, *Delaware Estuary News*, 1996, *6*(3), 6.
14. (a) P. Layman, *Chem. Eng. News*, Mar. 20, 1995, 16; (b) Anon., *Chem. Eng. News*, May 17, 1999, 23.
15. (a) J.H. Siegell, *Chem. Eng.*, 1996, *103*(6), 92; (b) N. Mukhopadhyay and E.C. Moretti, *Reducing and Controlling VOC. American Institute of Chemical Engineers*, New York, 1993.
16. (a) R.W. Baker, J. Kaschmekat, and J.G. Wijmans, *Chemtech.*, 1996, *26*(7), 37; (b) Anon., *Chem. Eng.*, 1996, *103*(9), 125.
17. *Presidential Green Chemistry Challenge*. EPA 744-K-96–001, Washington, DC, July, 1996, 23.
18. T.G. Kemper and T. Gum, *News Fats Oil, Relat. Mater*, 1994, *5*(8), 898, 902.
19. (a) The Solvents Council of the Chemical Manufacturers Association, *Mod. Paint Coat*, 1996, *86*(4), 22; (b) D.W. Hairston, *Chem. Eng.*, 1997, *104*(2), 55; (c) E.M. Kirschner, *Chem. Eng. News*, June 20, 1994, 13; (d) R.A. Sheldon, *Green Chem.*, 2005, *7*, 267; (e) W.M. Nelson, *Green Solvents For Chemistry—Perspectives and Practice*, Oxford University Press, New York, 2003; (f) D.J. Adams, P.J. Dyson, and S.J. Tavener, *Chemistry in Alternative Reaction Media*, Wiley, Hoboken, NJ, 2004; (g) J.M. deSimone, *Science*, 2002, *297*, 799; (h) K. Mikami, ed., *Green Reaction Media in Organic Synthesis*, Wiley, Hoboken, NJ, 2005; (i) F.M. Kerton, *Alternative Solvents for Green Chemistry*, Royal Society of Chemistry, Cambridge, UK, 2009.
20. A. Russell, J. Milford, M.S. Bergin, S. McBride, L. McNair, Y. Yang, W.R. Stockwell, and B. Croes, *Science*, 1995, *269*, 491.
21. Ferro Corporation advertisement, *Chem. Eng. News*, June 3, 1996, 48.
22. DuPont advertisement, *Chem. Eng. News*, Oct. 30, 1995, 33.
23. E.M. Kirschner, *Chem. Eng. News*, June 20, 1994, 13.
24. Eastman Chemical advertisement, *Chem. Eng. News*, July 24, 1995, 29.
25. S.C. Stinson, *Chem. Eng. News*, Nov. 25, 1996, 41.
26. A.M. Brownstein, *Chemtech.*, 1995, *25*(11), 34; R.A. Grey and D. Armstead U.S. patent 5,401,857, 1995.
27. W. Cabri, I. Candiani, M. Colombo, L. Franzoi, and A. Bedeschi, *Tetrahedron Lett.*, 1995, *36*, 949.
28. Anon., *Chem. Eng. News*, Sept. 11, 1995, 45.
29. Advertisement, *Chem. Eng. News*, Nov. 24, 1997, 92.
30. (a) D. Vinci, M. Donaldson, J.P. Hallet, E.A. John, P. Pollet, C.A. Thomas, J.D. Grilly, P.G. Jessop, C.L. Liotta, and C.A. Eckert, *Chem. Commun.*, 2007, 1427; (b) N. Jiang, D. Vinci, C.L. Liotta, C.A. Eckert, and A.J. Ragaukas, *Ind. Eng. Chem. Res.*, 2008, *47*, 627. (c) M.E. Donaldson, V.L. Mestre, D. Vinci, C.L. Liotta, and C.E. Eckert, *Ind. Eng. Chem. Res.*, 2009, *48*, 2541.
31. (a) S.M. Pereira, G.P. Savage, and G.W. Simpson, *Synth. Commun.*, 1995, *25*, 1023; (b) D.W. Zhu, *Macromolecules*, 1996, *29*, 2813.
32. H. Nakano and T. Kitazume, *Green Chem.*, 1999, *1*(1), 21.
33. (a) D.P. Curran and A. Ogawa, *J. Org. Chem.*, 1997, *62*, 450; (b) Anon., *Chem. Eng. News*, Feb. 24, 1997, 31.
34. E.M. Kirschner, *Chem. Eng. News*, June 20, 1994, 13.
35. (a) M. Freemantle, *Chem. Eng. News*, Mar. 30, 1998, 32; (b) M.J. Earle, P.B. McCormac, and K.R. Seddon, *Green Chem.*, 1999, *1*(1), 23; (c) N. Karodia, S. Guise, C. Newlands, and J.-A. Andersen, *Chem. Commun.*, 1998, 2341; (d) M.J. Earle, P.B. McCormac, and K.R. Seddon, *Chem. Commun.*, 1998, 2245; (e) P.J. Dyson, D.J. Ellis, D.G. Parker, and T. Welton, *Chem. Commun.*, 1999, 25; (f) M. Freemantle, *Chem. Eng. News*, Mar. 1, 1999, 11; May 10, 1999, 9; (g) B. Ellis, W. Keim, and P. Wasserscheid, *Chem. Commun.*, 1999, 337.
36. K. Weissermel and H.-J. Arpe, *Industrial Organic Chemistry*, 2nd ed., pp. 27–29, 35–37, VCH, Weinheim, 1993.
37. H.A. Wittcoff and B.G. Reuben, *Industrial Organic Chemicals*, Vol. 94, p. 152, Wiley, New York, 1996.
38. A. Rajathurai, *Chem. Ind.* (*London*), 1997, 758.
39. G. Samdani, *Chem. Eng.*, 1995, *102*(5), 17.
40. M. Selva, C.A. Marques, and P. Tundo, *J. Chem. Soc. Perkin 1*, 1994, 1323; Preprints A.C.S. *Div. Environ. Chem.*, 1994, *34*(2), 336.
41. (a) G.M. Kissinger, Preprints A.C.S. *Div. Environ. Chem.*, 1997, *37*(1), 360; (b) C.M. Caruana, *Chem. Eng. Prog.*, 2000, *96*(1), 11; (c) G.S. Ondrey, P. Silverberg, and T. Kamiya, *Chem. Eng.*, 2000, *107*(1), 3.
42. B.F. Haumann, *Int. News Fats Oils Relat. Mater*, 1997, *8*, 165.
43. N.W. Said, *Int. News Fats Oils Relat. Mater*, 1998, *9*(2), 139.
44. F. Cataldo, *J. Appl. Polym. Sci.*, 1995, *58*, 2063.
45. K. Weissermel and H.-J. Arpe, *Industrial Organic Chemistry*, 2nd ed., 388–391, VCH Weinheim, 1993.
46. K. Tanaka, *Solvent-Free Organic Synthesis*, Wiley-VCH, Weinheim, 2003.
47. (a) D.C. Dittmer, *Chem. Ind.* (*London*), 1997, 388–391; (b) J.O. Metzer, *Angew. Chem. Int. Ed.*, 1998, *37*, 2975.
48. R.S. Varma, *Green Chem.* 1999, *1*(1), 43.
49. H.M.S. Kumar, B.V. Subbareddy, S. Anjaneyulu, and J.S. Yadav, *Synth. Commun.*, 1998, *28*, 3811.
50. R. Ballini, G. Bosica, and P. Forconi, *Tetrahedron*, 1996, *52*, 1677.
51. U. Constantino, M. Curini, F. Marmottini, D. Rosati, and E. Pisani, *Chem. Lett.*, 1994, *23*, 2215.
52. (a) A. Soriente, A. Spinella, M. DeRosa, M. Giordano, and A. Scettri, *Tetrahedron Lett.*, 1997, *38*, 289; (b) A. Boruah, M. Baruah, D. Prajapati, and J.S. Sandhu, *Synth. Commun.*, 1998, *28*, 653.
53. Y. Higashio and T. Nakayama, *Catal. Today*, 1996, *28*, 127.
54. I. Almena, A. Diaz–Ortiz, E. Diez–Barra, A. de la Hoz, and A. Loupy, *Chem. Lett.*, 1996, *25*, 333.
55. (a) Z.V. Todres, *Organic Mechanochemistry and Its Practical Applications*, CRC Press, Boca Raton, FL, 2006; (b) Anon., *Chem. Eng. News*, Nov. 9, 2009, 41.

56. F. Toda, *Acc. Chem. Res.*, 1995, *28*, 480.
57. B. Yao, J. Bassus, and R. Lamartine, *New J. Chem.*, 1996, *20*, 913.
58. J. Antesberger, G.W.V. Cave, M.C. Ferrarelli, and J.L. Atwood, *Chem. Commun.*, 2005, 892.
59. A.L. Garay, A. Pichon, and S.L. James, *Chem. Soc. Rev.*, 2007, *36*, 846.
60. H. Hagiwara, S. Ohtsubo, and M. Kato, *Tetrahedron*, 1997, *53*, 2415.
61. (a) M. Ghiaci and J. Asghari, *Synth. Commun.*, 1998, *28*, 2213; (b) M. Ghiaci and G.H. Imanzadeh, *Synth. Commun.*, 1998, *28*, 2275.
62. Y. Toyoshi, T. Nakato, R. Tamura, H. Takahashi, H. Tsue, K.-I. Hirao, and T. Okuhara, *Chem. Lett.*, 1998, *27*, 135.
63. S.B. Umbarkar, A.V. Biradar, S.M. Mathew, S.B. Shelke, K.M. Malshe, P.T. Patil, S.P. Dadge, S.P. Niphadkar, and M.K. Dongare, *Green Chem.*, 2006, *8*, 488.
64. B. Rodriguez, T. Rantanen, and C. Bolm, *Angew. Chem. Int. Ed.*, 2006, *45*, 6924.
65. T.H. Grindstaff, *Chem. Eng. News*, Nov. 7, 1994, 3.
66. M. Hasegawa, Y. Akiho, and Y. Kanda, *J. Appl. Polym. Sci.*, 1995, *55*, 297.
67. C.D. Dimitrakopoulos, S.P. Kowalczyk, and K.-W. Lee, *Polymer*, 1995, *36*, 4983.
68. I.P. Parkin, *Chem. Ind.* (*London*), 1997, 725.
69. (a) A.A. Boyd-Boland, M. Chai, Y.Z. Luo, Z. Zhang, M.J. Yang, J.B. Pawilisxyn, and T. Gorecki, *Environ. Sci. Technol.*, 1994, *28*, 569A; (b) R. Eistert and K. Levsen, *J. Am. Soc. Mass Spect.*, 1995, *6*(11), 1119; (c) D.L. Heglud and D.C. Tilotta, *Environ. Sci. Technol.*, 1996, *30*, 1212; (d) T.K. Choudhury, K.O. Gerhardt, and T.P. Mawhinney, *Environ. Sci. Technol.*, 1996, *30*, 3259; (e) S.P. Thomas, R.S. Ranian, G.R.B. Webster, and L.P. Sarna, *Environ. Sci. Technol.*, 1996, *30*, 1521; (f) E.M. Thurman and M.S. Mills, *Solid-Phase Extraction—Principles and practice*, Wiley, New York, 1998; (g) J. Pawliszyn, *Solid Phase Microextraction—Theory and Practice*, Wiley, New York, 1997; (h) V. Comello, *R & D* (*Cahners*), 1999, *41*(2), 44; (i) R. Marsili, *R & D* (*Cahners*) 1994, *41*(2), 46.
70. H. McNab, *Aldrichchimica Acta*, 2004, *37*(1), 19.
71. (a) R.J.J. Jachuck, *Process Intensification in the Chemical Industry*, CRC Press, Boca Raton, FL, 2006; (b) T. Wirth, *Microreactors in Organic Synthesis and Catalysis*, Wiley-VCH, Weinheim, 2008; (c) Y. Wang and J.D. Holladay, *Microreactor Technology and Process Intensification*, ACS Symposium, 914, Oxford University Press, 2005; (d) B.P. Mason, K.E. Price, J.L. Steinbacher, A.R. Bogdan, and D.T. McQuade, *Chem. Rev.*, 2007, *107*, 2300; (e) M.M. Sharma, *Pure Appl. Chem.*, 2002, *74*, 2265; (f) Anon., *Catal. Today*, 2007, *125*, (1–2), (an issue of the journal microreactors.); (g) P. Mitchell, *Chem. Ind.* (*London*), Apr. 4, 2005, 16; (h) G.N. Doku, W. Verboom, D.N. Reinhardt, and A. van den Berg, *Tetrahedron*, 2005, *61*(11), 2733; (i) V. Hessel, S. Hardt, and H. Lowe, *Chemical Micro-Process Engineering—Fundamentals, Modeling and Reactions*, Wiley-VCH, Weinheim, 2004; (j) R.C. Costello, *Chem. Eng.*, 2004, *111*(4), 27; (k) M. Spear, *Chem. Ind.* (*London*), May 1, 2006, 16; (l) L. Kiwi-Minsker and A. Renken, *Catal. Today*, 2005, *110*, 2; (m) P. Watts and C. Wiles, *Chem. Commun.*, 2007, 443; (n) C. Tsouris and J. Vorcelli, *Chem. Eng. Prog.*, 2003, *99*(10), 50; (o) A.M. Rouhi, *Chem Eng. News*, July 5, 2004, 18; (p) S.V. Luis and E. Garcia-Verdugo, *Chemical Reactions and Processes Under Flow Conditions*, Royal Society of Chemisty, 2009; (q) M. Doble, *Chem. Eng. Prog.*, 2008, *104*(8), 33; (r) A. Weiler, *Chem. Ind. (Lond.)*, April 27, 2009, 19; (s) M.P. Dudukovic, *Science*, 2009, *325*, 698; (t) T. vanGerven and A. Stankiewicz, *Ind. Eng. Chem. Res.*, 2009, *48*, 2465; (u) T. Wirth, ed., *Microreactors in Organic Synthesis and Catalysis*, Wiley-VCH, Weinheim, 2008; (v) T.R. Dietrich, *Microchemical Engineering in Practice*, Wiley, Hoboken, NJ, 2009.
72. P.L. Short, *Chem. Eng. News*, Oct. 20, 2008, 37.
73. D.A. Waterkamp, M. Heiland, M. Schluter, J.C. Saluvlagelalu, F. Beyersdorff, and J. Thoming, *Green Chem.*, 2007, *9*, 1084.
74. Y. Wada, M.A. Schmidt, and K.F. Jensen, *Ind. Eng. Chem. Res*, 2006, *45*, 8036.
75. T. Gustafsson, F. Ponten, and P.H. Seeberger, *Chem. Commun.*, 2008, 1100.
76. D. Bekder, M. Ludwig, L.-W. Wang, and M.T. Reetz, *Angew. Chem. Int. Ed.*, 2006, *45*, 2463.
77. Y. Ushiogi, T. Hasle, Y. Linuma, A. Takata, and J.-i. Yoshida, *Chem. Commun.*, 2007, 2947.
78. (a) A. Nagaki, K. Kawamura, S. Suga, T. Ando, M. Sawamoto, and J.-i. Yoshida, *J. Am. Chem. Soc.*, 2004, *126*, 14702; (b) T. Iwasaki, A. Nagaki, and J.-i. Yoshida, *Chem. Commun.*, 2007, 1263.
79. H. Zhao, J.-X. Wang, Q.-A. Wang, J.-F. Chen, and J. Yun, *Ind. Eng. Chem. Res.*, 2007, *46*, 8229.
80. A.A. Hussan, O. Sandre, V. Cabuil, and P. Tabeling, *Chem. Commun.*, 2008, 1783.
81. L. Frenz, A. El Harrak, M. Pauly, S. Begin-Colin, A.D. Griffiths, and J.-C. Baret, *Angew. Chem. Int. Ed.*, 2008, *47*, 6817.
82. B.K.H. Yen, A. Gunther, M.A. Schmidt, K.F. Jensen, and M.G. Bawendi, *Angew. Chem. Int. Ed.*, 2005, *44*, 5447.
83. (a) E. Quevedo, J. Steinbacher, and D.T. McQuade, *J. Am. Chem. Soc.*, 2005, *127*, 10498; (b) S.J. Broadwater, S.L. Roth, K.E. Price, M. Kobaslija, and D.T. McQuade, *Org. Biomol. Chem.*, 2005, *3*, 2899; (c) S.L. Poe, M.A. Cummings, M.P. Haaf, and D.T. McQuade, *Angew. Chem. Int. Ed.*, 2006, *45*, 1545; (d) H. Song, D.L. Chen, and R.F. Ismagilov, *Angew. Chem. Int. Ed.*, 2006, *45*, 7336.
84. (a) M. Freemantle, *Chem. Eng. News*, Feb. 7, 2005, 11; (b) P.K. Plucinski, D.V. Bavykin, S.T. Kolaczkowski, and A.A. Lapkin, *Ind. Eng. Chem. Res.*, 2005, *44*, 9683.
85. R. Mullin, *Chem. Eng. News*, Sept. 24, 2007, 49.
86. G. Ondrey, *Chem. Eng.*, 2006, *113*(5), 13.
87. G. Ondrey, *Chem. Eng.*, 2008, *115*(1), 11.
88. D. Huang, Y.C. Lu, Y.J. Wang, L. Yang, and G.S. Luo, *Ind. Eng. Chem. Res.*, 2008, *47*, 3870.
89. R. Subramanian and L.D. Schmidt, *Angew. Chem. Int. Ed.*, 2005, *44*, 302.
90. B. Dunk and R. Jachuck, *Green Chem.*, 2000, *2*, G13.
91. P. Oxley, C. Brechtelsbauer, F. Ricard, N. Lewis, and C. Ramshaw, *Ind. Eng. Chem. Res.*, 2000, *39*, 2175.
92. C.Y. Tai, C.-T. Tai, M.-H. Chang, and H.-S. Liu, *Ind. Eng. Chem. Res.*, 2007, *46*, 5536.
93. G. Ondrey, *Chem. Eng.*, 2007, *114*(6), 10.
94. J.L. White, *Twin Screw Extrusion—Technology and Principles*, Hanser Gardner, Cincinnati, OH, 1990.

95. Y. Higuchi, S. Kuwahara, S. Hieda, and M. Kurokawa, *Chem. Abstr.*, 1995, *123*, 229, 395.
96. M. Sasai and H. Yamada, *Chem. Abstr.*, 1995, *122*, 188, 519.
97. G. Parkinson, *Chem. Eng.*, 1997, *104*(12), 25.
98. (a) J. Callari, *Plast. World*, 1993, *51*(9), 10; (b) E. Uhland and W. Wiedmann, *Macromol. Symp.*, 1994, *83*, 59; (c) A.D. Sagar and E.W. Merrill, *Polymer*, 1995, *36*, 1883.
99. DN.-S. Hon and W.Y. Chao, *J. Appl. Polym. Sci.*, 1993, *50*, 7.
100. (a) M. Xanthos, ed., *Reactive Extrusion: Principles and Practice*, 3rd ed. Hanser, Cincinnati, OH, 1992; (b) M. Lambla, *Macromol. Symp.*, 1994, *83*, 37; (c) H.G. Fritz, Q. Cai, U. Bolz, and R. Anderlik, *Macromol. Symp.*, 1994, *83*, 93; (d) P. Cassagnau and M. Taha, *J. Appl. Polym. Sci.*, 1996, 60, 1765; (e) D. Burlett and J.T. Lindt, *Rubber Chem. Technol.*, 1993, *66*, 411; (f) K. Kircher, *Chemical Reactions in Plastics Processing*, Hanser Gardner, Cincinnati, OH, 1987.
101. B.K. Kim and K.J. Kim, *Adv. Polym. Technol.*, 1993, *12*, 263.
102. (a) M. Ratzsch, *Prog. Polym. Sci.*, 1994, *19*, 1011; (b) G.-H. Hu, J.J. Flat, and M. Lambla, *Macromol. Chem. Macromol. Symp.*, 1993, *75*, 137.
103. (a) A.N. Wilkinson, D. Cole, and S.B. Tattum, *Polym. Bull*, 1995, 35, 751; (b) M. Fiorini, C. Berti, V. Ignatov, M. Toselli, and F. Pilati, *J. Appl. Polym. Sci.*, 1995, *55*, 1157; (c) K.-F. Su and K.-H. Wei, *J. Appl. Polym. Sci.*, 1995, *56*, 79.
104. M. Evstatiev, J.M. Schultz, S. Petrovich, G. Geolgiev, S. Fakirov, and K. Friedrich, *J. Appl. Polym. Sci.*, 1998, *67*, 723.
105. K.L.L. Eersels and G. Groeninckx, *J. Appl. Polym. Sci.*, 1997, *63*, 573.
106. W.M. Stevels, A. Bernard, P. van de Witte, P.J. Dijkstra, and J. Feijen, *J. Appl. Polym. Sci.*, 1996, *62*, 1295.
107. D. Lohse, *Polymeric Compatibilizers—Uses and Benefits in Polymer Blends*, Hanser Gardner, Cincinnati, OH, 1996.
108. A. Hesse and M. Ratzsch, *J. Macromol. Sci. Pure Appl. Chem.*, 1995, *A32*, 1305.
109. A. Hesse and M. Ratzsch, *J. Macromol. Sci. Pure Appl. Chem.*, 1994, *A31*, 1425.
110. G. Samay, T. Nagy, and J.L. White, *J. Appl. Polym. Sci.*, 1995, *56*, 1423.
111. R. Joachim, W. Holger, M. Philipp, S. Rudiger, W. Christof, F. Christian, K. Jorg, and M. Rolf, *Macromol. Symp.*, 1996, *102*, 241.
112. J. Rosch, R. Mulhaupt, and G.H. Michler, *Macromol. Symp.*, 1996, *112*, 141.
114. C. Maier and M. Lambla, *Angew. Macromol. Chem.*, 1995, *231*, 145.
114. (a) H. Xie, M. Seay, K. Oliphant, and W.E. Baker, *J. Appl. Polym. Sci.*, 1993, *48*, 1199; (b) K.E. Oliphant, K.E. Russell, and W.E. Baker, *Polymer*, 1995, *36*, 1597.
115. B. Turcsanyi, *Macromol. Rep.*, 1995, *A32*, 255.
116. L.P. Engle and K.B. Wagener, *J. Macromol. Sci. Rev. Macromol. Chem.*, 1993, *C33*, 239.
117. (a) K. Buehler and M. Gebauer, *Chem. Abstr.*, 1996, *124*, 290, 672; (b) Y.-J. Sun, G.-H. Hu, and M. Lambla, *Angew. Macromol. Chem.*, 1995, *229*, 1.
118. G.-H. Hu, Y.-J. Sun, and M. Lambla, *J. Appl. Polym. Sci.*, 1996, *61*, 1039.
119. (a) G.-H. Hu and H. Cartier, *J. Appl. Polym. Sci.*, 1999, *71*, 125; (b) H. Cartier and G.-H. Hu, *J. Polym. Chem. A Polym. Chem.*, 1998, *36*, 2763.
120. (a) J.H. Wang, V.P. Kurkov, L. Theard, and D. Rosendale, *Chem. Abstr.*, 1995, *123*, 287, 364; (b) I.M. Vermeesch, G. Groeninckx, and M.M. Coleman, *Macromolecules*, 1993, *26*, 6643; (c) I.M. Vermeesch and G. Groeninckx, *J. Appl. Polym. Sci.*, 1994, *53*, 1365.
121. A.M. Khan and J.A. Pojman, *Trends Polym. Sci.*, 1996, *4*, 253.
122. (a) P.R. Hornsby, J.F. Tung, and K. Taverdi, *J. Appl. Polym. Sci.*, 1994, *54*, 899; 1994, *53*, 891; (b) H. Kye and J.L. White, *J. Appl. Polym. Sci.*, 1994, *52*, 1249; (c) Z.-G. Yang and B. Lauke, *J. Appl. Polym. Sci.*, 1995, *57*, 679.
123. H. Wautier, *Chem. Abstr.*, 1995, *122*, 291, 756.
124. (a) W. Michaeli, H. Hocker, U. Berghaus, and W. Frings, *J. Appl. Polym. Sci.*, 1993, *48*, 871; (b) F.L. Binsbergen, W. Sjardijn, and U. Berghaus, *Chem. Abstr.*, 1993, *118*, 234, 707.
125. T. Hatsu, *Chem. Abstr.*, 1993, *119*, *28*, 864.
126. M. Okabe and T. Fukuoka, *Chem. Abstr.*, 1995, *123*, 314, 898.
127. T. Groth, W. Joentgen, N. Mueller, and U. Liesenfelder, *Chem. Abstr.*, 1995, *122*, 106, 855.
128. S. Coudray, J. Pascault, and M. Taha, *Polym. Bull.*, 1994, *32*, 605.
129. (a) D.S. Breslow, *Prog. Polym. Sci.*, 1993, *18*(6), 1141; (b) F. Stelzer, *J. Macromol. Sci. Pure Appl. Chem.*, 1996, *A33*, 941 (metathesis polymerization).
130. G.S. Samdani, *Chem. Eng.*, 1994, *101*(11), 27.
131. S.P. Yushanov, A.I. Isayev, and S.H. Kim, *Rubber Chem. Technol.*, 1998, *71*, 168.
132. R.S. Varma, *Green Chem.*, 2008, *10*, 1129; *Green Chem. Lett. Rev.*, 2007, *1*(1), 37.
133. M. Poliakoff, N.J. Meehan, and S.K. Ross, *Chem. Ind. (London)*, 1999, 750.
134. S. Lamare, B. Caillaud, K. Rouie, I. Goubet, and M.D. Legoy, *Biocatal. Biotransform.*, 2001, *19*, 361.
135. (a) E. Kiran and J.F. Brennecke, eds, *Supercritical Fluid Engineering Science: Fundamentals and Applications.* Am. Chem. Soc. Symp., 514, Washington, DC, 1993; (b) K.W. Hutchenson and N.R. Foster, eds, *Innovations in Supercritical Fluids: Science and Technology.* Am. Chem. Soc. Symp., 608, Washington, DC, 1995; (c) T.J. Bruno and J.F. Ely, eds, *Supercritical Fluid Technology*, p. 191, CRC Press, Boca Raton, FL; (d) R. van Eldik and C.D. Hubbard, *Chemistry Under Extreme or Non-classical Conditions*, Wiley, New York, 1996; (e) P.E. Savage, S. Gopolan, T.I. Mizan, C.J. Martino, and E.E. Brock, *AIChEJ* 1995, *41*, 1723–1778; (f) J.F. Brennecke, *Chem. Ind.*, 1996, 831; (g) M. Poliakoff and S. Howdle, *Chem. Br.*, 1995, *31*(2), 118; (h) T. Clifford and K. Bartle, *Chem. Ind.* (*London*), 1996, 449; (i) H. Black, *Environ. Sci. Technol.*, 1996, *30*, 125A; (j) R. Fernandez-Prini and M.L. Japas, *Chem. Soc. Rev.*, 1994, *23*, 155; (l) F. Stuber, A.M. Vazquez, M.A. Larrayoz, and F. Recasens, *Ind. Eng. Chem. Res.*, 1996, *35*, 3618; (m) K.A. Shaffer and J.M. deSimone, *Trends Polym. Sci.*, 1995, *3*, 146; (n) M.A. Abraham and A.K. Sunol, *Supercritical Fluids: Extraction and Pollution Prevention.* Am. Chem. Soc. Symp., 670, Washington, DC, 1997; (o) B. Bungert, G. Sadowski, and W. Arlt, *Ind. Eng. Chem. Res.*, 1998, *37*, 3208; (p) R. Noyori, ed., *Chem. Rev.*, 1999, *99*, 353–634; (q) K. Chin, C. Crabb, G. Ondrey, and T. Kamiya, *Chem. Eng.*, 1998, *105*(11), 32;

(r) A.S. Gopolan, C.M. Wai, and H.K. Jacobs, *Supercritical Carbon Dioxide—Separations and Processes*, Am. Chem. Soc. Symp. 860, Oxford University Press, New York, 2003; (s) T. Seki and A. Baiker, *Chem. Rev.*, 2009, *109*, 2409.
136. (a) M.A. McHugh and V.J. Krukonis, *Supercritical Fluid Extraction*, Butterworth–Heinemann, Boston, 1994; (b) L.T. Taylor, *Supercritical Fluid Extraction*, Wiley, New York, 1996; (c) A.N. Clifford, T. Lynch, R.J.H. Wilson, K.D. Bartle, C.M. Wai, J.W. King, G. Davidson, M. Kaplan, G. Liang, and D.A. Moyler, *J. Chem. Technol. Biotechnol.* 1996, *65*, 293.
137. L. Cooke, *Agric. Res.*, 1995, *43*(3), 18.
138. N.O. Maness, D. Chrz, T. Pierce, and G.H. Brusewitz, *J. Am. Oil Chem. Soc.*, 1995, *72*, 665.
139. E. Reverchon, G. Della Porta, and F. Senatore, *J. Agric. Food Chem.*, 1995, *43*, 1654.
140. S. Samdani, *Chem. Eng.*, 1995, *102*(8), 23.
141. B.C. Roy, M. Goto, and T. Hirose, *Ind. Eng. Chem. Res.*, 1996, *35*, 607.
142. R.L. Mendes, J.P. Coelho, H.L. Fernandes, I.J. Marrucho, J.M.S. Cabral, J.M. Novais, and A.F. Palavra, *J. Chem. Technol. Biotechnol.*, 1995, *62*, 53.
143. D.L. Taylor and D.K. Larick, *J. Agric. Food Chem.*, 1995, *43*, 2369.
144. I. Medina and J.L. Martinez, *J. Chem. Technol. Biotechnol.*, 1997, *68*, 14.
145. (a) M. Baker, *New Sci.*, Feb. 7, 1998, 22; (b) J.L. Leazer, Jr. S. Gant, A. Houck, W. Leonard and C.J. Welch, *Environ. Sci. Technol.*, 2009, *43*, 2018.
146. M.S. Kuk and R.J. Hron, *J. Am. Oil Chem. Soc.*, 1994, *71*, 1353.
147. M.J. Cocero and L. Calvo, *J. Am. Oil Chem. Soc.*, 1996, *73*, 1573.
148. N.T. Dunford and F. Temelli, *J. Am. Oil Chem. Soc.*, 1995, *72*, 1009.
149. T. Tsuda, K. Mizuno, K. Ohshima, S. Kawakishi, and T. Osawa, *J. Agric. Food Chem.*, 1995, *43*, 2803.
150. J. King and W.V. O'Farrell, *Int. News Fats Oils Relat. Mater.*, 1997, *8*, 1047.
151. P.J. Wan and R.J. Hron, *Int. News Fats Oils Relat. Mater.*, 1998, *9*, 707.
152. E. Sahle-Demessie, *Ind. Eng. Chem. Res.*, 1997, *36*, 4906.
153. F.P. Lucien and N.R. Foster, *Ind. Eng. Chem. Res.*, 1996, *35*, 4686.
154. (a) C.A. Blaney and S.U. Hossan, *Chemtech.*, 1997, *27*(2), 48; (b) E. Kiran, K. Malki, and H. Pohler, *Polym. Mater. Sci. Eng.*, 1996, *74*, 231.
155. G. Parkinson, *Chem. Eng.*, 1998, *105*(8), 17.
156. (a) G.A. Montero, T.D. Giorgio, and K.B. Schnelle, Jr., *Environ. Prog.*, 1996, *15*(2), 112; (b) J.H. Krieger, S. Borman, R.M. Baum, and M. Freemantle, *Chem. Eng. News*, Mar. 20, 1995, 55.
157. (a) J. Puiggene, A. Adivinacion, E. Velo, and F. Recasens, *Catal. Today*, 1994, *20*, 541; (b) K.F. Ng, N.K. Nair, K.Y. Liew, and A.M. Noor, *J. Am. Oil Chem. Soc.*, 1997, *74*, 963.
158. M.S. Roby and N. Totakura, U.S. patent 5,478,921, 1995.
159. K. Sakanishi, H. Obata, I. Mochida, and T. Sakaki, *Ind. Eng. Chem. Res.*, 1996, *35*, 335.
160. (a) L.A. Blanchard, D. Hancu, E.J. Beckman, and J.F. Brennecke, *Nature*, 1999, *399*, 28; (b) M. Freemantle, *Chem. Eng. News*, May 10, 1999, 9.
161. (a) P.G. Jessop and B. Subramaniam, *Chem. Rev.*, 2007, *107*, 2666; (b) E.M. Hill, J.M. Broering, J.P. Hallett, A.S. Bommarius, C.L. Liotta, and C.A. Eckert, *Green Chem.*, 2007, *9*, 888; (c) J.P. Hallett, C.L. Kitchens, R. Hernandez, C.L. Liotta, and C.A. Eckert, *Acc. Chem. Res.*, 2006, *39*, 531; (d) G.R. Akien and M. Poliakoff, *Green Chem.*, *11*, 1083.
162. (a) E.K. Wilson, *Chem. Eng. News*, Apr. 15, 1996, 27; (b) S. Wang and C.M. Wai, *Environ. Sci. Technol.*, 1996, *30*, 3111; (c) W. Cross, Jr., A. Akgerman, and C. Erkey, *Ind. Eng. Chem. Res.*, 1996, *35*, 1765; (d) N.G. Smart, T.E. Carleson, S. Elshani, S. Wang, and C.M. Wai, *Ind. Eng. Chem. Res.*, 1997, *36*, 1819.
163. (a) Y. Lin, N.G. Smart, and C.M. Wai, *Environ. Sci. Technol.*, 1995, *29*, 2706; (b) S. Iso, Y. Meguro, and Z. Yoshida, *Chem. Lett.*, 1995, 365; (c) G.S. Samdani, *Chem. Eng.*, 1995, *102*(6), 17; (d) N. Smart, C. Wai, and C. Phelps, *Chem. Br.*, 1998, *34*(8), 34; (e) M.J. Carrott, B.E. Waller, N.G. Smart, and C.M. Wai, *Chem. Commun.*, 1998, 373.
164. A.V. Yazdi and E.J. Beckman, *Ind. Eng. Chem. Res.*, 1997, *36*, 2368.
165. A.V. Yazdi and E.J. Beckman, *Ind. Eng. Chem. Res.*, 1996, *35*, 3644.
166. J.M. Murphy and C. Erkey, *Ind. Eng. Chem. Res.*, 1997, *36*, 537.
167. (a) E. Sahle-Demessie, K.L. Levien, and J.J. Morrell, *Chemtech.*, 1998, *28*(3), 12; (b) S.P. Nalawade, F. Pilcclhioni, and P.B.M. Jansssen, *Prog. Polym. Sci.*, 2006, *31*(1), 19; (c) L.J.M. Jacobs, M.F. Kemmere, and J.T.F. Keurentjes, *Green Chem.*, 2008, *10*, 731; (d) J.W.S. Lee, K. Wang, and B. Park, *Ind. Eng Chem. Res.*, 2005, *44*, 92; (e) M.F. Kemmere and T. Meyer, eds, *Supercritical Carbon Dioxide in Polymer Reaction Engineering*, Wiley-VCH, Weinheim, 2005; (f) M. Pagliaro, *Angew. Chem. Int. Ed.*, 2006, *45*, 6079.
168. Anon., *Chem. Eng. News*, Sept. 16, 1996, 42.
169. G. Gavend, B. Vulliermet, C. Perre, and M. Carles, U.S. patent 5,512,058, 1996.
170. J.J. Watkins and T.J. McCarthy, *Macromolecules*, 1994, *27*, 4845; 1995, *28*, 4067; (b) E. Kung, A.J. Lesser, and T.J. McCarthy, *Macromolecules*, 1998, *31*, 4160.
171. T. Kajitani, Y. Uosaki, and T. Moriyoshi, *J. Appl. Polym. Sci.*, 1995, *57*, 587.
172. (a) Y.-T. Shih, J.H. Su, G. Manivannan, P.H.C. Lee, S.P. Sawan, and W.D. Spall, *J. Appl. Polym. Sci.*, 1996, *59*, 695, 707; (b) K.A. Arora, A.J. Lesser, and T.J. McCarthy, *Macromolecules*, 1998, *31*, 4614.
173. S.J. Mulvaney and S.S.H. Rizvi, *Food Technol.*, 1993, *47*(12), 74.
174. G. Luna-Barcenas, S.K. Kanakia, I.C. Sanchez, and K.P. Johnston, *Polymer*, 1995, *36*, 3173.
175. J.-H. Kim, T.E. Paxton, and D.L. Tomasko, *Biotechnol. Prog.*, 1996, *12*, 650.
176. S. Mawson, K.P. Johnston, D.E. Betts, J.B. McClain, and J.M. DeSimone, *Macromolecules*, 1997, *30*, 71.
177. S. Mawson, S. Kanakia, and K.P. Johnston, *J. Appl. Polym. Sci.*, 1997, *64*, 2105.
178. E. Reverchon, G. Della Porta, A. diTrolio, and S. Pace, *Ind. Eng. Chem. Res.*, 1998, *37*, 952.
179. S. Moisan, J.-D. Marty, F. Cansell, and C. Aymonier, *Chem. Commun.*, 2008, 1428.
180. D. Alessi, A. Cortesi, F. Kikic, N.R. Foster, S.J. Macnaughton, and I. Colombo, *Ind. Eng. Chem. Res.*, 1996, *35*, 4718.

181. (a) J.B. McClain, D. Londono, J.R. Combes, T.J. Romack, D.A. Canelas, D.E. Betts, G.D. Wignall, E.T. Samulski, and J.M. de Simone, *J. Am. Chem. Soc.*, 1996, *118*, 917; (b) M.L. O'Neill, Q. Cao, M. Fang, K.P. Johnston, S.P. Wilkinson, C.D. Smith, J. Kerschner, and S.H. Jureller, *Ind. Eng. Chem. Res.*, 1998, *37*, 3067; (c) R.B. Gupta and J.-J. Shim, *Solubility In sc Carbon Dioxide*, CRC Press, Boca Raton, FL, 2007.
182. W.H. Tuminello, G.T. Dee, and M.A. McHugh, *Macromolecules*, 1995, *28*, 1506.
183. K.A. Kennedy, G.W. Roberts, and J.M. deSimone, *Adv. Polym. Sci.*, 2005, *175*, 329.
184. X. Zhao, R. Watkins, and S.W. Barton, *J. Appl. Polym. Sci.*, 1995, *55*, 773.
185. (a) E.J. Beckman, U.S. E.P.A. Presidential Green Chemistry Challenge award, 2002; (b) E.J. Beckman, *Ind. Eng. Chem. Res.*, 2003, *42*, 1598.
186. (a) D. Canelas and de Simone, *Chem. Br.*, 1998, *34*(8), 38; Preprints A.C.S. *Div. Environ. Chem.*, 1997, *37*(1), 352; (b) T.-M. Yong, W.P. Hems, J.L.M. van Nunen, A.B. Holmes, J.H.G. Steinke, P.L. Taylor, J.A. Segal, and D.A. Griffin, *Chem. Commun.*, 1997, 1811; (c) S. Zhou and B. Chu, *Macromolecules*, 1998, *31*, 7746; (d) A. Cooper, *J. Mater. Chem.*, 2000, *10*, 207.
187. (a) J.B. McClain, D.E. Betts, D.A. Canelas, E.T. Samulski, J.M. de Simone, J.D. Londono, H.D. Cochran, G.D. Wignall, D. Chillura-Martin, and R. Trioli, *Science*, 1996, *274*, 2049; (b) J. Kaiser, *Science*, 1996, *274*, 2013; (c) Z. Guam and J.M. de Simone, *Macromolecules*, 1994, *27*, 5527.
188. (a) J.M. de Simone, *Macromol. Symp.*, 1995, *98*, 795; (b) J.M. de Simone, E.E. Maury, Z. Guan, J.R. Combes, Y.Z. Menceloglu, M.R. Clark, J.B. McClain, T.J. Romack, and C.D. Mistele, Preprints A.C.S. *Div. Environ. Chem.*, 1994, *34*(2), 212; (c) D.A. Canelas and J.M. de Simone, *Adv. Polym. Sci.*, 1997, *133*, 103–140. (d) J.L. Kendall, D.A. Canelas, J.L. Young, and J.M. de Simone, *Chem. Rev.*, 1999, *99*, 543.
189. (a) K.A. Shaffer, T.A. Jones, D.A. Canelas, J.M. de Simone, and S.P. Wilkinson, *Macromolecules*, 1996, *29*, 2704; (b) K.C. Fox, *Science*, 1994, *265*, 321; (c) Y.L. Hsiao, E.E. Maury, J.M. de Simone, S. Mawson, and K.P. Johnston, *Macromolecules*, 1995, *28*, 8159; (d) T.J. Romack, J.M. de Simone, and T.A. Treat, *Macromolecules*, 1995, *28*, 8429; (e) J.M. de Simone, E.E. Maury, Y.Z. Menceloglu, J.B. McClain, T.J. Romack, and J.R. Combes, *Science*, 1994, *265*, 356; (f) Y.-L. Hsiao and J.M. de Simone, *J. Polym. Sci. A Polym. Chem.*, 1997, *35*, 2009; (g) C. Lepilleur and E.J. Beckman, *Macromolecules*, 1997, *30*, 745; (h) M.L. O'Neill, M.Z. Yates, K.P. Johnston, C.D. Smith, and S.P. Wilkinson, *Macromolecules*, 1998, *31*, 2838, 2848; (i) M.F. Kemmere, M.W.A. Kuijpers, and J.T.F. Keurentjes, *Macromol. Symp.*, 2007, *248*, 182; (j) A.M. Gregory, K.J. Thurecht, and S.M. Howdle, *Macromolecules*, 2008, *41*, 1215.
190. (a) T.J. Romack, E.E. Maury, and J.M. de Simone, *Macromolecules*, 1995, *28*, 912; (b) T.J. Romack, J.R. Combes, and J.M. De Simone, *Macromolecules*, 1995, *28*, 1724; (c) R. Leiberich, H.E. Gasche, and H. Waldman, *Chem. Abstr.*, 1996, *124*, 118,313; (d) R. Leiberich, H.-U. Dummersdorf, H. Waldman, H.E. Gasche, and Z. Kricsfalussy, *Chem. Abstr.*, 1996, *124*, 118,314; (e) G. Ondrey, *Chem. Eng.*, 2005, *112*(9), 15.
191. (a) R.S. Clough, C.L.S. Elsbernd, and J.E. Gozum, *Chem. Abstr.*, 1996, *124*, 262,009; (b) J.E. Gozum and M.C. Palazzotto, *Chem. Abstr.*, 1996, *124*, 262,010; (c) S. Beuermann, M. Buback, and C. Schmaltz, *Macromolecules*, 1998, *31*, 8069.
192. (a) J.M. de Simone, E.E. Maury, J.R. Combes, and Y.Z. Menceloglu, U.S. patent 5,312,882, 1994; (b) D.A. Canelas and J.M. de Simone, *Macromolecules*, 1997, *30*, 5673; (c) M.R. Clark, J.L. Kendall and J.M. de Simone, *Macromolecules*, 1997, *30*, 6011.
193. A.I. Cooper, W.P. Hems, and A.B. Holmes, *Macromol. Rapid Commun.*, 1998, *19*, 353.
194. D.A. Canelas, D.E. Betts, J.M. de Simone, M.Z. Yates, and K.P. Johnston, *Macromolecules*, 1998, *31*, 6794.
195. M.A. Quader, R. Snook, R.G. Gilbert, and J.M. de Simone, *Macromolecules*, 1997, *30*, 6015.
196. (a) J.P. Kennedy and T. Pernecker, U.S. Patent 5,376,744, 1994; *Polym. Bull*, 1994, *32*, 537; 1994, *33*, 13, 259; (b) W. Baade, R. Heinrich, G. Heinrich, G. Langstein, T. Mulder, and J. Puskas, *Chem. Abstr.*, 1996, *124*, 233, 442.
197. M.R. Clark and J.M. de Simone, *Macromolecules*, 1995, *28*, 3002; Presidential Green Chemistry Challenge, EPA 744-K-96-001, Washington, DC, July, 1996, 40.
198. D.J. Darensbourg, N.W. Stafford, and T. Katsurao, *J. Mol. Catal. A Chem.*, 1995, *104*, L1.
199. M. Super and E.J. Beckman, *Macromol. Symp.*, 1998, *127*, 89.
200. G. Parkinson, *Chem. Eng.*, 1999, *106*(3), 25.
201. K.K. Kapellen, C.D. Mistele, and J.M. De Simone, *Macromolecules*, 1996, *29*, 495.
202. (a) C.D. Mistele, H.H. Thorp, and J.M. de Simone, *J. Macromol. Sci. Pure Appl. Chem.*, 1996, *A33*, 953; (b) J.G. Hamilton, J.J. Rooney, J.M. de Simone, and C. Mistele, *Macromolecules*, 1998, *31*, 4387.
203. (a) S.M. Gross, R.D. Givens, M. Jikei, J.R. Royer, J.M. de Simone, R.G. Odell, and G.K. Hamer, *Macromolecules*, 1998, *31*, 9090; (b) S.M. Gross, D. Flowers, G. Roberts, D.J. Kiserow, and J.M. de Simone, *Macromolecules*, 1999, *32*, 3167.
204. (a) C.M. Wai, F. Hunt, M. Ji, and X. Chen, *J. Chem. Ed.*, 1998, *75*, 1641; (b) A. Baiker, *Chem. Rev.*, 1999, *99*, 453; (c) J.A. Darr and M. Poliakoff, *Chem. Rev.*, 1999, *99*, 495; (d) D.J. Cott, K.J. Ziegler, V.P. Owens, J.D. Glennon, A.E. Graham, and J.D. Holmes, *Green Chem.*, 2005, *7*, 105.
205. J.M. Tanko and J.F. Blackert, In: P.T. Anastas and C.A. Farries, eds, *Benign by Design: Alternative Synthetic Design*. Am. Chem. Soc. Symp. 577, Washington, DC, 1994, 98.
206. A.K. Dillow, S.L.J. Yun, D. Suleiman, D.L. Boatright, C.L. Liotta, and C.A. Eckert, *Ind. Eng. Chem. Res.*, 1996, *35*, 1801.
207. K. Chandler, C.W. Culup, D.R. Lamb, C.L. Liotta, and C.A. Eckert, *Ind. Eng. Chem. Res.*, 1998, *37*, 3252.
208. F. Temelli, J.W. King, and G.R. List, *J. Am. Oil Chem. Soc.*, 1996, *73*, 699.
209. P.G. Jessop, T. Ikariya, and R. Noyori, *Science*, 1995, *269*, 1065; *Chem. Rev.*, 1999, *99*, 475.
210. (a) P.G. Jessop, Y. Hsiao, T. Ikariya, and R. Noyori, *J. Am. Chem. Soc.*, 1996, *118*, 344; (b) Anon., *Chem. Ind.* (*London*), 1995, 406.
211. (a) O. Krocher, R.A. Koppel, and A. Baiker, *Chem. Commun.*, 1996, 1497; (b) O. Krocher, R.A. Koppel, M. Froba, and A. Baiker, *J. Catal.*, 1998, *178*, 284; (c) O. Krocher, R.A. Koppel, A. Baiker, *Chem. Commun.*, 1997, 453.

212. (a) M.J. Burk, S. Feng, M.F. Gross, and W. Tumas, *J. Am. Chem. Soc.*, 1995, *117*, 8277; (b) R.F. Service, *Science*, 1995, *269*, 1339; (c) Presidential Green Chemistry Challenge, EPA 744-K-96-011, July 1996, 38; (d) M. Schmidtkamp, D. Chen, W. Leitner, J. Klankermeyer, and G. Francio, *Chem. Commun.*, 2007, 4012; (e) K. Burgemeister, G. Francio, H. Hugl, and W. Leitner, *Chem. Commun.*, 2005, 6026.
213. J. Xiao, S.C.A. Nefkens, P.G. Jessop, T. Ikariya, and R. Noyori, *Tetrahedron Lett.*, 1996, *37*, 2813.
214. (a) M.G. Hitzler and M. Poliakoff, *Chem. Commun.*, 1997, 1667; (b) S.C. Stinson, *Chem. Eng. News*, July 14, 1997, 37; (c) J.A. Darr and M. Poliakoff, *Chem. Rev.*, 1999, *99*, 495; (d) M. Poliakoff, N.J. Meehan, and S.K. Ross, *Chem. Ind. (London)*, 1999, 750; (e) R. Liu, C. Wu, Q. Wang, J. Ming, Y. Hao, Y. Yu, and F. Zhao, *Green Chem.*, 2009, *11*, 979.
215. J. Kobayashi, Y. Mori, and S. Kobayashi, *Chem. Commun.*, 2005, 2567.
216. R.A. Bourne, J.G. Stevens, J. Ke, and M. Poliakoff, *Chem. Commun.*, 2007, 4632.
217. T. Seki, J.-D. Grunwaldt, and A.A. Backer, *Ind. Eng. Chem. Res.*, 2008, *47*, 4561.
218. (a) J.W. Rathke, R.J. Klingler, and T.R. Krause, *Organometallics*, 1991, *10*, 1350; (b) Y. Guo and A. Akgerman, *Ind. Eng. Chem. Res.*, 1997, *36*, 4581.
219. D.R. Pesiri, D.K. Morita, W. Glaze, and W. Tumas, *Chem Commun.*, 1998, 1015.
220. (a) R.S. Oakes, A.A. Clifford, K.D. Bartle, M.T. Pett, and C.M. Rayner, *Chem. Commun.*, 1999, 247; (b) M. Freemantle, *Chem. Eng. News*, Feb. 15, 1999, 12.
221. M.G. Hitzler, F.R. Smail, S.K. Ross, and M. Poliakoff, *Chem. Commun.*, 1998, 359.
222. (a) J.A. Bannister, P.D. Lee, and M. Poliakoff, *Organometallics*, 1995, *14*, 3876; (b) M. Poliakoff and S. Howdle, *Chem. Br.*, 1995, *31*, 118.
223. P.G. Jessop, T. Ikariya, and R. Noyori, *Organometallics*, 1995, 14, 1510.
224. (a) Y. Iwai, H. Uchida, Y. Mori, H. Higashi, T. Matsuki, T. Furuya, and Y. Arai, *Ind. Eng. Chem. Res.*, 1994, *33*, 2157; (b) H. Uchida, Y. Iwai, M. Amiya, and Y. Arai, *Ind. Eng. Chem. Res.*, 1997, *36*, 424; 1998, *37*, 595.
225. (a) S.L. Suib, *Chem. Rev.* 1993, *93*, 821; (b) M.L. Occelli and H.E. Robson, eds, *Synthesis of Microporous Materials, Expanded Clays and other Microporous Solids*, p. 57, van Nostrand–Reinhold, New York, 1992.
226. R. Ishii, H. Wada, and K. Ooi, *Chem. Commun.*, 1998, 1705.
227. D.A. Ward and E.I. Ko, *J. Catal.*, 1994, *150*, 18.
228. A.J. Mesiano, E.J. Beckman, and A.J. Russell, *Chem. Rev.*, 1999, *99*, 623.
229. (a) E. Catoni, E. Cernia, and C. Palocci, *J. Mol. Catal. A Chem.*, 1996, *105*, 79; (b) E. Cernia, C. Palocci, F. Gasparrini, D. Misiti, and N. Fagnano, *J. Mol. Catal.*, 1994, *89*, L11.
230. J.F. Martins, I.B. de Carvalho, T.C. de Sampaio, and S. Barreiros, *Enzyme Microb. Technol.*, 1994, *16*, 785.
231. H. Gunnlaugsdottir and B. Swik, *J. Am. Oil Chem. Soc.*, 1995, *72*, 399.
232. M. Habulin, V. Krmelj, and Z. Knez, *J. Agric. Food Chem.*, 1996, *44*, 338.
233. D. Lyons-Johnson, *Agric. Res.*, 1997, *45*(8), 22.
234. (a) A.J. Russel and E.J. Beckman, *American Chemical Society meeting*, New Orleans, March 1996, PMSE 039; (b) A.J. Russell, E.J. Beckman, D. Abderrahmare, and A.K. Chaudhary, U.S. patent 5,478,910, 1995.
235. (a) K.P. Johnston, K.L. Harrison, M.J. Clarke, S.M. Howdle, M.P. Heitz, F.V. Bright, C. Carlier, and T.W. Randolph, *Science*, 1996, *271*, 624; (b) E.J. Beckman, *Science*, 1996, *271*, 613; (c) M. Rouhi, *Chem. Eng. News*, Feb. 5, 1996, 8.
236. (a) M.J. Clarke, K.L. Harrison, K.P. Johnston, and S.M. Howdle, *J. Am. Chem. Soc.*, 1997, *119*, 6399; (b) M. Rouhi, *Chem. Eng. News*, Aug. 11, 1997, 40.
237. G.B. Jacobson, J. Brady, J. Watkin, and W. Tumas, Preprints A.C.S. *Div. Environ. Chem.*, 1999, *39*(1), 466.
238. (a) A.I. Cooper, J.D. Londono, G. Wignall, J.B. McClain, E.T. Samulski, J.S. Lin, A. Dobrynin, M. Rubenstein, A.L.C. Burke, J.M.J. Frechet, and J.M. de Simone, *Nature*, 1997, *389*, 368; (b) M. Rouhi, *Chem. Eng. News*, Sept. 29, 1997.
239. M.W.P.L. Bears, P.E. Froehling, and E.W. Meijer, *Chem. Commun.*, 1997, 1959.
240. M.D. Donohue and J.L. Geiger, Preprints A.C.S. *Div. Environ. Chem.*, 1994, *34*(2), 218.
241. (a) H. Black, *Environ. Sci. Technol.*, 1995, *29*, 497A; (b) Presidential Green Chemistry Challenge, EPA 744-K-96-001, July 1996, 25; (c) M. Ward, *Technol. Rev.*, 1997, *100*(8), 22; (d) M. McCoy, *Chem. Eng. News*, May 29, 2006, 7.
242. (a) W. Saus, D. Knittel, and E. Schollmeyer, *Text Res. J.*, 1993, *63*(3), 135; (b) D. Knittel, W. Saus, S. Holger, and E. Schollmeyer, *Angew. Makromol. Chem.*, 1994, *218*, 69; (c) B. Gebert, W. Saus, D. Knittel, H.-J. Buschmann, and E. Schollmeyer, *Text Res. J.*, 1994, *64*, 371; (d) M.J. Drews and C. Jordan, *Text Chem. Color*, 1998, *30*(6), 13; (e) B.L. West, S.G. Kazarian, M.F. Vincent, N.H. Brantley, and C.A. Eckert, *J. Appl. Polym. Sci.*, 1998, *69*, 911; (f) W.L.F. Santos, A. Moura, N.P. Povh, E.C. Muniz, and A.F. Rubira, *Macromol. Symp.*, 2006, *229*, 150; (g) Z.T. Liu, L. Zhang, Z. Liu, Z. Gao, W. Dong, H. Xiong, Y. Peng, and S. Tang, *Ind. Eng. Chem. Res.*, 206, *45*, 8932; (h) M.V.F. Cid, W. Buijs, and G.-J. Witkamp, *Ind. Eng. Chem. Res.*, 2007, *46*, 3941.
243. (a) L.E. Bowman, C.G. Caley, R.T. Hallen, and J.L. Fulton, *Text Res. J.*, 1996, *66*, 795; (b) L.E. Bowman, N.H. Reader, R.T. Hallen, and A. Butenhoff, *Text Res. J.*, 1998, *68*, 732.
244. Anon., *Chem. Eng. News*, Sept. 14, 1998, 33.
245. G. Parkinson, *Chem. Eng.*, 1998, *105*(7), 27.
246. R.R. Mattews and C.J. Muti, Jr., Preprints A.C.S. *Div. Environ. Chem.*, 1998, *38*(1), 79.
247. The Clean Technology Group [Leaflet], *School of Chemistry*, University of Nottingham, UK, 1998.
248. S.K. Ritter, *Chem. Eng. News*, July 9, 2007, 35.
249. G. Ondrey, *Chem. Eng.*, 1998, *105*(6), 19.
250. (a) L.J. Sealock, Jr., D.C. Elliott, E.G. Baker, A.G. Fassbender, and L.J. Silva, *Ind. Eng. Chem. Res.*, 1996, *35*, 4111; (b) Z.Y. Ding, M.A. Frisch, L. Li, and E.F. Gloyna, *Ind. Eng. Chem. Res.*, 1996, *35*, 3257; (c) E.F. Gloyna and L. Li, *Environ. Prog.*, 1995, *14*, 182; (d) G. Parkinson, *Chem. Eng.*, 1997, *104*(4), 23.
251. L.R. Ember, *Chem. Eng. News*, Mar. 24, 2008, 29.
252. G. Parkinson, *Chem. Eng.*, 1996, *103*(6), 25.
253. B.R. Foy, K. Waldthause, M.A. Sedillo, and S.J. Buelow, *Environ. Sci. Technol.*, 1996, *30*, 2790.
254. G. Parkinson, *Chem. Eng.*, 1998, *105*(7), 29.

255. (a) E.J. Parsons, *Chemtech.*, 1996, *26*(7), 30; (b) B. Kuhlmann, E.M. Arnett, and M. Siskin, *J. Org. Chem.*, 1994, *59*, 3098, 5377; (c) A.R. Katritzsky, S.M. Allin, and M. Siskin, *Acc. Chem. Res.*, 1996, *29*, 399.
256. G. Parkinson, *Chem. Eng.*, 1997, *104*(5), 35; (b) R.L. Holliday, J.W. King, and G.R. List, *Ind. Eng. Chem. Res.*, 1997, *36*, 932.
257. M. Buback, U. Elstner, F. Rindfleisch, and M.A. McHugh, *Macromol. Chem. Phys.*, 1997, *198*, 1189.
258. (a) P.A. Hamley, T. Llkenhans, J.M. Webster, E. Garcia-Verdugo, E. Venasrdou, M.J. Clarke, R. Auerbach, W.B. Thomas, K. Whiston, and M. Poliakoff, *Green Chem.*, 2002, *4*, 235; (b) J. Fraga-Dubreuill, E. Garcia-Verdugo, P.A. Hamley, E.M. Vaquero, L.M. Dudd, I. Pearson, D. Housley, W. Partenheimer, W.B. Thomas, K. Whiston, and M. Poliakoff, *Green Chem.*, 2007, *9*, 1238.
259. C. Yan, J. Fraga-Dubreuil, E. Garcia-Verdugo, P.A. Hamley, and M. Poliakoff, *Green Chem.*, 2008, *10*, 98.
260. K. Tajima, M. Uchida, K. Minami, M. Osada, K. Sue, T. Nonaka, H. Hattori, and K. Arai, *Environ. Sci. Technol.*, 2005, *39*, 9721.
261. M. Sato, K. Matsushima, H. Kawanami, and Y. Ikuhsima, *Angew. Chem. Int. Ed.*, 2007, *46*, 6284.
262. K. Sue, M. Suzuki, K. Arai, T. Ohashi, H. Ura, K. Matsui, Y. Hakuta, M. Watanabe, and T. Hiaki, *Green Chem.*, 2006, *8*, 634.
263. K. Sue, A. Suzuki, M. Suzuki, K. Arai, Y. Hakuta, H. Hayashi, and T. Hiaki, *Ind. Eng. Chem Res.*, 2006, *45*, 623.
264. K. Chandler, F. Deng, A.K. Dillow, C.L. Liotta, and C.A. Eckert, *Ind. Eng. Chem. Res.*, 1997, *36*, 5175.
265. (a) K.S. Jerome and E.J. Parsons, *Organometallics*, 1993, *12*, 29918; (b) D. Farrusseng, S. Aguado, and C. Pinel, *Angew. Chem. Int. Ed.*, 2009, *48*, 7502.
266. (a) P. Reardon, S. Metts, C. Critten, P. Dougherty, and E.J. Parsons, *Organometallics*, 1995, *14*, 3810, 4023; (b) J. Dimine, S. Metts, and E.J. Parsons, *Organometallics*, 1995, *14*, 2043.
267. N. Carlsson, C. Habenicht, L.C. Kan, M.J. Antal, Jr., N. Bian, R.J. Cunningham, and M. Jones, Jr., *Ind. Eng. Chem. Res.*, 1994, *33*, 1989.
268. B.M. Kabyemela, T. Adschiri, R.M. Malaluan, K. Arai, and H. Ohzeki, *Ind. Eng. Chem. Res.*, 1997, *36*, 5063.
269. Y. Yang, S.B. Hawthorne, and D.J. Miller, *Environ. Sci. Technol.*, 1997, *31*, 430.
270. (a) C.-J. Li, *Chem. Rev.*, 1993, *93*, 2023; (b) A. Lubineau, J. Auge, and Y. Queneau, *Synthesis*, 1994, 741; (c) A. Lubineau, *Chem. Ind. (London)*, 1996, 123; (d) P.G. Sammes and D.J. Weller, *Synthesis*, 1995, 1221; (e) J.B.F.N. Engberts, *Pure Appl. Chem.*, 1995, *67*, 823; (f) C.-J. Li and T.-H. Chan, *Comprehensive Organic Reactions in Aqueous Media*, 2nd ed., Wiley, New York, 2007; (g) P.E. Savage, *Chem. Rev.*, 1999, *99*, 603; (h) I.T. Horvath, ed., Catalysis in Water, *J. Mol. Catal. A Chem.*, 1997, *116*, 1–309; (i) K. Nomura, *J. Mol. Chem. A Chem.*, 1998, *130*, 1 (transition metal-catalyzed hydrogenations in water); (j) P.A. Grieco, ed., *Organic Syntheses in Water*, Blackie Academic & Professional, London, 1998; (k) C.-J. Li, *Chem. Rev.*, 2005, *105*, 3095; (l) C.-J. Li and L. Chen, *Chem. Soc. Rev.*, 2007, *36*, 68; (m) T. Dwars, E. Paetzold, and G. Oehme, *Angew. Chem. Int. Ed.*, 2005, *44*, 7174; (n) U.M. Lindstrom, ed., *Organic Reactions in Water*, Blackwell, Oxford, 2007; (o) U.M. Lindstrom and F. Andersson, *Angew. Chem. Int. Ed.*, 2006, *45*, 548; (p) H.C. Hailes, *Org. Proc. Res. Dev.*, 2007, *11*, 114; (q) C.I. Herrerias, X. Yao, Z. Li, and C.-J. Li, *Chem. Rev.*, 2007, *107*, 2546; (r) D. Dallinger and C.O. Kappe, *Chem. Rev.*, 2007, *107*, 2563 (microwave-assisted reactions in water); (s) S. Narayan, J. Muldoon, M.G. Finn, V.V. Fokin, H.C. Kolb, and K.B. Sharpless, *Angew. Chem Int. Ed.*, 2005, *44*, 3275.
271. J.W. Wijnen, R.A. Steiner, and J.B.F.N. Engberts, *Tetrahedron Lett.*, 1995, *36*, 5389.
272. P.T. Buonora, K.G. Rosauer, and L. Dai, *Tetrahedron Lett.*, 1995, *36*, 4009.
273. (a) F. Marken, R.G. Compton, S.D. Bull, and S.G. Davies, *Chem. Commun.*, 1997, 995; (b) M.-F. Ruasse, I.B. Blagoeva, R. Ciri, L. Garcia-Rio, J.R. Leis, A. Marques, J. Mejuto, and E. Monnier, *Pure Appl. Chem.*, 1997, *69*, 1923.
274. (a) J. Haggin, *Chem. Eng. News*, Oct. 10, 1994, 28; (b) P. Cintas, *Chem. Eng. News*, Mar. 20, 1995, 4; (c) B. Cornils and W.A. Herrmann, eds, *Aqueous-Phase Organometallic Catalysis—Concepts and Applications*, 2nd ed., Wiley-VCH, Weinheim, 2004; (d) C.-J. Li and T.-H. Chan, *Tetrahedron*, 1999, *55*, 11149; (e) B.H. Lipschutz and S. Ghorai, *Aldrichchimica Acta*, 2008, *41*(3), 59; (f) M. Carril, R. San Martin, and E. Dominguez, *Chem. Soc. Rev.*, 2008, *37*, 639; (g) G.L. Turner, J.A. Morris, and M.F. Greaney, *Angew. Chem. Int. Ed.*, 2007, *46*, 7996.
275. (a) C.-J. Li, *Tetrahedron*, 1996, *52*, 5643; (b) S.C. Stinson, *Chem. Eng. News*, Apr. 8, 1996, 39; (c) A. Durant, J.-L. Delpancke, R. Winand, and J. Reisse, *Tetrahedron Lett.*, 1995, *36*, 4257; (d) H.J. Lim, G. Keum, S.B. Kang, B.Y. Chung, and Y. Kim, *Tetrahedron Lett.*, 1998, *39*, 4367; (e) L. Wang, X. Sun, and Y. Zhang, *Synth. Commun.*, 1998, *28*, 3263; (f) X.-H. Yi, J.X. Haberman, and C.-J. Li, *Synth. Commun.*, 1998, *28*, 2999; (g) L.W. Bieber, M.F. da Silva, R.C. da Costa, and L.O.S. Silva, *Tetrahedron Lett.*, 1998, *39*, 3655; (h) T.-P. Loh and G.-L. Chua, *Chem. Commun.*, 2006, 2739.
276. Y. Masuyama, A. Ito, M. Fukuzawa, K. Terada, and Y. Kurusu, *Chem. Commun.*, 1998, 2025.
277. M.B. Isaac and T.-H. Chan, *Tetrahedron Lett.*, 1996, *36*, 8957.
278. L.A. Paquette, Presidential Green Chemistry Challenge, EPA 744-K-96-001, Washington, DC, July 1996, 8.
279. T.-P. Loh and L.-L. Wei, *Tetrahedron Lett.*, 1998, *39*, 323; (b) S. Kobayashi, T. Busujima, and S. Nagayama, *Tetrahedron Lett.*, 1998, *39*, 1579; (c) S. Kobayashi, S. Nagayama, and T. Busujima, *J. Am. Chem. Soc.*, 1998, *120*, 8287.
280. (a) M. Beller and B. Cornils, *J. Mol. Catal. A Chem.*, 1995, *104*, 32; (b) B. Cornils and E. Wiebus, *Chemtech.*, 1995, (1), 33; (c) G. Papadogianakis and R.A. Sheldon, *New J. Chem.*, 1996, *20*, 175; (d) B. Cornils and E.G. Kuntz, *J. Organomet. Chem.*, 1995, *502*, 177; (e) S. Lemaire-Audoire, M. Savignac, D. Dupuis, and J.P. Genet, *Tetrahedron Lett.*, 1996, *37*, 2003; (f) C. Amatore, E. Blart, J.P. Genet, A. Jutand, S. Lemaiare-Audoire, and M. Savignac, *J. Org. Chem.*, 1995, *60*, 6829.
281. S. Trinkhaus, J. Holz, R. Selki, and A. Borner, *Tetrahedron Lett.*, 1997, *38*, 807.

282. G. Fremy, Y. Castanet, R. Grzybek, E. Monflier, A. Mortreux, A.M. Trzeciak, and J.J. Ziolkowsko, *J. Organomet. Chem.*, 1995, *505*, 11.
283. (a) B.E. Hanson and J.R. Zoeller, *Catal. Today*, 1998, *42*, 371–471; (b) M.S. Goedheijt, P.C.J. Kamer, and P.W.N.M. van Leeuwen, *J. Mol. Catal. A Chem.*, 1998, *134*, 243; (c) M.S. Goedheijt, J.N.H. Reek, P.C.J. Kamer, and P.W.N.M. van Leeuwen, *Chem. Commun.*, 1998, 2431; (d) G. Wullner, H. Jansch, S. Kannenberg, F. Schubert, and G. Boche, *Chem. Commun.*, 1998, 1509; (e) H. Gulyaset, P. Arva, and J. Bakos, *Chem. Commun.*, 1997, 2385; (f) H. Bahrmann, H. Bach, C.D. Frohning, H.J. Kleiner, P. Lappe, D. Petere, D. Regnat, and W.A. Herrmann, *J. Mol. Catal. A Chem.*, 1997, *116*, 49; (g) H. Ding, J. Kang, B.E. Hanson, and C.W. Kohlpainter, *J. Mol. Catal. A Chem.*, 1997, *124*, 21; (h) G. Verspui, G. Papadogianakis, and R.A. Sheldon, *Chem. Commun.*, 1998, 401.
284. H. Dibowski and F.P. Schmidtchen, *Tetrahedron*, 1995, *51*, 2325.
285. (a) T. Malmstrom and C. Andersson, *Chem. Commun.*, 1996, 1135; (b) T. Malmstrom, H. Weigl, and C. Andersson, *Organometallics*, 1995 *14*, 2593.
286. (a) D.C. Mudalige and G.L. Rempel, *J. Mol. Catal. A Chem.*, 1997, *116*, 309; (b) A.N. Ajjou and H. Alper, *J. Am. Chem. Soc.*, 1998, *120*, 1466.
287. (a) X. Zheng, J. Jiang, X. Liu, and Z. Jin, *Catal. Today*, 1998, *44*, 175; (b) Z. Jin, X. Zheng, and B. Fell, *J. Mol. Chem. A Chem.*, 1997, *116*, 55; (c) E. Karakhanov, T. Filippova, A. Maximov, V. Predeina, and A. Restakyan, *Macromol. Symp.*, 1998, *131*, 87.
288. D.M. Lynn, B. Mohr, and R.H. Grubbs, *J. Am. Chem. Soc.*, 1998, *120*, 1621.
289. (a) W.A. Herrmann and C.W. Kohlpainter, *Angew. Chem. Int. Ed. Engl.*, 1993, *32*, 1524; (b) R.V. Chaudhari, A. Battacharya, and B.M. Bhanage, *Catal. Today*, 1995, *24*, 123.
290. I. Toth, I. Guo, and B.E. Hanson, *J. Mol. Catal. A Chem.*, 1997, *116*, 217.
291. G. Fremy, E. Montiflier, J.-F. Carpentier, Y. Castanet, and A. Mortreux, *Angew. Chem. Int. Ed.*, 1995, *34*, 1474.
292. B. Fell, C. Schobben, and G. Papadogianakis, *J. Mol. Catal. A Chem.*, 1995, *101*, 179.
293. (a) C.M. Starks, C.C. Liotta, and M. Halpern, *Phase Transfer Catalysis—Fundamentals, Applications and Industrial Perspectives*, Chapman Hall, London, 1994; (b) E.V. Dehmlow and S.S. Dehmlow, *Phase Transfer Catalysis*, 3rd ed., VCH, Weinheim, 1993.
294. F. Fringuelli, G. Pani, O. Piermatti, and F. Pizzo, *Tetrahedron*, 1994, *50*, 11499.
295. (a) N.A. Bumagin, V.V. Bykov, L.I. Sukhomlinova, T.P. Tolstaya, and I.P. Beletskaya, *Organomet. Chem.* 1995, *486*, 259; (b) T. Jeffery, *Tetrahedron Lett.*, 1994, *35*, 3051.
296. A.V. Cheprakov, N.V. Ponomareva, and I.P. Beletskaya, *Organomet. Chem.*, 1995, *486*, 297.
297. H.-Y. Shu, H.D. Perlmutter, and H. Shaw, *Ind. Eng. Chem. Res.*, 1995, *34*, 3761.
298. S.E. Friberg, J. Yang, and T. Huang, *Ind. Eng. Chem. Res.*, 1996, *35*, 2856.
299. E.J. Colonia, A.B. Dixit, and N.S. Tavare, *Ind. Eng. Chem. Res.*, 1998, *37*, 1956.
300. (a) V.G. Sadvilkar, S.D. Samant, and V.G. Gaikar, *J. Chem. Technol. Biotechnol.*, 1995, *62*, 405; (b) V.G. Sadvilkar, B.M. Khadilkar, and V.G. Gaikar, *J. Chem. Technol. Biotechnol.*, 1995, *63*, 33; (c) B.M. Khadilkar, V.G. Gaikar, and A.A. Chitnavis, *Tetrahedron Lett.*, 1995, *36*, 8083.
301. I. Vilotijevic and T.F. Jamison, *Science*, 2007, *317*, 1189.
302. K. Ding, Y. Wang, L. Zhang, Y. Wu, and T. Matsuura, *Tetrahedron*, 1996, *52*, 1005.
303. (a) B. Abribat and Y. LeBigot, *Tetrahedron*, 1997, *53*, 2119; (b) Y. LeBigot and A. Gaset, *Tetrahedron*, 1996, *52*, 8245; *Synth. Commun.*, 1994, *24*, 1773, 2091.
304. (a) U. Ritter, N. Winkhofer, H.-G. Schmidt, and H.W. Roesky, *Angew. Chem. Int. Ed.*, 1996, *35*, 524; (b) U. Ritter, N. Winkhofer, and H. Roesky, U.S. patent 5,840,993, 1998.
305. (a) R.D. Rogers and G.A. Voth, eds, *Acc. Chem. Res.*, 2007, *40*(11), 1079 (issue on ionic liquids); (b) H. Weingartner, *Angew. Chem. Int. Ed.*, 2008, *47*, 654; (c) R. Ranke, S. Stolte, R. Stormann, J. Arning, and B. Jastorff, *Chem. Rev.*, 2007, *107*, 2183.
306. A. Arce, M.J. Earle, S.P. Katare, H. Rodriguez, and K.R. Seddon, *Chem. Commun.*, 2006, 2548.
307. S.-G. Lee, *Chem. Commun.*, 2006, 1049.
308. (a) D. Fang, X.-L. Zhou, Z.-W. Ye, and Z.-L. Liu, *Ind. Eng. Chem. Res.*, 2006, *45*, 7982; (b) K. Fukumoto and H. Ohno, *Angew. Chem. Int. Ed.*, 2007, *46*, 1852, 549.
309. (a) P. Wasserscheid and T. Welton, *Ionic Liquids in Synthesis*, 2nd ed., Wiley-VCH, Weinheim, 2008; (b) N. Jain, A. Kumar, S. Chauhan, and S.M.S. Chauhan, *Tetrahedron*, 2005, *61*, 1015; (c) S.V. Mahlhotra, *Ionic Liquids in Organic Synthesis*, Am. Chem. Soc. Symp. 950, Oxford University Press, New York, 2007; (d) M.A.P. Martins, C.P. Frizzo, D.N. Moreira, N. Zanatta, and H.G. Bonacorso, *Chem. Rev.*, 2008, *108*, 2015; (e) V.I. Parvulescu and C. Hardacre, *Chem. Rev.*, 2007, *107*, 2615.
310. A. Zhu, T. Jiang, D. Wang, B. Han, L. Liu, J. Huang, J. Zhang, and D. Sun, *Green Chem.*, 2005, *7*, 514.
311. V. Conte, E. Elakkari, B. Floris, V. Mirruzzo, and P. Tagliatesta, *Chem. Commun.*, 2005, 1587.
312. W.-L. Wong, K.-C. Cheung, P.-H. Chan, Z.-Y. Zhou, K.-H. Lee, and K.-Y. Wong, *Chem. Commun.*, 2007, 2175.
313. Anon., *Chem. Eng. News*, Jan. 24, 2005, 35.
314. H. Zhang, F. Xu, X. Zhou, G. Zhang, and C. Wang, *Green Chem.*, 2007, *9*, 1208.
315. (a) Y.Y. Wang, M.M. Luo, Q. Lin, H. Chen, and X.J. Li, *Green Chem.*, 2006, *8*, 545; (b) S.L. Desset, D.J. Cole-Hamilton, and D.F. Foster, *Chem. Commun.*, 2007, 1933; (c) S.L. Dessel, S.W. Reeder, and D.J. Cole-Hamilton, *Green Chem.*, 2009, *11*, 630.
316. (a) Z. Zhang, Y. Xie, W. Li, S. Hu, J. Song, T. Jiang, and B. Han, *Angew. Chem. Int. Ed.*, 2008, *47*, 1127; (b) A. Perosa, P. Tundo, M. Selva, and P. Canton, *Chem. Commun.*, 2006, 4480; (c) V. Mevellec, B. Leger, M. Mauduit, and A. Roucoux, *Chem. Commun.*, 2005, 2838.
317. H. Jiang, C. Wang, H. Li, and Y. Wang, *Green Chem.*, 2006, *8*, 1076.
318. M.C.-Y. Tang, K.-Y. Wong, and T.H. Chan, *Chem. Commun.*, 2005, 1345.
319. A.L. Miller III and N.B. Bowden, *Chem. Commun.*, 2007, 2051.
320. (a) P.D. deMaria, *Angew. Chem. Int. Ed.*, 2008, *47*, 6960; (b) N.M.T. Lourenco, S. Parreiros, and C.A.M. Afonso, *Green Chem.*, 2007, 734.
321. (a) C.S. Brazel and R.D. Rogers, eds, *Ionic Liquids in Polymer Systems*, Am. Chem. Soc. Symp. 913, Oxford University Press, New York, 2005; (b) G. Schmidt-Naake, I. Woecht, and

A. Schmalfuss, *Macromol. Symp.*, 2007, *259*, 226; (c) V. Strehmel, *Macromol. Symp.*, 2007, *254*, 25; (d) C. Guerrero-Sanchez, R. Hoogenboom, and U.S. Schubert, *Chem. Commun.*, 2006, 3797; (e) Y.S. Vygodskii, A.S. Shaplov, E.I. Lozinskaya, O.A. Filippov, E.S. Shubina, R. Bandari, and M.R. Buchmeister, *Macromolecules*, 2006, *39*, 7821.

322. (a) M.C. Kroom, J. van Spronsen, C.J. Peters, R.A. Sheldon, and G.-J. Witkamp, *Green Chem.* 2006, *8*, 246; (b) L.C. Branco, A. Serbanovic, M.N. daPonte, and C.A.M. Afonso, *Chem. Commun.*, 2005, 107.
323. W. Miao and T.H. Chan, *Acc. Chem. Res.*, 2006, *39*, 897.
324. (a) U. Hintermair, G. Zhao, C.C. Santini, M.J. Muldoon, and D.J. Cole-Hamilton, *Chem. Commun.*, 2007, 1462; (b) A. Rusager, R. Fehrmann, S. Flicker, R. van Hal, M. Haumann, and P. Wasserscheid, *Angew. Chem. Int. Ed.*, 2005, *44*, 815; (c) T. Sakati, C. Zhong, M. Tada, and Y. Iwasawa, *Chem. Commun.*, 2005, 2506; (d) M.A. Neouze, J. le Bidau, F. Leroux, and A. Vioux, *Chem. Commun.*, 2005, 1082; (e) O. Jimenez, T.E. Muller, C. Sievers, A. Spirkl, and J.A. Lercher, *Chem. Commun.*, 2006, 2974; (f) A. Rusager, R. Fehrmann, M. Hsaumann, B.S.K. Gorle, and P. Wasserscheid, *Ind. Eng. Chem. Res.*, 2005, *44*, 9853.
325. J.-P. Mikkola, P. Virtanen, H. Karhu, T. Salmi, and D.Y. Murzin, *Green Chem.*, 2006, *8*, 197.
326. J. Baudoux, K. Perrigaud, P.-J. Madec, A.-C. Gaumont, and I. Dez, *Green Chem.*, 2007, *9*, 1346.
327. (a) X.D. Mu, J.-Q. Meng, Z.-C. Li, and Y. Kou, *J. Am. Chem. Soc.*, 2005, *127*, 9694; (b) H. Ohno, *Macromol. Symp.*, 2007, *249–250*, 551; (c) Y. Shen, Y. Zhang, Q. Zhang, L. Niu, T. You, and A. Ivaska, *Chem. Commun.*, 2005, 4193.
328. (a) R.D. Rogers and K.R. Seddon, *Ionic Liquids—Industrial Applications for Green Chemistry*, Am. Chem. Soc. Symp. 818, Oxford University Press, New York, 2002; (b) M. Freemantle, *Chem. Eng. News*, Aug. 1, 2005, 33; Jan. 1, 2007, 23; (c) P.L. Short, *Chem. Eng. News*, Apr. 24, 2006, 15; (d) S.K. Ritter, *Chem. Eng. News*, Sep. 29, 2008, 36; (e) G. Parkinson, *Chem. Eng. Prog.*, 2004, *100*(9), 7; (f) N.V. Plechkova and K.R. Seddon, *Chem. Soc. Rev.*, 2008, *37*, 123; (g) P. Walter, *Chem. Ind.* (*London*), Aug. 11, 2008, 12.
329. G. Ondrey, *Chem. Eng.*, 2005, *112*(8), 16.
330. (a) F. Hermanutz, F. Gahr, E. Uerdingen, F. Meister, and B. Kosan, *Macromol. Symp.*, 2008, *262*, 2310; (b) N. Sun, M. Rahman, Y. Qin, M.L. Maxim, H. Rodriguez and R.D. Rogers *Green Chem.*, 2009, *11*, 646.
331. (a) T. Heinze, S. Dorn, M. Schobitz, T. Liebert, S. Kohler, and F. Meister, *Macromol. Symp.*, 2008, *262*, 8; (b) S. Barthel and T. Heinze, *Green Chem.*, 2006, *8*, 301; (c) J.-P. Mikkola, A. Kirilin, J.-C.Tuuf, A. Pranovich, B. Holmbom, C.M. Kustov, D.Y. Murzin, and T. Salmi, *Green Chem.*, 2007, *9*, 1229.
332. D.M. Phillips, L.F. Drummy, D.G. Conrady, D.M. Fox, R.R. Nack, M.O. Stone, C. Trulove, H.C. deLong, and R.A. Mantz, *J. Am. Chem. Soc.*, 2004, *126*, 14350.
333. M. Jacoby, *Chem. Eng. News*, Jan. 7, 2008, 7.
334. E.R. Parnham and R.E. Morris, *Acc. Chem. Res.*, 2007, *40*, 1005.
335. (a) E. Kuhlmann, S. Himmler, H. Giebelhaus, and P. Wasserscheid, *Green Chem.*, 2007, *9*, 233; (b) M. Smiglak, J.D. Holbrey, S.T. Griffin, W.M. Reichert, R. Swatloski, A.R. Katritzky, H. Yang, D. Zhang, K. Kirichenko, and R.D. Rogers, *Green Chem.*, 2007, *9*, 90.
336. J. Kagimoto, K. Fukumoto, and H. Ohno, *Chem. Commun.*, 2006, 2254.
337. Y.Y. Jiang, G.-N. Wang, Z. Zhu, Y.-T. Wu, J .Geng, and Z.-B. Zhang, *Chem Commun.*, 2008, 305.
338. (a) C. Pretti, C. Chiappe, D. Pieraccini, M. Gregori, F. Abramo, G. Monni, and L. Intorre, *Green Chem.*, 2006, *8*, 238; (b) K.M. Docherty and C.F. Kulpa, Jr., *Green Chem.*, 2005, *7*, 185; (c) Anon., *Chem. Eng. News*, July 10, 2006, 22.
339. T. Welton, *Green Chem.* 2008, *10*, 483.
340. L. Phan, D.L. Chiu, D.J. Heldlebrant, H. Huttenhower, E. John, X. Li, P. Pollet, R. Wang, C.A. Eckert, C.L. Liotta, and P.G. Jessop, *Ind. Eng. Chem. Res.*, 2008, *47*, 539.
341. (a) M. Freemantle, *Chem. Eng. News*, Sept. 12, 2005, 36; (b) Y. Fukaya, Y. Lizuka, K. Sekikawa, and H. Ohno, *Green Chem.*, 2007, *9*, 1155; (c) A.P. Abbott, D. Boothby, G. Capper, D.L. Davies, and R.K. Rasheed, *J. Am. Chem. Soc.*, 2004, *126*, 9142; (d) A.P. Abbott, G. Capper, D.L. Davies, R.K. Rasheed, and P. Shikotra, *Inorg. Chem.*, 2005, *44*, 6497; (e) F. Ilgen and B. Konig, *Green Chem.*, 2009, *11*, 848; (f) J. Zhang, T. Wu, S. Chen, P. Feng and X. Bu, *Angew. Chem. Int. Ed.*, 2009, *48*, 3486.
342. (a) B. Boutevin, J.-J. Robin, B. Boyer, G. Lamaty, A. Leydet, J.-P. Roque, and O. Senhaji, *New J. Chem.*, 1996, *20*, 137; (b) M. Dreja and B. Tieke, *Macromol. Rapid Commun.*, 1996, *17*, 825; (c) A. Guyot and K. Tauer, *Adv. Polym. Sci.*, 1994, *111*, 45; (d) M. Dreja, W. Pychkhout-Hintzen, and B. Tieke, *Macromolecules*, 1998, *31*, 272.
343. (a) D.A. Jaeger, S.G.G. Russell, and H. Shinozaki, *J. Org. Chem.*, 1994, *59*, 7544; (b) D. Ono, S. Yamamura, M. Nakamura, T. Takeda, A. Masuyama, and Y. Nakatsuji, *J. Am. Oil Chem. Soc.*, 1995, *72*, 853; 1993, *70*, 29; (c) K.A. Wilk, A. Bieniecki, B. Burczyk, and A. Sokolowski, *J. Am. Oil Chem. Soc.*, 1994, *71*, 81; (d) G.-W. Wang, X.-G. Lei, and Y.-C. Liu, *J. Am. Oil Chem. Soc.*, 1993, *70*, 731; (e) G.-W. Wang, X.-Y. Yuan, Y.-C. Liu, and X.G. Lei, *J. Am. Oil Chem. Soc.*, 1994, *71*, 727; (f) G.-W. Wang, X.-Y. Yuan, Y.-C. Liu, X.-G. Lei, and Q.-X. Guo, *J. Am. Oil Chem. Soc.*, 1995, *72*, 83; (g) T. Kida, N. Morishima, A. Masuyama, and Y. Nakatsuji, *J. Am. Oil Chem. Soc.*, 1994, *71*, 705; (h) C. Yue, J.M. Harris, P.-E. Hellberg, and K. Bergstrom, *J. Am. Oil Chem. Soc.*, 1996, *73*, 841; (i) R.C. Hoy and A.F. Joseph, *Int. News Fats Oils Relat. Mater.*, 1996, *7*, 428.
344. (a) D. Hanson, *Chem. Eng. News*, Nov. 25, 1996, 12; (b) E.M. Kirschner, *Chem. Eng. News*, Jan. 27, 1997, 44; (c) Anon., *Chem. Eng. News*, Feb. 17, 1997, 35; (d) G. Parkinson, *Chem. Eng.*, 1996, *103*(12), 17; (e) Anon., *Chem. Eng. News*, Feb. 23, 1998, 47 (advertisement).
345. (a) O. Nuyken, K. Meindl, A. Wokaun, and T. Mezger, *Macromol. Rep.*, 1995, *A32*, 447; (b) I.R. Dunkin, A. Gittinger, D.C. Sherrington, and P. Whittaker, *J. Chem. Soc. Chem. Commun.*, 1994, 2245; (c) Y. Okamoto, H. Yoshida, and S. Takamuku, *Chem. Lett.*, 1988, *17*, 569.
346. Anon., *Chem. Eng. Prog.*, 2005, *101*(10), 7.
347. M. Antonietti and T. Nestl, *Macromol. Rapid Commun.*, 1994, *15*, 111.
348. (a) N.M. van Os, ed., *Nonionic Surfactants Organic Chemistry*, Dekker, New York, 1997; (b) T. Krawczyk, *Int. News Fats Oils Relat. Mater.*, 1998, *9*, 271; (c) K. Esuma and M. Ueno, eds, *Structure Performance Relationships in Surfactants*, Dekker, New York, 1997; (d) B. Jonsson, B.

Lindman, K. Holmberg, and B. Kronberg, *Surfactants and Polymers in Aqueous Solution*, Wiley, New York, 1998; (e) D.R. Karsa, ed., *New Products and Applications in Surfactant Technology*, CRC Press, Boca Raton, FL, 1998; (f) J.H. van Ee, O. Misset, E.J. Baas, eds, *Enzymes in Detergency*, Dekker, New York, 1997; (g) T. Krawczyk, *Int. News Fats Oils Relat. Mater.*, 1998, *9*, 924 (additives for detergents); (h) P.M. Morse, *Chem. Eng. News*, Feb. 1, 1999, 35.

349. (a) P.J. Chenier, *Survey of Industrial Chemistry*, 2nd ed., p. 476, VCH, Weinheim, 1992; (b) S.J. Ainsworth, *Chem. Eng. News*, Jan. 23, 1995, 30; (c) S.J. Ainsworth, *Chem. Eng. News*, Jan. 24, 1994, 34.
350. U.S.E.P.A., Guide to Cleaner Technologies—Alternatives to Chlorinated Solvents for Cleaning and Degreasing, EPA/625/R-93/016, Feb. 1994, Cincinnati, OH.
351. G.S. Samdani, *Chem. Eng.*, 1994, *101*(11), 23.
352. Anon., *Appl. Catal. B.*, 1994, *4*, N4.
353. (a) M. Rouhi, *Chem. Eng. News*, July 3, 1995, 22; (b) H. Stromberg, *Rubber World*, 1995, *212*(5), 14.
354. Anon., *Pollut. Equip. News*, 2008, *41*(1), 12.
355. Anon., *Int. News Fats Oils Relat. Mater.*, 1995, *6*, 769.
356. Anon., *Chem. Eng. News*, June 3, 196, 11.
357. D.R. Karsa, *Biodegradability of Surfactants*, Blackie, London, 1995.
358. B. Moss, *Chem. Ind.* (*London*), 1996, 407.
359. I. Nikov and K. Paev, *Catal. Today*, 1995, *24*, 41.
360. P. Zini, *Polymeric Additives for High Performance Detergents*, Technomic, Lancaster, PA, 1995.
361. (a) Anon., *Chem. Eng. News*, Jan. 20, 1997, 28; (b) Presidential Green Chemistry Challenge, EPA 744-K-96-001, July, 1996, 5.
362. S. Matsumara, K. Aoki, and K. Toshima, *J. Am. Oil Chem. Soc.*, 1994, *71*, 749.
363. J.H. Miller, D.A. Quebedeaux, and J.D. Sauer, *J. Am. Oil Chem. Soc.*, 1995, *72*, 857.
364. (a) P. Layman, *Chem. Eng. News*, Feb. 13, 1995, 21; June 12, 1995, 12; (b) Anon., *Chem. Ind. (London)*, 1995, *43*, 485.
365. Anon., *Chem. Eng.*, 1996, *103*(6), 80.
366. (a) Presidential Green Chemistry Challenge, EPA 744-K-96-001, July, 1996, 16; (b) F. Nilsson, *Int. News Fats Oils Relat. Mater.*, 1996, *7*, 490; (c) Y.H. Paik and G. Swift, *Chem. Ind.* (*London*), 1995, 55; (d) W. von Rybinski and K. Hill, *Angew. Chem. Int. Ed. Engl.*, 1998, *37*, 1329.
367. C. Panintrarux, S. Adachi, and R. Matsuno, *J. Mol. Catal. B Enzyme*, 1996, *1*, 165.
368. (a) D.B. Sarney and En. Vulfson, *Trends Biotechnol.*, 1995, *13*, 164; (b) A. Ducret, A. Giroux, M. Trani, and R. Lortie, *J. Am. Oil Chem. Soc.*, 1996, *73*, 109; (c) C. Scheckermann, A. Schlotterbeck, M. Schmidt, V. Wray, and S. Lang, *Enzyme Microb. Technol.*, 1995, *17*, 157.
369. (a) S.-C. Lin, *J. Chem. Technol. Biotechnol.*, 1996, *66*, 109; (b) Y. Ishigami, *Int. News Fats Oils Relat. Mater.*, 1993, *4*, 1156; (c) N. Kosaric, ed., *Biosurfactants*, Dekker, New York, 1993; (d) H. Razafindralambo, M. Paquot, A. Baniel, Y. Popineau, C. Hbid, P. Jacques, and P. Thonart, *J. Am. Oil Chem. Soc.*, 1996, *73*, 149.
370. K. Phillips, R. Coons, A. Scott, and K. Walsh, *Chem. Week*, Jan. 28, 2008, 23.
371. G. Ondrey, *Chem. Eng.*, 2008, *115*(7), 14.
372. F. Goursaud, M. Berchel, J. Guilbot, N. Legros, L. Lemiegre, J. Marcilloux, D. Plusquellec, and T. Benvegnu, *Green Chem.*, 2008, *10*, 310.
373. M. McCoy, *Chem. Eng. News*, Mar. 19, 2007, 29; Jan. 21, 2008, 15.
374. (a) A. Heller, *Acc. Chem. Res.*, 1995, *28*, 503; (b) G. Parkinson, *Chem. Eng.*, 1996, *103*(9), 23; (c) S. Sitkiewitz and A. Heller, *New J. Chem.*, 1996, *20*, 233; (d) S. Strauss, *Technol. Rev.*, 1996, *99*(2), 23; (e) J. Say, R. Bonnecaze, A. Heller, S. Sitkiewitz, E. Heller, and P. Haugsjaa, U.S. patent 5,790,934, 1998; (f) R. Wang, K. Hashimoto, A. Fujishima, M. Chikuni, E. Kojima, A. Kitamura, M. Shimohigoshi, and T. Watanabe, *Nature*, 1997, *388*, 431; (g) R. Dagani, *Chem. Eng. News*, Sept. 21, 1998, 70; July 27, 1998, 14; (h) V. Romes, P. Pichat, C. Guillard, T. Chopin, and C. Lehaut, *Ind. Eng. Chem. Res.*, 1999, *38*, 3878; (i) H.J. Lee and C. Willis, *Chem. Ind. (Lond.)*, April 13, 2009, 21.
375. (a) A. Hattori, M. Yamamoto, H. Tada, and S. Ito, *Chem. Lett.*, 1998, 707; (b) H. Kisch, L. Zang, C. Lange, W.F. Maier, C. Antonius, and D. Meissner, *Angew. Chem. Int. Ed.*, 1998, *37*, 3034.
376. T. Noguchi, A. Fujishima, P. Sawunyama, and K. Hashimoto, *Environ. Sci. Technol.*, 1998, *32*, 3831.
377. M. Freemantle, *Chem. Eng. News*, Nov. 30, 1998, 29.
378. (a) J. Martyka, *Investing Tomorrow (University of Minnesota)*, Fall 1998, 25; (b) Anon., *Chem. Eng. News*, Oct. 27, 1997, 36.
379. (a) A.R. Marrion, ed., *The Chemistry and Physics of Coatings*, R Soc. Chem. Cambridge, England, 1994; (b) S. Paul, *Surface Coatings—Science and Technology*, 2nd ed. Wiley, New York. 1996; (c) M.S. Reisch, *Chem. Eng. News*, Oct. 14, 1996, 44–64; Oct. 27, 1997, 34; (d) D. Stoye and W. Freitag, *Resins for Coatings—Chemistry, Properties and Applications*, Hanser Gardner, Cincinnati, OH, 1996; (e) A. Singh and J. Silverman, *Radiation Processing of Polymers*, Hanser Gardner, Cincinnati, OH, 1991; (f) P.M. Morse, *Chem. Eng. News*, Oct. 12, 1998, 42; (g) P.K. Datta and J.S. Burnell-Gray, *Fundamentals of Coatings*, Vol 1, Royal Soc. Chem. Special Publication 206, Cambridge, UK, 1997; (h) S.K. Ghosh, ed., *Functional Coatings by Polymer Microencapsulation*, Wiley-VCH, Weinheim, 2006.
380. (a) U.S.E.P.A., Guide to Cleaner Technologies—Organic Coating Replacements, EPA/625/R-94/006, Cincinnati, OH, Sept. 1994; (b) Z.W. Wicks, Jr., F.N. Jones, and S.P. Pappas, *Organic Coatings: Science and Technology*, 3rd ed., Wiley, New York, 2007; (c) D.S. Richart, K.J. Coeling, and Z.W. Wicks, Jr., *Kirk–Othmer Encyclopedia of Chemical Technology*, 4th ed., Wiley, New York, 1993, *6*, 635, 661, 669; (d) R.E. Morgan, *Chem. Eng. Prog.*, 1996, *92*(11), 54; (e) M.S. Reisch, *Chem. Eng. News*, Oct. 16, 2006, 13; June 2, 2008, 28; (f) C. Challener, J. Coatings, *Technol.*, July 2006, 28.
381. (a) K. Doren, W. Freitag, and D. Stoye, *Water-Borne Coatings—The Environmentally Friendly Alternative*, Hanser Gardner, Cincinnati, OH, 1994; (b) D.R. Karsa and W.D. Davies, *Waterborne Coatings and Adhesives*, Royal Soc. Chem, Cambridge, UK, 1995; (c) A. Hofland, *J. Coat. Technol.*, 1995, *67*(Sept.), 113; (d) T. Provder, M.A. Winnick, and M.W. Urban, eds, *Film Formation in Waterborne Coatings*, Am. Chem. Soc. Symp. 684, Washington, DC, 1996; (e) J.E. Glass, *Technology for Waterborne Coatings*, Am. Chem. Soc. Symp.

663, Washington, DC, 1997; (f) T. Provder, M.A. Winnik, and M.W. Urban, eds, *Film Formation in Waterborne Coatings*, Am. Chem. Soc. Symp. 642, Washington, DC, 1996; (g) G. Monaghan, *J. Coatings Technol.*, 2007, *4*(10), 30.
382. (a) Anon., *Environ. Sci. Technol.*, 1997, *31*, 396A; (b) B. Emelie, U. Schuster, S. Echkersley, and S. Eckersley, *Prog. Org. Coat*, 1998, *34*, 49.
383. J. Zhu and G.P. Bierwagen, *Prog. Org. Coat*, 1995, *26*, 87.
384. H.-J. Streitberger and K.-F. Dossel, *Automotive Paints and Coatings*, 2nd ed, Wiley-VCH, Weinheim, 2008.
385. T. Wright, *Coatings World*, 2008, *13*(3), 28.
386. (a) A.T. Chem, L.E. Katz, R.T. Wojcik, and J.M. O'Connor, *Mod. Paint Coat.*, 1996, *86*(5), 51; (b) A.T. Chen, L.E. Katz, R.T. Wojcik, and J.M. O'Connor, *Proc Waterborne High-Solids. Powder Coat. Symp.* 23rd, 1996, 103; *Chem. Abstr.*, 1996, *125*, 198, 596; (c) P.G. Becker, H.-P. Klein, and M. Schwab, *Mod. Paint Coat.*, 1996, *86*(3), 28; (d) M. Bock and J. Petzoldt, *Mod. Paint Coat.*, 1996, *86*(2), 22; (e) W.O. Buckley, *Mod. Paint Coat.*, 1996, *86*(10), 81; (f) B.K. Kim and J.C. Lee, *J. Polymer Sci. Part A: Polym. Chem.*, 1996, *34*, 1095; *J. Appl. Polym. Sci.*, 1995, *58*, 1117; (g) K.-L. Noble, *Prog. Org. Coat.*, 1997, *32*, 131; (h) R.G. Coogan, *Prog. Org. Coat.*, 1997, *32*, 51; (i) G.A. Howarth and H.L. Manock, *Surf. Coat. Int.*, 1997, *80*, 324.
387. C.H. Hare, *Mod. Paint Coat.*, 1996, *86*(2), 28.
388. (a) T. Poggio, D. Lenti, and L. Masini, *Surf. Coat. Int.*, 1995, *78*, 289; (b) G. Moore, D.-W. Zhu, G. Clark, M. Pellerite, C. Burton, D. Schmidt, and C. Coburn, *Surf. Coat. Int.*, 1995, *78*, 377.
389. X. Zhang, J.P. Bell, and M. Narkis, *J. Appl. Polym. Sci.*, 1996, *62*, 1303.
390. (a) T.A. Misev, *Powder Coatings*, Wiley, London, 1991; (b) M. Narkis and N. Rosenzweig, eds, *Polymer Powder Technology*, Wiley, New York, 1995; (c) T.A. Misev and R. van der Linde, *Prog. Org. Coat.*, 1998, *34*, 160; (d) H. Satoh, Y. Harada, and S. Libke, *Prog. Org. Coat.*, 1998, *34*, 193; (e) K. Gotoh, H. Masuda, and K. Higashitani, eds, *Powder Technology Handbook*, 2nd ed., Dekker, New York, 1997.
391. K.M. Biller and B. MacFadden, *Mod. Paint Coat.*, 1996, *86*(9), 34.
392. M. Johansson, H. Faolken, A. Frestedt, and A. Hult, *J. Coat. Technol.*, 1998, *70*(884), 57.
393. (a) Anon., *Chem. Eng. News*, Nov. 24, 1997, 56; (b) M.A. Reisch, *Chem. Eng. News*, Oct. 27, 1997, 35.
394. T. Sugama, R. Kawase, C.C. Berndt, and H. Herman, *Prog. Org. Coat.*, 1995, *25*, 205.
395. G.B. Blanchet, *Chemtech.*, 1996, *26*(6), 31.
396. (a) P.J. Watson, S.J. Walton, and A. Flinn, *Surf. Coat. Int.*, 1995, *78*, 520; (b) E.J. Marx and R.J. Pawlik, *Mod. Paint Coat.*, 1996, *86*(6), 28; (c) G. Parkinson, *Chem. Eng.*, 1996, *103*(5), 29; (d) F.M. Witte, C.D. Goemans, R. van der Linde, and D.A. Stanssens, *Prog. Org. Coat.*, 1997, *32*, 241.
397. (a) J.S. Witzeman, *Prog. Org. Coat.*, 1996, *27*, 269; (b) R.M. Guida, *Mod. Paint Coat.*, 1996, *86*(6), 34.
398. F. Schmitt, A. Wenning, and J.V. Weiss, *Prog. Org. Coat.*, 1998, *34*, 227.
399. (a) L.D. Venham, D.A. Wicks, and P.E. Yeske, Preprints A.C.S. *Div. Environ. Chem.*, 1994, *34*(2), 389; (b) B. Vogt-Birnbrich, *Prog. Org. Coat.*, 1996, *29*, 31; (c) K.H. Zabel, R.E. Boomgaard, G.E. Thompson, S. Turgoose, and H.A. Braun, *Prog. Org. Coat.*, 1998, *34*, 236.
400. (a) G.P. Guidetti, G.L. Rigosi, and R. Marzola, *Prog. Org. Coat.*, 1996, *27*, 79; (b) J.C. Duncan and B.C. Goff, *Surf. Coat. Int.*, 1996, *79*, 28.
401. (a) J.P. Fouassier and J.F. Rabek, eds, *Radiation Curing in Polymer Science and Technology*, Vol 1–4. Chapman & Hall, London, 1989; *Photoinitiation, Photopolymerization and Photocuring*, Hanser Gardner, Cincinnati, OH, 1995; (b) A.V. Rao, D.S. Kanitkar, and A.K. Parab, *Prog. Org. Coat.*, 1995, *25*, 221; (c) S. Jonsson, P.-E. Sundell, J. Hultgren, D. Sheng, and C.E. Hoyle, *Prog. Org. Coat.*, 1996, *27*, 107; (d) W. Cunningham, *Adhes. Age*, 1996, *39*(4), 24; (e) M. Uminski and L.M. Saija, *Surf. Coat. Int.*, 1995, *78*, 244; (f) C. Roffey, *Photogeneration of Reactive Species for Ultraviolet Curing*, Wiley, Chichester, 1997; (g) Anon., *PCI*, 2007, *23*(8), 38; (h) K. Dietliker, K. Misteli, T. Jung, P. Contich, J. Benkhoff, and S.E. Sitzmann, *Eur. Coatings J.*, 2005, *20*.
402. K.-D. Suh, Y.S. Chon, and J.Y. Kim, *Polym Bull*, 1997, *38*, 287.
403. L.N. Price, *J. Coat. Technol.*, 1995, *67*(849), 27.
404. (a) T. Tanabe, A. Torres-Filhoi, and D.C. Neckers, *J. Polym. Sci. A Polym. Chem.*, 1995, *33*, 1691; (b) W. Rutsch, K. Dietliker, D. Leppard, M. Kohler, L. Misev, U. Kolczak, and G. Rist, *Prog. Org. Coat.*, 1996, *27*, 227.
405. C.D. Decker and T.L. Bendaikha, *J. Appl. Polym. Soc.*, 1998, *70*, 2269.
406. J.V. Crivello, R. Narayan, and S.S. Sternstein, *J. Appl. Polym. Sci.*, 1997, *64*, 3073.
407. (a) M. Johansson and A. Hult, *J. Coat. Technol.*, 1995, *67*(849), 35; (b) W. Shi and B. Ranby, *J. Appl. Polym. Sci.*, 1996, *59*, 1951.
408. (a) D.A. Tomalia, *Macromol. Symp.*, 1996, *101*, 243; (b) J.M.J. Frechet, C.J. Hawker, I. Gitsov, and J.W. Leon, *J. Macromol. Sci. Pure Appl. Chem.*, 1996, *A33*, 1399; (c) R. Dagani, *Chem. Eng. News*, June 3, 1996, 30; (d) G.R. Newkome, C.N. Moorefield, and F. Vogtle, *Dendritic Molecules: Concepts, Syntheses, Perspectives*, VCH, Weinheim, 1996; (e) G.R. Newkome, *Advances in Dendritic Macromolecules*, JAI Press, Greenwich, CT, 1996; (f) F. Vogtle, ed., *Topics Curr. Chem.*, 1998, *197*, 1–229; (g) O.A. Mattthews, A.N. Shipway, and J.F. Stoddart, *Prog. Polym. Sci.*, 1998, *23*, 1–56; (h) H.-F. Chow, T.K.-K. Mong, M.F. Nongrum, and C.-W. Wan, *Tetrahedron*, 1998, *54*, 8543; (i) A. Archut and F. Vogtle, *Chem. Soc. Rev.*, 1998, *27*, 233; (j) Y.-H. Kim, *J. Polym. Sci. A Polym. Chem.*, 1998, *36*, 1685 (hyperbranched polymers).
409. (a) Z.J. Wang, J.A. Arceneaux, and J. Hall, *Mod. Paint Coat.*, 1996, *86*(8), 24; (b) J. Odeberg, J. Rassing, J.-E. Jonsson, and B. Wesslen, *J. Appl. Polym. Sci.*, 1996, *62*, 435.
410. S.-Y. Kwak, *J. Appl. Polym. Sci.*, 1995, *55*, 1683.
411. (a) M.S. Reisch, *Chem. Eng. News*, Oct. 14, 1996, 64; (b) T.M. Ellison, *Chemtech.*, 1995, *25*(5), 36.
412. Anon., *Chemtech.*, 1991, *21*(6) (back cover).
413. (a) V.D. McGinniss, *Prog. Org. Coat.*, 1996, *27*, 153; (b) W.J. van Ooij, D. Surman, and H.K. Yasud, *Prog. Org. Coat.*, 1995, 25, 319; (c) L. Agres, Y. Segui, R. Delsol, and P. Raynaud, *J. Appl. Polym. Sci.*, 1996, *61*, 2015; (d) J.L.C. Fonseca, S. Tasker, D.C. Apperley, and J.P.S. Badyal, *Macromolecules*, 1996, *29*, 1705; (e) E. Finson and J. Felts., *TAPPI J.* 1995, *78*(1), 161; (f) M. Inagaki, S. Tasaka, and M. Makino, *J. Appl. Polym. Sci.*, 1997, *64*, 1031.

414. D.J. Brennan, J.E. White, A.P. Haag, S.L. Kram, M.N. Mang, S. Pikulin, and C.N. Brown, *Macromolecules*, 1996, *29*, 3707.
415. (a) G. Parkinson, *Chem. Eng.*, 1996, *103*(8), 19; (b) Y. Watanabe, K. Suzuki, and T. Masukara, U.S. patent 5,460,738, 1995; (c) H. Reuter and M.T. Brandherm, *Angew. Chem. Int. Ed. Engl.*, 1995, *34*, 1578; (d) H.K. Schmidt, *Macromol. Symp.*, 1996, *101*, 333.
416. V.P. Gupta and G.R. Brown, *J. Appl. Polym. Sci.*, 1998, *69*, 1901.
417. G. Ondrey, *Chem. Eng.*, 2004, *111*(2), 16.
418. W. Lepkowski, *Chem. Eng. News*, Jan. 22, 1996, 7.
419. L.M. Han and R.B. Timmons, *J. Polym. Sci. A Polym. Chem.*, 1998, *36*, 3121.
420. F. Denes, *Trends Polym. Sci.*, 1997, *5*(1), 23.
421. (a) S. Pellicori and M. Colton, *R&D (Cahners)*, 1999, *41*(3), 39; (b) V. Comello, *R&D (Cahners)*, 1999, *41*(6), 55.
422. R.M. Burgess and M.M. Morrison, *Chem. Eng. Prog.*, 1995, *91*(9), 63.
423. (a) T.A. Reddy and S. Erhan, *J. Polymer Sci. A Polym. Chem.*, 1994, *32*, 557; (b) S.G. Hong and F.J. Boerio, *J. Appl. Polym. Sci.*, 1995, *55*, 437; (c) G. Cerisola, A. Barbucci, and M. Caretta, *Prog. Org. Coat.*, 1994, *24*, 21.
424. (a) Anon., *Chem. Eng. News*, Oct. 27, 1997, *43*, 53; (b) H.-P. Rink and B. Mayer, *Prog. Org. Coat.*, 1998, *34*, 175; (c) P.H. Lamers, B.K. Johnston, and W.H. Tyger, *Polym. Degrad. Stabil.*, 1997, *55*, 309; (d) W.H.L. Weber, D.J. Scholl, and J.L. Gerlock, *Polym. Degrad. Stabil.*, 1997, *57*, 339; (e) M.E. Nichols, C.A. Darr, C.A. Smith, M.D. Thouless, and E.R. Fischer, *Polym. Degrad. Stabil.*, 1998, *60*, 291; (f) V. Dudler, T. Bolle, and G. Rytz, *Polym. Degrad. Stabil.*, 1998, *60*, 351.
425. (a) DuPont Magazine, Nov. 1995, 5; (b) Anon., *Mod. Paint Coat.*, 1996, *86*(7), 22, 24; (c) B.V. Gregorovich and I. Hazan, *Prog. Org. Coat.*, 1994, *24*, 131; (d) D. Smock, *Plast Technol.*, 1995, *53*(9), 44.
426. R. Maisch, O. Stahlecker, and M. Kieser, *Prog. Org. Coat.*, 1996, *27*,145.
427. T. Mezger, *Macromol. Symp.* 1995, *100*, 101.
428. K. Cottril, *Mod. Paint Coat.*, 1996, *86*(12), 20.
429. (a) Anon., *R&D (Cahners)*, 1998, *40*(9),154; (b) H. Schmidt and D. Fink, *Surf. Coat., Int.*, 1996, *79*, 66.
430. J.A. Antonelli, D.J. Yang, H.K. Yasuda, and F.T. Wang, *Prog. Org. Coat.*, 1997, *31*, 351.
431. (a) R.W. Bassemir, A. Bean, O. Wasilewski, D. Kline, L.W. Hillis, C. Su, I.R. Steel, and W.E. Rusterholz, *Kirk–Othmer Encyclopedia Chemical Technology*, 4th ed., Wiley, New York, 1995, *14*, 482; (b) J.A.G. Drake, ed., *Chemical Technology in Printing and Imaging Systems*, Royal Soc. Chem., Cambridge, UK, 1993; (c) R. Kubler, *Ullmann's Encyclopedia of Industrial Chemistry*, 5th ed., VCH, Weinheim, 1993, *A22*, 143.
432. (a) *Environ. Sci. Technol.*, 1995, *29*, 407A, 548A; (b) EPA 744-F-93-003, July 1993.
433. E. Cunningham, *Am. Ink Maker*, 1995, *73*(3), 12.
434. H. Gaines, *Am. Ink Maker*, 1994, *72*(7), 20.
435. (a) C. Rooney and K. Reid, *Am. Ink Maker*, 1995, *73*(4), 71; (b) D.M. Gentile, *Mod. Paint Coat.*, 1996, *86*(6), 40.
436. D. Fishman and F. Shapiro, *Am. Ink Maker*, 1996, *74*(11), 34, 56.
437. (a) A.A. Kveglis, *Am. Ink Maker*, 1996, *74*(2), 32; (b) L. Hahn, *Am. Ink Maker*, 1996, *74*(1), 26; (c) D. Toth, *Am. Ink Maker*, 1996, *74*(8), 48.
438. E.L. Cussler, *Abstracts Green Chemistry and Chemical Engineering Conference*, Washington, DC, 1997, 6.
439. M. Colditz, E. Kunkel, and K.-H. Bohne, *Ullmann's Encyclopedia of Industrial Chemistry*, 5th ed., VCH, Weinheim, 1987, *A9*, 38–44.
440. (a) K.P. Caballero, Preprints A.C.S. *Div. Environ. Chem.*, 1994, *34*(2), 253; (b) Printing Project, EPA 744-F-93-015; EPA 74-F-93-003, July 1993.
441. (a) A.V. Pocius, *Kirk–Othmer Encyclopedia of Chemical Technology*, 4th ed., Wiley, New York, 1991, *1*, 461; *Adhesion and Adhesives Technology: An Introduction*, Hanser Gardner, Cincinnati, OH, 1996; (b) W.D. Arendt, *Adhes, Age*, 1996, *39*(9), 37; (c) J.J. Owens, *Adhes. Age*, 1993, *36*(12), 14.
442. (a) M.S. Reisch, *Chem. Eng. News*, Mar. 4, 1991, 23; (b) W. Desmarteau and J.M. Loutz, *Prog. Org. Coat.*, 1996, *27*, 33.
443. H.T. Oien, *Adhes. Age*, 1996, *39*(2), 30.
444. Goodyear Tire and Rubber Co. annual report, Mar. 1995.
445. J.M. Brandon, *Chemtech.*, 1994, *24*(12), 42.
446. *R&D (Cahners)*, 1999, *41*(3), 11.
447. M.F. de Pompei, Preprints A.C.S. *Div. Environ. Chem.*, 1994, *34*(2), 393.
448. (a) D.E. Nikles, A.M. Lane, S. Cheng, and H. Fan, Preprints A.C.S. *Div. Environ. Chem.*, 1994, *34*(2), 417; (b) S. Cheng, H. Fan, N. Gogineni, B. Jacobs, J.W. Harrell, I.A. Jefcoat, A.M. Lane, and D.E. Nikles, *Chemtech.*, 1995, *25*(10), 35; (c) D.E. Nikles, J.W. Harrell, I.A. Jefcoat, and A.M. Lane, Preprints A.C.S. *Div. Environ. Chem.*, 1998, *38*(1), 85; (d) Anon., *Chem. Ind. (London)*, 1998, 291.
449. (a) J. Guth, J. Russo, T. Kay, N. King, and R. Beaven, *Cosmet Toilet*, 1993, *108*(11), 97; (b) E. Walls and H.K. Krummel, *Cosmet Toilet*, 1993, *108*(3), 111.
450. Presidential Green Chemistry Challenge, EPA 744-K-96-001, July 1996, 43.

RECOMMENDED READING

1. T.H. Grindstaff, *Chem. Eng. News*, Nov. 7, 1994, 3 (syntheses without solvents).
2. F. Toda, *Acc. Chem. Res.*, 1995, *28*, 480 (solid-state organic chemistry).
3. A. Lubineau, *Chem. Ind.*, 1996, 123 (reactions in water).
4. Alternative metal finishes, EPA/625/R-94/007, Sept. 1994.
5. Organic coatings replacements, EPA/625/R-94/006, Sept. 1994.
6. Z.M. Wicks, Jr., F.N. Jones, and S.P. Pappas, *Organic Coatings: Science and Technology*, Vol. 2. Wiley, New York, 1994, 226–228, 249–251, 268–271.
7. J.M. Tanko and J.F. Blackert. In: P.T. Anastas and C.A. Farris, eds, *Benign by Design: Alternative Synthetic Design*. Am. Chem. Soc. Symp., 577, American Chemical Society, Washington, DC, 1994, 98 (reactions in sc CO_2).
8. M.M. Sharma, *Pure. Appl. Chem.*, 2002, *74*, 2265 (selectivity engineering and process intensification).
9. R.A. Sheldon, *Green Chem.*, 2005, *7*, 267 (green solvents).
10. P.L. Short, *Chem. Eng. News*, Oct. 20, 2008, 37 (microreactors).

11. A.M. Rouhi, *Chem. Eng. News*, July 5, 2004, 18 (microreactors).
12. J.M. deSimone, *Science*, 2002, *297*, 799 (green solvents).
13. M. Doble, *Chem. Eng. Prog.*, 2008, *104* (8), 33 (process intensification).

EXERCISES

1. Visit a laboratory or plant with an extruder.
2. Pick an organic reaction that is normally run in a solvent and tell how you might run it in an extruder. If possible, pick one that is used industrially.
3. Visit a local dry cleaner, metal parts fabricator, print shop, or paint shop to see how successful they have been in eliminating solvent emissions.
4. Check the *Toxics Release Inventory*, or a comparable compilation, to see what point sources of solvents are near where you live. Try to figure out alternative processes that will significantly reduce or eliminate these emissions.
5. If possible, visit a plant that uses powder coatings or radiation-cured coatings.
6. If there is an sc carbon dioxide chromatograph in an analytical laboratory near you, see how the material is handled.
7. Poll your friends to see how many of them prefer shiny cars, shoes, floors, and furniture. In many cases, substituting a plain finish for the glamour finish would reduce solvent use. It has been possible to eliminate the shiny metal bumpers and strips on cars.

CHAPTER 9

Biocatalysis and Biodiversity

9.1 BIOCATALYSIS

9.1.1 Advantages and Limitations

Chemical reactions are usually carried out under rigorous conditions. They often involve highly reactive ingredients, which may be toxic or carcinogenic, that are used in the organic solvents. High temperatures and pressures may be involved, especially in the production of commodity chemicals. Biocatalysis offers the possibility of making many of the products that we need in water at or near room temperature. It uses whole organisms, usually microorganisms, or the enzymes from them, to carry out the chemical transformations.[1] It is an aspect of biotechnology. Chemoenzymatic syntheses, that is, those sequences of reactions where both chemical and biological steps are involved, are becoming increasingly common.

Biocatalysis offers many advantages[2]:

1. Often run in water with no need for organic solvents.
2. Often run at or near room temperature at atmospheric pressure.
3. No toxic metal ions.
4. No carcinogens.
5. No noxious emissions, just carbon dioxide.
6. No toxic wastes: wastes can be used as animal feed or composted.
7. Often use organic wastes as starting materials.
8. Higher selectivities: it is often possible to react only one of two similar sites.[3]
9. Chiral compounds can be produced.
10. No protecting groups are needed for reactions.
11. Less energy is needed.

About 7% of petroleum is used in making chemicals. Biocatalysis offers a method to switch to a base of renewable materials for a sustainable future. By so doing, it will help moderate global warming; its inherently cleaner chemistry will reduce air and water pollution. It will reduce morbidity and mortality in the chemical industry since it is inherently safer. It will also reduce the possibility of attacks by terrorists.

There are limitations to biocatalysis that current research is endeavoring to overcome:

1. The reaction may be slow, especially if done in an organic medium.
2. The run may have to be quite dilute.
3. The enzyme may be expensive and difficult to recover for reuse.
4. The enzyme may lose activity too quickly, especially if used at elevated temperatures.
5. The recovery of the product from dilute solution may be complicated and expensive.
6. The product may inhibit its further formation.
7. It may be difficult to obtain a pure enzyme, free of contamination by other enzymes that might degrade the substrate or produce by-products.
8. It may be difficult to shift the equilibrium in the reaction so that 100% conversion to the desired product takes place.
9. Conditions for optimal growth of the organisms may be difficult to work out.
10. Production of the desired product may require the addition of an elicitor or altered growing conditions.
11. The organism may not secrete the product into the medium so that the cells have to be destroyed to recover the product.

9.1.2 Industrial Products Produced by Biocatalysis

In 1994, Hinman[4] listed the major products produced from fermentation as follows:

Product	Metric Tons/Year
Ethanol	15 million
Monosodium glutamate	1 million
Citric acid	400,000
Lysine	115,000
Gluconic acid	50,000
Penicillin[5]	15,000

In 2007, the ethanol produced in the United States as a biofuel was 50 million liters, caused largely by government subsidies.[6] These are usually made from glucose, starch, molasses, or other such compounds, with added nutrients from corn steep liquor, soybean meal, blood meal, and so on. In addition, there are many valuable compounds made in lower volume.[7] These include lactic acid, malic acid, succinic acid,[8] L-phenylalanine, L-aspartic acid, other amino acids,[9] flavors and fragrances,[10] steroids, vitamins, other antibiotics, astaxanthin for feeding to salmon to produce pink flesh,[11] xanthan gum, biosurfactants,[12] tissue plasminogen activator, human growth hormone, erythropoietin, and many enzymes.[13] L-Aspartate is made from fumarate using an aspartase from *Escherichia coli* and L-phenylalamine by addition of ammonia to *trans*-cinnamic acid using phenylalanine ammonia lyase from *Rhodotorula rubra*. They can be made simultaneously from fumaric acid and phenylpyruvic acid with ammonia by an engineered *E. coli*.[14] The two products are converted to a dipeptide, aspartame, which is a popular artificial sweetener (**9.1** Schematic).

L-Malate (**9.2** Schematic) can be made by the hydration of fumarate using a fumarase from *Brevibacterium flavum*. It can also be produced at up to 109 g/L by a strain of *Saccharomyces cerevisiae*.[15] (Enzymes are usually named after the reaction that they catalyze with the suffix *ase*[16].)

A plant for the production of riboflavin in Germany has reduced the cost of the current chemical syntheses by one-half.[17] The worldwide market for enzymes[18] is at least 1 billion dollars each year.[19] The major use is in laundry detergents where they save energy by allowing washing to proceed at lower temperatures.[20] They are also used extensively in food processing (e.g., production of high-fructose corn syrup and the use of pectinases to clarify fruit juices). Over 60% of the enzymes are recombinant products.[21]

Fermentation has been used in the production of foods and drinks since the days of antiquity. A recent application is the large-scale production of yogurt. Probiotics, where living bacteria are left in the yogurt, may be beneficial to health.[22] Before the advent of cheap petroleum, a variety of other commodity chemicals were produced by fermentation. Acetone and *n*-butyl alcohol were produced by *Clostridium acetobutylicum*.[23] Ethanol, which was entirely made from fermentation, is now made in part by the hydration of ethylene. Acetic acid is now made largely by the carbonylation of methanol (**9.3** Schematic) using a rhodium catalyst in the presence of iodide ion.[24] Acetic acid produced by fermentation does not involve toxic carbon monoxide or methanol, nor does it involve expensive rhodium. Products from petroleum or natural gas are cheaper because they do not include the cost of natural resource depletion or of global warming. (See Chapter 17 for a discussion of such costs.) It is also possible that the application of some of the newer techniques will improve the fermentations so that they give better yields and are more cost-competitive. This is one of the keys to a sustainable future.[25] Industrial research in this area is increasing.[26] Many books cover the use of enzymes in industry.[27]

9.1.3 Techniques of Biocatalysis

A variety of techniques are being studied in an effort to remove the limitations of biocatalysis.[28] Some of these will now be examined together with some examples of their use.

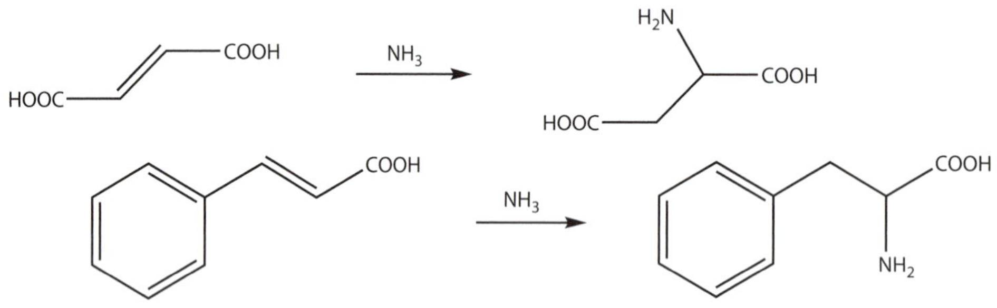

HOOC–CH=CH–COOH $\xrightarrow{NH_3}$ HOOC–CH$_2$–CH(NH$_2$)–COOH

C$_6$H$_5$–CH=CH–COOH $\xrightarrow{NH_3}$ C$_6$H$_5$–CH$_2$–CH(NH$_2$)–COOH

9.1 Schematic

HOOC–CH=CH–COOH $\xrightarrow{H_2O}$ HOOC–CH$_2$–CH(OH)–COOH

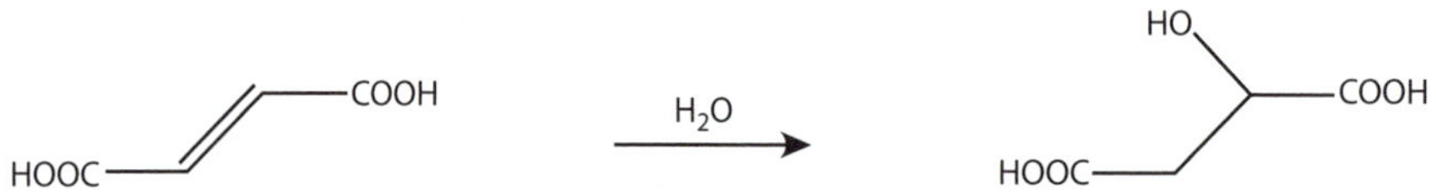

9.2 Schematic

9.1.3.1 Production of Ethanol

The methods used to prepare ethanol will be examined first, because ethanol is the chemical that is produced in largest volumes by fermentation.[29]

$$CH_3OH + CO \xrightarrow[I^{\ominus}]{\text{Rh catalyst}} CH_3COOH$$

9.3 Schematic

The *S. cerevisiae* often used to convert sugars into alcohol is inhibited by too much substrate and too much product. A typical final concentration of ethanol is 5.9%. By immobilizing the cells and adding more glucose and nutrients stepwise every few hours, ethanol was produced at up to 15.0%.[30] Another way to improve the process is to hunt for more tolerant organisms. A *Saccharomyces* used to make palm wine in Nigeria tolerated 50% sucrose solutions and led to a concentration of 21.5% ethanol.[31] Tropical yeasts can also tolerate somewhat higher temperatures (e.g., 40°C), which leads to faster rates. Thermophilic bacteria that can use both five- and six-carbon sugars make ethanol 10 times as fast as the usual fermentation with yeast.[32] Addition of zeolites to fermentation of molasses with *S. cerevisiae* increased the rate by 53%.[33] This is thought to involve the removal of inhibitors, such as toxic metal ions. Inhibitors may also include phenols, acetic acid, furfural, and 5-hydroxymethylfurfural.[34] Activated carbon and alumina have also been used for this purpose.[35] This allows the same medium to be used over a longer time.

Continuous removal of the ethanol by extraction with *n*-dodecanol has been used with an immobilized yeast in an 18-day run in an effort to lower the cost of production.[36] The main by-product was glycerol. Some companies have postponed trying to recover it.[37] The production of ethanol and its conversion to biodiesel fuel was carried out in one step using a two-phase system of oleic acid and an aqueous phase with *S. cerevisiae* plus a lipase from *Rhizomucor miehei*.[38] Continuous filtration of the cells with hollow-fiber membrane separation of the ethanol from the filtrate has also been used.[39] Ethanol was obtained from glucose in 90% yield by a system run for 150–185 days with continuous stripping of the ethanol by distillation.[40] Continuous removal of the ethanol by pervaporation through membranes was discussed in Chapter 7. Other methods of continuous removal include gas stripping, product removal by a solid phase, and the use of two aqueous layers.[41]

Costs might be lowered by combining steps and by starting with cheaper raw materials, such as wastes. Simultaneous saccharification and fermentation of cellulose (a polymer of glucose) has been accomplished by combining a cellulase and a β-glucosidase.[42] Cassava starch (another polymer of glucose) has been converted to 10.5% ethanol by a mixed culture of *Endomycopsis fibuligera* and *Zymomonas mobilis*.[43] The first organism converted the starch to sugar, which was then converted to ethanol by the second. The addition of 0.01% glucoamylase allowed the production of 13.2% ethanol in a yield of 98%. Cheaper substrates include stillage from sugar beets,[44] agricultural wastes, waste paper, and others.[45] To obtain favorable yields, both the cellulose and the hemicelluloses (polymers of five-carbon sugars) of lignocellulosic wastes must be used.[46] In addition, inhibitors, such as furfural, phenol, acetic acid, aromatic compounds, and metal ions, may have to be removed from lignocellulosic hydrolysates. In one case, acetic acid in hardwood spent sulfite liquor was removed by a mutant *S. cerevisiae* that grows on acetic acid, but not on xylose, glucose, mannose, or fructose.[47] *Candida parapsilosis* converted xylose to xylitol selectively in the presence of glucose and mannose in hydrolysate of aspen wood.[48] A mixed culture of *S. cerevisiae* and *Candida shehatae* produced both ethanol and xylitol simultaneously from a mixture of glucose and xylose.[49] Both products were also obtained from the latter organism growing on xylose alone.[50] Xylitol has also been produced from xylose by *Debaryomyces hansenii*.[51] Xylitol has a high sweetening power, does not cause dental caries, and is an insulin-independent carbohydrate for insulin-dependent diabetics. Its potential market is large, but not nearly as large as the potential market for ethanol used as fuel.

Klebsiella oxytoca was engineered to include genes to produce enzymes for the hydrolysis of both cellulose and hemicellulose.[52] It was able to produce ethanol in 83% yield from mixed office waste paper. The yield was 539 L/metric ton of waste. The paper had to be pretreated with water (with or without 1% sulfuric acid) at 140°C for 30 min. The need for the cellulase was reduced by recycling some of the residue to reduce costs. A plant in Louisiana uses this process on sugar cane wastes.[53] Other plants producing ethanol from cellulose are run by Iogen in Canada and by Abengoa Bioenergy in Spain.[54] The use of intermittent ultrasound in such systems can reduce the amount of enzyme needed by one-half.[55] The residue after converting corn to ethanol was a good ingredient for making food for *Tilapia* (a fish popular for aquaculture).[56] Credit for such by-products helps reduce the overall cost. Microwaves have speeded up other enzymatic reactions but have not been applied to the production of ethanol.[57] Magnets have increased the rate of ethanol production three- to four-fold.[58]

The U.S. National Energy Laboratory has used a genetically engineered *Z. mobilis* to convert agricultural residues and sawdust to ethanol in a process that uses both the glucose and xylose. It is estimated that this will reduce the price of ethanol from $1.20/gallon to $0.70/gallon.[59] (These prices are out-of-date but do show the potential cost savings.) Recombinant *E. coli* converted willow wood to ethanol, which is estimated to cost $0.48/L.[60] The market price in Sweden is $0.35/L. It is estimated that the price could be reduced by $0.063/L if the process started with waste and by another $0.063/L if the ethanol was not distilled. This makes a good case for starting with waste paper and using pervaporation to isolate the ethanol and then another membrane to dehydrate it. Critics contend that ethanol from corn is not a renewable energy source because it takes at least a barrel of petroleum to produce a barrel of ethanol.[61] This includes the energy needed to make and apply the fertilizer, tilling and harvesting the corn, hydrolyzing the corn starch, distilling the ethanol, and so on. The use of wastes and the use of membranes in the process will reduce the need for petroleum. It is also possible to use biodiesel fuel (obtained from vegetable oils by alcoholysis) as well as energy from the wind, the sun, and hydropower in the processing. The critics do not mention the costs of natural resource depletion and global warming. To put the problem into perspective, enormous savings of fuel would result if Americans lived in more compact cities and went from place to place by means other than driving alone in their cars.

9.1.3.2 Immobilization of Enzymes[62]

Altus Biologics purifies and crystallizes enzymes, which are then cross-linked with glutaraldehyde.[63] This results in marked improvements in stability to heat, extremes of pH, exogenous proteases, and organic solvents. Full activity remained for a product produced from thermolysin after 4 days at 65°C. More than 95% of the activity remained after 1 h in 50% aqueous tetrahydrofuran at 40°C, compared with only 36% for free thermolysin. The enzymes can be used in regioselective acylation of alcohols and amines, synthesis of peptides, resolution of acids and alcohols, chemoselective hydrolysis of esters and amides, modification of sugars, and in making modified penicillins. They can be filtered off and reused 10–20 times so that final productivities of 1:1000 to 1:100,000 catalyst per product can be obtained. Rates of hydrolyses catalyzed by these enzymes in organic solvents can equal or exceed those in water. Microporous cross-linked protein crystals ("bioorganic zeolites") have also been used to separate molecules by size, chemical structure, and chirality.[64] Roger Sheldon has modified the CLEC process so that the enzymes do not need to be crystallized.[65]

Whole cells are often immobilized in calcium alginate gel beads.[66] Lactic acid bacteria immobilized in this way have been used to recover the protein from cheese whey (the liquid left over from the making of cheese).[67] The lactic acid that is formed precipitates the proteins. Encapsulation in this way increased the stability of *Bacillus sphaericus* to heat and light for use in mosquito control.[68] An enzyme was immobilized on magnetite in this way.[69] This was used in a magnetic field that stabilized the fluidized bed in the hydrolysis of maltodextrin. Calcium alginate beads containing β-glucosidase were treated with tetramethoxysilane to encapsulate them with silica.[70] This gave excellent protein retention and thermal stability, with an operational life of several months. Chitin and chitosan-based materials have been used to immobilize enzymes.[71]

Immobilized biocatalysts[72] often have lower activity owing to diffusion and access problems. An enzyme immobilized on a macroporous monolith was more active than the one in beads, presumably because of better accessibility to the active sites.[73] Enzymes have also been encapsulated in porous thin-walled polymeric microcapusules.[74] Biocatalytic plastic made by acryloylation of subtilisin and α-chymotrypsin followed by free-radical-catalyzed polymerization with vinyl monomers, such as methyl methacrylate, styrene, and vinyl acetate, in the presence of a surfactant contains up to 50% enzyme.[75] They are 10 times less active than the free enzyme in water, but much more active in organic media than the free enzyme. There is no loss in activity after storage for several months at room temperature. This provides a way to put enzymes into tubing, membranes, coatings, and such. The method has been suggested for nonfouling paints for ships. Biocatalytic plastic has also been made by incorporating enzymes into polyurethane foams.[76] Vinyl acetate–alkyl acrylate copolymer latex paints have been used with sucrose and glycerol to immobilize nongrowing whole cells.[77] The products can be shipped dry and activated by adding water when desired. Microcapsules with several alternating layers of poly(allylamine hydrochloride) and poly(styrene sulfonate) have also been used to protect enzymes.[78]

There are some cases in which the immobilized enzyme has a higher activity than its free form. A lipase put on silica by the sol–gel method, using an alkyltrimethoxysilane, had twice the activity of the free enzyme.[79] It retained 63% of its activity after 3 months (i.e., much more than that of the free enzyme). It was stable enough mechanically for use in fluidized beds. Inclusion of Fe_3O_4 nanoparticles in the preparation made the enzyme easy to separate. Some enzymes trapped in silica by the sol–gel process are as active as the free enzymes and are much more stable thermally and to pH changes.[80]

An immobilized *Candida cylindracea* lipase retained 80% of its activity after 336 h of operation.[81] It was 37 times as stable as the native enzyme. If glucose oxidase was physisorbed on to silica, its activity dropped to one-sixth in 3 weeks.[82] If it was attached covalently by a triethoxysilane method, the activity increased over the 3-week period. Lipases in elastomeric silicone rubbers showed as much as 54 times the activity of the free enzyme.[83] Attachment at more than one point can enhance stability. A thermophilic esterase from *Bacillus strearothermophilus* was attached through its amino groups to a glyoxyl agarose gel.[84] Multipoint attachment increased the stability 30,000-fold compared with 600-fold for one-point attachment. Covalent multipoint attachment of carboxypeptidase to an agarose gel raised its stability a thousand fold.[85] Zeolites have been used as supports for cutinase[86] and trypsin.[87] A copolymer of *N*-isopropylacrylamide and *N*-acryloxysuccinimide was used to immobilize α-chymotrypsin as a material that dissolved below 35.5°C and precipitated at temperatures higher than that. It was 83.5% as active as the free enzyme initially. Eighty percent of this activity remained after nine precipitation–dissolution cycles.[88] Because the cost of immobilization can be a major factor in industrial catalysis, cheaper methods are being sought. In one method, β-glucuronidase immobilized on the oil bodies of *Brassica napus* was as active as the free enzyme.[89] Some organisms can form biofilms. These biofilms are being studied as a way of immobilization in enzymatic reactions.[90] (Other immobilization methods were described in Chapter 5.)

9.1.3.3 Biocatalysis at High Concentrations

One of the limitations of biocatalysis is that many reactions must be run in dilute media. Stepwise addition of the

OH → Aldehyde dehydrogenase → O

9.4 Schematic

substrate and continuous removal of the product can help, as mentioned earlier. However, sometimes this is unnecessary. The enzymatic hydrolysis of acrylonitrile produces acrylic acid at 390 g/L.[91] Some enzymes can be produced by solid-state fermentations.[92] *Kluyvereromyces lactis* produces β-galactosidase when grown on corn grits or wheat bran moistened with deproteinized milk whey. An acid protease has been produced by *Rhizopus oligosporus* grown on rice bran. Sugar beet pulp has also been used as a substrate for the production of enzymes. An entomopathogenic fungus for the control of insects, *Metarrhizium anisopliae*, has been grown on rice and rice husk. Xylanase has been made and used *in situ* by *Aspergillus oryzae* on wood chips for use in biobleaching.[93] Griseofulvin has been made by growing *Penicillium griseofulvin* on rice bran with a 50% water content.[94] Solid-state fermentation[95] offers superior simplicity of the enzyme productivity process, lower energy needs, lower waste output, and ease of recovery of the product. Occasionally the enzyme is just extracted with water. The process can be used to upgrade agricultural wastes, such as wheat bran, apple pomace, corn stalks, and mussel-processing wastes, to products such as ethanol and citric acid.[96] It is akin to raising edible fungi on an old log.

A lipase has been used to convert solid triolein (glycerol trioleate) to the monooleate by treatment with glycerol at 8°C.[97] Other lipases have been used in the hydrolysis and transestirification of oils, as well as in the esterification of fatty acids without solvents.[98]

Peptides can be produced from eutectic mixtures of amino acid derivatives with the addition of a small amount of solvent.[99] Immobilized subtilisin and thermolysin were used with 19–24% water or an alcohol to produce polypeptides. Subtilisin on celite (a diatomaceous earth) was used to convert a mixture of L-phenylalanine ethyl ester and L-leucinamide containing 10% triethyleneglycol dimethylether to L-phenylalanineleucinamide in 83% yield. Addition of 30% 2:1 ethanol/water reduced the time needed from 40 h to 4 h. The enzyme could be used three more times. These reaction mixtures contained 0.13–0.75 g peptide per gram reaction mixture compared with 0.015–0.035 when the reaction was carried out in solution. The small amount of solvent used could probably be recovered for recycle by vacuum distillation. This method may find wider use whenever the enzyme is not inhibited by the substrate and product. With the advent of enzymes with greater thermal stability more solids that melt on heating will become possible substrates. Peptides have also been made with thermolysin in solid-to-solid reactions in water, in over 90% yield.[100] For this to work, it may be necessary for the reactants to have a very slight solubility in the water. After removal of the dipeptide by filtration, the filtrate containing the enzyme might be used in another run.

Solid enzymes have been used as fixed bed catalysts with vaporized substances.[101] As an example, allyl alcohol has been oxidized to acrolein with aldehyde dehydrogenase (**9.4** Schematic). No solvent is needed. The dry enzyme is more stable to heat than it would be in water. Mass transfer is improved over a reaction in water. There is no need to immobilize the enzyme. The process can be continuous. Gaseous ethanol has been converted to acetaldehyde using an immobilized alcohol oxidase.[102] Esters have been made in a similar fashion.[103] (An example was given in Chapter 7 under process intensification.) As more thermally stable enzymes are found, this method will be applicable to more substrates. It may also be possible to operate under a partial vacuum to increase the range of substrates.

9.1.3.4 The Use of Enzymes in Organic Solvents

Our overall goal should be to reduce the use of organic solvents. However, their use can overcome some problems that are otherwise difficult to handle. The use of organic solvents with enzymes allows one to use water-sensitive compounds or substrates that are difficultly soluble in water.[104] Thermal stability is improved. Water-dependent side reactions are suppressed. There is a shift to synthesis over hydrolysis. There are some problems: Proteins are usually insoluble in many organic solvents. Organic solvents often inhibit the enzymes. Reaction rates may vary with the surface area, porosity, and the extent to which the protein is swollen by the solvent. Hydrogen bonding and the conformation of the polymer may be altered. The solvent, water, and the substrate may act as ligands that compete for the metal ion in the active site. Tetrahydrofuran and dimethylsulfoxide are strong ligands. When the reaction is run in sc carbon dioxide, carbamates may be formed from lysine and arginine. The rates of the enzymatic reactions usually drop dramatically in the shift from water to organic solvents. A trace of water, at least part of a monomolecular layer on the enzyme, must be present, although, in at least one case, methanol can be substituted for it.[105] Workers usually examine the amount of water needed for optimal results.

The use of cross-linked enzyme crystals to circumvent some of these problems was discussed earlier. The rate, when they are used, can be as high as in water. When these are used in aqueous acetonitrile, the rate can be increased 100-fold by adding a buffer of a suitable acid (e.g., *p*-nitrophenol, phenylboronic acid, or triphenyl-acetic acid) and its sodium salt.[106] A solid-state buffer of lysine plus lysine hydrochloride increased the activity of immobilized subtilisin in transesterifications run in hexane or toluene.[107] Subtilisin and chymotrypsin showed greater activity in polar organic solvents when they were treated with an aqueous buffer, silica, and 1-propanol and then used without drying.[108] The catalytic activity and enantioselectivity of *Candida antarctica* lipase in toluene improved when it was used with an amine, such as triethylamine.[109] The use of sc carbon dioxide as a medium, as described in Chapter 8, also overcomes some of the problems in the use of organic solvents.[110] Ionic liquids have also been used as solvents.[111]

A variety of other approaches are being tried to remove these limitations; however, before examining them, the use of two-phase systems in which only one is organic will be covered. Enzymes can be used in reverse micelles.[112] In one case, β-galactosidase was encapsulated by bis(2-ethylhexyl) sodium sulfosuccinate (a surfactant) and lecithin in isooctane. The lecithin kept the enzyme from going into the bulk aqueous phase. (Surfactants can be used to make enzymes soluble in organic solvents.) It was used in the hydrolysis of a galactose ester with the product being separated continuously with centrifugal partition chromatography to give an 85% conversion in 5 h. The reaction of menthol with propionic anhydride using *Candida rugosa* lipase in a microemulsion was run continuously with an utrafiltration step in the loop to harvest the product in the oil and reuse the enzyme.[113] Two-phase cultures can overcome the problem of toxicity of the product or substrate to the enzyme.[114] Styrene has been converted to its epoxide with 95% or more enantioselectivity using xylene oxygenase in *E. coli* in such a system. Such an extractive bioconversion has been used with animal cell cultures where one phase contains aqueous polyethylene glycol and the other aqueous dextran. The bacterium *Serratia macrescens* has been used to produce a chitinase in a two-phase system of 2% aqueous polyethylene glycol (molecular weight 20,000, top phase) and 5% dextran in water.[115] The cells remain in the top phase. The enzyme is harvested by replacing the lower phase to prevent back inhibition. β-Galactosidase has been produced by *E. coli* in a two-phase system of polyethylene glycol—aqueous phosphate.[116] Some of the cells were disrupted by ultrasound to release the enzyme to the top layer for harvest. The cells remained in the bottom layer. The culture was run intermittently by adding new top phases. A pectinase was produced in a two-phase system of aqueous polyethylene glycol 4000 and dextran by *Polyporus squamosa*. The fungal growth stayed in the bottom layer and the pectinase was harvested from the top layer.[117]

Bioconjugation of polyethylene glycol with enzymes increases their thermal stability at the same time that it increases their solubility and activity in organic solvents.[118] The attachment can be accomplished in a variety of ways. The reagents shown in **9.5** Schematic can react with amino groups in the proteins.

Another method is to attach a polyethylene glycol with an acrylate on one end to the enzyme. The resulting monomer is then polymerized with other acrylic monomers. Some of these conjugated enzymes can be used without solvent (e.g., in the transesterification of triglycerides at 58°C).

Acetylation of horseradish peroxidase increased its half-life at 65°C five-fold and made it more tolerant to dimethylformamide, tetrahydrofuran, and methanol.[119] Chymotrypsin is more active in organic media when immobilized than when it is just suspended.[120] The selectivity of enzymes in organic media can vary with the solvent.[121] The transesterification of racemic 1-phenylethanol with vinyl butyrate using subtilisin gave the best enantioselectivity in dioxane (61%) and the least in *N*-methylacetamide (3%). Pretreatment of a lipase from *C. rugosa* in isopropyl alcohol increased the activity slightly and the enantioselectivity by more than 10-fold in the resolution of ester (**9.6** Schematic.[122]) The *R*-isomer had 93.1% enantiomeric excess ee and the *S*-isomer had 94.4% ee.

A cyclodextrin glycoslytransferase favored α-cyclodextrin in ethanol, acetonitrile, and tetrahydrofuran, but favored up to 82% β-cyclodextrin in *tert*-butyl alcohol.[123]

Surfactants can be used to solubilize enzymes in organic media and increase their activity.[124] Ion pairing of α-chymotrypsin with bis(2-ethylhexyl) sodium sulfosuccinate

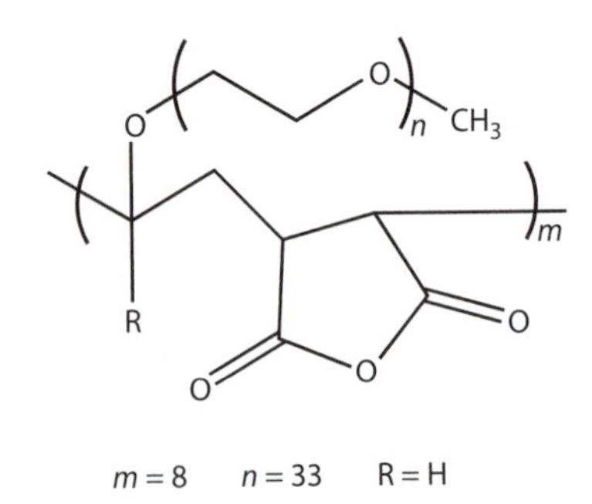

9.5 Schematic

9.6 Schematic

gave near-aqueous activity in a transesterification in isooctane.[125] Enzymes can also be coated with lipids to make them soluble in organic solvents.[126] The precipitate from mixing aqueous solutions of the two was lyophilized. It was soluble in benzene, ethyl acetate, isopropyl ether ethanol, and dimethylsulfoxide. The lipid used (a surfactant) is shown (**9.7** Schematic).

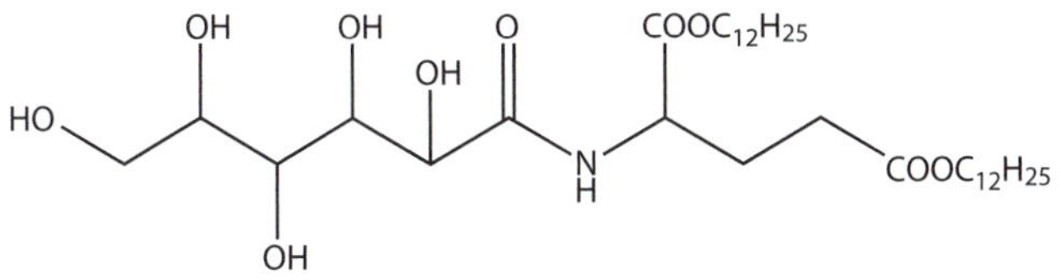

9.7 Schematic

About 150 lipid molecules were present per one enzyme. It was used to esterify 1-phenylethanol with lauric acid in isooctane, giving 95% conversion in 2 h. A lipid-coated lipase has also been used to esterify lauric acid with glycerol in sc carbon dioxide with 90% conversion.[127] Catalase with less than 5% of its surface amino groups modified by the nonionic surfactant **9.8** Schematic had a higher activity in 1,1,1-trichloroethane than when the enzyme was conjugated with polyethylene glycol.[128]

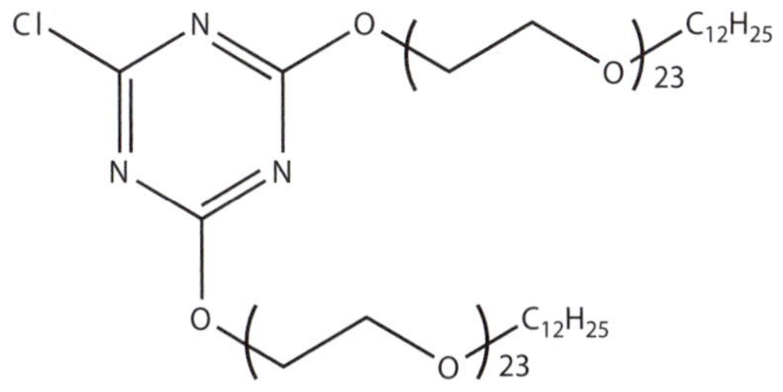

9.8 Schematic

Crown ethers also enhance the activity of enzymes in organic solvents.[129] The enzyme is lyophilized in the presence of the crown ether before use. α-Chymotrypsin treated in this way with 18-crown-6 was 640 times as active in transesterification of amino acid esters as the untreated enzyme. However, the rate was still 150 times lower than in water. The use of surfactants to improve activity in organic solvents seems preferable because of a greater rate increase. They also avoid the toxicity and expense of the crown ethers. The cross-linked enzyme crystals also offer a way to obtain reasonable rates in organic solvents.

Another way of improving on the rates in the presence of organic solvents has been pioneered by Arnold and coworkers.[130] They carried out the directed evolution of a *p*-nitrophenylesterase by random mutagenesis and screening plus pairwise gene recombination. The improvement in activity in one generation was not large. However, by carrying positive mutants through four generations, they developed an enzyme that performed as well in 30% aqueous dimethylformamide as the wild type did in water. The method can be used to evolve thermally stable enzymes and even to evolve some enzymes that can carry out reactions not found in nature. As few as four amino acid substitutions can convert an oleate 12-desaturase to a hydroxylase, and as few as six can carry out the opposite transformation.[131]

More recent work on directed evolution[132] has developed site-directed mutagenesis where the amino acids at the active site are replaced one by one.[133] Another technique is gene shuffling where DNA from an organism is chopped into pieces and then recombined with or without DNA pieces from a second organism.[134] This is especially useful since it can be used with organisms that cannot be cultured. (Only 0.1–1.0% of microorganisms can be cultured.)[135]

Such techniques have developed enzymes that are promiscuous, that is, will accept a variety of substrates.[136] In the case of a malonic acid decarboxylase, the wild-type enzyme gave one optical isomer and a mutant gave the other (**9.9** Schematic).[137]

In some cases, a mutant enzyme takes on a different function. For example, an alanine racemase from *Geobacillus thermophilus* lost this ability and became an aldolase when Tyr265 was replaced by alanine.[138]

9.1.3.5 Extremophiles

Extremophiles are organisms, usually Archaea, that grow normally under extreme conditions of heat, cold, salt, pressure, anoxia, or pH.[139] They live wherever there is water, to depths of 4.2 km and temperatures up 110°C. This includes deep sea hydrothermal vents, hot springs, the Antarctica, salt lakes, piles of sulfide minerals, sewage sludge, and such. Their enzymes are sometimes called extremozymes. The use of such enzymes tolerant of high temperatures could speed up many reactions. Salt tolerance may translate into better tolerance of organic solvents. Use of the high-temperature enzymes with foods would not require a separate sterilization step to remove unwanted bacteria. Use of enzymes adapted to cold[140] could allow processing of food to modify flavor and to tenderize meat without loss of nutrients.[141] Much energy might be saved if cold-adapted enzymes could be used in detergents to clean effectively in water at room temperature.[142] Proctor and Gamble markets "Tide Coldwater," which is said to save $65/year in energy.[143] An understanding of how these extremozymes tolerate such environments may lead to ways to stabilize ordinary enzymes to extreme conditions.[144] Several hyperthermophilic enzymes,[145] with optimal temperatures of 95–110°C, have been purified. These include hydrolases, oxidoreductases, dehydrogenases, hydrogenases, DNA polymerase, and glucose isomerase.

The upper limit on thermal stability of enzymes may be about 120°C.[146] Conformational stability and resistance to the usual denaturation mechanisms must be present. The aldehyde ferredoxin oxidoreductase from *Pyrococcus*

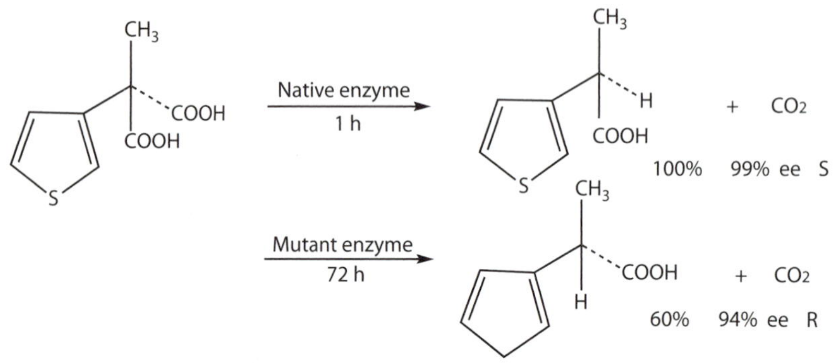

9.9 Schematic

furiosus (which has an optimal growth temperature of 100°C) has a lower surface area/volume ratio and greater than the usual number of salt bridges between acidic and basic amino acids.[147] The enzyme contains tungsten. A protease from *B. stearothermophilus* that had optimum activity at 60°C was converted to one operable at 100°C by introducing modifications that increased its rigidity.[148] This involved introduction of a disulfide bridge and substitution of six of the enzyme's amino acids by site-directed mutagenesis, including glycine for alanine and alanine for proline. It is amazing that protein monolayers applied to glass spheres or other surfaces by the Langmuir–Blodgett technique are stable to 150–200°C.[149]

The complete gene sequence of 1.7 million base pairs of the archaean *Methanococcus jannaschii* has been determined.[150] The microorganism grows at about 100°C, using only inorganic compounds and emitting the by-product methane. *Pyrolobus fumari* grows at 90–113°C at vents in the mid-Atlantic ridge using sulfur compounds and hydrogen.[151] It needs no organic matter or oxygen to live. A recombinant xylanase from *Thermatoga maritima* was active—over several hours at 100°C—in removing lignin and sugars at pH 3.5–10.[152] It may be useful in the enzymatic pulping of wood. Two enzymes from thermophilic bacteria convert glucose to gluconic acid and hydrogen with 100% efficiency.[153] No carbon dioxide is formed. This may be useful in producing hydrogen fuel, if uses for large amounts of gluconic acid can be found. Most hydrogen is produced from natural gas now, rather than by electrolysis of water. Hydrogen from glucose could be used in an ammonia plant.

The search for cold-adapted enzymes has involved bacteria from soils on mountain tops,[154] salmon intestines,[155] and whales rotting deep in the ocean.[156] Enzymes from these sources have greater flexibility and lower thermal stability. A bacterial lipase from *Psychrobacter immobilis* had very little arginine compared with lysine, a low content of proline, a small hydrophobic core and a very small content of salt bridges.[157] Comparison of the lactic dehydrogenases of Antarctic fish with other fish showed that the active site had not changed, but the surrounding structures were more flexible.[158]

Microorganisms that grow under the high pressures of the deep sea can be more resistant to organic solvents.[159] A strain of *Pseudomonas putida* from the deep sea tolerated 50% toluene. Microbial systems from deep basalt aquifers (1000 m deep) exist on hydrogen plus water containing some carbon dioxide.[160] Other microbes are found in sediments hundreds of meters below the floor of the Pacific Ocean and in granite several kilometers below the surface in Canada and Sweden. Halophilic archaeans accumulate inorganic ions at concentrations greater than those in the environment (in places such as the Great Salt Lake in Utah and the Dead Sea in Israel), whereas halophilic bacteria accumulate small osmoprotectants, such as glycerol.[161] Halophilic enzymes require high salt concentrations (1–4 M) for stability and activity. A salt-loving bacterium (an *Altermonas*) from a hot spring in western Utah is able to neutralize nerve gases without formation of any toxic by-products.[162] Extremophiles may be useful in bioremediation.

The bacterium *Deinococcus radiodurans* is able to resist intense ionizing radiation (500,000–3 million rad).[163] A human would die with less than 500 rad. This tolerance may have resulted from evolution of the ability to resist severe dehydration. Plants that have evolved to be dried down and then get revived by water often contain the sugar trehalose (α-D-glucopyranosyl-1,1′-α-D-glucopyranoside).[164] It protects insects, plants, and others from dessication, frost, and heat.[165] It produces a stable glassy state that enables the proteins to retain their native conformations and suppresses aggregation of denatured proteins. It can be made enzymatically from maltose[166] and starch.[167] Addition of trehalose before drying can stabilize invertase,[168] as well as enzymes that cut DNA.[169] Addition of trehalose can allow reactions to be run at higher temperatures (e.g., reverse transcriptase reactions that are 20 times faster at 60°C than at ambient temperature).[170] Combining this with thermophilic enzymes may allow long-term storage of dry enzymes without refrigeration. A combination of diethylaminoethyldextran and a polyol kept an alcohol oxidase active for 2 years in the dry state, whereas without these additives it lost activity in a month at room temperature.[171] A 2-M solution of xylitol made β-galactoside 10,000 more stable to high pressure.[172] Resistance to high pressure will become more important as more foods are sterilized in this manner.

A β-glucosidase from *P. furiosus* has been used to make glucoconjugates at 95°C for 15 h.[173] Some extremozymes are already in commercial use.[174] A DNA polymerase from *Thermus aquaticus*, found in a hot spring in Yellowstone Park in Wyoming, is used in the polymerase chain reaction. The reaction was not practical for widespread use until the thermophilic enzyme was found. ThemoGen, Inc. (now part of Medi Chem) (Chicago, Illinois) offers a series of esterases derived from thermophilic organisms, some of which have optimal activities at 65°C to more than 80°C.[175] They have a greater tolerance for organic solvents than the usual enzymes. Diversa (now Verennium, San Diego, California) uses a somewhat different approach to the recovery of extremozymes from extremophiles.[176] Thermophilic organisms can be difficult to culture. The company bypasses this step by isolating the DNA from a mixture of microorganisms and then cutting it into fragments, which are cloned in another bacterium, such as *E. coli*. The expression products of several clones are pooled and screened for activity robotically. A lead that is found can be optimized by random DNA mutagenesis. They now sell kits of glycosidases, lipases, aminotransferases, phosphatases, and cellulases, each of which may contain up to 10 recombinant enzymes. (For more on genetic engineering, see Section 9.1.6.) Many uses are envisioned for hyperthermophilic enzymes.[177]

9.1.3.6 Catalytic Antibodies

An antibody formed in an animal in response to a model compound for the transition state of a desired reaction may serve as a catalyst for that reaction. It is hoped that these catalytic antibodies will function in the same way as natural enzymes.[178] Reactions catalyzed by them include hydrolysis of esters, opening of epoxides by hydroxyl groups, Diels–Alder reactions, oxy-Cope reactions, cationic cyclization, elimination of hydrogen fluoride to form unsaturated ketones, and aldol reactions.[179] Antibody aldolases are much more tolerant of varied structures than natural enzymes.[180] One cyclization gave the disfavored product (**9.10** Schematic).[181]

Usually, the rate of the catalyzed reaction has been relatively slow compared with catalysis by enzymes. In one Diels–Alder reaction the rate enhancement was 2618 times. Occasionally, the rates do rival those of enzymes. A study of mutants of a catalytic antibody for the hydrolysis of *p*-nitrophenyl esters produced one that increased the rate by an additional 100-fold.[182] The most active antibodies are ones that increased the rate of hydrolysis of *p*-methylsulfonylphenyl esters by a factor of 100,000 and aldol reactions by a factor of 2.3×10^8.[183] The application of combinational chemistry to this field may provide substantial increases in rates.[184] The catalytic antibodies may be most useful in catalyzing reactions for which there are no suitable natural enzymes. An antibody produced in an animal can sometimes be transferred to a plant for production.[185]

9.1.3.7 Miscellaneous Techniques

Some oxidoreductases require NADH as a cofactor.[186] To use them in organic synthesis, as in the reaction of a ketone to an alcohol, it is necessary to have an efficient system to continuously regenerate them. A common way is to include in the same reaction formic acid and formate dehydrogenase, the by-product being carbon dioxide.[187] The regeneration of the cofactor can also be done electrochemically with or without the addition of a hydrogenase.[188] The use of whole organisms eliminates this need.

The ammonia that inhibited the growth of a mouse hybridoma cell line was reduced about 50% by a pH gradient across a hydrophobic microporous membrane.[189] A combined bioreaction and separation in centrifugal fields (10,000 rpm) was used to keep dextransucrase at the top of the supporting sucrose solution and the high-molecular-weight dextran at the rotor wall.[190] The rate of hydrolysis of lactose by *Lactobacillus bulgaricus* was enhanced by ultrasound, which caused some of the β-galactosidase to be released from the cells.[191] Ultrasonication somewhat increased the rate of hydrolysis of sucrose by invertase.[192]

Biocatalytic membrane reactors, which combine reaction and separation, are in widespread use.[193]

9.1.4 The Range of Biocatalysis

Enzymes and microbes can be used in a variety of ways in organic synthesis.[194] These include oxidation, reduction, acylation, halogenation, esterification, hydrolysis, formation of amides, nitriles, and others. A few examples from the literature will be given to illustrate the nature and scope of the reactions.

9.1.4.1 Oxidation and Reduction[195]

Aromatic compounds can be converted to vicinal glycols (**9.11** Schematic).[196]

9.10 Schematic

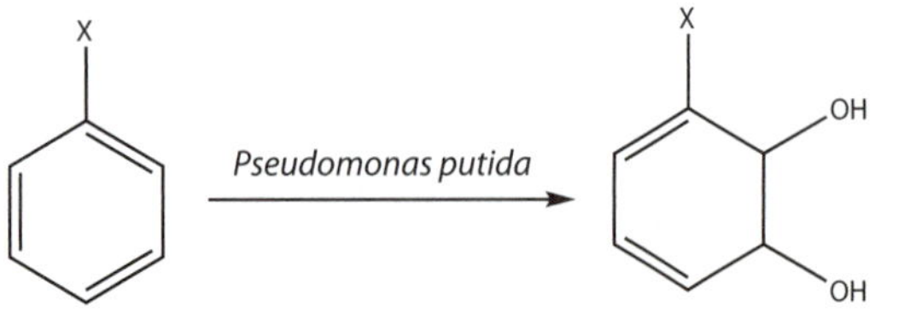

Pseudomonas putida

9.11 Schematic

By proper choice of conditions, various isomers can be produced at up to 88% ee. Some aromatic compounds can be hydroxylated without loss of aromaticity (**9.12** Schematic).[197]

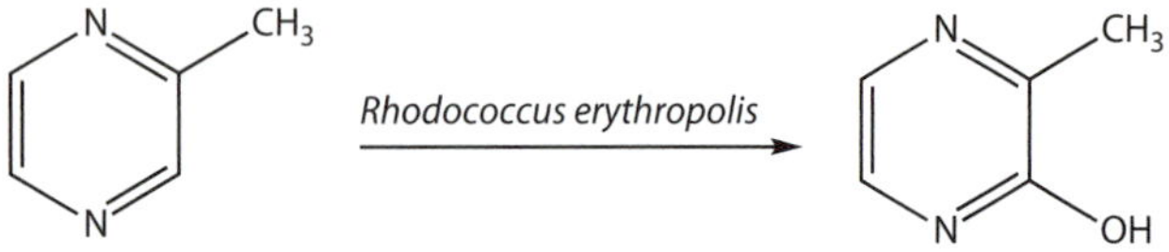

Rhodococcus erythropolis

9.12 Schematic

Phenols can be hydroxylated selectively (**9.13** Schematic), reactions that are hard to accomplish by chemical means.[198]

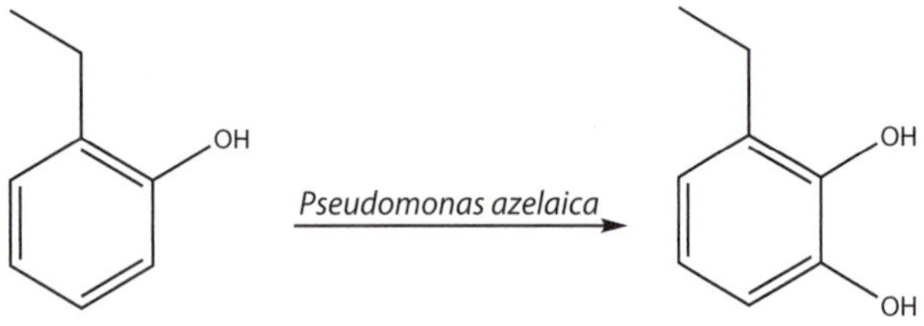

Pseudomonas azelaica

9.13 Schematic

Side chains can be selectively oxidized (**9.14** Schematic).

Laccase[199] can be used to catalyze the oxidation of aromatic methyl or methylol groups to the corresponding aldehydes (where $R = Cl$, alkoxy, nitro) by oxygen in the presence of a benzothiazoline compound (**9.15** Schematic).[200]

Such transformations can also use 1-hydroxy-1-*H*-benzotriazole as an oxidation mediator in place of the benzothiazoline.[201] An alkane hydroxylase has been used to convert alkanes to alcohols.[202] Cholesterol can be oxidized to the ketone, cholestenone, with cholesterol oxidase in reversed micelles of water in isooctane in 100% conversion.[203] Aliphatic alcohols can be oxidized to the corresponding carboxylic acid with *Acetobacter aceti* as long as the amount of substrate is kept below the inhibiting level.[204] The Baeyer–Villiger reaction of cyclic ketones[205] can be carried out with monooxygeneases from *P. putida* (**9.16** Schematic) or *Acetobacter calcoaceticus*, the former being preferred, because the latter is pathogenic.[206]

E-coniferyl alcohol is converted into (+)pinoresinol by an oxidase in the presence of a guiding protein (**9.17** Schematic). In the absence of the protein, the product is racemic.[207]

Cytochrome P450[208] can be used to introduce hydroxy groups into esters such as propyl phenylacetate with 99% selectivity.[209] Myristic acid gave 60% of the β-hydroxy acid and 40% of the α-hydroxyacid.[210] A pea peroxygenase has also been used to produce α-hydroxyacids.[211] *Bacillus megaterium* has been used to convert linoleic acid to monoepoxides.[212]

Enzymatic reductions are commonly used in asymmetrical synthesis[213] (see Chapter 10). One of the most common methods uses baker's yeast (*S. cerevisiae*). The complete DNA sequence of 12 million nucleotides and 6000 genes has been determined.[214] It can be used to reduce β-ketoesters in petrol plus a small amount of water to the *S* alcohol ester in 100% conversion with more than 98% ee.[215] When only water is used as the medium, the enantioselectivity is reduced greatly.[216] In contrast with baker's yeast and *Geotrichum candidum*, which produce the *S* alcohols from the reduction of ketones, *Yarrowia lipolytica* gives the *R*-isomer, but in only moderate yields.[217] *Pichia farinosa* also produces the *R*-isomer.[218] (For more on the preparation of single optical isomers, see Chapter 10.)

A dehydrogenase from *Rhodococcus ruber* was used in the transfer hydrogenation of ketones with isopropyl alcohol in 70–91% conversion and with >99% ee.[219] The enzyme was stable in 50% isopropyl alcohol and in 20% acetone. Reduction of ketones to give the *S* alcohols has also been carried out with an immobilized dehydrogenase from *Thermoanaerobacter ethanolicus* using isopropyl alcohol to regenerate the NADPH.[220]

Hexanal has been reduced to hexanol by passing the aldehyde over dried cells of *S. cerevisiae* at 65–85°C.[221]

Other reductions can also be carried out with baker's yeast, as shown in **9.18** Schematic for the quinoline oxide.[222]

Sodium salts of carboxylic acids can be reduced to the corresponding alcohols with *Colletotrichum gloeosporioides*

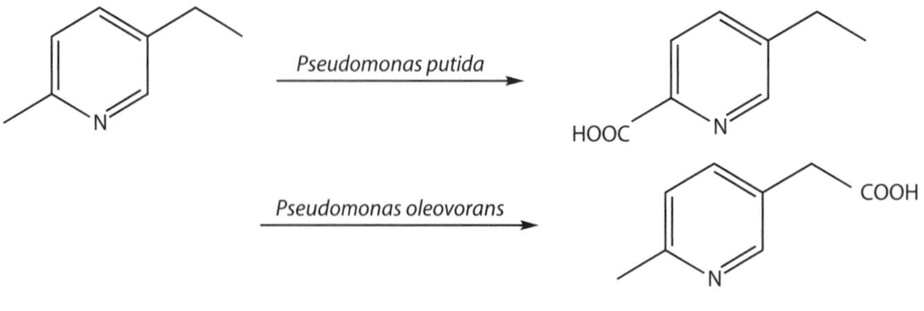

Pseudomonas putida

Pseudomonas oleovorans

9.14 Schematic

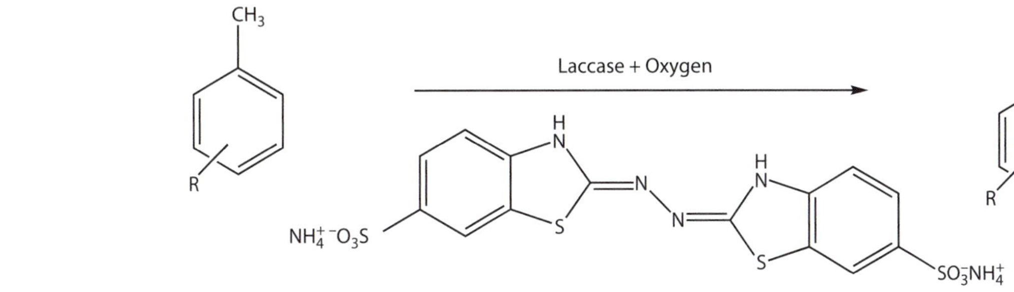

9.15 Schematic

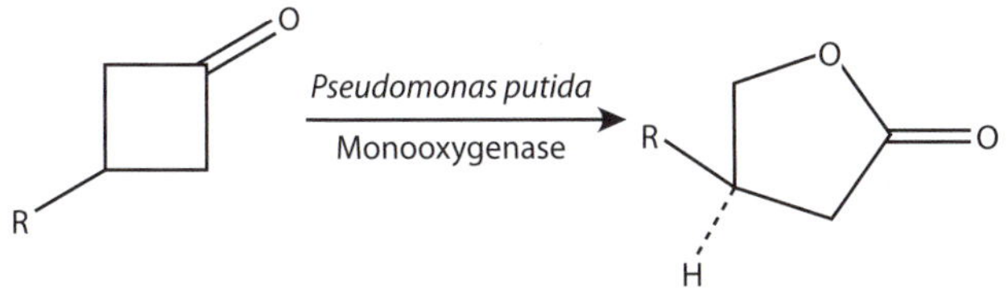

9.16 Schematic

in 70% yield.[223] This is a reduction that is difficult to carry out with the usual inorganic catalysts.

Lipases can be used to convert carboxylic acids to peracids with hydrogen peroxide. The peracids are used *in situ* for the oxidation of olefins, such as oleic acid, oleyl alcohol, and poly(1,3-butadiene).[224]

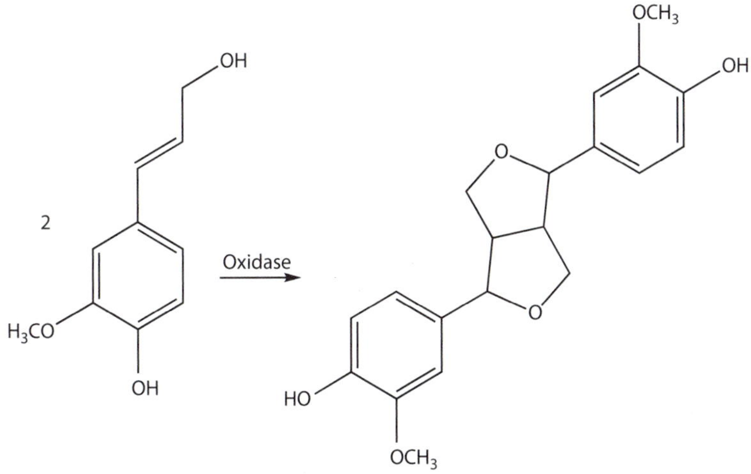

9.17 Schematic

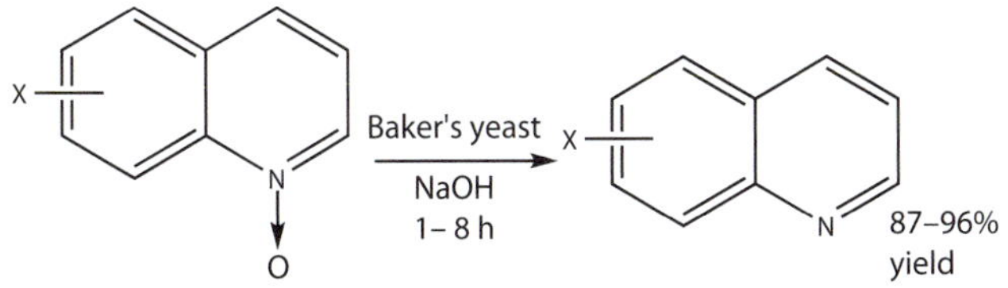

9.18 Schematic

9.1.4.2 Preparation and Hydrolysis of Esters

One of the largest uses for lipases[225] may be in the hydrolysis and transesterification of oils and fats.[226] The reactions with enzymes are much gentler than the ones used today, such as hydrolysis of a fat or oil in water at 150–260°C for 3–24 h. They are also less capital-intensive. Lipases can be used for the hydrolysis and subsequent reesterification with butyl alcohol of olive and rapeseed oil in 100% conversion. An *Aspergillus* lipase hydrolyzed various fats and oils in the presence of an aqueous buffer in 90–99% yields in 2–24 h.[227] Much of the lipase remained in the emulsion at the interface between the two layers. It could be recycled repeatedly with no loss of activity in a week. The ability to recycle is important because the cost of the enzyme is a large factor in using this method. Lipases from oat caryopses (grains)[228] and germinating rapeseed[229] have been suggested as being economically viable.

Biodiesel fuel can be made by the lipase-catalyzed alcoholysis of oils such as rapeseed oil in over 90% conversion.[230] (The United States produced 75 millon gallons of biodiesel fuel in 2005.[231]) (The reaction is shown in **9.19** Schematic with triolein and ethanol for convenience. Rapeseed oil also contains some other fatty acids.)

Methanol is often used, but ethanol would be better because it can be made by fermentation. There is a problem with the small amounts of stearates present crystallizing out in cold winters. The use of isopropyl alcohol reduces this problem. The other possibility is to crystallize it out at the plant before it reaches the market. A better solution might be to use a desaturase to introduce a double bond into the stearate. Transesterification of rapeseed oil with 14.6% lipase from *C. rugosa* (with no added solvent) took place in 1 h.[232] When the level of enzyme was dropped to 0.3%, the reaction took 10 h. If the enzyme was immobilized, the use of the large amount would not be a problem because the enzyme could easily be recycled. Removal of the glycerol formed from the alcoholysis of fats and oils can improve yields. When it was removed by adsorption on silica gel, the yield was 98%.[233]

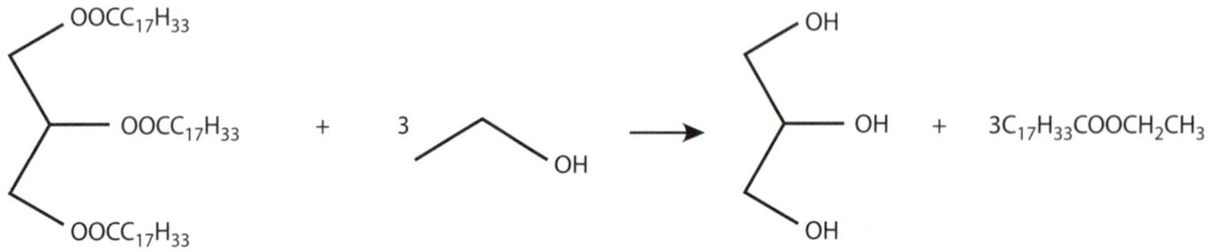

9.19 Schematic

One fraction of the lipase from *C. rugosa* gave a 98% yield in the esterification of oleic acid with 1-butanol.[234] Polyunsaturated fatty acids have been esterified with glycerol using an immobilized lipase to produce the triester in 95% yield.[235] Esterification of fatty acids using a silica-supported lipase gave 96–98% yields of ester when a 4-Å molecular sieve was used to remove the water formed in the reaction.[236] Another way to remove the water continuously is to operate in a partial vacuum (0.7 bar at 50°C), the yield being 96% in one case.[237] Lauryl oleate has been prepared in 85% yield without solvent using a fungal lipase on cellulosic biomass together with a molecular sieve to pick up the by-product water.[238] Immobilized lipases are used in the commercial production of isopropyl myristate, isopropyl palmitate, and 2-ethylhexyl palmitate.[239] Degussa (now Evonik) produces cetyl ricinolate and decyl cocoate enzymatically with less color and odor than when these compounds are made chemically.[240]

Hollow-fiber membrane reactors with immobilized lipases have been used for the continuous hydrolysis of triglycerides[241] and in the esterification of fatty acids.[242] There was no deactivation of the enzyme in the former case in 16 days. In a comparable run in solution, the enzyme lost 80% of its activity in 2 days of operation. The latter case used dodecanol and decanoic acid in hexane to give the ester in 97% yield. The half-life of the immobilized enzyme was 70 days. The integration of reaction and separation can decrease product inhibition, increase selectivity, shift equilibria, and reduce the number of downstream operations.[243]

Mono- and diglycerides are important food emulsifiers. They can be made enzymatically by the glycerolysis of fats and oils.[244] The reaction can be run continuously in a membrane reactor with an enzyme half-life of 3 weeks at 40°C. Higher concentrations of glycerol can be used if silica is present to prevent blockage of the enzyme by glycerol.[245] Depending on the conditions and lipase chosen for transesterifications with ethanol, the product can be the 2-monoglyceride,[246] the 1,2-diglyceride,[247] or the 1,3-diglyceride.[248] (The chemical route to monoglycerides involves temperatures of 240–260°C.) Biosurfactants can be made in this way.[249] They can also be made by the enzymatic reaction of long-chain fatty acids with sugars.[250]

Enzymes can be used to selectively hydrolyze only one of two ester groups in a molecule. An esterase from *P. putida* was used to hydrolyze dimethyl adipate to its half ester in more than 99% yield.[251] Pig liver esterase was used to hydrolyze dialkyl phthalates to the monoesters in 84–93% yields in 1–13 h.[252] A porcine pancreatic lipase was used to hydrolyze linear diol diacetates to the monoacetates in 79–95% yields.[253] Esters have been prepared from 12-hydroxystearic acid and C_8–C_{18} alcohols in 82–90% yields without esterification of the 12-hydroxy group, by using an immobilized lipase from *R. miehei*.[254] The L-enantiomer of octanoyl lactate has been prepared with a lipase in hexane (**9.20** Schematic).[255]

Ricinoleic acid, which differs only by having a *cis*-9-double bond, formed oligomers with an immobilized lipase from *C. rugosa*.[256] These were much less colored than the ones made by the commercial process, which is done at 200°C.

Since esterification is a reversible reaction, enzymatic esterifications in organic solvents often employ vinyl esters of fatty acids.[257] The by-product vinyl alcohol isomerizes to acetaldehyde (**9.21** Schematic), which cannot participate in the back reaction.

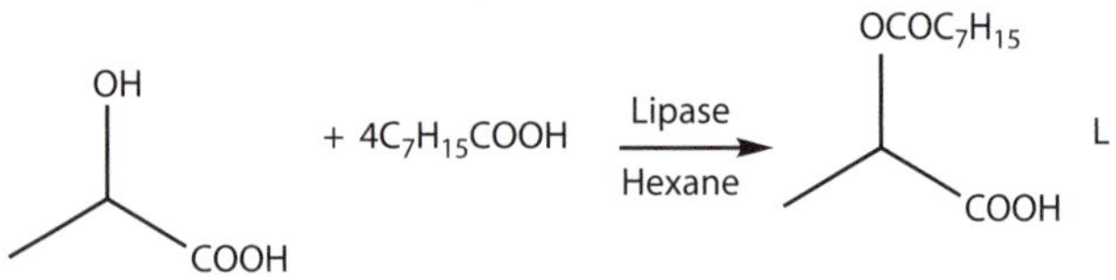

9.20 Schematic

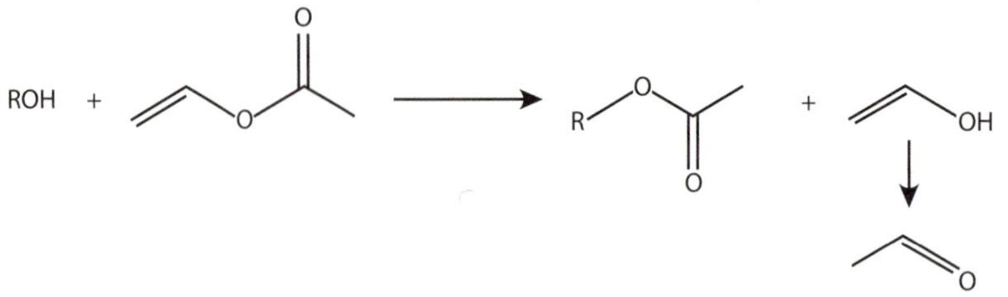

9.21 Schematic

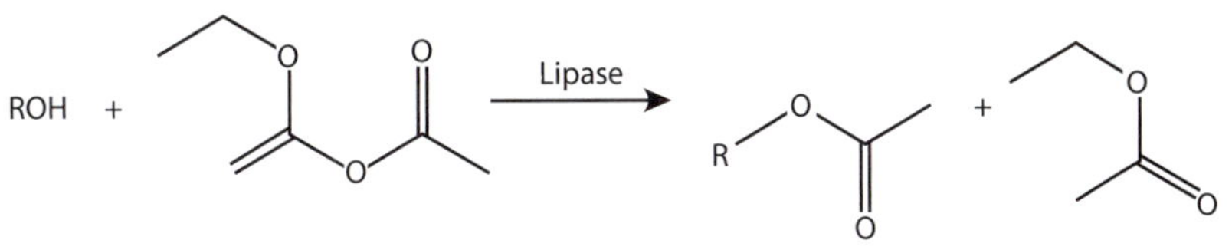

ROH + [vinyl ester] —Lipase→ [ester] + [ethyl acetate]

9.22 Schematic

Because some enzymes are deactivated by acetaldehyde, the variant in **9.22** Schematic in which the by-product is ethyl acetate has been developed as an improvement.[258] (The deactivation is due to formation of Schiff bases with the ε-amino group of lysine residues.[259]) Acyl carbonates can be used (**9.23** Schematic) in the same way.[260]

ROH + [acyl carbonate] —Lipase→ [ester] + CO_2 + [isopropanol]

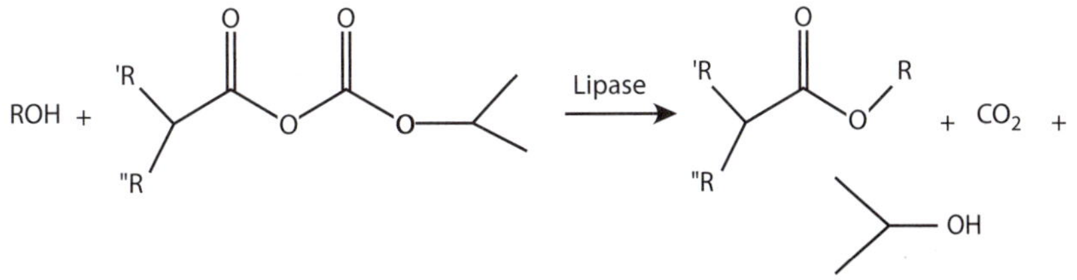

9.23 Schematic

2,2,2-Trifluoroethyl esters can also be used. In one case, *N* or *O* acylation depended on the solvent (**9.24** Schematic).[261]

Monoglycerides can be made by treatment of the vinyl ester of a fatty acid with glycerol (without solvent) in 100% conversion.[262] A lipase from *Penicillium roquefortii* has been used to react unsaturated fatty acid vinyl esters with glycerol to give more than 95% selectivity to 1,3-diglycerides in 85% yield.[263] Diols can be converted selectively to monoacetates.[264] An example is given in reaction **9.25**.

Enzymatic preparation of esters has also been done in the gas phase[265], in sc carbon dioxide[266] and in ionic liquids.[267]

9.1.4.3 Preparation of Acids, Amides, and Nitriles

The commercial preparations of acetic, citric, and malic acids were discussed earlier in this chapter. There are also a number of other carboxylic acids that can be produced by biocatalytic methods.

Genetically engineered *E. coli* has been used to convert glucose from corn into succinic acid (20–52 g/L).[268] Succinic acid can be converted, in turn, by known methods into 1,4-butanediol, tetrahydrofuran, and *N*-methylpyrrolidone, all of which are valuable commercial chemicals. The process for succinic acid is said to be 20–50% cheaper than other routes. A full-scale plant is in operation. Glycolic acid can be oxidized to glyoxylic acid microbially in 85% yield.[269] The catalyst could be recovered and reused 30 times. The product was used to prepare *N*-(phosphonomethyl)glycine (**9.26** Schematic) (an herbicide known as glyphosate).

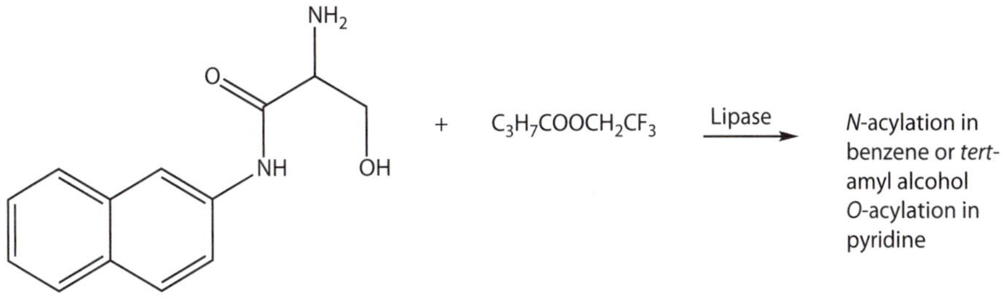

[amide] + $C_3H_7COOCH_2CF_3$ —Lipase→ *N*-acylation in benzene or *tert*-amyl alcohol; *O*-acylation in pyridine

9.24 Schematic

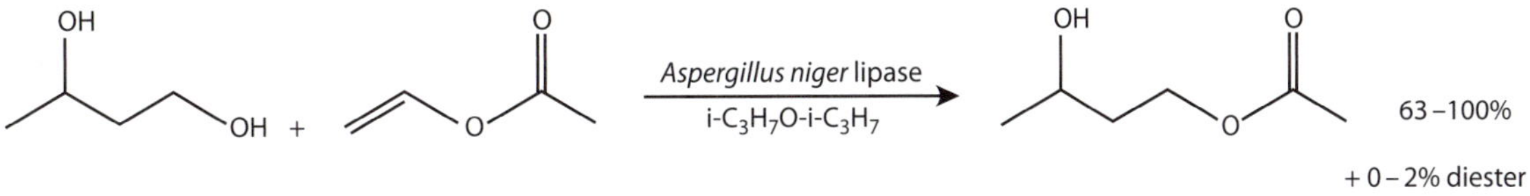

[diol] + [vinyl acetate] —*Aspergillus niger* lipase, $i\text{-}C_3H_7O\text{-}i\text{-}C_3H_7$→ [monoacetate] 63–100% + 0–2% diester

9.25 Schematic

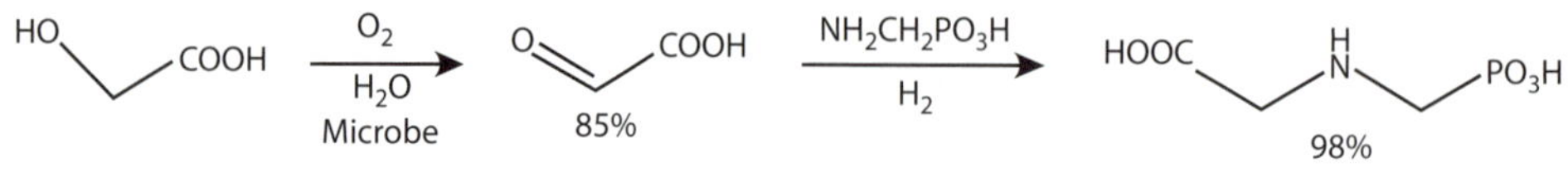

9.26 Schematic

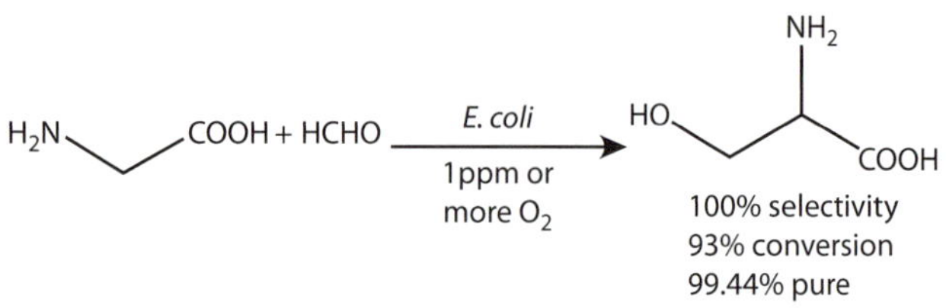

9.27 Schematic

Glycine has been converted into L-serine by reaction **9.27** Schematic with formaldehyde (a carcinogen) using *E. coli*.[270]

Limed straw and bagasse (the residue after the juice is pressed from sugar cane) can be fed to ruminant animals.[271] These can replace corn, which requires large amounts of fertilizer, herbicides, and insecticides, and this can lead to contamination of groundwater. Alternatively, it can be treated with rumen bacteria in an anaerobic fermentor to produce acetic, propionic, and butyric acids. Salts of these acids can be pyrolyzed to produce ketones. These can then be reduced to alcohols with hydrogen. Such syntheses allow the conversion of unwanted wastes, such as municipal solid waste, sewage sludge, and agricultural wastes, to valuable chemicals or fuels, instead of the wastes ending up in an incinerator or a landfill. The process replaces nonrenewable petroleum as a source of these chemicals.

Propionic acid can also be made from the hemicelluloses in wood.[272] These have few other uses. Steam-exploded *Populus tremuloides* was hydrolyzed enzymatically and then the hydrolysate was treated with *Propionibacterium acidipropinoici* to produce 18 g/L of propionic acid. Because this organism is inhibited by the product, various methods have been developed to remove it as formed. Use of ditridecylamine in oleyl alcohol with a hollow-fiber membrane gave a higher yield, final product concentration, and purity at the same time that the production of acetate and succinate was reduced in a run of 1.5 months.[273] The acid was recovered with aqueous sodium hydroxide using a second hollow-fiber membrane.

Long-chain unsaturated fatty acids, which are of interest for good health,[274] can also be produced biocatalytically.[275] Arachidonic acid (**9.28** Schematic) has been produced by growing the fungus, *Mortierella alpina*, on a mixture of glucose and potatoes.[276] γ-Linolenic acid (**9.29** Schematic) has been produced by growing *Cunninghamella echinulatun* on starch plus potassium nitrate or urea.[277] The eicosapentaenoic (**9.30** Schematic) and docosahexenoic (**9.31** Schematic) acids from fish oils help prevent heart attacks. Much effort has been directed toward preparing them more conveniently from algal, Antarctic sea ice bacteria, and fungal cultures.[278] The use of bacterial cultures may be the best, because they are often easier to culture than algae and tend to produce less complex mixtures of fatty acids.

Desaturases[279] are involved in the biosynthesis of these acids. It may be possible to use such enzymes for the commercial conversion of saturated fats to unsaturated ones. Surplus butterfat and tallow might be converted into more nutritious oils in this way. (Saturated fats in the diet have been implicated in heart problems.)

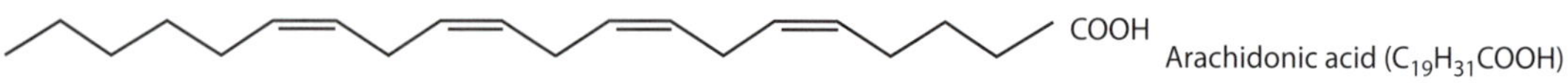

9.28 Schematic

9.29 Schematic

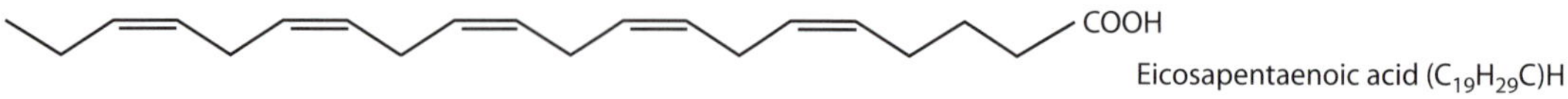

9.30 Schematic

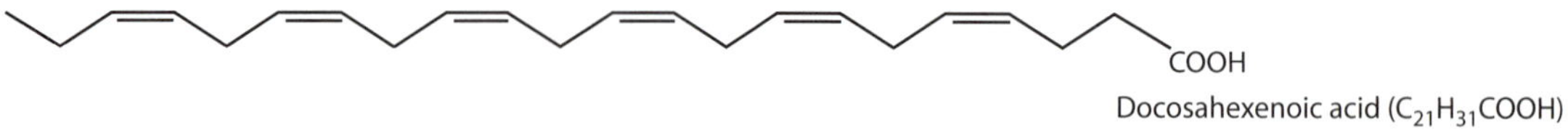

9.31 Schematic

A *Flavobacterium* sp. can be used to hydrate linoleic acid to 10-hydroxy-12(*Z*)octadecenoic acid, an analog of ricinoleic acid.[280] Ricinoleic acid (12-hydroxy-9(*Z*)octadecenoic acid) from castor oil is important in the preparation of paints and varnishes.[281] *Cryptococcus neoformans* can convert *n*-pentadecane to the corresponding α-ω-dicarboxylic acid ($HOOC(CH_2)_{13}COOH$).[282] *Candida tropicalis* has been used in the same way to produce the diacids commercially at up to 153 g/L.[283] Toluene can be converted to 4-hydroxybenzoic acid (**9.32** Schematic), a monomer used to prepare polyesters for use at high temperatures, with 99% selectivity by *P. putida*.[284] Further work is needed to speed up the reaction before it can be used commercially. Pyrrole-2-carboxylic acid can be made from pyrrole and carbon dioxide (**9.33** Schematic) using *B. megaterium*.[285] It should be possible to improve the 59% yield by continuous removal of the product as formed.

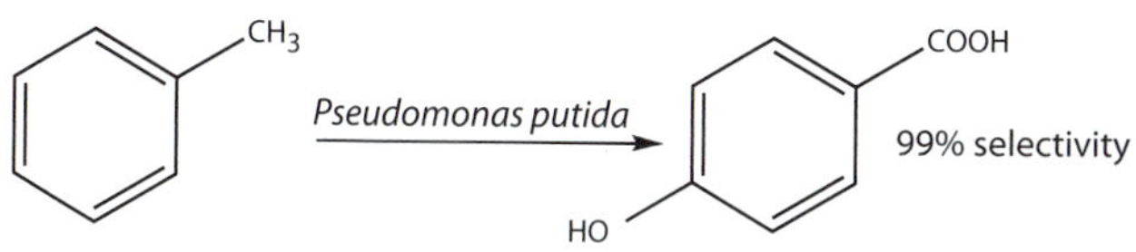

CH_3 — Pseudomonas putida → COOH, HO — 99% selectivity

9.32 Schematic

+ HCO_3^- — Bacillus megaterium ⇌ COOH 52%

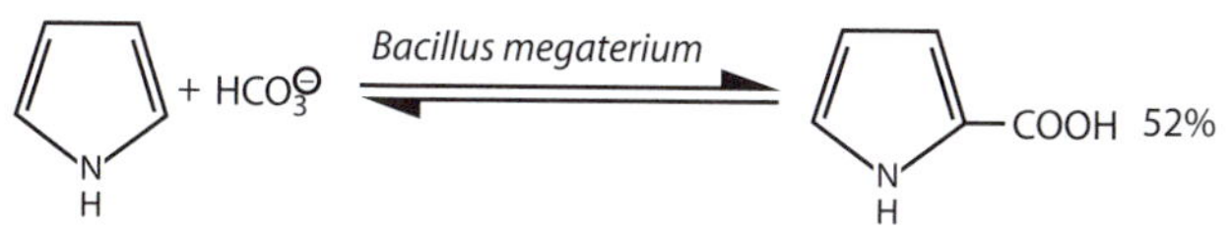

9.33 Schematic

Carbon dioxide has also been used with a pyruvate decarboxylase to convert acetaldehyde to lactic acid in 81% yield (**9.34** Schematic).[286]

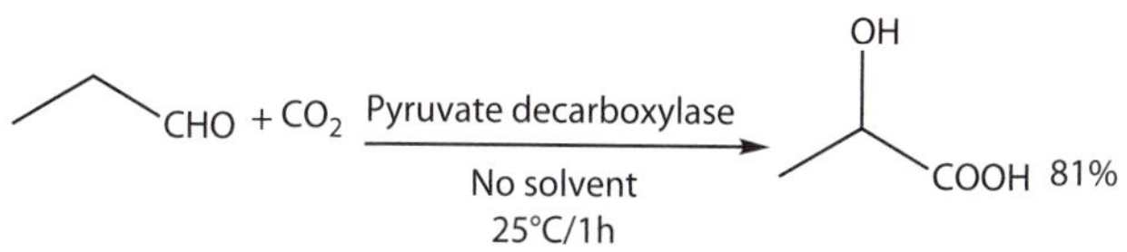

CHO + CO_2 — Pyruvate decarboxylase, No solvent, 25°C/1h → OH, COOH 81%

9.34 Schematic

Wacker Chemie produces cysteine from glucose using *E. coli* K12 (**9.35** Schematic).[287]

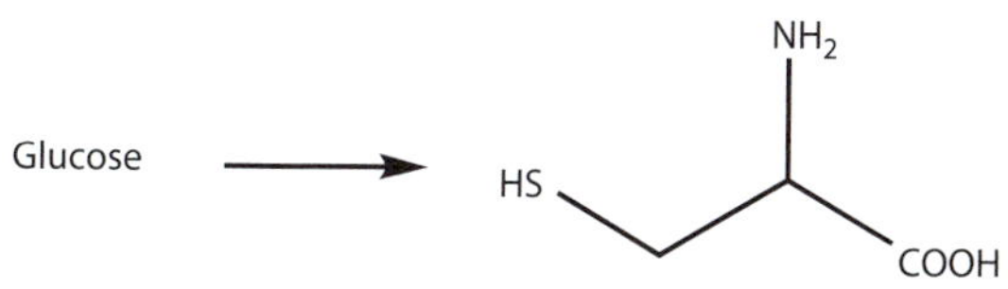

Glucose → HS, NH_2, COOH

9.35 Schematic

An enzymatic Friedel–Crafts reaction has been carried out with polyprenyl transferase (**9.36** Schematic) (where $n = 1, 12$).[288]

Triglycerides can be converted to the corresponding amides using *C. antarctica*, with ammonia in *tert*-butyl alcohol.[289] Olive oil gave a 90% yield of oleamide in 72 h at 60°C. Industrial reactions of this type are run at 200°C. The enzyme can also be used to convert oleic acid to oleamide in 90% yield in *tert*-amyl alcohol containing two equivalents of *n*-butyl alcohol. The intermediate butyl ester is formed *in situ*. *C. antarctica* lipase can also be used to prepare monoesteramides from dimethylsuccinate (**9.37** Schematic).[290] The corresponding imide can be formed if the medium is hexane. Amines can be protected as their allyl carbamates by formation with diallyl carbonate and a lipase (**9.38** Schematic).[291]

Amides can be made by the enzymatic hydrolysis of nitriles. Nitto Chemical Industry of Japan uses *Rhodococcus rhodocrous* to prepare acrylamide from acrylonitrile (**9.39** Schematic).[292] Lonza is using it for the hydrolysis of 3-cyanopyridine to niacinamide in a plant in China. Depending on the microorganism, the product of hydrolysis of a nitrile may be amide or the carboxylic acid (**9.40** Schematic).[293] *Rhodococcus* spp. have also been used to hydrolyze only one of two nitrile groups in aromatic dinitriles in 86% yield.[294] *Rhodococcus rhodochrous* has been used to hydrolyze aliphatic dinitriles, such as adiponitrile, to mixtures of mono- and dicarboxylic acids.[295] *Pseudomonas chlororaphis* has been used to convert adiponitrile to 5-cyanovaleramide with 96% selectivity at 97% conversion.[296] This is a significant improvement over the best chemical process.

The enzymatic hydrolysis of the amide in penicillin G (**9.41** Schematic) has replaced a cumbersome multistep chemical synthesis in methylene chloride that produced waste salts.[297] The reaction can be run in a polysulfone reactor with continuous removal of the product amine in the permeate.[298]

A new amide can be made using an immobilized penicillin G amidase in >98% yields.[299] The use of an expandase to convert penicillins to cephalosporins (**9.42** Schematic) also replaced a more expensive multistep chemical synthesis.[300] Penicillin acylase has also been used to remove phenylacetyl protecting groups from amines in the synthesis of oligonucleotides.[301]

Transaminases can convert α-keto acids to α-amino acids in one step compared to the usual multistep chemical procedure that involves reduction of the carbonyl to the alcohol, conversion to the mesylate, treatment with sodium azide, and reduction to the amino acid (**9.43** Schematic).[302]

Phenylalanine aminomutase can convert an α-amino acid to a β-amino acid.[303]

Hydroxynitrile lyases can be used to make cyanohydrins (**9.44** Schematic) in 80–100% yields.[304]

Proteases can be used to prepare dipeptides.[305] The addition of cyclodextrins can lead to higher activity in organic solvents.[306]

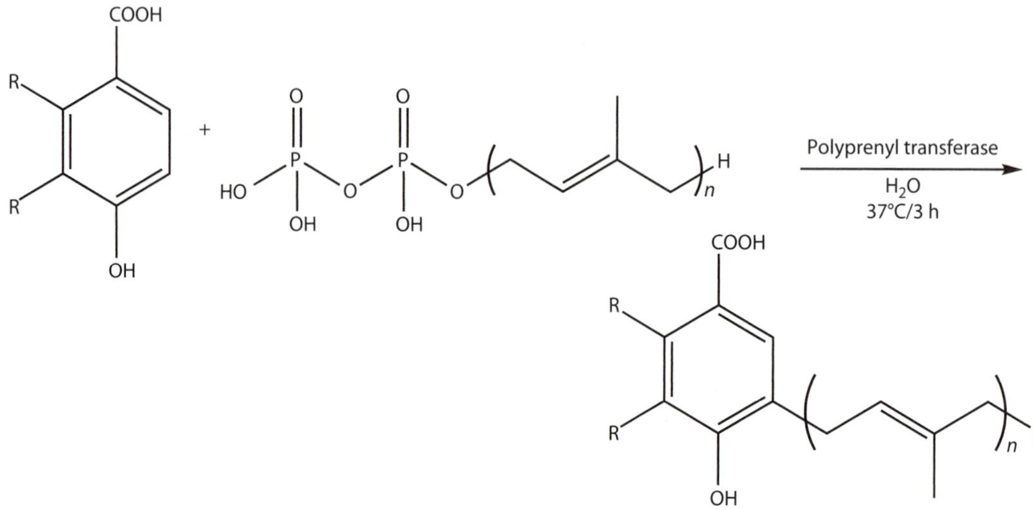

9.36 Schematic

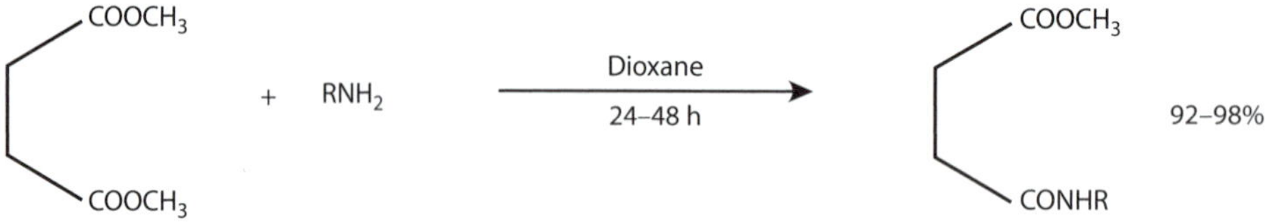

9.37 Schematic

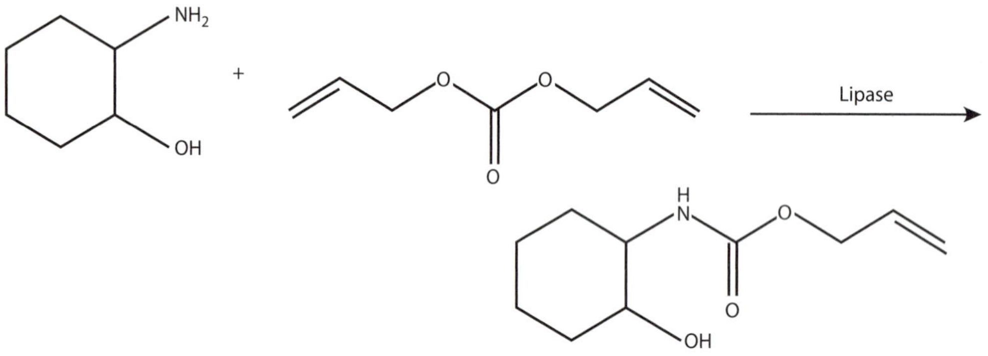

9.38 Schematic

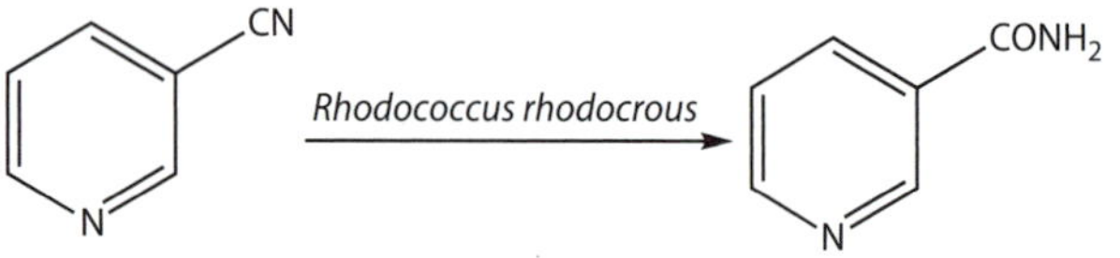

9.39 Schematic

9.1.4.4 Compounds with Two or More Hydroxyl Groups

Glycerol can be produced in 70–90% yields by a triose phosphate-deficient mutant of *S. cerevisiae*.[307] It has been produced commercially in China from glucose and urea using *Candida glycerinogenes*.[308] It can also be produced by raising halophytic microalgae. There is no need to use these routes due to the surplus of glycerol from the production of biodiesel fuel. Various groups are trying to find uses for it. A continuous culture of *C. acetobutylicum* growing on glycerol gave a 62% yield of butyl alcohol at 84% conversion.[309] It can be converted to propionic acid with a *Propionobaterium* in a membrane reactor without the formation of any acetic acid.[310] It has also been converted to 3-hydroxypropionaldehyde by *K. oxytoca*.[311] This product can be converted to an important monomer, acrylic acid, by dehydration and oxidation. Glycerol has also been converted to 1,3-propanediol, which can be used to make poly(trimethylene terephthalate).[312] The continuous fermentation with *Klebsiella pneumoniae* produced the product in 80–96% yield (35–48 g/L). The continuous culture was 2–3.5 times as productive as a batch culture. An analog of EDTA, which may be cheaper to produce, has

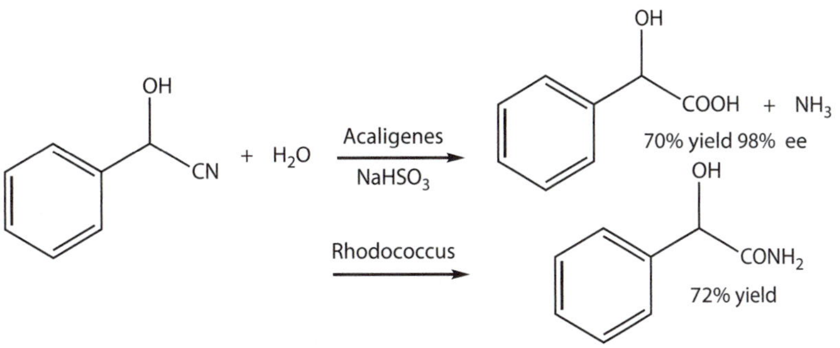

OH
CN + H_2O Acaligenes $NaHSO_3$ → OH COOH + NH_3 70% yield 98% ee
Rhodococcus → OH $CONH_2$ 72% yield

9.40 Schematic

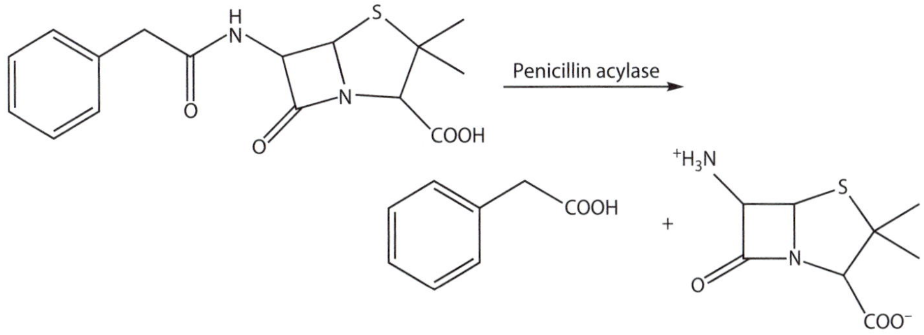

Penicillin acylase →
COOH + $^{+}H_3N$ … COO^-

9.41 Schematic

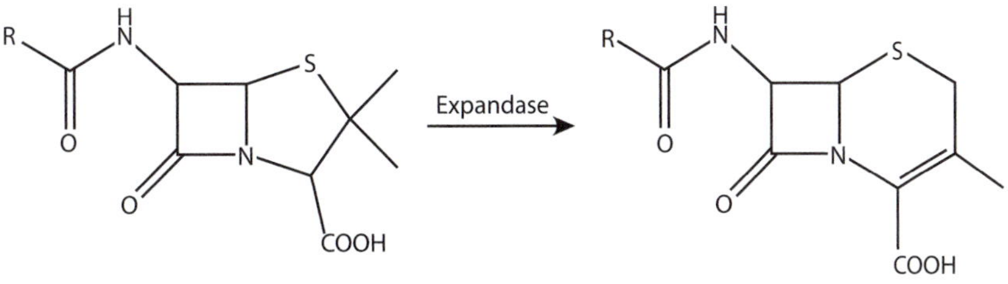

Expandase →

9.42 Schematic

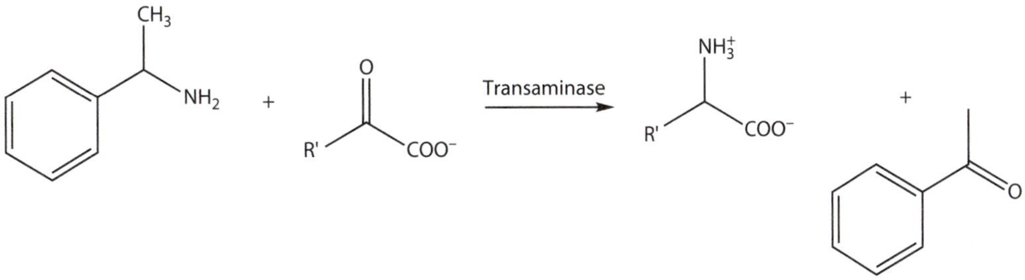

CH_3 NH_2 + R' O COO^- Transaminase → NH_3^+ R' COO^- + O

9.43 Schematic

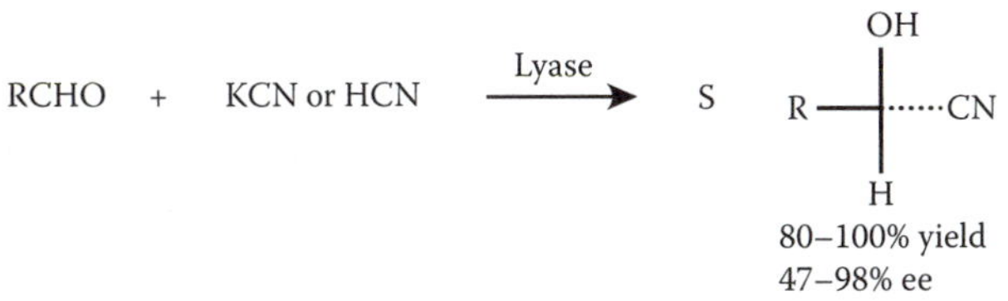

RCHO + KCN or HCN Lyase → S R—C(OH)(H)—CN
80–100% yield
47–98% ee

9.44 Schematic

been made by the action of *Amycolotopsis orientalis* on a mixture of glycerol and urea (**9.45** Schematic).[313]

The use of metabolic engineering to bolster some pathways and eliminate others has made it possible to produce both 1,3-propanediol (by *K. pneumoniae*) and 1,2-propanediol (by *Thermoanaerobacterium thermosaccharolyticum*) from sugars.[314] DuPont and Genecor (now Danisco) have worked

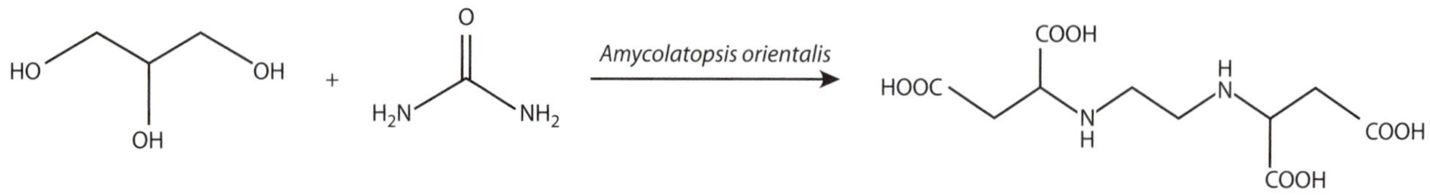

Amycolatopsis orientalis

9.45 Schematic

out a process to make 1,3-propanediol from hydrolyzed corn starch using a recombinant microorganism.[315] It has been known for some time that this polymer has good properties, but its commercial production has been delayed for want of a good inexpensive source of 1,3-propanediol. Degussa (now Evonik) made the diol by hydration of acrolein followed by reduction.[316] The same 3-hydroxypropionaldehyde can be obtained from glycerol by fermentation. Shell makes the diol by continuous ethylene oxide hydroformylation.[317] The raw materials will probably be derived from petroleum, although they could be derived from renewable sources. The ethylene could be produced by dehydration of ethanol. The synthesis gas could be made from biomass. The intermediate ethylene oxide is a carcinogen. Almost 3 billion lb of 1,2-propanediol is made each year from propylene oxide, which is made, in turn, from propylene derived from petroleum. It is used in unsaturated polyester resins, de-icing fluids, antifreeze, and others. It could be made from renewable resources such as sugars. Glycols can also be made from epoxides using epoxide hydrolases.[318]

Carbohydrate synthesis usually involves selective protection of some groups, reaction of the others and then deprotection. The enzymatic aldol reaction can produce some sugars without the need for any protecting groups.[319] An example is the preparation of 6-deoxy-L-sorbose (**9.46** Schematic).[320]

Transketolase

6-Deoxy-L-sorbose

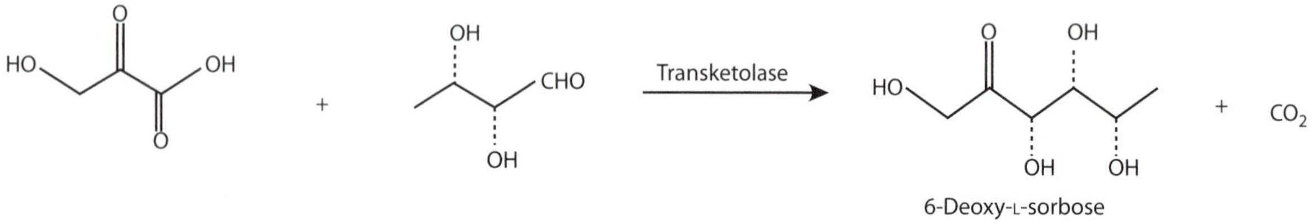

9.46 Schematic

Both starting materials were prepared enzymatically. Genetically engineered bacteria have been used to convert glucose to 2-keto-L-gulonic acid, eliminating four steps in the synthesis of ascorbic acid (vitamin C).[321]

Cyclodextrins (see Chapter 5) can be prepared in one step from starch using a recombinant enzyme based on a thermophilic bacterium, a species of *Thermoanaerobacter*.[322] Cyclodextrin glucanotransferase from *Bacillus macerans* and other organisms can also be used.[323] Cyclodextrins can reduce substrate and product inhibition, can act as chaparones for enzymes, and can increase the solubility of starting materials in water.[324]

Many polysaccharides, such as agar, agarose, alginic acid, and carrageen, are extracted from marine algae.[325] Controlled microalgal culture including genetic engineering, species and strain selection, timing and method of harvesting, drying, and storage conditions can give 15–20 times the biomass of natural systems.[326] These polysaccharides show up in practically every food market as thickeners and stabilizers for foods.[327] They are common in fat-free foods. L-ascorbic acid (vitamin C), β-carotene, glycerol, and docosahexaenoic acid (needed in infant nutrition) can also be produced by algae.[328] The microalgae *Chlorella* and *Spirulina* are sold as foods. Industrial-size photobioreactors that use optical fibers to direct solar light are being developed.[329] In Japan, macroalgae are cultured on nets for sale as foods.

9.1.4.5 Preparation of Aromatic Compounds

The carcinogen benzene (derived from petroleum) is the starting point for many important industrial chemicals. The syntheses often involve stringent conditions and may produce toxic by-products. The preparation of adipic acid for the manufacture of nylon is an example (**9.47** Schematic).

In contrast, biological systems produce aromatic compounds such as phenylalanine and tryptophan (**9.48** Schematic) under mild conditions from renewable resources without the formation of toxic by-products.

The pathway proceeds from glucose through shikimic acid.[330] This acid is used in making "Tamiflu," a drug for avian flu.[331] Frost and co-workers[332] have altered the pathways shown in Figure 9.1 to favor the desired aromatic compounds. This has been done by genetically engineering the microorganisms to cut out or reduce certain unwanted pathways while enhancing others.[333] It has been possible in this way to convert glucose from corn into catechol, hydroquinone, phloroglucinol, resorcinol, adipic acid, gallic acid (3,4,5-trihydroxybenzoicacid), and vanillin (4-hydroxy-3-methoxybenzaldehyde).[334]

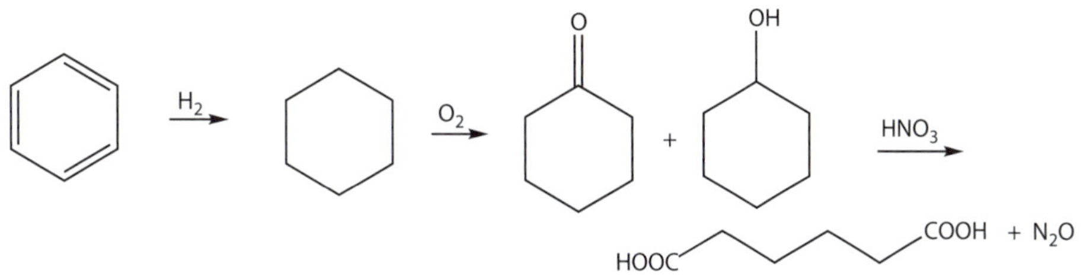

9.47 Schematic

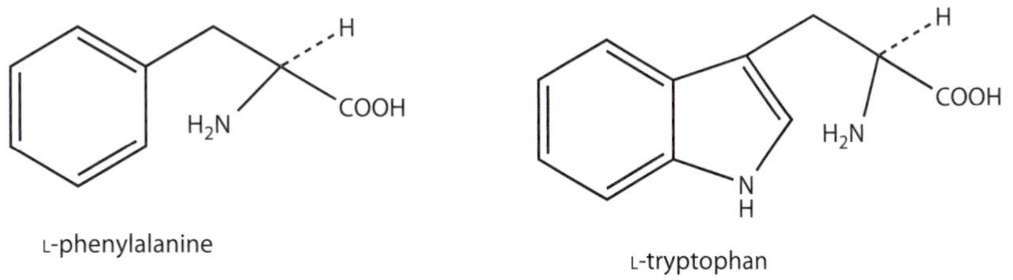

L-phenylalanine

L-tryptophan

9.48 Schematic

The yield of phenylalanine has been doubled. Some phosphoenol pyruvate is used to introduce the glucose into the cell. An *E. coli* mutant has been found that does not need this.[335] It gave increased yields of phenylalanine and tyrosine.

Indigo, the dye used to color blue jeans, can be made enzymatically by removal of the side chain of tryptophan to give indole, which can be dihydroxylated enzymatically, then oxidized with oxygen to indigo (**9.49** Schematic).

The present commercial route starts with the highly toxic aniline and produces waste salts. An even "greener" way to indigo would be to raise it as a crop as done in Colonial America. A new commercial strain produces five times more indigo than the usual plant.[336]

Glucose can be converted to caprolactam (the monomer for nylon 6) through lysine (**9.50** Schematic).[337]

Many of these processes cannot compete with the chemical routes used today. If the total environmental costs of global warming, natural resource depletion, waste disposal, and such were included, more of them would be practical today. Research continues to reduce their costs by increasing yields, starting with waste biomass where both five- and six-carbon sugars must be utilized and so forth.

9.1.4.6 Miscellaneous Applications

Most of the industrial applications of enzymes[338] mentioned earlier have dealt with the production of fine chemicals or drugs.[339] Several other large, or potentially large, uses are described in the following.

Enzymes are now being used by the pulp and paper industry.[340] Ligninolytic fungi can be used to remove lignin.[341] Mutants that lack cellulases are used. This can save 30% of the electrical energy and improve paper quality. *Phanerochaete chrysosporium* is the organism of choice. Lipases that hydrolyze triglycerides are being used to control pitch in mills.[342] Seasoning wood chips with *Ophiostoma piliferium* fungus for 2 weeks reduced the extractives by 25–40%.[343] Xylanases reduce the amount of chemicals needed for bleaching; in some cases, for a 25% saving in chemical costs. A thermophilic xylanase from *Dictyoglomus thermophilum* works optimally at 70–85°C, which makes it easier to fit into mill operations. Another from *Bacillus amyloliquefaciens* has optimal activity at 80°C, excellent stability at 50°C, and no cellulase activity.[344] Pulp fibers treated with a xylanase and then a laccase gave paper with 25% higher dry tensile strength and 46% more wet tensile strength. Cellulases used in controlled ways can smooth fibers by removing fine fibrils, enhance drainage, and promote ink removal in repulping.[345] Levanases can be used in combination with biocides for improved control of slime in paper mills.[346]

Biofinishing of cotton and woolen fabrics with cellulases, pectinases, and proteases is being used to remove protruding fibers and to avoid chemical finishing agents such as sodium hydroxide.[347] The resulting fabrics are smoother and softer, but do lose some tensile strength in the process. Cellulases[348] are also used in detergents to reduce pilling of cotton fabrics.[349] The wettability of poly(ethylene terephthalate) fabrics is better improved with lipases than with the conventional treatment with sodium hydroxide.[350] Fibers from kenaf, jute hemp, and flax are stronger when prepared with bacteria (30°C for 10 days) than when the unwanted parts of the plants are removed chemically.[351] Linen may make a comeback.[352] It does not need the warm climate and many agrochemicals required by cotton. The oil of the seeds is valuable. Polyvinyl alcohol that is used to stiffen cotton during weaving (a warp size) can be removed enzymatically in 1 h at 30–55°C, and the effluent can be fed to a conventional wastewater treatment plant.[353] Conventional desizing

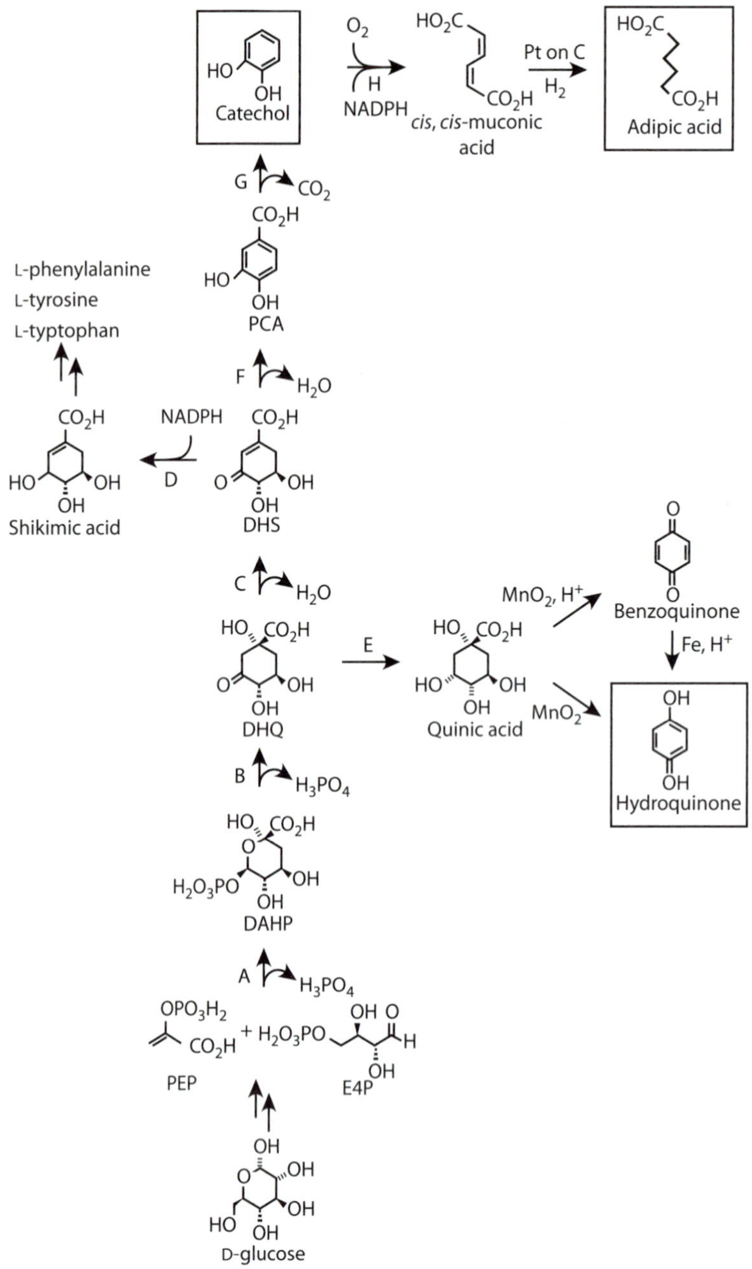

Figure 9.1 Altered pathways for production of aromatic compounds from glucose. (A) DAHP synthase; (B) DHQ synthase; (C) DHQ dehydratase; (D) shikimate dehydrogenase; (E) quinate dehydrogenase; (F) DHS dehydratase; (G) protocatechuate decarboxylase; (H) catechol 1,2-dioxygenase. (Reprinted with the permission of J.W. Frost.)

is done at 80°C for 30 min. It would be better to dissolve the size in water and then recover it for reuse by ultrafiltration. Wool bleached with a peroxide and then biopolished with a protease is less itchy, easier to dye and machine-washable.[354] This replaces processes that use chlorine.

Biomining involves the extraction of metal ions from their ores.[355] Low-grade ores of copper are piled on impervious aprons and treated with sulfuric acid. The *Thiobacillus ferrooxidans* that grows under these conditions converts the copper sulfide to copper sulfate that drains off with the sulfuric acid for recovery by electrowinning. Twenty-five percent of all copper is produced in this way. A combination of *T. ferrooxidans* and *Thiobacillus thiooxidans* leached metals from fly ash from municipal incinerators (recovering 81% Zn, 52% Al, 89% Cu, 64% Ni, 12% Cr, and 100% Cd).[356] Although this is a way to recover the metals and detoxify the ash, it would better to recycle the objects containing the metals rather than putting them in an incinerator. This system may also be applicable to the ash from coal-burning power plants. Thermophilic bacteria

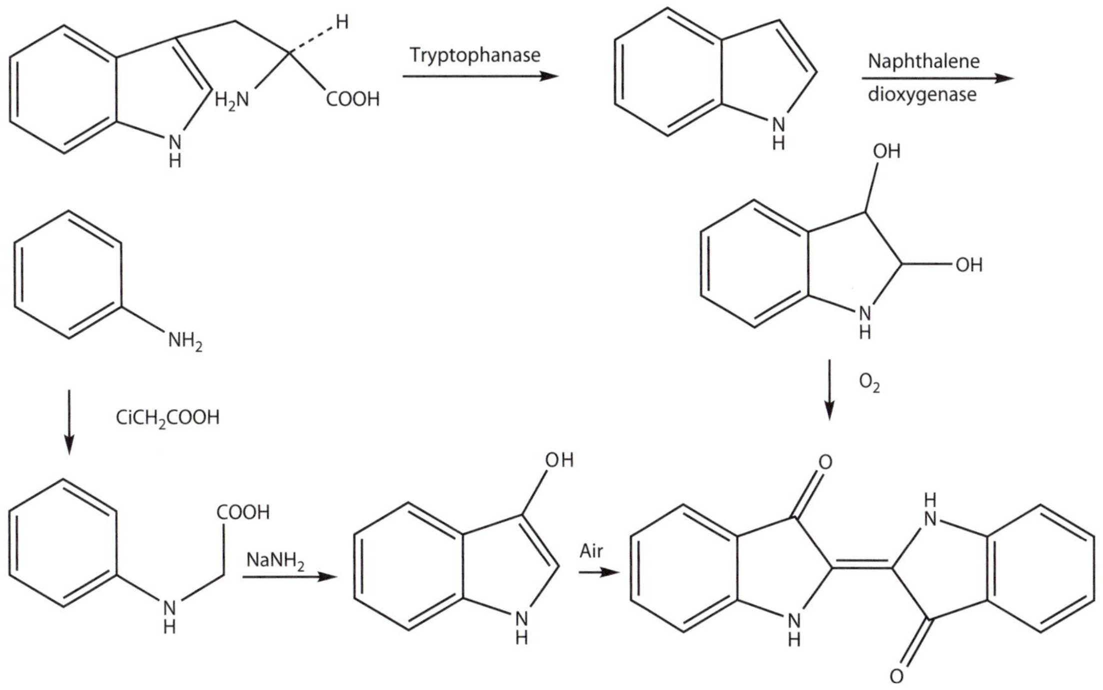

9.49 Schematic

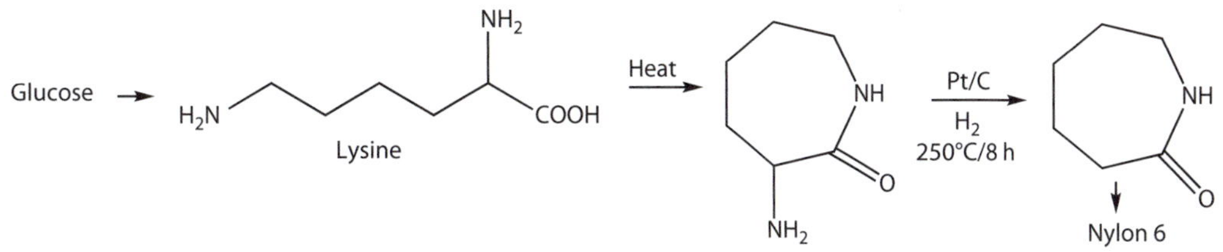

9.50 Schematic

have been used at 50°C to recover 98% of the gold from ores not amenable to cyanide leaching.[357] Phosphoric acid is usually obtained from calcium phosphate ore by treatment with sulfuric acid. It can also be recovered at room temperature using the gluconic acid and 2-ketogluconic acids produced by *Pseudomonas cepacia* and *Erwinia herbicola*. For practical use, a way to recover and reuse these acids must be found. A book on the environmental effects of mining is available.[358]

Biomineralization[359] involves proteins that guide the assembly of shells, diatoms, and nanoparticles[360] of Fe_3O_4. These include the silaffins that guide the formation of the silica in diatoms.[361] Such studies may lead to new ways to prepare silicon compounds, nanoparticles, and stronger composites.

Sulfur dioxide from electric power plants is a major source of acid rain. Although much of this can be taken out of the stack gas with the proper scrubbers, it would be desirable to remove the sulfur from the coal before it is shipped to the power plant. Inorganic sulfur, in the form of pyrite (FeS_2), can be removed by flotation of the ground coal. The sulfur present in organic compounds is harder to remove. Biodesulfurization of coal can be done with *R. rhodocrous* or *Agrobacterium* spp. without loss of carbon.[362] The sulfur in petroleum can be converted to sulfuric acid by *T. denitrificans* and then removed by washing with aqueous sodium hydroxide or ammonium hydroxide.[363] Energy BioSystems of Houston, Texas, has studied such a process in a pilot plant.[364] The process has been used in a 5000 barrel/day refinery in Alaska. The Petroleum Energy Center (Tokyo) is using a *Paenibacillus* sp. to reduce the sulfur to less than 50 ppm.[365] Current systems for the desulfurization of petroleum use cobalt and molybdenum catalysts with hydrogen at high temperatures. They do not remove all of the sulfur. What is left contributes to acid rain and can foul catalytic systems on the tail pipes of diesel engines. Sulfate-reducing bacteria, such as *Desulfovibrio desulfuricans*, have been used with anaerobic sewage sludge to reduce part of a sulfur dioxide stream to hydrogen sulfide. This could then be combined with the rest of the sulfur dioxide to produce elemental sulfur in a Claus plant.[366] *P. putida* cells, immobilized in calcium alginate, have been used to convert minor amounts of hydrogen sulfide in gas streams to elemental sulfur.[367] The use of bacteria to remove nitrogen from the nitrogen compounds in petroleum is being studied.[368] The hope is to obtain a set of bacteria that will remove both nitrogen and sulfur in a single treatment. One of the big advantages of renewable energy sources, such as solar, wind, tidal, and hydropower, is the absence of sulfur compounds.

The aqueous extraction of edible oils from oilseeds works better when enzymes are used to hydrolyze structural polysaccharides in the cell walls first.[369] It is possible to recover 90% of the oil by boiling the ground, treated seeds and then centrifuging and collecting the layer of oil. Further work is needed to make the process more efficient. Each oilseed species may require different enzymes. *Rhizopus oryzae* cells have also been used to recover the oil without the use of solvent.[370] The aqueous process may replace ones using hexane, which is flammable. Extraction with sc carbon dioxide was discussed in Chapter 8. A cellulase–pectinase combination helped the extraction of antioxidant phenols from grape pomace (the waste left from making grape juice).[371]

The fungus, *Monascus purpureus*, produces yellow, orange, and red pigments of the polyketide type, which have been used for coloring food in the Far East for centuries.[372] It can be grown on ethanol. These pigments inhibit the mutagenicity of the heterocyclic amines formed by cooking meals at high temperatures. This may offer an alternative to the use of synthetic food dyes, for which safety concerns have been expressed. Some synthetic food dyes have already been banned in the United States. One strain of the red yeast produces not only the color, but also a statin (a drug that lowers the level of cholesterol in the blood).[373]

Biofiltration can be used to remove volatile organic compounds, including odors, from waste air.[374] *Pseudomonas* spp. can remove toluene and *Bacillus* spp. can remove ethanol. A biofilter with mixed microbes removed toluene, xylenes, methylethylketone, butyl acetate, and methylisobutylketone satisfactorily for over one year.[375] The process is suitable for use on food-processing plants, public treatment works, and others, to control odors. The cost can be quite a bit lower than that of incineration or adsorption on carbon. A biofilter with *Nitrosomonas europaea* immobilized in calcium alginate removed 97.5% of ammonia from air.[376]

Bioremediation often uses bacteria[377] and fungi.[378] These are selected from contaminated sites or engineered to degrade common organic pollutants. Phytoremediation involves the use of hyperaccumulating flowering plants to remove toxic metal ions from the soil.[379] *Brassica juncea* can accumulate selenium, lead, chromium, cadmium, nickel, zinc, and copper. *Thlaspi caerulescens* can accumulate cadmium and zinc. It can accumulate up to 25 g of zinc per kilogram. It is being tested at the former zinc mine in Palmerton, Pennsylvania. Tetraethyllead used to be made by DuPont in New Jersey across the river from Wilmington, Delaware. The best plant found so far for removing the lead from the ground is *Ambrosia artemisiifolia* (ragweed). (Many people are allergic to the pollen of this plant.) More efficient plants are being sought. The plants containing the metals would be burned to recover the metals for reuse. A Japanese process employs anaerobic bacteria to reduce the level of selenium in industrial wastewater to less than 0.1 ppm.[380] The process may be suitable for selenium-containing water in the western United States where wildlife refuges have been contaminated, but the water would have to be deaerated first.

Municipal or food-processing sludges can be converted to single-cell protein for animal feed using thermophilic bacteria, the product being sterilized by the heat in the process.[381] It is possible to remove 99% of the biological oxygen demand, total dissolved solids, ammonia, nitrogen, and phosphorus from wastewater by growing duckweed on it. The plant can be harvested and sold for use as fertilizer. Microbial fuel cells can be used to reduce the oxygen demand of wastewater at the same time that they generate electricity to help run the plant.[382] (For more on fuel cells, see Chapter 15.) Pig waste is a problem in the Netherlands, Iowa, North Carolina, and other areas. By running it through a series of three reactors, anaerobic, aerobic (nitrification), and anaerobic (denitrification), the chemical oxygen demand was reduced by 90%, ammonium ion by 99.8%, and nitrate ion by 98.8%.[383] It is possible that methane could be generated at the same time and used for power at the farm. The same might be done with chicken waste in Delaware. Enzymes, such as xylanases and proteases, are being used to improve the digestibility of animal feeds.[384] A large market exists for the addition of phytase[385] to the feeds to reduce the phosphorus content of the animal waste by about 30%.

9.1.5 Enzymatic Polymerization

Enzymatic polymerization and oligomerization can be used to make polyesters, polypeptides, polysaccarides, polymers from phenols, polymers from anilines, and many others.[386] This approach could lead to fewer side reactions, higher regio- and stereoselectivity, under milder conditions. Oligomeric polymers can be prepared from lactones. Caprolactone can be polymerized in bulk with lipases (**9.51** Schematic) to polymers with molecular weights of 7000.[387]

Other lactones and propylene carbonate (1,3-dioxane-2-one) polymerized in the same way.[388] Lactide (3,6-dimethyl-1,4-dioxane-2,5-dione) can be polymerized in bulk with a

Lipase → $H\!-\!\!(O(CH_2)_nCO)_m\!\!-\!OH$

$n = 5, 10, 11, 14$

9.51 Schematic

lipase to a polyester with a molecular weight of 126,000.[389] This polymer, poly(lactic acid), is produced by The Nature Works (a joint venture of Cargill and Teijin) in a 300 million lb/year plant in Nebraska. Lyoprotectants, such as polyethylene glycol, glucose, sucrose, and cellobiose, added before lyophilization, enhanced the activity. When a small amount of vinyl methacrylate is present in the polymerization, the product is a macromonomer in which the hydroxyl group in the foregoing formula has been converted to a methacrylate. This underwent further polymerization with free radical initiators. The hydroxycarboxylic acid corresponding to $n = 10$ was polymerized with *C. cylindracea* lipase to a polymer with a weight-average molecular weight of 35,000.[390] A block copolymer was made by inclusion of hydroxy-terminated poly(1,3-butadiene) in the polymerization of caprolactone.[391] Block copolymers of caprolactone and methyl methacrylate have been made in sc carbon dioxide using atom transfer radical polymerization.[392]

Polymers have been prepared from diols such as 1,6-hexanediol with dimethylesters of succinic, adipic, fumaric, maleic, isophthalic, and terephthalic acids using lipases.[393] The by-product methanol was removed with a molecular sieve or with a nitrogen sparge. Number average molecular weights up to 12,900 were obtained. Similar polymerizations have been carried out using divinyl, bis(2-chloroethyl), and bis(2,2,2-trifluoroethyl) esters.[394] The highest weight-average molecular weight, 46,400, was obtained when the by-product alcohol was removed with a vacuum. The reaction of divinyladipate with triols, such as glycerol, in the presence of a lipase also led to linear polymers with molecular weights as high as 10,391.[395] Divinyl sebacate and glycerol gave a weight-average molecular weight of 19,000.[396] ω-Hydroxy esters have also been polymerized by lipases.[397] Oligomeric polyesters with molecular weights of only a few thousands are suitable for reactions with diisocyanates to produce polyurethanes. Glucose has been reacted with one end of divinyl sebacate in the presence of a lipase to produce a vinyl monomer, which gave poly(6–*O*-vinylsebacyl-D-glucose) on polymerization with 2,2′-azobisisobutyronitrile (an initiator for free radical polymerization), with a number average molecular weight of 11,000.[398] Maltitol and lactitol have been converted to polymers in the same way.[399] Sucrose has been used with divinyladipate in a similar sequence.[400] Sorbitol, adipic acid, and 1,9-nonanediol were polymerized in bulk at 90°C to a copolymer with a weight-average molecular weight of 44,000.[401]

Sucrose-1′-acrylate has been prepared by the reaction of vinyl acrylate with sucrose, using subtilisin.[402] It can be polymerized with free radical initiators in water to a polymer with a number average molecular weight of 2.4 million. It can also be copolymerized with acrylic acid. When copolymerized with a cross-linking diacrylate, it produces a copolymer that can pick up 1000 times its weight of water. Such polymers are useful in diapers and seed treatments. The usual superabsorbents derived from acrylamide (a neurotoxin) do not work as well in salt solutions. Because the sugar acrylate polymers have no ionic groups, they do not have this problem. Other sugar monomers are shown in **9.52** Schematic.[403] The first was made using *C. antarctica*.

Disaccharide fluorides have been treated with a cellulase to produce cellulose, amylose, and xylan.[404] A mutant strain of *Acetobacter xylinum* produced cellulose that imparted higher tensile strength and superior properties to paper than did cellulose from higher plants.[405] An economical mass production technique has been devised for it.

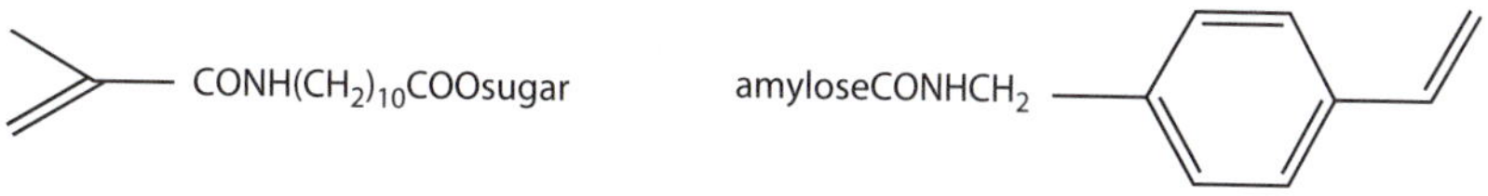

$CONH(CH_2)_{10}COOsugar$ amyloseCONHCH$_2$

9.52 Schematic

Polyhydroxyalkanoates are produced by bacteria for storage purposes.[406] Monsanto (formerly Zeneca) formerly produced 1000 tons/year of a random copolymer of 3-hydroxybutyrate and 3-hydroxyvalerate by growing *Alcaligenes eutrophus* on a mixture containing glucose and propionic acid. The high cost of $15/kg for this biodegradable polymer has limited its use. Metabolix and Archer Daniels Midland have improved the process and lowered the cost so that a 50,000 ton/year plant has been built in Iowa.[407] Growth of *Pseudomonas oleovorans* on a mixture on *n*-octane and 1-octene produces a poly(hydroxyalkanoate) that can be cross-linked with an electron beam to a biodegradable rubber.[408] Attempts are being made to reduce the costs of polyhydroxyalkanoates by the use of wastes as growth media and at the same time solving waste disposal problems. These include swine waste liquor,[409] palm oil waste,[410] and alcoholic distillery wastewater.[411] Another way to reduce costs would be to transfer the necessary genes to a crop plant that could be grown in a field. The polyhydroxyalkanoates have been produced in transgenic *Arabidopsis* and tobacco plants. They have also been produced on an experimental basis in poplar trees.[412] The polymer was extracted from the ground dry leaves with chloroform. It is desirable to find a less dangerous solvent for the extraction. It is possible that these methods will make it easier to replace some of the polyolefins that are in widespread use. It is also possible to produce polyethylene by polymerization of ethylene derived from ethanol from fermentation. Dow Chemical is building a plant in Brazil to do this. However, the polyethylene is not biodegradable the way the polyhydroxyalkanoates are. (For more on the production of polymers and other materials from renewable raw materials, see Chapter 12.)

9.53 Schematic

Phenols can be converted to polymers by the action of peroxidases and hydrogen peroxide (**9.53** Schematic).[413] The goal is to prepare phenolic resins without the use of formaldehyde (a carcinogen). Yields can be as high as 90–95%. The products from phenol and *p*-cresol are only partially soluble in dimethylformamide. Molecular weights can vary from a few thousands up to 800,000.[414] Enzymol International (Columbus, OH, USA) is commercializing the process using a peroxidase from soybeans.[415] Bisphenol A gives soluble polymers when the hydrogen peroxide is added in small portions every 15 min.[416]

It is not completely clear how these polymers will be fabricated into useful articles. Phenol–formaldehyde resins go through a soluble prepolymer stage that is cured to give the final product. It is possible that soluble oligomers made by the action of peroxidases on phenols could be cured in a melt with diisocyanates, diepoxides, dianhydrides, or bisoxazolines. A soluble monomer, prepared by the action of horseradish peroxidase on 4(2-methacrylatoethyl)phenol, was cured with either light or 2,2′-azobisisobutyronitrile.[417] A soluble oligomer with a degree of polymerization up to 12 has been prepared by polymerizing hydroquinone with one hydroxyl group blocked with a sugar and then removing the sugar by hydrolysis (**9.54** Schematic).[418]

9.54 Schematic

Natural phenols, such as syringic acid and cardanol, have also been polymerized with enzymes.[419] The soluble lignin from pulping wood in aqueous ethanol could be available in much larger amounts, if the industry adopts the process.

A poly(phenylene oxide) has been made by oxidation of 2,6-dimethylphenol with laccase and oxygen (**9.55** Schematic).[420] General Electric (now SABIC) polymerizes this phenol to a polymer of high molecular weight using oxygen with a copper salt and pyridine.

Aniline can also be polymerized by horseradish peroxidase and hydrogen peroxide to electrically conducting polymers.[421] If run in the presence of sulfonated polystyrene, this leads to a water-soluble doped polymer in a one-pot, benign process. Horseradish peroxidase and hydrogen peroxide convert *p*-aminobenzoic acid to a self-doped, water-soluble electrically conducting polymer.[422] The method also converts *o*-, *m*-, and *p*-phenylenediamines to polymers.[423]

Tetraalkoxysilanes can be converted to polysiloxanes in 1 h at 20°C, using a siloxane polymerase from a marine sponge.[424] Many living organisms use such enzymes to produce shape-controlled silica skeletons.

Enzymatic polymerization will have its greatest value when it uses renewable raw materials to make, under mild conditions, polymers that are biodegradable. It should avoid toxic chemicals and toxic catalysts. Some polymers that can be prepared in this way cannot be made by the usual chemical methods.

9.1.6 Genetic Engineering

It is evident from the earlier discussion that genetic engineering plays an integral role in the design of better biocatalysts. Over 60% of industrial enzymes are recombinant products.[425] The stability of subtilisin increased 1000 times when a calcium ion-binding loop was deleted, followed by directed mutagenesis and selection.[426] Blocked metabolic pathways can allow desired products to accumulate (e.g., the food ingredients L-glutamic acid and citric acid).[427] Analogs to the antibiotic, erythromycin, have been made in *Streptomyces coelicolor* containing a genetic block.[428] Putting the genes of one organism into another can produce a recombinant organism that can convert both glucose and xylose to ethanol, as mentioned earlier. Pregnenolone and progesterone have been produced from a combination of sugars and ethanol in recombinant *S. cerevisiae*.[429] The sweet noncariogenic protein, monellin, has been produced by recombinant *Candida utilis*.[430]

Plant biotechnology,[431] as applied to field crops, has a variety of goals. A few examples will illustrate some of these. Antisense inhibition of polyphenol oxidase gene expression prevents discoloration of potatoes after bruising.[432] A transgenic potato expressing a double-stranded RNA-specific ribonuclease (from yeast) was resistant to a potato spindle tuber viroid.[433] Decaffeinated coffee has been produced by silencing the gene for xanthosine N^7 methyltransferase, which is

9.55 Schematic

involved in the production of caffeine.[434] Proteins in *E. coli* can be enriched with essential amino acids, such as methionine, threonine, lysine, and leucine.[435] The work was done with cattle feed in mind, but it should also be applicable to cereal crops in developing nations where some people have protein-deficient diets. This supplements the efforts of conventional breeding to produce high-lysine corn for animal feed. Plant lipids can be altered by genetic engineering.[436] Lauric acid is usually obtained from coconut oil. By putting the thioesterase gene from the California bay tree (*Umbellularia californica*) into rapeseed, a transgenic plant producing an oil that is nearly 40% laurate has been obtained.[437] Another route involves the introduction of the gene for petroselenic acid (the δ-6 isomer of oleic acid) into rapeseed. Ozone oxidation of petroselenic acid gives lauric and adipic acids. The latter is used in making nylon. Trees genetically engineered to produce less lignin would be easier to pulp for paper.[438] Only field tests will tell whether they will be more susceptible to storm damage. They would have to be sterile so that they do not hybridize with the same or related species located nearby.

Efforts to modify the cotton plant aim to improve its resistance to insects, herbicides, viruses, nematodes, and stress.[439] Seeds free of gossypol are also desired. Cotton that expresses the natural insecticidal protein of *Bacillus thuringiensis* was used in 61% of the cotton crop in the United States in 2000.[440] In 2007, genetically engineered cotton plants had risen to 87%.[441] Calgene has engineered cotton to produce melanin so that the brown or black fiber will require little or no dyeing.[442] Further effort is being directed toward red and blue fibers. This method eliminates a chemical-dyeing operation that is often the source of contaminated wastewater. Agracetus has put bacterial genes for polyhydroxybutyrate into cotton so that a hybrid fiber is produced.[443] The resultant fiber is an all-natural equivalent of polyester–cotton blend fabrics.

Transgenic plants can lead to increased crop yields, reduced use of pesticides, and a shift to integrated pest management and conservation tillage.[444] Field trials of transgenic plants in the United States in 1994 involved 41% for herbicide resistance, 22% for insect resistance, and 18% for improved product quality.[445] Of the ones for herbicide tolerance, 45% were for corn and 24% for soybeans. For the period 1987–1998, crops were modified genetically 29% for herbicide tolerance, 24% for insect resistance, 20% for product quality, 10% for viral resistance, 5% for agronomic properties, 5% for fungal resistance, and 7% for other properties.[446] For 2007, 51% were for soybeans, 31% for maize, 13% for cotton, and 5% for canola.[447] In 2007, there were 114 million ha of genetically modified crops planted in the world, mainly soybean, maize, cotton, and canola.[448] Of this, 58 million ha was in the United States, 19 million ha in Argentina, 15 million ha in Brazil, and the rest in a number of other countries for a total of 23 countries. Half of the world's soybeans and one-third of its maize came from plants genetically engineered for herbicide or insect resistance.[449] Other field trials for maize involve a lower requirement for fertilizer and improved starch quality.[450] Maize with *B. thuringiensis* genes can be eaten in France, but not grown there.[451] Critics would prefer that food crops be genetically engineered for desirable traits other than just for resistance to insects and herbicides. There are now 15 species of weeds that are tolerant to the herbicide used on the genetically modified crops. There is concern that resistance developed to the antibiotics, such as ampicillin, used as a genetic marker in 95% of transgenic crops might jump to bacteria and make the antibiotics worthless in medicine.[452] Norway has banned all crops having antibiotic-resistant marker genes. It is now possible to produce transgenic corn without the ampicillin marker. Transgenic flax can be grown in herbicide-contaminated soil that would otherwise prevent crop rotation.[453]

Plants may be able to serve as bioreactors for biopharmaceuticals.[454] γ-Linolenic acid has been made in tobacco and *Arabidopsis* by inserting a gene coding for a δ-6-desaturase.[455] Human hemoglobin has been produced in a genetically engineered tobacco plant.[456] Therapeutic proteins have been made in duckweed.[457] If recombinant antibodies can be produced in the seeds of plants, large-scale production could be achieved at low cost. The product might be stored in the seeds instead of in the refrigerator. At present they are produced in microbial or mammalian cell cultures. An effective edible vaccine for an *E. coli* strain that causes diarrhea has been made in potatoes.[458] A vaccine for hepatitis B made in a genetically modified banana costs one cent per dose compared to $15 for a conventional one.[459] It does not need refrigeration. Vaccines made in plants avoid the problem of contaminants, such as diseases and viruses, in vaccines of animal origin.[460] Some human proteins have been produced in the milk of transgenic cattle, pigs, mice, goats, and rabbits.[461] Human antithrombin III and spider silk have both been produced in the milk of transgenic goats.[462] Human prolactin has been produced in the milk of mice at 30–50 times the levels achievable in cell culture.[463] Human growth hormone has been produced in the bladders of transgenic mice.[464] Although the concentration in the urine is low, the purification is simplified, because urine contains little protein and lipid. In contrast to making it in milk, both sexes of the animals can be used throughout their lives. Further work is needed on the use of larger transgenic animals. Transgenic chickens are desired for better disease resistance, altered sex ratio, and for production of immunoglobulins in the eggs.[465]

Each year about 500,000 children in southeast Asia go blind from a deficiency in vitamin A. The University of Freiburg, the Rockefeller Foundation, and Syngenta have developed "golden rice," which contains β-carotene that could cure this problem.[466]

There are some other risks in using agricultural biotechnology.[467] Gene transfer might occur from genetically engineered microorganisms to others in the enviroment.[468] Genetically modified *Pseudomonas fluorescens* on wheat was found up to 2 m away. The next year, it was on newly sown wheat plants and on nearby weeds.[469] Pollen from transgenic grass traveled up to 13 miles.[470] In sugar beets, there was no transfer of marker genes to indigenous microorganisms. Yet, it is well known that bacteria can exchange genes on occasion. If corn, cotton, potato, and soybean plants that incorporate *B. thuringiensis* genes for insect resistance hybridize with wild plants of the same or related species, the resulting wild plant may be resistant to herbivores that normally keep it in check. The result could be a superweed. To envision the effect of a superweed, one need only to think of some of the highly invasive noxious weeds that have been brought to one country from another. In the eastern United States, weeds such as *Alliaria officinalis*, *Rosa multiflora*, *Lonicera japonica*, *Celastrus orbiculatus*, *Lythrum salicaria*, and *Pueraria lobata* are threatening native plant communities. The insect toxin now transferred to the wild relative could kill off natural pollinators for the crop or other beneficial insects.[471] Corn pollen containing the *B. thuringiensis* toxin gene transferred to milkweed plants killed nearly half of the monarch butterfly caterpillars that fed on them.[472] Herbicide-resistant crops may lead to additional use of herbicides to control weeds that develop resistance to glyphosphate, with resultant contamination of surface and groundwater. Additional soil erosion may occur compared with a field with a surface cover crop under the main crop. There has already been a case of gene transfer from a crop of the mustard family to a wild relative. The use of *B. thuringiensis* spray on a regular calendar basis led to insect resistance in Hawaii. Not everyone is satisfied with proposals to avoid resistance by planting some nontransgenic plants in nearby refugia.[473] Labeling of transgenic crops has been suggested to protect consumers who may be allergenic to the foreign proteins present.[474] A soybean that incorporated a gene from Brazil nuts was not put on the market when testing showed that people allergic to Brazil nuts were also sensitive to the new soybean.

Transgenic fish can grow faster and use food more efficiently.[475] Some have already escaped where they might outcompete native fish.[476] Sterile fish could eliminate this problem. Another problem is keeping genetically modified crops apart from regular crops. Contamination has occurred in six cases.[477] If someone ate a crop with a drug in it, the result could be tragic. Plant-made pharmaceuticals are nearing the market.[478] The Union of Concerned Scientists has summarized the advantages and risks of the production of industrial chemicals and drugs in genetically modified crops.[479]

Advantages

Lauric acid from canola instead of from imported coconuts.
Higher value crops for farmers, for example, for tobacco farmers.
Cheaper drugs.
Vaccines that do not need to be stored in the refrigerator.
Vaccines that can be taken orally, instead of by needle.
Lower capital investment relative to cell cultures.
Prepare enzymes such as trypsin, laccase, and β-glucuronidase.
Prepare monoclonal antibodies and hormones.
Prrepare renewable plastics such as poly(β-hydroxyalkanoates).
Easy to scale up.
Avoid contamination by animal diseases or viruses.
Easier to handle and requiring less land than a herd of genetically modified animals.

Risks

Cross-pollination with regular crops may introduce drugs where they are not wanted. The same is true for wild relatives of the crops.
The food supply may become contaminated with drugs, if harvested crops are mixed by accident.
The seasonable nature of crop production may make it hard to keep a plant running all year long.

Possible strategies

Put genes in crops not used for food such as tobacco and guayule.
Physical isolation of drug and farm crops.
"Bioreactor crop" zones.
Produce sterile plants.
"Zero tolerance" for drugs in foods.
Harvest the product from a part of the plant that people do not eat, as long as none is in the part of the plant that is eaten.
The pros and cons of genetically modified crops are hotly debated.[480]

The Ecological Society of America advises the use of caution in the introduction of transgenic organisms into the field. The U.S. National Research Council advises caution with genetically modified animals.[481] (For more on the alternatives to traditional agriculture, see Chapter 11.)

9.2 BIODIVERSITY

9.2.1 The Value of Biodiversity

The discussion of biocatalysis has focused primarily on the chemistry of microorganisms. The advantages of preserving as many different ones in as many different habitats as possible should be apparent.[482] The traditional search for new types of activity has involved screening soil samples from many different places. This implies the need to preserve the varied places that the soils came from. Although not as popular with persons who donate money for conservation purposes as animals with fur or feathers, they are just as important. The range of applications is extended now that it is possible to take the genes from one organism and put them into another one.

Native plants and animals are sources of new foods, drugs, and other uses.[483] They have also provided ideas for new chemistry that would not have been thought of otherwise. Roughly 80% of the top 50 prescription drugs used today are of natural origin, or derivatives of them, or are based on leads from them.[484] The ability to return to the wild for new genetic material to breed into crops has saved a number of them. The classic case is the Irish potato famine of the nineteenth century when a blight struck the potato crop. One million people died and another 1 million emigrated out of a total population of 6 million. In 1969–1970, 85% of the hybrid corn was derived from a strain with a male sterility trait. A crisis developed when a Southern corn blight attacked the hybrid corn. Regular corn was not susceptible to the blight.[485] The disease resistance of a wild rice in India's forests saved much of Asia's rice from a pandemic blight.[486] In California, a uniform rootstock for grapes has been threatened by a virulent disease, so that the rootstocks are now being replaced by genetically varied ones.[487] A small wild patch of perennial corn in Mexico offers the possibility of changing the corn crop from annual to perennial, with savings in planting and soil erosion costs. An everblooming strawberry found in Utah offers the possibility of strawberries at any time of the warmer part of the year. *Pinus radiata*, originally found only in a small part of California, is now an important timber tree in many parts of the world. The thousands of plants and animals used as food in the past need to be reexamined for foods to supplement the few dozen species that provide the bulk of the world's food today. Some may prove to be more nutritious and more suitable for sustainable agriculture.

Many plants and animals have evolved chemical defense systems to keep themselves from being eaten or crowded out by other plants.[488] One way to look for new drug leads is to test those plants and animals in the fields, forests,[489] and the seas[490] that are conspicuous and look like they could be eaten easily. Another is to see what is used in tribal and herbal medicine.[491] Although some of these are surrounded by myth and superstition, others have proved to be valuable leads to new drugs. (The dangers of taking herbal supplements before they have been thoroughly tested for safety and efficacy were described in Chapter 1.) The hit rate for drugs from plants is 1 in 125, whereas it is 1 in 10,000 for chemically synthesized compounds.[492] As a drug, a compound does not necessarily have to be used for the same purpose that it is in the wild. Botulism toxin can be used in small amounts as a muscle relaxant.[493] Curare, an Amazonian Indian arrow poison, is useful as a skeletal muscle relaxant.[494] The protein contortrostatin, found in the venom of the copperhead snake, is able to reduce the rate of growth of tumors by 90%.[495] The deadly venom of the fish-eating snail, *Conus magus*, contains a 25-amino-acid polypeptide that is a pain killer.[496] Many compounds from plants and animals are being screened against diseases for which adequate drugs are not yet known, such as AIDS, cancer, and Alzheimer's disease.

Some of the more important substances found by screening of natural products will be described in the following. Their chemical structures are given in Figure 9.2. Before the discovery of colchicine, seven-membered rings of carbon atoms were thought to be rare. It contains two fused seven-membered rings. This discovery sparked the synthesis of many more. Before the discovery that reserpine, from the Indian vine *Rauwolfia serpentina*, could affect the brain, chemical treatment of mental illness was not thought to be possible. Today, many antidepressants are used. Artemisinin from *Artemisia annua* (which is used in Chinese herbal medicine) is effective against malaria that is resistant to the usual drug chloroquine.[497] It introduced the idea that transannular peroxides can be effective drugs. Simpler analogs also work.[498] Hervelines from the bark of *Hernandia voyronii* (used in herbal practice in Madagascar) are also effective against drug-resistant malaria.[499] The tree resists water well and has been used in dugout canoes and coffins so much that it is now among the most endangered species in the country. For it to be used as a source of the drug commercially, the tree would have to be farmed or the drug produced in tissue culture. Notice that its structure has no relation to that of artemisinin. Enediynes were not anticipated to be drugs, but calicheamicin, dynemicin, and others are effective against cancer.[500] This has spawned a great deal of activity in the synthesis of enediynes and in studies of their cyclization to aromatic structures.[501]

Azadirachtin from *Azadirachta indica* (the neem tree) is a potent insect antifeedant[502] that promises to replace harmful insecticides.[503] Polygodial (from *Polygonum hydropiper*) is another insect antifeedant.[504] Before it can be used widely, a way will have to be found to stabilize it against environmental degradation. Balanol from fungi is a potent inhibitor of protein kinase C enzymes and may lead to new drugs for diseases involving these.[505] The triterpene betulinic acid, which is derived from the bark of the white birch tree, is active against tumors.[506] Cycloartenol triterpenes from *Cimicifuga foetida* (used in traditional Chinese medicines) are anti-inflammatory and analgesic agents.[507] Bryostatin from a marine bryozoan is an anticancer agent.[508] Efforts are being made to farm the bryozoan to produce the drug. Camptothecin is a natural quinoline derivative that is used to treat ovarian cancer.[509] Compactin is effective in lowering cholesterol levels in the blood.[510] Discodermolides are marine natural products that exhibit immunosuppresive activity.[511]

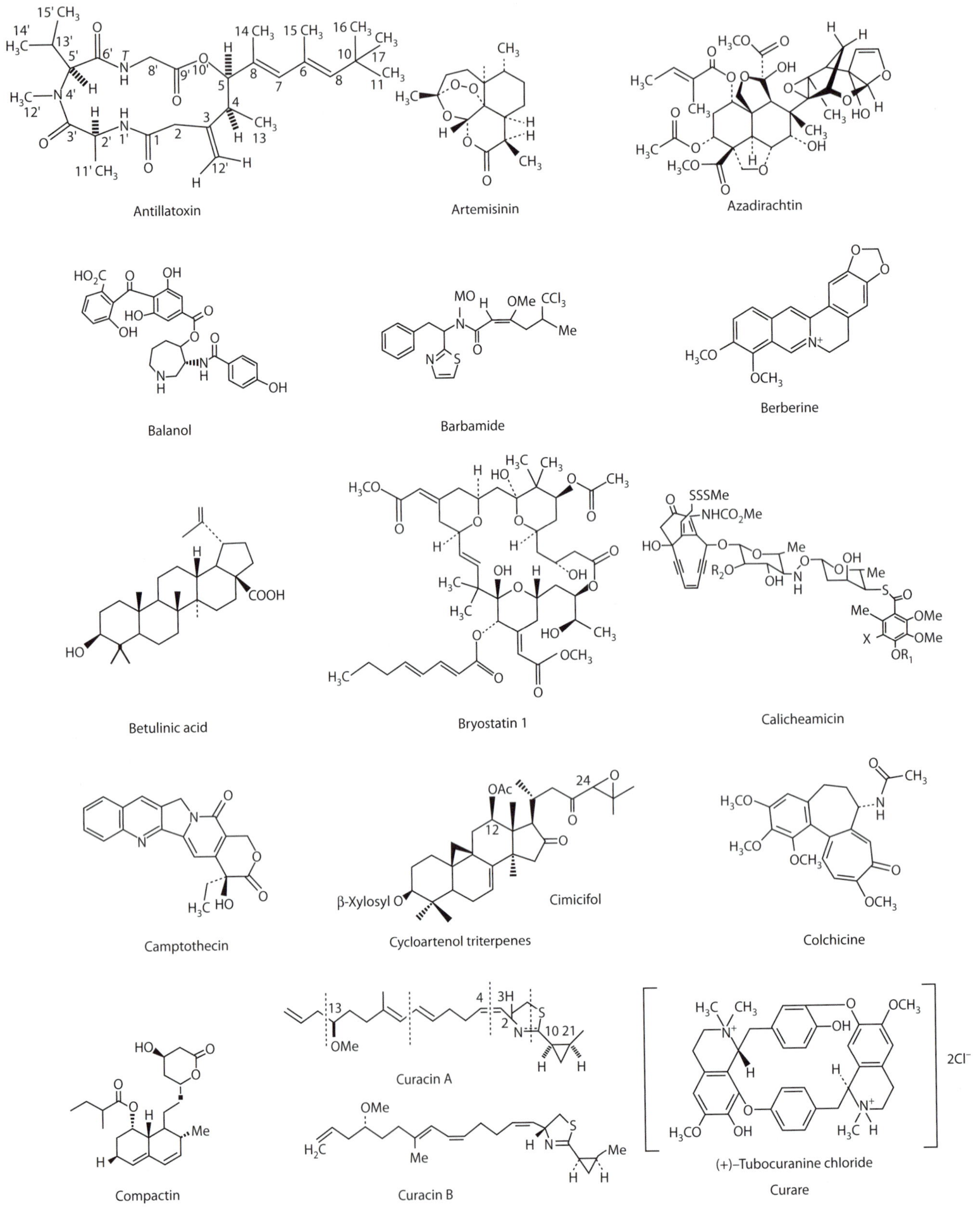

Antillatoxin

Artemisinin

Azadirachtin

Balanol

Barbamide

Berberine

Betulinic acid

Bryostatin 1

Calicheamicin

Camptothecin

Cimicifol

Cycloartenol triterpenes

Colchicine

Compactin

Curacin A

Curacin B

(+)-Tubocuranine chloride

Curare

Figure 9.2 Substances in some natural products.

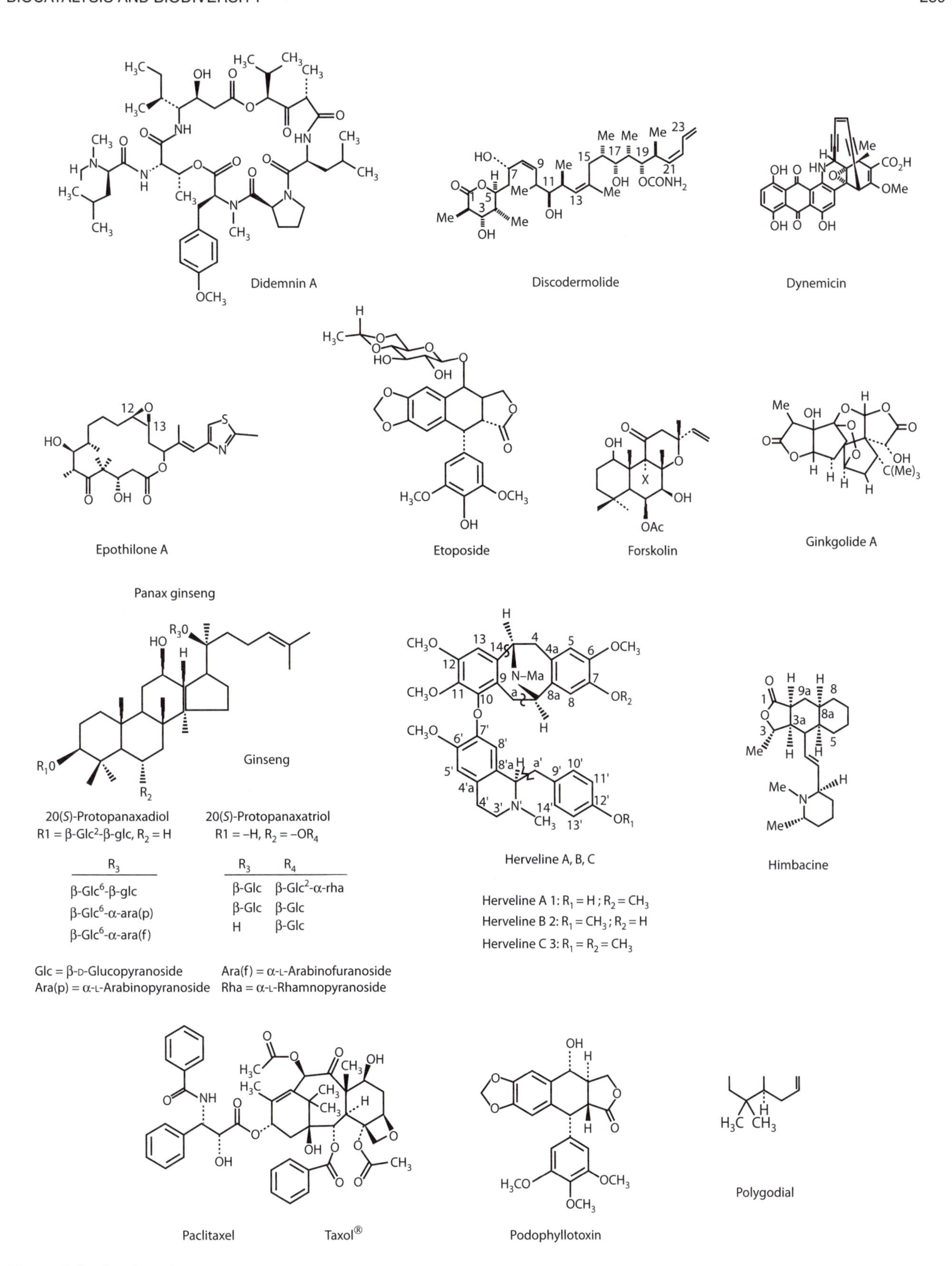

Figure 9.2 Continued

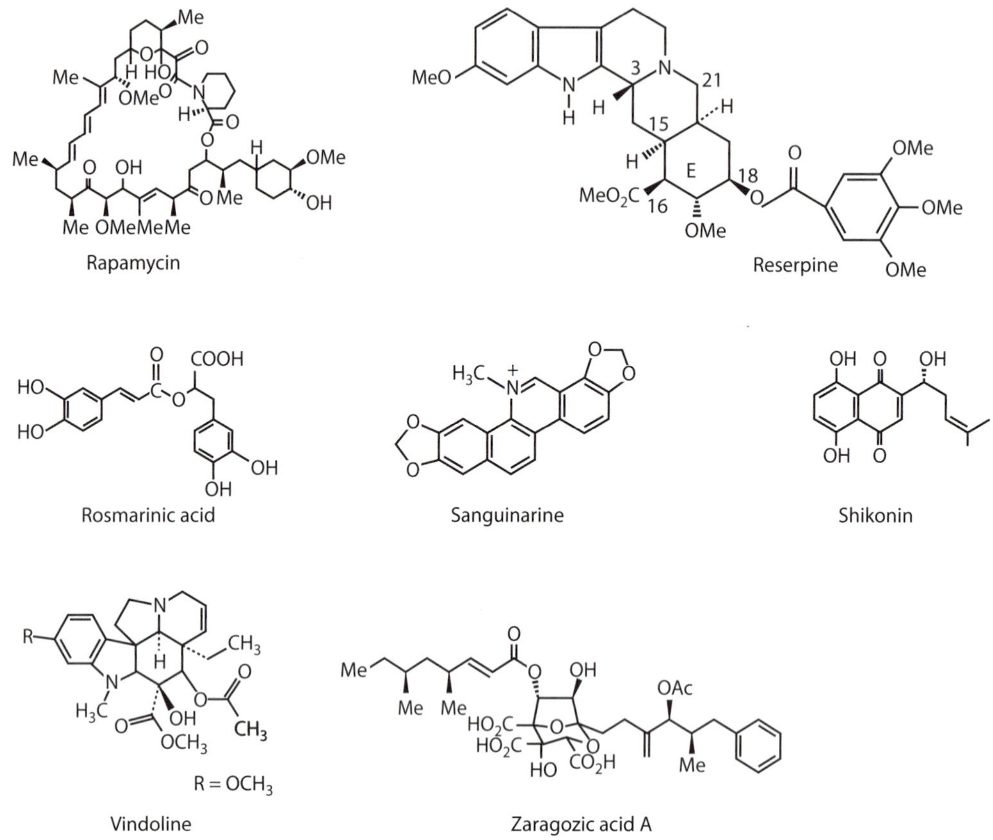

Rapamycin

Reserpine

Rosmarinic acid

Sanguinarine

Shikonin

Vindoline

$R = OCH_3$

Zaragozic acid A

Figure 9.2 Continued

Epibatidine (**9.56** Schematic) is a potent analgesic from the Ecuadorian poison frog *Epipedobates tricolor*.[512] Although the compound itself is too toxic to use as a drug, an analog developed by Abbott Laboratories (**9.57** Schematic) is less toxic, nonaddictive, and as effective as morphine.[513] It was still not good enough to commercialize.

Epothilones A and B are promising anticancer agents that bind to microtubules in the cell in the same way that paclitaxel (Taxol) does and are more potent.[514] Paclitaxel was obtained first from the bark of the Pacific yew (*Taxus brevifolia*).[515] Widespread extraction of the compound from this source could have wiped out the species. Fortunately, it can now be extracted from the needles of more common yews or produced in tissue culture. (See Section 9.2.3 on tissue culture.) Forskolin is a diterpene (from the roots of *Coleus forskohlii*) that lowers blood pressure.[516] Himbacine, from an Australian pine tree, offers a potential treatment for Alzheimer's disease.[517] Combretastatin, from the bark of the African bush willow, *Combretum caffrum*, cuts the flow of blood to tumors, causing 95% of the cancer cells to die in 24 h, but does not harm healthy blood vessels.[518]

Important drugs have been derived from two plants common in the woodlands of the eastern United States. One is etoposide (an anticancer drug) based on the podophyllotoxin found in *Podophyllum peltatum*.[519] The second is sanguinarine (from *Sanguinaria canadensis*), which is used for treatment of periodontal disease. A common fence lizard of the western United States, *Sceloporus occidentalis*, has

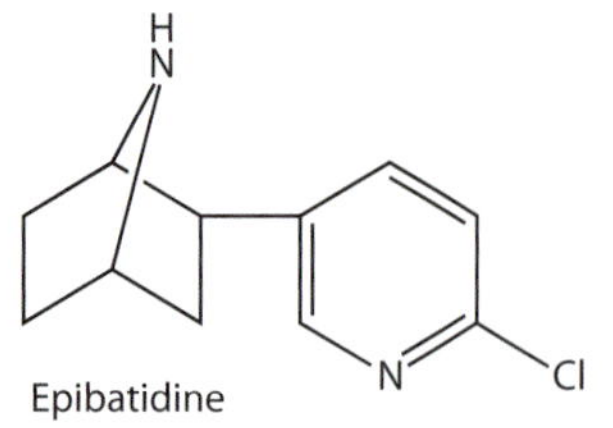

Epibatidine

9.56 Schematic

ABT 594

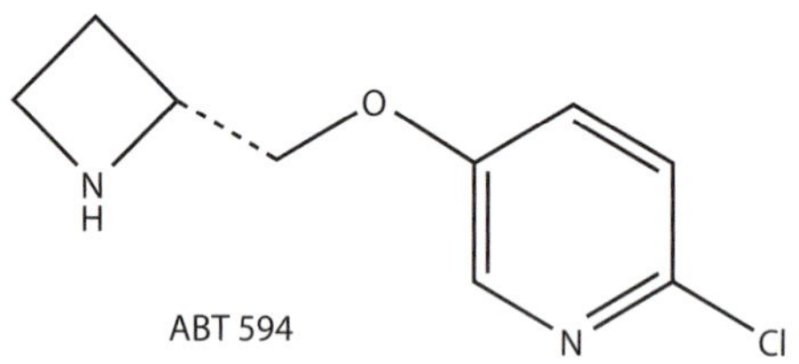

9.57 Schematic

something in the blood that kills the bacteria responsible for Lyme disease.[520]

Some of the organisms from which interesting drugs have been obtained are of very limited distribution. Rapamycin is produced by *Streptomyces hydroscopicus*, which is endemic (grows nowhere else) to Easter Island in the Pacific Ocean.[521] The oil from *Dicerandra frutescens*, a mint that grows only on a few hundred acres of Florida scrub, is a natural insecticide.[522] Curacin A, with potential use in cancer therapy, was extracted from the blue-green alga, *Lyngbya majuscula*, growing in the ocean on the western side of Curacao.[523] The same alga also contained antillatoxin, which was toxic to fish, and barbamide, which was toxic to snails. The same species growing at the northern tip of the island produced little curacin A, but produced kalkitoxin, a fish poison not found in the alga at the first site. This shows the importance of preserving subspecies and races as well as species of organisms. Many active compounds are produced, not by the species being studied, but by symbionts, which could differ in different locations. It may be possible to transfer the genes from the symbionts to *E. coli*, which would then turn out the desired compounds. Poison dart frogs get the poisons from the ants that they eat and the ants get them from the insects that they eat.[524]

9.2.2 Conservation of Natural Biodiversity

Humans' use of the land has brought many changes that have resulted in an increased rate of extinction of species.[525] The greatest threat is from habitat destruction, followed, in descending order, by competition with alien species, pollution, overexploitation, and disease.[526] One-fifth of all freshwater fish are extinct, threatened, or endangered.[527] Physical alterations of rivers, via contamination with sediment, waste, or exotic species, in the United States have put 67% of mussels, 64% of crayfish, 36% of fish, and 35% of amphibians in jeopardy or have made them extinct. The introduction of the Nile perch into Lake Victoria in Africa has caused the extinction or near extinction of many fishes. One-fourth of the mammals in the world are at risk of extinction: 46% of monkeys and apes; 36% of shrews and moles; 33% of pigs, antelopes, and cattle; 11% of birds; 25% of amphibians; and 34% of fishes.[528] There is a worldwide decline in amphibian populations for reasons that are just being identified. These may include fungal disease, parasitic worms, increases in ultraviolet light, and so on.[529] Of the species in the United States in 1900, 1.5% are now extinct (8.6% for freshwater mussels).[530]

The World Conservation Union reports that 34,000 plant species out of 270,000 are threatened.[531] This includes one-third of all lily, palm, and *Dipterocarpacae* species. Forests once covered 40% of the land on earth. This has now been reduced by one-third.[532] Tropical forests have been reduced by 50% in the last 50 years, half of this being due to slash and burn agriculture.[533] Forest certification is being used to combat illegal logging.[534] More than 2000 species of birds (about 15% of the world's total) have been lost in Polynesia since humans colonized the islands.[535] Losses have been highest in areas of high endemism. Some of this has been due to introduced animals, such as predatory snails, the brown tree snake on Guam, and the mongoose on Hawaii. Since 1778, Hawaii has lost 18 species of birds and 84 of the 980 species of plants. Another 133 of the species have populations of <100 plants. A big problem has been the introduction of grazing animals (i.e., goats and pigs) into an ecosystem that evolved without grazing animals. Introduction of highly invasive animals and weeds from other continents has led to a loss of biodiversity.[536] Sterilization of ballast water is being encouraged, and may be mandated, as a way to reduce invasions by aquatic species.[537] About 43% of the earth's surface has a reduced capacity to supply benefits to humankind owing to overgrazing, deforestation, and other such conditions.[538] Acid rain has damaged the forests of Scandinavia and the northeastern United States. Nitrogen from acid rain has caused a reduction in biodiversity in grasslands, as native grasses have been replaced with weed grasses.[539] Global warming may require a northern migration of species that cannot take place in many places because forest fragmentation has interspersed unsuitable habitat between forest fragments.[540] Warming may also be too rapid for migration to occur effectively.

There is a need for everyone to realize that biodiversity is valuable.[541] The goal is to save enough of each habitat in each stage of succession so that there will always be some.[542] No-take reserves are being set up in an effort to rebuild overexploited fish stocks, as well as to protect terrestrial species.[543] They will work only if they include the areas of highest biodiversity and are large enough.[544] Assigning catch shares also works.[545] Ecotourism is often suggested as a way to conserve biodiversity, but it must be managed carefully not to have a negative effect.[546] Human population growth rates will have to be slowed or even made negative to bring about a balance between population and the natural resource base. (The chemistry of family planning will be discussed in Chapter 16.) Overconsumption in the developed nations must be brought down to a sustainable level.[547] (Materials for a sustainable economy will be covered in Chapter 12; the chemistry of longer wear in Chapter 13; energy from renewable sources in Chapter 15; and the chemistry of recycling in Chapter 14.) The sustainable use of forests should be possible without the need to cut all of them.[548] It is also important to see how many degraded systems can be restored and to what extent.[549] Many books in conservation biology and restoration ecology are available for readers who wish to pursue the subjects further.

9.2.3 Plant Cell and Tissue Culture

Any animals or plants that are found to contain valuable drugs or other substances will have to be grown in farms or in cell culture to provide a reliable supply and to avoid overcollection from the wild.[550] In some cases where the growth requirements are difficult to satisfy, it may be easier to transfer the proper genes to other organisms for which cultural requirements are known. This assumes that the substance or a compound of equivalent efficacy cannot be made conveniently by chemical means.

Aquaculture is being used for organisms from the sea. CalBioMarine Technologies is raising the bryozoan *Bugula neritina* in closed culture to produce bryostatin 1. It may be possible to raise a sessile marine organism on rafts in bays, the way some oysters are raised in Japan.[551] At least one-third of the world's seafood is supplied by aquaculture.[552] It does have some problems that need to be solved, such as introduction of unwanted seaweeds, parasites, and pathogens, as well as destruction of mangrove forests and escapes of non-native fish. Australians have been able to get the endangered bluefin tuna to spawn in captivity.[553] This offers the possibility of raising the fish more cheaply than catching it from the wild, thus saving the species.

Plant cell and tissue culture[554] is a well-established technique that can be useful for drugs, colorants, flavors, fragrances, insect antifeedants, insecticides, and others. Ginseng, a favorite medicine in the Far East, is an example. The plant, *Panax ginseng*, became scarce in the mountains of the eastern United States. The next step was to cultivate it on farms in Wisconsin. The final step was to produce it in cell culture.[555] One large run in Japan produced 1 ton of cells, the same amount as 600 ha did in farms.

Plant cell and tissue culture has had some notable successes in producing shikonin (a pigment), berberine, and rosmarinic acid.[556] In some other cases, the cultures have not produced the same natural products in the same concentrations as in intact plants. There have also been occasional problems of cell line instability. Higher plants grow more slowly than microorganisms so that cultures may need to be grown for a few weeks and be kept sterile for this time. These problems are being overcome gradually as better knowledge of growth requirements, transport, and compartmentatlization of products, as well as biosynthetic pathways, accumulates.[557] Sometimes, providing the proper precursor can lead to increased yields.[558] The method has been used for chemical transformations, such as ring closures. For commercial use, cultures that overproduce relative to the wild plant are desired.

Various methods have been used to increase yields.[559] Varying the carbon/nitrogen ratio may help. It may be necessary to use a two-stage culture with a growth phase followed by a product-producing phase induced by a change in the medium. Alternating sucrose-rich and nitrate-rich media raised the yield of shikonin from *Lithosperum erythrorhizon* cell cultures.[560] Because recovery of the product may be 50–90% of the cost, it is desirable to select strains that excrete the product into the medium while the cells remain viable. If the product stays in the cells, they must be broken up, and are lost, to recover the product. Excretion may be helped by pH changes, heat, electroporation, and permeabilizers, such as dimethylsulfoxide.[561] Many of the compounds that are potentially useful as drugs are produced by plants as defenses against attack by fungi, insects, and larger herbivores. Treatment of cell cultures with elicitors, such as fungal extracts or various microbial polysaccharides, can often raise the level of the desired natural product.[562] Continuous extraction of the desired product can remove back inhibition and increase yields.[563] Sanguinarine can be collected with silicone oils. It can also be recovered from cultures of other cells with polymeric adsorbents.[564] Terpenes, which often are stored in oil glands and tend to be volatilized from cell cultures, can be recovered by extraction with a triglyceride of C_8–C_{10} fatty acids.[565]

Many secondary metabolites are produced in the roots of plants. Treatment of roots separated from the plants with *Agrobacterium rhizogenes* can cause them to grow rapidly as "hairy roots."[566] This can cause increased production of the desired compounds. Hairy root cultures of *Catharanthus roseus* made more vindoline than cell cultures.[567] Such cultures of *Papaver somniferum* produced up to 2–10% dry weight of sanguinarine.[568] Cultures of *Sesamum indicum* made 50 times as much of the antimicrobial compound **9.58** as the intact plant.[569]

A callus culture made by the transformation of *Trichosanthes kirilowii* roots with *A. rhizogenes* produced 1% by dry weight of a ribosome-inactivating protein compared with 0.01% in an ordinary dried root.[570] Artemisinin has been produced by hairy root cultures of *A. annua*.[571]

By using the various techniques outlined in the foregoing, overproduction of various compounds has been achieved. A few examples will be given. A cell suspension culture of *R. serpentina* produced 12 times as much raucaffricine as intact plants.[572] A callus cultures of *Dauca carota* produced up to 15% by dry weight of an anthocyanin.[573] Cultured cells of *Aralia cordata* contained 13% by dry weight of an anthocyanin after 3 weeks growth.[574] Anthocyanins are being studied as natural food colors. *Coptis japonica* cells can make up to 10% berberine.[575] Callus of *Eremophila* spp. can accumulate up to 20% of dry weight of verbascoside.[576] When *Mentha spicata* plantlets from shoot tip culture are grown on the right medium, the yield of monoterpenes can be higher than in the intact plant.[577] Considerable effort has been focused

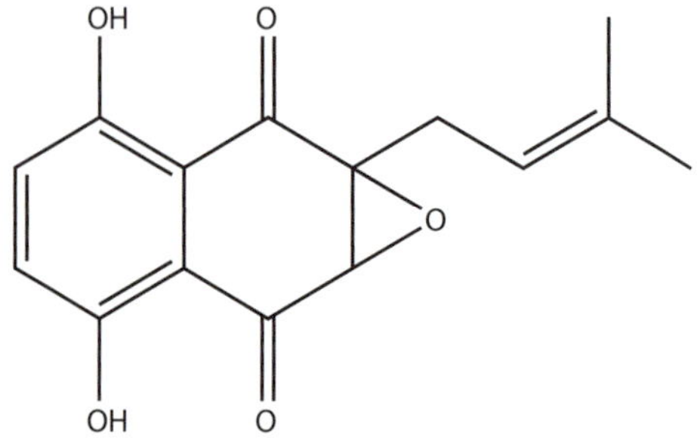

OH O

O

OH O

9.58 Schematic

O

$COOCH_3$

Methyl jasmonate

9.59 Schematic

cell culture as a way to prepare.[578] This is now the preferred method of producing the compound. Most of the compound is excreted to the medium. The best yields have been obtained by the use of methyl jasmonate as an elicitor.[579] *Taxus* media cells grown in suspension for 2 weeks produced 0.606% paclitaxel, compared with the usual 0.007% in dry yew bark. Methyl jasmonate (**9.59** Schematic) is produced by plants in response to wounding. It causes the synthesis of protease inhibitors.

Plant cell and tissue culture methods can be used for the mass propagation of plants. Cells (often from the growing tip) can be grown in a medium containing one hormone to form a callus or a suspension.

Changing to another hormone causes plantlets to form.[580] Each cell contains all the genetic information needed to form a whole plant. The process is called somatic embryogenesis. Much effort has been expended in extending the method to a large variety of families of plants, primarily those of economic interest. Practically all the house plants sold in the United States are propagated in this way in Florida. The method is simple enough so that it can be performed in the basement of one's home if one knows how. In Vietnam homes, it has been used to propagate potatoes, coffee, sugar cane, and orchids.[581] It is very useful with bamboo, which does not bloom very often, as well as with trees that may take many years to reach flowering size. The goal is to make it easier to propagate plants in this way than to collect them in the wild and thus prevent overcollecting. The method can also be used to propagate rare plants for reintroduction into their former ranges in the wild. To maintain genetic diversity, more than one source of plant material should be used.

Useful compounds, such as astaxanthin, lutein, β-carotene, ω −3 fatty acids, oil and phycocyanin, can be produced in cultures of microalgae.[582] New reactor designs are being developed to make these cultures more efficient and cheaper.[583]

REFERENCES

1. (a) H.E. Shoemaker, D. Mink, and M.G. Wubbolts, *Science*, 2003, *299*, 1694; (b) S. Lutz and U.T. Bornscheuer, eds, *Protein Engineering Handbook*, Wiley-VCH, Weinheim, 2008; (c) C. Ratlege and B. Kristiansen, eds, *Basic Biotechnology*, 2nd ed., Cambridge University Press, Cambridge, UK, 2001; (d) T. Palmer, *Enzymes, Biochemistry and Clinical Chemistry*, Horwood Publishing, Chichester, 2001; (e) K. Drauz and H. Waldmann, eds, *Enzyme Catalysis in Organic Synthesis*, Wiley-VCH, Weinheim, 2002; (f) H. Griengl, ed., *Biocatalysis*, Springer Verlag, New York, 2000; (g) M.J. Burk, W.-D. Fessner, and C.-H. Wong, eds, *Adv. Syn. Catal.*, 2003, *345*(6–7), 647–865 (issue on biocatalysis); (h) K. Buchholz, V. Kasche, and U.T. Bornscheuer, *Biocatalysts and Enzyme Telchnology*, Wiley-VCH, Weinheim, 2005; (i) A.S. Bommarius and B.R. Riebel, *Biocatalysis-Fundamentals and Applications*, Wiley-VCH, Weinheim, 2004; (j) R.D. Schmid, *Pocket Guide to Biotechnology and Genetic Engineering*, Wiley-VCH, Weinheim, 2003; (k) M. Doble, A.K. Kruthiventi, and V.G. Gaikar, eds, *Biotransformations and Bioprocesses*, CRC Press, Boca Raton, FL, 2004; (l) C. Ratledge and B. Kristiansen, eds, *Basic Biotechnology*, Cambridge University Press, Cambridge, UK, 2006; (m) W.-D. Fessner and T. Anthonsen, *Modern Biotechnology—Stereoselective and Environmentally Friendly Reactions*, Willey-VCH, Weinheim, 2009; (n) http://umbbd.ahc.umn.edu biocatalysis; (o) E. Garcia-Junceda, ed., *Multi-Step Enzyme Catalysis—Biotransformations and Chemoenzymatic Synthesis*, Wiley-VCH, Weinheim, 2008.
2. J. Lalonde, *Chem. Eng.*, 1997, *104*(9), 108.
3. D.J. Milner, *J. Chem. Technol. Biotechnol.*, 1995, *63*, 301.
4. R.L. Hinman, *Chemtech*, 1994, *24*(6), 45.
5. M.A. Penalva, R.T. Rowlands, and G. Turner, *Trends Biotechnol.*, 1998, *16*, 483.
6. R.F. Service, *Science*, 2008, *322*, 522.
7. (a) M.J. van der Werf, W.J.J. van der Tweel, J. Kamphuis, S. Hartmans, and J.A.M. deBont, *Trends Biotechnol.*, 1994, *12*(3), 95; (b) T. Nagasawa and H. Yamada, *Pure Appl. Chem.*, 1995, *67*, 1241; (c) K.B. Lee Jr. and L.S. Hu, *Chem. Ind. (Lond.)*, 1996, 334; (d) S.S. Sengha, *Kirk–Othmer Encyclopedia Chemical Technology*, 4th ed., Wiley, New York, 1993, *10*, 361.
8. (a) G. Ondrey, *Chem. Eng.*, 2007, *114*(4), 18; (b) Anon., *Chem. Eng. News*, Mar. 24, 2003, 12; (c) Anon., *Chem. Eng. News*, July 13, 2009, 17; (d) Anon., *Chem. Eng. News*, Nov. 2, 2009, 15.
9. (a) D. Rozzell, *Biocatalytic Production of Amino Acids and Derivatives*, Oxford University Press, Oxford, UK, 1992; (b) Anon., *Chem. Eng. News*, Oct. 2, 2006, 28.
10. (a) A. Gabelman, ed., *Bioprocess Production of Flavor, Fragrance and Color Ingredients*, Wiley, New York, 1994; (b) J. Sime, *Chem. Br.*, 1998, *34*(5), 26; (c) S. Serra, C. Fuganti, and E. Brenna, *Trends Biotechnol.*, 2005, *23*(4), 193; (d) A. Lomascolo, C. Stentelaire, M. Asther, and L. Lesage-Meessen, *Trends Biotechnol.*, 1999, *17*, 282.
11. (a) M. McCoy, *Chem. Eng. News*, Oct. 29, 2007, 22; (b) Anon., *Chem. Eng. News*, Mar. 24, 2003, 12; (c) T. Katsuda, A. Lababpour, K. Shimahara, and S. Katoh, *Enzyme Microb., Technol.*, 2004, *35*, 81.
12. S. Mukherjee, P. Das, and R. Sen, *Trends Biotechnol.*, 2006, *24*, 509.
13. (a) P.H. Nielsen, H. Malmos, T. Damhus, B. Diderichsen, H.K. Nielsen, M. Simonsen, H.E. Schiff, A. Oestergaard, H.S. Olsen, P. Eigtved, and T.K. Nielsen, *Kirk–Othmer*

Encyclopedia of Chemical Technology, 4th ed., Wiley, New York, 1994, *9*, 567–621; (b) K. Niranjan, M.R. Okos, and M. Renkowitz, eds, *Environmentally Responsible Food Processing*, AIChE Symp. 90, Washington, DC, 1994.
14. Y.-P. Chao, T.-E. Lo, and N.-S. Luo, *Enzyme Microb. Technol.*, 2000, *27*, 19.
15. X. Wang, C.S. Gong, and G.T. Tsao, *Biotechnol. Lett.*, 1996, *18*, 1441.
16. B.H. Dawson, J.W. Barton, and G.R. Petersen, *Biotechnol. Prog.*, 1997, *13*, 512.
17. Anon., *Chem. Eng. News*, May 4, 1998, 18.
18. H. Uhlig, *Industrial Enzymes and their Applications*, Wiley, New York, 1998.
19. A. Thayer, *Chem. Eng. News*, Apr. 25, 1994, 9.
20. (a) J.H. van Ee, O. Misset, and J. Baas, *Enzymes in Detergency*, Dekker, New York, 1997; (b) K.-E. Jaeger and M.T. Reetz, *Trends Biotechnol.*, 1998, *16*, 396.
21. (a) D. Cowan, *Trends Biotechnol.*, 1996, *14*(6), 177; (b) J. Hodgson, *Biotechnology*, 1994, *12*, 789.
22. Y.-K. Lee, K. Nomoto, S. Salminen, and S.L. Gorbach, *Handbook for Probiotics*, Wiley, New York, 1999.
23. (a) M.-C. Lai and R.W. Traxler, *Enzyme Microb. Technol.*, 1994, *16*, 1021; (b) D.R. Woods, *Trends Biotechnol.*, 1995, *13*, 259; (c) L. Girbal and P. Soucaille, *Trends Biotechnol.*, 1998, *16*, 11.
24. K. Weissermel and H.-J. Arpe, *Industrial Organic Chemistry*, 2nd ed., VCH, Weinheim, 1993, 167.
25. (a) G.S. Sayler, J. Sanseverino, and K.L. Davis, eds, *Biotechnology in the Sustainable Environment*, Plenum, New York, 1997; (b) Anon., *Biotechnology for Clean Industrial Products and Processes—Towards Industrial Sustainability*, Paris, 1998; (c) B. Zechendorf, *Trends Biotechnol.*, 1999, *17*, 219; (d) P. Lorenz and H. Zinke, *Trends Biotechnol.*, 2005, *23*(12), 570.
26. (a) J.A.M. van Balken, *Biotechnological Innovations in Chemical Synthesis*, Butterworth–Heinemann, Oxford, 1997; (b) M. McCoy, *Chem. Eng. News*, June 22, 1998, 13; (c) A. Shanley, *Chem. Eng.*, 1998, *105*(7), 63; (d) R. D'Aquino, *Chem. Eng.*, 1999, *106*(3), 37; (d) H. Danner and R. Braun, *Chem. Soc. Rev.*, 1999, *28*, 395.
27. (a) B. Kamm, P.R. Gruber, and M. Kamm, *Biorefineries—Industrial Processes and Products—Status Quo and Future Directions*, Wiley, New York, 2006; (b) W. Aehle, ed., *Enzymes in Industry—Production and Applications*, 3rd ed., Wiley VCH, Weinheim, 2004; (c) H. Bisswanger, *Practical Enzymology*, Wiley-VCH, Weinheim, 2004; (d) C.T. Hou, ed., *Handbook of Industrial Biocatalysis*, CRC Press, Boca Raton, FL, 2005; (e) B.C. Saha and D.C. Demirjian, *Applied Biocatalysis in Specialty Chemicals and Pharmaceuticals*, A.C.S. Symp. 776, Oxford University Press, Oxford, 2000; (f) A.L. Demain, *Trends Biotechnol.*, 2000, *18*, 26; (g) R.N. Patel, ed., *Biocatalysis in the Pharmaceutical and Biotechnological Industries*, Taylor and Francis, Boca Raton, FL, 2006; (h) A. Liese, K. Seelbach, and C. Wandrey, eds, *Industrial Biotransformations*, Wiley VCH, Weinheim, 2006; (i) G. Festel, J. Knoll, and H. Gotz, *Chem. Ind.* (*Lond.*), April 5, 2004, 21; (j) J. Tao, G-Q.L. Lin, and A. Liese, *Biocatalysis for Pharmaceutical Industry*, Wiley, Hoboken, NJ, 2008; (k) Anon., *Chem. Eng. Prog.*, 2005, *101*(10), 34–55; (l) S.K. Ritter, *Chem. Eng. News*, Apr. 3, 2006, 69.
28. (a) R.P. Chauchan and J.M. Woodley, *Chemtech*, 1997, *27*(6), 26; (b) S.M. Roberts, *J. Chem. Soc. Perkin Trans. I*, 2000, 611; (c) N.C. Price and L. Stevens, *Fundamentals of Enzymology*, 3rd ed., Oxford University Press, Oxford, 2000; (d) B.C. Saha, *Fermentation Biotechnology*, A.C.S. Symp. 862, Oxford University Press, Oxford, 2003; (e) D. Ringe and G.A. Petsko, *Science*, 2008, *320*, 1428.
29. C.E. Wyman, ed., *Handbook on Biotechnol: Production and Utilization*, Taylor & Francis, Washington, DC, 1996.
30. P. Xu, A.T. Thomas, and C.D. Gilson, *Biotechnol. Lett.*, 1996, *18*, 1439.
31. (a) L.I. Ezeogu and A.C. Emeruwa, *Biotechnol. Lett.*, 1993, *15*, 83; (b) P.B. Moracs, *Biotechnol. Lett.*, 1996, *18*, 1351.
32. G. Ondrey, *Chem. Eng.*, 1997, *104*(11), 23.
33. (a) M. Ergun, S.F. Mutlu, and O. Gorel, *J. Chem. Technol. Biotechnol.*, 1997, *68*, 147; (b) H. Siva Raman, A. Chandwadkar, S.A. Baliga, and A.A. Prabhune, *Enzyme Microb. Technol.*, 1994, *16*, 719.
34. (a) H. Zollner, *Handbook of Enzyme Inhibitors*, 3rd ed., Wiley-VCH, Weinheim, 1999; (b) C. Martin, M. Galbe, C.E. Wahlbom, B. Hahn-Hagerdahl, and L.J. Jonsson, *Enzyme Microb. Technol.*, 2002, *31*, 274; (c) I. deBari, E. Viola, D. Barisano, M. Cardinale, F. Nanna, F. Zimbardi, G. Cardinale, and G. Braccio, *Ind. Eng. Chem. Res.*, 2002, *41*, 1745; (d) H. Miyafuji, H. Danner, M. Neureiter, C. Thomassen, J. Bvochora, O. Szolar, and R. Braun, *Enzyme Microb. Technol.*, 2003, *32*, 396; (e) C. Martin and L.J. Jonsson, *Enzyme Microb. Technol.*, 2003, *32*, 386; (f) A. Martinez, M.E. Rodriguez, M.L. Wells, S.W. York, J.F. Preston, and L.O. Ingram, *Biotechnol. Prog.*, 2001, *17*, 287; (g) J.R. Weil, B. Dien, R. Bothast, R. Hendrickson, N.S. Mosier, and M.R. Ladisch, *Ind. Eng. Chem. Res.*, 2002, *41*, 6132.
35. (a) T. Ikegami, Y. Yamada, and H. Ando, *Biotechnol. Lett.*, 1998, *20*, 673; (b) M. Soupioni, E. Polichroniadou, M. Tokatlidou, M. Kanellaki, and A.A. Koutinas, *Biotechnol. Lett.*, 1998, *20*, 495.
36. M. Gyamerah and J. Glover, *J. Chem. Technol. Biotechnol.*, 1996, *66*, 145.
37. Anon., *Chem. Eng. News*, Apr. 16, 2001, 14.
38. A.C. Oliveira, M.F. Rosa, J.M.S. Cabral, and M.R. Aires-Barros, *J. Mol. Catal. B Enzymol.*, 1998, *5*, 29.
39. M. Kobayashi, K. Ishida, and K. Shimizu, *J. Chem. Technol. Biotechnol.*, 1995, *63*, 141.
40. F. Taylor, M.J. Kurantz, N. Goldberg, and J.C. Craig, Jr., *Biotechnol. Prog.*, 1995, *11*, 693; *Biotechnol. Lett.*, 1998, *20*, 67.
41. C.-H. Park and Q. Geng, *Sep. Purif. Methods*, 1992, *21*(2), 127.
42. G. Philippidis, T.K. Smith, and C.E. Wyman, *Biotechnol. Bioeng.*, 1993, *41*, 846.
43. O.V.S. Reddy and S.C. Basappa, *Biotechnol. Lett.*, 1996, *18*, 1315.
44. K.-Y. Lee and S.-T. Lee, *J. Chem. Technol. Biotechnol.*, 1996, *66*, 349.
45. (a) M.E. Himmel, J.O. Baker, and R.P. Overend, eds, *Enzymatic Conversion of Biomass for Fuels Production*, A.C.S. Symp. 566, Washington, DC, 1994; (b) J.N. Sadler and M.H. Penner, eds, *Enzymatic Degradation of Insoluble Carbohydrates*, A.C.S. Symp. 618, Washington, DC, 1995; (c) L. Olsson, B. Hahn-Hagerdal, *Enzyme Microb. Technol.*, 1996, *18*, 312; (d) M. McCoy, *Chem. Eng. News*, Dec. 7, 1998, 29.

46. (a) V. Meyrial, J.P. Delgenes, C. Romieu, R. Moletta, and A.M. Gounot, *Enzyme Microb. Technol.*, 1995, *17*, 535; (b) R. Eklund and G. Zacchi, *Enzyme Microb. Technol.*, 1995, *17*, 255.
47. H. Schneider, *Enzyme Microb. Technol.*, 1996, *19*, 94.
48. L. Preziosi-Belloy, V. Nolleau, and J.M. Navarro, *Enzyme Microb. Technol.*, 1997, *21*, 124.
49. T. Lebeau, T. Jouenne, and G.-A. Junter, *Enzyme Microb. Technol.*, 1997, *21*, 265.
50. S. Sanchez, V. Braup, E. Castro, A.J. Moya, and F. Camacho, *Enzyme Microb. Technol.*, 1997, *21*, 355.
51. (a) J.M. Dominguez, *Biotechnol. Lett.*, 1998, *20*, 53; (b) J.C. Parajo, H. Dominguez, and J.M. Dominguez, *Enzyme Microb. Technol.*, 1997, *21*, 18.
52. T.A. Brooks and L.O. Ingram, *Biotechnol. Prog.*, 1995, *11*, 619.
53. G. Parkinson, *Chem. Eng.*, 1998, *105*(3), 21.
54. G. Ondrey, *Chem. Eng.*, 2006, *113*(4), 27.
55. B.E. Wood, H.C. Aldrich, and L.O. Ingram, *Biotechnol. Prog.*, 1997, *13*, 232.
56. Y.V. Wu, et al., *J. Am. Oil Chem. Soc.*, 1994, *71*, 1041.
57. D.D. Young, J. Nichols, R.M. Kelly, and A. Dieters, *J. Am. Chem. Soc.*, 2008, *130*, 10048.
58. M.A. daMatta, J.B.F. Munez, A. Schuler, and M. daMotta, *Biotechnol. Prog.*, 2004, *20*, 393.
59. *Environ. Sci. Technol.*, 1994, *28*, 461A.
60. M. vonSivers, G. Zacchi, L. Olsson, and B. Hahn-Hugerdal, *Biotechnol. Prog.*, 1994, *10*, 555.
61. (a) A.S. Gordon, *Chem. Eng. News*, Oct. 17, 1994, 2; (b) S. Lauren, *Chem. Eng. News*, Jan. 23, 1995, 5; (c) G. Peaff, *Chem. Eng. News*, July 11, 1994, 4; (d) D. Pimentel, *Environ. Prog.*, 2001, *20*(3), O3; (e) D. Pimentel, *Natural Resoures Res.*, 2005, *14*, 65–76; (f) G. Hess, *Chem. Eng. News*, Sept. 12, 2005, 28.
62. (a) G.F. Bickerstaff, *Immobilization of Enzymes and Cells*, Humana Press, Totowa, NJ, 1996; (b) W. Tischer and V. Kasche, *Trends Biotechnol.*, 1999, *17*, 326; (c) U.T. Bornscheuer, *Angew. Chem. Int. Ed.*, 2003, *42*, 3336.
63. (a) Anon., *Chem. Eng. News*, June 3, 1996, 11; (b) N.L. St Clair and M.A. Navia, *J. Am. Chem. Soc.*, 1992, *114*, 7314; (c) A.L. Margolin, *Trends Biotechnol.*, 1996, *14*(7), 223; (d) J.J. Lalonde, *Chemtech*, 1997, *27*(2), 38; (e) Anon., *Chem. Eng. News*, July 11, 1994, 35; (f) R.A. Persichetti, N.L. St Clair, M.A. Navia, and A.L. Margolin, *J. Am. Chem. Soc.*, 1995, *117*, 2732; (g) N. Khalaf, C.P. Govardhan, J.J. Lalonde, R.A. Persichetti, Y.-F. Wang, and A.L. Margolin, *J. Am. Chem. Soc.*, 1996, *118*, 5494; (h) J.J. Lalonde, C.P. Govardhan, N. Khalaf, A.G. Martinez, K. Visuri, and A.L. Margolin, *J. Am. Chem. Soc.*, 1995, *117*, 6845; (i) D. Hairston, *Chem. Eng.*, 1995, *102*(5), 53; (j) L. Walters, *Chem. Ind. (Lond.)*, 1997, 412; (k) T. Zelinski and H. Waldmann, *Angew. Chem. Int. Ed.*, 1997, *36*, 722; (l) J.J. Roy and T.E. Abraham, *Chem. Rev.*, 2004, *104*, 3705; (m) A.L. Margolin and M.A. Navia, *Angew. Chem. Int. Ed.*, 2001, *40*, 2205.
64. L.Z. Vilenchik, J.P. Griffith, N.L. St Clair, M.A. Navia, and A.L. Margolin, *J. Am. Chem. Soc.*, 1998, *120*, 4290.
65. R.A. Sheldon, *Org. Lett.*, 2000, *2*, 1361.
66. A.M. O'Reilly and J.A. Scott, *Enzyme Microb. Technol.*, 1995, *17*, 636.
67. G.S. Samdani, *Chem. Eng.*, 1994, *101*(12), 23.
68. Y.M. Elcin, *Enzyme Microb. Technol.*, 1995, *17*, 587.
69. G.S. Samdani, *Chem. Eng.*, 1995, *102*(10), 17.
70. O. Heichal-Segal, S. Rappoport, and S. Braun, *Biotechnology*, 1995, *13*, 798.
71. B. Krajewska, *Enzyme Microb. Technol.*, 2004, *35*, 126.
72. (a) E. Kokufuta, *Prog. Polym. Sci.*, 1992, *17*, 647; (b) V.M. Balcao, A.L. Paiva, and F.X. Malcata, *Enzyme Microb. Technol.*, 1996, *18*, 392.
73. F. Svec, J.M.J. Frechet, *Science*, 1996, *273*, 205.
74. S. Borman, *Chem. Eng. News*, May 30, 1994, 17.
75. (a) P. Wang, M.V. Sergeeva, L. Lim, and J.S. Dordick, *Nat. Biotechnol.*, 1997, *15*, 789; (b) J.S. Dordick, S.J. Novick, and M.V. Sergeeva, *Chem. Ind. (Lond.)*, 1998, 17; (c) R. Peters and R. Sikorski, *Science*, 1997, *277*, 1849; (d) R. Rawls, *Chem. Eng. News*, Apr. 28, 1997, 27.
76. L. Ember, *Chem. Eng. News*, Sept. 15, 1997, 26.
77. M. Flickinger, J.L. Schottel, D.R. Bond, A. Aksan, and L.E. Scriven, *Biotechnol. Prog.*, 2007, *23*, 2.
78. (a) O. Kreft, M. Prevot, H. Mohwald, and G.B. Sukhorukov, *Angew. Chem. Int. Ed.*, 2007, *46*, 5605; (b) Y. Wang and F. Caruso, *Chem. Commun.*, 2004, 1528; (c) C. Gao, X. Liu, J. Shen, and H. Mohwald, *Chem. Commun.*, 2002, 1928.
79. (a) M.T. Reetz, A. Zonta, J. Simpelkamp, and W. Konen, *Chem. Commun.*, 1996, 1397; (b) M.T. Reetz, A. Zonta, V. Vijayakrishnan, and K. Schimossek, *J. Mol. Catal. A Chem.*, 1998, *134*, 251.
80. (a) I. Gill and A. Ballesteros, *J. Am. Chem. Soc.*, 1998, *120*, 8587; (b) A. Heller, Q. Chen, and L. Kenausis, *J. Am. Chem. Soc.*, 1998, *120*, 4586; (c) T.K. Jain, I. Roy, T.K. De, and A. Maitra, *J. Am. Chem. Soc.*, 1998, *120*, 11092; (d) I. Gill, E. Pastor, and A. Ballesteros, *J. Am. Chem. Soc.*, 1999, *121*, 9487; (e) M.T. Reetz, R. Wenkel, and D. Avnir, *Synthesis*, 2000, 781; (f) I. Gill and A. Ballesteros, *Trends Biotechnol.*, 2000, *18*, 469; (g) S. Hudson, J. Cooney, and E. Magner, *Angew. Chem. Int. Ed.*, 2008, *47*, 8582; (h) W. Fu, A. Yamaguchi, H. Kaneda, and N. Teramae, *Chem. Commun.*, 2008, 853; (i) R.R. Naik, M.M. Tomczak, H.R. Luckarift, J.C. Spain, and M.O. Stone, *Chem. Commun.*, 2004, 1684.
81. J.M. Moreno and J.V. Sinisterra, *J. Mol. Catal.*, 1994, *93*, 357; *J. Mol. Catal. A Chem.*, 1995, *98*, 171.
82. C. Bruning and J. Grobe, *J. Chem. Soc. Chem. Commun.*, 1995, 2323.
83. (a) I. Gill, E. Pastor, and A. Ballesteros, *J. Am. Chem. Soc.*, 1999, *121*, 9487; (b) A. Ragheb, M.A. Brook, and M. Hrynyk, *Chem. Commun.*, 2003, 2314.
84. (a) R. Fernandez–Lafuente, D.A. Cowan, and A.N.P. Wood, *Enzyme Microb. Technol.*, 1995, *17*, 366; (b) R. Fernandez-Lafuente, C.M. Rossell, V. Rodriguez, and J.M. Guisan, *Enzyme Microb. Technol.*, 1995, *17*, 517.
85. J. Pedroche, M.M. Yust, J. Giron-Guisan, and F. Millan, *Enzyme Microb. Technol.*, 2002, *31*, 711.
86. A.P.V. Goncalves, J.M. Lopes, F. Lemos, B. Enzyo, F.R. Ribeiro, D.M.F. Prazeres, J.M.S. Cabral, and M.R. Aires-Barros, *J. Mol. Catal. B Enzymol.*, 1996, *1*, 53.
87. J.F. Diaz and K.J. Balkus, Jr., *J. Mol. Catal. B Enzymol.*, 1997, *2*, 115.
88. J.-P. Chen and M.-S. Hsu, *J. Mol. Catal. B Enzymol.*, 1997, *2*, 233.
89. B. Kuhnel, L.A. Holbrook, M.M. Moloney, and G.J.H. van Rooijen, *J. Am. Oil. Chem. Soc.*, 1996, *73*, 1533.

90. X.Z. Li, J.S. Webb, S. Kjelleberg, and B. Rosche, *Appl. Environ. Microbiol.*, 2006, *72*, 5678.
91. T. Nagasawa and H. Yamada, *Pure Appl. Chem.*, 1995, 67, 1241.
92. (a) N. Roche, P.L. Berna, C. Desgranges, and A. Durand, *Enzyme Microb. Technol.*, 1995, *17*, 935; (b) B. Dorta, R.J. Ertola, and J. Arcas, *Enzyme Microb. Technol.*, 1996, *19*, 434; (c) M. Bercerra, M.I.G. Siso, *Enzyme Microb. Technol.*, 1996, *19*, 39; (d) L. Ikasari and D.A. Mitchell, *Enzyme Microb. Technol.*, 1996, *19*, 171; (e) A. Archana and T. Satyanarayana, *Enzyme Microb. Technol.*, 1997, *21*, 12; (f) S.A. Shaikh, J.M. Khire, and M. Khan, *J. Ind. Microb. Biotechnol.*, 1997, *19*, 239; (g) N.H.A. El-Nasser, S.M. Helmy, and A.A. EI-Gammal, *Polym. Degrad. Stabil.*, 1997, *55*, 249; (h) E.M. Silva and S.-T. Yang, *Biotechnol. Prog.*, 1998, *14*, 580; (i) E. Rosales, S.R. Couto, and A. Sanroman, *Biotechnol. Lett.*, 2002, *24*, 701; (j) P. Vats and U.C. Banerjee, *Enzyme Microb. Technol.*, 2004, *35*, 3.
93. J. Szendefy, G. Szakacs, and L. Christopher, *Enzyme Microb. Technol.*, 2006, *39*, 1354.
94. S.S. Saykhedkar and R.S. Singhal, *Biotechnol. Prog.*, 2004, *20*, 1280.
95. (a) L.P. Ooijkaas, F.J. Weber, R.M. Buitelaar, J. Tramper, and A. Rinzema, *Trends Biotechnol.*, 2000, *18*, 356; (b) N. Laurent, R. Haddoub, and S.L. Flitsch, *Trends Biotechnol.*, 2008, *26*, 328.
96. (a) G. Peiji, Q. Yinbo, Z. Xin, Z. Mingtian, and D. Yongcheng, *Enzyme Microb. Technol.*, 1997, *20*, 581; (b) M. Gutierrez-Correa and R.P. Tengerdy, *Biotechnol. Lett.*, 1998, *20*, 45; (c) J.P. Smits, A. Rinzema, J. Tramper, H.M. van Sonsbeek, J.C. Hage, A. Kaynak, and W. Knol, *Enzyme Microb. Technol.*, 1998, *22*, 50; (d) Z. Zhang and K. Shetty, *J. Agric. Food Chem.*, 1998, *46*, 786; (e) J. Pintado, A. Torrado, M.P. Gonzales, and M.A. Murado, *Enzyme Microb. Technol.*, 1998, *23*, 149; (f) M. Lu, J.D. Brooks, and I.S. Maddox, *Enzyme Microb. Technol.*, 1997, *21*, 392; (g) J. Cordova, M. Nemmaoui, M. Ismaili–Alaoui, A. Morin, S. Roussos, M. Raimbault, and B. Benjilali, *J. Mol. Catal. B Enzymol.*, 1998, *5*, 75.
97. U.T. Bornscheuer and T. Yamane, *Enzyme Microb. Technol.*, 1994, *16*, 864.
98. (a) Z. Knezevic, L. Mojovic, and B. Adnadjevic, *Enzyme Microb. Technol.*, 1998, *22*, 275; (b) M. Tuter, F. Arat, L. Dandik, and H.A. Aksog, *Biotechnol. Lett.*, 1998, *20*, 291; (c) N. Goma-Doncescu and M.D. Legoy, *J. Am. Oil Chem. Soc.*, 1997, *74*, 1137; (d) B. Selmi and D. Thomas, *J. Am. Oil Chem. Soc.*, 1998, *75*, 691; (e) J.A. Arcos, C. Otero, and C.G. Hill, Jr., *Biotechnol. Lett.*, 1998, *20*, 617; (f) A.M. Fureby, P. Adlercreutz, and B. Mattiasson, *J. Am. Oil Chem. Soc.*, 1996, *73*, 1489.
99. (a) X. Jorba, I. Gill, and E.N. Vulfson, *J. Agric. Food Chem.*, 1995, *43*, 2536; (b) I. Gill and E. Vulfson, *Trends Biotechnol.*, 1994, *12*(4), 118; (c) M. Erbeldinger, X. Ni, and P.J. Halling, *Enzyme Microb. Technol.*, 1998, *23*, 141.
100. (a) P.J. Halling, *Enzyme Microb. Technol.*, 1995, *17*, 601; (b) R.V. Ulijn and P.J. Halling, *Green Chem.*, 2004, *6*, 488.
101. (a) A.J. Russell, F.X. Yang, *Chemtech*, 1996, *26*(10), 24; (b) N. Hidaka and T. Matsumoto, *Ind. Eng. Chem. Res.*, 2000, *39*, 909.
102. N. Hidaka and T. Matsumoto, *Ind. Eng. Chem. Res.*, 2000, *39*, 909.
103. S. Lamare, M.-D. Legoy, and M. Graber, *Green Chem.*, 2004, *6*, 445–458.
104. (a) G. Bell, P.J. Halling, B.D. Moore, J. Partridge, and P.G. Rees, *Trends Biotechnol.*, 1995, *13*, 468; (b) L. Kvittingen, *Tetrahedron*, 1994, *50*, 8253; (c) M.H. Vermue and J. Tramper, *Pure Appl. Chem.*, 1995, *67*(2), 346; (d) A.M. Klibanov, *Trends Biotechnol.*, 1997, *15*, 97; (e) R. Leon, P. Fernandes, H.M. Pincheiro, and J.M.S. Calral, *Enzyme Microb. Technol.*, 1998, *23*, 483; (f) E.N. Vulfson, P.J. Halling, and H.L. Holland, eds, *Enzymes in Nonaqueous Solvents: Methods and Protocols*, Humana Press, Totowa, NJ, 2001; (g) A.L. Serdakowski and J.S. Dordick, *Trends Biotechnol.*, 2008, *26*(1), 48.
105. G.A. Hutcheon, M.C. Parker, A. James, and B.D. Moore, *Chem. Commun.*, 1997, 931.
106. K. Xu and A.M. Klibanov, *J. Am. Chem. Soc.*, 1996, *118*, 9815.
107. E. Zacharis, B.D. Moore, and P.J. Halling, *J. Am. Chem. Soc.*, 1997, *119*, 12396.
108. J. Patridge, P.J. Halling, and B.D. Moore, *Chem. Commun.*, 1998, 841.
109. M.-C. Parker, S.A. Brown, L. Robertson, and N.J. Turner, *Chem. Commun.*, 1998, 2247.
110. H.R. Hobbs and N.R. Thomas, *Chem. Rev.*, 2786.
111. (a) F. vanRantwijk and R.A. Sheldon, *Chem. Rev.*, 2007, *107*, 2757; (b) R.A. Sheldon, R.M. Lau, M.J. Sorgedrager, F. van-Rantwijk, and K.R. Seddon, *Green Chem.*, 2002, *4*, 147; (c) S. Cantone, U. Hanefeld, and A. Basso, *Green Chem.*, 2007, *9*, 954.
112. (a) Y. Yamada, R. Kuboi, and I. Komasawa, *Biotechnol. Prog.*, 1995, *11*, 682; (b) A.N. Rajagopal, *Enzyme Microb. Technol.*, 1996, *19*, 606.
113. B. Orlich and R. Schomacker, *Enzyme Microb. Technol.*, 2001, *28*, 42.
114. (a) G.M. Zijlstra, C.D. de Gooijer, L.A. van der Pol, and J. Tramper, *Enzyme Microb. Technol.*, 1996, *19*, 2; (b) M.G. Wubbolts, J. Hoven, B. Melgert, and B. Witholt, *Enzyme Microb. Technol.*, 1994, *16*, 887; (c) J.A.M. de Bont, *Trends Biotechnol.*, 1998, *16*, 493.
115. J.-P. Chen and M.-S. Lee, *Enzyme Microb. Technol.*, 1995, *17*, 1021.
116. R. Kuboi, H. Umakoshi, and I. Komasawa, *Biotechnol. Prog.*, 1995, *11*, 202.
117. M.G. Antov and D.M. Pericin, *Enzyme Microb. Technol.*, 2001, *28*, 467.
118. (a) Y. Inada, M. Furukawa, H. Sasaki, Y. Kodera, M. Hiroto, H. Nishimura, and A. Matsushima, *Trends Biotechnol.*, 1995, *13*, 86; (b) A. Matsushima, Y. Kodera, M. Hiroto, H. Nishimura, and Y. Inada, *J. Mol. Catal. B Enzymol.*, 1996, *2*, 1; (c) Z. Yang, M. Domach, R. Auger, F.X. Yang, and A.J. Russell, *Enzyme Microb. Technol.*, 1996, *18*, 82; (d) Z. Yang, A.J. Mesiano, S. Venkatasubramanian, S.H. Gross, J.M. Harris, and A.J. Russell, *J. Am. Chem. Soc.*, 1995, *117*, 4843; (e) H. Lee and T.G. Park, *Biotechnology*, 1998, *14*, 508; (f) Y. Kodera, K. Sakurai, Y. Satoh, T. Uemura, Y. Keneda, H. Nishimura, M. Hiroto, A. Matsushima, and Y. Inada, *Biotechnol. Lett.*, 1998, *20*, 177.
119. E. Miland, M.R. Symth, and C.O. Fagain, *Enzyme Microb. Technol.*, 1996, *19*, 63.

120. V.M. Suzawa, Y.L. Khmelnitsky, L. Giarto, J.S. Dordick, and D.S. Clark, *J. Am. Chem. Soc.*, 1995, *117*, 8435.
121. G. Carrea, G. Ottolina, and S. Riva, *Trends Biotechnol.*, 1995, *13*, 63.
122. I.J. Colton, S.N. Ahmed, and R.J. Kazlauskas, *J. Org. Chem.* 1995, *60*, 212.
123. A.D. Blackwood and C. Bucke, *Enzyme Microb. Technol.*, 2000, *27*, 704.
124. (a) S.Y. Huang, H.L. Chang, and M. Goto, *Enzyme Microb. Technol.*, 1998, *22*, 552; (b) S. Basher, M. Nakajima, and U. Cogan, *J. Am. Oil Chem. Soc.*, 1996, *73*, 1475; (c) A. Fishman, S. Basheer, S. Shatzmiller, and U. Cogan, *Biotechnol. Lett.*, 1998, *20*, 535.
125. V.M. Paradkar and J.S. Dordick, *J. Am. Chem. Soc.*, 1994, *116*, 5009.
126. (a) Y. Okahata and T. Mori, *Trends Biotechnol.*, 1997, *15*, 50; *J. Mol. Catal. B Enzymol.*, 1998, *5*, 119; (b) T. Mori and Y. Okahata, *Chem. Commun.*, 1998, 2215.
127. T. Mori, A. Kobayashi, and Y. Okahata, *Chem. Lett.*, 1998, 921.
128. Q. Jene, J.C. Pearson, and C.R. Lowe, *Enzyme Microb. Technol.*, 1997, *20*, 69.
129. (a) J. Broos, I.K. Sakodinskaya, J.F.J. Engbersen, W. Verboom, and D.N. Reinhoudt, *J. Chem. Soc. Chem. Commun.*, 1995, 255; (b) J. Broos, J.F.J. Engbersen, I.K. Sakodinskaya, W. Verboom, and D.N. Reinhoudt, *J. Chem. Soc. Perkin Trans.*, 1995, 2899; (c) J.F.J. Engbersen, J. Broos, W. Verboom, and D.N. Reinhoudt, *Pure Appl. Chem.*, 1996, *68*, 2171.
130. (a) J.C. Moore and F.H. Arnold, *Nat. Biotechnol.*, 1996, *14*, 458; (b) E.K. Wilson, *Chem. Eng. News*, Oct. 16, 1995, 22; (c) R.L. Rawls, *Chem. Eng. News*, Apr. 12, 1999, 38; (d) C. Schmidt-Dannert and F.H. Arnold, *Trends Biotechnol.*, 1999, *17*, 135; (e) B.L. Iverson and R.R. Breaker, *Trends Biotechnol.*, 1998, *16*, 52; (f) U.T. Bornscheuer, *Angew. Chem. Int. Ed.*, 1998, *37*, 3105; (g) J. Affholer and F.H. Arnold, *Chemtech*, 1999, *29*(9), 34; (h) J. Sutherland, *Chem. Ind.* (*Lond.*), 1999, 745.
131. P. Brown, J. Shanklin, E. Whittle, and C. Somerville, *Science*, 1998, *282*, 1315.
132. (a) S. Brakmann and K. Johnsson, ed., *Directed Molecular Evolution of Proteins*, Wiley-VCH, Weinheimm 2002; (b) S. Brakmann and A. Schwienhorst, *Evolutionary Methods in Biotechnology—CleverTricks for Directed Evolution*, Wiley-VCH. Weinheim, 2004; (c) A. Svendsen, ed., *Enzyme Functionality—Design, Engineering and Screening*, Marcel Dekker, New York, 2003; (d) R. Chatterjee and L. Yuan, *Trends Biotechnol.*, 2006, *24*(1), 28.
133. (a) M.T. Reetz, M. Bocola, J.D. Carballeira, D. Zha, and A. Vogel, *Angew. Chem. Int. Ed.*, 2005, *44*, 4192; (b) M.T. Reetz, L.-W. Wang, and M. Bocola, *Angew. Chem. Int. Ed.*, 2006, *45*, 1236; (c) M.T. Reetz, J.D. Carbelleira, and A. Vogel, *Angew. Chem. Int. Ed.*, 2006, *45*, 7745.
134. (a) W. Stemmer and B. Holland, *Am. Sci.*, 2003, *91*, 526; (b) K.A. Powell, S.W. Ramer, S.B. delCardayre, W.P.C. Stemmer, M.B. Tobin, P.F. Longhamp, and G.W. Huisman, *Angew. Chem. Int. Ed.*, 2001, *40*, 3948.
135. T.C. Galvao, W.W. Mohn, and V. deLorenzo, *Trends Biotehnol.*, 2005, *23*(10), 497.
136. K. Hult and P. Berglund, *Trends Biotechnol.*, 2007, *25*, 231.
137. Y. Ijima, K. Matoishi, Y. Terao, N. Doi, H. Yanagawa, and H. Ohta, *Chem. Commun.*, 2005, 877.
138. F.P. Seebeck and D. Hilvert, *J. Am. Chem Soc.*, 2003, *125*, 10158.
139. (a) M.W.W. Adams and R.M. Kelly, *Chem. Eng. News*, Dec. 18, 1995; (b) M.W.W. Adams and R.M. Kelly, *Biocatalysis at Extreme Temperatures*, A.C.S. Symp. 498, Washington, DC, 1992; (c) M.W.W. Adams, F.B. Perler, and R.M. Kelly, *Biotechnology*, 1995, *13*, 662; (d) J. Newell, *Chem. Br.*, 1995, *31*, 925; (e) R.M. Kelly, J.A. Baross, and M.W.W. Adams, *Chem. Br.*, 1994, *30*, 555; (f) E.C. deMacrio and A.J.L. Macario, *Trends Biotechnol.*, 1994, *12*, 512; (g) J. van der Oost, W.M. de Vos, and G. Antranikian, *Trends Biotechnol.*, 1996, *14*, 415; (h) C.P. Govardhan and A.L. Margolin, *Chem. Ind.* (*Lond.*), 1995, 689; (i) W.S. Fyfe, *Science*, 1996, *273*, 448; 1994, *265*, 471; (j) J. van der Oost, W.M. de Vos, and G. Antranikian, *Trends Biotechnol.*, 1996, *14*, 415; (k) M.C. Davis, *Trends Biotechnol.*, 1998, *16*, 102; (l) M. Gross, *Life on the Edge: Amazing Creatures Thriving in Extreme Environments*, Plenum, New York, 1998; (m) *Extremozymes and Commercially Important Extremophiles: The Next Wave of Industrial Manufacturing*, Wiley, New York, 1997; (n) D.A. Wharton, *Life at the Limits: Organisms in Extreme Environments*, Cambridge University Press, Cambridge, UK, 2002; (o) www.archaea.unsw.edu.au; (p) J. Laybown-Parry, *Science*, 2009, *324*, 1521; (q) A. Boetius and S. Joye, *Science*, 2009, *324*, 1523.
140. C.J. Marshall, *Trends Biotechnol.*, 1997, *15*, 359.
141. R.O. Bustos, C.R. Romo, and M.G. Healy, *J. Chem. Technol. Biotechnol.*, 1996, *65*, 193.
142. E. Pennisi, *Science*, 1997, *276*, 705.
143. R. Petkewich, *Environ. Sci. Technol.*, 2005, *39*, 478A.
144. C. Vieilli and J.G. Zeikus. *Trends Biotechnol.*, 1996, *14*(6), 183.
145. M.W.W. Adams and R.M. Kelly, *Trends Biotechnol.*, 1998, *16*, 329.
146. R.M. Daniel, *Enzyme Microb. Technol.*, 1996, *19*, 74.
147. (a) M.K. Chan, S. Muskund, A. Kletzin, M.W.W. Adams, and D.C. Rees, *Science*, 1995, *267*, 1463; (b) Anon., *Chem. Eng. News*, Mar 13, 1995, 33.
148. (a) B. van den Burg, G. Vriend, O.R. Veltman, G. Venema, and V.G.H. Eijsink, *Proc. Natl. Acad. Sci. USA*, 1998, *95*, 2056; (b) Anon., *Chem. Eng. News*, Mar. 9, 1998, 28.
149. C. Nicolini, *Trends Biotechnol.*, 1997, *15*, 395.
150. C. Holden, *Science*, 1996, *271*, 1061.
151. C. Holden, *Science*, 1997, *275*, 933.
152. C.-C. Chen, R. Adolphson, J.F.D. Dean, K.-E.L. Ericksson, M.W.W. Adams, J. Westpheling, *Enzyme Microb. Technol.*, 1997, *20*, 39.
153. Anon., *Chem. Ind.* (*Lond.*), 1996, 525.
154. Y. Morita, K. Kondoh, Q. Hasan T. Sakaguchi, Y. Murakami, K. Yokoyama, and E. Tamiya, *J. Am. Oil Chem. Soc.*, 1997, *74*, 1377.
155. Y. Morita, T. Nakamura, Q. Hasan, Y. Murakami, K. Yokoyama, and E. Tamiya, *J. Am. Oil Chem. Soc.*, 1997, *74*, 441.
156. A. Coghlan, *New Sci.*, 1998, Mar. 14, 24.
157. J.L. Arpigny, J. Lamotte, and C. Gerday, *J. Mol. Catal. B Enzymol.*, 1997, *3*, 29.
158. (a) P.A. Fields and G.N. Somero, *Proc. Natl. Acad. Sci. USA*, 1998, *95*, 11476; (b) Anon., *Chem. Eng. News*, Sept. 21, 1998, 56.
159. C. Kato, A. Inoue, and K. Horikoshi, *Trends Biotechnol.*, 1996, *14*, 6.

160. (a) T.O. Stevens and J.P. McKinley, *Science*, 1995, *270*, 450; (b) J. Kaiser, *Science*, 1995, *270*, 377; (c) E. Wilson, *Chem. Eng. News*, Oct 23, 1995, 8.
161. O. Dym, M. Mevarech, and J.L. Sussman, *Science*, 1995, *267*, 1344.
162. S.M. Edgington, *Biotechnology*, 1994, *12*, 1338.
163. (a) M.J. Daly, and K.W. Minton, *Science*, 1995, *270*, 1318; (b) J. Lin, R. Qi, C. Aston, J. Jing, T.S. Anantharaman, B. Mishra, O. White, M.J. Daly, K.W. Minton, J.C. Venter, and D.C. Schwartz, *Science*, 1999, *285*, 1558; (c) O.W. White, J.A. Eisen, J.F. Heidelberg, E.K. Hickey, et al., *Science*, 1999, *286*, 1571.
164. R. Roser and C. Colaco, *New Sci.*, 1993, May 15, 25.
165. M.A. Singer and S. Lindquist, *Trends Biotechnol.*, 1998, *16*, 460.
166. (a) Y. Yoshida, N. Nakamura, and K. Horikoshi, *Enzyme Microb. Technol.*, 1998, *22*, 71; (b) M. Kato, *J. Mol. Catal. B*, 1999, *6*, 223.
167. T. Higashiyana, *Pure Appl. Chem.*, *74*, 1263.
168. C. Schebor, L. Burin, M.P. Buera, J.M. Aquilera, and J. Chirife, *Biotechnology*, 1997, *13*, 857.
169. S. Rossi, M.P. Buera, S. Moreno, and J. Chirife, *Biotechnol. Prog.*, 1997, *13*, 609.
170. G. Parkinson, *Chem. Eng.*, 1999, *106*(3), 23.
171. G. Parkinson, *Chem. Eng.*, 1998, *105*(3), 21.
172. V. Athes and D. Combes, *Enzyme Microb. Technol.*, 1998, *22*, 532.
173. L. Fischer, R. Bromann, S.W.M. Kengen, W.M. de Vos, and F. Wagner, *Nat. Biotechnol.*, 1996, *14*, 88.
174. (a) R.M. Kelly, J.A. Baross, and M.W.W. Adams, *Chem. Br.*, 1994, *30*(7), 555; (b) C.C. Chester, *Environment*, 1996, *38*(8), 11.
175. (a) Anon., *Chem. Eng. News*, July 15, 1996, 66; (b) Thermo Gen. R & D Product Guide. 1996.
176. (a) M.B. Brennan, *Chem. Eng. News*, Oct 14, 1996, 31; (b) G. Parkinson, *Chem. Eng.*, 1996, *103*(3), 23; (c) S. Roberts and G. Ondrey, *Chem. Eng.*, 1996, *103*(8), 43; (d) Anon., *Chem. Eng. News*, Sept. 23, 1996: inside front cover. (e) Anon., *Chem. Eng. News*, Aug. 25, 1997, 13.
177. D.A. Comfort, S.R. Chhabra, S.B. Conners, C.-J. Chou, K.L. Epting, M.R. Johnson, K.L. Jones, A.C. Sehgal, and R.M. Kelly, *Green Chem.*, 2004, *6*, 459.
178. (a) J. Johnson, *Chem. Ind. (Lond.)*, 1995, 128; (b) P.G. Schultz and R.A. Lerner, *Science*, 1995, *269*, 1835; (c) G.M. Blackburn, A. Datta, and L.J. Patridge, *Pure Appl. Chem.*, 1996, *68*, 2009; (d) P. Kast and D. Hilvert. *Pure Appl. Chem.*, 1996, *68*, 2017; (e) M. Resmini, A.A.P. Meekel, and U.K. Pandit, *Pure Appl. Chem.*, 1996, *68*, 2025; (f) A.J. Kirby, *Angew. Chem. Int. Ed.*, 1996, *35*, 720; (g) E. Keinan, ed., *Catalytic Antibodies*, Wiley-VCH, Weinheim, 2004; (h) P.G. Schultz, J. Yin, and R.A. Lerner, *Angew. Chem. Int. Ed.*, 2002, *41*, 4427; (i) P. Wentworth, *Science*, 2002, *296*, 2247.
179. (a) J. Wagner, R.A. Lerner, and C.F. Barbas, III, *Science*, 1995, *270*, 1797; (b) S. Borman. *Chem. Eng. News*, Jan. 1, 1996, 25; (c) C.-H. Lin, T.Z. Hoffman, Y. Xie, P. Wirsching, and K.D. Janda, *Chem. Commun.*, 1998, 1075; (d) F.E. Romesberg, B. Spiller, P.G. Schultz, R.C. Stevens, *Science*, 1998, *279*, 1929; (e) T. Li, R.A. Lerner, and K.D. Janda, *Acc. Chem. Res.*, 1997, *30*, 115.
180. (a) C.F. Brands III, A. Heine, G. Zhong, T. Hoffmann, S. Gramatikova, R. Bjornestedt, B. List, J. Anderson, E.A. Stura, I.A. Wilson, and R.A. Lerner, *Science*, 1997, *278*, 2085; (b) G. Zhong, R.A. Lerner, and C.F. Barbas III, *Angew. Chem. Int. Ed.*, 1999, *38*, 3738.
181. R. Weinstain, R.A. Lerner, C.F. Barbas III, and D. Shabat, *J. Am. Chem. Soc.*, 2005, *127*, 13104.
182. (a) P.A. Patten, N.S. Gray, P.L. Yang, C.B. Marks, G.J. Wedemayer, J.J. Boniface, R.C. Stevens, and P.G. Schultz, *Science*, 1996, *271*, 1086; (b) J. Yu, S.Y. Choi, S. Lee, H.J. Yoon, S. Jeong, H. Mun, H. Park, and P.G. Schultz, *Chem. Commun.*, 1997, 1957.
183. C.-H. Lo, P. Wentworth, K.W. Jung, J. Yoon, J.A. Ashley, and K.D. Janda, *J. Am. Chem. Soc.*, 1997, *119*, 10251.
184. A. Persidis, *Nat. Biotechnol.*, 1997, *15*, 1313.
185. M.D. Smith, *Biotechnol. Adv.*, 1996, *14*, 267.
186. (a) T. Kometani, H. Yoshii, and R. Matsuno, *J. Mol. Catal. B Enzymol.*, 1996, *1*, 45; (b) R. Deveaux–Basseguy, A. Bergel, and M. Comtat, *Enzyme Microb. Technol.*, 1997, *20*, 248.
187. (a) K. Seelbach, B. Riebel, W. Hummel, M.-R. Kula, V.I. Tishkov, A.M. Egorov, C. Wandrey, and U. Kragl, *Tetrahedron Lett.*, 1996, *37*, 1377; (b) H.K. Kolbl, H. Hildebrand, N. Piel, T. Schroder, and W. Zitzmann, *Pure Appl. Chem.*, 1988, *60*, 825; (c) K. Seelbach and U. Kragl, *Enzyme Microb. Technol.*, 1997, *20*, 389.
188. (a) Q.-J. Chi and S.J. Dong, *J. Mol. Catal. B Enzymol.*, 1996, *1*, 193; (b) S. Kuwubata, K. Nishida, and H. Yoneyama, *Chem. Lett.*, 1994, 407; (c) S.B. Sobolov, M.D. Leonida, A. Bartoszko–Malik, K.I. Voivodov, F. McKinney, J. Kim, and A.J. Fry, *J. Org. Chem.*, 1996, *61*, 2125; (d) J. Cantet, A. Bergel, and M. Comtat, *Enzyme Microb. Technol.*, 1996, *18*, 72; (e) J.M. Obon, P. Casanova, A. Manjon, V.M. Fernandez, and J.L. Iborra, *Biotechnol. Prog.*, 1997, *13*, 557; (f) W.A.C. Somers, W. van Hartingsveld, E.C.A. Stigter, and J.P. van der Lugt, *Trends Biotechnol.*, 1997, *15*, 495. (g) M. Andersson, R. Otto, H. Holmberg, and P. Aldercreutz, *Biocatal Biotransform*, 1997, *15*, 281; (h) F. Hollmann, K. Hofstter, L.T. Habicher, B. Haurer, and A. Schmid, *J. Am. Chem Soc.*, 2005, *127*, 6540; (i) G. Ondrey, *Chem. Eng.*, 2005, *112*(9), 18; (j) L. Greiner, I. Schroeder, D.H. Muller, and A. Liese, *Green Chem.*, 2003, *5*, 697.
189. M. Schneider, M. El Alaoui, U. vonStockar, and I.W. Marison, *Enzyme Microb. Technol.*, 1997, *20*, 268.
190. S.J. Setford and P.E. Barker, *J. Chem. Technol. Biotechnol.*, 1994, *61*, 19.
191. D. Wang, M. Sakakibara, N. Kondoh, and K. Suzuki, *J. Chem. Technol. Biotechnol.*, 1996, *65*, 86.
192. M. Sakakibara, D. Wang, R. Takahashi, K. Takahashi, and S. Mori, *Enzyme Microb. Technol.*, 1996, *18*, 444.
193. (a) L. Giorno and E. Drioli, *Trends Biotechnol.*, 2000, *18*, 339; (b) G. Ondrey, *Chem. Eng.*, 2009, *115*(1), 14.
194. (a) S. Servi, ed., *Microbial Reagents in Organic Synthesis*, Kluwer Academic, Dordrecht, Netherlands, 1992; (b) G.A. Veldink, J.F.G. Vliegenthardt, eds, *Catal. Today*, 1994, *22*(3), 407–621; (c) J.R. Hanson, *An Introduction to Biotransformations in Organic Chemistry*, Oxford University Press, Oxford, 1995; (d) H. Danner and R. Braun, *Chem. Soc. Rev.*, 1999, *28*, 395; (e) M. Petersen and A. Kiener, *Green Chem.*, 1999, *1*, 99; (e) National Research Council, *Biobased Industrial Products—Priorities for Research and*

Commercialization, National Academy Press, Washington, DC, 2000; (f) A. Corma, S. Iborra, and A. Velty, *Chem. Rev.*, 2007, *107*, 2411; (g) R. Hatti-Kaul, U. Tornvall, L. Gustafsson, and P. Borjesson, *Trends Biotechnol.*, 2007, *25*, 119; (h) S.M. Roberts, ed., *Tetrahedron*, 2004, *60*(3), 483–806; (i) J. Tao, *Biocatalysis for Pharmaceutical Industry – Discover, Development and Manufacturing*, Wiley, New York, 2008; (j) D.J. Pollard and J.M. Woodley, *Trends Biotechnol.*, 2007, *25*, 66; (k) J.M. Woodley, *Trends Biotechnol.*, 2008, *26*, 321; (l) N. Ran, L. Zhao, Z. Chen, and J. Tao, *Green Chem.*, 2008, *10*, 361; (l) W.-D. Fessner, ed., *Top. Curr. Chem.*, 1999, *200*, 1–254; (m) R. Leon, P. Fernandes, H.M. Pinheiro, and J.M.S. Cabral, *Enzyme Microb. Technol.*, 1998, *23*, 483; (m) A. LIese, K. Seelbach, and L.C. Wandrey, *Industrial Biotransformations*, Wiley-VCH, Weinheim, 2000; (n) R.B. Silverman, *The Organic Chemistry of Enzyme-Catalyzed Reactions*, Academic Press, San Diego, 2000.

195. (a) H.L. Holland, *Organic Synthesis with Oxidative Enzymes*, VCH, Weinheim, 1992; (b) M.P.J. van Deurzen, F. van Rantwijk, and R.A. Sheldon, *Tetrahedron*, 1997, *53*, 13183; (c) S. Colonna, N. Gaggero, C. Richelmi, and P. Pasta, *Trends Biotechnol.*, 1999, *17*,163; (d) F. Hollman, K. Hofstetter, and A. Schmid, *Trends Biotechnol.*, 2006, *24*(4), 163.
196. (a) C.C.R. Allen, D.R. Boyd, H. Dalton, N.D. Sharma, I. Brannigan, N.A. Kerley, G.N. Sheldrake, and S.C. Taylor, *J. Chem. Soc. Chem. Commun.*, 1995, 117; (b) C.C.R. Allen, D.R. Boyd, H. Dalton, N.D. Sharma, S.A. Haughey, R.A.S. McMordie, B.T. McMurray, G.N. Sheldrake, and K. Sproule, *J. Chem. Soc. Chem. Commun.*, 1995, 119; (c) C.T. Marshall and J.M. Woodley, *Biotechnology*, 1995, *13*, 1072; (d) T. Hudlicky, D. Gonzalez, and D.T. Gibson, *Aldrichchimica Acta*, 1999, *32*(2), 35; (e) L.P. Wackett, *Enzyme Microb. Technol.*, 2002, *31*, 577.
197. (a) A. Kiener, *Chemtech*, 1995, *25*(9), 31; (b) M. Torimura, H. Yoshida, K. Kano, T. Ikeda, T. Nagasawa, and T. Ueda, *Chem. Lett.*, 1998, 295; (c) M. Petersen and A. Kiener, *Green Chem.*, 1999, *1*, 99; (d) M. Torimura, H. Yoshida, K. Kano, T. Ikeda, T. Yoshida, and T. Nagasawa, *J. Mol. Catal. B*, 2000, *8*, 265.
198. (a) M. Held, W. Suske, A. Schmid, K.-H. Engesser, H.-P.E. Kohler, B. Witholt, and M.C. Wubbolts, *J. Mol. Catal. B Enzymol.*, 1998, *5*, 87; (b) A. Schmid, H.-P.E. Kohler, K.-H. Engesser, *J. Mol. Catal. B Enzymol.*, 1998, *5*, 311.
199. (a) S. Riva, *Trends Biotechnol.*, 2006, *24*(5), 219; (b) S. Witayakran and A.J. Ragauskas, *Adv. Synth. Catal.*, 2009, *351*, 1187.
200. (a) A. Potthast, T. Rosenau, C.-L. Chen, and J.S. Gratzl, *J. Org. Chem.*, 1995, *60*, 4320; *J. Mol. Catal. A Chem.*, 1996, *108*, 5; (b) T. Rosenau, A. Potthast, C.-L. Chen, and J.S. Gratzl, *Synth Commun.*, 1996, *26*, 315.
201. E. Fritz-Landhals and B. Kunath, *Tetrahedron Lett.*, 1998, *39*, 5955.
202. J.B. van Beilen, J. Kingma, and B. Witholt, *Enzyme Microb. Technol.*, 1994, *16*, 905.
203. A. Gupta, R. Nagarajan, and A. Kilara, *Ind. Eng. Chem. Res.*, 1995, *34*, 2910.
204. D. Druaux, G. Mangeot, A. Endrizzi, and J.-M. Belin, *J. Chem. Technol. Biotechnol.*, 1997, *68*, 214.
205. D. Mandal, A. Ahmed, M.I. Khan, and R. Kumar, *J. Mol. Catal. A*, 2002, *181*, 237.
206. (a) A. Willetts, *Trends Biotechnol.*, 1997, *15*, 55; (b) R. Gagnon, G. Grogan, M.S. Levitt, S.M. Roberts, P.W.H. Wan, and A.J. Willetts, *J. Chem. Soc. Parkin Trans. 1*, 1994, 2537; (c) R. Gagnon, G. Grogan, S.M. Roberts, R. Villa, and A.J. Willetts, *J. Chem. Soc. Perkin Trans. 1*, 1995, 1505; (d) R. Gagnon, G. Grogan, E. Groussain, S. Pedragosa-Moreau, P.F. Richardson, S.M. Roberts, A.J. Wiletts, V. Alphand, J. Lebreton, and R. Furstoss, *J. Chem. Soc. Perkin Trans. 1*, 1995, 2527; (e) M.J. Lenn and C.J. Knowles, *Enzyme Microb. Technol.*, 1994, *16*, 964; (f) S.M. Roberts and P.W.H. Wan, *J. Mol. Catal. B Enzymol.*, 1998, *4*, 111; (g) J.D. Stewart, K.W. Reed, C.A. Martinez, J. Zhu, G. Chen, and M.M. Kayser, *J. Am. Chem. Soc.*, 1998, *120*, 3541.
207. (a) L.B. Davin, H.-B. Wang, A.L. Crowell, D.L. Bedgar, D.M. Martin, S. Sarkanen, and N.G. Lewis, *Science*, 1997, *275*, 362; (b) J. Kaiser, *Science*, 1997, *275*, 306.
208. (a) V.B. Urlachen and S. Eiben, *Trends Biotechnol.*, 2006, *24*(7), 324; (b) C.-H. Yun, K.-H. Kim, D.-Y. Kim, H.C. Jung, and J.-G. Pan, *Trends Biotechnol.*, 2007, *25*, 289.
209. M. Landwehn, L. Hochrein, C.R. Otey, A. Kasrayan, J.-E. Backvall, and F.H. Arnold, *J. Am. Chem. Soc.*, 2006, *128*, 6058.
210. O. Shoji, T. Fujishiro, H. Nakajima, M. Kim, S. Nagano, Y. Shiro, and Y. Watanabe, *Angew. Chem. Int. Ed.*, 2007, *46*, 3656.
211. W. Adam, M. Lazarus, C.R. Saha-Moller, and P. Schreir, *Acc. Chem. Res.*, 1999, *32*, 837.
212. C.T. Hou, *J. Am. Oil Chem. Soc.*, 2006, *83*, 677.
213. *Chem. Eng. News*, Apr. 1996, 29, 53.
214. M. North, *Tetrahedron Lett.*, 1996, 37, 1699.
215. O. Rotthaus, D. Kruger, M. Demuth, and K. Schaffner, *Tetrahedron*, 1997, *53*,935.
216. (a) K. Nakamura, Y. Inoue, and A. Ohno, *Tetrahedron Lett.*, 1995, *36*, 265; (b) K. Nakamura, K. Kitano, T. Matsuda, and A. Ohno, *Tetrahedron Lett.*, 1996, *37*, 1629.
217. G. Fantin, M. Fogagnolo, P.P. Giovannini, A. Medici, P. Pedrini, F. Gardini, and R. Lanciotti, *Tetrahedron*, 1996, 52, 3547.
218. H. Ikeda, E. Sato, T. Sugai, and H. Ohta, *Tetrahedron*, 1996, *52*, 8113, 8123.
219. W. Stampfer, B. Kosjek, C. Moitzi, W. Kroutil, and K. Faber, *Angew. Chem. Int. Ed.*, 2002, *41*, 1014.
220. M.M. Musa, K.I. Ziegelmann-Fjeld, C. Vieille, J.C. Zeckus, and R.S. Phillipe, *Angew. Chem. Int. Ed.*, 2007, *46*, 3091.
221. I. Goubet, T. Maugard, S. Lamare, and M.D. Legoy, *Enzyme Microb. Technol.*, 2002, *31*, 425.
222. W. Baik, D.I. Kim, S. Koo, J.U. Rhee, S.H. Shin, and B.H. Kim, *Tetrahedron Lett.*, 1997, *38*, 845.
223. G. Fronza, C. Fuganti, P. Grasselli, S. Servi, G. Zucchi, M. Barrens, and M. Villa, *J. Chem. Soc. Chem. Commun.*, 1995, 439.
224. (a) M.R. Gen Klaas and S. Warwel, *J. Am. Oil Chem. Soc.*, 1996, *73*, 1453; *J. Mol. Catal. A Chem.*, 1997, *117*, 311; *Synth. Commun.*, 1998, *28*, 251; (b) M. Paccar, J. Gross, E. Lubbert, S. Tolzer, S. Krauss, K.-H. van Pee, and A. Berkessel, *Angew. Chem. Int. Ed.*, 1997, *36*, 1196; (c) A.W.P. Jarvie, N. Overton, C.B. St Pourcain, *Chem. Commun.*, 1998, 177.
225. (a) K.-E. Jaeger and M.T. Reetz, *Trends Biotechnol.*, 1998, *16*, 396; (b) A.R.M. Yahya, W.A. Anderson, and M. Moo-Young, *Enzyme Microb. Technol.*, 1998, *23*, 438; (c) S.H. Krishna and N.G. Karanth, *Catal. Rev.-Sci. Eng.*, 2002, *44*,

499; (d) F. Hasan, A.A. Shah, and A. Hames, *Enzyme Microb. Technol.*, 2006, *39*, 235–251; (e) A.R.M. Yaha, W.A. Anderson, and M. Moo-Young, *Enzyme Microb. Technol.* 1998, *23*, 438; (f) N. Gandhi, N.S. Patil, S.B. Savant, and J.B. Joshi, *Catal. Rev.-Sci.Eng.*, 2000, *42*, 439; (g) P. Villeneuve, J.M. Muderhwa, J. Graille, and M.J. Haas, *J. Mol. Catal. B*, 2000, *9*, 113.

226. (a) A. Svendsen, *Int. News Fats Oils Relat. Mater.*, 1994, *5*, 619; (b) X.Y. Wu, S. Jaaskelainen, and Y.-Y. Linko, *Enzyme Microb. Technol.*, 1996, *19*, 226; (c) P. Villeneuve and T.A. Foglia, *Int. News Fats Oils Relat. Mater.*, 1997, *8*, 640; (d) N.N. Gandhi, *J. Am. Oil Chem. Soc.*, 1997, *74*, 621; (e) R.D. Schmid and R. Verger, *Angew. Chem. Int. Ed.*, 1998, *37*, 1608; (f) B. Rubin and E.A. Dennis, eds, *Methods in Enzymology*, vol. 284, Lipases: part A, Biotechnology. Academic Press, San Diego, 197; (g) R. Lortie, *Biotechnol. Adv.*, 1997, *15*, 1–15.
227. X. Fu, X. Zhu, K. Gao, and J. Duan, *J. Am. Oil Chem. Soc.*, 1995, *72*, 527.
228. S. Parmar and E.G. Hammond, *J. Am. Oil Chem. Soc.*, 1994, *71*, 881.
229. (a) I. Jachmanian, M. Perifanova–Nemska, M.-A. Grompone, and K.D. Mukherjee, *J. Agric. Food Chem.*, 1995, *43*, 2992; (b) I. Jachmanian and K.D. Mukherjee, *J. Agric. Food Chem.*, 1995, *43*, 2997.
230. (a) L.A. Nelson, T.A. Foglia, and W.N. Marmer, *J. Am. Oil Chem. Soc.*, 1996, *73*, 1191; (b) G. Parkinson, *Chem. Eng.*, 2002, *109*(8), 23; (c) H. Fukuda, S. Hama, S. Tamalampudi, and H. Noda, *Trends Biotechnol.*, 208, *26*, 668.
231. A. Tullo, *Chem. Eng. News*, July 31, 2006, 12.
232. Y.-Y. Linko, M. Lamsa, A. Huhtala, and P. Linko, *J. Am. Oil Chem. Soc.*, 1994, *71*, 1411.
233. D.E. Stevenson, R.A. Stanley, and R.H. Furneaux, *Enzyme Microb. Technol.*, 1994, *16*, 478.
234. Y.-Y. Linko and X.Y. Wu, *J. Chem. Technol. Biotechnol.*, 1996, *65*, 163.
235. Y. Kosugi and N. Azuma, *J. Am. Oil Chem. Soc.*, 1994, *71*, 1397.
236. P.E. Sonnet, G.P. McNeill, and W. Jun, *J. Am. Oil Chem. Soc.*, 1994, *71*, 1421.
237. P.E. Napier, H.M. Lacerda, C.M. Rosell, R.H. Valivety, A.M. Vaidya, and P.J. Halling, *Biotechnol. Prog.*, 1996, *12*, 47.
238. J.-P. Chen and J.-B. Wang, *Enzyme Microb. Technol.*, 1997, *20*, 615.
239. Y.P. Yong and B. Al-Duri, *J. Chem. Technol. Biotechnol.*, 1996, *65*, 239.
240. M.S. Reisch, *Chem. Eng. News*, Mar. 25, 2002, 21.
241. L. Giorno, R. Molinari, E. Direly, and D.B.P. Cesti, *J. Chem. Technol. Biotechnol.*, 1995, *64*, 345.
242. Z. Ujang, N. Al-Sharbati, and A.M. Vaidya, *Biotechnol. Prog.*, 1997, *13*, 39.
243. (a) A.L. Paiva and F.X. Malcata, *J. Mol. Catal. B Enzymol.*, 1997, *3*, 99–109; (b) R.P. Chauhan and J.M. Woodley, *Chemtech*, 1997, *27*(6), 26.
244. (a) U.T. Bornscheuer, *Enzyme Microb. Technol.*, 1995, *17*, 578; (b) S.J. Kwon, J.J. Han, and J.S. Rhee, *Enzyme Microb. Technol.*, 1995, *17*, 700; (c) R. Multzsch, W. Lokotsch, B. Steffen, S. Lang, J.O. Metzger, H.J. Schafer, S. Warwel, and F. Wagner, *J. Am. Oil Chem. Soc.*, 1994, *71*, 721; (d) F.J. Plou, M. Barandiaran, M.V. Calvo, A. Ballesteros, and E. Pastor, *Enzyme Microb. Technol.*, 1996, *18*, 66; (e) R. Rosu, Y. Uozaki, Y. Iwasaki, and T. Yamane, *J. Am. Oil Chem. Soc.*, 1997, *74*, 445; (f) H. Noureddini and S.E. Harmeier, *J. Am. Oil Chem. Soc.*, 1998, *75*, 1359; (g) M.A. Jackson and J.W. King, *J. Am. Oil Chem. Soc.*, 1997, *74*, 103.
245. (a) E. Castillo, V. Dossat, A. Marty, J.S. Condoret, and D. Combes, *J. Am. Oil Chem. Soc.*, 1997, *74*, 77; (b) E. Castillo, V. Dossat, C. Didier, and M. Alain, *J. Am. Oil Chem. Soc.*, 1998, *75*, 309.
246. A. Zaks and A.T. Gross, U.S. patent 5,316,927 (1994).
247. (a) A.M. Fureby, L. Tian, P. Adlercreutz, and B. Mattiasson, *Enzyme Microb. Technol.*, 1997, *20*, 198; (b) G.P. McNeill and P.E. Sonnet, *J. Am. Oil Chem. Soc.*, 1995, *72*, 213.
248. H.M. Ghazali, S. Hamidah, and Y.B.C. Man, *J. Am. Oil Chem. Soc.*, 1995, *72*, 633.
249. D.B. Sarney and E.N. Vulfson, *Trends Biotechnol.*, 1995, *13*, 164.
250. (a) O.P. Ward, J. Fang, and Z. Li, *Enzyme Microb. Technol.*, 1997, *20*, 52; (b) D. Charlemagne, M.D. Legoy, *J. Am. Oil Chem. Soc.*, 1995, *72*, 61; (c) Q.-H. Zhou and N. Kosaric, *J. Am. Oil Chem. Soc.*, 1995, *72*, 67; (d) P. Skagerlind, K. Larsson, M. Barfoed, and K. Hult, *J. Am. Oil Chem. Soc.*, 1997, *74*, 39.
251. E. Ozaki, T. Uragaki, K. Sakashita, and A. Sakimae, *Chem. Lett.*, 1995, 539.
252. D.J. Ager and I. Prakash, *Synth. Commun.*, 1995. *25*, 739.
253. O. Houille, T. Schmittberger, and D. Uguen, *Tetrahedron Lett.*, 1996, *37*, 625.
254. M. Ghosh and D.K. Battacharyya, *J. Am. Oil Chem. Soc.*, 1998, *75*, 1057.
255. C. Torres and C. Otero, *Enzyme Microb. Technol.*, 2001, *29*, 3.
256. Y. Yoshida, M. Kawase, C. Yamaguchi, and T. Yamane, *J. Am. Oil Chem. Soc.*, 1997, *74*, 261.
257. (a) K. Nakamura, M. Kinoshita, and A. Ohno, *Tetrahedron*, 1995, *51*, 8799; (b) G. Lin, H.-C. Liu, *Tetrahedron Lett.*, 1995, *36*, 6067; (c) U.T. Bornscheuer and T. Yamane, *J. Am. Oil Chem. Soc.*, 1995, *72*, 193; (d) F.F. Bruno, J.A. Akkara, M. Ayyagari, D.L. Kaplan, R. Gross, G. Swift, and J.S. Dordick, *Macromolecules*, 1995, *28*, 8881; (e) M. Indlekofer, M. Funke, W. Claasen, and M. Reuss, *Biotechnol. Prog.*, 1995, *11*, 436.
258. M. Schudok and G. Kretzschmar, *Tetrahedron Lett.*, 1997, *38*, 387.
259. H.K. Weber, J. Zuegg, K. Faber, and J. Pleiss, *J. Mol. Catal. B Enzymol.*, 1997, *3*, 131.
260. E. Guibe-Jampel, Z. Chalecki, M. Bassir, and M. Gelo-Pujic, *Tetrahedron*, 1996, *52*, 4397.
261. C. Ebert, L. Gardossi, P. Linda, and R. Vesnauer, *Tetrahedron*, 1996, *52*, 4867.
262. U.T. Bornscheuer and T. Yamane, *J. Am. Oil Chem. Soc.*, 1995, *72*, 193.
263. C. Waldinger and M. Schneider, *J. Am. Oil Chem. Soc.*, 1996, *73*, 1513.
264. (a) L. Heiss and H.-J. Gais, *Tetrahedron Lett.*, 1995, *36*, 3833, 3837; (b) K.-F. Hsiao, F.L. Yang, S.-H. Wu, and K.-T. Wang, *Biotechnol. Lett.*, 1996, *18*, 1277.
265. S. Lamare, B. Caillaud, K. Roule, I. Goubet, and M.D. Legoy, *Biocatal. Biotransformation*, 2001, *19*, 361.

266. (a) S. Srivastava, J. Modak, and G. Madras, *Ind. Eng. Chem. Res.*, 2002, *41*, 1940; (b) T. Matsuda, K. Watanabe, T. Harada, and K. Nakamura, *Catal. Today*, 2004, *96*, 103; (c) T. Matsuda, T. Harada, and K. Nakamura, *Green Chem.*, 2004, *6*, 440.
267. (a) K. Nakashima, T. Maruyama, N. Kamiya, and M. Goto, *Chem. Commun.*, 2005, 4297; (b) M. Eckstein, M. Sesing, U. Kragl, and P. Adlercreutz, *Biotechnol. Lett.*, 2002, *24*, 867.
268. (a) K. Fouhy, *Chem. Eng.*, 1997, *104*(2), 25; (b) R & D Mag, 1996, Dec. 64; (c) H. Song and S.Y. Lee, *Enzyme Microb. Technol.*, 2006, *39*, 352.
269. J.E. Gavagan, S.K. Fager, J.E. Seip, M.S. Payne, D.L. Anton, and R. DiCosimo, *J. Org. Chem.*, 1995, *60*, 3957.
270. D. Ura, T. Hashimukai, T. Matsumoto, and N. Fukuhara, U.S. patent 5,382,517 (1985).
271. M. Holtzapple, The Presidential Green Chemistry Challenge Awards Programs EPA 744-K-96-001, Washington, DC, 1996, July 7; Preprints A.C.S. *Div. Environ. Chem.*, 1997, *37*(1), 344.
272. J.A. Ramsay, M.-C. Aly Hassan, and B.A. Ramsay, *Enzyme Microb. Technol.*, 1998, *22*, 292.
273. Z. Jin and S.-T. Yang, *Biotechnol. Prog.*, 1998, *14*, 457.
274. (a) R.S. Lees and M. Karel, *Omega Minus Three Fatty Acids in Health and Disease*, Dekker, New York, 1990; (b) I.S. Newton, *Chem. Ind.* (*Lond.*), 1997, 302.
275. (a) O.P. Ward, *Int. News Fats Oils Relat. Mater.*, 1995, *6*, 683; (b) I. Gill and R. Valivety, *Trends Biotechnol.*, 1997, *15*, 401, 470; (c) T. Nakahara, T. Yokochi, T. Higashihara, S. Tanaka, T. Yaguchi, and D. Honda, *J. Am. Oil Chem. Soc.*, 1996, *73*, 1421; (d) M. Certik, E. Sakuradani, and S. Shimizu, *Trends Biotechnol.*, 1998, *16*, 500.
276. N. Totani and K. Oba, *Appl. Microbiol. Biotechnol.*, 1988, *28*(2), 135.
277. H.-C. Chen and C.-C. Chang, *Biotechnol. Prog.*, 1996, *12*, 338.
278. (a) P. Bajpai and P.K. Bajpai, *J. Biotechnol.*, 1993, *30*, 161; (b) N. Shirasaka and S. Shimizu, *J. Am. Oil Chem. Soc.*, 1995, *72*, 1545; (c) M. Cartens, E.M. Grima, A.R. Medina, A.G. Gimenez, J.I. Gonzalez, *J. Am.*, 1993, *70*, 119; (g) D.S. Nichols, P.D. Nichols, T.A. McMeekin, and C.W. Sullivan, Preprints A.C.S. *Div. Environ. Chem.*, 1996, *36*(2), 283; (h) Z. Cohen, *J. Am. Oil Chem. Soc.*, 1994, *71*, 941; (i) D.J. O'Brien and G.E. Senske, *J. Am. Oil Chem. Soc.*, 1994, *71*, 947; (j) B.F. Haumann, *Int. News Oils Relat. Meter.*, 1998, *9*(12), 1108.
279. (a) B. Behrouzian, P.-H. Buist, and J. Shanklin, *Chem. Commun.*, 2001, 401; (b) J.M. Dyer, D.C. Chapital, R.T. Mullen, J.-C.W. Kuan, and A.B. Pepperman, Jr., A.C.S. Meeting in New Orleans, Mar. 2003, AGFD 115.
280. C.T. Hou, *J. Am. Oil Chem. Soc.*, 1994, *71*, 975.
281. H.A. Wittcoff and B.G. Reuben, *Industrial Organic Chemicals Wiley*, New York, 1996, 373.
282. E.-C. Chan and J. Kuo, *Enzyme Microb. Technol.*, 1997, *20*, 585.
283. S. Liu, C. Le, X. Fang, and Z. Cao, *Enzyme Microb. Technol.*, 2004, *34*, 73.
284. E.S. Miller and S.W. Peretti, *Abstracts of 2nd Green Chemistry and Chemical Engineering Conference*, Washington, DC, 6/30/98–7/2/98, 3.
285. (a) M. Wieser, T. Yoshida, and T. Nagasawa, *Tetrahedron Lett.*, 1998, *39*, 4309; (b) T. Matsuda, Y. Ohashi, T. Harada, R. Yangihara, T. Nagasawa, and K. Nakamura, *Chem. Commun.*, 2001, 2194.
286. (a) M. Miyazaki, M. Shibue, K. Ogino, H. Nakamura, and H. Maeda, *Chem. Commun.*, 2001, 1800; (b) G. Parkinson, *Chem. Eng.*, 2001, *108*(11), 21.
287. G. Parkinson, *Chem. Eng.*, 2001, *108*(10), 19.
288. L. Wessjohann and B. Sontag, *Angew. Chem. Int. Ed.*, 1996, *35*, 1697.
289. M.C. de Zoete, A.C. Kock–van Dalen, F. van Rantwijk, and R.A. Sheldon, *J. Mol. Catal. B Enzymol.*, 1996, *1*, 109; 1996, *2*, 19, 141.
290. S. Puertas, F. Rebolledo, and V. Gotor, *Tetrahedron*, 1995, *51*, 1495.
291. B. Orsat, P.B. Alper, W. Moree, C.-P. Mak, and C.-H. Wong, *J. Am. Chem. Soc.*, 1996, *118*, 712.
292. S.T. Stinson, *Chem. Eng. News*, July 17, 1995, 16.
293. E. Endo, T. Yamagami, and K. Tamura, U.S. patent 5,326,702 (1994).
294. J. Crosby, J. Moilliet, J.S. Parratt, and N.J. Turner, *J. Chem. Soc. Perkin Trans. 1*, 1994, 1679.
295. O. Meth-Cohn and M.-X. Wang, *Chem. Commun.*, 1997, 1041.
296. R. di Cosimo, S.M. Hennessey, and J.E. Gavagan, *Abstracts 2nd Green Chemistry and Chemical Engineering Conference*, Washington, DC, 6/30/98–7/2/98, 3.
297. (a) R.A. Sheldon, *Chemtech*, 1994, *24*(3), 38; (b) H. Waldman and R. Reidel, *Angew. Chem. Int. Ed.*, 1997, *36*, 647.
298. J. Bryjak, M. Bryjak, and A. Noworyata, *Enzyme Microb. Technol.*, 1996, *19*, 196.
299. L. de Martin, C. Ebert, G. Garau. L. Gardossi, and P. Lina, *J. Mol. Catal. B*, 1999, *6*, 437.
300. (a) L. Crawford, A.M. Stepan, P.C. McAda, J.A. Rambosek, M.J. Conder, V.A. Vinci, and C.D. Reeves, *Biotechnology*, 1995, *13*(1), 58; (b) Anon., *Chem. Eng. News*, June 3, 1996, 33.
301. H. Waldmann and R, Reidel, *Angew. Chem. Int. Ed.*, 1997, *36*, 647.
302. (a) A. Iwasaki, Y. Yamada, N. Kizaki, Y. Ikenaka, and J. Hasegawa, *Appl. Microbiol. Biotechnol.*, 2006, *69*(5), 499; (b) A.M. Thayer, *Chem. Eng. News*, Feb. 18, 2008, 18.
303. K.L. Klettke, S. Sanyal, W. Mutatu, and K.D. Walker, *J. Am. Chem. Soc.*, 2007, *129*, 6988.
304. (a) S. Forster, J. Roos, F. Effenberger, H. Wajant, and A. Sprauer, *Angew. Chem. Int. Ed.*, 1996, *35*, 437; (b) M. Schmidt, S. Herve, N. Klempier, and H. Griengl, *Tetrahedron*, 1996, *52*, 7833; (c) H. Griengl, A. Hickel, D.V. Johnson, C. Kratky, M. Schmidt, and H. Schwab, *Chem. Commun.*, 1997, 1933.
305. (a) L. Hedstrom, ed., *Chem. Rev.*, 2002, *102*, 4429–4906.
306. M. Hasegawa, S. Yamamoto, M. Kobayashi, and H. Kise, *Enzyme Microb. Technol.*, 2003, *32*, 356.
307. C. Compagno, F. Boschi, and B.M. Ranzi, *Biotechnol. Prog.*, 1996, *12*, 581.
308. H. Jin, H. Fang, and J. Zhuge, *Biotechnol. Lett.*, 2003, *25*, 311.
309. J.C. Andrade and I. Vasconcelos, *Biotechnol. Lett.*, 2003, *25*, 121.
310. P. Boyaval, C. Corre, and M.-N. Madec, *Enzyme Microb. Technol.*, 1994, *16*, 883.

311. (a) Anon., *Chemtech*, 1994, *24*(7), 52; (b) P.J. Slininger, E. Vancauwenberge, and R.J. Rothhast, U.S. patent 4,962,027 (1990).
312. (a) D.C. Cameron, A.M. Held, M.-Y. Zhu, and F.A. Skraly, *Anaheim A.C.S. Meeting*, April 1995, *Biochem. Technol. Secretariat.*, 131; (b) K. Menzel, A.-P. Zeng, and W.D. Deckwer, *Enzyme Microb. Technol.*, 1997, *20*, 82.
313. N. Zwicker, U. Theobald, H. Zahner, and H.-P. Fiedler, *J. Ind. Microb. Biotechnol.*, 1997, *19*, 280.
314. (a) D.C. Cameron, N.E. Altaras and A.J. Shaw, *Biotechnol. Prog.*, 1998, *14*, 116; (b) J. Alper, *Science*, 1999, *283*, 1625.
315. (a) Anon., *R&D (Cahners)* 1997, Feb. 7; (b) Anon., *Chem. Eng. News*, Dec. 2, 1996, 8.
316. Anon., *Chem. Eng. News*, Dec. 14, 1998, 19.
317. (a) Anon., *Chem. Eng. News*, Jan. 13, 1997, 8; (b) Anon., *Chem. Eng. News*, March 30, 2009, 14.
318. (a) A. Archelas and R. Furstoss, *Trends Biotechnol.*, 1998, *16*, 108; (b) I.V.J. Archer, *Tetrahedron*, 1997, *53*, 15617; (c) S. Aguila, R. Vazquez-Duhalt, R. Tinoco, M. Rivera, G. Pecchi, and J.B. Alderete, *Green Chem.*, 2008, *10*, 647.
319. (a) H.J.M. Gijsen, L. Qiao, W. Fitz, and C.-H. Wong, *Chem. Rev.*, 1996, *96*, 443; (b) C.-H. Wong, R.L. Halcamb, Y. Ichikawa, T. Kajimoto, *Angew. Chem. Int. Ed.*, 1995, *34*, 412, 521; (c) S.-H. Jung, J.-H. Jeong, P. Miller, and C.-H. Wong, *J. Org. Chem.*, 1994, *59*, 7182; (d) H.J.M. Gijsen and C.-H. Wong, *J. Am. Chem. Soc.*, 1995, *117*, 2947; (e) G.-J. Boons, *Tetrahedron*, 1996, *52*, 1116; (f) A.J. Humphrey, N.J. Turner, R. McCague, and S.J.C. Taylor, *J. Chem. Soc. Chem. Commun.*, 1995, 2475.
320. L. Hecquet, J. Bolte, and C. Demuynck, *Tetrahedron*, 1996, *52*, 8223.
321. G. Parkinson, *Chem. Eng.*, 1998, *105*(10), 19.
322. (a) S. Pedersen, L. Dijkhuizen, B.W. Dijkstra, B.F. Jensen, and S.T. Jorgenson, *Chemtech*, 1995, *25*(12), 19; (b) T.-J. Kim, B.-C. Kim, and H.-S. Lee, *Enzyme Microb. Technol.*, 1997, *20*, 506.
323. A. Tonkova, *Enzyme Microb. Technol.*, 1998, *22*, 678.
324. R. Villalonga, R. Cao, and A. Fragoso, *Chem. Rev.*, 2007, *107*, 3088.
325. (a) R.J. Radmer, *Bioscience*, 1996, *46*(4), 263; (b) E.W. Becker, *Microalgae: Biotechnology and Microbiology*, Cambridge University Press, Cambridge, UK, 1994; (c) D. Renn, *Trends Biotechnol.*, 1997,*15*, 9.
326. (a) W. Harvey, *Biotechnology*, 1988, *6*, 487; (b) F. Chen, *Trends Biotechnol.*, 1996, *14*, 421.
327. M. Weber and J. Gradwohl, *Science*, 1995, *268*, 1514.
328. C. Vilchez, I. Garbayo, M.V. Lobato, and J.M. Vega, *Enzyme Microb. Technol.*, 1997, *20*, 562.
329. J.C. Ogbonna and H. Tanaka, *Chemtech*, 1997, *27*(7), 43.
330. (a) E. Haslam, *Shikimic Acid Metabolism and Metabolites*, Wiley, New York, 1993; (b) N.J. Grinter, *Chemtech*, 1998, *28*(7), 33; (c) K.M. Draths, D.R. Knop, and J.W. Frost, *J. Am. Chem. Soc.*, 1999, *121*, 1603.
331. (a) R. Carr, F. Ciccone, R. Gabel, M. Guinn, D. Johnston, J. Mastriona, T. Vandermeer, and M. Groaning, *Green Chem.*, 2008, *10*, 743; (b) J.-F. Tremblay, *Chem. Eng. News*, Apr. 10, 2006, 33.
332. (a) J. Frost, *EPA J.*, 1994, *20*(3–4), 22; (b) J.W. Frost and K.M. Draths, *Chem. Br.*, 1995, *31*, 206; (c) J.W. Frost and J. Lievense, *New J. Chem.*, 1994, *18*, 341; (d) K.M. Draths and J.W. Frost, *J. Am. Chem. Soc.*, 1995, *117*, 2395; (e) K.D. Snell, K.M. Draths, and J.W. Frost, *J. Am. Chem. Soc.*, 1996, *118*, 5605. (f) K.M. Draths and J.W. Frost, In: P.T. Anastas and C.A. Farris, eds, *Benign by Design—Alternative Synthetic Design for Pollution Prevention*, A.C.S. Symp. 577, Washington, DC, 1994, 32.
333. A. Berry, *Trends Biotechnol.*, 1996, *14*(7), 250.
334. (a) K. Li and J.W. Frost, *J. Am. Chem. Soc.*, 1998, *120*, 10545; (b) J.W. Frost, *Chem. Eng. News*, Nov. 30, 1998, 4; (c) C.A. Hansen and J.W. Frost, *J. Am. Chem. Soc.*, 2002, *124*, 5926; (d) W. Niu, K.M. Draths, and J.W. Frost, *Biotechnol. Prog.*, 2002, *18*, 201.
335. (a) N. Flores, J. Xiap, A. Berry, F. Bolivar, and F. Valle, *Nat. Biotechnol.*, 1996, *14*, 620; (b) A.L. Demain, *Nat. Biotechnol.*, 1996, *14*, 580.
336. Anon., *Chem. Ind.* (*Lond.*), 1999, 689.
337. J.W. Frost, U.S. 7,399,855 (2008).
338. (a) J. Ogawa and S. Shimizu, *Trends Biotechnol.*, 1999, *17*, 13; (b) D. Hairston, *Chem. Eng.*, 1997, *104*(12), 32; (c) B. Zechendorf, *Trends Biotechnol.*, 1999, *17*, 219.
339. (a) M. McCoy, *Chem. Eng. News*, Jan. 4, 1999, 10; (b) R.N. Patel, *J. Am. Oil Chem. Soc.*, 1996, *73*, 1363.
340. (a) T.K. Kirk, following papers, New Orleans A.C.S. Meeting, Mar. 1996. Cellulose papers 122, 137–146, 203–214, 228–234; (b) K.E. Hammel, *New J. Chem.*, 1996, *20*, 195; (c) P. Broda, *Paper Technol.*, 1995, *36*(10), 39; (d) T.W. Jeffries and L. Viikari, eds, *Enzymes for Pulp and Paper Processing*. A.C.S. Symp. 655, Washington, DC, 1996; (e) K.E.L. Eriksson, A. Cavaco-Paulo, eds, *Enzyme Applications in Fiber Processing*. A.C.S. Symp. 687, Washington, DC, 1998; (f) G. Ondrey and G. Parkinson, *Chem. Eng.*, 1997, *104*(4), 33; (g) M.B. Brennan, *Chem. Eng. News*, Mar. 23, 1998, 39.
341. G.M. Scott and R. Swaney, *TAPPI J.* 1998, *81*(12), 153.
342. K.-E. Jaeger and M.T. Reetz, *Trends Biotechnol.*, 1998, *16*, 396.
343. (a) P. Josefsson, F. Nilsson, L. Sundsrom, C. Noberg, E. Lie, M.B. Jansson, and G. Henriksson, *Ind. Eng. Chem. Res.*, 2006, *45*, 2374; (b) A. Gutleirrez, J.C. delRio, M.J. Martinez, and A.T. Martinez, *Trends Biotechnol.*, 2001, *19*, 340.
344. J.D. Breicia, F. Sineriz, M.D. Baigori, G.R. Castro, and R. Hatti.-Kaul, *Enzyme Microb. Technol.*, 1998, *22*, 43.
345. Anon., *Chem. Eng. News*, Aug. 1, 1994, 33.
346. A. Chaudhary, L.K. Gupta, J.K. Gupta, and U.C. Banerjee, *Biotechnol. Adv.*, 1998, *16*, 899.
347. (a) S.V. Chikkodi, *Text Chem. Color.*, 1996, *28*(3), 28, (b) A. Cavaco-Paulo, L. Almeida, *Text Chem. Color.*, 1996, *28*(6), 28; (c) S.V. Chikkodi, S. Khan, R.D. Mehti, *Text Res. J.*, 1995, *65*, 564; (d) M. Ueda, H. Koo, T. Wakida, and Y. Yoshimura, *Text Res. J.*, 1994, *64*, 615; (e) Y. Li and I.R. Hardin, *Text Res. J.*, 1998, *68*, 671; *Text Color Chem.*, 1998, *30*(9), 23; (f) M.M. Hartzell and Y.-L. Hsieh, *Text Res. J.*, 1998, *68*, 233; (g) C. Byrne, *Chem. Ind.* (*Lond.*), 1999, 343; (h) K.-E.L. Eriksson and A. Cavalco-Paulo, eds, *Enzyme Applications in Fiber Processing*, A.C.S. Symp. 687, Oxford University, Oxford, 1998; (i) V.G. Yachmenev, E.J. Blanchard, and A.H. Lambert, *Ind. Eng. Chem. Res.*, 2998, *37*, 3919; (j) T. Tzanov, M. Calafell, G.M. Guebitz, and A. Calvaco-Paulo, *Enzyme Microb. Technol.*, 2001, *29*, 357; (k) M. Calafell and P. Garriga, *Enzyme Microb., Technol.*, 2004, *34*, 326.
348. M.K. Bhat and S. Bhat, *Biotechnol. Adv.*, 1997, *15*, 583.

349. (a) T.S. Jakobsen, P. Lindegaard, and M. Chan, *Int. News Fats Oils Relat. Mater.*, 1998, *9*, 788. (b) T. Karwczyk, *Int. News Fats Oils Relat. Mater.*, 1997, *8*, 6.
350. (a) Y.-L. Hseih and L.A. Cram, *Text Res. J.*, 1998, *68*, 311; (b) G.M. Guelbitz and A. Cavaco-Paulo, *Trends Biotechnol.*, 2008, *26*(1), 32.
351. (a) G.N. Ramaswamy, C.G. Ruff, and C.R. Boyd, *Text Res. J.*, 1994, *64*, 305; (b) D.E. Akin, W.H. Morrison III, G.R. Gamble, L.L. Rigsby, G. Henriksson, and K.-E.L. Eriksson, *Text Res. J.*, 1997, *67*, 279.
352. M. Ossola and Y.M. Galante, *Enzyme Microb. Technol.*, 2004, *34*, 177.
353. T. Mori, M. Sakimoto, T. Kagi, and T. Saki, *J. Chem. Technol. Biotechnol.*, 1997, *68*, 151.
354. G. Ondrey, *Chem. Eng.*, 2005, *112*(6), 16.
355. (a) C.C. Gaylarde and H.A. Videla, eds, *Bioextraction and Biodeterioration of Metals*, Cambridge University Press, Cambridge, UK, 1995; (b) A.S. Moffat, *Science*, 1994, *264*, 778; (c) J. Haddadin, C. Dagot, and M. Fick, *Enzyme Microb. Technol.*, 1995, *17*, 290; (d) D.E. Rawlings, ed., *Biomining: Theory, Microbes and Industrial Processes*, Springer, Berlin, 1997; (e) D.E. Rawlings, *Pure Appl. Chem.*, 2004, *76*, 847.
356. S. Hadlington, *Chem. Br.*, 1998, *34*(6), 15.
357. (a) G.S. Samdani, *Chem. Eng.*, 1995, *102*(3), 25; (b) G. Parkinson, *Chem. Eng.*, 1997, *104*(4), 23; (c) J. Barrett and M. Hughes, *Chem. Br.*, 1997, *33*(6), 23.
358. E.A. Ripley, R.E. Redmann, A.A. Crowder, T.C. Ariano, C.A. Corrigan, and R.J. Farmer, *Environmental Effects of Mining*, St. Lucie Press, Delray Beach, FL, 1996.
359. (a) E. Bauerlein, P. Behrens, and M. Epple, eds, *Handbook of Biomineralization*, Wiley-VCH, Weinheim, Weinheim, 2007; (b) E. Bauerlein, ed., *Biomineralization*, 2nd ed., Willey-VCH, Weinheim. 2004; (c) P.M. Dove, J.J. deYoreo, and S. Weiner, eds, *Biomineralization*, Mineralogical Society of America and the Geochemical Society, Washington, DC, 2003; (d) S. Mann, *Biomineralization—Principles and Concepts in Bioinorganic Materials Chemistry*, Oxford University Press, Oxford, 2002; (e) L.A. Estroff, ed., *Chem. Rev.*, 2008, *108*, 4329–4978.
360. T. Matsunaga, T. Suzuki, M. Tanaka, and A. Arakaki, *Trends Biotechnol.*, 207, *25*(4), 182.
361. (a) S. Wenzl, R. Deutzmann, R. Hett, E. Hochmuth, and M. Sumper, *Angew. Chem. Int. Ed.*, 2004, *43*, 5933; (b) S.L. Rovner, *Chem. Eng. News*, Sept. 14, 2009, 30.
362. (a) J.L. Shennan, *J. Chem. Technol. Biotechnol.*, 1996, *67*, 109; (b) J.L. Fox, *Biotechnology*, 1993, *11*, 782; (c) J.J. Kilbane II and K. Jackowski, *Biotechnol. Bioeng.*, 1992, *40*, 1107; (d) M. Constanti, J. Giralt, and A. Bordons, *Enzymes Microb. Technol.*, 1996, *19*, 214; (e) K.A. Gray, O.S. Pogrbinsky, G.T. Mrachko, L. Xio, D.J. Monticello, and C.H. Squires, *Nat. Biotechnol.*, 1996, *14*, 1705.
363. (a) B. Rajganesh, K.L. Sublette, C. Camp, and M.R. Richardson, *Biotechnol. Prog.*, 1995, *11*, 228; (b) *Appl. Catal. A*, 1995, *130*, N10.
364. (a) Anon., *Chem. Eng. News*, Apr. 3, 1995, 9; (b) Anon., *Chem. Eng. News*, May 30, 1994, 16; (c) D.J. Monticello, *Chemtech*, 1998, *28*(7), 38; (d) G. Parkinson, *Chem. Eng.*, 1998, *105*(4), 23; (e) G. Parkinson, *Chem. Eng.*, 2000, *107*(3), 21.
365. G. Ondrey, *Chem. Eng.*, 1998, *105*(5), 17.
366. P.T. Selvaraj and K.L. Sublette, *Biotechnol. Prog.*, 1995, *11*, 153.
367. Y.-C. Chung, C. Huang, and C.-P. Tseng, *Biotechnol. Prog.*, 1996, *12*, 773.
368. (a) M.J. Benedik, P.R. Gibbs, R.R. Riddle, and R.C. Willson, *Trends Biotechnol.*, 1998, *16*, 390; (b) G. Parkinson, *Chem. Eng.*, 1997, *104*(6), 23.
369. (a) A. Rosenthal, D.L. Pyle, and K. Niranjan, *Enzyme Microb. Technol.*, 1996, *19*, 402; (b) D. Shankar, Y.C. Agrawal, B.C. Sarkar, and B.P.N. Singh, *J. Am. Oil Chem. Soc.*, 1997, *74*, 1543; (c) A. Ranelli and G. de Mattia, *J. Am. Oil Chem. Soc.*, 1997, *74*, 1105.
370. M. Torres, J.J. Mendez, V. Sanahuja, and R. Canela, *Biocatal. Biotransformation*, 2003, *21*, 129.
371. A. Meyer, S.M. Jepsen, and N.S. Sorensen, *J. Agric. Food Chem.*, 1998, *46*, 2439.
372. (a) P. Juzlova, L. Martinkova, J. Lozinski, and F. Machek, *Enzyme Microb. Technol.*, 1994, *16*, 996; (b) M.-H. Chen and M.R. Johns, *Enzyme Microb. Technol.*, 1994, *16*, 584; (c) S. Izawa, N. Harada, T. Watanabe, N. Kotokawa, A. Yamamoto, H. Haystsu, and S. Arimoto-Kobayashi, *J. Agric. Food Chem.*, 1997, *45*, 3980.
373. Anon., Johns Hopkins Med Lett Health after 50, 1999, *11*(5), 3.
374. (a) T.S. Webster, J.S. Devinny, E.M. Torres, and S.S. Basrai, *Environ. Prog.*, 1996, *15*(3), 141; (b) K. Kiared, L.L. Bibeau, R. Brzezinski, G. Viel, and M. Heitz, *Environ. Prog.*, 1996, *15*(3), 148; (c) Anon., *Chem. Eng.*, 1997, *104*(2), 86; (d) J. Luo and A. van Oostrom, *Pure Appl. Chem.*, 1997, *69*, 2403; (e) H.J. Campbell and M.A. Connor, *Pure Appl. Chem.*, 1997, *69*, 2411; (f) K.S. Betts, *Environ. Sci. Technol.*, 1997, *31*, 547A; (g) P.A. Gostomski, ed., *Environ. Prog.*, 1999, *18*, 151; (h) R. Iranpour, H.H.J. Cox, M.A. Deshusses, and E.D. Schroeder, *Environ. Prog.*, 2005, *24*(3), 254; (i) M. Doble, *Chem. Eng.*, 2006, *113*(6), 35–41; (j) J.S. Devinny, ed., *Environ. Prog.*, 2003, *22*(2), 79–144, J18; (k) I. Kim, *Chem. Eng. Prog.*, 2004, *100*(1), 8; (l) A. Govind, *Pollution Equipment News*, 2009, *42*(1), 10; 2009, *42*(3), 12.
375. J. Paca, E. Klapkova, M. Halecky, K. Jones, and T.S. Webster, *Environ. Prog.*, 2006, *25*(4), 365.
376. Y.-C. Chung, and C. Huang, *Environ. Prog.*, 1998, *17*(2), 70.
377. (a) R.M. Atlas, *Chem. Eng. News*, Apr. 3, 1995, 32; (b) R. Renner, *Environ. Sci. Technol.*, 1997, *31*, 188A.
378. H. Singh, *Mycoremediation*, Wiley, New York, 2006.
379. (a) A.M. Rouhi, *Chem. Eng. News*, Jan. 13, 1997, 21; (b) A.S. Moffat, *Science*, 1995, *269*, 303; (c) D.E. Salt and I. Chet, *Chet. Biotechnology*, 1995, *13*, 468; (d) I. Kim, *Chem. Eng.*, 1996, *103*(12), 39; (e) P.B.A.N. Kumar, V. Dushenkov, H. Motto, and I. Raskin, *Environ. Sci. Technol.*, 1995, *29*, 1232, 1239; (f) S.L. Brown, R.L. Chaney, J.S. Angle, and A.J.M. Baker, *Environ. Sci. Technol.*, 1995, *29*, 1581; (g) M.E. Watanabe, *Environ. Sci. Technol.*, 1997, *31*, 182A; (h) E.L. Kruger, T.A. Anderson, and J.R. Coats, eds, *Phytoremediation of Soil and Water Contaminants*. A.C.S. Symp. 664, Washington, DC, 1997; (i) S.C. McCutcheon and J.L. Schnoor, *Phytoremediation*, Wiley, New York, 2003.
380. G. Parkinson, *Chem. Eng.*, 1997, *104*(1), 21.
381. Anon., *Chem. Eng.*, 1997, *104*(2), 81.
382. T. Zhang, C. Cui, S. Chen, X. Ai, H. Yang, P. Shen, and Z. Peng, *Chem. Commun.*, 2006, 2257.

383. X. Font, N. Adroer, M. Poch, and T. Vicent, *J. Chem. Technol. Biotechnol.*, 1997, *68*, 75.
384. M. McCoy, *Chem. Eng. News*, May 4, 1998, 29.
385. (a) B.-L. Liu, A. Rafiq, Y.-M. Tzeng, and A. Rob, *Enzyme Microb. Technol.*, 1998, *22*, 415; (b) Anon., *Chem. Eng. News*, May 25, 1998, 26.
386. (a) S. Kobayashi, *Adv. Polym. Sci.*, 1995, *121*, 1; (b) G. Swift, D.L. Kaplan, R.A. Gross, eds, Preprints A.C.S. *Div. Polym. Mater. Sci. Eng.*, 1996, *74*, 1–4, 6, 32, 37, 37, 39, 67; (c) R.A. Gross, D.L. Kaplan, and G. Swift, *Enzymes in Polymer Synthesis*. A.C.S. Symp. 684, Washington, DC, 1997; (d) S.L. Aggarwal and S. Russo, *Compr. Polym. Sci. 2nd Suppl., Pergamon*, Oxford, 1996; (e) I.K. Varma, A.-C. Albertsson, R. Rajkhowa, and R.K. Srivastava, *Prog. Polym. Sci.*, 2005, *30*, 949; (f) R.A. Gross and H.N. Cheng, *Biocatalysis in Polymer Science*, A.C.S. Symp. 840, Oxford University Press, Oxford, 2003; (g) R.A. Gross, A. Kumar, and B. Kabra, *Chem. Rev.*, 2001, *101*, 2097; (h) S. Kobayashi, H. Uyama, and S. Kimura, *Chem. Rev.*, 2001, *101*, 3793.
387. (a) H. Uyama and S. Kobayashi, *Chem. Lett.*, 1993, *22*, 1149; (b) F. Binns and A. Taylor, *Tetrahedron*, 1995, *51*, 12929.
388. (a) H. Uyama, H. Kikuchi, K. Takeya, N. Hoshi, and S. Kobayashi, *Chem. Lett.*, 1996, 107; (b) H. Kikuchi and S. Kobayashi, *Chem. Lett.*, 1995, 1047; (c) K. Kullmer, H. Kikuchi, H. Uyama, and S. Kobayashi, *Macromol. Rapid Commun.*, 1998, *19*, 127; (d) S. Kobayashi, H. Uyama, S. Namekawa, and H. Hayakawa, *Macromolecules*, 1998, 31, 5655; (e) S. Kobayashi, K. Takeya, S. Suda, and H. Uyama, *Macromol. Chem. Phys.*, 1998, *199*, 1729; (f) S. Kobayashi, H. Kikuchi, and H. Uyama, *Macromol. Rapid Commun.*, 1997, *18*, 575; (g) S. Kobayashi and H. Uyama, *Macromol. Symp.*, 1999, *144*, 237; (h) S. Kobayashi, *Macromol. Rapid. Commun.*, 2009, *30*, 237–266.
389. (a) S. Matsumura, K. Mabuchi, and K. Toshima, *Macromol. Rapid Commun.*, 1997, *18*, 477; (b) P. vanHoek, A. Aristidou, J.J. Hahn, and A. Patist, *Chem. Eng. Prog.*, 2003, *99*(1), 37S.
390. D. O' Hagan and N.A. Zaida, *Polymer*, 1994, *35*, 3577.
391. A. Kumar, Y. Wang, M.A. Hillmyer, and R.A. Gross, *Polym. Preprints*, 2002, *43*(1), 264.
392. (a) C.J. Duxbury, W. Wang, M. deGeus, A. Heise, and S.M. Howdle, *J. Am. Chem. Soc.*, 2005, *127*, 2384; (b) S. Villarroya, K.J. Thurecht, A. Heise, and S.M. Howdle, *Chem. Commun.*, 2007, 3805.
393. (a) G. Mezoul, T. Lalot, M. Brigodiot, and E. Marechal, *Macromol. Chem. Phys.*, 1996, *197*, 3581; (b) G. Mezoul, T. Lalot, M. Brigodiot, and E. Marechal, *J. Polym. Sci. Part A Polym. Chem.*, 1995, *33*, 2691; *Polym. Bull.*, 1996, *36*, 541.
394. (a) H. Uyama and S. Kobayashi, *Chem. Lett.*, 1994, 1687; (b) Z.-L. Wang, K. Hiltunen, P. Orava, J. Seppala, and Y.-Y. Linko, *J. Macromol. Sci. Pure Appl. Chem.*, 1996, *A33*, 599; (c) A.J. Russell, E.J. Beckman, D. Abderrahmare, and A.K. Chaudhary, U.S. patent 5,478,910 (1995); (d) Y.-Y. Linko and J. Seppala, *Chemtech*, 1996, *26*(8), 25; (e) A.K. Chaudhary, J. Lopez, E.J. Beckman, and A.J. Russell, *Biotechnol. Prog.*, 1997, *13*, 318.
395. B.J. Kline, E.J. Beckman, and A.J. Russell, *J. Am. Chem. Soc.*, 1998, *120*, 9475.
396. H. Uyama, K. Inada, and S. Kobayashi, *Macromol. Rapid Commun.*, 1999, *20*, 171.
397. A.L. Gutman, D. Knaani, and T. Bravdo, *Macromol Symp.*, 1997, *122*, 39.
398. M. Kitagawa and Y. Tokiwa, *Biotechnol. Lett.*, 1998, *20*, 627.
399. Y. Miura, T. Ikeda, N. Wada, H. Sato, and K. Koyayashi, *Green Chem.*, 2003, *5*, 610.
400. (a) M.R. Borges and R. Balaban, *Macromol. Symp.*, 2007, *258*, 25; (b) L. Ferreira, M.A. Ramos, M.H. Gil, and J.S. Dordick, *Biotechnol. Prog.*, 2002, *18*, 986.
401. A.-S. Kulshrestha, A. Kumar, W. Gao, and R.A. Gross, *Polym. Preprints*, 2003, *44*(2), 585.
402. (a) J.S. Dordick, R.J. Linhardt, and D.G. Rethwisch, *Chemtech*, 1994, *24*(1), 33; (b) X. Chen, A. Johnson, J.S. Dordick, and D.G. Rethwisch, *Macromol. Chem. Phys.*, 1994, *195*, 3567; (c) Anon., *Chem. Ind.* (*Lond.*), 1996, 281; (d) G. Wulff, J. Schmid, T. Venhoff, *Macromol. Chem. Phys.*, 1996, *197*, 259; (e) X.-C. Liu and J.S. Dordick, *J. Polym. Sci. Part A Polym. Chem.*, 1999, *37*, 1665.
403. (a) U. Geyer, D. Klemm, K. Pavel, and H. Ritter, *Macromol. Rapid Commun.*, 1995, *16*, 337; (b) K. Kobayashi, S. Kamiya, and N. Enomoto, *Macromolecules*, 1996, *29*, 8670.
404. (a) J.H. Lee, R.M. Brown, Jr., S. Kuga, S.-I. Shoda, and S. Kobayashi, *Proc. Natl. Acad. Sci. USA.*, 1994, *91*, 7425; (b) S. Kobayashi, E. Okamoto, X. Wen, and S.-I. Shoda, *J. Macromol. Sci. Pure Appl. Chem.*, 1996, *A33*, 1375; (c) S. Kobayashi, X. Wen, and S.-I. Shoda, *Macromolecules*, 1996, *29*, 2698; (d) S.-I. Shoda and S. Kobayashi, *Trends Polym. Sci.*, 1997, *5*(4), 109.
405. (a) T. Tsuchida and F. Yoshinaga, *Pure Appl. Chem.*, 1997, *69*, 2453; (b) S. Bae and M. Shoda, *Biotechnol. Prog.*, 2004, *20*, 1366.
406. (a) D.P. Mobley, ed., *Microbial Synthesis of Polymers and Polymer Precursors*. Hanser-Gardner, Cincinnati, OH, 1994; (b) Y. Poirier, C. Nawrath, and C. Someville, *Biotechnology*, 1995, *13*, 142; (c) S.Y. Lee, *Trends Biotechnol.*, 1996, *14*, 431; *Nat. Biotechnol.*, 1997, *15*, 17; (d) S.J. Sim, K.D. Snell, S.A. Hogan, J.A. Stubbe, C. Rha, and A.J. Sinskey, *Nat. Biotechnol.*, 1997, *15*,63.
407. Anon., *Chem. Eng. News*, Mar. 20, 2006, 22.
408. G.J.M. de Koning, H.M.M. van Bilsen, P.J. Lemstra, W. Hazenberg, B. Witholt, H. Preusting, J.G. van der Galien, A. Schirmer and D. Jendrossek, *Polymer*, 1994, *35*, 2090.
409. K.-S. Cho, H.W. Ryu, C.-H. Park, and P.R.L. Goodrich, *Biotechnol. Lett.*, 1997, *19*, 7.
410. G. Ondrey, *Chem. Eng.*, 1998, *105*(1), 21.
411. H. Son, G. Park, and S. Lee, *Biotechnol. Lett.*, 1996, *18*, 1229.
412. (a) G. Parkinson, *Chem. Eng.*, 1997, *104*(1), 23; (b) A. Steinbuchel and B. Fuchtenbusch, *Trends Biotechnol.*, 1998, *16*, 419; (c) Anon., *Chem. Eng. News*, Nov. 2, 2009, 15.
413. (a) H. Uyama, H. Kurioka, I. Kaneko, and S. Kobayashi, *Chem. Lett.*, 1994, 423; (b) H. Kurioka, J. Sugihara, I. Komatsu, and S. Kobayashi, *Bull. Chem. Soc. Jpn.*, 1995, *68*, 3209; (c) H. Kurioka, J. Sugihara, and S. Kobayashi, *Bull. Chem. Soc. Jpn.*, 1996, *69*, 189; (d) S. Kobayashi, H. Kurioka, and H. Uyama, *Macromol. Rapid Commun.*, 1996, *17*, 503; (e) H. Kurioka, I. Komatsu, H. Uyama, and S. Kobayashi, *Macromol. Rapid Commun.*, 1994, *15*, 507; (f)

J.A. Akkara, M.S.R. Ayyagari, and F.F. Bruno, *Trends Biotechnol.*, 1999, *17*, 67; (g) H. Uyama, H. Kurioka, J. Sugihara, I. Komatsu, and S. Kobayashi, *J. Polym. Sci. Part A Polym. Chem.*, 1997, *35*, 1453; (h) K. Saito, G. Sun, and H. Hishido, *Green Chem. Lett. Rev.*, 2007, *1*(1), 47; (i) S. Kobayashi, H. Uyama, H. Tonami, T. Oguchi, H. Higashimura, R. Ikeda, and M. Kubota, *Macromol. Symp.*, 2001, *175*, 1; (j) Y.-J. Kim, H. Uyama, and S. Kobayashi, *Macromolecules*, 2003, *36*, 5058.
414. I. Kaneko, S. Kobayashi, H. Uyama, and H. Kurioka, *Chem. Abstr.*, 1995, *123*, 314, 799.
415. D. Hairston, *Chem. Eng.*, 1995, *102*(5), 53.
416. S. Kobayashi, H. Uyama, T. Ushiwata, T. Uchiyama, J. Sugihara, and H. Kurioka, *Macromol. Chem. Phys.*, 1998, *199*, 777.
417. H. Uyama, C. Lohavisavapanich, R. Ikeda, and S. Kobayashi, *Macromolecules*, 1998, *31*, 554.
418. P. Wang, B.D. Martin, S. Parida, D.G. Rethwisch, and J.S. Dordick, *J. Am. Chem. Soc.*, 1995, *117*, 12885.
419. H. Uyama and S. Kobashi, *Chemtech.*, 1999, *29*(10), 22.
420. (a) R. Ikeda, H. Uyama, and S. Kobayashi, *Macromolecules*, 1996, *29*, 3053; (b) R. Ikeda, J. Sugihara, H. Uyama, S. Kobayashi, *Macromolecules*, 1996, *29*, 8702; (c) S. Kobayashi and H. Higashimura, *Prog. Polym. Sci.*, 2003, *28*, 1015.
421. (a) H. Zemel and J.F. Quinn, *Chem. Abstr.*, 1995, *123*, *199*, 816; (b) L.A. Samuelson, A. Anagnostopoulos, K.S. Alva, J. Kumar, and S.K. Tripathy, *Macromolecules*, 1998, *31*, 4376; (c) P. Zurer, *Chem. Eng. News*, Feb. 15, 1999, 68.
422. K.S. Alva, K.A. Marx, J. Kumar, and S.K. Tripathy, *Macromol. Rapid Commun.*, 1996 *17*, 859.
423. (a) D. Ichinohe, T. Muranaka, T. Sasaki, M. Kobayashi, and H. Kise, *J. Polym. Sci. Part A Polym. Chem.*, 1998, *36*, 2593; (b) D. Ichinohe, T. Muranaka, and H. Kise, *J. Appl. Polym. Sci.*, 1998, *70*, 717.
424. (a) Y. Zhou, K. Shimizu, J.N. Cha, G.D. Stucky, and D.E. Morse, *Angew. Chem. Int. Ed.*, 1999, *38*, 870; (b) D.E. Morse, *Trends Biotechnol.*, 1999, *17*, 230.
425. (a) D. Cowan, *Trends Biotechnol.*, 1996, *14*, 177; (b) J. Hodgson, *Biotechnology*, 1994, *12*, 789.
426. S.L. Strausberg, P.A. Alexander, D.T. Gallagher, G.L. Gilliland, B.L. Barnett, P.N. Bryan, *Biotechnology*, 1995, *13*, 669.
427. P.S.J. Cheetham, *Chem. Ind. (Lond.)*, 1995, 265.
428. (a) J.R. Jacobsen, C.R. Hutchinson, D.E. Cane, and C. Khosla, *Science*, 1997, *277*, 367; (b) R.F. Service, *Science*, 1997, *277*, 319.
429. C. Duport, R. Spagnoli, E. Degryse, and D. Pompon, *Nat. Biotechnol.*, 1998, *16*,186.
430. (a) K. Kondo, Y. Miura, H. Sone, K. Kobayashi, and H. Iijima, *Nat. Biotechnol.*, 1997, *15*, 543; (b) R. Dansby, *Nat. Biotechnol.*, 1997, *15*, 419.
431. R.B. Flavell, *Trends Biotechnol.*, 1995, *13*, 317–398.
432. C.W.B. Bachem, *Biotechnology*, 1994, *12*, 1101.
433. T. Sano, A. Nagayama, T. Ogawa, I. Ishida, and Y. Okada, *Nat. Biotechnol.*, 1997, *15*, 1290.
434. A. Coghlan, *New Sci.*, 1998, Mar. 21, 11.
435. (a) M. Beauregard, C. Dupont, R.M. Teather, and M.A. Hefford, *Biotechnology*, 1995, *13*, 974; (b) S.C. Falco, T. Guida, M. Locke, J. Mauvais, C. Sanders, R.T. Ward, and P. Webber, *Biotechnology*, 1995, *13*, 577; (c) R.J. Wallace and A. Chesson, *Biotechnology in Animal Feeds and Animal Feeding*. VCH, Weinheim, 1995.
436. (a) F.P. Wolter, *Int. New Fats Oils Relat. Mater.*, 1993, *4*(1), 93; (b) A.S. Moffat, *Science*, 1995, *268*, 658; (c) R.T. Topfer, N. Martini, and J. Schell, *Science*, 1995, *268*, 681.
437. (a) A.J. del Vecchio, *Int. Fats Oils Relat. Mater.*, 1996, *7*(3), 230; (b) D.J. Murphy, *Trends Biotechnol.*, 1996, *14*(6), 206; (c) Anon., *Chem. Ind. (Lond.)*, 1994, 841; (d) H.M. Devies and F.J. Flider, *Chemtech*, 1994, *24*(4), 33.
438. (a) G.K. Podila and D.F. Karnosky, *Chem. Ind.*, 1996, 976; (b) T. Tzfira, A. Zuker, and A. Altman, *Trends Biotechnol.*, 1998, *16*, 439; (c) C.C. Mann and M.L. Plummer, *Science*, 2002, *295*, 1626.
439. M.E. John, *Chem. Ind.*, 1994, 676; *Chemtech*, 1994, *24*(10), 27.
440. (a) R.T. Fraley, *Science*, 1996, *574*, 1994; (b) A.M. Thayer, *Chem. Eng. News*, Apr. 28, 1997, 15; (c) Anon., *Chem. Eng. News*, July 10, 2000, 15.
441. ERS/US Dept. Agric. 2007.
442. Anon., *Chem. Eng. News*, July 29, 1996, 18.
443. M. Brennan, *Chem. Eng. News*, Nov. 18, 1996, 9.
444. (a) Anon., *Chem. Eng. News*, Mar. 22, 1999, 22; (b) D. Ferber, *Science*, 1999, *286*, 1662.
445. M. Ward, *Biotechnology*, 1994, *12*, 967.
446. J. Grisham, *Chem. Eng. News*, Mar. 15, 1999, 40.
447. A. McKeown, 12/4/08, www.worldwatch.org/node/5950.
448. L.M. Zahn, P.J. Hines, E. Pennisi, J. Travis, eds., *Science*, 2008, *320*, 465–497.
449. A.S. Moffat, *Science*, 1998, *282*, 2176.
450. J. Dunwell, *Chem. Ind. (Lond.)*, 1995, 730.
451. M. Balter, *Science*, 1997, *275*, 1063.
452. (a) B. Hileman, *Chem. Eng. News*, 1999, Mar. 15, 36; 1999, May 24, 7; (b) P. Wyner, *Chem. Ind. (Lond.)*, 1998, 422.
453. Anon., *Nat. Biotechnol.*, 1996, *14*, 685.
454. (a) L. Miele, *Trends Biotechnol.*, 1997, *15*, 45; (b) J.K.-C. Ma and M.B. Hein, *Trends Biotechnol.*, 1995, *13*, 522; (c) D. Rotman, *Technol. Rev.*, 1998, *101*(5), 34; (d) S.N. Chapman and T.M.A. Wilson, *Chem. Ind. (Lond.)*, 1997, 550.
455. G. Freyssinet and T. Thomas, *Pure Appl. Chem.*, 1998, *70*, 61.
456. Anon., *Chem. Ind. (Lond.)*, 1997, 200.
457. Anon., *Chem. Ind. (Lond.)*, 1998, 1005.
458. (a) R. Rawls, *Chem. Eng. News*, May 4, 1998, 12; (b) C. Potera, *Technol. Rev.*, 1998, *101*(5), 66.
459. Anon., *Environment*, 2001, *43*(9), 6.
460. E. Dorey, *Chem. Ind. (Lond.)*, Feb. 20, 2006, 8.
461. (a) D. Drunkard, I. Cottingham, I. Garner, S. Bruce, M. Dalrymple, G. Lasser, P. Bishop, and D. Foster, *Nat. Biotechnol.*, 1996, *14*, 867; (b) A. Thayer, *Chem. Eng. News*, Aug. 26, 1996, 23; (c) Anon., *Chem. Eng. News*, Mar. 9, 1998, 9; (d) R. Rawls, *Chem. Eng. News*, Oct. 6, 1997, 33; (e) L.M. Houdebine, *Transgenic Animals—Generation and Use*, Harwood Academic, Amsterdam, 1997.
462. (a) C. Holden, *Science*, 1999, *284*, 903; (b) Anon., *Chem. Eng. News*, May 3, 1999, 15; (c) Baguishi, E. Behboodi, D.T. Melican, J.S. Pollock, M.M. Destrempes, C. Cammuso, J.L. Williams, S.D. Nims, C.A. Porter, P. Midura, M.J. Palacios, S.L. Ayres, R.S. Denniston, M.L. Hayes, C.A. Ziomek, H.M. Meade, R.A. Godke, W.G. Gavin, E.W. Overstrom, and Y. Echelard, *Nat. Biotechnol.*, 1999, *17*, 456.

463. A. Marshall, *Nat. Biotechnol.*, 1998, *16*, 8.
464. (a) H. Meade and C. Ziomek, *Nat. Biotechnol.*, 1998, *16*, 21; (b) D.E. Kerr, F. Liang, K.R. Bondioli, H. Zhao, G. Krebich, R.J. Wall, and T.-T. Sun, *Nat. Biotechnol.*, 1998, *16*, 75.
465. H. Sang, *Trends Biotechnol.*, 1994, *12*, 415.
466. B. Hileman, *Chem. Eng. News*, May 16, 2005, 29.
467. (a) P. Kareiva and J. Stark, *Chem. Ind.* (*Lond.*), 1994, 52; (b) M. Williamson, *Trends Biotechnol.*, 1996, *14*, 449; (c) J. Rissler and M. Mellon, *The Ecological Risks of Engineered Crops.*, MIT Press, Cambridge, MA, 1996.
468. J.R. Stephenson and A. Warnes, *J. Chem. Technol. Biotechnol.*, 1996, *65*, 5.
469. (a) F.A.A.M. de Leij, *Biotechnology*, 1995, *13*, 1488; (b) I.P. Thompson, A.K. Lilley, R.J. Ellis, P.A. Bramwell, and M.J. Bailey, *Biotechnology*, 1995, *13*, 1493; (c) P.A. Goy and J.H. Duesing, *Nat. Biotechnol.*, 1996, *14*, 39.
470. B. Hileman, *Chem. Eng. News*, Sept. 27, 2004, 5.
471. (a) T.H. Schuler, G.M. Poppy, B.R. Kerry, and I. Denholm, *Trends Biotechnol.*, 1999, *17*, 210; (b) B. Halweil, *World Watch*, 1999, *12*(1), 9.
472. B. Hileman, *Chem. Eng. News*, May 24, 1999, 7; 1999, May 31, 5.
473. M. Mellon, *Science*, 1996, *274*, 703.
474. (a) Anon., *Chem. Ind.* (*Lond.*), 1998, 245; (b) H.J. Miller, *Science*, 1999, *284*, 1471.
475. R.H. Devlin, L.F. Sunstrom, and W.M. Muir, *Trends Biotechnol.*, 2006, *24*(2), 89.
476. E. Stokstad, *Science*, 2002, *297*, 1797.
477. Anon., *Chem. Eng. News*, Dec. 15, 2008, 24.
478. A. Spok, *Trends Biotechnol.*, 2007, *25*, 74.
479. (a) M. Mellon, *Catalyst*, (Union of Concerned Scientists) 2002, *1*(2), 3; (b) J. Rissler, *Catalyst*, (Union of Concerned Scientists) 2005, *4*(1), 2.
480. (a) G. Conko, J. Shave, and G. Azeez, *Chem. Eng. Prog.*, 2005, *101*(9), 4, 6; (b) A. McHughen, *Pandora's Picnic Basket—The Potential and Hazards of Genetically-Modified Crops*, Oxford University Press, Oxford, 2000.
481. E. Stokstad, *Science*, 2002, *297*, 1257.
482. (a) A.L. Demain, *Nat. Biotechnol.*, 1998, *16*, 3; (b) N.R. Pace, *Science*, 1997, *276*, 734; (c) E.F. de Long, *Trends Biotechnol*, 1997, *15*, 203; (c) E.K. Wilson, *Chem. Eng. News*, Jan. 20, 2003, 37; (d) A.M. Rouhi, *Chem. Eng. News*, Apr. 10, 2006, 67; (e) X. Li and L. Quing, *Trends Biotechnol.*, 2005, *23*(11), 539; (f) L.M. Jarvis, *Chem. Eng. News*, Dec. 8, 2007, 22; (g) M. Fischbach and C.T.L. Walsh, *Science*, 2009, *325*, 1089.
483. (a) M. Hayes, S. Wrigley, R. Thomas, and E.J.T. Chrystal, eds, *Phytochemical Diversity: A Source of New Industrial Products*. Royal Society of Chemistry, Special Publication 200, Cambridge, UK, 1997; (b) F. Grifo and J. Rosenthal, *Biodiversity and Human Health*, Island Press, Covelo, CA, 1997; (c) S. Grabley and R. Thiericke, *Drug Discovery from Nature*, Springer-Verlag, Heidelberg, 1999; (d) P.B. Kaufman, L.J. Cseke, S. Warber, J.A. Duke, and H.L. Brielmann, *Natural Products from Plants*, CRC Press, Boca Raton, FL, 1999; (e) Atta-ur-Rahman and M.I. Choudhary, *Pure Appl. Chem.*, 2002, *74*, 511; (f) E. Chivian and A. Bernstein, *Sustaining Life: How Human Health Depends on Biodiversity*, Oxford University Press, Oxford, 2008; (g) M. Burke, *Chem. Br.*, 2003, *39*(12), 27; (h) I. Raskin, N. Borisjuk, A. Brinker, D.A. Moreno, C. Ripoli, N. Yacoby, J.M. O'Neil, T. Comwell, and E. Fridlender, *Trends Biotechnol.*, 2002, *20*, 522; (i) M.J. Plotkin, *Medicine Quest: In Search of Nature's Healing Secrets*, Viking, 2000; (j) H. Mooney, G. Mace, J.D. Sachs, E.M. Baillee, W.J. Sutherland, P.R. Annsworth, N. Ash et al., *Science*, 2009, *325*, 1473, 1502, 1503; (k) E. Chivian and A. Bernstein, eds., *Sustaining Life—How Human Health Depends on Biodiversity*, Oxford University Press, New York, 2008; (l) A.D. Kinghorn, E.J.C. deBlanco, B. Chai, J. Orjala, N.R. Farnsworth, D.D. Soejarto et al., *Pure Appl. Chem.*, 2009, *81*, 1051; (m) H.B. Bode, *Angew. Chem. Ind. Ed.*, 2009, *48*, 6394.
484. (a) Conservation International, Washington, DC, Fact Sheet, Feb. 1996; (b) I. Paterson and E.A. Anderson, *Science*, 2005, *310*, 451.
485. C.S. Levings III, *Science*, 1996, *272*, 1279.
486. M. Meyers, *Science*, 1995, *269*, 358.
487. Anon., *Chem. Ind.*, 1996, *317*, 320.
488. (a) W. Agosta, *Bombardier Beetles and Fever Trees: A Close-up Look at Chemical Warfare and Signals in Animals and Plants*, Addison–Wesley, New York, 1995; (b) T. Eisner, M. Eisner, and M. Siegler, *Secret Weapons—Defenses of Insects, Spiders, Scorpions and Other Many-Legged Creatures*, Harvard University Press, Cambridge, MA, 2005.
489. (a) G.A. Cordell, C.K. Angerhofer, and J.M. Pezzuto, *Pure Appl. Chem.*, 1994, *66*, 2283; (b) A.D. Kinghorn and E.-K. Seo, *Chemtech*, 1996, *26*(7), 46; (c) P.R. Seidl, O.R. Gottleib, and M.A. Kaplan, eds, *Chemistry of the Amazon: Biodiversity, Natural Products and Environmental Issues*. A.C.S. Symp. 588, Washington, DC, 1995, 2–7; (d) A.D. Kinghorn and M.F. Balandrin, *Human Medicinal Agents from Plants*. A.C.S. Symp. 534, Washington, DC, 1993; (e) W.V. Reid, ed., *Biodiversity Prospecting*, World Resources Institute, Washington, DC, 1993; (f) N.J. Walton, *Chem. Br.*, 1992, 525; (g) A.M. Rouhi, *Chem. Eng. News*, Apr. 7, 1997, 14; (h) P.L. Short, *Chem. Eng. News*, Feb. 26, 2007, 27; (i) A. Crozier, M.N. Clifford, and H. Ashihara, eds, *Plant Secondary Metabolites—Occurrence, Structure and Role in the Human Diet*, Blackwell Publishing, Oxford, 2006; (j) J.E. Smith, N.J. Rowan, and R. Sullivan, *Biotechnol. Lett.*, 2002, *24*, 1839.
490. (a) R.R. Colwell, *Science*, 1995, *267*, 1611; (b) D.J. Faulkner, *Chem. Br.*, 1995, *31*, 680; (c) F. Flam, *Science*, 1994, *266*, 1324; (d) L. Bongiorna and F. Pietra, *Chem. Ind.* (*Lond.*), 1996, 54; (e) B. Baker, L. Tangley, G. Liles, A.N.C. Morse, D.E. Morse, R.J. Radmer, and B.K. Carte, *Bioscience*, 1996, *46*(4), 240–271; (f) D.A. Cowan, *Trends Biotechnol.*, 1997, *15*, 129; (g) W. Fenical, *Trends Biotechnol*. 1997, *15*, 339.
491. (a) K. Hostettmann, A. Marston, M. Maillard, and M. Hamburger, eds, *Phytochemistry of Plants Used in Traditional Medicine*, Oxford University Press, Oxford, UK, 1995; (b) Ciba Foundation, *Ethnobotany and the Search for New Drugs*, Wiley, New York, 1994; (c) R. Rawls, *Chem. Eng. News*, Sept. 23, 1996, 53; (d) J. Mervis, *Science*, 1996, *273*, 578; (e) L.D. Lawson and R. Bauer, eds, *Phytomedicines of Europe: Chemistry and Biological Activity*, A.C.S. Symp. 691, Washington, DC, 1998; (f) V.E. Tyler, *Chemtech*, 1997, *27*(5), 52; (g) P.J. Houghton, *Chem. Ind.* (*Lond.*), 1999, 15.
492. J.L. Fox, *Biotechnology*, 1995, *13*, 544.

493. A.L. Harvey, *Chem. Ind. (Lond.)*, 1995, 914.
494. S. Budavari, M.J. O'Neil, A. Smith, P.E. Heckelman, and J.F. Kinneary, eds, *Merck Index*, 12th ed., Merck & Co., White-house Station, NJ, 1996, 1671.
495. Anon., *Chem. Br.*, 1998, *34*(10), 17.
496. (a) Anon., *Nat. Biotechnol.*, 1997, *15*, 5; (b) G. Stix, *Sci. Am.*, April 2005, 88.
497. (a) P.T. Lansbury and D.M. Nowak, *Tetrahedron Lett.*, 1992, *33*, 1029; (b) D.S. Torok and H. Ziffer, *Tetrahedron Lett.*, 1995, *36*, 828; (c) R.K. Haynes and S.C. Vonwiller, *Acc. Chem. Res.*, 1997, *30*, 73; (d) N.J. White, *Science*, 2008, *320*, 330; (e) R.K. Haynes, W.-Y. Ho, H.-W. Chan, B. Fugmann, J. Stetter S.L. Croft, L. Vivas, W. Peters, and B.L. Robinson, *Angew. Chem. Int. Ed.*, 2004, *43*, 1381.
498. (a) G.H. Posner, *Tetrahedron Lett.*, 1996, *37*, 815; (b) J.A. Vroman, H.N. El Sohly, and M.A. Avery, *Synth. Commun.*, 1998, *28*, 1555; (c) P.M. O'Neill, N.L. Searle, K.J. Raynes, J.L. Maggs, S.A. Ward, R.C. Storr, B.K. Park, and G.H. Posner, *Tetrahedron Lett.*, 1998, *39*, 6065; (d) M. Brennan, *Chem. Eng. News*, Mar. 2, 1998, 38.
499. P. Rasoanaivo, S. Ratsimamanga-Urverg, C. Galeffi, M. Nicoletti, F. Frappier, and M.-T. Martin, *Tetrahedron*, 1995, *51*, 1221.
500. (a) K.C. Nicolaou, *Chem. Br.*, 1994, *30*(1), 33; (b) K.C. Nicolaou, E.A. Theodorakis, and C.F. Claiborne, *Pure Appl. Chem.*, 1996, *68*, 2129; (c) I. Dancy, T. Skrydstrup, C. Crevisy, and J.-M. Beau, *J. Chem. Soc. Chem. Commun.*, 1995, 799.
501. (a) M. Mladenova, M. Alami, and G. Linstrumelle, *Synth. Commun.*, 1995, *25*, 1401; (b) J.W. Grissom and D. Huang, *Angew. Chem. Int. Ed.*, 1995, *34*, 2037; (c) S. Borman, *Chem. Eng. News*, Aug. 28, 1995, 28; (d) J.W. Grissom, G.U. Gunawardena, D. Klingberg, and D. Huang, *Tetrahedron*, 1996, *52*, 6453; (e) R. Gleiter and R. Ritter, *Angew. Chem. Int. Ed.*, 1994, *33*, 2470.
502. S.V. Ley and P.L. Toogood, *Chem. Br.*, 1990, (1), 31.
503. (a) S. Johnson, E.D. Morgan, I.D. Wilson, M. Spraul, and M. Hofmann, *J. Chem. Soc. Perkin Trans. I*, 1994, 1499; (b) W.-J. Koot and S.V. Ley, *Tetrahedron*, 1995, *51*, 2077; (c) A.A. Denholm, L. Jennens, S.V. Ley, and A. Wood, *Tetrahedron*, 1995, *51*, 6591; (d) G.E. Veitch, E. Beckmann, B.J. Burke, A. Boyer, S.L. Maslen, and S.V. Ley, *Angew. Chem. Int. Ed.*, 2007, *46*, 7629; (e) G.E. Veitch, A. Boyer, and S.V. Ley, *Angew. Chem. Int. Ed.*, 2008, *47*, 9402.
504. A.F. Barrero, E.A. Manzaneda, J. Altaregos, S. Salido, J.M. Ramos, M.S.J. Simmonds, and W.N. Blaney, *Tetrahedron*, 1995, *51*, 7435.
505. S. Borman, *Chem. Eng. News*, Sept. 12, 1994, 6.
506. (a) E. Pisha, H. Chai, I.-S. Lee, T.E. Chagwedera, N.R. Farnsworth, G.A. Cordell, C.W.W. Beecher, H.H.S. Fong, A.D. Kinghorn, D.M. Brown, M.C. Wani, M.E. Wall, T.J. Hieken, T.K. DasGupta, and J.M. Pezzuto, *Nat. Med.*, 1995, *1*, 1046; (b) Anon., *Chem. Eng. News*, Oct. 9, 1995, 32.
507. S. Kadota, J.X. Li, K. Tanaka, and T. Namba, *Tetrahedron*, 1995, *51*, 1143.
508. (a) J. de Brabander and M. Vandewalle, *Pure Appl. Chem.*, 1996, *68*, 715; (b) S. Pain, *New Sci.*, 1996, Sept. 14, 38; (c) P.A. Wender, J. De Brabander, P.G. Harran, J.-M. Jimenez, M.F.T. Koehler, B. Lippa, C.-M. Park, and M. Shiozaki, *J. Am. Chem. Soc.*, 1998, *120*, 4534; (d) D.A. Evans, P.H. Carter, E.M. Carreira, J.A. Prunet, A.B. Charette, and M. Lautens, *Angew. Chem. Int. Ed.*, 1998, *37*, 3254; (e) *New Sci.*, 1998, Feb. 7, 11; (f) P.A. Wender, B.A. deChristopher, and A.J. Schrier, *J. Am. Chem. Soc.*, 2008, *130*, 6658; (g) G.E. Keck, M.B. Kraft, A. Truong, W. Li, C.C. Sanchez, N. Kedei, N.E. Lewin, and P.M. Blumberg, *J. Am. Chem. Soc.*, 2008, *130*, 6660.
509. M.A. Ciufolini and F. Roschangar, *Angew. Chem. Int. Ed.*, 1996, *35*, 1692.
510. (a) T.H. Hiyama, G.B. Reddy, T. Minami, and T. Hanamoto, *Bull. Chem. Soc. Jpn.*, 1995, *68*, 350; (b) T. Hiyama, T. Minami, and K. Takahashi, *Bull. Chem. Soc. Jpn.*, 1995, *68*, 364.
511. (a) D.T. Hung, J.B. Nerenberg, and S.L. Schreiber, *J. Am. Chem. Soc.*, 1996, *118*, 11054; (b) C.A. Bewley and D.J. Faulkner, *Angew. Chem. Int. Ed.*, 1998, *37*, 2163; (c) E. deLemos, F.-H. Poree, A. Commercon, J.F. Betzer, A. Pancrazi, and J. Ardisson, *Angew. Chem. Int. Ed.*, 2007, *46*, 1917; (d) M. Freemantle, *Chem. Eng. News*, Mar. 1, 2004, 33.
512. (a) R. Xu, G. Chu, and D. Bai, *Tetrahedron Lett.*, 1996, *37*, 1463; (b) M.L. Trudell and C. Zhang, *J. Org. Chem.*, 1996, *61*, 7189; (c) C.D. Jones and N.S. Simpkins, *Tetrahedron Lett.*, 1998, *39*, 1021, 1023; (d) N.S. Sirisoma and C.R. Johnson, *Tetrahedron Lett.*, 1998, *39*, 2059.
513. (a) E. Strauss, *Science*, 1998, *279*, 32; (b) A.W. Bannon, M.W. Decker, M.W. Hollalday, P. Curzon, D. Donnelly-Roberts, P.S. Puttfarcken, R.S. Bitner, A. Diaz, A.H. Dickenson, R.H. Porsolt, M. Williams, and S.P. Arneric, *Science*, 1998, *279*, 77.
514. (a) S. Borman, *Chem. Eng. News*, 1996, Dec. 23, 24; (b) K.C. Nicolaou, Y. He, D. Vourloumis, H. Vallberg, Z. Yang, *Angew. Chem. Int. Ed.*, 1996, *35*, 2399; (c) Z. Yang, Y. He, D. Vourloumis, H. Vallberg, and K.C. Nicolaou, *Angew. Chem. Int. Ed.*, 1997, *36*, 166; (d) D.-S. Su, D. Meng, P. Bertinato, A. Balog, E.J. Sorensen, S.J. Danishefsky, Y.-H. Zheng, T.-C. Chou, L. He, and S.B. Horwitz, *Angew. Chem. Int. Ed.*, 1997, *36*, 757; (e) K.C. Nicolaou, F. Roschangar, and D. Vourloumis, *Angew. Chem. Int. Ed.*, 1998, *37*, 2014; (f) A. Balog, C. Harris, K. Savin, X.-G. Zhang, T.C. Chou, and S.J. Danishefsky, *Angew. Chem. Int. Ed.*, 1998, *37*, 2675; (g) R. Finlay, *Chem. Ind. (Lond.)*, 1997, 991; (h) K.C. Nicolaou, A. Ritzen, and K. Namoto, *Chem. Commun.*, 2001, 1523; (i) U. Klar, B. Buchmann, W. Schwede, W. Skuballa, J. Hoffmann, and R.B. Lichmer, *Angew. Chem. Int. Ed.*, 2006, *45*, 7942; (j) F. Cachoux, T. Isarno, M. Wartman, and K.-H. Altmann, *Angew. Chem. Int. Ed.*, 2005, *44*, 7469; (k) F. Feyen, F. Cachoux, J. Gertsch, M. Wartman, and K.-H. Altmann, *Acc. Chem. Res.*, 2008, *41*(1), 21.
515. (a) G.I. Georg, T.T. Chen, I. Ojima, and D.M. Wyas, eds, *Taxane Anticancer Agents: Basic Science and Current Status*. A.C.S. Symp. 583, Washington, DC, 1995; (b) P. Jenkins, *Chem. Br.*, 1996, *32*(11), 43; (c) S. Borman, *Chem. Eng. News*, Apr. 26, 1999, 35; (d) J. Andrako, *Chem. Eng. News*, June 16, 1997, 7; (e) D.G.I. Kingston, *Chem. Commun.*, 2001, 867.
516. (a) B. Delpech, D. Calvo, and R. Lett, *Tetrahedron Lett.*, 1996, *37*, 1015, 1019; (b) R. Calvo, M. Port, B. Delpech, and R. Lett, *Tetrahedron Lett.*, 1996, *37*, 1023; (c) S. Zimmermann, S. Bick, P. Welzel, H. Meuer, and W.S. Sheldrick, *Tetrahedron*, 1995, *51*, 2947.
517. S. Chackalamannil, R.J. Davies, T. Asberom, D. Doller, and D. Leone, *J. Am. Chem. Soc.*, 1996, *118*, 9812.

518. Anon., *Chem. Ind. (Lond.)*, 1997, 406.
519. (a) D.E. Bogucki and J.L. Charlton, *J. Org. Chem.*, 1995, *60*, 588; (b) D. Stadler and T. Bach, *Angew. Chem. Int. Ed.*, 2008, *47*, 7557.
520. W. Stolzenburg, *Nat. Conserv.*, 1999, *49*(1), 7.
521. (a) A.B. Smith III, S.M. Condon, J.A. McCauley, J.L. Leazer, Jr., J.W. Leahy, R.E. Maleczka, Jr., *J. Am. Chem. Soc.*, 1997, *119*, 947; (b) M.L. Maddess, M.N. Tackett, H. Watanabe, P.E. Brennan, C.D. Spilling, J.S. Scott, D.P. Osborn, and S.V. Ley, *Angew. Chem. Intl. Ed.*, 2007, *46*, 591.
522. L. Aylsworth, *Nat. Conserv.*, 1998, *48*(2), 6.
523. (a) J.D. White, T.-S. Kim, and M. Nambu, *J. Am. Chem. Soc.*, 1995, *117*, 5612; (b) W.H. Gerwick, P.J. Proteau, D.G. Nagle, E. Hamel, A. Blokhin, and D.L. Slate, *J. Org. Chem.*, 1994, *59*, 1243; (c) E.W. Schmidt, *Trends Biotechnol.*, 2005, *23*(9), 437; (d) Anon., *Chem. Eng. News*, Oct. 14, 2002, 4.
524. Anon., *Chem. Eng. News*, Aug. 15, 2005, 38.
525. (a) E.O. Wilson, ed., *Biodiversity*. National Academy Press, Washington, DC, 1988; (b) E.T. LaRoe, *Our Living Resources*. National Biological Service, Sup. Documents, Pittsburgh, PA, 1995; (c) A.T. Durning, *World Watch*, 1996, *9*(6), 28; 1997, *10*(1), 25; (d) C. Bright, *World Watch*, 1999, *12*(3), 12; (e) F.S. Chapin, O.E. Sala, I.C. Burke, J.P. Grime, D.U. Hooper, W.K. Lauenroth, A. Lombard, H.A. Mooney, A.R. Mosier, S. Naeem, S.W. Pacala, J. Roy, W.L.S. Teffen, D. Tilman, *BioScience*, 1998, *48*(1), 45; (f) S.L. Pimm and J.H. Lawton, *Science*, 1998, *279*, 2068; (g) M.A. Cochrane, A. Alencar, M.D. Schulze, C.M. Souza, Jr., D.C. Nepstad, P. Lefebvre, E.A. Davidson, *Science*, 1999, *284*, 1832; (h) J. Jernvall and P.C. Wright, *Proc. Natl. Acad. Sci. USA*, 1998, *95*, 11279; (i) P. Kareiva, S. Watts, R. McDonald, and T. Boucher, *Science*, 2007, *316*, 1866; (j) E.W. Sanderson, M. Jaiteh, M.A. Levy, K.H. Redford, and A.V. Wannebo, *BioScience*, 2002, *52*(10), 891.
526. (a) D.S. Wilcove, D. Rothstein, J. Dubow, A. Phillips, and E. Losos, *BioScience*, 1998, *48*, 607; (b) B. Czech, P.R. Krausman, and P.K. Devers, *BioScience*, 2000, *50*, 593; (c) D. Pimentel, L. Lach, R. Zuniga, and D. Morrison, *BioScience*, 2000, *50*, 53.
527. (a) J.N. Abramovitz, *World Watch*, 1995, *8*(5), 27; (b) B.A. Stein and S.R. Flack, *Environment*, 1997, *39*(4), 6.
528. (a) C. Holden, *Science*, 1996, *274*, 183; (b) J. Schipper, J.S. Chanson, F. Chiozza, N.A. Cox, M. Hoffmann, V. Katariya, J.C. Venter et al., *Science*, 2008, *322*, 225; (c) B.A. Stein, L.S. Kutner, and J.S. Adams, eds, *Precious Heritage—The Status of Biodiversity in the United States*, Oxford Univerisity Press, 2000; (d) B. Groombridge and M.D. Jenkins, *World Atlas of Biodiversity*, University of California Press, Berkeley, 2002; (e) R.E. Hester and R.M. Harrison, eds, *Biodiversity Under Threat*, Royal Soc. Chem, Cambridge, UK, 2007.
529. (a) P.T.J. Johnson, K.B. Lunde, E.G. Ritchie, and A.E. Launer, *Science*, 1999, *284*, 802; (b) V. Morell, *Science*, 1999, *284*, 728; (c) J. Kaiser, *Science*, 1998, *281*, 23; (d) J. Pelley, *Environ. Sci. Technol.*, 1998, *32*, 352A; (d) K.R. Lips, J.D. Diffendorfer, J.R. Mendelson III, and M.W. Sears, *PLoS Biol.*, 2008, *6*(3), e72; (e) www.globalamphibians.org.
530. T. Eisner, J. Lubchenko, E.O. Wilson, D.S. Wilcove, and M.J. Bean, *Science*, 1995, *269*, 1231.
531. C. Holden, *Science*, 1998, *280*, 385.
532. (a) N. Meyers, *Science*, 1995, *268*, 823; (b) Anon., *Environ. Sci. Technol.*, 1998, *32*, 225A; (c) N. Williams, *Science*, 1998, *281*, 1426; (d) A.S. Moffatt, *Science*, 1998, *282*, 1253; (e) I.R. Noble and R. Dirzo, *Science*, 1997, *277*, 522.
533. (a) M.E. Soule and M.A. Sanjayan, *Science*, 1998, *279*, 2060; (b) R.L. Chazdon, *Science*, 1998, *281*, 1295; (c) M. Williams, *Deforesting the Earth*, University of Chicago Press, Chicago, 2003; (d) F. Achard, H.D. Eva, H.-J. Stibig, P. Maynaux, J. Gallego, T. Richards, and J.-P. Malingreau, *Science*, 2002, *297*, 999; (e) geodata.grid.unep.ch.
534. K.A. Vogt, B.C. Larson, J.C. Gordon, D.J. Vogt, and A. Fanzeres, *Forest Certification—Roots, Issues, Challenges and Benefits*, CRC Press, Boca Raton, FL, 1999.
535. (a) S.L. Pimm, G.J. Russell, J.L. Gittleman, and T.M. Brooks, *Science*, 1995, *269*, 347; (b) G.W. Cox, *Alien Species in North America and Hawaii—Impacts on Natural Ecosystems*, Island Press, Washington, DC, 1999.
536. (a) D. Tenenbaum, *Technol. Rev.*, 1996, *99*(6), 32; (b) C. Bright, *World Watch*, 1995, *8*(4), 10; 1999, *12*(1), 22; (c) R.G. Westbrooks, *Invasive Plants, Changing the Landscape of America*. Federal Interagency Committee for the Management of Noxious and Exotic Weeds., Washington, DC, 1998; (d) R. Devine, *Alien Invasion: America's Battle with Non-Native Animals and Plants*. National Geographic Society, Washington, DC, 1998; (e) M. Williamson, *Biological Invasions*, Chapman & Hall, London, 1996; (f) N. Shigesada and K. Kawasaki, *Biological Invasions: Theory and Practice*. Oxford University Press, Oxford, 1997; (g) H.A. Mooney and R.J. Hobbs, eds, *Invasive Species in a Changing World*, Island Press, Washington, DC, 2000; (h) D.F. Sax, J.J. Stachowiz, and S.D. Gaines, eds, *Species Invasions—Insights Into Ecology, Evolution and Biogeography*, Sinauer, Sunderland, MA, 2005; (i) J. vanDriesche and R. vanDriesche, *Nature Out of Place: Biological Invasions in the Global Age*, Island Press, Washington, DC, 2001; (j) J.A. McNeely, H.A. Mooney, L.E. Neville, P.J. Schei, and J.K. Waage, eds, *Global Strategy On Invasive Alien Species*, Island Press, Washington, DC, 2001. (k) Y. Baskin, *A. Plague of Rats and Rubbervines*, Island Press, Washington, DC, 2002.
537. (a) K. Christen, *Environ. Sci. Technol.*, 2001, *35*, 14A; (b) C.D. Hunt, D.C. Tanis, T.G. Stevens, R.M. Frederick, and R.A. Everett, *Environ. Sci. Technol.*, 2005, *39*, 321A.
538. (a) G.C. Daily, *Science*, 1995, *269*, 350; (b) C. Sekercioglu, G. Daily, and P. Ehrlich, *Proc. Nat. Acad. Sci. USA*, 2004, *101*(52), 18042; (c) E. Stokstad, *Science*, 2005, *308*, 41.
539. (a) D.A. Wedin and D. Tilman, *Science*, 1996, *274*, 1720; (b) C.J. Stevens, N.B. Dise, J.O. Mountford, and D.J. Gowing, *Science*, 2004, *303*, 1876.
540. (a) G. Taubes, *Science*, 1995, *267*, 1595; (b) T.E. Lovejoy and L. Hannah, eds, *Climate Change and Biodiversity*, Yale University Press, New Haven, 2005; (c) D.A. Lindenmeyer and J. Fischer, *Habitat Fragmentation and Landscape Change*, Island Press, Washington, DC, 2006.
541. (a) D. Pearce and D. Moran, *The Economic Value of Biodiversity*. Earthscan, London, 1994; (b) M. Milstein, *Science*, 1995, *270*, 226; (c) P.R. Andrews, R. Borris, E. Dange, M.P. Gupta, L.A. Mitscher, A. Monge, N.J. de Sourz, and J.G. Topliss, *Pure Appl. Chem.*, 1996, *68*, 2325, 2333; (d) K. Raustiala and D.G. Victor, *Environment*, 1996, *38*(4), 17; (e) World Resources Institute. Global Biodiversity Strategy.

Washington, DC, 1992; (f) S.R. Kellert, *The Value of Life—Biological Diversity and Human Society*. Island Press, Washington, DC, 1996; (g) Y.L. Baskin, *The Work of Nature—How the Diversity of Life Sustains Us*. Island Press, Washington, DC, 1997; (h) M.L. Reaka-Kudla, D.E. Wilson, and E.O. Wilson, eds, *Biodiversity II: Understanding and Protecting Our Biological Resources*. Joseph Henry Press, Washington, DC, 1997; (i) P. Bachman, M. Kohl, and R. Pavinen, eds, *Assessment of Biodiversity for Improving Forest Planning*, Kluwer Academic, Dordrecht, Netherlands, 1998; (j) F. Powledge, *BioScience*, 1998, *48*, 347; (k) M. Loreau, S. Naeem, and P. Inchaustl, *Biodiversity and Ecosystem Functioning*, Oxford University Press, Oxford, 2002.

542. (a) T.C. Foin, *BioScience*, 1998, *48*, 177; (b) W.V. Reid, *Environment*, 1997, *39*(7), 16; (c) I.M. Goklany, *BioScience*, 998, *48*, 941; (d) B. Thorne-Miller, *The Living Ocean—Understanding and Protecting Marine Biodiversity*, 2nd ed., Island Press, Washington, DC, 1998; (e) O.H. Frankel, A.H.D. Brown, and J.J. Burdon, *The Conservation of Plant Biodiversity*. Cambridge University Press, New York, 1995; (f) A.G. Johns, *Timber Production and Biodiversity Conservation in Tropical Rain Forests*. Cambridge University Press, New York, 1997; (g) L.R.K. Baydack, H. Campa III, and J.B. Haufler, eds, *Practical Approaches to the Conservation of Biological Diversity*, Island Press, Washington, DC, 1998; (h) S. Peck, *Planning for Biodiversity*, Island Press, Washington, DC, 1998; (i) D.L. Dekker-Robertson and W.J. Libby, *BioScience*, 1998, *48*, 471; (j) R.F.G. Ormond, J.D. Gage, M.V. Angle, and C. Tickell, eds, *Marine Biodiversity—Patterns and Processes*, Cambridge University Press, New York, 1997; (k) V.H. Heywood and R.T. Watson, eds, *Global Biodiversity Assessment*, Cambridge University Press, New York, 1996.
543. (a) L.W. Botsford, J.C. Castilla, and C.H. Peterson, *Science*, 1997, *277*, 509; (b) A. Hastings and L.W. Botsford, *Science*, 1999, *284*, 1537; (c) C. Holden, *Science*, 1999, *283*, 1631; (d) L. Watling and E.A. Norse, eds, *Conserv. Biol.*, 1998, *12*(6), 1178–1240; (e) D. Malakoff, *Science*, 1998, *282*, 2168; (f) M. Batisse, *Environment*, 1997, *39*(5), 6; (g) L.K. Bergen and N.H. Carr, *Environment*, 2003, *45*(2), 8; (h) K. McLeod and B.H. Leslie, eds., *Ecosystem-Based Management for the Oceans*, Island Press, Washington, DC, 2009.
544. (a) J.J. Armesto, R. Rozzi, C. Smith-Ramirez, and M.T.K. Arroyo, *Science*, 1998, *282*, 1271; (b) C.H. Flather, M.S. Knowles, and I.A. Kendall, *BioScience*, 1998, *48*, 365.
545. C. Castelllo, S.D. Gaines, and J. Lynham, *Science*, 2008, *321*, 1678.
546. (a) M.S. Honey, *Environment*, 1999, *41*(5), 4; (b) W.M. Getz, L. Fortmann, D. Cumming, J. du Toit, J. Hilty, R. Martin, M. Murphree, N. Owen-Smith, A.M. Starfield, and M.I. Westphal, *Science*, 1999, *283*, 1855; (c) A. Inamdar, H. de Jode, K. Lidsay, and S. Cobb, *Science*, 1999, *284*, 1856.
547. P.R. Ehrlich and A.H. Ehrlich, *One With Neneveh—Politics, Consumption and the Human Future*, Island Press, Washington, DC, 2004.
548. (a) D.A. Taylor, *Environment*, 1997, *39*(1), 6; (b) C. Maser, B.T. Bormann, M.H. Brookes, A.R. Kiester, and J.F. Weigand, *Sustainable Forestry—Philosophy, Science and Economics*. St. Lucie Press, Delray Breach, FL, 1994.
549. (a) L.J. Sauer, *The Once and Future Forest—A Guide to Forest Restoration Strategies*. Island Press, Washington, DC, 1998; (b) D.A. Falk, C.I. Millar, and M. Olwell, eds, *Restoring Diversity—Strategies for Reintroduction of Endangered Plants*. Island Press, Washington, DC, 1996; (c) D.A. Falk, M.A. Palmer, and J.B. Zedler, eds, *Foundations of Restoration Ecology*, Island Press, Washington, DC, 2006; (d) R.B. Primack, *A Primer of Conservation Biology*, 4th ed., Sinauer Associates, Sunderland, MA, 2008; (e) L. Roberts, R. Stone, and A. Sugden, eds., *Science*, 2009, *325*, 555–576; (f) J.M.R. Benayas, A.C. Newton, A. Diaz, and J.M. Bullock, *Science*, 2009, *325*, 1121.
550. (a) A.M. Rouhi, *Chem. Eng. News*, Nov. 20, 1995, 42; (b) P.H. Canter, H. Thomas, and E. Ernst, *Trends Biotechnol.*, 2005, *23*(4), 180.
551. R. Osinga, J. Tramper, and R.H. Wijffels, *Trends Biotechnol.*, 1998, *16*, 130.
552. (a) R.L. Naylor, S.L. Williams, and D.R. Strong, *Science*, 2001, *294*, 1655; (b) R. Subasinghe, D. Soto, and J. Jia, *Rev., Aquaculture*, 2009, *1*(1), 2.
553. (a) R. Ellis, National Public Radio, 1/2/09; (b) R. Ellis, *Tuna—A Love Story*, Alfred A.Knopf, New York, 2008.
554. (a) A.H. Scragg, ed., *J. Biotechnol.*, 1992, *26*, 1–99; (b) F. Pierik, *In Vitro Culture of Higher Plants*, 3rd ed., Kluwer, Dordrecht, Netherlands, 1987; (c) M.A., L.J. Borowitzka, *Microalgal. Biotechnology*, Cambridge University Press, Cambridge, U.K., 1988; (d) *Richmond. Handbook of Microalgal Mass Culture*, CRC Press, Boca Raton, FL, 1986; (e) R.H. Smith, *Plant Tissue Culture*. Academic Press, San Diego, CA, 1992; (f) A. Stafford and G. Warren, *Plant Cell and Tissue Culture*. Wiley, New York, 1991; (g) R.A. Dixon, and R.A. Gonzales, *Plant Cell Culture—A Practical Approach*, 2nd ed., IRL Press, Oxford, UK, 1994; (h) S.S. Bhojwani, *Plant Tissue Culture—Applications and Limitations*. Elsevier, Amsterdam, 1990; (i) F. Canstabel and I.K. Vasil, *Phytochemicals in Plant Cell Cultures*. Academic Press, San Diego, CA, 1988; (j) O.L. Gamborg and G.C. Phillips, eds, *Plant Cell, Tissue, and Organ Culture—Fundamental Methods*. Springer-Verlag, New York, 1995; (k) G. Payne, V. Bringi, C. Prince, and M.L. Shuler, *Plant Cell and Tissue Culture in Liquid Systems*. Hanser, Munich, 1991, 177–223; (l) J.W. Pollard and J.M. Walker, *Plant Cell and Tissue Culture Techniques*. Humna, Totowa, NJ, 1990; (m) I.K. Vasil and T.A. Thorpe, *Plant Cell and Tissue Culture*. Kluwer Academic, Dordrecht, Netherlands, 1994; (n) W.R. Sharp, Z. Chen, D.A. Evans, P.V. Ammirato, M.R. Sondahl, Y.-P.S. Bajaj, and Y. Yamada, eds, *Handbook Plant Cell Culture*, vols. 1–6. McGraw-Hill, New York, 1983–1989; (o) D. Knorr, C. Caster, H. Dorneburg, R. Dorn, S. Graf, D. Havkin-Frenkel, A. Podstolski, and U. Werrman, *Food Technol.*, 1993, *47*(12), 57; (p) *Biotechnology in Agriculture and Forestry*, vol. 20—High Technology and Micropropagation, Springer, Berlin, 1992; (q) J.P. Kutney, *Pure Appl. Chem.*, 1997, *69*, 431; (r) T.-J. Fu, *Chemtech*, 1998, *28*(1), 40; (s) A.H. Shulman, *Trends Biotechnol.*, 1998, *16*, 1; (t) *Cuppett. Int. News Fats Oils Relat. Mater.*, 1998, *9*, 588; (u) H. Dornenburg and D. Knorr, *Food Technol.*, 1997, *51*(11), 47 and following papers, pp. 56–72; (v) J. Coleman, D. Evans, and A. Kearns, *Plant Cell Culture*, Taylor & Francis, Boca Raton, FL, 2003.

555. (a) Y.-H. Zhang and J.-J. Yu, *Biotechnol. Prog.*, 1996, *12*, 567; (b) A. Scheidegger, *Trends Biotechnol.*, 1990, *8*, 197; (c) O. Sticher, *Chemtech*, 1998, *28*(4), 26; (d) Y.-H. Zhang and J.-J. Zhong, *Enzyme Microb. Technol.*, 1997, *21*, 59; (e) Z.-Y. Zhang and J.-J. Zhong, *Biotechnol. Prog.*, 2004, *20*, 1076.
556. T.M. Kutchan, A. Bock, and H. Dittrich, *Phytochemistry*, 1994, *35*, 353.
557. J.P. Kutney, *ACC. Chem. Res.*, 1993, *26*, 559; *Pure Appl. Chem.*, 1999, *71*, 1025.
558. J.P. Kutney, *Pure Appl. Chem.*, 1996, *68*, 2073; 1994, *66*, 2243.
559. (a) R.M. Buitelaar and J. Tramper, *J. Biotechnol.*, 1992, *23*, 111; (b) R. Tom, B. Jardin, C. Chavarie, and J. Archambault, *J. Biotechnol.*, 1991, *21*, 1; (c) R. Tom, B. Jardin, C. Chavarie, D. Rho, and J. Archambault, *J. Biotechnol.*, 1991, *21*, 21; (d) B. Jardin, R. Tom, C. Chavarie, D. Rho, and J. Archambault, *J. Biotechnol.*, 1991, *21*, 43; (e) R. Ballica, D.D.Y. Rhy, and C.I. Kado, *Biotechnol. Bioeng.*, 1993, *41*, 1075; (f) D.I. Kim, H. Pedersen, and C.-K. Chin, *J. Biotechnol.*, 1991, *21*, 201; (g) A.K. Lipsky, *J. Biotechnol.*, 1992, *26*, 83; (h) H. Dornenburg and D. Knorr, *Enzyme Microb. Technol.*, 1995, *17*, 674; (i) P. Komaraiah, M. Navratil, M. Carlsson, P. Jeffers, M. Brodelius, P.E. brodelius, M. Kieran, and C.-F. Mandenius, *Biotechnol. Prog.*, 2004, *20*, 1245.
560. V. Srinivasan and D.D.Y. Ryu, *Biotechnol. Bioeng.*, 1993, *42*,793.
561. C.-H. Park and B.C. Martinez, *Biotechnol. Bioeng.*, 1992, *40*, 459.
562. (a) I.M. Whitehead and D.R. Threlfall, *J. Biotechnol.*, 1992, *26*, 63; (b) N.P. Malpathak and S.B. David, *Biotechnol. Lett.*, 1992, *14*, 965; (c) M. Tani, K. Takeda, K. Yazaki, and M. Tabata *Phytochemistry*, 1993, *34*, 1285; (d) H. Dorneburg and D. Knorr, *J. Agric. Food Chem.*, 1994, *42*, 1048; (e) S.K. Rijhwani and J.V. Shanks, *Biotechnol. Prog.*, 1998, *14*, 442; (f) A. Namdeo, S. Patil, and D.P. Fulzele, *Biotechnol. Prog.*, 2002, *18*, 159.
563. P. Brodelius and H. Pedersen, *Trends Biotechnol.*, 1993, *11*, 30.
564. R.D. Williams, N. Chauret, C. Bedard, and J. Archambault, *Biotechnol. Bioeng.*, 1992, *40*, 971.
565. A.J. Parr, *J. Biotechnol.*, 1989, *10*, 1.
566. (a) H.E. Flores, *Chem. Ind.* (*Lond.*), 1992, 274; (b) S.A. McKelvery, J.A. Gehrig, K.A. Hollar, and W.R. Curtis, *Biotechnol. Prog.*, 1993, *9*, 317; (c) P.M. Doran, ed., *Hairy Roots—Culture and Applications*. Harwood Academic, Amsterdam, 1997; (d) S. Guillo, J. Tremouillaux-Guiller, P.K. Pati, M. Rideau, and P. Gantey, *Trends Biotechnol.*, 2006, *24*(9), 403.
567. (a) R. Bhadra, S. Vani, and J.V. Shanks, *Biotechnol. Bioeng.*, 1993, *41*, 581; (b) S.K. Rijhwani and J.V. Shanks, *Biotechnol. Prog.*, 1998, *14*, 442.
568. R.D. Williams and B.E. Ellis, *Phytochemistry*, 1993, *32*, 719.
569. T. Ogasawara, K. Chiba, and M. Tada, *Phytochemistry*, 1993, *33*, 1095.
570. J.E. Thorup, K.A. McDonald, A.P. Jackman, N. Bhatia, and A.M. Dendekar, *Biotechnol. Prog.*, 1994, *10*, 345.
571. (a) C.Z. Liu, Y.C. Wang, F. Ouyang, H.C. Ye, and G.F. Li, *Biotechnol. Lett.*, 1998, *20*, 265; (b) J.-W. Wang and R.X. Tan, *Biotechnol. Lett.*, 2002, *24*, 1153.
572. D.H. Schubel, C.M. Ruyter, J. Stockiat, *Phytochemistry*, 1989, *28*, 491.
573. L. Rajendran, G.A. Ravishankar, L.V. Venktaraman, and K.R. Prathiba, *Biotechnol. Lett.*, 1992, *14*, 707.
574. K. Sakamoto, K. Iida, K. Sawamura, K. Hajiro, Y. Asada, T. Yoshikawa, and T. Furuya, *Phytochemistry*, 1993, *33*, 357.
575. K. Matsubara, S. Kitani, T. Yoshioka, T. Morimoto, Y. Fujita and Y. Yamada, *J. Chem. Technol. Biotechnol.*, 1989, *46*, 61.
576. B. Dell, C.L. Elsegood, and E.L. Ghisalberta, *Phytochemistry*, 1989, *28*, 1871.
577. T. Hirata, S. Murakami, K. Ogihara, and T. Suga, *Phytochemistry*, 1990, *29*, 493.
578. (a) M. Seki and S. Furusaki, *Chemtech*, 1996, *26*(3), 41; (b) L.J. Pestchanker, S.C. Roberts, and M.L. Shuler, *Enzyme Microb. Technol.*, 1996, *19*, 256; (c) N. Mirjalili and J.C. Linden, *Biotechnol. Prog.*, 1996, *12*, 110; (d) J. Luo, X.G. Mei, L. Liu, and D.W. Hu, *Biotechnol. Lett.*, 2002, *24*, 561; (e) R.M. Cusido, J. Palazon, M. Bonfill, A. Navia-Osorio, C. Morales, and M.T. Pinol, *Biotechnol. Prog.*, 2002, *18*, 418.
579. (a) Y. Yukimune, H. Tabata, Y. Higashi, and Y. Hara, *Nat. Biotechnol.*, 1996, *14*, 1129; (b) J. Pezzuto, *Nat. Biotechnol.*, 1996, *14*, 1083.
580. P.D. Denchev, A.I. Kulkin, and A.H. Scragg, *J. Biotechnol.*, 1992, *26*, 99.
581. M. Timm, *Biotechnology*, 1989, *7*, 118.
582. (a) X.-M. Shi, Y. Jiang, and F. Chen, *Biotechnol. Prog.*, 2002, *18*, 723; (b) A.P. Carvalho, L.A. Meireles, and F.X. Malcata, *Biotechnol. Prog.*, 2006, *22*, 1490; (c) R.H. Wijffels, *Trends Biotechnol.*, 2008, *26*(1), 26.
583. (a) J.W. Kram, *Biodiesel.*, 2007, *4*(11), 40; (b) M. Phelan, *Chem. Eng.*, 2008, *115*(9), 22; (c) G. Ondrey, *Chem. Eng.*, 2008, *115*(7), 16; (d) G. Ondrey, *Chem. Eng.*, 2008, *115*(1), 15.

RECOMMENDED READING

1. J.W. Frost and J. Lievense. *New J. Chem.* 1994, *18*, 341; or J.W. Frost and K.M. Draths, *Chem. Br.* 1995, *31*, 206.
2. A.D. Kinghorn, M.F. Balandrin, eds, *Human Medicinal Agents from Plants*, A.C.S. Symp. 534, Washington, DC, 1993, pp. 2–10.
3. F. Flam, *Science*, 1994, *226*, 1324.
4. M.W.W. Adams and R.M. Kelly, *Chem. Eng. News*, Dec. 18, 1995, 32.
5. A.J. Kirby, *Angew Chem. Int. Ed. Engl.*, 1996, *35*, 720–722.
6. J. Lalonde, *Chem. Eng.* 1997, *104*(9), 108.
7. A.L. Paiva and F.X. Malcata, *J. Mol. Catal. B Enzymol.*, 1997, *3*, 99–106.
8. H.E. Shoemaker, D. Mink, and B.G. Wulbbolts, *Science*, *299*, 1694 (biocatalysis).
9. D. Ringe and G.A. Petsko, *Science*, 2008, *320*, 1428 ("How enzymes work").
10. G. Conko, J. Shave, and G. Azeez, *Chem. Eng. Prog.*, 2005, *101*(9), 4–6 (pro and con of genetically-modified crops).
11. P. Karieva, S. Watts, R. McDonald, and T. Boucher, *Science*, 2007, *316*, 1866 (man's global footprint).

EXERCISES

1. Visit a microbiology laboratory near you to see how fermentation is done.
2. Check the library for books on fermentation that were written in 1910–1930 to see what compounds were made by the method then that are made from petroleum or natural gas today. What modern techniques would you use to improve the processes to make them competitive today?
3. Pick a present-day process for a large volume chemical that is made by a process that is energy-intensive or involves the use of toxic reagents or catalysis. How would you go about finding an organism that would make the same chemical from renewable resources? You may need to use a simple chemical step in addition to the biocatalytic ones.
4. Are there any food-processing plants in your area that are having waste disposal problems? If so, see if you can devise a biocatalytic system to make something useful and saleable from the wastes.
5. Check with your state Natural Heritage Program (or an equivalent program if you live in a country other than the United States) to see what percentage of the native flora and fauna are extinct, endangered, or threatened. See if you can find out why these populations are in trouble?
6. Check with your department of agriculture to see what exotic plants and animals are in your area and what threats they pose to the native flora and fauna. Then go into the field to verify this.
7. Pick a reduction by baker's yeast in a recent journal or in *Organic Syntheses*. Using this procedure try reducing a different compound by this method.
8. Visit a nearby wastewater treatment plant. See if it is using the best available technology.

CHAPTER 10

Stereochemistry

10.1 IMPORTANCE OF OPTICAL ISOMERS

Most organic molecules in living organisms contain asymmetrical centers (i.e., they are chiral). (The terminology of stereochemistry has been reviewed by Moss.[1]) For example, amino acids that are incorporated into proteins are L and sugars in carbohydrates are D. It is understandable that the three-dimensional structures of the receptors in proteins for small molecules will favor only one optical isomer (i.e., the one that fits sterically, hydrogen bonds properly, and so on). Most compounds made for use by plants and animals will have to be single optical isomers. These include agricultural chemicals, drugs, flavors, food additives, fragrances, and such. This is especially important with drugs,[2] for which the unwanted isomer may produce toxic side effects.

There are several possibilities: (a) The unwanted isomer may be inert, in which case its synthesis is a waste, for it must be discarded or recycled, if this is possible. (b) The unwanted isomer is useful for another entirely different purpose. (c) The unwanted isomer can ruin the effects of the desired one and may prove to be toxic.[3] (d) The isomers may racemize in the body so that the racemate may be used. Sheldon has listed some examples (**10.1** Schematic) that illustrate these cases.[4]

The reader will appreciate the difference in flavors of the two terpene isomers. Aspartame is the popular artificial sweetener found in many soft drinks and other foods. Dramatic differences can also be found in the isomers of drugs (**10.2** Schematic).

Some drugs may not need to be single isomers because they racemize in the body. The analgesic ibuprofen (**10.3** Schematic) is an example. Even so, Merck planned to market the *S*-isomer as faster acting (12 min versus 30 min for the racemic drug).[2d] (The isomerization presumably

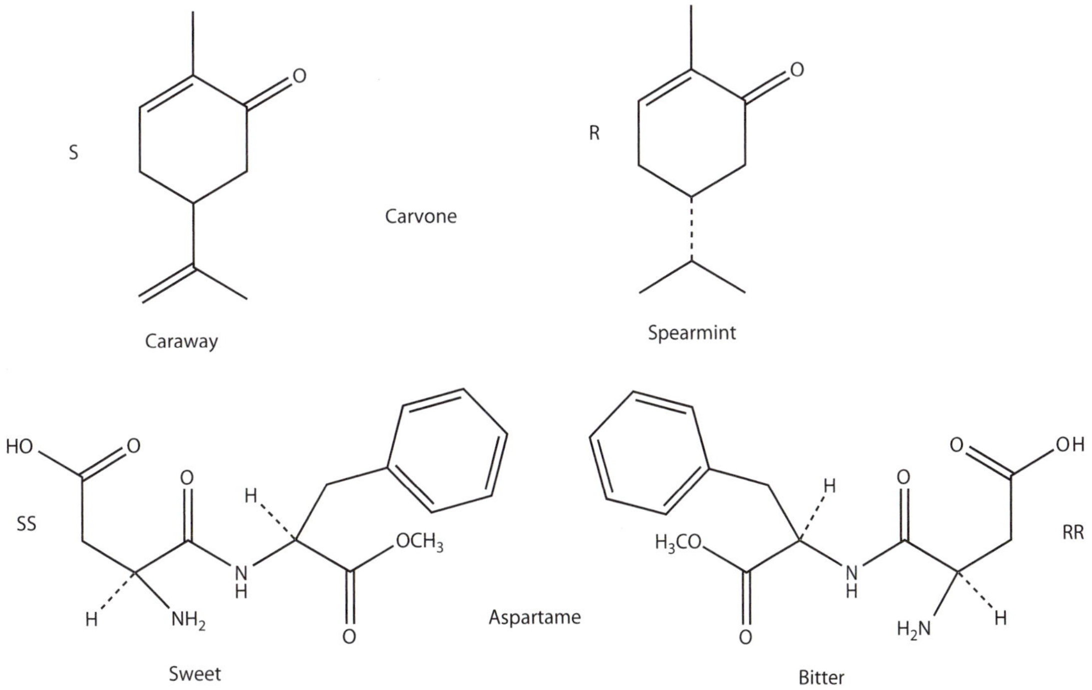

10.1 Schematic

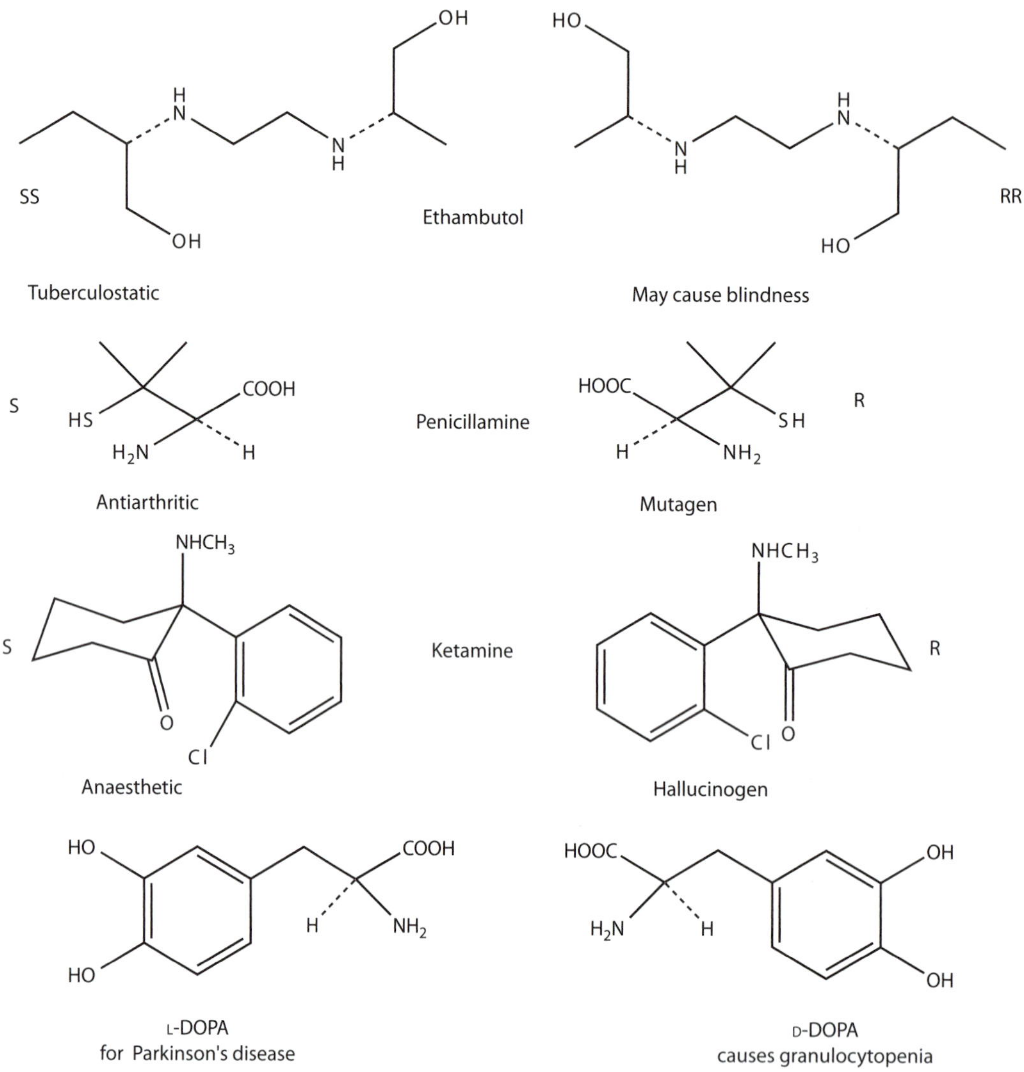

SS Ethambutol RR

Tuberculostatic | May cause blindness

S Penicillamine R

Antiarthritic | Mutagen

S Ketamine R

Anaesthetic | Hallucinogen

L-DOPA for Parkinson's disease | D-DOPA causes granulocytopenia

10.2 Schematic

involves removal of a proton from the carbon atom α to the carboxylic acid.)

Whether this is applicable to other drugs that isomerize in the body depends on the rate of isomerization and the rate at which the desired effect is wanted.

Optical isomers can also have different effects on plants (**10.4** Schematic).

10.2 CHIRAL POOL

The desired optical isomer can sometimes be produced as a natural product. The caraway and spearmint flavors mentioned earlier (see **10.1** Schematic) are extracted from crop plants. Other single isomers can be produced by enzymatic reactions, as described in Chapter 9. Thus, the

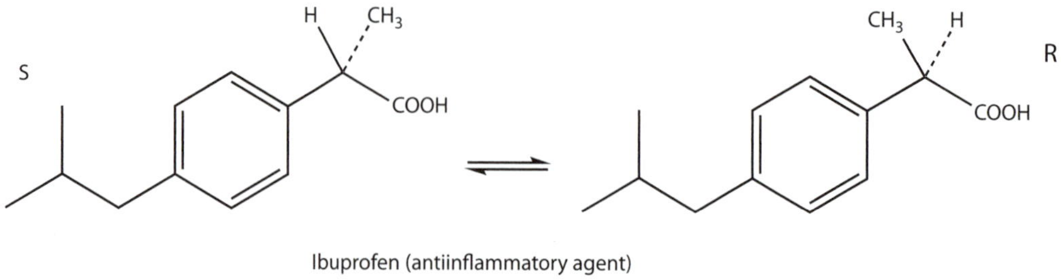

S ⇌ R

Ibuprofen (antiinflammatory agent)

10.3 Schematic

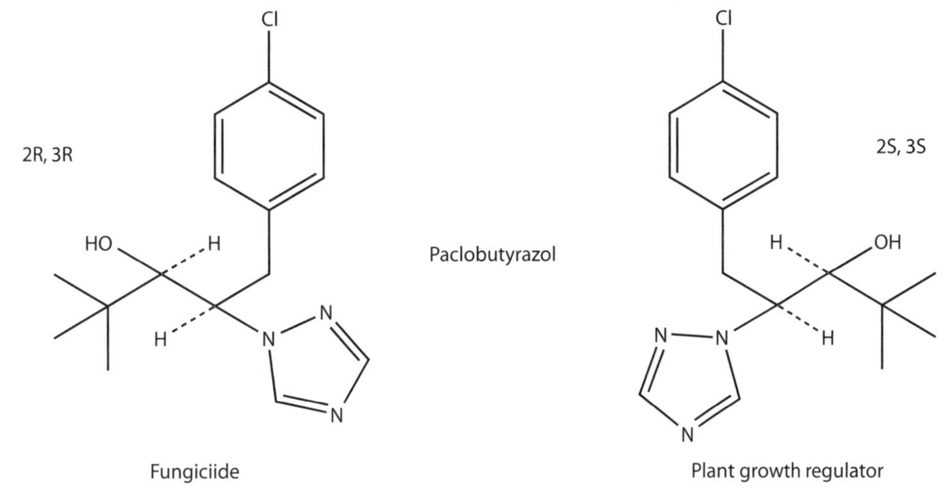

10.4 Schematic

L-aspartic acid and L-phenylalanine needed to make aspartame could be obtained in this way. In some instances, chemical synthesis may be better. In others, it may be the only way to obtain the compounds.

Almost all the syntheses require a natural product or a derivative. These are drawn from a chiral pool of alkaloids, amino acids, hydroxy acids, sugars, terpenes, and others. Typical alkaloids include ephedrine, cinchonidine, quinidine, and sparteine (**10.5** Schematic).

Typical L-amino acids are glutamic acid, lysine, phenylalanine, and proline (**10.6** Schematic).

There are many other α-amino acids. These have been selected to show typical diversity, one with an extra carboxylic acid group, one with an extra amino group, an aromatic one, and one with a heterocyclic ring. Derivatives of α-amino acids, such as the amino alcohols obtained by reduction of the acids or their esters, are also used.

L-α-Hydroxyacids such as those in **10.7** Schematic can be used.[5]

Many derivatives of tartaric acid are used, including esters and the diols resulting from the addition of Grignard reagents to their ketals.

D-Sugars such as glucose, lactose, and sucrose (**10.8** Schematic) are also used.[6]

Derivatives obtained by oxidation, reduction, and other means are also used (**10.9** Schematic).

Terpenes used include 3-carene, camphor, limonene, menthol, and α- and β-pinenes, as well as their derivatives (**10.10** Schematic).

Addition of borane to α-pinene produces a useful chiral reducing agent. 10-camphorsulfonic acid is a useful resolving agent for amines.

Chiral chemistry is a very active field covered by numerous books[7] and reviews.[8] The treatment here will emphasize the themes of earlier chapters.

1. Minimize waste, especially toxic wastes. Pick processes that do not discard the 50% of unwanted isomer. Avoid waste salts.
2. Avoid solvents, especially toxic ones, when possible.
3. Avoid toxic reagents and harsh or energy-intensive conditions.
4. Keep the number of steps low. Avoid systems that require putting on a chiral auxiliary, running the desired reaction, and then removing the auxiliary.
5. Provide for the simple recovery and recycle of chiral ligands.
6. Find conditions that give close to quantitative yields and enantioselectivity to avoid further processing that may reduce yields. For example, losses can occur when crystallization is necessary to increase the enantiomeric excess (ee).
7. Find conditions that produce the desired compound in a reasonable time and with a simple workup.

10.3 RESOLUTION OF RACEMIC MIXTURES

A guanidine resolved spontaneously from dichloromethane (**10.11** Schematic).[9]

In a few cases it is possible to crystallize the desired optical isomer by seeding a supersaturated solution of the racemate with the desired isomer. This has been used with a sulfide ester alcohol intermediate in the synthesis of the drug diltiazem (**10.12** Schematic).[2d]

Repeated crystallization of the racemate of the compound in **10.13** Schematic led to the preferential enrichment of enantiomers (up to as much as 100% ee) in the mother liquor.[10]

A phosphorus compound was resolved by formation of a palladium chelate (**10.14** Schematic).[11]

Other spontaneous resolutions have occurred under the influence of achiral compounds.[12] Sublimation of L-serine

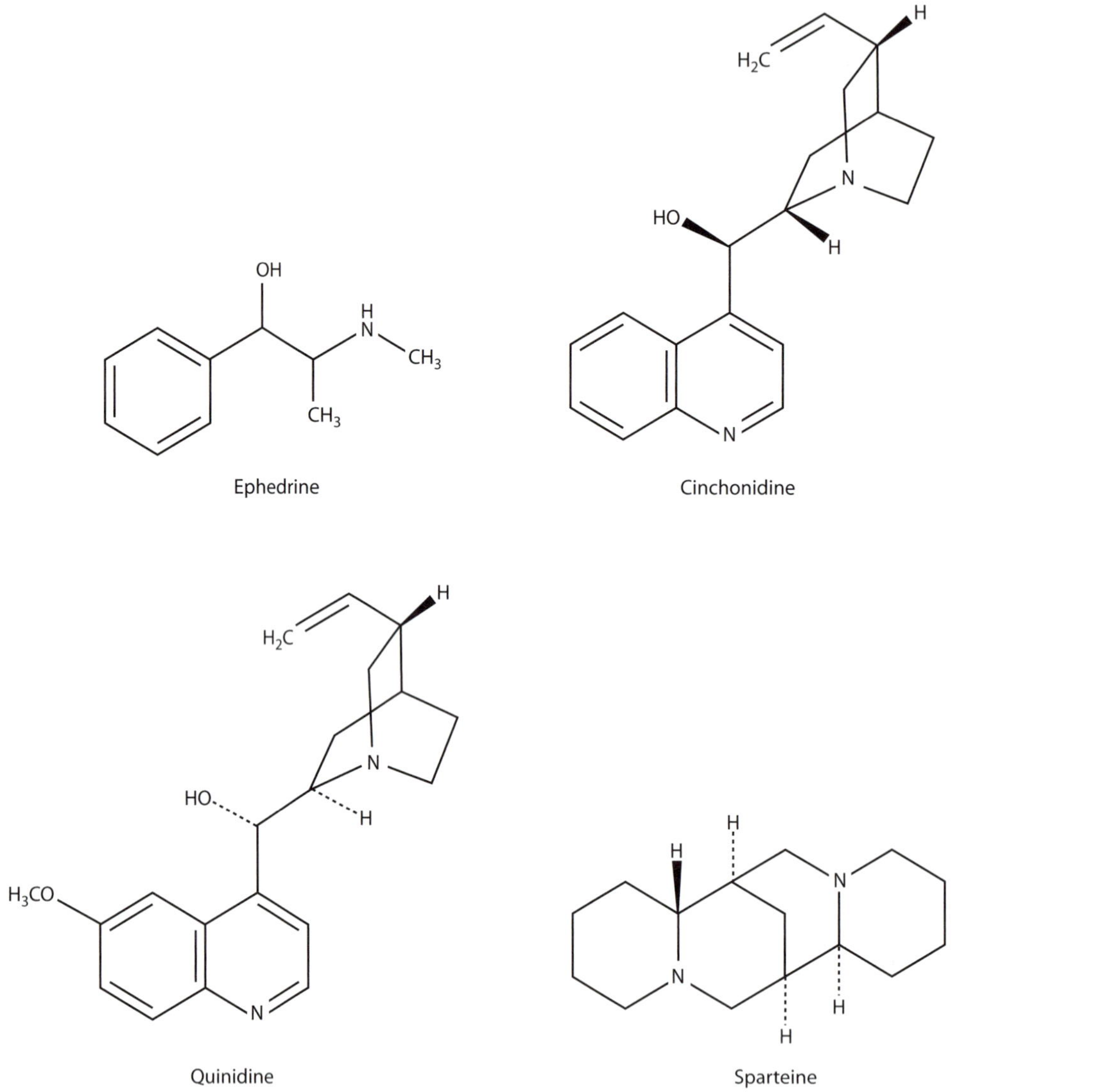

Ephedrine

Cinchonidine

Quinidine

Sparteine

10.5 Schematic

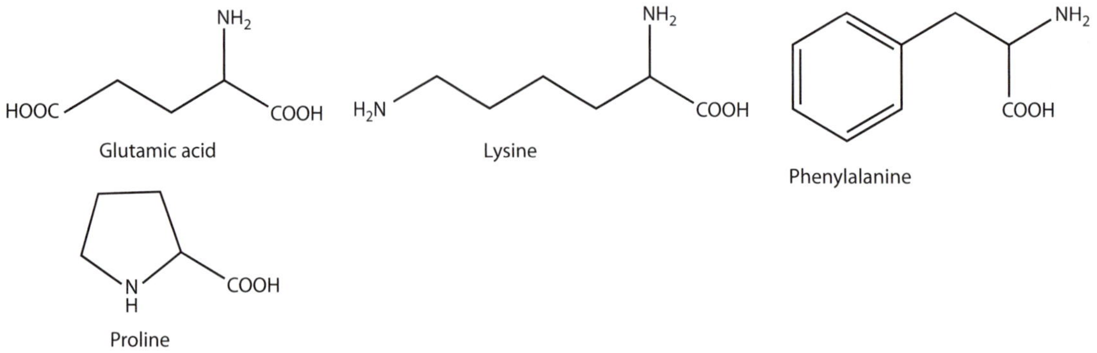

Glutamic acid

Lysine

Phenylalanine

Proline

10.6 Schematic

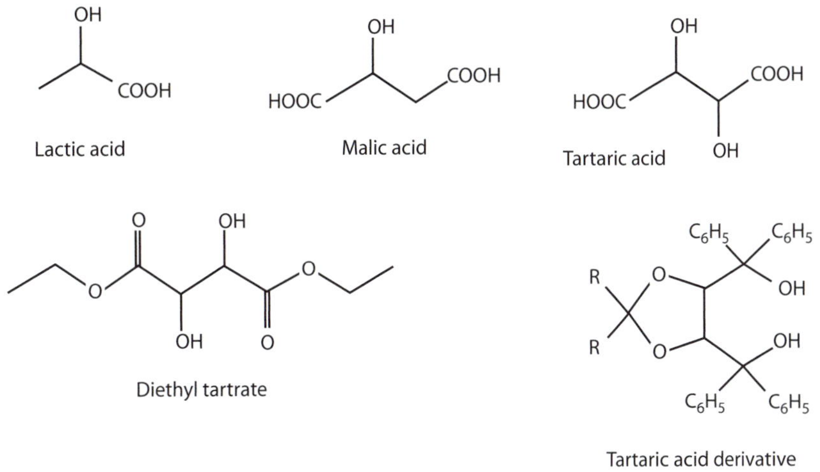

Lactic acid Malic acid Tartaric acid

Diethyl tartrate Tartaric acid derivative

10.7 Schematic

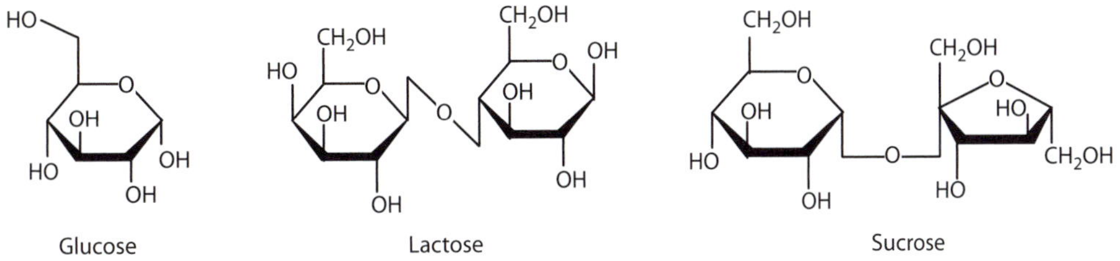

Glucose Lactose Sucrose

10.8 Schematic

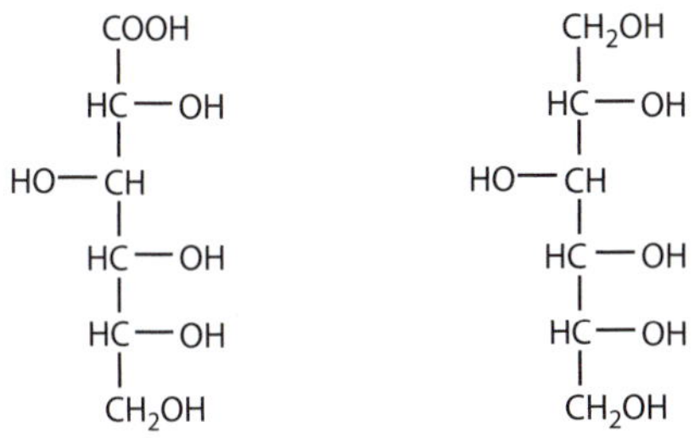

10.9 Schematic

with 3% ee at 205°C/2 hours raised the ee to 69%, but destroyed half of the compound.[13] High-intensity circularly polarized light has been used to destroy the unwanted isomer of racemic tartaric acid.[14] L-Tartaric acid was obtained in only 11% ee. Considerable development work will be needed to raise this to a useful level. A disadvantage of the method is that it wastes 50% of the material.

The molecular recognition of chiral guests by enantiomerically pure hosts to form inclusion compounds can be used to separate racemates[15] (as mentioned in Chapter 7). Thus, 1,1′-bi(2-naphthol) (BINOL; **10.15** Schematic) has been used to separate a variety of nitrogen, phosphorus, and sulfur compounds, including amine oxides and sulfoxides. The procedure can be as simple as mixing the guest with the host and then distilling off the unwanted isomer. The desired isomer is then obtained by stronger heating under vacuum.

A diol derived from tartaric acid (TADDOL) has been used to obtain alcohols and epoxides with more than 97% ee (**10.16** Schematic).[16]

The mixture to be resolved is stirred with a suspension of the host diol, followed by filtration of the inclusion compound. The guest is released by heating under vacuum. The host can be used over again. The method is described as simple, inexpensive, and environmentally safe. There is the disadvantage of having to locate a suitable host, especially for larger and more complicated molecules. BINOL can be resolved by formation of a 1:1 inclusion compound (**10.17** Schematic) with a quaternary ammonium salt derived from the amino acid, leucine.[17] The quaternary ammonium salt can be recovered for reuse.

2-Alkylcyclohexanones have been resolved in 100% yield (74–94% ee) by formation of an inclusion compound with the cyclohexanone ketal analog of TADDOL.[18]

A methylated α-cyclodextrin has been used as a host to resolve a ketal (**10.18** Schematic) in water.[19] The inclusion compound of the *R*-isomer precipitates from water in 77% yield. The yield is higher than 50% because there is some equilibration of the isomers in water. To prevent isomerization in storage, it is stored as the host–guest complex. The compound is a sex pheromone of the olive fly.

As mentioned in Chapter 7, molecular imprinting of polymers[20] can also be used to prepare hosts for the removal of the desired optical isomer from racemates. Foam flotation

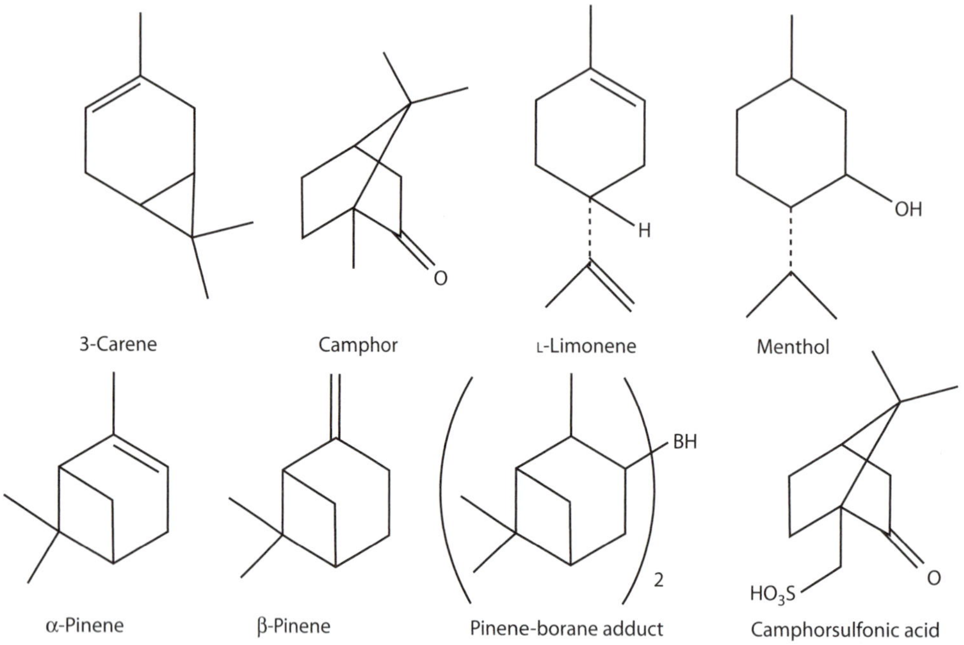

3-Carene Camphor L-Limonene Menthol

α-Pinene β-Pinene Pinene-borane adduct Camphorsulfonic acid

10.10 Schematic

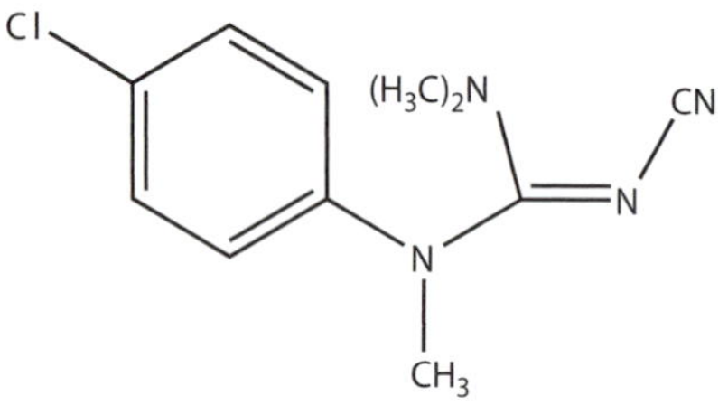

10.11 Schematic

with imprinted polymers has been used in chiral separations.[21] Flotation is often used to separate minerals from their ores. The polymers can be reused repeatedly. If good separation factors can be obtained, the imprinted polymers could be used in membranes. This would eliminate the need to hydrolyze a derivative made in a kinetic resolution. This may turn out to be the lowest-cost method of resolution.

Diastereoisomeric resolution is a standard method that is still used in the preparation of 65% of single-isomer drugs.[21] This involves reacting the racemic mixture with an optically active resolving agent to form a mixture of diastereoisomers that can be separated by crystallization. Lactic, mandelic, or camphorsulfonic acid might be used with an amine. Chiral crown ethers can also be used with amines.[22] An optically active α-methylbenzylamine, ephredrine, or another alkaloid might be used with a carboxylic acid. The use of families of structurally related resolving agents has led to more rapid separations.[23] DSM is using this method on a scale of hundreds of kilograms. An example is the use of a single enantiomer of tartaric acid, acylated with benzoic, *p*-toluic, and *p*-anisic acids, to resolve *O*-chloro-α-phenethylamine. One of the resolving agents is thought to inhibit nucleation. The use of expensive alkaloids can be avoided in the resolution of α-hydroxy and α-alkoxyacids through the use of calcium *O,O´*-dibenzoyltartrate.[24]

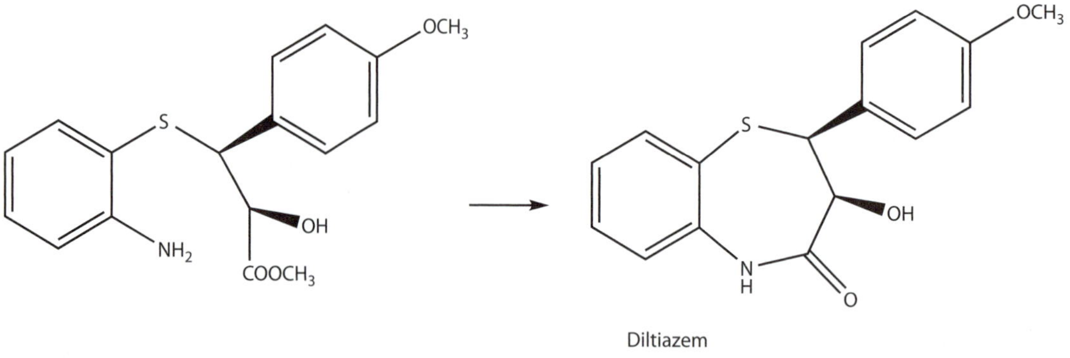

Diltiazem

10.12 Schematic

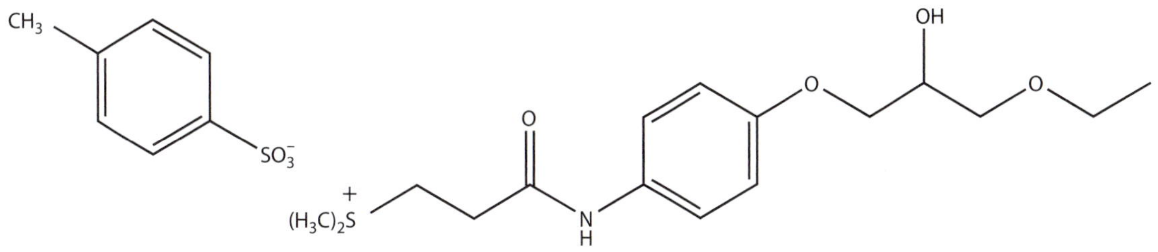

10.13 Schematic

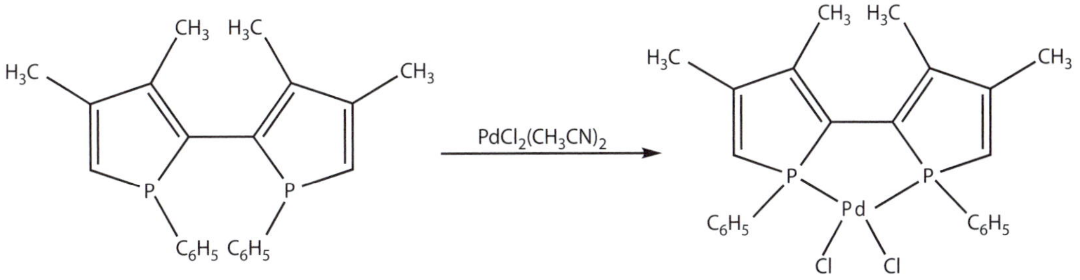

10.14 Schematic

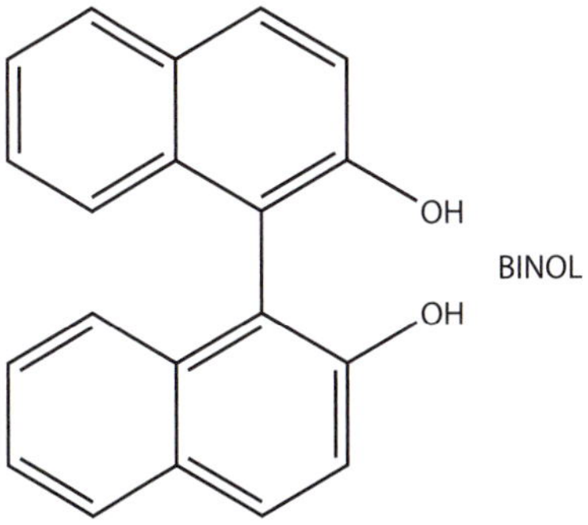

10.15 Schematic

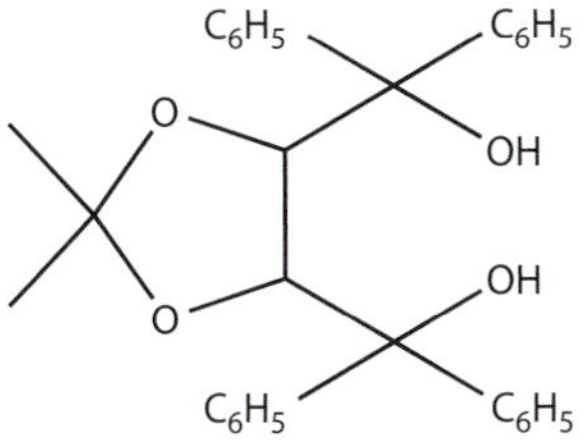

10.16 Schematic

10.17 Schematic

An alcohol could be converted to an ester, which could be hydrolyzed back to the alcohol after the separation, provided conditions that would not favor racemization during the hydrolysis were selected. This might be done by a trans-esterification under acidic conditions, such as is done in the conversion of poly(vinyl acetate) to poly(vinyl alcohol) with methanol. The problems with this method are that the maximum yield will be less than 50% if there are any losses in crystallization, and waste salts will be formed when the desired isomer is liberated from its salt or ester. The resolving agent, which may be expensive, must be recovered in high yield for recycling to the next run.

Chiral columns are used routinely for the analytical separation of optical isomers.[25] Typical columns use tribenzoates, or tris(phenyl carbamates), of cellulose or amylose as well as cyclodextrins. Flash chiral chromatography has also been performed on a cellulose carbamate on silica.[26] Preparative-scale chiral chromatography is being adapted to larger amounts that are sufficient for the preparation of some drugs: 10,000 kg/year of (*R*)-3-chloro-1-phenylpropanol is separated in this way.[2d] Simulated moving-bed chromatography, in which a series of columns is linked in a cycle with the solution being pumped around repeatedly,

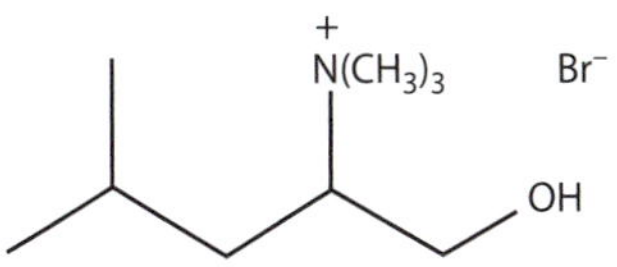

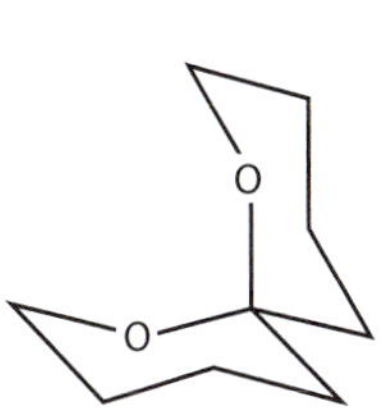

10.18 Schematic

and the head and tail of the band are taken off in each cycle, is used. This uses less solvent, but more packing, than the usual column chromatography. For some drugs that are not needed in large amounts, this can be the cheapest and fastest way to proceed.[27]

Chiral membranes have also been used to separate isomers. A cellulose tris(3,5-dimethylphenyl carbamate) has been used to separate an amino alcohol β-blocker, although the *S*-isomer was obtained in only 55% ee and the *R*-isomer in 23% ee.[28] On the other hand, a chiral membrane of a poly(L-glutamate) derivative completely resolved racemic tryptophan.[29] Apoenzymes (enzymes lacking their usual cofactor) have been placed in the pores of a membrane consisting of a gold film covered with polypyrrole.[30] In one of these, D-phenylalanine passed through five times faster than the L-isomer. (For more details on membrane separations, see Chapter 7.) This method holds considerable promise for catalytic membrane reactors, especially if the unwanted isomer can be racemized *in situ* and the % ee can be improved.

Enzymes are often used in kinetic resolutions, when one isomer reacts much faster than the other.[31] This can be done on an industrial scale. Zeneca uses a dehalogenase on racemic 2-chloropropionic acid to obtain about 2000 ton/year of the L-isomer for use in making a herbicide.[32] Enzymatic resolutions may involve ring opening of epoxides,[33] formation or hydrolysis of esters and amides, reduction of carbonyl compounds, and so on. Lipases are often used for this purpose.[34] A typical example is reaction **10.19**.[35]

Various factors affect the enantioselectivity in such hydrolyses. These include the concentration and the choice of the organic solvent used.[36] In another, pretreatment of the lipase with 2-propanol raised the enantioselectivity 25-fold to 93% ee.[37] In an esterification of menthol, coating the lipase with a surfactant raised the activity 100-fold in isooctane.[38] Entrapment of the lipase in a gel made by the hydrolysis of a mixture of tetramethoxysilane and *n*-propyltrimethoxysilane (**10.20** Schematic) improved the activity severalfold and raised the enantioselectivity. The catalyst could be reused many times, although there was a 15–20% loss of activity in the first two or three cycles, after which the activity became constant.[39]

Microwave radiation increased the rate 1–14-fold and the ee 3–9-fold (up to 92% ee for one isomer and 96% ee for the other) in the lipase-catalyzed esterification of an alcohol with vinyl acetate.[40] Ultrasound increased the rate 7–83-fold. Directed evolution can occasionally be used to improve the enantioselectivity of lipases. A lipase for the hydrolysis of racemic *p*-nitrophenyl 2-methyldecanoate that gave 2% ee was run through four generations of error-prone polymerase chain reactions to raise the selectivity to 81% ee.[41]

Lipases have also been used to prepare enantiomerically pure amines (**10.21** Schematic).[42]

L-Methionine has been prepared from *N*-acetyl-D,L-methionine by deacylation with an aminoacylase.[43] The lipase was supported on an ion-exchange resin. The reaction was driven to completion by chromatographic ion-exchange removal of the by-product acetic acid and unreacted substrate, the racemic mixture being fed in pulses to the column. The cross-liked enzyme crystals marketed by Altus Biologics are very useful in enzymatic resolutions, because of their enhanced stability to heat and organic solvents and because they can be reused many times. (For more details on enantiospecific reactions with enzymes, see Chapter 9.)

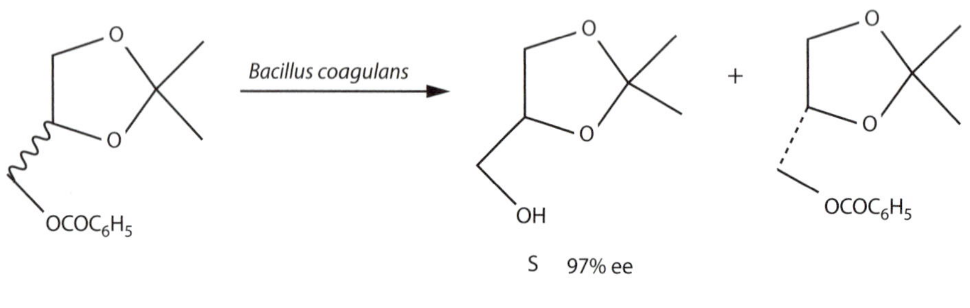

10.19 Schematic

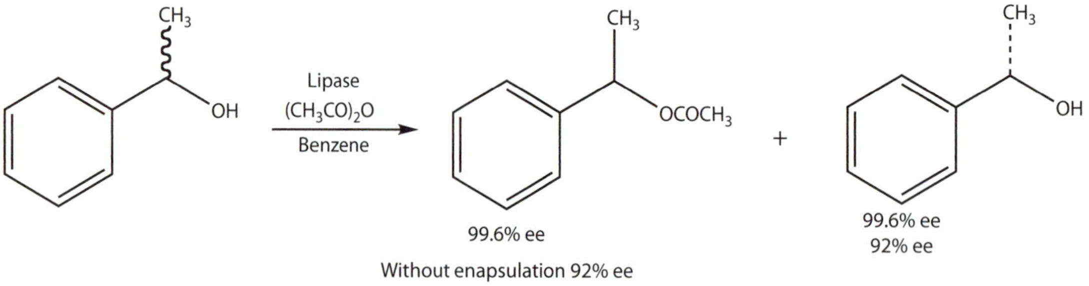

10.20 Schematic

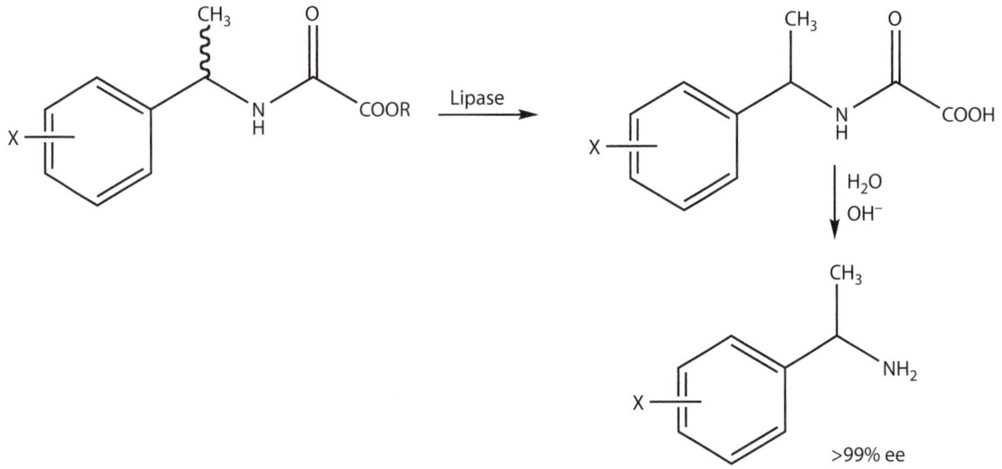

10.21 Schematic

The problem with most of the foregoing methods is that the maximum yield is 50%. The unwanted isomer is often discarded as waste. In other instances, it is racemized[44] by acid, base, heat, a racemase,[45] or other means and then recycled to the process. Base works well with any carbonyl compound that has a hydrogen atom on the α-carbon atom. A convenient way to racemize amines is to make a Schiff base with benzaldehyde, which can then isomerize by way of the isomeric ketamine (**10.22** Schematic).

When resolution and racemization are carried out in the same vessel at the same time, the process has been termed deracemization or dynamic kinetic resolution.[46] The maximum yield then becomes 100%, with no need to recycle an unwanted isomer. The process involves an enzyme that converts the desired isomer to a derivative plus a racemase or a metal salt that isomerizes the unwanted isomer. The process can be run continuously.[47] Some examples will be given to illustrate the scope of this approach.

A combination of a ruthenium catalyst and a lipase was used to convert an amine to an acetamide in up to 95% yield and up to 99% ee.[48] The same procedure can be used to convert alcohols to their acetates (**10.23** Schematic).[49]

This method has been used to prepare chiral polyesters (**10.24** Schematic).[50]

Racemic alanine amides were converted to the D-amino acid using a D-aminopeptidase and α-amino-ε-caprolactam racemase in >99.7% yield.[51]

The *R*,*R*-isomer of α-bromoamide (**10.25** Schematic) reacts faster than the *S*,*R*-isomer.[52]

Biocatalytic deracemization can be accomplished where the racemizing agent is either biological or chemical.[53] Two organisms have been used together to hydrolyze styrene oxide. *Aspergillus niger* works only on the *R*-isomer with retention of configuration. *Beauvaria bassiana* uses only the *S*-isomer with inversion. The result is that a single product is formed (**10.26** Schematic).

Geotrichum candidum has been used to deracemize 1-arylethanols (**10.27** Schematic).[54]

At least two enzymes are thought to be involved in a process that must go through oxidation to the ketone and then reduction to the alcohol. An alcalase, in the presence of pyridoxal-5-phosphate as an isomerizing agent, has been used to convert racemic amino acid esters to the L-enantiomers with 90–98% ee in 87–95% yield.[55]

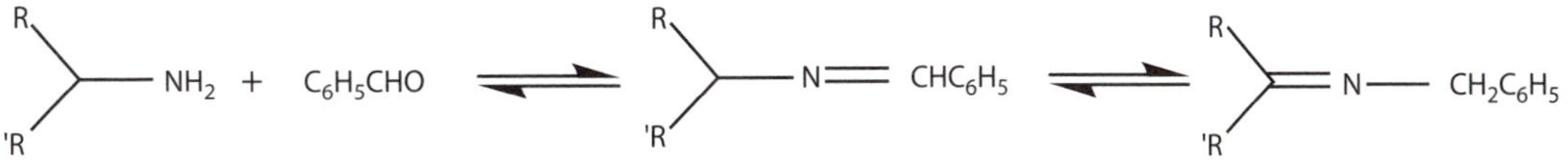

10.22 Schematic

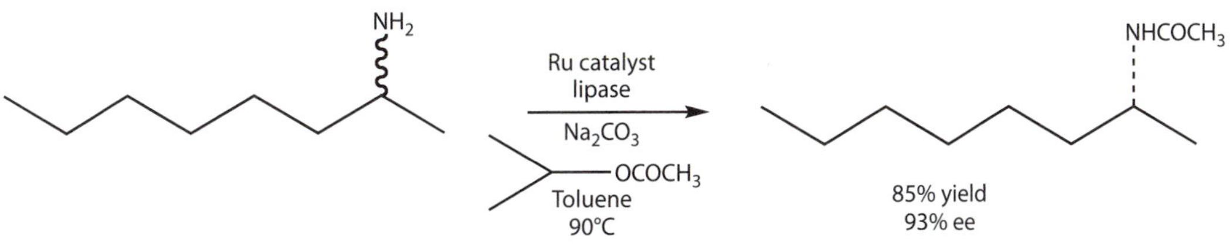

10.23 Schematic

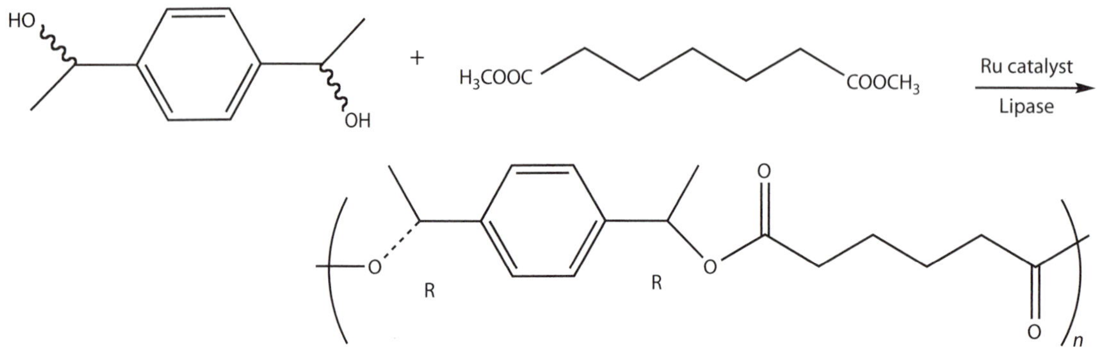

10.24 Schematic

Silica has been used to promote hemithioacetal epimerization by dissociation–recombination in an enzymatic esterification (**10.28** Schematic).[56]

A palladium salt has been used to isomerize an allylic acetate during an enzymatic hydrolysis (**10.29** Schematic).[57]

An allylic alcohol isomerized during an enzymatic esterification (**10.30** Schematic).[58]

Triethylamine has been used as a racemizing agent in the hydrolysis of thioesters (**10.31** Schematic).[59]

Other dynamic resolutions involve *in situ* racemization by reversible Michael reactions,[60] retroaldol reactions,[61] enolization of esters,[62] and Ru(II)-catalyzed isomerization of alcohols.[63]

These examples are a step in the right direction, but improvements are still needed. Some of the reactions take place over days. The times might be shortened by using enzymes that are active at higher temperatures. Microwaves might help. When the enzymatic reactions are carried out in organic solvents, the times might be shortened by using a surfactant or by using cross-linked enzymes crystals (as described in Chapter 9). The use of immobilized enzymes and supported catalysts may help in the workups and in the recycling for the next run. There are still some by-products to be removed. Frequently, there is still the need to prepare an ester in a separate step before obtaining the desired acid or alcohol from the resolution. Two of the foregoing examples are especially instructive. The resolution of racemic 1-arylalkanols with *G. candidum* in water (see **10.27** Schematic) gave up to 99% ee without the need to make any derivatives or use any chemical reagents. The quantitative resolution of racemic tryptophan by a poly(L-glutamate) membrane used no derivatives and no chemical reagents. These suggest the use of a deracemization catalyst, chemical or biological, in conjunction with a chiral membrane in a continuous reactor. The catalyst could be in solution or in the membrane. This could be the quickest and least-expensive way to make the alcohols, amines, carboxylic acids, and such, needed for single-isomer drugs, flavors, agricultural chemicals, and the like. It would also produce the least waste. Use of a membrane reactor in the continuous reduction of an emulsion of

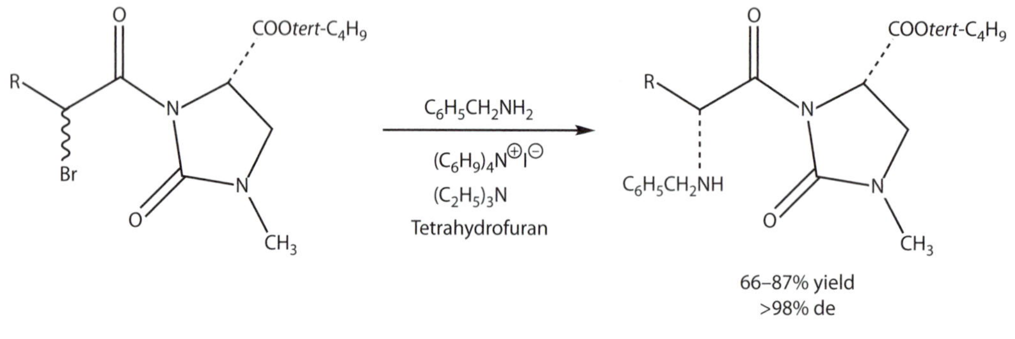

10.25 Schematic

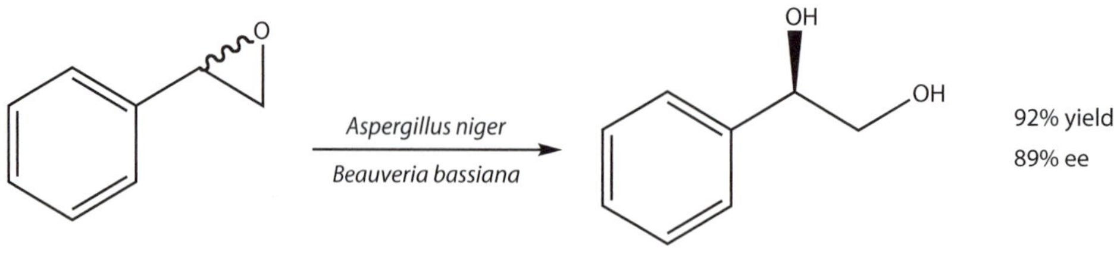

10.26 Schematic

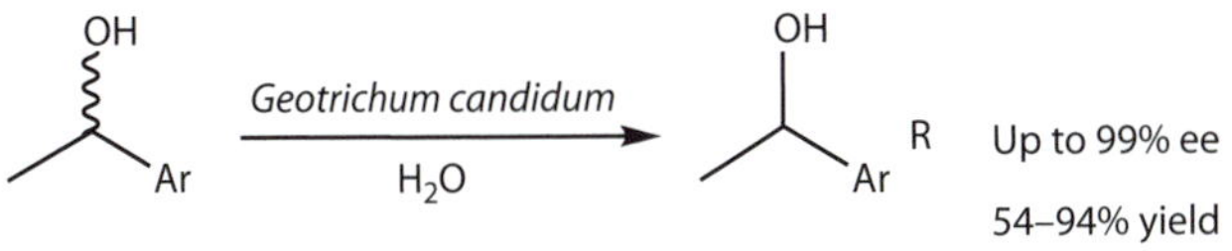

OH Geotrichum candidum H_2O Ar OH R Ar Up to 99% ee 54–94% yield

10.27 Schematic

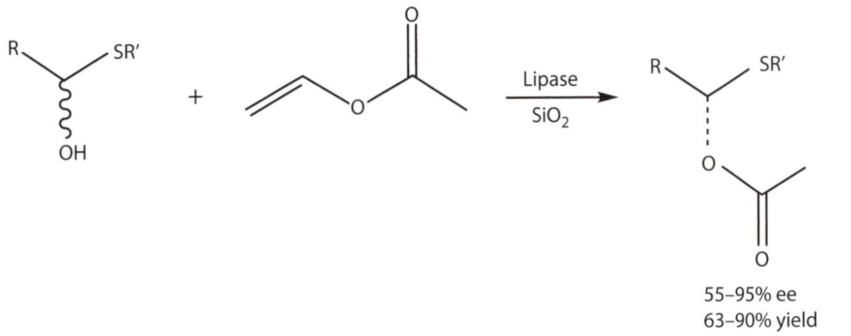

R SR' OH + O O Lipase SiO_2 R SR' O O 55–95% ee 63–90% yield

10.28 Schematic

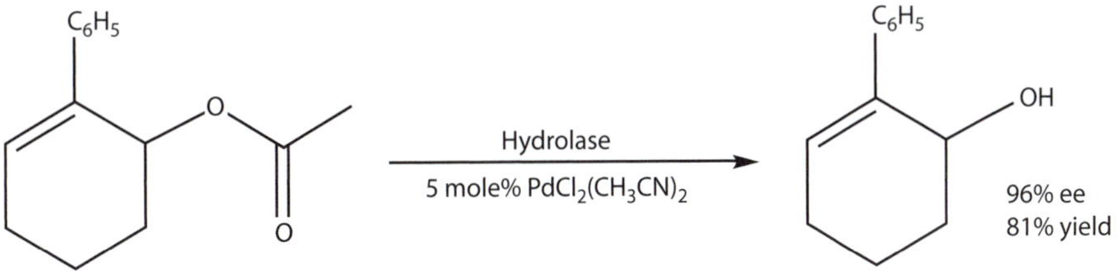

C_6H_5 O O Hydrolase 5 mole% $PdCl_2(CH_3CN)_2$ C_6H_5 OH 96% ee 81% yield

10.29 Schematic

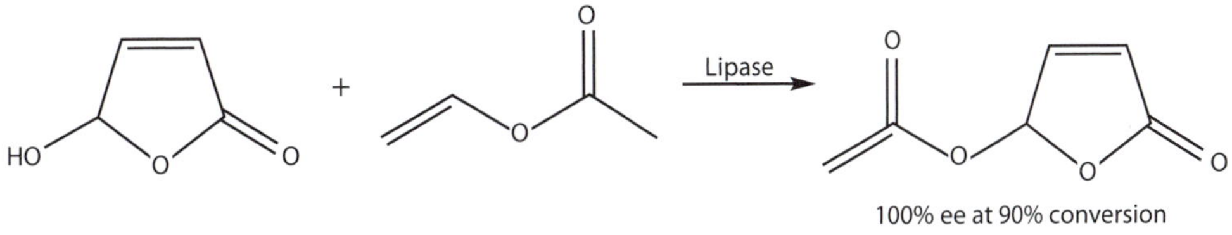

HO O O + O O Lipase O O O O 100% ee at 90% conversion

10.30 Schematic

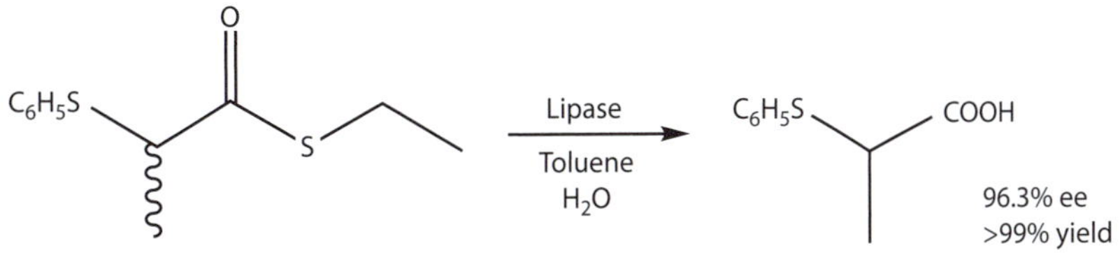

C_6H_5S O S Lipase Toluene H_2O C_6H_5S COOH 96.3% ee >99% yield

10.31 Schematic

2-octanone to *S*-2-octanol (>99.5% ee) with a *Candida parapsilosis* reductase increased the turnover number by ninefold, compared with a classic reactor.[64] (For more on membrane separations, see Chapter 7.)

Catalytic antibodies have been used to resolve aldols by the retroaldol reaction to obtain 95–99% ee.[65]

Benzyl alcohols have been resolved in parallel kinetic resolutions that use two reagents with selectivities for opposite isomers to make carbonates in simultaneous reactions (**10.32** Schematic).[66]

Some racemic compounds might be separated without the need to make chemical derivatives. Chiral diols bond selectively (*SS*-isomer favored over the *RR*-isomers by a factor of 6 or 7) by hydrogen bonding to the amidine (**10.33** Schematic).[67]

10.4 ASYMMETRICAL SYNTHESIS

The best approach is to start with a natural optically active compound and modify it by reactions that will not racemize the product or intermediates. Unfortunately, this is possible in only a relatively few cases. Some of the problems mentioned in the foregoing might be solved if one could devise a catalytic asymmetrical synthesis for a desired compound.[68] Activity in this field is intense. The many small

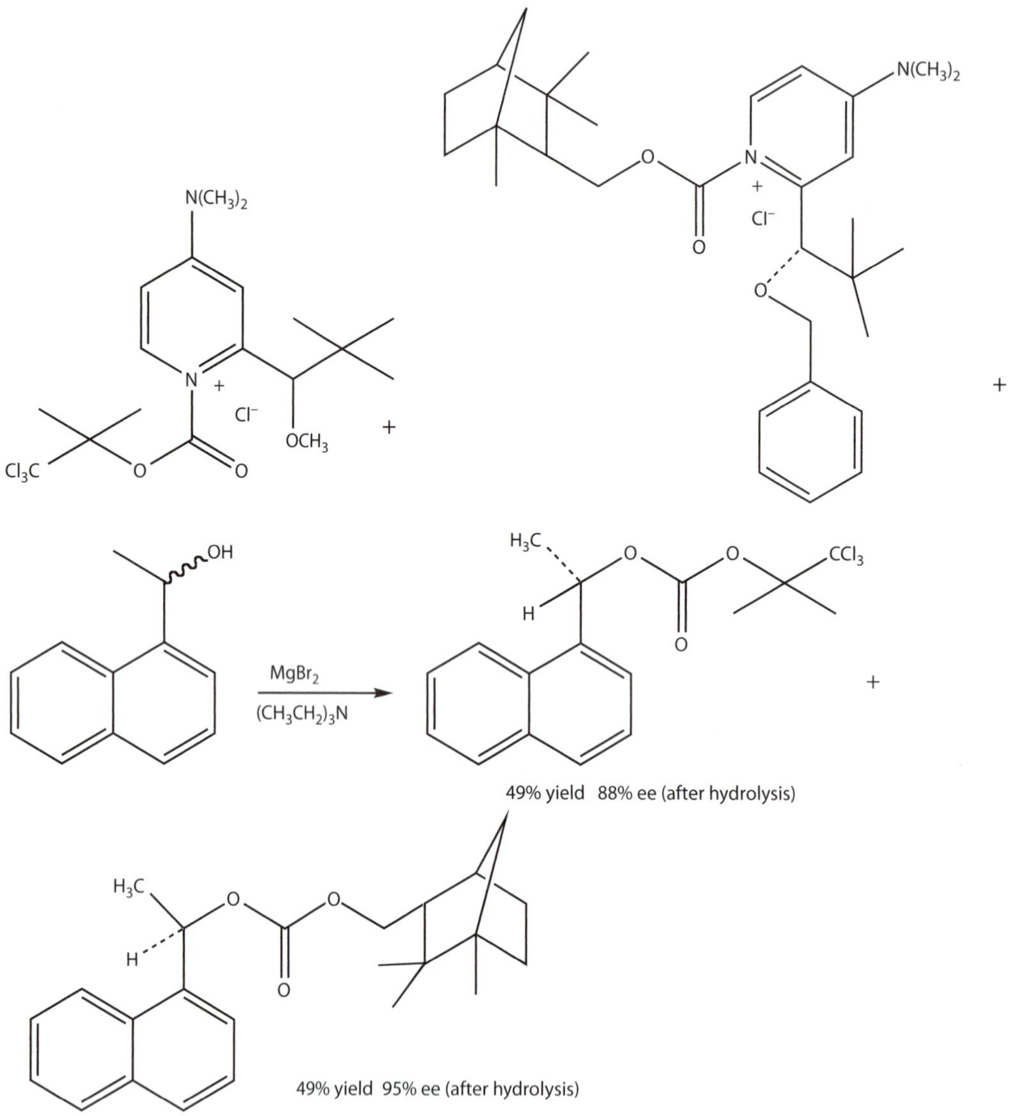

$N(CH_3)_2$ Cl^- OCH_3 Cl_3C O O + $N(CH_3)_2$ O N + Cl^- O O +

OH $MgBr_2$ $(CH_3CH_2)_3N$ H_3C H O O CCl_3 O +

49% yield 88% ee (after hydrolysis)

H_3C H O O O

49% yield 95% ee (after hydrolysis)

10.32 Schematic

companies in this field (plus some of their suppliers) include the following:

Altus Biologics, Cambridge, Massachusetts: enzymes for syntheses

Austin Chemical, Buffalo Grove, Illinois: chiral chromatography

Cambrex, Warren, New Jersey[69]

Celltech and Chirotech, Cambridge, England[70]

Chemi, Milan, Italy

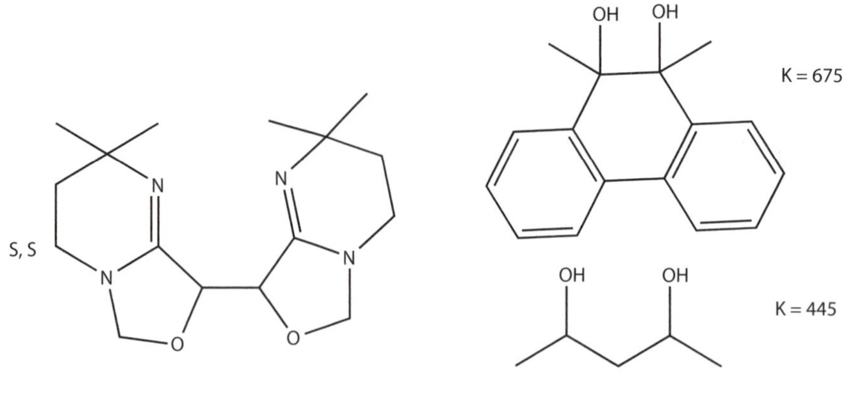

S, S N N N N O O

OH OH K = 675

OH OH K = 445

10.33 Schematic

Chiragene Warren, New Jersey
Chiral Quest, New Jersey
Chiral Technologies, West Chester, Pennsylvania: chiral chromatography
Chirogen, Australia
Daiso, Osaka, Japan
Evotec, Abingdon, England
Rhodia Chi Rex, England
SepraChem and Sepracor, Marlborough, Massachusetts[38]: single-isomer drugs
Solvias, Switzerland
SynChem, Chicago, Illinois
Syncom, Groningen, the Netherlands
Synthon Chiragenics, Lansing, Michigan: chiral reagents
Takasago, Rockleigh, New Jersey
Zeeland, Zeeland, Michigan.

Ideally, the product would be made in 100% ee so that no crystallization would be needed and no losses would occur. In practice, this seldom happens. To be useful commercially, the ee should be over 90% so that a minimal number of crystallizations will provide the pure optical isomer. Many of the reactions reported in the literature do not meet this standard. However, they may be important mileposts on the way to high enantioselectivity. Despite the successes of computational chemistry in many areas, it is still not possible to model the transition states of the reactions well enough to select catalyst that will provide high stereoselectivity. Most of the good catalysts are found empirically and raised to high selectivity by systematic variation of the substituents. (Some variations are done to get around existing patents.) The application of combinatorial chemistry may help. Rather than finding a general catalyst that gives high enantioselectivity and yields for a whole class of compounds, it may well be necessary to tailor the steric and electronic effects in each one to a specific substrate and set of conditions, Reasonable turnover rates are also required for the catalysts. Some of the better methods of asymmetrical synthesis will be described in the following, together with their advantages and limitations.

The use of enzymes in enantioselective reactions[71] was covered in Chapter 9 and earlier in this chapter. Most single-isomer products are still made by chemical methods. Chemists are becoming more familiar with biocatalysis and it can be expected to supplant some of the chemical methods.

Chiral auxiliaries are often used on a stoichiometric basis. They are often derived from natural amino acids and the amino alcohols obtained by their reduction.[72] A typical alkylation reaction (**10.34** Schematic) carried out in this way is shown.[73] (Similar alkylations have been carried out using polymeric analogs, the ee varying with the resin used [up to 90% ee when Wang resin was used].)[74]

The starting compound was also condensed with an aldehyde in 82–84% yield with more than 98% ee. A corresponding amide from an α,β-unsaturated acid underwent the Diels–Alder reaction in 86–90% yield with more than 99% ee. The original amino alcohol was probably treated with phosgene to close the oxazolidone ring.[75] The less toxic dimethyl carbonate might be a suitable replacement for the phosgene. The use of another chiral auxiliary for alkylation is shown in **10.35** Schematic.[76]

Terpene derivatives are also used as chiral auxiliaries, as in example **10.36** Schematic.[77]

Although these syntheses give high enantioselectivity, they do have associated problems. It is necessary to make the chiral auxiliary in more than one step, to use it on a

Base, RBr → Hydrolysis → 61–86% yield, >99% ee

10.34 Schematic

RX, Base → Pd, H_2 → H_2N–CH(R)–COOH; 60% yield, >99% ee

10.35 Schematic

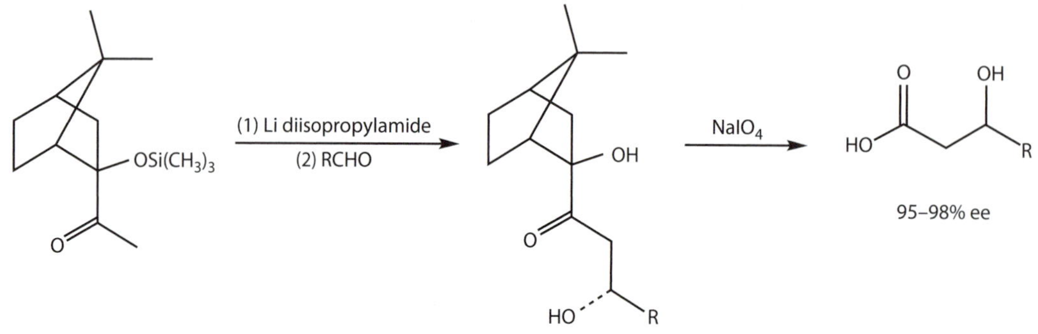

(1) Li diisopropylamide
(2) RCHO
$NaIO_4$
95–98% ee

10.36 Schematic

stoichiometric basis, and to recover it at the end for recycling. Some of the reagents may be toxic; hence separating the chiral auxiliary or its remnants from the product may complicate the workup. Waste salts are produced in the process, and therefore the process can be expensive.

Asymmetrical reduction of ketones to alcohols[78] can be done in a number of ways. Reductions with microorganisms[79] were described in Chapter 9. By choosing appropriate microbes both *R*- and *S*-isomers were obtained. The inclusion of dimethylsulfide in a reduction of β-ketoesters with Baker's yeast raised the enantioselectivity from 94% to 98%.[80] One dehydrogenase reduced both *R*- and *S*-isomers of perillaldehyde to a single product (**10.37** Schematic).[81]

A stoichiometric amount of a chiral borane, such as diisopinocampheylchloroborane, which is made by addition of borane to α-pinene, can be used to achieve ee's as high as 91–98%.[82] Achiral boranes can be used in stoichiometric amounts with a chiral catalyst (**10.38** Schematic).[83] The catalyst was designed using a transition state model. The ee was lower in methylene chloride than in toluene. Other workers have used this type of catalyst in a polymeric form to simplify the workup (**10.39** Schematic).[84]

The dimethylsulfide adduct of borane was the reducing agent, giving the alcohol in up to 89% ee in 95% conversion in 20 min. The catalyst could be reused at least three times. The reduction has also been carried out by transfer hydrogenation using 2-propanol as the hydrogen donor and ruthenium complexes as catalysts with ee's up to 98%.[85] A variety of catalysts were used, two of which are shown in **10.40** Schematic.

Catalytic reductions produce far less waste or material to be recycled than methods that use stoichiometric amounts of reducing agents. They usually give higher volume yields and productivities than microbial reductions.[86] α-Ketoesters have been reduced with hydrogen using a cinchona alkaloid-modified platinum catalyst to give up to 95% ee.[87] Methyl pyruvate was reduced to 98.8% *R*-methyl lactate and 1.2% *S*-methyl lactate using this method with a poly(vinyl pyrrolidone)-stabilized catalyst.[88]

One of the most popular ligands is 2,2′-binaphthyl (BINAP),[89] which can be made from BINOL through its ditriflate[90] (**10.41** Schematic). (The synthesis of BINOL from 2-naphthol by oxidation with air was given in Chapter 4.) Resolution of the BINOL can be accomplished with the inclusion compound described earlier in this chapter, or with *N*-benzyl-cinchonidium chloride,[91] or through a cyclic menthyl phosphite.[92] A polymer-supported BINAP, available commercially, can be recovered for reuse by filtration.[93]

A wide variety of analogs of BINAP[94] and BINOL[95] have been prepared. These include substituents on the rings,

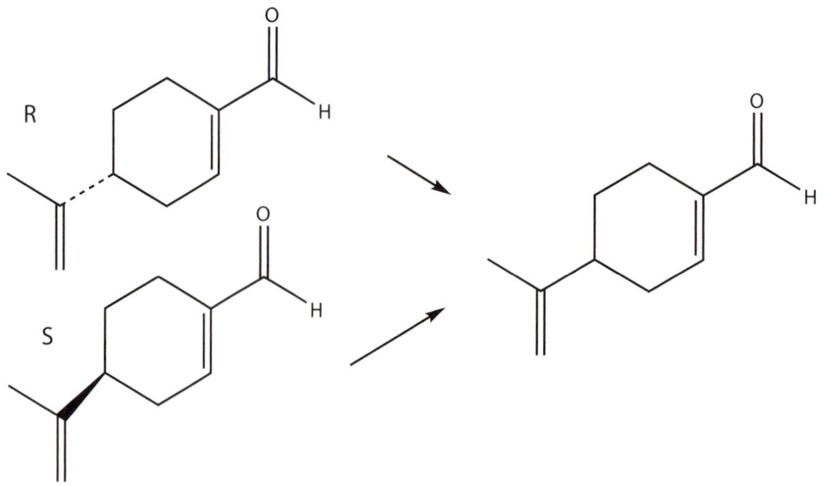

R
S

10.37 Schematic

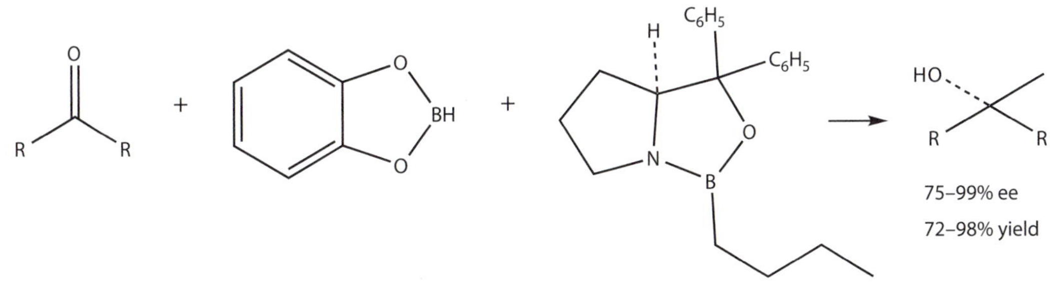

10.38 Schematic

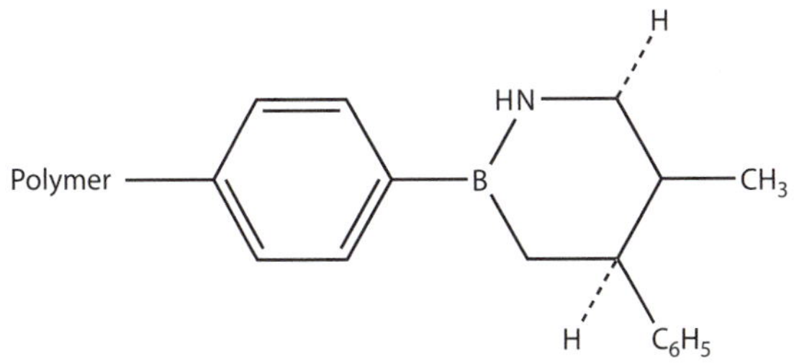

10.39 Schematic

on phosphorus, and on oxygen with or without the addition of metal ions. Cyclic phosphorus compounds have been made from BINOL.

A tetrakis (sodium sulfonate) from BINAP has been used with a ruthenium catalyst in a thin film of ethylene glycol on porous glass in the asymmetrical reduction of an unsaturated carboxylic acid (**10.42** Schematic).[96]

The catalyst could be reused several times, an important factor considering the high cost of ruthenium and the ligand. No ruthenium was lost by leaching. Even so, the ruthenium would have to be recovered chemically after the useful life of the catalyst. Chiral ruthenium catalysts can also be used to reduce cyclic Schiff bases in 82% to more than 99% yield with 77–97% ee.[97]

Optimization of the phosphine ligands has led to high enantioselectivity in the hydrogenation of α-acetamidoacrylic esters with rhodium catalysts and of β-ketoesters with ruthenium catalysts.[98] Two of the diphosphine ligands are shown in **10.43** Schematic.

The diphosphine on the left was used with a rhodium catalyst and hydrogen to reduce the dehydroamino acid derivatives. The same reduction could also be carried out with sugar-derived phosphinites, the product being obtained in 97–99% ee.[99] The diphosphine on the right, or better its counterpart, in which isopropyl has replaced methyl, was used with a ruthenium catalyst and hydrogen or 2-propanol in the reductions of β-ketoesters. 1-Acetylnaphthalene was reduced with a ruthenium catalyst using the ligand on the left with 2-propanol (**10.44** Schematic) in more than 99% yield and 97% ee.

The top reaction (where R = phenyl, R' = R'' = methyl) was run with a different diphosphine using a polymeric surfactant as the anion for rhodium to give 99–92% ee.[100] Optimization of the phosphine structure would probably raise this to the levels shown in **10.44** Schematic. The polymeric surfactant was derived from a copolymer of acrylamide with maleic anhydride (**10.45** Schematic).

The catalyst could be filtered off and reused up to nine times. This shows the advantages of supported catalysts in

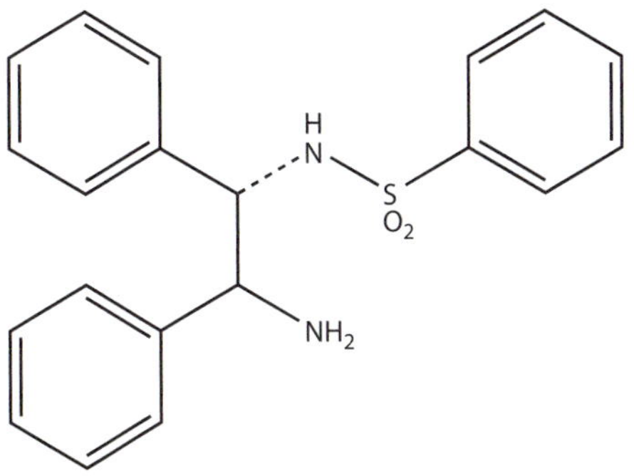

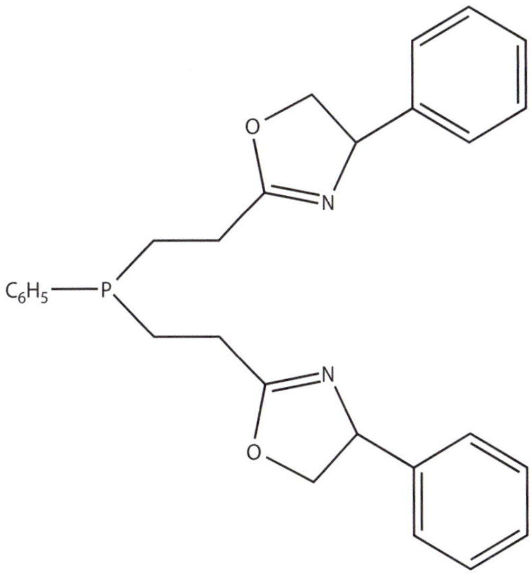

10.40 Schematic

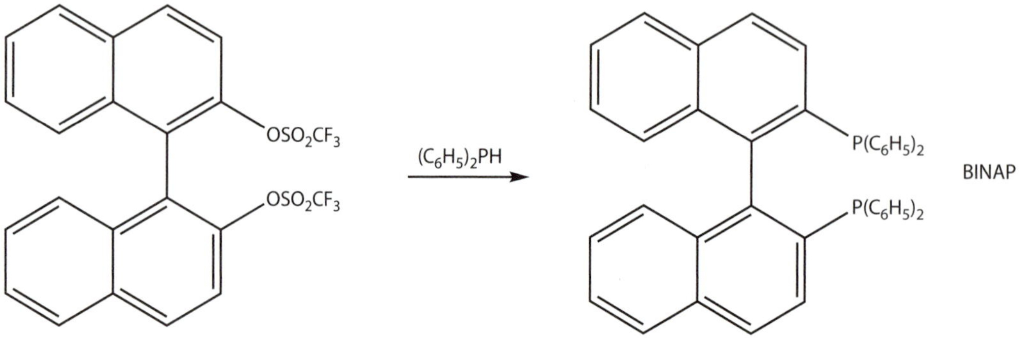

OSO_2CF_3
OSO_2CF_3
$(C_6H_5)_2PH$
$P(C_6H_5)_2$
$P(C_6H_5)_2$
BINAP

10.41 Schematic

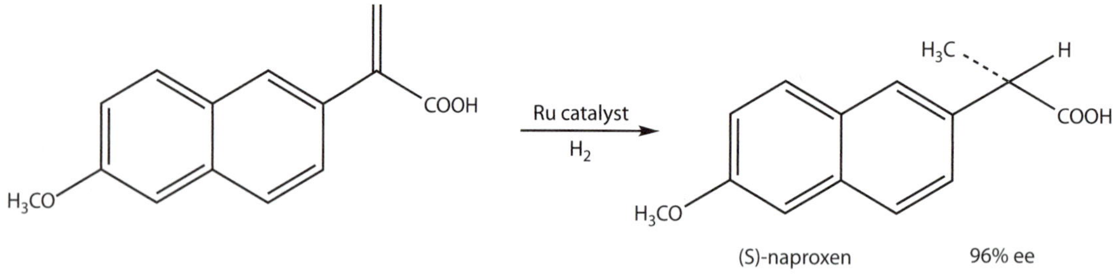

COOH
H_3CO
Ru catalyst
H_2
H_3C
H
COOH
H_3CO
(S)-naproxen 96% ee

10.42 Schematic

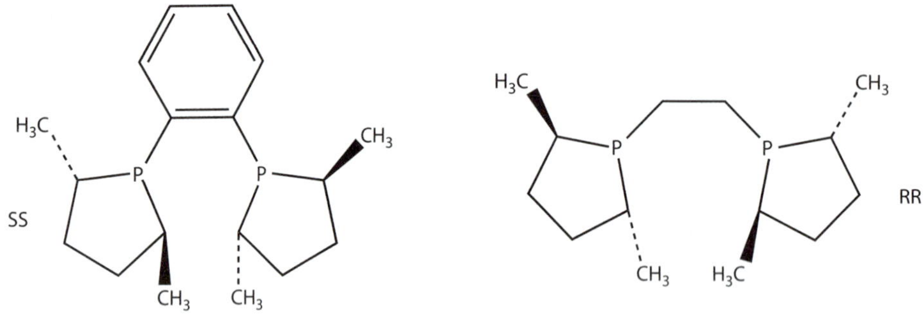

H_3C
CH_3
SS
P
P
CH_3
CH_3
H_3C
CH_3
P
P
RR
CH_3
H_3C

10.43 Schematic

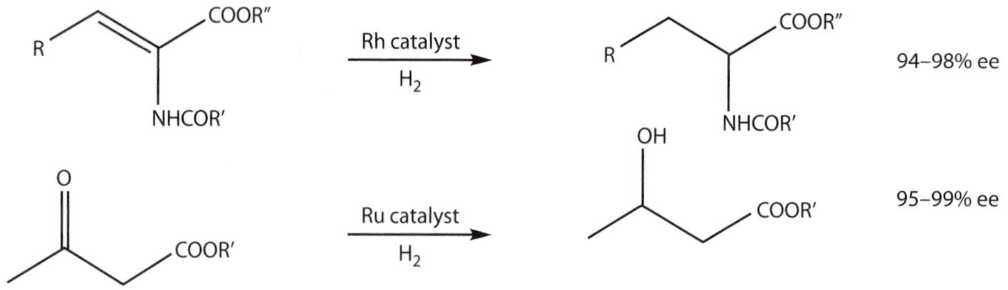

R
COOR″
NHCOR′
Rh catalyst
H_2
R
COOR″
NHCOR′
94–98% ee

O
COOR′
Ru catalyst
H_2
OH
COOR′
95–99% ee

10.44 Schematic

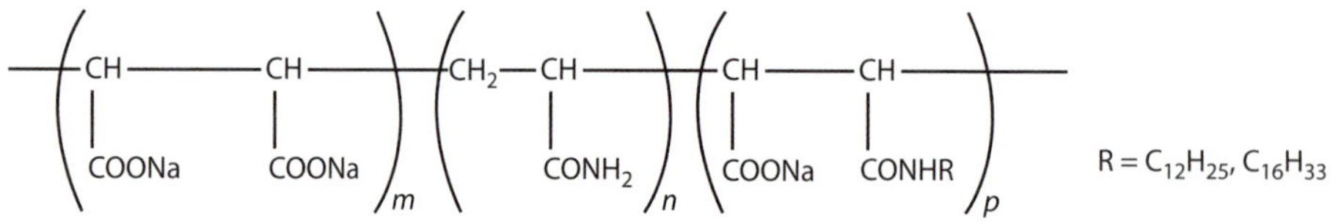

CH CH CH_2 CH CH CH
COONa COONa $CONH_2$ COONa CONHR
m n p
$R = C_{12}H_{25}, C_{16}H_{33}$

10.45 Schematic

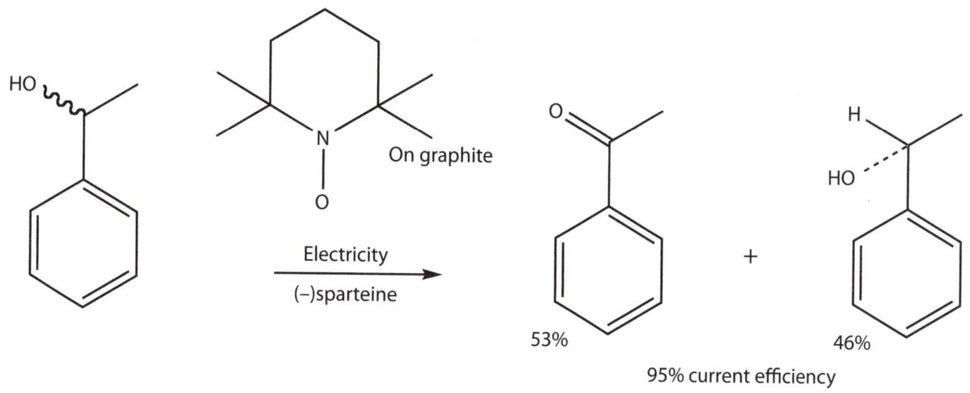

HO On graphite Electricity (–)sparteine 53% + HO H 46% 95% current efficiency

10.46 Schematic

simplifying workups and in the conservation of expensive materials.

BINOL has also been used in many other catalysts for asymmetrical synthesis. A ruthenium BINOL complex in a poly(dimethylsiloxane) membrane was used to reduce methyl acetoacetate with hydrogen to the corresponding alcohol ester with 92–93% ee.[101] The activity was superior to that of a comparable homogeneous system. The membrane could be reused. Salts of BINOL with titanium, ytterbium, and zinc have been used as chiral acid catalysts.[102] Heterobimetallic alkoxides derived from two molecules of BINOL, one trivalent metal ion, such as a rare earth, and one alkali metal ion are chiral basic catalysts.[103]

Chiral oxidations are important in synthesis.[104] An example that illustrates the use of electricity to replace a reagent, and an insoluble catalyst for ease of workup, is shown in **10.46** Schematic.[105]

The use of enantioselective epoxidation is common.[106] Chloroperoxidases have been used with hydrogen peroxide or *tert*-butylhydroperoxide to epoxidize olefins (**10.47** Schematic) to epoxides with 88–96% ee.[107]

The use of hydrogen peroxide with a poly(D-leucine) catalyst (**10.48** Schematic) offers a simple workup.[108] The catalyst can be reused.

Use of poly(L-leucine) gives the other isomer. Sodium perborate can also be used as the oxidant. This type of starting material can be oxidized using hydrogen peroxide or *tert*-butylhydroperoxide with La, Y, or Yb-BINOL catalysts in the presence of a molecular sieve to give comparable stereospecificity without the poly(leucine), but recovery of the catalyst for reuse is more difficult.[109]

Jacobsen and co-workers have devised highly enantiospecific manganese complexes for epoxidations of olefins with sodium hypochlorite.[110] A variety of other metal ions have been used with the same ligands in various reactions.[111] Peracids and amine *N*-oxides can also be used as oxidants. A typical catalyst is shown in **10.49** Schematic.

(The *trans*-1,2-cyclohexanediamine used in this complex has also been used in a variety of chiral reagents and ligands for asymmetrical cyclopropanation, alkylation, dihydroxylation, and such[112]). Merck and Sepracor have developed a drug synthesis based on such a system (**10.50** Schematic).[113]

Jacobsen and co-workers have used similar catalyst in the enantioselective opening of epoxide rings. Stereospecific hydrolysis with a cobalt acetate chelate can be used to resolve racemic epoxides.[114] Propylene oxide was opened with trimethylsilylazide in the presence of an *RR*-chromium

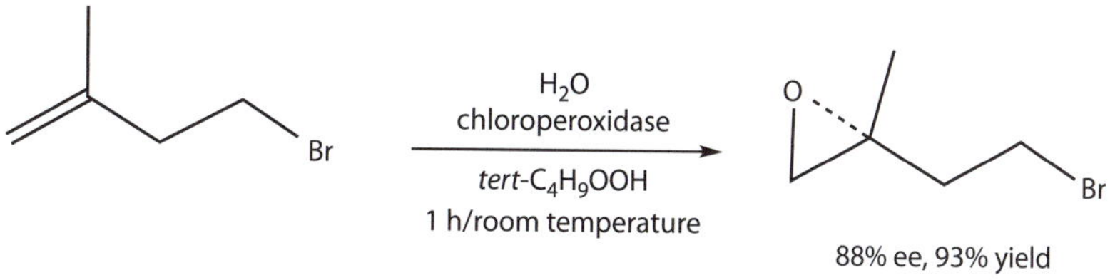

Br H_2O chloroperoxidase *tert*-C_4H_9OOH 1 h/room temperature O Br 88% ee, 93% yield

10.47 Schematic

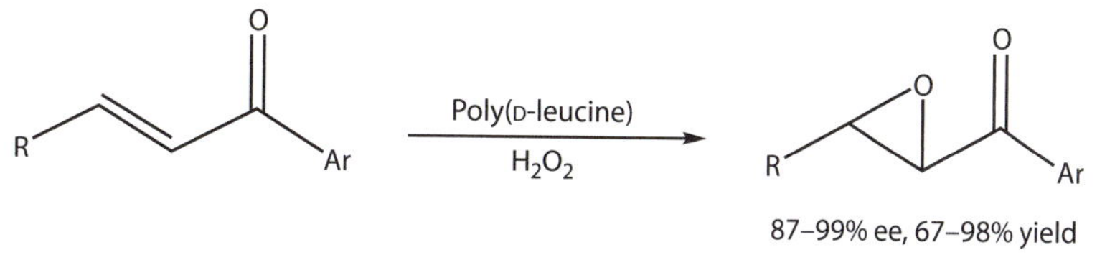

O R Ar Poly(D-leucine) H_2O_2 O O R Ar 87–99% ee, 67–98% yield

10.48 Schematic

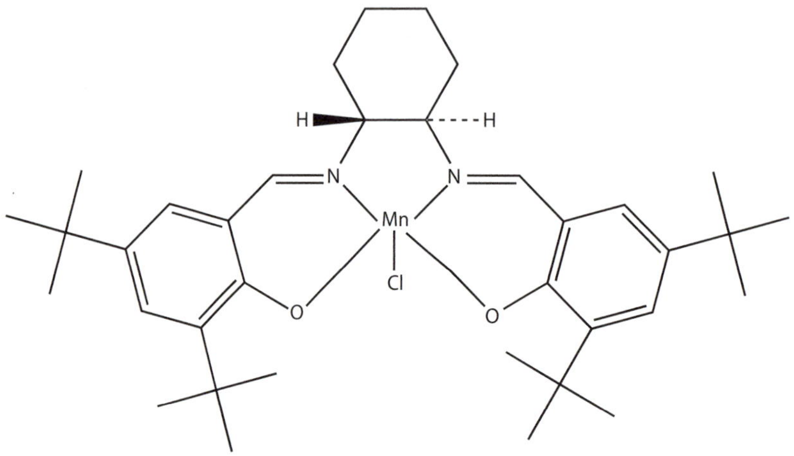

10.49 Schematic

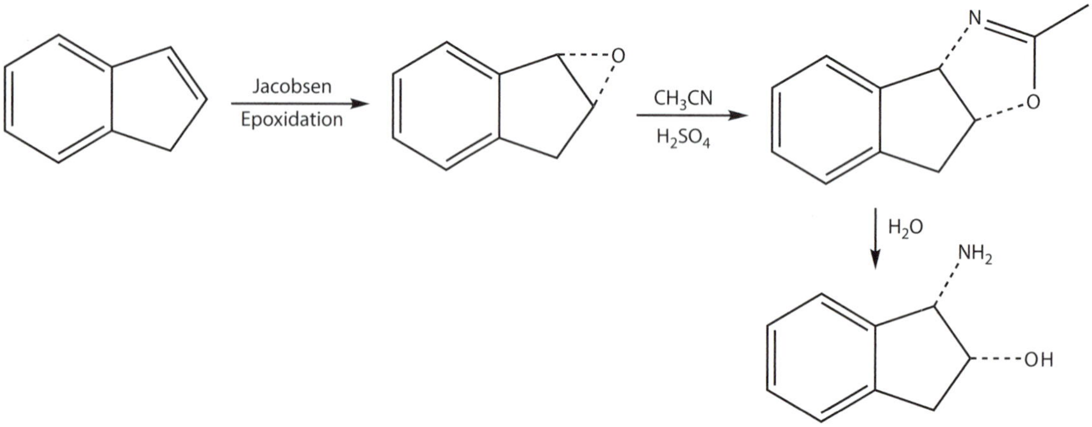

10.50 Schematic

azide chelate catalyst to produce (*S*)-1-azido-2-trimethylsiloxypropane in quantitative yield with 97% ee.[115] Cyclohexene oxide was opened with benzoic acid in the presence of 1 mol% cobalt chelate catalyst to give the hydroxybenzoate in 98% yield with 77% ee.[116]

Jacobsen epoxidations have also been carried out with the manganese chelate being made in zeolites.[117] No manganese leached out. These reactions have been somewhat slower and have often tended to give somewhat lower ee's. Polymeric versions of the Jacobsen catalyst have also been used, but have often given lower ee's.[118] Use of the Jacobsen catalyst occluded in a cross-linked poly(dimethylsiloxane) membrane reactor gave 78% conversion to a product with 57% ee in 8 h.[119] Thus, further work is needed to improve these systems, which offer easy recovery of the catalyst for reuse.

A disadvantage of the Jacobsen methods is that they are energy intensive because low temperatures (0°C to as low as –78°C) are often required for optimal stereoselectivity. As pointed out in Chapter 4, the ligands in metal chelate catalyst often undergo oxidation themselves, shortening the useful lifetimes of the catalysts, although encapsulation in a zeolite can slow this oxidation. It would be interesting to see

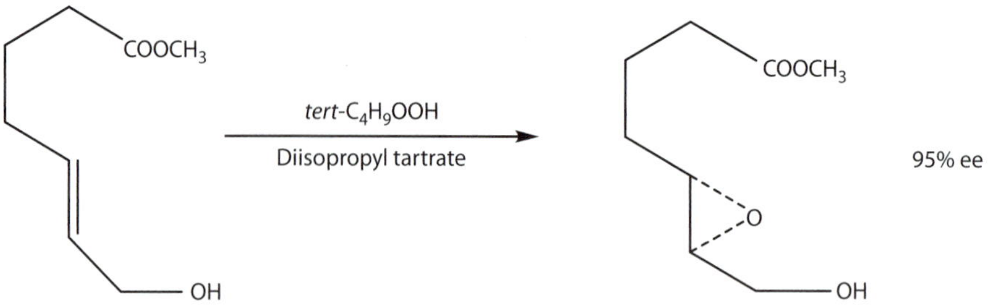

10.51 Schematic

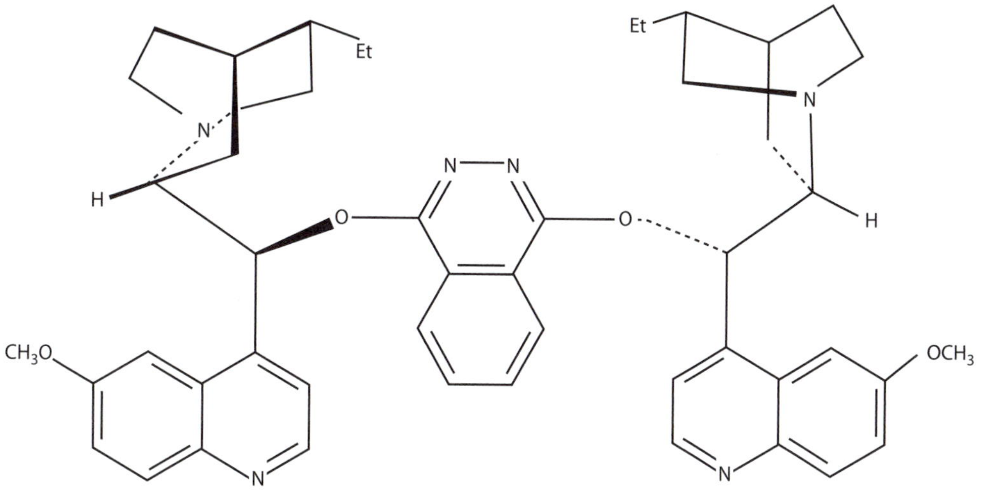
Et
Et
N
N
N—N
H
H
O
O
CH3O
OCH3
N
N

10.52 Schematic

if a completely inorganic catalyst could be devised for the epoxidations. A manganese derivative of heteropoly-acid might work.

Sharpless and co-workers have devised an enantiospecific oxidation of allylic alcohols to the epoxides using *tert*-butylhydroperoxide in the presence of titanium(IV) isopropoxide and dialkyl tartrates (**10.51** Schematic).[120]

The Sharpless group is best known for development of a method for the asymmetrical dihydroxylation of olefins.[121] A typical oxidation uses 1 mol% of potassium osmate dihydrate, 1 mol% ligand, potassium carbonate, potassium ferricyanide, and methylsulfonamide in aqueous *tert*-butyl alcohol. Other oxidants such as *N*-methylmorpholine-*N*-oxide can be used in place of the potassium ferricyanide. One of the preferred ligands is a bis(dihydroquinidine ether) of phthalazine (**10.52** Schematic)

Reaction **10.53** was run with the amine oxide.

The method gives higher enantioselectivity with *trans*- than with *cis*-olefins.[122] Catalytic asymmetrical aminohydroxylation of double bonds (**10.54** Schematic) can be done by a modification of the method using chloramine-T.[123]

Frequently, the hydroxysulfonamide crystallizes out of the reaction mixture so that it is easily recovered by filtration. Substitution of benzyl-OCONClNa for the chloramine-T raised the yield to as high as 80% and the ee to as high as 99%.[124]

Because the molecular weight of the ligand is relatively large, 1 mol% of it amounts to quite a bit on a weight basis. As it is expensive, a convenient method of recovering it for recycle is needed. The dihydroxylation of an alkyl cinnamate was carried out in 81% yield and more than 99% ee with a phthalazine-alkaloid catalyst attached to silica.[125] Another silica-supported catalyst gave 92% yield and 97% ee in the dihydroxylation of styrene.[126] The catalyst could be

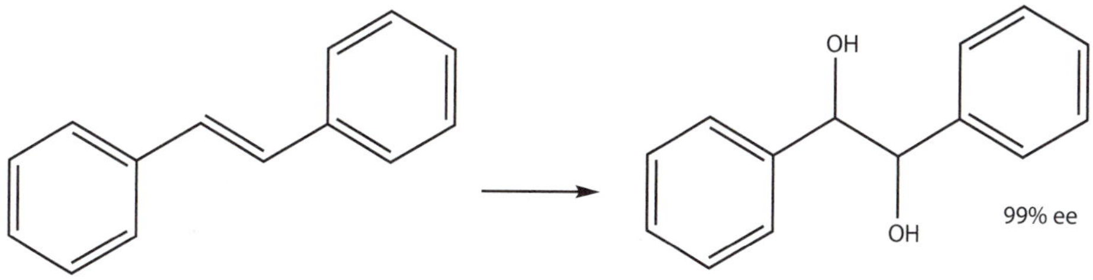
OH
OH
99% ee

10.53 Schematic

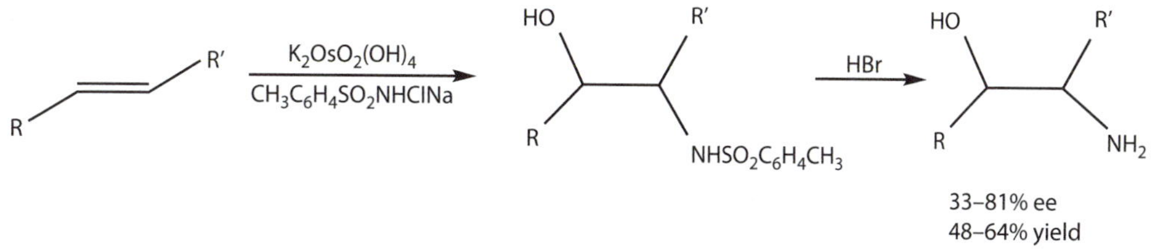
R'
R
K2OsO2(OH)4
CH3C6H4SO2NHClNa
HO
R'
R
NHSO2C6H4CH3
HBr
HO
R'
R
NH2
33–81% ee
48–64% yield

10.54 Schematic

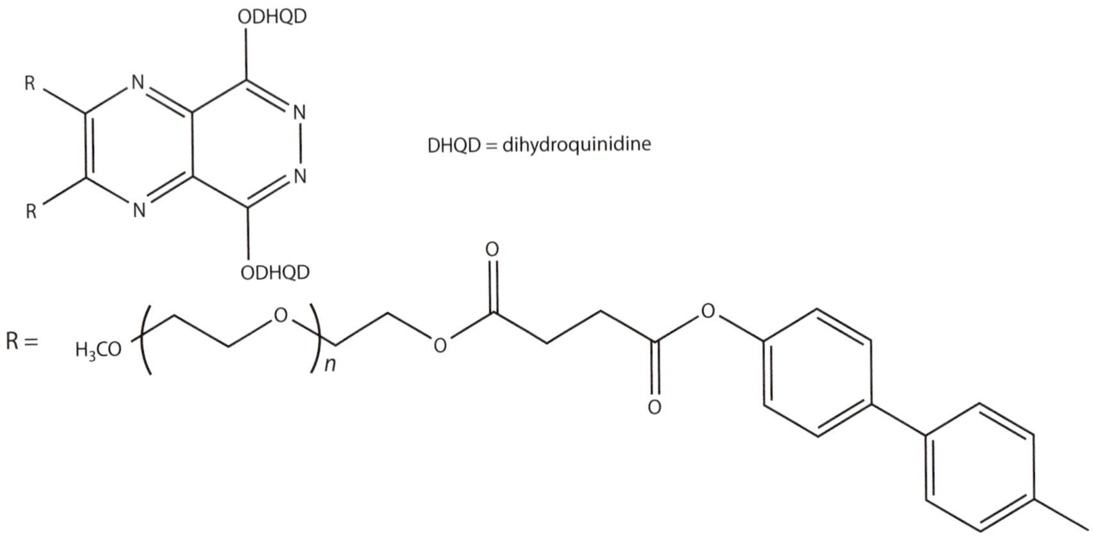

ODHQD
R N N N R N ODHQD
DHQD = dihydroquinidine
R = H_3CO (O)n O O O O

10.55 Schematic

reused seven times with no loss of ee, but small amounts of potassium osmate had to be added with each cycle. Polymeric derivatives of cinchona alkaloids have also been used for ease of recovery by filtration.[127] One of these catalysts could be reused three times with no loss in activity. The ee was 88%. Another gave an 86% yield and 65–97% ee in the dihydroxylation of styrene.[128] It could be used five times more if a little more osmium was added at each cycle. A better system that does not sacrifice any enantioselectivity uses a soluble polymer and gives up to 99% ee with 86–92% yields.[129] The ligand is attached to polyethylene glycol monomethyl ether (**10.55** Schematic). At the end of the reaction, addition of *tert*-butyl methyl ether precipitated the polymer, which was removed by filtration, the product remaining in solution. The recovery of the polymer was 98%. It could be used repeatedly with almost no change in enantioselectivity or yield. (The use of ultrafiltration to separate the polymer would eliminate the need for the *tert*-butyl methyl ether.) A similar dihydroxylation in which the cinnamate substrate was part of an insoluble polymer gave 98% ee at 98% conversion.[130] The phthalazine-alkaloid catalyst was attached to a soluble polymer. Ideally, the osmium compound should remain on the polymer with no loss, or if it is attached to a soluble polymer, be easy to separate for reuse by ultrafiltration. Microencapsulated osmium tetroxide is now available.[131]

Methods using osmium catalysts should be compared with those where the double bond is epoxidized, followed by ring opening with an epoxide hydrolase, that is systems that use no osmium.

The Sharpless epoxidation of allylic alcohols with *tert*-butylhydroperoxide and titanium(IV) isoproxide was carried out with a polymeric tartrate made from tartaric acid and 1,8-octanediol.[132] The yield of epoxide (92%) was comparable with that when dimethyl tartrate was used (91%), but the 79% ee was lower than the 98% ee found with the dimethyl tartrate. It may be possible to raise the ee by further variations in the structure of the polymer. The heterogenization of a catalyst for a homogeneous reaction often requires optimization to obtain comparable or higher yields and stereospecificity.

The toxicity of the osmium compounds is worrisome. They are also expensive. Ruthenium has been used in place of osmium, but its compounds can also be toxic and expensive.[133] Because potassium permanganate is also used in the dihydroxylation of double bonds, manganese compounds should be tried with asymmetrical ligands. This may be a good place to use combinatorial chemistry. In view of the recent use of titanium silicates in oxidation reactions (see Chapter 4), titanium compounds should also be screened. The ideal would be to find an asymmetrical completely inorganic compound that would do the job. The cost of the oxidizing agent can be minimized by regenerating it electrochemically.[134]

Many asymmetrical reactions are carried out to form carbon–carbon single bonds. Addition of triethylaluminum to benzaldehyde in the presence of (*S*)-octahydroBINOL gave the alcohol addition product (**10.56** Schematic) in 100% yield and 96.4% ee.[135]

An interesting specialized case involves asymmetrical autocatalysis in which the product of the reaction is the ligand (**10.57** Schematic).[136] The reaction is seeded with 98.8% ee product.

If the reaction is seeded with leucine of 2% ee the reaction product has 21–26% ee, and if seeded with 2-butanol of 0.1% ee the reaction product has 73–76% ee. Using either of these products as seed for a repeat run raises the ee to more than 95%. Similar chiral amplification is seen in the addition of diethylzinc to benzaldehyde in the presence of a

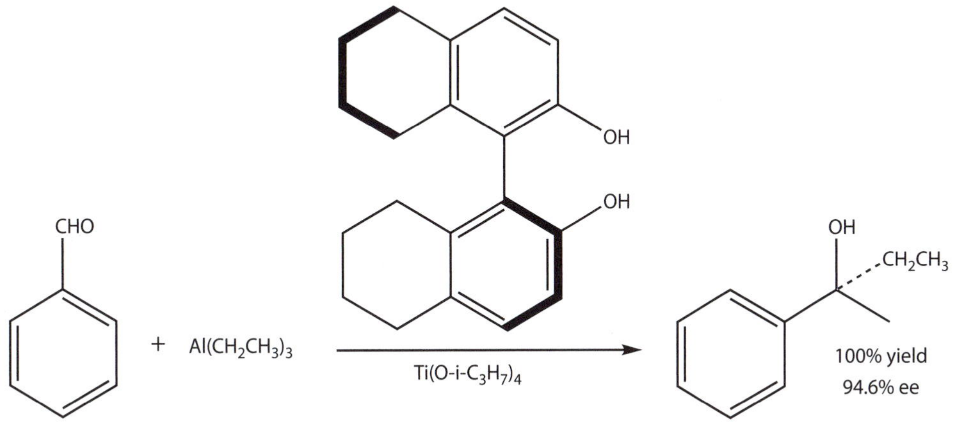

10.56 Schematic

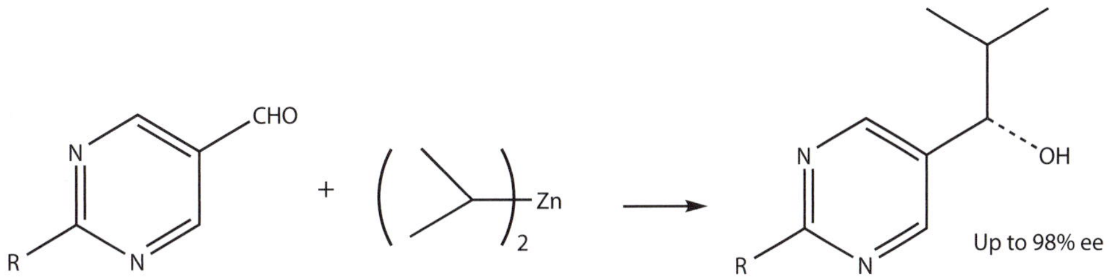

10.57 Schematic

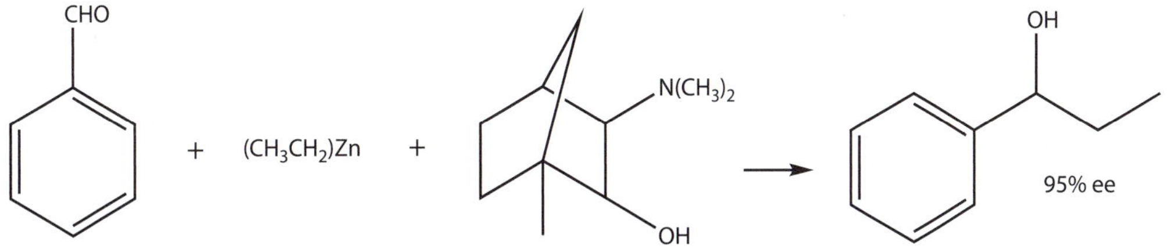

10.58 Schematic

diaminoalcohol (**10.58** Schematic), with 15% ee where the product formed has an ee of 95%.[137] (Such effects have also been seen in several other reactions.)

Chiral Lewis acids derived from tartaric acid have been used in the asymmetrical addition of dialkylzinc compounds to aldehydes[138] and nitroolefins (**10.59** Schematic).[139] Another chiral Lewis acid has been used to catalyze the reaction of aldehydes with allyltrimethylsilane (**10.59** Schematic).[140]

An analog of the second catalyst, where F is replaced by O-*i*-C_3H_7, has been used to catalyze the reaction of benzaldehyde with a silyl enol ether (**10.60** Schematic).[141]

The enantioselectivity was much poorer in the noncoordinating solvents, methylene chloride and toluene. Thus, it is important to remember that solvents can coordinate with a metal center in the transition state and play an important role in determining selectivity.

Bisoxazoline copper complexes have also been used as chiral Lewis acids that give high yields and enantioselectivities (**10.61** Schematic).[142]

The enantioselectivity can vary with the anion. With a related copper chelate, the hexafluoroantimonate was a more enantioselective catalyst than the trifluoromethanesulfonate. This shows that anions are more than innocent bystanders. Copper bisoxazoline complexes have also been used in asymmetrical Diels–Alder reactions[143] and ene reactions.[144] The activity is said to approach that of enzymes. When runs were made in hexane in the presence of florisil, the catalyst could be recovered by filtration and reused with only a slight reduction of activity. Various Lewis acids of magnesium,[145] tin,[146] titanium,[147] and boron[148] have also be en used as catalysts for such reactions. An asymmetrical Diels–Alder reaction with an aluminum Lewis acid is shown in **10.62** Schematic.[149]

Chiral bases are also used in asymmetrical synthesis. An alkylation is shown in **10.63** Schematic.[150]

Another chiral base that has been used is potassium *N*-methylephredinate (**10.64** Schematic).[151]

Chiral bases made by treatment of (*R*)-BINOL with a combination of lanthanum isopropoxide and sodium *tert*-butoxide

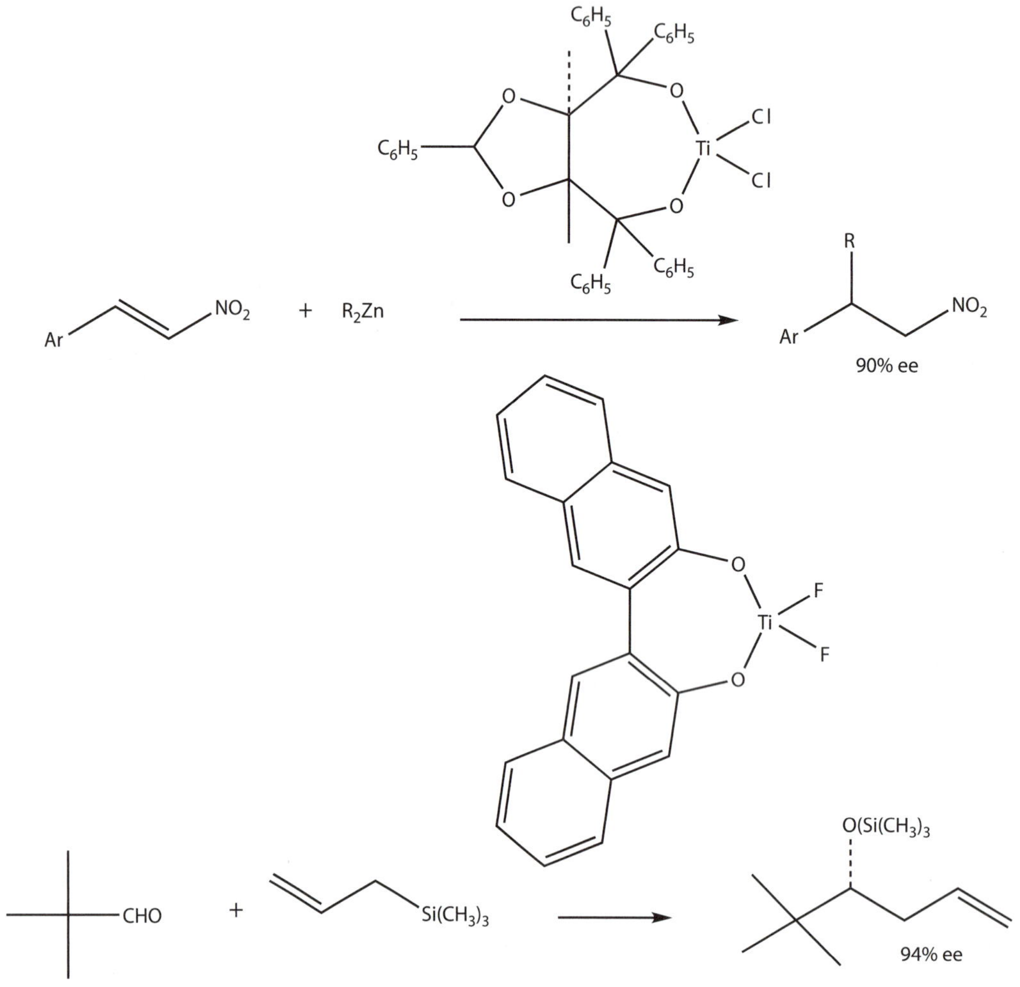

C_6H_5 C_6H_5 O O O C_6H_5 Ti Cl Cl O O C_6H_5 C_6H_5

Ar NO_2 + R_2Zn → R Ar NO_2

90% ee

O Ti F F O

CHO + $Si(CH_3)_3$ → $O(Si(CH_3)_3$

94% ee

10.59 Schematic

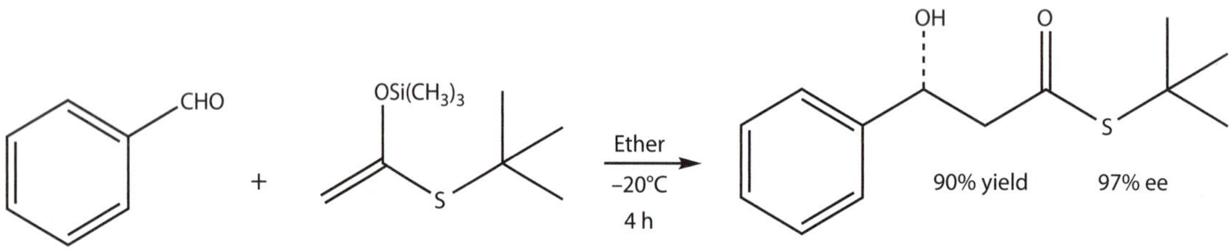

CHO + $OSi(CH_3)_3$ S

Ether

−20°C

4 h

OH O S

90% yield 97% ee

10.60 Schematic

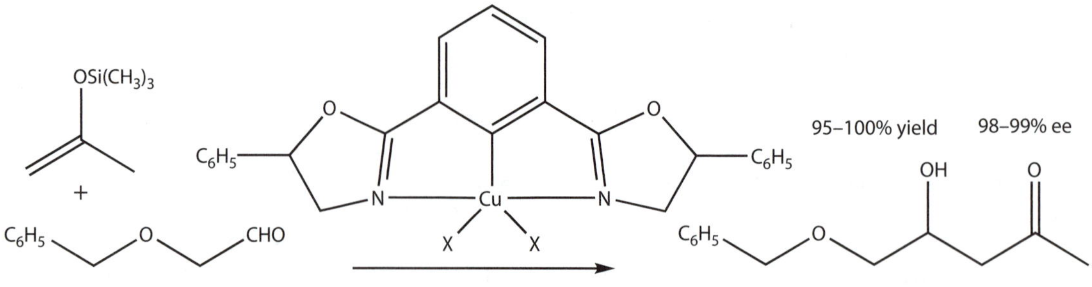

$OSi(CH_3)_3$

+

C_6H_5 O CHO

C_6H_5 O N Cu N O C_6H_5

X X

95–100% yield 98–99% ee

C_6H_5 O OH O

10.61 Schematic

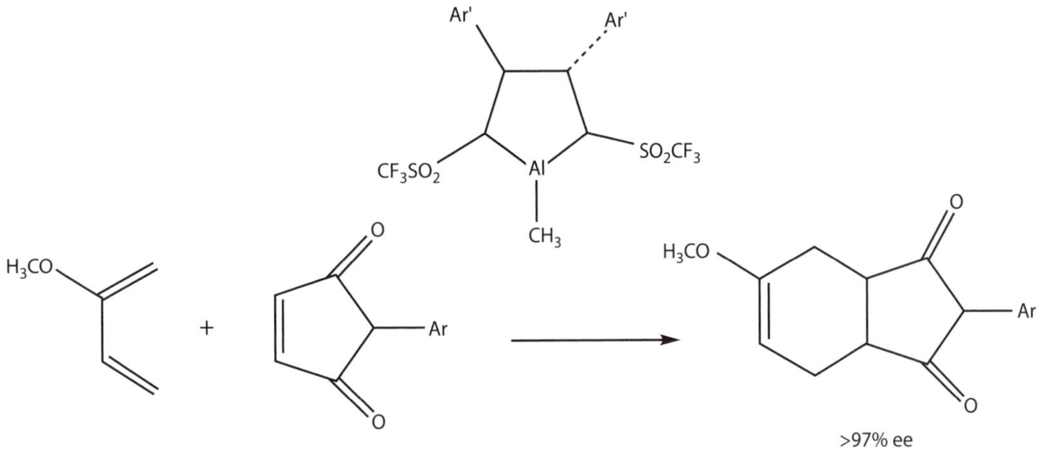

10.62 Schematic

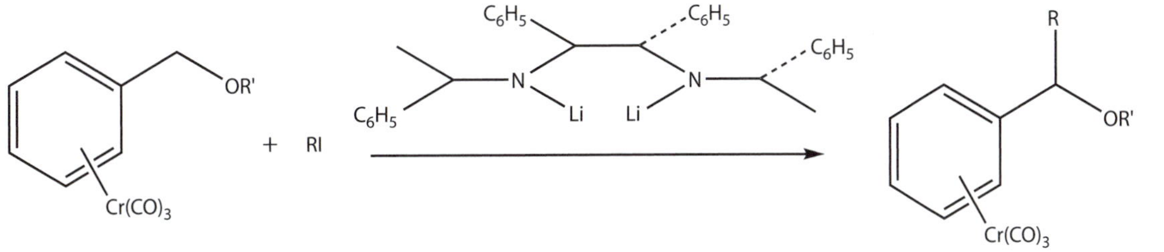

10.63 Schematic

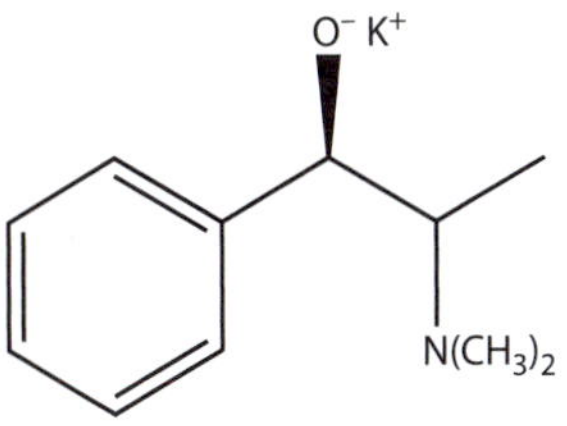

10.64 Schematic

have been used in asymmetrical Michael reactions to produce products with 92% ee.[152]

Chiral phase transfer catalysts have been used in the synthesis of amino acids (**10.65** Schematic).[153] (The amino acid is obtained by hydrolysis of the product shown.) This type of catalyst has also been used successfully in the epoxidation of α,β-unsaturated ketones in toluene with aqueous sodium hypochlorite.[154]

Various amines derived from 1,1′-binaphthalenes, such as the one below, have also been used as phase transfer catalysts (**10.66** Schematic).[155]

Efforts are being made to simplify the workups and the recovery of chiral ligands for recycling by the use of polymeric reagents. In some cases this works well, but in others optimization is needed to make the results as good as those in solution. Asymmetrical protonation with a polymer-supported chiral proton donor (**10.67** Schematic) has given better results than the comparable reaction in solution.[156] The polymer puts additional steric constraints on the system, which in this case are favorable. The product is obtained in 95% yield and 94% ee. In the counterpart in solution, the ee was only 21–50%. Polymeric chiral catalysts have also been used in the addition of zinc alkyls to aldehydes. Use of a proline-based copolymer in a continuous asymmetrical synthesis with an ultrafiltration membrane gave 80% ee (**10.68** Schematic).[157] There was no deactivation in 7 days. A boron-containing polymer (**10.69** Schematic) gave

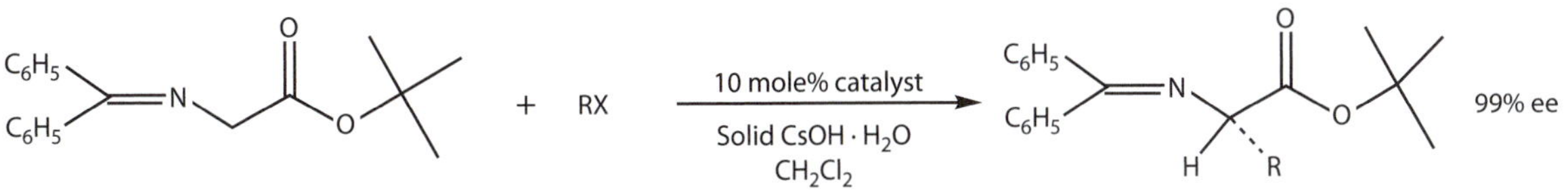

10.65 Schematic

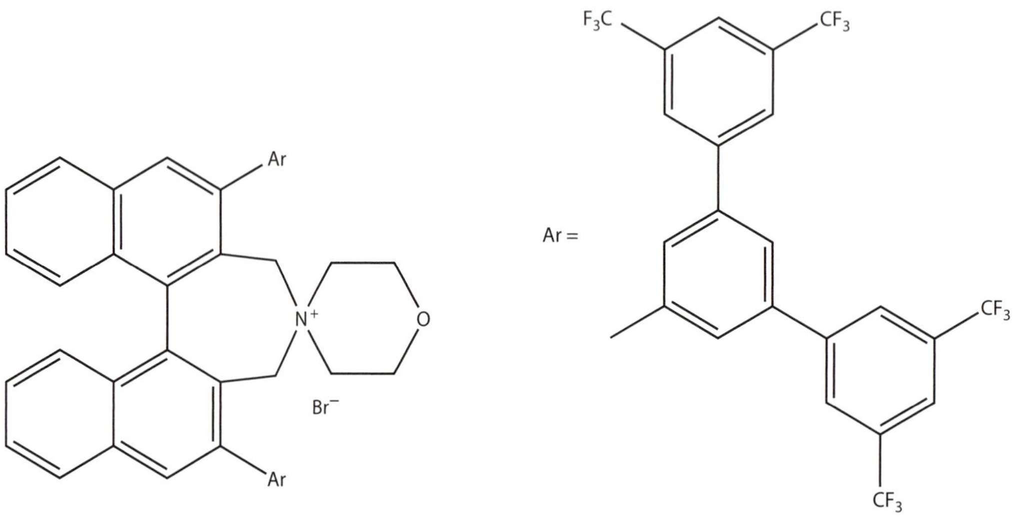

10.66 Schematic

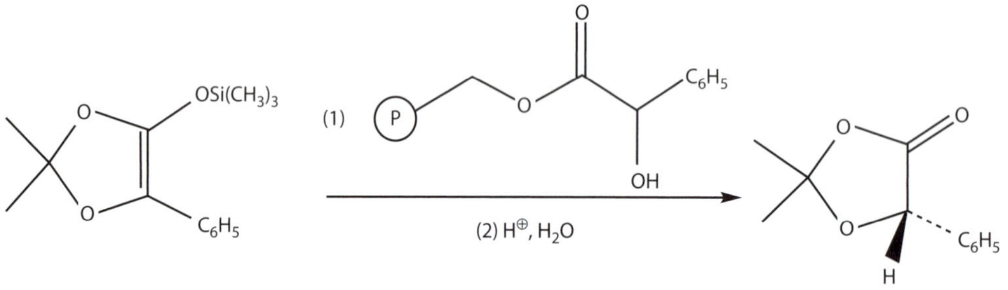

10.67 Schematic

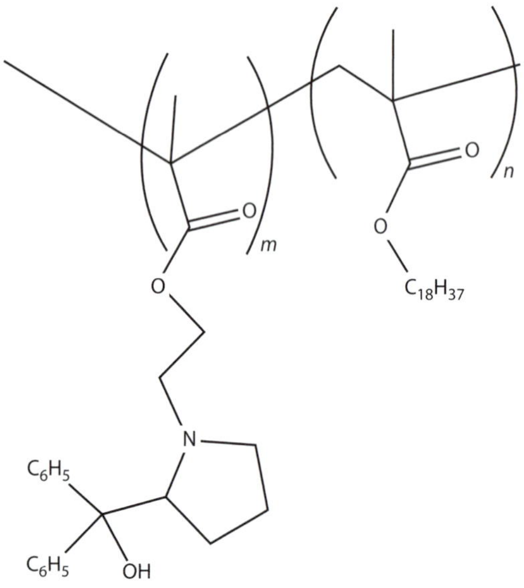

10.68 Schematic

only 28–51% ee compared with the 65–75% ee found with model compounds in solution.[158]

Further work with supported catalysts in membrane reactors is needed to raise the enantioselectivity to commercially useful levels. Care should be taken to pick supports, polymers or otherwise, that are very stable under the conditions of the reaction so that they will be long-lived.

Organocatalysis,[159] usually with proline or its derivatives, provides a way of making drugs and other compounds without the cost of noble metal catalysts and without leaving traces of them in the products. An example is shown below (**10.70** Schematic).[160]

The mechanism involves an aminium salt and an enamine (**10.71** Schematic). Imidazolidinones, such as the one below, have also been used as catalysts for such reactions (**10.72** Schematic).[161]

Poly(leucine) immobilized on polystyrene catalyzed an epoxidation with hydrogen peroxide (**10.73** Schematic).[162]

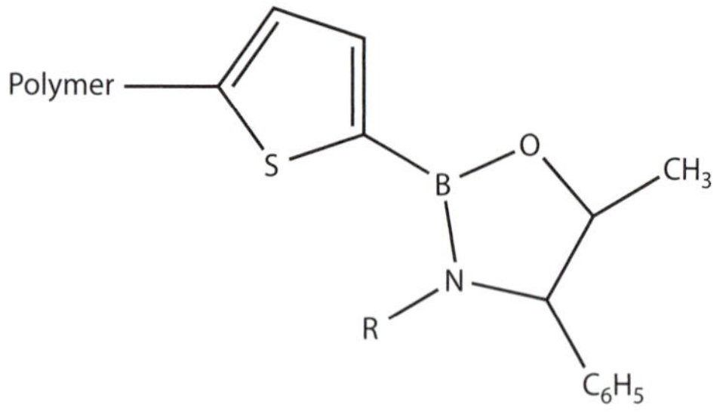

10.69 Schematic

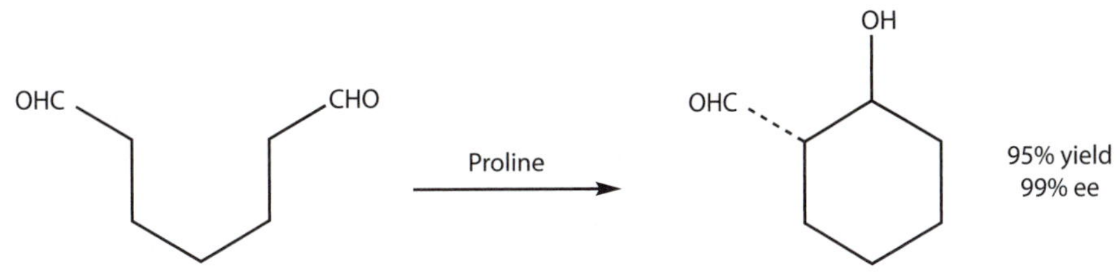

OHC CHO Proline OHC OH 95% yield 99% ee

10.70 Schematic

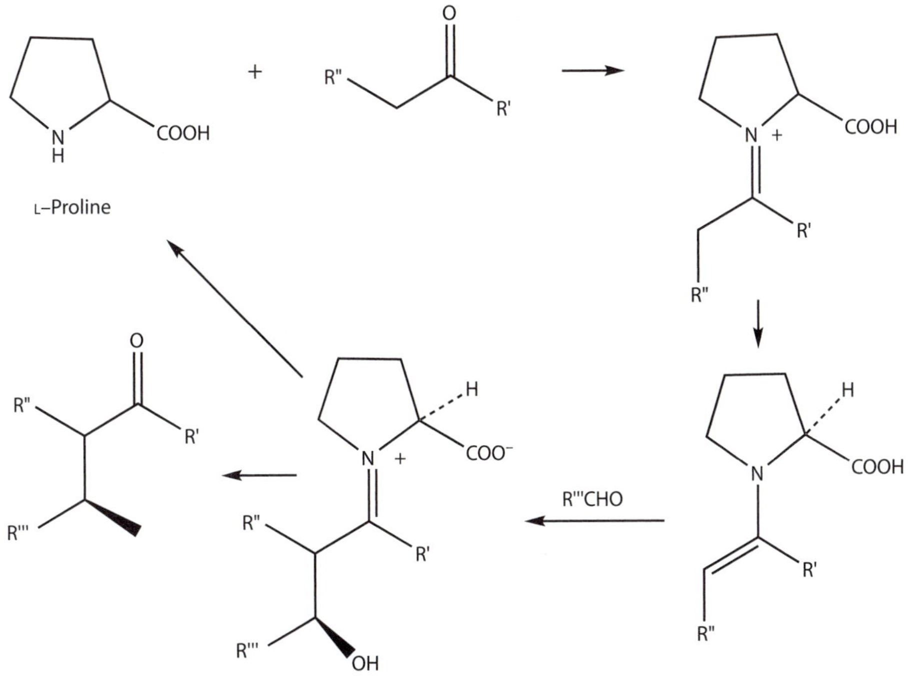

L-Proline COOH R'' R' O COOH R'' R' H COOH R'''CHO H COO^- R'' R' R''' OH R'' R' R'''

10.71 Schematic

R' R R'' CHO O CH_3 C_6H_5 H_2 X^- 20 mole% R' R'' CHO R

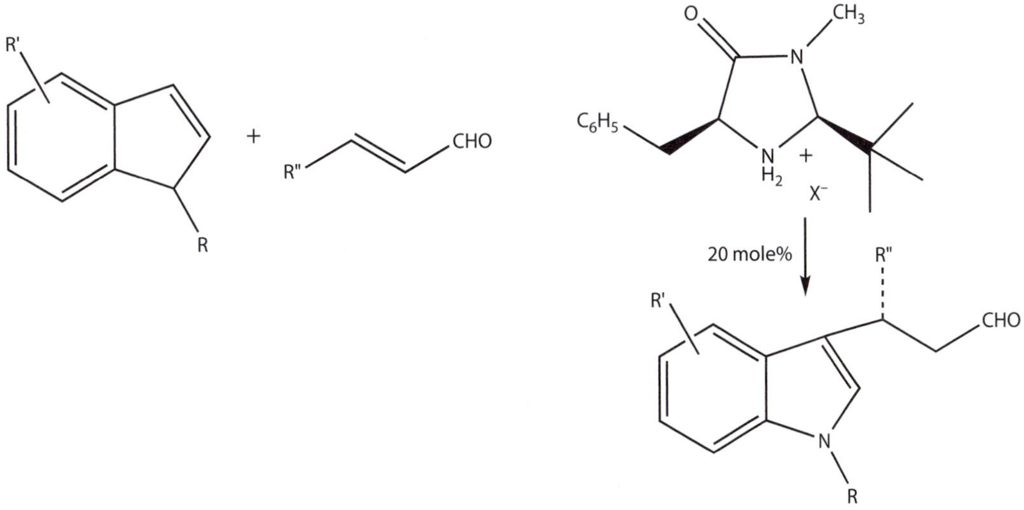

10.72 Schematic

Poly(L-leucine)/polystyrene
$Na_2CO_3 \cdot H_2O_2$
$CH_3CH_2CH_2OCH_3$
20 min
99% conversion
96% ee

10.73 Schematic

(1) Rh(BINAP)$^{\oplus}$
(2) H_2O
$N(C_2H_5)_2$
CHO
Citronellal
$ZnBr_2$
(–)Isopulegol
H_2
Ni
OH
(–)Menthol

10.74 Schematic

Many other types of reactions have been used in asymmetrical syntheses. A rhodium-catalyzed allylic rearrangement (**10.74** Schematic) has been used to make 1000 metric tons of (–)menthol per year.[163]

The rhodium catalyst gives 400,000 turnovers. A drawback of the route is the need to recrystallize the isopulegol at –50°C to achieve 100% ee. Other asymmetrical reactions include hydroformylation, hydrocyanation, cyclopropanation, aziridination, and so forth. The various suggestions for improvement of the reactions discussed in this chapter can be applied to these also. It would be preferable, if possible, to avoid the use of toxic hydrogen cyanide altogether. Extensive work has gone into fine-tuning the metallocene catalysts used for the stereospecific polymerization of olefins.[164] Perhaps some of the principles used there can be applied here, too.

REFERENCES

1. (a) G.P. Moss, *Pure Appl. Chem.*, 1996, 68, 2193; (b) E.L. Eliel, S.M. Wilen, and M.P. Doyle, *Basic Organic Stereochemistry*, Wiley-Interscience, New York, 2001.
2. (a) J. Caldwell., *Chem. Ind. (Lond.)*, 1995, 176; (b) M.J. Cannarsa., *Chem. Ind. (Lond.)*, 1996, 374; (c) I.W. Davies and P.J. Reider., *Chem. Ind. (Lond.)*, 1996, 374.; (d) S.J. Stinson., *Chem. Eng. News*, Sept. 19, 1994, 38; Oct. 9, 1997, 44; Oct. 20, 1998, 38; Sept. 19, 1999, 83; (e) H.Y. Aboul-Enein and I. Wainer., *The Impact of Stereochemistry on Drug Development and Use*, Wiley, New York, 1997; (f) A. Richards and R. McCague, *Chem. Ind. (Lond.)*, 1997, 422; (g) S.C. Stinson, *Chem. Eng. News*, Oct. 11, 1999, 101; Nov. 22, 1999, 57; May 8, 2000, 59; (h) A.M. Rouhi, *Chem. Eng. News*, June 10, 2002, 43; June 14, 2004, 47; (i) S.C. Stinson, *Chem. Eng. News*, Oct. 23, 2000, 55; May 14, 2001, 45.
3. S.C. Stinson., *Chem. Eng. News*. June 2, 1997, 28.
4. R.A. Sheldon., *Chirotechnology—Industrial Synthesis of Optically Active Compounds*, Dekker, New York, 1993.
5. (a) G.M. Coppola and H.F. Schuster, *Alpha-Hydroxyacids in Enantioselective Synthesis*, Wiley-VCH, Weinheim, 1997; (b) J. Gawrowski and K. Gawrowski., *Tartaric and Malic Acids in Synthesis: A Source Book of Building Blocks, Ligands, Auxiliaries, and Resolving Agents*, Wiley, New York, 1999.
6. P. G. Hultin, M.A. Earle, and M. Sudharshan., *Tetrahedron*, 1997, *53*, 14823.
7. (a) D.J. Ager, ed., *Handbook of Chiral Chemicals*, 2nd ed., CRC Press, Boca Raton, FL, 2005; (b) H.U. Blaser and E. Schmidt, eds, *Asymmetric Catalysis on Industrial Scale*, Wiley-VCH, Weinheim, 2004; (c) A. Berkessel and H. Groger, eds, *Asymmetric Organocatalysis*, Wiley-VCH, Weinheim, 2005; (d) M. Christmann and S. Brase, *Asymmetric Synthesis—The Essentials*, Wiley-VCH, Weinheim, 2007; (e) D.E. deVos, I.F.J.L.V. Lankelecom, and P.A. Jacobs, eds, *Chiral Catalyst Immobilization and Recycling*, Wiley-VCH, Weinheim, 2000; (f) G.Q. Lin, Y.-M. Li, and A.S.C. Chan, *Principles and Applications of Asymmetric Synthesis*, Wiley, New York, 2001; (g) I. Ojima, ed., *Catalytic Asymmetric Synthesis*, 2nd ed., Wiley, New York, 2000; (h) D. Enders and K.-E. Jaeger, eds, *Asymmetric Synthesis with Chemical and Biological Methods*, Wiley-VCH, Weinheim, 2007; (i) P.A. Dalko, ed., *Enantioselective Organocatalysis—Reactions and Experimental Procedures*, Willey-VCH, Weinheim, 2006; (j) K. Mikami and M. Lautens, eds, *New Frontiers in Asymmetric Catalysis*, Wiley, New York, 2007; (k) L.A. Paquette, ed., *Chiral Reagents for Asymmetric Synthesis*, Wiley, New York, 2003; (l) W. Lough and I. Wainer, eds, *Chirality in Natural and Applied Science*, Blackwell Science, Oxford, 2002; (m) G. Subramanian, ed., *Chiral Separation Techniques—A Practical Approach*, Wiley-VCH, Weinheim, 2000; (n) E.N. Jacobsen, A. Pfaltz, and

H. Yamamoto, *Comprehensive Asymmetric Catalysis, 3*, Springer-Verlag, Berlin, 1999; (o) J.M.J. Williams, *Catalysis in Asymmetric Synthesis*, Academic, San Diego, 1999; (p) T.-L. Ho, *Stereoselectivity in Synthesis*, Wiley, New York, 1999; (q) V. Gotor, J. Alfonso, and E. Garsia-Uridiales, eds., *Asymmetric Organic Synthesis with Enzymes*, Wiley-VCH, 2008; (r) H. Amouri and M. Gruselle, *Chirality in Transition Metal Chemistry*, Wiley, Hoboken, NJ, 2008.

8. (a) B.W. Gung, B. leNoble, eds, *Chem. Rev.*, 1999, *99*, 1067; (b) H.-U. Blaser, *Chem. Commun.*, 2003, 293 (enantioselective catalysis for fine chemicals); (c) M. Breuer, K. Ditrich, T. Habicher, B. Hauer, M. Kessler, R. Sturmer, and T. Zelenski, *Angew. Chem Int. Ed.*, 2004, *43*, 788–824 (industrial enantioselectivity); (d) C. Bolm and J.A. Gladysz, eds, *Chem. Rev.*, 2003, *103*(8), 2761–3400 (enantioselective catalysis); (e) J.-A. Ma and D. Cahard, *Angew. Chem. Int. Ed.*, 2004, *43*, 4566 (asymmetric catalysis).
9. I. Cunningham, S.J. Coles, and M.B. Hursthouse, *Chem. Commun.*, 2000, 61.
10. R. Tamura, H. Takahashi., K. Hirotsu, Y. Nakajima, T. Ushio, F. Toda., *Angew. Chem. Int. Ed.*, 1998, *37*, 2876.
11. O. Tissot, M. Gouygou, F. Dallemer, J.C. Daran, and G.G.A. Balavoine, *Angew. Chem. Int. Ed.*, 2001, *40*, 1076.
12. L. Perez-Garcia and D.B. Amabilino, *Chem. Soc. Rev.*, 2002, *31*, 342.
13. R.H. Perry, C. Wu, M. Neflu, and R.G. Cooks, *Chem. Commun.*, 2007, 1071.
14. (a) Y. Shimizu and S. Kawanishi, *Chem. Commun.*, 1996, 1333; (b) B.L. Feringa and R.A. van Delden, *Angew. Chem. Int. Ed.*, 1999, *38*, 3419.
15. (a) G. Kaupp, *Angew. Chem. Int. Ed.*, 1994, *33*, 728; (b) F. Toda, *Acc Chem. Res.*, 1995, *28*, 483.
16. (a) D. Seebach, A.K. Beck, and A. Heckel, *Angew. Chem. Int. Ed.*, 2001, *40*, 93; (b) I. Sereewatthanawut, A.T. Boam, and A.G. Livingston, *Macromol. Symmp.*, 2008, *264*, 184.
17. F. Toda and K. Tanaka, *Chem. Commun.*, 1997, 1087.
18. T. Tsunoda, H. Kaku, M. Nagaku, and E. Okuyama, *Tetrahedron. Lett.*, 1997, *38*, 7759.
19. K. Yannakopoulou, D. Mentzafos, I.M. Mavridis, and K. Dandika, *Angew. Chem. Int. Ed.*, 1996, *35*, 2480.
20. (a) I.A. Nicholls, L.I. Andersson, and K. Mosbach, *Trends. Biotechnol.*, 1995, *13*(2), 47; (b) M. Yoshida, K. Uezu, M. Goto, S. Furusaki, M. Takagi, *Chem. Lett.*, 1998, *27*, 925.
21. D.W. Armstrong, J.M. Schneiderheinze, Y.-S. Hwang, and B. Sellerger, *Anal. Chem.*, 1998, *70*, 3717.
22. X.X. Zhang, J.S. Bradshaw, and R.M. Izatt., *Chem. Rev.*, 1997, *97*, 3313.
23. (a) T. Vries, H. Wynberg, M. Kellogg, Q.B. Broxterman, A. Minnaard, B. Kaptein, S. van der Sluis, L. Hulshof, and J. Kooistra, *Angew. Chem. Int. Ed.*, 1998, *37*, 2387; (b) A. Collet, *Angew. Chem. Int. Ed.*, 1998, *37*, 3239; (c) S. Stinson, *Chem. Eng. News*, May 25, 1998, 9; (d) J.W. Nieuwenhuijzen, R.F.P. Grimbergen, C. Koopman, R.M. Kellogg, T.R. Vries, K. Pouwer, E. van Echten, B. Kaptein, L.A. Hulshof, and Q.B. Broxterman, *Angew. Chem. Int. Ed.*, 2002, *41*, 4281; (e) M. Leeman, G. Brasile, E. Gelens, T. Vries, B. Kaptein, and R. Kellogg, *Angew. Chem. Int. Ed.*, 2008, *47*, 1287.
24. A. Mravik, Z. Bocskei, Z. Katona, I. Markovits, E. Fogassy, *Angew. Chem. Int. Ed.*, 1997, 36, 1534.
25. (a) P. Camilleri, V. deBiasi, A. Hutt, *Chem. Br.*, 1994, *30*(1), 43; (b) E. Yashima and Y. Okamoto, *Bull. Chem. Soc. Jpn.*, 1995, *68*, 3289; (c) K. Lorenz, E. Yashima, and Y. Okamoto, *Angew. Chem. Int. Ed.*, 1998, *37*, 1922; (d) Y. Okamoto and E. Yashima, *Angew. Chem. Int. Ed.*, 1998, *37*, 1021.
26. S.A. Matlin, S.J. Grieb, and A.M. Belenguer, *J. Chem. Soc. Chem. Commun.*, 1995, 301.
27. (a) M.M. Coy, *Chem. Eng. News*, June 19, 2000, 17; (b) C. Wibowo and L.O'Young, *Chem. Eng. Prog*, 2005, *101*(11), 22; (c) M. Juza, M. Mazzotti, and M. Morbidelli, *Trends Biotechnol.*, 2000, *18*, 108.
28. E. Yashima, J. Noguchi, and Y. Okamoto, *J. Appl. Polym. Sci.*, 1994, *54*, 1087.
29. N. Ogata., *Macromol. Symp.*, 1995, *98*, 543.
30. (a) B.B. Lakshmi, C.R. Martin, *Nature*, 1997, *388*, 758; (b) S. Borman, *Chem. Eng. News*, Aug. 25 1997, 10.
31. (a) E. Schoffers, A. Golebiowski, and C.R. Johnson., *Tetrahedron*, 1996, *52*, 3769; (b) R. Noyori, M. Tokunaga, and M. Kitamura, *Bull. Chem. Soc. Jpn.*, 1995, *68*, 36; (c) A.J. Carnell, *Chem. Br.*, 1997, *33*(5), 49; (d) C.R. Johnson, *Acc. Chem. Res.*, 1998, 31, 333.
32. S. Taylor, *Chem. Br.*, 1998, *34*(5), 23.
33. (a) I.V.J. Archer, *Tetrachedron*, 1997, *53*, 15617; (b) A. Archelas and R. Furstoss, *Trends Biotechnol.*, 1998, *16*, 108; (c) A.L. Botes, C.A.G.M. Weijers, and M.S. van Dyk, *Biotechnol. Lett.*, 1998, *20*, 421–427.
34. F. Thiel, *Chem. Rev.*, 1995, *95*, 2203.
35. F. Molinari, O. Brenna, M. Valenti, and F. Aragozzini, *Enzyme. Microb. Technol.*, 1996, *19*, 551.
36. M. Kinoshita and A. Ohno., *Tetrahedron* 1996, *52*, 5397.
37. I.J. Colton, S.N. Ahmed, and R.J. Kazlauskas, *J. Org. Chem.*, 1995, *60*, 212.
38. N. Kamiya, M. Goto, and F. Nakashio, *Biotechnol. Prog.*, 1995, *11*, 270.
39. M.T. Reetz, A. Zonta, and J. Simpel Kamp, *Angew. Chem. Int. Ed.*, 1995, *34*, 301.
40. G. Lin and W.-Y. Lin, *Tetrahedron. Lett.*, 1998, *39*, 4333.
41. M.T. Reetz, A. Zonta, K. Schimossek, K. Liebeton, and K.-E. Jaeger, *Angew. Chem. Int. Ed.*, 1997, *36*, 2830.
42. D.T. Chapman, D.H.G. Crout, M. Mahmoudian, D.I.C. Scopes, and P.W. Smith., *Chem. Commun.*, 1996, 2415.
43. Anon., *Chem. Eng. News*, Apr. 15, 1996.
44. (a) E.J. Ebbers, G.J.A. Ariaans, J.P.M. Houbiers, A. Bruggink, and B. Zwanenburg., *Tetrahedron*, 1997, *53*, 9417; (b) J.H. Koh, H.M. Jeong, and J. Park, *Tetrahedron Lett.*, 1998, *39*, 5545.
45. R. Dagani, *Chem. Eng. News*,, Oct. 9, 1995.
46. (a) S. Caddick and K. Jenkins, *Chem. Soc. Rev.*, 1996, *25*, 447; (b) F.F. Huerta, A.B.E. Minidis, and J.-E. Backvall, *Chem. Soc. Rev.*, 2001, *30*, 321; (c) H. Pellissier, *Tetrahedron*, 2003, *59*, 8291; (d) O. Plamies and J.-E. Backvall, *Trends. Biotechnol.*, 2004, *22*, 123; (e) B.M. Trost and D.R. Fandrick, *Aldrichim. Acta*, 2007, *40*(3), 59; (f) K. Benaissi, M. Poliakoff, and N.R. Thomas, *Green Chem.*, 2009, *11*, 617; (g) P. Lozano, T. deDiego, C. Mira, K. Montague, M. Vaultier, and J.L. Iborra, *Green Chem.*, 2009, *11*, 538.
47. C. Roengpithya, D.A. Patterson, A.G. Livingston, P.C. Taylor, J.L. Irwin, and M.R. Parrett, *Chem. Commun.*, 2007, 3462.

48. J. Paetzold and J.-E. Backvall, *J. Am. Chem. Soc.*, 2005, *127*, 17620.
49. J.B. Martin-Matute, M. Edin, K. Bogar, and J.-E. Backvall, *Angew. Chem. Int. Ed.*, 2004, *43*, 6535.
50. I. Hilker, G. Rabani, G.K.M. Verzijl, A.R.A. Palmans, and A. Heise, *Angew. Chem. Int. Ed.*, 2006, *45*, 2130.
51. Y. Asano and S. Yamaguchi, *J. Am. Chem. Soc.*, 2005, *127*, 7696.
52. J.A. O'Meara, M. Jung, and T. Durst, *Tetrahedron Lett.*, 1995, *36*, 2559.
53. H. Stecher and K. Faber, *Synthesis*, 1997, 1.
54. K. Nakamura, Y. Inoue, T. Matsuda, and A. Ohno, *Tetrahedron Lett.*, 1995, *36*, 6263.
55. S.-T. Chen, W.-H. Huang, and K.-T. Wang, *J. Org. Chem.*, 1994, *59*, 7580.
56. S. Brand, M.F. Jones, and C.M. Rayner. *Tetrahedron Lett.*, 1995, *36*, 8493.
57. (a) J.V. Allen and J.M.J. Williams, *Tetrahedron Lett.*, 1996, *37*, 1859; (b) R. Sturmer., *Angew. Chem. Int. Ed.*, 1997, *36*, 1173.
58. H. van der Deen, A.D. Cuiper, R.P. Hof, A. van Oeveren, B.L. Feringa, and R.M. Kellogg, *J. Am. Chem. Soc.*, 1996, *118*, 3801.
59. D.S. Tan, M.M. Gunter, and D.G. Drueckhammer, *J. Am. Chem. Soc.*, 1995, *117*, 9093.
60. S.C. Stinson, *Chem. Eng. News*, Jan. 18, 1999, 86.
61. P.R. Carlier, W.W.-F. Lam, N.C. Wan, and I.D. Williams, *Angew. Chem. Int. Ed.*, 1998, *37*, 2252.
62. (a) P.-J. Um and D.G. Drueckhammer, *J. Am. Chem. Soc.*, 1998, *120*, 5605; (b) M.M. Jones and J.M.J. Williams, *Chem. Commun.*, 1998, 2519.
63. A.L.E. Larsson, B.A. Persson, and J.-E. Backvall, *Angew. Chem. Int. Ed.*, 1997, *36*, 1211.
64. A. Liese, T. Zelinski, M-R. Kula, H. Kierkels, M. Karutz, U. Kragl, and C. Wandrey, *J. Mol. Catal. B. Enzymol.*, 1998, *4*, 91.
65. G. Zhong, D. Shabat, B. List, J. Anderson, S.C. Sinha, R.A. Lerner, and C.F. Barbas III, *Angew. Chem. Int. Ed.*, 1998, *37*, 2481.
66. (a) E. Vedejs and X. Chen, *J. Am. Chem. Soc.*, 1997, *119*, 2584; (b) P. Somfai, *Angew. Chem. Int. Ed.*, 1997, *36*, 2731; (c) J. Eames, *Angew. Chem. Int. Ed.*, 2000, *39*, 885.
67. Y. Dobashi, K. Kobayashi, and A. Dobashi, *Tetrahedron Lett.*, 1998, *39*, 93.
68. (a) B.W. Gung and B. le Noble, eds, *Chem. Rev.*, 1999, *99*, 1067–1480; (b) P.J. Walsh and M.C. Kozlowski, *Fundamentals of Asymmetric Catalysis*, University Science Books, Sausalito, CA, 2008.
69. D.L. Illman, *Chem. Eng. News*, Feb. 13, 1995, 44.
70. P. Layman, *Chem. Eng. News*, 1995, Sept. 25, 1995, 17.
71. (a) V. Gotor, I. Alfonso, and E. Garcia-Urdiales, eds, *Asymmetric Organic Synthesis with Enzymes*, Wiley-VCH, Weinheim, 2008; (b) R.N. Patel, ed., *Stereoselective Biocatalysis*, Marcel Dekker, New York, 1999; (c) A.M. Thayer, *Chem. Eng. News*, Aug. 14, 2006, *15*, 29; (d) R.N. Patel, *Enzyme Microb. Technol.*, 2002, *31*, 804.
72. (a) A. Studer, *Synthesis*, 1996, 793; (b) D.J. Ager, I. Prakash, and D.R. Schaad, *Chem. Rev.*, 1996, *96*, 835; (c) J. Seyden-Penne, *Chiral Auxiliaries and Ligands in Asymmetric Synthesis*, Wiley, New York, 1995; (d) K. Ruck-Braun and H. Kunz, *Chiral Auxiliaries in Cycloadditions*, Wiley-VCH, Weinheim, 1999.
73. (a) M.P. Sibi, P.K. Deshpande and J. Ji, *Tetrahedron Lett.*, 1995, *36*, 8965; (b) D.J. Ager, I. Prakash, and D.R. Schaad, *Aldrichim. Acta*, 1997, *30*(1), 3 (review of use of chiral oxazolidinones in asymmetrical synthesis).
74. K. Burgess and D. Lim, *Chem. Commun.*, 1997, 785.
75. S.C. Stinson, *Chem. Eng. News*, Oct. 24, 1994, 32.
76. M.A. Schuerman, K.I. Keverline, and R.G. Hiskey, *Tetrahedron. Lett.*, 1995, *36*, 825.
77. C. Palomo, A. Gonzalez, J.M. Garcia, C. Landa, M. Oiarbide, S. Rodriguez, and A. Linden, *Angew. Chem. Int. Ed.*, 1998, *37*, 180.
78. (a) M. Wills, *Chem. Soc. Rev.*, 1995, *24*, 177; (b) R. Noyori and T. Ohkuma, *Pure. Appl. Chem.*, 1999, *71*, 1493; (c) R. Noyori and T. Ohkuma, *Angew. Chem. Int. Ed.*, 2001, *40*, 40; (d) O.I. Kolodiazhnyi, *Tetrahedron*, 2003, *59*, 5953; (e) X. Cui and K. Burgess, *Chem. Rev.*, 2005, *105*, 3272; (f) M.J. Kirsche and Y. Sun, eds, *Acc. Chem. Res.*, 2007, *40*, 1237–1419.
79. R.N. Patel, R.L. Hanson, A. Banerjee, and L.J. Szarka, *J. Am. Oil. Chem. Soc.*, 1997, *74*, 1345.
80. R. Hayakawa, K. Nozawa, M. Shimizu, and T. Fujisawa, *Tetrahedron Lett.*, 1998, *39*, 67.
81. Anon., *Chem. Eng. News*, June 2, 2008, 44.
82. (a) R.K. Dhar, *Aldrichchim Acta*, 1994, *27*, 43; (b) P.V. Ramachandran, B. Gong, and H.C. Brown, *J. Org. Chem.*, 1995, *60*, 41.
83. E.J. Corey and C.J. Helal, *Tetrahedron Lett.*, 1995, *36*, 9153.
84. C. Caze, N.E. Moualij, C.J. Lock, and J. Ma., *J. Chem. Soc. Perkin. Trans.*, 1995, *1*, 345.
85. (a) T. Langer and G. Helmchen, *Tetrahedron Lett.*, 1996, *37*, 1381; (b) S. Hashiguchi, A. Fujii, J. Takehara, T. Ikariya, and R. Noyori, *J. Am. Chem. Soc.*, 1995, *177*, 7562; (c) J.-X. Gao, T. Ikariya, and R. Noyori, *Organometallics*, 1996, *15*, 1087; (d) Y. Jiang, Q. Jiang, G. Zhu, and X. Zhang, *Tetrahedron Lett.*, 1997, *38*, 215.
86. E.I. Klabunovskii and R.A. Sheldon, *Cattech*, 1997, *1*(2), 153.
87. (a) H.-U. Blaser, H.-P. Jalett, M. Muller, and M. Studer, *Catal. Today*, 1997, *37*, 441; (b) C. Le Blond, J. Wang, J. Liu, A.T. Andrews, and Y.-K. Sun, *J. Am. Chem. Soc.*, 1999, *121*, 4920.
88. X. Zuo, H. Liu, and M. Liu, *Tetrahedron Lett.*, 1998, *39*, 1941.
89. L. Pu, *Chem, Rev.*, 1998, *98*, 2405.
90. (a) T. Ohkuma, H. Ooka, S. Hashiguchi, T. Ikariya, and R. Noyori, *J. Am. Chem. Soc.*, 1995, *117*, 2675; (b) T. Ohta, T. Miyake, N. Seido, H. Kumobayashi, and H. Takaya, *J. Org. Chem.*, 1995, *60*, 357; (c) S. Akutagawa., *Appl. Catal. A*, 1995, *128*, 171; (d) K. Tani, J.-I. Onouchi, T. Yamagata, and Y. Kataoka, *Chem. Lett.*, 1995, *24*, 955; (e) Anon., *Chem. Eng. News*, Feb. 5, 1996, 46; (f) S.J. Stinson, *Chem. Eng. News*, Sept. 19, 1994, 38; Oct. 9, 1995, 44.
91. D. Cai, D.L. Hughes, T.R. Verhoeven, and P.J. Reider, *Tetrahedron Lett.*, 1995, *36*, 7991.
92. S.J. Stinson, *Chem. Eng. News*, Sept. 19, 1994, 38; Oct. 9, 1995, 44.
93. Anon., *Chem. Eng. News*, Jan. 26, 1998, 13.
94. (a) H. Shimizu, I. Nagasaki, and T. Saito, *Tetrahedron*, 2005, *61*, 5405; (b) T.P. Yoon and E. Jacobsen, *Science*, 2003, *299*, 1691; (c) M. Barthod, G. Mignani, G. Woodward, and M. Lemaire, *Chem. Rev.*, 2005, *105*, 1801.

95. (a) M. Shibasaski, M. Kanai, and S. Matsunaga, *Aldrichchim. Acta*, 2006, *39*(2), 31; (b) Anon., *ChemFiles* (*Aldrich*), 2006, *6*(10), 5; (c) M. Shibasaki and N. Yoshikawa, *Chem. Rev.*, 2002, *102*, 2187; (d) J. Inanaga, H. Furuno, and T. Hayano, *Chem. Rev.*, 2002, *102*, 2211; (e) J.M. Brunel, Ch*em. Rev.*, 2005, *105*, 857; (f) B. List and J.W. Yang, *Science*, 2006, *313*, 1584.
96. K.T. Wan and M.E. Davis, *J. Catal.*, 1995, *152*, 25; *148*, 1; *Nature* 1994, *370*, 449.
97. N. Uematsu, A. Fujii, S. Hashiguchi, T. Ikariya, and R. Noyori, *J. Am. Chem. Soc.*, 1996, *118*, 4916.
98. (a) M.J. Burk, M.F. Gross, T.G.P. Harper, C.S. Kalberg, J.R. Lee, and J.P. Martinez, *Pure Appl. Chem.*, 1996, *68*(1), 37; (b) M.J. Burk, M.F. Gross, and J.P. Martinez, *J. Am. Chem. Soc.*, 1995, *177*, 9375; (c) M.J. Burk, Y.M. Wang, and J.R. Lee, *J. Am. Chem. Soc.*, 1996, *118*, 5142; (d) J. Albrecht and U. Nagel, *Angew. Chem. Int. Ed.*, 1996, *35*, 407; (e) F.-Y. Zhang, C.-C. Pai, and A.S.C. Chan, *J. Am. Chem. Soc.*, 1998, *120*, 5808; (f) J. Kang. J.H. Lee, S.H Ahn, and J.S. Choi, *Tetrahedron Lett.*, 1998, *39*, 5523; (g) A. Tungler and K. Fodor, *Catal. Today*, 1997, *37*, 191; (h) B.R. James, *Catal. Today*, 1997, *37*, 209; (i) M.T. Reetz, A. Gosberg, R. Goddard, and S.-H. Kyung, *Chem. Commun.*, 1998, 2077; (j) M.J. Bruk, C.S. Kalberg, and A. Pizzano, *J. Am. Chem. Soc.*, 1998, *120*, 4345; (k) T. Naota, H. Takaya, and S.-I. Murahashi, *Chem. Rev.*, 1998, *98*, 2599.
99. (a) T.V. Rajan Babu, T.A. Ayers, and A.L. Casalnuovo, *J. Am. Chem. Soc.*, 1994, *116*, 4101; (b) M. Dieguez, O. Pamies, and C. Claver, *Chem. Rev.*, 2004, *104*, 3189.
100. H.N. Flach, I. Grassert, and G. Oehme, *Macromol. Chem. Phys.*, 1994, *195*, 3289.
101. D. Tas, C. Thoelen, I.F.J. Vankelecom, and P.A. Jacobs, *Chem. Commun.*, 1997, 2323.
102. (a) S. Kobayashi., *Pure Appl. Chem.*, 1998, *70*, 1019; (b) K. Ishihara and H. Yamamoto, *Cattech*, 1997, *1*(1), 51; (c) C.M. Yu, H.-S. Choi, W.-H. Jung, H.-J. Kim, and J. Shin, *Chem. Commun.*, 1997, 763; (d) C.M. Yu, S.-K. Yoon, H.-S. Choi, and K. Baek, *Chem. Commun.*, 1997, 761; (e) K. Mikami, S. Matsukawa, T. Volk, and M. Terada, *Angew. Chem. Int. Ed.*, 1997, *36*, 2768
103. (a) M. Shigasaki, H. Sasai, and T. Arai., *Angew. Chem. Int. Ed.*, 1997, *36*, 1236; (b) M. Shibasaki, H. Sasai, T. Arai, and T. Ida, *Pure Appl. Chem.*, 1998, *70*, 1027; (c) T. Arai, H. Sasai, K. Yamaguchi, and M. Shibasaki, *J. Am. Chem. Soc.*, 1998, *120*, 441; (d) L. Pu, *Chem. Rev.*, 1998, *98*, 2405; (e) Y.M.A. Yamada, N. Yoshikawa, H. Sasai, and M. Shibasaki, *Angew. Chem. Int. Ed.*, 1997, *36*, 1871; (f) M. Shibasaki, M. Kanai, S. Matsunaga, and N. Kumagai, *Acc. Chem. Res.*, 2009, *42*, 1117.
104. T. Katsuki, ed., *Asymmetric Oxidation Reactions*, Oxford University Press, Oxford, 2001.
105. Y. Kashiwagi, Y. Yanagisawa, F. Kurashima, J.-I. Anzai, T. Osa, and J.M. Bobbitt, *Chem. Commun.*, 1996, 2745.
106. (a) E. Hoft., *Top. Curr. Chem.*, 1993, *164*, 63; (b) E.M. McGarrigle, D.G. Gilheany, Q.-H. Xia, H.-Q. Ge, C.-P. Ye, Z.-M. Liu, and K.-X. Su, *Chem. Rev.*, 2005, *105*, 1563–1915.
107. (a) L.P. Hager and E.J. Allain, U.S. patent 5,358,860, (1994); (b) A.F. Dexter, E.J. Lakner, R.A. Campbell, and L.P. Hager, *J. Am. Chem. Soc.* 1995, 117, 6412; (c) F.J. Lakner. R.A. Campbell, and L.P. Hager, *J. Am. Chem. Soc.*, 1995, *117*, 6412; (c) F.J. Lakner, K.P. Cain, and L.P. Hager, *J. Am. Chem. Soc.*, 1997, *119*, 443.
108. (a) M.E.L. Sanchez and S.M. Roberts, *J. Chem. Soc. Perkin. Trans.*, 1995, *1*, 1467; (b) W. Kroutil, P. Mayon, M.E. Lasterra-Sanchez, S.J. Maddrell, S.M. Roberts, S.R. Thornton, C.J. Todd, and M. Tuter, *Chem. Commun.*, 1996, 845; (c) M.W. Cappi, W.-P. Chen, R.W. Flood, Y.-W. Liao, S.M. Roberts, J. Skidmore, J.A. Smith, and N.M. Williamson., *Chem. Commun.*, 1998, 1159; (d) P.A. Bentley, S. Bergeron, M.W. Cappi, D.E. Hibbs, M.B. Hursthouse, T.C. Nugent, R. Pulido, S.M. Roberts, and L.E. Wu. *Chem. Commun.*, 1997, 739; (e) J.V. Allen, K.-H. Drauz, R.W. Flood, S.M. Roberts, and J. Skidmore., *Tetrahedron Lett.*, 1999, *40*, 5417; (f) L. Carde, H. Davies, T.P. Geller, S.M. Roberts., *Tetrahedron Lett.*, 1999, *40*, 5421.
109. (a) K. Daikai, M. Kamaura, and J. Inanaga, *Tetrahedron Lett.*, 1998, *39*, 7321; (b) S. Watanabe, Y. Kobayashi, T. Arai, H. Sasai, M. Bougauchi, and M. Shibasaki, *Tetrahedron Lett.*, 1998, *39*, 7353; (c) M. Bougauchi, S. Watanabe, T. Arai, H. Sasai, and M. Shibasaki, *J. Am. Chem. Soc.*, 1997, *119*, 2329.
110. (a) B. Brandes and E.N. Jacobsen, *Tetrahedron Lett.*, 1995, *36*, 5123; (b) M. Palucki, G.J. McCormick, and E.N. Jacobsen, *Tetrahedron Lett.*, 1995, *36*, 5457; (c) Y. Noguchi, R. Irie, T. Fukuda, and T. Katsuki, *Tetrahedron Lett.*, 1996, *37*, 4533; (d) M. Palucki, N.S. Finney, P.J. Pospisil, M.L. Guler, T. Ishida, and E.N. Jacobsen, *J. Am. Chem. Soc.*, 1998, *120*, 948.
111. (a) P.G. Cozzi, *Chem. Soc. Rev.*, 2004, *33*, 410; (b) S. Stinson, *Chem. Eng. News*, July 23, 2001, 9.
112. (a) Y.L. Bennani and S. Hanessian, *Chem. Rev.*, 1997, *97*, 3161; (b) C. Halm and M.J. Kurth, *Angew. Chem. Int. Ed.*, 1998, *37*, 510 (polymer-supported catalyst).
113. S.C. Stinson, *Chem. Eng. News*, May, 1994, *16*, 6.
114. M. Tokunaga. J.F. Larrow, F. Kakiuchi, and E.N. Jacobsen, *Science*, 1997, *277*, 936.
115. (a) J.F. Larrow, S.E. Schaus, and E.N. Jacobsen, *J, Am. Chem. Soc.*, 1996, *118*, 7420; (b) L.E. Martinez, J.L. Leighton., D.H. Carsten, E.N. Jacobsen, *J. Am. Chem. Soc.*, 1995, *117*, 5897; (c) J.L. Leighton and E.N. Jacobsen, *J. Org. Chem.*, 1996, *61*, 389; (d) M. Rouhi, *Chem. Eng. News*, Aug. 12, 1996, 10.
116. E.N. Jacobsen, F. Kakiuchi, R.G. Konsler, J.F. Larrow, and M. Tokunaga, *Tetrahedron Lett.*, 1997, *38*, 773.
117. (a) S.B. Ogunwumi and T. Bein, *Chem. Commun.*, 1997, 901; (b) M.J. Sabater, A. Corma, A.L. Domenech, V. Fornes, and H. Garcia, *Chem. Commun.*, 1997, 1285.
118. (a) B.B. De, B.B. Lohray, S. Sivaram, and P.K. Dhal, *J. Polym. Sci. A. Polym. Chem.*, 1997, *35*, 1809; (b) L. Canali and D.C. Sherington, *Chem. Sov. Rev.*, 1999, *28*, 85.
119. I.F.J. Van Kelcom, D. Tas, R.F. Parton, V.V. Vyver, and P.A. Jacobs, *Angew. Chem. Int. Ed.*, 1994, *35*, 1346.
120. S. Kotha, *Tetrahedron*, 1994, *50*, 3639.
121. (a) H.C. Kolb, M.S. Van Nieuwenhze, and K.B. Sharpless, *Chem. Rev.*, 1994, *94*, 2483; (b) D.S. Berrisford, C. Bolm, and K.B. Sharpless, *Angew. Chem. Int. Ed.*, 1995, *34*, 1059; (c) H. Becker and K.B. Sharpless, *Angew. Chem. Int. Ed.*, 1996, *35*, 449.
122. S.C. Stinson, *Chem. Eng. News*, Apr. 28, 1997, 26.
123. (a) H. Becker, G. Li, H.-T. Chang, K.B. Sharpless, *Angew. Chem. Int. Ed.*, 1996, *35*, 451; (b) O. Reiser, *Angew. Chem.*

Int. Ed., 1996, *35*, 1308; (c) Anon., *Chem. Eng. News*, Feb. 19, 1996, 6; (d) P. O.'Brien, *Angew. Chem. Int. Ed.*, 1999, *38*, 326.
124. K.L. Reddy and K.B. Sharpless, *J. Am. Chem. Soc.*, 1998, *120*, 1207.
125. C.E. Song, C.R. Oh, S.W. Lee, S.-G. Lee, L. Canali, and D.C. Sherrington, *Chem. Commun.*, 1998, 2435.
126. C. Bolm, A. Maischak, and A. Gerlach, *Chem. Commun.*, 1997, 2353.
127. A. Petri, D. Pini, and P. Salvadori, *Tetrahedron Lett.*, 1995, *36*, 1549.
128. P. Salvadori, D. Pini, and A. Petri, *J. Am. Chem. Soc.*, 1997, *119*, 6929.
129. (a) C. Bolm and A. Gerlach, *Angew. Chem. Int. Ed.*, 1997, *36*, 741; (b) S. Borman, *Chem. Eng. News*, Aug. 26, 1996, 35.
130. H. Han and K.D. Janda, *Angew. Chem. Int. Ed.*, 1997, *36*, 1731.
131. Anon., *Chem. Eng. News*, Dec. 13, 1999, 35.
132. L. Canali, J.K. Karjalainen, D.C. Sherrington, and O. Hormi, *Chem. Commun.*, 1997, 123.
133. T.K.M. Shing, E.K.W. Tam, V.W.-F. Tai, I.H.F. Chung, and O. Jiang, *Chem. Eur. J.*, 1996, *2*, 50.
134. S.C. Stinson, *Chem. Eng. News*, Apr. 24, 1995, 37.
135. A.S.C. Chan, F.-Y. Zhang, and C-W. Yip, *J. Am. Chem. Soc.*, 1997, *119*, 4080.
136. (a) T. Shibata, K. Choji, T. Hayase, Y. Aizu, and K. Soai, *Chem. Commun.*, 1996, *751*, 1235; (b) T. Shibata, H. Morioka, T. Hayase, K. Choji, and K. Soai, *J. Am. Chem. Soc.*, 1996, *118*, 471; (c) C. Bolm, F. Bienewald, and A. Seger, *Angew. Chem. Int. Ed.*, 1996, *35*, 1657; (d) P. Zurer, *Chem. Eng. News*, Nov. 30, 1998, 9; (e) K. Soai, T. Shibata, H. Morioka, and K. Choji, *Nature* 1995, *378*, 767; (f) T. Shibata, J. Yamamoto, N. Matsumoto, S. Yonekubo, S. Osanai, and K. Soai, *J. Am. Chem. Soc.*, 1998, *120*, 12157; (g) M. Avalos, R. Babiano, P. Cintas, J.L. Jimenez, and J.C. Palacios, *Chem. Commun.*, 2000, 887; (h) M.H. Todd, *Chem. Soc. Rev.*, 2002, *31*, 211; (i) I. Sato, H. Urabe, S. Ishiguro, T. Shibata, and K. Soai, *Angew. Chem. Int. Ed.*, 2003, *42*, 315; (j) J. Podlech and T. Gehring, *Angew. Chem. Int. Ed.*, 2005, *44*, 5776.
137. (a) M. Kitamura, S. Suga, H. Oka, and R. Noyori, *J. Am. Chem. Soc.*, 1998, *120*, 9800; (b) C. Girard and H.B. Kagan, *Angew. Chem. Int. Ed.*, 1998, *37*, 2922.
138. M. Braun, *Angew. Chem. Int. Ed.*, 1996, *35*, 519.
139. H. Schafer and D. Seebach, *Tetrahedron*, 1995, *51*, 2305.
140. R.O. Duthaler and A. Hafner, *Angew. Chem. Int. Ed.*, 1997, *36*, 43.
141. G.E. Keck and D. Krishnamurthy, *J. Am. Chem. Soc.*, 1995, *117*, 2363.
142. (a) D.A. Evans, J.A. Murry, and M.C. Kozlowski, *J. Am. Chem. Soc.*, 1996, *118*, 5814; (b) S.C. Stinson, *Eng. News*, April, 1995, *24*, 37; (c) D.A. Evans, T. Rovis, and J.S. Johnson, *Pure Appl. Chem.*, 1999, *71*, 1407.
143. (a) D.A. Evans, J.A. Murry, P. von Matt, R.D. Norcross, and S.J. Miller, *Angew. Chem. Int. Ed.*, 1995, *34*, 798; (b) I.W. Davies, L. Gerena, D. Cai, R.D. Larsen, T.R. Verhoeven, and P.J. Reider, *Tetrahedron Lett.*, 1997, *38*, 1145; (c) M. Johannsen, S. Yao, and K.A. Jorgensen, *Chem. Commun.*, 1997, 2169; (d) S. Yao, M. Hohannsen, H. Audrain, R.G. Hazell, and K.A. Jorgensen, *J. Am. Chem. Soc.*, 1998, *120*, 8599; (e) D.A. Evans. E.J. Olhava, J.S. Johnson, J.M. Janey., *Angew. Chem. Int. Ed.*, 1998, *37*, 3372.
144. D.A. Evans, C.S. Burgey, N.A. Paras, T. Vojkovsky, and S.W. Tregay, *J. Am. Chem. Soc.*, 1998, *120*, 5824.
145. S. Kanemasa, Y, Oderaotoshi, S.-I, Sakaguchi, H. Yamamoto, J. Tanaka, E. Wadd, and D.P. Curran., *J. Am. Chem. Soc.*, 1998, *120*, 3074.
146. D.A. Evans, D.W.C. MacMillan, and K.R. Campos., *J. Am. Chem. Soc.*, 1997, *119*, 10859.
147. (a) D. Seebach, R. Dahinden, R.E. Marti, A.K. Beck, D.A. Plattner, and F.N.M. Kuhnle., *J. Org. Chem.* 1995, *60*, 1788; (b) C. Haase, C.R. Sarko, and M. DiMare., *J. Org. Chem.*, 1995, *60*, 1777; (c) E. Wada, W. Pei, and S. Kanemasa., *Chem. Lett.*, 1994, 2345.
148. (a) K. Ishihara. H. Kurihara, and H. Yamamoto., *J. Am. Chem. Soc.*, 1996, *118*, 3049; (b) Y. Hayashi, J.J. Rhode, and E.J. Corey., *J. Am. Chem. Soc.*, 1996, *118*, 5502.
149. E.J. Corey, and M.A., Letavic., *J. Am. Chem. Soc.*, 1995, *117*, 9616.
150. E.L.M. Cowton, S.E. Gibson, M.J. Schneider, and M.H. Smith., *Chem. Commun.*, 1996, 839.
151. M. Amadji, J. Vadecard, J.-C. Plaquevent, L. Duhamel, and P. Duhamel, *J. Am. Chem. Soc.*, 1996, *118*, 12483.
152. (a) H. Sasai, T. Arai, Y. Satow, K.N. Houk, and M. Shibasaki, *J. Am. Chem. Soc.*, 1995, *117*, 6194; (b) H. Steinhagen and G. Helmchen, *Angew. Chem. Int. Ed.*, 1996, *35*, 2339.
153. (a) E.J. Corey, F. Xu, and M.C. Noe, *J. Am. Chem. Soc.*, 1997, *119*, 12414; (b) E.J. Corey, M.C. Noe, and F. Xu, *Tetrahedron Lett.*, 1998, *39*, 5347; (c) K. Marouka, ed., *Asymmetric Phase Transfer Catalysis*, Wiley-VCH, Weinheim, 2008; (d) T. Ooi and K. Msaruoka, *Angew. Chem. Int. Ed.*, 2007, *46*, 4222.
154. B. Lygo and P.G. Wainwright, *Tetrahedron Lett.*, 1998, *39*, 1599.
155. X. Wang, M. Kitamura, and K. Maruoka, *J. Am. Chem. Soc.*, 2007, *129*, 1038.
156. F. Cavelier, S. Gomez, R. Jacquier, and J. Verducci, *Tetrahedron Lett.*, 1994, *35*, 2891.
157. (a) C. Dreisbach, G. Wischnewski, U. Kragl, and C. Wandrey, *J. Chem. Soc. Perkin. Trans., 1*, 1995, 875; (b) U. Kragl and C. Dreisbach, *Angew. Chem. Int. Ed.*, 1996, *35*, 642.
158. N.E. Moualij and C. Caze, *Eur. Polym. J.*, 1995, *31*, 193.
159. (a) A. Berkessel and H. Groger, *Organocatalysis in Asymmetric Synthesis*, Wiley, New York, 2004; (b) P.I. Dalko, *Enantioselective Organoatalysis—Reactions and Experimental Procedures*, Wiley-VCH, Weinheim, 2007; (c) K.K. Houk and B. List, eds, *Acc. Chem. Res.*, 2004, *37*, 487–631. (d) R.M. Kellogg, *Angew. Chem. Int. Ed.*, 2007, *46*, 494; (e) P. Kocosky and A.V. Malkov, *Tetrahedron*, 206, *62*, 243; (f) Anon., *ChemFiles*. (*Aldrich*), Milwaukee, WI, 2006, 3.
160. C. Pidathala, L. Hoang, N. Vignola, and B. List, *Angew. Chem. Int. Ed.*, 2003, *42*, 2785.
161. J.F. Austin and D.W.C. MacMillan lII. *J. Am. Chem. Soc.*, 2002, *124*, 1172.
162. J.V. Allen, K.-H. Drauz, R.W. Flood, S.M. Roberts, and J. Skidmore, *Tetrahedron Lett.*, 1999, *40*, 5417.
163. (a) I. Ojima, ed., *Catalytic Asymmetric Synthesis*, 2nd ed., Wiley, New York, 2000; (b) S.C. Stinson, *Chem. Eng. News*, July, 1996, *15*, 35.
164. (a) A.E. Hamielec and J.B.P. Soares, *Prog. Polym. Sci.* 1996, *21*, 651; (b) M.R. Ribeiro, A. Deffieux and M.F. Portella, *Ind. Eng. Chem. Res.*, 1997, *36*, 1224.

RECOMMENDED READING

1. S. Kotha, *Tetrahedron*, 1994, *50*, 3639.
2. S.C. Stinson, *Chem. Eng. News*, Sept. 21, 1998, 83.
3. G. Kauup, *Angew. Chem. Int. Ed.*, 1994, *33*, 728–729.
4. T.P. Yoon and E.N. Jacobsen, *Science*, 2003, *299*, 1691.
5. B. List and J.W. Yang, *Science*, 2006, *313*, 1584.

EXERCISES

1. See what chiral chemistry is being done at your school or industrial laboratory. What methods are being used? Are the chemists aware of some of the less wasteful approaches mentioned in this chapter?
2. Look up chiral inorganic compounds that do not contain carbon. How would you make such a compound that would be an oxidation catalyst with a long lifetime? (Many oxidation catalysts fail because of oxidation of the organic ligands on the metal ion.)
3. Devise a system for a Sharpless dihydroxylation of an olefin in which no osmium or alkaloid-derived ligand is lost on repeated reuse. You may want to refer to Chapters 5 and 7 for hints.
4. Devise an efficient membrane reactor which will allow one optical isomer to pass, but which will keep the catalyst in the reactor so that none is lost. This will eliminate the need for further processing to release the desired isomer from a derivative.

CHAPTER 11

Agrochemicals

11.1 THE NATURE AND USE OF AGROCHEMICALS

Agrochemicals include fertilizer, fungicides, herbicides, and insecticides. The last three are pesticides.[1] The world used 2.1 billion kg of pesticides in 1995,[2] with about a fourth of this being used in the United States. The cost of the insecticides and fungicides used in the United States in 2005 was $2.4 billion.[3] Out of this world total, 10–25% is used on one crop, namely cotton.[4]

	U.S. Market	World Market
	(in millions of kg of active ingredient)	
Herbicides	301	1002
Insecticides	153	767
Fungicides	74	256
Other	40	107
Total	568	2132

The U.S. total in 1995 was a record high.[5] Pesticides are used on 900,000 farms and by 69 million households in the United States.[6] The herbicide, atrazine, was the one used in the largest amount, 21 million kg.

Early insecticides were often inorganic arsenic, copper, lead, and sulfur compounds.[7] An example is Bordeaux mixture, which contains copper sulfate and calcium hydroxide. Copper chromium arsenate was used as a wood preservative, but has been phased out (see Chapter 4). Organic natural products, such as pyrethrin and rotenone (**11.1** Schematic), have also been used. Rotenone is used today for killing undesirable fish, such as carp, before restocking ponds with game fish. Many synthetic pyrethrins are used today because of their relative safety to humans. However, both rotenone and pyrethrin[8] are broad-spectrum insecticides that can be very toxic to beneficial insects.[9]

DDT [1,1-bis(4-chlorophenyl)-2,2,2-trichloroethane: (**11.2** Schematic)], also called dichlorodiphenyltrichloroethane, was

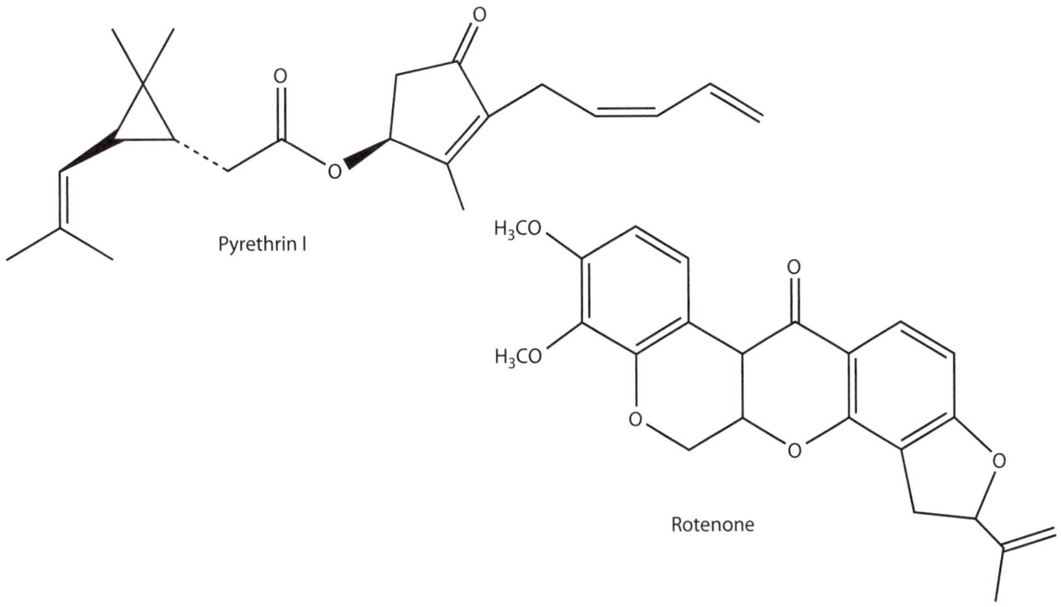

Pyrethrin I

H_3CO

H_3CO

Rotenone

11.1 · Schematic

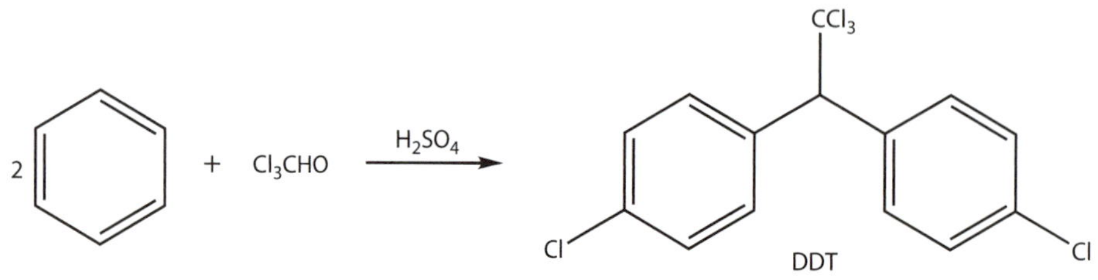

2 benzene + Cl_3CHO $\xrightarrow{H_2SO_4}$ DDT

11.2 Schematic

introduced about the time of World War II. This was followed by a flood of many other highly chlorinated hydrocarbons, including toxaphene (**11.3** Schematic).

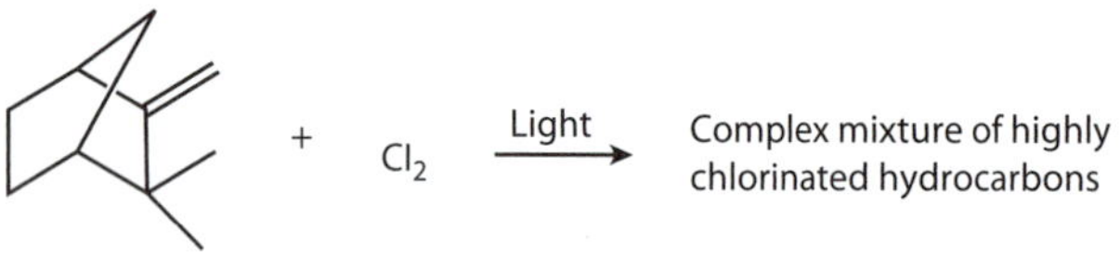

+ Cl_2 $\xrightarrow{\text{Light}}$ Complex mixture of highly chlorinated hydrocarbons

11.3 Schematic

Most of these have since been banned in the United States and many other countries, although they are still used in some.[10] They caused reproductive failure in birds, such as the bald eagle, osprey, pelican, and peregrine falcon, owing to eggshell thinning; DDT is a human carcinogen. DDT is still used in Africa in the control of malaria, along with insecticide-treated bed nets and elimination of standing water.[11] Toxaphene was very toxic to fish, the median lethal concentration (LC_{50}) for trout and blue gill is 0.003–0.005 ppm, and for birds, it is 40–71 mg/kg.

The organophosphate insecticides, such as malathion (**11.4** Schematic), came next. These are less persistent in the environment, but more toxic to humans. Carbamates, such as carbaryl (**11.5** Schematic), were also introduced.

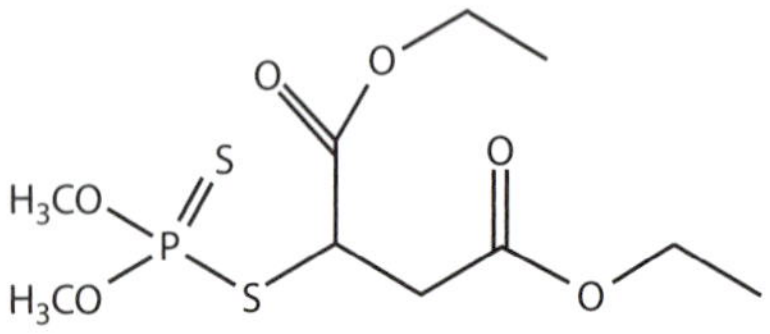

11.4 Schematic

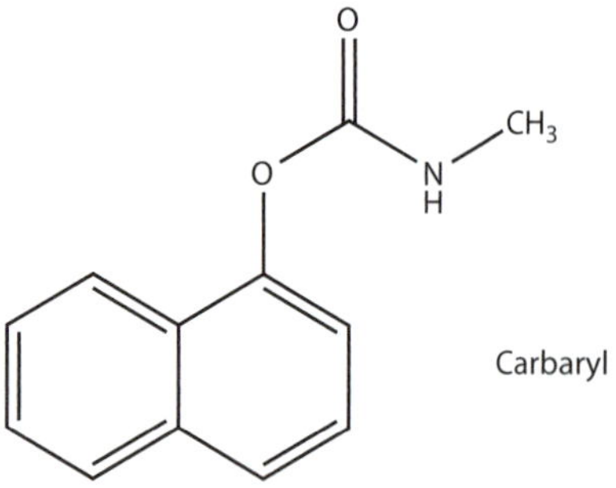

Carbaryl

11.5 Schematic

Carbaryl is made by the reaction of methylisocyanate with 2-naphthol. This is the product made at the plant in Bhopal, India, where the disaster mentioned in Chapter 1 occurred.

Newer insecticides being introduced today may be more selective and less toxic to nontarget organisms.[12] For example, the insecticide pirimicarb (**11.6** Schematic) is highly selective for aphids and has little effects on biological control organisms.[13]

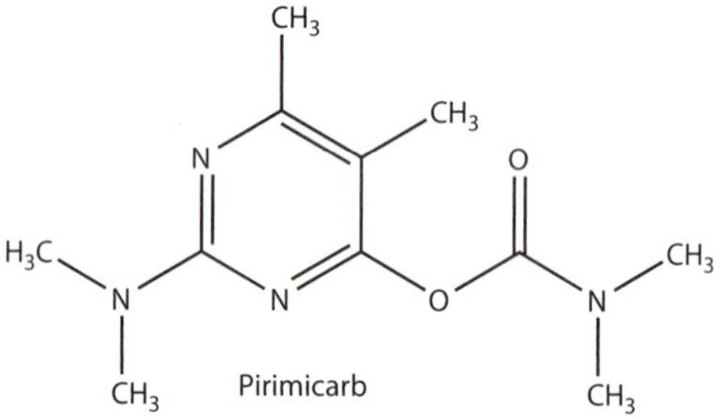

Pirimicarb

11.6 Schematic

The first synthetic herbicide[14] was 2,4-dichlorophenoxyacetic acid (**11.7** Schematic). The one used in largest volume today is atrazine, which is typical of a group of triazine herbicides. Glyphosate (very low mammalian toxicity) is used extensively to kill vegetation nonselectively.[15] Alachlor is a typical amide herbicide. Trifluralin is representative of the dinitroanilines. Paraquat (very toxic to humans) is used extensively to kill vegetation before planting a new crop in no-till agriculture.[16] This practice reduces soil erosion greatly.

Two new classes of herbicides are active at much lower dosages. The sulfonylureas (**11.8** Schematic)[17] are used at 2–4 g/ha. The sulfonylureas have low mammalian toxicity. Only 1.4 oz/acre of the imidazolinones (see **11.8** Schematic)[18] is needed.

The herbicide flumiclorac (**11.9** Schematic), for use on maize and soybeans, is neither a teratogen nor a carcinogen. It hydrolyzes too quickly to contaminate groundwater.[19] (This will only be effective as long as the hydrolysis products are nontoxic.)

Typical fungicides (**11.10** Schematic) include sulfur, Bordeaux mixture, dithiocarbamates, captan, and benomyl.[20]

Fertilizers are usually simpler compounds that supply nitrogen, potassium, and phosphorus to plants.[21] Typical compounds are potassium chloride, ammonium nitrate,

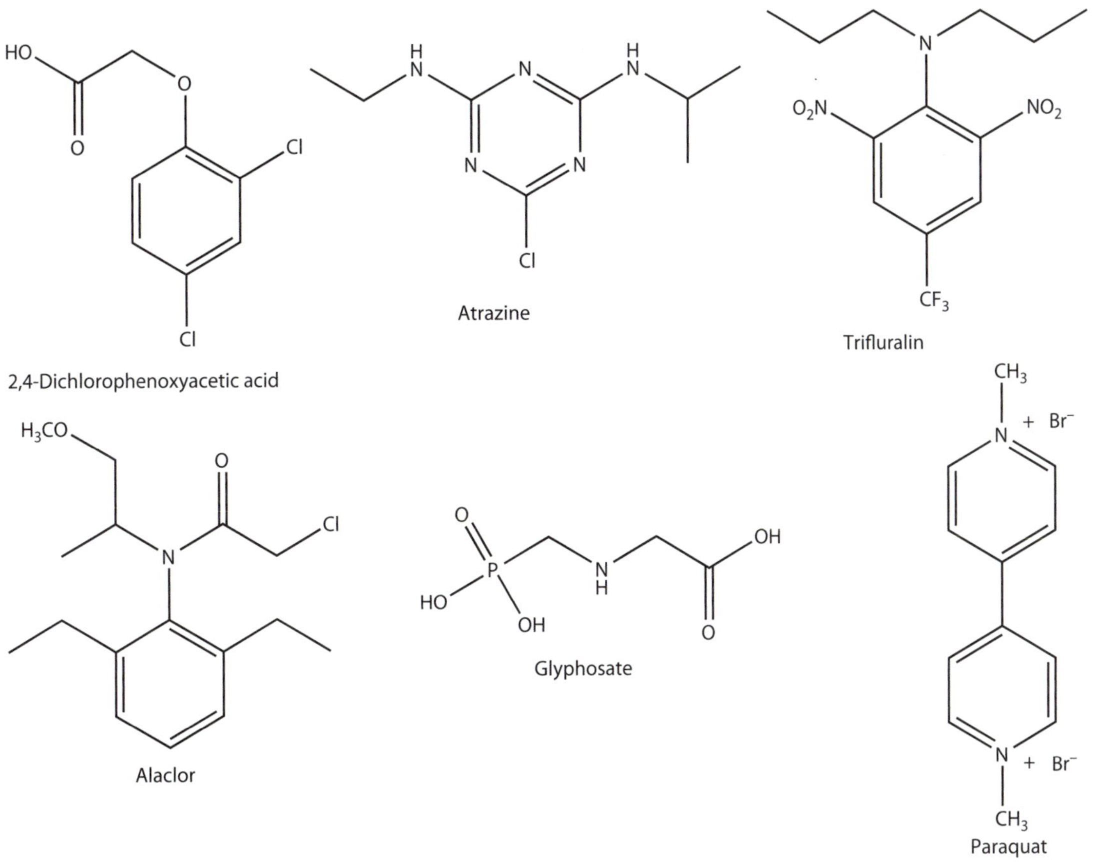

2,4-Dichlorophenoxyacetic acid

Atrazine

Trifluralin

Alaclor

Glyphosate

Paraquat

11.7 Schematic

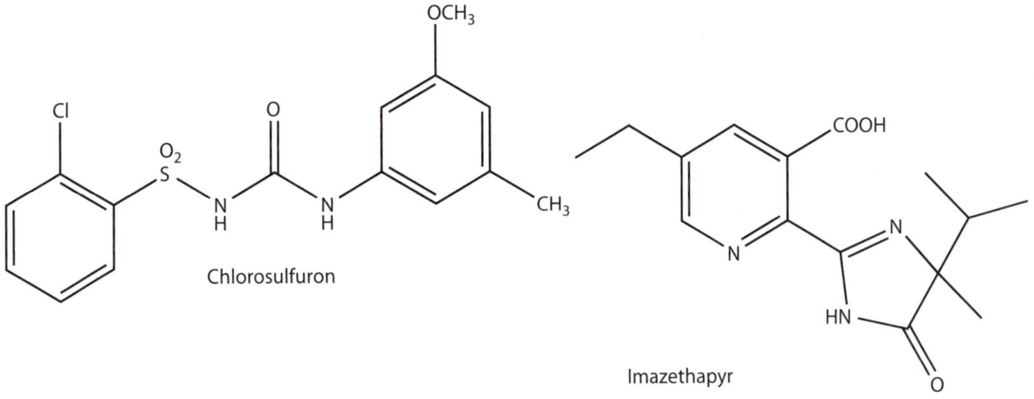

Chlorosulfuron

Imazethapyr

11.8 Schematic

diammonium phosphate, and urea–formaldehyde condensates. The last are for slow release.

Controlled release of agrochemicals (e.g., by hydrolysis of a polymeric ester) can offer the advantages of constant level, smaller dose, reduced evaporation loss, lower toxicity, longer life, decreased environmental pollution, and reduced effect on nontarget species by wind or runoff.[22] Systemic pesticides are preferred. This ensures protection of the growing tip of the plant. For weed killers, this means that it is not necessary to hit every leaf of the plant with the herbicide. Pesticides are often applied on a spray schedule according to the calendar as a prophylactic measure. Chemical pesticides are used 98% of the time.

11.2 PROBLEMS WITH AGROCHEMICALS[23]

Pesticides are an essential part of food and fiber production, but must be used judiciously in integrated pest

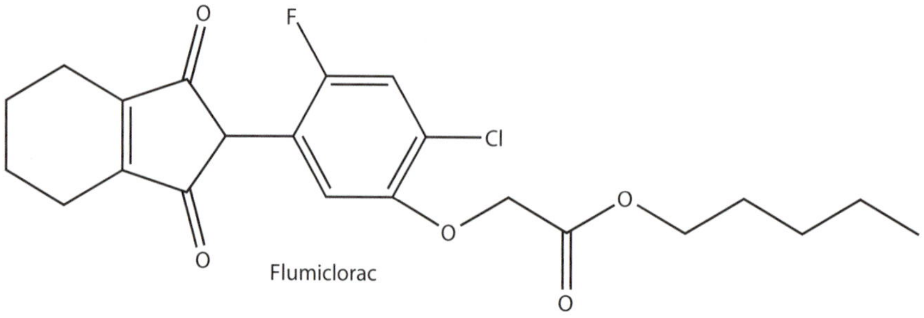

11.9 Schematic

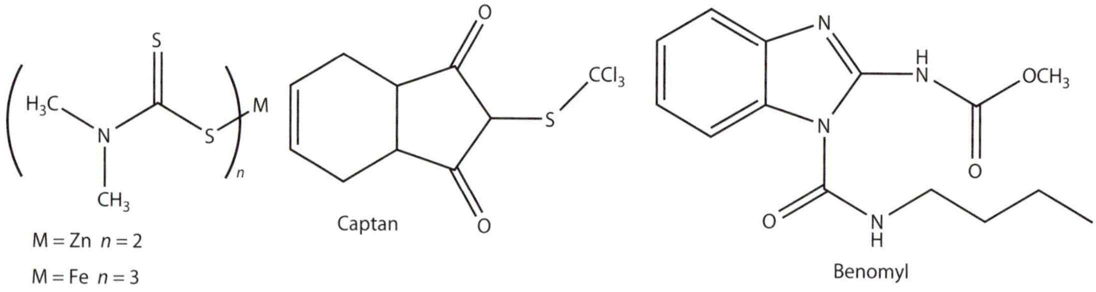

11.10 Schematic

management to avoid problems. Some, like DDT, do enough harm so that they are being banned or their use restricted.

11.2.1 Resistance to Pesticides

The selection of organisms to withstand unusual environments was discussed in Chapter 9. The organism was challenged with the reagent or new conditions. The few percent that survived were challenged again. The process was repeated until the organism developed resistance to the challenge.[24] This is exactly what has happened in the use of pesticides. There are now about 500 species of insects and mites, 100 species of weeds, and 150 plant pathogens that are resistant to pesticides that formerly controlled them.[25] When resistance develops in a pest species the farmer's first response is to increase the dosage. When this no longer works, he switches to another pesticide or mixture of pesticides. At some point, this no longer works, and the crop can no longer be raised in the location in question. (American farmers now have to apply two to five times as much pesticide to do what one application did in the early 1970s.)

The mechanisms of resistance can vary. In bacterial resistance to antibiotics, these include pumping the antibiotic out of the cell, changing the cell wall to avoid the antibiotic, enzymatic destruction of the antibiotic, and the development of substitute proteins not targeted by the antibiotic.[26] It may take longer to develop resistance in insects and plants than in microorganisms simply because the generation times are longer.

The best way to avoid the development of resistance is to devise other methods of controlling the pest, where possible, and to use the pesticide as a last resort, when nothing else works. When possible, a variety of approaches that involve different mechanisms can be rotated.

11.2.2 Effects of Pesticides on Nontarget Organisms

The ideal pesticide would affect only one pest species and nothing else. In practice, this is almost never possible. Only a few insect viruses (baculoviruses)[27] are this specific. In designing a pesticide, one looks for a difference between the metabolism of the pest and everything else. An insecticide that attacks a metabolic route in insects that is not present in humans or other mammals should be less toxic to the latter. An herbicide that interferes with a plant process not found in animals (e.g., photosynthesis) should be less toxic to animals. In practice, many of these differences have not been enough to protect humans and other animals from the toxicity.[28] Rachel Carson was among the first to point this out.[29] Current problems and practices have been summarized in a recent popular book.[30]

It is estimated that there are 1 million pesticide poisonings in the world each year.[31] These are more likely to happen in the developing world where fewer precautions are taken when spraying. China had 10,000 deaths in 1993. The United States exports many pesticides that are banned or restricted in this country.[32] About three-fourths of these exports lacked chemical-specific information. The U.S.

EPA received 1500 reports of adverse effects of pesticides covering 9200 incidents in 1996.[33] The number rose to 46,000 in 1999.[34] Organophosphate and carbamate insecticides inhibit cholinesterase, which is present in humans as well as in insects. They can cause headaches, nausea, dizziness, vision problems, and other discomforts.[35] The carbamate insecticide, carbaryl, may be linked to some cancers in the United Kingdom.[36] An epidemic of nearsightedness in Japan in the late 1950s to the early 1970s may be linked to the use of organophosphate insecticides.[37] Some farmers want organophosphate sheep dips to be banned, because of their neuropsychological effects on humans.[38] Organophosphate and carbamate insecticides are under review by the EPA to see if additional restrictions should be placed on their use.[39]

Certain tumors are more common among farmers.[40] Cancer has been correlated with the maize production in the Midwest and with cotton and vegetable growing in the southeastern United States. Farmers have an increased incidence of non-Hodgkin's lymphoma, an often fatal form of cancer.[41] There is also evidence that some widely used pesticides can suppress the immune system and lead to more illnesses.[42] Parkinson's disease has been associated with the use of pesticides.[43] This includes the natural pesticide rotenone.[44] They may also increase the incidence of diabetes.[45] Indirect detrimental effects of pesticides may go unnoticed or be hard to pin down. Pesticides are now being evaluated as endocrine disrupters,[46] as mentioned in Chapter 3. The natural insecticides sometimes used in organic farming can also be harmful. Ingestion or inhalation of sabadilla can cause gastrointestinal symptoms and hypotension (low blood pressure).[47] The problems of nicotine are well known.

Chlorpyrifos, diazonon, *cis*-permethrin, and *trans*-permethrin were detected in >67% of a sample of 168 child care centers in the United States.[48] There were more than 2500 illnesses caused by pesticide use in or near schools in 1998–2002.[49] Chlorpyrifos[50] and diazinon[51] have now been banned. Most uses of the carbamate insecticide, carbofuran, have been cancelled.[52]

Misuse of pesticides can occur. Methyl parathion (**11.11** Schematic) is an insecticide that is permitted for use on crops outdoors, but is banned from indoor use. It was used for control of cockroaches in 2700 houses in Mississippi and surrounding states in 1996. About 330 families had to be evacuated from their homes. Decontamination is expected to cost 50 million dollars.[53] Ironically, some people in Louisiana deliberately sprayed their homes with methyl parathion to obtain "free" repairs. As mentioned in Chapter 1, pesticide residues on foods are monitored and are not normally a problem.[54] Some fruit in the United Kingdom had residues that exceeded the legal limits.[55] However, there have been some instances of pesticides used illegally on food crops.[56] This tends to happen more often on foods imported into the United States and Canada from developing nations.[57]

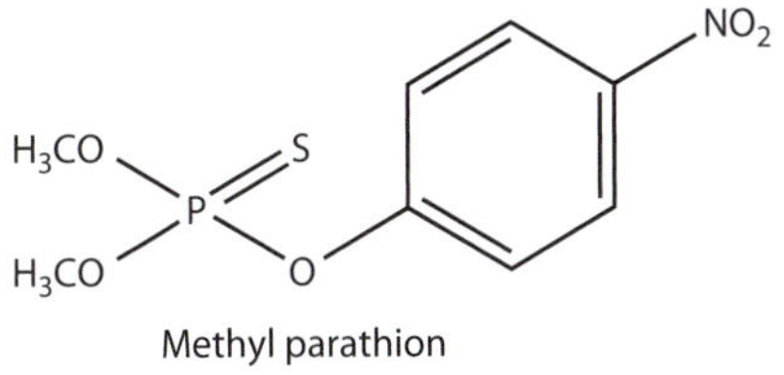

Methyl parathion

11.11 Schematic

Pesticides harm or kill many nontarget organisms. These include bats, birds, and fish, which often eat insects. The use of diclofenac to treat sick cattle in India has been banned since the practice has caused their near extinction of several species of vultures.[58] Meloxicam has been suggested as a substitute that will not harm the vultures.

Agricultural chemicals, such as atrazine and glyphosate, may also be part of the reason for the recent decline in amphibian populations, either directly or indirectly, by making the weakened animals more susceptible to disease and parasites.[59] Beneficial insects that may prey on the pest species are also killed. The killing of zooplankton by low levels of pesticides in agricultural runoff is thought to be responsible for some algal blooms in the Netherlands.[60] There is now a shortage of natural pollinators in some places as a result of pesticide use.[61] Some plants no longer have pollinators. The answer is not just to bring in more hives of honeybees. They too have been harmed by pesticides. France has suspended the marketing of imidacloprid, which kills honeybees.[62] Clothianidin, which is used as a seed treatment for maize, is a possible cause of "colony collapse disorder." They are not efficient pollinators for many plants, including crops such as alfalfa and lima beans. The killing of earthworms reduces soil tilth, including the ability of air and water to penetrate the soil. The use of herbicides also destroys soil tilth and can promote loss of topsoil through erosion. They can reduce erosion if used in no-till farming, which makes plowing unnecessary (**11.12** Schematic).

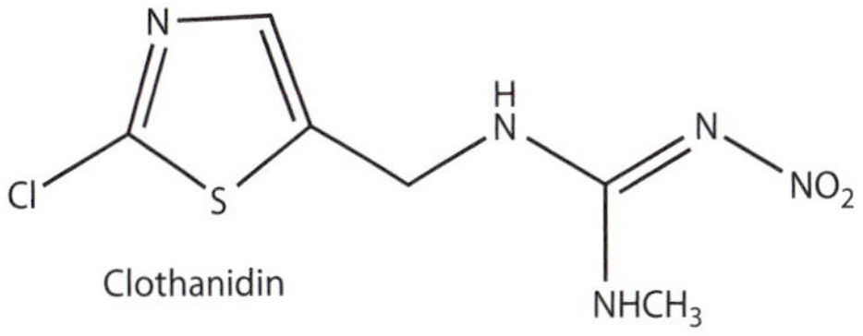

Clothanidin

11.12 Schematic

Antibiotics are used routinely in feeds for animals, such as cattle, hogs, and poultry, to promote growth and prevent disease.[63] Fifty million pounds of antibiotics are used each year in the United States, with over half of it being used in animal feed. Politics has been a factor in the continued use of antibiotics in animal feed.[64] The FDA wanted to eliminate this practice 30 years ago. Fluoroquinolines were approved for use in chicken feed in

1996 over the objections of the Centers for Disease Control and Prevention. By 1999, 18–30% of the Campylobacter isolated from humans showed resistance to fluoroquinolines. Cipro for humans is marketed as Baytril for chicken feed. The FDA has now banned the use of Baytril.[65] The problem is that this type of "directed evolution" has produced strains of bacteria that cause human diseases that are resistant to antibiotics (**11.13** Schematic).[66]

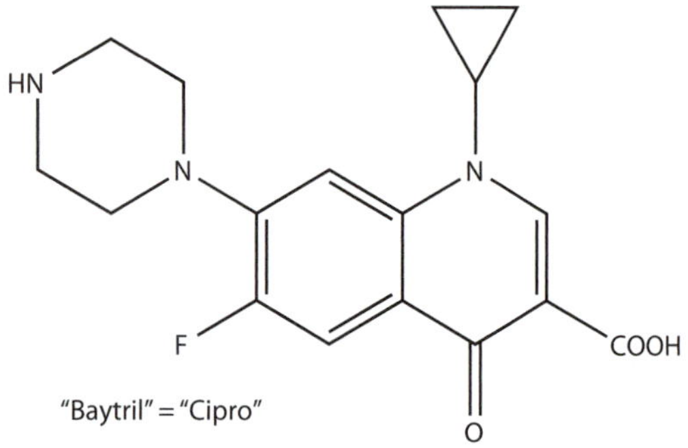

"Baytril" = "Cipro"

11.13 Schematic

Denmark, Finland, and Sweden have eliminated the antibiotics in animal feed. Denmark, the largest exporter of pork in the world, has done so with a 1% increase in costs to the farmer. Antibiotic use in the country has dropped by 54%.[67] *E. coli* O157:H7, which is present in 2% of cattle, causes 200,000 cases of illness, including 250 deaths, in the Untied States each year.[68] A vaccine is being developed for it. A simpler way to avoid this strain is to feed the cattle with hay instead of grain for the last 5 days before slaughter.[69] (The acids formed in the colon of the cattle from digestion of the starch in the grain causes the development of strains that are resistant to the stomach acid of humans.) The EU has banned virginiamycin, spiramycin, tylosin phosphate, and bacitracin zinc from animal feeds[70] Sweden banned antibiotics from animal feeds in the 1980s. It has substituted better hygiene on the farm for the antibiotics. Bacteriophages (viruses of bacteria) have been suggested as replacements for antibiotics in agriculture.[71]

11.2.3 Pesticides and Fertilizers in Air and Water

Pesticides do not always go where they are intended to go and, if they do, they do not always stay there. In aerial spraying, 50–75% of the spray can miss the target.[72] If the insecticide lands on areas adjacent to the field, it can harm natural insect pollinators and predators. If an herbicide lands on adjacent areas, it can kill vegetation that is the home of natural predators and pollinators. Sensitive crops in adjacent fields can be killed. If the crop is young, most of an insecticidal spray ends up on the ground.

Rain may also wash some of the spray off the leaves. Most pesticides are not compounds with high molecular weights. They can evaporate in the heat of the sun. Once airborne, they can travel to remote parts of the world, where rain can bring them to the surface again.[73] This long-range aerial transport explains how remote lakes in the Arctic and remote islands in the oceans have become contaminated. The pre-emergence herbicide, acetochlor (**11.14** Schematic), was detected in 29% of rain samples at four sites in Iowa and in 17% of stream samples from 51 sites in nine states.[74]

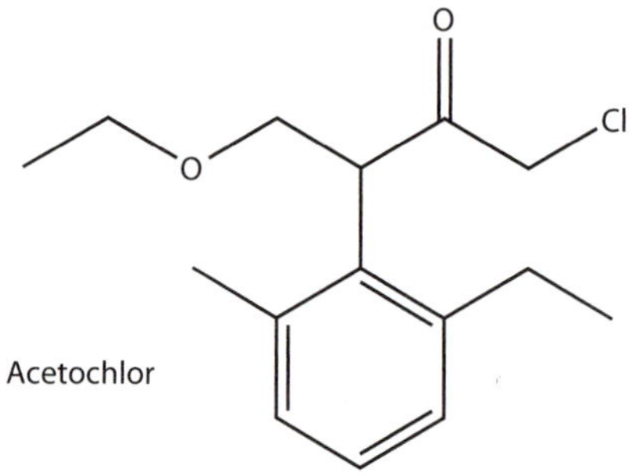

Acetochlor

11.14 Schematic

A study of ambient air in California showed that concentrations of 1,3-dichloropropene and methyl bromide (used as soil fumigants) were high enough in some places to be of concern.[75] Some of the herbicides put on the lawn can be tracked into the house, where it can become carpet dust.[76]

The pesticide can enter the ground with rains. It can remain intact or be degraded to metabolites that may or may not be toxic. Rain can also carry some of it into the nearest lake or river.[77] Fertilizers can be moved in the same way. They can cause eutrophication of lakes and rivers. Ground-water contamination[78] by farm or ranch runoff has been reported in 44 states in the United States.[79] Half of the wells for drinking water in the United States contain nitrates. (Nitrates can come from fertilizers or animal wastes. In sufficient amounts, they can cause depletion of oxygen in the blood of infants.) The average annual loss of soil from farmland is 5 tons/acre. Some 14 million Americans are routinely exposed to weed killers in their drinking water.[80] Atrazine is the most commonly used herbicide. Although only 1–2% may be lost from fields into rivers, this is enough to kill some aquatic plants.[81] At high doses, atrazine triggers mammary tumors in susceptible mice.[82] It can cause chromosomal breaks in Chinese hamster ovary cells at a level of 3 ppb.[83] Chlorotriazines,[84] such as atrazine, have been shown to be endocrine disrupters. Atrazine is not removed from drinking water by conventional water treatment.[85] Polymeric analogs of atrazine, copolymers of acrylamide and allyltriazines, solve the problem of evaporation into the air and may be less likely to get into groundwater and surface water. One of them is shown in **11.15** Schematic. They are effective herbicides.[86]

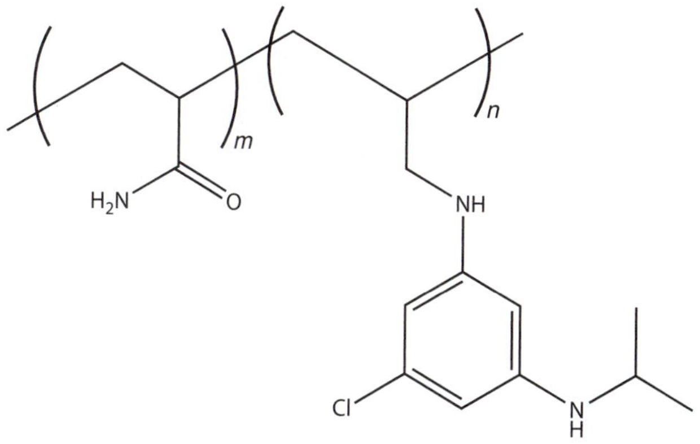

11.15 Schematic

It is common for the metabolite of the pesticide to be found more frequently than the pesticide itself.[87] The level of atrazine in groundwater was half that of its metabolites. Metabolites found in a stream included hydroxyatrazine, deethylhydroxyatrazine, and deisopropylhydroxyatrazine (**11.16** Schematic).[88] The half life of atrazine in the soil to give nonphytotoxic products varies from 60 days to more than a year.[89] The herbicide alachlor is present at low levels in groundwater in 16 states. It is classed as a "probable human carcinogen." It is degraded to a variety of products in soil (**11.17** Schematic). Some of these are weak mutagens.[90]

Atrazine has been banned in the EU but not in the United States.[91]

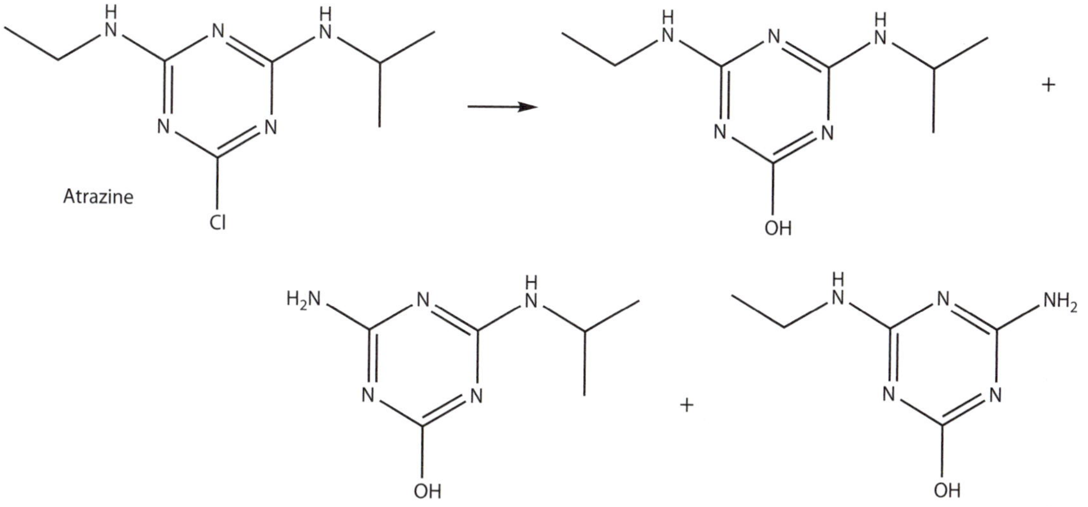

Atrazine

11.16 Schematic

Soil

Ar = 2,6-diethylphenyl

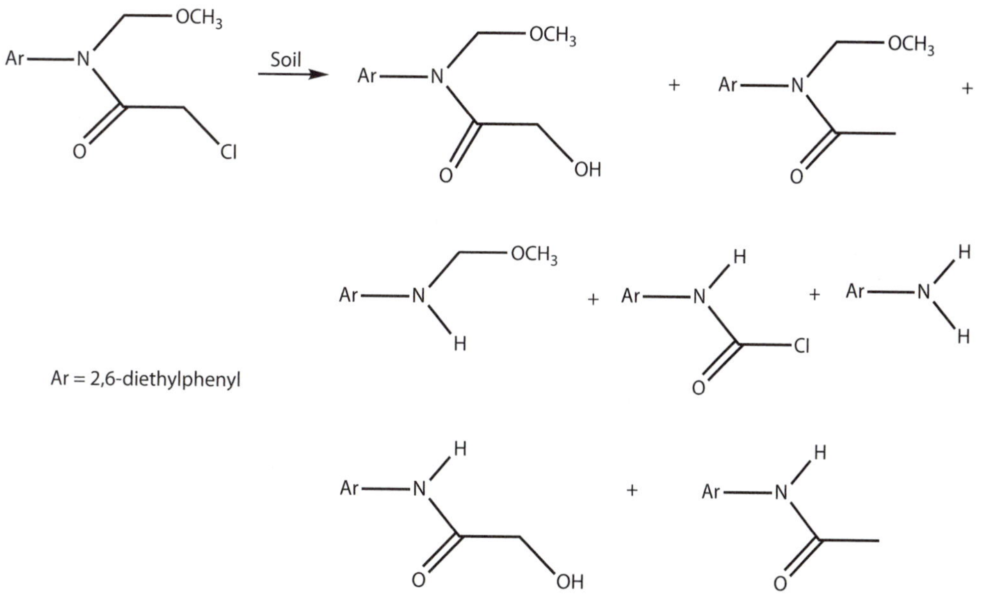

11.17 Schematic

11.2.4 Sustainable Agriculture

Sustainable agriculture[92] can be defined as a system that indefinitely meets the demands for agricultural output at socially acceptable economic and environmental costs.[93] Another group says that sustainable agriculture involves a system for food and fiber production that can maintain high levels of production with minimal environmental impact and can support viable rural communities.[94] It conserves natural resources. The present system of agriculture in the United States is unsustainable. It is the most energy-intensive system in the world.[95] Fuel is needed not only for the machines that plow the ground, sow the seed, cultivate, spray, and harvest the crop, and carry it an average of 1400 miles to the consumer, but also to make the fertilizer and pesticides used on the farm. For example, the synthesis of ammonia (**11.18** Schematic) is energy-intensive; it is based on fossil fuels and produces carbon dioxide, which is involved in global warming. The hydrogen for the process is derived from natural gas or petroleum.[96] Natural gas accounts for 70% of the cost of making ammonia.[97]

$$CH_4 + H_2O \xrightarrow[700\text{–}800°C]{\text{Ni catalyst}} CO + 3H_2$$

$$CO + H_2O \xrightarrow[425°C]{\text{Fe catalyst}} CO_2 + H_2 \quad \text{85–90\% yield at 20–22\% conversion}$$

$$N_2 + 3H_2 \xrightarrow[400\text{–}600°C]{\text{Fe catalyst}} 2NH_3$$

11.18 Schematic

One reason for the high energy consumption is that the exit gases from the ammonia synthesis must be cooled to –10 to –20°C to condense out the ammonia and must then be reheated for recycling. Sustainable agriculture should have a minimum of external inputs.

Crop losses caused by insects went up from 7 to 13% in the period from 1945 to 1989, despite the use of 10 times as much insecticide in 1989.[98] The reasons probably involve the shift to intensive monoculture and the loss of natural predators. As mentioned earlier, the lack of specificity and the toxicity of insecticides are problems that need to be solved in a system of sustainable agriculture.

The loss of topsoil must be reduced greatly so that the agricultural productivity does not decline. The cost of soil erosion in the Untied States has been estimated to be $44 billion/year.[99] This includes the cost of nutrients that have to be replaced with more fertilizer and the damage caused by the silt in the rivers. It cost $28/yard^3 to dredge the harbor of Green Bay, Wisconsin. There was some thought of selling this topsoil back to the farmers. This idea was abandoned when it was found that putting 35–66-ft buffers of grass on each side of the stream could keep the soil out of the harbor for $2/yard^3.[100] The U.S. Department of Agriculture (USDA) has a new policy of paying farmers to put in such buffers to reduce the amount of soil, fertilizers, and pesticides entering streams.[101] Fertilizer runoff from Midwestern farms in the United States has caused oxygen depletion (<2 mg/L dissolved oxygen) in 18,000 km^2 of the Gulf of Mexico.[102] (This dead zone is exceeded in size only by dead areas in the Baltic and Black Seas.[103]) This has an effect on the Gulf's fisheries. Fertilizer runoff is also a problem in restoring the health of Chesapeake Bay in Maryland.[104] Soils in natural areas are often rich in organic matter. They have a porous structure that allows rain to infiltrate readily, decreasing runoff. This soil structure is lacking in conventional agriculture using herbicides and synthetic fertilizers. It is present when organic farming is used.[105] (This involves returning all crop residues and animal waste to the land.)

Humans have used a great many species of plants for food over the millennia. However, most agriculture in the United States today is based on about two dozen crops. Many of these older food crops need to be reinvestigated to look for more nutritious ones that can thrive with little inputs of fertilizer, pesticide, and water.[106] Many of our major crops are annuals, so that plowing (minimized in the no-till system) and reseeding are needed every year. Some groups are investigating perennial crops to give sustainable yields with minimal soil erosion.[107] Current efforts are focused on breeding for higher yields.

The EPA has joint programs with the USDA and some growers to minimize the use of pesticides.[108]

One billion people worldwide rely on fish as their main source of protein. The United States now imports 80% of its seafood. As the world's fisheries have declined, attention has turned to aquaculture. For it to be sustainable in the future, current problems of replacement of mangroves by ponds, use of wild fish as feed, pollution from wastes, escapes of exotic species or varieties, and the short lifetimes of ponds must be solved.[109] Shrimp and salmon farming have many of these problems. It would be better to farm herbivorous fish.

11.3 ALTERNATIVE AGRICULTURE

The field of alternative agriculture is covered by several books and reviews,[110] from which much of the following discussion has been taken. Many standard agricultural methods disrupt the natural processes that suppress pests.[111] Crop monocultures tend to invite pests, especially when there is no alternative food and when their natural enemies no longer have places to live. The increased mechanization of agriculture has resulted in the removal of many hedgerows where these natural enemies as well as natural pollinators could live. The use of herbicides and synthetic fertilizers instead of cover crops and organic fertilizers, together with application of fungicides and insecticides, has resulted in much less life in the soil, as well as to soil erosion. Some of the organisms in a healthy soil[112] can be very beneficial in controlling pest species. The key to alternative agriculture is a thorough understanding of the ecology of the crop, the pests, and their natural enemies. This requires a great deal of study and monitoring of populations. The various approaches for the control of pest populations are grouped, in the following, under the categories of insects, weeds, fungi, and such. In some cases, an approach can apply to more than one group. In addition, many new crop protection agents or leads for them can come from studies of natural products.[113] "Spinosad" and "Spinetoram" for insect control have been found by Dow Agrosciences in this way.[114]

11.3.1 Insects and Other Animals

11.3.1.1 Mechanical and Cultural Methods

The first approach should be to plant a variety of the crop that has been bred to be resistant to the pest. Unfortunately, a general confidence in chemical methods of control has slowed the breeding of resistant varieties. Some insects can be removed mechanically. Hand picking is done in some home gardens. A vacuum cleaner can remove insects from lettuce.[115] Unfortunately, this system has not proved effective with other crops. Not all mechanical traps are effective. A popular type for the control of biting flies uses a UV light to attract insects to an electrified grid. A study in northern Delaware found only 31 biting flies (0.22%) in a total of 13,789 insects from such traps.[116] The nontarget insects were from 104 insect families. There were 1868 natural predators and parasites[117] present in the sample. Many mosquitoes are not attracted to light, but are to the carbon dioxide emitted by an animal. Not only is the trap ineffective, but it is a threat to insect biodiversity. A trap with carbon dioxide plus octenol (a component of cow's breath) was much more effective.[118] The mosquitoes were killed by the synthetic pyrethroid insecticide in the trap. (Carbon dioxide has also been used to lure sand flies to a trap where they are caught on fabric panels coated with mineral oil.[119]) Mosquito populations can be reduced by eliminating standing water where they can breed (e.g., gutters that do not drain well) and ponds and marshes, making them more accessible to small mosquito-eating fish can help. Animals that eat mosquitoes should be encouraged (e.g., bats, nighthawks, swallows, frogs, and dragonflies).[120]

Insects in grain can be killed by microwaves as the grain enters the grain elevator.[121] Because there is more water in the insects than in the grain, they heat up first. It can also be done with an e-beam at the rate of 40 tons/h.[122] Passing in air at less than 17°C also deters the insect feeding.[123] Radiofrequency radiation has been used to kill insects on stored fruit.[124] Alternating vacuum and a carbon dioxide atmosphere can be used to sterilize fruits and vegetables.[125]

Barn owls have been used for effective rat control in palm oil plantations in Malaysia.[126] Putting up nest boxes helps to attract the one barn owl needed for each 10 ha. Putting up nest boxes for the Great Tit (*Parus major*) has been used to reduce the insects on fruit trees in the Netherlands.[127] The introduction of mongoose into Hawaii to control rats was a disaster. The mongoose is diurnal and the rat nocturnal, so that the two do not meet. However, the mongoose has decimated some of the bird species on the islands.[128]

Crop rotations can be a valuable way to disrupt a pest's life cycle. (They also increase profitability.[129]) Diseases of potatoes in Maine can be controlled by rotation with canola.[130] The soybean nematode is a problem if the farmer tries to grow the same crop in the same field year after year. If corn is planted in alternative years, the nematode eggs die out and there is no problem.[131] A *Pasteuria* bacterium, added to the soil, can also help control this nematode.[132] In England, control of a cyst nematode was disrupted when the farmer applied a fungicide to his crop and killed the soil fungi that controlled the nematode. Other strategies for controlling the soybean nematode involve use of compounds (**11.19** Schematic) that inhibit the hatching of the eggs or which stimulate them to hatch when there is no crop for them to feed on. The compound on the left gave 87% control of hatching at 54 μg/L. That on the right stimulated hatching at as low as 10^{-12} g/L. A potato cyst nematode, normally controlled by carbamates, organophosphates, or methyl bromide, can be controlled by *trans*-1,3-diphenylpropenone (**11.20** Schematic), which kills the nematodes as well as inhibits hatching. It has low oral toxicity (to mice) and low phytotoxicity.[133]

It also helps destroy the insect or its eggs when it is overwintering. Crop residues are sometimes burned to do this. A better system, which avoids the air pollution associated with burning, is to compost the crop residues.[134] If done properly, the temperature in the compost pile is enough to destroy insects, fungi, and weed seeds The compost can then be put on the fields to increase the organic matter of the soil.

Intercropping is a useful technique.[135] Trap crops can lure the pest away from the main crop. Alfalfa has been used to keep lygus bugs off cotton in this way. Squash has been used to keep squash bugs and cucumber beetles off watermelon and cantaloupe. After the insects collect on the squash they can be sprayed without harming the beneficial insects in the area.[136] Weeds can take nutrients and sunlight intended for the crop. However, the presence of weeds cannot be judged to be detrimental automatically. There are some instances where they are beneficial to the crop and should not be mown until they clearly interfere with the crop yield. Several mechanisms are operative. Morning glories controlled the *Argus* tortoise

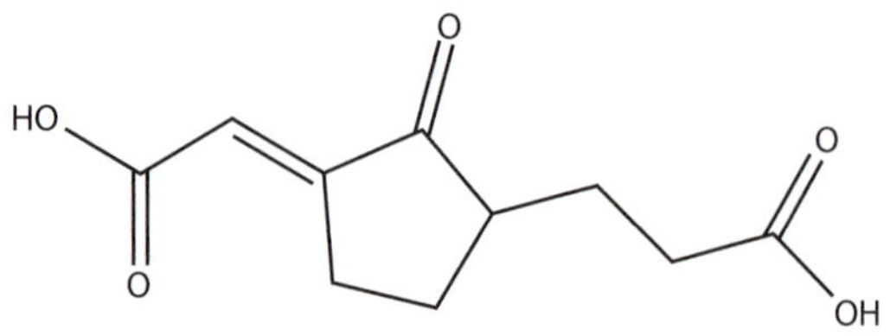

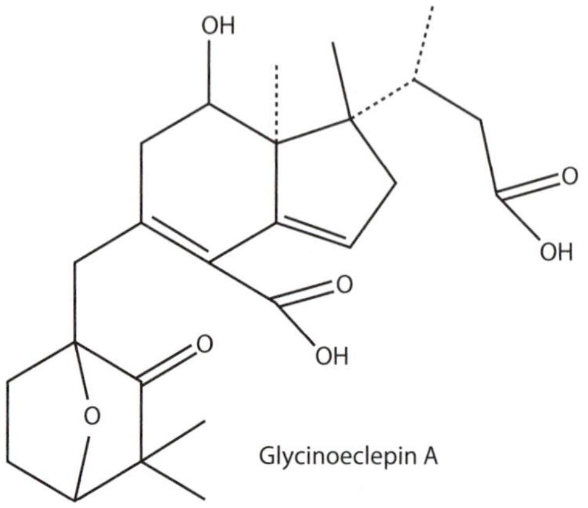

Glycinoeclepin A

11.19 Schematic

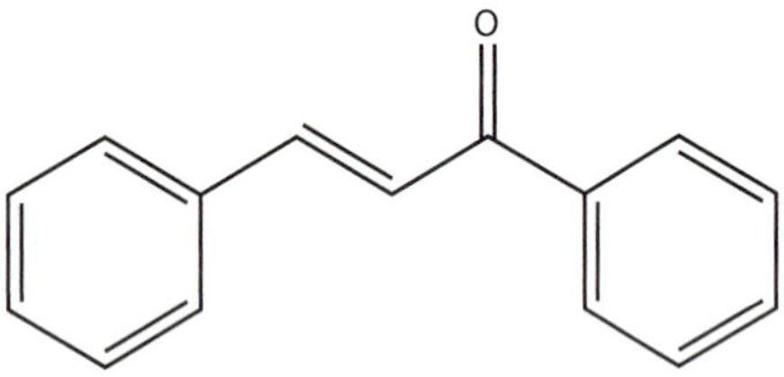

11.20 Schematic

beetle in sweet potatoes by serving as an alternative host for a parasite for the beetle. Wild mustard in broccoli served as trap crop for *Phyllotrea cruciferae. Eleusine indica* and *Leptochioa filliformis* kept leaf hoppers off beans by chemical repellency or masking. When beans are intercropped with wheat, there is impairment in the searching behavior of aphids. When maize is grown with beans, there is an increase in the numbers of beneficial insects and interference with colonization. When peanuts and field beans are raised together, the aphids become trapped on the epidermal hairs of the beans. Intercropping cabbage with tomato and tobacco results in feeding inhibition of flea beetles by the odors of the nonhost plants. Strongly aromatic plants, such as onions and tomatoes, can disturb the mechanics of orientation of insects. Extracts of marigolds have been used as larvicides for the mosquito, *Aedes aegypti*.[137]

Chemical defoliants for cotton can be replaced by a propane flame.[138]

Zebra mussel larvae can be killed by pumping the water through 0.15–0.4 mm tapered slots.[139] Low-frequency radio waves can cause zebra mussels to close their shells causing half of them to die in 20 days.[140] Encapsulated potassium chloride can kill the mussels when they ingest it without hurting fish.[141]

11.3.1.2 Predators, Parasites, and Pathogens

Natural enemies often keep insect populations under control. Some of the biggest problems have come from the introduction of animals and plants to new continents without their natural enemies. There was a tremendous population explosion of the Japanese beetle in the eastern United States in the 1930s. It was brought under control by the importation of a fungus (milky spore disease). The state of Delaware was marked off in a grid pattern. Every square with grass in it was inoculated with the fungus. Today, the problem with the beetle is minor. Most introductions to new continents do not become problems.[142] Of intentional introductions 25–68% have become established and only 2% have become pests. Five percent of unintentional introductions have become established, but only 7% of these have become pests. Nine of 20 species of introduced domestic animals have become serious environmental problems (e.g., pigs and goats in Hawaii). Eleven of the world's worst weeds were crop plants in other regions. Guava and ginger are noxious weeds in Hawaii. When the natural enemies are brought in for controlling foreign insects and weeds, the return on the investment is 30-fold. Great care must be taken to ensure that the natural enemies will not choose to attack any other species. *Compsilura concinnati* introduced into the United States in 1906–1986 to control gypsy moths has decimated the populations of *Cecropia* and *Promethea* moths.[143] Sterile male release can be used to wipe out a relatively small population of a new introduction. Large numbers of sterile males must be released to stamp out the wild ones. This technique was successful in eliminating the screw worm fly from the southeastern United States. It has also been used to eliminate the tsetse fly from Zanzibar.[144]

Many predators, parasites, and diseases of problem species[145] can be encouraged by providing them with refuges in which to live. In areas where these local populations may not exert sufficient control, mass rearing (often using synthetic diets[146]) and timely release into the fields may solve the problem. A few of the many examples will be given.[147] Beneficial mites are used to control the two-spotted spider mite on 50–70% of California strawberries. The protozoan *Nosema locustae* is used to control grasshoppers. *B. thuringiensis* is used to control Lepidoptera (butterflies and moths). It must not be overused because resistance to it can develop. Resistance in the diamondback moth is an example.[148] It can also be a problem if it gets on nontarget Lepidoptera. An area in Oregon where it was used showed a loss of two-thirds of the species present and a 95–99% decline in the biomass of caterpillars.[149] This same problem may result from the use of "Spinosad," a mixture of spinosyns A and D (**11.21** Schematic) isolated from the soil bacterium *Saccharopolyspora spinosa*, if it is used in similar situations. It won a Presidential Green Chemistry Challenge Award in 1999.[150]

maize.[159] The beetle, *Rodiola cardinalis*, has been used to control the cottony cushion scale of citrus.[160] Many parasitic organisms (e.g., wasps that lay their eggs on insects) are highly host and habitat specific. The yellow citrus ant is used to keep harmful insects off oranges in China.[161]

Many baculoviruses[162] (i.e., insect viruses) are species specific. They have been used for the successful control of the gypsy moth, velvet bean caterpillar (on 5.9 million ha of soybeans in Brazil), cotton bollworm, codling moth (on apples, pears, and walnuts), rhinoceros beetle, potato tuber moth, sawflies, and porina moth.[163] Resistance has developed in the codling moth in Germany.[164] They are also effective against mosquitoes.[165] The gypsy moth was introduced into the United States in Massachusetts from Europe in 1869. It defoliates many trees if there are no checks on its population. The "war" against it has often used diflubenzuron ("dimilin") (**11.22** Schematic), which affects many nontarget species.[166]

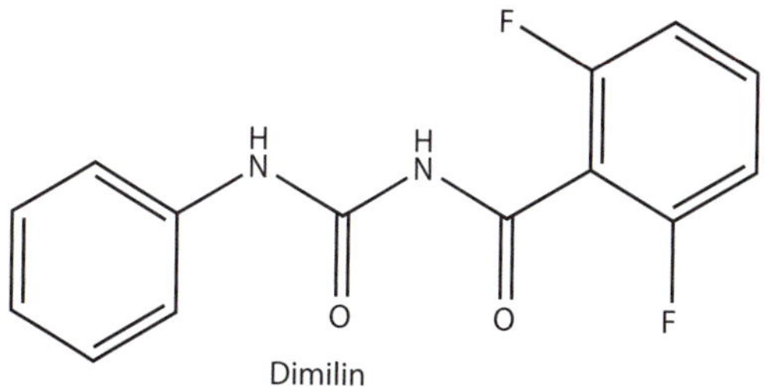

Spinosyn A (R = H)
Spinosyn D (R = CH_3)

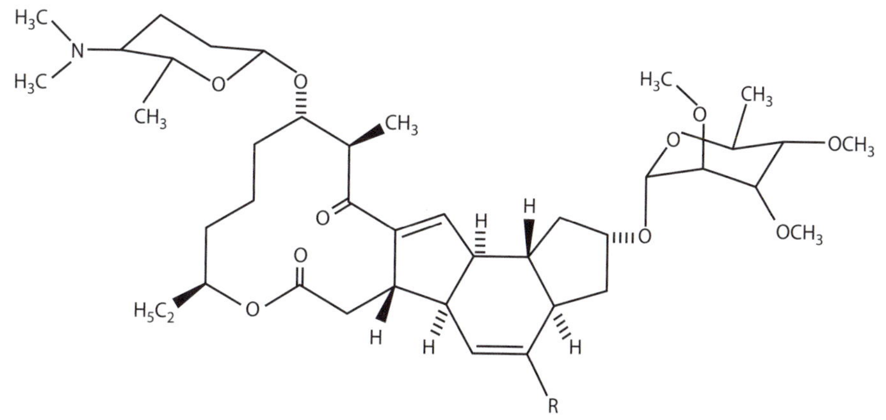

11.21 Schematic

It does not harm most beneficial insects, has a low toxicity to mammals and birds, does not bioaccumulate, breaks down quickly in sunlight, adsorbs to, and does not leach from soils, but still has a broad spectrum of activity against Lepidoptera. Strains of *Bacillus thuringiensis* that control Coleoptera (beetles) have also been found. *Bacillus sphaericus* works better on mosquitoes because it resists ultraviolet light and degradation by proteases, and can be prepared by low-cost methods.[151] Care should be taken not to overuse it, as *Culex* mosquitoes can become resistant to it.

The fungus *Beauveria bassiana* is used to control insects in crops and turf as well as flies in cattle sheds[152] This includes the diamond back moth and Indian meal moths that have become resistant to *Bacillus thuringiensis*. This and another fungus (*Paecilomyces fumosoroseus*) kill white flies that are resistant to chemical pesticides.[153] When freeze-dried and stored, there is 70% survival after 5 months. The fungus *Metarhizum anisoplia* controls locusts, sugar beet root maggots, cockroaches, carpenter ants, and termites.[154] Recent work has made it more stable in storage and more effective in the field.[155] The entomopathogenic nematode genera, *Steinernema, Psammomermis*, and *Heterorhabditis*, have also been used on soil insects.[156] The actual killing of the host is done by bacteria that are symbiotic with the nematodes.[157] Other nematodes have been used to control houseflies[158] and the sap beetle on

Dimilin

11.22 Schematic

This type of spraying is thought to have eliminated about two dozen species of desirable moths and butterflies from New Jersey. *B. thuringiensis* is more expensive and more selective, but still harms other Lepidoptera. The virus ("Gypchek") is the most selective. It is just coming into use for control of the moth. A fungus imported to control the moth also works in some cases. Small mammals, such as mice, and shrews also help by eating the larvae. There are also some less specific insect viruses that can control up to 30 species. Some of the problems in the use of baculoviruses are limited stability to ultraviolet light, the high pH of some leaf surfaces, and the slowness of death of the moths. The first can be improved by the use of ultraviolet screening agents, such as fluorescent brighteners, sometimes used with a feeding stimulant.[167] Antioxidants also help. Encapsulation with starch helps.[168] Trehalose might also help stabilize it. Faster kills of maize earworms are being obtained by incorporating an appetite-stopping hormone.[169] Large companies often consider such niche markets too small to warrant their attention.

11.3.1.3 Antifeedants

It is not necessary to kill an insect pest to protect the crop. Sometimes, repellents will do the job.[170] Many plants have evolved with alkaloids, such as quinine, quinidine, cinchonidine, strychnine, and brucine (**11.23** Schematic), that deter feeding.[171]

Two of the best studied insect antifeedants are azadirachin from the neem tree (*Azadirachta indica*)[172] and polygodial from *Polygonum hydropiper* (**11.24** Schematic).[173] Parts of the neem tree have been used in India for food protection for many years. Neem extracts deter feeding in 135 species and disrupt mating. These include the cockroach, Mediterranean fruit fly, gypsy moth, and locusts. Azadirachtin hydrolyzes readily so that it will not be persistent in soils. Its use may be limited somewhat by its sensitivity to acid, base, and ultraviolet light. Its use might be extended by encapsulation using an ultraviolet screening agent in the coating. (Abamectin, a natural product pesticide that degrades rapidly in air and sunlight, has been stabilized by encapsulation with the protein zein.[174]) Neem extracts are available commercially. Polygodial has marked antifeedant activity against insects and fish. Its environmental stability is not very good. It has a very hot taste to humans.[175] The optical isomer where the aldehyde group has the opposite configuration is tasteless. Other antifeedants, such as quinine, taste bitter to humans. Whether or not other species perceive these in the same way is uncertain. Numerous analogs have been made in the hope of remedying some of the deficiencies, enhancing activity, and obtaining a compound that can be made economically on a commercial scale.

Numerous other insect antifeedants have been isolated from a variety of sources. One way to search for these is to look for a plant that would appear to be good food for nearby insects but that is not eaten. Two compounds that are about as effective as azadirachtin are shown in **11.25** Schematic.[176] Mycotoxins can be as effective as azadirachtin in some cases, as in example, **11.26** Schematic.[177] It would be instructive to try this technique on some of the highly invasive exotic

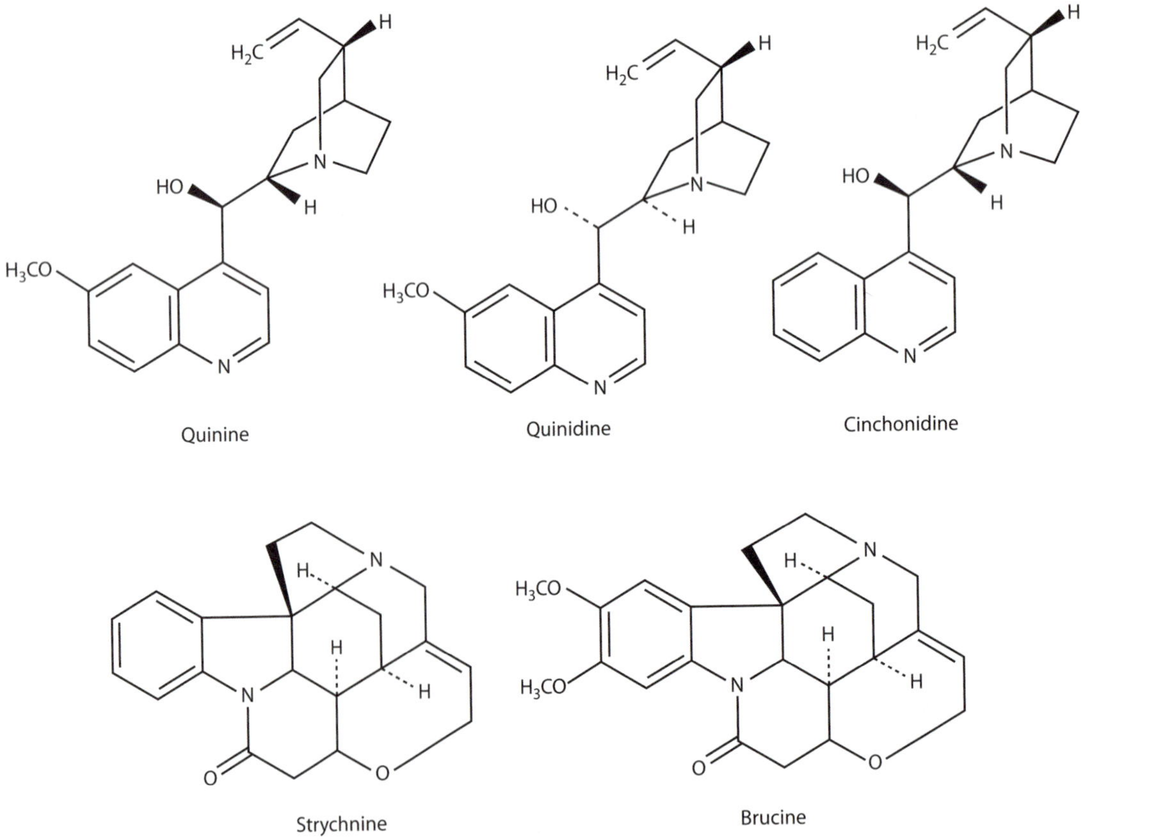

Quinine

Quinidine

Cinchonidine

Strychnine

Brucine

11.23 Schematic

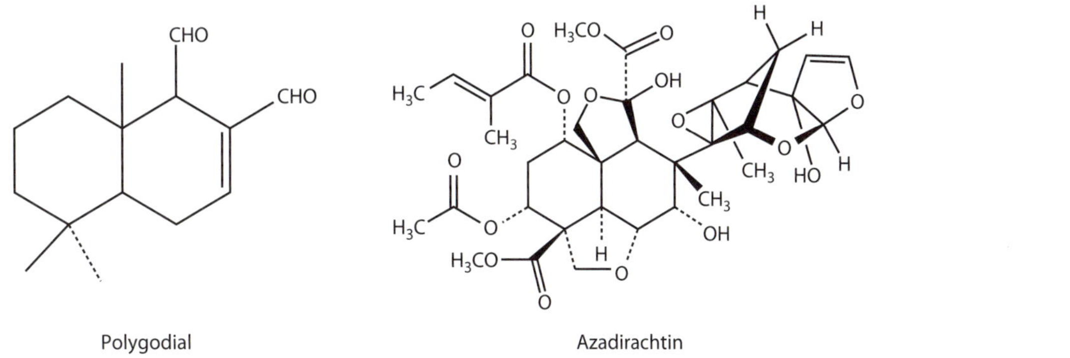

Polygodial

Azadirachtin

11.24 Schematic

weeds found in the eastern United States. It is possible that weeds such as kudzu, *Lonicera japonica, Alliaria petiolata*, and *Celastrus orbiculatus* contain antifeedants that might be used on crops in the areas where they are pests. Their harvest for this purpose might help contain their spread and preserve natural biodiversity.

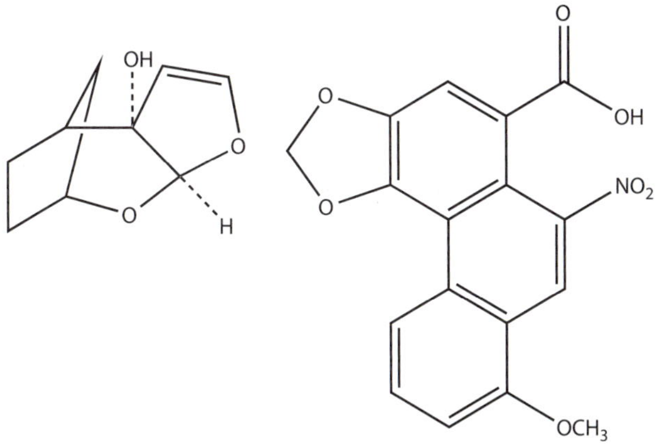

11.25 Schematic

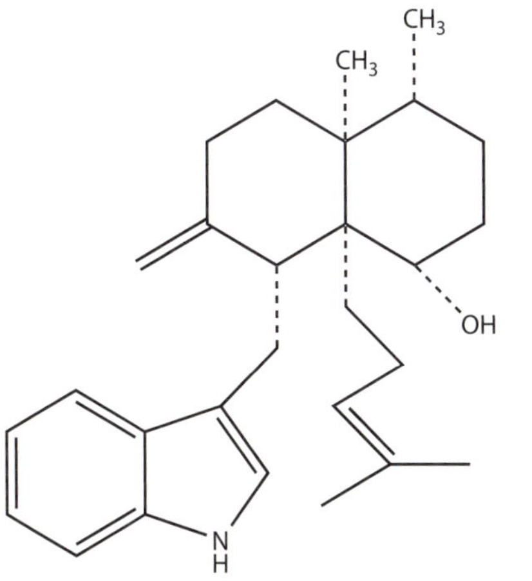

11.26 Schematic

Most antifeedants are not as active as the available pesticides, so that they would have to be used at higher levels. They would work best if they were systemic to protect the growing tip of the plant. The supply from natural sources might not be large enough for crop use in many cases. The antifeedant derived from betulin from the bark of birch trees (*Betula papyrifera*), shown in **11.27** Schematic, is potentially available in large supply.[178]

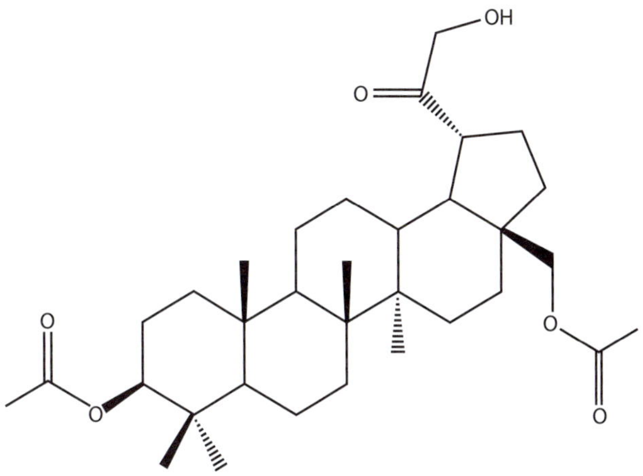

11.27 Schematic

Another approach is to breed plants with large amounts of antifeedant in them. Breeders have succeeded in producing cotton containing high levels of gossypol in the leaves but not in the seeds[179] This deters the tobacco budworm (*Heliothis virescens*) while still allowing the seeds to be used as a source of edible cottonseed oil.

The gaur, a wild ox from Asia, produces an antifeedant[180] (**11.28** Schematic) and a caterpillar produces an antifeedant[181] (**11.29** Schematic) on their hairs.

N,N-diethyl-*m*-toluamide (DEET) (**11.30** Schematic) has been used as an insect repellent by people.[182] A bicyclic C_{16} aldehyde from *Callicarpa americana* works just as well.[183] DuPont has developed nepetalactone analogs that work even better[184] (**11.30** Schematic).

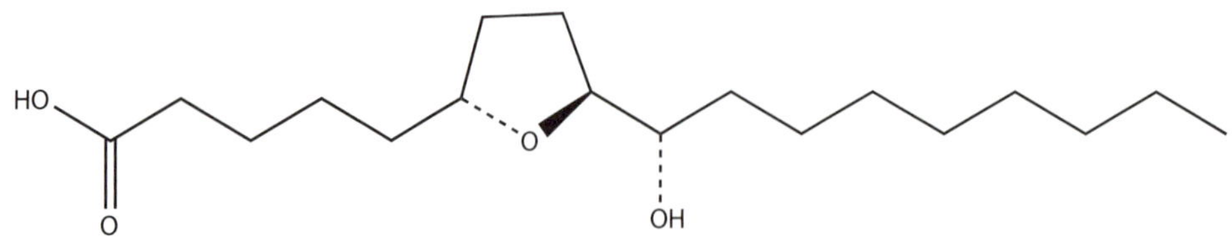

11.28 Schematic

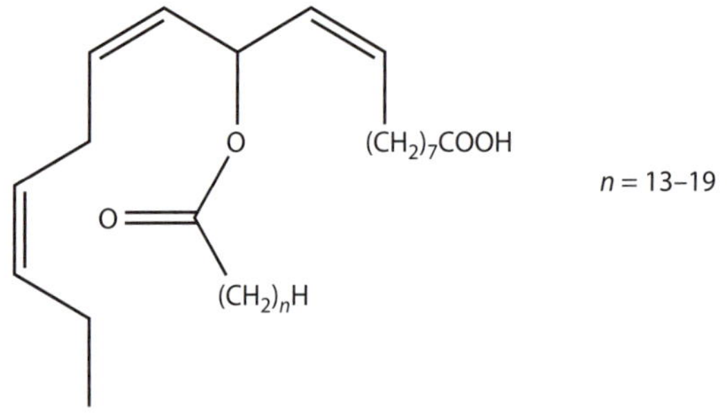

$n = 13–19$

11.29 Schematic

DEET

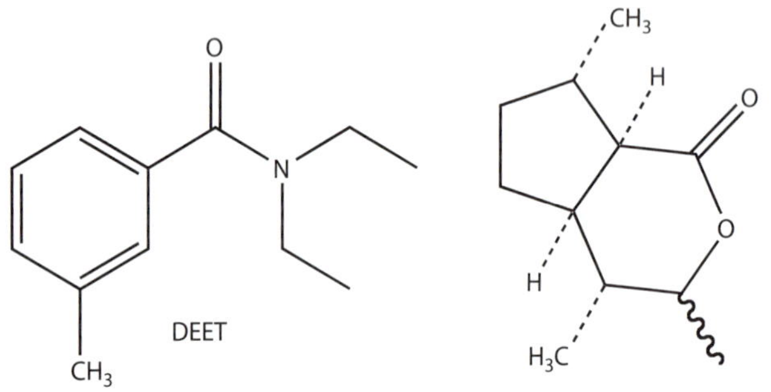

11.30 Schematic

Marine organisms contain a variety of antifeedants that protect them from fish, sea urchins, and such. (Some are also active against insects.[185]) Three are shown in **11.31** Schematic. The first is from a gastropod,[186] the second from a nudibranch,[187] and the third from bryozoan.[188]

The observation that birds do not eat Concord grapes led to the isolation of methyl anthranilate (**11.32** Schematic), which is now used to keep gulls off airports and Canada geese off lawns.[189] Before this use, 15,000 gulls had to be shot in 1991 at John F. Kennedy Airport in New York City.

11.3.1.4 Pheromones

The lives of insects depend on chemical cues that are used to locate food,[190] a mate, a place to lay eggs, and other activities. Those produced by the insect are called pheromones.[191] There are alarm, aggregation,[192] sex,[193] and other kinds of pheromones. Such chemical cues can be used to control insect populations by luring them to traps or causing them to be repelled from the crop. Putting the chemicals of the host plant that enable the insect to find it, the trail pheromone, the sex pheromone, or the aggregation pheromone in the trap can capture the insects. Putting these same chemicals in many

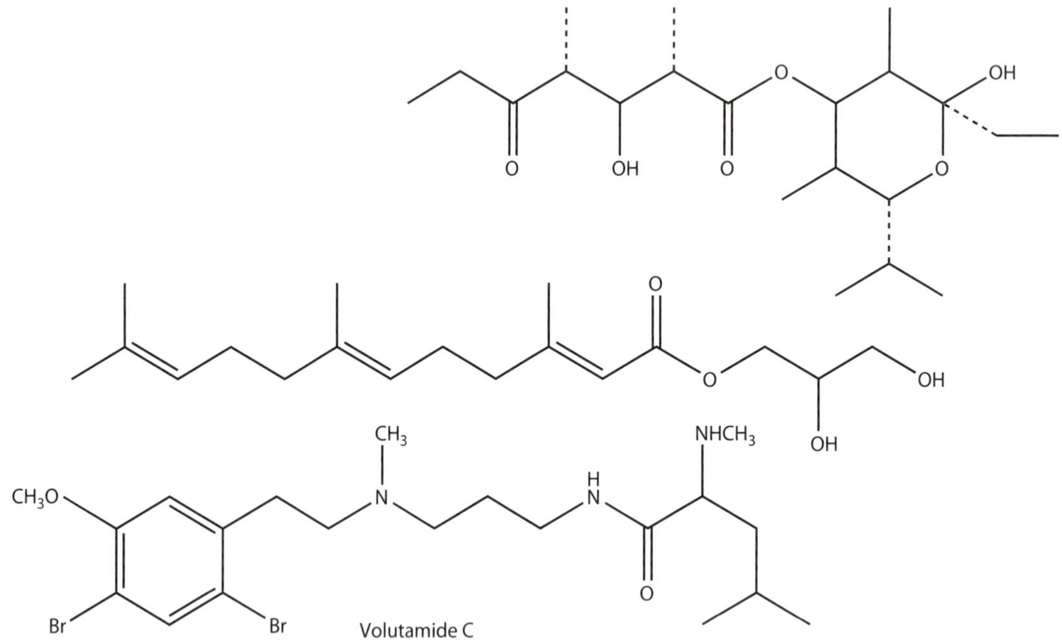

Volutamide C

11.31 Schematic

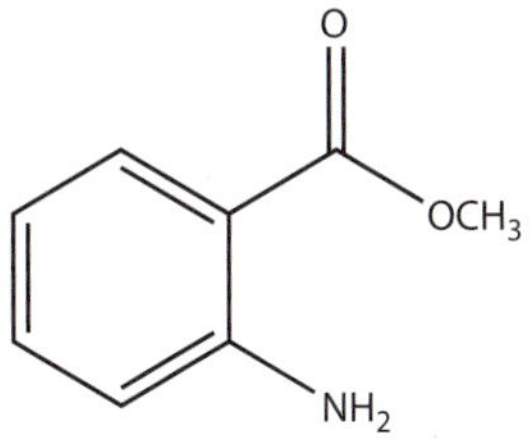

11.32 Schematic

places other than the crop can confuse the insects and disrupt their normal processes.[194] This technique of mating disruption with sex pheromones is used to control cotton bollworms, codling moths on apples and pears, and other insects. About 70% of Brazil's tomato crop is sprayed with a chemical insecticide every year to control a caterpillar pest. The sex pheromone of the moth, *Scrobipalpuloides absoluta*, has been identified (**11.33** Schematic) and used very effectively to trap the male moths.[195] Its general use may be able to eliminate 90% of the spraying. This method is species specific; hence, it will not harm nontarget organisms.

A few examples of pheromones will be shown to indicate the diversity of chemical structures. Many sex pheromones of Lepidoptera tend to be long-chain aliphatic aldehydes, acetates, or alcohols. The sex pheromone of a rice borer is Z-11-hexadecenal (**11.34** Schematic).[196] The sex pheromone of the Israeli pine blast scale is shown in **11.35** Schematic.[197]

Almost all the sex pheromones are mixtures. They have to be enough for the thousands of insect species. The mixture used to disrupt the mating of the codling moth consists principally of (*E,E*)-8,10-dodecadien-1-ol, with minor amounts of dodecan-1-ol, (*E*)-9-dodecen-1-ol, and tetradecan-1-ol (**11.36** Schematic).[198]

Mating has also been disrupted by analogs, in which a cyclopropene group has been used in place of the double bond (**11.37** Schematic), in the natural sex pheromone[199] (e.g., for the female housefly).

There are also many other types of compounds (**11.38** Schematic) used by insects other than Lepidoptera as sex pheromones. The methyl esters of L-valine and L-isoleucine are used by the cranberry white grub,[200] periplanone B by cockroaches, ipsdienol by bark beetles,[201] and olean by the olvie fruit fly.[202] The isocoumarin is part of the trail pheromone of the ant, *Lasius fuliginosus*.[203]

The sugarcane rootstalk borer weevil is attracted to aggregate when it is attracted by food, others of its kind, and

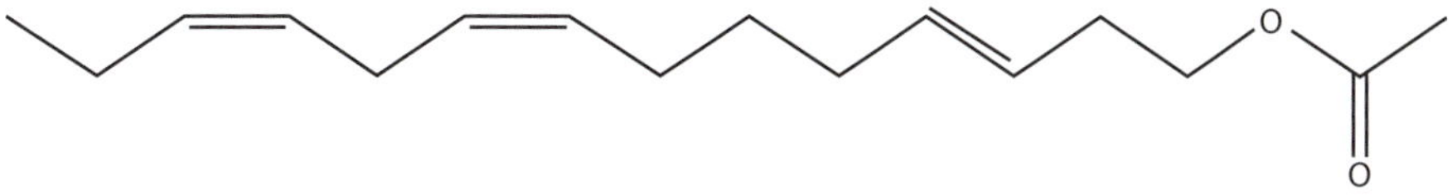

11.33 Schematic

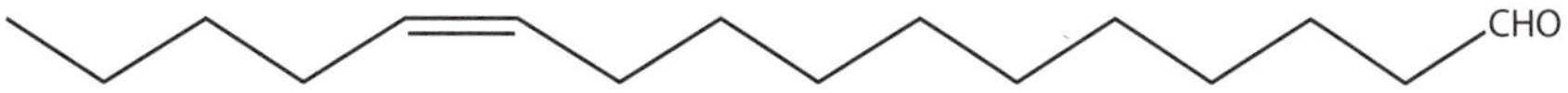

11.34 Schematic

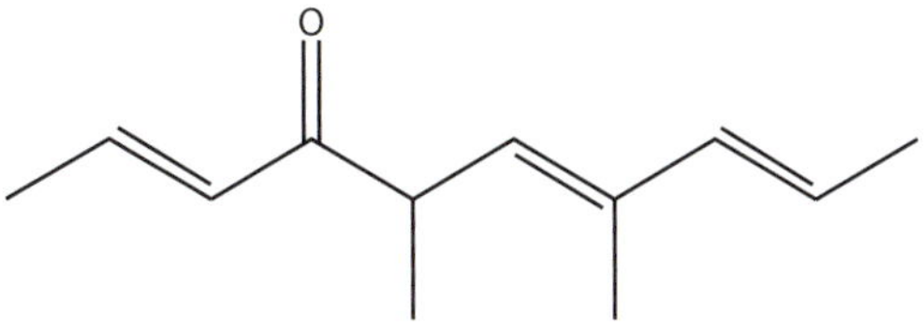

11.35 Schematic

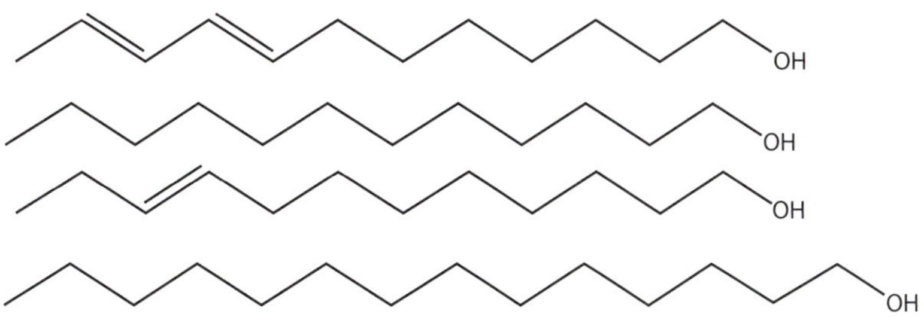

11.36 Schematic

11.37 Schematic

L-Valine methyl ester

L-Isoleucine methyl ester

Ipsdienol

Olean

Periplanone B

11.38 Schematic

by their frass.[204] If the pheromones involved can be identified, they can be used in traps. Insects interact with the terpenes in plants in a variety of ways.[205] The insect may find the host plant in this way, either to eat it or to lay eggs in it. Terpenes such as pulegone (**11.39** Schematic),[206] thujone, camphor, and citronellal repel insects (**11.40** Schematic). Linalool has been used as an insecticide on house plants.

Pulegone

11.39 Schematic

Blends of oils from cloves, eucalyptus, lemon grass, cinnamon, and others that block the insect neurotransmitter octopamine are now available commercially for use as insecticides.[207]

Because these are relatively volatile compounds, their use as lures may require slow-release formulations that would release the proper amount of terpene as a function of time. Such formulations are used with the sex pheromone traps. Sometimes it is as simple as impregnating a piece of fiberboard with the liquid. Sex pheromones have also been incorporated into plasticized polyvinyl chloride and in laminated polymer films.[208]

Dioryctria abietivorella oviposits on white pine in response to volatile emissions of myrcene and 3-carene (**11.41** Schematic).[209] Terpenes can both deter and stimulate

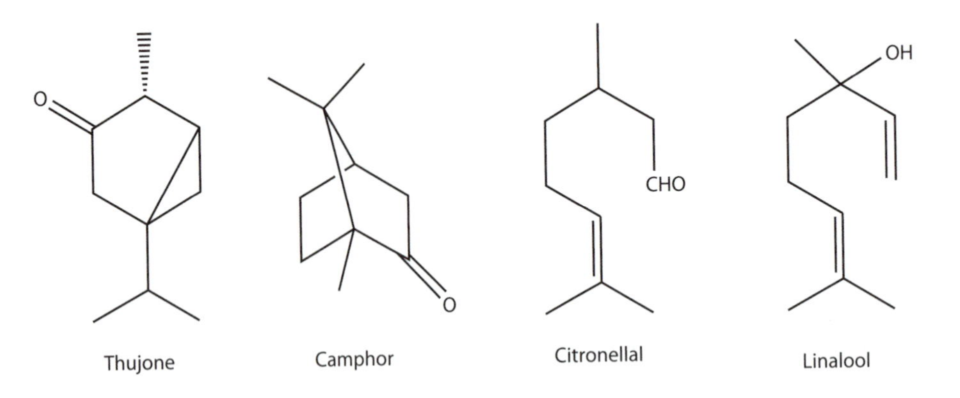

11.40 Schematic

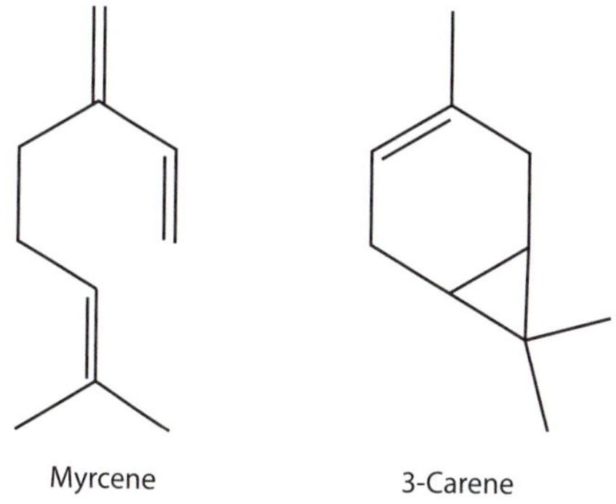

Myrcene 3-Carene

11.41 Schematic

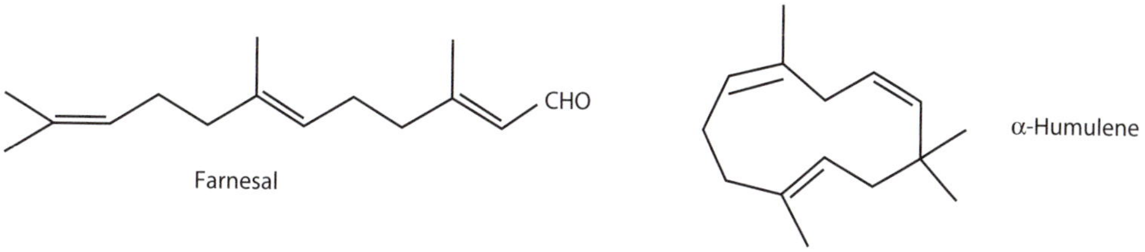

Farnesal α-Humulene

11.42 Schematic

11.43 Schematic

11.44 Schematic

oviposition of the European corn borer. In a series that was screened, the best deterrent was farnesal and the best stimulant α-humulene (**11.42** Schematic).[210]

An egg parasitoid uses the sex pheromone of the tussock moth to find the eggs, which are deposited shortly after mating.[211] An oviposition attractant for the mosquito, *Culex pipiens fatigans*, is shown in **11.43** Schematic.[212]

The use of such materials to induce the mosquitoes to lay eggs in locations unsuitable for their development could reduce the amount of pesticide needed to control populations. The European cherry fruit fly lays only one egg in a cherry and then marks the cherry with an ovipositing deterrent (**11.44** Schematic).[213] Spraying the tree with this deterrent reduced the infestation dramatically.

The bark from pines infested with the southern pine beetle attracts two species of parasitoids.[214] If the compounds involved can be identified, they might be put on the pine trees at the first sign of a beetle. It might be necessary to augment the supply of parasitoids by releasing some reared in the laboratory.

In some insects, a male marks the female after mating so that she will not mate again. If this pheromone can be used on the females before any male arrives, any eggs laid will be sterile.

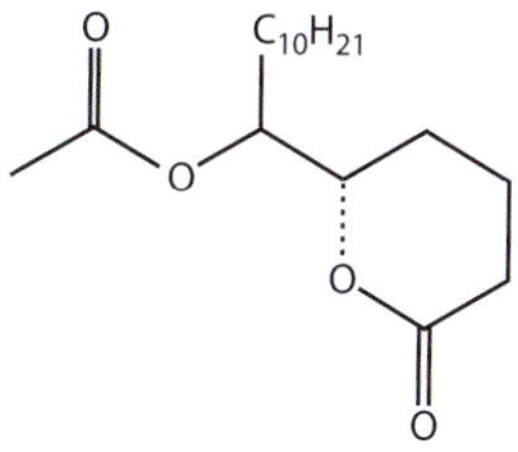

A bait attractive to termites lures them into a trap containing the insecticide hexafluomuron, which kills them.[215] This system uses much less insecticide than the previous system of treating the soil (**11.45** Schematic).

There are also other traps that may not use pheromones. A biodegradable decoy for an apple orchard contains high-fructose corn syrup with an insecticide plus cayenne pepper to keep other animals away. It replaces three sprayings of insecticide to keep the apple maggot fly under control.[216]

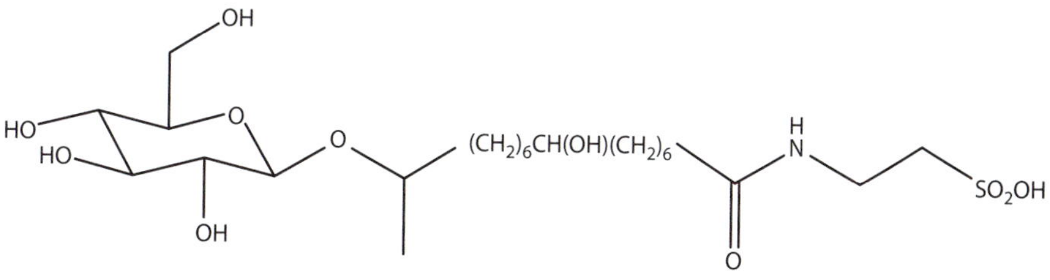

Hexaflumuron

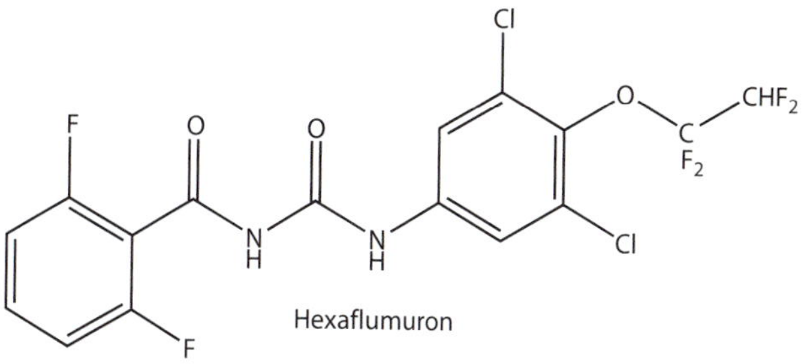

11.45 Schematic

This type can also be used on blueberries. Cloth cows in Zimbabwe with a black center and blue ends attract tsetse flies to an insecticide in the cloth. This technique has reduced the cases of sleeping sickness in cattle from 10,000 to 50 per year over 17 years.[217]

11.3.1.5 Insect Growth Regulators

Insects have exoskeletons. To grow, they must molt. This process is controlled by 20-hydroxyecdysone (**11.46** Schematic), a molting hormone. Insects can be destroyed by preventing them from molting or causing them to molt prematurely. Some plants use a defense of making phytoecdysteroids to interfere with insect growth regulation. *Ajuga reptans* does this.[218] Four phytoecdysteroids (**11.47** Schematic)—20-hydroxyecdysone, norcyasterone, cyasterone, and isocyasterone—are produced at 0.5% in hairy root cultures.[219]

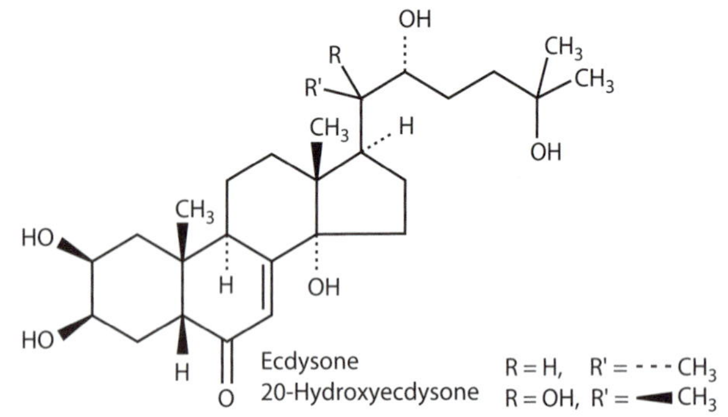

Neotenin (**11.48** Schematic) is an insect juvenile hormone.[220] The level of juvenile hormone must drop for metamorphosis to take place. A carbamate (**11.49** Schematic) causes fatal morphological changes in the insect if applied when the inherent juvenile hormone is low.[221] It is nontoxic to warm-blooded animals and fish. It could still be toxic to many beneficial insects.

Dow has devised diacylhydrazine agonists of ecdysone (**11.50** Schematic) for caterpillar control that induce a premature lethal molt in a few hours.[222] It does not harm beneficial insects, such as bees, ladybugs, and wasps. However, it could be toxic to other butterflies and moths, at least some of which are natural pollinators. Fleas on animals can be controlled by methoprene (**11.51** Schematic), a juvenile hormone mimic that prevents the conversion of larvae into pupae.[223] Another chemical, lufenuron (**11.52** Schematic), is given to the animal orally. Any flea that bites the animal becomes sterile by inhibition of the development of chitin, a vital polymer in the exoskeleton. These chemicals should have minimal effects on other species.

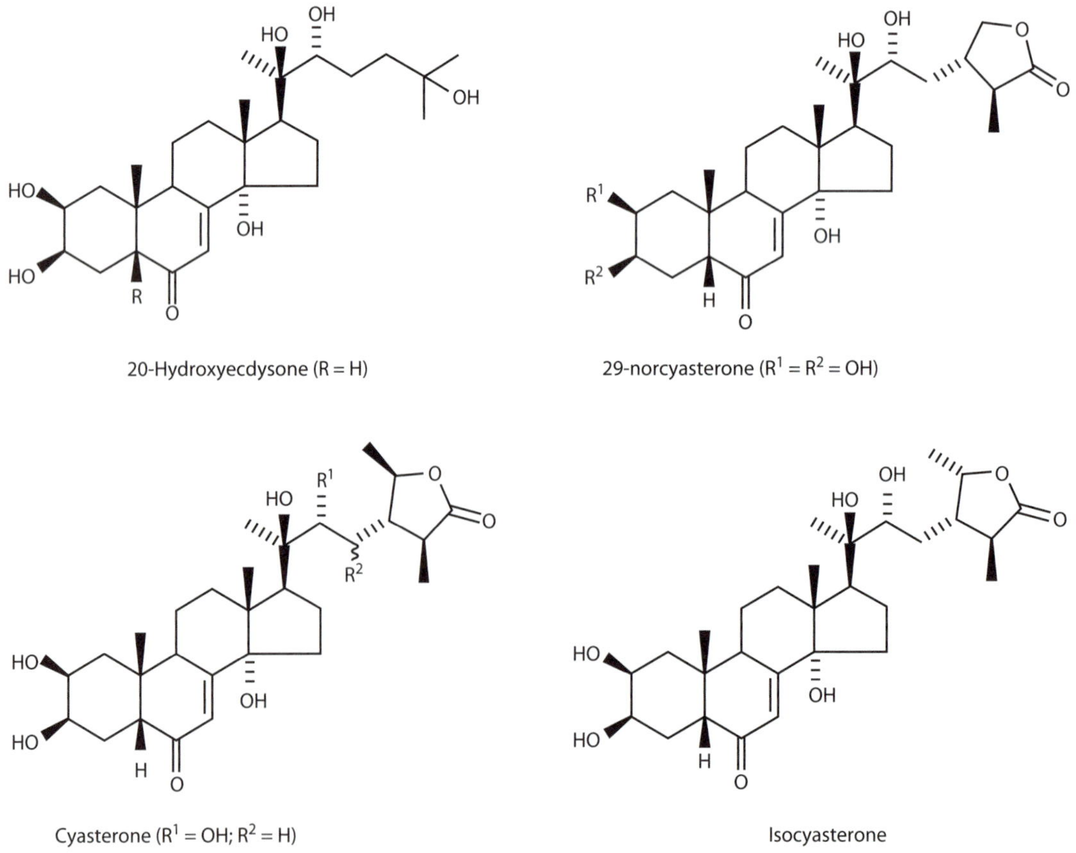

Ecdysone R = H, R' = - - - CH_3
20-Hydroxyecdysone R = OH, R' = ◄ CH_3

11.46 Schematic

20-Hydroxyecdysone (R = H)

29-norcyasterone ($R^1 = R^2 = OH$)

Cyasterone ($R^1 = OH$; $R^2 = H$)

Isocyasterone

11.47 Schematic

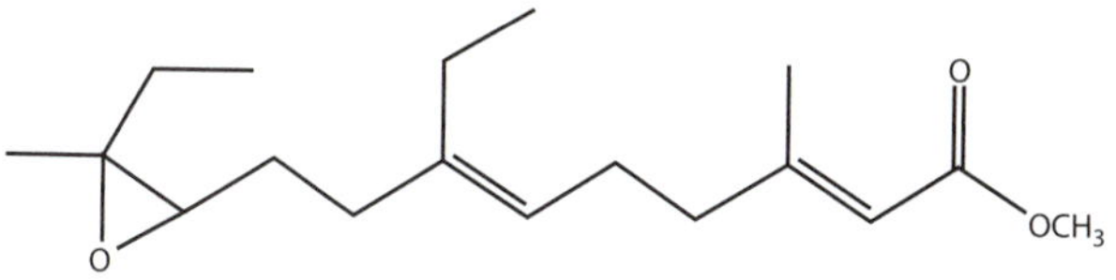

11.48 Schematic

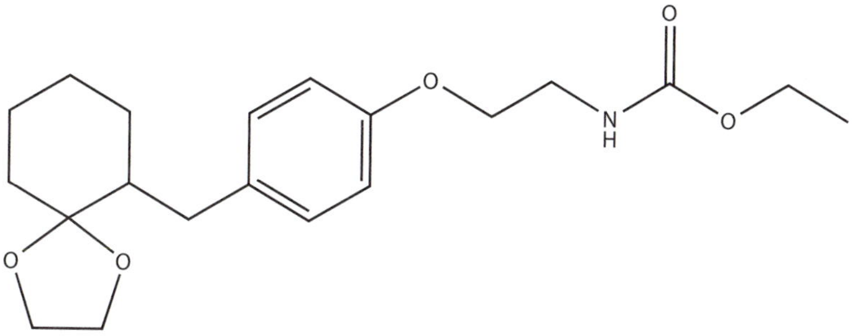

11.3.1.6 Phytoalexins

The coevolution of plants with bacteria, fungi, insects, nematodes, viruses, and others that attack them has led to various defenses in plants.[225] Chemicals produced in response to the attack are called phytoalexins. They include poisons, feeding deterrents, proteins that block the insect's digestive enzymes (such as protease inhibitors), chitinase (for destroying the fungal cell wall or the insect's exoskeleton), enzymes for the production of callose to wall off the area of attack, volatile compounds that attract parasitoids to the insect, phytoecdysteroids to interfere with molting, accumulation of phenols, and other mechanisms. In some cases, the response is systemic and protects the plant for days. Localized cell and tissue death at the site of infection may deter the enemy.[226] Insects may, in turn, detoxify, sequester, or bypass these defenses. When the beet army worm feeds on maize seedlings, volicitin (**11.53** Schematic) in the worm's oral secretions causes the plant to produce volatile compounds, such as terpenes, that attract wasps that parasitize the worm.[227]

Jasmonic acid (also derived by biosynthesis from 18-carbon unsaturated fatty acids; **11.54** Schematic) is often involved in the signaling in the plant to respond to the threat.[228]

Salicylic acid (**11.55** Schematic) is another signaling molecule in the plant's response. The amount of jasmonic acid rises in 5 min after the insect feeds on the tobacco plant. The acid reaches the roots in 2 h and a flush of nicotine reaches the leaves (up to 120 mg/g leaf) in 7 h. Both of these acids can trigger the release of volatile compounds from the plant. These include not only terpenes, but also methyl jasmonate and methyl salicylate, which can signal nearly plants to mount their defenses against attack.[229] Methyl jasmonate also inhibits the sprouting of potatoes.[230]

Cyclic hydroxamic acids (**11.56** Schematic) stored in corn and wheat seedlings as their glucosides are released by a glucosidase when the plant is attacked.[231] These resist insect and microbial attacks.

It may be possible to protect plants by treating them with elicitors than cause the synthesis of phytoalexins before there are very many insects or disease organisms around.[232]

11.49 Schematic

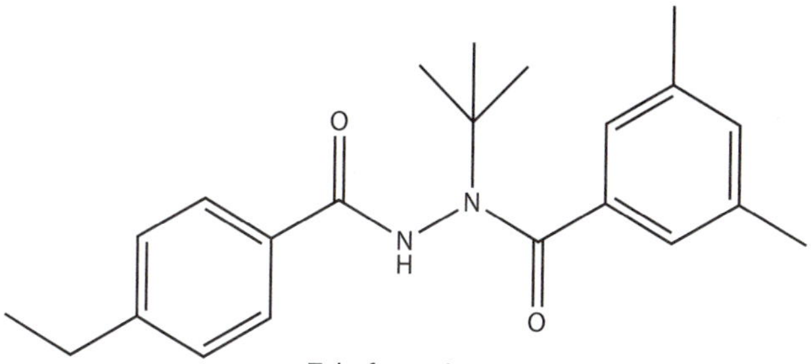

Tebufenozide

11.50 Schematic

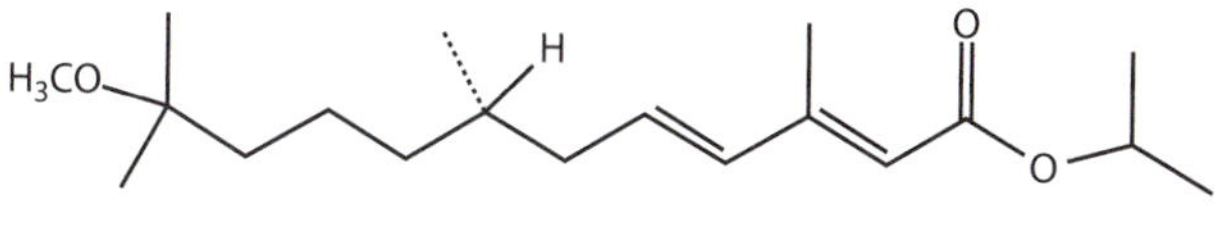

(S)-Methoprene

11.51 Schematic

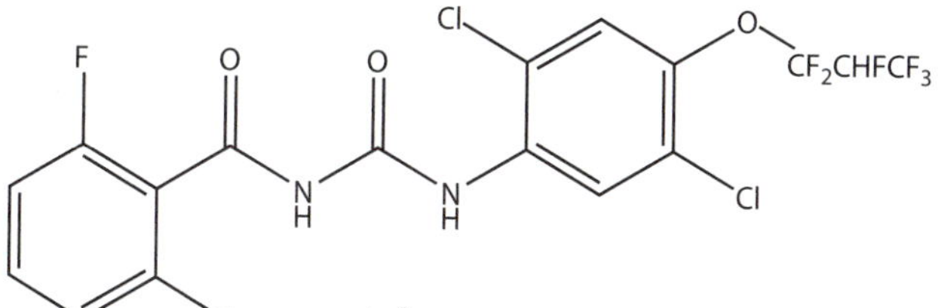

Lufenuron

11.52 Schematic

The prothoracicotropic hormones that activate the synthesis of ecdysone in Lepidoptera are highly species specific, unlike other insect neuropeptides. If something can be found that blocks the receptor for these hormones, it may be a highly specific insect control agent.[224]

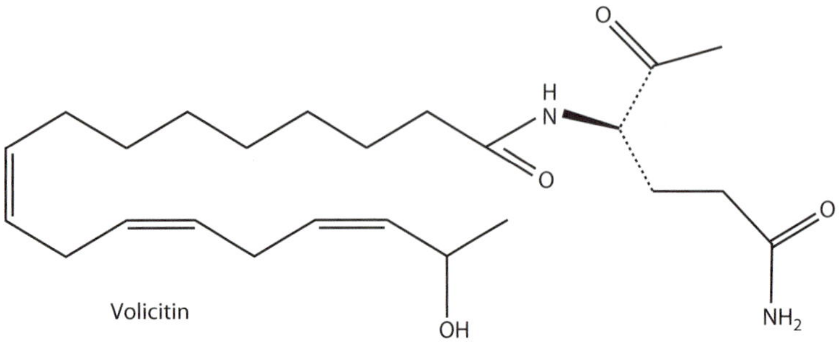

Volicitin

11.53 Schematic

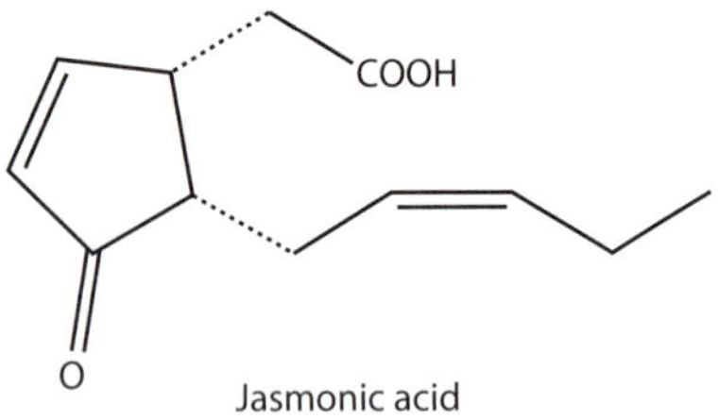

Jasmonic acid

11.54 Schematic

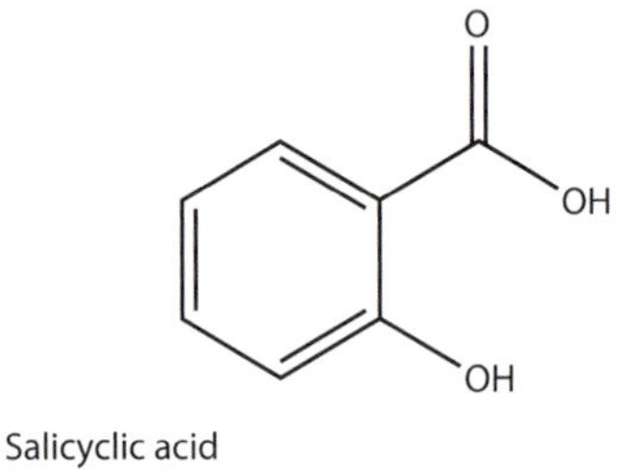

Salicyclic acid

11.55 Schematic

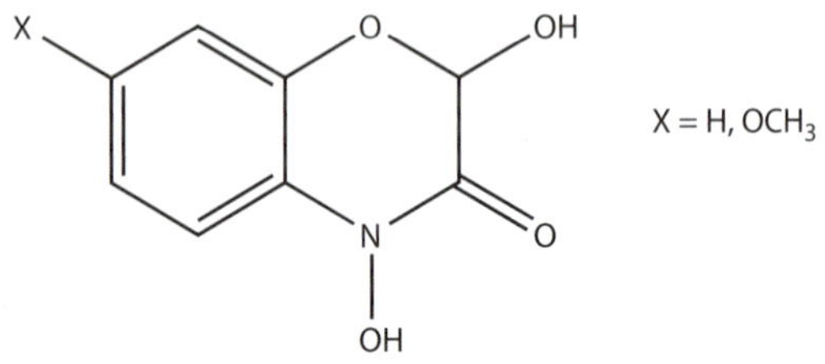

X = H, OCH_3

11.56 Schematic

Chitosan from the deacetylation of chitin (an acetylated polymer from glucosamine found in crab shells, fungi, and insect exoskeletons) can elicit such a response. So can oligogalacturonides (degree of polymerization 8–15). Fungal endogalacturonases cause the release of such oligomers from pectin in plant cell walls in 15 min. Hydrogen peroxide, as such, or in some slow-release bound form, might be a suitable elicitor for sweet potatoes, beans, and sugar beets, where attack by *Fusarium* fungi produces hydrogen peroxide that stimulates the production of phytoalexins.[233] Aluminum chloride has been shown to be an elicitor for resveratrol (**11.57** Schematic) in grapevines.[234]

This phytoalexin helps the plant resist attack by gray mold (*Botrytis cinerea*). Another possibility is to use a phytoalexin made by a chemical synthesis apart from the plant. Novartis sells a benzothiadiazole (Actigard) that elicits systemic resistance to bacterial and fungal pathogens.[235]

11.3.1.7 Miscellaneous Methods of Insect Control

Fumigants are used to kill organisms in the soil, in stored fruits, and grains, houses, libraries, and so on. The debate about the use of methyl bromide, which can cause ozone depletion in the stratosphere, was described in Chapter 3. Methyl iodide has been proposed as a replacement because it photolyzes in the troposphere and is unlikely to cause ozone depletion.[236] However, it is a highly toxic cancer-suspect agent. Chopped broccoli has been worked into the soil before planting cauliflower to eliminate a wilt caused by *Verticillium dahliae*.[237] This replaced methyl bromide. Irradiation of food can be used to replace fumigation for insects.[238] It has been used on spices, strawberries, tomatoes, grapefruit, poultry, and others. Chilling fruits to 34°F (1°C) or less for 12 days will kill fly eggs and maggots.[239] Forcing hot air through papayas and citrus eliminates fruit flies.[240] Repellents such as azadirachtin and the homoglynolide A can be used to keep insects out of stored grain.[241]

The pirate bug has been used to control insects such as the Indian meal moth in stored grain.[242] Stored grain can also be protected from insects by an atmosphere of carbon dioxide.[243] Insects in library and museum materials can be killed by keeping the materials under nitrogen for 8–10 days.[244] Another

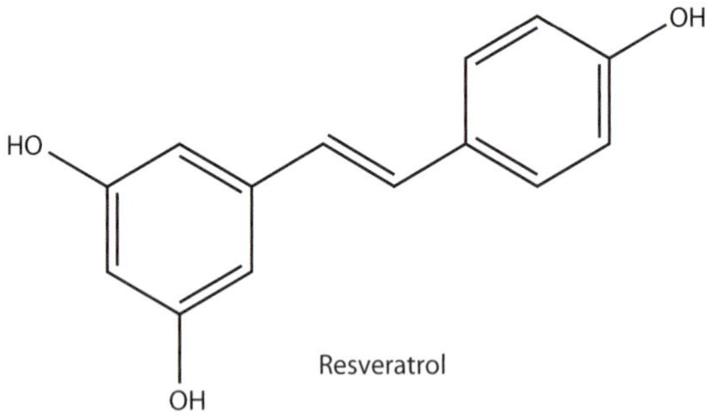

Resveratrol

11.57 Schematic

method is to heat to 140°F (60°C) for 2–4 h. Freezing to –20 to –40°C works if the temperature is reduced within 24 h.

Cockroaches can be eliminated from homes by putting boric acid or silica gel in cracks and other places where they may live.[245] These materials are thought to work by destruction of the waxy coating on the insect's body. They may also be ingested by grooming. Diatomaeceous earth killed insects in stored grain as well as organophosphate insecticides did.[246] Coating apples and pear trees with kaolin clay controlled pear psylla insects.[247] This treatment also works on grapes and citrus.

Flies that have translucent guts can be killed by eating dyes that will form singlet oxygen in the light.[248] Phloxine B (**11.58** Schematic) and uranine (**11.59** Schematic) are suitable dyes. They are used in food and cosmetics. When these two were ingested by fruit flies, followed by light, there was 100% mortality in 2–12 h. The method is suitable for use with pheromone traps that may or may not contain an insecticide. Some traps contain sticky panels for the insects. Others just have a transparent trap with a small hole that the insects can get into but cannot seem to figure out how to get out. This is an alternative to phosphorothioate insecticides, such as malathion. Because no light enters the guts of mammals, birds, and fish, they are unaffected by such treatments. As more than 100 photosensitizers have been isolated from higher plants, this may be another mode of natural plant defense.

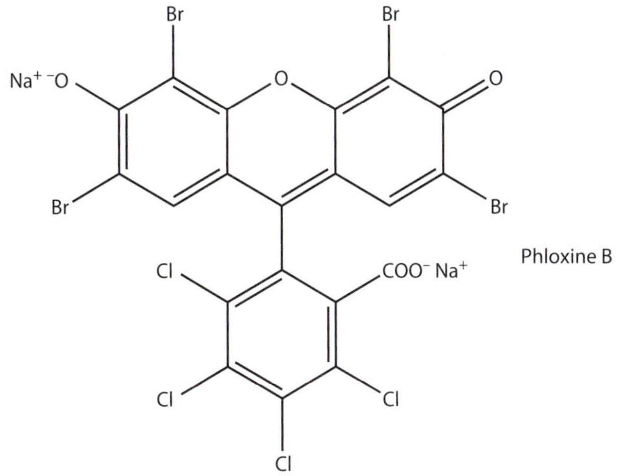

Phloxine B

11.58 Schematic

Soft-bodied insects, such as aphids, can often be controlled by relatively pure soaps of fatty acids. Acid mine drainage involves the oxidation of pyrite and other metal sulfides by *Thiobacillus ferrooxidans*. It can be controlled by detergents that wash away the surface fatty film on the bacteria that normally protects them from the acid.[249] The glycolipids on the surface of the leaves of petunias have insecticidal properties. The preparation of analogs has shown the diheptanoyl and dioctanoyl esters of sucrose to be potent insecticides for tobacco aphids and sweet potato white

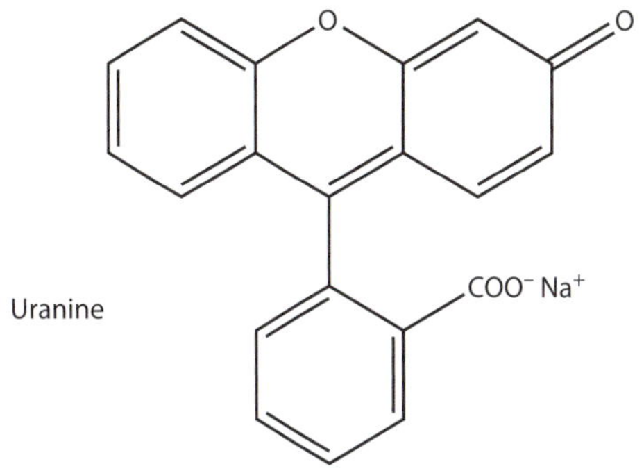

Uranine

11.59 Schematic

flies.[250] Sugar esters extracted from *Nicotiaum glutinosa* controlled white flies.[251] (These glycolipids break down the outer coating of soft-bodied insects, but are relatively nontoxic to hard-bodied insects, such as lady beetles.) Soft-bodied insets can also be controlled by applying horticultural oils (clean petroleum fractions), either during the growing season or in the off season.[252]

11.3.2 Weed Control

11.3.2.1 Mechanical and Cultural Methods

Before the advent of herbicides, weeds were removed by mechanical cultivation. Herbicides were adopted because they made weeding cheaper without any mechanical damage to the crop. Both methods involve more erosion of soil than is desirable. There are some other ways to cut down on the crop losses to weeds[253]

The seeds of the crop should be as free of weed seeds as possible. Proper composting of animal wastes before their use as fertilizer will involve a temperature high enough to kill any weed seeds present. Weeds are primarily pioneer plants of disturbed habitats. The seeds can remain dormant in the soil seed bank for many years. The seed germination is triggered by more than a minimal diurnal variation of temperature, or in some cases by light. In Iowa, tillage at night reduced the incidence of weeds by 50% or more.[254] If a no-till system, with a relatively harmless herbicide such as a glyphosate, is used, the seeds will not be brought to the surface and will be less likely to germinate. For those that do germinate, a second pass with glyphosate can be used. (Although glyphosate[255] is relatively harmless, it is made with hydrogen cyanide, which is very toxic.) Monsanto has developed a process for making the intermediate iminodiacetic acid, which eliminates the need for the HCN,[256] but has built new plants based on both routes.[257] Even the new route is not free of toxic chemicals, because the diethanolamine is made by reacting the carcinogen ethylene oxide with ammonia) (**11.60** Schematic).

Ridge tillage uses an herbicide only under the 6–8-in.-high ridge that serves as the seedbed for the crop.[258] The ridge is

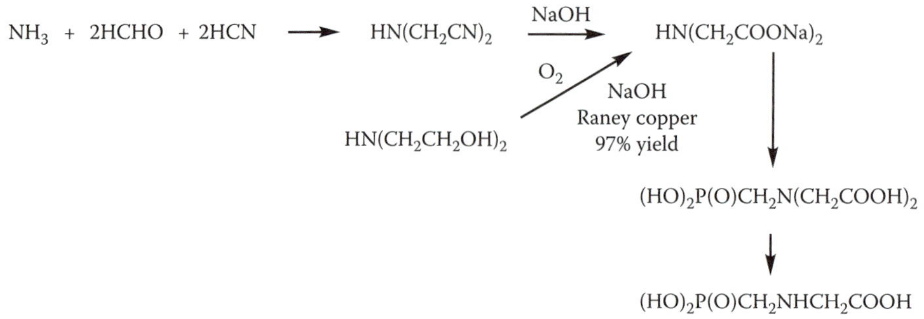

$$NH_3 + 2HCHO + 2HCN \longrightarrow HN(CH_2CN)_2 \xrightarrow{NaOH} HN(CH_2COONa)_2$$

$$HN(CH_2CH_2OH)_2 \xrightarrow[\text{NaOH, Raney copper, 97\% yield}]{O_2} HN(CH_2COONa)_2$$

$$HN(CH_2COONa)_2 \longrightarrow (HO)_2P(O)CH_2N(CH_2COOH)_2 \longrightarrow (HO)_2P(O)CH_2NHCH_2COOH$$

11.60 Schematic

warmer so that the seed germinates before any seeds between the rows. Use of a string trimmer for the weeds that did develop between the rows gave a system that reduced herbicide use by 60% over conventional spraying.[259] The fertilizer is applied to the ridge with the first cultivation, the amount being limited to what the crop can actually use. Putting the fertilizer in a narrow subsurface band under the seed with conservation tillage is better than broadcasting the fertilizer.[260] These methods reduce soil loss by 90% over that in a plowed field. If the weeds between the rows are mown, less soil will be lost than if the soil is actually disturbed to take out the weeds

Small succulent weeds between rows of taller woodier plants can be removed with a tractor-drawn flamer. The aim is not to burn up the weeds but just to achieve cell rupture. This method has been used to weed alfalfa, maize, cotton, sugarcane, and soybeans. High-technology sprayers that determine the precise locations of weeds in soybeans by light and computer chips have been used to reduce the amount of herbicide needed by using only short bursts to spray the weeds.[261] If this technique can be adapted to flame spraying or to the use of mechanical cutters, the herbicide can be eliminated. Weeds often grow in patches. Such spraying systems have also been used for the application of water and fertilizer.[262]

Planting crops in narrow rows (0.25–038 m) can help suppress weeds by shading them.[263] When this was used, with some cultivation, soybean and sunflower yields did not go down. Maize yield was 10% less than that in conventional practice with herbicides. Weed grasses in wheat were reduced by a combination of seeding 50% more per acre, choosing a taller wheat variety, and applying the fertilizer 5 months before planting so that the deeper roots (versus the weed grasses) of the wheat reached it first.[264] Crop rotation also helps, because some weeds tend to associate with certain crops. Intercropping can suppress weeds.[265] Intercropping maize and velvet beans has worked well in Honduras. This has doubled or tripled the yield of maize. Legumes are often used as cover crops, which may not be harvested.[266] This living mulch reduces soil erosion, suppresses weeds, and puts nitrogen into the soil, so that less or no nitrogen fertilizer has to be added.[267] Weed suppression works well if the cover crop is also mown or, even better, if it is rolled and crimped with the new crop being planted through the old one.[268] In one form of conservation tillage, a tall forage soybean was grown and cut, after which broccoli seedlings were planted. This method cut the use of chemicals on the brocculi in half.[269] Nonliving mulches can also suppress weeds. They are used with strawberries, pineapples, sugarcane, and some vegetables. They are very common in home landscaping. Crop residues can often be used as mulches. For a sustainable agriculture, it would be better not to use a plastic derived from petroleum as a mulch.

11.3.2.2 Predators and Pathogens for Weed Control

Chickens and geese have been used to remove weeds from an apple orchard intercropped with potatoes.[270] This does require frequent moving of the animals and the fences used to contain them. Sterile grass carp have been used to remove aquatic weeds.[271] Many of the worst weeds have been imported from other parts of the world without their natural enemies. In these cases, biological control can be very effective, if one is absolutely certain that control agents are species specific and will not shift to other hosts. Meeting this requirement means that finding a suitable control agent may take quite a bit of time. An example is the use of moths and weevils for water hyacinth, which is a problem in the southeastern United States.[272] The larvae of the soybean looper have been used to control kudzu in North Carlolina.[273] Each larva is injected with the egg of a stingless wasp that kills the looper before it becomes a moth. The egg makes the larva eat at 1.5 times the normal rate.

Fungal pathogens can also be useful in controlling other weeds[274] *C. gloeosporioides* controls northern joint vetch. *Phytopthora palmivora* has been used to control strangler vine in citrus groves. *Puccinia canaliculata* is registered for control of a species of sedge, *Cyperus esculenta*. *Colletotrichum truncatum* is effective on hemp sesbania, a common weed in cotton and soybean fields.[275] The bacterium *Xanthomonas campestris* can control *Poa annua* in turf, but the treatment is slow, requiring 50–60 days.[276] Velvetleaf can be controlled in this way.[277] The insect that feeds on the seeds can spread the fungi. Fungal spores of *Myrothecium verrucaria* in oil killed Bermua grass, pigweed, and bindweed as effectively as atrazine and 2,4-D.[278] The fungus *Sphacelotheca holci* attacks Johnson grass without hurting sugarcane.[279] Pantocin B from *Erwinia herbicola* controls the fire blight caused by *Erwinia amylovora* (**11.61** Schematic).[280]

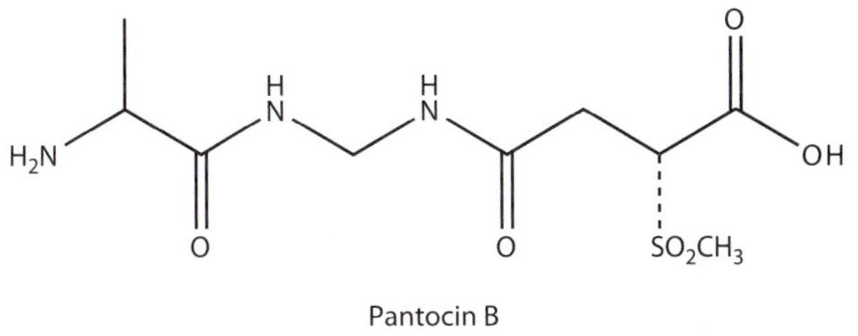

Pantocin B

11.61 Schematic

Thus, it is possible to use native fungi, which are simply applied at higher levels when needed. These need to be easy to grow in quantity and to remain viable after prolonged storage. The right formulation can enhance activity and improve consistency of treatment. They need to be safe for nontarget organisms.

It is also possible to use toxins isolated from fungi. The cyclic tetrapeptide, tentoxin, from *Alternaria* species causes severe chlorosis in many problem weeds without hurting maize or soybeans.[281] Another cyclic peptide (**11.62** Schematic) from *Alternaria alternate* kills the weed, *Centaurea maculosa*.[282] The toxin is host specific.

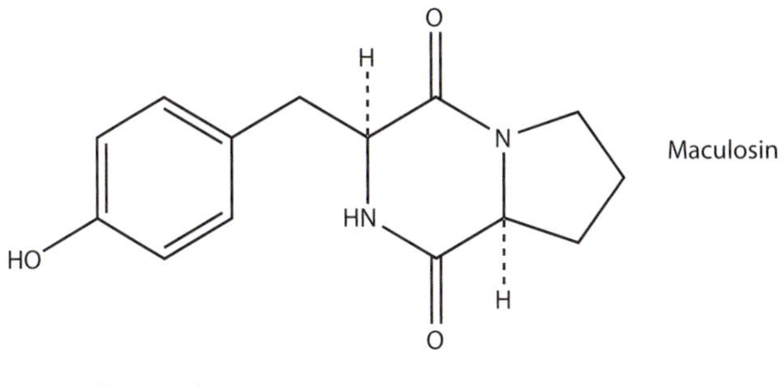

Maculosin

11.62 Schematic

Materials from higher plants can also be useful. Corn gluten meal is used as a pre-emergence herbicide.[283] Cutting Edge Formulations sells emulsions of DL-limomene that kill weeds as well as glyphosate and paraquat do.[284] EDEN Bioscience Corporation markets a hairpin protein developed at Cornell University that triggers a plant's natural defense system against pests and disease.[285]

Algae can be controlled with ultrasound.[286]

11.3.2.3 Allelopathy

Many plants elaborate chemicals (allelochemicals) that interfere with the growth of others around them.[287] The black walnut tree is an example. It is possible for some of these materials to be of use in controlling weeds.[288] Genetic transfer of allelopathy to control weeds in crops has been proposed.[289] (Intercropping the allelopathic species with a tolerant crop is a possibility for weed control.) A plant growth inhibitor (**11.63** Schematic) from *Ailanthus altissima* inhibits the growth of several weeds.[290] (This tree is a weed in the eastern United States.)

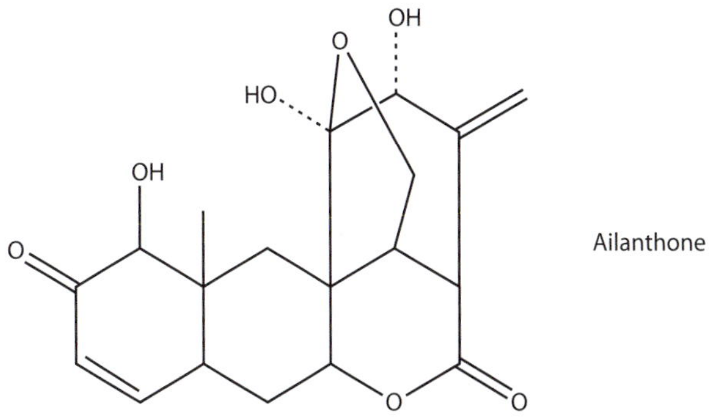

Ailanthone

11.63 Schematic

The volatiles from defatted seed meal of *Brassica napus* reduced the germination of lettuce seed from 95 to 10–15%.[291] Perhaps, the meal could be applied to a field to inhibit germination of weeds. An allelopathic substance from the rhizomes of *Elymus repens* inhibited the germination of alfalfa at a concentration of 10^{-6} g/mL.[292] A root culture of *Menispermum dauricum* produced germination stimulants[293] for *Striga hermonthica*, a root parasite weed.[294] Such stimulants might be used to cause the weed seed to germinate in the absence of the crop, so that the seedlings would die out. Many germination stimulants are not very stable in the field. Perhaps, they can be stabilized with antioxidants, light stabilizers, encapsulation, and such.

Antibiotics can also be used for weed control. Monsanto isolated one from a *Streptomyces* culture (**11.64** Schematic) that showed over 90% phytotoxicity to several broad-leaved weeds, without hurting wheat at a level of 7 g/acre.[295]

11.3.2.4 Killing Weeds with Light

Porphyrins are formed biosynthetically from δ-aminolevulinic acid (**11.65** Schematic). Application of this material can lead to porphyrins in the dark. When the plant is exposed to light, the excess porphyrins produce singlet oxygen that kills the tissues.[296] This method has some problems besides having to spray in the dark. There is no cheap chemical synthesis for the compound that gives high yields. This may change now that a method for converting waste paper to levulinic acid in 70% yield has been developed.[297] It has been made by microorganisms commercially in Japan.[298] The culture broth killed several dicotyledonous weeds. This may be cheap enough for commercial use. (δ-Aminolevulinic

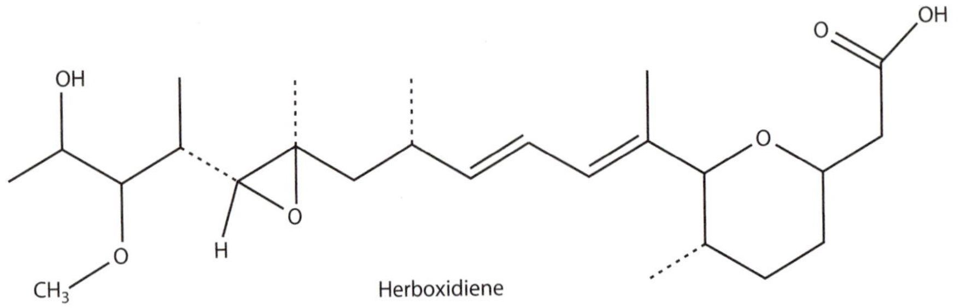

Herboxidiene

11.64 Schematic

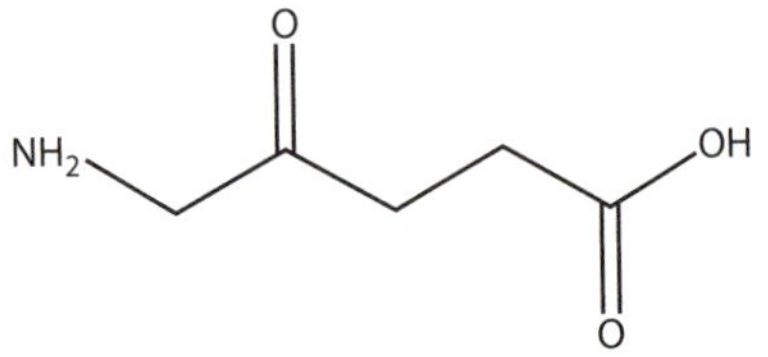

11.65 Schematic

acid and various porphyrins have also been used with light in the treatment of skin diseases.[299])

11.3.3 Control of Plant and Animal Pathogens

It is important to be able to control plant diseases caused by bacteria, fungi, and viruses in an environmentally sound way,[300] as far as possible without the use of chemical pesticides. The classic method has been to breed resistant crops using genetic material from resistant wild relatives. Now that the molecular genetics of plant disease resistance is starting to be understood and disease resistance is starting to be understood and disease resistance genes characterized,[301] it may be possible to use genetic engineering to develop resistant cultivars more quickly. The problem is that some breeders stopped breeding for disease resistance because they felt that cheap chemicals would take care of the disease. They are beginning to breed for resistance again now that blights have developed that chemical fungicides cannot handle. "Because late blight (of potatoes) had been controlled with chemicals since the middle of the 20th century, breeding for resistance to the disease was not a top priority in the United States."[302] An alternative program for control of fungal diseases of wheat and barley in the Pacific Northwest of the United States involves the use of resistant cultivars, seed treatments with fungicides, and planting dates that are least favorable for infection.[303] For root pathogens, crop rotations with nonhost crops can also help.

Bananas in the stores may be a thing of the past soon, because a blight that is not controlled by chemical fungicides is developing in Asia and may get to the plantations of Central and South America.[304] Global trade and travel have become more common and diseases can move around faster.[305] Because bananas have very few seeds, it is hard to breed them.

There is an effort to exploit microbe–microbe interactions to control plant pathogens.[306] The beneficial microbes may control the pathogens in a variety of ways.[307] These include competition for living space, competition for key nutrients, production of siderophores that capture nutrient metal ions more effectively, production of plant hormones, induction of resistance mechanisms in the host, production of antimicrobial substance, making cell-wall-degrading enzymes, and so on. Genetic engineering is being used in an effort to improve these beneficial microbes.[308] The problems in the use of beneficial microbes involve finding inexpensive methods of production[309] and storage as well as obtaining consistent results under field conditions. The latter means that the variables involved need to be better understood. The storage stability of beneficial fungi and bacteria has been improved from 6 weeks to 6 months at room temperature, and 2 years in the refrigerator, by adding a sugar to stabilize cell membranes.[310] Some biocontrol methods work better in greenhouses than in open fields simply because the environment in the greenhouse is easier to control. A few of the more promising examples will be given.

Pseudomonas fluorescens controls fire blight by outcompeting with it for nutrients.[311] The soil fungus *Trichoderma* sp. controls late blight of potatoes caused by another fungus, *Phytopthora infestans*, as well as some species of *Pythium* that cause damping off (a fungal disease).[312] The *Trichoderma* can be applied as a seed treatment. Another soil fungus, *Gliocladium virens*, controls damping off caused by *Pythium ultimum* and *Rhizoctonia solani*. Both of these beneficial fungi can be grown in fermenters. Seed treatment with a mixture of bacteria, *Enterobacter cloacae* and *Burkholderia cepacia*, prevented damping off caused by *Pythium* and *Fusarium* species.[313] The use of a methyl cellulose coating improves some of these seed treatments.[314] *Fusarium* wilt has also been prevented by mixed cropping *Allium fistulosum* with bottle gourd to provide the *Pseudomonas* bacterium that controlled the *Fusarium*.[315] When used on cotton, *Aspergillus flavus* strain AF36 does not produce aflatoxin and outcompetes other strains that do so that the seed can be used as animal feed.[316] AgraQuest has developed a strain of the soil bacterium *Bacillus subtilis* that controls a variety of fungal and bacterial diseases.[317] It colonizes the leaf surface and competes with the pathogen for space and nutrients, as well as preventing attachment physically. It also produces metabolites that help destroy the pathogen's germ tubes and membranes. The nematode *Steinerema riobrave* controls the curculio beetle on peaches.[318]

The fungus *Ampelomyces quisqualis* is marketed for the control of powdery mildew on grapes.[319] Mild strains of bacteria, fungi, and viruses can prevent disease by outcompeting more virulent strains.[320] A mild strain of a citrus virus is used in Brazil for this purpose.[321] This method is also used to prevent chestnut blight in Europe.[322] The American chestnuts were one-third of the trees of the oak–hickory–chestnut association in the eastern United States until an imported blight wiped them out, except for a few stump sprouts. The challenge is to find enough strains of the hypovirulent fungus to reintroduce the chestnut.

Lactobacillus bacteria can destroy gut pathogens in farm animals for faster growth.[323] This offers an alternative to the use of antibiotics. A mixture of 29 types of live bacteria is being marketed for the elimination of *Salmonella* from chickens.[324] These are sprayed as a mist over newly hatched chicks. A similar technique works on pigs.[325] Screening 25,000 isolates from the poultry production environment produced several species of bacteria whose bacteriocins (polypeptides) reduced *Campylobacter* to nearly undetectable levels in chickens.[326] Chickens fed xylanase grow faster and lose 99% of the *Campylobacter jejuni* in their guts.[327] A combination of six bacteriophages and *Gluconobacter asaii*, a bacterium normally found on apples and pears, reduced *Listeria monocytogenes* by 99.99% on melons.[328] These are three of the most important organisms in foodborne illness.[329] A brief burst of 290°F steam kills *Salmonella* on chickens.[330] This is an alternative to irradiation or treatment with bactericides.

Postharvest diseases can also be controlled with other organisms.[331] Naturally occurring *Pseudomonas* bacteria can control rotting of apples, citrus, and pears while they are in storage.[332] One possibility is to put the bacteria in a fruit coating. A combination of *Candida saitoana* modified chitosan and lysozyme is used in this way commercially.[333] 1-Methylcyclopropene in α-cyclodextrin is used commercially to delay ripening and to prevent mold on apples in cold storage.[334] Molding of kiwi fruit by *Botrytis* can be controlled by an extract of *Trichoderma* fungus that contains 6-pentyl-α-pyrone (**11.66** Schematic).[335] It is also effective against several fungi. Because it is expensive to make chemically, its growth on moistened ground maize is being studied, but yields are still low at 2 g/kg.[336]

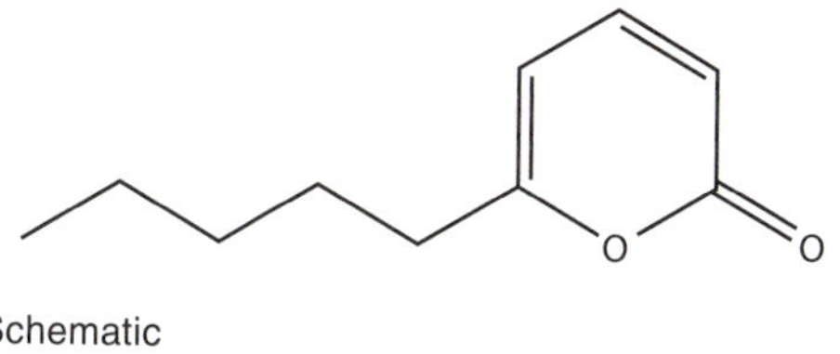

11.66 Schematic

The essential oil of *Salvia fructicosa*, which contains the terpenes 1,8-cineole, α- and β-thujone, and camphor, kills herpes simplex virus and several strains of bacteria.[337] Antimicrobial peptides of plants and animals[338] that resist infection of the host can now be produced in recombinant bacteria.[339]

Many fungicides are being isolated from a variety of natural sources. Some of them may be useful as such. More often, variations on their structure are made to enhance activity; to reduce toxicity to the crop plant; to increase their stability to hydrolysis, light, and enzymes; and to make their chemical syntheses commercially attractive. Strobilurin (**11.67** Schematic) is such a case. It was isolated from a wood-rotting mushroom. This lead has resulted in a whole new class of fungicides, the β-methoxyacrylates.[340] A related compound is found in another mushroom.

A variety of other fungicides with many different types of structures are being isolated from natural sources. Some of the structures are simple enough so that analogs could be made relatively easily. Others are complicated enough that total synthesis would be impractical on a commercial scale. In such cases, the hope is that simpler analogs of a part of the structure will be active. If not, the material will have to be isolated from natural sources, perhaps with some synthetic modification later. Microorganisms can often be grown in large fermenters, providing that a proper medium can be found. *Coniothyrium minitas* for biocontrol of *Sclerotinia sclerotiorum* has been produced by solid-state fermentation.[341]

A few examples can illustrate the wide diversity of active compounds. An antibiotic from a culture of *Streptomyces flaveus* (**11.68** Schematic) showed strong activity against four fungi with little toxicity to the host plants.[342] Polycyclopropanes (**11.69** Schematic) from *Streptomyces* fermentation broths are active against filamentous fungi.[343] A diketone (**11.70** Schematic), from the fungus *Trichoderma viride*, is a fungal growth inhibitor for the biocontrol of soilborne pathogens.[344]

The rice blast fungus, *Magnaporthe grisea*, uses a special cell for pathogenesis. A polypeptide from the yeast, *Saccharomyces cerevisiae*, blocks the formation of this cell and prevents infection by some strains of the fungus.[345] The authors plan to make a variety of analogs, perhaps by combinatorial chemistry, to test on this and other strains. Chitinase has also been used to control fungi.[346] Germinating seeds can produce compounds that protect the seeds from pathogens. Malted barley is active against plant pathogenic fungi and bacteria.[347] Milk has been used to control powdery mildew on squash and cucumbers in Brazil.[348] These should be much less expensive and easier to produce than many of the other compounds mentioned in the foregoing.

Extracts of the heartwood of western red cedar killed spores and inhibited the growth of *Phytopthora ramorum*, the cause of sudden oak death in California.[349] The wood (*Thuja dolabrata*) in a 1200-year-old pagoda in Japan shows no signs of rot or termites.[350] The most active compound in the wood against such invading organisms is carvacrol (**11.71** Schematic).

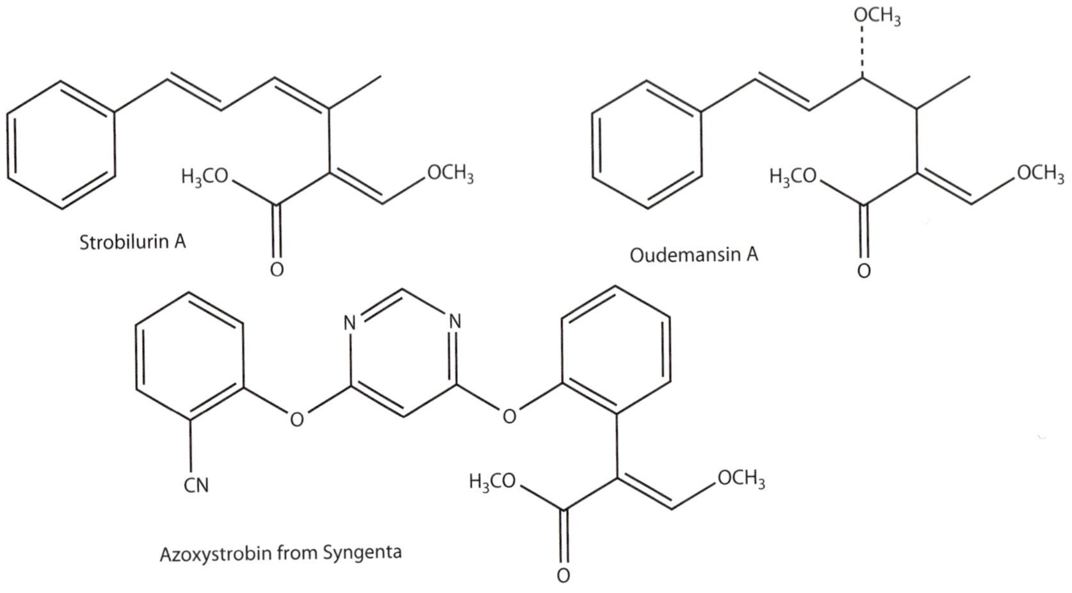
OCH3
H3CO
OCH3
Strobilurin A
O
H3CO
OCH3
Oudemansin A
O
N
N
O
O
CN
H3CO
OCH3
Azoxystrobin from Syngenta
O

11.67 Schematic

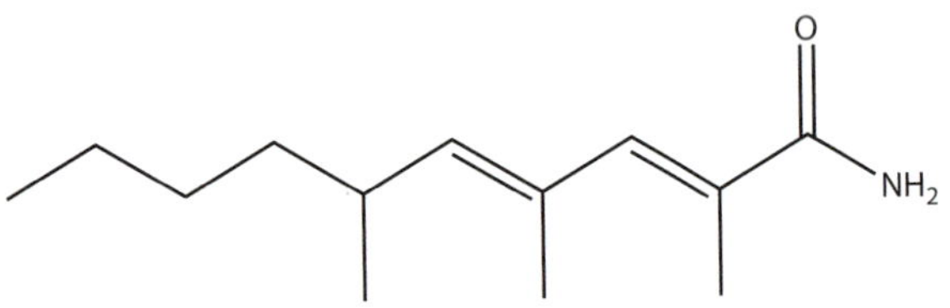
O
NH2

11.68 Schematic

It is possible that this or a similar compound could be used to replace some of the treatments now used on wood. Carvacrol is fairly volatile, bp 237°C. It would be interesting to see how fast it would be released from wood treated with it under pressure. If it is released too fast for purposes of preservation, the wood could be impregnated with a monomer derived from carvacrol that might then be polymerized in the wood by heat or irradiation. An acrylate from carvacrol is one possibility. An alternative would be to make such a

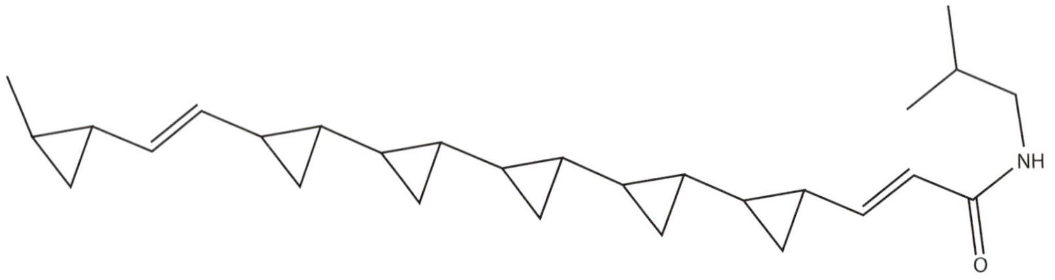
NH
O

11.69 Schematic

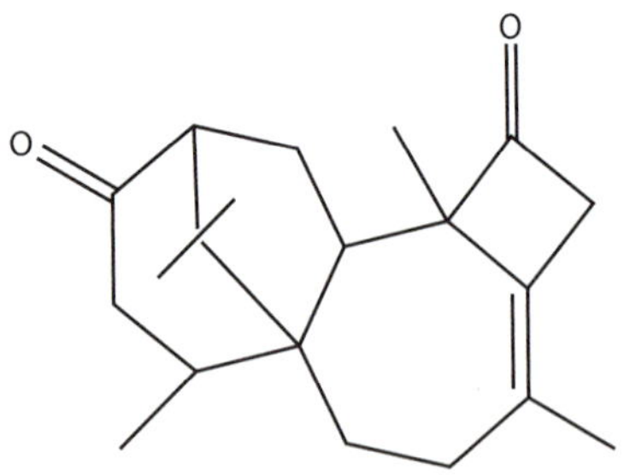
O
O

11.70 Schematic

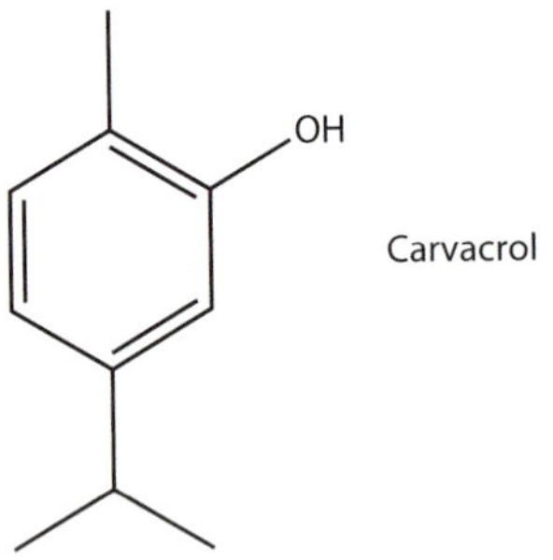
OH
Carvacrol

11.71 Schematic

polymer and then impregnate the wood with it. The idea is to obtain slow release by hydrolysis or by some action of the invading organism. Another possibility would be to use a carvacrol-impermeable sealer on the impregnated wood. A disadvantage is that any new surface, such as a saw cut, might have to be sealed anew. If wood preservatives are considered as pesticides, they are more than one-third of the total used.[351] The use of steel or concrete utility poles has been suggested as a way of reducing this usage.

11.3.4 Fertilizers

There is a global nitrogen overload resulting from fertilizer use, fossil fuel emissions, and the growing of nitrogen-fixing crops, such as soybeans.[352] The world uses more than 140 million tons of fertilizer each year, with maize taking the most, followed by wheat and soybeans.[353] The United States used 33 million tons in 2007.[354] About 20% of this leaves the watershed in the rivers. To this must be added the nitrogen and phosphorus that wash off or leach through the soil when animal waste[355] is applied to the land in amounts too large for the crop to assimilate.[356] For example, the nitrogen input to the Baltic Sea has quadrupled since 1900. The result has been algal blooms with subsequent dead zones in coastal areas worldwide, including the Adriatic, Baltic, and Mediterranean Seas, Hong Kong, Japan, the Gulf of Mexico, and Chesapeake Bay in the United States. The third largest, the one in the Gulf of Mexico, is the size of the state of New Jersey and is due to 1.8 million metric tons of nitrogen that comes down the Mississippi River each year, three times the amount that came down 40 years ago.

Experiments with irrigated wheat in Mexico demonstrated that with less fertilizer and better timing (one-third at the time of planting and two-thirds 6 weeks later), the yield and quality were just as substantial and after-tax profits were 12–17% more.[357] The use of legumes as cover crops and intercrops to put nitrogen into the soil was discussed earlier. If practiced widely, this would greatly reduce the need for synthetic nitrogenous fertilizers. The example of intercropping beans and maize in Honduras that was mentioned earlier is especially instructive, because maize accounts for about 45% of fertilizer consumption in the United States.[358] In some cases, inoculation of the legumes with nitrogen-fixing bacteria can help.[359] The recycling of all organic materials on the farm will help prevent loss of nutrients. It will also help cut erosion and soil loss. Phosphorous is often lost with the soil that is lost by soil erosion. Municipal organic wastes, including treated sewage sludge, as well as agricultural wastes, can also be used to build up the organic matter in the soil and add nutrients to it.[360] More than 8000 farms in the United States are now composting dead animals, manure, and crop residues.[361] This also increases the water infiltration rate and water-holding capacity of the soil. As long as chemical insecticides that harm them are not applied, *Lumbricus terrestris* earthworms can help incorporate the organic matter into the soil.[362] The worms feed by dragging surface matter into their burrows where nitrogen compounds can be released into the soil. Water infiltration, root penetration, and air exchange are also helped by this process.

Outbreaks of the microorganism, *Pfiesteria piscicida*, in Maryland and North Carolina, which caused fish kills and some human illness, have been linked to the nutrient pollution from hog and chicken farms, municipal sewage, and aquaculture impoundments.[363] Several states have enacted legislation to control pollution from animal waste and more has been proposed for the 450,000 feed lots in the United States.[364] The Netherlands has the most comprehensive legislation of this type. The main barrier to the wider use of such wastes, as well as composted waste of swine and chickens, is the cost of transportation. This, apparently, is not a problem in the home-gardening market. Composted sewage sludge from Milwaukee, Wisconsin (Milorganite), and composted cow manure from Vermont (Moo Doo) are available in a local gardening store in Newark, Delaware. Other uses that are technically feasible, but may not be economic unless a dollar value is assigned to the harm caused by the nutrient pollution, include combustion to produce electricity, production of biogas, algae, duckweed, single-cell protein, and mushrooms.[365] Biogas produced from poultry waste could be used to heat and cool the chicken houses. Duckweed, single-cell protein, and fly maggots raised on the waste could be fed back to the chickens. The use of low phytate corn or adding phytase to the feed can reduce the phosphorus in the waste by about 38%.[366] The use of aluminum sulfate to reduce the phosphorus runoff from the land by 70%[367] may result in an undesirable buildup of aluminum and acidification of the soil.

Pig manure in North Carolina has been treated to remove 99.7% biological oxygen demand, 98.7% nitrogen, 95% phosphorus, and 99% zinc.[368] The calcium phosphate produced is used as a fertilizer. Composted rock phosphate is said to be as good as superphosphate.[369] This avoids the need for sulfuric acid in making phosphate fertilizer. The need for phosphate fertilizers can be reduced by the use of symbiotic fungi (mycorrhizae) that live in or attached to the roots or plants. They are essential for good growth of many native trees and herbaceous plants. It may be helpful to use them more on crop plants. They will grow as long as no fungicides are applied to the crop. Among the companies selling them is Plant Health Care of Pittsburgh, Pennsylvania. The hunt for improved strains is ongoing.

A different approach to using less fertilizer and fewer pesticides is to shift to different crops. Industrial hemp requires no pesticides and little fertilizer compared with cotton, which is typically grown with lots of pesticides.[370] "If you smoke industrial hemp, all you'll get is a headache." Hemp fiber has been used to make cloth for hundreds of years.

11.4 LAWNS

Lawns are given a special place in this discussion because of the many problems associated with their ubiquitous use.

Lawns have come into widespread acceptance since the invention of the lawn mower in 1830.[371] There are 40 million acres of lawn in the United States cared for at a cost of $10.4 billion/year.[372] They require about 27,000 gallons of water per acre each week to keep them green in the summer. This use of water could be avoided if the exotic cool season grasses that make up the lawn were allowed to go dormant (and brown) as they do naturally. In Novato, California, in water-short Marin County, just north of San Francisco, California, where 33% of the water supply goes on to lawns, the water district will pay up to 310 dollars per home to rip out the lawn and replace it with plants adapted to the semi-arid climate.[373]

The average lawn receives 5–10 lb of pesticides per acre per year. This is the heaviest pesticide application of any land area in the United States. Off the 285 million lb of pesticides used annually in the United States, 67 million lb are applied to lawns.[374] In New Jersey, more pesticides are used on lawns and golf courses than for agriculture.[375] More than 2 million lb are applied to lawns in the state each year. More fungicides are used on turfgrass each year in the United States than on any other crop.[376] Golf courses in New York and Japan use seven to eight times as much pesticide per acre (18 lb/year on Long Island, New York) as nearby agricultural land.[377] (The pollution from golf courses apparently varies with the locality and the methods of management used, being higher than agricultural land in some cases,[378] and lower in others.[379])

There are 61 million lawn mowers in the United States, most of them powered by gasoline.[380] For a unit of fuel, they emit 50 times as much air pollution as an automobile. Very few people use hand-powered mowers, which cause little or no air and noise pollution. Municipal solid waste contains about 18% of yard waste (see Chapter 14). In the summer, most of this consists of grass clippings and, in the fall, leaves raked off the lawn to keep the grass from smothering. It is better to let the clippings stay on the lawn and to put the leaves in a compost pile, either in the back yard, a neighborhood pile, or a town pile. The nutrients can then be reused. A modified gibberellin has been found that reduces the growth of grass by threefold, so that mowing can be less frequent.[381] It is not yet on the market.

In 1995 a popular four-step lawn care program from Scott used the following:

- Step 1: 1.51 lb/acre of pre-emergence herbicide (pendimethalin).
- Step 2: herbicides for broad-leaved plants (1.51 lb/acre 2,4-dichlorophenoxyaceic acid plus 1.51 lb/ acre 2(2-methyl-4-chlorophenoxy)propionic acid).
- Step 3: 0.461 lb/acre insecticide (diazinon).
- Step 4: no pesticide.

Each step included fertilizer (30–3–10 in step 1, 32–4–3 in step 2, 28–6–4 in step 3, and 32–3–10 in step 4 [at 123 lb/acre]). (The numbers refer to the amounts of nitrogen, phosphorus, and potassium in fertilizers.) Part of the nitrogen in the fertilizer was available to plants immediately (ammonium dihydrogenphosphate) and part of it after slow release (urea plus urea–formaldehyde reaction products). The high nitrogen content of the fertilizer makes the grass grow rapidly and would not be used on an agricultural crop. In 2008, the product had 2,4-dichlorophenoxyacetic acid, 1-(3-chlorophenyl)piperazine (mecoprop), and 2-methoxy-3,6-dichlorolbenzoic acid (dicamba) in step 2. In step 4, diazinon had been replaced by bifenthrin (a synthetic pyrethrin).

The use of clover in lawns to fix nitrogen was common until, as one author puts it, a turf company declared clover to be a weed and sold a chemical to kill it.[382] Lawns of bent, foxtail, and fescue grasses need less potassium and have fewer dandelions, because both the dandelions and Kentucky blue grass require relatively large amounts of potassium.[383] (Efforts are underway to see if this principle can be applied to weeds in other crops.) Scott also markets moss control granules for lawns containing a mixture of iron(II), magnesium, and potassium sulfates to be applied at 122 lb/acre. Diazinon is a cholinesterase inhibitor that is toxic to birds and earthworms. As a result of a number of bird killings, its use on lawns and golf courses has been banned.

The biggest problem is that the materials applied to the lawn do not necessarily stay there when it rains or the lawn is overwatered.[384] They can go into groundwater or into streams. The result can be loss of aquatic vegetation and animal life in streams and lakes from the herbicides and insecticides, as well as eutrophication, which often involves growth of undesirable species, from the fertilizer. The result may well be a loss in biodiversity. Biodiversity can also be lost if native wild flowers and shrubs are replaced by lawn, as often happens. (The lawn is a biological desert of a few species of Eurasian grasses.[385])

Lawns prevent soil erosion, and serve as a fire break and as a place to play. There is not much playing on lawns. Where there are fine lawns, the children usually play on the driveway or in the street. People usually sit on their deck instead of on the lawn. Lawns are largely a matter of custom. Although they can be made less harmful by the methods of alternative agriculture discussed in the foregoing, the questions should be: Are they needed? Can the same objectives be accomplished in other less harmful ways?[386]

The first step is to keep the amount of lawn to a minimum using species and cultivars resistant to drought and pathogens. Alternatives to lawns include (a) ground covers, such as species of *Vinca*, *Pachysandra*, *Hedera*, or others; (b) wildflower meadows; (c) mixtures of flowering shrubs, herbaceous perennials, and annuals with paths of gravel, bark, brick, flagstone, or other materials in them; (d) the natural leaf litter and flowers of forest floor, (e) moss, where applicable, and (f) xerophytic (i.e., dry) plantings of cacti, yuccas, and such.[387] Meadow plantings offer lower maintenance, lower long-term costs, colorful flowers, and a home for beneficial insects, birds, and butterflies.[387] They do not have to be watered or fertilized. Some corporations are now using such meadows in lieu of lawns to help their bottom lines.

11.5 GENETIC ENGINEERING

Genetic engineering offers great potential for improving crops (as mentioned in Chapter 9).[388] The goals include resistance to drought, disease, frost, and salt; less need for fertilizer, higher protein content; and higher levels of limiting amino acids in proteins. It can often produce the desired plant in less time than conventional breeding and achieve results not available by conventional breeding, such as producing 12-carbon fatty acid derivatives in rape. Strategies to improve the resistance of plants to bacterial diseases by genetic engineering include producing antibacterial proteins, inhibiting bacterial pathogenicity or virulence factors, enhancing natural plant defenses, inducing cell death at the site of infection, and enhancing production of elicitors for disease response.[389] Modification of the DNA of nitrogen-fixing *Rhizobium* has made it more competitive in nodule formation in legumes.[390] However, the major uses today of imparting resistance to insects and herbicides have been questioned.[391]

Putting foreign genes in crops can outwit insects and fungi. Losses of stored peas to weevils have been reduced by putting in the gene for an α-amylase inhibitor from a common bean.[392] Putting a chitinase gene into rape made it resistant to three fungal pathogens.[393] Growth of insect larvae was suppressed when a chitinase gene was put into rice and tobacco plants.[394] Transgenic rice with an introduced protease inhibitor from potato was resistant to the pink stem borer.[395] This inhibitor is destroyed when the potato is cooked so that no one will eat it. Tobacco plants were made more resistant to soilborne fungi by the addition of an antiviral protein from pokeweed (*Phytolacca americana*).[396] When mosquitoes eat *Chlorella* that has been engineered to contain a gene for a mosquito digestive control hormone that prevents the digestion of food, they die.[397] The freeze-dried alga can be stored for long periods. Clearly, the technique works. The problem is one of the invading organisms developing resistance.

There are some alternatives to plants containing *Bacillus thuringiensis* for reducing the damage to maize by insects. The fungus *Beauveria bassiana* can live in the maize plant and not hurt it, while at the same time being lethal to maize borers.[398] Maize rootworms (which mature to beetles) are attracted to baits containing a curcurbitacin E glycoside.[399] The baits can contain a red dye that kills the worms when photoactivated by sunlight. Crop rotation also helps, although some insects are becoming resistant to it.[400]

A tomato was engineered to be more tolerant to salt.[401] Plants have also been engineered to be resistant to drought.[402] Genetically engineered flood-tolerant rice can be flooded to kill weeds in a simple, cheap, and organic way. One suggestion is to combine these with organic farming.[403]

Soybeans have been genetically engineered to produce so little polyunsaturated oil that hydrogenation is unnecessary and no *trans* fats are made.[404] Alternatively, the oil from one USDA soybean does not need it either.[405] The pigeon pea is the main source of protein for 1 billion people.[406] Conventional breeding over 30 years produced plants that matured in 3 months with 50% higher yields. The hybrid plants require neither nitrogen fertilizer nor irrigation.

Gene silencing by RNA interference in transgenic plants is being studied for the control of insects in cases where the present methods do not work well.[407] It needs further study under real world conditions.

Critics feel that genetically modified crops merely reinforce the farmer's reliance on pesticides. They would rather see integrated weed management involving crop rotation, choosing crops that outcompete weeds, intercropping, planting at higher densities, and other mechanical and cultural practices described earlier in this chapter.[408]

11.6 INTEGRATED PEST MANAGEMENT

Integrated pest management[409] requires an understanding of the ecology of the crop, the pest, and other plants and animals in the system. It relies heavily on monitoring the pest population. Ecologists and entomologists may be used as consultants. When the level of the pest passes an economic threshold[410] that cannot be controlled by mechanical and cultural means, additional special steps must be taken. These may include augmenting the supply of predatory insects, nematodes, fungi, or other organisms. Spraying is a last resort. Then, the least harmful and most species-specific material is used in spot spraying of areas where the pest population is high. *Bacillus thuringiensis* and other biopesticides[411] are often used in place of chemical insecticides.

There have been some notable successes. In 1986, the brown leafhopper had become a major pest of rice in Indonesia. Before the use of insecticides began, it had been a secondary pest. The government banned 57 out of 66 insecticides for rice and removed subsidies from the rest. It gave 10–12-week training courses in integrated pest management to 200,000 farmers. As a result, pesticide applications have been cut to 18% or their earlier amount, rice yields have gone up 15%, and the government has saved 1 billion dollars.[412] The use of pesticides on rice in California's Central Valley has been reduced by 99% since 1985.[413] It is possible to grow cotton without pesticides and without a transgenic variety. Tiny parasitic wasps and lacewing predators are used along with pheromone traps.[414] Organically grown cotton sells for at least 40% more than regular cotton. The amount of pesticides used on cotton in Texas has been reduced by 90% since 1966.[415] In Massachusetts, integrated pest management has reduced pesticide use on apples, strawberries, cranberries, maize, and potatoes by as much as 40–60%.[416] It has been estimated that alternative agriculture in Missouri could reduce soil loss by 70%, energy use by 22%, herbicide use by up to 40%, nitrogen fertilizer use by up to 30%, and production costs by 17%.[417] The U.S. National Arboretum in Washington, District of Columbia, has used monitoring and spot spraying to reduce its pesticide volume by three-fourths since 1992.[418] It has also shifted to more horticultural oils, insecticidal soaps, and insect growth regulators as alternatives to chemical pesticides. Growers of apples and pears in the Pacific Northwest are relying more and more on pheromone mating disruption, sterile male and parasite releases, plus spot spraying to control the codling moth.[419] Almond and walnut growers in California have eliminated pesticides by planting cover crops that attract beneficial predators.[420] Area-wide integrated pest management in the United States in 1997 included 7700 acres of apples and pears in the Pacific Northwest, 40,000 acres for the maize rootworm in the Corn Belt, 46,000 acres for leafy spurge, and 200,000 acres for the cotton bollworm and tobacco budworm in Mississippi.[421]

An interesting experiment on low-input agriculture is happening in Cuba today.[422] Since Russian aid ended in 1991, the inputs of fertilizer, pesticides, and petroleum have dropped by more than 50%. Its pest-monitoring system is said to be among the most comprehensive in the world. There are many stations where mass rearing of beneficial insects is done. Home gardens have proliferated. It is still too early to tell how well this will work out, but in 2008 60% of Cuba's food was imported.[423] This may have something to do with Cuba's system of government.

Low-input, sustainable agriculture sometimes results in reduced yields per acre, with 7–15% declines being common.[424] This is offset, at least partially, by reduced costs for fertilizers and pesticides. Low-input systems reduce risk and variability in profits. A 22-year study in Pennsylvania of farming without synthetic fertilizers and pesticides using (a) legumes, grass, and cattle; and (b) maize and soybeans with all crop residues and manure returned to the soil found that yields and profits from maize were comparable with those of a conventional system.[425] In years of drought, yields from the organic farm were 20–40% higher.[426] There was greater soil retention of carbon and nitrogen, together with a 50% reduction of energy use. The additional carbon in the soil led to easier infiltration and retention of water. This is one way to sequester carbon to moderate global warming.[427] Organic farming led to three times as many earthworms, twice as many insects, 40% more mycorrhizal fungi, and 84% more bat foraging.[428]

Organic farming[429] often caters to customers who will pay more for a product. It has grown more than 20% a year for the last 6 years.[430] As mentioned in Chapter 1, food scientists agree that there is no need to buy organic produce in an effort to avoid pesticide residues. Only in a few cases has the produce been tastier or more nutritious.[431] Some organic apples were sweeter and less tart.[432] Organically grown kiwis contained more vitamin C and minerals.[433] Organic peppers grown in Spain contained more vitamin C and carotenoids.[434] Significantly, 63% of organic farmers are conducting their own on-farm research. In one area of California, the top yield of rice is 100 bags of 100 lb each per acre. The organic farmer obtained only 85 bags, but sold his rice for twice as much.[435] In the Pacific Northwest, integrated pest management has led to higher profits and lower economic risk.[436] Ontario organic dairy farms were 60% more profitable than conventional ones.[437] In India, there was no difference in income between conventional and sustainable agriculture.[438] In developing countries, such as those in Africa, where labor is cheaper than fertilizer and pesticides, organic farming can be profitable and can lead to 20–90% increases in production.[439]

A 21-year study of organic fruit production in Switzerland found that it gave a 20% lower yield than conventional methods.[440] The trial used 56% less energy, had 40% more roots with mycorrhizae, three times as many earthworms, twice as many spiders, and used 97% less conventional pesticides. An organic vegetable-livestock farm in Sweden operated since 1982 sells its cattle and pigs at a 20% premium.[441] In contrast to the lower yields usually found in organic farming, organic pecans outyielded conventional ones at 44 lb compared to 26 lb/ acre.[442] This may have happened due to the poultry litter, compost, rock minerals, and mycorrrhizal fungi used in the organic plots. This suggests that if organic farmers added a little conventional fertilizer to compensate for the loss of nutrients with removal of the crop, yields of the two methods would be comparable. Comparison of 18 tomato farms in California, half conventional and half organic, showed that pesticides offered no economic advantage.[443]

The USDA has set standards for organic food. The "USDA Organic" label means that the food has been produced without the use of irradiation, pesticides, genetically modified crops, sewage sludge, synthetic fertilizer, antibiotics, and growth hormones.[444] Setting the standards involved considerable wrangling since agribusiness wanted laxer standards so that it could capture part of the fast-growing market for organic foods.

The current system of farm subsidies in the United States is often mentioned as a barrier to low-input agriculture.[445] To remain in a price-support system, a farmer may have to continue a monoculture with heavy inputs of fertilizers and pesticides and without crop rotation. Farm subsidies in the United States are $8.5 billion/year.[446] Conventional economics assigns no cost to soil erosion, water pollution, loss of biodiversity, loss of natural pollinators, and global warming.[447] If these costs were included, sustainable agriculture would come out ahead of conventional agriculture.

In some cases, misconceptions led to the unnecessary use of pesticides. Cotton farmers in the Imperial Valley of California saw *Lygus hesperus* on the cotton and sprayed, even though work at the nearby experimental station showed the insects did not lower the yield of cotton.[448] By ignoring this and other recommendations based on local research, the industry almost collapsed before it was revived by the adoption of integrated pest management. Vietnamese rice farmers sprayed leaf-folders until they were shown by demonstration plots that the insects made no difference on the yield.[449]

Consumer expectations may be unrealistically high. Of the pesticides used on oranges, 60–80% are just to improve the cosmetic appearance of the skin.[450] Homeowners may grab the spray can instead of picking an insect off a plant or just pulling up a weed by hand. People who are fearful of pesticide residues in their food may still use pesticides freely in their yards and in their homes.

Large companies have been slow to enter the niche markets required by integrated pest management. They prefer broad-spectrum products with large volumes, which are patentable and provide some effective advantage over the competition. The products must have acceptable toxicological and environmental effects.[451] They feel that the costs of developing and registering a biopesticide that is highly specific cannot be recouped in the marketplace. Thus, government agricultural experimental stations may have to develop a product and then hunt for a small company to produce and distribute it, or handle the whole operation themselves. Unfortunately, funding for such programs is declining.[452] Progressive chemical companies might look at the business more broadly and develop an agricultural service for farmers that involved monitoring services, beneficial insects when needed, pheromones, or other services together with ecologists and entomologists to help the farmer. Several large companies are already in the business of supplying a wide variety of small-volume products to customers (e.g., dyes and pigments, flavors and fragrances, stabilizers for plastics, and others). Studies have shown that small companies tend to be more innovative than larger ones.

REFERENCES

1. (a) G.W.A. Milne, ed., *CRC Handbook of Pesticides*, CRC Press, Boca Raton, FL, 1995; (b) L.G. Copping and H.G. Hewitt, *Chemistry and Mode of Action of Crop Protection Agents.*, Royal Society of Chemistry, Cambridge, 1998; (c) H. Ohkawa, H. Mayagawa, and P.W. Lee, *Pesticide Chemistry—Crop Protection, Public Health and Environmental Safety*, Willey-VCH, Weinheim, 2007; (d) F. Muller, *Agrochemicals, Composition, Production, Toxicology and Applications*, Wiley, New York, 2000; (e) National Research Council, *The Future Role of Pesticides in U.S. Agriculture*, National Academy Press, Washington, DC, 2001; (f) D. Baker and N.K. Umetsu, eds, *The Agrochemical Discovery: Insect, Weed and Fungal Control*, Oxford University Press, Oxford, 201; (g) N.N. Ragsdale and J.N. Seiber, *Pesticides: Managing Risks and Optimizing Benefits*, A.C.S. Symp. 734, Oxford University Press, New York, 1999; (h) W. Kramer and U. Schirmer, eds, *Modern Crop Protection Compounds*, Wiley-VCH, Weinheim, 2007; (i) G. Voss and G. Ramos, eds, *Chemistry of Crop Protection*, Wiley-VCH, Weinheim, 2002.
2. P.C. Kearney, D.R. Shelton, and W.C. Koskinen, *Kirk–Othmer Encyclopedia of Chemical Technology*, 4th ed., Wiley, New York, 1997, *22*, 419.
3. D. Rich, *On Earth* (Natural Resources Defense Council), 2006, *28*(1), 6.
4. B. Burdett, *Chem. Ind. (Lond.)*, 1996, 882.
5. (a) *Anon.*, *Chem. Ind. (Lond.)*, 1996, 440; (b) *Anon.*, *Chem. Eng. News*, June 3, 1996, 21.
6. *Anon.*, *Environ. Sci. Technol.*, 1994, *28*, 355A.
7. (a) P.J. Chenier, *Survey of Industrial Chemistry*, 2nd ed., VCH, Weinheim, 1992; (b) R.L. Metcalf, *Kirk–Othmer Encyclopedia of Chemical Technology*, 4th ed., Wiley, New York, 1995, *14*, 524.
8. (a) J.E. Casida and G.B. Quistad, *Pyrethrum Flowers: Production, Chemistry, Toxicology and Uses*, Oxford University Press, Oxford, 1995.
9. National Research Council Board on Agriculture, *Ecologically Based Pest Management: New Solutions for a New Century*, National Academy Press, Washington, DC, 1996, 47.
10. *Anon.*, *Chem. Eng. News*, June 23, 1997, 23.
11. (a) N. Lubick, *Environ. Sci. Technol.*, 2007, *41*, 6323; (b) J.-F. Tremblay, *Chem. Eng. News*, May 29, 2006, 19.
12. (a) G.G. Briggs, ed., *Advances in the Chemistry of Insect Control III*, Royal Society of Chemistry, Cambridge, 1994; (b) G.A. Best and A.D. Ruthven, *Pesticides—Developments, Impacts and Controls*, Royal Society of Chemistry, Cambridge, 1995; (c) N.N. Ragsdale, P.C. Kearney, and J.R. Plimmerm, eds, *Eighth International Congress of Pesticide Chemistry: Options 2000*, American Chemical Society, Washington, DC, 1995.
13. National Research Council Board on Agriculture, *Ecologically Based Pest Management: New Solutions for a New Century*, National Academy Press, Washington, DC, 1996, 47.
14. J.M. Bradow, C.P. Dionigi, R.M. Johnson, and S. Wojkowski, *Kirk–Othmer Encyclopedia of Chemical Technology*, 4th ed., Wiley, New York, 1995, *13*, 73.
15. J.E. Franz, M.K. Mao, and J.A. Sikorski, eds, *Glyphosate: A Unique Glboal Herbicide*, A.C.S. Monogr 189, American Chemical Society, Washington, DC, 1997.
16. C.J. Baker, K.E. Saxton, and W.R. Ritchie, *No-Tillage Seeding—Science and Practice*, CAB International, Wallingford, CT, 1996.
17. (a) *Anon.*, *Environ. Sci. Technol.*, 1996, *30*, 425A; (b) G. Dinelli, R. Vicari, A. Bonetti, and P. Catizone, *J. Agric. Food Chem.*, 1997, *45*, 1940.
18. (a) H.G. Jackson, *Chemtech*, 1995, *25*(1), 48; (b) S.J. Ainsworth, *Chem. Eng. News*, Apr. 29, 1996, 35; (c) S.J. Stout, A.R. daCunha, G.L. Picard, and M.M. Safarpour, *J. Agric. Food Chem.*, 1996, *44*, 2182.
19. Anon., *Environ. Sci. Technol.*, 1995, *29*, 67A.
20. (a) B.A. Dreikorn and W.J. Owen, *Kirk–Othmer Encyclopedia of Chemical Technology*, 4th ed., Wiley, New York, 12, 204; (b) D. Hutson and J. Miyamoto, *Fungicidal Activity—Chemical and Biological Approaches to Plant Protection*, Wiley, New York, 1998; (c), L. Zirngibl, *Antifungal Azoles: A Comprehensive Survey of Their Structures and Properties*, Wiley-VCH, Weinheim, 1998.
21. G. Hoffmeister, *Kirk–Othmer Encyclopedia of Chemical Technology*, 4th ed., Wiley, New York, 1993, *10*, 433.
22. (a) E.-R. Kenawy and J. Macromol, *Sci. Rev. Macromol. Chem.*, 1998, *C38*, 365; (b) Anon., *Green Chem.*, 1999, *1*, G46.

23. (a) R.E. Hester and R.M. Harrison, eds, *Agricultural Chemicals and the Environment*, Royal Society of Chemistry, Cambridge, 1996; (b) P.A. Matson, W.J. Parton, A.G. Power, and M.J. Swift, *Science*, 1997, *277*, 504j; (c) D. Pimentel, ed., *CRC Handbook of Pest Management in Agriculture*, 2nd ed., CRC Press, Boca Raton, FL, 1990; (d) Hough, *The Global Politics of Pesticides*, Earthscan, London, 1999; (e) N.N. Ragsdale and J.N. Seiber, eds, *Pesticides: Managing Risks and Optimizing Benefits*, A.C.S. Symp. 734, Oxford University Press, New York, 1999; (f) A. Trewavas, *Chem. Ind. (Lond.)*, Jan. 19, 2004, 12; (g) T. Marrs and B. Balantyne, eds, *Pesticide Toxicology and International Regulation*, Willey, Chichester, 2004; (h) A. Ravishankara, J.S. Daniel, and R.W. Portmann, *Science*, 2009, *326*, 56; (i) Anon., *Chem. Eng. News*, Sept. 21, 2009, 29.
24. (a) F. Flam, *Science*, 1994, *265*, 1032; (b) H.A. Wichman, R.R. Badgett, L.A. Scott, C.M. Boulianne, and J.J. Bull, *Science*, 1995, *285*, 422; (c) J.N. Thompson, *Science*, 1999, *284*, 2116.
25. (a) P.R. Ehrlich and A.H. Ehrlich, *Technol. Rev.*, 1997, *100*(1), 38; (b) S.B. Powles and J.A.M. Holtum, *Herbicide Resistance in Plants*, Lewis, Boca Raton, FL, 1994; (c) N. Holmes, J. Chen, and R. Rivera, *Amicus J.* (National Resources Defense Council), 2000, *21*(4), 22; (d) I. Denholm and M.S. Willianson, *Science*, 2002, *297*, 2222.; (e) J. Msarshall-Claark and I. Yamaguchi, *Agrochemical Resistance: Extent, Mechanism and Detection*, Oxford University Press, Oxford, 2002.
26. G.J. Chin and J. Marx, eds, *Science*, 1994, *264*, 360.
27. F.R. Hutner-Fujita, *Insect Viruses and Pest Management*, Wiley, New York, 1998.
28. (a) J. Wargo, *Our Children's Toxic Legacy*, 2nd ed, Yale University Press, New Haven, CT, 1998; (b) M.A. Kamrin, *Pesticide Profiles: Toxicity, Environmental Impact and Fate*, CRC Press, Boca Raton, FL, 1997; (c) A. Newman, *Environ. Sci. Technol.*, 1995, *29*, 324A; (d) J.J. Johnston, ed., *Pesticides and Wildlife*, A.C.S. Symp. 771, Oxford University Press, Oxford, 2000.
29. R. Carson, *Silent Spring*, Houghton–Mifflin, Boston, 1962.
30. M. Pollan, *The Omnivore's Dilemma: A Natural History of Four Meals*, Penguin Press, New York, 2006.
31. G. Gardner, In: L. Starke, ed., *State of the World 1996*, W.W. Norton, New York, 1996, 90.
32. B. Hileman, *Chem. Eng. News*, Mar. 4, 1996, 6.
33. Anon., *Chem. Eng. News*, Jan 6, 1997, 19; Jan 3, 2000, 17.
34. Anon., *Chem. Eng. News*, Jan. 3, 2000, 17.
35. S. Schubert, T. Shistar, and J. Feldman, *Pesticides and You*, 1996, 16(1–2), special insert pp. 13–16; *National Coalition Against the Misuse of Pesticides*, Washington, DC.
36. Anon., *Chem. Ind. (Lond.)*, 1995, 907.
37. J. Kaiser, *Science*, 1997, *276*, 201.
38. (a) J. Monk, *Chem. Ind. (Lond.)*, 1996, 108; (b) C. Shannon, *Chem. Ind. (Lond.)*, 1998, 1008; (c) M. Mar, *Chem. Ind. (Lond.)*, 1998, 408.
39. (a) D. Hanson, *Chem. Eng. News*, May 4, 1998, 38; (b) B. Hileman, *Chem. Eng. News*, Feb. 4, 2002, 23.
40. National Research Council Board on Agriculture, *Ecologically Based Pest Management: New Solutions for a New Century*, National Academy Press, Washington, DC, 1996, 40.
41. R. Broderick, *Univ. Minn. Alumni News*, Spring, 1996, *1*(1), 9.
42. (a) B. Hileman, Anon., *Chem. Eng. News*, Mar. 18, 1996, 23; (b) Anon., *Chem. Ind.*, 1996, 196.
43. A. King, *Chem. Ind. (Lond.)*, June 11, 2007, 9; (b) M. Murphy, *Chem. Ind. (Lond.)*, Apr. 7, 2008, 5; (c) P. Walter, *Chem. Ind. (Lond.)*, May 7, 2007, 5.
44. L. Helmuth, *Science*, 2000, *290*, 1068.
45. Anon., *Chem. Ind. (Lond.)*, June 23, 2008, 9.
46. T. Colburn, D. Dumanoski, and J.P. Myers, *Our Stolen Future*, Dutton, New York, 1996.
47. X. Zang, E.K. Fukuda, and J.D. Rosen, *J. Agric. Food Chem.*, 1997, *45*, 1758; 1998, *46*, 2206.
48. N.S. Tulve, P.A. Jones, M.G. Nishioka, R.C. Fortmann, C.W. Groghan, J.Y. Zhou, A. Fraser, C. Cave, and W. Friedman, *Environ. Sci. Tehnol.*, 2006, *40*, 6269.
49. Anon., *Chem. Eng. News*, Aug. 1, 2005, 26.
50. (a) Anon., *Chem. Eng. News*, Aug. 18, 2008, 26; B. Hileman, *Chem. Eng, News*, June 12, 2000, 11.
51. (a) Anon., *Chem. Eng. News*, Aug. 23, 2004, 7; (b) P.J. Phillips, S.W. Ator, and E.A. Nystrom, *Environ Sci. Technol.*, 2007, *41*, 4246.
52. (a) Anon., *Chem. Eng. News*, Jan. 5, 2009, 18; (b) Anon., *Chem. Eng. News*, Nov. 9, 2009, 36.
53. (a) B. Hileman, *Chem. Eng. News*, Dec. 16, 1996, 28; Jan. 27, 1997, 22; (b) Anon., *Chem. Eng. News*, Jan. 13, 1997, 16; (c) *Chem. Eng. News*, Mar. 17, 1997.
54. D. Hamilton, *Pesticide Residues in Food and Drinking Water*, Wiley, New York, 2001.
55. Anon., *Chem. Ind. (Lond.)*, June 21, 2004, 8.
56. B. Hileman, *Chem. Eng. News*, Feb. 27, 1995, 8.
57. Anon., *Chem. Ind. (Lond.)*, 1998, 637.
58. (a) C. O'Driscoll, *Chem. Ind. (Lond.)*, Sept. l8, 2008, 10; (b) C. Holden, *Science*, 2006, *311*, 587.
59. (a) B. Hileman, *Chem. Eng. News*, Mar. 29, 1999, 22; (b) J.J. LaClair, J.A. Bantle, and J. Dumont, *Environ. Sci. Technol.*, 1998, *32*, 1453; (c) B. Hileman, *Chem. Eng. News*, Apr. 22, 2002. 9; (d) R.A. Relyea, *Ecol. Applications*, 2005, *15*, 618.
60. D. MacKenzie, *New Sci.*, Mar. 28, 1998, 13.
61. (a) S.L. Buchmann and G.P. Nabhan, *Forgotten Pollinators*, Island Press, Washington, DC, 1996; (b) G.P. Nabhan and S.L. Buchmann, Pollination services: biodiversity's direct link to world food stability, In: G. Daly, ed., *Ecosystem Services*, Island Press, Washington, DC, 1996; (c) G. Allen-Wardell, P. Bernhardt, R. Bitner, A. Burquez, S. Buchmann, J. Cane, P.A. Cox, V. Dalton, P. Feinsinger, D. Inouye, M. Ingram, C.E. Jones, K. Kennedy, P. Kevan, H. Koopowitz,, R. Medellin, S. Medellin-Morales, G.P. Nabhan, B. Pavlik, V. Tepedino, P. Torchio, and S. Walker, *Conserv. Biol.*, 1998, *12*(1), 8.
62. (a) Anon., *Chem. Eng. News*, May 31, 2004, 10; (b) S. Everts, *Chem. Eng. News*, May 26, 2008, 10; (c) J.K. Kaplan, *Agric. Res.*, 2008, *5*, 8.
63. R.J. Goldberg, *Environ. Defense Fund Lett.*, 1999, *30*(3), 4; (b) D. Ferber, *Science*, 2000, *288*, 792.
64. S. Falkow and D. Kennedy, *Science*, 2001, *291*, 397.
65. B. Hileman, *Chem. Eng. News*, Aug. 8, 2005, 16.
66. (a) R.L. Dorit, *Am. Sci.*, 2009, *97*(1), 20; (b) M. Shnayerson and M.J. Plotkin, *The Killers Within: The Deadly Rise of Drug-Resistant Bacteria*, Little, Brown, New York, 2002.
67. D. Ferber, *Science*, 2003, *301*, 1027.
68. R.A. Lovett, *Science*, 1998, *282*, 1404.
69. (a) F. Diez-Gonzalez, T.R. Callaway, M.G. Kizoulis, and J.B. Russell, *Science*, 1998, 281, 1666; (b) J. Couzin, *Science*, 1998, *281*, 1578.

70. (a) Anon., *Chem. Eng. News*, Dec. 21, 1998, 12; (b) Anon, *Chem. Ind. (Lond.)*, 1999, *7*, 824.
71. N.K. Petty, T.J. Evans, P.C. Fineran, and G.P.C. Salmond, *Trends Biotechnol.*, 2007, *25*(1), 7.
72. (a) G. Gardner, In: L. Starke, ed., *State of the World 1996*, WW Norton, News York, 1996, 90; (b) P.R. Ehrlich and A.H. Ehrlich, *Technol. Rev.*, 1997, *100*(1), 38.
73. D.A. Kurtz, ed., *Long Range Transport of Pesticides*, Lewis, Boca Raton, FL, 1990.
74. (a) D.W. Kolpin, B.K. Nations, D.A. Goolsby, and E.M. Thurman, *Environ. Sci. Technol.*, 1996, *30*, 1459; (b) R.J. Gilliom, *Environ. Sci. Technol.*, 2007, *41*, 3408.
75. L.W. Baker, D.L. Fitzell, J.N. Seiber, T.R. Parker, T. Shibamoto, M.W. Poore, K.E. Longley, R.P. Tomlin, R. Propper, and D.W. Duncan, *Environ. Sci. Technol.*, 1996, *30*, 1365.
76. M.G. Nishioka, H.M. Burkholder, M.C. Brinkman, S.M. Gordon, and R.G. Lewis, *Environ. Sci. Technol.*, 1996, *30*, 3313.
77. (a) G.R. Conway and J.N. Pretty, *Unwelcome Harvest: Agriculture and Pollution*, Earthscan, London, 1991; (b) R.E. Hester and R.M. Harrison, eds, *Agricultural Chemicals and the Environment*, Royal Society of Chemistry, Cambridge, 1996; (c) R.D. Wauchope, R.L. Graney, S. Cryder, C. Eadsforth, A.W. Klein, and K.D. Racke, *Pure Appl. Chem.*, 1995, *67*, 2089; (d) S.Z. Cohen, R.D. Wauchope, A.W. Klein, C.V. Eadsforth, and R. Graney, *Pure Appl. Chem.*, 1995, *67*, 2109; (e) D.A. Goolsby, E.M. Thurman, M.L. Pomes, M.T. Meyer, and W.A. Battaglin, *Environ. Sci. Technol.*, 1997, *31*, 1325; (f) W.T. Foreman, M.S. Majewski, D.A. Goolsby, R.H. Coupe, and F.W. Wiebe, Preprints A.C.S. *Div. Environ. Chem.*, 1999, *39*(1), 440.
78. (a) D.C. Adriano, A.K. Iskander, and I.P. Murarka, eds, *Contamination of Groundwaters*, St. Lucie Press, Delray Beach, FL, 1994; (b) D.W. Kolpin, J.E. Barbash, and R.J. Gilliom, *Environ. Sci. Technol.*, 1998, *32*, 558.
79. J.G. Mitchell, *Natl. Geogr. Mag.*, 1996, *189*(2), 106.
80. (a) J. Long, *Chem. Eng. News*, Oct. 24, 1994, 8; (b) B. Hileman, *Chem. Eng. News*, June 12, 1995, 5; (c) B. Hileman, *Chem. Eng. News*, Mar. 5, 1990, 26; (d) Anon., *Chem. Eng. News*, Aug. 21, 1995, 26.
81. (a) S.P. Schottler, S.J. Elsenreich, and P.D. Capel, *Environ. Sci. Technol.*, 1994, *28*, 1079; (b) R. Renner, *Environ. Sci. Technol.*, 1996, *30*, 110A.
82. R. Stone, *Science*, 1995, *268*, 368.
83. A. Newman, *Environ. Sci. Technol.*, 1995, *29*, 450.
84. L.G. Ballantine, J.E. McFarland, and D.S. Hackett, eds, *Triazine Herbicides: Risk Assessment*, A.C.S. Symp. 683, Washington, DC, 1998.
85. R.P. Richards, D.B. Packer, B.R. Christensen, and D.P. Tierney, *Environ. Sci. Technol.*, 1995, *29*, 406.
86. T. Konstantinova, L. Metzova, and H. Konstantinov, *J. Appl. Polym. Sci.*, 1994, *54*, 2187.
87. (a) D.W. Kolpin, E.M. Thurman, and D.A. Goolsby, *Environ. Sci. Technol.*, 1996, *30*, 335; (b) M.T. Meyer and E.M. Thurman, eds, *Herbicide Metabolities in Surface Water and Groundwater*. A.C.S. Symp. 630, Washington, DC, 1996; (c) D.W. Kolpin, E.M. Thurman, and S.M. Linhart, *Eviron. Sci. Technol.*, 2001, *35*, 1217.
88. (a) R.N. Lerch, W.W. Donald, Y.-X. Li, and E.E. Alberts, *Environ. Sci. Technol.*, 1995, *29*, 2759; (b) E.M. Thurman, M.T. Meyer, M.S. Mills, L.R. Zimmerman, C.A. Perry, and D.A. Goolsby, *Environ. Sci. Technol.*, 1994, *28*, 2267; (c) N. Shapir and R.T. Mandelbvaum, *J. Agric. Food Chem.*, 1997, *45*, 4481.
89. S. Alvey and D.E. Crowley, *Environ. Sci. Technol.*, 1996, *30*, 1596.
90. (a) D.N. Tessier and J.M. Clark, *J. Agric. Food Chem.*, 1995, *43*, 2504; (b) S. Mangiapan, E. Benfenati, P. Grasso, M. Terreni, M. Pregnalato, G. Pagani, and D. Barcelo, *Environ. Sci. Technol.*, 1997, *31*, 3637; (c) D.S. Aga, S. Heberle, D. Rentsch, R. Hany, and S.R. Muller, *Environ. Sci. Technol.*, 1999, *33*, 3452; (d) W.H. Graham, D.W. Graham, F. Denoyelles, Jr., V.H. Smith, C.K. Larive, and E.M. Thurman, *Environ. Sci. Technol.*, 1999, *33*, 4471; (e) K. Christen, *Environ. Sci. Technol.*, 1999, *33*, 230A; (f) M.L. Hladik, J.J. Hsiao, and A.L. Roberts, *Environ. Sci. Technol.*, 2005, *39*, 6561.
91. (a) M. Murphy, *Chem. Ind (Lond.)*, Feb. 7, 2005, 10; (b) Anon., *Chem. Eng. News*, Oct. 12, 2009, 36.
92. (a) G.C. Conway and E.B. Barbier, *After the Green Revolution: Sustainable Agriculture for Development*, Earthscan, London, 1990; (b) C.A. Edwards, R. Celesete, and F.E. Hutchinson, eds, *Sustainable Agricultural System*, St. Lucie Press, Delray Beach, FL, 1990; (c) J.N. Pretty, *Regenerating Agriculture: Polices and Practice for Sustainability and Self-Reliance*, Joseph Henry Press, Washington, DC, 1995; (d) S. Corey, D.J. Dahl, and W.M. Milne, eds, *Pest Control and Sustainable Agriculture*, CSIRO, East Melbourne, Australia, 1993; (e) L.A. Thrupp, *New Partnerships for Sustainable Agriculture*, World Resources Institute, Washington, DC, 1996; (f) Faeth, R.C. Repetto, K. Kroll, Q. Dai, and G. Helmers, *Paying the Farm Bill: U.S. Agricultural Policy and the Transition to Sustainable Agriculture*, World Resources Institute, Washington, DC, 1991; (g) B. Vorley and d. Keeney, *Bugs in the System—Reinventing the Pesticide Industry for Sustainable Agriculture*, Earthscan, London, 1998; (h) C.F. Jordan, *Working with Nature—Resource Management for Sustainability*, Harwood Academic, Amsterdam, 1998; (i) G. Conway, *The Doubly Green Revolution: Food for All in the 21st Century*, Cornell University Press, Cornell, NY, 1998; *Environment*, 2000 *42*(1), 8; (j) R. Hester and R. Harrison, eds, *Sustainability in Agriculture*, Royal Society of Chemistry, Cambridge, 2005; (k) D.B. Lindenmeyer and J.F. Franklin, eds, *Towards Forest Sustainability*, Island Press, Washington, DC, 2004; (l) E. Stokstad, *Science*, 2008, *319*, 1474, "International Assessment of Agricultural Science and Technology"; (m) S.K. Ritter, *Chem. Eng. News*, Feb. 16, 2009, 13.
93. (a) P.R. Crosson, *Resources for the Future*, Fall, 1994, (117), 10; (b) C.O. Stockle, R.I. Papendick, K.E. Saxton, G.S. Campbell, and F.K. van Evert, *Am. J. Alt. Agric.*, 1994, *9*(1–2), 45; (c) N. Uri, *Agriculture and the Environment*, Nova Science, Commack, New York, 1999.
94. M. Mellon, J. Rissler, and F. McCamant, *Union of Concerned Scientists briefing paper*, May 1995.
95. (a) B. Bergland, *Am. J. Alt. Agric.*, 1993, *8*(4), 160; (b) A. Kimbrill, ed., *Fatal Harvest—The Tragedy of Industrial Agriculture*, Island Press, Washington, DC, 2002.
96. P.J. Chenier, *Survey of Industrial Chemistry*, 2nd ed., VCH, Weinheim, 1992, *60*, 70.
97. Anon., *Chem. Eng. News*, Sept. 12, 1994, 9.

98. (a) G. Gardner, In: L. Starke, ed., *State of the World 1996*, WW Norton, New York, 1996, 89; (b) C.M. Benbrook and E. Growth, *Pest Management at the Crossroads*, Consumers Union, Yonkers, NY, 1996.
99. (a) J. Glanz, *Science*, 1995, *267*, 1088; (b) D. Pimentel, C. Harvey, P. Resosudarmo, K. L. Sinclair, D. Kurz, M. McNair, S. Crist, L. Shpritz, L. Fitton, R. Saffouri, and R. Blair, *Science*, 1995, *267*, 1117.
100. C. Rathmann, *Land Water*, 1997, *41*(2), 10.
101. Anon., *Environ. Defense Fund Lett.*, 1997, *28*(3), 1.
102. (a) J. Kaiser, *Science*, 1996, *274*, 3315; (b) C. Houge, *Chem. Eng. News*, Oct. 5, 2009, 33.
103. (a) C. Holden, *Science*, 1999, *285*, 661; (b) R.J. Diaz and R. Rosenberg, *Science*, 2008, *321*, 926;(c) C. Hogue, *Chem. Eng. News*, Oct. 2, 2006; Oct. 8, 2007, 11.
104. E.L. Appleton, *Environ. Sci. Technol.*, 1995, *29*, 550A.
105. (a) L.E. Drinkwater, P. Wagoner, and M. Sarrantonio, *Nature*, 1998, *396*, 262; (b) D. Tilman, *Nature*, 1998, *396*, 211; (c) W.H. Schlesinger, *Science*, 1999, *284*, 2095; (d) P.E. Rasmussen, K.W.T. Goulding, J.R. Brown, P.R. Grace, H.H. Janzen, and M. Korschens, *Science*, 1998, *282*, 893.
106. (a) J. Janick, ed., *Progress in New Crops*, ASHS Press, Alexandria, VA, 1996; (b) *Lost Crops of Africa*, vol. 1., Grains, 1996; vol. 2, Vegetables, 2006; vol. 3, Fruits, 2008, National Academic Press, Washington, DC.
107. (a) A.S. Moffat, *Science*, 1996, *274*, 1469; (b) P.L. Scheinost, D.L. Lammer, X. Cai, T.D. Murray, and S.S. Jones, *Am. J. Alt. Agric.*, 2003, *16*(4), 147.
108. (a) Anon., *Chem. Eng. News*, Aug. 22, 1994, 17; Jan. 2, 1995, 16; (b) G. Parkinson, *Chem. Eng.*, 1996, *103*(5), 29.
109. (a) J.E. Bardach, *Sustainable Aquaculture*, Wiley, New York, 1997; (b) A.P. McGinn, *World Watch*, 1998, *11*(2), 10; (c) R.L. Naylor, R.J. Goldberg, H. Mooney, M. Beveridge, J. Clay, C. Folke, N. Kautsky, J. Lubchenko, J. Primavera, and M. Williams, *Science*, 1998, *282*, 883; (d) R.R. Stickney and J.P. McVey, eds, *Responsible Marine Aquaculture*, Oxford University Press, Oxford, 2002; (e) B. Costa-Pierce, A. Desbonnett, P. Edwards, and D. Baker, eds, *Urban Aquaculture*, CABI Publishing, Wallingford, UK, 2005; (f) P. Edwards, D. Little, and H. Demaine, eds, *Rural Aquaculture*, CABI Publishing, 2002; (g) E. Engelhaupt, *Environ. Sci. Technol.*, 2007, *41*, 4188.
110. (a) National Research Council Board on Agriculture, *Ecologically Based Pest Management: New Solutions for a New Century*, National Academy Press, Washington, DC, 1996; (b) Office of Technology Assessment, *Biologically Based Technologies for Pest Control*, U.S. Superintendent of Documents, Washington, DC, 1995; (c) M.A. Altieri, *Biodiversity and Pest Management in Agroecosystems*, Food Products Press, New York, 1994; (d) C.R.N. Godfrey, ed., *Agrochemicals from Natural Products*, Dekker, New York, 1995; (e) L.G. Copping, ed., *Crop Protection Agents from Nature: Natural Products and Analogues*, Royal Society of Chemistry, Cambridge, 1996; (f) R.L. Metcalf, *Kirk–Othmer Encyclopedia of Chemical Technology*, 4th ed., Wiley, New York, 1995, *14*, 524; (g) D. Pimentel, ed., *Techniques for Reducing Pesticide Use: Economic and Environmental Benefits*, Wiley, New York, 1997.
111. National Research Council Board on Agriculture, *Ecologically Based Pest Management: New Solutions for a New Century*, National Academy Press, Washington, DC, 1996, 2.
112. (a) D. Coleman and D.A. Crossley, Jr., *Fundamentals of Soil Ecology*, Academic, San Diego, 1996; (b) H.P. Collins, G.P. Robertson, and M.J. Klug, eds, *The Significance and Regulation of Soil Biodiversity*, Kluwer Academic, Dordrecht, the Netherlands, 1995; (c) R.I. Papendick and J.F. Parr, eds, *Am. J. Alt. Agric.*, 1992, *7*(1–2), 2–68; (d) A. H. Fitter, ed., *Ecological Interactions in Soil—Plants, Microbes and Animals*, Blackwell Science, Oxford, 1985.
113. (a) L.L.G. Copping, ed., *Crop Protection Agents from Nature: Natural Products and Analogues*, Crit. Rep. Appl. Chem. Vol. 35, Royal Society of Chemistry, Cambridge, 1996; (b) P.A. Hedin, R.M. Hollingsworth, E.P. Masler, J. Miyamoto, and D.G. Thompson, eds, *Phytochemicals for Pest Control*, A.C.S. Symp. 658, Washington, DC, 1997; (c) G.D. Crouse, *Chemtech*, 1998, *28*(1), 36; (c) S.K. Wrigley, M.A. Hayes, R. Thomas, E.J.T. Chrystal, and N. Nicholson, *New Leads for the Pharmaceutical and Agrochemical Industries*, Special publication 257, Royal Society of Chemistry, Cambridge, 2000; (d) L.G. Copping, and B.P.S. Kambag, eds, *Pest Management Science*, 2000, *56*(8), 649–722.
114. Dow Agrosciencies, Presidential Green Chemistry Challenge Awards, 1999, 2008.
115. Office of Technology Assessment, *Biologically Based Technologies for Pest Control*, U.S. Superintendent of Documents, Washington, DC, 1995, 11.
116. T.B. Frick and D.W. Tallamy, *Entomol. News*, 1996, *107*(2), 77.
117. M.J. Crawley, ed., *The Population Biology of Predators, Parasites and Diseases*, Blackwell Science, Oxford, 1992.
118. (a) S. Adams, *Agric. Res.*, 1996, *44*(3), 12; (b) D.L. Kline, *Agric. Res.*, 1999, *47*(11), 23.
119. Anon., *Technol. Rev.*, 1998, *101*(5), 15.
120. M.A. Howland, *Land Water*, 1996, *40*(5), 45.
121. L. Cooke, *Agric. Res.*, 1996 *44*(5), 21.
122. D.A. Cleghorn, D.N. Ferro, W. Flinn, and S.V. Nablo, *Boston ACS Meeting*, Aug. 2002, AGFD 62.
123. G.C. Ziobro, *Chem. Ind. (Lond.)*, 1998, 428.
124. A. Flores, J. Suszkiw, and M. Wood, *Agric. Res.*, 2003, *51*(2), 15.
125. L. Richards, *Chem. Ind. (Lond.)*, July 3, 2006, 8.
126. Anon., *Int. News Fats Oils Relat. Mater.*, 1995, *6*, 888.
127. C.M.M. Mols and M.E. Visser, *J. Appl. Ecol.*, 2002, *39*, 888.
128. Office of Technology Assessment, *Biologically Based Technologies for Pest Control*, U.S. Superintendent of Documents, Washington, DC, 1995, 85.
129. L.D. King and D.L. Hoag, *Am. J. Alt. Agric.*, 1998, *13*(1), 12.
130. E. Peabody, *Agric. Res.*, 2007, *55*(6), 9.
131. (a) National Research Council Board on Agriculture, *Ecologically Based Pest Management: New Solutions for a New Century*, National Academy Press, Washington, DC, 1996, 20; (b) G.A. Kraus, B. Johnson, A. Kongsjahju, and G.L. Tylka, *J. Agric. Food Chem.*, 1994, *42*, CT 1839; (c) G.A. Kraus, S.J. Vander Louw, G.L. Tylka, and D.H. Soh, *J. Agric. Food Chem.*, 1996, *44*, 1548; (d) A.G. Whitehead, *Plant Nematode Control*, CAB International, Wallingford, CT, 1998.
132. D. Lyons-Johnson, *Agric Res.*, 1997, *45*(9), 7.
133. J.A. Gonzalez and A. Estevez-Braun, *J. Agric. Food Chem.*, 1998, *46*, 1163.
134. (a) M. deBartoldi, ed., *The Science of Composting*, Chapman & Hall, New York, 1996; (b) D.B. Churchill, W.R. Horwath, L.F. Elliott, and D.M. Bilsland, *Am. J. Alt. Agric.*, 1996, *11*, 7.

135. (a) M.A. Altieri, *Biodiversity and Pest Management in Agroecosystems*, Food Products Press, New York, 1994; (b) D.U. Hooper and P.M. Vitousek, *Science*, 1997, *277*, 1302.
136. J. Suszkiw, *Agric. Res.*, 1997, *45*(9), 16.
137. M.M. Green, *Science*, 1997, *277*, 623.
138. D.Comis, *Agric. Res.*, 2005, *53*(2), 18.
139. G. Ondrey, *Chem. Eng.*, 2004, *111*(3), 17.
140. C. Zandonella, *New Sci.*, Sept. 18, 2001, 14.
141. D.L. Aldridge, P. Elliott, and G.D. Mogridge, *Environ. Sci. Technol.*, 2006, *40*, 975.
142. (a) H.M.T. Hokkanen, *Am. J. Alt. Agric.*, 1995, *10*(2), 64; (b) H.M.T. Hokkanen and J.M. Lynch, eds, *Biological Control: Risks and Benefits*, Cambridge University Press, New York, 1995; (c) F. Fenner and Bernardini Fantini, *Biological Control of Vertebrate Pests*, CABI Publishing, Wallingford, UK, 1999; (d) E. Wajnberg, J.K. Scott, and P.C. Quimby, eds, *Evaluating Indirect Ecological Effects of Biological Control*, Oxford University Press, Oxford, 2000.
143. M.N. Jensen, *Science*, 2000, *290*, 2230.
144. (a) D.H. Kinley III, *Environment*, 1998, *40*(7), 14; (b) *New Sci.*, Oct. 29, 1997, 12.
145. (a) P.B. McEvoy, J.W. Kloepper, B.A. Federici, J.V. Maddox, M.R. Strand, J.J. Obrycki, D.W. Onstad, M.I. McManus, H.R. Smith, C.L. Remington, D. Secord, and P. Kareiva, *Bioscience*, 1996, *46*(6), 398–436 (whole issue); (b) C. Thies and T. Tscharntbe, *Science*, 1999, *285*, 893; (c) M.E. Hochberg and A.R. Ives, eds, *Parasitoid Population Biology*, Princeton University Press, Princeton, NJ, 2000; (d) J. Suszkiw, *Agric. Res.*, 2004, *52*(1), 12.
146. D. Senft, *Agric. Res.*, 1997, *45*(6), 4; (b) T.N. Ananthakrishnan, ed., *Technology in Biological Control*, Science Publishers, Enfield, NH.
147. (a) L.G. Copping, ed., *Crop Protection Agents from Nature—Natural Products and Analogues*, Royal Society of Chemistry, Cambridge, 1996; (b) National Research Council Board on Agriculture, *Ecologically Based Pest Management: New Solutions for a New Century*, National Academy Press, Washington, DC, 1996, 46.
148. (a) Office of Technology Assessment, *Biologically Based Technologies for Pest Control*, U.S. Superintendent of Documents, Washington, DC, 1995, 33, 51, 77, 150; (b) T. Weaver, *Agric. Res.*, 1999, *47*(3), 26.
149. Anon., *Environ. Sci. Technol.*, 1994, *28*, 461A.
150. (a) R. Dagani, *Chem. Eng. News*, July 5, 1999, 30; (b) S.D. West, *J. Agric. Food Chem.*, 1997, *45*, 3107.
151. (a) J.-W. Liu, W.H. Yap, T. Thanabalu, and A.G. Porter, *Nat. Biotechnol.*, 1996, *14*, 343; (b) C.F. Curtis, *Nat. Biotechnol.*, 1996, *14*, 265.
152. (a) R. Mestel, *New Sci.*, June 10, 1995, 7; (b) R.J. St. Leger and D.W. Roberts, *Trends Biotechnol.*, 1997, *15*, 83; (c) H. Becker, *Agric. Res.*, 1999, *47*(9), 20; 2000, *48*(11), 12.
153. (a) G. Parkinson, *Chem. Eng.*, 1996, *103*(3), 27; (b) J. deOuattro, D. Senft, and M. Wood, *Agric. Res.*, 1997, *45*(2), 4; (c) S.M. Hays, *Agric. Res.*, 1998, *46*(8), 2.
154. (a) S.K. Ritter, *Chem. Eng. News*, Dec. 4, 2006, 92; (b) G. Ondrey, *Chem. Eng.*, 2006, *113*(3), 18; (c) A.K. Raina and M.S. Wright, *Agric. Res.*, 2005, *53*(11), 23; (d) E. Peabody, *Agric. Res.*, 2006, *54*(9), 16.
155. J. Suszkiv, *Agric. Res.*, 2008, *56*(8), 4.
156. (a) F.R. Hall and J.W. Barry, eds, *Biorational Pest Control 595, Agents—Formulation and Delivery*. A.C.S. Symp. 595, Washington, DC, 1995, 197; (b) D. Lyons-Johnson, *Agric. Res.*, 1997, *45*(4), 10; (c) D. Stanley, *Agric. Res.*, 1996, *44*(11), 22.
157. (a) P.S. Grewal, E.E. Lewis, and R. Gaugler, *J. Chem. Ecol.*, 1997, *23*, 503; (b) D. Bowen, T.A. Rocheleau, M. Blackburn, O. Andreev, E. Golubeva, R. Bhartia, and R.H. French-Constant, *Science*, 1998, *280*, 2129; (c) E. Strauss, *Science*, 1998, *280j*, 2050; (d) *Nat. Biotechnol.*, 1998, *16*, 124.
158. T. Weaver, *Agric. Res.*, 1998, *46*(10), 19.
159. D. Lyons-Johnson, *Agric. Res.*, 1997, *45*(4), 10.
160. National Research Council Board on Agriculture, *Ecologically Based Pest Management; New Solutions for a New Century*, National Academy Press, Washington, DC, 1996, *46*, 51.
161. S. Pain, *New Sci.*, Apr. 14, 2001, 46.
162. (a) L.E. Volkman, *Science*, 1995, *269*, 1834; (b) F.R. Hunter-Fujita, *Insect Viruses and Pest Management*, Wiley, New York, 1998.
163. (a) G.J. Persley, ed., *Biotechnology and Integrated Pest Management*, CAB International, Wallingford, CT, 1996; (b) S.M. Thiem, *Am. J. Alt. Agric.*, 1995, *10*(2), 57; (c) L.F. Elliott and J.M. Lynch, *Am. J. Alt. Agric.*, 1995, *10*(2), 67; (d) S.O. Duke, J.J. Menn, and J.R. Plimmer, eds, *Pest Control with Enhanced Environmental Safety*, A.C.S. Symp. 524, Washington, DC, 1993, 239.
164. S. Asser-Kaiser, E. Fritsch, K. Undorf-Spahn, J. Kienzle, K.E. Eberle, N.A. Gund, A. Reineke, C.P.W. Zebitz, D.G. Heckel, J. Huber, and J.E. Jehle, *Science*, 2007, *317*, 1916.
165. (a) T. Weaver-Missick, *Agric. Res.*, 2001, *49*(4), 29; (b) J.J. Becnel, *Agric. Res.*, 2003, *51*(10), 23.
166. M.J. Walters, *Nat. Conserv.*, 1995, *45*(3), 8.
167. (a) F.R. Hall and J.W. Barry, eds, *Biorational Pest Control Agents—Formulation and Delivery*, A.C.S. Symp. 595, Washington, DC, 1995, 153; (b) R. Farrar, Jr., *Agric. Res.*, 1999, *47*(2), 23; (c) F.R. Hall and J.J. Menn, eds, *Biopesticides: Use and Delivery*, Humana Press, Totowa, NJ, 1999.
168. B. Hardin, *Agric. Res.*, 1998, *46*(6), 20.
169. J. Suszkiw, *Agric. Res.*, 1998, *46*(3), 25.
170. (a) D.C. Carlson, *Kirk–Othmer Encyclopedia of Chemical Technology*, 4th ed., Wiley, New York, 1997, *21*, 236; (b) S.V. Ley and P.L. Toogood, *Chem. Br.*, 1990, *26*(1), 31; (c) A. Gonzalez–Coloma, C. Gutierrez, R. Cabrera, and M. Reina, *J. Agric. Food Chem.*, 1997, *45*, 946; (d) A. Gonzalez-Coloma, A. Guadrano, C. Gutierrez, R. Cabrera, E. de la Pena, G. de la Fuente, and M. Reina, *J. Agric. Food Chem.*, 1998, *46*, 286.
171. H. Eichenseer and C.A. Mullin, *J. Chem. Ecol.*, 1997, *23*(1), 71.
172. (a) H. Schmutter, ed., *The Neem Tree: Source of Unique Natural Products for Integrated Pest Management*, Medicinal, Industrial and Other Purposes, VCH, New York, 1995; (b) A.A. Denholm, L. Jennens, S.V. Ley, and A. Wood, *Tetrahedron*, 1995, *51*, 6591; (c) M. Nakatani, R.C. Huang, H. Okamura, T. Iwagawa, K. Tadera, and H. Naoki, *Tetrahedron*, 1995, *51*, 11731; (d) J. Kumar and B.S. Parmar, *J. Agric. Food Chem.*, 1996, *44*, 2137; (e) J.D. Stark and J.F. Walter, *J. Agric. Food Chem.*, 1995, *43*, 507; (f) S.Y. Szeto and M.T. Wan, *J. Agric. Food Chem.*, 1996, *44*, 1160;

(g) S. Johnson, E.D. Morgan, I.D. Wilson, M. Spraul, and M. Hofmann, *J. Chem. Soc., Perkin Trans. 1*, 1994, 1499; (h) D. Tenenbaum, *Technol. Rev.*, 1996, *99*(2), 21; (i) S.R. Yakkundi, R. Thejavathi, and B. Ravindranath, *J. Agric. Food Chem.*, 1995, *43*, 2517; (j) S.V. Ley, *Pure Appl. Chem.*, 1994, *66*, 2099.

173. (a) A.F. Farrero, E.A. Manzaneda, J. Altarejos, S. Salido, J.M. Ramos, M.S.J. Simmonds, and W.M. Blaney, *Tetrahedron*, 1995, *51*, 7435; (b) J.G. Urones, I.S. Marcos, B.G. Perez, A.M. Lithgow, D. Diez, P.M. Gomez, P. Basabe, and N.M. Garrido, *Tetrahedron*, 1995, *51*, 1845; (c) M. Jonassohn, R. Davidsson, P. Kahnberg, and O. Sterner, *Tetrahedron*, 1997, *53*, 237.
174. R.J. Demchak and R.A. Dybas, *J. Agric. Food Chem.*, 1997, *45*, 260.
175. G. Cimino, A. Spinella, and G Sodano, *Tetrahedron Lett.*, 1984, *25*, 4151.
176. (a) G.M.R. Tombo and D. Bellus, *Angew. Chem. Int. Ed.*, 1991, *30*, 1208; (b) L. Lajide, P. Escoubas, and J. Mizutani, *J. Agric. Food Chem.*, 1993, *41*, 669.
177. J.B. Gloer, *Acc. Chem. Res.*, 1995, *28*, 343.
178. F.-Y. Huang, B.Y. Chung, M.D. Bentley, and A.R. Alford, *J. Agric. Food Chem.*, 1995, *43*, 2513.
179. J.C. McCarty, Jr., P.A. Hedin, and R.D. Stipanovic, *J. Agric. Food Chem.*, 1996, *44*, 613.
180. Anon., *Chem. Eng. News*, Apr. 14, 2003, 49.
181. Anon., *Chem. Eng. News*, May 13, 2002, 34.
182. M. Jacoby, *Chem. Eng News*, Aug. 5, 2002, 35.
183. N. Eisberg, *Chem. Ind.* (*Lond.*), Apr. 9, 2007, 8.
184. Anon., *Chem. Eng. News*, Sept. 13, 2004, 30.
185. K.A. El Sayed, D.C. Dunbar, T.L. Perry, S.P. Wilkins, and M.T. Hamann, *J. Agric. Food Chem.*, 1997, *45*, 2735.
186. M.L. Ciavatti, M. Gavagnin, R. Puliti, G. Cimino, E. Martinez, J. Ortea, and C.A. Mattia, *Tetrahedron*, 1996, *52*, 12831.
187. E.I. Graziani, R.J. Andersen, P.J. Krug, and D.J. Faulkner, *Tetrahedron*, 1996, *52*, 6869.
188. A.M. Montanari, W. Fenical, N. Lindquist, A.Y. Lee, and J. Clardy, *Tetrahedron*, 1996, *52*, 5371.
189. (a) R.A. Kerr, *Science*, 1994, *266*, 5463; (b) K.M. Reese, *Chem. Eng. News*, Mar. 20, 1995, 84.
190. (a) R. Teranishi, R.G. Buttery, and H. Sugisawa, eds, *Bioactive Volatile Compounds from Plants*, A.C.S. Symp. 525, Washington, DC, 1993; (b) E.A. Bernays and J.A.A. Renwick, *Host–Plant Selection by Phytophagous Insects*, Chapman & Hall, New York, 1994; (c) T. Eisner, *For Love of Insects*, Harvard University Press, Cambridge, MA 2003; (d) A.M. Rouhi, *Chem. Eng. News*, Nov. 13, 2000, 33; (e) W. Agosta, *Thieves, Deceiver and Killers—Tales of Chemistry in Nature*, Princeton University Press, Princeton, 2001; (f) H.F. Nijhout, *Insect Hormones*, Princeton University Press, Princeton, 1998.
191. (a) E.D. Morgan, *Chem. Ind.* (*Lond.*), 1994, 370; (b) K. Mori, *Chem. Commun.*, 1997, 1153; *Acc. Chem. Res.*, 2000, *33*, 102; (c) J. Hardie and A.K. Minks, *Pheromones of Non-Lepidopteran Insects Associated With Agricultural Plants*, Oxford Universisy Press, Oxford, 1999; (d) www.pherobase.com.
192. (a) H.M. Putambekar and D.G. Naik, *Synth. Commun.*, 1998, *28*, 2399; (b) R.J. Petroski and D. Weisleder, *J. Agric. Food Chem.*, 1997, *45*, 943.
193. (a) H. Arn, M. Toth, and E. Priesner, *List of Sex Pheromones of Lepidoptera and Related Attractants*, 2nd ed., IOBC, Montavet, France, 1992; (b) M.S. Mayer and J.R. McLaughlin, *Handbook of Insect Pheromones and Sex Attractants*, CRC Press, Boca Raton, FL, 1991.
194. V. Glaser, *Biotechnology*, 1995, *13*, 219.
195. (a) A.B. Atygalle, G.N. Jham, A. Svatos, R.T.S. Frighetto, J. Meinwald E.F. Vilela, F.A. Ferrara, and M.A. Uchoa–Fernandes, *Tetrahedron Lett.*, 1995, *36*, 5471; (b) Anon., *Chem. Ind.* (*Lond.*), 1995, 638.
196. S. Gil, M.A. Lazaro, R. Mestres, F. Millan, and M. Parra, *Synth. Commun.*, 1995j, *25*, 351.
197. K. Mori and M. Amaike, *J. Chem. Soc., Perkin Trans. 1*, 1994, 2727.
198. A.-C. Backman, M. Bengtsson, and P. Witzgall, *J. Chem. Ecol.*, 1997, *23*, 807.
199. (a) J.R. Al Dulayymi, M.S. Baird, M.J. Simpson, S. Nyman, and G.R. Port, *Tetrahedron*, 1996, *52*, 12509; (b) T. Ando, R. Ohno, K. Ikemoto, and M. Yamamoto, *J. Agric. Food Chem.*, 1996, *44*, 3350.
200. A. Zhang, P.S. Robbins, W.S. Leal, C.E. Linn, Jr., M.G. Villani, and W.L. Roelofs, *J. Ecol.*, 1997, *23*, 231.
201. T. Eisner and J. Meinwald, eds, *Chemical Ecology: The Chemistry of Biotic Interaction*, National Academy Press, Washington, DC, 1995, 106.
202. K. Mori, *Pure Appl. Chem.*, 1996, *68*, 2111; 1994, *66*, 1991.
203. (a) F. Kern, R.W. Klein, E. Janssen, H.-J. Bestmann, A.B. Attygalle, D. Schafer, and U. Maschwitz, *J. Chem. Ecol.*, 1997, *23*, 779; (b) H.-J. Bestmann, E. Ubler, and B. Holldobler, *Angew. Chem. Int. Ed.*, 1997, *36*, 395.
204. A.R. Harari and P.J. Landolt, *J. Chem. Ecol.*, 1997, *23*, 857.
205. R. Tsao and J.R. Coats, *Chemtech*, 1995, *25*(7), 23.
206. G. Franzios, M. Mirotsou, E. Hatziapostolou, J. Kral, Z.G. Scouras, and P. Mavragani–Tsipidou, *J. Agric. Food Chem.*, 1997, *45*, 2690.
207. (a) Anon., *Chem. Eng. News*, July 5, 1999, 18; (b) G Ondrey, *Chem. Eng.*, 1999, *106*(8), 17; (c) P.L. Short, *Chem. Eng. News*, Oct. 1, 2007, 24.
208. (a) D. Shailaja, S. Merajudin, S.M. Ahmed, and M. Yaseen, *J. Appl. Polym. Sci.*, 1997, *64*, 1373; (b) S.L. Wilkinson, *Chem. Eng. News*, Sept. 11, 2000, 25.
209. S. Shu, G.G. Grant, D. Langevin, D.A. Lombardo, and L. MacDonald, *J. Chem. Ecol.*, 1997, *23*, 35.
210. B.F. Binder and J.C. Robbins, *J. Agric. Food Chem.*, 1997, *45*, 980.
211. N. Arakaki, S. Wakamura, T. Yasuda, and K. Yamagishi, *J. Chem. Ecol.*, 1997, *23*, 153.
212. (a) C. Gravier-Pelletier, Y. LeMerrer, and J.-C. Depezay, *Tetrahedron*, 1995, *51*, 1663; (b) C. Bonini, M. Checconi, G. Righi, and L. Rossi, *Tetrahedron*, 1995, *51*, 4111.
213. K. Mori and Z.-H. Qian, *Ann.*, 1994, 291.
214. B.T. Sullivan, C.W. Berisford, and M.J. Dalusky, *J. Chem. Ecol.*, 1997, *23*, 837.
215. U.S. EPA Presidential Green Chemistry Challenge Award, 2000.
216. (a) Anon., *Environment*, 2000, *42*(5), 6; (b) B. Hardin, *Agric. Res.*, 2000, *48*(1), 12.
217. J. Gewolb, *Science*, 2001, *294*, 45.
218. M.-P. Calcagno, F. Camps, J. Coll, E. Mele, and F. Sanchez-Baeza, *Tetrahedron*, 1995, *51*, 12119.

219. F. Camps and J. Coll, *Phytochemistry*, 1993, *32*, 1361.
220. (a) R.L. Metcalf, *Kirk–Othmer Encyclopedia of Chemical Technology*, 4th ed., Wiley, New York, 1995, *14*, 569; (b) C.R.A. Godfrey, ed., *Agrochemicals from Natural Products*, Dekker, New York, 1995, 147.
221. Z. Wimmer, M. Rejzek, M. Zarevuck, A.J. Kuldova, I. Hrdy, V. Nemec, and M. Romanuk, *J. Chem. Ecol.*, 1997, *23*, 605.
222. (a) M. Thirugnanam and G.R. Carlson, Preprints A.C.S. *Div. Environ. Chem.*, 1997, *37*(1), 346; (b) L.R. Raber, *Chem. Eng. News*, July 6, 1998, 25.
223. E. Wilson, *Chem. Eng. News*, July 31, 1995, 16.
224. A.M. Rouhi, *Chem. Eng News*, Jan. 8, 1996, 23.
225. (a) R.S. Fritz and E.L. Simms, eds, *Plant Resistance to Herbivores and Pathogens*, University of Chicago Press, Chicago, 1992; (b) N. Benhamou, *Trends Plant Sci.*, 1996, *1*, 233; (c) b. Baker, P. Zambryski, B. Staskawicz, and S.P. Dinesh–Kumar, Science, 1997, *276*, 726; (d) E.E. Farmer, *Science*, 1997, *276*, 912; (e) H.T. Alborn, T.C.J. Turlings, T.H. Jones, G. Stenhagen, J.H. Loughrin, and J.H. Tumlinson, *Science*, 1997, *276*, 945; (f) E. Blee, *Int. News Fats Oils Relat. Mater.*, 1995, *6*, 852; (g) P.J. O'Donnell, C. Calvert, R. Atzorn, C. Wasternack, H.M. O'Leyser, and D.J. Bowles, *Science*, 1996, *274*, 1914; (h) A. Toumadke and W.C. Johnson, Jr., *J. Am. Chem. Soc.*, 1995, *117*, 7023; (i) J.L. Bi, J.B. Murphy, and G.W. Felton, *J. Chem. Ecol.*, 1997, *23*, 97; (j) D.W. Tallamy and M.J. Raupp, eds, *Phytochemical Induction by Herbivores*, Wiley, New York, 1991.
226. (a) G. Strittmatter, J. Janssens, C. Opsomer, and J. Botterman, *Biotechnology*, 1995, *13*, 1085; (b) N.K. Clay, A.M. Adio, C. Denoux, G. Jander, and F.M. Ausubel, *Science*, 2009, *323*, 95; (c) E.W. Weiler, *Angew. Chem Int. Ed.*, 2003, *42*, 392; (d) M.T. Nishimura, M. Stein, B.-H. Hou, J.P. Vogel, H. Edwards, and S.C. Somerville, *Science*, 2003, *301*, 969; (e) A. Kessler and I.T. Baldwin, *Science*, 2001, *291*, 2141; (f) G. Pohnert, *Angew. Chem. Int. Ed.*, 2000, *39*, 4352; (g) Anon., *Chem. Eng. News*, May 14, 2001, 80; (i) R. Tollrian and C.D. Harvell, eds, *The Ecology and Evolution of Inducible Defenses*, Princeton University Press, Princeton, 1999.
227. (a) H.T. Alborn, T.C.J. Turlings, T.H. Jones, G. Stenhagen, J.H. Loughrin, and J.H. Tumlinson, *Science*, 1997, *276*, 945; (b) P.W. Pare, H.T. Alborn, and J.H. Tumlinson, *Proc. Natl. Acad. Sci. USA*, 1998, *95*, 13971; (c) F. Schroeder, *Angew. Chem. Int. Ed.*, 1998, *37*, 1213; (d) G. Pohnert, T. Koch, and W. Boland, *Chem. Commun.*, 1999, 1087.
228. (a) S. Pain, *New Sci.*, Mar. 4, 1995, 13; (b) W. Boland, J. Hopke, J. Donath, J. Nuske, and F. Bublitz, *Angew. Chem. Int. Ed.*, 1995, *34*, 1600; (c) R. Stevenson, *Chem. Br.*, 1998, *34*(4), 29.
229. (a) T.P. Delaney, S. Uknes, B. Vernooji, L. Friedrich, K. Weymann, D. Negrotto, T. Gaffney, M. Gut-Rella, H. Kessmann, E. Ward, and J. Ryals, *Science*, 1994, *266*, 1247; (b) V. Shulaev, P. Silverman, and I. Raskin, *Nature*, 1997, *385*, 718; (c) T. Taapken, S. Blechert, E.W. Weiler, and M.H. Zenk, *J. Chem. Soc., Perkin Trans. 1*, 1994, 1439; (d) S.L. Wilkinson, *Chem. Eng. News*, July 30. 2001, 42; (e) S. Xaio, S. Ellwod, O. Calis, E. Patrick, T. Li, M. Coleman, and J.G. Turner, *Science*, 2001, *291*, 118; (f) R. Liechti and E.E. Farmer, *Science*, 2002, *296*, 1649; (g) D. Wang, N.D. Weaver, M. Kesarwami, and X. Dong, *Science*, 2005. *308*, 1036; (h) S.-W. Park, E. Kaimoyo, D. Kumar, S. Moshesr, and D.F. Kiessig, *Science*, 2007, *318*, 113; (i) A. Gfeller and E.E. Farmer, *Science*, 2004, *306*, 1515.
230. C. Martin, *Chem. Br.*, 1995, *31*, 663.
231. (a) C.A. Escobar, M. Kluge, and D. Sicker, *Tetrahedron Lett.*, 1997, *38*, 1017; (b) S.R. Desai, P. Kumar, and W.S. Chilton, *Chem. Commun.*, 1996, 1321; (c) P. Kumar and W.S. Chilton, *Tetrahedron Lett.*, 1994, *35*, 3247; (d) M. Frey, P. Chomet, E. Glawischnig, C. Stettner, S. Grun, A. Winklmair, W. Eisenreich, A Bacher, R.B. Meeley, S.P. Briggs, K. Simcox, and A. Gierl, *Science*, 1997, *277*, 696.
232. (a) R. Hammerschmidt and J. Kuc, *Induced Resistance to Disease in Plants*, Kluwer Academic, Dordrecht, 1995.;(b) G. Parkinson, *Chem. Eng.*, 2000, *107*(5), 23; (c) J. Engelberth, T. Koch, F. Kuhnemann, and W. Boland, *Angew. Chem. Int. Ed.*, 2000, *39*, 1860.
233. A. Murai, K. Sato, and T. Hasegawa, *Chem. Lett.*, 1995, 883.
234. M. Adrian, P. Jeandet, R. Bessis, and J.M. Joubert, *J. Agric. Food Chem.*, 1996, *44*, 1979.
235. M. Jacobs, *Chem. Eng. News*, Apr. 19, 1999, 5.
236. J. Gan and S.R. Yates, *J. Agric. Food Chem.*, 1996, *44*, 4001.
237. Anon., *Environment*, 1996, *38*(8), 23.
238. (a) P. Loaharanu, *Food Technol.*, 1994, *48*(5), 124; (b) D.W. Thayer, *Food Technol.*, 1994, *48*(5), 132; (c) M.H. Stevenson, *Food Technol.*, 1994, *48*(5), 141; (d) C.H. McMurray, M.F. Patterson, and E.M. Stewart, *Chem. Ind.* (*Lond.*), 1998, 433.
239. M. Wood, *Agric. Res.*, 1995, *43*(3), 22.
240. B. Hardin, *Agric. Res.*, 1998, *46*(1), 4.
241. K. Mori and Y. Matsushima, *Synthesis*, 1995, 845.
242. L. Cooke, *Agric. Res.*, 1995, *43*(3), 19.
243. J.D. Harmon, *Integrated Pest Management in Museum*, Library and Archival Facilities, Harmon Preservation Pest Management, Indianapolis, IN, 1993, 92.
244. J.D. Harmon, *Integrated Pest Management in Museum*, Library and Archival Facilities, Harmon Preservation Pest Management, Indianapolis, IN, 1993, pp. 87, 93.
245. (a) J.D. Harmon, *Integrated Pest Management in Museum*, Library and Archival Facilities, Harmon Preservation Pest Management, Indianapolis, IN, 1993, p. 101. (b) H.D. Klein and A.M. Wenner, *Tiny Game Hunting: Environmentally Healthy Ways to Trap and Kill the Pests in Your Home and Garden*, University of California Press, Berkeley, CA, 2001.
246. Anon., *Chem. Ind.* (*Lond.*), 2001, 362.
247. J. McBride, *Agric. Res.*, 2000, *48*(11), 14; (b) J. Garcia, *Agric. Res.*, 2001, *49*(3), 18.
248. (a) J.R. Heitz and K.R. Downum, eds, *Light-Activated Pest Control*, A.C.S. Symp. 616, Washington, DC, 1995; (b) R.F. Service, *Science*, 1995, *268*, 806; (c) M. Wood, *Agric. Res.*, 1996j, *44*(1), 20; (d) J.P. Alcantara-Licudine, N.L. Bui, M.K. Kawate, and Q.X. Li, *J. Agric. Food Chem.*, 1998, *46*, 1005.
249. Anon., *Land Water*, 1996, *40*(6), 50.
250. (a) O.T. Chortyk, J.G. Pomonis, and A.W. Johnson, *J. Agric. Food Chem.*, 1996, *44*, 1551; (b) O.T. Chortyk, S.J. Kays, and Q. Teng, *J. Agric. Food Chem.*, 1997, *45*, 270; (c) G. Parkinson, *Chem. Eng.*, 2000, *107*(7), 19; (d) G.J. Puterka, *Agric. Res.*, 2000, *48*(6), 23; (e) R.M. Bliss, *Agric. Res.*, 2005, *53*(6), 8.
251. H. Becker, *Agric. Res.*, 1998, *46*(7), 14.
252. R.L. Metcalf, *Kirk–Othmer Encyclopedia of Chemical Technology*, 4th ed., Wiley, New York, 1995, *14*, 575.

253. (a) S. Radosevich, *Weed Ecology*, 2nd ed., Wiley, New York, 1997, 350–388; (b) M.K. Upadhyaya and R.E. Blackshaw, *Non-Chemical Weed Management*, Oxford University Press, Oxford, 2007; (c) R. van Driesche M. Hoddle, and T. Center, *Control of Pests and Weeds by Natural Enemies*, Blackwell Publishing, Oxford, 2008.
254. H. Becker, *Agric. Res.*, 1995, *43*(12), 10.
255. J.E. Franz, M.K. Mao, and J.A. Sikorski, *Glyphosate—A Unique Global Herbicide*, A.C.S. Monogr. 189, Washington, DC, 1997.
256. (a) Presidential Green Chemistry Challenge Award Recipients, U.S. EPA 744-K-97-003, Sept. 1997, 16; (b) J.H. Clark, *Green Chem.*, 1999, *1*, 1.
257. (a) Anon., *Chem. Eng. News*, Mar. 8, 1999, 15; (b) M. McCoy, *Chem. Eng. News*, Mar. 22, 1999, 17.
258. D.R. Lighthall, *Am. J. Alt. Agric.*, 1996, *11*, 168.
259. L. Cooke, *Agric. Res.*, 1997, *45*(7), 20.
260. P.E. Rasmussen, *Am. J. Alt. Agric.*, 1996, *11*, 108.
261. (a) L. Cooke, *Agric. Res.*, 1996, *44*(4), 15; (b) J.V. Stafford, *Chem. Ind.* (*Lond.*), 2000, 98; (c) J. Lee and T. Weaver-Missick, *Agric. Res.*, 1999, *47*(6), 4; (d) L. McGinnis, *Agric. Res.*, 2006, *54*(8), 18.
262. D. Elstein and E. Peabody, *Agric. Res.*, 2005, *53*(10), 14; (b) J.V. Stafford, *Chem. Ind.* (*Lond.*), 2000, 98.
263. F. Forcella, M.E. Westgate, and D.D. Warnes, *Am. J. Alt. Agric.*, 1992, *7*(4), 161.
264. (a) D. Senft, *Agric. Res.*, 1997, *45*(5), 19; (b) J. Weiner, H.-W. Griepentrog, and L. Kristensen, *J. Appl. Ecol.*, 2001, *38*, 784,
265. (a) R.L. Zimdahl, *Am. J. Alt. Agric.*, 1995, *10*(3), 138; (b) *Am. J. Alt. Agric.*, 1995, *10*(3), 107; (c) D. Tenenbaum, *Technol. Rev.*, 1996, *99*(2), 21.
266. (a) K.S. Smallwood, *Am. J. Alt. Agric.*, 1996, *11*, 155; (b) M.A. Altieri, *Biodiversity and Pest Management in Agroecosystems*, Food Products, New York, 1994, 83; (c) D. Elstein and D. Comis, *Agric. Res.*, 2005, *53*(6), 14; (d) K. Thorpe, *Agric. Res.*, 2000, *48*(9), 23; (e) H. Becker, *Agric. Res.*, 2001, *49*(2), 14; (f) D. Comis and A. Perry, *Agric. Res.*, 2009, 57(8), 7; (g) D. O'Brien, *Agric. Res.*, 2009, *57*(9), 19.
267. (a) V. Klinkenborg, *Natl. Geogr. Mag.*, 1995, *188*(6), 69; (b) E.B. Mallory, J.I. Posner, and J.O. Baldcock, *Am. J. Alt. Agric.*, 1998, *13*(1), 2; (c) D. Comis, *Agric. Res.*, 2008, *56*(6), 18.
268. (a) L.G. McGinnis, *Agric. Res.*, 2008, *56*(8), 6; (b) D. Comis, *Agric. Res.*, 2008, *56*(4), 18.
269. D. Stanley, *Agric. Res.*, 1997, *45*(3), 12.
270. M.S. Clark, S.H. Gage, L.B. DeLind, and M. Lennington, *Am. J. Alt. Agric.*, 1995, *10*(3), 114.
271. National Research Council Board on Agriculture, *Ecologically Based Pest Management: New Solutions for a New Century*, National Academy Press, Washington, DC, 1996, 36.
272. J. Lee, *Agric. Res.*, 1996, *44*(5), 14.
273. R. Bragg, *New York Times*, Sept. 7, 1997, 12.
274. F.R. Hall and J.W. Barry, *Biorational Pest Control Agents—Formulation and Delivery*, A.C.S. Symp. 595, Washington, DC, 1995, 238, 252.
275. Anon., *Environment*, 1995, *37*(7), 22.
276. S.A. Duke, J.J. Menn, and J.R. Plimmer, eds, *Pest Control with Enhanced Environmental Safety*, A.C.S. Symp. 524, Washington, DC, 1994, 79, 117.
277. B. Hardin, *Agric. Res.*, 1998, *46*(10), 14.
278. G. Parkinson, *Chem. Eng.*, 2001, *108*(8), 21.
279. J. Suszkiw, *Agric. Res.*, 1999, *47*(5), 21.
280. A.E. Sutton and J. Clardy, *Org. Lett.*, 2000, *2*, 319.
281. P.A. Hedin, J.J. Menn, and R.M. Hollingsworth, eds, *Natural and Engineered Pest Management Agents*, A.C.S. Symp. 551, Washington, DC, 1992, 268.
282. M.M. Bobylev, L.I. Bobyleva, and G.A. Strobel, *J. Agric. Food Chem.*, 1996, *44*, 3960.
283. M.C. McDade and N.E. Christians, *Am. Alt. Agric.*, 2000, *15*, 189.
284. U.S. EPA Presidential Green Chemistry Challenge Award, 2008, p. 22.
285. (a) U.S. EPA Presidential Green Chemistry Challenge Award, 2001, p. 14; (b) G. Parkinson, *Chem. Eng.*, 2000, *107*(5), 23.
286. South Santee Aquaculture Inc., Charleston, SC, 2007.
287. Inderjit, K.M.M. Dakshini, and F.A. Einhellig, eds, *Allelopathy: Organisms, Processes and Applications*, A.C.S. Symp. 582, Washington, DC, 1994.
288. J.R. Vyvyan, *Tetrahedron*, 2002, *58*, 163.
289. L. Pons, *Agric. Res.*, 2005, *53*(5), 18.
290. L.-J. Lin, G. Peiser, B.-P. Ying, K. Mathias, F. Karasina, Z. Wang, J. Itatani, L. Green, and Y.-S. Hwang, *J. Agric. Food Chem.*, 1995, *43*, 1708.
291. P.D. Brown and M.J. Morra, *J. Agric. Food Chem.*, 1995, *43*, 3070.
292. S.A. Korhammer and E. Haslinger, *J. Agric. Food Chem.*, 1994, *42*, 2048.
293. (a) J.W.J.F. Thuring, N.W.J.T. Heinsman, R.W.A.W.M. Jacobs, G.H.L. Nefkens, and B. Zwanenburg, *J. Agric. Food Chem.*, 1997, *45*, 507; (b) J.W.J.F. Thuring, G.H.L. Nefkens, and B. Zwanenburg, *J. Agric. Food Chem.*, 1997, *45*, 1409; (c) B. Zwanenburg and J.W.J.F. Thuring, *Pure Appl. Chem.*, 1997, *69*, 651; (d) P. Welzel, S. Rohrig, and Z. Milkova, *Chem. Commun.*, 1999, 2017; (e) K. Yoneyama, D. Sato, Y. Takeuchi, H. Sekimoto, and T. Yokota, Anaheim A.C.S. Meeting, Mar. 2004, AGFD 26.
294. (a) Y. Ma, A.G.T. Babiker, I.A. Ali, Y. Sugimoto, and S. Inanaga, *J. Agric. Food Chem.*, 1996, *44*, 3355; (b) P. Welzel, S. Rohrig, and Z. Milkova, *Chem. Commun.*, 1999, 2017; (c) B. Halweil, *World Watch*, 2001, *14*(5), 26.
295. (a) A.J.F. Edmonds, W. Trueb, W. Oppolzer, and P. Cowley, *Tetrahedron*, 1997, *53*, 2785; (b) N.D. Smith, P.J. Kocienski, and S.D.A. Street, *Synthesis*, 1996, 652; (c) M.G. Banwell, C.T. Bui, G.W. Simpson, and K.G. Watson, *Chem. Commun.*, 1996, 723.
296. S.A. Duke and C.A. Rebeiz, eds, *Porphyric Pesticides: Chemistry*, Toxicology and Pharmaceutical Applications, A.C.S. Symp. 559, Washington, DC, 1994, 1, 65, 206.
297. R. Dagani, *Chem. Eng. News*, July 5, 1999, 30.
298. (a) C. Sasikala, C.V. Ramana, and P.R. Rao, *Biotechnol. Prog.*, 1994, *10*, 451; (b) G. Ondrey, *Chem. Eng.*, 2004, *111*(6), 17.
299. (a) A.M. Rouhi, *Chem. Eng. News*, Nov. 2, 1998, 22; (b) R.K. Pandey and C.K. Herman, *Chem. Ind.* (*Lond.*), 1998, 739.
300. (a) R.N. Strange, *Plant Disease Control: Toward Environmentally Acceptable Methods*, Chapman & Hall, London, 1993; (b) N.A. Rechcigl and J.E. Rechcigl, eds, *Environmentally Safe Approaches to Crop Disease Control*, CRC Lewis Publishers, Boca Raton, FL, 1997.

301. (a) B.J. Staskawicz, F.M. Ausubel, B.J. Baker, J.G. Ellis, and J.D.G. Jones, *Science*, 1995, *268*, 661; (b) F. Ratcliff, B.D. Harrison, and D.C. Baulcombe, *Science*, 1997, *276*, 1558; (c) W. Ligterink, T. Kroj, U. Zur Nieden, H. Hirt, and D. Scheel, *Science*, 1997, *276*, 2054; (d) J.E. Galan and A. Collmer, *Science*, 1999, *284*, 1322; (e) N.S. Al–Kaff, et al., *Science*, 1998, *279*. 2113.
302. D. Stanley, *Agric. Res.*, 1997, *45*(5), 10.
303. R.W. Smiley, *Am. J. Alt. Agric.*, 1996, *11*, 95.
304. (a) F. Pearce, *Conserv. Mag.*, 2008, *9*(4), 28; (b) D. Grimm, *Science*, 2008, *322*, 1046.
305. P.J. Hines and JMarx, eds, *Science*, 2001, *292*, 2269–2276.
306. (a) J.W. Kloeper, *Bioscience*, 1996, *46*, 406; (b) P.H. Abelson, *Science*, 1995, *269*, 1027.
307. (a) D.N. Dowling and F. O'Gara, *Trends Biotechnol.*, 1994, *12j*, 133; (b) G.H. Goldman, C. Hayes, and G.E. Harman, *Trends Biotechnol.*, 1994, *12*, 478.
308. B.R. Glick and Y. Bashan, *Biotechnol. Adv.*, 1997, *15*, 353.
309. L.P. Ooijhas, et al., *Enzyme Microb. Technol.*, 1998, *22*, 480.
310. K.B. Stelljes, *Agric. Res.*, 1997, *45*(10), 22.
311. K.B. Stelljes and D. Senft, *Agric. Res.*, 1998, *46*(1), 14.
312. (a) H. Becker, *Agric. Res.*, 1996, *44*(2), 22; (b) F.R. Hall and J.W. Barry, eds, *Biorational Pest Control Agents—Formulation and Delivery*, A.C.S. 595, Washington, DC, 1995, 166–173.
313. (a) J. Suszkiw, *Agric. Res.*, 1996, *44*(6), 16; (b) R. Lumsden, *Agric. Res.*, 1997, *45*(4), 23.
314. A. Coghlan, *New Sci.*, Feb. 10, 1996, 19.
315. P.A. Hedin, *Naturally Occurring Pest Bioregulators*, A.C.S. Symp. 449, Washington, DC, 1991, 407, 417.
316. E. Peabody, *Agric. Res.*, 2007, *55*(7), 20.
317. S. Durham, *Agric. Res.*, 2008, *54*(3), 14.
318. (a) Anon., *Chem. Eng. News*, Nov. 5, 2001, 47; (b) R. Winder, *Chem. Ind.* (*Lond.*), Jan. 23, 2006, 20.
319. (a) M. Burke, *Chem. Ind.* (*Lond.*), 1996, 910; (b) *Environ. Sci. Technol.*, 1994, *28*, 357A; (c) J. Hodgson, *Biotechnology*, 1994, *12*, 481.
320. (a) G. Ji, R. Beavis, and R.P. Novick, *Science*, 1997, *276*, 2027; (b) B. Sneh, *Biotechnol. Adv.*, 1998, *16*(1), 1–32.
321. National Research Council Board on Agriculture, *Ecologically Based Pest Management: New Solutions for a New Century*, National Academy Press, Washington, DC, 1996, 51.
322. B. Chen, *Science*, 1994, *264*, 1762.
323. Anon., *New Sci.*, Mar. 14, 1998, 17.
324. (a) V. Glaser, *Nat. Biotechnol.*, 1998, *16*, 413; (b) J. Core, *Agric. Res.*, 2004, *52*(1), 20.
325. A. Flores, *Agric. Res.*, 2004, *52*(3), 9.
326. S. Durham, *Agric. Res.*, 2005, *53*(11), 8.
327. N. Jones, *New Sci.*, Mar. 24, 2001, 6.
328. R.M. Bliss, *Agric. Res.*, 2007, *55*(3), 22.
329. J.M. Fonseca and S. Ravishankar, *Am. Sci.*, 2007, *95*, 494.
330. D. Stanley, *Agric. Res.*, 1997, *45*(10), 16.
331. C.L. Wilson and M.E. Wisniewski, *Biocontrol of Postharvest Diseases, Theory and Practice*, CRC Press, Boca Raton, FL, 1994.
332. (a) Anon., *Chem. Eng. News*, Mar. 13, 1995, 26; (b) National Research Council Board on Agriculture, *Ecologically Based Pest Management: New Solutions for a New Century*, National Academy Press, Washington, DC, 1996, 51.
333. R. Bliss, *Agric. Res.*, 2002, *50*(7), 12.
334. (a) S. Everts, *Chem. Eng. News*, Oct. 29, 2007, 10; (b) J. Suszkiw, *Agric. Res.*, 2007, *55*(9), 18.
335. (a) C. Holden, *Science*, 1995, *270*, 1443; (b) S.R. Parker, H.G. Cutler, J.M. Jacyno, and R.A. Hill, *J. Agric. Food Chem.*, 1997, *45*, 2774; (c) M. Rito-Palomares, A. Negrete, L. Miranda, C. Flores, E. Galindo, and Serranno-Carreon, *Enzyme Microb. Technol.*, 2001, *28*, 625.
336. J.M. Cooney, D.R. Lauren, D.J. Jensen, and L.J. Perry–Meyer, *J. Agric. Food Chem.*, 1997, *45*, 531, 2802.
337. A. Sivropoulou, C. Nikolaou, E. Papanikolaou, S. Kokkini, T. Lanaras, and M. Arsenakis, *J. Agric. Food Chem.*, 1997, *45*, 3197.
338. (a) J. Suszkiw, *Agric. Res.*, 1998, *46*(6), 23; (b) V.M. Gomes, M.-I. Mossqueda, A. Blanco–Labra, M.P. Sales, K.V.S. Fernandes, R.A. Corderio, and J. Xavier–Filho, *J. Agric. Food Chem.*, 1997, *45*, 4110; (c) K. Seetharaman, E. Whitehead, N.P. Keller, R.D. Vaniska, and A.W. Rooney, *J. Agric. Food Chem.*, 1997, *45j*, 3666; (d) D. Yang, O. Chertov, S.N. Bykoskaia, Q. Chen, M.J. Buffo, J. Shogan, M. Anderson, J.M. Schroder, J.M. Wang, O.M.Z. Howard, and J.J. Oppenheim, *Science*, 1999, *286*, 525.
339. R.E.W. Hancock and R. Lehrer, *Trends Biotechnol.*, 1998, *16*, 82.
340. (a) J.M. Clough and C.R.A. Godfrey, *Chem. Br.*, 1995, *31*, 466; (b) A. Miller, *Chem. Ind.* (*Lond.*), 1997, 7; (c) S. Zapf, A. Werle, T. Anke, D. Klostermeyer, B. Steffan, and W. Steglich, *Angew. Chem. Int. Ed.*, 1995, *34*, 196; (d) Anon., *Chem. Eng. News*, Nov. 25, 1996, 16.
341. E.E. Jones, F.J. Weber, J. Oostra, A. Rinzema, A. Mead, and J.M. Whipps, *Enzyme Microb. Technol.*, 2004, *34*, 196.
342. B.K. Hwang, J.Y. Lee, B.S. Kim, and S.S. Moon, *J. Agric. Food Chem.*, 1996, *44*, 3653.
343. A.B. Charette and H. Lebel, *J. Am. Chem. Soc.*, 1996, *118*, 10327.
344. L. Mannina, A.L. Segre, A. Ritieni, V. Fogliano, F. Vinale, G. Randazzo, L. Maddau, and A. Bottalico, *Tetrahedron*, 1997, *53*, 3135.
345. J.L. Beckerman, F. Naider, and D.J. Ebbole, *Science*, 1997, *276*, 1116.
346. R.S. Patil, V. Ghormade, and M.V. Deshpande, *Enzyme Microb. Technol.*, 2000, *26*, 473.
347. Anon., *R&D* (*Cahners*), 1998, *40*(10), 126.
348. Anon., *Chem. Ind.* (*Lond.*), 1999, 905.
349. L. McGinnis, *Agric. Res.*, 2008, *56*(4), 10.
350. (a) C. Putnam, *New Sci.*, June 24, 1995, 24; (b) Y.-J. Ahn, S.-B. Lee, T. Okubo, and M. Kim, *J. Chem. Ecol.*, 1995, *21*, 263.
351. *Environ. Sci. Technol.*, 1997, *31*, 180A.
352. (a) A.S. Moffat, *Science*, 1998, *279*, 988; (b) D. Malakoff, *Science*, 1998, *281*, 190; (c) J. Pelley, *Environ. Sci. Technol.*, 1998, *32*, 462A; (d) P. Vitousek, J.D. Aber, R.W. Howarth, G.E. Likens, P.A. Matson, D.W. Schindler, W.H. Schlesinger, and D.G. Tilman, *Ecol. Appl.*, 1997, *7*(3), 737; (e) J.N. Galloway, A.R. Townsend, J.W. Erisman, M. Bekunda, Z. Cai, J.R. Freney, L.A. Martinelli, S.P. Seitzinger, and M.A. Sutton, *Science*, 2008, *320*, 889; (f) E.W. Boyer and R.A. Howarth, eds, *The Nitrogen Cycle at Regional to Global Scales*, Kluwer, Norwell, MA, 2002; (g) L. McGinnis, *Agric. Res.*, 2009, *57*(2), 20; (h) P.M. Vitousek, R. Naylor, T. Crews, M.B. David, L.E. Drinkwater, E. Holland et al., *Science*, 2009, *324*, 1519; 2009, *326*, 665.

353. A. Thayer, *Chem. Eng. News*, Mar. 23, 1998, 7.
354. Economic Research Service, U.S. Dept. of Agriculture, Washington, DC.
355. (a) J.R. Porter, N. Chirinda, C. Felby, and J.E. Olesen, *Science*, 2008, *320*, 1421; (b) S. Jutzi and LEAD Team, *Livestock's Long Shadow*, FAO, Rome, 2006; (c) M.A. Mallin, *Am. Sci.*, 2000, *88*(1), 26.
356. (a) Anon., *Environ. Sci. Technol.*, 1998, *32*, 355A; (b) D.C. Taylor and D.H. Rickerl, *Am. J. Alt. Agric.*, 1998, *13*(2), 61.
357. P.A. Matson, R. Naylor, and I. Ortiz-Monasterio, *Science*, 1998, *280*, 112.
358. E. Kirschner, *Chem. Eng. News*, Mar. 18, 1996, 16.
359. (a) Y. Bashan, *Biotechnol. Adv.*, 1998, *16*, 729; (b) C. Amiet-Charpentier, J.-P Benoit, D. LeMeurlay, P. Gadille, and J. Richard, *Macromal. Symp.*, 2000, *151*, 611.
360. (a) T.A. Obreza, *Am. J. Alt. Agric.*, 1996, *11*, 147; (b) S.S. Andres and L. Lohr, *Am. J. Alt. Agric.*, 1996, *11*, 147; (c) K.H. Tan, *Principles of Soil Chemistry*, 3rd ed., Dekker, New York, 1998; (d) M.H.B. Hayes and W.S. Wilson, *Humic Substances, Peats and Sludges—Health and Environmental Aspects*, Special Publication 172, Royal Society of Chemistry, Cambridge, 1997; (e) J.S. Gaffney, N.A. Marley, and S.B. Clark, eds, *Humic and Fulvic Acids: Isolation, Structure and Environmental Role*, A.C.S. Symp. 651, Washington, DC, 1996.
361. R.M. Kashmanian and R.F. Rynk, *Am. J. Alt. Agric.*, 1998, *13*, 40.
362. M.R. Werner, *Am. J. Alt. Agric.*, 1996, *11*, 186.
363. (a) J. Pelley, *Environ. Sci. Technol.*, 1998, *32*, 26A; (b) B. Hileman, *Chem. Eng. News*, Oct. 13, 1997, 14.
364. (a) J. Pelley, *Environ, Sci. Technol.*, 1998, *32*, 305A; (b) K.S. Betts, *Environ. Sci. Technol.*, 1998, *32*, 535A; (c) B. Baker, *Bioscience*, 1998, *48*, 996.
365. (a) T.W. Bruulsema, Boston A.C.S. Meeting, FERT 005, Aug. 1998; (b) L.L. Rectenwald and R.W. Drenner, *Environ. Sci. Technol.*, 2000, *34*, 522; (c) M.V. Jimenez-Perez, P. Sanchez-Castillo, O. Romera, D. Fernandez-Moreno, and C. Perez-Martinez, *Enzyme Microb. Technol.*, 2004, *34*, 392.
366. (a) P.H. Abelson, *Science*, 1999, *283*, 2015; (b) P. Vats, M.S. Bhattacharyya, and U.C. Banerjee, *Critical Rev. Environ. Sci. Technol.*, 2005, *35*, 469.
367. (a) T. Weaver, *Agric. Res.*, 1998, *46*(6), 12; (b) K.H. Nahm, *Critical Rev. Environ. Sci. Technol.*, 2005, *35*, 487.
368. (a) G. Ondrey, *Chem. Eng.*, 2005, *112*(4), 17; (b) L. Pons, *Agric. Res.*, 2005, *53*(3), 14; 2007, *55*(5), 19.
369. (a) C.P Singh and A. Amberger, *Am. J. Alt. Agric.*, 1995, *10*(2), 82; (g) G. Ondrey, *Chem. Eng.*, 1999, *106*(6), 19; (c) J.F. Fernandez-Bertran, *Pure Appl. Chem.*, 1997, *71*, 581.
370. J. Makower, *Wilmington Delaware News J.*, Aug. 3, 1995, D3.
371. (a) W. Schultz, *The Chemical-Free Lawn*, Rodale Press, Emmaus, PA, 1989; (b) V.S. Jenkins, *The Lawn: A History of an American Obsession*, Smithsonian Institution Press, Washington, DC, 1994; (c) K.M. Reese, *Chem. Eng. News*, Mar. 16, 1998, 80; (d) T. Steinberg, *Amerian Green—The Obsessive Quest for the Perfect Lawn*, W.W. Norton, New York, 2005.
372. P.D. Thacker, *Environ. Sci. Technol.*, 2005, *39*, 98A.
373. M. Amato, *Garbage*, July/Aug. 1990, 34.
374. (a) *University of California at Berkeley Wellness Lett.*, 1995, *11*(11), 6; (b) S.C. Templeton, D. Zilberman, and S.J. Yoo, *Environ. Sci. Technol.*, 1998, *32*, 416A; (c) A.S. Moffat, *Science*, 1999, *284*, 1249.
375. J. Nogaki, *Delaware Estuary News*, 1995, *5*(4), 12.
376. J.M. Vargas, Jr., *Management of Turfgrass Diseases*, 2nd ed., Lewis, Boca Raton, FL, 1993.
377. (a) A.E. Platt, *World Watch*, 1994, *7*(3), 27; (b) Anon., *Environment*, 1994, *36*(6), 21.
378. T. Suzuki, H. Kondo, K. Yaguchi, T. Maki, and T. Suga, *Environ. Sci. Technol.*, 1998, *32*, 920.
379. (a) Anon., *Chem. Eng. News*, Apr. 20, 1998, 71; (b) Turf Grass Chemicals Symposium, Boston A.C.S. Meeting, AGRO 059-167, Aug. 1998.
380. J. Makower, *Wilmington Delaware News J.*, Apr. 21, 1994.
381. (a) C. Holden, *Science*, 1995, *268*, 1571; (b) K.M. Reese, *Chem. Eng. News*, July 29, 1996, 80.
382. W. Schultz, *The Chemical-Free Law*, Rodale Press, Emmaus, PA, 1989, 112.
383. E.A. Tilman, D. Tilman, M.J. Crawley, and A.E. Johnston, *Ecol. Appl.*, 1999, *9*, 103.
384. J.M. Clark and M.P. Kenna, *Fate and Management of Turfgrass Chemicals*, A.C.S. Symp. 743, Washington, DC, 2000.
385. S. Apfelbaum and J. Broughton, *Land Water*, 1998, *42*(1), 6.
386. (a) F.H. Bohrmann, D. Balmori. G.T. Geballe, and L. Vernegaard, *Redesigning the American Lawn: A Search for Environmental Harmony*, Yale University Press, New Haven, CT, 1993. (b) C. Uhl, *Conserv. Biol.*, 1998, *12*, 1175; (c) K.M. Reese, *Chem. Eng. News*, May 4, 1998, 88; (d) S. Daniels, *The Wild Lawn Handbook—Alternatives to the Traditional Front Lawn*, Macmillan, New York, 1995.
387. (a) Anon., *Native Plants in the Landscape Conference*, 6/12/97, Millersville University, Millersville, PA; (b) N. Diboll, *Prairie Nursery*, Westfield, WI; (c) *Bowman's Hill Wildflower Preserve*, New Hope, PA.
388. (a) J.D. deVault, K.J. Hughes, O.A. Johnson, and S.K. Narwang, *Biotechnology*, 1996, *14*, 46. (b) D. Hanson, *Chem. Eng. News*, May 13, 1996, 22; (c) A.M. Thayer, *Chem. Eng. News*, Apr. 19, 1999, 21; (d) P.H. Abelson and P.J. Hines, *Science*, 1999, *285*, 367 and following reviews, 368–389; (e) N. Federoff and N.M. Brown, *Mendel in the Kitchen—A Scientist's View of Genetciallly-Modified Foods*, Joseph Henry Press, Washington, DC, 2004; (f) K. Rajasekaran, T. Jacks and J. Finley, eds, *Crop Biotechnology*, A.C.S. Symp. 829, Washington, DC, 2002; (g) D.K. Letourneau and B.E. Burrows, *Genetically-Engineered Organisms: Environmental and Human Health Effects*, CRC Press, Boca Raton, FL, 2002; (h) N. Halford, ed., *Plant Biotechnology—Current and Future applications of Genetically-Modified Crops*, Wiley, New York, 2006; (i) www.PewAgBiotech.org; (j) P.F. Urquin *High Tech Harvest—Understanding Genetically-Modified Food Plants*, Westview (Perseus), Boulder, CO, 2002.
389. F. Mourgues, M.-N. Brisset, and E. Chevreau, *Trends Biotechnol.*, 1998, *16*, 203.
390. (a) R. Haselkorn, *Nat. Biotechnol.*, 1997, *15*, 511; (b) P. Mavingui, M. Flores, D. Romero, E. Martinez–Romero, and R. Palacios, *Nat. Biotechnol.*, 1997, *15*, 564.
391. (a) J. Rissler and M. Mellon, *The Ecological Risks of Engineered Crops*, MIT Press, Cambridge, MA., 1996; (b) B. Hileman, *Chem. Eng. News*, Aug. 21, 1995, 8.

392. R.E. Shade, H.E. Schroeder, J.J. Pueyo, L.M. Tabe, L.L. Murdock, T.J.V. Higgins, and M.J. Crispeels, *Biotechnology*, 1994, *12*, 793.
393. R. Grison, B. Grezes–Besset, M. Schneider, N. Lucante, L. Olsen, J.-M. Leguay, and A. Toppan, *Nat. Biotechnol.*, 1996, *14*, 643.
394. G. Parkinson, *Chem. Eng.*, 1999, *106*(3), 27.
395. X. Duan, X. Li, Q. Xue, M. Abo—El-Saad, D. Xu, and R. Wu, *Nat. Biotechnol.*, 1996, *14*, 494.
396. O. Zoubenko, F. Uckun, Y. Hur. I. Chat, and N. Turner, *Nat. Biotechnol.*, 1997, *15*, 992.
397. (a) Anon., *Bioscience*, 1998, *48*, 240; (b) Anon., *New Sci.*, Dec. 13, 1997, 17.
398. D. Lyon-Johnson, *Agric. Res.*, 1997, *45*(11), 12.
399. (a) J. Lee, *Agric. Res.*, 1999, *47*(4), 16; (b) Anon., *Chem. Eng. News*, May 4, 1998, 40; (c) A.B. DeMilo, C.-J. Lee, R.F.W. Schroder, W.F. Schmidt, and D.J. Harrison, *J. Entomol. Sci.*, 1998, *33*, 343; (d) R.F.W. Schroder, A.B. DeMilo, C.-J. Lee, and P.A.W. Martin, *J. Entomol. Sci.*, 1998, *33*, 355; (e) D. Comis, *Agric. Res.*, 1997, *45*(10), 4; (f) D. Pimentel and M. Pimentel, *Science*, 2000. *288*, 1966.
400. D. Ferber, *Science*, 2000, *287*, 1390.
401. Anon., *Chem. Ind.* (*Lond.*), 2001, 4918; (c) M. Jacoby, *Chem. Eng. News*, Sept. 28, 2009, 60.
402. (a) P. Walter, *Chem. Ind.* (*Lond.*), July 18, 2005, 18; (b) E. Pennisi, *Science*, 2008, *320*, 171; (c) M. Jacoby, *Chem. Eng. News*, Sept. 28, 2009, 60.
403. (a) P. Ronald, *Conserv. Mag.*, 2008, *9*(3), 35; (b) P.C. Ronald and R.W. Adamchak, *Tomorrow's Table—Organic Farming, Genetics and Future of Food*, Oxford University Press, New York, 2008.
404. D. Comis and R.F. Wilson, *Agric. Res.*, 2005, *53*(7), 2, 4.
405. R.M. Bliss, *Agric. Res.*, 2004, *52*(2), 22.
406. E. Stokstad, *Science*, 2007, *316*, 196.
407. (a) V. Vance and H. Vaucheret, *Science* 2001, *292*, 2277; (b) D.R.G. Price and J.A. Gatehouse, *Trends Biotechnol.*, 2008, *26*(7), 393; (c) M. Scharf, *Chem. Ind.* (*Lond.*), Aug. 11, 2008, 20.
408. (a) R. Wrubel, *Technol. Rev.*, 1994, *97*(4), 57; (b) B. Halweil, *World Watch*, 1999, *12*(4), 219; (c) D. Gurian-Sherman and E. Robinson, *Catalyst* (Union of Concerned Scientist) Summer 2009, 11.
409. (a) H.F. van Emden and D.B. Pealkall, *Beyond Silent Spring*, Chapman & Hall, London, 1996; (b) A.R. Leslie and G.W. Cuperus, eds, *Successful Implementation of Integrated Pest Management for Agricultural Crops*, Lewis, Boca Raton, FL, 1993; (c) A.R. Leslie, *Handbook of Integrated Pest Management for Turf and Ornamentals*, Lewis, Boca Raton, FL, 1994; (d) G. Gardner, *World Watch*, 1996, *9*(2), 20; (e) D. Dent, *Integrated Pest Management*, Chapman & Hall, London, 1995; (f) D.J. Chadwick and J. Marsh, eds, *Crop Protection and Sustainable Agriculture*, Ciba Foundation Symp. 177, Wiley, New York, 1993; (g) N. Uri, *Agriculture and the Environment*, Nova Science, Commack, NY, 1999; (h) P.A. Matson, W.J. Parton, A.G. Power, and M.J. Swift, *Science*, 1997, *277*, 504; (i) A. Shani, *Chemtech*, 1998, *28*(3), 30; (j) D. Evans, *Chem. Br.*, 1998, *34*(7), 20; (k) P. Crowley, H. Fischer, and A. Devonshire, *Chem. Br.*, 1998, *34*(7), 25; (l) M. Luszniak and J. Pickett, *Chem. Br.*, 1998, *34*(7), 29; (m) J. Stetter and F. Lieb, *Angew. Chem. Int. Ed.*, 2000, *39*, 1725; (n) S.R. Gliessman, *Agroecology: Ecological Processes in Sustainable Agriculture*, Ann Arbor Press, Chelsea, MI, 1998; (o) M. Wilson, ed., *Optimizing Pesticide Use*, Wiley, Chichester, 2003; (p) B.L. Gardner, *American Agriculture in the 20th Century—How It Fluourished and What It Cost*, Harvard University Press, Cambridge, MA, 2002; (r) D.L. Jackson and L.L. Jackson, *The Farm As A Natural Habitat - Reconnecting Food Systems With Ecosystems*, Island Press, Washington, DC, 2002; (s) www.ipmworld.umn.edu; (t) L.L. Jackson, *Ecol. Restoration*, 2002, *20*(2), 96; (u) S.J. Scherr and J.A. McNeely, *Farming With Nature—The Science and Practice of Ecoagriculture*, Island Press, Washington, DC, 2007.
410. L.G. Highley and L.P. Pedigo, eds, *Economic Thresholds for Integrated Pest Management*, University of Nebraska Press, Lincoln, NE, 1996.
411. (a) J. Wood, *Chem. Ind.* (*Lond.*), Oct. 27, 2008, 19; (b) O. Lopez, J.G. Fernanez-Bolanos, and M.V. Gil, *Green Chem.*, 2005, *7*, 431.
412. (a) G. Gardner, *World Watch*, 1996, *9*(2), 25; (b) P.R. Ehrlich and A.H. Ehrlich, *Technol. Rev.*, 1997, *100*(1), 38.
413. J. Emory, *Nat. Conserv.*, 1994, *44*(6), 14.
414. J. Pleydell-Bouverie, *New Sci.*, Sept. 24, 1994, 26.
415. G. Gardner, In: L. Strake, ed., *State of the World*, W.W. Norton, New York, 1996, 88.
416. (a) C.S. Holligsworth, M.J. Paschall, N.L. Cohen, and W.M. Coli, *Am. J. Alt. Agric.*, 1993, *8*(2), 78; (b) C.S. Hollingsworth and W.M. Coli, *Am. J. Alt. Agric.*, 2001, *16*(4), 177.
417. J. Ikerd, *Am. J. Alt. Agric.*, 1992, *7*(3), 121.
418. D. Comis, *Agric. Res.*, 1996, *44*(1), 12.
419. (a) Anon., *Am. J. Alt. Agric.*, 1995, *10*(3), 121;(b) D. Senft, *Agric. Res.*, 1997, *45*(5), 4.
420. J. Pelley, *Environ. Sci. Technol.*, 2000, *34*, 373A.
421 (a) R.M. Faust, *Agric. Res.*, 1997, *45*(10), 2; (b) J.K. Kaplan, *Agric. Res.*, 2003, *51*(2), 4; (c) R.M. Faust, *Agric. Res.*, 2004, *52*(11), 1.
422. (a) I. Perfecto, *Am. J. Alt. Agric.*, 1994, *9*(3), 98; (b) J. Simon, *Amicus J.* (Natural Resources Defense Council), 1997, *18*(4), 35; (c) C. Flavin, *World Watch*, 1997, *10*(3), 37; (d) F. Funes, ed., *Sustainable Agriculture and Resistance: Transforming Food Production Production in Cuba*, Food First Books, Milford, CT; (e) csanr.wsu.edu/Cuba/CubaTripReport2003-09-09.pdf.
423 a) WHYY, *National Public Radio*, Jan. 2009; (b) www.ipsnews.net/news.asp?idnews = 42347.
424. (a) L.K. Lee, *Am. J. Alt. Agric.*, 1992, *7*(1–2), 82; (b) W.S. Roberts and S.M. Swinton, *Am. J. Alt. Agric.*, 1996, *11*, 10; (c) J.D. Smolik, T.L. Dobbs, and D.H. Rickerl, *Am. J. Alt. Agric.*, 1995, *10*(1), 25; (d) K.M. Painter, D.L. Young, D.M. Granatstein, and D.J. Mulla, *Am. J. Alt. Agric.*, 1995, *10*(2), 88.
425. (a) L.E. Drinkwater, P. Wagoner, and M. Sarrantonio, *Nature*, 1998, *396*, 262; (b) D. Tilman, *Nature*, 1998, *396*, 211.
426. C. Macilwain, Q. Schiermeier, L. Nelson, J. Giles, and V. Gewin, *Nature*, 2004, *428*, 796.
427. (a) G. Sposito, *The Chemistry of Soils*, Oxford University Press, Oxford, 2008; (b) D. Comis, *Agric Res.*, 2007, *55*(6), 4.
428. D. Nierenberg, *World Watch*, 2005, *18*(1), 22.
429. (a) J. Fernandez–Cornejo, C. Greene, R. Penn, and D. Newton, *Am. J. Alt. Agric.*, 1998, *13*, 69; (b) E. McCann, S. Sullivan, D. Erickson, and R. DeYoung, *Environ. Manage.*, 1997, *21*, 747; (c) B. Hileman, *Chem. Eng. News*, Mar. 13, 2000, 11. (d) OECD, *Organic Agriculture—Sustainability, Markets*

and Policies, CABI Publishing, Wallingford, UK, 2003; (e) C. Badgeley, J. Moghtader, E. Quintero, E. Zakin, M.J. Chappell, K. Aviles-Vazquez, A. Samulon, and I. Perfecto, *Renewable Agriculture and Food Systems*, 2007, *22*, 86.
430. Anon., *Am. J. Alt. Agric.*, 1996, *11*, 131.
431. www.organic-center.org/science.nutri.php.
432. E. Samuel, *New Sci.*, Apr. 21, 2001, 15.
433. L. Richards, *Chem. Ind.* (*Lond.*), Mar. 26, 2007, 8.568.
434. Anon., *Chem. Ind.* (*Lond.*), Jun. 11, 2007, 8.
435. V. Klinkenborg, *Natl. Geographic Mag.*, 1995, *188*(6), 61.
436. F.L. Young, A.G. Ogg, Jr., and R.I. Papendick, *Am. J. Alt. Agric.*, 1994, *9*(1–2), 52.
437. Y.O. Ogini D.P. Stonehouse, and E.A. Clark, *Am. J. Alt. Agric.*, 1999, *14*, 59.
438. E. van der Werf, *Am. J. Alt. Agric.*, 1993, *8*(4), 185.
439. (a) A. Herro, *World Watch*, 2006, *19*(5), 5; (b) J. Pretty, *Environment*, 2003, *45*(9), 8. (c) F. Pearce, *New Sci.*, Feb. 3, 2001, 16; (d) P.A. Sanchez, *Science*, 2002, *295*, 2019; (e) N. Parrott and T. Marsden, *The Real Green Revolution, Organic and Agroecological Farming in the South*, Greenpeace Environmental Trust, U.K.
440. (a) F.P. Weibel, *Am. J. Alt. Agric.*, 2001, *16*(4), 191; (b) P. Mader, A. Fliessbach, D. Dubois, L. Gunst, P. Fried, and U. Niggli, *Science*, 2002, *296*, 1694.
441. W. Lockeretz, *Am. J. Alt. Agric.*, 1999, *14*, 37.
442. A. Flores, *Agric. Res.*, 2008, *56*(10), 4.
443. D.K. Letourneau and B. Goldstein, *J. Apl. Ecol.*, 2001, *38*, 557.
444. (a) Anon., *University of California Berkeley Wellness Lett.*, 2003, *19*(5), 2; (b) B. Hileman, *Chem. Eng. News*, Jan. 8, 2001, 24; Mar. 13, 2000, 11.
445. (a) P.L. Diebel, D.B. Taylor, and S.S. Batie, *Am. J. Alt. Agric.*, 1993, 8(3), 120; (b) P.H. Abelson, *Science*, 1995, *267*, 943; (c) C.H. Cummings, *World Watch*, 2006, *19*(4), 38.
446. D. Hanson, *Chem. Eng. News*, Jan. 2, 1995, 18.
447. P.R. Ehrlich and A.H. Ehrlich, *Technol. Rev.*, 1997, *100*(1), 46.
448. D. Dent, *Integrated Pest Management*, Chapman & Hall, London, 1995, 282.
449. (a) D. McKenzie, *New Sci.*, Aug. 26, 1995, 6; (b) A.M. Rouhi, *Chem. Eng. News*, Oct. 7, 1996, 10.
450. T. Colburn, D. Dumanoski, and J.P. Meyers, *Our Stolen Future*, Dutton, New York, 1996, 299.
451. F.R. Hall and J.W. Barry, eds, *Biorational Pest Control Agents—Formulation and Delivery*, A.C.S. Symp. 595, Washington, DC, 1995, *5*, 178.
452. D. Normile, *Science*, 2008, *320*, 303.

RECOMMENDED READING

1. B. Hileman, *Chem. Eng. News*, Mar. 18, 1996, 23 (immune system suppression caused by pesticides).
2. T. Eisner and J. Meinwald, eds, *Chemical Ecology: The Chemistry of Biotic Interaction*, National Academy Press, Washington, DC, 1995, 103–111 (chemistry of sex attraction in insects).
3. G. Gardner, In: L. Strake, ed., *State of the World 1996*, W.W. Norton, New Work, 1996, 78–94 (preserving agricultural resources).
4. (Alternate for Ref. 3). M. Mellon, J. Rissler, and F. McCamant, *Union of Concerned Scientists Briefing Paper*, "Sustainable Agriculture," 1995.
5. I. Perfecto, *Am. J. Alt. Agric.*, 1994, *9*(3), 98 (integrated pest management in Cuba).
6. R. Wrubel, *Technol. Rev.*, 1994, *97*(4), 57 (herbicide-resistant crops).
7. P.A. Matson, W.J. Parton, A.G. Power, and M.J. Swift, *Science*, 1997, *277*, 504 (problems with agrochemicals).
8. B. Hileman, *Chem. Eng. News*, July 19, 1999, 42 (transgenic crops).
9. A.S. Moffat, *Science*, 1998, *279*, 988 (global nitrogen overload).
10. N. Halford, *Chem. Ind.* (*Lond.*), 2001, 505 (transgenic crops).

EXERCISES

1. Visit a garden supply store to see what chemicals are being sold for use around the home. Look up the chemicals in the *Merck Index* and other suitable sources to see what they are and what their toxicities are. Include Scott's four-step lawn care program in the survey. If possible compare the levels of application to those recommended by your local agricultural extension agent for use on farm crops.
2. See if any of your neighbors use electric insect traps and "Bag-a-Bug" traps. Neither is effective.
3. Compare prices of organic farm produce and conventional produce in a local store.
4. See how many compost piles there are in your neighborhood.
5. Check with your university agricultural department to see that research is being conducted on sustainable agriculture.
6. Is there any contamination of groundwater or surface water from agricultural or lawn runoff near where you live? If so, what are the sources? Propose some solutions for any problems that may have resulted from it.
7. Land use can change with the passing of time. Are there any former agricultural fields in your area that are now being used for other purposes that contain copper, arsenic and lead? Is remediation required?
8. What is the status of bees in the area where you live, both honeybees and other bees? Are crops and flowers being pollinated adequately?
9. Is there any fish farming near where you live? If so, what have they done to eliminate the environmental problems of aquaculture?

CHAPTER 12

Materials for a Sustainable Economy

12.1 INTRODUCTION

Most organic chemicals in use today are derived from nonrenewable petroleum and natural gas, with some still being made from coal. After use, the bulk of these ultimately end up as carbon dioxide, the main greenhouse gas that is causing global warming. For a sustainable future,[1] these must be based on renewable resources[2] from fields and forests. Before the advent of cheap petroleum and natural gas, many of them were. It is instructive to look back at these older methods to see if they might be economical today if the cost of global warming were included in the prices of the organic chemicals that we use. Further optimization of the processes with the knowledge that has accumulated since they were in use should increase yields and lower costs. Biocatalysis (see Chapter 9) is an example. Formerly it was used more for the production of commodity chemicals than it is today. Some of these earlier fermentation methods have not been studied much lately. Because energy is often a large factor in producing chemicals, renewable sources (see Chapter 15) will have to replace the fossil fuels used today. Biocatalysis with renewable raw materials has become more popular with the recent escalation of prices for oil and gas. The adoption of such sources of energy and raw materials will cut the amount of acid rain substantially, for there is not much sulfur in either the raw materials or the fuels.

The following sections will explore routes to the basic chemicals that are now in use, as well as some new possibilities. They should be read in conjunction with one of the standard texts[3] on industrial organic chemistry. (Specific references will not be given for well-known processes in these texts.) Carcinogens and very toxic chemicals will be avoided where possible. Energy-intensive processes will be avoided, if there is more than one possible route. Some methods of converting oil and natural gas to commodity chemicals use as much as one-third of the starting material for energy to fuel the process. They often involve high temperatures, with low conversions, and much recycling. Sometimes, it is necessary to quench a hot stream to obtain the desired product. After the product is separated, the stream must be reheated to the high temperature for the next cycle. Many of today's processes involve oxidation of the starting hydrocarbon to move oxygen into the molecule. Renewable raw materials have oxygen in them, so that converting them to the desired products involves different methods, which may not require such extreme temperatures and so much energy. Removal of oxygen atoms may be involved. Although biogas (methane) derived from anaerobic fermentation of organic matter could be used with today's processes for natural gas, there would not be enough of it.

12.2 COMMODITY CHEMICALS FROM RENEWABLE RAW MATERIALS

The 50 largest-volume chemicals contain many derived from fossil carbon sources. Their 1995 volumes in billions of pounds produced in the United States[4] are ethylene (46.97), ammonia (35.60), propylene (25.69), methyl *tert*-butyl ether (17.62), ethylene dichloride (17.26), nitric acid (17.24), ammonium nitrate (15.99), benzene (15.97), urea (15.59), vinyl chloride (14.98), ethylbenzene (13.66), styrene (11.39), methanol (11.29), carbon dioxide (10.89), xylene (9.37), formaldehyde (8.11), terephthalic acid (7.95), ethylene oxide (7.62), toluene (6.73), *p*-xylene (6.34), cumene (5.63), ethylene glycol (5.23), acetic acid (4.68), phenol (4.16), propylene oxide (4.00), butadiene (3.68), carbon black (3.32), isobutylene (3.23), acrylonitrile (3.21), vinyl acetate (2.89), acetone (2.76), butyraldehyde (2.68), adipic acid (1.80), nitrobenzene (1.65), and bisphenol A (1.62). Ammonia is included because the hydrogen used in its synthesis is derived from methane or another fossil hydrocarbon. Many of these compounds are intermediates in the synthesis of polymers. Polyethylene is the world's most common plastic (41% of the total), with others including polyvinyl chloride (23%), polypropylene (21%), polystyrene (11%), and acrylonitrile-butadiene-styrene plus styrene-acrylonitrile copolymers (4%).[5] The use of poly(ethylene terephthalate) is growing. The United States produced 18 billion tons of polyethylene, 8.7 billion tons of polypropylene, 2.7 billion tons of polystyrene, 591 million tons of polyamides, 6.6 billion tons of poly(vinyl chloride), 3.5 billion tons of resins made with formaldehyde, and 296 million tons of epoxy resins in 2007.[6]

12.2.1 Olefins and Chemicals Derived from Them

Ethylene for polymerization to the most widely used fermentation (**12.1** Schematic).[7] Dow Chemical and Braskem are building a plant in Brazil to make ethylene and polyethylene from ethanol from sugar cane.[8] Solvay Indupa is planning a plant in Brazil to convert such ethylene to vinyl chloride. Ethanol can be converted to ethylene in 100% yield with silicotungstic acid in MCM-411 zeolite catalyst or with iron(III) phosphate hydrate in 97.6% yield.[9] The ethanol used need not be anhydrous. Dehydration of 20% aqueous ethanol over HZSM-5 zeolite gave 76–83% ethylene, 2% ethane, 6.6% propylene, 2% propane, 4% butenes, and 3% *n*-butane.[10] Presumably, the paraffins could be dehydrogenated catalytically after separation from the olefins.[11] Ethylene can be dimerized to 1-butene with a nickel catalyst.[12] It can be trimerized to 1-hexene with a chromium catalyst and a phosphine with a 98% yield.[13] Ethylene is often copolymerized with 1-hexene to produce linear low-density polyethylene. The trimerization and copolymerization can be done in one vessel.[14] Brookhart and co-workers have developed iron, cobalt, nickel, and palladium dimine catalysts that produce similar branched polyethylene from ethylene alone.[15] Sasol makes 1-octene commercially by a similar process with 70% selectivity.[16] A similar chromium phosphine system has been used to convert isoprene to a trimer in >90% yield with up to 87% of the trimers being linear.[17] Mixed higher olefins can be made by reaction of ethylene with triethylaluminum or by the Shell higher olefins process, which employs a nickel phosphine catalyst.

The catalytic dehydrogenation of 1-butene gives 1,3-butadiene,[18] which can be dimerized to 4-vinyl-1-cyclohexene by copper(I)-containing zeolite, with over 99% selectivity.[19] This material is converted to styrene by catalytic oxidation with oxygen in the presence of steam with 92% selectivity. This can provide the styrene for conversion to polystyrene and to styrene-butadiene rubber. Isobutene to make polyisobutylene rubber, used as the liner of tires, can be made from 1-butene by acid-catalyzed isomerization.

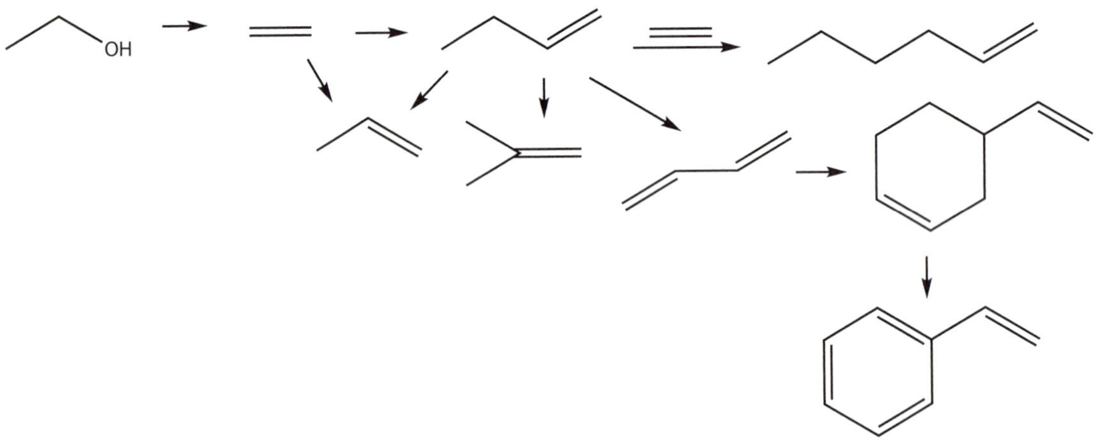

OH

12.1 Schematic

Propylene can be produced by the metathesis of ethylene and 2-butene, obtained by the catalytic isomerization of 1-butene.[20] Tungsten- or molybdenum-containing catalysts are used. Because the reaction is reversible, an excess of ethylene will help drive the reaction to propylene. A mixture of acetone and 1-butanol can be made by the fermentation of sugars. Propylene could be produced by catalytic hydrogenation, followed by dehydration of the intermediary isopropyl alcohol. The 1-butanol could be dehydrated to 1-butene, which could be isomerized to 2-butene for metathesis with ethylene to produce more propylene. Thus, two fermentation plants could convert sugars from the hydrolysis of carbohydrates to ethylene and propylene. These two monomers are used to make large volumes of plastics, especially because they are inexpensive. Copolymers of ethylene and propylene are used as rubbers. Elastomeric polypropylene can be prepared from propylene alone using metallocene catalysts.[21] Ethylene–propylene copolymers with the properties of plasticized polyvinyl chloride can be prepared with metallocene catalysts.[22] A single site catalyst has been used to make a synthetic oil with the monomer cost being one-third of that of the 1-decene that is used currently.[23]

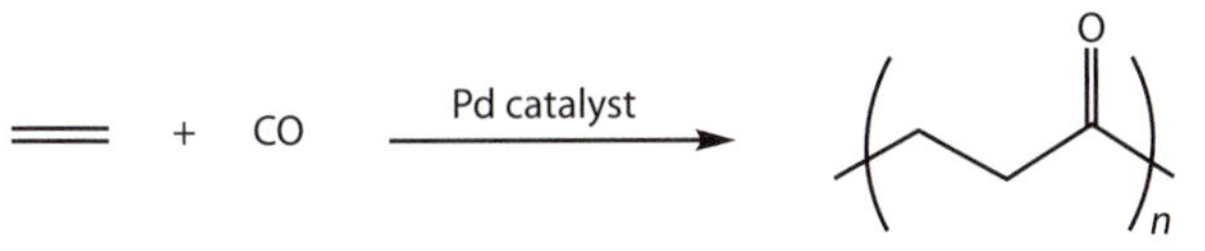

Pd catalyst

12.2 Schematic

Thermoplastic engineering resins are being made by the copolymerization of ethylene with carbon monoxide (**12.2** Schematic) with a small amount of propylene present in the mixture.[24] The process also works with higher olefins.

A third possible route to propylene converts sugars to propionic acid by fermentation. (Propionic acid has also been made by the hydroxycarbonylation of ethylene using a molybdenum-containing catalyst.[25]) The propionic acid would be hydrogenated catalytically as such, or as an ester, to 1-propanol (**12.3** Schematic), which could be dehydrated to propylene.

silicalite (see Chapter 4). Vinyl chloride (a carcinogen) is produced by adding chlorine to ethylene, followed by cracking the resultant ethylene dichloride. Vinyl acetate is made by the oxidative addition of acetic acid to ethylene using a palladium catalyst. A corresponding addition of water to ethylene gives acetaldehyde, which might be made more directly by the oxidation of ethanol from fermentation. It was formerly made by copper-catalyzed dehydrogenation of ethanol or by silver-catalyzed oxidation of ethanol with air.

Acrylic acid for acrylic ester monomers is made by the air oxidation of propylene to acrolein, then to acrylic acid (**12.5** Schematic).[26] Methacrylic acid for methacrylates can be made in the same way from isobutylene.[27]

Another route to methyl methacrylate makes methyl propionate from ethylene, carbon monoxide and methanol, which is then reacted with formaldehyde (carcinogenic) and finally dehydrated to give the product (**12.6** Schematic).[28]

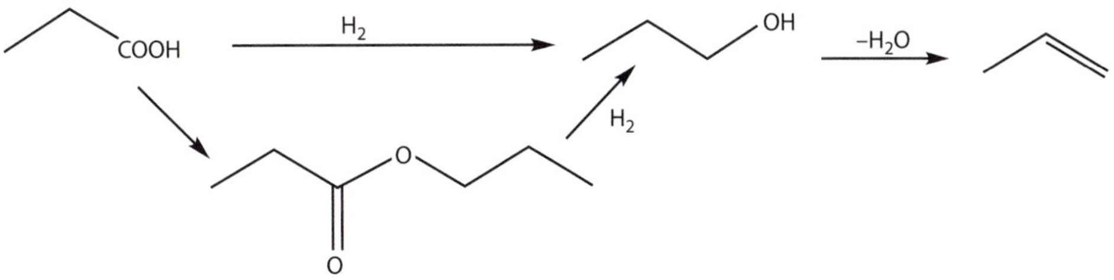

12.3 Schematic

Numerous other large-volume compounds can be made from ethylene and propylene by standard petrochemical methods. Ethylene can be oxidized to ethylene oxide using oxygen with a silver catalyst (**12.4** Schematic). Propylene can be oxidized to propylene oxide with hydrogen peroxide and titanium

Citric acid derived from sugars by fermentation can be converted to methacrylic acid with hot water (as mentioned in Chapter 8). Acrylonitrile (a carcinogenic monomer) is made by the oxidation of propylene with a mixture of ammonia and oxygen.[29]

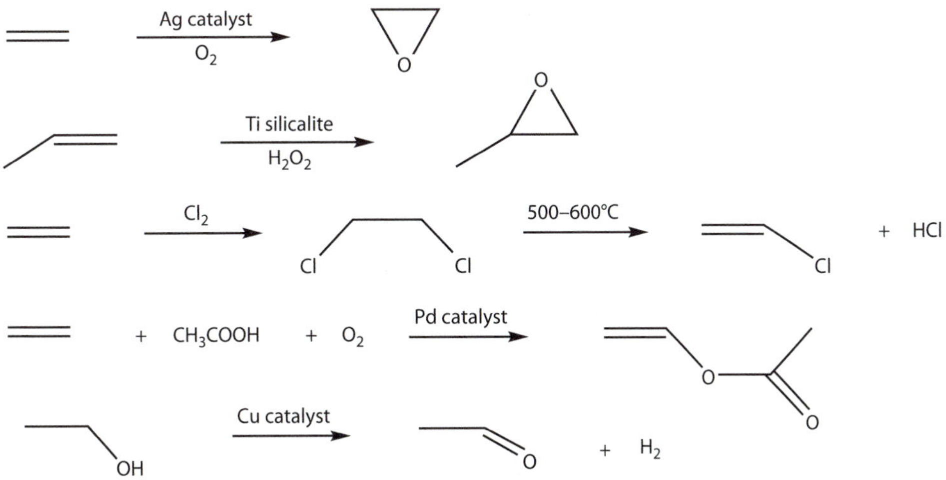

12.4 Schematic

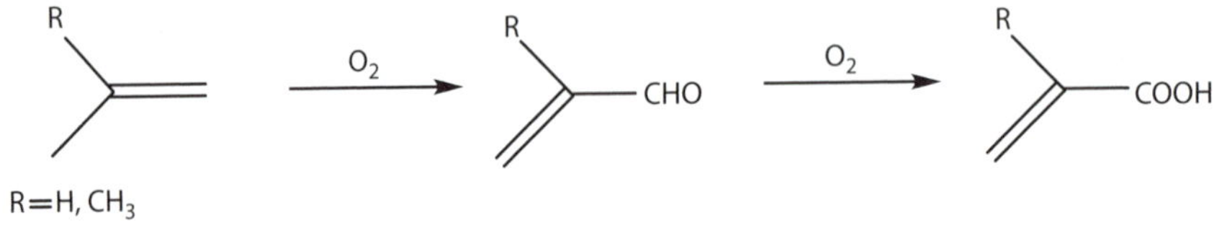

12.5 Schematic

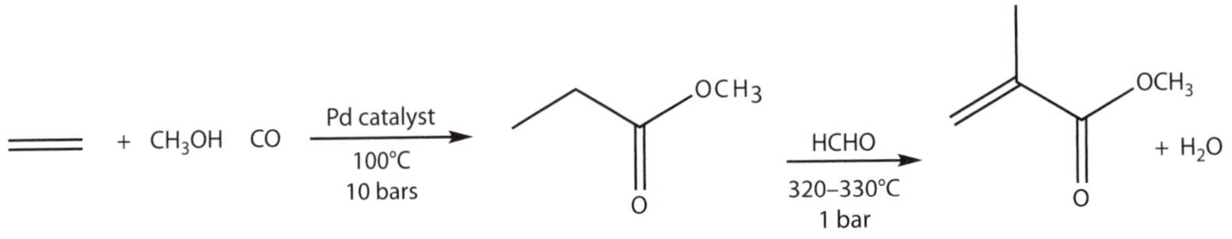

12.6 Schematic

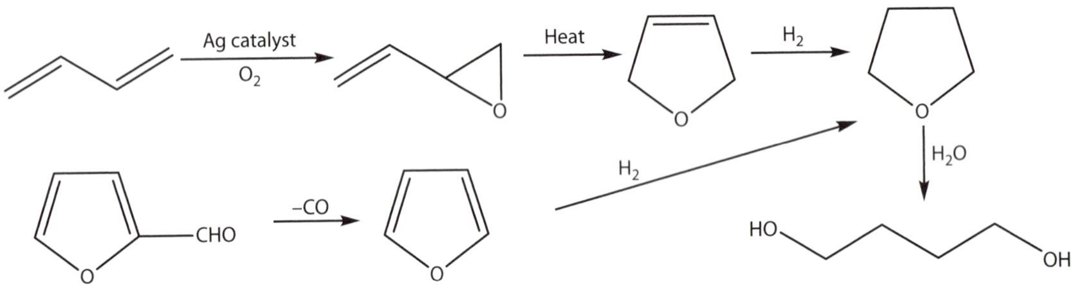

12.7 Schematic

Eastman Chemical has devised a way to convert 1,3-butadiene to various commodity chemicals by way of 3,4-epoxy-1-butene (**12.7** Schematic).[30]

Another route to tetrahydrofuran starts with furfural obtained by the acidic dehydration of pentose sugars found in oat hulls, maize cobs, and other vegetable wastes. Catalytic decarbonylation gives furan, which is reduced to tetrahydrofuran. Tetrahydrofuran can be polymerized to glycols used in making polyurethanes by reaction with diisocyanates. The 1,4-butanediol is used in making poly(butylene terephthalate). Dehydrogenative cyclization of 1,4-butanediol yields butyrolactone, which can be converted to pyrrolidones by treatment with ammonia or amines such as ethanolamine (**12.8** Schematic). The monomer, *N*-vinyl-2-pyrrolidone, is made by vinylation of 2-pyrrolidone with acetylene. The vinylation might also be done by reaction with ethylene oxide or ethylene carbonate followed by dehydration.[31] It might be possible to carry out the vinylation with acetaldehyde instead. This monomer is used in radiation-cured coatings (**12.8** Schematic).

12.2.2 Alcohols and Carboxylic Acids

The preparation of 1,3-propanediol from glucose or glycerol by biocatalysis was described in Chapter 9. It is used in making poly(trimethylene terephthalate). DuPont advertises this polymer as being made from renewable raw materials. Only the third that is the diol is. A new route to terephthalic acid is needed for the other two-thirds. Ethylene glycol and 1,2-propanediol are obtained by the hydrolysis of the corresponding epoxides. A mixture of ethylene, propylene, and butylene glycols, together with a small amount of glycerol, has been produced in 94% conversion by the hydrogenolysis of sorbitol with an Ru/Sn on carbon catalyst in the presence of sodium hydroxide.[32] Sorbitol is obtained by the reduction of glucose. (See also work by Wang et al.[33])

Ethyl acetate can be obtained by the dehydrogenation of ethanol in 95.7% yield.[34] The process may involve a disproportionation of the intermediate acetaldehyde. It can also be produced by addition of acetic acid to ethylene using a heteropolyacid catalyst.[35] In another process that probably involves acetaldehyde as an intermediate, bioethanol can be converted to biobutanol over a calcium apatite catalyst at 300°C/1.78 s with 76% selectivity, the rest of the products being higher alcohols.[36] Direct esterification of ethanol using acetic acid made by fermentation is another route. Acetone was prepared in 97–98% yield by treatment of acetic acid with cerium(IV) oxide on silica[37] or manganese nodules from the Indian Ocean.[38] Zirconium oxide/metal ion catalysts have been used for 99% yields.[39] This is an alternative to the fermentation that produces it along with 1-butanol and

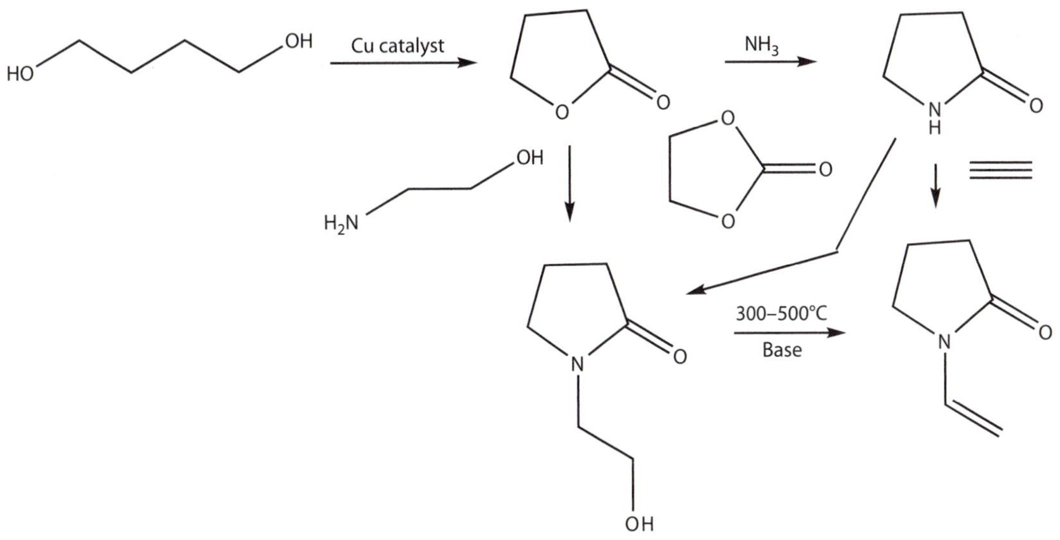

12.8 Schematic

might be preferable, in that any process that produces two coproducts must find adequate outlets in the market for both of them. There is a loss of one carbon atom when acetic acid is converted to acetone.

Succinic acid has been produced in 90% yield by the anaerobic fermentation of carbohydrates.[40] It was heated to form succinic anhydride, which was then oxidized to maleic anhydride with oxygen, using an iron phosphate catalyst (**12.9** Schematic) in more than 95% selectivity at up to 58% conversion.

Maleic anhydride can also be made by the air oxidation of butenes. Maleic anhydride is used in the preparation of unsaturated polyester resins. It can also be reduced to butyrolactone and to 1,4-butanediol. It also offers a possible synthesis of phthalic anhydride by a Diels–Alder reaction with 1,3-butadiene (**12.10** Schematic).

Phthalic anhydride is used in the preparation of alkyd and other resins, as well as in the preparation of esters (e.g., di-2-ethylhexyl phthalate), which serve as plasticizers for polyvinyl chloride. (Citrate esters[41], such as acetyltri-*n*-butylcitrate, are safer plasticizers, but cost more.) It can also be converted to terephthalic acid for the preparation of poly(ethylene terephthalate) through potassium phthalate using the Henkel process which employs a zinc–cadmium catalyst at high temperatures.

Lactic acid can be produced by the fermentation of starch wastes or cheese whey.[42] It has also been prepared from ethylene, carbon dioxide, and water using a $Pt/Sn/SiO_2$

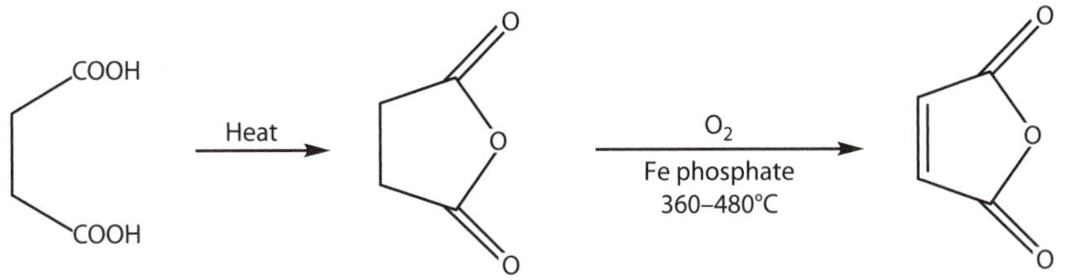

12.9 Schematic

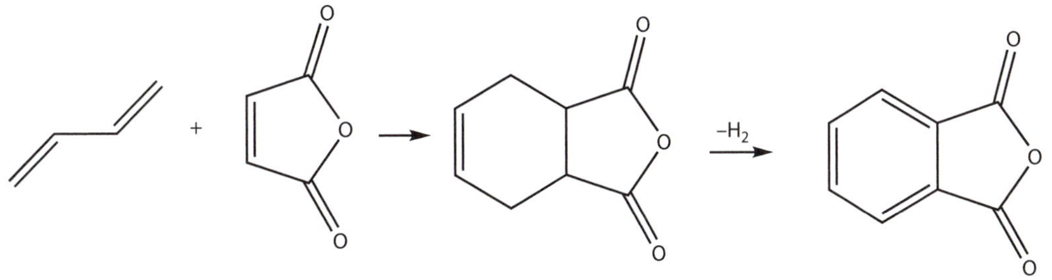

12.10 Schematic

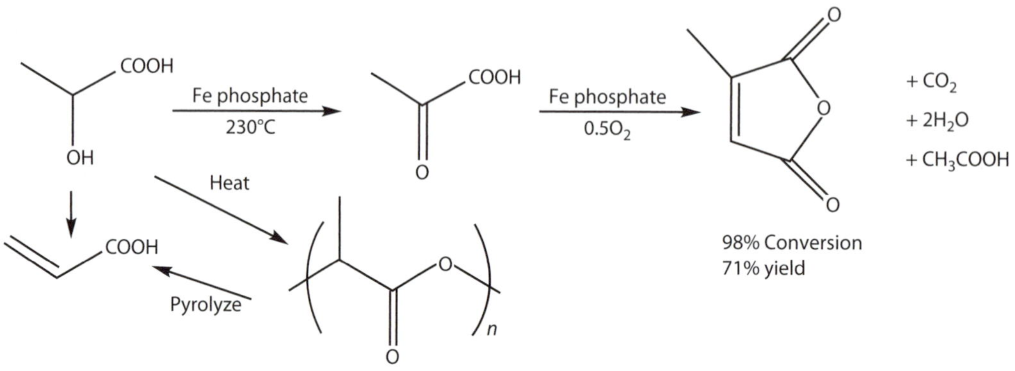

12.11 Schematic

catalyst.[43] It can be converted to an analogue of maleic anhydride using a two-step oxidation (**12.11** Schematic).[44]

This anhydride may be suitable for some of the uses of maleic anhydride in the preparation of polymers. Dehydration of lactic acid to the important monomer, acrylic acid, gave only a 17% yield.[45] Pyrolysis of the acetate of methyl lactate gave methyl acrylate in 90% yield, but was reported to be too expensive to use commercially. A better route being developed by Codexis, Novozymes, and Cargill dehydrates 3-hydroxyproionic acid made by fermentation.[46] Oligomers of lactic acid can be prepared by simple heating.[47] It may be possible to pyrolyze them to acrylic acid in high yields.

The development of biodiesel fuel has produced a surplus of cheap glycerol that companies are starting to use to make other chemicals.[48] Solvay and Dow are converting it to epichlorohydrin in a process that still discards chlorine as sodium chloride (**12.12** Schematic).[49]

Archr Daniels Midland, Ashland, Cargill, Dow Chemical, Huntsman, Snergy and Virent have all proposed or built plants to convert glycerol to 1,2-propanediol (1,2-propylene glycol) by a process devised by Galen Suppes.[50] The intermediate 1-hydroxy-2-propanone (acetol) is removed by reactive distillation, then reduced to the diol with copper chromite and hydrogen (**12.13** Schematic).

1,2-Propylene glycol is less toxic than ethylene glycol and can replace it in antifreeze and deicing fluids. Glycerol can be converted to 1,3-propanediol by fermentation. This diol can be oxidized with air in the presence of methanol using an Au/Fe_2O_3 catalyst to methyl 3-hydroxypropionate with 90% selectivity at 94% conversion.[51] Dehydration leads to methyl acrylate, a common monomer. Acrolein can be produced with 75% selectivity at 50% conversion with a zinc sulfate catalyst in water at 360°C/10–60 s or in 67% yield and 84% selectivity with a phosphoric acid on carbon catalyst at 260°C.[52] The same transformation can be carried out with a phosphotungstic acid on alumina catalyst in 68% selectivity at 69% conversion.[53] Acrolein can be oxidized to the monomer, acrylic acid. Glycerol can be converted to diglycerol[54] and other ethers. A 1:1 acetal forms on treatment of a mixture of sucrose and glycerol with a sucrose phosphorolase.[55] Mono- and diglycerides of long chain fatty acids are used as food emulsifiers. Acetals and a cyclic carbonate can also be made from glycerol.

Adipic acid for the preparation of nylon is made starting with benzene. It can be made from glucose via catechol, as

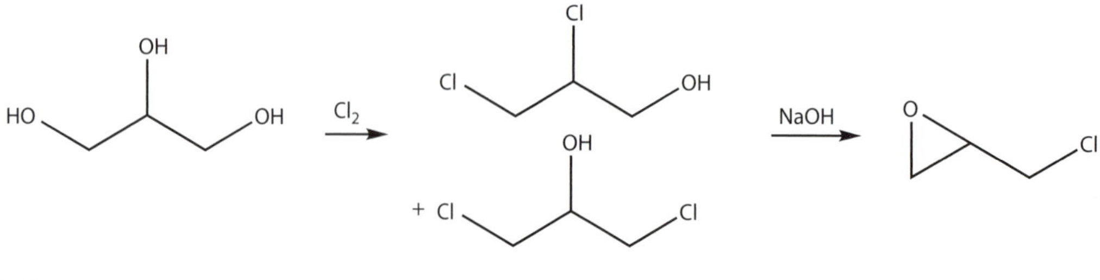

12.12 Schematic

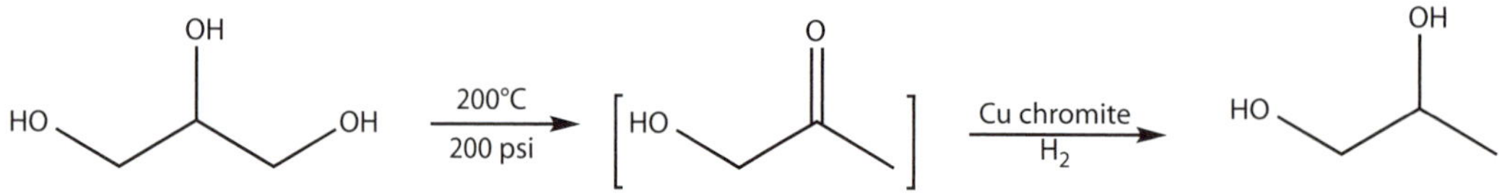

12.13 Schematic

$$CH_3(CH_2)_{10}CH{=}CH(CH_2)_4COOH \xrightarrow{O_3} CH_3(CH_2)_{10}COOH + HOOC(CH_2)_4COOH$$

12.14 Schematic

shown by Frost and co-workers (see Chapter 9). It can also be made by the oxidation of petroselenic acid with ozone, the companion product lauric acid also being valuable (**12.14** Schematic).[56] Petroselenic acid constitutes 70–80% of the seed oil of many species of Umbelliferae. This requires growing a new crop or putting the appropriate genes into a microorganism or higher plant (such as rapeseed).

It may be possible to prepare adipic acid from glucose by a chemical route. Glucose can be oxidized to an α–ω dicarboxylic acid (glucaric acid) by air with a platinum catalyst.[57] The next step would be to replace all of the hydroxyl groups with hydrogen. This has been done with tartaric acid using hydrogen iodide (**12.15** Schematic).[58] The challenge is to make it catalytic in iodide or to find another system of reduction.

Another way to prepare dibasic acids for the preparation of polyamides would be to oxidize cyclohexene from the Diels–Alder reaction of 1,3-butadiene and ethylene or cyclooctene made from 1,3-butadiene by cyclic dimerization, followed by a reduction that might involve conjugation of the double bonds *in situ* (**12.16** Schematic). The latter may be preferable because the former requires forcing conditions.[59] It may be possible to run the former reaction under high pressure or with ultrasound or with a metal complex catalyst (such as a metal triflate) to reduce the electron density of the diene by complexation.

The diamines needed for the polyamides can be made by conversion of the acids to their ammonium salts followed by heating to form the dinitriles, which can then be reduced. Adipic acid can be reduced to 1,6-hexanediol (for use in

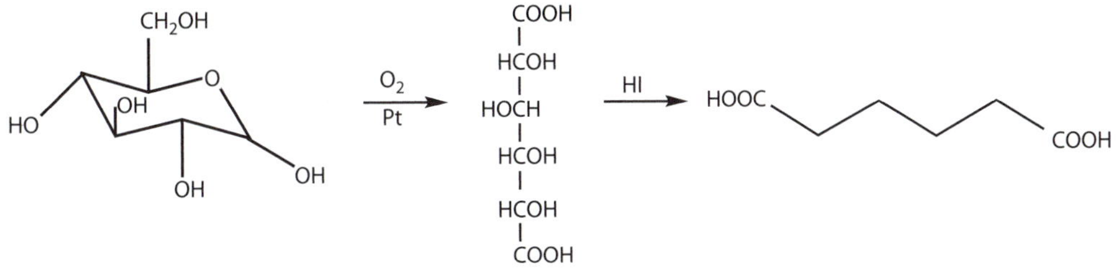

CH2OH, O, OH, HO, OH, OH; O_2 / Pt; COOH, HCOH, HOCH, HCOH, HCOH, COOH; HI; HOOC, COOH

12.15 Schematic

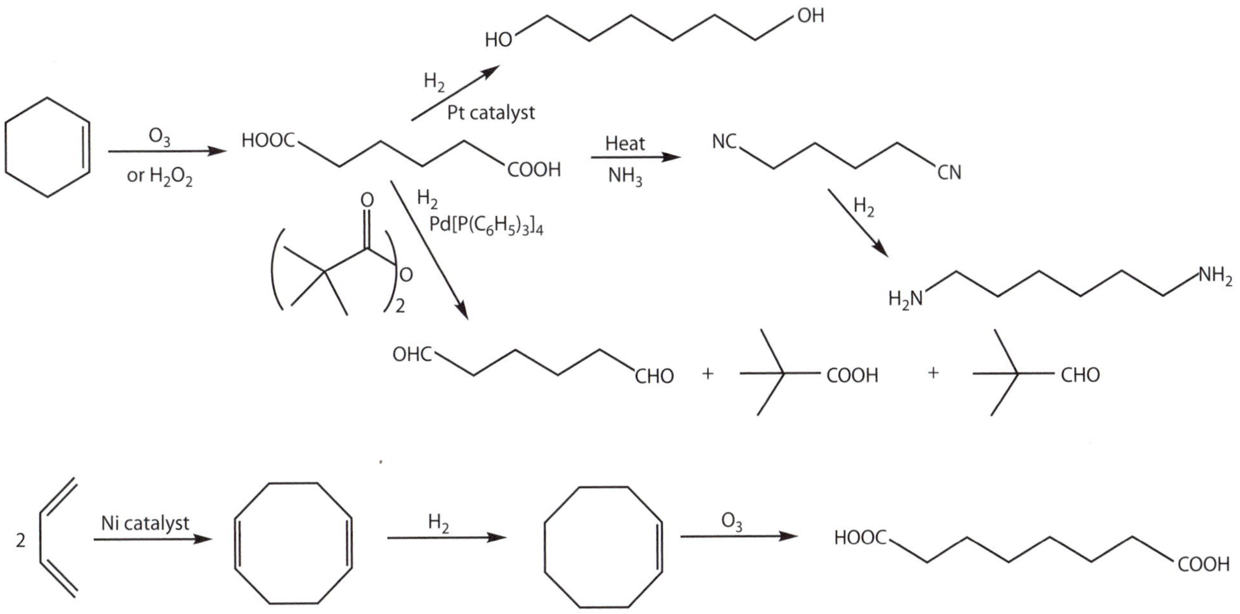

HO, OH; H_2 / Pt catalyst; O_3 / or H_2O_2; HOOC, COOH; Heat / NH_3; NC, CN; H_2; H_2N, NH_2; O, O, 2; H_2 / $Pd[P(C_6H_5)_3]_4$; OHC, CHO + COOH + CHO

2; Ni catalyst; H_2; O_3; HOOC, COOH

12.16 Schematic

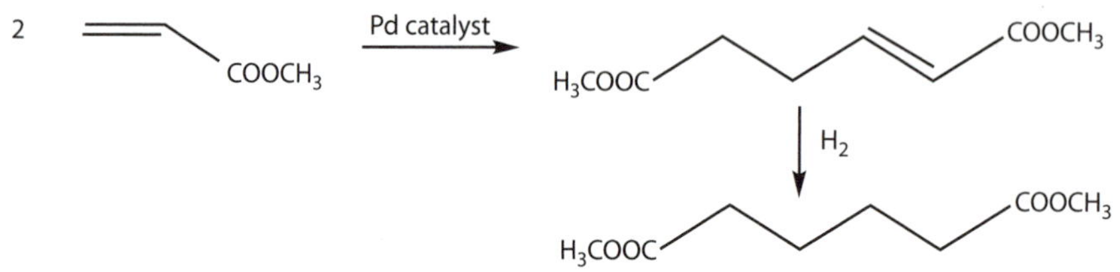

2 $CH_2=CHCOOCH_3$ —Pd catalyst→ H_3COOC–CH₂CH₂CH=CH–COOCH₃ —H_2→ $H_3COOC(CH_2)_4COOCH_3$

12.17 Schematic

polyurethanes) with hydrogen in a 95% yield using a modified platinum catalyst.[60] Earlier processes, which went through the dimethyl ester, gave a 90% yield. Adipic acid can be reduced with hydrogen in the presence of excess pivalic anhydride with a palladium catalyst to produce adipaldehyde in 94–99% yield.[61] The pivalic acid formed in the reaction is reconverted to pivalic anhydride for the next run. The by-product pivaldehyde can be oxidized to pivalic acid for reuse. Adipaldehyde might be used to cross-link proteins (e.g., in the immobilization of enzymes or in the tanning of leather). Dialkyl adipates can also be made by the tail-to-tail dimerization of alkyl acrylates using palladium or other metal catalysts (**12.17** Schematic).[62] The alkyl acrylates could be made from glucose, as described above.

Paper mill sludge (as well as other forms of cellulose) can be converted to levulinic acid in 70–90% yield by hydrolysis to glucose followed by controlled heating (**12.18** Schematic) in a process developed by Biofine, Inc.[63] Levulinic acid can be converted by known processes to the solvents, tetrahydrofuran and 2-methylfuran, to succinic acid, to 5-aminolevulinic acid for weed control (see Chapter 11), and to diphenolic acid. Diphenolic acid may be useful in waterborne epoxy coatings and other resins. If a suitable acid catalyst that can be left in the resin can be found, it may be possible to convert diphenolic acid, by self-acylation, to an analogue of phenol–formaldehyde resins, thus eliminating the use of the carcinogenic formaldehyde. The by-product formic acid and furfural may also be valuable.

12.2.3 Epoxides and Aromatic Compounds

Many epoxy resins are based on the diglycidyl ether of bisphenol A, made by the reaction of epichlorohydrin (very toxic) with the diol in the presence of base (**12.19** Schematic).

A better system that avoids the toxic reagent and the by-product salts would be to use epoxides derived from 4-vinyl-1-cyclohexene, 1,5-cyclooctadiene, oligomeric poly(butadiene), unsaturated oils, terpenes, and such, that come from renewable sources or can be derived from them as outlined in the foregoing section. This will require considerable redesigning of the resins. Epoxidized vegetable oils mixed with diepoxides, glass fiber, and a photocationic initiator have been polymerized with sunlight to give materials suitable for roofing, culverts, boats, medical casts, and pipes.[64] The starting resins are nontoxic, have an infinite shelf life in the dark, and need no refrigeration. Bisphenol A is also used in the preparation of polycarbonate resins. Its preparation requires a renewable source of phenol. Styrene can be reduced to ethylbenzene which can be oxidized to the hydroperoxide, followed by cleavage to phenol and acetaldehyde with an acid catalyst (**12.20** Schematic). The styrene might be oxidized to benzoic acid, for which conversion to phenol is known.[65] Benzoic acid can also be converted to terephthalic acid by the Henkel process.

The benzoic acid might also be made by the Diels–Alder reaction of 1,3-butadiene with acrylic acid followed by catalytic dehydrogenation. Treatment of phenol with ammonia at

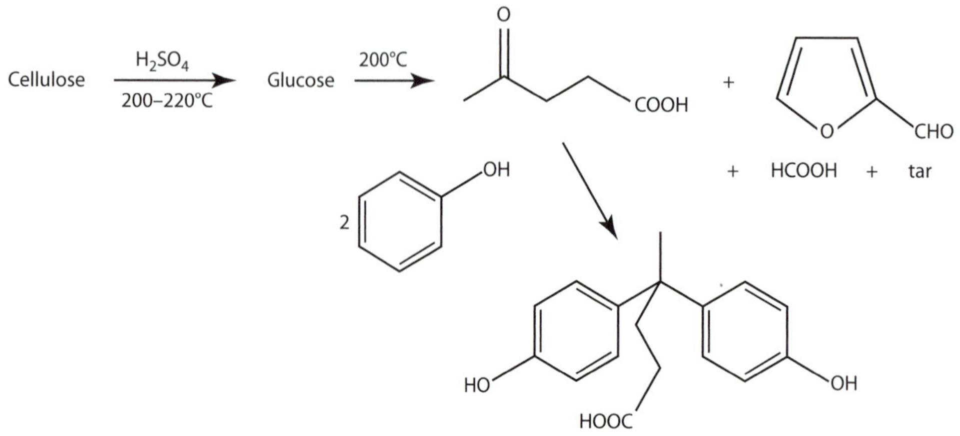

Cellulose —H_2SO_4, 200–220°C→ Glucose —200°C→ levulinic acid (COOH) + furfural (CHO) + HCOOH + tar

levulinic acid + 2 phenol (OH) → diphenolic acid (HO, OH, HOOC)

12.18 Schematic

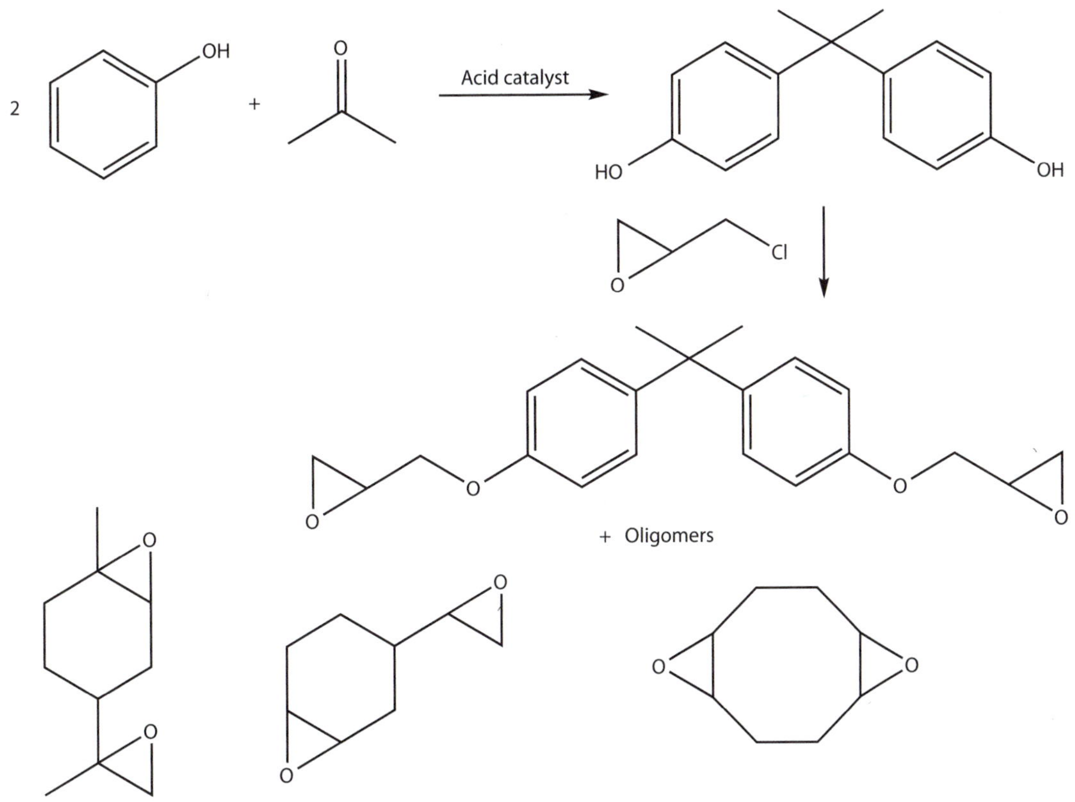

12.19 Schematic

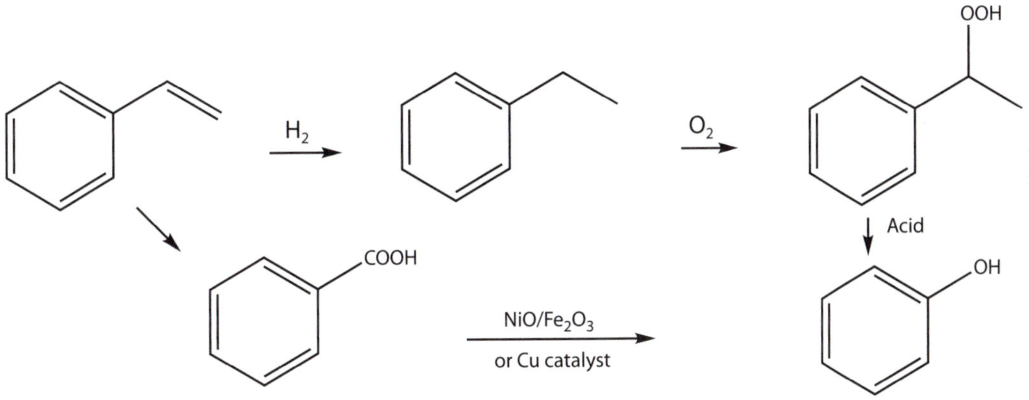

12.20 Schematic

high temperatures produces aniline, as mentioned in Chapter 2. Ethylbenzene can be rearranged to xylenes with zeolite catalysts. Thus, it could serve as a source of phthalic, isophthalic, and terephthalic acids by the oxidation of *o*-, *m*-, and *p*-xylenes. (The xylenes and other aromatic hydrocarbons can also be made by the dehydrocyclization of ethylene, propylene, and butenes, or their corresponding alkanes.[66] Benzene can also be made from methane.[67])

p-Cresol can be made from *p*-cymene derived from terpenes by dehydrogenation (**12.21** Schematic).

It might then be oxidized enzymatically with hydrogen peroxide to produce a polymer, as described in Chapter 9. A final cure with diisocyanates or bisoxazolines would yield the equivalent of a phenol–formaldehyde resin without the use of the carcinogenic formaldehyde. Terephthalic acid for poly(ethylene terephthalate) might be made by the oxidation of *p*-cymene, although two carbon atoms would be lost. The amount of terpenes available may not be enough for this use.

Catechol, hydroquinone, resorcinol, and pyrogallol can be produced from glucose by the enzymatic methods of Frost and co-workers (see Chapter 9). Catechol can also be obtained by the demethylation and cleavage of lignin, the polymer that holds wood together.[68] Frost and co-workers have shown how to convert catechol to adipic acid. Thus, the

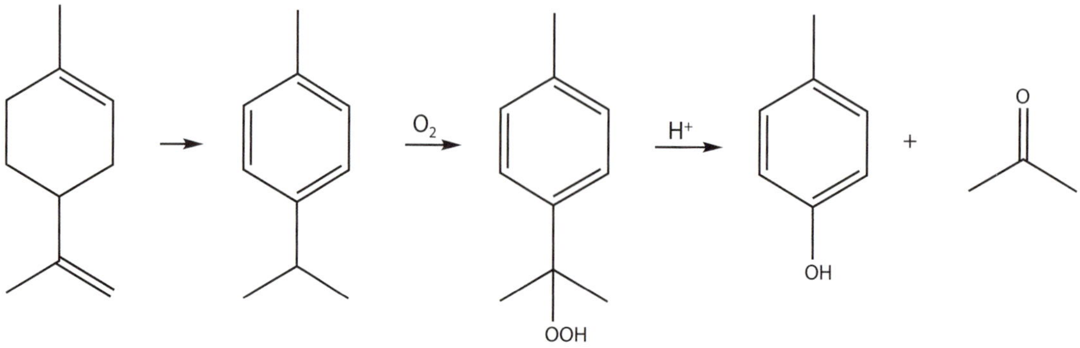

12.21 Schematic

lignin from wood, which is now underutilized, might be converted into nylon. The intermediate cic,cis-muconic acid in Frost's procedure, or its esters, might be converted to terephthalic acid or its esters by the Diels–Alder reaction with ethylene under high pressure, or with acetylene, possibly with a cobalt catalyst, or with acrylic acid to cyclic adducts, which could be dehydrogenated catalytically, as done in the Dow process for styrene from vinylcyclohexene. If acrylic acid were used, a final selective decarboxylation would be required.

A major use of phenol is in phenol–formaldehyde adhesives for wood. Tannins are polyphenols from plants, as in bark,[69] that may be able to replace phenol in such adhesives. A typical repeating unit is shown in **12.22** Schematic. These repeating units are often linked by C_4–C_6 or C_4–C_8 bonds. The atoms in the ring are numbered. Typical dehydrodimers are connected through the number 4 carbon atom of one ring to number 8 of the other, and in the second case number 4 to number 6 of the other.

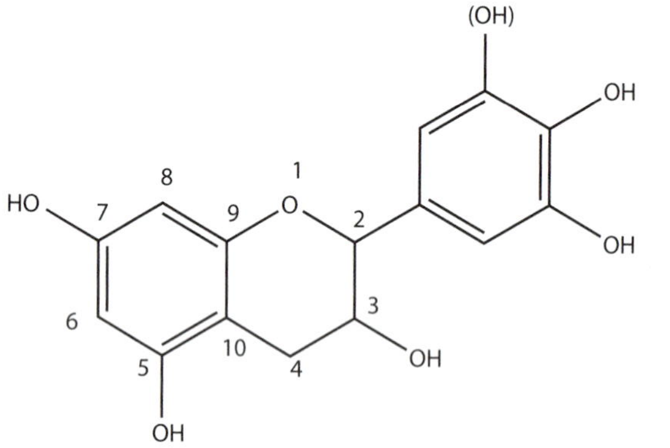

12.22 Schematic

Some tannins can be hydrolyzed to gallic acid (**12.23** Schematic). The pyrogallol was reacted with formaldehyde to give resins comparable with those made from phenol.[70] Perhaps furfural can be used instead of formaldehyde. The most interesting adhesives from tannins use no formaldehyde and have no toxicity. They are made by the self-condensation of tannins in the presence of silica and sodium hydroxide, or other catalysts. The heterocyclic ring opens to yield a reactive 2-position that substitutes open positions in the aromatic rings of the adjacent chains of tannins.[71] This is suitable for interior use. Further study or the addition of organic cross-linking agents such as diisocyanates might make it suitable for boards for exterior use. The use of tannins is attractive because the waste tree bark (from which they are often obtained) is frequently located next to the plywood or

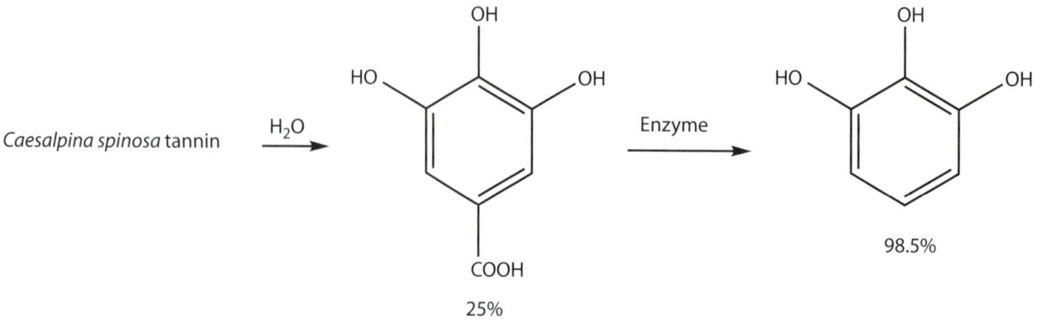

Caesalpina spinosa tannin $\xrightarrow{H_2O}$ gallic acid (25%) $\xrightarrow{\text{Enzyme}}$ pyrogallol (98.5%)

12.23 Schematic

particle board mill. Systems such as this may also be useful in tanning leather without the use of chromium. Lignins, which are also polyphenols, can also be used as adhesives for wood without formaldehyde. The soluble lignin can be cross-linked oxidatively with a hydrogen peroxide plus sulfur dioxide redox system or with diepoxides, diisocyanates, and so on.[72] It should be possible to use a nonvolatile reducing agent instead of sulfur dioxide.

Plywood can also be held together with a combination of soy flour and a polymeric azetidinium salt normally used as a wet strength resin for paper.[73] Columbia Forest Products now uses this system exclusively.

12.2.4 Chemicals from Fats and Oils[74]

Fats and oils (triglycerides) from plants and animals are renewable sources of chemicals, but the amounts of the chemicals made from them are small compared with those made from petroleum and natural gas. This may change as biodiesel fuel (e.g., ethyl oleate) made by the alcoholysis of oils becomes common. Such esters may be useful as environmentally friendly solvents.[75] Unsaturated oils, such as linseed oil, are the basis of oil-based paints, which cure by cross-linking through oxidation by air. Soaps are the potassium or sodium salts of long-chain fatty acids obtained by the hydrolysis of triglycerides. The dibasic dimer fatty acids obtained by the dimerization of oleic and linoleic acids (both C_{18} acids) are made into oligomeric fatty amides, which are used to cure epoxy resins. The unsaturated oils are also epoxidized to epoxides for use in epoxy resins and as stabilizers for plastics. There are also efforts to raise *Vernonia galamensis* that naturally contains epoxy oils.[76] Polymers have been made from epoxidized soybean oil and di- or polyamines with fillers.[77]

Hydrogenation of fats leads to long-chain alcohols that can be converted to sulfate detergents, such as sodium lauryl sulfate. These alcohols can also be converted to acetals of sugars, the alkyl glycoside detergents.[78] (See Chapter 3, Section 3.7 and Chapter 8, Section 8.6 for more on "green" detergents.) Fatty amines are obtained by reduction of fatty nitriles, which are made by heating the ammonium salts of the fatty acids. These can be made into quaternary ammonium salts for use as detergents, phase transfer catalysts, and germicides. Castor oil contains 90% ricinoleic acid residues, and hence it has a hydroxyl functionality of about 2.7 (**12.24** Schematic). The reaction of castor oil with diisocyanates has been used to make polyurethanes.[79] In one case, hydroxyethyl-methacrylate was included to form a prepolymer with a terminal methacrylate group that could be cured with benzoyl peroxide.

$CH_3(CH_2)_5CH(OH)CH_2CH{=}CH(CH)_7COOH$ Ricinoleic acid

12.24 Schematic

Heating a mixture of ricinoleic, maleic, and sebacic acids in a melt produced a polyanhydride with a molecular weight of over 100,000, which could be used as a biodegradable polymer for slow drug release.[80]

Vegetable oils can also be used as food-grade lubricants after the addition of antioxidants.[81] A genetically engineered soybean oil from DuPont contains 80% monounsaturated acid residues and 6% polyunsaturated ones.[82] It can be used as a hydraulic fluid or as a food-grade lubricant. Biodegradable ester hydraulic fluids with triple the stability of mineral oil-based ones can be made by treating unsaturated fatty acids from vegetable oils with sulfuric acid (to cause addition of the carboxylic acid group to the double bond).[83] The use of various fatty acid esters as cutting oils for metal working is covered in Chapter 13. Propylene glycol monolinoleate is a coalescing agent for latex paints that cures into the paint film, reduces odor, and increases scrub resistance (**12.25** Schematic).[84]

$CH_3(CH_2)_4CH{=}CHCH_2CH{=}CH(CH_2)_7COOCH_2CH(OH)CH_3$

12.25 Schematic

Both fatty and resin acids (diterpene acids) are obtained as a by-product of the pulping of wood for paper. The mixture is known as tall oil. The separated resin acids are known as rosin. Resin acids such as abietic acid (**12.26** Schematic) can also be obtained as exudates of pine trees.

Paper is often sized by forming the aluminum salt of rosin on it. This allows printing on paper by keeping the ink from soaking into it. Such paper tends to degrade over the years, and hence archival paper in books and journals is made with alkaline sizing instead. Esters of resin with alcohols such as glycerol and pentaerythritol are used as tackifying resins for adhesives. These "hard resins" compete with oligomeric resins from petroleum fractions.[85]

Monoterpenes,[86] such as limonene from citrus peel and pinenes from pine trees, are also valuable. A diepoxide from limonene for epoxide resins was mentioned before. Cationic polymerization of β-pinene followed by reduction of the

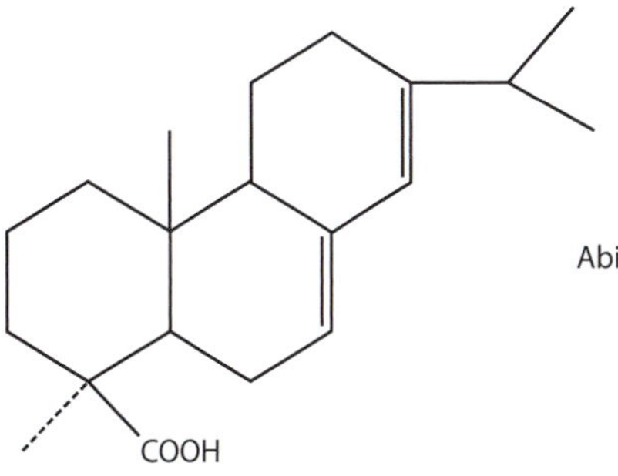

Abietic acid

COOH

12.26 Schematic

polymer produced a polymer with good transparency with an M_w >50,000 and a T_g of 130°C (**12.27** Schematic).[87]

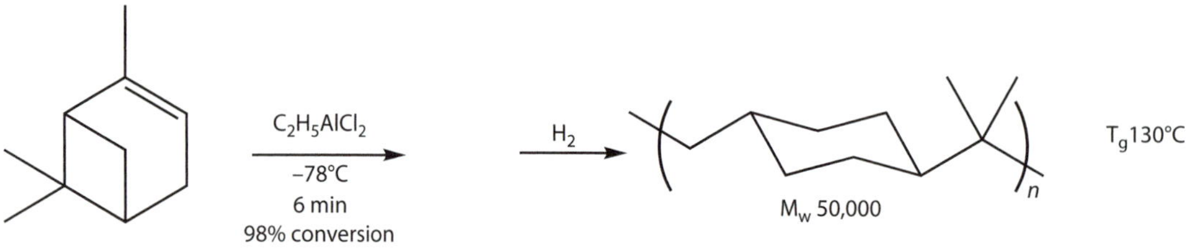

$C_2H_5AlCl_2$, −78°C, 6 min, 98% conversion; H_2; M_w 50,000; T_g 130°C

12.27 Schematic

aliphatics, 35.9% xylenes, and 0.1% aromatics of more than C_{11}.[97] This is one way to produce the xylene needed to make terephthalate polymers. When canola oil was used instead of methanol in this process, a maximum yield of C_2–C_4 olefins of 26% was obtained.[98] Aromatic compounds were also produced at the same time. An aluminum phosphate catalyst converted methanol (77% conversion) to a mixture in which more than two-thirds of the products were ethylene and propylene.[99] The selectivity for ethylene was 58% and that for propylene was 21%. Total, Sud Chemie, and Mitsui have plants for the conversion to ethylene and propylene.[100]

Ammonia is produced by the reaction of nitrogen with hydrogen derived from petroleum or natural gas. This process uses 60% of all the hydrogen produced. Hydrogen from the electrolysis of waster costs quite a bit more. The hydrogen could be made from biomass, as shown in **12.28** Schematic. Ammonia can be reacted with carbon dioxide to form urea, much of which ends up in slow-release fertilizer. The use of leguminous cover crops to put nitrogen into the soil could reduce the demand for fertilizers based on ammonia or urea (as described in Chapter 11). Melamine for melamine–formaldehyde resins is made by heating urea with an aluminum oxide catalyst (**12.29** Schematic). These resins are used in counter tops and plastic dinnerware. It should be possible to replace the carcinogenic formaldehyde with various cross-linking agents, such as glyoxal, bisanhydrides, bisoxazolines, diisocyanates, and the like.

12.2.5 The Use of Synthesis Gas from Biomass

Synthesis gas is usually made from natural gas or petroleum. It can also be made from biomass[88] and from municipal solid waste.[89] A recent method used flash evaporation over a rhodium/cerium catalyst at about 800°C with a contact time of <50 ms.[90] This mixture of carbon monoxide and hydrogen can then be converted to a variety of materials by standard petrochemical methods (**12.28** Schematic) using a variety of catalysts.[91]

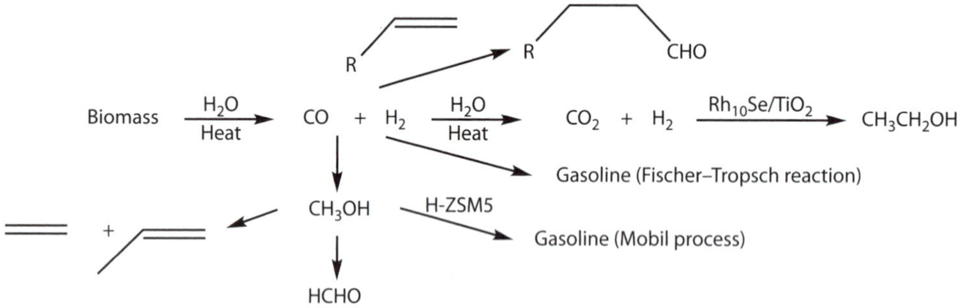

R; R CHO; Biomass $\xrightarrow[\text{Heat}]{H_2O}$ CO + H_2 $\xrightarrow[\text{Heat}]{H_2O}$ CO_2 + H_2 $\xrightarrow{Rh_{10}Se/TiO_2}$ CH_3CH_2OH; Gasoline (Fischer–Tropsch reaction); CH_3OH; H-ZSM5; Gasoline (Mobil process); +; HCHO

12.28 Schematic

The water–gas shift reaction is used to convert the carbon monoxide to carbon dioxide and more hydrogen. These, when reacted under the influence of the special catalyst, form ethanol with 83% selectivity.[92] The hydrogen can be used to reduce the carbon monoxide to methanol. The synthesis gas can also be used in the hydroformylation of olefins to form aldehydes.[93] It can be converted to gasoline via the Fischer–Tropsch reaction.[94] Synthesis gas from surplus glycerol can also be used in this reaction.[95] If the iron catalyst is supported on MgO/SiO_2, the yield of C_2–C_4 olefins is increased to a ratio of 0.65 olefins to paraffins.[96] The Mobil process converts methanol to gasoline over a zeolite. If the process is run at 538°C, a higher temperature than usual, the product consists of 60.5% C_1–C_4 aliphatics, 3.5% higher

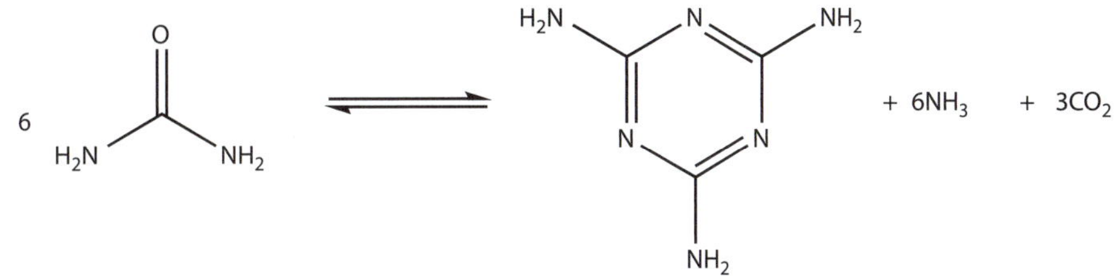

$$6\ H_2NC(O)NH_2 \rightleftharpoons C_3N_3(NH_2)_3 + 6NH_3 + 3CO_2$$

12.29 Schematic

12.3 USE OF NATURAL POLYMERS

Before the advent of cheap petroleum, many natural polymers were being used or being studied for use in plastics. Since then, synthetic polymers have dominated the marketplace. It should be possible to make the plastics that we need from cellulose, hemicellulose, starch, chitin, lignin, proteins, or other natural products.[101] Ideally, the plastic or the other product should have properties similar to or better than the one derived from petroleum that it is designed to replace. It should neither require more energy to make nor cost more.

The majority of these polymers and their derivatives will be biodegradable and compostable. Biodegradability[102] is advantageous when the useful life is done and in agricultural mulch as well as in sutures that do not have to be removed surgically.[103] The strength of the bioplastics, as well as some conventional plastics, can often be increased by natural fibers, at least some of which can come from agricultural by-products such as corn shucks and chicken fibers.[104] The challenge is to keep the bioplastics out of the conventional ones when the latter are recycled.

12.3.1 Cellulose

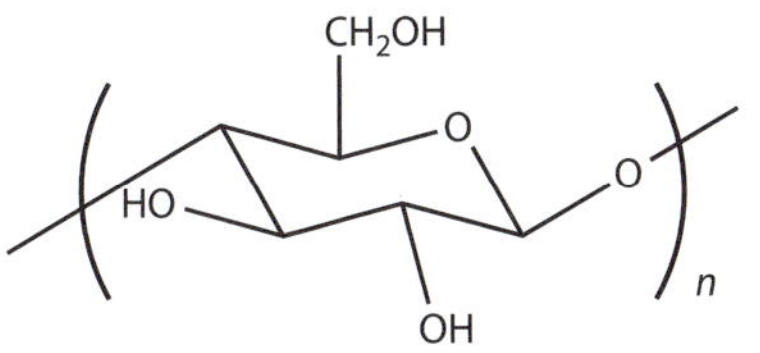

Cellulose (**12.30** Schematic)[105] is found in wood, along with about 24% hemicellulose and 22% lignin, and in cotton, sisal, jute, linen, hemp, and other similar products. It is a polymer of glucose.

It can also be made from sucrose by *Acetobacter xylinum*.[106] This bacterial cellulose has high mechanical strength and might become an important material if the cost can be brought down. Regenerated cellulose fibers[107] are known as rayon and regenerated films as cellophane. They are usually made by the xanthate process, which uses sodium hydroxide and carbon disulfide (**12.31** Schematic). The xanthate solution is converted to fiber or film by passage into a bath of acid. This relatively polluting process is being replaced by one in which a solution of cellulose in *N*-methylmorpholine-*N*-oxide is passed into water.[108] The solvent is recovered and recycled. This method produces stronger materials, because the resulting molecular weight is higher. The product costs twice as much as cotton, a price that does not take into account the environmental effects of the large amount of irrigation water, fertilizers, and pesticides required by the crop. (The drying up of the Aral Sea is one result.) When cellulose from trees is used in this process, these environmental problems are reduced. Trees also have the advantage in that they can be harvested any time of the year, in contrast with the seasonal harvesting of an annual crop.[109] Steam-exploded wheat straw can be used as a source of cellulose for making rayon.[110] BASF Corp. is studying a process of making rayon and derivatives of cellulose in ionic liquids.[111]

Kenaf also avoids some of these problems, but is an annual crop.[112] Its 50:50 blends with cotton perform as well in textiles as 100% cotton, while having better luster, texture, dye uptake, and colorfastness.[113] Linen (from flax which does not require as much water to grow as cotton) has handle, comfort, strength, and water absorption properties that are superior to cotton, but it has less than 1% of the market. Efforts are being made to make it easier to use by improving the processes by which the fibers are separated from the rest of the plant. This retting is usually done by leaving the plants in a wet place in the field for a few weeks. The new methods involve heating with sodium dodecylsulfate and oxalic acid in water for 2 h at 75°C[114] or steam explosion.[115] Retting by enzymes is also being studied.[116] The tenacity of flax can be improved by 50% and that of jute by 110% by grafting on some methyl methacrylate or 2-hydroxyethyl methacrylate.[117] The flax plant can serve as a source of both fiber for clothing and oil for paints, epoxides, and the like. Hemp has been used in textiles for centuries. It requires less water, fertilizers, and pesticides than cotton and can be grown in a cold climate.[118] It can be grown

Cellulose: $[\text{–O–}C_6H_7O_2(OH)(CH_2OH)\text{–}]_n$ (structure labels: CH_2OH, O, HO, O, OH, n)

12.30 Schematic

$$\text{CellOH} \xrightarrow{NaOH} \text{CellO}^- Na^+ \xrightarrow{CS_2} \text{CellOC(S)S}^- Na^+ \xrightarrow{H^+} \text{CellOH} \quad \text{Rayon}$$

12.31 Schematic

in Canada, but not in the United States, since the same species is smoked for marijuana.

A possible way to lower the cost of fibers and films of regenerated cellulose would be to run cellulose through a twin-screw ultrasonic extruder with a minimum of the solvent and passing the exudate through a stream of hot air to recover the solvent for reuse. This stronger cellophane could be used in place of many plastic films used today. The high moisture vapor transmission of cellophane is reduced by a coating of nitrocellulose or poly(vinylidene chloride), which also makes the film heat-sealable.[119] Cellulose, in the forms of wood flour, bamboo fiber, and the like, has also been used as a filler or reinforcing agent for plastics.[120]

A great many derivatives of cellulose, principally esters and ethers, have been made.[121] Methyl, ethyl, carboxymethyl, hydroxyethyl, and hydroxypropyl ethers are made commercially today. These are used as water-soluble polymers, except for ethylcellulose which is a tough plastic used in screw-driver handles, and so on. Methylcellulose plasticized with polyethyleneglycol has been tested as an edible coating for preventing water loss from fruit.[122] An alternate method uses microemulsions of beeswax, candelilla wax, or carnauba wax to deposit the edible film on fruit.[123] Cellulose acetate and cellulose acetate butyrate are also commercial plastics. The main factor inhibiting the larger use of these cellulose derivatives is their higher cost relative to polyolefins and other plastics derived from petroleum and natural gas. If simpler, cheaper methods for the preparation of the cellulose derivatives can be devised, their competitive situation should improve.

It may be possible to prepare many cellulose derivatives in extruders, with or without the addition of ultrasound and microwaves. If acetic anhydride were used, the by-product acetic acid could be removed through a vent. If a cyclic anhydride, such as maleic or succinic anhydride, were used, there would be no need for the vent. The product should dissolve in aqueous base and might be useful in situations in which sodium carboxymethylcellulose is used today. Polymeric anhydrides might also be used. Cellulose acetate has been reacted with cyclic lactones in an extruder.[124] Cellulose acetate containing a citrate plasticizer can be extruded.[125] Methylcellulose might be made with dimethyl carbonate in the presence of a base. It should be possible to design reagents to disrupt the hydrogen bonds of cellulose, then react with it in the extruder. Solvents for cellulose such as *N*-methylmorpholine-*N*-oxide, lithium chloride/*N,N*-dimethylacetamide,[126] and ammonia/*N,N*-dimethylformamide,[127] can serve as a starting point for ideas. Cellulose aerogels with densities as low as 0.06 g/cc have been made using 7 sodium hydroxide/12 urea/81 water.[128] The simplest approach would be to make a cellulose carbamate using urea. Grafting on *N,N*-dimethylacrylamide, *N*-acryloylmorpholine, *N*-vinylpyrrolidone, and the like, with or without *N,N*-methylenebisacrylamide as a cross-linking agent, may also be possible. The amide could also be made *in situ* by reaction of the cellulose with isocyanates or oxazolines. Formaldehyde is used with some solvents to dissolve cellulose. Perhaps an aldehyde, such as butyraldehyde or furfural, could be used in the extruder along with other reagents that would react with the cellulose, but not with the aldehyde. A final possibility would be to see if a little aldehyde plus water could be extruded, with a vent for the water at the end. An acid catalyst may be necessary to form the acetals in these reactions.

Trimethylsilycellulose has been prepared from cellulose and 1,1,1,3,3,3-hexamethyldisilazane in liquid ammonia in an autoclave.[129] There were no by-product salts, no tedious purification, and no degradation of the polymer chain. Such a process might be run in an extruder. The cost of the reagent may be a drawback to widespread use.

12.3.2 Starch

Starch is also a polymer of glucose, but with α-glucosidyl linkages instead of the β ones in cellulose. It consists of a linear polymer, amylose, and a branched polymer of a much higher molecular weight, amylopectin.[130] Starch fibers extruded from a mixture of high-amylose cornstarch containing 30% water have been suggested for use in paper and textiles.[131] Before they can be used, their resistance to water needs to be increased by some sort of cross-linking, perhaps with glyoxal or adipaldehyde. Compression-molded high-amylose starch containing water and/or glycerol is tough.[132] It is stronger when cellulose microfibrils are added.[133] Because plasticizers can evaporate, migrate, or leach out over time, it might be wise to try some of a higher molecular weight, such as polyglycerol or polyethylene oxide. Plates, eating utensils, and other items have been molded from starch containing some water or glycerol.[134] It is best to add a coating to increase the resistance to water. Perhaps cutin, the outer covering of leaves, or waxes from the outsides of fruits, can be recovered from food-processing wastes and used for this purpose. A sucrose oleate might act both as a plasticizer and as a hydrophobic agent.

A composite of limestone, potato starch, and a small amount of cellulose fiber is being tested for containers used in fast-food restaurants.[135] Foamed "peanuts" of starch for packing purposes are available commercially for use in place of those made of polystyrene.[136] Colored versions are sold for children to wet and stick together.[137]

Tough leathery plastics have been made by extrusion of a starch-methyl acrylate graft copolymer.[138] Grafting can also be done in the extruder, as done with acrylamide with ammonium persulfate as a catalyst at the rate of 109–325 g/min.[139] Because starch is more tractable than cellulose, it is possible to prepare many derivatives by carboxymethylation, acylation, and amination in minutes with the aid of microwaves.[140] Use of some hydrophobic reagents might remedy some of the problems of water resistance. Another way to improve the resistance to water is by using polymer blends. Starch has been blended with a copolymer of ethylene and vinyl alcohol for this prupose.[141] Foamed plates have been prepared by baking a mixture of starch and aqueous polyvinyl alcohol.[142] Water or glycerol was still required as a plasticizer in both. Plates, knives, forks, and the like have been made by molding 30% of a synthetic polyester with 70% starch or gluten.[143] (The polyester probably attaches to the starch or gluten by formation of an ester or amide.) Poly(3-hydroxybutyrate), polylactic acid, and polycaprolactone also work.[144] Foamed polyurethanes have been made from starch, polycaprolactone, water, and an aliphatic diisocyanate.[145]

Rapidly dissolving water-soluble films have been made from mixtures of pectin, starch, and glycerol.[146] Water-soluble films are useful in packaging materials that are to be dissolved in water, such as food ingredients, detergents, and pesticides. Dissolving the whole package eliminates the problem of disposing of the package.

The use of materials for single-use throw-away plates, bowls, and tableware, and their disposal, would be reduced greatly if dishes of china and tableware of stainless steel were used and washed on the site.

12.3.3 Hemicellulose and Chitin

The hemicellulose[147] from the pulping of trees for paper is an underused resource. A small amount is being hydrolyzed to xylose for hydrogenation to the sweetener, xylitol.[148] A good use is as a substrate for fermentation, as is being done in the conversion of biomass to ethanol. (See Chapters 9 and 15.) A combination of sorbitol and arabinoxylan from corn hulls has been used as a moisture barrier on grapes to prevent weight loss on storage.[149] Apple waste consisting of 14% pectin, 12% hemicellulose, and 41% lignocellulose can be injection molded.[150]

Chitin[151] is another underused resource. It is the second most abundant polymer on earth, found in crab shells, shrimp shells, and so on. It is a polymer of an acetylated glucosamine (where the 2-OH is replaced by 2-$NHCOCH_3$). The amine from the hydrolysis of chitin is called chitosan. Although many uses have been explored, neither polymer has found widespread use.[152] Chitin can be converted to fibers by making a dibutyryl derivative.[153] Such a derivatization might also be tried in an extruder. Fibers have been spun from chitosan.[154] Chitosan that has been partially acylated to introduce acetyl, butyryl, hexanoyl, or lauroyl groups has been spun from aqueous solution.[155] Subsequent heating to higher than 180°C gave fibers with high-tensile strength and modulus that were insoluble in water and formic acid. A chitin/cellulose blend is marketed as a textile fiber under the name of "Crabyon."[156] The use of chitosan as a coagulant for the wastewater from making tofu has been described as using one waste to treat another.[157] The product is suitable for animal food. A water-resistant adhesive has been prepared from dihydroxyphenylalanine, tyrosinase, and chitosan.[158] Nanoparticles of copper carboxymethyl chitosan killed 99% of *Staphylococcus aureus.*[159] Chitosan has been modified for use in gene therapy to replace viruses as vectors.[160]

12.3.4 Lignin

Most lignin[161] from the pulping of wood for paper is burned for fuel. Many efforts have been made to find higher-value uses for it. Nearly all of it is removed from wood in a Kraft process that puts sulfonate groups into it. (The structure varies with the species. A schematic structure for spruce lignin is shown in **12.32** Schematic.[162] More phenolic groups are formed in the pulping process.) A process, which may not yet be commercial, removes the lignin from wood with aqueous ethanol.[163] This "organosolv" lignin has a number-average molecular weight of less than 1000 and a polydispersity of 2.4–6.3. It is soluble in some organic solvents and in dilute alkali. Pulping of wheat straw has been done with aqueous acetone instead of ethanol.[164] Such lignin may find greater use in adhesives and resins than Kraft lignin. It has been used as a reinforcing filler for poly(ethylene-*co*-vinylacetate.)[165] Any reaction that can use the phenol group is of interest. One possibility is to demethylate lignin to a catechol, which can then be reacted with a polyamine, possibly an amine-rich protein, using an oxidative enzyme to form the equivalent of mussel glue, the adhesive that holds the mussel to the rock. This adhesive could cure under water, for surgery or for boats. Demethylated lignin might also be useful in the tanning of leather. The molecular weight of the demethylated organosolv lignin will need to be raised for it to be strong. This might be done by enzymatic polymerization, by reaction with furfural or a bisoxazoline. (Phenol-furfural resins are used to bond abrasive wheels and brake linings.) Waste phenol–formaldehyde resin has been liquified to liquid phenols in more than 95% yield by hydrogenolysis in hot tetralin.[166] This process should be tried on organosolv lignin. Another route to "green phenols" is the pyrolysis of softwood bark at 500°C, at 1.5 psi.[167] This gave 30% oil (70% phenolic content), 29% charcoal, 10% gas, and 31% water.

Lignin has been tested in blends with several different polymers, including silicones, acrylics, polyurethanes, epoxy resins, and polyvinyl chloride.[168] Graft copolymers of lignin with styrene are thermoplastic.[169] IBM is using lignin in its circuit boards. Sarkanen and co-workers have prepared plastics containing 85% of nonderivatized Kraft lignin by including poly(vinyl acetate), diethyleneglycol dibenzoate, and indene.[170] They feel that such a mixture could be extruded. They also report thermoplastic formulations of 95–100% alkylated Kraft lignin. Earlier workers have been able to put only 25–40% lignin into polyurethanes, phenol–formaldehyde resins, epoxy resins, acrylic polymers, and so on. Lignin makes lignin-starch composites more hydrophobic than starch alone.[171] A thermoplastic material devised by the Fraunhofer Institute and manufactured by Technaro consists of lignin, cellulose fibers, and "natural additives."[172] It can be used in car parts, computer housings, toys, and so on.

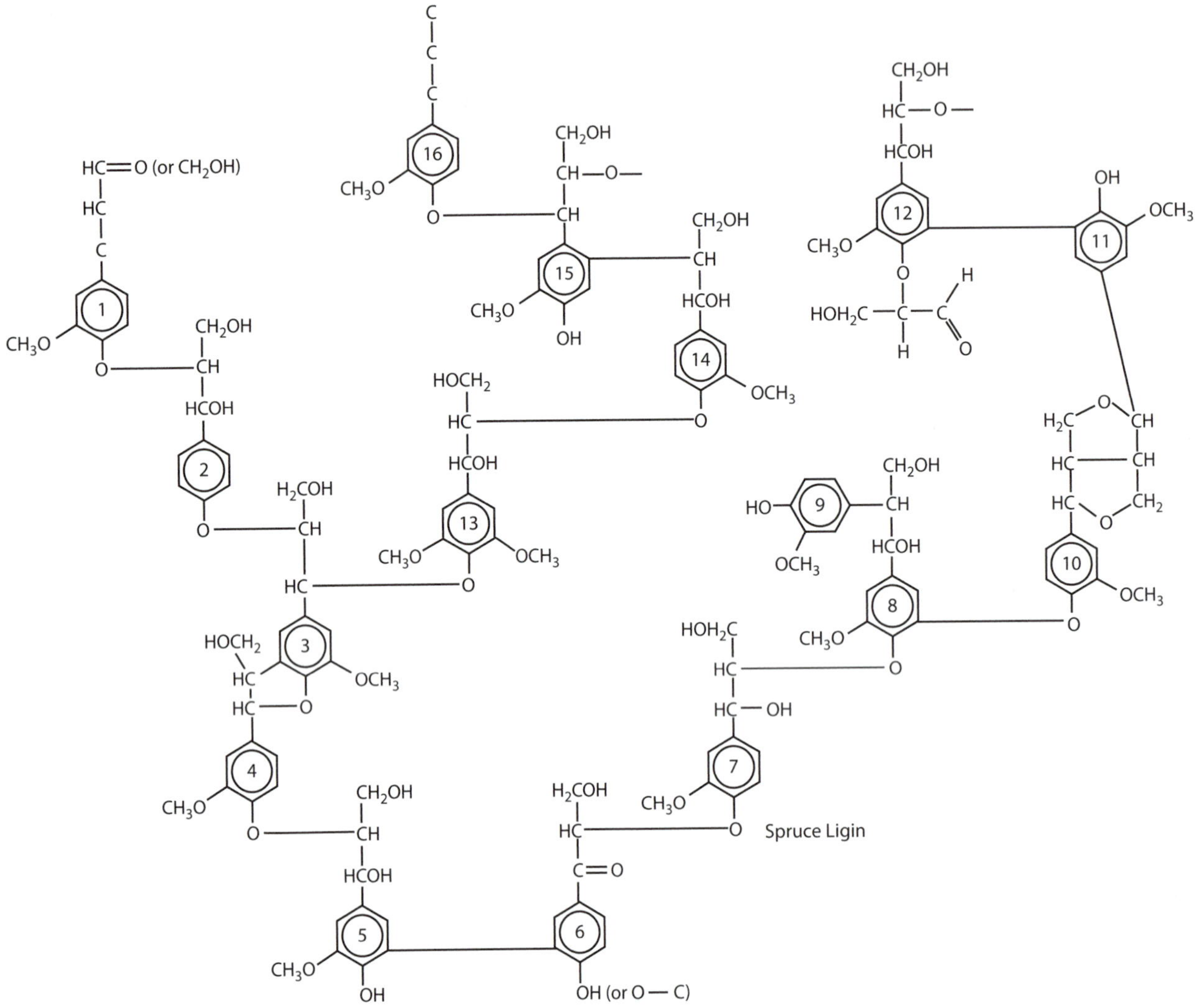

12.32 Schematic (Reprinted from E. Adler, *Wood Sci. Technol.*, 1977, 11, 169–218. With permission of Springer Verlag and Chalmers Tekniska Hogskola.)

Steam-exploded wood has been molded into composites without any adhesive binder.[173] Wood can be modified to improve its properties.[174] Some cheaper woods can be made equivalent to rare tropical woods, which may help save the latter. Impregnation of beechwood with allyl glycidyl ether and methyl methacrylate followed by polymerization with γ-radiation increases its strength and protects it against aging.[175]

The United States printed 56 million newspapers each day in 2004.[176] An enormous number of trees would be saved if these newspapers were read using a wireless personal digital assistant instead of the traditional way. This would produce 32–140 times less carbon dioxide, much less sulfur dioxide and nitrogen oxides, and use 26–67 times less water.

12.3.5 Proteins

Efforts to use proteins in plastic date back to the work of Henry Ford on soybeans.[177] The problems of these and other protein-based films, fibers, and plastics include brittleness, too much moisture absorption, and the need to stabilize them against microbial degradation. Brittleness can be due to very low molecular weight, very high hydrogen bonding between chains, or very high cross-linking. Plasticizers can be added to disrupt the hydrogen bonding. Preferably, they should be polymeric, even grafted to the protein, to avoid problems of leaching, migration, or volatilization. The addition of rubbers can reduce brittleness, provided that the phase separation of the rubbery domains is in the proper form. The best ones involve a light grafting of the rubber to the parent polymer. A polyurethane rubber, perhaps based on an oligomeric ester diol, might be suitable for a protein. Strength can also be increased by the addition of reinforcing fibers and fillers. Cellulose or glass fibers might be suitable. The molecular weight can be increased by some cross-linking. The water absorption can be reduced by hydrophobic groups added by grafting on a hydrophobic monomer, blending in a more hydrophobic polymer, a hydrophobic surface coating on the finished article, or other measures. These methods are being applied to proteins with some success. In the best cases, they involve additives derived from renewable resources. Occasionally these methods can be used to produce edible films that extend food life.[178]

Many potential new uses for soy protein and oil are being studied.[179] Soy protein plastics plasticized with polyhydric alcohols, such as glycerol, ethylene glycol, and propylene glycol, are said to have the potential to compete with engineering polymers if they can be kept dry.[180] (Polyvinyl alcohol was not suitable as a plasticizer.[181]) Films with tensile strengths of 2000 psi and 500% elongation can be blown from extrusion-blended soy protein and 30–40% of aliphatic polyesters.[182] Although they are brown and have a caramel-like odor, they may be suitable for trash bags and mulching film. (For comparison, low-density polyethylene trash bags have tensile strengths of 2000–3000 psi and elongations to break of 300–600%.)

Mixtures of zein (corn protein) or soy protein with starch for extrusion or molding were cross-linked to increase their tensile strength and water resistance.[183] (Zein treated with hot flax oil[184] or oleic acid[185] is also resistant to water.) There is a surplus of corn protein as a by-product of the conversion of corn to high-fructose corn syrup.[186] Blends of plasticized corn gluten meal and poly(caprolactone) have been suggested for use as biodegradable packaging.[187] The tensile strength of zein films plasticized with glycerol, with or without polypropylene glycol, was increased two to threefold by cross-linking with formaldehyde or with glutaraldehyde.[188] The water vapor permeability went up when the cross-linking agent was dialdehyde starch (made by oxidation of some of the pyranose rings with periodate). Zein fiber was cross-linked with citric acid or 1,2,3,4-butanetetracarboxylic acid using a sodium dihydrogenphosphate catalyst.[189] It had good stability to aqueous acetic acid. Soy protein plasticized with glycerol and water was spun, then finished with a combination of glutaraldehyde and acetic anhydride and oriented by stretching. It had a water uptake of only 1.15%.[190] Gelatin has been converted to tough plastic films by cross-linking with 3% hexamethylenediisocyanate, followed by orientation by stretching.[191] The films swelled in water. A hydrophobic surface treatment or coating would be needed for general use. (Gelatin is obtained from the collagen in bones and hides.)

Useful plastics have been made from the milk protein, casein.[192] The best ones were made using formaldehyde for cross-linking. A more benign cross-linking agent would be desirable. Sterilized edible films, with potential use in food packaging, have been made by gamma-irradiation of casein plasticized with propylene glycol or, better, triethyleneglycol.[193] An aqueous dispersion of modified, plasticized wheat gluten was found to be a good film former.[194] Films of wheat gluten may be suitable for controlled atmosphere packaging of foods because they have a high selectivity for carbon dioxide over oxygen (28 at 24°C and 100% relative humidity).[195] The puncture resistance of films of cottonseed protein plasticized with glycerol increased on cross-linking with formaldehyde, glyoxal, or glutaraldehyde.[196]

The grafting of vinyl monomers to proteins as a way to improve toughness and resistance to water does not seem to have been explored. Nor has the addition of rubbery particles been studied. The brittleness of polypeptides has been exploited in a positive way in one case. The polypeptides can be used in a chewing gum that can be easily removed from hard surfaces because it becomes brittle on drying.[197]

Soy protein treated with alkali or the enzyme, trypsin, is useful in adhesives.[198] Plywood made with new-modified soy protein glues performs just as well as that made with conventional phenol–formaldehyde resins.[199]

Chicken feathers can be converted to fibers for use in diapers.[200] Each chicken produces 0.5–1.0 lb of feathers. Feathers, hides, hoofs, and hair all contain proteins containing cystine cross-links. Cured rubber containing disulfide and polysulfide cross-links can be devulcanized by heating and shearing with rubber chemicals, by ultrasound and by microwaves.[201] It can then be revulcanized. It should be possible to prepare useful fibers and films from feathers, hides, hoofs, and hair with the use of an ultrasonic extruder perhaps with a volatile plasticizer, possibly sc carbon dioxide, which can be recovered through a vent. This would lower the temperature needed for the extrusion. Molded objects might be made in a similar fashion. The water resistance should be adequate, except perhaps with feathers, where they could be made hydrophobic by grafting or blending. If done properly, this might also provide moth resistance for shirts made from chicken feathers. Birds oil their feathers by preening using oil from an oil gland. Chicken feathers have also been used for filters[202], molded plastic,[203] and circuit boards.[204]

"Pleather," a leather–fiber composite, can be made from waste scraps of leather[205] and hides by hot pressing.[206] Spider silk is very strong and resilient.[207] Dragline silk is 5 times as strong as steel (on a weight basis), twice as elastic as nylon, and waterproof. It is being studied to see if polyamides with similar properties can be made. A mutant *Bacillus subtilis* self-assembles to form threads as long as 1 m that are as strong as steel on a weight basis.[208] If this can be scaled up, a promising biodegradable fiber may be available for general use.

12.3.6 Rubber

In a sustainable future, natural rubbers[209] will be used more and synthetic rubbers less. The problem of allergies to the protein in Hevea rubber can be avoided by using "Yulex" rubber from guayule.[210] The epoxidation of natural rubber can be used to extend the applications of rubber and improve bonding.[211]

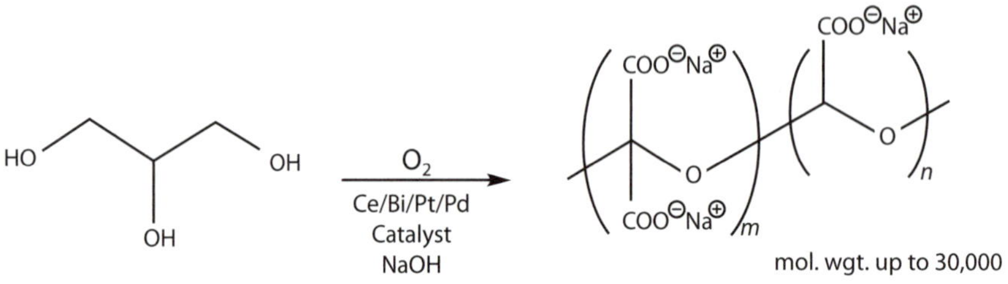

12.33 Schematic

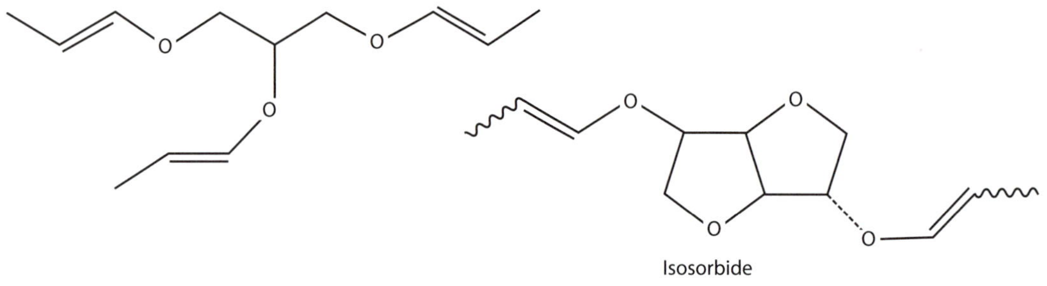

12.34 Schematic

12.4 POLYMERS FROM RENEWABLE RAW MATERIALS[212]

12.4.1 Products from Polyols

Several approaches, both chemical and biological, are being taken to convert sugars, amino acids, and related compounds to new polymers for commercial use.[213] Glycerol has been oxidized to a polymeric carboxylic acid (**12.33** Schematic) that may find use in detergents to complex calcium ions.[214]

Propenyl ethers from glycerol and isosorbide (**12.34** Schematic) are useful in photoinitiated cationic polymerization (e.g., for use in coatings).[215]

The polymerization is said to be fast with a low energy input, is not inhibited by air, uses no solvents, is nonpolluting, and uses inexpensive materials. The photopolymerization of various epoxidized oils and terpenes also does this. The propenyl ethers were made by isomerization of the corresponding allyl ethers with a ruthenium catalyst. The allyl ethers were made from the hydroxy compounds with allyl bromide and base, a process that produces waste salts. It is possible that monomers could be made by the reaction of the hydroxy compounds with propylene oxide followed by dehydration such that no waste salts would be formed.

Isosorbide is made by the acid-catalyzed dehydration of sorbitol, in turn, obtained by the reduction of glucose.[216] It has been reacted with acid chlorides of dibasic acids to form polyesters (**12.35** Schematic) with molecular weights as high as 34,000 and melting points as high as 181°C.[217] Many have lower molecular weights.

For use in plastics, this might require chain extension with diisocyanates, bisoxazolines, carbodiimides, bis(anhydrides), and so on. One use of the oligomers would be to form

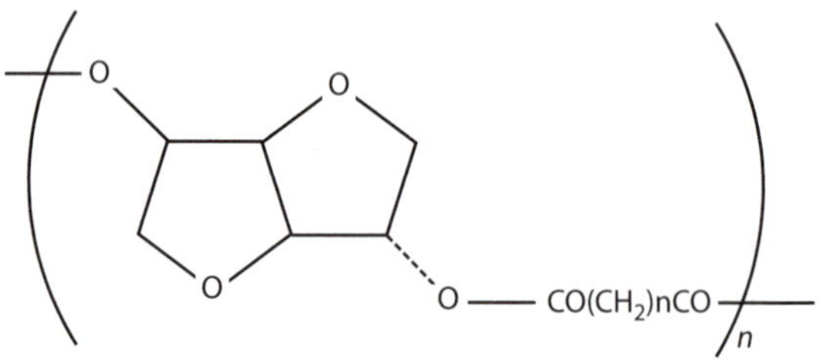

12.35 Schematic

polyurethanes by reaction with isocyanates. The use of acid chlorides can be avoided if the polymers are made by ester exchange or made enzymatically, with compounds such as divinyl adipate. Poly(butylene sebacate) with a molecular weight of 46,400 has been made from bis(2,2,2-trifluoroethyl)sebacate and 1,4-butanediol.[218] One polyester with a molecular weight of 24,000 has been made by ester exchange from isosorbide and a dimethyl ester (**12.36** Schematic), which can be derived from renewable sources (i.e., furfural and acetaldehyde).[219] (Similar polymers have been made using ethylene glycol as the diol.[220]) Polycarbonates and polyurethanes have also been prepared from isosorbide and other "sugar diols."[221] Fatty esters of isosorbide have been used as plasticizers.[222]

A variety of vinyl sugar monomers have been prepared and polymerized.[223] Many of the syntheses involve various protecting groups. Probably, the most practical and direct syntheses are those involving enzymatic formation of acrylates (as described in Chapter 9). This has been used to produce sucrose acrylate, as well as a monomer (**12.37** Schematic) made from α-methylgalactose and vinyl acrylate. A copolymer of this monomer with *N,N*-methylenebis(acrylamide) imbibed 100 times its weight of water. This swelling was not affected by

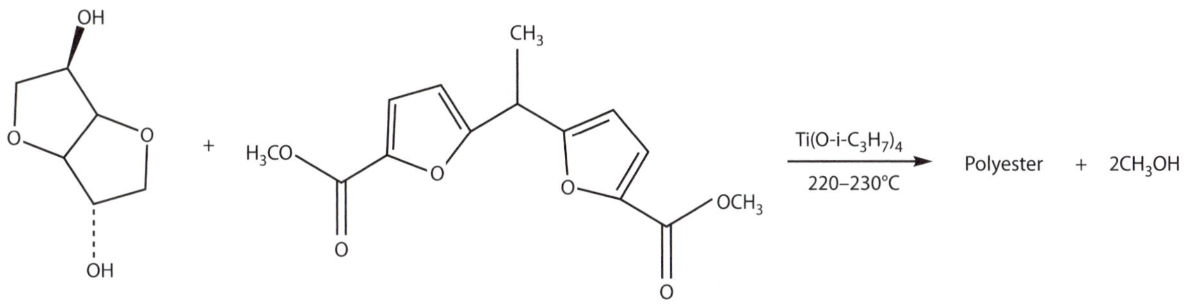

$$\text{Isosorbide} + \text{dimethyl ester (furan)} \xrightarrow[220\text{–}230°C]{Ti(O\text{-}i\text{-}C_3H_7)_4} \text{Polyester} + 2CH_3OH$$

12.36 Schematic

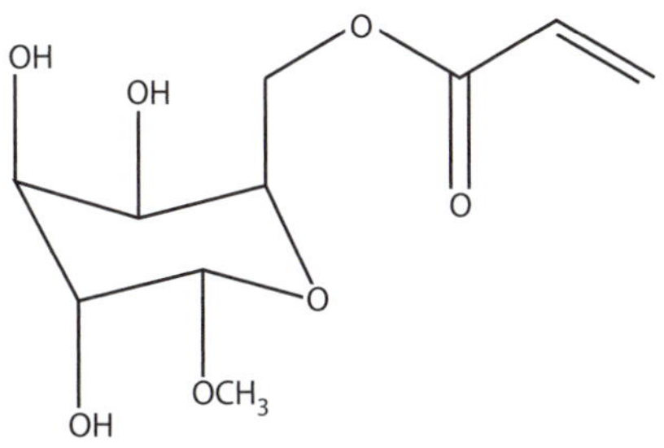

12.37 Schematic

sodium chloride in the water. Most "super-slurpers" swell less in the presence of dissolved salts. Thus, the new copolymer may be of use in collecting urine in diapers. Another polymeric hydrogel was made by polymerization of sucrose-1′-acrylate using sucrose-6,1′-diacrylate (prepared by enzymatic transesterification of sucrose and vinyl acrylate) as the cross-linking agent.[224]

The monomer in **12.38** Schematic was made from glycerol, acetone, and acetylene. The alternating copolymer with maleic anhydride cross-links on standing in a moist atmosphere.[225] (The carboxylic acid that results from the action of water on the anhydride catalyzes the cleavage of the acetal. The remaining anhydride then reacts with the newly formed hydroxyl groups.)

Epoxy resin adhesives for aluminum, glass, and steel have been prepared by converting octaallyl or octacrotylsucrose to the corresponding epoxides which were cured with diethylenetriamine.[226] Several plastics, resins, and adhesives have been based on soybean oil.[227] Vinyl esters of the fatty acids can be used as monomers. The oil can be epoxidized, then hydrolyzed to glycols, which can be converted to acrylates or maleates for polymerization. Natural fibers, such as hemp, can be used with such materials to form inexpensive composites.

Galactose-1,6-dialdehyde has been used as a substitute for glutaraldehyde in cross-linking proteins.[228]

A methacrylate made from gluconolactone and 2-aminoethylmethacrylate has been polymerized without blocking any hydroxyl groups (**12.39** Schematic).

12.4.2 Furfural

Many efforts have been made to base polymers on furfural made from pentoses.[229] The polymers may be useful, but tend to have lower thermal stability than the usual synthetic polymers. Polyesters based on furfural were mentioned earlier. The acid-catalyzed polymerization of furfuryl alcohol is used in foundry cores.[230] Furfural that has been decarbonylated, then hydrogenated to tetrahydrofuran polymerizes to a polymeric diol used in making polyurethanes. Furfural has been condensed with cardanol (*m*-pentadecadienylphenol) from cashew nut shell oil in the presence of other phenols to produce polymeric resins.[231] Cardanol and hydrogenated cardanol have been polymerized with horseradish peroxidase to soluble polymers in up to 85% yield.[232] Plasticizers that

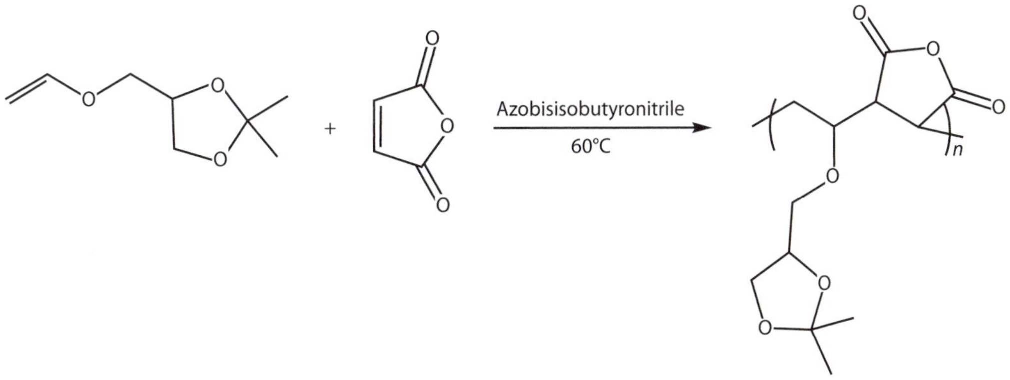

$$\text{vinyl ether acetal} + \text{maleic anhydride} \xrightarrow[60°C]{\text{Azobisisobutyronitrile}} \text{alternating copolymer}_n$$

12.38 Schematic

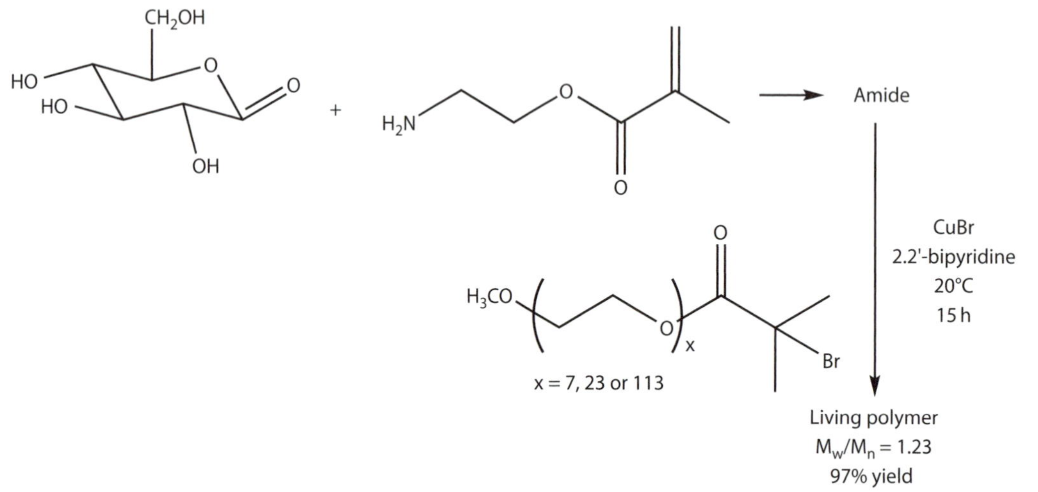

12.39 Schematic

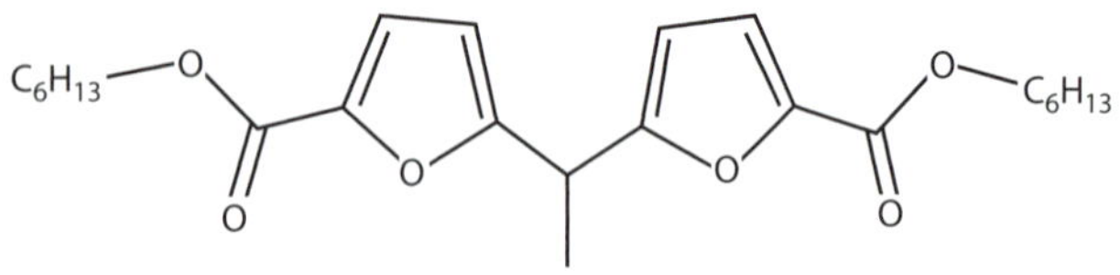

12.40 Schematic

are effective in polyvinyl chloride, such as **12.40** Schematic, have been made from furfural.[233]

Hexoses can be converted to 5-hydroxymethylfurfural by acids, such as hydrogen chloride or ion exchange resins, in water/solvent mixtures.[234] For example, the reaction with fructose can be 85% selective at 90% conversion. Pentoses give furfural.[235] Another system combines the hydrolysis of cellulose by concentrated hydrochloric acid with production of the furan derivative (**12.41** Schematic).[236]

The aldehydes can react with phenols to make resins with both the aldehyde and methylol groups reacting.[237] The phenols could be derived from tannins or lignin. The 5-hydroxymethylfurfural can be reduced to a diol or oxidized to a dicarboxylic acid by oxygen with an Au/TiO_2 catalyst in a 98% yield.[238] These could be converted to polyesters or polyamides. A copolyester from methyl furoate, isosorbide, acetaldehyde, and a methyl ester of a dibasic acid is shown in **12.42** Schematic.[239]

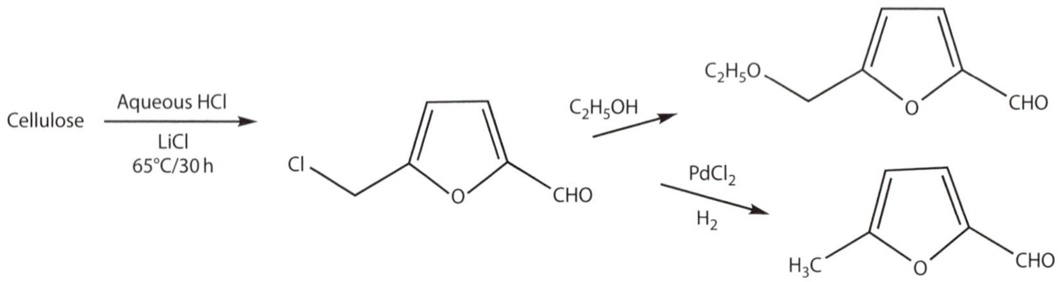

12.41 Schematic

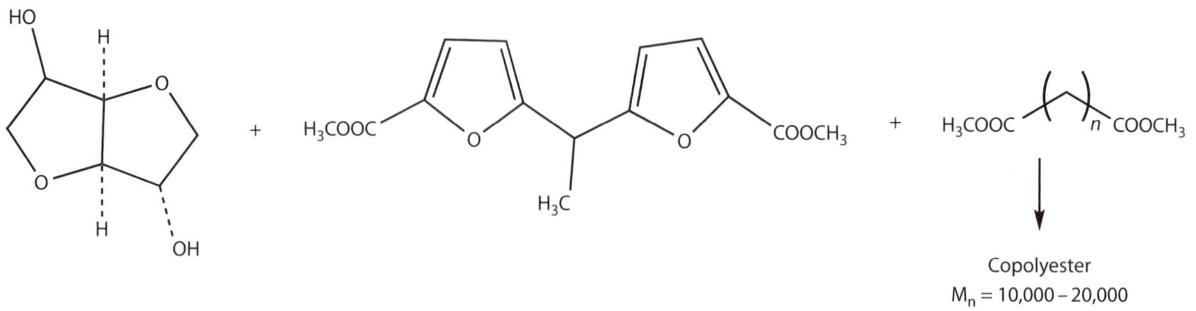

12.42 Schematic

12.4.3 Polyurethanes

Renewable materials can be used in the preparation of polyurethanes. Glycosides of polytetrahydrofuran have been used with diisoyanates.[240] Castor oil (which contains about 2.7 OH per molecule) has also been used.[241] Polyols derived from epoxidized soybean oil have been used to make polyurethanes.[242] They can be made by hydrolysis or methanolysis of the epoxides.[243] Another route is through hydroformylation, followed by reduction.[244] They can replace up to 30% of the polyols derived from petroleum. Lactic acid oligomers can be used.[245] "Wheat Board" is made from wheat straw and isocyanates.[246] It is said to be lighter and as strong as and more resistant to moisture than conventional particle board. No formaldehyde is needed for this building material.

12.4.4 Polyesters[247]

The production of poly(hydroxyalkanoates)[248] by bacteria was described in Chapter 9. They are being used in cups, cosmetic bottles, water filters, and credit cards,[249] and large-volume applications are expected now that their prices are more competitive with those of polyolefins. (Monsanto abandoned development of them in 1998 as the costs remained 25–50% above those of polyolefins.[250] Metabolix and Archer Daniels Midland are building a 50,000 metric ton/year plant in Iowa to produce the polymer.[251] Metabollix has genetically modified switchgrass to produce up to 3.7% poly (hydroxyalkanoate), but feels that 5.0–7.5% is necessary to make the method commercially feasible.[252] Inclusion of wood fiber makes them cheaper and causes nucleation that allows faster processing.[253] They may find use in eco-packaging because their barrier properties to carbon dioxide and water are reasonably good.[254] One of their big selling points is their biodegradability.[255] It allows them to be recycled by composting.

Polylactic acid[256] is also biodegradable, decomposing relatively quickly under conditions in the body. This allows it and related copolymers with glycolic and other acids to be used in surgery, where something implanted in the body does not have to be removed surgically.[257] It is produced by The Nature Works (owned by Cargill) in a 300 million lbs/year plant in Nebraska. A wide variety of catalysts has been used for the polymerization.[258] The polymer is said to be suitable as commodity plastic for packaging, being similar to polystyrene in many respects.[259] The strength is increased by orientation. When plasticized with varying amounts of lactide (the cyclic dimer) or other oligomers of lactic acid, its properties can mimic those of flexible polyvinyl chloride, low-density polyethylene, linear low-density polyethylene, polypropylene, and polystyrene. It can also be plasticized with 20% acetyltributyl citrate.[260]

Kureha is building a commercial plant to make poly (glycolic acid) in West Virginia.[261] This polymer has better gas barrier properties than poly(ethylene-co-vinyl alcohol).

Copolymers of carbon dioxide with propylene oxide,[262] cyclohexene oxide,[263] and limonene oxide[264] are being studied. A new company, Novomer, has been set up to make them.[265]

12.4.5 Polyamides

Polyamides can be made from renewable resources in several ways. Soybean-based dimer C_{36} fatty acids have been polymerized with a variety of diamines, including 1,4-phenylenediamine.[266] Oxazolines, made from fatty acids and ethanolamine (**12.43** Schematic), have been polymerized up to a weight-average molecular weight of 148,000 with a polydispersity of 3.4.[267] Other chain lengths can also be used. Copolymers of ones with long chains and ones with short chains have been made. Some of these may find use as compatibilizers for polyamides. Poly(γ-stearyl-L-glutamate) has been made by polymerization of the cyclic anhydride (**12.44**

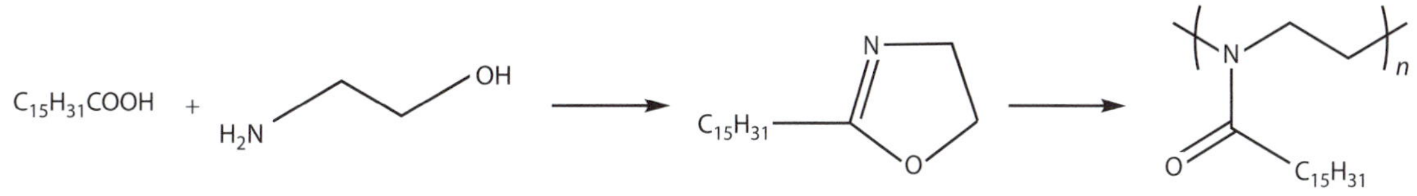

12.43 Schematic

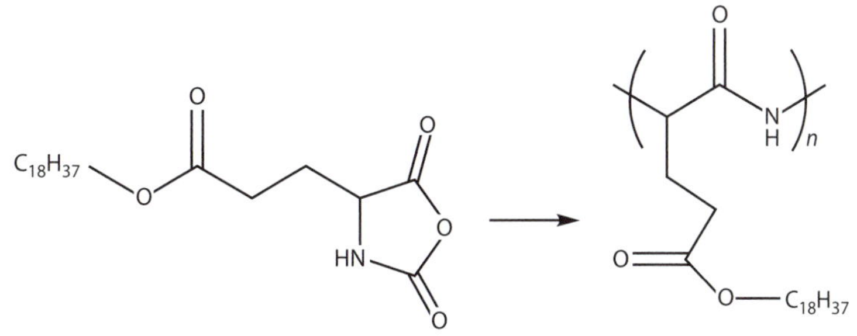

12.44 Schematic

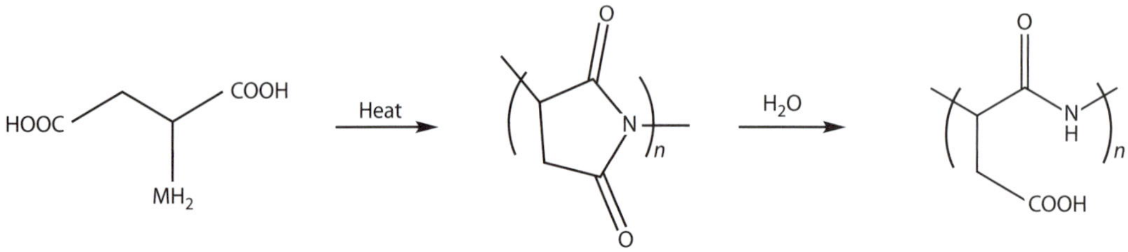

HOOC COOH MH$_2$ Heat O N n O H_2O O N H n COOH

12.45 Schematic

Schematic).[268] These amino acid anhydrides are usually made with phosgene. The use of dimethyl carbonate should be tested in the preparation of the monomers.

Polyaspartic acid can be made by heating aspartic acid without solvent to form a polysuccinimide, which is then hydrolyzed (**12.45** Schematic).[269] It can also be made directly from maleic anhydride and ammonia. It promises to be useful as a scale inhibitor in water, an antiredeposition agent in detergents, and the like. Gamma-irradiation converts it to a biodegradable superabsorbent material that takes up 3400 g water per gram of dry polymer.[270] (Most superabsorbent polymers are based on acrylamide, a neurotoxin.)

Poly(γ-glutamic acid), an analogue of poly(aspartic acid) with another methylene group in the side chain (**12.46** Schematic), is produced by *Bacillus subtilis*.[271] (Notice that polyglutamic acid made by chemical synthesis involves the other carboxylic group.) Esterification improves the processibility and the solubility in organic solvents.

Adipic acid and hexamethylenediamine have been copolymerized with 37.5–47.5 mol% of proline, alanine, or glycine to produce polyamides with number-average molecular weights of about 20,000.[272] They have been suggested for use as biodegradable food packaging. Their cost could be reduced if mixed amino acids from the hydrolysis of waste protein could be used. It might also be possible to pass a mixture of adipic acid, hexamethylenediamine, and waste protein through a vented extruder to make such products. Polyamides with molecular weights up to 26,700 have been made from the benzyl ester of lysine and activated diesters of dicarboxylic acids.[273] Tyrosine-based polycarbonates have also been made (**12.47** Schematic).[274]

Thermally reversible polymers might make recycling easier. Some brittle ones have been based on sorbic acid, lysine, glycine, and maleic anhydride (**12.48** Schematic).[275]

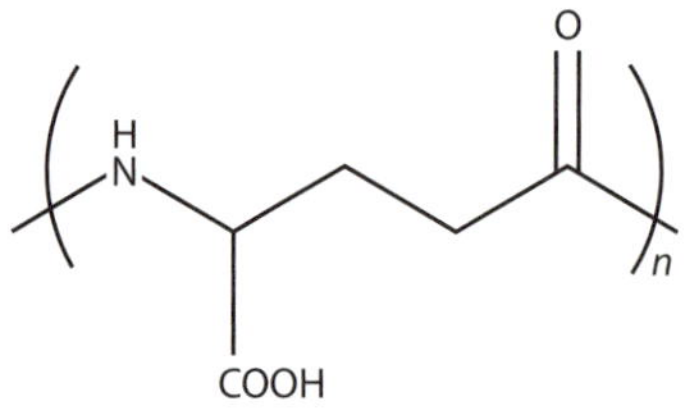

H N O n COOH

12.46 Schematic

The polymers form by the Diels–Alder reaction. A furan ester has been used in a similar way at 60°C with another bismaleimide (**12.49** Schematic) to make polymers with molecular weights as high as 30,000.[276] The problem is that Diels–Alder adducts of furans reverse at relatively low temperatures, in this case at 90°C or more. Such polymers would have to be used at room temperature or not much higher.

12.5 CONCLUSIONS AND RECOMMENDATIONS

There are ways to make many of the major polymers used today from renewable raw materials. Many of the syntheses will require further study and optimization before they can be used commercially. The economics will change with the shift to new raw materials and what is inexpensive now may not be so after the change. It may be less expensive to use the natural polymers again. As many of these have been given little attention since the advent of cheap petroleum and natural gas, further study using the additional knowledge and methods available today is indicated. This should include new methods of fabrication, new chemical derivatives of

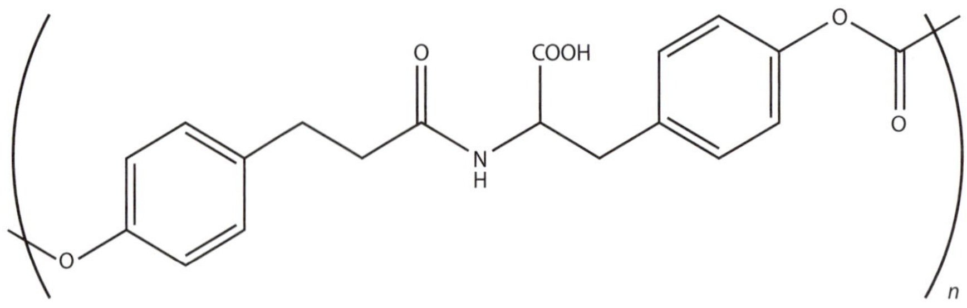

O COOH O N H O O n

12.47 Schematic

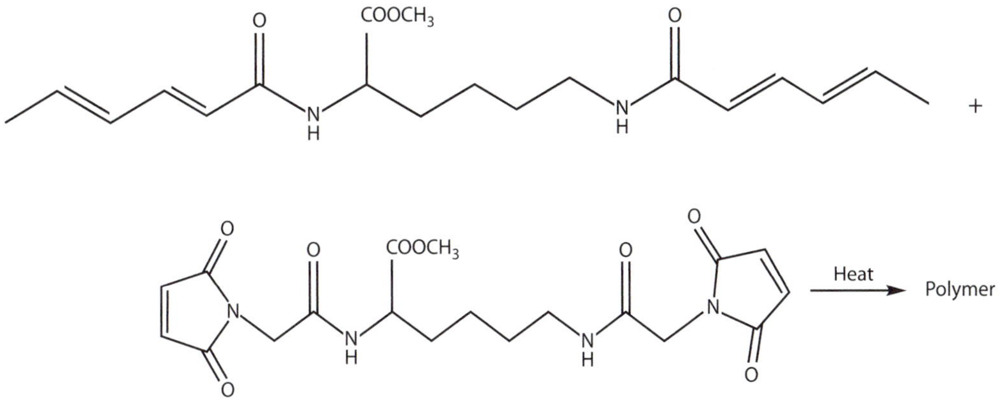

12.48 Schematic

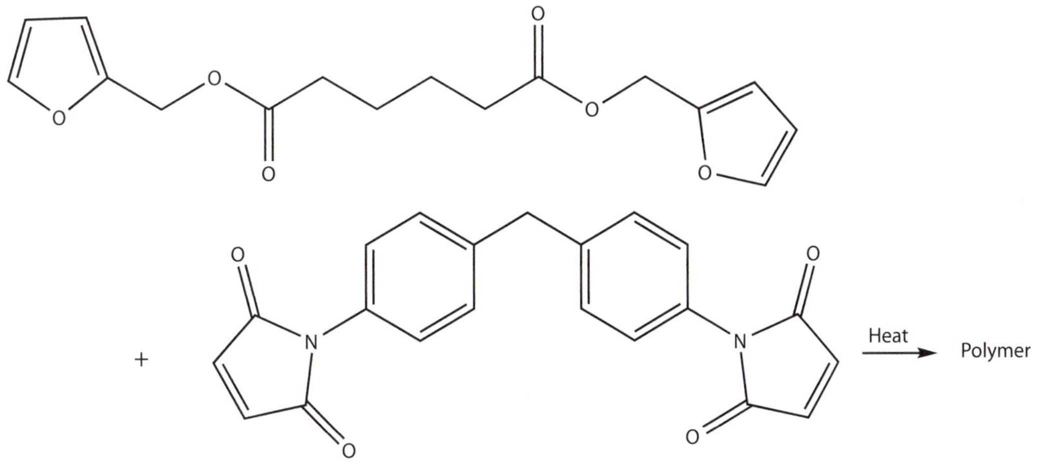

12.49 Schematic

natural polymers, bonding agents for fillers, compatibilizers for polymer blends, polymeric plasticizers, surface coatings as barriers for gases and liquids and for scratch or hardness resistance, methods of graft polymerization, and ways of upgrading present-day wastes to the status of valuable coproducts. By gaining a better understanding of natural polymers, such as spider silk and mussel glue, it may be possible to prepare valuable new materials from renewable raw materials. Such research may hasten the change from products based on petroleum and natural gas to those based on materials from forests and fields, especially if methods can be found to make products from the latter less expensive.

The biggest unsolved need is to find a way to curb the overconsumption of material goods by developed nations.[277] There must be a way to shift back from single-use, throwaway items to reusable ones. Chapter 13 will cover ways to make things last longer. Ways must be found to recycle materials in items that are worn out over and over again, instead of just burning them for their energy content. Recycling will be covered in Chapter 14. All of this must be done with much less energy than is used today, because it will have to come from renewable sources, as described in Chapter 15.

REFERENCES

1. (a) National Commission on the Environment, *The Report of the National Commission on the Environment, Choosing a Sustainable Future*, Island Press, Washington, DC, 1993; (b) J. Dewulf and H. vanLangenhove, eds, *Renwables–Based Technology: Sustainability Assessment*, Wiley, NY, 2006; (c) K. Geiser, *Materials Matter: Toward a Sustainable Materials Policy*, MIT Press, Cambridge, MA, 2001; (d) R. Lankey and P.A. Anastas, eds, *Advancing Sustainability Through Green Chemistry and Engineering*, ACS Symp. 823, Washington, DC, 2002; (e) F. Cavani, G. Centi, S. Perathoner, and F. Trifiro, eds., *Sustainable Industrial Chemistry*, Wiley-VCH, Weinheim, 2010.
2. (a) H. Zoebelein, ed., *Dictionary of Renewable Resources*, 2nd ed., Wiley-VCH, Weinheim, 2001; (b) B.A. Tokay, *Ullmann's Encyclopedia of Industrial Chemistry*, 5th ed.,

VCH, Weinheim, 1985, *A4*, 99; (c) D. Gaskell, *Chem. Br.*, 1998, *34*(2), 49; (d) A. Louwrier, *Biotechnol. Appl. Biochem.*, 1998, *27*, 1; (e) G. Fuller, T. A. McKeon, and D.D. Bills, eds, *Agricultural Materials as Renewable Resources; Nonfood and Industrial Applications*, ACS Symp. 647, Washington, DC, 1996; (f) J.E. Recheigl and H.C. MacKinnon, *Agricultural Uses of By-Products and Wastes*, ACS Symp. 668, Washington, DC, 1997; (g) B.C. Saha and J. Woodward, *Fuels and Chemicals from Biomass*, ACS Symp. 666, Washington, DC, 1997; (h) C. Crabb, *Chem. Eng.*, 2000, *107*(10), 31; (i) A. Thayer, *Chem. Eng. News*, May 29, 2000, 40; (j) J.J. Bozell and M.K. Patel, *Feedstocks for the Future—Renewables for the Production of Chemicals and Materials*, ACS Symp. 921, 2006; (k) G. Centi and R.A. vanSanten, eds, *Catalysis for Renewables, From Feedstock to Energy Production*, Wiley-VCH, Weinheim, 2007; (l) J.J. Bozell, *Chemicals and Materials from Renewable Resources*, ACS Symp. 784, Washington, DC, 2001; (m) M. Graziani and P. Formasiero, eds, *Renewable Resources and Renewable Energy*, CRC/Taylor & Francis, Boca Raton, FL, 2007; (n) I. Law, M. Smallwood, and W. Smith, *Chem. Ind. (Lond.)*, July 19, 2004, 18; (o) J.H. Clark and F. Deswarte, eds., *Introduction to Chemicals from Bismass*, Wiley, Hoboken, NJ, 2008.
3. (a) K. Weissermel and H.-J. Arpe, *Industrial Organic Chemistry*, 4th ed., Wiley-VCH, Weinheim, 2003; (b) H.A. Wittcoff, B.G. Reuben, and J.S. Plotkin, *Industrial Organic Chemicals*, Wiley-Interscience, Hoboken, NJ, 2004; (c) P.J. Chenier, *Survey of Industrial Chemistry*, Kluwer Academic Publishers, Norwell, MA, 2002; (d) G.A. Olah and A. Molnar, *Hydrocarbon Chemistry*, Wiley, NY, 1995.
4. E.M. Kirschner, *Chem. Eng. News*, Apr. 8, 1996, 17.
5. M.S. Reisch, *Chem. Eng. News*, May 26, 1997, 14.
6. Staff, *Chem. Eng. News*, July 7, 2008, 67.
7. C.B. Philips and R. Datta, *Ind. Eng. Chem. Res.*, 1997, *36*, 4466.
8. Anon., *Chem. Eng. News*, June 16, 2008, 33.
9. (a) D. Varili, T. Dogu, and G. Dogu, *Ind. Eng. Chem. Res.*, 2008, *47*, 4071; (b) M. Inaba, K. Murata, M. Saito, and I. Takahara, *Green Chem.*, 2007, *9*, 638.
10. A.K. Talukdar, K.G. Bhattacharyya, and S. Sivasanker, *Appl. Catal. A*, 1997, *148*, 357.
11. M. Jacoby, *Chem. Eng. News*, Aug. 2, 1999, 6.
12. (a) P. Andrews, J.M. Corker, J. Evans, and M. Webster, *J. Chem. Soc. Dalton Trans.*, 1994, 1337; (b) H.K. Hall, Jr., *Chemtech*, 1997, *27*(6), 20; (c) G.R. Lappin, L.H. Nemec, J.D. Sauer, and J.D. Wagner, *Kirk–Othmer Encyclopedia of Chemical Technology*, 4th ed., Wiley, New York, 1996, *17*, 839.
13. (a) D.S. McGuinness, P. Wasserscheid, W. Keim, C. Hu, U. Englert, J.T. Dixon, and C. Grove, *Chem. Commun.*, 2003, 334; (b) R.D. Kohn, *Angew. Chem. Int. Ed.*, 2008, *47*, 245.
14. Z. Ye, F. AlObadi, and S. Zhu, *Macromol. Rapid Commun.*, 2004, *25*, 647.
15. (a) L.K. Johnson, S. Mecking, and M. Brookhart, *J. Am. Chem. Soc.*, 1996, *118*, 267; (b) C.M. Killian, D.J. Tempel, L.K. Johnson, and M. Brookhart, *J. Am. Chem. Soc.*, 1996, *118*, 11664; (c) L.K. Johnson, C.M. Killian, and M. Brookhart, *J. Am. Chem. Soc.*, 1995, *117*, 6414; (d) M. Brookhart, J.M. DeSimone, B.E. Grant, and M.J. Tanner, *Macromolecules*, 1995, *28*, 5378; (e) M. Freemantle, *Chem. Eng. News*, Apr. 13, 1998, *120*, 7143; (g) Z. Guan, P.M. Cotts, E.F. McCord, and S.J. McLain, *Science*, 1999, *283*, 2059; (h) R.J. Piccolini and J.S. Plotkin, *Chemtech*, 1999, *29*(5), 39.
16. (a) A. Bollman, K. Blann, J.T. Dixon, F.M. Hess, E. Killian, H. Maumela, et al., *J. Am. Chem. Soc.*, 2004, *126*, 14712; (b) Anon., *Chem. Eng. News*, Nov. 1, 2004, 24.
17. K.E. Bowen, M. Charensuk, and D.F. Wass, *Chem. Commun.*, 2007, 2835.
18. M.L. Morgan, *Chem. Ind.* (*Lond.*), 1998, 90.
19. (a) J. Haggin, *Chem. Eng. News*, June 20, 1994, 34; (b) R.W. Diesen, U.S. patent 5,276,257 (1994); (c) R.W. Diesen, K.A. Burdett, R.S. Dixit, and S.S.T. King, U.S. patent 5,329,057 (1994).
20. (a) K.J. Ivin and J.C. Mol, *Olefin Metathesis and Metathesis Polymerization*, Academic, San Diego, 1997, 93; (b) A. Gil and M. Montes, *Ind. Eng. Chem. Res.*, 1997, *36*, 1431; (c) A.H. Tullo, *Chem. Eng. News*, Mar. 17, 2003, 21; (d) A.H. Tullo, *Chem. Eng. News*, Apr. 23, 2007, 28.
21. (a) M.D. Bruce, G.W. Coates, E. Hauptman, R.M. Waymouth, and J.W. Ziller, *J. Am. Chem. Soc.*, 1997, *119*, 11174; (b) R. Baum, *Chem. Eng. News*, Jan. 16, 1995, 6.
22. (a) D.F. Oxley, *Chem. Ind.* (*Lond.*), 1998, 305; (b) D. Rotman, *Chem. Week*, May 21, 1997, 23; (c) K. Mashima, Y. Nakayama, and A. Nakamura, *Adv. Polym. Sci.*, 1997, *133*, 1; (d) W. Kaminsky, *Pure Appl. Chem.*, 1998, *70*, 1229; (e) M.R. Ribeiro, A. Deffieux, and M.F. Portela, *Ind. Eng. Chem. Res.*, 1997, *36*, 1224.
23. Anon., *R & D Cahners*, 2001, *43*(9), 45.
24. (a) Anon., *Chem. Eng. News*, Nov. 4, 1996, 17; (b) Y. Koide and A.R. Barron, *Macromolecules*, 1996, *29*, 1110; (c) E. Drent and P.H.M. Budzelaar, *Chem. Rev.*, 1996, *96*, 663; (d) A.S. Abu-Surrah, R. Wursche, and B. Rieger, *Macromol. Chem. Phys.*, 1997, *198*, 1197; (e) A. Sommazzi and F. Garbassi, *Prog. Polym. Sci.*, 1997, *22*, 1547; (f) A. Gray, *Chem. Br.*, 1998, *34*(3), 44.
25. M. Brennan, *Chem. Eng. News*, Apr. 1, 1996, 8.
26. N. Nojiri, Y. Sakai, and Y. Watanabe, *Catal. Rev. Sci. Eng.*, 1995, *37*(1), 145.
27. (a) G. Parkinson, *Chem. Eng.*, 1997, *104*(6), 21; (b) T. Jinbo, T. Pondo, M. Murakami, T. Matsushisa, K. Kawahara, and N. Horiuchi, *Chem. Abstr.*, 1994, *121*, 281422; (c) T. Kuroda and T. Shiotani, *Chem. Abstr.*, 1994, *121*, 281426–7.
28. G. Parkinson, *Chem. Eng.*, 2000, *107*(13), 21.
29. S.J. Ainsworth, *Chem. Eng. News*, May 22, 1995, 13.
30. (a) S. Stinson, *Chem. Eng. News*, Aug. 21, 1995, 7; (b) C. Ondrey, *Chem. Eng.*, 1995, *102*(9), 17; (c) M.L. Morgan, *Chem. Ind.* (*Lond.*), 1997, 166; (d) T.J. Remans, P.A. Jacobs, J. Martens, D.A.G. van Oeffelen, and M.H.G. Steijns, U.S. patent 5,811,601 (1998).
31. (a) H.A. Wittcoff, B.G. Reuben, and J.S. Plotkin, *Industrial Organic Chemicals*, 2nd ed., Wiley-Interscience, Hoboken, NJ, 2004; (b) Anon., *J. Chem. Ind. Eng. (China)*, 2005, *22*(7), 545.
32. G. Gubitosa and B. Casale, U.S. patent 5,354,914 (1994).
33. (a) K. Wang, M.C. Hawley, and T.D. Furney, *Ind. Eng. Chem. Res.*, 1995, *34*, 3766; (b) H. Li, W. Wang, and J.F. Deng, *J. Catal.*, 2000, *191*, 257.
34. (a) J. Ryczkowski, *Appl. Catal. A*, 1995, *125*, N18; (b) Anon., *Chem. Eng. News*, Feb. 1, 1999, 14; (c) N.P. Chopey, *Chem. Eng.*, 2003, *110*(12), 44.

35. I.D. Dobson, *Green Chem.*, 2003, *5*, G78.
36. T. Tsuchida, S. Sakuma, T. Takeguchi, and W. Ueda, *Ind. Eng. Chem. Res.*, 2006, *45*, 8634.
37. M. Glinski, J. Kijenski, and A. Jakubowski, *Appl. Catal. A*, 1995, *128*, 209.
38. K.M. Parida, A. Samal, and N.N. Das, *Appl. Catal. A*, 1998, *166*, 201.
39. K. Parida and H.K. Mishra, *J. Mol. Catal. A*, 1999, *139*, 73.
40. (a) S.K. Yedur, J. Dulebohn, T. Werpy, and K.A. Berglund, *Ind. Eng. Chem. Res.*, 1996, *35*, 663; (b) Anon., *Chem. Eng. News*, Mar. 24, 2003, 12.
41. Anon., *Chem. Eng. News*, Jan. 17, 2005, 67.
42. (a) I.-K. Yoo, H.N. Chang, E.G. Lee, Y.K. Chang, and S.-H. Moon, *Biotechnol. Lett.*, 1997, *19*, 237; (b) D. Porro, L. Brambilla, B.M. Ranzi, E. Martegani, and L. Alberghina, *Biotechnol. Prog.*, 1995, *11*, 294; (c) Anon., *Chemtech*, 1994, *24*(10), 38; (d) M.G. Adsul, A.J. Varma, and D.V. Gokhale, *Green Chem.*, 2007, *9*, 58.
43. J. Llorca, P.R. de la Piscina, J.-L.G. Fierro, J. Sales, and N. Homs, *J. Mol. Catal. A. Chem.*, 1997, *118*, 101.
44. M. Ai and K. Ohdan, *Chem. Lett.*, 1996, *25*, 247.
45. (a) D.C. Wadley, M.S. Tam, P.B. Kokitkar, J.E. Jackson, and D.J. Miller, *J. Catal.*, 1997, *165*, 162; (b) G.C. Gunter, D.J. Miller, and J.E. Jackson, *J. Catal.*, 1994, *148*, 252.
46. (a) A. Tullo, *Chem. Eng. News*, June 3, 27, 2005, 11; (b) Anon., *Chem. Eng. News*, Jan. 21, 2008, 25.
47. K. Enomoto, M. Ajioka, and A. Yamaguchi, U.S. patent 5, 310865 (1994).
48. (a) M. Pagliaro and M. Rossi, *The Future of Glycerol, New Usages for a Versatile Raw Material*, Royal Society of Chemistry, Cambridge, U.K., 2008; (b) F. Jerome, Y. Pouilloux, and J. Barrault, *Chem. Sus. Chem.*, 2008, *1*, 586; (c) R. D'Aquino and G. Ondrey, *Chem. Eng.*, 2007, *114*(9), 31; (d) M. McCoy, *Chem. Eng. News*, June 1, 2009, 16; (e) M. Pagliaro, *The Future of Glyceral*, 2nd ed., *Royal Society of Chemistry*, Cambridge, UK.
49. (a) G. Ondrey, *Chem. Eng.*, 2007, *114*(5), 17; (b) M.S. Reisch and A.H. Tullo, *Chem. Eng. News*, Dec. 18, 2006, 30.
50. (a) U.S.EPA, *Presidential Green Chemistry Challenge Awards*, June 2006, to Galen J. Suppes; (b) S. Shelley, *Chem. Eng. Prog.*, 2007, *103*(8), 6; (c) M.G. Musilono, L.A., Scarpino, F. Maurello, and R. Pietropaolo, *Green Chem.*, 2009, *11*, 1511; (d) A-Y. Yin, X-Y. Guo, W-L. Dai, and K-N. Fan, *Green Chem.*, 2009, *11*, 1514.
51. E. Taarning, A.T. Madssen, J.M. Marchetti, K. Egeblad, and C.H. Christensen, *Green Chem.*, 2008, *10*, 408.
52. (a) L. Ott, M. Bicker, and H. Vogel, *Green Chem.*, 2006, *8*, 214; (b) W. Yan and G.J. Suppes, *Ind. Eng. Chem. Res.*, 2009, *48*, 3279.
53. S.-H. Chai, H.-P. Wang, Y. Liang, and B.-Q. Xu, *Green Chem.*, 2007, *9*, 1130.
54. J. Barrault, Y. Pouilloux, J.M. Clacens, C. Vanhove, and S. Bancquart, *Catal. Today*, 2002, *75*, 177.
55. C. Goedel, T. Sawangwan, M. Mueler, A. Schwarz, and B. Nidetzsky, *Angew. Chem. Int. Ed.*, 2008, *47*, 10086.
56. D.J. Murphy, *Chem. Br.*, 1995, *31*(4), 300.
57. C.L. Mehltretter, C.E. Rist, and B.H. Alexander, U.S. patent 2,472,168 (1949).
58. S. Coffey, ed., *Rodd's Chemistry of Carbon Compounds*, 2nd ed., Elsevier, Amsterdam, 1976, I G, 21.
59. F. Fringuelli and A. Taticchi, *Dienes in the Diels–Alder Reaction*, Wiley, NY, 1990, 65.
60. G. Parkinson, *Chem. Eng.*, 1998, *105*(11), 19.
61. K. Nagayama, I. Shimizu, and A. Yamamoto, *Chem. Lett.*, 1998, *27*, 1143.
62. (a) G.L. Tembe, A.R. Bandyopadhyay, P.A. Ganeshpure, and S. Satish, *Catal. Rev. Sci. Eng.*, 1996, *38*, 299; (b) G.M. di Renzo, P.S. White, and M. Brookhart, *J. Am. Chem. Soc.*, 1996, *118*, 6225; (c) T. Hayashi, *Chem. Abstr.*, 1995, *122*, 291736.
63. (a) R. Dagani, *Chem. Eng. News*, July 5, 1999, 30; (b) G.M. Kirchhoff and T. Williamson, *Green Chem.*, 1999, *1*, G124.
64. J.V. Crivello, R. Narayan, and S.S. Sternstein, *J. Appl. Polym. Sci.*, 1997, *64*, 2073.
65. J. Miki, M. Asanuma, Y. Tachibana, and T. Shikada, *J. Chem. Soc. Chem. Commun.*, 1994, 1685.
66. (a) P. Meriaudeau and C. Naccache, *Catal. Rev. Sci. Eng.*, 1997, *39*, 5; (b) V.R. Choudhary and P. Devadas, *J. Catal.*, 1997, *172*, 475.
67. Y. Xu and L. Lin, *Appl. Catal. A*, 1999, *188*, 53.
68. S.Y. Lin and C.W. Dence, eds, *Methods in Lignin Chemistry*, Springer, Berlin, 1992, 374.
69. (a) R.W. Hemingway and P.E. Laks, eds, *Plant Polyphenols: Synthesis, Properties, Significance*, Plenum, NY, 1992; (b) R. W. Hemingway and A.H. Conner, eds, *Adhesives from Renewable Resource*, ACS Symp. 385, Washington, DC, 1989; (c) E. Haslam, *Plant Polyphenols: Vegetable Tannins Revisited*, Cambridge University Press, Cambridge, 1989.
70. J.M. Garro-Galvez and B. Riedl, *J. Appl. Polym. Sci.*, 1997, *65*, 399.
71. (a) A. Pizzi, ed., *Advanced Wood Adhesives Technology*, Dekker, NY, 1994, 149; (b) E. Masson, A. Pizzi, and M. Merlin, *J. Appl. Polym. Sci.*, 1997, *64*, 243; (c) A. Pizzi, N. Meikleham, and A. Stephanou, *J. Appl. Polym. Sci.*, 1995, *55*, 929; (d) A. Pizzi and N. Meikleham, *J. Appl. Polym Sci.*, 1995, *55*, 1265; (e) R. Garcia and A. Pizzi, *J. Appl. Polym. Sci.*, 1998, *70*, 1083, 1093.
72. A. Pizzi, ed., *Advanced Wood Adhesives Technology*, Dekker, NY, 1994, 219.
73. (a) U.S.EPA, *Presidential Green Chemistry Challenge Awards*, July, 2007, to K. Li and Hercules (now Ashland); (b) K. Li, S. Peshkova, and X. Geng, *J. Am. Chem. Soc.*, 2004, *81*, 487.
74. (a) F. Gunstone and R. Hamilton, *Oleochemical Manufacture and Applications*, Sheffield Academic Press, Sheffield, U.K., 2001; (b) U. Biermann, W. Friedt, S. Lang, W. Luhs, G. Machmuller, J.O. Metzger, M.R.G. Klaas, H.J. Schafer, and M.P. Schneider, *Angew. Chem. Int. Ed.*, 2000, *39*, 2206; (c) P.L. Nayak, *J. Macromol. Sci. Rev. Macromol. Chem.*, 2000, *C40*, 1.
75. (a) United Soy Board, *Feedstocks*, 1994, *4*(4), 4; (b) Anon., *Chem. Eng. News*, Nov. 24, 1997, 92 [advertisement]; (c) S.G. Wildes, *Chem. Innovation*, 2001, *31*(5), 22.
76. D.J. Murphy, *Chem. Br.*, 1995, *31*(4), 300.
77. Z.S. Liu, S. Erhan, J. Xu, and P.D. Calvert, *J. Appl. Polym. Sci.*, 2002, *85*, 2100.
78. (a) W. von Rybinski and K. Hill, *Angew. Chem. Int. Ed.*, 1998, *37*, 1329; (b) I. Hama, *Int. News Fats Oils Relat. Mater.*, 1997, *8*, 628; (c) R. Tsushima, *Int. News Fats Oils Relat. Mater.*, 1997,

8, 362; (d) J.H. Houston, *Int. News Fats Oils Relat. Mater.*, 1997, *8*, 928; (e) A.S. Shanley, *Chem. Eng.*, 2000, *107*(5), 55; (f) Anon., *Green Chem.*, 1999, *1*, G8.

79. (a) L. Zhang and H. Ding, *J. Appl. Polym. Sci.*, 1997, *64*, 1393; (b) P. Nayak, D.K. Mishra, D. Parida, K.C. Sahoo, M. Nanda, S. Lenka, and P.L. Nayak, *J. Appl. Polym. Sci.*, 1997, *63*, 671; (c) P. Gong and L. Zhang, *Ind. Eng. Chem. Res.*, 1998, *37*, 2681.
80. A.J. Domb and R. Nudelman, *J. Polym. Sci. A Polym. Chem.*, 1995, *33*, 717.
81. (a) D. Hairston, *Chem. Eng.*, 1994, *101*(8), 65; (b) A.L. Boehman, W.H. Swain, D.E. Weller, and J.M. Perez, Preprints A.C.S. *Div. Environ. Chem.*, 1998, *38*(2), 157; (c) P. Gwynne, A.C.S. Chemistry, Winter 1999, 15.
82. G. Parkinson, *Chem. Eng.*, 1996, *103*(11), 21.
83. (a) G. Ondrey, *Chem. Eng.*, 1998, *105*(13), 19; (b) S. Durham and M. Wood, *Agric. Res.*, 2002, *50*(4), 22.
84. U.S.EPA, *Presidential Green Chemistry Challenge Awards*, 2005 (to Archer Daniels Midland).
85. (a) I. Puskas, *Chemtech*, 1995, *25*(12), 43; (b) T. Hayakawa, H. Harihara, A.G. Anderson, A.P.E. York, K. Suzuki, H. Yasuda, and K. Takehira, *Angew. Chem. Int. Ed.*, 1996, *35*, 192; (c) R. Mildenlerg, M. Zander, and G. Collin, *Hydrocarbon Resins*, Wiley-VCH, Weinheim, 1997; (d) M.J. Zohuriaan and H.O. Midian, *J. Macromol. Sci.-Rev. Macromol. Chem.*, 2000, *C40*, 23.
86. E. Breitmaier, *Terpenes—Flavors, Fragrances, Pharmaca, Pheromones*, Wiley-VCH, Weinheim, 2006.
87. K. Satoh, H. Sugiyama, and M. Kamigaito, *Green Chem.*, 2006, *8*, 878.
88. (a) D. Wang, S. Czernik, D. Montane, M. Mann, and E. Chornet, *Ind. Eng. Chem. Res.*, 1997, *36*, 1507; (b) X. Xu, Y. Matsumura, J. Stenberg, and M.J. Antal, Jr., *Ind. Eng. Chem. Res.*, 1996, *39*, 893; (c) C. Okkerse and H. van Bekkum, *Green Chem.*, 1997, *1*, 107; (d) G. Parkinson, *Chem Eng.*, 2000, *107*(10), 23; (e) G. Ondrey, *Chem. Eng.*, 1999, *106*(6), 21; (f) M. Asadullah, S.-i. Ito, K. Kunimori, M. Yamada, and K. Tomishige, *Environ. Sci. Technol.*, 2002, *36*, 4476; (g) M. Asadullah, K. Fujimoto, and K. Tomishige, *Ind. Eng. Chem. Res.*, 2001, *40*, 8594; (h) F. Vogel, M.H. Waldner, A.A. Rouff, and S. Rabe, *Green Chem.*, 2007, *9*, 616.
89. (a) G. Parkinson, *Chem. Eng.*, 1998, *105*(10), 23; (b) G. Ondrey, *Chem. Eng.*, 2005, *112*(8), 17.
90. J.R. Salge, B.J. Dreyer, P.J. Dauenhauer, and L.D., Schmidt, *Science*, 2006, *314*, 801.
91. G. Braca, *Oxygenates by Homologation* or *Carbon Monoxide Hydrogenation with Metal Complexes*, Kluwer Academic, Dordrecht, 1994.
92. H. Kurakata, Y. Izumi, and K.-I. Aika, *Chem. Commun.*, 1996, 389.
93. M. Beller, B. Cornils, C.D. Frohning, and C.W. Kohlpaintner, *J. Mol. Catal.*, 1995, *104*, 17.
94. (a) A.A. Adesina, *Appl. Catal. A*, 1996, *138*, 345; (b) M. Freemantle, *Chem. Eng. News*, Aug. 12, 1996, 31; (c) D.B. Bukur and X. Lang, *Ind. Eng. Chem. Res.*, 1999, *38*, 3270; (d) B. Eisenberg, R.A. Fiato, T.G. Kaufmann, and R.F. Bauman, *Chem.Tech.*, 1999, *29*(10), 32.
95. (a) D.A. Simonetti, J. Rass-Hansen, E.L. Kunkes, R.R. Soares, and J.A. Dumesic, *Green Chem.*, 2007, *9*, 1073; (b) R.R. Soares, D.A. Simonetti, and J.A. Dumesic, *Angew. Chem. Int. Ed.*, 2006, *45*, 3982.
96. N.G. Gallegos, A. M. Alvarez, M.V. Cagnoli, J.F. Bengoa, S.G. Marchetti, R.C. Mercader, and A.A. Yeramian, *J. Catal.*, 1996, *161*, 132.
97. H.A. Wittcoff and B.G. Reuben, *Industrial Organic Chemicals*, Wiley, NY, 1996, 325.
98. S.P.R. Katikaneni, J.D. Adjaye, R.O. Idem, and N.N. Bakhshi, *Ind. Eng. Chem Res.*, 1996, *35*, 3332.
99. P.A. Barnett, R.H. Jones, J.M. Thomas, G. Sankar, I.J. Shannon, and C.R.A. Catlow, *Chem. Commun.*, 1996, 2001.
100. (a) Anon., *Chem. Eng. News*, Oct. 13, 2008, 21; (b) Anon., *Chem Eng. News*, Dec. 3, 2007, 37; (c) J.-F. Tremblay, *Chem. Eng. News*, Sept. 1, 2008, 13.
101. (a) L.H. Sperling and C.E. Carraher, *Encyclopedia of Polymer Science and Engineering*, 2nd ed., 1988, *12*, 658; (b) C.G. Gebelein and C.E. Carraher, *Industrial Biotechnological Polymers*, Technomic, Lancaster, PA, 1995; (c) A. Gandini, *Compr. Polym. Sci.*, 1st Suppl., Pergamon, Oxford, 1992, 528; (d) N. Mundigler, B. Herbinger, E. Berghofer, G. Schleining, and B. Grzeskowiak, *Carbohydr. Polym.*, 1995, *26*, 271; (e) D.L. Kaplan, *Biopolymers from Renewable Resources*, Springer, NY, 1998; (f) P.L. Nayak, *J. Macromol. Sci.-Rev. Macromol. Chem.*, 1999, *C39*, 481.
102. (a) E.S. Stevens, *Green Plastics: An Introduction to the New Science of Biodegrdable Plastics*, Princeton University Press, NJ, 2001; (b) A. benBrahim, *Chem. Br.*, 2002, *38*(4), 40; (c) S.H. Iman, R.V. Greene, and B.R. Zaidi, eds, *Biopolymers – Utilizing Nature's Advanced Materials*, ACS Symp. 723, Washington, DC, 1999; (d) A. Rouilly and L. Rigal, *J. Macromol. Sci., Polym. Rev.*, 2002, *C42*(4), 441; (e) Anon., *Biopolymers, 10*, Wiley-VCH, Weinheim.
103. S. Slomkowki, S. Blazwicz, S. Pielka, and A. Nadolny, eds, *Biomaterials in Regenerative Medicine*, Macromol. Symp., 2007, *253*, 1–190.
104. (a) A.K. Mohanty, M. Misra, and L.T. Drzal, *Natural Fibers, Biopolymers and Biocomposites*, CRC Press, Boca Raton, FL, 2005; (b) R.P. Wool and X.S. Sun, *Bio-Based Polymers and Composites*, Elsevier, Amsterdam, 2005; (c) N. Reddy and Y. Yang, *Trends Biotechnol.*, 2005, *23*(1), 22 (biofibers from agricultural byproducts).
105. (a) A.D. French, N.R. Bertoniere, O.A. Battista, J.A. Cuculo, and D.G. Gray, *Kirk–Othmer Encyclopedia of Chemical Technology*, 4th ed., Wiley, New York, 1993, *5*, 476; (b) R. Gilbert, *Cellulosic Polymers-Blends and Composites*, Hanser Gardner, Cincinnati, OH, 1994; (c) K.J. Saunders, *Organic Polymer Chemistry*, 2nd ed., Chapman & Hall, London, 1988; (d) T. Heinze and W.G. Glasser, eds, *Cellulose Derivatives: Modification, Characterization and Nanostructures*, ACS 688, Washington, DC, 1998; (e) D. Klemm, B. Philipp, and T. Heinze, *Comprehensive Cellulose Chemistry, vol. 1: FundamentalsandAnalyticalMethods: vol. 2: Functionalization of Cellulose*, Wiley-VCH, Weinheim, 1998; (f) O.L. Bobleter, *Prog. Polym. Sci.*, 1994, *19*, 797; (g) H.A. Krassig, *Cellulose: Structure, Accessibility and Reactivity*, Gordon & Breach, Philadelphia, PA, 1993; (h) D.N.-S. Hon and N. Shiraishi, eds, *Wood and Cellulose Chemistry*, 2nd ed., Dekker, NY, 2001; (i) T.K. Lindhorst, *Essentials of Carbohydrate Chemistry and Biochemistry*, Wiley-VCH, Weinheim, 2007.
106. (a) K. Tajima, M. Fujiwara, and M. Takai, *Macromol. Symp.*, 1995, *99*, 149; (b) L.W. Dalton, *Chem. Eng. News*, Apr. 26, 2004, 24532.

107. (a) C. Woodings, *Regenerated Cellulose Fibers*, CRC Press, 2001; (b) L. Zhang, D. Ruan, and J. Zhou, *Ind. Eng. Chem. Res.*, 2001, *40*, 5923.
108. (a) S.A. Mortimer and A.A. Peguy, *J. Appl. Poym. Sci.*, 1996, *60*, 305, 1747; (b) C. O'Driscoll, *Chem. Br.*, 1996, *32*(12), 27; (c) S. Dobson, *Chem. Ind. (Lond.)*, 1995, 870; (d) G. Parkinson, *Chem. Eng.*, 1995, *103*(10); (e) M. Hirami, *J. Macromol. Sci. Pure Appl. Chem.*, 1996, *A33*, 1825; (e) F. Wendler, A. Konkia, and T. Heinze, *Macromol. Symp.*, 2008, *262*, 72.
109. R.A. Sedjo and D. Botkin, *Environment*, 1997, *39*(10), 14.
110. B. Focher, A. Marzetti, E. Marsano, G. Conio, A. Tealdi, A. Cosani, and M. Terbojevich, *J. Appl. Polym. Sci.*, 1998, *67*, 961.
111. P.L. Short, *Chem. Eng. News*, Dec. 11, 2006, 22.
112. L. McGraw, *Agric. Res.*, 2000, *48*(8), 14.
113. G.N. Ramaswamy and E.P. Easter, *Text. Res. J.*, 1997, *67*, 803.
114. G. Henriksson, K.-E.L. Eriksson, L. Kimmel, and D.E. Akin, *Text. Res. J.*, 1998, *68*, 942.
115. R.W. Kessler and R. Kohler, *Chemtech*, 1996, *26*(12), 34.
116. R.M. Bliss, *Agric. Res.*, 2005, *53*(11), 12.
117. (a) M.A. Ali, M.A. Khan, K.M.I. Ali, and G. Hinrichsen, *J. Appl. Polym. Sci.*, 1998, *70*, 843; (b) K.M.I. Ali, M.A. Khan, M.A. Ali, and K.S. Akhunzada, *J. Appl. Polym. Sci.*, 1999, *71*, 841.
118. R. Melamede, *Chem. Ind. (Lond.)*, 2001, 724.
119. L. Zhang and Q. Zhou, *Ind. Eng. Chem. Res.*, 1997, *36*, 2651.
120. Y. Mi, X. Chen, and Q. Guo, *J. Appl. Polym Sci.*, 1997, *64*, 1267.
121. (a) K.J. Edgar, C.M. Buchanan, J.S. Debenham, P.A. Rundquist, B.D. Seiler, M.C. Shelton, and D. Tindall, *Prog. Polym. Sci.*, 2001, *26*, 1605; (b) T. Heinze and T. Liebert, *Prog. Polym. Sci.*, 2001, *26*, 1689–1762; (c) T. Heinze and K. Fischer, eds, *Macromol. Symp.*, 2007, *244*, 1–203.
122. F. Debeaufort and A. Voiley, *J. Agric. Food Chem.*, 1997, *45*, 685.
123. R.D. Hagenmaier and R.A. Baker, *J. Agric. Food Chem.*, 1997, *45*, 349.
124. H. Warth, R. Mulhaupt, and J. Schatzle, *J. Appl. Polym. Sci.*, 1997, *64*, 231.
125. (a) A.K. Mohanty, L.T. Drzal, A. Wibowo, and M. Misra, *Polym. Preprints*, 2002, *43*(1), 336; (b) A.C. Wibowo, A.K. Mohanty, M. Misra, and L.T. Drzal, *Ind. Eng. Chem. Res.*, 2004, *43*, 4883.
126. (a) B. Morgenstern and H.W. Kammer, *Trends Polym. Sci.*, 1996, *4*(3), 87; (b) C. Vaca-Garcia, S. Thibaud, M.E. Borredon, and G. Gozzelion, *J. Am. Oil Chem. Soc.*, 1998, *75*, 315; (d) J.N. Nayak, Y. Chen, and J. Kim, *Ind. Eng. Chem. Res.*, 2008, *47*, 1702.
127. D. Klemm, T. Heinze, A Stein, and T. Liebert, *Macromol. Symp.*, 1995, *99*, 129; (b) H. Henze-Wethkamp, P. Zugenmaier, A. Stein, and D. Klemm, *Macromol. Symp.*, 1995, *99*, 245.
128. (a) J. Cai, S. Kimura, M. Wada, S. Kuga, and L. Zhang, *Chem Sus Chem*, 2008, *1*, 149; (b) M. Pinnow, H.-P. Fink, C. Fanter, and J. Kunze, *Macromol. Symp.*, 208, *262*, 129.
129. W. Mormann and T. Wagner, *Macromol. Rapid Commun.*, 1997, *18*, 515.
130. P.J. Frazier, A. Donald, and P. Richmond, *Starch: Structure and Functionality*, Royal Society of Chemistry Special Publication 205, Cambridge, 1997.
131. D. Eagles, D. Lesnoy, and S. Barlow, *Text. Res. J.*, 1996, *66*, 277.
132. (a) J.J.G. van Soest and D.B. Borger, *J. Appl. Polym. Sci.*, 1997, *64*, 631; (b) J.J.G. van Soest and F.G. Vliegenthart, *Trends Biotechnol.*, 1997, *15*, 208; (c) A. Rindlav-Westling, M. Standing, and P. Gatenholm, *Carbohydr. Polym.*, 1998, *36*(2–3), 217; (d) F. Ania, M. Dunkel, R.K. Bayer, and F.J. Balta-Calleja, *J. Appl. Polym. Sci.*, 2002, *85*, 1246; (e) R.F.T. Stepto, *Macromol. Symp.* 2009, *279*, 163; (f) L. Janssen and L. Moscicki, *Thermoplastic Starch—A Green Material for Various Industries*, Wiley-VCH, Weinheim, 2009.
133. (a) A. Dufresne and M.R. Vignon, *Macromolecules*, 1998, *31*, 2693; (b) D.-H. Kim, S.-K. Na, and J.-S. Park, *J. Appl. Polym. Sci.*, 2003, *88*, 2100.
134. (a) P. Forssell, J. Mikkila, T. Suortti, J. Seppalla, and K. Poutanen, *J. Macromol. Sci. Pure Appl. Chem.*, 1996, *A33*, 703; (b) J.J.G. van Soest and N. Knooren, *J. Appl. Polym. Sci.*, 1997, *64*, 1411; (c) K. Poutanen and P. Forssell, *Trends Polym. Sci.*, 1996, *4*, 128; (d) F. Meuser, D.J. Manners, and W. Seibel, eds, *Plant Polymeric Carbohydrates*, Royal Society of Chemistry Special Publication 134. Cambridge, 1993; (e) G. Fuller, T.A. McKeon, and D.D. Bills, eds, *Agricultural Materials as Renewable Resources: Nonfood and Industrial Applications*, ACS Symp. 647, Washington, DC, 1996; (f) M.L. Fishman, R.B. Friedman, and S.J. Huang, eds, *Polymers from Agricultural Coproducts*, ACS Symp. 575, Washington, DC, 1994, 50; (g) B.J. Domingo and S.A. Morris, *J. Appl. Polym. Sci.*, 1999, *71*, 2147; (h) R.F.T. Stepto, *Macromol. Symp.*, 200, *152*, 74.
135. (a) Anon., *Chem. Eng. News*, Oct. 28, 1996, 29; (b) R. Dagani, *Chem. Eng. News*, July 27, 1998, 14; (c) P. Millner, *Agric. Res.*, 2001, *49*(1), 23.
136. D. Dallabrida, *Wilmington Delaware News J.*, June 3, 1997, B5.
137. T. Colonnese, B. Bodio, and J. Hardy, *Green Chem.*, 2002, *4*, G50.
138. G.F. Fanta and R.L. Shogren, *J. Appl. Polym. Sci.*, 1997, *65*, 1021.
139. J.L. Willett, ACS Meeting, New Orleans, IEC 204, Mar. 2003.
140. S.P. McCarthy and B. Koroskenyi, *Polym. Mat. Sci. Eng.*, 2002, *86*, 350.
141. P.J. Stenhouse, J.A. Ratto, and N.S. Schneider, *J. Appl. Polym. Sci.*, 1997, *64*, 2613.
142. (a) R.L. Shogren, J.W. Lawton, K.F. Tiefenbacher, and L. Chen, *J. Appl. Polym. Sci.*, 1998, *68*, 2129; (b) G.A.R. Nobes, W.J. Orts, G.M. Glenn, and M.V. Harper, Chicago ACS Meeting IEC98, Aug. 2001.
143. (a) V. Voller, K. Stetson, and M. Bhattacharya, *University of Minnesota Institute of Technology*, Inventing Tomorrow, Minneapolis, Minnesota, 1998, *23*(2), 30; (b) L. Averous, N. Fauconnier, L. Moro, and C.F. Fringant, *J. Appl. Polym. Sci.*, 2000, *76*, 1117.
144. L. Averous, *J. Macromol. Sci., Polym. Rev.*, 2004, *C44*, 231–274.
145. R. Alfani, S. Iannace, and L. Nicholas, *J. Appl. Polym. Sci.*, 1998, *68*, 739.
146. M.L. Fishman, D.R. Coffin, J.J. Unruh, and T. Ly, *J. Macromol. Sci. Pure Appl. Chem.*, 1996, *A33*, 639.
147. (a) N.S. Thompson, *Kirk–Othmer Encyclopedia of Chemical Technology*, 4th ed., Wiley, New York, 1995, *13*, 13; (b) A.

Ebringerova and T. Heinze, *Macromol. Rapid Commun.*, 2000, *21*, 542.
148. Anon., *Chem. Eng. News*, Nov. 13, 1995, 14.
149. P. Zhang and R.L. Whistler, *J. Appl. Polym. Sci.*, 2004, *93*, 2896.
150. N. Mundigler, B. Herbinger, E. Berghofer, G. Schleining, and B. Grzeskowiak, *Carbohydr. Polym.*, 1995, *26*, 271.
151. (a) S. Tokura, S.-I. Nishimura, N. Sakairi, and N. Nishi, *Macromol., Symp.*, 1996, *101*, 389; (b) K. Kurita, S. Ishii, K. Tomita, S.-I. Nishimura, and K. Shimoda, *J. Polym. Sci. A Poym. Chem.*, 1994, *32*, 1027; (c) G. Skjak-Braek, T. Anthonsen, and P.A. Sandford, eds, *Chitin and Chitosan; Sources, Chemistry, Biochemistry, Physical Properties and Applications*, Elsevier Applied Science, London, 1989; (d) S. Salmon and S.M. Hudson, *J. Macromol. Sci. Pure Appl. Sci.*, 1997, *C37*, 199; (e) D.K. Singh and A.R. Ray, *J. Macromol. Sci.-Rev. Macromol. Chem.*, 2000, *C40*, 69; (f) K. Kurita, *Prog. Polym. Sci.*, 2001, *26*, 1921; (g) P.K. Dutta, M.N.V. Ravikumar, and J. Dutta, *J. Macromol. Sci. – Polym. Rev.*, 2002, *C42*, 307.
152. (a) K.C. Gupta and M.N.V. Kumar, *J. Macromol. Sci. – Polym. Rev.*, 2000, *C40*, 273; (b) M. Rinaudo, *Prog. Polym. Sci.*, 2006, *31*(7), 603; (c) H. Sashiwa and S. Aiba, *Prog. Polym. Sci.*, 2004, *29*(9), 887; (d) E. Guibal, *Prog. Polym. Sci.*, 2005, *30*, 71; (e) D.J. Macquarrie and J.J. Hardy, *Ind. Eng. Chem. Res.*, 2005, *44*, 8499–8520.
153. C. Urbanczyk, B. Lipp-Symonowicz, I. Szosland, A. Jeziorny, W. Urbaniak-Domagala, K. Dorau, H. Wrzosek, S. Stajnowski, S. Kowalska, and E. Sztajnert, *J. Appl. Polym Sci.*, 1997, *65*, 807.
154. J.Z. Knaul and K.A.M. Creber, *J. Appl. Polym. Sci.*, 1997, *66*, 117.
155. C.L. Yue, R. Kumar, and S.P. McCarthy, Boston A.C.S. Meeting Abstracts, Aug. 1998, POLY 251; *Polym. Prepr. (Am. Chem. Soc. Div. Polym. Chem.)*, 1998, *39*(2), 132.
156. S. Hirano, *Macromol. Symp.*, 2001, *168*, 21.
157. H.K. Jun, J.S. Kim, H.K. No, and S.P. Myers, *J. Agric. Food Chem.*, 1994, *42*, 1834.
158. K. Yamada, T. Chen, G. Kumar, O. Vesnovsky, L.D.T. Topoleski, and G.F. Payne, *Biomacromolecules*, 2000, *1*, 252.
159. C. Gu, B. Sun, W. Wu, F. Wang, and M. Zhu, *Macromol. Symp.*, 2007, *254*, 160.
160. T.-H. Kim, H.-L. Jiang, D. Jere, I.-K. Park, M.-H. Cho, J.-W. Nah, Y.-J. Choi, T. Akaike, and C.-S. Cho, *Prog. Polym. Sci.*, 2007, *32*, 726.
161. (a) S.Y. Lin and S.E. Lebo, Jr., *Kirk–Othmer Encyclopedia of Chemical Technology*, 4th ed., Wiley, New York, 1995, *15*, 268; (b) S.Y. Lin and I.S. Lin, *Ullmann's Encyclopedia of Industrial Chemistry*, 5th ed., VCH, Weinheim, 1990, *A15*, 305; (c) N.G. Lewis and S. Sarkanen, eds, *Lignin and Lignan Biosynthesis*, ACS Symp. 697, Washington, DC, 1998; (d) W.G. Glasser, R. Northey, and T.P. Schultz, eds, *Lignin Properties and Materials*, ACS Symp. 742, Washington, DC, 1999; (e) A.M. Rouhi, *Chem. Eng. News*, Nov. 13, 2000, 29; Apr. 2, 2001, 2001, 52; (f) A. Steinbuchel and M. Hofrichter, eds, *Biopolymers – Biology, Chemistry, Biotechnology, Applications. Vol. 1; Lignin, Humic Acid, Coal*, Wiley-VCH, Weinheim, 2001.
162. S.Y. Lin and S.E. Lebo, Jr., *Kirk–Othmer Encyclopedia of Chemical Technology*, 4th ed., Wiley, New York, 1995, *15*, 270.
163. (a) J.J. Meister, *J. Macromol. Sci., Polym. Rev.*, 2002, *C42*, 235; (b) S. Caparros, J. Ariza, G. Garrotte, F. Lopez, and M.J. Diaz, *Ind. Eng. Chem. Res.*, 2007, *46*, 623.
164. L. Jimenez, J.C. Garcia, I. Perez, J. Ariza, and F. Lopez, *Ind. Eng. Chem. Res.*, 2001, *40*, 6201.
165. J. Rosch and R. Mulhaupt, *Polym. Bull.*, 1994, *32*, 361.
166. G. Parkinson, *Chem. Eng.*, 1997, *104*(8), 21; 1999, *106*(3), 27.
167. D. Feldman, D. Banu, M. Lacasse, J. Wang, and C. Luchian, *J. Macromol. Sci. Pure Appl. Sci.*, 1995, *A32*, 1643.
168. M.-J. Chen, D.W. Gunnells, D.J. Gardner, O. Milstein, R. Gersonde, H.J. Feine, A. Huttermann, R. Frund, H.D. Ludemann, and J.J. Meister, *Macromolecules*, 1996, *29*, 1389.
169. J.D. Gelorme and L. Kosbar, *Green Chemistry and Engineering Conference*, Washington, DC, June 23–25, 1997.
170. Y. Li, J. Mlynar, and S. Sarkanen, *J. Polym. Sci. B Polym. Phys.*, 1997, *35*, 1899.
171. S. Baumberger, C. Lapierre, and B. Monties, *J. Agric. Food Chem.*, 1998, *46*, 2234.
172. Anon., *R&D (Cahners)*, 1999, *41*(13), 11.
173. M.N. Angles, J. Reguant, D. Montane, F. Ferrando, X. Farriol, and J. Salvado, *J. Appl. Polym. Sci.*, 1999, *73*, 2485.
174. C.A.S. Hill, *Wood Modification – Chemical, Thermal and Other Processes*, Wiley, NY, 2006.
175. D. Solpan and O. Guven, *J. Appl. Polym. Sci.*, 1999, *71*, 1515.
176. M.W. Toffel and A. Horvath, *Environ. Sci. Technol.*, 2004, *38*, 2961.
177. B.Y. Tao, *Chem. Ind. (Lond.)*, 1994, 906.
178. (a) F. Callegarin, J.-A.Q. Gallo, F. Debeaufort, and A. Voilley, *J. Am. Oil Chem. Soc.*, 1997, *74*, 1183; (b) A. Gennadios, C.L. Weller, and M.A. Hanna, *Int. News Fats Oils Relat. Mater.*, 1997, *8*, 622; (c) R.D. Hagemaier and R.A. Baker, *J. Agric. Food Chem.*, 1997, *45*, 349; (d) F. Debeaufort and A. Voilley, *J. Agric. Food Chem.*, 1997, *45*, 685; (e) S.L. Wilkinson, *Chem. Eng. News*, June 15, 1998, 26; (f) G. Parkinson, *Chem. Eng.*, 2000, *107*(8), 17; (g) J. Irissin-Mangata, B. Boutevin, and G. Bauduin, *Polym. Bull.*, 1999, *43*, 441; (h) J. Irissin-Mangata, G. Bauduon, and B. Boutevin, *Polym. Bull.*, 2000, *44*, 409.
179. (a) K. Smith, *Int. News Fats Oils Relat. Mater.*, 1996, *7*, 1212; 1998, *9*, 844; (b) N. Rangavajhyala, J. Ghorpade, and M. Hanna, *J. Agric. Food Chem.*, 1997, *45*, 4204.
180. (a) S. Wang, H.-J. Sue, and J. Jane, *J. Macromol. Sci. Pure App. Sci.*, 1996, *A33*, 557; (b) X. Mo, X.S. Sun, and Y. Wang, *J. Appl. Polym. Sci.*, 1999, *73*, 2595.
181. Y. Zhang, S. Ghasemzadeh, A.M. Kotliar, S. Kumar, and S. Presnell, *J. Appl. Polym. Sci.*, 1999, *71*, 11.
182. (a) R. Narayan, *United Soybean Board News Release*, St. Louis, Missouri, May 11, 1998; (b) K. Smith, *Int. News Fats Oils Relat. Mater.*, 1998, *9*, 844; (c) Anon., *Chem. Eng. News*, Nov. 9, 1998, 89.
183. M.L. Fishman, R.B. Friedman, and S.J. Huang, eds, *Polymers from Agricultural Corproducts*, ACS Symp. 575, Washington, DC, 1994, 92.
184. (a) Anon., *Chem. Ind. (Lond.)*, 1997, 294; (b) G.W. Padua, *Cereal Chem.*, 1997, *83*, 74.
185. H.-M. Lai, P.H. Geil, and G. W. Padua, *J. Appl. Polym. Sci.*, 1999, *71*, 1267.

186. (a) Y.V. Wu, R.R. Rosati, D.J. Sessa, and P.B. Brown, *J. Agric. Food Chem.*, 1995, *43*, 1585; (b) Q. Wu, H. Sakabe, and S. Isobe, *Ind. Eng. Chem. Res.*, 2003, *42*, 6765.
187 D. Aithani and A.K. Mohanty, *Ind. Eng. Chem. Res.*, 2006, *45*, 6147.
188. N. Parris and D.R. Coffin, *J. Agric. Food Chem.*, 1997, *45*, 1596.
189. Y. Yang, L. Wang, and S. Li, *J. Appl. Polym Sci.*, 1996, *59*, 433.
190. H.C. Huang, E.G. Hammond, C.A. Reitmeier, and D.J. Myers, *J. Am. Oil Chem. Soc.*, 1995, *72*, 1453.
191. W. Zhao, A. Kloczkowski, J.E. Mark, B. Erman, and I. Bahar, *Chemtech*, 1996, *26*(3), 32; *J. Macromol. Sci. Pure Appl. Sci.*, 1996, *A33*, 525.
192. (a) R.D. Deanin, I. Skeist, and P.G. Hereld, eds, *Renewable Resources for Plastics Growth and Change in adhesives.* A.C.S. *Div. Chemical Marketing and Economics*, Washington, DC, 1975, 157; (b) L. McGinnis, *Agric. Res.*, Burlington, NJ, 2007, *55*(5), 16; (c) Literature, American Casein Company, 2000.
193. (a) D. Brault, G. D'Aprano, and M. Lacroix, *J. Agric. Food Chem.*, 1997, *45*, 2964; (b) E. Mezgheni, G. D'Aprano, and M. Lacroix, *J. Agric. Food Chem.*, 1998, *46*, 318.
194. J.T.P. Derksen, F.P. Cuperus, and P. Kolster, *Prog. Org. Coat.*, 1996, *27*, 45.
195. H. Mujica-Paz and N. Gontard, *J. Agric. Food Chem.*, 1997, *45*, 4101.
196. C. Marquie, A.-M. Tessier, C. Aymard, and S. Guilbert, *J. Agric. Food Chem.*, 1997, *45*, 922.
197. S. Hartman, U.S. patent 5,580,590 (1996).
198. (a) U. Kalapathy, N.S. Hettiarachchy, D.J. Myers, and K.C. Rhee, *J. Am. Oil Chem. Soc.*, 1995, *72*, 1461; 1996, *73*, 1063; (b) U. Kalapathy, N.S. Hettiarachchy, D. Myers, and M.A. Hanna, *J. Am. Oil Chem. Soc.*, 1995, *72*, 507.
199. (a) Anon., *Chem. Br.*, 1998, *34*(10), 18; (b) *Technol. Rev.*, 1998, *101*(6), 21; (c) U.S.EPA, *Presidential Green Chemistry Challenge Award*, 2007 (to K. Li, Hercules and Columbia Forest Products).
200. (a) M. Rouhi, *Chem. Eng. News*, Feb. 23, 1998, 66; (b) G. Parkinson, *Chem. Eng.*, 1998, *105*(3), 21.
201. (a) W.C. Warner, *Rubber Chem. Technol.*, 1994, *67*, 559; (b) A.I. Isayev, J. Chen, and A. Tukachinsky, *Rubber Chem. Technol.*, 1995, *68*, 267; (c) A.I. Isayev, S.P. Yushanov, and J. Chen, *J. Appl. Polym. Sci.*, 1996, *59*, 803, 815; (d) G.S. Samdani, *Chem. Eng.*, 1995, *102*(1), 17.
202. W.F. Schmidt, *Agric. Res.*, 2005, *53*(1), 23.
203. Anon., *Agric. Res.*, 2006, *54*(8), 16.
204. B. Halford, *Chem. Eng. News*, Sept. 6, 2004, 36.
205. (a) H.P. German, *Science and Technology for Leather into the Next Millennium*, McGraw-Hill, New Delhi, 1999; (b) J.-F. Tremblay, *Chem. Eng. News*, Feb. 2. 2009, 18.
206. B. de Castro, M.A. Ferreira, R.T. Markus, and A. Wyler, *J. Macromol. Sci. Pure Appl. Chem.*, 1997, *A34*, 109.
207. (a) C. Holden, *Science*, 1995, *270*, 739; (b) M.A. Colgin and R.V. Lewis, *Chem. Ind. (Lond.)*, 1995, 1009; (c) A.H. Simmons, C.A. Michal, and L.W. Jelinski, *Science*, 1995, *271*, 84; (d) C. Jackson and J.P. O'Brien, *Macromolecules*, 1995, *28*, 5975; (e) S.B. Warner, M. Polk, and K. Jacob, *J. Macromol. Sci.-Rev. Macromol. Chem.*, 1999, *C39*. 643; (f) V. Gilman, *Chem. Eng. News*, June 16, 2003, 27; (g) Z. Shao, F. Vollrath, Y. Yang, and H.C. Thogersen, *Macromolecules*, 2003, *36*, 1157; (i) J.H. Exler, D. Hummerich, and T. Scheibel, *Angew. Chem. Int. Ed.*, 2007, *46*, 3559; (j) W. Kim and V.P. Conticello, *J. Macromol. Sci., Polym. Rev.*, 2007, *C47*, 119; (k) M. Heim, D. Keerl, and T. Scheibel, *Angew. Chem. Int. Ed.*, 2009, *48*, 3584.
208. C. Potera, *Science*, 1997, *276*, 1499.
209. J.B. vanBeilen and Y. Poirier, *Trends Biotechnol.*, 2007, *25*, 522.
210. (a) E.K. Abraham and P. Ramesh, *J. Macromol. Sci., Polym. Rev.*, 2002, *C42*, 185; (b) Anon., *Chem. Eng. News*, May 5, 2008, 23.
211. V.B. Nguyen and C.H. Chu, *J. Macromol. Sci. Pure Appl. Sci.*, 1996, *A33*, 1949.
212. (a) R. Gross, C. Scholza, and G. Leatham, eds, *Polymers from Renewable Resources*, A.C.S. Symp. 764, Washington, DC, 2000; (b) S.H. Imam, R.V. Greene, and B.R. Zaida, eds, *Biopolymers Utilizing Nature's Advanced Materials*, ACS Symp 723, Washington, DC, 1999; (c) J. Evans, *Chem. Br.*, 2000, *36*(2), 41.
213. (a) L.A. Kleintjens, ed., *Macromol. Symp.*, 1998, *127*, 1–273; (b) O. Akaranta, *Surf. Coat. Int.*, 1996, *79*, 152; (c) D.L. Kaplan, *Biopolymers from Renewable Resources*, Springer, NY, 1998.
214. H. Kimura, *J. Polym. Sci. A Polym. Chem.*, 1996, *34*, 3595, 3607, 3615; 1998, *36*, 189, 195.
215. (a) J.V. Crivello, R. Narayan, S.A. Bratslavsky, and B. Yang, *Macromol. Symp.*, 1996, *107*, 75; (b) J.V. Crivello and S.A. Bratslavsky, *J. Macromol. Sci. Pure Appl. Chem.*, 1994, *A31*, 1927; (c) J.V. Crivello and S.K. Rajaraman, *J. Polym. Sci. A Polym, Chem.*, 1997, *35*, 1579, 1593.
216. K.K. Bhatia, U.S. patent 6,864,378 (2005) (to DuPont).
217. (a) M. Okada, Y. Okada, A. Tao, and K. Aoki, *J. Appl. Polym Sci.*, 1996, *62*, 2257; (b) H.R. Kricheldrof and N. Probst, *Macromol. Rapid Commun.*, 1995, *16*, 231; (c) M. Majdoub, A. Loupy, and G. Fleche, *Eur. Polym. J.*, 1994, *30*, 1431; (d) R. Storbeck and M. Ballauff, *J. Appl. Polym. Sci.*, 1996, *59*, 1199.
218. Y.Y. Linko and J. Seppala, *Chemtech*, 1996, *26*(8), 25.
219. M. Okada, K. Tachikawa, and K. Aoi, *J. Polym. Sci. A Polym. Chem.*, 1997, *35*, 2729; *J. Appl. Polym. Sci.*, 1999, *74*, 3342.
220. A. Khrouf, S. Boufi, R. El Gharbi, N.M. Belgacem, and A. Gandini, *Polym. Bull.*, 1996, *37*, 589.
221. (a) H.R. Kricheldrof, *J. Macromol. Sci. Rev. Macromol. Chem.*, 1997, *C37*, 599; (b) H.R. Kricheldrof, S.-J. Sun, C.-P. Chen, and T.-C. Chang, *J. Polym. Sci. A Polym. Chem.*, 1997, *35*, 1611.
222. J. vanHaveren, E.A. Oostveen, and D.S. vanEs, Anaheim ACS Meeting, Mar. 2004, CELL 17.
223. (a) X. Chen, J.S. Dordick, and D.G. Rethwisch, *Macromolecules*, 1995, *28*, 6014; *Carbohydr. Polym.*, 1995, *28*, 15; (b) E.-J. Yaacoub, B. Skeries, and K. Buchholz, *Macromol. Chem. Phys.*, 1997, *198*, 899; (c) G. Wulff, J. Schmid, and T. Venhoff, *Macromol. Chem. Phys.*, 1996, *197*, 159.
224. N.S. Patil, Y. Li, D.G. Rethwisch, and J.S. Dordick, *J. Polym. Sci. A Polym. Chem.*, 1997, *35*, 2221.
225. B. Rerna and S. Ramakrishnan, *Polym. Bull.*, 1997, *38*, 537.
226. N.D. Sachinvala, D.L. Winsor, R.K. Menescal, I. Ganjian, W.P. Niemczura, and M.H. Litt, *J. Polym Sci. A Polym. Chem.*, 1998, *36*, 2397.

227. (a) R.P. Wool, *Chemtech*, 1999, *29*(6), 44; (b) Anon., *Green Chem.*, 1999, *1*, G3.
228. R. Schoevaart and T. Kieboom, *Carbohydrate Res.*, 2001, *334*, 1.
229. (a) A. Gandini, *Encyclopedia of Polymer Science Engineering*, 2nd ed., Wiley, New York, 1987, *7*, 454; (b) A. Gandini and M.N. Belgacem, *Prog. Polym. Sci.*, 1997, *22*, 1203–1379; (c) C. Gousse and A. Gandini, *Polym. Bull.*, 1998, *40*, 389; (d) K.J. Zeitsch, *The Chemistry & Technology of Furfural and Its Many Byproducts*, Elsevier, Amsterdam, 2000; (e) M. Jacoby, *Chem. Eng. News*, July 6, 2009, 26.
230. M. Choura, N.M. Belgacem, and A. Gandini, *Macromolecules*, 1996, *29*, 3839.
231. S.K. Swain, S. Sahoo, D.K. Mohapatra, B.K. Mishra, S. Lenka, and P.L. Nayak, *J. Appl. Polym. Sci.*, 1994, *54*, 1413.
232. K.S. Alva, P.L. Nayak, J. Kumar, and S.K. Tripathy, *J. Macromol. Sci. Pure Appl. Sci.*, 1997, *A34*, 665.
233. R.D. Sanderson, D.F. Schneider, and I. Schreuder, *J. Appl. Polym. Sci.*, 1995, *55*, 1837.
234. (a) Y. Roman-Leshkov, J.N. Chheda, and J.A. Dumesic, *Science*, 2006, *312*, 1933; (b) J.N. Chheda, G.W. Huber, and J.A. Dumesic, *Angew. Chem. Int. Ed.*, 2007, *46*, 7164–7183; (c) J.N. Chheda, Y. Roman-Leshkov, and J.A. Dumesic, *Green Chem.*, 2007, *9*, 342; (d) H. Zhao, J.E. Holladay, H. Brown, and Z.C. Zhang, *Science*, 2007, *316*, 1597; (e) G. Yong, Y. Zhang, and J.Y. Ying, *Angew. Chem. Int. Ed.*, 2008, *47*, 9345; (f) G. Ondrey, *Chem. Eng.*, 2009, *116*(6), 14.
235. K.J. Zeitsch, *Chem. Innov.*, 2001, *31*(1), 41.
236. M. Mascal and E.B. Nikitin, *Angew. Chem. Int. Ed.*, 2008, *47*, 7924.
237. F.W. Lichtenthaler, *Acc. Chem. Res.*, 2002, *35*, 728.
238. E. Taening, I.S. Nielsen, K. Egeblad, R. Madsen, and C.H. Christensen, *ChemSusChem*, 2008, *1*, 75.
239. M. Okada, K. Tachikawa, and K. Aoi, *J. Appl. Polym. Sci.*, 1999, *74*, 3342.
240. M.J. Donnelly, *Polym. Int.*, 1995, *37*, 1, 297.
241. (a) D.K. Mishra, D. Parida, S.S. Nayak, S. Lenka, and P.L. Nayak, *Macromol. Rep.*, 1995, *A32*, 499; (b) S.S. Nayak, D.K. Mishra, P.L. Nayak, and S. Lenka, *Macromol. Rep.*, 1995, *A32*, 511.
242. United Soybean Board, *Feedstocks*, St. Lowis, Missori, 1999, *4*(1), 3.
243. (a) U.S.EPA, *Presidential Green Chemistry Challenge Award*, 2007 (to Cargill); (b) L. Crandall, *Inform*, 2002, *13*, 626.
244. P. Kandanarachichi, A. Guo, D. Demydov, and Z. Petrovic, *J. Am. Oil Chem. Soc.*, 2003, *79*, 1221.
245. (a) K. Hiltunen, J.V. Seppala, and M. Harkunen, *J. Appl. Polym. Sci.*, 1997, *64*, 865; (b) J. Seppala. J.F. Selin, and T. Su, *Chem. Abstr.*, 1994, *121*, 180504; (c) K. Hiltunen and J.V. Seppala, *J. Appl. Polym. Sci.*, 1998, *67*, 1011, 1017.
246. C. Aregood, *Wilmington Delaware News J.*, Sept. 18, 1996, B5.
247. Y. Ikada and H. Tsuji, *Macromol. Rapid Commun.*, 2000, *21*, 117.
248. (a) S.Y. Lee, *Trends Biotechnol.*, 1996, *14*, 431; (b) D.P. Mobley, *Plastics from Microbes–Microbial Synthesis of Polymers and Polymer Precursors*, Hanser Gardner, Cincinnati, OH, 1994; (c) Y. Poirier, C. Nawrath, and C. Somerville, *Biotechnology*, 1995, *13*, 142; (d) R. Sharma and A.R. Ray, *J. Macromol. Sci. Rev. Macromol. Chem.*, 1995, *C35*, 327; (e) H.-M. Muller and D. Seebach, *Angew. Chem. Int. Ed.*, 1993, *32*, 477; (f) R.H. Marchessault, *Trends Polym. Sci.*, 1996, *4*, 163; (g) S. Williams and O. Peoples, *Chem. Br.*, 1997, *33*(12), 29; (h) R.M. Weiner, *Trends Biotechnol.* 1997, *15*, 390; (i) H. Daniell and C. Guda, *Chem. Ind. (Lond.)*, 1997, 555; (j) I. Kim, G. Ondrey, and T. Kamiya, *Chem. Eng.*, 1998, *105*(7), 43; (k) K. Sudesh, H. Abe, and Y. Doi, *Prog. Polym. Sci.*, 2000, *25*, 1503; (k) S.S. Im, Y.H. Kim, J.S. Yoon, and I.-J. Chin, eds, *Macromol. Symp.*, 2005, *224*, 1–70.
249. Anon., *Chem. Ind. (Lond.)*, 1997, 374.
250. Anon., *Eur. Chem. News*, 1998, *69*(1839), 63.
251. G. Ondrey, *Chem. Eng.*, 2006, *113*(3), 18.
252. (a) A.H. Tullo, *Chem. Eng. News*, Sept. 29, 2008, 21; (b) Anon., *Chem. Eng. News*, Aug. 18, 2008, 13.
253. V.E. Reinsch and S.S. Kelley, *J. Appl. Polym. Sci.*, 1997, *64*, 1785.
254. (a) O. Miguel, M.J. Fernandez-Berridi, and J.J. Iruin, *J. Appl. Polym Sci.*, 1997, *64*, 1849; (b) O. Miguel, T.A. Barbari, and J.J. Irwin, *J. Appl. Polym. Sci.*, 1999, *71*, 2391.
255. G.J.L. Griffin, *Chemistry and Technology of Biodegradable Polymers*. Chapman & Hall, London, 1994; (b) M.J.R. Blackman, *Environmentally Degradable of Polymers*, Patent Office, London, 1993.
256. (a) M. Vert, G. Schwarch, and J. Coudane, *J. Macromol. Sci. Pure Appl. Chem.*, 1995, *A32*, 787; (b) S. Owen, M. Masaoka, R. Kawamura, and N. Sakota, *J. Macromol. Sci. Pure Appl. Chem.*, 1995, *A32*, 843; (c) M. Spinu, C. Jackson, M.Y. Keating, and K.H. Gardner, *J. Macromol. Sci. Pure Appl. Chem.*, 1996, *A33*, 1497; (d) A. Thayer, *Chem. Eng. News*, Dec. 8, 1997, 14; (e) J.-C. Bogaert and P. Coszach, *Macromol. Symp.*, 2000, *153*, 287; (f) S. Stinson, *Chem. Eng. News*, Feb. 28, 2000, 14; (g) Anon., *Green Chem.*, 2000, *2*, G37; (h) G. Parkinson, *Chem. Eng.*, 2000, *107*(2), 19; (i) A. Sodergard and M. Stoll, *Prog. Polym. Sci.*, 2002, *27*, 1123; (j) R. Mehta, V. Kumar, H. Bhunia, and S.N. Upadhyay, *J. Macromol. Sci., Polym. Rev.*, 2005, *C45*, 325; (k) B. Gupta, N. Revagade, and J. Hilborn, *Prog. Polym. Sci.*, 2007, *32*, 455; (l) C.K. Williams, *Chem. Soc. Rev.*, 2007, *36*, 1573; (m) S.S. Im, Y.H. Kim, J.S. Yoon, and I.-J. Chin, eds, *Macromol. Symp.*, 2005, *224*, 71–180.
257. J.D. Bronzino, ed., *The Bioengineering Handbook*, CRC Press, Boca Raton, FL, 1995, 611.
258. R.H. Platel, L.M. Hodgson, and C.K. Williams, *J. Macromol. Sci., Polym. Rev.*, 2008, *48*(1), 11–63.
259. R.G. Sinclair, *J. Macromol. Sci. Pure Appl. Chem.*, 1996, *A33*, 585.
260. L.V. Labrecque, R.A. Kumar, V. Dave, R.A. Gross, and S.P. McCarthy, *J. Appl. Polym. Sci.*, 1997, *66*, 1507.
261. (a) M. McCoy, *Chem. Eng. News*, Apr. 28, 2008, 28; (b) G. Ondrey, *Chem. Eng.*, 2008, *115*(4), 14.
262. (a) S.D. Allen, D.R. Moore, E.B. Lobkovsky, and G.W. Coates, *J. Am. Chem. Soc.*, 2002, *124*, 14284; (b) X.-B. Lu and Y. Wang, *Angew. Chem. Int. Ed.*, 2004, *43*, 3574; (c) G.W. Coates and D.R. Moore, *Angew. Chem. Int. Ed.*, 2004, *43*, 6618–6639.
263. (a) W.J. vanMeerendonk, R. Duchateau, C.E. Konig, and G.-J.M. Gruter, *Macromol. Rapid Commun.*, 2004, *25*, 382; (b) D.J. Darensbourg, R.M. Mackiewicz, A.L. Phelps, and D.R. Billodeaux, *Acc. Chem. Res.*, 2004, *37*, 836; (c) T.H. Zevaco, A. Janssen, J. Sypien, and E. Dinjus, *Green Chem.*, 2005, *7*, 659.

264. C.M. Byrne, S.D. Allen, E.B. Lobkovsky, and G.W. Coates, *J. Am. Chem. Soc.*, 2004, *126*, 11404.
265. Anon., *Chem. Eng. News*, Nov. 12, 2007, 25.
266. (a) X.-D. Fan, Y. Deng, J. Waterhouse, and P. Pfromm, *J. Appl. Polym. Sci.*, 1998, *68*, 305; (b) Y. Deng, X.D. Fan, and J. Waterhouse, *J. Appl. Polym. Sci.*, 1999, *73*, 1081.
267. M. Beck, P. Birnbrich, U. Eicken, H. Fischer, W.E. Fristad, B. Hase, and H.-J. Krause, *Angew. Makromol. Chem.*, 1994, *223*, 217.
268. D.S. Poche, W.H. Daly, and P.S. Russo, *Macromolecules*, 1995, *28*, 6745.
269. (a) The Presidential Green Challenge Awards Program, U.S. Environmental Agency, EPA744-K-96-001, Washington, DC, July 1996, 5; (b) S.H. Ramsey, U.S. patent 5,449,748 (1995); (c) J.E. Donachy and C.S. Sikes, *J. Polym. Sci. A Polym. Chem.*, 1994, *32*, 789; (d) G. Parkinson, *Chem. Eng.*, 1997, *104*(8), 25; (e) L. Wood and G. Calton, U.S. patent 5,659,008 (1997).
270. M. Tomida, M. Yabe, and Y. Arakawa, *Polymer*, 1997, *38*, 2791.
271. H. Kubota, Y. Nambu, and T. Endo, *J. Polym. Sci. A Polym. Chem.*, 1995, *33*, 85.
272. (a) I. Arvanitoyannis, E. Psomiadou, N. Yamamoto, E. Nikolaou, and J.M.V. Blanchard, *Polymer*, 1995, *36*, 2957; (b) I. Arvanitoyannis, E. Nikolaou, and N. Yamamoto, *Angew. Makromol. Chem.*, 1994, *221*, 67.
273. I. Gachard, B. Coutin, and H. Sekiguchi, *Macromol. Chem. Phys.*, 1997, *198*, 1375.
274. K. James and J. Kohn, *Trends Polym. Sci.*, 1996, *4*, 394.
275. M. Reinicke and H. Ritter, *Macromol. Chem. Phys.*, 1994, *195*, 2445.
276. N. Kuramoto, K. Hayashi, and K. Nagai, *J. Polym. Sci. A Polym. Chem.*, 1994, *32*, 2501.
277. (a) E.O. Wilson, *Science*, 1998, *279*, 2048; (b) T. Princen, M. Maniates, and K. Cona, eds, *Confronting Consumption*, Island Press, Washington, DC, 2002; (c) T. Princen, *The Logic of Sufficiency*, MIT Press, Washington, DC, 2005; (d) S.K. Sikdar, *Ind. Eng. Chem. Res.*, 2007, *46*, 4727–4733.
278. E. Adler, *Wood Sci. Technol.*, 1977, 11, 169–218.

RECOMMENDED READING

1. B.Y. Tao, *Chem. Ind. (Lond.)*, 1994, 906 [industrial products from soybeans].
2. D.J. Murphy, *Chem. Br.*, 1995, *31*(4), 300 [oleochemicals].
3. J.N. BeMiller, *Kirk–Othmer Encyclopedia of Chemical Technology*, 4th ed., Wiley, New York, 1992, *4*, 932–938 [cellulose, hemicellulose, and starch].
4. M.A. Colgin and R.V. Lewis, *Chem. Ind. (Lond.)*, 1995, 1009 [spider silk].
5. E.M. Kirschner, *Chem. Eng. News*, Apr. 8, 1996 [top 50 chemicals].
6. C.-H. Zhou, J.N. Beltramini, Y.-X. Fan, and G.Q. Lu, *Chem. Soc. Rev.*, 2008, *37*, 527 [commodity chemicals from glycerol].
7. A. Behr, J. Eilting, K. Irawadi, J. Leschinski, and F. Linder, *Green Chem. A*, 2008, *10*, 13 [chemicals from glycerol].
8. C.H. Christensen, J. Rass-Hansen, C.C. Marsden, E. Taarning, and K. Egeblad, *ChemSusChem*, 2008, *1*, 283 [chemicals from biomass].

EXERCISES

1. Devise a way to make benzene from renewable materials. Look up the major uses of benzene. Because it is a carcinogen, can we do without it in a sustainable future?
2. What chemical reactions could be run on soy or corn protein to decrease brittleness and water absorption? Can you devise a system to do this with reagents from renewable sources for use in a twins-crew exruder?
3. Choose a food-processing waste. How could something useful be made from it? If you cannot think of one, try chicken feathers.
4. Cutin is a material on the surfaces of leaves and fruits, which keeps the water in and the rain out. What food-processing wastes might be used as its source for putting a hydrophobic coating on molded protein products? Outline the steps to do this.
5. What reactions could be run on the lignin from the pulping of wood with aqueous ethanol to upgrade it for large-scale applications.
6. Knowing that sulfonation of aromatic compounds is a reversible process, can you think of a way to remove the sulfonate groups from Kraft lignin? The product would be an analogue of "organosolv" lignin.

CHAPTER 13

Chemistry of Long Wear

13.1 WHY THINGS WEAR OUT

The first step in achieving longer wear is to have reusable ones that can be used many times before they wear out or break and have to be recycled. If the process of wear or breakage is understood, it may be possible to intervene by altering the design or manner of use so that a longer service life is obtained.[1]

Metals can rust or corrode. Rubber can harden and crack with age. Plastic may embrittle with age. Colors may fade in the sunlight. These are due to oxidation. Efforts to minimize these processes involve isolation of the object from the environment, by a coating or other means, or by the use of additives that prevent the oxidation. Concrete may deteriorate under the influence of deicing salts. Other objects wear out mechanically. A cutting edge becomes dull. The knees of the pants wear through from abrasion. Stockings snag and trip. Teeth come out of zippers. Plastics may scratch or stain. The rung on the chair may break. Glass and china objects may break if they are dropped or hit. Some plastics are brittle on impact. Biological factors may also be involved. The clothes moth likes to eat woolen garments. Water-based paints, inks, cutting fluids, and such may serve as substrates for the growth of bacteria and fungi. Termites and carpenter ants may eat the wood in houses. Understanding the ecology of these organisms can sometimes provide means to prevent their entry. Otherwise, biocides are used.

There is often a weak link in an appliance or piece of equipment. A plain steel screw in a stainless steel pot will fail first. The screw is used because stainless steel is harder to machine. A plain steel knife keeps its cutting edge better than one made of stainless steel. A small spring may fatigue and break. A nut may be lost from a bolt. Many nuts are locked on by putting a monomer on the threads, which then polymerizes when no more oxygen is present.[2] This is an anaerobic adhesive. The important thing is for the equipment to be designed for ease of disassembly for easy replacement of the weak link. This will require more screws and fewer rivets, welds, and metal parts encased in plastic. It may require more standardization of sizes of screws or other parts. The second factor is that the spare part should be easy to obtain, preferably locally. A generation ago it was possible to buy the one screw needed at the neighborhood hardware store. Today, one may have to drive to a mall and buy a prepackaged assortment of two dozen screws. It may be best if the owner can replace the part himself. The next best option would be a local repair service. For more complicated equipment, shipment to a central repair facility might be necessary. Generators for automobiles are handled in this manner. Other weak links include the elastic thread in underwear that fails long before the cloth, and the holes that develop in pockets long before they form anywhere else in the pants. Ideally, the weak link would be redesigned so that it is no longer weak. This may require chemists to devise new and better materials.

Some of the failures can be due to improper use of the object or lack of preventive maintenance. Running without coolant will damage some machinery (e.g., a car). Storing a steel bicycle or shovel outdoors will encourage rusting. An item of polyethylene would melt if put in the oven. Plain glass might crack in the oven, whereas Pyrex glass would not. Glasses may etch in the dishwasher over time if the detergent is too alkaline. Making the china mug thicker than the delicate teacup reduces breakage. Putting wet objects directly on wooden furniture may ruin the finish. This can be avoided by using furniture with tops of melamine–formaldehyde resin laminates. Shovel handles may break if used for prying. The cutting edge on the saw will be dulled if one tries to saw through wood with nails in it. A small rung on a chair may break if someone steps on it. Torque wrenches were devised to prevent the application of too much force to nuts and bolts.

Some of the failures are due to the improper choice of materials or poor designs. The cutting edges on saws, planes, chisels, knives, and such will stay sharp longer if they are made of a good tool steel. They should be easy to resharpen. The spring will last longer if it is made of a good tool steel and not overextended in use. It is easy to cast "white metal," but the castings tend to be somewhat brittle and often break with time. A household strainer of stainless steel will last longer than one made of plain steel. Stainless steel knives, forks, and spoons will hold up better than those plated with silver, and there is no tarnish to be taken off. Chutes and trucks carrying rocks will last longer if they are lined with rubber or with ultrahigh molecular weight polyethylene. Levees at New Orleans failed because the force of Hurricane Katrina exceeded that expected by the designers.[3] In nearby Mississippi, houses worth $400,000 collapsed because they lacked $20 cross braces. The area has now adopted a building code.

A ceramic may not have been fired hot enough. Furniture, plywood, and particle board not made with waterproof glue should not be put into exterior use. The galvanized coating on wire or gutters may not be thick enough to weather well. Gutters of aluminum, copper, stainless steel, and plastic avoid this problem. Furniture drawers made with dovetailed joints will hold together longer than plain ones just nailed or stapled together. A floor made of oak will wear less than one made of pine.

Coatings containing pinholes fail sooner than others. A better coalescing agent for water-based coatings may be needed. Coatings may not adhere well to substrates that have not been sanded or cleaned. The coating may not adhere well if it has not been designed for the substrate in question. This can be a problem for inks for polyolefins, which lack functional groups. The usual practice is to treat polypropylene film with a corona discharge to put some oxygen-containing groups on the surface. Paper sized under acidic conditions embrittles with age, but paper sized under neutral conditions lasts much longer and is specified for archival use. Neutral sizing also permits better recycling of water in the plant. A natural wood finish on a home may have to be redone every couple of years. An oil-based paint or one of the newer latex paints for exterior use would last two to three times longer. If the home were built with aluminum or plastic siding, or finished in stucco, brick, or stone, this periodic recoating of the surface would be unnecessary. Roofs made of slate or tile last longer than those made of asphalt shingles. Dyes and pigments that are not resistant to light should not be used in objects to be used outdoors or even next to the window in the living room.

Stockings and panty hose would last longer if they were made of rip-stop nylon, a woven fabric. Making them thicker would also help. Rotating the rug or moving the furniture once in a while, so that the wear will be in different places, will make the rug last longer.

A great many useful objects are made of plastic. The preferred methods of fabrication of many plastics involve melting the plastic and putting it into a mold or forcing it through a spinneret to form a fiber. If this process is done at too high a temperature, the polymer or the additives in it may be degraded. The extra heat is usually applied in an effort to speed up the line. As the molecular weight increases, so does the viscosity and the strength. To obtain the maximum strength, the molecular weight should be high enough to be on a plateau where no further increase in one will affect the other. The usual molding conditions are a compromise between optimal strength and a viscosity that can be handled in the equipment. Adding ultrasound to an extruder can raise the molecular weight above that which can be processed without it. Ultrahigh molecular weight polyethylene (molecular weights of 1 million or more) cannot be processed by these methods. However, when fiber is prepared by a solution method, the fiber is strong enough to be used in bulletproof vests. Further research on cross-linking or chain extension during fabrication may lead to stronger, tougher parts. On the other hand, too much cross-linking can lead to hard, brittle objects. The glass transition temperature is also important. If a rubbery polymer is used below its t_g, it will be rigid instead of rubbery. If a plastic is used above its t_g, it will gradually change its size or shape, a process termed creep.

Additives to the polymers make a world of difference.[4] A brittle polymer, such as polystyrene, can be made to resist impact by the inclusion of little rubber balls a few microns in size. The rubber diffuses and dissipates the force that hits the objects. The trick is to find a way to make the rubber separate in the proper manner during fabrication. The best method involves light grafting to the preformed rubber during polymerization of the other monomers. Preformed rubber particles of the core–shell type have been used to toughen acrylic polymers.[5] Powdered, cross-linked 500 nm rubber particles have been added to polypropylene and ethylene–propylene copolymers to give thermoplastic elastomers.[6] The same principle applies to the macroscopic particles of rubber that make asphalt pavements last longer (see Chapter 14). Reinforcing fibers and fillers can increase strength. The lifetimes of automobile tires have been increased greatly in the past 30 years by switching to a radial design of the belts and using reinforcing silica fillers in poly(1,3-butadiene) treads.[7] The carcass is often made of a different rubber (e.g., a copolymer of styrene and 1,3-butadiene). Shoe heels are a different story. They are made of highly filled reclaimed rubber. If they were made of the same rubber as tire tread, wear would not be a problem.

The useful lives of some polymers are determined by how long the plasticizers and stabilizers remain in them. Losses by volatilization, migration, and leaching are greatest in thin sections, such as fibers and films. The plasticizer in the poly(vinyl chloride) seat covers in a car volatilizes and condenses on the inside of the windshield. It is responsible for the smell of a new car. The seat covers become less flexible as the plasticizer leaves. The use of plasticizing comonomers, such as vinyl stearate, or polymeric plasticizers can obviate the problem of plasticizer loss. The use of newer polyolefins, with properties of flexible poly(vinyl chloride) instead of poly(vinyl chloride) itself, will eliminate the problem of plasticizer loss (see Chapter 12). The problem with the elastic natural rubber "cut" thread in underwear may be twofold. Stabilizers may come out in the wash, allowing the unsaturated polymer to oxidize more readily. Grafting of the stabilizer may also be possible. This may be the wrong rubber to use. The choice of a less unsaturated rubber, such as polyisobutylene or a polyurethane, might solve the problem of the elastic thread in underwear. Some plastics are not stabilized adequately. Because the stabilizers usually cost several times as much as the polymer, there is a tendency to skimp on the amounts used. The "general-purpose grade" should not be used for any demanding applications.

13.2 STABILIZERS FOR POLYMERS

Plastics fail for a variety of reasons.[8] Cellulose nitrate film in museum collections is failing owing to the release of nitric acid.[9] The degree of degradation is related to the amount of sulfate left in the sample when it was made. Old paper can be sterilized and coated by a plasma containing hexamethyldisiloxane to improve its life and tensile strength.[10] Poly(vinyl chloride) objects fail from the loss of both plasticizer and hydrogen chloride. Polymers with ester backbones may be subject to hydrolysis in use. The most common type of degradation is oxidation, both thermal and light-catalyzed. Any of these processes that lower the molecular weight below that needed for adequate strength will limit the useful life of the polymer. On the other hand, there are a few places for which controlled degradation is desired (e.g., agricultural mulch, litter, and surgical sutures). Transition metal dithiocarbamates can be added to the first two to give time-controlled photolysis.[11] Low-density polyethylene containing nanoparticles of titanium dioxide has also been used for agricultural mulch.[12] (For polyesters used in the third application, see Chapter 12.)

Polymer stability varies greatly. Poly(methyl methacrylate), bisphenol polycarbonate, and poly(vinyl fluoride) last for many years of outdoor exposure. The first two can be used for unbreakable windows in buses, schools, and elsewhere. The problem is that they become scratched. A highly cross-linked scratch-resistant coating is put on automobile headlights of these polymers by radiation curing. Poly(vinyl fluoride) can be used to coat the exteriors of buildings. Clear exterior finishes for wood only last a year or two, much less than pigmented finishes.[13] Polyurethanes made with aliphatic isocyanates are more stable outdoors than those made with aromatic isocyanates. The latter absorb more ultraviolet light. Polyolefin polymers are subject to oxidation. Polypropylene and poly(1,3-butadiene) could not be used commercially without the addition of stabilizers. A trace of antioxidant is added even before the workup of these polymers when they are made. The radicals from these polymers are stabilized by being tertiary in the first case and allylic in the second case (**13.1** Schematic).

Polyethylene, which lacks the many branches, is more stable, but still requires the addition of stabilizers. Polyisobutylene, which is a saturated rubber, is more stable than the unsaturated rubbers. Poly(methyl methacrylate), in which oxidation would have to produce a primary radical, is much more stable. Polyamide degradation begins with abstraction of a hydrogen from the methylene group next to the nitrogen atom, giving a radical that can then react with oxygen.[14]

13.1 Schematic

A wide variety of stabilizers are used in plastics and rubbers.[15] It is common for more than one to be used in a given polymer. The total amount added, for example, to polypropylene, is usually under 1%. Because stabilizers are used up in protecting the polymer, more may have to be added when used items are recycled by remolding.[16] Oxidation of a hydrocarbon is a chain reaction (**13.2** Schematic).[17]

$$R^\cdot + O_2 \longrightarrow ROO^\cdot \xrightarrow{RH} ROOH + R^\cdot$$

$$ROOH + M^{n+} \longrightarrow ROO^\cdot + M^{(n-1)+} + H^+$$

$$ROOH + M^{(n-1)+} \longrightarrow RO^\cdot + M^{n+} + OH^-$$

$$RO^\cdot + RH \longrightarrow ROH + R^\cdot$$

13.2 Schematic

The initiating radical may be formed by mechanochemical scission during molding[18] or by ozone in the air.[19] If a metal ion that can have multiple valences is present (e.g., a trace of iron from the extruder), the oxidation can be accelerated. In the case of polypropylene, a new radical and, hence, a new hydroperoxide can be formed by backbiting (**13.3** Schematic).

An alkoxy radical, formed as previously mentioned, can cleave the chain to form a ketone. The primary radical would promptly abstract a hydrogen atom from something to form a tertiary radical. The extent of degradation of polypropylene is sometimes estimated by measuring the carbonyl content by the infrared spectrum of the sample. If light is present, the ketone can undergo further cleavage.

Antioxidants[20] interrupt these chains by forming radicals that are too weak to abstract hydrogen atoms from carbon atoms. These radical scavengers contain hydrogen atoms that are easily abstracted. The most active are aromatic amines[21] such as those in **13.4** Schematic. The radicals formed are stabilized by electron delocalization in a number of resonance structures, some of which are shown in **13.5** Schematic. These are staining antioxidants owing to the imine structures that can form. Abstraction of a second hydrogen atom would lead to highly colored quinone imines. The dihydroquinoline cannot do this and is classed as semistaining. Staining antioxidants can be used when the color does not matter, as in gasoline and black rubber tires. Ozone causes cracking in rubber under stress. The cleavage of carbon–carbon double bonds by ozone is well known. Antiozonants tend to be of the aminodiphenylamine type.[22] Waxes that bloom to the surface are used to limit the entry of ozone.

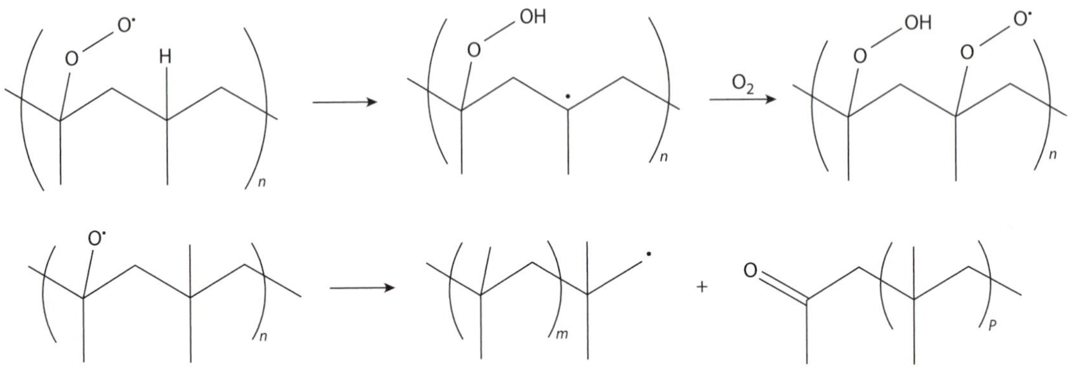

13.3 Schematic

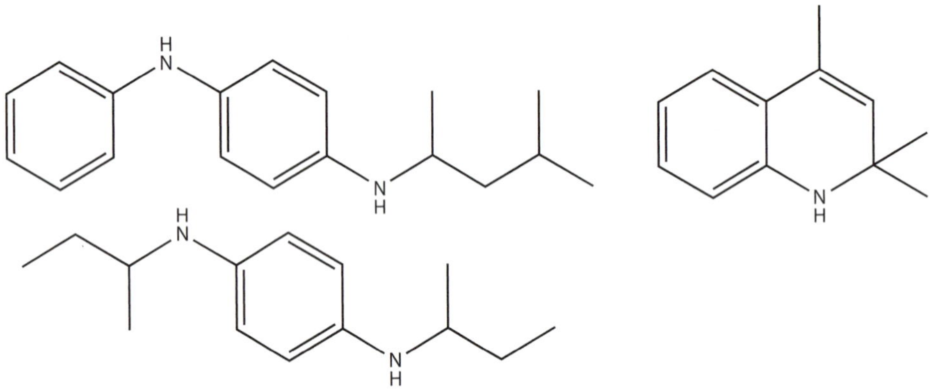

13.4 Schematic

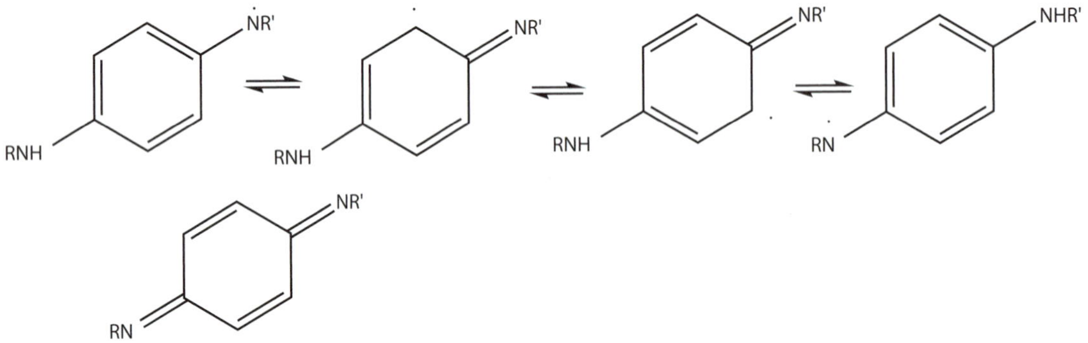

13.5 Schematic

The less active hindered phenols are used in most plastics to avoid the staining. The simplest, and perhaps the cheapest, is 2,6-di-*tert*-butyl-4-methylphenol. Abstraction of hydrogen leads to a resonance-stabilized radical (**13.6** Schematic) that cannot abstract a hydrogen from a carbon atom owing to steric hindrance and low reactivity. Although this phenol can be used in gasoline, it is too volatile for use in most plastics, especially in thin sections, as in fibers and films. For these uses, analogs of higher molecular weight, such as the ester of pentaerythritol (**13.7** Schematic), are used.

Nitrogen-containing phenols that break two chains are more effective antioxidants than other phenolic antioxidants (**13.8** Schematic).

Dinitriles have also shown promise as antioxidants (**13.9** Schematic).[24] The hydroperoxides formed in the oxidation can be decomposed by the use of S(II), S(IV), or P(III) compounds.[25] A few peroxide decomposers are shown in **13.10** Schematic. The cyclic sulfite is derived from a hindered phenol.[26] Because the antioxidants and peroxide decomposers operate by different mechanisms, they are synergistic and are often used together (**13.10** Schematic).

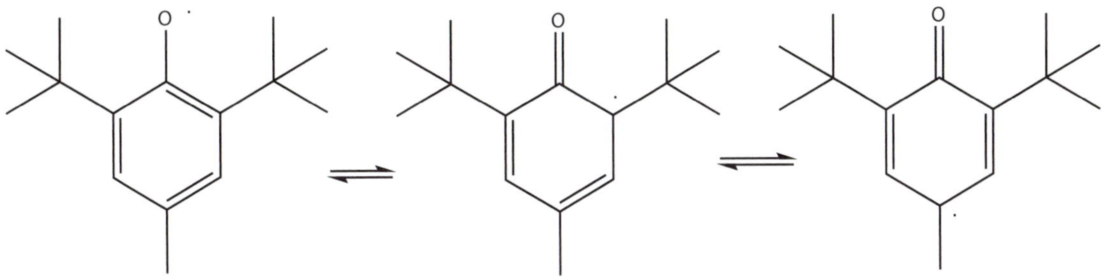

13.6 Schematic

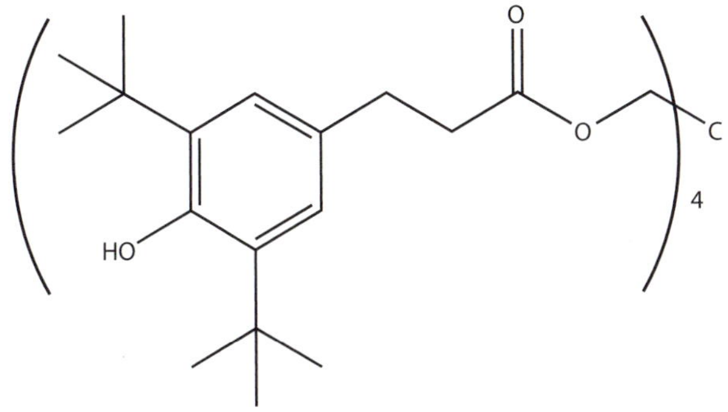

13.7 Schematic

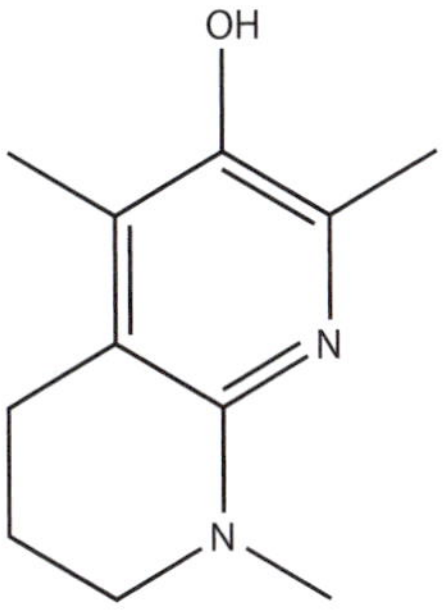

13.8 Schematic

The hydroperoxide oxidizes the sulfur or phosphorus (**13.11** Schematic).

The rate of degradation increases when polyolefins are used around copper wire. Copper is also undesirable in gasoline. Metal deactivators are used in these cases. The oxalylhydrazone probably forms a metal complex that is too insoluble to act as a catalyst for oxidation. The propylenediamine derivative is used as a metal deactivator for gasoline. It may work by changing the redox potential of the copper. Talc accelerates the rate of loss of mechanical properties when used as a nucleating agent for polypropylene.[27] This may be due to traces of transition metal ions in the mineral. Metal-containing pigments sometimes behave in the same way. It might also be due to adsorption of stabilizers onto the talc.[28] Citric acid and EDTA are food-grade metal deactivators for use in systems more polar than hydrocarbons (**13.12** Schematic). Citric acid is good for deactivating iron (III).

Additional measures have to be taken when light is involved in the degradation of the polymer.[29] Electroluminescent polymers have been encapsulated in resins to keep oxygen and water out and prevent degradation by singlet oxygen.[30] Use of carbon black as a pigment also keeps the light out and prevents photodegradation.[31] Ultraviolet screening agents can also be used to keep the light out. The two types work by photo-enolization (**13.13** Schematic), which allows the energy to be dissipated as heat without harming the polymer.[32] Typically, R and R′ are long alkyl chains that make the compounds more soluble in polyolefins. Excited states can be quenched by metal chelates, such as those shown in **13.14** Schematic (which has a light green color).

The best light stabilizers are the hindered amines. A typical one is shown in **13.15** Schematic. They are used along with the antioxidants and peroxide decomposers. Their exact mechanism seems to be a matter of debate. One mechanism is shown in **13.15** Schematic.[33]

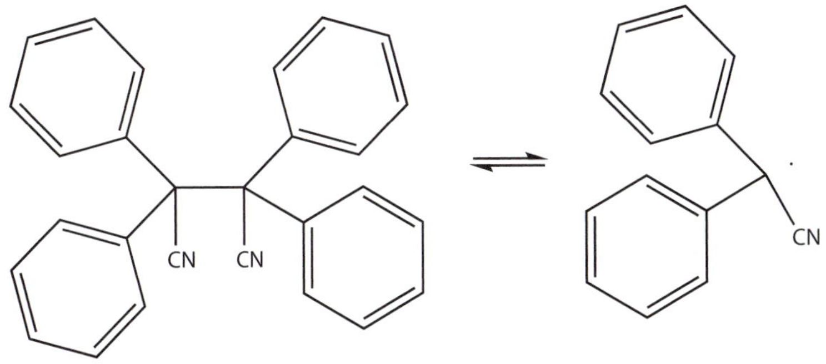

Thus, each molecule destroys more than one radical chain. A variety of analogs of higher molecular weight, often oligomeric polymers, are commercially available for use in fibers and films, where the one shown in **13.15** Schematic might be too volatile.[34] Another method that has been used to reduce volatility is to attach a stabilizer to the polymer that is to be protected.[35] For example, a hindered piperidinol or aminopiperidine was reacted with a graft copolymer of maleic

13.9 Schematic

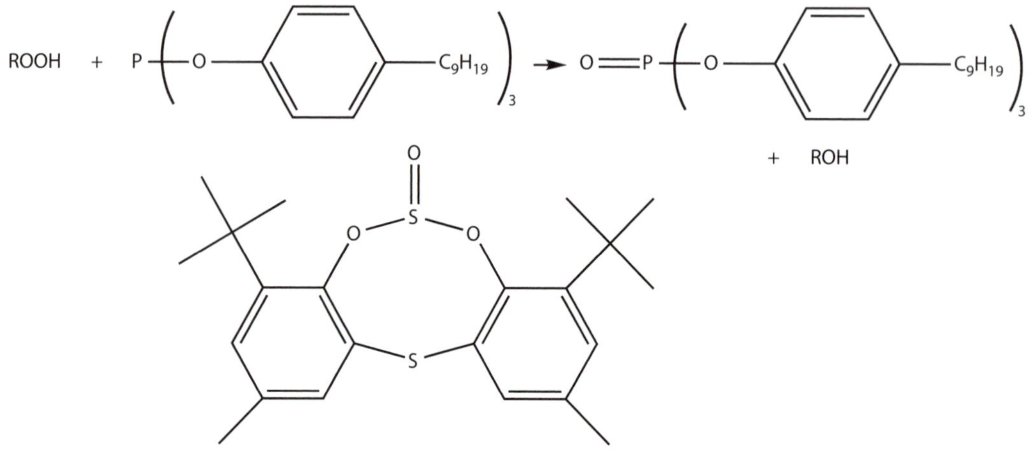

$$ROOH + P(O-C_6H_4-C_9H_{19})_3 \rightarrow O=P(O-C_6H_4-C_9H_{19})_3 + ROH$$

13.10 Schematic

$$ROOH + S(CH_2CH_2COOC_{18}H_{37})_2 \longrightarrow ROH + O=S(CH_2CH_2COOC_{18}H_{37})_2$$

13.11 Schematic

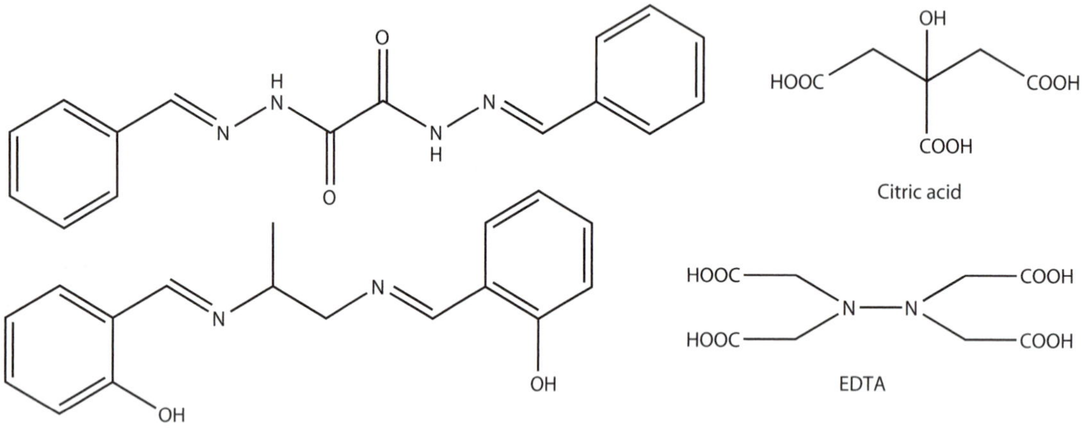

Citric acid

EDTA

13.12 Schematic

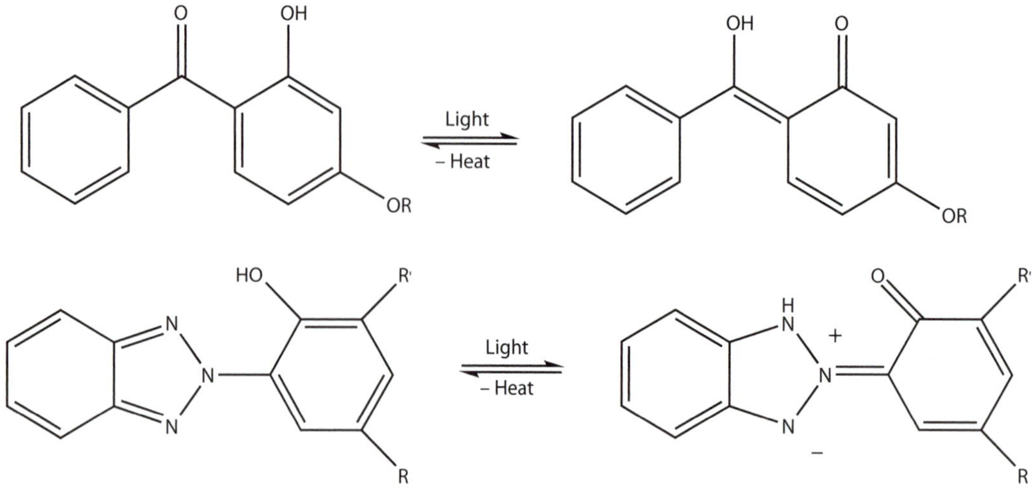

Light
– Heat

Light
– Heat

13.13 Schematic

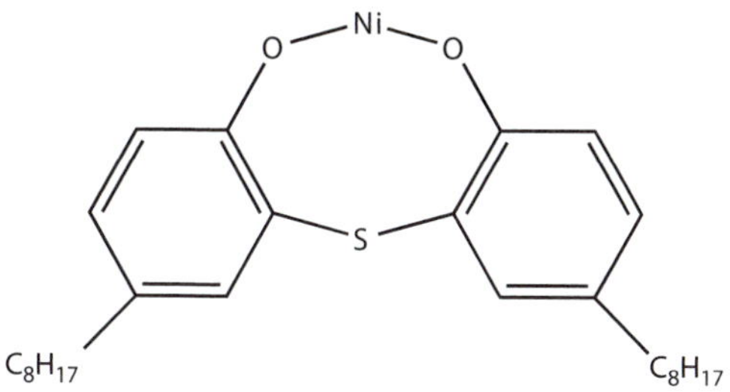

13.14 Schematic

anhydride on polypropylene to form an ester and amide, respectively (**13.16** Schematic).[36]

A hindered amine monomer was grafted to polypropylene in the presence of a peroxide at 200°C.[37] Grafting the hydroxybenzophenone monomer (**13.17** Schematic) to wood stabilized the surface to light, whereas a monomeric equivalent did not.[38]

The hindered amine and a peroxide decomposer have also been combined in one molecule (**13.18** Schematic)[39]

A product made by grafting a mixture of styrene and vinyltrimethoxysilane onto ethylene–propylene–diene monomer terpolymer rubber (EPDM) was able to replace acrylonitrile-1,3-butadiene–styrene terpolymer (ABS).[40] The product showed improved heat and light stability by removal of the unsaturation. The need to use the carcinogenic acrylonitrile was eliminated.

Acid acceptors are used when acids may be involved in degradation of the polymer. Calcium stearate is often used with polypropylene to neutralize any mineral acids, such as hydrochloric acid, that might result from the reaction of water with catalyst residues. Various mixed metal laurates are used as stabilizers in poly(vinyl chloride). Efforts are underway to replace the cadmium in some of these mixtures (see Chapter 4). Epoxidized soybean oil has been used as the acid acceptor, along with dioctyltinbis(thioglycolate), to improve the heat stability of poly(vinyl chloride).[41]

Synthetic antioxidants, such as 2,6-di-*tert*-butyl-4-methylphenol, are not allowed in foods in Japan. In the United States they can be used in the wrapper around the cereal in small amounts, but not in the cereal itself. There is great interest in natural antioxidants that can be used to protect foods.[42] That these natural phenols from plants may scavenge free radicals in the body and thereby reduce the incidence of heart attacks and cancer makes them especially important. Vitamin E (**13.19** Schematic) is the one that we must all have in our diets.

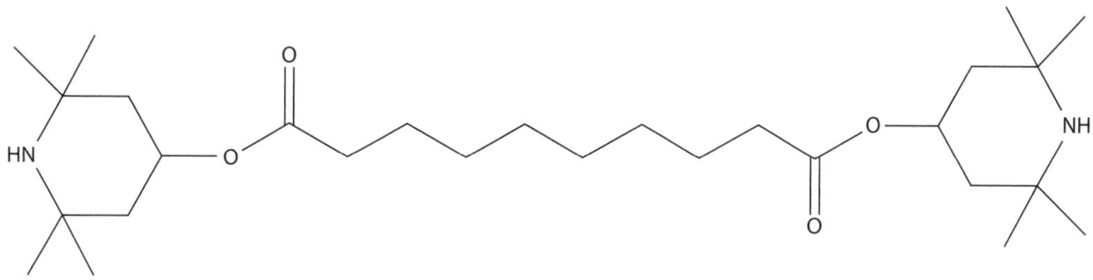

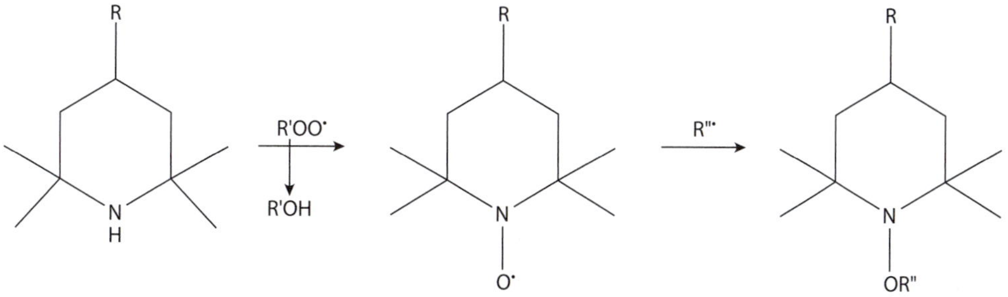

13.15 Schematic

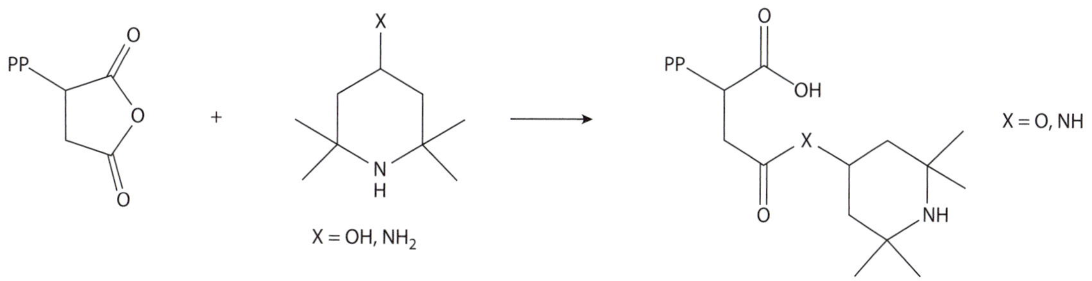

13.16 Schematic

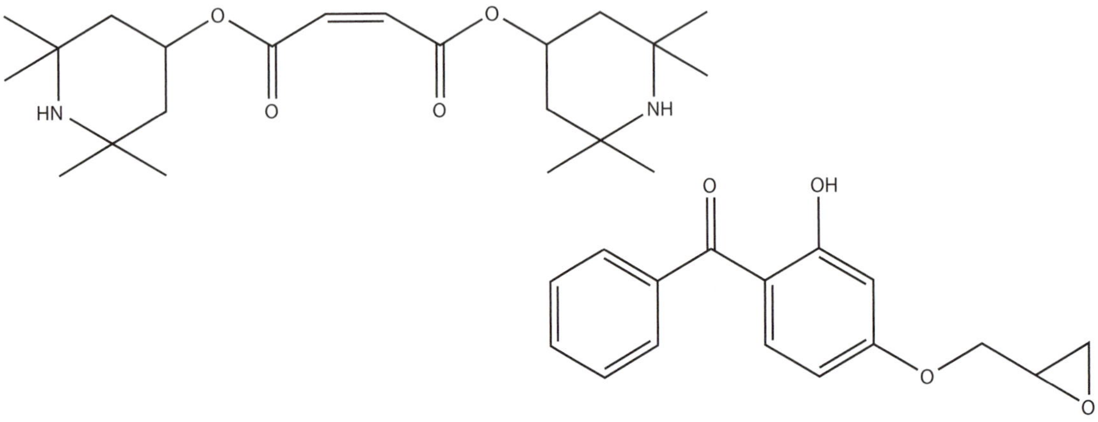

13.17 Schematic

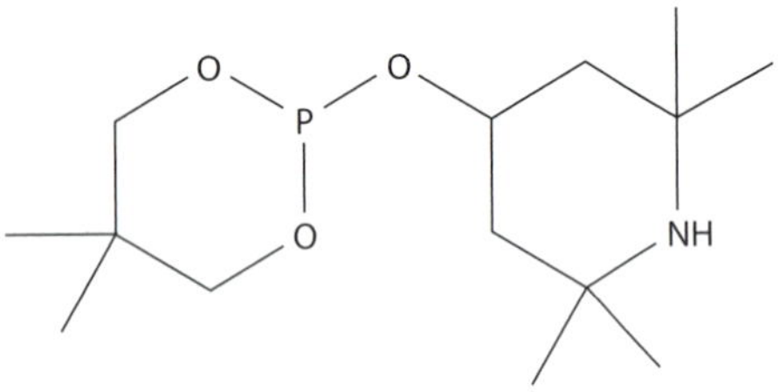

13.18 Schematic

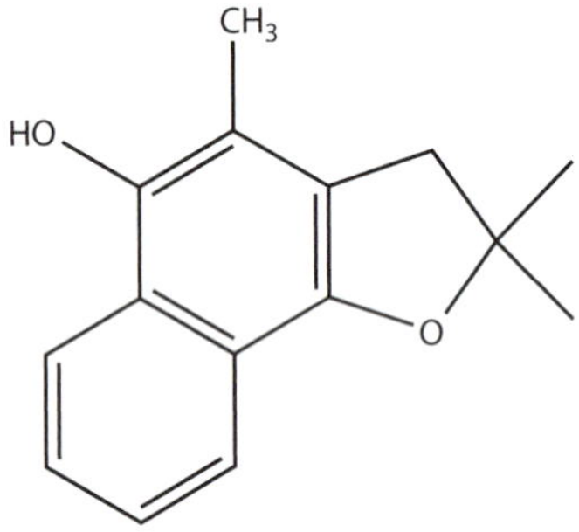

Some synthetic analogs (**13.20** Schematic) show higher activity than vitamin E.[43] The naphthalene unit presumably allows enhanced electron delocalization to form a more stable radical. Plant flavanoids can be good antioxidants. Luteolin is a better antioxidant than 2,6-di-*tert*-butyl-4-methylphenol.[44] Green tea is a good source of antioxidant flavanoids, one of the most active being epigallocatechin gallate (**13.21** Schematic).[45] It may act by inhibiting the enzyme urokinase (in cases where it causes the reduction of the size of tumors).[46]

A wide variety of extracts of plants and spices have also shown antioxidant activity, often comparable with or better than that of 2,6,-di-*tert*-butyl-4-methylphenol.[47] These include the phenols in extracts of rosemary,[48] black pepper,[49] buckwheat,[50] peanut hulls,[51] and others.[52]

Additives in disposable plasticware may interfere with laboratory experiments. In one case, a quaternary ammonium salt biocide and a slip agent, oleamide, came out.[53]

13.20 Schematic

13.3 LUBRICATION, WEAR, AND RELATED SUBJECTS

If two objects are rubbed together, they may wear by abrasion. The heat developed in the process may cause them to deteriorate by oxidation or loss of strength, or they may melt. In fact, this is one way to weld two pieces of plastics together. They are rotated against each other until the surfaces melt together and then removed from the apparatus to cool. A lubricant[54] lowers the coefficient of friction, allowing one surface to slide over the other more easily without, or with less, abrasion. It can also, when passed through continuously, cool the surfaces and remove any particles that form.

There are some special cases for which it is unnecessary to add a lubricant. Magnetic bearings and bearings that are

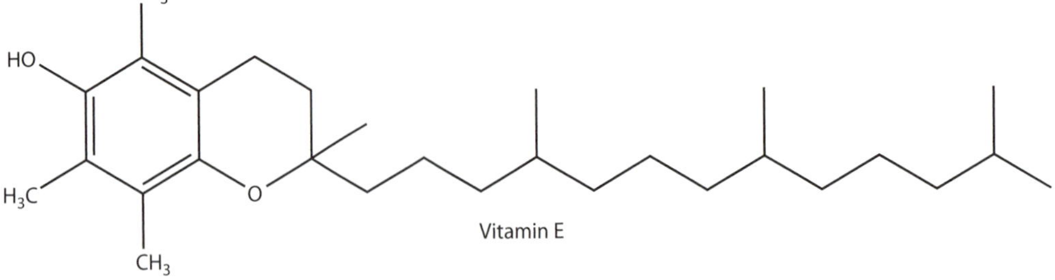

Vitamin E

13.19 Schematic

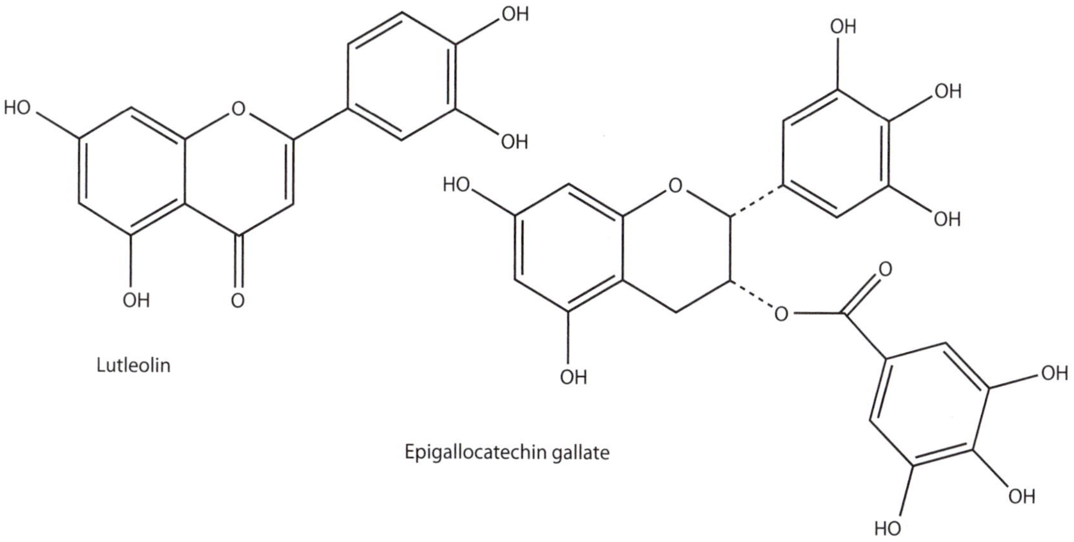

Lutleolin

Epigallocatechin gallate

13.21 Schematic

suspended by a pressurized gas have no metal-to-metal contact so that wear and heat buildup are avoided.[55] Several companies sell oil-free vacuum pumps that eliminate the trouble and expense of changing the oil, disposal costs of used oil and contamination of process chambers, which is especially important in the manufacture of semiconductors and liquid crystal displays.[56] Plastic gears of nylon or polytetrafluoroethylene are self-lubricating and need no added lubricant. They reduce the noise associated with machinery. They may be made with graphite or molybdenum disulfide in them as additional built-in lubricants. These are layered substances where the layers tend to slide over one another easily. Carbon-graphite normally has about 15% porosity. The pores can be filled with organic resins, Cu, Sb, bronze, or Ni–Cr to improve lubricity so that no liquid lubricant is needed.[57] One impregnated with Sb running continuously against a polished SiC counter-face lasted 3 years. Copper particles,[58] glass fibers, carbon fibers, and such in polytetrafluoroethylene also reduce wear.[59] Cross-linking the polymer with an electron beam raised its abrasion resistance 10^3–10^4-fold and its creep resistance two- to three-fold.[60] A small amount of silicone oil (1.5–2.0%) in a thermoplastic polyurethane improved its wear characteristics.[61] Polyetheretherketone has a lower coefficient of friction and wear rate when 7.5%-nm-sized zirconia is included in it.[62] Films of carbon deposited from a radiofrequency plasma of argon have a coefficient of friction of 0.001, which is 20 times lower than that of molybdenum disulfide and 40 times lower than that of polytetrafluoroethylene.[63] If a way can be found to apply them to large objects, they should be tried to prevent the fouling of pipes and ships by marine organisms.

Hollow[64], nested,[65] and spherical[66] nanoparticles of tungsten disulfide, as well as onions of molybdenum disulfide,[67] are better lubricants than the usual molybdenum disulfide. They may roll like ball bearings.

As mentioned earlier in this chapter, wear[68] can sometimes be reduced by putting a coating of polyurethane rubber or of ultrahigh molecular weight polyethylene on a metal surface. Sometimes, wear can also be reduced by making the surface harder. Cross-linked coatings on the surfaces of plastics can do this. The wear of high-density polyethylene was reduced by a thin surface film made by plasma polymerizing a mixture of acetylene, hydrogen, and silane.[69] Harder surfaces were produced on poly(methyl methacrylate) by transamidation with diamines.[70] Another way to do it is to deposit a transparent layer of silica or titania by the sol–gel method. High-energy ion irradiation can be used to make the surfaces of common polymers, such as polyethylene, poly(ethylene terephthalate), polystyrene, nylon, and others, harder than steel.[71] This might provide a way of eliminating the scratching and staining of plastic dinnerware made of melamine–formaldehyde plastics. Another possibility would be to apply a diamond coating by the three-laser method.[72] Diamonex Co. offers amorphous and polycrystalline diamond coatings (prepared by chemical vapor deposition[73]) that increase the wear resistance of plastics, glass, and metals.[74] Diamond-like coatings are hard and slick with a coefficient of friction as low as 0.001 compared to 0.15 for graphite and 0.04 for polytetrafluoroethylene so that lubricants are not needed with them.[75] Plastic dishes coated in this way offer advantages over glass and ceramic dinnerware in not scratching when in use and in not breaking when dropped. Mitsubishi uses acetylene with a radiofrequency plasma to coat poly(ethylene terephthalate) beverage bottles with a diamond-like coating to improve their barrier properties to oxygen and carbon dioxide up to 10-fold.[76] The bottles can be reused.

Face materials for seals can be made of carbon-graphite, silicon carbide, tungsten carbide, or alumina.[77] These require lubrication. Cutting tools with tungsten carbide tips or diamond coatings are common. Metal surfaces may also be hardened by carburizing, nitriding, or metallizing with a metal with greater lubricity or hardness (see Chapter 4). Thin films of titanium nitride, titanium carbide, and alumina (5 μm), which extend tool life 4–20 times, have been deposited by chemical vapor deposition,[78] magnetron sputtering, and pulsed direct current sputtering.[79] Multilayered coatings of TiN/VN and TiN/NbN were even harder. A tool coated with $Ti_{1-x}Al_xN/a\text{-}Si_3N_4$ is stable enough thermally so that no cutting oil is needed in machining.[80] Clear multilayered films of Al_2O_3/ZrO_2 should be able to protect a wide variety of objects without the use of any toxic metal ions. Small amounts of silicon and titanium make an alloy of aluminum, magnesium, and boron slightly harder than boron nitride at one third the cost.[81] It has only 5% of the wear of tungsten carbide (WC) and reduces power use up to 10%.[82] Rhenium diboride is as hard as BN and can be used for cutting where diamond cannot be used.[83] Superhard steel that outperforms tungsten carbide at one third the price is made by quenching atomized molten steel and then processing it into coatings.[84] Addition of nanoparticles of aluminum oxide to aluminum metal can impart the wear resistance of bearing steel.[85]

The hardness and resistance to wear of aluminum alloy quasicrystals, such as $Al_{65}Cu_{23}Fe_{12}$, may make them suitable as protective coatings for engine parts and other applications during which sliding motions can cause abrasion.[86] They can be toughened by inclusion of some iron aluminide.

Ceramics,[87] which are normally brittle, can be made less so if their grain size is reduced to the nanometer range[88] or if layered structures are used.[89] The nacre ($CaCO_3$) in shells is sturdy although pure aragonite is brittle. A mimic made by controlled unidirectional freezing of a colloidal suspension, freeze drying, and filling with an epoxy resin or metal was as strong as bone.[90] An ice-templated structure of layers of alumina and poly(methyl methacrylate) had 300 times the toughness of the ingredients.[91] Another mimic of nacre is a composite of platelets of alumina in chitosan in alternating layers.[92] A ceramic that can be stretched 10–50% is made of 40% yttria-stabilized zirconia, 30% magnesium aluminate, and 30% α-alumina.[93] Further work is needed to suppress cavitation during deformation. Layered Ti_3SiC_2 is ductile, stiff, lightweight, and machinable. It can be used in a high-speed tool bit without lubrication.[94]

Epoxy resins containing 10% of 32-nm particles of titanium dioxide were more resistant to cracking under strain and to scratching than either the unfilled resin or when the resin was filled with the usual microparticles of titanium dioxide. Nanocomposites of nylon containing a few percent of 1-nm-thick platelets of clay had a much higher heat distortion temperature and were stiffer, stronger, and just as tough as the unfilled nylon.[95] The technique has been used in a variety of polymers with similar but sometimes lesser effects.[96] Polyorganosilsesquioxanes[97] and carbon nanotubes[98] have also been used to improve polymer properties.

Ceramics can be made tougher by the inclusion of ceramic fibers.[99] Fibers of amorphous Si–Al–C–O material[100] and $SiBN_3C$[101] are suitable for use in air up to 1500–1600°C, whereas those of crystalline SiC cannot be used above about 1300°C because of the loss of mechanical strength above this temperature. These fibers are prepared by firing fibers spun from polymeric precursors, as in **13.22** Schematic.

Silicon carbide fibers are made from polydimethyldisilane (**13.23** Schematic).[102]

Boron nitride fibers are made through a spinnable intermediate (**13.24** Schematic).[103]

Ceramics reinforced with such fibers may increase the energy efficiency of engines by allowing them to run at higher temperatures. A ductile ceramic consisting of an interpenetrating network of single crystals of alumina and single crystals of gadolinium aluminate ($GdAlO_3$) has been made by unidirectional solidification of a eutectic mixture of the two materials. It shows high strength and metal-like pliability up to 1600°C.[104] Sodium aluminosilicate glass can be made up to five times stronger by exchanging potassium ions for sodium ions in the surface and then back-exchanging with sodium ions.[105] Samples fail safely by multiple cracking. (The safety glass in automobiles, which cracks in a similar fashion, consists of two sheets of glass with a layer of poly(vinyl butyral) between them.) Thin ceramic thermal

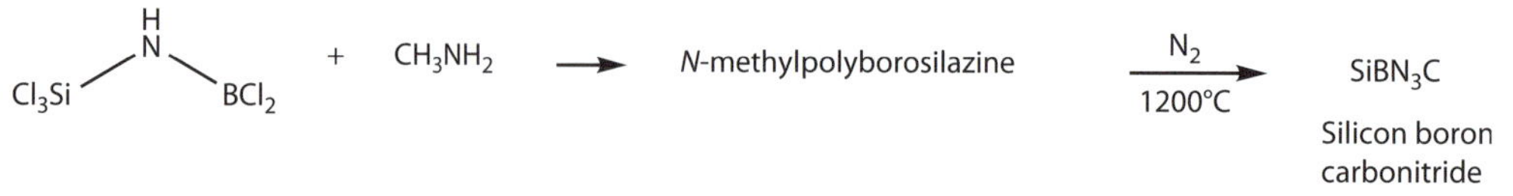

$Cl_3Si{-}NH{-}BCl_2$ + CH_3NH_2 ⟶ *N*-methylpolyborosilazine $\xrightarrow[1200°C]{N_2}$ $SiBN_3C$ (Silicon boron carbonitride)

13.22 Schematic

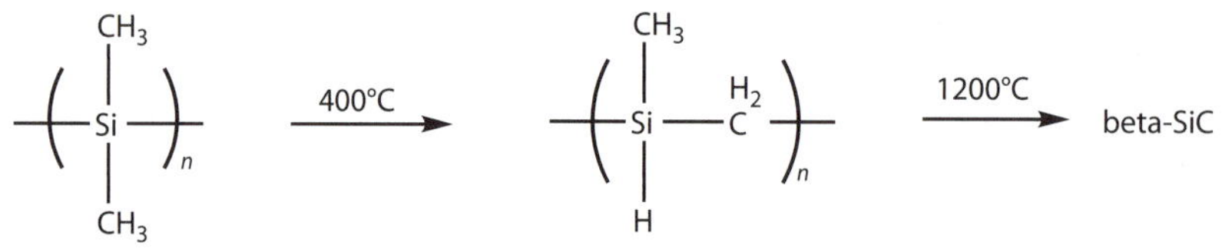

$-(Si(CH_3)_2)_n-$ $\xrightarrow{400°C}$ $-(SiH(CH_3){-}CH_2)_n-$ $\xrightarrow{1200°C}$ beta-SiC

13.23 Schematic

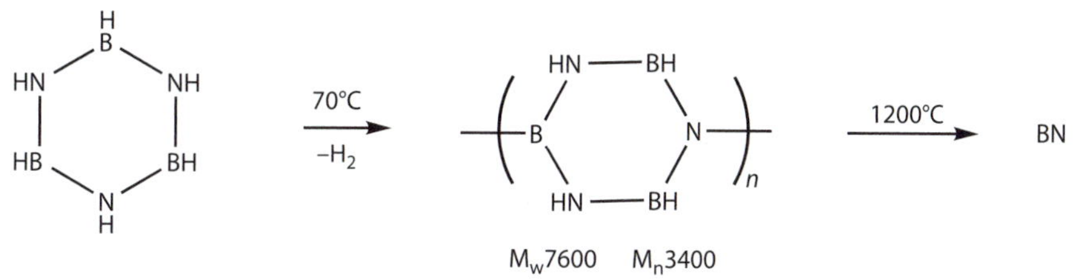

13.24 Schematic

barrier coatings on gas turbine blades allow them to run under conditions of higher temperature where the engines are more efficient.[106]

Most lubricating oils are based on petroleum.[107] For optimum results, numerous additives are required. These include antifoaming agents (typically silicone oils), antioxidants, corrosion inhibitors, detergents or dispersants, metal deactivators (such as zinc dithiocarbamates or dithiophosphates), pour-point depressants [such as a poly(alkyl methacrylate)], viscosity index improvers [such as ethylene–propylene copolymers, hydrogenated diene–styrene copolymers, or a poly(alkyl methacrylate) (from C_4,C_{12},C_{18} methacrylates)]. Pour-point depressants lower the temperature at which the oil can be used (e.g., needed in engine oil in parked cars in the winter). They prevent wax from crystallizing out in a form that would interfere with the functioning of the engine. Viscosity index improvers[108] reduce the amount of the decrease in viscosity as the oil heats up. Extreme pressure lubricants, as needed for gears, use phosphorus or sulfur compounds, such as sulfurized polybutene and zinc dialkyldithiophosphates. As one part slides over another, these compounds produce lubricating layers of metal sulfides or phosphides on the metal surfaces.[109] The zinc dialkyldithiophosphate also functions as an inhibitor for corrosion and oxidation. Greases are made by adding 4–20% of a metal (usually lithium or calcium) fatty acid soap to an oil to thicken it. Inorganic thickeners, such as silica and montmorillonite (a clay), can also be used. A stable suspension of boric acid in motor oil can reduce friction by two thirds and result in a 4–5% reduction in fuel consumption.[110]

Used engine oils can be reused. Sometimes, all that is needed is removal of particles by filtration or centrifugation. In others, it may be necessary to remove volatile acids and water by heating, followed by treatment with sulfuric acid, then lime, and, sometimes, bleaching clay. A final distillation under vacuum completes the rerefining.[111] Another system uses propane at ambient temperature in a continuous process.[112] The additives and impurities precipitate and settle out, after which the propane is flashed off and the oil distilled. A new package of additives is usually required before reuse. A company in Florida reports an "electric mobile oil refiner" that can extend the life of motor oil "indefinitely."[113] Texaco (now Chevron) offers an on-site mobile reclamation and refortification service for lubricants.[114] Safety-Kleen has a similar service, but has discontinued its service on used antifreeze.

Synthetic oils[115] are used for more demanding applications, such as use at higher temperatures. Esters of pentaerythritol [$C(CH_2OCOR)_4$, where R averages about C_7] are often used at higher temperatures. Phosphate esters, such as tri(*o*-tolyl)phosphate, and silicone oils are also used at higher temperatures. Silicone grease is made by dissolving a lithium soap in a silicone oil. Solid lubricants, such as the layered materials graphite and molybdenum disulfide,[116] are used for extreme conditions of temperature or chemical resistance. Hollow nanoparticles of tungsten disulfide outperform these materials and may come into common use.[117] They are made by treating hollow nanoparticles of tungsten (VI) oxide with hydrogen sulfide and hydrogen at 800–850°C.

High-oleic vegetable oils can be used in place of mineral oils[118] (see Chapter 12). They can also be used as hydraulic fluids. They can be used as food-grade lubricants, are from renewable sources, and are biodegradable, which helps if a spill occurs. Additives are also required for these oils. Ethanolysis of such oils would produce ethyl oleate for use as fuel with the same diesel engine that uses the oils for lubricant and in its hydraulic lines. Oligomeric esters of oleic acid, formed by self-addition of the carboxylic acid groups to the double bond in the presence of a sulfuric acid catalyst, can also be used as biodegradable lubricants.[119] Some ionic liquids are lubricants (**13.25** Schematic).[120]

Cutting oils are used in metal working to cool the workpiece and the tool, to lubricate, and to remove the cuttings. These can be water-based, as in an emulsion of 2–5% mineral oil with added antifoaming agent, biocide, and corrosion inhibitor. They can also be based on polyalkylene glycols plus fatty acids, stabilizers, biocide, corrosion inhibitor, and such, without any mineral oil.[121] Triethanolamine salts of ricinoleic acid polymers or of the half esters of C_{36} dimer fatty acids are suitable corrosion inhibitors for water-based cutting oils.[122] The emulsions containing oil can be recycled on site in a unit that can be mobile.[123] The process involves

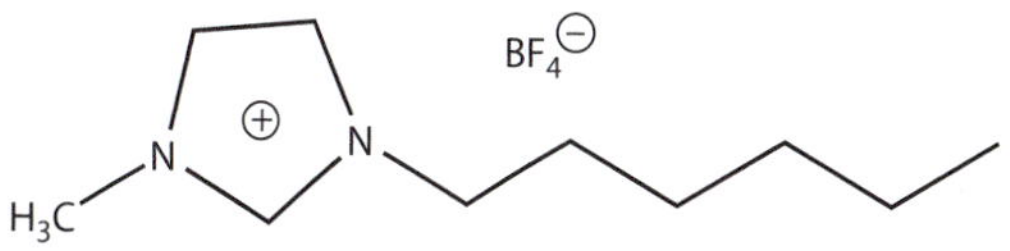

13.25 Schematic

filtration, pasteurizing, centrifugation, and addition of more additives. The use of hollow-fiber membranes to transport the oil and reject the water in cutting oils has also been studied.[124] Monsanto has developed a protein-based metal-working fluid.[125] It has been found that coating a tungsten carbide cutting tool with a nanolayer of tungsten plus tungsten disulfide eliminates the need for a cutting oil. It also allows the tool to be designed with sharper edges, which reduces the force on it by a factor of two to three.[126] This helps reduce the amount of heat generated. The TiAlN one mentioned above works even better since it is more heat-stable. (Laser machining also eliminates the need for a cutting oil.[127]) Diamond coatings can be used on tools also, but not for cutting steel, since the formation of iron carbides destroys the coatings.[128] The problem of a thermal mismatch (i.e., the coating expands at a different rate than the substrate on heating) that can produce delamination and cracking of the diamond film has been solved by creating a "three dimensionally thermally and compositionally graded interface."[129]

Grinding fluids may be a 2% emulsion of oil in water. Without such a coolant, enough heat may be generated to damage the workpiece. Studies are being conducted to eliminate the fluids by the use of liquid nitrogen as the coolant.[130] Another system uses minute amounts of a nonhazardous ester oil applied by a microlubrication system plus a jet of cold air.[131] Perhaps, the particles in grinding wheels can be redesigned with tungsten–tungsten disulfide coatings, mentioned earlier, for tools. Sharper particles would reduce the heat generated in grinding.

Air-cooled heat exchangers eliminate the need for water treatment and the need for disposal or recycling of solutions of antifreeze.[132] The Volkswagen "beetle" uses an air-cooled engine. Texaco (now Chevron) is marketing an "Extended Life Anti-Freeze/Coolant" for cars and light trucks that lasts for 100,000 miles. Engine coolants can be recycled profitably.[133]

13.4 INHIBITION OF CORROSION

Despite man's best efforts, many objects have their lifetimes shortened by corrosion.[134] One third of chemical plant failures in the United States are due to corrosion.[135] Some notable examples are the Three Mile Island nuclear reactor, a nuclear reactor in Ohio, the Alaska pipeline, and the Motiva (now Valero) oil refinery in Delaware City, DE. The cost of corrosion in the United States is $300 billion.[136] Corrosion should be monitored to be sure that it is not excessive.[137] Batelle has devised a system with a new layer between the base coat and the topcoat that fluoresces under a scanning device if corrosion has started. This can reveal it before it is visible to the naked eye.[138] The corrosion of metals is an electrochemical process (**13.26** Schematic).

$$M \rightleftharpoons M^{n+} + ne^-$$

13.26 Schematic

The forward reaction takes place at the anode, the reverse at the cathode. An electrically conducting pathway is needed to complete the circuit. Corrosion is often localized, forming pits and crevices wherever the passivating layer of oxide is broken by aggressive anions. Galvanic corrosion can occur when two dissimilar metals are joined, one becoming the anode and the other the cathode. Magnesium and zinc have been used as sacrificial anodes to protect steel from rusting in pipelines, ships, hot water heaters, and other equipment. A sprayed zinc anode has been used to protect the steel bars in concrete highways from the effects of deicing salts.[139] Environmental effects such as pH, nature and concentration of anions, concentration of oxygen, and temperature influence the corrosion process. Living organisms, such as bacteria, sometimes produce metabolites that cause corrosion.[140] This is also true of buildings and monuments made of stone, on which biofilms can increase the deposition of pollutants as well as produce metabolic acids.[141] Acid rain is particularly hard on limestone, marble, and sandstone, which is held together by carbonates. Genetically engineered aerobic bacteria form a protective film on surfaces in cooling water systems that cuts the corrosion of steel, stainless steel, copper, and brass by 35- to 40-fold, by thwarting the growth of the usual corrosion-causing bacteria.[142] This could be cheaper and less damaging to the environment than treatment with biocides.

The recovery of artifacts from the sunken ship, Titanic, shows the results of corrosion over 89 years at 0°C and pH 8.2 under 2.5 miles of water.[143] Aided by bacteria and other agents, steel had corroded to give FeO(OH), Fe_2O_3, and $FeCO_3$. Lead carbonate and lead sulfide were derived from the paint. Brass had corroded to Cu(OH)Cl. There was a slow conversion of silver to silver chloride.

Where possible, materials that will not corrode should be chosen.[144] These include glass, plastics,[145] and ceramics.[146] The first and last are brittle so that when they are used in the chemical process industry, it is usually as a lining for a steel vessel. New ceramic composites consisting of silicon carbide or alumina matrices reinforced with nicalon ceramic fibers have reduced brittleness and resist heat, erosion, and chemicals. (See the earlier discussion for improved ceramic fibers and less brittle ceramics.) Glass linings and piping may be used more in the food and pharmaceutical industries, which require high purity with a minimum of contamination. Plastics reinforced with glass or graphite fibers (i.e., composites) may achieve much wider use in the future for noncorroding bridge decks, utility poles, and other exposed surfaces. This would circumvent the problem of concrete deteriorating under the influence of deicing salts. The use of sand, instead of deicing salts, would also accomplish this. The use of the more expensive calcium–magnesium acetates in place of the sodium chloride might also do it with less corrosion, contamination of water supplies with sodium ion, and death of vegetation along roads. The use of recycled poly(ethylene terephthalate) to protect the reinforcing steel rods in concrete from corrosion has been demonstrated.[147] Zinc glucuronate migrates through the concrete to give 99.2% protection to the steel bars in concrete.[148] Great Barrier Systems sells CONDURx, which protects the steel bars by reducing the amount of water that penetrates the concrete.[149] Dow Chemical sells glass-filled polyurethane rods to reinforce concrete.[150] Polytetrafluoroethylene resists corrosion and chemical attack well, but creeps, so that it, too, is restricted to equipment linings.[151]

Titanium resists corrosion well, but is expensive.[152] Tantalum is even more resistant but is also expensive. Tantaline (Denmark) uses chemical vapor deposition to put a 50 μm layer of tantalum on the surface of stainless steel to improve its resistance to corrosion.[153] Stainless steel and other nickel alloys are cheaper for use in equipment and resist reagents such as formic acid.[154] The commonly used Monel metal is an alloy of 66.6% nickel and 31.5% copper. One of the best steel alloys contains 20% chromium, 29% nickel, 2.5% molybdenum, and 3.5% copper. Its rate of corrosion is 28 μm/yr. A stainless steel containing 6% molybdenum plus about 25% nickel and 20% chromium lost <1 mil/yr in saturated sodium chloride. Lasershot peening can be used to treat the surfaces of metals to improve resistance to corrosion and extend the lifetime up to five times.[155]

A variety of reagents are used to inhibit corrosion.[156] The shift from solvent-based to water-based paints, cleaning solutions, inks, cutting oils, and such will require increased use of such inhibitors. Corrosion can be inhibited in various ways. A reagent may complex with the metal surface and alter the redox potential of the metal or prevent adsorption of aggressive ions. A passivating layer may be formed and stabilized. The diffusion of oxygen to the surface may be inhibited or the oxygen in the surrounding medium may be scavenged. The surface may also be isolated by application of a protective coating of another metal or a polymer.

Chromate ion is an anodic inhibitor for steel that passivates the surface. Because chromate is toxic, it is being replaced[157] with molybdates. Chromate works in the absence of oxygen. The replacements, molybdate and tungstate, require oxygen to be effective. Cerium and magnesium[158] compounds are alternatives for chromate for the corrosion protection of aluminum. Alkylphosphonic acids perform as well as chromate and have superior adhesion.[159] Silanes, such as 1,2-ethylenebis(triethoxysilane), are also being studied as replacements for chromate.[160] A trimethylsilane plasma has been used to put a base coat on aluminum with no need for chromium.[161] Layer-by-layer deposition of MCM 22/silica composite films gave corrosion resistance to aluminum comparable to chromate.[162] Corrosion-inhibiting pigments may contain chromate or lead.[163] Red lead, Pb_3O_4, was a popular pigment in corrosion-inhibiting primers for steel. These are being replaced by calcium and zinc molybdates or phosphomolybdates. Selective precipitation of calcium or magnesium carbonates on cathodic areas or formation of phosphate films restricts access of oxygen to the metal surface. Hydrazine and sodium sulfite are used to scavenge the oxygen from boiler feedwater.[164]

Organic inhibitors may absorb on the whole surface. In hydrophobic fatty amines, the amine coordinates with the metal and the long carbon chains stick out from the surface. Carboxymethylated fatty amines (i.e., *N*-substituted glycines) chelate the metal on the surface. If a chelating agent forms an insoluble surface coating, it may be a good corrosion inhibitor. If it forms a soluble chelate, corrosion may be enhanced. Alkylcatechols also inhibit the corrosion of steel. [Catechol forms strong chelates with Fe (III).] Benzotriazole protects copper from corrosion by formation of a copper (I) chelate. Ethynylcyclohexanol absorbs to steel and then polymerizes (**13.27** Schematic).

Volatile corrosion inhibitors can be used with objects stored in closed containers, such as tools during shipment. A typical one is a dicyclohexylamine salt that dissociates to produce the amine (**13.28** Schematic).

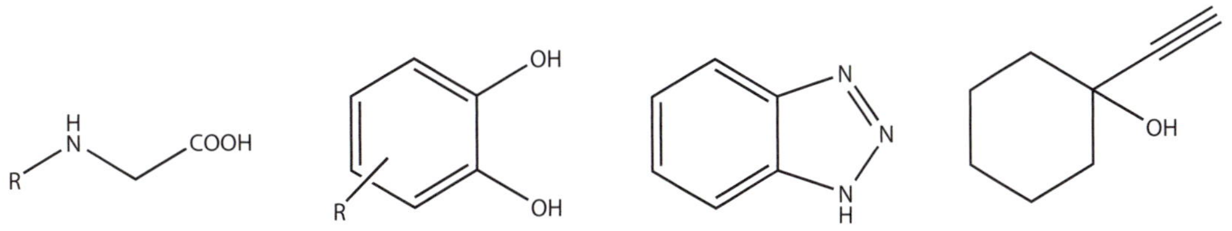

13.27 Schematic

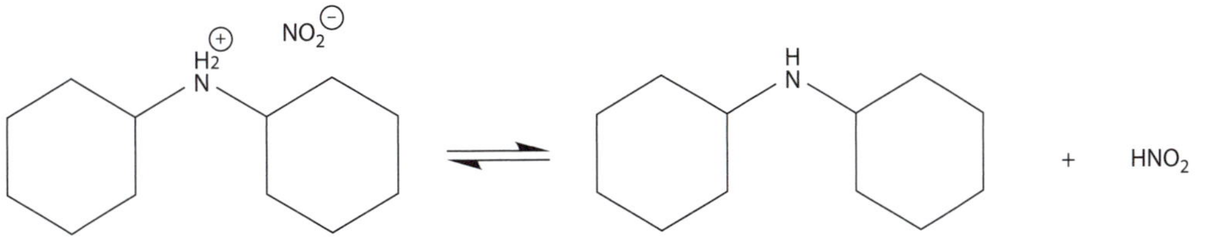

13.28 Schematic

Low molecular weight copolymers of styrene with maleic or acrylic acid have been used as their *N*,*N*-dimethylethanolamine salts to prevent the corrosion of zinc and aluminum in coatings applied from water.[165] A mixture of styrene and 4-fluorophenylmaleimide polymerizes spontaneously on aluminum, iron, and copper to provide corrosion protection.[166] The iron in a steel surface has been used in a redox polymerization of a mixture of styrene and 4-carboxyphenylmaleimide at room temperature to produce uniform pinhole-free films 1 μm thick (**13.29** Schematic).[167]

Corrosion-resistant binders for magnetic tapes have been prepared from diaminoquinones (**13.30** Schematic).[168] This diol was mixed with other diols and then reacted with diisocyanates to form polyurethanes.

The electrically conducting polymers, polyaniline and polypyrrole, prevent corrosion of steel.[169] One of the best formulations uses polyaniline with zinc nitrate, which is then covered with an epoxy resin topcoat. Polyorganosiloxanes have been grafted to starch using the sol–gel method with alkoxysilanes. These materials have been complexed with cerium ions to provide corrosion protection for aircraft, which is expected to be 50% cheaper than conventional coatings.[170]

Chemical process plants often use epoxy, acrylic, polyurethane, and other coatings to prevent corrosion.[171] Food cans often have epoxy resin linings. A coating made by the plasma polymerization of a 1:1 perfluorobutane–hydrogen mixture protected copper from aqueous corrosion.[172] Auto makers often use electrogalvanized steel sheet with a 4- to 14-μm coating of zinc.[173] Automobile frames are given a phosphatizing treatment before painting to help protect against road salts. This produces a surface of $Zn_3(PO_4)_2$, $Zn_2Fe(PO_4)_2$, and $Fe_3(PO_4)_2$. Zinc phosphate has been used as a pigment in vinyl paints to replace zinc chromate for the prevention of corrosion of steel.[174] Phenylphosphonic acid has been used (without zinc) in solvent-borne paints for *in situ* phosphatizing.[175] Steel coated by the polymerization of trimethylsilane in an argon–hydrogen plasma outperformed the current phosphated galvanized steel system.[176] If this can be done for a whole automobile, it will reduce the problem of contamination of wastewater with zinc and phosphate.

The hydroxyl groups on the surface of steel have been treated with 3-aminopropyltrimethoxysilane to produce an amine surface for reaction into an epoxy resin for corrosion protection.[177] Terpolymers of maleic anhydride, styrene, and C_4–C_{18} alkyl acrylates have been used as corrosion inhibitors for aluminum flake pigments.[178] (The anhydride may react with hydroxyl groups on the surface.) The use of lamellar mica in organic waterborne coatings for steel retards corrosion by limiting the ingress of water.[179] This works even better if the lamellar mica is coated with iron (III) oxide.[180]

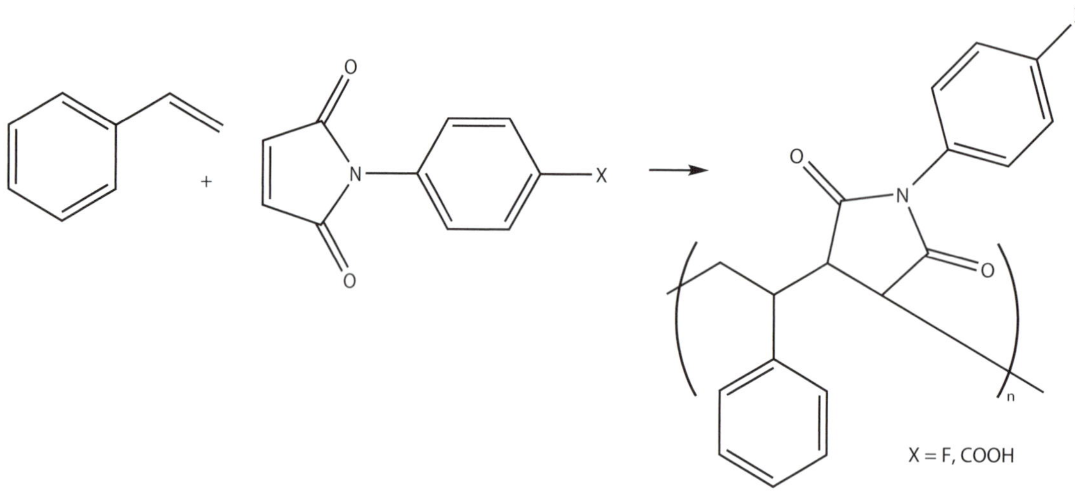

X = F, COOH

13.29 Schematic

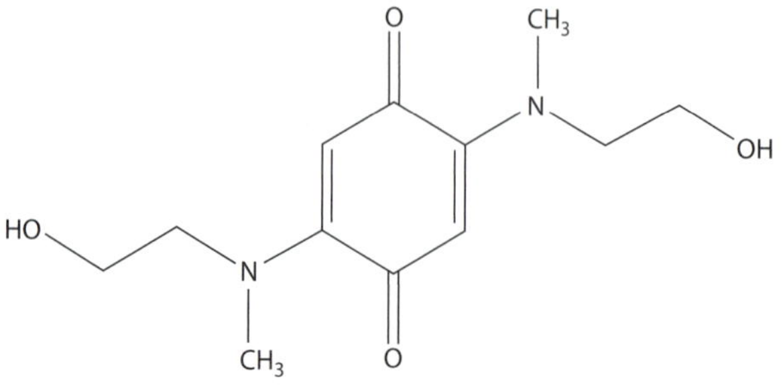

13.30 Schematic

Porcelain enamel is also used to protect steel. Kitchen pots of this material were in common use before the widespread adoption of aluminum and stainless steel pots. The main problem with enameled steel is that the coatings can chip on impact. Coatings of Portland cement are used to cover steel pipes. The use of electroplating, flame spraying, explosion cladding,[181] hot-dipping, chemical vapor deposition, ion implantation, and others to provide surfaces of metals more resistant to corrosion was covered in Chapter 4.

A coating of the zeolite SAPO 11 put on in an ionic liquid with microwaves can protect aluminum without any chromate.[182]

Films of oxides can protect underlying metals. Magnesium and aluminum form surfaces that protect them in this way. To prevent the formation of a dull powdery surface, some aluminum surfaces, such as those for use on the exteriors of buildings, are anodized electrochemically in sulfuric acid, which produces a shiny film of alumina containing some sulfate and water. It is no longer necessary to use chromate in anodizing.[183] The anodized surface of aluminum can be protected by the addition of chelating agents, such as 8-hydroxyquinoline and benzotriazoles, in a sealing coating of isostearic acid.

Cooling water treatment[184] involves not only corrosion control but also prevention of deposition of scale in pipes. Polyaspartic acid (prepared as in Chapter 12) can be used as a dispersant and inhibitor of mineral scale.[185] Magnetic fields applied outside the pipes can reduce calcium carbonate buildup or produce a softer less tenacious scale of aragonite instead of the usual crystal form, calcite.[186] A pulsed field can keep some of the crystals suspended instead of depositing. Magnetic fields also do this for carbon scale on stainless steel equipment in oil refineries and for sodium fluorosilicate scale in the manufacture of phosphoric acid. This should be simpler and cheaper than adding a chemical to do the job.

13.5 MENDING

Mending and patching can make many items last longer so that they do not need to be replaced as soon. The chemistry may have to be different when this has to be done at room temperature on something that was done at elevated temperature in the factory. A scratch or chip in the baked-on finish on a car or an appliance may have to be touched up with a different coating dissolved in a solvent. Epoxy body solder for auto body repairs has to cure at room temperature. The challenge, of course, is to make the repair as strong and durable as the original material, and to do so even if the surface preparation has been less than perfect. Plastic wood and patching plaster are familiar to many persons. Self-healing polymers have been devised.[187] An epoxy resin contained a ruthenium ring-opening catalyst and microcapsules of dicyclopentadiene, which polymerized when broken by a crack.[188] Another contained a microencapsulated mercaptan and a microencapsulated glycidyl ester (**13.31** Schematic).[189]

$C(CH_2OCOCH_2CH_2SH)_4$ + [diglycidyl tetrahydrophthalate] → Healed polymer

13.31 Schematic

The iron-on patch was a big step forward in repairing clothing. The hot-melt adhesive used on it must adhere well to the fabric initially and through repeated laundering or dry cleaning, must not soften in the clothes drier, must not be damaged by repeated flexing in use, and must be applicable with a home iron under conditions that may not be well controlled. One material that is used is a terpolyamide composed of 6, 6/36, and 6/12 nylons that soften at 100–150°C.[190] (The numbers refer to the number of carbon atoms in the ingredients, with those in the diamine being given first.) Such hot-melt adhesives are also available as fusible webs that can be used to reinforce fabrics.[191] Although these adhesives are not altered chemically when they are applied, it should be possible to develop some that will cure to tougher polymers on application. A mixture of microencapsulated diepoxide and curing agent would cure when the capsules were ruptured by heat or pressure. Some adhesives, such as that sold for installing Velcro fasteners, cure by evaporation of solvent. Velcro fasteners on stickyback tape (that contains no solvent) are now available commercially. "Shoe Goo II" applied to worn spots on shoes can extend the life of shoes appreciably. It appears to be a thick solution of a rubber in a mixture of propyl acetate and petroleum distillate. Similar solutions of polymers are available for mending broken china and other items. Adhesive tapes on release paper can be removed and used to hem garments (i.e., to replace sewing). Glue sticks are also available for trimming, and putting in hems, zippers, and what not. Some of these techniques should be applicable to putting replacement liners in pant pockets.

The weak links in clothing include knees, elbows, pockets, and zippers. Ideally, these should be made more durable in the first place. Zippers with plastic (properly stabilized) teeth large enough so as not to catch on the adjacent fabric, on a strong fabric backing, should not wear out before the rest of the garment. Stronger and thicker, if necessary, cloth could be used for the pockets. Areas of wear such as knees, elbows, and seats could be reinforced with strong fabric. When the sharp points of shirt collars are badly worn, the collars can be removed and the shirts worn without them. Clothing that wears out beyond the ability to patch it can sometimes be used for other purposes.[192] Long pants with worn-out knees can be converted to shorts. A long-sleeved sweater with worn-out elbows can be converted to a short-sleeved one. A dress can be cut off to a blouse. Worn-out sweaters can be converted to mittens. Portions of one worn-out garment that are still good can be used to patch another. Sections of adult clothing that are still good can be converted to clothing for children. In earlier years, scraps of clothing went into patchwork quilts or braided rugs. Old rags are also useful for cleaning around the house.

13.6 MISCELLANEOUS

The biocides needed to protect wood and many other items were covered in Chapter 3. Flame retardants[193] are sometimes used in fabrics and in plastics. In one sense, they may be thought of as compounds that make objects last longer. Typically, they are compounds that contain high levels of chlorine, bromine, or phosphorus. Antimony oxide is often used with the halogen compounds as a synergist. Hydrated aluminum hydroxide is also in common use. To be effective, these materials have to be used at such high levels, often 10–40%, that the physical properties of the polymer are lowered appreciably. The nanocomposites of clays in polymers mentioned earlier burn less readily than the unfilled polymers, without losing strength from the filler. The use of halogen compounds conflicts with efforts to reduce their use, as discussed in Chapter 3.

The use of polymers that are inherently resistant to ignition seems preferable to trying to protect polymers, such as polyolefins, that are much more flammable. Aromatic polyamides and phenol–formaldehyde resins resist ignition and tend to char in a flame. They have been used in fire suits. Efforts should be made to minimize possible sources of ignition. This includes removal of smoking materials, charcoal lighter fluid, candles, paint thinner, kerosene heaters, or others from the household. An electric stove is less likely to ignite clothing than a gas stove. A screen in front of the fireplace also helps. Smoke detectors and automatic sprinkler systems also help and are required by law in many localities.

13.7 THE FUTURE

Chemistry can provide stronger, more durable materials that will last longer. Studies of the ways in which polymers and other materials degrade and wear out may lead to new ways of intercepting the processes. This may mean the design of new stabilizer systems, new methods of construction, or new polymers.

The problem of overconsumption by the materials-oriented societies of industrialized nations is a difficult one that must be solved for a sustainable future.[194] A possible approach is to make the real cost of an item apparent to the purchaser at the time of purchase. This might encourage the purchaser to buy the item based on the lowest cost per unit of performance instead of the lowest initial cost. The label might contain items such as cost of the container, cost of the product, cost for each use, expected lifetime, cost per month or year, amount of use to be expected before resharpening is needed, and more. This would show that the cost of an item packaged in a throwaway container is higher than that of one packaged in a reusable container. A customer would have a choice of paying 10 cents for a throwaway glass bottle or 1 cent per use for a refillable glass bottle. A saw of poor steel that had to be sharpened every few months would be less desirable than one of good steel that lasted 10 years before resharpening was needed. A label-rating system might also be used. This might be a list of features of good construction with those in the item for purchase being checked. The presence of dovetailed joints on drawers in furniture or double seams on clothing are examples.

There are examples of such a required rating system in use today, at least in the United States. New cars are required to have the expected miles per gallon of fuel posted on them. Electrical appliances, such as hot water heaters, are required to have energy-efficiency ratings on them. Foods in the grocery store must be labeled with the price per unit weight as well as the price of the package. The source of the ratings might have to be a government agency, such as the U.S. National Institute of Standards and Technology. Private testing laboratories are also suitable, as long as their ratings are thorough and objective. Underwriters Laboratories does it for electrical equipment. Consumers' Union does it for many consumer goods. Materials scientists may have to help in devising additional accelerated tests, if none exist today. Frequently, the standard methods already certified by the American Society for Testing and Materials will suffice.

The jobs that will be lost in the manufacture of throwaway items will be counterbalanced by an increased number of jobs in maintenance and repair. Many items are discarded today because it is cheaper to buy a new one than to repair the old one. This calls for research on new automated methods of repair as well as on new designs that make repair simple. Increasing the cost of disposal may also help. (See Chapter 17 for more on environmental economics and Chapter 18 for greening of consumers.)

REFERENCES

1. (a) M.E. Eberhart, *Why Things Break*, Harmony Books, New York, 2003; (b) D.J. Wulpi, *Understanding How Components Fail*, 2nd ed., ASM International, Materials Park, OH, 1999; (c) A.F. Liu, *Mechanics and Mechanisms of Fracture: An Introduction*, ASM International, Materials Park, OH, 2005.
2. S.C. Temin, *Encyclopedia of Polymer Science and Engineering*, 2nd ed., Wiley, New York, 1985, *1*, 570.
3. E. Kintisch, *Science*, 2005, *310*, 953.
4. (a) M.P. Stevens, *J. Chem. Educ.*, 1993, *70*, 445, 535, 713; (b) R. Gachter and H. Muller, *Plastics Additives Handbook*, 4th ed., Hanser–Gardner, Cincinnati, OH, 1993; (c) S. Al-Malaiba, A. Golovoy, and C.A. Wilbie, eds, *Chemistry and Technology of Polymer Additives*, Blackwell Science, Oxford, 1999.
5. P.A. Lovell, *Trends Polym. Sci.*, 1996, *4*, 264.
6. X. Zhang, G. Wei, Y. Liu, J. Gao, Y. Zhu, Z. Song, F. Huang, M. Zhang and J. Qiao, *Macromol. Symp.*, 2003, *193*, 261.

7. (a) U. Goerl, A. Hunsche, A. Mueller, and H.G. Koban, *Rubber Chem. Technol.*, 1997, *70*, 608; (b) A.S. Hashim, B. Azahari, Y. Ikeda, and S. Kohjiya, *Rubber Chem. Technol.*, 1998, *71*, 289; (c) J.T. Byers, *Rubber World*, 1998, *218*(6), 38, 40, 44, 46.
8. (a) W. Brostow and R.D. Corneliussen, *Failure of Plastics*, Hanser–Gardner, Cincinnati, OH, 1986; (b) W. Schnabel, *Polymer Degradation—Principles and Practical Applications*, Hanser–Gardner, Cincinnati, OH, 1982; (c) R.L. Clough, N.C. Billingham, and K.T. Gillen, eds, *Polymer Durability—Degradation, Stabilization and Lifetime Prediction*, ACS Adv. Chem. Ser. 249, Washington, DC, 1996; (d) N.S. Allen, M. Edge, T. Corrales, A. Childs, C.M. Liauw, F. Catalina, C. Peinado, A. Minihan, and D. Aldcroft, *Polym. Degrad. Stabil.*, 1998, *61*, 183; (e) A. Rivaton, J.-L. Gardette, B. Mailhot, and S. Morlat-Therlas, *Macromol. Symp.*, 2005, *225*, 129–146.
9. (a) S.C. Stinson, *Chem. Eng. News*, Sept. 9, 1996, 34; (b) A. Quye, *Chem. Ind.* (*Lond.*), 1998, 599; (c) Y. Shashoua, *Macromol. Symp.*, 2006, *238*, 67.
10. E. Vassallo, L. Laguardia, D. Ricci, and G. Bonizzoni, *Macromol. Symp.*, 2006, *238*, 46.
11. (a) G. Scott, *Trends Polym. Sci.*, 1997, *5*, 361; (b) G. Scott, *Macromol. Symp.*, 199, *144*, 113.
12. L. Zan, W. Fa, and S. Wang, *Environ. Sci. Technol.*, 2006, *40*, 1681.
13. T. Daniel, M.S. Hirsch, K. McClelland, A.S. Ross, and R.S. Williams, *JCT-Coatings Technol.*, 2004, *1*(9), 42.
14. (a) B. Lanska, *Angew. Makromol. Chem.*, 1997, *252*, 139; (b) P.N. Thanki and R.P. Singh, *J. Macromol. Sci. Rev. Macromol. Chem.*, 1998, *C38*, 595.
15. (a) M.P. Stevens, *J. Chem. Educ.*, 1993, *70*, 535; (b) M. Dexter, *Kirk–Othmer Encyclopedia of Chemical Technology*, 4th ed., Wiley, New York, 1992, *3*, 424; (c) R.L. Clough, N.C. Billingham, and K.T. Gillen, eds, *Polymer Durability: Degradation, Stabilization and Lifetime Prediction*, ACS Adv. Chem. Ser. 249, Washington, DC, 1996; (d) G.E. Zaikov, ed., *Degradation and Stabilization of Polymers – Theory and Practice*, Nova Science, Commack, NY, 1995; (e) S.H. Hamid, M.B. Amin, and A.G. Maadhah, eds, *Handbook of Polymer Degradation and Stabilization*, Dekker, New York, 1992; (f) N.S. Allen and M. Edge, *Fundamentals of Polymer Degradation and Stabilization*, Elsevier, London, 1992; (g) G. Scott, *Mechanism of Polymer Degradation and Stabilization*, Elsevier, London, 1990; *Atmospheric Oxidation and Antioxidants*, Elsevier, Amsterdam, 1993; (h) J. Pospisil and S. Nespurek, *Polym. Degrad. Stabil.*, 1995, *49*, 99; *Macromol. Symp.*, 1997, *115*, 143; (i) H. Zweifel, *Macromol. Symp.*, 1997, *115*, 181; (j) W.D. Habicher, I. Bauer, K. Scheim, C. Rautenberg, A. Lossack, and K. Yamaguchi, *Macromol. Symp.*, 1997, *115*, 93; (k) J.A. Kuczkowski, *Rubber World*, 1995, *212*(5), 19; (l) G. Scott, *Antioxidants in Science, Technology, Medicine and Nutrition*, Albion, Chichester, 1997; (m) G. Pritchard, ed., *Plastics Additives: An A-Z Reference*, Chapman & Hall, New York, 1998; (n) M. Bolgar, J. Hubball, S. Meronek, and J. Groeger, *Handbook for Chemical Analysis of Plastic and Polymer Additives*, CRC Press, Boca Raton, FL, 2008.
16. (a) R. Pfaendner, H. Herbst, K. Hoffmann, and F. Sitek, *Angew. Makromol. Chem.*, 1995, *232*, 193; (b) J. Pospisil, F.A. Sitek, and R. Pfaendner, *Polym. Degrad. Stabil.*, 1995, *48*, 351.
17. E.T. Denisov and V.V. Azatyan, *Inhibition of Chain Reactions*, Gordon and Breach, London, 2000.
18. G. Scott, *Polym. Degrad. Stabil.*, 1995, *48*, 315.
19. F. Gugumus, *Polym. Degrad. Stab.*, 1998, *62*, 403.
20. (a) Y.A. Shlyapnikov, S.G. Kiryushkin, and A.P. Mar' in, *Antioxidative Stability of Polymers*, Taylor & Francis, London, 1996; (b) D.W. Hairston, *Chem. Eng.*, 1996, *103*(5), 71; (c) E.T. Denisov and I.B. Afanasev, *Oxidation and Antioxidants in Organic Chemistry and Biochemistry*, CRC Press, Boca Raton, FL, 2005.
21. (a) B. Lanska, *Polym. Degrad. Stabil.*, 1996, *53*, 89, 99; (b) J. Pospisil, *Adv. Polym. Sci.*, 1995, *124*, 87.
22. R.P. Lattimer, C.K. Rhee, and R.W. Layer, *Kirk–Othmer Encyclopedia of Chemical Technology*, 4th ed., Wiley, New York, 1992, *3*, 448.
23. M. Wijtmans, D.A. Pratt, L. Valgimigli, G.A. diLabio, G.F. Pedulli, and N.A. Porter, *Angew. Chem. Int. Ed.*, 2003, *42*, 4370.
24. (a) M. Frenette, P.D. MacLean, L.R.C. Barclay, and J.C. Scaiano, *J. Am. Chem. Soc.*, 2006, *128*, 16432; (b) H.-G. Korth, *Angew. Chem. Int. Ed.*, 2007, *46*, 5274.
25. W.D. Habicher, I. Bauer, and J. Pospisil, *Macromol. Symp.*, 2005, *225*, 147.
26. A. Gunther, T. Konig, W.D. Habicher, and K. Schwetlick, *Polym. Degrad. Stabil.*, 1997, *55*, 209.
27. M.S. Rabello and J.R. White, *J. Appl. Polym. Sci.*, 1997, *64*, 2505.
28. D. Vaillant, J. Lacoste, and J. Lemaire, *J. Appl. Polym. Sci.*, 1997, *65*, 609.
29. (a) J.F. Rabek, *Photodegradation of Polymer*, Springer, Heidelberg, 1996; (b) J.F. Rabek, *Polymer Photodegradation—Mechanisms and Experimental Methods*, Chapman & Hall, London, 1995; (c) R.L. Clough and S.W. Shalaby, eds, *Irradiation of Polymers: Fundamentals and Technological Applications*, ACS Symp. 620, Washington, DC, 1996; (d) V.Y. Shlyapintokh, *Photochemical Conversion and Stabilization of Polymers*, Hanser–Gardner, Cincinnati, OH, 1985; (e) A. Faucitano, A. Buttafava, G. Camino, and L. Greci, *Trends Polym. Sci.*, 1996, *4*(3), 92; (f) F. Gugumus, *Polym. Degrad. Stabil.*, 1994, *44*, 273, 299; 1995, *50*, 101; (g) H.B. Olayan, H.S. Hamid, and E.D. Owen, *J. Macromol. Sci. Rev. Macromol. Chem.*, 1996, *C36*, 671; (h) A. Tidjani, *J. App. Polym. Sci.*, 1997, *64*, 2497; (i) C. Decker, Biry, and K. Zahouily, *Polym. Degrad. Stabil.*, 1995, *49*, 111; (j) K. Kikkawa, *Polym. Degrad. Stabil.*, 1995, *49*, 135; (k) A. Rivaton, *Polym. Degrad. Stabil.*, 1995, *49*, 163; (l) M.C. Celina and R.A. Assink, *Polymer Durability and Radiation Effects*, ACS Symp., 978, Oxford University Press, New York, 2007.
30. (a) B.H. Cumpston and K.F. Jensen, *Trends Polym. Sci.*, 1996, *4*(5), 151; (b) L. Peeters and H.J. Giese, *Trends Polym. Sci.*, 1997, *5*(5), 161.
31. (a) J. Mwila, M. Miraftab, and A.R. Horrocks, *Polym. Degrad. Stabil.*, 1994, *44*, 351; (b) A.R. Horrocks and M. Liu, *Macromol. Symp.*, 2003, *202*, 199.
32. M. Zayat, P. Garcia-Parejo, and D. Levy, *Chem. Soc. Rev.*, 2007, *36*, 1270.
33. J.-J. Lin, M. Cuscurida, and H.G. Waddill, *Ind. Eng. Chem. Res.*, 1997, *36*, 1944.
34. F. Gugumus, *Polym. Degrad. Stabil.*, 1998. *60*, 99, 119; 1999, *66*, 133.

35. J. Pospisil, W.D. Habicher, and S. Nespurek, *Macromol. Symp.*, 2001, *164*, 389.
36. R. Mani, N. Sarkar, R.P. Singh, and S. Sivaram, *J. Macromol. Sci. Pure Appl. Chem.*, 1996, *A33*, 1217.
37. J. Malik, G. Ligner, and L. Avar, *Polym. Degrad. Stabil.*, 1998, *60*, 205–213.
38. M. Kiguchi and P.D. Evans, *Polym. Degrad. Stabil.*, 1998, *61*, 33.
39. K. Schweltlick and W.D. Habicher, *Angew. Makromol. Chem.*, 1995, *232*, 239.
40. D.-J. Park, C.-S. Ha, and W.-J. Cho, *J. Appl. Polym. Sci.*, 1998, *67*, 1345.
41. C. Garrigues, A. Guyot, and V.H. Tran, *Polym. Degrad. Stabil.*, 1994, *45*, 103.
42. (a) J.T. Kumpulainen and J.T. Salonen, *Natural Antioxidants and Food Quality in Atherosclerosis and Cancer Prevention*, Royal Society of Chemistry, Cambridge, 1997; (b) F. Shahidi, ed., *Natural Antioxidants: Chemistry, Health Effects and Applications*, American Oil Chemists Society, Champaign, IL, 1996; (c) G. Scott, *Chem. Br.*, 1995, *31*(11), 879; (d) I. Kubo, *Chemtech*, 1999, *29*(8), 37; (e) E. Haslam, *Practical Polyphenolics: from Structure to Molecular Recognition and Physiological Action*, Cambridge University Press, Cambridge, 1998.
43. L.R.C. Barclay, C.D. Edwards, K. Mukai, Y. Egawa, and T. Nishi, *J. Org. Chem.*, 1995, *60*, 2739.
44. (a) G.O. Igile, W. Oleszek, M. Jurzysta, S. Burda, M. Fafunso, and A.A. Fasanmade, *J. Agric. Food Chem.*, 1994, *42*, 2445; (b) K. Samejima, K. Kanazawa, H. Ashida, and G.-I. Danno, *J. Agric. Food Chem.*, 1995, *43*, 410; (c) H. Yamamoto, J. Sakakibara, A. Nagatsu, and K. Sekiya, *J. Agric. Food Chem.*, 1998, *46*, 862.
45. (a) J.A. Vinson, J. Jang, Y.A. Dabbagh, M.M. Serry, and S. Cai, *J. Agric. Food Chem.*, 1995, *43*, 2798; (b) J.A. Vinson, Y.A. Dabbagh, M.M. Serry, and J. Jang, *J. Agric. Food Chem.*, 1995, *43*, 2800; (c) G.-C. Yen and H.-Y. Chen, *J. Agric. Food Chem.*, 1995, *43*, 27; (d) U.N. Wanasundara and F. Shahidi, *J. Am. Oil Chem. Soc.*, 1996, *73*, 1183.; (e) S.-W. Huang and E.N. Frankel, *J. Agric. Food Chem.*, 1997, *45*, 3033; (f) Y. He and F. Shahidi, *J. Agric. Food Chem.*, 1997, *45*, 4262; (g) G. Ondrey, *Chem. Eng.*, 2003, *110*(4), 21; (h) M. Kurisawa, J.E. Chung, H. Uyama, and S. Kobayashi, *Chem. Commun.*, 2004, 94; (i) B. Zhou and Z.-L. Liu, *Pure Appl. Chem.*, 2005, *77*, 1887.
46. Anon., *Tufts University Health and Nutrition Letter*, Medford, MA, 1997, *15*(6), 2.
47. (a) S.J. Risch and C.-T. Ho, eds, *Spices: Flavor Chemistry and Antioxidant Properties*, ACS Symp. 660, Washington, DC, 1997; (b) O.I. Aruoma, *Int. News Fats Oils Relat. Mater.*, 1997, *8*, 1236.
48. D.A. Pearson, E.N. Frankel, R. Aeschbach, and J.B. German, *J. Agric. Food Chem.*, 1997, *45*, 578.
49. N. Tipsrisukond, L.N. Fernando, and A.D. Clarke, *J. Agric. Food Chem.*, 1998, *46*, 4329.
50. (a) M. Watanabe, *J. Agric. Food Chem.*, 1997, *45*, 839; (b) M. Watanabe, Y. Ohshita, and T. Tsushida, *J. Agric. Food Chem.*, 1997, *45*, 1039.
51. P.-D. Duh and G.-C. Yen, *J. Am. Oil Chem. Soc.*, 1997, *74*, 745.
52. (a) C. Ganthavorn and J.S. Hughes, *J. Am. Oil Chem. Soc.*, 1997, *74*, 1025; (b) M.T. Satue–Gracia, M. Heinonen, and E.N. Frankel, *J. Agric. Food Chem.*, 1997, *45*, 3362; (c) G.O. Adegoke and A.G.G. Krishna, *J. Am. Oil Chem. Soc.*, 1998, *75*, 1047; (d) T. Miyake and T. Shibamoto, *J. Agric. Food Chem.*, 1997, *45*, 1819; (e) D.L. Madhavi, M.A.L. Smith, and G. Mitiku, *J. Agric. Food Chem.*, 1997, *45*, 1506; (f) J.H. Chen and C.-T. Ho, *J. Agric. Food Chem.*, 1997, *45*, 2374.
53. G.R. McDonald, A.L. Hudson, S.M.J. Dunn, H. You, G.E. Baker, R.M. Whittal, J.W. Martin, A. Jha, D.E. Emondson, and A. Holt, *Science*, 2008, *322*, 917.
54. (a) E.R. Booser, ed., *Handbook of Lubrication and Tribology*, CRC Press, Boca Raton, FL, 1994; (b) E. Rabinowicz, *Friction and Wear of Materials*, 2nd ed., Wiley, New York, 1995; (c) K.C. Ludema, Friction, *Wear and Lubrication–A Textbook in Tribology*, CRC Press, Boca Raton, FL, 1996; (d) *ASM Handbook 18, Friction, Lubrication and Wear Technology*, ASM International, Materials Park, OH, 1992; (e) D. Klamann, *Ullmann's Encyclopedia Industrial Chemistry*, 5th ed., VCH, Weinheim, 1990, *A15*, 423–518; (f) E.R. Booser, *Kirk–Othmer Encyclopedia of Chemical Technology*, 4th ed., Wiley, New York, 1995, *15*, 463; (g) R. Glyde, *Chem. Br.*, 1997, *37*(7), 39; (h) T. Mang and W. Dresell, eds, *Lubricants and Lubrication*, Wiley-VCH, Weinheim, 2001; (i) A. Shanley, *Chem. Eng.*, 2000, *107*(7), 33; (j) A. Fisher and K. Robzin, *Wear and Wear Protection*, Wiley-VCH, Weinheim, 2009.
55. G. Ondrey, *Chem. Eng.*, 2002, *109*(11), 35.
56. (a) E. Bez, *R & D (Cahners)*, 1998, *40*(7), 73; (b) Anon., *R & D (Cahners)*, 1998, *40*(1), 87; *40*(11), 18, 62; (c) V. Comello, *R & D (Cahners)*, 1999, *41*(5), 59; (d) C. Crabb, *Chem. Eng.*, 2000, *107*(2), 37.
57. G.H. Phelps, *Chem. Eng.*, 2009, *115*(1), 46.
58. Z.-Z. Zhang, Q.-J. Xue, W.-M. Liu, and W.-C. Shen, *J. Appl. Polym. Sci.*, 1998, *70*, 1455; 1999, *72*, 361, 751; 1999, *73*, 2611; 1999, *74*, 797.
59. Q.-J. Xue, Z.Z. Zhang, W.-M. Liu, and W.-C. Shen, *J. Appl. Polym. Sci.*, 1998, *69*, 1393.
60. G. Parkinson, *Chem. Eng.*, 2003, *110*(1), 19.
61. (a) T. Bremner, D.J.T. Hill, M.I. Killeen, J.H. O'Donnell, P.J. Pomery, D. St. John, and A.K. Whittaker, *J. Appl. Polym. Sci.*, 1997, *65*, 939; (b) D.J.T. Hill, M.I. Killeen, J.H. O'Donnell, P.J. Pomery, D. St. John, and A.K. Whittaker, *J. Appl. Polym. Sci.*, 1996, *61*, 1757.
62. Q. Wang, Q. Xue, W. Shen, and J. Zhang, *J. Appl. Polym. Sci.*, 1998, *69*, 135.
63. Anon., (a) *R & D (Cahners)*, 1997, *39*(13), 72; 1998, *40*(10), 126; (b) D. Bradley, *Chem. Br.*, 1997, *33*(12), 18; (c) G. Ondrey, *Chem. Eng.*, 1997, *104*(11), 31.
64. M. Jacoby, *Chem. Eng. News*, September 3, 2001, 35.
65. B. Halford, *Chem. Eng. News*, August 29, 2005, 30.
66. M. Jacoby, *Chem. Eng. News*, Oct. 24, 2005, 51.
67. Anon., *Chem. Eng. News*, Sept. 18, 2000, 47.
68. R. Chattopadhyay, *Surface Wear: Analysis, Treatment and Prevention*, ASM International, Materials Park, OH, 2001.
69. X. Zou, X. Yi, and Z. Fang, *J. Appl. Polym. Sci.*, 1998, *70*, 1561.
70. J. Pavlinec, M. Lazar, and I. Janigova, *J. Macromol. Sci. Pure Appl. Chem.*, 1997, *A34*, 81.
71. (a) R. Dangani, *Chem. Eng. News*, Jan. 9, 1995, 24; (b) E.H. Lee, L.G.R. Rao, and L.K. Mansur, *Trends Polym. Sci.*, 1996, *4*, 229.
72. (a) R. Dagani, *Chem. Eng. News*, Dec. 15, 1997, 11; (b) R.F. Service, *Science*, 1997, *278*, 2057.

73. M.N.R. Ashfold and P.W. May, *Chem. Ind.* (*Lond.*), 1997, 505.
74. (a) Anon., *R & D* (*Cahners*), 1998, *40*(10), 180; (b) Diamonex Performance Products, Allentown, PA, August 1998.
75. Anon., *Chem. Eng. News*, Jan. 18, 2007, 48.
76. G. Ondrey, *Chem. Eng.*, 2004, *111*(2), 16.
77. L.J. Thorwart, *Chem. Eng. Prog.*, 1996, *92*(4), 80.
78. W.S. Rees, Jr., *Chemical Vapor Deposition of Nonmetals*, VCH, Weinheim, 1996.
79. (a) W.D. Sproul, *Science*, 1996, *273*, 889; (b) J.I. Brauman and P. Szuromi, *Science*, 1996, *273*, 855; (c) D.H. Lowndes, D. Geohegan, A.A. Puretzky, D.P. Norton, and C.M. Rouleau, *Science*, 1996, *273*, 898; (d) K. Legg and B. Taylor, *R & D* (*Cahners*), 1998, *40*(10), 4UK.
80. S. Veprek and M. Jilek, *Pure Appl. Chem.*, 2002, *74*, 475.
81. G. Ondrey, *Chem. Eng.*, 1999, *106*(13), 19.
82. G. Ondrey, *Chem. Eng.*, 2009, *115*(1). 17.
83. H.Y. Chung, M.B. Weinberger, J.B. Levine, A. Kavner, J.-M. Yang, S.H. Tolbert, and R.B. Kaner, *Science*, 2007, *316*, 436.
84. (a) D. Branagan and J.V. Burch, U.S. patent 6,767,419 (2004); (b) D. Branagan and Y. Tang, Composites Part A, *Appl. Sci. Eng.*, 2002, *33*(6), 855; (c) Idaho National Engineering and Environmental Laboratory, Idaho Falls, Idaho.
85. W. Schulz, *Chem. Eng. News*, Oct. 16, 2000, 39.
86. M. Jacoby, *Chem. Eng. News*, Mar. 15, 1999, 44; Dec. 4, 2000, 14; Jan. 15, 2007, 16.
87. R. Riedel, ed., *Handbook of Ceramic Hard Materials*, 2 vol., Wiley-VCH, Weinheim, 2000.
88. R. Dagani, *Chem. Eng. News*, June 7, 1999, 25.
89. (a) W.G. Clegg, *Science*, 1999, *286*, 1097; (b) M.P. Rao, A.J. Sanchez-Herencia, G.E. Beltz, R.M. McMeeking, and F.F. Lange, *Science*, 1999, *286*, 102.
90. S. Deville, E. Saiz, R.K. Nalla, and A.P. Tomsia, *Science*, 2006, *311*, 515.
91. E. Munch, M.E. Launey, D.H. Alsem, E. Salz, A.P. Tomisa, and R.O. Ritchie, *Science*, 2008, *322*, 1516.
92. L.J. Bonderer, A.R. Studart, and L.J. Gauckler, *Science*, 2008, *319*, 1069.
93. B.-N. Kim, K. Hiraga, K. Morita, and Y. Sakka, *Nature*, 2001, *413*, 288.
94. M.W. Barsoum and T. El-Raghy, *Am. Sci.*, 2001, *89*, 334.
95. G. Lawton, *Chem. Ind.* (*Lond.*), 2001, 174.
96. (a) A.B. Morgan, *Material Matters* (*Aldrich Chemical Co.*), Milwaukee, WI, 2007, *2*(1), 20; (b) B. Chen, J.R.G. Evans, H. Greenwell, P. Boulet, P.V. Coveney, A.A. Bowden, and A. Whiting, *Chem. Soc. Rev.*, 2008, *37*, 568; (c) A. Usuki and M. Kato, *Adv. Polym. Sci.*, 2005, *179*, 1; (d) S.S. Ray and M. Okamoto, *Prog. Polym Sci.*, 2003, *28*, 1539–1627; (e) L.A. Goettler, K.Y. Lee, and H. Thakkar, *J. Macromol. Sci. – Polym. Rev.*, 2007, *47*, 291.
97. (a) M. Joshi and B.S. Butola, *J. Macromol. Sci. – Polym. Rev.*, 2004, *C44*, 389; (b) J.E. Mark, *Acc. Chem. Res.*, 2006, *39*, 881.
98. S. Shelley, *Chem. Eng.*, 2003, *110*(1), 29; (b) J. Yang, Z. Zhanag, K. Friedrich, and A.K. Schwarb, *Macromol. Rapid Commun.*, 2007, *28*, 955.
99. W.K. Tredway, *Science*, 1998, *282*, 1275.
100. T. Ishikawa, S. Kajii, K. Matsunaga, T. Hogami, Y. Kohtoku, and T. Nagasawa, *Science*, 1998, *282*, 1295.
101. P. Baldus, M. Jansen, and D. Sporn, *Science*, 1999, *285*, 699.
102. G. Odian, *Principles of Polymerization*, 4th ed., Wiley-Interscience, Hoboken, N.J., 2004, 174.
103. L.G. Sneddon, M.G. LMirabelli, A.T. Lynch, P.J. Fazen, K. Su, and J.S. Beck, *Pure Appl. Chem.*, 1991, *63*, 407.
104. (a) Y. Waku, N. Nakagawa, T. Wakamoto, H.J. Ohtsubo, K. Shimizu, and Y. Kohtoku, *Nature*, 1997, *389*, 49; (b) E. Wilson, *Chem. Eng. News*, Sept. 8, 1997, 12.
105. D.J. Green, R. Tandon, and V.M. Sglawo, *Science*, 1999, *283*, 1295.
106. (a) Y. Tamarin, *Protective Coatings for Turbine Blades*, ASM International, Materials Park, OH, 2002; (b) N.P. Padture, M. Gell, and E.H. Jordan, *Science*, 2002, *296*, 280.
107. (a) S. Ritter, *Chem. Eng. News*, Mar. 13, 2006, 38; (b) A. Shanley, *Chem. Eng.*, 2000, *107*(7), 33.
108. M.K. Mishra and R.G. Saxton, *Chemtech*, 1995, *25*(4), 35.
109. G. Smith, *Chem. Br.*, 2000, *36*(4), 38.
110. M. Whitfield, *Chem. Ind.* (*Lond.*), Aug. 27, 2007, 10.
111. (a) D.W. Brinkman and J.R. Dickson, *Environ. Sci. Technol.*, 1995, *29*, 81; (b) J.R. Dickson and D. Wilkinson, *Environ. Sci. Technol.*, 1995, *29*, 87; (c) J.P. Martins, *Ind. Eng. Chem. Eng.*, 1998, *105*(8), 17.
112. G. Parkinson, *Chem. Eng.*, 1996, *103*(8), 19.
113. J. Johnson, *Environ. Sci. Technol.*, 1995, *29*, 72A.
114. Texaco Inc., *Environment, Health and Safety Review*, Harrison, NY, 1994.
115. (a) P.M. Morse, *Chem. Eng. News*, Sept. 7, 1998, 21; (b) L. Rudnick and R. Shubkin, *Synthetic Lubricants and High-Performance Functional Fluids*, Marcel Dekker, New York, 1999.
116. Anon., *Chem. Eng. News*, Sept. 18, 2000, 47.
117. (a) R. Dagani, *Chem. Eng. News*, June 23, 1997, 10; (b) L. Rapoport, S.Y. Bilik, Y. Feldman, M. Momyonfer, S.R. Cohen, and R. Tenne, *Nature*, 1997, *387*, 791.
118. (a) K. Lal, Preprints ACS *Div. Environ. Chem.*, 1994, *34*(2), 474; (b) Anon., *R & D (Cahners)*, 1996, *38*(13), 9; (c) D. Hariston, *Chem. Eng.*, 1994, *101*(8), 65; (d) G. Parkinson, *Chem. Eng.*, 1996, *103*(11), 21; (e) S.Z. Erhan and J.M. Perez, *Biobased Industrial Fluids and Lubricants*, Am. Oil Chem. Soc., Champaign, IL, 2002; (f) S. Boyde, *Green Chem.*, 2002, *4*, 293–307.
119. (a) T.A. Isbell, H.B. Frykman, T.P. Abbott, J.E. Lohr, and J.C. Drozd, *J. Am. Oil Chem. Soc.*, 1997, *74*, 473; (b) C. Cermak, *Agric. Res.* 2007, *55*(4), 23.
120. (a) C. Ye, W. Liu, and L. Yu, *Chem. Commun.*, 2001, 2244; (b) F. Zhou, Y. Liang, and W. Liu, *Chem. Soc. Rev.*, 209, 38, 2590.
121. D.S. Shukla and V.K. Jain, *Chemtech*, 1997, *27*(5), 32.
122. (a) S. Watanabe, H. Nakagawa, Y. Ohmori, T. Fujita, M. Sakamoto, and T. Haga, *J. Am. Oil Chem. Soc.*, 1996, *73*, 807; (b) S. Watanabe, Y. Ohmori, T. Fujita, and M. Sakamoto, *J. Am. Oil Chem. Soc.*, 1994, *71*, 1003.
123. A.R. Gavaskar, R.F. Olfenbuttel, and J.A. Jones, *U.S. Environmental Protection Agency*, EPA/600/SR-93/114, Cincinnati, OH, August 1993.
124. M.J. Semmens, *National Center for Clean Industrial and Treatment Technologies—Activities Report*, Houghton, MI, 10/95–12/96, 52.
125. J. Pelley, *Environ. Sci. Technol.*, 1997, *31*, 138A.
126. R. Komanduri, *Green Chemistry Conference*, Washington, DC, June 23, 1997.
127. Anon., *R & D* (*Cahners*), 1997, *39*(10), 41.

128. (a) R.H. Wentorf, Jr., *Kirk–Othmer Encyclopedia of Chemical Technology*, 4th ed., Wiley, New York, 1992, *4*, 1082; (b) K.E. Spear and J.P. Dismukes, *Synthetic Diamond–Emerging CVD Science and Technology*, Wiley, New York, 1994; (c) W. Lepkowski, *Chem. Eng. News*, Jan. 22, 1996, 7.
129. R.K. Singh, D.R. Gilbert, J. Fitz-Gerald, S. Harkness, and D.G. Lee, *Science*, 1996, *272*, 396.
130. S. Malkin, *National Environmental Technology for Waste Prevention Institute*, University of Massachusetts at Amherst, 1997.
131. S. Malkin, *N.E.T.I News* (*National Environmental Technology of Waste Prevention Institute, University of Massachusetts*), 1998, *1*(4).
132. R. Mukherjee, *Chem. Eng. Prog.*, 1997, *93*(2), 26.
133. A.R. Gavaskar, R.E. Olfenbuttel, and J.A. Jones, U.S. E.P.A./600/R-92/024; *Chem. Abstr.*, 1993, *119*, 228, 756.
134. (a) P.J. Moran and P.M. Natishan, *Kirk–Othmer Encyclopedia of Chemical Technology*, 4th ed., Wiley, New York, 1993, *7*, 548; (b) D.A. Jones, *Principles and Prevention of Corrosion*, Macmillan, New York, 1992; (c) F. Mansfeld, ed., *Corrosion Mechanisms*, Dekker, New York, 1987; (d) P. Marcus and J. Oudar, *Corrosion Mechanisms in Theory and Practice*, Dekker, New York, 1995; (e) *ASM Handbook 13, Corrosion*, ASM International, Materials Park, OH, 1987; (f) D. Talbot and J. Talbot, *Corrosion Science and Technology*, CRC Press, Boca Raton, FL, 1997; (g) P.E. Elliott, *Chem. Eng. Prog.*, 1998, *94*(5), 33; (h) T.W. Swaddle, *Inorganic Chemistry—An Industrial and Environmental Perspective*, Academic, San Diego, CA, 1996; (i) T.E. Graedel and C. Leygraf, *Atmospheric Corrosion*, Wiley, New York, 2000; (j) Z. Ahmad, *Principles of Corrosion Engineering and Corrosion Control*, Butterworth-Heineman, Burlington, MA, 2006; (k) J.R. Davis, *Corrosion: Understanding the Basics*, ASM International, Materials Park, OH 2000; (l) A.S. Khanna, *Introduction to High Temperature Oxidation and Corrosion*, ASM International, Materials Park, OH, 2002; (m) C. Shargay and C. Spurrell, *Chem. Eng.*, 2003, *110*(5), 42; (n) S.A. Bradford, *Corrosion Control*, 2nd ed., ASM International, Materials Park, OH, 2001; (o) A.H. Tullo, *Chem. Eng. News*, Sept. 17, 2007, 20; (p) R.W. Revie, *Corrosion and Corrosion Control*, 4th ed., Wiley, Hoboken, NJ, 2008.
135. C. Punckt, M. Bolscher, H.H. Rotermund, A.S. Mikhailov, L. Organ, N. Budiansky, J.R. Scully, and J.L. Hudson, *Science*, 2004, *305*, 1133.
136. R. Marshall, *Chem. Eng.*, 2008, *115*(8), 5.
137. R.D. Kane, *Chem. Eng.*, 2007, *114*(6), 34.
138. (a) Anon., *Chem. Eng. News*, Feb. 2, 2009, 16; (b) Anon., *Chem. Eng. News, Oct*. 26, 2009, 24.
139. (a) J.E. Bennett, *Chem. Eng. Prog.*, 1998, *94*(7), 77; (b) T.H. Lewis, Jr., *Chem. Eng. Prog.*, 1999, *95*(6), 55.
140. (a) Anon., *Chem. Eng. News*, Mar. 17, 1997, 29; (b) B.J. Little, R.I. Ray, and P.A. Wagner, *Chem. Eng. Prog.*, 1998, *94*(9), 51; (c) J. Jass and H.M. Lappin–Scott, *Chem. Ind.* (*Lond.*), 1997, 682; (d) S.J. Yuan, S.O. Pehkonen, Y.P. Ting, E.T. Kang, and K.G. Neoh, *Ind. Eng. Chem. Res.*, 2008, *47*, 3008.
141. (a) P. Young, *Environ. Sci. Technol.*, 1996, *30*, 206A; *New Sci.*, 1996, *152*(2054), 36; (b) G. Allen and J. Beavis, *Chem. Br.*, 1996, *32*(9), 24.
142. G. Parkinson, *Chem. Eng.*, 1998, *105*(11), 21.
143. (a) M. Freemantle, *Chem. Eng. News*, Oct. 17, 1994, 49; (b) P. Stoffyn–Egli and D.E. Buckley, *Chem. Br.*, 1995, *31*(7), 551.
144. (a) P. Schweizer, *Corrosion Resistance Tables*, 4th ed., Dekker, New York, 1995; (b) B. Craig and D. Anderson, *Handbook of Corrosion Data*, 2nd ed., ASM International, Materials Park, OH, 1995; (c) S.L. Chawla and R.K. Gupta, *Materials Selection for Corrosion Control*, ASM International, Materials Park, OH, 1993; (d) P. Schweizer, *Corrosion–Resistant Piping Systems*, Dekker, New York, 1994.
145. (a) P. Khaladkar, *Chem. Eng.*, 1995, *102*(10), 94; (b) G. Parkinson, *Chem. Eng.*, 1995, *102*(4), 115.
146. (a) J.K. Wessel, *Chem. Eng.*, 1996, *103*(10), 80; (b) N. Paxton, *Chem. Eng.*, 1996, *103*(10), 84.
147. F. Bourges–Fricoteaux, *J. Coat. Technol.*, 1998, *70*(884), 63.
148. C. Chandler, M. Kharshan, and A. Furman, *5th Green Chemistry Conference*, Washington, DC, June, 2001, 31.
149. M. Rouhi, *Chem. Eng. News*, Dec. 17, 2001, 53.
150. A.H. Tullo, *Chem. Eng. News*, Jan. 28, 2002, 22.
151. Anon., *Chem. Eng.*, 1997, *104*(2), 99.
152. J.E. Deily, *Chem. Eng. Prog.*, 1997, *93*(6), 50.
153. G. Ondrey, *Chem. Eng.*, 2008, *115*(11), 16.
154. (a) C.M. Schillmoller, *Chem. Eng. Prog.*, 1997, *93*(2), 66; (b) E.L. Hibner and D.S. Fende, *Chem. Eng. Prog.*, 1999, *95*(4), 69; (c) M. Picciotti and F. Picciotti, *Chem. Eng. Prog.*, 2006, *102*(12), 45; (d) P. Crook, *Chem. Eng. Prog.*, 2007, *103*(5), 45.
155. Anon., *R & D* (*Cahners*), 1998, *40*(10), 154.
156. (a) B.G. Clubley, ed., *Chemical Inhibitors for Corrosion Control*, R. Soc. Chem. Spec. Publ. 71, Cambridge, 1990; (b) D.W. Hairston, *Chem. Eng.*, 1996, *103*(3), 65; (c) Y.I. Kuznetsov, *Organic Inhibitors of Corrosion of Metals*, Plenum, New York, 1996.
157. (a) R.L. Twite and G.P. Bierwagen, *Prog. Org. Coat.*, 1998, *33*(2), 91; (b) A. Barbucci, M. Delucchi, and G. Cerisola, *Prog. Org. Coat.*, 1998, *33*(2), 131.
158. G. Ondrey, *Chem. Eng.*, 2006, *113*(12), 13.
159. (a) I. Maege, E. Jaehne, A. Henke, H.-J.P. Adler, C. Braun, C. Jung, and M. Stratmann, *Macromol. Symp.*, 1997, *126*, 7; (b) K.S. Betts, *Environ. Sci. Technol.*, 1999, *33*, 87A.
160. W.J. van Ooij and T. Child, *Chemtech*, 1998, *28*(2), 26.
161. H.K. Yasuda, Q.S. Yu, C.M. Reddy, C.E. Moffitt, and D.M. Wieliczka, *J. Appl. Polym. Sci.*, 2002, *85*, 1387, 1443.
162. J. Choi, Z. Lai, S. Ghosh, D.E. Beving, Y. Yan, and M. Tsapatsis, *Ind. Eng. Chem. Res.*, 2007, *46*, 7096.
163. C. Simpson, *Chemtech*, 1997, *27*(4), 40.
164. B. Buecker, *Chem. Eng.*, 2008, *115*(2), 30.
165. (a) B. Muller and G. Imblo, *J. Appl. Polym. Sci.*, 1996, *59*, 57; (b) B. Muller and D. Mebarek, *Angew. Makromol. Chem.*, 1994, *221*, 177.
166. (a) R. Agarwal and J.P. Bell, *J. Appl. Polym. Sci.*, 2000, *76*, 875; (b) H. Zheng, K. Nainani, and J.P. Bell, *J. Appl. Polym. Sci.*, 2002, *85*, 1749.
167. X. Zhang and J.P. Bell, *J. Appl. Polym Sci.*, 1997, *66*, 1667.
168. (a) D.E. Nikles and J.L. Liang, U.S. Patent 5,475,066, 1995; (b) S.B. Hait and D.E. Nikles, *Polym. Preprints*, 2001, *42*(2), 674; (c) E. Vaccaro and D.A. Scola, *Chemtech*, 1999, *29*(7), 15.
169. (a) Y. Wei, J. Wang, X. Jia, J.-M. Yeh, and P. Spellane, *Polymer*, 1995, *36*, 4535; (b) G. Parkinson, *Chem. Eng.*, 1996, *103*(7), 17; (c) T.P. McAndrew, *Trends Polym. Sci.*, 1997, *5*(1), 7; (d). W. Su and J.O. Iroh, *J. Appl. Polym. Sci.*, 1997, *65*, 417, 617; 1999, *71*, 1293; (e) B.N. Grgur, N.V. Krstajic, M.V. Vojnovic, C. Lacnjevac, and L. Gajic–Krstajic,

Prog. Org. Coat., 1998, *33*, 1; (f) T. Schauer, A. Joos, L. Dulog, and C.D. Eisenbach, *Prog. Org. Coat.*, 1998, *33*, 20; (g) J.O. Iroh and W. Su, *J. Appl. Polym. Sci.*, 1997, *66*, 2433; (h) N. Gospodinova and L. Terlemzyan, *Prog. Polym. Sci.*, 1998, *23*, 1443; (i) L.-M. Liu and K. Levon, *J. Appl. Polym. Sci.*, 1999, *73*, 2849; (j) B. Wessling, *Chem. Innov.*, 2001, *31*(1), 35; (k) K.G. Shah, G.S. Akundy, and J.O. Iroh, *J. Appl. Polym. Sci.*, 2002, *85*, 1669; (l) Y.S. Negi and P.V. Adhyapak, *J. Macromol. Sci.-Polym. Rev.*, 2002, *C42*, 35.
170. K. Fouhy, *Chem. Eng.*, 1997, *104*(2), 19.
171. (a) E.V. Bowry, *Chem. Eng. Prog.*, 1996, *92*(8), 81; (b) G.P. Bierwagen, ed., *Organic Coatings for Corrosion Control*, ACS Symp. 689, Washington, DC, 1998; (c) K. Pianoforte, *Coatings World*, Feb. 2008, 34.
172. Y. Lin and H. Yasuda, *J. Appl. Polym. Sci.*, 1996, *60*, 543.
173. R. Williams, *Chemtech*, 1995, *25*(4), 49.
174. B. del Amo, R. Romagnoli, V.F. Vetere, and L.S. Hernandez, *Prog. Org. Coat.*, 1998, *33*, 28.
175. (a) Y. Yu and C.-T. Lin, *Ind. Eng. Chem. Res.*, 1997, *36*, 368; (b) M.C. Whitten, Y.-Y. Chuang, and C.-T. Lin, *Ind. Eng. Chem. Res.*, 2002, *42*, 5232.
176. T.J. Lin, J.A. Antonelli, D.J. Yang, H.K. Yasuda, and F.T. Wang, *Prog. Org. Coat.*, 1997, *31*, 351.
177. J. Jang and E.K. Kim, *J. Appl. Polym. Sci.*, 1999, *71*, 585.
178. B. Muller, A. Paulus, B. Lettmann, and U. Poth, *J. Appl. Polym. Sci.*, 1998, *69*, 2169.
179. S. Gee, *Surf. Coat. Int.*, 1997, *80*, 316.
180. P. Kalenda, A. Kolendova, V. Stengl, P. Antos, J. Subrt, Z. Kvaca, and S. Bakardjoeva, *Prog. Org. Coatings*, 2004, *49*, 137.
181. J.G. Banker, *Chem. Eng. Prog.*, 1996, *92*(7), 40.
182. R. Cai, M. Sun, Z. Chen, R. Munoz, C. O'Neill, D.E. Beving, and Y.S. Yan, *Angew. Chem. Int. Ed.*, 2008, *47*, 525.
183. (a) G.P. Shulman, A.J. Bauman, and T.D. Brown, Preprints ACS *Div. Environ. Chem.*, 1994. *34*(2), 420; (b) R.L. Twite and G.P. Bierwagen, *Prog. Org. Coat.*, 1998, *33*, 91; (c) K.H. Zabel, R.E. Boomgaard, G.E. Thompson, S. Turgoose, and H.A. Braun, *Prog. Org. Coat.*, 1998, *34*, 236.
184. (a) C. Frayner, *Modern Cooling Water Treatment Practice*, Chemical Publishing, New York, 1998; (b) C.R. Ascolese and D.I. Bain, *Chem. Eng. Prog.*, 1998, *94*(3), 49; (c) M. McCoy, *Chem. Eng News*, Dec. 21, 1998, 21; (d) K.S. Belts, *Environ. Sci. Technol.*, 1999, *33*, 87A.
185. (a) R.J. Ross and L.P. Koskan, Preprints ACS *Div. Environ. Chem.*, 1997, *37*(1), 343; (b) G. Fan, L.P. Koskan, and R.J. Ross, Preprints ACS. *Div. Environ. Chem.*, 1998, *38*(1), 80.
186. (a) R.F. Benson, *Chemtech*, 1997, *27*(4), 34; (b) Anon., *Paper Technol.*, 1995, *36*(10), 11; (c) M. Burke, *Chem. Ind. (Lond.)*, 1996, 244; (d) A. Shanley, *Chem. Eng.*, 1998, *105*(8), 59; (e) G. Parkinson, *Chem. Eng.*, 1998, *105*(8), 25.
187. (a) D.Y. Wu, S. Meure, and D. Solomon, *Prog. Polym. Sci.*, 2008, *33*(5), 479; (b) B. Ghosh and M.W. Urban, *Science*, 2009, 323, 1458; (c) Anon., Chem. Eng. News, June 1, 2009, 14.
188. S. Stinson, *Chem. Eng. News*, Feb. 19, 2001, 13.
189. Y.C. Yuan, M.Z. Rong, M.Q. Zhang, J. Chen, G.C. Yang, and X.M. Li, *Macromolecules*, 2008, *41*, 5197.
190. W. Haller, G. Tauber, W. Dierichs, J. Herold, and W. Brockmann, *Ullmann's Encyclopedia of Industrial Chemistry*, 5th ed., VCH, Weinheim, 1985, *A1*, 260.
191. P. Brewbaker, R. Dignan, and S. Meyers, eds, *Singer Clothing Care and Repair*, C. DeCosse, Minnetonka, MN, 1985, *54*, 93–94.
192. L.A. Liddell, *Clothes and Your Appearance*, Good-heart–Willcox, South Holland, IL, 1988, 175, 185.
193. (a) D. Price, ed., *Polym. Degrad. Stabil.*, 1996, *54*, 117–416; (b) M.S. Reisch, *Chem. Eng. News*, Feb. 24, 1997, 19; (c) C. Martin, *Chem. Br.*, 1998, *34*(6), 20; (d) J.Q. Wang and W.K. Chow, *J. Appl. Polym. Sci.*, 2005, *97*, 366; (e) A.R. Horrocks and D. Price, *Fire Retardant Materials*, CRC Press, Boca Raton, FL, 2001; (f) S.Y. Lu and I. Hamerton, *Prog. Polym. Sci.*, 2002, *27*, 1661.
194. E.O. Wilson, *Science*, 1998, *279*, 2048.

RECOMMENDED READING

1. D. Klamann, *Ullmann's Encyclopedia of Industrial Chemistry*, 5th ed., VCH, Weinheim, 1990, *A15*, 505–506 [recycling oil].
2. M.P. Stevens, *J. Chem. Ed.*, 1993, *70*, 445, 535, 713 [additives and stabilizers for polymers].
3. R. Dagani, *Chem. Eng. News*, Dec. 15, 1997, 11 [three-laser method for making diamond coatings].
4. Y. Waku, N. Nakagawa, T. Wakamoto, H. Ohtsubo, K. Shimizu, and Y. Kohtoku, *Nature*, 1997, *389*, 49 [ductile ceramic].
5. A. Shanley, *Chem. Eng.*, 2000, *107*(7), 33. [lubricants]
6. S. Ritter, *Chem. Eng. News*, Mar. 13, 2006, 38. [motor oil]
7. A.H. Tullo, *Chem. Eng. News*, Sept. 17, 2007, 20. [corrosion]
8. C. Shargay and C. Spurrell, *Chem. Eng.*, 2003, *110*(5), 42. [corrosion].

EXERCISES

1. Make a list of weak links in products that you use. How might these be replaced to make products that would last longer?
2. List some things that you have found difficult to repair. How would you redesign them to make them easy to repair?
3. Do you have anything in your house that has been made from scraps that might have otherwise been thrown out?
4. Visit a fabric store to see what chemistry is there.
5. Devise a method to protect buildings of sandstone, limestone, and marble from acid rain.
6. What items of plastic in your home have worn out recently? Have there been any premature failures? What could be done to make such products more durable?
7. Deicing salts can cause concrete to deteriorate. What could be put in or on the concrete to prevent this? How might one polymerize a monomer in the concrete to make it stronger and more durable?
8. The pink polyvinyl chloride strapping on the beach chairs developed brown splotches in one season, despite periodic treatment with a liquid supplied by the supplier. Diagnose the problem and suggest a solution.

CHAPTER 14

Chemistry of Recycling

14.1 WASTE

The minimization of waste by the chemical industry was discussed in Chapter 1. The waste to be discussed here is municipal solid waste. This is the trash discarded from homes, businesses, construction sites, schools, or others. For a sustainable future, it will be necessary to recycle as much of this as possible. The amount of waste varies with the country (Table 14.1), with the United States leading the list as the world's foremost throw-away nation.[1]

Table 14.1 Amount of Waste Generated by Various Countries

Country	Waste per Person per Year (tons)
United States	0.88
Australia	0.74
Canada	0.50
Denmark	0.40
Japan	0.35
Germany	0.34
Switzerland	0.32
Sweden	0.28
France	0.26
United Kingdom	0.25
Italy	0.24
Spain	0.22

The quality of life does not follow this order. Several of the countries have higher incomes per capita and life expectancies than the United States. Several of the countries that compete with the United States in trade discard less waste per person. The overconsumption that leads to so much waste has implications beyond the boundaries of the country in question. For example, the United States obtains chromium and platinum from South Africa, calcium fluoride from Greenland, and tungsten from China. The developed nations import beef, woods, bananas, cocoa, coffee, pineapples, shrimp, and other consumables from tropical countries. The developed nations are partly responsible for any environmental degradation involved in the production of these crops. The United States uses one-third of materials in the world.[2] Europe, Japan, and the United States have 16% of the world's population but account for 80% of the materials used.

The United States generated 208 million tons of municipal solid waste in 1995.[3] This amounts of 4.3 lb/person/day. It consisted of 37.6% paper and paperboard, 15.9% yard trimmings, 9.3% plastics, 8.3% metals, 6.6% food, 6.6% glass, 6.6% wood, and 9.0% of other materials by weight. Containers and packaging accounted for 34.1% of the total. (In 1996, 209 million tons of municipal solid waste was generated, together with 136 million tons of construction and demolition waste.[4]) A study in Clark County, Washington, showed that 2.5% of the material discarded was reusable as it was.[5] The reusable items included food in the original containers, furniture, cosmetics, nuts, bolts, appliances, toys, clothing, carpet, wood building materials (41% of the total), plastic buckets, and ornamental glass. The average business in the United States throws away 250 lb of paper per person per year.[6] The advent of computers and photocopiers has not reduced the amount of paper used in offices.[7] Seventy percent of the municipal solid waste in the United States ends up in landfills.[8] These have an average remaining life of 10 years.

Only 27% of the municipal solid waste generated in the United States in 1995 was recycled. In 2006, 82 million tons of waste was recycled, which was 32.5%.[9] More than 700 curbside recycling programs and more than 9000 drop-off centers were in operation in 1995. The latter are energy-intensive, because people usually drive to them in their cars. By 2008, there were 8500 curbside recycling programs in the United States.[10] In 1995, the recycled material included 40% of paper and paperboard, 38% of containers and packaging, 52% of aluminum packaging, 54% of steel packaging, 52% of paper and paperboard packaging, 27% of glass packaging, 14% of wood packaging, and 10% of plastic containers and packaging.[3] Office paper was recycled at the rate of 44%, magazines at 28%, clothing at 16%, tires at 17%, nonferrous metals at 70% (largely owing to recovery of lead batteries), and ferrous metals at 31%. (By 1997, the recycling rate for paper and paperboard had grown to 45%.[11]) The United States exports 10 million tons of scrap iron each year.[12] It also exports a million containers of used clothing to developing nations each year.

The drop-off centers for voluntary recycling in Delaware collect only 3.5% of residential waste in the state.[13] Industrial recycling and commercial recycling raise the average for the state to 15.3%, the lowest among the northeastern states in the United States. The highest recycling rate among the states, 54.6%, is in New Jersey, where curbside collection of recyclables is mandatory. Communities that have over 50% recovery of recyclables, such as San José, California; Seattle, Washington; and Portland, Oregon, have either mandatory programs or pay-as-you-throw systems (where the cost of waste disposal to the resident is per volume or weight).[14] (Minnesota has 1843 communities with pay-as-you-throw systems.[15]) Recycling has been successful for apartments and condominiums as well as for individual homes.[16] For high-rise buildings, one system uses a chute that goes to a turntable and compactor.[17] A person punches in the type of material on the floor control panel and drops the material in the slot.

A model study on Long Island, New York, a few years ago, demonstrated that it was possible to recycle 84% of municipal solid waste. It is instructive to look at what some industries have done to approach zero waste.[18] Mad River Brewing Co. in Blue Lakes, California, has a 97% recycling rate. In addition to recycling the standard commodities, such as cans and bottles, paper is printed on both sides, office paper is shredded for packing material, spent grain is sold to local composting companies and farmers, employees repair and use wooden pallets, and wood scraps become firewood. The former Chrysler auto assembly plant in Newark, Delaware, recycled 95% of waste plastic and cardboard, 85% of office paper, 100% of scrap metal, and 20% of wooden pallets.[19] In one year, this avoided disposal costs of 349,760 dollars for hauling and disposal at a landfill and earned 212,288 dollars when sold. It should be possible to reuse more of the wooden pallets or to replace them with more durable, reusable plastic pallets.

Landfills received 57% of the waste, and another 16% was burned, usually with recovery of energy. (In 2005, landfills received 70%.[8]) Both these methods must be used carefully to prevent environmental contamination. Landfills must be lined with liners that will not be punctured by the heavy equipment running over them, and must have provisions for leachate treatment and monitoring wells to be sure that groundwater does not become contaminated. The methane from them must be used for fuel or vented to the atmosphere properly to avoid fires and explosions. Many Superfund sites are old landfills at which these precautions were not taken. Several old landfills that replaced valuable wetlands can be seen from the train as one passes through New York City. The former Fresh Kills landfill, then the largest in the United States, is in a former marsh on Staten Island in New York City. It is now closed. Incinerators[20] also have their limitations. The emissions must be scrubbed to remove hydrogen chloride [from poly(vinyl chloride) and other chlorinated materials], nitrogen oxides generated during combustion, mercury, dioxins, particulates, and such. The remaining ash may contain toxic, heavy metal compounds. Although incineration can recover some of the energy in waste, it does not recover nearly as much as would be saved if reusable items were used instead of throwaway ones. For example, using and discarding 1000 plastic teaspoons consumes 10 times more energy than it takes to make one stainless steel teaspoon and wash it 1000 times.[21] The tragedy is that both of these disposal methods throw away useful, valuable materials. Replacement of these puts additional pressure on the natural resource base.

14.2 RECYCLING

14.2.1 Introduction

Recycling[22] is simplified if the material is free of contaminants. This requires as much separation as possible at the source. The process is complicated because many consumer items are made of more than one material. A steel beverage may have an aluminum top. The aseptic drink box consists of several layers. Polypropylene film for foods is usually coated with a barrier of poly(vinylidene chloride). It must be remembered that the paper, steel, aluminum, glass, plastic, and such that need to be recycled come in many different grades and compositions. These have been optimized for specific uses, and they may not do well in others. Thus, recycle may have to be to a less-demanding use. Paper can be made by heating and grinding (thermomechanical pulp) or by treating the wood with chemicals to remove the lignin.[23] It may or may not be bleached. Various fillers, sizes, wet strength resins, and other additives may be used on it. Some paper may be dyed or waxed. Steel, aluminum, copper, and other metals are available in many different alloys. Plain steel and stainless steel would not be recycled together. Many different alloys are used in stainless steel. A different alloy is used in the lid (e.g., 0.35% Mn and 4.5% Mg) of the aluminum can than in the body (e.g., 1.1% Mn and 11% Mg) of the can.[24] Deep drawing is important for the body and ease of working the pull tab is important in the lid. Aluminum pans are made of other alloys. Some scrap dealers will not even take aluminum pans. (Aluminum is also alloyed for various uses with iron, copper, silicon, lithium, chromium, lead, zinc, and zirconium.) Window glass has a different composition than that in containers and must be kept separate. Pyrex glass has yet another composition.

The composition of plastics can vary even more. Various polyethylenes vary in molecular weight, molecular weight distribution, degree of branching, length of branching, crystallinity, and so on. For polypropylene, variables of tacticity and crystallinity must be included. Copolymers can be random, block, or graft, and many sometimes contain some of the homopolymers. The additive packages used in the various polymers may vary. These include stabilizers, impact modifiers, flame retardants, and dyes, among others (as described in Chapter 13). Coatings and laminates can cause additional problems. Clothing often consists of mixtures of fibers, as in cotton–polyester fabrics. Automobile tires may have one rubber in the tread, another in the carcass, and a polyisobutylene liner inside as a gas barrier. In addition, the degree of cross-linking in rubbers and other polymers may vary.

Some curbside recycling programs allow the comingling of recyclable materials that are relatively easy to separate to save work on the part of the consumer. For example, steel cans can be separated from aluminum cans with a magnet. Separation of a mixture of containers is more difficult, but can be done. If the containers can be passed down a line one at a time, on-line, and then sensed by infrared, visible, or ultraviolet light, by x-ray, or by other rapid means, blasts of air can remove them into the proper bins. Carpets and plastics can be separated with a hand-held spectrometer.[25] Metals and their alloys can be separated with a hand-held x-ray fluorescence analyzer.[26] Separation of mixed municipal solid waste, as done in Delaware in past years, has been difficult.[27] Aluminum separated by eddy currents created in the metal includes both beverage cans and pie tins, which are different alloys and do not recycle well together. The glass recovered has a mixture of colors of much less value than glass of separate colors. The presence of ceramics in the glass interferes with its recycling. Air classification can remove light objects such as paper and plastic film. The plastic film interferes with the repulping of the paper. A film of polyethylene terephthalate will come out in a different place than a bottle of the same material. Sink–float techniques can also be used on mixtures after removal of the paper and plastic films. This may require shredding of metal and plastic containers, as well as breaking the glass, so that containers with air inside do not float. Compost made from the organic content of the waste is contaminated with slivers of glass, as well as with any toxic heavy metals that may be present in the pigments on the paper. Steam classification has been used to separate metals plastics, glass, and paper (as cellulose pulp) to reduce the amount of solid waste by 80–90%.[28]

The recycling techniques applied to the various clean streams of materials will now be covered.

14.2.2 Paper

Asia is the world's largest consumer of paper using 58 million tons/year followed by Europe at 44 million tons and the United States at 41 million tons, but the United States leads in per capita consumption at 350 kg per year followed by China at 25 kg and India at 5 kg.[29]

14.2.2.1 Recycling versus Incineration

The recycling of paper[30] has advantages over incineration to recover energy from it,[31] despite claims to the contrary by some authors.[32] Recycling saves more energy than that obtained by incineration, because making new paper requires energy to harvest trees and so on. Recycling creates 3 times as many jobs as incineration. In England, at least, recycling helps the balance of payments and avoids landfill costs. Those who advocate burning rather than recycling assume that fossil fuels will be used to harvest and pulp the trees. Fossil energy may be used, but it does not have to be in the future. People who recycle paper are working toward a goal of 100% recycled fiber content, with zero wastewater discharge from the plant. In a few instances, this is possible today. Containerboard is being made in Germany[33] and corrugated cardboard in New York[34] from 100% postconsumer material. Newsprint and tissue are being made in a German mill from mixed newspaper and magazines.[35]As the techniques of recycling gradually improve, the postconsumer content of recycled paper is climbing.

14.2.2.2 Deinking

Some paper can be recycled without deinking. The resulting sheet will be somewhat gray, but can still be reprinted in a legible fashion. For some uses, such as toilet tissue, there should be no need for deinking or bleaching. Some recycled fiber that has not been deinked can be added with a surfactant in the making of new newsprint.[36] Usually deinking is required.[37] Not all uses require the same level of deinking. Sorted office paper can be converted to tissue, white linerboard, and other grades (containing 100% recycled fiber), with less stringent optical requirements than printing and writing grades.[38] The paper is usually repulped in the presence of a base, a surfactant, and other additives at 45–60°C for 4–60 min. A typical run may contain 0.8–1.5% sodium hydroxide, 1–3% sodium silicate, 0.25–1.5% surfactant, 0.5–2.0% hydrogen peroxide, and 0.15–0.4% diethylenetriaminepentaacetic acid. The fiber is swollen by the sodium hydroxide at pH 8 to 10, reducing the adhesion of ink to the fibers. The sodium silicate prevents the redeposition of ink, and the hydrogen peroxide counteracts yellowing. The diethylenetriaminepentaacetic acid ties up any metal ions present that might cause the hydrogen peroxide to decompose. The ink[39] particles are then removed by washing the pulp or by flotation.[40] Washing is best for small particles of ink (2–20 μm) and flotation for larger particles (40–250 μm).[41] The presence of clay filler particles from magazines improves the efficiency of flotation deinking, so that newspapers and magazines are often repulped together.[42] Pressurized flotation cells are better than those operating at atmospheric pressure.[43] Column flotation is superior to that done in a tank.[44] In some flotation deinking, calcium chloride is used with a long-chain fatty acid to make the ink particles hydrophobic.[45]

Mechanical means are also used to enhance the removal of ink from the fibers. Passing the fibers between rotating disks can help.[46] When ultrasound is used, there is less need for chemicals, and the resultant pulp is stronger.[47] Steam explosion, during which paper containing 50% moisture is passed through a hot coaxial feeder, reduces the need for chemicals and eliminates the need for flotation.[48] This process can also handle laser and xerographic inks, which can be difficult to remove by other means. There is some loss in tensile strength.

The removal of toners (the inks) used in photocopiers can offer problems.[49] Toners are low-melting resins (e.g., styrene-1,3-butadiene copolymers, acrylic polymers, or polyesters) containing carbon black.[50] Various techniques are used to facilitate their removal. One way is to melt them during the deinking process so that they agglomerate, cool the mixture to harden them, and then remove them by screens.[51] Another is to pass the paper slurry through a twin-shaft kneader at room temperature, which shifts the particle size to one in which flotation can be used to remove them.[52] Ultrasound can be used to break up toner particles for easier removal.[53] 1-Octadecanol enhances the detachment of toners and increases flotation efficiency.[54] Planetary mixing starting at 90°C and ending at 25°C can help.[55] Cationic surfactants, such as cetyltrimethylammonium bromide and poly(dimethyldiallylammonium chloride), can act as both the frothing agent and the collector in the flotation deinking of photocopier paper.[56] Ricoh has devised a way of removing toners from sheets of copy paper by treating with a hot solution and then drying.[57] Xerox has devised toners that dissolve in base in 1 h at 50°C when the paper is recycled.[58] Toners based on alkali-degradable polymers can be removed (97% removal) by flash hydrolysis in 0.5% alkali in 200–300 s at 210°C.[59] Soy-based toners swell in the repulping medium, creating stresses that dislodge them from the fiber.[60] Toshiba has devised toners that can be decolorized by heat or solvent so that the paper can be reused.[61] Photoerasable toners have been made based on copolymers of methacrylophenone.[62] It should also be possible to base them on dyes that bleach in ultraviolet light. Electrophoretic inks, in which a pigment in a microcapsule responds to an electric field, may also be useful in reusable paper.[63] Thus, the sheet of paper can be used again in the same office without repulping. Xerox and others have devised systems that erase themselves in 16–24 h of printing or by passing the paper through the printer again immediately.[64] The paper can be used as many as 50 times. "Scott LimpiaMax" paper towels can be reused.[65]

14.2.2.3 Stickies

Paper may contain a variety of other materials that must be removed during repulping.[66] These include sizing agents, retention aids, wet and dry strength additives, coatings, fillers, and others. In addition, recycled paper may contain staples, paper clips, pressure-sensitive adhesives (as on U.S. stamps and return address labels), hot-melt adhesive (as in book bindings), wax, plastic tapes, polystyrene windows in envelopes, and so on.[67] Pressure-sensitive adhesives may contain various rubbers, such as styrene-1,3-butadiene copolymers, polyisoprene, natural rubber, and poly(1,3-butadiene), plus tackifying resins, such as esters of rosin. Hot-melt adhesives[68] may contain polyethylene, copolymers of ethylene and vinyl alcohol, wax, and such. Coating binders may be styrene-1,3-butadiene copolymer rubber, polyvinyl acetate, poly(vinyl chloride), vinyl acrylates, or others. Barrier films may be made of wax, polyethylene, polypropylene, poly(ethylene terephthalate), or nitrocellulose. Radiation-cured inks may be based on epoxyacrylates, polyol acrylates, urethane acrylates, polyester acrylates, and such. Although some of these materials can be removed by screens, others tend to plug screens and may form spots in the recycled paper.[69] Collectively, they are referred to as *stickies*. One way of removing them is by treatment with sc carbon dioxide, containing a lower alkane, such as pentane.[70]

Air Products has devised a process using oxygen, sodium hydroxide, sodium silicate, and the like, at 180–200°C at 60 psig, followed by treatment with a reducing agent, such as sodium hydrosulfite, that removes dyes, inks, wet strength resins, and detackifies adhesives in mixed office wastepaper.[71] Another system uses 1% mineral spirits, 1.22% sodium silicate, and 3% sodium hydroxide in water at 50°C.[72] The polymers present become scavengers for the ink, forming particles that can be removed by screening. In more common processes, the stickies that cannot be removed by screening are detackified, agglomerated, or dispersed.[73] Talc is sometimes added to reduce tackiness. Dispersants, such as fatty alcohol ethoxylates and naphthalene sulfonates, can keep some of them in a colloidal state in which they will not interfere. Ultrasound can help in dispersing them.[74] Flocculation with high molecular weight cationic polymers can also be used. The method used will depend on the nature of the stickies and the equipment in the mill. Some pressure-sensitive adhesives become more spherical on heating, so that they can be removed by centrifugation or screening techniques.[75] Most hot-melt adhesives come out on screens as strips and fragments.[76]

Some work has been done to redesign the materials so that the stickies problems do not occur. Waxes used on 3–5% of corrugated cardboard cartons (for food items) complicate recycling,[77] although there is one system that is said to be able to reduce the wax from 30% to less than 1% mechanically.[78] Waxes made with a low percentage of stearic acid and a few tenths of a percent of a fatty alcohol ethoxylate redisperse during alkaline repulping.[79] All of the wax on old corrugated cardboard can be removed with sc carbon dioxide at 100°C, so that the wax can be reused.[80] Paper coated with a combination of zein (from maize) and paraffin wax can be dewaxed with the enzyme, α-chymotrypsin.[81] Acrylic polymers, which may contain carboxylic acid groups, can be used to replace wax on corrugated boards.[82] Polycaprolactone–paper composites have good water resistance and biodegradability.[83] These might be able to replace objects made of waxed paper. The conditions of alkaline repulping would have to be such that the polyester was hydrolyzed.

Repulpable hot-melt adhesive can be made with softening points high enough that they will not soften during repulping, or they can be made dispersible in aqueous base.[84] Binders for deinkable inks have been made by the reaction of phthalic anhydride and neopentyl glycol (2,2-dimethyl-1,3-propanediol) or trimethylolpropane to form polyesters.[85] Dyes that can be bleached by reduction during repulping may offer advantages.[86] The use of dyed paper could also be reduced. Most stickies are polymers. Substituting copolymers made with acrylic acid, methacrylic acid, or maleic anhydride would make them soluble in the repulping medium.[87] Such polymers are widely used in water-based coatings and adhesives (see Chapter 8). Another possibility is to use polyesters from the polyglycol diacids (**14.1** Schematic) sold by Hoechst Celanese.[88] These could be made to be soluble in water.

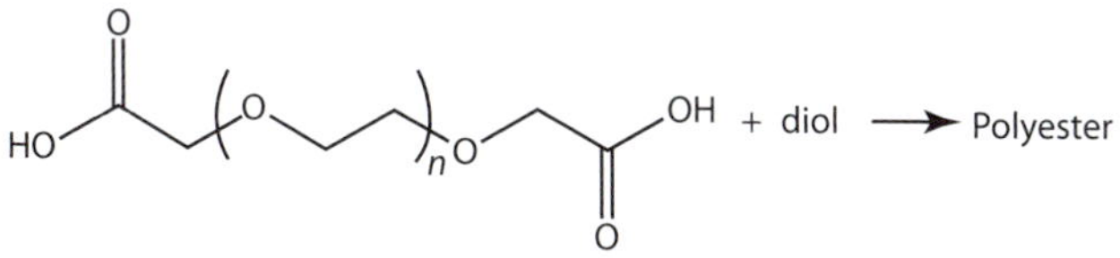

14.1 Schematic

The problem of polystyrene windows in envelopes can be solved by reverting to the use of glassine paper windows. Glassine paper is made by a combination of heat and pressure. Its cost might be reduced if it can be made directly in the paper to be folded into the envelope by pressure and microwave heating. Ultrasound might help in the process. Windows that are just cutouts are now in use and seem to function satisfactorily. The U.S. Postal Service is to change the pressure-sensitive adhesive on stamps to one that will repulp more easily. Going back to the water-moistenable ones used in past years would solve the problem.

14.2.2.4 Enzymatic Repulping

No deinking chemicals are necessary when old newspapers are repulped with a cellulase.[89] The larger uninked fibers are retained by a plastic mesh screen, whereas the smaller inked ones go through to a separate vessel. Enzymes facilitate toner removal, increase brightness, improve pulp drainage, preserve fiber integrity, and lower chemical costs in the repulping of mixed office wastepaper.[90] The drainage is best when the enzyme is combined with a polyacrylamide containing quaternary ammonium groups.[91] Xylanases improved drainability without hurting tensile strength when repulping old paperboard containers.[92] Buckman Labs markets an esterase to cleave esters in adhesives on wastepaper during repulping.[93] The deinked pulp can be made into fine writing and printing papers. Old newspapers have been converted to supersoft tissue using enzymatic repulping.[94] Enzymes should be useful in removing soy-based inks.

14.2.2.5 Changes in Fibers in Reuse

Recycled paper contains more fines and shorter fibers. Some strength is lost in repeated cycling of thermomechanical fibers (as in newspapers).[95] The surface becomes rougher and the ratio of oxygen to carbon goes up, perhaps by oxidation. There is a substantial change in the fiber properties in the first cycle, but most physical properties stabilize after about four cycles.[96] This is thought to be due to a drop in interfiber bonding and not to a loss of strength of individual fibers.[97] There is an irreversible hornification of some mechanical pulps on drying that reduces fiber bonding.[98] Beating (i.e., mechanical action to produce more fibrils on the surfaces of the fibers) can restore the strength properties lost on drying and rewetting.[99] The minimum mechanical force necessary to obtain the desired result should be used throughout the repulping process. This may mean that some of the action of the machines used with virgin pulp may need to be made gentler.

In one study, newspaper was recycled without the use of water.[100] The paper was fiberized mechanically, and the fibers were air-laid to produce a new sheet. It was necessary to treat the fibers first with ozone and then with ammonia to achieve good interfiber bonding when the air-laid sheet was pressed. The tensile strength of the resulting paper approached that of the original newsprint. Such a process would eliminate aqueous effluents.

14.2.2.6 Extent of Recycling of Paper

Europe recovered 48 million tons of paper, nearly 50% of the total used, in 2001, of which 40 million tons went back into paper and paperboard.[101]

By using the techniques outlined in the foregoing for aqueous repulping, some mills have been able to obtain very high rates of recycle of postconsumer paper. U.K. Paper uses deinked pulp of 75–100% recycled content.[102] Bridgewater Paper, in England, produces newsprint that averages 95% recycled content.[103] Some urban minimills[104] can make recycled linerboard from 100% old corrugated containers.[105] A mill next to the former Fresh Kills landfill on Staten Island, New York, makes 100% recycled containerboard.[106] It "harvests an urban forest." New Jersey had 13 paper mills running only on wastepaper and 8 steel minimills using scrap steel in 1999. The total annual output of these two industries was more than 1 billion dollars.[107] Some of these minimills are able to operate closed systems with no effluent.[108] Zero liquid effluent is now practical for corrugating medium and linerboard mills.[109] A few other mills have achieved it.[110] A Kimberly–Clark mill in Australia will pulp thinnings from *Pinus radiata* plantations with magnesium bisulfite, bleach with hydrogen peroxide, recover all spent sulfite liquor, capture all gases from cooking, and either compost all fines or use them for fuel.[111] Further study of the use of reverse osmosis, ultrafiltration, biological oxidation, and such may make it possible for other mills to have zero effluents. Paper mill effluents have been notorious for their contamination of streams.[112] Even the natural resin and fatty acids in the wood can be toxic to fish.[113] It would be better to recover these acids for industrial use, as done in many mills.

14.2.2.7 Nonpaper Uses for Wastepaper

Mixed wastepaper can also be used for uses other than paper.[114] Extensive cleaning may not be needed in something such as a wood-based composite that will be painted or overlaid. Various composites have been made. These include (a) molding of paper fibers mixed with polyethylene or polypropylene to form door panels, trunk liners for cars, and plastic lumber; (b) nonwoven mats of up to 90% paper fibers held together by synthetic fibers by entanglement or thermal bonding; (c) wet laid boards made with binders, such as resins, asphalt, and such, for use as acoustical insulating board; (d) composites of wood, paper fiber, and gypsum or Portland cement; and (e) structural paper products made from wet-formed waffle-like sections. A board from 50:50 wood cement is easy to saw and nail. Such a board might be made from the calcium sulfate, derived from flue gas desulfurization, and sludge from the recycling of paper. It could replace the present drywall. Putting two waffle-like sections of re-formed, recycled paper together with an adhesive produces a sandwich that is 30–200% stronger than conventional corrugated fiberboard. Polypropylene, containing 50% waste newspaper flour, is sold in Japan.[115] Maleated polypropylene can be used as a compatibilizer. Old telephone books and plastic milk jugs have been made into flat panels and moldings normally made of wood in Australia.[116] They are twice as strong and thrice as stiff as wood. They repel water and are free of knots and defects. Wastepaper shredded into 2–3-in flakes is used in hydroseeding in Iowa.[117] It breaks down into finer pieces in the hydroseeder tank. Shredded wastepaper is used as animal bedding instead of straw in Chester County, Pennsylvania. It can also be used as packing around objects being shipped in boxes in place of polystyrene "peanuts." With the proper flame retardant added, it can be used as insulation in houses. Pencils can also be made of recycled paper.[118]

14.2.2.8 Uses for Paper Mill Sludge

Many uses have been explored for paper mill sludge.[119] (More than 4 billion kg are produced in the United States each year.[120]) It can be used in making boards, tiles, plastics; as animal litter; or even fed to cattle. One of the best uses for the sludge from repulping mixed office wastepaper is to mix it into newsprint to lower costs while producing a higher-quality pulp.[121] Phenolic resin composites have been made with paper sludge.[122] Bacteria or fungi can convert paper mill sludge to single cell protein. Yeasts can convert it to ethanol. Methane can be produced from it by anaerobic fermentation. Other disposal methods include composting, land application, and incineration. (A record 200,000 tons of nonrecyclable paper was converted to compost in the United States in 1998.[123]) The best approach is to minimize the amount of sludge formed by use of an acrylamide copolymer as a retention aid to keep more of the fines in the paper being made.[124] The uses of sludges from the repulping of wastepaper will be limited somewhat by the presence of ink particles, fillers, and such. They should still be suitable for many of the foregoing uses. The residue from incineration of sludges might be used again as a filler for paper. It has been made into glass aggregate used in roofing shingles, asphalt, and sandblasting grit.[125] Sludges should be too valuable to dispose of them in a landfill.

One proposal for the use of scrap wood that is not suited for making paper is to raise termites on it.[126] The methane evolved can be collected and used, and the termites can be harvested for animal feed.

14.2.2.9 Nonwood Fibers for Paper

There is much concern about the loss of biodiversity as the old-growth forests of the Pacific Northwest of the United States and Canada are cut. Old-growth boreal forests supply 15% of the world's paper pulp, and old-growth temperate forests supply 1%.[127] Efforts are being made to find other sources for paper so that these forests can be preserved. In Canada, sawmill waste is the source of 60% of the fiber for paper.[128] This might also be used to make chipboard for construction. Canada is the largest producer of newsprint in the world.

Bags made from bamboo and sisal in Brazil were more resistant to bursting than those made from pulp from pine trees.[129] Paper can also be made from hemp, kenaf, wheat straw, rice straw, and bagasse (the residue left after the juice is pressed out of sugarcane).[130] If done properly, nonwood can replace both softwood and hardwood pulps with few problems and little loss in performance.[131] Kenaf (*Hibiscus cannabinus*)[132] and hemp can be grown in one season compared with a harvesting cycle of 15–25 years for southern pine in the United States of America and of 6–8 years for *Eucalyptus* in Brazil.[133] Hemp produces 3–6 tons/acre/year and kenaf produces 6–10 tons/acre/year, twice as much as southern pine.[134] Newsprint from kenaf is as strong as that made from trees, is brighter, slower to yellow, and less susceptible to ink run-off and smudging.[135] However, its price will not become competitive until larger amounts are used. The chemical pulping of hemp, kenaf, and jute gives high-quality pulp that costs too much for bulk applications at present. Extrusion pulping with aqueous sodium hydroxide, sodium sulfide, and such has been used to make pulp that should be competitive with wood pulp in a variety of paper applications.[136] The quality is comparable with softwood chemical pulp. Enzymatic pretreatment of nonwoody plants yields paper, with as good or better physical properties as conventional pulping, with much energy saved.[137] Spain saves wood by pulping wheat straw with 65% aqueous methanol at 175°C for 75 min.[138] Fibers pulped from grasses, such as *Phragmites communis* (an invasive weed in marshes of the eastern United States), with performic acid were weaker, but could be used to replace 30–70% of birch wood pulp.[139] A mill to produce paper from flax straw and maize stalks was opened in Nebraska in 1997.[140] It produces currency paper, writing paper, newsprint, and cardboard. Spent grain from a brewery has been fractionated to produce a fraction containing 1–2% protein, 30% cellulose, 60% hemicellulose, and several percent silica that can be used in paper at a level of 10%.[141] Greater production of paper from agricultural wastes will decrease the pressure to cut forests. This might be encouraged by removing the subsidies and charging more for logging in national forests in the United States. It might also be encouraged by a tax deduction. Even more important is to figure out how the consumption of paper can be reduced. (The United States with 4.7% of the world's population uses 31% of the world's paper and paperboard.[142])

U.S. shoppers receive 18 billion catalogs each year.[143] This is about 60 catalogs per capita and consumes 3.3 million tons of paper. Customers could ask to have their names removed from the mailing lists or receive an electronic catalog. Companies could be greener by printing the catalogs on 100% postconsumer paper.

14.2.3 Plastics

World consumers used 215 billion lb of the five most commonly used plastics in 1996[144] and 250 billion lb in 2007.[145] This included 41% polyethylene, 23% poly(vinyl chloride), 21% polypropylene, 11% polystyrene, and 4% styrene–acrylonitrile copolymers. The world production of polyethylene terephthalate in 1996 was 9.8 billion lb. The United States' plastics use in 1995 was 71.2 billion lb, of which 27% was used for packaging, 20% in building, 13% in consumer products, and the rest in a miscellany of uses. The United States' plastics use in 2007 was 84 billion lb, of which 18.2 billion lb was high-density polyethylene and 8.7 billion lb was poly(ethylene terephthalate).[146] The use of plastics continues to rise each year. The consumption of poly(ethylene terephthalate) is growing at the rate of 15–20% per year, with much of this going into bottles.

14.2.3.1 Recycling Problems

The recycling of plastics[147] is not growing rapidly. In the United States, less than 10% of plastics packaging is recycled. Only 4% of plastic grocery bags is recycled (e.g., to decking lumber).[148] In 1993, only about 2% of all plastics produced was recycled.[149] Some recycling facilities are being sold to smaller companies by the large chemical companies or are closing.[150] One problem is obtaining an assured stream of clean material to recycle. The few success stories include poly(ethylene terephthalate) bottles, many of which are returned under deposit legislation.[151] In 1996, 485,000 metric tons was collected in the world for recycling. (Of this, 330,000 metric tons was in North America.) There is even a report of a shortage of these bottles for recycling.[152] These are used to produce foamed insulation,[153] fibers for various applications, shower curtains, paint brushes, and others. However, the recycling rate for polyethylene terephthalate bottles dropped from 27.8% in 1996 to 25.4% in 1997 to 16% in 2007, whereas total use was rising.[154] The recycling rate for high-density polyethylene was 6% in 2007. During the same time period, the recycling rate for all plastic containers dropped from 24.5% to 23.7%. In 2005, 2.1 billion lb of plastic bottles were recycled in the United States, which was 24.6% of the total.[155] This includes both poly(ethylene terephthalate) and high-density polyethylene bottles, which made up 96% of all plastic bottles. Small amounts of poly(vinyl chloride) with the polyester can seriously contaminate the batch.[156]

Coca-Cola is building the world's largest poly(ethylene terephthalate) recycling plant in England.[157] The company has also bought into the Recycle Bank in Philadelphia. This Bank has increased the recycling rate in the city from 15% to 50% by giving participants credits that can be used at local businesses. Unfortunately, these moves may delay the return of refillable bottles.

High-density polyethylene milk jugs are being recycled into other containers, usually for nonfood uses, as well as into drainage pipes, trash cans, grocery bags, traffic barrier cones, and such. If they go into new food containers, they must be in an inner layer in a laminate with virgin material on the surface. The polypropylene cases of car batteries are also recycled well. Although poly(vinyl chloride) can be recycled into bottles, pipes, and window frames, it is recycled less than other plastics.[158] Solvay recycles poly(vinyl chloride) in plants in Italy and Japan.[159] Ground plastic is dissolved in methyl ethyl ketone at 100–140°C, the solution is filtered, and then the polymer is precipitated. Metals, poly(ethylene terephthalate), and nylon do not dissolve.

The plastics that are in other throwaway packaging[160] are difficult to handle. They may be in thin films or contaminated with food residues, and may be in a great many trash cans in a great many places. The high costs of collecting and sorting raise the prices of recycled plastics to a point that most manufacturers would rather use virgin plastics. (The cost of virgin resin does not include global warming and natural resource depletion.) If the pieces of plastic can be sent down a line one at a time, an on-line spectroscopic analyzer can sort them. For example, polyethylene and polypropylene can be separated by pulsed laser photoacoustic technology.[161] Hand-held devices that work by Raman spectroscopy are also available.[162] If the materials to be sorted are crumpled films, laminates, core-sheath fibers,[163] or polymer blends, such systems will not work well. Foamed cups and "peanuts" of polystyrene are voluminous and require some form of densification to make transportation to the recycling center economical.[164]

Many plastic objects are marked with a number inside three chasing arrows: 1 is polyethylene terephthalate; 2 is high-density polyethylene; 3 is poly(vinyl chloride); 4 is low-density polyethylene; 5 is polypropylene; 6 is polystyrene; and 7 is other (including a multilayer container). Such a system allows a consumer to place the used object in the proper bin, if such bins are provided. The consumer must remember that the lid or cap is often a different plastic from the rest of the container and must be conscientious in looking for the numbers on every piece. Curbside recycling is not going to provide that many bins. There is also a trend of all recyclables being comingled in one bin at the curbside. Manual sorting at a recycling center might work, but would be time-consuming. This means that some automated system is required for the separation of the various plastics. Some companies that make plastics have sought to encourage the recycling of their products by takeback programs. DuPont makes its spun-bonded polyethylene mailing bags with 25% postconsumer high-density polyethylene from milk and water jugs. The company collects the used bags from its customers for recycling into plastic lumber.[165] As these mailing bags look very much like their paper counterparts, they get into paper recycling and cause problems as stickies.[166] Waste polyethylene can be used to hold together other waste materials to make new products, as mentioned earlier under the recycling of paper. Waste cedar fibers can be used with up to 60% waste polyethylene to make plastic lumber.[167]

14.2.3.2 Recycling Methods

Recycling is easiest when only one polymer is present and no separation is needed. Polypropylene fertilizer bags can be reprocessed without additional stabilizers if they were properly stabilized the first time.[168] Because of the degradation in aging and remolding, it is often necessary to add additional stabilizers when a plastic is recycled into a new plastic object.[169] Carpets are often made with one polymer in the facing and another in the backing. Hoechst Celanese has devised one made entirely from polyester that should be easier to recycle at the end of its useful life.[170] This also simplifies what one is to do with scraps from the manufacturing and installation processes. Interface Inc. does not sell carpets. It leases them, services them, and replaces them when needed. Thus, it knows the composition of what it recycles.[171] Nylon-6 is easier to recycle than nylon-6,6 because only one compound has to be recovered instead of two after hydrolysis. The company diverted >38 million kg from landfills since 1994. Profits are up, plant effluents are down, costs are down, and employees are happy.

The poly(ethylene terephthalate) soda bottle is a mixture in that it has a label and cap of polypropylene. Redesign has eliminated the polyethylene base cup used formerly. It may be possible to prepare an all-polyester soda bottle.[172] The label might be replaced by ink jet printing. A shrink label of polyester is also possible, but care would be required in its installation because its softening point is about 100°C higher than that of the current label of polypropylene. A cap of polyester may also be possible if crystallization of the polymer can be speeded up to shorten molding cycles. Sodium bicarbonate and carbonate have been used as nucleating agents to shorten the molding cycles of recycled poly(ethylene terephthalate).[173] (Bottle makers often add a crystallization inhibitor so that the bottles remain clear. A cap would not have to be clear.) The seal of the cap liner might have to be made of a different polymer, which in the small amount involved, one hopes, would not interfere with recycling of the bottles. DuPont is marketing a glass fiber-reinforced molding compound made from used soda bottles and x-ray film from which the emulsion has been removed.[174] The key to this reuse is to prevent degradation of the molecular weight by keeping the polyester dry to prevent hydrolysis, minimizing oxidative degradation during recycle, and keeping rogue materials, such as polyethylene and poly(vinyl chloride) out of the system.[175]

Foamed polystyrene waste (15–35%) has been dissolved in a mixture of styrene and divinylbenzene for copolymerization.[176] This was followed by sulfonation to produce an ion-exchange resin. High temperatures in an oxidative environment tend to decrease the physical properties of recycled nylon-6,6.[177] Poly(vinyl chloride) from bottles, appliances, and pipes can be remolded into pipes, with the addition of fresh stabilizer, with or without the addition of virgin poly(vinyl chloride), without loss of tensile strength or modululs.[178] French auto makers are studying ways of recycling poly(vinyl chloride) from cars at the end of their useful lives.[179] The material is used in seat covers, dashboards, and trim. Greenpeace has accused the German plastics industry of misleading the public on the amount of poly(vinyl chloride) that is being recycled.[180] According to Greenpeace, 98% of poly(vinyl chloride) waste is dumped or burned. An ingenious way has been found to remove the paint from old car bumpers of polypropylene before recycling.[181] The bumper is cut into pieces and then run through a series of four rolls at 100°C, with the second roll running slightly faster than the others. The polypropylene stretches, but the paint does not so it flakes off.

14.2.3.3 Use of Compatibilizers and Handling of Mixtures

Much effort is being expended on polymer blends to reduce the need for separation and to improve the properties of recycled materials. Because most polymers are incompatible with one another, this usually requires the addition of a compatibilizer[182] or the preparation of one *in situ*. (Compatibilizers are often block or graft copolymers, with parts similar to each of the polymers that are to be made compatible.) A few examples will be given. A polypropylene grafted with maleic anhydride has been used to compatibilize blends of polypropylene and nylon-6,6.[183] (The amino end groups in the polyamide react with the anhydride groups in the graft copolymer to form a new amide.) A mixture of poly(ethylene terephthalate) and linear low-density polyethylene can be compatibilized using 10% of the sodium salt of poly(ethylene-*co*-methacrylic acid).[184] Perhaps, this could be used on a mixture of soda bottles and milk jugs. Reactive extrusion of a mixture of poly(vinyl chloride) and polyethylene with a peroxide and triallylisocyanurate gave a compatible mixture.[185] In this case, the compatibilizer was formed *in situ* by grafting. The product was a cross-linked copolymer. Moldings of a mixed waste of polystyrene and polyethylene had increased strength.[186] For this, the mixture was not compatible. The improvement in properties was due to the separation of the less than 30% polystyrene as fibers. Small amounts of hydrogenated styrene–1,3-butadiene block copolymers were used to compatibilize mixtures of polystyrene and polyethylene.[187] Blends of polypropylene and polystyrene were compatibilized with 2.5% polystyrene-block-poly(ethylene-*co*-propylene).[188]

Efforts are being made to handle mixtures typical of waste streams. A mixture of polyethylene, poly(ethylene terephthalate), and poly(vinyl chloride) bottles, typical of those used in Italy, was molded with an elastomeric block copolymer of styrene, ethylene, and 1,3-butadiene to produce improvements in impact strength and elongation at break.[189] Solid-state, high-shear extrusion of mixed waste linear low-density polyethylenes produced block and graft copolymers *in situ* to achieve improved compatibility.[190] For this, free radicals were generated mechanically rather than through the addition of a peroxide. The resulting mixture was then remolded. The concept is expected to be applicable to various mixtures of waste plastics to eliminate the need for sorting them.[191] This might be suitable for the "gnarled fence posts" made from recycled materials that are available at some garden stores.

A different approach taken by Rensselaer Polytechnic Institute is to separate the polyolefin plastics in the domestic waste stream.[192] It is said to be able to reclaim 98% of the plastics. The polystyrene is dissolved in mixed xylenes at room temperature, low-density polyethylene at 75°C, high-density polyethylene at 105°C, and polypropylene at 130°C. There may be problems with this scheme. Prior separation of poly(vinyl chloride) and poly(ethylene terephthalate) may be required. Polypropylene film is often coated with poly(vinylidene chloride) as a barrier coating. In addition, many polymers dissolve very slowly, even over days, to give highly viscous solutions. (The higher the molecular weights of the polymers, the greater these problems.) After flash evaporation of the solvent, some of it will remain in the polymer until the polymer is passed through a vented extruder. Traces of solvent are difficult to remove from viscous matrices.

14.2.3.4 Use of Chemical Reactions in Recycling[193]

The concern about possible contamination of food containers means that they can be recycled to only nonfood contact uses in the United States. To reuse poly(ethylene terephthalate) in contact with food, it must be broken down, purified (to remove metal compounds, colors, and such), and resynthesized (**14.2** Schematic). This has been done by hydrolysis, methanolysis, and glycolysis.[194] (The solvolysis can be complete in 4–10 min when microwaves are used with a zinc acetate catalyst.[195] It can be done in 10 min at 380°C.[196]) (For nonfood uses, such as the use of poly(ethylene terephthalate) in magnetic tapes, it may be sufficient to melt the polymer and filter out the chromium or iron oxides.[197])

14.2 Schematic

The technique is also applicable to cross-linked polyesters made from unsaturated polyesters and styrene, as in the sheet-molding compounds used in cars and boats.[198] The ester linkages were broken by treatment with ethanolic potassium hydroxide or by ethanolamine. The latter was preferred, because no waste salts were formed by neutralization, as they were with the potassium hydroxide. The filler and glass fibers could be removed at this stage, if desired. The alcohol-soluble product was suitable for use in new bulk-molding compounds. A method that allows the cross-linked polyester to recycle back into the same use involves heating with propylene glycol, re-esterification with additional maleic anhydride, and adding more styrene.[199] An alternative way of recycling the sheet-molding compound would be to pulverize it and use it as a filler in new sheet-molding compound. Waste polycarbonates can be broken down by phenolysis.[200] The bisphenol in it can be recovered in 80–90% yield by methanolysis[201] or hydrolysis.[202] (Direct reuse by remolding would be preferable when possible.) Polytetrahydrofuran can be converted to monomer by a hot treatment with hydrochloric acid on kaolin.[203] The solid catalyst was still active after repeated use.

Any condensation polymer containing ester, carbonate, amide, imide, urethane, or similar groups can be treated in this way to recover monomers or oligomers for making the same or other polymers. Soluble products from cross-linked polymers will usually have to be used in an application other than the original one. Nylon-6 wastes from carpets, or other, have been hydrolyzed to the monomer caprolactam (**14.3** Schematic).[204] (The carpet backing left behind has been tested as a filler in roofing shingles.) Phosphoric acid can be used as a catalyst for the reaction.

95% yield by hydrogenolysis in hot tetralin (the source of the hydrogen).[210] The mixture of liquid phenols (75% were cresols and xylenols) could be used to make new phenolic resins. (When simple pyrolysis was used, the yield of phenols was only 30–50%.[211]) This illustrates a way to recycle a thermoset (i.e., cross-linked by heat) polymer chemically, instead of just using it as a filler in a new polymer. Supercritical water has been used to break down polymers. Waste phenolic resin gave a mixture of phenols in 28% yield.[212] Polystyrene gave 92.4% selectivity for styrene, 5.1% for toluene, and 2.4% for benzene.[213] The distillation residue from the preparation of 2,4-toluenediisocyanate allowed recovery of 80 mol% of the diisocyanate as 2,4-diaminotoluene.[214] Styrene–butadiene rubber with water and hydrogen peroxide at 450°C gave a mixture of benzene, toluene, ethylbenzene, benzaldehyde, phenol, acetophenone, and benzoic acid in over 90% yield.[215] scWater at 385–600°C for 30–90 s is being used to burn off carbon compounds so that inorganic materials can be recovered.[216] This includes calcium carbonate from the ink sludge from recycling paper and precious metal catalysts.

Oxidation of waste plastics, such as polyethylene, polypropylene, polystyrene, poly(alkyl acrylates), and nylon-6,6, with NO/O_2 for 16 h at 170°C, led to mixtures of carboxylic acids, for which uses would have to be developed if this method was applied to large volumes of waste plastics.[217] Hydrogenation of polyethylene at 150°C for 10 h over silica/alumina-supported zirconium hydride catalysts gave a 100%

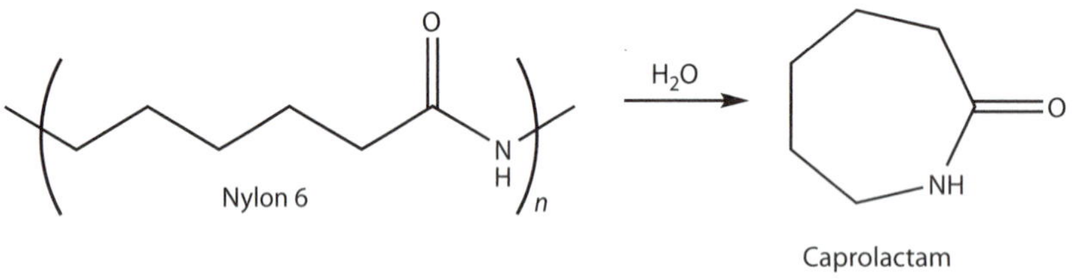

Nylon 6 $\xrightarrow{H_2O}$ Caprolactam

14.3 Schematic

Waste nylon carpet has also been pyrolyzed in a fluidized bed at 350°C over KOH/Al_2O_3 to recover 85% of the caprolactam.[205] Nylon-6 waste has been upgraded by reactive extrusion with trimellitic anhydride (without solvent) to produce polyamideimides, with number average molecular weights of 9000–30,000.[206] Polyamides have also been converted to monomers by treatment with ammonia.[207] Polyurethanes can be degraded by aminolysis or transesterification to liquid products useful in adhesives, foams, and composites[208] (e.g., as hardeners for epoxy resins). General Electric has devised a way to recover 4,4′-hexafluoroisopropylidenebis (phthalic anhydride) from scrap polyimides using lithium hydroxide in *N*-methylpyrrolidone at 225°C.[209] The glass fiber, diamine, and bisanhydride are recovered. The solvent and lithium hydroxide are recycled back to the process. Waste phenolic resin has been liquefied in more than

conversion to saturated oligomers.[218] Polypropylene gave a 40% conversion to lower alkanes at 190°C for 15 h. It is not clear what use these materials would have, other than serving as feedstock for a petroleum refinery.

Ketal structures have been built into cross-linked polymers for cleavage by acid to permit recycling. Compound **14.4** has been used to make cleavable epoxy resins.[219]

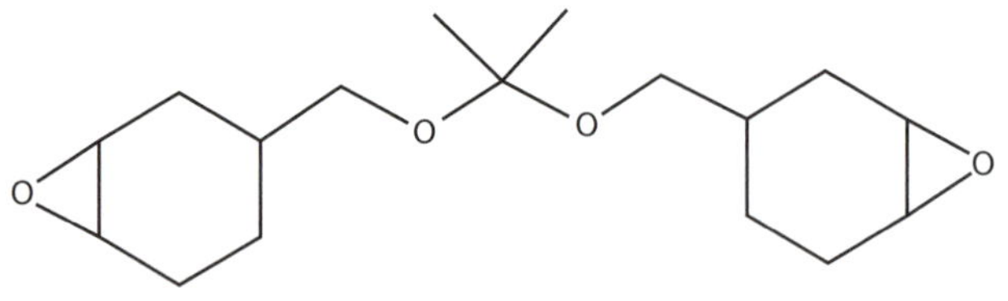

14.4 Schematic

The cured resins can be cleaved by methanesulfonic or *p*-toluenesulfonic acids in ethanol. An orthoester has been cross-linked with a dithiol (**14.5** Schematic), and then cleaved to a difunctional monomer suitable for recycling.[220] Systems involving ketal and orthoester cleavages may be too expensive for anything other than specialized applications.

Azo initiator
+ HSRSH → Cross-linked polymer
5 mol 3% CF_3COOH

14.5 Schematic

14.2.3.5 *Pyrolysis of Plastics*

A few polymers can be converted back to their monomers for purification and repolymerization. Polymers formed by ring-opening polymerization fall in this class, as shown in the foregoing by the conversion of nylon-6 back into caprolactam. When ethyl cyanoacrylate is used as a binder for metal and ceramic powders, it can be recovered for reuse by pyrolysis at 180°C.[221] The monomer can be obtained by pyrolysis of polymethyl methacrylate in 92–100% yield,[222] poly(α-methylstyrene) in 95–100% yield, and polytetrafluoroethylene in 97–100% yied.[223] Polystyrene can be depolymerized to styrene containing some styrene dimer by heating solid acids or bases at 350–400°C.[224] It is possible that the dimer could be recycled to the next run to produce more monomer. The best yield (>99% styrene) was obtained by passing polystyrene through a fluidized bed of a solid catalyst at 400–700°C, with a contact time of more than 60 s.[225]

Other polyolefins do not depolymerize cleanly to olefins.[226] Sometimes, the pyrolyzate is added back to the feedstock for a petroleum refinery.[227] A pilot plant in Portugal converts plastic waste to fuels and feedstock for a refinery.[228] A municipal plant in Japan converts 6000 metric tons/year of waste plastic to 2400–3600 metric tons/year of heavy oil.[229] Many pyrolyses are designed to obtain a mixture of liquid and gaseous fuels.[230] This[231] is little better than recovery of energy by incineration of mixed waste. It is not recycling in the sense of using the organic material over and over again to make the same or a similar product.

Pyrolysis of polyethylene and polypropylene for 1 s at 750°C gives mixtures of ethylene, propylene, and butenes in 91–92% yield.[232] Pyrolysis of polypropylene over HY zeolite gives a mixture of C_4–C_9 hydrocarbons, mostly monoolefins.[233] Cracking of polyethylene over H-ZSM-5 or sulfated zirconia can favor propene and isoalkenes at high catalyst levels.[234] Butenes and pentenes are also formed. Pyrolysis of mixed plastics in a fluidized bed yielded 25–45% gas plus 30–50% oil rich in aromatics, including benzene, toluene, and xylenes.[235] Poly(vinyl chloride) interferes with this process by sticking together in the bed and plugging tubes. Plastics have also been cracked in vented extruders, the existing gas being washed to remove hydrogen chloride.[236] Catalytic degradation of polyethylene with solid acid catalysts at 400°C produced 8.4% gas and 90.5% liquid.[237] Degradation of polyethylene, polypropylene, polyisobutylene, and squalane over the zeolite H-ZSM-5 produced products rich in gasoline-range chemicals.[238] Feeding molten low-density polyethylene to aH-Ga-silicate at 525°C gave a more than 70% yield of aromatic hydrocarbons.[239] Heating low-density polyethylene with MCM-41 zeolite produced a fuel similar to gasoline in 96% yield.[240] An iron-active carbon catalyst converted polypropylene in the presence of a small amount of hydrogen sulfide to 98% colorless distillate at 380°C.[241] Poly(vinyl chloride) and poly(ethylene terephthalate) must be removed from the entering waste.

Heating poly(butylene terephthalate) with 3 mol% of a tin catalyst in *o*-dichlorobenzene converted it to cyclic oligomers, which might be of value in reaction injection molding.[242]

14.2.3.6 *Biodegradable Polymers*

Biodegradable polymers[243] have been advocated by some to solve litter problems and to offer composting as a way of disposing of waste plastic. Merely being in a landfill will not ensure the decomposition of plastic. During the excavation of landfills, layers have been dated by the dates on recovered newspapers.[244] Truly biodegradable polymers, with groups that enzymes can recognize, were described in Chapter 12. Synthetic polymers such as polyethylene were made to degrade in sunlight or fall apart by bacterial action in several ways.[245] One is to put some carbonyl groups into polyethylene by including a vinyl ketone in the polymerization. Another is to accelerate photodegradation by inclusion of a metal salt [e.g., an iron (III) dithiocarbamate], which may be coated with a material that has to be removed by biodegradation before the photodegradation begins. Time control is obtained by an appropriate choice of metal compound and its concentration. Starch-filled polyolefins are degradable in that the bacteria can remove the starch, but the polyolefin itself remains. (In the United States the law requires that the plastic rings that hold six beverage cans together must degrade in sunlight to avoid animals becoming caught and dying in the rings.)

Polycaprolactone–paper composites have good resistance to water and are biodegradable.[246] Garbage bags of 50:50 potato starch/polycaprolactone are available commercially at about 3 times the price of polyethylene.[247] Larger volume is expected to reduce the cost to twice that of polyethylene. The Michigan Biotechnology Institute is developing maize-based coating for paperboard to replace polyethylene and wax coating on paper.[248] DuPont is producing Biomax polyesters, which are modified polyethylene terephthalates that degrade in composting in about 8 weeks.[249] Up to three "aliphatic monomers" are incorporated into the polyester to provide the linkages for bacterial degradation. The polyesters are expected to find use as coatings for paper plates and cups, bags for yard waste, and such. Eastman sells a biodegradable polyester for use in film that is made from adipic acid, terephthalic acid, and 1,4-butanediol.[250] A biodegradable polyamide that can be melt-processed has been made by copolymerization of aspartic and 4-aminobenzoic acids (**14.6** Schematic).[251]

If a way can be found to make it from ammonia, maleic anhydride, and 4-aminobenzoic acid, it may be relatively inexpensive and useful. It could be the biodegradable equivalent of nylon-6 and nylon-6,6. Chain extension with bisoxazolines or other reagents in the extruder may be needed to increase the molecular weight to obtain optimum physical properties.

Biodegradable polymers certainly have a place in the future, but with some limitations. If they are on paper that is being recycled, they will have to be designed to come off in the process. If they get into streams of plastics being recycled, the recycling could be ruined. For example, starch would decompose below the molding temperatures of nylon or polyester. If photodegradants enter the waste plastic stream by this route, the lifetime of the remolded plastic may be short. It will be difficult to devise a sorting and labeling system that will keep the usual synthetic plastic and biodegradable polymers separate in the waste stream. (A school system in Virginia decided to use biodegradable cutlery and trash bags so that they could be composted together with food waste.[252]) The best approach may be to cut back on the use of throwaway items in favor of those that are durable and last a long time in repeated use.[253]

14.2.4 Rubber

14.2.4.1 Scope of the Problem

The recycling of rubber[254] is largely a problem of what to do with used automobile and truck tires, for they constitute the bulk of rubber products that are produced. In 1996, about 250 million tires were discarded in the United States.[255] The "reuse/recovery rate," which includes burning for energy, reached 82%, with another 8% being retreaded.[256] Roughly the same number of tires were discarded in 1994 and 1995,[257] despite the fact that Americans are driving more miles each year. (Fewer were burned in these earlier years.) The leveling off in the number discarded may be due to the increased use of silica with a silane coupling agent as a reinforcement that improves tread life, tear strength, cut-growth resistance, and ozone aging, while decreasing fuel consumption.[258] Tires can now last for 100,000 miles, twice as long as they did 30 years ago. The average scrap tire weighing 20 lb contains 14% fabric, 16% steel, and 70% rubber.[259] The rubber contains carbon black as a reinforcing agent and is cured with sulfur.[260]

Rubber tires tend to work to the surface in landfills and serve as breeding places for mosquitoes when water gets

H_3PO_4, 180°C/4.5 h; M_w 29,000; T_g 91°C; T_m 282°C

14.6 Schematic

into them.[261] There are about 800 million used tires in stockpiles. Some stockpiles have caught fire and burned for months, as at Winchester, Virginia. The best use of a used tire is for retreading. The first retreading increases the life of the tire by 20,000–40,000 more miles.[262] Bus and truck tires can be retreaded 2 or 3 times, and airplane tires as many as 12 times.[263] In 1995, 10% of discarded automobile tires were retreaded compared with 55–70% of bus and truck tires. A new material that helps prevent delamination of tire treads from carcasses may spur more retreading.[264] About 12 million tires are used each year in civil engineering applications, such as retaining walls, artificial reefs, crash barriers, and floor mats. Use in reefs has proven to be an ecological blunder with tires washing up on shore.[265] Drainage pipes that are cheaper than those made of concrete or steel have been made by strapping tires with steel bands to form the pipes.[266] Old tires split around the middle have been buried under golf greens to retain irrigation water longer.[267]

14.2.4.2 Uses for Powdered Rubber Tires

About 6 million tires are ground each year for a variety of uses. Coarsely chopped ones are often used in children's playgrounds to minimize the impact of falls. They might also be used in garden paths, sidewalks, patios, and such. They have also been used as agricultural mulch that does not break down so that it does not have to be replaced annually.[268] Ground tire rubber has been used to replace sand or anthracite in filters for wastewater.[269] Finely ground materials are produced by cryogenic grinding, the fiber and steel being removed in the process.[270] Ford and Michelin are planning to use powdered rubber from tires in new tires.[271] A 1991 U.S. law specifies the use of powdered rubber from tires in federally funded asphalt highway paving.[272] This is done in Arizona, California, Florida, and Texas, but other states have yet to follow suit. The increased cost of the asphalt–rubber mixture may be the source of some of the protests. The price per unit of performance may actually be lower in the long run. Asphalt–rubber mixtures impart improved low-temperature crack resistance to roads.[273] This could translate into fewer potholes in roads in the spring of each year and less money spent on patching them. A road made with 18% rubber and 82% asphalt lasted longer and created less noise, 5 dB for cars and 8 dB for trucks.[274] A study at the University of Detroit suggests that 6–7% rubber is optimal.[275] For best results, the rubber should be ground to less than 80 mesh. Powdered rubber has also been tested in concrete.[276] Styrene block copolymer rubbers can be used in asphalt to extend the life of the asphalt up to 5 times.[277]

Composite Particles, Inc. has used mixtures of reactive gases, such as fluorine and oxygen, to introduce an oxygen functionality, such as hydroxyl and carboxylic acid groups, on the surface of powdered rubber tires.[278] The treated particles can be used at up to 40% in polyurethanes with no loss in physical properties. The product is cheaper because the treated rubber particles cost much less than the raw materials for the polyurethane. The particles can also be used in epoxy resins,[279] polysulfides, polyesters, and other polymers. The company also uses the surface activation process on polyethylene. Perhaps, this can be used in the recycling of polyolefin plastics. Another method of surface activation of rubber particles uses a coating that can cross-link on curing in a rubber matrix to improve the tensile strength.[280] These rubber particles can be used at up to 100% in making conveyor belts, mats, and damping sheets. Energy and equipment costs are reduced because mixing and calendaring steps are not needed.

14.2.4.3 Devulcanization

Scrap tires can be devulcanized in a variety of ways.[281] The powdered rubber can be softened with terpenes, and then masticated with chemicals that break the disulfide bonds.[282] Good physical properties were obtained when the devulcanized rubber was used at 50–70 parts per hundred of virgin natural rubber in revulcanization. Similar devulcanization can be done in an open mill with high shearing.[283] However, the recycled rubber had only 50–85% of the original rubber's tensile strength. Waste rubber can be devulcanized in heated ultrasonic screw extruders in a continuous process.[284] The product can be reprocessed, shaped, and revulcanized similar to virgin rubber. The retention of mechanical properties is said to be good, the resulting rubber having a tensile strength of 14.2 MPa (about 70% that of a vulcanizate of natural rubber) and 670% elongation at break. Interchain disulfides may change to cyclic sulfur compounds in the process. Scrap tires can be recyled by heating at 200°C/1250 psig.[285] When phthalic anhydride is included in the process, 75% of the original mechanical properties can be obtained. Silicone rubber can also be devulcanized ultrasonically.[286] Reclaimed rubber can be used in a variety of products, including vehicle mudguards, carpet underlays, adhesives, sound insulation, and floor mats. It was formerly used more with virgin rubber in the preparation of new tires, but the process gives insufficient crack resistance in radial tires.[287] The crack resistance of such mixtures might be improved by the use of treated silica fillers. The sulfur in finely ground scrap tires can also be removed by sulfur-using bacteria in water over a period of several days.[288] The product can be used with virgin rubber to make new tires with improved properties, which are cheaper than a tire of all-new rubber. Devulcanized rubber might be used to bind coarsely ground rubber into roofing shingles, insulating board, acoustic board, floors that are easy on the feet, and other construction uses.

14.2.4.4 Tires to Energy

Scrap tires can be pyrolyzed to a mixture of gas, oil, and activated carbon.[289] The zinc and the sulfur stay in the char.

The process is being used on a commercial scale in Hokkaido, Japan, but appears to be uneconomical in the United States. (Mixtures of scrap tires with waste plastics have also been pyrolyzed.[290]) The recovered carbon black can be used in new tires.[291] It appears to be cheaper in the United States to burn the tires for energy at lime kilns, paper mills, and such.[292] The sulfur in the tires must be trapped by the lime or by other means. The combustion conditions must be such that the usual thick black smoke of burning rubber is not present. Recycling the rubber in the tires to more items of rubber would be preferable to just burning them as fuel. The real key to fewer tires to recycle is to devise better land use systems that require fewer automobiles (see Chapter 15).

14.2.5 Metals

14.2.5.1 Methods of Recycling

Aluminum, lead, and steel are the principal metals that are recycled.[293] The aluminum comes mostly from beverage cans. Of the 130 billion metal cans used in the United States each year,[294] 99 billion are made of aluminum.[295] In 1997, 59.1% of the aluminum beverage cans made in the United States were recycled.[296] In 2007, 54 billion cans were recycled, 53.8% of the total.[297] Americans discarded 32 billion soda cans containing 435,000 tons of aluminum in 2002.[298] Cars are the largest single source of scrap iron.[299] The lead is from car batteries. (In the United States, 96.5% of lead–acid batteries were recycled in 1996.[300]) Clean streams of metals recycle best. Non-can aluminum consists of different alloys than the cans and should not be mixed with them for recycle.

Physical methods of separation are preferred to chemical ones.[301] Magnets are used to separate steel cans from municipal solid waste. After the battery, gas tank, radiator, and reusable parts are removed from a used automobile, it is shredded. Air classification removes the fabrics and light plastics. The steel is taken out by a magnet. The aluminum is separated from the copper, zinc, and lead by density using a sink–float technique and then from the remaining materials in an eddy current separator. If the polyurethane foam, which can be recycled,[302] is removed mechanically, the remaining plastics can be separated by dissolution in various solvents, acetone for acrylonitrile–1,3-butadiene-styrene terpolymers, tetrahydrofuran for poly(vinyl chloride), and xylene or cyclohexanone for polypropylene. This solution process probably suffers from the disadvantages mentioned earlier in connection with the use of solvent to separate polyolefins in solid waste. Much thought is being given to the preparation of cars that will be easier to recycle at the end of their useful lives.[303] DuPont has made a prototype "greener" car in which poly(ethylene terephthalate) is used to replace many of the other plastics found in the usual car. Polyester cushions, seat covers, and headrests were used. DuPont says that the polyester part could be recycled as a unit or remolded in the same way that soda bottles are.

Steel food containers should be detinned before melting down in order to recover the valuable tin and to avoid contamination of the batch of steel. Impurity zinc can be removed from aluminum by distilling it out at 900°C under argon in 1 min.[304] Zinc can be recovered from galvanized steel by distillation at 397°C at 10 Pa.[305] It can also be recovered in 99% yield by extraction with 40% aqueous sodium hydroxide at 200°F, followed by electrodeposition of the zinc, which recycles the sodium hydroxide to the process.[306]

Americans discarded 2.5 billion steel clothes hangers (195 million tons) in 2006.[307] A substitute is an "Eco Hanger" made of 100% recycled and recyclable paper, which lasts about 2 months in normal closet use. Steel minimills can operate with nearly 100% scrap steel to make reinforcing bars for concrete and other items; in 1999, there were eight steel minimills in New Jersey.[308] Scrap provided 56% of the steel in the United States in 1999. The making of steel cans from recycled steel saves 60–70% of the energy needed to make them from iron ore. Recycling aluminum cans saves 95% of the energy needed to make them from bauxite ore. Aluminum recycling reduces air pollution by 95% and water pollution by 97%, compared with making cans from bauxite. The emissions of fluorocarbons from the aluminum fluoride used in the electrolytic preparation of aluminum are eliminated. The paper labels and inner and outer coatings are burned off when the metal is remelted. In contrast to organic materials, such as paper and plastics, the metals can be used over and over again with little or no loss.

Precious metals are recovered from catalysts after burning off any organic materials present.[309] Quantitative recycling of Raney nickel catalyst can be done on a lab scale by attaching it to the stirrer bar.[310]

14.2.5.2 Recycling By-Products of the Manufacture of Metals

When steel is poured from one ladle to another, a certain amount of oxidation of the molten metal occurs. The residual "kish" can be leached with acid to recover flakes of graphite, which could supply 6% of the market in the United States.[311] If this leaching is done with hydrochloric acid, the leachate can be combined with pickling liquor that is obtained by dissolving the scale and rust off metal sheets. One plant sprays this liquid down a tower 100 ft high at 640°C (1200°F) to recover iron oxide and hydrogen chloride for recycle to the process. Zinc and iron have been recovered from pickling steel with 20% hydrochloric acid.[312] Iron, chromium, and nickel have been recovered from pickling liquor of stainless steel pickled with hydrofluoric and nitric acids.[313] The sludge from air pollution control of a blast furnace has been treated with hydrochloric acid and sodium hypochlorite to recover iron, carbon, zinc, and lead.[314] Other metal oxide wastes have been reduced to the metals with coal or coke to recover the metals.[315]

Some aluminum oxide and nitride form when aluminum is melted and cast. This "dross" can be heated to 3000–5000°C in an electric arc furnace under argon to recover 80–90% of the aluminum in it.[316] The process is said to be 30% cheaper than the current recovery method. Less "dross" is produced if the argon plasma is used in remelting aluminum scrap.[317] Aluminum recycling produces a salt slag as waste.[318] Half of this is a salt flux that can be recovered and recycled. The rest is impure aluminum oxide that might possibly be combined with bauxite for the preparation of aluminum. Spent pot liners in the aluminum refinery can be processed to recover aluminum and calcium fluorides for recycle to the process and carbon for use as a fuel.[319]

14.2.5.3 Recovery of Metals from Petroleum Residues

Nickel and vanadium in the fly ash from the burning of heavy oil can be recovered by chlorination and distillation, with 67% recovery of the nickel and 100% recovery of the vanadium.[320] Another method is to leach out the nickel with aqueous ammonium chloride, followed by treatment with hydrogen sulfide to recover nickel sulfide (in 87% yield), which could be refined in the usual way.[321] The vanadium was recovered next in 78% yield by solvent extraction with trioctylamine, followed by treatment of the extract with aqueous sodium carbonate to take the vanadium back into water, and finally ammonium chloride to produce ammonium vanadate.

14.2.6 Glass

Most glass recycling[322] is done with jars and bottles that are melted down and made into more jars and bottles. As with metals, this process can be done over and over again. Other types of glass, such as windows and light bulbs, as well as ceramic tableware, have different chemical compositions and must be kept out of the container-making process. They too could be recycled into more window glass and light bulbs, if there was enough of them in one place at any given time. The containers must be sorted by color (clear, green, and brown). If the containers are whole, this can be done automatically by an on-line color sensor. Broken glass of mixed colors is a problem, especially if it is in small fragments, and might have to be diverted to another secondary use. After sorting by color, the glass is pulverized. Paper labels can be removed by screening or suction. Steel caps are removed with a magnet. The pulverized glass (called cullet) can be melted down and made into new containers. More commonly, additional sand, limestone, and sodium carbonate are added in making a new batch of glass. The addition of the cullet lowers the melting point, saves energy, and prolongs furnace life.

The United States produces 40 billion glass containers (10.6 million tons) each year. (In 2008, 25% of the glass jars and containers made in the United States were recycled, down from 35% in 1997 and 37.9% in 1996.[323]) In addition, 800,000 tons are imported. Because many of the imported containers are green wine bottles, too many used green containers enter the waste stream for effective recycling back into green containers. Thus, they may have to be diverted to secondary uses, such as fiberglass and road aggregate. About 60% of the streets of Baltimore, Maryland, are paved with glassphalt, a mixture of glass and asphalt. Up to 40% glass can be used in glassphalt. Reflector beads for roads can be made from recycled clear glass. Small porous glass spheres made from recycled glass are sold as lightweight fillers and insulators.[324] Auto glass can be converted into bathroom tiles.[325] Discarded fluorescent light bulbs can be crushed, the mercury and fluorescent material removed, and then be converted to fiberglass insulation.[326] This can keep some mercury out of landfills, but requires an effective collection system for used lamps. The U.S. Congress has passed resolutions calling for the EPA to regulate the disposal of mercury-containing lamps.[327] Powdered glass of mixed colors can be converted to rhodesite, a double-layer silicate, which is an ion-exchange resin and chemical adsorbent, by heating with sodium hydroxide or sodium carbonate solutions at 100–150°C for 1–7 days.[328] Pulverized glass has also been mixed with a reactive binder and compacted into blocks for the construction of walls.[329] The blocks serve as insulation that reduces the energy needed for heating and air-conditioning.

14.2.7 Miscellaneous Recycling

14.2.7.1 Composting

Yard wastes (grass, leaves, and tree limbs) constitute 15.9% of municipal solid waste, which can easily be recycled by composting (see also Chapter 11). The best method is to eliminate the grass clippings by cutting the lawn frequently and leaving the clipping on the lawn where they fall. The leaves can be composted on site, in the backyard, which eliminates the cost of collection. With the proper moistening and turning of the pile, odor will not be a problem. Food wastes can be included if the composter has a screen to keep out dogs, raccoons, rats, and such.[330] Under proper composting conditions, the decomposition process will produce temperatures of 56–60°C, which will kill any pathogens present. Composting can also be done on a neighborhood or on a municipal basis, although there will be a cost for collection.[331] In Newark, Delaware, leaves are raked to the curbside by the homeowner and then sucked into a city truck for transport to the municipal composting facility. The resulting compost is available to residents of the city. Many states and cities now prohibit yard waste in trash. Texas has 50 large-scale composting operations.[332] Brush disposal in Texas landfills has dropped 75% since 1992 owing to this. More than 8000 American farms now have composting operations.[333] African cities are also composting.[334]

Compost can also be produced from a variety of other organic wastes. Undeliverable bulk mail has been mixed with food and yard waste for composting.[335] Pulp and paper mill sludge can be composted instead of being put into a landfill.[336] Animal waste from chickens, cattle, hogs, or other farm animals can be composted directly or used in an anaerobic system to produce biogas with compost as a residue.[337] (Contamination of water by pig and poultry farms is a problem in Delaware, Maryland, North Carolina, and the Netherlands.) Sewage sludge is often handled in the latter way. Compost from municipal solid waste has not worked well, because it may contain toxic heavy metal ions, slivers of glass, and pieces of plastic.[338]

Compost consists of material that does not rot readily. The humic and fulvic acids in it are ill-defined, alkali-soluble materials derived from lignin.[339] They are aromatic compounds containing carboxylic acid, hydroxyl (both phenolic and alcoholic), ketone, and quinone groups. Incorporation of compost into the soil recycles nutrients, provides ion-exchange capacity for calcium, magnesium, and other ions, and promotes aggregation of soil particles, water infiltration, and water retention. Conventional agriculture depletes soil organic matter to the extent of 50% in 25 years in the temperate zone and in 5 years in the tropics.[340] The application of compost from yard waste, crop wastes, and sewage sludge can enhance the fertility of the soil.

14.2.7.2 Uses for Food-Processing Wastes

Food-processing wastes are often used as food for animals. Corn gluten meal, left from the preparation of high-fructose corn syrup, can be used as food for fish in aquaculture.[341] So can the distillers dry grains left from the fermentation to produce ethanol. Dietary fiber for consumption by humans can be recovered from pear and kiwi wastes.[342] Tomato skins can be removed by steam and abrasion in a process that eliminates the traditional use of sodium hydroxide for peeling. The waste can be fed to animals. Brewer's spent grain has been used to make bread.[343] A plant in Missouri converts turkey scraps into an oil resembling biodiesel at 100–200 barrels/day.[344]

14.2.7.3 Recycling Wood and Construction Wastes

More wood needs to be recycled. Wooden pallets need to be reused over and over again. When broken beyond repair, they can be chipped for use in chipboard[345] or for use in paper. They can be combined with wood waste from construction or demolition for this purpose. (Frigidaire is now using reusable polyethylene pallets, which may last longer.[346]) New Haven, Connecticut, shreds a combination of stumps, trees, and other wood waste to produce mulch. Problems have arisen from burying such waste on construction sites. Sinkholes have formed and methane has leaked into basements.

Residential construction waste is made up of 42% wood, 26% drywall, 11% masonry, 4% cardboard, 2% metals, and 15% other materials.[347] Roughly 80% of this waste can be recycled. The key is in segregating the wastes on site. This can be done by the contractor, the subcontractor, or a cleanup company. Some scrap lumber can be used as braces in the new dwelling. The rest can be handled as in the foregoing for chipboard. One way that has been used to keep it out of the landfill is to put it in a pile at the edge of the site with a "free" sign on it. Scrap drywall can be remade into new drywall. Pulverized drywall can be used as a soil amendment. The ideal is to refurbish existing buildings for new uses instead of knocking them down and building new ones. In some cases, wooden buildings are burned down by local fire departments as the cheapest way to get rid of them. The problem is that it costs a lot for hand labor to take apart a building so that the materials in it can be salvaged. New automated methods for doing this need to be explored. Perhaps, roofing shingles could be removed by suction using a mobile crane ("cherry picker") or by an automated chisel that would go under the shingles. The old shingles could be recycled into new shingles or into asphalt. After cutting one hole in the roof, hooks on a crane might pull off a beam or a panel at a time. The nails in the downed beams could be removed by saws in a machine that sized up the board by sensors. Masonry might be removed by a mechanical chisel. It is possible that high-pressure jets of water could separate the mortar from the bricks after the walls were taken down. Another possibility would be to use automatic chisels followed by a planing machine guided by automatic sensors. Used brick was popular in home construction in the United States at one time. The mortar could be used in making new mortar, in concrete, or on agricultural soil.

There are eight companies in the United States that take apart houses and sell the salvaged materials at a discount.[348]

In the United States, 90% of single-family homes are framed in wood, usually on site ("stick-built").[349] Steel framing was as cheap as wood, but was used in only 13,000 houses out of the 1.3 million built in the United States in 1993.[350] The use of steel could help save old-growth forests, but its production does require a lot of energy. Preparation of steel framing for houses from scrap steel is a possibility. In Japan and Sweden, houses are often made as wall panels in a factory and then lifted into place on site by a crane. Such a system allows for designs that minimize waste and centralize what is waste for ease of recycling. Commercial buildings in the United States are usually made of steel, concrete, and masonry. Because concrete made from fly ash from electric power plants is stronger than that made with sand, this offers an outlet for waste fly ash.[351] Fly ash can also go into the concrete blocks used in construction. Heating it with coal at 1300°C under nitrogen produces $Si_3Al_3O_3N_5$ ceramics.[352] "Red Mud," the residue from the extraction of bauxite for aluminum manufacture, has been used with calcium oxide and ferrous slag to make tiles and bricks.[353] (One plant in Louisiana produces over 1 million tons of "Red Mud" each year.)

Wood is also being saved by substituting concrete cross ties for wooden ones on railroads. However, production of cement for concrete is an energy-intensive operation and trees are a renewable resource when managed properly. Railroad ties made of 100% recycled plastic are being tested.[354] Japan uses 25 billion disposable wooden chopsticks per year, 95% of which are imported.[355] There is a campaign to switch to reusable stainless steel ones. China has the same problem.

14.2.7.4 Reusing Carpets and Clothing

Textiles are about 5% of municipal solid waste. Their recycling rate is 25%. In the United States, 66 million tons is landfilled each year.[356] Scrap carpet and used carpet of polyester or of nylon can be converted back to monomer for reuse as described earlier. (Seven billion pounds of carpet is discarded each year in the United States.[357]) They can also be used as reinforcing fibers for concrete, nonwoven, and lower-melting resins, such as polypropylene and polyethylene.[358] This might require a surface treatment of the fiber (or with the polyolefins, the matrix) to increase polarity. (In former years, textile fibers were used to produce high rag content paper.) This might be done to produce strong paper for currency, mailing envelopes, and such. Carpet waste has also been converted into needle-punched thermoformable trunk liners, door, and roof panels for cars, and into panels that can replace plywood in some applications, such as floors.[359] Collins and Aikman (Georgia) takes back vinyl-backed carpet from customers and puts it into recycled backing for modular tile products.[360]

Similar uses might be found for worn-out clothing. Clothing that is still useable is collected by charities for resale.[361] The part that cannot be sold in the United States is shipped to developing nations at the rate of 1 million containers a year.[362] It is then used as such or resewn into other garments. Recycling textiles generates 37 times as many jobs as landfills and incinerators.[363] Clothing made of cotton/polyethylene terephthalate can also be recycled by selectively hydrolyzing the polyester with sodium hydroxide, and then dissolving the cotton remaining in *N*-methylmorpholine–*N*-oxide for spinning into rayon.[364] Presumably, the terephthalic acid could also be recovered for reuse.

Used mattresses can be resold in 15 states in the United States after sterilization by dry baking or chemical disinfection.[365] In other cases, they are sometimes dismantled and the components sold.

14.2.7.5 Handling Used Oil, Antifreeze, Ink, and Paint

Methods of recycling oil[366] were discussed in Chapter 13. Unfortunately, not all of the oil goes into the recycling containers. About half of the used oil from by motorists who change their own oil goes down storm drains. In an effort to curtail this pollution, some cities are putting labels on storm drains pointing out that anything poured down the drain goes into the nearest river or lake. The oil can harm fish and other aquatic organisms. A year after the Exxon Valdez oil spill, the effects could still be seen in the cytochrome P4501A of the fish *Anoplarchus purpuresens*.[367] Improperly discarded oil filters release 3 times the amount of the Exxon Valdez oil spill into the environment of the United States each year.[368] Oil filters can be crushed and heated to recover both the steel and the oil. In Delaware, this oil (4500 gallons/year) is recycled at the oil refinery at Delaware City.[369] Liquid carbon dioxide is being used to take motor oil off high-density polyethylene containers, so that both can be recycled.[370] Safety-Kleen Corp. offers nationwide collection for used oil from cars and other sources.[371] Mobile on-site services are also available for removing toxic compounds and acid degradation products from ethylene glycol antifreeze, with the addition of new corrosion inhibitors, so that the antifreeze can be reused.[372] Organoclays have been used to remove oils from water.[373]

Ink recycled with the aid of mobile equipment costs 15–35% less than new ink.[374] About 60–75% of reconstituted ink is from waste ink, with the remainder being new materials to rebuild ink. Each year U.S. households discard 68 million gallons of unused paint.[375] Another 79 million gallons of architectural paint is discarded. This has been combined with waste polyethylene to produce a tough plastic that can be sold. A company in Massachusetts recycles leftover paint by straining it and mixing it with new paint.[376] The cost of recycled paint is $<50\%$ that of new paint.[377] Paint is the largest volume item in household hazardous waste collections. A company in Toledo, Ohio, heats paint sludge to 400–580°F and then uses it to make caulks, adhesives, and sealants.[378] Gauze plus paint sludge from painting cars can be used to seal roofs, a process that is cheaper than a new roof.[379]

14.2.7.6 Toner Cartridges

Some manufacturers have taken back used toner cartridges (used in laser printing), cleaned and inspected them, replaced worn parts, if needed, refilled them with toner, and sold them.[380]

14.2.7.7 Use of Baths of Molten Metal and Plasma Arcs

When no other way of recycling is available, organic wastes can be dropped into a bath of molten iron, at 1650°C, to produce carbon monoxide, hydrogen, and hydrogen chloride (if chlorine is present).[381] Lime and silica fluxes may be used to form a slag layer. Heavy metals collect in the bath.

No nitrogen oxides, sulfur oxides, or dioxins are evolved. Destruction of organic compounds is more than 99.999%. The synthesis gas formed can be used to make methanol or for hydroformylation. Any hydrogen chloride formed would have to be removed by a scrubber. This is said to be cheaper than incineration.

A plasma arc torch, operating at 3000°F, converts wastes to hydrogen, carbon monoxide, metal, and a glassy slag, so that nothing is wasted and nothing goes to the landfill.[382] The slag can be used in asphalt or concrete. Air emissions are one tenth those of an incinerator. The process could make landfills obsolete. Fort Pierce, Florida, planned to build a $425 million plant to process 8000 tons of municipal solid waste per day plus what is in the existing landfill but has scaled back the size of the plant.[383] Two plants are operating in Japan. Others are in Washington state and Hawaii. Startech Environmental Corporation markets the technology.[384] Aseptic juice containers consist of layers of paper and aluminum plus adhesive layers. Tetra Pak and Alcoa recycle these with plasma jet technology in Brazil.[385]

14.2.7.8 Uses for Inorganic Wastes (Other than Glass and Metals)

A few examples will be given of efforts to recycle inorganic wastes. Only 22% of coal fly ash is utilized, with outlets in cement and concrete products, filler in asphalt, grit for snow and ice control, road base stabilization, and others.[386] Concrete made with fly ash is stronger and is up to 30–50% cheaper.[387] Building blocks can be made from fly ash, waste calcium sulfate from flue gas desulfurization, and lime or Portland cement.[388] More could be used if we built our houses of concrete blocks instead of wood. Efforts have also been made to recover the aluminum, iron, and other metals in fly ash. This would reduce the need to mine so much new ore. As the world switches from fossil fuels to renewable sources of energy, the problem of what to do with fly ash will disappear. The ash from incinerated sewage sludge has been mixed with silica, alumina, and lime for conversion to tiles and floor panels in Japan.[389] Spent bleaching earth from the treatment of edible fats and oils presents a disposal problem.[390] The best approach is to regenerate it for reuse. Extraction with isopropyl alcohol or hot water under pressure removes adhering oil so that the solid can be reused as a filter cake. It can also be put into a cement kiln or used in land application. Contaminated sulfuric acid can be cleaned up by cracking at 1000°C, followed by conversion of the exit gas to fresh sulfuric acid.[391] The process is in use at three plants in Europe and at six in the United States. Zero discharge of wastewater is the goal of industry. This involves biological treatments, reverse osmosis, and so on. The color of spent dyebath water can be removed by treatment with ozone so that the water can be reused.[392]

14.2.7.9 Electronic Waste

Large numbers of television sets, computers, cell phones, and such are discarded each year by developed nations because they no longer work or the owner wants a newer model.[393] Those in working condition could be reused by someone else. Others could be reused after simple repairs. Many end up in landfills, where lead may leach out of the cathode ray tubes.[394] Others end up in China or Africa, where they may be taken apart for salvage under very unsafe conditions[395] Governments are searching for better systems.[396] NEC Solutions America sells computers that do not contain the usual lead, barium, boron, cobalt, and bromine.[397] It run on one third the energy of a typical computer. There is no effective system in the United States for recycling batteries.

14.3 METHODS AND INCENTIVES FOR SOURCE REDUCTION

14.3.1 Range of Approaches

The following hierarchy of methods of use goes from those with the most waste to those producing the least waste. (An example is given in most cases.)

Use once and then bury it in a landfill (throwaway packaging).
Use once and then compost it (biodegradable packaging).
Use once and then burn it with recovery of energy (throwaway packaging).
Use once and then make it into a different object, which after use can be burned or buried (high-density polyethylene milk jug made into a container for motor oil).
Use once and then make it into a different object that will last for many years (a polyethylene terephthalate soft-drink bottle made into fiber for clothing or a sleeping bag).
Use once, and then melt and reshape it to original use (recycled aluminum cans or single-use glass bottles).
Use once, and then melt, and reshape it to a different long-term use (steel can be made into a reinforcing bar for concrete).
Use once and then hydrolyze the polymer to recover the raw materials, which are then converted into a duplicate of the original object (a polyethylene terephthalate soft-drink bottle made into another soft-drink bottle).
Use 10–60 times and then remake into a duplicate object (a refillable glass or a poly(ethylene terephthalate) soft-drink bottle).
Use 10–60 times and then remake into a different object with a long life (a polycarbonate milk bottle recycled into a part for an automobile).
Use repeatedly, sharpen as needed and replace small parts that wear out (saw and safety razor).
Use indefinitely (china mug or drinking glass).
Is the object or its content needed at all (bottled water, neckties)?

14.3.2 Making More with Less Material

The best approaches are to reduce, reuse, or recycle, listed in order of decreasing importance. To a manufacturer, source reduction[398] often means making the same number of objects with less material. This may be a design change, such as substituting an indented bottom for the polyethylene base cup formerly used on a poly(ethylene terephthalate) soft-drink bottle. The use of a stronger plastic resin, perhaps with a higher molecular weight or a better molecular weight distribution, may allow the bottle to be made thinner. Optimization of biaxial orientation may lead to increased strength so that less material is needed. In 1978, a 2-L soft-drink bottle of poly(ethylene terephthalate) weighed 100 g and cost 0.48 dollars. Today, it weighs 50 g and costs 0.10–0.12 dollars.[399] The milk jug of high-density polyethylene of the early 1970s weighed 95 g. In 1989 it weighed 60 g.[400] The 12-fluid-oz aluminum beverage can has become thinner over the years until now 32 cans weigh 1 lb.[401] This represents a 50% reduction in weight. The pressure in the can helps it to hold its shape. Another method of reduction is to achieve longer wear (see Chapter 13). Because of improvements in the materials and methods of construction, automobile tires now last twice as many miles as they did 30 years ago.

14.3.3 Throwaway Items and the Consumer

Source reduction for the consumer means obtaining products that he or she needs with a minimum of waste to be discarded. The most important approach is to substitute reusable objects for single-use, throwaway ones. It is instructive to compare what is done today in the United States with what was done 50 years ago and with what is still done in many other countries[402] (Table 14.2).

Too many items are discarded because a person wants something new; a fad passes and an item goes out of fashion; the person gains or loses weight; the child outgrows it; it was stored improperly; a minor repair or refinishing is needed; and so on. It should be possible to reverse some of these trends to minimize waste, energy consumption, and natural resource depletion without adversely affecting the health or longevity of people. One third of Americans today are obese, in part because of all of the "labor-saving" devices to which they have become accustomed.

There is some question over why some of the items in Table 14.2 need to be disposable. Two billion disposable razors are used each year in the United States. Reusable safety razors work, with the need to dispose of only used blades. This author has minimized his use of razors and shaving cream, while saving 10 min every day, by wearing a

Table 14.2 Approach to Substitution of Reusable Objects for Throwaways

Today	What Was Formerly Done
Paper napkin	Cloth napkin
Paper tissue	Handkerchief
Plastic bag for waste	Wash-out garbage can
Discard broken shovel and buy a new one	Replace broken handle on shovel
Single-use soft-drink container	Glass bottle with 0.02 dollars deposit
Single-use polyethylene milk jug	Milk delivered to door in refillable glass bottles
Single-use glass cider jug	Take own bottle to farm for cider
Assortment of two dozen screws in plastic container	Buy one screw at local hardware store
Discard faulty toaster	Local repair of toaster
Two disposable mouse traps in plastic package	Buy one mouse trap and use it for years
Bottled water	Canteen
Disposable polystyrene cup	China cup or drinking glass
Disposable diapers	Cloth diapers collected from home and cleaned by diaper service
Disposable camera	Use camera indefinitely, just buy new film as need
Disposable razor	Safety razor used indefinitely, just buy new blades
Aluminum pan under frozen TV dinner	Use own pan repeatedly
Cling wrap and aluminum foil to wrap food	Wax paper
Plastic grocery bags	Paper grocery bags or market basket
Weed whacker or herbicide	Hand clippers
Recreation vehicle	Tent
Disposable needles for syringes	Sterilize needles in autoclave
Electric pencil sharpener	Turn sharpener by hand
Photocopier	Copy by hand or blue print
One car per driver	Ride bus or street car
Disposable ballpoint pen	Fountain pen
Leaf blower	Rake

beard. Kodak advertises its single-use camera as 86% recyclable. A regular camera lasts so long that recycling is a minor issue. It requires only new film. No film is needed in digital cameras. The use of wooden pencils has resulted in overexploitation of red cedar (*Juniperus virginiana*) trees and more recently of *Libocedrus decurrens*.[403] Pencil board can be made today from recycled paper.[404] A mechanical pencil requires only replacement lead, which is used completely. A fountain pen requires only replacement ink. A cut-out hole eliminates the need for a plastic window in an envelope. A rubber stamp or a fountain pen eliminates the need for an adhesive return address label. A moist sponge can eliminate the need for the strip of adhesive tape sometimes used to seal envelopes. If the material inside is not personal, the flap can just be tucked in. An air-cooled engine requires no engine coolant.

Alliance Medical Corporation cleans used surgical instruments, sterilizes them, and sells them to hospitals at a discount.[405] Before the 1970s, hospitals did this themselves. A few stores sell secondhand building materials, such as doors, kitchen cabinets, and carpets.[406] Since 1992, a church in Wilmington, Delaware, has run a warehouse for used building materials, such as kitchen sinks, that are donated by local businesses, contractors, and families remodeling their homes. It provides help to low-income families that are improving their homes.[407]

Some wasteful practices are part of local customs. There is no need to buy shoes for a child who cannot walk. Many stores in the United States do half their annual business in the month before Christmas. Many of the presents given are ill-chosen and unwanted. Some people consider Christmas—a religious observance—to be overcommercialized. The same is true of many wedding presents. Would a gift certificate at a local grocery store be a suitable replacement? Synthetic Christmas trees can be used year after year without having to cut a tree in the forest. This is probably more environmentally sound than "recycling" a real tree by chipping it for use as mulch. One waste disposal firm in Delaware distributed 1000 free plastic bags for handling the used trees, thus adding plastic to the waste stream.[408] Some stores replace their storefronts well before they are worn out because this brings in more customers and sales go up.[409]

Instead of having your own of everything, could you share a tool with neighbors, borrow it from a civic association, rent it when needed, or hire a person with one to do the job? As an example, a person with a half acre lot probably does not need his own chain saw. There is a small movement in this direction. There are 200 sharing villages in Denmark, where 10% of all new housing is in cohousing clusters, which use half as much land as the usual suburb.[410] Some cohousing communities share tools, such as staple guns, sewing machines, leaf rakes, carpentry tools, toilet plungers, backpacks, or tents. A system of car sharing is developing in Europe. This has spread to the United States, where a person can join a group such as Philly Car Share or Zip Car. Could you spend a little more for a washable wall paint so that you would not have to repaint so often? Is the deck really necessary, or could you use a lawn chair? Houses being built in the United States today are larger and average fewer occupants in them than those built a generation ago. Millions of people, including 14 million widows, live alone in houses in the United States. Is there a way that we can alter our social system so that more of these singles can share a house?

14.3.4 Containers and Packaging

Containers and packaging[411] constitute 34% of municipal solid waste by volume. There are many ways in which this can be reduced without lowering the standards of sanitation. More businesses are shifting to multitrip containers for delivering their products.[412] In one case (Nalco), the use of 96,000 15–800 gallons stainless steel containers has eliminated the need to dispose of 3 million drums and 30 million lb of chemical waste since the program began.[413] Use of returnable, refillable, pesticide containers reduces the likelihood of contamination of the environment, while saving time and money.[414] Manufacturers ship more than 90% of all products in the United States in corrugated cardboard boxes.[415] Multitrip containers for the delivery of food to supermarkets could save large amounts of corrugated cardboard. This calls for some innovation to produce easy-to-open and -close containers that nest or can be collapsed for reshipment to the manufacturer. Standardization of package sizes or at least crate sizes could allow them to be used by many different companies, with credit going to the company that provided them in each specific case. A German manufacturer is studying the use of collapsible polypropylene crates for use with its products. These might be made of plastic or other material. 3M has a reusable, collapsible-packaging system that saves the company 4 million dollars/year.[416] Frigidaire has a similar program.[417]

There are also many opportunities for reduction of packaging at the consumer level. The Giant chain of supermarkets in the eastern United States offers over 100 different foods and other items in bulk (in addition to the usual fruits and vegetables). The customer uses a scoop to fill a plastic bag, and then puts a sticker and a plastic tie on it. The items include cereals, candy, dried fruits, nuts, pasta, snacks, and pet supplies. They are usually cheaper than those that are prepackaged. Unfortunately, this is only a small portion of the items sold in the store. It should be possible to add beans, detergents, coffee, and other items. Beverages might be dispensed into the customer's own container, as done with milk in parts of Germany. The only requirement may be that the cashier at the checkout counter be able to recognize the item inside the bag or jar. Cereal in large boxes, instead of in individual variety packs, is 78% less expensive and generates 54% less waste.[418] Buying fruit juice in frozen concentrate, rather than in shrink-wrapped packs, cuts the cost by

41% and the waste by 61%. The least packaging for breakfast cereal uses the form-and-fill package made of plastic film and uses no outer box. The box of rolled oats used to be made of all paper. Now it comes with a plastic top, which makes it harder to recycle. Some bread is double wrapped, even though a single layer is sufficient. Some fabric softeners are now available as liquid concentrates. The customer buys the original container, which can then be refilled with concentrates as needed. There are also examples in which small items are put into oversized packages so that the package can carry an advertisement. It would be better to put the advertisement on the bin where they are kept.

Beverage containers account for over 5% of municipal solid waste. As mentioned earlier in this chapter, the United States uses 99 billion cans and 46 billion glass containers each year. David Saphire of Inform, Inc. has done a careful analysis of the use of refillable containers for this purpose.[419] Refillable glass or polyethylene terephthalate bottles making 25 trips use 91–93% less material and lead to 96% fewer bottles as waste. Even at eight trips, refillable glass bottles use 78% less glass than one-way bottles carrying the same amount of beverage. At 25 trips, glass beer bottles will consume 93% less energy than would be required to make the single-use bottles to carry the same amount of beverage. The energy used in washing the refillable bottles is more than offset by the energy that would be required to make new bottles. A 12-fluid-oz refillable bottle used 10 times takes 75% less energy per use as a recycled glass or aluminum container and 84–91% less than a throwaway one.[420] Consumers might be surprised if they knew what they throw away in the cost of packaging for single-use containers. The cost of the 2-L polyester bottle of a soft drink is 0.10–0.12 dollars, which amounts to 14% of the 0.78 price per bottle. (The price will vary with the store and with the brand. This was the lowest price found in a supermarket in Newark, Delaware, in September 1997.) A typical aluminum beverage can cost 0.057 dollars,[421] which was 29% of the 0.20 dollars price of the container filled with a soft drink. The cost of the glass jar for baby food may amount to one third of the purchase price.[422] An alternative that saves money and reduces waste is to mash the food yourself or to puree it in a kitchen blender.

Poly(ethylene terephthalate) bottles are not refilled with beverages in the United States for fear of contamination. After soaking with benzene, butyric acid, malathion, and lindane for 2 weeks at 40°C, only part of the contaminants could be removed by 8% aqueous ethanol.[423] Sophisticated gas chromatographic headspace analysis is used to reject contaminated bottles in Europe where they are refilled with more beverage. The refillable polyester bottle is heavier than the single-use one. Use of a barrier coating of silica[424] or a diamond-like material, put on by plasma polymerization, to prevent contaminants from contacting the polyester might solve this problem for the United States (see later). Brief treatment of the interior of the bottle with fluorine at the time of manufacture might work. This method is used with polyethylene tanks for gasoline that are used on automobiles. The problem of scuff marks from repeated use that detract from the appearance of the container might be solved by a hard-surface coating, such as the cross-linked acrylic coatings used on plastic automobile headlights. The ideal coatings would not interfere with the recycling of the bottle at the end of its useful life. Poly(ethylene naphthalate) softens at a higher temperature is harder to scratch and is a better oxygen barrier, which extends the shelf life for beer.[425] Its use in blends with poly(ethylene terephthalate) in beer bottles could complicate recycling, unless a good marking, deposit, and refund system is developed. Coating the inside of the polyethylene terephthalate bottle with a 0.1-μm layer of hydrogenated carbon, with or without nitrogen doping, using a plasma at less than 50°C reduces oxygen transport by 30-fold.[426] This may eliminate the need for the poly(ethylene naphthalate) beer bottle. Such a coating might make it possible to refill poly(ethylene terephthalate) bottles for soft drinks repeatedly in the United States. Polyester bottles must be washed with less alkaline detergents than those used with glass bottles to avoid hydrolysis of the polymer. The use of enzyme-containing detergents at lower temperatures should help.

It may be possible to devise a refillable aluminum can for beverages. It could be shaped like the aluminum bottles that laboratory chromatographic alumina is shipped in.

Milk and orange juice are also sold in half gallon bisphenol A polycarbonate jugs. A jug may cost 1.10 dollars, but the cost per trip is low when spread over 40–60 trips.[427] A deposit of 0.25 dollars ensures a high return rate. Some 8-fluid-oz school milk bottles of polycarbonate realize over 100 trips. Just as with the polyester bottles, these also have to be washed under mild conditions to avoid hydrolysis. (Bisphenol A is an endocrine disrupter, which may be detrimental to the health of fetuses and young children.) At the end of their useful life, the manufacturer (formerly General Electric, now SABIC) takes back the bottles for remolding into bottle crates and such. This system has been used in the United States, Sweden, and Switzerland for over 10 years. Most milk is sold in paper cartons coated with polyethylene, or in jugs made of high-density polyethylene. The system for milk requiring the least packaging is probably powdered skim milk equivalent to 22.7 L (24 qt) of liquid milk in a cardboard box covered with aluminum foil. It can be stored at room temperature and requires refrigeration only after it is made up as liquid milk. The polycarbonate bottle cannot be used for soft drinks because it is not a suitable barrier to carbon dioxide. A challenge is to find a suitable barrier that will not interfere with recycling the container at the end of its useful life as a bottle. An inorganic barrier coating, such as silica, might work.

The average American consumes 52 gallons of soft drinks each year. This is more than the amount of water that he or she drinks. For comparison, the average American drank 26 gallons of coffee, 24 gallons of milk, and 32 gallons of beer[428] and also consumed 158 lb of sugar each year. About one third of this comes from soft drinks, which use sucrose or high-fructose corn syrup or both. Both promote dental caries.[429] A typical soft drink contains 10–14% sugar, 0.37% flavoring, and 0.185–0.74% citric acid in water saturated with carbon dioxide.[430] It may also contain color, caffeine, and preservatives such as sodium benzoate. Phosphoric acid and other acids may be used instead of citric acid. The pH before carbonation is 2.35–2.66. It contains no vitamins, minerals, protein (usually), fiber, or complex carbohydrates. Its consumption at such levels raises serious nutritional questions. If other beverages were substituted for it, container waste would drop.

Bottled water is popular in the United States today. Some people feel that it tastes better, has fewer impurities, and may confer higher social status than tap water.[431] The analysis of 37 brands in 1991 found that 24 had one or more values that were not in compliance with the drinking water standards of the United States.[432] Some were just bottled municipal water. Testing of 1000 bottles of 103 brands in 1999 found that about "one third of the brands had at least one sample containing levels of contamination that exceeded state or industry limits or guidelines."[433] Bottled water does not contain sufficient fluoride ion for good dental health.[434] It would be better to drink at the nearest water fountain or to fill your reusable bottle there. This would eliminate some plastic container waste. (Bottled water does have a place in some of the less-developed countries in which the standards of sanitation may be inadequate to protect the public water supply.) One company uses refillable bottles of poly(ethylene terephthalate) for its mineral water in Holland and Belgium.[435] The headspace of returned bottles is checked by gas chromatography–mass spectrometry at the rate of 0.1 s/bottle. The label is cut off with a high-pressure water jet at the rate of 550 labels/min. The sleeve waste is recycled. The bottles are cleaned at 167°F, inspected physically, filled, and a new shrink label of polyethylene applied. The crate of six 1.5-L bottles has a deposit of 7.20 dollars, half for the crate and half for the bottles (i.e., 0.60 dollars/bottle). The return rate is 94%. Americans drank 24 gallons/person at a cost of 26 billion dollars.[436] This is such a waste of plastics, energy, and dollars that some cities, such as San Francisco, California, have banned their sale.

14.3.5 Using Less Paper

The United States consumes enormous amounts of paper each year, 1.5 tons/household/year.[437] It takes 5 billion ft^3 of trees plus 35 million tons of recycled wastepaper to make this paper. Each year Americans receive 4 million tons of junk mail, 47% of which is discarded by the recipient without looking at it.[438] It took 1.5 trees for the junk mail received at each household. Each U.S. office worker discarded 225 lb of paper in 1988.[439] Various methods have been suggested for reducing office waste, such as using both sides of the page, using smaller sheets for short memos, using e-mail, circulating memos instead of making copies, eliminating coversheets and wide margins, or making fewer photocopies. New incentives are needed for making fewer photocopies and reducing the number of computer printouts. (Charging individuals more for each one might help.) Various predictions of a cashless society, where transactions would be done by computer and e-mail, have been made in the past. This is starting to happen. Use of e-mail for bank statements and other investment information could reduce the volume of mail. If every home with a television set could use it for this purpose, the volume of paper used in mail would be reduced. This may require further developments in technology. A shift from paper packaging back to dishes of china at fast-food restaurants could reduce the use of paper. The restaurants would have to install more dishwashers. A surcharge on paper dishes and wrappings for take-out food might encourage more customers to eat on site and might reduce litter along roads.

Because newspaper is such a large component of household paper, it deserves special consideration. It can be replaced by a system that transmits the information in the newspaper to the customer over the telephone lines during the night. In the morning, the customer queries the terminal to obtain the table of contents. He might also print out the portions of interest. When such systems were test-marketed earlier, they flopped. One problem may have been that they cannot be taken to the office to be read at lunchtime. A standard command on the terminal to print out the front page, the sports scores, and the stock prices, or whatever else was of special interest, might have taken care of this. It is also possible that the system was tried before people became used to the World Wide Web and similar systems. Updating the news sections hourly might be a selling point with some customers. Most scientific journals are now on-line. Popular magazines might be handled in the same way. This would eliminate the need for finding space for them in the home and the need for finding a place to recycle them. It is also possible to read them at a central library, either in person or through a web site. Sharing with friends is a good way to use magazines that contain no time-sensitive material.

When the price of newsprint has risen, newspapers have gone to lighter, thinner paper, and have reduced the sizes of margins. Newsprint is a relatively weak paper made by grinding the tree and bleaching the product. This is a thermomechanical pulp, in contrast with writing paper where the lignin is removed by chemical means. The challenge is to find a dry strength additive for paper that will produce a thinner, stronger paper that will not cost any more than current newsprint. This may also involve finding an inexpensive additive to increase opacity, because thinner paper might allow print on one side to be seen from the other. Smaller headlines, type, and margins would reduce the number of pages in the daily newspaper. However, smaller type is harder to read. Neither the newspaper nor its readers may want to go to a more concise writing style to save paper and the cost of it. Limiting the size of advertisements could save a lot of space. This could be done as a matter of policy or by increasing the rates for the larger ads. It might also be done by government regulation or a tax on the larger ads. The problem is that advertisements support most of the newspaper. Advertisers might seek other media under these conditions. Customers would probably not buy many newspapers without advertisements because of the much higher price. If the cost of the newspaper was determined by the number of sections bought, people might buy fewer. Some might opt for only the international and national news, plus the sports section. Others might be content with just national and local news without the sports, business, real estate, and automotive sections. This could reduce the price and amount of paper used by as much as 50%. It would require the development of more complicated vending machines for newspapers. Newspapers are already facing stiff competition for advertising from electronic and other media. It would be instructive for Americans to see what is done in countries where newsprint is less plentiful and more expensive.

14.3.6 Life-Cycle Analyses

Life-cycle analyses[440] are becoming increasingly popular (see Chapter 17 for more details). They consider the amounts of energy and materials used, together with the waste and pollution produced, in taking a product from cradle to grave. ("Cradle to cradle" is a better term since it implies use of as much waste as possible for new useful material.[441]) Unfortunately, they are not all done in the same way or with the same assumptions. Most assume the use of energy from fossil fuels rather than from renewable sources. Very few consider the costs of global warming and natural resource depletion, not to mention the loss of biodiversity. It is important to see what assumptions have been made and who paid for the study, for it may not be one that is completely objective. The use of paper versus polystyrene for cups may tend to favor the foamed polystyrene one unless a china mug or a drinking glass is included.[442] The reusable one will win because washing it is not very expensive and does not involve the energy and materials to make a new one. Paper versus plastic is also debated for grocery bags. The winner here is the string bag or market basket that the customer brings to the store. Some paper bags have a statement on them that they are recyclable and are made from a renewable resource. Some also say, "reuse this bag for 2 cents credit." Apparently, this is not enough of an incentive for customers, for very few of them do it. (Customers could also bring back for reuse the thin polyethylene film bags used to carry and store fruits and vegetables.) A local supermarket has used bags of high-density polyethylene with labels that say, "Save a tree! By using plastic bags, you help save the trees and forests that are cut each year to make paper bags. Recycle! Help the environment." This is encouraging the use of a nonrenewable resource. It also ignores the fact that the recycling of plastic films is not working well. A thin plastic bag closed by heat-sealing, a clip, or a paper or plastic tie might replace the shrink-wrapped polystyrene tray. Trays and egg cartons can be made of paper, a renewable resource, by pulp molding, the way they used to be. One company has been "earning greenie points" by packaging its toaster ovens in a molded pulp made from 100% used newspaper.[443] This replaced foamed polystyrene.

Plastic bags from stores have become a litter problem. Ireland put a 0.15 euro tax on each one, which reduced usage by 90%.[444] Bangladesh, Taiwan, and South Africa have banned the bags.[445] Denmark and Japan tax them. The United States recycled 812 million lb of plastic bags and film in 2006, a small fraction of that produced.[446] The largest recycler, Trex, makes decks of 1.5 billion grocery bags plus sawdust and wood scraps.

14.3.7 Ecolabels

Ecolabels[447] designed to influence consumers' choices are popular. A label attached by a pressure-sensitive adhesive to a plastic film around papers in the mail said, "This plastic cover (polyethylene) is not damaging to the environment. It can be recycled, is nontoxic, and requires no more energy to produce than recycled paper. It helps both to reduce costs and the number of returns." Some advertisements have arrived in white, low-density polyethylene envelopes with a label attached with a pressure-sensitive adhesive, together with the chasing arrows sign and the message, "This bag is recyclable." This implies that the adhesives and the paper in the label will not hurt the ability to recycle it. This is questionable, but may depend on the process used. If the paper is left in, this may produce weak spots in the new plastic, because it will not adhere well to the polyethylene and may decompose at the temperature of remolding. The reuse is limited to pigmented materials and cannot be used to make clear plastic. Even the American Chemical Society has shipped two magazines together in clear low-density polyethylene with the chasing arrows marked with "100% recyclable." While the film was heat-sealed around the magazines, a paper label was held on by a pressure-sensitive adhesive. One journal has arrived at the author's home with a separate sheet of paper inside the heat-sealed envelope. The sign on the paper sheet said, "Recyclable plastic. Please recycle." The address was on a paper label attached to the sheet of paper by a water-moistenable adhesive. Another such bag even specified in the label on the plastic, "minimum 30% industrial content—minimum 15% post consumer content." Another magazine arrived with the sheet of paper inside a biodegradable plastic bag. The address was on a paper label attached by a water-moistenable adhesive. The message on the piece of paper was, "Printed with soy ink. Printed on recycled paper." There are simpler ways to handle magazines in the mail. The journal *Science* uses a paper label attached by a water-based adhesive at one corner of the front. The system using the least material, and among the easiest to recycle, was found on a May 1997 issue of the magazine *Horticulture*. The address was printed on a white rectangular space at the bottom of the cover using an ink-jet printer. It might be possible to use ink-jet printing with a nontoxic edible ink to mark oranges, bananas, and other foods instead of putting a plastic sticker on each one. Ink-jet printing can also be used to label bottles to replace the shrink-fit labels of a different plastic than that in the bottle. Further development may be needed to make them as colorful as the current shrink labels. Refillable bottles can also use painted or embossed labels so that no new one is required when the bottle is reused. The liberal use of the term "recycled" can be misleading. It has always been good manufacturing practice to put scrap made in the plant back into the process. Today, a lot of it is being called recycled. It would be better to reserve the term for postconsumer material that has been reused. Very few products are made from 100% postconsumer waste.

14.3.8 Ecoparks

An ecopark is envisioned as a place where everything that goes into the system emerges as useful products. The waste of one industrial plant is used as starting material for another. The outstanding example is in Kalundborg, Denmark.[448] The electric power plant sends steam to the Statoil refinery, to the Novo Nordisk plant, and to the town for heating buildings. It also burns some gas that the refinery would otherwise flare. Excess heat also goes to a fish farm that produces 250 tons of fish each year. Ninety percent of the energy in the coal is used. Sulfur dioxide from the scrubber on the power plant is converted to calcium sulfate for the production of drywall. The drywall plant also uses gas from the refinery that would otherwise be flared. Sulfur removed from oil at the refinery is converted to sulfuric acid for use at a plant 50 km away. The fly ash and residual ash from the power plant are used in cement and roads. Sludges from the fish farm and the enzyme plant of Novo Nordisk are used as fertilizers. All of this collaboration occurred without government action, unless one includes the effects of government regulations. There is also an industrial partnership in southern Ontario, where calcium sulfate from scrubbing the flue gas of a power plant is made into drywall and the waste heat from generating the power is used for growing vegetables in eight acres of greenhouses.[449] An ecofarming project on Fiji is also trying to use everything.[450] Brewery waste, which used to be dumped into the harbor (and damaged the corals), is mixed with rice straw, sawdust, or shredded newspaper for use as a substrate for the growth of mushrooms. The residue from the mushroom culture is fed to chickens. The chicken waste is used to produce biogas. The residue from the biogas generator becomes fertilizer for the fishpond. Vegetables are grown hydroponically on the top of the fishpond.

14.3.9 Economics of Recycling

Some authors feel that recycling does not make sense economically.[451] They cite high collecting and sorting costs. They also point out that a pricing mechanism that reflects the true social cost, including resource depletion and environmental damage, is needed. Others feel that recycling is economically viable today.[452] Recycling creates more jobs than landfilling or incineration.[453] In Seattle, the average cost to dispose of a ton of nonrecyclable municipal waste in 1995 was 105 dollars. The average cost to collect and process a ton of recyclables was 28 dollars. One way of reducing the amount of municipal solid waste is unit pricing, a "pay as you throw" policy.[454] This can take the form of stickers that have to be purchased and placed on each bag of trash (as in Ithaca, New York) or stickers on each bag more than one. (For more on environmental economics, see Chapter 17.)

14.3.10 Role of Government in Reducing Consumption

It is apparent that source reduction and higher recycling are unlikely to happen without some form of government intervention.[455] This can take the form of laws banning certain uses, taxing undesired practices, subsidizing desirable practices, labeling laws showing true costs, or other such. Thus, it is possible to consider taxes on throwaway items and those based on nonrenewable raw materials, such as petroleum, natural gas, and coal. Denmark's tax on waste, started in 1987, is the highest in Europe.[456] Austria, Belgium, Finland, France, the Netherlands, and the United Kingdom also tax waste. Sweden and Norway are considering such taxes. The U.S. Congress is studying the removal of the hundreds of billions of dollars of subsidies given each year to virgin resource producers.[457] These subsidies to the mining, timber, and petroleum industries are considered to be impediments to recycling. Maine has a law requiring disposal fees paid at the time of purchase of major new appliances, furniture, bathtubs, and mattresses.[458] There are also laws that require a minimum recycle content in new items.[459] Sweden and Germany have banned unsolicited direct mailings.[460] If the U.S. Postal Service would require first-class postage, the amount used on personal letters, for junk mail, the amount of such mail would decrease.

The 65% recycling rate for aluminum beverage containers in the United States in 1994 (54% in 2007) is considered high.[461] However, if one looks at the billions of cans that are lost from the recycling loop each year, it is easy to see that a better system is needed. There is a proven way to obtain more of them back. Saphire has reviewed the role of deposits in retrieving used beverage containers.[462] Maine redemption rates for beer and soft-drink containers are 92%, for wine containers 79%, for juices 75%. In the other ten states with deposit-refund laws, only beer and soft-drink containers are covered. New York has the lowest redemption rate of the states with bottle bills, with a redemption rate of 66% for soft drinks and 79% for beer. Delaware's bottle bill is relatively ineffective, because aluminum containers and 2-L plastic containers are exempted. British Columbia is considering expansion of its 27-year-old bottle bill to cover sports drinks, bottled water, fruit and vegetable juice, wine, and spirits.[463] Higher rates of return can be obtained with larger deposits, as shown by the use of 0.25 dollars per container for some milk containers, as mentioned earlier in this chapter. A nationwide bottle bill with the coverage of that in Maine and with higher deposits than those now used in bottle bill states would increase recycling greatly and provide more jobs at the same time.

Michigan's bottle bill resulted in 4684 new jobs in the state. A tiered system with the payback of the full deposit for a refillable container, half back for a one-way container that can be recycled, and none for a one-way container that cannot be recycled has been recommended. Requiring the use of standardized (generic bottles), as done in Germany, also helps. Any bottler can use any other bottler's bottles just by putting on a new label. Efforts can be made to make the recycling of containers as convenient as possible by the use of reverse vending machines. Deposits on jam jars, detergent bottles, or others may also reduce waste. Denmark and Prince Edward Island have outlawed nonrefillable containers for beer and soft drinks. The return rate in Denmark is 98%. However, beverages exported by Denmark are in throwaway containers. In the grocery stores in the United States, an item for purchase must be marked not only with the price of the packaged item, but also with the cost per unit weight or unit volume. It might be appropriate to also require the cost of the container per use. This would demonstrate the economy of using refillables.

Germany is further along than other countries in making manufacturers responsible for the ultimate recycling or disposal of what they produce. A 1991 law requires that companies either collect their own waste packaging for recycling or pay fees to Duales System Deutschland to do it for them under the "Green Dot" system.[464] In 1994, the recycling fees were highest for plastic (Table 14.3).

The system is working, but not without some growing pains. In 1997, 5.6 million metric tons of used packaging materials were collected, which was 89% of that used. The recycling rates were glass 89%, corrugated cardboard 93%, plastics 69%, composites 78%, aluminum 86%, and tinplate 84%.[465] Other European countries have complained that they have been inundated with material for recycling from Germany. Pyrolysis of plastic waste to produce oil and gas for fuel has been used, but really should not be called recycling.[466]

Table 14.3 Recycling Fees for Containers in Germany

Component	$/kg
Plastics	1.97
Composites	1.40
Beverage cartons	1.13
Aluminum	1.00
Tinplate	0.37
Paper	0.33
Natural materials	0.13
Glass	0.10

The European Commission has proposed a takeback law covering electronic equipment, including appliances, office machinery, and telecommunications equipment.[467] (For more on how public policy can influence the environment, see Chapter 18.)

14.4 OVERALL PICTURE

For a sustainable future,[468] more things will have to be made from renewable resources and fewer from nonrenewable resources. This means using paper instead of plastic wherever possible, unless the plastic is based on a renewable source (as described in Chapter 12). The throwaway habit must be thrown away in favor of reusable objects, designed for long life, easy repair, and ease of recycling of the materials in them. Objects made of 100% postconsumer waste must become common, instead of being rare as they are today.

No one has found a way to curb the overconsumption of industrialized nations. (E. O. Wilson pointed out that if all nations used materials at the same rate as the United States, it would take the natural resources of two more Planet Earths.[469]) First, society must agree that this is a necessary objective. Next, some system of incentives to reduce and disincentives to use lavishly must be devised. This is the "carrot–stick" approach. This will involve a culture change.[470] Persons who live in the United States are constantly bombarded with messages to buy. The advertising to do this is 500 dollars/person/year. Credit card companies deluge the average person with offers of cards that make buying on credit easy. The message is "Buy now—Pay later–Why wait to have what you want?" This makes the objects purchased cost significantly more. One can think of restrictions on advertising or taxes on advertising and credit cards, but these could be very unpopular. An approach that has increased the savings rate in some countries is similar to the individual retirement account (IRA) used in the United States. Money is set aside before taxes for purchase of a first home or for a college education. There are no taxes until the money is withdrawn for the intended purpose. Premature withdrawal for another purpose imposes a tax penalty. (The individual savings rate in the United States is significantly lower than that in most other developed nations.)

In a finite world, there will be limits to what people can expect to use.[471] This means that the constant growth in the sales of a product, that some executives seem to expect, is unrealistic. One can see these unrealistic expectations expressed in almost any issue of *Chemical Engineering News*. The time may come when closed landfills are mined for the materials in them.

REFERENCES

1. (a) G.M. Levy, ed., *Packaging in the Environment*, Blackie Academic and Professional, London, 1993, 150; (b) G. Matos and L. Wagner, *Ann. Rev. Energy Environ.*, 1998, *23*, 107.
2. N. Holmes, J. Chen, and R. Rivera, *Amicus J.* (Natural Resources Defense Council), 2000, *21*(4), 22.
3. U.S. Environmental Protection Agency, Characterization of Municipal Solid Waste in the United States: 1994 Update. EPA 530-S-94-042, Nov. 1994 and 1996 update.
4. (a) A. Levey and E. Yermoli, *Waste Age's Recycling Times*, 1998, *10*(24), 3; (b) K. Egan, *Waste Age's Recycling Times*, 1998, *10*(11), 8.
5. K.M. White, *Waste Age's Recycling Times*, 1997, *9*(7), 7.
6. (a) S. Cassel, *Technol. Rev.*, 1992, *95*(6), 20; (b) ID2 Communications, Victoria, British Columbia, Canada gives 100–200 lb per worker per year, 2008.
7. P. Calmels and R. Harris, *Pulp Pap. Int.*, 1994, *36*(12), 47.
8. Anon., *R&D Cahners*, 2005, *47*(12), 32.
9. L. Zhito, www.americanprofile.com.
10. M. Chester, E. Martin, and N. Sathaye, *Environ. Sci. Technol.*, 2008, *42*, 2142.
11. J.M. Heumann, *Waste Age's Recycling Times*, 1998, *10*(7), 2.
12. D.J. Hanson, *Chem. Eng. News*, Feb. 26, 1996, 22.
13. M. Murray, *Wilmington Delaware Sunday News J.*, Apr. 18, 1999, A1.
14. (a) J.M. Heumann, *Waste Age's Recycling Times*, 1997, *9*(18), 7; (b) Anon., *Environment*, 1998, *4*(4), 24.
15. K.M. White, *Waste Age's Recycling Times*, 1999, *11*(1), 14.
16. K. Egan, *Waste Age's Recycling Times*, 1998, *10*(19), 13.
17. J.M. Heumann, *Waste Age's Recycling Times*, 1997, *9*(25), 13.
18. K.A. O'Connell, *Waste Age's Recycling Times*, 1998, *10*(17), 6.
19. Delaware Economic Development Office, *Delaware Econom. Dev.*, Dover, DE, 1996, *1*(3), 8.
20. (a) R.E. Hester and R.M. Harrison, eds, *Waste Incineration and the Environment*, Royal Society of Chemistry, Cambridge, 1994. (b) G. Parkinson, *Chem. Eng.*, 1997, *104*(3), 17; *104*(4), 23; (c) T. Pavone, *Chem. Process (Chicago)*, 1997; *60*(4), 32; (d) J. Swithenbank, V. Nasserzadeh, B.C.R. Ewan, I. Delay, D. Lawarence, and B. Jones, *Environ. Prog.*, 1997, *16*(1), 65; (e) Anon., *Chem. Eng.*, 1996, *103*(5), 51; (f) J. Carlin, *Chem. Ind. (Lond.)*, 1998, 928; (g) B. Piasecki, D. Rainey, and K. Fletcher, *Am. Sci.*, 1998, *86*, 364; (h) R.E. Hester and R.M. Harriison, eds, *Environmental and Health Impact of Solid Waste Management Activities*, Royal Society of Chemistry, Cambridge, 2003; (i) Z. Wang, H. Richter, J.B. Howard, J. Jordan, J. Carlson, and Y.A. Levendis, *Ind. Eng. Chem. Res.*, 2004, *43*, 2873.
21. R. Desmond, *EDF Lett. (Environmental Defense Fund)*, 1998, *27*(3), 7.
22. (a) J.T. Aquino, ed., *Waste Age/Recycling Times Recycling Handbook*, Lewis Publishers, Boca Raton, FL., 1995; (b) C. Boener and K. Chilton, *Kirk–Othmer Encyclopedia of Chemical Technology*, 4th ed., Wiley, New York, 1996, *20*, 1075; (c) A.G.R. Manser and A.A. Keeling, *Practical Handbook of Processing and Recycling Municipal Waste*. CRC Press, Boca Raton, FL, 1996; (d) C.P. Rader, ed., *Plastics, Rubber and Paper Recycling: A Pragmatic Approach*. ACS Symp. 609, Washington, DC, 1995; (e) A.K.M. Rainbow, ed., *Why Recycle?* A.A. Balkema, Rotterdam, 1994; (f) C.R. Rhyner, L.J. Schwartz, R.B. Wenger, M.G. Kohrell, *Waste Management and Resource Recovery*, Lewis Publishers, Boca Raton, FL, 1995; (g) S.M. Turner, *Recycling*, American Chemical Society, Washington, DC, Dec. 1993; (h) R. Waite, *Household Waste Recycling*. Earthscan, London, 1995; (i) J. Carless, *Taking Out the Trash: A No-Nonsense Guide to Recycling*. Island Press, Washington, DC, 1992; (j) H.F. Lund, ed., *The McGraw-Hill Recycling Handbook*, 2nd ed., Hightstown, NJ, 2000.
23. J.C. Roberts, *The Chemistry of Paper*. Royal Society of Chemistry, Cambridge, 1996.
24. J.T. Staley and W. Hairpin, *Kirk–Othmer Encyclopedia of Chemical Technology*, 4th ed., Wiley, New York, 1992, *2*, 212.
25. M.S. Reisch, *Chem. Eng. News*, Dec. 10, 2007, 24.
26. www.innov-xsys.com.
27. G. Parkinson, An automated plant using similar techniques started in Hanover, Germany in 2000, *Chem. Eng.*, 1999, *106*(7), 19.
28. (a) Anon., *Pap. Ind.*, 2007, *24*(1), 13; (b) J.G. Leahy, K.D. Tracy, and M.H. Eley, *Biotechnol. Lett.*, 2003, *25*, 479; (c) M.H. Eley and C.C. Holloway, *Appl. Biochem. and Biotechnol.*, 1988, *17*(1–3), 125.
29. M.E. Marley, *Pap. Technol.*, 2002, *43*(6), 2.
30. (a) F. Berman, *Trash to Cash*, St. Lucie Press, Delray Beach, FL, 1996; (b) C.G. Thompson, *Recycled Papers—The Essential Guide*, MIT, Press, Cambridge, MA, 1992.
31. B. Bateman, *Pap. Technol.*, 1996, *37*(1), 15.
32. (a) A. Karma, J. Engstrom, and T. Kutinlahti, *Pulp. Pap. Can.*, 1994, *95*(11), 38; (b) U. Arena, M.L. Mastellone, F. Perugini, and R. Clift, *Ind. Eng. Chem. Res.*, 2004, *43*, 5702.
33. B. Fransson and I. Emanuelsson, *Pap. Technol.*, 2002, *43*(7), 40.
34. M. Shaw, *Pulp and Paper*, Nov. 2002, 29.
35. (a) T. Pfitzner, *Pulp Pap. Int.*, 2003, *45*(4), 18; (b) J. Toland, *Pulp Pap. Int.*, 2003, *45*(4), 25.
36. T. Blain and J. Grant, *Pulp Can.*, 1994, *95*(5), 43.
37. (a) J.K. Borchard, *Kirk–Othmer Encyclopedia of Chemical Technology*, 4th ed., Wiley, New York, 1997, *21*, 10; *Chem. Ind. (Lond.)*, 1993, 273; *Prog. Pap. Recycl.*, 1994, *3*(4), 47; (b) L.D. Ferguson, *TAPPI J.*, 1992, *75*(7), 75; *75*(8), 49; (c) F.R. Hamilton, *PIMA Mag.*, Jan. 1992, 20; (d) T. Woodward, *PIMA Mag.*, Jan. 1992, 34; (e) A. Johnson, *Pap. Technol.*, 1992, *33*(11), 20; (f) C. Silverman, *Am. Ink Maker*, 1991, *69*(11), 28; (g) R.M. Rowell, T.L. Laufenberg, and J.K. Rowell, eds, *Materials Interactions Relevant to Recycling Wood-Based Materials*. Materials Research Society, Pittsburgh, PA, 1992; (h) F.J. Sutman, M.B.K. Letscher, and R.J. Dexter, *TAPPI J.*, 1996, *79*(3), 177; (i) R.C. Thompson, *Surf. Coat Int.*, 1998, *81*(5), 230; (j) M. Pescantin, *Pap. Technol.*, 2002, *43*(9), 47.
38. J.K. Borchardt, D.W. Matalamabi, V.G. Lott, and D.B. Grimes, *TAPPI J.*, 1997, *80*(10), 269.
39. S. Ritter, *Chem. Eng. News*, Nov. 16, 1998, 35.
40. P. Somasundaran, L. Zhang, S. Krishnakumar, and R. Slepetys, *Prog. Pap. Recycl.*, 1999, *8*(3), 22.
41. G. Sun and Y. Deng, *Prog. Pap. Recycl.*, 2000, *10*(1), 13.
42. M.S. Mahagaonkar, K.R. Stack, and P.W. Banham, *TAPPI J.*, 1998, *81*(12), 101.
43. J. Milliken, *TAPPI J.*, 1997, *80*(9), 79.
44. S. Dessureault, P. Carabin, A. Thom, J. Kleuser, and P. Gitzen, *Prog. Pap. Recycl.*, 1998, *8*(1), 23.

45. M.M. Sain, C. Daneualt, and M. Lapoint, *Am. Ink Maker*, 1996, *74*(4), 26.
46. G. Rangamannar and L. Silveri, *TAPPI J.*, 1990, *73*(7), 188.
47. (a) Anon., *Prog. Pap. Recycl.*, 1996, *6*(1), 73; (b) J. Bredael, N.J. Sell, and J.C. Norman, *Prog. Pap. Recycl.*, 1996, *6*(1), 24.
48. (a) J.D. Taylor and E.K.C. Yu, *Chemtech.*, 1995, *25*(2), 38; (b) A.K. Sharma, W.K. Forester, and E.H. Shriver, *TAPPI J.*, 1996, *79*(5), 211; (c) F. Ruzinsky and B.V. Kokta, *Prog. Pap. Recycl.*, 1998, *7*(3), 47; 2000, *9*(2), 30.
49. (a) J.K. Borchardt, D.W. Matalamaki, V.G. Lott, and J.H. Rask, *Prog. Pap. Recycl.*, 1994, *3*(4), 47; (b) E.-M. Debzi, R.H. Marchessault, G. Excoffier, and H. Chanzy, *Macromol. Symp.*, 1999, *143*, 243; (c) X.-D. Fan, Y. Deng, J. Waterhouse, P. Pfromm, and W.W. Carr, *J. Appl. Polym. Sci.*, 1999, *74*, 1563.
50. D.A. Johnson and E.V. Thomson, *TAPPI J.*, 1995, *78*(2), 41.
51. (a) C.R. Olson, J.D. Hall, and I.J. Philippe, *Prog. Pap. Recycl.*, 1993, *2*(2), 24; (b) W. Hilberg, *Pulp Pap.*, 1994, Sept. 157; (c) L.D. Ferguson and D.H. McBride, *Pulp Pap. Can.*, 1994, *95*(2), 62.
52. L.D. Ferguson and D.H. McBride, *Pulp Pap. Can.*, 1994, *95*(2), 62.
53. (a) L.G. Offill and R.A. Venditti, *Prog. Pap. Recycl.*, 1994, *3*(4), 64; (b) R. Levandoski, J. Norman, G. Pepelnjak, and T. Drnovsek, *Prog. Pap. Recycl.*, 1999, *8*(3), 53.
54. (a) Q.-M. Chen and H.-H. Chang, *Prog. Pap. Recycl.*, 1998, *7*(4), 58; (b) J. Zheng, R.A. Vendetti, and H.G. Olf, *Prog. Pap. Recycl.*, 1999, *9*(1), 30.
55. E. Tremblay, M.M. Sain, C. Daneault, and M. Lapointe, *Prog. Pap. Recycl.*, 1998, *8*(1), 45.
56. G. Wu, Y. Deng, and J. Zhu, *Prog. Pap. Recycl.*, 1998, *7*(4), 20; 2000, *10*(1), 13.
57. Thoresen, *Waste Age's Recycling Times*, Sept. 21, 1993, 12.
58. G. Sacripanti and S. Kittellberger, *Chem. Br.*, 2001, *37*(4), 52.
59. E.-M. Debzi, R.H. Marchessault, G. Excoffier, and H. Chanzy, *Macromol. Symp.*, 1999, *143*, 243.
60. (a) X.-D. Fan, Y. Deng, J. Waterhouse, P. Pfromm, and W.W. Carr, *J. Appl. Polym. Sci.*, 1999, *74*, 1563; (b) Anon., *R&D (Cahners)*, 2003, *45*(9), 49.
61. (a) S. Machida, S. Takayama, N. Ikeda, T.I. Urano, and K. Sano, Preprints ACS, *Div. Environ. Chem.*, 1999, *39*(2), 220; (b) E. Yermoli, *Waste Age's Recycling Times*, 1998, *10*(2), 3; (c) Toshiba, U.S. patent 6,203, 2001. (d) Anon., *Green Chem.*, 1999, *1*, G163.
62. K. Sugita, *Prog. Org. Coat*, 1997, *31*, 87.
63. (a) P. Zurer, *Chem. Eng. News*, July 20, 1998, 12; (b) K.D. Vincent, U.S. Patent 6,045,955, 2000; (c) R. Dagani, *Chem. Eng. News*, Jan. 15, 2001, 40; (d) J.A. Rogers, *Science*, 2001, *291*, 1502; (e) M. Gross, *Chem. Br.*, 2001, *37*(7), 22; (f) M. Granmar and A. Cho, *Science*, 2005, *308*, 785; (g) E. Katz and I. Willner, *Chem. Commun.*, 2005, 5641.
64. (a) F. Hayden, *Chem. Eng. News*, Jan. 8, 2007, 72; (b) R. Klajin, P.J. Wesson, K.J.M. Bishop, and B.A. Grzybowski, *Angew. Chem. Int. Ed.*, 2009, *48*, 7035.
65. Kimberly-Clark Annual Report for 2002, Dallas, TX, 15.
66. (a) J.C. Roberts, *The Chemistry of Paper*. Royal Society of Chemistry, Cambridge, 1996; (b) G.M. Scott, J.K. Borchardt, L.D. Ferguson, M.R. Doshi, and J.M, Dyer, *Prog. Pap. Recycl.*, 1998, *7*(4), 63–82.
67. (a) T. Friberg, *Prog. Pap. Recycl.*, 1996, *6*(1), 70; (b) E. Kiran, K. Malki, and H. Pohler, *Polym. Mater Sci. Eng.*, 1996, *74*, 231; (c) N.N.-C. Hsu, *Prog. Pap. Recycl.*, 1996, *6*(1), 63; (d) M. Muvundamina, *TAPPI J.*, 1997, *80*(9), 129; (e) J. Dyer, *Prog. Pap. Recycl.*, 1998, *8*(1), 69; (f) R. McKinney, *Pulp Pap. Int.*, 1998, *40*(6), 45.
68. A.V. Pocius, *Kirk–Othmer Encyclopedia of Chemical Technology*, 4th ed., Wiley, New York, 1991, *1*, 461.
69. (a) M. Prein, *Prog. Pap. Recycl.*, 1992, *1*(4), 59; (b) J.K. Borchardt and M.R. Doshi, *Prog. Pap. Recycl.*, 1992, *1*(4), 70.
70. E. Kiran, K. Malki, and H. Pohler, *Polym. Mater Sci. Eng.*, 1996, *74*, 231.
71. (a) G.S. Samdani, *Chem. Eng.*, 1995, *102*(11), 17; (b) Anon., *Chem. Eng.*, 1995, *102*(12), 66.
72. M. Muvundamina and J. Liu, *TAPPI J.*, 1997, *80*(11), 172.
73. T.M. Woodward, *Pap. Technol.*, 1995, *36*(7), 27.
74. J. Lane and A. Manning, *Prog. Pap. Recycl.*, 1998, *7*(4), 56.
75. F.J. Saint Amand, B. Perrin, and P. deLuca, *Prog. Pap. Recycl.*, 1998, *7*(4), 39.
76. T. Math, *Pulp Pap. Can.*, 1994, *95*(4), 52.
77. (a) H. Ridgley, *Waste Age's Recycling Times*, 1998, *10*(6), 13; (b) M.J. Nowak, S.J. Severtson, M.J. Coffey, and D.H. Slinkman. In: B.N. Brogdon, ed., *Fundamental Advances & Innovations in the Pulp & Paper Industry*, AIChE Symp., New York, 1999, *95*(322), 144.
78. Anon., *Waste Age's Recycling Times*, 1998, *10*(7), 13.
79. (a) E. Back, *Pulp Pap. Int.*, 1994, *36*(10), 47; (b) B. Drehmer and E.L. Back, *Pap. Technol.*, 1995, *36*(3), 36.
80. T.C. Stauffer, R.A. Venditti, R.D. Gilbert, J.F. Kadla, Y. Chernyak, and G.A. Montero, *J. Appl. Polym. Sci.*, 2001, *81*, 1107.
81. N. Parris, P.J. Vergano, L.C. Dickey, P.H. Cooke, J.C. Craig, *J. Agric. Food Chem.*, 1998, *46*, 4056.
82. R.L. Park, *Paperboard Packag*, 1995, *80*(2), 5.
83. S.-I. Akahori and Z. Osawa, *Polym. Degrad. Stabil.*, 1994, *45*, 261.
84. J.J. Owens, *Adhes. Age*, 1993, *36*(12), 14.
85. (a) M.J.C. Cudero, M.M. Lopez-Gonzales, and J.M. Barrales-Rienda, *J. Polym. Sci. A Polym. Chem.*, 1996, *34*, 1059; (b) M.J.C. Cudero, M.M.C. Lopez-Gonzalez, E.C. Chico, and J.M. Barrales-Rienda, *J. Appl. Polym. Sci.*, 1997, *66*, 2409; (c) M.M. Lopez-Gonzalez, M.J.C. Cudero, and J.M. Barrales-Rienda, *J. Polym. Sci. A Polym. Chem.*, 1997, *35*, 3409.
86. R. Klein and H. Grossman, *Pap. Technol.*, 1995, *36*(8), 61.
87. J. Garrett, P.A. Lovell A.J. Shae, and R.D. Viney, *Macromol. Symp.*, 2000, *151*, 487.
88. Anon., *Chem. Eng. News*, June 24, 1996, 18.
89. (a) J. Woodward, *Biotechnology*, 1994, *12*, 905; (b) R. Dinus and T. Welt, *Prog. Pap. Recyl.*, 1994, *3*(4), 63, 64.
90. (a) O.U. Heise, J.P. Unwin, J.H. Klungness, W.G. Fineran, Jr., M. Sykes, and S. Abubakr, *TAPPI J.*, 1996, *79*(3), 207; (b) P. Bajpai and P.K. Bajpai, *TAPPI J.*, 1998, *81*(12), 111; (c) A. Roring and R.D. Haynes, *Prog. Pap. Recycl.*, 1998, *7*(3), 73; (d) J.M. Jobbins and N.E. Franks, *TAPPI J.*, 1997, *80*(9), 73; (e) A.L. Morkbak and W. Zimmermann, *Prog. Pap. Recycl.*, 1998, *7*(3), 33; (f) S. Vyas and A. Lachke, *Enzyme Microb. Technol.*, 2003, *32*, 236.
91. G. Stork, H. Pereira, T.M. Wood, E.M. Dusterhoft, A. Toft, and J. Puls, *TAPPI J.*, 1995, *78*(2), 79, 89.
92. H. Pala, M.A. Lemos, M. Mota, and F.M. Gama, *Enzyme Microb. Technol.*, 2001, *29*, 274.
93. G. Parkinson, *Chem. Eng.*, 2002, *100*(11), 19.

94. N.W. Lazorisak, J.F. Schmitt, and R. Smith, U.S. Patent 5, 620, 565, 1997.
95. I. Eriksson, P. Lunabba, A. Pettersson, and G. Carlsson, *TAPPI J.*, 1997, *80*(7), 151.
96. J. Clewley and N. Wiseman, *Pap. Technol.*, 1995, *36*(8), 51.
97. J. Frenzel, *Pap. Technol.*, 1995, *36*(8), 40.
98. (a) K.N. Law, J.L. Valade, and J. Quan, *TAPPI J.*, 1996, *79*(3), 167; (b) U. Weise and H. Paulapuro, *Prog. Pap. Recycl.*, 1998, *7*(3), 14.
99. J. Bailey, *Pulp Pap. Can.*, 1996, *97*(1), 10.
100. J.L. Minor and R.H. Atalla. Strength loss in recycled fibers & methods of restoration, In: R.M. Rowell, T.L. Laufenberg, and J.K. Rowell, eds, *Materials Interactions Relevant to Recycling Wood-Based Materials.* Materials Research Society, Pittsburgh, PA, 1992, 215.
101. T. Friberg and G. Brelsford, *Solutions (from TAPPI and PIMA)*, Aug. 2002, 27.
102. K. Cathie, *Pap. Technol.*, 1997, *38*(5), 34.
103. Anon., *Pap. Technol.*, 1995, *36*(7), 13.
104. K.L. Patrick, *Pulp Pap.*, 1994, *68*(13), 75.
105. J. Schultz, *Paperboard Packag.*, 1995, *80*(2), 16.
106. C.L. Reynolds, *TAPPI J.*, 1997, *80*(9), 102.
107. L.R. Brown, *World Watch*, 1999, *12*(2), 13.
108. S. Westergard, *Pap. Technol.*, 1995, *36*(8), 28.
109. N. Wiseman and G. Ogden, *Pap. Technol.*, 1996, *37*(1), 31.
110. (a) D.G. Meadows, *TAPPI J.*, 1996, *79*(1), 63; (b) R.T. Klinker, *TAPPI J.*, 1996, *79*(1), 97.
111. J. Osborne, *Pulp Pap. Int.*, 1994, *36*(4), 100.
112. M.R. Servos, K.R. Munbittrick, J.H. Carey, and G.J. Van Der Krick, eds, *Environmental Fate and Effects of Pulp and Paper Mill Effluents*. St. Lucie Press, Delray Beach, FL, 1996.
113. H.-W. Liu, S.-N. Lo, and H.-C. Lavallee, *TAPPI J.*, 1996, *79*(5), 145.
114. (a) B.W. English, Recycling paper for nonpaper uses, In: C.L. Verrill, ed., *Advances in Forest Products: Environmental and Process Engineering*. AIChE Symp. 304, New York, 1994, *90*, 1; (b) B.W. English, R.A. Young, J.K. Rowell. In: R.M. Rowell, T.L. Laufenberg, and J.K. Rowell, eds, *Paper and Composites from Agrobased Resources*. CRC Press, Boca Raton, FL, 1997, 269.
115. X. Yuan, Y. Zhang, and X. Zhang, *J. Appl. Polym. Sci.*, 1999, *71*, 333.
116. Anon., *New Sci.*, Feb. 7, 1998, 15.
117. D. Dahl, *Land Water*, 1995, *39* (Jan./Feb.), 50.
118. R.L. Maine, *Prog. Pap. Recycl.*, 1993, *2*(3), 76.
119. (a) L. Webb, *Pulp Pap. Int.*, 1994, *36*(10), 18; (b) J. Glenn, *Prog. Pap. Recycl.*, 1998, *7*(3), 54.
120. J.J. Gregg, A.K. Zander, and T.L. Theis, *TAPPI J.*, 1997, *80*(9), 157.
121. C.W. Simpson and R. Lam, *TAPPI J.*, 1997, *80*(9), 67.
122. J. Jang, H. Chung, M. Kim, and Y. Kim, *J. Appl. Polym. Sci.*, 1998, *69*, 2043.
123. E. Yermoli, *Waste Age's Recycling Times*, 1999, *11*(1), 12.
124. H. Xiao, R. Pelton, and A. Hamielec, *TAPPI J.*, 1996, *79*(4), 129.
125. K.M. White, *Waste Age's Recycling Times*, 1998, *10*(8), 4.
126. R. Tranpour, M. Stenstrom, G. Tchobanoglous, D. Miller, J. Wright, and M.V. Vossoughi, *Science*, 1999, *285*, 706.
127. A.T. Mattoon, *World Watch*, 1998, *11*(2), 20.
128. Anon., *Pap. Technol.*, 1995, *36*(4), 32.
129. P. Knight, *Pulp Pap. Int.*, 1994, *36*(9), 71.
130. (a) B. Barber, *Waste Age's Recycling Time*, 1997, *9*(7), 13; (b) E. Ayres, *World Watch*, 1993, *6*(5), 5.
131. (a) C.F. Baker, *Pap. Technol.*, 1998, *39*(4), 27; (b) L. Paavilainen, *Pulp Pap. Int.*, 1998, *40*(6), 61.
132. (a) W. Tao, T.A. Calamari, F.F. Shih, and C. Cao, *TAPPI J.*, 1997, *80*(12), 162; (b) T.A. Calamari, W. Tao, and W.R. Goynesz, *TAPPI J.*, 1997, *80*(8), 149.
133. A.T. Mattoon, *World Watch*, 1998, *11*(2), 20.
134. W.H. Morrison III, D.E. Akin, G. Ramaswamy, and B. Baldwin, *Text Res. J.*, 1996, *66*, 651.
135. Anon., *Environment*, 1996, *38*(4), 24.
136. G. van Roekel, *Pap. Technol.*, 1997, *38*(5), 37.
137. G. Giovannozzi-Sermanni, P.L. Cappelletto, A.D.'Annibale, and C. Perani, *TAPPI J.*, 1997, *80*(6), 139.
138. L. Jimenez, F. Maestre, M. J. de la Torre, and I. Perez, *TAPPI J.*, 1997, *80*(12), 148.
139. (a) A. Seisto and K. Poppius-Levlin, *TAPPI J.*, 1997, *80*(9), 215; (b) A. Seisto, K. Poppius-Levin, and T. Jousimaa, *TAPPI J.*, 1997, *80*(10), 235.
140. Anon., *R&D (Cahners)*, 1997, *39*(3), 11.
141. G.S. Samadani, *Chem. Eng.*, 1994, *101*(11), 25.
142. A.T. Mattoon, *World Watch*, 1998, *11*(2), 20.
143. (a) A. Garzon, *On Earth* (Natural Resources Defense Council), 2008, *29*(4), 61; (b) Environmental Defense Solutions (Environmental Defense fund), *Environ. Defense Solutions*, 2004, *35*(6), 9.
144. (a) M.S. Reisch, *Chem. Eng. News*, May 26, 1997, 14, 17; (b) S.J. Ainsworth, *Chem. Eng. News*, Apr. 18, 1994, 11.
145. P.J. Lemstra, *Science*, 2009, *323*, 725.
146. (a) Staff, *Chem. Eng. News*, July 7, 2008, 67; (b) M. McCoy, *Chem. Eng. News*, Mar. 16, 2009, 30.
147. (a) J. Brandrup, M. Bittner, W. Michaeli, and G. Meges, eds, *Recycling and Recovery of Plastics*. Hanser–Gardner, Cincinnati, OH, 1996; (b) R.J. Ehrig, *Plastics Recycling*, Hanser–Gardner, Cincinnati, OH, 1992; (c) A.L. Bisio and M.Xanthos, *How to Manage Plastics Waste*. Hanser–Gardner, Cincinnati, OH, 1994; (d) W. Heitz, ed., *Macromol. Symp.*, 1992, *57*, 1–395; (e) N. Basta, G. Ondrey, R. Rakagopal, and T. Kamiya, *Chem. Eng.*, 1997, *104*(6), 43; (f) League of Women Voters. *The Plastics Waste Primer: A Handbook for Citizens*. Washington, DC, 1993; (g) G.L. Nelson, *Chemtech*, 1995, *25*(12), 50; (h) J. Scheirs, *Polymer Recycling: Science, Technology and Applications*. Wiley, New York, 1998; (i) J. Pospisil, S. Nespurek, R. Pfaendner, and H. Zweifel, *Trends in Polym. Sci.*, 1997, *5*, 294; (j) N.T. Dintcheva, N. Jilov, and F.P. LaMantia, *Polym. Degrad. Stabil.*, 1997, *57*, 191; (k) C. Crabb, *Chem. Eng.*, 2000, *107*(6), 41; (l) J. Brandrup, *Macromol. Symp.*, 1999, *144*, 439; (m) J. Kahovec, ed., *Macromol. Symp.*, 1998, *135*, 1–373; (n) A.L. Andrady, *Plastics and the Environment*, Wiley, New York, 2003; (o) A. Azapagic, A. Elmsley, and I. Hamerton, *Polymers: The Environmental and Sustainable Development*, Wiley, Chichester, 2003; (p) K.C. Frisch, D. Klempner, and G. Prentice, *Advances in Plastics Recycling*, 2 vols, CRC Press, Boca Raton, FL, 2001.
148. H. Ridgley, *Waste Age's Recycling Times*, 1997, *9*(17), 6.
149. (a) M.S. Reisch, *Chem. Eng. News*, May 22, 1995, 41; (b) K. Egan, *Waste Age's Recycling Times*, 1997, *9*(23), 1; (c) R.A. Denison, *Something to Hide: The Sorry State of Plastics*

Recycling, Environmental Defense Fund, Washington, DC, Oct. 21, 1997; (d) *EDF Lett. (Environmental Defense Fund)*, 1998, *28*(1), 2.

150. (a) Anon., *Chem. Eng. News*, July 22, 1996, 26; (b) C. Rovelo, *Waste Age's Recycling Times*, 1997, *9*(8), 15; (c) Anon., *Chem. Eng. News*, Dec. 9, 1996, 18; (d) E.M. Kirschner, *Chem. Eng. News*, Nov. 4, 1996, 19; (e) Anon., *Chem. Eng. News*, June 23, 1997, 15; (f) J.M. Heumann, *Waste Age's Recycling Times*, 1997, *9*(13), 14; (g) J.L. Bast, *Waste Age's Recycling Times*, 1997, *9*(13), 15; (h) D.L. NaQuin, *Waste Age's Recycling Times*, 1997, *9*(18), 13; (i) Anon., *Chem. Eng. News*, Sept. 28, 1998, 14; (i) Anon., *Chem. Eng. News*, Sept. 3, 2001, 15.
151. (a) W. Weizer, *Chem. Ind. (Lond.)*, 1995, 1013; (b) R.A. Denison, *EDF Lett. (Environmental Defense Fund)*, New York, 1993, *24*(6), 4. (c) H. Hansler, *Chem. Ind. (Lond.)*, 2000, 429.
152. (a) G.P. Karayannidis, D.E. Kokkalas, and D.N. Biliaris, *J. Appl. Polym. Sci.*, 1995, *56*, 405; (b) A.H. Tullo, *Chem. Eng. News*, Oct. 15, 2007, 15; (c) H. Hansler, *Chem. Ind. (Lond.)*, July 10, 2000, 429.
153. D. O'Sullivan, *Chem. Eng. News*, Feb. 12, 1990, 25.
154. (a) J.M. Heumann, *Waste Age's Recycling Times*, 1998, *10*(17), 23; (b), M. McCoy, *Chem. Eng. News*, Mar. 16, 2009, 30.
155. Anon., *American Chemistry*, Mar./Apr. 2007, 26.
156. (a) K. Egan, *Waste Age's Recycling Times*, 1998, *10*(9), 1; (b) M. Paci and F.P. LaMantia, *Polym. Degard. Stab.*, 1999, *63*, 11.
157. (a) J.-E. Johansson, *Chem. Ind. (Lond.)*, Mar. 24, 2008, 16; (b) Anon., *Chem. Eng. News*, Sept. 10, 2007, 21.
158. (a) J.M. Heumann, *Waste Age's Recycling Times*, 1998, 10(9), 1; (b) S. Ulutan, *J. Appl. Polym. Sci.*, 1998, *69*, 865; (c) D. Braun, *Prog. Polym. Sci.*, 2002, *27*, 2171, 2183.
159. (a) Anon., *Chem. Eng. News*, Mar. 4, 2002, 15; Jan. 19, 2004, 17; (b) A. Scott, *Chem. Week*, Mar. 6, 2002, 31.
160. (a) S.E. Selke, *Packaging and the Environment: Alternatives, Trends and Solutions*, 2nd ed., Technomic, Lancaster, PA, 1994; (b) A. Brody and K. Marsh, *The Wiley Encyclopedia of Packaging Technology*, 2nd ed., Wiley, New York, 1997; (c) Anon., *Food Technol.*, 1990, *44*(7), 98; (d) K. Dotson, *Int. News Fats Oils Relat. Mater*, 1991, *2*, 854.
161. J.K.S. Wan, K.P. Vepsalainen, and M.S. Ioffe, *J. Appl. Polym. Sci.*, 1994, *54*, 25.
162. (a) Anon., *R&D (Cahners)*, 1998, *40*(10), 130; (b) M. Jacoby, *Chem. Eng. News*, Aug. 16, 1999, 38.
163. Anon., *Chem. Eng. News*, Feb. 13, 1995, 11.
164. H. Ridgley, *Waste Age's Recycling Times*, 1997, *9*(19).
165. S.J. Ainsworth and A.M. Thayer, *Chem. Eng. News*, Oct. 17, 1994, 10.
166. K. O'Connell, *Waste Age's Recycling Times*, 1998, *10*(2), 1.
167. J. Ogando, *Plast Technol.*, 1992, *38*(3), 23.
168. D. Staicu, G. Banica, and S. Stoica, *Polym. Degrad. Stabil.*, 1994, *46*, 259.
169. (a) R. Pfaendner, H. Herbst, K. Koffmann, and F. Sitek, *Angew. Makromol. Chem.*, 1995, *232*, 193; (b) M.K. Loultchea, M. Proietto, N. Jilov, and F.P. la Mantia, *Polym. Degrad. Stabil.*, 1997, *57*, 77; *Macromol. Symp.*, 2000, *152*, 201; (c) J. Pospisil, S. Nespurek, R. Pfaenner, and H. Zweifel, *Trends Polym. Sci.*, 1997, *5*, 294; (d) F.P. LaMantia, *Macromol. Symp.*, 2000, *152*, 201; (e) C.N. Kartallis, C.D. Papaspyrides, and R. Pfaendner, *J. Appl. Polym. Sci.*, 2003, *88*, 3033.
170. L. Willis, *Chemtech.*, 1994, *24*(2), 51.
171. Anon., *World Watch*, 2007, *20*(3), 29.
172. H. Ridgley, *Waste Age's Recycling Times*, 1997, *9*(20), 7.
173. M. Xanthos, B.C. Baltzis, and P.P. Hsu, *J. Apply. Polym. Sci.*, 1997, *64*, 1423.
174. Anon., *Chem. Br.*, 1996, *32*(3), 11.
175. M. Paci and F.P. la Mantia, *Polym. Degrad. Stabil.*, 1998, *61*, 417.
176. J. Simitzis and D. Fountas, *J. Appl. Polym. Sci.*, 1995, *55*, 879.
177. P.-A. Eriksson, P. Boydell, K. Eriksson, J.-A.E. Manson, and A.-C. Albertsson, *J. Apply. Polym. Sci.*, 1997, *65*, 1619.
178. (a) M. Wenguang and F.P. la Mantia, *J. App. Polym. Sci.*, 1996, *59*, 759; (b) F.P. la Mantia, ed., *Recycling of Poly(vinyl chloride) and Mixed Waste*. ChemTec Publishing, Toronto, 1995; (c) S. Ulatan, *J. Appl. Polym. Sci.*, 1998, *69*, 865.
179. Anon., *Chem. Eng. News*, Feb. 24, 1997, 16.
180. Anon., *Chem. Ind. (Lond.)*, 1997, 369.
181. G. Parkinson, *Chem. Eng.*, 1996, *103*(12), 23.
182. S. Dutta and D. Lohse, *Polymeric Compatibilizers: Uses and Benefits in Polymer Blends*, Hanser–Gardner, Cincinnati, OH, 1996.
183. (a) J. Duvall, C. Selliti, V. Topolkaraev, A. Hiltner, E. Baer, and C. Myers, *Polymer*, 1994, *35*, 3948; (b) G.H. Kim, S.S. Hwang, B.G. Cho, and S.M. Hong, *Macromol. Symp.*, 2007, *249-250*, 485.
184. N.K. Kalfoglou, D.S. Skafidas, and D.D. Sotiropoulou, *Polymer*, 1994, *35*, 3624.
185. P. van Ballegovic and A. Rudin, *J. Apply. Polym. Sci.*, 1990, *39*, 2097.
186. T.J. Nosker, D.R. Morrow, R.W. Renfree, K.E. Van Ness, and J.J. Donaghy, *Nature*, 1991, *350*, 563.
187. (a) E. Kroeze, G. ten Brinke, and G. Hadziioannou, *Polym. Bull.*, 1997, *38*, 203; (b) S.C. Tjong and S.A. Xu, *J. Appl. Polym. Sci.*, 1998, *68*, 1099.
188. G. Radonji and C. Musil, *Angew. Makromol. Chem.*, 1997, *251*, 141.
189. F.P. la Mantia, *Polym. Degard. Stabil.*, 1993, *42*, 213.
190. (a) G. Parkinson, *Chem. Eng.*, 1996, *103*(12), 21; (b) A.R. Nesarikar, S.H. Carr, N. Khait, and F.M. Mirabella, *J. Appl. Polym. Sci.*, 1997, *63*, 1179.
191. ReSyk, Inc. literature, Brigham City, Utah, 2003.
192. S. Samdani, *Chem. Eng.*, 1995, *102*(4), 19.
193. W. Hoyle and D.R. Karsa, *Chemical Aspects of Plastic Recycling*, Special Publication 199. Royal Society of Chemistry, Cambridge, 1997.
194. (a) G.K. Wallace, *TAPPI J.*, 1996, *79*(3), 215; (b) D. Paszun and T. Spychaj, *Ind. Eng. Chem. Res.*, 1997, *36*, 1373; (c) D.-C. Wang, L.-W. Chem, and W.-Y. Chiu, *Angew. Makromol. Chem.*, 1995, *230*, 47; (d) G. Parkinson, *Chem. Eng.*, 1996, *103*(3), 19; (e) G. Samadani, *Chem. Eng.*, 1995, *102*(5), 15; (f) F. Johnson, D.L. Sikkenga, K. Danawala, and B.I. Rosen, U.S. Patent 5,473, 102, 1995; (g) G. Guclu, A. Kasgoz, S. Ozbudak, S. Ozgumus, and M. Orbay, *J. App. Polym. Sci.*, 1998, *69*, 2311; (h) G. Parkinson, *Chem. Eng.*, 1999, *106*(12), 21; (i) J.-W. Chen and L.-W. Chen, *J. Apply. Polym. Sci.*, 199, *73*, 35; (j) B.-K. Kim, G.-C. Hwang, S.-Y. Bai, S.C. Yi, and H. Kumazawa, *J. Appl. Polym. Sci.*, 2001, *81*, 2102; (k) G.P. Karayannidis and D.S. Axilias, *Macromol. Mater. Eng.*, 2007, *292*, 128–146.

195. A. Krazn, *J. App. Polym. Sci.*, 1998, *69*, 1115.
196. H. Wang, L. Chen, X. Liu, Y. Zhang, Z. Wu, and Y. Zhou, *Polym. Preprints*, 2002, *43*(1), 473.
197. K.A. O'Connell, *Waste Age's Recycling Times*, 1997, *9*(19), 15.
198. H. Winter, H.A.M. Mostert, P.J.H.M. Smeets, and G. Paas, *J. Appl. Polym. Sci.*, 1995, *57*, 1409.
199. K.H. Yoon, A.T. DiBenedetto, and S.J. Huang, *Polymer*, 1997, *38*, 2281.
200. S.J. Shafer, U.S. Patent 5,336,814, 1993.
201. R. Pinero, J. Garcia, and M.J. Cocero, *Green Chem.*, 2005, *7*, 380.
202. G. Ondrey, *Chem. Eng.*, 2004, *111*(13), 14.
203. H. Mueller, *Chem. Abstr.*, 1996, *124*, 9639.
204. (a) P. Bassler and M. Kopietz, U.S. Patent 5,495,015, 1996; (b) E. Fuchs and T. Witzil. U.S. Patent 5,495,014, 1996; (c) H. Liehr, U. Schollar, and G. Keunecke, *Chem. Abstr.*, 1995, *122*, 266265; (d) H.L. Fuchs, J. Ritz, and G. Neubauer, *Chem. Abstr.*, 1995, *123*, 314992; (e) S. Sifniades and A.B. Levy, *Chem. Abstr.*, 1995, *123*, 341272; (f) M. Kopietz, U. Kalck, S. Jones, P. Bassler, and C.-U. Priester, U.S. Patent 5,455,346, 1995. (g) G. Ondrey, *Chem. Eng.*, 1998, *105*(1), 21; (h) S.J. Zhou, I. Brubaker, and R. Sedath, 2nd *Green Chemistry and Engineering Conference Abstracts*, Washington, DC, June 30–July 2, 1998, 35; (i) K.-S. Betts, *Environ. Sci. Technol.*, 1999, *33*, 87A; (j) C. Crabb, *Chem. Eng.*, 1999, *106*(6), 54; (k) A.H. Tullo, *Chem. Eng.*, 1997, *104*(10), 25.
205. G. Parkinson, *Chem. Eng.*, 1997, *104*(10), 25.
206. K.-J. Eichhorn, D. Lehmann, and D. Voigt, *J. Appl. Polym. Sci.*, 1996, *62*, 2053.
207. R.J. McKinney, U.S. Patent 5,395,974, 1995.
208. (a) J.E. Kresta, H.X. Xiao, J. Kytner, and I. Cejpek, *Abstr. ACS Meet.* Washington, DC, Aug. 1994, *Macromol. Sec.* 0003; Soc. Automot. Eng. Special Publication SP–1099, 1995, 179; *Chem. Abstr.*, 1995, *123*, 342, 389; (b) X. Xue, M. Omoto, T. Kidai, and Y. Imai, *J. Appl. Polym. Sci.*, 1995, *56*, 127.
209. Anon., *Chem. Eng. News*, May 5, 1997, 59.
210. G. Parkinson, *Chem. Eng.*, 1997, *104*(8), 21.
211. G. Parkinson, *Chem. Eng.*, 1997, *104*(10), 21.
212. H. Tagaya, Y.-I. Suzuki, T. Asou, J.-I. Kadokawa, and K. Chiba, *Chem. Lett.*, 1998, 937.
213. W.D. Lilac and S. Lee, *Preprints Pap. Natl. Meet. Am. Chem. Soc. Div. Environ. Chem.*, 1998, *105*(7), 29.
214. G. Parkinson, *Chem. Eng.*, 1998, *105*(7), 29.
215. Y. Park, J.N. Hool, C.W. Curtis, and B. Roberts, *Ind. Eng. Chem. Res.*, 2001, *40*, 756.
216. Anon., *Chematur Engineering*, Karlskoga, Sweden, 2006.
217. (a) A. Pifer and A. Sen, *Angew. Chem. Int. Ed.* 1998, *37*, 3306; (b) J.E. Remias, T.A. Pavlosky, and A. Sen, Preprints ACS, *Div. Environ. Chem.*, 2000, *40*(2), 69.
218. V. Dufaud and J.M. Basset, *Angew. Chem. Int. Ed.* 1998, *37*, 806.
219. S.L. Buchwalter and L.L. Kosbar, *J. Polym. Sci. A Polym. Chem.*, 1996, *34*, 249.
220. T. Endo, T. Suzuki, F. Sanda, and T. Takata, *Macromolecules*, 1996, *29*, 3315.
221. C. Birkinshaw, M. Buggy, and A. O'Neill, *J. Chem. Technol. Biotechnol.*, 1996, *66*, 19.
222. (a) G. Ondrey, *Chem. Eng.*, 1998, *105*(10), 29; (b) G. Parkinson, *Chem. Eng.*, 1997, *104*(12), 25. (c) Anon., *Chem. Eng. News*, Oct. 12, 1998, 21.
223. H. Sawada, *Encyclopedia of Polymer Science Engineering*, 2nd ed., Wiley, New York, 1986, *4*, 733.
224. (a) R. Lin and R.L. White, *J. Appl. Polym. Sci.*, 1997, *63*, 1287; (b) Z. Zhang, T. Hirose, S. Nishio, Y. Morioka, N. Azuma, A. Ueno, H. Okhita, and M. Okada, *Ind. Eng. Chem. Res.*, 1995, *34*, 4514; (c) P. Carniti, A. Gervasini, P.L. Beltrame, A. Audisio, and F. Bertini, *Appl. Catal. A*, 1995, *127*, 139; (d) G.S. Samadani, *Chem. Eng.*, 1995, *102*(10), 23; (e) Y.D.M. Simard, M.R. Kamal, and D.G. Cooper, *J. Appl. Polym. Sci.*, 1995, *58*, 843; (f) T.E. Ponsford and H.T. Ponsford. U.S. Patent 5,406,010, 1995; (g) T. Sawaguchi and M. Seno, *J. Polym. Sci. A Polym. Chem.*, 1998, *36*, 209; (h) O.S. Woo and L.J. Broadbelt, *Catal. Today*, 1998, *40*, 121.
225. A. Northemann, *Chem. Abstr.*, 1995, *123*, 229–258.
226. (a) Z. Zhibo, S. Nishio, Y. Morioka, A. Ueno, H. Okhita, Y. Tochihara, T. Mizushima, and N. Kakuta, *Catal. Today*, 1996, *29*, 303; (b) S. Lovett, F. Berruti, and L.A. Behie, *Ind. Eng. Chem. Res.*, 1997, *36*, 4436; (c) J.M. Arandes, I. Abajo, D. Lopez-Valerio, I. Fernandez, M.J. Azkoiti, M. Olazar, and J. Bilbao, *Ind. Eng. Chem. Res.*, 1997, *36*, 4523; (d) R.R. Guddeti, R. Knight, and E.D. Grossmann, *Ind. Eng. Chem. Res.*, 2000, *39*, 1171; (e) W. Kaminsky and F. Hartmann, *Angew. Chem. Int. Ed.* 2000, *39*, 331.
227. (a) Anon., *Chem. Ind. (Lond.)*, 1994, 847; (b) G. de La Puente, J.M. Arandes, and U.A. Sedran, *Ind. Eng. Chem. Res.*, 1997, *36*, 4530; (c) G. Ondrey, *Chem. Eng.*, 1998, *105*(1), 21; (d) J. Aguado and D. Serrano, *Feedstock Recycling of Plastics Waste*, Royal Society of Chemistry, Cambridge, 1999; (e) M.A. Uddin, Y. Sabata, A. Muto, and K. Murata, *Ind. Eng. Chem. Res.*, 1999, *38*, 1406.
228. G. Parkinson, *Chem. Eng.*, 1996, *103*(3), 17.
229. T. Kamiya, *Chem. Eng.*, 1997, *104*(2), 42.
230. (a) M.J. Mcintosh, G.G. Arzoumanidis, and F.E. Brockmeier, *Environ. Prog.*, 1998, *17*(1), 19; (b) G. Parkinson, *Chem. Eng.*, 1998, *105*(4), 23.
231. J. Aguado and D. Serrano, *Feedstock Recycling of Plastic Wastes*, Royal Society of Chemistry, Cambridge, 1999.
232. (a) R.W.J. Westerhout, J.A.M. Kuipers, and W.P.M. van Swaaij, *Ind. Eng. Chem. Res.*, 1998, *37*, 841; (b) R.W.J. Westerhout, J. Waanders, J.A.M. Kuipers, and W.P.M. van Swaaij, *Ind. Eng. Chem. Res.*, 1998, *37*, 2293, 2316.
233. W. Zhao, S. Hasegawa, J. Fujita, F. Yoshii, T. Sasaki, K. Makuuchi, J. Sun, and S.-I. Nichimoto, *Polym. Degrad. Stabil.*, 1996, *53*, 129, 199.
234. (a) R. Lin and R.L. White, *J. Appl. Polym. Sci.*, 1995, *58*, 1151; (b) P.N. Sharratt, Y.-H. Lin, A.A. Garforth, and J. Dwyer, *Ind. Eng. Chem. Res.*, 1997, *36*, 5118.
235. (a) W. Kaminsky, B. Schlesselmann, and C.M. Simon, *Polym. Degard. Stabil.*, 1996, *53*, 189; (b) W. Kaminsky, *Angew. Makromol. Chem.*, 1995, *232*, 151.
236. W. Michaeli and V. Lackner, *Angew. Makromol. Chem.*, 1995, *232*, 167.
237. P.L. Beltrame, P.J. Carniti, G. Audisio, and F. Bertini, *Polym. Degrad. Stabil.*, 1989, *26*, 209.

238. (a) R.C. Mordi, J. Dwyer, and R. Fields, *Polym. Degrad. Stabil.*, 1994, *46*, 57; (b) D.L. Negelein, R. Lin, and R.L. White, *J. Appl. Polym. Sci.*, 1998, *67*, 341; (c) Y. Uemichi, J. Nakamura, T. Itoh, M. Sugioka, A. R. Garforth, and J. Dwyer, *Ind. Eng. Chem. Res.*, 1999, *38*, 385.
239. (a) Y. Uemichi, K. Takuma, and A. Ayame, *Chem. Commun.*, 1998, 1975; (b) K. Takuma, Y. Uemichi, and A. Ayame, *Appl. Catal. A*, *192*, 273.
240. J. Aguado, D.P. Serrano, M.D. Romero, and J.M. Escola, *Chem. Commun.*, 1996, 725.
241. I. Nakamura and K. Fujimoto, *Catal. Today*, 1996, *27*, 175.
242. P.A. Hubbard, W.J. Brittain, W.L. Mattice, and D.J. Brunelle, *Macromolecules*, 1998, *31*, 1518.
243. (a) G. Scott and D. Gilead, eds, *Degradable Polymers: Principles and Applications*, Chapman & Hall, New York, 1996; (b) G.J.L. Griffin, *Chemistry and Technology of Biodegradable Polymers*. Chapman & Hall, London, 1994; (c) S.J. Huang, ed., *Polym. Degrad. Stabil.*, 1994, *45*(2), 165–249; (d) A.-C. Albertsson and S. Karlsson, eds, *Macromol. Symp.*, 1998, *130*, 1–410; (e) J. Kahovec, ed., *Macromol. Symp.*, 1997, *123*, 1–249; (f) R. Chandra and R. Rustgi, *Prog. Polym. Sci.*, 1998, *23*, 1273; (g) R.A. Gross and B. Kalra, *Science*, 2002, *297*, 803; (h) E.S. Stevens, *Green Plastics*, Princeton University Press, Princeton, 2002.
244. W. Rathje and C. Murphy, *Rubbish: The Archaeology of Garbage*, Harper–Collins, New York, 1992.
245. (a) R. Arnaud, P. Dabin, J. Lemaire, S. Al-Malaika, S. Chohan, M. Coker, G. Scott, A. Fauve, and A. Maaroufi, *Polym. Degrad. Stabil.*, 1994, *46*, 211; (b) B.S. Yoon, M.H. Suh, S.H. Cheong, J.E. Yie, S.H. Yoon, and S.H. Lee, *J. Appl. Polym. Sci.*, 1996, *60*, 1677; (c) B.G. Kang, S.H. Yoon, S.H. Lee, J.E. Yie, B.S. Yoon, and M.H. Shu, *J. Appl. Polym. Sci.*, 1996, *60*, 1977.
246. S.-I. Akahori and Z. Osawa, *Polym. Degrad. Stabil.*, 1994, *45*, 261.
247. G.S. Samadani, *Chem. Eng.*, 1995, *102*(3), 23.
248. R. Narayan, Polym. Preprints (ACS), *Div. Polym. Chem.*, 1993, *34*(1), 917.
249. (a) Anon., *R&D (Cahners)*, 1997, *39* (Jan.) 7; (b) M.S. Reisch, *Chem. Eng. News*, July 7, 1997 26.
250. (a) G. Parkinson, *Chem. Eng.*, 1997, *104*(12), 27; (b) Anon., *Chem. Eng. News*, Sept. 13, 2004, 11.
251. T. Nakato, M. Tomida, A. Kusuno, M. Shibata, and T. Kakuchi, *Polym. Bull.*, 1998, *40*, 647.
252. Anon., *Waste Age's Recycling Times*, 1998, *10*(7), 5.
253. J. Tritten, *Environ. Sci. Technol*, 1999, *33*, 194A.
254. (a) J.P. Paul, *Kirk–Othmer Encyclopedia of Chemical Technology*, 4th ed., Wiley, New York, 1997, *21*, 22; (b) V.K. Sharma, M. Mincrini, F. Fortuna, F. Cogini, and G. Cornacchia, *Energy Convers. Manage.*, 1998, *39*, 511; (c) W. Klingensmith and K. Baranwal, *Rubber World*, 1998, *218*(3), 41; (d) S.K. De, A.I. Isayev, and K. Khait, eds, *Rubber Recycling*, CRC Press, Boca Raton, FL, 2005; (e) B. Adhikari, D. De, and S. Maiti, *Prog. Polym. Sci.*, 2000, *25*, 909.
255. J.M. Heumann, *Waste Age's Recycling Times*, 1997, *9*(7), 3.
256. Goodyear Tire and Rubber Co. Annual Report, Akron, Ohio, 1996, 20.
257. (a) Anon., *Rubber World*, 1996, *213*(4), 12; (b) Abstr. ACS Meeting, Washington, DC, Aug. 1994, MACR 23.
258. (a) A.M. Thayer, *Chem. Eng. News*, July 17, 1995, 38; (b) L.R. Evans, J.C. Hope, and W.H. Waddell, *Rubber World*, 1995, *212*(3), 21; (c) Goodyear Tire and Rubber Co. Quarterly Report, Akron, Ohio, June 12, 1996; (d) J.-A.E. Bice, S.D. Paktar, and T.A. Okel, *Rubber World*, 1997, *217*(1), 58.
259. J.D. Osborn, *Rubber World*, 1995, *212*(2), 34.
260. (a) J.E. Mark, B. Erman, and F.R. Eirich, *Science and Technology of Rubber*, Academic, San Diego, 1994; (b) J.L. White, *Rubber Processing: Technology, Materials and Principles*, Hanser–Gardner, Cincinnati, OH, 1995; (c) W. Hofmann, *Rubber Technology Handbook*, Hanser–Gardner, Cincinnati, OH, 1989; (d) N.P. Cheremisinoff, *Elastomer Technology Handbook*, CRC Press, Boca Raton, FL, 1993; (e) K. Nadgi, *Rubber as an Engineering Material: Guidelines for Users*. Hanser–Gardner, Cincinnati, OH, 1992.
261. K. Reese, *Today's Chem. Work*, 1995, *4*(2), 75.
262. A.K. Bhowmick, M.M. Hall, and H.A. Benarey, eds, *Rubber Products Manufactures Technology*, Dekker, New York, 1994, 899.
263. C.P. Rader, *Plastics, Rubber and Paper Recycling: A Pragmatic Approach*, ACS Symp. 609, Washington, DC, 1995, 237.
264. (a) M. Reisch, *Chem. Eng. News*, Dec. 19, 1994, 6; (b) M.S. Reisch, *Chem. Eng. News*, Aug. 29, 1994, 14.
265. B. Skoloff, *Wilmington, DE News J.*, Feb. 17, 2007, A12.
266. E. Kersten, *Land and Water*, 1997, *41*(2), 52.
267. R. Mestel, *New Sci.*, Apr. 22, 1995, 6.
268. (a) Anon., *R&D (Cahners)*, 1997, *39*(1), 11; (b) J.M. Heumann, *Waste Age's Recycling Times*, 1997, *9*(11), 6.
269. G. Ondrey, *Chem. Eng.*, 2007, *113*(13), 13.
270. G. Ondrey, *Chem. Eng.*, 1999, *106*(6), 19.
271. (a) M.W. Rouse, *Rubber World*, 1995, *212*(2), 23; (b) C.P. Rader, ed., *Plastics, Rubber and Paper Recycling: A Pragmatic Approach*, ACS Symp. 609, Washington, DC, 1995, 207; (c) K.M. White, *Waste Age's Recycling Times*, 1997, *9*(26), 5.
272. L.H. Lewandowski, *Rubber Chem. Technol.*, 1994, *67*, 447.
273. (a) A.K. Bhowmick, M.M. Hall, and H.A. Benarey, eds, *Rubber Products Manufacturing Technology*, Dekker, New York, 1994, 901; (b) A. Coomarasamy and S.A.M. Hesp, *Rubber World*, 998, *218*(2), 26; (c) I. Gawel, R. Stephowski, and F. Czechowski, *Ind. Eng. Chem. Res.*, 2006, *45*, 3044.
274. Anon., University of Detroit Mercy Annual Report, Michigan, 1993–1994, 13.
275. M.S. Reisch, *Chem. Eng. News*, Aug. 5, 1996, 13.
276. K.-H. Chung and Y.-K. Hong, *J. Appl. Polym. Sci.*, 1999, *72*, 35.
277. B.D. Baumann, *Rubber World*, 1993, *208*(1), 16; 1995, *212*(2), 30.
278. M.J. Boynton and A. Lee, *J. Appl. Polym. Sci.*, 1997, *66*, 271.
279. W. Dierkes, *Rubber World*, 1996, *214*(2), 25.
280. W. Dierkes, *Rubber World*, 1996, *214*(2), 25.
281. (a) W.C. Warner, *Rubber Chem. Technol.*, 1994, *67*(3), 559; In: C. P. Rader, ed. *Plastics, Rubber and Paper Recycling: A Pragmatic Approach*, ACS Symp. 609, Washington, DC, 1995, 245; (b) V.Y. Levin, S.H. Kim, and A.I. Isayev, *Rubber Chem. Technol.*, 1997, *70*, 641.
282. M.J. Myhre and D.A. McKillop, *Rubber World*, 1996 *214*(May), 42.

283. (a) G.S. Samdani, *Chem. Eng.*, 1995, *102*(1), 17; (b) C.J. Brown and W.F. Watson, *Rubber World*, 1998, *218*(2), 34.
284. (a) A.I. Isayev, J. Chen, and A. Tukachinsky, *Rubber Chem. Technol.*, 1995, *68*, 267; (b) A.I. Isayev, S.P. Yushanov, and J. Chen, *J. Appl. Polymer Sci.*, 1996, *59*, 803, 815; (c) A. Tukachinsky, D. Schaorm, A.I. Isayev, *Rubber Chem. Technol.*, 1996, *69*(1), 92; (d) V.Y. Levin, S.H. Kim, A.I. Isayev, J. Massey, and E. von Meerwell, *Rubber Chem. Technol.*, 1996, *69*(1), 104; (e) M. Tapale and A.I. Isayev, *J. Appl. Polym. Sci.*, 1998, *70*, 2007; (f) S.P. Yushanov, A.I. Isayev, and S.H. Kim, *Rubber Chem. Technol.*, 1998, *71*, 168; (g) P. Walter, *Chem. Ind. (Lond.)*, Aug. 21, 2006, 10.
285. A.R. Tripathy, J.E. Morin, D.E. Williams, S.J. Eyles, and R.J. Farris, *Macromolecules*, 2002, *35*, 4616.
286. (a) B. Diao, A.I. Isayev, V.Y. Levin, and S.H. Kim, *J. Appl. Polym. Sci.*, 1998, *69*, 2691; (b) B. Diao, A.I. Isayev, and V.Y. Levin, *Rubber Chem. Technol.*, 1999, *72*, 152; (c) S.E. Shim, V.V. Yashin, and A.I. Isayev, *Green Chem.*, 2004, *6*, 291.
287. J.P. Paul, *Encyclopedia of Polymer Science Engineering*, 2nd ed., Wiley, New York, 1988, *14*, 787.
288. (a) G.S. Samdani, *Chem. Eng.*, 1995, *102*(1), 23; (b) M. Christiansson, B. Stenberg, L.R. Wallenberg, and O. Holst, *Biotechnol. Lett.*, 1998, *20*, 637; (c) Anon., *R&D (Cahners)*, 1997, (10), 28.
289. (a) H. Teng, M.A. Serio, M.A. Wojtowicz, R. Bassilakis, and P.R. Solomon, *Ind. Eng. Chem. Res.*, 1995, *34*, 3102; (b) Anon., *Chem. Eng. News*, Sept. 18, 1995, 29; (c) M.A. Wojtowicz and M.A. Serio, *Chemtech.*, 1996, *26*(10), 48; (d) G. Samdani, *Chem. Eng.*, 1994, *101*(6), 17; (e) G.P. Huffman, Preprints ACS. *Div. Environ. Chem.*, 1994, *34*(2), 478; (f) G. Parkinson, *Chem. Eng.*, 1997, *104*(9), 25; (g) G. Ondrey, *Chem. Eng.*, 1999, *106*(1), 25; 1999, *106*(6), 17; (h) G. San Miguel, G.D. Fowler, and C.J. Sollars, *Ind. Eng. Chem. Res.*, 1998, *37*, 2430; (i) G. Ondrey, *Chem. Eng.*, 1999, *106*(6), 17; (j) P. Williams, *Green Chem.*, 2003, *5*, G20; (k) J.A. Consea, I. Martin-Gullon, R. Font, and J. Jauhiainen, *Environ. Sci. Technol.*, 2004, *38*, 3189; (l) K. Unapumnuk M. Lu, and T.C. Keener, *Ind. Eng. Chem. Res.*, 2006, *45*, 8757.
290. G.P. Huffman and N. Shah, *Chemtech.*, 1998, *26*(12), 34.
291. (a) S. Bhadra, P.P. De, N. Mondal, R. Mukhapadhyaya, and S. DasGupta, *J. Appl. Polym.*, 2003, *89*, 465; (b), G. Ondrey, *Chem. Eng.*, 2007, *114*(13), 16.
292. Anon., *Chem. Eng. News*, Aug. 4, 1997, 82–83.
293. (a) H.V. Makar, *Kirk–Othmer Encyclopedia of Chemical Technology*, 4th ed., Wiley, New York, 1996, *20*, 1092; (b) W.H. Richardson, *Kirk–Othmer Encyclopedia of Chemical Technology*, 4th ed., Wiley, New York, 1996, *20*, 1106; (c) I.K. Wernick and N.J. Themeils, *Ann. Rev. Energy Environ.*, 1998, *23*, 465; (d) S.K. Ritter, *Chem. Eng. News*, June 8, 2009, 53.
294. A.L. Brody, *Kirk–Othmer Encyclopedia of Chemical Technology*, 4th ed., Wiley, New York, 1994, *11*, 834.
295. K. Egan, *Waste Age's Recycling Times*, 1997, *9*(8), 7.
296. K.M. White, *Waste Age's Recycling Times*, 1998, *10*(8), 1.
297. Anon., Business Wire, July 21, 2008.
298. Anon., *World Watch*, 2004, *17*(1), back cover.
299. B.J. Jody, E.J. Daniels, and N.F. Brockmeier, *Chemtech.*, 1994, *24*(11), 41.
300. Anon., *Waste Age's Recycling Times*, 1998, *10*(24), 5.
301. T.J. Veasey, R.J. Wilson, and D.M. Squires, *The Physical Separation and Recovery of Metals from Wastes*. Gordon & Breach, Newark, NJ, 1993.
302. (a) K.C. Frisch, D. Klempner, and G. Prentice, *Recycling of Polyurethanes*, CRC Press, Boca Raton, FL, 1999; (b) W. Rabhofer and E. Weigand, *Automotive Polyurethanes*, CRC Press, Boca Raton, FL, 2001.
303. M. Freemantle, *Chem. Eng. News*, Nov. 27, 1995, 25.
304. G. Ondrey, *Chem. Eng.*, 2000, *107*(11), 21.
305. G.S. Samdani, *Chem. Eng.*, 1995, *102*(6), 17.
306. G. Parkinson, *Chem. Eng.*, 1998, *105*(11), 23.
307. R. Petkewich, *Chem. Eng. News*, Apr. 23, 2007, 72.
308. L.R. Brown, *World Watch*, 1999, *12*(2), 13.
309. R.T. Jacobsen, *Chem. Eng. Prog.*, 2005, *101*(2), 20; *101*(3), 22; *101*(4), 41.
310. W.M. Czaplik, J.-M. Neudorfl, and A.T. vonWangelin, *Green Chem.*, 2007, *9*, 1163.
311. J. Szekely and G. Trapaga, *Technol. Rev.*, 1995, *98*(1), 30.
312. I. Miesiai, *Ind. Eng. Chem. Res.*, 2005, *44*, 1004.
313. J. Hermoso, J. Dufour, J.L. Galvez, C. Negro, and F. Lopez-Mateos, *Ind. Eng. Chem. Res.*, 2005, *44*, 5750.
314. P. vanHerck, C. Vandecasteele, R. Swennen, and R. Mortier, *Environ. Sci. Technol.*, 2000, *34*, 3802.
315. G. Ondrey, *Chem. Eng.*, 2006, *113*(5), 14.
316. G. Parkinson, *Chem. Eng.*, 1997, *104*(5), 31.
317. Anon., *R&D (Cahners)*, 1997, *39*(10), 53.
318. Anon., *Chemtech.*, 1994, *24*(11), 31.
319. G. Parkinson, *Chem. Eng.*, 1996, *103*(3), 23.
320. K. Murase, K.-I. Nishikawa, K.-I. Machida, and G.-Y. Adachi, *Chem. Lett.*, 1994, *23*, 1845.
321. S. Akita, T. Maeda, and H. Takeuchi, *J. Chem. Technol. Biotechnol.*, 1995, *62*, 345.
322. (a) C.P. Ross, *Kirk–Othmer Encyclopedia of Chemical Technology*, 4th ed., Wiley, New York, 1996, *20*, 1127; (b) M. Gruver, *Wilmington Delaware News J.*, Oct. 3, 2009, A8.
323. K. Egan, *Waste Age's Recycling Times*, 1998, *10*(11), 35.
324. Anon., *Chem. Eng. News*, Feb. 7, 2003, 52.
325. B. Hughes, *Waste Age's Recycling Times*, 1998, *10*(17), 10.
326. (a) G. Parkinson, *Chem. Eng.*, 1996, *103*(2), 21; (b) K. Egan, *Waste Age's Recycling Times*, 1998, *10*(24), 4.
327. K.A. O'Connell, *Waste Age's Recycling Times*, 1997, *9*(13), 6.
328. M.W. Grutzeck and J.A. Marks, *Environ. Sci. Technol.*, 1999, *33*, 312.
329. The Presidential Green Chemistry Challenge Awards Program, U.S. Environmental Protection Agency, EPA 744–K-96–001, Washington, DC, July 1996, 14.
330. J.T.B. Tripp and A.P. Cooley, *EDF Lett. (Environmental Defense Fund)*, New York, 1996, *27*(6), 4.
331. R. Simon, *Forbes*, May 28, 1990, 136.
332. K.A. O'Connell, *Waste Age's Recycling Times*, 1998, *10*(14), 13.
333. R.M. Kashmanian and R.F. Rynk, *Am. J. Alt. Agric.*, 1998, *13*, 40.
334. P. Drechsel and D. Kunze, eds, *Waste Composting for Urban and Peri-Urban Agriculture*, Oxford University Press, Oxford, 2001.
335. K.A. O'Connell, *Waste Age's Recycling Times*, 1997, *9*(9), 7.
336. M.J. Jackson and M.A. Lane, *J. Agric. Food Chem.*, 1997, *45*, 2354.

337. Anon., *Environment*, 1995, *37*(2), 22.
338. L.F. Diaz, G.R. Savage, L.L. Eggerth, and C.G. Golueke, *Composting and Recycling Municipal Solid Waste*, Lewis Publishers, Boca Raton, FL, 1993.
339. (a) F.J. Stevenson, *Humus Chemistry: Genesis, Composition, Reactions*, Wiley, New York, 1994; (b) K.H. Tan, *Principles of Soil Chemistry*, Dekker, New York, 1982, 48–82; (c) H.L. Bohn, B.L. McNeal, and G.A. O'Connor, *Soil Chemistry*, Wiley, New York, 1979, 92; (d) W. Kordel, M. Dassenakis., J. Iintelmann, and S. Padberg, *Pure Appl. Chem*., 1997, *69*(7), 1571; (e) J.S. Gaffney, N.A. Marley, and S.B. Clark, *Humic and Fulvic Acids: Isolation, Structure and Environmental Role*, ACS Symp. 651, Washington, DC, 1996; (f) M.H.B. Hayes and W.S. Wilson, *Humic Substances, Peats and Sludges—Health and Environmental Aspects*, Special Publication 172, Royal Society of Chemistry, Cambridge, 1997; (g) U.S. Environmental Protection Agency, *Composting Yard and Municipal Waste*, Technomic, Lancaster, PA., 1995; (h) E. Ghabbour and D. Davies, eds, *Humic Substances: Structures, Models and Functions*, Special Publication 259, Royal Society of Chemistry, Cambridge, 2000; (i) E. Tipping, *Cation Binding by Humic Substances*, Cambridge University Press, Cambridge, 2002.
340. P.A. Matson, W.J. Parton, A.G. Power, and M.J. Swift, *Science*, 1997, *277*, 506.
341. D.J. Sessa and Y.V. Wu, *Int. News Fats Oils Relat. Mater*, 1996, *7*(3), 274.
342. M.A. Martin-Cabrejas, R.M. Esteban, F.J. Lopez-Andreu, K. Waldron, and R.P. Selvendran, *J. Agric. Food Chem*., 1995, *43*, 662.
343. Anon., *Environment*, 1994, *36*(10), 22.
344. (a) Anon., *Environ. Sci. Technol*., 2003, *37*, 389A; 2004, *38*, 265A; (b) T. Staedler, *Technol. Rev*., 2003, *106*(5), 73.
345. Anon., *Land Water*, 1993, *37*(July/Aug.), 49.
346. H. Ridgley, *Waste Age's Recycling Times*, 1997, *9*(23), 4.
347. P. Yost and E. Lund, *Residential Construction Waste Management—A Builder's Field Guide–How to Save Money and Landfill Space*, National Association of Home Builders Research Center, Upper Marlboro, MD, 1997.
348. (a) J. Harkinson, *The Environmental Magazine*, 2000, *11*(2), 44; (b) C. Szczepanski, *On Earth* (Natural Resources Defense Council), 2002, *24*(3), 9; (c) Anon., *Wilmington DE News J*., Apr. 30, 2003, B7; (d) Anon., *Wilmington DE News J*., Oct. 22, 2004, B7; (e) J. Gershon, *On Earth*, 2005, *27*(2), 11; (f) Anon., *Environment*, 2001, *43*(2), 4.
349. S. Shulman, *Technol. Rev*., 1995, *98*(1), 18.
350. C. Yost, *Chemtech*., 1995. *25*(6), 51.
351. G. Ondrey, *Chem. Eng*., 2003, *110*(12), 15.
352. Q. Qiu, V. Hlavacek, and S. Prochazka, *Ind. Eng. Chem. Res*., 2005, *44*, 2469, 2477.
353. V. Mymrin, H. deA.Ponte, O.F. Lopes, and A.V. Vaamonde, *Green Chem*., 2003, *5*, 357.
354. H. Ridgley, *Waste Age's Recycling Times*, 1997, *9*(21), 6.
355. Anon., *World Watch*, 2006, *19*(1), inside front cover.
356. H. Ridgley, *Waste Age's Recycling Times*, 1998, *10*(8), 6; 1998, *10*(14), 5.
357. K.A. O'Connell, *Waste Age's Recycling Times*, 1998, *10*(24), 12.
358. M.S. Reisch, *Chem. Eng. News*, Nov. 7, 1994, 12.
359. (a) A. Hoyle, *Nonwovens Ind*., 1995, *26*(2), 32; 1994, *25*(12), 68; (b) M. Rice, *Wilmington Delaware News J*., Jan. 7, 1995, E4.
360. K.A. Sager, *Waste Age's Recycling Times*, 1997, *9*(17).
361. K. Egan, *Waste Age's Recycling Times*, 1997, *9*(9), 16.
362. (a) D.J. Hanson, *Chem. Eng. News*, Feb. 26, 1996, 23; (b) H. Ridgley, *Waste Age's Recycling Times*, 1997, *9*(22), 3.
363. S. Halsted, *Waste Age's Recycling Times*, 1997, *9*(11), 5.
364. I.I. Negulescu, H. Kwon, B.J. Collier, J.R. Collier, and A. Pendse, *Text Chem. Color*, 198, *30*(6), 31.
365. B. Hughes, *Waste Age's Recycling Times*, 1998, *10*(14), 12.
366. D.A. Becker and D.W. Brinkman, *Kirk–Othmer Encyclopedia of Chemical Technology*, 4th ed., Wiley, New York, 1997, *21*, 1.
367. B.R. Woodin, R.M. Smolowitz, and J.J. Stegeman, *Environ. Sci. Technol*., 1997, *31*, 1198.
368. Anon., *Pollut. Equip. News*, June 1997, 45.
369. M. Murray, *Wilmington Delaware News J*., Dec. 8, 1995.
370. G. Ondrey, *Chem. Eng*., 1998, *105*(5), 21.
371. Anon., *Pollut. Equip. News*, June, 1997, 36.
372. Anon., *Chem. Eng. News*, Aug. 14, 1995, 33.
373. G. Alther, *Pollut. Equip. News*, June, 1997, 58.
374. K.A. O'Connell, *Waste Age's Recycling Times*, 1998, *10*(14), 1, 10.
375. P. Walter, *Chem. Ind. (Lond.)*, Apr. 9, 2007, 9.
376. K. Egan, *Waste Age's Recycling Times*, 1997, *9*(21), 5.
377. Anon., *J. Coatings Technol*., 2007, *4*(10), 28.
378. K.A. Sager, *Waste Age's Recycling Times*, 1997, *9*(20), 16.
379. Anon., *Environ. Prog*., 2005, *24*(1), 3.
380. Information from Brookrock Corp. Newark, DE.
381. (a) S. Ottwell, *Chem. Eng. (Rugy, Engl.)*, 1996, (Feb.), 15; (b) C.J. Nagel, C.A. Chanechuk, E.W. Wong, and R.D. Bach, *Environ. Sci. Technol*., 1996, *30*, 2155; (c) E. Rafferty, *Chem. Eng*., 1997, *104*(11), 43; (d) J.J. Cudahy, *Environ. Prog*., 1999, *18*, 285; (e) Molten Metal Technologies, Waltham, MA.
382. (a) D. NaQuin, *Waste Age's Recycling Times*, 1998, *10*(14), 14; (b) K. Bullis, *Technol. Rev*., Mar./Apr. 2007, 25; (c) C.G. Daye, *Chicago AC Meeting*, Aug. 2001, SCHB 4; (d) D. Talbot, *Technol. Rev*., 2004, *107*(1), 24; (e) P. McKenna, *New Sci*., April 25, 2009, 33.
383. "Plasma Arc Technology", www.google.com, 2008.
384. Anon., *Chem. Eng. Prog*., 2004, *100*(1), 15.
385. G. Parkinson, *Chem. Eng. Prog*., 2005, *101*(7), 13.
386. (a) S. Tyson and T. Blackstock, *Paper Pap.-Am. Chem. Soc., Div. Fuel Chem*., 1996, *41*(2), 587; C.A., 1996, *124*, 209, 949; (b) R.E. Hughes, G.S. Dreher, M. Rostam-Abadi, D.M. Moore, and P.J. DeMaris, *Paper. Pap.-Am. Chem. Soc., Div. Fuel Chem*., 1996, *41*(2), 597; C.A., 1996, *124*, 239, 889.
387. D. NaQuin, *Waste Age's Recycling Time*, 1997, *9*(26), 14.
388. J. Beretka, R. Cioffi, L. Santoro, and G.L. Valenti, *J. Chem. Technol. Biotechnol*., 1994, *59*, 243.
389. G. Parkinson, *Chem. Eng*., 1997, *104*(5), 29.
390. (a) W. Zschau, *Int. News Fats Oils Relat. Mater*., 1994, *5*(12), 1375; (b) C. Chung and V. Eidman, *Int. News Fats Oils Relat. Mater*, 1997, *8*, 739.
391. J.H. Krieger and M. Fremantle, *Chem. Eng. News*, July 7, 1997, 10.
392. J.H. Krieger and M. Freemantle, *Chem. Eng. News*, July 7, 1997, 10.

393. (a) E. Grossman, *High Tech Trash—Digital Devices, Hidden Toxics and Human Health*, Island Press, Washington, DC, 2006; (b) C. Carroll, *Nat. Geo. Mag.*, 2008, *213*(1), 64; (c) J. Johnson, *Chem. Eng. News*, May 26, 2008, 32; Sept. 29, 2008, 30; (d) K. Torzewski, *Chem. Eng.* 2009, *116*(4), 24; (e) L. Zhan and Z. Xu, *Environ. Sci. Technol.*, 2009, *43*, 7074; (f) O.A. Ogunseitan, J.M. Schoenung, J-D.M. Saphores, and A.A. Shapiro, *Science*, 2009, *326*, 670.
394. A.J. Saterlay, S.J. Wilkins, and R.G. Compton, *Green Chem.*, 2001, *3*, 149.
395. (a) B. Hileman, *Chem. Eng. News*, July 1, 2002, 15; (b) M.S. Reisch, *Chem. Eng. News*, June 28, 2004, 28; (c) Anon., *Chem. Eng. News*, Oct. 31, 2005, 21.
396. (a) B. Hileman, *Chem. Eng. News*, Jan. 2, 2006, 18; (b) R.E. Hester and R.M. Harrison, *Electronic Waste Management*, Royal Society of Chemistry, Cambridge, 2008; (c) L. Mastny, *World Watch*, 2006, *19*(3), 12; (d) K. Betts, *Environ. Sci. Technol.*, 2008, *42*, 1393.
397. (a) K.S. Betts, *Environ. Sci. Technol.*, 2002, *36*, 370A; (b) Anon., *Environment*, 2002, *44*(10), 5.
398. J.D. Underwood, *Chem. Int. (Lond.)*, 1994, 18.
399. Anon., *Package News*, 1997, May, 6.
400. *The Solid Waste Management Problem*, 2nd ed., The Council on Solid Waste Solutions, Washington, DC, 1989, 11.
401. H. Ridgley, *Waste Age's Recycling Times*, 1998, *10*(7), 1.
402. S. Strasser, *Waste and Want: A Social History of Trash*, Metropolitan Books, New York, 1999.
403. H. Petroski, *The Pencil—A History of Design and Circumstance*, Knopf, New York, 1990.
404. R.L. Maine, *Prog. Pap. Recycl.*, 1993, *2*(3), 76.
405. D. NaQuin, *Waste Age's Recycling Times*, 1998, *10*(19), 3.
406. Anon., *Washington Delaware Sunday News J.*, Dec. 20, 1998, B7.
407. M. Billington, *Wilmington Delaware News J.*, Mar. 14, 1999, B1.
408. Hockessin (Delaware) *Community News*, Dec. 21, 1995, 26.
409. P. Gopal, *Wilmington Delaware News J.*, Oct. 26, 1998, B1.
410. G. Gardner, *World Watch*, 1999, *12*(4), 10.
411. (a) S.E.M. Selke, *Packaging and the Environment*. Technomic, Lancaster, PA, 1990; 2nd ed., 1994; (b) E.J. Stilwell, R.C. Canty, P.W. Kopf, and A.M. Montrone, *Packaging for the Environment—A Partnership for Progress*, American Management Association, New York, 1991; (c) G.M. Levy, ed., *Packaging in the Environment*, Blackie Academic and Professional, London, 1993.
412. D. Saphire, *Benefits of Reusable Shipping Containers—Delivering the Goods*, Inform New York, 1995.
413. The Presidential Green Chemistry Challenge Awards Program, U.S. Environmental Protection Agency, Washington, DC, EPA 744-K-96–001, 1996, July 33.
414. Anon., *Pollout. Equip. News*, 1997, *30*(5), 4.
415. K.A. O'Connell, *Waste Age's Recycling Times*, 1997, *9*(24), 13.
416. K. Holler, *Waste Age's Recycling Times*, 1997, *9*(24), 10.
417. H. Ridgley, *Waste Age's Recycling Times*, 1997, *9*(23), 4.
418. Anon., *Environment*, 1997, *39*(3), 23.
419. D. Saphire, *Case Reopened—Reassessing Refillable Bottles.* Inform, New York, 1994.
420. H.E. French, *World Watch*, 1993, *6*(6), 32.
421. B. Regan, *Am. Metal Market*, Mar. 10, 1997, *105*(47).
422. S.T. Firkaly, *Into the Mouths of Babes—A Natural Food Nutrition and Feeding Guide for Infants and Toddlers*, rev. ed., Betterway Books, Cincinnati, OH, 1995.
423. (a) V. Komolprasertand and A.R. Lawson, *J. Agric. Food Chem.*, 1997, *45*, 444; (b) M. Inagaki, S. Tasaka, and M. Makino, *J. Appl. Polym. Sci.*, 1997, *64*, 1031; (c) N. Inagaki, S. Tasaka, and H. Hiramatsu, *J. Appl. Polym. Sci.*, 1999, *71*, 2091; (d) F. Garbassi and E. Occhiello, *Macromol. Symp.*, 1999, *139*, 107; (e) Anon., *Mod. Plastics*, 2000, *77*(1), 12; 2000, *77*(6), 12; (f) A. Tullo, *Chem. Eng. News*, May 22, 2000, 25.
424. (a) M. Creatore, F. Palumbo, and R. d'Agostilno, *Pure Appl. Chem.*, 2002, *74*, 407; (b) J. Hyun, M. Pope J. Smith, M. Park, and J.J. Cuomo, *J. Appl. Polym. Sci.*, 2000, *75*, 1158.
425. *Package News*, 1997, May 6.
426. (a) G. Ondrey, *Chem. Eng.*, 1999, *106*(6), 21; (b) M. Defosse, *Mod. Plastics*, 2000, *77*(3), 26; 2000, *77*(8), 23; (c) G. Ondrey, *Chem. Eng.*, 2004, *111*(2), 16.
427. (a) Letter. G. T. Kimbrough of Ashland Superamerica to A.S. Matlack, 2/15/96. (b) *Packaging*, 1995, *66*(705), 13.
428. (a) Anon., *Chem. Eng. News*, Aug. 25, 1997, 68; (b) Anon., *University of California Berkeley Wellness Lett.* 1999, *15*(5), 8; 2007, *23*(5), 2; (c) Anon., *Tufts Univ. Health Nutr. Lett.*, 2003, *21*(3), 1; Apr. 2003, 3.
429. T.H. Grenby, *Pure Appl. Chem.*, 1997, *69*(4), 709.
430. (a) H. Bennett, *Chemical Formulary*, Chemical Publishing, New York, 1976, *19*, 308; (b) P.K. Ashurst, *Chemistry and Technology of Soft Drinks and Juices*, Sheffield Academic Press, London, 1998.
431. K.M. Reese, *Chem. Eng. News*, June 8, 1998, 64.
432. H.E. Allen, M.A. Henderson, and C.N. Haas, *Chemtech.*, 1991, *21*(12), 738.
433. Anon., *Amicus J.* (Natural Resources Defense Council), 2000, *22*(2), 7.
434. Anon., *Tufts Univ. Health Nutr. Lett.*, 1997, *15*(7), 3.
435. Anon., *Package Dig.*, 1995, *32*(2), 68.
436. (a) E. Royte, *Bottlemania: How Water Went on Sale and We Bought It*, Bloomsburg, New York, 2008; (b) J. Motavalli, *EDF Solutions*, 2008, *39*(4) 10; (c) E. Arnold and J. Larsen, *Earth Policy Institute*, Feb. 2, 2006; (d) Anon., *Tufts Univ. Health Nutr. Lett.*, 2007, *25*(3), 7.
437. M. Smith, *The U.S. Paper Industry and Sustainable Production*. MIT Press, Cambridge, MA, 1997.
438. J. Williams, *Wilmington Delaware News J.*, Apr. 25, 1996, D1.
439. R. Graff and B. Fishbein, *Reducing Office Paper Waste*, Inform, New York, 1991.
440. (a) G.L. Nelson, *Chemtech.*, 1995, *25*(12), 50; (b) R.A. Denison, *Ann. Rev. Energy Environ.*, 1996, *21*, 191.
441. W. McDonough and M. Braungart, *Cradle to Cradle: Remaking the Way That We Make Things*, North Point Press, New York, 2002.
442. M.B. Hocking, *Science*, 1991, *251*, 504.
443. Anon., *Packag Dig.*, 1995, Nov. 42.
444. K. Christen, *Environ. Sci. Technol.*, 2002, *36*, 406A.
445. (a) Anon., *Pap. Technol.*, 2002, *43*(8), 7; (b) J.-F. Tremblay, *Chem. Eng. News*, July 29, 2002, 19; Oct. 27, 2003, 28.
446. Anon., *Chem. Eng. News*, Mar. 31, 2008, 16.
447. P. Chatton, *Surface Coat. Int.*, 1995, *78*(11), 490.
448. (a) N. Gertler and J.R. Ehrenfeld, *Technol. Rev.*, 1996, *99*(2), 48; (b) R.A. Frosch, *Environment*, 1995, *37*(10), 16; (c) M. Dwortzon, *Technol. Rev.*, 1998, *100*(9), 18; (d) H. Grann,

The industrial symbiosis at Kalundborg, Denmark, In: D.J. Richards, ed., *The Industrial Green Game. Implications for Environmental Design & Management*, pp. 117–123, Royal Society of Chemistry, Cambridge, 1997; (e) R. vanBerkel, T. Fujita, S. Hashimoto, and M. Fujil, *Environ. Sci. Technol.*, 2009, *43*, 1271.
449. M. Nisbet, G.J. Venta, and M. Klein, *Environ. Prog.*, 1998, *17*(2), 96.
450. H. Kane, *World Watch*, 1997, *10*(4), 29.
451. (a) C. Hendrickson, L. Lave, and F. McMichael, *Chemtech.*, 1995, *25*(5), 56; (b) R.A. Denison and J.F. Ruston, *EDF Lett. (Environmental Defense Fund)*, New York, 1996, *27*(5), 7; (c) C. Boerner and K. Chilton, *Environment*, 1994, *36*(1), 7.
452. (a) *World Watch*, 1995, *8*(4), 39; (b) K.M. White and C. Miller, *Waste Age's Recycling Times*, 1993, Mar 9, 7; (c) C. Hanisch, *Environ. Sci. Technol.*, 2000, *34*, 170A. (d) F. Ackerman, *Why Do We Recycle? Markets, Values & Public Policy*, Island Press, Washington, DC, 1997.
453. J.M. Heumann, *Waste Age's Recycling Times*, 1998, *10*, 15.
454. Anon., *Environment*, 1996, *38*(9), 22.
455. F. Ackerman, *Why Do We Recycle? Markets, Values and Public Policy*. Island Press, Washington, DC, 1997.
456. M.S. Andersen, *Environment*, 1998, *40*(4), 10.
457. K.A. O'Connell, *Waste Age's Recycling Times*, 1997, *9*(5), 1; *9*(6), 7.
458. F. Ackerman, *Environment*, 1992, *34*(5), 2.
459. R.F. Stone, A.D. Sagar, and N.A. Ashford, *Technol. Rev.*, 1992, *95*(5), 48.
460. S. Odendahl, *Pulp Pap. Can.*, 1994, *95*(4), 30.
461. *Environ. Sci. Technol.*, 1995, *29*(9), 407A.
462. D. Saphire, *Case Reopened—Reassessing Refillable Bottles*, Inform, New York, 1994, 222, 233, 244, 264.
463. K.M. White, *Waste Age's Recycling Times*, 1997, *9*(10), 6.
464. (a) Anon., *Chem. Eng. News.*, July 7, 1997, 23; (b) P. Layman, *Chem. Eng. News*, Oct. 31, 1994, 10; (c) L. Gottsching, *Pap. Technol.*, 1996, *37*(1), 47; (d) M. Ryan, *World Watch*, 1994, *7*(4), 9; (e) M. Ryan, *World Watch*, 1993; *6*(5), 28; (f) B.K. Fishbein, *Germany, Garbage and the Green Dot: Challenging the Throwaway Society*. Inform, New York, 1994.
465. (a) J.M. Heumann, *Waste Age's Recycling Times*, 1998, *10*(11), 29; (b) K.M. White, *Waste Age's Recycling Times*, 1998, *10*(14), 4.
466. S. Samdani, *Chem. Eng.*, 1995, *102*(4), 21.
467. K.A. O'Connell, *Waste Age's Recycling Times*, 1997, *9*(26), 7.
468. (a) R.W. Kates, *Environment*, 2000, *42*(3), 10; (b) P.F. Barlett and G.W. Chase, eds, *Sustainability on Campus—Stories and Strategies for Change*, Island Press, Washington, DC, 2004. (c) T.E. Graedel and R.J. Klee, *Environ. Sci. Technol.*, 2002, *36*, 523; (d) S. Dresner, *The Principles of Sustainability*, Stylus Publishing, Herndon, VA, 2002.
469. E.O. Wilson, *Science*, 1998, *279*, 2048.
470. (a) G. Gardner and P. Sampat, *Mind Over Matter: Recasting the Role of Materials in Our Lives*. Worldwatch Paper, 144, World Watch Institute, Washington, DC, 1998; (b) J.B. Schor, *The Overspent American—Why We Want What We Don't Need*, Harper Perennial, New York, 1999.
471. (a) D.H. Meadows, D.L. Meadows, and J. Randers, *Beyond the Limits*, Chelsea Green Publishers, Post Mills, VT, 1992; (b) A. Durning, *How Much Is Enough? The Consumer Society and the Future of the Earth*, Norton, New York, 1992; (c) R.W. Kates, *Environment*, 2000, *42*(3), 10; (d) T. Jackson, *The Earthscan Reader on Sustainable Consumption*, Stylus Publishing, Herndon, VA, 2006.

RECOMMENDED READING

1. M.S. Turner, *Recycling—Information Pamphlet*, American Chemical Society, Washington, DC, 1993.
2. H. Kane, In: L. Starke, ed., *Shifting to Sustainable Industries. State of the World 1996*, pp. 152–167, Worldwatch Institute, Washington, DC, 1996.
3. J.E. Young and A. Sachs. In: L. Starke, ed., *Creating a Sustainable Materials Economy. State of the World 1995*, pp. 76–94, Worldwatch Institute, Washington, DC, 1995.
4. U.S. Consumption and The Environment, Union of Concerned Scientists, Cambridge, MA, Feb. 1994.
5. P. Harrison, Sharing the Blame. *Environ. Action*, 1994, *26*(2), 17; *The Third Revolution: Environment, Population and A Sustainable World*, Chapter 18, p. 255, I. B. Tauris, New York, 1992.
6. G.L. Nelson., Ecology and plastics, *Chemtech.*, 1995, *25*(12), 50.

EXERCISES

1. Watch your neighbors at home and at school to see what they discard that is still usable or easily repaired. One way to do this is to walk down the street on trash day before the trash man arrives.
2. Ask a person from an earlier generation what was used before throwaway items came into common use. Could we safely revert to these?
3. Check the items in the Goodwill or another thrift store to see what fraction of them are new or practically new.
4. See what items are sold in bulk in the supermarket. Figure out how other items might be sold in this way and thus reduce packaging. Do the same thing for a hardware store.
5. Look at the ecolabels in a store to see if any might be misleading.
6. Spun-bonded polyethylene (e.g., Tyvek) is used in strong mailing envelopes. How should these be marked so that they don't cause complications?
7. How could a household waste-exchange program be set up in your area?
8. The use of computers and e-mails was expected to reduce the use of paper. Why hasn't it? Can you devise a system so that these do reduce the amount of paper that is used?
9. Look at the contents of the trash or recycling containers next to the photocopiers in your building. What types of mistakes are people making when they copy? How would you redesign the photocopier to avoid these?
10. Are there any groups in your area that share cars or other equipment?
11. Do a life-cycle analysis on artificial turf versus the usual lawn for a park or a campus.

CHAPTER 15

Energy and the Environment

15.1 ENERGY-RELATED PROBLEMS[1]

Energy is needed for heating, cooling, and lighting homes, offices, and manufacturing plants; cooking; transportation; farming; manufacturing of goods; and a variety of other uses. Buildings in the United States use 36% of the total energy and 65% of electricity, produce 30% of greenhouse emissions, use 30% of raw materials, produce 30% of waste and consume 12% of potable water.[2] Homes in the United States use 46% of their energy for space heating, 15% for heating water, 10% for food storage, 9% for space cooling, 7% for lighting, and 13% for other uses.[3] The corresponding figures for commercial buildings are 31%, 4%, 5%, 16%, 28%, and 16%. Transportation accounts for one-fourth of the energy consumption and two-thirds of the oil usage in the United States. Industry uses 37% of the energy in the United States. Agriculture, including the transportation and processing of foods uses about 17% of the energy used in the United States. Industrialized societies tend to be energy-intensive. Most of this energy is derived from fossil fuels. In 1993, oil provided 39% of the world's energy, natural gas 22%, and coal 26%, whereas hydropower and other renewable energy sources provided only 8%.[4] Nuclear power accounted for 5% of the total. In 2004, the world obtained 39% from oil, 24% from coal, 23% from natural gas, 7% from nuclear sources and 7% from renewable sources of which 90% was hydropower.[5] This energy consumption has produced various problems, some of which are quite serious. Even nuclear energy and renewable energy have problems that need to be solved. Although technology can solve some of these problems, others may require changes in our lifestyles, habits, value systems, and outlooks. This review will focus primarily on the role that chemistry can play in solving some of the problems.[6] It will also mention areas in which social factors play important roles.

15.1.1 The Generation of Electric Power[7]

The burning of coal is used to generate much of the electric power in the United States. For example, Delmarva Power obtains 55.9% of its power from coal, 7.3% from gas, 34.2% from nuclear sources, 0.5% from oil, and 2.1% from renewable sources.[8] Strip mining of coal turns the earth upside down. Although reclamation is now required, the original diverse natural system is not replaced. Instead, a more restricted selection of plants is used. The soil may be quite acidic owing to oxidation of iron sulfide pyrite brought to the surface during the mining. The same process can result in acid mine drainage into local streams, resulting in a drastic reduction in their biota. Tailings ponds must be constructed so that no fine coal enters the stream and the dam is strong enough to withstand heavy rains and possible floods. Mining underground coal is the most hazardous occupation in the United States. Miners may be exposed to roof falls, dust explosions, methane explosions, black lung disease from breathing coal dust, and other problems.

Burning coal in the power plant produces sulfur dioxide, nitrogen oxides, fine particulate matter, and emissions of toxic heavy metals, such as mercury (see in Chapter 4) Electric power generators are responsible for 78% of the sulfur dioxide, 64% of the nitrogen oxides, 40% of fine particulates, 21% of the airborne mercury, and 35% of the carbon dioxide released in the United States.[9] Many old plants in existence when the Clean Air Act was passed were exempted from the regulations. It has been estimated that if the 559 dirtiest plants met the current source emissions standards, it would cut the emission of nitrogen oxides by 69% and that of sulfur dioxide by 77%.[10] Although emissions of sulfur dioxide, carbon monoxide, and lead have fallen since 1988, ground-level ozone has remained about the same, and nitrogen oxides are increasing.[11] Annual crop damage from ground-level ozone in the United States is about 3 billion dollars.[12]

The sulfur dioxide comes from sulfur in the coal. Although inorganic sulfur in pyrite can be removed by flotation of fine coal, organically bound sulfur is much harder to remove, and usually is not. The sulfur dioxide in stack gas is often captured with lime to form calcium sulfite, which can be oxidized to calcium sulfate (gypsum). In places such as Denmark and Japan, this gypsum is used in making wallboard for houses. Until recently, in the United States only 5% of this gypsum was used in wallboard because it could not compete with natural gypsum.[13] The result was that more than 12 million tons of flue gas gypsum was landfilled each year in the United States. (About three-fourths of it was still being landfilled in 2009.[14]) Ways of removing the sulfur dioxide with regenerable adsorbents have been studied, but do not seem to be in widespread use.[15]

Nitrogen oxides are produced when fuels are burned at high temperatures. The amount formed is less at lower temperatures. Catalytic combustion using a palladium catalyst has been studied as a way of burning the fuel at a lower temperature.[16] Nitrogen oxides can also be removed by reduction of the stack gases with ammonia, hydrogen, carbon monoxide, or alkanes using various catalysts to form nitrogen.[17] Nitrogen oxides formed during combustion react with hydrocarbons in the presence of light to form ground-level ozone. The U.S. EPA has proposed to require deep cuts (up to 85%) in the amount of nitrogen oxides emitted by power plants in the eastern United States as a way of reducing ground-level ozone.[18] This may require more than just burning at a lower temperature. The Edison Electric Institute, a trade association of utilities, calls the proposal "unfair, expensive and misdirected." The American Chemistry Council (formerly the Chemical Manufacturers Association) is also concerned about the proposal. Motor vehicles produce about half of the nitrogen oxides in the air. Power plants produce much of the rest. Ships at sea account for 14% of the total emissions of nitrogen oxides, from the burning of fossil fuels, and 16% of the sulfur dioxide arising from the burning of petroleum. Their residual oil fuel may contain as much as 4.5% sulfur.[19] Some of these emissions drift over and fall out on land in places such as Scandinavia. Acid rain from sulfur dioxide and nitrogen oxides emissions[20] has wiped out the fish populations in 9500 lakes in Norway with another 5300 remaining at risk.[21] Much of the emissions arise in other countries, such as the United Kingdom and drift in with the winds.[22] (A study by the British government concluded that up to 24,000 persons die prematurely in the United Kingdom each year following short-term air pollution episodes involving ozone, particulates, and sulfur dioxide.[23]) Simply reducing the supply of acid rain now may not be enough to restore these lakes, because the buffer capacity of the soil of the watershed may have been exhausted. Such depletion of calcium ions in the soil has been documented in the mountains of New Hampshire in the United States.[24] Acid rain also causes reductions in crop and timber yields, as well as corrosion of metals in cars, power lines, and the like, and of buildings made of limestone, marble, and sandstone. It is also partially responsible for the eutrophication of water in places such as the Chesapeake Bay in Maryland.

Power plants are also among the sources of the fine particulates (under 2.5 μm) that are beginning to be regulated by the U.S. EPA as a way to better health. Better electrostatic precipitators or scrubbing devices may be needed to control these. (Other sources include motor vehicles and the burning of wood. In Los Angeles, 14% of the fine carbon-containing particles come from residential fireplaces, with the value going up to 30% in winter.[25]) This, together with lowering the ozone standard from 0.12 to 0.08 ppm is expected to prevent 15,000 premature deaths and up to 350,000 cases of aggravated asthma each year in the United States.[26] These new standards are unpopular with the American industry.[27]

The fly ash formed in coal combustion also represents a disposal problem[14] (see also Chapter 14). Although there are some uses such as in concrete and bricks, soil stabilization, soil conditioner, and landfill cover, more need to be found.[28] Additional uses in wallboard, concrete blocks, and other construction materials should be possible. Other ashes include bottom ash and boiler slag. Experiments have been run on the recovery of iron, aluminum, and other metals from the ashes, but the processes may not be economical at this time. This could reduce the need to mine for these other materials. Coal-fired power plants produce over 100 million tons of ash annually in the United States. Coal fly ash is routinely mixed with water and put into settling basins. This process extracts some arsenic, cadmium, mercury, selenium, and strontium into water, which can then cause abnormalities in amphibians and can contaminate drinking water.[29] On December 22, 2008, a tailings pond dam in Kingston, Tennessee, broke, releasing 5.4 million cubic yards of ash into the valley covering 275 acres with 6 ft of ash.

About 34% of the energy in the coal is recovered as electricity in the usual power plant.[30] The rest goes out in the cooling water. Water drawn in from a river or lake contains fish, fish larvae, fish eggs, and a variety of other organisms that are often killed in the process. Fish screens have been designed to divert fish, but do nothing for smaller organisms. These are killed when the cooling water is chlorinated to prevent biofouling of the tubes. The exit water must then be dechlorinated by reducing the chlorine to chloride ion with sulfur dioxide, sodium sulfite, or sodium metabisulfite to avoid killing more organisms in the receiving body of water.[31] The thermal pollution from the released water can help fish and shellfish grow more in the winter, but can be deadly to them in the summer. Substitution of cooling towers for the once-through cooling water would reduce the losses of organisms, but would be more expensive.[32] Air cooling is also possible but may be even more expensive. There is an urgent need to find uses for this waste heat. One of the best ways is to use it to heat and cool buildings in a process called district heating. (The waste heat can also be recovered as steam for industrial plant use, the process being called cogeneration.) This can recover 80–85% or more of the energy in the fuel[33] and is often used in Europe. District heating amounts to 50% of the power in Denmark, 38% in the Netherlands, 36% in Finland, and 31% in Russia, but only 8% in the United States. (The waste heat from the power plant just north of Wilmington, Delaware, could be used for this purpose in buildings in downtown Wilmington, but it is not.) Perhaps, further research on cheap, efficient insulation for the pipes is needed. Drag-reducing surfactants, such as *N*-acetylsarcosinate ($C_{16}H_{33}N(CH_3)CH_2CONa^+$), can reduce drag in these systems as much as 70%.[34] More research is needed on methods of transforming this waste heat to a stored form that can be shipped as a liquid or in a slurry in a pipe for reclaiming at more distant locations. (Heat storage systems are discussed later.)

15.1.2 Transportation

The 5% of the world's population living in the United States consumes 25% of its oil.[35] (The United States has only 3% of the world's oil reserves.[36]) The transportation sector uses 65% of this oil. Almost all (97%) transportation in the United States relies on oil. Personal automobiles consume 50% of the oil.[37] Transportation accounted for one-third of the carbon dioxide emissions of the United States in 2003.[38] Gasoline consumption was almost 114 billion gal in 1996.[39] America's drivers burned an average of 336 million gal of gasoline per day for the first 8 months of 1997.[40] Half of this oil was imported. In June 1992, the cost of this oil plus the importation of cars in which it could be used was 6.6 billion dollars.[41] Partly as a result of this, the United States is now the world's largest debtor nation.[42] Europe imports 76% of its petroleum.[43] Transportation produces 20% of all carbon dioxide in the world and 67% of the pollutant emissions in the world's cities.

Numerous environmental problems are associated with the production and use of this oil. Oil spills were discussed in Chapter 1. Extensive studies have been made on the fate of the oil spilled in Alaskan waters from the *Exxon Valdez*.[44] Leaks in pipelines have also been problems.[45] Twenty-five years after a spill in Alberta, Canada, large amounts of oil were still present at depths of 10–40 cm. Motorists who change their own oil often dump the used oil down a storm drain, from which it finds its way to the nearest river. Efforts to eliminate this practice include laws, providing convenient collection stations for used oil, and putting warning signs on the pavement next to storm drains. Oil can also get into water from the exhausts of older outboard motors which burn a mixture of 1 qt of oil to 1 gal of gasoline. There is also the problem of what to do with used drilling muds and oil well brines, especially on offshore drilling platforms. The methyl *tert*-butyl ether added to gasoline to reduce air pollution has shown up in shallow wells and springs in several urban areas.[46] This may have resulted both from spills and by evaporation or leakage from tanks. Most of it in Donner Lake, California, came from recreational boating.[47] The additive is not used any longer.[48] Ethanol is used instead.

Motor vehicles are responsible for much of the ground-level ozone. They produce half of the nitrogen oxides in the air. Heavy-duty diesel engines account for a quarter of this.[49] There are about 5 million heavy-duty diesel engines and 2 million light-duty diesel engines on roads in the United States. Short (50 ns) electrical bursts can eliminate not only the nitrogen oxides, but also the soot in the diesel exhaust. In addition, motor vehicles emit unburned hydrocarbons and carbon monoxide. The cost of air pollution caused by cars and trucks in the United States has been estimated at \$30–\$200 billion/yr.[50]

The soot from diesel engines and wood smoke carry mutagenic polycyclic aromatic hydrocarbons and their nitro derivatives.[51] Some typical ones are shown in **15.1** Schematic, the first being a potent mutagen in the Ames *Salmonella* test. Catalytic afterburners (containing platinum) are now required on new wood stoves sold in parts of the United States to reduce air pollution from them.

Motor vehicle transport accounts for half of the world's oil consumption and generates one-fifth of all greenhouse emissions. Tailpipe emissions are the largest source of air pollution in nearly half of the cities of the world, surpassing wood fires, coal-burning plants, and chemical manufacturing.[52] New rules for lowering the maximum sulfur content in gasoline and diesel fuel, both in the United States[53] and in Europe,[54] are designed to reduce this pollution by enhancing the performance of catalytic converters. (The average sulfur content in U.S. gasoline in February 1999 was 330 ppm.) Several methods have been used to lower the sulfur content to <1 ppm.[55]

Urban sprawl is a big part of the problem in the United States. Americans in cities drive five times as far as persons

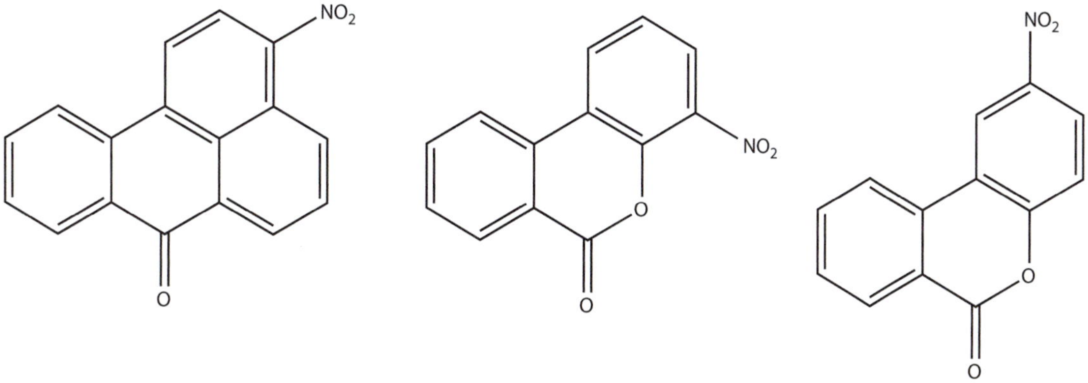

NO_2 NO_2 NO_2
O O O O O

15.1 Schematic

who live in European cities.[56] The United States has more cars per capita than any other developed nation.[57] An Australian has described the situation as "totally out of control." Only 3% of travel in the U.S. cities is by public transport. One-fourth of the travel in European cities is by this method, whereas it is two-thirds in Hong Kong and Tokyo. Close to half of all urban space in the United States is for cars. The average time to commute to work in the United States is 22 min.[58] Only 4% travel to work by public transportation. Only 3% walk to work.[59] Only 22% of car miles are to work, whereas about a third are to pick up children and to go shopping.[60] Many children are effectively immobilized by unsafe roads and the lack of sidewalks in suburbia, which prevent them from walking. Children old enough to drive cars often go to shopping malls to find a place to walk and meet their friends. Some of them also have a curious energy-intensive pastime of cruising their cars round and round a loop in the center of Newark, Delaware, on Friday night, as they watch their friends in other cars. Widening roads accounts for 44% of highway costs. Subsidies to roads are seven times those for public transportation. The difference between work and welfare in the inner city may exist, in part, because of inadequate public transportation.

The problem in the United States is becoming worse.[61] The population in Delaware grew by 11% in a 10-year period, but the number of vehicle miles went up by 55%.[62] There are now more cars than licensed drivers.[63] Delaware cars average 17 miles/gal. The problem is that the price of gasoline is too low. At \$1.17 gal^{-1} in 1999, it was cheaper in inflation-adjusted dollars than it was in 1960.[64] At \$1.10 gal^{-1}, it was the cheapest in inflation-adjusted dollars that it had ever been in the 79-year history of recorded pump prices.[65] It was cheaper than milk, apple juice, and even, some, mineral water. The price rose as high as \$4.00 gal^{-1} in 2008, but dropped to \$1.83 gal^{-1} in March 2009. Americans are choosing less fuel-efficient cars today. Sport utility vehicles, vans, and light trucks, which are subject to lower government fuel efficiency standards,[66] are now almost half of the total fleet. Americans tend to choose larger cars over smaller ones[67] because they think they are safer. The fuel efficiency of cars was 21 mpg in 1994.[68] The U.S. Congress raised this to 27.5 mpg in 2007.[69] They also choose options such as lower mileage, the ability to accelerate rapidly, four-wheel drive, antiskid brakes, automatic transmission, air-conditioning, and the like. Advertisements for new cars emphasize more power, more pickup, more ability to drive off the road, and more luxury.[70] The raising of speed limits above 55 mph has not helped because fuel efficiency declines above this speed.

The cost of mobility in the United States is about 20% of the family income. In Europe, the figure is closer to 7%. The problem is that much of the cost of automobile transportation is hidden. For each dollar that the average American associates with his car, there is a 2-dollar hidden subsidy in the form of road construction and repairs, street maintenance, traffic management, and parking enforcement. It has even been suggested that the price of the Persian Gulf War be paid for at the gas pump. Taxes on gasoline are relatively low in the United States, but are 64% in the European Union.[71] Suggestions for putting the cost where the motorist can see it include road-use taxes, elimination of free parking except in very small lots, linking the cost of car insurance and license fees to the miles traveled, elimination of free company cars, taxes on gas guzzlers, and so on. Taxes and subsidies might be rearranged such that it is expensive and inconvenient to drive one person per car, but inexpensive and convenient to take public transportation or to walk or bike. A double-track rail line can carry as many passengers as a 16-lane highway can. A subway uses one-sixth as much energy per passenger as a car carrying one person. Air travel is more energy-intensive than travel by rail, suggesting that trips of 100–200 miles should be made by rail. Supersonic aircraft[72] require twice as much fuel per passenger mile as regular aircraft does.

Alternatives to sprawl include clustered housing and public transit-oriented development.[73] A return to more compact living with village-like living has been proposed.[74] People in compact neighborhoods in San Francisco made 42% fewer trips by car than suburbanites.[75] Mixed-use developments cut auto use. Portland, Oregon, has urban growth boundaries and other laws that are leading to a more livable city.[76] Curitiba, Brazil, has an integrated public transportation system that functions well. The combination of cheap gasoline, few safe places to walk in suburbia, and abundant food has made one-third of Americans obese (i.e., 20% or more over optimum weight). World oil production is predicted to peak in 2010.[77] (U.S. oil production peaked in 1970.) This should lead to higher gasoline prices. To moderate the problems associated with global warming, it would be desirable to change to a less energy-intensive system before 2010.

15.1.3 Miscellaneous Problems

There has been a tendency to mechanize all sorts of tools and gadgets in the name of saving labor or living a life of luxury. Gasoline-powered lawn mowers and chain saws emit just as much pollution as the outboard motors mentioned in the foregoing section.[78] This is many times what comes out of the tailpipe of an automobile for the same amount of fuel consumed. Lawn mowers and saws that are run by hand are still available, although not very common. They give the operator more exercise. There is some question as to whether the average home owner needs a powered tooth brush, pencil sharpener, carving knife, can opener, lawn edger, or leaf blower. Some of the larger items may even be purchased as status symbols.

Single-use throwaway items require more energy to make than reusable ones on a per-use basis, as discussed in Chapter 14. The production of ammonia for fertilizers is energy-intensive. Much of it could be replaced by leguminous cover crops, less lawn, as mentioned in Chapter 11. Runoff of excess fertilizer has caused a dead zone of 6000 $miles^2$ to form in the Gulf of Mexico.[79] It has also led to *Pfiesteria*-caused fish kills in North Carolina and Maryland.[80]

Ammonia manufacture from nitrogen and hydrogen is an energy-intensive process that consumes about 5% of the world's production of natural gas (for fuel and as a source of hydrogen).[81] The reaction is usually run at about 450°C over an iron catalyst at about 10% conversion per pass. Chemists have long sought a milder, better way of doing this. Recent research suggests that such a route may be possible. Molybdenum amides have been used to split nitrogen to form molybdenum nitrides (**15.2** Schematic).[82] Hydrogen alone does not reduce these to ammonia, but catalysts that intermediately form metal hydrides might. These include catalysts containing iron, nickel, cobalt, palladium, platinum, and so on. A manganese nitride has been used as a source of ammonia in a reaction with a double bond (**15.2** Schematic).[83] The challenge is to figure out the mechanism of this reaction so that modifications can lead to the preparation of ammonia in the form of more useful compounds.

Other experiments on nitrogen fixation have used transition metal complexes of nitrogen.[84] One used a tungsten–nitrogen complex with a ruthenium–hydrogen complex at 55°C to produce a 55% yield (based on tungsten) of ammonia.[85] Intermetallic compounds of iron and titanium have been used with ruthenium on alumina to make some ammonia, at 450–500°C, via interstitial metal nitrides.[86] Schrock has been able to prepare ammonia from nitrogen and hydrogen, but the process is not suitable for industrial use (**15.3** Schematic).[87]

Although these leads are suggestive, considerable additional work may be needed to devise a practical commercial system of making ammonia.

The use of nuclear energy to generate power produces no greenhouse gases, but it can cause pollution.[88] Mining the ore exposes the miners to radon, which can cause cancer. In one area in Colorado, houses were built on radioactive tailings piles. Nuclear plants, such as the one in Salem, New Jersey, require massive amounts of cooling water. The use of cooling towers reduces this. Even if, in normal operations, negligible amounts of radioactive materials are emitted, there is the possibility of accidents, as seen at Chernobyl and Three Mile Island, as described in Chapter 1. Accidents of this type can expose many more people over a much larger area than an accident in an ordinary chemical or power plant. There is also the unresolved problem of long-term storage of the used fuel. An underground repository for nuclear weapons-related waste opened near Carlsbad, New Mexico, in 1999.[89]

The cost of electricity produced in this way is higher than that produced from conventional fuels. France produces 75% of its electricity in nuclear plants, Sweden 51%, Switzerland 42%, Spain 38%, Japan 28%, Germany 27%, and the United States 22%. There are 104 nuclear power plants in the United States.[90] Sweden has voted to phase out nuclear energy. The German government wants to phase out the nuclear power industry.[91]

Nuclear power has been proposed as part of the solution to global warming.[88] Safety is still a problem.[92] The head of a reactor in Ohio almost corroded through, 6 inches of steel having been dissolved. Newer designs, such as the pebble-bed reactor, are said to be safer.[93] Toshiba has a 10 MW

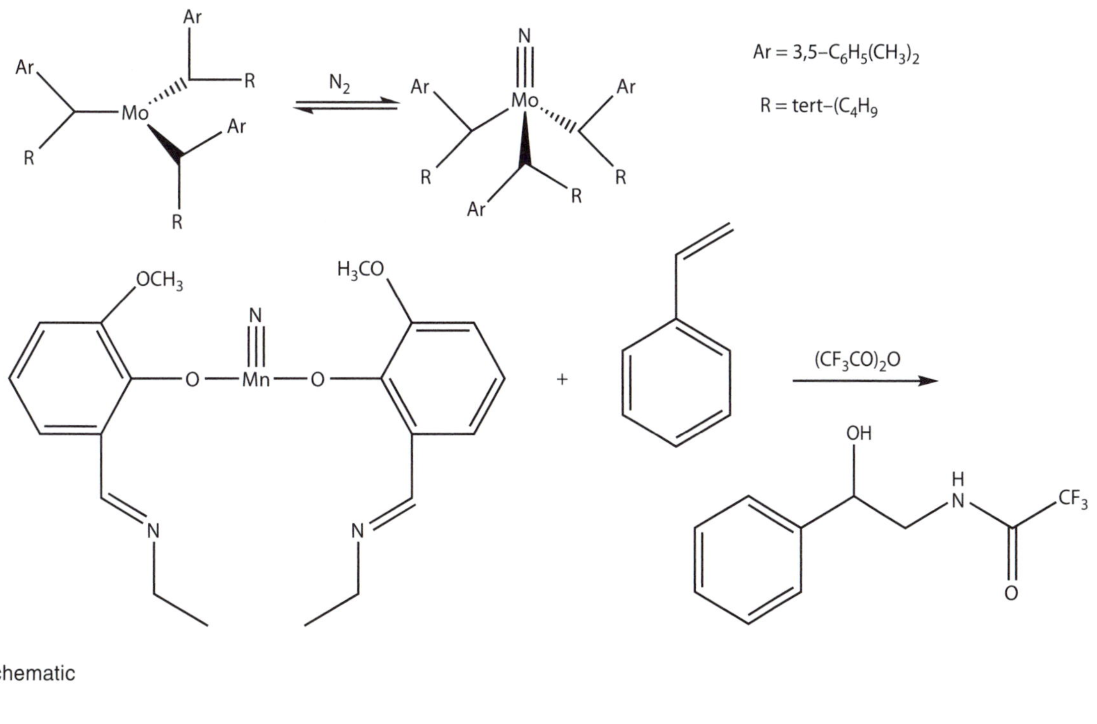

Ar = 3,5-$C_6H_5(CH_3)_2$

R = tert-(C_4H_9)

N_2; $(CF_3CO)_2O$

15.2 Schematic

HIPT N HIPT N Mo N HIPT N N

[2,6-lutidiniumH][B(C_6F_5)$_4$]

(Cp^*)$_2$Cr

heptane

HIPT NH_3 HIPT N Mo N HIPT N N

N_2 → NH_3

where Cp* = pentamethylcyclopentadienyl

where HIPT =

15.3 Schematic

reactor for remote locations that is said to be "super-safe, small and simple."[94]

Environmental problems can also be present when renewable sources of energy are used. Condensed steam from geothermal plants may have to be cleaned up before it can be discharged. Hydrothermal gases in Yellowstone National Park contain nickel, copper, zinc, silicon, lead, arsenic, and antimony.[95] Reinjection into the reservoir avoids this problem. A geothermal system in Idaho uses a heat exchanger with isopentane as a working fluid in a Rankine cycle.[96] Hydropower requires dams that remove the niches in running streams that foster biodiversity. Cool, well-aerated running water favors different species more than lakes do. A shaded stream can have a different flora and fauna that may be more diverse than that in the sunny lake behind the dam. The dams also interfere with the spawning of migratory fishes, such as salmon, shad, and eel. Further research on fish ladders is needed to develop more effective designs.

The use of biomass for energy can also involve problems.[97] Half of the wood harvested in the world is used for fuel for heating and cooking. In many developing nations, this is resulting in increasingly long trips for obtaining wood and in deforestation. If other ways can be found to provide the needed energy (e.g., by tapping the sun's energy), reforestation may be easier to accomplish. Biogas generators at home will also help.[98] A biomass-to-energy plant on the Big Island of Hawaii destroyed about half of the remaining primary forest before it was closed. (Hawaii has a very high incidence of endemic plants that evolved because of its isolation.) Harvesting of peat for fuel in northern Europe is threatening the bog flora. The harvesting of whales for oil for lamps in the nineteenth century may have started their decline. Biofuels have some major problems.[99] Ethanol from corn uses as much fossil energy to produce as it contains. It would take all the corn and soybeans grown in the United States to supply just 12% of the gasoline demand and 6% of the diesel fuel demand in the United States. The situation is similar in Europe. Use of biofuels has increased food prices. Ethanol from cellulose will require large amounts of land, will deplete nutrients in the soil, and may enhance erosion. It has been suggested that preservation of forests and reforestation would be a better way to mitigate emissions of carbon dioxide. Growing algae on the carbon dioxide emissions of power plants is being studied, but will require more efficient photobioreactors to be economical.[100]

Very high particulate levels result from cooking on unvented biomass stoves in rural Mexico.[101] Typically, tortilla cooking is done 3–4 days a week, with 2–4 h each day. The use of liquified petroleum gas eliminates the particulates, but produces another problem in cities. In Mexico City, leaks of this gas accounted for one-third to one-half of the hydrocarbons in the air.[102] (Mexico City is notorious for its severe air pollution. Ozone levels often reach 0.20–0.30 ppm.) Traditional tortilla making uses a lot of water and energy and results in a discharge of water containing a lot of lime. A careful study of this process has reduced the waste greatly by using exact amounts of lime and water together with infrared cooking in a cheaper process.[103]

15.1.4 Global Warming[104]

The world is now 0.8°C warmer than it was in the nineteenth century, according to the United Nations Intergovernmental Panel on Climate Change.[105] Eleven of the past 16 years have been the hottest of this century. The year 1998 was the warmest year up until that time since record keeping began in 1856. It has now been surpassed by 2006 as the hottest year.[106] Based on tree rings, pollen, sediments, gases trapped in glaciers, and corals, it was the hottest year in the last 1000 years.[107] This is the greatest rate of temperature change in the last 10,000 years. Global minimum temperatures are rising faster than the maxima.[108] Temperatures in polar regions have risen faster than the global average (e.g., as much as 6°F in Alaska).[109]

Although the presence of aerosols from volcanoes and air pollution have complicated the study,[110] it is now believed to be due to human activities. It is due to putting more greenhouse gases, such as carbon dioxide, nitrous oxide, chlorofluorocarbons, hydrofluorocarbons, perfluorocarbons, and sulfur hexafluoride, into the atmosphere.[111] Sulfur hexafluoride is the most potent, being 23,900 times as potent as carbon dioxide.[112] Efforts are being made to reduce its use by the electrical and other industries, or at least to use systems to capture it for recycling. Carbon dioxide is by far the largest, amounting to 23 billion tons each year.[113] The preindustrial level of 280 ppm in the air has risen to 380 ppm today, largely from the burning of fossil fuels.[114] This value could reach 500 ppm by the year 2100 and result in a rise of global temperature of 1.0–3.5°C. This warming was first predicted by Arrhenius 100 years ago.[115]

The United States with 5% of the world's population produced about one-fourth of the carbon dioxide emitted each year, the maximum for any developed nation (Table 15.1). In 1990, the biggest producers of carbon dioxide after the United States were the former Soviet Union, China, and India.[116] China was the second largest emitter after the United States in 1996.[117] China has now surpassed the United States by 8%.[118] Burning or clearing tropical forests accounts for about 20% of carbon dioxide emissions.[119]

Table 15.1 Carbon Dioxide Emissions by Country in 1990

Nation	CO_2 Emitted in Millions of Metric Tons
United States	4957
Japan	1173
Germany	1012
United Kingdom	584
Canada	457
Australia	289
Spain	261
the Netherlands	168
Czech Republic	170
Austria	59
Sweden	61
Denmark	52
Switzerland	44
Norway	36
New Zealand	2

Along with the increase in temperature, by 2100 there will be a rise in sea level of 15–95 cm.[120] Glaciers and islands will shrink. More frequent extreme weather events such as droughts and floods are expected. A 1-m rise in sea level would threaten half of Japan's industrial areas. An estimated 118 million people may be at risk from the rising sea level.[121] Many species of plants and animals will have to migrate poleward, a process that may be difficult, considering the extensive habitat fragmentation in many countries. The responses to global warming can be seen in some species already. A gradual warming of the California current by 1.2–1.6°C since the 1950s has resulted in an 80% decline in the zooplankton, a 60% decline in Cassin's auklet, and a 90% decline in the sooty shearwater.[122] More winters without sea ice in Antarctica have caused a drop in the krill population.[123] Birds are laying their eggs earlier in the season in the United Kingdom.[124] The bleaching of corals on tropical reefs is considered by some to be an early warning sign of global warming.[125] The range of Edith's checkerspot butterfly in the western United States, Mexico, and Canada is moving northward.[126] Absorption of carbon dioxide by the oceans is lowering the pH to a point where it may interfere with the calcification of animals.[127] There is also the fear that global warming could shift ocean currents, with the result that Europe could become much colder.[128]

Agriculture may expand northward, but crop yields in existing agricultural areas could suffer if they become drier. Countries in the western United States and in the Andes rely on snow in the mountains for water in the growing season. The snow is decreasing, which means there is less water when it is needed.[129] Growers promote faster growth of some greenhouse crops by adding carbon dioxide, but the effect is not general. Photosynthesis may rise in C_3 plants, with less rise in C_4 plants.[130] If carbon dioxide is not the limiting nutrient, no increase in growth may occur. The amount of detrital carbon is three times that held in living plants.[131] Elevated carbon dioxide levels on some plants can result in changes in the soil fungi and the arthropods that feed on them.[132] A rise in temperature may cause more rapid decomposition of carbon in the soil by bacteria and fungi, with the result that the total system becomes a net emitter of carbon dioxide, rather than a sink. Arctic tundra is now a source of carbon emissions, rather than a sink.[133]

Warmer temperatures may cause vectors for tropical diseases to move northward.[134] These include malaria, schistosomiasis, dengue fever, cholera, and others. Malaria and dengue fever are already occurring at higher altitudes than they used to. More heat waves will mean more heat-related deaths, as happened in Chicago and Paris.

Various suggestions have been made for sequestering some of the carbon dioxide. Planting trees (e.g., on degraded land) could remove 12–15% of fossil fuel emissions during 1995–2050.[135] Tropical forests can sequester up to 200 metric tons of carbon dioxide per hectare. Organic farming with cover crops can put more carbon back into the soil (see Chapter 11). Carbon dioxide can be made into useful products by chemical reactions.[136] One reaction is the treatment of carbon dioxide with hydrogen to produce methanol or dimethyl ether for use as fuels.[137] The hydrogen must be obtained from renewable sources (e.g., electrolysis of water using electricity from hydropower, wind energy, or solar energy). If it is made in the usual way by reaction of methane with water, there is nothing to be gained. Since much more petroleum and natural gas are used for fuels than for the production of chemicals, the significant reduction of carbon dioxide levels by this method would require a major expansion of the chemical industry. Because the chemical industry consumes 22% of the energy used by American industry, it needs to reduce emissions of carbon dioxide.[138] The production of cement generates 10% of the carbon dioxide produced by humans.[139] The worldwide use of energy for refineries is 5 million barrels of oil per day.

Other suggestions are to put the carbon dioxide into saline aquifers at least 800-m (2600-ft) deep or into the ocean below 3500 m (11,500 ft).[140] Although this may be a way of handling some of it, there are some unanswered questions. Deep-well injection of chemicals has had some undesirable side effects and is being curtailed. Some artificial earthquakes have been produced in this way. A natural earthquake that released the gas suddenly could kill people in the vicinity, just as the carbon dioxide released by an overturning lake did in the Cameroon. It would also be difficult to collect the carbon dioxide from mobile sources, such as cars and trucks. However, if they were all electric, the carbon dioxide could be collected at a central power plant. In addition, the large amounts of carbon dioxide emitted would require a large number of injection wells. The hydrates that form when liquid carbon dioxide is put into the deep ocean spread out in pools on the ocean floor and then dissolve in water that would be moved by ocean currents.[141] It seems much better to reduce our consumption through increased energy efficiency and to switch to renewable sources of energy.[142]

The extent of concern about global climate change varies. Insurance companies are concerned because more storms in recent years have caused them to pay out more money than expected to settle claims.[143] Low-lying island nations are concerned, for they stand to lose more with rising sea levels. The Union of Concerned Scientists issued a statement endorsed by 1500 prominent scientists, including 98 Nobel Prize winners and dozens of officials of national academies, calling for definite action to reduce emissions to below 1990 levels.[144] The Kyoto Treaty of 1997 calls for nations to reduce their greenhouse emissions,[145] but some observers feel that the reductions agreed on will not be nearly enough to stabilize the world climate.[146] There is also much bickering among nations over which country should do how much and in what way.[147]

Companies and countries that profit from the sale of oil seem less concerned.[148] The head of Exxon said:

> A number of countries, including the United States, appear to be on a policy course that could cause steep reductions in the consumption of oil and other fossil fuels, on the unproven theory that their use significantly affects Earth's climate. Such a policy could have a severe economic impact and force drastic changes in lifestyles and standards of living all over the world.[149]

The Global Climate Coalition, the U.S. fossil fuel sector's lobbying arm, said: "It's really not necessary to begin reducing emissions by 2005."[150] British Petroleum, which has a division that makes solar cells, feels that the time to begin to reduce emissions is now.[151] Shell, General Motors, Monsanto, Sunoco, and DuPont feel that it is time to begin reducing emissions, but the American Chemistry Council (formerly the Chemical Manufacturers Association) is opposed to taking the actions required by the Kyoto agreement.[152]

Estimates of the cost to prevent further global warming vary widely.[153] A United Nations report on global warming says that the effects of global warming could be mitigated at surprisingly little cost.[154] A study by the Pew Center on Global Climate Change concluded that emissions in developing nations can be reduced without slowing down the countries' economic growth.[155] A group of 2400 scientists feel that the total benefits from reducing emissions will outweigh the cost.[156] They feel that the change can be made without harming employment or living standards. The cost of reducing emissions using some of the advanced technologies now available or on the horizon will be virtually zero, according to two workers in the field.[157] The U.S. Department of Energy (DOE) calculates the net economic cost to stabilize emissions as near zero.[158] The World Resources Institute has estimated that stabilizing carbon dioxide emissions at 1990 levels by 2020 will have a minimal effect on the U.S.

economy.[159] The costs of the new technologies will be offset by savings in the fuel that will not have to be purchased. Sir Nichholas Stern calculates the cost to stabilize the concentration of carbon dioxide at a reasonable level by 2050 to be 1% of the gross domestic product.[160]

On the other hand, the "Global Climate Coalition," a lobbying group for the fossil fuel industry, concluded that stabilizing emissions would cost each household $2061 each year. Charles River Associates estimated the cost of complying with the Kyoto Protocol at $100 billion/yr.[161] The Argonne National Laboratory says that limiting emissions could devastate energy-intensive industries, such as the chemical industry and cost many jobs.[162] Critics of this study say that it is not possible to approach the problem by a single-sector way. Jobs lost in the oil industry might be replaced by new ones in the photovoltaic cell industry. Dale Jorgenson, an economics professor at Harvard University, feels that trying to stabilize greenhouse gases in the atmosphere "would be very, very damaging to the world economy."[163] A report from the Conference Board and McKinsey & Co. estimates that a third to a half of U.S. greenhouse emissions could be eliminated at a small cost to the economy by new industrial, building, and appliance efficiencies.[164]

Because 5% of global carbon emissions come from U.S. transportation (more than any other sector in any other country), this is a good place to start reductions.[165] Improved fuel efficiency standards for American motor vehicles have been suggested as one way of reducing emissions. However, a group called the Coalition for Vehicle Choice opposes such standards.[166] Fuel efficiency standards were raised by some in 2008. Hybrid vehicles with electric motors and gasoline or diesel engines that have twice the fuel efficiency of current cars are now available.[167] Carbon taxes, which have been proposed as a way of limiting emissions, are very unpopular with the chemical industry.[168] Such taxes or cap and trade systems are starting.[169] It should also be possible to eliminate the $150–250 billion dollars of annual subsidies in the world for fossil fuels.[170] One can conclude from the general foregoing discussion that when major changes are required, a lot of politics can be involved. The success of the Montreal Treaty in protecting the ozone layer suggests that it should be possible to use a similar treaty (the Kyoto Treaty or some modified version) for global warming.[171] There should also be new business opportunities for those who are willing to go into new areas.[172] Unfortunately, the research and development activity of the U.S. energy sector in 1999 was extremely low.[173]

15.2 HEATING, COOLING, AND LIGHTING BUILDINGS

Buildings produce 39% of the total carbon dioxide emissions in the United States. Sixty percent of the energy used by a building is for heating and cooling, with 40% for lighting and appliances. Reducing the energy used in buildings offers the largest cost-effective way of reducing carbon dioxide emissions.[174] Putting up a new building is best, but retrofits are also good. Before a building can be sold or rented in Europe, its energy efficiency must be calculated and provided to the potential buyers or renters.[175]

15.2.1 Use of Trees and Light Surfaces

The large savings in energy that can result from living closer to work in more compact cities[176] was discussed earlier. The saving of energy by using waste heat from the generation of electricity for heating and cooling of buildings was also mentioned. Cities are islands of heat not only due to the artificial lighting used in buildings, but also due to the many black surfaces that absorb heat from the sun.[177] One-sixth of the electricity used in the United States is used to cool buildings. A combination of lighter roofs, walls, and road surfaces, together with the planting of shade trees in appropriate places, can reduce the need for air-conditioning by 18–30%.[178] (The range on houses may be as much as 10–70% depending on factors such as the amount of insulation in the attic and others.) It could make Los Angeles 5°F cooler in the summer. A black roof reflects 5% of solar energy, whereas a white roof reflects 85%. As a white roof becomes dirty, its effectiveness can be reduced by 20%.[179] Washing with soap and water returns its maximum effectiveness. A self-cleansing roof surface might be made with a surface layer of titanium dioxide catalyst for light-catalyzed destruction of the dirt[180] (see Chapter 8). Another alternative is a surface of a perfluorinated polymer to which the dirt would not adhere well, so that rain could wash it off. Concrete roads absorb less heat from the sun than ones of black asphalt. Parking lots can be shaded by trees in traffic islands. For individual homes, deciduous trees can be planted on the western and southern sides to cast maximum shadows on the house. A tree on the east can be used to shade the air-conditioner. Shrubs around the foundation will reduce the temperature of the walls and the adjacent soil. Coniferous trees can be planted on the northern side to break winter winds.

Green roofs (roofs with plants such as *Sedums* growing on them) insulate the building, delay the rate of runoff of rain, reduce the energy needed for air-conditioning, and make the roof last longer.[181] No watering is needed unless there is a prolonged drought. The building must be built stronger to support the extra weight.

15.2.2 Solar Heating and Cooling

Passive solar heating and cooling[182] can save 50% of the energy needed for a home at the latitude of Philadelphia, Pennsylvania. (A building in temperate, semiarid Argentina that used this technique saved 80% of its heating costs.[183]) This involves putting the long face of the building toward the south. Living spaces can be put on the south and storage spaces and garages on the north. Large windows are put on the south and small windows on the north. Other windows are located to allow cross-ventilation in the summer. Overhanging eaves are set to keep out the summer sun, but to allow the lower winter sun to enter the building. The windows should be double or triple glazed with thermal breaks (i.e., insulation) on any metal frames. The building must be well insulated and weather-stripped and follow LEED (Leadership in Energy and Environmental Design, U.S. Green Building Council) specifications. A large mass of concrete, brick, or stone is put in a floor, interior wall, or chimney to store heat overnight. A vent on the roof can be arranged so that heat rising in a stack effect can be used to bring cooler air into the building in the summer. Small turbines used in this way are driven by the rising air or by the wind. Ventilation at night will cool down the building, which, if closed in the morning, will take quite a while to warm up. Heat exchangers can be used with exhaust air to recover some of its heat during the winter.

Additional possibilities include a green roof; putting a solar water heater on the roof and drying the laundry on a line outside, instead of drying in a clothes dryer. Heat pumps can be used to upgrade low-grade heat, such as that found in used cooling water and wastewater.[184] A heat pump that draws its heat from the ground, rather than the air, will be more efficient in cool weather.[185] A water loop geothermal heat pump that circulated water around a building's hot and cool sides and that replaced a boiler saved 54% of the energy under optimum conditions.[186] An electric utility company in Delaware suggests vertical loops 10–15-ft apart that go to a depth of 80–180 ft.[187] (The temperature of the ground below about 30 ft is constant the year round in Delaware.) A groundwater system heat pump in Portland, Oregon, is still good after 50 years of operation.[188] Air drawn in through underground pipes can be used for cooling in the summer and is partially preheated for use in winter. Before the large-scale adoption of air-conditioning, some Chicago hotels were cooled by air drawn in from the city's system of tunnels. A cinema in India has been cooled by a combination of evaporative cooling, a wind tower, and an earth air tunnel.[189] There is a possibility of locating a refrigerator in the home such that it could be cooled in part by outside winter air. Buildings can be heated by putting a transparent plate in front of a wall, preferably blackened, with an air gap in between.[190] The air that heats up in the gap can be moved into the building with a small fan. Such a method can be used to retrofit existing buildings. Solar heat pipe panels have also been used to transfer heat from outside to the inside of the building.[191] The water left in the evacuated tube evaporates on the hot outside end and condenses on the cooler end on the inside of the building. Solar roofs have been used to heat air in Northern Ireland.[192] Solar energy has also been used for district heating with the aid of parabolic concentrators that focus the sun's rays on a pipe containing water.[193]

Efforts have been made to optimize the absorption of heat by the black surface. One paint consists of a mixed oxide of iron, manganese, and copper with phenoxy and silicone resins.[194] Stainless steel–aluminum nitride coatings, made by magnetron sputtering, have a solar absorptance of 0.93–0.96.[195] Alumina pigmented with nickel particles is in common use in solar collectors. It has a solar absorptance of 0.96 and a thermal emittance of 0.12–0.17 at 60–100°C.[196] A black chrome consisting of $CuCrO_4 + CuO + Cu_2O$ absorbed more than 0.9 units in the visible region with less than 0.1 thermal emittance in the infrared region.[197] Black nickel is said to be as good as black chrome.[198] To maximize the absorption of heat, one system uses a U-shaped apparatus with a porous matrix in the back channel (Figure 15.1).[199] Another uses a transparent honeycomb insulation in the top and bottom covers.[200]

A new solar cooker uses an aluminized plastic sheet parabolic reflector to direct the sun's heat to a storage block beneath the pot inside the house.[201] In another, vacuum tube heat pipes are hooked to the oven.[202] Solar cookers offer a way of compensating for fuelwood shortages in developing countries and of relieving pressure on forests. Wood and crops can be dried faster in a solar tunnel than in the open air.[203] Solar energy has been used to heat swimming pools.[204] Solar-powered water sterilizers used in Hawaii and other places heat the water to 65°C.[205] It is possible to desalinate water with solar energy[206] at half the price of evaporative desalting with fossil fuels or by reverse osmosis.[207]

There is virtually no use of passive solar heating and cooling in Delaware. Although the small additional cost in building this way is offset by savings in fuel costs after a few years, many buyers may opt for the lowest initial cost (but higher long-term cost). Some buyers will spend money for extras, such as decks and hot tubs, but may know nothing about the advantages of passive solar heating and cooling. In contrast, Davis, California, has a building code that almost requires passive solar heating and cooling. It is also a place where it is safe to ride bicycles. Each weekday morning, children put on their safety helmets and ride their bikes to

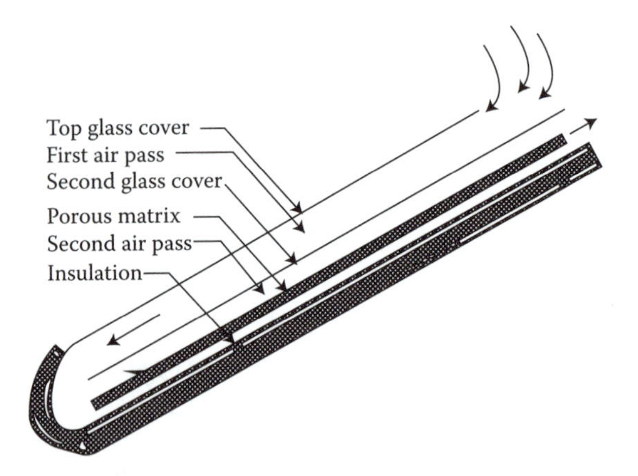

Figure 15.1 Counterflow solar air heater with porous matrix. (Reprinted from A.A. Mahamad, *Sol. Energy*, *60*, 71–76. Copyright 1997 with permission of Elsevier Science.)

school. The California Energy Commission now requires passive solar heating and cooling in all new houses in the state.[208]

15.2.3 Refrigeration with Solar Energy

Cooling by absorption refrigeration was discussed in Chapter 3. Solar energy can be used as the heat needed to power the system. (Biogas or waste heat from a power plant can also do so.) The best system seems to be one using lithium bromide–water, although it is limited to 0°C as the lowest temperature.[209] The ammonia–water system can go to lower temperatures, but is less efficient. Numerous other materials are being tested in the search for cheap, efficient systems. These include other lithium salts, with or without zinc salts,[210] polyethylene glycol dimethyl ethers with chlorofluorocarbons,[211] charcoal with methanol,[212] active carbon and water with a zeolite,[213] granular calcium chloride with ammonia,[214] and strontium chloride with ammonia.[215] The reactions in the last system are shown in **15.4** Schematic.

$$SrCl_2 \cdot NH_3 + 7 \text{ liquid } NH_3 \underset{\text{Day}}{\overset{\text{Night}}{\rightleftharpoons}} SrCl_2 \cdot 8NH_3 + \text{Heat}$$

15.4 Schematic

A solar ammonia–carbon ice maker produced 500 kg ice per day.[216] Electricity from photovoltaic cells can also be used to run refrigerators.[217] Photovoltaic cells were also used to power a thermoelectric device using the Peltier effect (which involves passing a current through a junction of two dissimilar metals) to reach –3°C.[218] The system has no moving parts and is small, lightweight, reliable, noiseless, portable, and is potentially low cost when mass-produced. The best current material for the Peltier effect, Bi_2Te_3/Sb_2Te_3, with a figure of merit of 2.4 (ZT), still requires too much energy to make this system competitive with most compressor-driven refrigeration.[219] The search is on for materials with a ZT > 3.[220] If these can be found, the conventional refrigerator may become a thing of the past.

15.2.4 Heat Storage

The intermittent nature of sunlight requires that there be some means of storing heat at night and on cloudy days. A mass of concrete or masonry is used for that purpose in passive solar homes. The heat can also be stored in beds of rock,[221] in the soil,[222] in tanks of water,[223] or in aquifers.[224] Some of these are for seasonal storage. In one case, 30-m vertical heat exchangers placed in the ground achieved storage efficiencies of 70%. Water pits of concrete lined with stainless steel or high-density polyethylene and insulated with mineral wool have been used in Germany for the seasonal storage of solar heat. One being built in Hamburg, with 3200-m² collectors and 4500 m³ of water, is expected to supply 60% of the heat needed in the winter by 120 houses. Another is being built to service 570 apartments. The average temperature for one in Italy on July 31, 1996 was 69.3°C and was expected to reach 80°C by the start of the heating season.[225]

For indoor storage, it is desirable to have storage materials that take up less space, but may cost more. This can be accomplished by materials that undergo phase changes, dehydration, and so on.[226] The compound must change state in a useful temperature range, have a relatively high latent heat per unit volume, and be cheap. Inorganic ones are nonflammable, but may be corrosive. Organic compounds probably would not be corrosive, but could burn. For example, calcium chloride hexahydrate melts at 25°C; thus, it melts by day and freezes by night. But it, along with other hydrates such as magnesium chloride hexahydrate and magnesium nitrate hexahydrate, tends to supercool.[227] The use of 0.5% sodium metaborate dihydrate with a eutectic mixture of magnesium chloride hexahydrate and magnesium nitrate hexahydrate reduced the supercooling from 16°C to 2°C. Another problem is that the anhydrous salt may settle out and not rehydrate properly. This can be minimized by using thickeners such as 3% acrylic acid copolymer with sodium sulfate decahydrate containing 3% borax to prevent supercooling.[228] Carboxymethylcellulose (2–4%) was also used with sodium acetate trihydrate containing 2% potassium sulfate.

The hydrates or phase-change materials may also be incorporated in other materials. Calcium chloride hexahydrate has been used to fill pores in silica, alumina, carbon, and metals.[229] This may be a way of eliminating problems of the settling of anhydrous salts. If a phase-change material is incorporated into the walls or floors of a building, the need for a separate storage vessel is eliminated. A mixture of butyl palmitate and butyl stearate, melting at 21°C, was put into wallboard at a level of 23%.[230] In another case, a mixture of lauric and myristic acids was used in wallboard.[231] Methyl palmitate, containing some methyl stearate, has also been used.[232] BASF[233] and DuPont[234] sell such wallboard. BASF has built energy-efficient demonstration houses at Patterson, New Jersey, and the University of Nottingham.[235] Neopentyl glycol, which undergoes an endothermic transition from a monoclinic to face-centered cubic crystal at 39°C, has been used in floors.[236] A little graphite (<1%) is used to help nucleate the process.

When higher-temperature heat is available, other systems are possible. A school in Munich, Germany, dehydrated 7 metric tons of zeolites using 130°C steam in the district heating system during off-peak hours, then passed moist air over them in the day to recover the heat.[237] Pellets of calcium hydroxide containing zinc, aluminum, and copper additives were dehydrated to calcium oxide using solar heat from a solar concentrator; then the reaction was reversed to recover the heat.[238] Zeolite 13X has been used to store carbon dioxide obtained from the decomposition of calcium carbonate at 825°C.[239] Such temperatures are available with solar furnaces (where a whole field of mirrors focus on the reaction vessel) (**15.5** Schematic).

$$CaCO_3 \xrightleftharpoons{825^\circ C} CaO + CO_2$$

15.5 Schematic

The reaction is free of toxicity and corrosiveness, has no side reactions, and the raw material is cheap. More work is needed in this area to find reactions that work well with low-grade waste heat and with heat from flat plate or parabolic solar collectors.

15.2.5 Lighting

Compact fluorescent lamps use about one-fourth as much energy as the incandescent bulbs that they replace.[240] Such lamp replacement has been a common part of many energy-conservation campaigns. For outdoor lighting, low-pressure sodium lamps are eight times as efficient as incandescent lamps, three times as efficient as mercury lamps, and 40% better than high-pressure sodium lamps.[241] The yellow color of the outdoor light should not interfere with most outdoor activities. Astronomers concerned about light pollution recommend "shoebox"-type lighting fixtures that direct the light downward to where it is needed and save energy at the same time.[242] The Czech Republic requires fully shielded light fixtures.[243] A sulfur bulb (the size of a golf ball), which gives off four times as much light as a mercury lamp at one-third the cost, is being tested by the U.S. DOE.[244] The light produced by irradiating sulfur in argon with microwaves is similar to sunlight, but it has very little energy emitted in the infrared and ultraviolet ranges. It puts out as much light as two hundred and fifty 100-W incandescent bulbs. When used with light pipes, it could illuminate large areas of stores, offices, and outdoor areas. The cost of illuminating shopping centers and factories in the United States is 9 billion dollars each year.

New luminescent materials are being studied for lighting and for displays.[245] A xenon plasma can be up to 65% efficient, compared with low-pressure mercury lamps, which are about 30% efficient. The ultraviolet light from the plasma can be converted to red light with a $LiGdF_4{:}Eu^{3+}$ phosphor in a process during which one photon of vacuum ultraviolet radiation is converted to two visible photons.[246] If suitable phosphors can be found to produce visible light, it could result in the elimination of all ordinary fluorescent lamps that contain mercury (see Chapter 4). It would increase the efficiency of current xenon lamps used in simulating sunlight in accelerated testing of materials and as sources of light for fiber optic illumination.[247] Workers at Los Alamos Laboratory have made fluorescent lamps in which the mercury has been replaced with a carbon field emitter of electrons.[248] Blue light-emitting diodes (LEDs) based on gallium nitride can be used to produce white light.[249] The combinations of phosphors needed for this are being optimized. They are more energy-efficient than halogen lamps. LEDs are used in traffic lights and tail lights of automobiles. Those that produce white light are available for use in lamps in homes and stores but are still expensive.[250] They offer longer lifetimes and higher efficiencies than conventional light sources. As they become cheaper, it may be possible to light the room with LEDs in the wall. Organic LEDs are also being studied.[251] They have high efficiencies and long lifetimes (in colors other than blue), but the cost of fabrication is high. They are available commercially. These might remove the need for gallium, which is not a very abundant element. The need for the mercury, used in fluorescent lamps, may be eliminated.

People prefer to work under natural lighting, rather than under artificial light. They also like to be able to see out of the windows. "Daylighting" can bring natural light to more areas in a home or office. It involves the use of skylights, atria, and the like, to bring light to more indoor area.[252] Windows in interior walls also help. These can be empty or of transparent or translucent plastic or glass. In former years, skylights were more common on factories. Skylights with vertical panes of glass situated on homes are sometimes called clerestories. Light from skylights can be brought to various parts of the home through light pipes.[253] In some ways, they may resemble a periscope on a submarine. The pipes (typically 0.25 m in diameter) can be hollow plastic tubes of a polymer, such as poly(methyl methacrylate). This polymer is used in optical fibers for carrying light short distances from a central bulb to numerous outlets, as in the instrument panel of an automobile. The principle involved is that of total internal reflection. Light striking the surface at less than the critical angle is reflected. Vacuum metallization of glass or plastic with aluminum, or other methods of making plastic mirrors, might allow a variety of plastic substrates to be used.[254] (Light pipes might allow the lighting of buildings with fewer, more powerful, but more efficient, bulbs.) Mirrors placed at the upper section of a window can be used to bring more light to the ceiling and to corners of rooms.[255] Daylighting reduces the need for air-conditioning by eliminating the heat from unnecessary lamps. In one office building in Brussels, this technique reduced energy consumption by 60%.[256]

Windows can let in too much heat in the summer and let too much escape at night and in the winter. These problems are being approached in several ways.[257] One is the use of low-emissivity coatings that allow visible light to pass but reflect infrared light. These thin coatings consist of metal oxides or noble metals put on in a variety of ways.[258] A 150-nm (one-fourth the wavelength of the average visible light) film can be used for this purpose. One of the most common coatings consists of indium tin oxide. It can be put on the glass in a variety of ways. One way is to spray a solution of tin(IV) chloride, indium(III) chloride, and some fluoride ion on to a glass surface at 400–700°C. The trace of fluoride eliminates haze. Dual nozzles can also be used with tin(IV) chloride and nitrogen in one and water plus nitrogen in the other. A widely used system uses a layer of silver between two layers of tin oxide. One way to make this is by magnetron sputtering, using three sputtering chambers. This involves a glow discharge through argon containing some oxygen at 10^{-3}–10^{-2} mbar, with some source of tin or silver and permanent magnets below the source plate. Vacuum evaporation, with or without an electron beam, is another technique for putting on the coatings.[259] Dip coating, followed by firing, is also possible. Mixed alkoxides plus water can be used in this method by the sol–gel approach. Coatings of zinc sulfide (15 nm) over silver (20 nm) have also been used.[260] The silver can be replaced with aluminum (10 nm) with some loss of efficiency. A possible problem with the use of indium oxide-containing coatings is the limited amount of indium available. About 120 tons of indium is produced in the world each year, as a by-product of zinc mining. Carbon nanotubes may replace indium tin oxide where it is used as a transparent electrode.[261]

Efforts are underway to produce "smart windows" that will react with sunlight to keep out excess light. Photochromic[262] sun glasses can be made by heating the lens in a bath of molten silver nitrate plus silver chloride at 400°C for 4 h, followed by further heating at up to 660°C. The silver chloride put in this way is converted to silver metal in sunlight and then reconverted to silver chloride when the light is removed. It is possible that this method could be extended to windows. As mentioned in Chapter 7, some polymers show inverse solubility in water. Thus, a polymer blend of polypropylene oxide and poly(styrene-*co*-hydroxyethyl methacrylate) has been used in a window.[263] When sunlight heats up the window, the polymers precipitate preventing entry of too much light. A commercial material of this type called Cloud Gel is under development.[264] Systems are being developed that will not freeze in the winter.[265]

A great deal of effort is going into developing electrochromic[266] windows that can be darkened by application of an electric current. Such systems are commercial in car mirrors now. The electrochromic material is put on one sheet of glass separated from the counterelectrode on another sheet of glass by a solution of electrolyte. Current efforts are directed toward increasing the durability over many cycles and decreasing the cost and the response time. Two of the most used electrochromic materials are tungsten(VI) oxide and Prussian blue (a ferric ferrocyanide).[267] The former works by reduction to the intensely blue tungsten(V). The latter forms the intensely blue $Fe(II)[Fe(III)(CN)_6]^-$ from colorless $Fe(II)[Fe(II)(CN)_6]^{2-}$. The tungsten oxide varies in efficiency with the method of preparation. It can be put on the glass by evaporation, sputtering, chemical vapor deposition, spraying, the sol–gel method, or others, with the resulting film varying in porosity, density, and morphology.[268] Tungsten(VI) oxoalkoxides and β-diketonates have been used in chemical vapor deposition.[269] Inclusion of some fluoride ion in it leads to rapid dynamics and good durability.[270] Another method involves dipping an indium–tin oxide-coated glass in an aqueous solution of sodium tungstate and hydrogen peroxide, followed by heating.[271] This improved the response time. However, heating it too much caused crystallization, which reduced the performance. Putting in some lithium (Li/W = 0.14) by evaporation in an electron beam gave a film with good reversibility and coloration efficiency.[272] The transmission of light dropped from 80% to 20% when an electric current was applied. Polyaniline and polyalkylthiophenes have been tested with tungsten oxide.[273]

Other materials being investigated include ferrocene with a bipyridinium salt,[274] niobium oxide,[275] nickel oxohydroxide,[276] and cobalt oxohydroxide.[277] The cobalt oxohydroxide is pale yellow in the reduced state and dark gray in the oxidized state. Vanadium dioxide doped with W, Mo, or Nb blocks infrared light above 29°C, but is yellow-green.[278] A typical electrolyte is lithium perchlorate in propylene carbonate. Solid electrolytes, such as a lithium salt (perchlorate, tetrafluoroborate, or triflate), in a polyepoxide[279] or in a polyvinyl chloride gel in ethylene carbonate–propylene carbonate,[280] lithium iodide in polyvinyl butyral,[281] and Nafion H (a polymeric perfluorocarbon–sulfonic acid),[282] have also been tested. Some other systems use suspended particles between two panes of glass.[283] When the particles are aligned by an electric field, the window becomes transparent. Combination photovoltaic–electrochromic devices are under study.[284]

Much heat is lost through glass windows, even those that are double-glazed. An improvement is the double-glazed window with a vacuum inside.[285] Low-emissivity glass was used with minute pillars inside to prevent collapse. The system was still good after 1 year.[286] Other systems use transparent insulation in the windows.[287] Some of the best use silica aerogels (very light silica gel) are made by the sol–gel method from tetraalkoxysilane (**15.6** Schematic).[288]

$$Si(OCH_3)_4 + 4H_2O \longrightarrow 4CH_3OH + Si(OH)_4 \longrightarrow (SiO_2)_n$$

15.6 Schematic

In one case, the gel was washed with methanol to remove the water, then with amyl acetate, and finally dried with supercritical carbon dioxide. Drying in an oven caused the gel to collapse to a denser xerogel. Inclusion of some fluoride in the process gave excellent transparency. When 1.5–2.0-cm layers of these aerogels, which contained up to 98% porosity, were placed in a double-glazed window with some vacuum, thermal losses were six times less than in the usual double-glazed window. This makes them as good as the usual insulated wall in reducing heat losses. Translucent aerogel granules have been made by BASF using a double-spray nozzle with aqueous solutions of sodium silicate and sulfuric acid. Although these starting materials are cheaper, the drying with supercritical fluids is still expensive. More recent work has found an additive that allows an aerogel, dried under normal conditions of temperature and pressure, to spring back to its original shape after drying.[289] Reaction with chlorotrimethylsilane, methyltrimethoxysilane, or aryltrimethoxysilane has been used to eliminate the need for supercritical drying. Hoechst was developing an aerogel that does not require drying with a supercritical fluid[290] and has sold the process to Cabot.[291] The wet gel made from sodium silicate is treated with an organic compound before drying at atmospheric pressure. The gel has permanent water-repellent properties. Its thermal conductivity is about one-third that of conventional insulation, such as polyurethane and polystyrene foams. The brittleness and moisture pickup of silica aerogels can be reduced by starting with an alkyltrimethoxysilane instead of a tetraalkoxysilane.[292] Silica aerogels can be made 100 times stronger by treatment with a polyisocyanate.[293] These do not require drying with sc fluids.

Aerojet has made organic aerogels with densities of 0.1–0.2 g/mL^3 for use in insulators and capacitors.[294] Transparent macroporous monoliths have also been made by polymerization of trimethylolpropanetrimethacrylate using toluene, ethyl formate, and other liquids as porogens.[295] Poly(alkyl methacrylates) have good resistance to outdoor weathering and can be used in windows.

Glass capillaries mounted perpendicular to the panes of glass have also been used as transparent insulation.[296] Fifty years ago, hollow glass bricks were used to let light into buildings. For places where it is not necessary to see out or to have ventilation, they would seem to offer less loss of heat than the usual windows.

The goal of many of these systems is to have automatic control, so that a person does not have to adjust anything with the time of the day or the season of the year. Low-emissivity coatings for glass keep the heat from the summer sun out of the house, and keep heat in the house in winter. Yet there are times when letting the heat in would help in heating the building. Plain glass plus curtains controlled by photocells and temperature sensors might be used. A light roof is needed to reflect heat in the summer, but a dark one may be needed to absorb heat in winter. Automating this transition would require reversible panels or a device that would be black, yet would be covered with a white pigment slurry in the summer. To avoid snow and ice covering photovoltaic cell arrays in the winter, the panels may be mounted on the vertical walls of the building. There may also be conflicts between shade from trees to reduce the heat on a building in the summer and photovoltaic cell arrays to generate electricity for air-conditioning.

15.3 RENEWABLE ENERGY FOR ELECTRICITY AND TRANSPORT[297]

15.3.1 Alternative Fuels[298]

Various alternative fuels for motor vehicles have been suggested as a way of reducing air pollution. Those that are still derived from coal, natural gas, or petroleum will do nothing to mitigate the release of greenhouse gases that lead to global warming. A bus powered by natural gas is less polluting than one using diesel fuel. An electric car[299] transfers the pollution to the power plant where the electricity is generated and where it may be easier to control. If the electricity is made in one of the power plants in the mid-western United States that was grandfathered by the Clean Air Act and not required to clean up the stack gases, then pollution might be higher with the electric car than with current ones burning gasoline. If the electric cars use lead–acid batteries, lead loss in mining and processing, as well as recycling, must be kept low.[300] If the electricity is made by hydropower, wind power, photovoltaic cells, or other renewable means, then the electric car would reduce both air pollution and global warming. The use of photovoltaic cells on the roof and hood of the car has been suggested.[301]

Methyl ether produces less soot and nitrogen oxides when used as a fuel in diesel engines.[302] It may be a way for trucks with diesel engines to meet the particulate standards in the United States. It is made from synthesis gas (**15.7** Schematic).

$$CO + 2H_2 \longrightarrow CH_3OH \xrightarrow[\text{HMnW}_{12}\text{PO}_{40}\text{ under CO}]{225°\text{C}} \underset{96\%}{CH_3OCH_3} + \underset{4\%}{CH_3COOCH_3}$$

15.7 Schematic

The synthesis gas is usually made by treating methane or petroleum with steam at high temperatures. It can also be made from biomass; thus it could help reduce global warming. The methanol made from biomass in this way is a possible fuel for cars. The problem is that at the current level of use of cars and trucks, it would take huge amounts of biomass.

Some of the carbon dioxide produced by burning fuels could be converted into ethanol for use as a fuel (**15.8** Schematic).[303] To be effective in reducing climate change, the hydrogen would have to be produced by a renewable route (e.g., electrolysis of water with electricity from a renewable source). Almost all of it is made from synthesis gas derived from fossil hydrocarbons today. Ethanol is also made by the fermentation of sugars derived from corn and other sources (as discussed in Chapter 9).[304] It is cost-competitive in the United States only because of a government subsidy, which some persons consider to be "corporate welfare."[305] Current pricing does not take into account the costs of environmental degradation. Much current research is directed toward improving the process to make it competitive, even under the current pricing system. The preparation of diesel fuel by the

$$CO_2 + H_2 \xrightarrow[350°\text{C}]{Rh_{10}Se/TiO_2} CH_3CH_2OH \quad 83\%\ \text{selectivity}$$

15.8 Schematic

alcoholysis of plant oils[306] to give, for example, ethyl oleate from rapeseed oil, was also described in Chapter 9. This biodiesel fuel produces less particulate matter on combustion than diesel fuel from petroleum.[307] By the current pricing system, this biodiesel fuel is two to three times as expensive as diesel fuel derived from petroleum. Both ethanol and ethyl oleate produced by these routes are almost free of sulfur. They are relatively nontoxic and, if spilled, biodegrade quickly. High oleic vegetable oils can be used as lubricants in place of mineral oils.[308]

Methane can be produced by anaerobic fermentation. This is often performed in sewage-treatment plants where the methane is used as fuel to dry the sewage sludge. This biogas can also be produced from plant and animal wastes on the farm or at home. China has 5 million such digesters and India has 1–2 million.[309] These provide fuel for cooking and electric power generation. The residue from generation of the biogas is used as fertilizer. The pollution of streams and lakes from the runoff of fields with too much manure spread on them could be eliminated by the use of biogas generators. (In Delaware and Maryland, fish kills by toxins produced by *Pfiesteria*, a dinoflagellate, have been traced to overenrichment of streams by agricultural runoff, including that from poultry farms. In North Carolina, the problem was runoff from hog farms. Hog farms in the Netherlands also have problems with runoff.) It has been suggested that these wastes be used to produce biogas, which could then be used to heat the chicken houses in winter. The biogas could also be used for cooking and for heating and cooling the farmer's house and nearby houses. If too much gas was produced in some seasons, it might be compressed into a truck for transport and sale elsewhere. Biogas can also be used to fuel the truck.[310] These animal wastes might also be used in substrates on which edible mushrooms are grown. Horse manure is used for this purpose in the mushroom-growing areas of southeastern Pennsylvania.

Landfills in the United States produce 8.4 million metric tons of methane each year. The gas from 750 landfills could be recovered economically, but only in 120 was it actually collected in 1995.[311] By 1998, the number had grown to 259.[312] (The landfill in Wilmington, Delaware, will supply gas to the electric power plant a mile away for 20 years.[313]) Any carbon dioxide and hydrogen sulfide impurities in landfill gas are removed before the gas is burned for fuel.[314]

Biomass can also be used for energy.[315] In 1999, biomass supplied 3% of the energy used in the United States.[316] Over 2 billion persons in the world rely on biomass fuels.[317] Energy plantations, preferably on land unsuited for other crops, could provide 600 billion kWh of electricity at \$0.04–0.05 kWh^{-1}, according to one estimate. (The prices of electricity quoted here and elsewhere in this chapter come from references of different dates and may not be easy to compare.) Most experiments use fast-growing poplars and willows or switchgrass (*Panicum virgatum*). Switchgrass needs little fertilizer or herbicides and can be harvested twice a year.[318] The European system of coppicing may be suitable for harvesting the biomass from the trees. The tops are cut off every few years, leaving the stumps to resprout. Cutting similar to a hedge might also be suitable. Industrial coppicing for wood for British power plants involves machine harvesting every 3 years.[319] To prevent loss of soil fertility, the ashes from the power plant would have to be spread on the ground of the plantation at intervals. Burlington, Vermont, has a 200 tons/day biogasification plant that can use wood, crop residues, yard waste, and energy crops to produce electricity at \$0.055 kWh^{-1}.[320] The gas produced by pyrolysis at 1500°F is a mixture of methane, carbon monoxide, and hydrogen. The residual char is used to provide the heat for the pyrolysis. The gases and liquids obtained from biomass by pyrolysis or other means can be upgraded by a variety of catalytic methods, including cracking with zeolites.[321]

15.3.2 A Hydrogen Economy

Hydrogen has been proposed as a clean fuel that produces only water on combustion.[322] Also, some members of the U.S. Congress have pushed for more research in this area.[323] Almost all hydrogen produced today (96%) is made by steam reforming of coal, natural gas, or petroleum.[324] For a sustainable future, the hydrogen must come from renewable sources.[325] The steam reforming reaction can also be applied to the methane in biogas or in the gas obtained by the pyrolysis of biomass. The water–gas shift reaction can then be run on the carbon monoxide in the pyrolysis gas to produce more hydrogen (**15.9** Schematic).

$$CO + H_2O \xrightarrow[350\text{–}400°C]{\text{Fe catalyst}} CO_2 + H_2$$

15.9 Schematic

In addition to the method of pyrolysis used in Burlington, Vermont, there are other methods. One produces a bio-oil that is then subjected to steam reforming to produce hydrogen.[326] Wet biomass, such as water hyacinths and cattails grown in tertiary wastewater treatment ponds, can be gasified in supercritical water using a carbon catalyst to produce a mixture of hydrogen, carbon monoxide, carbon dioxide, and methane containing a little ethane and propane.[327] Treatment of wet cellulose with a nickel catalyst at 350°C for 1 h recovered 70% of the hydrogen in the cellulose, together with carbon dioxide and some methane.[328]

Only 4% of the hydrogen used today is produced by the electrolysis of water.[329] Hydrogen produced in this way costs four times more than that made from petroleum and natural gas. Cheap electricity from renewable sources is needed to encourage the production by electrolysis. The electricity generated by photovoltaic cells can be used to electrolyze water to produce hydrogen. It might cost less and use less equipment if both steps could be run in one apparatus.

Photoelectrochemical generation of hydrogen from water[330] has been performed with up to 18% efficiency.[331] It has been carried out with a stacked gallium arsenide–galium indium phosphide cell with 12.4% efficiency using sunlight.[332] Systems under study include niobates ($A_4Nb_4O_{17}$, where A is calcium, strontium, lanthanum, potassium, or rubidium) used with cadmium sulfide or nickel,[333] cerium(IV) oxide[334] or copper(I) oxide,[335] and titanium dioxide.[336] Dye-sensitized solar cells, using a ruthenium bipyridyl complex with tin and zinc oxides, gave about 8% efficiency in direct sunlight.[337] A seaside power plant in Japan has used solar energy to produce hydrogen from water, then used the hydrogen to reduce carbon dioxide to methane over a nickel–zirconium catalyst at 300°C.[338] The methane was burned to generate more electricity. It would be more direct to generate the electricity with the photocell.

Microorganisms contain hydrogenases that allow them to produce hydrogen.[339] *Anabaena variabilis* has been used in a photobioreactor to produce hydrogen.[340] This system has been proposed as a way of utilizing carbon dioxide from power plants. The biomass produced could be converted to more hydrogen through synthesis gas, as described earlier. *Chlamydomonas reinhardtii* has also been used to produce hydrogen.[341] The search is on for more efficient strains or other organisms. A combination of enzymes from thermophilic bacteria (*Thermoplasma acidophilum* and *Pyroccoccus furiosus*) has been used to convert glucose to gluconolactone, which hydrolyzes to gluconic acid (**15.10** Schematic), with 100% selectivity.[342]

$$CH_3CH_2OH \xrightarrow[\text{Light}]{H_4SiW_{12}O_{40}} CH_3CHO \quad 100\%$$

15.10 Schematic

A mixed culture of *Clostridium butyricum* and *Enterobacter aerogenes* makes more than 2 mol of hydrogen from glucose, the other products being carbon dioxide and a mixture of acetic, butyric, and lactic acids.[343] The glucose might be obtained from the hydrolysis of cellulose or starch. The photodehydrogenation of ethanol to acetaldehyde over a heteropolyacid also gives 100% selectivity.[344] If either of these is used on a large scale, additional uses will have to be found for the gluconic acid and acetaldehyde, or they can be converted to carbon dioxide and more hydrogen via synthesis gas.

Storage of the hydrogen fuel poses some problems.[345] The gas can be compressed for storage at fixed sites. For use in vehicles, this would add considerable weight in the form of steel tanks. Various methods that would allow more hydrogen to be stored in lighter-weight tanks are being studied. Graphite fiber-reinforced epoxy resin tanks may be suitable. The U.S. DOE target is a material that can hold 6 wt% hydrogen at ambient temperature and deliver it in a fully reversible manner over thousands of cycles with short recharging times, preferably without the use of noble metal catalysts.[346] The main classes being studied include metal hydrides, such as ammonia borane (NH_3BH_3),[347] lithium nitrides,[348] and metal organic frameworks, such as zinc dicarboxylates,[349] carbon nanotubes,[350] and aromatic hydrocarbons.[351]

Other systems propose to store hydrogen by reversible chemical reactions. Ammonia has been suggested as a fuel,[352] either burned as such or thermally dissociated to recover the hydrogen. A hydrazine fuel cell uses nickel and cobalt electrodes instead of platinum and carbon.[353] The hydrazine is stored as a hydrazone. Benzene could be hydrogenated to cyclohexane, which could be shipped to a destination where it would be dehydrogenated to recover the hydrogen.[354] Because benzene is a carcinogen, it might be better to use toluene or naphthalene. Asemblon is marketing Hydronol™, a proprietary mixture of this type which could fit into existing gas stations and minimize the risk of explosions from hydrogen.[351] Hydrogenation of fullerenes for hydrogen storage has also been proposed.[355] Methanol has been suggested for the storage of hydrogen. It can be dehydrogenated to methyl formate, which can be converted back to methanol by hydrogen produced by the electrolysis of aqueous sodium chloride in the presence of the formate.[356] The process is said to be 20% cheaper than transporting liquid hydrogen. There is a "crossover" problem of methanol passing through the membrane which might be solved by using a zeolite 3A membrane. Formic acid may be useful in small fuel cells for computers.[357] There is a problem of the carbon monoxide formed when reforming a hydrocarbon on board poisoning the platinum catalyst. It can be removed under fuel cell operating conditions by a small amount of oxygen using a copper(II)/ceria catalyst.[358]

The first hydrogen filling station has been set up in Hamburg, Germany.[359] Hydrogen-fueled vehicles can be 70% efficient compared with about 23% efficiency for ones with a conventional fuel and engine.

15.3.3 Fuel Cells[360]

Hydrogen would be burned in a fuel cell to power an electric vehicle. The only product would be water. This is the reverse of the electrolysis of water to produce hydrogen and oxygen. The system is not subject to the Carnot cycle, which limits efficiency in the usual electric power plants,[361] and 60% electrical efficiency can be obtained. Demonstration cars and buses have been built.[362] The fuel cells fit in the same space as the present diesel engine in the bus. An aircraft powered by hydrogen is being planned.[363] Hydrogen is lighter than the present jet fuel, but requires four times as much storage space. The lighter fuel may permit more payload if the tanks that store the hydrogen do not weigh too much. The prototype minivan used tanks made of carbon fiber-reinforced plastic mounted on the roof. It was able to travel 150 miles between refueling. The use of methanol, ethanol, natural gas, or gasoline will require a reformer on board to convert the fuel to hydrogen. Such systems are being studied.[364] Such cars may achieve twice as many miles per gallon as cars with internal combustion engines. A problem that is being overcome is that the traces of carbon monoxide in the hydrogen from reforming poison the catalyst. Fuel cells can now be run directly on butane, ethane, or propane.[365]

The principal types of fuel cells being developed for various uses are (a) molten carbonate, (b) solid oxide, (c) polymer electrolyte, and (d) phosphoric acid. A typical molten carbonate is a mixture of lithium and potassium carbonates at 650°C running at 60% efficiency. If by-product heat can be utilized, then the total energy efficiency will be higher. Fuel cells can run on landfill gas even if carbon dioxide is present. They can have single-cell lifetimes of 30,000 h.[366] Many cells are stacked in series to obtain the desired voltage. Solid oxide systems, such as zirconium and cerium oxide mixtures,[367] can be run at 600°C or higher. One 25-kW prototype ran 13,194 h before a scheduled shutdown. No reformer is needed for hydrocarbon fuels. The electrodes do not contain noble metals. A perfluorinated sulfonic acid, such as Nafion, is used as a solid electrolyte in some fuel cells that are the most likely to be used in vehicles.[368] No liquid electrolyte is present. A platinum catalyst for the reaction is used. Concentrated phosphoric acid at 200°C is used in some fuel cells. One in Tokyo ran continuously for 9477 h before being shut down for scheduled service. Stationary fuel cells have advantages for electric power production. They can be nonpolluting. Transmission losses are reduced over those of conventional systems when the fuel cells are located near to where the power will be used. Fuel cells come in modules, and so it is easier to expand the production of power than it is with a conventional electric plant using fossil fuels. Energy Conversion Devices has developed a fuel cell based on a magnesium alloy with neither noble metal catalysts nor a membrane.[369] Microbial fuel cells generate electricity at the same time in which they clean up wastewater.[370] They do not require electrodes of noble metals.

Before fuel cells come into widespread use, their cost must be reduced. There is concern that, just as in the case of biofuels, the use of fuel cells has been oversold. Critics feel that greenhouse gas emissions can be controlled more quickly and easily by improving the energy efficiency of vehicles and buildings.[371] The challenge is to get over the idea that each driver has to have his own car and drive alone in it.

15.3.4 Solar Thermal Systems[372]

Heat from the sun can be concentrated by parabolic troughs, parabolic dishes, or by mirrors to produce steam for generating electrical power. These can be used to heat a fluid at the focal point of the parabola from which the fluid is then pumped to the steam generator. The heat can also be aimed at a central receiver. The central receiver allows higher temperatures to be reached. These systems can be made to track the sun for greater efficiency. A 750-ha plant in the California desert using parabolic troughs has operated for 20 years with more than 94% power availability. It produces electricity for as little as \$0.10 kWh^{-1}. Further improvements, such as vacuum insulation for the heat-carrying pipes, better absorbers, antireflective coatings, and such, are expected to lower the cost of the electricity to a level at which it can be competitive with that from conventional plants that burn fossil fuel.

Solar absorbers of Mo/Al_2O_3 can be deposited on copper-coated glass using two separate electron beams.[373] These offer 0.955 absorbance at 350°C, when topped with an antireflective coating of alumina. Another offers 90% solar absorbance with low thermal emittance through the use of a 50-nm Al_2O_3/10–12-nm AlCuFe/70-nm Al_2O_3-layered coating.[374] The AlCuFe coating is applied by ion beam sputtering. Cathodic arc chemical vapor deposition of absorbant surfaces is reported to be faster than other methods.[375] The stainless steel–aluminum nitride coatings put on by magnetron sputtering (mentioned earlier) are stable to 330–400°C.[376] Black cobalt oxide absorbers applied by electrodeposition are stable to 300°C, but not to 400°C.[377] Further research is directed at absorbers that will be stable at 500°C.

Parabolic dishes can produce higher-temperature steam than parabolic troughs. The central receiver for a field of mirrors (a heliostat field) usually contains a nitrate salt with blackened particles used at up to 900°C. A plant in the California desert uses 2000 motorized mirrors aimed at a 91-m metal tower containing molten nitrate salts up to 565°C. The salt stays hot for 2–12 h after the sun goes down so that some additional electricity can be generated after dark.[378] Higher temperatures, up to 1600°C or more, are possible using magnesium oxide powder in argon as a heat-transport medium in a black-lined receiver.[379] This should increase the efficiency of the system. Molten nitrate salts and ceramic bricks have been used for heat storage so that some additional electricity can be generated after the sun sets.

Solar furnaces have been suggested for thermochemical cycles for producing hydrogen (**15.11** Schematic).[380]

Reaction in a solar furnace over a Fe_2O_3/Al_2O_3 catalyst at 1100–1200 K for 20 s

$$I_2 + SO_2 + 2H_2O \longrightarrow 2HI + H_2SO_4$$

$$2HI \longrightarrow H_2 + I_2$$

$$H_2SO_4 \longrightarrow SO_2 + H_2O + 0.5O_2$$

15.11 Schematic

A solar chimney was used to generate electricity in Spain for several years. The 200-m chimney with a 45,000-m^2

collector area at the base was shaped like an inverted funnel. Heat rising through the chimney produced wind that ran the generator. The residual heat in the soil allowed some additional electricity to be generated at night. A solar tower that will produce 200 MW is being built in Australia.[381] Solar ponds trap the sun's heat in salt solutions. The temperature gradient in the pond is used to run a heat engine (Stirling engine) to generate electricity.[382] They are relatively inefficient. A temperature gradient between warm surface water and cold deep water (of at least 20°C) in the ocean can be used to drive a heat engine with ammonia or another material as the working fluid.[383] This has been done in Hawaii where deep water is close to land. The nutrient-laden deep water brought to the surface can be used for aquaculture to derive additional benefits.

Beaming power to earth from solar power satellites has also been suggested.[384] Problems could occur if the incoming beam missed the target receiver.

15.3.5 Photovoltaic Cells[385]

Photovoltaic cells convert light to electricity at the place where it is to be used. There is no need for a central facility with a concentration of equipment or for long transmission lines (as well as the losses of electricity in them). The peak in the electricity generation may well occur on the hot summer afternoon when the electrical system of the community is strained and expensive supplemental generators must be brought into service. Thus, its cost should be compared with the cost of this supplemental electricity in this situation. (In 2009, electricity from the best photovoltaic systems cost about twice as much as peak power from a conventional plant.) Photovoltaic systems are cost-competitive now in places remote from power lines. Further research is needed to make them cost-competitive in areas where an electric network is already in place. The current problem is that the most efficient photovoltaic cells are expensive to produce and the cheaper ones are not efficient enough. Photovoltaic systems use no toxic fuel, use an inexhaustible fuel, produce no emissions, require little if any maintenance, and have no moving parts. In 1998, the United States produced 44% of the world's solar cells, Japan 24%, and the rest of the world 11%.[386] In 2004, the United States produced 12%, Japan 50%, Europe 26%, and the rest of the world 12%.[387] The market for photovoltaic cells was growing at the rate of 40%/yr in 2005.[388]

High-efficiency solar cells can be made from silicon wafers cut from single crystals grown from a melt.[389] One of the most efficient systems is made by a series of steps.[390] Its efficiency is 23.5% under 1 sun. (1 sun means that no mirrors are used to concentrate the light. Because mirrors are cheaper than solar cells they are sometimes used.)

1. Diffuse boron tribromide into the back to produce p-type silicon (i.e., with a deficiency of electrons sometimes referred to as holes).
2. Texturize the front by selective etching through a mask to produce a surface of inverted pyramids (to maximize absorption by allowing light to hit twice).
3. Diffuse phosphorus oxytrichloride into the front to form n-type silicon (i.e., with an excess of electrons).
4. Passivate by thermal oxidation to form silicon dioxide on the surface.
5. Metallize the front by vacuum evaporation to form a Ti/Pd/Ag layer (to minimize the reflection of light).
6. Metallize the back by vacuum evaporation of aluminum (to reflect light within the cell to maximize absorbance).
7. Put on buried contacts in grooves cut by a laser (to minimize losses due to shading by the grid).
8. For use outdoors, cells are usually encapsulated in an ethylene–vinyl acetate copolymer.
9. Attach foundation, frame, cover glass, wires, and an inverter to convert the direct current produced by the cell to alternating current.

The number of steps needs to be reduced to reduce the cost.[391] One proposal is to coat both sides at once.[392]

Uncoated, untextured silicon may reflect as much as 36% of the incident light. This can be reduced to about 12% by a single coat of titanium put on by the sol–gel method, and to 3% by a double titanium dioxide coating.[393] Other materials for antireflective layers include silicon nitride,[394] tantalum oxide,[395] radiofrequency-sputtered indium tin oxide,[396] and zinc oxide.[397] Aluminum nitride topped with three layers of angled titanium dioxide nanorods produced a coating that reflected almost no light.[398] Nanoporous poly(methyl methacrylate) allows 99.3% of the incident light to pass through.[399] A sulfur hexafluoride–oxygen plasma has also been used to texturize silicon surfaces.[400] Abrasive cutting wheels have also been used.[401] The discoloration of the ethylene–vinyl acetate encapsulant can be reduced by stabilizers or by the use of a glass that filters out the ultraviolet light.[402]

Cutting the silicon wafer into thin slivers increases the surface area 13-fold to produce a cell that is about 20% efficient.[403] The energy to produce the cell can be recouped in 1.5 years compared with 3 years for a conventional cell. Polycrystalline silicon is cheaper than single-crystal silicon. It can be made by slow casting of an ingot of silicon. An 18.6% efficiency has been obtained in a cell made with it.[404] It is made by reduction of trichlorosilane with hydrogen.[405] Amorphous silicon has a higher absorbance than crystalline silicon and can be used in thinner layers, down to less than 1 μm.[406] It can be crystallized with a laser, but this might require application of several layers to obtain the thickness needed for a solar cell.[407] Amorphous hydrogenated silicon can be made in a plasma from silane and methane.[408] Such thin-film cells[409] can reach efficiencies of 9–10%. Rear reflectors increase efficiency by keeping the light in.[410] Unfortunately, these thin-film cells have a lower efficiency after a few years owing to diffusion of aluminum from the contacts into the silicon. Some of the loss can be restored by annealing at 150°C. Some of the loss can also be avoided by using thermal insulation to keep the cell hotter.[411] If a triple junction cell (i.e., one containing the equivalent of three stacked cells) is used, the efficiency can be as high as 12.8%.[412] Flexible plastic cells of amorphous silicon have been made by automated roll-to-roll chemical vapor deposition processes.[413] Amorphous silicon on crystalline silicon has been used in an effort to decrease the thickness of crystalline silicon needed.[414] The cell efficiency was 14.7–16%.

There are some other thin-film devices that have equal or higher efficiencies.[415] Cadmium telluride in 1–2-μm films is 15–16% efficient.[416] The most efficient (about 17–19.9%) thin-film cells have been made from Cu(In, Ga)Se_2.[417] Cheaper routes that use fewer vacuum processes are being explored, but the resulting cells are not as efficient.[418] These include roll-to-roll fabrication of flexible cells using inkjet printing. The efficiency is limited by any crystal defects that may be present.[419] Various methods of preparation are being studied in an effort to prepare films with fewer defects.[420] For the preparation of cadmium telluride, these include sublimation, evaporation, chemical vapor deposition, spraying, and electrodeposition. A film free of defects has been obtained by spraying cadmium and tellurium solutions containing phosphines on to a substrate at temperatures higher than 250°C.[421] Chemical vapor deposition is often used to make gallium arsenide and other materials for solar cells.[422] An example is the pyrolysis of the single precursor, Ga[(As-*tert*-$C_4H_9)_2]_3$, to form gallium arsenide. It affords a safer preparation than routes that use pyrophoric or very toxic substances such as trimethylgallium, arsine, and so on (**15.12** Schematic).

$$GaS(CH_3)_3 + AsH_3 \xrightarrow{600\text{–}800°C} GaS + 3CH_4$$

15.12 Schematic

InP has been prepared in a similar fashion at 167°C.[423] A way of avoiding the use of poisonous hydrogen selenide is to treat the copper–indium alloy with selenium vapor at 400°C.[424]

The band gap of a semiconductor is the energy needed to promote an electron from a valence band to a conduction band (i.e., to make an electron mobile).[425] Typical band gaps are silicon 1.2 eV, gallium arsenide 1.4 eV, cadmium selenide 1.7 eV, cadmium sulfide 2.5 eV, and $CuInS_2$ 1.53 eV. The energy in visible light is 1.4–1.5 eV. For a good photovoltaic cell, it is desirable to choose a material with a band gap in or close to this range. A europium-doped calcium fluoride crystal has been used to shift some of the shorter wavelengths in sunlight to longer wavelengths more suitable for silicon-based cells.[426]

Gallium arsenide is more efficient than most other semiconductors,[427] especially when used in tandem with another cell material that absorbs in another part of the solar spectrum, so that more of the light can be used.[428] Mechanically stacked GaSb/GaAs cells were 34% efficient at 100 suns and 30% at 40 suns.[429] A tandem solar cell of InGaP/GaAs had a 30% efficiency at 1 sun.[430] The possible problem of a mismatch of the two different crystals can be solved by inclusion of a flexible layer.[431] A holographic system focuses light, diverts the unwanted infrared radiations, splits the light, and focuses the two bands on different solar cells, one of gallium arsenide and the other of silicon (Figure 15.2).[432] A 20-fold concentration of the light is obtained. It will require a device to track the sun. It is described as a "major breakthrough" that will make solar power competitive with that from fossil fuels. The electricity will be produced at a cost of \$0.057 kWh^{-1} compared with a cost of \$0.075 kWh^{-1} in a conventional plant using pulverized coal as a fuel. (Care should be used in comparing prices, because they may vary with the time and place.)

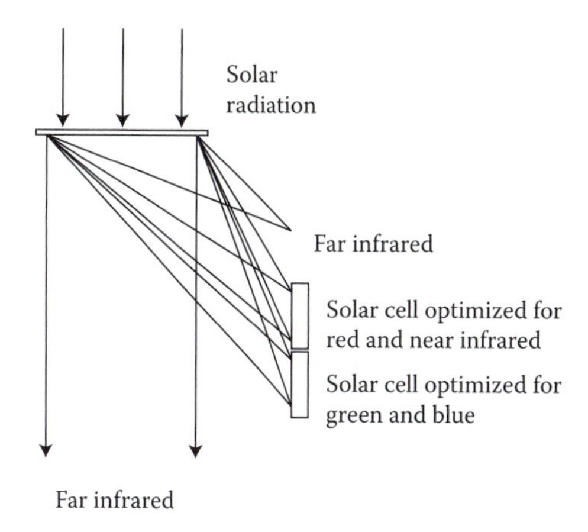

Figure 15.2 Schematic of a holographic systems to focus light in solar cells. (Reprinted from J.E. Ludman, J. Riccobono, I.V. Semenova, et al., *Sol. Energy*, *60*, 1–9. Copyright 1997 with permission of Elsevier Science.)

The most efficient photovoltaic cell (42.8%) uses multiple junctions of indium gallium phosphide, gallium arsenide, silicon, and indium gallium arsenide.[433] The next best (40.8%) uses a triple junction of gallium indium phosphide, gallium arsenide, and germanium at 326 suns.[434]

A group in Switzerland has developed a photovoltaic cell that can function as a window for a building.[435] In one example, a ruthenium pyridine complex photosensitizer is attached to the titanium dioxide semiconductor by a phosphate. An iodine-based electrolyte (KI_3 dissolved in 50:50 ethylene carbonate/propylene carbonate) is between the panes. All the films are so thin that they are transparent. The efficiency is 10–11%. Other dyes have been tested to avoid the use of expensive ruthenium.[436] The best cell efficiencies are 8.6–9.1%.[437] The dye in the best cell is shown in **15.13** Schematic.

Variants on such cells have included fullerenes[438] and polypyrrole.[439] Poly(thiophene)–fullerene cells are 5% efficient.[440] Since the colors of quantum dots vary with their size, it is hoped that they can be used in tandem solar cells. The best cells using them are only 2% efficient.[441] The use of solid electrolytes will avoid problems that might occur if a seal on a liquid electrolyte leaked. Photovoltaic cells have also been in the form of nanowires, which could harvest more of the light.[442] Mirrors to concentrate the sunlight are

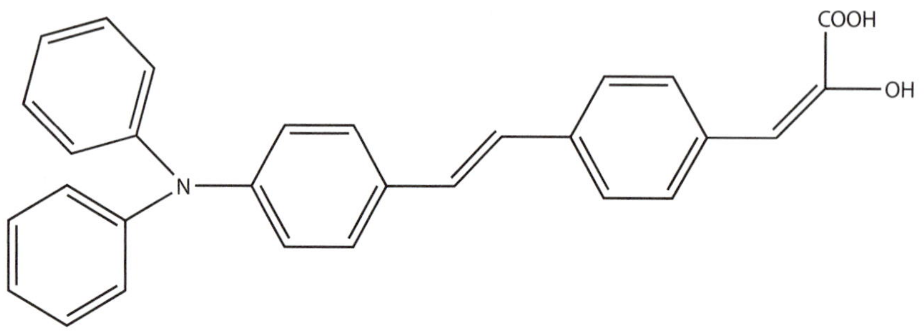

15.13 Schematic

cheaper than solar cells but the extra heat lowers the efficiency. A new system that may solve this problem puts the mirrors reflecting various wavelengths of light on the edges of the cell.[443] A second new system uses mirrors with tracking the sun for up to 35% efficiency with III–V cells.

Silicon has the advantage of being abundant and nontoxic. It is used in 90% of solar cells.[385] Most other cells contain toxic metals or elements that are not abundant and probably not cheap. If materials other than silicon come into common use, some sort of collection and recycling system will be needed to keep discarded cells from ending up in incinerators and landfills where they may be sources of pollution.[444] (Silicon Valley, California, has 29 Superfund sites.[445])

Solar cells are being used for many purposes in many places around the world: for heating and lighting homes, lighting bus stops, pumping water, and many others.[446] They can be used in place of the usual cladding or roofing of a new building, which avoids some of the costs for the usual materials used for these purposes.[447] They may work best if cleaned and serviced at intervals by an independent serviceman or one connected with the local utility.[448] Individual home owners may forget to clean them. Self-cleaning modules based on the photocatalytic action of thin layers of titanium dioxide may be possible[449] (see also Chapter 8). A home in Osaka, Japan, fed 77% of its photovoltaic power into the grid between 10:00 a.m. and 2:00 p.m. on a clear day in May.[450] This suggests that 3 kW of photovoltaic power on a house at this location would satisfy most of the needs for the house and alleviate peak power demand. A house in Florida with photovoltaic cells and white roof tiles required 83% less energy to air-condition than a conventional house.[451] With more solar cells, it might not require any energy from the electrical grid. It is also possible to utilize both the heat and light of the sun with the solar array.[452]

A test house built in Freiburg, Germany, operated for 3 years without being connected with the local electrical grid.[453] Electricity from photovoltaic cells was used to produce hydrogen and oxygen by the electrolysis of water. The hydrogen was stored at 28 bar. Electricity was stored in lead–acid batteries or was obtained by reaction of the hydrogen and oxygen in a fuel cell. Catalytic combustion of the hydrogen was used for cooking and space heating. The house used passive solar heating as well as transparent insulation in the windows. Its annual energy consumption was 8.7 kWh/m^2 compared with 110–140 kWh/m^2 for a standard German house. The only problem was that the living room was a little cool. The builders of the house would like a better method for long-term high-energy storage.

15.3.6 Other Sources of Renewable Energy

Other sources include hydropower, tidal energy, wave energy, geothermal energy, and wind.[454] Further development of hydropower[455] is possible, for many existing dams do not involve the generation of electricity. This may be partly due to the small heads of water involved. The erection of new dams for this purpose involves tradeoffs, because natural biotic communities as well as agricultural fields, houses, and towns are often lost in the process. The wind was used more as a source of power in the past than it is today. Before the days of rural electrification in the United States, it was common practice to use windmills to pump water for the farm. The Netherlands is famous for its windmills of former years that were used to grind grain, and so on. Windpower is cost-competitive today in some areas.[456] Part of the cost is offset because the land under the turbines can be used for other purposes, such as farming and grazing. Some new projects are producing electricity at a wholesale price of \$0.045 kmh^{-1}.[457] Wind farms are located in places where the wind blows fairly steadily, such as mountain passes and at the seashore. The Great Plains of the western United States has been suggested as a place for wind farms. A 100 MW wind farm is planned for Minnesota and a 50 MW one for Iowa. The power of the wind goes up as the cube of the wind speed. In 1994, California had 15,000 wind turbines, generating enough electricity to supply a city of the size of San Francisco. They operated 95% of the time. The wind turbines in Denmark supplied 20% of the country's electricity in 2006.[458] Hydropower from Norway and Sweden is used when the wind does not blow. The island of Samso, Denmark, has been used as a demonstration of "life after oil."[459] It used a combination of windpower, solar heating for houses, and a wood-fired power plant for district heating, but refused to use biogas from the manure from the pigs.

Delaware is planning the first offshore wind farm in the United States.[460] Globally, wind energy installations grew an average of 22%/yr from 1990 to 1999.[461] In 1998, the wind energy industry was valued at 2 billion dollars. Both wind and sunlight are intermittent sources of energy. A facility that uses both will be more reliable than one using either source alone.[462] An Australian sail boat combines the two.[463] Interconnected grid systems will be more reliable than single units.

Electricity is generated from steam in the ground in places such as New Zealand, California, Nevada, Utah, and Hawaii.[464] Geothermal energy is also used to heat 80% of the homes in Iceland. The steam and hot water from underground reservoirs contain gases, such as hydrogen sulfide and ammonia, as well as dissolved minerals that may contain arsenic, mercury, vanadium, nickel, or others. The geothermal plants can be polluting if these are discharged at the surface. A better system is to reinject the used water, not only to avoid the need to treat the water to remove pollutants, but also to form more steam and to prevent land subsidence. A plant is being built in the Imperial Valley of California to recover 30,000 metric tons of zinc annually before the reinjection of the used water.[465] The plants at the geysers near Santa Rosa, California have been seeing steam production drop 10%/yr since the mid-1980s, despite reinjection of 31% of the water produced.[466] A better system is to use a closed loop with a working fluid other than water (e.g., isobutene or pentane). This may also be better in utilizing low-grade heat. Studies have also been made of using the heat of hot, dry rock to produce steam to generate electricity. The temperature at the bottom of a 12,000-ft borehole in New Mexico was 177°C (350°F). For production of steam, it will probably be necessary to shatter the rock at the bottom of the hole with explosives. Oil well-fracturing techniques can also used. A 1 MW system of this type is being built in Australia.[467]

Tidal energy can be used to generate power in places where the tidal range is fairly large.[468] Power is generated as the water goes in over a dam and again as the water comes out over the dam. A plant in France uses this method. Proposals have been made to use the exceptionally high tides in the Bay of Fundy in Canada for this purpose. The waves in the ocean are a much more general source of power, which is largely untapped. The advantages of waves over offshore wind power include waves more of the time than wind, no damage to birds and bats and minimal visual impact for people to complain about. A few prototypes have been tested for the generation of power. The best may be those with the fewest moving parts. One system uses sand-filled ballast tanks to keep the apparatus in place.[469] As the wave passes, air is expelled upward through turbines that generate electricity. As the water level drops, another air turbine will generate electricity. A wind turbine can be mounted on top, if desired. The electricity from such a unit in Japan costs about twice as much as that from an oil-fired power plant.[470] Other schemes use pistons actuated by the water to drive a generator. The up-and-down motion of a floating device anchored to the bottom (a "bobbing duck") could also be used to generate power. An underwater system utilizes the pressure differences caused by the passing of the wave. The air in two mushroom-shaped floaters that are each partially filled with water is transferred between the two as the wave passes.[471] Two commercial wave power systems were in operation in 2009,[472] an oscillating water column system off the coast of the Island of Islay, Scotland, and a segmented pontoon system off the coast of Portugal. Sync Wave Energy is building a wave energy harvester off the coast of the British Columbia which it feels will produce electricity competitive with that of wind power.[473] Underwater turbines have been installed in the East River off Manhattan, New York, and in the Severn River in England.[474] They might be used wherever strong currents are available. Another system will use the motions of the sea to generate electricity from piezoelectric polymers, such as poly(vinylidene fluoride).[475] (Pressure on a piezoelectric material generates an electric current.) Such polymers can also be used to produce electricity from low-grade heat (e.g., water at <100°C).[476] If such systems can be scaled up economically, they could increase the efficiency of conventional power plants and reduce the thermal pollution of streams.

The challenge is to reduce the costs of the various forms of renewable energy[477] to where they can compete with fossil fuels today. If the various environmental and health costs were included in the price, the cost of electricity from an unscrubbed coal-burning power plant would about double. This new form of accounting would make it easier to replace fossil fuels with renewable forms of energy.

15.3.7 Energy Storage

Because some of the sources of renewable energy are intermittent, it will be necessary to combine them with some that are not, or find ways to store enough energy until the source is again available. It might be possible to supplement energy from the wind or the sun with energy from biogas, biomass, or hydropower. This might require building an additional plant and the expense that goes with it. At present, solar and wind power are supplemented with that from natural gas and other fossil fuels or with hydropower.

Energy can be stored using pumped storage reservoirs, compressed air in tanks, flywheels, or hot water.[478] Flywheels with electromagnetic bearings can be up to 90% efficient in storing and releasing energy.[479] The possibility of storing energy in highly extendible materials, such as rubbers, is being studied.[480] The storage of energy as heat, hydrogen, or in other chemical compounds was discussed earlier. An additional possibility is to use light to form a higher-energy compound that can later be reconverted to the original compound with release of the energy. The reaction must be 100% reversible over many cycles, should work with sunlight, and have a high quantum yield. The most studied reactions of this type involve the conversion of norbornadienes to quadricyclanes, which may be small molecules or attached to polymers.[481] A typical example is shown in **15.14** Schematic.

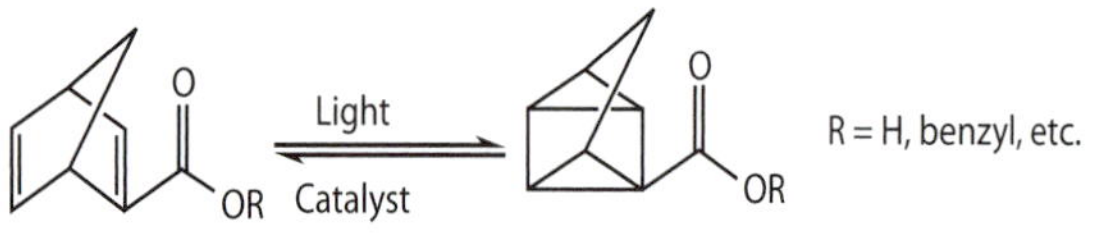

Light / Catalyst — R = H, benzyl, etc.

15.14 Schematic

Storage of energy in batteries[482] is common. The search is on for batteries of common nontoxic materials that combine light weight and compactness with high-energy density. The most common goal is a better battery for widespread use in motor vehicles.[483] For this use, it must be fully reversible over many cycles and be relatively easy to recharge quickly. The major candidates for electric vehicle batteries are shown in Table 15.2.

The first three are commercially available today. The nickel–cadmium battery has the disadvantage of using toxic cadmium. The lead–acid batteries used in cars today weigh too much and wear out too soon for widespread vehicles. The nickel–metal hydride battery appears to be available today for general use.[484] It is essentially a nickel–hydrogen cell with a hydrogen-storing alloy, such as $LaNi_5$, as the anode with aqueous potassium hydroxide containing some sodium hydroxide and lithium hydroxide as the electrolyte.[485] The active material in the cathode is nickel hydroxide. The overall reaction in the battery is shown in **15.15** Schematic (where M represents the hydrogen-storing alloy).

$$Ni(OH)_2 + M \rightleftharpoons NiOOH + MH$$

15.15 Schematic

Lithium batteries are being studied intensively owing to their light weight and high-energy density.[486] Modifications to prevent the occasional fire when overcharging laptops are being made. Most lithium batteries use a lithium-intercalated carbon anode and a cathode of a lithium metal oxide made of cobalt, nickel, or manganese. Lithium iron phosphate is thought to be safer. It is being used in portable power tools and in a car battery made by A123 for an electric car maker in Norway.[487] The electrolyte is a lithium salt, such as perchlorate, hexafluorophosphate, hexafluoroarsenate, or triflate, in solvents such as ethylene carbonate and propylene carbonate. The electrolyte can also be a solid containing the salt, such as polyethylene oxide or polypropylene oxide, which yields an all-solid battery.[488] A plasticizer can be added to increase conductivity in the solid electrolyte. The challenge is to increase the conductivity to the desired range of 10^2–10^1 S/cm from the more usual 10^{-4}–10^5 S/cm. An electrolyte film of a poly(epichlorohydrin-*co*-oxirane) and poly(acrylonitrile-*co*-1,3-butadiene) blend, swollen with a solution of 40% lithium perchlorate in propylene carbonate, has a conductivity of more than 10^3 S/cm.[489] A combination of polymer (**15.16** Schematic) with lithium chlorate has a conductivity of 10^3 S/cm.[490]

15.16 Schematic

Table 15.2 Candidates for Electric Vehicle Batteries

Battery	Range/Charge (miles)	Life (years)	Specific Energy (W-h/kg)
Nickel–cadmium	65–140	8	45–55
Lead–acid	50–70	2–3	35–40
Nickel–metal hydride	100–120	8	High 60s–70
Lithium ion	150–300	5–10	100–150

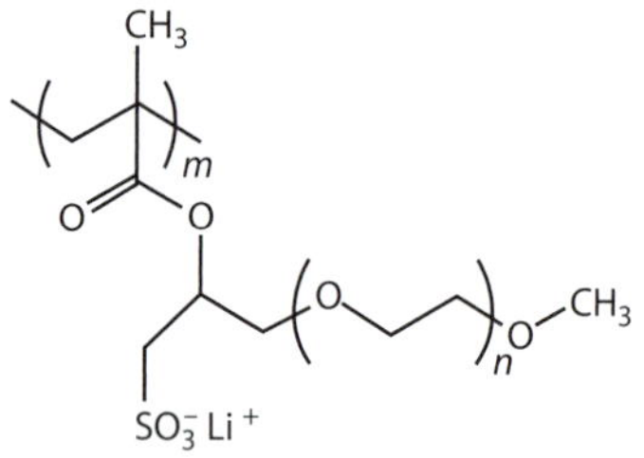

Zinc-air batteries are also being studied for use in vehicles.[491] The anode consists of particles of zinc in aqueous potassium hydroxide. The energy storage is about 160 Wh/kg. The range of the vehicle is expected to be about 250–300 miles. A drawback is that the zinc must be separated from the air electrode for recharging. The plan is to exchange the discharged battery for a charged one at a recharging station. An exchange system may be the best for many vehicle batteries. It circumvents the problems connected with charges that are too rapid. The system may be easier to apply to buses than to private cars. An alternative for the latter is to plug in the car while it is parked at a lot or in the garage at home at night. A more difficult problem to solve is abuse of batteries by drivers who accelerate and decelerate rapidly and repeatedly, or who drive until the battery is almost dead.

Each year, the world used 60 billion batteries made with a zinc anode and a manganese dioxide cathode with an electrolyte of potassium hydroxide. Substituting potassium or barium ferrate for the manganese dioxide permits a three-electron change instead of a two-electron one, increasing the battery's capacity by 50% or more.[492] The Fe(VI) battery is rechargeable, based on abundant starting materials, and environmentally benign. Large sodium–sulfur batteries are being tested at stationery sites for load leveling.[493]

15.4 USE OF LESS COMMON FORMS OF ENERGY FOR CHEMICAL REACTIONS

Many chemical reactions must be heated to obtain adequate rates. This can consume as much as one-third of the energy in the starting hydrocarbon in some reactions used in refining petroleum. Reactions that must be run at high temperatures with short contact times followed by quick quenching are especially difficult. These are often run to low conversion so that much recycling is needed.

The use of electricity, light, ultrasound, and microwaves has advantages for many chemical reactions. The total energy input may be reduced. Selectivity may be improved. The time required for the reaction can occasionally be shortened dramatically, which would permit greater throughput in a plant. Some of the reactions may not be possible at all with simple heating. These forms of energy are clean, meaning that unlike the usual catalysts, they require no separation at the end of the reaction, and there is no waste from them to dispose of. These methods are underused, probably because the usual chemists and chemical engineers are unfamiliar with them. They should not be neglected with the assumption that they will be more expensive. They may actually be cheaper.

15.4.1 Electricity

Electricity is used on a large scale in the production of inorganic compounds such as chlorine and sodium hydroxide. It is used to refine metals such as aluminum and copper and can also be applied to titanium.[494] However, the only large-volume application in organic chemistry appears to be the hydrodimerization of acrylonitrile (**15.17** Schematic) to form adiponitrile (2,000,000 tons/yr), which is then reduced to the diamine or hydrolyzed to adipic acid for the preparation of nylon-6,6.[495]

CN $\xrightarrow[H_2O]{e^{\ominus}}$ NC CN

15.17 Schematic

Electrochemistry[496] can be used to regenerate an expensive or toxic reagent *in situ*. An example is the electrical regeneration of periodate that is used for the oxidation of vicinal diols such as sugars. When used with osmium oxidations, it can keep the level of the toxic metal reagent quite low. In Chapter 4, an example of the conversion of a naphthoquinone to its epoxide by electrolysis of aqueous sodium iodide was given. The sodium hydroxide and iodine produced by electrolysis react to form hypoiodite that adds to the double bond to form the hydroxyiodide, which then eliminates sodium iodide by the action of the sodium hydroxide to reform the sodium iodide.[497] Electricity can be used in oxidations and reductions instead of reagents. When iodine is reduced to hydriodic acid in this way, no waste products are formed.[498] Phosphorus-containing wastes are produced in the conventional process using phosphorus. The new process is also cheaper. Ketones can be converted to alcohols in up to 98% yield by electrocatalytic hydrogenation using rhodium-modified electrodes.[499] There is no catalyst to be separated at the end of the reaction. The electrochemical hydrogenation of water-immiscible olefins and acetylenes is enhanced by concurrent ultrasonication.[500] Hydrogenation of edible oils using a cell with a Nafion membrane with a ruthenium anode on one side and a platinum or palladium black cathode on the other produced less of the undesirable *trans*-isomers than conventional hydrogenations.[501]

An oxidation using a nickel hydroxide electrode is shown in **15.18** Schematic.[502] Electrochemistry is also a way to produce radicals and anions. The hydrodimerization of acrylonitrile to adiponitrile just mentioned may involve the coupling of free radicals. The coupling of carbonyl compounds, such as *p*-tolualdehyde, to form pinacols with up to 100% selectivity, by way of free radicals, can be done electrically.[503] Anions can also be formed electrochemically and used *in situ*, as in example (**15.19** Schematic).[504]

H_3C CHO N N Cl $\xrightarrow{e}$ H_3C COOH N N Cl 80–85%

15.18 Schematic

Electrochemistry can also be used in a variety of other "clean technologies," including separations by electrodialysis, recovery of metal ions from waste streams, and destruction of cyanide ion and organic compounds in waste streams.[505]

O CH_3O $\xrightarrow[CO_2]{e}$ HO COOH CH_3O

98% conversion 98% yield

15.19 Schematic

15.4.2 Light

Light is environmentally benign, leaving no residues to be removed in the workup of a reaction. It can catalyze some reactions that are difficult or impossible to run in other ways (e.g., some cycloadditions).[506] Most photochemistry is carried out using light with shorter wavelengths than those found in sunlight. The challenge is to use sunlight as is, or when concentrated. The degree of concentration can vary with the optical system. The large concentration obtained with solar furnaces has been suggested for isomerization, cycloadditions, catalytic cyclizations, and purification of water.[507]

A few examples will be given to illustrate current trends in research. Dithianes, benzyl ethers, and related compounds have been cleaved by the use of visible light with a dye (**15.20** Schematic).[508] (To use more of the light, mixtures of dyes can be used.)

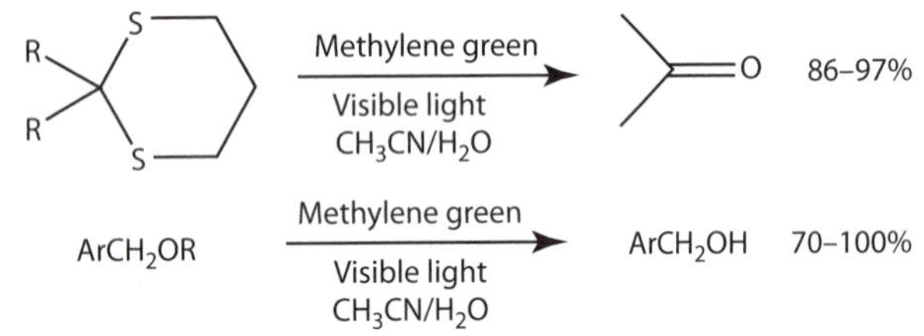

Methylene green; Visible light; CH_3CN/H_2O — 86–97%

$ArCH_2OR$ → Methylene green; Visible light; CH_3CN/H_2O → $ArCH_2OH$ 70–100%

15.20 Schematic

Benzonitrile in methanolic potassium hydroxide can be hydrolyzed to benzamide in 96% yield using an oxophosphoporphyrin catalyst with >420-nm light at 20°C.[509] A combination of visible light and water was used in a cyclization (**15.21** Schematic) to produce substituted pyridines with almost no by-products.[510]

Pinacols have been made with sunlight (**15.22** Schematic).[511]

Epoxides can be polymerized with visible light.[512] Oxidation of hydrocarbons in zeolites with blue light gives improved selectivity.[513] Isobutane can be converted to *tert*-butylhydroperoxide with 98% selectivity. Benzaldehyde is produced from toluene, acrolein from propylene, and acetone from propane.

Photoredox reactions can be carried out with semiconductors. Irradiation of nitrobenzene in the presence of zinc oxide particles in alcohol produced phenylhydroxylamine in 73% yield.[514] Irradiation of 4-cyanophenylazide in ethanol containing titanium dioxide gave 4-cyanoaniline in 98% yield. Photooxidation of cyclohexane in the presence of titanium dioxide gave 85.4% cyclohexanone, 2.6% cyclohexanol, and 12% carbon dioxide.[515] To be of use commercially as part of a route to nylon-6,6, the amount of carbon dioxide must be reduced. This method has also been studied for the detoxification of wastewater containing organic contaminants, such as trichloroethylene.[516]

Acylhydroquinones can be produced from 1,4-benzoquinone and an aldehyde (**15.23** Schematic) with light from a sunlamp.[517] In contrast to the usual acylation with an acylchloride, this process produces no by-product salts. The analogous reaction has been carried out in 90% yield in sunlight with naphthoquinone and *n*-butyraldehyde.[518]

15.4.3 Ultrasound

Many applications of ultrasound[519] to the synthesis of organic compounds have been given in earlier chapters. This technique is especially useful when a solid in a liquid has to react. This may be a matter of the surface being cleaned continuously by the cavitation. (Cavitation produces high temperature and pressures for very short times.) It can help when there are two immiscible liquids. In this case, mechanical stirring for macromixing may be combined with

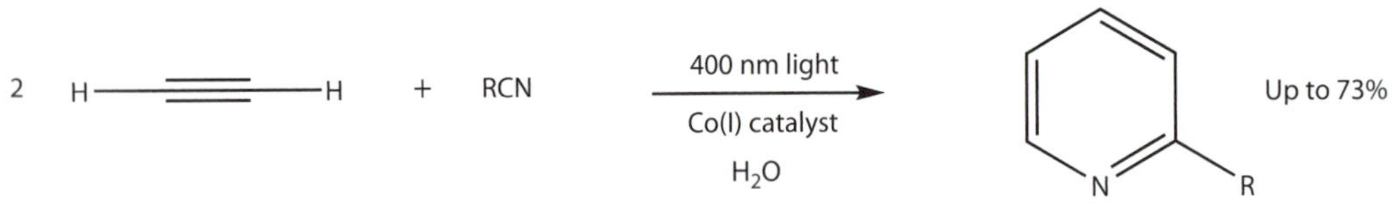

2 H—≡—H + RCN → 400 nm light; Co(I) catalyst; H_2O → Up to 73%

15.21 Schematic

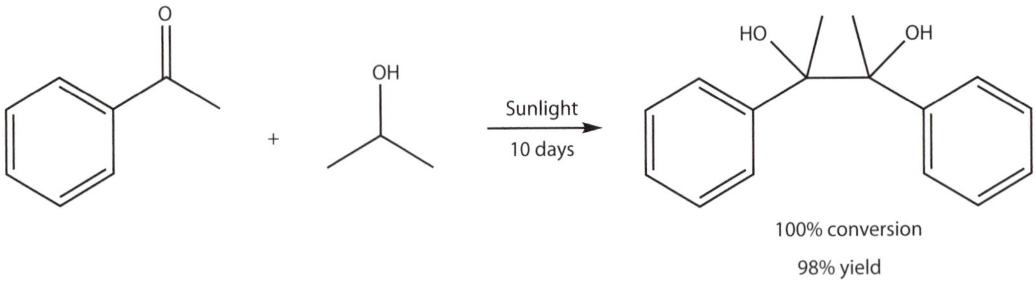

Sunlight; 10 days — 100% conversion; 98% yield

15.22 Schematic

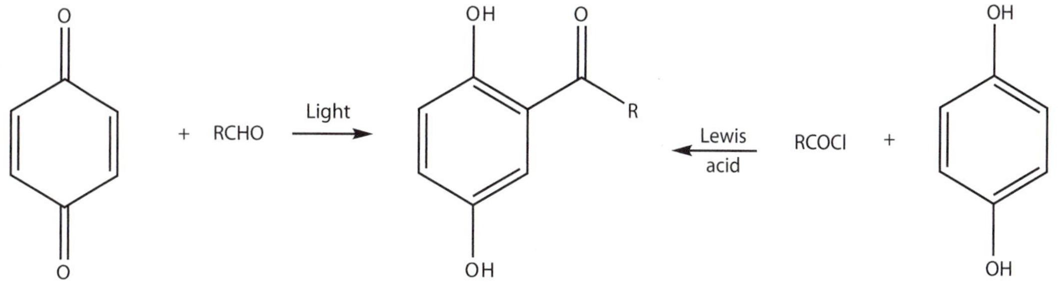

15.23 Schematic

ultrasound for micromixing. The instantaneous high pressures obtained can be used to accelerate some reactions when the products occupy a smaller volume than the starting materials (e.g., the Diels–Alder reaction).

The cost of the ultrasonic energy needed for these mixing and cleaning processes in reactions is not high because only relatively small amounts of energy are needed. In contrast, the energy needed for doing chemistry with ultrasound is much higher (e.g., generating hydroxyl radical in water). If the hydroxyl radical is to be used to initiate a polymerization, only a little may be needed, and so it may not cost too much. The cost is still not prohibitive for reactions such as lowering the molecular weight of polymers, if relatively few chain scissions are needed. A couple of examples from the field of polymers will illustrate some of what can be done. Block copolymers of polyethylene and acrylamide have been made by ultrasonic cleavage of the polyethylene in the presence of acrylamide.[520] Maleic anhydride grafted to polypropylene in 93% yield when the two were ultrasonicated at 60°C in the presence of benzoyl peroxide.[521] Nitration of phenol with 9 wt% aqueous nitric acid can be done in 2 h with a 94% conversion to a 70:27 *p*-nitrophenol/*o*-nitrophenol mixture with ultrasound but the conversion is only 30% to a 49:48 *p*-nitrophenol/*o*-nitrophenol mixture in 48 h without it.[522]

Ultrasound is used to control crystallization to improve filtration and reduce processing time, as for table sugar.[523] It can be used to solder metals and ceramics without a flux.[524] It can also be used to weld metals and plastics, as in holding a nonwoven fabric together.

Ultrasound can be scaled up to large volumes in batch or continuous flow systems. It can also be combined with an extruder, where it speeds up extrusion and allows the processing of polymers of higher molecular weight that may have improved tensile strength. It is a technique that deserves to be used more often, for it can often save money by reducing reaction times.

15.4.4 Microwaves

As illustrated by examples in earlier chapters, the use of microwave energy,[525] instead of conventional heating, often results in good yields in very short times. Ovens with homogeneous fields of focused microwaves, if temperature control and power modulation are present, can lead to improved yields compared with conventional ovens. Such ovens are supplied by Biotage, CEM, Milestone, and Anton Paar. There must be something present that absorbs the microwaves.

In the synthesis of zeolites, it is necessary to hold the gel at elevated temperatures for extended periods to induce the zeolite to crystallize. Zeolites can be made from fly ash in 30 min when microwaves are used, compared with 24–48 h when the mixture is just heated.[526] In the synthesis of zeolite MCM-41, heating for 80 min at 150°C is replaced by the use of microwaves for 1 min at 160°C.[527] Varma et al. have run many organic reactions with clay catalysts without solvent in open vessels for 1–3 min to obtain good yields of products.[528] These include the preparation of imines and enamines from ketones and amines in 95–98% yields, oxidation of alcohols to ketones with iron(III) nitrate on clay, cleavage of thioketals, and others. The rearrangement of a cyclic oxime to a lactam has been carried out in quantitative yield with microwaves in 5 min in the presence of the zeolite AlMCM-41 (**15.24** Schematic).[529]

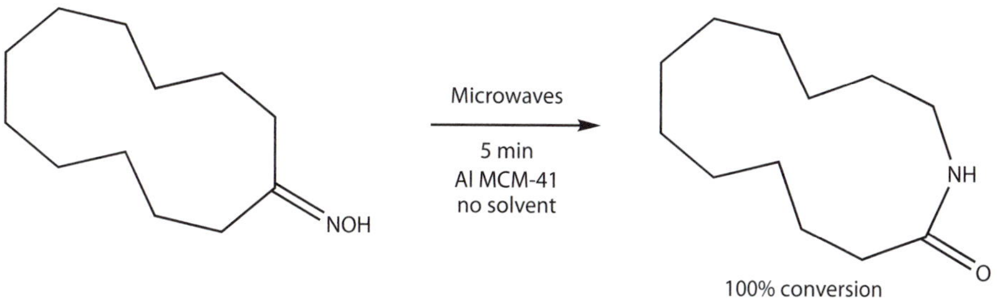

15.24 Schematic

That these yields cannot be obtained in a comparable period of simple heating make these workers feel that there is a special microwave effect. It is possible that the effect is due to instantaneous high temperatures at the surface of the solid. It may also be that microwaves facilitate contact between the solid and liquid reagents. If so, a combination of ultrasound and microwaves might be even better for reactions that take somewhat longer to run with microwaves than those just cited. Others feel that there is no such effect.[530] While the debate about the special microwave effect is still going on, it appears to have been settled for at least this one class of reactions.

The application of microwave energy as an alternative to conventional heating needs to be tried wherever microwave-absorbing (i.e., polar) materials are present. These include not only inorganic compounds, but also water, alcohols, amides, and many other oxygen-containing compounds, but not hydrocarbons, such as hexane and toluene. Xylene can be heated with microwaves by dispersing 1–2% of cobalt or magnetite, 5–2-nm nanoparticles, in it.[531] A magnet is used to recover these for reuse in the next run. Microwaves enhance the solid-state polymerization of polyethylene terephthalate at 236°C and nylon at 182–226°C by enhancing the diffusion of small molecules.[532] There is considerable current enhancement when microwaves are used with platinum microelectrodes.[533] Microwaves have been used with metal powders at high temperature to produce parts that are superior to those made by the usual sintering in powder metallurgy.[534] They have also been used in the production of ceramics.[535] Regeneration of zeolites that had been used to trap volatile organic compounds with microwaves, required only one-third as much energy as conventional heating.[536]

Microwaves should be well suited for continuous tubular reactors, where they might reduce the amount of energy needed, reduce costs, and increase the output of a plant. If water were inert, it might be used to absorb the microwave radiation, the reagents being in solution or in an emulsion. Further research will undoubtedly find many more applications for this technique.

REFERENCES

1. (a) J. Randolph and G.M. Masters, *Energy for Sustainability—Technology, Planning, Pollicy*, Island Press, Washington, DC, 2007; (b) J. Goldemberg, ed., *World Energy Assessment—Energy and The Challenges of Sustainability*, United Nations Development Programme, New York, 2001; (c) E. Kintisch, *Science*, 2007, *318*, 547; (d) J. Holdren, ed., *Science*, 2007, *315*, 737–812 (issue on energy); (e) N. Armaroli and V. Balzani, *Angew. Chem. Int. Ed.*, 2007, *46*, 52–66; (f) L.R. Brown, *Futurist*, 2006, *40*(4), 18; (g) L.R. Brown, *Plan B 4.0, Mobilizing to Save Civilization*, W.W. Norton, New York, 2008; (h) Worldwatch Institute, *American Energy—The Renewable Path to Energy Security*, Washington, DC, 2007; (i) Union of Concerned Scientists, www.ucsusa.org/clean_energy; (j) R. Dell and D. Rand, *Clean Energy*, Royal Society of Chemistry, Cambridge, 2004.
2. A. Vijayan and A. Kumar, *Environ. Prog.*, 2005, *24*(2), 125.
3. A. Bisio and S.R. Boots, *Energy Technology and the Environment*, Wiley, New York, 1995, 1131.
4. Anon., *Chem. Eng. News.*, 1996, May 17, 23.
5. R. Winder, *Chem. Ind. (Lond.)*, July 4, 2005, 14.
6. (a) C.A.C. Sequeira and J.B. Moffat, eds, *Chemistry, Energy and the Environment*, Royal Society of Chemistry, Cambridge, 1998; (b) J. de S. Arons, H. van der Kooi, and K. Sankaranarayanan, *Efficiency and Sustainability in the Energy and Chemical Industries*, Marcel Dekker, New York, 2004; (c) W.J. Storck, *Chem. Eng. News*, Feb. 5, 2001, 19.
7. (a) R.E. Hester and R.M. Harrison, eds, *Environmental Impact of Power Generation*, Royal Society of Chemistry, Cambridge, 1999; (b) R.F. Hirsch and A.H. Serchuk, *Environment*, 1999, *41*(7), 4.
8. Anon., *Wilmington Delaware News J.* 1996, Aug. 13, A5.
9. J.I. Levy, J.K. Hammitt, Y. Yanagisawa, and J.D. Spengler, *Environ. Sci. Technol.*, 1999, *33*, 4364.
10. Anon., *Environ. Sci. Technol.*, 1998, *32*, 399A.
11. Anon., *Chem. Eng. News*, 1999, Jan. 11, 22.
12. R. Tyson, *Environ. Sci. Technol.* 1997, *31*, 508A.
13. E.N. Kaufman, M.H. Little, and P.T. Selvarej, *J. Chem. Technol. Biotechnol.*, 1996, *66*, 365.
14. J. Johnson, *Chem. Eng. News*, Feb. 23, 2009, 44.
15. (a) M. Capone, *Kirk–Othmer Encyclopedia of Chemical Technology*, 4th ed., Wiley, New York, 1997, *23*, 446; (b) L.S. Jae, H.K. Jun, S.Y. Jung, T.J. Lee, C.K. Ryu, and J.C. Kim, *Ind. Eng. Chem. Res.*, 2005, *44*, 9973.
16. J.A. Cusumano, *J. Chem. Ed.*, 1995, *72*, 959.
17. R.L. Berglund, *Kirk–Othmer Encyclopedia of Chemical Technology*, 4th ed., Wiley, New York, 1994, *9*, 1022.
18. (a) L. Raber, *Chem. Eng. News*, 1997, Oct. 20, 14; 1998, Oct. 5, 11; (b) W.V. Cicha, *Chem. Eng. News*, 1997, Dec. 15, 6; (c) G. Parkinson, *Chem. Eng.*, 2000, *1072*(2), 27.
19. J.J. Corbett and P. Fischbeck, *Science*, 1997, *278*, 823.
20. J. McCormick, *Acid Earth—The Politics of Acid Pollution*, 3rd ed., Earthscan, London, 1997.
21. Anon., *Environ. Sci. Technol.*, 1997, *31*, 459A.
22. G. Davison and C.N. Hewitt, *Air Pollution in the United Kingdom*, Royal Society of Chemistry, Cambridge, Special Publication 210, 1997.
23. Anon., *Environ. Sci. Technol.*, 1998, *32*, 171A.
24. (a) G.E. Likens, C.T. Driscoll, and D.C. Buso, *Science*, 1996, *272*, 244; (b) G.E. Likens, K.C. Weathers, T.J. Butler, and D.C. Buso, *Science*, 1998, *282*, 1991; (c) J.M. Gunn, W. Keller (and following papers), *Restor. Ecol.*, 1998, *6*, 316–390.
25. (a) W.F. Rogge, L.M. Hildeman, M.A. Mazurek, G.R. Cass, and B.R.T. Simoneit, *Environ. Sci. Technol.*, 1998, *32*, 13; (b) J.D. McDonald, B. Zielinska, E.M. Fujita, J.C. Sagebiel, J.C. Chow, and J.G. Watson, *Environ. Sci. Technol.*, 2000, *34*, 2080.
26. (a) J. Goodill, *Hockessin Community News*, 1997, Oct. 7, 4; (b) J. Kaiser, *Science*, 2000, *289*, 22, 711; (c) C. Hogue, *Chem. Eng. News*, July 3, 2000, 6.
27. (a) L.R. Raber, *Chem. Eng. News*, 1997, Aug. 11, 26; 1998, Sept. 7, 27; (b) J. Kaiser, *Science*, 1998, *280*, 193; (c) A.S. Lefohn, D.S. Shadwick, and S.D. Ziman, *Environ. Sci. Technol.* 1998, *32*, 276A; (d) J. Grisham, *Chem. Eng. News*,

1999, May 24, 5; (e) R.E. Benedick, *Ozone Diplomacy—New Directions in Safeguarding the Planet*, 2nd ed., Harvard University Press, Cambridge, MA, 1998.
28. Connectiv, Pollution Prevention and Energy Reduction Conference, Newark, DE, Oct. 7, 1997.
29. R. Tyson, *Environ. Sci. Technol.*, 1997, *31*, 408A.
30. J. Romm, M. Levine, M. Brown, and E. Petersen, *Science*, 1998, *279*, 669.
31. E.D. Weil and S.R. Andler, *Kirk–Othmer Encyclopedia of Chemical Technology*, 4th ed., Wiley, New York, 1997, *23*, 311.
32. D. Mescher, *The Review*, University of Delaware, Newark, DE, 1997, Sept. 16, A2.
33. (a) A. Johansson, *Clean Technology*, Lewis Publishers, Boca Raton, FL, 1992, 142; (b) S. Honeyman, *Chem. Ind.* (*Lond.*) 1998, *920*; (c) K. Christen, *Environ. Sci. Technol.*, 1998, *32*, 531A; (d) T.R. Castlen and P.F. Scheve, *Am. Sci.*, 2009, *97*(1), 26.
34. I. Harwigsson and M. Hellsten, *J. Am. Oil. Chem. Soc.*, 1996, *73*, 921.
35. Environmental Action Fact Sheet, Washington, DC, Feb. 1991.
36. Environmental Defense Fund, Jan. 2002.
37. (a) Anon., *Chem. Eng. News*, 1994, Sept. 19, 20; (b) C. Steiner, *$20 Per Gallon*, Grand Central Publishing, New York, 2009; (c) A. Schafer, H.D. Jacoby, J.B. Heywood, and I.A. Waltz, *Am. Sci.*, 2009, 97, 476; (d) D. Sperling and D. Gordon, *Two Billion Cars: Driving Toward Sustainability*, Oxford University Press, New York, 2008.
38. J.J. Romm, *Hell and High Water: Global Warming—The Solution and the Politics—What We Should Do*, William Morrow, New York, 2007.
39. G. Peaff, *Chem. Eng. News*, 1997, Aug. 4, 20.
40. H.J. Hebert, *Wilmington Delaware News J.*, 1997, Oct. 18, F5.
41. P.H. Abelson, *Science*, 1992, *257*, 1459.
42. P.H. Abelson, *Science*, 1997, *227*, 1587.
43. T. Krawczyk, *Int. News Fats Oils Relat. Mater.*, 1996, *7*, 1320.
44. (a) T.R. Loughlin, ed., *Marine Mammals and the Exxon Valdez*, Academic, San Diego, 1994; (b) D.A. Wolfe, M.J. Hameedi, J.A. Galt, G. Watabayashi, J. Short, C. O'Claire, S. Rice, J. Michel, J.R. Payne, J. Braddock, S. Hanna, and D. Sale, *Environ. Sci. Technol.*, 1994, *28*, 560A; (c) A. Weiner, C. Berg, T. Gerlach, J. Grunblatt, K. Holbrook, and M. Kuwada, *Restor. Ecol.*, 1997, *5*, 44; (d) S.F. Sugai, J.E. Lindstrom, and J.F. Braddock, *Environ. Sci. Technol.*, 1997, *31*, 1564; (e) S.D. Rice, R.B. Spies, D.A. Wolfe, and B.A. Wright, eds, *Proceedings of the Exxon Valdez Oil Spill*, American Fisheries Society, Bethesda, MD, 1996; (f) *Chem. Eng. News*, 1994, Aug. 1, 17; (g) R.M. Atlas, *Chem. Br.*, 1996, *32*(5), 42; (h) J. Kaiser, *Science*, 1999, *284*, 247; (i) T.A. Birkland, *Environment*, 1998, *40*(7), 4; (j) C. Holden, *Science*, 1998, *280*, 1697.
45. Z. Wang, M. Fingas, S. Blenkinsopp, G. Sergy, M. Landriault, L. Sigouin, and P. Lambert, *Environ. Sci. Technol.*, 1998, *32*, 2222.
46. P.J. Squillace, J.S. Zogorski, W.G. Wilber, and C.V. Price, *Environ. Sci. Technol.*, 1996, *30*, 1721.
47. J.E. Reuter, C.F. Harrington, W.R. Cullen, D.A. Bright, and K.J. Reimer, *Environ. Sci. Technol.*, 1998, *32*, 3666.
48. (a) J. Grisham, *Chem. Eng. News*, 1999, May 17, 10; (b) P.M. Morse, *Chem. Eng. News*, 1999, Apr. 12, 26; (c) M. McCoy, *Chem. Eng. News*, 1999, Apr. 5, 9; (d) C. Hogue, *Chem. Eng. News*, Mar. 27, 2000, 6.
49. L.J. Penvenne, *Technol. Rev.*, 1997, *100*(4), 12.
50. Anon., *Chem. Eng. News*, 1997, Sept. 15, 21.
51. (a) T. Enya, H. Suzuki, T. Watanabe, T. Hirayama, and Y. Hisamatsu, *Environ. Sci. Technol.* 1997, *31*, 2772; (b) B.J. Finlayson-Pitts and J.N. Pitts, *Science*, 1997, *276*, 1045.
52. O. Tunali, *World Watch*, 1996, *9*(1), 26.
53. (a) C.M. Cooney, *Environ. Sci. Technol.*, 1998, *32*, 272A; (b) Anon., *Chem. Eng. News*, 1999, May 10, 6.
54. G. Parkinson, *Chem. Eng.*, 1998, *105*(8), 27.
55. (a) Anon., *Chem. Eng. Prog.*, 2006, *102*(4), 12; (b) A.-J. Hernandez-Maldnado and R.T. Yang, *Cat. Rev. Sci. Eng.*, 2004, *46*, 111; *Ind. Eng. Chem. Res.*, 2003, *42*, 3103; (c) E. Ito and J.A.R. van Veen, *Catal. Today*, 2006, *116*(4), 446.
56. M. Hamer, *New Sci.*, Sept. 14, 1996, 5.
57. (a) D. Hemenway, *N. Eng. J. Med.*, 1998, *339*, 843; (b) L. Miller, *Wilmington DE News J.*, Aug. 30, 2003, A3.
58. Anon., *Univ. Calif. Berkeley Wellness Lett.*, 1997, *13*(4), 8.
59. Bureau of Transportation Statistics, Washington, DC, Mar. 28, 2003.
60. J.H. Kay, *Asphalt Nation*, Crown, New York, 1997; (b) *Technol. Rev.*, 1997, *100*(5), 53.
61. W. Harrington and V. McConnell, *Environment*, 2003, *45*(9), 22.
62. Anon., *Wilmington Delaware Sunday News*, 1997, Mar. 2, H1.
63. (a) V. Helmbreck, *Wilmington Delaware News J.*, 1996, Nov. *12*, D1; (b) L. Miller, *Wilmington Delaware News J.*, Aug. 30, 2003, A3.
64. Anon., *Univ. Calif. Berkley Wellness Lett.*, 1996, *12*(10), 1.
65. Anon., *Chevron Letter to Shareholders*, 2nd quarter, San Francisco 1998, 13.
66. (a) Union of Concerned Scientists, *Motor-Vehicle Fuel Efficiency and Global Warming*, Cambridge, MA, May, 1991; (b) R. Hwang, *Nucleus* (Union of Concerned Scientists), 1998, *20*(2), 1.
67. F.R. Field, III and J.P. Clark, *Technol. Rev.*, 1997, *100*(1), 28.
68. Anon., *Environ. Sci. Tehnol.*, 2006, *40*, 5167.
69. J.M. Broder and M. Maynard, *New York Times*, Dec. 1, 2007.
70. D.B. Aulenbach, *Chem. Eng. News*, 1997, June 2, 4.
71. R. Lukman, *Chem. Ind.* (*Lond.*), 1995, 940.
72. D.W. Fahey, E.R. Keim, K.A. Boering, C.A. Brock, J.C. Wilson, H.H. Jonsson, S. Anthony, T.F. Hanisco, P.O. Wennberg, R.C. Miake-Lye, R.J. Salawitch, N. Louisnard, E.L. Woodbridge, R.S. Gao, S.G. Donnelly, R.C. Wamsley, L.A. del Negro, S. Solomon, B.C. Daube, S.C. Wofsy, C.R. Webster, R.D. May, K.K. Kelly, M. Loewwnstein, J.R. Podolske, and K.R. Chan, *Science*, 1995, *270*, 70.
73. (a) D. Young. *Alternatives to Sprawl*, Lincoln Institute of Land Policy, Cambridge, MA, 1995; (b) P. Newman and J. Kenworthy, *Sustainability and Cities—Overcoming Automobile Dependence*, Island Press, Washington, DC, 1998; (c) R. Cervero, *The Transit Metropolis—A Global Inquiry*, Island Press, Washington, DC, 1998; (d) J. Pelley, *Environ. Sci. Technol.*, 2000, *34*, 13A.
74. T. Hylton, *Save Our Lands, Save Our Towns: A Plan for Pennsylvania*, RB Books, Harrisburg, PA, 1995.
75. C. Wieman, *Technol. Rev.*, 1996, *99*(4), 48.
76. M. O'Meara, *World Watch*, 1998, *11*(5), 8.

77. (a) R.A. Kerr, *Science*, 1998, *281*, 1128; (b) K.S. Deffeyes, *Hubbert's Peak—The Impending World Oil Shortage*, Princeton University Press, Princeton, 2001.
78. Anon., *Environment*, 1999, *41*(3), 23.
79. R. Tyson, *Environ. Sci. Technol.*, 1997, *31*, 454A.
80. (a) B. Hileman, *Chem. Eng. News*, 1997, Oct. 13, 14; (b) T. Stuckey, *Wilmington Delaware News J.*, 1997, Oct. 18, B6.
81. H.A. Wittcoff and B.G. Rueben, *Industrial Organic Chemicals*, Wiley, New York, 1996, 313.
82. (a) R. Baum, *Chem. Eng. News*, 1995, May 15, 5; (b) R.L. Rawls, *Chem. Eng. News*, 1998, June 22, 29; (c) C.C. Cummins, *Chem. Commun.*, 1998, 1777.
83. J. duBois, C.S. Tomooka, J. Hong, E.M. Carreira, and M.W. Day, *Angew. Chem. Int. Ed.*, 1997, *36*, 1645.
84. F. Tuczek and N. Lehnert, *Angew. Chem. Int. Ed.*, 1998, *37*, 2636.
85. (a) Y. Nishibayashi, S. Iwai, and M. Hidai, *Science*, 1998, *279*, 540; (b) G.J. Leigh, *Science*, 1998, *279*, 506.
86. K.-I. Machida, M. Ito, K. Hirose, and G.-Y. Adachi, *Chem. Lett.*, 1998, 1117.
87. R. Schrock, *Acc. Chem. Res.*, 2005, *38*(12), 955.
88. (a) B. Hileman, *Chem. Eng. News*, 1996, Oct. 28, 24; (b) W.C. Sailor, D. Bodansky, C. Braun, S. Fetter, and B. van der Zwaan, *Science*, 2000, *288*, 1177; (c) J. Johnson, *Chem. Eng. News*, Oct. 2, 2000, 39; (d) T. Zoelnner, *Uranium—War, Energy and the Rock That Shaped the World*, Viking Press, Penguin Group, New York, 2009; (e) M. Jacoby, *Chem. Eng. News*, Nov. 16, 2009, 44; (f) http://www.nuclear.gov
89. J. Johnson, *Chem. Eng. News*, 1999, Apr. 12, 34.
90. J. Johnson, *Chem. Eng. News*, 1997, May 12, 14.
91. R. Koenig, *Science*, 1998, *282*, 1967.
92. J. Johnson, *Chem. Eng. News*, May 12, 2003, 27.
93. (a) G.H. Marcus and A.E. Levin, *Phys. Today*, 2002, *55*(4), 54; (b) D. Hairston, N. Franz, and T. Kamiya, *Chem. Eng.*, 2001, *108*(6), 33.
94. M. Ryan, *Am. Sci.*, 2009, *97*(2), 112.
95. B. Planer-Friedrich and B.J. Merkel, *Environ. Sci. Technol.*, 2006, *40*, 3181.
96. R. Peltier, *Power*, 2007, *151*(12), 46.
97. (a) D.L. Klass, *Biomass for Renewable Energy, Fuels and Chemicals*, Academic Press, San Diego, 1998; (b) *Chem. Eng. Prog*, 2008, *104*(8), S2–S23.
98. Letter, S.J. McCormick (Nature Conservancy in Action), and A.S. Matlack, Jan. 10, 2003.
99. (a) D. Schneider, *Am. Sci.*, 2006, *94*(5), 408; (b) R. Righelato and D.V. Spracklen, *Science*, 2007, *317*, 902; (c) P. Zurer, *Chem. Eng. News*, July 24, 2006, editorial page; (d) J. Johnson, *Chem. Eng. News*, Dec. 4, 2006, 57; Jan. 1, 2007, 19; (e) F. Rosillo-Calle, S. Hemstock, P. deGroot, and J. Woods, eds, *The Biomass Assessment Handbook—Bioenergy for a Sustainable Environment*, Stylus Publishing, Herndon, VA, 2006; (f) G.W. Huber and B.E. Dale, *Sci. Am.*, 2009, *301*(1), 52; (g) M. Voith, *Chem. Eng. News*, April 27, 2009, 10; (h) R.F. Service, *Science*, 2009, *326*, 516; (i) T.F. McGowan, ed., *Biomass and Alternate Fuel Systems—An Engineering and Economic Guide*, Wiley-AIChe, Hoboken, NJ, 2009; (j) D. Charles, *Science*, 2009, *324*, 587; (k) D. Keeney, *Environ. Sci. Technol.*, 2009, *43*, 8; (l) D. Malakoff, *Conservation Magazine*, 2009, *10*(2), 20; (m) P. Kumar, D.M. Barrett, M.J. Delwiche, and P. Stroeve, *Ind. Eng. Chem. Res.*, 2009, *48*, 3713; (n) T.R. Sinclair, Am. Sci., 2009, *97*, 400; (o) D. Tilman, R. Socolow, J.A. Foley, J. Hill, E. Larson, L. Lynd, et al., *Science*, 2009, *325*, 270.
100. (a) M. Voith, *Chem. Eng. News*, Jan. 26, 2009, 22 4; (b) R.F. Service, *Science*, 2009, 325, 379.
101. M. Brauer, K. Barlett, J. Regalado-Pineda, and R. Perez-Padilla, *Environ. Sci. Technol.*, 1996, *30*, 104.
102. R. Baum, *Chem. Eng. News*, 1995, Apr. 10, 8.
103. J. Cohen, *Science*, 1995, *267*, 824.
104. (a) A. Gore, *An Inconvenient Truth: The Planetary Emergency of Global Warming and What We Can Do About It*, Melcher/Rodale, 2006; (b) Anon., *Am. Sci.*, 2007, *95*(3), after p. 282; (c) *Environment*, 2004, *46*(10) [issue on global climate change]; (d) B. Hileman, *Chem. Eng. News*, Oct. 6, 2006, 33; Mar. 21, 2005, 47; (e) C. Hogue, *Chem. Eng. News*, Feb. 2, 2009, 11; (b) C. Houge, *Chem. Eng. News*, June 22, 2009, 10.
105. (a) D.R. Easterling, B. Horton, P.D. Jones, T.C. Peterson, T.R. Karl, D.E. Parker, M.J. Salinger, V. Razuvayev, N. Plummer, P. Jamason, and C.K. Folland, *Science*, 1997, *277*, 364; (b) K. Hasselman, *Science*, 1997, *276*, 914; (c) R.A. Kerr, *Science*, 1996, *273*, 34; (d) B. Hileman, *Chem. Eng. News*, 1999, Aug. 9, 16; Apr. 24, 2000, 9; (e) R.A. Kerr, *Science*, 2000, *288*, 589; (f) D.R. Easterling, G.A. Meehl, C. Permasan, S.A. Changnon, T.R. Karl, and L.O. Mearns, *Science*, 2000, *289*, 2068; (g) R.T. Watson, ed., *Climate Change 2001 Synthesis Report*, Cambridge, University Press, Cambridge, 2001; (h) D. Hanson, *Chem. Eng. News*, Nov. 26, 2007, 7; (i) Intergovernmental Panel on Climate Change Report, Fourth Assessment, Cambridge University Press, Cambridge, UK, Nov. 17, 2007.
106. R.A. Kerr, *Science*, 2007, *317*, 182.
107. K.R. Briffa, T.J. Osborn, *Science*, 1999, *284*, 926.
108. R.D. Alward, *Science*, 1991, *283*, 229.
109. E.W. Lempinen, *Science*, 2006, *314*, 609.
110. R.A. Kerr, *Science*, 1995, *270*, 1565; 1995, *268*, 802.
111. (a) Anon., *Chem. Eng. News*, 1996, Apr. 22, 29; (b) K.P. Shine and W.T. Sturges, *Science*, 2007, *315*, 1804.
112. (a) K.S. Betts, *Environ. Sci. Technol.*, 1998, *32*, 487A; (b) M. Maiss, C.A.M. Brenninkmeijer, *Environ. Sci. Technol.*, 1998, *32*, 3077.
113. (a) B. Hileman, *Chem. Eng. News*, 1997, May 5, 37; (b) P.E. Kauppi, *Science*, 1995, *270*, 1454.
114. (a) P.S. Zurer, *Chem. Eng. News*, 1995, Mar. 13, 27; (b) R.A. Feely, C.L. Sabine, K. Lee, W. Barelson, J. Kleypass, V.J. Fabry, and F.J. Millero, *Science*, 2004, *305*, 362.
115. J. Uppenbrink, *Science*, 1996, *272*, 1122.
116. Anon., *Science*, 1990, *248*, 1308.
117. Anon., *Chem. Eng. News*, 1997, Jan. 13. www.scidev.net/en/news/china-is-second-biggest-greenhouse-gas-emitter.html
118. Anon., *Chem. Ind.* (*Lond.*), July 9, 2007, 5.
119. R.E. Gullison, P.C. Frumhoff, J.G. Canadell, C.B. Field, D.C. Nepsltad, K. Hayhoe, R. Avissar, L.M. Curran, P. Friedlingstein, C.D. Jones, and C. Nobre, *Science*, 2007, *316*, 985.
120. J. Kaiser, *Science*, 1997, *278*, 217.
121. Anon., *Chem. Ind.* (*Lond.*), 1995, 1001.
122. (a) D.K. Hill, *Science*, 1995, *267*, 1911; (b) S.H. Schneider and T.L. Root, eds, *Wildlife Responses to Climate Change—North American Case Studies*, Island Press, Washington, DC, 2001.

123. (a) V. Loeb, V. Siegel, O. Holn-Hansen, R. Hewitt, W. Fraser, W. Trivelpiece, and S. Trivelpiece, *Nature*, 1997, *387*, 897; (b) *Chem. Eng. News*, *1997*, July 7, 37.
124. (a) H.Q.P. Crick, C. Dudley, D.E. Glue, and D.L. Thomson, *Nature*, 1997, *388*, 526; (b) K.M. Reese, *Chem. Eng. News*, 1997, Sept. 8, 64.
125. B. Hileman, *Chem. Eng. News*, 1995, Nov. 27, 18.
126. R.L. Hotz, *Wilmington Delaware News J.*, 1996, Aug. 29, A13.
127. R. Petkewich, *Chem. Eng. News*, Feb. 23, 2009, 56.
128. (a) R.A. Kerr, *Science*, 1998, *281*, 156; (b) W.S. Broecker, *Science*, 1997, *278*, 1582.
129. (a) R.A. Kerr, *Science*, 2007, *318*, 1859; (b) R.S. Bradley, M. Vuille, H.F. Diaz, and W. Vergara, *Science*, 2006, *312*, 1755; (c) R.F. Service, *Science*, 2004, *303*, 1124.
130. (a) J.H. Lawton, *Science*, 1995, *269*, 328; (b) E. Culotta, *Science*, 1995, *268*, 654; (c) A. Raschi, P. Van Gardingen, F. Miglietta, and R. Tognetti, eds, *Plant Responses to Elevated CO_2—Evidence from Natural Springs*, Cambridge University Press, New York, 1997; (d) E.H. De Luca, J.G. Hamilton, S.L. Naidu, R.B. Thomas, J.A. Andrews, A. Finzi, M. Lavine, R. Matamala, J.E. Mohan, G.R. Hendrey, and W.H. Schlesinger, *Science*, 1999, *248*, 1177.
131. (a) J. Cebrian, C.M. Duarte, *Science*, 1995, 268, 1606; (b) Anon., *Environment*, 1999, *41*(3), 21.
132. T.H. Jones, L.J. Thompson, J.H. Lawton, T.M. Bezemer, R.D. Bardgett, T.M. Blackburn, K.D. Bruce, P.F. Cannon, G.S. Hall, S.E. Hartley, G. Howson, C.G. Jones, C. Kampichler, E. Kandeler, and D.A. Ritchie, *Science*, 1998, *280*, 441.
133. B. Hileman, *Chem. Eng. News*, 1999, Aug. 9, 16.
134. (a) B. Hileman, *Chem. Eng. News*, 1996, Jan. 22, 9; 1995, Oct. 2, 19; (b) P. Zurer, *Chem. Eng. News*, 1996, July 15, 12; (c) R. Stone, *Science*, 1995, *267*, 957; (d) G.C. Daily and P.R. Ehrlich, *Annu. Rev. Energy Environ.*, 1996, *21*, 125; (e) P.R. Epstein, *Science*, 1999, *285*, 347; (f) K.J. Linthicum, A. Anyamba, C.J. Tucker, P.W. Kelley, M.F. Myers, and C.J. Peters, *Science*, 1999, *285*, 397; (g) C.D. Harwell, K. Kim, J.M. Burkholder, R.R. Colwell, P.R. Epstein, D.J. Grimes, E.E. Hofmann, E.K. Lipp, A.D.M.E. Osterhaus, R.M. Overstreet, J.W. Porter, G.W. Smith, and G.R. Vasta, *Science*, 1999, *285*, 1505; (h) C.D. Harvell, C.E. Mitchell, J.R. Ward, S. Altizer, A.P. Dobson, R.S. Ostfeld, and M.D. Samuel, *Science*, 2002, *296*, 2158.
135. (a) A.S. Moffat, *Science*, 1997, *277*, 315; (b) D.H. Janzen, *Science*, 1997, *277*, 883.
136. (a) M.M. Halmann, *Chemical Fixation of Carbon Dioxide—Methods for Recycling Carbon Dioxide into Useful Products*, CRC Press, Boca Raton, FL, 1994; (b) M.M. Halmann and M. Steinberg, *Greenhouse Gas Carbon Dioxide Mitigation: Science and Technology*, Lewis Publishers, Boca Raton, FL, 1999; (c) K.M.K. Yu, I. Curcic, J. Gabriel, and S.C.E. Tsang, *ChemSusChem*, 2008, *1*, 893; (d) S.K. Ritter, *Chem. Eng. News*, Apr. 30, 2007, 11.
137. (a) S. Wilkinson, *Chem. Eng. News*, 1997, Oct. 13, 6; (b) E.E. Benson, C.P. Kubiak, A.J. Sathrum, and J.M. Smieja, *Chem. Soc. Rev.*, 2009, *38*, 89.
138. B. Hileman, *Chem. Eng. News*, 1997, June 30, 30.
139. D. Kreuze, *Technol. Rev.*, 1998, *101*(3), 30.
140. (a) J. Johnson, *Chem. Eng. News*, Apr. 2, 2007, 48; Oct. 29, 25; (b) B. Hileman, *Chem. Eng. News*, Sept. 24, *207*, 74; (c) P. Webster, *Science*, 2005, *309*, 2145; (d) D. Normile, ed., *Science*, 2009, *325*, 1642–1658; (e) P. Barbaro and C. Bianchini, *Catalysis for Sustainable Energy Production*, Wiley-VCH, Weinheim, 2009.
141. (a) P.G. Brewer, G. Friederich, E.T. Peltzer, and F.M. Orr, Jr., *Science*, 1999, *284*, 943; (b) A. Yamasaki and H. Teng, Preprints Papers Natl. Meet. *ACS Div. Environ. Chem.* 1998, *38*(1), 22; (c) R. Ohmura and Y.H. Mori, *Environ. Sci. Technol.*, 1998, *32*, 1120.
142. (a) J.A. Turner, *Science*, 1999, *285*, 687; (b) M.A. Brown, M.D. Levine, J.P. Romm, A.H. Rosenfeld, and J.G. Koomey, *Annu. Rev. Energy Environ.*, 1998, *23*, 287; (c) Anon., *Chem. Eng. News*, 1999, Aug. 16, 237; (d) D.J.C. MacKay, *Sustainable Energy Without the Hot Air*, U.I.T. Cambridge, Cambridge, UK, 2009.
143. (a) C. Flavin, *World Watch*, 1997, *10*(1), 10; (b) B. Hileman, *Chem. Eng. News*, 1997, Apr. 14, 28.
144. Anon., *Science*, 1997, *278*, 23.
145. (a) B. Hileman, *Chem. Eng. News*, 1997, Dec. 15, 9; 1997, Dec. 22, 20; (b) B. Bolin, *Science*, 1998, *279*, 330.
146. D. Malakoff, *Science*, 1997, *278*, 2048.
147. B. Hileman, *Chem. Eng. News*, 1998, Nov. 16, 26; 1998, Nov. 23, 10.
148. (a) J.D. Mahlman, *Annu. Rev. Energy Environ.*, 1998, *23*, 83; (b) T.M. Parris, *Environment*, 1998, *40*(9), 3; (c) R.J. Sutherland, *Science*, 1998, *281*, 647; (d) J. Romm, M. Levine, M. Brown, and E. Petersen, *Science*, 1998, *281*, 648.
149. L.R. Raymond, *Exxon Corp. Annu. Rep.*, Houston, TX, 1996.
150. B. Hileman, *Chem. Eng. News*, 1996, Dec. 23, 5.
151. B. Hileman, *Chem. Eng. News*, 1997, June 9, 24.
152. (a) B. Hileman, *Chem. Eng. News*, 1998, Nov. 23, 10; (b) D. Hanson, *Chem. Eng. News*, 1998, Nov. 2, 10; (c) Anon., *Chem. Ind. (Lond.)*, 1998, 867; (d) J. Romm, *Cool Companies: How the Best Business Boost Profits and Productivity by Cutting Greenhouse Emissions*, Island Press, Washington, DC, 1999.
153. (a) R. Repetto and D. Austin, *The Costs of Climate Protection: A Guide for the Perplexed*, World Resources Institute, Washington, DC, 1997; (b) J.J. Romm, *Cool Companies: How the Best Businesses Boost Profits and Productivity by Cutting Greenhouse Gas Emissions*, Island Press, Washington, DC, 1999.
154. R.A. Kerr, *Science*, 1995, *270*, 731.
155. Anon., *Chem. Eng. News*, 1999, June 21, 24.
156. (a) *Chem. Eng. News*, 1997, June 23, 23; (b) *Environ. Sci. Technol.*, 1997, *31*, 180A.
157. Anon., *Chem. Eng. News*, 1997, Apr. 28, 20.
158. B. Hileman, *Chem. Eng. News*, 1997, Oct. 6, 10.
159. Anon., *Chem. Eng. News*, 1997, June 16, 24.
160. Anon., *Economist*, Nov. 4, 2006, 14.
161. C. Cooper, *Chem. Eng.*, 1998, *105*(8), 46.
162. B. Hileman, *Chem. Eng. News*, 1997, July 21, 11.
163. B. Hileman, *Chem. Eng. News*, 1997, Feb. 17, 9.
164. Anon., *Chem. Eng. News*, Dec. 10, 2007, 27.
165. (a) M. Fergusson and I. Skinner, *Environment*, 1999, *41*(1), 24; (b) Transportation Research Board, National Research Council, *Toward a Sustainable Future: Addressing the Long-Term Effects of Motor Vehicle Transportation on Climate and Ecology*, National Academy of Sciences Press, Washington, DC, 1997.
166. B. Hileman, *Chem. Eng. News*, 1997, Jan. 15, 14.

167. (a) D. Malakoff, *Science*, 1999, *285*, 680; (b) P. Gwynne, *R&D (Cahners)*, 1998, *40*(40), 26; (c) Anon., *R&D (Cahners)*, 1998, *40*(10), 127.
168. (a) P. Layman, *Chem. Eng. News*, 1997, June 16, 22; 1995, Oct. 2, 17.
169. (a) D.G. Victor, J.C. House, and S. Joy, *Science*, 2005, *309*, 1820; (b) W.H. Schlesinger, *Science*, 2006, *314*, 1217; (c) M. Todd, *Chem. Ind.* (*Lond.*), Aug. 15, 20.
170. (a) B. Hileman, *Chem. Eng. News*, 1997, June 23, 26; (b) Anon., *World Watch*, 2006, *19*(1), 32.
171. B. Hileman, *Chem. Eng. News*, 1997, Sept. 15, 24; 1997, Sept. 22, 9.
172. B. Hileman, *Chem. Eng. News*, 1997, Sept. 22, 22.
173. R.M. Margolis and D.M. Kammen, *Science*, 1999, *285*, 690.
174. M. Voith, *Chem. Eng. News*, Nov. 17, 2008, 15; (b) E.L. Peterson, M. Beger, and Z.T. Richards, *Science*, 2008, *319*, 1759.
175. Anon., *Modern Plastics*, Dec. 2005, 32.
176. Organization for Economic Cooperation and Development, *Urban Energy Handbook—Good Local Practice*, Paris, 1995.
177. (a) A.H. Rosenfeld, J.J. Romm, H. Akbari, and A.C. Lloyd, *Technol. Rev.*, 1997, *100*(2), 52; (b) H. Akbari, S. Davis, S. Dorsano, J. Huang, and S. Winnett, eds, *Cooling Our Communities—A Guidebook on Tree Planting and Light-Colored Surfacing*, U.S. Environmental Protection Agency, PM-221, 22P-2001, GPO Document 055-000-00371-8, Jan. 1992; (c) G. Moll and S. Ebenreck, *Shading Our Cities: A Resource Guide for Urban and Community Forests*, Island Press, Washington, DC, 1989; (d) Green Living Staff, A.S. Moffat, and M. Schiler, *Energy-Efficient and Environmental Landscaping*, Appropriate Solutions Press, South Newfane, VT, 1994.
178. A. Meier, ed., *Energy Buildings*, 1997, *25*, 99–177.
179. S.E. Bretz and H. Akbari, *Energy Buildings*, 1997, *25*, 159.
180. A. Heller, *Acc. Chem. Res.*, 1995, *28*, 503.
181. (a) Anon., *Environment*, 2000, *42*(10), 7; (b) N.B. Solomon, *Architect. Rec.*, 2003, *191*(3), 149; (c) B. Booth, *Environ. Sci. Technol.*, 2006, *40*, 4046; (d) S. Saiz, C. Kennedy, B. Bass, and K. Pressnail, *Environ. Sci. Technol.*, 2006, *40*, 4312; (e) V. Klinkenborg, *Nat. Geo. Mag.*, May 2009, 215(5), 84.
182. (a) J.D. Balcomb, *Passive Solar Buildings*, MIT Press, Cambridge, MA., 1992; (b) P. O'Sullivan, *Passive Solar Energy in Buildings*, Elsevier Applied Science, New York, 1988; (c) C. Carter and J. de Villiere, *Principles of Passive Solar Building Design*, Pergamon, New York, 1987; (d) A. Bisio and S.R. Boots, *Energy Technology and the Environment*, Wiley, New York, 1995, 468; (e) Union of Concerned Scientists, *Put Renewable Energy to Work in Buildings*, Cambridge, MA, Jan. 1993; (f) J. Nieminen, *Energy Buildings*, 1994, *21*, 187; (g) N. Lenssen and D.M. Roodman, *State of the World 1995*, Worldwatch Institute, W.W. Norton, New York, 1995, 95; (h) F. Sick and J. Leppanen, *Sol. Energy*, 1994, *53*, 379; (i) Steven Winter Associates, *The Passive Solar Design and Construction Handbook*, Wiley, New York, 1997; (j) www.ucsusa.org/clean_energy; (k) www.eere/energy.gov/RE/passive_passive.html; (l) www.nrdc.org/buildinggreen/strategies/energy.asp; (m) Clinton Foundation, "Energy Efficiency Building Retrofit Program," Fact sheet, May 2007, New York; (n) A.K. Athienitis and M. Santamouris, *Thermal analysis and Design of Passive Solar Buildings*, Stylus Publishing, Herndon, VA, 2002.
183. C. Filippin, A. Beascochea, A. Esteves, C. de Rosa, L. Cortegoso, and D. Estelrich, *Sol. Energy*, 1998, *63*, 105.
184. (a) J.J. Gelegenis, N.G. Koumoutsos, *Chem. Eng. Prog.*, 1996, *92*(2), 42; (b) G. Walker, *The Stirling Alternative: Power Systems, Refrigerants and Head Pumps*, Gordon & Breach, Langhorne, PA, 1994; (c) D. Macmichael, *Heat Pumps*, 2nd ed., Pergamon, Oxford, 1988.
185. A. Trombe, L. Serres, *Energy Buildings*, 1994, *21*, 155.
186. L. Xinguo, *Energy Convers. Manage.*, 1998, *39*, 295.
187. Anon., *Delmarva Power Energy News*, 1996, Sept. 1.
188. M.S. Hatten and W.B. Morrison, *ASHRAE J.*, 1995, *37*(7), 45.
189. A.K. Singh, G.N. Tiwari, N. Lugani, and H.P. Garg, *Energy Convers. Manage.*, 1996, *37*, 531.
190. (a) G. Rockendorf, S. Janssen, and H. Felten, *Sol. Energy*, 1996, *58*, 33; (b) H. Gunnewick, E. Brundrett, and K.G.T. Hollands, *Sol. Energy*, 1996, *58*, 227.
191. G. Oliveti and N. Arcuri, *Sol. Energy*, 1996, *57*, 345.
192. S.N.G. Lo and B. Norton, *Sol. Energy*, 1996, *56*, 143.
193. M. Ronnelid and B. Karlsson, *Sol. Energy*, 1996, *57*, 93.
194. Z.C. Orel and B. Orel, *Sol. Energy Mater. Sol. Cells*, 1995, *36*, 11.
195. Q.-C. Zhang, *Sol. Energy Mater. Sol. Cells*, 1998, *52*, 95.
196. (a) T. Tesfamichael, W.E. Vargas, E. Wackelgard, and G.A. Niklasson, *Sol. Energy Mater. Sol. Cells*, 1998, *55*, 251; (b) S. Suzer, F. Kadirgan, H.M. Sohmen, A.J. Wetherilt, and I.E. Ture, *Sol. Energy Mater. Sol. Cells*, 1998, *52*, 55.
197. P.J. Sebastian, J. Quintana, and F. Avila, *Sol. Energy Mater. Sol. Cells*, 1997, *45*, 57.
198. M. Koltun, G. Gukhman, and A. Gavrilina, *Sol. Energy Mater. Sol. Cells*, 1994, *33*, 41.
199. A.A. Mohamad, *Sol. Energy*, 1997, *60*, 71.
200. N.D. Kaushika and M. Arulanantham, *Sol. Energy Mater. Sol. Cells*, 1996, *44*, 383.
201. P. Kariukinyahoro, R.R. Johnson, and J. Edwards, *Sol. Energy*, 1997, *59*, 11.
202. A. Balzar, P. Stumpf, S. Eckhoff, H. Ackermann, and M. Grupp, *Sol. Energy*, 1996, *58*, 63.
203. (a) M. Reuss, *Sol. Energy*, 1997, *59*, 259; (b) R.J. Fuller and W.W.S. Charters, *Sol. Energy*, 1997, *59*, 151; (c) C. Palaniappan and S.V. Subramanian, *Sol. Energy*, 1998, *63*, 31.
204. R. Croy and F.A. Peuser, *Sol. Energy*, 1994, *53*(1), 47.
205. (a) N. Chege, *World Watch*, 1995, *8*(5), 8; (b) J.D. Burch and K.E. Thomas, *Sol. Energy*, 1998, *64*, 87.
206. (a) G. Mink, M.M. Aboabboud, and E. Karmazsin, *Sol. Energy*, 1998, *62*, 309; (b) www.solar-desalination.com/4.html.
207. G.S. Samdani, *Chem. Eng.*, 1995, *102*(3), 19.
208. California Energy Commission, Title 24, Part 6, 2005 standards, "California Energy Standards for Residential and Nonresidential Buildings," Sacramento, CA.
209. (a) M.A. Siddiqui, *Energy Convers. Manage.*, 1997, *38*, 889, 905; (b) A.T. Bulgan, *Energy Convers. Manage.*, 1997, *38*, 1431; (c) M. Niang, T. Cachot, and P. Le Goff, *Energy Convers. Manage.*, 1997, *38*, 1701; (d) M. Izquierido, F. Hernandez, and E. Martin, *Sol. Energy*, 1994, *53*, 431; (e) T.A. Ameel, K.G. Gee, and B.D. Wood, *Sol. Energy*, 1995, *54*, 65; (f) M. Hammad and Y. Zurigat, *Sol. Energy*, 1998, *62*, 79.

210. (a) Saghiruddin and M.A. Sissiqui, *Energy Convers. Manage.*, 1996, *37*, 421, 433; (b) R. Yang and P.-L. Yang, *Sol. Energy*, 1995, *54*, 13, 19; (c) T.A. Ameel, K.G. Gee, and B.D. Wood, *Sol. Energy*, 1995, *54*, 65.
211. (a) I. Dincer, M. Edin, and I.G. Ture, *Energy Convers. Manage.*, 1996, *37*, 51; (b) K. Das and A. Mani, *Energy Convers. Manage.*, 1996, *37*, 87.
212. O. St C. Headley, *Sol. Energy*, 1994, *53*, 191.
213. L. Zhenyan, L. Yunzhuang, and Z. Jiaxin, *Sol. Energy Sol. Cells*, 1998, *52*, 45.
214. S.O. Enibe and O.C. Iloeje, *Sol. Energy*, 1997, *60*, 77.
215. (a) A. Erhard and E. Hahne, *Sol. Energy*, 1997, *59*, 155; (b) N.K. Bansal, J. Blumenberg, H.J. Kavasch, and T. Roettinger, *Sol. Energy*, 1997, *61*, 127.
216. M. Nieman, J. Kreuzburg, K.R. Schreitmuller, and L. Leppers, *Sol. Energy*, 1997, *59*, 67.
217. T.A. Kattakayam and K. Srinivasan, *Sol. Energy*, 1996, *56*, 543.
218. H. Sofrata, *Energy Convers. Manage.*, 1996, *37*, 269.
219. A. Kraft, *Chem. Ind. (Lond.)*, Dec. 3, 2001, 771.
220. (a) F.J. DiSalvo, *Science*, 1999, *285*, 703; (b) T.M. Tritt, *Science*, 1999, *283*, 804; (c) W. Wang, X. Lu, T. Zhang, G. Zhang, W. Jiang, and X. Li, *J. Am. Chem. Soc.*, 2007, *129*, 6702; (d) B. Halford, *Chem. Eng. News*, Jan. 14, 2008, 12.
221. (a) M. Santamouris, C.A. Balaras, E. Dascalaki, and M. Vallindras, *Sol. Energy*, 1994, *53*, 411; (b) I.A. Abbud, G.O.G. Lof, and D.C. Hittle, *Sol. Energy*, 1995, *54*, 75.
222. (a) K. Morino and T. Oka, *Energy and Buildings*, 1994, *21*, 65; (b) M. Reuss, M. Beck, and J.P. Muller, *Sol. Energy*, 1997, *59*, 247; (c) D.S. Breger, J.E. Hubbell, H. El Hasnaqui, and J.E. Sunderland, *Sol. Energy*, 1996, *56*, 493.
223. (a) R. Kubler, N. Fisch, and E. Hahne, *Sol. Energy*, 1997, *61*, 97; (b) G. Oliveti and N. Arcuri, *Sol. Energy*, 1995, *54*, 85.
224. M.T. Kangas and P.D. Lund, *Sol. Energy*, 1994, *53*, 237.
225. G. Oliveti, N. Arcuri, and S. Ruffolo, *Sol. Energy*, 1998, *62*, 281.
226. (a) P. Brousseau and M. Lacroix, *Energy Convers. Manage.*, 1996, *37*, 599; (b) D.W. Hawes and D. Feldman, *Sol. Energy Mater. Sol. Cells*, 1992, *27*, 91, 103.
227. G.A. Lane, *Sol. Energy Mater. Sol. Cells*, 1992, *27*, 135.
228. H.W. Ryu, S.W. Woo, B.C. Shin, and S.D. Kim, *Sol. Energy Mater. Sol. Cells*, 1992, *27*, 161.
229. E.A. Levitsky, *Sol Energy Mater Sol Cells*, 1996, *44*, 219.
230. S. Scalat, D. Banu, D. Hawes, J. Paris, F. Haghighata, and D. Feldman, *Sol. Energy Mater. Sol. Cells*, 1996, *44*, 49.
231. H. Kaasinen, *Sol. Energy Mater. Sol. Cells*, 1992, *27*, 181.
232. D. Feldman, D. Banu, and D.W. Hawes, *Sol. Energy*, 1995, *36*, 147.
233. G. Ondrey, *Chem. Eng.*, 2005, *112*(9), 16.
234. Anon., *Chem. Eng. News*, May 7, 2007, 33.
235. (a) A. Brice, *ICIS Chem. Business*, Feb. 4–10, 2008, 24; (b) N. Eisberg, *Chem. Ind. (Lond.)*, Nov. 20, 2006, 20.
236. M. Barrio, J. Font, D.O. Lopez, J. Muntasell, J.L. Tamarit, *Sol. Energy Mater. Sol. Cells*, 1992, *27*, 127.
237. G. Parkinson, *Chem. Eng.*, 1996, *103*(3), 19.
238. I. Fujii, M. Ishino, S. Akiyama, M.S. Murthy, and K.S. Rajanandam, *Sol. Energy*, 1994, *53*, 329.
239. K. Kyaw, T. Shibata, F. Watanabe, and H. Matsuda, *Energy Convers. Manage.*, 1997, *38*, 1025.
240. (a) A.H. Rosenfeld and D. Hafemeister, *Sci. Am.*, 1988, Apr. 78; (b) J. Johnson, *Chem. Eng. News*, Dec. 3, 2007, 46.
241. N. Henbest, *New Sci.*, 1989, Feb. 11, 44.
242. G. Weaver, private communication.
243. Anon., *Chem. Eng. News*, Apr. 22, 2002, 56.
244. (a) Anon., *Environment*, 1994, *36*(10), 24; (b) *Chem. Ind. (Lond.)*, 1994, 844.
245. T. Justel, H. Nikol, and C. Ronda, *Angew. Chem. Int. Ed.*, 1998, *37*, 3084.
246. (a) M. Rouhi, *Chem. Eng. News*, 1999, Feb. 8, 37; (b) R.T. Wegh, H. Donker, K.D. Oskam, and A. Meijerink, *Science*, 1999, *283*, 663.
247. Cermax Lamp Engineering Guide ILC Technology, Sunnyvale, CA, 1998.
248. Anon., *R&D (Cahners)* 1998, *40*(4), 100.
249. (a) Anon., *R&D (Cahners)*, 1998, *40*(4), 7; (b) B. Gill, ed., *Group III Nitride Semiconductor Compounds: Physics and Applications*, Oxford University Press, New York, 1998; (c) D. Wilson, *Technol. Rev.*, 2000, *103*(3), 27; (d) Y.S.L. Chung, M.Y. Jeon, and C.K. Kim, *Ind. Eng. Chem. Res.*, 2009, 48, 740; (e) Anon., *Technol. Rev.*, 2009, 112(4), 16; (f) H.A. Hoppe, *Angew. Chem. Ind. Ed.*, 2009, 48, 3572.
250. (a) J. Walker, *R&D (Cahners)*, 2004, *46*(5), 56; (b) D. Talbot, *Technol. Rev.*, 2003, *106*(4), 30; (c) J. Johnson, *Chem. Eng. News*, Dec. 3, 2007, 46.
251. (a) J.D. Anderson, E.M. McDonald, P.A. Lee, M.L. Anderson, E.L. Ritchie, H.K. Hall, T. Hopkins, E.A. Mash, J. Wang, A. Padias, S. Thayumanavan, S. Barlow, S.R. Marder, G.E. Jabbour, S. Shaheen, B. Kippelen, N. Peyghambarian, R.M. Wightman, and N.R. Armstrong, *J. Am. Chem. Soc.*, 1998, *120*, 9646; (b) W.-L. Yu, J. Pei, Y. Cao, W. Huang, and A.J. Heeger, *Chem. Commun.*, 1999, 1837; (c) P. Livingstone, *R&D (Cahners)*, 2008, *50*(1), 43; (b) K. Mullen and U. Scherf, *Organic Light Emitting Devices: Synthesis, Properties and Applications*, Wiley-VCH, Weinheim, 2005; (d) E. Polikparov and M.E. Thompson, *Mater. Matters (Aldrich)*, Milwaukee, WI, 2007, *3*(2), 21.
252. (a) N. Lenssen and D.M. Roodman, *State of the World 1995*, World Watch Institute, W.W. Norton, New York, 1995, 104; (b) J. Wienold, ed., *Solar Energy*, 2002, *73*, 75–135 [issue on "daylighting"].
253. (a) P.D. Swift and G.B. Smith, *Sol. Energy Mater. Sol. Cells*, 1995, *36*, 159; (b) W.C. Chen, C.C. Chang, and N.H. Wang, *J. Appl. Polym. Sci.*, 1997, *66*, 2103.
254. (a) P. Schissel, G. Jorgensen, C. Kennedy, and R. Goggin, *Sol. Energy Mater. Sol. Cells*, 1994, *33*, 183; (b) G. Correa, R. Almanza, I. Martinea, M. Mazari, and J.C. Cheang, *Sol. Energy Mater. Sol. Cells*, 1998, *52*, 231.
255. T. Thwaites, *New Sci.*, 1995, Jan. 7, 20.
256. A. Viljoen, J. Dubiel, M. Wilson, and M. Fontoynont, *Sol. Energy*, 1997, *59*, 179.
257. (a) K. Frost, D. Arasteh, and J. Eto, *Savings from Energy Efficiency Windows: Current and Future Savings from New Fenestration Technologies in the Residential Market*, Lawrence Berkeley, Berkeley, CA, Apr., 1993; (b) M.D. Hutchins, ed., *Sol. Energy*, 1998, *62*, 145–245; (c) S. Koichi, Y. Nagashima, and S. Nakagaki, U.S. patent, 5,958,811 (1999).
258. (a) C.S. Rondestvedt, Jr., *Kirk–Othmer Encyclopedia of Chemical Technology*, 3rd ed., New York, 1983, *23*, 222; (b) B.H.W.S. de Jong, *Ullmann's Encyclopedia Industrial*

Chemistry, 5th ed., VCH, Weinheim, 1989, *A12*, 425; (c) H. Dislich. In: D.R. Uhlmann and N.J. Kreidl, *Glass Science and Technology*, vol. 2, Academic, San Diego, 1984; 252–271; (d) R. Kirsch, ed., *Metals in Glassmaking*, Elsevier, Amsterdam, 1993, 372–384; (e) H. Rawson, *Glasses and Their Applications*, Institute of Metals, London, 1991, 85–91.

259. S.M.A. Durrani, E.E. Khawaja, J. Shirokoff, M.A. Daous, G.D. Khatak, M.A. Salim, and M.S. Hussain, *Sol. Energy Mater. Sol. Cells*, 1996, *44*, 37.
260. X. Zhang, S. Yu, and M. Ma, *Sol. Energy Mater. Sol. Cells*, 1996, *44*, 279.
261. (a) A.M. Thayer, *Chem. Eng. News*, Nov. 12, 2007, 29; (b) B. Halford, *Chem. Eng. News*, Feb. 20, 2006, 34.
262. (a) A.V. Dotsenko, L.B. Glebov, and V.A. Tsekhomskii, *Physics and Chemistry of Photochromic Glasses*, CRC Press, Boca Raton, FL, 1997; (b) M.A. Sobhan, R.T. Kivaisi, B. Stjerna, and C.G. Graqvist, *Sol. Energy Mater. Sol. Cells*, 1996, *44*, 451; (c) M. Ire, ed., *Chem. Rev.*, 2000, *100*(5), 1683–1890.
263. W. Eck and H. Wilson, *Adv. Mater.*, 1995, *7*, 800.
264. R. Dagani, *Chem. Eng. News*, 1997, June 9, 26.
265. A. Beck, W. Korner, H. Scheller, J. Fricke, W.J. Platzer, and V. Wittwer, *Sol. Energy Mater. Sol. Cells*, 1995, *36*, 339.
266. (a) P.M.S. Monk, R.J. Mortimer, and D.R. Rosseinsky, *Electrochromism: Fundamentals and Applications*, VCH, Deerfield Beach, FL, 1995; (b) P.M.S. Monk, R.J. Mortimer, and D.R. Rosseinsky, *Chem. Br.*, 1995, *31*, 380; (c) M. Green, *Chem. Ind. (Lond.)*, 1996, 641; (d) M. Kitao, K. Izawa, and S. Yamada, *Sol. Energy Mater. Sol. Cells*, 1995, *39*, 115–203; (e) C.M. Lampert, *Sol. Energy Mater. Sol. Cells*, 1998, *52*, 207; (f) A. Georg, W. Graf, D. Schweiger, V. Wittwer, P. Nitz, and H.R. Wilson, *Sol. Energy*, 1998, *62*, 215; (g) R.J. Mortimer, *Chem. Soc. Rev.*, 1997, *26*, 147; (h) C.G. Granqvist, *Pure Appl. Chem.*, 2008, *80*, 2489.
267. (a) J. Jayalakshmi, H. Gomathi, and G.P. Rao, *Sol. Energy Mater. Sol. Cells*, 1997, *45*, 201; (b) I.D. Brotherston, Z. Cao, G. Thomas, P. Weglicki, and J.R. Owen, *Sol. Energy Mater. Sol. Cells*, 1995, *39*, 257; (c) E. Masetti, D. Dini, and F. Decker, *Sol. Energy Mater. Sol. Cells*, 1995, *39*, 301.
268. (a) S.A. Agnihotry, Rashmi, R. Ramchandran, and S. Chandra, *Sol. Energy Mater. Sol. Cells*, 1995, *36*, 289; (b) P.V. Kamat, *Chemtech*, 1995, *25*(6), 22.
269. D.V. Baxter, M.H. Chisolm, S. Doherty, and N.E. Gruhn, *Chem. Commun.*, 1996, 1129.
270. A. Azens and C. Granqvist, *Sol. Energy Mater. Sol. Cells*, 1996, *44*, 333.
271. H. Wang, M. Zhang, S. Yang, L. Zhao, and L. Ding, *Sol. Energy Mater. Sol. Cells*, 1996, *43*, 345.
272. Y. Yonghong, Z. Jiayu, G. Peifu, and T. Jinfa, *Sol. Energy Mater. Sol. Cells*, 1997, *46*, 349.
273. (a) C. Arbizzani, M. Mastragostino, and A. Zanelli, *Sol. Energy Mater. Sol. Cell*, 1995, *39*, 213; (b) S. Panero, B. Scrosati, M. Baret, B. Cecchini, and E. Masetti, *Sol. Energy Mater. Sol. Cells*, 1995, *39*, 239.
274. R. Cinnsealach, G. Boschloo, S.N. Rao, and D. Fitzmaurice, *Sol. Energy Mater. Sol. Cells*, 1998, *55*, 215.
275. N. Ozer, T. Barreto, T. Buyuklimanli, and C.M. Lampert, *Sol. Energy Mater. Sol. Cells*, 1995, *36*, 433.
276. (a) Y.Z. Xu, M.Q. Qiu, S.C. Qiu, J. Dai, G.J. Cao, H.H. He, and J.Y. Wang, *Sol. Energy Mater. Sol. Cells*, 1997, *45*, 105; (b) L.D. Kadam and P.S. Patil, *Sol. Energy Mater. Sol. Cells*, 2001, *69*, 361.
277. C.N. Polo da Fonseca, M.-A. De Paoli, and A. Gorenstein, *Sol. Energy Mater. Sol. Cells*, 1994, *33*, 73.
278. G. Ondrey, *Chem. Eng.*, 2004, *111*(9), 20.
279. (a) M. Andrei, A. Roggero, L. Marchese, and S. Passerini, *Polymer*, 1994, *35*, 3592; (b) M.-H. Cui, J.-S. Guo, H.-Q. Xie, Z.-H. Wu, and S.-C. Qui, *J. Appl. Polym. Sci.*, 1997, *65*, 1739; (c) H.Q. Xie, D. Xie, and Y. Liu, *J. Appl. Polym. Sci.*, 1998, *70*, 2417.
280. L. Su, J. Fang, and Z. Lu, *J. Appl. Polym. Sci.*, 1998, *70*, 1955.
281. S. Gopal, R. Ramchandran, and R.S.A. Agnihotry, *Sol. Energy Mater. Sol. Cells*, 1997, *45*, 17.
282. A. Pennisi and F. Simone, *Sol. Energy Mater. Sol. Cells*, 1995, *39*, 333.
283. Anon., *R&D (Cahners)*, 1998, *40*(7), 110.
284. (a) D.K. Benson and H.M. Branz, *Sol. Energy Mater. Sol. Cells*, 1995, *39*, 203; (b) P. Zurer, *Chem. Eng. News*, 1996, Oct. 21, 10.
285. R.E. Collins and T.M. Simko, *Sol. Energy*, 1998, *62*, 189.
286. (a) J.D. Garrison and R.E. Collins, *Sol. Energy*, 1995, *55*(3), 151; (b) M. Lenzen and R.E. Collins, *Sol. Energy*, 1997, *61*, 11.
287. (a) M. Saito and M. Shukuya, *Sol. Energy*, 1996, *58*, 247; (b) A.G. Lien, A.G. Hestnes, and O. Aschehoug, *Sol. Energy*, 1997, *59*, 27.
288. (a) Technical Insights, Inc. (Englewood/Fort Lee, NJ), *Aerogels and Xerogels: Growth and Opportunities for the Early 21st Century*, Wiley, New York, 1996; (b) L.C. Klein, ed., *Sol–Gel Technology*, Noyes, Park Ridge, NJ, 1988, 226; (c) Y.A. Attia, *Sol–Gel Processing and Applications*, Plenum, New York, 1994; (d) M. Schneider and A. Baiker, *Catal. Rev. Sci. Eng.*, 1995, *37*, 515; (e) Y.-Y. Wang, Y.-B. Gao, Y.-H. Sun, and S.-Y. Chen, *Catal. Today*, 1996, *30*, 171; (f) L.L. Hench and R. Orefice, *Kirk–Othmer Encyclopedia of Chemical Technology*, 4th ed., New York, 1997, *22*, 497; (g) N. Husing and U. Schubert, *Angew. Chem. Int. Ed.*, 1998, *37*, 22; (h) A.C. Pierre and G.M. Pajonk, *Chem. Rev.*, 2002, *102*, 4243–4254.
289. G.S. Samdani, *Chem. Eng.*, 1995; *102*(6), 23.
290. (a) G. Ondrey, *Chem. Eng.*, 1997, *104*(7), 21; (b) J.H. Krieger and M. Freemantle, *Chem. Eng. News*, 1997, July 7, 16.
291. Anon., *Chem. Eng. News*, 1998, Aug. 10.
292. U. Schubert, *J. Chem. Soc. Dalton Trans.*, 1996, 3347.
293. Anon., *Chem. Eng. News*, Sept. 16, 2002, 8.
294. D.L. Illman, *Chem. Eng. News*, 1995, Feb. 13, 64.
295. J.H.G. Steinke, I.R. Dunkin, and D.C. Sherrington, *Macromolecules*, 1996, *29*, 5826.
296. B. Peuportier and J. Michel, *Sol. Energy*, 1995, *54*(1), 13.
297. (a) A. Johansson, *Clean Technology*, Lewis Publishers, Boca Raton, FL, 1992, 74–94; (b) T.B. Johansson, H. Kelly, A.K.N. Reddy, and R.H. Williams, eds, *Renewable Energy—Sources for Fuels and Electricity*, Island Press, Washington, DC, 1993; (c) E.A. Torrero, *Kirk–Othmer Encyclopedia of Chemical Technology*, 4th ed., New York, 1997, *21*, 218–236; (d) J.A.G. Drake, ed., *Electrochemistry and Clean Energy*, Royal Society of Chemistry, Cambridge, 1994; (e) K.S. Brown, *Science*, 1999, *285*, 678; (f) V. Comello, *R&D (Cahners)*, 1998, *40*(5), 22; (g) L. Freris and D. Infield, *Renewable Energy in Power Systems*, Wiley, New York, 2008; (h) EUREC Agency, *The Future of Renewable Energy 2, Prospects and Directions*,

Stylus Publishing, Herndon, VA, 2002; (i) D. Assmann, U. Laumanns, and D. Uh, eds, *Renewable Energy—A Global Review of Technologies, Policies and Markets*, Stylus Publishing, Herndon, VA, 2006.
298. (a) F.S. Sterrett, ed., *Alternate Fuels and the Environment*, Lewis Publishers, Boca Raton, FL, 1994; (b) W.M. Kreucher, *Chem. Ind. (Lond.)*, 1995, 601; (c) D.C. Carslaw and N. Fricker, *Chem. Ind. (Lond.)*, 1995, 593; (d) Union of Concerned Scientists, *Alternative Transportation Fuels*, Cambridge, MA, Dec. 1991.
299. (a) D. Sperling, *Chem. Ind. (Lond.)*, 1995, 609; (b) J.J. MacKenzie, *The Keys to the Car: Electric and Hydrogen Vehicles for the 21st Century*, World Resources Institute, Washington, DC, 1994.
300. L.B. Lave, C.T. Handrickson, and F.C. McMichael, *Science*, 1995, *268*, 993.
301. K. Sasaki, M. Yokota, H. Nagayoshi, and K. Kamisako, *Sol. Energy Mater. Sol. Cells*, 1997, *47*, 259.
302. (a) A.M. Rouhi, *Chem. Eng. News*, 1995, May 29, 37; (b) S. Samdani, *Chem. Eng.*, 1995, *102*(4), 17; (c) R.W. Wegman, *J. Chem. Soc. Chem. Commun.*, 1994, 947; (d) Q. Ge, Y. Huang, F. Qiu, and S. Li, *Appl. Catal. A*, 1998, *167*, 23; (e) G. Parkinson, *Chem. Eng.*, 1999, *106*(10), 19.
303. H. Kurakata, Y. Izumi, and K.-I. Aika, *Chem. Commun.*, 1996, 389.
304. (a) P.H. Abelson, *Science*, 1995, *268*, 955; (b) M.E. Himmel, J.O. Baker, and R.P. Overend, eds, *Enzymatic Conversion of Biomass Fuels Production*, A.C.S. Symp. 566, Washington, DC, 1994; (c) B.C. Saha and J. Woodward, *Fuels and Chemicals from Biomass*, A.C.S. Symp. 666, Washington, DC, 1997.
305. S. Moore, *Wilmington Delaware News J.*, 1997, July 12, A7.
306. (a) I. Lee, I.A. Johnson, and E.G. Hammond, *J. Am. Oil Chem. Soc.*, 1996, *73*, 631; (b) T. Krawczyk, *Int. News Fats, Oils Relat. Mater.*, 1996, *7*, 800–815; (c) R. Varese and M. Varese, *Int. News Fats, Oils Relat. Mater.*, 1996, *7*, 816–824; (d) P. Bondioli, A. Gasparoli, A. Lanzani, E. Fedeli, S. Veronese, and M. Sala, *J. Am. Oil Chem. Soc.*, 1995, *72*, 699; (e) H.S. Sii, H. Masjuki, and A.M. Zaki, *J. Am. Oil Chem. Soc.*, 1995, *72*, 905; (f) R.O. Dunn, M.W. Shockley, and M.O. Bagby, *J. Am. Oil Chem Soc.*, 1996, *73*, 1719; (g) L.C. McGraw, *Agric. Res.*, 1998, *46*(4), 21; (h) R. Cascone, R. Sierra, A. Smith, C. Granda, M.T. Holtzapple, J.C. Liao, and W. Higashide, *Chem. Eng. Prog.*, 2008, *104*(8), S2–S23; (i) A. Swasamy, K.Y. Cheah, P. Fornasiero, F. Kemasuor, S. Zinoviev, and S. Miertus, *ChemSumChem*, 2009, 2, 278–300; (j) R. Jothiramalingam, and M.K. Wang, *Ind. Eng. Chem. Res*, 2009, 48, 6162.
307. D.Y.Z. Chang, J.H. Van Gerpen, I. Lee, L.A. Johnson, E.G. Hammond, and S.J. Marley, *J. Am. Oil Chem. Soc.*, 1996, *73*, 1549.
308. K. Lal, Preprint papers, *Natl. Meet. ACS Div. Environ. Chem.*, 1994, *34*(2), 474.
309. (a) D.O. Hall and J.I. House, *Sol. Energy Mater. Sol. Cells*, 1995, *37*, 521; (b) P. Sampat, *World Watch*, 1995, *8*(6), 21.
310. P. Borjesson and B. Mattiasson, *Trends Biotechnol.*, 2008, *26*(1), 7.
311. (a) Anon., *Environ. Sci. Technol.*, 1965, *29*, 67A; (b) *Environment*, 1998, *40*(1), 22.
312. C.A. Grinder, *Waste News*, 1999, *5*(12), 1.
313. M. Murray, *Wilmington Delaware News J.*, 1995 July 12, B1.
314. (a) K. Fouhy and S. Shelley, *Chem. Eng.*, 1997, *104*(5), 55; (b) C. He, D.J. Herman, R.G. Minet, and T.-T. Tsotsis, *Ind. Eng. Chem. Res.*, 1997, *36*, 4100; (c) B. Eklund, E.P. Anderson, B.L. Walker, and D.B. Burrows, *Environ. Sci. Technol.*, 1998, *32*, 2233.
315. (a) A. Kendall, A. McDonald, and A. Williams, *Chem. Ind. (Lond.)*, 1997, 342; (b) D.P.L. Murphy, ed., *Energy from Crops*, Semundo, Cambridge, 1996; (c) V.R. Tolbert and A. Schiller, *Am. J. Alt. Agric.*, 1996, *11*, 148–149; (d) W. Patterson, *Power from Plants: The Global Implications of New Technologies for Electricity from Biomass*, Earthscan, London, 1994; (e) M.R. Khan, ed., *Clean Energy from Waste and Coal*, A.C.S. Symp. 515, Washington, DC, 1992; (f) T.B. Johansson, H. Kelly, A.K.N. Reddy, and R.H. Williams, eds, *Renewable Energy: Sources for Fuels and Electricity*, Island Press, Washington, DC, 1993, *38*, 56; (g) A. Faaij, *Energy from Biomass and Waste*, Utrecht University, Utrecht, 1997; (h) A.V. Bridgwater and D.G.B.L. Boocock, eds, *Developments in Thermochemical Biomass Conversion*, Chapman & Hall, London, 1997; (i) T.E. Bull, *Science*, 1999, *285*, 1209; (j) J.A. Turner, *Science*, 1999, *285*, 1209.
316. Anon., *Chem. Eng. News*, 1999, Aug. 23, 30.
317. D.M. Kammen, *Environment*, 1999, *41*(5), 10.
318. Anon., *Environment*, 1999, *41*(1), 20.
319. F. Pearce, *New Sci.*, 1995, Jan. 14, 12.
320. (a) G.S. Samdani, *Chem. Eng.*, 1994, *101*(11), 19; (b) G. Parkinson, *Chem. Eng.*, 1997, *104*(10), 23.
321. (a) A.V. Bridgwater, *Appl. Catal.*, 1994, *116*, 5; (b) C. Zhao, Y. Kou, A.A. Lemonidou, X. Li, and J.A. Lercher, *Angew. Chem. Int. Ed.*, 2009, *48*, 3987.
322. (a) J.M. Ogden and R.H. Williams, *Solar Hydrogen—Moving Beyond Fossil Fuels*, World Resources Institute, Washington, DC, 1989; (b) J.C. Cannon, Harnessing Hydrogen: The Key to Sustainable Transportation, *Int. News Fats, Oils Relat. Mater.*, New York, 1995; (c) T.B. Johansson, H. Kelley, A.K.N. Reddy, and R.H. Williams, eds, *Renewable Energy: Sources for Fuels and Electricity*, Island Press, Washington, DC, 1993, 925; (d) D.A.J. Rand and R.M. Dell, Hydrogen economy: Challenges and prospects, *Royal Soc. Chem. Publishing*, Cambridge, UK, 2007.
323. W. Lepkowski, *Chem. Eng. News*, 1995, Mar. 6, 22; 1996, Oct. 21, 31.
324. (a) K.V. Kordesch and G.R. Simader, *Chem. Rev.*, 1995, *95*, 191; (b) J.D. Holladay, J. Hu, D.L. King, and Y. Wang, *Catal. Today*, 2009, *139*, 244.
325. V.N. Parmon, *Catal. Today*, 1997, *35*, 153.
326. D. Wang, S. Czernik, D. Montane, M. Mann, and E. Chornet, *Ind. Eng. Chem. Res.*, 1997, *36*, 1507.
327. (a) X. Xu, Y. Matsumura, J. Stenberg, and M.J. Antal, Jr., *Ind. Eng. Chem. Res.*, 1996, *35*, 2522; (b) X. Xu and M.J. Antal, Jr., *Environ. Prog.*, 1998, *17*, 215.
328. T. Minowa, T. Ogi, and S.-Y. Yokoyama, *Chem. Lett.*, 1995, *285*, 937.
329. (a) S.E. Nilsen and K. Andreassen, *Int. News Fats, Oils Relat. Mater.*, 1996, *7*, 1120; (b) B.E. Logan, *Environ. Sci. Technol.*, 2004, *38*, 160A.
330. (a) T. Takata, A. Tanaka, M. Hara, J.N. Kondo, and K. Domen, *Catal. Today*, 1998, *44*, 17; (b) J.R. Durrant, *Chem. Ind.*

(*Lond.*), 1998, 838; (c) J. Jacoby, *Chem. Eng. News.*, August 10, 2009, 7.
331. (a) A. Heller, *Acc. Chem. Res.*, 1995, *28*, 503; (b) J.R. Bolton, *Sol. Energy*, 1996, *57*, 37; (c) Anon., *Chem. Eng. News*, Oct. 2, 2000, 45.
332. (a) O. Khaselev and J.A. Turner, *Science*, 1998, *280*, 425; (b) S.S. Kocha, D. Montgomery, M.W. Peterson, and J.A. Turner, *Sol. Energy Mater. Sol. Cells*, 1998, *52*, 389.
333. (a) K. Sayama, H. Arakawa, and K. Domen, *Catal. Today*, 1996, *28*, 175; (b) J. Yoshimura, A. Tanaka, J.N. Kondo, and K. Domen, *Bull. Chem. Soc. Jpn.*, 1995, *68*, 2439; (c) M. Freemantle, *Chem. Eng. News*, 1999, June 28, 10; (d) H.G. Kim, D.W. Hwang, J. Kim, Y.G. Kim, and J.S. Lee, *Chem. Commun.*, 1999, 1077; (e) W. Shangguan, K. Inoue, and A. Yoshida, *Chem. Commun.*, 1998, 779; (f) D.W. Hwang, H.G. Kim, J. Kim, K.Y. Cho, Y.G. Kim, and J.S. Lee, *J. Catal.*, 2000, *193*, 40.
334. K.-H. Chung and D.-C. Park, *Catal. Today*, 1996, *30*, 157.
335. (a) S. Ikeda, T. Takata, T. Kondo, G. Hitoki, M. Hara, J.N. Kondo, K. Domen, H. Hosono, H. Kawazoe, and A. Tanaka, *Chem. Commun.*, 1998, 2185; (b) M. Hara, T. Kondo, M. Komoda, S. Ikeda, K. Shinohara, A. Tanaka, J.N. Kondo, and K. Domen, *Chem. Commun.*, 1998, 357.
336. S. Ichikawa and R. Doi, *Catal. Today*, 1996, *27*, 271.
337. K. Tennakone, G.R.R.A. Kumara, I.R.M. Kottegoda, and V.P.S. Perera, *Chem. Commun.*, 1999, 15.
338. G. Parkinson, *Chem. Eng.*, 1998, *105*(10), 21.
339. (a) J.C. Fontecilla-Camps, A. Volbeda, and M. Frey, *Trends Biotechnol.* 1996, *14*(11), 417; (b) J. Benemann, *Nat. Biotechnol.*, 1996, *14*, 1101; (c) A. Fiechter, ed., *Advances in Biochemical Engineering and Biotechnology, vol. 52: Microbial and Enzymatic Byproducts*, Springer, New York, 1995; (d) M.W.W. Adams and E.I. Stiefel, *Science*, 1998, *282*, 1842; (e) J.W. Peters, W.N. Lanzilotta, B.J. Lemon, and L.C. Seefeldt, *Science*, 1998, *282*, 1853.
340. S.A. Markov, M.J. Bazin, and D.O. Hall, *Enzyme Microb. Technol.*, 1995, *17*, 306.
341. (a) J. Kaiser, *Science*, 2000, *287*, 1581; (b) M.L. Ghirardi, L. Zhang, J.W. Lee, T. Flynn, M. Seibert, E. Greenbaum, and A. Melis, *Trends Biotechnol.*, 2000, *18*, 506; (c) M. Freemantle, *Chem. Eng. News*, July 22, 2002, 35.
342. (a) J. Woodward, S.M. Mattingly, M. Danson, D. Hough, N. Ward, and M. Adams, *Nat. Biotechnol.*, 1966, *14*, 872; (b) J. Haggin, *Chem. Eng. News*, 1996, July 8, 5; (c) Anon., *Chem. Ind.* (*Lond.*), 1996, 525.
343. H. Yokoi, T. Tokushige, J. Hirose, S. Hayashi, and Y. Takasaki, *Biotechnol. Lett.*, 1998, *20*, 143.
344. J. Haggin, *Chem. Eng. News*, 1996, Aug. 5, 26.
345. (a) J. Graetz, *Chem. Soc. Rev.*, 2009, *38*, 73; (b) A.W.C. van den Berg and C.O. Arean, *Chem. Commun.*, 2008, 668; (c) M. Jacoby, *Chem. Eng. News*, Jan. 28, 2008, 67; (d) U. Eberle, M. Felderhoff, and F. Schuth, *Angew. Chem. Int. Ed.*, 2009, 48, 6608.
346. C. Read, G. Thomas, G. Ordaz, and S. Satyapal, *Mater. Matters* (*Aldrich*), Milwaukee, WI, 2007, *2*(2), 3.
347. (a) A. Karkambar, C. Ardahl, T. Autrey, and G.L. Soloveichik, *Mater. Matters* (*Aldrich*), Milwaukee, WI, 2007, *2*(2), 6, 11; (b) T.B. Marder, *Angew. Chem. Int. Ed.*, 2007, *46*, 8116.
348. (a) Y.H. Hu and E. Ruckenstein, *Ind. Eng. Chem. Res.*, 2008, *47*, 48; (b) J. Yang, A. Sudik, D.J. Siegel, D. Halliday, A. Drews, R.O. Carter, III, C. Wolverton, G.J. Lewis, J.W.A. Sachtler, J.J. Low, S.A. Faheem, D.A. Lesch, and V. Ozolins, *Angew. Chem. Int. Ed.*, 2008, *47*, 882.
349. (a) J.L.C. Rowsell and O.M. Yaghi, *Angew. Chem. Int. Ed.*, 2005, *44*, 4670; (b) M. Jacoby, *Chem. Eng. News*, May 19, 2003, 11.
350. L. Schlapbach and A. Zuttel, *Nature*, 2001, *414*, 353.
351. (a) E.D. Naeemi, D. Graham, and B.F. Norton, *Mater. Matters* (*Aldrich*), Milwaukee, WI, 2007, *2*(2), 23; (b) N. Kariya, A. Fukuoka, and M. Ichikawa, *Chem. Commun.*, 2003, 690; (c) P. Ferreira-Aparicio, I. Rodriguez-Ramos, and A. Guerrero-Ruoz, *Chem. Commun.*, 2002, 2082; (d) G. Parkinson, *Chem. Eng.*, 2003, *110*(3), 21.
352. (a) P. Mitchell, *Chem. Eng. News*, 1997, Sept. 22, 6; (b) K. Lovegrove and A. Luzzi, *Sol. Energy*, 1996, *56*, 361; (c) R.M. Steele, *Chemtech*, 1999, *29*(8), 28; (d) C.H. Christensen, T. Johannessen, R.Z. Sorenson, and J.K. Norskov, *Catal. Today*, 2006, *111*, 140.
353. (a) G. Ondrey, *Chem. Eng.*, 2008, *115*(2), 11; (b) K. Asazawa, K. Yamida, H. Tanaka, A. Oka, M. Taniguchi, and T. Kobayashi, *Angew. Chem. Int. Ed.*, 2007, *46*, 8024; (d) S.K. Singh, X-B. Zhang, and Q. Xu, *J. Am. Chem. Soc.*, 2009, *131*, 9894.
354. (a) G.S. Samdani, *Chem. Eng.*, 1994, *101*(12), 19; (b) W.-W. Xu, G.P. Rosini, M. Gupta, C.M. Jensen, W.C. Kaska, K. Krogh-Jespersen, and A.S. Goodman, *Chem. Commun.*, 1997, 2273; (d) S. Yolcular and O. Olgun, *Catal. Today*, 2008, *138*, 198.
355. Anon., *Chem. Ind.* (*Lond.*), 1994, 326.
356. (a) G. Ondrey, *Chem. Eng.*, 1999, *106*(2), 23; (b) G.A. Olah, *Angew. Chem. Int. Ed.*, 2005, *44*, 2636; (c) E. Willcocks, *Chem. Br.*, 2003, *39*(1), 27.
357. (a) G. Ondrey, *Chem. Eng.*, 2006, *113*(4), 24; (b) S. Uhm, H.J. Lee, Y. Kurn, and J. Lee, *Angew. Chem. Int. Ed.*, 2008, *47*, 10163; (c) Y. Park, S.-S. Kim, and O.H. Han, *Angew. Chem. Int. Ed.*, 2008, *47*, 94; (d) M. Ojeda and E. Iglesia, *Angew. Chem. Int. Ed.*, 2009, *48*, 4800.
358. M. Tada, R. Bal, X. Mu, R. Coquet, S. Namba, and Y. Iwasawa, *Chem. Commun.*, 2007, 4689.
359. Anon., *Environment*, 1999, *41*(1999), *41*(3), 21.
360. (a) W. Vielstich, H. Gasteiger, and A. Lamm, *Handbook of Fuel Cells—Fundamentals, Technology and Applications*, Wiley-VCH, Weinheim, 2003, *4*; (b) F. de Bruijn, *Green Chem.*, 2005, *7*(3), 132; (c) E. Dorey, *Chem. Ind.* (*Lond.*), Feb. 12, 2007, 10; (d) H.W. Cooper, *Chem. Eng. Prog.*, 2007, *103*(11), 14; (e) R. Datta, *Fuel Cell Principles, Components and Assemblies*, Wiley-Blackwell, Hoboken, NJ, 2009.
361. E.J. Hoffman, *Power Cycles and Energy Efficiency*, Academic, San Diego, 1996.
362. (a) M. Miller, *Hydrogen Fuel Cell Vehicles*, Union of Concerned Scientists, Cambridge, MA, Feb. 1995; (b) T. Newton, *Chem. Br.*, 1997, *33*(1), 29; (c) K. Fouhy and G. Ondrey, 1996, *103*(8), 46; (d) R. Edwards, *New Sci.*, 1996, *152*(2057), 40; (e) *Chem. Eng. News*, 1996, May 20, 38; (f) T.R. Ralph and G.A. Hards, *Chem. Ind.* (*Lond.*), 1998, 337; (g) V. Raman, *Chem. Ind.* (*Lond.*), 1997, 771; (h) J. Yamaguchi, *Automot. Eng.*, 1998, *106*(7), 54.
363. Anon., *Appl. Catal. B*, 1997, *12*, N14.
364. (a) Anon., *Chem. Eng. News*, 1997, Oct. 27; (b) B. Hileman, *Chem. Eng. News*, 1997, Apr. 21, 31; (c) Anon., *Chem. Eng. News*, 1997, Jan. 13, 20; (d) Anon., *Chem. Eng. News*, 1997,

Aug. 4, 16; (e) A. Hamnett, *Catal. Today*, 1997, *38*, 445; (f) C. Sloboda, *Chem. Eng.*, 1997, *104*(11), 49; (g) M. Jacoby, *Chem. Eng. News*, 1999, Aug. 16, 7; (h) Anon., *Chem. Eng. News*, 1999, Jan. 11, 22; (i) S. Velu, K. Suzuki, and T. Osaki, *Chem. Commun.*, 1999, 2341; (j) R. Corfield, *Chem. Ind. (Lond.)*, Aug. 21, 2006, 22.
365. (a) S. Park, J.M. Vohs, and R.J. Gorte, *Nature*, 2000, *404*, 265; (b) T. Hibino, A. Hashimoto, T. Inoue, J.-I. Tokumo, S.-I. Yoshida, and M. Sano, *Science*, 2000, *288*, 2031.
366. Y. Hishinuma and M. Kunikata, *Energy Convers. Manage.*, 1997, *38*, 1237.
367. (a) Anon., *Chem. Ind. (Lond.)*, 1996, 823; (b) D. Hart, *Chem. Ind. (Lond.)*, 1998, 344; (c) V. Thangadurai, A.K. Shukla, and J. Gopalakrishnan, *Chem. Commun.*, 1998, 2647; (d) D.J.L. Brett, A. Atkinson, N.P. Brandon, and S.J. Skinner, *Chem. Soc. Rev.*, 2008, *37*, 1568; (e) S. Park, J.M. Vohs, and R.J. Gorte, *Nature*, 2000, *404*, 265; (f) L. Yang, S. Wang, K. Blinn, M. Liu, Z. Liu, Z. Cheng, and M. Liu, *Science*, 2009, *326*, 126; (g) T. Suzuki, Z. Hasan, Y. Funahashi, T. Yamaguchi, Y. Fujishiro, and M. Awano, *Science*, 2009, *325*, 126.
368. K.R. Williams and G.T. Burstein, *Catal. Today*, 1997, *38*, 401.
369. M. Reisch, *Chem. Eng. News*, Oct. 22, 2001, 22.
370. (a) K. Rabaey and W. Verstraete, *Trends Biotechnol.*, 2005, *23*, 291; (b) M.J. Moehlenbrock and S.D. Minteer, *Chem. Soc. Rev.*, 2008, *37*, 1188; (c) B.E. Logan, ed., *Eviron. Sci. Technol.*, 2006, *40*, 5172–5218 [issue on microbial fuel cells]; (d) M. Rosenbaum, F. Zhao, U. Shroder, and F. Scholz, *Angew. Chem. Int. Ed.*, 2006, *45*, 6658; (e) X. Cao, X. Huang, P. Liang, K. Xiao, Y. Zhou, X. Zhang, and B.E. Logan, *Environ. Sci. Technol.*, 2009, *43*, 7148.
371. (a) J.J. Romm, *The Hype about Hydrogen: Fact and Fiction in the Race to Save the Planet*, Island Press, Washington, DC.
372. (a) D.A. Beattie, *History and Overview of Solar Heat Technologies*, MIT Press, Cambridge, MA, 1997; (b) S. Moran and J.T. McKinnon, *World Watch*, 2008, *21*(2), 26; (c) Anon., *Technol. Rev.*, 2007, *110*(5), S1; (d) R. Peltier, *Power*, 2007, *151*(12), 40; (e) E. Corcoran, *Forbes*, Sept. 3, 2007, 80–92; (f) D. Rotman, *Technol. Rev.*, 2009, *112*(4), 47; (g) G. Ondrey, *Chem. Eng.*, 2009, *116*(2), 11; (h) G. Johnson, *Nat. Geo. Mag.*, 2009, *215*(9), 28.
373. Q.-C. Zhang, Y. Yin, and D.R. Mills, *Sol. Energy Mater. Sol. Cells*, 1996, *40*, 43.
374. T. Eisenhammer, *Sol. Energy Mater. Sol. Cells*, 1997, *46*, 53.
375. Y. Yin, D.R. McKenzie, and W.D. McFall, *Sol. Energy Mater. Sol. Cells*, 1996, *44*, 69.
376. Q.-C. Zhang, K. Zhao, B.-C. Zhang, L.-F. Wang, Z.-L. Shen, Z.-J. Zhou, D.-L. Xie, and B.-F. Li, *Sol. Energy*, 1998, *64*, 109.
377. E. Barrera, I. Gonzalez, and T. Viveros, *Sol. Energy Mater. Sol. Cells*, 1998, *51*, 69.
378. Anon., *Science*, 1996, *271*, 1061.
379. V. Verlotski, M. Schaus, and M. Pohl, *Sol. Energy Mater. Sol. Cells*, 1997, *45*, 227.
380. S. Brutti, G. de Maria, G. Cerri, A. Giovannelli, B. Brunette, P. Cafarelli, E. Semprin, V. Barbarossa, and A. Ceroli, *Ind. Eng. Chem. Res.*, 2007, *46*, 6393.
381. M. Burke, *Environ. Sci. Technol.*, 2002, *36*, 403A.
382. A. Johansson, *Clean Technology*, Lewis Publishers, Boca Raton, FL, 1992, 146.
383. W.H. Avery and C. Wu, *Renewable Energy from the Ocean, A Guide to OTEC*, Oxford University Press, Oxford, 1994.
384. P.E. Glaser, F.P. Davidson, and K.I. Csigi, eds, *Solar Power Satellites—A Space Energy System for Earth*, Wiley, New York, 1998.
385. (a) D.Y. Coswami, ed., *Advances in Solar Energy*, Stylus Publishing, Herndon, VA, 2005 [an annual publication]; (b) H. Scheer, *The Solar Economy*, Stylus Publishing, Herndon, VA, 2004; (c) German Solar Energy Society, *Planning and Installing Photovoltaic Systems, A Guide for Installers, Architects and Engineers*, Stylus Publishing, Herndon, VA, 2005; (d) R. Eisenberg, ed., *Inorg. Chem.*, 2005, *44*, 6799–6911; (e) A. Luque and S. Hegedus, eds, *Handbook of Photovoltaic Science and Engineering*, Wiley, New York, 2003; (f) A.H. Tullo, *Chem. Eng. News*, Nov. 20, 2006, 25; (g) J. Johnson, *Chem. Eng. News*, Oct. 20, 2008, 40; (h) M. Behar, *On Earth (Natural Resources Defense Council)*, 2009, *31*(1), 26.
386. (a) Anon., *Chem. Eng. News*, 1997, July 7, 30; (b) J. Johnson, *Chem. Eng. News*, 1998, Mar. 30, 24.
387. P.D. Maycock, *P.V. News*, Sept.–Oct., 18.
388. (a) M. Rynn, *Amicus J. (Natural Resources Defense Council)*, 1999, *21*(2), 15; (b) C. Flavin and M. O'Meara, *World Watch*, 1998, *11*(5), 23; (c) J. Johnson, *Chem. Eng. News*, Oct. 9, 2000, 45; (d) R.F. Service, *Science*, 2008, *319*, 718.
389. (a) A.S. Bouazzi, M. Abaab, and B. Rezig, *Sol. Energy Mater. Sol. Cells*, 1997, *46*, 29; (b) J. Zhao, A. Wang, P. Altermat, S.R. Wenham, and M.A. Green, *Sol. Energy Mater. Sol. Cells*, 1996, *41/42*, 87.
390. W. Wettling, *Sol. Energy Mater. Sol. Cells*, 1995, *38*, 487, 494.
391. (a) A. Hubner, C. Hampe, and A.G. Aberle, *Sol. Energy Mater. Sol. Cells*, 1997, *46*, 67; (b) M. Ghannam, S. Sivoththaman, J. Poortmans, J. Szlufcik, J. Nijs, R. Mertens, and R. van Overstraeten, *Sol. Energy*, 1997, *59*, 101; (c) M. Pauli, T. Reindl, W. Kruhler, F. Holmberg, and J. Muller, *Sol. Energy Mater. Sol. Cells*, 1996, *41/42*, 119; (d) A.U. Ebong, C.B. Honsberg, and S.R. Wenham, *Sol. Energy Mater. Sol. Cells*, 1996, *44*, 271; (e) F.C. Marques, J. Urdanivia, and I. Chambouleyron, *Sol. Energy Mater. Sol. Cells*, 1998, *52*, 285.
392. P. Doshi, A. Rohatgi, M. Ropp, Z. Chen, D. Ruby, and D.L. Meier, *Sol. Energy Mater. Sol. Cells*, 1996, *41/42*, 31.
393. L.C. Klein, ed., *Sol–Gel Technology*, Noyes, Park Ridge, NJ, 1988, 80.
394. Y. Yazawa, T. Kitatani, J. Minemura, K. Tamura, K. Mochizuki, and T. Warabisako, *Sol. Energy Mater. Sol. Cells*, 1994, *35*, 39.
395. F.Z. Tepehan, F.E. Ghodsi, N. Ozer, and G.G. Tepehan, *Sol. Energy Mater. Sol. Cells*, 1997, *46*, 311.
396. E. Aperathitis, Z. Hatzopoulos, M. Androulidaki, V. Foukaraki, A. Kondilis, C.G. Scott, D. Sands, and P. Panayotatos, *Sol. Energy Mater. Sol. Cells*, 1997, *45*, 161.
397. (a) F. Zhu, T. Fuyuki, H. Matsunami, and J. Singh, *Sol. Energy Mater. Sol. Cells*, 1995, *39*, 1; (b) M. Martinez, J. Herrero, and M.T. Gutierrez, *Sol. Energy Mater. Sol. Cells*, 1997, *45*, 75.
398. Anon., *Chem. Eng. News*, Mar. 5, 2007, 52.
399. S. Walheim, E. Schaffer, J. Mlynek, and U. Steiner, *Science*, 1999, *283*, 520.

400. S. Winderbaum, O. Reinhold, and F. Yun, *Sol. Energy Mater. Sol. Cells*, 1997, *46*, 239.
401. (a) P. Faith, C. Borst, C. Zechner, E. Bucher, G. Willeki, and S. Narayana, *Sol. Energy Mater. Sol. Cells*, 1997, *48*, 229; (b) T. Machida, A. Miyazawa, Y. Yokosawa, H. Nakaya, S. Tanaka, T. Nunoi, H. Kumada, M. Murakami, and T. Tomita, *Sol. Energy Mater. Sol. Cells*, 1997, *48*, 243.
402. (a) F.J. Pern, *Sol. Energy Mater. Sol. Cells*, 1996, *41/42*, 587; (b) A.W. Czanderna and F.J. Pern, *Sol. Energy Mater. Sol. Cells*, 1996, *43*, 101; (c) P. Klemchuk, M. Ezrun, G. Lavigne, W. Holley, J. Galica, and S. Agro, *Polym. Degrad. Stabil.*, 1997, *55*, 347; (d) F.J. Pern, *Angew. Makromol. Chem.*, 1997, *252*, 195.
403. P. Walter, *Chem. Ind. (Lond.)*, Dec. 18, 2006, 11.
404. (a) A. Rohatgi and S. Narasimha, *Sol. Energy Mater. Sol. Cells*, 1997, *48*, 187; (b) J. Nijs, S. Sivoththaman, J. Szlufcik, K. De Clercq, F. Duerinckx, E. Van Kerschaever, R. Einhaus, J. Poortmans, T. Vermeulen, and R. Mertens, *Sol. Energy Mater. Sol. Cells*, 1997, *48*, 199; (c) Z. Yuwen, L. Zhongming, M. Chundong, H. Shaoqi, L. Zhiming, Y. Yuan, and C. Zhiyun, *Sol. Energy Mater. Sol. Cells*, 1997, *48*, 167; (d) A.V. Shah, R. Platz, and H. Keppner, *Sol. Energy Mater. Sol. Cells*, 1995, *38*, 501; (e) M. Reisch, *Chem. Eng. News*, July 30, 2007, 15.
405. G. Ondrey, *Chem. Eng.*, 2004, *111*(13), 15; 2007, *114*(8), 14.
406. (a) A.V. Shah, R. Platz, and H. Keppner, *Sol. Energy Mater. Sol. Cells*, 1995, *38*, 505; (b) E.A. Schiff, A. Matsuda, M. Hack, S.J. Wagner, and R. Schropp, eds, *Amorphous Silicon Technology: 1995, 1996, and 1997*, Materials Research Society, Pittsburgh, PA, 1995–1997; (c) C. Eberspacher, H.W. Schock, D.S. Ginley, T. Catalano, and T. Wada, eds, *Thin Films for Photovoltatic and Related Device Applications*, Materials Research Society, Pittsburgh, PA, 1996, *426*; (d) B. Jagannathan, W.A. Anderson, and J. Coleman, *Sol. Energy Mater. Sol. Cells*, 1997, *46*, 289.
407. R.F. Service, *Science*, 1998, *279*, 1300.
408. (a) F. Demichelis, G. Crovini, C.F. Pirri, E. Tresso, R. Galloni, R. Rizzoli, C. Summonte, and P. Rava, *Sol. Energy Mater. Sol. Cells*, 1995, *37*, 315; (b) H. Nishiwaki, K. Uchihashi, K. Takaoka, M. Nakagawa, H. Inoue, A. Takeoka, S. Tsuda, and M. Ohnishi, *Sol. Energy Mater. Sol. Cells*, 1995, *37*, 295; (c) C.R. Wronski, *Sol. Energy Mater. Sol. Cells*, 1996, *41/42*, 427; (d) Y. Hishikawa, K. Ninomiya, E. Maruyama, S. Kuroda, A. Terakawa, K. Sayama, H. Tarui, M. Sasaki, S. Tsuda, and S. Nakano, *Sol. Energy Mater. Sol. Cells*, 1996, *41/42*, 441; (e) B. Rech, C. Beneking, and H. Wagner, *Sol. Energy Mater. Sol. Cells*, 1996, *41/42*, 475; (f) M. Kubon, E. Boehmer, F. Siebke, B. Rech, C. Beneking, and H. Wagner, *Sol. Energy Mater. Sol. Cells*, 1996, *41/42*, 485; (g) M.S. Haque, H.A. Naseem, and W.D. Brown, *Sol. Energy Mater. Sol. Cells*, 1996, *41/42*, 543; (h) N. Gonzalez, J.J. Gandia, J. Carabe, and M.T. Gutierrez, *Sol. Energy Mater. Sol. Cells*, 1997, *45*, 175; (i) R. Martins, *Sol. Energy Mater. Sol. Cells*, 1997, *45*, 1.
409. A. Shah, P. Torres, R. Tscharner, N. Wyrsch, and H. Keppner, *Science*, 1999, *285*, 692.
410. K. Winz, C.M. Fortman, T. Eickhoff, C. Beneking, H. Wagner, H. Fujiwara, and I. Shimizu, *Sol. Energy Mater. Sol. Cells*, 1997, *49*, 195.
411. (a) T. Yamawaki, S. Mizukami, A. Yamazaki, and H. Takahashi, *Sol. Energy Mater. Sol. Cells*, 1997, *47*, 125; (b) M. Kondo, H. Nishio, S. Kurata, K. Hayashi, A. Takenaka, A. Ishikawa, K. Nishimura, H. Yamagishi, and Y. Tawada, *Sol. Energy Mater. Sol. Cells*, 1997, *49*, 1.
412. S. Guha, J. Yang, A. Banerjee, T. Glatfelter, and S. Sugiyama, *Sol. Energy Mater. Sol. Cells*, 1997, *48*, 365.
413. T. Yoshida, S. Fujikake, S. Kato, M. Tanda, K. Tabuchi, A. Takano, Y. Ichikawa, and H. Sakai, *Sol. Energy Mater. Sol. Cells*, 1997, *48*, 383.
414. (a) F. Roca, G. Sinno, G. DiFrancia, P. Prosini, G. Pascarella, and D. Della Sala, *Sol. Energy Mater. Sol. Cells*, 1997, *48*, 15; (b) G. Ballhorn, K.J. Weber, S. Armand, M.J. Stocks, and A.W. Blakers, *Sol. Energy Mater. Sol. Cells*, 1998, *52*, 61.
415. (a) A.M. Hermann, *Sol. Energy Mater. Sol. Cells*, 1998, *55*, 75; (b) O. Savadogo, *Sol. Energy Mater. Sol. Cells*, 1998, *52*, 361; (c) A. Shah, P. Torres, R. Tscharner, N. Wyrsch, and H. Keppner, *Science*, 1999, *285*, 692; (d) M. Pagliaro, R. Ciriminna, and G. Palmisano, *ChemSusChem*, 2008, *1*, 880; (e) A. Nieswand, *R&D (Cahners)*, 2006, *48*(12), 13; (f) J.S. Moore, *J. Am. Chem. Soc.*, 2008, *130*, 12201.
416. (a) C. Eberspacher, C.F. Gay, and P.D. Oskowitz, *Sol. Energy Mater. Sol. Cells*, 1996, *41/42*, 637; (b) S. Kumazawa, S. Shibutani, T. Nishio, T. Aramoto, H. Higuchi, T. Arita, A. Hanafusa, K. Omura, M. Murozono, and H. Takakawa, *Sol. Energy Mater. Sol. Cells*, 1997, *49*, 205.
417. (a) M.A. Contreras, J. Tuttle, A. Gabor, A. Tennant, K. Ramanathan, S. Asher, A. Franz, J. Keane, L. Wang, and R. Noufi, *Sol. Energy Mater. Sol. Cells*, 1996, *41/42*, 231; (b) A. Catalano, *Sol. Energy Mater. Sol. Cells*, 1996, *41/42*, 205; (c) C. O'Driscoll, *Chem. Br.*, 1998, *34*(4), 40; (d) Anon., *R&D (Cahners)*, 1999, *41*(5), 7; (e) A.M. Herman, R. Westfall, and R. coind, *Sol. Energy Mater. Sol. Cells*, 1998, *52*, 355; (f) A.M. Fernandez, *Sol. Energy Mater. Sol. Cells*, 1998, *52*, 423; (g) K. Ramanathan, R. Noufi, J. Granata, J. Webb, and J. Keane, *Sol. Energy Mater. Sol. Cells*, 1998, *55*, 15; (h) R.N. Bhattacharya, W. Batchelor, J.E. Granata, F. Hasoon, H. Wiesner, K. Ramanathan, J. Keane, and R.N. Noufi, *Sol. Energy Mater. Sol. Cells*, 1998, *55*, 83.
418. (a) A.M. Hermann, R. Westfall, and R. Wind, *Sol. Energy Matter. Sol. Cells*, 1998, *52*, 355; (b) A.M. Fenandez, M.E. Calixto, P.J. Sebastian, S.A. Gamboa, A.M. Hermann, and R.N. Noufi, *Sol. Energy Mater. Sol. Cells*, 1998, *52*, 423; (c) P.K.V. Pillai and K.P. Vijayakumar, *Sol. Energy Mater. Sol. Cells*, 1998, *51*, 47.
419. E. Ahmed, A. Zegadi, A.E. Hill, R.D. Pikington, R.D. Tomlinson, A.A. Dost, W. Ahmed, S. Leppavuori, J. Levoska, and O. Kusmartseva, *Sol. Energy Mater. Sol. Cells*, 1995, *36*, 227.
420. A.C. Rastogi and K.S. Balakrishman, *Sol. Energy*, 1995, *36*, 121.
421. R.F. Service, *Science*, 1996, *271*, 922.
422. (a) A.C. Jones, *Chem. Br.*, 1995, *31*(5), 389; *Chem. Soc. Rev.*, 1997, *26*, 101; (b) S.M. Gates, *Chem. Rev.*, 1996, *96*, 1519; (c) D.A. Atwood, V.O. Atwood, A.H. Cowley, R.A. Jones, J.L. Atwood, and S.G. Bott, *Inorg. Chem.*, 1994, *33*, 3251; (d) A. Adachi, A. Kudo, and T. Sakata, *Bull. Chem. Soc. Jpn.*, 1995, *68*, 3283.
423. M. Green and P. O'Brien, *Chem. Commun.*, 1998, 2459.
424. P.J. Sebastian, A.M. Fernandez, and A. Sanchez, *Sol. Energy Matter. Sol. Cells*, 1995, *39*, 55.
425. S. Sze, ed., *Modern Semiconductor Device Physics*, Wiley, New York, 1997.

426. K. Kawano, K. Arai, H. Yamada, N. Hashimoto, and R. Nakata, *Sol. Energy Mater. Sol. Cells*, 1997, *48*, 35.
427. K. Takahashi, S. Yamada, T. Unno, and S. Kuma, *Sol. Energy Mater. Sol. Cells*, 1998, *50*, 169.
428. H. Matsubara, T. Tanabe, A. Moto, Y. Mine, and S. Takagishi, *Sol. Energy Mater. Sol. Cells*, 1998, *50*, 177.
429. V.K. Sharma, A. Colangelo, and G. Spagna, *Energy Convers. Manage.*, 1995, *36*, 161, 239.
430. J.A. Hutchby, M.L. Timmons, R. Venkatasubramanian, P. Sharps, and R.A. Whisnant, *Sol. Energy Mater. Sol. Cells*, 1994, *35*, 9.
431. R.F. Service, *Science*, 1997, *276*, 356.
432. J.E. Ludman, J. Riccobono, I.V. Semenova, N.O. Reinhand, W. Tai, X. Lo, G. Syphers, E. Rahls, G. Sliker, and J. Martin, *Sol. Energy*, 1997, *60*, 1.
433. E. Kintisch, *Science*, 2007, *317*, 583.
434. (a) K.S. Betts, *Envion. Sci. Technol.*, 2007, *41*, 3038; (b) Anon., *Chem. Eng. Prog.*, 2008, *104*(9), 18.
435. (a) P. Pechy, F.P. Rotzinger, M.K. Nazeeruddin, O. Kohle, S.M. Zakeeruddin, R. Humphry-Baker, and M. Gratzel, *J. Chem. Soc. Chem. Commun.*, 1995, 65; (b) A. Coghlan, *New Sci.*, 1995, Jan. 28, 22; (c) M. Freemantle, *Chem. Eng. News*, 1998, Oct. 26, 44; (d) M.K. Nazeeruddin, R. Humphry-Baker, M. Gratzel, and B.A. Murrer, *Chem. Commun.*, 1998, 719; (e) H. Sugihara, L.P. Singh, K. Seyama, H. Arakawa, M.K. Nazeeruddin, and M. Gratzel, *Chem. Lett.*, 1998, *27*, 1005; (f) U. Bach, D. Lupo, P. Comte, J.E. Moser, F. Weissortel, J. Salbeck, H. Sprietzer, and M. Gratzel, *Nature*, 1998, *395*, 583; (g) W.C. Sinke and M.M. Wienk, *Nature*, 1998, *395*, 544; (h) A. Hagfeldt and M. Gratzel, *Acc. Chem. Res.* 2000, *33*, 269; (i) N.S. Lewis, *ChemSusChem*, 2009, *2*, 383.
436. (a) N. Robertson, *Angew. Chem. Int. Ed.*, 2006, *45*, 2338; (b) J.-H. Yum, D.P. Hagberg, S.-J. Moon, K.M. Karlson, T. Marinado, L. Sun, A. Hagfeldt, M.K. Nazeeruddin, and M. Gratzel, *Angew. Chem. Int. Ed.*, 2009, *48*, 1576; (c) A. Mishra, M.K.L.R. Fischer and P. Baurele, *Angew. Chem. Int. Ed.*, 2009, *48*, 2474.
437. (a) H. Choi, C. Baik, S.O. Kang, J. Ko, M.-S. Kang, M.K. Nazeeruddin, and M. Gratzel, *Angew. Chem. Int. Ed.*, 2008, *47*, 327; (b) S. Hwang, J.C. Lee, C. Park, H. Lee, C. Kim, C. Park, M.-H. Lee, W. Lee, J. Park, K. Kim, N.-G. Park, and C. Ki, *Chem. Commun.*, 2007, 4887.
438. (a) M. Freemantle, *Chem. Eng. News*, 1999, Apr. 12, 13; (b) J.-F. Nierengarten, J.-F. Eckert, J.-F. Nicoud, L. Ouali, V. Krasnikov, and G. Hadziioannou, *Chem. Commun.*, 1999, 617.
439. (a) K. Murakoshi, R. Kogure, Y. Wada, and S. Yanagida, *Sol. Energy Mater. Sol. Cells*, 1998, *55*, 113; (b) G.G. Wallace, P.C. Dastoor, D.L. Officer, and C.O. Too, *Chem. Innovation*, 2000, *30*(4), 14.
440. (a) B.C. Thompson and J.M.J. Frechet, *Angew. Chem. Int. Ed.*, 2008, *47*, 58–77; (b) D. Kronholm and J.C. Hummelen, *Mater. Matters (Aldrich)*, Milwaukee, WI, 2007, *2*(3), 16; (c) Y. Liang, Y. Wu, D. Feng, S-T. Tsai, H-J. Son, G. Li, and L. Yu, *J. Am. Chem. Soc.*, 2009, *131*, 56.
441. D. Talbot, *Technol. Rev.*, 2007, *110*(2), 49.
442. B. Tian, T.J. Kempa, and C.M. Lieber, *Chem. Soc. Rev.*, 2009, *38*, 16.
443. (a) K. Bullis, *Technol. Rev.* 2008, *111*(5), 104; (b) M.J. Currie, J.K. Mapel, T.D. Heidel, S. Groffri, and M.C. Baldo, *Science*, 2008, *321*, 226; (c) SolFocus, Inc., www.solfocus.com.; (d) J.M. Higgins, *Futurist*, 2009, *43*(3), 25.
444. R.F. Service, *Science*, 1996, *272*, 1744.
445. Anon., *Environment*, 1999, *41*(5), 22.
446. C. Flavin and M. O'Meara, *World Watch*, 1997, *10*(3), 28.
447. (a) J.P. Louineau, F. Crick, B. McNelis, R.D.W. Scott, B.E. Lord, R. Noble, D. Anderson, R. Hill, and N.M. Pearsall, *Sol. Energy Mater. Sol. Cells*, 1994, *35*, 461; (b) T. Yagiura, M. Morizane, K. Murata, K.T. Yagiura, K. Uchihashi, S. Tsuda, S. Nakano, T. Ito, S. Omoto, Y. Yamashita, H. Yamakawa, and T. Fujiwara, *Sol. Energy Mater. Sol. Cells*, 1997, *47*, 2227; (c) M. Yoshino, T. Mori, M. Mori, M. Takahashi, S.-I. Oshida, and K. Shirasawa, *Sol. Energy Mater. Sol. Cells*, 1997, *47*, 235; (d) S.R. Wenham, S. Bowden, M. Dickinson, R. Largent, N. Shaw, C.B. Honsberg, M.A. Green, and P. Smith, *Sol. Energy Mater. Sol. Cells*, 1997, *47*, 325.
448. F.R. Goodman, E.A. De Meo, and R.M. Zavadil, *Sol. Energy Mater. Sol. Cells*, 1994, *35*, 385.
449. (a) Anon., *R&D (Cahners)*, 1999, *41*(5), 9; (b) R. Dagani, *Chem. Eng. News*, 1998, July 27, 14; (c) R. Wang, K. Hashimoto, A. Fujishima, M. Chikuni, E. Kojima, A. Kitamura, M. Shimohigoshi, and T. Watanabe, *Nature*, 1997, *388*, 431.
450. N. Tsujino, T. Ishida, A. Takeoka, Y. Mukino, E. Sakoguchi, M. Ohsumi, M. Ohinishi, S. Nakano, and Y. Kuwani, *Sol. Energy Mater. Sol. Cells*, 1994, *35*, 497.
451. Anon., *Environment*, 1998, *40*(8), 21.
452. (a) M.M. Koltun, *Sol. Energy Mater. Sol. Cells*, 1994, *35*, 31; (b) T. Fujisawa and T. Tani, *Sol. Energy Mater. Sol. Cells*, 1997, *47*, 135; (c) M.D. Bazilian, F. Leenders, B.G.C. van der Ree, and D. Prasad, *Solar Energy*, 2001, *71*, 57.
453. K. Voss, A. Goetzberger, G. Bopp, A. Haberle, A. Heinzel, and H. Lehmberg, *Sol. Energy*, 1996, *58*, 17.
454. (a) M. Brower, *Cool Energy: Renewable Solutions to Environmental Problems*, MIT Press, Cambridge, MA, 1992; *Environmental Impacts of Renewable Energy Technologeis*, Union of Concerned Scientists, Cambridge, MA, Aug. 1992; (b) K.S. Betts, *Environ. Sci. Technol.*, 2000, *34*, 306A.
455. R. Hunt and J. Hunt, *Chem. Ind. (Lond.)*, 1998, 227.
456. (a) R. Gasch and J. Twele, ed., *Wind Power Plants—Fundamentals, Design, Construction and Operation*, Stylus Publishing, Herndon, VA, 2004; (b) J.F. Manwell, J.G. McGowan, and A.L. Rogers, *Wind Power Explained—Theory, Design and Application*, 2nd ed., Wiley, New York, 2009; (c) T. Ackermann, ed., *Wind Power in Power Systems*, Wiley, New York, 2005; (d) S. Heier, *Grid Integration of Wind Energy Conversion Systems*, 2nd ed., Wiley, New York, 2006; (e) J. Deyette, *Catalyst* (Union of Concerned Scientists), 2006, *5*(2), 2.
457. J. Topping, *Chem. Eng. News*, 1997, Sept. 22, 24.
458. D. Schneider, *Am. Sci.*, 2007, *95*, 490.
459. R. Mandelbaum, *Discover*, 2004, *25*(6), 48.
460. Anon., *Bluewater Wind*, Hoboken, NJ.
461. (a) K.S. Brown, *Science*, 1999, *285*, 678; (b) C. Flavin, *World Watch*, 2000, *13*(2), 8; (c) Anon., *Environment*, 2000, *42*(5), 5.
462. T. Markvart, *Sol. Energy*, 1996, *57*, 277.
463. D. Drollette, *Technol. Rev.*, 1998, *101*(4), 54.
464. (a) D. Tenenbaum, *Technol. Rev.*, 1995, *98*(1), 38; (b) P.M. Wright, *Chem. Ind. (Lond.)*, 1998, 208; (c) J.E. Mock, J.W. Tester, and P.M. Wright, *Annu. Rev. Energy Environ.*, 1997, *22*, 305; (d) J. Johnson, *Chem. Eng. News*, August 17, 2009, 30.

465. G. Ondrey, *Chem. Eng.*, 1999, *106*(2), 19.
466. G. Vogel, *Science*, 1997, *275*, 761.
467. A. Macdonald-Smith, *Wilmington DE News J.*, Feb. 22, 2009, C5.
468. T.B. Johansson, H. Kelly, A.K.N. Reddy, and R.H. Williams, eds, *Renewable Energy: Sources for Fuels and Electricity*, Island Press, Washington, DC, 1993, 515, 530.
469. J. Webb, *New Sci.*, 1995, July 29, 6.
470. Anon., *Science*, 1998, *280*, 1843.
471. H. Blankesteijn, *New Sci.*, 1996, Sept. 14, 20.
472. (a) B.E. Erickson, *Environ. Sci. Technol.*, 2001, *35*, 80A; (b) E. Robinson and J. Rogers, *Catalyst* (Union of Concerned Scientists), 2008, *7*(1), 18; (c) A. von Jouanne, *Mech. Eng.*, 2006, *128*(12), 24; (d) N. Lubick, *Environ. Sci. Technol.*, 2009, *43*, 2204; (e) J. Scruggs and P. Jacob, *Science*, 2009, *323*, 1176.
473. Anon., *Chem. Eng. Prog.*, 2006, *102*(12), 14.
474. (a) J. Johnson, *Chem. Eng. News*, Oct. 4, 2004, 23; (b) E. Jeffries, *Chem. Ind.* (*Lond.*), Jan. 29, 2007, 22; (c) R. Stone, *Science*, 2003, *299*, 339.
475. J. Geary, *New Sci.*, 1995, Jan. 14, 19.
476. G. Parkinson, *Chem Eng.*, 1998, *105*(7), 29.
477. C. Flavin and N. Lenssen, *Technol. Rev.*, 1995, *98*(4), 42.
478. (a) T.B. Johansson, H. Kelly, A.K.N. Reddy, and R.H. Williams, eds, *Renewable Energy: Sources for Fuels and Electricity*, Island Press, Washington, DC, 1993, 171, 1042; (b) S.M. Schoening, J.M. Eyer, J.J. Iannucci, and S.A. Horgan, *Annu. Rev. Energy Environ.*, 1996, *21*, 347; (c) J. Fahey, *Forbes*, 2008, *182*(11), 122; (d) C. Forsberg and M. Kazimi, *Science*, 2009, *325*, 32; (e) D. Charles, Science, 2009, *324*, 172 ; (f) B.S. Lee and D.E. Gushee, *Chem. Eng. Prog.*, 2009, *105*(4), 22.
479. C. Flavin, *World Watch*, 1996, *9*(1), 15.
480. R.J. Farris, University of Massachusetts at Amherst, private communication, June, 1997.
481. (a) H. Cao, Y. Akimoto, Y. Fujiwara, Y. Tanimoto, L.-P. Zhang, and C.-H. Tung, *Bull. Chem. Soc. Jpn.*, 1995, *68*, 3411; (b) M. Maafi, C. Lion, and J.-J. Aaron, *New J. Chem.*, 1996, *20*, 559; (c) I. Nishimura, A. Kameyama, T. Sakurai, and T. Nishikubo, *Macromolecules*, 1996, *29*, 3818.
482. (a) J. Besenhard, *Handbook of Battery Materials*, Wiley-VCH, Weinheim, 1999; (b) R. Dell, *Chem. Br.*, 2000, *36*(3), 34; (c) M. Tsuchida, ed., *Macromol. Symp.*, 2000, *156*, 171, 179, 187, 195, 203, 223; (d) M.S. Whittingham, R.F. Savinell, and T. Zawodzinski, eds, *Chem. Rev.*, 2004, *104*, 4243–4886; (e) R.M. Dell and D.A.J. Rand, *Understanding Batteries*, Royal Society of Chemistry, Cambridge, 2001; (f) R. Armstrong and A. Robertson, *Chem. Br.*, 2002, *38*(2), 38; (g) T.W. Walker, *Chem. Eng. Prog*, 2008, *104*(3), S23; (h) M.R. Palacin, *Chem. Soc. Rev.*, 2009, *38*, 2565; (i) J. Chen and F. Cheng, *Acc. Chem. Res.*, 2009, *42*, 713.
483. (a) D. Sperling, *Electric Vehicles and Sustainable Transportation*, Island Press, Washington, DC, 1995; (b) S. Dunn, *World Watch*, 1997, *10*(2), 19; (c) S. Wilkinson, *Chem. Eng. News*, 1997, Oct. 13, 18; (d) J. Glanz, *New Sci.*, 1995, Apr. 15, 32; (e) G.L. Henriksen, W.H. DeLuca, and D.R. Vissers, *Chemtech*, 1994, *24*(11), 32; (f) For a magnesium battery see M. Freemantle, *Chem. Eng. News*, Oct. 16, 2000, 8.
484. N.L.C. Steele, and D.T. Allen, *Environ. Sci. Technol.*, 1998, *32*(1), 40A.
485. (a) F.-S. Cai, G.-Y. Zhang, J. Chen, X.-L. Gou, H.-K. Liu, and S.-X. Dou, *Angew. Chem. Int. Ed.*, 2004, *43*, 4212.
486. (a) J.R. Owen, *Chem. Soc. Rev.*, 1997, *26*, 259; (b) M. Jacoby, *Chem. Eng. News*, Dec. 27, 2007, 26; (c) M. Yoshi, R.J. Brodd, and A. Kozawa, *Lithion Ion Batteries: Science and Technologies*, Springer, New York, 2009.
487. (a) Anon., *Chem. Eng. News*, Jan. 12, 2009, 26; (b) N. Shirouzu, *Wall St. J.*, Jan. 11, 2008, A1.
488. (a) A. Coghlan, *New Sci.*, Apr. 1, 1995, 25; (b) J.-Y. Sanchez, F. Alloin, and D. Benraban, *Macromol. Symp.*, 1997, *114*, 85; (c) C. Roux, W. Gorecki, J.Y. Sanchez, and E. Belorizky, *Macromol. Symp.*, 1997, *114*, 211; (d) J.-F. Moulin, P. Damman, and M. Dosiere, *Macromol. Symp.*, 1997, *114*, 237; (e) F.M. Gray, *Polymer Electrolytes*, Royal Society of Chemistry, Cambridge, UK, 1997; (f) V. Chandrasebhar, *Adv. Polym. Sci.*, 1998, *135*, 139; (g) G.S. MacGlashan, Y.G. Andrew, and P.G. Bruce, *Nature*, 1999, *398*, 792.
489. J.-L. Ju, Q.-C. Gu, H.-S. Xu, and C.-Z. Yang, *J. Appl. Polym. Sci.*, 1998, *70*, 353.
490. S.-W. Hu and S.-B. Fang, *Macromol. Rapid Commun.*, 1998, *19*, 539.
491. (a) G. Parkinson, *Chem. Eng.*, 1997, *104*(8), 29; 2000, *107*(12), 23; (b) G.S. Samdani, *Chem. Eng.*, 1995, *102*(1), 21; (c) *R&D* (*Cahners*) 1998, *40*(13), 9; (d) Anon., *Technol. Rev.*, *104*(7), 86.
492. (a) S. Licht, B. Wang, and S. Ghosh, *Science*, 1999, *285*, 1039; (b) M. Freemantle, *Chem. Eng. News*, Aug. 16, 1999, 4; (c) S. Licht and R. Tel-Vered, *Chem. Commun.*, 2004, 628; (d) S. Licht, X. Yu, and D. Ou, *Chem. Commun.*, 2007, 2753.
493. (a) G. Parkinson, *Chem. Eng.*, 2002, *109*(3), 19; (b) P. Fairley, *Technol. Rev.*, 2003, *106*(2), 50; (c) D. Pletcher and F. Walsh, *Chem. Ind.* (*Lond.*), 2001, 564.
494. M. Freemantle, *Chem. Eng. News*, Sept. 25, 2000, 12.
495. (a) J.D. Genders and D. Pletcher, *Chem. Ind.* (*Lond.*), 1996, 682; (b) H. Wendt, S. Rausche, and T. Borucinski, *Adv. Catal.*, 1994, *40*, 87.
496. (a) K.D. Moeller, *Tetrahedron*, 2000, *56*, 9527; (b) J. Grimshaw, *Electrochemical Reactions and Mechanisms in Organic Chemistry*, Elsevier, Amsterdam, 2001; (c) H. Lund and O. Hammerich, eds, *Organic Electrochemistry*, 4th ed., Dekker, New York, 2001; (e) K. Izutsu, *Electrochemistry in Non-Aqueous Solutions*, Wiley-VCH, Weinheim, 2002.
497. N. Takano, M. Ogata, and N. Takeno, *Chem. Lett.*, 1996, *25*, 85.
498. G. Parkinson, *Chem. Eng.*, 1996, *103*(5), 23.
499. S. Chardon-Noblat, I.M.F. DeOliveira, J.C. Moutet, and S. Tingry, *J. Mol. Catal. A: Chem*, 1995, *99*, 13.
500. M. Fremantle, *Chem. Eng. News*, June 9, 1997, 11.
501. W. An, J.K. Hong, P.N. Pintauro, K. Warner, and W. Neff, *J. Am. Oil Chem. Soc.*, 1998, *75*, 917.
502. M. Abdel-Azzem and M. Zahran, *Bull. Chem. Soc. Jpn.*, 1994, *67*, 1879.
503. K. Matsuda, M. Atobe, and T. Nonaka, *Chem. Lett.*, 1994, *23*, 1619.
504. A.S.C. Chan, T.T. Huang, J.H. Wagenknecht, and R.E. Miller, *J. Org. Chem.*, 1995, *60*, 742.
505. (a) K. Scott, *Electrochemical Processes for Clean Technology*, Royal Society of Chemistry, Cambridge, UK, 1995; (b) E.R. Beaver, and following papers, *Chemtech*, 1996, *26*(4), 9–58; (c) G. Parkinson and G. Ondrey, *Chem. Eng.*, 1996, *103*(5),

37; (d) S. Mazrou, H. Kerdjoudji, A.T. Cherif, A. Elmidaoui, and J. Molenat, *New J. Chem.*, 1998, *22*, 355; (e) D. Simonsson, *Chem. Soc. Rev.*, 1997, *26*, 181; (f) N.P. Cheremisinoff, *Electrotechnology: Industrial and Environmental Applications*, Noyes Publications, Westwood, NJ, 1996.

506. (a) V. Ramamurthy and K.S. Shanze, *Organic Photochemistry*, Dekker, New York, 1997; (b) C.E. Wayne and R.P. Wayne, *Photochemistry*, Oxford University Press, Oxford, 1996; (c) D.C. Neckers, D.H. Volmam, and G. von Bunau, *Adv. Photochem.*, 1999, *25*, and earlier volumes; (d) A. Bhattacharya, *Prog. Polym. Sci.*, 2000, *25*, 371; (e) V. Balzani, A. Credi, and M. Venturi, *ChemSusChem.*, 2008, *1*, 26; (f) V.M. Parmon, D. Kozlov, and P. Smirniotis, *Photocatalysis—Catalysts, Kinetics and Reactors*, Wiley-VCH, Weinheim, 2009; (g) P. Klan and J. Wirz, *Photochemistry of Organic Compounds*, Wiley, Hoboken, NJ, 2008.
507. J.H. Krieger and M. Freemantle, *Chem. Eng. News*, 1997, July 7, 15.
508. G.A. Epling and Q. Wang. In: P.T. Anastas and C.T. Farris, eds, *Benign by Design*, A.C.S. Symp. 577, Washington, DC, 1994, 64.
509. T. Igarashi, K. Konishi, and T. Aida, *Chem. Lett.*, 1998, 1039.
510. B. Heller and G. Oehme, *J. Chem. Soc. Chem. Commun.*, 1995, 179.
511. J.-T. Li, J.-H. Yang, and T.-S. Li, *Green Chem.*, 2003, *5*, 433.
512. J.V. Crivello and M. Sangermann, *Polym. Preprints*, 2001, *42*(2), 783.
513. P. Zurer, *Chem. Eng. News*, Apr. 1, 1996, 5.
514. R. Bao, W.S. Dodson, T.C. Bruton, and Y. Li, Preprints papers *Natl. Meet. ACS Div. Environ. Chem.*, 1994, *34*(2), 411.
515. P. Pichat, *Catal. Today*, 1994, *19*, 313.
516. (a) R. Memming, *Top. Curr. Chem.*, 1994, *169*, 171; (b) Y. Zhang, *Chem. Ind. (Lond.)*, 1994, 714.
517. G.A. Kraus, M. Kirihara, and Y. Wu. In: P.T. Anastas and C.T. Farris, eds, *Benign by Design*, ACS Symp. 577, Washington, DC, 1994, 76.
518. C. Schiel, M. Oelgemoller, and J. Mattay, *Synthesis*, 2001, 1275.
519. (a) J.-L. Luche, *Synthetic Organic Sonochemistry*, Plenum, New York, 1998; (b) L.H. Thompson and L.K. Doraiswamy, *Ind. Eng. Chem. Res.*, 1999, *38*, 1215–1249; (c) P. Cintas and J.-L. Luche, *Green Chem.*, 1999, *1*, 115; (d) A.B. Pandit, P.S. Kumar, and M.S. Kumar, *Chem. Eng. Prog.*, 1999, *95*(5), 43; (e) T.J. Mason, *Sonochemistry*, ACS Oxford Chemistry Primers 70, Oxford University Press, Oxford, 2000; (f) T.J. Mason and D. Peters, *Practical Sonochemistry: Power Ultrasound Uses and Applications*, 2nd ed., Horwood Publishing, Chichester, 2002; (g) Y.G. Adewicyi, *Ind. Eng. Chem Res.*, 2001, *40*, 4681–4715; (h) D. Cheeke, *Fundamentals and Applications of Ultrasonic Waves*, CRC Press, Boca Raton, FL, 2002.
520. H. Fujiwara, J. Tanaka, and A. Horiuchi, *Polym. Bull.*, 1996, *36*, 723.
521. E.A.G. Gonzalez de los Santos, M.J.L. Gonzalez, and M.C. Gonzalez, *J. Appl. Polym. Sci.*, 1998, *68*, 45.
522. N.S. Nandurkar, M.D. Bhor, S.D. Samant, and B.M. Bhanage, *Ind. Eng. Chem. Res.*, 2007, *46*, 8590.
523. G. Parkinson, *Chem. Eng.*, 2002, *109*(12), 15.
524. Wenesco, Inc., Chicago, IL.
525. (a) A. Loupy, *Microwaves in Organic Synthesis*, 2nd ed., Wiley-VCH, Weinheim, 2002; (b) V. Polshettir and R.A. Varma, *Chem. Soc. Rev.*, 2008, *37*, 1546; (c) B.L. Hayes, *Microwave Synthesis—Chemistry at the Speed of Light*, CEM Publishing, Matthews, NC; (d) C.O. Kappe, *Chem. Soc. Rev.*, 2008, *37*, 1127; (e) M. Nuchter, B. Ondruschka, W. Bonrath, and A. Gum, *Green Chem.*, 2004, *6*, 128–141; (f) B.L. Hayes, *Aldrichchim. Acta*, 2004, *37*(2), 66; (g) V. Marx, *Chem. Eng. News*, Dec. 13, 2005, 14; (g) C.O. Kappe, D. Dallinger, and S. Murphee, eds, *Practical Microwave Synthesis for Organic Chemists—Strategies, Instruments and Protocols*, Wiley-VCH, Weinheim, 2008.
526. X. Querol, A. Alastuey, A.L. Soler, F. Plana, J.M. Andres, R. Juan, P. Ferrer, and C.R. Ruiz, *Environ. Sci. Technol.*, 1997, *31*, 2527.
527. C.-G. Wu and T. Bein, *Chem. Commun.*, 1996, 925.
528. (a) R.S. Varma, *Green Chem.*, 1999, *1*(1), 43; (b) R.S. Varma, K.P. Naicker, D. Kumar, R. Dahiya, and P.J. Liesen, *J. Microw. Power Electromagn. Energy*, 1999, *54*(2), 113.
529. T.D. Conesa, J.M. Campelo, J.H. Clark, R. Luque, D.J. Macquarrie, and A.A. Romero, *Green Chem.*, 2007, *9*, 1109.
530. T. Razzaq and C.O. Kappe, *ChemSusChem*, 2008, *1*(1–2), 123.
531. A. Holzwarth, J. Lou, T.A. Hatton, and P.E. Laibinis, *Ind. Eng. Chem. Res.*, 1998, *37*, 2701.
532. F.K. Mallon and W.H. Ray, *J. Appl. Polym. Sci.*, 1998, *69*, 1203.
533. R.G. Compton, B.A. Coles, and F. Marken, *Chem. Commun.*, 1998, 2595.
534. R. Roy, D. Agrawal, J. Cheng, and S. Gedevanishvili, *Nature*, 1999, *399*, 668.
535. D.E. Clark, W.H. Sutton, and D.A. Lewis, eds, *Microwaves—Theory and Application in Materials Processing IV: Microwave and RF Technology: From Science to Application*, Ceramic Trans., vol. 80, American Ceramic Society, Westerville, OH, 1997.
536. W.C. Conner and R. Laurence, Department of Chemical Engineering, University of Massachusetts at Amherst, private communication, June 1997.

RECOMMENDED READING

1. J. Kaiser, *Science*, 1997, *278*, 217 [global warming].
2. Union of Concerned Scientists, *Assessing the Hidden Costs of Fossil Fuels*, Cambridge, MA, Jan. 1993.
3. L. Starke, ed., *State of the World 1995*, W.W. Norton, New York, 1995, 58–75 [energy from sun and wind]; 95–112 [better buildings].
4. Union of Concerned Scientists, Putting Renewable Energy to Work in Buildings, Cambridge, MA, Jan. 1993.
5. M. Brower, Environmental Impacts of Renewable Energy Technologies, Union of Concerned Scientists, Cambridge, MA, Aug. 1992.
6. K. Voss, A. Goetzberger, G. Bopp, A. Haberle, A. Heinzel, and H. Lehmberg, *Sol. Energy*, 1996, *58*, 17 [self-sufficient solar house in Freiburg, Germany].
7. J.E. Ludman, J. Riccobono, I.V. Semenova, N.O. Reinhand, W. Tai, X. Lo, G. Syphers, E. Rallis, G. Sliker, and J. Martin, *Sol. Energy*, 1997, *60*, 1 [holographic system for photovoltaic cells].
8. W. Wettling, *Sol. Energy Mater. Sol. Cells*, 1995, *38*, 494 [efficient photovoltaic cells from crystalline silicon].

9. J.A. Turner, *Science*, 1999, *285*, 687.
10. L.R. Brown, *Futurist*, 2006, *40*(4), 18 [energy and the planet].
11. L.R. Brown, *Plan B 4.0: Mobilizing to Save Civilization*, W.W. Norton, New York, 2009.
12. N. Aramaroli and V. Balzani, *Angew. Chem. Int. Ed*., 2007, *46*, 52–66 [future energy supplies].
13. R.F. Service, *Science*, 2008, *319*, 718 [solar cells].
14. H.W. Cooper, *Chem. Eng. Prog*., 2007, *103*(11), 34 [fuel cells].
15. E. Robinson and J. Rogers, *Catalyst* (*Union of Concerned Scientists*), Spring 2008, 19 [wave power].

EXERCISES

1. Keep track of the single-use throwaway items that you use for a week. What reusable items could be used instead?
2. Monitor the cars on a road during rush hour to see how many cars have more than one person in them. Also count walkers and bikers.
3. See if you can find the following in your area:
   Passive solar heating and cooling
   Active solar heating
   Photovoltaic cells
   Windmill
   A greenway
   Use of biogas from a landfill
   District heating
   An electric car
   Geothermal system
   District heating
   When you find them, see if the owner or operator will be willing to share operating data with you. Does your university do research on any of these?
4. Devise a system of incentives and disincentives to help solve the problem of global warming in your area.
5. Hydrogen can be stored by hydrogenating benzene to cyclohexane, the recovered later by dehydrogenation. Benzene is a carcinogen. Figure out a better chemical way to store hydrogen that does not involve a carcinogen.
6. Light energy can be stored by converting norbornadienes to quadricyclanes, and the energy released later by reversing the reaction. A cheaper, better way is needed that has a high-energy density and that is reversible over a great many cycles. Can you devise one?
7. Waste heat might be recovered for use later at the same or a different location by conversion of one chemical compound to another. Can you figure out a system that will do this with minimal losses over many cycles. The chemical should be abundant, cheap and, preferably, suitable for transport through a pipeline. It will be even better if the system can store excess summer heat for use in the winter. (Nature does this by converting carbon dioxide in the air to wood in trees.)
8. Is there as nuclear power plant near you? Has it had any safety problems? Visit it to see how it works.

CHAPTER 16

Population and the Environment

16.1 THE PROBLEMS

Aging farmers often divide their farmlands among their children. This is not a problem if there are no more than two children. However, many farm families have many children to help in the work on the farm. In this case, the children each have less land than the parents. If this is done over a few generations, the final farm size may not be enough to provide sufficient food for the family.[1] Sociologists studied this problem in the Rio Grande Valley in New Mexico in the 1930s. Rwanda affords a more recent illustration. The population grew from 2.5 to 8.8 million from 1950 to 1994 as the average woman had eight children, the highest rate in the world.[2] The nation had the highest population density in Africa. The average family was trying to support itself on only 0.7 ha in the 1990s.[3] This may have been one of the reasons for the bitter civil war to have erupted in that nation.

There are various responses to this shortage of land. One of them is to move ever higher on the mountainside and cut down the forest so that crops can be raised causing severe erosion. Such a farmland does not last very long; hence, it is necessary to move fairly often and repeat the process. Some live in flood-prone areas because they have no other place to live, as in Bangladesh. The floods become worse as other farmers in mountains upstream cut down more forests to create cropland. Another response is to send the husband to work in another country to earn money to support the family. This leaves the wife to look after the children, farm the fields, gather the firewood, get the water, and do all the other chores. In many places in the developing world, firewood and water are in short supply; hence, longer and longer trips are needed to obtain them. This is one reason for deforestation. Many families, unable to obtain more land, move to the cities in search of jobs. Here, they may live in shantytowns lacking adequate water, sanitation, and other services. They may do menial jobs or be unemployed. The number of urban poor in Latin America was 44 million in 1970. In 1990, the number had reached 115 million.[4] The number was 113 million in 2005.[5] These were 60% of the total poor.

The nutrients in the tropical forests are largely in the vegetation. Slash and burn agriculture is one way of obtaining them. The forest is cut down and burned to release the nutrients, which are then available for growing crops. After a few years, this system is depleted and the family moves on to repeat the process in another location. It takes an extended time period for the land to become forest again and recover. If there are too many people doing this, the whole forest can be depleted and destroyed. In some countries, efforts have been made to resettle surplus population by clearing tropical forests. The soils are fragile and remain fertile for only a few years. This has been true in parts of Brazil. In some parts of this country, soils treated in this way undergo an irreversible change in structure, a hardening, that prevents any future use. These changes can occur on land used for grazing cattle as well as on land used for sowing crops.

The overall effects of these processes can lead to desertification, deforestation; increased flooding, less water in the ground and in rivers; depletion of underground aquifers[6]; loss of biodiversity; loss of topsoil; more disease,[7] famine, and unemployment; as well as political instability.[8] The world has undergone a population explosion, with the time to add each new billion becoming shorter (Table 16.1).[9] On March 8, 2009, the population of the world was 6,765,317,000.[10]

Even though fertility is declining (down from six children per woman to three in the developing world), it is still above the replacement level of 2.1.[11] Seventy countries (largely in the developed world) have achieved stable or negative growth.[12] In the developing world, where 98% of future growth is expected to have occured,[13] most countries have not reached this level. There is also a large population momentum effect as large numbers of persons reach reproductive

Table 16.1 Population Growth Over Time

Date	Population (in billions)
1830	1
1930	2
1960	3
1974	4
1987	5
1999	6

age. In most countries in Africa, 45% of the people are younger than 15 years. In India 60% of the population is below of 25 years.

Of the world's 6.7 billion people, 1.3 billion live in absolute poverty. Approximately 30% of the world's labor force of 2.8 billion people are either unemployed or underemployed (i.e., not earning enough to lift themselves and their families out of poverty).[14] At the present rate of population growth, 1 billion new jobs are needed. Roughly, 1 billion people go to bed hungry each night.[15] There are food shortages in 86 countries. Africa produced 30% less food per capita in 1998 than it did in 1967.[16] Half of the children in Guatemala under 5 years of age are undernourished.[17] In Mali, 23% of the children are too thin. More than 1.5 billion do not have clean drinking water or sanitation. Women have a disproportionate share of the problems. They represent 66% of the illiterates and 70% of the world's poor. The world's poorest country in 1993 was Mozambique. In 2008, the gross domestic product (GDP) per capita per year was 700 dollars.[18] Its population of 16 million is growing at the rate of 2.7%/year. At this rate, the population will double in 26 years. In 2008, Malawi, Somalia, Comoros, and the Solomon Islands had annual GDP per capita incomes of only $600.[19] Ethiopia, with an annual GDP per capita of $841 in 2006 will see its population double in 22 years. (This is the country where a major famine a few years ago required international help.) Tanzania with a GDP per capita per year of 800 dollars will double its population in 21 years. In contrast, the figure for the world's richest country, Norway is $43,400 followed by Switzerland with $40,680. These are followed in descending order by the United States, Japan, Denmark, Sweden, Finland, Ireland, and Austria. Infant mortality is highest in Afghanistan and Mozambique at 162:1000 live births. It is lowest in Japan, Singapore, Iceland, and Sweden at 5:1000. Life expectancy is lowest in Mozambique at 42.1 years and highest in Macau at 84.4 years, in Andorra at 82.7 years, and in Japan at 82.1 years.[20] Many of the countries with the lowest life expectancies are in subSaharan Africa where adult prevalence rates of HIV/AIDS vary from 10% to 39%.

The Legatum Institute in Dubai United Arab Emirates rated 140 countries in its 2008 "Prosperity Index."[21] Australia was at the top and Yemen at the bottom. A woman in Yemen will have 6.2 children in her lifetime. The annual income per capita was $903 in 2006. The United Nations Human Development Index for 2007 rated 177 countries on life expectancy, income, and education. Iceland was on top having displaced Norway, which had the top place for the previous six years.[22] Niger is a semi-arid country on the southern edge of the Sahara Desert. It has a problem of desertification. The literacy rate is 17%. The average life span is 46 years.[23] Medecins Sans Frontieres has teams in the country providing "Plumpy Nut," a mixture of peanut paste, vegetable oil, vitamins, minerals, milk powder, and sugar, to severely malnourished children.[24] A woman in Niger will average 7.1 children in her lifetime. Her lifetime chance of dying of a pregnancy or childbirth complication is one in seven, the highest in the world.[25] (In more developed regions, the risk is one in 7300.) Niger's per capita per year income in 2006 was $901.

The United Nations Millennium Development Goals call for significant reductions in infant, child and maternal mortality, poverty, and hunger in the world by 2015.[26] It is unlikely that these goals will be met.[27]

Among the 85 nations that believe that their birth rates are too high are Zaire (6.7 children per woman), Saudi Arabia (6.4), Somalia (6.8), Cote d'Ivoire (5.7), and Syria (4.7). Somalia required an international team to try to restore order after its government collapsed. This effort failed. At the other extreme are Hong Kong (1.2), Spain (1.2), Germany (1.3), Italy (1.3), Greece (1.4), Austria (1.5), Bulgaria (1.5), Slovenia (1.5), and Romania (1.5).[28]

Various proposals have been made for ways to combat these problems. One is to resettle surplus population in new lands. The problem is that most of the suitable land is already occupied. Efforts to put people into the rain forests of Brazil and Indonesia have not worked well. The land was not very suitable for conventional farming. Migration to other countries has been used as a safety valve in the past, as in the case of Ireland. There is an ethical question over whether it is right to expect a nation that has kept its population growth rate under control to absorb the surplus from ones that have not.

Plant breeders carried out a "green revolution" by producing more productive varieties of wheat, rice, and other crops. These varieties required more water or fertilizer, which some of the very poor could not afford. In addition, they decreased the genetic base of these crops, perhaps eliminating some less productive varieties that are more tolerant of drought, poor soil, and disease. From the foregoing figures, it can be seen that another green revolution is needed if the food supply is to keep up with the number of people. With current technology, it is not clear how this might be accomplished.[29] Breeding plants that produce food in drier and more saline soils, or that could be irrigated by seawater, is being studied. A genetically engineered *Arabidopsis thaliana* was able to grow in 0.2 M sodium chloride.[30] Further work is needed to apply this method to crop plants. We now live in a human-dominated landscape, where one-third to one-half of the earth's surface has been transformed and where over half of the accessible freshwater has been put to use.[31] This makes it difficult to develop new irrigated lands. The loss of agricultural land to urban sprawl each year will make it that much more difficult to produce enough food to feed everyone adequately. Sixty-six percent of the world's marine fisheries are either fully exploited or overexploited. Africa now needs 14 million tons more grain annually than it produces.[32]

There are food surpluses in Europe and the United States, partially as a result of government subsidies. It has been suggested that these be used to feed hungry people in the developing world. This might be used to buy time to confront the real problem of overpopulation. There are questions on who would pay the farmers for the food, if those who need it the most are too poor to pay for it. There is also the question of whether these surpluses were produced using sustainable methods (i.e., without excessive loss of topsoil or depletion of underground aquifers, or salinization of irrigated lands). Even with its modern food distribution system, the United States still has some hungry people in it. Another suggestion to release more food for human use is to adopt vegetarian diets. It takes about 9 lb of feed to produce 1 lb of beef. It takes only 2 lb of feed to produce 1 lb of chicken or catfish. If people in affluent countries eat less meat, there would be more food to feed those who are hungry. Some meat is produced on rangeland that is unsuitable for growing crops. It has also been suggested in China that people eat the grain, rather than using it into make beer. Another suggestion is to have fewer animals as pets, so that food now being fed to pets could be used to feed humans instead.

The United States is one of the fastest-growing nations in the industrialized world.[33] It is adding 2.4 million people each year, one-third of whom come by immigration. (Immigration and births to immigrant women accounted for 70% of the increase in the population in the 1990s.[34]) It is the third most heavily populated nation in the world after China and India, with a population of 264 million in 1996.[35] The population on March 8, 2009 was 305,967,529.[36] Its incidence of teenage pregnancy is much higher than that in other industrialized nations. This is four to six times that in western Europe and nine times that in Japan and the Netherlands.[37] It cost U.S. taxpayers $9 billion in 2004 in health care, foster care, child welfare, food stamps, prison, and lower earnings. The abstinence-only sex education used in the United States from 2000 to 2008 has been shown to be no better than no sex education in preventing teen pregnancy.[38] Students who receive comprehensive sex education have half as many teen pregnancies as the others.

The affluence and overconsumption of the Americans mean that the 24 million people born in the last decade will consume more resources than the entire population of Africa.[39] America's annual population growth causes 2.5 times the global warming of Africa.[40] A child born in the United States will have twice the impact of one born in Sweden, 3 times one in Italy, 13 times one in Brazil, 35 times one in India, 140 times one in Bangladesh, and 280 times one in Nepal.[41] The average person in a developed nation uses 5 times more fertilizer, 12 times more oil, and 24 times more natural gas as his or her counterpart in a developing nation.[42] He or she produces 40 times more industrial waste and 52 times more industrial effluents. The United States leads the world in the per capita emissions of carbon dioxide.[43] The influence of the consumption in developed nations is felt beyond their borders because they need to import many raw materials. As an example, tropical forests in Costa Rica were cut and replaced with grass so that cattle could be grown on it to supply America's fast-food restaurants with hamburgers. Tropical forests have also been cut to supply hardwood for furniture, paneling, and so on, for use in developed nations. Many minerals used by the United States are mined in developing nations.

How many people can the world support with present-day technology?[44] Opinions vary, but it is clear that one cannot expect endless population growth and endless increases in food production.[45] It is important to support these people in a sustainable manner that does not foreclose any options for future generations. This means that practices that permanently damage land by loss of topsoil, nutrients, depletion of groundwater, or other, must be stopped. The loss of biodiversity must be stopped. Our behavior must be changed so that no further loss of the ozone layer occurs and global warming is moderated.[46] It is also clear that if the population becomes too large, the standard of living will decline in more places than it has at present. If the population grows to the point where the natural resource base can no longer support it, it may be controlled by famine, disease, and war.[47] A more humane way of approaching the problem is through wider use of family planning on a large scale. The rest of this chapter will deal with possible ways of doing this, as well as point out additional research that needs to be done.

16.2 CHEMISTRY OF HUMAN REPRODUCTION

The reproductive process is controlled by several hormones, as shown in **Chart 16.1**.[48] The biological effects and the interaction of hormones in females are shown on the left and those in males are shown on the right.

The hypothalamus produces gonadotropin-releasing hormone (LHRH or GnRH), which stimulates the pituitary gland to produce follicle-stimulating hormone (FSH), and luteinzing hormone (LH). LHRH is a decapeptide. FSH and LH are heterodimeric glycoproteins (with α- and β-subunits) with molecular weights of 34 and 28.5 kDa, respectively. FSH and LH stimulate an oocyte (an immature ovum) to develop. (The female is born with all the oocytes that she will ever have.) At the same time, the ovarian follicles synthesize the estrogen, estradiol, the biosynthesis starting from cholesterol (**16.1** Schematic). Small amounts are also made elsewhere. Estrone and estriol are active metabolites of estradiol. (Note that these estrogens have aromatic A rings and hence they are phenols.)

One ovum matures in the follicle in each 28-day cycle. At the same time, the endometrium (the lining of the uterus) proliferates to receive the ovum if it is fertilized. During the middle and late portions of this follicular phase, the concentration of FSH falls and the concentration of LH rises (in response to an increasing level of estrogen). This triggers ovulation by rupture of the follicle at about the 14th day. The follicle then becomes the corpus luteum, which produces progesterone. The estrogen and progesterone inhibit FSH

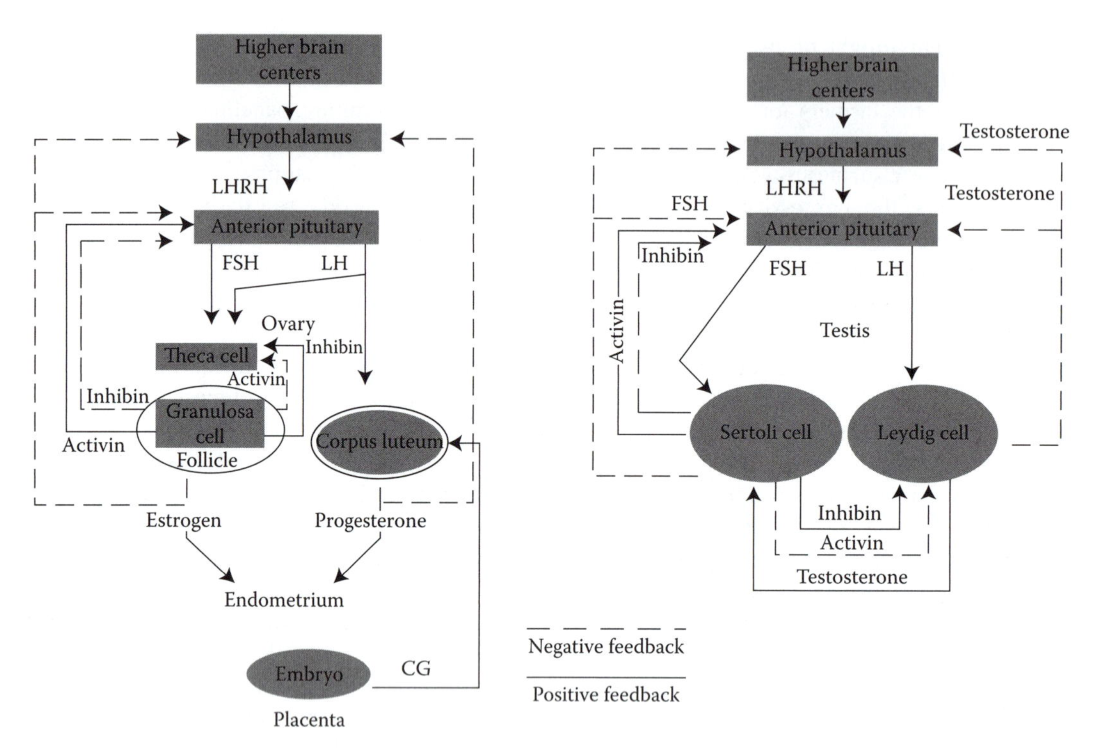

Chart 16.1 Chemistry and control of reproduction. (Reprinted from W.J. DeKoning, G.A. Walsh, A.S. Wrynn, and D.R. Headon, *Biotechnol.*, 1994, *12*, 988. With permission of Nature Biotechnology and the authors.)

Cholesterol

Estrone

Estradiol

Estriol

16.1 Schematic

and LH. The life of the corpus luteum is about 14 days. After ovulation, the uterus has sulfated carbohydrates that bind the L-selectin protein that arrives with the implanting blastocyst.[49] If fertilization[50] takes place, the lifetime of the corpus luteum is extended by the action of chorionic gonadotropin secreted by the implanting blastocyst (which results by cell division of the fertilized ovum). Human chorionic gonadotropin (hCG) is another dimeric glycoprotein, with a molecular weight of 36.6 kDa. When fertilization occurs, the zygote puts up a protective barrier, which is cross-linked with hydrogen peroxide, to keep out other sperm.[51] Inhibin and activin are glycoproteins that are involved in feedback

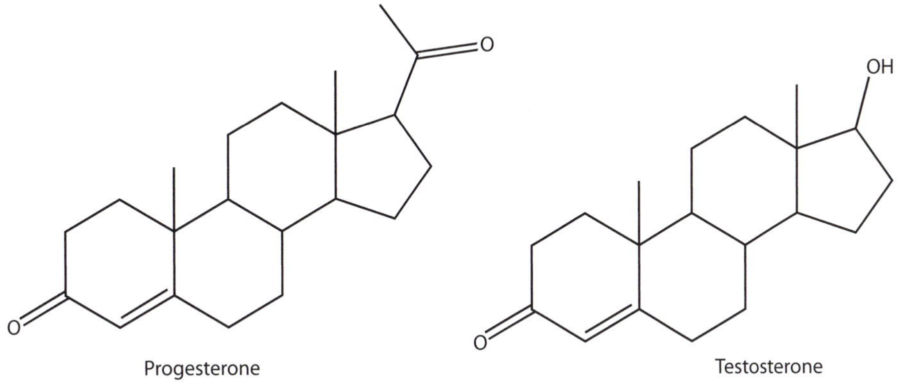

Progesterone

Testosterone

16.2 Schematic

loops. During pregnancy, the production of progesterone (**16.2** Schematic) by the placenta is about 10 times that of the ovary. If fertilization does not occur, two superficial layers of the endometrium are shed in menstrual bleeding, and the cycle begins again.

In the male, the FSH and LH cause the testes to produce androgens (male sex hormones), of which testosterone (see **16.2** Schematic) is typical. (Both progesterone and testosterone are also derived by biosynthesis from cholesterol. A microbiological process for the conversion of cholesterol to testosterone for therapeutic use has been developed using a *Mycobacterium*.[52]) Sperm are infertile when they leave the testis. They mature in their 2-week trip through the epididymis, which leads to the vas deferens and the penis. The gene septin 4 and the protein ARTS are necessary to remove excess cytoplasm from the sperm in order to make it fertile.[53]

The use of androstenedione (**16.3** Schematic), as a metabolic precursor of testosterone, to increase athletic prowess is unproven and may lead to detrimental effects, such as elevated blood lipid levels, liver problems, and testicular atrophy.[54]

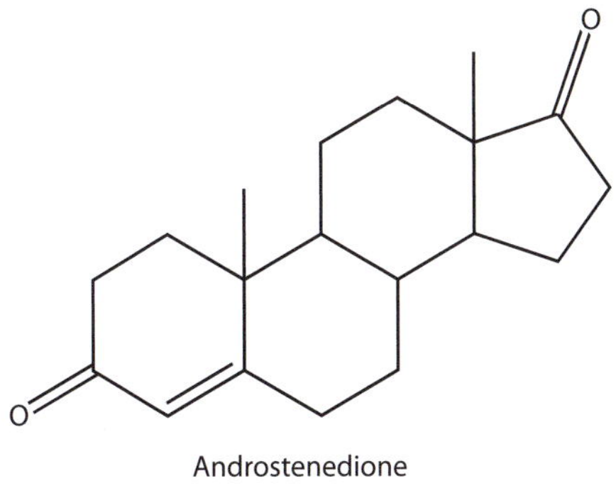

Androstenedione

16.3 Schematic

16.3 THE CHEMISTRY OF FAMILY PLANNING

In the United States, 60% of the pregnancies are unintended.[55] One in 20 American women has an unintended pregnancy each year.[56] Unwanted children often receive less, or less caring, attention and can become social problems. A world fertility survey indicated that 500 million women do not want any more children, did not want their last child, or want to control their fertility, but lack access to the necessary family-planning knowledge or services.[57] The majority of people, 67%, favor family planning and contraception.[58] Thus, there is a major need for better application of the methods known to be effective today, as well as research to lead to newer, and possibly cheaper, methods with fewer side effects. (Publicly funded family planning programs prevent nearly 2 million unintended pregnancies and more than 800,000 abortions in the United States each year.[59]) Two reviews give good coverage of the subject.[60]

In theory, knowledge of the woman's menstrual cycle might allow a couple to restrict intercourse to periods where fertilization would not occur. In practice, the method is unreliable. The time of ovulation can vary from month to month. The ovum must be fertilized within 12–24 h after ovulation. However, sperm can remain viable in the female reproductive tract for 5–7 days and possibly more. Other more reliable methods must be considered.[61] It is important to remember that almost all drugs have side effects, which may be minor in most persons, but may be major in some. Thus, any method that is used must have minimal side effects. Doctors select among the different methods to achieve the best one for any given individual.

16.3.1 The Use of Sex Hormones

Progestins can be used alone as oral contraceptives,[62] but they are less effective and have more side effects, such as intermenstrual or breakthrough bleeding and spotting, than when used in combination with estrogens. These inhibit ovulation by negative feedback on the hypothalamus and the pituitary gland, which decreases the secretion of FSH and LH. Newer oral contraceptives use a quarter of the estrogen and one-tenth of the progestin used in former years.[63] They often vary the amounts and proportions of estrogen and progestin through the phases of the menstrual cycle to minimize the amounts needed and the side effects. For this purpose, the natural estrogens and progestins would have to be used in large amounts, because they are not absorbed well from the gastrointestinal tract and remain active for only a short time. Synthetic analogues have overcome these drawbacks, such that only 30–35 μg of estrogen and 0.15–1.00 mg progestin need to be used. Two typical ones are shown in **16.4** Schematic.

Implantation of analogues of the sex hormones in slow-release formulations can offer protection for 5 years. Norplant consists of six match-like cylinders of levonorgestrel (see **16.4** Schematic) in a silastic matrix (made from a copolymer containing dimethylsiloxane and methylvinylsiloxane units). The implant can be removed surgically any time when the woman no longer desires protection. Occasionally, there have been problems with the removal.

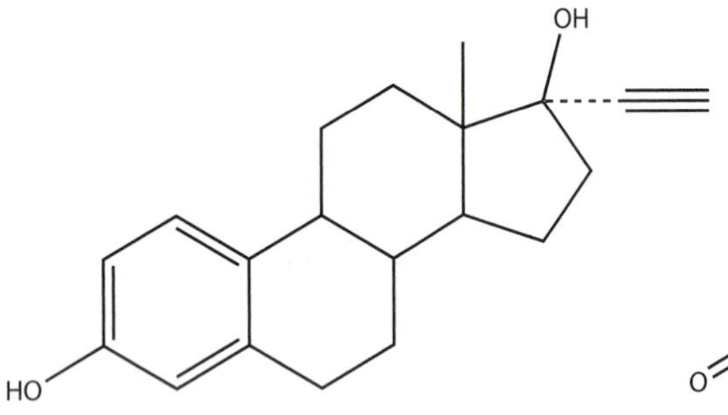

OH HO Ethynylestradiol

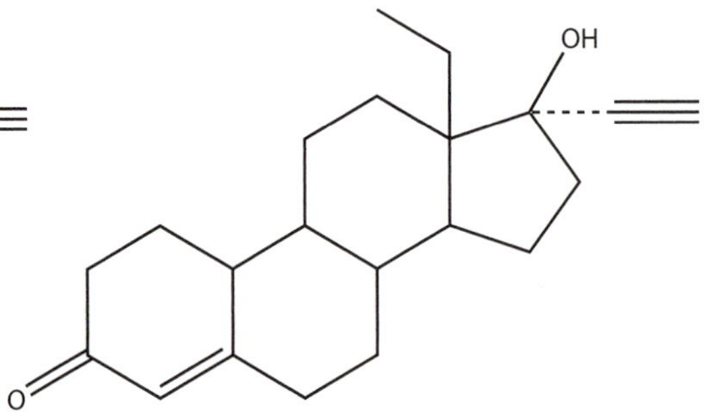

OH O Levonorgestrel

16.4 Schematic

For this reason, slow-release formulations of levonorgestrel in biodegradable polymers, such as poly(caprolactone), are being studied. The effective life is expected to be 12 months. There is a problem if the woman wants it to be removed surgically in a shorter time during which the polymer may have degraded appreciably.

The methyl ether of ethinyl estradiol (a monoether of the phenol group) can also be used as the ether is cleaved in the body. A variety of synthetic progestins are used.

The health benefits of oral contraceptives go beyond contraception and outlast the reproductive years.[63] These include reduced incidence of ovarian and endometrial cancers. Oral contraceptives do not increase the risk of breast cancer, heart disease, or blood clots.[64]

Long-acting injectable analogues of the sex hormones (**16.5** Schematic) are also used. Depo Provera is injected every 3 months as a microcrystalline aqueous suspension. The active compound, medroxyprogesterone acetate, dissolves slowly as needed. Another compound that is injected once a month is an ester that probably hydrolyzes slowly.

Contraceptive vaginal rings have been made that release the steroid slowly. (The walls of the vagina are easily penetrated by the hormones.) This can allow the use of drugs that have low oral activity. For example, an estrogen–progestin mixture can be incorporated in a silastic matrix. The ring has the advantage that it can be removed any time without surgery. Mirena™ is 99% effective and lasts for 5 years. It has been used in Europe for 17 years.[65]

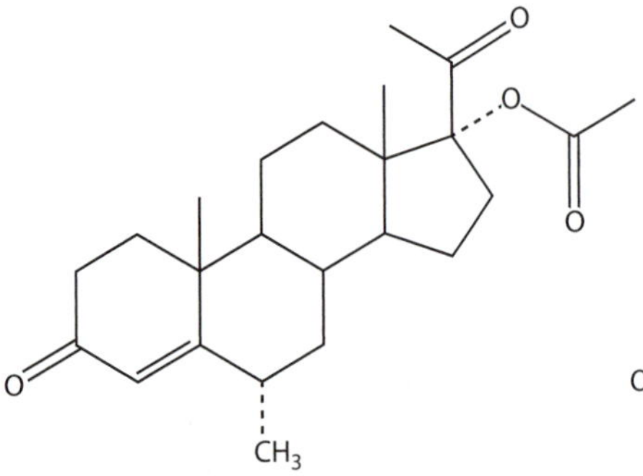

O O O O CH_3 Medroxyprogesterone acetate

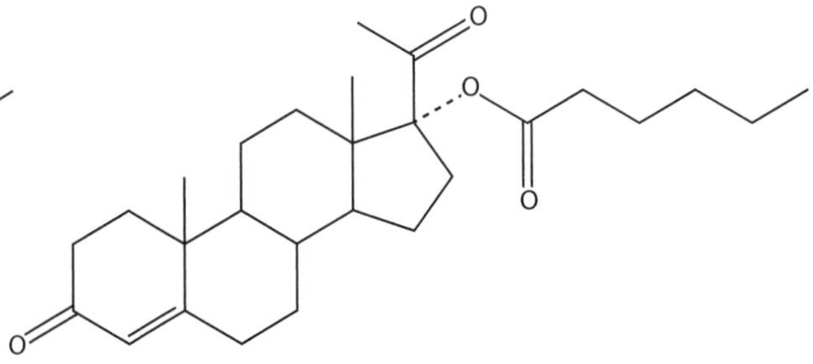

O O O O 17-α-hydroxyprogesterone caproate

16.5 Schematic

Contraceptives are also used to control the numbers of pigeons, geese, and wild horses.[66] Nicarbazin disables the hatching of birds. It is cheaper to just shoot surplus white-tailed deer.

16.3.2 Intrauterine Devices

Intrauterine devices (IUDs) can also contain steroids. One type contains 38 mg of progesterone, which is released slowly over 1 year. The one containing levonorgestrel is under development. Unmedicated rings appear to work just as effectively, although the details of their action are not well known. They appear to create an environment unsuitable for sperm and ovum and inhibit implantation. They can be made of polyethylene, stainless steel, or copper, with the last being the most common. Some can be left in for 10 years.[67]

16.3.3 Morning-After Treatments and Abortifacients

The chance of pregnancy can be reduced by 75% by a high dose of oral contraceptives taken within 72 h (under a doctor's supervision) after unprotected intercourse.[68] This usually consists of taking one set of pills followed by a second set 12 h later. The use of a product containing only levonorgestrel has cut the side effect of nausea by over 50%.[69] Additional work is needed to improve the efficacy and to reduce the side effect of nausea further. Inclusion of an antinausea drug may help. Danazol (**16.6** Schematic) may be the answer. It is said to be highly effective with few side effects.

The antiprogestin, mifepristone (RU-486; see **16.6** Schematic), can be used in conjunction with a prostaglandin as a morning-after pill.[70] It induced menstruation in 98% of women who were up to ten days late with their periods. Mifepristone can also be used to terminate pregnancy in the first trimester in a nonsurgical abortion.[71] When the mifepristone is followed with a prostaglandin, such as misoprostol[72] (**16.7** Schematic), given orally 36–48 h later, the abortion rate is greater than 95%. It is approved for use until up to 49 days of pregnancy in France and China, and for up to 63 days in Sweden and the United Kingdom. More than 200,000 women in France and the United Kingdom have used this method. It was approved for use in the United States in September 2000.[73] There were no side effects in 98% of the cases in the United States.[74] Mifepristone is also being studied for the treatment of Cushing's syndrome (caused by overactive adrenal glands), infertility, breast cancer, endometriosis, uterine fibroid tumors, brain tumors, and for the management of labor at the end of a normal pregnancy.

Mifepristone binds to the receptor that normally binds progesterone, preventing transcription of the genes normally triggered by progesterone. Other routes by which abortifacients may work include inhibition of progesterone biosynthesis, increased metabolism of progesterone, and inhibition of the chorionic gonadotropin that supports the corpus luteum. Prostaglandins serve many functions in the body, including a role in childbirth.[75] The body produces them from arachidonic acid (**16.7** Schematic), which is obtained from the diet or is produced, in turn, from linoleic acid, an essential fatty acid.

A combination of two prescription drugs, misoprostol (used for treatment of gastric ulcers; see **16.7** Schematic) and methotrexate (used for cancer, arthritis, and psoriasis; **16.8** Schematic), can also be used to terminate pregnancy. Of the 178 women in the early weeks of pregnancy receiving this

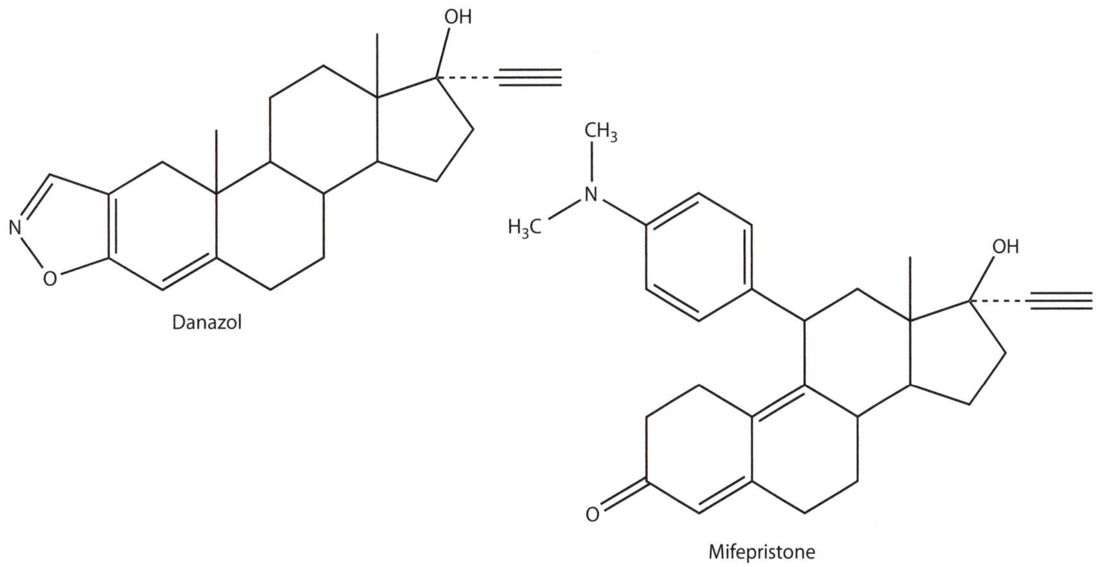

OH
N
O
Danazol

CH_3
N
H_3C
OH
O
Mifepristone

16.6 Schematic

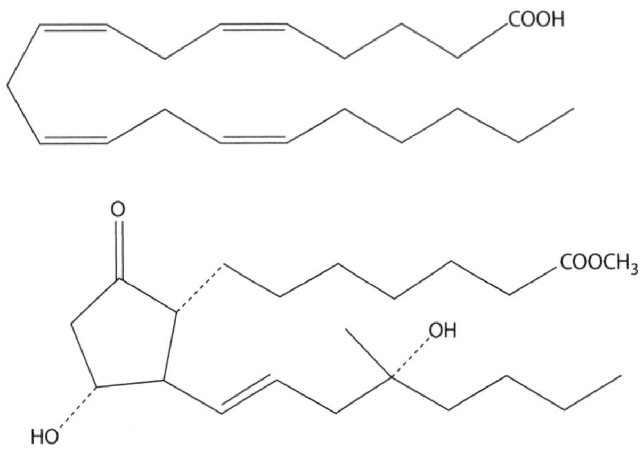

Arachidonic acid

Misoprostol

16.7 Schematic

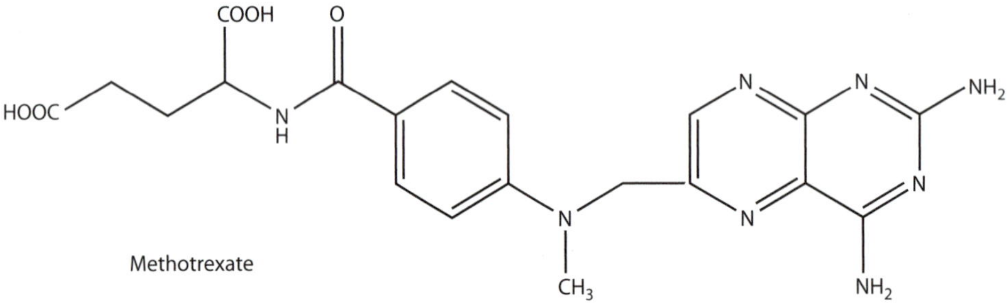

Methotrexate

16.8 Schematic

treatment, 171 terminated. More than 650 women have now used this method with good results[76]. In countries without access to mifepristone, each drug has also been used alone for medical abortions.[77]

The isocupressic acid (**16.9** Schematic) in the needles of *Pinus ponderosa* can cause abortions in the cattle that eat the needles.[78] This may be a possible lead for new abortifacients for humans.

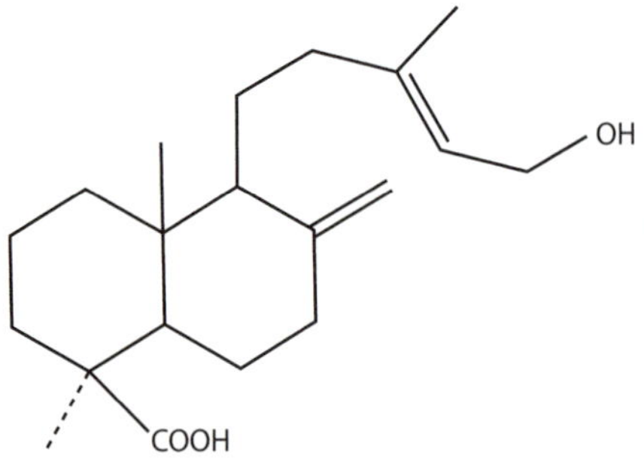

Isocupressic acid

16.9 Schematic

16.3.4 Physical Barriers

The physical barriers include diaphragms and cervical caps made of rubber that fit in the vagina over the cervix. They are mechanical barriers to block the passage of sperm and are often used with spermicidal agents, such as nonoxynol (a surfactant; **16.10** Schematic).

These must fit well, be inserted correctly, and remain in place long enough for the spermicide to act. A vaginal sponge is less effective. India is testing oil from the neem tree (*Azadirachta indica*), with and without other plant extracts, as a vaginal cream or suppository for contraception.[79] (This is the tree that produces the insect repellent

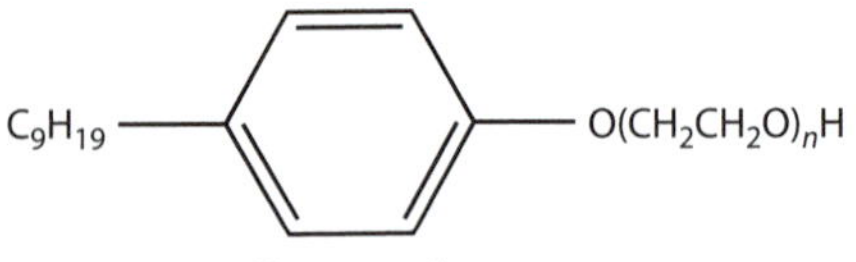

C_9H_{19}—C_6H_4—$O(CH_2CH_2O)_nH$

Nonoxynol

16.10 Sehematic

discussed in Chapter 11.) One problem that must be overcome is the sharp unpleasant odor of the oil.

The condom is the only method of contraception that prevents the transmission of diseases at the same time.[80] Sexually transmitted diseases include those caused by *Chlamydia trachomatis, Neisseria gonorrhoeae, Trichomonas vaginalis, Treponema pallidum*, herpes simplex virus II, human papilloma virus, and HIV. Chlamydia I infection[81] is the most prevalent sexually transmitted disease in the United States, with up to 4 million cases per year. It can cause pelvic inflammatory disease, tubular pregnancies, and infertility in women. It responds to treatment with antibotics.[82] Its complete genome has been sequenced. (Gonorrhea is caused by *Neisseria* and syphilis by *Treponema*. The United States has the highest rate of sexually transmitted diseases in the industrialized world.[83]) Although it may be possible to develop vaccines for these in the future, almost none is available today.[84] Two companies sell vaccines for human papilloma virus, the cause of cervical cancer.[85] Genital herpes has been prevented by controlled release of antibodies from poly(ethylene-*co*-vinyl acetate) disks.[86] Circumcision reduces the risk of HIV by 65%.[87] Genital *Herpes simplex* increases the chance of HIV transmission.

It is now commonplace for teenagers and college students in the United States to "hook up" (have a casual sexual encounter with no intention of a long-term relationship).[88] While most of them use contraceptives, this practice may contribute to some early, unintended pregnancies.

Condoms work well if they are of good quality, are not too old, and stay on. They must be used with water or glycol-based lubricants, for petroleum-based lubricants cause them to lose strength and possibly fail. Condoms are made by dipping a form coated with a coagulant into a latex of natural rubber. After drying, a second dip is done. The problem is that there are flaws in many of the condoms. About 10–15% of the rubber used in making condoms and rubber latex gloves is wasted in rejects.[89] Former testing did not detect holes smaller than 10 μm.[90] The problem is that the 5-μm holes that are present are large enough to allow the HIV to pass through. The virus particles have a diameter of 0.1 μm, compared with 6.0 μm for the bacteria causing syphilis and 45 μm for the sperm. Hepatitis B virus is about 20 nm in size. Use of a condom reduces the chance of transmitting HIV tenfold, but not to zero. Newer testing methods using high-voltage electricity can decrease the measurable size to 1 μm or less.[91] These are said to find the small holes through which the viruses could pass. Condom quality improved in the 4 years up 1990.[92] The Johns Hopkins School of Public Health reports that that HIV cannot penetrate an intact latex condom.[93] (In 1998, there were 33 million cases of AIDS in the world, including 5.8 million new cases that year.[94] There are 22 million cases of HIV/AIDS in subSaharan Africa.)

There is a need for vaginal creams that are not only spermicidal but which also inactivate bacteria and viruses. Among those being developed is an emulsion of oil, water, and two common laboratory detergents that kills fungi, bacteria, viruses, sperm, and red blood cells, but no other animal cells.[95] Another is a cellulose ether quaternary ammonium salt that coats mucosa of the reproductive tract and slows down the sperm.[96] It is used with dextran sulfate, a known virucide, and a surfactant that dissolves the outer shell of the sperm. Despite high hopes and much research, only one of many has worked and it ("PRO 2000") gave only a 30% reduction in a small trial.[97] A larger trial was in progress in 2009.

Another problem is that some persons are allergic to the proteins often found in natural rubber. This can be avoided by the use of a polyurethane condom. Unfortunately, as now made, these are more likely to tear (7–9% of the time) than the rubber latex condoms (2% of the time).[98] It has been suggested that they be made of guayule rubber latex, which isfree of the allergens.[99] Thermoplastic elastomers, such as polystyrene-*b*-(ethylene-*co*-butene)-*b*-polystyrene have been used to make gloves that are more resistant to ozone than those made of natural rubber.[100] These should resist aging better than those made of natural rubber. It should be possible to mold a condom without holes from a thermoplastic rubber.

16.3.5 Contraceptives for Males

Additional contraceptives for males are being developed to supplement the condom. A study in Minnesota has shown that birth rates correlate with sperm count in the range of 46.5–123 million/mL.[101] Weekly injections of testosterone in 399 men reduced the sperm count for 98.6% of them.[102] A sperm count of 3 million or fewer almost guaranteed no conception. Only four pregnancies occurred in the trial. The testosterone can also be administered by skin patches, a method more convenient than injections.[103] A combination of a skin patch of testosterone and a progesterone pill reduced men's sperm counts to zero.[104] When the treatment was discontinued, the sperm counts gradually returned to normal. A male contraceptive pill containing testosterone and desogestrel was 100% effective in clinical trials with no side effects.[105] The Population Council and Sano are working with a synthetic androgen that is 10 times as active as testosterone. It maintains normal sexual functions and muscle mass without overstimulating the growth of the prostate gland. Organon is testing an androgen that is taken orally in the pill form.[106] Indenopyridines, which disrupt the development of sperm, are being tested by the Research Triangle Institute in North Carolina.[107] Any method that relies on inhibiting spermatogenesis may require 2–3 months to become effective. This time is needed for normal sperm originally present to pass through and be eliminated from the system. Spermatogenesis is controlled by FSH and LH. It may be possible to inhibit them, but testosterone supplements may have to be given at the same time to maintain libido. Despite these interesting leads, a male contraceptive is still not available commercially.[108]

Some male contraceptives tested in the past have been too toxic or have not been reversible when the substance was no longer administered or have interfered with libido. Gossypol (**16.11** Schematic) from cottonseed was used in China, but was found to have the first two problems.

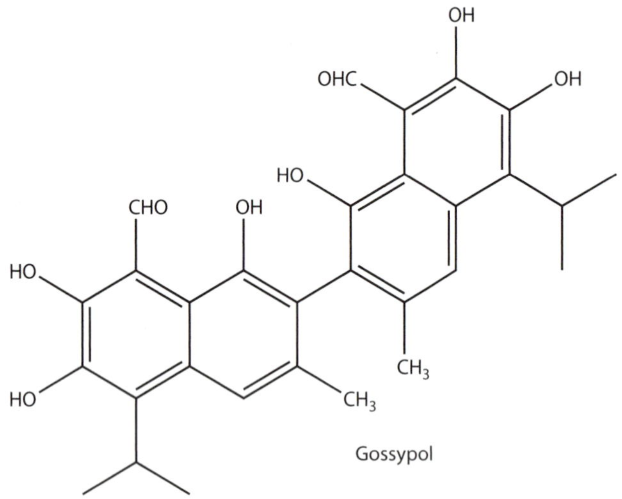

Gossypol

16.11 Schematic

Phenoxybenzamine

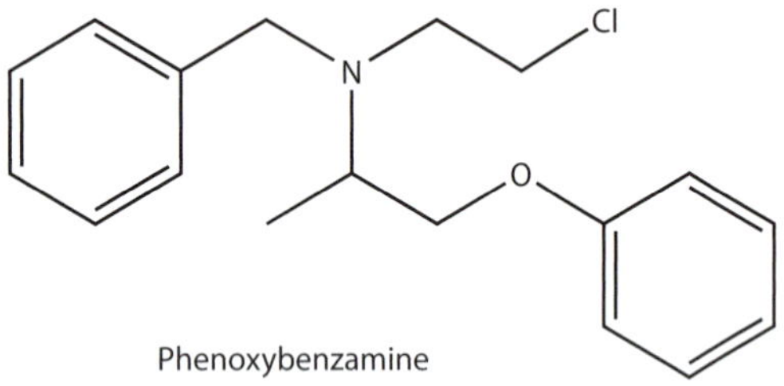

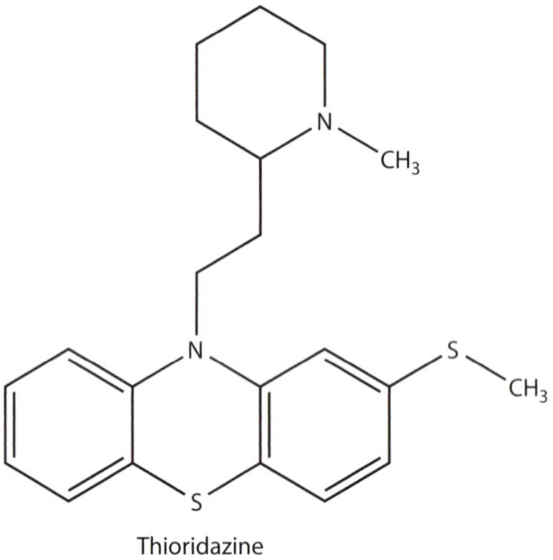

16.12 Schematic

Extracts of the Chinese vine, *Tripterygium wilfordii*, have also shown some promise. This male contraceptive affects sperm maturation in the epididymis. Two other drugs work by another mechanism. They prevent the contractions of the muscles that run along the vas deferens so that seminal fluid and sperm are not pushed along. This results in a "dry" orgasm with no ejaculation. The drugs are phenoxybenzamine (**16.12** Schematic), which is normally used to control hypertension, and thioridazine (see **16.12** Schematic), which is an antispychotic.[109] The lack of ejaculation was discovered as a side effect of the normal uses of the drugs.

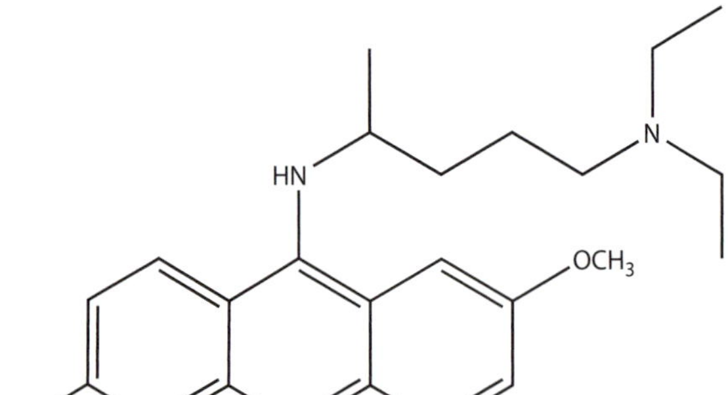

16.3.6 Sterilization

Surgical sterilization is a widely used method of contraception for both men and women. The problem is that it is very difficult to reverse if a couple wants to have more children, as when a child dies. It may also be too expensive for very poor women in developing countries. A nonsurgical approach sterilizes a woman by insertion of quinacrine (**16.13** Schematic) into the uterus by a trained paramedic.[110] It produces scar tissue in the Fallopian tubes that prevents pregnancy. This drug has been used over a long time period as an antimalarial. The cost of the treatment is estimated at less than 1.00–2.50 dollars compared with 12 dollars for laparoscopic sterilization (45 dollars if the cost of the whole team is included). Insertion of small springs into the Fallopian tubes may provide permanent birth control without surgery.[111] Vasectomy for men is already inexpensive.

A nonsurgical vasectomy has been investigated in India. A styrene-maleic anhydride copolymer was injected into the vas deferens to block the duct.[112] If removal is desired later, the polymer can be dissolved out with dimethysulfoxide. Reversible vas occlusion has been administered successfully

Thioridazine

Quinacrine

16.13 Schematic

to over 300,000 males. It involves no surgery, is reversible, and costs less than $5.00 in developing countries.

16.3.7 Newer Methods under Study[113]

The relative effectiveness of the various methods is given by the approximate numbers in Table 16.2[114]

A thorough knowledge of the process of human reproduction may lead to other approaches. The gamete recognition proteins in sperm–egg interactions in animals are just beginning to be understood.[115] The process involves five steps:

1. Molecules released by the egg attract the sperm.
2. The protein integrin on the egg binds to a small piece of the sperm protein fertilin.
3. The acrosomal vesicle on the head of the sperm opens.
4. A protease creates a hole in the egg envelope for the sperm to pass.
5. The two cells fuse.

A 13-amino-acid polypeptide that mimicks the binding region of fertilin in step 2 inhibited fertilization in *in vitro* trials.

Efforts have been made to develop vaccines that will prevent pregnancy. Ideally, these would target some antigens that are associated only with the male (e.g., the enzymes present in the head of the sperm that hydrolyze part of the zona pellucida, a glycoprotein coating, that the sperm must penetrate to reach the ovum). Sperm membrane proteins might also be suitable antigens.[116] Vaccines have also been made to hCG, which is essential for implantation of the blastocyst. A subunit of this protein was attached to a disabled toxoid as a carrier to make the antigen. Vaccines for males have been directed toward antibodies that neutralize gonadotropin-releasing hormone, which is needed for the production of sperm. Although the vaccines work, they only last a few months, so that booster shots are required.

Another approach is to find compounds that block the enzymes that allow the sperm to make a hole through the zona pellucida.[117] One product that does this was 92% effective in preventing pregnancy in rats, indicating a need for further optimization. It is possible that combinatorial synthesis of polypeptides can lead to compounds that will competitively block the FSH, LH, and hCG at their receptor sites. Recombinant analogues of FSH, LH, hCG, and LHRH are being reported.[118] Some of these might be useful in contraception. Others might be useful in treating infertility and other reproductive deficiencies. Each stage in human reproduction needs further study to see if synthetic analogues of the natural compounds can block the receptors or otherwise disrupt the process without undesirable side effects.

Table 16.2 Effectiveness of Various Contraceptive Method

Method	% Pregnant in 1 Year
Rhythm method	8.5
Spermicides	7
Female condom	5
Male condom	2
Oral contraceptive	1
Sterilization	0.5
Norplant	0.3

16.4 SUMMARY OF THE PROBLEM

Relatively few drug companies carry out research on contraception. Of the large pharmaceutical companies in the world, only Ortho and Wyeth are doing contraceptive research and development.[119] Most companies are concerned about potential liability, poor profits, and high development costs.[120] Vaccine research boomed after a 1986 U.S. law shielded vaccine makers from liability.[121] The committee of the U.S. Institute of Medicine investigating the problem recommends that U.S. product liability law be reformed, that health insurance cover contraceptives, and that international health agencies pool their funds to create a large market. Many of the poor people of the world cannot afford to pay for the contraceptives. Over a 5-year period, an American woman may spend 1784 dollars for oral contraceptives, or 1290 dollars for Depo Provera, or 850 dollars for Norplant.[122] A woman in a developing nation may obtain these for a much lower price if they were supplied by an international health agency. There has been a campaign for contraceptive equity in the United States. Health insurance that covers Viagra, a drug for male impotence, should also cover contraceptives for women, but sometimes it does not.

Even though the poor may not be able to afford to pay for family-planning services, it is important to provide them. In the United States, every public dollar spent on such services saves the taxpayers almost $4.50.[123] On the international scene, it will be cheaper to subsidize the distribution of family-planning services than to provide famine relief, refugee camps, money for defense, military intervention, and so on. Less than 1% of developmental aid goes to family planning. The United States ranks last among 21 industrialized nations in providing developmental aid to other countries (as a percentage of gross national product). It spends 0.10%, compared with the 0.97% of Denmark, the highest in the list.[124]

The state of Kerala in southern India has shown how to control population growth, even though it is not wealthy.[125] Its per capita income ($619/yr) in 2005 was 1/65 that of the United States. (In the rest of India in 2005, the per capita income was $514.) Its population of 30 million people lives in land about the size of Vancouver Island, British Columbia. Life expectancy is high. Infant mortality is low. The birth rate is lower, and the literacy rate is higher than that in the United States. The state spends 60% of its budget on health and education. Women and men are treated equally under this system. It also has old-age pensions. The secret to the control of population in developing nations appears to be to improve the social status of women and to focus on the satisfaction of basic human needs.[126] Secure land tenure also helps.

Tunisia has had reproductive health clinics in every city and village since 1957 when it gained independence.[127] Its population has grown from 4 to 9 million since then. Eighty percent of the people are middle class, 85% own their own homes, and every boy and girl goes to school. The per capita income per year was $8210 in 2006. Life expectancy is 75.3 years. In contrast, neighboring Algeria's population has grown from 4 to 38 million and the country had a civil war. The annual per capita income was $7082 in 2006. Life expectancy is 73.5 years.

Pakistan has been at the other extreme in government expenditures. In 1990, it spent more than twice as much on the military as on health and education combined.[128] Its population of 146 million in 1999 grew to 172,800,048 by July 2008. The fertility rate of 5.8 children per woman in 1999 had dropped to 4.1 by 2008. The Pakistani Minister of Population Welfare visited Washington, DC in 1997 to ask for financial aid to deal with this problem. At that time, 25% of the married women in the country used contraceptives.[129] Pakistan's health services are said to be improving.[130] The semiautonomous Gilgit region of northern Pakistan is an exception, having attained a stable population.[131] The process of making the sexes equal began in 1984. Infant mortality has fallen by 80% and birth rates have been reduced by half.

The cost of providing family planning and reproductive health services for a year to every couple in the world that needs and wants them has been estimated at 17 billion dollars. This is what the world spends each week on armaments.[132] This same amount is spent for pet food in the United States and Europe each year.

Fertility declines in East Asia since 1950 have been dramatic, dropping to about replacement levels or below in Japan, Taiwan, Korea, Iran, and China.[133] In Thailand, fertility dropped from 6 children per woman in the 1960s, to 3.7 in 1980, and to the replacement level of 2.1 in 1991.[134] Adult women in Thailand have a literacy rate of 90% compared with 96% for men. In Latin America, Costa Rica has achieved a stable population with an annual per capita income of $10,964 in 2006.

Contraceptive use varies widely among countries. The highest use is in France and Belgium, with rates of 80% and 79%, respectively. The United States ranks tenth at 74%. The lowest rate, 1%, is in Somalia. In Ethiopia, the rate is 2%. In the poorest country in the world in 1993, Mozambique (80 dollars per capita), the use rate was 4%. Military intervention was necessary in Somalia. Ethiopia required extensive famine relief. Postwar reconstruction in Mozambique has required hundreds of millions of dollars in foreign aid.[135]

Conservative religious groups are making it more difficult to provide family planning services and supplies to those nations that need them.[136]

GH Brundtland, former prime minister of Norway, said, "Population growth is one of the most serious obstacles to world prosperity and sustainable development."[137] The biologist Norman Myers says, "Family planning could bring more benefits to more people at less cost than any other single technology now available to the human race."[138] Lester R. Brown says "Filling the family planning gap may be the most urgent item on the global agenda. The benefits are enormous and the cost minimal."[139] Another author feels, "Population growth is just one agent in the degradation of our environment. Unsustainable consumption patterns in industrialized nations, poverty, unequal access to resources and the low status of women in many parts of the world are all implicated."[140] The example of Kerala shows how to slow population growth rates. No one is quite sure how to reduce consumption to sustainable levels, although this will have to be done at some point.

REFERENCES

1. L.R. Brown and B. Halweil, *World Watch*, 1999, *12*(5), 20.
2. (a) L.R. Brown, In: L. Starke, ed., *State of the World 1995*, W.W. Norton, NY, 1995, 14; (b) J. Gasana, *World Watch*, 2002, *15*(5), 24.
3. P. Uvin, *Environment*, 1996, *38*(3), 6.
4. Anon., *Science*, 1996, *272*, 657.
5. www.worldbank.org/urban/symposium2005/papers/fay.pdf.
6. S.L. Postel, *Bioscience*, 1998, *48*, 629; S.L. Postel, *World Watch*, 1999, *12*(5), 30.
7. (a) D. Pimentel, M. Tort, L. D'Anna,A. Krawic, J. Berger, J. Rossman, F. Mugo, N. Doon, M. Shriberg, E. Howard, S. Lee, and Talbot., *Bioscience*, 1998, *48*, 817; (b) G. Mock and P. Steele, *Environment*, 2006, *48*(1), 8; (c) J.K. McKee, *Sparing Nature–The Conflict Between Human Population Growth and Earth's Biodiversity*, Rutgers University Press, New Brunswick, NJ, 2003.
8. (a) M. Painter and W.H. Durham, *The Social Causes of Environmental Destruction in Latin America.* University of Michigan Press, Ann Arbor, 1995; (b) N. Myers, *Ultimate Security: The Environmental Basic of Political Stability*, W.W. Norton, NY, 1993; (c) L. Lassonde, *Coping with Population Challenges*, Earthscan, London, 1997; (d) G.D. Ness and M.V. Golay, *Population and Strategies for National Sustainable Development*, Earthscan, London, 1997; (e) G. Hardin, *Living Within Limits: Ecology, Economics and Population Taboos*, Oxford University Press, Oxford, 1993; (f) M.C. Cruz, C.A. Meyer, R. Repetto, and R. Woodward, *Population Growth, Poverty and Environmental Stress: Frontier Migration in the Philippines and Costa Rica*, World Resources Institute, Washington, DC, 1992; (g) B. Cartledge., *Population and the Environment—The Linacre Lectures 1993–1994*, Oxford University Press, Oxford, 1995; (h) A. Najam and M. Brower, *Poverty, Development and the Environment*, Union of Concerned Scientists, Cambridge,

MA, May 1994; (i) N. Polunin, ed., *Population and Global Security*, Cambridge University Press, Cambridge, 1998; (j) N. Birdsall, A.C. Kelley, and S.W. Sinding, eds, *Demographic Change, Economic Growth and Poverty in the Developing World*, Oxford University Press, NY, 2001; (k) A. Goudie, *The Human Impact on the Natural Environment*, 6th ed., Blackwell Publishing, Malden, MA, 2005.

9. (a) Anon., *Popline*, July/Aug. 1998, *20*, 3 (Population Institute, Washington, DC); (b) O. Elgun., *Popline*, July/Aug. 1999, *21*, 1 (Population Institute, Washington, DC).
10. (a) www.optimumpopulation.org; (b) www.census.gov/main/www/popclock.html.
11. J. Bongaarts, *Science*, 1998, *282*, 419.
12. Anon., *Popline*, Jan./Feb. 1999, *21*, 3 (Population Institute, Washington, DC).
13. (a) Anon., *Popline*, Mar./Apr. 1999, *21*, 4 (Population Institute, Washington, DC); (b) C.N.R. Rao, *Science*, 2009, *325*, 126.
14. Anon., *Popline*, Mar./Apr. 1995, *17*, 1 (Population Institute, Washington, DC).
15. G. Daily, P. Dasgupta, B. Bolin, P. Crosson, J. duGuerny, P. Ehrlich, C. Folke, A.M. Jansson, B.-O. Jansson, N. Kautsky, A. Kinzig, S. Levin, K.-G. Maler, P. Pinstrup-Andersen, S. Siniscalco, and L.B. Walker, *Science*, 1998, *281*, 1291.
16. (a) Anon., *Popline*, Mar./Apr. 1998, *20*, 3 (Population Institute, Washington, DC); (b) G. Conay and G. Toenniessen, *Science*, 2003, *299*, 1187.
17. Anon., *Popline*, Sept./Oct. 1997, *19*, 2–4 (Population Institute, Washington, DC).
18. (a) Anon., *Popline*, May/June 1995, *17*, 1 (Population Institute, Washington, DC); (b) www.airninja.com/worldfacts/countries/Mozambique/gdppercapita.htm.
19. Central Intelligence Agency (CIA) World Factbook (World Bank Data), Washington, DC.
20. en.wilkipedia.org/wiki/List_of_countries_by_life_expectancy (from CIA World Factbook 2008 estimates).
21. C. Holden, *Science*, 2008, *322*, 509.
22. Anon., *Science*, 2007, *318*, 1701.
23. Anon., *Environment*, 2005, *47*(10), 6.
24. M. Enserink, *Science*, 2008, *322*, 37.
25. (a) Anon., *Popline* (Population Institute, Washington, DC), Nov./Dec., 2007; (b) www.mrdowling.com/800gdppercapita.html.
26. Anon., *Ecosystems and Human Well-Being: Synthesis Report–Millennium Ecosystem Assessment*, Island Press, Washington, DC, 2005.
27. M. Campbell, J. Cleland, A. Ezeh, and N. Brata, *Science*, 2007, *315*, 1501.
28. R. Johnston, *Science*, 1997, *276*, 664.
29. (a) C. Mann, *Science*, 1997, *277*, 1038; (b) L.T. Evans, *Feeding the Ten Billion: Plants and Population Growth*, Cambridge University Press, Cambridge, 1998.
30. (a) M.P. Apse, G.S. Aharon, W.A. Snedden, and E. Blumwald, *Science*, 1999, *285*, 1256; (b) W.B. Frommer, U. Ludewig, and D. Rentsch, *Science* 1999, *285*, 1222.
31. (a) P.M. Vitousek, H.A. Mooney, J. Lubchenko, and J.M. Melillo, *Science*, 1997, *277*, 494; (b) S.L. Postel, G.C. Daily, and P.R. Ehrlich, *Science*, 1996, *271*, 785.
32. K.M. Reese, *Chem. Eng. News*, 1996 Mar., *18*, 88.
33. (a) Anon., *Popline*, May/June 1997, *19*, 6 (Population Institute, Washington, DC); (b) Anon., *Science*, 2008, *321*, 1139.
34. Letter from W. Fornos, Population Institute, Washington, DC, Mar. 1999.
35. R.E. Schmid, *Wilmington Delaware News J.*, 1996 Mar., *14*, A11.
36. (a) www.census.gov/ipc/www/idbnew.html; (b) www.census.gov/main/www/popclock.html.
37. (a) Anon., *Popline*, Sept./Oct. 1998, *20*, 3 (Population Institute, Washington, DC); (b) www.teenpregnancy.org/costs.
38. P.K. Kohler, L.E. Manhart, and W.E. Lafferty, *J. Adolescent Health*, 2008, *42*(4), 344.
39. (a) *New Sci*., Sept. 1994, *17*, 7; (b) R.W. Kates, *Environment*, 2000, *42*(3), 10.
40. (a) P. Harrison, *Amicus. J.* (Natural Resources Defense Council), 1994, *15*(4), 16; (b) Anon., *Chem. Ind.* (*Lond.*), July 9, 2007, 5.
41. R. Goodland and H. Daly, *Ecol. Appl.*, 1996, *6*, 1002, 1011.
42. P. Harrison, *Environ. Action*, 1994, *26*(2), 17.
43. C. Lynch, *Amicus J.* (Natural Resources Defense Council), 1998, *19*(4), 15.
44. (a) D.H. Meadows, *Beyond the Limits: Confronting Global Collapse, Envisioning a Sustainable Future*, Chelsea Green Publishing, Mills, VT, 1992; (b) L.R. Brown and H. Kane, *Full House: Reassessing the Earth's Carrying Capacity*, W.W. Norton, NY, 1994; (c) J.E. Cohen, *How Many People Can the Earth Support?* W.W. Norton, NY, 1995; (d) P. Harrison, *The Third Revolution*, Penguin Books, London, 1993; (e) L.A. Mazur, ed., *Beyond the Numbers—A Reader on Population, Consumption, and the Environment*, Island Press, Washington, DC, 1994; (f) J.E. Cohen, *Science*, 1995, *269*, 341; (g) G.D. Moffet, *Critical Masses*, Viking Penguin, NY, 1994; (h) www.optimumpopulation.org.
45. A.A. Bartlett, *Science*, 1997, *277*, 1746.
46. B.C. O'Neill, F.L. MacKellar, and W. Lutz, *Population and Climate Change*, Cambridge University Press, NY, 2001.
47. T.F. Homer-Dixon, *Environment, Scarcity and Violence*, Princeton University Press, Princeton, NJ, 2001.
48. (a) W.J. deKoning, G.A. Walsh, A.S. Wrynn, and D.R. Headon, *Biotechnology*, 1994, *12*, 988; (b) I. Pollard, *A Guide to Reproduction: Social Issues and Human Concerns*, Cambridge University Press, Cambridge, 1994; (c) J.W. Gunnet and L.A. Dixon, *Kirk–Othmer Encyclopedia of Chemical Technology*, 4th ed., Wiley, New York, 1995, *13*, 433; (d) J. Sandow, G. Neef, K. Prezewowsky, and U. Stache, *Ullmann's Encyclopedia of Industrial Chemistry*, 5th ed., VCH, Weinheim, 1989; *A13*, 101; (e) B.P. Morgan and M.S. Moynihan, *Kirk–Othmer Encyclopedia of Chemical Technology*, 4th ed., Wiley, New York, 1997, *22*, 858; (f) P. Primakoff and D.G. Myles, *Science*, 2002, *296*, 2183;(g) H-Y. Fan, Z. Liu, M. Shimada, E.L. Sterneck, P.F. Johnson, S.M. Hedrick, and J.S. Richards, *Science*, 2009, *324*, 890.
49. (a) O.D. Genbacev, A. Prakobphol, R.A. Foulk, A.R. Krtolica, D. Ilic, M.S. Singer, Z-Q. Yang, L.L. Kiessling, S.D. Rosen, and S.J. Fisher, *Science*, 2003, *299*, 405; (b) S.A. Robertson, W.V. Ingman, S. O'Leary, D.J. Sharkey, and K.P. Tremellen, *J. Reproductive Immunol.*, 2002, *57*, 109.
50. M.L. Stitzel and G. Seydoux, *Science*, 2007, *316*, 407.
51. Anon., *Chem. Eng. News*, Dec. 13, 2004, 28.
52. W.-H. Liu and C.-K. Lo, *J. Ind. Microbiol. Biotechnol.*, 1997, *19*, 269.

53. N. Delaney, *Chem. Ind.* (*Lond.*), Apr. 4, 2005, 8.
54. (a) P. Zurer, *Chem. Eng. News*, Sept. 1998, *28*, 37; (b) S.K. Ritter, *Chem. Eng. News*, Nov. 17, 2003, 66.
55. Anon., *Popline*, July/Aug. 1997, *19*, 1 (Population Institute, Washington, DC).
56. C. Richards, *Planned Parenthood Federation of America*, Nov. 2006.
57. A.A. Leiserowitz, R.W. Kates, and T.M. Parris, *Environment*, 2005, *47*(9), 22.
58. W. Fornos., *San Diego Union-Tribune*, May 19, 1996.
59. R.B. Gold, A. Sonfield, C.L. Richards, and J.J. Frost, *Alan Guttmacher Institute Report*, Feb. 24, 2009.
60. (a) D.W. Hahn, *Kirk–Othmer Encyclopedia of Chemical Technology*, 4th ed., Wiley, New York, 1993, *7*, 219; (b) P.F. Harrison and A. Rosenfield, *Contraceptive Research and Development—Looking to the Future*, National Academy of Sciences Press, Washington, DC, 1996.
61. (a) J.A. Robertson, *Children of Choice—Freedom and the New Reproductive Technologies*, Princeton University Press, Princeton, NJ, 1994; (b) P. Senanayake and M. Potts, eds, 2nd ed., *Atlas of Contraception*, Informa Health Care, Taylor & Francis, Boca Raton, FL, 2008; (c) J. Guillebaud, *Contraception Today: A Pocketbook for General Practioners and Practice Nurses*, Taylor & Francis, Boca Raton, FL, 2007.
62. (a) C. Djerassi, *From the Lab Into the World: A Pill for People, Pets and Bugs*, American Chemical Society, Washington, DC, 1994; (b) J.Z. Guo, D.W. Hahn, and M.P. Wachter, *Kirk–Othmer Encyclopedia of Chemical Technology*, 4th ed., Wiley, New York, 1995, *13*, 481; (c) C. Djerassi, *This Man's Pill- Reflections on the 50th Anniversary of the Pill*, Oxford University Press, Oxford, 2001; (d) L.V. Marks, *Sexual Chemistry: A History of the Contraception Pill*, Yale University Press, New Haven, CT, 2001.
63. Anon., *Univ. Calif. Berkeley Wellness Lett.*, 1999, *15*(12), 5.
64. Anon., *Chem. Ind.* (*Lond.*), 1999, 44.
65. Letter, *Planned Parenthood of Delaware*, Fall, 2001.
66. (a) E. Wolf, *Conservation*, 2008, *9*(1), 43; (b) D. Fox, *Conservation*, 2007, *4*, 20.
67. Anon., *Univ. Calif. Berkeley Wellness Lett.*, 2006, *22*(12), 8.
68. (a) Anon., *Univ. Calif. Berkeley Wellness Lett.*, 1996, *12*(6), 8; (b) Anon., *Univ. Calif. Berkeley Wellness Lett.*, 1996, *13*(1), 1; (c) Anon., *Univ. Calif. Berkeley Wellness Lett.*, 2000; *17*(2), 8; (d) Anon., *Nat. Biotechnol.*, 1996, *14*, 930.
69. (a) Anon., *Popline*, July/Aug. 1999, *21*, 2 (Population Institute, Washington, DC); (b) Anon., *J. Couzin, Science*, 2005, *310*, 38; (c) www.not-2-late.com.; (d) www.backupyour birthcontrol.org.; (e) Anon., *Univ. Calif. Berkeley Wellness Lett.*, 2002, *18*(11), 6.
70. R.F. Service, *Science*, 1994, *266*, 1480.
71. (a) I.M. Spitz and I. Agranat, *Chem. Ind.* (*Lond.*), 1995, 89; (b) E.-E. Baulieu, *The Abortion Pill: RU-486, A Woman's Choice*, Simon & Shuster, NY, 1991; (c) L. Lauder, *RU-486: The Pill That Could End the Abortion Wars Against Your Family and Why Women Don't Have It*, Addison Wesley, Reading, MA, 1991.
72. (a) R.W. Hale and S. Zinberg, *New Eng. J. Med.*, 2001, *344*(1), 59; (b) A.B. Goldberg, M.B. Greenberg, and P.D. Darney, *New Eng J. Med.*, 2001, *344*(1), 38.
73. (a) M. Balter, *Science*, 2000, *290*, 39; (b) P. Zurer, *Chem. Eng. News*, Oct. 30, 2000, 37; (c) www.earlyoptionpill.com.
74. (a) S. Silver, *Wilmington Delaware News J.*, July 3, 1997, A14; (b) M. Balter, *Science*, 2000, *290*, 39.
75. (a) P.W. Collins, *Kirk–Othmer Encyclopedia of Chemical Technology*, 5th ed., Wiley, New York, 1996, *20*, 302; (b) V. Ullrich and R. Brugger, *Angew. Chem. Int. Ed.*, 1994, *33*, 1911; (c) Y. Sugimoto, A. Yamasaki, E. Segi, K.T. Tsuboi, Y. Aze, T. Nishimura, et al., *Science*, 1997, *277*, 681; (d) C.D. Funk, *Science*, 2001, *294*, 1871.
76. (a) Letter, *Planned Parenthood Federation of America*, New York, Oct. 17, 1997; (b) K. Armstrong, *Wilmington, Delaware News J.*, Oct. 19, 1994, *17*, A8; (c) D. Brown, *Wilmington, Delaware News J.*, Oct. 1994, *19*, A7.
77. W.R. Ewart and B. Winikoff, *Science*, 1998, *281*, 520.
78. (a) D.R. Gardner, R.J. Molyneux, L.F. James, K.E. Panter, and B.L. Stegelmeier, *J. Agric. Food Chem.*, 1994, *42*, 756; (b) K.E. Panter, R.J. Molyneux, L.F. James, and B.L. Stegelmeier, *J. Agric. Food Chem.*, 1996, *44*, 3257; (c) M.S. Al-Mahmoud, S.P. Ford, R.E. Short, D.B. Farley, L. Christenson, and J.P.N. Rosazza, *J Agric. Food Chem.*, 1995, *43*, 2154.
79. Anon., *Science*, 1997, *278*, 1233.
80. Anon., *Univ. Calif. Berkeley Wellness Lett.*, 2002, *18*(8), 4.
81. (a) P. Bavoli and P. Wyrick, Chlamydia: Genomics and Pathogenesis, Horizon *Biosscience*, Norwich, UK, 2006; (b) Anon., *Univ. Calif. Berkeley Wellness Lett.*, 2006, *22*(11), 4.
82. S. Ebrahim, *"Sexually Transmitted Infections," Centers for Disease Control and Prevention*, Feb. 2005.
83. (a) Anon., *Univ. Calif. Berkeley Wellness Lett.*, 1998, *15*(1), 4; (b) R.S. Stephens, S. Kalman, C. Lammel, J. Fan, R. Marathe, L. Aravina, W. Mitchell, L. Olinger, R.L. Tatusov, Q. Zhao, E.V. Koonin, and R.W. Davis, *Science*, 1998, *282*, 754; (c) T. Hatch, *Science*, 1998, *282*, 638; (d) T. Nordenberg, *F.D.A. Consumer*, 1999, *33*(4), 24.
84. B. Dixon, *Biotechnology*, 1994, *12*, 1314.
85. J. Kaiser, *Science*, 2008, *320*, 860.
86. J.K. Sherwood, L. Zeitlin, K.J. Whaley, R.A. Cone, and M. Saltzman, *Nat. Biotechnol.*, 1996, *14*, 468.
87. J. Cohen, *Science*, 2005, *309*, 1002.
88. (a) K. Bogle, *Hooking Up: Sex, Dating and Relationships on Campus*, New York University Press, New York, 2008; (b) T. Wolfe, *Hooking Up*, Farrar, Straus and Giroux, 2000.
89. G. Mathew, R.P. Singh, R. Lakshminarayanan, and S. Thomas, *J. Appl. Polym Sci.*, 1996, *61*, 2035.
90. C.M. Ronald, *Rubber World*, 1993, *208*(3), 15.
91. Anon., *R & D* (*Cahners*), 1997, *39*(10), 61.
92. Anon., *Consumer Rep.* 1999, *64*(6), 46.
93. Anon., *Popline*, May/June 1992, *21*, 1 (Population Institute, Washington, DC).
94. M. Balter, *Science*, 1998, *282*, 1790.
95. J. Alper, *Science*, 1999, *284*, 1754.
96. A.M. Rouh, *Chem. Eng. News*, June 1999, *28*, 24.
97. (a) A.M. Thayer, *Chem. Eng. News*, Sept. 22, 2008, 17; (b) J. Cohen, *Science*, 2009, *323*, 996; J. Cohen, *Science*, 2008, *319*, 1026; J. Cohen, *Science*, 2008, *321*, 512.
98. Anon., *Univ. Calif. Berkeley Wellness Lett.*, 1995, *12*(2), 8.
99. (a) K.M. Reese, *Chem. Eng. News*, Apr., 1995, *24*, 82; (b) M. Wood, *Agric. Res.*, 1999, *47*(5), 18.
100. N.R. Legge and R.C. Mowbray, *Rubber World*, 1994, *211*(1), 43.
101. Anon., *Chem. Ind.* (*Lond.*), 1997, 157.

102. Anon., *Popline*, Mar/Apr. 1996, *18*, 4 (Population Institute, Washington, DC).
103. L. Gopinath, *Chem. Br.*, 1997, *33*(9), 32.
104. Anon., *Popline*, May/June 1999, *21*, 3 (Population Institute, Washington, DC).
105. (a) Anon., *Chem. Ind. (Lond.)*, 2000, 484; (b) Anon., *Popline*, Washington, DC, July–August, 2000, *22*, 2.
106. Anon., Wilmington, *Delaware News J.*, Oct. 7, 1997, A2.
107. Anon., *Chem. Eng. News*. Sept. 1994, *5*, 35.
108. K. Megget, *Chem. Ind. (Lond.)*, May 5, 2008, 20.
109. P. Aldhous, *New Sci.*, Mar. 1995, *4*, 16.
110. (a) T. Black, *Popline*, Jan/Feb. 1996, *18*, 3 (Population Institute, Washington, DC); (b) Anon., *Popline*, Mar/Apr. 1999, *21*, 1 (Population Institute, Washington, DC); (c) http://www.quinacrine.com; (d) J. Lippes, *Science*, 2002, *297*, 1121; (e) C.M. Dobson, *Science*, 2004, *304*, 1259.
111. Anon., *Chem. Ind. (Lond.)*, 2002 (6), 6.
112. (a) P.F. Harrison and A. Rosenfield, *Contraceptive Research and Development—Looking to the Future*, National Academy Press, Washington, DC, 1996, 119; (b) Anon., *Institute for Development Training*, 212 E. Rosemary St., Chapel Hill, NC.
113. S.J. Nass and J.F. Strauss III, *New Frontiers in Contraceptive Research: A Blueprint for Action*, National Academies Press, Washington, DC, 2004.
114. R.A. Hatcher, ed., *Contraceptive Technology*, 17th ed., Ardent Media, NY, 1998.
115. (a) V.D. Vacquier, *Science*, 1998, *281*, 1995; (b) S. Tanaka, T. Kunath, A.-K. Hadiantonakis, A. Nagy, and J. Rossant, *Science*, 1998, *282*, 2072; (c) E. Strauss, *Science*, 1999, *283*, 169; (d) Y. Chen, M.J. Cann, T.N. Litrin, V. Iowigenko, M.L. Sinclair, L.R. Levin, and J. Buck, *Science*, 2000, *289*, 625; (e) M. Miyado, G. Yamada, S. Yamada, H. Hasuwa, Y. Nakamura, F. Ryu, K. Suzuki, K. Kosai, K. Inoue, A. Ogura, M. Okabe, and E. Mekada, *Science*, 2000, *287*, 321.
116. P. Aldhous, *Science*, 1994, *266*, 1484.
117. (a) S. Wilkinson, *Chem. Eng. News*, June 1997, *30*, 9; (b) D.J. Burks, R. Carballada, H.D.M. Moore, and P.M. Saling, *Science*, 1995, *269*, 83; (c) R.J. Aitkin, *Science*, 1995, *269*, 39; (d) L.H. Bookbinder, A. Cheng., and J.D. Bleil, *Science*, 1995, *269*, 86.
118. (a) W.J. deKoning, G.A. Walsh, A.S. Wrynn, and D.R. Headon, *Biotechnology*, 1994, *12*, 988; (b) V. Garcia-Campayo, A. Sato, B. Hirsch, T. Sugahara, M. Muyan, A.J.W. Hsueh, and I. Boime, *Nat. Biotechnol.*, 1997, *15*, 663; (c) B. Kutscher, M. Bernd, T. Beckers, E.E. Polymeropoulus, and J. Engel, *Angew. Chem. Int. Ed.*, 1997, *36*, 2149.
119. C. Djerassi, *Science*, 1996, *272*, 1858.
120. (a) R.F. Service, *Science* 1994, *266*, 1480, 1489; (b) S. Borman, *Chem. Eng. News*, June, 1996, *3*, 6; (c) R.F. Service, *Science*, 1996, *272*, 1258.
121. Anon., *Science*, 1995, *268*, 791.
122. K. Michelman, *Lett. NARAL Members*, Oct. 1997.
123. G. Feldt, *Letter to Planned Parenthood Federation of America Members*, Aug. 1997.
124. J. Cohen, *Science*, 1997, *277*, 760.
125. (a) K. Mahadevan and M. Sumangala, *Social Development, Cultural Change and Fertility Decline: A Study of Fertility Change in Kerala*, Sage, New Delhi, 1989; (b) B. McKibben, *Hope, Human and Wild: True Stories of Living Lightly on the Earth*, Hungry Mind, St Paul, MN, 1995; (c) R. Jeffrey, *Politics, Women and Well-Being: How Kerala Became a "Model,"* Macmillan Academic & Professional, London, 1992; (d) B.A. Prakash, *Kerala's Economy, Performance, Problems, Prospects*, Sage, New Delhi, 1994; (e) R.W. Franke and B.H. Chasin, *Radical Reform as Development in an Indian State*, 2nd ed., Institute for Food Development Policy, Oakland, CA, 1994; (f) www.pppinindia.com/state-economy-kerala.asp.
126. K.M. Leisinger and K. Schmitt, *All Our People—Population Policy with a Human Face*, Island Press, Washington, DC, 1994.
127. (a) G.A. Geyer, *Tunisia: The Story of a Country That Works*, Stacey International, London, 2004; (b) Anon., *Popline*, Sept.–Oct. 2004, 2 (Population Institute, Washington, DC); (c) www.mrdowling.com/800gdppercapita.html.
128. (a) Anon., *Union of Concerned Scientists, Poverty, Development and the Environment*, Cambridge, MA, May 1994; (b) Anon., *Popline*, July/Aug. 1999, *21*, 2 (Population Institute, Washington, DC).
129. Anon., *Popline*, 1997, July/Aug. 3, *19* (Population Institute, Washington, DC).
130. Anon., *Popline*. May–June 2007, 2, *29* (Population Institute, Washington, DC)
131. C. Hertzman, *Am. Sci.*, 2001, *89*(6), 538.
132. (a) Letter from W. Fornos (Population Institute, Washington, DC), Sept. 1999; (b) Anon., *World Watch* 1999, *12*(1), 39; (c) U.N. Development Programme, *Human Development Report 1998*, Oxford University Press, NY, 1998, 30–37.
133. G. Feeney, *Science*, 1994, *266*, 1518.
134. Anon., *Popline*, 1995; May/June, 2, *17* (Population Institute, Washington, DC).
135. M. Renner., In: L. Starke, ed., *State of the World 1995*, W.W. Norton, NY, 1995, 150.
136. (a) A. Glasser, A.M. Gulmezoglu, G.P. Schmid, C.G. Moreno, and P.F.A. van Look, *Lancet*, 2006, *368*, 1595; (b) J. Clelland, S. Bernstein, A. Ezeh, A. Glasier, and J. Innis, *Lancet*, 2006, *368*, 1810.
137. G.H. Brundtland, *Environment*, 1994, *36*(10), 16.
138. N. Myers, *People Planet*, 1993, *2*(2), 37.
139. W. Fornos, *Letter, Population Institute*, Washington, DC, May 2006.
140. K. Krane, *Amicus J.* (Natural Resources Defense Council), 1994, *16*(3), 45.

RECOMMENDED READING

1. D.W. Hahn, McGuire, and G. Bialy, *Kirk–Othmer Encyclopedia of Chemical Technology*, 4th ed., Wiley, New York, 1993, *7*, 219 [contraceptives].
2. L.R. Brown. In: L. Starke, ed., *State of the World 1995*, W.W. Norton, NY, 3 ["Nature's Limits"].
3. P. Harrison, *The Third Revolution: Population and a Sustainable World*, I.B. Tauris, NY, 1992, 255 [differences in consumption between the rich and the poor].
4. J. Bongaarts, *Science*, 1998, *282*, 419 [declining fertility and future population trends in the world].
5. L.R. Brown and B. Halweil, *World Watch*, 1999, *12*(5), 20 [effects of increasing populations on the environment].

6. M. Campbell, J. Cleland, A. Ezeh and NN. Brata, *Science*, 2007, *315*, 1501 [population growth and the Millennium Development Goals].

EXERCISES

1. What family-planning services are available in your community for both rich and poor?
2. Check with your local department of health to see what the rates of teenage pregnancy; infant mortality; AIDS, and other sexually transmitted diseases; rape; and births are in the area where you live or go to school.
3. Was sex education in your home and school thorough and realistic?
4. Do you know some people with large families? If it is possible to do so tactfully, see what their attitudes are toward zero population growth.

CHAPTER 17

Environmental Economics

17.1 INTRODUCTION

The previous chapters have covered many ways of reducing pollution by prevention. This is often cheaper than putting something on the end of the pipe or at the top of the smokestack. For example, substitution of an aqueous cleaning process for one using trichloroethylene means that no solvent has to be purchased, recycled, or disposed of. No solvent emissions result from the new process, and none can be spilled and seep into the groundwater. The used water can be treated on site or at the local wastewater treatment plant. Processes that will help society attain a sustainable future have also been discussed. According to Ed Wasserman, President of the American Chemical Society for 1999, "Green chemistry is effective, profitable and it is the right thing to do for our health and that of our planet."[1] The problem is that companies and ordinary citizens have been slow to adopt these new processes.[2] This chapter will explore what role money plays in this and what other social and political factors may be involved.

Under conventional economics, market forces can destroy or profoundly affect the life support system on which we depend. This destruction is now evident in worldwide problems, such as stratospheric ozone depletion and global warming. On a smaller scale, a commercially important species can be driven to near extinction by ordinary market forces.[3] The striped bass fishery on the East Coast of the United States provided fish for the restaurant trade. As the supply of fish was reduced by overfishing, the restaurants charged more and customers were willing to pay the price. Finally, when the population was seriously depleted, the federal government initiated a moratorium on fishing for the species. After a few years without harvesting, the population began to recover.

A new hybrid discipline of environmental economics (also called ecological economics) tries to include the value of natural support systems in a consideration of profit and loss. Ecologists and economists are coming to know each other. The International Society for Ecological Economics publishes the journal *Ecological Economics*. A number of reviews of the subject are available.[4]

17.2 NATURE'S SERVICES

Nature provides a number of important services[5] free of charge. These are seldom included in economic calculations. Robert Costanza et al. estimated their total value at $33 trillion/yr in 1997.[6] (The authors emphasize that these figures are preliminary and will require further refinement in the future.) For comparison, the world's gross national product per year was about 18 trillion dollars. The gross annual national product of the United States was 6.9 trillion dollars. The 17 services were estimated for 16 different biomes, including forest, grassland, wetlands, ocean, and so on. They included

- Gas regulation: composition of the atmosphere, ozone layer, etc.
- Climate regulation: temperature, precipitation, etc.
- Disturbance regulation: resilience to storms, droughts, etc.
- Water regulation: the hydrologic cycle
- Water supply: storage and retention of water, aquifers, etc.
- Erosion and sedimentation control
- Soil formation
- Nutrient cycling: nitrogen, phosphorus, etc.
- Waste treatment: recovery of nutrients
- Pollination
- Biological pest control
- Refugia: nursery grounds, overwintering grounds, etc.
- Food production: fish, game, etc.
- Raw materials: lumber, fuel, fodder, etc.
- Genetic resources: medicines, genes for resistance to pests, etc.
- Recreation: ecotourism, sport fishing, etc.
- Cultural: aesthetic, artistic, educational, or scientific value

The largest value, 17 trillion dollars, is assigned to nutrient cycling. Because ecosystem services are often ignored or undervalued, the social costs of a project may exceed the benefits. As an example, clear-cutting a forest will produce income for the landowner, the timber company, and the workers. It will also decrease the value of nearby homes. The sediment that washes off in the process may make the streams unsuitable for spawning fish, fill up reservoirs, and kill the coral reef in the ocean where the stream enters it. When it rains, more water will run off and less will soak into the ground. Floods will occur more often. Another result may be that the stream now dries up in the summer so that the adjacent town has no water to drink or to use in irrigating crops. If these services could be replicated at all, the cost would be high. The water might be brought in by pipes from another watershed. The fish might be imported from another part of the globe. The electricity from the lost hydropower might be supplied by a new coal-burning plant.

Such natural services may be interrupted in other ways. If the chemical or sewage plant upstream puts something toxic into the stream that the water treatment plant of the city downstream cannot remove by standard treatments, a new water supply or a new treatment method will have to be found at increased cost. In earlier years before the toxicity of the polychlorinated biphenyls was fully appreciated, General Electric released enough of them into the Hudson River north of Albany, New York, so that the whole Hudson River from Hudson Falls to New York City is now a Superfund site.[7] Fishermen are advised not to eat the fish they catch. It is not always easy to calculate the cost of a fishery lost to toxic heavy-metal ions or acids draining out of a mine site.[8] One settlement, involving a salmon fishery in a river in Idaho, was for 60 million dollars. The Exxon Valdez oil spill in Alaska cost Exxon 3 billion dollars. The persons whose wells become contaminated by leachate from the nearby landfill will face the costs of bringing water from a distance. This was a cost that was not included when the landfill was built. The U.S. National Research Council has recommended that the U.S. Department of Commerce resume development of a method to better measure environmental costs.[9] There is a book on the law and ecosystem services.[10]

The economic value of biodiversity has also been calculated by others.[11] Pimentel et al.[12] estimate the value of biodiversity at $3 trillion/yr. Their list of services includes the following:

- Soil formation
- Biological nitrogen fixation
- Crop and livestock genetics
- Biological pest control
- Plant pollination
- Pharmaceuticals

They point out that biodiversity is necessary for the sustainability of agricultural, forest, and natural ecosystems on which humans depend. Roughly 99% of pests are controlled by natural enemies and by plant resistance. Loss of key pollinators may mean loss of a crop. Honeybees can pollinate some crops, but not all of them. There are already species for which no pollinators are left.[13] Current extinction rates are 1000–10,000 times the natural rate. The concern is that keystone species (i.e., species without which the ecosystem cannot function) may be lost. These are more likely to be soil microbes or insects rather than pandas or tigers, despite the popular appeal of the latter.

Just the extent of human alteration of the natural cycles is causing problems. Soil microbes and lightning fix about 90–140 million tons of nitrogen each year. Humans add 140 million more tons as fertilizer made by the chemical industry to increase crop yields.[14] This has accelerated the loss of biodiversity. Nitrogen is a key element in controlling species composition, diversity, and dynamics in both terrestrial and aquatic systems. Many of the original native plants function best at low levels of nitrogen. Many soils are now so saturated with nitrate that vital nutrients, such as calcium and potassium ions, are being lost to groundwater and streams. Algal blooms and toxic *Pfiesteria* outbreaks are more common. This has contributed to long-term declines in coastal fisheries. There is now, as a result of fertilizer runoff from Midwestern farms, a large area in the Gulf of Mexico with too little dissolved oxygen to support the fishery. Some of the nitrogen is released to the air as N_2O, a greenhouse gas that helps to deplete the ozone layer. (Other ways of handling nitrogen for agricultural purposes are given in Chapter 11.)

According to one source,[15] "degradation of our natural environment continues accelerating in scope and impact" under the pretext of the economic necessity "to balance budgets" and to provide opportunities for "development." Another author feels that growth and trade are increasingly incompatible with environmental protection.[16] Others even question the advantages of continuing economic growth in developed nations that are already overconsuming.[17] Yet, almost every issue of *Chemical Engineering News* mentions the desire to grow, to produce, and to sell more.

The chemical industry relies on natural capital,[18] which must be used in a sustainable way. This means that the present generation should fulfill its needs in a way that does not jeopardize the prospects of future generations.[19] It means that natural ecosystems must not be exploited to a point at which they lack the resilience to recover.[20] In 1961, 70% of the earth's renewable resources were being used. In 1999, the figure was 120%.[21] Traditional economics views natural capital as endless or easy to substitute for, so that one can use it and move on to more of it somewhere else. All that is needed to exploit it is labor and money. During most of man's existence on earth, this may have seemed true. Today, the limits of the earth[22] are apparent, and the frontier philosophy must be replaced by an ethic of looking out for what we have, so that it will last indefinitely. Extractive industries, such as timbering and mining, must produce products no faster than the trees can grow back or renewable substitutes can be found for the minerals.[23] Each person living in Germany, Japan, the Netherlands, and the United States consumes 45–85 metric tons of natural resources each year.[24] In the subsidized mining of coal in Germany, 29 metric tons of earth per person in Germany is moved each year. Most of this is waste. In the United States, 15 metric tons of topsoil per person per year is lost as part of agricultural production. These hidden costs do not appear in the usual economic statistics. With the continuing decline in the concentration of desired metal or mineral in the rock, the time may come when the landfills of today are mined for metals. The figure for gross national product often contains money from products harvested unsustainably, remediation of past mistakes, and reduction of pollution.

17.3 ENVIRONMENTAL ACCOUNTING

17.3.1 The Ecological Footprint

Each year, 6 million ha of land in the world undergo desertification, 17 million ha are deforested, and soil erosion exceeds soil formation by 26 billion tons. The ecological footprint is an accounting tool devised by William Rees to measure the productive land area needed to supply the resource consumption and waste assimilation of a given human population.[25] Twenty percent of the world's population, largely in industrialized nations, consumes 80% of its resources. Some typical values of the productive land or water needed to support a person at a given material standard indefinitely are given in Table 17.1.

Many developed nations do not have enough productive land or water to support the current material consumption. Typical deficits are given in Table 17.2.

Table 17.1 Typical Values of Land or Water Needed to Support a Person

Country	Land or Water Needed (ha/person)
United States	5.1
Canada	4.3
the Netherlands	3.3
India	0.4
World	1.8

Table 17.2 Typical Land and Water Deficits by Country

Country	Deficits
Japan	1.76
Korea	1.81
Austria	2.15
Belgium	2.8
United Kingdom	2.65
Denmark	2.38
France	2.22
German	2.66
the Netherlands	2.85
Switzerland	2.56
United States	2.29

The water footprint of common consumer goods varies widely with typical values such as 140 L per cup of coffee, 70 L per apple, 2700 L per cotton shirt, 40 L per slice of bread, 15,500 L per kg of beef, and 8000 L per pair of leather shoes.[26]

Materials to make up the deficits are supplied by other nations. Thus, there may be natural resource exploitation that is unsustainable, pollution, and waste in the supplier nation that does show up on the balance sheet of the consuming nation.

Calculation of the ecological footprints of cities around the Baltic Sea showed that the area required for the assimilation of their waste nitrogen, phosphorus, and carbon dioxide would be 390–975 times the land area of the cities.[27] The amount of forest land needed to assimilate carbon dioxide from all the world's cities would be three times that available on earth. The World Wildlife Fund "Living Planet Index" tracks the effect of human populations on biodiversity.[28] In 2004, the biggest consumers of nonrenewable resources per capita were the United Arab Emirates, the United States, Kuwait, Australia, and Sweden.[28]

17.3.2 Life Cycle Analyses

Environmental life cycle analyses follow a product from cradle to grave.[29] (Following it from cradle to cradle would be better.) They may not be easy to do and they may be expensive. They have to take into account the environmental effect at each stage: the raw materials and the way they were obtained; the manufacturing process; any transportation involved; the influence of use; and how the object will be disposed of at the end of its useful life. When no hard data from detailed studies are available, simplifying assumptions may have to be made. It is not surprising that different persons evaluating the same product can come to different conclusions. The first thing to do in looking at an analysis is to see if the group that funded it might have a vested interest in the outcome.[30] The second thing to do is to see what assumptions have been made.

Several studies have compared the environmental effects of disposable diapers with cloth diapers.[31] (The world uses 450 billion disposable diapers per year, of which 93 billion are used in India.) A 1988 analysis by the cloth diaper industry favored cloth as having a smaller environmental effect. A 1990 study made for Procter & Gamble, makers of disposable diapers, favored disposable diapers. A 1990 study by the American Paper Institute found the two kinds to be equivalent. A 1991 study by the cloth diaper industry found cloth to be superior.

A disposable diaper is often made with a nonwoven polypropylene fiber face sheet and a transfer layer next to the skin backed by a layer of cellulose and a superabsorbent polymer, with a polyethylene cover on the outside. Over the years, they have become thinner. The polyethylene and polypropylene are unlikely to have been made from renewable raw materials, although it is conceivable that they might be in the distant future. The superabsorbent may or may not have been made at least partially from starch. Thus, one must assess the possibility of oil spills in drilling for and recovering the oil. What wastes were associated with this? Was any forest destroyed in the process? What pollution or wastes came out of the oil refinery? What type of forest was cut to produce the cellulose? Was old-growth forest cut? Did the logging result in sediment damaging a fishery or in other loss of biodiversity? What pollution and wastes were involved in making the superabsorbent? The acrylonitrile that may have been grafted to starch to produce the superabsorbent is a carcinogen. What was the energy needed to make the diaper and to transport it to the user? The energy probably came from fossil fuel rather than from a renewable source. What was the fuel efficiency of the equipment used in the various steps of acquiring the raw materials, making the diaper, and transporting it? After use, the diaper was probably buried in a landfill where even the cellulose part may not decompose. (Landfill cores have often been dated by the dates of the newspapers found in them.)

The cotton diaper also involves the use of energy that probably comes from fossil fuels. Again, what was the fuel efficiency of the equipment used in the various steps in the life cycle? The cotton may have been raised with subsidized irrigation water and with pesticides and fertilizers, some of which may have seeped into the groundwater or the nearest stream. Were the workers in the cotton mill protected from dust? Because the diaper can be used many times, the main environmental effect will involve making the detergent, the washing machine, and heating the water. What wastes and pollution may have accompanied the making of the detergent? Is the energy to heat the water from a fossil fuel or a renewable source? What is the energy efficiency of the water heater? How much hot water is used? Swedish clothes washers may take only a small fraction of the water used in an automatic clothes washer in the United States. How hot is the water? Is the detergent used in warm water instead of hot water? Detergents for use in unheated water are now available commercially. Is the diaper dried on a line in the sun or in a drier powered by gas or electricity? Are two worn diapers sewn together to obtain more life from them? When the child no longer needs the diapers, will they be given to a friend who will use them until they wear out? At the end of their useful life, will the diapers be used as wiping clothes, in rag content paper, or just sent to the landfill? How do you depreciate the washing machine that is used to wash other clothes as well?

One of the studies of cloth diapers assumed a life of 92 uses (a study funded by the disposable industry), whereas the other assumed 167 uses (a study funded by the cloth diaper industry). A study funded by the disposable diaper industry assumed an energy cost for transporting the cotton to China, where the diapers were made. One would have to check to see what fraction of cloth diapers is made this way. Disposable diaper makers have pushed composting as a method of disposal. Most cities do not compost this type of waste. The polyolefins would not degrade well in this process, even though the lack of stabilizers in the polypropylene might allow it to become powder.

Cotton towels last through about 100 washes. Hotels can make them last longer by posting signs that suggest that guests not ask that they be washed everyday.[32]

This comparison of diapers shows how many assumptions and judgment calls have to be made in typical life cycle analyses. To perform a comprehensive analysis would require collecting a large amount of data that might be expensive to obtain. It also points out the possible misuse of data. A diaper that involved no use of nonrenewable materials would be preferred.

It is also likely that if a study does not reach the conclusion desired by the sponsor, the data may never be released. Analysis of 111 studies on soft drinks, fruit juice, and milk found that studies funded by the beverage industry were four to eight times more likely to be favorable to the funders than those from independent research.[33] In some other comparisons, longer-lived alternatives are neglected. Attempts to compare the environmental effects of throwaway paper and plastic cups do not include the china mug. The mug will take more energy to make, but it can last for decades. During its life, it could be washed using soap or detergents and energy from renewable sources. It would have the least effect on the environment. Comparisons of paper and plastic grocery bags overlook the possibility of bringing your own basket or string bag. For shopping in a supermarket several string bags would do the job, with the least effect on the environment.

Life cycle assessment is a technique that is still evolving. The U.S. EPA is studying the technique and trying to standardize it enough to allow meaningful ecolabels to be put on consumer items.[34] Among the effects being considered are ozone depletion, global warming, smog formation, human toxicity, noise, energy use, and nonrenewable resource depletion. Tradeoffs may sometimes have to be made (e.g., choosing the lesser influence of toxic chemical use and global warming). The process endeavors to quantify "energy and raw material requirements, atmospheric emissions, waterborne emissions, solid wastes, and other release for the entire life cycle of a product, package, process, material, or activity." A factor that should be included is whether or not the product has been designed for easy disassembly for recycle at the end of its useful life. Many other countries have government-sanctioned ecolabeling programs, including the European Community, Canada, France, Germany, Austria, the Netherlands, Singapore, New Zealand, and Japan.[35] Although these may differ in detail, the goal is to promote products that have reduced environmental effects during their entire life cycles. They should also help consumers make informed choices in their purchases. This will help avoid goods that unintentionally promote a dysfunctional economy.

Ecolabels have varying standards of stringency. Forest Stewardship Council and Marine Stewardship Council labels are considered to be good. Consumers Union has classified them into "meaningful and verified," such as "Green Seal," "Rainforest Alliance Certified," and "USDA Organic," and into "not meaningful and unverified," such as "cruelty free," "environmentally friendly," "no chemicals," and "natural."[36]

17.3.3 Cost–Benefit Analyses

Cost–benefit analyses[37] can also involve uncertainties and tradeoffs. Some of the biggest problems involve health and safety. What value should be assigned to a human life? Is it the same for a 30-year-old and a 75-year-old? Will this value be the same for a person in a developing nation as it is in a developed one? How can the cost of sickness be measured? One can add up the costs of doctor's visits, medicines, and income lost from missing work. What dollar value can be put on time lost from school or the misery of just feeling lousy? Lead in the environment can lower the IQ of children. What dollar value should be assigned to this? What lost value is assigned to a dirty lake? Does one add up the cost of buying fish instead of pulling them out of the lake, the cost of driving elsewhere to fish or swim, and the cost of cleaning up the water for drinking?

Cost–benefit analyses have been popular with some members of the U.S. Congress and polluters, who hope that new regulations can really be justified economically.[38] However, that the benefits are significantly larger than the cost[39] has not eliminated the grumbling when the cost of the change is high. This was true with the lowering of the ozone standard from 0.12 to 0.08 ppm and reducing the size of the particular matter to be regulated.[40] The new regulations were estimated to prevent 20,000 premature deaths per year and 250,000 sick days. The cost would be $6.5–8.5 billion/yr versus benefits of 120 billion dollars. The new particulate standards could require redesign of diesel engines in trucks and buses, as well as putting new catalytic converters on wood stoves. Changes in coal-burning power plants may also be required.

The U.S. EPA has been credited with doing a good job with cost–benefit analyses.[41] The analyses are said to be explicit and transparent, with all the assumptions stated. The cost of each analysis for lead in gasoline, asbestos, ozone depletion, and such ran from 731,000 to 1,035,000 dollars. When the underlying science is good, the analyses can be helpful. If it is weak, there may be no point in running the analysis. According to EPA calculations, huge benefits have resulted from the Clean Air Act over the period of 1970–1990.[42] The cost was 523 billion dollars and the benefits 6–50 trillion dollars. Curbing acid rain in the United States under the Clean Air Amendments of 1990 produced 12 dollars in benefits, of reduced illnesses and deaths and better visibility, for every 1 dollar in compliance costs.[43]

Many cost–benefit analyses are of limited value. In addition to the problems of putting a value on human health, there may be simplifying assumptions and comparisons that are not valid. This has led to their value being questioned.[44]

17.3.4 Green Accounting

Pollution and waste represent inefficiencies of production,[45] which might be eliminated through pollution prevention.[46] For example, if the United States were as energy-efficient as Sweden or Japan, it could save $200 billion/yr. The chemical industry in the United States spends over 4 billion dollars each year on pollution control, an amount equal to 2.7% of sales.[47] An additional 2.1% of sales is spent on related capital expenditures.

Antiquated accounting systems are one hindrance to the identification of opportunities for pollution prevention. The costs of waste may be hidden in overheads, research and development, marketing, product registration, and other unrelated accounts.[48] The waste and all costs associated with it should be assigned to the process and product that produces it. This may turn out to be as much as 20% of the total production cost of the product. Amoco's Yorktown, Virginia, oil refinery thought its environmental costs were 3% of operations, until it checked carefully and found them to be 22%.[49] DuPont found a case where the environmental costs were 19% of the manufacturing costs. When the true cost is properly assigned, management may have a greater incentive to restudy the whole process in the search for a better one. The pollution prevention alternative may offer opportunities for increased productivity and reduced cost. It can be cheaper than any end-of-the-pipe treatment. It may eliminate the need for waste disposal, especially disposal of hazardous waste, scrubbers on exit air, need to report toxic releases, future liabilities, and contingencies, such as fines or need to remediate spills. For example, if the solvent is replaced by water, there will be no need for an air scrubber. No solvent will need to be purchased. If no toxic reagents are used, then none will be left in exit air or water. If a use can be found for the waste, if will generate revenue rather than require a disposal fee. If no waste is sent to a landfill, then none will leach out in the event of a failure of the liner and there will be no long-term liability. In addition, a clean process may help the corporate image, which may help sales of other products as well. Capital requests for pollution prevention projects will fare better in competition with other capital requests if all these factors are taken into account. Plants with green-accounting systems have roughly three times as many pollution prevention projects as those with no cost-accounting systems. The projects yielded an average savings of 3.49 dollars for every dollar spent.[50] A study by the Business Roundtable (a group of top executives of large corporations) concluded that pollution prevention efforts can be successful when they are incorporated into existing business procedures, rather than being treated in separate plans, and when specific approaches are not mandated.[51]

Many plants have associated costs that are not borne by the plant, but which are externalized to others. These are also referred to as societal costs. These are not included in the usual accounting system. Businesses may not be legally accountable for them. Consider, for example, the power plant that burns high-sulfur coal. Some such plants have scrubbers that take out most of the sulfur dioxide from the stack gas. Other plants were in existence when the Clean Air Act went into effect and are grandfathered so that no scrubber is required.[52] In one place along the Ohio River, an old plant puts out six times more sulfur dioxide per unit of electricity generated as a nearby newer plant. The air receiving the sulfur dioxide drifts eastward to New York and New England, where it decreases forest and crop yields and can kill some forests. Tall stacks put on some plants in the Midwest reduce the concentration of sulfur dioxide at ground level locally, but increase long-range transport. The sulfur dioxide is converted to sulfuric acid in the air. This causes corrosion of objects made of metal, limestone, marble, and sandstone. These include automobiles, which have to have an acid-resistant finish, power lines, buildings, statues, and others. In former years, sulfur oxides from metal smelters often killed all vegetation within miles of the smelter, as at INCO's smelter in Sudbury, Ontario. (After resisting clean-up for many years, INCO found that implementing a redesigned smelting process not only reduced emissions but also reduced costs, so that it made a return on its investment.[53] It sells the sulfuric acid made from the sulfur dioxide at a profit and now markets the technology to others.) High concentrations of sulfur oxides in smog in London and in Donora, Pennsylvania, killed many persons in the 1950s. There is an incentive to keep the grandfathered plants running as long as possible because they are cheaper. The cost of adding scrubbers would raise the price of electricity to customers in Ohio, but not to the point that their electricity would be more expensive than that in New England.

In former years, the city of Philadelphia used what it thought was the most cost-effective system of handling the city's wastewater. This involved only primary treatment (i.e., simple filtration to remove solids). The solids were transported to the ocean by barges for dumping. The filtrate was dumped directly into the Delaware River. The result was that a section of the Delaware River contained too little dissolved oxygen for fish to survive. Migratory fish, such as shad, could not go up the river past this block. Swimming was no longer possible at the former beaches near New Castle, Delaware. In one summer, these waste solids, together with wastewater from New Jersey rivers, caused an area of 2000 square miles of ocean to have too little dissolved oxygen to support life. The ocean could no longer be considered an infinite sink for wastes. This system was probably cheaper for residents of Philadelphia, but it was not for its neighbors downstream. It took years of litigation and a suit by the states of Delaware and New Jersey plus several environmental organizations to persuade Philadelphia to put in plants for secondary treatment of the wastewater. The federal government put up the bulk of the money to build the new plants.

Individuals can also export some of their costs to their neighbors. Agricultural runoff of soil, pesticides, and fertilizer can lead to algal blooms, *Pfiesteria* outbreaks, loss of fish, and loss of biodiversity. Leaky septic tanks can contaminate groundwater and surface water. Soil erosion from construction sites can plug culverts, fill up reservoirs and harbors, and such. Because the downstream costs from construction fall on government, more and more regulations are being put into place to try to keep the costs on the construction site. Development of a site often leads to greater runoff of water in storms, which can lead to stream-bank erosion. Regulations are now being aimed at water retention basins that will keep the excess water and costs on site. Individuals also operate cars with tailpipe emissions that can cause smog and contribute to acid rain. The costs of these are borne by the general public, especially those with respiratory problems. Although efforts are being made by government to regulate this, many states still violate the standards for ground-level ozone several days each summer. The cost of picking up litter along roads falls on the taxpayers or on the people who live along the road.

One problem in some of these examples is the diffuse nature of the effects. The public may not be able to associate the problems with the real source, which may be far away. Because the effects can be diffuse, it may be hard to obtain enough persons together to mount a strong lobby for relief. Some lobbies are not as effective as others. Based on what has happened in the Pacific Northwest of the United States, the timbering lobby appears to be much stronger than that of the salmon fisherman. Campaign contributions may be part of the problem. A logging company may be able to afford larger contributions than individual fishermen. Some environmentalists feel very strongly about this problem. Donella Meadows of Dartmouth has said, "The first priority is campaign reforms. We're lost until we get democracy back."[54] Accounting methods that reflect the true cost of these various actions to society are needed.

17.4 CORPORATIONS

17.4.1 Additional Reasons Why Pollution Prevention May Not Be Adopted

Incomplete accounting is not the only reason for the slowness of businesses to adopt pollution prevention.[55] Several pollution prevention opportunities were identified at a Dow plant in Texas that manufactures isocyanates.[56] The savings would have been $1 million/yr with payback periods ranging from 15 months to 5 years. The new processes would have eliminated 500,000 lb of waste each year and allowed the hazardous waste incinerator to be close. The changes were not made because the business center thought that more money could be made by other investments. A 2-year study of Dow's Midland, Michigan, plant by a team of community activists and company project engineers (organized by the Natural Resources Defense Council) revealed opportunities for a 37% reduction in waste and a 43% reduction in emissions at a capital expense of 3.1 million dollars and a potential savings of $5.4 million/yr.[57] These changes were adopted. A Dow plant in Louisiana has annual waste reduction contests that have produced a stream of improvements over several years.[58] Problems that often prevent implementation at other places were said to be absent at this plant. These include (a) a fear of losing control over funds, (b) giving prevention a low priority, and (c) a belief that the new process would not work.

Paul Tebo of DuPont has outlined his company's intention to reduce emissions and waste with the goal of zero emissions.[59] He says that proactive companies are finding that what is good for the environment is good for business and can lead to a competitive advantage. An example he gives is the company's nylon plant at Chattanooga, Tennessee, which now converts 99.8% of the raw materials to saleable products. As a result, the site's wastewater plant will be shut down and the remaining wastewater sent to the city's treatment plant, which will save $250,000 yr^{-1}. This has been done by finding uses for by-products that were discarded. (By-products of DuPont's manufacture of nylon were are discussed in Chapter 1.) A second example is a herbicide plant in Indonesia that will produce only one or two bags of ash each week from its solid and liquid waste. New plants are more likely to incorporate pollution prevention than old ones that are fully depreciated, especially if the products of the old ones are considered to be mature and are not growing in sales. Ecoparks, where one company's waste is another's raw material, may become more common.[60] An example is the one at Kalundborg, Denmark[61] (described in Chapter 14).

OSi Specialties, a subsidiary of Witco (now Chemtura), was nudged into action by the U.S. EPA.[62] The problem was loss of methyl and ethyl chloride in the wastewater from a process of making ethers from polyethylene oxide and polypropylene oxide. A complete study of all of the waste streams at the plant resulted in a solution that cost 600,000 dollars and saved 800,000 dollars. This was done by adding a unit that converted the excess methyl chloride from the process to methanol, which could be sold. An attorney for the company says, "But if it hadn't made economic sense, we wouldn't have done it." If dimethyl carbonate could be used for the end-capping, there would be no waste salts produced and almost no wastewater.

A study done by INFORM found that the "obstacles to pollution prevention were not regulatory, technological or even economic, but primarily institutional."[63] Most companies assumed that their processes were efficient. Some pollution control departments had little knowledge of the processes that produced the wastes that they handled. Another study found that environmental managers often lack the whole-hearted support of the company's business managers, so that it is hard for them to get things done.[64]

The U.S. DOE has found that seven industrial sectors—chemicals, aluminum, steel, metal casting, glass, petroleum refining, and forest products—consume 80% of the energy used in manufacturing in the United States and are responsible for 80% of all waste and pollution from manufacturing.[65] Research and development averages about 2.8% of sales for all manufacturing in the country. Aside from the chemical industry, the six other segments spend about one-third of this. These industries have relatively mature markets and technologies. Their scale requires large capital expenditures for change, and payback periods may be long. They are characterized by conservative investment strategies and a reluctance to innovate. This has led to high pollution, high energy use,[66] and little incentive to improve. According to the DOE, the chemical industry spends more on research and development than the average manufacturer. However, most of this is said to be on product development, rather than on basic research or research to improve processes. The DOE has put up some federal money to help companies in these industries to innovate in areas where the companies are unlikely to innovate on their own, but for which the payoffs may be large for the whole country.

Individuals may not act in the best interests of the company if personal incentives are not right.[67] Managers are often averse to risk. Because pollution prevention projects can involve both effort and risk, the managers must be compensated accordingly, or some system must be devised to shift the risk from the managers to the firm. The wholehearted support of top management for pollution prevention also helps. Without it, such projects may become low-priority ones. A study of 12 large manufacturers showed that small projects were subjected to higher hurdles than large ones by middle management.[68] This suggests grouping several small projects into a large one in order to achieve the attention of top management.

A company that has downsized to the point where many employees have too much work to do may have little time to do more than service existing product lines. Management may have little incentive to invest money or research in product lines that are not growing and that it considers to be mature. Too much emphasis on the bottom line next quarter may lead to a demand for unrealistically short payback times for pollution prevention projects. There have been cases where a plant manager looked good on the balance sheet by spending very little money for maintenance. That this approach may end up costing more in the long run than planned preventive maintenance may not become apparent until the plant manager has transferred to another location or has retired. For example, a small leak in the roof will do little damage if fixed promptly. If it is allowed to remain until it becomes large, the repairs may include not only new roofing but also the beams that support the roof, the ceiling, and floors in the room below, and any equipment in the rooms that would be damaged by water. Management may also be wary of a pollution prevention project that it feels falls under someone else's patent. It may feel that any royalty charged might be enough to negate any cost savings that might result from the project. It may also look at end-of-the-pipe treatments as simpler and easier than re-examining their whole process of find ways to improving it. There may also be the fear that the new process will not pay for itself in savings, with the result that competitors who do not clean up will have an advantage in the marketplace. If management feels that a better environmental technology is just around the corner, it may fear locking into a process that might be obsolete in a few years.

A drop-in replacement that requires no new equipment is more likely to be adopted than one that requires major purchases of new equipment. If more equipment has to be added to space that is not in the present plant, then the cost of expanding the building must be considered. Some process changes can involve major changes to equipment unfamiliar to the current personnel. For example, the discharge of heavy-metal ions in the wastewater from an electroplating plant can be avoided by vacuum metallization, chemical vapor deposition, and other such processes. The equipment will be quite different and expensive, and plant personnel may distrust it because they are not familiar with it. Older engineers may recall problems with newer methods in former years, before the techniques were improved to current levels. The use of ultrasound and microwaves in organic reactions may fall into this category. Aqueous degreasing to replace chlorinated solvents may be distrusted for fear of rusting. Prompt removal of water with a jet of air, or inclusion of corrosion inhibitors, can prevent the formation of rust. A synthetic organic chemist may not know how to run a fermentation, even though doing so might eliminate toxic reagents and hazardous wastes as well as organic solvents.[69]

A municipality may not improve its wastewater treatment plant because its officials were elected on the promise of no new taxes. The electorate may not vote for any new bonds to be issued either.

17.4.2 Other Aspects of Corporate Finance

The high cost of accidents involving toxic chemicals was mentioned in Chapter 1. In a fire or explosion, there is the cost of rebuilding the plant, the possible loss of business to competitors during the period when your plant cannot supply, hospital and rehabilitation bills for the injured, government fines for safety violations, awards by juries for damage to the community, and loss of a good company image that may have helped to sell your products. This should provide a powerful incentive for substituting greener processes that use less toxic chemicals. In addition to the reasons for accidents in Chapter 1, there may also be the feeling that "It can't happen here. We've been doing it this way for years and have never had a problem."

Corporations are often said to place their interests over those of the community, the state, and the environment.[70] It is to their financial advantage to externalize as many costs as possible rather than internalize them. After the logging company removes the timber in some countries, replanting may be left to the government. This is in addition to the downstream loss of fisheries mentioned earlier. A generation or two ago in the United States, beer and soft drinks were sold in refillable bottles with deposits on them. Since then, the bottlers have turned to single-use throwaway containers. The washing lines have been dismantled and their crews laid off or shifted to other jobs. The consuming public pays more now for the packaging and for its disposal. The local government usually provides the landfill or incinerator to deal with the packaging after it is used (see also Chapter 14). Firms may tend to overexploit costless inputs such as air and water, in what Hardin called the "tragedy of the commons."

Firms are in business to make money. This sometimes results in abuses, such as unsustainable use of raw materials, or contamination of air and water. In a typical situation, the more the firm sells, the more the money it makes. This hardly encourages conservation. Ways are needed by which corporation can make more money by selling less material. A few ways have been found to do this. DuPont has a contract with a major auto company to paint the new cars. This provides an incentive to minimize the amount of paint used. If it just provided the paint in the usual way, the way to maximize the profit would be to sell more paint per car. Businesses may use less material, but make just as much money, by leasing, servicing, and taking back equipment, as they could by selling it and letting the customer take care of its disposal at the end of its useful life. If done properly, the leasing and servicing system could build up customer loyalty. Some of the parts in obsolete equipment might be reused in updated new models. Interface, a carpet company in Georgia, leases its flooring products to customers who return them when it is worn out so that Interface can recycle it into new flooring materials.[71] Xerox saves hundreds of millions of dollars each year with its programs to reuse and recycle photocopiers.[72] Old machines are updated by replacing modules in the same housing. Laws that require the manufacturer to repair and take back the product at the end of its useful life can create incentives for recovery remanufacture and recycling.[73]

Little work is done on drugs for major tropical diseases because the pharmaceutical industry fears that the persons who contract the illnesses will be too poor to purchase the drugs.[74] However, some drug companies have sold drugs of questionable benefit in developing countries.

Many firms that make money by supplying fossil fuels are reluctant to diversify into energy-supply companies. They could also make money by selling and servicing equipment and techniques of energy conservation or by selling energy from renewable sources. However, the techniques for obtaining energy from renewable sources may be unfamiliar to them. Electric power plants could sell more energy if they used their waste heat for district heating and cooling instead of putting it into a river (see also Chapter 15). Some public utilities commissions in New England have worked out systems for rewarding utilities for selling energy conservation to their customers so that less electricity needs to be generated.

A firm in the business of selling agricultural chemicals might diversify into integrated pest management. This would mean fewer employees in chemical plants, but many more employees as ecologists, entomologists, soil scientists, and such to monitor insect populations, crop damage, fertilizer needs, or others. Such firms could sell consulting services. They might also obtain patents on insect traps, pheromones, or insect repellents, on which they could obtain royalties on the use by others. (Agricultural chemicals are covered in Chapter 11.)

These examples show a switch from pure manufacturing to broader companies that provide a variety of services as well as products. Unfortunately, large firms are less innovative than small ones and are often reluctant to make the kinds of changes suggested here. The changes would involve the retraining of workers no longer needed in production, but the new company structure would require more total employees. As far as consumer products are concerned, it often takes more labor to repair one than to produce a new one by mass production methods.[75] This overlooks the cost of disposal.

American companies are becoming greener as more of them embrace the "triple bottom line" of profit, environment, and social responsibility.[76] Companies with strict standards are finding that their stock market valuation has improved and that it is easier to get loans and insurance.[77] Socially responsible investing is growing.[78] Companies are being rated on their "triple bottom line" by Innovest Strategic Value Advisors, Sustainable Asset Management, and the Equator Principles Financial Institutions.[79]

17.5 ENVIRONMENTAL ECONOMICS OF INDIVIDUALS

17.5.1 Making Choices

Individuals try to minimize their expenses, but they do not always buy or do what is cheapest, at least not what is the cheapest per unit of performance, or cheapest over the long term. They sometimes justify this in terms of increased "convenience." Some of this is a matter of what dollar value they place on their time that is not spent at work earning money. The leaf blower is much more expensive and uses more materials and fossil fuel than a rake that is used by hand. The rake can also provide needed exercise. Yet, they seem to be increasing in popularity. The use of single-use throwaway beverage containers, which are more expensive than refillable ones on a per trip basis, was mentioned earlier. A similar situation exists in fast-food restaurants. Throwaway dishes and eating utensils cost more per use than washable, reusable ones. They are popular with the restaurant management because no equipment, labor, or space is required for washing them. Consumers often buy less efficient items, such as fuel-inefficient vans, justifying their purchases by "convenience" and "I can afford it." They may also buy for style and fashion, rather than longer life and better wear.

There are often hidden costs not seen by the consumer when he or she makes his or her choice. Stores in America create the illusion of abundance and perpetuate the myth of plenty. Few consumers see the natural resource depletion and unsustainable nature of this. The environmental abuse that created the object may be well out of sight, thousands of miles away. Buying a teak bench or tropical hardwood paneling may involve destruction of tropical forests by unsustainable harvesting. Wild flowers are popular with home gardeners. Buying them may contribute to the decimation of natural populations if the plants are collected from the wild. The situation is not helped by nurserymen who collect from the wild and then grow the plants in their nurseries for a year and label them as "nursery grown." Ecolabels will help consumers make informed choices only if the labels are honest. This may require certification by an unimpeachable agency, possibly by government.

The purchase of food may involve similar hidden costs. The food may have been raised on land that formerly was a diverse forest or prairie. It may have involved the heavy use of pesticides and soil loss. Because the average food item consumed in the United States has been shipped over a thousand miles, the energy cost in distributing it is high. Buying local produce would reduce this energy cost. Without informative labels, how can the shopper tell the difference? People may inadvertently contribute to the loss of local farm products by buying a new house on what may have been some of the best farmland in the state. This will eventually raise food prices because the yields from marginal land will be lower or food will have to come from farther away. People who buy waterfront lots, complete with a boat slip at the back of the house, may not think about the productive wetland that was replaced, but may be disappointed at the decline in the fish populations that they came to catch. The abuser may be far away from the problem that he or she helped to create. A farmer in Lancaster County, Pennsylvania, who uses a lot of fertilizer in the hope of maximizing his crop yields may not realize that he is contributing to the decline of life in the Chesapeake Bay downstream (i.e., unless he goes to the Bay to fish or sail in his off time). The motorists who change their own oil and pour it down the storm drain or sewer may not realize that they are degrading the river below, where they like to fish on weekends. Nor are they thinking of the extra cost to the next town below to purify the river water for drinking. The amount that goes into the waterways of the United States in this way each year is about 430 million L, compared with the 45 million L spilled by the Exxon Valdez in Alaska some years ago.[80] These are examples of "social traps", in which short-term goals and narrow interests lead to outcomes that no one wants.[81]

There is a tendency of Americans to buy by the lowest initial price rather than by the lowest price per unit of time. Passive solar heating and cooling raises the initial cost of a home, but lowers the cost of home ownership over a period of years. Outdoor paint that will last for 3 years will probably be cheaper than paint that will last for 5 years. However, it is not cheaper per unit of performance when the time, labor, and money involved in repainting every 3 years is considered.

People may think that a problem they cannot see is not real, as in the case of global warming or depletion of the ozone layer. They may also give no thought to what happens to their trash after they put it at the curb to be collected or what happens after they flush the toilet. Landfills, incinerators, and wastewater treatment plants have all been big problems for some communities. Some Superfund sites are former landfills.

Green consumers can take effective action against businesses that are not environmentally responsible in some cases. If the business markets a product directly to consumers, a consumer boycott of its products may be effective in persuading the company to change. Such was the case when a European company promoted its baby formula over breast feeding in Africa. If the offending firm markets only to other firms that then make the consumer products, the boycotts may be less effective. Putting a correcting resolution on the ballot for the next annual meeting of the shareholders can call attention to the problem. Such resolutions seldom win, but sometimes the company would prefer to settle the complaint than have a major confrontation at the annual meeting.

17.5.2 Ecotourism

Ecotourism is becoming big business.[82] Two journals are devoted to this subject.[83] If done properly, it can be an important way of preserving biodiversity and bringing income to the area. It is the top source of income for Costa Rica, which is doing it in a sustainable fashion.[84] It is the first developing country to halt and then reverse its deforestation. The country uses a tax on fossil fuels to pay rural land owners \$50 ha/yr to protect their forests and \$400 ha^{-1}/yr to farm ecologically. It has also achieved zero population growth at a population of 4 million.

However, there are limits. The actions of tourists must not be so much that they degrade the area, or tourists will stop coming.[85] (Certification by a government or another reputable agency helps to minimize these problems.[86]) In Sri Lanka, the coast was degraded by the lack of proper disposal means for garbage and wastewater.[87] Trekking in the Himalayan Mountains has resulted in deforestation and litter accumulation along trails.[88] Tourism in Bali has resulted in the fragmentation and degradation of coastal ecosystems of coral reefs, seagrass, and mangroves.[89] The message in all of these cases is that proper service facilities must be provided for the tourists who come. The amount and extent of resource use should be planned in advance by the community and the businesses there. Recreational diving in eastern Australia involved the breakage of 0.6–1.9 corals per 30-min dive.[90] The bulk of the damage was done by a small number of inexperienced divers. Predive briefing helped reduce the problem. Because the most damaging activity of all was the anchoring of the boats that brought the divers, putting in established mornings should help.

A big fish on the reef in front of underwater photographers will bring in much more money than the same fish at the end of a fishing line.[91] Many more people will see it on the reef. The small coral reef fish are more valuable for tourism than they would be in someone's home aquarium thousands of miles away. There can be other abuses as well. A man whose family is hungry may poach the wild game to eat or to sell to outsiders who will pay a lot for it. A person who uses poisons or blast fishing in Indonesian coral reefs destroys the reef for short-term personal gain at the expense of long-term gain for the whole community.[92] These types of problems can be minimized by systems that channel at least some of the money from tourism to the people who live there, instead of it all going to the trekking company or the hotel. It is estimated that if the coral reefs in Milne Bay, Papua New Guinea, are left in pristine condition, they will be worth more in tourism revenues than revenues from all other local industries combined.[93] Bird watching at Cape May, New Jersey, on the Atlantic migration flyway, is a big business, bringing 31 million dollars to the area each year.[94] Wild animal enthusiasts spent \$1.24 billion on hotel rooms, food, and equipment in New Jersey in 2001, more than the \$987 spent by hunters and fishermen in the state.[95] Each year 38.4 billion dollars are spent on viewing wildlife in the United States.

Networks of no-take marine refuges have been proposed as a way of protecting biodiversity as well as tourism. If suitably placed, these refuges provide fish and fish eggs and larvae for other areas where fish have been overharvested.[96] The concept is good, but it is proving difficult to put into practice. The key impediment is the long tradition of free access to fish in the ocean commons in spite of the evidence of damage due to overfishing. This is another example of the "tragedy of the commons." This has caused the proposed no-take part of the Florida Keys National Marine Sanctuary to shrink from 10% of the total to less than 0.5%, an amount that may be too small to be effective.

17.6 GOVERNMENT ACTIONS AFFECTING ENVIRONMENTAL ECONOMICS

17.6.1 Role of Government

The government of Bhutan has suggested replacing the usual gross domestic product (GDP) per capita per year with a gross national happiness (GNH).[97] The rating is based on (1) promotion of equitable and sustainable socioeconomic development, (2) preservation and promotion of cultural values, and (3) conservation of natural resources and establishment of good governance. Other rating systems that proposed a replacement for GDP per capita per year are Genuine Progress Indicator, Happy Planet Index, and Human Development Index.

The interplay of the marketplace can allow results to happen that are not in the long-term interest of society. For example, it can drive a species to extinction. It can so degrade the "commons" of air and water that they can no longer provide the free natural services that they once did. Government can provide a series of incentives to encourage individuals and businesses to manage these resources in a sustainable way.[98] It can set up accreditation systems to certify that a product has been produced in an environmentally sound fashion (e.g., a green label indicating that the tropical wood has been harvested in a sustainable manner).[99] Germany has used a government-regulated "Blue Angel" logo on environmentally sound products, such as mercury-free thermometers, for the past 20 years.[100] Voluntary labeling of wood grown on a sustainable basis and crops produced by organic-farming methods is being tried.[101] It is endorsed by environmental organizations. Such systems need independent verification of compliance. Such certification may offer some competitive advantage.

Most of the current regulations in the Untied States are based on considerations of health and safety.[102] In some cases, it has been necessary to use a precautionary principle and act when the problem is definite, but before the final scientific evidence is in, as in the Montreal Protocol for substances that deplete the ozone layer.[103] As problems of ozone depletion, global warming, and loss of biodiversity have reached global dimensions, international treaties among many nations have become necessary to address these problems.[104] International institutions, such as the World Bank, as well as governments are working harder to prevent further damage to our life support system, the earth.[105]

The State Environment Protection Inspectorate of Poland is trying to improve Poland's poor environmental record by a "name and shame" policy of publishing a list of the country's polluters each year.[106] Since 1990, the list has dropped from 80 to 69.

Government can establish laws to restrict abuses. If something is terribly bad, it can be banned completely, as occurred in the United States in the banning of seed treatments involving organomercury compounds. Mercury is involved in enough fish advisories that setting a limit of zero on wastewater may be desirable. This would cause a shift of any remaining chlorine plants using mercury electrodes to the well-proved method using Nafion-containing cells. (Taiwan had seven plants still using the mercury process in 1999. One of them made the news in 1999 when Formosa Plastics shipped 3000 metric tons of mercury-containing waste to Cambodia, where it was rejected.[107] Attempts to have it treated in California failed, so that the waste went back to Taiwan, where the people do not want it either. Taiwan has been slow to adopt adequate controls on environmental pollution.) Putting a zero limit on toxic heavy-metal ions in wastewater might allow sewage sludge to change from something that is hard to get rid of to a profitable soil amendment. A law proposed in Vermont would require labeling of consumer products containing mercury to help in recycling them.[108] Manufacturers of fluorescent lamps oppose this law.

Undesirable practices can be taxed and the money can be used to subsidize desirable ones.[109] An example would be a tax on an automobile that had poor fuel efficiency, with the subsidy going to one with high fuel efficiency. A $0.13 tax per plastic bag in Ireland reduced its usage by 90%.[110] A tax on low-nutrient snack food has been proposed.[111] Government can collect and distribute information on pollution prevention. It can require that information for consumers be included on product labels (e.g., nutritional labels on food packages). It can use seed money to support research, sometimes jointly with businesses. Preserves can be set up to protect biodiversity.[112]

Some of these systems have worked well, but many have been criticized by those affected and others. Some of the regulatory systems are cumbersome and need streamlining. It is not always easy to tell whether the law, or its interpretation, is flawed, or those being regulated are just grumbling because they would prefer not to be regulated. The system is still evolving, sometimes in different ways in different countries. We still lack effective systems for dealing with some of the problems of the "commons." All involved need to work toward more effective systems that will lead all of us into a sustainable future.

17.6.2 Regulations

Regulations[113] have proved to be quite valuable in curbing some of the worst pollution of air and water from point sources. They have been less effective in handling pollution from nonpoint sources. Just the threat of regulation can cause companies to study ways of reducing their pollution. Many efficient, profitable ways of reducing pollution tend to surface after regulations come into effect.[114] Regulations can reduce a firm's resistance to change. They can also provide a level playing field where a company that cleans up at some cost is not penalized in the marketplace by losing market share to one that has not. Penalties for noncompliance must be high enough to hurt the company's bottom line, so that they are not just considered as part of doing business. (Some businesses that would prefer to avoid the costs of complying with regulations quote the costs without mentioning the benefits and call for "regulatory reform."[115])

Industry in the United States has argued for greater flexibility in correcting the problems than is allowed in some of the command-and-control laws. It says that this will allow it to seek the least-cost solutions to problems. This has started to happen. The case of OSi Specialties cited earlier involved the U.S. EPA and the company in a plant-wide approach. The U.S. OSHA has now invited each of the companies with the worst safety records to implement an improved safety and health program, with the guarantee that there will be significant worker involvement.[116] A pilot study of this method in Maine lowered the injury rate by 30%. However, on a national scale, the program has been halted by a court ruling in a suit brought by the National Association of Manufacturers.[117] The court ruled that OSHA did not follow the proper procedures in implementing the plan.

"Cap and trade" systems allow the polluter more flexibility in finding the least-cost way to comply.[118] The government puts a cap on the amount of a pollutant that can be emitted (at a lower level than found at the time). For sulfur dioxide emissions in the United States, this was set at half of the former level. A company that cleaned up to more than this level was given credits to sell to one that could not meet the 50% level. This led to the desired reduction of emissions 30% ahead of schedule at one-tenth the cost of some industry predictions.[119] This involved the use of low-sulfur coal instead of adding scrubbers, and should offer an additional incentive to improve methods for reducing the sulfur levels in mined coal. This method has been proposed for tackling the problem of global warming.[120] An alternative is a carbon tax on emissions. A "cap and trade" system is being used in northern New Jersey to limit the amount of mercury entering the Pasaic River.[121] A proposal has been made to lump all the pollutants from one site in a single trading credit instead of having separate credits for each one.[122] These and other market-based environmental tools have been endorsed widely.[123] For the protection of biodiversity, these include disincentives such as access fees, user fees, and noncompliance fees, as well as cost sharing of reserves, individual fishing quotas, trading rights, and such.

Deposit–refund systems work, as shown in 10 states in the United States and in many other countries. They are effective in eliminating improper disposal as well as in reclaiming a sorted, relatively uncontaminated stream for recycling. About 54% of the aluminum beverage cans used in the United States are recycled. However, in Vermont the deposit returns over 90% of them. Copying the Vermont example with a federal law covering the entire country would eliminate an enormous amount of waste and natural resource depletion. Deposit–refund systems for lead–acid batteries and lubricating oils have been suggested. Although most lead–acid batteries are recycled, some still find their way into landfills and incinerators, where they can cause pollution. The problem of improper disposal of lubricating oil was discussed earlier. Recovering it would allow it to be cleaned up for reuse. Extended product responsibility is being used in Germany. More takeback requirements for manufacturers could lead to greater repair and reuse as well as to less waste and more recycle.[124] A similar system might be used on standardized nestable or collapsible shipping containers for a variety of goods. Governments can help industry standardize sizes so that items can be reused by just changing the label to that of the new company. Beer and wine bottles are handled this way in Germany.

Right-to-know laws have been successful in reducing pollution in the United States. They were enacted in the aftermath of the disaster at Bhopal, India. The Toxic Release Inventory, begun in 1986, requires polluters to quantify their emissions annually[125] (see also Chapter 1). Frequently, top-level managers were unaware of the extent of emissions. The law has been a powerful incentive to reduce emissions. Companies are afraid of losing business if their corporate image is tarnished. Being listed in the local paper as the biggest polluter in the area or being put on a "dirty dozen" list is undesirable from their standpoint. Ten firms with "grandfathered" plants in Texas have announced that they will voluntarily reduce emissions by 10,000 tons annually.[126]

Massachusetts and New Jersey have gone a step further and require information about the movement of toxic chemicals within plants.[127] The laws also require the plants to produce proposals that reduce the amounts of hazardous chemicals used. In Massachusetts during 1990–1995, byproducts were reduced by 30% and toxic chemical use by 20%. Cost savings were reported by 67% of the nearly 600 companies involved. The benefits to the state are estimated to be 91 million dollars, which exceeded costs by 77 million dollars, even when excluding human health and ecological benefits, which are hard to monetize. Only 12% of the companies wanted the program eliminated. This included the Massachusetts Chemical Technology Alliance, a trade association, in conjunction with the American Chemistry Council (formerly the Chemical Manufacturers Association).[128] The New Jersey law directs the companies to prepare a pollution prevention plan. This costs an average of 26,000 dollars. For every dollar spent on the process, facilities predict a net savings of 5–8 dollars with a payback period of 1–2 years.

California has a law, enacted in 1986, Proposition 65, that requires manufacturers to list on the label of their products any significant amounts of compounds in them that might cause cancer or have reproductive toxicity.[129] This has prompted companies to reformulate their products to remove such materials.

17.6.3 Jobs and Regulations

Hueting[130] mentions three myths in the environmental debate that must be reversed for progress to be made: (a) the environment conflicts with employment; (b) production must grow to create financing; and (c) it is too expensive for society to save the environment. There is no evidence that environmental regulations in the United States have harmed the economy.[131] Deposit–refund legislation in 10 states in the United States has created jobs, for example, 4684 in

Michigan, 3800 in New York, 1800 in Massachusetts, and 350 in Vermont.[132] States with stronger environmental programs outperformed states with weaker programs. In the period 1965–1990, levels of carbon monoxide, sulfur dioxide, and lead were reduced in southern California. During this same period, employment grew 50% above the national average, wages were above the national average, and so was the local economy. Nations with the most stringent environmental regulations show the best economic performance. Plants with poor environmental records are no more profitable than cleaner plants in the same industry.[133] There is no evidence that superior environmental performance puts a company at a disadvantage in the marketplace. Only about 0.1% of layoffs are due to environmental regulation. The relocation of jobs offshore is mainly due to lower labor costs, not to environmental regulation. A shift to a sustainable economy is expected to create more jobs in energy efficiency, recycling, and public transportation than will be lost in the oil and coal industries, car manufacturing, and waste disposal. Other jobs will appear in wind energy, photovoltaic cells, bike path construction, and so on.[134]

Spending on environmental technologies creates jobs in part because the business is labor-intensive and often uses capital goods that are produced domestically.[135] There are more than 100,000 environmental technology firms in the United States.[136] They employ 1.3 million people and make 180 billon dollars each year. Their exports are $16 billion/yr and provide a 9.3 billion dollar trade surplus. World markets for the environmental technology industries were 295 billion dollars in 1992.[137] The industry is driven by regulation. Its growth may slow down, at least in the industrialized nations, as major sources of pollution are brought under control. The next step will be to devise new processes that meet the environmental and productivity needs of the industry.[138] As an example, this includes jobs created when a new business is set up to make products from recycled materials.[139] It can also include a new machine that does a better job (e.g., the Maytag front-loading washing machine that handles larger loads in a gentler fashion than the usual top-loading machine, while saving water and electricity).[140] Further work is needed to reduce the price, which at present is twice that of the top loader.

As for the myth that production must grow to create financing, most growth comes from the most environmentally damaging activities. This includes natural resource depletion, carbon dioxide emissions, and other such. The myth that it is too expensive to save the environment is wrong. Using a hand rake instead of the leaf blower, or drinking tap water instead of bottled water, saves materials, energy, and dollars. Use of a bicycle instead of driving alone in your car shows similar savings. Raising two children instead of six is much cheaper and impinges much less on the environment. The combination of economic growth and conservation will be possible in a sustainable economy that does not deplete natural capital and in which clean technologies are no more expensive than current ones when all the costs are considered.

17.6.4 Subsidies

The market is distorted by the many subsidies provided by governments. Subsidies more than $0.95–1.95 trillion/yr are given to environmentally destructive activities.[141] Many of these subsidies shore up declining extractive industries. Coal mining in Germany is an example. In the United States, these include below-cost timber sales in the national forests, cheap mining claims, irrigation water provided by dams built at public expense, grazing fees for public lands that are too low for sustainable use, price supports for crops raised in monoculture, highway construction, and policing costs not assigned to the automobile.[142] Those who use natural capital should be expected to pay the full cost of it,[143] as suggested in a report on the United States written by the U.N. Organization for Economic Cooperation and Development. If they did, they might use less of it, and recycled materials could compete better in the markets. The National Recycling Coalition in the United States urges the elimination of subsidies so that recycled materials can compete better with virgin materials.[144] The price of petroleum in the United States is artificially low, which encourages its wasteful use (as described in Chapter 15). Worldwide subsidies to fossil fuels amount to $58–300 billion/yr.[145] The subsidy to fossil fuels in the United States is $25 billion/yr.[146] The subsidy for water for western agriculture in the United States is $4.4 billion/yr.[147] The U.S. Department of Agriculture also pays farmers to keep 28 million acres of marginal crop land out of production in a Conservation Reserve Program.[148] Agriculture in developed nations receives subsidies of $311 billion/yr.[149] Subsidy-driven overcapitalization of the world's fishing fleets has resulted in drastic declines in many fish populations.[150] National subsidies to fishing in the North Atlantic Ocean have been $2.5 billion/yr.

Some authors feel that governments, whether Brazil or the United States, tend to perpetuate the status quo, complete with favors to special interests.[151] Subsidies allow inefficiencies to remain in the system. Subsidies, if used, should be taken off virgin materials and put on to recycled ones. They should be removed from energy-inefficient forms of transportation and given to the energy-efficient ones. They might also be used to jump-start new technologies for a sustainable future. This was done in California for wind energy. Subsidies in the form of tax relief or joint research projects with industry might be used to put pollution prevention processes into plants.

17.6.5 Taxes

Green taxes may be the least costly way to a sustainable future.[152] Taxes on waste, on pollution, and on environmentally damaging products, such as pesticides, fertilizers, motor vehicle fuels, as well as unsustainable natural resource depletion, have been proposed.[153] Such ecotaxes are being tested in Europe.[154] Many of these taxes are revenue-neutral, the money being put back into the same industries in the form of reductions in other taxes. The amount put back can favor those industries with the most efficient processes and producing the least pollution. These include taxes on carbon dioxide, sulfur dioxide, and nitrogen oxide emissions; lead in gasoline, fertilizer, and batteries in Sweden; toxic waste and water pollution in Germany; water pollution and household waste in the Netherlands; carbon dioxide in Norway; and water pollution in France. Other taxes are also becoming more common. Denmark taxes cars and trucks by weight, which is often related to fuel efficiency. It also has a tax on disposable beverage containers, plates, cups, and cutlery. There are also taxes on paper and plastic carrier bags, pesticides in containers smaller than 1 kg, nickel–cadmium batteries, and incandescent light bulbs. Belgium has a tax on disposable razors that are not recycled, as well as one on disposable beer bottles. In Australia, products made entirely from recycled paper are exempt from sales tax. All of these are designed to shift taxes away from labor and capital and on to energy and the environment. The taxes appear to be working.[155] Taxes on heavy-metal emissions in the Netherlands have cut emissions of mercury by 97%. Iowa has reduced fertilizer use by a tax on it.[156] Similar taxes have been proposed to cut fertilizer use in Europe.[157] Denmark's waste tax reduced waste by 26% from 1987 to 1997, with recycling going up at the same time.[158] England is studying the use of ecotaxes to improve the environment.[159] There is a movement to make the ecotaxes uniform over the European Union.[160] China has proposed taxes on wastewater, noise, solid waste, and low-level radioactive waste.[161]

Regulatory taxation of fossil fuels is favored as a means of combating global warming.[162] However, the European chemical industry is against energy taxes for fear that these could hinder its ability to compete with companies in other countries (e.g., the United States) where energy is cheaper.[163] A study by the World Resources Institute favors taxing pollution, congestion, and waste instead of payrolls, incomes, and profits.[164] The taxes would include tolls on congested urban highways, taxes on the carbon content of fuels, and fees for collecting household waste. A study by the Worldwatch Institute proposes a shift of taxes from wages and profits to pollution and resource depletion.[165] The study predicts that this would create jobs and boost living standards without damaging industrial competitiveness and without increasing the overall tax burden. The American Chemistry Council is opposed to the idea. Some of the common objections to pollution taxes, which may not be valid, are that they are a license to pollute, they may be accepted as a cost of doing business, and they may be harshest on smaller firms.[166]

Green taxes must not be dodged by firms bringing in products made in other countries that do not have these taxes. Tariffs on imported products could neutralize this.[167] Tariffs on clean technology and on products produced with such technology could be removed. One proposal is to handle his by giving a country most sustainable nation status, in the same way that the United States gives most favored nation status to others for trading purposes. This could translate into low or no tariffs for a country using sustainable harvesting of resources.

Another change in government rules might eliminate a practice of some firms that try to delay cleaning up by litigation, feeling that legal fees are cheaper than cleaning up.[168] The proposal is to have the U.S. Securities and Exchange Commission require companies to report hazardous waste sites as financial liabilities. This would show up in the company's bottom line until the site was cleaned up.

REFERENCES

1. R. Dagani, *Chem. Eng. News*, July 5, 1999, 38.
2. (a) L. Ember, *Chem. Eng. News*, Mar. 20, 1995, 6; (b) B. Dalal-Clayton, *Getting to Grips with Green Plans—National Level Experience in Industrial Countries*, Earthscan, London, 1996.
3. (a) C.H. Freese, *Wild Species as Commodities—Managing Markets, and Ecosystems for Sustainability*, Island Press, Washington, DC, 1998; (b) T.M. Swansoon, ed., *The Economics and Ecology of Biodiversity Decline: The Forces Driving Global Change*, Cambridge University Press, New York, 1998; (c) S. Iudicello, M. Weber, and R. Wieland, *Fish, Markets and Fishermen*, Island Press, Washington, DC, 1999; (d) E. Bulte, R. Damania, L. Gillson, and K. Lindsay, *Science*, 2004, *306*, 420.
4. (a) C.D. Kolstad, *Environmental Economics*, Oxford University Press, Oxford, 1999; (b) S.L. Pimm, *The World According to Pimm: A Scientist Audits the Earth*, McGraw-Hill, New York, 2001; (c) R.N. Stavins, *Economics of the Environmennt: Selected Readings*, 4th ed., W.W. Norton, New York, 2000; (d) J.J. Rao, *Chem. Eng. Prog.*, 2001, *97*(11), 38; (e) E.A. Davidson, *You Can't Eat GNP—Economics As If Ecology Mattered*, Perseus Publishing, Cambridge, MA, 2000; (f) A. Gilpin, *Environmental Economics: A Critical Overview*, Wiley, New York, 2000; (g) D. Pearce and E.B. Barbier, *Blueprint for a Sustainable Economy*, Earthscan, London, 2000; (h) H.E. Daly and J. Farley, *Ecological Economics—Principles and Applications*, Island Press, Washington, DC, 2003; (i) N.O. Keohane and S.M. Olmstead, *Markets and the Environment*, Island Press, Washington, DC, 2007; (j) C.S. Russell, *Applying Economics to the Environment*, Oxford University Press, Oxford, 2001; (k) L.R. Brown, *The Earth Policy Reader*, Earth Policy Institute, Washington, DC, 2002 (www.earth-policy.org/Books/index.htm); (l) T.L. Cherry, S. Kroll, and J.F. Shogren, *Environmental Economics—Experimental Methods*, Taylor and Francis, Boca Raton, FL, 2007.
5. (a) G.C. Daily, ed., *Nature's Services—Societal Dependence on Natural Ecosystems*, Island Press, Washington, DC, 1997; (b) F. Hinterberger, E.F. Granek, S. Polasky, S. Aswani, L.A. Cramer, D.M. Stoms, et al. *Ecol. Econ.*, 1997, *23*(1), 1; (c) G.C. Daily and K. Ellison, *The New Economy of Nature, The Quest to Make Conservation Profitable*, Island Press, Washington,

DC, 2002; (d) G. Heal, *Nature and the Marketplace, Capturing the Value of Ecosystem Services*, Island Press, Washington, DC, 2000; (e) R. Costanza, H. Daly, C. Folke, P. Hawken, C.S. Holling, A.J. McMichael, D. Pimentel, and D. Rapport, *BioScience*, 2000, *50*(2), 149; (f) K. Ellison, *Conservation in Practice*, 2005, *6*(3), 38; (g) J. Withgott, Science, 2004, *305*, 1100; (h) R.S. Farrow, C.B. Goldberg, and M.J. Small, eds, *Environ. Sci. Technol.*, 2000, *34*, 1381–1461; (i) C. Kremen, J.O. Niles, M.G. Dalton, G.C. Daily, P.R. Ehrlich, J.P. Fay, D. Grewal, and R.P. Guillery, *Science*, 2000, *288*, 1828; (j) E.B. Barbier, E.W. Koch, B.R. Silliman, S.D. Hacker, E. Wolanski, J. Primavera, et al., *Science*, 2008, *319*, 321; (k) M. Palmer, E. Bernhardt, E. Chornesky, S. Collins, A. Dobson, C. Duke, B. Gold, et al., *Science*, 2004, *304*, 1251.

6. (a) R. Costanza, R. d'Agre, R. deGroot, S. Farber, M. Grasso, B. Hannon, K. Limburg, S. Naeem, R.V. O.Neill, J. Paruelo, R.G. Raskin, P. Sutton, and M. van den Belt, *Nature*, 1997, *387*, 253; (b) J.N. Abramovitz, *World Watch*, 1997, *10*(5), 9; (c) W. Roush, *Science*, 1997, *276*, 1029; (d) S.L. Pimm, *Nature*, 1997, *387*, 231; (e) M. Rouhi, *Chem. Eng. News*, June 30, 1997, 38; (f) D. Pearce, *Environment*, 1998, *40*(2), 23.
7. (a) R. Hoyle, *Nat. Biotechnol.*, 1997, *15*, 1227; (b) Anon., *Environ. Sci. Technol.*, 1998, *32*, 492A.
8. R. Renner, *Environ. Sci. Technol.*, 1998, *32*, 86A.
9. Anon., *Chem. Eng. News*, July 26, 1999, 26.
10. J.B. Ruhl, S.E. Kraft, and C.L. Lant, *The Law and Policy of Ecosystem Services*, Island Press, Washington, DC, 2007.
11. (a) D. Pearce and D. Moran, *The Economic Value of Biodiversity*, Earthscan, London, 1994; (b) E.B. Barbier, J.C. Burgess, and C. Folke, *Paradise Lost? The Ecological Economics of Biodiversity*, Earthscan, London, 1994; (c) T.M. Swanson, ed., *The Economics and Ecology of Biodiversity Decline*, Cambridge University Press, Cambridge, 1995; (d) R. Baker, *Saving All the Parts—Reconciling Economics and the Endangered Species Act*, Island Press, Washington, DC, 1993; (e) D.M. Roodman, *The Natural Wealth of Nations: Harnessing the Market for the Environment*, W.W. Norton, New York, 1998; (f) A. Duraiappah, Policy Brief on "Putting the Right Price on Nature: Environmental Economics," U.N. Environment Program, Geneva, Switzerland 2006, www.scidev.net/biodiversity.
12. (a) D. Pimentel, C. Wilson, C. McCulllum, R. Huang, P. Dwen, J. Flack, Q. Tran, T. Saltman, and B. Cliff, *Bioscience*, 1997, *47*, 747; (b) D. Pimentel, L. Lach, R. Zuniga, and D. Morrison, *BioScience*, 2000, *50*(1), 53 [costs of non-native species in the U.S.A.].
13. S.L. Buchmann and G.P. Nabhan, *The Forgotten Pollinators*, Island Press, Washington, DC, 1996.
14. P.M. Vitousek, J.D. Aber, R.W. Howarth, G.E. Likens, P.A. Matson, D.W. Schinder, W.H. Schlesinger, and D.G. Tilman, *Ecol. Appl.*, 1997, *7*, 737–750.
15. T. Heyd, *Environ. Ethics*, 1997, *19*, 437.
16. R.U. Ayres, *Ecol. Econ.*, 1996, *19*, 117.
17. (a) R. Douthwaite, *The Growth Illusion*, Council Oaks Books, Gabriola Island, British Columbia, Canada, 1993; (b) R. Ayers, *Turning Point—The End of the Growth Paradigm*, Earthscan, London, 1998; (c) M. Carley and P. Spapens, *Sharing the World—Sustainable Living and Global Equity in the 21th Century*, Earthscan, London, 1998; (d) A. Durning, *How Much is Enough—The Consumer Society and the Future of the Earth*, Norton, New York, 1992; (e) R.U. Ayres, *Environ. Sci. Technol.*, 1998, *32*, 361A.
18. (a) T. Prugh, *National Capital and Human Economic Survival*, International Society of Ecological Economics, ISEE Press, Solomons, MD, 1995; (b) A.M. Jansson, M. Hammer, C. Folke, and R. Costanza, eds, *Investing in Natural Capital: The Ecological Economics Approach to Sustainability*, Island Press, Washington, DC, 1994; (c) A. Clewell and N. Lopoukhine, *Restor. Manage. Notes*, 1996, *14*, 151; (d) C. Folke, C.S. Holling, and C. Perrings, *Ecol. Appl.*, 1996, *6*, 1018.
19. (a) A. Durning, *Technol. Rev.*, 1991, *94*(4), 57; (b) H.E. Daly, *Beyond Growth: The Economics of Sustainable Development*, Beacon Press, Boston, 1996; (c) J.C.J.M. van den Berghand and J. van der Straaten, *Toward Sustainable Development—Concepts, Methods and Policy*, Island Press, Washington, DC, 1994; (d) J. Carew-Reid, *Strategies for National Sustainable Development—A Handbook for Their Preparation and Implementation*, ICUN, Earthscan, London, 1995; (e) D. Pearce, ed., *Models of Sustainable Development*, Edward Elgar, Cheltenham, UK, 1996; (f) B. Nath, L. Hens, and D. Devuyst, eds, *Textbook on Sustainable Development*, VUB University Press, Brussels, 1996; (g) K. Ginther, E. Denters, and P.J.I.M. de Waart, eds, *Sustainable Development and Good Governance*, M. Nijhoff, Dordrecht, 1995; (h) D.W. Pearce, *Sustainable Development: Economics and Environment in the Third World*, Gower, Brookfield, VT, 1990; (i) M. Jacobs, *Sustainable Development: Greening the Economy*, Fabian Society, London, 1990; (j) C.A.C. Nieto, *Sustainable Development in Theory and Practice: A Costa Rican Case Study*, PhD Dissertation, University of Delaware, Newark, DE, 1997; (k) Anon., *Sustainable Development: OECD Policy Approaches for the 21st Century*, OECD, Paris, 1997; (1) C. Raviaoli and P. Elkins, *Economists and the Environment*, Zed Books, London, 1995, 161–200; (m) President's Council on Sustainable Development, *Sustainable Developments*, Washington, DC, 1995; (n) R. Goodland and H. Daly, *Ecol. Appl.*, 1996, 6, 1002; (o) H.E. Daly and J.B. Cobb, Jr., *For the Common Good*, Beacon Press, Boston, 1994; (p) T. O'Riordan and H. Voisey, eds, *The Transition to Sustainability—the Politics of Agenda 21 in Europe*, Earthscan, London, 1998; (q) R. Beaton and C. Maser, *Reuniting Economy and Ecology in Sustainable Development*, Lewis Publishers, Boca Raton, FL, 1999; (r) L.F.D. Muschett, ed., *Principles of Sustainable Development*, St Lucie Press, Delray Beach, FL, 1997; (s) P. Hawken, A. Lovins, and H. Lovins, *Natural Capitalism: Creating the Next Industrial Revolution*, Little Brown, New York, 1999; (t) C. Maser, *Ecological Diversity in Sustainable Development—The Vital and Forgotten Dimension*, Lewis Publishers, Boca Raton, FL, 1999; (u) C.J. Kibert, ed., *Reshaping the Built Environment—Ecology, Ethics and Economics*, Island Press, Washington, DC, 1999; (v) G. Atkinson, R. Dubourg, K. Hamilton, M. Munasinghe, D. Pearce, and C. Young, *Measuring Sustainable Development: Macroeconomics and the Environment*, Edward Elgar, Cheltenham, UK, 1997; (w) J. Pelley, *Environ. Sci. Technol.*, 2000, *34*, 206A; (x) R. Costanza, H. Daly, C. Folke, P. Hawken, C.S. Holling, A.J. McMichael, D. Pimentel, and D. Rapport, *BioScience*, 2000, *50*(2), 149.
20. C. Folke, C.S. Holling, and C. Perrings, *Ecol. Appl.*, 1996, *6*, 1018.

21. M. Wackernagel, N.B. Schulz, D. Deumling, A.C. Lanares, M. Jenkins, V. Kapos, C. Monfreda, J. Loh, N. Meyers, R. Norgaard, and J. Randers, *Proc. Natl. Acad. Sci. USA*, 2002, *99*, 9266.
22. (a) G. Hardin, *Living within Limits: Ecology, Economics and Population Taboos*, Oxford University Press, Oxford, 1993; (b) L.R. Brown, *State of the World 1995*, W.W. Norton, New York, 1995, 3; (c) R. Costanza, ed., *Ecol. Econ.*, 1995, *15*(2), 89–157; (d) E. Ostrom, J. Burger, C.B. Field, R. B. Norgaard, and D. Policansky, *Science*, 1999, *284*, 278.
23. (a) T.M. Power, *Extraction and the Environment—The Economic Battle to Control Our Natural Landscapes*, Island Press, Washington, DC, 1996; (b) T.M. Power, *Lost Landscapes and Failed Economics—The Search for a Value of Place*, Island Press, Washington, DC, 1996; (c) G. Chichilnisky, *Amicus. J. (Natural Resources Defense Council)*, 1998, *20*(2), 13.
24. C. Runyan, *World Watch*, 1997, *10*(5), 9.
25. (a) W. Rees and M. Wackernagel, *Our Ecological Footprint: Reducing Human Impact on the Earth*, New Society, Philadelphia, 1996; (b) M. Weckernagel and W.E. Rees, *Ecol. Econ.*, 1997, *20*, 3.
26. M. Voith, *Chem. Eng. News*, Oct. 6, 2008, 12.
27. C. Folke, A. Jansson, J. Larsson, and R. Costanza, *Ambio*, 1997, *26*(3), 167.
28. www.panda.org.
29. (a) B.W. Vigon, D.A. Tolle, B.W. Cornaby, H.C. Latham, C.L. Harrison, T.L. Boguski, R.G. Hunt, and J.D. Sellers, *U.S. EPA Risk Reduction Engineering Laboratory. Life Cycle Assessment—Inventory Guidelines and Principles*, Lewis, Boca Raton, FL 1994; (b) G.A. Keoleian, D. Menerey, B.W. Vigon, D. A. Tolle, B.W. Cornaby, H.C. Latham, C.L. Harrison, T. Boguski, R.G. Hunt, and J.D. Sellers, *Product Life Cycle Assessment to Reduce Health Risks and Environmental Impact*, Noyes, Park Ridge, NJ, 1994; (c) S. van der Ryn and S. Cowan, *Ecological Design*, Island Press, Washington, DC, 1995, 90–98; (d) J.S. Hirschhorn, *Chemtech*, 1995, *25*(4), 6; (e) J. Nash and M.D. Stoughton, *Environ. Sci. Technol.*, 1994, *28*, 236A; (f) T.E. Graedel, *Streamlined Life-Cycle Assessment*, Prentice-Hall, Paramus, NJ, 1998; (g) D. Ciambrone, *Environmental Life Cycle Analysis*, Lewis, Boca Raton, FL, 1997; (h) T. Krawczyk, *Int. News Fats Oils Relat. Mater.*, 1997, *8*, 266; (i) J. Kaiser, *Science*, 1999, *285*, 685; (j) C. Hendrickson, A. Horvath, S. Joshi, and L. Lave, *Environ. Sci. Technol.*, 1998, *32*, 184A; (k) S.T. Chubbs and B.A. Steiner, *Environ. Prog.*, 1998, *17*(2), 92; (l) J.J. Marano and S. Rogers, *Environ. Prog.*, 1999, *18*, 267; (m) M.A. Curran, ed., *Environ. Prog.*, 2000, *19*, 61–145; (n) M.Z. Hauschild, *Environ. Technol.*, 2005, *39*, 81A; (o) A. Tukker, *Environ. Sci. Technol.*, 2002, *36*, 71A.
30. D. Hanson, *Chem. Eng. News*, Dec. 21, 1998, 26.
31. (a) C. Crossen, *Tainted Truth: The Manipulation of Fact in America*, Simon and Schuster, New York, 1994, 140–143; (b) Anon., *World Watch*, 2007, *20*(2), inside front cover.
32. R.B. Blackburn and J. Payne, *Green Chem.*, 2004, *6*, G59.
33. Anon., *Tufts Univ. Health Nutr. Lett.*, 2007, *25*(1), 3.
34. (a) G.A. Davis, *The Use of Life Cycle Assessment in Environmental Labeling Programs*, EPA/742-R-93-003, Washington, DC, Sept. 1993; (b) M.A. Curran and T.J .Skone, *Environ. Prog.*, 2003, *22*(1), 1.
35. (a) B. Allen, *Green Chem.*, 2000, *2*, G19; (b) L.H. Gulbrandsen, *Environment*, 2005, *47*(5), 8.
36. (a) Anon., *Univ. Calif. Berkeley Wellness Lett.*, 2005, *21*(10), 6; (b) Anon., *The Consumer Union Guide to Environmental Labels*, www.eco-labels.org, Yonkers, NY.
37. (a) J.A. Dixon, L.F. Scura, R.A. Carpenter, and P.B. Sherman, *Economic Analysis of Environmental Impacts*, 2nd ed., Earthscan, London, 1994; (b) B.B. Marriott, *Environmental Impact Assessment: A Practical Guide*, McGraw-Hill, New York, 1997; (c) E.J. Calabrese and L.A. Baldwin, *Performing Ecological Risk Assessments*, St. Lucie Press, Delray Beach, FL, 1993; (d) P.A. Erickson, *A Practical Guide to Environmental Impact Assessment*, Academic, San Diego, CA, 1994; (e) S. Farrow and M. Toman, *Environment*, 1999, *41*(2), 12.
38. (a) D.J. Hanson, *Chem. Eng. News*, July 17, 1995, 45; (b) G.W. Suter, II, *Ecological Risk Assessment*, 2nd ed., CRC Press, Boca Raton, FL, 2007; (c) R.L. Revesz and M.A. Livermore, *Rethinking Rationality—How Cost-Benefit Analysis Can Better Protect the Environment and Our Health*, Oxford University Press, New York, 2008.
39. J. Johnson, *Chem. Eng. News*, Aug. 24, 1998, 14.
40. (a) L.R. Raber, *Chem. Eng. News*, Feb. 3, 1997, 28; (b) C.M. Cooney, *Environ. Sci. Technol.*, 1997, *31*, 14A; (c) T. Agres, *R&D (Cahners)*, 1998, *40*(5), 15.
41. (a) R.D. Morgenstern, ed., *Economics Analyses at EPA Assessing Regulatory Impact*, Resources for the Future, Washington, DC, 1997; (b) J.D. Graham and J.K. Hartwell, *The Greening of Industry: A Risk Management Approach*, Harvard University Press, Cambridge, MA, 1997.
42. (a) Anon., *Chem. Eng. News*, Oct. 27, 1997, 18; (b) Anon., *Environ. Sci. Technol.*, 1998, *32*, 14A; (c) Anon., *Chem. Eng. News*, Nov. 22, 1999, 45 [benefits of Clean Air Act 1990 Amendments]; (d) C.M. Cooney, *Environ. Sci. Technol.*, 2000, *34*, 67A [benefits of Clean Air Act 1990 Amendments]; (e) C.M. Cooney, *Environ. Sci. Technol.*, 2000, *34*, 67A.
43. Anon., *Environment*, 1997, *39*(10), 23.
44. (a) M. O'Brian, *Making Better Environmental Decisions: An Alternative to Risk Assessment*, MIT Press, Cambridge, MA, 2000; (b) Q. Zhang, J.C. Crittenden, and J.R. Mihelcic, *Environ. Sci. Technol.*, 2001, *35*, 1282; (c) H. Gavaghan, *Science*, 2000, *290*, 911; (d) F. Ackerman and L. Heinzerling, *Priceless—On Knowing the Price of Everything and the Value of Nothing*, The New Press, New York, 2004; (e) O. Pilkey, *Science*, 2008, *320*, 1423; (f) C. Hogue, *Chem. Eng. News*, Jan. 29, 2007, 32; (g) C.D. Brewer, *Science*, 2009, *325*, 1075.
45. P. Hawken, *The Ecology of Commerce, A Declaration of Sustainability*, Harper-Collins, New York, 1993, 177.
46. (a) J.A. Cichowicz, *How to Control Costs in Your Pollution Prevention Program*, Wiley, New York, 1997; (b) J.H. Clark, *Chemistry of Waste Minimisation*, Chapman & Hall, London, 1995; (c) S.T. Thomas, *Facility Manager's Guide to Pollution Prevention and Waste Minimization*, BNA Books, Washington, DC, 1995; (d) J.R. Aldrich, *Pollution Prevention Economics: Financial Impact on Business and Industry*, McGraw-Hill, New York, 1996; (e) R.C. Kirkwood and A.J. Longley, *Clean Technology and the Environment*, Blackie Academic, London, 1995; (f) P. Sharratt and M. Sparshott, *Case Studies in Environmental Technology*, IChemE Rugby, UK, 1996; (g) N.P. Cheremisinoff and A. Bendavid-Val, *Green Profits*, Butterworth-Heinemann, Woburn, MA, 2001; (h) J.D. Underwood, *Chem. Ind., (Lond.)*, Jan. 3, 1994, 18.
47. A. Thayer, *Chem. Eng. News*, July 3, 1995, 10.

48. (a) *Green Ledgers: Case Studies in Corporate Environmental Accounting*, World Resources Institute, Washington, DC, 1995; (b) M. Spitzer and H. Elwood, *An Introduction to Environmental Accounting as a Business Management Tool: Key Concepts and Terms*, U.S. Environmental Protection Agency, EPA 742-R-95-001, June 1995.
49. D. Shannon, *Environ. Sci. Technol.*, 1995, *29*, 309A.
50. M. Dorfman, C. Miller, and W. Muir, *Environmental Dividends: Cutting More Chemical Wastes*, Inform, New York, 1992.
51. Anon., *Environ. Sci. Technol.*, 1994, *28*, 214A.
52. (a) A. Thayen, *Chem. Eng. News*, May 5, 1997, 28; (b) J. Johnson, *Chem. Eng. News*, Nov. 2, 2009, 24.
53. D. Munton, *Environment*, 1998, *40*(6), 4.
54. J. Gersh, *Amicus. J. (Natural Resources Defense Council)*, 1999, *21*(1), 37.
55. (a) R. Gottlieb, *Reducing Toxics—A New Approach to Policy and Industrial Decision-making*, Island Press, Washington, DC, 1995; (b) T. Vinas and J.S. McClenahen, *Industry Week*, 2004, *253*(7), 53.
56. (a) L. Greer and C. van L. Sels, *Environ. Sci. Technol.*, 1997, *31*, 418A; (b) T. Chapman, *Chem. Ind.* (*Lond.*), 1998, 475; (c) Anon., *Environ. Sci. Technol.*, 1998, *32*, 307A.
57. (a) E. Robbins, *Amicus. J. (Natural Resources Defense Council)*, 1999, *21*(3), 8; (b) J. Johnson, *Chem. Eng. News*, Aug. 17, 1998, 34; Sept. 13, 1999, 22.
58. K. Nelson. In: R. Socolow, C. Andrews, F. Berkhout, and V. Thomas, eds, *Industrial Ecology and Global Change*, Cambridge University Press, Cambridge, 1994, 371.
59. (a) S.L. Wilkinson, *Chem. Eng. News*, Aug. 4, 1997, 35; (b) Anon., *Green Chemistry and Engineering Conference*, Washington, DC, June 23–25, 1997; (c) P.V. Tebo., *Chemtech*, 1998, *28*(3), 8.
60. (a) K. Schmidt, *New Sci.*, 1996, *150*(2032), 32; (b) B. Hileman, *Chem. Eng. News*, May 29, 1995, 34; (c) S.M. Edgington, *Biotechnology*, 1995 *13*, 33.
61. H. Grann. In: D.J. Richards, ed., *The Industrial Green Game—Implications for Environmental Design and Management*, Royal Society of Chemistry, Cambridge, UK, 1997, 117–123.
62. D.J. Hanson, *Chem. Eng. News*, Dec. 8, 1997, 18.
63. J.D. Underwood, *Chem. Ind.*, (*Lond.*), 1994, 18.
64. E. Kirschner, *Chem. Eng. News*, Feb. 26, 1996, 19.
65. J. Johnson, *Chem. Eng. News*, Sept. 29, 1997, 19.
66. J.J. Romm, *Cool Companies: How the Best Businesses Boost Profits and Productivity by Cutting Greenhouse Gas Emissions*, Island Press, Washington, DC, 1999.
67. R. Socolow, C. Andrews, F. Berkhout, and V. Thomas, eds, *Industrial Ecology and Global Change*, Cambridge University Press, Cambridge, 1994, 384, 390, 413.
68. M. Ross, *Financial Manage.*, 1986, *15*(4), 15.
69. Organization for Economic Cooperation and Development, *Biotechnology for Clean Industrial Products and Processes*, Organization for Economic Cooperation and Development, Paris, 1998, 58.
70. (a) P. Hawken, *The Ecology of Commerce, A Declaration of Sustainability*, Harper-Collins, New York, 1993, 116; (b) D.C. Korten, *The Post-Corporate World: Life After Capitalism*, Berrett-Koehler, San Francisco, 1999; (c) T. Clarke, *Ecologist*, 1999, *29*(3), 158.
71. P.M. Morse, *Chem. Eng. News*, July 26, 1999, 19.
72. J. Johnson, *Chem. Eng. News*, June 21, 1999, 25.
73. E.B. Bennett and T.E. Graedel, *Environ. Sci. Technol*, 2000, *34*, 541.
74. (a) P. Wilmshurst, *Chem. Ind.* (*Lond.*), 1997, 706; (b) M. Jacobs, *Chem. Eng. News*, May 17, 1999, 5; (c) S. Garrattini, *Science*, 1997, *275*, 287.
75. R.U. Ayres, *Environ. Sci. Technol.*, 1998, *32*, 366A.
76. (a) W. Greider, *On Earth (Natural Resources Defense Council)*, Fall, 2003, 20; (b) A.W. Savitz and K. Weber, *The Triple Bottom Line—How Today's Best-Run Companies are Achieving Economic, Social and Environmental Success*, Jossey-Bass, New York, 2006.
77. (a) Anon., *Environ. Sci. Technol.*, 2000, *34*, 459A; (b) Anon., *Environment*, Nov., 2000, 6.
78. (a) J. Motavelli, *Environmental Defense Solutions*, 2004, *35*(1), 10; (b) V. Dunn, *Chem. Ind.* (*Lond.*), Feb. 2, 2004, 16.
79. (a) K. Elllison, *Nat. Conserv. Mag.*, 2002, *52*(4), 44; (b) www.innovestgroup.com/home.htm; (c) www.equator-principles.com/principles.shtml; (d) Environmental valuation & cost-benefit news, *Sustainable Asset Management Sustainability Yearbook 2008*, Apr. 7, 2008; (e) Anon., *Chem.Eng. News*, Sept. 28, 2009, 33.
80. M. Webster, T. O'Brien, and R. Collins, *Chem. Ind.* (*Lond.*), 1997, 810.
81. J.C.J.M. van den Bergh and J. van der Straaten, *Toward Sustainable Development*, Island Press, Washington, DC, 1994, 104.
82. (a) T. Whelan, *Nature Tourism—Managing for the Environment*, Island Press, Washington, DC, 1991; (b) G. Neale, ed., *The Green Travel Guide*, 2nd ed., Earthscan, London, 1999; (c) M. Honey, *Ecotourism and Sustainable Development—Who Owns Paradise?* Island Press, Washington, DC, 1998; (d) R.B. Primack, D. Bray, H.A. Galletti, and I. Ponciano, eds, *Timber, Tourists and Temples—Conservation and Development in the Maya Forest of Belize, Guatemala and Mexico*, Island Press, Washington, DC, 1998; (e) L. France, *The Earthscan Reader in Sustainable Tourism*, Earthscan, London, 1997; (f) H. Youth, *World Watch*, 2000, *13*(3), 120; (g) Anon., *Environment*, 1999, *41*(10), 7; (g) D.B. Weaver, ed., *The Encyclopedia of Ecotourism*, Oxford University Press, 2001; (h) G. Neale, ed., *The Green Travel Guide*, 2nd ed., Earthscan, London, 1999; (i) D.A. Fennell, *Ecotourism Program Planning*, Oxford University Press, UK, 2002.
83. *J. Ecotourism* and *J. Sustainable Tourism*, both from Channel View Publications, Bristol, UK.
84. (a) J. Tidwell, *Conservation Front Lines*, Conservation International, Washington, DC, 2006, *6.1*, 6; (b) P.J. Ferraro and A. Kiss, *Science*, 2002, *298*, 1718; (c) G.C. Daily, T. Soderquist, S. Aniyar, K. Arrow, P. Dasgupta, P.R. Ehrlich, et al., *Science*, 2000, *289*, 395.
85. (a) R. Buckley, *Case Studies in Ecotourism*, Oxford University Press, UK, 2003; (b) R. Buckley, C. Pickering, and D.B. Weaver, eds, *Nature-Based Tourism, Environment and Land Management*, Oxford University Press, UK, 2004; (c) L. France, *The Earthscan Reader in Sustainable Tourism*, Stylus Publishing, Herndon, VA, 1997; (d) P.W. McRandle, *World Watch*, 2006, *19*(4), 5; (e) P.F.J. Eagles, S.F. McCool, and C.D. Haynes, *Sustainable Tourism in Protected Areas—Guidelines for Planning and Management*, Island Press, Washington, DC, 2002; (f) A. Ananthaswamy, *New Sci.*, Mar. 6, 2004, 6.

86. M. Honey, *Environment*, 2003, *45*(6), 8.
87. A.T. White, V. Barker, and G. Tantrigama, *Ambio*, 1997, *26*, 335.
88. S.C. Rai and R.C. Sundriyal, *Ambio*, 1997, *26*, 235.
89. D. Knigh, B. Mitchell, and G. Wall, *Ambio*, 1997, *26*, 90.
90. V.J. Harriott, D. Davis, and S.A. Banks, *Ambio*, 1997, *26*, 173.
91. H. Youth, *L.World Watch*, 2000, *13*(3), 12.
92. H. Cesar, C.G. Lundin, S. Bettencourt, and J. Dixon, *Ambio*, 1997, *26*, 345.
93. G. Kula, *Conserv. Int.* "*Cl. News from the Front*," 1997, *3*(1).
94. P. Dunne, *Wilmington Delaware News J.*, July 29, 1997, B2.
95. (a) National Survey of Fishing, Hunting an Wildlife Association, *Wilmington Delaware News J.*, Oct. 21, 2002, B2; (b) Anon., *Outdoor Delaware*, Fall, 2002, 29.
96. (a) J.C. Ogden, *Science*, 1997, *278*, 1414; (b) R. Fujita and J. Lubchenko, *KDF Lett.* (Environmental Defense Fund), 1998, *29*(1), 7; (c) Anon., *Chem. Eng. News*, Jan. 12, 1998, 23.
97. (a) A. Esty, *Am. Sci.*, 2004, *92*(6), 513; (b) Anon., *Conserv. Mag.*, 2009, *10*(1), 22.
98. R. Socolow, C. Andrews, F. Berkhout, and V. Thomas, eds, *Industrial Ecology and Global Change*, Cambridge University Press, Cambridge, 1994, 406.
99. (a) J. Barrett, *Pulp Pap. Int.*, 1994, *36*(12), 53; (b) M. Wagner, *Amicus J. (Natural Resources Defense Council)*, 1997, *19*(1), 17; (c) *Amicus. J. (Natural Resources Defense Council)*, 1998, *20*(2), 46; (d) B. Allen, *Green Chem.*, 2000, *2*, G19.
100. J. Gersh, *Amicus. J. (Natural Resources Defense Council)*, 1999, *21*(1), 40.
101. N. Dudley, C. Elliott, and S. Stolton, *Environment*, 1997, *39*(6), 16.
102. M.S. Reisch, *Chem. Eng. News*, Jan. 12, 1998, 98.
103. (a) B. Hileman, *Chem. Eng. News*, Feb. 9, 1998, 16; (b) J.A. Tickner, ed., *Precaution: Environmental Science, and Preventive Public Policy*, Island Press, Washington, DC, 2002.
104. (a) D.D. Nelson, *International Environmental Auditing*, Government Institutes, Rockville, MD, 1998; (b) P. Sands, *Principles of International Environmental Law*, 2nd ed., Cambridge University Press, Cambridge, 2003; (c) P. Sands and P. Galizzi, eds, *Documents in International Evironmental Law*, 2nd ed., Cambridge University Press, Cambridge, 2004.
105. (a) B. Hileman, *Chem. Eng. News*, June 23, 1997, 26; (b) E. Papadakis, *Environmental Politics and Institutional Change*, Cambridge University Press, New York, 1997.
106. Anon., *Chem. Ind.* (*Lond.*), 1998, 376.
107. J.-F. Tremblay, *Chem. Eng. News*, May 31, 1999, 19.
108. M.G. Malloy, *Waste Age's Recycling Times*, 1998, *10*(6), 6.
109. www.oecd.org/env/policies/taxes/index.htm.
110. Anon., *Chem. Ind.* (*Lond.*), 2002, (6), 5.
111. (a) F. Katz, *Food Process.*, Apr. 2005, 42; (b) U.S.D.A. Economic Research Service, Bull. 747-08, Washington, DC.
112. J.A. Dixon and P.B. Sherman, *Economics of Protected Areas*, Island Press, Washington, DC, 1990.
113. (a) A. Gouldson and J. Murphy, *Regulatory Realities—the Implementation and Impact of Industrial Environmental Regulation*, Earthscan, London, 1998; (b) Anon., *Chem. Eng. News*, Nov. 9, 1998, 18; (c) J. Johnson, *Chem. Eng. News*, Jan. 24, 2000, 14.
114. R. Socolow, C. Andrews, F. Berkhout, and V. Thomas, eds, *Industrial Ecology and Global Change*, Cambridge University Press, Cambridge, 1994, 384.
115. (a) Anon., *Environ. Sci. Technol.*, 1998, *32*, 213A; (b) B. Hileman, *Chem. Eng. News*, Mar. 23, 1998, 30; (c) P.R. Portney, *Environment*, 1998, *40*(2), 14; (d) C. Hogue, *Chem. Eng. News*, Sept. 27, 2003, 27; Feb. 26, 2001, 32; Aug. 23, 2004, 26.
116. Anon., *Chem. Eng. News*, Dec. 1, 1997, 14.
117. Anon., *Chem. Eng. News*, Apr. 19, 1999, 38.
118. (a) F. Krupp, *EDF Lett.* (Environmental Defense Fund), 1998, *29*(1), 3; (b) G.T. Svendsen, *Public Choice and Environmental Regulation—Tradable Permit Systems in the United States and Carbon Dioxide Taxation in Europe*, Edward Elgar, Cheltenham, UK, 1998; (c) S. Sorrell and J. Skea, *Pollution for Sale: Emissions Trading and Joint Implementation*, Edward Elgar, Cheltenham, UK, 1999.
119. (a) R.A. Kerr and J. Kaiser, *Science*, 1998, *282*, 1024; (b) B. Hileman, *Chem. Eng. News*, Mar. 2, 1998, 28; (c) J. Boyd, D. Burstraw, A. Krupnick, V. McConnell, R.G. Newell, K. Palmer, J.N. Sanchiro, and M. Walls, *Environ. Sci. Technol.*, 2003, *37*, 216A.
120. (a) M. Webster, *On Earth*, 2009, *31*(1), 58; (b) G.T. Svendsen, *Public Choice and Environmental Regulation—Tradable Permit Systems in the United States and Carbon Dioxide Taxation in Europe*, Edgar Elger, Cheltenham, UK, 1998; (c) S. Sorrell and J. Shea, *Pollution for Sale: Emissions Trading and Joint Implementation*, Edgar Elger, Cheltenham, UK, 1999; (d) P.C. Fusaro and M. Yuen, *Green Trading Markets: Developing the Second Wave*, Elsevier Science, Amsterdam, 2005.
121. *New Jersey Discharger* (New Jersey Department of Environmental Protection), Trenton, NJ, 1997, *5*(1).
122. S. Schaltegger and T. Thomas, *Ecol. Econ.*, 1996, *19*, 35.
123. (a) J.B. Hockenstein, R.N. Stavins, and B.W. Whitehead, *Environment*, 1997, *19*(4), 13; (b) J.R. Aldrich, *Pollution Prevention Economics: Financial Impacts on Business and Industry*, McGraw-Hill, New York, 1996, 24; (c) *The Distributive Effects of Economic Instruments for Environmental Policy*, Organization for Economic Cooperation and Development, Paris, 1994; (d) *Managing the Environment; the Role of Economic Instruments*, Organization for Economic Cooperation and Development, Paris, 1994; (e) *Environmental Policy; How to Apply Economics Instruments*, Organization for Economic Cooperation and Development, Paris, 1991; (f) Anon., *Economic Instruments for Environmental Protection*, Organization for Economic Cooperation and Development, Paris, 1989; (g) Anon., *Renewable Natural Resources: Economic Incentives for Improved Management*, Organization for Economic Cooperation and Development, 1989; (h) Anon., *Saving Biological Diversity: Economic Incentive*, Organization for Economic Cooperation and Development, Paris, 1997.
124. G.A. Davis, C.A. Wilt, and J.N. Barkenbus, *Environment*, 1997, *39*(7), 10.
125. (a) M. Webster, T. O'Brien, and R. Collins, *Chem. Ind.* (*Lond.*), 1997, 810; (b) S.A. Hearne, *Environment*, 1996, *38*(6), 4.
126. (a) Anon., *Chem. Eng. News*, Nov. 24, 1997, 19; (b) A. Thayer, *Chem. Eng. News*, May 5, 1997, 28.
127. (a) M. Becker, K. Geiser, and C. Keenan, *Environ. Sci. Technol.*, 1997, *31*(12), 560A; (b) D. Beardsley, T. Davies, and R. Hersh, *Environment*, 1997, *39*(7), 6; (c) B. Allen, *Green Chem.*, 1999, *1*, G23.

128. (a) J. Holtzman and M. de Vito, *Chem. Eng. News*, Oct. 13, 1997, 2; (b) Anon., *Chem. Eng. News*, Dec. 22, 1997, 19; (c) M.J. de Vito, *Chem. Ind. (Lond.)*, 1998, 36.
129. (a) R.A. Lovett, *Environ. Sci. Technol.*, 1997, *31*, 368A; (b) *Amicus. J. (Natural Resources Defense Council)*, 1998, *19*(4), 46.
130. R. Hueting, *Ecol. Econ.*, 1996, *18*, 81.
131. (a) Anon., *Chem. Eng. News*, Mar. 20, 1995, 17; (b) R.H. Bezdek, *Ambio*, 1989, *18*, 274; (c) M. Renner, Jobs in a sustainable economy, Worldwatch paper 104, Sept. 1991, Worldwatch Institute, Washington, DC; (d) Anon., *Futurist*; Mar.–Apr., 1992, *32*, 49; (e) Anon., *Chem. Eng. News*, Nov. 16, 1992, 12; (f) J. Johnson, *Environ. Sci. Technol.*, 1995, *29*, 19A; (g) R.H. Bezdek, *Environment*, 1993, *35*(7), 7; (h) E. Goldstein, *Environ. Manage.*, 1996, *20*, 313; (i) *Environ. Sci. Technol.*, 1995, *29*, 310A; (j) L. Ember, *Chem. Eng. News*, Jan. 23, 1995, 19; (k) E.S. Goodstein, *The Trade-Off Myth: Fact and Fiction about Jobs and the Environment*, Island Press, Washington, DC, 1999.
132. www.bottlebillhawaii.org/factls.htm.
133. Anon., *Chem. Eng. News*, Mar. 20, 1995, 17.
134. (a) Anon., *Environment*, 2000, *42*(5), 6; (b) Anon., *Chem. Eng. News*, Sept. 25, 2000, 33.
135. (a) Anon., *Environ. Sci. Technol.*, 1995, *29*, 173A; (b) *Chem. Ind. (Lond.)*, 1995, 361.
136. (a) J. Johnson, *Chem. Eng. News*, Dec. 1, 1997, 15; (b) D.R. Berg and G. Ferrier, *Chemtech*, 1999, *29*(3), 45; (c) Anon., *Chem. Eng. News*, June 22, 1998, 22; (d) K.S. Betts, *Environ. Sci. Technol.*, 1998, *32*, 353A.
137. Anon., *Chem. Ind. (Lond.)*, 1994, 398.
138. Anon., *Chem. Eng. News*, Nov. 2, 1998, 16.
139. (a) L. Jarvis, *Amicus. J. (Natural Resources Defense Council)*, 1998, *20*(2), 24; (b) J. Makower, *Good, Green Jobs*, California Department of Conservation, Sacramento, CA, 1995.
140. (a) W. Nixon, *Amicus. J. (Natural Resources Defense Council)*, 1998, *20*(2), 16; (b) M. Reisner, *Amicus. J. (Natural Resources Defense Council)*, 1998, *20*(2), 19.
141. (a) Anon., *Environ. Sci. Technol.*, 1997, *31*, 82A; (b) N. Myers and J. Kent, *Perverse Subsides: Tax $s Undercutting Our Economies and Environments Alike*, International Institute for Sustainable Development, Winnipeg, 1998; (c) J. Gersh, *Amicus. J. (Natural Resources Defense Council)*, 1999, *21*(1), 37; (d) D.M. Roodman, *World Watch*, 1995, *8*(5), 13; (e) N. Myers, *Science*, 2000, *287*, 2419; (f) N. Myers and J. Kent, *Perverse Subsidies—How Misused Tax Dollars Harm the Environment and the Economy*, Island Press, Washington, DC, 2001; (g) Worldwatch Institute, *State of the World 2008*, W.W. Norton, New York, 2008; (h) A. Balmford, A. Bruner, P. Cooper, R. Costanza, S. Farber, R.E. Green M. Jenkins, et al., *Science*, 2002, *297*, 950; (i) P.R. Ehrlich and A.H. Ehrlich, *One With Nineveh—Politics, Consumption and the Human Future*, Island Press, Washington, DC, 2004; (j) C. Pye-Smith, *The Subsidy Scandal—How Your Government Wastes Your Money to Wreck Your Environment*, Stylus Publishing, Herndon, VA, 2002.
142. (a) T. Prugh, *Natural Capital and Human Economic Survival*, ISEE Press, Solomons, MD, 1995, *114*, 140; (b) M. Khanna and D. Zilberman, *Ecol. Econ.*, 1997, *23*(1), 25.
143. E. Rodenburg, *Environment*, 1997, *39*(4), 25.
144. J.M. Heumann, *Waste Age's Recycling Times*, 1997, *9*(24), 2.
145. (a) B. Hileman, *Chem. Eng. News*, June 23, 1997, 26; (b) C. Lynch, *Amicus. J. (Natural Resources Defense Council)*, 1998, *19*(4), 15.
146. R. Gelbspan, *Amicus. J. (Natural Resources Defense Council)*, 1998, *19*(4), 22.
147. D. Pimentel, C. Wilson, C. McCullum, R. Huang, P. Dwen, J. Flack, Q. Tran, T. Saltman, and B. Cliff, *Bioscience*, 1997, *47*, 754.
148. C. Anderson, *Wilmington Delaware News J.*, Dec. 13, 1997, F5.
149. C. Coles, *Futurist*, 2003, *37*(3), 13.
150. (a) D. Pauly, V. Christensen, J. Dalsgaard, R. Froese, and F. Torres, Jr., *Science*, 1998, *279*, 860; (b) N. Williams, *Science*, 1998, *279*, 809 (c) C. Ash, *Science*, 2004, *305*, 1242.
151. (a) C. Uhl, P. Barreto, A. Verissimo, E. Vidal, P. Amaral, A.C. Barros, C. Souza, Jr., J. Johns, and J. Gerwing, *Bioscience*, 1997, *47*, 160; (b) J. Gersh, *Amicus. J. (Natural Resources Defense Council)*, 1999, *21*(1), 37.
152. (a) B.S. Dunkiel, *Environment*, 1996, *38*(10), 16; (b) M. Hyman, *Chem. Ind. (Lond.)*, 1999, 528; (c) Green Tax Database, http://www.oecd.org/env/policies/taxes/index.htm.
153. (a) *Taxation and the Environment*, Organization for Economic Cooperation and Development, Paris, 1993; (b) S. Bernow, R. Costanza, H. Daly, R. DeGennaro, D. Erlandson, D. Ferris, P. Hawken, J.A. Hoerner, J. Lancelot, T. Marx, D. Norland, I. Peters, D.L. Roodman, C. Schneider, P. Shyamsundar, and J. Woodwell, *BioScience*, 1998, *48*, 193; (c) T. O'Riordan, ed., *Ecotaxation*, Earthscan, London, 1997; (d) D.M. Roodman, *World Watch*, 1995, *8*(5), 13.
154. (a) Anon., *Environment and Taxation: The Cases of the Netherlands, Sweden, and the United States*, Organization for Economic Cooperation and Development, Paris, 1994; (b) M. Burke, *Environ. Sci. Technol.*, 1997, *31*, 84A; (c) Anon., *Chem. Ind. (Lond.)*, 1996, *38*(3), 16; (d) F. Muller, *Environment.*, 1996, *38*(2), 12; (e) D. Cansier and R. Krumm, *Ecol. Econ.*, 1997, *23*(1), 59; (f) C. Hanisch, *Environ. Sci. Technol.*, 1998, *32*, 540A; (g) P. Layman, *Chem. Eng. News*, June 14, 1999, 14.
155. L. Ember, *Chem. Eng. News*, June 2, 1997, 8.
156. R. Tyson, *Environ. Sci. Technol.*, 1997, *31*, 454A.
157. Anon., *Environ. Sci. Technol.*, 1997, *31*, 407A.
158. C. Hanisch, *Environ. Sci. Technol.*, 1998, *32*, 540A.
159. Anon., *Chem. Ind. (Lond.)*, 1998, 248.
160. C. Martin, *Chem. Br.*, 1997, *33*(12), 33.
161. R.A. Bohm, C. Ge, M. Russell, J. Wang, and J. Yang, *Environment*, 1998, *40*(7), 10.
162. D.J. Wolfson and D.C. Koopmans, *Ecol. Econ.*, 1996, *19*, 55.
163. (a) M. Reisch, *Chem. Eng. News*, June 28, 1999, 17; (b) T. Chapman, *Chem. Ind. (Lond.)*, 1999, 85; (c) Anon., *Chem. Ind. (Lond.)*, 1999, 203.
164. Anon., *Chem. Eng. News*, Nov. 1992, *23*, 13.
165. (a) Anon., *Chem. Ind. (Lond.)*, 1997, 370; (b) D.M. Roodman. In: L. Starke, ed., *State of the World 1996*, W.W. Norton, New York, 1996, 168.
166. J.C.J.M. van den Bergh and J. van der Straaten, eds, *Toward Sustainable Development*, Island Press, Washington, DC, 1994, 268.

167. (a) M. Khanna and D. Zilberman, *Ecol. Econ.*, 1997, *23*(1), 25; (b) P. Hawken, *The Ecology of Commerce. A Declaration of Sustainability*, Harper-Collins, New York, 1993, 198; (c) K. Gallagher and J. Werksman, eds, *The Earthscan Reader on International Trade and Sustainable Development*, Stylus Publishing, Sterling, VA, 2002.
168. (a) R. Begley, *Environ. Sci. Technol.*, 1997, *31*, 355A; (b) Anon., *Chem. Eng. News*, Feb. 5, 2001, 20.

RECOMMENDED READING

1. D.M. Roodman. In: L. Starke, ed., *State of the World 1996*, W.W. Norton, New York, 1996, 168 [taxing pollution and waste].
2. S. Nathan, *Chem. Ind. (Lond.)*, 1995, 824 [green taxes in Europe].
3. Anon., *Chem. Eng. News*, Mar. 20, 1995, 17 [slowness of industries to adopt pollution prevention].
4. D. Pimentel, C. Wilson, C. McCullum, R. Huang, P. Dwen, J. Flack, Q. Tran, T. Saltman, and B. Cliff, *Bioscience*, 1997, *47*, 747 [value of biodiversity].
5. C. Crossen, *Tainted Truth: The Manipulation of Fact in America*, Simon & Schuster, New York, 1994, 140–143 [life-cycle analyses of diapers].
6. R.U. Ayres, *Environ. Sci. Technol.*, 1998, *32*, 366A [toward a zero-emissions economy].
7. M. Palmer, E. Bernhardt, E. Chornesky, S. Collins, A. Dobson, C. Duke, B. Gold, et al., *Science*, 2004, *304*, 1251.

EXERCISES

1. How would you revise the tax system to favor a clean environment and a sustainable future?
2. What does your state or country subsidize?
3. Does your state or country offer any flexible, market-based alternatives to the "command-and-control" system of reducing pollution?
4. Do a life cycle analysis of some commonly used item (e.g., a paper napkin versus a linen napkin or a natural versus an artificial Christmas tree).
5. What factors would have to be taken into account in a cost–benefit analysis of a regulation to outlaw the release of mercury into water?
6. List some individuals or groups or firms in your areas that externalize some of their costs at the expense of the public.

CHAPTER 18

Greening

18.1 INTRODUCTION

Chapter 17 described some economic and institutional factors that slow down the adoption of pollution prevention. This chapter continues this theme by considering the extent to which individuals, governments, and businesses have embraced the concepts of a clean environment and a sustainable future.

Rachel Carson pointed out the dangers of the indiscriminate use of pesticides in her book *Silent Spring*, published in 1962.[1] Gaylord Nelson, a former senator from Wisconsin, originated Earth Day, the first one being held in 1970.[2] These are two of the major events that gave rise to the environmental movement of today in the United States. The United States Congress responded to this movement by enacting the Clean Air Act, the Clean Water Act, and the Endangered Species Act, as well as other laws. What has happened since then? One poll conducted in the United States in 1994 found that 23% of the respondents considered themselves to be active environmentalists, and 56% were sympathetic to the environmental cause.[3] More than half felt that environmental laws had not gone far enough, with only 16% feeling that the laws had gone too far. Environmental groups were considered favorably by 74% of the respondents.

Progress has been made in some areas since 1970. Emissions of pollutants from point sources into air and water have decreased. Toxic releases are decreasing. Some Superfund sites have been cleaned up. Businesses would no longer think of dumping a barrel of waste solvent on the ground at the landfill site so that the barrel could be used again for the same purpose. Control of pollutants from nonpoint sources is still a problem. There is now more international cooperation and discussion of global problems, such as ozone depletion by chlorofluorocarbons (CFCs) and the effect of population growth on the environment. Several nations have achieved zero population growth rates. Population growth rates are still high in some of the developing nations least able to deal with them. A great many of the world's fisheries are overfished. Biodiversity continues to decline. More consumer goods are marked with the postconsumer recycled content. Many of these problems have been dealt with in earlier chapters. By analyzing the greening that has occurred during this period, it may be possible to find ways to speed up the process of attacking the remaining problems.

18.2 INDIVIDUALS

Many persons who say that they are green may be so only in limited ways, owing to problems that they do not associate themselves with or do not understand.[4] Use the quiz below to see how green you are:

1. What do you put on your lawn?
2. What do you take off your lawn and what do you do with it?
3. What do you use to carry the groceries in after you buy them?
4. What toxic chemicals do you have in your home and garage?
5. Do you use any single-use, throwaway items?
6. How many cars of what type are there in your family?
7. How long has it been since you rode on a train?
8. What do you do with the used oil taken from your car?
9. How do you dry the washed clothes?
10. How often do you mend an object instead of discarding it and buying a new one?
11. What do you do with clothes or toys that the children have outgrown?
12. Do you have anything in your home or office that was made using toxic chemicals?
13. Do you own anything made of teak or mahogany?
14. What do you recycle?
15. How many children do you have or intend to have?
16. When did you last contact your congressman?
17. Does your home or office have active or passive solar heating and cooling? Could these be added?
18. Does the wastewater from your toilet receive tertiary treatment?
19. How many suits or dresses do you own?
20. How far do you live from work?
21. Do you obey the speed limits on highways?

Readers will recognize what many of these questions are getting at from the material given in earlier chapters. Runoff from lawns, golf courses, and farms contaminates many streams, decreasing the biodiversity in them. About 18% of municipal solid waste is yard waste. If you put grass clippings or leaves in the trash can, you are contributing to the landfill problem. Several of the questions are directed at using less material and less energy. Groceries can be carried in string bags, or in your own basket. There are many times where you can substitute reusable items for single-use ones. Mending something can substitute for buying a new one. Passing things that you no longer need on to friends can eliminate their need to buy something. Living close to where you work may allow you to get there by walking or cycling. Taking the train instead of your car or an airplane can save energy. So can drying the wash on an outdoor line. There is nothing wrong with changing your own oil in your car. The problem is that about half of it ends up going down storm drains to contaminate streams.

You may have a surprising number of chemicals considered to be hazardous in your home. These include pesticides, lye for cleaning drains, hydrocarbon solvents for cleaning, paints, varnishes, and others. Check the list of chemicals that household hazardous waste drives try to collect. Figuring out what you own that was made with toxic chemicals will be more of a job. You can probably figure out what formaldehyde was used to make, but beyond that it may be difficult. Do not forget that the gasoline in your car was probably made using an alkylation step catalyzed by hydrogen fluoride or sulfuric acid. The leather in your shoes was probably tanned with a chromium salt. For the person with a chemical background, a search of industrial encyclopedias and books on industrial chemistry may help you decide what toxic chemicals were used. It is also true that some toxic chemicals are so useful that it will be a long time before they are displaced. Acrylonitrile and ethylene oxide fall in this class. If the company that made the item that you are using has a consumer services telephone number, you can ask them what hazardous chemicals were used in its manufacture. They may or may not tell you.

Some problems are not recognized because they are out of sight. Most people probably do not know whether or not the local wastewater treatment plant has tertiary treatment. It may not have, especially if residents defeated the last referendum for a bond issue, to build such a facility. They may not know that their septic tank is leaking if the leak is directly into a stream or lake. They may not know what landfill or incinerator their trash goes to. They may not realize that the teak and mahogany that they value probably come from overharvested tropical forests. Destruction of natural forest may also be taking place on their own property if they buy a new home on land that was forest or if they remove all native shrubs and wild flowers from a wooded lot so that they can plant grass. There is also a feeling that the environmental problem is caused by someone else. People may blame the ozone alerts in the summer on emissions from local manufacturing plants, rather than from too many people driving in too many cars, as they drive alone in their cars. They may be forced to drive because they lobbied against pedestrian ways between developments, against sidewalks along roads, and against retail businesses anywhere near their homes. It may be easy to recognize a well-publicized problem and miss others. A home near the author's has a bumper sticker on the car to support recycling. (Recycling of aluminum beverage cans dropped from 60% a few years earlier to 54% in 2007.) The home also has an underground piping system for spray irrigation of the lawn. This keeps the lawn green during the hot, dry summer when the grass would normally go dormant. At the same time, the county has tried to decide whether to destroy a productive wetland or a wooded valley for a reservoir to meet the needs for more water for the future. Most people in the United States drive faster than the posted speed limits. Driving at the speed limit saves lives and contributes less to global warming, since fuel efficiency decreases as speeds increase.

Consumer choices can make a difference.[5] If no one took the plastic grocery bags at the store, they would no longer be offered. Americans like shiny cars with glamour finishes. If they could settle for a finish more like that on a truck or lawn mower, solvent would no longer be needed to apply the finish coat. In Delaware, this would have meant the elimination of over 1 million lb of solvent emissions per year from the two automobile assembly plants in the state. (Boyh plants were closed by mid 2009, although one may reopen to produce electric cars.) Energy efficiency could be increased if consumers heeded the energy efficiency rating on new cars and appliances. Energy would be saved and tailpipe emissions reduced if they bought small cars instead of vans, sport utility wagons, and pickup trucks. Wood is a renewable form of energy, but if a person buys and uses a wood stove without a catalytic combustion aid in the stack, the stove will be a source of air pollution. It is difficult for consumers to test different brands on their own. However, they may be able to obtain the information needed to make an intelligent decision from a consumer-testing group such as Consumer Reports. It takes effort to be really green. It may mean more time in the library and more time shopping around for a reusable item instead of the throwaway version that is more common in the stores. It also takes extra effort to read the green logo that the manufacturer has put on a product to determine that it is really meaningful and that important points, which might be negative, have not been left out. Green consumerism is said to be making a difference and is pushing manufacturers into offering products that are friendlier to the environment.[6]

There are times when consumers work against the better environmental choice. This usually happens when the fear of the unknown and a lack of seeing the overall picture produces a "not in my backyard" response. Piggyback rail service, where the truck trailers are hauled long distances on trains, and then unloaded for local delivery, saves energy. However, when a proposal was made to add such service to an existing railyard in Elsmere, Delaware, all the local residents could think of was more trucks on local roads. "Stop Chessie" bumper stickers appeared and the facility was never put in. A good use for treated wastewater is to spray it onto land to increase crop yields. Although research over two or more decades has shown the method to be sound, some neighbors of the project near Odessa, Delaware, were still afraid. The facility is now in operation. Land application of sewage sludge as a soil amendment is also hard for some people to accept. Although fluoridation of water is the best method known for preventing dental caries, some groups still prevent its use in some cities. Although everyone produces the trash, no one wants the landfill near where one lives.

A single individual or family may feel helpless in trying to green the country or the world. One person can make a difference. Rachel Carson did it with her book *Silent Spring* (published in 1962) that described the overuse of persistent pesticides. When the book was published it received both favorable[7] and unfavorable[8] reviews. She was criticized by companies that made insecticides and by government agencies favoring their use. Her writing was said to present a "one-sided picture" and to contain "highly dramatized presentations." A 1987 review presented a more moderate appraisal.[9] A 1997 biography points out that she checked her facts meticulously and that in many points, she was right.[10] A 1997 letter still says that there are inaccuracies in the book.[11] Another writer feels that she was basically correct in what she wrote.[12] Theo Colburn has called attention to the environmental threats posed by hormone-disrupting chemicals.[13] She and her co-workers have called attention to what may prove to be an important problem, certainly one that justifies further research. Her book has also been condemned by "book should be ignored" and "science is missing." This is almost a replay of the 1962 reviews of *Silent Spring*. It is not always easy to be an environmental pioneer challenging the status quo, but if your facts are right you should eventually be accepted.

Delaware still has the extensive salt marshes needed to support one of the largest shorebird migrations on the continent, largely because of the work of one man, former governor Russell Peterson, and the Coastal Zone Management Act that he convinced the legislature to pass. The act still stands and works after many years, despite repeated attacks on it by people who would like to industrialize the area. The Earth Day started by Gaylord Nelson is observed each year.

The number of schools including environmental education in their curricula is increasing, although not as fast as desired.[14] In some cases, student environmental groups have prodded sluggish administrators into launching environmental projects that saved the colleges significant amounts of money.[15] More than 280 U.S. college and university presidents have signed the American College and University Presidents Climate Commitment to prepare a program to combat global warming.[16] The Talloires Declaration for greening campuses has been signed by more than 270 college and university presidents in 43 countries on five continents.[17]

If people think through their causes carefully and explain them to their elected officials on a local, state, or national scale, good legislation may result that corrects a problem. Calling the problem to the attention of others through letters to the editor and op-ed pieces may encourage others to join the cause and contact their legislators also. If you encourage candidates for office to explain their positions and then vote carefully, you may also be able to encourage the correction of problems. Unfortunately, too many voters make their decisions based on the incomplete data given on television news, without understanding the full situation. Individuals have sometimes been able to influence corporations to do better with less pollution by buying some stock and then putting a proposal on the ballot for the annual shareholders meeting. Although most such proposals fail, the threat of a confrontation at the annual meeting may sometimes force the company to adopt the proposal.[18] The Interfaith Center for Corporate Responsibility of New York City pioneered this approach. It submits 80 or more resolutions a year. Individuals can also help monitor the health of local streams in Stream Watch programs. If the stream's health seems to be declining, it can be called to the attention of the local authorities, who will see if anyone upstream is violating a discharge permit.

18.3 NONGOVERNMENTAL ORGANIZATIONS

Individuals can also join environmental organizations that can carry out detailed studies of problems and propose possible solutions to them.[19] The organization may also function as an alerting service on pending legislation, and such. One of the oldest is the Sierra Club, which played a role in the establishment of the national park and national monument system in the United States.[20] One of the world's largest environmental organizations, The Nature Conservancy, was founded in the 1950s by a group of ecologists from the Ecologists Union (which had split from the Ecological Society of America in 1946) for the purpose of preserving species.[21] Its program is laid out by scientists and implemented in a nonconfrontational manner. Its staff consists not only of ecologists, but also of realtors and tax specialists who help in the acquisition of land. In contrast, the much smaller Earth First (founded in 1980) often engages in dramatic, and sometimes violent, confrontational civil disobedience. Greenpeace also uses the dramatic confrontational approach.[22] An example is the use of a small boat to try to prevent the dumping of toxic wastes into the ocean. This attracts the attention of the media, which can bring the problem before millions of people who may have been unaware of it. Such tactics were successful in preventing the sinking of an oil storage buoy in the North Sea, the cheapest option for Shell, which owned the buoy. Greenpeace feared that this would contaminate the sea. The buoy was towed to shore for dismantling and, hopefully, recycling of the materials in it. Greenpeace was also instrumental in placing hydrocarbon refrigerants into European refrigerators to replace CFCs (see Chapter 3).

Environmental Defense Fund[23] and the Nature Resources Defense Council consist of lawyers supplemented by staffs of scientists. They defend the environment by bringing lawsuits under existing laws. For example, a project that would destroy old-growth forest or wetlands may not have been thought out carefully. They can ask the court to delay the project until an environmental impact statement has been prepared and submitted. These groups also propose new legislation and devise new economic instruments for the control of polluting activities. California's Proposition 65, which requires that consumer products containing carcinogens must list them on the label, is such a law. This law has prompted many manufacturers to reformulate their products to remove the carcinogens. Occasionally, the organizations work with industries to find solutions to problems.[24] Environmental Defense Fund showed Wal-Mart, the world's retailer, how to increase efficiency and profits by greening. It also asked the 100 largest chemical companies in the United States to run basic toxicity tests on 71% of the 3000 chemicals used in largest volume in the United States for which the data are lacking.[25] As of January 1998, 11 had agreed to do it, and six had declined to do it. The American Chemistry Council [formerly the Chemical Manufacturers Association (CMA)] and the U.S. EPA had worked out an agreement for the testing by 1999. Companies have pledged to provide data on more than 2200 of the 2800 high production volume chemicals in commerce.[26] No company had volunteered to provide data on 17 substances by 2006. The collection of data was still incomplete by the end of 2008.

The Center for Science in the Public Interest, founded in 1971, has focused on improving the nutritional status of the food that Americans eat.[27] Among its successes is the ban on the use of sulfites (to which some people are sensitive) on produce. Its newsletter, *Nutrition Action*, has 800,000 subscriptions.

The Rainforest Action Network has been able to persuade many forest products companies, such as Home Depot, Lowe's, and FedEx Kinko's, to stop buying products obtained by logging endangered old growth forests by publicizing abuses and using boycotts.[28]

Membership in environmental organizations, such as The Nature Conservancy and the World Wildlife Fund, has grown steadily over the years.[29] The former had 1 million members in 1999.[30] Greenpeace peaked in 1990 with 2.35 million members. Some others have had rising and falling membership, depending on what confidence people have had in the government administration in Washington, DC, at varying times. It is the middle class that makes up most of the membership. Members of Friends of the Earth in England were 74% from the middle class. Professional or technical people made up 84% of the membership.

National environmental organizations frequently operate by "mail order." They buy mailing lists, and they try to target the most likely prospective members. The mailing can shape concern for tropical rain forests, whales, and others. They may exaggerate the extent of the crises, and they must propose solutions. Pictures of wildlife covered with oil from a spill can motivate people. Some organizations seem to have a monthly crisis that they use in trying to solicit additional funds from members.

Not everyone appreciates the efforts of environmental groups.[31] The Environmental Working Group put together the data from the Toxic Release Inventory under the title of *Dishonorable Discharge: Toxic Pollution of America's Waters*. One reporter wrote this up under the title, *Information Can Be Quite Uninformative*,[32] and proceeded to ridicule the originators. Readers were quick to point out that the thesis of the report did have some merit and should not have been written up in this way.[33] One pointed out that the next page in the magazine covered the rising levels of unexplainable deformities in frogs, which may turn out to be related to what substances are dumped into lakes and streams.

Not all environmental groups take the positions advocated by modern science. Jeremy Rifkin of The Foundation on Economic Trends has been described as a man "who's never met an engineered gene he liked."[34] (The value of such genes was described in Chapter 9.) There are also organizations whose names make them sound as if they are proenvironment, but that are set up to prevent environmental improvement. The National Wetlands Coalition (which uses a symbol of a duck flying over a swamp) was set up by a group of oil and gas companies, together with real estate developers, to ease restrictions on converting wetlands to drilling sites and shopping malls.[35] Keep America Beautiful was founded by the bottling industry to organize antilitter campaigns. Its sponsors fight every new deposit–refund legislation that is proposed. Consumer Alert was set up to fight any new regulations relating to consumer safety. The Council for Responsible Nutrition is a group of manufacturers of dietary supplements the purpose of which is to prevent regulation of their products as foods by the U.S. FDA.[36]

There has also been some backlash to the environmental movement.[37] A biotech industry consultant said, "You have some professional protest groups that make money by peddling fear."[38] It is apparent from the letters to the editor of *Chemical Engineering News* on the subject that at least some chemists are not in sympathy with the environmental movement. The ecologists, Anne and Paul Ehrlich, have written a rebuttal to three such books pointing out inaccuracies and misstatements.[39] Another book describes the "environmentalist's extreme, sensationalist view" of toxic chemicals in the environment.[40] The author feels that *Silent Spring* is a polemic filled with alarmist statements. She is employed by the American Council on Science and Health, an industry-sponsored organization. *Green Watch* is a twice-monthly newsletter covering the activities of environmental groups.[41] Its advertisement seeking subscriptions says

> You should know what environmental activists are plotting and how they could severely damage your reputation and profits this year. New hidden threats may be targeting the environmental policies of your company, your association or your clients.

18.4 GOVERNMENT

Governments are supposed to take the long view that will do the most good for the greatest number of people in the long term. Ecopolitics leading to a green, sustainable society should be part of this.[42] This trend seemed to be true from 1970 up to 1995 with the passage of numerous environmental laws in the United States. The air and water of the country became cleaner during this period. Then the antienvironmental backlash hit as the 104th Congress convened in 1995. This proved to be the worst Congress on environmental matters in 25 years, as rated by the League of Conservation Voters.[43] With a maximum score of 100, 111 representatives and 24 senators received zeros. It was an environmental backslide[44] as efforts were made to weaken major environmental laws and cut funding for renewable energy studies, the EPA, national parks, and others in the name of "regulatory reform," removal of "burdensome regulations," and "balancing the budget."[45] A number of the attempts showed up as riders on unrelated appropriation bills, presumably to avoid floor debate and with the hope that they would pass because other congressmen would not want to delay an important appropriations bill. This happened despite the overwhelming support of the public for strong environmental and public health standards. That the efforts failed in large part was probably due to this public, which made its feelings known to the congress and to the administration. The administration included the staunch environmentalist Vice President Al Gore.[46] The Office of Technology Assessment, which won praise from many people for its reports over 23 years, was a casualty of the 104th Congress.[47] Many environmentalists would like to see it started again.

The American public had greened over the years, but many of the special interest groups had not. Part of the problem with the 104th Congress was the power of campaign contributions. It takes a lot of money to run for election to congress. Congressmen had voted three out of four times for bills (to relax federal environmental protection) that were supported by groups that had given them money through political action committees.[48] These groups included agribusiness, chemical manufacturers, and natural resource extraction companies. Many chemical companies, as well as the American Chemistry Council, also give "soft" money to the major political parties.[49] Many congressmen were still not green in 1999. The U.S. House of Representatives passed the Mandates Information Act (H.R. 350), which requires estimates by the Congressional Budget Office of the costs imposed on the private sector by a regulation, but requires no estimates of its benefits.[50]

This process was repeated in the 8 years of the Bush Administration[51] (2000–2008) as efforts were made to water down environmental regulations in response to campaign contributions from industry.[52] Most of these efforts failed as environmental groups took their cases to the courts.[53]

Delaware has cleaned up 12 Superfund sites (out of the 20 most-polluted sites in the state) to the point that they are virtually free of toxins.[54] About 445 had been cleaned up nationwide by August 1997. The Delaware Department of Natural Resources and Environmental Control has shown a willingness to work with polluters who violate their discharge permits to help them clean up.[55] The department feels that this is a better approach than the use of fines, but this approach is not supported by some local environmental groups. The compliance rate for the period 1995–September 1996 was 99%, with an average of only 63 violations per month. The department points out that runoff from roads, parking lots, overfertilized yards, and farms produced 10 times more pollution than that from the 75 permitted discharge pipes.

Europe has greened as well over the years. (There are green parties in at least 13 countries in Europe.[56]) The Green Party in Germany has been active. One result is the system making manufacturers responsible for the disposal of what they produce at the end of its useful life. Several countries in Europe are incorporating ecotaxes into their tax systems, as described in Chapter 17. Thirteen countries in Europe have achieved zero population growth. Most companies in the United Kingdom do not find environmental laws to be onerous.[57] The EU has greened more in some ways than the United States, so that there have been conflicts over climate change, genetically engineered crops, and the use of hormones in raising beef.[58]

Under the guidance of the United Nations, more and more attention is being given to the earth's environmental problems,[59] including that of environmentally sustainable development. Many international conferences sponsored by the United Nations have dealt with world problems. The nations of the world have frequently recognized the problems. Actions taken (Montreal Protocol of 1987) have been effective in the phase-out of the CFCs that threaten the ozone layer.[60] The 1997 conference, in Kyoto, Japan, made a start in mitigating global warming (i.e., if the nations can actually implement the reductions of greenhouse gases agreed upon). In other instances, the promise of improvement remains largely unfulfilled.[61] Ten years after the Brundtland report[62] on the need for a sustainable future and 5 years after the Earth Summit at Rio de Janeiro,[63] most of the problems identified were still there in 1997 or had worsened. The same was true after the meeting in Johannesburg in 2002.[64] (There have also been some problems in persuading some countries to comply with the treaties, such as the convention on International Trade in Endangered Species and the ban on CFCs.[65]) The United States has tried to eliminate a black market in CFCs.[66]

Nongovernmental organizations (NGOs) can send observers, but not delegates, to these international conferences. Despite this second-rate status, they have influenced the process of sustainable development.[67] While at conferences, they can disseminate information and offer new ideas and suggestions. The idea of sustainable development has become part of mainstream thinking, much more so than 20 years ago.

International lending organizations have been criticized by environmental groups for supporting projects that damage the environment. Pressure from environmental NGOs is forcing them to change. The U.S. Agency for International Development has changed the most, to the point that it focuses on sustainable forms of development and even supports biodiversity projects.[68] The World Bank has made less progress, but has changed and is still changing.[69] Its initial claims (1994) for a change to a "green agenda" were viewed with skepticism by environmental groups. It funds projects using fossil fuels that emit much more carbon dioxide than the alternative energy projects that it funds.[70] It cut back its family-planning aid in 1996.[71] In 1998, it wanted to lift its 1991 ban on investments in logging primary tropical forests.[72] There have been complaints that the power of its inspection panel has been weakened.[73] In 1996, it had 200 environmental specialists in its staff of 11,000. Robert Goodland of the World Bank is in the forefront of environmental economics.[74] The International Monetary Fund[75] lends money to countries that have problems with balance of payments. In return, it expects the country to show fiscal austerity. This structural adjustment lending is exempt from environmental assessments. The result can be excessive exportation of natural resources to the detriment of the environment. It is said to have greened the least of the three lending agencies.

The World Trade Organization settles disputes among nations when their individual trade laws conflict. There are complaints that international trade is winning out over the environment in some of the decisions. For example, it ruled in 1998 that a U.S. law requiring turtle excluder devices to protect endangered species of sea turtles from being trapped in the nets of shrimp fishermen was really a trade barrier and need not be enforced.[76] The North American Free Trade Agreement (NAFTA) rules have also been abused to impact the environment negatively.[77]

18.5 CORPORATIONS

Many businesses have become greener since the founding of the U.S. EPA in 1970[78] and some more than others.[79] R. E. Chandler of Monsanto describes the response of the chemical industry to environmentalism as consisting of three overlapping stages.[80] The first phase was the denial phase. Industry tended to deny that major problems existed. It often predicted dramatically high costs, job losses, and plant closures if the regulations were enforced. This exaggeration caused it to lose credibility with both legislators and the general public. The second phase was the "risk–benefit" one. Major U.S. chemical companies launched expensive public relations campaigns to convince the public that its emotions and fears of the unknown should be replaced by a more industry-like view of facts and reason in which the risk was small. Monsanto's Chemical Facts of Life of the late 1970s and early 1980s was such a program. The third phase involves the "public right to know," which may have been a reaction to the disaster at Bhopal, India. It includes a proactive openness with the community. The goal is to go beyond compliance with all laws to reduce all toxic and hazardous emissions, ultimately to zero. This phase incorporates pollution prevention. Some companies view environmental concerns as profit-drivers (i.e., through higher sales of cleaner products). The actions of the 104th Congress in the United States suggest that there are many companies that are still in the first phase.

Another author describes the first phase as an inherently defensive one, where management believes that it is being treated unjustly by the regulatory authority.[81] The third stage is one during which there is a sustainable, interactive program of environmental management within the company. Another author classifies corporations as red, yellow, or green, depending on the extent of their conversion to a long-term, opportunity-seeking, company-wide, cradle-to-grave performance system, which is the green one.[82] He and others feel that the green phase should offer a competitive advantage[83] and be part of a sustainable future.[84]

It is possible that the companies that grumble the most about environmental regulations are still in the first phase. It is said that, "No major piece of environmental legislation has ever been supported by corporate America."[85] A few examples will illustrate the complaining often coming from companies, or trade association that represent them. The extent of the complaining may go up as the cost of compliance becomes high, or as the amount of change required to comply becomes large. Earnest W. Deavenport, Jr., of Eastman Chemical and the American Chemistry Council have pushed for regulatory reform.[86] An analysis sponsored by Pfizer concluded that the laws disregard reality, that the regulatory framework is impossible to administer properly and that they foster excessive litigation.[87] The U.S. EPA, which administers the laws enacted by the U.S. Congress, is subjected to a lot of criticism. Its expansion of the number of chemicals to be reported in the Toxic Release Inventory was objected to by the National Council of Chemical Distributors (which has 333 members).[88] The American Petroleum Institute, representing refinery operators, says the cost of the new rule for reducing the emission of toxic chemicals from refineries will exceed the benefit.[89] The reduction would be 227,000 tons annually. It also objected to the reduction of sulfur in gasoline.[90] The National Association of Manufacturers says that "There are some serious problems with our environmental regulatory system."[91] The association did find that over half of the members that voluntarily changed manufacturing processes to reduce waste or emissions saved money as a result. Industrial groups have fought the "credible evidence rule," which says that any evidence deemed credible can be used to determine whether a facility is violating emissions standards.[92] Electrical utilities have sued the EPA over a rule that would reduce the emission of nitrogen oxides by 900,000 tons/yr.[93]

The new standards for ozone (0.08 ppm) and small particulates are expected to prevent 15,000–20,000 premature deaths and as many as 350,000 cases of aggravated asthma each year[94] (see also Chapter 17). They are supported by the American Lung Association and other environmental groups. They have been opposed by the American Trucking Association, the U.S. Chamber of Commerce, the National Association of Manufacturers, the American Chemistry Council, the Air Quality Standards Association (a group of electric power generators, oil companies, and vehicle manufacturers), the Synthetic Organic Chemical Manufacturers Association (now Society of Chemical Manufacturers and Affiliates), and the National Coalition of Petroleum Retailers.[95] Opponents of the rule say that the cost of compliance would cripple them. The EPA estimates the annual cost of compliance at 6.5–8.5 billion dollars and the Reason Foundation (an industry-supported think tank in Los Angeles) at 90–150 billion dollars ($250 per capita).[96] S. Buccino of the Natural Resources Defense Council pointed out that past estimates of the cost of air quality improvements have been high.[97] In 1970, Lee Lacocca predicted that the Clean Air Act "could prevent continued production of automobiles." In the 1980s, utilities predicted the cost of control of acid rain at $1500/ton NO_x, yet the cost of a credit to emit a ton today is 100 dollars. The oil companies predicted that the cost of cleaner gasoline would be up to $0.25/gal. It turned out to be only $0.03–0.05/gal more. The Global Climate Coalition (a lobbying group for coal, oil, auto, and chemical industries) and the CMA (now American Chemistry Council) have predicted severe economic impacts as a result of the Kyoto agreement by which the United States must reduce its emissions of greenhouse emission below 1990 levels.[98] The coalition predicts unemployment levels not seen since the Great Depression. The coalition has also attacked the science behind global warming.[99] The Competitive Enterprise Institute and several members of the U.S. Congress sued President Clinton to block the release of a congressionally mandated report on how climate would affect U.S. taxpayers.[100] They called it "junk science." Environmental groups predict no additional cost, because the cost of energy efficiency will be offset by the savings from not having to purchase as much energy. Because energy efficiency and renewable energy are more labor-intensive than current oil and car production, jobs may have to change, but there will be little, if any, loss. A Business Roundtable (a group of top corporate executives) study predicted job losses of 200,000–2 million from the 1990 amendments to the Clean Air Act.[101] The actual number, as of June 1994, was 2363. The U.S. Chamber of Commerce has requested the U.S. EPA to "correct faulty scientific data."[102] (The numbers can vary due to site-specific conditions and varying methods of measurement.) "Critics of federal regulatory policies often plead for 'sound science', a cryptic rallying cry for those who really want to discourage regulation."[103]

It is important to remember where regulations come from. Many are the result of tragedies, such as the Toxic Substances Control Act, which resulted from the accident at Bhopal, India. Others try to correct abuses that impair the health and safety of citizens that individuals and companies are not taking care of voluntarily. Speed limits on highways are intended to lower accidents, but many motorists exceed them. The Motiva (now Valero) refinery in Delaware deferred maintenance on a tank. The managers instructed operators not to fill the tank above the hole. A welder on the tank ignited the hydrocarbons in the sulfuric acid in the tank, which resulted in a fire and explosion that killed the welder and injured seven others, and spilled one million gallons of concentrated sulfuric acid. Now Delaware has a law that all tanks of 25,000 gallons or more must be registered and inspected by the state. The paperwork that goes with regulations is burdensome but no one has found a good substitute for it.

Many U.S. industries are becoming greener based on what they say at meetings and in print.[104] However, it is not possible to tell how green they are by reading their annual environmental reports.[105] The level of disclosure is said to vary greatly. There is a high degree of subjectivity and selectivity in what is reported, often with a lack of quantification of environmental impacts. A survey of 54 large U.S. corporations, which expressed environmental concern through such reports, found that only one-third of the companies would switch to a less toxic material if it added 1% to the cost of the product, but only two said they would be willing to raise the cost by 5%.[106] Public relations firms can be hired to write such reports, as well as prepare advertisements, and even provide "citizen" letter-writing campaigns.[107] The advertisements may picture green forests and wildlife next to the plant. At the same time, the company, through its membership in a trade association, may be lobbying congress to relax the laws that regulate it. One report says that businesses in Great Britain are paying little more than lip service to the environment.[108] Skeptics have been harsh in their criticisms of these practices.[109.] They accuse firms of trying to spread doubt and uncertainty.

Advertising is big business. In 1989, its cost was 120 dollars for every person in the world. The average child in the United States sees 20,000 commercials in a year each with the message, "Buy this."[110] This is one cause of the overconsumption in the developed nations. Corporations are also flooding the schools with teaching aids and study materials that may contain information that is misleading or inaccurate, as well as advertising.[111] The U.S. EPA suggests caution in accepting "green" advertising claims. Its suggestions include looking for claims that are specific, being wary of claims that are overly broad or vague, realizing that many materials do not degrade in landfills, or elsewhere.[112] Many earth-friendly claims have been called "phony baloney."[113] The U.S. Federal Trade Commission fined Bayer $3.2 million for claiming that adding a green tea extract to a multivitamin capsule could reduce a person's weight without the need for exercise.[114] It is investigating the green assertions of carbon-offset claims. The sign on the bottom of the six-pack of Coors beer says, "Do not litter highways or public grounds. Keep our country beautiful." The consumer would be much more likely to see this message if it were on the side of the six-pack holder instead of on the bottom.

The CMA (now the American Chemistry Council) was 125 years old in 1997. Its over 190 member companies make over 90% of the chemicals that are produced in the United States.[115] Its future involves embracing sustainable development, establishing a health and environmental research program and bringing greater flexibility into government regulations.[116] It is collaborating with the U.S. EPA in the testing of high-volume chemicals for toxicity and as endocrine disruptors.[117] (The Society of Chemical Manufacturers and Affiliates and the American Chemistry Council objects to the "design" of the program, fearing that the cost of testing will be a hardship on its members.[118] It also criticizes the U.S. EPA for ignoring the "unique needs of small-batch and custom chemical producers" and "its opposition to promising suggestions."[119]) The Chemical Industry Institute of Toxicology is funded by the American Chemistry Council.[120] The Council is proud of the record of the chemical industry in reducing releases of chemicals on the Toxics Release Inventory by more than 60% in 6 years.[121] However, it opposed expansion of the Toxics Release Inventory in the courts, losing its suit in the appeals court.[122] Its request to have ethylene glycol removed from the Inventory was denied by the EPA.[123] It does not support additional testing of the health effects of biphenyl, carbonyl sulfide, chlorobenzene, ethylene glycol, methylisobutylketone, phenol, and trichlorobenzene, as suggested by the EPA.[124] It also opposes mandatory reporting of toxic chemical use, although such systems appear to be working well in Massachusetts and New Jersey[125] (see also Chapter 17). The Synthetic Organic Chemical Manufacturers Association also objects to this.[126] Such inventory reporting should offer a maximum of flexibility for those companies who dislike "command-and-control" regulation. The Toxics Release Inventory requires no reduction in emissions at all, but companies often choose to make reductions after they realize how much they are losing in material and dollars.[127] They may also be motivated by wanting a good public image. The American Chemistry Council has spent 40 million dollars over 5 years on an advertising campaign to create a better public image for the chemical industry, but has discontinued the campaign as a result of its failure to change public attitudes.[128] The advertising campaign started again in 2005.[129]

Advertising may not be able to change an image created by accidents and some illegal practices. Many firms have been fined for exploiting the public by price fixing. The following partial list has been compiled from reports in *Chemical Engineering News* and *Chemistry and Industry*:

Paraffin wax ENI, Exxon-Mobil, Sasol, Total
Chloroacetic and organic peroxides Arkema, Degussa (now Evonik), Akzo-Nobel, Hoechst
Vitamins Roche, BASF, Aventis, Takeda, Merck, Solvay
Citric acid Roche, ADM, Bayer, Cerestar
Sodium gluconate ADM, Akzo-Nobel
Rubber Chemicals Crompton (now Chemtura), Flexsys, Bayer
Sorbates Hoechst, Eastman, Chisso, Ueno
Methionine Degussa (now Evonik), Aventis, Nippon Soda
Industrial gases Linde, Air Liquide, Air Products, BOC
Nucleotide flavor enhancers Ajinomoto
Polyester staple fiber Nan Ya Plastics, Ko Sa
Rubber DuPont-Dow Elastomers, Crompton (now Chemtura), Bayer, DSM, ENI, Shell, Dow, Bayer, Zeon
Sodium perchlorate and inorganic peroxides Kemira, Arkema, Akzo-Nobel
Poly(methyl methacrylate), ICI (now Akzo-Nobel), Arkema, Degussa (now Evonik), BASF, Lucite
Adipic-based polyester polyols Bayer
Choline chloride Akzo-Nobel, BASF, UCB
Microcrystalline cellulose Asahi Kasei, FMC
Carbon fiber Toray Industries
Transatlantic shipment of chemicals Odjfell Seachem, Jo Tankers
Calcium carbide and magnesium Akzo Nobel, Evonik
Tin and heat stabilizers Ciba (now BASF), Akzo Nobel, Arkema, Reagens, Chemson Group, Baerlocher

In several cases, company executives have been sent to jail. Several W.R. Grace executives were on trial for continuing to mine asbestos-containing vermiculite knowing that it was making many of the residents of Libby, Montana, ill.[130] They were acquitted.

Pharmaceutical companies have also received fines for efforts to keep their profits up:

- Abbott Labs's six months expulsion for inappropriate advertising to health professionals, such as trips to greyhound tracks and tennis tournaments.
- Astra Zeneca made "false or misleading safety claims on Crestor," deceived the Patent Office into extending patent protection on Losec, defrauded Medicaid on drug pricing, paid Ranbaxy to delay the introduction of a generic version of Nexium.
- Aventis conspired with Andrx to keep a generic form of Cardizem off the market for a year.
- Bayer overcharged Medicaid and interfered with first responders and the Chemical Safety and Hazard Investigation Board investigations into the August 28, 2008, fire and explosion that killed two workers at its West Virginia plant by being overly secretive[131].
- Bristol Meyers Squibb lied to the Federal Trade Commission about dealings with a generic drug maker on Plavix.

- Cephalon paid a generic company to delay competition with a patented drug.
- Eli Lilly's illegal promotion of drug for unapproved use and spreading false rumor of shortage of drug to increase sales.
- Glaxo Smith Kline failed to disclose negative as well as favorable clinical trial results, overcharged Medicaid, owes $2.7 billion in back taxes, and is engaged in price fixing.
- Merck's misleading ads for Vioxx, withholding negative data on clinical trials.
- Pfizer's Clean Air Act violations; promoting Bextra for uses, the FDA had disapproved ($2.3 billion settlement).
- Ranbaxy's falsified data on applications to FDA.
- Schering Plough violated Good Manufacturing Practice, made agreements with generic companies to delay the introduction of generic drugs, and withheld data on clinical trials.
- Serono's deceptive mechanical device to boost the sales of AIDS drugs.

Drug companies have also filed frivolous patent applications to try to extend patent coverage on drugs. The companies claim that they need to maintain good profits to cover the high costs of research and development. This bothers the public when it finds out that the companies have been spending two to three times as much money on drug salesmen as on research and development.[132]

The showpiece of the American Chemistry Council is its "Responsible Care" program.[133] The program originated in Canada in 1985 and was adopted by the American Chemistry Council in 1988. It has spread to South America, Europe, Africa, Asia,[134] and Australia, and is now in 40 countries. The British Chemical Distributors and Traders Association is encouraging all of its members to sign up for the program.[135] Only 92 of its 117 members had done so by 1998. Companies pledge to adhere to 10 guiding principles:

1. Recognize and respond to community concerns.
2. Develop chemicals that are safe to make, transport, use, and dispose of.
3. Give priorities to environmental, health, and safety in planning products and processes.
4. Give information on chemical-related health and safety hazards promptly to officials, workers and the public, and to recommend protective measures.
5. Advise customers on the safe use and handling of products.
6. Operate plants in a way that protects the environment and the health and safety of workers and the public.
7. Conduct research on environmental, health, and safety aspects of products, processes, and waste.
8. Resolve problems created by the past handling and disposal of hazardous materials.
9. Participate with government and others to create responsible regulations to safeguard the community, workplace, and the environment.
10. Offer assistance to others who produce, use, transport, and dispose of chemicals.

Skeptics, who felt at first that the program was just a public relations ploy, now consider it to be real and worthwhile. They do point out that it may be written too broadly and that it may not be quantitative enough. The president of the American Chemistry Council has said, "We are acutely aware that the initiative loses credibility if we say one thing with Responsible Care and say another thing when we deal with Congress or the regulatory agencies." Independent third-party verification is starting to be used by some companies.[136] The United Nations Commission on Sustainable Development plans to "examine voluntary initiatives and agreements," but the American Chemistry Council does not want "a formal U.N. mechanism to oversee these programs."[137] Efforts need to be increased to bring labor into the program, for it has felt left out.[138] If this approach and similar ones prove workable and verifiable, this may be a good alternative to enactment of further regulations.[139]

There are some remaining skeptics. The U.S. Public Interest Research Group Education Fund found that more than 75% of American Chemistry Council members were unable or unwilling to answer seven questions about toxic chemical releases, accidents, and storage and transportation safety.[140] Some scientists doing research on the toxicities of industrial chemicals have felt that industry has been overly secretive when asked to provide data.[141] Part of this may be due to the litigious society of the United States. The Coalition for Effective Environmental Information (an industry group) and the American Chemistry Council have been reluctant to disclose the "worst-case scenarios" required under the Clean Air Act's 1990 amendments to the general public for fear that it might help terrorists.[142] The Working Group on Community Right-to-Know (a coalition of 150 organizations) feels that public disclosure is the best way to minimize risk.[143] Public disclosure might provide a strong incentive for a company to replace its processes with greener ones using less toxic chemicals. The public was openly critical at a meeting in Texas to explain companies' risk-management plans.[144] A way must be found to improve the dialogue between chemical plants and the communities around them.

Europe adopted a Registration, Evaluation, Authorisation and Restriction of Chemicals (REACH) law in 2008.[145] The American Chemistry Council believes that this law will stifle chemical innovation.[146] However, American companies that export chemicals to Europe will have to comply with it.

The American Chemistry Council was instrumental in having the comments of a respected toxicologist on polybrominated diphenyl ethers removed from an EPA assessment.[147] A regional EPA administrator was forced to resign for wanting Dow to remove the dioxins from two rivers in Michigan.[148] These events have led to a U.S. Congressional investigation of the influence of the American Chemistry Council on the EPA.[149] A bill to measure synthetic chemicals in the bodies of Californians, supported by health and environmental groups, but opposed by the American Chemistry Council, was vetoed by the governor.[150]

In 1988, The Coalition for Environmentally Responsible Economics (a group of investors and environmentalists) put together a code (first called the Valdez Principles, now the CERES Principles) for corporations to follow.[151] It includes

1. Reducing releases into the biosphere and protecting biodiversity.
2. Sustainable use of natural resources.
3. Minimizing waste and disposing of it properly.
4. Sustainable use of energy.
5. Minimizing environmental, health, and safety risks to employees and the public.
6. Elimination of products that cause environmental damage.
7. Correcting any damage that may have been caused to people and to the environment.
8. Informing those involved of any potential hazards in a timely fashion.
9. Environmental interests must be represented on the Board of Directors.
10. An annual self-audit will be made public.

This code is broader than Responsible Care. It includes the concept of a sustainable future and requires that a self-audit be made public. Large companies have been slow to embrace it. Management usually recommends a vote against the many shareholder resolutions that propose it in the proxies for the annual meetings (e.g., GE in 1993 and Goodyear Tire and Rubber in 1996). Those that have adopted it include Sun Company, General Motors,[152] H. B. Fuller, Polaroid, and Arizona Public Service.[153]

The International Organization for Standardization (ISO) was formed in 1946 to facilitate standardization as a means of promoting international trade. It is located in Geneva, Switzerland. Its ISO 9000 series, which deals with product quality, has become almost a requirement for companies to do business.[154] The series is being extended to an ISO 14,000 series dealing with environmental management.[155] These standards will include formalization of corporate policies and audits, evaluation of performance, labeling, and life cycle assessments. The standard will be worldwide and will require mandatory third-party verification. The goal is to support environmental protection in balance with socioeconomic needs. If these standards are adopted as widely as the ISO 9000 series has been, they may become a powerful force in cleaning up the environment. Companies buying the chemicals may insist that their suppliers adhere to the principles. This may be especially helpful in nations that lack an effective regulatory system.[156]

It is not easy to tell just how green a company is.[157] Stock analysts do not consider environmental performance a major factor in evaluating companies.[84,158] The U.S. National Academy of Engineering is developing methods to do this.[159] The Investor Responsibility Research Center evaluates companies based on Superfund sites, spills, total emissions, enforcement actions, and penalties.[160] (Such data can usually be obtained from state environmental departments in the United States.) The Hamburg Environmental Institute evaluated the environmental friendliness of nearly 70 of the largest chemical and pharmaceutical corporations in its annual Top 50 report.[161] The top 10 were deemed "proactive." The rankings of the seven U.S. corporations in this category in the report for 1995 were Johnson & Johnson first, 3M third, Procter & Gamble fourth, Dow fifth, Baxter International sixth, Bristol Meyers Squibb ninth, and DuPont tenth. Henkel (Germany) was second, Ciba–Geigy (Switzerland) seventh, and Unilever (Netherlands) eighth. Some of the rankings change with the year. DuPont was 27th in 1994. Unilever was 22nd in 1994. Six U.S. companies were among the 22 classified as "active," the next lower level. The remaining 18 companies were classified as "reactive," the next lower level, or "passive," a still lower level. Monsanto was 19th in the report. The U.S. firms classified as "passive" were Merck 43rd, Colgate–Palmolive 47th, Occidental 48th, and GE Plastics 49th. The lowest ranking, "negative" was given to firms that could not be ranked "due to insufficient communication of their environmental performance." Firms in this category included American Home Products, Exxon Chemical, Pfizer, Schering Plough, Warner–Lambert, and Rhone–Poulenc. Overall, companies were weakest when judged on the "sustainability" of their products.

DuPont won the highest score among chemical companies in the Dow Jones Sustainability Index in 2001.[162] Akzo-Nobel was first in 2007. The Global Reporting Initiative evaluated about 700 companies for sustainability and the triple bottom line.[163] The Council of Institutional Investors rates companies based on their proxy statements.[164] Covalence (Swiss) ranked companies for their ethics.[165] In 2006, GlaxoSmithKline, Merck, and Bristol Meyers Squibb were the top drug companies; BASF, Air Products, DuPont, and Bayer were the top chemical companies; and BP Statoil was the top oil company. Government Metrics International rated 2100 companies for accountability giving perfect grades to Air Products and Chemicals, DuPont, General Electric, Pfizer, Praxair, and 3M.[166] The Sustainability Yearbook from SAM and Price-Waterhouse-Coopers gave the electricity company Endresa the best grade.[167] Of the 1000 companies rated, 67 were placed in the gold class. B Corporation rates companies on social responsibility.[168] SEE Change rated 3M the best in savings by pollution prevention.[169] Lists of the "world's most admired companies" was by peer group rating.[170] BASF and DuPont were the top chemical companies, Johnson & Johnson and Novartis topped the pharmaceutical companies, while Exxon-Mobil and Chevron were the most admired oil companies with BP being fourth. Some of these rating systems apparently do not take into account the accidents described in Chapter 1 and the fines for deceptive practices described earlier in this chapter. The Environmental Sustainability Index from Yale and Columbia Universities rates countries, placing Finland, Norway, and Uruguay at the top and North Korea, Taiwan, Turkmenistan, Iraq, and Uzbekistan at the bottom.[171] The United States is number 45 on the list.

The Council on Economic Priorities (a New York NGO) produces an annual list of the worst polluters in the United States.[172] These ratings are based on toxic releases, regulatory compliance, waste, cleanups, and worker health and safety of more than 100 companies. The 1994 list included Union Carbide (now Dow) (which had three times as many spills as the next worst competitor, as well as a higher than average frequency of health and safety violation), Exxon (now Exxon-Mobil), Texaco (now Chevron Texaco), International Paper, Southern Company, Westinghouse, Westvaco, and Maxxam (unsustainable harvesting of redwood forests). The 1995 list included Formosa Plastics, Exxon (now Exxon-Mobil), Maxxam, and Southern Company (industry's top emitter of carbon dioxide). Texaco (now Chevron-Texaco), Union Carbide, Westinghouse, International Paper, and Westvaco were not on the 1995 list. "Businesses on the list call it biased, distorted or just plain wrong." Along with the list, the Council recommends ways in which the companies can improve.

Some companies that are not very green make the news. Condea Vista was found guilty of "wanton and reckless disregard of public safety" by a Louisiana jury for a leak of 1,2-dichloroethane over an 8-month period.[173] Georgia Pacific was fined for failure to install pollution control equipment for volatile organic compounds (VOCs) at 11 of its plants in the southeastern United States.[174] Installation of the control equipment would reduce emissions by 5000 tons/yr. A Formosa Plastics poly(vinyl chloride) plant in Texas agreed to try to implement a zero-discharge system for wastewater to a bay rich in shrimp and oysters.[175] This followed years of lawsuits, environmental violations, and millions of dollars of fines. A ship carrying about 3000 metric tons of mercury-containing waste from a Formosa Plastics Taiwan plant to Cambodia was sent back to Taiwan by the Cambodian government.[176] ICI (now Akzo Nobel) was reprimanded by the Environment Agency of the British government for 1996 spills of 1,2-dichloroethane, vinylidene chloride, chloroform, trichloroethylene, and naphtha.[177] When a titanium tetrachloride leak caused a road to be closed and a spill of light oil into an estuary occurred 2 weeks after ICI (now Akzo Nobel) promised to clean up its act, the Environment Agency shut down the titanium dioxide plant. The U.S. Department of Justice took Borden Chemical and Plastics to court in an effort to force it to clean up groundwater contaminated with vinyl chloride and 1,2-dichloroethane at its poly(vinyl chloride) plant in Louisiana.[178]

At the other extreme is 3M, which has saved 1.4 billion dollars in its Pollution Prevention Pays campaign through 3450 projects in the period 1975–1991.[179] Pioneer Hi-Bred International developed a superior soybean for use as an animal food by inserting a gene from a Brazil nut.[180] When company testing showed that humans allergic to Brazil nuts would also be allergic to the soybean, it decided not to release the new plant. Even though the soybean was meant to be fed to animals, there was a possibility of its getting into food for humans.

Many companies appear to be somewhere between the extremes of "proactive" and "reactive," but the general direction is toward becoming greener. In addition to the criteria for judging companies already mentioned, there are some for which the necessary information is difficult to obtain. For example, it is hard to know what role a company plays in an antienvironmental stand taken by a trade association that it belongs to. Companies have also been criticized for being greener in things that are relatively inexpensive than in ones involving a lot of money, but that may have more environmental impact. There is also a legitimate debate about whether or not some company actions are green or not. For example, a new insecticide that contaminated air and water less than its predecessors would be an improvement, but not as much as less pesticide use through integrated pest management (see Chapter 2 for more details). The records of a few companies will be given to illustrate the range of greening. Some of these can be compared with the ratings given in the foregoing by environmental organizations. The two companies cited that make agricultural chemicals have shown little interest in alternative agriculture, just as the oil companies cited have shown little interest in renewable energy.

Ashland Oil reerected an oil tank on the banks of the Monongahela River some years ago. When filled with oil in 1988, the tank ruptured suddenly so that about 1 million gal of diesel oil slashed over the dike meant to contain it and went into the river, contaminating it for miles.[181] Since then, starting in 1989, Ashland has cosponsored The Ohio River Sweep, an annual riverbank cleanup of trash. The company's refinery in Catlettsburg, Kentucky, was granted an operating permit by the state in 1995, following two decades of emissions violations caused by equipment malfunctions and deficient operating procedures.[182] The emissions included sulfur dioxide and calcium oxide dust. Ashland has "made impressive progress toward excellence in the areas of environment, health, and safety."[183] "It recognizes its responsibility to protect and maintain the quality of the environment." It has reduced its toxic chemical releases by more than 52% since 1990. Through the Wildlife Habitat Council (an industry group), the company has set up three sites on its properties as wildlife areas. It also contributes to The Nature Conservancy, The Appalachian Trail Club, The American Trust, The Conservation Fund, and the Izaak Walton League. It has held household hazardous chemical waste collections for several years. Its service-station-stores in Saint Paul, Minnesota, use polycarbonate milk bottles that can be refilled 60 times. The company's service stations have facilities for collecting used motor oil for recycling. Ashland

Chemical was the first chemical company to link its corporate Toxic Release Inventory, Superfund, solid waste, and wastewater report to the U.S. EPA's Envirofacts data warehouse, which makes them available to anyone with access to the World Wide Web.[184]

Chevron had two oil spills and a release of silica–alumina catalyst in 1991–1993.[185] Its oil spills have been reduced by 97% since 1989. Losses owing to fire in 1993 were 37 million dollars. There was a fire and explosion in the hydrocracking unit in Richmond, California, in 1999.[186] Fines of 1.4 million dollars were paid in 1993. A fine of $7 million was levied for leaks at its El Segundo marine terminal under the Clean Air Act.[187] One week later, the state of California fined it $2 million for a leak in a pipeline carrying jet fuel.[188] The company is adding double-hulled tankers to its fleet as a way to reduce the possibility of oil spills.[189] The company has been on trial in Ecuador for the many oil spills in the Amazon that have not been cleaned up.[190] On the other hand, Jared Diamond has commended the company for its oil field operations in Papua New Guinea in a rain forest.[191] Since 1989, and up to 1994, it had contributed about 7 million dollars "to hundreds of conservation, wildlife preservation and environmental research and educational organizations." The company has won awards for its work with wildlife habitat in Wyoming, Colorado, and other places.[192] Environmental groups have accused it of green-washing in its "People Do" and other advertising.[193] They claim that the company is frequently doing only what the law requires, while at the same time its lobbyists (through its membership in the American Petroleum Institute) are trying to convince the Congress to weaken the laws. Chevron's chairman has called for regulatory reform because, "Our regulatory system does not have a principle that says 'enough is enough.'" He says that the regulatory process has grown "bloated on an unbalanced diet of zero-risk thinking."[194] Company management recommended that shareholders vote against shareholder proposals at the annual meeting asking that the company not drill for oil in the Arctic National Wildlife Refuge, that the company publish a report on the environmental and safety hazards for communities surrounding its plants, and that the company publish a report on the contribution of its products to global warming, as well as efforts to moderate this.[195]

In the early days of the U.S. EPA, the agency asked permission to inspect a Dow plant. When the company refused in 1978, the EPA inspected the plant by flying an airplane above it. Dow sued the EPA for illegal search and seizure, but lost in a case that reached the U.S. Supreme Court.[196] Dow is a greener company today. It planned to reduce waste by 50%, losses from spills and leaks by 90%, release of dioxins by 90%, and of toxic chemicals by 75%, by the year 2005.[197] A joint project with the Natural Resources Defense Council at its Midland, Michigan, plant reduced emissions and waste with a savings of $2.3 million per year.[198] A 20-year efficiency campaign aims to cut energy use in half by 2015.[199] Its president describes the third wave of environmentalism, which will be led by industry, because it can link business reality and environmental progress. (He says that the first wave was driven by environmental groups and the second by the command-and-control methods of government.) According to him, "Poor environmental, health and safety performance represents a waste of precious resources." David T. Buzzelli, a vice president of Dow, cochaired the President's Council on Sustainable Development.[200] Dow received a Presidential Green Chemistry Challenge Award for the use of carbon dioxide in blowing polystyrene foam.[201] Its second entry was the use of microemulsions to replace traditional organic solvents. However, the company is still pushing the use of methylene chloride, trichloroethylene, and perchloroethylene in its advertising.[202] As mentioned in Chapter 17, Dow's Louisiana plant has an effective pollution prevention program, but a plant in La Porte, Texas, has not put in pollution prevention that would save $1 million/yr. The plant in Louisiana was fined $2.4 million for air and water violations in 2003.[203] The opportunities identified for the reduction of waste and emissions that save dollars at its Midland, Michigan, plant have been implemented.[204] Dow's contributions to congressmen through its political action committees have been more than double those of DuPont and Monsanto.[205] Dow promised to take corrective measures to remove the dioxins in the river below its Midland, Michigan, plant in 2003, but had not started by 2008.[206] A report by Innovest Strategic Value Advisors was critical of the company on dioxins and other issues.[207] Dow and Cargill were producing poly(lactic acid) from starch, for use as a replacement for some polymers that are based on petroleum.[208] Dow has since left the project, the plant now being owned by The Nature Works. As mentioned earlier, DuPont-Dow Elastomers was fined for fixing prices on poly(chloroprene). Dow has established the Dow Chair of Sustainable Chemistry at the University of California, Berkeley.[209]

When Delaware's bottle bill and Coastal Zone Management Act were proposed in the 1970s, the DuPont Company testified against them in public hearings. Since then, the company has become greener. Paul Tebo of DuPont points to the company's Chattanooga, Tennessee, nylon plant which now achieves a 99.8% yield of saleable products, a Lycra process that recovers 99.95% of its spinning solvent, a biodegradable oil for two-cycle engines, shipment of neoprene rubber in a Surlyn bag that can be compounded right into the rubber, and an herbicide plant in Java that has close to zero emissions.[210] According to him, "We're saying the reason we're making all these changes is not just that it is good for the environment, it's good for business." He is against implementation of the Kyoto Protocol.[211] The company is striving for zero emissions. DuPont now makes methyl isocyanate from *N*-methylformamide on demand on site and no longer ships it (as discussed in Chapter 2). It also makes phosgene as needed, so that the amount on site is small. This is described as an interim measure until a good phosgene-free route to isocyanates can be developed. The company has now affiliated with the Pew Center on Global Climate Change, which supports the Kyoto Protocol for reduction of greenhouse gases.[212]

DuPont's former chairman, Edgar S. Woolard, said, "As long as environmental protection is in a special category assigned to certain people instead of a part of a mental check-list with which every person approaches every task, then our environmental accomplishments will remain reactive and corrective rather than proactive and innovative."[213] He also favored regulatory reform of the "current regulatory morass."[214] The company presents grants to groups that protect open space and natural areas. It has put a conservation easement on its farm near Chestertown, Maryland.[215] A plant in Spain, where historic buildings have been preserved and the grass is cut by rare or endangered local breeds of cattle, horses, sheep, and donkeys, instead of lawn mowers, has been described as a model for environmental and community relations in Europe.[216] The company is not without its critics. The U.S. EPA fined the company 1.89 million dollars for mislabeling herbicides (for neglecting to state that eye protection was needed while spraying).[217] It settled for 38.5 million-dollar lawsuits concerning contamination of land adjacent to its Pompton Lakes, New Jersey, plant with lead and mercury.[218] It has been fined for the release of a cloud of fuming sulfuric acid at a plant in Kentucky in 1995[219] and for releases at other plants.[220] A Texas shrimper has asked DuPont to use its reverse osmosis membranes to provide zero discharge from its Victoria, Texas, plant, to protect nearby shrimp beds.[221] Environmentalists opposed a DuPont proposal to strip-mine titanium minerals adjacent to the Okefenokee National Wildlife Refuge, for fear that mining would alter the swamp's ground and surface water levels.[222] They also objected to the company drilling for oil in the Grand Staircase–Escalante National Monument in Utah.[223] It was fined for wastewater violations at its Edgemoor, Delaware, titanium dioxide plant.[224] The Delaware Department of Natural Resources and Environmental Control has ordered the company to clean up the dioxins in piles of waste at this site.[225] This had not been done by 2008. The U.S. EPA sued DuPont to recover the cost of cleanup of the Superfund site next to its former pigments plant in Newport, Delaware.[226] The company was fined $16.5 million for failing to report the presence of perfluorooctanoic acid in wells near its plant in West Virginia.[227] The company was sued for trying to prevent the appointment of an environmentalist to a committee that advised the Office of the U.S. Trade Representative.[228] (DuPont Pharmaceutical paid a fine of $44.5 million for misleading marketing and promotion of Coumadin.[229])

Former chairman of DuPont, Chad Holliday, Jr., pledged the company to a sustainable future.[230] Critics accuse the company of trying to "wrap itself in a green flag."

Exxon is famous for the 1989 oil spill from its tanker, *Exxon Valdez*, in Alaska, which resulted in the largest environmental fines in U.S. history.[231] In the period 1990–1995, it spilled only 250 gal of oil.[232] The release of toxic chemicals has declined over 50% since 1987. The release of VOCs from chemical operations has been reduced by 50% since 1990. Hazardous waste disposal has been reduced by 75% since 1990. "The company is committed to continuous efforts to improve environmental performance throughout its activities." The company has "developed procedures to prevent seismic activity during periods of fish spawning." The company provides financial support for projects to save the jaguar and the tiger, as well as for the Bermuda Biological Station for Research, the Royal Society for Nature Conservation in the United Kingdom, and others. The company belongs to the American Chemistry Council and the American Petroleum Institute. The company's environmental coordinator says, "Government and private research programs cannot afford to continue the adversarial approaches taken in the past."[233] The company objected to the proposal of the Texas Natural Resource Conservation Commission to reduce emissions of NO_x by 90% by 2007, calling the proposal "draconian reductions, for which technology has not been demonstrated."[234] The company is against the new EPA regulations for ozone and fine particulates. It feels that, "The government is rushing to impose these extreme new measures based on questionable science. While the benefits are uncertain, no one will escape the enormous costs."[235] It takes a similar position on global warming.[236] "The debate over global warming has been clouded by assertions based on incomplete and uncertain since … The public debate continues to be inaccurate, incomplete and misleading … The climate proposals announced by President Clinton on October 22 would reduce the U.S. gross domestic product by more than $200 billion." The Royal Society of Chemistry in England has censured the company for misinforming the public about the science of climate change.[237] Its "disinformation strategy parallels the tobacco industry's campaign to confuse the public about the dangers of smoking."[238]

Monsanto (which split into Monsanto and Solutia in 1997) has devised ways to make isocyanates without the use of phosgene (see Chapter 2). It has also found a way to make *p*-phenylenediamine antioxidants without the use of chlorine (see Chapter 3). The company pledge of 1990 includes the reduction of all toxic and hazardous releases and emissions, working toward the ultimate goal of zero effect.[239] It pledges also to "work to achieve sustainable agriculture through new technology and practices." It has offered two 1 million-dollar challenges to anyone who can help it find a cost-effective and commercially practical way to recover useful chemicals from its wastewater, at the same time reducing their effect on the environment. Both have been won by SRI International.[240] Monsanto also won a Presidential Green Chemistry Challenge Award in 1996 for a new method to make disodium iminodiacetate, an intermediate in the synthesis of the herbicide glyphosate.[241] The intermediate is made by the copper-catalyzed dehydrogenation of diethanolamine. The new route eliminates the use of hydrogen cyanide and formaldehyde, as well as eliminating

1 kg of waste for every 7-kg product. This "zero-waste" route gives a higher overall yield with fewer process steps. The company has produced Alachlor, an herbicide often found as a contaminant in groundwater, since 1969.[242] The company also markets insect-resistant corn and cotton, which incorporate genes for *Bacillus thuringiensis* toxins, as well as cotton and soybeans that are resistant to its herbicide glyphosate.[243] Environmentalists are concerned about the rate of build-up of resistance that will take place in insects and weeds (as mentioned in Chapter 11). There is also the question of the extent of soil erosion with soil kept bare except for the crop. Alternative agriculture could be a more sustainable alternative.

BP has had many problems. It was fined $9.5 million for violations of the Clean Air Act at eight refineries in 2000.[244] It was fined $1.5 million for safety breaches at its oil refinery in Grangemouth, Scotland, in 2000.[245] Deaths resulted from an explosion at its engineering polymers plant in Augusta, Georgia, in March 2001.[246] Its Texas City, Texas, refinery has had 41 deaths in 33 years.[247] (Fifteen of these deaths were from a fire and explosion in 2005.) Its Prudhoe Bay, Alaska, pipeline failure could have been avoided by more adequate inspection.[248] Among the causes of the deaths in Texas City and the pipeline leak were budget and production pressures, flawed communication of lessons learned, excessive decentralization of safety functions, and high management turnover. BP lobbied against environmental controls of VOCs that would have prevented deaths.[249] From 1997 to 2000, the company contributed to the election campaigns of two-thirds of the U.S. senators and representatives who voted against key green measures before the U.S. Congress.[250] BP recognizes climate change and has a solar subsidiary.[251] It has funded work on alternative energy at the Universities of California Berkley and Illinois Urbana.[252]

Motiva Enterprises (once Star Enterprise, now Valero) was a joint venture of Texaco, Shell, and the Saudi Arabian Oil Company. Its operations include a refinery, located near Delaware City, Delaware, which has had difficulties in meeting environmental standards over the years. (The following dates refer to items from the *Wilmington Delaware News Journal*, except as referenced otherwise.)

1/31/85: 19,000 gal of sulfuric acid spilled, contaminating Red Lion Creek
2/19/87: 60,000-dollar fine for emissions of sulfur dioxide
6/10/88: large emissions of carbon monoxide and coke dust
9/26/89: 425,000-dollar fine for air pollution and mishandling of hazardous wastes
8/14/92: 1.68 million-dollar fine for discharging toxic chemicals into the Delaware River
3/11/95: state of Delaware sues Star Enterprise for 19 odor violations since 9/16/94
4/12/95: odors attributed to a malfunctioning sulfur recovery unit—unmarked pipes misconnected, bypassing the wastewater treatment unit—suit claiming 120 violations of federal water rules in January 1994, settled out of court for 360,000 dollars
6/8/96: blower failure bathes homes in soot
6/11/96: odor problems with the sulfur recovery unit—fined $175,000 dollars for odor problems from September 1994 to December 1, 1995
7/2/96: odor problems due to starting operation without a scrubber
6/11/97: state of Delaware permit "caps the amount of sulfur that can be recovered and burned off during processing"
9/10/97 and 9/12/97: U.S. EPA fines Star Enterprise 125,000 dollars for violating the Clean Water Act
10/31/97: power failure causes refinery to emit a thick black smoke
12/15/97: four employees treated for caustic burns at the local hospital after an explosion
2/19/98: sulfur dioxide and carbon monoxide released after a fire disables a pollution control system
4/22/98: fined 125,000 dollars for polluting the river between 1993 and 1997
7/7/98: sues the Delaware Department of Natural Resources and Environmental Control to prevent required reductions in emissions of nitrogen oxides
7/12/98: fined 75,000 dollars for breaking a federal court ban on new wastewater violations between 1995 and 1997
9/2/98: federal judge orders monitoring studies of the damage that its pollution has caused to the Delaware River
5/25/99: a federal appeals court upholds the decision of the judge above[253]
10/28/99: barge spills oil into the Delaware River at Motiva dock
1/8/00: crude oil fuels fireball at plant
1/9/00: fined $210.000 for toxic emissions and pollution law violations
3/22/00: fined $390,000 for waste water violations
4/19/00: fined $146,000 for improper storage of hazardous waste and violating reporting and training rules
5/21/00: flash fire injures an employee
12/29/00: storage tank collapses under vacuum
4/4/01: leaked NaOH into a creek on February 27 and March 28
7/18/01: fire and explosion with 1 million gallons of concentrated sulfuric acid spilled—one dead, eight injured
1/23/02: fire due to a mechanical failure
3/31/02: vanadium found in a discharge to a river
4/27/02: workers must wear respirators or air masks near boilers due to leaks of carbon monoxide and sulfur dioxide
10/24/02: acid in boiler water causes corrosion and shutdown of boilers
10/29/02: smoke pours from refinery during incinerator shutdown due to corrosion
2/21/03: state rejects refinery's plan to convert sulfur dioxide to sodium sulfate for disposal into river, rather than install a scrubber
5/6/03: sulfuric acid spilled from storage tank
3/12/04: fined $200,000 for valve leaking carbon monoxide, hydrogen cyanide, dust and ammonia for over a month

3/18/04: employee scalded by steam; two workers scalded by steam major release of hydrogen sulfide and methane from compressor and 6/9/09: problems in the shutdown of the plant check valve failure[254]
6/9/09: problems in the shutdown of the plant
1/28/07: fire[255]
5/19/07: two deaths from asphyxiation under nitrogen in tank
9/8/07: millions of pounds of nitrates put into river in previous years not reported
10/8/07: Texaco and Motiva fined $2.25 million for polluting the river at the end of 20 years in federal court[256]
2/18/08: loss of sulfur dioxide, hydrogen sulfide, nitrogen oxides, soot, and ammonia after a power failure
2/26/09: steam and boiler problems leading to pollutant flaring and the first-ever total shutdown of refinery
The refinery has also maintained a productive natural marsh on its property for many years. As mentioned in Chapter 1, its toxic emissions are very small compared with those emitted by automobiles in northern New Castle County. The refinery is now closed permanently.
Texaco has had other problems as well:
The Company paid fines of 2,807,375 dollars in 1992, most of the fines being for an oil spill in Fairfax, Virginia.[257]
2/22/93: The State of Washington fined the company 9.4 million dollars for an oil spill.[258]
5/19/97: more than 10,000 gal of oil spilled from a ruptured pipeline in Louisiana.
Texaco ended operations in Ecuador in 1992. Oil spills and overflowing waste pits had contaminated the land and the drinking water. The company says that it complied with all the rules in force at the time. It attributes many of the spills to natural disasters. The jury at the International Water Tribunal in Amsterdam found that large quantities of hazardous waste entered the soil and water with at most superficial measures to minimize spills and contamination.[259]
3/26/97 and 1/4/97: Texaco settles a racial discrimination suit for 176 million dollars. The company has announced a comprehensive plan to promote employment and business opportunities for minorities.[260]

Texaco (now Chevron Texaco) supported the radio broadcasts of the Metropolitan Opera for many years. The company has made donations to the Marine Biological Laboratory in Woods Hole, Massachusetts, the New York Botanical Garden, and the Louisiana Nature Conservancy. It also supports an urban tree-planting program.[261] It donated 3000 acres of cypress–tupelo swamp in Louisiana to the Nature Conservancy for a bird refuge in 1994.[262] It has also donated 10,000 acres of land in Utah for a wildlife habitat in 1995.[263] The company also refers to "continued cost-inefficient government regulations" in the same report.

The president of Texaco stated in the "Environment, Health, and Safety Review for 1992," "The drive to lead is an intrinsic part of the Texaco way of life, and our work in the areas of environment, health and safety is no exception." In the corresponding report for 1996, the new chairman of the board said, "Excellence in environmental, health and safety performance must be an integral part of all our business operations."

REFERENCES

1. (a) R. Carson, *Silent Spring*, Houghton Mifflin, Boston, 1962; (b) R.M. Baum, *Chem. Eng. News*, June 4, 2007, 5.
2. (a) B. Ruben, *Environ. Action*, 1995, *27*(1), 11; (b) M. Jacobs, *Chem. Eng. News*, Apr. 24, 2000, 5.
3. B. Baker, *Environ. Action*, 1995, *27*(1), 9.
4. (a) G.T. Gardner and P.C. Stern, *Environmental Problems and Human Behavior*, Allyn & Bacon, Boston, 1996; (b) C.G. Herndl and S.C. Brown, eds, *Green Culture—Environmental Rhetoric in Contemporary America*, University of Wisconsin, Madison, WI, 1996; (c) K.I. Noorman and T.S. Uiterkamp, eds, *Green Households? Domestic Consumers, the Environment and Sustainability*, Earthscan, London, 1997; (d) G Gardner and P. Sampat, *Mind Over Matter: Recasting the Role of Materials in Our Lives*, World Watch Institute paper 144, Washington, DC, 1998; (e) R. Rosenblatt, ed., *Consuming Desires—Consumption, Culture and the Pursuit of Happiness*, Island Press/Shearwater Books, Washington, DC, 1999.
5. John, Javna, *Simple Things You Can Do To Save The Earth.* Earthworks Press, Berkeley, CA, 1989. (Some typical actions that a consumer can take.)
6. J. Makower, *Wilmington Delaware News J.*, 1995, D3.
7. (a) I.L. Baldwin, *Science*, 1962, *137*, 1042; (b) A.N. Hirshfield, M.F. Hirschfield, and J.A. Flaws, *Science*, 1996; *272*, 1444.
8. (a) Anon., *Chem. Eng. News*, Aug. 1962, 23; (b) W.J. Darby, *Chem. Eng. News*, Oct. 1, 1962, 60; (c) M. Heylin, *Chem. Eng. News*, Jan. 12, 1998, 134.
9. G.J. Macro, R.M. Hollingsworth, and W. Durham, eds, *Silent Spring Revisited.* American Chemical Society, Washington, DC, 1987.
10. (a) L. Lear, *Rachel Carson, Witness for Nature.* Holt, New York, 1997; (b) M. Zuk, *Science*, 1997, *278*, 1897; (c) J.M. DeMassa, *Chem. Eng. News*, Apr. 13, 1998, 8.
11. J.G. Edwards, *Chem. Eng. News*, Dec. 8, 1997, 2.
12. D. McGarvey, *Chem. Eng. News*, Sept. 8, 1997, 6.
13. (a) T. Colburn, D. Dunanoski, and J.P. Meyers, *Our Stolen Future.* Dutton, New York, 1996; (b) A.N. Hirschfield, M.F. Hirschfield, and J.A. Flaws, *Science*, 1996, *272*, 1444.
14. (a) J. Collet and S. Karakashian, eds, *Greening the College Curriculum—A Guide to Environmental Teaching in the Liberal Arts*, Island Press, Washington, DC, 1995; (b) F.S. Vandervoort and E.N. Anderson, *Science*, 1997, *275*, 909; (c) J. Blewitt and C. Cullingford, eds, *The Sustainability Curriculum–The Challenge for Higher Education*, Stylus Publishing, Herndon, VA, 2004.
15. (a) W.H. Mansfield III, *World Watch*, 1998 *11*(3), 24; (b) D. Eagon and J. Keniry, *Green Investment, Green Return—How Practical Conservation Projects Save Millions on America's Campuses*. National Wildlife Federation, Vienna, VA, 1998; (c) P.F. Bartlett and G.W. Chase, eds, *Sustainability on Campus–Stories and Strategies for Change*, MIT Press, Cambridge, MA, 2004.
16. Anon., *Chem. Eng. News*, June 18, 2007, 62.

17. K.S. Betts, *Environ. Sci. Technol.*, 2001, *35*, 198A.
18. (a) S.J. Bennett, R. Freierman, and S. George, *Corporate Realities and Environmental Truths—Strategies for Leading Your Business in the Environmental Era.* Wiley, New York, 1993; (b) A. Thayer, *Chem. Eng. News*, May 8, 2000, 13; (c) G. Heal, *When Principles Pay*, Columbia University Press, New York, 2008; (d) R. Mulin, *Chem. Eng. News*, April 23, 2007, 30.
19. P. Hawken, *Blessed Unrest–How the Largest Movement in the World Came Into Being and No One Saw It Coming*, Viking Press, New York, 2007.
20. U. Hjelmar, *The Political Practice of Environmental Organizations*, Avebury Studies in Green Research, Aldershot, UK, 1996, 72–120.
21. (a) C.G. Herndl and S.C. Brown, *Green Culture—Environmental Rhetoric in Contemporary America*, University of Wisconsin Press, Madison, WI, 1996, 236–260; (b) B. Birchard, *Nature's Keepers: The Remarkable Story of How The Nature Conservancy Became the Largest Environmental Organization in the World*, Jossey-Bass, New York, 2005.
22. (a) R. Hunt, *Chem. Br.*, 1995, *31*, 874; (b) D. Clery, *Science*, 1996, *272*, 1258; (c) P. Layman, *Chem. Eng. News*, Oct. 20, 1997, 24.
23. L.R. Crutchfield and H.McC. Grant, *Forces for Good–The Six Practices of High-Impact Nonprofits*, Wiley, New York, 2008.
24. (a) E. Fastiggi, *Catalyzing Environmental Results—Lesson in Advocacy Organization-Business Partnerships*, The Alliance for Environmental Innovation (Environmental Defense Fund and The Pew Charitable Trusts), Washington, DC, 1999; (b) J. Bendell, *Terms for Endearment: NGOs and Sustainable Development*, Greenleaf Publishing, Sheffield, UK, 2000; (c) Anon., *Chem. Eng. News*, Nov. 3, 2008, 17; (d) M. Boleat, *Chem. Ind. (Lond.)*, 2004, 14; (e) M. Gunther, *Fortune*, 2006, *154*(3), 42.
25. (a) J. Johnson, *Chem. Eng. News*, Sept. 8, 1997, 27; (b) J. Johnson, *Chem. Eng. News*, Dec. 15, 1997, 42; (c) *EDF Lett.;* 1998, *29*(1), 8.
26. (a) C. Hogue, *Chem. Eng. News*, Oct. 25, 2004, 45; (b) Anon., *Chem. Eng. News*, Mar. 27, 2006, 32; (c) C. Hogue, *Chem. Eng. News*, Jan. 8, 2007, 40.
27. M. Heylin, *Chem. Eng. News*, Feb. 26, 1996, 25.
28. (a) M. Gunther, *Fortune*, May 31, 2004, 158; (b) Letter, *Rainforest Action Network*, San Francisco, CA, May 2007.
29. (a) G. Jordan, and W.A. Maloney, *The Protest Business? Mobilizing Campaign Groups.* Manchester University Press, Manchester, UK, 1997; (b) L. M. Salamon and H.K. Anheier, *The Emerging Non-Profit Sector: An Overview*, Manchester University Press, Manchester, UK, 1996; (c) C. Runyan, *World Watch*, 1999, *12*(6), 12.
30. C. Marzec, *Nat. Conserv.*, 1999, *49*(5), 6.
31. J.R.E. Bliese, *The Greening of Conservative America*, Westview Press, Boulder, CO, 2001.
32. L. Raber, *Chem. Eng. News*, Nov. 16, 1996, 23.
33. R. Shogren and A. Browning, *Chem. Eng. News*, Jan. 6, 1997, 4.
34. E. Marshall, *Science*, 1996, *272*, 1094.
35. D.C. Korten, *When Corporations Rule the World.* Kumarian Press, West Hartford, CT, 1995, 143.
36. R. Rawls, *Chem. Eng. News*, Dec. 23, 1996, 27.
37. (a) R. Rowell, *Green Backlash: Subversion of the Environmental Movement.* Routledge, New York, 1996; (b) G. Easterbrook, *A Moment on the Earth: The Coming Age of Environmental Optimism.* Viking Penguin, New York, 1995; (c) S. Budiansky. *Nature's Keepers: The New Science of Nature Management.* Free Press, New York, 1995; (d) C.C. Mannand and M.L. Plummer, *Noah's Choice—the Future of Endangered Species.* Knopf, New York, 1995; (e) J.R. Dunn and J.E. Kinney, *Conservative Environmentalism: Reassessing the Means, Redefining the Ends.* Quorum Books, Westport, CT, 1996; (f) J. Lehr, ed., *Rational Readings on Environmental Concerns.* van Nostrand Reinhold, New York, 1992; (g) P.J. Wingate and J.D. Honeycutt, *Chem. Eng. News*, Oct. 2, 1995, 5; (h) J. Wortham, M. Gough, and A. Underdown, *Chem. Eng. News*, May 5, 1997, 5; (i) S. Milloy, *Science Without Sense.* Cato Institute, Washington, DC, 1996; (j) K.M. Reese, *Chem. Eng. News*, Jan. 15, 1996, 48; (k) Y. Xue, *Chem. Eng. News*, Aug. 4, 1997, 6.
38. B.E. Erickson, *Chem. Eng. News*, June 2, 2008, 36.
39. P.R. Ehrlich and A.H. Ehrlich, *Betrayal of Science and Reason—How Anti-Environmental Rhetoric Threatens Our Future*, Island Press, Washington, DC, 1996.
40. E.M. Whelan, *Toxic Terror: The Truth Behind the Cancer Scares.* Prometheus, Buffalo, NY, 1993.
41. *Letter from Green Watch*, Fairfax, VA, August 1997.
42. (a) D.A. Coleman, *Ecopolitics: Building a Green Society*, Rutgers University Press, New Brunswick, NJ, 1994; (b) S.P. Hays, *A History of Environmental Politics Since 1945*, University of Pittsburgh Press, Pittsburgh, 1999.
43. Anon., *Environ. Sci. Technol.*, 1996, *30*, 160A.
44. R.M. White, *Technol. Rev.*, 1996, *99*(2), 56.
45. (a) J. Davis, *Amicus J. (Natural Resources Defense Council)*, 1995, *17*(2), 18; (b) *Amicus J. (Natural Resources Defense Council)*, 1995, *17*(2), 3; (c) K. Durbin, *Amicus J. (Natural Resources Defense Council)*, 1995, *17*(3), 29; (d) D.N. Foley and K.D. Lassila, *Amicus J. (Natural Resources Defense Council)*, 1995, *17*(3); (e) Anon., *Amicus J. (Natural Resources Defense Council)*, 1995, *17*(3), 3; (f) D. Meadows, *Amicus J. (Natural Resources Defense Council)*, 1996, *18*(1), 12; (g) Anon., *Amicus J. (Natural Resources Defense Council)*, 1996, *18*(1), 55; (h) D.J. Hanson, *Chem. Eng. News*, July 24, 1995, 32; (i) P. Goldman, *Environment*, 1996, *38*(3), 41; (j) B. Hileman, *Chem. Eng. News*, Mar. 11, 1996, 8.
46. A. Gore. *Earth in the Balance—Ecology and the Human Spirit.* Houghton Mifflin, Boston, 1992.
47. (a) M. Jacobs, *Chem. Eng. News*, Oct. 9, 1995, 5; (b) B. Hileman, *Chem. Eng. News*, June 19, 1995, 21; (c) R.M. Baum, *Chem. Eng. News*, Aug. 31, 20-09, 3
48. D.N. Foley and K.D. Sassila, *Environ. Sci. Technol.*, 1996, *30*, 13.
49. G. Peaff, *Chem. Eng. News*, July 7, 1997, 27.
50. B. Hileman, *Chem. Eng. News*, Feb. 22, 1999, 23.
51. (a) R.M. Baum, *Chem. Eng. News*, May 26, 2008, 3; (b) C. Hogue, *Chem. Eng. News*, Dec. 22, 2008, 27.
52. C. Hogue, *Chem. Eng. News*, Feb. 12, 2001, 29.
53. C. Hogue, *Chem. Eng. News*, May 5, 2008, 34.
54. D. Thompson Jr., *Wilmington Delaware News J.*, Aug. 15, 1997, A1.
55. (a) C. Weiser, *Wilmington Delaware News J.*, June 11, 1997, A4; (b) C.A.G. Tulou, *Wilmington Delaware News J.*, Sept.

29, 1996, K4; (c) C.A.G. Tulou, *Wilmington Delaware News J.*, Aug. 10, 1998, A7.
56. W. Rudig, *Environment*, 2002, *44*(3), 20.
57. Anon., *Chem. Ind.* (*Lond.*), 1996 397; 2000, 49; 1999, 123.
58. B. Hileman, *Chem. Eng. News*, June 14, 1999, 21.
59. (a) J. Werksmann, *Greeting International Institutions.* Earthscan, London, 1995; (b) O.R. Young, *Environment*, 1999, *41*(8), 20.
60. D. Hinrichsen, *Amicus J. (Natural Resources Defense Council)*, 1996, *18*(3), 35.
61. W.C. Clark, *Environment*, 1997, *39*(7), 1.
62. World Commission on Environment and Development, *Our Common Future.* Oxford University Press, Oxford, 1987.
63. (a) J.B. Callicott and F.J.R. da Rocha, eds, *Earth Summit Ethics: Toward a Reconstructive Postmodern Philosophy of Environmental Education*, State University of New York, Albany, NY, 1996; (b) T. Heyd, *Environ. Ethics*, 1997, *19*, 434; (c) M. McCoy and P. McCully, In: I. Tellam and P. Chatterjee, eds, *The Road from Rio: An NGO Action Guide to Environment and Development*, International Books, Utrecht, 1993.
64. B. Hileman, *Chem. Eng. News*, Sept. 2, 2002, 11.
65. (a) E.B. Weiss and H.K. Jacobson, *Environment*, 1999, *41*(6), 16; (b) E.B. Weiss and H.K. Jackson, eds, *Engaging Counties: Strengthening Compliance with International Environmental Accords*, MIT Press, Cambridge, MA, 1998.
66. B. Hileman, *Chem. Eng. News*, Sept. 21, 1998, 41.
67. (a) T.G. Weiss and L. Gordenker, eds, *NGOs, the U.N. and Global Governance*, L. Rienner, Boulder, CO, 1996; (b) T. Prince and M. Finger, *Environmental NGOs in World Politics: Linking the Local and the Global*, Routledge, London, 1994; (c) P. Willetts, ed., *Conscience of the World: The Influence of NGOs in the UN System*, Hurst, London, 1996; (d) A. Fowler, *Strikling a Balance—A Guide to Enhancing the Effectiveness of Non-Governmental Organizations in International Development*, Earthscan, London, 1997.
68. T. Lovejoy, *Amicus J. (Natural Resources Defense Council)*, 1996, *17*(4), 40.
69. (a) P.J. Nelson, *The World Bank and Non-Governmental Organizations: The Limits of Apolitical Development*, Macmillan, Basingstoke, UK, 1995; (b) B. Hileman, *Chem. Eng. News*, June 23, 1997, 26; (c) J.J. Warford, *The Greening of Economic Policy Reform*, World Bank, Washington, DC, 1997; (d) J. Werksman, *Greening International Institutions*, Earthscan, London, 1996, 131–146; (e) J.A. Fox and L.D. Brown, *The Struggle for Accountability: The World Bank, NGOs and Grassroots Movements*, MIT Press, Cambridge, MA, 1998. (f) World Bank, *Greening Industry: New Roles for Communities, Markets and Governments*, Oxford University Press, New York, 1999, www.worldbank.org/nipr; (g) H.O. Bergesen and L. Lunde, *Dinosaurs or Dynamos? The United Nations and the World Bank at the Turn of the Century*, Earthscan, London, 1999.
70. C. Flavin, *World Watch*, 1997, *10*(6), 25.
71. Anon., *Popline* (*Population Institute*), Jan/Feb. 20, 1998, 3.
72. I.A. Bowles, R.E. Rice, R.A. Mittermeier, and G.A.B. da Fonseca, *Science*, 1998, *280*, 1899.
73. K. Christen, *Environ. Sci. Technol.*, 1999, *33*, 11A.
74. R. Goodland and H. Daly, *Ecol. Appl.*, 1996, *6*, 1002.
75. J. Gold, *Interpretation: The IMF and International Law*, Kluwer Law International, London, 1996.
76. (a) B. Hileman, *Chem. Eng. News*, Nov. 2, 1998, 17; (b) H. French, *World Watch*, 1999, *12*(6), 22; (c) H. French, *Vanishing Borders–Protecting the Planet in the Age of Globilization*, W.W. Norton, New York, 2000; (d) K. Danaher and R. Burbach, eds, *Globalize This–The Battle Against the World Trade Organization & Corporate Rule*, Common Courage Press, Monroe, ME, 2000; (e) C. Hogue, *Chem. Eng. News*, Dec. 10, 2001, 23.
77. C. Hogue, *Chem. Eng. News*, Apr. 4, 2003, 31; Dec. 27, 2003, 36.
78. (a) R.M. Baum, *Chem. Eng. News*, Oct. 30, 1995, 5; (b) J. Darabaris, *Corporate Environmental Management*, CRC Press, Boca Raton, Florida, 2008.
79. (a) D. Rejeski, *Technol. Rev.*, 1997, *100*(1), 56; (b) R. Breslow, *Chem. Eng. News*, Aug. 26, 1996, 72; (c) B.R. Allenby and D.J. Richards, *The Greening of Industrial Ecosystems*, National Academy Press, Washington, DC, 1994, (d) F. Cairncross, *Green Inc.—A Guide a Business and the Environment.* Earthscan, London, 1995; (e) J.D. Graham and J.K. Hartwell, *The Greening of Industry—A Risk Management Approach*, Harvard University Press, Cambridge, MA, 1997; (f) P. Groenewegen, K. Fischer, E.G. Jenkins, and J. Schot, eds, *The Greening of Industry Resource Guide and Bibliography*, Island Press, Washington, DC, 1995; (g) R.A. David. *The Greening of Business*, Gower, Brookfield, VT, 1991. (h) K.M. Reese, *Chem. Eng. News*, Oct. 2, 2000, 200. (i) P.G. Derr and E.M. McNamara, *Case Studies in Environmental Ethics*, Rowman and Littlefield Publishers, Lanham, MD, 2003.
80. R.E. Chandler, In: K. Martin and T.W. Bastock, *Waste Minimisation: A Chemist's Approach*, Royal Society of Chemistry, Cambridge, 1994, 62–74.
81. M.D. Rogers. *Business and the Environment*, Macmillan, Basingstoke, England, 1995, 3.
82. V.N. Bhat, *The Green Corporation—the Next Competitive Advantage.* Quorum Books, Westport, CO, 1996.
83. (a) K. Fischer and J. Schot, eds, *Environmental Strategies for Industry*, Island Press, Washington, DC, 1993; (b) T. Saunders and L. McGovern, *The Bottom Line of Green is Black*, Harper, San Francisco, 1994.
84. (a) D.J. Schell, *A Green Plan for Industry: 16 Steps to Environmental Excellence*, Government Institutes, Rockville, MD, 1998; (b) T. Chapman, *Chem. Ind.* (*Lond.*), 1998, 834; (c) L.D. DeSimone and F. Popoff, *Eco-Efficiency: The Business Link to Sustainable Development*, MIT Press, Cambridge, MA, 1997; (d) N.J. Roome, *Sustainability Strategies for Industry—the Future of Corporate Practice*, Island Press, Washington, DC, 1998; (e) K. Fischer and J. Schot, *Environmental Strategies for Industry—International Perspectives on Research Needs and Policy Implications*, Island Press, Washington, DC, 1993; (f) R. Howes, J. Skea, and B. Whelan, *Clean and Competitive? Motivating Environmental Performance in Industry*, Earthscan, London, 1998; (g) P.M. Morse, *Chem. Eng. News*, Aug. 3, 1998, 13; (h) D. Gaskell, *Chem. Br.*, 200, *36*(6), 36; (i) A.L. White, *Environment*, 1999, *41*(8), 30; (i) A.J. Hoffmann, *Environment*, 2000, *42*(5), 22; (k) M. Arnold, *Chem. Ind.* (*Lond.*), 1999, 921; (l) N.J. Roome, *Sustainability Strategies—The Future of Corporate Practice*, Island Press, Washington, DC, 1998; (m) C. Laszlo, *The Sustainable Company: How To Create Lasting Value Through Social and Environmental Performance*, Island Press, Washington, DC, 2003.

85. P. Hawken, *The Ecology of Commerce—a Declaration of Sustainability*, Harper, San Francisco, 1993, 31.
86. (a) D. Hanson, *Chem. Eng. News*, May 8, 1995, 24; (b) *Chem. Eng. News*, May 15, 1995, 17; (c) W. Storck, *Chem. Eng. News*, Oct. 5, 1998, 26.
87. Anon., *Chem. Eng. News*, May 15, 1995, 22.
88. D. Hanson, *Chem. Eng. News*, Apr. 7, 1997, 12.
89. (a) D. Hanson, *Chem. Eng. News*, Aug. 7, 1995, 8; (b) T. Agres, *R&D (Cahners)*, Dec. 1996, *38*, 29.
90. (a) J.L. Grisham, *Chem. Eng. News*, June 7, 1999, 21; (b) G. Parkinson, *Chem. Eng*, 2001, *108*(10), 25.
91. D. Hanson, *Chem. Eng. News*, Apr. 26, 1999, 10.
92. (a) Anon., *Chem. Eng. News*, Feb. 24, 1997, 25; (b) Anon., *Chem. Eng. News*, Sept. 14, 1998, 21.
93. Anon., *Environ. Sci. Technol.*, 1997, *31*, 124A.
94. L. Raber, *Chem. Eng. News*, Apr. 6, 1998, 12; Aug. 24, 1998, 26.
95. (a) Anon., *Chem. Eng. News*, July 28, 1997, 27; June 23, 1997, 23; (b) M. Jacobs, *Chem. Eng. News*, Apr. 14, 1997, 5; (c) D. Hanson, *Chem. Eng. News*, Dec. 16, 1996, 27; (d) L. Raber, *Chem. Eng. News*, Apr. 14, 1997, 10; May 12, 1997, 26; Aug. 11, 1997, 2; (e) Anon., *Chem. Ind. (Lond.)*, 1996, 965; (f) K. Chin, *Chem. Eng.*, 1997, *104*(1), 42; (g) J. Kaiser, *Science*, 1997, *277*, 466; (h) J. Grisham, *Chem. Eng. News*, May 24, 1999, 5.
96. (a) L. Raber, *Chem. Eng. News*, June 23, 1997, 11; July 7, 1997, 31; (b) B. Hileman, *Chem. Eng. News*, June 2, 1997, 7; (c) Anon., *Environ. Sci. Technol.*, 2000, *34*, 252A.
97. S. Buccino, *Chem. Ind. (Lond.)*, 1997, 666.
98. B. Hileman, *Chem. Eng. News*, Dec. 22, 1997, 20; April 24, 2000, 31.
99. P. Weiss, *Science*, 1996, *272*, 1734.
100. Anon., *Science*, 2000, *290*, 423.
101. L. Ember, *Chem. Eng. News*, Jan. 23, 1995, 19.
102. Anon., *Chem. Eng. News*, Jan. 23, 2006, 29.
103. S.C. Trombulak, D.S. Wilcove, and T.D. Male, *Science*, 2006, *312*, 973.
104. (a) B. Smart, ed., *Beyond Compliance: A New Industry View of the Environment*, World Resources Institute, Washington, DC, 1992; (b) P. Layman, *Chem. Eng. News*, Oct. 18, 1999, 33; (c) C. Holliday, S. Schmidheiny, and P. Watts, *Walking The Talk*, Greenleaf Publishing, Sheffield, UK, 2002.
105. (a) Anon., *Chem. Ind. (Lond.)*, 1995, *164*, 906; (b) B. Dale, *Chem. Ind. (Lond.)*, 1994, 976.
106. L.B. Lave and H.S. Matthews, *Technol. Rev.*, 1996, *99*(8), 68.
107. (a) D.C. Korten, *When Corporations Rule the World*, Kumarian Press, West Hartford, CO, 1995, 146–156; (b) D. Helvarg, *Amicus J. (Natural Resources Defense Council)*, 1996, *18*(2), 13; (c) D. Edward, *Ecologist*, 1999, *18*(2), 13; (c) D. Edward, *Ecologist*, 1999, *29*(3), 172; (d) F. Simon and M. Woodell, In: D.J. Richards, ed., *The Industrial Green Game—Implications for Environmental Design and Management*, Royal Society of Chemistry, Cambridge, UK, 1997, 212–224.
108. Anon., *Chem. Ind. (Lond.)*, 1998, 473.
109. (a) D. Fagin and M. Lavelle, *Toxic Deception: How the Chemical Industry Manipulates Science, Bends the Law and Endangers Your Health*, Common Courage Press, Monroe, ME, 1999; (b) G. Markowitz and D. Rosner, *Deceit and Denial—The Deadly Politics of Industrial Pollution*, University of California Press, Berkeley, CA & Milbank Memorial Fund, New York, 2002; (c) D. Michaels, *Doubt Is Their Product—How Industry's Assault on Science Threatens Your Health*, Oxford University Press, New York, 2008; (d) W. Freudenberg, R. Gramling, and D. Davidson, *World Watch*, 2008, *21*(3), 7.
110. D. Leonhardt, K. Kerwin, and L. Armstrong, *Business Week*, June 30, 1997, 62.
111. P. Wechsler, *Business Week*, June 30, 1997, 68.
112. "Green" Advertising Claims, EPA530-F-92-024, Oct. 1992.
113. K. Wiedemann, *Tufts Nutrition*, Spring 2008, 16.
114. (a) Anon., *Tufts University Health and Nutrition Lett.*, 2007, *25*(10), 3; (b) C. Hogue, *Chem. Eng. News*, Feb. 4, 2008, 24.
115. M. Heylin, *Chem. Eng. News*, June 2, 1997, 12; May 31, 1999, 15.
116. (a) G. Peaff, *Chem. Eng. News*, Nov. 25, 1996, 23; (b) J.L. Wilson, *Chem. Eng. News*, June 3, 1996, 5; (c) A.R. Hirsig, *Chem. Eng. News*, Sept. 8, 1997, 5.
117. (a) F.L. Webber, *Chem. Eng. News*, Nov. 9, 1998, 5; (b) J. Johnson, *Chem. Eng. News*, Nov. 2, 1998, 19; Mar. 8, 1999, 9.
118. R.S. Rogers, *Chem. Eng. News*, Apr. 12, 1999, 30.
119. E.H. Fording Jr., *Chem. Eng. News*, May 4, 1998, 5.
120. (a) M. McCoy, *Chem. Eng. News*, Aug. 2, 1999, 16; (b) A.M. Thayer, *Chem. Eng. News*, June 8, 1998, 17.
121. D.J. Hanson, *Chem. Eng. News*, July 15, 1996, 29.
122. Anon., *Chem. Eng. News*, Aug. 18, 1997, 33.
123. Anon., *Chem. Eng. News*, May 19, 1997, 29.
124. G. Parkinson, *Chem. Eng.*, 1996, *103*(11), 27.
125. (a) J. Johnson, *Chem. Eng. News*, July 28, 1997, 30; June 30, 1997, 26; (b) Anon., *Chem. Eng. News*, Dec. 9, 1996, 23; June 30, 1997, 25; Dec. 22, 1997, 19; (c) Anon., *Environ. Sci. Technol.*, 1998, *32*, 129A.
126. Anon., *Chem. Eng. News*, Mar. 10, 1997, 33.
127. A. Thayer, *Chem. Eng. News*, Apr. 13, 1998, 32.
128. (a) M. Jacobs, *Chem. Eng. News*, Nov. 24, 1997, 5; (b) Anon., *Chem. Ind. (Lond.)*, 1996, 961.
129. M. McCoy, *Chem. Eng. News*, Sept. 26, 2005, 10; June 26, 2006, 27.
130. (a) B. Anez, *Wilmington Delaware News J.*, Feb. 8, 2005, A6; (b) M. Reisch, *Chem. Eng. News*, May 18, 2009, 7.
131. J. Johnson, *Chem. Eng. News*, Mar. 16, 2009, 40.
132. P. Barry, *AARP Bull.*, Washington, DC, June 2002, 8.
133. (a) E. Kirchner, *Chem. Eng. News*, Apr. 22, 1996, 23; (b) L.R. Ember, *Chem. Eng. News*, May 29, 1995, 10; (c) M. Heylin, *Chem. Eng. News*, May 29, 1995, 5; (d) A. Shanley, G. Ondrey, and J. Chowdhury, *Chem. Eng.*, 1997, *104*(3), 39; (e) M.S. Reisch, *Chem. Eng. News*, Jan. 12, 1998, 104; May 11, 1998, 13; Oct. 26, 1998, 15; May 24, 1999, 18; Sept. 4, 2000, 21; (f) D. Hunter, ed., *Chem. Week*, July 5/12, 2000, 38–85; (g) R. Stevenson, *Chem. Br.*, 1999, *35*(5), 27; (h) F.M. Lynn, G. Busenberg, N. Cohen, and C. Chess, *Environ. Sci. Technol.*, 2000, *34*, 1881.
134. J.-F. Tremblay, *Chem. Eng. News*, Oct. 21, 1996, 21; June 23, 1997, 20.
135. Anon., *Chem. Ind. (Lond.)*, 1998, 872.
136. (a) M.S. Reisch, *Chem. Eng. News*, Apr. 14, 1997, 21; (b) M.S. Reisch, *Chem. Eng. News*, May 26, 2003, 15; (c) P. Short, *Chem. Eng. News*, May 26, 2003, 19.
137. K.S. Betts, *Environ. Sci. Technol.*, 1998, *32*, 303A.

138. (a) P. Layman, *Chem. Eng. News*, Dec. 22, 1997, 9; Jan. 5, 1998, 17; (b) Anon., *Chem. Ind.* (*Lond.*), 1998, 3; (c) Anon., *Chem. Eng. News*, Mar. 1, 1999, 20.
139. (a) S.J. Bennett, R. Freierman, and S. George, *Corporate Realities and Environmental Truths—Strategies for: Leading Your Business in the Environmental Era*, Wiley, New York, 1993, 6, 19, 22, 24; (b) J. Nash and J. Ehrenfeld, *Environment*, 1996, *38*(10), 16.
140. (a) Anon., *Environ. Sci. Technol.*, 1998, *32*, 171A; (b) Anon., *Chem. Eng. News*, Feb. 2, 1998, 18; (c) T.P. Silverste, *Chem. Eng. News*, Mar. 9, 1998, 2.
141. B. Hileman, *Chem. Eng. News*, Aug. 17, 1998, 36.
142. (a) D. Hanson, *Chem. Eng. News*, Sept. 14, 1998, 10; (b) Anon., *Chem. Eng. News*, May 10, 1999, 22; May 24, 1999, 25; (c) B. Hileman, *Chem. Eng. News*, Mar. 16, 1998, 26; (d) Anon., *Chem. Ind. (Lond.*), 1998, 464; (e) J. Johnson, *Chem. Eng. News*, May 17, 1999, 11.
143. Anon., *Chem. Ind.* (*Lond.*), 1998, 383.
144. J. Johnson, *Chem. Eng. News*, Feb. 15, 1999, 9; Mar. 1, 1999, 35.
145. G. Hess, *Chem. Eng. News*, Jan. 15, 2009, 19.
146. C. Hogue, M.P. Walls, and J. Tickner, *Chem. Eng. News*, Jan. 8, 2007, 34.
147. (a) C. Hogue, *Chem. Eng. News*, Sept. 22, 2008, 12; Oct. 6, 2008, 42; (b) M. Murphy, *Chem. Ind.* (*Lond.*), May 5, 2008, 5.
148. C. Hogue, *Chem. Eng. News*, May 12, 2008, 9; (b) Anon., *Chem. Eng. News*, July 28, 2008, 41; (c) C. Hogue, *Chem. Eng. News*, June 1, 2009, 4.
149. (a) J. Johnson, *Chem. Eng. News*, Apr. 7, 2008, 10; (b) C. Hogue and J. Johnson, *Chem. Eng. News*, Apr. 14, 2008, 35.
150. Anon., *Chem. Eng. News*, Oct. 17, 2008, 27.
151. (a) Anon., *Amicus J. (Natural Resources Defense Council)*, 1990, *12*(2), 3; (b) V.N. Bhat, *The Green Corporation—the Next Competitive Advantage*, Quorum Books, Westport, CO, 1996, 163; (c) http://www.ceres.org. (d) K. Betts, *Environ. Sci. Technol.*, 1999, *33*, 189A.
152. T.F. Walton, *Environ. Prog.*, 1996, *15*(1), 1.
153. S.A. Fenn, *Technol. Rev.*, 1995, *98*(5), 62.
154. (a) D. Green, *ISO 9000 Quality Systems Auditing*, Gower, Aldershot, UK, 1997; (b) W.A. Golomski, A.J.M. Pallett, J.G. Surak, and K.E. Simpson, *Food Technol.*, 1994, *48*(12), 57, 60, 63; (c) A. Badiru, *Industry's Guide to ISO 9000*, Wiley, New York, 1995.
155. (a) D. Hunt and C. Johnson, *Environmental Management Systems—Principles and Practice*, McGraw-Hill, New York, 1995; (b) J. Cascio, G. Woodside, and P. Mitchell, *ISO 14000 Guide*: *The New International Management Standards*, McGraw-Hill, New York, 1996; (c) J.M. Diller, *Chem. Eng. Prog*, 1997, *93*(11), 36; (d) R. Begley, *Environ. Sci. Technol.*, 1996, *30*(7), 298A; 1997, *31*(8), 364A; (e) G.S. Samdani, S. Moore, and G. Ondrey, *Chem. Eng.*, 1995, *102*(6), 41; (f) A.M. Thayer, *Chem. Eng. News*, Apr. 1, 1996, 11; Sept. 30, 1996, 27; (g) B. Rothery, *ISO 14000 and ISO 9000*, Gower Publishing, Aldershot, UK, 1995; (h) W.M. von Zharen, *ISO 14000*: *Understanding the Environmental Standards*, Government Institutes, Rockville, MD, 1996; (i) J. Cascio and J.S. Shideler, *Chemtech*, 1998, *28*(5), 49; (j) A. Schoffman and A. Tordini, *ISO 14001: A Practical Approach*, American Chemical Society, Washington, DC, 2000; (k) P.J. Knox, *Chem. Process* (*Chicago*), 1999, *62*(2), 26; (l) R. Krut and H. Gleckman, *ISO 14001—The Missed Opportunities*, Earthscan, London, 1998; (m) N.I. McClelland and B. St. John, *Environ. Prog.*, 1999, *18*(1), S3.
156. Anon., *Chem. Eng. News*, Jan. 29, 1996, 15.
157. J.W. Houck and O.F. Williams, eds, *Is the Good Corporation Dead?*, Rowman & Littlefield, London, 1996.
158. A. Thayer, *Chem. Eng. News*, May 4, 1998, 31.
159. (a) Anon., *Chem. Eng. News*, Aug. 30, 1999, 39; (b) Anon., *Industrial Environmental Performance Metrics: Challenges and Opportunities*, National Academy of Engineering, Washington, DC, 1999; http://national-academies.org.
160. V.N. Bhat, *The Green Corporation—The Next Competitive Advantage*. Quorum Books, Westport, CT, 1996, 165.
161. M. Freemantle, *Chem. Eng. News*, May 20, 1996, 30.
162. (a) A. Scott, *Chem. Week*, Oct. 24, 2007, 23; (b) Anon., *Chem. Eng. News*, Sept. 9, 2002, 15; (c) www.mallenbaker.net/esr/CSRfiles/djsgli.html; (d) Anon., *Chem. Eng. News*, Sept. 12, 2005, 14.
163. Anon., *Conservation in Practice*, 2005, *6* (4), 38.
164. www.cii.org.
165. E. Dorey, *Chem. Ind.* (*Lond.*), Feb. 20, 2006, 7.
166. Anon., *Chem. Eng. News*, Feb. 16, 2004, 12.
167. M. Murphy, *Chem. Ind.* (*Lond.*), Mar. 10, 2008, 7.
168. (a) K. Patttison, *On Earth (Natural Resources Defense Council)*, 2009, *30*(4), 20; (b) Anon., *World Watch*, 2008, *21*(2), 5.
169. E. Assadourin, *World Watch*, 2006, *19*(2), 16.
170. Anon., *Fortune*, 2009, *159*(5), 81.
171. (a) Anon., *Environment*, 2005, *47*(2), 7; (b) www.yale.edu/esis.
172. (a) Anon., *Chem. Ind.* (*Lond.*), 1994, 928; (b) Council on Economic Priorities, New York, 1995.
173. Anon., *Chem. Ind.* (*Lond.*), 1997, 855.
174. Anon., *Chem. Eng. News*, July 29, 1996, 33.
175. (a) S. Ainsworth, *Chem. Eng. News*, July 18, 1994, 9; (b) J.F. Tremblay, *Chem. Eng. News*, Feb. 24, 1997, 22.
176. Anon., *Chem. Eng. News*, Apr. 5, 1999, 11.
177. (a) Anon., *Chem. Ind.* (*Lond.*), 1997, 457; 1998, 289; (b) M. Reisch, *Chem. Eng. News*, June 23, 1997, 22; (c) G. Ondrey, *Chem. Eng.*, 1999, *106*(4), 29.
178. Anon., *Chem. Eng. News*, Nov. 7, 1994, 14.
179. S.J. Bennett, R. Freierman and S. George, *Corporate Realities and Environmental Truths—Strategies for Leading Your Business in the Environmental Era*, Wiley, New York, 1993, 156.
180. F. Krupp, *EDF Lett.* (*Environmental Defense Fund*), 1996, *27*(4), 4.
181. (a) M.R. Lee, *A Review of Oil Pollution Prevention Regulations after the Monongahela River Spill*, Congressional Research Service, Washington, DC, 1988; (b) U.S. Congress Hearing, *Oil Spill on the Monongahela and Ohio Rivers*, Washington, DC,1988.
182. W. Lepkowski, *Chem. Eng. News*, July 10, 1995, 7.
183. Ashland Inc, 1995 Annual Report on Environment, Health, and Safety, Ashland, Covington, KY.
184. Anon., *Chem. Eng. News*, July 21, 1997, 22.
185. Chevron, *Measuring Progress—A Report on Chevron's Environmental Performance*, San Francisco, CA, Oct. 1994.
186. Anon., *Chem. Eng.*, 1999, *106*(4), 64.
187. G. Ondrey, *Chem. Eng.*, 2000, *107*(11), 27.
188. R. Rivera, *Amicus J. (Natural Resources Defense Council)*, 2001, *2*(4), 10.

189. Chevron, 1997, Annual Report, San Francisco, CA.
190. K. Koenig, *World Watch*, 2004, *17*(1), 10.
191. J. Diamond, *Conserv. Pract.*, 2005, *6*(4), 12.
192. Chevron, Quarterly reports to shareholders for the third quarters of 1996 and 1997, San Francisco, CA.
193. (a) Earth Day 2000, *The Don't Be Fooled Report—the Top Ten Greenwashers of 1994*, San Francisco, CA, 1995; (b) D. Helvarg, *Amicus J. (Natural Resources Defense Council)*, 1996, *18*(2), 16.
194. Chevron, Report to shareholders for the second quarter of 1995.
195. (a) Chevron, 1995 meeting report, San Francisco, CA, June 10, 1995; (b) Chevron, Report to shareholders, San Francisco, CA, first quarter 1997; (c) Chevron, Proxy statement for its annual shareholders meeting, San Francisco, CA, Apr. 1999.
196. E.N. Brandt, *Growth Company—Dow Chemical's First Century*, Michigan State Press, East Lansing, 1997, 526.
197. L. Raber, *Chem. Eng. News*, May 6, 1996, 7.
198. (a) L. Greer, *Environ. Sci. Technol.*, 2000, *34*, 254A; (b) J. Johnson, *Chem. Eng. News*, Sept. 13, 1999, 22.
199. J. Johnson, *Chem. Eng. News*, Oct. 9, 2006, 28.
200. Anon., *Chem. Eng. News*, Aug. 19, 1996, 13.
201. The Presidential Green Chemistry Challenge Awards program—Summary of 1996 Award Entries and Recipients, EPA 744-K-96–110, Washington, DC, July 1996, *3*, 25.
202. Anon., *Chem. Eng. News*, Mar. 10, 1997, 1.
203. Anon., *Chem. Eng. News*, April 21, 2003, 42.
204. (a) J. Johnson, *Chem. Eng. News*, Aug. 17, 1998, 34; Sept. 13, 1999, 22; (b) L. Greer, *Environ. Sci. Technol.*, 2000, *34*, 254A.
205. D. Fagin and M. Lavelle, *Toxic Deception—How the Chemical Industry Manipulates Science, Bends the Law and Endangers Your Health*, Carol Publishing, Secaucus, NJ, 1996, *114*, 123.
206. Anon., *Chem. Eng. News*, July 16, 2007, 24; Jan. 14, 2008, 36.
207. Anon., *Chem. Eng. News*, Apr. 26, 2004, 12.
208. Anon., *Chem. Eng. News*, Dec. 1, 1997, 7
209. Anon., *Chem. Eng. News*, Oct. 15, 2008, 20.
210. (a) J.H. Krieger, *Chem. Eng. News*, July 8, 1996, 13; (b) Green Chemistry and Engineering Conference, Washington, DC, June 23–25, 1997; (c) P.V. Tebo, *Chemtech*, 1998, *28*(3), 8.
211. B. Hileman, *Chem. Eng. News*, July 30, 2001, 13.
212. B. Hileman, *Chem. Eng. News*, Nov. 23, 1998, 10.
213. S.J. Bennett, R.L. Freierman, and S. George, *Corporate Realties and Environmental Truths—Strategies for Leading Your Business in the Environmental Era*, Wiley, New York, 1993, 44.
214. E.S. Woolard, *Chem. Ind.* (*Lond.*), 1995, *436*, 969.
215. (a) M. Murray, *Wilmington Delaware News J.*, Apr. 11, 1996, B1; (b) J. Brooks, *Wilmington Delaware News J.*, Sept. 12, 1997, B7.
216. P.L. Layman, *Chem. Eng. News*, Mar. 29, 1999, 16.
217. G. Parkinson, *Chem. Eng.*, 1998, *105*(7), 31.
218. (a) C. Aregood, *Wilmington Delaware News J.*, June 17, 1997, B7; (b) A.M. Thayer, *Chem. Eng. News*, Dec. 16, 2002, 17; (c) M. Reisch, *Chem. Eng. News*, June 17, 2002, 6.
219. (a) R. Raber, *Wilmington Delaware News J.*, Sept. 3, 1997, B7; (b) C. Hogue, *Chem. Eng. News*, Aug. 7, 2000, 12.
220. Anon., *Chem. Eng. News*, July 30, 2007, 44.
221. (a) W. Lepkowski, *Chem. Eng. News*, July 15, 1996, 24; (b) Anon., *Environ. Action.*, 1996, *28*(1–2), 39.
222. (a) M. Reisch, *Chem. Eng. News*, Apr. 14, 1997, 9; (b) A.M. Thayer, *Chem. Eng. News*, Mar. 9, 1998, 10.
223. J. Gerstenzang, *Wilmington Delaware News J.*, Sept. 9, 1997, A4.
224. J. Montgomery, *Wilmington Delaware News J.*, June 26, 2000, B3.
225. J. Montgomery, *Wilmington Delaware News J.*, Mar. 25, 2002, A1.
226. M. Milford, *Wilmington Delaware News J.*, Sept. 14, 2002, B1.
227. C. Hogue, *Chem. Eng. News*, Dec. 19, 2005, 10.
228. C. Hogue, *Chem. Eng. News*, Mar. 26, 2001, 19; Sept. 3, 2001, 27.
229. Anon., *Chem. Week*, Aug. 15, 2001, 12.
230. (a) C. Holliday, S. Schmidheiny, and P. Watts, *Walking the Talk*, Greenleaf Publishing, Sheffield, UK, 2002; (b) R. Baum, *Chem. Eng. News*, Oct. 16, 2006, 10; (c) M.S. Reisch, *Chem. Eng. News*, Sept. 3, 2001, 17; (d) J.F. Tremblay, *Chem. Eng. News*, Mar. 19, 2007, 22.
231. (a) "Not Walking the Talk: duPont's Untold Safety Failures", Sept. 2005, www.dupontsafetyrevealed.org; (b) J. Montgomery, *Wilmington Delaware News J.*, Oct. 11, 2006, B6.
232. Valdez Litigation Update. Exxon Perspectives (a quarterly newsletter for shareholders), Houston, TX, Sept. 1996, 5.
233. Exxon. Environment, Health and Safety Progress Report for 1995, Houston, TX.
234. B. Hileman, *Chem. Eng. News*, May 27, 1996, 27.
235. Exxon Perspectives (a quarterly newsletter to shareholders), Houston, TX, Mar. 1997.
236. L. Raymond, et al. Exxon Perspectives (a quarterly newsletter to shareholders), Houston, TX, Dec. 1997.
237. E. Marshall, *Science*, 2006, *313*, 1871.
238. E. Robinson, *Catalyst* (*Union of Concerned Scientists*), Spring 2007, 2.
239. R.E. Chandler, In: K. Martin, T.W. Bastock, eds, *Waste Minimisation—A Chemist's Approach*, Royal Society of Chemistry, Cambridge, 1994, 71.
240. (a) Anon., *Chem. Eng. News*, Aug. 29, 1994, 21; (b) G. Samdani, *Chem. Eng.*, 1995, *102*(5), 19; (c) Anon., *Chem. Ind.* (*Lond.*), 1994, 668; (d) Anon., *Chem. Eng. News*, Jan. 9, 1995, 9; (e) Anon., *Chem. Eng. News*, Apr. 8, 1996, 36; (f) Anon., *Chem. Eng. News*, July 22, 1996, 36; (g) E.R. Beaver, *Environ. Prog.*, 1997, *16*(3), F3.
241. The Presidential Green Chemistry Challenge Awards Program—Summary of 1996 Award Entries and Recipients. EPA 744-K-96-001, U. S. Environmental Protection Agency, Washington, DC, July 1996, 2.
242. D. Fagin and M. Lavelle, *Toxic Deception; How the Chemical Industry Manipulates Science, Bends the Law and Endangers Your Health*, Carol Publishing, Secaucus, NJ, 1996.
243. (a) A. Thayer, *Chem. Eng. News*, Dec. 22, 1997, 18; (b) M.S. Reisch, *Chem. Eng. News*, Jan. 12, 1998, 106.
244. G. Parkinson, *Chem. Eng.*, 2001, *108*(2), 21.
245. Anon., *Chem. Eng. News*, Jan. 28, 2002, 18.
246. (a) A.M. Thayer, *Chem. Eng. News*, Dec. 24, 2001, 13; (b) M. Reisch and C. Hogue, *Chem. Eng. News*, Dec. 21, 2001, 13.

247. Anon., *Chem. Eng. News*, Jan. 28, 2008, 50; Feb. 11, 2008, 36.
248. N. Eisberg, *Chem. Ind.* (*Lond.*), May 28, 2007, 8.
249. M. Murphy and N. Eisberg, *Chem. Ind.* (*Lond.*), Apr. 7, 2009, 12, 13.
250. Anon., *Amicus J. (Natural Resources Defense Council)*, 2001, *22*(4), 9.
251. E. Assadourian, *World Watch.*, 2004, *17*(3), 28.
252. E. Wilson, *Chem. Eng. News*, Feb. 5, 2007, 9.
253. Anon., *Amicus J. (Natural Resources Defense Council)*, 1999, *21*(3), 46.
254. A. Denio, *Del-Chem. Bull.*, Wilmington, DE, November-December 2004, 12.
255. Delaware City Forum hearing at Delaware City, DE.
256. Anon., *On Earth*, 2008, *29*(4), 59.
257. Texaco Inc. Environment, *Health Safety Rev.*, Texaco, 1994, 13.
258. Anon., *Chem. Eng. News*, Feb. 22, 1993, 14.
259. (a) J. Kane, *Savages.* Knopf, New York, 1995, 190; (b) *Environment*, 1995, *37*(1), 21; (c) Texaco, Annual Report for 1994, Texaco, 13; (d) Texaco, Annual Report for 1995, Texaco, 22; (e) Report on the 1995 Annual Meeting of Texaco, Texaco, June 8, 1995; (f) Texaco Inc. Environment, *Health Safety Rev.,* Texaco, 1994, 21; (g) Texaco Inc. Environment, *Health Safety Rev.,* Texaco, 1996, 28.
260. (a) Anon., New York Times Dec. 22, 1996, E8; (b) P.I. Bijur, Letter, to shareholders, Texaco, Nov, 16, 1996.
261. Texaco Inc. Environment, *Health Safety Rev.,* Texaco, 1994, 22.
262. B. Anderson, *Nat. Conserv.*, 1994, *44*(2), 30.
263. Texaco, Annual Report for 1995, Texaco, 22.

RECOMMENDED READING

1. R.E. Chandler, In: K. Martin, T.W. Bastock, eds, *Waste Minimisation: A Chemist's Approach*, Royal Society of Chemistry, Cambridge, UK, 1994, 63 [stages of greening of companies].
2. J. Nash and J. Ehrenfeld, *Environment*, 1996, *38*(1), 16 [voluntary agreements for self-regulation].
3. D.C. Korten, *When Corporations Rule the World*, Kumarian Press, West Hartford, CT, 1995, 329 [the June 12, 1992 "Proactive Agenda for the Future" adopted by NGOs at Rio de Janeiro].
4. B. Hileman, *Chem. Eng. News*, Dec. 22, 1997, 20 [Kyoto conference on global warming].

EXERCISES

1. Pick a company in your geographic area and see how green it is. Check its Toxic Release Inventory, spills, Superfund sites, enforcement actions, penalties, annual report, items about it in the local newspaper, and what trade associations it belongs to, together with their actions. (If you cannot find all of these, use what you can. If you are in a country other than the United States, try to find comparable data.)
2. See what NGOs are active where you live. What do they do? How could they be more effective?
3. Consider the regulations in effect where you live. Are there too many or too few? Are they effective? How might they be improved? Have you discussed them with your elected officials?

Index

F

G

H

N

O

P

Q

R

S

T